A CONCISE HANDBOOK OF
MATHEMATICS, PHYSICS, AND ENGINEERING SCIENCES

A CONCISE HANDBOOK OF
MATHEMATICS, PHYSICS, AND ENGINEERING SCIENCES

Edited by

Andrei D. Polyanin

Alexei I. Chernoutsan

Authors

A.I. Chernoutsan, A.V. Egorov, A.V. Manzhirov, A.D. Polyanin,
V.D. Polyanin, V.A. Popov, B.V. Putyatin, Yu.V. Repina,
V.M. Safrai, A.I. Zhurov

CRC Press
Taylor & Francis Group
Boca Raton   London   New York

CRC Press is an imprint of the
Taylor & Francis Group an **informa** business
A CHAPMAN & HALL BOOK

CRC Press
Taylor & Francis Group
6000 Broken Sound Parkway NW, Suite 300
Boca Raton, FL 33487-2742

© 2011 by Taylor and Francis Group, LLC
CRC Press is an imprint of Taylor & Francis Group, an Informa business

No claim to original U.S. Government works

Printed in the United States of America on acid-free paper
10 9 8 7 6 5 4 3 2 1

International Standard Book Number: 978-1-4398-0639-5 (Hardback)

Library of Congress Cataloging-in-Publication Data

A concise handbook of mathematics, physics, and engineering sciences / editors, Andrei D. Polyanin,
    Alexei I. Chernoutsan.
        p. cm.
    Includes bibliographical references and index.
    ISBN 978-1-4398-0639-5 (hardback : alk. paper)
    1. Mathematics--Handbooks, manuals, etc. 2. Physics--Handbooks, manuals, etc. 3.
Engineering--Handbooks, manuals, etc. I. Polianin, A. D. (Andrei Dmitrievich) II. Chernoutsan, A. I.
III. Title.

QA40.C65 2011
500.2--dc22   2010021872

Visit the Taylor & Francis Web site at
http://www.taylorandfrancis.com

and the CRC Press Web site at
http://www.crcpress.com

CONTENTS

PREFACE

This is a concise multi-subject handbook, which consists of three major parts: mathematics, physics, and applied and engineering sciences. It presents basic notions, formulas, equations, problems, theorems, methods, and laws on each of the subjects in brief form. The absence of proofs and a concise presentation has permitted combining a substantial amount of reference material in a single volume. The handbook is intended for a wide audience of engineers and researchers (not specialized in mathematics or theoretical physics) as well as graduate and postgraduate students.

- The first part of the book contains chapters on arithmetics, elementary and analytic geometry, algebra, differential and integral calculus, functions of complex variable, integral transforms, ordinary and partial differential equations, special functions, probability theory, etc.

- The second part of the book contains chapters on molecular physics and thermodynamics, electricity and magnetism, oscillations and waves, optics, special relativity, quantum mechanics, atomic physics, etc.

- The third part of the book contains chapters on dimensional analysis and similarity, mechanics of point masses and rigid bodies, strength of materials, hydrodynamics, mass and heat transfer, electrical engineering, and methods for constructing empirical and engineering formulas.

A compact and clear presentation of the material allows the reader to get quick help on (or revise) the desired topic. Special attention is paid to issues that many engineers and students may find difficult to understand.

When selecting the material, the authors have given a pronounced preference to practical aspects; namely, to formulas, problems, methods, and laws that most frequently occur in sciences and engineering applications and university education. Many results are represented in tabular form.

For the convenience of a wider audience with different mathematical backgrounds, the authors tried to avoid special terminology whenever possible. Therefore, some of the topics and methods are outlined in a schematic and somewhat simplified manner, which is sufficient for them to be used successfully in most cases. Many sections were written so that they could be read independently. The material within subsections is arranged in increasing order of complexity. This allows the reader to get to the heart of the matter quickly.

The material of the reference book can be roughly categorized into the following three groups according to meaning:

1. The main text containing a concise, coherent survey of the most important definitions, formulas, equations, methods, theorems, and laws.

2. For the reader's better understanding of the topics and methods under study, numerous examples are given throughout the book.

3. Discussion of additional issues of interest, given in the form of remarks in small print.

For the reader's convenience, several long mathematical tables—indefinite and definite integrals, direct and inverse integral transforms (Laplace, Mellin, and Fourier transforms), and exact solutions of differential equations—which contain a large amount of information, are presented in the supplement of the book. Also included are some physical tables and the periodic table of the chemical elements.

This handbook consists of parts, chapters, sections, and subsections. Figures and tables are numbered separately in each section, while formulas (equations) and examples are numbered separately in each subsection. When citing a formula, we use notation like (M3.1.2.5), which means formula 5 in Subsection M3.1.2. For the reader's convenience, each citation number is preceded by a letter to indicate one of the major parts: mathematics (M), physics (P), engineering sciences (E), or supplements (S). At the end of each chapter, we present a list of main and additional literature sources containing more detailed information about topics of interest to the reader.

Special font highlighting in the text, cross-references, an extensive table of contents, and a detailed index help the reader to find the desired information.

Chapters M1, M2, and M6–M9 were written by V. M. Safrai and A. I. Zhurov, Chapters M3–M5, M10, and M14 by A. V. Manzhirov and V. A. Popov, Chapters M11–M13, E1, E4, E5, E7, and S1–S5 by A. D. Polyanin, Chapters P1–P8 by A. I. Chernoutsan, Chapter E2 by V. D. Polyanin, Chapter E3 by B. V. Putyatin, Chapter E6 by A. V. Egorov and Yu. V. Repina, and Chapters S6 and S7 by A. I. Chernoutsan and A. I. Zhurov. Part M was edited by A. D. Polyanin and parts E and S were edited by A. D. Polyanin and A. I. Chernoutsan.

We would like to express our deep gratitude to Vladimir Nazaikinskii for translating several chapters of this handbook.

The authors hope that this book will be helpful for a wide range of engineers, scientists, university teachers, and students engaged in the fields of physics, mechanics, engineering sciences, chemistry, biology, ecology, medicine as well as social and economical sciences.

Andrei D. Polyanin
Alexei I. Chernoutsan

EDITORS

 **Andrei D. Polyanin, D.Sc., Ph.D.,** is a well-known scientist of broad interests and is active in various areas of mathematics, mechanics, and chemical engineering sciences. He is one of the most prominent authors in the field of reference literature on mathematics.

Professor Polyanin graduated with honors from the Department of Mechanics and Mathematics of Moscow State University in 1974. He received his Ph.D. degree in 1981 and D.Sc. degree in 1986 at the Institute for Problems in Mechanics of the Russian (former USSR) Academy of Sciences. Since 1975, Professor Polyanin has been working at the Institute for Problems in Mechanics of the Russian Academy of Sciences; he is also Professor of Mathematics at Bauman Moscow State Technical University. He is a member of the Russian National Committee on Theoretical and Applied Mechanics and of the Mathematics and Mechanics Expert Council of the Higher Certification Committee of the Russian Federation.

Professor Polyanin is an author of more than 30 books in English, Russian, German, and Bulgarian as well as over 140 research papers and three patents. He has written a number of fundamental handbooks, including A. D. Polyanin and V. F. Zaitsev, *Handbook of Exact Solutions for Ordinary Differential Equations*, CRC Press, 1995 and 2003; A. D. Polyanin and A. V. Manzhirov, *Handbook of Integral Equations*, CRC Press, 1998 and 2008; A. D. Polyanin, *Handbook of Linear Partial Differential Equations for Engineers and Scientists*, Chapman & Hall/CRC Press, 2002; A. D. Polyanin, V. F. Zaitsev, and A. Moussiaux, *Handbook of First Order Partial Differential Equations*, Taylor & Francis, 2002; A. D. Polyanin and V. F. Zaitsev, *Handbook of Nonlinear Partial Differential Equations*, Chapman & Hall/CRC Press, 2004, and A. D. Polyanin and A. V. Manzhirov, *Handbook of Mathematics for Engineers and Scientists*, Chapman & Hall/CRC Press, 2007.

Professor Polyanin is Editor-in-Chief of the international scientific-educational Website *EqWorld — The World of Mathematical Equations* (http://eqworld.ipmnet.ru), which is visited by several thousands of users a day worldwide. He is also Editor of the book series *Differential and Integral Equations and Their Applications*, Chapman & Hall/CRC Press, London/Boca Raton. Professor Polyanin is a member of the Editorial Board of the journal *Theoretical Foundations of Chemical Engineering*.

In 1991, Professor Polyanin was awarded a Chaplygin Prize of the Russian Academy of Sciences for his research in mechanics. In 2001, he received an award from the Ministry of Education of the Russian Federation.

Address: Institute for Problems in Mechanics, 101 Vernadsky Ave., Bldg. 1, 119526 Moscow, Russia.

Home page: http://eqworld.ipmnet.ru/polyanin-ew.htm

Alexei I. Chernoutsan, Ph.D., is a prominent scientist in the fields of statistical physics, fluctuation kinetics, phase transitions, and percolation theory. He graduated from the Faculty of Physics of the Moscow State University in 1975. He received his Ph.D. degree in 1980 at the Landau Institute for Theoretical Physics of the Russian (former USSR) Academy of Sciences. Professor Chernoutsan has been a member of staff of the Gubkin Russian State University of Oil and Gas since 1980, Dean of the Faculty of Natural Sciences Education since 2002, and Head of the Department of Physics since 2009.

An experienced and multi-skilled lecturer in general physics, Professor Chernoutsan has taught students of the Gubkin Russian State University of Oil and Gas for many years. He is an active author in the field of popular science and since 1991 he has been Deputy Editor-in-Chief of *Kvant*, a famous Russian educational magazine for students. He is an author of 3 books and more than 80 scientific and educational publications.

Address: Gubkin Russian State University of Oil and Gas, 65 Leninskii Ave., 119991 Moscow, Russia.
E-mail: acher@gubkin.ru

Part I
Mathematics

Chapter M1
Arithmetic and Elementary Algebra

M1.1. Real Numbers

M1.1.1. Integer Numbers

▶ **Natural, integer, even, and odd numbers.** *Natural numbers*: 1, 2, 3, ... (all positive whole numbers).

Integer numbers (or simply *integers*): 0, ±1, ±2, ±3, ...

Even numbers: 0, 2, 4, ... (all nonnegative integers that can be divided evenly by 2). An even number can generally be represented as $n = 2k$, where $k = 0, 1, 2, ...$

Remark 1. Sometimes all integers that are multiples of 2, such as 0, ±2, ±4, ..., are considered to be even numbers.

Odd numbers: 1, 3, 5, ... (all natural numbers that cannot be divided evenly by 2). An odd number can generally be represented as $n = 2k + 1$, where $k = 0, 1, 2, ...$

Remark 2. Sometimes all integers that are not multiples of 2, such as ±1, ±3, ±5, ..., are considered to be odd numbers.

All integers as well as even numbers and odd numbers form *infinite countable sets*, which means that the elements of these sets can be enumerated using the natural numbers 1, 2, 3, ...

▶ **Prime and composite numbers.** A *prime number* is a positive integer that is greater than 1 and has no positive integer divisors other than 1 and itself. The prime numbers form an infinite countable set. The first ten prime numbers are: 2, 3, 5, 7, 11, 13, 17, 19, 23, 29, ...

A *composite number* is a positive integer that is greater than 1 and is not prime, i.e., has factors other than 1 and itself. Any composite number can be uniquely factored into a product of prime numbers. The following numbers are composite: $4 = 2 \times 2$, $6 = 2 \times 3$, $8 = 2^3$, $9 = 3^2$, $10 = 2 \times 5$, $12 = 2^2 \times 3$, ...

The number 1 is a special case that is considered to be neither composite nor prime.

▶ **Divisibility tests.** Below are some simple rules helping to determine if an integer is divisible by another integer.

All integers are divisible by 1.

Divisibility by 2: last digit is divisible by 2.

Divisibility by 3: sum of digits is divisible by 3.

Divisibility by 4: two last digits form a number divisible by 4.

Divisibility by 5: last digit is either 0 or 5.

Divisibility by 6: divisible by both 2 and 3.

Divisibility by 9: sum of digits is divisible by 9.

Divisibility by 10: last digit is 0.

Divisibility by 11: the difference between the sum of the odd-numbered digits (1st, 3rd, 5th, etc.) and the sum of the even-numbered digits (2nd, 4th, etc.) is divisible by 11.

Example 1. Let us show that the number 80729 is divisible by 11.

The sum of the odd-numbered digits is $\Sigma_1 = 8 + 7 + 9 = 24$. The sum of the even-numbered digits is $\Sigma_2 = 0 + 2 = 2$. The difference between them is $\Sigma_1 - \Sigma_2 = 22$ and is divisible by 11. Consequently, the original number is also divisible by 11.

▶ **Greatest common divisor and least common multiple.**

$1°$. The *greatest common divisor* of natural numbers $a_1, a_2, \ldots, a_n$ is the largest natural number, b, which is a common divisor to $a_1, \ldots, a_n$.

Suppose some positive numbers $a_1, a_2, \ldots, a_n$ are factored into products of primes so that

$$a_1 = p_1^{k_{11}} p_2^{k_{12}} \cdots p_m^{k_{1m}}, \quad a_2 = p_1^{k_{21}} p_2^{k_{22}} \cdots p_m^{k_{2m}}, \quad \ldots, \quad a_n = p_1^{k_{n1}} p_2^{k_{n2}} \cdots p_m^{k_{nm}},$$

where $p_1, p_2, \ldots, p_m$ are different prime numbers and the k_{ij} are nonnegative integers ($i = 1, 2, \ldots, n$; $j = 1, 2, \ldots, m$). Then the greatest common divisor b of $a_1, a_2, \ldots, a_n$ is calculated as

$$b = p_1^{\sigma_1} p_2^{\sigma_2} \cdots p_m^{\sigma_m}, \quad \sigma_j = \min_{1 \le i \le n} k_{ij}.$$

Example 2. The greatest common divisor of 180 and 280 is $2^2 \times 5 = 20$ due to the following factorization:

$$180 = 2^2 \times 3^2 \times 5 = 2^2 \times 3^2 \times 5^1 \times 7^0,$$

$$280 = 2^3 \times 5 \times 7 = 2^3 \times 3^0 \times 5^1 \times 7^1.$$

$2°$. The *least common multiple* of n natural numbers $a_1, a_2, \ldots, a_n$ is the smallest natural number, A, that is a multiple of all the a_k.

Suppose some natural numbers $a_1, \ldots, a_n$ are factored into products of primes just as in Item $1°$. Then the least common multiple of all the a_k is calculated as

$$A = p_1^{\nu_1} p_2^{\nu_2} \cdots p_m^{\nu_m}, \quad \nu_j = \max_{1 \le i \le n} k_{ij}.$$

Example 3. The least common multiple of 180 and 280 is equal to $2^3 \times 3^2 \times 5^1 \times 7^1 = 2520$ due to the factorization given in Example 2.

M1.1.2. Real, Rational, and Irrational Numbers

▶ **Real numbers.** The *real numbers* are all the positive numbers, negative numbers, and zero. Any real number can be represented by a *decimal fraction* (or simply *decimal*), finite or infinite. The set of all real numbers is denoted by $\mathbb{R}$.

All real numbers are categorized into two classes: the *rational* numbers and *irrational* numbers.

▶ **Rational numbers.** A *rational number* is a real number that can be written as a fraction (ratio) p/q with integer p and q ($q \ne 0$). It is only the rational numbers that can be written in the form of finite (terminating) or periodic (recurring) decimals (e.g., $1/8 = 0.125$ and $1/6 = 0.16666\ldots$). Any integer is a rational number.

The rational numbers form an infinite countable set. The set of all rational numbers is everywhere dense. This means that, for any two distinct rational numbers a and b such that $a < b$, there exists at least one more rational number c such that $a < c < b$, and hence there are infinitely many rational numbers between a and b. (Between any two rational numbers, there always exist irrational numbers.)

▶ **Irrational numbers.** An *irrational number* is a real number that is not rational; no irrational number can be written as a fraction p/q with integer p and q ($q \ne 0$). To the irrational numbers there correspond nonperiodic (nonrepeating) decimals. Here are examples of irrational numbers: $\sqrt{3} = 1.73205\ldots$, $\pi = 3.14159\ldots$

The set of irrational numbers is everywhere dense, which means that between any two distinct irrational numbers, there are both rational and irrational numbers. The set of irrational numbers is uncountable.

M1.2. Equalities and Inequalities. Arithmetic Operations. Absolute Value

M1.2.1. Equalities and Inequalities

Throughout Subsection 1.2.1, it is assumed that a, b, c, d are real numbers.

▶ **Basic properties of equalities.**

1. If $a = b$, then $b = a$.
2. If $a = b$, then $a + c = b + c$, where c is any real number; furthermore, if $a + c = b + c$, then $a = b$.
3. If $a = b$, then $ac = bc$, where c is any real number; furthermore, if $ac = bc$ and $c \neq 0$, then $a = b$.
4. If $a = b$ and $b = c$, then $a = c$.
5. If $ab = 0$, then either $a = 0$ or $b = 0$; furthermore, if $ab \neq 0$, then $a \neq 0$ and $b \neq 0$.

▶ **Basic properties of inequalities.**

1. If $a < b$, then $b > a$.
2. If $a \leq b$ and $b \leq a$, then $a = b$.
3. If $a \leq b$ and $b \leq c$, then $a \leq c$.
4. If $a < b$ and $b \leq c$ (or $a \leq b$ and $b < c$), then $a < c$.
5. If $a < b$ and $c < d$ (or $c = d$), then $a + c < b + d$.
6. If $a \leq b$ and $c > 0$, then $ac \leq bc$.
7. If $a \leq b$ and $c < 0$, then $ac \geq bc$.
8. If $0 < a \leq b$ (or $a \leq b < 0$), then $1/a \geq 1/b$.

M1.2.2. Addition and Multiplication of Numbers

▶ **Addition of real numbers.** The sum of real numbers is a real number.
Properties of addition:

$$
\begin{aligned}
a + 0 &= a &&\text{(property of zero)}, \\
a + b &= b + a &&\text{(addition is commutative)}, \\
a + (b + c) &= (a + b) + c = a + b + c &&\text{(addition is associative)},
\end{aligned}
$$

where a, b, c are arbitrary real numbers.

For any real number a, there exists its unique *additive inverse*, or its *opposite*, denoted by $-a$, such that

$$a + (-a) = a - a = 0.$$

▶ **Multiplication of real numbers.** The product of real numbers is a real number.
Properties of multiplication:

$$
\begin{aligned}
a \times 0 &= 0 &&\text{(property of zero)}, \\
ab &= ba &&\text{(multiplication is commutative)}, \\
a(bc) &= (ab)c = abc &&\text{(multiplication is associative)}, \\
a \times 1 &= 1 \times a = a &&\text{(multiplication by unity)}, \\
a(b + c) &= ab + ac &&\text{(multiplication is distributive)},
\end{aligned}
$$

where a, b, c are arbitrary real numbers.

For any nonzero real number a, there exists its unique *multiplicative inverse*, or its *reciprocal*, denoted by a^{-1} or $1/a$, such that

$$aa^{-1} = 1 \quad (a \neq 0).$$

M1.2.3. Ratios and Proportions

▶ **Operations with fractions and properties of fractions.** Ratios are written as fractions: $a : b = a/b$. The number a is called the *numerator* and the number b ($b \neq 0$) is called the *denominator* of a fraction.

Properties of fractions and operations with fractions:

$$\frac{a}{1} = a, \qquad \frac{a}{b} = \frac{ab}{bc} = \frac{a : c}{b : c} \qquad \text{(simplest properties of fractions);}$$

$$\frac{a}{b} \pm \frac{c}{b} = \frac{a \pm c}{b}, \qquad \frac{a}{b} \pm \frac{c}{d} = \frac{ad \pm bc}{bd} \qquad \text{(addition and subtraction of fractions);}$$

$$\frac{a}{b} \times c = \frac{ac}{b}, \qquad \frac{a}{b} \times \frac{c}{d} = \frac{ac}{bc} \qquad \text{(multiplication by a number and by a fraction);}$$

$$\frac{a}{b} : c = \frac{a}{bc}, \qquad \frac{a}{b} : \frac{c}{d} = \frac{ad}{bc} \qquad \text{(division by a number and by a fraction).}$$

▶ **Proportions. Simplest relations. Derivative proportions.** A proportion is an equation with a ratio on each side. A proportion is denoted by $a/b = c/d$ or $a : b = c : d$.

$1°$. The following simplest relations follow from $a/b = c/d$:

$$ad = bc, \qquad \frac{a}{c} = \frac{b}{d}, \qquad a = \frac{bc}{d}, \qquad b = \frac{ad}{c}.$$

$2°$. The following derivative proportions follow from $a/b = c/d$:

$$\frac{ma + nb}{pa + qb} = \frac{mc + nd}{pc + qd},$$

$$\frac{ma + nc}{pa + qc} = \frac{mb + nd}{pb + qd},$$

where m, n, p, q are arbitrary real numbers.

Some special cases of the above formulas:

$$\frac{a \pm b}{b} = \frac{c \pm d}{d}, \qquad \frac{a - b}{a + b} = \frac{c - d}{c + d}.$$

M1.2.4. Percentage

▶ **Definition. Main percentage problems.** A *percentage* is a way of expressing a ratio or a fraction as a whole number, by using 100 as the denominator. One *percent* is one per one hundred, or one hundredth of a whole number; notation: 1%.

Below are the statements of main percentage problems and their solutions.

$1°$. Find the number b that makes up $p\%$ of a number a. Answer: $b = \frac{ap}{100}$.

$2°$. Find the number a whose $p\%$ is equal to a number b. Answer: $a = \frac{100\,b}{p}$.

$3°$. What percentage does a number b make up of a number a? Answer: $p = \frac{100\,b}{a}\%$.

▶ **Simple and compound percentage.**

$1°$. *Simple percentage.* Suppose a cash deposit is increased yearly by the same amount defined as a percentage, $p\%$, of the initial deposit, a. Then the amount accumulated after t years is calculated by the simple percentage formula

$$x = a\left(1 + \frac{pt}{100}\right).$$

$2°$. *Compound percentage.* Suppose a cash deposit is increased yearly by an amount defined as a percentage, $p\%$, of the deposit in the previous year. If a is the initial deposit, then the amount accumulated after t years is calculated by the compound percentage formula

$$x = a\left(1 + \frac{p}{100}\right)^t.$$

M1.2.5. Absolute Value of a Number (Modulus of a Number)

▶ **Definition.** The absolute value of a real number a, denoted by $|a|$, is defined by the formula

$$|a| = \begin{cases} a & \text{if } a \geq 0, \\ -a & \text{if } a < 0. \end{cases}$$

An important property: $|a| \geq 0$.

▶ **Some formulas and inequalities.**

$1°$. The following relations hold true:

$$|a| = |-a| = \sqrt{a^2}, \quad a \leq |a|,$$
$$\big||a| - |b|\big| \leq |a + b| \leq |a| + |b|,$$
$$\big||a| - |b|\big| \leq |a - b| \leq |a| + |b|,$$
$$|ab| = |a|\,|b|, \quad |a/b| = |a|/|b|.$$

$2°$. From the inequalities $|a| \leq A$ and $|b| \leq B$ it follows that $|a + b| \leq A + B$ and $|ab| \leq AB$.

M1.3. Powers and Logarithms
M1.3.1. Powers and Roots

▶ **Powers and roots: the main definitions.** Given a positive real number a and a positive integer n, the nth power of a, written as a^n, is defined as the multiplication of a by itself repeated n times:

$$a^n = \underbrace{a \times a \times a \times \cdots \times a}_{n \text{ multipliers}}.$$

The number a is called the *base* and n is called the *exponent*.

Obvious properties: $0^n = 0$, $1^n = 1$, $a^1 = a$.

Raising to the zeroth power: $a^0 = 1$, where $a \neq 0$.

Raising to a negative power: $a^{-n} = \dfrac{1}{a^n}$, where n is a positive integer.

If a is a positive real number and n is a positive integer, then the *nth arithmetic root* or *radical* of a, written as $\sqrt[n]{a}$, is the unique positive real number b such that $b^n = a$. In the case of $n = 2$, the brief notation $\sqrt{a}$ is used to denote $\sqrt[2]{a}$.

The following relations hold:

$$\sqrt[n]{0} = 0, \qquad \sqrt[n]{1} = 1, \qquad \left(\sqrt[n]{a}\right)^n = a.$$

Raising to a fractional power $p = m/n$, where m and n are natural numbers:

$$a^p = a^{m/n} = \sqrt[n]{a^m}, \qquad a \geq 0.$$

▶ **Operations with powers and roots.** The properties given below are valid for any real exponents p and q ($a > 0$, $b > 0$):

$$a^{-p} = \frac{1}{a^p}, \qquad a^p a^q = a^{p+q}, \qquad \frac{a^p}{a^q} = a^{p-q},$$

$$(ab)^p = a^p b^p, \qquad \left(\frac{a}{b}\right)^q = \frac{a^q}{b^q}, \qquad (a^p)^q = a^{pq}.$$

In operations with roots (radicals) the following properties are used:

$$\sqrt[n]{ab} = \sqrt[n]{a}\,\sqrt[n]{b}, \qquad \sqrt[n]{\frac{a}{b}} = \frac{\sqrt[n]{a}}{\sqrt[n]{b}}, \qquad \sqrt[n]{a^m} = \left(\sqrt[n]{a}\right)^m, \qquad \sqrt[n]{\sqrt[m]{a}} = \sqrt[mn]{a}.$$

Remark. It often pays to represent roots as powers with rational exponents and apply the properties of operations with powers.

M1.3.2. Logarithms

▶ **Definition. The main logarithmic identity.** The *logarithm of a positive number b to a given base a* is the exponent of the power c to which the base a must be raised to produce b. It is written as $\log_a b = c$.

Equivalent representations:

$$\log_a b = c \quad \Longleftrightarrow \quad a^c = b,$$

where $a > 0$, $a \neq 1$, and $b > 0$.

Main logarithmic identity:

$$a^{\log_a b} = b.$$

Simple properties:

$$\log_a 1 = 0, \qquad \log_a a = 1.$$

▶ **Properties of logarithms. The common and natural logarithms.** Properties of logarithms:

$$\log_a(bc) = \log_a b + \log_a c, \qquad \log_a\left(\frac{b}{c}\right) = \log_a b - \log_a c,$$

$$\log_a(b^k) = k \log_a b, \qquad \log_{a^k} b = \frac{1}{k} \log_a b \quad (k \neq 0),$$

$$\log_a b = \frac{1}{\log_b a} \quad (b \neq 1), \qquad \log_a b = \frac{\log_c b}{\log_c a} \quad (c \neq 1),$$

where $a > 0$, $a \neq 1$, $b > 0$, $c > 0$, and k is any number.

The logarithm to the base 10 is called the *common* or *decadic logarithm* and written as

$$\log_{10} b = \log b \quad \text{or sometimes} \quad \log_{10} b = \lg b.$$

The logarithm to the base e (the *base of natural logarithms*) is called the *natural logarithm* and written as

$$\log_e b = \ln b,$$

where $e = \lim_{n \to \infty} \left(1 + \frac{1}{n}\right)^n = 2.718281\ldots$

The following relations hold:

$$\ln b \approx 2.30259 \lg b, \qquad \lg b \approx 0.43429 \ln b$$

M1.4. Binomial Theorem and Related Formulas

M1.4.1. Factorials. Binomial Coefficients. Binomial Theorem

▶ **Factorials. Binomial coefficients.**
Factorial:

$$0! = 1! = 1,$$
$$n! = 1 \times 2 \times 3 \times \cdots \times (n-1) \times n, \quad n = 2, \ 3, \ 4, \ \ldots$$

Double factorial:

$$0!! = 1!! = 1,$$
$$n!! = \begin{cases} (2k)!! & \text{if } n = 2k, \\ (2k+1)!! & \text{if } n = 2k+1, \end{cases}$$
$$(2k)!! = 2 \times 4 \times 6 \times \cdots \times (2k-2) \times (2k) = 2^k k!,$$
$$(2k+1)!! = 1 \times 3 \times 5 \times \cdots \times (2k-1) \times (2k+1),$$

where n and k are natural numbers.
Binomial coefficients:

$$C_n^k = \binom{n}{k} = \frac{n!}{k!\,(n-k)!} = \frac{n(n-1)\ldots(n-k+1)}{k!}, \quad k = 1, \ 2, \ 3, \ \ldots, \ n;$$
$$C_a^k = \frac{a(a-1)\ldots(a-k+1)}{k!}, \quad \text{where} \quad k = 1, \ 2, \ 3, \ \ldots,$$

where n is a natural number and a is any number.

▶ **Binomial theorem.** Let a, b, and c be real (or complex) numbers. The following formulas hold true:

$$(a \pm b)^2 = a^2 \pm 2ab + b^2,$$
$$(a \pm b)^3 = a^3 \pm 3a^2b + 3ab^2 \pm b^3,$$
$$(a \pm b)^4 = a^4 \pm 4a^3b + 6a^2b^2 \pm 4ab^3 + b^4,$$
$$\ldots\ldots\ldots\ldots\ldots\ldots\ldots\ldots\ldots\ldots\ldots\ldots$$
$$(a + b)^n = \sum_{k=0}^{n} C_n^k a^{n-k} b^k, \quad n = 1, \ 2, \ \ldots$$

The last formula is known as the *binomial theorem*, where the C_n^k are binomial coefficients.

M1.4.2. Related Formulas

▶ **Formulas involving powers ≤ 4.**

$$a^2 - b^2 = (a-b)(a+b),$$
$$a^3 + b^3 = (a+b)(a^2 - ab + b^2),$$
$$a^3 - b^3 = (a-b)(a^2 + ab + b^2),$$
$$a^4 - b^4 = (a-b)(a+b)(a^2 + b^2),$$
$$(a+b+c)^2 = a^2 + b^2 + c^2 + 2ab + 2ac + 2bc,$$
$$a^4 + a^2b^2 + b^4 = (a^2 + ab + b^2)(a^2 - ab + b^2).$$

▶ **Formulas involving arbitrary powers.** Let n be any positive integer. Then

$$a^n - b^n = (a - b)(a^{n-1} + a^{n-2}b + \cdots + ab^{n-2} + b^{n-1}).$$

If n is a positive even number, then

$$a^n - b^n = (a + b)(a^{n-1} - a^{n-2}b + \cdots + ab^{n-2} - b^{n-1})$$
$$= (a - b)(a + b)(a^{n-2} + a^{n-4}b^2 + \cdots + a^2b^{n-4} + b^{n-2}).$$

If n is a positive odd number, then

$$a^n + b^n = (a + b)(a^{n-1} - a^{n-2}b + \cdots - ab^{n-2} + b^{n-1}).$$

M1.5. Progressions

M1.5.1. Arithmetic Progression

$1°$. An *arithmetic progression*, or *arithmetic sequence*, is a sequence of real numbers for which each term, starting from the second, is the previous term plus a constant d, called the *common difference*, so that $a_{n+1} = a_n + d$, $n = 1, 2, 3, \ldots$ In general, the terms of an arithmetic progression are expressed as

$$a_n = a_1 + (n - 1)d, \qquad n = 1,\ 2,\ 3,\ \ldots,$$

where a_1 is the first term of the progression. An arithmetic progression is called *increasing* if $d > 0$ and *decreasing* if $d < 0$.

$2°$. An arithmetic progression has the property

$$a_n = \tfrac{1}{2}(a_{n-1} + a_{n+1}).$$

$3°$. The sum of n first terms of an arithmetic progression is calculated as

$$S_n = a_1 + \cdots + a_n = \tfrac{1}{2}(a_1 + a_n)n = \tfrac{1}{2}[2a_1 + (n-1)d]n.$$

M1.5.2. Geometric Progression

$1°$. A *geometric progression*, or *geometric sequence*, is a sequence of real numbers for which each term, starting from the second, is the previous term multiplied by a constant q, called the *common ratio*, so that $a_{n+1} = a_n q$, $n = 1, 2, 3, \ldots$ In general, the terms of a geometric progression are expressed as

$$a_n = a_1 q^{n-1}, \qquad n = 1,\ 2,\ 3,\ \ldots,$$

where a_1 is the first term of the progression.

$2°$. A geometric progression with positive terms has the property

$$a_n = \sqrt{a_{n-1} a_{n+1}}.$$

$3°$. The sum of n first terms of a geometric progression is calculated as ($q \neq 1$)

$$S_n = a_1 + \cdots + a_n = a_1 \frac{1 - q^n}{1 - q}.$$

M1.6. Mean Values and Some Inequalities

M1.6.1. Arithmetic Mean, Geometric Mean, and Other Mean Values

The *arithmetic mean* of a set of n real numbers $a_1, a_2, \ldots, a_n$ is defined as

$$m_{\mathrm{a}} = \frac{a_1 + a_2 + \cdots + a_n}{n}. \tag{1.6.1.1}$$

Geometric mean of n positive numbers $a_1, a_2, \ldots, a_n$:

$$m_{\mathrm{g}} = (a_1 a_2 \ldots a_n)^{1/n}. \tag{1.6.1.2}$$

Harmonic mean of n real numbers $a_1, a_2, \ldots, a_n$:

$$m_{\mathrm{h}} = \frac{n}{(1/a_1) + (1/a_2) + \cdots + (1/a_n)}, \qquad a_k \neq 0. \tag{1.6.1.3}$$

Quadratic mean (or *root mean square*) of n real numbers $a_1, a_2, \ldots, a_n$:

$$m_{\mathrm{q}} = \sqrt{\frac{a_1^2 + a_2^2 + \cdots + a_n^2}{n}}. \tag{1.6.1.4}$$

M1.6.2. Inequalities for Mean Values

Given n positive numbers $a_1, a_2, \ldots, a_n$, the following inequalities hold true:

$$m_{\mathrm{h}} \leq m_{\mathrm{g}} \leq m_{\mathrm{a}} \leq m_{\mathrm{q}}, \tag{1.6.1.5}$$

where the mean values are defined above by (1.6.1.1)–(1.6.1.4). The equalities in (1.6.1.5) are attained only if $a_1 = a_2 = \cdots = a_n$.

To make it easier to remember, let us rewrite inequalities (1.6.1.5) in words as

$$\boxed{\text{harmonic mean}} \leq \boxed{\text{geometric mean}} \leq \boxed{\text{arithmetic mean}} \leq \boxed{\text{quadratic mean}}.$$

M1.6.3. Some Inequalities of General Form

Let a_k and b_k be real numbers with $k = 1, 2, \ldots, n$.

Generalized triangle inequality:

$$\left| \sum_{k=1}^{n} a_k \right| \leq \sum_{k=1}^{n} |a_k|.$$

Cauchy's inequality (also known as the *Cauchy–Bunyakovsky inequality* or *Cauchy–Schwarz–Bunyakovsky inequality*):

$$\left(\sum_{k=1}^{n} a_k b_k \right)^2 \leq \left(\sum_{k=1}^{n} a_k^2 \right) \left(\sum_{k=1}^{n} b_k^2 \right).$$

Minkowski's inequality:

$$\left(\sum_{k=1}^{n} |a_k + b_k|^p \right)^{\frac{1}{p}} \leq \left(\sum_{k=1}^{n} |a_k|^p \right)^{\frac{1}{p}} + \left(\sum_{k=1}^{n} |b_k|^p \right)^{\frac{1}{p}}, \qquad p \geq 1.$$

M1.7. Some Mathematical Methods

M1.7.1. Proof by Contradiction

Proof by contradiction (also known as *reductio ad absurdum*) is an indirect method of mathematical proof. It is based on the law of non-contradiction (a statement cannot be true and false at the same time) and includes the following reasoning:

    1. Suppose one has to prove some statement S.

    2. One assumes that the opposite of S is true.

    3. Based on known axioms, definitions, theorems, formulas, and the assumption of Item 2, one arrives at a contradiction (deduces some obviously false statement).

    4. One concludes that the assumption of Item 2 is false and hence the original statement S is true, which was to be proved.

    Example. (Euclid's proof of the irrationality of the square root of 2 by contradiction.)

    1. It is required to prove that $\sqrt{2}$ is an irrational number, that is, a real number that cannot be represented as a fraction p/q, where p and q are both integers.

    2. Assume the opposite: $\sqrt{2}$ is a rational number. This means that $\sqrt{2}$ can be represented as a fraction

$$\sqrt{2} = p/q. \tag{1.7.1.1}$$

Without loss of generality the fraction p/q is assumed to be irreducible, implying that p and q are mutually prime (have no common factor other than 1).

    3. Square both sides of (1.7.1.1) and then multiply by q^2 to obtain

$$2q^2 = p^2. \tag{1.7.1.2}$$

The left-hand side is divisible by 2. Then the right-hand side, p^2, and hence p is also divisible by 2. Consequently, p is an even number so that

$$p = 2n, \tag{1.7.1.3}$$

where n is an integer. Substituting (1.7.1.3) into (1.7.1.2) and then dividing by 2 yields

$$q^2 = 2p^2. \tag{1.7.1.4}$$

Now it can be concluded, just as above, that q^2 and hence q must be divisible by 2. Consequently, q is an even number so that

$$q = 2m, \tag{1.7.1.5}$$

where m is an integer.

    It is now apparent from (1.7.1.3) and (1.7.1.5) that the fraction p/q is not simple, since p and q have a common factor 2. This contradicts the assumption made in Item 2.

    4. It follows from the results of Item 3 that the representation of $\sqrt{2}$ in the form of a fraction (1.7.1.1) is false, which means that $\sqrt{2}$ is irrational.

M1.7.2. Mathematical Induction

The method of proof by (complete) *mathematical induction* is based on the following reasoning:

    1. Let $A(n)$ be a statement dependent on n with $n = 1, 2, \ldots$ (A is a hypothesis at this stage).

    2. Base case. Suppose the initial statement $A(1)$ is true. This is usually established by direct substitution $n = 1$.

    3. Induction step. Assume that $A(n)$ is true for any n and then, based on this assumption, prove that $A(n + 1)$ is also true.

    4. Principle of mathematical induction. From the results of Items 2–3 it is concluded that the statement $A(n)$ is true for any n.

Example.

1. Prove the formula for the sum of odd numbers

$$1 + 3 + 5 + \cdots + (2n - 1) = n^2 \tag{1.7.2.1}$$

for any natural n.

2. For $n = 1$, we have an obvious identity: $1 = 1$.

3. Let us assume that formula (1.7.2.1) holds for any n. To consider the case of $n + 1$, let us add the next term, $(2n + 1)$, to both sides of (1.7.2.1) to obtain

$$1 + 3 + 5 + \cdots + (2n - 1) + (2n + 1) = n^2 + (2n + 1) = (n + 1)^2.$$

Thus, from the assumption of the validity of formula (1.7.2.1) for any n it follows that (1.7.2.1) is also valid for $n + 1$.

4. According to the principle of mathematical induction, this proves formula (1.7.2.1).

Remark. The first step, the formulation of an original hypothesis, is the most difficult part of the method of mathematical induction. This step is often omitted from the method.

M1.7.3. Proof by Counterexample

A *counterexample* is an example which is used to prove that a statement (proposition) is false. Counterexamples play an important role in mathematics. Whereas a complicated proof may be the only way to demonstrate the validity of a particular theorem, a single counterexample is all that is needed to refute the validity of a proposed theorem.

In general, the scheme of a proof by counterexample is as follows:

1. Given a proposition: all elements a that belong to a set A also belong to a set (possess a property) B.

2. Refutation of the proposition: one specifies an element a_* (counterexample) that belongs to A but does not belong to B.

Example. *Proposition:* Numbers in the form $2^{2^n} + 1$, where n is a positive integer, were once thought to be prime.

These numbers are prime for $n = 1, 2, 3, 4$. But for $n = 5$, we have a counterexample, since

$$2^{2^5} + 1 = 4294967297 = 641 \times 6700417;$$

it is a composite number.

Conclusion: When faced with a number in the form $2^{2^n} + 1$, we are not allowed to assume it is either prime or composite, unless we know for sure for some other reason.

Bibliography for Chapter M1

Bronshtein, I. N. and Semendyayev, K. A., *Handbook of Mathematics, 4th Edition*, Springer-Verlag, Berlin, 2004.

Courant, R. and Robbins, H., *What Is Mathematics?: An Elementary Approach to Ideas and Methods, 2nd Edition*, Oxford University Press, Oxford, 1996.

Franklin, J. and Daoud, A., *Introduction to Proofs in Mathematics*, Prentice Hall, New York, 1988.

Garnier, R. and Taylor, J., *100% Mathematical Proof*, John Wiley & Sons, New York, 1996.

Gelbaum, B. R. and Olmsted, J. M. H., *Theorems and Counterexamples in Mathematics*, Springer-Verlag, New York, 1990.

Jordan, B. E. and Palow, W. P., *Integrated Arithmetic & Algebra*, Addison-Wesley, Boston, 1999.

Krantz, S. G., *Dictionary of Algebra, Arithmetic, and Trigonometry*, CRC Press, Boca Raton, Florida, 2001.

Pólya, G., *Mathematics and Plausible Reasoning, Vol. 1: Induction and Analogy in Mathematics*, Princeton University Press, Princeton, 1990.

Rossi, R. J., *Theorems, Corollaries, Lemmas, and Methods of Proof*, Wiley-Interscience, Hoboken, N.J., 2006.

Thompson, J. E., *Arithmetic for the Practical Man*, Van Nostrand Reinhold, New York, 1973.

Weisstein, E. W., *CRC Concise Encyclopedia of Mathematics, 2nd Edition*, CRC Press, Boca Raton, Florida, 2003.

Zwillinger, D., *CRC Standard Mathematical Tables and Formulae, 31st Edition*, CRC Press, Boca Raton, Florida, 2002.

Chapter M2
Elementary Functions

Basic elementary functions: power, exponential, logarithmic, trigonometric, and inverse trigonometric (arc-trigonometric or antitrigonometric) functions. All other elementary functions are obtained from the basic elementary functions and constants by means of the four arithmetic operations (addition, subtraction, multiplication, and division) and the operation of composition (composite functions).

The graphs and the main properties of the basic as well as some other frequently occurring elementary functions of the real variable are described below.

M2.1. Power, Exponential, and Logarithmic Functions

M2.1.1. Power Function: $y = x^\alpha$ (α is an Arbitrary Real Number)

▶ **Graphs of the power function.** General properties of the graphs: the point $(1, 1)$ belongs to all the graphs, and $y > 0$ for $x > 0$. For $\alpha > 0$, the graphs pass through the origin $(0, 0)$; for $\alpha < 0$, the graphs have the vertical asymptote $x = 0$ ($y \to +\infty$ as $x \to 0$). For $\alpha = 0$, the graph is a straight line parallel to the x-axis.

Consider more closely the following cases.

Case 1: $y = x^{2n}$, where n is a positive integer ($n = 1, 2, \dots$). This function is defined for all real x and its range consists of all $y \geq 0$. This function is even, nonperiodic, and unbounded. It crosses the axis Oy and is tangential to the axis Ox at the origin $x = 0$, $y = 0$. On the interval $(-\infty, 0)$ this function decreases, and it increases on the interval $(0, +\infty)$. It attains its minimum value $y = 0$ at $x = 0$. The graph of the function $y = x^2$ (parabola) is given in Fig. M2.1 *a*.

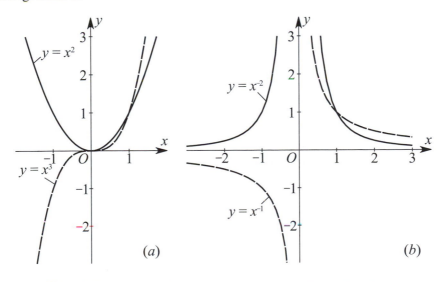

Figure M2.1. Graphs of the power function $y = x^n$, where n is an integer.

15

Case 2: $y = x^{2n+1}$, where n is a positive integer. This function is defined on the entire x-axis and its range coincides with the y-axis. This function is odd, nonperiodic, and unbounded. It crosses the x-axis and the y-axis at the origin $x = 0$, $y = 0$. It is an increasing function on the entire axis with no points of extremum, the origin being its inflection point. The graph of the function $y = x^3$ (cubic parabola) is shown in Fig. M2.1 *a*.

Case 3: $y = x^{-2n}$, where n is a positive integer. This function is defined for all $x \neq 0$, and its range is the semiaxis $y > 0$. It is an even, nonperiodic, unbounded function having no intersection with the coordinate axes. It increases on the interval $(-\infty, 0)$, decreases on the interval $(0, +\infty)$, and has no points of extremum. The graph of the function has a vertical asymptote $x = 0$. The graph of the function $y = x^{-2}$ is given in Fig. M2.1 *b*.

Case 4: $y = x^{-2n+1}$, where n is a positive integer. This function is defined for all $x \neq 0$, and its range is the entire y-axis. It is an odd, nonperiodic, unbounded function with no intersections with the coordinate axes. This is a decreasing function on the entire axis with no points of extremum. It has a vertical asymptote $x = 0$. The graph of the function $y = x^{-1}$ is given in Fig. M2.1 *b*.

Case 5: $y = x^{\alpha}$ with a noninteger $\alpha > 0$. This function is defined for all* $x \geq 0$ and its range is the semiaxis $y \geq 0$. This function is neither odd nor even and it is nonperiodic and unbounded. It crosses the axes Ox and Oy at the origin $x = 0$, $y = 0$ and increases everywhere in its domain, taking its smallest value at the point $x = 0$, $y = 0$. The graph of the function $y = x^{1/2}$ is given in Fig. M2.2.

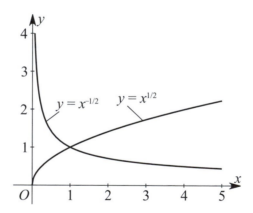

Figure M2.2. Graphs of the power function $y = x^{\alpha}$, where α is a noninteger.

Case 6: $y = x^{\alpha}$ with a noninteger $\alpha < 0$. This function is defined for all $x > 0$ and its range is the semiaxis $y > 0$. This function is neither odd nor even, it is nonperiodic and unbounded, and it has no intersections with the coordinate axes, which coincide with its horizontal and vertical asymptotes. This function is decreasing on its entire domain and has no points of extremum. The graph of the function $y = x^{-1/2}$ is given in Fig. M2.2.

▶ **Properties of the power function.** Basic properties of the power function:

$$x^{\alpha} x^{\beta} = x^{\alpha+\beta}, \quad (x_1 x_2)^{\alpha} = x_1^{\alpha} x_2^{\alpha}, \quad (x^{\alpha})^{\beta} = x^{\alpha\beta},$$

for any α and β, where $x > 0$, $x_1 > 0$, $x_2 > 0$.

---

* In fact, the power function $y = x^{1/n}$ with an odd integer n is also defined for all $x < 0$. Here, however, it is always assumed that $x \geq 0$. A similar assumption is made with regard to the functions of the form $y = x^{m/n}$, where m is a positive integer and m/n is an irreducible fraction.

Differentiation and integration formulas:

$$(x^\alpha)' = \alpha x^{\alpha-1}, \qquad \int x^\alpha \, dx = \begin{cases} \dfrac{x^{\alpha+1}}{\alpha+1} + C & \text{if } \alpha \neq -1, \\ \ln|x| + C & \text{if } \alpha = -1. \end{cases}$$

The Taylor series expansion in a neighborhood of an arbitrary point x_0:

$$x^\alpha = \sum_{n=0}^{\infty} C_\alpha^n x_0^{\alpha-n}(x-x_0)^n \quad \text{for} \quad |x-x_0| < |x_0|,$$

where $C_\alpha^n = \dfrac{\alpha(\alpha-1)\ldots(\alpha-n+1)}{n!}$ are binomial coefficients.

M2.1.2. Exponential Function: $y = a^x$ $(a > 0, \ a \neq 1)$

▶ **Graphs of the exponential function.** This function is defined for all x and its range is the semiaxis $y > 0$. This function is neither odd nor even, it is nonperiodic and unbounded, and it crosses the axis Oy at $y = 1$ and does not cross the axis Ox. For $a > 1$, it is an increasing function on the entire x-axis; for $0 < a < 1$, it is a decreasing function. This function has no extremal points; the axis Ox is its horizontal asymptote. The graphs of these functions have the following common property: they pass through the point $(0, 1)$. The graph of $y = a^x$ is symmetrical to the graph of $y = (1/a)^x$ with respect to the y-axis. For $a > 1$, the function a^x grows faster than any power of x as $x \to +\infty$, and it decays faster than any power of $1/x$ as $x \to -\infty$. The graphs of the functions $y = 2^x$ and $y = (1/2)^x$ are given in Fig. M2.3.

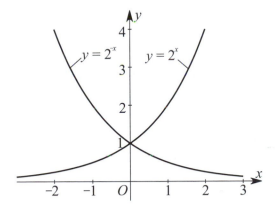

Figure M2.3. Graphs of the exponential function.

▶ **Properties of the exponential function.** Basic properties of the exponential function:

$$a^{x_1} a^{x_2} = a^{x_1+x_2}, \qquad a^x b^x = (ab)^x, \qquad (a^{x_1})^{x_2} = a^{x_1 x_2}.$$

Number e, *base of natural (Napierian) logarithms*, and the function e^x:

$$e = \lim_{n\to\infty}\left(1 + \frac{1}{n}\right)^n = 2.718281\ldots, \qquad e^x = \lim_{n\to\infty}\left(1 + \frac{x}{n}\right)^n.$$

The formula for passing from an arbitrary base a to the base e of natural logarithms:

$$a^x = e^{x \ln a}.$$

The inequality

$$a^{x_1} > a^{x_2} \quad \Longleftrightarrow \quad \begin{cases} x_1 > x_2 & \text{if } a > 1, \\ x_1 < x_2 & \text{if } 0 < a < 1. \end{cases}$$

The limit relations for any $a > 1$ and $b > 0$:

$$\lim_{x \to +\infty} \frac{a^x}{|x|^b} = \infty, \qquad \lim_{x \to -\infty} a^x |x|^b = 0.$$

Differentiation and integration formulas:

$$(e^x)' = e^x, \qquad \int e^x \, dx = e^x + C;$$

$$(a^x)' = a^x \ln a, \qquad \int a^x \, dx = \frac{a^x}{\ln a} + C.$$

Power series expansion:

$$e^x = 1 + \frac{x}{1!} + \frac{x^2}{2!} + \frac{x^3}{3!} + \cdots + \frac{x^n}{n!} + \cdots = \sum_{k=0}^{\infty} \frac{x^k}{k!}.$$

M2.1.3. Logarithmic Function: $y = \log_a x$ $(a > 0, a \neq 1)$

▶ **Graphs of the logarithmic function.** This function is defined for all $x > 0$ and its range is the entire y-axis. The function is neither odd nor even; it is nonperiodic and unbounded; it crosses the axis Ox at $x = 1$ and does not cross the axis Oy. For $a > 1$, this function is increasing, and for $0 < a < 1$, it is a decreasing function; it has no extremal points, and the axis Oy is its vertical asymptote. The common property of the graphs of such functions is that they all pass through the point $(1, 0)$. The graph of the function $y = \log_a x$ is symmetric to that of $y = \log_{1/a} x$ with respect to the x-axis. The modulus of the logarithmic function tends to infinity slower than any power of x as $x \to +\infty$ and slower than any power of $1/x$ as $x \to +0$. The graphs of the functions $y = \log_2 x$ and $y = \log_{1/2} x$ are shown in Fig. M2.4.

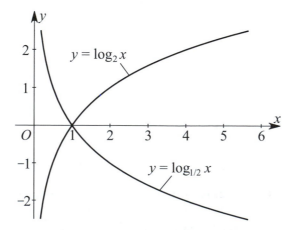

Figure M2.4. Graphs of the logarithmic function.

▶ **Properties of the logarithmic function.** By definition, the logarithmic function is the inverse of the exponential function. The following equivalence relation holds:

$$y = \log_a x \quad \Longleftrightarrow \quad x = a^y,$$

where $a > 0$, $a \neq 1$.

Basic properties of the logarithmic function:

$$a^{\log_a x} = x, \qquad\qquad \log_a(x_1 x_2) = \log_a x_1 + \log_a x_2,$$

$$\log_a(x^k) = k \log_a x, \qquad \log_a x = \frac{\log_b x}{\log_b a},$$

where $x > 0$, $x_1 > 0$, $x_2 > 0$, $a > 0$, $a \neq 1$, $b > 0$, $b \neq 1$.

The simplest inequality:

$$\log_a x_1 > \log_a x_2 \quad \Longleftrightarrow \quad \begin{cases} x_1 > x_2 & \text{if } a > 1, \\ x_1 < x_2 & \text{if } 0 < a < 1. \end{cases}$$

For any $b > 0$, the following limit relations hold:

$$\lim_{x \to +\infty} \frac{\log_a x}{x^b} = 0, \qquad \lim_{x \to +0} x^b \log_a x = 0.$$

The logarithmic function with the base e (*base of natural logarithms, Napierian base*) is denoted by

$$\log_e x = \ln x,$$

where $e = \lim_{n \to \infty} \left(1 + \frac{1}{n} \right)^n = 2.718281\ldots$

Formulas for passing from an arbitrary base a to the Napierian base e:

$$\log_a x = \frac{\ln x}{\ln a}.$$

Differentiation and integration formulas:

$$(\ln x)' = \frac{1}{x}, \qquad \int \ln x \, dx = x \ln x - x + C.$$

Power series expansion:

$$\ln(1 + x) = x - \frac{x^2}{2} + \frac{x^3}{3} - \cdots + (-1)^{n-1} \frac{x^n}{n} + \cdots = \sum_{k=1}^{\infty} (-1)^{k-1} \frac{x^k}{k}, \quad -1 < x \leq 1.$$

M2.2. Trigonometric Functions

M2.2.1. Trigonometric Circle. Definition of Trigonometric Functions

▶ **Trigonometric circle. Degrees and radians.** *Trigonometric circle* is the circle of unit radius with center at the origin of an orthogonal coordinate system Oxy. The coordinate axes divide the circle into four quarters (*quadrants*); see Fig. M2.5. Consider rotation of the polar radius issuing from the origin O and ending at a point M of the trigonometric circle.

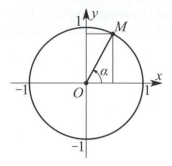

Figure M2.5. Trigonometric circle.

Let α be the angle between the x-axis and the polar radius OM measured from the positive direction of the x-axis. This angle is assumed positive in the case of counterclockwise rotation and negative in the case of clockwise rotation.

Angles are measured either in radians or in degrees. One radian is the angle at the vertex of the sector of the trigonometric circle supported by its arc of unit length. One degree is the angle at the vertex of the sector of the trigonometric circle supported by its arc of length $\pi/180$. The radians are related to the degrees by the formulas

$$1 \text{ radian} = \frac{180°}{\pi}; \qquad 1° = \frac{\pi}{180}.$$

▶ **Definition of trigonometric functions.** The *sine* of α is the ordinate (the projection to the axis Oy) of the point on the trigonometric circle corresponding to the angle of α radians. The *cosine* of α is the abscissa (projection to the axis Ox) of that point (see Fig. M2.5). The sine and the cosine are basic trigonometric functions and are denoted, respectively, by $\sin \alpha$ and $\cos \alpha$.

Other trigonometric functions are *tangent, cotangent, secant,* and *cosecant.* These are derived from the basic trigonometric functions, sine and cosine, as follows:

$$\tan \alpha = \frac{\sin \alpha}{\cos \alpha}, \qquad \cot \alpha = \frac{\cos \alpha}{\sin \alpha}, \qquad \sec \alpha = \frac{1}{\cos \alpha}, \qquad \operatorname{cosec} \alpha = \frac{1}{\sin \alpha}.$$

Table M2.1 gives the signs of the trigonometric functions in different quadrants. The signs and the values of $\sin \alpha$ and $\cos \alpha$ do not change if the argument α is incremented by $\pm 2\pi n$, where $n = 1, 2, \ldots$ The signs and the values of $\tan \alpha$ and $\cot \alpha$ do not change if the argument α is incremented by $\pm \pi n$, where $n = 1, 2, \ldots$

TABLE M2.1
Signs of trigonometric functions in different quarters.

Quarter	Angle in radians	$\sin \alpha$	$\cos \alpha$	$\tan \alpha$	$\cot \alpha$	$\sec \alpha$	$\operatorname{cosec} \alpha$
I	$0 < \alpha < \frac{\pi}{2}$	+	+	+	+	+	+
II	$\frac{\pi}{2} < \alpha < \pi$	+	−	−	−	−	+
III	$\pi < \alpha < \frac{3\pi}{2}$	−	−	+	+	−	−
IV	$\frac{3\pi}{2} < \alpha < 2\pi$	−	+	−	−	+	−

TABLE M2.2
Numerical values of trigonometric functions for some angles α (in radians).

Angle α	0	$\frac{\pi}{6}$	$\frac{\pi}{4}$	$\frac{\pi}{3}$	$\frac{\pi}{2}$	$\frac{2\pi}{3}$	$\frac{3\pi}{4}$	$\frac{5\pi}{6}$	π
$\sin\alpha$	0	$\frac{1}{2}$	$\frac{\sqrt{2}}{2}$	$\frac{\sqrt{3}}{2}$	1	$\frac{\sqrt{3}}{2}$	$\frac{\sqrt{2}}{2}$	$\frac{1}{2}$	0
$\cos\alpha$	1	$\frac{\sqrt{3}}{2}$	$\frac{\sqrt{2}}{2}$	$\frac{1}{2}$	0	$-\frac{1}{2}$	$-\frac{\sqrt{2}}{2}$	$-\frac{\sqrt{3}}{2}$	-1
$\tan\alpha$	0	$\frac{\sqrt{3}}{3}$	1	$\sqrt{3}$	∞	$-\sqrt{3}$	-1	$-\frac{\sqrt{3}}{3}$	0
$\cot\alpha$	∞	$\sqrt{3}$	1	$\frac{\sqrt{3}}{3}$	0	$-\frac{\sqrt{3}}{3}$	-1	$-\sqrt{3}$	∞

Table M2.2 gives the values of trigonometric functions for some values of their argument (the symbol ∞ means that the function is undefined for the corresponding value of its argument).

M2.2.2. Graphs of Trigonometric Functions

▶ **Sine: $y = \sin x$.** This function is defined for all x and its range is $y \in [-1, 1]$. The sine is an odd, bounded, periodic function (with period 2π). It crosses the axis Oy at the point $y = 0$ and crosses the axis Ox at the points $x = \pi n$, $n = 0, \pm1, \pm2, \ldots$ The sine is an increasing function on every segment $[-\frac{\pi}{2} + 2\pi n, \frac{\pi}{2} + 2\pi n]$ and is a decreasing function on every segment $[\frac{\pi}{2} + 2\pi n, \frac{3}{2}\pi + 2\pi n]$. For $x = \frac{\pi}{2} + 2\pi n$, it attains its maximal value ($y = 1$), and for $x = -\frac{\pi}{2} + 2\pi n$ it attains its minimal value ($y = -1$). The graph of the function $y = \sin x$ is called the *sinusoid* or *sine curve* and is shown in Fig. M2.6.

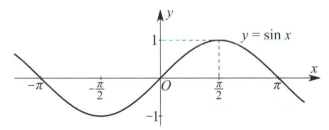

Figure M2.6. Graph of the function $y = \sin x$.

▶ **Cosine: $y = \cos x$.** This function is defined for all x and its range is $y \in [-1, 1]$. The cosine is a bounded, even, periodic function (with period 2π). It crosses the axis Oy at the point $y = 1$, and crosses the axis Ox at the points $x = \frac{\pi}{2} + \pi n$. The cosine is an increasing function on every segment $[-\pi + 2\pi n, 2\pi n]$ and is a decreasing function on every segment $[2\pi n, \pi + 2\pi n]$, $n = 0, \pm1, \pm2, \ldots$ For $x = 2\pi n$ it attains its maximal value ($y = 1$), and for $x = \pi + 2\pi n$ it attains its minimal value ($y = -1$). The graph of the function $y = \cos x$ is a sinusoid obtained by shifting the graph of the function $y = \sin x$ by $\frac{\pi}{2}$ to the left along the axis Ox (see Fig. M2.7).

▶ **Tangent: $y = \tan x$.** This function is defined for all $x \neq \frac{\pi}{2} + \pi n$, $n = 0, \pm1, \pm2, \ldots$, and its range is the entire y-axis. The tangent is an unbounded, odd, periodic function (with period π). It crosses the axis Oy at the point $y = 0$ and crosses the axis Ox at the points $x = \pi n$. This is an increasing function on every interval $(-\frac{\pi}{2} + \pi n, \frac{\pi}{2} + \pi n)$. This function has no points of extremum and has vertical asymptotes at $x = \frac{\pi}{2} + \pi n$, $n = 0, \pm1, \pm2, \ldots$ The graph of the function $y = \tan x$ is given in Fig. M2.8.

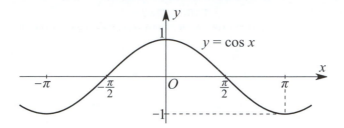

Figure M2.7. Graph of the function $y = \cos x$.

▶ **Cotangent:** $y = \cot x$. This function is defined for all $x \neq \pi n$, $n = 0, \pm 1, \pm 2, \dots$, and its range is the entire y-axis. The cotangent is an unbounded, odd, periodic function (with period π). It crosses the axis Ox at the points $x = \frac{\pi}{2} + \pi n$, and does not cross the axis Oy. This is a decreasing function on every interval $(\pi n, \pi + \pi n)$. This function has no extremal points and has vertical asymptotes at $x = \pi n$, $n = 0, \pm 1, \pm 2, \dots$ The graph of the function $y = \cot x$ is given in Fig. M2.9.

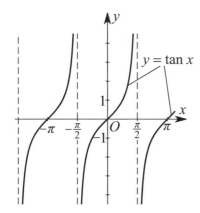

Figure M2.8. Graph of the function $y = \tan x$.   **Figure M2.9.** Graph of the function $y = \cot x$.

M2.2.3. Properties of Trigonometric Functions

▶ **Simplest relations.**

$$\sin^2 x + \cos^2 x = 1, \qquad \tan x \cot x = 1,$$
$$\sin(-x) = -\sin x, \qquad \cos(-x) = \cos x,$$
$$\tan x = \frac{\sin x}{\cos x}, \qquad \cot x = \frac{\cos x}{\sin x},$$
$$\tan(-x) = -\tan x, \qquad \cot(-x) = -\cot x,$$
$$1 + \tan^2 x = \frac{1}{\cos^2 x}, \qquad 1 + \cot^2 x = \frac{1}{\sin^2 x}.$$

▶ **Reduction formulas.**

$$\sin(x \pm 2n\pi) = \sin x, \qquad \cos(x \pm 2n\pi) = \cos x,$$
$$\sin(x \pm n\pi) = (-1)^n \sin x, \qquad \cos(x \pm n\pi) = (-1)^n \cos x,$$
$$\sin\left(x \pm \frac{2n+1}{2}\pi\right) = \pm(-1)^n \cos x, \qquad \cos\left(x \pm \frac{2n+1}{2}\pi\right) = \mp(-1)^n \sin x,$$

$$\sin\left(x \pm \frac{\pi}{4}\right) = \frac{\sqrt{2}}{2}(\sin x \pm \cos x), \qquad \cos\left(x \pm \frac{\pi}{4}\right) = \frac{\sqrt{2}}{2}(\cos x \mp \sin x),$$

$$\tan(x \pm n\pi) = \tan x, \qquad\qquad\qquad \cot(x \pm n\pi) = \cot x,$$

$$\tan\left(x \pm \frac{2n+1}{2}\pi\right) = -\cot x, \qquad \cot\left(x \pm \frac{2n+1}{2}\pi\right) = -\tan x,$$

$$\tan\left(x \pm \frac{\pi}{4}\right) = \frac{\tan x \pm 1}{1 \mp \tan x}, \qquad\qquad \cot\left(x \pm \frac{\pi}{4}\right) = \frac{\cot x \mp 1}{1 \pm \cot x},$$

where $n = 1, 2, \ldots$

▶ **Relations between trigonometric functions of single argument.**

$$\sin x = \pm\sqrt{1 - \cos^2 x} = \pm\frac{\tan x}{\sqrt{1 + \tan^2 x}} = \pm\frac{1}{\sqrt{1 + \cot^2 x}},$$

$$\cos x = \pm\sqrt{1 - \sin^2 x} = \pm\frac{1}{\sqrt{1 + \tan^2 x}} = \pm\frac{\cot x}{\sqrt{1 + \cot^2 x}},$$

$$\tan x = \pm\frac{\sin x}{\sqrt{1 - \sin^2 x}} = \pm\frac{\sqrt{1 - \cos^2 x}}{\cos x} = \frac{1}{\cot x},$$

$$\cot x = \pm\frac{\sqrt{1 - \sin^2 x}}{\sin x} = \pm\frac{\cos x}{\sqrt{1 - \cos^2 x}} = \frac{1}{\tan x}.$$

The sign before the radical is determined by the quarter in which the argument takes its values.

▶ **Addition and subtraction of trigonometric functions.**

$$\sin x + \sin y = 2\sin\left(\frac{x+y}{2}\right)\cos\left(\frac{x-y}{2}\right),$$

$$\sin x - \sin y = 2\sin\left(\frac{x-y}{2}\right)\cos\left(\frac{x+y}{2}\right),$$

$$\cos x + \cos y = 2\cos\left(\frac{x+y}{2}\right)\cos\left(\frac{x-y}{2}\right),$$

$$\cos x - \cos y = -2\sin\left(\frac{x+y}{2}\right)\sin\left(\frac{x-y}{2}\right),$$

$$\sin^2 x - \sin^2 y = \cos^2 y - \cos^2 x = \sin(x+y)\sin(x-y),$$

$$\sin^2 x - \cos^2 y = -\cos(x+y)\cos(x-y),$$

$$\tan x \pm \tan y = \frac{\sin(x \pm y)}{\cos x \cos y}, \qquad \cot x \pm \cot y = \frac{\sin(y \pm x)}{\sin x \sin y},$$

$$a\cos x + b\sin x = r\sin(x + \varphi) = r\cos(x - \psi).$$

Here, $r = \sqrt{a^2 + b^2}$, $\sin\varphi = a/r$, $\cos\varphi = b/r$, $\sin\psi = b/r$, and $\cos\psi = a/r$.

▶ **Products of trigonometric functions.**

$$\sin x \sin y = \tfrac{1}{2}[\cos(x-y) - \cos(x+y)],$$

$$\cos x \cos y = \tfrac{1}{2}[\cos(x-y) + \cos(x+y)],$$

$$\sin x \cos y = \tfrac{1}{2}[\sin(x-y) + \sin(x+y)].$$

▶ **Powers of trigonometric functions.**

$\cos^2 x = \frac{1}{2}\cos 2x + \frac{1}{2}$,

$\sin^2 x = -\frac{1}{2}\cos 2x + \frac{1}{2}$,

$\cos^3 x = \frac{1}{4}\cos 3x + \frac{3}{4}\cos x$,

$\sin^3 x = -\frac{1}{4}\sin 3x + \frac{3}{4}\sin x$,

$\cos^4 x = \frac{1}{8}\cos 4x + \frac{1}{2}\cos 2x + \frac{3}{8}$,

$\sin^4 x = \frac{1}{8}\cos 4x - \frac{1}{2}\cos 2x + \frac{3}{8}$,

$\cos^5 x = \frac{1}{16}\cos 5x + \frac{5}{16}\cos 3x + \frac{5}{8}\cos x$,

$\sin^5 x = \frac{1}{16}\sin 5x - \frac{5}{16}\sin 3x + \frac{5}{8}\sin x$,

$$\cos^{2n} x = \frac{1}{2^{2n-1}}\sum_{k=0}^{n-1} C_{2n}^k \cos[2(n-k)x] + \frac{1}{2^{2n}}C_{2n}^n,$$

$$\cos^{2n+1} x = \frac{1}{2^{2n}}\sum_{k=0}^{n} C_{2n+1}^k \cos[(2n-2k+1)x],$$

$$\sin^{2n} x = \frac{1}{2^{2n-1}}\sum_{k=0}^{n-1}(-1)^{n-k} C_{2n}^k \cos[2(n-k)x] + \frac{1}{2^{2n}}C_{2n}^n,$$

$$\sin^{2n+1} x = \frac{1}{2^{2n}}\sum_{k=0}^{n}(-1)^{n-k} C_{2n+1}^k \sin[(2n-2k+1)x].$$

Here, $n = 1, 2, \ldots$ and $C_m^k = \dfrac{m!}{k!\,(m-k)!}$ are binomial coefficients $(0! = 1)$.

▶ **Addition formulas.**

$$\sin(x \pm y) = \sin x \cos y \pm \cos x \sin y, \qquad \cos(x \pm y) = \cos x \cos y \mp \sin x \sin y,$$

$$\tan(x \pm y) = \frac{\tan x \pm \tan y}{1 \mp \tan x \tan y}, \qquad \cot(x \pm y) = \frac{1 \mp \tan x \tan y}{\tan x \pm \tan y}.$$

▶ **Trigonometric functions of multiple arguments.**

$\cos 2x = 2\cos^2 x - 1 = 1 - 2\sin^2 x$,

$\sin 2x = 2\sin x \cos x$,

$\cos 3x = -3\cos x + 4\cos^3 x$,

$\sin 3x = 3\sin x - 4\sin^3 x$,

$\cos 4x = 1 - 8\cos^2 x + 8\cos^4 x$,

$\sin 4x = 4\cos x\,(\sin x - 2\sin^3 x)$,

$\cos 5x = 5\cos x - 20\cos^3 x + 16\cos^5 x$,

$\sin 5x = 5\sin x - 20\sin^3 x + 16\sin^5 x$,

$$\cos(2nx) = 1 + \sum_{k=1}^{n}(-1)^k \frac{n^2(n^2-1)\ldots[n^2-(k-1)^2]}{(2k)!} 4^k \sin^{2k} x,$$

$$\cos[(2n+1)x] = \cos x \left\{ 1 + \sum_{k=1}^{n}(-1)^k \frac{[(2n+1)^2-1][(2n+1)^2-3^2]\ldots[(2n+1)^2-(2k-1)^2]}{(2k)!} \sin^{2k} x \right\},$$

$$\sin(2nx) = 2n\cos x \left[\sin x + \sum_{k=1}^{n}(-4)^k \frac{(n^2-1)(n^2-2^2)\ldots(n^2-k^2)}{(2k-1)!} \sin^{2k-1} x \right],$$

$$\sin[(2n+1)x] = (2n+1)\left\{ \sin x + \sum_{k=1}^{n}(-1)^k \frac{[(2n+1)^2-1][(2n+1)^2-3^2]\ldots[(2n+1)^2-(2k-1)^2]}{(2k+1)!} \sin^{2k+1} x \right\},$$

$$\tan 2x = \frac{2\tan x}{1-\tan^2 x}, \qquad \tan 3x = \frac{3\tan x - \tan^3 x}{1-3\tan^2 x}, \qquad \tan 4x = \frac{4\tan x - 4\tan^3 x}{1-6\tan^2 x + \tan^4 x},$$

where $n = 1, 2, \ldots$

▶ **Trigonometric functions of half argument.**

$$\sin^2\frac{x}{2} = \frac{1-\cos x}{2}, \qquad \cos^2\frac{x}{2} = \frac{1+\cos x}{2},$$

$$\tan\frac{x}{2} = \frac{\sin x}{1+\cos x} = \frac{1-\cos x}{\sin x}, \qquad \cot\frac{x}{2} = \frac{\sin x}{1-\cos x} = \frac{1+\cos x}{\sin x},$$

$$\sin x = \frac{2\tan\frac{x}{2}}{1+\tan^2\frac{x}{2}}, \qquad \cos x = \frac{1-\tan^2\frac{x}{2}}{1+\tan^2\frac{x}{2}}, \qquad \tan x = \frac{2\tan\frac{x}{2}}{1-\tan^2\frac{x}{2}}.$$

▶ **Differentiation formulas.**

$$\frac{d\sin x}{dx} = \cos x, \qquad \frac{d\cos x}{dx} = -\sin x, \qquad \frac{d\tan x}{dx} = \frac{1}{\cos^2 x}, \qquad \frac{d\cot x}{dx} = -\frac{1}{\sin^2 x}.$$

▶ **Integration formulas.**

$$\int \sin x\, dx = -\cos x + C, \qquad \int \cos x\, dx = \sin x + C,$$

$$\int \tan x\, dx = -\ln|\cos x| + C, \qquad \int \cot x\, dx = \ln|\sin x| + C,$$

where C is an arbitrary constant.

▶ **Power series expansions.**

$$\cos x = 1 - \frac{x^2}{2!} + \frac{x^4}{4!} - \frac{x^6}{6!} + \cdots + (-1)^n\frac{x^{2n}}{(2n)!} + \cdots \qquad (|x| < \infty),$$

$$\sin x = x - \frac{x^3}{3!} + \frac{x^5}{5!} - \frac{x^7}{7!} + \cdots + (-1)^n\frac{x^{2n+1}}{(2n+1)!} + \cdots \qquad (|x| < \infty),$$

$$\tan x = x + \frac{x^3}{3} + \frac{2x^5}{15} + \frac{17x^7}{315} + \cdots + \frac{2^{2n}(2^{2n}-1)|B_{2n}|}{(2n)!}x^{2n-1} + \cdots \qquad (|x| < \pi/2),$$

$$\cot x = \frac{1}{x} - \left(\frac{x}{3} + \frac{x^3}{45} + \frac{2x^5}{945} + \cdots + \frac{2^{2n}|B_{2n}|}{(2n)!}x^{2n-1} + \cdots\right) \qquad (0 < |x| < \pi),$$

where B_n are Bernoulli numbers (see Subsection M13.1.2).

▶ **Representation in the form of infinite products.**

$$\sin x = x\left(1 - \frac{x^2}{\pi^2}\right)\left(1 - \frac{x^2}{4\pi^2}\right)\left(1 - \frac{x^2}{9\pi^2}\right)\cdots\left(1 - \frac{x^2}{n^2\pi^2}\right)\cdots$$

$$\cos x = \left(1 - \frac{4x^2}{\pi^2}\right)\left(1 - \frac{4x^2}{9\pi^2}\right)\left(1 - \frac{4x^2}{25\pi^2}\right)\cdots\left(1 - \frac{4x^2}{(2n+1)^2\pi^2}\right)\cdots$$

▶ **Euler and de Moivre formulas. Relation to hyperbolic functions.**

$$e^{y+ix} = e^y(\cos x + i\sin x), \qquad (\cos x + i\sin x)^n = \cos(nx) + i\sin(nx), \qquad i^2 = -1,$$

$$\sin(ix) = i\sinh x, \qquad \cos(ix) = \cosh x, \qquad \tan(ix) = i\tanh x, \qquad \cot(ix) = -i\coth x.$$

M2.3. Inverse Trigonometric Functions

M2.3.1. Definitions. Graphs of Inverse Trigonometric Functions

▶ **Definitions of inverse trigonometric functions.** *Inverse trigonometric functions (arc functions)* are the functions that are inverse to the trigonometric functions. Since the trigonometric functions $\sin x$, $\cos x$, $\tan x$, $\cot x$ are periodic, the corresponding inverse functions, denoted by $\text{Arcsin}\, x$, $\text{Arccos}\, x$, $\text{Arctan}\, x$, $\text{Arccot}\, x$, are multi-valued. The following relations define the multi-valued inverse trigonometric functions:

$$\sin(\text{Arcsin}\, x) = x, \quad \cos(\text{Arccos}\, x) = x,$$
$$\tan(\text{Arctan}\, x) = x, \quad \cot(\text{Arccot}\, x) = x.$$

These functions admit the following verbal definitions: $\text{Arcsin}\, x$ is the angle whose sine is equal to x; $\text{Arccos}\, x$ is the angle whose cosine is equal to x; $\text{Arctan}\, x$ is the angle whose tangent is equal to x; $\text{Arccot}\, x$ is the angle whose cotangent is equal to x.

The principal (single-valued) branches of the inverse trigonometric functions are denoted by

$$\arcsin x \equiv \sin^{-1} x \quad \text{(arcsine is the inverse of sine)},$$
$$\arccos x \equiv \cos^{-1} x \quad \text{(arccosine is the inverse of cosine)},$$
$$\arctan x \equiv \tan^{-1} x \quad \text{(arctangent is the inverse of tangent)},$$
$$\text{arccot}\, x \equiv \cot^{-1} x \quad \text{(arccotangent is the inverse of cotangent)}$$

and are determined by the inequalities

$$-\frac{\pi}{2} \le \arcsin x \le \frac{\pi}{2}, \quad 0 \le \arccos x \le \pi \quad (-1 \le x \le 1);$$
$$-\frac{\pi}{2} < \arctan x < \frac{\pi}{2}, \quad 0 < \text{arccot}\, x < \pi \quad (-\infty < x < \infty).$$

The following equivalent relations can be taken as definitions of single-valued inverse trigonometric functions:

$$y = \arcsin x, \quad -1 \le x \le 1 \quad \Longleftrightarrow \quad x = \sin y, \quad -\frac{\pi}{2} \le y \le \frac{\pi}{2};$$
$$y = \arccos x, \quad -1 \le x \le 1 \quad \Longleftrightarrow \quad x = \cos y, \quad 0 \le y \le \pi;$$
$$y = \arctan x, \quad -\infty < x < +\infty \quad \Longleftrightarrow \quad x = \tan y, \quad -\frac{\pi}{2} < y < \frac{\pi}{2};$$
$$y = \text{arccot}\, x, \quad -\infty < x < +\infty \quad \Longleftrightarrow \quad x = \cot y, \quad 0 < y < \pi.$$

The multi-valued and the single-valued inverse trigonometric functions are related by the formulas

$$\text{Arcsin}\, x = (-1)^n \arcsin x + \pi n,$$
$$\text{Arccos}\, x = \pm \arccos x + 2\pi n,$$
$$\text{Arctan}\, x = \arctan x + \pi n,$$
$$\text{Arccot}\, x = \text{arccot}\, x + \pi n,$$

where $n = 0, \pm 1, \pm 2, \ldots$

The graphs of inverse trigonometric functions are obtained from the graphs of the corresponding trigonometric functions by mirror reflection with respect to the straight line $y = x$ (with the domain of each function being taken into account).

▶ **Arcsine:** $y = \arcsin x$. This function is defined for all $x \in [-1, 1]$ and its range is $y \in [-\frac{\pi}{2}, \frac{\pi}{2}]$. The arcsine is an odd, nonperiodic, bounded function that crosses the axes Ox and Oy at the origin $x = 0$, $y = 0$. This is an increasing function in its domain, and it takes its smallest value $y = -\frac{\pi}{2}$ at the point $x = -1$; it takes its largest value $y = \frac{\pi}{2}$ at the point $x = 1$. The graph of the function $y = \arcsin x$ is given in Fig. M2.10.

▶ **Arccosine:** $y = \arccos x$. This function is defined for all $x \in [-1, 1]$ and its range is $y \in [0, \pi]$. It is neither odd nor even. It is a nonperiodic, bounded function that crosses the axis Oy at the point $y = \frac{\pi}{2}$ and crosses the axis Ox at the point $x = 1$. This is a decreasing function in its domain, and at the point $x = -1$ it takes its largest value $y = \pi$; at the point $x = 1$ it takes its smallest value $y = 0$. For all x in its domain, the following relation holds: $\arccos x = \frac{\pi}{2} - \arcsin x$. The graph of the function $y = \arccos x$ is given in Fig. M2.11.

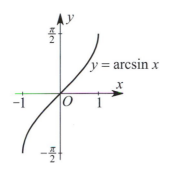

Figure M2.10. Graph of the function $y = \arcsin x$.

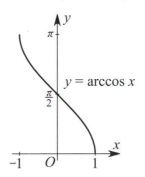

Figure M2.11. Graph of the function $y = \arccos x$.

▶ **Arctangent:** $y = \arctan x$. This function is defined for all x, and its range is $y \in (-\frac{\pi}{2}, \frac{\pi}{2})$. The arctangent is an odd, nonperiodic, bounded function that crosses the coordinate axes at the origin $x = 0$, $y = 0$. This is an increasing function with no points of extremum. It has two horizontal asymptotes: $y = -\frac{\pi}{2}$ (as $x \to -\infty$) and $y = \frac{\pi}{2}$ (as $x \to +\infty$). The graph of the function $y = \arctan x$ is given in Fig. M2.12.

▶ **Arccotangent:** $y = \text{arccot}\, x$. This function is defined for all x, and its range is $y \in (0, \pi)$. The arccotangent is neither odd nor even. It is a nonperiodic, bounded function that crosses the axis Oy at the point $y = \frac{\pi}{2}$ and does not cross the axis Ox. This is a decreasing function on the entire x-axis with no points of extremum. It has two horizontal asymptotes $y = 0$ (as $x \to +\infty$) and $y = \pi$ (as $x \to -\infty$). For all x, the following relation holds: $\text{arccot}\, x = \frac{\pi}{2} - \arctan x$. The graph of the function $y = \text{arccot}\, x$ is given in Fig. M2.13.

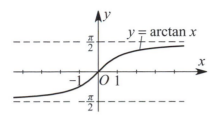

Figure M2.12. Graph of the function $y = \arctan x$.

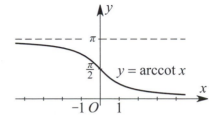

Figure M2.13. Graph of the function $y = \text{arccot}\, x$.

M2.3.2. Properties of Inverse Trigonometric Functions

▶ **Simplest formulas.**

$$\sin(\arcsin x) = x, \quad \cos(\arccos x) = x,$$
$$\tan(\arctan x) = x, \quad \cot(\text{arccot}\, x) = x.$$

▶ **Some properties.**

$$\arcsin(-x) = -\arcsin x, \qquad \arccos(-x) = \pi - \arccos x,$$
$$\arctan(-x) = -\arctan x, \qquad \operatorname{arccot}(-x) = \pi - \operatorname{arccot} x,$$

$$\arcsin(\sin x) = \begin{cases} x - 2n\pi & \text{if } 2n\pi - \frac{\pi}{2} \le x \le 2n\pi + \frac{\pi}{2}, \\ -x + 2(n+1)\pi & \text{if } (2n+1)\pi - \frac{\pi}{2} \le x \le 2(n+1)\pi + \frac{\pi}{2}, \end{cases}$$

$$\arccos(\cos x) = \begin{cases} x - 2n\pi & \text{if } 2n\pi \le x \le (2n+1)\pi, \\ -x + 2(n+1)\pi & \text{if } (2n+1)\pi \le x \le 2(n+1)\pi, \end{cases}$$

$$\arctan(\tan x) = x - n\pi \quad \text{if} \quad n\pi - \frac{\pi}{2} < x < n\pi + \frac{\pi}{2},$$
$$\operatorname{arccot}(\cot x) = x - n\pi \quad \text{if} \quad n\pi < x < (n+1)\pi.$$

▶ **Relations between inverse trigonometric functions.**

$$\arcsin x + \arccos x = \frac{\pi}{2}, \qquad \arctan x + \operatorname{arccot} x = \frac{\pi}{2};$$

$$\arcsin x = \begin{cases} \arccos \sqrt{1-x^2} & \text{if } 0 \le x \le 1, \\ -\arccos \sqrt{1-x^2} & \text{if } -1 \le x \le 0, \\ \arctan \dfrac{x}{\sqrt{1-x^2}} & \text{if } -1 < x < 1, \\ \operatorname{arccot} \dfrac{\sqrt{1-x^2}}{x} - \pi & \text{if } -1 \le x < 0; \end{cases} \qquad \arccos x = \begin{cases} \arcsin \sqrt{1-x^2} & \text{if } 0 \le x \le 1, \\ \pi - \arcsin \sqrt{1-x^2} & \text{if } -1 \le x \le 0, \\ \arctan \dfrac{\sqrt{1-x^2}}{x} & \text{if } 0 < x \le 1, \\ \operatorname{arccot} \dfrac{x}{\sqrt{1-x^2}} & \text{if } -1 < x < 1; \end{cases}$$

$$\arctan x = \begin{cases} \arcsin \dfrac{x}{\sqrt{1+x^2}} & \text{for any } x, \\ \arccos \dfrac{1}{\sqrt{1+x^2}} & \text{if } x \ge 0, \\ -\arccos \dfrac{1}{\sqrt{1+x^2}} & \text{if } x \le 0, \\ \operatorname{arccot} \dfrac{1}{x} & \text{if } x > 0; \end{cases} \qquad \operatorname{arccot} x = \begin{cases} \arcsin \dfrac{1}{\sqrt{1+x^2}} & \text{if } x \ge 0, \\ \pi - \arcsin \dfrac{1}{\sqrt{1+x^2}} & \text{if } x \le 0, \\ \arctan \dfrac{1}{x} & \text{if } x > 0, \\ \pi + \arctan \dfrac{1}{x} & \text{if } x < 0. \end{cases}$$

▶ **Addition and subtraction of inverse trigonometric functions.**

$$\arcsin x + \arcsin y = \arcsin\left(x\sqrt{1-y^2} + y\sqrt{1-x^2}\right) \quad \text{for} \quad x^2 + y^2 \le 1,$$
$$\arccos x \pm \arccos y = \pm\arccos\left[xy \mp \sqrt{(1-x^2)(1-y^2)}\right] \quad \text{for} \quad x \pm y \ge 0,$$
$$\arctan x + \arctan y = \arctan \frac{x+y}{1-xy} \quad \text{for} \quad xy < 1,$$
$$\arctan x - \arctan y = \arctan \frac{x-y}{1+xy} \quad \text{for} \quad xy > -1.$$

▶ **Differentiation formulas.**

$$\frac{d}{dx} \arcsin x = \frac{1}{\sqrt{1-x^2}}, \qquad \frac{d}{dx} \arccos x = -\frac{1}{\sqrt{1-x^2}},$$
$$\frac{d}{dx} \arctan x = \frac{1}{1+x^2}, \qquad \frac{d}{dx} \operatorname{arccot} x = -\frac{1}{1+x^2}.$$

$$\coth x = \frac{\sqrt{\sinh^2 x + 1}}{\sinh x} = \frac{\cosh x}{\sqrt{\cosh^2 x - 1}} = \frac{1}{\tanh x}.$$

▶ **Addition formulas.**

$$\sinh(x \pm y) = \sinh x \cosh y \pm \sinh y \cosh x, \qquad \cosh(x \pm y) = \cosh x \cosh y \pm \sinh x \sinh y,$$

$$\tanh(x \pm y) = \frac{\tanh x \pm \tanh y}{1 \pm \tanh x \tanh y}, \qquad \coth(x \pm y) = \frac{\coth x \coth y \pm 1}{\coth y \pm \coth x}.$$

▶ **Addition and subtraction of hyperbolic functions.**

$$\sinh x \pm \sinh y = 2 \sinh\left(\frac{x \pm y}{2}\right) \cosh\left(\frac{x \mp y}{2}\right),$$

$$\cosh x + \cosh y = 2 \cosh\left(\frac{x + y}{2}\right) \cosh\left(\frac{x - y}{2}\right),$$

$$\cosh x - \cosh y = 2 \sinh\left(\frac{x + y}{2}\right) \sinh\left(\frac{x - y}{2}\right),$$

$$\sinh^2 x - \sinh^2 y = \cosh^2 x - \cosh^2 y = \sinh(x + y) \sinh(x - y),$$

$$\sinh^2 x + \cosh^2 y = \cosh(x + y) \cosh(x - y),$$

$$(\cosh x \pm \sinh x)^n = \cosh(nx) \pm \sinh(nx),$$

$$\tanh x \pm \tanh y = \frac{\sinh(x \pm y)}{\cosh x \cosh y}, \qquad \coth x \pm \coth y = \pm\frac{\sinh(x \pm y)}{\sinh x \sinh y},$$

where $n = 0, \pm 1, \pm 2, \ldots$

▶ **Products of hyperbolic functions.**

$$\sinh x \sinh y = \tfrac{1}{2}[\cosh(x + y) - \cosh(x - y)],$$

$$\cosh x \cosh y = \tfrac{1}{2}[\cosh(x + y) + \cosh(x - y)],$$

$$\sinh x \cosh y = \tfrac{1}{2}[\sinh(x + y) + \sinh(x - y)].$$

▶ **Powers of hyperbolic functions.**

$$\cosh^2 x = \tfrac{1}{2}\cosh 2x + \tfrac{1}{2}, \qquad\qquad \sinh^2 x = \tfrac{1}{2}\cosh 2x - \tfrac{1}{2},$$

$$\cosh^3 x = \tfrac{1}{4}\cosh 3x + \tfrac{3}{4}\cosh x, \qquad\qquad \sinh^3 x = \tfrac{1}{4}\sinh 3x - \tfrac{3}{4}\sinh x,$$

$$\cosh^4 x = \tfrac{1}{8}\cosh 4x + \tfrac{1}{2}\cosh 2x + \tfrac{3}{8}, \qquad \sinh^4 x = \tfrac{1}{8}\cosh 4x - \tfrac{1}{2}\cosh 2x + \tfrac{3}{8},$$

$$\cosh^5 x = \tfrac{1}{16}\cosh 5x + \tfrac{5}{16}\cosh 3x + \tfrac{5}{8}\cosh x, \quad \sinh^5 x = \tfrac{1}{16}\sinh 5x - \tfrac{5}{16}\sinh 3x + \tfrac{5}{8}\sinh x,$$

$$\cosh^{2n} x = \frac{1}{2^{2n-1}} \sum_{k=0}^{n-1} C_{2n}^k \cosh[2(n-k)x] + \frac{1}{2^{2n}} C_{2n}^n,$$

$$\cosh^{2n+1} x = \frac{1}{2^{2n}} \sum_{k=0}^{n} C_{2n+1}^k \cosh[(2n-2k+1)x],$$

$$\sinh^{2n} x = \frac{1}{2^{2n-1}} \sum_{k=0}^{n-1} (-1)^k C_{2n}^k \cosh[2(n-k)x] + \frac{(-1)^n}{2^{2n}} C_{2n}^n,$$

$$\sinh^{2n+1} x = \frac{1}{2^{2n}} \sum_{k=0}^{n} (-1)^k C_{2n+1}^k \sinh[(2n-2k+1)x].$$

Here, $n = 1, 2, \ldots$ and C_m^k are binomial coefficients.

▶ **Hyperbolic functions of multiple argument.**

$$\cosh 2x = 2\cosh^2 x - 1, \qquad\qquad \sinh 2x = 2\sinh x \cosh x,$$

$$\cosh 3x = -3\cosh x + 4\cosh^3 x, \qquad \sinh 3x = 3\sinh x + 4\sinh^3 x,$$

$$\cosh 4x = 1 - 8\cosh^2 x + 8\cosh^4 x, \qquad \sinh 4x = 4\cosh x(\sinh x + 2\sinh^3 x),$$

$$\cosh 5x = 5\cosh x - 20\cosh^3 x + 16\cosh^5 x, \quad \sinh 5x = 5\sinh x + 20\sinh^3 x + 16\sinh^5 x.$$

$$\cosh(nx) = 2^{n-1}\cosh^n x + \frac{n}{2}\sum_{k=0}^{[n/2]}\frac{(-1)^{k+1}}{k+1}C_{n-k-2}^{k-2}2^{n-2k-2}(\cosh x)^{n-2k-2},$$

$$\sinh(nx) = \sinh x\sum_{k=0}^{[(n-1)/2]}2^{n-k-1}C_{n-k-1}^{k}(\cosh x)^{n-2k-1}.$$

Here, C_m^k are binomial coefficients and $[A]$ stands for the integer part of the number A.

▶ **Hyperbolic functions of half argument.**

$$\sinh\frac{x}{2} = \operatorname{sign} x\sqrt{\frac{\cosh x - 1}{2}}, \qquad \cosh\frac{x}{2} = \sqrt{\frac{\cosh x + 1}{2}},$$

$$\tanh\frac{x}{2} = \frac{\sinh x}{\cosh x + 1} = \frac{\cosh x - 1}{\sinh x}, \qquad \coth\frac{x}{2} = \frac{\sinh x}{\cosh x - 1} = \frac{\cosh x + 1}{\sinh x}.$$

▶ **Differentiation formulas.**

$$\frac{d\sinh x}{dx} = \cosh x, \qquad \frac{d\cosh x}{dx} = \sinh x,$$

$$\frac{d\tanh x}{dx} = \frac{1}{\cosh^2 x}, \qquad \frac{d\coth x}{dx} = -\frac{1}{\sinh^2 x}.$$

▶ **Integration formulas.**

$$\int \sinh x\,dx = \cosh x + C, \qquad \int \cosh x\,dx = \sinh x + C,$$

$$\int \tanh x\,dx = \ln\cosh x + C, \qquad \int \coth x\,dx = \ln|\sinh x| + C,$$

where C is an arbitrary constant.

▶ **Power series expansions.**

$$\cosh x = 1 + \frac{x^2}{2!} + \frac{x^4}{4!} + \frac{x^6}{6!} + \cdots + \frac{x^{2n}}{(2n)!} + \cdots \qquad (|x| < \infty),$$

$$\sinh x = x + \frac{x^3}{3!} + \frac{x^5}{5!} + \frac{x^7}{7!} + \cdots + \frac{x^{2n+1}}{(2n+1)!} + \cdots \qquad (|x| < \infty),$$

$$\tanh x = x - \frac{x^3}{3} + \frac{2x^5}{15} - \frac{17x^7}{315} + \cdots + (-1)^{n-1}\frac{2^{2n}(2^{2n}-1)|B_{2n}|x^{2n-1}}{(2n)!} + \cdots \qquad (|x| < \pi/2),$$

$$\coth x = \frac{1}{x} + \frac{x}{3} - \frac{x^3}{45} + \frac{2x^5}{945} - \cdots + (-1)^{n-1}\frac{2^{2n}|B_{2n}|x^{2n-1}}{(2n)!} + \cdots \qquad (|x| < \pi),$$

where B_n are Bernoulli numbers (see Subsection M13.1.2).

▶ **Relation to trigonometric functions.**

$$\sinh(ix) = i \sin x, \quad \cosh(ix) = \cos x, \quad \tanh(ix) = i \tan x, \quad \coth(ix) = -i \cot x, \qquad i^2 = -1.$$

M2.5. Inverse Hyperbolic Functions

M2.5.1. Definitions. Graphs of Inverse Hyperbolic Functions

▶ **Definitions of inverse hyperbolic functions.** The *inverse hyperbolic functions* (also known as the *area hyperbolic functions*) are the inverses of the respective hyperbolic functions. The following notation is used for inverse hyperbolic functions:

$$\operatorname{arcsinh} x \equiv \operatorname{arsinh} x \equiv \sinh^{-1} x \quad \text{(inverse of hyperbolic sine),}$$
$$\operatorname{arccosh} x \equiv \operatorname{arcosh} x \equiv \cosh^{-1} x \quad \text{(inverse of hyperbolic cosine),}$$
$$\operatorname{arctanh} x \equiv \operatorname{artanh} x \equiv \tanh^{-1} x \quad \text{(inverse of hyperbolic tangent),}$$
$$\operatorname{arccoth} x \equiv \operatorname{arcoth} x \equiv \coth^{-1} x \quad \text{(inverse of hyperbolic cotangent).}$$

Inverse hyperbolic functions can be expressed in terms of logarithmic functions:

$$\operatorname{arcsinh} x = \ln\left(x + \sqrt{x^2 + 1}\right) \quad (x \text{ is any}); \qquad \operatorname{arccosh} x = \ln\left(x + \sqrt{x^2 - 1}\right) \quad (x \geq 1);$$
$$\operatorname{arctanh} x = \frac{1}{2} \ln \frac{1 + x}{1 - x} \quad (|x| < 1); \qquad \operatorname{arccoth} x = \frac{1}{2} \ln \frac{x + 1}{x - 1} \quad (|x| > 1).$$

Here, only one (principal) branch of the function $\operatorname{arccosh} x$ is listed, the function itself being double-valued. In order to write out both branches of $\operatorname{arccosh} x$, the symbol $\pm$ should be placed before the logarithm on the right-hand side of the formula.

The graphs of the inverse hyperbolic functions are given below. These are obtained from the graphs of the corresponding hyperbolic functions by mirror reflection with respect to the straight line $y = x$ (with the domain of each function taken into account).

▶ **Inverse hyperbolic sine: $y = \operatorname{arcsinh} x$.** This function is defined for all x, and its range coincides with the y-axis. The $\operatorname{arcsinh} x$ is an odd, nonperiodic, unbounded function that crosses the axes Ox and Oy at the origin $x = 0$, $y = 0$. This is an increasing function on the entire x-axis with no points of extremum. The graph of the function $y = \operatorname{arcsinh} x$ is given in Fig. M2.18.

▶ **Inverse hyperbolic cosine: $y = \operatorname{arccosh} x$.** This function is defined for all $x \in [1, +\infty)$, and its range consists of $y \in [0, +\infty)$. The $\operatorname{arccosh} x$ is neither odd nor even; it is nonperiodic and unbounded. It does not cross the axis Oy and crosses the axis Ox at the point $x = 1$. It is an increasing function in its domain with the minimal value $y = 0$ at $x = 1$. The graph of the function $y = \operatorname{arccosh} x$ is given in Fig. M2.19.

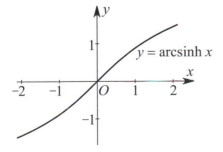

Figure M2.18. Graph of the function $y = \operatorname{arcsinh} x$.

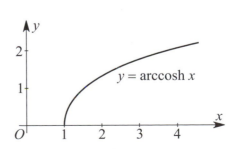

Figure M2.19. Graph of the function $y = \operatorname{arccosh} x$.

▶ **Inverse hyperbolic tangent:** $y = \operatorname{arctanh} x$. This function is defined for all $x \in (-1, 1)$, and its range consists of all y. The $\operatorname{arctanh} x$ is an odd, nonperiodic, unbounded function that crosses the coordinate axes at the origin $x = 0$, $y = 0$. This is an increasing function in its domain with no points of extremum and an inflection point at the origin. It has two vertical asymptotes: $x = \pm 1$. The graph of the function $y = \operatorname{arctanh} x$ is given in Fig. M2.20.

▶ **Inverse hyperbolic cotangent:** $y = \operatorname{arccoth} x$. This function is defined for $x \in (-\infty, -1)$ and $x \in (1, +\infty)$. Its range consists of all $y \neq 0$. The $\operatorname{arccoth} x$ is an odd, nonperiodic, unbounded function that does not cross the coordinate axes. It is a decreasing function on each of the semiaxes of its domain. This function has no points of extremum and has one horizontal asymptote $y = 0$ and two vertical asymptotes $x = \pm 1$. The graph of the function $y = \operatorname{arccoth} x$ is given in Fig. M2.21.

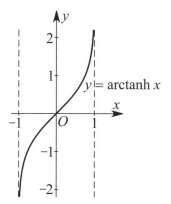

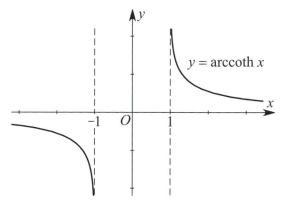

Figure M2.20. Graph of the function $y = \operatorname{arctanh} x$.   **Figure M2.21.** Graph of the function $y = \operatorname{arccoth} x$.

M2.5.2. Properties of Inverse Hyperbolic Functions

▶ **Simplest relations.**

$$\operatorname{arcsinh}(-x) = -\operatorname{arcsinh} x, \quad \operatorname{arctanh}(-x) = -\operatorname{arctanh} x, \quad \operatorname{arccoth}(-x) = -\operatorname{arccoth} x.$$

▶ **Relations between inverse hyperbolic functions.**

$$\operatorname{arcsinh} x = \operatorname{arccosh} \sqrt{x^2 + 1} = \operatorname{arctanh} \frac{x}{\sqrt{x^2 + 1}},$$

$$\operatorname{arccosh} x = \operatorname{arcsinh} \sqrt{x^2 - 1} = \operatorname{arctanh} \frac{\sqrt{x^2 - 1}}{x},$$

$$\operatorname{arctanh} x = \operatorname{arcsinh} \frac{x}{\sqrt{1 - x^2}} = \operatorname{arccosh} \frac{1}{\sqrt{1 - x^2}} = \operatorname{arccoth} \frac{1}{x}.$$

▶ **Addition and subtraction of inverse hyperbolic functions.**

$$\operatorname{arcsinh} x \pm \operatorname{arcsinh} y = \operatorname{arcsinh} \left(x \sqrt{1 + y^2} \pm y \sqrt{1 + x^2} \right),$$

$$\operatorname{arccosh} x \pm \operatorname{arccosh} y = \operatorname{arccosh} \left[xy \pm \sqrt{(x^2 - 1)(y^2 - 1)} \right],$$

$$\operatorname{arcsinh} x \pm \operatorname{arccosh} y = \operatorname{arcsinh} \left[xy \pm \sqrt{(x^2 + 1)(y^2 - 1)} \right],$$

$$\operatorname{arctanh} x \pm \operatorname{arctanh} y = \operatorname{arctanh} \frac{x \pm y}{1 \pm xy}, \quad \operatorname{arctanh} x \pm \operatorname{arccoth} y = \operatorname{arctanh} \frac{xy \pm 1}{y \pm x}.$$

▶ **Differentiation formulas.**

$$\frac{d}{dx}\operatorname{arcsinh} x = \frac{1}{\sqrt{x^2+1}}, \qquad \frac{d}{dx}\operatorname{arccosh} x = \frac{1}{\sqrt{x^2-1}},$$

$$\frac{d}{dx}\operatorname{arctanh} x = \frac{1}{1-x^2} \quad (x^2 < 1), \qquad \frac{d}{dx}\operatorname{arccoth} x = \frac{1}{1-x^2} \quad (x^2 > 1).$$

▶ **Integration formulas.**

$$\int \operatorname{arcsinh} x \, dx = x\operatorname{arcsinh} x - \sqrt{1+x^2} + C,$$

$$\int \operatorname{arccosh} x \, dx = x\operatorname{arccosh} x - \sqrt{x^2-1} + C,$$

$$\int \operatorname{arctanh} x \, dx = x\operatorname{arctanh} x + \frac{1}{2}\ln(1-x^2) + C,$$

$$\int \operatorname{arccoth} x \, dx = x\operatorname{arccoth} x + \frac{1}{2}\ln(x^2-1) + C,$$

where C is an arbitrary constant.

▶ **Power series expansions.**

$$\operatorname{arcsinh} x = x - \frac{1}{2}\frac{x^3}{3} + \frac{1\times 3}{2\times 4}\frac{x^5}{5} - \cdots + (-1)^n \frac{1\times 3\times\cdots\times(2n-1)}{2\times 4\times\cdots\times(2n)}\frac{x^{2n+1}}{2n+1} + \cdots \qquad (|x| < 1),$$

$$\operatorname{arcsinh} x = \ln(2x) + \frac{1}{2}\frac{1}{2x^2} + \frac{1\times 3}{2\times 4}\frac{1}{4x^4} + \cdots + \frac{1\times 3\times\cdots\times(2n-1)}{2\times 4\times\cdots\times(2n)}\frac{1}{2nx^{2n}} + \cdots \qquad (|x| > 1),$$

$$\operatorname{arccosh} x = \ln(2x) - \frac{1}{2}\frac{1}{2x^2} - \frac{1\times 3}{2\times 4}\frac{1}{4x^4} - \cdots - \frac{1\times 3\times\cdots\times(2n-1)}{2\times 4\times\cdots\times(2n)}\frac{1}{2nx^{2n}} - \cdots \qquad (|x| > 1),$$

$$\operatorname{arctanh} x = x + \frac{x^3}{3} + \frac{x^5}{5} + \frac{x^7}{7} + \cdots + \frac{x^{2n+1}}{2n+1} + \cdots \qquad (|x| < 1),$$

$$\operatorname{arccoth} x = \frac{1}{x} + \frac{1}{3x^3} + \frac{1}{5x^5} + \frac{1}{7x^7} + \cdots + \frac{1}{(2n+1)x^{2n+1}} + \cdots \qquad (|x| > 1).$$

Bibliography for Chapter M2

Abramowitz, M. and Stegun, I. A. (Editors), *Handbook of Mathematical Functions with Formulas, Graphs and Mathematical Tables*, National Bureau of Standards Applied Mathematics, Washington, D.C., 1964.

Adams, R., *Calculus: A Complete Course, 6th Edition*, Pearson Education, Toronto, 2006.

Anton, H., Bivens, I., and Davis, S., *Calculus: Early Transcendentals Single and Multivariable, 8th Edition*, John Wiley & Sons, New York, 2005.

Bronshtein, I. N. and Semendyayev, K. A., *Handbook of Mathematics, 4th Edition*, Springer-Verlag, Berlin, 2004.

Courant, R. and John, F., *Introduction to Calculus and Analysis, Vol. 1*, Springer-Verlag, New York, 1999.

Edwards, C. H., and Penney, D., *Calculus, 6th Edition*, Pearson Education, Toronto, 2002.

Gradshteyn, I. S. and Ryzhik, I. M., *Tables of Integrals, Series, and Products, 6th Edition*, Academic Press, New York, 2000.

Kline, M., *Calculus: An Intuitive and Physical Approach, 2nd Edition*, Dover Publications, New York, 1998.

Korn, G. A. and Korn, T. M., *Mathematical Handbook for Scientists and Engineers, 2nd Edition*, Dover Publications, New York, 2000.

Prudnikov, A. P., Brychkov, Yu. A., and Marichev, O. I., *Integrals and Series, Vol. 1, Elementary Functions*, Gordon & Breach, New York, 1986.

Sullivan, M., *Trigonometry, 7th Edition*, Prentice Hall, Englewood Cliffs, New Jersey, 2004.

Thomas, G. B. and Finney, R. L., *Calculus and Analytic Geometry, 9th Edition*, Addison Wesley, Reading, Massachusetts, 1996.

Weisstein, E. W., *CRC Concise Encyclopedia of Mathematics, 2nd Edition*, CRC Press, Boca Raton, Florida, 2003.

Zill, D. G. and Dewar, J. M., *Trigonometry, 2nd Edition*, McGraw-Hill, New York, 1990.

Zwillinger, D., *CRC Standard Mathematical Tables and Formulae, 31st Edition*, CRC Press, Boca Raton, Florida, 2002.

Chapter M3
Elementary Geometry

M3.1. Plane Geometry

M3.1.1. Triangles

▶ **Plane triangle and its properties.**

$1°$. A *plane triangle*, or simply a *triangle*, is a plane figure bounded by three straight line segments (*sides*) connecting three noncollinear points (*vertices*) (Fig. M3.1a). The smaller angle between the two rays issuing from a vertex and passing through the other two vertices is called an (*interior*) *angle* of the triangle. The angle adjacent to an interior angle is called an *external angle* of the triangle. An external angle is equal to the sum of the two interior angles to which it is not adjacent.

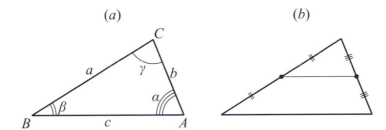

Figure M3.1. Plane triangle (*a*). Midline of a triangle (*b*).

A triangle is uniquely determined by any of the following sets of its parts:

1. Two angles and their included side.
2. Two sides and their included angle.
3. Three sides.

Depending on the angles, a triangle is said to be:

1. *Acute* if all three angles are acute.
2. *Right* (or *right-angled*) if one of the angles is right.
3. *Obtuse* if one of the angles is obtuse.

Depending on the relation between the side lengths, a triangle is said to be:

1. *Regular* (or *equilateral*) if all sides have the same length.
2. *Isosceles* if two of the sides are of equal length.
3. *Scalene* if all sides have different lengths.

$2°$. *Congruence tests for triangles:*

1. If two sides of a triangle and their included angle are congruent to the corresponding parts of another triangle, then the triangles are congruent.
2. If two angles of a triangle and their included side are congruent to the corresponding parts of another triangle, then the triangles are congruent.

3. If three sides of a triangle are congruent to the corresponding sides of another triangle, then the triangles are congruent.

3°. Triangles are said to be *similar* if their corresponding angles are equal and their corresponding sides are proportional.

Similarity tests for triangles:

1. If all three pairs of corresponding sides in a pair of triangles are in proportion, then the triangles are similar.
2. If two pairs of corresponding angles in a pair of triangles are congruent, then the triangles are similar.
3. If two pairs of corresponding sides in a pair of triangles are in proportion and the included angles are congruent, then the triangles are similar.

The areas of similar triangles are proportional to the squares of the corresponding linear parts (such as sides, altitudes, medians, etc.).

4°. The line connecting the midpoints of two sides of a triangle is called a *midline* of the triangle. The midline is parallel to and half as long as the third side (Fig. M3.1*b*).

Let a, b, and c be the lengths of the sides of a triangle; let α, β, and γ be the respective opposite angles (Fig. M3.1*a*); let R and r be the circumradius and the inradius, respectively; and let $p = \frac{1}{2}(a + b + c)$ be the semiperimeter.

Table M3.1 represents the basic properties and relations characterizing triangles.

TABLE M3.1
Basic properties and relations characterizing plane triangles.

No.	The name of property	Properties and relations
1	Triangle inequality	The length of any side of a triangle does not exceed the sum of lengths of the other two sides
2	Sum of angles of a triangle	$\alpha + \beta + \gamma = 180°$
3	Law of sines	$\dfrac{a}{\sin \alpha} = \dfrac{b}{\sin \beta} = \dfrac{c}{\sin \gamma} = 2R$
4	Law of cosines	$c^2 = a^2 + b^2 - 2ab \cos \gamma$
5	Law of tangents	$\dfrac{a + b}{a - b} = \dfrac{\tan\left[\frac{1}{2}(\alpha + \beta)\right]}{\tan\left[\frac{1}{2}(\alpha - \beta)\right]} = \dfrac{\cot\left(\frac{1}{2}\gamma\right)}{\tan\left[\frac{1}{2}(\alpha - \beta)\right]}$
6	Theorem on projections (law of cosines)	$c = a \cos \beta + b \cos \alpha$
7	Trigonometric angle formulas	$\sin \dfrac{\gamma}{2} = \sqrt{\dfrac{(p - a)(p - b)}{ab}}, \quad \cos \dfrac{\gamma}{2} = \sqrt{\dfrac{p(p - c)}{ab}},$ $\tan \dfrac{\gamma}{2} = \sqrt{\dfrac{(p - a)(p - b)}{p(p - c)}}, \quad \sin \gamma = \dfrac{2}{ab} \sqrt{p(p - a)(p - b)(p - c)}$
8	Law of tangents	$\tan \gamma = \dfrac{c \sin \alpha}{b - c \cos \alpha} = \dfrac{c \sin \beta}{a - c \cos \beta}$
9	Mollweide's formulas	$\dfrac{a + b}{c} = \dfrac{\cos\left[\frac{1}{2}(\alpha - \beta)\right]}{\sin\left(\frac{1}{2}\gamma\right)} = \dfrac{\cos\left[\frac{1}{2}(\alpha - \beta)\right]}{\cos\left[\frac{1}{2}(\alpha + \beta)\right]},$ $\dfrac{a - b}{c} = \dfrac{\sin\left[\frac{1}{2}(\alpha - \beta)\right]}{\cos\left(\frac{1}{2}\gamma\right)} = \dfrac{\sin\left[\frac{1}{2}(\alpha - \beta)\right]}{\sin\left[\frac{1}{2}(\alpha + \beta)\right]}$

Table M3.2 permits one to find the sides and angles of an arbitrary triangle if three appropriately chosen sides and/or angles are given. From the relations given in Tables M3.1 and M3.2, one can derive all missing relations by cyclic permutations of the sides a, b, and c and the angles α, β, and γ.

TABLE M3.2
Solution of plane triangles.

No.	Three parts specified	Formulas for the remaining parts
1	Three sides a, b, c	*First method.* One of the angles is determined by the law of cosines, $\cos\alpha = \dfrac{b^2 + c^2 - a^2}{2bc}$. Then either the law of sines or the law of cosines is applied. *Second method.* One of the angles is determined by trigonometric angle formulas. Further proceed in a similar way. Remark. The sum of lengths of any two sides must be greater than the length of the third side.
2	Two sides a, b and the included angle γ	*First method.* The side c is determined by the law of cosines, $c = \sqrt{a^2 + b^2 - 2ab\cos\gamma}$. The angle α is determined by either the law of cosines or the law of sines. The angle β is determined from the sum of angles in triangle, $\beta = 180° - \alpha - \gamma$. *Second method.* $\alpha + \beta$ is found from the sum of angles in triangle, $\alpha + \beta = 180° - \gamma$; $\alpha - \beta$ is found from the law of tangents, $\tan\dfrac{\alpha-\beta}{2} = \dfrac{a-b}{a+b}\cot\dfrac{\gamma}{2}$. Then α and β can be found. The third side c is determined by either the law of cosines or the law of sines.
3	A side c and the two angles α, β adjacent to it	The third angle γ is found from the sum of angles in triangle, $\gamma = 180° - \alpha - \beta$. Sides a and b are determined by the law of sines.
4	Two sides a, b and the angle α opposite one of them	The second angle is determined by the law of sines, $\sin\beta = \dfrac{b}{a}\sin\alpha$. The third angle is $\gamma = 180° - \alpha - \beta$. The third side is determined by the law of sines, $c = a\dfrac{\sin\gamma}{\sin\alpha}$. Remark. Five cases are possible: 1. $a > b$; i.e., the angle is opposite the greater side. Then $\alpha > \beta$, $\beta < 90°$ (the larger angle is opposite the larger side), and the triangle is determined uniquely. 2. $a = b$; i.e., the triangle is isosceles and is determined uniquely. 3. $a < b$ and $b\sin\alpha < a$. Then there are two solutions, $\beta_1 + \beta_2 = 180°$. 4. $a < b$ and $b\sin\alpha = a$. Then the solution is unique, $\beta = 90°$. 5. $a < b$ and $b\sin\alpha > a$. Then there are no solutions.

▶ **Medians, angle bisectors, and altitudes of a triangle.** A straight line through a vertex of a triangle and the midpoint of the opposite side is called a *median* of the triangle (Fig. M3.2*a*). The three medians of a triangle intersect in a single point lying strictly inside the triangle, which is called the *centroid* or *center of gravity* of the triangle. This point cuts the medians in the ratio 2 : 1 (counting from the corresponding vertices).

The length of the median m_a to the side a is equal to

$$m_a = \frac{1}{2}\sqrt{2(b^2 + c^2) - a^2} = \frac{1}{2}\sqrt{a^2 + 4b^2 - 4ab\cos\gamma}. \tag{M3.1.1.1}$$

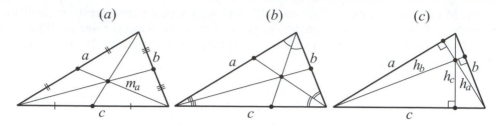

Figure M3.2. Medians (*a*), angle bisectors (*b*), and altitudes (*c*) of a triangle.

An *angle bisector* of a triangle is a line segment between a vertex and a point of the opposite side and dividing the angle at that vertex into two equal parts (Fig. M3.2*b*). The three angle bisectors intersect in a single point lying strictly inside the triangle. This point is equidistant from all sides and is called the *incenter* (the center of the *incircle* of the triangle). The angle bisector through a vertex cuts the opposite side in ratio proportional to the adjacent sides of the triangle.

The length of the angle bisector l_a drawn to the side a is given by the formulas

$$l_a = \frac{\sqrt{bc[(b+c)^2 - a^2]}}{b+c} = \frac{\sqrt{4p(p-a)bc}}{b+c},$$

$$l_a = \frac{2cb\cos\left(\frac{1}{2}\alpha\right)}{b+c} = 2R\frac{\sin\beta\sin\gamma}{\cos\left[\frac{1}{2}(\beta-\gamma)\right]} = 2p\frac{\sin\left(\frac{1}{2}\beta\right)\sin\left(\frac{1}{2}\gamma\right)}{\sin\beta + \sin\gamma},$$

(M3.1.1.2)

where R is the circumradius (see below).

An *altitude* of a triangle is a straight line passing through a vertex and perpendicular to the straight line containing the opposite side (Fig. M3.2*c*). The three altitudes of a triangle intersect in a single point, called the *orthocenter* of the triangle.

The length of the altitude h_a to the side a is given by the formulas

$$h_a = b\sin\gamma = c\sin\beta = \frac{bc}{2R},$$

$$h_a = 2(p-a)\cos\frac{\alpha}{2}\cos\frac{\beta}{2}\cos\frac{\gamma}{2} = 2(p-b)\sin\frac{\alpha}{2}\sin\frac{\beta}{2}\cos\frac{\gamma}{2}.$$

(M3.1.1.3)

The lengths of the altitude, the angle bisector, and the median through the same vertex satisfy the inequality $h_a \le l_a \le m_a$. If $h_a = l_a = m_a$, then the triangle is isosceles; moreover, the first equality implies the second, and vice versa.

▶ **Circumcircle and incircle.** A straight line passing through the midpoint of a segment and perpendicular to it is called the *perpendicular bisector* of the segment. The circle passing through the vertices of a triangle is called the *circumcircle* of the triangle. The center O_1 of the circumcircle, called the *circumcenter*, is the point where the perpendicular bisectors of the sides of the triangle meet (Fig. M3.3*a*). The feet of the perpendiculars drawn from a point Q on the circumcircle to the three sides of the triangle lie on the same straight line called the *Simpson line* of Q with respect to the triangle (Fig. M3.3*b*). The circumcenter, the orthocenter, and the centroid lie on a single line, called the *Euler line* (Fig. M3.3*c*).

The circle tangent to the three sides of a triangle and lying inside the triangle is called the *incircle* of the triangle. The center O_2 of the incircle (the *incenter*) is the point where the angle bisectors meet (Fig. M3.4*a*). The straight lines connecting the vertices of a triangle with the points at which the incircle is tangent to the respective opposite sides intersect in a single point G called the *Gergonne point* (Fig. M3.4*b*).

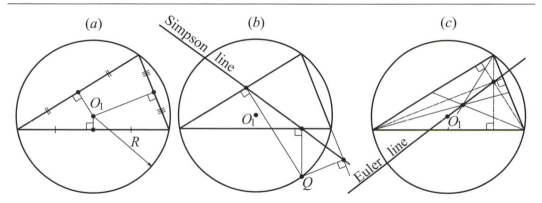

Figure M3.3. The circumcircle of a triangle. The circumcenter (*a*), the Simpson line (*b*), and the Euler line (*c*).

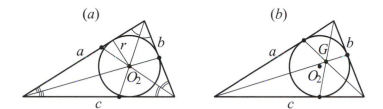

Figure M3.4. The incircle of a triangle (*a*). The incenter and the Gergonne point (*b*).

The inradius r and the circumradius R satisfy the relations

$$r = \sqrt{\frac{(p-a)(p-b)(p-c)}{p}} = p\tan\frac{\alpha}{2}\tan\frac{\beta}{2}\tan\frac{\gamma}{2} = (p-c)\tan\frac{\gamma}{2}, \quad \text{(M3.1.1.4)}$$

$$R = \frac{a}{2\sin\alpha} = \frac{b}{2\sin\beta} = \frac{c}{2\sin\gamma} = \frac{p}{4\cos\left(\frac{1}{2}\alpha\right)\cos\left(\frac{1}{2}\beta\right)\cos\left(\frac{1}{2}\gamma\right)}. \quad \text{(M3.1.1.5)}$$

The distance d between the circumcenter and the incenter is given by the expression

$$d = \sqrt{R^2 - 2Rr}. \quad \text{(M3.1.1.6)}$$

▶ **Area of a triangle.** The area S of a triangle is given by the formulas

$$S = \tfrac{1}{2}ah_a = \tfrac{1}{2}ab\sin\gamma = rp, \quad \text{(M3.1.1.7)}$$

$$S = \sqrt{p(p-a)(p-b)(p-c)} \quad \text{(Heron's formula)}, \quad \text{(M3.1.1.8)}$$

$$S = \frac{abc}{4R} = 2R^2\sin\alpha\sin\beta\sin\gamma, \quad \text{(M3.1.1.9)}$$

$$S = c^2\frac{\sin\alpha\sin\beta}{2\sin\gamma} = c^2\frac{\sin\alpha\sin\beta}{2\sin(\alpha+\beta)}. \quad \text{(M3.1.1.10)}$$

▶ **Right (right-angled) triangles.** A *right* triangle is a triangle with a right angle. The side opposite the right angle is called the *hypotenuse*, and the other two sides are called the *legs* (Fig. M3.5).

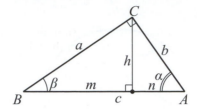

Figure M3.5. A right triangle.

The hypotenuse c, the legs a and b, and the angles α and β opposite the legs satisfy the following relations:

$$\alpha + \beta = 90°;$$

$$\sin \alpha = \cos \beta = \frac{a}{c}, \quad \sin \beta = \cos \alpha = \frac{b}{c},$$
$$\tan \alpha = \cot \beta = \frac{a}{b}, \quad \tan \beta = \cot \alpha = \frac{b}{a}. \qquad (\text{M3.1.1.11})$$

One also has

$$a^2 + b^2 = c^2 \quad (\text{PYTHAGOREAN THEOREM}), \qquad (\text{M3.1.1.12})$$
$$h^2 = mn, \quad a^2 = mc, \quad b^2 = nc, \qquad (\text{M3.1.1.13})$$

where h is the length of the altitude drawn to the hypotenuse; moreover, the altitude cuts the hypotenuse into segments of lengths m and n.

In a right triangle, the length of the median m_c drawn from the vertex of the right angle coincides with the circumradius R and is equal to half the length of the hypotenuse c, $m_c = R = \frac{1}{2}c$. The inradius is given by the formula $r = \frac{1}{2}(a + b - c)$. The area of the right triangle is $S = \frac{1}{2}ah_a = \frac{1}{2}ab$ (see also formulas (M3.1.1.4), (M3.1.1.5), and (M3.1.1.9)).

▶ **Isosceles and equilateral triangles.**

1°. An *isosceles* triangle is a triangle with two equal sides. These sides are called the *legs*, and the third side is called the *base* (Fig. M3.6a).

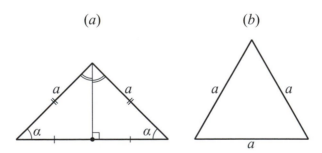

Figure M3.6. An isosceles triangle (*a*). An equilateral triangle (*b*).

Properties of isosceles triangles:

1. In an isosceles triangle, the angles adjacent to the base are equal.
2. In an isosceles triangle, the median drawn to the base is the angle bisector and the altitude.
3. In an isosceles triangle, the sum of distances from a point of the base to the legs is constant.

Criteria for a triangle to be isosceles:

1. If two angles in a triangle are equal, then the triangle is isosceles.
2. If a median in a triangle is also an altitude, then the triangle is isosceles.
3. If a bisector in a triangle is also an altitude, then the triangle is isosceles.

$2°$. An *equilateral* (or *regular*) triangle is a triangle with all three sides equal (Fig. M3.6b). All angles of an equilateral triangle are equal to $60°$. In an equilateral triangle, the circumradius R and the inradius r satisfy the relation $R = 2r$.

For an equilateral triangle with side length a, the circumradius and the inradius are given by the formulas $R = \frac{\sqrt{3}}{3}a$ and $r = \frac{\sqrt{3}}{6}a$, and the area is equal to $S = \frac{\sqrt{3}}{4}a^2$.

M3.1.2. Polygons

▶ **Polygons. Basic information.** A *polygon* is a plane figure bounded by a closed broken line. The straight line segments forming a polygon are called its *sides* (or *edges*). The points at which two sides meet are called the *vertices* (or *corners*) of the polygon. Two sides sharing a vertex, as well as two successive vertices (the endpoints of the same edge), are said to be adjacent. A polygon is said to be *convex* if it lies on one side of any straight line passing through two neighboring vertices. In what follows, we consider only simple convex polygons.

An *(interior) angle* of a convex polygon is the angle between two sides meeting in a vertex. A convex polygon is said to be *inscribed* in a circle if all of its vertices lie on the circle. A polygon is said to be *circumscribed* about a circle if all of its sides are tangent to the circle.

For a convex polygon with n sides, the sum of interior angles is equal to $180°(n-2)$. One can find the area of an arbitrary polygon by dividing it into triangles.

▶ **Properties of quadrilaterals.**
1. The diagonals of a convex quadrilateral meet.
2. The sum of interior angles of a convex quadrilateral equals $360°$ (Figs. M3.7a and b).
3. The lengths of the sides a, b, c, and d, the diagonals d_1 and d_2, and the segment m connecting the midpoints of the diagonals satisfy the relation $a^2 + b^2 + c^2 + d^2 = d_1^2 + d_2^2 + 4m^2$.
4. A convex quadrilateral is circumscribed if and only if $a + c = b + d$.
5. A convex quadrilateral is inscribed if and only if $\alpha + \gamma = \beta + \delta$.
6. The relation $ac + bd = d_1 d_2$ holds for inscribed quadrilaterals (PTOLEMY'S THEOREM).

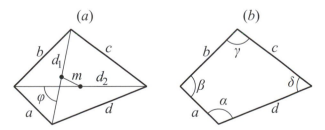

Figure M3.7. Quadrilaterals.

▶ **Areas of quadrilaterals.** The area of a convex quadrilateral is equal to

$$S = \frac{1}{2}d_1 d_2 \sin \varphi = \sqrt{p(p-a)(p-b)(p-c)(p-d) - abcd \cos^2 \frac{\beta+\delta}{2}}, \qquad \text{(M3.1.2.1)}$$

where φ is the angle between the diagonals d_1 and d_2 and $p = \frac{1}{2}(a+b+c+d)$.

The area of an inscribed quadrilateral is

$$S = \sqrt{p(p-a)(p-b)(p-c)(p-d)}. \qquad \text{(M3.1.2.2)}$$

The area of a circumscribed quadrilateral is

$$S = \sqrt{abcd \sin^2 \frac{\beta + \delta}{2}}. \qquad \text{(M3.1.2.3)}$$

▶ **Basic quadrilaterals.**

$1°$. A *parallelogram* is a quadrilateral such that both pairs of opposite sides are parallel (Fig. M3.8a).

(a) $\qquad\qquad\qquad\qquad\qquad (b)$

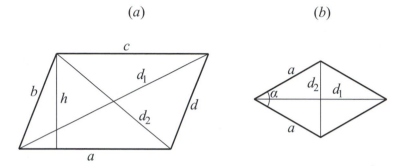

Figure M3.8. A parallelogram (a) and a rhombus (b).

Attributes of parallelograms (a quadrilateral is a parallelogram if):
1. Both pairs of opposite sides have equal length.
2. Both pairs of opposite angles are equal.
3. Two opposite sides are parallel and have equal length.

Properties of parallelograms:
1. The diagonals meet and bisect each other.
2. Opposite sides have equal length, and opposite angles are equal.
3. The diagonals and the sides satisfy the relation $d_1^2 + d_2^2 = 2(a^2 + b^2)$.
4. The area of a parallelogram is $S = ah$, where h is the altitude.

$2°$. A *rhombus* is a parallelogram in which all sides are of equal length (Fig. M3.8b).

Properties of rhombi:
1. The diagonals are perpendicular.
2. The diagonals are angle bisectors.
3. The area of a rhombus is $S = ah = a^2 \sin \alpha = \frac{1}{2} d_1 d_2$.

$3°$. A *rectangle* is a parallelogram in which all angles are right angles (Fig. M3.9a).

(a) $\qquad\qquad\qquad\qquad\qquad (b)$

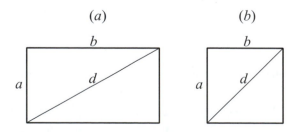

Figure M3.9. A rectangle (a) and a square (b).

Properties of rectangles:
1. The diagonals have equal lengths.
2. The area of a rectangle is $S = ab$.

4°. A *square* is a rectangle in which all sides have equal lengths (Fig. M3.9b). A square is also a rhombus with right angles.

Properties of squares:
1. All angles are right angles.
2. The diagonals are equal to $d = a\sqrt{2}$.
3. The diagonals meet at a right angle and are angle bisectors.
4. The area of a square is equal to $S = a^2 = \frac{1}{2}d^2$.

5°. A *trapezoid* is a quadrilateral in which two sides are parallel and the other two sides are nonparallel (Fig. M3.10). The parallel sides a and b are called the *bases* of the trapezoid, and the other two sides are called the *legs*. In an *isosceles* trapezoid, the legs are of equal length. The line segment connecting the midpoints of the legs is called the *median* of the trapezoid. The length of the median is equal to half the sum of the lengths of the bases, $m = \frac{1}{2}(a + b)$.

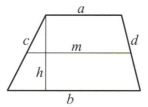

Figure M3.10. A trapezoid.

The perpendicular distance between the bases is called the *altitude* of a trapezoid.

Properties of trapezoids:
1. A trapezoid is circumscribed if and only if $a + b = c + d$.
2. A trapezoid is inscribed if and only if it is isosceles.
3. The area of a trapezoid is $S = \frac{1}{2}(a + b)h = mh = \frac{1}{2}d_1 d_2 \sin \varphi$, where φ is the angle between the diagonals d_1 and d_2.
4. The segment connecting the midpoints of the diagonals is parallel to the bases and has the length $\frac{1}{2}(b - a)$.

▶ **Regular polygons.** A convex polygon is said to be *regular* if all of its sides have the same length and all of its interior angles are equal. A convex n-gon is regular if and only if it is taken to itself by the rotation by an angle of $2\pi/n$ about some point O. The point O is called the *center* of the regular polygon. The angle between two rays issuing from the center and passing through two neighboring vertices is called the *central angle* (Fig. M3.11).

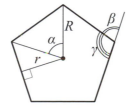

Figure M3.11. A regular polygon.

Properties of regular polygons:

1. The center is equidistant from all vertices as well as from all sides of a regular polygon.
2. A regular polygon is simultaneously inscribed and circumscribed; the centers of the circumcircle and the incircle coincide with the center of the polygon itself.
3. In a regular polygon, the central angle is $\alpha = 360°/n$, the external angle is $\beta = 360°/n$, and the interior angle is $\gamma = 180° - \beta$.
4. The circumradius R, the inradius r, and the side length a of a regular polygon satisfy the relations

$$a = 2\sqrt{R^2 - r^2} = 2R\sin\frac{\alpha}{2} = 2r\tan\frac{\alpha}{2}. \qquad \text{(M3.1.2.4)}$$

5. The area S of a regular n-gon is given by the formula

$$S = \frac{arn}{2} = nr^2\tan\frac{\alpha}{2} = nR^2\sin\frac{\alpha}{2} = \frac{1}{4}na^2\cot\frac{\alpha}{2}. \qquad \text{(M3.1.2.5)}$$

Table M3.3 presents several useful formulas for regular polygons.

TABLE M3.3
Regular polygons (a is the side length).

No.	Name	Inradius r	Circumradius R	Area S
1	Regular polygon	$\dfrac{a}{2\tan\frac{\pi}{n}}$	$\dfrac{a}{2\sin\frac{\pi}{n}}$	$\dfrac{1}{2}arn$
2	Triangle	$\dfrac{\sqrt{3}}{6}a$	$\dfrac{\sqrt{3}}{3}a$	$\dfrac{\sqrt{3}}{4}a^2$
3	Square	$\dfrac{1}{2}a$	$\dfrac{1}{\sqrt{2}}a$	a^2
4	Pentagon	$\sqrt{\dfrac{5+2\sqrt{5}}{20}}\,a$	$\sqrt{\dfrac{5+\sqrt{5}}{10}}\,a$	$\dfrac{\sqrt{25+10\sqrt{5}}}{4}a^2$
5	Hexagon	$\dfrac{\sqrt{3}}{2}a$	a	$\dfrac{3\sqrt{3}}{2}a^2$
6	Octagon	$\dfrac{1+\sqrt{2}}{2}a$	$\dfrac{\sqrt{2+\sqrt{2}}}{2}a$	$2(1+\sqrt{2})a^2$
7	Enneagon	$\dfrac{5+2\sqrt{5}}{2}a$	$\dfrac{1+\sqrt{5}}{2}a$	$\dfrac{\sqrt{5+2\sqrt{5}}}{2}a^2$
8	Dodecagon	$\dfrac{2+\sqrt{3}}{2}a$	$\dfrac{3+\sqrt{3}}{\sqrt{6}}a$	$3(2+\sqrt{3})a^2$

M3.1.3. Circle

▶ **Some definitions and formulas.** A *circle* is the set of all points in the plane that are the same fixed distance R from a fixed point O (Fig. M3.12a). The distance R is called the *radius* of the circle and the point O is called its *center*. A plane figure bounded by a circle, including its interior, is called a *disk*. A segment connecting two points on a circle is called a *chord*. A chord passing through the center of a circle is called a *diameter* of the circle (Fig. M3.12b). The length of a diameter is $d = 2R$. A straight line that touches a circle at a single point is called a *tangent*, and the common point is called the *point of tangency* (Fig. M3.12c). A straight line that cuts a circle at two points, an extended chord, is called a

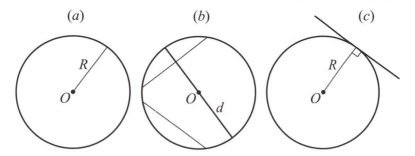

Figure M3.12. A circle (*a*), a diameter (*b*) and a tangent (*c*) of a circle.

secant. The angle formed by two radii is called a *central angle.* The angle formed by two chords with a common endpoint is called an *inscribed angle.*

Properties of circles and disks:

1. The circumference is $L = 2\pi R = \pi d = 2\sqrt{\pi S}$.
2. The area of a disk is $S = \pi R^2 = \frac{1}{4}\pi d^2 = \frac{1}{4}Ld$.
3. The diameter of a circle is a longest chord.
4. The diameter passing through the midpoint of a chord is perpendicular to the chord.
5. The radius drawn to the point of tangency is perpendicular to the tangent.
6. An inscribed angle is half the central angle subtended by the same chord, $\alpha = \frac{1}{2}\angle BOC$ (Fig. M3.13*a*).

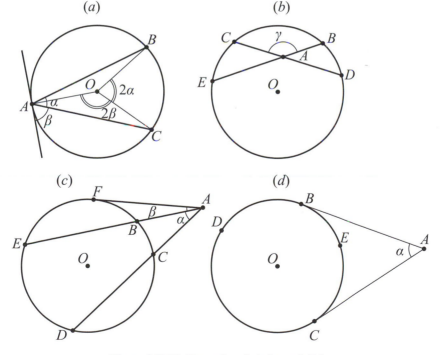

Figure M3.13. Properties of circles and disks.

7. The angle between a chord, AC, and the tangent to the circle at an endpoint, A, of the chord is equal to $\beta = \frac{1}{2}\angle AOC$ (Fig. M3.13*a*).
8. The angle between two chords, BE and CD, is $\gamma = \frac{1}{2}(\overset{\smile}{BC} + \overset{\smile}{ED})$ (Fig. M3.13*b*).

9. The angle between two secants, AD and AE, is $\alpha = \frac{1}{2}(D\overset{\smile}{E} - B\overset{\smile}{C})$ (Fig. M3.13c).
10. The angle between a secant, AE, and the tangent, AF, to the circle at a point, F, is equal to $\beta = \frac{1}{2}(F\overset{\smile}{E} - B\overset{\smile}{F})$ (Fig. M3.13c).
11. The angle between two tangents, AB and AC, is $\alpha = \frac{1}{2}(B\overset{\smile}{D}C - B\overset{\smile}{E}C)$ (Fig. M3.13d).
12. If two chords, BE and CD, meet, then $AC \cdot AD = AB \cdot AE = R^2 - m^2$, $m = OA$ (Fig. M3.13b).
13. For secants, AE and AD, the relations $AC \cdot AD = AB \cdot AE = m^2 - R^2$ hold (Fig. M3.13c).
14. For a tangent, AF, and a secant, AD, the relation $AF^2 = AC \cdot AD$ holds (Fig. M3.13c).

▶ **Segment and sector.** A plane figure bounded by two radii and one of the subtended arcs is called a *(circular) sector*. A plane figure bounded by an arc and the corresponding chord is called a *segment* (Fig. M3.14a). If R is the radius of the circle, l is the arc length,

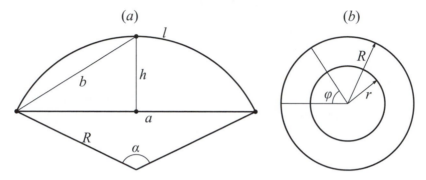

Figure M3.14. A segment (*a*) and an annulus (*b*).

a is the chord length, α is the central angle (in degrees), and h is the height of the segment, then the following relations hold:

$$a = 2\sqrt{2hR - h^2} = 2R\sin\frac{\alpha}{2},$$

$$h = R - \sqrt{R^2 - \frac{a^2}{4}} = R\left(1 - \cos\frac{\alpha}{2}\right) = \frac{a}{2}\tan\frac{\alpha}{4}, \tag{M3.1.3.1}$$

$$l = \frac{2\pi R\alpha}{360} \approx 0.01745\, R\alpha.$$

The area of a circular sector is given by the formula

$$S = \frac{lR}{2} = \frac{\pi R^2 \alpha}{360} \approx 0.00873\, R^2 \alpha, \tag{M3.1.3.2}$$

and the area of a segment not equal to a half-disk is given by

$$S_1 = \frac{\pi R^2 \alpha}{360} \pm S_\Delta, \tag{M3.1.3.3}$$

where S_Δ is the area of the triangle with vertices at the center of the disk and at the endpoints of the radii bounding the corresponding sector. One takes the minus sign for $\alpha < 180$ and the plus sign for $\alpha > 180$.

The arc length and the area of a segment can be found by the approximate formulas

$$l \approx \frac{8b - a}{3}, \quad l \approx \sqrt{a^2 + \frac{16h^2}{3}},$$
$$S_1 \approx \frac{h(6a + 8b)}{15}, \tag{M3.1.3.4}$$

where b is the chord of the half-segment (see Fig. M3.14a).

▶ **Annulus.** An *annulus* is a plane figure bounded by two concentric circles of distinct radii (Fig. M3.14b). Let R be the outer radius of an annulus (the radius of the outer bounding circle), and let r be the inner radius (the radius of the inner bounding circle). Then the area of the annulus is given by the formula

$$S = \pi(R^2 - r^2) = \frac{\pi}{4}(D^2 - d^2) = 2\pi\rho\delta, \tag{M3.1.3.5}$$

where $D = 2R$ and $d = 2r$ are the outer and inner diameters, $\rho = \frac{1}{2}(R + r)$ is the midradius, and $\delta = R - r$ is the width of the annulus.

The area of the part of the annulus contained in a sector of central angle φ, given in degrees (see Fig. M3.14b), is given by the formula

$$S = \frac{\pi\varphi}{360}(R^2 - r^2) = \frac{\pi\varphi}{1440}(D^2 - d^2) = \frac{\pi\varphi}{180}\rho\delta. \tag{M3.1.3.6}$$

M3.2. Solid Geometry

M3.2.1. Straight Lines, Planes, and Angles in Space

▶ **Mutual arrangement of straight lines and planes.**

1°. Two distinct straight lines lying in a single plane either have exactly one point of intersection or do not meet at all. In the latter case, they are said to be *parallel*. If two straight lines do not lie in a single plane, then they are called *skew lines*.

The angle between skew lines is determined as the angle between lines parallel to them and lying in a single plane (Fig. M3.15a). The distance between skew lines is the length of the straight line segment that meets both lines and is perpendicular to them.

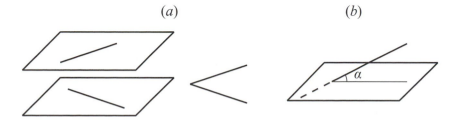

(a)   *(b)*

Figure M3.15. The angle between skew lines (a). The angle between a line and a plane (b).

2°. Two distinct planes either intersect in a straight line or do not have common points. In the latter case, they are said to be *parallel*. Coinciding planes are also assumed to be parallel. If two planes are perpendicular to a single straight line or each of them contains a pair of intersecting straight lines parallel to the corresponding lines in the other pair, then the planes are parallel.

$3°$. A straight line either entirely lies in the plane, meets the plane at a single point, or has no common points with the plane. In the last case, the line is said to be *parallel* to the plane.

The angle between a straight line and a plane is equal to the angle between the line and its projection onto the plane (Fig. M3.15*b*). If a straight line is perpendicular to two intersecting straight lines on a plane, then it is perpendicular to each line on the plane, i.e., *perpendicular to the plane*.

▶ **Polyhedral angles.**

$1°$. A *dihedral angle* is a figure in space formed by two half-planes issuing from a single straight line as well as the part of space bounded by these half-planes. The half-planes are called the *faces* of the dihedral angle, and their common straight line is called the *edge*. A dihedral angle is measured by its linear angle ABC (Fig. M3.16*a*), i.e., by the angle between the perpendiculars raised to the edge DE of the dihedral angle in both planes (*faces*) at the same point.

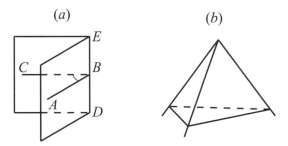

Figure M3.16. A dihedral (*a*) and a trihedral (*b*) angle.

$2°$. A part of space bounded by an infinite triangular pyramid is called a *trihedral angle* (Fig. M3.16*b*). The faces of this pyramid are called the *faces* of the trihedral angle, and the vertex of the pyramid is called the *vertex of a trihedral angle*. The rays in which the faces intersect are called the *edges* of a trihedral angle. The edges form *face angles*, and the faces form the dihedral angles of the trihedral angle. As a rule, one considers trihedral angles with dihedral angles less than π (or $180°$), i.e., convex trihedral angles. Each face angle of a convex trihedral angle is less than the sum of the other two face angles and greater than their difference.

Two trihedral angles are equal if one of the following conditions is satisfied:

1. Two face angles, together with the included dihedral angle, of the first trihedral angle are equal to the respective parts (arranged in the same order) of the second trihedral angle.
2. Two dihedral angles, together with the included face angle, of the first trihedral angle are equal to the respective parts (arranged in the same order) of the second trihedral angle.
3. The three face angles of the first trihedral angle are equal to the respective face angles (arranged in the same order) of the second trihedral angle.
4. The three dihedral angles of the first trihedral angle are equal to the respective dihedral angles (arranged in the same order) of the second trihedral angle.

$3°$. A *polyhedral angle* $OABCDE$ (Fig. M3.17*a*) is formed by several planes (*faces*) having a single common point (the *vertex*) and successively intersecting along straight lines $OA, OB, \ldots, OE$ (the *edges*). Two edges belonging to the same face form a *face angle* of the polyhedral angle, and two neighboring faces form a *dihedral angle*.

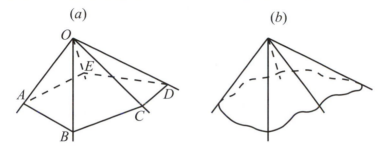

Figure M3.17. A polyhedral (*a*) and a solid (*b*) angle.

Polyhedral angles are equal (*congruent*) if one can be transformed into the other by translations and rotations. For polyhedral angles to be congruent, the corresponding parts (face and dihedral angles) must be equal.

A convex polyhedral angle lies entirely on one side of each of its faces. The sum $\angle AOB + \angle BOC + \cdots + \angle EOA$ of face angles (Fig. M3.17*a*) of any convex polyhedral angle is less that 2π (or $360°$).

4°. A *solid angle* is a part of space bounded by straight lines issuing from a single point (vertex) to all points of some closed curve (Fig. M3.17*b*). Trihedral and polyhedral angles are special cases of solid angles. A solid angle is measured by the area cut by the solid angle on the sphere of unit radius centered at the vertex. Solid angles are measured in steradians. The entire sphere forms a solid angle of 4π steradians.

M3.2.2. Polyhedra

▶ **General concepts.** A *polyhedron* is a closed object formed by intersecting planes. In other words, a polyhedron is a set of finitely many plane polygons satisfying the following conditions:

1. Each side of each polygon is simultaneously a side of a unique other polygon, which is said to be adjacent to the first polygon (via this side).
2. From each of the polygons forming a polyhedron, one can reach any other polygon by successively passing to adjacent polygons.

These polygons are called the *faces*, their sides are called the *edges*, and their vertices are called the *vertices* of a polyhedron.

A polyhedron is said to be *convex* if it lies entirely on one side of the plane of any of its faces; if a polyhedron is convex, then so are its faces.

EULER'S THEOREM. *If the number of vertices in a convex polyhedron is v, the number of edges is e, and the number of faces is f, then $v + f - e = 2$.*

▶ **Prism. Parallelepiped.**

1°. An n-sided *prism* is a polyhedron in which two faces are equal n-gons (the *base faces*) that lie on parallel planes and have respectively parallel sides, and the remaining n faces (*joining* or *lateral faces*) are parallelograms; see Fig. M3.18*a*. A *right* prism is a prism in which the lateral faces are perpendicular to the base faces; otherwise it is an *oblique* prism. A right prism is said to be *regular* if its base faces are regular polygons.

If l is the lateral edge length, S is the area of the base face, H is the height of the prism (perpendicular distance between the planes of the bases), P_{sec} is the perimeter of a normal

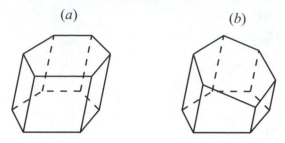

Figure M3.18. A prism (*a*) and a truncated prism (*b*).

section, one perpendicular to a lateral edge, and $S_{\sec}$ is the area of a normal section, then the area of the lateral surface S_{lat} and the volume V of the prism are given by the formulas

$$
\begin{aligned}
S_{\text{lat}} &= P_{\sec} l \\
V &= SH = S_{\sec} l.
\end{aligned}
\tag{M3.2.2.1}
$$

The portion of a prism between one of the bases and a plane nonparallel to it is called a *truncated prism* (Fig. M3.18*b*). The volume of a truncated prism is

$$
V = LS_1,
\tag{M3.2.2.2}
$$

where L is the length of the segment connecting the centroids of the base faces and S_1 is the area of the section of the prism by a plane perpendicular to this segment.

The volume of a truncated regular prism (its base being a regular n-gon) is expressed as

$$
V = S_{\sec} \frac{l_1 + l_2 + \cdots + l_n}{n},
$$

where $S_{\sec}$ is the area of a normal section and $l_1, l_2, \ldots, l_n$ are the lengths of the lateral edges.

2°. A prism whose bases are parallelograms is called a *parallelepiped*. All four diagonals in a parallelepiped intersect at a single point and bisect each other (Fig. M3.19*a*). A parallelepiped is said to be *rectangular* if it is a right prism and its base faces are rectangles. In a rectangular parallelepiped, all diagonals are equal (Fig. M3.19*b*).

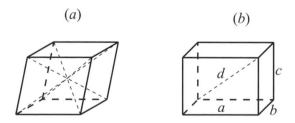

Figure M3.19. A parallelepiped (*a*) and a rectangular parallelepiped (*b*).

If a, b, and c are the lengths of the edges of a rectangular parallelepiped, then the diagonal d can be determined by the formula $d^2 = a^2 + b^2 + c^2$. The volume of a rectangular parallelepiped is given by the formula $V = abc$, and the lateral surface area is $S_{\text{lat}} = PH$, where P is the perimeter of the base face.

3°. A rectangular parallelepiped all of whose edges are equal ($a = b = c$) is called a *cube*. The diagonal of a cube is given by the formula $d^2 = 3a^2$. The volume of the cube is $V = a^3$, and the lateral surface area is $S_{\text{lat}} = 4a^2$.

▶ **Pyramid, obelisk, and wedge.**

1°. A *pyramid* is a polyhedron in which one face (the *base* of the pyramid) is an arbitrary polygon and the other (*lateral*) faces are triangles with a common vertex, called the *apex* of the pyramid (Fig. M3.20a). The base of an *n-sided* pyramid is an *n*-gon. The perpendicular through the apex to the base of a pyramid is called the *altitude* (*height*) of the pyramid.

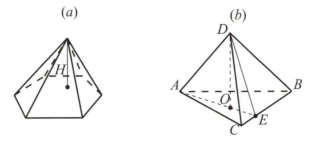

Figure M3.20. A pyramid (*a*). The altitude DO, the plane DAE, and the side BC in a triangular pyramid (*b*).

The volume of a pyramid is given by the formula

$$V = \frac{1}{3}SH, \tag{M3.2.2.3}$$

where S is the area of the base and H is the altitude of the pyramid.

If DO is the altitude of the pyramid $ABCD$ and $DA \perp BC$, then the plane DAE is perpendicular to BC (Fig. M3.20b).

If the pyramid is cut by a plane (Fig. M3.21a) parallel to the base, then

$$\frac{SA_1}{A_1A} = \frac{SB_1}{B_1B} = \cdots = \frac{SO_1}{O_1O},$$

$$\frac{S_{ABCDEF}}{S_{A_1B_1C_1D_1E_1F_1}} = \left(\frac{SO}{SO_1}\right)^2, \tag{M3.2.2.4}$$

where SO is the altitude of the pyramid.

The altitude of a *triangular pyramid* passes through the orthocenter of its base if and only if all pairs of opposite edges of the pyramid are perpendicular.

Given the length of the edges, $DA = a$, $DB = b$, $DC = c$, $BC = p$, $AC = q$, and $AB = r$, of a triangular pyramid (Fig. M3.21b), its volume can be found from the relation

$$V^2 = \frac{1}{288} \begin{vmatrix} 0 & r^2 & q^2 & a^2 & 1 \\ r^2 & 0 & p^2 & b^2 & 1 \\ q^2 & p^2 & 0 & c^2 & 1 \\ a^2 & b^2 & c^2 & 0 & 1 \\ 1 & 1 & 1 & 1 & 0 \end{vmatrix}, \tag{M3.2.2.5}$$

where the right-hand side contains a determinant.

A pyramid is said to be *regular* if its base is a regular *n*-gon and the altitude passes through the center of the base. The altitude (dropped from the apex) of a lateral face is called the *apothem* of a regular pyramid. For a regular pyramid, the lateral surface area is

$$S_{\text{lat}} = \frac{1}{2}Pl, \tag{M3.2.2.6}$$

where P is the perimeter of the base and l is the apothem.

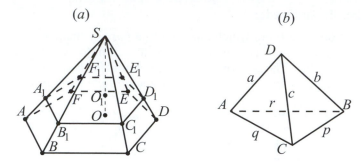

Figure M3.21. The original pyramid and a pyramid cut off by a plane (*a*). A triangular pyramid (*b*).

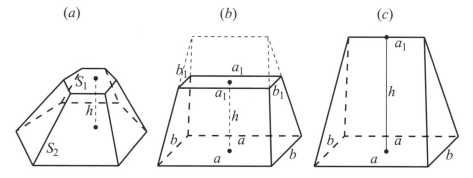

Figure M3.22. A frustum of a pyramid (*a*), an obelisk (*b*), and a wedge (*c*).

2°. If a pyramid is cut by a plane parallel to the base, then it splits into two parts: a pyramid similar to the original pyramid and the *frustum* (Fig. M3.22a). The volume of the frustum is

$$V = \frac{1}{3}h(S_1 + S_2 + \sqrt{S_1 S_2}) = \frac{1}{3}hS_2\left(1 + \frac{a}{A} + \frac{a^2}{A^2}\right), \qquad (M3.2.2.7)$$

where S_1 and S_2 are the areas of the bases, a and A are two respective sides of the bases, and h is the altitude (the perpendicular distance between the bases).

For a regular frustum, the lateral surface area is

$$S_{\text{lat}} = \frac{1}{2}(P_1 + P_2)l, \qquad (M3.2.2.8)$$

where P_1 and P_2 are the perimeters of the bases and l is the altitude of the lateral face.

3°. A hexahedron whose bases are rectangles lying in parallel planes and whose lateral faces form equal angles with the base, but do not meet at a single point, is called an *obelisk* (Fig. M3.22b). If a, b and a_1, b_1 are the sides of the bases and h is the altitude, then the volume of the hexahedron is

$$V = \frac{h}{6}[(2a + a_1)b + (2a_1 + a)b_1]. \qquad (M3.2.2.9)$$

4°. A pentahedron whose base is a rectangle and whose lateral faces are isosceles triangles and isosceles trapezoids is called a *wedge* (Fig. M3.22c). The volume of the wedge is

$$V = \frac{h}{6}(2a + a_1)b. \qquad (M3.2.2.10)$$

▶ **Regular polyhedra.** A polyhedron is said to be *regular* if all of its faces are equal regular polygons and all polyhedral angles are equal to each other. There exist five regular polyhedra (Fig. M3.23), whose properties are given in Table M3.4.

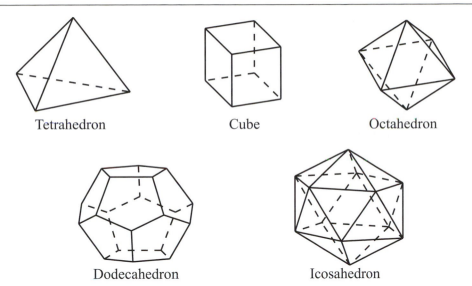

Tetrahedron   Cube   Octahedron

Dodecahedron   Icosahedron

Figure M3.23. Five regular polyhedra.

TABLE M3.4
Regular polyhedra (a is the edge length).

No.	Name	Number of faces and their shapes	Number of vertices	Number of edges	Total surface area	Volume
1	Tetrahedron	4 triangles	4	6	$a^2\sqrt{3}$	$\dfrac{a^3\sqrt{2}}{12}$
2	Cube	6 squares	8	12	$6a^2$	a^3
3	Octahedron	8 triangles	6	12	$2a^2\sqrt{3}$	$\dfrac{a^3\sqrt{2}}{3}$
4	Dodecahedron	12 pentagons	20	30	$3a^2\sqrt{25+10\sqrt{5}}$	$\dfrac{a^3}{4}(15+7\sqrt{5})$
5	Icosahedron	20 triangles	12	30	$3a^2\sqrt{3}$	$\dfrac{5a^3}{12}(3+\sqrt{5})$

M3.2.3. Solids Formed by Revolution of Lines

▶ **Cylinder.** A *cylindrical surface* is a surface in space swept by a straight line (the *generator*) moving parallel to a given direction along some curve (the *directrix*) (Fig. M3.24a).

$1°$. A solid bounded by a closed cylindrical surface and two planes is called a *cylinder*; the planes are called the *bases* of the cylinder (Fig. M3.24b).

If P is the perimeter of the base, P_{sec} is the perimeter of the section perpendicular to the generator, S_{sec} is the area of this section, S_{bas} is the area of the base, and l is the length of the generator, then the lateral surface area S_{lat} and the volume V of the cylinder are given by the formulas

$$S_{\text{lat}} = PH = P_{\text{sec}}l,$$
$$V = S_{\text{bas}}H = S_{\text{sec}}l. \tag{M3.2.3.1}$$

In a *right* cylinder, the bases are perpendicular to the generator. In particular, if the bases are disks, then one speaks of a *right circular cylinder*. The volume, the lateral surface

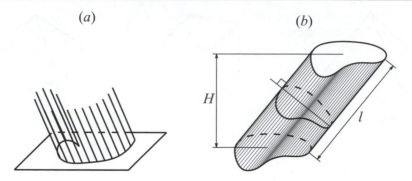

Figure M3.24. A cylindrical surface (*a*). A cylinder (*b*).

area, and the total surface area of a right circular cylinder are given by the formulas

$$
\begin{aligned}
V &= \pi R^2 H, \\
S_{\text{lat}} &= 2\pi R H, \\
S &= 2\pi R(R + H),
\end{aligned}
\tag{M3.2.3.2}
$$

where R is the radius of the base.

A right circular cylinder is also called a *round cylinder*, or simply a *cylinder*.

$2°$. The part of a cylinder cut by a plane nonparallel to the base is called a *truncated cylinder* (Fig. M3.25*a*).

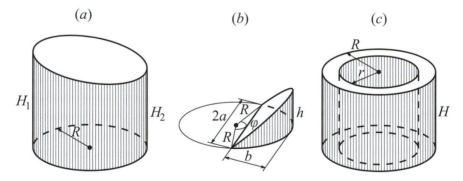

Figure M3.25. A truncated cylinder (*a*), a "hoof" (*b*), and a cylindrical tube (*c*).

The volume, the lateral surface area, and the total surface area of a truncated cylinder are given by the formulas

$$
\begin{aligned}
V &= \pi R^2 \frac{H_1 + H_2}{2}, \\
S_{\text{lat}} &= \pi R(H_1 + H_2), \\
S &= \pi R\left[H_1 + H_2 + R + \sqrt{R^2 + \left(\frac{H_2 - H_1}{2}\right)^2} \right],
\end{aligned}
\tag{M3.2.3.3}
$$

where H_1 and H_2 are the maximal and minimal generators.

$3°$. A *segment of a round cylinder (a "hoof")* is a portion of the cylinder cut by a plane that is nonparallel to the base and intersects it. If R is the radius of the cylindrical segment, h is the height of the "hoof," and b is its width (for the other notation, see Fig. M3.25b), then the volume V and the lateral surface area S_{lat} of the "hoof" can be determined by the formulas

$$V = \frac{h}{3b}\left[a(3R^2 - a^2) + 3R^2(b-R)\alpha\right] = \frac{hR^3}{b}\left(\sin\alpha - \frac{\sin^3\alpha}{3} - \alpha\cos\alpha\right),$$

$$S_{\text{lat}} = \frac{2\pi R}{b}[(b-R)\alpha + a], \tag{M3.2.3.4}$$

where $\alpha = \frac{1}{2}\varphi$ is measured in radians.

$4°$. A solid bounded by two closed cylindrical surfaces and two planes is called a *cylindrical tube*; the planes are called the bases of the tube. The volume of a round cylindrical tube (Fig. M3.25c) is

$$V = \pi H(R^2 - r^2) = \pi H\delta(2R - r) = \pi H\delta(2r + \delta) = 2\pi H\delta\rho, \tag{M3.2.3.5}$$

where R and r are the outer and inner radii, $\delta = R - r$ is the thickness, $\rho = \frac{1}{2}(R + r)$ is the midradius, and H is the height of the pipe.

▶ **Conical surface. Cone. Frustum of cone.** A *conical surface* is the union of straight lines (*generators*) passing through a fixed point (the *apex*) in space and any point of some space curve (the *directrix*) (Fig. M3.26a).

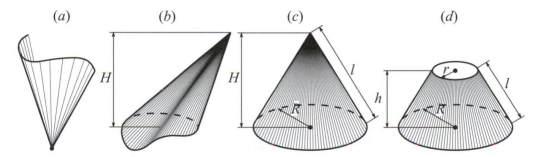

| (a) | (b) | (c) | (d) |

Figure M3.26. Conical surface (a). A cone (b), a right circular cone (c), and a frustum of a cone (d).

$1°$. A solid bounded by a conical surface with closed directrix and a plane is called a *cone*; the plane is the base of the cone (Fig. M3.26b). The volume of an arbitrary cone is given by the formula

$$V = \frac{1}{3}HS_{\text{bas}}, \tag{M3.2.3.6}$$

where H is the altitude of the cone and S_{bas} is the area of the base.

A *right circular cone* (Fig. M3.26c) has a disk as the base, and its vertex is projected onto the center of the disk. If l is the length of the generator and R is the radius of the base, then the volume, the lateral surface area, and the total surface area of the right circular cone are given by the formulas

$$V = \frac{1}{3}\pi R^2 H,$$

$$S_{\text{lat}} = \pi Rl = \pi R\sqrt{R^2 + H^2}, \tag{M3.2.3.7}$$

$$S = \pi R(R + l).$$

$2°$. If a cone is cut by a plane parallel to the base, then we obtain a *frustum of a cone* (Fig. M3.26d). The length l of the generator, the volume V, the lateral surface area S_{lat}, and the total surface area S of the frustum of a right circular cone are given by the formulas

$$l = \sqrt{h^2 + (R - r)^2},$$

$$V = \frac{\pi h}{3}(R^2 + r^2 + Rr),$$

$$S_{\text{lat}} = \pi l(R + r),$$

$$S = \pi[l(R + r) + R^2 + r^2],$$

(M3.2.3.8)

where r is the radius of the upper base and h is the altitude of the frustum of a cone.

▶ **Sphere. Spherical parts. The torus.**

$1°$. A *sphere* is the set of all points in space that are the same distance R from a fixed point O (Fig. M3.27a). The distance R is called the *radius* of the sphere and the point O is called its *center*. A straight line segment that passes through the center of a sphere and whose endpoints are on the sphere is called a *diameter* of the sphere. A solid formed by a sphere together with its interior is called a *ball*. Any section of the sphere by a plane is a circle. The section of the sphere by a plane passing through its center is called a *great circle* of radius R. There exists exactly one great circle passing through two arbitrary points on the sphere that are not antipodal (i.e., are not the opposite endpoints of a diameter); the smaller arc of this great circle is the shortest distance on the sphere between these points. The surface area S of a sphere and the volume V of the ball bounded by the sphere are given by

$$S = 4\pi R^2 = \pi D^2 = \sqrt[3]{36\pi V^2},$$

$$V = \frac{4\pi R^3}{3} = \frac{\pi D^3}{6} = \frac{1}{6}\sqrt{\frac{S^3}{\pi}},$$

(M3.2.3.9)

where $D = 2R$ is the diameter of the sphere.

(a)   (b)   (c)

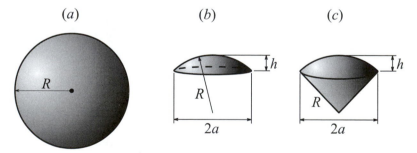

Figure M3.27. A sphere (*a*), a spherical cap (*b*), and a spherical sector (*c*).

$2°$. A portion of a ball cut from it by a plane is called a *spherical cap* (Fig. M3.27b). The width a (base radius), the area S_{lat} of the curved surface, the total surface area S, and the volume V of a spherical cap can be found from the formulas

$$a^2 = h(2R - h),$$

$$S_{\text{lat}} = 2\pi Rh = \pi(a^2 + h^2),$$

$$S = S_{\text{lat}} + \pi a^2 = \pi(2Rh + a^2) = \pi(h^2 + 2a^2),$$

(M3.2.3.10)

$$V = \frac{\pi h}{6}(3a^2 + h^2) = \frac{\pi h^2}{3}(3R - h),$$

where R and h are the radius and the height of the spherical cap.

$3°$. A portion of a ball bounded by the curved surface of a spherical cap and the conical surface whose base is the base of the cap and whose vertex is the center of the ball is called a *spherical sector* (Fig. M3.27c). The total surface area S and the volume V of a spherical sector are given by the formulas

$$S = \pi R(2h + a),$$
$$V = \frac{2}{3}\pi R^2 h, \qquad\qquad\qquad (\text{M3.2.3.11})$$

where a is the width of the spherical cap, h is its height, and R is the radius of the sector.

$4°$. A portion of a ball contained between two parallel plane secants is called a *spherical segment* (Fig. M3.28a). The curved surface of a spherical segment is called a *spherical zone*, and the plane circular surfaces are the *bases* of a spherical segment. The radius R of the ball, the radii a and b of the bases, and the height h of a spherical segment satisfy the relation

$$R^2 = a^2 + \left(\frac{a^2 - b^2 - h^2}{2h}\right)^2. \qquad\qquad (\text{M3.2.3.12})$$

The curved surface area S_{lat}, the total surface area S, and the volume V of a spherical segment are given by the formulas

$$S_{\text{lat}} = 2\pi Rh,$$
$$S = S_{\text{lat}} + \pi(a^2 + b^2) = \pi(2Rh + a^2 + b^2), \qquad (\text{M3.2.3.13})$$
$$V = \frac{\pi h}{6}(3a^2 + 3b^2 + h^2).$$

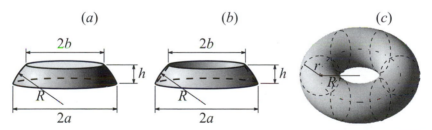

Figure M3.28. A spherical segment (a) and a spherical segment without the truncated cone inscribed in it (b). A torus (c).

If V_1 is the volume of the truncated cone inscribed in a spherical segment (Fig. M3.28b) and l is the length of its generator, then $V - V_1 = \frac{\pi h l^2}{6}$.

$5°$. A *torus* is a surface generated by revolving a circle about an axis coplanar with the circle but not intersecting it (Fig. M3.28c). If r is the radius of the circle being rotated and R is the distance from its center to the axis of revolution ($R > r$), then the surface area and the volume of the torus are given by

$$S = 4\pi^2 Rr = \pi^2 Dd,$$
$$V = 2\pi^2 Rr^2 = \frac{\pi^2 Dd^2}{4},$$

where $d = 2r$ and $D = 2R$ are the diameters of the generating circle and the circle of revolution.

Bibliography for Chapter M3

Alexander, D. C. and Koeberlein, G. M., *Elementary Geometry for College Students, 3rd Edition*, Houghton Mifflin Company, Boston, Massachusetts, 2002.

Gustafson, R. D. and Frisk, P. D., *Elementary Geometry, 3rd Edition*, Wiley, New York, 1991.

Moise, E., *Elementary Geometry from an Advanced Standpoint, 3rd Edition*, Addison Wesley, Boston, Massachusetts, 1990.

Musser, G. L. and Trimpe, L. E., *College Geometry: A Problem Solving Approach with Applications*, Prentice Hall, Englewood Cliffs, New Jersey, 1994.

Schultze, A. and Sevenoak, F. L., *Plane and Solid Geometry*, Adamant Media Corporation, Boston, Massachusetts, 2004.

Tussy, A. S. and Gustafson, R. D., *Basic Geometry for College Students: An Overview of the Fundamental Concepts of Geometry*, Thomson-Brooks/Cole, Pacific Grove, California, 2002.

Chapter M4
Analytic Geometry

M4.1. Points, Segments, and Coordinate Plane

M4.1.1. Cartesian and Polar Coordinates on Plane

▶ **Rectangular Cartesian coordinates in the plane.** A *rectangular Cartesian coordinate system* consists of two mutually perpendicular directed lines, called *coordinate axes*, each treated as a number line (see Subsection M6.1.1). The point of intersection of the axes is called the *origin* and usually labeled with the letter O, while the axes themselves are called the *coordinate axes*. As a rule, one of the coordinate axes is horizontal, directed from left to right, and called the *abscissa axis*. The other axis is vertical, directed upwards, and called the *ordinate axis*. The two axes are usually denoted by X or OX and Y or OY, respectively, and the coordinate system itself is denoted by XY or OXY. The two coordinate axes divide the plane into four parts, which are called *quadrants* and numbered I, II, III, and IV counterclockwise as shown in Fig. M4.1.

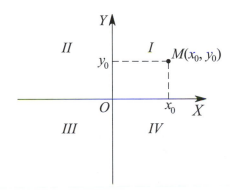

Figure M4.1. A rectangular Cartesian coordinate system.

Each point M in the plane is uniquely defined by a pair of real numbers $(x_0, \; y_0)$, called its *coordinates*, which specify its projections onto the X- and Y-axes. The numbers x_0 and y_0 are called, respectively, the *abscissa* and the *ordinate* of the point M.

Remark. Strictly speaking, the coordinate system introduced above is a *right rectangular Cartesian coordinate system*. A *left rectangular Cartesian coordinate system* can, for example, be obtained by changing the direction of one of the axes. A right rectangular Cartesian coordinate system is usually called simply a Cartesian coordinate system.

If A and B are two points in the plane, then the length of the segment AB will be denoted $|AB|$.

▶ **Polar coordinates.** A *polar coordinate system* is determined by a point O called the *pole*, a ray OA issuing from this point, which is called the *polar axis*, a scale segment for measuring lengths, and the positive sense of rotation around the pole. Usually, the counterclockwise sense is assumed to be positive (see Fig. M4.2a).

The position of each point M on the plane is determined by two *polar coordinates*, the *polar radius* $\rho = |OM|$ and the *polar angle* $\theta = \angle AOM$ (the values of the angle θ are defined up to an additive term $2\pi n$, where n is an integer). To be definite, one usually assumes that $0 \le \theta \le 2\pi$ or $-\pi \le \theta \le \pi$. The polar radius of the pole is zero, and its polar angle does not have any definite value.

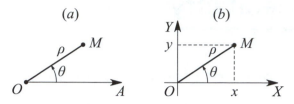

Figure M4.2. A polar coordinate system (a). Relationship between Cartesian and polar coordinates (b).

▶ **Relationship between Cartesian and polar coordinates.** Suppose that M is an arbitrary point in the plane, (x, y) are its rectangular Cartesian coordinates, and (ρ, θ) are its polar coordinates (see Fig. M4.2b). The transformation from one coordinate system to the other is expressed by the formulas

$$x = \rho \cos\theta, \qquad \text{or} \qquad \rho = \sqrt{x^2 + y^2}, \qquad \text{(M4.1.1.1)}$$
$$y = \rho \sin\theta \qquad\qquad\qquad \tan\theta = y/x,$$

where the polar angle θ is determined with regard to the quadrant where the point M lies.

Example. Let us find the polar coordinates ρ, θ ($0 \le \theta \le 2\pi$) of the point M whose Cartesian coordinates are $x = -3$, $y = -3$.

From formulas (M4.1.1.1), we obtain $\rho = \sqrt{(-3)^2 + (-3)^2} = 3\sqrt{2}$ and $\tan\theta = \frac{-3}{-3} = 1$. Since the point M lies in the third quadrant, we have $\theta = \arctan 1 + \pi = \frac{5}{4}\pi$.

M4.1.2. Distance Between Points. Division of Segment in Given Ratio. Area of a Polygon

▶ **Distance between points on plane.** The distance d between two arbitrary points $A_1(x_1, y_1)$ and $A_2(x_2, y_2)$ on the plane is given by the formula

$$d = \sqrt{(x_2 - x_1)^2 + (y_2 - y_1)^2},$$

where x and y with the respective subscripts are the Cartesian coordinates of these points, and by the formula

$$d = \sqrt{\rho_1^2 + \rho_2^2 - 2\rho_1\rho_2 \cos(\theta_2 - \theta_1)},$$

where ρ and θ with the respective subscripts are the polar coordinates of these points.

▶ **Angles between segments.** The angle β between arbitrary segments A_1A_2 and A_3A_4 joining the points $A_1(x_1, y_1)$, $A_2(x_2, y_2)$ and $A_3(x_3, y_3)$, $A_4(x_4, y_4)$, respectively, can be found from the relation

$$\cos\beta = \frac{(x_2 - x_1)(x_4 - x_3) + (y_2 - y_1)(y_4 - y_3)}{\sqrt{(x_2 - x_1)^2 + (y_2 - y_1)^2}\sqrt{(x_4 - x_3)^2 + (y_4 - y_3)^2}}.$$

▶ **Division of a segment in a given ratio.** Given two points, $A(x_1, y_1)$ and $B(x_2, y_2)$, and a number λ, the coordinates of the point $M(x, y)$ dividing the segment AB in the ratio $\lambda = |AM| : |MB|$ are expressed as

$$x = \frac{x_1 + \lambda x_2}{1 + \lambda}, \qquad y = \frac{y_1 + \lambda y_2}{1 + \lambda}. \tag{M4.1.2.1}$$

Example. Find the coordinates of the midpoint of a segment AB.

The midpoint of the segment corresponds to $\lambda = 1$. Substituting this value into (M4.1.2.1) gives $x = \dfrac{x_1 + x_2}{2}$, $y = \dfrac{y_1 + y_2}{2}$.

▶ **Area of a triangle.** The area S_3 of the triangle with vertices A_1, A_2, and A_3 is given by the formula

$$\pm S_3 = \frac{1}{2}[(x_2 - x_1)(y_3 - y_1) - (x_3 - x_1)(y_2 - y_1)]$$

$$= \frac{1}{2} \begin{vmatrix} x_2 - x_1 & y_2 - y_1 \\ x_3 - x_1 & y_3 - y_1 \end{vmatrix} = \frac{1}{2} \begin{vmatrix} x_1 & y_1 & 1 \\ x_2 & y_2 & 1 \\ x_3 & y_3 & 1 \end{vmatrix}, \tag{M4.1.2.2}$$

where x and y with respective subscripts are the Cartesian coordinates of the vertices, and by the formula

$$\pm S_3 = \frac{1}{2}[\rho_1\rho_2 \sin(\theta_2 - \theta_1) + \rho_2\rho_3 \sin(\theta_3 - \theta_2) + \rho_3\rho_1 \sin(\theta_1 - \theta_3)], \tag{M4.1.2.3}$$

where ρ and θ with respective subscripts are the polar coordinates of the vertices. In formulas (M4.1.2.2) and (M4.1.2.3), one takes the plus sign if the vertices are numbered counterclockwise (see Fig. M4.3a) and the minus sign otherwise.

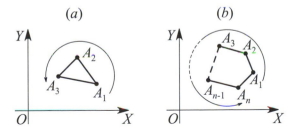

Figure M4.3. Area of a triangle (a) and of a polygon (b).

▶ **Area of a polygon.** The area S_n of the polygon with vertices $A_1, \dots, A_n$ is given by the formula

$$\pm S_n = \frac{1}{2}[(x_1 - x_2)(y_1 + y_2) + (x_2 - x_3)(y_2 + y_3) + \cdots + (x_n - x_1)(y_n + y_1)], \tag{M4.1.2.4}$$

where x and y with respective subscripts are the Cartesian coordinates of the vertices, and by the formula

$$\pm S_n = \frac{1}{2}[\rho_1\rho_2 \sin(\theta_2 - \theta_1) + \rho_2\rho_3 \sin(\theta_3 - \theta_2) + \cdots + \rho_n\rho_1 \sin(\theta_1 - \theta_n)], \tag{M4.1.2.5}$$

where ρ and θ with respective subscripts are the polar coordinates of the vertices. In formulas (M4.1.2.4) and (M4.1.2.5), one takes the plus sign if the vertices are numbered counterclockwise (see Fig. M4.3b) and the minus sign otherwise.

Remark. One often says that formulas (M4.1.2.2)–(M4.1.2.5) express the *oriented area* of the corresponding figures.

M4.2. Straight Lines and Points on Plane

M4.2.1. Equations of Straight Lines on Plane

▶ **Equation of a curve.** Given a coordinate system, the set of all points in the plane can be treated as the set of various pairs of numbers x, y. Relations imposed upon x and y define subsets of the plane.

A line in the plane is usually defined using an equation relating the Cartesian coordinates x and y. An equation

$$F(x, y) = 0$$

is an *equation of a curve in the plane* if the coordinates of all the points lying on the curve satisfy the equation and the coordinates of all those points that do not lie on the curve do not satisfy it.

Example 1. Derive an equation of the line all of whose points are equidistant from the points $A(0, 2)$ and $B(4, -2)$.

Let $M(x, y)$ be a point that belongs to the line. For the distances to A and B, we have

$$\rho(A, M) = \sqrt{x^2 + (y - 2)^2}, \qquad \rho(B, M) = \sqrt{(x - 4)^2 + (y + 2)^2}.$$

It follows from the relation $\rho(A, M) = \rho(B, M)$ that $x^2 + (y-2)^2 = (x-4)^2 + (y+2)^2$. Expanding and collecting similar terms yields $y = x - 2$, which is an equation that determines a straight line.

Parametric equations of a curve on the plane have the form

$$x = \varphi(t), \qquad y = \psi(t),$$

where x and y are treated as the coordinates of some point M for each value of the *variable parameter* t.

▶ **Slope-intercept equation of a straight line.** The *slope-intercept equation of a straight line* in the rectangular Cartesian coordinate system OXY has the form

$$y = kx + b, \tag{M4.2.1.1}$$

where $k = \tan \varphi$ is the slope of the line and b is the y-intercept of the line, i.e., the signed distance from the point of intersection of the line with the ordinate axis to the origin. Equation (M4.2.1.1) is meaningful for any straight line that is not perpendicular to the abscissa axis (see Fig. M4.4a).

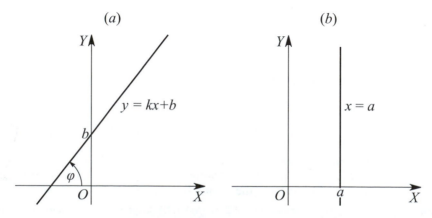

Figure M4.4. Straight lines on plane.

If a straight line is not perpendicular to the OX-axis, then its equation can be written as (M4.2.1.1), but if a straight line is perpendicular to the OX-axis, then its equation can be written as

$$x = a, \qquad \text{(M4.2.1.2)}$$

where a is the abscissa of the point of intersection of this line with the OX-axis (see Fig. M4.4b).

For the slope of a straight line, we also have the formula

$$k = \frac{y_2 - y_1}{x_2 - x_1}, \qquad \text{(M4.2.1.3)}$$

where $A_1(x_1, y_1)$ and $A_2(x_2, y_2)$ are two arbitrary points of the line.

▶ **Point-slope equation of a straight line.** In the rectangular Cartesian coordinate system OXY, the equation of a straight line with slope k passing through a point $A(x_1, y_1)$ has the form

$$y - y_1 = k(x - x_1). \qquad \text{(M4.2.1.4)}$$

If we set $x_1 = 0$ and $y_1 = b$ in equation (M4.2.1.4), then we obtain equation (M4.2.1.1).

▶ **Equation of a straight line passing through two given points.** The equation of a straight line passing through two distinct points $A_1(x_1, y_1)$ and $A_2(x_2, y_2)$ has the form

$$\frac{x - x_1}{x_2 - x_1} = \frac{y - y_1}{y_2 - y_1} \qquad (x_1 \neq x_2,\ y_1 \neq y_2). \qquad \text{(M4.2.1.5)}$$

If $x_1 = x_2$, this equation degenerates into $x = x_1$. If $y_1 = y_2$, the equation becomes $y = y_1$.

Example 2. Let us derive the equation of the straight line passing through the points $A_1(5, 1)$ and $A_2(7, 3)$. Substituting the coordinates of these points into formula (M4.2.1.5), we obtain

$$\frac{x - 5}{2} = \frac{y - 1}{2} \qquad \text{or} \qquad y = x - 4.$$

▶ **General equation of a straight line.** A linear equation of the form

$$Ax + By + C = 0 \qquad (A^2 + B^2 \neq 0) \qquad \text{(M4.2.1.6)}$$

is called the *general equation of a straight line* in the rectangular Cartesian coordinate system OXY. In rectangular Cartesian coordinates, each straight line is determined by an equation of degree 1, and, conversely, each equation of degree 1 determines a straight line.

If $B \neq 0$, then equation (M4.2.1.6) can be written as (M4.2.1.1), where $k = -A/B$ and $b = -C/B$.

Special cases of equation (M4.2.1.6):
1. If $A = 0$ and $B \neq 0$, then the equation becomes $y = -C/B$ and determines a straight line parallel to the axis OX.
2. If $B = 0$ and $A \neq 0$, then the equation becomes $x = -C/A$ and determines a straight line parallel to the axis OY.
3. If $C = 0$, then the equation becomes $Ax + By = 0$ and determines a straight line passing through the origin.

▶ **General equation of a straight line passing through a given point.** In the rectangular Cartesian coordinate system OXY, the general equation of a straight line passing through a point $M(x_1, y_1)$ in the plane has the form

$$A(x - x_1) + B(y - y_1) = 0. \qquad \text{(M4.2.1.7)}$$

▶ **Parametric equations of a straight line.** In the rectangular Cartesian coordinate system OXY, a straight line passing through a point $M(x_1, y_1)$ in the plane can be represented by the parametric equations

$$x = x_1 + At, \quad y = y_1 + Bt, \tag{M4.2.1.8}$$

where A and B are constants and t is a variable parameter.

A straight line passing through two points, $M_1(x_1, y_1)$ and $M_2(x_2, y_2)$, can be represented by the parametric equations

$$\begin{aligned} x &= x_1(1-t) + x_2 t, \\ y &= y_1(1-t) + y_2 t. \end{aligned} \tag{M4.2.1.9}$$

▶ **Intercept-intercept equation of a straight line.** The general equation of a straight line can be rewritten in the form

$$\frac{x}{a} + \frac{y}{b} = 1,$$

which is called the *intercept-intercept equation of a straight line*. The numbers a and b are the x- and y-intercepts of the straight line, i.e., the signed distances from the origin to the points at which the straight line crosses the coordinate axes (see Fig. M4.5).

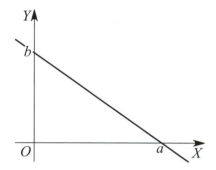

Figure M4.5. A straight line with intercept-intercept equation.

▶ **Equation of a pencil of straight lines.** The set of all straight lines passing through a fixed point M in the plane is called a *pencil of straight lines*, and the point M itself is called the *center of the pencil*. The equation determining all straight lines in the pencil is called the *equation of the pencil*.

$1°$. Given the Cartesian coordinates of the pencil center $M(x_1, y_1)$, then the equation of any straight line in the pencil has the form (M4.2.1.7), where A and B are arbitrary constants.

$2°$. If the equations of two straight lines in the pencil are known, $A_1 x + B_1 y + C_1 = 0$ and $A_2 x + B_2 y + C_2 = 0$, then the equation of the pencil can be written as

$$\alpha(A_1 x + B_1 y + C_1) + \beta(A_2 x + B_2 y + C_2) = 0,$$

where α and β are any numbers that are not simultaneously zero.

M4.2.2. Mutual Arrangement of Points and Straight Lines

▶ **Condition for three points to be collinear.** Suppose there are three distinct points, $M_1(x_1, y_1)$, $M_2(x_2, y_2)$, and $M_3(x_3, y_3)$, given in the Cartesian coordinate system OXY on the plane. They are collinear (lie on the same straight line) if and only if

$$\begin{vmatrix} x_1 & y_1 & 1 \\ x_2 & y_2 & 1 \\ x_3 & y_3 & 1 \end{vmatrix} = x_1 y_2 + x_2 y_3 + x_3 y_1 - x_1 y_3 - x_2 y_1 - x_3 y_2 = 0.$$

This condition reflects the fact that the area of the triangle with vertices at the above points is zero; cf. Eq. (M4.1.2.2).

▶ **Distance from a point to a straight line.** The distance d from a point $M(x_0, y_0)$ to a straight line given by the general equation $Ax + By + C = 0$ can be calculated as

$$d = \frac{|Ax_0 + By_0 + C|}{\sqrt{A^2 + B^2}}. \tag{M4.2.2.1}$$

Example 1. Let us find the distance from the point $M(2, 1)$ to the straight line $3x - 4y + 8 = 0$. We use formula (M4.2.2.1) to obtain

$$d = \frac{|3 \cdot 2 - 4 \cdot 1 + 8|}{\sqrt{3^2 + 4^2}} = \frac{10}{5} = 2.$$

▶ **Angle between two straight lines.**

$1°$. If two straight lines are given by the equations

$$y = k_1 x + b_1,$$
$$y = k_2 x + b_2,$$

where $k_1 = \tan \varphi_1$ and $k_2 = \tan \varphi_2$ are the slopes of the respective lines (see Fig. M4.6), the angle α between these straight lines is determined by the formula

$$\tan \alpha = \frac{k_2 - k_1}{1 + k_1 k_2} \qquad (k_1 k_2 \neq -1).$$

If $k_1 = k_2$, the straight lines are *parallel* ($\alpha = 0$).
If $k_1 k_2 = -1$, the straight lines are *perpendicular* ($\alpha = \frac{1}{2}\pi$).

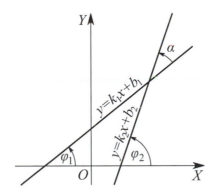

Figure M4.6. Angle between two straight lines.

Example 2. Given a triangle with vertices $A(-2, 0)$, $B(2, 4)$, and $C(4, 0)$, derive the equations of the side BC and the altitude AH.

Using (M4.2.1.5), one finds an equation for the side BC: $\dfrac{x-2}{4-2} = \dfrac{y-4}{0-4}$, or $y = -2x + 8$. It follows that the slope of this straight line is $k_{BC} = -2$. The above condition for two straight lines to be perpendicular gives $k_{AH} = -\dfrac{1}{k_{BC}} = \dfrac{1}{2}$. Using equation (M4.2.1.4), one obtains the equation for the altitude AH: $y - 0 = \frac{1}{2}(x + 2)$ or $y = \frac{1}{2}x + 1$.

$2°$. If two straight lines are defined by the general equations

$$A_1 x + B_1 y + C_1 = 0,$$
$$A_2 x + B_2 y + C_2 = 0,$$
(M4.2.2.2)

the angle α between them can be calculated from

$$\tan \alpha = \frac{A_1 B_2 - A_2 B_1}{A_1 A_2 + B_1 B_2} \qquad (A_1 A_2 + B_1 B_2 \neq 0).$$

If $A_1 B_2 - A_2 B_1 = 0$ (or $A_1/A_2 = B_1/B_2$), the straight lines are *parallel*.
If $A_1 A_2 + B_1 B_2 = 0$, the straight lines are *perpendicular*.

▶ **Point of intersection of straight lines.** Suppose that two straight lines are defined by general equations in the form (M4.2.2.2). Each common solution of equations (M4.2.2.2) determines a common point of the two lines.

If the determinant of system (M4.2.2.2) is not zero, i.e.,

$$\begin{vmatrix} A_1 & B_1 \\ A_2 & B_2 \end{vmatrix} = A_1 B_2 - A_2 B_1 \neq 0,$$

then the system is consistent and has a unique solution; hence, these straight lines are distinct and nonparallel and meet at the point $A(x_0, y_0)$, where

$$x_0 = \frac{B_1 C_2 - B_2 C_1}{A_1 B_2 - A_2 B_1}, \qquad y_0 = \frac{C_1 A_2 - C_2 A_1}{A_1 B_2 - A_2 B_1}.$$

▶ **Distance between parallel lines.** The distance between the parallel lines given by equations

$$A_1 x + B_1 y + C_1 = 0 \quad \text{and} \quad A_1 x + B_1 y + C_2 = 0$$

can be found using the formula

$$d = \frac{|C_1 - C_2|}{\sqrt{A_1^2 + B_1^2}}.$$

M4.3. Quadratic Curves

M4.3.1. Circle

▶ **Equations of a circle in the Cartesian coordinate system.** The *canonical equation of a circle* in a rectangular Cartesian coordinate system OXY has the form

$$x^2 + y^2 = a^2,$$
(M4.3.1.1)

where the point $O(0, 0)$ is the center of the circle and $a > 0$ is its radius (see Fig. M4.7a). The circle defined by equation (M4.3.1.1) is the locus of points equidistant (lying at the distance a) from its center.

The circle with radius a and center $A(x_0, y_0)$ is defined by the equation

$$(x - x_0)^2 + (y - y_0)^2 = a^2.$$
(M4.3.1.2)

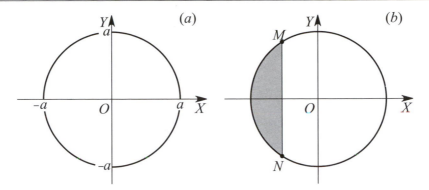

Figure M4.7. Circle.

The circle that passes through three noncollinear points $A_1(x_1, y_1)$, $A_2(x_2, y_2)$, and $A_3(x_3, y_3)$ can be described by the determinant equation

$$\begin{vmatrix} x^2 + y^2 & x & y & 1 \\ x_1^2 + y_1^2 & x_1 & y_1 & 1 \\ x_2^2 + y_2^2 & x_2 & y_2 & 1 \\ x_3^2 + y_3^2 & x_3 & y_3 & 1 \end{vmatrix} = 0.$$

The area of the *disk* bounded by a circle of radius a is given by the formula $S = \pi a^2$. The circumference of this circle is $L = 2\pi a$. The area of the figure bounded by the circle and a chord with endpoints $M(x_0, y_0)$ and $N(x_0, -y_0)$, shaded in Fig. 4.7b, is expressed as

$$S = \frac{\pi a^2}{2} + x_0 \sqrt{a^2 - x_0^2} + a^2 \arcsin \frac{x_0}{a}.$$

See also Subsection M3.1.3.

▶ **Other equations of a circle.** The equation of the circle (M4.3.1.1) can be represented in parametric form as

$$x = a \cos \theta, \quad y = a \sin \theta,$$

where the polar angle θ plays the role of the variable parameter.

In the polar coordinate system, the equation of the circle (M4.3.1.1) becomes

$$\rho = a.$$

Note that it does not contain the polar angle θ.

M4.3.2. Ellipse

▶ **Definition and the canonical equation of an ellipse.** An *ellipse* is the locus of points in the plane the sum of whose distances to two points, F_1 and F_2, is a constant quantity, denoted $2a$; see Fig. M4.8 a. Either of the points F_1 and F_2 is called a *focus* of the ellipse and the distance between them, $\rho(F_1, F_2) = 2c$, is called the *focal distance*.

In the rectangular Cartesian coordinate system where the X-axis is the straight line passing through the foci, the origin O coincides with the midpoint of the segment $F_1 F_2$, and the Y-axis passes through O and is perpendicular to the X-axis, as shown in Fig. M4.8 a, the equation of the ellipse has the simplest form

$$\frac{x^2}{a^2} + \frac{y^2}{b^2} = 1. \tag{M4.3.2.1}$$

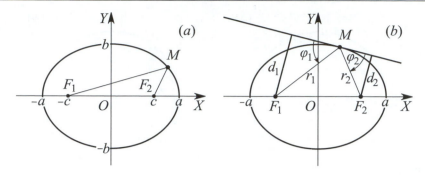

Figure M4.8. Ellipse (*a*). Tangent to an ellipse and the optical property of an ellipse (*b*).

Equation (M4.3.2.1) is called the *canonical equation of the ellipse.* Here, it is assumed that $a \geq b > 0$.

The positive numbers a and b are called, respectively, the *semimajor axis* and *semiminor axis* of the ellipse, with $b = \sqrt{a^2 - c^2}$. The number $c = \sqrt{a^2 - b^2}$ is called the *linear eccentricity* of the ellipse. The number $e = c/a = \sqrt{1 - b^2/a^2}$ is called the *eccentricity* or the *numerical eccentricity* of the ellipse, with $0 \leq e < 1$. The number $p = b^2/a$ is called the *focal parameter* (or simply the *parameter*) of the ellipse.

The point $O(0,0)$ is called the *center* of the ellipse. The points of intersection of the ellipse with the axes of symmetry, $A_1(-a, 0)$, $A_2(a, 0)$ and $B_1(0, -b)$, $B_2(0, b)$, are called its *vertices*. The straight line passing through the foci of an ellipse is known as its major axis and is sometimes called its *focal axis.* Either of the straight lines $x = \pm a/e$ ($e \neq 0$) is called a *directrix* of the ellipse. The focus $F_2(c, 0)$ and the directrix $x = a/e$ are said to be *right*, and the focus $F_1(-c, 0)$ and the directrix $x = -a/e$ are said to be *left*.

Remark. For $a = b$ ($c = 0$), equation (M4.3.2.1) becomes $x^2 + y^2 = a^2$ and determines a circle.

The area of the figure bounded by the ellipse is given by the formula $S = \pi ab$. The length of the ellipse can be calculated approximately by the formula $L \approx \pi \left[1.5 \, (a + b) - \sqrt{ab} \, \right]$.

▶ **Focal and focus-directrix properties of an ellipse.** The segments joining a point $M(x, y)$ of an ellipse with the foci $F_1(-c, 0)$ and $F_2(c, 0)$ are called the *left* and *right focal radii* of this point. We denote the lengths of the left and right focal radii by $r_1 = |F_1 M|$ and $r_2 = |F_2 M|$, respectively (see Fig. M4.8 *b*). By the definition of an ellipse,

$$r_1 + r_2 = 2a,$$

where r_1 and r_2 satisfy the relations

$$r_1 = \sqrt{(x + c)^2 + y^2} = a + ex, \qquad r_2 = \sqrt{(x - c)^2 + y^2} = a - ex.$$

The ellipse determined by equation (M4.3.2.1) on the plane is the locus of points for which the ratio of distances to a focus and the like directrix is equal to e:

$$r_1 \left| x + \frac{a}{e} \right|^{-1} = e, \qquad r_2 \left| x - \frac{a}{e} \right|^{-1} = e.$$

(A focus and a directrix are said to be *like* if both of them are right or left simultaneously.)
▶ **Equation of a tangent and the optical property of an ellipse.** The tangent to the ellipse (M4.3.2.1) at an arbitrary point $M_0(x_0, y_0)$ is given by the equation

$$\frac{x_0 x}{a^2} + \frac{y_0 y}{b^2} = 1.$$

The distances d_1 and d_2 from the foci $F_1(-c, 0)$ and $F_2(c, 0)$ to the tangent to the ellipse at the point $M_0(x_0, y_0)$ are expressed as

$$d_1 = \frac{1}{aN} |x_0 e + a| = \frac{r_1}{aN},$$
$$d_2 = \frac{1}{aN} |x_0 e - a| = \frac{r_2}{aN}, \qquad N = \sqrt{\left(\frac{x_0}{a^2}\right)^2 + \left(\frac{y_0}{b^2}\right)^2},$$

where $r_1 = r_1(M_0)$ and $r_2 = r_2(M_0)$ are the lengths of the focal radii of M_0 (see Fig. M4.8 b, where $M = M_0$).

The tangent at an arbitrary point $M_0(x_0, y_0)$ of an ellipse forms acute angles φ_1 and φ_2 with the focal radii of the point of tangency, and

$$\sin \varphi_1 = \frac{d_1}{r_1} = \frac{1}{aN}, \qquad \sin \varphi_2 = \frac{d_2}{r_2} = \frac{1}{aN}.$$

This fact, written as

$$\varphi_1 = \varphi_2,$$

is known as the *optical property of an ellipse*. It means that a light ray issued from one focus of the ellipse will reflect to the other focus (see Fig. M4.8 b, where $M = M_0$).

▶ **Equations of an ellipse in polar coordinates and parametric equations.** In polar coordinates (ρ, φ), with the pole coinciding with the right focus and the polar axis directed along the X-axis, the equation of an ellipse has the form

$$\rho = \frac{p}{1 + e \cos \varphi},$$

where $0 \leq \varphi \leq 2\pi$, $p = b^2/a$, and $e = \sqrt{1 - b^2/a^2}$. If the pole is taken at the left focus, this equation becomes

$$\rho = \frac{p}{1 - e \cos \varphi}.$$

The equation of an ellipse can also be represented in the parametric form

$$x = a \cos t, \quad y = b \sin t,$$

with the parameter t assuming any values from 0 to 2π.

M4.3.3. Hyperbola

▶ **Definition and the canonical equation of a hyperbola.** A *hyperbola* is the locus of points in the plane the absolute difference of whose distances to two points, F_1 and F_2, is a constant quantity, denoted $2a$; see Fig. M4.9 a. Either of the points F_1 and F_2 is called a *focus* of the hyperbola and the distance between them, $\rho(F_1, F_2) = 2c$, is called the *focal distance*.

In the rectangular Cartesian coordinate system where the X-axis is the straight line passing through the foci, the origin O coincides with the midpoint of the segment $F_1 F_2$, and the Y-axis passes through O and is perpendicular to the X-axis, as shown in Fig. M4.8 a, the equation of the hyperbola has the simplest form

$$\frac{x^2}{a^2} - \frac{y^2}{b^2} = 1, \qquad (M4.3.3.1)$$

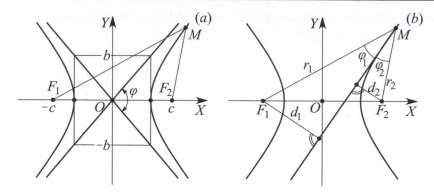

Figure M4.9. Hyperbola (*a*). The tangent to the hyperbola and optical property of a hyperbola (*b*).

and is called the *canonical equation of the hyperbola*.

The number a is called the *real semiaxis* of the hyperbola and the number b is called its *imaginary semiaxis*, with $b = \sqrt{c^2 - a^2}$. The number $c = \sqrt{a^2 + b^2}$ is known as the *linear eccentricity* of the hyperbola and the number $e = c/a = \sqrt{1 + b^2/a^2}$ is its *eccentricity*, $e > 1$. The number $p = b^2/a$ is called the *focal parameter* (or simply the *parameter*) of the hyperbola.

The point $O(0,0)$ is called the *center* of the hyperbola. The points $(-a, 0)$ and $(a, 0)$ at which the hyperbola crosses the X-axis are known as the *vertices* of the hyperbola. Either of the straight lines $x = \pm a/e$ is called a *directrix* of the hyperbola. The focus $F_2(c, 0)$ and the directrix $x = a/e$ are said to be *right*, while the focus $F_1(-c, 0)$ and the directrix $x = -a/e$ are said to be *left*.

A hyperbola consists of two parts, called its *branches*, lying in the domains $x \geq a$ and $x \leq -a$. It has two *asymptotes*, straight lines the hyperbola approaches at large distances from its center, which are given by

$$y = \frac{b}{a}x \quad \text{and} \quad y = -\frac{b}{a}x.$$

The branches of a hyperbola lie within two vertical angles formed by the asymptotes and are called its *left* and *right* branches. The angle φ between the asymptotes of a hyperbola is determined by the equation

$$\tan \frac{\varphi}{2} = \frac{b}{a}.$$

Remark. If $a = b$ ($e = \sqrt{2}$), then $\varphi = \frac{1}{2}\pi$. In this case, the hyperbola is said to be *equilateral* or *rectangular* and its asymptotes are mutually perpendicular. The equation of an equilateral hyperbola has the form $x^2 - y^2 = a^2$. If the asymptotes are taken to be the coordinate axes, then the equation of the hyperbola becomes $xy = a^2/2$; i.e., an equilateral hyperbola represents an inverse proportionality dependence.

▶ **Focal and focus-directrix properties of a hyperbola.** The segments joining a point $M(x, y)$ of the hyperbola with the foci $F_1(-c, 0)$ and $F_2(c, 0)$ are called the *left* and *right* *focal radii* of this point. We denote the lengths of the left and right focal radii by $r_1 = |F_1 M|$ and $r_2 = |F_2 M|$, respectively (see Fig. M4.9 *b*). By the definition of a hyperbola,

$$|r_1 - r_2| = 2a,$$

where r_1 and r_2 satisfy the relations

$$r_1 = \sqrt{(x + c)^2 + y^2} = \begin{cases} a + ex & \text{for } x > 0, \\ -a - ex & \text{for } x < 0, \end{cases}$$

$$r_2 = \sqrt{(x - c)^2 + y^2} = \begin{cases} -a + ex & \text{for } x > 0, \\ a - ex & \text{for } x < 0. \end{cases}$$

The hyperbola defined by equation (M4.3.3.1) on the plane is the locus of points for which the ratio of distances to a focus and the like directrix is equal to e:

$$r_1\left|x + \frac{a}{e}\right|^{-1} = e, \qquad r_2\left|x - \frac{a}{e}\right|^{-1} = e.$$

(A focus and a directrix are said to be *like* if both of them are right or left simultaneously.)

▶ **Equation of a tangent and the optical property of a hyperbola.** The tangent to the hyperbola (M4.3.3.1) at an arbitrary point $M_0(x_0, y_0)$ is given by the equation

$$\frac{x_0 x}{a^2} - \frac{y_0 y}{b^2} = 1.$$

The distances d_1 and d_2 from the foci $F_1(-c, 0)$ and $F_2(0, c)$ to the tangent to the hyperbola at the point $M_0(x_0, y_0)$ are expressed as

$$d_1 = \frac{|x_0 e + a|}{aN} = \frac{r_1}{aN},$$

$$d_2 = \frac{|x_0 e - a|}{aN} = \frac{r_2}{aN}, \qquad N = \sqrt{\left(\frac{x_0}{a^2}\right)^2 + \left(\frac{y_0}{b^2}\right)^2},$$

where r_1 and r_2 are the lengths of the focal radii of the point M_0 (see Fig. M4.9 b, where $M = M_0$).

The tangent at any point $M_0(x_0, y_0)$ of the hyperbola forms acute angles φ_1 and φ_2 with the focal radii of the point of tangency, and

$$\sin \varphi_1 = \frac{d_1}{r_1} = \frac{1}{aN}, \qquad \sin \varphi_2 = \frac{d_2}{r_2} = \frac{1}{aN}.$$

This fact, written as

$$\varphi_1 = \varphi_2,$$

is known as the *optical property of a hyperbola*. It means that a light ray issued from a focus of the hyperbola will reflect so as to appear as though issued from the other focus (see Fig. M4.9 b, where $M = M_0$).

The tangent to a hyperbola at any point bisects the angles between the straight lines joining this point with the foci. The tangent to a hyperbola at either of its vertices intersects the asymptotes at two points such that the distance between them is equal to $2b$.

▶ **Equations of a hyperbola in polar coordinates and parametric equations.** In polar coordinates (ρ, φ), with the pole coinciding with the right focus and the polar axis directed along the X-axis, the equation of the hyperbola has the form

$$\rho = \frac{p}{1 - e \cos \varphi},$$

where $0 \leq \varphi \leq 2\pi$, $p = b^2/a$, and $e = \sqrt{1 + b^2/a^2}$. If the pole is taken at the left focus, the equation of the hyperbola becomes

$$\rho = \frac{p}{1 + e \cos \varphi}.$$

A parametric representation for the right branch of a hyperbola is given by the equations

$$x = a \cosh t, \qquad y = b \sinh t,$$

with the parameter t assuming any real values.

A parametric representation that covers both branches of a hyperbola is given by the equations

$$x = a \sec t, \qquad y = b \tan t,$$

with $-\pi \leq t \leq \pi$ and $t \neq \pm\frac{1}{2}\pi$.

M4.3.4. Parabola

▶ **Definition and the canonical equation of a parabola.** A *parabola* is the locus of all points in the plane equidistant from a given point F and a given straight line l, with $F \notin l$; see Fig. M4.10 a. The point F is called the *focus* of the parabola and the straight line l is its *directrix*.

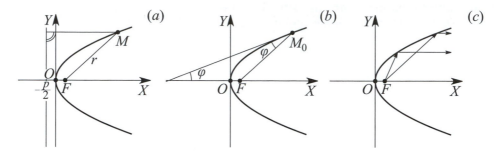

Figure M4.10. Parabola (a). Tangent to a parabola (b). Optical property of a parabola (c).

Let us draw a straight line through the focus F and perpendicularly to the directrix and denote the point at which this line crosses the directrix by C. Introduce the following Cartesian coordinate system: take the above line to be the X-axis (directed from C to F), the midpoint of the segment CF to be the origin O, and the perpendicular line through O to be the Y-axis. In this coordinate system, the parabola is determined by the equation

$$y^2 = 2px, \tag{M4.3.4.1}$$

where $p = |FC| > 0$. The number p is called *focal parameter* and equation (M4.3.4.1) is called the *canonical equation of the parabola*.

A parabola consists of an infinite branch symmetric about the X-axis. The point $O(0,0)$ is called the *vertex* of the parabola. The directrix of the parabola is given by the equation $x = -p/2$. The number $p/2$ is known as the *focal distance*. The segment joining a point $M(x,y)$ on the parabola with the focus $F(p/2,0)$ is called the *focal radius* of the point.

▶ **Focal properties of a parabola.** If r denotes the length of the focal radius FM, then by the definition of a parabola,

$$r = x + \frac{p}{2}.$$

As is apparent from Fig. M4.10a, the number r also satisfies the relation

$$r = \sqrt{\left(x - \frac{p}{2}\right)^2 + y^2}.$$

▶ **Equation of a tangent and the optical property of a parabola.** The tangent to the parabola (M4.3.4.1) at an arbitrary point $M_0(x_0, y_0)$ is given by the equation

$$yy_0 = p(x + x_0). \tag{M4.3.4.2}$$

The angle φ between the tangent to the parabola at a point $M_0(x_0, y_0)$ and the focal radius FM_0 is determined by

$$\cos \varphi = \frac{y_0}{\sqrt{y_0^2 + p^2}}.$$

The same relation holds for the angle between the tangent (M4.3.4.2) and the X-axis. This property of a parabola is called its *optical property*: a light ray issued from the focus reflects off the parabola in the direction parallel to the parabola axis (see Fig. M4.10 c).

▶ **Equation of a parabola in polar coordinates and parametric equations.** In polar coordinates (ρ, φ), with the pole at the focus of the parabola and the polar axis directed along the parabola axis, the equation of the parabola has the form

$$\rho = \frac{p}{1 - \cos \varphi},$$

where $0 < \varphi < 2\pi$.

Parametric equations of a parabola are

$$x = \tfrac{1}{2}pt^2, \quad y = pt,$$

with the parameter t assuming any real values.

M4.3.5. Transformation of Quadratic Curves to Canonical Form

▶ **General equation of a quadratic curve. Translation and rotation.** A set of points in the plane whose coordinates in the rectangular Cartesian coordinate system satisfy the general second-order algebraic equation

$$a_{11}x^2 + 2a_{12}xy + a_{22}y^2 + 2a_{13}x + 2a_{23}y + a_{33} = 0 \qquad (\text{M4.3.5.1})$$

is called a (bivariate) *quadratic curve* (or just *quadratic*); it is also known as a *second-order curve*. If equation (M4.3.5.1) does not determine a real geometric object, this equation is said to determine an *imaginary quadratic curve*.

Equation (M4.3.5.1) may be simplified using the following transformations of the Cartesian coordinate system:

1. Translation:
$$x = \bar{x} + x_0, \quad y = \bar{y} + y_0. \qquad (\text{M4.3.5.2})$$

It means that the origin $O(0,0)$ is transferred to the point $\bar{O}(x_0, y_0)$ and the coordinate axes are moved parallel to the original ones; $\bar{x}$ and $\bar{y}$ denote the new coordinates.

2. Rotation:
$$x = \hat{x} \cos \varphi - \hat{y} \sin \varphi, \quad y = \hat{x} \sin \varphi + \hat{y} \cos \varphi. \qquad (\text{M4.3.5.3})$$

The coordinate axes are rotated about the origin, which does not move, by the angle φ counterclockwise; $\hat{x}$ and $\hat{y}$ denote the new coordinates.

▶ **Canonical equations of quadratic curves. The classification table.** With transformations (M4.3.5.2)–(M4.3.5.3), equation (M4.3.5.1) can be reduced to one of the *nine canonical forms* classified in Table M4.1. The first five curves, with $\delta \neq 0$, are *nondegenerate* (their canonical equations contain two quadratic terms proportional to x^2 and y^2). The last four curves, with $\delta = 0$, are *degenerate* (their canonical equations contain only one quadratic term, x^2 or y^2). Curves 3, 5, 7, 8, and 9, with $\Delta = 0$, split into straight lines; their equations can be represented as the product of two factors linear in the coordinates, each having the form $(\alpha_n x + \beta_n y + \gamma_n)$, on the left-hand side and zero on the right-hand side.

TABLE M4.1
Classification of quadratic curves.

No.	Curve name	Canonical equation	Conditions for invariants
1	Ellipse	$\dfrac{x^2}{a^2} + \dfrac{y^2}{b^2} = 1$	$\delta > 0,\ I\Delta < 0$
2	Imaginary ellipse (no real points)	$\dfrac{x^2}{a^2} + \dfrac{y^2}{b^2} = -1$	$\delta > 0,\ I\Delta > 0$
3	Pair of imaginary straight lines intersecting at a real point	$\dfrac{x^2}{a^2} + \dfrac{y^2}{b^2} = 0$	$\delta > 0,\ \Delta = 0$
4	Hyperbola	$\dfrac{x^2}{a^2} - \dfrac{y^2}{b^2} = 1$	$\delta < 0,\ \Delta \neq 0$
5	Pair of intersecting straight lines (degenerate hyperbola)	$\dfrac{x^2}{a^2} - \dfrac{y^2}{b^2} = 0$	$\delta < 0,\ \Delta = 0$
6	Parabola	$y^2 = 2px$	$\delta = 0,\ \Delta \neq 0$
7	Pair of parallel straight lines	$x^2 - a^2 = 0$	$\delta = \Delta = 0,\ \sigma < 0$
8	Pair of imaginary parallel straight lines (no real points)	$x^2 + a^2 = 0$	$\delta = \Delta = 0,\ \sigma > 0$
9	Pair of coinciding straight lines	$x^2 = 0$	$\delta = \Delta = \sigma = 0$

▶ **Invariants of quadratic curves.** Quadratic curves can be studied using the three *invariants*

$$I = a_{11} + a_{22}, \quad \delta = \begin{vmatrix} a_{11} & a_{12} \\ a_{12} & a_{22} \end{vmatrix}, \quad \Delta = \begin{vmatrix} a_{11} & a_{12} & a_{13} \\ a_{12} & a_{22} & a_{23} \\ a_{13} & a_{23} & a_{33} \end{vmatrix}, \tag{M4.3.5.4}$$

whose values do not change under parallel translations and rotations of the coordinate axes, and the sign of the quantity

$$\sigma = \begin{vmatrix} a_{11} & a_{13} \\ a_{13} & a_{33} \end{vmatrix} + \begin{vmatrix} a_{22} & a_{23} \\ a_{23} & a_{33} \end{vmatrix}. \tag{M4.3.5.5}$$

The invariant Δ is called the *large discriminant* of equation (M4.3.5.1). The invariant δ is called the *small discriminant*.

The quadratic curves can be classified based on the values of the invariants, specified in the last column in Table M4.1.

▶ **Characteristic equation of quadratic curves.** The properties of quadratic curves can be studied using the *characteristic equation*

$$\begin{vmatrix} a_{11} - \lambda & a_{12} \\ a_{12} & a_{22} - \lambda \end{vmatrix} = 0 \quad \text{or} \quad \lambda^2 - I\lambda + \delta = 0. \tag{M4.3.5.6}$$

The roots λ_1 and λ_2 of the characteristic equation (M4.3.5.6) are eigenvalues of a real symmetric matrix, $[a_{ij}]$, and hence are real.

The invariants I and δ are expressed in terms of the roots λ_1 and λ_2 as follows:

$$I = \lambda_1 + \lambda_2, \quad \delta = \lambda_1 \lambda_2. \tag{M4.3.5.7}$$

▶ **Nondegenerate case $\delta \neq 0$. Reduction of quadratic curves to canonical form.** First, by applying the translation transformation (M4.3.5.2) with

$$x_0 = -\frac{1}{\delta} \begin{vmatrix} a_{13} & a_{12} \\ a_{23} & a_{22} \end{vmatrix}, \quad y_0 = -\frac{1}{\delta} \begin{vmatrix} a_{11} & a_{13} \\ a_{12} & a_{23} \end{vmatrix},$$

one reduces equation (M4.3.5.1) to the form

$$a_{11}\bar{x}^2 + 2a_{12}\bar{x}\bar{y} + a_{22}\bar{y}^2 + \frac{\Delta}{\delta} = 0. \qquad \text{(M4.3.5.8)}$$

Then, with the rotation transformation (M4.3.5.3) where $\bar{x}$ and $\bar{y}$ are substituted for x and y and the angle φ determined by

$$\tan 2\varphi = \frac{2a_{12}}{a_{11} - a_{22}} \qquad \left(\text{if } a_{11} = a_{22}, \text{ then } \varphi = \frac{\pi}{4} \right),$$

equation (M4.3.5.8) is transformed into

$$\lambda_1 \hat{x}^2 + \lambda_2 \hat{y}^2 + \frac{\Delta}{\delta} = 0,$$

where λ_1 and λ_2 are the roots of the characteristic equation (M4.3.5.6).

Note the following formulas for an ellipse:

$$a^2 = -\frac{1}{\lambda_2}\frac{\Delta}{\delta} = -\frac{\Delta}{\lambda_1 \lambda_2^2}, \qquad b^2 = -\frac{1}{\lambda_1}\frac{\Delta}{\delta} = -\frac{\Delta}{\lambda_1^2 \lambda_2} \qquad (\lambda_1 \geq \lambda_2),$$

where a and b are the semimajor and semiminor axes of the ellipse.

Similar formulas for a hyperbola have the form

$$a^2 = -\frac{1}{\lambda_1}\frac{\Delta}{\delta} = -\frac{\Delta}{\lambda_1^2 \lambda_2}, \qquad b^2 = \frac{1}{\lambda_1}\frac{\Delta}{\delta} = \frac{\Delta}{\lambda_1^2 \lambda_2} \qquad (\lambda_1 \geq \lambda_2).$$

▶ **Degenerate case $\delta = 0$. Reduction of quadratic curves to canonical form.** If $\delta = 0$, equation (M4.3.5.1) can be rewritten as

$$(\alpha x + \beta y)^2 + 2a_{13}x + 2a_{23}y + a_{33} = 0. \qquad \text{(M4.3.5.9)}$$

If the coefficients a_{13} and a_{23} are respectively proportional to α and β, i.e., $a_{13} = k\alpha$ and $a_{23} = k\beta$, then equation (M4.3.5.9) becomes $(\alpha x + \beta y)^2 + 2k(\alpha x + \beta y) + a_{33} = 0$, and hence

$$\alpha x + \beta y = -k \pm \sqrt{k^2 - a_{33}},$$

which determines a pair of real (or imaginary) parallel straight lines.

If a_{13} and a_{23} are not proportional to α and β, then equation (M4.3.5.9) can be rewritten as

$$(\alpha x + \beta y + \gamma)^2 + 2k(\beta x - \alpha x + q) = 0. \qquad \text{(M4.3.5.10)}$$

The parameters k, γ, and q can be determined by comparing the coefficients in equations (M4.3.5.9) and (M4.3.5.10). If the line $\alpha x + \beta y + \gamma = 0$ is treated as the axis $\bar{O}\bar{X}$ and the line $\beta x - \alpha x + q = 0$ as the axis $\bar{O}\bar{Y}$ and the new coordinates are expressed as

$$\hat{x} = \frac{\beta x - \alpha x + q}{\pm\sqrt{\alpha^2 + \beta^2}}, \qquad \hat{y} = \frac{\alpha x + \beta y + \gamma}{\pm\sqrt{\alpha^2 + \beta^2}},$$

then equation (M4.3.5.10) acquires the form

$$\hat{y}^2 = 2p\hat{x},$$

where $p = |k|/\sqrt{\alpha^2 + \beta^2}$. The axis $\bar{O}\bar{X}$ points to the half-plane where the sign of $\beta x - \alpha x + q$ is opposite to that of k.

The focal parameter p of a parabola is expressed in terms of the invariants I, δ, and Δ and the roots λ_1 and λ_2 ($\lambda_1 \geq \lambda_2$) of the characteristic equation (M4.3.5.6) as follows:

$$p = \frac{1}{I}\sqrt{-\frac{\Delta}{I}} = \frac{1}{\lambda_1}\sqrt{-\frac{\Delta}{\lambda_1}} > 0, \qquad \lambda_2 = 0.$$

M4.4. Coordinates, Vectors, Curves, and Surfaces in Space

M4.4.1. Vectors and Their Properties

▶ **Notion of a vector.** A directed line segment connecting an initial point A and a terminal point B (see Fig. M4.11) is called a *vector* and denoted $\overrightarrow{AB}$. The nonnegative number equal to the length of the segment AB is called the *length* (or *magnitude*) of the *vector* $\overrightarrow{AB}$ and denoted $|\overrightarrow{AB}|$. The vector $\overrightarrow{BA}$ is said to be *opposite* to the vector $\overrightarrow{AB}$; it has the same magnitude but opposite direction. Vectors are usually denoted by a single lowercase letter, either with an arrow above (e.g., $\vec{a}$) or without (e.g., **a**); the latter is the most common notation for vectors, in which case a boldface lowercase letter is used.

Figure M4.11. Vector $\overrightarrow{AB}$.

Two vectors are said to be *collinear* (*parallel*) if they lie on the same straight line or on parallel lines. Three vectors are said to be *coplanar* if they lie in the same plane or in parallel planes. A vector 0 whose initial and terminal points coincide is called the *zero vector* (or *null vector*); its length is zero ($|0| = 0$) and its direction is assumed to be arbitrary. A vector **e** of length one is called a *unit vector*.

Two vectors are called *equal* is they are collinear and have the same magnitude and direction. It follows that, for any vector **a** and any point A, there exists a unique vector $\overrightarrow{AB}$ with its start point at A that is equal to **a**. For this reason, vectors in analytical geometry are defined up to their position, so that all vectors obtained from each other by parallel transport are considered to be the same.

▶ **Sum and difference of vectors.** The *sum* **a** + **b** *of vectors* **a** *and* **b** is defined as the vector directed from the initial point of **a** to the terminal point of **b** where the start of **b** is placed at the tip of **a**. This method of the addition of vectors is called the *triangle rule* (see Fig. M4.12 *a*). The sum **a** + **b** can also be found using the *parallelogram rule* as shown in Fig. M4.12 *b*. The *difference* **a** − **b** *of vectors* **a** *and* **b** is defined as the vector that must be added to **b** to get **a**: **b** + (**a** − **b**) = **a** (see Fig. M4.12 *c*).

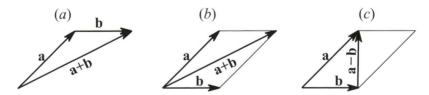

Figure M4.12. The sum of vectors: triangle rule (*a*) and parallelogram rule (*b*). The difference of vectors (*c*).

The *product* λ**a** *of a vector* **a** *by a number* λ is defined as the vector whose magnitude is equal to $|\lambda \mathbf{a}| = |\lambda||\mathbf{a}|$ and direction coincides with that of **a** if $\lambda > 0$ or is opposite to it if $\lambda < 0$.

Remark. If **a** = 0 or $\lambda = 0$, then the resulting product is the zero vector. In this case, the direction of the product λ**a** is undetermined.

▶ **Main properties of operations with vectors.**
1. $\mathbf{a} + \mathbf{b} = \mathbf{b} + \mathbf{a}$ (commutativity).
2. $\mathbf{a} + (\mathbf{b} + \mathbf{c}) = (\mathbf{a} + \mathbf{b}) + \mathbf{c}$ (associativity of addition).
3. $\mathbf{a} + 0 = \mathbf{a}$ (existence of zero vector).
4. $\mathbf{a} + (-\mathbf{a}) = 0$ (existence of opposite vector).
5. $\lambda(\mathbf{a} + \mathbf{b}) = \lambda\mathbf{a} + \lambda\mathbf{b}$ (distributivity with respect to addition of vectors).
6. $(\lambda + \mu)\mathbf{a} = \lambda\mathbf{a} + \mu\mathbf{a}$ (distributivity with respect to addition of constants).
7. $\lambda(\mu\mathbf{a}) = (\lambda\mu)\mathbf{a}$ (associativity of product).
8. $1\mathbf{a} = \mathbf{a}$ (multiplication by unity).

M4.4.2. Coordinate Systems

▶ **Cartesian coordinate system. Some useful formulas.** A *rectangular Cartesian coordinate system* (also called just *rectangular coordinate system* or *Cartesian coordinate system*) is defined by three pairwise perpendicular directed straight lines OX, OY, and OZ (the *coordinate axes*) concurrent at a single point O (the *origin*).

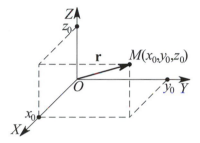

Figure M4.13. A point in a rectangular Cartesian coordinate system.

For an arbitrary point M in space, let us draw through it three planes parallel to the planes OYZ, OXZ, and OXY. These planes will intersect the coordinate axes OX, OY, and OZ at three points. Denote by x_0, y_0, and z_0 the distances from these points to the origin O (see Fig. M4.13). The numbers x_0, y_0, and z_0 are, respectively, called the x-coordinate (or *abscissa*), the y-coordinate (or *ordinate*), and the z-coordinate of the point M. One usually uses the notation $M(x_0, y_0, z_0)$ to specify that the point M has the coordinates (x_0, y_0, z_0).

Planes parallel to the coordinate planes are coordinate surfaces on which one of the coordinates is constant. Straight lines parallel to the coordinate axes are coordinate lines along which only one coordinate varies and the other two remain constant. Coordinate surfaces meet at coordinate lines.

Each point M in three-dimensional space uniquely defines a vector $\overrightarrow{OM}$, which is called the *position vector* of the point M. The coordinates of the position vector coincide with those of M and one usually writes $\mathbf{r} = \overrightarrow{OM} = (x_0, y_0, z_0)$.

The distance between two points, $M_1(x_1, y_1, z_1)$ and $M_2(x_2, y_2, z_2)$, is given by the formula

$$d = \sqrt{(x_2 - x_1)^2 + (y_2 - y_1)^2 + (z_2 - z_1)^2} = |\mathbf{r}_2 - \mathbf{r}_1|, \qquad (M4.4.2.1)$$

where $\mathbf{r}_2 = \overrightarrow{OM_2}$ and $\mathbf{r}_1 = \overrightarrow{OM_1}$ are the position vectors of the points M_1 and M_2, respectively (see Fig. M4.14).

Any triple of numbers (x, y, z) can be identified with a point P and a position vector $\overrightarrow{OP}$, whose coordinates are these numbers. An arbitrary vector (x, y, z) can be represented as

$$(x, y, z) = x\mathbf{i} + y\mathbf{j} + z\mathbf{k},$$

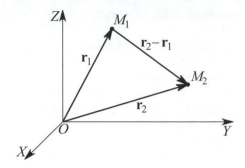

Figure M4.14. Distance between points.

where $\mathbf{i} = (1, 0, 0)$, $\mathbf{j} = (0, 1, 0)$, and $\mathbf{k} = (0, 0, 1)$ are the unit vectors with the same directions as the coordinate axes OX, OY, and OZ (*basis vectors*).

Two vectors $\mathbf{r}_1 = (x_1, y_1, z_1)$ and $\mathbf{r}_2 = (x_2, y_2, z_2)$ are equal to each other if and only if the relations

$$x_1 = x_2, \quad y_1 = y_2, \quad z_1 = z_2$$

hold simultaneously. The coordinates of the sum or difference of vectors and the product of a vector by a scalar are calculated as

$$(x_1, y_1, z_1) \pm (x_2, y_2, z_2) = (x_1 \pm x_2, y_1 \pm y_2, z_1 \pm z_2),$$
$$\alpha (x, y, z) = (\alpha x, \alpha y, \alpha z).$$

If a point M divides a directed segment $\overrightarrow{M_1 M_2}$ in a ratio λ, then the coordinates of this point are given by

$$x = \frac{x_1 + \lambda x_2}{1 + \lambda}, \quad y = \frac{y_1 + \lambda y_2}{1 + \lambda}, \quad z = \frac{z_1 + \lambda z_2}{1 + \lambda} \quad \text{or} \quad \mathbf{r} = \frac{\mathbf{r}_1 + \lambda \mathbf{r}_2}{1 + \lambda}, \quad \text{(M4.4.2.2)}$$

where $\lambda = |\overrightarrow{M_1 M}| / |\overrightarrow{M M_2}|$. The special case where M is the midpoint of $\overrightarrow{M_1 M_2}$ corresponds to $\lambda = 1$.

The angles α, β, and γ between $\overrightarrow{M_1 M_2}$ and the coordinate axes OX, OY, and OZ are determined by

$$\cos \alpha = \frac{x_2 - x_1}{|\mathbf{r}_2 - \mathbf{r}_1|}, \quad \cos \beta = \frac{y_2 - y_1}{|\mathbf{r}_2 - \mathbf{r}_1|}, \quad \cos \gamma = \frac{z_2 - z_1}{|\mathbf{r}_2 - \mathbf{r}_1|},$$

with

$$\cos^2 \alpha + \cos^2 \beta + \cos^2 \gamma = 1.$$

The numbers $\cos \alpha$, $\cos \beta$, and $\cos \gamma$ are called the *direction cosines* of the vector $\overrightarrow{M_1 M_2}$.

The angle φ between two vectors $\overrightarrow{M_1 M_2}$ and $\overrightarrow{M_3 M_4}$ defined by the points $M_1 (x_1, y_1, z_1)$, $M_2 (x_2, y_2, z_2)$, $M_3 (x_3, y_3, z_3)$, and $M_4 (x_4, y_4, z_4)$ can be found from

$$\cos \varphi = \frac{(x_2 - x_1)(x_4 - x_3) + (y_2 - y_1)(y_4 - y_3) + (z_2 - z_1)(z_4 - z_3)}{|\mathbf{r}_2 - \mathbf{r}_1| \, |\mathbf{r}_4 - \mathbf{r}_3|}.$$

The area of the triangle with vertices M_1, M_2, and M_3 is given by the formula

$$S = \frac{1}{4} \sqrt{\begin{vmatrix} y_1 & z_1 & 1 \\ y_2 & z_2 & 1 \\ y_3 & z_3 & 1 \end{vmatrix}^2 + \begin{vmatrix} z_1 & x_1 & 1 \\ z_2 & x_2 & 1 \\ z_3 & x_3 & 1 \end{vmatrix}^2 + \begin{vmatrix} x_1 & y_1 & 1 \\ x_2 & y_2 & 1 \\ x_3 & y_3 & 1 \end{vmatrix}^2}.$$

The volume of the pyramid with vertices M_1, M_2, M_3, and M_4 is equal to

$$V = \frac{1}{6}|D|, \qquad D = \begin{vmatrix} x_2 - x_1 & y_2 - y_1 & z_2 - z_1 \\ x_3 - x_1 & y_3 - y_1 & z_3 - z_1 \\ x_4 - x_1 & y_4 - y_1 & z_4 - z_1 \end{vmatrix} = \begin{vmatrix} 1 & x_1 & y_1 & z_1 \\ 1 & x_2 & y_2 & z_2 \\ 1 & x_3 & y_3 & z_3 \\ 1 & x_4 & y_4 & z_4 \end{vmatrix},$$

and the volume of the parallelepiped spanned by vectors $\overrightarrow{M_1 M_2}$, $\overrightarrow{M_1 M_3}$, and $\overrightarrow{M_1 M_4}$ is equal to

$$V = |D|.$$

▶ **Cylindrical coordinates.** *Cylindrical coordinates* are a generalization of polar coordinates (see Subsection M4.1.1) that adds a third dimension. If a point M is specified by its cylindrical coordinates, they are the polar coordinates ρ and φ of the projection of M onto a base plane (usually OXY) and the distance (usually z) of M from this base plane (see Fig. M4.15 a). It is usually assumed that $0 \le \varphi \le 2\pi$ (or $-\pi \le \varphi \le \pi$). For cylindrical coordinates, the coordinate surfaces are planes $z = $ const perpendicular to the axis OZ, half-planes $\varphi = $ const bounded by the axis OZ, and cylindrical surfaces $\rho = $ const with axis OZ.

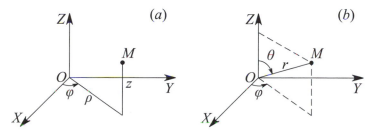

Figure M4.15. Point in cylindrical (*a*) and spherical (*b*) coordinates.

Let M be an arbitrary point in space with its Cartesian coordinates (x, y, z) and cylindrical coordinates (ρ, φ, z). The conversion formulas from Cartesian to cylindrical and from cylindrical to Cartesian coordinates are as follows:

$$\begin{aligned} x &= \rho \cos \varphi, & \rho &= \sqrt{x^2 + y^2}, \\ y &= \rho \sin \varphi, & \tan \varphi &= y/x, \\ z &= z, & z &= z, \end{aligned}$$

where the polar angle φ is taken with regard to the quadrant in which the projection of the point M onto the base plane lies.

▶ **Spherical coordinates.** The *spherical coordinates* of a point M are defined as the length $r = |\overrightarrow{OM}|$ of its position vector, the *azimuthal angle* φ from the positive direction of the axis OX to the projection of M onto the plane OXY, and the *zenithal angle* θ from the positive direction of the axis OZ to M (see Fig. M4.15 b). It is usually assumed that $0 \le \varphi \le 2\pi$ and $0 \le \theta \le \pi$ (or $-\pi \le \varphi \le \pi$ and $0 \le \theta \le \pi$). For spherical coordinates, the coordinate surfaces are spheres $r = $ const centered at the origin, half-planes $\varphi = $ const bounded by the axis OZ, and cones $\theta = $ const with vertex O and axis OZ.

The conversion formulas from the Cartesian coordinates (x, y, z) to the spherical coordinates (r, φ, θ) and back are as follows:

$$\begin{aligned} x &= r \sin \theta \cos \varphi, & r &= \sqrt{x^2 + y^2 + z^2}, \\ y &= r \sin \theta \sin \varphi, & \tan \varphi &= y/x, \\ z &= r \cos \theta, & \tan \theta &= \sqrt{x^2 + y^2}/z, \end{aligned}$$

where the angle φ is determined from the same considerations as in the case of cylindrical coordinates.

M4.4.3. Scalar, Cross, and Scalar Triple Products of Vectors

▶ **Scalar product of two vectors.** The *scalar product* (also known as the *dot product*) of two vectors **a** and **b**, is defined as the product of their magnitudes by the cosine of the angle between the vectors (see Fig. M4.16),

$$\mathbf{a} \cdot \mathbf{b} = |\mathbf{a}||\mathbf{b}| \cos \varphi.$$

It follows that $\mathbf{a} \cdot \mathbf{b} > 0$ if the angle between **a** and **b** is acute, $\mathbf{a} \cdot \mathbf{b} < 0$ if it is obtuse, and $\mathbf{a} \cdot \mathbf{b} = 0$ if it is right.

Remark. The scalar product of a vector **a** by a vector **b** is also denoted by $(\mathbf{a} \cdot \mathbf{b})$, $(\mathbf{a}, \mathbf{b})$, and **ab**.

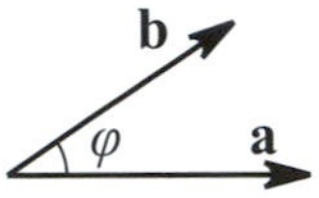

Figure M4.16. Scalar product of two vectors.

Properties of the scalar product:
1. $\mathbf{a} \cdot \mathbf{b} = \mathbf{b} \cdot \mathbf{a}$ (commutativity).
2. $\mathbf{a} \cdot (\mathbf{b} + \mathbf{c}) = \mathbf{a} \cdot \mathbf{b} + \mathbf{a} \cdot \mathbf{c}$ (distributivity with respect to addition of vectors). This property holds for any number of summands.
3. If **a** and **b** are collinear, then $\mathbf{a} \cdot \mathbf{b} = \pm|\mathbf{a}||\mathbf{b}|$. (The plus sign is taken if **a** and **b** have the same direction, and the minus sign is taken if they have opposite directions.)
4. $(\lambda \mathbf{a}) \cdot \mathbf{b} = \lambda (\mathbf{a} \cdot \mathbf{b})$ (associativity with respect to a scalar factor).
5. $\mathbf{a} \cdot \mathbf{a} = |\mathbf{a}|^2$. The scalar product $\mathbf{a} \cdot \mathbf{a}$ is denoted by $\mathbf{a}^2$ (the *scalar square* of the vector **a**).
6. The magnitude of a vector is expressed via the scalar product as

$$|\mathbf{a}| = \sqrt{\mathbf{a} \cdot \mathbf{a}} = \sqrt{\mathbf{a}^2}.$$

7. Two nonzero vectors **a** and **b** are perpendicular if and only if $\mathbf{a} \cdot \mathbf{b} = 0$.
8. The scalar products of the basis vectors are

$$\mathbf{i} \cdot \mathbf{j} = \mathbf{i} \cdot \mathbf{k} = \mathbf{j} \cdot \mathbf{k} = 0, \quad \mathbf{i} \cdot \mathbf{i} = \mathbf{j} \cdot \mathbf{j} = \mathbf{k} \cdot \mathbf{k} = 1.$$

9. If vectors are given by their coordinates, $\mathbf{a} = (a_x, a_y, a_z)$ and $\mathbf{b} = (b_x, b_y, b_z)$, then

$$\mathbf{a} \cdot \mathbf{b} = (a_x\mathbf{i} + a_y\mathbf{j} + a_z\mathbf{k})(b_x\mathbf{i} + b_y\mathbf{j} + b_z\mathbf{k}) = a_xb_x + a_yb_y + a_zb_z.$$

10. The *Cauchy–Schwarz inequality*

$$|\mathbf{a} \cdot \mathbf{b}| \le |\mathbf{a}||\mathbf{b}|.$$

11. The *Minkowski inequality*

$$|\mathbf{a} + \mathbf{b}| \le |\mathbf{a}| + |\mathbf{b}|.$$

12. The *angle φ between vectors* **a** and **b** is determined by the formula

$$\cos \varphi = \frac{\mathbf{a} \cdot \mathbf{b}}{|\mathbf{a}||\mathbf{b}|} = \frac{a_xb_x + a_yb_y + a_zb_z}{\sqrt{a_x^2 + a_y^2 + a_z^2}\sqrt{b_x^2 + b_y^2 + b_z^2}}.$$

▶ **Cross product of two vectors.** The *cross product* of a vector **a** by a vector **b** is defined as the vector **c** denoted by $\mathbf{a} \times \mathbf{b}$ (see Fig. M4.17), satisfying the following three conditions:

1. Its absolute value (magnitude) is equal to the area of the parallelogram spanned by the vectors **a** and **b**; i.e.,

$$|\mathbf{c}| = |\mathbf{a} \times \mathbf{b}| = |\mathbf{a}||\mathbf{b}| \sin \varphi.$$

2. It is perpendicular to the plane of the parallelogram; i.e., $\mathbf{c} \perp \mathbf{a}$ and $\mathbf{c} \perp \mathbf{b}$.
3. The vectors **a**, **b**, and **c** form a *right-handed trihedral*; i.e., the vector **c** points to the side from which the sense of the shortest rotation from **a** to **b** is counterclockwise.

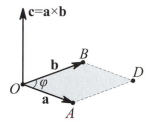

Figure M4.17. Cross product of two vectors.

Remark. The cross product of a vector **a** by a vector **b** is also denoted by $\mathbf{c} = [\mathbf{a}, \mathbf{b}]$.

Properties of cross product:

1. $\mathbf{a} \times \mathbf{b} = -\mathbf{b} \times \mathbf{a}$ (anticommutativity).
2. $\mathbf{a} \times (\mathbf{b} + \mathbf{c}) = \mathbf{a} \times \mathbf{b} + \mathbf{a} \times \mathbf{c}$ (distributivity with respect to the addition of vectors). This property holds for any number of summands.
3. Vectors **a** and **b** are collinear if and only if $\mathbf{a} \times \mathbf{b} = 0$. In particular, $\mathbf{a} \times \mathbf{a} = 0$.
4. $(\lambda\mathbf{a}) \times \mathbf{b} = \mathbf{a} \times (\lambda\mathbf{b}) = \lambda(\mathbf{a} \times \mathbf{b})$ (associativity with respect to a scalar factor).
5. The cross products of basis vectors are

$$\mathbf{i} \times \mathbf{i} = \mathbf{j} \times \mathbf{j} = \mathbf{k} \times \mathbf{k} = 0, \quad \mathbf{i} \times \mathbf{j} = \mathbf{k}, \quad \mathbf{j} \times \mathbf{k} = \mathbf{i}, \quad \mathbf{k} \times \mathbf{i} = \mathbf{j}.$$

6. If the vectors are given by their coordinates $\mathbf{a} = (a_x, a_y, a_z)$ and $\mathbf{b} = (b_x, b_y, b_z)$, then

$$\mathbf{a} \times \mathbf{b} = \begin{vmatrix} \mathbf{i} & \mathbf{j} & \mathbf{k} \\ a_x & a_y & a_z \\ b_x & b_y & b_z \end{vmatrix} = (a_y b_z - a_z b_y)\mathbf{i} + (a_z b_x - a_x b_z)\mathbf{j} + (a_x b_y - a_y b_x)\mathbf{k}.$$

7. The area of the parallelogram spanned by vectors **a** and **b** is equal to

$$S = |\mathbf{a} \times \mathbf{b}| = \sqrt{\begin{vmatrix} a_y & a_z \\ b_y & b_z \end{vmatrix}^2 + \begin{vmatrix} a_x & a_z \\ b_x & b_z \end{vmatrix}^2 + \begin{vmatrix} a_x & a_y \\ b_x & b_y \end{vmatrix}^2}.$$

8. The area of the triangle spanned by vectors **a** and **b** is equal to

$$S = \frac{1}{2}|\mathbf{a} \times \mathbf{b}| = \frac{1}{2}\sqrt{\begin{vmatrix} a_y & a_z \\ b_y & b_z \end{vmatrix}^2 + \begin{vmatrix} a_x & a_z \\ b_x & b_z \end{vmatrix}^2 + \begin{vmatrix} a_x & a_y \\ b_x & b_y \end{vmatrix}^2}.$$

▶ **Conditions for vectors to be parallel or perpendicular.**
A vector **a** is collinear to a vector **b** if

$$\mathbf{b} = \lambda\mathbf{a} \quad \text{or} \quad \mathbf{a} \times \mathbf{b} = 0.$$

A vector **a** is perpendicular to a vector **b** if

$$\mathbf{a} \cdot \mathbf{b} = 0.$$

Remark. In general, the condition $\mathbf{a} \cdot \mathbf{b} = 0$ implies that the vectors **a** and **b** are perpendicular or one of them is the zero vector. The zero vector can be viewed to be perpendicular to any other vector.

▶ **Scalar triple product of three vectors.** The *scalar triple product* of vectors **a**, **b**, and **c** is defined as the scalar product of **a** by the cross product of **b** and **c**:

$$[\mathbf{abc}] = \mathbf{a} \cdot (\mathbf{b} \times \mathbf{c}).$$

Remark. The scalar triple product of three vectors **a**, **b**, and **c** is also denoted by **abc**.

Properties of scalar triple product:
1. $[\mathbf{abc}] = [\mathbf{bca}] = [\mathbf{cab}] = -[\mathbf{bac}] = -[\mathbf{cba}] = -[\mathbf{acb}]$.
2. $[\mathbf{aab}] = [\mathbf{bab}] = 0$ or $\mathbf{a} \cdot (\mathbf{a} \times \mathbf{b}) = \mathbf{b} \cdot (\mathbf{a} \times \mathbf{b}) = 0$.
3. $[(\mathbf{a} + \mathbf{b})\mathbf{cd}] = [\mathbf{acd}] + [\mathbf{bcd}]$ (distributivity with respect to addition of vectors). This property holds for any number of summands.
4. $[\lambda\mathbf{abc}] = \lambda[\mathbf{abc}]$ (associativity with respect to a scalar factor).
5. If the vectors are given by their coordinates $\mathbf{a} = (a_x, a_y, a_z)$, $\mathbf{b} = (b_x, b_y, b_z)$, and $\mathbf{c} = (c_x, c_y, c_z)$, then

$$[\mathbf{abc}] = \begin{vmatrix} a_x & a_y & a_z \\ b_x & b_y & b_z \\ c_x & c_y & c_z \end{vmatrix}.$$

6. The scalar triple product $[\mathbf{abc}]$ is equal to the volume V of the parallelepiped spanned by the vectors **a**, **b**, and **c** taken with the sign + if the vectors **a**, **b**, and **c** form a right-handed trihedral and the sign − if the vectors form a left-handed trihedral,

$$[\mathbf{abc}] = \pm V.$$

7. Three nonzero vectors **a**, **b**, and **c** are coplanar if and only if $[\mathbf{abc}] = 0$. In this case, the vectors **a**, **b**, and **c** are linearly dependent; they satisfy a relation of the form $\alpha\mathbf{a} + \beta\mathbf{b} + \gamma\mathbf{c} = 0$.

M4.5. Line and Plane in Space

M4.5.1. Plane in Space

▶ **General equation of a plane.** In a Cartesian coordinate system, a plane is given by a first-order algebraic equation.

The *general* (*complete*) *equation of a plane* has the form

$$Ax + By + Cz + D = 0, \tag{M4.5.1.1}$$

where $A^2 + B^2 + C^2 \neq 0$.
1. For $D = 0$, the equation defines a plane passing through the origin.
2. For $A = 0$ (respectively, $B = 0$ or $C = 0$), the equation defines a plane parallel to the axis OX (respectively, OY or OZ).
3. For $A = D = 0$ (respectively, $B = D = 0$ or $C = D = 0$), the equation defines a plane passing through the axis OX (respectively, OY or OZ).
4. For $A = B = 0$ (respectively, $A = C = 0$ or $B = C = 0$), the equation defines a plane parallel to the plane OXY (respectively, OXZ or OYZ).

▶ **Intercept equation of a plane.** A plane $Ax + By + Cz + D = 0$ that is not parallel to the axis OX (i.e., $A \neq 0$) meets this axis at a (signed) distance $a = -D/A$ from the origin (see Fig. M4.18). The number a is called the x-intercept of the plane. Similarly, one defines the y-intercepts $b = -D/B$ (for $B \neq 0$) and the z-intercept $c = -D/C$ (for $C \neq 0$). Then such a plane can be defined by the equation

$$\frac{x}{a} + \frac{y}{b} + \frac{z}{c} = 1,$$

which is called the *intercept equation of the plane*.

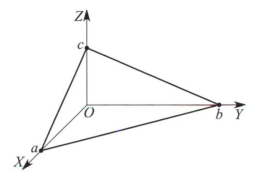

Figure M4.18. A plane with intercept equation.

Remark. A plane parallel to the axis OX but nonparallel to the other two axes is defined by the equation $y/b + z/c = 1$, where b and c are the y- and z-intercepts of the plane. A plane simultaneously parallel to the axes OX and OY can be represented in the form $z/c = 1$.

▶ **Equation of the plane passing through a point M_0 and perpendicular to a vector N.** The equation of the plane passing through a point $M_0(x_0, y_0, z_0)$ and perpendicular to a vector $\mathbf{N} = (A, B, C)$ has the form

$$A(x - x_0) + B(y - y_0) + C(z - z_0) = 0, \qquad \text{or} \qquad (\mathbf{r} - \mathbf{r}_0) \cdot \mathbf{N} = 0, \qquad \text{(M4.5.1.2)}$$

where $\mathbf{r}$ and $\mathbf{r}_0$ are the position vectors of the points $M(x, y, z)$ and $M_0(x_0, y_0, z_0)$, respectively (see Fig. M4.19). The vector $\mathbf{N}$ is called a *normal vector*. Its direction cosines are

$$\cos \alpha = \frac{A}{\sqrt{A^2 + B^2 + C^2}}, \quad \cos \beta = \frac{B}{\sqrt{A^2 + B^2 + C^2}}, \quad \cos \gamma = \frac{C}{\sqrt{A^2 + B^2 + C^2}}.$$

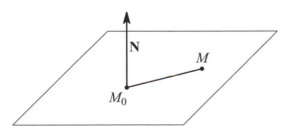

Figure M4.19. Plane passing through a point M_0 and perpendicular to a vector $\mathbf{N}$.

▶ **Equation of the plane passing through a point and parallel to another plane.** The plane that passes through a point $M_0(x_0, y_0, z_0)$ and is parallel to a plane $Ax+By+Cz+D=0$ is given by equation (M4.5.1.2).

▶ **Equation of the plane passing through three points.** The plane passing through three points $M_1(x_1, y_1, z_1)$, $M_2(x_2, y_2, z_2)$, and $M_3(x_3, y_3, z_3)$ (see Fig. M4.20) is described by the equation

$$\begin{vmatrix} x - x_1 & y - y_1 & z - z_1 \\ x_2 - x_1 & y_2 - y_1 & z_2 - z_1 \\ x_3 - x_1 & y_3 - y_1 & z_3 - z_1 \end{vmatrix} = 0, \quad \text{or} \quad \left[(\mathbf{r} - \mathbf{r}_1)(\mathbf{r}_2 - \mathbf{r}_1)(\mathbf{r}_3 - \mathbf{r}_1)\right] = 0, \quad \text{(M4.5.1.3)}$$

where $\mathbf{r}$, $\mathbf{r}_1$, $\mathbf{r}_2$, and $\mathbf{r}_3$ are the position vectors of the points $M(x, y, z)$, $M_1(x_1, y_1, z_1)$, $M_2(x_2, y_2, z_2)$, and $M_3(x_3, y_3, z_3)$, respectively.

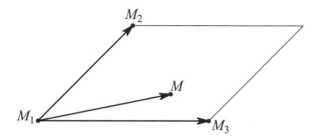

Figure M4.20. Plane passing through three points.

Remark 1. Equation (M4.5.1.3) means that the vectors $\overrightarrow{M_1 M}$, $\overrightarrow{M_1 M_2}$, and $\overrightarrow{M_1 M_3}$ are coplanar.

Remark 2. If the three points $M_1(x_1, y_1, z_1)$, $M_2(x_2, y_2, z_2)$, and $M_3(x_3, y_3, z_3)$ are collinear, then equation (M4.5.1.3) is satisfied identically.

Example 1. Let us construct an equation of the plane passing through the three points $M_1(1, 1, 1)$, $M_2(2, 2, 1)$, and $M_3(1, 2, 2)$.

Obviously, the points M_1, M_2, and M_3 are not collinear, since the vectors $\overrightarrow{M_1 M_2} = (1, 1, 0)$ and $\overrightarrow{M_1 M_3} = (0, 1, 1)$ are not collinear. According to (M4.5.1.3), we have

$$\begin{vmatrix} x - 1 & y - 1 & z - 1 \\ 1 & 1 & 0 \\ 0 & 1 & 1 \end{vmatrix} = 0,$$

whence the desired equation is $x - y + z - 1 = 0$.

▶ **Equation of the plane passing through two points and parallel to a straight line.** The plane passing through two points $M_1(x_1, y_1, z_1)$ and $M_2(x_2, y_2, z_2)$ and parallel to a straight line with direction vector $\mathbf{R} = (l, m, n)$ (see Fig. M4.21) is given by the equation

$$\begin{vmatrix} x - x_1 & y - y_1 & z - z_1 \\ x_2 - x_1 & y_2 - y_1 & z_2 - z_1 \\ l & m & n \end{vmatrix} = 0, \quad \text{or} \quad \left[(\mathbf{r} - \mathbf{r}_1)(\mathbf{r}_2 - \mathbf{r}_1)\mathbf{R}\right] = 0, \quad \text{(M4.5.1.4)}$$

where $\mathbf{r}$, $\mathbf{r}_1$, and $\mathbf{r}_2$ are the position vectors of the points $M(x, y, z)$, $M_1(x_1, y_1, z_1)$, and $M_2(x_2, y_2, z_2)$, respectively.

Remark. If the vectors $\overrightarrow{M_1 M_2}$ and $\mathbf{R}$ are collinear, then equations (M4.5.1.4) become identities.

Example 2. Find an equation of the plane passing through the points $M_1(0, 1, 0)$ and $M_2(1, 1, 1)$ and parallel to the straight line with direction vector $\mathbf{R} = (0, 1, 1)$.

According to (M4.5.1.4), we have

$$\begin{vmatrix} x - 0 & y - 1 & z - 0 \\ 1 - 0 & 1 - 1 & 1 - 0 \\ 0 & 1 & 1 \end{vmatrix} = 0,$$

whence the desired equation is $-x - y + z + 1 = 0$.

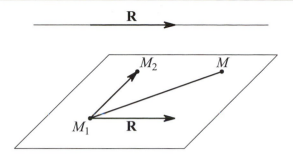

Figure M4.21. Plane passing through two points and parallel to a line.

▶ **Equation of the plane passing through a point and parallel to two straight lines.**
The plane passing through a point $M_0(x_0, y_0, z_0)$ and parallel to two straight lines with
direction vectors $\mathbf{R}_1 = (l_1, m_1, n_1)$ and $\mathbf{R}_2 = (l_2, m_2, n_2)$ is given by the equation

$$\begin{vmatrix} x - x_0 & y - y_0 & z - z_0 \\ l_1 & m_1 & n_1 \\ l_2 & m_2 & n_2 \end{vmatrix} = 0, \qquad \text{or} \qquad \big[(\mathbf{r} - \mathbf{r}_0)\mathbf{R}_1\mathbf{R}_2\big] = 0,$$

where $\mathbf{r}$ and $\mathbf{r}_0$ are the position vectors of the points $M(x, y, z)$ and $M_0(x_0, y_0, z_0)$, respectively.

The equation of the plane passing through a point $M_0(x_0, y_0, z_0)$ and parallel to two
noncollinear vectors $\mathbf{R}_1 = (l_1, m_1, n_1)$ and $\mathbf{R}_2 = (l_2, m_2, n_2)$ can be represented in the form
(M4.5.1.2) with A, B, and C being the coordinates of the vector $\mathbf{R} = \mathbf{R}_1 \times \mathbf{R}_2$.

Example 3. Find an equation of the plane P that passes through the point $M_0(2, -1, 1)$ and is perpendicular
to the planes P_1 and P_2 defined by $3x + 2y - z + 4 = 0$ and $x + y + z - 3 = 0$.

The vectors $\mathbf{N}_1 = (3, 2, -1)$ and $\mathbf{N}_2 = (1, 1, 1)$ are normal to P_1 and P_2 and parallel to P. Their cross product
is

$$\mathbf{N} = \mathbf{N}_1 \times \mathbf{N}_2 = \begin{vmatrix} \mathbf{i} & \mathbf{j} & \mathbf{k} \\ 3 & 2 & -1 \\ 1 & 1 & 1 \end{vmatrix} = 3\mathbf{i} - 4\mathbf{j} + 1\mathbf{k}.$$

The vector $\mathbf{N}$ is perpendicular to the desired plane P, which therefore satisfies the equation

$$3(x - 2) - 4(y + 1) + (z - 1) = 0 \qquad \text{or} \qquad 3x - 4y + z - 11 = 0.$$

▶ **Equation of the plane passing through two points and perpendicular to a given
plane.** The plane passing through two points $M_1(x_1, y_1, z_1)$ and $M_2(x_2, y_2, z_2)$ and perpendicular to the plane $Ax + By + Cz + D = 0$ (see Fig. M4.22) is determined by the
equation

$$\begin{vmatrix} x - x_1 & y - y_1 & z - z_1 \\ x_2 - x_1 & y_2 - y_1 & z_2 - z_1 \\ A & B & C \end{vmatrix} = 0, \qquad \text{or} \qquad \big[(\mathbf{r} - \mathbf{r}_1)(\mathbf{r}_2 - \mathbf{r}_1)\mathbf{N}\big] = 0, \qquad \text{(M4.5.1.5)}$$

where $\mathbf{r}$, $\mathbf{r}_1$, and $\mathbf{r}_2$ are the position vectors of the points $M(x, y, z)$, $M_1(x_1, y_1, z_1)$, and
$M_2(x_2, y_2, z_2)$, respectively.

Remark. If the straight line passing through the points $M_1(x_1, y_1, z_1)$ and $M_2(x_2, y_2, z_2)$ is perpendicular
to the original plane, then the desired plane is undetermined and equations (M4.5.1.5) become identities.

▶ **Equation of the plane passing through a point and perpendicular to two planes.**
The plane passing through a point $M_1(x_1, y_1, z_1)$ and perpendicular to two (nonparallel)

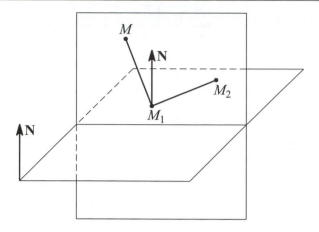

Figure M4.22. Plane passing through two points and perpendicular to a given plane.

planes $A_1x + B_1y + C_1z + D_1 = 0$ and $A_2x + B_2y + C_2z + D_2 = 0$ (see Fig. M4.23) is given by the equation

$$\begin{vmatrix} x - x_1 & y - y_1 & z - z_1 \\ A_1 & B_1 & C_1 \\ A_2 & B_2 & C_2 \end{vmatrix} = 0, \qquad \text{or} \qquad \big[(\mathbf{r} - \mathbf{r}_1)\mathbf{N}_1\mathbf{N}_2\big] = 0, \tag{M4.5.1.6}$$

where $\mathbf{N}_1 = (A_1, B_1, C_1)$ and $\mathbf{N}_2 = (A_2, B_2, C_2)$ are normals to the given planes and $\mathbf{r}$ and $\mathbf{r}_1$ are the position vectors of the points $M(x, y, z)$ and $M_1(x_1, y_1, z_1)$, respectively.

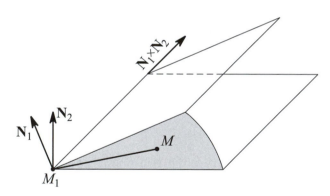

Figure M4.23. Plane passing through a point and perpendicular to two planes.

Remark 1. Equations (M4.5.1.6) mean that the vectors $\overrightarrow{M_1M}$, $\mathbf{N}_1$, and $\mathbf{N}_2$ are coplanar.

Remark 2. If the original planes are parallel, then the desired plane is undetermined. In this case, equations (M4.5.1.6) become identities.

Example 4. Let us find an equation of the plane passing through the point $M_1(0, 1, 2)$ and perpendicular to the planes $x - y + z - 3 = 0$ and $-x + y + z + 4 = 0$.

According to (M4.5.1.6), we have

$$\begin{vmatrix} x - 0 & y - 1 & z - 2 \\ 1 & -1 & 1 \\ -1 & 1 & 1 \end{vmatrix} = 0,$$

whence the desired equation is $x + y - 1 = 0$.

▶ **Equation of planes passing through the line of intersection of planes.** The planes passing through the line of intersection of the planes $A_1x + B_1y + C_1z + D_1 = 0$ and $A_2x + B_2y + C_2z + D_2 = 0$ are given by the equation

$$\alpha(A_1x + B_1y + C_1z + D_1) + \beta(A_2x + B_2y + C_2z + D_2) = 0,$$

which is called the *equation of a pencil of planes.* Here α and β are arbitrary parameters $(\alpha^2 + \beta^2 \neq 0)$.

M4.5.2. Line in Space

▶ **Parametric equations of a straight line.** The *parametric equations of the line* that passes through a point $M_1(x_1, y_1, z_1)$ and is parallel to a direction vector $\mathbf{R} = (l, m, n)$ (see Fig. M4.24) are

$$x = x_1 + lt, \quad y = y_1 + mt, \quad z = z_1 + nt, \qquad \text{or} \qquad \mathbf{r} = \mathbf{r}_1 + t\mathbf{R}, \qquad \text{(M4.5.2.1)}$$

where $\mathbf{r} = \overrightarrow{OM}$ and $\mathbf{r}_1 = \overrightarrow{OM_1}$. As the parameter t varies from $-\infty$ to $+\infty$, the point M with position vector $\mathbf{r} = (x, y, z)$ determined by formula (M4.5.2.1) runs over the entire straight line in question. It is convenient to use parametric equations (M4.5.2.1) if one needs to find the point of intersection of a straight line with a plane.

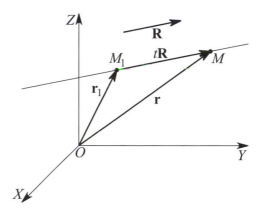

Figure M4.24. Straight line passing through a point and parallel to a direction vector.

The numbers l, m, and n characterize the direction of the straight line in space; they are called the *direction coefficients* of the straight line. For a unit vector $\mathbf{R} = \mathbf{R}^0$, the coefficients l, m, n are the cosines of the angles α, β, and γ formed by this straight line (the direction vector $\mathbf{R}^0$) with the coordinate axes OX, OY, and OZ. These cosines can be expressed via the coordinates of the direction vector $\mathbf{R}$ as

$$\cos\alpha = \frac{l}{\sqrt{l^2 + m^2 + n^2}}, \quad \cos\beta = \frac{m}{\sqrt{l^2 + m^2 + n^2}}, \quad \cos\gamma = \frac{n}{\sqrt{l^2 + m^2 + n^2}}.$$

▶ **Canonical equations of a straight line.** The equations

$$\frac{x - x_1}{l} = \frac{y - y_1}{m} = \frac{z - z_1}{n}, \qquad \text{or} \qquad (\mathbf{r} - \mathbf{r}_1) \times \mathbf{R} = 0, \qquad \text{(M4.5.2.2)}$$

are called the *canonical equations of the straight line* through the point $M_1(x_1, y_1, z_1)$ with the position vector $\mathbf{r}_1 = (x_1, y_1, z_1)$ and parallel to the direction vector $\mathbf{R} = (l, m, n)$.

Remark 1. One can obtain canonical equations (M4.5.2.2) from parametric equations (M4.5.2.1) by eliminating the parameter t.

Remark 2. In the canonical equations, all coefficients l, m, and n cannot be zero simultaneously, since $|\mathbf{R}| \neq 0$. But some of them may be zero. If one of the denominators in equations (M4.5.2.2) is zero, this means that the corresponding numerator is also zero.

▶ **General equation of a straight line.** The *general equation of a straight line in space* defines it as the line of intersection of two planes (see Fig. M4.25) and is given analytically by a system of two linear equations

$$A_1 x + B_1 y + C_1 z + D_1 = 0,$$
$$A_2 x + B_2 y + C_2 z + D_2 = 0. \qquad \text{(M4.5.2.3)}$$

The normals to the planes are $\mathbf{N}_1 = (A_1, B_1, C_1)$ and $\mathbf{N}_2 = (A_2, B_2, C_2)$. The direction vector $\mathbf{R}$ is equal to the cross product of the normals $\mathbf{N}_1$ and $\mathbf{N}_2$; i.e.,

$$\mathbf{R} = \mathbf{N}_1 \times \mathbf{N}_2, \qquad \text{(M4.5.2.4)}$$

and its coordinates l, m, and n can be obtained by the formulas

$$l = \begin{vmatrix} B_1 & C_1 \\ B_2 & C_2 \end{vmatrix}, \qquad m = \begin{vmatrix} C_1 & A_1 \\ C_2 & A_2 \end{vmatrix}, \qquad n = \begin{vmatrix} A_1 & B_1 \\ A_2 & B_2 \end{vmatrix}.$$

Remark 1. Simultaneous equations of the form (M4.5.2.3) define a straight line if and only if the coefficients A_1, B_1, and C_1 in one of them are not proportional to the respective coefficients A_2, B_2, and C_2 in the other.

Remark 2. For $D_1 = D_2 = 0$ (and only in this case), the line passes through the origin.

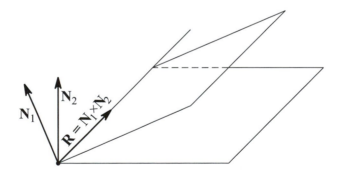

Figure M4.25. Straight line as intersection of two planes.

Example. Let us reduce the equation of the straight line

$$x + 2y - z + 1 = 0, \qquad x - y + z + 3 = 0$$

to canonical form.

We choose one of the coordinates arbitrarily; say, $x = 0$. Then

$$2y - z + 1 = 0, \qquad -y + z + 3 = 0,$$

and hence $y = -4$, $z = -7$. Thus the desired line contains the point $M(0, -4, -7)$. We find the cross product of the vectors $\mathbf{N}_1 = (1, 2, -1)$ and $\mathbf{N}_2 = (1, -1, 1)$ and, according to (M4.5.2.4), obtain the direction vector $\mathbf{R} = (1, -2, -3)$ of the desired line. Therefore, with (M4.5.2.2) taken into account, the equations of the line become

$$\frac{x}{1} = \frac{y + 4}{-2} = \frac{z + 7}{-3}.$$

▶ **Equations of a straight line passing through two points.** The canonical equations of the straight line (see Fig. M4.26) passing through two points $M_1(x_1, y_1, z_1)$ and $M_2(x_2, y_2, z_2)$ are

$$\frac{x - x_1}{x_2 - x_1} = \frac{y - y_1}{y_2 - y_1} = \frac{z - z_1}{z_2 - z_1}, \qquad \text{or} \qquad (\mathbf{r} - \mathbf{r}_1) \times (\mathbf{r}_2 - \mathbf{r}_1) = 0, \qquad (\text{M4.5.2.5})$$

where $\mathbf{r}$, $\mathbf{r}_1$, and $\mathbf{r}_2$ are the position vectors of the points $M(x, y, z)$, $M_1(x_1, y_1, z_1)$, and $M_2(x_2, y_2, z_2)$, respectively.

The parametric equations of this line are

$$\begin{aligned} x &= x_1(1 - t) + x_2 t, \\ y &= y_1(1 - t) + y_2 t, \qquad \text{or} \qquad \mathbf{r} = (1 - t)\mathbf{r}_1 + t\mathbf{r}_2. \\ z &= z_1(1 - t) + z_2 t, \end{aligned} \qquad (\text{M4.5.2.6})$$

Remark. Eliminating the parameter t from equations (M4.5.2.6), we obtain equations (M4.5.2.5).

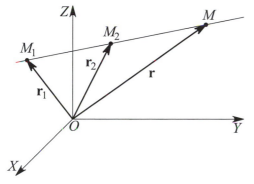

Figure M4.26. Straight line passing through two points.

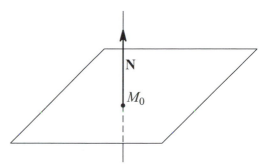

Figure M4.27. Straight line passing through a point and perpendicular to a plane.

▶ **Equations of a straight line passing through a point and perpendicular to a plane.** The equations of the straight line passing through a point $M_0(x_0, y_0, z_0)$ and perpendicular to the plane given by the equation $Ax + By + Cz + D = 0$ (see Fig. M4.27) are

$$\frac{x - x_0}{A} = \frac{y - y_0}{B} = \frac{z - z_0}{C}.$$

M4.5.3. Mutual Arrangement of Points, Lines, and Planes

▶ **Angles between lines in space.** Consider two straight lines determined by vector parametric equations $\mathbf{r} = \mathbf{r}_1 + t\mathbf{R}_1$ and $\mathbf{r} = \mathbf{r}_2 + t\mathbf{R}_2$. The angle φ between these lines (see Fig. M4.28) can be obtained from the formulas

$$\cos \varphi = \frac{\mathbf{R}_1 \cdot \mathbf{R}_2}{|\mathbf{R}_1||\mathbf{R}_2|}, \qquad \sin \varphi = \frac{|\mathbf{R}_1 \times \mathbf{R}_2|}{|\mathbf{R}_1||\mathbf{R}_2|}.$$

If the lines are given by the canonical equations

$$\frac{x - x_1}{l_1} = \frac{y - y_1}{m_1} = \frac{z - z_1}{n_1} \quad \text{and} \quad \frac{x - x_2}{l_2} = \frac{y - y_2}{m_2} = \frac{z - z_2}{n_2}, \qquad (\text{M4.5.3.1})$$

then the angle φ between the lines can be found from the formulas

$$\cos\varphi = \frac{l_1 l_2 + m_1 m_2 + n_1 n_2}{\sqrt{l_1^2 + m_1^2 + n_1^2}\sqrt{l_2^2 + m_2^2 + n_2^2}},$$

$$\sin\varphi = \frac{\sqrt{\begin{vmatrix} m_1 & n_1 \\ m_2 & n_2 \end{vmatrix}^2 + \begin{vmatrix} n_1 & l_1 \\ n_2 & l_2 \end{vmatrix}^2 + \begin{vmatrix} l_1 & m_1 \\ l_2 & m_2 \end{vmatrix}^2}}{\sqrt{l_1^2 + m_1^2 + n_1^2}\sqrt{l_2^2 + m_2^2 + n_2^2}}. \tag{M4.5.3.2}$$

Example 1. Let us find the angle between the lines

$$\frac{x}{1} = \frac{y-2}{2} = \frac{z+1}{2} \quad \text{and} \quad \frac{x}{0} = \frac{y-2}{3} = \frac{z+1}{4}.$$

Using the first formula in (M4.5.3.2), we obtain

$$\cos\varphi = \frac{1\cdot 0 + 2\cdot 3 + 2\cdot 4}{\sqrt{1^2 + 2^2 + 2^2}\,\sqrt{0^2 + 3^2 + 4^2}} = \frac{14}{15},$$

and hence $\varphi \approx 0.3672$ rad.

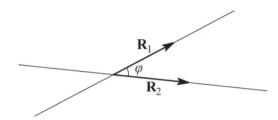

Figure M4.28. Angle between two lines in space.

▶ **Conditions for two lines to be parallel.** Two straight lines given by vector parametric equations $\mathbf{r} = \mathbf{r}_1 + t\mathbf{R}_1$ and $\mathbf{r} = \mathbf{r}_2 + t\mathbf{R}_2$ are parallel if

$$\mathbf{R}_2 = \lambda\mathbf{R}_1 \quad \text{or} \quad \mathbf{R}_2 \times \mathbf{R}_1 = 0,$$

i.e., if their direction vectors $\mathbf{R}_1$ and $\mathbf{R}_2$ are collinear. This can be written as

$$\frac{l_1}{l_2} = \frac{m_1}{m_2} = \frac{n_1}{n_2}.$$

Remark. If parallel lines have a common point (i.e., $\mathbf{r}_1 = \mathbf{r}_2$ in parametric equations), then they coincide.

▶ **Conditions for two lines to be perpendicular.** Two straight lines given by vector parametric equations $\mathbf{r} = \mathbf{r}_1 + t\mathbf{R}_1$ and $\mathbf{r} = \mathbf{r}_2 + t\mathbf{R}_2$ are perpendicular if

$$\mathbf{R}_1 \cdot \mathbf{R}_2 = 0. \tag{M4.5.3.3}$$

This condition can be written as

$$l_1 l_2 + m_1 m_2 + n_1 n_2 = 0. \tag{M4.5.3.4}$$

Example 2. Let us show that the lines

$$\frac{x-1}{2} = \frac{y-3}{1} = \frac{z}{2} \quad \text{and} \quad \frac{x-2}{1} = \frac{y+1}{2} = \frac{z}{-2}$$

are perpendicular.

Indeed, condition (M4.5.3.4) is satisfied,

$$2\cdot 1 + 1\cdot 2 + 2\cdot(-2) = 0,$$

and hence the lines are perpendicular.

▶ **Theorem on the arrangement of two lines in space.** *Two straight lines in space can:*
a) *be skew;*
b) *lie in the same plane and not meet each other, i.e., be parallel;*
c) *meet at a point;*
d) *coincide.*

A general characteristic of all four cases is the determinant of the matrix

$$\begin{pmatrix} x_2 - x_1 & y_2 - y_1 & z_2 - z_1 \\ l_1 & m_1 & n_1 \\ l_2 & m_2 & n_2 \end{pmatrix}, \qquad \text{(M4.5.3.5)}$$

whose entries are taken from the canonical equations of the lines (M4.5.3.1).

In cases *a–d* of the theorem, for the matrix (M4.5.3.5) we have, respectively:
a) the determinant is nonzero;
b) the last two rows are proportional to each other but are not proportional to the first row;
c) the last two rows are not proportional, and the first row is their linear combination;
d) all rows are proportional.

In cases *b–d* the determinant is zero.

▶ **Angle between planes.** Consider two planes given by the general equations

$$\begin{aligned} A_1 x + B_1 y + C_1 z + D_1 &= 0, \\ A_2 x + B_2 y + C_2 z + D_2 &= 0. \end{aligned} \qquad \text{(M4.5.3.6)}$$

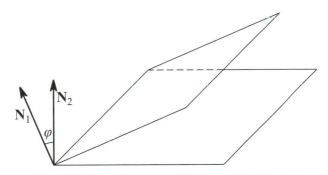

Figure M4.29. Angle between two planes.

The angle between two planes (see Fig. M4.29) is defined as any of the two adjacent dihedral angles formed by the planes (if the planes are parallel, then the angle between them is by definition equal to 0 or π). One of these dihedral angles is equal to the angle φ between the normal vectors $\mathbf{N}_1 = (A_1, B_1, C_1)$ and $\mathbf{N}_2 = (A_2, B_2, C_2)$ to the planes, which can be determined by the formula

$$\cos \varphi = \frac{A_1 A_2 + B_1 B_2 + C_1 C_2}{\sqrt{A_1^2 + B_1^2 + C_1^2}\sqrt{A_2^2 + B_2^2 + C_2^2}} = \frac{\mathbf{N}_1 \cdot \mathbf{N}_2}{|\mathbf{N}_1||\mathbf{N}_2|}.$$

▶ **Conditions for two planes to be parallel.** Two planes given by the general equations (M4.5.3.6) are parallel if and only if the following *condition for the planes to be parallel* is satisfied:

$$\frac{A_1}{A_2} = \frac{B_1}{B_2} = \frac{C_1}{C_2} \neq \frac{D_1}{D_2};$$

in this case, the planes do not coincide.

Two planes coincide if they are parallel and have a common point. Two planes given by the general equations (M4.5.3.6) coincide if and only if the following *condition for the planes to coincide* is satisfied:

$$\frac{A_1}{A_2} = \frac{B_1}{B_2} = \frac{C_1}{C_2} = \frac{D_1}{D_2}.$$

▶ **Conditions for two planes to be perpendicular.** Planes are perpendicular if their normals are perpendicular. Two planes determined by the general equations (M4.5.3.6) are perpendicular if and only if the following *condition for the planes to be perpendicular* is satisfied:

$$A_1 A_2 + B_1 B_2 + C_1 C_2 = 0 \qquad \text{or} \qquad \mathbf{N}_1 \cdot \mathbf{N}_2 = 0, \qquad (M4.5.3.7)$$

where $\mathbf{N}_1 = (A_1, B_1, C_1)$ and $\mathbf{N}_2 = (A_2, B_2, C_2)$ are the normals to the planes.

> **Example 3.** Let us show that the planes $x - y + z = 0$ and $x - y - 2z + 5 = 0$ are perpendicular. Since condition (M4.5.3.7) is satisfied,
>
> $$1 \cdot 1 + (-1) \cdot (-1) + 1 \cdot (-2) = 0,$$

we see that the planes are perpendicular.

▶ **Angle between a straight line and a plane.** Consider a plane given by the general equation

$$Ax + By + Cz + D = 0 \qquad (M4.5.3.8)$$

and a line given by the canonical equations

$$\frac{x - x_1}{l} = \frac{y - y_1}{m} = \frac{z - z_1}{n}. \qquad (M4.5.3.9)$$

The angle between the line and the plane (see Fig. M4.30) is defined as the complementary angle θ of the angle φ between the direction vector $\mathbf{R} = (l, m, n)$ of the line and the normal $\mathbf{N} = (A, B, C)$ to the plane. For this angle, one has the formula

$$\sin \theta = |\cos \varphi| = \frac{|Al + Bm + Cn|}{\sqrt{A^2 + B^2 + C^2}\sqrt{l^2 + m^2 + n^2}} = \frac{|\mathbf{N} \cdot \mathbf{R}|}{|\mathbf{N}|\,|\mathbf{R}|}.$$

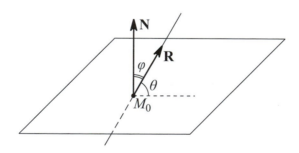

Figure M4.30. Angle between a straight line and a plane.

▶ **Conditions for a straight line and a plane to be parallel.** A plane given by the general equation (M4.5.3.8) and a line given by canonical equations (M4.5.3.9) are parallel if the following two conditions hold:

$$Al + Bm + Cn = 0,$$
$$Ax_1 + By_1 + Cz_1 + D \neq 0.$$

The first condition means that the direction vector of the straight line is perpendicular to the normal to the plane and the second condition means that the line is not contained in the plane.

▶ **Condition for a straight line and a plane to be perpendicular.** A line given by canonical equations (M4.5.3.9) and a plane given by the general equation (M4.5.3.8) are perpendicular if the line is collinear to the normal to the plane (is a normal itself), i.e., if

$$\frac{A}{l} = \frac{B}{m} = \frac{C}{n}, \qquad \text{or} \qquad \mathbf{N} = \lambda \mathbf{R}.$$

▶ **Intersection of a straight line and a plane.** Consider a plane given by the general equation (M4.5.3.8) and a straight line given by parametric equations

$$x = x_1 + lt, \quad y = y_1 + mt, \quad z = z_1 + nt.$$

The coordinates of the point $M_0(x_0, y_0, z_0)$ of intersection of the line with the plane (see Fig. M4.30), if the point exists at all, are determined by the formulas

$$x_0 = x_1 + lt_0, \quad y_0 = y_1 + mt_0, \quad z_0 = z_1 + nt_0,$$

where

$$t_0 = -\frac{Ax_1 + By_1 + Cz_1 + D}{Al + Bm + Cn}.$$

▶ **Distance from a point to a plane.** The distance from a point $M_0(x_0, y_0, z_0)$ to a plane given by the general equation (M4.5.1.1) is determined by the formula

$$d = \frac{|Ax_0 + By_0 + Cz_0 + D|}{\sqrt{A^2 + B^2 + C^2}}.$$

▶ **Distance between two parallel planes.** We consider two parallel planes given by the general equations $Ax + By + Cz + D_1 = 0$ and $Ax + By + Cz + D_2 = 0$. The distance between them is

$$d = \frac{|D_1 - D_2|}{\sqrt{A^2 + B^2 + C^2}}.$$

▶ **Distance from a point to a straight line.** The distance from a point $M_0(x_0, y_0, z_0)$ to a line given by canonical equations (M4.5.2.2) is determined by the formula

$$d = \frac{\sqrt{\begin{vmatrix} m & n \\ y_1 - y_0 & z_1 - z_0 \end{vmatrix}^2 + \begin{vmatrix} n & l \\ z_1 - z_0 & x_1 - x_0 \end{vmatrix}^2 + \begin{vmatrix} l & m \\ x_1 - x_0 & y_1 - y_0 \end{vmatrix}^2}}{\sqrt{l^2 + m^2 + n^2}}.$$

▶ **Distance between straight lines.** Consider two nonparallel lines given in the canonical form

$$\frac{x - x_1}{l_1} = \frac{y - y_1}{m_1} = \frac{z - z_1}{n_1},$$
$$\frac{x - x_2}{l_2} = \frac{y - y_2}{m_2} = \frac{z - z_2}{n_2}.$$

The distance between them can be calculated by the formula

$$d = \frac{\pm \begin{vmatrix} x_1 - x_2 & y_1 - y_2 & z_1 - z_2 \\ l_1 & m_1 & n_1 \\ l_2 & m_2 & n_2 \end{vmatrix}}{\sqrt{\begin{vmatrix} l_1 & m_1 \\ l_2 & m_2 \end{vmatrix}^2 + \begin{vmatrix} m_1 & n_1 \\ m_2 & n_2 \end{vmatrix}^2 + \begin{vmatrix} n_1 & l_1 \\ n_2 & l_2 \end{vmatrix}^2}} \tag{M4.5.3.10}$$

(minus sign should be taken if the determinant is negative). The condition that the determinant in the numerator in (M4.5.3.10) is zero is the *condition for the two lines in space to meet*.

M4.6. Quadric Surfaces (Quadrics)

M4.6.1. Quadrics and Their Canonical Equations

▶ **Central surfaces.** A segment joining two points of a surface is called a *chord*. If there exists a point in space, not necessarily lying on the surface, that bisects all chords passing through it, then the surface is said to be *central* and the point is called the *center* of the surface.

The equations listed below for central surfaces are given in *canonical form*; i.e., the center of a surface is at the origin, and the surface symmetry axes are the coordinate axes. Moreover, the coordinate planes are symmetry planes.

▶ **Ellipsoid.** An *ellipsoid* is a central surface defined by the equation

$$\frac{x^2}{a^2} + \frac{y^2}{b^2} + \frac{z^2}{c^2} = 1, \tag{M4.6.1.1}$$

where the numbers a, b, and c are the lengths of the segments called the *semiaxes* of the ellipsoid (see Fig. M4.31 *a*). The coordinates of all points of the ellipsoid satisfy the inequalities $-a \le x \le a$, $-b \le y \le b$, and $-c \le z \le c$.

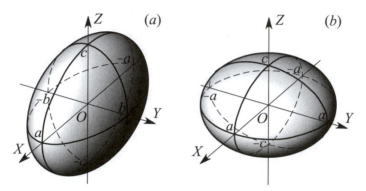

Figure M4.31. Triaxial ellipsoid (*a*) and spheroid (*b*).

If $a \ne b \ne c$, then the ellipsoid is said to be *triaxial*, or *scalene*. If $a = b \ne c$, then the ellipsoid is called a *spheroid*; it can be obtained by rotating the ellipse $x^2/a^2 + z^2/c^2 = 1$, $y = 0$ lying in the plane OXZ about the axis OZ (see Fig. M4.31 *b*). If $a = b > c$, then the ellipsoid is an *oblate spheroid*, and if $a = b < c$, then the ellipsoid is a *prolate spheroid*. If $a = b = c$, then the ellipsoid is the sphere of radius a given by the equation $x^2 + y^2 + z^2 = a^2$.

An arbitrary plane section of an ellipsoid is an ellipse (or, in a special case, a circle). The volume of an ellipsoid is equal to $V = \frac{4}{3}\pi abc$.

Remark. About the sphere, see also Subsection M3.2.3.

▶ **Hyperboloids.** A *one-sheeted hyperboloid* is a central surface defined by the equation

$$\frac{x^2}{a^2} + \frac{y^2}{b^2} - \frac{z^2}{c^2} = 1, \tag{M4.6.1.2}$$

where a and b are the *real semiaxes* and c is the *imaginary semiaxis* (see Fig. M4.32 *a*).

A *two-sheeted hyperboloid* is a central surface defined by the equation

$$\frac{x^2}{a^2} + \frac{y^2}{b^2} - \frac{z^2}{c^2} = -1, \tag{M4.6.1.3}$$

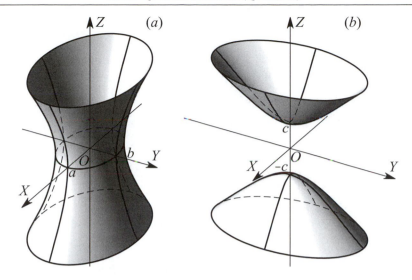

Figure M4.32. One-sheeted (a) and two-sheeted (b) hyperboloids.

where c is the *real semiaxis* and a and b are the *imaginary semiaxes* (see Fig. M4.32 b). A two-sheeted hyperboloid consists of two parts whose points lie at $z \leq -c$ and $z \geq c$.

A hyperboloid approaches the surface

$$\frac{x^2}{a^2} + \frac{y^2}{b^2} - \frac{z^2}{c^2} = 0,$$

which is called an *asymptotic cone*, infinitely closely.

A plane passing through the axis OZ intersects each of the hyperboloids (M4.6.1.2) and (M4.6.1.3) in two hyperbolas and the asymptotic cone in two straight lines, which are the asymptotes of these hyperbolas. The section of a hyperboloid by a plane parallel to OXY is an ellipse. The section of a one-sheeted hyperboloid by the plane $z = 0$ is an ellipse, which is called the *gorge* or *throat* ellipse.

For $a = b$, we deal with the *hyperboloid of revolution* obtained by rotating a hyperbola with semiaxes a and c about its focal axis $2c$ (which is an imaginary axis for a one-sheeted hyperboloid and a real axis for a two-sheeted hyperboloid). If $a = b = c$, then the hyperboloid of revolution is said to be *right*, and its sections by the planes OXZ and OYZ are equilateral hyperbolas.

▶ **Cone.** A *cone* is a central surface defined by the equation

$$\frac{x^2}{a^2} + \frac{y^2}{b^2} - \frac{z^2}{c^2} = 0. \tag{M4.6.1.4}$$

The cone (see Fig. M4.33) defined by (M4.6.1.4) has vertex at the origin, and for its base we can take the ellipse with semiaxes a and b in the plane perpendicular to the axis OZ at the distance c from the origin. This cone is the asymptotic cone for the hyperboloids (M4.6.1.2) and (M4.6.1.3). For $a = b$, we obtain a *right circular cone*.

Remark. About the cone, see also Subsection M3.2.3.

▶ **Paraboloids.** In contrast to the surfaces considered above, paraboloids are not central surfaces. For the equations listed below, the vertex of a paraboloid lies at the origin, the axis OZ is the symmetry axis, and the planes OXZ and OYZ are symmetry planes.

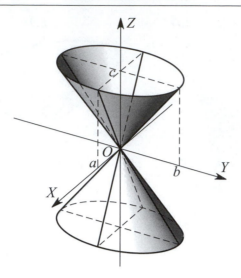

Figure M4.33. A cone.

An *elliptic paraboloid* (see Fig. M4.34 *a*) is a noncentral surface defined by the equation

$$\frac{x^2}{p} + \frac{y^2}{q} = 2z, \qquad\qquad (M4.6.1.5)$$

where $p > 0$ and $q > 0$ are parameters. All points of an elliptic paraboloid lie in the domain $z \geq 0$.

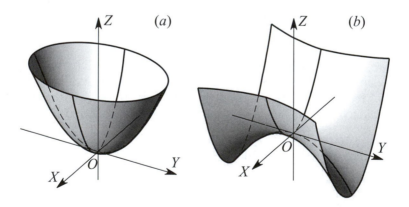

Figure M4.34. Elliptic (*a*) and hyperbolic (*b*) paraboloids.

The sections of an elliptic paraboloid by planes parallel to the axis OZ are parabolas, and the sections by planes parallel to the plane OXY are ellipses. If $p = q$, then we have a *paraboloid of revolution*, which is obtained by rotating the parabola $2pz = x^2$ lying in the plane OXZ about its axis.

The volume of the part of an elliptic paraboloid cut by the plane perpendicular to its axis at a height h is equal to $V = \frac{1}{2}\pi abh$, i.e., half the volume of the elliptic cylinder with the same base and altitude.

A *hyperbolic paraboloid* (see Fig. M4.34 *b*) is a noncentral surface defined by the equation

$$\frac{x^2}{p} - \frac{y^2}{q} = 2z, \tag{M4.6.1.6}$$

where $p > 0$ and $q > 0$ are parameters.

The sections of a hyperbolic paraboloid by planes parallel to the axis OZ are parabolas, and the sections by planes parallel to the plane OXY are hyperbolas.

M4.6.2. Quadrics (General Theory)

▶ **General equation of a quadric. Translation and rotation.** A *quadric* is a set of points in three-dimensional space whose coordinates in the rectangular Cartesian coordinate system satisfy a second-order algebraic equation

$$a_{11}x^2 + a_{22}y^2 + a_{33}z^2 + 2a_{12}xy + 2a_{13}xz + 2a_{23}yz + 2a_{14}x + 2a_{24}y + 2a_{34}z + a_{44} = 0, \tag{M4.6.2.1}$$

or

$$(a_{11}x + a_{12}y + a_{13}z + a_{14})x + (a_{21}x + a_{22}y + a_{23}z + a_{24})y$$
$$+ (a_{31}x + a_{32}y + a_{33}z + a_{34})z + a_{41}x + a_{42}y + a_{43}z + a_{44} = 0,$$

with symmetric coefficients, $a_{ij} = a_{ji}$ ($i, j = 1, 2, 3, 4$); the factors 2 appearing in some terms are introduced for further convenience. If equation (M4.6.2.1) does not define a real geometric object, then one says that this equation defines an *imaginary quadric*.

Equation (M4.6.2.1) can be simplified using the transformations of translation and rotation.

1. Translation:
$$x = \bar{x} + x_0, \quad y = \bar{y} + y_0, \quad z = \bar{z} + z_0. \tag{M4.6.2.2}$$

This transformation means that the origin $O(0, 0, 0)$ is translated to the point $\bar{O}(x_0, y_0, z_0)$ with the new axes of coordinates remaining parallel to the original ones; $\bar{x}$, $\bar{y}$, and $\bar{z}$ are the new coordinates.

2. Rotation:

$$x = e_{11}\hat{x} + e_{12}\hat{y} + e_{13}\hat{z}, \quad y = e_{21}\hat{x} + e_{22}\hat{y} + e_{23}\hat{z}, \quad z = e_{31}\hat{x} + e_{32}\hat{y} + e_{33}\hat{z}. \tag{M4.6.2.3}$$

This transformation means that all points are rotated about the origin O, with e_{11}, e_{21}, e_{31} being the direction cosines of the axis $O\widehat{X}$, e_{12}, e_{22}, e_{32} those of the axis $O\widehat{Y}$, and e_{13}, e_{23}, e_{33} those of the axis $O\widehat{Z}$ in the initial coordinate system $OXYZ$.

▶ **Classification of quadrics.** With successive application of transformations (M4.6.2.2)–(M4.6.2.3), equation (M4.6.2.1) can be reduced to one of the following 17 *canonical forms*, each of which is associated with a certain class of quadrics (see Table M4.2). The first six surfaces, with $\delta \neq 0$, are *nondegenerate*; their canonical equations contain three quadratic terms proportional to x^2, y^2, and z^2. The other surfaces, 7–17, with $\delta = 0$, are *degenerate*; their canonical equations contain only two (proportional to x^2 and y^2) or even one (x^2) quadratic term. The last five surfaces, 13–17, disintegrate into planes (real or imaginary); their equations can be represented as the product of two factors linear in coordinates on the left-hand side and zero on the right-hand side.

TABLE M4.2
Classification of quadrics.

No.	Surface	Canonical equation	Conditions for invariants
1	Ellipsoid	$\dfrac{x^2}{a^2} + \dfrac{y^2}{b^2} + \dfrac{z^2}{c^2} = 1$	$\delta \neq 0,\ \Delta < 0,\ S\delta > 0,\ T > 0$
2	Imaginary ellipsoid	$\dfrac{x^2}{a^2} + \dfrac{y^2}{b^2} + \dfrac{z^2}{c^2} = -1$	$\delta \neq 0,\ \Delta > 0,\ S\delta > 0,\ T > 0$
3	Imaginary cone with real vertex	$\dfrac{x^2}{a^2} + \dfrac{y^2}{b^2} + \dfrac{z^2}{c^2} = 0$	$\delta \neq 0,\ \Delta = 0,\ S\delta > 0,\ T > 0$
4	One-sheeted hyperboloid	$\dfrac{x^2}{a^2} + \dfrac{y^2}{b^2} - \dfrac{z^2}{c^2} = 1$	$\delta \neq 0,\ \Delta > 0,$ $S\delta > 0$ or $T > 0$ (not both > 0)
5	Two-sheeted hyperboloid	$\dfrac{x^2}{a^2} + \dfrac{y^2}{b^2} - \dfrac{z^2}{c^2} = -1$	$\delta \neq 0,\ \Delta < 0,$ $S\delta > 0$ or $T > 0$ (not both > 0)
6	Real cone	$\dfrac{x^2}{a^2} + \dfrac{y^2}{b^2} - \dfrac{z^2}{c^2} = 0$	$\delta \neq 0,\ \Delta = 0,$ $S\delta > 0$ or $T > 0$ (not both > 0)
7	Elliptic paraboloid	$\dfrac{x^2}{p} + \dfrac{y^2}{q} = 2z \quad (p, q > 0)$	$\delta = 0,\ \Delta < 0,\ T > 0$
8	Hyperbolic paraboloid	$\dfrac{x^2}{p} - \dfrac{y^2}{q} = 2z \quad (p, q > 0)$	$\delta = 0,\ \Delta > 0,\ T < 0$
9	Elliptic cylinder	$\dfrac{x^2}{a^2} + \dfrac{y^2}{b^2} = 1$	$\delta = \Delta = 0,\ T > 0,\ S\sigma < 0$
10	Imaginary elliptic cylinder	$\dfrac{x^2}{a^2} + \dfrac{y^2}{b^2} = -1$	$\delta = \Delta = 0,\ T > 0,\ S\sigma > 0$
11	Hyperbolic cylinder	$\dfrac{x^2}{a^2} - \dfrac{y^2}{b^2} = 1$	$\delta = \Delta = 0,\ T < 0,\ \sigma \neq 0$
12	Parabolic cylinder	$y^2 = 2px$	$\delta = \Delta = 0,\ T = 0$
13	Pair of real intersecting planes	$\dfrac{x^2}{a^2} - \dfrac{y^2}{b^2} = 0$	$\delta = \Delta = 0,\ T < 0,\ \sigma = 0$
14	Pair of imaginary planes intersecting in a real straight line	$\dfrac{x^2}{a^2} + \dfrac{y^2}{b^2} = 0$	$\delta = \Delta = 0,\ T > 0,\ \sigma = 0$
15	Pair of real parallel planes	$x^2 = a^2$	$\delta = \Delta = T = 0,\ \Sigma < 0$
16	Pair of imaginary parallel planes	$x^2 = -a^2$	$\delta = \Delta = T = 0,\ \Sigma > 0$
17	Pair of real coinciding planes	$x^2 = 0$	$\delta = \Delta = T = 0,\ \Sigma = 0$

▶ **Invariants of quadrics.** The shape of a quadric can be identified using four *invariants* and two *semi-invariants* without reducing equation (M4.6.2.1) to canonical form.

The four main invariants are

$$S = a_{11} + a_{22} + a_{33}, \tag{M4.6.2.4}$$

$$T = \begin{vmatrix} a_{11} & a_{12} \\ a_{21} & a_{22} \end{vmatrix} + \begin{vmatrix} a_{11} & a_{13} \\ a_{31} & a_{33} \end{vmatrix} + \begin{vmatrix} a_{22} & a_{23} \\ a_{32} & a_{33} \end{vmatrix}, \tag{M4.6.2.5}$$

$$\delta = \begin{vmatrix} a_{11} & a_{12} & a_{13} \\ a_{12} & a_{22} & a_{23} \\ a_{13} & a_{23} & a_{33} \end{vmatrix}, \tag{M4.6.2.6}$$

$$\Delta = \begin{vmatrix} a_{11} & a_{12} & a_{13} & a_{14} \\ a_{12} & a_{22} & a_{23} & a_{24} \\ a_{13} & a_{23} & a_{33} & a_{34} \\ a_{14} & a_{24} & a_{34} & a_{44} \end{vmatrix}, \tag{M4.6.2.7}$$

whose values are preserved under parallel translations and rotations of the coordinate axes.
The semi-invariants are

$$\sigma = \Delta_{11} + \Delta_{22} + \Delta_{33},$$

$$\Sigma = \begin{vmatrix} a_{11} & a_{14} \\ a_{41} & a_{44} \end{vmatrix} + \begin{vmatrix} a_{22} & a_{24} \\ a_{42} & a_{44} \end{vmatrix} + \begin{vmatrix} a_{33} & a_{34} \\ a_{43} & a_{44} \end{vmatrix},$$

whose values are preserved only under rotations of the coordinate axes, with Δ_{ij} being the cofactor of the entry a_{ij} in Δ.

The last column in Table M4.2 allows the classification of the quadrics in accordance with the values of the invariants S, T, δ, and Δ and semi-invariants σ and Σ.

▶ **Characteristic quadratic form of a quadric.** The *characteristic quadratic form*

$$F(x, y, z) = a_{11}x^2 + a_{22}y^2 + a_{33}z^2 + 2a_{12}xy + 2a_{13}xz + a_{23}yz$$

corresponding to equation (M4.6.2.1) and its *characteristic equation*

$$\begin{vmatrix} a_{11} - \lambda & a_{12} & a_{13} \\ a_{12} & a_{22} - \lambda & a_{23} \\ a_{13} & a_{23} & a_{33} - \lambda \end{vmatrix} = 0, \quad \text{or} \quad \lambda^3 - S\lambda^2 + T\lambda - \delta = 0 \qquad \text{(M4.6.2.8)}$$

permit studying the main properties of quadrics.

The roots λ_1, λ_2, and λ_3 of the characteristic equation (M4.6.2.8) are the eigenvalues of the real symmetric matrix $[a_{ij}]$ and hence are always real. The invariants S, T, and δ can be expressed in terms of the roots λ_1, λ_2, and λ_3 as follows:

$$S = \lambda_1 + \lambda_2 + \lambda_3, \quad T = \lambda_1\lambda_2 + \lambda_1\lambda_3 + \lambda_2\lambda_3, \quad \delta = \lambda_1\lambda_2\lambda_3.$$

The expressions of the parameters of the main quadrics via the invariants T, δ, and Δ and the roots λ_1, λ_2, and λ_3 of the characteristic equation are listed in Table M4.3.

TABLE M4.3
Expressions of the parameters of the main quadrics via the invariants
(M4.6.2.4)–(M4.6.2.7) and the roots of the characteristic equation (M4.6.2.8).

Surface	Canonical equation	Parameters of the quadrics	Remarks
Ellipsoid	$\dfrac{x^2}{a^2} + \dfrac{y^2}{b^2} + \dfrac{z^2}{c^2} = 1$	$a^2 = -\dfrac{1}{\lambda_3}\dfrac{\Delta}{\delta},\quad b^2 = -\dfrac{1}{\lambda_2}\dfrac{\Delta}{\delta},$ $c^2 = -\dfrac{1}{\lambda_1}\dfrac{\Delta}{\delta},\quad \delta = \lambda_1\lambda_2\lambda_3$	$a \geq b \geq c,$ $\lambda_1 \geq \lambda_2 \geq \lambda_3 > 0$
One-sheeted hyperboloid	$\dfrac{x^2}{a^2} + \dfrac{y^2}{b^2} - \dfrac{z^2}{c^2} = 1$	$a^2 = -\dfrac{1}{\lambda_2}\dfrac{\Delta}{\delta},\quad b^2 = -\dfrac{1}{\lambda_1}\dfrac{\Delta}{\delta},$ $c^2 = \dfrac{1}{\lambda_3}\dfrac{\Delta}{\delta},\quad \delta = \lambda_1\lambda_2\lambda_3$	$a \geq b,$ $\lambda_1 \geq \lambda_2 > 0 > \lambda_3$
Two-sheeted hyperboloid	$\dfrac{x^2}{a^2} + \dfrac{y^2}{b^2} - \dfrac{z^2}{c^2} = -1$	$a^2 = \dfrac{1}{\lambda_3}\dfrac{\Delta}{\delta},\quad b^2 = \dfrac{1}{\lambda_2}\dfrac{\Delta}{\delta},$ $c^2 = -\dfrac{1}{\lambda_1}\dfrac{\Delta}{\delta},\quad \delta = \lambda_1\lambda_2\lambda_3$	$a \geq b,$ $\lambda_1 > 0 > \lambda_2 \geq \lambda_3$
Elliptic paraboloid	$\dfrac{x^2}{p} + \dfrac{y^2}{q} = 2z$	$p = \dfrac{1}{\lambda_2}\sqrt{-\dfrac{\Delta}{T}},\quad q = \dfrac{1}{\lambda_1}\sqrt{-\dfrac{\Delta}{T}},\quad T = \lambda_1\lambda_2$	$p > 0,\quad q > 0,$ $\lambda_1 \geq \lambda_2 > \lambda_3 = 0$
Hyperbolic paraboloid	$\dfrac{x^2}{p} - \dfrac{y^2}{q} = 2z$	$p = \dfrac{1}{\lambda_1}\sqrt{-\dfrac{\Delta}{T}},\quad q = -\dfrac{1}{\lambda_3}\sqrt{-\dfrac{\Delta}{T}},\quad T = \lambda_1\lambda_3$	$p > 0,\quad q > 0,$ $\lambda_1 > \lambda_2 = 0 > \lambda_3$

Bibliography for Chapter M4

Alexander, D. C. and Koeberlein, G. M., *Elementary Geometry for College Students, 3rd Edition,* Houghton Mifflin Company, Boston, Massachusetts, 2002.

Blau, H. I., *Foundations of Plane Geometry,* Prentice Hall, Englewood Cliffs, New Jersey, 2002.

Chauvenet, W., *A Treatise on Elementary Geometry,* Adamant Media Corporation, Boston, Massachusetts, 2001.

Efimov, N. V., *Higher Geometry,* Mir Publishers, Moscow, 1980.

Fogiel, M. (Editor), *High School Geometry Tutor, 2nd Edition,* Research & Education Association, Englewood Cliffs, New Jersey, 2003.

Fuller, G. and Tarwater, D., *Analytic Geometry, 7th Edition,* Addison Wesley, Boston, Massachusetts, 1993.

Gustafson, R. D. and Frisk, P. D., *Elementary Geometry, 3rd Edition,* Wiley, New York, 1991.

Hartshorne, R., *Geometry: Euclid and Beyond,* Springer, New York, 2005.

Jacobs, H. R., *Geometry, 2nd Edition,* W. H. Freeman & Company, New York, 1987.

Jacobs, H. R., *Geometry: Seeing, Doing, Understanding, 3rd Edition,* W. H. Freeman & Company, New York, 2003.

Jurgensen, R. and Brown, R. G., *Geometry,* McDougal Littell/Houghton Mifflin, Boston, Massachusetts, 2000.

Kay, D., *College Geometry: A Discovery Approach, 2nd Edition,* Addison Wesley, Boston, Massachusetts, 2000.

Kletenik, D. V., *Problems in Analytic Geometry,* University Press of the Pacific, Honolulu, Hawaii, 2002.

Kostrikin, A. I. (Editor), *Exercises in Algebra: A Collection of Exercises in Algebra, Linear Algebra and Geometry,* Gordon & Breach, New York, 1996.

Kostrikin, A. I. and Shafarevich, I. R. (Editors), *Algebra I: Basic Notions of Algebra,* Springer-Verlag, Berlin, 1990.

Leff, L. S., *Geometry the Easy Way, 3rd Edition,* Barron's Educational Series, Hauppauge, New York, 1997.

Moise, E., *Elementary Geometry from an Advanced Standpoint, 3rd Edition,* Addison Wesley, Boston, Massachusetts, 1990.

Musser, G. L., Burger, W. F., and Peterson, B. E., *Mathematics for Elementary Teachers: A Contemporary Approach, 6th Edition,* Wiley, New York, 2002.

Musser, G. L., Burger, W. F., and Peterson, B. E., *Essentials of Mathematics for Elementary Teachers: A Contemporary Approach, 6th Edition,* Wiley, New York, 2003.

Musser, G. L. and Trimpe, L. E., *College Geometry: A Problem Solving Approach with Applications,* Prentice Hall, Englewood Cliffs, New Jersey, 1994.

Postnikov, M. M., *Lectures in Geometry, Semester I, Analytic Geometry,* Mir Publishers, Moscow, 1982.

Privalov, I. I., *Analytic Geometry, 32nd Edition [in Russian],* Lan, Moscow, 2003.

Riddle, D. R., *Analytic Geometry, 6th Edition,* Brooks Cole, Stamford, 1995.

Roe, J., *Elementary Geometry,* Oxford University Press, Oxford, 1993.

Schultze, A. and Sevenoak, F. L., *Plane and Solid Geometry,* Adamant Media Corporation, Boston, Massachusetts, 2004.

Suetin, P. K., Kostrikin, A. I., and Manin, Yu. I., *Linear Algebra and Geometry,* Gordon & Breach, New York, 1997.

Tussy, A. S. and Gustafson, R. D., *Basic Geometry for College Students: An Overview of the Fundamental Concepts of Geometry,* Thomson-Brooks/Cole, Pacific Grove, California, 2002.

Vygodskii, M. Ya., *Mathematical Handbook: Higher Mathematics,* Mir Publishers, Moscow, 1971.

Weeks, A. W. and Adkins, J. B., *A Course in Geometry: Plane and Solid,* Bates Publishing Company, Sandwich, Massachusetts, 1982.

Woods, F. S., *Higher Geometry: An Introduction to Advanced Methods in Analytic Geometry, Phoenix Edition,* Dover Publications, New York, 2005.

Chapter M5

Algebra

M5.1. Polynomials and Algebraic Equations

M5.1.1. Polynomials and Their Properties

▶ **Definition of a polynomial.** A *polynomial of degree* n of a scalar variable x is an expression of the form

$$f(x) \equiv a_n x^n + a_{n-1} x^{n-1} + \cdots + a_1 x + a_0 \quad (a_n \neq 0), \tag{M5.1.1.1}$$

where $a_0, \ldots, a_n$ are real or complex numbers ($n = 0, 1, 2, \ldots$). Polynomials of degree zero are nonzero numbers.

Two polynomials are *equal* if they have the same coefficients of like powers of the variable.

▶ **Main operations over polynomials.**

1°. The *sum* (*difference*) of two polynomials $f(x)$ of degree n and $g(x)$ of degree m is the polynomial of degree $l \leq \max\{n, m\}$ whose coefficient of each power of x is equal to the sum (difference) of the coefficients of the same power of x in $f(x)$ and $g(x)$, i.e., if

$$g(x) \equiv b_m x^m + b_{m-1} x^{m-1} + \cdots + b_1 x + b_0, \tag{M5.1.1.2}$$

then the sum (difference) of polynomials (M5.1.1.1) and (M5.1.1.2) is

$$f(x) \pm g(x) = c_l x^l + c_{l-1} x^{l-1} + \cdots + c_1 x + c_0, \quad \text{where} \quad c_k = a_k \pm b_k \quad (k = 0, 1, \ldots, l).$$

If $n > m$ then $b_{m+1} = \cdots = b_n = 0$; if $n < m$ then $a_{n+1} = \cdots = a_m = 0$.

2°. To *multiply* a polynomial $f(x)$ of degree n by a polynomial $g(x)$ of degree m, one should multiply each term in $f(x)$ by each term in $g(x)$, add the products, and collect similar terms. The degree of the resulting polynomial is $n + m$. The product of polynomials (M5.1.1.1) and (M5.1.1.2) is

$$f(x)g(x) = c_{n+m} x^{n+m} + c_{n+m-1} x^{n+m-1} + \cdots + c_1 x + c_0, \quad c_k = \sum_{i,j=0}^{i+j=k} a_i b_j,$$

where $k = 0, 1, \ldots, n + m$.

3°. Each polynomial $f(x)$ of degree n can be *divided* by any other polynomial $p(x)$ of degree m ($p(x) \neq 0$) *with remainder*, i.e., uniquely represented in the form $f(x) = p(x)q(x) + r(x)$, where $q(x)$ is a polynomial of degree $n - m$ (for $m \leq n$) or $q(x) = 0$ (for $m > n$), referred to as the *quotient*, and $r(x)$ is a polynomial of degree $l < m$ or $r(x) = 0$, referred to as the *remainder*.

If $r(x) = 0$, then $f(x)$ is said to be *divisible* by $p(x)$ (without remainder).

If $m > n$, then $q(x) = 0$ and $r(x) \equiv f(x)$.

▶ **Methods for finding quotient and remainder.**

$1°$. *Horner's scheme.* To divide a polynomial $f(x)$ of degree n (see (M5.1.1.1)) by the polynomial $p(x) = x - b$, one uses Horner's scheme: the coefficients of $f(x)$ are written out in a row, starting from a_n; b is written on the left; then one writes the number a_n under a_n, the number $a_n b + a_{n-1} = b_{n-1}$ under a_{n-1}, the number $b_{n-1}b + a_{n-2} = b_{n-2}$ under a_{n-2}, $\ldots$, the number $b_1 b + a_0 = b_0$ under a_0. The number b_0 is the remainder in the division of $f(x)$ by $p(x)$, and $a_n, b_{n-1}, \ldots, b_1$ are the coefficients of the quotient.

Remark. To divide $f(x)$ by $p(x) = ax + b$ ($a \neq 0$) with remainder, one first uses Horner's scheme to divide by $p_1(x) = x - (-\frac{b}{a})$; now if $q_1(x)$ and r_1 are the quotient and remainder in the division of $f(x)$ by $p_1(x)$, then $q(x) = \frac{1}{a}q_1(x)$ and $r = r_1$ are the quotient and remainder in the division of $f(x)$ by $p(x)$.

Example 1. Let us divide $f(x) = x^3 - 2x^2 - 10x + 3$ by $p(x) = 2x + 5$.
We use Horner's scheme to divide $f(x)$ by $p_1(x) = x + 5/2$:

$$
\begin{array}{r|cccc}
 & 1 & -2 & -10 & 3 \\
\hline
-\dfrac{5}{2} \Big| & 1 & -\dfrac{9}{2} & \dfrac{5}{4} & -\dfrac{1}{8}
\end{array}
$$

Thus $f(x) = p(x)q(x) + r(x)$, where

$$q(x) = \frac{1}{2}\left(x^2 - \frac{9}{2}x + \frac{5}{4}\right) = \frac{1}{2}x^2 - \frac{9}{4}x + \frac{5}{8}, \quad r = -\frac{1}{8}.$$

POLYNOMIAL REMAINDER THEOREM. *The remainder in the division of a polynomial $f(x)$ by the polynomial $p(x) = x - b$ is the number equal to the value of the polynomial $f(x)$ at $x = b$.*

$2°$. *Long division.* To divide a polynomial $f(x)$ of degree n by a polynomial $p(x)$ of degree $m \leq n$, one can use *long division*.

Example 2. Let us divide $f(x) = x^3 + 8x^2 + 14x - 5$ by $p(x) = x^2 + 3x - 1$.
We use long division:

$$
\begin{array}{r}
\underline{x^3 + 8x^2 + 14x - 5}\,\Big|\,\underline{x^2 + 3x - 1} \\
\end{array}
$$

$$
\begin{array}{rl}
x^3 + 8x^2 + 14x - 5 & \big|\, x^2 + 3x - 1 \\
\underline{x^3 + 3x^2 \quad - x} & \big|\, x + 5 \\
5x^2 + 15x - 5 & \\
\underline{5x^2 + 15x - 5} & \\
0 &
\end{array}
$$

Thus $f(x) = p(x)q(x) + r(x)$, where $q(x) = x + 5$ and $r(x) = 0$; i.e., $f(x)$ is divisible by $p(x)$.

Example 3. Let us divide $f(x) = x^3 - 4x^2 + x + 1$ by $p(x) = x^2 + 1$.
We use long division:

$$
\begin{array}{rl}
x^3 - 4x^2 + x + 1 & \big|\, x^2 + 1 \\
\underline{x^3 \qquad\quad + x} & \big|\, x - 4 \\
-4x^2 \qquad + 1 & \\
\underline{-4x^2 \qquad - 4} & \\
5 &
\end{array}
$$

Thus $f(x) = p(x)q(x) + r(x)$, where $q(x) = x - 4$ and $r(x) = 5$.

▶ **Expansion of polynomials in powers of a linear binomial.** For each polynomial $f(x)$ given by equation (M5.1.1.1) and any number c, one can write out the *expansion of $f(x)$ in powers of $x - c$*:

$$f(x) = b_n(x - c)^n + b_{n-1}(x - c)^{n-1} + \cdots + b_1(x - c) + b_0.$$

To find the coefficients $b_0, \ldots, b_n$ of this expansion, one first divides $f(x)$ by $x - c$ with remainder. The remainder is b_0, and the quotient is some polynomial $g_0(x)$. Then one divides

$g_0(x)$ by $x - c$ with remainder. The remainder is b_1, and the quotient is some polynomial $g_1(x)$. Then one divides $g_1(x)$ by $x - c$, obtaining the coefficient b_2 as the remainder, etc. It is convenient to perform the computations by Horner's scheme (see above).

The coefficients in the expansion of a polynomial $f(x)$ in powers of the difference $x - c$ are related to the values of the polynomial and its derivatives at $x = c$ by the formulas

$$b_0 = f(c), \quad b_1 = \frac{f'_x(c)}{1!}, \quad b_2 = \frac{f''_{xx}(c)}{2!}, \quad \dots, \quad b_n = \frac{f_x^{(n)}(c)}{n!},$$

where the *derivative of a polynomial* $f(x) = a_n x^n + a_{n-1} x^{n-1} + \dots + a_1 x + x_0$ with real or complex coefficients $a_0, \dots, a_n$ is the polynomial $f'_x(x) = n a_n x^{n-1} + (n-1) a_{n-1} x^{n-2} + \dots + a_1$, $f''_{xx}(x) = [f'_x(x)]'_x$, etc. (see Subsection M6.2.1).

M5.1.2. Linear and Quadratic Equations

▶ **Linear equations.** The linear equation

$$ax + b = 0 \quad (a \neq 0)$$

has the solution

$$x = -\frac{b}{a}.$$

▶ **Quadratic equations.** The quadratic equation

$$ax^2 + bx + c = 0 \quad (a \neq 0) \tag{M5.1.2.1}$$

has the roots

$$x_{1,2} = \frac{-b \pm \sqrt{b^2 - 4ac}}{2a}.$$

The existence of real or complex roots is determined by the sign of the discriminant $D = b^2 - 4ac$:

Case $D > 0$. There are two distinct real roots.

Case $D < 0$. There are two distinct complex conjugate roots.

Case $D = 0$. There are two equal real roots.

VIÈTE THEOREM. *The roots of a quadratic equation (M5.1.2.1) satisfy the following relations:*

$$x_1 + x_2 = -\frac{b}{a}, \quad x_1 x_2 = \frac{c}{a}.$$

M5.1.3. Cubic Equations

▶ **Incomplete cubic equation.**

$1°$. *Cardano's solution.* The roots of the incomplete cubic equation

$$y^3 + py + q = 0 \tag{M5.1.3.1}$$

have the form

$$y_1 = A + B, \quad y_{2,3} = -\frac{1}{2}(A + B) \pm i \frac{\sqrt{3}}{2}(A - B),$$

where

$$A = \left(-\frac{q}{2} + \sqrt{D}\right)^{1/3}, \quad B = \left(-\frac{q}{2} - \sqrt{D}\right)^{1/3}, \quad D = \left(\frac{p}{3}\right)^3 + \left(\frac{q}{2}\right)^2, \quad i^2 = -1,$$

and A, B are arbitrary values of the cubic roots such that $AB = -\frac{1}{3}p$.

The number of real roots of a cubic equation depends on the sign of the discriminant D:

Case $D > 0$. There is one real and two complex conjugate roots.

Case $D < 0$. There are three real roots.

Case $D = 0$. There is one real root and another real root of double multiplicity.

2°. *Trigonometric solution.* If an incomplete cubic equation (M5.1.3.1) has real coefficients p and q, then its solutions can be found with the help of the trigonometric formulas given below.

(*a*) Let $p < 0$ and $D < 0$. Then

$$y_1 = 2\sqrt{-\frac{p}{3}}\cos\frac{\alpha}{3}, \quad y_{2,3} = -2\sqrt{-\frac{p}{3}}\cos\left(\frac{\alpha}{3} \pm \frac{\pi}{3}\right),$$

where the values of the trigonometric functions are calculated from the relation

$$\cos\alpha = -\frac{q}{2\sqrt{-(p/3)^3}}.$$

(*b*) Let $p > 0$ and $D \geq 0$. Then

$$y_1 = 2\sqrt{\frac{p}{3}}\cot(2\alpha), \quad y_{2,3} = \sqrt{\frac{p}{3}}\left[\cot(2\alpha) \pm i\frac{\sqrt{3}}{\sin(2\alpha)}\right],$$

where the values of the trigonometric functions are calculated from the relations

$$\tan\alpha = \left(\tan\frac{\beta}{2}\right)^{1/3}, \quad \tan\beta = \frac{2}{q}\left(\frac{p}{3}\right)^{3/2}, \quad |\alpha| \leq \frac{\pi}{4}, \quad |\beta| \leq \frac{\pi}{2}.$$

(*c*) Let $p < 0$ and $D \geq 0$. Then

$$y_1 = -2\sqrt{-\frac{p}{3}}\frac{1}{\sin(2\alpha)}, \quad y_{2,3} = \sqrt{-\frac{p}{3}}\left[\frac{1}{\sin(2\alpha)} \pm i\sqrt{3}\cot(2\alpha)\right],$$

where the values of the trigonometric functions are calculated from the relations

$$\tan\alpha = \left(\tan\frac{\beta}{2}\right)^{1/3}, \quad \sin\beta = \frac{2}{q}\left(-\frac{p}{3}\right)^{3/2}, \quad |\alpha| \leq \frac{\pi}{4}, \quad |\beta| \leq \frac{\pi}{2}.$$

In the above three cases, the real value of the cubic root should be taken.

▶ **Complete cubic equation.** The roots of a complete cubic equation

$$ax^3 + bx^2 + cx + d = 0 \quad (a \neq 0) \tag{M5.1.3.2}$$

are calculated by the formulas

$$x_k = y_k - \frac{b}{3a}, \quad k = 1, 2, 3,$$

where y_k are the roots of the incomplete cubic equation (M5.1.3.1) with the coefficients

$$p = -\frac{1}{3}\left(\frac{b}{a}\right)^2 + \frac{c}{a}, \quad q = \frac{2}{27}\left(\frac{b}{a}\right)^3 - \frac{bc}{3a^2} + \frac{d}{a}.$$

VIÈTE THEOREM. *The roots of a complete cubic equation (M5.1.3.2) satisfy the following relations:*

$$x_1 + x_2 + x_3 = -\frac{b}{a}, \quad x_1 x_2 + x_1 x_3 + x_2 x_3 = \frac{c}{a}, \quad x_1 x_2 x_3 = -\frac{d}{a}.$$

M5.1.4. Fourth-Degree Equation

▶ **Special cases of fourth-degree equations.**

$1°$. The *biquadratic equation*

$$ax^4 + bx^2 + c = 0$$

can be reduced to a quadratic equation (M5.1.2.1) by the substitution $\xi = x^2$. Therefore, the roots of the biquadratic equations are given by

$$x_{1,2} = \pm\sqrt{\frac{-b + \sqrt{b^2 - 4ac}}{2a}}, \qquad x_{3,4} = \pm\sqrt{\frac{-b - \sqrt{b^2 - 4ac}}{2a}}.$$

$2°$. The *reciprocal (algebraic) equation*

$$ax^4 + bx^3 + cx^2 + bx + a = 0$$

can be reduced to a quadratic equation by the substitution

$$y = x + \frac{1}{x}.$$

The resulting quadratic equation has the form

$$ay^2 + by + c - 2a = 0.$$

$3°$. The *generalized reciprocal equation*

$$ax^4 + bx^3 + cx^2 + \lambda bx + \lambda^2 a = 0$$

can be reduced to a quadratic equation by the substitution

$$y = x + \frac{\lambda}{x}.$$

The resulting quadratic equation has the form

$$ay^2 + by + c - 2a\lambda = 0.$$

▶ **General fourth-degree equation.**

$1°$. *Reduction of a general fourth-degree equation to an incomplete equation.* The general fourth-degree equation

$$ax^4 + bx^3 + cx^2 + dx + e = 0 \quad (a \neq 0)$$

can be reduced to an incomplete equation of the form

$$y^4 + py^2 + qy + r = 0 \tag{M5.1.4.1}$$

by the substitution

$$x = y - \frac{b}{4a}.$$

$2°$. *Descartes–Euler solution.* The roots of the incomplete equation (M5.1.4.1) are given by the formulas

$$\begin{aligned} y_1 &= \tfrac{1}{2}\left(\sqrt{z_1} + \sqrt{z_2} + \sqrt{z_3}\right), & y_2 &= \tfrac{1}{2}\left(\sqrt{z_1} - \sqrt{z_2} - \sqrt{z_3}\right), \\ y_3 &= \tfrac{1}{2}\left(-\sqrt{z_1} + \sqrt{z_2} - \sqrt{z_3}\right), & y_4 &= \tfrac{1}{2}\left(-\sqrt{z_1} - \sqrt{z_2} + \sqrt{z_3}\right), \end{aligned} \tag{M5.1.4.2}$$

where z_1, z_2, z_3 are the roots of the cubic equation (cubic resolvent of equation (M5.1.4.1))

$$z^3 + 2pz^2 + (p^2 - 4r)z - q^2 = 0. \tag{M5.1.4.3}$$

The signs of the roots in (M5.1.4.2) are chosen from the condition

$$\sqrt{z_1}\sqrt{z_2}\sqrt{z_3} = -q.$$

The roots of the fourth-degree equation (M5.1.4.1) are determined by the roots of the cubic resolvent (M5.1.4.3); see Table M5.1.

TABLE M5.1
Relations between the roots of an incomplete equation of fourth-degree and the roots of its cubic resolvent.

Cubic resolvent (M5.1.4.3)	Fourth-degree equation (M5.1.4.1)
All roots are real and positive*	Four real roots
All roots are real: one is positive and two are negative*	Two pairs of complex conjugate roots
One real root and two complex conjugate roots	Two real roots and two complex conjugate roots

* By the Viète theorem, the product of the roots z_1, z_2, z_3 is equal to $q^2 \geq 0$.

$3°$. *Ferrari solution.* Let z_0 be any of the roots of the auxiliary cubic equation (M5.1.4.3). Then the four roots of the incomplete equation (M5.1.4.1) are found by solving the following two quadratic equations:

$$y^2 - \sqrt{z_0}\, y + \frac{p + z_0}{2} + \frac{q}{2\sqrt{z_0}} = 0,$$

$$y^2 + \sqrt{z_0}\, y + \frac{p + z_0}{2} - \frac{q}{2\sqrt{z_0}} = 0.$$

M5.1.5. Algebraic Equations of Arbitrary Degree and Their Properties

▶ **Simplest equations of degree n and their solutions.**

$1°$. The *binomial algebraic equation*

$$x^n - a = 0 \quad (a \neq 0)$$

has the solutions

$$x_{k+1} = \begin{cases} a^{1/n}\left(\cos \dfrac{2k\pi}{n} + i \sin \dfrac{2k\pi}{n} \right) & \text{for } a > 0, \\[3mm] |a|^{1/n}\left(\cos \dfrac{(2k+1)\pi}{n} + i \sin \dfrac{(2k+1)\pi}{n} \right) & \text{for } a < 0, \end{cases}$$

where $k = 0, 1, \ldots, n - 1$ and $i^2 = -1$.

$2°$. Equations of the form

$$x^{2n} + ax^n + b = 0,$$
$$x^{3n} + ax^{2n} + bx^n + c = 0,$$
$$x^{4n} + ax^{3n} + bx^{2n} + cx^n + d = 0$$

are reduced by the substitution $y = x^n$ to a quadratic, cubic, and fourth-degree equation, respectively, whose solution can be expressed by radicals (see Subsections M5.1.2–M5.1.4).

Remark. In the above equations, n can be noninteger.

$3°$. The *generalized reciprocal (algebraic) equation*

$$a_0 x^{2n} + a_1 x^{2n-1} + \cdots + a_{n-1} x^{n+1} + a_n x^n$$
$$+ \lambda a_{n-1} x^{n-1} + \lambda^2 a_{n-2} x^{n-2} + \cdots + \lambda^{n-1} a_1 x + \lambda^n a_0 = 0 \qquad (a_0 \neq 0).$$

can be reduced to an equation of degree n by the substitution

$$y = x + \frac{\lambda}{x}.$$

Example 1. The equation

$$ax^6 + bx^5 + cx^4 + dx^3 + cx^2 + bx + a = 0,$$

which is a special case of the reciprocal equation with $n = 3$ and $\lambda = 1$, can be reduced to the cubic equation

$$ay^3 + by^2 + (c - 3a)y + d - 2b = 0$$

by the substitution $y = x + 1/x$.

▶ **Equations of general form and their properties.** An *algebraic equation of degree n* has the form

$$a_n x^n + a_{n-1} x^{n-1} + \cdots + a_1 x + a_0 = 0 \quad (a_n \neq 0), \tag{M5.1.5.1}$$

where a_k are real or complex coefficients. Denote the polynomial of degree n on the left-hand side in equation (M5.1.5.1) by

$$P_n(x) \equiv a_n x^n + a_{n-1} x^{n-1} + \cdots + a_1 x + a_0 \quad (a_n \neq 0). \tag{M5.1.5.2}$$

A value $x = x_1$ such that $P_n(x_1) = 0$ is called a *root* of equation (M5.1.5.1) (and also a root of the polynomial $P_n(x)$). A value $x = x_1$ is called a *root of multiplicity m* if $P_n(x) = (x - x_1)^m Q_{n-m}(x)$, where m is an integer ($1 \leq m \leq n$), and $Q_{n-m}(x)$ is a polynomial of degree $n - m$ such that $Q_{n-m}(x_1) \neq 0$.

THEOREM 1 (FUNDAMENTAL THEOREM OF ALGEBRA). *Any algebraic equation of degree n has exactly n roots (real or complex), each root counted according to its multiplicity.*

Thus, the left-hand side of equation (M5.1.5.1) with roots $x_1, x_2, \ldots, x_s$ of the respective multiplicities $k_1, k_2, \ldots, k_s$ ($k_1 + k_2 + \cdots + k_s = n$) can be factorized as follows:

$$P_n(x) = a_n (x - x_1)^{k_1} (x - x_2)^{k_2} \ldots (x - x_s)^{k_s}.$$

THEOREM 2. *Any algebraic equation of an odd degree with real coefficients has at least one real root.*

THEOREM 3. *Suppose that equation (M5.1.5.1) with real coefficients has a complex root $x_1 = \alpha + i\beta$. Then this equation has the complex conjugate root $x_2 = \alpha - i\beta$, and the roots x_1, x_2 have the same multiplicity.*

THEOREM 4. *Any rational root of equation (M5.1.5.1) with integer coefficients a_k is an irreducible fraction of the form p/q, where p is a divisor of a_0 and q is a divisor of a_n. If $a_n = 1$, then all rational roots of equation (M5.1.5.1) (if they exist) are integer divisors of the free term.*

THEOREM 5 (ABEL–RUFFINI THEOREM). *Any equation (M5.1.5.1) of degree $n \leq 4$ is solvable by radicals, i.e., its roots can be expressed via its coefficients by the operations of addition, subtraction, multiplication, division, and taking roots (see Subsections M5.1.2– M5.1.4). In general, equation (M5.1.5.1) of degree $n > 4$ cannot be solved by radicals.*

▶ **Relations between roots and coefficients. Discriminant of an equation.**

VIÈTE THEOREM. *The roots of equation (M5.1.5.1) (counted according to their multiplicity) and its coefficients satisfy the following relations:*

$$(-1)^k \frac{a_{n-k}}{a_n} = S_k \quad (k = 1, 2, \ldots, n),$$

where S_k are elementary symmetric functions of $x_1, x_2, \ldots, x_n$:

$$S_1 = \sum_{i=1}^{n} x_i, \quad S_2 = \sum_{1 \leq i < j}^{n} x_i x_j, \quad S_3 = \sum_{1 \leq i < j < k}^{n} x_i x_j x_k, \quad \ldots, \quad S_n = x_1 x_2 \ldots x_n.$$

Note also the following relations:

$$(n-k)a_{n-k} + \sum_{j=1}^{k} a_{n-(k-j)} s_j = 0 \quad (k = 1, 2, \ldots, n)$$

with symmetric functions $s_j = \sum_{i=1}^{n} x_i^j$.

The *discriminant D of an algebraic equation* is the product of a_n^{2n-2} and the squared Vandermonde determinant $\Delta(x_1, x_2, \ldots, x_n)$ of its roots:

$$D = a_n^{2n-2} [\Delta(x_1, x_2, \ldots, x_n)]^2 = a_n^{2n-2} \prod_{1 \leq j < i \leq n} (x_i - x_j)^2.$$

The discriminant D is a symmetric function of the roots $x_1, x_2, \ldots, x_n$, and is equal to zero if and only if the polynomial $P_n(x)$ has at least one multiple root.

▶ **Bounds for the roots of algebraic equations with real coefficients.**

1°. All roots of equation (M5.1.5.1) in absolute value do not exceed

$$N = 1 + \frac{A}{|a_n|}, \tag{M5.1.5.3}$$

where A is the largest of $|a_0|, |a_1|, \ldots, |a_{n-1}|$.

The last result admits the following generalization: all roots of equation (M5.1.5.1) in absolute value do not exceed

$$N_1 = \rho + \frac{A_1}{|a_n|}, \tag{M5.1.5.4}$$

where $\rho > 0$ is arbitrary and A_1 is the largest of

$$|a_{n-1}|, \quad \frac{|a_{n-2}|}{\rho}, \quad \frac{|a_{n-3}|}{\rho^2}, \quad \ldots, \quad \frac{|a_0|}{\rho^{n-1}}.$$

For $\rho = 1$, formula (M5.1.5.4) turns into (M5.1.5.3).

Remark. Formulas (M5.1.5.3) and (M5.1.5.4) can also be used for equations with complex coefficients.

Example 2. Consider the following equation of degree 4:

$$P_4(x) = 9x^4 - 9x^2 - 36x + 1.$$

Formula (M5.1.5.3) for $n = 4$, $|a_n| = 9$, $A = 36$ yields a fairly rough estimate $N = 5$, i.e., the roots of the equation belong to the interval $[-5, 5]$. Formula (M5.1.5.4) for $\rho = 2$, $n = 4$, $|a_n| = 9$, $A_1 = 9$ yields a better estimate for the bounds of the roots of this polynomial, $N_1 = 3$.

$2°$. A constant K is called an upper bound for the real roots of equation (M5.1.5.1) or the polynomial $P_n(x)$ if equation (M5.1.5.1) has no real roots greater than or equal to K; in a similar way, one defines a lower and an upper bound for positive and negative roots of an equation or the corresponding polynomial.

Let

K_1 be an upper bound for the positive roots of the polynomial $P_n(x)$,

K_2 be an upper bound for the positive roots of the polynomial $P_n(-x)$,

$K_3 > 0$ be an upper bound for the positive roots of the polynomial $x^n P_n(1/x)$,

$K_4 > 0$ be an upper bound for the positive roots of the polynomial $x^n P_n(-1/x)$.

Then all nonzero real roots of the polynomial $P_n(x)$ (if they exist) belong to the intervals $(-K_2, -1/K_4)$ and $(1/K_3, K_1)$.

Next, we describe three methods for finding upper bounds for positive roots of a polynomial.

Maclaurin method. Suppose that the first m leading coefficients of the polynomial (M5.1.5.2) are nonnegative, i.e., $a_n > 0$, $a_{n-1} \geq 0$, ..., $a_{n-m+1} \geq 0$, and the next coefficient is negative, $a_{n-m} < 0$. Then

$$K = 1 + \left(\frac{B}{a_n}\right)^{1/m} \tag{M5.1.5.5}$$

is an upper bound for the positive roots of this polynomial, where B is the largest of the absolute values of negative coefficients of $P_n(x)$.

Example 3. Consider the fourth-degree equation from Example 2. In this case, $m = 2$, $B = 36$ and formula (M5.1.5.5) yields $K = K_1 = 1 + (36/9)^{1/2} = 3$. Now, consider the polynomial $P_4(-x) = 9x^4 - 9x^2 + 36x + 1$. Its positive roots has the upper bound $K_2 = 1 + (9/9)^{1/2} = 2$. For the polynomial $x^4 P_4(1/x) = x^4 - 36x^3 - 9x^2 + 9$, we have $m = 1$, $K_3 = 1 + 36 = 37$. Finally, for the polynomial $x^4 P_4(-1/x) = x^4 + 36x^3 - 9x^2 + 9$, we have $m = 2$, $k_4 = 1 + 9^{1/2} = 4$. Thus if $P_4(x)$ has real roots, they must belong to the intervals $(-2, -1/4)$ and $(1/37, 3)$.

Newton method. Suppose that the polynomial $P_n(x)$ and all its derivatives $P_n'(x)$, ..., $P_n^{(n)}(x)$ take positive values for $x = c$. Then c is an upper bound for the positive roots of $P_n(x)$.

Example 4. Consider the polynomial from Example 2 and calculate the derivatives

$$P_4(x) = 9x^4 - 9x^2 - 36x + 1, \quad P_4'(x) = 36x^3 - 18x - 36, \quad P_4''(x) = 108x^2 - 18, \quad P_4'''(x) = 216x, \quad P_4''''(x) = 216.$$

It is easy to check that for $x = 2$ this polynomial and all its derivatives take positive values, and therefore $c = 2$ is an upper bound for its positive roots.

▶ **Theorems on the number of real roots of polynomials.** The number of all negative roots of a polynomial $P_n(x)$ is equal to the number of all positive roots of the polynomial $P_n(-x)$.

$1°$. The exact number of positive roots of a polynomial whose coefficients form a sequence that does not change sign or changes sign only once can be found with the help of the Descartes theorem (rule of signs).

DESCARTES THEOREM. *The number of positive roots (counted considering their multiplicity) of a polynomial $P_n(x)$ with real coefficients is either equal to the number of sign alterations between consecutive nonzero coefficients or is less than it by a multiple of 2.*

Applying the Descartes theorem to $P_n(-x)$, we obtain a similar theorem for the negative roots of the polynomial $P_n(x)$.

Example 5. Consider the cubic polynomial

$$P_3(x) = x^3 - 3x + a^2 \quad (a \neq 0).$$

Its coefficients have the signs $+ - +$, and therefore we have two alterations of sign. Therefore, the number of positive roots of $P_3(x)$ is equal either to 2 or to 0. Now, consider the polynomial $P_3(-x) = -x^3 + 3x + a^2$. The sequence of its coefficients changes sign only once. Therefore, the original equation has one negative root.

$2°$. *A stronger version of the Descartes theorem.* Suppose that all roots of a polynomial $P_n(x)$ are real*; then the number of positive roots of $P_n(x)$ is equal to the number of sign alterations in the sequence of its coefficients, and the number of its negative roots is equal to the number of sign alterations in the sequence of coefficients of the polynomial $P_n(-x)$.

Example 6. Consider the characteristic polynomial of the symmetric matrix

$$P_3(x) = \begin{vmatrix} -2-x & 1 & 1 \\ 1 & 1-x & 3 \\ 1 & 3 & 1-x \end{vmatrix} = -x^3 + 14x + 20,$$

which has only real roots. The sequence of its coefficients changes sign only once, and therefore it has a single positive root. The number of its negative roots is equal to two, since this polynomial has three nonzero real roots and only one of them can be positive.

$3°$. If two neighboring coefficients of a polynomial $P_n(x)$ are equal to zero, then the roots of the polynomial cannot be all real (in this case, the stronger version of the Descartes theorem cannot be used).

$4°$. The number of real roots of a polynomial $P_n(x)$ greater than a fixed c is either equal to the number of sign alterations in the sequence $P_n(c), \ldots, P_n^{(n)}(c)$ or is by an even number less. If all roots of $P_n(x)$ are real, then the number of its roots greater than c coincides with the number of sign alterations in the sequence $P_n(c), \ldots, P_n^{(n)}(c)$.

Example 7. Consider the polynomial

$$P_4(x) = x^4 - 3x^3 + 2x^2 - 2a^2x + a^2.$$

For $x = 1$, we have $P_4(1) = -a^2$, $P_4'(1) = -1 - 2a^2$, $P_4''(1) = -2$, $P_4'''(1) = 6$, $P_4''''(1) = 24$. Thus, there is a single sign alteration, and therefore the polynomial has a single real root greater than unity.

M5.2. Determinants and Matrices

M5.2.1. Determinants

▶ **Second-order, third-order, and nth-order determinants.**

$1°$. The *second-order determinant* is a number Δ associated with 4 scalar quantities $a_{11}, a_{12}, a_{21}, a_{22}$, arranged in a 2×2 square table. It is denoted and calculated as

$$\Delta = \begin{vmatrix} a_{11} & a_{12} \\ a_{21} & a_{22} \end{vmatrix} = a_{11}a_{22} - a_{12}a_{21}.$$

The numbers a_{11}, a_{12}, a_{21}, and a_{22} are called elements of the determinant Δ.

$2°$. The *third-order determinant* is a number Δ associated with a 3×3 square table of 9 scalar quantities; it is denoted and calculated as

$$\Delta = \begin{vmatrix} a_{11} & a_{12} & a_{13} \\ a_{21} & a_{22} & a_{23} \\ a_{31} & a_{32} & a_{33} \end{vmatrix}$$

$$= a_{11}a_{22}a_{33} + a_{12}a_{23}a_{31} + a_{13}a_{21}a_{32} - a_{13}a_{22}a_{31} - a_{12}a_{21}a_{33} - a_{11}a_{23}a_{32}.$$

---

* This is the case, for instance, if we are dealing with the characteristic polynomial of a symmetric matrix.

This expression is obtained by the *triangle rule* (*Sarrus scheme*), illustrated by the following diagrams, where entries occurring in the same product with a given sign are joined by segments:

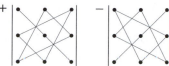

3°. The *nth-order determinant* is a number Δ associated with an $n \times n$ square table of n^2 scalar quantities; it is denoted by

$$
\Delta = \begin{vmatrix}
a_{11} & a_{12} & \cdots & a_{1n} \\
a_{21} & a_{22} & \cdots & a_{2n} \\
\vdots & \vdots & \ddots & \vdots \\
a_{n1} & a_{n2} & \cdots & a_{nn}
\end{vmatrix}.
\tag{M5.2.1.1}
$$

The numbers a_{ij} are elements of the determinant Δ.

The determinant (M5.2.1.1) is calculated using the formulas

$$
\begin{aligned}
\Delta &= a_{i1}A_{i1} + a_{i2}A_{i2} + \cdots + a_{in}A_{in} \\
&= a_{1j}A_{1j} + a_{2j}A_{2j} + \cdots + a_{nj}A_{nj}
\end{aligned}
\tag{M5.2.1.2}
$$

for any ith row and jth column. Here, A_{ij} is the *cofactor* of the element a_{ij}, which is defined as $A_{ij} = (-1)^{i+j}M_{ij}$, where M_{ij} is the *minor* corresponding to a_{ij}. The minor M_{ij} is defined as the $(n-1)$st-order determinant of size $(n-1) \times (n-1)$ obtained from the original determinant by removing the ith row and the jth column (i.e., the row and the column that intersect at a_{ij}). It follows from (M5.2.1.2) that the calculation of an nth-order determinant is reduced to the calculation of n determinants of order $n-1$.

The first formula in (M5.2.1.2) is called the *cofactor expansion of the determinant along row i* and the other one is called the *cofactor expansion of the determinant along column j*.

▶ **Properties of determinants.**
1. If a determinant contains a row (column) consisting of all zeroes, then this determinant is equal to zero.
2. If a determinant has two proportional rows (columns), then the determinant is zero.
3. If a determinant has a row (column) that is a linear combination of its other rows (columns), then the determinant is zero.
4. If two rows (columns) are interchanged, the determinant changes its sign.
5. If each element of a row (column) is divisible by a common number, this number can be factored out of the determinant.
6. The determinant does not change if a linear combination of some of its rows (columns) is added to another row (column).

Remark. The determinant is equal to zero if and only if its rows (columns) are linearly dependent.

▶ **Calculation of determinants.**

1°. Determinants can be calculated using the above properties.

Example 1. Find the determinant

$$
\Delta = \begin{vmatrix}
-13 & 25 & 17 \\
26 & -34 & -26 \\
36 & -33 & -24
\end{vmatrix}.
$$

We first factor out the common divisor 2 of the elements in the second row, then add the resulting second row to the first one and then add the second row multiplied by –2 to the third one to obtain

$$\Delta = 2 \begin{vmatrix} -13 & 25 & 17 \\ 13 & -17 & -13 \\ 36 & -33 & -24 \end{vmatrix} = 2 \begin{vmatrix} 0 & 8 & 4 \\ 13 & -17 & -13 \\ 10 & 1 & 2 \end{vmatrix}.$$

In the last determinant, by adding the third column multiplied by –2 to the second one and by using the cofactor expansion along the first row, one obtains

$$\Delta = 2 \begin{vmatrix} 0 & 0 & 4 \\ 13 & 9 & -13 \\ 10 & -3 & 2 \end{vmatrix} = 2 \times 4 \begin{vmatrix} 13 & 9 \\ 10 & -3 \end{vmatrix} = 8\,(-39 - 90) = -1032.$$

2°. Determinants are often calculated using the cofactor expansion formulas (M5.2.1.2). To this end, its is convenient to take a row or a column that contains many zero elements.

Example 2. Find the third-order determinant

$$A = \begin{vmatrix} 1 & -1 & 2 \\ 6 & 1 & 5 \\ 2 & -1 & -4 \end{vmatrix}.$$

We perform the cofactor expansion along the second column:

$$\det A = \sum_{k=1}^{3} (-1)^{k+2} a_{k2} M_{k2} = (-1)^{1+2} \times (-1) \times \begin{vmatrix} 6 & 5 \\ 2 & -4 \end{vmatrix} + (-1)^{2+2} \times 1 \times \begin{vmatrix} 1 & 2 \\ 2 & -4 \end{vmatrix} + (-1)^{3+2} \times (-1) \times \begin{vmatrix} 1 & 2 \\ 6 & 5 \end{vmatrix}$$

$$= 1 \times [6 \times (-4) - 5 \times 2] + 1 \times [1 \times (-4) - 2 \times 2] + 1 \times [1 \times 5 - 2 \times 6] = -49.$$

M5.2.2. Matrices. Types of Matrices. Operations with Matrices

▶ **Definition of a matrix. Types of matrices.** A *matrix* of *size* (or *dimension*) $m \times n$ is a rectangular table with *entries* a_{ij} ($i = 1, 2, \ldots, m; j = 1, 2, \ldots, n$) arranged in m rows and n columns:

$$A \equiv \begin{pmatrix} a_{11} & a_{12} & \cdots & a_{1n} \\ a_{21} & a_{22} & \cdots & a_{2n} \\ \vdots & \vdots & \ddots & \vdots \\ a_{m1} & a_{m2} & \cdots & a_{mn} \end{pmatrix}. \tag{M5.2.2.1}$$

Note that, for each entry a_{ij}, the index i refers to the ith row and the index j to the jth column. Matrices are briefly denoted by uppercase letters (for instance, A, as here), or by the symbol $[a_{ij}]$, sometimes with more details: $A \equiv [a_{ij}]$ ($i = 1, 2, \ldots, m; j = 1, 2, \ldots, n$). The numbers m and n are called the *dimensions* of the matrix.

The *null* or *zero matrix* is a matrix whose entries are all equal to zero: $a_{ij} = 0$ ($i = 1, 2, \ldots, m, j = 1, 2, \ldots, n$).

A *column vector* or *column* is a matrix of size $m \times 1$. A *row vector* or *row* is a matrix of size $1 \times n$. Both column and row vectors are often simply called *vectors*.

A *square matrix* is a matrix of size $n \times n$, and n is called the dimension of this square matrix. The *main diagonal* of a square matrix is its diagonal from the top left corner to the bottom right corner with the entries $a_{11}\ a_{22}\ \ldots\ a_{nn}$. Table M5.2 lists the main types of square matrices.

▶ **Basic operations with matrices.** Two matrices are *equal* if they are of the same size and their respective entries are equal.

The *sum* of two matrices $A \equiv [a_{ij}]$ and $B \equiv [b_{ij}]$ of the same size $m \times n$ is the matrix $C \equiv [c_{ij}]$ of size $m \times n$ with the entries

$$c_{ij} = a_{ij} + b_{ij}.$$

TABLE M5.2
Some types of square matrices.

Type of square matrix $[a_{ij}]$	Entries
Unit (identity) $I = [\delta_{ij}]$	$a_{ij} = \delta_{ij} = \begin{cases} 1, & i = j, \\ 0, & i \neq j \end{cases}$ (δ_{ij} is the Kronecker delta)
Diagonal	$a_{ij} = \begin{cases} \text{any}, & i = j, \\ 0, & i \neq j \end{cases}$
Upper triangular (superdiagonal)	$a_{ij} = \begin{cases} \text{any}, & i \leq j, \\ 0, & i > j \end{cases}$
Lower triangular (subdiagonal)	$a_{ij} = \begin{cases} \text{any}, & i \geq j, \\ 0, & i < j \end{cases}$
Symmetric	$a_{ij} = a_{ji}$
Skew-symmetric (antisymmetric)	$a_{ij} = -a_{ji}$
Hermitian (self-adjoint)	$a_{ij} = \bar{a}_{ji}$ ($\bar{a}_{ji}$ is the complex conjugate of a number a_{ji})

The sum of two matrices is denoted by $C = A + B$, and the operation is called *addition of matrices*.

Properties of addition of matrices:

$$A + O = A \qquad \text{(property of zero matrix)},$$
$$A + B = B + A \qquad \text{(commutativity)},$$
$$(A + B) + C = A + (B + C) \qquad \text{(associativity)},$$

where matrices A, B, C, and zero matrix O have the same size.

The *difference* of two matrices $A \equiv [a_{ij}]$ and $B \equiv [b_{ij}]$ of the same size $m \times n$ is the matrix $C \equiv [c_{ij}]$ of size $m \times n$ with entries

$$c_{ij} = a_{ij} - b_{ij} \quad (i = 1, 2, \ldots, m; \; j = 1, 2, \ldots, n).$$

The difference of two matrices is denoted by $C = A - B$, and the operation is called *subtraction of matrices*.

The *product* of a matrix $A \equiv [a_{ij}]$ of size $m \times n$ by a scalar λ is the matrix $C \equiv [c_{ij}]$ of size $m \times n$ with entries

$$c_{ij} = \lambda a_{ij} \quad (i = 1, 2, \ldots, m; \; j = 1, 2, \ldots, n).$$

The product of a matrix by a scalar is denoted by $C = \lambda A$, and the operation is called *multiplication of a matrix by a scalar*.

Properties of multiplication of a matrix by a scalar:

$$0A = O \qquad \text{(property of zero)},$$
$$(\lambda\mu)A = \lambda(\mu A) \qquad \text{(associativity with respect to a scalar factor)},$$
$$\lambda(A + B) = \lambda A + \lambda B \qquad \text{(distributivity with respect to addition of matrices)},$$
$$(\lambda + \mu)A = \lambda A + \mu A \qquad \text{(distributivity with respect to addition of scalars)},$$

where λ and μ are scalars, matrices A, B, C, and zero matrix O have the same size.

The *product* of a matrix $A \equiv [a_{ij}]$ of size $m \times p$ and a matrix $B \equiv [b_{ij}]$ of size $p \times n$ is the matrix $C \equiv [c_{ij}]$ of size $m \times n$ with entries

$$c_{ij} = \sum_{k=1}^{p} a_{ik}b_{kj} \quad (i = 1, 2, \ldots, m; \; j = 1, 2, \ldots, n);$$

i.e., the entry c_{ij} in the ith row and jth column of the matrix C is equal to the sum of products of the respective entries in the ith row of A and the jth column of B. Note that the product is defined for matrices of *compatible size*; i.e., the number of columns in the first matrix should be equal to the number of rows in the second matrix. The product of two matrices A and B is denoted by $C = AB$, and the operation is called *multiplication of matrices*.

Example 1. Consider two matrices

$$A = \begin{pmatrix} 1 & 2 \\ 6 & -3 \end{pmatrix} \quad \text{and} \quad B = \begin{pmatrix} 0 & 10 & 1 \\ -6 & -0.5 & 20 \end{pmatrix}.$$

The product of the matrix A and the matrix B is the matrix

$$C = AB = \begin{pmatrix} 1 & 2 \\ 6 & -3 \end{pmatrix} \begin{pmatrix} 0 & 10 & 1 \\ -6 & -0.5 & 20 \end{pmatrix}$$
$$= \begin{pmatrix} 1 \times 0 + 2 \times (-6) & 1 \times 10 + 2 \times (-0.5) & 1 \times 1 + 2 \times 20 \\ 6 \times 0 + (-3) \times (-6) & 6 \times 10 + (-3) \times (-0.5) & 6 \times 1 + (-3) \times 20 \end{pmatrix} = \begin{pmatrix} -12 & 9 & 41 \\ 18 & 61.5 & -54 \end{pmatrix}.$$

Properties of multiplication of matrices:

$AO = O_1$ (property of zero matrix),

$(AB)C = A(BC)$ (associativity of the product of three matrices),

$AI = A$ (multiplication by unit matrix),

$A(B + C) = AB + AC$ (distributivity with respect to a sum of two matrices),

$\lambda(AB) = (\lambda A)B = A(\lambda B)$ (associativity of the product of a scalar and two matrices),

$SD = DS$ (commutativity for any square and any diagonal matrices),

where λ is a scalar, matrices A, B, C, square matrix S, diagonal matrix D, zero matrices O and O_1, and unit matrix I have the compatible sizes.

Two square matrices A and B are said to *commute* if $AB = BA$, i.e., if their multiplication is subject to the commutative law (in general, this is not the case).

▶ **Transpose, orthogonal, and adjoint matrix.** The *transpose* of a matrix $A \equiv [a_{ij}]$ of size $m \times n$ is the matrix $C \equiv [c_{ij}]$ of size $n \times m$ with entries

$$c_{ij} = a_{ji} \quad (i = 1, 2, \ldots, n; \; j = 1, 2, \ldots, m).$$

The transpose is denoted by $C = A^T$.

Example 2. If $A = (a_1, a_2)$ then $A^T = \begin{pmatrix} a_1 \\ a_2 \end{pmatrix}$.

Properties of transposes:

$$(A + B)^T = A^T + B^T, \quad (\lambda A)^T = \lambda A^T, \quad (A^T)^T = A,$$
$$(AC)^T = C^T A^T, \quad\quad\quad O^T = O_1, \quad\quad\quad I^T = I,$$

where λ is a scalar; matrices A, B, and zero matrix O have size $m \times n$; matrix C has size $n \times l$; zero matrix O_1 has size $n \times m$.

THEOREM (DECOMPOSITION OF MATRICES). *For any square matrix A, the matrix $S_1 = \frac{1}{2}(A + A^T)$ is symmetric and the matrix $S_2 = \frac{1}{2}(A - A^T)$ is skew-symmetric. The representation of A as the sum of symmetric and skew-symmetric matrices is unique: $A = S_1 + S_2$.*

A square matrix A is said to be *orthogonal* if $A^T A = AA^T = I$, i.e., $A^T = A^{-1}$, where A^{-1} is the inverse of A (see Subsection M5.2.3).

Properties of orthogonal matrices:
1. If A is an orthogonal matrix, then A^T is also orthogonal.
2. The product of two orthogonal matrices is an orthogonal matrix.
3. Any symmetric orthogonal matrix is involutive, i.e., $AA = I$.

The *complex conjugate* of a matrix $A \equiv [a_{ij}]$ of size $m \times n$ is the matrix $C \equiv [c_{ij}]$ of size $m \times n$ with entries

$$c_{ij} = \bar{a}_{ij} \quad (i = 1, 2, \ldots, m; \; j = 1, 2, \ldots, n),$$

where $\bar{a}_{ij}$ is the complex conjugate of a_{ij}. The complex conjugate matrix is denoted by $C = \overline{A}$.

The *adjoint matrix* of a matrix $A \equiv [a_{ij}]$ of size $m \times n$ is the matrix $C \equiv [c_{ij}]$ of size $n \times m$ with entries

$$c_{ij} = \bar{a}_{ji} \quad (i = 1, 2, \ldots, n; \; j = 1, 2, \ldots, m).$$

The adjoint matrix is denoted by $C = A^*$.

Properties of adjoint matrices:

$$(A + B)^* = A^* + B^*, \quad (\lambda A)^* = \bar{\lambda} A^*, \quad (A^*)^* = A,$$
$$(AC)^* = C^* A^*, \quad\quad O^* = O_1, \quad\quad I^* = I,$$

where λ is a scalar; matrices A, B, and zero matrix O have size $m \times n$; matrix C has size $n \times l$; zero matrix O_1 has size $n \times m$.

Remark. If a matrix is *real* (i.e., all its entries are real), then the corresponding transpose and the adjoint matrix coincide.

A square matrix A is said to be *normal* if $A^* A = AA^*$. A normal matrix A is said to be *unitary* if $A^* A = AA^* = I$, i.e., $A^* = A^{-1}$, where A^{-1} is the inverse of A (see Subsection M5.2.3).

▶ **Trace of a matrix.** The *trace* of a square matrix $A \equiv [a_{ij}]$ of size $n \times n$ is the sum of its diagonal entries,

$$\mathrm{Tr}(A) = \sum_{i=1}^{n} a_{ii}.$$

If λ is a scalar and square matrices A and B have the same size, then

$$\mathrm{Tr}(A + B) = \mathrm{Tr}(A) + \mathrm{Tr}(B), \quad \mathrm{Tr}(\lambda A) = \lambda \mathrm{Tr}(A), \quad \mathrm{Tr}(AB) = \mathrm{Tr}(BA),$$

▶ **Minors. Rank and defect of a matrix.** In a square or rectangular matrix, let us select k arbitrary rows and k arbitrary columns to make up a square submatrix. The kth-order determinant formed by the entries where the selected rows and columns intersect is called a kth-order *minor* of the matrix.

The *rank of a matrix* A is the maximum order of nonzero minors of A. The rank of a matrix A is denoted rank(A). If all entries of a matrix are zero, the rank of the matrix is taken to be zero.

Properties of the rank of a matrix:

1. The rank of a matrix does not change if: a row (column) whose entries are all zero is deleted; some rows (columns) are interchanged; a row (column) is multiplied by a nonzero number; the entries of one row (column) multiplied by any number are added to the respective entries of another row (column); and the rows are substituted by columns while the columns are substituted by the respective rows (for square matrices).

Example 3. Find the rank of the matrix $A = \begin{pmatrix} 1 & 2 & 3 & 6 \\ 2 & 3 & 1 & 6 \\ 3 & 1 & 2 & 6 \end{pmatrix}$.

Subtract the sum of the first three columns from column 4 and then delete the resulting column, whose entries are now all zero, to obtain the matrix $A_1 = \begin{pmatrix} 1 & 2 & 3 \\ 2 & 3 & 1 \\ 3 & 1 & 2 \end{pmatrix}$, which has the same rank as A. Since $\det(A_1) = -18 \neq 0$, we have rank(A_1) = 3, and hence rank(A) = 3.

2. For any matrices A and B of the same size the following inequality holds:

$$\text{rank}(A + B) \leq \text{rank}(A) + \text{rank}(B).$$

3. For a matrix A of size $m \times n$ and a matrix B of size $n \times k$, the *Sylvester inequalities* hold:

$$\text{rank}(A) + \text{rank}(B) - n \leq \text{rank}(AB) \leq \min\{\text{rank}(A), \text{rank}(B)\}.$$

For a square matrix A of size $n \times n$, the value $d = n - \text{rank}(A)$ is called the *defect* of the matrix A, and A is called a *d-fold degenerate matrix*. The rank of a nondegenerate square matrix $A \equiv [a_{ij}]$ of size $n \times n$ is equal to n.

4. Let r be the rank of a matrix; basic minor of this matrix is its nonzero minor of the order r. Basic rows (columns) of the matrix are the rows (columns) forming the basic minor.

THEOREM ON BASIC MINOR. *Basic rows (resp., basic columns) of a matrix are linearly independent. Any row (resp., any column) of a matrix is a linear combination of its basic rows (resp., columns).*

▶ **Linear dependence of row vectors (column vectors).** A row vector (column vector) B is a *linear combination* of row vectors (column vectors) $A_1, \ldots, A_k$ if there exist scalars $\alpha_1, \ldots, \alpha_k$ such that

$$B = \alpha_1 A_1 + \cdots + \alpha_k A_k.$$

Row vectors (column vectors) $A_1, \ldots, A_k$ are said to be *linearly dependent* if there exist scalars $\alpha_1, \ldots, \alpha_k$ ($\alpha_1^2 + \cdots + \alpha_k^2 \neq 0$) such that

$$\alpha_1 A_1 + \cdots + \alpha_k A_k = O,$$

where O is the zero row vector (column vector).

Row vectors (column vectors) $A_1, \ldots, A_k$ are said to be *linearly independent* if, for any $\alpha_1, \ldots, \alpha_k$ ($\alpha_1^2 + \cdots + \alpha_k^2 \neq 0$) we have

$$\alpha_1 A_1 + \cdots + \alpha_k A_k \neq O.$$

Remark. Row vectors (column vectors) $A_1, \ldots, A_k$ are linearly dependent if and only if at least one of them is a linear combination of the others.

▶ **Determinant of a matrix.** For any square matrix A of the form (M5.2.2.1), one can calculate its determinant (M5.2.1.1), denoted $\det A$, $\det(A)$, or $|A|$.

The determinant of a matrix has the following properties:

1. The determinant of a triangular (upper or lower) and a diagonal matrices is equal to the product of its entries on the main diagonal. In particular, the determinant of the unit matrix is equal to 1.

2. The determinant of the product of two matrices A and B of the same size is equal to the product of their determinants,

$$\det(AB) = \det A \det B.$$

3. The determinant is invariant under matrix transposition:

$$\det A = \det A^T.$$

M5.2.3. Inverse Matrix. Functions of Matrices

▶ **Inverse matrices.** Let A be an $n \times n$ square matrix and let I be the unit matrix of the same size.

A square matrix A is called *nonsingular* or *nondegenerate* if $\det A \neq 0$.

THEOREM. *A square matrix is nondegenerate if and only if its rows (columns) are linearly independent.*

A square matrix A is called *invertible* if one can find a matrix B such that $AB = BA = I$. The matrix B is called the *inverse* of A and denoted A^{-1}. An invertible matrix A has a unique inverse.

THEOREM. *A square matrix A is invertible if and only if its determinant is nonzero (i.e., A is nonsingular).*

If the matrix A is defined by the table (M5.2.2.1), then its inverse is calculated as

$$A^{-1} = \begin{pmatrix} \dfrac{A_{11}}{\det A} & \dfrac{A_{21}}{\det A} & \cdots & \dfrac{A_{n1}}{\det A} \\ \dfrac{A_{12}}{\det A} & \dfrac{A_{22}}{\det A} & \cdots & \dfrac{A_{n2}}{\det A} \\ \vdots & \vdots & \ddots & \vdots \\ \dfrac{A_{1n}}{\det A} & \dfrac{A_{2n}}{\det A} & \cdots & \dfrac{A_{nn}}{\det A} \end{pmatrix}. \tag{M5.2.3.1}$$

where A_{ij} is the cofactor of the element a_{ij} of the determinant of A; the definition of A_{ij} can be found after formula (M5.2.1.2).

Properties of the inverse of a matrix:

$$(AB)^{-1} = B^{-1}A^{-1}, \quad (\lambda A)^{-1} = \frac{1}{\lambda}A^{-1},$$
$$(A^{-1})^{-1} = A, \quad (A^{-1})^T = (A^T)^{-1}, \quad (A^{-1})^* = (A^*)^{-1},$$

where the square matrices A and B are assumed to be nonsingular and the scalar λ to be nonzero.

▶ **Powers of square matrices.** The product of several identical square matrices A can be written as a *positive integer power* of the matrix A: $AA = A^2$, $AAA = A^2A = A^3$, etc. For a positive integer k, one defines $A^k = A^{k-1}A$ as the kth power of A. For a nondegenerate matrix A, one defines $A^0 = AA^{-1} = I$, $A^{-k} = (A^{-1})^k$. Powers of a matrix have the following properties:

$$A^p A^q = A^{p+q}, \quad (A^p)^q = A^{pq},$$

where p and q are arbitrary positive integers and A is an arbitrary square matrix; or p and q are arbitrary integers and A is an arbitrary nondegenerate matrix.

There exist matrices A^k whose positive integer power is equal to the zero matrix, even if $A \neq O$. If $A^k = O$ for some integer $k > 1$, then A is called a *nilpotent matrix*.

A matrix A is said to be *involutive* if it coincides with its inverse: $A = A^{-1}$ or $A^2 = I$.

▶ **Polynomials and functions of matrices.** A *polynomial with matrix argument* is the expression obtained from a scalar polynomial $f(x)$ by replacing the scalar argument x with a square matrix X:

$$f(X) = a_0 I + a_1 X + a_2 X^2 + \cdots,$$

where a_i ($i = 0, 1, 2, \ldots$) are real or complex coefficients. The polynomial $f(X)$ is a square matrix of the same size as X.

The exponential function of a square matrix X can be represented as the following convergent series:

$$e^X = 1 + X + \frac{X^2}{2!} + \frac{X^3}{3!} + \cdots = \sum_{k=0}^{\infty} \frac{X^k}{k!}.$$

The inverse matrix has the form

$$(e^X)^{-1} = e^{-X} = 1 - X + \frac{X^2}{2!} - \frac{X^3}{3!} + \cdots = \sum_{k=0}^{\infty} (-1)^k \frac{X^k}{k!}.$$

Remark. Note that $e^X e^Y \neq e^Y e^X$, in general. The relation $e^X e^Y = e^{X+Y}$ holds only for commuting matrices X and Y.

Some other functions of matrices can be expressed in terms of the exponential function:

$$\sin X = \frac{1}{2i}(e^{iX} - e^{-iX}), \quad \cos X = \frac{1}{2}(e^{iX} + e^{-iX}),$$

$$\sinh X = \frac{1}{2}(e^X - e^{-X}), \quad \cosh X = \frac{1}{2}(e^X + e^{-X}).$$

M5.2.4. Eigenvalues and Characteristic Equation of a Matrix. The Cayley–Hamilton Theorem

▶ **Eigenvalues and spectra of square matrices.** An *eigenvalue* of a square matrix A is any real or complex λ for which the matrix $F(\lambda) \equiv A - \lambda I$ is degenerate. The set of all eigenvalues of a matrix A is called its *spectrum*, and $F(\lambda)$ is called its *characteristic matrix*. The inverse of an eigenvalue, $\mu = 1/\lambda$, is called a *characteristic value*.

A square matrix is nondegenerate if and only if all its eigenvalues are different from zero.

A nonzero (column) vector X satisfying the condition

$$AX = \lambda X$$

is called an *eigenvector of the matrix A* corresponding to the eigenvalue λ. Eigenvectors corresponding to distinct eigenvalues of A are linearly independent.

► **Characteristic equation of a matrix.** The algebraic equation of degree n

$$f_A(\lambda) \equiv \det(A - \lambda I) \equiv \det\left[a_{ij} - \lambda\delta_{ij}\right] \equiv \begin{vmatrix} a_{11} - \lambda & a_{12} & \cdots & a_{1n} \\ a_{21} & a_{22} - \lambda & \cdots & a_{2n} \\ \vdots & \vdots & \ddots & \vdots \\ a_{n1} & a_{n2} & \cdots & a_{nn} - \lambda \end{vmatrix} = 0$$

is called the *characteristic equation* of the matrix A of size $n \times n$, and $f_A(\lambda)$ is called its *characteristic polynomial*. The spectrum of the matrix A (i.e., the set of all its eigenvalues) coincides with the set of all roots of its characteristic equation.

Example 1. The characteristic equation of the matrix

$$A = \begin{pmatrix} 4 & -8 & 1 \\ 5 & -9 & 1 \\ 4 & -6 & -1 \end{pmatrix}$$

has the form

$$f_A(\lambda) \equiv \det \begin{pmatrix} 4 - \lambda & -8 & 1 \\ 5 & -9 - \lambda & 1 \\ 4 & -6 & -1 - \lambda \end{pmatrix} = -\lambda^3 - 6\lambda^2 - 11\lambda - 6 = -(\lambda + 1)(\lambda + 2)(\lambda + 3) = 0.$$

Therefore the spectrum of the matrix A consists of three eigenvalues: $\lambda_1 = -1$, $\lambda_2 = -2$, and $\lambda_3 = -3$.

Let λ_j be an eigenvalue of a square matrix A. Then:
1) $\alpha\lambda_j$ is an eigenvalue of the matrix αA for any scalar α;
2) λ_j^p is an eigenvalue of the matrix A^p ($p = 0, \pm 1, \ldots, \pm N$ for a nondegenerate A; otherwise, $p = 0, 1, \ldots, N$), where N is a natural number;
3) a polynomial $f(A)$ of the matrix A has the eigenvalue $f(\lambda)$.

The matrix power series $\sum_{k=0}^{\infty} \alpha_k A^k$ is convergent if and only if the power series $\sum_{k=0}^{\infty} \alpha_k \lambda_j^k$ is convergent for each eigenvalue λ_j of A.

Regarding bounds for eigenvalues, see Subsection M5.1.5.

Let the positive integer s_i be the *multiplicity* of the eigenvalue λ_i of the characteristic equation of the matrix A of size $n \times n$. Note that $\sum_i s_i = n$.

The determinant $\det A$ is equal to the product of all eigenvalues of A, each eigenvalue counted according to its multiplicity, i.e.,

$$\det A = \prod_i \lambda_i^{s_i}.$$

The trace $\mathrm{Tr}(A)$ is equal to the sum of all eigenvalues of A, each eigenvalue counted according to its multiplicity, i.e.,

$$\mathrm{Tr}(A) = \sum_i s_i \lambda_i.$$

► **Cayley–Hamilton theorem. Sylvester theorem.**

CAYLEY–HAMILTON THEOREM. *Each square matrix A satisfies its own characteristic equation; i.e., $f_A(A) = 0$.*

Example 2. Let us illustrate the Cayley–Hamilton theorem by the matrix in Example 1:

$$f_A(A) = -A^3 - 6A^2 - 11A - 6I$$

$$= -\begin{pmatrix} 70 & -116 & 19 \\ 71 & -117 & 19 \\ 64 & -102 & 11 \end{pmatrix} - 6\begin{pmatrix} -20 & 34 & -5 \\ -21 & 35 & -5 \\ -18 & 28 & -1 \end{pmatrix} - 11\begin{pmatrix} 4 & -8 & 1 \\ 5 & -9 & 1 \\ 4 & -6 & -1 \end{pmatrix} - 6\begin{pmatrix} 1 & 0 & 0 \\ 0 & 1 & 0 \\ 0 & 0 & 1 \end{pmatrix} = O.$$

A scalar polynomial $p(\lambda)$ is called an *annihilating polynomial* of a square matrix A if $p(A) = 0$. For example, the characteristic polynomial $f_A(\lambda)$ is an annihilating polynomial of A. The unique monic annihilating polynomial of least degree is called the *minimal polynomial* of A and is denoted by $\psi(\lambda)$. The minimal polynomial is a divisor of every annihilating polynomial.

By dividing an arbitrary polynomial $f(\lambda)$ of degree n by an annihilating polynomial $p(\lambda)$ of degree m ($p(\lambda) \neq 0$), one obtains the representation

$$f(\lambda) = p(\lambda)q(\lambda) + r(\lambda),$$

where $q(\lambda)$ is a polynomial of degree $n - m$ (if $m \leq n$) or $q(\lambda) = 0$ (if $m > n$) and $r(\lambda)$ is a polynomial of degree $l < m$ or $r(\lambda) = 0$. Hence

$$f(A) = p(A)q(A) + r(A),$$

where $p(A) = 0$ and $f(A) = r(A)$. The polynomial $r(\lambda)$ in this representation is called the *interpolation polynomial* of A.

Example 3. Let

$$f(A) = A^4 + 4A^3 + 2A^2 - 12A - 10I,$$

where the matrix A is defined in Example 1. Dividing $f(\lambda)$ by the characteristic polynomial $f_A(\lambda) = -\lambda^3 - 6\lambda^2 - 11\lambda - 6$, we obtain the remainder $r(\lambda) = 3\lambda^2 + 4\lambda + 2$. Consequently,

$$f(A) = r(A) = 3A^2 + 4A + 2I.$$

The Cayley–Hamilton theorem can also be used to find the powers and the inverse of a matrix A (since if $f_A(A) = 0$, then $A^k f_A(A) = 0$ for any positive integer k).

Example 4. For the matrix in Examples 1–3, one has

$$f_A(A) = -A^3 - 6A^2 - 11A - 6I = 0.$$

Hence we obtain

$$A^3 = -6A^2 - 11A - 6I.$$

By multiplying this expression by A, we obtain

$$A^4 = -6A^3 - 11A^2 - 6A.$$

Now we use the representation of the cube of A via lower powers of A and eventually arrive at the formula

$$A^4 = 25A^2 + 60A + 36I.$$

For the inverse matrix, by analogy with the preceding, we obtain

$$A^{-1} f_A(A) = A^{-1}(-A^3 - 6A^2 - 11A - 6I) = -A^2 - 6A - 11I - 6A^{-1} = 0.$$

The definitive result is

$$A^{-1} = -\frac{1}{6}(A^2 + 6A + 11I).$$

THEOREM. *Every analytic function of a square $n \times n$ matrix A can be represented as a polynomial of the same matrix,*

$$f(A) = \frac{1}{\Delta(\lambda_1, \lambda_2, \ldots, \lambda_n)} \sum_{k=1}^{n} \Delta_{n-k} A^{n-k},$$

where $\Delta(\lambda_1, \lambda_2, \ldots, \lambda_n)$ is the Vandermonde determinant

$$\Delta(\lambda_1, \lambda_2, \ldots, \lambda_n) \equiv \begin{vmatrix} 1 & 1 & \cdots & 1 \\ \lambda_1 & \lambda_2 & \cdots & \lambda_n \\ \lambda_1^2 & \lambda_2^2 & \cdots & \lambda_n^2 \\ \vdots & \vdots & \ddots & \vdots \\ \lambda_1^{n-1} & \lambda_2^{n-1} & \cdots & \lambda_n^{n-1} \end{vmatrix} = \prod_{1 \leq j < i \leq n} (\lambda_i - \lambda_j)$$

and Δ_i is obtained from Δ by replacing the $(i+1)$st row by $(f(\lambda_1), f(\lambda_2), \dots, f(\lambda_n))$.

In some cases, an analytic function of a matrix A can be computed by a formula in the following theorem.

SYLVESTER'S THEOREM. *If all eigenvalues of a matrix A are distinct, then*

$$f(A) = \sum_{k=1}^{n} f(\lambda_k) Z_k, \quad Z_k = \frac{\prod_{i \neq k}(A - \lambda_i I)}{\prod_{i \neq k}(\lambda_k - \lambda_i)}.$$

M5.3. Systems of Linear Algebraic Equations

M5.3.1. Consistency Condition for a Linear System

▶ **Notion of a system of linear algebraic equations.** A *system of m linear equations with n unknown quantities* has the form

$$
\begin{aligned}
a_{11}x_1 + a_{12}x_2 + \cdots + a_{1k}x_k + \cdots + a_{1n}x_n &= b_1, \\
a_{21}x_1 + a_{22}x_2 + \cdots + a_{2k}x_k + \cdots + a_{2n}x_n &= b_2, \\
&\cdots\cdots\cdots\cdots\cdots\cdots\cdots\cdots\cdots\cdots\cdots\cdots\cdots \\
a_{m1}x_1 + a_{m2}x_2 + \cdots + a_{mk}x_k + \cdots + a_{mn}x_n &= b_m,
\end{aligned}
\tag{M5.3.1.1}
$$

where $a_{11}, a_{12}, \dots, a_{mn}$ are the *coefficients of the system*; $b_1, b_2, \dots, b_m$ are its *constant terms*; and $x_1, x_2, \dots, x_n$ are the unknown quantities.

System (M5.3.1.1) is said to be *homogeneous* if all its constant terms are equal to zero. Otherwise (i.e., if there is at least one nonzero free term) the system is called *nonhomogeneous*.

If the number of equations is equal to that of the unknown quantities ($m = n$), system (M5.3.1.1) is called a *square system*.

A *solution* of system (M5.3.1.1) is a set of n numbers $x_1, x_2, \dots, x_n$ satisfying the equations of the system. A system is said to be *consistent* if it admits at least one solution. If a system has no solutions, it is said to be *inconsistent*. A consistent system of the form (M5.3.1.1) is called a *determined system* if it has a unique solution. A consistent system with more than one solution is said to be *undetermined*.

It is convenient to use matrix notation for systems of the form (M5.3.1.1),

$$AX = B, \tag{M5.3.1.2}$$

where $A \equiv [a_{ij}]$ is a matrix of size $m \times n$ called the *basic matrix* of the system; $X \equiv [x_i]$ is a column vector of size n; $B \equiv [b_i]$ is a column vector of size m.

▶ **Consistency condition for a general linear system.** System (M5.3.1.1) or (M5.3.1.2) is associated with two matrices: the basic matrix A of size $m \times n$ and the *augmented matrix* A_1 of size $m \times (n+1)$ formed by the matrix A supplemented with the column of the constant terms, i.e.,

$$
A = \begin{pmatrix} a_{11} & a_{12} & \cdots & a_{1n} \\ a_{21} & a_{22} & \cdots & a_{2n} \\ \vdots & \vdots & \ddots & \vdots \\ a_{m1} & a_{m2} & \cdots & a_{mn} \end{pmatrix}, \quad A_1 = \begin{pmatrix} a_{11} & a_{12} & \cdots & a_{1n} & b_1 \\ a_{21} & a_{22} & \cdots & a_{2n} & b_2 \\ \vdots & \vdots & \ddots & \vdots & \vdots \\ a_{m1} & a_{m2} & \cdots & a_{mn} & b_m \end{pmatrix}. \tag{M5.3.1.3}
$$

KRONECKER–CAPELLI THEOREM. *A linear system (M5.3.1.1) or (M5.3.1.2) is consistent if and only if its basic matrix and its augmented matrix (M5.3.1.3) have the same rank, i.e.,* $\mathrm{rank}(A_1) = \mathrm{rank}(A)$.

▶ **Equivalent systems of equations. The elementary transformations.** Two systems are said to be *equivalent* if their sets of solutions coincide.

Systems of linear equations can be simplified using the following three types of elementary transformations:

1. Interchange of two equations (or the corresponding rows of the augmented matrix).
2. Multiplication of both sides of one equation (or the corresponding row of the augmented matrix) by a nonzero constant.
3. Adding to both sides of one equation both sides of another equation multiplied by a constant (adding to some row of the augmented matrix its other row multiplied by a constant).

Under the above elementary transformations, a system of linear equations reduces to an equivalent system of equations.

M5.3.2. Finding Solutions of a System of Linear Equations

▶ **System of two equations with two unknown quantities.** A system of two equations with two unknown quantities has the form

$$a_1 x + b_1 y = c_1,$$
$$a_2 x + b_2 y = c_2. \tag{M5.3.2.1}$$

Depending on the coefficients a_k, b_k, c_k, the following three cases are possible:

$1°$. If $\Delta = a_1 b_2 - a_2 b_1 \neq 0$, then system (M5.3.2.1) has a unique solution,

$$x = \frac{c_1 b_2 - c_2 b_1}{a_1 b_2 - a_2 b_1}, \quad y = \frac{a_1 c_2 - a_2 c_1}{a_1 b_2 - a_2 b_1}.$$

$2°$. If $\Delta = a_1 b_2 - a_2 b_1 = 0$ and $a_1 c_2 - a_2 c_1 = 0$ (the case of proportional coefficients), then system (M5.3.2.1) has infinitely many solutions described by the formulas

$$x = t, \quad y = \frac{c_1 - a_1 t}{b_1} \quad (b_1 \neq 0),$$

where t is arbitrary.

$3°$. If $\Delta = a_1 b_2 - a_2 b_1 = 0$ and $a_1 c_2 - a_2 c_1 \neq 0$, then system (M5.3.2.1) has no solutions.

▶ **General square system of linear equations with $m = n$.** A square system of linear equations has the form (M5.3.1.1) with $m = n$.

$1°$. *Cramer's rule.* If the determinant of the matrix of system (M5.3.1.1) with $m = n$ is different from zero, i.e., $\Delta = \det A \neq 0$, then the system admits a unique solution, which is expressed by formulas

$$x_1 = \frac{\Delta_1}{\Delta}, \quad x_2 = \frac{\Delta_2}{\Delta}, \quad \dots, \quad x_n = \frac{\Delta_n}{\Delta}, \tag{M5.3.2.2}$$

where Δ_k $(k = 1, 2, \dots, n)$ is the determinant of the matrix obtained from A by replacing its kth column with the column of constant terms:

$$\Delta_k = \begin{vmatrix} a_{11} & a_{12} & \cdots & a_{1\,k-1} & b_1 & a_{1\,k+1} & \cdots & a_{1n} \\ a_{21} & a_{22} & \cdots & a_{2\,k-1} & b_2 & a_{2\,k+1} & \cdots & a_{2n} \\ \vdots & \vdots & \ddots & \vdots & \vdots & \vdots & \ddots & \vdots \\ a_{n1} & a_{n2} & \cdots & a_{n\,k-1} & b_n & a_{n\,k+1} & \cdots & a_{nn} \end{vmatrix}.$$

Example 1. Using Cramer's rule, let us find the solution of the system of linear equations

$$2x_1 + x_2 + 4x_3 = 16,$$
$$3x_1 + 2x_2 + x_3 = 10,$$
$$x_1 + 3x_2 + 3x_3 = 16.$$

The determinant of its basic matrix is different from zero,

$$\Delta = \begin{vmatrix} 2 & 1 & 4 \\ 3 & 2 & 1 \\ 1 & 3 & 3 \end{vmatrix} = 26 \neq 0,$$

and we have

$$\Delta_1 = \begin{vmatrix} 16 & 1 & 4 \\ 10 & 2 & 1 \\ 16 & 3 & 3 \end{vmatrix} = 26, \quad \Delta_2 = \begin{vmatrix} 2 & 16 & 4 \\ 3 & 10 & 1 \\ 1 & 16 & 3 \end{vmatrix} = 52, \quad \Delta_3 = \begin{vmatrix} 2 & 1 & 16 \\ 3 & 2 & 10 \\ 1 & 3 & 16 \end{vmatrix} = 78.$$

Therefore, by Cramer's rule (M5.3.2.2), the only solution of the system has the form

$$x_1 = \frac{\Delta_1}{\Delta} = \frac{26}{26} = 1, \quad x_2 = \frac{\Delta_2}{\Delta} = \frac{52}{26} = 2, \quad x_3 = \frac{\Delta_3}{\Delta} = \frac{78}{26} = 3.$$

2°. System (M5.3.1.1) with $m = n$ can be treated in the matrix form (M5.3.1.2) where A is a square matrix. If $\det A \neq 0$, the system has a unique solution

$$X = A^{-1}B,$$

expressed in terms of the inverse A^{-1}, which can be found by formula (M5.2.3.1).

3°. *Reduction of a system to a triangular form* (*Gaussian method*). Suppose that $\det A \neq 0$. The Gaussian method is based on elementary transformations (see Subsection M5.3.1) used for the reduction of a given system to an equivalent system having the triangular form

$$x_1 + \alpha_{12}x_2 + \alpha_{13}x_3 + \cdots + \alpha_{1n}x_n = \beta_1,$$
$$x_2 + \alpha_{23}x_3 + \cdots + \alpha_{2n}x_n = \beta_2,$$
$$\dots\dots\dots\dots\dots\dots\dots\dots$$
$$x_{n-1} + \alpha_{n-1,n}x_n = \beta_{n-1},$$
$$x_n = \beta_n.$$

This system can be easily solved: inserting $x_n = \beta_n$ (from the last equation) into the preceding $(n-1)$st equation, one finds x_{n-1}. Then, inserting the values obtained for x_n, x_{n-1} into the $(n-2)$nd equation, one finds x_{n-2}. Proceeding in this way, one finally finds x_1. This back substitution process is described by the formulas

$$x_k = \beta_k - \sum_{s=k+1}^{n} \alpha_{ks}x_s \quad (k = n-1, n-2, \dots, 1).$$

Example 2. Solve the system

$$x_1 + x_2 - 2x_3 = -2,$$
$$2x_1 + 3x_2 + x_3 = 9,$$
$$3x_1 + 2x_2 + 2x_3 = 7.$$

Multiply the first equation by -2 and add to the second one. Multiply the first equation by -3 and add to the third one. As a result, the first equation together with the two obtained make up the equivalent system

$$x_1 + x_2 - 2x_3 = -2,$$
$$x_2 + 5x_3 = 13,$$
$$-x_2 + 8x_3 = 13.$$

Adding together the last two equations and dividing the result by 13, one arrives at the triangular system

$$x_1 + x_2 - 2x_3 = -2,$$
$$x_2 + 5x_3 = 13,$$
$$x_3 = 2.$$

Solving this system from bottom to top, one finds that

$$x_3 = 2, \quad x_2 = 13 - 5x_3 = 3, \quad x_1 = -2 - x_2 + 2x_3 = -1.$$

$4°$. *The Jordan–Gauss method.* Let us introduce some definitions. An unknown x_i is called *resolved* or *basic* if it enters only in one equation of the system with coefficient 1 and is not contained in the other equations.

If each equation of the system contains a resolved unknown, this system is called *resolved*. The unknowns of the system that are not basic are called *free*.

In order to find all solutions of a consistent system of linear equations, it suffices to find an equivalent resolved system. If all the unknowns happen to be basic, the resolved system gives the values of these unknowns. Otherwise, the basic unknowns are expressed in terms of the free ones.

Description of the method. Let us write down the system of linear equations (M5.3.1.1) as the table

x_1	$\cdots$	x_k	$\cdots$	x_n	
a_{11}	$\cdots$	a_{1k}	$\cdots$	a_{1n}	b_1
$\cdots$	$\cdots$	$\cdots$	$\cdots$	$\cdots$	$\cdots$
a_{r1}	$\cdots$	a_{rk}	$\cdots$	a_{rn}	b_r
$\cdots$	$\cdots$	$\cdots$	$\cdots$	$\cdots$	$\cdots$
a_{m1}	$\cdots$	a_{mk}	$\cdots$	a_{mn}	b_m

For a resolving entry $a_{rk} \neq 0$, the following procedure is called the *Jordan transformation*:
1) multiply the rth row of the table by $1/a_{rk}$;
2) add the resulting rth row multiplied by $-a_{1k}$ to the first row;
3) add the rth row multiplied by $-a_{2k}$ to the second row; and so on for all remaining rows.

After that, the unknown x_k becomes resolved, with all entries of the kth column equal to zero except that $a_{rk} = 1$.

By choosing other resolving entries in different rows and performing the respective Jordan transformations, one arrives at a resolved system equivalent to the original one.

If, at some point, the coefficients of the unknowns in a row become all zero and the free term of that row is nonzero, then the system of equations is inconsistent. If all entries of a row, including the free term, become zero, then this row is crossed out from the table.

Example 3. Solve the system of equations

$$2x_1 - 3x_2 + 5x_3 = 1,$$
$$x_1 + 2x_2 - 3x_3 = -7,$$
$$2x_1 + 5x_3 = 4.$$

Rewrite this system as a table and reduce it to a resolved form in six steps:

x_1	x_2	x_3	
2	−3	5	1
[1]	2	−3	−7
2	0	5	4

$\Longrightarrow$

x_1	x_2	x_3	
0	−7	11	15
1	2	−3	−7
0	−4	11	18

$\Longrightarrow$

x_1	x_2	x_3	
0	−7	11	15
1	2	−3	−7
0	3	0	3

$\Longrightarrow$

x_1	x_2	x_3	
0	−7	11	15
1	2	−3	−7
0	[1]	0	1

$\Longrightarrow$

x_1	x_2	x_3	
0	0	11	22
1	0	−3	−9
0	1	0	1

$\Longrightarrow$

x_1	x_2	x_3	
0	0	[1]	2
1	0	−3	−9
0	1	0	1

$\Longrightarrow$

x_1	x_2	x_3	
0	0	1	2
1	0	0	−3
0	1	0	1

In the tables above, the resolving entries are boxed. The following sequence of actions has been performed: (1) double the second row has been subtracted from the first and third ones, (2) the first row has been subtracted from the third one, (3) the third row has been divided by 3, (4) the third row multiplied by 7 (resp., by -2) has been added to the first (resp., second) one, (5) the first row has been divided by 11, and (6) the first row multiplied by 3 has been added to the second one. Thus, the original system acquires the resolved form

$$\begin{cases} 0 \cdot x_1 + 0 \cdot x_2 + 1 \cdot x_3 = 2, \\ 1 \cdot x_1 + 0 \cdot x_2 + 0 \cdot x_3 = -3, \\ 0 \cdot x_1 + 1 \cdot x_2 + 0 \cdot x_3 = 1. \end{cases}$$

The resulting solution is $x_1 = -3$, $x_2 = 1$, $x_3 = 2$.

Example 4. Solve the system of equations

$$\begin{cases} 2x_1 + 7x_2 + 3x_3 + x_4 = 6, \\ 3x_1 + 5x_2 + 2x_3 + 2x_4 = 4, \\ 9x_1 + 4x_2 + x_3 + 7x_4 = 2. \end{cases}$$

With Jordan transformations, this system is reduced to the resolved form

$$\begin{cases} -11x_2 - 5x_3 + x_4 = -10, \\ x_1 + 9x_2 + 4x_3 = 8. \end{cases}$$

Hence, the set of all solutions to the original system is given by

$$x_1 = 8 - 9x_2 - 4x_3, \quad x_4 = -10 + 11x_2 + 5x_3,$$

with x_2 and x_3 assuming any real values.

▶ **General system of m linear equations with n unknown quantities.** Suppose that system (M5.3.1.1) is consistent and its basic matrix A has rank r. First, in the matrix A, one finds a submatrix of size $r \times r$ with a nonzero rth-order determinant and drops the $m - r$ equations whose coefficients do not belong to this submatrix (the dropped equations follow from the remaining ones and can, therefore, be neglected). In the remaining equations, the $n - r$ unknown quantities (free unknown quantities) that are not involved in the said submatrix should be transferred to the right-hand sides. Thus, one obtains a system of r equations with r unknown quantities, which can be solved by any of the methods described above in the current subsection.

Remark. If the rank r of the basic matrix and the rank of the augmented matrix of system (M5.3.1.1) are equal to the number of the unknown quantities n, then the system has a unique solution.

▶ **Existence of nontrivial solutions of a homogeneous system.** Consider the homogeneous system (M5.3.1.1), with $b_1 = b_2 = \cdots = b_m = 0$. This system is always consistent, since it always has the so-called *trivial solution* $x_1 = x_2 = \cdots = x_n = 0$.

THEOREM. *A homogeneous system has a nontrivial solution if and only if the rank of the matrix A is less than the number of the unknown quantities n.*

It follows that a square homogeneous system has a nontrivial solution if and only if the determinant of its matrix of coefficients is equal to zero, $\det A = 0$.

M5.4. Quadratic Forms

M5.4.1. Quadratic Forms and Their Transformations

▶ **Quadratic form with n variables.** A real quadratic form is a homogeneous polynomial of degree 2 in n variables $x_1, x_2, \ldots, x_n$ of the form

$$A_n(x_1, x_2, \ldots, x_n) \equiv \sum_{i,j=1}^{n} a_{ij} x_i x_j, \tag{M5.4.1.1}$$

with real coefficients a_{ij} satisfying the symmetry condition $a_{ij} = a_{ji}$.

The quadratic form (M5.4.1.1) can be conveniently written in short matrix notation

$$A_n(X) = X^T A X, \qquad (M5.4.1.2)$$

where $X \equiv [x_i]$ is a column vector consisting of n elements, X^T is its transpose, and $A \equiv [a_{ij}]$ is an $n \times n$ symmetric matrix, called the *matrix of the quadratic form*.

A real-valued quadratic form $A_n(X)$ is said to be:
a) *positive definite* (resp., *negative definite*) if $A_n(X) > 0$ (resp., $A_n(X) < 0$) for any $X \neq 0$;
b) *indefinite* if there exist vectors X and Y such that $A_n(X) > 0$ and $A_n(Y) < 0$;
c) *nonnegative* (resp., *nonpositive*) if $A_n(X) \geq 0$ (resp., $A_n(X) \leq 0$) for all $X \neq 0$.

The determinant $\det A$ of the matrix A is called the *discriminant of the quadratic form* $A_n(X)$. A quadratic form is called *degenerate* if its discriminant is zero.

▶ **Criteria of positive and negative definiteness of a quadratic form.**

$1°$. A real quadratic form $A_n(X)$ is positive definite, negative definite, indefinite, nonnegative, nonpositive if the eigenvalues λ_i of its matrix $A \equiv [a_{ij}]$ are all positive, are all negative, some are positive and some negative, are all nonnegative, are all nonpositive, respectively.

$2°$. *Sylvester criterion.* A real quadratic form $A_n(X)$ is positive definite if and only if it satisfies the conditions

$$\Delta_1 \equiv a_{11} > 0, \quad \Delta_2 \equiv \begin{vmatrix} a_{11} & a_{12} \\ a_{21} & a_{22} \end{vmatrix} > 0, \quad \ldots, \quad \Delta_n \equiv \det A > 0.$$

If the signs of the minor determinants alternate,

$$\Delta_1 < 0, \quad \Delta_2 > 0, \quad \Delta_3 < 0, \ldots,$$

then the quadratic form is negative definite.

▶ **Transformations of a real quadratic form.** Let us find out how the coefficient matrix changes under a linear transformation of the variables

$$x_i = \sum_{k=1}^{n} b_{ik} y_k \qquad (i = 1, 2, \ldots, n), \qquad (M5.4.1.3)$$

where b_{ik} are real numbers. In matrix notation, transformation (M5.4.1.3) becomes

$$X = BY, \qquad (M5.4.1.4)$$

where $Y \equiv [y_i]$ is a column vector of size n and $B \equiv [b_{ij}]$ is a transformation matrix of size $n \times n$.

Substituting (M5.4.1.4) into (M5.4.1.2) gives

$$A_n(X) = Y^T B^T A B Y = Y^T \widetilde{A} Y = \widetilde{A}_n(Y),$$

where

$$\widetilde{A} = B^T A B. \qquad (M5.4.1.5)$$

It follows that the discriminant of a quadratic form changes according to the rule

$$\det \widetilde{A} = \det A \, (\det B)^2.$$

In what follows, only nondegenerate transformations of variables are considered, i.e., those with $\det B \neq 0$. The rank of the coefficient matrix remains unchanged under such transformations. The rank of the coefficient matrix is usually said to be the *rank of the quadratic form*.

M5.4.2. Canonical and Normal Representations of a Quadratic Form

▶ **Canonical representation of a quadratic form.** Any real quadratic form (M5.4.1.1) can be reduced to the form

$$A_n(X) = \sum_{i=1}^{r} \lambda_i y_i^2 \equiv A_r(Y) \tag{M5.4.2.1}$$

using an appropriate nondegenerate linear transformation (M5.4.1.3).

This representation is called a *canonical representation* of the quadratic form, the real coefficients $\lambda_1, \ldots, \lambda_r$ are called the *canonical coefficients*.

Reduction of the quadratic form (M5.4.1.1) to the canonical form (M5.4.2.1) is not unique and can be performed using various linear transformations of the form (M5.4.1.3).

LAW OF INERTIA OF QUADRATIC FORMS. *The number of terms with positive coefficients and the number of terms with negative coefficients in any canonical representation of a real quadratic form does not depend on the method used to obtain such a representation.*

The *index of inertia* of a real quadratic form is the integer r equal to the number of nonzero coefficients in its canonical representation (this number coincides with the rank of the quadratic form). Its *positive index of inertia* is the integer p equal to the number of positive coefficients in the canonical representation of the form, and its *negative index of inertia* is the integer q equal to the number of its negative canonical coefficients. The integer $s = p - q$ is called the *signature* of the quadratic form.

A real quadratic form $A_n(X)$ is

a) positive definite (resp., negative definite) if $p = n$ (resp., $q = n$);
b) indefinite if $p \neq 0$ and $q \neq 0$;
c) nonnegative (resp., nonpositive) if $q = 0$, $p < n$ (resp., $p = 0$, $q < n$).

THEOREM. *For any real symmetric quadratic form $A_n(X)$ there exists a real orthogonal transformation (M5.4.1.3), whose matrix B possesses the property $B^T B = B B^T = I$, that reduces the quadratic form to the canonical form (M5.4.2.1). The canonical coefficients $\lambda_1, \ldots, \lambda_n$ are eigenvalues of the quadratic form matrix A.*

▶ **Lagrange's method of reduction of a quadratic form to a canonical form.** For the canonical form (M5.4.1.1), consider the following two cases.

Case 1. Suppose that $a_{mm} \neq 0$ for some m ($1 \leq m \leq n$). By letting

$$A_n(X) = \frac{1}{a_{mm}} \left(\sum_{k=1}^{n} a_{mk} x_k \right)^2 + A_{n-1}(X), \tag{M5.4.2.2}$$

one can easily verify that the quadratic form $A_{n-1}(X)$ does not contain the variable x_m (it contains $n - 1$ variables or fewer). This method of isolating a perfect square in a quadratic form can always be applied if the matrix $[a_{ij}]$ ($i, j = 1, 2, \ldots, n$) contains nonzero diagonal elements.

Case 2. Suppose that $a_{mm} = a_{ss} = 0$, but $a_{ms} \neq 0$. In this case, the quadratic form can be represented as

$$A_n(X) = \frac{1}{2a_{ms}} \left[\sum_{k=1}^{n} (a_{mk} + a_{sk}) x_k \right]^2 - \frac{1}{2a_{ms}} \left[\sum_{k=1}^{n} (a_{mk} - a_{sk}) x_k \right]^2 + A_{n-2}(X), \tag{M5.4.2.3}$$

where $A_{n-2}(X)$ does not contain the variables x_m and x_s (it contains $n-2$ variables), and the linear forms in square brackets are linearly independent (and therefore can be taken as new independent variables or coordinates).

By combining the above two procedures, the quadratic form $A_n(X)$ can always be represented in terms of squared linear forms; these forms are linearly independent, since each contains a variable which is absent from the other linear forms. By taking the linear forms to be new independent variables, one obtains the canonical representation of the quadratic form (M5.4.2.1).

Note that the main formulas (M5.4.2.2) and (M5.4.2.3) can be rewritten as

$$A_n(X) = \frac{1}{4a_{mm}}\left(\frac{\partial A_n}{\partial x_m}\right)^2 + A_{n-1}(X), \tag{M5.4.2.2a}$$

$$A_n(X) = \frac{1}{8a_{ms}}\left[\left(\frac{\partial A_n}{\partial x_m} + \frac{\partial A_n}{\partial x_s}\right)^2 - \left(\frac{\partial A_n}{\partial x_m} - \frac{\partial A_n}{\partial x_s}\right)^2\right] + A_{n-2}(X). \tag{M5.4.2.3a}$$

Example. Reduce the quadratic form

$$A_3(X) = 4x_1^2 + x_2^2 + x_3^2 - 4x_1x_2 - 4x_1x_3 + 4x_2x_3$$

to a canonical form.

Using formula (M5.4.2.2a) with $m = 1$, we get

$$A_3(X) = \tfrac{1}{16}(8x_1 - 4x_2 - 4x_3)^2 + 2x_2x_3 = (2x_1 - x_2 - x_3)^2 + A_2(X).$$

Further applying formula (M5.4.2.3a) with $m = 2$ and $s = 3$ to $A_2(X) = 2x_2x_3$, we obtain

$$A_2(X) = 2x_2x_3 = \tfrac{1}{8}(2x_2 + 2x_3)^2 - \tfrac{1}{8}(2x_3 - 2x_2)^2 = \tfrac{1}{2}(x_2 + x_3)^2 - \tfrac{1}{2}(x_2 - x_3)^2.$$

The two formulas just obtained yield a canonical representation of the original form:

$$A_3(X) = y_1^2 + \tfrac{1}{2}y_2^2 - \tfrac{1}{2}y_3^2,$$

where

$$y_1 = 2x_1 - x_2 - x_3, \qquad y_2 = x_2 + x_3, \qquad y_3 = x_2 - x_3.$$

▶ **Jacobi's formula.** Introduce the following notation:

$$A\begin{pmatrix} x_1 & x_2 & \cdots & x_k \\ y_1 & y_2 & \cdots & y_k \end{pmatrix} = \sum_{i,j=1}^{k} a_{ij}x_ix_j$$

Let

$$D_k = A\begin{pmatrix} 1 & 2 & \cdots & k \\ 1 & 2 & \cdots & k \end{pmatrix} \neq 0 \qquad (k = 1, 2, \ldots, r),$$

where r is the rank of the quadratic form (M5.4.1.1). Then the form (M5.4.1.1) admits the canonical representation

$$A_n(X) = \frac{1}{a_{11}}\,y_1^2 + \sum_{k=2}^{r} \frac{1}{D_{k-1}D_k}\,y_k^2,$$

where

$$y_k = c_{kk}x_k + c_{k,k+1}x_{k+1} + \cdots + c_{kn}x_n, \qquad c_{kq} = A\begin{pmatrix} 1 & 2 & \cdots & k-1 & k \\ 1 & 2 & \cdots & k-1 & q \end{pmatrix};$$

$$k = 1, 2, \ldots, r; \quad q = k, k+1, \ldots, n.$$

▶ **Normal representation of a real quadratic form.** Any real quadratic form (M5.4.1.1) admits the *normal representation*

$$A_n(X) = \sum_{i=1}^{n} \varepsilon_i z_i^2,$$

where $z_1, \ldots, z_n$ are the new variables and $\varepsilon_1, \ldots, \varepsilon_n$ are coefficients taking the values $-1, 0, 1$.

A normal representation can be obtained by the following transformations:

1. One obtains the canonical representation (M5.4.2.1), for example, by Lagrange's method.

2. With the nondegenerate coordinate transformation

$$y_i = \begin{cases} \dfrac{1}{\sqrt{\lambda_i}}\, z_i & \text{for } \lambda_i > 0, \\[2mm] \dfrac{1}{\sqrt{-\lambda_i}}\, z_i & \text{for } \lambda_i < 0, \\[2mm] z_i & \text{for } \lambda_i = 0, \end{cases}$$

the canonical representation can be converted to a normal representation.

▶ **Simultaneous reduction of two quadratic forms to sums of squares.**

THEOREM. *Let $A_n(X)$ and $B_n(X)$ be real symmetric quadratic forms in n variables and let $B_n(X)$ be positive definite. Then there exists a real transformation (M5.4.1.3) that reduces the two forms to*

$$A(X) = \sum_{k=1}^{n} \lambda_k y_k^2, \quad B(X) = \sum_{k=1}^{n} y_k^2,$$

where y_k are new variables. The set of real $\lambda_1, \ldots, \lambda_n$ coincides with the spectrum of eigenvalues of the matrix $B^{-1}A$; this set consists of the roots of the algebraic equation

$$\det(A - \lambda B) = 0.$$

M5.5. Linear Spaces

M5.5.1. Concept of a Linear Space. Its Basis and Dimension

▶ **Definition of a linear space.** A *linear space* or a *vector space* over a field of *scalars* (usually, the field of real numbers or the field of complex numbers) is a set $\mathcal{V}$ of elements $\mathbf{x}, \mathbf{y}, \mathbf{z}, \ldots$ (also called *vectors*) of any nature for which the following conditions hold:

I. There is a rule that establishes correspondence between any pair of elements $\mathbf{x}, \mathbf{y} \in \mathcal{V}$ and a third element $\mathbf{z} \in \mathcal{V}$, called the *sum* of the elements $\mathbf{x}, \mathbf{y}$ and denoted by $\mathbf{z} = \mathbf{x} + \mathbf{y}$.

II. There is a rule that establishes correspondence between any pair $\mathbf{x}, \lambda$, where $\mathbf{x}$ is an element of $\mathcal{V}$ and λ is a scalar, and an element $\mathbf{u} \in \mathcal{V}$, called the *product of a scalar λ and a vector* $\mathbf{x}$ and denoted by $\mathbf{u} = \lambda\mathbf{x}$.

III. The following eight axioms are assumed for the above two operations:

1. Commutativity of the sum: $\mathbf{x} + \mathbf{y} = \mathbf{y} + \mathbf{x}$.
2. Associativity of the sum: $(\mathbf{x} + \mathbf{y}) + \mathbf{z} = \mathbf{x} + (\mathbf{y} + \mathbf{z})$.
3. There is a zero element 0 such that $\mathbf{x} + 0 = \mathbf{x}$ for any $\mathbf{x}$.

4. For any element $\mathbf{x}$ there is an opposite element $\mathbf{x}'$ such that $\mathbf{x} + \mathbf{x}' = 0$.
5. A special role of the unit scalar 1: $1 \cdot \mathbf{x} = \mathbf{x}$ for any element $\mathbf{x}$.
6. Associativity of the multiplication by scalars: $\lambda(\mu\mathbf{x}) = (\lambda\mu)\mathbf{x}$.
7. Distributivity with respect to the addition of scalars: $(\lambda + \mu)\mathbf{x} = \lambda\mathbf{x} + \mu\mathbf{x}$.
8. Distributivity with respect to a sum of vectors: $\lambda(\mathbf{x} + \mathbf{y}) = \lambda\mathbf{x} + \lambda\mathbf{y}$.

This is the definition of an abstract linear space. We obtain a *specific linear space* if the nature of the elements and the operations of addition and multiplication by scalars are concretized.

Example 1. Consider the set of all free vectors in three-dimensional space. If addition of these vectors and their multiplication by scalars are defined as in analytic geometry (see Subsection M4.5.1), this set becomes a linear space denoted by B_3.

Example 2. Consider the *n-dimensional coordinate space* $\mathbb{R}^n$, whose elements are ordered sets of n arbitrary real numbers $(x_1, \ldots, x_n)$. The generic element of this space is denoted by $\mathbf{x}$, i.e., $\mathbf{x} = (x_1, \ldots, x_n)$, and the reals $x_1, \ldots, x_n$ are called the *coordinates* of the element $\mathbf{x}$. From the algebraic standpoint, the set $\mathbb{R}^n$ may be regarded as the set of all row vectors with n real components.

The operations of addition of elements of $\mathbb{R}^n$ and their multiplication by scalars are defined by the following rules:

$$(x_1, \ldots, x_n) + (y_1, \ldots, y_n) = (x_1 + y_1, \ldots, x_n + y_n),$$
$$\lambda(x_1, \ldots, x_n) = (\lambda x_1, \ldots, \lambda x_n).$$

Remark. If the field of scalars λ, μ, ... in the above definition is the field of all real numbers, the corresponding linear spaces are called *real linear spaces*. If the field of scalars is that of all complex numbers, the corresponding space is called a *complex linear space*. In many situations, it is clear from the context which field of scalars is meant.

The above axioms imply the following properties of an arbitrary linear space:

1. The zero vector is unique, and for any element $\mathbf{x}$ the opposite element is unique.
2. The zero vector 0 is equal to the product of any element $\mathbf{x}$ by the scalar 0.
3. For any element $\mathbf{x}$, the opposite element is equal to the product of $\mathbf{x}$ by the scalar -1.
4. The *difference* of two elements $\mathbf{x}$ and $\mathbf{y}$, i.e., the element $\mathbf{z}$ such that $\mathbf{z} + \mathbf{y} = \mathbf{x}$, is unique.

▶ **Basis and dimension of a linear space. Isomorphism of linear spaces.** An element $\mathbf{y}$ is called a *linear combination* of elements $\mathbf{x}_1, \ldots, \mathbf{x}_k$ of a linear space $\mathcal{V}$ if there exist scalars $\alpha_1, \ldots, \alpha_k$ such that

$$\mathbf{y} = \alpha_1\mathbf{x}_1 + \cdots + \alpha_k\mathbf{x}_k.$$

Elements $\mathbf{x}_1, \ldots, \mathbf{x}_k$ of the space $\mathcal{V}$ are said to be *linearly dependent* if there exist scalars $\alpha_1, \ldots, \alpha_k$ such that $|\alpha_1|^2 + \cdots + |\alpha_k|^2 \neq 0$ and

$$\alpha_1\mathbf{x}_1 + \cdots + \alpha_k\mathbf{x}_k = 0,$$

where 0 is the zero element of $\mathcal{V}$.

Elements $\mathbf{x}_1, \ldots, \mathbf{x}_k$ of the space $\mathcal{V}$ are said to be *linearly independent* if for any scalars $\alpha_1, \ldots, \alpha_k$ such that $|\alpha_1|^2 + \cdots + |\alpha_k|^2 \neq 0$, we have

$$\alpha_1\mathbf{x}_1 + \cdots + \alpha_k\mathbf{x}_k \neq 0.$$

Remark 1. Elements $\mathbf{x}_1, \ldots, \mathbf{x}_k$ of a linear space $\mathcal{V}$ are linearly dependent if and only if at least one of them is a linear combination of the others.

Remark 2. If at least one of the elements $\mathbf{x}_1, \ldots, \mathbf{x}_k$ is equal to zero, then these elements are linearly dependent. If some of the elements $\mathbf{x}_1, \ldots, \mathbf{x}_k$ are linearly dependent, then all these elements are linearly dependent.

Example 3. The elements $\mathbf{i}_1 = (1, 0, \ldots, 0)$, $\mathbf{i}_2 = (0, 1, \ldots, 0)$, $\ldots$, $\mathbf{i}_n = (0, 0, \ldots, 1)$ of the space $\mathbb{R}^n$ (see Example 2) are linearly independent. For any $\mathbf{x} = (x_1, \ldots, x_n) \in \mathbb{R}^n$, the vectors $\mathbf{x}, \mathbf{i}_1, \ldots, \mathbf{i}_n$ are linearly dependent.

A *basis* of a linear space $\mathcal{V}$ is defined as any system of linearly independent vectors $\mathbf{e}_1, \ldots, \mathbf{e}_n$ such that for any element $\mathbf{x}$ of the space $\mathcal{V}$ there exist scalars $x_1, \ldots, x_n$ such that

$$\mathbf{x} = x_1 \mathbf{e}_1 + \cdots + x_n \mathbf{e}_n.$$

This relation is called the *representation of an element* $\mathbf{x}$ *in terms of the basis* $\mathbf{e}_1, \ldots, \mathbf{e}_n$, and the scalars $x_1, \ldots, x_n$ are called the *coordinates* of the element $\mathbf{x}$ in that basis.

UNIQUENESS THEOREM. *The representation of any element* $\mathbf{x} \in \mathcal{V}$ *in terms of a given basis* $\mathbf{e}_1, \ldots, \mathbf{e}_n$ *is unique.*

Let $\mathbf{e}_1, \ldots, \mathbf{e}_n$ be any basis in $\mathcal{V}$ and vectors $\mathbf{x}$ and $\mathbf{y}$ have the coordinates $x_1, \ldots, x_n$ and $y_1, \ldots, y_n$ in that basis. Then the coordinates of the vector $\mathbf{x} + \mathbf{y}$ in that basis are $x_1 + y_1$, $\ldots, x_n + y_n$, and the coordinates of the vector $\lambda \mathbf{x}$ are $\lambda x_1, \ldots, \lambda x_n$ for any scalar λ.

Example 4. Any three noncoplanar vectors form a basis in the linear space B_3 of all free vectors. The n elements $\mathbf{i}_1 = (1, 0, \ldots, 0)$, $\mathbf{i}_2 = (0, 1, \ldots, 0)$, $\ldots$, $\mathbf{i}_n = (0, 0, \ldots, 1)$ form a basis in the linear space $\mathbb{R}^n$.

A linear space $\mathcal{V}$ is said to be n-*dimensional* if it contains n linearly independent elements and any $n + 1$ elements are linearly dependent. The number n is called the *dimension* of that space, $n = \dim \mathcal{V}$.

A linear space $\mathcal{V}$ is said to be *infinite-dimensional* ($\dim \mathcal{V} = \infty$) if for any positive integer N it contains N linearly independent elements.

THEOREM 1. *If* $\mathcal{V}$ *is a linear space of dimension* n, *then any* n *linearly independent elements of that space form its basis.*

THEOREM 2. *If a linear space* $\mathcal{V}$ *has a basis consisting of* n *elements, then* $\dim \mathcal{V} = n$.

Example 5. The dimension of the space B_3 of all free vectors is equal to 3. The dimension of the space $\mathbb{R}^n$ is equal to n.

Two linear spaces $\mathcal{V}$ and $\mathcal{V}'$ over the same field of scalars are said to be *isomorphic* if there is a one-to-one correspondence between the elements of these spaces such that if elements $\mathbf{x}$ and $\mathbf{y}$ from $\mathcal{V}$ correspond to elements $\mathbf{x}'$ and $\mathbf{y}'$ from $\mathcal{V}'$, then the element $\mathbf{x} + \mathbf{y}$ corresponds to $\mathbf{x}' + \mathbf{y}'$ and the element $\lambda \mathbf{x}$ corresponds to $\lambda \mathbf{x}'$ for any scalar λ.

Remark. If linear spaces $\mathcal{V}$ and $\mathcal{V}'$ are isomorphic, then the zero element of one space corresponds to the zero element of the other.

THEOREM. *Any two* n-*dimensional real (or complex) spaces* $\mathcal{V}$ *and* $\mathcal{V}'$ *are isomorphic.*

▶ **Affine space.** An *affine space* is a nonempty set $\mathcal{A}$ that consists of elements of any nature, called *points*, for which the following conditions hold:
I. There is a given linear (vector) space $\mathcal{V}$, called the *associated linear space*.
II. There is a rule by which any ordered pair of points $A, B \in \mathcal{A}$ is associated with an element (vector) from $\mathcal{V}$; this vector is denoted by $\overrightarrow{AB}$ and is called the *vector issuing from the point* A *with endpoint at* B.
III. The following conditions (called *axioms of affine space*) hold:
1. For any point $A \in \mathcal{A}$ and any vector $\mathbf{a} \in \mathcal{V}$, there is a unique point $B \in \mathcal{A}$ such that $\overrightarrow{AB} = \mathbf{a}$.
2. $\overrightarrow{AB} + \overrightarrow{BC} = \overrightarrow{AC}$ for any three points $A, B, C \in \mathcal{A}$.

By definition, the *dimension of an affine space* $\mathcal{A}$ is the dimension of the associated linear space $\mathcal{V}$, $\dim \mathcal{A} = \dim \mathcal{V}$.

Any linear space may be regarded as an affine space.

In particular, the space $\mathbb{R}^n$ can be naturally considered as an affine space. Thus if $A = (a_1, \ldots, a_n)$ and $B = (b_1, \ldots, b_n)$ are points of the affine space $\mathbb{R}^n$, then the corresponding vector $\overrightarrow{AB}$ from the linear space $\mathbb{R}^n$ is defined by $\overrightarrow{AB} = (b_1 - a_1, \ldots, b_n - a_n)$.

Let $\mathcal{A}$ be an n-dimensional affine space with the associated linear space $\mathcal{V}$. A *coordinate system* in the affine space $\mathcal{A}$ is a fixed point $O \in \mathcal{A}$, together with a fixed basis $\mathbf{e}_1, \ldots, \mathbf{e}_n \in \mathcal{V}$. The point O is called the *origin* of this coordinate system.

Let M be a point of an affine space $\mathcal{A}$ with a coordinate system $O\mathbf{e}_1 \ldots \mathbf{e}_n$. One says that the point M has *affine coordinates* (or simply coordinates) $x_1, \ldots, x_n$ in this coordinate system, and one writes $M = (x_1, \ldots, x_n)$ if $x_1, \ldots x_n$ are the coordinates of the radius-vector $\overrightarrow{OM}$ in the basis $\mathbf{e}_1, \ldots, \mathbf{e}_n$, i.e., $\overrightarrow{OM} = x_1\mathbf{e}_1 + \cdots + x_n\mathbf{e}_n$.

M5.5.2. Subspaces of Linear Spaces

▶ **Concept of a linear subspace and a linear span.** A subset $\mathcal{L}$ of a linear space $\mathcal{V}$ is called a *linear subspace* of $\mathcal{V}$ if the following conditions hold:

1. If $\mathbf{x}$ and $\mathbf{y}$ belong to $\mathcal{L}$, then the sum $\mathbf{x} + \mathbf{y}$ belongs to $\mathcal{L}$.
2. If $\mathbf{x}$ belongs to $\mathcal{L}$ and λ is an arbitrary scalar, then the element $\lambda\mathbf{x}$ belongs to $\mathcal{L}$.

The *null subspace* in a linear space $\mathcal{V}$ is its subset consisting of the single element zero. The space $\mathcal{V}$ itself can be regarded as its own subspace. These two subspaces are called *improper subspaces*. All other subspaces are called *proper subspaces*.

Example 1. A subset B_2 consisting of all free vectors parallel to a given plane is a subspace in the linear space B_3 of all free vectors.

The *linear span* $L(\mathbf{x}_1, \ldots, \mathbf{x}_m)$ of vectors $\mathbf{x}_1, \ldots, \mathbf{x}_m$ in a linear space $\mathcal{V}$ is, by definition, the set of all linear combinations of these vectors, i.e., the set of all vectors of the form

$$\alpha_1\mathbf{x}_1 + \cdots + \alpha_m\mathbf{x}_m,$$

where $\alpha_1, \ldots, \alpha_m$ are arbitrary scalars. The linear span $L(\mathbf{x}_1, \ldots, \mathbf{x}_m)$ is the least subspace of $\mathcal{V}$ containing the elements $\mathbf{x}_1, \ldots, \mathbf{x}_m$.

If a subspace $\mathcal{L}$ of an n-dimensional space $\mathcal{V}$ does not coincide with $\mathcal{V}$, then $\dim \mathcal{L} < n = \dim \mathcal{V}$.

Let elements $\mathbf{e}_1, \ldots, \mathbf{e}_k$ form a basis in a k-dimensional subspace of an n-dimensional linear space $\mathcal{V}$. Then this basis can be supplemented by elements $\mathbf{e}_{k+1}, \ldots, \mathbf{e}_n$ of the space $\mathcal{V}$, so that the system $\mathbf{e}_1, \ldots, \mathbf{e}_k, \mathbf{e}_{k+1}, \ldots, \mathbf{e}_n$ forms a basis in the space $\mathcal{V}$.

THEOREM ON THE DIMENSION OF A LINEAR SPAN. *The dimension of a linear span* $L(\mathbf{x}_1, \ldots, \mathbf{x}_m)$ *of elements* $\mathbf{x}_1, \ldots, \mathbf{x}_m$ *is equal to the maximal number of linearly independent vectors in the system* $\mathbf{x}_1, \ldots, \mathbf{x}_m$.

▶ **Sum and intersection of subspaces.** The *intersection* of subspaces $\mathcal{L}_1$ and $\mathcal{L}_2$ of one and the same linear space $\mathcal{V}$ is, by definition, the set of all elements $\mathbf{x}$ of $\mathcal{V}$ that belong simultaneously to both spaces $\mathcal{L}_1$ and $\mathcal{L}_2$. Such elements form a subspace of $\mathcal{V}$.

The *sum* of subspaces $\mathcal{L}_1$ and $\mathcal{L}_2$ of one and the same linear space $\mathcal{V}$ is, by definition, the set of all elements of $\mathcal{V}$ that can be represented in the form $\mathbf{y} + \mathbf{z}$, where $\mathbf{y}$ is an element of $\mathcal{V}_1$ and $\mathbf{z}$ is an element of $\mathcal{L}_2$. The sum of subspaces is also a subspace of $\mathcal{V}$.

THEOREM. *The sum of dimensions of arbitrary subspaces* $\mathcal{L}_1$ *and* $\mathcal{L}_2$ *of a finite-dimensional space* $\mathcal{V}$ *is equal to the sum of the dimension of their intersection and the dimension of their sum.*

Example 2. Let B_3 be the linear space of all free vectors (in three-dimensional space). Denote by $\mathcal{L}_1$ the subspace of all free vectors parallel to the plane OXY, and by $\mathcal{L}_2$ the subspace of all free vectors parallel to the plane OXZ. Then the sum of the subspaces $\mathcal{L}_1$ and $\mathcal{L}_2$ coincides with B_3, and their intersection consists of all free vectors parallel to the axis OX.

The dimension of each space $\mathcal{L}_1$ and $\mathcal{L}_2$ is equal to two, the dimension of their sum is equal to three, and the dimension of their intersection is equal to unity.

M5.5.3. Coordinate Transformations Corresponding to Basis Transformations in a Linear Space

▶ **Basis transformation and its inverse.** Let $\mathbf{e}_1, \ldots, \mathbf{e}_n$ and $\widetilde{\mathbf{e}}_1, \ldots, \widetilde{\mathbf{e}}_n$ be two arbitrary bases of an n-dimensional linear space $\mathcal{V}$. Suppose that the elements $\widetilde{\mathbf{e}}_1, \ldots, \widetilde{\mathbf{e}}_n$ are expressed via $\mathbf{e}_1, \ldots, \mathbf{e}_n$ by the formulas

$$
\begin{aligned}
\widetilde{\mathbf{e}}_1 &= a_{11}\mathbf{e}_1 + a_{12}\mathbf{e}_2 + \cdots + a_{1n}\mathbf{e}_n, \\
\widetilde{\mathbf{e}}_2 &= a_{21}\mathbf{e}_1 + a_{22}\mathbf{e}_2 + \cdots + a_{2n}\mathbf{e}_n, \\
&\cdots\cdots\cdots\cdots\cdots\cdots\cdots\cdots\cdots\cdots \\
\widetilde{\mathbf{e}}_n &= a_{n1}\mathbf{e}_1 + a_{n2}\mathbf{e}_2 + \cdots + a_{nn}\mathbf{e}_n.
\end{aligned}
$$

Thus, the transition from the basis $\mathbf{e}_1, \ldots, \mathbf{e}_n$ to the basis $\widetilde{\mathbf{e}}_1, \ldots, \widetilde{\mathbf{e}}_n$ is determined by the matrix

$$
A \equiv \begin{pmatrix}
a_{11} & a_{12} & \cdots & a_{1n} \\
a_{21} & a_{22} & \cdots & a_{2n} \\
\vdots & \vdots & \ddots & \vdots \\
a_{n1} & a_{n2} & \cdots & a_{nn}
\end{pmatrix}.
$$

Note that $\det A \neq 0$, i.e., the matrix A is nondegenerate.

The transition from the basis $\widetilde{\mathbf{e}}_1, \ldots, \widetilde{\mathbf{e}}_n$ to the basis $\mathbf{e}_1, \ldots, \mathbf{e}_n$ is determined by the matrix $B \equiv [b_{ij}] = A^{-1}$. Thus, we can write

$$
\widetilde{\mathbf{e}}_i = \sum_{j=1}^{n} a_{ij}\mathbf{e}_j, \qquad \mathbf{e}_k = \sum_{j=1}^{n} b_{kj}\widetilde{\mathbf{e}}_j \qquad (i, k = 1, 2, \ldots, n). \tag{M5.5.3.1}
$$

▶ **Relations between coordinate transformations and basis transformations.** Suppose that in a linear n-dimensional space $\mathcal{V}$, the transition from its basis $\mathbf{e}_1, \ldots, \mathbf{e}_n$ to another basis $\widetilde{\mathbf{e}}_1, \ldots, \widetilde{\mathbf{e}}_n$ is determined by the matrix A (see above). Let $\mathbf{x}$ be any element of the space $\mathcal{V}$ with the coordinates $(x_1, \ldots, x_n)$ in the basis $\mathbf{e}_1, \ldots, \mathbf{e}_n$ and the coordinates $(\widetilde{x}_1, \ldots, \widetilde{x}_n)$ in the basis $\widetilde{\mathbf{e}}_1, \ldots, \widetilde{\mathbf{e}}_n$, i.e.,

$$
\mathbf{x} = x_1\mathbf{e}_1 + \cdots + x_n\mathbf{e}_n = \widetilde{x}_1\widetilde{\mathbf{e}}_1 + \cdots + \widetilde{x}_n\widetilde{\mathbf{e}}_n.
$$

Then using formulas (M5.5.3.1), we obtain the following relations between these coordinates:

$$
x_j = \sum_{i=1}^{n} \widetilde{x}_i a_{ij}, \qquad \widetilde{x}_k = \sum_{l=1}^{n} x_l b_{lk}, \qquad j, k = 1, \ldots, n.
$$

In terms of matrices and row vectors, these relations can be written as follows:

$$
(x_1, \ldots, x_n) = (\widetilde{x}_1, \ldots, \widetilde{x}_n)A, \qquad (\widetilde{x}_1, \ldots, \widetilde{x}_n) = (x_1, \ldots, x_n)A^{-1}
$$

or, in terms of column vectors,

$$
(x_1, \ldots, x_n)^T = A^T(\widetilde{x}_1, \ldots, \widetilde{x}_n)^T, \qquad (\widetilde{x}_1, \ldots, \widetilde{x}_n)^T = (A^{-1})^T(x_1, \ldots, x_n)^T,
$$

where the superscript T indicates the transpose of a matrix.

M5.5.4. Euclidean Space

▶ **Definition and properties of a Euclidean space.** A *real Euclidean space* (or simply, *Euclidean space*) is a real linear space $\mathcal{V}$ endowed with a *scalar product* (also known as *inner product* and *dot product*), which is a real-valued function of two arguments $\mathbf{x} \in \mathcal{V}$, $\mathbf{y} \in \mathcal{V}$ denoted by $\mathbf{x} \cdot \mathbf{y}$, and satisfying the following conditions (axioms of the scalar product):

1. Symmetry: $\mathbf{x} \cdot \mathbf{y} = \mathbf{y} \cdot \mathbf{x}$.
2. Distributivity: $(\mathbf{x}_1 + \mathbf{x}_2) \cdot \mathbf{y} = \mathbf{x}_1 \cdot \mathbf{y} + \mathbf{x}_2 \cdot \mathbf{y}$.
3. Homogeneity: $(\lambda \mathbf{x}) \cdot \mathbf{y} = \lambda(\mathbf{x} \cdot \mathbf{y})$ for any real λ.
4. Positive definiteness: $\mathbf{x} \cdot \mathbf{x} \geq 0$ for any $\mathbf{x}$, and $\mathbf{x} \cdot \mathbf{x} = 0$ if and only if $\mathbf{x} = 0$.

Example 1. Consider the linear space B_3 of all free vectors in three-dimensional space. The space B_3 becomes a Euclidean space if the scalar product is introduced as in analytic geometry (see Subsection M4.5.3):

$$\mathbf{x} \cdot \mathbf{y} = |\mathbf{x}|\,|\mathbf{y}| \cos \varphi,$$

where φ is the angle between the vectors $\mathbf{x}$ and $\mathbf{y}$.

Example 2. Consider the n-dimensional coordinate space $\mathbb{R}^n$ whose elements are ordered systems of n arbitrary real numbers, $\mathbf{x} = (x_1, \ldots, x_n)$. Endowing this space with the scalar product

$$\mathbf{x} \cdot \mathbf{y} = x_1 y_1 + \cdots + x_n y_n,$$

we obtain a Euclidean space.

THEOREM. *For any two elements $\mathbf{x}$ and $\mathbf{y}$ of a Euclidean space, the Cauchy–Schwarz inequality holds:*

$$(\mathbf{x} \cdot \mathbf{y})^2 \leq (\mathbf{x} \cdot \mathbf{x})(\mathbf{y} \cdot \mathbf{y}).$$

Here equality holds if and only if one of the vectors is 0 or one vector is a multiple of the other.

A linear space $\mathcal{V}$ is called a *normed space* if it is endowed with a *norm*, which is a real-valued function of $\mathbf{x} \in \mathcal{V}$, denoted by $\|\mathbf{x}\|$ and satisfying the following conditions:

1. Homogeneity: $\|\lambda \mathbf{x}\| = |\lambda|\,\|\mathbf{x}\|$ for any real λ.
2. Positive definiteness: $\|\mathbf{x}\| \geq 0$ and $\|\mathbf{x}\| = 0$ if and only if $\mathbf{x} = 0$.
3. The *triangle inequality* (also called the *Minkowski inequality*) holds for all elements $\mathbf{x}$ and $\mathbf{y}$:

$$\|\mathbf{x} + \mathbf{y}\| \leq \|\mathbf{x}\| + \|\mathbf{y}\|. \tag{M5.5.4.1}$$

The value $\|\mathbf{x}\|$ is called the *norm of an element* $\mathbf{x}$ or its *length*.

THEOREM. *Any Euclidean space becomes a normed space if the norm is introduced by*

$$\|\mathbf{x}\| = \sqrt{\mathbf{x} \cdot \mathbf{x}}. \tag{M5.5.4.2}$$

COROLLARY. *In any Euclidean space with the norm (M5.5.4.2), the triangle inequality (M5.5.4.1) holds for all its elements $\mathbf{x}$ and $\mathbf{y}$.*

The *distance between elements* $\mathbf{x}$ and $\mathbf{y}$ of a Euclidean space is defined by

$$d(\mathbf{x}, \mathbf{y}) = \|\mathbf{x} - \mathbf{y}\|.$$

One says that φ is the *angle* between two elements $\mathbf{x}$ and $\mathbf{y}$ of a Euclidean space if

$$\cos \varphi = \frac{\mathbf{x} \cdot \mathbf{y}}{\|\mathbf{x}\|\,\|\mathbf{y}\|}.$$

Two elements $\mathbf{x}$ and $\mathbf{y}$ of a Euclidean space are said to be *orthogonal* if their scalar product is equal to zero, $\mathbf{x} \cdot \mathbf{y} = 0$.

PYTHAGOREAN THEOREM. *Let $\mathbf{x}_1, \ldots \mathbf{x}_m$ be mutually orthogonal elements of a Euclidean space, i.e., $\mathbf{x}_i \cdot \mathbf{x}_j = 0$ for $i \neq j$. Then*

$$\|\mathbf{x}_1 + \cdots + \mathbf{x}_m\|^2 = \|\mathbf{x}_1\|^2 + \cdots + \|\mathbf{x}_m\|^2.$$

Example 3. In the Euclidean space B_3 of free vectors with the usual scalar product (see Example 1), the following relations hold:

$$\|\mathbf{a}\| = |\mathbf{a}|, \quad (\mathbf{a} \cdot \mathbf{b})^2 \le |\mathbf{a}|^2 |\mathbf{b}|^2, \quad |\mathbf{a} + \mathbf{b}| \le |\mathbf{a}| + |\mathbf{b}|.$$

In the Euclidean space $\mathbb{R}^n$ of ordered systems of n numbers with the scalar product defined in Example 2, the following relations hold:

$$\|\mathbf{x}\| = \sqrt{x_1^2 + \cdots + x_n^2},$$

$$(x_1 y_1 + \cdots + x_n y_n)^2 \le (x_1^2 + \cdots + x_n^2)(y_1^2 + \cdots + y_n^2),$$

$$\sqrt{(x_1 + y_1)^2 + \cdots + (x_n + y_n)^2} \le \sqrt{x_1^2 + \cdots + x_n^2} + \sqrt{y_1^2 + \cdots + y_n^2}.$$

▶ **Orthonormal basis in a finite-dimensional Euclidean space.** For elements $\mathbf{x}_1, \dots, \mathbf{x}_m$ of a Euclidean space, the mth-order determinant $\det[\mathbf{x}_i \cdot \mathbf{x}_j]$ is called their *Gram determinant*. These elements are linearly independent if and only if their Gram determinant is different from zero.

One says that n elements $\mathbf{i}_1, \dots, \mathbf{i}_n$ of an n-dimensional Euclidean space $\mathcal{V}$ form its *orthonormal basis* if these elements have unit norm and are mutually orthogonal, i.e.,

$$\mathbf{i}_i \cdot \mathbf{i}_j = \begin{cases} 1 & \text{for } i = j, \\ 0 & \text{for } i \ne j. \end{cases}$$

THEOREM. *In any n-dimensional Euclidean space $\mathcal{V}$, there exists an orthonormal basis.*

Orthogonalization of linearly independent elements:

Let $\mathbf{e}_1, \dots, \mathbf{e}_n$ be n linearly independent vectors of an n-dimensional Euclidean space $\mathcal{V}$. From these vectors, one can construct an orthonormal basis of $\mathcal{V}$ using the following algorithm (called *Gram–Schmidt orthogonalization*):

$$\mathbf{i}_i = \frac{\mathbf{g}_i}{\sqrt{\mathbf{g}_i \cdot \mathbf{g}_i}}, \quad \text{where} \quad \mathbf{g}_i = \mathbf{e}_i - \sum_{j=1}^{i} (\mathbf{e}_i \cdot \mathbf{i}_j) \mathbf{i}_j \quad (i = 1, 2, \dots, n). \tag{M5.5.4.3}$$

Remark. In any n-dimensional ($n > 1$) Euclidean space $\mathcal{V}$, there exist infinitely many orthonormal bases.

Properties of an orthonormal basis of a Euclidean space:

1. Let $\mathbf{i}_1, \dots, \mathbf{i}_n$ be an orthonormal basis of a Euclidean space $\mathcal{V}$. Then the scalar product of two elements $\mathbf{x} = x_1 \mathbf{i}_1 + \cdots + x_n \mathbf{i}_n$ and $\mathbf{y} = y_1 \mathbf{i}_1 + \cdots + y_n \mathbf{i}_n$ is equal to the sum of products of their respective coordinates:

$$\mathbf{x} \cdot \mathbf{y} = x_1 y_1 + \cdots + x_n y_n.$$

2. The coordinates of any vector $\mathbf{x}$ in an orthonormal basis $\mathbf{i}_1, \dots, \mathbf{i}_n$ are equal to the scalar product of $\mathbf{x}$ and the corresponding vector of the basis (or the projection of the element $\mathbf{x}$ on the axis in the direction of the corresponding vector of the basis):

$$x_k = \mathbf{x} \cdot \mathbf{i}_k \quad (k = 1, 2, \dots, n).$$

Remark. In an arbitrary basis $\mathbf{e}_1, \dots, \mathbf{e}_n$ of a Euclidean space, the scalar product of two elements $\mathbf{x} = x_1 \mathbf{e}_1 + \cdots + x_n \mathbf{e}_n$ and $\mathbf{y} = y_1 \mathbf{e}_1 + \cdots + y_n \mathbf{e}_n$ has the form

$$\mathbf{x} \cdot \mathbf{y} = \sum_{i=1}^{n} \sum_{j=1}^{n} a_{ij} x_i y_j,$$

where $a_{ij} = \mathbf{e}_i \cdot \mathbf{e}_j$ $(i, j = 1, 2, \dots, n)$.

Two Euclidean spaces $\mathcal{V}$ and $\widetilde{\mathcal{V}}$ are said to be *isomorphic* if one can establish a one-to-one correspondence between the elements of these spaces satisfying the following conditions: if elements $\mathbf{x}$ and $\mathbf{y}$ of $\mathcal{V}$ correspond to elements $\widetilde{\mathbf{x}}$ and $\widetilde{\mathbf{y}}$ of $\widetilde{\mathcal{V}}$, then the element $\mathbf{x} + \mathbf{y}$ corresponds to $\widetilde{\mathbf{x}} + \widetilde{\mathbf{y}}$; the element $\lambda \mathbf{x}$ corresponds to $\lambda \widetilde{\mathbf{x}}$ for any λ; the scalar product $(\mathbf{x} \cdot \mathbf{y})_\mathcal{V}$ is equal to the scalar product $(\widetilde{\mathbf{x}} \cdot \widetilde{\mathbf{y}})_{\widetilde{\mathcal{V}}}$.

THEOREM. *Any two n-dimensional Euclidean spaces $\mathcal{V}$ and $\widetilde{\mathcal{V}}$ are isomorphic.*

Bibliography for Chapter M5

Anton, H., *Elementary Linear Algebra, 8th Edition*, Wiley, New York, 2000.

Beecher, J. A., Penna, J. A., and Bittinger, M. L., *College Algebra, 2nd Edition*, Addison Wesley, Boston, Massachusetts, 2004.

Bellman, R. E., *Introduction to Matrix Analysis* (McGraw-Hill Series in Matrix Theory), McGraw-Hill, New York, 1960.

Bernstein, D. S., *Matrix Mathematics: Theory, Facts, and Formulas with Application to Linear Systems Theory*, Princeton University Press, Princeton, 2005.

Blitzer, R. F., *College Algebra, 3rd Edition*, Prentice Hall, Englewood Cliffs, New Jersey, 2003.

Cullen, C. G., *Matrices and Linear Transformations, 2nd Edition*, Dover Publications, New York, 1990.

Dugopolski, M., *College Algebra, 3rd Edition*, Addison Wesley, Boston, 2002.

Eves, H., *Elementary Matrix Theory*, Dover Publications, New York, 1980.

Franklin, J. N., *Matrix Theory*, Dover Publications, New York, 2000.

Gantmacher, F. R., *Matrix Theory, Vol. 1, 2nd Edition*, American Mathematical Society, Providence, Rhode Island, 1990.

Gantmacher, F. R., *Applications of the Theory of Matrices*, Dover Publications, New York, 2005.

Gelfand, I. M., *Lectures on Linear Algebra*, Dover Publications, New York, 1989.

Gelfand, I. M. and Shen, A., *Algebra*, Birkhauser, Boston, Massachusetts, 2003.

Gilbert, J. and Gilbert, L., *Linear Algebra and Matrix Theory, 2nd Edition*, Brooks Cole, Stamford, Connecticut, 2004.

Hazewinkel, M. (Editor), *Handbook of Algebra, Vol. 1*, North Holland, Amsterdam, 1996.

Hazewinkel, M. (Editor), *Handbook of Algebra, Vol. 2*, North Holland, Amsterdam, 2000.

Hazewinkel, M. (Editor), *Handbook of Algebra, Vol. 3*, North Holland, Amsterdam, 2003.

Horn, R. A. and Johnson, C. R., *Matrix Analysis*, Cambridge University Press, Cambridge, England, 1990.

Korn, G. A. and Korn, T. M., *Mathematical Handbook for Scientists and Engineers, 2nd Edition*, Dover Publications, New York, 2000.

Kostrikin, A. I. and Manin, Yu. I., *Linear Algebra and Geometry*, Gordon & Breach, New York, 1997.

Kostrikin, A. I. and Shafarevich, I. R., *Algebra I: Basic Notions of Algebra*, Springer-Verlag, Berlin, 1990.

Kurosh, A. G., *Lectures on General Algebra*, Chelsea Publishing, New York, 1965.

Kurosh, A. G., *Algebraic Equations of Arbitrary Degrees*, Firebird Publishing, New York, 1977.

Lancaster, P. and Tismenetsky, M., *The Theory of Matrices, Second Edition: With Applications* (Computer Science and Scientific Computing), Academic Press, Boston, Massachusetts, 1985.

Lay, D. C., *Linear Algebra and Its Applications, 3rd Edition*, Addison Wesley, Boston, Massachusetts, 2002.

Lial, M. L., Hornsby, J., and Schneider, D. I., *College Algebra, 9th Edition*, Addison Wesley, Boston, Massachusetts, 2004.

Lipschutz, S. and Lipson, M., *Schaum's Outline of Linear Algebra, 3rd Edition*, McGraw-Hill, New York, 2000.

MacDuffee, C. C., *The Theory of Matrices*, Dover Publications, New York, 2004.

Meyer, C. D., *Matrix Analysis and Applied Linear Algebra, Package Edition*, Society for Industrial & Applied Mathematics, University City Science Center, Philadelphia, 2001.

Mikhalev, A. V. and Pilz, G., *The Concise Handbook of Algebra*, Kluwer Academic, Dordrecht, Boston, 2002.

Perlis, S., *Theory of Matrices*, Dover Publications, New York, 1991.

Schneider, H. and Barker, G. Ph., *Matrices and Linear Algebra, 2nd Edition*, Dover Publications, New York, 1989.

Shilov, G. E., *Linear Algebra*, Dover Publications, New York, 1977.

Strang, G., *Introduction to Linear Algebra, 3rd Edition*, Wellesley Cambridge Pr., Wellesley, 2003.

Strang, G., *Linear Algebra and Its Applications, 4th Edition*, Brooks Cole, Stamford, Connecticut, 2005.

Sullivan, M., *College Algebra, 7th Edition*, Prentice Hall, Englewood Cliffs, New Jersey, 2004.

Tobey, J. and Slater, J., *Intermediate Algebra, 5th Edition*, Prentice Hall, Englewood Cliffs, New Jersey, 2005.

Turnbull, H. W. and Aitken, A. C., *An Introduction to the Theory of Canonical Matrices, Phoenix Edition*, Dover Publications, New York, 2004.

Zhang, F., *Matrix Theory*, Springer, New York, 1999.

Chapter M6

Limits and Derivatives

M6.1. Basic Concepts of Mathematical Analysis

M6.1.1. Number Sets. Functions of Real Variable

▶ **Real axis, intervals, and segments.** The *real axis* is a straight line with a point O chosen as the origin, a positive direction, and a scale unit.

There is a one-to-one correspondence between the set of all real numbers $\mathbb{R}$ and the set of all points of the real axis, with each real x being represented by a point on the real axis separated from O by the distance $|x|$ and lying to the right of O for $x > 0$, or to the left of O for $x < 0$.

One often has to deal with the following number sets (sets of real numbers or sets on the real axis).

1. Sets of the form (a, b), $(-\infty, b)$, $(a, +\infty)$, and $(-\infty, +\infty)$ consisting, respectively, of all $x \in \mathbb{R}$ such that $a < x < b$, $x < b$, $x > a$, and x is arbitrary are called *open intervals* (sometimes simply *intervals*).

2. Sets of the form $[a, b]$ consisting of all $x \in \mathbb{R}$ such that $a \le x \le b$ are called *closed intervals* or *segments*.

3. Sets of the form $(a, b]$, $[a, b)$, $(-\infty, b]$, $[a, +\infty)$ consisting of all x such that $a < x \le b$, $a \le x < b$, $x \le b$, $x \ge a$ are called *half-open intervals*.

A *neighborhood of a point* $x_0 \in \mathbb{R}$ is defined as any open interval (a, b) containing x_0 $(a < x_0 < b)$. A neighborhood of the "point" $+\infty$, $-\infty$, or ∞ is defined, respectively, as any set of the form $(b, +\infty)$, $(-\infty, c)$ or $(-\infty, -a) \cup (a, +\infty)$ (here, $a \ge 0$).

▶ **Lower and upper bound of a set on a straight line.** The *upper bound* of a set of real numbers is the least number that bounds the set from above. The *lower bound* of a set of real numbers is the largest number that bounds the set from below.

In more details: let a set of real numbers $X \in \mathbb{R}$ be given. A number β is called its upper bound and denoted $\sup X$ if for any $x \in X$ the inequality $x \le \beta$ holds and for any $\beta_1 < \beta$ there exists an $x_1 \in X$ such that $x_1 > \beta_1$. A number α is called the lower bound of X and denoted $\inf X$ if for any $x \in X$ the inequality $x \ge \alpha$ holds and for any $\alpha_1 > \alpha$ there exists an $x_1 \in X$ such that $x_1 < \alpha_1$.

Example 1. For a set X consisting of two numbers a and b $(a < b)$, we have

$$\inf X = a, \quad \sup X = b.$$

Example 2. For intervals (open, closed, and half-open), we have

$$\inf(a, b) = \inf[a, b] = \inf(a, b] = \inf[a, b) = a,$$
$$\sup(a, b) = \sup[a, b] = \sup(a, b] = \sup[a, b) = b.$$

One can see that the upper and lower bounds may belong to a given set (e.g., for closed intervals) and may not (e.g., for open intervals).

The symbol $+\infty$ (resp., $-\infty$) is called the upper (resp., lower) bound of a set unbounded from above (resp., from below).

▶ **Real-valued functions of real variable. Methods of defining a function.**

$1°$. Let D and E be two sets of real numbers. Suppose that there is a relation between the points of D and E such that to each $x \in D$ there corresponds some $y \in E$, denoted by $y = f(x)$. In this case, one speaks of a *function* f defined on the set D and taking its values in the set E. The set D is called the *domain of the function* f, and the subset of E consisting of all elements $f(x)$ is called the *range of the function* f. This functional relation is often denoted by $y = f(x)$, $f : D \to E$, $f : x \mapsto y$.

The following terms are also used: x is the *independent variable* or the *argument*; y is the *dependent variable*.

$2°$. The most common and convenient way to define a function is the *analytic method*: the function is defined explicitly by means of a formula (or several formulas) depending on the argument x; for instance, $y = 2 \sin x + 1$.

Implicit definition of a function consists of using an equation of the form $F(x, y) = 0$, from which one calculates the value y for any fixed value of the argument x.

Parametric definition of a function consists of defining the values of the independent variable x and the dependent variable y by a pair of formulas depending on an auxiliary variable t (parameter): $x = p(t)$, $y = q(t)$.

Quite often functions are defined in terms of convergent series or by means of tables or graphs. There are some other methods of defining functions.

$3°$. The *graph of a function* is the representation of a function $y = f(x)$ as a line on the plane with orthogonal coordinates x, y, the points of the line having the coordinates $x, y = f(x)$, where x is an arbitrary point from the domain of the function.

▶ **Single-valued, periodic, odd and even functions.**

$1°$. A function is *single-valued* if each value of its argument corresponds to a unique value of the function. A function is *multi-valued* if there is at least one value of its argument corresponding to two or more values of the function. In what follows, we consider only single-valued functions, unless indicated otherwise.

$2°$. A function $f(x)$ is called *periodic* with period T (or T-*periodic*) if $f(x + T) = f(x)$ for any x.

$3°$. A function $f(x)$ is called *even* if it satisfies the condition $f(x) = f(-x)$ for any x. A function $f(x)$ is called *odd* if it satisfies the condition $f(x) = -f(-x)$ for any x.

▶ **Decreasing, increasing, monotone, and bounded functions.**

$1°$. A function $f(x)$ is called *increasing or strictly increasing* (resp., *nondecreasing*) on a set $D \subset \mathbb{R}$ if for any $x_1, x_2 \in D$ such that $x_1 > x_2$, we have $f(x_1) > f(x_2)$ (resp., $f(x_1) \geq f(x_2)$). A function $f(x)$ is called *decreasing or strictly decreasing* (resp., *nonincreasing*) on a set D if for all $x_1, x_2 \in D$ such that $x_1 > x_2$, we have $f(x_1) < f(x_2)$ (resp., $f(x_1) \leq f(x_2)$). All such functions are called *monotone functions*. Strictly increasing or decreasing functions are called *strictly monotone*.

$2°$. A function $f(x)$ is called *bounded* on a set D if $|f(x)| < M$ for all $x \in D$, where M is a finite constant. A function $f(x)$ is called *bounded from above* (*bounded from below*) on a set D if $f(x) < M$ ($M < f(x)$) for all $x \in D$, where M is a real constant.

▶ **Composite and inverse functions.**

$1°$. Consider a function $u = u(x)$, $x \in D$, with values $u \in E$, and let $y = f(u)$ be a function defined on E. Then the function $y = f\big(u(x)\big)$, $x \in D$, is called a *composite function* or the *superposition* of the functions f and u.

$2°$. Consider a function $y = f(x)$ that maps $x \in D$ into $y \in E$. The *inverse function* of $y = f(x)$ is a function $x = g(y)$ defined on E and such that $x = g(f(x))$ for all $x \in D$. The inverse function is often denoted by $g = f^{-1}$.

For strictly monotone functions $f(x)$, the inverse function always exists. In order to construct the inverse function $g(y)$, one should use the relation $y = f(x)$ to express x through y. The function $g(y)$ is monotonically increasing or decreasing together with $f(x)$.

M6.1.2. Limit of a Sequence

▶ **Some definitions.** Suppose that there is a correspondence between each positive integer n and some (real or complex) number denoted, for instance, by x_n. In this case, one says that a *numerical sequence* (or, simply, a *sequence*) $x_1, x_2, \ldots, x_n, \ldots$ is defined. Such a sequence is often denoted by $\{x_n\}$; x_n is called the *generic term* of the sequence.

Example 1. For the sequence $\{n^2 - 2\}$, we have $x_1 = -1$, $x_2 = 2$, $x_3 = 7$, $x_4 = 14$, etc.

A sequence is called *bounded* (bounded from above, bounded from below) if there is a constant M such that $|x_n| < M$ (respectively, $x_n < M$, $x_n > M$) for all $n = 1, 2, \ldots$

▶ **Limit of a sequence.** A number b is called the *limit of a sequence* $x_1, x_2, \ldots, x_n, \ldots$ if for any $\varepsilon > 0$ there is $N = N(\varepsilon)$ such that $|x_n - b| < \varepsilon$ for all $n > N$.

If b is the limit of the sequence $\{x_n\}$, one writes $\lim\limits_{n \to \infty} x_n = b$ or $x_n \to b$ as $n \to \infty$.

The limit of a constant sequence $\{x_n = c\}$ exists and is equal to c, i.e., $\lim\limits_{n \to \infty} c = c$. In this case, the inequality $|x_n - c| < \varepsilon$ takes the form $0 < \varepsilon$ and holds for all n.

Example 2. Let us show that $\lim\limits_{n \to \infty} \dfrac{n}{n+1} = 1$.

Consider the difference $\left| \dfrac{n}{n+1} - 1 \right| = \dfrac{1}{n+1}$. The inequality $\dfrac{1}{n+1} < \varepsilon$ holds for all $n > \dfrac{1}{\varepsilon} - 1 = N(\varepsilon)$. Therefore, for any positive ε there exists an $N = \dfrac{1}{\varepsilon} - 1$ such that for $n > N$ we have $\left| \dfrac{n}{n+1} - 1 \right| < \varepsilon$.

It may happen that a sequence $\{x_n\}$ has no limit at all. For example, this is the case for the sequence $\{x_n\} = \{(-1)^n\}$. A sequence that has a finite limit is called *convergent*.

THEOREM (BOLZANO–CAUCHY). *A sequence x_n has a finite limit if and only if for any $\varepsilon > 0$, there is N such that the inequality*

$$|x_n - x_m| < \varepsilon$$

holds for all $n > N$ and $m > N$.

▶ **Properties of convergent sequences.**

1. Any convergent sequence can have only one limit.

2. Any convergent sequence is bounded. From any bounded sequence one can extract a convergent subsequence.*

3. If a sequence converges to b, then any of its subsequences also converges to b.

4. If $\{x_n\}$, $\{y_n\}$ are two convergent sequences, then the sequences $\{x_n \pm y_n\}$, $\{x_n \cdot y_n\}$, and $\{x_n/y_n\}$ (in this ratio, it is assumed that $y_n \neq 0$ and $\lim\limits_{n \to \infty} y_n \neq 0$) are also convergent

---

* Let $\{x_n\}$ be a given sequence and let $\{n_k\}$ be a strictly increasing sequence with k and n_k being natural numbers. The sequence $\{x_{n_k}\}$ is called a *subsequence* of the sequence $\{x_n\}$.

and

$$\lim_{n\to\infty} (x_n \pm y_n) = \lim_{n\to\infty} x_n \pm \lim_{n\to\infty} y_n;$$

$$\lim_{n\to\infty} (cx_n) = c \lim_{n\to\infty} x_n \qquad (c = \text{const});$$

$$\lim_{n\to\infty} (x_n \cdot y_n) = \lim_{n\to\infty} x_n \cdot \lim_{n\to\infty} y_n;$$

$$\lim_{n\to\infty} \frac{x_n}{y_n} = \frac{\lim_{n\to\infty} x_n}{\lim_{n\to\infty} y_n}.$$

5. If $\{x_n\}$, $\{y_n\}$ are convergent sequences and the inequality $x_n \leq y_n$ holds for all n, then $\lim_{n\to\infty} x_n \leq \lim_{n\to\infty} y_n$.

6. If the inequalities $x_n \leq y_n \leq z_n$ hold for all n and $\lim_{n\to\infty} x_n = \lim_{n\to\infty} z_n = b$, then $\lim_{n\to\infty} y_n = b$.

▶ **Increasing, decreasing, and monotone sequences.** A sequence $\{x_n\}$ is called *increasing or strictly increasing* (resp., *nondecreasing*) if the inequality $x_{n+1} > x_n$ (resp., $x_{n+1} \geq x_n$) holds for all n. A sequence $\{x_n\}$ is called *decreasing or strictly decreasing* (resp., *nonincreasing*) if the inequality $x_{n+1} < x_n$ (resp., $x_{n+1} \leq x_n$) holds for all n. All such sequences are called *monotone sequences*. Strictly increasing or decreasing sequences are called *strictly monotone*.

THEOREM. *Any monotone bounded sequence has a finite limit.*

Example 3. It can be shown that the sequence $\left\{ \left(1 + \dfrac{1}{n}\right)^n \right\}$ is bounded and increasing. Therefore, it is convergent. Its limit is denoted by the letter e:

$$e = \lim_{n\to\infty} \left(1 + \frac{1}{n}\right)^n \qquad (e \approx 2.71828).$$

Logarithms with the base e are called *natural* or *Napierian*, and $\log_e x$ is denoted by $\ln x$.

▶ **Properties of positive sequences.**

$1°$. If a sequence x_n ($x_n > 0$) has a limit (finite or infinite), then the sequence

$$y_n = \sqrt[n]{x_1 \cdot x_2 \dots x_n}$$

has the same limit.

$2°$. From property $1°$ for the sequence

$$x_1, \quad \frac{x_2}{x_1}, \quad \frac{x_3}{x_2}, \quad \dots, \quad \frac{x_n}{x_{n-1}}, \quad \frac{x_{n+1}}{x_n}, \quad \dots,$$

we obtain a useful corollary

$$\lim_{n\to\infty} \sqrt[n]{x_n} = \lim_{n\to\infty} \frac{x_{n+1}}{x_n},$$

under the assumption that the second limit exists.

Example 4. Let us show that $\lim\limits_{n\to\infty} \dfrac{n}{\sqrt[n]{n!}} = e$.

Taking $x_n = \dfrac{n^n}{n!}$ and using property $2°$, we get

$$\lim_{n\to\infty} \frac{n}{\sqrt[n]{n!}} = \lim_{n\to\infty} \frac{x_{n+1}}{x_n} = \lim_{n\to\infty} \left(1 + \frac{1}{n}\right)^n = e.$$

▶ **Infinitely small and infinitely large sequences.** A sequence x_n converging to zero is called *infinitely small* or *infinitesimal*.

A sequence x_n whose terms infinitely grow in absolute values with the growth of n is called *infinitely large* or *"tending to infinity."* In this case, the following notation is used: $\lim\limits_{n\to\infty} x_n = \infty$. If, in addition, all terms of the sequence starting from some number are positive (negative), then one says that the sequence x_n converges to "plus (minus) infinity," and one writes $\lim\limits_{n\to\infty} x_n = +\infty$ $\left(\lim\limits_{n\to\infty} x_n = -\infty\right)$. For instance, $\lim\limits_{n\to\infty} (-1)^n n^2 = \infty$, $\lim\limits_{n\to\infty} \sqrt{n} = +\infty$, $\lim\limits_{n\to\infty} (-n) = -\infty$.

Theorem (Stolz). *Let x_n and y_n be two infinitely large sequences, $y_n \to +\infty$, and y_n increases with the growth of n (at least for sufficiently large n): $y_{n+1} > y_n$. Then*

$$\lim_{n\to\infty} \frac{x_n}{y_n} = \lim_{n\to\infty} \frac{x_n - x_{n-1}}{y_n - y_{n-1}},$$

provided that the right limit exists (finite or infinite).

Example 5. Let us find the limit of the sequence

$$z_n = \frac{1^k + 2^k + \cdots + n^k}{n^{k+1}}.$$

Taking $x_n = 1^k + 2^k + \cdots + n^k$ and $y_n = n^{k+1}$ in the Stolz theorem, we get

$$\lim_{n\to\infty} z_n = \lim_{n\to\infty} \frac{n^k}{n^{k+1} - (n-1)^{k+1}}.$$

Since $(n-1)^{k+1} = n^{k+1} - (k+1)n^k + \cdots$, we have $n^{k+1} - (n-1)^{k+1} = (k+1)n^k + \cdots$, and therefore

$$\lim_{n\to\infty} z_n = \lim_{n\to\infty} \frac{n^k}{(k+1)n^k + \cdots} = \frac{1}{k+1}.$$

▶ **Upper and lower limits of a sequence.** The limit (finite or infinite) of a subsequence of a given sequence x_n is called a *partial limit* of x_n. In the set of all partial limits of any sequence of real numbers, there always exists the largest and the least (finite or infinite). The largest (resp., least) partial limit of a sequence is called its *upper* (resp., *lower*) *limit*. The upper and lower limits of a sequence x_n are denoted, respectively,

$$\overline{\lim_{n\to\infty}}\, x_n, \qquad \underline{\lim_{n\to\infty}}\, x_n.$$

Example 6. The upper and lower limits of the sequence $x_n = (-1)^n$ are, respectively,

$$\overline{\lim_{n\to\infty}}\, x_n = 1, \qquad \underline{\lim_{n\to\infty}}\, x_n = -1.$$

A sequence x_n has a limit (finite or infinite) if and only if its upper limit coincides with its lower limit:

$$\lim_{n\to\infty} x_n = \overline{\lim_{n\to\infty}}\, x_n = \underline{\lim_{n\to\infty}}\, x_n.$$

M6.1.3. Limit of a Function. Asymptotes

▶ **Definition of the limit of a function. One-sided limits.**

$1°$. One says that b is the *limit of a function* $f(x)$ as x tends to a if for any $\varepsilon > 0$ there is $\delta = \delta(\varepsilon) > 0$ such that $|f(x) - b| < \varepsilon$ for all x such that $0 < |x - a| < \delta$.

Notation: $\lim\limits_{x \to a} f(x) = b$ or $f(x) \to b$ as $x \to a$.

One says that b is the limit of a function $f(x)$ as x tends to $+\infty$ if for any $\varepsilon > 0$ there is $N = N(\varepsilon) > 0$ such that $|f(x) - b| < \varepsilon$ for all $x > N$.

Notation: $\lim\limits_{x \to +\infty} f(x) = b$ or $f(x) \to b$ as $x \to +\infty$.

In a similar way, one defines the limits for $x \to -\infty$ or $x \to \infty$.

THEOREM (BOLZANO–CAUCHY 1). *A function $f(x)$ has a finite limit as x tends to a (a is assumed finite) if and only if for any $\varepsilon > 0$ there is $\delta > 0$ such that the inequality*

$$|f(x_1) - f(x_2)| < \varepsilon \tag{M6.1.3.1}$$

holds for all x_1, x_2 such that $|x_1 - a| < \delta$ and $|x_2 - a| < \delta$.

THEOREM (BOLZANO–CAUCHY 2). *A function $f(x)$ has a finite limit as x tends to $+\infty$ if and only if for any $\varepsilon > 0$ there is $\Delta > 0$ such that the inequality (M6.1.3.1) holds for all $x_1 > \Delta$ and $x_2 > \Delta$.*

$2°$. One says that b is the *left-hand limit* (resp., *right-hand limit*) of a function $f(x)$ as x tends to a if for any $\varepsilon > 0$ there is $\delta = \delta(\varepsilon) > 0$ such that $|f(x) - b| < \varepsilon$ for $a - \delta < x < a$ (resp., for $a < x < a + \delta$).

Notation: $\lim\limits_{x \to a-0} f(x) = b$ or $f(a - 0) = b$ (resp., $\lim\limits_{x \to a+0} f(x) = b$ or $f(a + 0) = b$).

▶ **Properties of limits.** Let a be a number or any of the symbols ∞, $+\infty$, $-\infty$.

1. If a function has a limit at some point, this limit is unique.

2. If c is a constant function of x, then $\lim\limits_{x \to a} c = c$.

3. If there exist $\lim\limits_{x \to a} f(x)$ and $\lim\limits_{x \to a} g(x)$, then

$$\lim_{x \to a} \left[f(x) \pm g(x) \right] = \lim_{x \to a} f(x) \pm \lim_{x \to a} g(x);$$

$$\lim_{x \to a} c f(x) = c \lim_{x \to a} f(x) \quad (c = \text{const});$$

$$\lim_{x \to a} f(x) \cdot g(x) = \lim_{x \to a} f(x) \cdot \lim_{x \to a} g(x);$$

$$\lim_{x \to a} \frac{f(x)}{g(x)} = \frac{\lim\limits_{x \to a} f(x)}{\lim\limits_{x \to a} g(x)} \quad \left(\text{if } g(x) \neq 0, \ \lim_{x \to a} g(x) \neq 0 \right).$$

4. Let $f(x) \leq g(x)$ in a neighborhood of a point a ($x \neq a$). Then $\lim\limits_{x \to a} f(x) \leq \lim\limits_{x \to a} g(x)$, provided that these limits exist.

5. If $f(x) \leq g(x) \leq h(x)$ in a neighborhood of a point a and $\lim\limits_{x \to a} f(x) = \lim\limits_{x \to a} h(x) = b$, then $\lim\limits_{x \to a} g(x) = b$.

These properties hold also for one-sided limits.

▶ **Limits of some functions.**

$$\textit{First noteworthy limit:} \qquad \lim_{x \to 0} \frac{\sin x}{x} = 1.$$

$$\textit{Second noteworthy limit:} \qquad \lim_{x \to \infty} \left(1 + \frac{1}{x} \right)^x = e.$$

Some other frequently used limits:

$$\lim_{x\to 0}\frac{(1+x)^n-1}{x}=n,\qquad \lim_{x\to\infty}\frac{a_n x^n+a_{n-1}x^{n-1}+\cdots+a_1 x+a_0}{b_n x^n+b_{n-1}x^{n-1}+\cdots+b_1 x+b_0}=\frac{a_n}{b_n},$$

$$\lim_{x\to 0}\frac{1-\cos x}{x^2}=\frac{1}{2},\quad \lim_{x\to 0}\frac{\tan x}{x}=1,\quad \lim_{x\to 0}\frac{\arcsin x}{x}=1,\quad \lim_{x\to 0}\frac{\arctan x}{x}=1,$$

$$\lim_{x\to 0}\frac{e^x-1}{x}=1,\quad \lim_{x\to 0}\frac{a^x-1}{x}=\ln a,\quad \lim_{x\to 0}\frac{\ln(1+x)}{x}=1,\quad \lim_{x\to 0}\frac{\log_a(1+x)}{x}=\log_a e,$$

$$\lim_{x\to 0}\frac{\sinh x}{x}=1,\quad \lim_{x\to 0}\frac{\tanh x}{x}=1,\quad \lim_{x\to 0}\frac{\operatorname{arcsinh} x}{x}=1,\quad \lim_{x\to 0}\frac{\operatorname{arctanh} x}{x}=1,$$

$$\lim_{x\to +0}x^a\ln x=0,\quad \lim_{x\to +\infty}x^{-a}\ln x=0,\quad \lim_{x\to +\infty}x^a e^{-x}=0,\quad \lim_{x\to +0}x^x=1,$$

where $a>0$ and $b_n\neq 0$.

▶ See Subsection M6.2.3, where L'Hospital rules for calculating limits with the help of derivatives are given.

▶ **Asymptotes of the graph of a function.** An *asymptote* of the graph of a function $y=f(x)$ is a straight line whose distance from a point (x,y) on the graph of $y=f(x)$ tends to zero if at least one of the coordinates (x,y) tends to infinity.

The line $x=a$ is a *vertical asymptote* of the graph of the function $y=f(x)$ if at least one of the one-sided limits of $f(x)$ as $x\to a\pm 0$ is equal to $+\infty$ or $-\infty$.

The line $y=kx+b$ is an *oblique asymptote* of the graph of $y=f(x)$ if at least one of the limit relations holds:

$$\lim_{x\to +\infty}[f(x)-kx-b]=0\quad\text{or}\quad \lim_{x\to -\infty}[f(x)-kx-b]=0.$$

If there exist finite limits

$$\lim_{x\to +\infty}\frac{f(x)}{x}=k,\qquad \lim_{x\to +\infty}[f(x)-kx]=b,\qquad\qquad\text{(M6.1.3.2)}$$

then the line $y=kx+b$ is an oblique asymptote of the graph for $x\to +\infty$ (in a similar way, one defines an asymptote for $x\to -\infty$).

Example. Let us find the asymptotes of the graph of the function $y=\dfrac{x^2}{x-1}$.

1°. The graph has a vertical asymptote $x=1$, since $\lim\limits_{x\to 1}\dfrac{x^2}{x-1}=\infty$.

2°. Moreover, for $x\to\pm\infty$, there is an oblique asymptote $y=kx+b$ whose coefficients are determined by the formulas (M6.1.3.2):

$$k=\lim_{x\to\pm\infty}\frac{x}{x-1}=1,\quad b=\lim_{x\to\pm\infty}\left(\frac{x^2}{x-1}-x\right)=\lim_{x\to\pm\infty}\frac{x}{x-1}=1.$$

Thus, the equation of the oblique asymptote has the form $y=x+1$. Fig. M6.1 shows the graph of the function under consideration and its asymptotes.

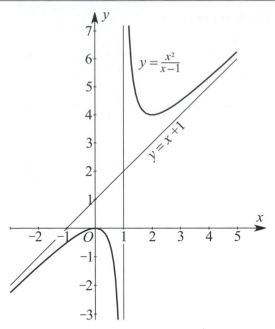

Figure M6.1. The graph of the function $y = \dfrac{x^2}{x-1}$ and its asymptotes.

M6.1.4. Infinitely Small and Infinitely Large Functions

▶ **Definitions.** A function $f(x)$ is called *infinitely small* for $x \to a$ if $\lim\limits_{x \to a} f(x) = 0$.

A function $f(x)$ is said to be *infinitely large* for $x \to a$ if for any $K > 0$ the inequality $|f(x)| > K$ holds for all $x \neq a$ in a small neighborhood of the point a. In this case, one writes $f(x) \to \infty$ as $x \to a$ or $\lim\limits_{x \to a} f(x) = \infty$. (In these definitions, a is a finite number or any of the symbols ∞, $+\infty$, $-\infty$.) If $f(x)$ is infinitely large for $x \to a$ and $f(x) > 0$ ($f(x) < 0$) in a neighborhood of a (for $x \neq a$), one writes $\lim\limits_{x \to a} f(x) = +\infty$ (resp., $\lim\limits_{x \to a} f(x) = -\infty$).

▶ **Properties of infinitely small and infinitely large functions.**

1. The sum and the product of finitely many infinitely small functions for $x \to a$ is an infinitely small function.

2. The product of an infinitely small function $f(x)$ for $x \to a$ and a function $g(x)$ which is bounded in a neighborhood U of the point a (i.e., $|g(x)| < M$ for all $x \in U$, where $M > 0$ is a constant) is an infinitely small function.

3. $\lim\limits_{x \to a} f(x) = b$ if and only if $f(x) = b + g(x)$, where $g(x)$ is infinitely small for $x \to a$.

4. A function $f(x)$ is infinitely large at some point if and only if the function $g(x) = 1/f(x)$ is infinitely small at the same point.

▶ **Comparison of infinitely small quantities. Symbols of the order: O and o.** Functions $f(x)$ and $g(x)$ that are infinitely small for $x \to a$ are called *equivalent* near a if $\lim\limits_{x \to a} \dfrac{f(x)}{g(x)} = 1$. In this case one writes $f(x) \sim g(x)$.

Examples of equivalent infinitely small functions:

$$(1 + \varepsilon)^n - 1 \sim n\varepsilon, \qquad a^\varepsilon - 1 \sim \varepsilon \ln a, \qquad \log_a(1 + \varepsilon) \sim \varepsilon \log_a e,$$

$$\sin \varepsilon \sim \varepsilon, \qquad \tan \varepsilon \sim \varepsilon, \qquad 1 - \cos \varepsilon \sim \tfrac{1}{2}\varepsilon^2, \qquad \arcsin \varepsilon \sim \varepsilon, \qquad \arctan \varepsilon \sim \varepsilon,$$

where $\varepsilon = \varepsilon(x)$ is infinitely small for $x \to a$.

Functions $f(x)$ and $g(x)$ are said to be of the *same order* for $x \to a$, and one writes $f(x) = O\big(g(x)\big)$ if $\lim\limits_{x \to a} \dfrac{f(x)}{g(x)} = K$, $0 < |K| < \infty$.*

A function $f(x)$ is of a *higher order of smallness* compared with $g(x)$ for $x \to a$ if $\lim\limits_{x \to a} \dfrac{f(x)}{g(x)} = 0$, and in this case, one writes $f(x) = o\big(g(x)\big)$.

M6.1.5. Continuous Functions. Discontinuities of the First and the Second Kind

▶ **Continuous functions.** A function $f(x)$ is called *continuous* at a point $x = a$ if it is defined at that point and its neighborhood and $\lim\limits_{x \to a} f(x) = f(a)$.

For continuous functions, a small variation of their argument $\Delta x = x - a$ results in a small variation of the function $\Delta y = f(x) - f(a)$, i.e., $\Delta y \to 0$ as $\Delta x \to 0$. (This property is often used as a definition of continuity.)

A function $f(x)$ is called *right-continuous* at a point $x = a$ if it is defined at that point (and to its right) and $\lim\limits_{x \to a+0} f(x) = f(a)$. A function $f(x)$ is called *left-continuous* at a point $x = a$ if it is defined at that point (and to its left) and $\lim\limits_{x \to a-0} f(x) = f(a)$.

▶ **Properties of continuous functions.**

1. Suppose that functions $f(x)$ and $g(x)$ are continuous at some point a. Then the functions $f(x) \pm g(x)$, $cf(x)$, $f(x)g(x)$, $\dfrac{f(x)}{g(x)}$ $(g(a) \neq 0)$ are also continuous at a.

2. Suppose that a function $f(x)$ is continuous on the segment $[a, b]$ and takes values of different signs at its endpoints, i.e., $f(a)f(b) < 0$. Then there is a point c between a and b at which $f(x)$ vanishes:
$$f(c) = 0 \quad (a < c < b).$$

3. If $f(x)$ is continuous at a point a and $f(a) > 0$ (resp., $f(a) < 0$), then there is $\delta > 0$ such that $f(x) > 0$ (resp., $f(x) < 0$) for all $x \in (a - \delta, a + \delta)$.

4. Any function $f(x)$ that is continuous at each point of a segment $[a, b]$ attains its largest and its smallest values, M and m, on that segment.

5. A function $f(x)$ that is continuous on a segment $[a, b]$ takes any value $c \in [m, M]$ on that segment, where m and M are, respectively, its smallest and its largest values on $[a, b]$.

6. If $f(x)$ is continuous and increasing (resp., decreasing) on a segment $[a, b]$, then on the segment $\big[f(a), f(b)\big]$ (resp., $\big[f(b), f(a)\big]$) the inverse function $x = g(y)$ exists, and is continuous and increasing (resp., decreasing).

7. If $u(x)$ is continuous at a point a and $f(u)$ is continuous at $b = u(a)$, then the composite function $f\big(u(x)\big)$ is continuous at a.

Remark. Any elementary function is continuous at each point of its domain.

▶ **Points of discontinuity of a function.** A point a is called a *point of discontinuity of the first kind* for a function $f(x)$ if there exist finite one-sided limits $f(a + 0)$ and $f(a - 0)$, but the relations $\lim\limits_{x \to a+0} f(x) = \lim\limits_{x \to a-0} f(x) = f(a)$ do not hold. The value $|f(a + 0) - f(a - 0)|$ is called the *jump* of the function at the point a. In particular, if $f(a + 0) = f(a - 0) \neq f(a)$, then a is called a *point of removable discontinuity*.

---

* There is another definition of the symbol O. Namely, $f(x) = O\big(g(x)\big)$ for $x \to a$ if the inequality $|f(x)| \leq K|g(x)|$, $K = \text{const}$, holds in some neighborhood of the point a (for $x \neq a$).

Examples of functions with discontinuities of the first kind.

1. The function $f(x) = \begin{cases} 0 & \text{for } x < 0 \\ 1 & \text{for } x \geq 0 \end{cases}$ has a jump equal to 1 at the discontinuity point $x = 0$.

2. The function $f(x) = \begin{cases} 0 & \text{for } x \neq 0 \\ 1 & \text{for } x = 0 \end{cases}$ has a removable discontinuity at the point $x = 0$.

A point a is called a *point of discontinuity of the second kind* if at least one of the one-sided limits $f(a + 0)$ or $f(a - 0)$ does not exist or is equal to infinity.

Examples of functions with discontinuities of the second kind.

1. The function $f(x) = \sin \dfrac{1}{x}$ has a second-kind discontinuity at the point $x = 0$ (since this function has no one-sided limits as $x \to \pm 0$).

2. The function $f(x) = 1/x$ has an infinite limit as $x \to 0$, so it has a second-kind discontinuity at the point $x = 0$.

M6.1.6. Convex and Concave Functions

▶ **Definition of convex and concave functions.**

$1°$. A function $f(x)$ defined and continuous on a segment $[a, b]$ is called *convex* (or *convex downward*) if for any x_1, x_2 in $[a, b]$, the *Jensen inequality* holds:

$$f\left(\frac{x_1 + x_2}{2}\right) \leq \frac{f(x_1) + f(x_2)}{2}. \tag{M6.1.6.1}$$

The geometrical meaning of convexity is that all points of the graph curve between two graph points lie below or on the rectilinear segment joining the two graph points (see Fig. M6.2 a).

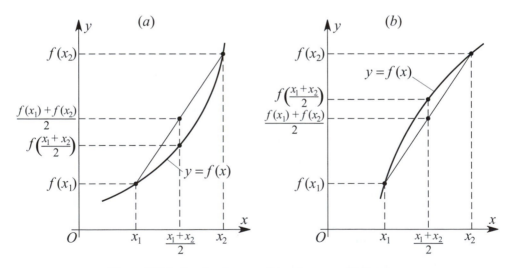

Figure M6.2. Graphs of convex (a) and concave (b) functions.

If for $x_1 \neq x_2$, condition (M6.1.6.1) holds with $<$ instead of $\leq$, then the function $f(x)$ is called *strictly convex*.

$2°$. A function $f(x)$ defined and continuous on a segment $[a, b]$ is called *concave* (or *convex upward*) if for any x_1, x_2 in $[a, b]$ the following inequality holds:

$$f\left(\frac{x_1 + x_2}{2}\right) \geq \frac{f(x_1) + f(x_2)}{2}. \tag{M6.1.6.2}$$

The geometrical meaning of concavity is that all points of the graph curve between two graph points lie above or on the rectilinear segment joining the two graph points (see Fig. M6.2 b).

If for $x_1 \neq x_2$, condition (M6.1.6.2) holds with $>$ instead of $\geq$, then the function $f(x)$ is called *strictly concave*.

▶ **Some properties of convex and concave functions.**

1. The product of a convex (concave) function and a positive constant is a convex (concave) function.

2. The sum of two or more convex (concave) functions is a convex (concave) function.

3. A non-constant convex (resp., concave) function $f(x)$ on a segment $[a, b]$ cannot attain its largest (resp., smallest) value inside the segment.

4. A function $f(x)$ that is continuous on a segment $[a, b]$ and twice differentiable on the interval (a, b) is convex downward (resp., convex upward) if and only if $f''(x) \geq 0$ (resp., $f''(x) \leq 0$) on that interval.

M6.1.7. Convergence of Functions

▶ **Pointwise, uniform, and nonuniform convergence of functions.** Let $\{f_n(x)\}$ be a sequence of functions defined on a set $X \subset \mathbb{R}$. The sequence $\{f_n(x)\}$ is said to be *pointwise convergent* to $f(x)$ as $n \to \infty$ if for any fixed $x \in X$, the numerical sequence $\{f_n(x)\}$ converges to $f(x)$. The sequence $\{f_n(x)\}$ is said to be *uniformly convergent* to a function $f(x)$ on X as $n \to \infty$ if for any $\varepsilon > 0$ there is an integer $N = N(\varepsilon)$ such that for all $n > N$ and all $x \in X$, the following inequality holds:

$$|f_n(x) - f(x)| < \varepsilon. \tag{M6.1.7.1}$$

Note that in this definition, N is independent of x. For a sequence $\{f_n(x)\}$ pointwise convergent to $f(x)$ as $n \to \infty$, by definition, for any $\varepsilon > 0$ and any $x \in X$, there is $N = N(\varepsilon, x)$ such that (M6.1.7.1) holds for all $n > N(\varepsilon, x)$.

If one cannot find such an N independent of x and depending only on ε (i.e., one cannot ensure (M6.1.7.1) uniformly; to be more precise, there is an $\varepsilon > 0$ such that for any $N > 0$ there is a $k_N > N$ and $x_N \in X$ such that $|f_{k_N}(x_N) - f(x_N)| \geq \varepsilon$), then one says that the sequence $\{f_n(x)\}$ *converges nonuniformly* to $f(x)$ on the set X.

▶ **Basic theorems.** Let X be an interval on the real axis.

THEOREM. *Let $f_n(x)$ be a sequence of continuous functions uniformly convergent to $f(x)$ on X. Then $f(x)$ is continuous on X.*

COROLLARY. *If the limit function $f(x)$ of a pointwise convergent sequence of continuous functions $\{f_n(x)\}$ is discontinuous, then the convergence of the sequence $\{f_n(x)\}$ is nonuniform.*

Example. The sequence $\{f_n(x)\} = \{x^n\}$ converges to $f(x) \equiv 0$ as $n \to \infty$ uniformly on each segment $[0, a]$, $0 < a < 1$. However, on the segment $[0, 1]$ this sequence converges nonuniformly to the discontinuous function $f(x) = \begin{cases} 0 & \text{for } 0 \leq x < 1, \\ 1 & \text{for } x = 1. \end{cases}$

CAUCHY CRITERION. *A sequence of functions $\{f_n(x)\}$ defined on a set $X \in \mathbb{R}$ uniformly converges to $f(x)$ as $n \to \infty$ if and only if for any $\varepsilon > 0$ there is an integer $N = N(\varepsilon) > 0$ such that for all $n > N$ and $m > N$, the inequality $|f_n(x) - f_m(x)| < \varepsilon$ holds for all $x \in X$.*

▶ **Geometrical meaning of uniform convergence.** Let $f_n(x)$ be continuous functions on the segment $[a, b]$ and suppose that $\{f_n(x)\}$ uniformly converges to a continuous function $f(x)$ as $n \to \infty$. Then all curves $y = f_n(x)$, for sufficiently large $n > N$, belong to the strip between the two curves $y = f(x) - \varepsilon$ and $y = f(x) + \varepsilon$ (see Fig. M6.3).

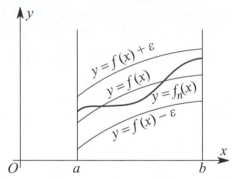

Figure M6.3. Geometrical meaning of uniform convergence of a sequence of functions $\{f_n(x)\}$ to a continuous function $f(x)$.

M6.2. Differential Calculus for Functions of a Single Variable

M6.2.1. Derivative and Differential: Their Geometrical and Physical Meaning

▶ **Definition of derivative and differential.** The *derivative* of a function $y = f(x)$ at a point x is the limit of the ratio

$$y' = \lim_{\Delta x \to 0} \frac{\Delta y}{\Delta x} = \lim_{\Delta x \to 0} \frac{f(x + \Delta x) - f(x)}{\Delta x},$$

where $\Delta y = f(x + \Delta x) - f(x)$ is the increment of the function corresponding to the increment of the argument Δx. The derivative y' is also denoted by y'_x, $\dot{y}$, $\dfrac{dy}{dx}$, $f'(x)$, $\dfrac{df(x)}{dx}$.

Example 1. Let us calculate the derivative of the function $f(x) = x^2$.
By definition, we have

$$f'(x) = \lim_{\Delta x \to 0} \frac{(x + \Delta x)^2 - x^2}{\Delta x} = \lim_{\Delta x \to 0} (2x + \Delta x) = 2x.$$

The increment Δx is also called the differential of the independent variable x and is denoted by dx.

A function $f(x)$ that has a derivative at a point x is called *differentiable* at that point. The differentiability of $f(x)$ at a point x is equivalent to the condition that the increment of the function, $\Delta y = f(x + dx) - f(x)$, at that point can be represented in the form $\Delta y = f'(x)\, dx + o(dx)$ (the second term is an infinitely small quantity compared with dx as $dx \to 0$; see Subsection M6.1.4).

A function differentiable at some point x is continuous at that point. The converse is not true, in general; continuity does not always imply differentiability.

A function $f(x)$ is called *differentiable* on a set D (interval, segment, etc.) if for any $x \in D$ there exists the derivative $f'(x)$. A function $f(x)$ is called *continuously differentiable* on D if it has the derivative $f'(x)$ at each point $x \in D$ and $f'(x)$ is a continuous function on D.

The *differential dy* of a function $y = f(x)$ is the principal linear part of its increment Δy at the point x, so that $dy = f'(x)dx$, $\Delta y = dy + o(dx)$.

The approximate relation $\Delta y \approx dy$ or $f(x + \Delta x) \approx f(x) + f'(x)\Delta x$ (for small Δx) is often used in numerical analysis.

▶ **Physical and geometrical meaning of the derivative. Tangent line.**

$1°$. Let $y = f(x)$ be the function describing the path y traversed by a body by the time x. Then the derivative $f'(x)$ is the velocity of the body at the instant x.

$2°$. The *tangent line* or simply the *tangent* to the graph of the function $y = f(x)$ at a point $M(x_0, y_0)$, where $y_0 = f(x_0)$, is defined as the straight line determined by the limit position of the secant MN as the point N tends to M along the graph. If α is the angle between the x-axis and the tangent line, then $f'(x_0) = \tan \alpha$ is the slope ratio of the tangential line (Fig. M6.4).

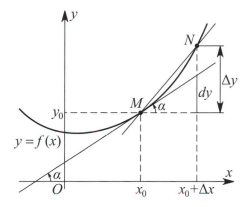

Figure M6.4. The tangent to the graph of a function $y = f(x)$ at a point (x_0, y_0).

Equation of the tangent line to the graph of a function $y = f(x)$ *at a point* (x_0, y_0):

$$y - y_0 = f'(x_0)(x - x_0).$$

Equation of the normal to the graph of a function $y = f(x)$ *at a point* (x_0, y_0):

$$y - y_0 = -\frac{1}{f'(x_0)}(x - x_0).$$

M6.2.2. Table of Derivatives and Differentiation Rules

The derivative of any elementary function can be calculated with the help of derivatives of basic elementary functions and differentiation rules.

▶ **Table of derivatives of basic elementary functions (a = const).**

$$(a)' = 0, \qquad\qquad (x^a)' = ax^{a-1},$$
$$(e^x)' = e^x, \qquad\qquad (a^x)' = a^x \ln a,$$
$$(\ln x)' = \frac{1}{x}, \qquad\qquad (\log_a x)' = \frac{1}{x \ln a},$$
$$(\sin x)' = \cos x, \qquad\qquad (\cos x)' = -\sin x,$$
$$(\tan x)' = \frac{1}{\cos^2 x}, \qquad\qquad (\cot x)' = -\frac{1}{\sin^2 x},$$
$$(\arcsin x)' = \frac{1}{\sqrt{1 - x^2}}, \qquad\qquad (\arccos x)' = -\frac{1}{\sqrt{1 - x^2}},$$
$$(\arctan x)' = \frac{1}{1 + x^2}, \qquad\qquad (\text{arccot}\, x)' = -\frac{1}{1 + x^2},$$
$$(\sinh x)' = \cosh x, \qquad\qquad (\cosh x)' = \sinh x,$$

$$(\tanh x)' = \frac{1}{\cosh^2 x}, \qquad (\coth x)' = -\frac{1}{\sinh^2 x},$$

$$(\operatorname{arcsinh} x)' = \frac{1}{\sqrt{1 + x^2}}, \qquad (\operatorname{arccosh} x)' = \frac{1}{\sqrt{x^2 - 1}},$$

$$(\operatorname{arctanh} x)' = \frac{1}{1 - x^2} \quad (x^2 < 1), \qquad (\operatorname{arccoth} x)' = \frac{1}{1 - x^2} \quad (x^2 > 1).$$

▶ **Differentiation rules.**

1. *Derivative of the sum (difference) of functions:*

$$[u(x) \pm v(x)]' = u'(x) \pm v'(x).$$

2. *Derivative of the product of a function and a constant:*

$$[au(x)]' = au'(x) \qquad (a = \text{const}).$$

3. *Derivative of the product of functions:*

$$[u(x)v(x)]' = u'(x)v(x) + u(x)v'(x).$$

4. *Derivative of the quotient of functions:*

$$\left[\frac{u(x)}{v(x)}\right]' = \frac{u'(x)v(x) - u(x)v'(x)}{v^2(x)}.$$

5. *Derivative of a composite function:*

$$\left[f(u(x))\right]' = f'_u(u)u'(x).$$

6. *Derivative of a parametrically defined function $x = x(t)$, $y = y(t)$:*

$$y'_x = \frac{y'_t}{x'_t}.$$

7. *Derivative of an implicit function defined by the equation $F(x, y) = 0$:*

$$y'_x = -\frac{F_x}{F_y} \qquad (F_x \text{ and } F_y \text{ are partial derivatives}).$$

8. *Derivative of the inverse function $x = x(y)$ (for details see footnote*):*

$$x'_y = \frac{1}{y'_x}.$$

9. *Derivative of a composite exponential function:*

$$[u(x)^{v(x)}]' = u'vu^{v-1} + v'u^v \ln u = u^v \left(u'\frac{v}{u} + v' \ln u\right).$$

---

* Let $y = f(x)$ be a differentiable monotone function on the interval (a, b) and $f'(x_0) \neq 0$, where $x_0 \in (a, b)$. Then the inverse function $x = g(y)$ is differentiable at the point $y_0 = f(x_0)$ and $g'(y_0) = \dfrac{1}{f'(x_0)}$.

10. *Derivative of a composite function of two arguments*:

$$\left[f\big(u(x), v(x)\big)\right]' = f_u(u, v)u' + f_v(u, v)v' \qquad (f_u \text{ and } f_v \text{ are partial derivatives}).$$

11. *Logarithmic derivative*:

$$[\ln u(x)]' = \frac{u'(x)}{u(x)}.$$

Example 1. Let us calculate the derivative of the function $\dfrac{x^2}{2x + 1}$.
Using the rule of differentiating the ratio of two functions, we obtain

$$\left(\frac{x^2}{2x+1}\right)' = \frac{(x^2)'(2x+1) - x^2(2x+1)'}{(2x+1)^2} = \frac{2x(2x+1) - 2x^2}{(2x+1)^2} = \frac{2x^2 + 2x}{(2x+1)^2}.$$

Example 2. Let us calculate the derivative of the function $\ln \cos x$.
Using the chain rule or the logarithmic derivative formula, we get

$$(\ln \cos x)' = \frac{1}{\cos x}(\cos x)' = -\tan x.$$

Example 3. Let us calculate the derivative of the function x^x. Using the rule of differentiating the composite exponential function with $u(x) = v(x) = x$, we have

$$(x^x)' = x^x \ln x + x x^{x-1} = x^x (\ln x + 1).$$

M6.2.3. Theorems about Differentiable Functions. L'Hospital Rule

▶ **Main theorems about differentiable functions.**

ROLLE THEOREM. *If the function $y = f(x)$ is continuous on the segment $[a, b]$, differentiable on the interval (a, b), and $f(a) = f(b)$, then there is a point $c \in (a, b)$ such that $f'(c) = 0$.*

LAGRANGE THEOREM. *If the function $y = f(x)$ is continuous on the segment $[a, b]$ and differentiable on the interval (a, b), then there is a point $c \in (a, b)$ such that*

$$f(b) - f(a) = f'(c)(b - a).$$

This relation is called the *formula of finite increments*.

CAUCHY THEOREM. *Let $f(x)$ and $g(x)$ be two functions that are continuous on the segment $[a, b]$, differentiable on the interval (a, b), and $g'(x) \neq 0$ for all $x \in (a, b)$. Then there is a point $c \in (a, b)$ such that*

$$\frac{f(b) - f(a)}{g(b) - g(a)} = \frac{f'(c)}{g'(c)}.$$

▶ **L'Hospital's rules on indeterminate expressions of the form $0/0$ and ∞/∞.**

THEOREM 1. *Let $f(x)$ and $g(x)$ be two functions defined in a neighborhood of a point a, vanishing at this point, $f(a) = g(a) = 0$, and having the derivatives $f'(a)$ and $g'(a)$, with $g'(a) \neq 0$. Then*

$$\lim_{x \to a} \frac{f(x)}{g(x)} = \frac{f'(a)}{g'(a)}.$$

Example 1. Let us calculate the limit $\lim\limits_{x \to 0} \dfrac{\sin x}{1 - e^{-2x}}$.
Here, both the numerator and the denominator vanish for $x = 0$. Let us calculate the derivatives

$$f'(x) = (\sin x)' = \cos x \qquad \Longrightarrow \qquad f'(0) = 1,$$
$$g'(x) = (1 - e^{-2x})' = 2e^{-2x} \qquad \Longrightarrow \qquad g'(0) = 2 \neq 0.$$

By the L'Hospital rule, we find that

$$\lim_{x \to 0} \frac{\sin x}{1 - e^{-2x}} = \frac{f'(0)}{g'(0)} = \frac{1}{2}.$$

THEOREM 2. *Let $f(x)$ and $g(x)$ be two functions defined in a neighborhood of a point a, vanishing at a, together with their derivatives up to the order $n-1$ inclusively. Suppose also that the derivatives $f^{(n)}(a)$ and $g^{(n)}(a)$ exist and are finite, $g^{(n)}(a) \neq 0$. Then*

$$\lim_{x \to a} \frac{f(x)}{g(x)} = \frac{f^{(n)}(a)}{g^{(n)}(a)}.$$

THEOREM 3. *Let $f(x)$ and $g(x)$ be differentiable functions and $g'(x) \neq 0$ in a neighborhood of a point a ($x \neq a$). If $f(x)$ and $g(x)$ are infinitely small or infinitely large functions for $x \to a$, i.e., the ratio $\dfrac{f(x)}{g(x)}$ at the point a is an indeterminate expression of the form $\dfrac{0}{0}$ or $\dfrac{\infty}{\infty}$, then*

$$\lim_{x \to a} \frac{f(x)}{g(x)} = \lim_{x \to a} \frac{f'(x)}{g'(x)}$$

(provided that there exists a finite or infinite limit of the ratio of the derivatives).

Remark. The L'Hospital rule 3 is applicable also in the case of a being one of the symbols ∞, $+\infty$, $-\infty$.

▶ **Methods for interpreting other indeterminate expressions.**

$1°$. Expressions of the form $0 \cdot \infty$ and $\infty - \infty$ can be reduced to indeterminate expressions $\dfrac{0}{0}$ or $\dfrac{\infty}{\infty}$ by means of algebraic transformations, for instance:

$$u(x)\,v(x) = \frac{u(x)}{1/v(x)} \qquad\qquad \text{transformation rule} \quad 0 \cdot \infty \Longrightarrow \frac{0}{0},$$

$$u(x) - v(x) = \left(\frac{1}{v(x)} - \frac{1}{u(x)} \right) : \frac{1}{u(x)v(x)} \quad \text{transformation rule} \quad \infty - \infty \Longrightarrow \frac{0}{0}.$$

$2°$. Indeterminate expressions of the form 1^∞, ∞^0, 0^0 can be reduced to expressions of the form $\dfrac{0}{0}$ or $\dfrac{\infty}{\infty}$ by taking logarithm and using the formulas $\ln u^v = v \ln u = \dfrac{\ln u}{1/v}$.

Example 2. Let us calculate the limit $\lim\limits_{x \to 0} (\cos x)^{1/x^2}$.
We have the indeterminate expression 1^∞. We find that

$$\ln \lim_{x \to 0} (\cos x)^{1/x^2} = \lim_{x \to 0} \ln(\cos x)^{1/x^2} = \lim_{x \to 0} \frac{\ln \cos x}{x^2} = \lim_{x \to 0} \frac{(\ln \cos x)'}{(x^2)'} = \lim_{x \to 0} \frac{(-\tan x)}{2x} = -\frac{1}{2}.$$

Therefore, $\lim\limits_{x \to 0} (\cos x)^{1/x^2} = e^{-1/2} = \dfrac{1}{\sqrt{e}}$.

M6.2.4. Higher-Order Derivatives and Differentials. Taylor's Formula

▶ **Derivatives and differentials of higher orders.** The *second-order derivative* or the second derivative of a function $y = f(x)$ is the derivative of the derivative $f'(x)$. The second derivative is denoted by y'' and also by y''_{xx}, $\dfrac{d^2 y}{dx^2}$, $f''(x)$.

The derivative of the second derivative of a function $y = f(x)$ is called the third-order derivative, $y''' = (y'')'$. The nth-order derivative of the function $y = f(x)$ is defined as the derivative of its $(n-1)$th derivative:

$$y^{(n)} = (y^{(n-1)})'.$$

The nth-order derivative is also denoted by $y_x^{(n)}$, $\dfrac{d^n y}{dx^n}$, $f^{(n)}(x)$.

The *second-order differential* is the differential of the first-order differential, $d^2 y = d(dy)$. If x is the independent variable, then $d^2 y = y'' \cdot (dx)^2$. In a similar way, one defines differentials of higher orders.

▶ **Table of higher-order derivatives of some elementary functions.**

$$(x^a)^{(n)} = a(a-1)\ldots(a-n+1)x^{a-n}, \qquad (a^x)^{(n)} = (\ln a)^n a^x,$$

$$(\ln x)^{(n)} = (-1)^{n-1}(n-1)!\,\frac{1}{x^n}, \qquad (\log_a x)^{(n)} = (-1)^{n-1}\frac{(n-1)!}{\ln a}\,\frac{1}{x^n},$$

$$(\sin x)^{(n)} = \sin\left(x + \frac{\pi n}{2}\right), \qquad (\cos x)^{(n)} = \cos\left(x + \frac{\pi n}{2}\right),$$

$$(\sinh x)^{(n)} = \begin{cases} \cosh x & \text{if } n \text{ is odd,} \\ \sinh x & \text{if } n \text{ is even,} \end{cases} \qquad (\cosh x)^{(n)} = \begin{cases} \cosh x & \text{if } n \text{ is even,} \\ \sinh x & \text{if } n \text{ is odd.} \end{cases}$$

▶ **Rules for calculating higher-order derivatives.**

1. *Derivative of a sum (difference) of functions*:

$$[u(x) \pm v(x)]^{(n)} = u^{(n)}(x) \pm v^{(n)}(x).$$

2. *Derivatives of a function multiplied by a constant*:

$$[au(x)]^{(n)} = au^{(n)}(x) \qquad (a = \text{const}).$$

3. *Derivatives of a product*:

$$[u(x)v(x)]'' = u''(x)v(x) + 2u'(x)v'(x) + u(x)v''(x),$$

$$[u(x)v(x)]''' = u'''(x)v(x) + 3u''(x)v'(x) + 3u'(x)v''(x) + u(x)v'''(x),$$

$$[u(x)v(x)]^{(n)} = \sum_{k=0}^{n} C_n^k u^{(k)}(x)v^{(n-k)}(x) \qquad \text{(Leibniz formula)},$$

where C_n^k are binomial coefficients, $u^{(0)}(x) = u(x)$, $v^{(0)}(x) = v(x)$.

4. *Derivatives of a composite function*:

$$\left[f(u(x))\right]'' = f_{uu}''(u_x')^2 + f_u' u_{xx}'',$$

$$\left[f(u(x))\right]''' = f_{uuu}'''(u_x')^3 + 3f_{uu}'' u_x' u_{xx}'' + f_u' u_{xxx}'''.$$

5. *Derivatives of a parametrically defined function* $x = x(t)$, $y = y(t)$:

$$y'' = \frac{x_t' y_{tt}'' - y_t' x_{tt}''}{(x_t')^3}, \qquad y''' = \frac{(x_t')^2 y_{ttt}''' - 3x_t' x_{tt}'' y_{tt}'' + 3y_t'(x_{tt}'')^2 - x_t' y_t' x_{ttt}'''}{(x_t')^5}, \qquad y^{(n)} = \frac{(y^{(n-1)})_t'}{x_t'}.$$

6. *Derivatives of an implicit function defined by the equation* $F(x, y) = 0$:

$$y'' = \frac{1}{F_y^3}\left(-F_y^2 F_{xx} + 2F_x F_y F_{xy} - F_x^2 F_{yy}\right),$$

$$y''' = \frac{1}{F_y^5}\Big(-F_y^4 F_{xxx} + 3F_x F_y^3 F_{xxy} - 3F_x^2 F_y^3 F_{xyy} + F_x^3 F_y F_{yyy} + 3F_y^3 F_{xx} F_{xy}$$

$$- 3F_x F_y^2 F_{xx} F_{yy} - 6F_x F_y^2 F_{xy}^2 - 3F_x^3 F_{yy}^2 + 9F_x^2 F_y F_{xy} F_{yy}\Big),$$

where the subscripts denote the corresponding partial derivatives.

7. *Derivatives of the inverse function* $x = x(y)$:

$$x''_{yy} = -\frac{y''_{xx}}{(y'_x)^3}, \qquad x'''_{yyy} = -\frac{y'''_{xxx}}{(y'_x)^4} + 3\frac{(y''_{xx})^2}{(y'_x)^5}, \qquad x_y^{(n)} = \frac{1}{y'_x}[x_y^{(n-1)}]'_x.$$

▶ **Taylor's formula.** Suppose that in a neighborhood of a point $x = a$, the function $y = f(x)$ has derivatives up to the order $(n + 1)$ inclusively. Then for all x in that neighborhood, the following representation holds:

$$f(x) = f(a) + \frac{f'(a)}{1!}(x - a) + \frac{f''(a)}{2!}(x - a)^2 + \cdots + \frac{f^{(n)}(a)}{n!}(x - a)^n + R_n(x), \quad \text{(M6.2.4.1)}$$

where $R_n(x)$ is the *remainder term* in Taylor's formula.

The remainder term can be represented in different forms:

$$R_n(x) = o[(x - a)^n] \qquad\qquad\qquad\qquad\qquad \text{(Peano)},$$

$$R_n(x) = \frac{f^{(n+1)}\big(a + k(x - a)\big)}{(n + 1)!}(x - a)^{n+1} \qquad\qquad \text{(Lagrange)},$$

$$R_n(x) = \frac{f^{(n+1)}\big(a + k(x - a)\big)}{n!}(1 - k)^n (x - a)^{n+1} \qquad \text{(Cauchy)},$$

$$R_n(x) = \frac{f^{(n+1)}\big(a + k(x - a)\big)}{n!\,p}(1 - k)^{n+1-p}(x - a)^{n+1} \quad \text{(Schlömilch and Roche)},$$

$$R_n(x) = \frac{1}{n!}\int_a^x f^{(n+1)}(t)(x - t)^n\, dt \qquad\qquad \text{(integral form)},$$

where $0 < k < 1$ and $p > 0$; k depends on x, n, and the structure of the remainder term. The remainders in the form of Lagrange and Cauchy can be obtained as special cases of the Schlömilch formula with $p = n + 1$ and $p = 1$, respectively.

For $a = 0$, the Taylor's formula (M6.2.4.1) turns into

$$f(x) = f(0) + \frac{f'(0)}{1!}x + \frac{f''(0)}{2!}x^2 + \cdots + \frac{f^{(n)}(0)}{n!}x^n + R_n(x)$$

and is called the *Maclaurin formula*.

The Maclaurin formula for some functions:

$$e^x = 1 + \frac{x}{1!} + \frac{x^2}{2!} + \frac{x^3}{3!} + \cdots + \frac{x^n}{n!} + R_n(x),$$

$$\sin x = x - \frac{x^3}{3!} + \frac{x^5}{5!} - \frac{x^7}{7!} + \cdots + (-1)^n \frac{x^{2n+1}}{(2n + 1)!} + R_{2n+1}(x),$$

$$\cos x = 1 - \frac{x^2}{2!} + \frac{x^4}{4!} - \frac{x^6}{6!} + \cdots + (-1)^n \frac{x^{2n}}{(2n)!} + R_{2n}(x).$$

M6.2.5. Extremal Points. Points of Inflection

▶ **Maximum and minimum. Points of extremum.** Let $f(x)$ be a differentiable function on the interval (a, b) and $f'(x) > 0$ (resp., $f'(x) < 0$) on (a, b). Then $f(x)$ is an *increasing* (resp., *decreasing*) function on that interval*.

Suppose that there is a neighborhood of a point x_0 such that for all $x \neq x_0$ in that neighborhood we have $f(x) > f(x_0)$ (resp., $f(x) < f(x_0)$). Then x_0 is called a point of *local minimum* (resp., *local maximum*) of the function $f(x)$.

Points of local minimum or maximum are called points of *extremum*.

---

* At some isolated points of the interval, the derivative may vanish.

▶ **Necessary and sufficient conditions for the existence of extremum.** Suppose that $f(x)$ is continuous in some neighborhood $(x_0 - \delta, \ x_0 + \delta)$ of a point x_0 and differentiable at all points of the neighborhood except, possibly, x_0.

NECESSARY CONDITION OF EXTREMUM. *A function $f(x)$ can have an extremum only at points in which its derivative either vanishes or does not exist (or is infinite).*

FIRST SUFFICIENT CONDITION OF EXTREMUM. *If $f'(x) > 0$ for $x \in (x_0 - \delta, \ x_0)$ and $f'(x) < 0$ for $x \in (x_0, \ x_0 + \delta)$, then x_0 is a point of local maximum of this function. If $f'(x) < 0$ for $x \in (x_0 - \delta, \ x_0)$ and $f'(x) > 0$ for $x \in (x_0, \ x_0 + \delta)$, then x_0 is a point of local minimum of this function.*

If $f'(x)$ is of the same sign for all $x \ne x_0$, $x \in (x_0 - \delta, \ x_0 + \delta)$, then x_0 cannot be a point of extremum.

SECOND SUFFICIENT CONDITION OF EXTREMUM. *Let $f(x)$ be a twice differentiable function in a neighborhood of x_0. Then the following statements hold:*

(i) $f'(x_0) = 0$ and $f''(x_0) < 0$ $\implies$ *$f(x)$ has a local maximum at the point x_0;*

(ii) $f'(x_0) = 0$ and $f''(x_0) > 0$ $\implies$ *$f(x)$ has a local minimum at the point x_0.*

THIRD SUFFICIENT CONDITION OF EXTREMUM. *Let $f(x)$ be a function that is n times differentiable in a neighborhood of a point x_0 and $f'(x_0) = f''(x_0) = \cdots = f^{(n-1)}(x_0) = 0$, but $f^{(n)}(x_0) \ne 0$. Then the following statements hold:*

(i)   *n is even and $f^{(n)}(x_0) < 0$* $\implies$ *$f(x)$ has a local maximum at the point x_0;*

(ii) *n is even and $f^{(n)}(x_0) > 0$* $\implies$ *$f(x)$ has a local minimum at the point x_0.*

If n is odd, then x_0 cannot be a point of extremum.

▶ **Largest and the smallest values of a function.** Let $y = f(x)$ be continuous on the segment $[a, b]$ and differentiable at all points of this segment except, possibly, finitely many points. Then the largest and the smallest values of $f(x)$ on $[a, b]$ belong to the set consisting of $f(a)$, $f(b)$, and the values $f(x_i)$, where $x_i \in (a, b)$ are the points at which $f'(x)$ is either equal to zero or does not exist (is infinite).

▶ **Direction of the convexity of the graph of a function.** The graph of a differentiable function $y = f(x)$ is said to be *convex upward* (resp., *convex downward*) on the interval (a, b) if for each point of this interval, the graph lies below (resp., above) the tangent line at that point.

If the function $y = f(x)$ is twice differentiable on the interval (a, b) and $f''(x) < 0$ (resp., $f''(x) > 0$), then its graph is convex upward (resp., downward) on that interval. (At some isolated points of the interval, the second derivative may vanish.)

Thus, in order to find the intervals on which the graph of a twice differentiable function $f(x)$ is convex upward (resp., downward), one should solve the inequality $f''(x) < 0$ (resp., $f''(x) > 0$).

▶ **Inflection points.** An *inflection point* on the graph of a function $y = f(x)$ is defined as a point $(x_0, \ f(x_0))$ at which the graph passes from one side of its tangent line to another. At an inflection point, the graph changes the direction of its convexity.

Suppose that the function $y = f(x)$ has a continuous second derivative $f''(x)$ in some neighborhood of a point x_0. If $f''(x_0) = 0$ and $f''(x)$ changes sign as x passes through the point x_0, then $(x_0, \ f(x_0))$ is an inflection point.

M6.2.6. Qualitative Analysis of Functions and Construction of Graphs

▶ **General scheme of analysis of a function and construction of its graph.**
1. Determine the domain in which the function is defined.
2. Determine whether the function is odd or even and whether it is periodic.
3. Find the points at which the graph crosses the coordinate axes.
4. Find the asymptotes of the graph.
5. Find extremal points and intervals of monotonicity.
6. Determine the directions of convexity of the graph and its inflection points.
7. Draw the graph, using the properties 1 to 6.

Example. Let us examine the function $y = \dfrac{\ln x}{x}$ and construct its graph.

We use the above general scheme.
1. This function is defined for all x such that $0 < x < +\infty$.
2. This function is neither odd nor even, since it is defined only for $x > 0$ and the relations $f(-x) = f(x)$ or $f(-x) = -f(x)$ cannot hold. Obviously, this function is nonperiodic.
3. The graph of this function does not cross the y-axis, since for $x = 0$ the function is undefined. Further, $y = 0$ only if $x = 1$, i.e., the graph crosses the x-axis only at the point $(1, 0)$.
4. The straight line $x = 0$ is a vertical asymptote, since $\lim\limits_{x \to +0} \dfrac{\ln x}{x} = -\infty$. We find the oblique asymptotes:

$$k = \lim_{x \to +\infty} \frac{y}{x} = 0, \quad b = \lim_{x \to +\infty} (y - kx) = 0.$$

Therefore, the line $y = 0$ is a horizontal asymptote of the graph.
5. The derivative $y' = \dfrac{1 - \ln x}{x^2}$ vanishes for $\ln x = 1$. Therefore, the function may have an extremum at $x = e$. For $x \in (0, e)$, we have $y' > 0$, i.e., the function is increasing on this interval. For $x \in (e, +\infty)$, we have $y' < 0$, and therefore the function is decreasing on this interval. At $x = e$ the function attains its maximal value $y_{\max} = \dfrac{1}{e}$.

One should also examine the points at which the derivative does not exist. There is only one such point, $x = 0$, and it corresponds to the vertical asymptote (see Item 4).
6. The second derivative $y'' = \dfrac{2 \ln x - 3}{x^3}$ vanishes for $x = e^{3/2}$. On the interval $(0, e^{3/2})$, we have $y'' < 0$, and therefore the graph is convex upward on this interval. For $x \in (e^{3/2}, +\infty)$, we have $y'' > 0$, and therefore the graph is convex downward on this interval. The value $x = e^{3/2}$ corresponds to an inflection point of the graph, with the ordinate $y = \frac{3}{2} e^{-3/2}$.
7. Using the above results, we construct the graph (Fig. M6.5).

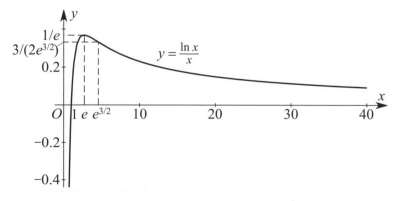

Figure M6.5. Graph of the function $y = \frac{\ln x}{x}$.

▶ **Transformations of graphs of functions.** Let us describe some methods which in many cases allow us to construct the graph of a function if we have the graph of a simpler function.

1. The graph of the function $y = f(x) + a$ is obtained from that of $y = f(x)$ by shifting the latter along the axis Oy by the distance $|a|$. For $a > 0$ the shift is upward, and for $a < 0$ downward (see Fig. M6.6 a).

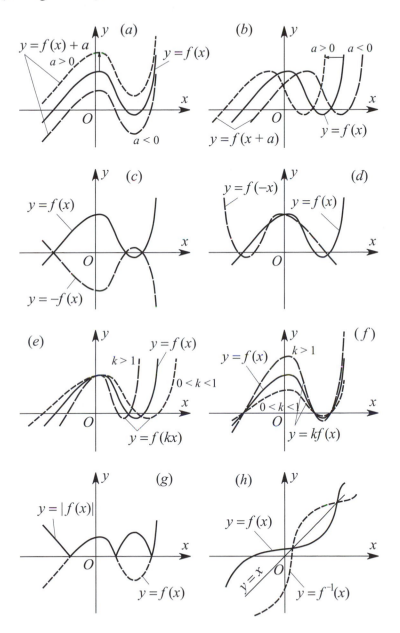

Figure M6.6. Transformations of graphs of functions.

2. The graph of the function $y = f(x + a)$ is obtained from that of $y = f(x)$ by shifting the latter along the Ox by the distance $|a|$. For $a > 0$ the shift is to the left, and for $a < 0$ to the right (see Fig. M6.6 b).

3. The graph of the function $y = -f(x)$ is obtained from that of $y = f(x)$ by symmetric reflection with respect to the axis Ox (see Fig. M6.6 c).

4. The graph of the function $y = f(-x)$ is obtained from that of $y = f(x)$ by symmetric reflection with respect to the axis Oy (see Fig. M6.6 d).

5. The graph of the function $y = f(kx)$ for $k > 1$ is obtained from that of $y = f(x)$ by contracting the latter k times to the axis Oy, and for $0 < k < 1$ by extending the latter $1/k$ times from the axis Oy. The points at which the graph crosses the axis Oy remain unchanged (see Fig. M6.6 e).

6. The graph of the function $y = kf(x)$ for $k > 1$ is obtained from that of $y = f(x)$ by extending the latter k times from the axis Ox, and for $0 < k < 1$ by contracting the latter $1/k$ times to the axis Ox. The points at which the graph crosses the axis Ox remain unchanged (see Fig. M6.6 f).

7. The graph of the function $y = |f(x)|$ is obtained from that of $y = f(x)$ by preserving the parts of the latter for which $f(x) \geq 0$ and symmetric reflection, with respect to the axis Ox, of the parts for which $f(x) < 0$ (see Fig. M6.6 g).

8. The graph of the inverse function $y = f^{-1}(x)$ is obtained from that of $y = f(x)$ by symmetric reflection with respect to the straight line $y = x$ (see Fig. M6.6 h).

M6.2.7. Approximate Solution of Equations (Root-Finding Algorithms for Continuous Functions)

▶ **Preliminaries.** For a vast majority of algebraic (transcendental) equations of the form

$$f(x) = 0, \tag{M6.2.7.1}$$

where $f(x)$ is a continuous function, there are no exact formulas for the roots.

When solving the equation approximately, the first step is to bracket the roots, i.e., find sufficiently small intervals containing exactly one root each. Such an interval $[a, b]$, where the numbers a and b satisfy the condition $f(a)f(b) < 0$ (which is assumed to hold in what follows), can be found, say, graphically.

The second step is to compute successive approximations $x_n \in [a, b]$ $(n = 1, 2, \ldots)$ to the desired root $c = \lim_{n \to \infty} x_n$, usually by one of the following methods.

▶ **Bisection method.** To find the root of equation (M6.2.7.1) on the interval $[a, b]$, we bisect the interval. If $f\left(\dfrac{a+b}{2}\right) = 0$, then $c = \dfrac{a+b}{2}$ is the desired root. If $f\left(\dfrac{a+b}{2}\right) \neq 0$, then of the two intervals $\left[a, \dfrac{a+b}{2}\right]$ and $\left[\dfrac{a+b}{2}, b\right]$ we take the one at whose endpoints the function $f(x)$ has opposite signs. Now we bisect the new, smaller interval, etc. As a result, we obtain either an exact root of equation (M6.2.7.1) at some step or an infinite sequence of nested intervals $[a_1, b_1], [a_2, b_2], \ldots$ such that $f(a_n)f(b_n) < 0$. The root is given by the formula $c = \lim_{n \to \infty} a_n = \lim_{n \to \infty} b_n$, and the estimate

$$0 \leq c - a_n \leq \frac{1}{2^n}(b - a)$$

is valid.

The following two methods are more efficient.

▶ **Regula falsi method (false position method).** Suppose that the derivatives $f'(x)$ and $f''(x)$ exist on the interval $[a, b]$ and the inequalities $f'(x) \neq 0$ and $f''(x) \neq 0$ hold for all $x \in [a, b]$.

If $f'(a)f''(a) > 0$, then we take $x_0 = a$ for the zero approximation; the subsequent approximations are given by the formulas

$$x_{n+1} = x_n - \frac{f(x_n)}{f(b) - f(x_n)}(b - x_n), \qquad n = 0, 1, \ldots$$

If $f'(a)f''(a) < 0$, then we take $x_0 = b$ for the zero approximation; the subsequent approximations are given by the formulas

$$x_{n+1} = x_n - \frac{f(x_n)}{f(a) - f(x_n)}(a - x_n), \qquad n = 0, 1, \ldots$$

The regula falsi method has the first order of local convergence as $n \to \infty$, that is,

$$|x_{n+1} - c| \le k|x_n - c|,$$

where k is a constant depending on $f(x)$ and c is the root of equation (M6.2.7.1).

The regula falsi method has a simple geometric interpretation. The straight line (secant) passing through the points $(a, f(a))$ and $(b, f(b))$ of the curve $y = f(x)$ meets the abscissa axis at the point x_1; the value x_{n+1} is the abscissa of the point where the line passing through the points $(x_0, f(x_0))$ and $(x_n, f(x_n))$ meets the x-axis (see Fig. M6.7 a).

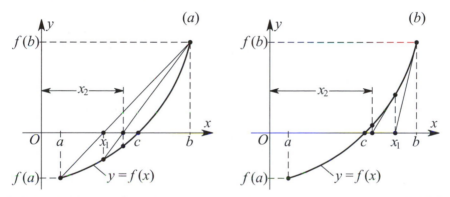

Figure M6.7. Graphical construction of successive approximations to the root of equation (M6.2.7.1) by the regula falsi method (a) and the Newton–Raphson method (b).

▶ **Newton–Raphson method.** Suppose that the derivatives $f'(x)$ and $f''(x)$ exist on the interval $[a, b]$ and the inequalities $f'(x) \ne 0$ and $f''(x) \ne 0$ hold for all $x \in [a, b]$.

If $f(a)f''(a) > 0$, then we take $x_0 = a$ for the zero approximation; if $f(b)f''(b) > 0$, then $x_0 = b$. The subsequent approximations are computed by the formulas

$$x_{n+1} = x_n - \frac{f(x_n)}{f'(x_n)}, \qquad n = 0, 1, \ldots$$

If the initial approximation x_0 is sufficiently close to the desired root c, then the Newton–Raphson method exhibits quadratic convergence:

$$|x_{n+1} - c| \le \frac{M}{2m}|x_n - c|^2,$$

where $M = \max\limits_{a \le x \le b} |f''(x)|$ and $m = \min\limits_{a \le x \le b} |f'(x)|$.

The Newton–Raphson method has a simple geometric interpretation. The tangent to the curve $y = f(x)$ through the point $(x_n, f(x_n))$ meets the abscissa axis at the point x_{n+1} (see Fig. M6.7 b).

The Newton–Raphson method has a higher order of convergence than the regula falsi method. Hence the former is more often used in practice.

M6.3. Functions of Several Variables. Partial Derivatives

M6.3.1. Point Sets. Functions. Limits and Continuity

▶ **Sets on the plane and in space.** The distance between two points A and B on the plane and in space can be defined as follows:

$$\rho(A, B) = \sqrt{(x_A - x_B)^2 + (y_A - y_B)^2} \qquad \text{(on the plane),}$$

$$\rho(A, B) = \sqrt{(x_A - x_B)^2 + (y_A - y_B)^2 + (z_A - z_B)^2} \quad \text{(in three-dimensional space),}$$

$$\rho(A, B) = \sqrt{(x_{1A} - x_{1B})^2 + \cdots + (x_{nA} - x_{nB})^2} \qquad \text{(in n-dimensional space),}$$

where x_A, y_A and x_B, y_B, and x_A, y_A, z_A and x_B, y_B, z_B, and x_{1A}, ..., x_{nA} and x_{1B}, ..., x_{nB} are Cartesian coordinates of the respective points.

An ε-*neighborhood of a point* M_0 (on the plane or in space) is the set consisting of all points M (resp., on the plane or in space) such that $\rho(M, M_0) < \varepsilon$, where it is assumed that $\varepsilon > 0$. An ε-*neighborhood of a set* K (on the plane or in space) is the set consisting of all points M (resp., on the plane or in space) such that $\inf_{M_0 \in K} \rho(M, M_0) < \varepsilon$, where it is assumed that $\varepsilon > 0$.

An *interior point* of a set D is a point belonging to D, together with some neighborhood of that point. An *open set* is a set containing only interior points. A *boundary point* of a set D is a point such that any of its neighborhoods contains points both inside and outside D. A *closed set* is a set containing all its boundary points. A set D is called a *bounded set* if $\rho(A, B) < C$ for any points $A, B \in D$, where C is a constant independent of A, B. Otherwise (i.e., if there is no such constant), the set D is called *unbounded*.

▶ **Functions of two or three variables.** A (numerical) *function* on a set D is, by definition, a relation that sets up a correspondence between each point $M \in D$ and a unique numerical value. If D is a plane set, then each point $M \in D$ is determined by two coordinates x, y, and a function $z = f(M) = f(x, y)$ is called a *function of two variables*. If D belongs to a three-dimensional space, then one speaks of a *function of three variables*. The set D on which the function is defined is called the *domain* of the function. For instance, the function $z = \sqrt{1 - x^2 - y^2}$ is defined on the closed circle $x^2 + y^2 \leq 1$, which is its domain.

The *graph* of a function $z = f(x, y)$ is the surface formed by the points $(x, y, f(x, y))$ in three-dimensional space. For instance, the graph of the function $z = ax + by + c$ is a plane, and the graph of the function $z = \sqrt{1 - x^2 - y^2}$ is a semisphere.

A *level line* of a function $z = f(x, y)$ is a line on the plane x, y with the following property: the function takes one and the same value $z = c$ at all points of that line. Thus, the equation of a level line has the form $f(x, y) = c$. A *level surface* of a function $u = f(x, y, z)$ is a surface on which the function takes a constant value, $u = c$; the equation of a level surface has the form $f(x, y, z) = c$.

A function $f(M)$ is called *bounded* on a set D if there is a constant C such that $|f(M)| \leq C$ for all $M \in D$.

▶ **Limit of a function at a point and its continuity.** Let M be a point that comes infinitely close to some point M_0, i.e., $\rho = \rho(M_0, M) \to 0$. It is possible that the values $f(M)$ come close to some constant b.

One says that b is the *limit of the function* $f(M)$ at the point M_0 if for any (arbitrarily small) $\varepsilon > 0$, there is $\delta > 0$ such that for all points M belonging to the domain of the function and satisfying the inequality $0 < \rho(M_0, M) < \delta$, we have $|f(M) - b| < \varepsilon$. In this case, one writes $\lim_{\rho(M, M_0) \to 0} f(M) = b$.

A function $f(M)$ is called *continuous* at a point M_0 if $\lim\limits_{\rho(M,M_0)\to 0} f(M) = f(M_0)$. A function is called *continuous on a set* D if it is continuous at each point of D. Any continuous function $f(M)$ on a closed bounded set is bounded on that set and attains its smallest and its largest values on that set.

M6.3.2. Differentiation of Functions of Several Variables

For the sake of brevity, we consider the case of a function of two variables. However, all statements can be easily extended to the case of n variables.

▶ **Total and partial increments of a function. Partial derivatives.** A *total increment* of a function $z = f(x, y)$ at a point (x, y) is

$$\Delta z = f(x + \Delta x, y + \Delta y) - f(x, y),$$

where Δx, Δy are increments of the independent variables. *Partial increments* in x and in y are, respectively,

$$\Delta_x z = f(x + \Delta x, y) - f(x, y),$$
$$\Delta_y z = f(x, y + \Delta y) - f(x, y).$$

Partial derivatives of a function z with respect to x and to y at a point (x, y) are defined as follows:

$$\frac{\partial z}{\partial x} = \lim_{\Delta x \to 0} \frac{\Delta_x z}{\Delta x}, \quad \frac{\partial z}{\partial y} = \lim_{\Delta y \to 0} \frac{\Delta_y z}{\Delta y}$$

(provided that these limits exist). Partial derivatives are also denoted by z_x and z_y, $\partial_x z$ and $\partial_y z$, or $f_x(x, y)$ and $f_y(x, y)$.

▶ **Differentiable functions. Differential.** A function $z = f(x, y)$ is called *differentiable* at a point (x, y) if its increment at that point can be represented in the form

$$\Delta z = A(x, y)\Delta x + B(x, y)\Delta y + o(\rho), \qquad \rho = \sqrt{(\Delta x)^2 + (\Delta y)^2},$$

where $o(\rho)$ is a quantity of a higher order of smallness compared with ρ as $\rho \to 0$ (i.e., $o(\rho)/\rho \to 0$ as $\rho \to 0$). In this case, there exist partial derivatives at the point (x, y), and $z_x = A(x, y)$, $z_y = B(x, y)$.

A function that has continuous partial derivatives at a point (x, y) is differentiable at that point.

The *differential* dz *of a function* $z = f(x, y)$ is defined as follows:

$$dz = f_x(x, y)\Delta x + f_y(x, y)\Delta y.$$

Taking the differentials dx and dy of the independent variables equal to Δx and Δy, respectively, one can also write $dz = f_x(x, y)\, dx + f_y(x, y)\, dy$.

The relation $\Delta z = dz + o(\rho)$ for small Δx and Δy is widely used for approximate calculations, in particular, for finding errors in numerical calculations of values of a function.

Example 1. Suppose that the values of the arguments of the function $z = x^2 y^5$ are known with the error $x = 2 \pm 0.01$, $y = 1 \pm 0.01$. Let us calculate the approximate value of the function.

We find the increment of the function z at the point $x = 2$, $y = 1$ for $\Delta x = \Delta y = 0.01$, using the formula $\Delta z \approx dz = 2 \cdot 2 \cdot 1^5 \cdot 0.01 + 5 \cdot 2^2 \cdot 1^4 \cdot 0.01 = 0.24$. Therefore, we can accept the approximation $z = 4 \pm 0.24$.

If a function $z = f(x, y)$ is differentiable at a point (x_0, y_0), then

$$f(x, y) = f(x_0, y_0) + f_x(x_0, y_0)(x - x_0) + f_y(x_0, y_0)(y - y_0) + o(\rho).$$

Hence, for small ρ (i.e., for $x \approx x_0$, $y \approx y_0$), we obtain the approximate formula

$$f(x, y) \approx f(x_0, y_0) + f_x(x_0, y_0)(x - x_0) + f_y(x_0, y_0)(y - y_0).$$

The replacement of a function by this linear expression near a given point is called *linearization*.

▶ **Composite function.** Consider a function $z = f(x, y)$ and let $x = x(u, v)$, $y = y(u, v)$. Suppose that for $(u, v) \in D$, the functions $x(u, v)$, $y(u, v)$ take values for which the function $z = f(x, y)$ is defined. In this way, one defines a *composite function* on the set D, namely, $z(u, v) = f\big(x(u, v), y(u, v)\big)$. In this situation, $f(x, y)$ is called the outer function and $x(u, v)$, $y(u, v)$ are called the inner functions.

Partial derivatives of a composite function are expressed by

$$\frac{\partial z}{\partial u} = \frac{\partial f}{\partial x} \frac{\partial x}{\partial u} + \frac{\partial f}{\partial y} \frac{\partial y}{\partial u},$$

$$\frac{\partial z}{\partial v} = \frac{\partial f}{\partial x} \frac{\partial x}{\partial v} + \frac{\partial f}{\partial y} \frac{\partial y}{\partial v}.$$

For $z = z(t, x, y)$, let $x = x(t)$, $y = y(t)$. Thus, z is actually a function of only one variable t. The derivative $\frac{dz}{dt}$ is calculated by

$$\frac{dz}{dt} = \frac{\partial z}{\partial t} + \frac{\partial z}{\partial x} \frac{dx}{dt} + \frac{\partial z}{\partial y} \frac{dy}{dt}.$$

This derivative, in contrast to the partial derivative $\frac{\partial z}{\partial t}$, is called a *total derivative*.

▶ **Second partial derivatives and second differentials.** The *second partial derivatives* of a function $z = f(x, y)$ are defined as the derivatives of its first partial derivatives and are denoted as follows:

$$\frac{\partial^2 z}{\partial x^2} = z_{xx} \equiv (z_x)_x, \qquad \frac{\partial^2 z}{\partial x\, \partial y} = z_{xy} \equiv (z_x)_y,$$

$$\frac{\partial^2 z}{\partial y\, \partial x} = z_{yx} \equiv (z_y)_x, \qquad \frac{\partial^2 z}{\partial y^2} = z_{yy} \equiv (z_y)_y.$$

The derivatives z_{xy} and z_{yx} are called *mixed derivatives*. If the mixed derivatives are continuous at some point, then they coincide at that point, $z_{xy} = z_{yx}$.

In a similar way, one defines higher-order partial derivatives.

The *second differential* of a function $z = f(x, y)$ is the expression

$$d^2 z = d(dz) = (dz)_x \Delta x + (dz)_y \Delta y = z_{xx}(\Delta x)^2 + 2 z_{xy} \Delta x \Delta y + z_{yy}(\Delta y)^2.$$

In a similar way, one defines $d^3 z$, $d^4 z$, etc.

▶ **Implicit functions and their differentiation.** Consider the equation $F(x, y) = 0$ with a solution (x_0, y_0). Suppose that the derivative $F_y(x, y)$ is continuous in a neighborhood of the point (x_0, y_0) and $F_y(x, y) \neq 0$ in that neighborhood. Then the equation $F(x, y) = 0$ defines a continuous function $y = y(x)$ (called an *implicit function*) of the variable x in a neighborhood of the point x_0. Moreover, if in a neighborhood of (x_0, y_0) there exists a continuous derivative F_x, then the implicit function $y = y(x)$ has a continuous derivative expressed by $\dfrac{dy}{dx} = -\dfrac{F_x}{F_y}$.

Consider the equation $F(x, y, z) = 0$ that establishes a relation between the variables x, y, z. If $F(x_0, y_0, z_0) = 0$ and in a neighborhood of the point (x_0, y_0, z_0) there exist continuous partial derivatives F_x, F_y, F_z such that $F_z(x_0, y_0, z_0) \neq 0$, then equation $F(x, y, z) = 0$, in a neighborhood of (x_0, y_0), has a unique solution $z = \varphi(x, y)$ such that $\varphi(x_0, y_0) = z_0$;

moreover, the function $z = \varphi(x, y)$ is continuous and has continuous partial derivatives expressed by

$$\frac{\partial z}{\partial x} = -\frac{F_x}{F_z}, \qquad \frac{\partial z}{\partial y} = -\frac{F_y}{F_z}.$$

Example 2. For the equation $x \sin y + z + e^z = 0$ we have $F_z = 1 + e^z \neq 0$. Therefore, this equation defines an implicit function $z = \varphi(x, y)$ on the entire plane, and its derivatives have the form $\dfrac{\partial z}{\partial x} = -\dfrac{\sin y}{1 + e^z}$, $\dfrac{\partial z}{\partial y} = -\dfrac{x \cos y}{1 + e^z}$.

▶ **Jacobian. Dependent and independent functions. Invertible transformations.**

$1°$. Two functions $f(x, y)$ and $g(x, y)$ are called *dependent* if there is a function $\Phi(z)$ such that $g(x, y) = \Phi(f(x, y))$; otherwise, the functions $f(x, y)$ and $g(x, y)$ are called *independent*.

The *Jacobian* is the determinant of the matrix whose elements are the first partial derivatives of the functions $f(x, y)$ and $g(x, y)$:

$$\frac{\partial(f, g)}{\partial(x, y)} \equiv \begin{vmatrix} \dfrac{\partial f}{\partial x} & \dfrac{\partial f}{\partial y} \\ \dfrac{\partial g}{\partial x} & \dfrac{\partial g}{\partial y} \end{vmatrix}. \tag{M6.3.2.1}$$

1) If the Jacobian (M6.3.2.1) in a domain D is identically equal to zero, then the functions $f(x, y)$ and $g(x, y)$ are dependent in D.

2) If the Jacobian (M6.3.2.1) is nonzero in D, then the functions $f(x, y)$ and $g(x, y)$ are independent in D.

$2°$. Functions $f_k(x_1, x_2, \ldots, x_n)$, $k = 1, 2, \ldots, n$, are called dependent in a domain D if there is a function $\Phi(z_1, z_2, \ldots, z_n)$ such that

$$\Phi\big(f_1(x_1, x_2, \ldots, x_n), f_2(x_1, x_2, \ldots, x_n), \ldots, f_n(x_1, x_2, \ldots, x_n)\big) = 0 \qquad \text{(in } D);$$

otherwise, these functions are called independent.

The *Jacobian* is the determinant of the matrix whose elements are the first partial derivatives:

$$\frac{\partial(f_1, f_2, \ldots, f_n)}{\partial(x_1, x_2, \ldots, x_n)} \equiv \det\left(\frac{\partial f_i}{\partial x_j}\right). \tag{M6.3.2.2}$$

The functions $f_k(x_1, x_2, \ldots, x_n)$ are dependent in a domain D if the Jacobian (M6.3.2.2) is identically equal to zero in D. The functions $f_k(x_1, x_2, \ldots, x_n)$ are independent in D if the Jacobian (M6.3.2.2) does not vanish in D.

$3°$. Consider the transformation

$$y_k = f_k(x_1, x_2, \ldots, x_n), \qquad k = 1, 2, \ldots, n. \tag{M6.3.2.3}$$

Suppose that the functions f_k are continuously differentiable and the Jacobian (M6.3.2.2) differs from zero at a point $(x_1^\circ, x_2^\circ, \ldots, x_n^\circ)$. Then, in a sufficiently small neighborhood of this point, equations (M6.3.2.3) specify a one-to-one correspondence between the points of that neighborhood and the set of points $(y_1, y_2, \ldots, y_n)$ consisting of the values of the functions (M6.3.2.3) in the corresponding neighborhood of the point $(y_1^\circ, y_2^\circ, \ldots, y_n^\circ)$. This means that the system (M6.3.2.3) is locally solvable in a neighborhood of the point $(x_1^\circ, x_2^\circ, \ldots, x_n^\circ)$, i.e., the following representation holds:

$$x_k = g_k(y_1, y_2, \ldots, y_n), \qquad k = 1, 2, \ldots, n,$$

where g_k are continuously differentiable functions in the corresponding neighborhood of the point $(y_1^\circ, y_2^\circ, \ldots, y_n^\circ)$.

M6.3.3. Directional Derivative. Gradient. Geometrical Applications

▶ **Directional derivative.** One says that a *scalar field* is defined in a domain D if any point $M(x, y)$ of that domain is associated with a certain value $z = f(M) = f(x, y)$. Thus, a thermal field and a pressure field are examples of scalar fields. A *level line* of a scalar field is a level line of the function that specifies the field (see Subsection M6.3.1). Thus, isothermal and isobaric curves are, respectively, level lines of thermal and pressure fields.

In order to examine the behavior of a field $z = f(x, y)$ at a point $M_0(x_0, y_0)$ in the direction of a vector $\mathbf{a} = \{a_1, a_2\}$, one should construct a straight line passing through M_0 in the direction of the vector $\mathbf{a}$ (this line can be specified by the parametric equations $x = x_0 + a_1 t$, $y = y_0 + a_2 t$) and study the function $z(t) = f(x_0 + a_1 t, \ y_0 + a_2 t)$. The derivative of the function $z(t)$ at the point M_0 (i.e., for $t = 0$) characterizes the change rate of the field at that point in the direction $\mathbf{a}$. Dividing $z'(0)$ by $|\mathbf{a}| = \sqrt{a_1^2 + a_2^2}$, we obtain the so-called *derivative in the direction* $\mathbf{a}$ of the given field at the given point:

$$\frac{\partial f}{\partial \mathbf{a}} = \frac{1}{|\mathbf{a}|} \big[a_1 f_x(x_0, y_0) + a_2 f_y(x_0, y_0) \big].$$

The *gradient* of the scalar field $z = f(x, y)$ is, by definition, the vector-valued function

$$\operatorname{grad} f = f_x(x, y)\mathbf{i} + f_y(x, y)\mathbf{j},$$

where $\mathbf{i}$ and $\mathbf{j}$ are unit vectors along the coordinate axes x and y. At each point, the gradient of a scalar field is orthogonal to the level line passing through that point. The gradient indicates the direction of maximal growth of the field. In terms of the gradient, the directional derivative can be expressed as follows:

$$\frac{\partial f}{\partial \mathbf{a}} = \frac{\mathbf{a}}{|\mathbf{a}|} \operatorname{grad} f.$$

The gradient is also denoted by $\nabla f = \operatorname{grad} f$.

Remark. The above facts for a plane scalar field obviously can be extended to the case of a spatial scalar field.

▶ **Geometrical applications of the theory of functions of several variables.**

1. The *equation of the tangent plane* to the surface $z = f(x, y)$ at a point (x_0, y_0, z_0), where $z_0 = f(x_0, y_0)$, has the form

$$z = f(x_0, y_0) + f_x(x_0, y_0)(x - x_0) + f_y(x_0, y_0)(y - y_0).$$

The vector of the normal to the surface at that point is

$$\mathbf{n} = \big\{ -f_x(x_0, y_0), \ -f_y(x_0, y_0), \ 1 \big\}.$$

2. If a surface is defined by the equation $\Phi(x, y, z) = 0$, then the equation of its tangent plane at the point (x_0, y_0, z_0) has the form

$$\Phi_x(x_0, y_0, z_0)(x - x_0) + \Phi_y(x_0, y_0, z_0)(y - y_0) + \Phi_z(x_0, y_0, z_0)(z - z_0) = 0.$$

A normal vector to the surface at this point is

$$\mathbf{n} = \big\{ \Phi_x(x_0, y_0, z_0), \ \Phi_y(x_0, y_0, z_0), \ \Phi_z(x_0, y_0, z_0) \big\}.$$

3. Consider a surface defined by the parametric equations

$$x = x(u, v), \quad y = y(u, v), \quad z = z(u, v)$$

or, in vector form, $\mathbf{r} = \mathbf{r}(u, v)$, where $\mathbf{r} = \{x, y, z\}$, and let $M_0\big(x(u_0, v_0), \ y(u_0, v_0), \ z(u_0, v_0)\big)$ be the point of the surface corresponding to the parameter values $u = u_0$, $v = v_0$. Then the vector of the normal to the surface at the point M_0 can be expressed by

$$\mathbf{n}(u, v) = \frac{\partial \mathbf{r}}{\partial u} \times \frac{\partial \mathbf{r}}{\partial v} = \begin{vmatrix} \mathbf{i} & \mathbf{j} & \mathbf{k} \\ x_u & y_u & z_u \\ x_v & y_v & z_v \end{vmatrix},$$

where all partial derivatives are calculated at the point M_0.

M6.3.4. Extremal Points of Functions of Several Variables

▶ **Conditions of extremum of a function of two variables.**

1°. *Points of minimum, maximum, or extremum.* A point (x_0, y_0) is called a *point of local minimum* (resp., *maximum*) of a function $z = f(x, y)$ if there is a neighborhood of (x_0, y_0) in which the function is defined and satisfies the inequality $f(x, y) > f(x_0, y_0)$ (resp., $f(x, y) < f(x_0, y_0)$). Points of maximum or minimum are called *points of extremum.*

2°. A *necessary condition of extremum.* If a function has the first partial derivatives at a point of its extremum, these derivatives must vanish at that point. It follows that in order to find points of extremum of such a function $z = f(x, y)$, one should find solutions of the system of equations

$$f_x(x, y) = 0, \quad f_y(x, y) = 0.$$

The points whose coordinates satisfy this system are called *stationary points.* Any point of extremum of a differentiable function is its stationary point, but not every stationary point is a point of its extremum.

3°. *Sufficient conditions of extremum* are used for the identification of points of extremum among stationary points. Some conditions of this type are given below.

Suppose that the function $z = f(x, y)$ has continuous second derivatives at a stationary point. Let us calculate the following quantity at this point:

$$\Delta = f_{xx} f_{yy} - f_{xy}^2.$$

The following statements hold:

1) If $\Delta > 0$, $f_{xx} > 0$, then the stationary point is a point of local minimum;
2) If $\Delta > 0$, $f_{xx} < 0$, then the stationary point is a point of local maximum;
3) If $\Delta < 0$, then the stationary point cannot be a point of extremum.

In the degenerate case, $\Delta = 0$, a more delicate analysis of a stationary point is required. In this case, a stationary point may happen to be a point of extremum and may not.

Remark. In order to find points of extremum, one should check not only stationary points, but also points at which the first derivatives do not exist or are infinite.

4°. *The smallest and the largest values of a function.* Let $f(x, y)$ be a continuous function in a closed bounded domain D. Any such function takes its smallest and its largest values in D.

If the function has partial derivatives in D, except at some points, then the following method can be helpful for determining the coordinates of the points (x_{min}, y_{min}) and (x_{max}, y_{max}) at which the function attains its minimum and maximum, respectively. One should find all internal stationary points and all points at which the derivatives are infinite or do not exist. Then one should calculate the values of the function at these points and compare these with its values at the boundary points of the domain, and then choose the largest and the smallest values.

▶ **Extremal points of functions of three variables.** For functions of three variables, points of extremum are defined in exactly the same way as for functions of two variables. Let us briefly describe the scheme of finding extremal points of a function $u = \Phi(x, y, z)$. Finding solutions of the system of equations

$$\Phi_x(x, y, z) = 0, \quad \Phi_y(x, y, z) = 0, \quad \Phi_z(x, y, z) = 0,$$

we determine stationary points. For each stationary point, we calculate the values of

$$\Delta_1 = \Phi_{xx}, \quad \Delta_2 = \begin{vmatrix} \Phi_{xx} & \Phi_{xy} \\ \Phi_{xy} & \Phi_{yy} \end{vmatrix}, \quad \Delta_3 = \begin{vmatrix} \Phi_{xx} & \Phi_{xy} & \Phi_{xz} \\ \Phi_{xy} & \Phi_{yy} & \Phi_{yz} \\ \Phi_{xz} & \Phi_{yz} & \Phi_{zz} \end{vmatrix}.$$

The following statements hold:

1) If $\Delta_1 > 0$, $\Delta_2 > 0$, $\Delta_3 > 0$, then the stationary point is a point of local minimum;

2) If $\Delta_1 < 0$, $\Delta_2 > 0$, $\Delta_3 < 0$, then the stationary point is a point of local maximum.

▶ **Conditional extremum of a function of two variables. Lagrange function.** A point (x_0, y_0) is called a *point of conditional or constrained minimum* (resp., *maximum*) of a function

$$z = f(x, y) \tag{M6.3.4.1}$$

under the additional condition*

$$\varphi(x, y) = 0 \tag{M6.3.4.2}$$

if there is a neighborhood of the point (x_0, y_0) in which $f(x, y) > f(x_0, y_0)$ (resp., $f(x, y) < f(x_0, y_0)$) for all points (x, y) satisfying the condition (M6.3.4.2).

For the determination of points of conditional extremum, it is common to use the *Lagrange function*

$$\Phi(x, y, \lambda) = f(x, y) + \lambda \varphi(x, y),$$

where λ is the so-called *Lagrange multiplier*. Solving the system of three equations (the last equation coincides with the condition (M6.3.4.2))

$$\frac{\partial \Phi}{\partial x} = 0, \quad \frac{\partial \Phi}{\partial y} = 0, \quad \frac{\partial \Phi}{\partial \lambda} = 0,$$

one finds stationary points of the Lagrange function (and also the value of the multiplier λ). The stationary points may happen to be points of extremum. The above system yields only necessary conditions of extremum, but these conditions may be insufficient; it may happen that there is no extremum at some stationary points. However, with the help of other properties of the function under consideration, it is often possible to establish the character of a critical point.

Example 1. Let us find an extremum of the function

$$z = x^n y, \tag{M6.3.4.3}$$

under the condition

$$x + y = a \qquad (a > 0, \quad n > 0, \quad x \geq 0, \quad y \geq 0). \tag{M6.3.4.4}$$

Taking $\varphi(x, y) = x + y - a$, we construct the Lagrange function

$$\Phi(x, y, \lambda) = x^n y + \lambda(x + y - a).$$

Solving the system of equations

$$\Phi_x \equiv n x^{n-1} y + \lambda = 0,$$
$$\Phi_y \equiv x^n + \lambda = 0,$$
$$\Phi_\lambda \equiv x + y - a = 0,$$

we find the coordinates of a unique stationary point,

$$x_\circ = \frac{an}{n+1}, \quad y_\circ = \frac{a}{n+1}, \quad \lambda_\circ = -\left(\frac{an}{n+1}\right)^n,$$

which corresponds to the conditional maximum of the given function, $z_{\max} = \dfrac{a^{n+1} n^n}{(n+1)^{n+1}}$.

---

* This condition is also called a *constraint*.

Remark. In order to find points of conditional extremum of functions of two variables, it is often convenient to express the variable y through x (or vice versa) from the additional equation (M6.3.4.2) and substitute the resulting expression into the right-hand side of (M6.3.4.1). In this way, the original problem is reduced to the problem of extremum for a function of a single variable.

Example 2. Consider again the extremum problem of Example 1 for the function of two variables (M6.3.4.3) with the constraint (M6.3.4.4). After the elimination of the variable y from (M6.3.4.3)–(M6.3.4.4), the original problem is reduced to the extremum problem for the function $z = x^n(a - x)$ of one variable.

▶ **Conditional extrema of functions of several variables.** Consider a function $u = f(x_1, \ldots, x_n)$ of n variables under the condition that x_1, $\ldots$, x_n satisfy m equations ($m < n$):

$$
\begin{cases}
\varphi_1(x_1, \ldots, x_n) = 0, \\
\varphi_2(x_1, \ldots, x_n) = 0, \\
\cdots\cdots\cdots\cdots\cdots, \\
\varphi_m(x_1, \ldots, x_n) = 0.
\end{cases}
$$

In order to find the values of x_1, $\ldots$, x_n for which f may have a conditional maximum or minimum, one should construct the Lagrange function

$$
\Phi(x_1, \ldots, x_n; \lambda_1, \ldots, \lambda_m) = f + \lambda_1\varphi_1 + \lambda_2\varphi_2 + \cdots + \lambda_m\varphi_m
$$

and equate to zero its first partial derivatives with respect to the variables x_1, $\ldots$, x_n and the parameters λ_1, $\ldots$, λ_m. From the resulting $n + m$ equations, one finds x_1, $\ldots$, x_n (and also the values of the unknown Lagrange multipliers λ_1, $\ldots$, λ_m). As in the case of functions of two variables, the question whether the given function has points of conditional extremum can be answered on the basis of additional investigation.

Example 3. Consider the problem of finding the shortest distance from the point (x_0, y_0, z_0) to the plane

$$
Ax + By + Cz + D = 0. \tag{M6.3.4.5}
$$

The squared distance between the points (x_0, y_0, z_0) and (x, y, z) is equal to

$$
R^2 = (x - x_0)^2 + (y - y_0)^2 + (z - z_0)^2. \tag{M6.3.4.6}
$$

In our case, the coordinates (x, y, z) should satisfy equation (M6.3.4.5) (this point should belong to the plane). Thus, our problem is to find the minimum of the expression (M6.3.4.6) under the condition (M6.3.4.5). The Lagrange function has the form

$$
\Phi = (x - x_0)^2 + (y - y_0)^2 + (z - z_0)^2 + \lambda(Ax + By + Cz + D).
$$

Equating to zero the derivatives of Φ with respect to x, y, z, and λ, we obtain the following system of algebraic equations:

$$
2(x - x_0) + A\lambda = 0, \quad 2(y - y_0) + B\lambda = 0, \quad 2(z - z_0) + C\lambda = 0, \quad Ax + By + Cz + D = 0.
$$

Its solution has the form

$$
x = x_0 - \frac{1}{2}A\lambda, \quad y = y_0 - \frac{1}{2}B\lambda, \quad z = z_0 - \frac{1}{2}C\lambda, \quad \lambda = \frac{2(Ax_0 + By_0 + Cz_0 + D)}{A^2 + B^2 + C^2}. \tag{M6.3.4.7}
$$

Thus we have a unique answer, and since the distance between a given point and the plane can be realized at a single point (x, y, z), the values obtained should correspond to that distance. Substituting the values (M6.3.4.7) into (M6.3.4.6), we find the squared distance

$$
R^2 = \frac{(Ax_0 + By_0 + Cz_0 + D)^2}{A^2 + B^2 + C^2}.
$$

M6.3.5. Differential Operators of the Field Theory

▶ **Hamilton's operator and first-order differential operators.** Hamilton's operator, commonly known as the *nabla vector operator* or the *gradient operator*, is the symbolic vector

$$\nabla = \mathbf{i}\,\frac{\partial}{\partial x} + \mathbf{j}\,\frac{\partial}{\partial y} + \mathbf{k}\,\frac{\partial}{\partial z}.$$

This vector can be used for expressing the following differential operators:

1) gradient of a scalar function $u(x, y, z)$:

$$\operatorname{grad} u = \mathbf{i}\,\frac{\partial u}{\partial x} + \mathbf{j}\,\frac{\partial u}{\partial y} + \mathbf{k}\,\frac{\partial u}{\partial z} = \nabla u;$$

2) divergence of a vector field $\mathbf{a} = P\,\mathbf{i} + Q\,\mathbf{j} + R\,\mathbf{k}$:

$$\operatorname{div} \mathbf{a} = \frac{\partial P}{\partial x} + \frac{\partial Q}{\partial y} + \frac{\partial R}{\partial z} = \nabla \cdot \mathbf{a}$$

(scalar product of the nabla vector and the vector $\mathbf{a}$);

3) rotation of a vector field $\mathbf{a} = P\,\mathbf{i} + Q\,\mathbf{j} + R\,\mathbf{k}$:

$$\operatorname{curl} \mathbf{a} = \begin{vmatrix} \mathbf{i} & \mathbf{j} & \mathbf{k} \\ \frac{\partial}{\partial x} & \frac{\partial}{\partial y} & \frac{\partial}{\partial z} \\ P & Q & R \end{vmatrix} = \nabla \times \mathbf{a}$$

(vector product of the nabla vector and the vector $\mathbf{a}$).

Each scalar field $u(x, y, z)$ generates a vector field $\operatorname{grad} u$. A vector field $\mathbf{a}(x, y, z)$ generates two fields: the scalar field $\operatorname{div} \mathbf{a}$ and the vector field $\operatorname{curl} \mathbf{a}$.

▶ **Second-order differential operators.** The following differential identities hold:

$$
\begin{aligned}
&1) \quad \operatorname{curl}\operatorname{grad} u = \mathbf{0} \quad \text{or} \quad (\nabla \times \nabla)\,u = \mathbf{0},\\
&2) \quad \operatorname{div}\operatorname{curl} \mathbf{a} = 0 \quad \text{or} \quad \nabla \cdot (\nabla \times \mathbf{a}) = 0.
\end{aligned}
$$

The following differential relations hold:

$$1) \quad \operatorname{div}\operatorname{grad} u = \Delta u = \frac{\partial^2 u}{\partial x^2} + \frac{\partial^2 u}{\partial y^2} + \frac{\partial^2 u}{\partial z^2},$$

$$2) \quad \operatorname{curl}\operatorname{curl} \mathbf{a} = \operatorname{grad}\operatorname{div} \mathbf{a} - \Delta\mathbf{a},$$

where Δ is the *Laplace operator*, $\Delta u = \nabla \cdot (\nabla u) = \nabla^2 u$.

Bibliography for Chapter M6

Adams, R., *Calculus: A Complete Course, 6th Edition*, Pearson Education, Toronto, 2006.

Boyer, C. B., *The History of the Calculus and Its Conceptual Development*, Dover Publications, New York, 1989.

Brannan, D., *A First Course in Mathematical Analysis*, Cambridge University Press, Cambridge, England, 2006.

Browder, A., *Mathematical Analysis: An Introduction*, Springer-Verlag, New York, 1996.

Courant, R. and John, F., *Introduction to Calculus and Analysis, Vol. 1*, Springer-Verlag, New York, 1999.

Edwards, C. H., and Penney, D., *Calculus, 6th Edition*, Pearson Education, Toronto, 2002.

Kline, M., *Calculus: An Intuitive and Physical Approach, 2nd Edition*, Dover Publications, New York, 1998.

Landau, E., *Differential and Integral Calculus*, American Mathematical Society, Providence, Rhode Island, 2001.

Silverman, R. A., *Essential Calculus with Applications*, Dover Publications, New York, 1989.

Zorich, V. A., *Mathematical Analysis*, Springer-Verlag, Berlin, 2004.

Chapter M7

Integrals

M7.1. Indefinite Integral

M7.1.1. Antiderivative. Indefinite Integral and Its Properties

▶ **Antiderivative.** An *antiderivative* (or *primitive function*) of a given function $f(x)$ on an interval (a, b) is a differentiable function $F(x)$ such that its derivative is equal to $f(x)$ for all $x \in (a, b)$:

$$F'(x) = f(x).$$

Example 1. Let $f(x) = 2x$. Then the functions $F(x) = x^2$ and $F_1(x) = x^2 - 1$ are antiderivatives of $f(x)$, since $(x^2)' = 2x$ and $(x^2 - 1)' = 2x$.

THEOREM. *Any function $f(x)$ continuous on an interval (a, b) has infinitely many continuous antiderivatives on (a, b). If $F(x)$ is one of them, then any other antiderivative has the form $F(x) + C$, where C is a constant.*

▶ **Indefinite integral.** The *indefinite integral* of a function $f(x)$ is the set, $F(x) + C$, of all its antiderivatives. This fact is written as

$$\int f(x)\, dx = F(x) + C.$$

Here, $f(x)$ is called the *integrand* (or the *integrand function*). The process of finding an integral is called integration. The differential dx indicates that the integration is carried out with respect to x.

Example 2. $\displaystyle\int 6x^2\, dx = 2x^3 + C$, since $(2x^3)' = 6x^2$.

▶ **Most important corollaries of the definition of the indefinite integral.** *Differentiation is the inverse of integration*:

$$\frac{d}{dx}\left(\int f(x)\, dx \right) = f(x).$$

Integration is the inverse of differentiation:*

$$\int f'(x)\, dx = f(x) + C.$$

The latter formula serves to make up tables of indefinite integrals. The procedure is often reversed here: an integral is first given in explicit form (i.e., the function $f(x)$ on the right-hand side is prescribed), and then the integrand is obtained by differentiation.

---

* Integration recovers the function from its derivative, to an additive constant.

M7.1.2. Table of Basic Integrals. Properties of the Indefinite Integral. Examples of Integration

▶ **Table of basic integrals.** Listed below are most common indefinite integrals, which are important for the integration of more complicated expressions:

$$\int x^a \, dx = \frac{x^{a+1}}{a+1} + C \quad (a \neq -1),$$

$$\int \frac{dx}{x} = \ln|x| + C,$$

$$\int \frac{dx}{x^2 + a^2} = \frac{1}{a} \arctan \frac{x}{a} + C,$$

$$\int \frac{dx}{x^2 - a^2} = \frac{1}{2a} \ln\left|\frac{x-a}{x+a}\right| + C,$$

$$\int \frac{dx}{\sqrt{a^2 - x^2}} = \arcsin \frac{x}{a} + C,$$

$$\int \frac{dx}{\sqrt{x^2 + a}} = \ln\left|x + \sqrt{x^2 + a}\right| + C,$$

$$\int e^x \, dx = e^x + C,$$

$$\int a^x \, dx = \frac{a^x}{\ln a} + C,$$

$$\int \ln x \, dx = x \ln x - x + C,$$

$$\int \ln ax \, dx = x \ln ax - x + C,$$

$$\int \sin x \, dx = -\cos x + C,$$

$$\int \cos x \, dx = \sin x + C,$$

$$\int \tan x \, dx = -\ln|\cos x| + C,$$

$$\int \cot x \, dx = \ln|\sin x| + C,$$

$$\int \frac{dx}{\sin x} = \ln\left|\tan \frac{x}{2}\right| + C,$$

$$\int \frac{dx}{\cos x} = \ln\left|\tan\left(\frac{x}{2} + \frac{\pi}{4}\right)\right| + C,$$

$$\int \frac{dx}{\sin^2 x} = -\cot x + C,$$

$$\int \frac{dx}{\cos^2 x} = \tan x + C,$$

$$\int \arcsin x \, dx = x \arcsin x + \sqrt{1 - x^2} + C,$$

$$\int \arccos x \, dx = x \arccos x - \sqrt{1 - x^2} + C,$$

$$\int \arctan x \, dx = x \arctan x - \frac{1}{2} \ln(1 + x^2) + C,$$

$$\int \operatorname{arccot} x \, dx = x \operatorname{arccot} x + \frac{1}{2} \ln(1 + x^2) + C,$$

$$\int \sinh x \, dx = \cosh x + C,$$

$$\int \cosh x \, dx = \sinh x + C,$$

$$\int \tanh x \, dx = \ln \cosh x + C,$$

$$\int \coth x \, dx = \ln|\sinh x| + C,$$

$$\int \frac{dx}{\sinh x} = \ln\left|\tanh \frac{x}{2}\right| + C,$$

$$\int \frac{dx}{\cosh x} = 2 \arctan e^x + C,$$

$$\int \frac{dx}{\sinh^2 x} = -\coth x + C,$$

$$\int \frac{dx}{\cosh^2 x} = \tanh x + C,$$

$$\int \operatorname{arcsinh} x \, dx = x \operatorname{arcsinh} x - \sqrt{1 + x^2} + C,$$

$$\int \operatorname{arccosh} x \, dx = x \operatorname{arccosh} x - \sqrt{x^2 - 1} + C,$$

$$\int \operatorname{arctanh} x \, dx = x \operatorname{arctanh} x + \frac{1}{2} \ln(1 - x^2) + C,$$

$$\int \operatorname{arccoth} x \, dx = x \operatorname{arccoth} x + \frac{1}{2} \ln(x^2 - 1) + C,$$

where C is an arbitrary constant.

A more extensive table of indefinite integrals can be found in Section S1.1.

▶ **Properties of the indefinite integral.**

1. *A constant factor can be taken outside the integral sign:*

$$\int a f(x) \, dx = a \int f(x) \, dx \quad (a = \text{const}).$$

2. *Integral of the sum or difference of functions (additivity):*

$$\int [f(x) \pm g(x)] \, dx = \int f(x) \, dx \pm \int g(x) \, dx.$$

3. *Integration by parts*:

$$\int f(x)g'(x)\,dx = f(x)g(x) - \int f'(x)g(x)\,dx.$$

4. *Repeated integration by parts* (generalization of the previous formula):

$$\int f(x)g^{(n+1)}(x)\,dx = f(x)g^{(n)}(x) - f'(x)g^{(n-1)}(x) + \cdots + (-1)^n f^{(n)}(x)g(x)$$

$$+ (-1)^{n+1} \int f^{(n+1)}(x)g(x)\,dx, \qquad n = 0, 1, \ldots$$

5. *Change of variable* (*integration by substitution*):

$$\int f(x)\,dx = \int f(\varphi(t))\,\varphi'(t)\,dt, \qquad x = \varphi(t).$$

On computing the integral using the change of variable $x = \varphi(t)$, one should rewrite the resulting expression in terms of the original variable x using the inverse substitution $t = \varphi^{-1}(x)$.

▶ **Examples of direct integration of elementary functions.**

$1°$. With simple algebraic manipulation and the properties listed above, the integration may often be reduced to tabulated integrals.

Example 1. $\displaystyle \int \frac{2x-1}{\sqrt{x}}\,dx = \int \left(2\sqrt{x} - \frac{1}{\sqrt{x}}\right)dx = 2\int x^{1/2}\,dx - \int x^{-1/2}\,dx = \frac{4}{3}x^{3/2} - 2x^{1/2} + C.$

$2°$. Tabulated integrals can also be used where any function $\varphi(x)$ appears in place of x; for example,

$$\int e^x\,dx = e^x + C \quad \Longrightarrow \quad \int e^{\varphi(x)}\,d\varphi(x) = e^{\varphi(x)} + C;$$

$$\int \frac{dx}{x} = \ln|x| + C \quad \Longrightarrow \quad \int \frac{d\sin x}{\sin x} = \ln|\sin x| + C.$$

The reduction of an integral to a tabulated one may often be achieved by taking some function inside the differential sign.

Example 2. $\displaystyle \int \tan x\,dx = \int \frac{\sin x\,dx}{\cos x} = \int \frac{-d\cos x}{\cos x} = -\int \frac{d\cos x}{\cos x} = -\ln|\cos x| + C.$

$3°$. Integrals of the form $\displaystyle \int \frac{dx}{ax^2 + bx + c},\, \int \frac{dx}{\sqrt{ax^2 + bx + c}}$ can be computed by making a perfect square:

$$ax^2 + bx + c = a\left(x + \frac{b}{2a}\right)^2 - \frac{b^2}{4a} + c.$$

Then one should replace dx with the equal differential $d\left(x + \frac{b}{2a}\right)$ and use one of the four formulas in the second and third rows in the table of integrals given at the beginning of the current subsection.

Example 3. $\displaystyle \int \frac{dx}{\sqrt{2x - x^2}} = \int \frac{dx}{\sqrt{1 - (x-1)^2}} = \int \frac{d(x-1)}{\sqrt{1 - (x-1)^2}} = \arcsin(x - 1) + C.$

$4°$. The integration of a polynomial multiplied by an exponential function can be accomplished by using the formula of integration by parts (or repeated integration by parts) given above.

Example 4. Compute the integral $\int (3x + 1) e^{2x}\, dx$.

Taking $f(x) = 3x + 1$ and $g'(x) = e^{2x}$, one finds that $f'(x) = 3$ and $g(x) = \frac{1}{2} e^{2x}$. On substituting these expressions into the formula of integration by parts, one obtains

$$\int (3x + 1) e^{2x}\, dx = \frac{1}{2}(3x + 1) e^{2x} - \frac{3}{2} \int e^{2x}\, dx = \frac{1}{2}(3x + 1) e^{2x} - \frac{3}{4} e^{2x} + C = \left(\frac{3}{2} x - \frac{1}{4} \right) e^{2x} + C.$$

Remark 1. More complex examples of the application of integration by parts or repeated integration by parts can be found in Subsection M7.1.6.

Remark 2. Examples of using a change of variable (see Property 5 above) for the computation of integrals can be found in Subsections M7.1.4 and M7.1.5.

M7.1.3. Integration of Rational Functions

▶ **Partial fraction decomposition of a rational function.** A *rational function* (also known as a *rational polynomial function*) is a quotient of polynomials:

$$R(x) = \frac{P_n(x)}{Q_m(x)}, \tag{M7.1.3.1}$$

where

$$P_n(x) = a_n x^n + \cdots + a_1 x + a_0,$$
$$Q_m(x) = b_m x^m + \cdots + b_1 x + b_0.$$

The fraction (M7.1.3.1) is called *proper* if $m > n$ and *improper* if $m \leq n$.

Every proper fraction (M7.1.3.1) can be decomposed into a sum of partial fractions. To this end, one should factorize the denominator $Q_m(x)$ into irreducible multipliers of the form

$$(x - \alpha_i)^{p_i}, \qquad\qquad i = 1, 2, \ldots, k; \tag{M7.1.3.2a}$$
$$(x^2 + \beta_j x + \gamma_j)^{q_j}, \qquad j = 1, 2, \ldots, s, \tag{M7.1.3.2b}$$

where the p_i and q_j are positive integers satisfying the condition $p_1 + \cdots + p_k + 2(q_1 + \cdots + q_s) = m$; $\beta_j^2 - 4\gamma_j < 0$. The rational function (M7.1.3.1) can be represented as a sum of irreducibles and to each irreducible of the form (M7.1.3.2) there correspond as many terms as the power p_i or q_i:

$$\frac{A_{i,1}}{x - \alpha_i} + \frac{A_{i,2}}{(x - \alpha_i)^2} + \cdots + \frac{A_{i,p_i}}{(x - \alpha_i)^{p_i}}; \tag{M7.1.3.3a}$$
$$\frac{B_{j,1} x + D_{j,1}}{x^2 + \beta_j x + \gamma_j} + \frac{B_{j,2} x + D_{j,2}}{(x^2 + \beta_j x + \gamma_j)^2} + \cdots + \frac{B_{j,q_j} x + D_{j,q_j}}{(x^2 + \beta_j x + \gamma_j)^{q_j}}. \tag{M7.1.3.3b}$$

The constants $A_{i,l}$, $B_{j,r}$, $D_{j,r}$ are found by the *method of undetermined coefficients*. To that end, one should equate the original rational fraction (M7.1.3.1) with the sum of the above partial fractions (M7.1.3.3) and reduce both sides of the resulting equation to a common denominator. Then, one collects the coefficients of like powers of x and equates them with zero, thus arriving at a system of linear algebraic equations for the $A_{i,l}$, $B_{j,r}$, and $D_{j,r}$.

Example 1. This is an illustration of how a proper fraction can be decomposed into partial fractions:

$$\frac{b_5 x^5 + b_4 x^4 + b_3 x^3 + b_2 x^2 + b_1 x + b_0}{(x+a)(x+c)^3(x^2+k^2)} = \frac{A_{1,1}}{x+a} + \frac{A_{2,1}}{x+c} + \frac{A_{2,2}}{(x+c)^2} + \frac{A_{2,3}}{(x+c)^3} + \frac{Bx+D}{x^2+k^2}.$$

▶ **Integration of a proper fraction.**

$1°$. To integrate a proper fraction, one should first rewrite the integrand (M7.1.3.1) in the form of a sum of partial fractions. Below are the integrals of most common partial fractions (M7.1.3.3a) and (M7.1.3.3b) (with $q_j = 1$):

$$\int \frac{A}{x-\alpha}\, dx = A\ln|x-\alpha|, \qquad \int \frac{A}{(x-\alpha)^p}\, dx = -\frac{A}{(p-1)(x-\alpha)^{p-1}},$$

$$\int \frac{Bx+D}{x^2+\beta x+\gamma}\, dx = \frac{B}{2}\ln(x^2+\beta x+\gamma) + \frac{2D-B\beta}{\sqrt{4\gamma-\beta^2}}\arctan\frac{2x+\beta}{\sqrt{4\gamma-\beta^2}}. \qquad \text{(M7.1.3.4)}$$

The constant of integration C has been omitted here. More complex integrals of partial fractions (M7.1.3.3b) with $q_j > 1$ can be computed using the formula

$$\int \frac{Bx+D}{(x^2+\beta x+\gamma)^q}\, dx = \frac{P(x)}{(x^2+\beta x+\gamma)^{q-1}} + \lambda \int \frac{dx}{x^2+\beta x+\gamma}, \qquad \text{(M7.1.3.5)}$$

where $P(x)$ is a polynomial of degree $2q-3$. The coefficients of $P(x)$ and the constant λ can be found by the method of undetermined coefficients by differentiating formula (M7.1.3.5).

Remark. The following recurrence relation may be used in order to compute the integrals on the left-hand side in (M7.1.3.5):

$$\int \frac{Bx+D}{(x^2+\beta x+\gamma)^q}\, dx = \frac{(2D-B\beta)x + D\beta - 2B\gamma}{(q-1)(4\gamma-\beta^2)(x^2+\beta x+\gamma)^{q-1}} + \frac{(2q-3)(2D-B\beta)}{(q-1)(4\gamma-\beta^2)}\int \frac{dx}{(x^2+\beta x+\gamma)^{q-1}}.$$

Example 2. Compute the integral $\int \frac{3x^2-x-2}{x^3+8}\, dx$.

Let us factor the denominator of the integrand, $x^3+8 = (x+2)(x^2-2x+4)$, and perform the partial fraction decomposition:

$$\frac{3x^2-x-2}{(x+2)(x^2-2x+4)} = \frac{A}{x+2} + \frac{Bx+D}{x^2-2x+4}.$$

Multiplying both sides by the common denominator and collecting the coefficients of like powers of x, we obtain

$$(A+B-3)x^2 + (-2A+2B+D+1)x + 4A+2D+2 = 0.$$

Now equating the coefficients of the different powers of x with zero, we arrive at a system of algebraic equations for A, B, and D:

$$A+B-3 = 0, \quad -2A+2B+D+1 = 0, \quad 4A+2D+2 = 0.$$

Its solution is: $A = 1$, $B = 2$, $D = -3$. Hence, we have

$$\int \frac{3x^2-x-2}{x^3+8}\, dx = \int \frac{1}{x+2}\, dx + \int \frac{2x-3}{x^2-2x+4}\, dx$$

$$= \ln|x+2| + \ln(x^2-2x+4) - \frac{1}{\sqrt{3}}\arctan\frac{x-1}{\sqrt{3}} + C.$$

Here, the last integral of (M7.1.3.4) has been used.

$2°$. The integrals of proper rational functions defined as the ratio of a polynomial to a power function $(x - \alpha)^m$ are given by the formulas

$$\int \frac{P_n(x)}{(x-\alpha)^m} \, dx = -\sum_{k=0}^{n} \frac{P_n^{(k)}(\alpha)}{k!\,(m-k-1)(x-\alpha)^{m-k-1}} + C, \qquad m > n+1;$$

$$\int \frac{P_n(x)}{(x-\alpha)^{n+1}} \, dx = -\sum_{k=0}^{n-1} \frac{P_n^{(k)}(\alpha)}{k!\,(n-k)(x-\alpha)^{n-k}} + \frac{P_n^{(n)}(\alpha)}{n!} \ln|x-\alpha| + C,$$

where $P_n(x)$ is a polynomial of degree n and $P_n^{(k)}(\alpha)$ is its kth derivative at $x = \alpha$.

$3°$. Suppose the roots in the factorization of the denominator of the fraction (M7.1.3.1) are all real and distinct:

$$Q_m(x) = b_m x^m + \cdots + b_1 x + b_0 = b_m(x-\alpha_1)(x-\alpha_2)\ldots(x-\alpha_m), \qquad \alpha_i \neq \alpha_j.$$

Then the following formula holds:

$$\int \frac{P_n(x)}{Q_m(x)} \, dx = \sum_{k=1}^{m} \frac{P_n(\alpha_k)}{Q'_m(\alpha_k)} \ln|x-\alpha_k| + C,$$

where $m > n$ and the prime denotes a derivative.

▶ **Integration of improper fractions.**

$1°$. In order to integrate an improper fraction, one should first isolate a proper fraction by division with remainder. As a result, the improper fraction is represented as the sum of a polynomial and a proper fraction,

$$\frac{a_n x^n + \cdots + a_1 x + a_0}{b_m x^m + \cdots + b_1 x + b_0} = c_m x^{n-m} + \cdots + c_1 x + c_0 + \frac{s_{m-1} x^{m-1} + \cdots + s_1 x + s_0}{b_m x^m + \cdots + b_1 x + b_0} \qquad (n \geq m),$$

which are then integrated separately.

Example 3. Evaluate the integral $I = \displaystyle\int \frac{x^2}{x-1} \, dx$.

Let us rewrite the integrand (improper fraction) as the sum of a polynomial and a proper fraction: $\dfrac{x^2}{x-1} = x + 1 + \dfrac{1}{x-1}$. Hence, $I = \displaystyle\int \left(x + 1 + \frac{1}{x-1} \right) dx = \frac{1}{2}x^2 + x + \ln|x-1| + C$.

$2°$. The integrals of improper rational functions defined as the ratio of a polynomial to a simple power function $(x - \alpha)^m$ are evaluated by the formula

$$\int \frac{P_n(x)}{(x-\alpha)^m} \, dx = \sum_{k=m}^{n} \frac{P_n^{(k)}(\alpha)}{k!\,(k-m+1)}(x-\alpha)^{k-m+1} + \frac{P_n^{(m-1)}(\alpha)}{(m-1)!} \ln|x-\alpha|$$

$$- \sum_{k=0}^{m-2} \frac{P_n^{(k)}(\alpha)}{k!\,(m-k-1)(x-\alpha)^{m-k-1}} + C,$$

where $n \geq m$.

Remark 1. The indefinite integrals of rational functions are always expressed in terms of elementary functions.

Remark 2. Some of the integrals reducible to integrals of rational functions are considered in Subsections M7.1.5 and M7.1.6.

M7.1.4. Integration of Irrational Functions

The integration of some irrational functions can be reduced to that of rational functions using a suitable change of variables. In what follows, the functions $R(x, y)$ and $R(x_1, \ldots, x_k)$ are assumed to be rational functions in each of the arguments.

▶ **Integration of expressions involving radicals of linear-fractional functions.**

$1°$. The integrals with roots of linear functions

$$\int R\left(x, \sqrt[n]{ax + b}\right) dx$$

are reduced to integrals of rational functions by the change of variable $z = \sqrt[n]{ax + b}$.

Example 1. Evaluate the integral $I = \int x\sqrt{1 - x}\, dx$.

With the change of variable $\sqrt{1 - x} = z$, we have $x = 1 - z^2$ and $dx = -2z\, dz$. Substituting these expressions into the integral yields

$$I = -2\int (1 - z^2)z^2\, dz = -\frac{2}{3}z^3 + \frac{2}{5}z^5 + C = -\frac{2}{3}\sqrt{(1 - x)^3} + \frac{2}{5}\sqrt{(1 - x)^5} + C.$$

$2°$. The integrals with roots of linear-fractional functions

$$\int R\left(x, \sqrt[n]{\frac{ax + b}{cx + d}}\right) dx$$

are reduced to integrals of rational functions by the substitution $z = \sqrt[n]{\dfrac{ax + b}{cx + d}}$.

$3°$. Integrals containing the product of a polynomial by a simple power function of the form $(x - a)^\beta$ are evaluated by the formula

$$\int P_n(x)(x - a)^\beta\, dx = \sum_{k=0}^{n} \frac{P_n^{(k)}(a)}{k!\,(k + \beta + 1)}(x - a)^{k+\beta+1},$$

where $P_n(x)$ is a polynomial of degree n, $P_n^{(k)}(a)$ is its kth derivative at $x = a$, and β is any positive or negative proper fraction (to be more precise, $\beta \neq -1, -2, \ldots, -n - 1$).

▶ **Euler substitutions. Trigonometric substitutions.** We will be considering integrals involving the radical of a quadratic trinomial:

$$\int R\left(x, \sqrt{ax^2 + bx + c}\right) dx,$$

where $b^2 \neq 4ac$. Such integrals are expressible in terms of elementary functions.

$1°$. *Euler substitutions.* The given integral is reduced to the integral of a rational fraction by one of the following three Euler substitutions:

1) $\sqrt{ax^2 + bx + c} = t \mp x\sqrt{a}$ if $a > 0$;

2) $\sqrt{ax^2 + bx + c} = xt \pm \sqrt{c}$ if $c > 0$;

3) $\sqrt{ax^2 + bx + c} = t(x - x_1)$ if $4ac - b^2 < 0$,

where x_1 is a root of the quadratic equation $ax^2 + bx + c = 0$. In all three cases, the variable x and the radical $\sqrt{ax^2 + bx + c}$ are expressible in terms of the new variable t as (the formulas correspond to the upper signs in the substitutions):

1) $x = \dfrac{t^2 - c}{2\sqrt{a}\,t + b}$, $\quad \sqrt{ax^2 + bx + c} = \dfrac{\sqrt{a}\,t^2 + bt + c\sqrt{a}}{2\sqrt{a}\,t + b}$, $\quad dx = 2\dfrac{\sqrt{a}\,t^2 + bt + c\sqrt{a}}{(2\sqrt{a}\,t + b)^2}\,dt$;

2) $x = \dfrac{2\sqrt{c}\,t - b}{a - t^2}$, $\quad \sqrt{ax^2 + bx + c} = \dfrac{\sqrt{c}\,t^2 - bt + c\sqrt{a}}{a - t^2}$, $\quad dx = 2\dfrac{\sqrt{c}\,t^2 - bt + c\sqrt{a}}{(a - t^2)^2}\,dt$;

3) $x = \dfrac{(t^2 + a)x_1 + b}{t^2 - a}$, $\quad \sqrt{ax^2 + bx + c} = \dfrac{(2ax_1 + b)t}{t^2 - a}$, $\quad dx = -2\dfrac{(2ax_1 + b)t}{(t^2 - a)^2}\,dt$.

$2°$. *Trigonometric substitutions.* The function $\sqrt{ax^2 + bx + c}$ can be reduced, by making a perfect square in the radicand, to one of the three forms:

$$1) \quad \sqrt{a}\sqrt{(x-p)^2 + q^2} \quad \text{if} \quad a > 0;$$

$$2) \quad \sqrt{a}\sqrt{(x-p)^2 - q^2} \quad \text{if} \quad a > 0;$$

$$3) \quad \sqrt{-a}\sqrt{q^2 - (x-p)^2} \quad \text{if} \quad a < 0,$$

where $p = -\frac{1}{2}b/a$. Different trigonometric substitutions are further used in each case to evaluate the integral:

$$1) \quad x - p = q\tan t, \quad \sqrt{(x-p)^2 + q^2} = \frac{q}{\cos t}, \quad dx = \frac{q\,dt}{\cos^2 t};$$

$$2) \quad x - p = \frac{q}{\cos t}, \quad \sqrt{(x-p)^2 - q^2} = q\tan t, \quad dx = \frac{q\sin t\,dt}{\cos^2 t};$$

$$3) \quad x - p = q\sin t, \quad \sqrt{q^2 - (x-p)^2} = q\cos t, \quad dx = q\cos t\,dt.$$

Example 2. Evaluate the integral $\int \sqrt{6 + 4x - 2x^2}\,dx$.

This integral corresponds to case 3 with $a = -2$, $p = 1$, and $q = 2$. The integrand can be rewritten in the form:
$$\sqrt{6 + 4x - 2x^2} = \sqrt{2}\sqrt{3 + 2x - x^2} = \sqrt{2}\sqrt{4 - (x-1)^2}.$$

Using the trigonometric substitution $x - 1 = 2\sin t$ and the formulas $\sqrt{3 + 2x - x^2} = 2\cos t$ and $dx = 2\cos t\,dt$, we obtain

$$\int \sqrt{6 + 4x - 2x^2}\,dx = 4\sqrt{2}\int \cos^2 t\,dt = 2\sqrt{2}\int (1 + \cos 2t)\,dt$$

$$= 2\sqrt{2}\,t + \sqrt{2}\sin 2t + C = 2\sqrt{2}\arcsin\frac{x-1}{2} + \sqrt{2}\sin\left(2\arcsin\frac{x-1}{2}\right) + C$$

$$= 2\sqrt{2}\arcsin\frac{x-1}{2} + \frac{\sqrt{2}}{2}(x-1)\sqrt{4 - (x-1)^2} + C.$$

▶ **Integral of a differential binomial.** The *integral of a differential binomial*,

$$\int x^m(a + bx^n)^p\,dx,$$

where a and b are constants, and n, m, p are rational numbers, is expressible in terms of elementary functions in the following three cases only:

1) If p is an integer. For $p \geq 0$, removing the brackets gives the sum of power functions. For $p < 0$, the substitution $x = t^r$, where r is the common denominator of the fractions m and n, leads to the integral of a rational function.

2) If $\frac{m+1}{n}$ is an integer. One uses the substitution $a + bx^n = t^k$, where k is the denominator of the fraction p.

3) If $\frac{m+1}{n} + p$ is an integer. One uses the substitution $ax^{-n} + b = t^k$, where k is the denominator of the fraction p.

Remark. In cases 2 and 3, the substitution $z = x^n$ leads to integrals of the form $3°$ above.

M7.1.5. Integration of Exponential and Trigonometric Functions

▶ **Integration of exponential and hyperbolic functions.**

1. Integrals of the form $\displaystyle\int R(e^{px}, e^{qx})\, dx$, where $R(x, y)$ is a rational function of its arguments and p, q are rational numbers, may be evaluated using the substitution $z^m = e^x$, where m is the common denominator of the fractions p and q. In the special case of integer p and q, we have $m = 1$, and the substitution becomes $z = e^x$.

Example 1. Evaluate the integral $\displaystyle\int \frac{e^{3x}\, dx}{e^x + 2}$.

This integral corresponds to integer p and q: $p = 1$ and $q = 3$. So we use the substitution $z = e^x$. Then $x = \ln z$ and $dx = \dfrac{dz}{z}$. Therefore,

$$\int \frac{e^{3x}\, dx}{e^x + 2} = \int \frac{z^2 dz}{z + 2} = \int \left(z - 2 + \frac{4}{z+2}\right) dz = \frac{1}{2} z^2 - 2z + 4\ln|z + 2| + C = \frac{1}{2} e^{2x} - 2e^x + 4\ln(e^x + 2) + C.$$

2. Integrals of the form $\displaystyle\int R(\sinh ax,\, \cosh ax)\, dx$ are evaluated by converting the hyperbolic functions to exponentials, using the formulas $\sinh ax = \frac{1}{2}(e^{ax} - e^{-ax})$ and $\cosh ax = \frac{1}{2}(e^{ax} + e^{-ax})$, and performing the substitution $z = e^{ax}$. Then

$$\int R(\sinh ax,\, \cosh ax)\, dx = \frac{1}{a} \int R\left(\frac{z^2 - 1}{2z},\, \frac{z^2 + 1}{2z}\right) \frac{dz}{z}.$$

Alternatively, the substitution $t = \tanh\left(\dfrac{ax}{2}\right)$ can also be used to evaluate integrals of the above form. Then

$$\int R(\sinh ax,\, \cosh ax)\, dx = \frac{2}{a} \int R\left(\frac{2t}{1 - t^2},\, \frac{1 + t^2}{1 - t^2}\right) \frac{dt}{1 - t^2}.$$

▶ **Integration of trigonometric functions.**

1. Integrals of the form $\displaystyle\int R(\sin ax,\, \cos ax)\, dx$ can be converted to integrals of rational functions using the fundamental trigonometric substitution $t = \tan\left(\dfrac{ax}{2}\right)$:

$$\int R(\sin ax,\, \cos ax)\, dx = \frac{2}{a} \int R\left(\frac{2t}{1 + t^2},\, \frac{1 - t^2}{1 + t^2}\right) \frac{dt}{1 + t^2}.$$

Example 2. Evaluate the integral $\int \dfrac{dx}{2 + \sin x}$.

Using the fundamental trigonometric substitution $t = \tan \dfrac{x}{2}$, we have

$$\int \frac{dx}{2 + \sin x} = 2 \int \frac{dt}{\left(2 + \dfrac{2t}{1 + t^2}\right)(1 + t^2)} = \int \frac{dt}{t^2 + t + 1} = 2 \int \frac{d(2t + 1)}{(2t + 1)^2 + 3}$$

$$= \frac{2}{\sqrt{3}} \arctan \frac{2t + 1}{\sqrt{3}} + C = \frac{2}{\sqrt{3}} \arctan \left(\frac{2}{\sqrt{3}} \tan \frac{x}{2} + \frac{1}{\sqrt{3}}\right) + C.$$

2. Integrals of the form $\int R(\sin^2 ax, \cos^2 ax, \tan ax)\, dx$ are converted to integrals of rational functions with the change of variable $z = \tan ax$:

$$\int R(\sin^2 ax, \cos^2 ax, \tan ax)\, dx = \frac{1}{a} \int R\left(\frac{z^2}{1 + z^2}, \frac{1}{1 + z^2}, z\right) \frac{dz}{1 + z^2}.$$

3. Integrals of the form

$$\int \sin ax \cos bx\, dx, \qquad \int \cos ax \cos bx\, dx, \qquad \int \sin ax \sin bx\, dx$$

are evaluated using the formulas

$$\sin \alpha \cos \beta = \tfrac{1}{2}[\sin(\alpha + \beta) + \sin(\alpha - \beta)],$$
$$\cos \alpha \cos \beta = \tfrac{1}{2}[\cos(\alpha + \beta) + \cos(\alpha - \beta)],$$
$$\sin \alpha \sin \beta = \tfrac{1}{2}[\cos(\alpha - \beta) - \cos(\alpha + \beta)].$$

4. Integrals of the form $\int \sin^m x \cos^n x\, dx$, where m and n are integers, are evaluated as follows:
    (a) if m is odd, one uses the change of variable $\cos x = z$, with $\sin x\, dx = -dz$;
    (b) if n is odd, one uses the change of variable $\sin x = z$, with $\cos x\, dx = dz$;
    (c) if m and n are both even nonnegative integers, one should use the degree reduction formulas

$$\sin^2 x = \tfrac{1}{2}(1 - \cos 2x), \qquad \cos^2 x = \tfrac{1}{2}(1 + \cos 2x), \qquad \sin x \cos x = \tfrac{1}{2} \sin 2x.$$

Example 3. Evaluate the integral $\int \sin^5 x\, dx$.

This integral corresponds to odd m: $m = 5$. With simple rearrangement and the change of variable $\cos x = z$, we have

$$\int \sin^5 x\, dx = \int (\sin^2 x)^2 \sin x\, dx = -\int (1 - \cos^2 x)^2\, d\cos x = -\int (1 - z^2)^2\, dz$$

$$= \tfrac{2}{3} z^3 - \tfrac{1}{5} z^5 - z + C = \tfrac{2}{3} \cos^3 x - \tfrac{1}{5} \cos^5 x - \cos x + C.$$

Remark. In general, the integrals $\int \sin^p x \cos^q x\, dx$ are reduced to the integral of a differential binomial by the substitution $y = \sin x$.

M7.1.6. Integration of Polynomials Multiplied by Elementary Functions

Throughout this section, $P_n(x)$ designates a polynomial of degree n.

▶ **Integration of the product of a polynomial by exponential functions.** General formulas:

$$\int P_n(x)e^{ax}\,dx = e^{ax}\left[\frac{P_n(x)}{a} - \frac{P'_n(x)}{a^2} + \cdots + (-1)^n\frac{P_n^{(n)}(x)}{a^{n+1}}\right] + C,$$

$$\int P_n(x)\cosh(ax)\,dx = \sinh(ax)\left[\frac{P_n(x)}{a} + \frac{P''_n(x)}{a^3} + \cdots\right] - \cosh(ax)\left[\frac{P'_n(x)}{a^2} + \frac{P'''_n(x)}{a^4} + \cdots\right] + C,$$

$$\int P_n(x)\sinh(ax)\,dx = \cosh(ax)\left[\frac{P_n(x)}{a} + \frac{P''_n(x)}{a^3} + \cdots\right] - \sinh(ax)\left[\frac{P'_n(x)}{a^2} + \frac{P'''_n(x)}{a^4} + \cdots\right] + C.$$

These formulas are obtained by repeated integration by parts; see Property 4 from Subsection M7.1.2 with $f(x) = P_n(x)$ for $g^{(n+1)}(x) = e^{ax}$, $g^{(n+1)}(x) = \cosh(ax)$, and $g^{(n+1)}(x) = \sinh(ax)$, respectively.

In the special case $P_n(x) = x^n$, the first formula gives

$$\int x^n e^{ax}\,dx = e^{ax}\sum_{k=0}^{n}\frac{(-1)^{n-k}}{a^{n+1-k}}\frac{n!}{k!}x^k + C. \qquad \text{(M7.1.6.1)}$$

▶ **Integration of the product of a polynomial by a trigonometric function.**

1°. General formulas:

$$\int P_n(x)\cos(ax)\,dx = \sin(ax)\left[\frac{P_n(x)}{a} - \frac{P''_n(x)}{a^3} + \cdots\right] + \cos(ax)\left[\frac{P'_n(x)}{a^2} - \frac{P'''_n(x)}{a^4} + \cdots\right] + C,$$

$$\int P_n(x)\sin(ax)\,dx = \sin(ax)\left[\frac{P'_n(x)}{a^2} - \frac{P'''_n(x)}{a^4} + \cdots\right] - \cos(ax)\left[\frac{P_n(x)}{a} - \frac{P''_n(x)}{a^3} + \cdots\right] + C.$$

These formulas are obtained by repeated integration by parts; see Property 4 from Subsection M7.1.2 with $f(x) = P_n(x)$ for $g^{(n+1)}(x) = \cos(ax)$ and $g^{(n+1)}(x) = \sin(ax)$, respectively.

2°. To evaluate integrals of the form

$$\int P_n(x)\cos^m(ax)\,dx, \qquad \int P_n(x)\sin^m(ax)\,dx,$$

with $m = 2, 3, \ldots$, one should first use the trigonometric formulas

$$\cos^{2k}(ax) = \frac{1}{2^{2k-1}}\sum_{i=0}^{k-1}C_{2k}^i\cos[2(k-i)ax] + \frac{1}{2^{2k}}C_{2k}^k \qquad (m = 2k),$$

$$\cos^{2k+1}(ax) = \frac{1}{2^{2k}}\sum_{i=0}^{k}C_{2k+1}^i\cos[(2k-2i+1)ax] \qquad (m = 2k+1),$$

$$\sin^{2k}(ax) = \frac{1}{2^{2k-1}}\sum_{i=0}^{k-1}(-1)^{k-i}C_{2k}^i\cos[2(k-i)ax] + \frac{1}{2^{2k}}C_{2k}^k \qquad (m = 2k),$$

$$\sin^{2k+1}(ax) = \frac{1}{2^{2k}}\sum_{i=0}^{k}(-1)^{k-i}C_{2k+1}^i\sin[(2k-2i+1)ax] \qquad (m = 2k+1),$$

thus reducing the above integrals to those considered in Item 1°.

3°. Integrals of the form

$$\int P_n(x)\,e^{ax}\sin(bx)\,dx, \qquad \int P_n(x)\,e^{ax}\cos(bx)\,dx$$

can be evaluated by repeated integration by parts.

In particular,

$$\int x^n e^{ax}\sin(bx) = e^{ax}\sum_{k=1}^{n+1}\frac{(-1)^{k+1}n!}{(n-k+1)!\,(a^2+b^2)^{k/2}}x^{n-k+1}\sin(bx+k\theta)+C,$$

$$\int x^n e^{ax}\cos(bx) = e^{ax}\sum_{k=1}^{n+1}\frac{(-1)^{k+1}n!}{(n-k+1)!\,(a^2+b^2)^{k/2}}x^{n-k+1}\cos(bx+k\theta)+C,$$

$(*)$

where

$$\sin\theta = -\frac{b}{\sqrt{a^2+b^2}}, \qquad \cos\theta = \frac{a}{\sqrt{a^2+b^2}}.$$

▶ **Integrals involving power and logarithmic functions.**

1°. The formula of integration by parts with $g'(x) = P_n(x)$ is effective in the evaluation of integrals of the form

$$\int P_n(x)\ln(ax)\,dx = Q_{n+1}(x)\ln(ax) - a\int\frac{Q_{n+1}(x)}{x}\,dx,$$

where $Q_{n+1}(x) = \int P_n(x)\,dx$ is a polynomial of degree $n+1$. The integral on the right-hand side is easy to take, since the integrand is the sum of power functions.

Example. Evaluate the integral $\int \ln x\,dx$.

Setting $f(x) = \ln x$ and $g'(x) = 1$, we find $f'(x) = \dfrac{1}{x}$ and $g(x) = x$. Substituting these expressions into the formula of integration by parts, we obtain $\int \ln x\,dx = x\ln x - \int dx = x\ln x - x + C$.

2°. The easiest way to evaluate integrals of the more general form

$$I = \int\sum_{i=0}^{n}\ln^i(ax)\left(\sum_{j=0}^{m}b_{ij}x^{\beta_{ij}}\right)dx,$$

where the β_{ij} are arbitrary numbers, is to use the substitution $z = \ln(ax)$, so that

$$I = \int\sum_{i=0}^{n}z^i\left(\sum_{j=0}^{m}\frac{b_{ij}}{a^{\beta_{ij}+1}}e^{(\beta_{ij}+1)z}\right)dz.$$

By removing the brackets, one obtains a sum of integrals like $\int x^n e^{ax}\,dx$, which are easy to evaluate by formula (M7.1.6.1).

M7.2. Definite Integral

M7.2.1. Basic Definitions. Classes of Integrable Functions. Geometrical Meaning of the Definite Integral

▶ **Basic definitions.** Let $y = f(x)$ be a bounded function defined on a finite closed interval $[a, b]$. Let us partition this interval into n elementary subintervals defined by a set of points $\{x_0, x_1, \ldots, x_n\}$ such that $a = x_0 < x_1 < \cdots < x_n = b$. Each subinterval $[x_{k-1}, x_k]$ will be characterized by its length $\Delta x_k = x_k - x_{k-1}$ and an arbitrarily chosen point $\xi_k \in [x_{k-1}, x_k]$. Let us make up an *integral sum* (a *Cauchy–Riemann sum*, also known as a *Riemann sum*)

$$s_n = \sum_{k=1}^{n} f(\xi_k)\Delta x_k \qquad (x_{k-1} \le \xi_k \le x_k).$$

If, as $\Delta x_k \to 0$ for all k and, accordingly, $n \to \infty$, there exists a finite limit of the integral sums s_n and it depends on neither the way the interval $[a, b]$ was partitioned, nor the selection of the points ξ_k, then this limit is denoted $\int_a^b f(x)\,dx$ and is called the *definite integral* (also the *Riemann integral*) of the function $y = f(x)$ over the interval $[a, b]$:

$$\int_a^b f(x)\,dx = \lim_{n \to \infty} s_n \qquad \left(\max_{1 \le k \le n} \Delta x_k \to 0\right).$$

In this case, the function $f(x)$ is called *integrable* on the interval $[a, b]$.

▶ **Classes of integrable functions.**

1. If a function $f(x)$ is continuous on an interval $[a, b]$, then it is integrable on this interval.

2. If a bounded function $f(x)$ has finitely many jump discontinuities on $[a, b]$, then it is integrable on $[a, b]$.

3. A monotonic bounded function $f(x)$ on $[a, b]$ is always integrable on $[a, b]$.

▶ **Geometric meaning of the definite integral.** If $f(x) \ge 0$ on $[a, b]$, then the integral $\int_a^b f(x)\,dx$ is equal to the area of the domain $D = \{a \le x \le b,\ 0 \le y \le f(x)\}$ (the area of the curvilinear trapezoid shown in Fig. M7.1).

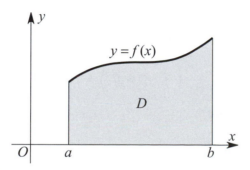

Figure M7.1. The integral of a nonnegative function $f(x)$ on an interval $[a, b]$ is equal to the area of the shaded region.

M7.2.2. Properties of Definite Integrals and Useful Formulas

▶ **Qualitative properties of integrals.**

1. If a function $f(x)$ is integrable on $[a, b]$, then the functions $cf(x)$, with $c = $ const, and $|f(x)|$ are also integrable on $[a, b]$.

2. If two functions $f(x)$ and $g(x)$ are integrable on $[a, b]$, then their sum, difference, and product are also integrable on $[a, b]$.

3. If a function $f(x)$ is integrable on $[a, b]$ and its values lie within an interval $[c, d]$, where a function $g(y)$ is defined and continuous, then the composite function $g(f(x))$ is also integrable on $[a, b]$.

4. If a function $f(x)$ is integrable on $[a, b]$, then it is also integrable and on any subinterval $[\alpha, \beta] \subset [a, b]$. Conversely, if an interval $[a, b]$ is partitioned into a number of subintervals and $f(x)$ is integrable on each of the subintervals, then it is integrable on the whole interval $[a, b]$.

5. If the values of a function are changed at finitely many points, this will not affect the integrability of the function and will not change the value of the integral.

▶ **Properties of integrals in terms of identities.**

1. *The integral over a zero-length interval is zero*:

$$\int_a^a f(x)\,dx = 0.$$

2. *Antisymmetry under the swap of the integration limits*:

$$\int_a^b f(x)\,dx = -\int_b^a f(x)\,dx.$$

3. *Linearity.* If functions $f(x)$ and $g(x)$ are integrable on an interval $[a, b]$, then

$$\int_a^b \left[Af(x) \pm Bg(x) \right] dx = A \int_a^b f(x)\,dx \pm B \int_a^b g(x)\,dx$$

for any numbers A and B.

4. *Additivity.* If $c \in [a, b]$ and $f(x)$ is integrable on $[a, b]$, then

$$\int_a^b f(x)\,dx = \int_a^c f(x)\,dx + \int_c^b f(x)\,dx.$$

Remark. This property is also valid in the case where $c \notin [a, b]$.

5. *Differentiation with respect to a variable upper limit.* If $f(x)$ is continuous on $[a, b]$, then the function $\Phi(x) = \int_a^x f(t)\,dt$ is differentiable on $[a, b]$, and $\Phi'(x) = f(x)$. This fact can be written as

$$\frac{d}{dx}\left(\int_a^x f(t)\,dt \right) = f(x).$$

6. *Newton–Leibniz formula*:

$$\int_a^b f(x)\,dx = F(x)\,\Big|_a^b = F(b) - F(a),$$

where $F(x)$ is an antiderivative of $f(x)$ on $[a, b]$.

7. *Integration by parts.* If functions $f(x)$ and $g(x)$ have continuous derivatives on $[a, b]$, then

$$\int_a^b f(x)g'(x)\,dx = \left[f(x)g(x)\right]\Big|_a^b - \int_a^b f'(x)g(x)\,dx.$$

8. *Repeated integration by parts:*

$$\int_a^b f(x)g^{(n+1)}(x)\,dx = \left[f(x)g^{(n)}(x) - f'(x)g^{(n-1)}(x) + \cdots + (-1)^n f^{(n)}(x)g(x)\right]_a^b$$

$$+ (-1)^{n+1} \int_a^b f^{(n+1)}(x)g(x)\,dx, \qquad n = 0, 1, \ldots$$

9. *Change of variable (substitution) in a definite integral.* Let $f(x)$ be a continuous function on $[a, b]$ and let $x(t)$ be a continuously differentiable function on $[\alpha, \beta]$. Suppose also that the range of values of $x(t)$ coincides with $[a, b]$, with $x(\alpha) = a$ and $x(\beta) = b$. Then

$$\int_a^b f(x)\,dx = \int_\alpha^\beta f\big(x(t)\big)\,x'(t)\,dt.$$

Example. Evaluate the integral $\displaystyle\int_0^3 \frac{dx}{(x-8)\sqrt{x+1}}$.

Perform the substitution $x + 1 = t^2$, with $dx = 2t\,dt$. We have $t = 1$ at $x = 0$ and $t = 2$ at $x = 3$. Therefore,

$$\int_0^3 \frac{dx}{(x-8)\sqrt{x+1}} = \int_1^2 \frac{2t\,dt}{(t^2-9)t} = 2\int_1^2 \frac{dt}{t^2-9} = \frac{1}{3}\ln\left|\frac{t-3}{t+3}\right|\Big|_1^2 = \frac{1}{3}\ln\frac{2}{5}.$$

M7.2.3. Asymptotic Formulas for the Calculation of Integrals

Below are some general formulas, involving arbitrary functions and parameters, that may be helpful for obtaining asymptotics of integrals.

▶ **Asymptotic formulas for integrals with weak singularity as $\varepsilon \to 0$.** We will consider integrals of the form

$$I(\varepsilon) = \int_0^a \frac{x^{\beta-1} f(x)}{(x+\varepsilon)^\alpha}\,dx,$$

where $0 < a < \infty$, $\beta > 0$, $f(0) \neq 0$, and $\varepsilon > 0$ is a small parameter.

The integral diverges as $\varepsilon \to 0$ for $\alpha \geq \beta$, that is, $\lim\limits_{\varepsilon \to 0} I(\varepsilon) = \infty$. In this case, the leading term of the asymptotic expansion of the integral $I(\varepsilon)$ is given by

$$I(\varepsilon) = \frac{\Gamma(\beta)\Gamma(\alpha-\beta)}{\Gamma(\alpha)} f(0)\varepsilon^{\beta-\alpha} + O(\varepsilon^\sigma) \quad \text{if} \quad \alpha > \beta,$$

$$I(\varepsilon) = -f(0)\ln\varepsilon + O(1) \qquad\qquad \text{if} \quad \alpha = \beta,$$

where $\Gamma(\beta)$ is the gamma function and $\sigma = \min[\beta - \alpha + 1, 0]$.

▶ **Asymptotic formulas for Laplace integrals of special form as $\lambda \to +\infty$.** Consider a *Laplace integral of the special form*

$$I(\lambda) = \int_0^a x^{\beta-1} \exp\big(-\lambda x^\alpha\big) f(x)\,dx,$$

where $0 < a < \infty$, $\alpha > 0$, and $\beta > 0$.

The following formula, called *Watson's asymptotic formula*, holds as $\lambda \to +\infty$:

$$I(\lambda) = \frac{1}{\alpha} \sum_{k=0}^{n} \frac{f^{(k)}(0)}{k!} \Gamma\left(\frac{k+\beta}{\alpha}\right) \lambda^{-(k+\beta)/\alpha} + O\left(\lambda^{-(n+\beta+1)/\alpha}\right).$$

Remark. Watson's formula also holds for improper integrals with $a = \infty$ if the original integral converges absolutely for some $\lambda_0 > 0$.

▶ **Asymptotic formulas for Laplace integrals of general form as $\lambda \to +\infty$.** Consider a *Laplace integral of the general form*

$$I(\lambda) = \int_a^b f(x) \exp[\lambda g(x)] \, dx, \tag{M7.2.3.1}$$

where $[a, b]$ is a finite interval and $f(x)$, $g(x)$ are continuous functions.

Leading term of the asymptotic expansion of the integral (M7.2.3.1) as $\lambda \to +\infty$. Suppose the function $g(x)$ attains a maximum on $[a, b]$ at only one point $x_0 \in [a, b]$ and is differentiable in a neighborhood of x_0, with $g'(x_0) = 0$, $g''(x_0) \neq 0$, and $f(x_0) \neq 0$. Then the leading term of the asymptotic expansion of the integral (M7.2.3.1), as $\lambda \to +\infty$, is given by

$$\begin{aligned} I(\lambda) &= f(x_0)\sqrt{-\frac{2\pi}{\lambda g''(x_0)}} \exp[\lambda g(x_0)] && \text{if } a < x_0 < b, \\ I(\lambda) &= \frac{1}{2} f(x_0)\sqrt{-\frac{2\pi}{\lambda g''(x_0)}} \exp[\lambda g(x_0)] && \text{if } x_0 = a \text{ or } x_0 = b. \end{aligned} \tag{M7.2.3.2}$$

Note that the latter formula differs from the former by the factor $1/2$ only.

Under the same conditions, if $g(x)$ attains a maximum at either endpoint, $x_0 = a$ or $x_0 = b$, but $g'(x_0) \neq 0$, then the leading asymptotic term of the integral, as $\lambda \to +\infty$, is

$$I(\lambda) = \frac{f(x_0)}{|g'(x_0)|} \frac{1}{\lambda} \exp[\lambda g(x_0)], \qquad \text{where } x_0 = a \text{ or } x_0 = b. \tag{M7.2.3.3}$$

For more accurate asymptotic estimates for the Laplace integral (M7.2.3.1), see below.

▶ **Asymptotic formulas for a power Laplace integral.** Consider the *power Laplace integral*, which is obtained from the exponential Laplace integral (M7.2.3.1) by substituting $\ln g(x)$ for $g(x)$:

$$I(\lambda) = \int_a^b f(x)[g(x)]^\lambda \, dx, \tag{M7.2.3.6}$$

where $[a, b]$ is a finite closed interval and $g(x) > 0$. It is assumed that the functions $f(x)$ and $g(x)$ appearing in the integral (M7.2.3.6) are continuous; $g(x)$ is assumed to attain a maximum at only one point $x_0 = [a, b]$ and to be differentiable in a neighborhood of $x = x_0$, with $g'(x_0) = 0$, $g''(x_0) \neq 0$, and $f(x_0) \neq 0$. Then the leading asymptotic term of the integral, as $\lambda \to +\infty$, is expressed as

$$I(\lambda) = f(x_0)\sqrt{-\frac{2\pi}{\lambda g''(x_0)}} [g(x_0)]^{\lambda+1/2} \qquad \text{if } a < x_0 < b,$$

$$I(\lambda) = \frac{1}{2} f(x_0)\sqrt{-\frac{2\pi}{\lambda g''(x_0)}} [g(x_0)]^{\lambda+1/2} \quad \text{if } x_0 = a \text{ or } x_0 = b.$$

Note that the latter formula differs from the former by the factor $1/2$ only.

Under the same conditions, if $g(x)$ attains a maximum at either endpoint, $x_0 = a$ or $x_0 = b$, but $g'(x_0) \neq 0$, then the leading asymptotic term of the integral, as $\lambda \to +\infty$, is

$$I(\lambda) = \frac{f(x_0)}{|g'(x_0)|} \frac{1}{\lambda} [g(x_0)]^{\lambda+1/2}, \qquad \text{where} \quad x_0 = a \quad \text{or} \quad x_0 = b.$$

M7.2.4. Mean Value Theorems. Properties of Integrals in Terms of Inequalities

▶ **Mean value theorems.**

THEOREM 1. *If $f(x)$ is a continuous function on $[a, b]$, there exists at least one point $c \in (a, b)$ such that*

$$\int_a^b f(x)\,dx = f(c)(b-a).$$

The number $f(c)$ is called the mean value of the function $f(x)$ on $[a, b]$.

THEOREM 2. *If $f(x)$ is a continuous function on $[a, b]$, and $g(x)$ is integrable and of constant sign ($g(x) \geq 0$ or $g(x) \leq 0$) on $[a, b]$, then there exists at least one point $c \in (a, b)$ such that*

$$\int_a^b f(x)g(x)\,dx = f(c) \int_a^b g(x)\,dx.$$

▶ **Properties of integrals in terms of inequalities.**

1. *Estimation theorem.* If $m \leq f(x) \leq M$ on $[a, b]$, then

$$m(b-a) \leq \int_a^b f(x)\,dx \leq M(b-a).$$

2. *Inequality integration theorem.* If $\varphi(x) \leq f(x) \leq g(x)$ on $[a, b]$, then

$$\int_a^b \varphi(x)\,dx \leq \int_a^b f(x)\,dx \leq \int_a^b g(x)\,dx.$$

In particular, if $f(x) \geq 0$ on $[a, b]$, then $\int_a^b f(x)\,dx \geq 0$.

Further on, it is assumed that the integrals on the right-hand sides of the inequalities of Items 3–6 exist.

3. *Absolute value theorem* (*integral analogue of the triangle inequality*):

$$\left| \int_a^b f(x)\,dx \right| \leq \int_a^b |f(x)|\,dx.$$

4. *Bunyakovsky's inequality* (*Cauchy–Schwarz–Bunyakovsky inequality*):

$$\left(\int_a^b f(x)g(x)\,dx \right)^2 \leq \int_a^b f^2(x)\,dx \int_a^b g^2(x)\,dx.$$

5. *Cauchy's inequality*:

$$\left(\int_a^b [f(x) + g(x)]^2\,dx \right)^{1/2} \leq \left(\int_a^b f^2(x)\,dx \right)^{1/2} + \left(\int_a^b g^2(x)\,dx \right)^{1/2}.$$

6. *Minkowski's inequality* (generalization of Cauchy's inequality):

$$\left(\int_a^b |f(x) + g(x)|^p\,dx \right)^{\frac{1}{p}} \leq \left(\int_a^b |f(x)|^p\,dx \right)^{\frac{1}{p}} + \left(\int_a^b |g(x)|^p\,dx \right)^{\frac{1}{p}}, \qquad p \geq 1.$$

M7.2.5. Geometric and Physical Applications of the Definite Integral

▶ **Geometric applications of the definite integral.**

1. The *area of a domain* D bounded by curves
$y = f(x)$ and $y = g(x)$ and straight lines $x = a$ and $x = b$ in the xy-plane (see Fig. M7.2 a) is calculated by the formula

$$S = \int_a^b \left[f(x) - g(x) \right] dx.$$

If $g(x) \equiv 0$, this formula gives the area of a curvilinear trapezoid bounded by the x-axis, the curve $y = f(x)$, and the straight lines $x = a$ and $x = b$.

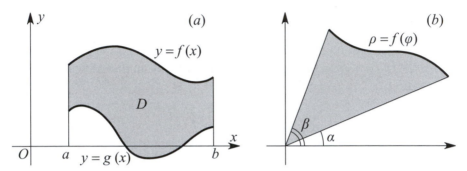

Figure M7.2. (a) A domain D bounded by two curves $y = f(x)$ and $y = g(x)$ on an interval $[a, b]$; (b) a curvilinear sector.

2. *Area of a domain* D. Let $x = x(t)$ and $y = y(t)$, with $t_1 \leq t \leq t_2$, be parametric equations of a piecewise-smooth simple closed curve bounding on its left (traced counterclockwise) a domain D with area S. Then

$$S = -\int_{t_1}^{t_2} y(t) x'(t) \, dt = \int_{t_1}^{t_2} x(t) y'(t) \, dt = \frac{1}{2} \int_{t_1}^{t_2} \left[x(t) y'(t) - y(t) x'(t) \right] dt.$$

3. *Area of a curvilinear sector.* Let a curve $\rho = f(\varphi)$, with $\varphi \in [\alpha, \beta]$, be defined in the polar coordinates ρ, φ. Then the area of the curvilinear sector $\{\alpha \leq \varphi \leq \beta; \ 0 \leq \rho \leq f(\varphi)\}$ (see Fig. M7.2 b) is calculated by the formula

$$S = \frac{1}{2} \int_\alpha^\beta [f(\varphi)]^2 \, d\varphi.$$

4. *Area of a surface of revolution.* Let a surface of revolution be generated by rotating a curve $y = f(x) \geq 0$, $x \in [a, b]$, about the x-axis; see Fig. M7.3. The area of this surface is calculated as

$$S = 2\pi \int_a^b f(x) \sqrt{1 + [f'(x)]^2} \, dx.$$

5. *Volume of a body of revolution.* Let a body of revolution be obtained by rotating about the x-axis a curvilinear trapezoid bounded by a curve $y = f(x)$, the x-axis, and straight lines $x = a$ and $x = b$; see Fig. M7.3. Then the volume of this body is calculated as

$$V = \pi \int_a^b [f(x)]^2 \, dx.$$

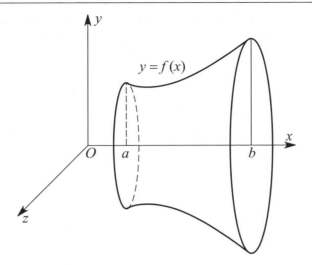

Figure M7.3. A surface of revolution generated by rotating a curve $y = f(x)$.

6. *Arc length of a plane curve defined in different ways.*
(*a*) If a curve is the graph of a continuously differentiable function $y = f(x)$, $x \in [a, b]$, then its length is determined as

$$L = \int_a^b \sqrt{1 + [f'(x)]^2} \, dx.$$

(*b*) If a plane curve is defined parametrically by equations $x = x(t)$ and $y = y(t)$, with $t \in [\alpha, \beta]$ and $x(t)$ and $y(t)$ being continuously differentiable functions, then its length is calculated by

$$L = \int_\alpha^\beta \sqrt{[x'(t)]^2 + [y'(t)]^2} \, dt.$$

(*c*) If a curve is defined in the polar coordinates ρ, φ by an equation $\rho = \rho(\varphi)$, with $\varphi \in [\alpha, \beta]$, then its length is found as

$$L = \int_\alpha^\beta \sqrt{\rho^2(\varphi) + [\rho'(\varphi)]^2} \, d\varphi.$$

7. The *arc length of a spatial curve* defined parametrically by equations $x = x(t)$, $y = y(t)$, and $z = z(t)$, with $t \in [\alpha, \beta]$ and $x(t)$, $y(t)$, and $z(t)$ being continuously differentiable functions, is calculated by

$$L = \int_\alpha^\beta \sqrt{[x'(t)]^2 + [y'(t)]^2 + [z'(t)]^2} \, dt.$$

▶ **Physical applications of the integral.**
1. *Work of a variable force.* Suppose a point mass moves along the x-axis from a point $x = a$ to a point $x = b$ under the action of a variable force $F(x)$ directed along the x-axis. The mechanical work of this force is equal to

$$A = \int_a^b F(x) \, dx.$$

2. *Mass of a rectilinear rod of variable density.* Suppose a rod with a constant cross-sectional area S occupies an interval $[0, l]$ on the x-axis and the density of the rod material is a function of x: $\rho = \rho(x)$. The mass of this rod is calculated as

$$m = S \int_0^l \rho(x)\, dx.$$

3. *Mass of a curvilinear rod of variable density.* Let the shape of a plane curvilinear rod with a constant cross-sectional area S be defined by an equation $y = f(x)$, with $a \le x \le b$, and let the density of the material be coordinate dependent: $\rho = \rho(x, y)$. The mass of this rod is calculated as

$$m = S \int_a^b \rho\big(x, f(x)\big) \sqrt{1 + [y'(x)]^2}\, dx.$$

If the shape of the rod is defined parametrically by $x = x(t)$ and $y = y(t)$, then its mass is found as

$$m = S \int_a^b \rho\big(x(t), y(t)\big) \sqrt{[x'(t)]^2 + [y'(t)]^2}\, dt.$$

4. The *coordinates of the center of mass of a plane homogeneous material curve* whose shape is defined by an equation $y = f(x)$, with $a \le x \le b$, are calculated by the formulas

$$x_c = \frac{1}{L} \int_a^b x \sqrt{1 + [y'(x)]^2}\, dx, \qquad y_c = \frac{1}{L} \int_a^b f(x) \sqrt{1 + [y'(x)]^2}\, dx,$$

where L is the length of the curve.

If the shape of a plane homogeneous material curve is defined parametrically by $x = x(t)$ and $y = y(t)$, then the coordinates of its center of mass are obtained as

$$x_c = \frac{1}{L} \int_a^b x(t) \sqrt{[x'(t)]^2 + [y'(t)]^2}\, dt, \qquad y_c = \frac{1}{L} \int_a^b y(t) \sqrt{[x'(t)]^2 + [y'(t)]^2}\, dt.$$

5. The *coordinates of the center of mass of a homogeneous curvilinear trapezoid* bounded by a curve $y = f(x)$, the x-axis, and the straight lines $x = a$ and $x = b$ (see Fig. M7.1) are given by

$$x_c = \frac{1}{S} \int_a^b x f(x)\, dx, \qquad y_c = \frac{1}{2S} \int_a^b [f(x)]^2\, dx, \qquad S = \int_a^b f(x)\, dx,$$

where S is the area of the trapezoid.

M7.2.6. Improper Integrals with Infinite Integration Limits

An improper integral is an integral with an infinite limit (limits) of integration or an integral of an unbounded function.

▶ Integrals with infinite limits.

$1°$. Let $y = f(x)$ be a function defined and continuous on an infinite interval $a \le x < \infty$. If there exists a finite limit $\lim\limits_{b \to \infty} \int_a^b f(x)\, dx$, then it is called a (convergent) *improper integral* of $f(x)$ on the interval $[a, \infty)$ and is denoted

$\int_a^\infty f(x)\,dx$. Thus, by definition

$$\int_a^\infty f(x)\,dx = \lim_{b\to\infty} \int_a^b f(x)\,dx. \qquad \text{(M7.2.6.1)}$$

If the limit is infinite or does not exist, the improper integral is called *divergent*.

The *geometric meaning of an improper integral* is that the integral $\int_a^\infty f(x)\,dx$, with $f(x) \geq 0$, is equal to the area of the unbounded domain between the curve $y = f(x)$, its asymptote $y = 0$, and the straight line $x = a$ on the left.

2°. Suppose an antiderivative $F(x)$ of the integrand function $f(x)$ is known. Then the improper integral (M7.2.6.1) is

(i) *convergent* if there exists a finite limit $\lim\limits_{x\to\infty} F(x) = F(\infty)$;

(ii) *divergent* if the limit is infinite or does not exist.

In case (i), we have

$$\int_a^\infty f(x)\,dx = F(x)\Big|_a^\infty = F(\infty) - F(a).$$

Example 1. Let us investigate the improper integral $I = \int_a^\infty \dfrac{dx}{x^\lambda}$, $a > 0$.

The integrand $f(x) = x^{-\lambda}$ has an antiderivative $F(x) = \dfrac{1}{1-\lambda} x^{1-\lambda}$ if $\lambda \neq 1$. Depending on the value of the parameter λ, we have

$$\lim_{x\to\infty} F(x) = \frac{1}{1-\lambda} \lim_{x\to\infty} x^{1-\lambda} = \begin{cases} 0 & \text{if } \lambda > 1, \\ \infty & \text{if } \lambda < 1. \end{cases}$$

Therefore, if $\lambda > 1$, the integral is convergent and is equal to $I = F(\infty) - F(a) = \dfrac{a^{1-\lambda}}{\lambda - 1}$, and if $\lambda < 1$, the integral is divergent. It is easy to show that the integral is also divergent if $\lambda = 1$.

3°. Improper integrals for other infinite intervals are defined in a similar way:

$$\int_{-\infty}^b f(x)\,dx = \lim_{a\to-\infty} \int_a^b f(x)\,dx,$$

$$\int_{-\infty}^\infty f(x)\,dx = \int_{-\infty}^c f(x)\,dx + \int_c^\infty f(x)\,dx,$$

where c is an arbitrary number. Note that if either improper integral on the right-hand side of the latter relation is convergent, then, by definition, the integral on the left-hand side is also convergent. If at least one of the integrals on the right-hand side is divergent, the integral on the left is called divergent.

4°. Properties 2–4 and 6–9 from Subsection M7.2.2, where a can be equal to $-\infty$ and b can be ∞, apply to improper integrals as well; it is assumed that all quantities on the right-hand sides exist (the integrals are convergent).

▶ **Sufficient conditions for convergence of improper integrals.** In many problems, it suffices to establish whether a given improper integral is convergent or not and, if yes, evaluate it. The theorems presented below can be useful in doing so.

THEOREM 1 (CAUCHY'S CONVERGENCE CRITERION). *For the integral (M7.2.6.1) to be convergent it is necessary and sufficient that for any $\varepsilon > 0$ there exist a number R such that the inequality*

$$\left| \int_{\alpha}^{\beta} f(x)\, dx \right| < \varepsilon$$

holds for any $\beta > \alpha > R$.

THEOREM 2. *If $0 \le f(x) \le g(x)$ for $x \ge a$, then the convergence of the integral $\int_{a}^{\infty} g(x)\, dx$ implies the convergence of the integral $\int_{a}^{\infty} f(x)\, dx$; moreover, $\int_{a}^{\infty} f(x)\, dx \le \int_{a}^{\infty} g(x)\, dx$. If the integral $\int_{a}^{\infty} f(x)\, dx$ is divergent, then the integral $\int_{a}^{\infty} g(x)\, dx$ is also divergent.*

THEOREM 3. *If the integral $\int_{a}^{\infty} |f(x)|\, dx$ is convergent, then the integral $\int_{a}^{\infty} f(x)\, dx$ is also convergent; in this case, the latter integral is called absolutely convergent.*

Example 2. The improper integral $\int_{1}^{\infty} \dfrac{\sin x}{x^2}\, dx$ is absolutely convergent, since $\left| \dfrac{\sin x}{x^2} \right| \le \dfrac{1}{x^2}$ and the integral $\int_{1}^{\infty} \dfrac{1}{x^2}\, dx$ is convergent (see Example 1).

THEOREM 4. *Let $f(x)$ and $g(x)$ be integrable functions on any finite interval $a \le x \le b$ and let there exist a limit, finite or infinite,*

$$\lim_{x \to \infty} \frac{f(x)}{g(x)} = K.$$

Then the following assertions hold:
1. If $0 < K < \infty$, both integrals

$$\int_{a}^{\infty} f(x)\, dx, \qquad \int_{a}^{\infty} g(x)\, dx \qquad\qquad (\text{M7.2.6.2})$$

are convergent or divergent simultaneously.
2. If $0 \le K < \infty$, the convergence of the latter integral in (M7.2.6.2) implies the convergence of the former integral.
3. If $0 < K \le \infty$, the divergence of the latter integral in (M7.2.6.2) implies the divergence of the former integral.

THEOREM 5 (COROLLARY OF THEOREM 4). *Given a function $f(x)$, let its asymptotics for sufficiently large x have the form*

$$f(x) = \frac{\varphi(x)}{x^{\lambda}} \qquad (\lambda > 0).$$

Then: (i) if $\lambda > 1$ and $\varphi(x) \le c < \infty$, then the integral $\int_{a}^{\infty} f(x)\, dx$ is convergent; (ii) if $\lambda \le 1$ and $\varphi(x) \ge c > 0$, then the integral is divergent.

THEOREM 6. *Let $f(x)$ be an absolutely integrable function on an interval $[a, \infty)$ and let $g(x)$ be a bounded function on $[a, \infty)$. Then the product $f(x)g(x)$ is an absolutely integrable function on $[a, \infty)$.*

THEOREM 7 (ANALOGUE OF ABEL'S TEST FOR CONVERGENCE OF INFINITE SERIES). *Let $f(x)$ be an integrable function on an interval $[a, \infty)$ such that the integral (M7.2.6.1) is* convergent *(maybe not absolutely) and let $g(x)$ be a monotonic and bounded function on* $[a, \infty)$. *Then the integral*

$$\int_a^\infty f(x) g(x)\, dx \qquad (M7.2.6.3)$$

is convergent.

THEOREM 8 (ANALOGUE OF DIRICHLET'S TEST FOR CONVERGENCE OF INFINITE SERIES). *Let (i) $f(x)$ be an integrable function on any finite interval $[a, A]$ and*

$$\left| \int_a^A f(x)\, dx \right| \le K < \infty \qquad (a \le A < \infty);$$

(ii) $g(x)$ be a function tending to zero monotonically as $x \to \infty$: $\lim_{x \to \infty} g(x) = 0$. Then the integral (M7.2.6.3) is convergent.

Example 3. Let us show that the improper integral $\displaystyle\int_a^\infty \frac{\sin x}{x^\lambda}\, dx$ is convergent for $a > 0$ and $\lambda > 0$.

Set $f(x) = \sin x$ and $g(x) = x^{-\lambda}$ and verify conditions (i) and (ii) of Theorem 8. We have

(i) $\quad \left| \displaystyle\int_a^A \sin x\, dx \right| = |\cos a - \cos A| \le 2$;

(ii) since $\lambda > 0$, the function $x^{-\lambda}$ is monotonically decreasing and tends to zero as $x \to \infty$.

So both conditions of Theorem 8 are met, and therefore the given improper integral is convergent.

M7.2.7. Improper Integrals of Unbounded Functions

▶ **Basic definitions.**

$1°$. Let a function $f(x)$ be defined and continuous for $a \le x < b$, but $\lim_{x \to b-0} f(x) = \infty$. If there exists a finite limit $\lim_{\lambda \to b-0} \displaystyle\int_a^\lambda f(x)\, dx$, it is called the (convergent) *improper integral of the unbounded function $f(x)$ over the interval $[a, b]$.* Thus, by definition

$$\int_a^b f(x)\, dx = \lim_{\lambda \to b-0} \int_a^\lambda f(x)\, dx. \qquad (M7.2.7.1)$$

If no finite limit exists, the integral is called *divergent.*

If $\lim_{x \to a+0} f(x) = \infty$, then, by definition, it is assumed that

$$\int_a^b f(x)\, dx = \lim_{\gamma \to a+0} \int_\gamma^b f(x)\, dx.$$

Finally, if $f(x)$ is unbounded near a point $c \in (a, b)$ and both integrals $\displaystyle\int_a^c f(x)\, dx$ and $\displaystyle\int_c^b f(x)\, dx$ are convergent, then, by definition,

$$\int_a^b f(x)\, dx = \int_a^c f(x)\, dx + \int_c^b f(x)\, dx.$$

If at least one of the integrals on the right-hand side is divergent, the integral on the left-hand side is called divergent.

$2°$. The geometric meaning of an improper integral of an unbounded function and also sufficient conditions for convergence of such integrals are similar to those for improper integrals with infinite limit(s).

▶ **Convergence tests for improper integrals of unbounded functions.** Presented below are theorems for the case where the only singular point of the integrand function is the right endpoint of the interval $[a, b]$.

THEOREM 1 (CAUCHY'S CONVERGENCE CRITERION). *For the integral (M7.2.7.1) to be convergent it is necessary and sufficient that for any $\varepsilon > 0$ there exist a number $\delta > 0$ such that for any δ_1 and δ_2 satisfying $0 < \delta_1 < \delta$ and $0 < \delta_2 < \delta$ the following inequality holds:*

$$\left| \int_{b-\delta_1}^{b-\delta_2} f(x)\,dx \right| < \varepsilon.$$

THEOREM 2. *If $0 \le f(x) \le g(x)$ for $a \le x < b$, then the convergence of the integral $\int_a^b g(x)\,dx$ implies the convergence of the integral $\int_a^b f(x)\,dx$, with $\int_a^b f(x)\,dx \le \int_a^b g(x)\,dx$. If the integral $\int_a^b f(x)\,dx$ is divergent, then the integral $\int_a^b g(x)\,dx$ is also divergent.*

Example. For any continuous function $\varphi(x)$ such that $\varphi(1) = 0$, the improper integral $\int_0^1 \dfrac{dx}{\varphi^2(x) + \sqrt{1-x}}$ is convergent and does not exceed 2, since $\dfrac{1}{\varphi^2(x) + \sqrt{1-x}} < \dfrac{1}{\sqrt{1-x}}$, while the integral $\int_0^1 \dfrac{dx}{\sqrt{1-x}}$ is convergent and is equal to 2.

THEOREM 3. *Let $f(x)$ and $g(x)$ be continuous functions on $[a, b)$ and let the following limit exist:*

$$\lim_{x \to b} \frac{f(x)}{g(x)} = K \qquad (0 < K < \infty).$$

Then both integrals

$$\int_a^b f(x)\,dx, \qquad \int_a^b g(x)\,dx$$

are either convergent or divergent simultaneously.

THEOREM 4. *Let a function $f(x)$ be representable in the form*

$$f(x) = \frac{\varphi(x)}{(b-x)^\lambda} \qquad (\lambda > 0),$$

where $\varphi(x)$ is continuous on $[a, b]$ and the condition $\varphi(b) \ne 0$ holds.

Then: (i) if $\lambda < 1$ and $\varphi(x) \le c < \infty$, then the integral $\int_a^b f(x)\,dx$ is convergent; (ii) if $\lambda \ge 1$ and $\varphi(x) \ge c > 0$, then this integral is divergent.

M7.2.8. Approximate (Numerical) Methods for Computation of Definite Integrals

For approximate computation of an integral like $\int_a^b f(x)\,dx$, let us break up the interval $[a, b]$ into n equal subintervals with length $h = \dfrac{b-a}{n}$. Introduce the notation: $x_0 = a, x_1, \dots, x_n = b$ (the partition points), $y_i = f(x_i)$, $i = 0, 1, \dots, n$.

1°. *Rectangle rules*:

$$\int_a^b f(x)\,dx \approx h(y_0 + y_1 + \cdots + y_{n-1}),$$

$$\int_a^b f(x)\,dx \approx h(y_1 + y_2 + \cdots + y_n).$$

The error of these formulas, R_n, is proportional to h and is estimated using the inequality

$$|R_n| \le \tfrac{1}{2}h(b-a)M_1, \qquad M_1 = \max_{a \le x \le b}\left|f'(x)\right|.$$

2°. *Trapezoidal rule*:

$$\int_a^b f(x)\,dx \approx h\left(\frac{y_0 + y_n}{2} + y_1 + y_2 + \cdots + y_{n-1}\right).$$

The error of this formula is proportional to h^2 and is estimated as

$$|R_n| \le \tfrac{1}{12}h^2(b-a)M_2, \qquad M_2 = \max_{a \le x \le b}\left|f''(x)\right|.$$

3°. *Simpson's rule*:

$$\int_a^b f(x)\,dx \approx \tfrac{1}{3}h[y_0 + y_n + 4(y_1 + y_3 + \cdots + y_{n-1}) + 2(y_2 + y_4 + \cdots + y_{n-2})],$$

where n is even. The error of approximation by Simpson's rule is proportional to h^4:

$$|R_n| \le \tfrac{1}{180}h^4(b-a)M_4, \qquad M_4 = \max_{a \le x \le b}\left|f^{(4)}(x)\right|.$$

Simpson's rule yields exact results for the case where the integrand function is a polynomial of degree two or three.

M7.3. Double and Triple Integrals

M7.3.1. Definition and Properties of the Double Integral

▶ **Definition and properties of the double integral.** Suppose there is a bounded set of points defined on the plane, so that it can be placed in a minimal enclosing circle. The diameter of this circle is called the *diameter of the set*. Consider a domain D in the xy-plane. Let us partition D into n nonintersecting subdomains (cells). The largest of the cell diameters is called the *partition diameter* and is denoted $\lambda = \lambda(\mathcal{D}_n)$, where $\mathcal{D}_n$ stands for the partition of the domain D into cells. Let a function $z = f(x,y)$ be defined in D. Select an arbitrary point in each cell (x_i, y_i), $i = 1, 2, \ldots, n$, and make up an *integral sum*,

$$s_n = \sum_{i=1}^n f(x_i, y_i)\,\Delta S_i,$$

where ΔS_i is the area of the ith subdomain.

If there exists a finite limit, $\mathcal{J}$, of the sums s_n as $\lambda \to 0$ and it depends on neither the partition $\mathcal{D}_n$ nor the selection of the points (x_i, y_i), this limit is denoted $\iint_D f(x,y)\,dx\,dy$ and is called the double integral of the function $f(x,y)$ over the domain D:

$$\iint_D f(x,y)\,dx\,dy = \lim_{\lambda \to 0} s_n.$$

This means that for any $\varepsilon > 0$ there exists a $\delta > 0$ such that for all partitions $\mathcal{D}_n$ such that $\lambda(\mathcal{D}_n) < \delta$ and for any selection of the points (x_i, y_i), the inequality $|s_n - \mathcal{J}| < \varepsilon$ holds. In this case, the function $f(x,y)$ is called integrable over the domain D.

▶ **Classes of integrable functions.** Further on, it is assumed that D is a closed bounded domain.

1. If $f(x, y)$ is continuous in D, then the double integral $\iint_D f(x, y) \, dx \, dy$ exists.

2. If $f(x, y)$ is bounded and the set of points of discontinuity of $f(x, y)$ has a zero area (e.g., the points of discontinuity lie on finitely many continuous curves in the xy-plane), then the double integral of $f(x, y)$ over the domain D exists.

▶ **Properties of the double integral.**

1. *Linearity.* If functions $f(x, y)$ and $g(x, y)$ are integrable in D, then

$$\iint_D \left[a f(x, y) \pm b g(x, y) \right] dx \, dy = a \iint_D f(x, y) \, dx \, dy \pm b \iint_D g(x, y) \, dx \, dy,$$

where a and b are any numbers.

2. *Additivity.* If the domain D is split into two subdomains D_1 and D_2 that do not have common internal points and if the function $f(x, y)$ is integrable in either subdomain, then

$$\iint_D f(x, y) \, dx \, dy = \iint_{D_1} f(x, y) \, dx \, dy + \iint_{D_2} f(x, y) \, dx \, dy.$$

3. *Estimation theorem.* If $m \le f(x, y) \le M$ in D, then

$$mS \le \iint_D f(x, y) \, dx \, dy \le MS,$$

where S is the area of the domain D.

4. *Mean value theorem.* If $f(x, y)$ is continuous in D, then there exists at least one internal point $(\bar{x}, \bar{y}) \in D$ such that

$$\iint_D f(x, y) \, dx \, dy = f(\bar{x}, \bar{y}) \, S.$$

The number $f(\bar{x}, \bar{y})$ is called the *mean value of the function* $f(x, y)$ in D.

5. *Integration of inequalities.* If $\varphi(x, y) \le f(x, y) \le g(x, y)$ in D, then

$$\iint_D \varphi(x, y) \, dx \, dy \le \iint_D f(x, y) \, dx \, dy \le \iint_D g(x, y) \, dx \, dy.$$

In particular, if $f(x, y) \ge 0$ in D, then $\iint_D f(x, y) \, dx \, dy \ge 0$.

6. *Absolute value theorem*

$$\left| \iint_D f(x, y) \, dx \, dy \right| \le \iint_D |f(x, y)| \, dx \, dy.$$

▶ **Geometric meaning of the double integral.** Let a function $f(x, y)$ be nonnegative in D. Then the double integral $\iint_D f(x, y) \, dx \, dy$ is equal to the volume of a cylindrical body with base D in the plane $z = 0$ and bounded from above by the surface $z = f(x, y)$; see Fig. M7.4.

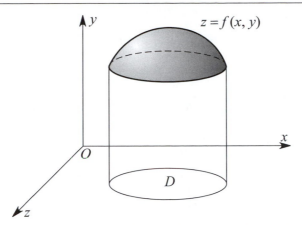

Figure M7.4. A double integral of a nonnegative function $f(x, y)$ over a domain D is equal to the volume of a cylindrical body with base D in the plane $z = 0$ and bounded from above by the surface $z = f(x, y)$.

M7.3.2. Computation of the Double Integral

▶ **Use of iterated integrals.**

$1°$. If a domain D is defined in the xy-plane by the inequalities $a \leq x \leq b$ and $y_1(x) \leq y \leq y_2(x)$ (see Fig. M7.5 a), then*

$$\iint_D f(x, y)\, dx\, dy = \int_a^b dx \int_{y_1(x)}^{y_2(x)} f(x, y)\, dy. \qquad (\text{M7.3.2.1})$$

The expression on the right-hand side is called an *iterated integral*. Note that the variable x in the inner integral is considered constant when integrating.

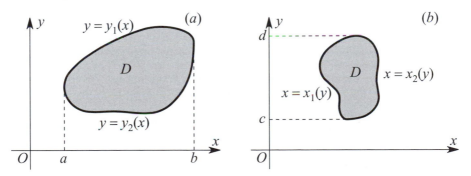

Figure M7.5. Computation of a double integral using iterated integrals: (a) illustration to formula (M7.3.2.1), (b) illustration to formula (M7.3.2.2).

$2°$. If $D = \{c \leq y \leq d,\ x_1(y) \leq x \leq x_2(y)\}$ (see Fig. M7.5 b), then

$$\iint_D f(x, y)\, dx\, dy = \int_c^d dy \int_{x_1(y)}^{x_2(y)} f(x, y)\, dx. \qquad (\text{M7.3.2.2})$$

Example 1. Compute the integral

$$I = \iint_D \frac{dx\, dy}{(ax + by)^2},$$

where $D = \{0 \leq x \leq 1,\ 1 \leq y \leq 3\}$ is a rectangle, $a > 0$, and $b > 0$.

---

* It is assumed that in (M7.3.2.1) and (M7.3.2.2) the double integral on the left-hand side and the inner integral on the right-hand side exist.

Using formula (M7.3.2.2), we get

$$\iint_D \frac{dx\,dy}{(ax+by)^2} = \int_1^3 dy \int_0^1 \frac{dx}{(ax+by)^2}.$$

Compute the inner integral:

$$\int_0^1 \frac{dx}{(ax+by)^2} = -\frac{1}{a(ax+by)}\Big|_{x=0}^{x=1} = \frac{1}{a}\left(\frac{1}{by} - \frac{1}{by+a}\right).$$

It follows that

$$I = \frac{1}{a}\int_1^3 \left(\frac{1}{by} - \frac{1}{by+a}\right) dy = \frac{1}{ab}\ln\frac{3(a+b)}{a+3b}.$$

$3°$. Consider a domain D inscribed in a rectangle $\{a \le x \le b,\ c \le y \le d\}$. Let the boundary of D, within the rectangle, be intersected by straight lines parallel to the coordinate axes at two points only, as shown in Fig. M7.6 a. Then, by comparing formulas (M7.3.2.1) and (M7.3.2.2), we arrive at the relation

$$\int_a^b dx \int_{y_1(x)}^{y_2(x)} f(x,y)\,dy = \int_c^d dy \int_{x_1(y)}^{x_2(y)} f(x,y)\,dx,$$

which shows how the order of integration can be changed.

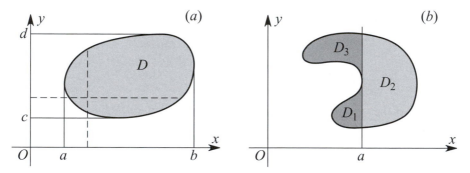

Figure M7.6. Illustrations to the computation of a double integral in a simple (a) and a complex (b) domain.

$4°$. In the general case, the domain D is first split into subdomains considered in Items $1°$ and $2°$, and then the property of additivity of the double integral is used. For example, the domain D shown in Fig. M7.6 b is divided by the straight line $x = a$ into three subdomains D_1, D_2, and D_3. Then the integral over D is represented as the sum of three integrals over the resulting subdomains.

▶ **Change of variables in the double integral.**

$1°$. Let $x = x(u,v)$ and $y = y(u,v)$ be continuously differentiable functions that map one-to-one a domain D_1 in the uv-plane onto a domain D in the xy-plane, and let $f(x,y)$ be a continuous function in D. Then

$$\iint_D f(x,y)\,dx\,dy = \iint_{D_1} f\big(x(u,v),\,y(u,v)\big)|J(u,v)|\,du\,dv,$$

where $J(u,v)$ is the *Jacobian* (or *Jacobian determinant*) of the mapping of D_1 onto D:

$$J(u,v) = \frac{\partial(x,y)}{\partial(u,v)} = \begin{vmatrix} \frac{\partial x}{\partial u} & \frac{\partial x}{\partial v} \\ \frac{\partial y}{\partial u} & \frac{\partial y}{\partial v} \end{vmatrix} = \frac{\partial x}{\partial u}\frac{\partial y}{\partial v} - \frac{\partial x}{\partial v}\frac{\partial y}{\partial u}.$$

The fraction before the determinant is a common notation for a Jacobian.

The absolute value of the Jacobian characterizes the extension (contraction) of an infinitesimal area element when passing from x, y to u, v.

$2°$. The Jacobian of the mapping defining the change from the Cartesian coordinates x, y to the polar coordinates ρ, φ,

$$x = \rho \cos \varphi, \qquad y = \rho \sin \varphi, \tag{M7.3.2.3}$$

is equal to

$$J(\rho, \varphi) = \rho. \tag{M7.3.2.4}$$

Example 2. Given a sphere of radius R and a right circular cylinder of radius $a < R$ whose axis passes through the sphere center, find the volume of the figure the cylinder cuts out of the sphere.

The volume of this figure is calculated as

$$V = 2 \iint_{x^2+y^2 \leq a^2} \sqrt{R^2 - x^2 - y^2} \, dx \, dy.$$

Passing in the integral from x, y to the polar coordinates (M7.3.2.3) and taking into account (M7.3.2.4), we obtain

$$V = 2 \int_0^{2\pi} \int_0^a \sqrt{R^2 - \rho^2} \, \rho \, d\rho \, d\varphi = \frac{4\pi}{3} \left[R^3 - (R^2 - a^2)^{3/2} \right].$$

M7.3.3. Geometric and Physical Applications of the Double Integral

▶ **Geometric applications of the double integral.**

1. *Area of a domain D in the xy-plane*:

$$S = \iint_D dx \, dy.$$

2. *Area of a surface defined by an equation $z = f(x, y)$ with $(x, y) \in D$ (the surface is projected onto a domain D in the xy-plane)*:

$$S = \iint_D \sqrt{\left(\frac{\partial f}{\partial x}\right)^2 + \left(\frac{\partial f}{\partial y}\right)^2 + 1} \, dx \, dy.$$

3. *Calculation of volumes.* If a domain U of the three-dimensional space is defined by $\{(x, y) \in D, \ f(x, y) \leq z \leq g(x, y)\}$, where D is a domain in the xy-plane, the volume of U is calculated as

$$V = \iint_D \left[g(x, y) - f(x, y) \right] dx \, dy.$$

The three-dimensional domain U is a cylinder with base D bounded by the surface $z = f(x, y)$ from below and the surface $z = g(x, y)$ from above. The lateral surface of this body consists of segments of straight lines parallel to the z-axis.

4. *Area of a surface defined parametrically by equations $x = x(u, v)$, $y = y(u, v)$, $z = z(u, v)$, with $(u, v) \in D_1$*:

$$S = \iint_{D_1} \sqrt{EG - F^2} \, du \, dv.$$

Notation used:

$$E = \left(\frac{\partial x}{\partial u}\right)^2 + \left(\frac{\partial y}{\partial u}\right)^2 + \left(\frac{\partial z}{\partial u}\right)^2,$$

$$G = \left(\frac{\partial x}{\partial v}\right)^2 + \left(\frac{\partial y}{\partial v}\right)^2 + \left(\frac{\partial z}{\partial v}\right)^2,$$

$$F = \frac{\partial x}{\partial u} \frac{\partial x}{\partial v} + \frac{\partial y}{\partial u} \frac{\partial y}{\partial v} + \frac{\partial z}{\partial u} \frac{\partial z}{\partial v}.$$

5. *Area of a surface defined by a vector equation* $\mathbf{r} = \mathbf{r}(u, v) = x(u, v)\mathbf{i} + y(u, v)\mathbf{j} + z(u, v)\mathbf{k}$, *with* $(u, v) \in D_1$:

$$S = \iint_{D_1} |\mathbf{n}(u, v)|\, du\, dv,$$

where $\mathbf{n}(u, v) = \mathbf{r}_u \times \mathbf{r}_v$ is a normal vector to the surface; the subscripts u and v denote the respective partial derivatives.

Remark. The formulas from Items 4 and 5 are equivalent—they define one and the same surface in two forms, scalar and vector, respectively.

▶ **Physical applications of the double integral.** Consider a flat plate that occupies a domain D in the xy-plane. Let $\gamma(x, y)$ be the surface density of the plate material (the case $\gamma = \text{const}$ corresponds to a homogeneous plate).

1. *Mass of a flat plate*:

$$m = \iint_D \gamma(x, y)\, dx\, dy.$$

2. *Coordinates of the center of mass of a flat plate*:

$$x_c = \frac{1}{m} \iint_D x\gamma(x, y)\, dx\, dy, \quad y_c = \frac{1}{m} \iint_D y\gamma(x, y)\, dx\, dy,$$

where m is the mass of the plate.

3. *Moments of inertia of a flat plate* about the coordinate axes:

$$I_x = \iint_D y^2\gamma(x, y)\, dx\, dy, \quad I_y = \iint_D x^2\gamma(x, y)\, dx\, dy.$$

The moment of inertia of the plate about the origin of coordinates is calculated as $I_0 = I_x + I_y$.

M7.3.4. Definition and Properties of the Triple Integral

▶ **Definition of the triple integral.** Let a function $f(x, y, z)$ be defined in a domain U of the three-dimensional space. Let us break up U into n subdomains (cells) that do not have common internal points. Denote by $\lambda = \lambda(\mathcal{U}_n)$ the *diameter* of the resulting partition $\mathcal{U}_n$, i.e., the maximum of the cell diameters (the diameter of a domain in space is the diameter of the minimal sphere enclosing the domain). Select an arbitrary point, (x_i, y_i, z_i), $i = 1, 2, \ldots, n$, in each cell and make up an *integral sum*

$$s_n = \sum_{i=1}^{n} f(x_i, y_i, z_i)\, \Delta V_i,$$

where ΔV_i is the volume of the ith cell. If there exists a finite limit of the sums s_n as $\lambda(\mathcal{U}_n) \to 0$ that depends on neither the partition $\mathcal{U}_n$ nor the selection of the points (x_i, y_i, z_i), then it is called the triple integral of the function $f(x, y, z)$ over the domain U and is denoted

$$\iiint_U f(x, y, z)\, dx\, dy\, dz = \lim_{\lambda \to 0} s_n.$$

▶ **Properties of the triple integral.** The properties of triple integrals are similar to those of double integrals.

1. *Linearity.* If functions $f(x, y, z)$ and $g(x, y, z)$ are integrable in a domain U, then

$$\iiint_U \left[a f(x, y, z) \pm b g(x, y, z) \right] dx \, dy \, dz$$

$$= a \iiint_U f(x, y, z) \, dx \, dy \, dz \pm b \iiint_U g(x, y, z) \, dx \, dy \, dz,$$

where a and b are any numbers.

2. *Additivity.* If a domain U is split into two subdomains, U_1 and U_2, that do not have common internal points and if a function $f(x, y, z)$ is integrable in either subdomain, then

$$\iiint_U f(x, y, z) \, dx \, dy \, dz = \iiint_{U_1} f(x, y, z) \, dx \, dy \, dz + \iiint_{U_2} f(x, y, z) \, dx \, dy \, dz.$$

3. *Estimation theorem.* If $m \leq f(x, y, z) \leq M$ in a domain U, then

$$mV \leq \iiint_U f(x, y, z) \, dx \, dy \, dz \leq MV,$$

where V is the volume of U.

4. *Mean value theorem.* If $f(x, y, z)$ is continuous in U, then there exists at least one internal point $(\bar{x}, \bar{y}, \bar{z}) \in U$ such that

$$\iiint_U f(x, y, z) \, dx \, dy \, dz = f(\bar{x}, \bar{y}, \bar{z}) \, V.$$

The number $f(\bar{x}, \bar{y}, \bar{z})$ is called the *mean value of the function f* in the domain U.

5. *Integration of inequalities.* If $\varphi(x, y, z) \leq f(x, y, z) \leq g(x, y, z)$ in a domain U, then

$$\iiint_U \varphi(x, y, z) \, dx \, dy \, dz \leq \iiint_U f(x, y, z) \, dx \, dy \, dz \leq \iiint_U g(x, y, z) \, dx \, dy \, dz.$$

6. *Absolute value theorem*:

$$\left| \iiint_U f(x, y, z) \, dx \, dy \, dz \right| \leq \iiint_U |f(x, y, z)| \, dx \, dy \, dz.$$

M7.3.5. Computation of the Triple Integral. Some Applications. Iterated Integrals and Asymptotic Formulas

▶ **Use of iterated integrals.**

1°. Consider a three-dimensional body U bounded by a surface $z = g(x, y)$ from above and a surface $z = h(x, y)$ from below, with a domain D being the projection of the body onto the xy-plane. In other words, the domain U is defined as $\{(x, y) \in D : h(x, y) \leq z \leq g(x, y)\}$. Then

$$\iiint_U f(x, y, z) \, dx \, dy \, dz = \iint_D dx \, dy \int_{h(x,y)}^{g(x,y)} f(x, y, z) \, dz.$$

2°. If, under the same conditions as in Item 1°, the domain D of the xy-plane is defined as $\{a \leq x \leq b, \, y_1(x) \leq y \leq y_2(x)\}$, then

$$\iiint_U f(x, y, z) \, dx \, dy \, dz = \int_a^b dx \int_{y_1(x)}^{y_2(x)} dy \int_{h(x,y)}^{g(x,y)} f(x, y, z) \, dz.$$

▶ **Change of variables in the triple integral.**

1°. Let $x = x(u, v, w)$, $y = y(u, v, w)$, and $z = z(u, v, w)$ be continuously differentiable functions that map, one to one, a domain Ω of the u, v, w space onto a domain U of the x, y, z space, and let a function $f(x, y, z)$ be continuous in U. Then

$$\iiint_U f(x, y, z)\, dx\, dy\, dz = \iiint_\Omega f\big(x(u, v, w),\, y(u, v, w),\, z(u, v, w)\big)\, |J(u, v, w)|\, du\, dv\, dw,$$

where $J(u, v, w)$ is the *Jacobian* of the mapping of Ω onto U:

$$J(u, v, w) = \frac{\partial(x, y, z)}{\partial(u, v, w)} = \begin{vmatrix} \dfrac{\partial x}{\partial u} & \dfrac{\partial x}{\partial v} & \dfrac{\partial x}{\partial w} \\[2mm] \dfrac{\partial y}{\partial u} & \dfrac{\partial y}{\partial v} & \dfrac{\partial y}{\partial w} \\[2mm] \dfrac{\partial z}{\partial u} & \dfrac{\partial z}{\partial v} & \dfrac{\partial z}{\partial w} \end{vmatrix}.$$

The expression in the middle is a common notation for a Jacobian.

The absolute value of the Jacobian characterizes the expansion (or contraction) of an infinitesimal volume element when passing from x, y, z to u, v, w.

2°. The Jacobians of some common transformations in space are listed in Table M7.2.

TABLE M7.2
Some curvilinear coordinates in space and the respective Jacobians.

Name of coordinates	Transformation	Jacobian, J
Cylindrical coordinates ρ, φ, z	$x = \rho \cos \varphi, \ y = \rho \sin \varphi, \ z = z$	ρ
Generalized cylindrical coordinates ρ, φ, z	$x = a\rho \cos \varphi, \ y = b\rho \sin \varphi, \ z = z$	$ab\rho$
Spherical coordinates r, φ, θ	$x = r \cos \varphi \sin \theta, \ y = r \sin \varphi \sin \theta, z = r \cos \theta$	$r^2 \sin \theta$
Generalized spherical coordinates r, φ, θ	$x = ar \cos \varphi \sin \theta, \ y = br \sin \varphi \sin \theta,$ $z = cr \cos \theta$	$abcr^2 \sin \theta$
Parabolic cylinder coordinates σ, τ, z	$x = \sigma\tau, \ y = \frac{1}{2}(\tau^2 - \sigma^2), \ z = z$	$\sigma^2 + \tau^2$

▶ **Some geometric and physical applications of the triple integral.**

1. *Volume of a domain U*:

$$V = \iiint_U dx\, dy\, dz.$$

2. *Mass of a body of variable density $\gamma = \gamma(x, y, z)$ occupying a domain U*:

$$m = \iiint_U \gamma\, dx\, dy\, dz.$$

3. *Coordinates of the center of mass*:

$$x_c = \frac{1}{m} \iiint_U x\gamma\, dx\, dy\, dz, \quad y_c = \frac{1}{m} \iiint_U y\gamma\, dx\, dy\, dz, \quad z_c = \frac{1}{m} \iiint_U z\gamma\, dx\, dy\, dz.$$

4. *Moments of inertia* about the coordinate axes:

$$I_x = \iiint_U \rho_{yz}^2 \gamma\, dx\, dy\, dz, \quad I_y = \iiint_U \rho_{xz}^2 \gamma\, dx\, dy\, dz, \quad I_z = \iiint_U \rho_{xy}^2 \gamma\, dx\, dy\, dz,$$

where $\rho_{yz}^2 = y^2 + z^2$, $\rho_{xz}^2 = x^2 + z^2$, and $\rho_{xy}^2 = x^2 + y^2$.

If the body is homogeneous, then $\gamma = \mathrm{const}$.

Example. Given a bounded homogeneous elliptic cylinder,

$$\frac{x^2}{a^2} + \frac{y^2}{b^2} = 1, \quad 0 \le z \le h,$$

find its moment of inertia about the z-axis.

Using the generalized cylindrical coordinates (see the second row in Table M7.2), we obtain

$$I_x = \gamma \iiint_U (x^2 + y^2)\, dx\, dy\, dz = \gamma \int_0^h \int_0^{2\pi} \int_0^1 \rho^2(a^2 \cos^2 \varphi + b^2 \sin^2 \varphi) ab\rho\, d\rho\, d\varphi\, dz$$

$$= \frac{1}{4} ab\gamma \int_0^h \int_0^{2\pi} (a^2 \cos^2 \varphi + b^2 \sin^2 \varphi)\, d\varphi\, dz = \frac{1}{4} ab\gamma \int_0^{2\pi} \int_0^h (a^2 \cos^2 \varphi + b^2 \sin^2 \varphi)\, dz\, d\varphi$$

$$= \frac{1}{4} abh\gamma \int_0^{2\pi} (a^2 \cos^2 \varphi + b^2 \sin^2 \varphi)\, d\varphi = \frac{1}{4} \pi ab(a^2 + b^2)h\gamma.$$

5. *Potential of the gravitational field* of a body U at a point (x, y, z):

$$\Phi(x, y, z) = \iiint_U \gamma(\xi, \eta, \zeta)\frac{d\xi\, d\eta\, d\zeta}{r}, \quad r = \sqrt{(x - \xi)^2 + (y - \eta)^2 + (z - \zeta)^2},$$

where $\gamma = \gamma(\xi, \eta, \zeta)$ is the body density. A material point of mass m is pulled by the gravitating body U with a force $\mathbf{F}$. The projections of $\mathbf{F}$ onto the x-, y-, and z-axes are given, respectively, by

$$F_x = km\frac{\partial \Phi}{\partial x} = km \iiint_U \gamma(\xi, \eta, \zeta)\frac{\xi - x}{r^3}\, d\xi\, d\eta\, d\zeta,$$

$$F_y = km\frac{\partial \Phi}{\partial y} = km \iiint_U \gamma(\xi, \eta, \zeta)\frac{\eta - y}{r^3}\, d\xi\, d\eta\, d\zeta,$$

$$F_z = km\frac{\partial \Phi}{\partial z} = km \iiint_U \gamma(\xi, \eta, \zeta)\frac{\zeta - z}{r^3}\, d\xi\, d\eta\, d\zeta,$$

where k is the gravitational constant.

M7.4. Line and Surface Integrals

M7.4.1. Line Integral of the First Kind

▶ **Definition of the line integral of the first kind.** Let a function $f(x, y, z)$ be defined on a piecewise smooth curve $\overset{\frown}{AB}$ in the three-dimensional space $\mathbb{R}^3$. Let the curve $\overset{\frown}{AB}$ be divided into n subcurves by points $A = M_0, M_1, M_2, \dots, M_n = B$, thus defining a partition $\mathcal{L}_n$. The longest of the chords $M_0M_1, M_1M_2, \dots, M_{n-1}M_n$ is called the *diameter of the partition* $\mathcal{L}_n$ and is denoted $\lambda = \lambda(\mathcal{L}_n)$. Let us select on each arc $\overset{\frown}{M_{i-1}M_i}$ an arbitrary point (x_i, y_i, z_i), $i = 1, 2, \dots, n$, and make up an *integral sum*

$$s_n = \sum_{i=1}^n f(x_i, y_i, z_i)\, \Delta l_i,$$

where Δl_i is the length of $\overset{\frown}{M_{i-1}M_i}$.

If there exists a finite limit of the sums s_n as $\lambda(\mathcal{L}_n) \to 0$ that depends on neither the partition $\mathcal{L}_n$ nor the selection of the points (x_i, y_i, z_i), then it is called the *line integral of the first kind* of the function $f(x, y, z)$ over the curve $\overset{\frown}{AB}$ and is denoted

$$\int_{AB} f(x, y, z)\, dl = \lim_{\lambda \to 0} s_n.$$

A line integral is also called a *curvilinear integral* or a *path integral*.

If the function $f(x, y, z)$ is continuous, then the line integral exists. The line integral of the first kind does not depend of the direction the path $\overset{\frown}{AB}$ is traced; its properties are similar to those of the definite integral.

▶ **Computation of the line integral of the first kind.**

1. If a plane curve is defined in the form $y = y(x)$, with $x \in [a, b]$, then

$$\int_{AB} f(x, y)\, dl = \int_a^b f\big(x, y(x)\big) \sqrt{1 + (y_x')^2}\, dx.$$

2. If a curve $\overset{\smile}{AB}$ is defined in parametric form by equations $x = x(t)$, $y = y(t)$, and $z = z(t)$, with $t \in [\alpha, \beta]$, then

$$\int_{AB} f(x, y, z)\, dl = \int_\alpha^\beta f\big(x(t), y(t), z(t)\big) \sqrt{(x_t')^2 + (y_t')^2 + (z_t')^2}\, dt. \qquad \text{(M7.4.1.1)}$$

If a function $f(x, y)$ is defined on a plane curve $x = x(t)$, $y = y(t)$, with $t \in [\alpha, \beta]$, one should set $z_t' = 0$ in (M7.4.1.1).

Example. Evaluate the integral $\displaystyle\int_{AB} xy\, dl$, where $\overset{\smile}{AB}$ is a quarter of an ellipse with semiaxes a and b.

Let us write out the equations of the ellipse for the first quadrant in parametric form:

$$x = a \cos t, \qquad y = b \sin t \qquad (0 \le t \le \pi/2).$$

We have $\sqrt{(x_t')^2 + (y_t')^2} = \sqrt{a^2 \sin^2 t + b^2 \cos^2 t}$. To evaluate the integral, we use formula (M7.4.1.1) with $z_t' = 0$:

$$\int_{AB} xy\, dl = \int_0^{\pi/2} (a \cos t)(b \sin t)\sqrt{a^2 \sin^2 t + b^2 \cos^2 t}\, dt$$

$$= \frac{ab}{2} \int_0^{\pi/2} \sin 2t \sqrt{\frac{a^2}{2}(1 - \cos 2t) + \frac{b^2}{2}(1 + \cos 2t)}\, dt = \frac{ab}{4} \int_{-1}^1 \sqrt{\frac{a^2 + b^2}{2} + \frac{b^2 - a^2}{2}u}\, u\, du$$

$$= \frac{ab}{4} \frac{2}{b^2 - a^2} \frac{2}{3}\left(\frac{a^2 + b^2}{2} + \frac{b^2 - a^2}{2}u\right)^{3/2}\Bigg|_{-1}^1 = \frac{ab}{3} \frac{a^2 + ab + b^2}{a + b}.$$

▶ **Applications of the line integral of the first kind.**

1. *Length of a curve $\overset{\smile}{AB}$:*

$$L = \int_{AB} dl.$$

2. *Mass of a material curve $\overset{\smile}{AB}$ with a given line density $\gamma = \gamma(x, y, z)$:*

$$m = \int_{AB} \gamma\, dl.$$

3. *Coordinates of the center of mass of a material curve $\overset{\smile}{AB}$:*

$$x_c = \frac{1}{m} \int_{AB} x\gamma\, dl, \qquad y_c = \frac{1}{m} \int_{AB} y\gamma\, dl, \qquad z_c = \frac{1}{m} \int_{AB} z\gamma\, dl.$$

To a material line with uniform density there corresponds $\gamma = \text{const}$.

M7.4.2. Line Integral of the Second Kind

▶ **Definition of the line integral of the second kind.** Let a vector field

$$\mathbf{a}(x, y, z) = P(x, y, z)\,\mathbf{i} + Q(x, y, z)\,\mathbf{j} + R(x, y, z)\,\mathbf{k}$$

and a piecewise smooth curve $\overset{\smile}{AB}$ be defined in some domain in $\mathbb{R}^3$. By dividing the curve by points $A = M_0, M_1, M_2, \dots, M_n = B$ into n subcurves, we obtain a partition $\mathcal{L}_n$. Let us select on each arc $\overset{\smile}{M_{i-1}M_i}$ an arbitrary point (x_i, y_i, z_i), $i = 1, 2, \dots, n$, and make up a sum of dot products

$$s_n = \sum_{i=1}^{n} \mathbf{a}(x_i, y_i, z_i) \cdot \overrightarrow{M_{i-1}M_i},$$

called an *integral sum*.

If there exists a finite limit of the sums s_n as $\lambda(\mathcal{L}_n) \to 0$ (λ is the diameter of the partition; see Subsection M7.4.1) that depends on neither the partition $\mathcal{L}_n$ nor the selection of the points (x_i, y_i, z_i), then it is called the *line integral of the second kind* of the vector field $\mathbf{a}(x, y, z)$ along the curve $\overset{\smile}{AB}$ and is denoted

$$\int_{AB} \mathbf{a} \cdot d\mathbf{r}, \quad \text{or} \quad \int_{AB} P\,dx + Q\,dy + R\,dz.$$

The line integral of the second kind depends on the direction the path is traced, so that

$$\int_{AB} \mathbf{a} \cdot d\mathbf{r} = -\int_{BA} \mathbf{a} \cdot d\mathbf{r}.$$

A line integral over a closed contour $\mathcal{C}$ is called a *closed path integral* (or a *circulation*) of a vector field $\mathbf{a}$ around $\mathcal{C}$ and is denoted

$$\oint_{\mathcal{C}} \mathbf{a} \cdot d\mathbf{r}.$$

Physical meaning of the line integral of the second kind: $\int_{AB} \mathbf{a} \cdot d\mathbf{r}$ determines the work done by the vector field $\mathbf{a}(x, y, z)$ on a particle of unit mass when it travels along the arc $\overset{\smile}{AB}$.

▶ **Computation of the line integral of the second kind.**

$1°$. For a plane curve $\overset{\smile}{AB}$ defined as $y = y(x)$, with $x \in [a, b]$, and a plane vector field $\mathbf{a}$, we have

$$\int_{AB} \mathbf{a} \cdot d\mathbf{r} = \int_{a}^{b} \left[P\big(x, y(x)\big) + Q\big(x, y(x)\big) y_x'(x) \right] dx.$$

$2°$. Let $\overset{\smile}{AB}$ be defined by a vector equation $\mathbf{r} = \mathbf{r}(t) = x(t)\mathbf{i} + y(t)\mathbf{j} + z(t)\mathbf{k}$, with $t \in [\alpha, \beta]$. Then

$$\int_{AB} \mathbf{a} \cdot d\mathbf{r} = \int_{AB} P\,dx + Q\,dy + R\,dz$$

$$= \int_{\alpha}^{\beta} \left[P\big(x(t), y(t), z(t)\big) x_t'(t) + Q\big(x(t), y(t), z(t)\big) y_t'(t) + R\big(x(t), y(t), z(t)\big) z_t'(t) \right] dt. \quad \text{(M7.4.2.1)}$$

For a plane curve $\overset{\smile}{AB}$ and a plane vector field $\mathbf{a}$, one should set $z'(t) = 0$ in (M7.4.2.1).

▶ **Potential and curl of a vector field.**

1°. A vector field $\mathbf{a} = \mathbf{a}(x, y, z)$ is called *potential* if there exists a function $\Phi(x, y, z)$ such that

$$\mathbf{a} = \operatorname{grad} \Phi, \quad \text{or} \quad \mathbf{a} = \frac{\partial \Phi}{\partial x}\mathbf{i} + \frac{\partial \Phi}{\partial y}\mathbf{j} + \frac{\partial \Phi}{\partial z}\mathbf{k}.$$

The function $\Phi(x, y, z)$ is called a *potential* of the vector field $\mathbf{a}$. The line integral of the second kind of a potential vector field along a path $\overset{\frown}{AB}$ is equal to the increment of the potential along the path:

$$\int_{AB} \mathbf{a} \cdot d\mathbf{r} = \Phi\big|_B - \Phi\big|_A.$$

2°. The *curl* of a vector field $\mathbf{a}(x, y, z) = P\mathbf{i} + Q\mathbf{j} + R\mathbf{k}$ is the vector defined as

$$\operatorname{curl} \mathbf{a} = \left(\frac{\partial R}{\partial y} - \frac{\partial Q}{\partial z}\right)\mathbf{i} + \left(\frac{\partial P}{\partial z} - \frac{\partial R}{\partial x}\right)\mathbf{j} + \left(\frac{\partial Q}{\partial x} - \frac{\partial P}{\partial y}\right)\mathbf{k} = \begin{vmatrix} \mathbf{i} & \mathbf{j} & \mathbf{k} \\ \frac{\partial}{\partial x} & \frac{\partial}{\partial y} & \frac{\partial}{\partial z} \\ P & Q & R \end{vmatrix}.$$

The vector $\operatorname{curl} \mathbf{a}$ characterizes the rate of rotation of $\mathbf{a}$ and can also be described as the circulation density of $\mathbf{a}$. Alternative notations: $\operatorname{curl} \mathbf{a} \equiv \nabla \times \mathbf{a} \equiv \operatorname{curl} \mathbf{a}$.

▶ **Necessary and sufficient conditions for a vector field to be potential.** Let U be a simply connected domain in $\mathbb{R}^3$ (i.e., a domain in which any closed contour can be deformed to a point without leaving U) and let $\mathbf{a}(x, y, z)$ be a vector field in U. Then the following four assertions are equivalent to each other:

(1) the vector field $\mathbf{a}$ is potential;

(2) $\operatorname{curl} \mathbf{a} \equiv \mathbf{0}$;

(3) the circulation of $\mathbf{a}$ around any closed contour $C \in U$ is zero, or, equivalently, $\oint_C \mathbf{a} \cdot d\mathbf{r} = 0$;

(4) the integral $\int_{AB} \mathbf{a} \cdot d\mathbf{r}$ is independent of the shape of $\overset{\frown}{AB} \in U$ (it depends only on the initial and final points).

M7.4.3. Surface Integral of the First Kind

▶ **Definition of the surface integral of the first kind.** Let a function $f(x, y, z)$ be defined on a smooth surface D. Let us break up this surface into n elements (cells) that do not have common internal points and let us denote this partition by $\mathcal{D}_n$. The *diameter*, $\lambda(\mathcal{D}_n)$, of a partition $\mathcal{D}_n$ is the largest of the diameters of the cells (see Paragraph M7.3.4-1). Let us select in each cell an arbitrary point (x_i, y_i, z_i), $i = 1, 2, \ldots, n$, and make up an *integral sum*

$$s_n = \sum_{i=1}^{n} f(x_i, y_i, z_i)\,\Delta S_i,$$

where ΔS_i is the area of the ith element.

If there exists a finite limit of the sums s_n as $\lambda(\mathcal{D}_n) \to 0$ that depends on neither the partition $\mathcal{D}_n$ nor the selection of the points (x_i, y_i, z_i), then it is called the *surface integral of the first kind* of the function $f(x, y, z)$ and is denoted $\iint_D f(x, y, z)\,dS$.

▶ **Computation of the surface integral of the first kind.**

1°. If a surface D is defined by an equation $z = z(x, y)$, with $(x, y) \in D_1$, then

$$\iint_D f(x, y, z)\, dS = \iint_{D_1} f\left(x, y, z(x, y)\right) \sqrt{1 + (z'_x)^2 + (z'_y)^2}\, dx\, dy.$$

2°. If a surface D is defined by a vector equation $\mathbf{r} = \mathbf{r}(x, y, z) = x(u, v)\,\mathbf{i} + y(u, v)\,\mathbf{j} + z(u, v)\,\mathbf{k}$, where $(u, v) \in D_2$, then

$$\iint_D f(x, y, z)\, dS = \iint_{D_2} f\left(x(u, v), y(u, v), z(u, v)\right) |\mathbf{n}(u, v)|\, du\, dv,$$

where $\mathbf{n}(u, v) = \mathbf{r}_u \times \mathbf{r}_v$ is a normal to the surface D; the subscripts u and v denote the respective partial derivatives.

▶ **Applications of the surface integral of the first kind.**

1°. Area of a surface D:

$$S_D = \iint_D dS.$$

2°. Mass of a material surface D with a surface density $\gamma = \gamma(x, y, z)$:

$$m = \iint_D \gamma(x, y, z)\, dS.$$

3°. Coordinates of the center of mass of a material surface D:

$$x_c = \frac{1}{m} \iint_D x\gamma\, dS, \quad y_c = \frac{1}{m} \iint_D y\gamma\, dS, \quad z_c = \frac{1}{m} \iint_D z\gamma\, dS.$$

To the uniform surface density there corresponds $\gamma = \text{const}$.

M7.4.4. Surface Integral of the Second Kind

▶ **Definition of the surface integral of the second kind.** Let a vector field $\mathbf{a}(x, y, z) = P\,\mathbf{i} + Q\,\mathbf{j} + R\,\mathbf{k}$ be defined on a smooth oriented surface D. Let us perform a partition, $\mathcal{D}_n$, of the surface D into n elements (cells) that do not have common internal points. Also select an arbitrary point $M_i(x_i, y_i, z_i)$, $i = 1, 2, \ldots, n$, for each cell and make up an *integral sum* $s_n = \sum_{i=1}^{n} \mathbf{a}(x_i, y_i, z_i) \cdot \mathbf{n}_i^\circ\, \Delta S_i$, where ΔS_i is area of the ith cell and $\mathbf{n}_i^\circ$ is the unit normal to the surface at the point M_i, the orientation of which coincides with that of the surface.

If there exists a finite limit of the sums s_n as $\lambda(\mathcal{D}_n) \to 0$ (λ is the diameter of the partition, see Subsection M7.4.3) that depends on neither the partition $\mathcal{D}_n$ nor the selection of the points $M_i(x_i, y_i, z_i)$, then it is called the *surface integral of the second kind* (or the *flux of the vector field* $\mathbf{a}$ *across the oriented surface D*) and is denoted

$$\iint_D \mathbf{a}(x, y, z) \cdot \overrightarrow{dS}, \quad \text{or} \quad \iint_D P\, dy\, dz + Q\, dx\, dz + R\, dx\, dy.$$

Note that the surface integral of the second kind changes its sign when the orientation of the surface is reversed.

▶ **Computation of the surface integral of the second kind.**

$1°$. If a surface D is defined by an equation $z = z(x, y)$, with $(x, y) \in D_2$, then the normal $\mathbf{n}(x, y) = \mathbf{r}_x \times \mathbf{r}_y = -z_x \mathbf{i} - z_y \mathbf{j} + \mathbf{k}$ orients the surface D "upward," in the positive direction of the z-axis; the subscripts x and y denote the respective partial derivatives. Then

$$\iint_D \mathbf{a} \cdot \overrightarrow{dS} = \pm \iint_{D_2} \left(-z_x P - z_y Q + R\right) dx \, dy,$$

where $P = P\big(x, y, z(x, y)\big)$, $Q = Q\big(x, y, z(x, y)\big)$, and $R = R\big(x, y, z(x, y)\big)$. The plus sign is taken if the surface has the "upward" orientation, and the minus sign is chosen in the opposite case.

M7.4.5. Integral Formulas of Vector Calculus

▶ **Ostrogradsky–Gauss theorem (divergence theorem).** Let a vector field $\mathbf{a}(x, y, z) = P(x, y, z)\mathbf{i} + Q(x, y, z)\mathbf{j} + R(x, y, z)\mathbf{k}$ be continuously differentiable in a finite simply connected domain $V \subset \mathbb{R}^3$ and let S denote the surface of V oriented by an outward normal. Then the *Ostrogradsky–Gauss theorem* (or the *divergence theorem*) holds:

$$\iint_S \mathbf{a} \cdot \overrightarrow{dS} = \iiint_V \operatorname{div} \mathbf{a} \, dx \, dy \, dz,$$

where $\operatorname{div} \mathbf{a}$ is the *divergence* of the vector $\mathbf{a}$, which is defined as follows:

$$\operatorname{div} \mathbf{a} = \frac{\partial P}{\partial x} + \frac{\partial Q}{\partial y} + \frac{\partial R}{\partial z}.$$

Thus, the flux of a vector field across a closed surface in the outward direction is equal to the triple integral of the divergence of the vector field over the volume bounded by the surface. In coordinate form, the Ostrogradsky–Gauss theorem reads

$$\iint_S P \, dy \, dz + Q \, dx \, dz + R \, dx \, dy = \iiint_V \left(\frac{\partial P}{\partial x} + \frac{\partial Q}{\partial y} + \frac{\partial R}{\partial z}\right) dx \, dy \, dz.$$

▶ **Stokes's theorem (curl theorem).**

$1°$. Let a vector field $\mathbf{a}(x, y, z)$ be continuously differentiable in a domain of the three-dimensional space $\mathbb{R}^3$ that contains an oriented surface D. The orientation of a surface uniquely defines the direction in which the boundary of the surface is traced; specifically, the boundary is traced counterclockwise when looked at from the direction of the normal to the surface. Then the circulation of the vector field around the boundary $\mathcal{C}$ of the surface D is equal to the flux of the vector $\operatorname{curl} \mathbf{a}$ across D:

$$\oint_{\mathcal{C}} \mathbf{a} \cdot d\mathbf{r} = \iint_D \operatorname{curl} \mathbf{a} \cdot \overrightarrow{dS}.$$

In coordinate notation, *Stokes's theorem* reads

$$\oint_{\mathcal{C}} P \, dx + Q \, dy + R \, dz = \iint_D \left(\frac{\partial R}{\partial y} - \frac{\partial Q}{\partial z}\right) dy \, dz + \left(\frac{\partial P}{\partial z} - \frac{\partial R}{\partial x}\right) dx \, dz + \left(\frac{\partial Q}{\partial x} - \frac{\partial P}{\partial y}\right) dx \, dy.$$

$2°$. For a plane vector field $\mathbf{a}(x, y) = P(x, y)\mathbf{i} + Q(x, y)\mathbf{j}$, Stokes's theorem reduces to *Green's theorem*:

$$\oint_{\mathcal{C}} P \, dx + Q \, dy = \iint_D \left(\frac{\partial Q}{\partial x} - \frac{\partial P}{\partial y}\right) dx \, dy,$$

where the contour $\mathcal{C}$ of the domain D on the xy-plane is traced counterclockwise.

Bibliography for Chapter M7

Adams, R., *Calculus: A Complete Course, 6th Edition*, Pearson Education, Toronto, 2006.

Anton, H., *Calculus: A New Horizon, 6th Edition*, Wiley, New York, 1999.

Anton, H., Bivens, I., and Davis, S., *Calculus: Early Transcendental Single Variable, 8th Edition*, John Wiley & Sons, New York, 2005.

Aramanovich, I. G., Guter, R. S., et al., *Mathematical Analysis (Differentiation and Integration)*, Fizmatlit Publishers, Moscow, 1961.

Borden, R. S., *A Course in Advanced Calculus*, Dover Publications, New York, 1998.

Brannan, D., *A First Course in Mathematical Analysis*, Cambridge University Press, Cambridge, England, 2006.

Bronshtein, I. N. and Semendyayev, K. A., *Handbook of Mathematics, 4th Edition*, Springer-Verlag, Berlin, 2004.

Browder, A., *Mathematical Analysis: An Introduction*, Springer-Verlag, New York, 1996.

Clark, D. N., *Dictionary of Analysis, Calculus, and Differential Equations*, CRC Press, Boca Raton, Florida, 2000.

Courant, R. and John, F., *Introduction to Calculus and Analysis, Vol. 1*, Springer-Verlag, New York, 1999.

Danilov, V. L., Ivanova, A. N., et al., *Mathematical Analysis (Functions, Limits, Series, Continued Fractions)* [in Russian], Fizmatlit Publishers, Moscow, 1961.

Dwight, H. B., *Tables of Integrals and Other Mathematical Data*, Macmillan, New York, 1961.

Edwards, C. H., and Penney, D., *Calculus, 6th Edition*, Pearson Education, Toronto, 2002.

Fedoryuk, M. V., *Asymptotics, Integrals and Series* [in Russian], Nauka Publishers, Moscow, 1987.

Fikhtengol'ts, G. M., *Fundamentals of Mathematical Analysis, Vol. 2*, Pergamon Press, London, 1965.

Fikhtengol'ts, G. M., *A Course of Differential and Integral Calculus, Vol. 2* [in Russian], Nauka Publishers, Moscow, 1969.

Gradshteyn, I. S. and Ryzhik, I. M., *Tables of Integrals, Series and Products, 6th Edition*, Academic Press, New York, 2000.

Kaplan, W., *Advanced Calculus, 5th Edition*, Addison Wesley, Reading, Massachusetts, 2002.

Kline, M., *Calculus: An Intuitive and Physical Approach, 2nd Edition*, Dover Publications, New York, 1998.

Landau, E., *Differential and Integral Calculus*, American Mathematical Society, Providence, Rhode Island, 2001.

Marsden, J. E. and Weinstein, A., *Calculus, 2nd Edition*, Springer-Verlag, New York, 1985.

Mendelson, E., *3000 Solved Problems in Calculus*, McGraw-Hill, New York, 1988.

Polyanin, A. D., Polyanin, V. D., et al., *Handbook for Engineers and Students. Higher Mathematics. Physics. Theoretical Mechanics. Strength of Materials, 3rd Edition* [in Russian], AST/Astrel, Moscow, 2005.

Prudnikov, A. P., Brychkov, Yu. A., and Marichev, O. I., *Integrals and Series, Vol. 1, Elementary Functions*, Gordon & Breach, New York, 1986.

Silverman, R. A., *Essential Calculus with Applications*, Dover Publications, New York, 1989.

Strang, G., *Calculus*, Wellesley-Cambridge Press, Massachusetts, 1991.

Taylor, A. E. and Mann, W. R., *Advanced Calculus, 3rd Edition*, John Wiley, New York, 1983.

Thomas, G. B. and Finney, R. L., *Calculus and Analytic Geometry, 9th Edition*, Addison Wesley, Reading, Massachusetts, 1996.

Widder, D. V., *Advanced Calculus, 2nd Edition*, Dover Publications, New York, 1989.

Zorich, V. A., *Mathematical Analysis*, Springer-Verlag, Berlin, 2004.

Zwillinger, D., *CRC Standard Mathematical Tables and Formulae, 31st Edition*, CRC Press, Boca Raton, Florida, 2002.

Chapter M8
Series

M8.1. Numerical Series and Infinite Products
M8.1.1. Convergent Numerical Series and Their Properties. Cauchy's Criterion

▶ **Basic definitions.** Let $\{a_n\}$ be a numerical sequence. The expression

$$a_1 + a_2 + \cdots + a_n + \cdots = \sum_{n=1}^{\infty} a_n$$

is called a *numerical series* (*infinite sum, infinite numerical series*), a_n is the *generic term of the series*, and

$$s_n = a_1 + a_2 + \cdots + a_n = \sum_{k=1}^{n} a_k$$

is the *nth partial sum of the series*. If there exists a finite limit $\lim\limits_{n \to \infty} s_n = S$, the series is called *convergent*, and S is called the *sum of the series*. In this case, one writes $\sum\limits_{n=1}^{\infty} a_n = S$. If $\lim\limits_{n \to \infty} s_n$ does not exist (or is infinite), the series is called *divergent*. The series $a_{n+1} + a_{n+2} + a_{n+3} + \cdots$ is called the *nth remainder of the series*.

Example 1. Consider the series $\sum\limits_{n=1}^{\infty} aq^{n-1} = a + aq + aq^2 + \cdots$ whose terms form a *geometric progression* with ratio q. This series is convergent for $|q| < 1$ (its sum has the form $S = \frac{a}{1-q}$) and is divergent for $|q| \geq 1$.

▶ **Necessary condition for a series to be convergent. Cauchy's criterion.**

1. *A necessary condition for a series to be convergent.* For a convergent series $\sum\limits_{n=1}^{\infty} a_n$, the generic term must tend to zero, $\lim\limits_{n \to \infty} a_n = 0$. If $\lim\limits_{n \to \infty} a_n \neq 0$, then the series is divergent.

Example 2. The series $\sum\limits_{n=1}^{\infty} \cos \dfrac{1}{n}$ is divergent, since its generic term $a_n = \cos \dfrac{1}{n}$ does not tend to zero as $n \to \infty$.

The above necessary condition is insufficient for the convergence of a series.

Example 3. Consider the series $\sum\limits_{n=1}^{\infty} \dfrac{1}{\sqrt{n}}$. Its generic term tends to zero, $\lim\limits_{n \to \infty} \dfrac{1}{\sqrt{n}} = 0$, but the series $\sum\limits_{n=1}^{\infty} \dfrac{1}{\sqrt{n}}$ is divergent because its partial sums are unbounded,

$$s_n = \frac{1}{\sqrt{1}} + \frac{1}{\sqrt{2}} + \cdots + \frac{1}{\sqrt{n}} > n\frac{1}{\sqrt{n}} = \sqrt{n} \to \infty \quad \text{as} \quad n \to \infty.$$

2. *Cauchy's criterion of convergence of a series.* A series $\sum\limits_{n=1}^{\infty} a_n$ is convergent if and only if for any $\varepsilon > 0$ there exists an $N = N(\varepsilon)$ such that for all $n > N$ and any positive integer k, the following inequality holds: $|a_{n+1} + \cdots + a_{n+k}| < \varepsilon$.

▶ **Properties of convergent series.**

1. If a series is convergent, then any of its remainders is convergent. Removal or addition of finitely many terms does not affect the convergence of a series.

2. If all terms of a series are multiplied by a nonzero constant, the resulting series preserves the property of convergence or divergence (its sum is multiplied by that constant).

3. If the series $\sum_{n=1}^{\infty} a_n$ and $\sum_{n=1}^{\infty} b_n$ are convergent and their sums are equal to S_1 and S_2,

respectively, then the series $\sum_{n=1}^{\infty}(a_n \pm b_n)$ are convergent and their sums are equal to $S_1 \pm S_2$.

4. Terms of a convergent series can be grouped in successive order; the resulting series has the same sum. In other words, one can insert brackets inside a series in an arbitrary order. The inverse operation of opening brackets is not always admissible. Thus, the series $(1-1)+(1-1)+\cdots$ is convergent (its sum is equal to zero), but, after removing the brackets, we obtain the divergent series $1 - 1 + 1 - 1 + \cdots$ (its generic term does not tend to zero).

M8.1.2. Convergence Criteria for Series with Positive (Nonnegative) Terms

▶ **Basic convergence (divergence) criteria for series with positive terms.**

1. *The first comparison criterion.* If $0 \le a_n \le b_n$ (starting from some n), then the convergence of the series $\sum_{n=1}^{\infty} b_n$ implies the convergence of $\sum_{n=1}^{\infty} a_n$; and the divergence of the series $\sum_{n=1}^{\infty} a_n$ implies the divergence of $\sum_{n=1}^{\infty} b_n$.

2. *The second convergence criterion.* Suppose that there exists a finite limit

$$\lim_{n\to\infty} \frac{a_n}{b_n} = \sigma,$$

where $0 < \sigma < \infty$. Then $\sum_{n=1}^{\infty} a_n$ is convergent (resp., divergent) if and only if $\sum_{n=1}^{\infty} b_n$ is convergent (resp., divergent).

Corollary. Suppose that $a_{n+1}/a_n \le b_{n+1}/b_n$ starting from some N (i.e., for $n > N$). Then convergence of the series $\sum_{n=1}^{\infty} b_n$ implies convergence of $\sum_{n=1}^{\infty} a_n$, and divergence of $\sum_{n=1}^{\infty} a_n$ implies divergence of $\sum_{n=1}^{\infty} b_n$.

3. *D'Alembert criterion.* Suppose that there exists the limit (finite or infinite)

$$\lim_{n\to\infty} \frac{a_{n+1}}{a_n} = D.$$

If $D < 1$, then the series $\sum_{n=1}^{\infty} a_n$ is convergent. If $D > 1$, then the series is divergent. For $D = 1$, the d'Alembert criterion cannot be used for deciding whether the series is convergent or divergent.

Example 1. Let us examine the convergence of the series $\sum_{n=1}^{\infty} n^k x^n$ with $x > 0$, using the d'Alembert criterion. Taking $a_n = n^k x^n$, we get

$$\frac{a_{n+1}}{a_n} = \left(1 + \frac{1}{n}\right)^k x \to x \quad \text{as} \quad n \to \infty.$$

Therefore, $D = x$. It follows that the series is convergent for $x < 1$ and divergent for $x > 1$. The series is divergent for $x = 1$, since a_n does not tend to zero as $n \to \infty$.

4. *Cauchy criterion.* Suppose that there exists the limit (finite or infinite)

$$\lim_{n \to \infty} \sqrt[n]{a_n} = K.$$

For $K < 1$, the series $\sum_{n=1}^{\infty} a_n$ is convergent; for $K > 1$, the series is divergent. For $K = 1$, the Cauchy criterion cannot be used to establish convergence of a series.

Remark. The Cauchy criterion is stronger than the d'Alembert criterion, but the latter is, in many cases, simpler than the former.

5. *Gauss's criterion.* Suppose that the ratio of two consecutive terms of a series can be represented in the form

$$\frac{a_n}{a_{n+1}} = \lambda + \frac{\mu}{n} + o\left(\frac{1}{n}\right) \quad \text{as} \quad n \to \infty.$$

The series $\sum_{n=1}^{\infty} a_n$ is convergent if $\lambda > 1$ or if $\lambda = 1$ and $\mu > 1$. The series is divergent if $\lambda < 1$ or if $\lambda = 1$ and $\mu \leq 1$.

6. *Maclaurin–Cauchy integral criterion.* Let $f(x)$ be a nonnegative nonincreasing continuous function on the interval $1 \leq x < \infty$. Let $f(1) = a_1$, $f(2) = a_2$, ..., $f(n) = a_n$, ... Then the series $\sum_{n=1}^{\infty} a_n$ is convergent if and only if the improper integral $\int_1^{\infty} f(x)\,dx$ is convergent.

Example 2. The *harmonic series* $\sum_{n=1}^{\infty} \frac{1}{n} = 1 + \frac{1}{2} + \frac{1}{3} + \cdots$ is divergent, since the integral $\int_1^{\infty} \frac{1}{x}\,dx$ is divergent. In a similar way, one finds that the series $\sum_{n=1}^{\infty} \frac{1}{n^\alpha}$ is convergent for $\alpha > 1$ and divergent for $\alpha \leq 1$.

M8.1.3. Convergence Criteria for Arbitrary Numerical Series. Absolute and Conditional Convergence

▶ **Arbitrary series. Leibniz, Abel, and Dirichlet convergence criteria.**

1. *Leibniz criterion.* Suppose that the terms a_n of a series $\sum_{n=1}^{\infty} a_n$ have alternating signs, their absolute values form a nonincreasing sequence, and $a_n \to 0$ as $n \to \infty$. Then this "alternating" series is convergent. If S is the sum of the series and s_n is its nth partial sum, then the following inequality holds for the error $|S - s_n| \leq |a_{n+1}|$.

Example 1. The series $1 - \frac{1}{2^2} + \frac{1}{3^3} - \frac{1}{4^4} + \frac{1}{5^5} - \cdots$ is convergent by the Leibniz criterion. Taking $S \approx s_4 = 1 - \frac{1}{2^2} + \frac{1}{3^3} - \frac{1}{4^4}$, we obtain the error less than $a_5 = \frac{1}{5^5} = 0.00032$.

2. *Abel criterion.* Consider the series

$$\sum_{n=1}^{\infty} a_n b_n = a_1 b_1 + a_2 b_2 + \cdots + a_n b_n + \cdots, \qquad \text{(M8.1.3.1)}$$

where a_n and b_n are two sequences or real numbers.

Series (M8.1.3.1) is convergent if the series

$$\sum_{n=1}^{\infty} b_n = b_1 + b_2 + \cdots + b_n + \cdots \qquad \text{(M8.1.3.2)}$$

is convergent and the a_n form a bounded monotone sequence ($|a_n| < K$).

3. *Dirichlet criterion.* Series (M8.1.3.1) is convergent if partial sums of series (M8.1.3.2) are bounded uniformly in n,

$$\left| \sum_{k=1}^{n} b_k \right| \leq M \qquad (n = 1, 2, \ldots),$$

and the sequence $a_n \to 0$ is monotone.

Example 2. Consider the series $\sum_{n=1}^{\infty} a_n \sin(nx)$, where $a_n \to 0$ is a monotonically decreasing sequence.

Taking $b_n = \sin(nx)$ and using a well-known identity, we find the partial sum

$$s_n = \sum_{k=1}^{n} \sin(kx) = \frac{\cos\left(\frac{1}{2}x\right) - \cos\left[\left(n + \frac{1}{2}\right)x\right]}{2\sin\left(\frac{1}{2}x\right)} \qquad (x \neq 2m\pi; \quad m = 0, \pm 1, \pm 2, \ldots).$$

This sum is bounded for $x \neq 2m\pi$:

$$|s_n| \leq \frac{1}{\left|\sin\left(\frac{1}{2}x\right)\right|}.$$

Therefore, by the Dirichlet criterion, the series $\sum_{n=1}^{\infty} a_n \sin(nx)$ is convergent for any $x \neq 2m\pi$. Direct verification shows that this series is also convergent for $x = 2m\pi$ (since all its terms at these points are equal to zero).

Remark. The Leibniz and the Abel criteria can be deduced from the Dirichlet criterion.

▶ **Absolute and conditional convergence.**

1. *Absolutely convergent series.* A series $\sum_{n=1}^{\infty} a_n$ (with terms of arbitrary sign) is called *absolutely convergent* if the series $\sum_{n=1}^{\infty} |a_n|$ is convergent.

Any absolutely convergent series is convergent. In order to establish absolute convergence of a series, one can use all convergence criteria for series with nonnegative terms given in Subsection M8.1.2 (in these criteria, a_n should be replaced by $|a_n|$).

Example 3. The series $1 + \frac{1}{2^2} - \frac{1}{3^2} - \frac{1}{4^2} + \frac{1}{5^2} + \frac{1}{6^2} - \cdots$ is absolutely convergent, since the series with the absolute values of its terms, $\sum_{n=1}^{\infty} \frac{1}{n^2}$, is convergent (see the second series in Example 2 of Subsection M8.1.2 for $\alpha = 2$).

2. *Conditionally convergent series.* A convergent series $\sum_{n=1}^{\infty} a_n$ is called *conditionally convergent* if the series $\sum_{n=1}^{\infty} |a_n|$ is divergent.

Example 4. The series $1 - \frac{1}{2} + \frac{1}{3} - \frac{1}{4} + \cdots$ is conditionally convergent, since it is convergent (by the Leibniz criterion), but the series with absolute values of its terms is divergent (it is a harmonic series; see Example 2 in Subsection M8.1.2).

Any rearrangement of the terms of an absolutely convergent series (in particular, a convergent series with nonnegative terms) neither violates its absolute convergence nor changes its sum. Conditionally convergent series do not possess this property: the terms of a conditionally convergent series can be rearranged in such order that the sum of the new series becomes equal to any given value; its terms can also be rearranged so as to result in a divergent series.

M8.1.4. Multiplication of Series. Some Inequalities

▶ **Multiplication of series. Cauchy, Mertens, and Abel theorems.** A product of two infinite series $\sum\limits_{n=0}^{\infty} a_n$ and $\sum\limits_{n=0}^{\infty} b_n$ is understood as a series whose terms have the form $a_n b_m$ (n, $m = 0, 1, \ldots$). The products $a_n b_m$ can be ordered to form a series in many different ways. The following theorems allow us to decide whether it is possible to multiply series.

CAUCHY THEOREM. *Suppose that the series* $\sum\limits_{n=0}^{\infty} a_n$ *and* $\sum\limits_{n=0}^{\infty} b_n$ *are absolutely convergent and their sums are equal to A and B, respectively. Then any product of these series is an absolutely convergent series and its sum is equal to AB. The following Cauchy multiplication formula holds:*

$$\left(\sum_{n=0}^{\infty} a_n\right)\left(\sum_{n=0}^{\infty} b_n\right) = \sum_{n=0}^{\infty}\left(\sum_{m=0}^{n} a_m b_{n-m}\right). \tag{M8.1.4.1}$$

MERTENS THEOREM. *The Cauchy multiplication formula (M8.1.4.1) is also valid if one of the series,* $\sum\limits_{n=0}^{\infty} a_n$ *or* $\sum\limits_{n=0}^{\infty} b_n$, *is absolutely convergent and the other is (conditionally) convergent. In this case, the product is a convergent series, possibly, not absolutely convergent.*

ABEL THEOREM. *Consider two convergent series with sums A and B. Suppose that the product of these series in the form of Cauchy (M8.1.4.1) is a convergent series with sum C. Then $C = AB$.*

▶ **Inequalities.**

1. *Generalized triangle inequality:*

$$\left|\sum_{n=1}^{\infty} a_n\right| \le \sum_{n=1}^{\infty} |a_n|.$$

2. *Cauchy inequality (Cauchy–Schwarz–Bunyakovsky inequality):*

$$\left(\sum_{n=1}^{\infty} a_n b_n\right)^2 \le \left(\sum_{n=1}^{\infty} a_n^2\right)\left(\sum_{n=1}^{\infty} b_n^2\right).$$

3. *Minkowski inequality:*

$$\left(\sum_{n=1}^{\infty} |a_n + b_n|^p\right)^{\frac{1}{p}} \le \left(\sum_{n=1}^{\infty} |a_n|^p\right)^{\frac{1}{p}} + \left(\sum_{n=1}^{\infty} |b_n|^p\right)^{\frac{1}{p}}, \qquad p \ge 1.$$

In all these inequalities it is assumed that the series on the right-hand sides are convergent.

M8.2. Function Series

M8.2.1. Pointwise and Uniform Convergence of Function Series

▶ **Convergence of a function series at a point. Convergence domain.** A *function series* is a series of the form

$$u_1(x) + u_2(x) + \cdots + u_n(x) + \cdots = \sum_{n=1}^{\infty} u_n(x),$$

where $u_n(x)$ are functions defined on a set $X \subset \mathbb{R}$. The series $\sum\limits_{n=1}^{\infty} u_n(x)$ is called *convergent*

at a point $x_0 \in X$ if the numerical series $\sum\limits_{n=1}^{\infty} u_n(x_0)$ is convergent. The set of all $x \in X$ for which the function series is convergent is called its *convergence domain*. The sum of the series is a function of x defined on its convergence domain.

In order to find the convergence domain for a function series, one can use the convergence criteria for numerical series described in Subsections M8.1.2 and M8.1.3 (with the variable x regarded as a parameter).

A series $\sum\limits_{n=1}^{\infty} u_n(x)$ is called *absolutely convergent on a set* X if the series $\sum\limits_{n=1}^{\infty} |u_n(x)|$ is convergent on this set.

Example. The function series

$$1 + x + x^2 + x^3 + \cdots$$

is convergent for $-1 < x < 1$ (see Example 1 in Subsection M8.1.1). Its sum is defined on this interval, $S = \dfrac{1}{1-x}$.

The series $\sum\limits_{k=n+1}^{\infty} u_k(x)$ is called the *remainder of a function series* $\sum\limits_{n=1}^{\infty} u_n(x)$. For a series convergent on a set X, the relation $S(x) = s_n(x) + r_n(x)$, where $s_n(x)$ is the partial sum of the series and $r_n(x)$ is the sum of its remainder, implies that $\lim\limits_{n\to\infty} r_n(x) = 0$ for $x \in X$.

▶ **Uniformly convergent series. Condition of uniform convergence.** A function series is called *uniformly convergent* on a set X if for any $\varepsilon > 0$ there exists an N (dependent on ε but not on x) such that for all $n > N$, the inequality

$$\left| \sum_{k=n+1}^{\infty} u_k(x) \right| < \varepsilon$$

holds for all $x \in X$.

A *necessary and sufficient condition of uniform convergence of a series*. A series $\sum\limits_{n=1}^{\infty} u_n(x)$ is uniformly convergent on a set X if and only if for any $\varepsilon > 0$ there exists an N (independent of x) such that for all $n > N$ and all $m = 1, 2, \ldots$, the inequality

$$\left| \sum_{k=n+1}^{n+m} u_k(x) \right| < \varepsilon$$

holds for all $x \in X$.

M8.2.2. Basic Criteria of Uniform Convergence. Properties of Uniformly Convergent Series

▶ **Criteria of uniform convergence of series.**

1. *Weierstrass criterion of uniform convergence.* A function series $\sum\limits_{n=1}^{\infty} u_n(x)$ is uniformly convergent on a set $X \subset \mathbb{R}$ if there is a convergent number series $\sum\limits_{n=1}^{\infty} a_n$ with

nonnegative terms such that $|u_n(x)| \le a_n$ for all sufficiently large n and all $x \in X$. The series $\sum\limits_{n=1}^{\infty} a_n$ is called a *majorant series* for $\sum\limits_{n=1}^{\infty} u_n(x)$.

Example. The series $\sum\limits_{n=1}^{\infty} (-1)^n \dfrac{\sin nx}{n^2}$ is uniformly convergent for $-\infty < x < \infty$, since $\left| (-1)^n \dfrac{\sin nx}{n^2} \right| \le \dfrac{1}{n^2}$, and the numerical series $\sum\limits_{n=1}^{\infty} \dfrac{1}{n^2}$ is convergent (see the second series in Example 2 in Subsection M8.1.2).

2. *Abel criterion of uniform convergence of function series.* Consider a function series

$$\sum_{n=1}^{\infty} u_n(x)v_n(x) = u_1(x)v_1(x) + u_2(x)v_2(x) + \cdots + u_n(x)v_n(x) + \cdots , \qquad \text{(M8.2.2.1)}$$

where $u_n(x)$ and $v_n(x)$ are sequences of functions of the real variable $x \in [a, b]$.

Series (M8.2.2.1) is uniformly convergent on the interval $[a, b]$ if the series

$$\sum_{n=1}^{\infty} v_n(x) = v_1(x) + v_2(x) + \cdots + v_n(x) + \cdots \qquad \text{(M8.2.2.2)}$$

is uniformly convergent on $[a, b]$ and the functions $u_n(x)$ form a monotone sequence for each x and are uniformly bounded (i.e., $|u_n(x)| \le K$ with a constant K independent of n, x).

3. *Dirichlet criterion of uniform convergence of function series.* Series (M8.2.2.1) is uniformly convergent on the interval $[a, b]$ if the partial sums of the series (M8.2.2.2) are uniformly bounded, i.e.,

$$\left| \sum_{k=1}^{n} v_k(x) \right| \le M = \text{const} \qquad (x \in [a, b], \quad n = 1, 2, \ldots),$$

and the functions $u_n(x)$ form a monotone sequence (for each x) that uniformly converges to zero on $[a, b]$ as $n \to \infty$.

▶ **Properties of uniformly convergent series.** Let $\sum\limits_{n=1}^{\infty} u_n(x)$ be a function series that is uniformly convergent on a segment $[a, b]$, and let $S(x)$ be its sum. Then the following statements hold.

THEOREM 1. *If all terms $u_n(x)$ of the series are continuous at a point $x_0 \in [a, b]$, then the sum $S(x)$ is continuous at that point.*

THEOREM 2. *If the terms $u_n(x)$ are continuous on $[a, b]$, then the series admits term-by-term integration:*

$$\int_a^b S(x)\,dx = \int_a^b \left(\sum_{n=1}^{\infty} u_n(x) \right) dx = \sum_{n=1}^{\infty} \int_a^b u_n(x)\,dx.$$

Remark. The condition of continuity of the functions $u_n(x)$ on $[a, b]$ can be replaced by a weaker condition of their integrability on $[a, b]$.

THEOREM 3. *If all terms of the series have continuous derivatives and the function series* $\sum\limits_{n=1}^{\infty} u'_n(x)$ *is uniformly convergent on $[a, b]$, then the sum $S(x)$ is continuously differentiable on $[a, b]$ and*

$$S'(x) = \left(\sum_{n=1}^{\infty} u_n(x) \right)' = \sum_{n=1}^{\infty} u'_n(x)$$

(i.e., the series admits term-by-term differentiation).

M8.3. Power Series

M8.3.1. Radius of Convergence of Power Series. Properties of Power Series

▶ **Abel theorem. Convergence radius of a power series.** A *power series* is a function series of the form

$$\sum_{n=0}^{\infty} a_n x^n = a_0 + a_1 x + a_2 x^2 + a_3 x^3 + \cdots \tag{M8.3.1.1}$$

(the constants $a_0, a_1, \ldots$ are called the *coefficients* of the power series), and also a series of a more general form

$$\sum_{n=0}^{\infty} a_n (x - x_0)^n = a_0 + a_1(x - x_0) + a_2(x - x_0)^2 + a_3(x - x_0)^3 + \cdots,$$

where x_0 is a fixed point. Below, we consider power series of the first form, since the second series can be transformed into the first by the replacement $\bar{x} = x - x_0$.

ABEL THEOREM. *A power series $\sum_{n=0}^{\infty} a_n x^n$ that is convergent for some $x = x_1$ is absolutely convergent for all x such that $|x| < |x_1|$. A power series that is divergent for some $x = x_2$ is divergent for all x such that $|x| > |x_2|$.*

Remark. There exist series convergent for all x, for instance, $\sum_{n=1}^{\infty} \dfrac{x^n}{n!}$. There are series convergent only for $x = 0$, for instance, $\sum_{n=1}^{\infty} n!\, x^n$.

For a given power series (M8.3.1.1), let R be the least upper bound of all $|x|$ such that the series (M8.3.1.1) is convergent at point x. Thus, by the Abel theorem, the series is (absolutely) convergent for all $|x| < R$, and the series is divergent for all $|x| > R$. The constant R is called the *radius of convergence* of the power series, and the interval $(-R, R)$ is called its *interval of convergence*. The problem of convergence of a power series at the endpoints of its convergence interval has to be studied separately in each specific case. If a series is convergent only for $x = 0$, the convergence interval degenerates into a point (and $R = 0$); if a series is convergent for all x, then, obviously, $R = \infty$.

▶ **Formulas for the radius of convergence of power series.**

$1°$. The radius of convergence of a power series (M8.3.1.1) with finitely many zero terms can be calculated by the formulas

$$R = \lim_{n \to \infty} \left| \frac{a_n}{a_{n+1}} \right| \quad \text{(obtained from the d'Alembert criterion for numerical series)},$$

$$R = \lim_{n \to \infty} \frac{1}{\sqrt[n]{|a_n|}} \quad \text{(obtained from the Cauchy criterion for numerical series)}.$$

Example 1. For the power series $\sum_{n=1}^{\infty} \dfrac{3^n}{n} x^n$, using the first formula for the radius of convergence, we get

$$R = \lim_{n \to \infty} \left| \frac{a_n}{a_{n+1}} \right| = \lim_{n \to \infty} \left| \frac{n+1}{3n} \right| = \frac{1}{3}.$$

Therefore, the series is absolutely convergent on the interval $-\frac{1}{3} < x < \frac{1}{3}$ and is divergent outside that interval.

At the left endpoint of the interval, for $x = -\frac{1}{3}$, we have the conditionally convergent series $\sum_{n=1}^{\infty} \dfrac{(-1)^n}{n}$, and at the right endpoint, for $x = \frac{1}{3}$, we have the divergent numerical series $\sum\limits_{n=1}^{\infty} \frac{1}{n}$. Thus, the series under consideration is convergent on the semi-open interval $\left[-\frac{1}{3}, \frac{1}{3}\right)$.

If the number of zero coefficients in a power series is infinite, the above formulas for R are inapplicable. In such cases, one can directly apply the d'Alembert or Cauchy criteria to the series.

Example 2. For the power series $\sum\limits_{n=1}^{\infty} \frac{4^n}{n} x^{2n}$, the d'Alembert criterion gives

$$\lim_{n\to\infty} \left| \frac{a_{n+1} x^{2(n+1)}}{a_n x^{2n}} \right| = \lim_{n\to\infty} \left| \frac{4nx^2}{n+1} \right| = 4x^2.$$

Then the given series is absolutely convergent if $4x^2 < 1$, or on the interval $-\frac{1}{2} < x < \frac{1}{2}$, and hence $R = \frac{1}{2}$. (It is easily seen that the original series diverges at both endpoints of the interval of convergence.)

2°. Suppose that a power series (M8.3.1.1) is convergent at a boundary point of its convergence interval, say, for $x = R$. Then its sum is left-continuous at that point,

$$\lim_{x\to R-0} \sum_{n=0}^{\infty} a_n x^n = \sum_{n=0}^{\infty} a_n R^n.$$

Example 3. Having the expansion

$$\ln(1 + x) = x - \frac{x^2}{2} + \frac{x^3}{3} - \cdots + (-1)^{n+1} \frac{x^n}{n} + \cdots \qquad (R = 1)$$

in the domain $-1 < x < 1$ and knowing that the series

$$1 - \frac{1}{2} + \frac{1}{3} - \cdots + (-1)^{n+1} \frac{1}{n} + \cdots$$

is convergent (by the Leibniz criterion for series with terms of alternating sign), we conclude that the sum of the last series is equal to $\ln 2$.

▶ **Properties of power series.** On any closed segment belonging to the (open) convergence interval of a power series, the series is uniformly convergent. Therefore, on any such segment, the series has all the properties of uniformly convergent series described in Subsection M8.2.2. Therefore, the following statements hold:

1. A power series (M8.3.1.1) admits term-by-term integration on any segment $[0, x]$ for $|x| < R$,

$$\int_0^x \left(\sum_{n=0}^{\infty} a_n x^n \right) dx = \sum_{n=0}^{\infty} \frac{a_n}{n+1} x^{n+1}$$

$$= a_0 x + \frac{a_1}{2} x^2 + \frac{a_2}{3} x^3 + \cdots + \frac{a_n}{n+1} x^{n+1} + \cdots .$$

Remark 1. The value of x in this formula may coincide with an endpoint of the convergence interval ($x = -R$ and/or $x = R$), provided that series (M8.3.1.1) is convergent at that point.

Remark 2. The convergence radii of the original series and the series obtained by its term-by-term integration on the segment $[0, x]$ coincide.

2. Inside the convergence interval (for $|x| < R$), the series admits term-by-term differentiation of any order, in particular,

$$\frac{d}{dx} \left(\sum_{n=0}^{\infty} a_n x^n \right) = \sum_{n=1}^{\infty} n a_n x^{n-1}$$

$$= a_1 + 2a_2 x + 3a_3 x^2 + \cdots + n a_n x^{n-1} + \cdots .$$

Remark 1. The sum of a power series is a function that has derivatives of any order inside the interval of convergence.

Remark 2. The convergence radii of the original series and the series obtained by its term-by-term differentiation coincide.

M8.3.2. Taylor and Maclaurin Power Series

▶ **Basic definitions.** Let $f(x)$ be an infinitely differentiable function at a point x_0. The *Taylor series* for this function is the power series

$$\sum_{n=0}^{\infty} \frac{1}{n!} f^{(n)}(x_0)(x - x_0)^n = f(x_0) + f'(x_0)(x - x_0) + \frac{1}{2} f''(x_0)(x - x_0)^2 + \cdots,$$

where $0! = 1$ and $f^{(0)}(x_0) = f(x_0)$.

A special case of the Taylor series (for $x_0 = 0$) is the *Maclaurin series*:

$$\sum_{n=0}^{\infty} \frac{1}{n!} f^{(n)}(0) x^n = f(0) + f'(0)x + \frac{1}{2} f''(0)x^2 + \cdots.$$

A formal Taylor series (Maclaurin series) for a function $f(x)$ may be:

1) divergent for $x \neq x_0$,
2) convergent in a neighborhood of x_0 to a function different from $f(x)$,
3) convergent in a neighborhood of x_0 to the function $f(x)$.

In the last case, one says that $f(x)$ is *expandable in a Taylor series* in the said neighborhood, and one writes

$$f(x) = \sum_{n=0}^{\infty} \frac{1}{n!} f^{(n)}(x_0)(x - x_0)^n.$$

▶ **Conditions of expansion in Taylor series.** A *necessary and sufficient condition for a function $f(x)$ to be represented by its Taylor series* in a neighborhood of a point x_0 is that the remainder term in the Taylor formula* should tend to zero as $n \to \infty$ in this neighborhood of x_0.

In order that $f(x)$ could be represented by its Taylor series in a neighborhood of x_0, it suffices that all its derivatives in that neighborhood be bounded by the same constant, $|f^{(n)}(x)| \leq M$ for all n.

Uniqueness of the Taylor series expansion. If a function $f(x)$ is representable by the sum of a power series, the coefficients of this series are determined uniquely (since this series is the Taylor series of $f(x)$ and its coefficients have the form $\dfrac{f^{(n)}(x_0)}{n!}$, where $n = 0, 1, 2, \dots$). Therefore, in problems of representing a function by a power series, the answer does not depend on the method adopted for this purpose.

▶ **Representation of some functions by the Maclaurin series.** The following representations of elementary functions by Maclaurin series are often used in applications:

$$e^x = 1 + x + \frac{x^2}{2!} + \frac{x^3}{3!} + \cdots + \frac{x^n}{n!} + \cdots;$$

$$\sin x = x - \frac{x^3}{3!} + \frac{x^5}{5!} - \cdots + (-1)^{n-1} \frac{x^{2n-1}}{(2n-1)!} + \cdots;$$

---

* Different representations of the remainder in the Taylor formula are given in Subsection M6.2.4.

$$\cos x = 1 - \frac{x^2}{2!} + \frac{x^4}{4!} - \cdots + (-1)^n \frac{x^{2n}}{(2n)!} + \cdots \;;$$

$$\sinh x = x + \frac{x^3}{3!} + \frac{x^5}{5!} + \cdots + \frac{x^{2n-1}}{(2n-1)!} + \cdots \;;$$

$$\cosh x = 1 + \frac{x^2}{2!} + \frac{x^4}{4!} + \cdots + \frac{x^{2n}}{(2n)!} + \cdots \;;$$

$$(1+x)^\alpha = 1 + \alpha x + \frac{\alpha(\alpha-1)}{2!}x^2 + \cdots + \frac{\alpha(\alpha-1)\ldots(\alpha-n+1)}{n!}x^n + \cdots \;;$$

$$\ln(1+x) = x - \frac{x^2}{2} + \frac{x^3}{3} - \cdots + (-1)^{n+1}\frac{x^n}{n} + \cdots \;;$$

$$\arctan x = x - \frac{x^3}{3} + \frac{x^5}{5} - \cdots + (-1)^{n+1}\frac{x^{2n-1}}{2n-1} + \cdots .$$

The first five series are convergent for $-\infty < x < \infty$ ($R = \infty$), and the other series have unit radius of convergence, $R = 1$.

M8.3.3. Operations with Power Series. Summation Formulas for Power Series

▶ **Addition, subtraction, multiplication, and division of power series.**

1. *Addition and subtraction of power series.* Two series $\sum_{n=0}^{\infty} a_n x^n$ and $\sum_{n=0}^{\infty} b_n x^n$ with convergence radii R_a and R_b, respectively, admit term-by-term addition and subtraction on the intersection of their convergence intervals:

$$\sum_{n=0}^{\infty} a_n x^n \pm \sum_{n=0}^{\infty} b_n x^n = \sum_{n=0}^{\infty} c_n x^n, \qquad c_n = a_n \pm b_n.$$

The radius of convergence of the resulting series satisfies the inequality $R_c \geq \min[R_a, R_b]$.

2. *Multiplication of power series.* Two series $\sum_{n=0}^{\infty} a_n x^n$ and $\sum_{n=0}^{\infty} b_n x^n$, with the respective convergence radii R_a and R_b, can be multiplied on the intersection of their convergence intervals, and their product has the form

$$\left(\sum_{n=0}^{\infty} a_n x^n\right)\left(\sum_{n=0}^{\infty} b_n x^n\right) = \sum_{n=0}^{\infty} c_n x^n, \qquad c_n = \sum_{k=0}^{n} a_k b_{n-k}.$$

The convergence radius of the product satisfies the inequality $R_c \geq \min[R_a, R_b]$.

3. *Division of power series.* The ratio of two power series $\sum_{n=0}^{\infty} a_n x^n$ and $\sum_{n=0}^{\infty} b_n x^n$, $b_0 \neq 0$, with convergence radii R_a and R_b can be represented as a power series

$$\frac{\sum_{n=0}^{\infty} a_n x^n}{\sum_{n=0}^{\infty} b_n x^n} = c_0 + c_1 x + c_2 x^2 + \cdots = \sum_{n=0}^{\infty} c_n x^n, \tag{M8.3.3.1}$$

whose coefficients can be found, by the method of indefinite coefficients, from the relation

$$(a_0 + a_1 x + a_2 x^2 + \cdots) = (b_0 + b_1 x + b_2 x^2 + \cdots)(c_0 + c_1 x + c_2 x^2 + \cdots).$$

Thus, for the unknown c_n, we obtain a triangular system of linear algebraic equations

$$a_n = \sum_{k=0}^{n} b_k c_{n-k}, \qquad n = 0, 1, \ldots,$$

which is solved consecutively, starting from the first equation:

$$c_0 = \frac{a_0}{b_0}, \quad c_1 = \frac{a_1 b_0 - a_0 b_1}{b_0^2}, \quad c_n = \frac{a_n}{b_0} - \frac{1}{b_0} \sum_{k=1}^{n} b_k c_{n-k}, \quad n = 2, 3, \ldots$$

The convergence radius of the series (M8.3.3.1) is determined by the formula

$$R_1 = \min\left[R_a, \frac{\rho}{M+1}\right],$$

where ρ is any constant such that $0 < \rho < R_b$; ρ can be chosen arbitrarily close to R_b; and M is the least upper bound of the quantities $|b_m/b_0|\rho^m$ ($m = 1, 2, \ldots$), so that $|b_m/b_0|\rho^m \leq M$ for all m.

▶ **Composition of functions representable by power series.** Consider a power series

$$z = f(y) = a_0 + a_1 y + a_2 y^2 + \cdots = \sum_{n=0}^{\infty} a_n y^n \qquad \text{(M8.3.3.2)}$$

with convergence radius R. Let the variable y be a function of x that can be represented by a power series

$$y = \varphi(x) = b_0 + b_1 x + a_2 x^2 + \cdots = \sum_{n=0}^{\infty} b_n x^n \qquad \text{(M8.3.3.3)}$$

with convergence radius r. It is required to represent z as a power series of x and find the convergence radius of this series.

Formal substitution of (M8.3.3.3) into (M8.3.3.2) yields

$$z = f\big(\varphi(x)\big) = \sum_{n=0}^{\infty} a_n \left(\sum_{k=0}^{\infty} b_k x^k \right)^n = A_0 + A_1 x + A_2 x^2 + \cdots = \sum_{n=0}^{\infty} A_n x^n, \quad \text{(M8.3.3.4)}$$

where

$$A_0 = a_0 + a_1 b_0 + a_2 b_0^2 + \cdots,$$
$$A_1 = a_1 b_1 + 2 a_2 b_0 b_1 + 3 a_3 b_0^2 b_1 + \cdots,$$
$$A_2 = a_1 b_2 + a_2 (b_1^2 + 2 b_0 b_2) + 3 a_3 (b_0 b_1^2 + b_0^2 b_2) + \cdots,$$

$$\cdots \cdots \cdots \cdots \cdots \cdots \cdots \cdots \cdots \cdots \cdots \cdots \cdots \cdots \cdots \cdots \cdots$$

THEOREM ON CONVERGENCE OF SERIES (M8.3.3.4).

(i) *If series (M8.3.3.2) is convergent for all y (i.e., $R = \infty$), then the convergence radius of series (M8.3.3.4) coincides with the convergence radius r of series (M8.3.3.3).*

(ii) *If $0 \leq |b_0| < R$, then series (M8.3.3.4) is convergent on the interval $(-R_1, R_1)$, where*

$$R_1 = \frac{(R - |b_0|)\rho}{M + R - |b_0|},$$

and ρ is an arbitrary constant such that $0 < \rho < r$; ρ can be chosen arbitrarily close to r; and M is the least upper bound of the quantities $|b_m|\rho^m$ ($m = 1, 2, \ldots$), so that $|b_m|\rho^m \leq M$ for all m.

(iii) *If $|b_0| > R$, then series (M8.3.3.4) is divergent.*

Remark. Case (i) is realized if, for instance, (M8.3.3.2) has finitely many terms.

▶ **Simplest summation formulas for power series.** Suppose that the sum of a power series is known,

$$\sum_{k=0}^{\infty} a_k x^k = S(x). \qquad \text{(M8.3.3.6)}$$

Then, using term-by-term integration (on the convergence interval), one can find the following sums:

$$\sum_{k=0}^{\infty} a_k k^m x^k = \left(x \frac{d}{dx} \right)^m S(x);$$

$$\sum_{k=0}^{\infty} a_k (nk + m) x^{nk+m-1} = \frac{d}{dx} \left[x^m S(x^n) \right];$$

$$\sum_{k=0}^{\infty} \frac{a_k}{nk+m} x^{nk+m} = \int_0^x x^{m-1} S(x^n)\, dx, \qquad n > 0, \ m > 0; \qquad \text{(M8.3.3.7)}$$

$$\sum_{k=0}^{\infty} a_k \frac{nk+s}{nk+m} x^{nk+s} = x \frac{d}{dx} \left[x^{s-m} \int_0^x x^{m-1} S(x^n)\, dx \right], \qquad n > 0, \ m > 0;$$

$$\sum_{k=0}^{\infty} a_k \frac{nk+m}{nk+s} x^{nk+s} = \int_0^x x^{s-m} \frac{d}{dx} \left[x^m S(x^n) \right] dx, \qquad n > 0, \ s > 0.$$

Example 2. Let us find the sum of the series $\sum_{k=0}^{\infty} k x^{k-1}$.

We start with the well-known formula for the sum of an infinite geometrical progression:

$$\sum_{k=0}^{\infty} x^k = \frac{1}{1-x} \qquad (|x| < 1).$$

This series is a special case of (M8.3.3.6) with $a_k = 1$, $S(x) = 1/(1-x)$. The series $\sum_{k=0}^{\infty} k x^{k-1}$ can be obtained from the left-hand side of the second formula in (M8.3.3.7) for $m = 0$ and $n = 1$. Substituting $S(x) = 1/(1-x)$ into the right-hand side of that formula, we get

$$\sum_{k=0}^{\infty} k x^{k-1} = \frac{d}{dx} \frac{1}{1-x} = \frac{1}{(1-x)^2} \qquad (|x| < 1).$$

M8.4. Fourier Series

M8.4.1. Representation of 2π-Periodic Functions by Fourier Series. Main Results

▶ **Dirichlet theorem on representation of a function by Fourier series.** A function $f(x)$ is said to satisfy the *Dirichlet conditions* on an interval (a, b) if:

1) this interval can be divided into finitely many intervals on which $f(x)$ is monotone and continuous;

2) at any discontinuity point x_0 of the function, there exist finite one-sided limits $f(x_0 + 0)$ and $f(x_0 - 0)$.

DIRICHLET THEOREM. *Any 2π-periodic function that satisfies the Dirichlet conditions on the interval $(-\pi, \pi)$ can be represented by its Fourier series*

$$f(x) = \frac{a_0}{2} + \sum_{n=1}^{\infty} \left(a_n \cos nx + b_n \sin nx\right) \qquad \text{(M8.4.1.1)}$$

whose coefficients are defined by the Euler–Fourier formulas

$$
\begin{aligned}
a_n &= \frac{1}{\pi} \int_{-\pi}^{\pi} f(x) \cos nx\, dx, \qquad n = 0,\ 1,\ 2,\ldots, \\
b_n &= \frac{1}{\pi} \int_{-\pi}^{\pi} f(x) \sin nx\, dx, \qquad n = 1,\ 2,\ 3,\ldots
\end{aligned}
\qquad \text{(M8.4.1.2)}
$$

At the points of continuity of $f(x)$, the Fourier series converges to $f(x)$, and at any discontinuity point x_0, the series converges to $\frac{1}{2}[f(x_0 + 0) + f(x_0 - 0)]$.
The coefficients a_n and b_n of the series (M8.4.1.1) are called the *Fourier coefficients*.

Remark. Instead of the integration limits $-\pi$ and π in (M8.4.1.2), one can take c and $c + 2\pi$, where c is an arbitrary constant.

▶ **Lipschitz and Dirichlet–Jordan convergence criteria for Fourier series.** LIPSCHITZ CRITERION. *Suppose that $f(x)$ is continuous at a point x_0 and for sufficiently small $\varepsilon > 0$ satisfies the inequality $|f(x_0 \pm \varepsilon) - f(x_0)| \le L\varepsilon^{\sigma}$, where L and σ are constants, $0 < \sigma \le 1$. Then the representation (M8.4.1.1)–(M8.4.1.2) holds at $x = x_0$.*

In particular, the conditions of the Lipschitz criterion hold for continuous piecewise differentiable functions.

Remark. The Fourier series of a continuous periodic function with no additional conditions (for instance, of its regularity) may happen to be divergent at infinitely many (even uncountably many) points.

DIRICHLET–JORDAN CRITERION. *Suppose that $f(x)$ is a function of bounded variation on some interval $(x_0 - h, x_0 + h) \in (-\pi, \pi)$ (i.e., $f(x)$ can be represented as a difference of two monotonically increasing functions). Then the Fourier series (M8.4.1.1)–(M8.4.1.2) of the function $f(x)$ at the point x_0 converges to the value $\frac{1}{2}[f(x_0 + 0) + f(x_0 - 0)]$.*

▶ **Asymptotic properties of Fourier coefficients.**

$1°$. Fourier coefficients of an absolutely integrable function tend to zero as n goes to infinity: $a_n \to 0$ and $b_n \to 0$ as $n \to \infty$.

$2°$. Fourier coefficients of a continuous 2π-periodic function have the following limit properties:

$$\lim_{n \to \infty} (na_n) = 0, \qquad \lim_{n \to \infty} (nb_n) = 0,$$

i.e., $a_n = o(1/n)$ and $b_n = o(1/n)$.

$3°$. If a continuous periodic function is continuously differentiable up to the order $m - 1$ inclusively, then its Fourier coefficients have the following limit properties:

$$\lim_{n \to \infty} (n^m a_n) = 0, \qquad \lim_{n \to \infty} (n^m b_n) = 0,$$

i.e., $a_n = o\left(n^{-m}\right)$ and $b_n = o\left(n^{-m}\right)$.

M8.4.2. Fourier Expansions of Periodic, Nonperiodic, Even, and Odd Functions

▶ **Expansion of $2l$-periodic and nonperiodic functions in Fourier series.**

$1°$. The case of $2l$-periodic functions can be easily reduced to that of 2π-periodic functions by changing the variable x to $z = \dfrac{\pi x}{l}$. In this way, all the results described above for 2π-periodic functions can be easily extended to $2l$-periodic functions.

The Fourier expansion of a $2l$-periodic function $f(x)$ has the form

$$f(x) = \frac{a_0}{2} + \sum_{n=1}^{\infty} \left(a_n \cos \frac{n\pi x}{l} + b_n \sin \frac{n\pi x}{l} \right), \tag{M8.4.2.1}$$

where

$$a_n = \frac{1}{l} \int_{-l}^{l} f(x) \cos \frac{n\pi x}{l}\, dx, \qquad b_n = \frac{1}{l} \int_{-l}^{l} f(x) \sin \frac{n\pi x}{l}\, dx. \tag{M8.4.2.2}$$

$2°$. A nonperiodic (aperiodic) function $f(x)$ defined on the interval $(-l, l)$ can also be represented by a Fourier series (M8.4.2.1)–(M8.4.2.2); however, outside that interval, the sum of that series $S(x)$ may differ from $f(x)$*.

▶ **Fourier expansion of even and odd functions.**

$1°$. Let $f(x)$ be an even function, i.e., $f(-x) = f(x)$. Then the Fourier expansion of $f(x)$ on the interval $(-l, l)$ has the form of the *cosine Fourier series*:

$$f(x) = \frac{a_0}{2} + \sum_{n=1}^{\infty} a_n \cos \frac{n\pi x}{l},$$

where the Fourier coefficients have the form

$$a_n = \frac{2}{l} \int_{0}^{l} f(x) \cos \frac{n\pi x}{l}\, dx \qquad (b_n = 0).$$

$2°$. Let $f(x)$ be an odd function, i.e., $f(-x) = -f(x)$. Then the Fourier expansion of $f(x)$ on the interval $(-l, l)$ has the form of the *sine Fourier series*:

$$f(x) = \sum_{n=1}^{\infty} b_n \sin \frac{n\pi x}{l},$$

where the Fourier coefficients have the form

$$b_n = \frac{2}{l} \int_{0}^{l} f(x) \sin \frac{n\pi x}{l}\, dx \qquad (a_n = 0).$$

---

* The sum $S(x)$ is a $2l$-periodic function defined for all x, but $f(x)$ may happen to be nonperiodic or even undefined outside the interval $(-l, l)$.

Example. Let us find the Fourier expansion of the function $f(x) = x$ on the interval $(-\pi, \pi)$.
Taking $l = \pi$ and $f(x) = x$ in the formula for the Fourier coefficients and integrating by parts, we obtain

$$b_n = \frac{2}{\pi} \int_0^\pi x \sin(nx)\, dx = \frac{2}{\pi} \left(-\frac{1}{n} x \cos(nx) \Big|_0^\pi + \frac{1}{n} \int_0^\pi \cos(nx)\, dx \right) = -\frac{2}{n} \cos(n\pi) = (-1)^{n+1} \frac{2}{n}.$$

Therefore, the Fourier expansion of $f(x) = x$ has the form

$$f(x) = 2 \sum_{n=1}^{\infty} (-1)^{n+1} \frac{\sin(nx)}{n} \qquad (-\pi < x < \pi).$$

$3°$. If $f(x)$ is defined on the interval $(0, l)$ and satisfies the Dirichlet conditions, it can be represented by the cosine Fourier series, as well as the sine Fourier series (with the help of the above formulas). The cosine Fourier expansion of $f(x)$ on the interval $(0, l)$ corresponds to the extension of $f(x)$ to the interval $(-l, 0)$ as an even function: $f(-x) = f(x)$. The sine Fourier expansion of $f(x)$ on $(0, l)$ corresponds to the extension of $f(x)$ to the interval $(-l, 0)$ as an odd function: $f(-x) = -f(x)$. Both series on the interval $(0, l)$ give the values of $f(x)$ at points of its continuity and the value $\frac{1}{2}[f(x_0+0)+f(x_0-0)]$ at points of its discontinuity; outside the interval $(0, l)$, these two series represent different functions.

▶ **Fourier series in complex form.** The complex Fourier expansion of a function $f(x)$ on an interval $(-l, l)$ has the form

$$f(x) = \sum_{n=-\infty}^{\infty} c_n e^{i\omega_n x},$$

where

$$\omega_n = \frac{n\pi}{l}, \qquad c_n = \frac{1}{2l} \int_{-l}^{l} f(x) e^{-i\omega_n x}\, dx; \qquad n = 0, \pm 1, \pm 2, \ldots$$

The expressions $e^{i\omega_n x}$ are called *complex harmonics*, the coefficients c_n are *complex amplitudes*, ω_n are *wave numbers* of the function $f(x)$, and the set of all wave numbers $\{\omega_n\}$ is called the *discrete spectrum* of the function.

M8.4.3. Criteria of Uniform and Mean-Square Convergence of Fourier Series

▶ **Criteria of uniform convergence of Fourier series.**
LIPSCHITZ CRITERION. *The Fourier series of a function $f(x)$ converges uniformly to that function on an interval $[-l, l]$ if on a wider interval $[-L, L]$ $(-L < -l < l < L)$ the following inequality holds:*

$$|f(x_1) - f(x_2)| \le K |x_1 - x_2|^\sigma \quad \text{for all} \quad x_1, x_2 \in [-L, L],$$

where K and σ are constants, $0 < \sigma \le 1$.

Corollary. The Fourier series of a continuous function $f(x)$ converges uniformly to that function on an interval $[-l, l]$ if on a wider interval the function $f(x)$ has a bounded derivative $f'(x)$.

For any continuously differentiable $2l$-periodic function $f(x)$, its Fourier series [defined by formulas (M8.4.2.1)–(M8.4.2.2)] is uniformly convergent to $f(x)$.

▶ **Fourier series of square-integrable functions. Parseval identity.**

1°. For a continuous 2π-periodic function $f(x)$, its Fourier series (M8.4.1.1)–(M8.4.1.2) converges to $f(x)$ in mean square, i.e.,

$$\int_{-\pi}^{\pi} [f(x) - f_n(x)]^2 \, dx \to 0 \quad \text{as} \quad n \to \infty,$$

where $f_n(x) = \frac{1}{2}a_0 + \sum_{k=1}^{n}(a_k \cos kx + b_k \sin kx)$ is a partial sum of the Fourier series.

2°. If $f(x)$ is integrable on the segment $[-\pi, \pi]$ and the integral $\int_{-\pi}^{\pi} f^2(x) \, dx$ exists as an improper integral with finitely many singularities, then the Fourier series (M8.4.1.1)–(M8.4.1.2) is mean-square convergent to $f(x)$.

3°. Let $f(x) \in L^2[-\pi, \pi]$ be a square-integrable function on the segment $[-\pi, \pi]$. Then its Fourier series (M8.4.1.1)–(M8.4.1.2) is mean-square convergent to $f(x)$, and the *Parseval identity* holds:

$$\frac{a_0^2}{2} + \sum_{n=1}^{\infty}(a_n^2 + b_n^2) = \frac{1}{\pi} \int_{-\pi}^{\pi} f^2(x) \, dx,$$

where a_n, b_n are defined by (M8.4.1.2). Note that the functions considered in Items 1° and 2° belong to $L^2[-\pi, \pi]$.

Bibliography for Chapter M8

Brannan, D., *A First Course in Mathematical Analysis*, Cambridge University Press, Cambridge, England, 2006.

Bromwich, T. J. I., *Introduction to the Theory of Infinite Series*, American Mathematical Society, Providence, Rhode Island, 2005.

Bronshtein, I. N. and Semendyayev, K. A., *Handbook of Mathematics, 4th Edition*, Springer-Verlag, Berlin, 2004.

Danilov, V. L., Ivanova, A. N., et al., *Mathematical Analysis (Functions, Limits, Series, Continued Fractions)* [in Russian], Fizmatlit Publishers, Moscow, 1961.

Davis, H. F., *Fourier Series and Orthogonal Functions*, Dover Publications, New York, 1989.

Fedoryuk, M. V., *Asymptotics, Integrals, Series* [in Russian], Nauka, Moscow, 1987.

Fichtenholz, G. M., *Functional Series*, Taylor & Francis, London, 1970.

Fikhtengol'ts (Fichtenholz), G. M., *A Course in Differential and Integral Calculus, Vols. 2 and 3* [in Russian], Nauka Publishers, Moscow, 1969.

Gradshteyn, I. S. and Ryzhik, I. M., *Tables of Integrals, Series, and Products, 6th Edition*, Academic Press, New York, 2000.

Hansen, E. R., *A Table of Series and Products*, Prentice Hall, Englewood Cliffs, London, 1975.

Hirschman, I., *Infinite Series*, Greenwood Press, New York, 1978.

Hyslop, J. M., *Infinite Series*, Dover Publications, New York, 2006.

Jolley, L. B. W., *Summation of Series*, Dover Publications, New York, 1961.

Knopp, K., *Infinite Sequences and Series*, Dover Publications, New York, 1956.

Knopp, K., *Theory and Applications of Infinite Series*, Dover Publications, New York, 1990.

Mangulis, V., *Handbook of Series for Scientists and Engineers*, Academic Press, New York, 1965.

Pinkus, A. and Zafrany, S., *Fourier Series and Integral Transforms*, Cambridge University Press, Cambridge, England, 1997.

Prudnikov, A. P., Brychkov, Yu. A., and Marichev, O. I., *Integrals and Series, Vol. 1, Elementary Functions*, Gordon & Breach, New York, 1986.

Zhizhiashvili, L.V., *Trigonometric Fourier Series and Their Conjugates*, Kluwer Academic, Dordrecht, 1996.

Zygmund, A., *Trigonometric Series, 3rd Edition*, Cambridge University Press, Cambridge, England, 2003.

Chapter M9
Functions of Complex Variable

M9.1. Complex Numbers

M9.1.1. Definition of a Complex Number. Arithmetic Operations with Complex Numbers

▶ **Definition of a complex number. Geometric interpretation.** The set of complex numbers is an extension of the set of real numbers. An expression of the form $z = x + iy$, where x and y are real numbers, is called a *complex number*, and the symbol i is called the *imaginary unit*, which possesses the property $i^2 = -1$. The numbers x and y are called, respectively, the *real* and *imaginary parts* of z and denoted by

$$x = \operatorname{Re} z \quad \text{and} \quad y = \operatorname{Im} z.$$

The complex number $x + i0$ is identified with real number x, and the number $0 + iy$ is denoted by iy and is said to be *pure imaginary*. Two complex numbers $z_1 = x_1 + iy_1$ and $z_2 = x_2 + iy_2$ are *equal* if $x_1 = x_2$ and $y_1 = y_2$.

The complex number $\bar{z} = x - iy$ is said to be *conjugate* to the number z.

A complex number $z = x + iy$ can be conveniently represented as a point (x, y) in a two-dimensional Cartesian coordinate system (see Fig. M9.1). The axes OX and OY are called the *real* and *imaginary* axis, respectively, and the plane OXY is called the *complex plane*. The notions of a complex number and a point on the complex plane are identical.

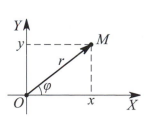

Figure M9.1. Geometric interpretation of a complex number.

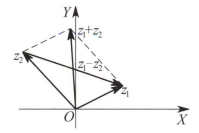

Figure M9.2. The sum and difference of complex numbers.

▶ **Addition, subtraction, multiplication, and division of complex numbers.** The *sum or difference* of complex numbers $z_1 = x_1 + iy_1$ and $z_2 = x_2 + iy_2$ is defined as the number

$$z_1 \pm z_2 = x_1 \pm x_2 + i(y_1 \pm y_2).$$

The *geometric meaning of the operations of addition and subtraction of complex numbers* is as follows: the sum and the difference of complex numbers z_1 and z_2 are the vectors equal to the directed diagonals of the parallelogram spanned by the vectors z_1 and z_2 (Fig. M9.2). The following inequalities hold (Fig. M9.2):

$$|z_1 + z_2| \leq |z_1| + |z_2|, \quad |z_1 - z_2| \geq \big||z_1| - |z_2|\big|. \tag{M9.1.2.1}$$

Inequalities (M9.1.2.1) become equalities if and only if the arguments of the complex numbers z_1 and z_2 coincide (i.e., $\arg z_1 = \arg z_2$; see Subsection M9.1.2) or one of the numbers is zero.

The *product* $z_1 z_2$ of complex numbers $z_1 = x_1 + iy_1$ and $z_2 = x_2 + iy_2$ is defined to be the number

$$z_1 z_2 = (x_1 x_2 - y_1 y_2) + i(x_1 y_2 + x_2 y_1).$$

The product of a complex number $z = x + iy$ by its conjugate is always nonnegative:

$$z\bar{z} = x^2 + y^2.$$

If $z_2 \neq 0$, then the *quotient* of z_1 and z_2 is defined as

$$\frac{z_1}{z_2} = \frac{x_1 x_2 + y_1 y_2}{x_2^2 + y_2^2} + i\frac{x_2 y_1 - x_1 y_2}{x_2^2 + y_2^2}. \tag{M9.1.1.1}$$

Relation (M9.1.1.1) can be obtained by multiplying the numerator and the denominator of the fraction z_1/z_2 by $\bar{z}_2$.

M9.1.2. Trigonometric Form of Complex Numbers. Powers and Radicals

▶ **Modulus and argument of a complex number.** There is a one-to-one correspondence between complex numbers $z = x + iy$ and points M with coordinates (x, y) on the plane with a Cartesian rectangular coordinate system OXY or with vectors $\overrightarrow{OM}$ connecting the origin O with M (Fig. M9.1). The length r of the vector $\overrightarrow{OM}$ is called the *modulus* (also *magnitude* and *absolute value*) of the number z and is denoted by $r = |z|$, and the angle φ formed by the vector $\overrightarrow{OM}$ and the positive direction of the OX-axis is called the *argument* (also *phase*) of the number z and is denoted by $\varphi = \text{Arg}\, z$.

The modulus of a complex number is determined by the formula

$$|z| = \sqrt{x^2 + y^2}.$$

The argument $\text{Arg}\, z$ is determined up to a multiple of 2π, $\text{Arg}\, z = \arg z + 2k\pi$, where k is an arbitrary integer and $\arg z$ is the *principal value* of $\text{Arg}\, z$ determined by the condition $-\pi < \arg z \leq \pi$. The principal value $\arg z$ is given by the formula

$$\arg z = \begin{cases} \arctan(y/x) & \text{for } x > 0, \\ \pi + \arctan(y/x) & \text{for } x < 0,\, y \geq 0, \\ -\pi + \arctan(y/x) & \text{for } x < 0,\, y < 0, \\ \pi/2 & \text{for } x = 0,\, y > 0, \\ -\pi/2 & \text{for } x = 0,\, y < 0. \end{cases}$$

For $z = 0$, $\text{Arg}\, z$ is undefined.

▶ **Trigonometric form of complex numbers.** Since $x = r\cos\varphi$ and $y = r\sin\varphi$, it follows that the complex number can be written in the *trigonometric* (or *polar*) *form*

$$z = x + iy = r(\cos\varphi + i\sin\varphi).$$

For two complex numbers written in trigonometric form, $z_1 = r_1(\cos\varphi_1 + i\sin\varphi_1)$ and $z_2 = r_2(\cos\varphi_2 + i\sin\varphi_2)$, the following arithmetic rules are valid:

$$z_1 z_2 = r_1 r_2\big[\cos(\varphi_1 + \varphi_2) + i\sin(\varphi_1 + \varphi_2)\big], \qquad \frac{z_1}{z_2} = \frac{r_1}{r_2}\big[\cos(\varphi_1 - \varphi_2) + i\sin(\varphi_1 - \varphi_2)\big].$$

In the latter formula, it is assumed that $z_2 \neq 0$.

▶ **Powers and radicals.** For any positive integer n, the nth power of z is calculated by *de Moivre's formula*

$$z^n = r^n(\cos n\varphi + i \sin n\varphi),$$

For $z \neq 0$ and positive integer n, there are exactly n distinct values of the nth root of $z = r(\cos\varphi + i\sin\varphi)$, which are determined by

$$\sqrt[n]{z} = z^{1/n} = \sqrt[n]{r}\left(\cos\frac{\varphi + 2k\pi}{n} + i\sin\frac{\varphi + 2k\pi}{n}\right) \quad (k = 0, 1, 2, \ldots, n-1).$$

Example. Let us find all values of $\sqrt[3]{i}$.

Let us represent the complex number $z = i$ in trigonometric form. We have $r = |z| = 1$ and $\varphi = \arg z = \frac{\pi}{2}$. The distinct values of the cube root are calculated by the formula

$$\omega_k = \sqrt[3]{1}\left(\cos\frac{\frac{\pi}{2} + 2\pi k}{3} + i\sin\frac{\frac{\pi}{2} + 2\pi k}{3}\right) \quad (k = 0, 1, 2),$$

so that

$$\omega_0 = \cos\frac{\pi}{6} + i\sin\frac{\pi}{6} = \frac{\sqrt{3}}{2} + i\frac{1}{2},$$

$$\omega_1 = \cos\frac{5\pi}{6} + i\sin\frac{5\pi}{6} = -\frac{\sqrt{3}}{2} + i\frac{1}{2},$$

$$\omega_2 = \cos\frac{3\pi}{2} + i\sin\frac{3\pi}{2} = -i.$$

The roots are shown in Fig. M9.3.

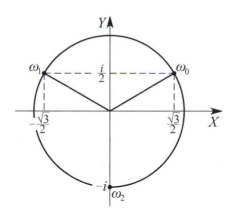

Figure M9.3. The roots of $\sqrt[3]{i}$.

M9.2. Functions of Complex Variables

M9.2.1. Basic Concepts. Differentiation of a Function of a Complex Variable

▶ **Some concepts and definitions.** A subset D of the complex plane such that each point of D has a neighborhood contained in D (i.e., D is open) and two arbitrary points of D can be connected by a broken line lying in D (i.e., D is connected) is called a *domain* in the complex plane. Each point of D is its *interior point*. A point that does not lie in D but whose arbitrary neighborhood contains points of D is called a *boundary point* of D. The set of all boundary points of D is called the *boundary* of D. The union of a domain D with its boundary is called a *closed domain* and denoted by $\overline{D}$. The boundary of a domain can consist of finitely many closed curves, segments, and points; the curves and cuts are assumed to be piecewise smooth.

The simplest examples of domains are neighborhoods of points on the complex plane. A *neighborhood* of a point a on the complex plane is understood as the set of points z such that $|z - a| < R$, i.e., the interior of the disk of radius $R > 0$ centered at the point a. The *extended complex plane* is obtained by augmenting the complex plane with the fictitious *point at infinity*. A *neighborhood of the point at infinity* is understood as the set of points z such that $|z| > R$ (including the point at infinity itself).

If to each point z of a domain D there corresponds a single point w (resp., a number of points w), then one says that there is a *single-valued* (resp., *multi-valued*) *function* $w = f(z)$ defined on the domain D. If we set $z = x + iy$ and $w = u + iv$, then defining a function $w = f(z)$ of the complex variable z is equivalent to defining two functions $\operatorname{Re} f = u = u(x, y)$ and $\operatorname{Im} f = v = v(x, y)$ of two real variables. If the function $w = f(z)$ is single-valued on D and the images of distinct points of D are distinct, then the mapping determined by this function is said to be *schlicht*. The notions of boundedness, limit, and continuity for single-valued functions of a complex variable do not differ from the corresponding notions for real functions of two real variables.

▶ **Differentiability. The Cauchy–Riemann conditions.** Let a single-valued function $w = f(z)$ be defined in a neighborhood of a point z. If there exists a limit

$$\lim_{h \to 0} \frac{f(z + h) - f(z)}{h} = f'_z(z),$$

then the function $w = f(z)$ is said to be *differentiable* at the point z and $f'_z(z)$ is called its *derivative* at the point z.

Cauchy–Riemann conditions. If the functions $u(x, y) = \operatorname{Re} f(z)$ and $v(x, y) = \operatorname{Im} f(z)$ are differentiable at a point (x, y), then the *Cauchy–Riemann conditions*

$$\frac{\partial u}{\partial x} = \frac{\partial v}{\partial y}, \qquad \frac{\partial u}{\partial y} = -\frac{\partial v}{\partial x} \qquad \text{(M9.2.1.1)}$$

are necessary and sufficient for the function $w = f(z)$ to be differentiable at the point $z = x + iy$.

If the function $w = f(z)$ is differentiable, then

$$w'_z = u_x + iv_x = v_y - iu_y = u_x - iu_y = v_y + iv_x,$$

where the subscripts x and y indicate the corresponding partial derivatives.

Remark. The Cauchy–Riemann conditions are sometimes also called the d'Alembert–Euler conditions.

The rules for arithmetic operations on the derivatives and those for taking the derivative of a composite function and the inverse function (if it exists) have exactly the same form as in the case of functions of a real variable:

1. $\left[\alpha f_1(z) \pm \beta f_2(z)\right]'_z = \alpha [f_1(z)]'_z \pm \beta [f_2(z)]'_z$, where α and β are arbitrary complex constants.
2. $\left[f_1(z) f_2(z)\right]'_z = [f_1(z)]'_z f_2(z) + f_1(z)[f_2(z)]'_z$.
3. $\left[\dfrac{f_1(z)}{f_2(z)}\right]'_z = \dfrac{[f_1(z)]'_z f_2(z) - f_1(z)[f_2(z)]'_z}{f_2^2(z)} \qquad (f_2(z) \neq 0)$.
4. If a function $w = f(z)$ is differentiable at a point z and a function $W = F(w)$ is differentiable at the point $w = f(z)$, then the composite function $W = F(f(z))$ is differentiable at the point z and $W'_z = [F(f(z))]'_z = F'_f(f)f'_z(z)$.

5. If a function $w = f(z)$ is differentiable at a point z and the inverse $z = g(w) \equiv f^{-1}(w)$ exists and is differentiable at the point w, then

$$[f^{-1}(w)]'_w = \frac{1}{f'_z(z)|_{z=f^{-1}(w)}} \qquad (f'_z(z) \neq 0).$$

▶ **Analyticity. The maximum modulus principle. The Liouville's theorem.** A single-valued function differentiable in some neighborhood of a point z_0 is said to be *analytic* (*regular, holomorphic*) at this point.

A function $w = f(z)$ is analytic at a point z_0 if and only if it can be represented by a power series

$$f(z) = \sum_{k=0}^{\infty} c_k (z - z_0)^k$$

converging in some neighborhood of z_0.

A function analytic at each point of the domain D is said to be *analytic* in D.

A function $w = f(z)$ is said to be *analytic at the point at infinity* if the function $F(z) = f(1/z)$ is analytic at the point $z = 0$.

A function $w = f(z)$ is analytic at the point at infinity if and only if this function can be represented by a power series

$$f(z) = \sum_{k=0}^{\infty} b_k z^{-k}$$

converging for sufficiently large $|z|$.

If a function $w = f(z)$ is analytic at a point z_0 and $f'_z(z_0) \neq 0$, then $f(z)$ has an analytic inverse function $z(w)$ defined in a neighborhood of the point $w_0 = f(z_0)$. If a function $w = f(z)$ is analytic at a point z_0 and the function $W = F(w)$ is analytic at the point $w_0 = f(z_0)$, then the composite function $W = F[f(z)]$ is analytic at the point z_0. If a function is analytic in a domain D and continuous in $\overline{D}$, then its value at any interior point of the domain is uniquely determined by its values on the boundary of the domain. The analyticity of a function at a point implies the existence and analyticity of its derivatives of arbitrary order at this point.

Single-valued functions, as well as single-valued branches of multi-valued functions, are analytic everywhere on the domains where they are defined. It follows from (M9.2.1.1) that the real and imaginary parts $u(x, y)$ and $v(x, y)$ of a function analytic in a domain are *harmonic* in this domain, i.e., satisfy the Laplace equation

$$\Delta f = f_{xx} + f_{yy} = 0$$

in this domain.

Remark. If $u(x, y)$ and $v(x, y)$ are two arbitrary harmonic functions, then the function $f(z) = u(x, y) + iv(x, y)$ is not necessarily analytic, since for the analyticity of $f(z)$ the functions $u(x, y)$ and $v(x, y)$ must satisfy the Cauchy–Riemann conditions.

Example 1. The function $w = z^2$ is analytic.
Indeed, since $z = x + iy$, we have $w = (x + iy)^2 = x^2 - y^2 + i2xy$, $u(x, y) = x^2 - y^2$, and $v(x, y) = 2xy$. The Cauchy–Riemann conditions

$$u_x = v_y = 2x, \quad u_y = -v_x = -2y$$

are satisfied at all points of the complex plane, so the function $w = z^2$ is analytic.

Example 2. The function $w = \bar{z}$ is not analytic.

Indeed, since $z = x + iy$, we have $w = x - iy$, $u(x, y) = x$, $v(x, y) = -y$. The Cauchy–Riemann conditions are not satisfied,

$$u_x = 1 \ne -1 = v_y, \qquad u_y = -v_x = 0,$$

so the function $w = \bar{z}$ is not analytic.

MAXIMUM MODULUS PRINCIPLE. *If a function $w = f(z)$ that is not identically constant is analytic in a domain D and continuous in $\overline{D}$, then its modulus cannot attain a maximum at an interior point of D.*

LIOUVILLE'S THEOREM. *If a function $w = f(z)$ is analytic and bounded in the entire complex plane, then it is constant.*

Remark. The Liouville theorem can be stated in the following form:
if a function $w = f(z)$ is analytic in the extended complex plane, then it is constant.

▶ **Geometric meaning of the derivative.** *Geometric meaning of the absolute value of the derivative.* Suppose that a function $w = f(z)$ is analytic at a point z_0 and $f'_z(z_0) \ne 0$. Then the value $|f'_z(z_0)|$ determines the dilatation (similarity) coefficient at the point z_0 under the mapping $w = f(z)$. The value $|f'_z(z_0)|$ is called the *dilatation ratio* if $|f'_z(z_0)| > 1$ and the *contraction ratio* if $|f'_z(z_0)| < 1$.

Geometric meaning of the argument of the derivative. The argument of the derivative $f'_z(z_0)$ is equal to the angle by which the tangent at the point z_0 to any curve passing through z_0 should be rotated to give the tangent to the image of the curve at the point $w_0 = f(z_0)$. For $\varphi = \arg f'_z(z) > 0$, the rotation is counterclockwise, and for $\varphi = \arg f'_z(z) < 0$, the rotation is clockwise.

▶ **Elementary functions.**

$1°$. Consider the *functions $w = z^n$ and $w = \sqrt[n]{z}$* for positive integer n. The function

$$w = z^n$$

is single-valued. It is schlicht in the sectors $2\pi k/n < \varphi < 2\pi(k + 1)/n$, $k = 0, 1, 2, \ldots$, each of which is transformed by the mapping $w = z^n$ onto the plane w with a cut on the positive real semiaxis.

The function

$$w = \sqrt[n]{z}$$

is an n-valued function for $z \ne 0$, and its value is determined by the value of the argument chosen for the point z. If a closed curve C does not surround the point $z = 0$, then, as the point z goes around the entire curve C, the point $w = \sqrt[n]{z}$ for a chosen value of the root also moves along a closed curve and returns to the initial value of the argument. But if the curve C surrounds the origin, then, as the point z goes around the entire curve C in the *positive sense* (in the *counterclockwise* direction), the argument of z increases by 2π and the corresponding point $w = \sqrt[n]{z}$ does not return to the initial position. It will return there only after the point z goes n times around the entire curve C. If a domain D does not contain a closed curve surrounding the point $z = 0$, then one can single out n continuous single-valued functions, each of which takes only one of the values $w = \sqrt[n]{z}$; these functions are called the *branches* of the multi-valued function $w = \sqrt[n]{z}$. One cannot single out n separate branches of the function $w = \sqrt[n]{z}$ in any neighborhood of the point $z = 0$; accordingly, the point $z = 0$ is called a *branch point* of this function.

$2°$. The *Zhukovskii function*

$$w = \frac{1}{2}\left(z + \frac{1}{z}\right)$$

is defined and single-valued for all $z \neq 0$; it is schlicht in any domain that does not simultaneously contain any points z_1 and z_2 such that $z_1 z_2 = 1$.

3°. The *exponential function* $w = e^z$ is defined by the formula

$$w = e^z = e^{x+iy} = e^x(\cos y + i \sin y).$$

The function $w = e^z$ is analytic everywhere. For the exponential function, the usual differentiation rule is preserved:

$$(e^z)'_z = e^z.$$

The basic property of the exponential function is also preserved:

$$e^{z_1} e^{z_2} = e^{z_1 + z_2}.$$

For $x = 0$ and $y = \varphi$, the definition of the exponential function implies the *Euler formula* $e^{i\varphi} = \cos\varphi + i \sin\varphi$, which permits one to write any complex number with modulus r and argument φ in the *exponential form*

$$z = r(\cos\varphi + i \sin\varphi) = re^{i\varphi}.$$

The exponential function is periodic with imaginary period $2\pi i$, and the mapping $w = e^z$ is schlicht in the strip $0 \le y < 2\pi$.

4°. The *logarithm* is defined as the inverse of the exponential function: if $e^w = z$, then

$$w = \operatorname{Ln} z.$$

This function is defined for $z \neq 0$. The logarithm satisfies the following relations:

$$\operatorname{Ln} z_1 + \operatorname{Ln} z_2 = \operatorname{Ln}(z_1 z_2), \quad \operatorname{Ln} z_1 - \operatorname{Ln} z_2 = \operatorname{Ln}\frac{z_1}{z_2},$$

$$\operatorname{Ln}(z^n) = n \operatorname{Ln} z, \quad \operatorname{Ln} \sqrt[n]{z} = \frac{1}{n}\operatorname{Ln} z.$$

The exponential form of complex numbers readily shows that the logarithm is infinite-valued:

$$\operatorname{Ln} z = \ln|z| + i \operatorname{Arg} z = \ln|z| + i \arg z + 2\pi ki, \quad k = 0, \pm 1, \pm 2, \ldots \quad \text{(M9.2.1.2)}$$

The quantity $\ln z = \ln|z| + i \arg z$ is taken to be the principal value of this function. Just as with the function $w = \sqrt[n]{z}$, we see that if the point $z = 0$ is surrounded by a closed curve C, then the point $w = \operatorname{Ln} z$ does not return to its initial position after z goes around C in the positive sense, since the argument of w increases by $2\pi i$. Thus if a domain D does not contain a closed curve surrounding the point $z = 0$, then in D one can single out infinitely many continuous and single-valued branches of the multi-valued function $w = \operatorname{Ln} z$; the differences between the values of these branches at each point of the domain are equal to $2\pi ki$, where k is an integer. This cannot be done in an arbitrary neighborhood of the point $z = 0$, and this point is called a *branch point* of the logarithm.

5°. *Trigonometric functions* are defined in terms of the exponential function as follows:

$$\cos z = \frac{e^{iz} + e^{-iz}}{2}, \qquad \sin z = \frac{e^{iz} - e^{-iz}}{2i},$$

$$\tan z = \frac{\sin z}{\cos z} = -i\frac{e^{iz} - e^{-iz}}{e^{iz} + e^{-iz}}, \qquad \cot z = \frac{\cos z}{\sin z} = i\frac{e^{iz} + e^{-iz}}{e^{iz} - e^{-iz}}.$$

These are properties of the functions $\cos z$ and $\sin z$:
1. They are analytic for any z.
2. The usual differentiation rules are valid:

$$(\sin z)'_z = \cos z, \quad (\cos z)'_z = -\sin z.$$

3. They are periodic with real period $T = 2\pi$.
4. $\sin z$ is an odd function, and $\cos z$ is an even function.
5. In the complex plane, they are unbounded.
6. The usual trigonometric relations hold:

$$\cos^2 z + \sin^2 z = 1, \quad \cos 2z = \cos^2 z - \sin^2 z, \quad \text{etc.}$$

The function $\tan z$ is analytic everywhere except for the points

$$z_k = \frac{\pi}{2} + k\pi, \qquad k = 0, \pm 1, \pm 2, \ldots,$$

and the function $\cot z$ is analytic everywhere except for the points

$$z_k = k\pi, \qquad k = 0, \pm 1, \pm 2, \ldots$$

The functions $\tan z$ and $\cot z$ are periodic with real period $T = \pi$.

$6°$. *Hyperbolic functions* are defined by the formulas

$$\cosh z = \frac{e^z + e^{-z}}{2}, \qquad\qquad \sinh z = \frac{e^z - e^{-z}}{2},$$

$$\tanh z = \frac{\sinh z}{\cosh z} = \frac{e^z - e^{-z}}{e^z + e^{-z}}, \qquad \coth z = \frac{\cosh z}{\sinh z} = \frac{e^z + e^{-z}}{e^z - e^{-z}}.$$

For real values of the argument, each of these functions coincides with the corresponding real function. Hyperbolic and trigonometric functions are related by the formulas

$$\cosh z = \cos iz, \quad \sinh z = -i \sin iz, \quad \tanh z = -i \tan iz, \quad \coth z = i \cot iz.$$

$7°$. *Inverse trigonometric and hyperbolic functions* are expressed via the logarithm and hence are infinite-valued:

$$\text{Arccos } z = -i \operatorname{Ln}(z + \sqrt{z^2 - 1}), \qquad \text{Arcsin } z = -i \operatorname{Ln}(iz + \sqrt{1 - z^2}),$$

$$\text{Arctan } z = -\frac{i}{2} \operatorname{Ln} \frac{1 + iz}{1 - iz}, \qquad \text{Arccot } z = -\frac{i}{2} \operatorname{Ln} \frac{z + i}{z - i},$$

$$\text{arccosh } z = \operatorname{Ln}(z + \sqrt{z^2 - 1}), \qquad \text{arcsinh } z = \operatorname{Ln}(z + \sqrt{z^2 - 1}),$$

$$\text{arctanh } z = \frac{1}{2} \operatorname{Ln} \frac{1 + z}{1 - z}, \qquad \text{arccoth } z = \frac{1}{2} \operatorname{Ln} \frac{z + 1}{z - 1}.$$

The principal value of each of these functions is obtained by choosing the principal value of the corresponding logarithmic function.

$8°$. The *power function* $w = z^\gamma$ is defined by the relation

$$z^\gamma = e^{\gamma \operatorname{Ln} z}, \tag{M9.2.1.3}$$

where $\gamma = \alpha + i\beta$ is an arbitrary complex number. Substituting $z = re^{i\varphi}$ into (M9.2.1.3) yields

$$z^\gamma = e^{\alpha \ln r - \beta(\varphi + 2k\pi)} e^{i\alpha(\varphi + 2k\pi) + i\beta \ln r}, \qquad k = 0, \pm 1, \pm 2, \ldots \tag{M9.2.1.4}$$

It follows from relation (M9.2.1.4) that the function $w = z^\gamma$ has infinitely many values for $\beta \neq 0$.

$9°$. The *general exponential function* is defined by

$$w = \gamma^z = e^{z\,\mathrm{Ln}\,\gamma} = e^{z\ln|\gamma|}e^{zi\,\mathrm{Arg}\,\gamma}, \qquad (\text{M9.2.1.5})$$

where $\gamma = \alpha + i\beta$ is an arbitrary nonzero complex number. The function (M9.2.1.5) is a set of separate mutually independent single-valued functions that differ from one another by the factors $e^{2k\pi iz}$, $k = 0, \pm1, \pm2, \ldots$

Example 3. Let us calculate the values of some elementary functions at specific points:

1. $\cos 2i = \frac{1}{2}(e^{2ii} + e^{-2ii}) = \frac{1}{2}(e^2 + e^{-2}) = \cosh 2 \approx 3.7622.$
2. $\ln(-2) = \ln 2 + i\pi$, since $|-2| = 2$ and the principal value of the argument is equal to π.
3. $\mathrm{Ln}(-2)$ is calculated by formula (M9.2.1.2):

$$\mathrm{Ln}(-2) = \ln 2 + i\pi + 2\pi k i = \ln 2 + (1 + 2k)i\pi \quad (k = 0, \pm1, \pm2, \ldots).$$

4. $i^i = e^{i\,\mathrm{Ln}\,i} = e^{i(i\frac{\pi}{2}+2\pi ki)} = e^{-\frac{\pi}{2}-2\pi k} \quad (k = 0, \pm1, \pm2, \ldots).$

The main elementary functions $w = f(z) = u(x, y) + iv(x, y)$ of the complex variable $z = x + iy$ are given in Table M9.1.

TABLE M9.1
Main elementary functions $w = f(z) = u(x, y) + iv(x, y)$ of the complex variable $z = x + iy$.

No.	Complex function $w = f(z)$	Algebraic form $f(z) = u(x, y) + iv(x, y)$	Zeros of nth order	Singularities
1	z	$x + iy$	$z = 0,\ n = 1$	$z = \infty$ is a first-order pole
2	z^2	$x^2 - y^2 + i\,2xy$	$z = 0,\ n = 2$	$z = \infty$ is a second-order pole
3	$\dfrac{1}{z-(x_0+iy_0)}$ (x_0, y_0 are real numbers)	$\dfrac{x-x_0}{(x-x_0)^2 + (y-y_0)^2} + i\,\dfrac{-(y-y_0)}{(x-x_0)^2 + (y-y_0)^2}$	$z = \infty,\ n = 1$	$z = x_0 + iy_0$ is a first-order pole
4	$\dfrac{1}{z^2}$	$\dfrac{x^2 - y^2}{(x^2 + y^2)^2} + i\,\dfrac{-2xy}{(x^2 + y^2)^2}$	$z = \infty,\ n = 2$	$z = 0$ is a second-order pole
5	$\sqrt{z}$	$\pm\left[\left(\dfrac{x+\sqrt{x^2+y^2}}{2}\right)^{1/2} + i\left(\dfrac{-x+\sqrt{x^2+y^2}}{2}\right)^{1/2}\right]$	$z = 0$ is a branch point	$z = 0$ is a first-order branch point $z = \infty$ is a first-order branch point
6	e^z	$e^x \cos y + ie^x \sin y$	no zeros	$z = \infty$ is an essential singular point
7	$\mathrm{Ln}\,z$	$\ln|z| + i(\arg z + 2k\pi),$ $k = 0, \pm1, \pm2, \ldots$	$z = 1,\ n = 1$ (for the branch corresponding to $k = 0$)	Logarithmic branch points for $z = 0,\ z = \infty$
8	$\sin z$	$\sin x \cosh y + i \cos x \sinh y$	$z = \pi k,\ n = 1$ $(k = 0, \pm1, \pm2, \ldots)$	$z = \infty$ is an essential singular point
9	$\cos z$	$\cos x \cosh y + i(-\sin x \sinh y)$	$z = \frac{1}{2}\pi + \pi k,\ n = 1$ $(k = 0, \pm1, \pm2, \ldots)$	$z = \infty$ is an essential singular point
10	$\tan z$	$\dfrac{\sin 2x}{\cos 2x + \cosh 2y} + i\,\dfrac{\sinh 2y}{\cos 2x + \cosh 2y}$	$z = \pi k,\ n = 1$ $(k = 0, \pm1, \pm2, \ldots)$	$z = \frac{1}{2}\pi + \pi k$ $(k = 0, \pm1, \pm2, \ldots)$ are first-order poles

▶ **Analytic continuation.** Let two domains D_1 and D_2 have a common part γ of the boundary, and let single-valued analytic functions $f_1(z)$ and $f_2(z)$, respectively, be given in these domains. The function $f_2(z)$ is called a *direct analytic continuation* of $f_1(z)$ into the domain D_2 if there exists a function $f(z)$ analytic in the domain $D_1 \cup \gamma \cup D_2$ and satisfying the condition

$$f(z) = \begin{cases} f_1(z) & \text{for } z \in D_1, \\ f_2(z) & \text{for } z \in D_2. \end{cases} \tag{M9.2.1.6}$$

If such a continuation is possible, then the function $f(z)$ is uniquely determined. If the domains are simply connected and the functions $f_1(z)$ and $f_2(z)$ are continuous in $D_1 \cup \gamma$ and $D_2 \cup \gamma$, respectively, and coincide on γ, then $f_2(z)$ is the direct analytic continuation of $f_1(z)$ into the domain D_2. In addition, suppose that the domains D_1 and D_2 are allowed to have common interior points. A function $f_2(z)$ is called a direct analytic continuation of $f_1(z)$ through γ if $f_1(z)$ and $f_2(z)$ are continuous in $D_1 \cup \gamma$ and $D_2 \cup \gamma$, respectively, and their values on γ coincide. At the common interior points of D_1 and D_2, the function determined by relation (M9.2.1.6) can be double-valued.

M9.2.2. Integration of Functions of Complex Variables

▶ **Definition and properties of the integral of a function of a complex variable.** Suppose that an oriented curve C connecting points $z = a$ and $z = b$ is given on the complex plane and a function $w = f(z)$ of the complex variable z is defined on the curve. We divide the curve C into n parts, $a = z_0, z_1, \ldots, z_{n-1}, z_n = b$, arbitrarily choose $\xi_k \in [z_k, z_{k+1}]$, and compose the integral sum

$$\sum_{k=0}^{n-1} f(\xi_k)(z_{k+1} - z_k).$$

If there exists a limit of this sum as $\max |z_{k+1} - z_k| \to 0$, independent of the way it is partitioned and the choice of the points ξ_k, then this limit is called the *integral* of the function $w = f(z)$ over the curve C and is denoted by

$$\int_C f(z)\, dz. \tag{M9.2.2.1}$$

Properties of the integral of a function of a complex variable:

1. If α, β are arbitrary constants, then $\int_C [\alpha f(z) + \beta g(z)]\, dz = \alpha \int_C f(z)\, dz + \beta \int_C g(z)\, dz$.
2. If $\widetilde{C}$ is the same curve as C but with the opposite sense, then $\int_{\widetilde{C}} f(z)\, dz = -\int_C f(z)\, dz$.
3. If $C = C_1 \cup \cdots \cup C_n$, then $\int_C f(z)\, dz = \int_{C_1} f(z)\, dz + \cdots + \int_{C_n} f(z)\, dz$.
4. If $|f(z)| \le M$ at all points of the curve C, then the following *estimate of the absolute value of the integral* holds: $\left| \int_C f(z)\, dz \right| \le Ml$, where l is the length of the curve C.

If C is a piecewise smooth curve and $f(z)$ is bounded and piecewise continuous, then the integral (M9.2.2.1) exists. If $z = x + iy$ and $w = u(x, y) + iv(x, y)$, then the computation of the integral (M9.2.2.1) is reduced to finding two ordinary curvilinear integrals:

$$\int_C f(z)\, dz = \int_C u(x, y)\, dx - v(x, y)\, dy + i \int_C v(x, y)\, dx + u(x, y)\, dy. \tag{M9.2.2.2}$$

Remark. Formula (M9.2.2.2) can also be written in a form convenient for memorizing:

$$\int_C f(z)\, dz = \int_C (u + iv)(dx + i\, dy).$$

If the curve C is given by the parametric equations $x = x(t)$, $y = y(t)$ ($t_1 \le t \le t_2$), then

$$\int_C f(z)\,dz = \int_{t_1}^{t_2} f(z(t))z'_t(t)\,dt,$$

where $z = z(t) = x(t) + iy(t)$ is the complex parametric equation of the curve C.

If $f(z)$ is an analytic function in a simply connected domain D containing the points $z = a$ and $z = b$, then the *Newton–Leibniz formula* holds:

$$\int_a^b f(z)\,dz = F(b) - F(a),$$

where $F(z)$ is a primitive of the function $f(z)$, i.e., $F'_z(z) = f(z)$ in the domain D.

If $f(z)$ and $g(z)$ are analytic functions in a simply connected domain D and $z = a$ and $z = b$ are arbitrary points of the domain D, then the *formula of integration by parts* holds:

$$\int_a^b f(z)\,dg(z) = f(b)g(b) - f(a)g(a) - \int_a^b g(z)\,df(z).$$

If an analytic function $z = g(w)$ determines a single-valued mapping of a curve $\widetilde{C}$ onto a curve C, then

$$\int_C f(z)\,dz = \int_{\widetilde{C}} f(g(w))g'_w(w)\,dw.$$

▶ **Cauchy's theorems.**

CAUCHY'S THEOREM FOR A SIMPLY CONNECTED DOMAIN. *If a function $f(z)$ is analytic in a simply connected domain D bounded by a contour C and is continuous in $\overline{D}$, then $\int_C f(z)\,dz = 0$.*

CAUCHY'S THEOREM FOR A MULTIPLY CONNECTED DOMAIN. *If a function $f(z)$ is analytic in a multiply connected domain D bounded by a contour C consisting of several closed curves and is continuous in $\overline{D}$, then $\int_C f(z)\,dz = 0$ provided that the sense of all curves forming C is chosen in such a way that the domain D lies to the same side of the contour.*

▶ **Cauchy's integral and related integrals. Morera's theorem.** If a function $f(z)$ is analytic in an n-connected domain D and continuous in $\overline{D}$, and C is the boundary of D, then for any interior point z of this domain the *Cauchy integral formula* holds:

$$f(z) = \frac{1}{2\pi i} \int_C \frac{f(\xi)}{\xi - z}\,d\xi. \tag{M9.2.2.3}$$

(Here integration is carried out in the positive sense of C; i.e., the contour C is traced so that the domain D lies to the left of the contour.) Under the same assumptions as above, the derivatives of arbitrary order of the function $f(z)$ at any interior point z of the domain are expressed as

$$f_z^{(n)}(z) = \frac{n!}{2\pi i} \int_C \frac{f(\xi)}{(\xi - z)^{n+1}}\,d\xi \qquad (n = 1, 2, \dots). \tag{M9.2.2.4}$$

For an arbitrary smooth curve C, not necessarily closed, and for a function $f(\xi)$ everywhere continuous on C, possibly except for finitely many points at which this function has an integrable discontinuity, the right-hand side of formula (M9.2.2.3) defines a *Cauchy-type integral.* The function $F(z)$ determined by a Cauchy-type integral is analytic at any point that does not belong to C. If C divides the plane into several domains, then the Cauchy-type integral generally determines different analytic functions in these domains.

Formulas (M9.2.2.3) and (M9.2.2.4) allow one to calculate the integrals

$$\int_C \frac{f(\xi)}{\xi - z} \, d\xi = 2\pi i f(z), \quad \int_C \frac{f(\xi)}{(\xi - z)^{n+1}} \, d\xi = \frac{2\pi i}{n!} f_z^{(n)}(z). \tag{M9.2.2.5}$$

Example 1. Let us calculate the integral

$$\int_C \operatorname{Im} z \, dz,$$

where C is the semicircle $|z| = 1$, $0 \le \arg z \le \pi$ (Fig. M9.4).

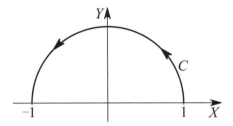

Figure M9.4. The semicircle $|z| = 1$, $0 \le \arg z \le \pi$.

Using formula (M9.2.2.2), we obtain

$$\int_C \operatorname{Im} z \, dz = \int_C y(dx + i \, dy) = \int_C y \, dx + i \int_C y \, dy = \int_1^{-1} \sqrt{1 - x^2} \, dx - i \cdot 0 = -\frac{\pi}{2}.$$

Example 2. Let us calculate the integral

$$\int_C \frac{dz}{z - z_0},$$

where C is the circle of radius R centered at a point z_0 with counterclockwise sense.

Using the integral formula (M9.2.2.5), we obtain

$$\int_C \frac{1}{z - z_0} \, dz = 2\pi i.$$

Example 3. Let us calculate the integral

$$\int_C \frac{dz}{z^2 + 1},$$

where C is the circle of unit radius centered at the point i with counterclockwise sense.

To apply the Cauchy integral formula (M9.2.2.3), we transform the integrand as follows:

$$\frac{1}{1 + z^2} = \frac{1}{(z - i)(z + i)} = \frac{1}{z + i} \frac{1}{z - i} = \frac{f(z)}{z - i}, \quad f(z) = \frac{1}{z + i}.$$

The function $f(z) = 1/(z + i)$ is analytic in the interior of the domain under study and on its boundary; hence the Cauchy integral formula (M9.2.2.3) and the first of formulas (M9.2.2.5) hold. From the latter formula, we obtain

$$\int_C \frac{dz}{z^2 + 1} = \int_C \frac{f(z)}{z - i} \, dz = 2\pi i f(i) = 2\pi i \frac{1}{2i} = \pi.$$

Formulas (M9.2.2.3) and (M9.2.2.4) imply the *Cauchy inequalities*

$$|f_z^{(n)}(z)| \le \frac{n!}{2\pi} \left| \int_C \frac{f(\xi)}{(\xi - z)^{n+1}} \, d\xi \right| \le \frac{n!Ml}{2\pi R^{n+1}},$$

where $M = \max\limits_{z \in D} |f(z)|$ is the maximum modulus of the function $f(z)$ in the domain D, R is the distance from the point z to the boundary C, and l is the length of the boundary C.

If, in particular, $f(z)$ is analytic in the disk $D = |z - z_0| < R$, and bounded in $\bar{D}$, then we obtain the inequality

$$|f_z^{(n)}(z_0)| \leq \frac{n!M}{R^n} \quad (n = 0, 1, 2, \ldots).$$

MORERA'S THEOREM. *If a function $f(z)$ is continuous in a simply connected domain D and $\int_C f(z)\,dz = 0$ for any closed curve C lying in D, then $f(z)$ is analytic in the domain D.*

M9.2.3. Taylor and Laurent Series

▶ **Taylor series.** If a series

$$\sum_{n=0}^{\infty} f_n(z) \tag{M9.2.3.1}$$

of analytic functions in a simply connected domain D converges uniformly in this domain, then its sum is analytic in the domain D.

If a series (M9.2.3.1) of functions analytic in a domain D and continuous in $\overline{D}$ converges uniformly in $\overline{D}$, then it can be differentiated termwise any number of times and can be integrated termwise over any piecewise smooth curve C lying in D.

ABEL'S THEOREM. *If the power series*

$$\sum_{n=0}^{\infty} c_n(z - a)^n \tag{M9.2.3.2}$$

converges at a point z_0, then it also converges at any point z satisfying the condition $|z - a| < |z_0 - a|$. Moreover, the series converges uniformly in any disk $|z - a| \leq q|z_0 - a|$, where $0 < q < 1$.

It follows from Abel's theorem that the domain of convergence of a power series is an open disk centered at the point a; moreover, this disk can fill the entire plane. The radius of this disk is called the *radius of convergence* of a power series. The sum of the power series inside the disk of convergence is an analytic function.

Remark. The radius of convergence R can be found by the *Cauchy–Hadamard formula*

$$\frac{1}{R} = \overline{\lim_{n \to \infty}} \sqrt[n]{|c_n|},$$

where $\overline{\lim}$ denotes the upper limit.

If a function $f(z)$ is analytic in the open disk D of radius R centered at a point $z = a$, then this function can be represented in this disk by its *Taylor series*

$$f(z) = \sum_{n=0}^{\infty} c_n(z - a)^n,$$

whose coefficients are determined by the formulas

$$c_n = \frac{f_z^{(n)}(a)}{n!} = \frac{1}{2\pi i} \int_C \frac{f(\xi)}{(\xi - a)^{n+1}}\,d\xi \quad (n = 0, 1, 2, \ldots),$$

where C is the circle $|z - a| = qR, 0 < q < 1$. In any closed domain belonging to the disk D, the Taylor series converges uniformly. Any power series expansion of an analytic function is its Taylor expansion. The Taylor series expansions of some elementary functions in powers of z are as follows:

$$e^z = 1 + z + \frac{z^2}{2!} + \frac{z^3}{3!} + \dots \quad (|z| < \infty), \tag{M9.2.3.3}$$

$$\cos z = 1 - \frac{z^2}{2!} + \frac{z^4}{4!} - \dots, \quad \sin z = z - \frac{z^3}{3!} + \frac{z^5}{5!} - \dots \quad (|z| < \infty), \tag{M9.2.3.4}$$

$$\cosh z = 1 + \frac{z^2}{2!} + \frac{z^4}{4!} + \dots, \quad \sinh z = z + \frac{z^3}{3!} + \frac{z^5}{5!} + \dots \quad (|z| < \infty), \tag{M9.2.3.5}$$

$$\ln(1 + z) = z - \frac{z^2}{2} + \frac{z^3}{3} - \dots \quad (|z| < 1), \tag{M9.2.3.6}$$

$$(1 + z)^a = 1 + az + \frac{a(a - 1)}{2!} z^2 + \frac{a(a - 1)(a - 2)}{3!} z^3 + \dots \quad (|z| < 1). \tag{M9.2.3.7}$$

The last two expansions are valid for the single-valued branches for which the values of the functions for $z = 0$ are equal to 0 and 1, respectively.

Remark. Series expansions (M9.2.3.3)–(M9.2.3.7) coincide with analogous expansions of the corresponding elementary functions of the real variable (see Subsection M8.3.2).

To obtain the Taylor series for other branches of the multi-valued function $\mathrm{Ln}(1 + z)$, one has to add the numbers $2k\pi i, k = \pm 1, \pm 2, \dots$ to the expression in the right-hand side:

$$\mathrm{Ln}(1 + z) = 2k\pi i + z - \frac{z^2}{2} + \frac{z^3}{3} - \dots.$$

▶ **Laurent series.** The domain of convergence of the function series $\sum\limits_{n=-\infty}^{\infty} c_n(z - a)^n$ is a circular annulus $K : r < |z - a| < R$, where $0 \leq r < R \leq \infty$. The sum of the series is an analytic function in the annulus of convergence. Conversely, in any annulus K where the function $f(z)$ is analytic, this function can be represented by the *Laurent series* expansion

$$f(z) = \sum_{n=-\infty}^{\infty} c_n(z - a)^n$$

with coefficients determined by the formulas

$$c_n = \frac{1}{2\pi i} \int_\gamma \frac{f(\xi)}{(\xi - a)^{n+1}} \, d\xi \quad (n = 0, \pm 1, \pm 2, \dots), \tag{M9.2.3.8}$$

where γ is the circle $|z - a| = \rho, r < \rho < R$. In any closed domain contained in the annulus K, the Laurent series converges uniformly.

The part of the Laurent series with negative numbers,

$$\sum_{n=-\infty}^{-1} c_n(z - a)^n = \sum_{n=1}^{\infty} \frac{c_{-n}}{(z - a)^n},$$

is called its *principal part*, and the part with nonnegative numbers,

$$\sum_{n=0}^{\infty} c_n(z - a)^n,$$

is called the *regular part*. Any expansion of an analytic function in positive and negative powers of $z - a$ is its Laurent expansion.

Example 1. Let us consider Laurent series expansions of the function

$$f(z) = \frac{1}{z(1-z)}$$

in a Laurent series in the domain $0 < |z| < 1$. This function is analytic in the annulus $0 < |z| < 1$ and hence can be expanded in the corresponding Laurent series. We write this function as the sum of elementary fractions:

$$f(z) = \frac{1}{z(1-z)} = \frac{1}{z} + \frac{1}{1-z}.$$

Since $|z| < 1$, we can use formula (M9.2.3.7) and obtain the expansion

$$\frac{1}{z(1-z)} = \frac{1}{z} + 1 + z + z^2 + \cdots$$

Example 2. Let us consider Laurent series expansions of the function

$$f(z) = e^{1/z}$$

in a Laurent series in a neighborhood of the point $a = 0$. To this end, we use the expansion (M9.2.3.3), where we should replace z by $1/z$. Thus we obtain

$$e^{1/z} = 1 + \frac{1}{1! \, z} + \frac{1}{2! \, z^2} + \cdots + \frac{1}{n! \, z^n} + \cdots \qquad (z \neq 0).$$

M9.2.4. Zeros and Isolated Singularities of Analytic Functions

▶ **Zeros of analytic functions.** A point $z = a$ is called a *zero* of a function $f(z)$ if $f(a) = 0$. If $f(z)$ is analytic at the point a and is not zero identically, then the least order of nonzero coefficients in the Taylor expansion of $f(z)$ centered at a, in other words, the number n of the first nonzero derivative $f^{(n)}(a)$, is called the *order of zero* of this function. In a neighborhood of a zero a of order n, the Taylor expansion of $f(z)$ has the form

$$f(z) = c_n(z-a)^n + c_{n+1}(z-a)^{n+1} + \ldots \qquad (c_n \neq 0, \; n \geq 1).$$

In this case, $f(z) = c_n(z-a)^n g(z)$, where the function $g(z)$ is analytic at the point a and $g(a) \neq 0$. A first-order zero is said to be *simple*. The point $z = \infty$ is a zero of order n for a function $f(z)$ if $z = 0$ is a zero of order n for $F(z) = f(1/z)$.

If a function $f(z)$ is analytic in a neighborhood of its zero a and is not identically zero in any neighborhood of a, then there exists a neighborhood of a in which $f(z)$ does not have any zeros other than a.

UNIQUENESS THEOREM. *If functions $f(z)$ and $g(z)$ are analytic in a domain D and their values coincide on some sequence a_k of points converging to an interior point a of the domain D, then $f(z) \equiv g(z)$ everywhere in D.*

ROUCHÉ'S THEOREM. *If functions $f(z)$ and $g(z)$ are analytic in a simply connected domain D bounded by a curve C, are continuous in $\overline{D}$, and satisfy the inequality $|f(z)| > |g(z)|$ on C, then the functions $f(z)$ and $f(z) + g(z)$ have the same number of zeros in D.*

▶ **Isolated singularities of analytic functions.** A point a is called an *isolated singularity* of a single-valued analytic function $f(z)$ if there exists a neighborhood of this point in which $f(z)$ is analytic everywhere except for the point a itself. The point a is called

1. A *removable singularity* if $\lim\limits_{z \to a} f(z)$ exists and is finite.
2. A *pole* if $\lim\limits_{z \to a} f(z) = \infty$.
3. An *essential singularity* if $\lim\limits_{z \to a} f(z)$ does not exist.

A necessary and sufficient condition for a point a to be a removable singularity of a function $f(z)$ is that the Laurent expansion of $f(z)$ around a does not contain the principal part.

If a function $f(z)$ is bounded in a neighborhood of an isolated singularity a, then a is a removable singularity of this function.

A necessary and sufficient condition for a point a to be a pole of a function $f(z)$ is that the principal part of the Laurent expansion of $f(z)$ around a contains finitely many terms:

$$f(z) = \frac{c_{-n}}{(z-a)^n} + \ldots + \frac{c_{-1}}{(z-a)} + \sum_{k=0}^{\infty} c_k (z-a)^k. \tag{M9.2.4.1}$$

The *order of a pole* a of a function $f(z)$ is defined to be the order of the zero of the function $F(z) = 1/f(z)$. If $c_{-n} \neq 0$ in expansion (M9.2.4.1), then the order of the pole a of the function $f(z)$ is equal to n. For $n = 1$, we have a *simple pole*.

A necessary and sufficient condition for a point a to be an essential singularity of a function $f(z)$ is that the principal part of the Laurent expansion of $f(z)$ around a contains infinitely many nonzero terms.

SOKHOTSKI'S THEOREM. *If a is an essential singularity of a function $f(z)$, then for each complex number A there exists a sequence of points $z_k \to a$ such that $f(z_k) \to A$.*

Example. Let us consider some functions with singular points of different kind.

1°. The function $f(z) = (1 - \cos z)/z^2$ has a removable singularity at the origin, since its Laurent expansion about the origin,

$$\frac{1 - \cos z}{z^2} = \frac{1}{2} - \frac{z^2}{24} + \frac{z^4}{720} - \ldots,$$

does not contain the principal part.

2°. The function $f(z) = 1/(1 + e^{z^2})$ has infinitely many poles at the points $z = \pm\sqrt{(2k+1)\pi i}$ ($k = 0, \pm 1, \pm 2, \ldots$). All these poles are simple poles, since the function $1/f(z) = 1 + e^{z^2}$ has simple zeros at these points. (Its derivative is nonzero at these points.)

3°. The function $f(z) = \sin(1/z)$ has an essential singularity at the origin, since the principal part of its Laurent expansion

$$\sin\frac{1}{z} = \frac{1}{z} - \frac{1}{z^3 3!} + \ldots$$

contains infinitely many terms.

The following two simplest classes of single-valued analytic functions are distinguished according to the character of singular points.

1. *Entire functions.* A function $f(z)$ is said to be *entire* if it does not have singular points in the finite part of the complex plane. An entire function can be represented by an everywhere convergent power series

$$f(z) = \sum_{n=0}^{\infty} c_n z^n.$$

An entire function can have only one singular point at $z = \infty$. If this singularity is a pole of order n, then $f(z)$ is a polynomial of degree n. If $z = \infty$ is an essential singularity, then $f(z)$ is called an *entire transcendental function*. If $z = \infty$ is a regular point (i.e., $f(z)$ is analytic for all z), then $f(z)$ is constant (*Liouville's theorem*). All polynomials, the exponential function, $\sin z$, $\cos z$, etc. are examples of entire functions. Sums, differences, and products of entire functions are themselves entire functions.

2. *Meromorphic functions.* A function $f(z)$ is said to be *meromorphic* if it does not have any singularities except for poles. The number of these poles in each finite closed domain D is always finite.

Suppose that a function $f(z)$ is analytic in a neighborhood of the point at infinity. The definition of singular points can be generalized to this function without any changes. But the criteria for the type of a singular point at infinity related to the Laurent expansion are different.

THEOREM. *In the case of a removable singularity at the point at infinity, the Laurent expansion of a function $f(z)$ in a neighborhood of this point does not contain positive powers of z. In the case of a pole, it contains finitely many positive powers of z. In the case of an essential singularity, it contains infinitely many powers of z.*

Let $f(z)$ be a multi-valued function defined in a neighborhood D of a point $z = a$ except possibly for the point a itself, and let $f_1(z)$, $f_2(z)$, ... be its branches, which are single-valued continuous functions in the domain where they are defined. The point a is called a *branch point (ramification point)* of the function $f(z)$ if $f(z)$ passes from one branch to another as the point z goes along a closed curve around the point z in a neighborhood of D. If the original branch is reached again after going around this curve m times (in the same sense), then the number $m - 1$ is called the *order* of the branch point, and the point a itself is called a *branch point of order* $m - 1$.

If all branches $f_k(z)$ tend to the same finite or infinite limit as $z \to a$, then the point a is called an *algebraic* branch point. (For example, the point $z = 0$ is an algebraic branch point of the function $f(z) = \sqrt[m]{z}$.) In this case, the single-valued function

$$F(z) = f(z^m + a)$$

has a regular point or a pole for $z = 0$.

If the limit of $f_k(z)$ as $z \to a$ does not exist, then the point a is called a *transcendental* branch point. For example, the point $z = 0$ is a transcendental branch point of the function $f(z) = \exp(\sqrt[m]{1/z})$.

In a neighborhood of a branch point a of finite order, the function $f(z)$ can be expanded in a *fractional power series* (*Puiseux series*)

$$f(z) = \sum_{k=-\infty}^{\infty} c_k (z - a)^{k/m}. \tag{M9.2.4.2}$$

If a new branch is obtained each time after going around this curve (in the same sense), then the point a is called a *branch point of infinite order (a logarithmic branch point)*. For example, the points $z = 0$ and $z = \infty$ are logarithmic branch points of the multi-valued function $w = \operatorname{Ln} z$. A logarithmic branch point is classified as a transcendental branch point.

For $a \neq \infty$, the expansion (M9.2.4.2) contains finitely many terms with negative k (infinitely many in the case of a transcendental point).

M9.2.5. Residues. Calculation of Definite Integrals

▶ **Residue of an analytic function at an isolated singular point.** The *residue* res $f(a)$ of a function $f(z)$ at an isolated singularity a is defined as the number

$$\operatorname{res} f(a) = \frac{1}{2\pi i} \oint_C f(z)\,dz, \tag{M9.2.5.1}$$

where the integral is taken in the positive sense over a contour C surrounding the point a and containing no other singularities of $f(z)$ in the interior.

Remark. Residues are sometimes denoted by $\mathrm{res}[f(z); a]$ or $\mathrm{res}_{z=a} f(z)$.

The residue $\mathrm{res}\, f(a)$ of a function $f(z)$ at a singularity a is equal to the coefficient of $(z - a)^{-1}$ in the Laurent expansion of $f(z)$ in a neighborhood of the point a,

$$\mathrm{res}\, f(a) = \frac{1}{2\pi i} \oint_C f(z)\, dz = c_{-1}.$$

Basic rules for finding the residues:

1. The residue of a function at a removable singularity is zero.
2. If a is a pole of order n, then

$$\mathrm{res}\, f(a) = \frac{1}{(n-1)!} \lim_{z \to a} \frac{d^{n-1}}{dz^{n-1}} \left[f(z)(z-a)^n \right]. \tag{M9.2.5.2}$$

3. For a simple pole ($n = 1$),

$$\mathrm{res}\, f(a) = \lim_{z \to a} \left[f(z)(z-a) \right].$$

4. If $f(z)$ is the quotient of two analytic functions,

$$f(z) = \frac{\varphi(z)}{\psi(z)},$$

in a neighborhood of a point a and $\varphi(a) \neq 0$, $\psi(a) = 0$, but $\psi_z'(a) \neq 0$ (i.e., a is a simple pole of $f(z)$), then

$$\mathrm{res}\, f(a) = \frac{\varphi(a)}{\psi_z'(a)}. \tag{M9.2.5.3}$$

5. If a is an essential singularity of $f(z)$, then to obtain $\mathrm{res}\, f(a)$, one has to find the coefficient c_{-1} in the Laurent expansion of $f(z)$ in a neighborhood of a.

▶ **Basic theorems on residues.**

A function $f(z)$ is said to be continuous on the boundary C of the domain D if for each boundary point z_0 there exists a limit $\lim_{z \to z_0} f(z) = f(z_0)$ as $z \to z_0$, $z \in D$.

CAUCHY'S RESIDUE THEOREM. *Let $f(z)$ be a function continuous on the boundary C of a domain D and analytic in the interior of D everywhere except for finitely many points $a_1, \ldots, a_n$. Then*

$$\int_C f(z)\, dz = 2\pi i \sum_{k=1}^{n} \mathrm{res}\, f(a_k), \tag{M9.2.5.4}$$

where the integral is taken in the positive sense of C.

The *logarithmic residue* of a function $f(z)$ at a point a is by definition the residue of its logarithmic derivative

$$\left[\ln f(z) \right]_z' = \frac{f_z'(z)}{f(z)}.$$

THEOREM. *The logarithmic derivative $f_z'(z)/f(z)$ has first-order poles at the zeros and poles of $f(z)$. Moreover, the logarithmic residue of $f(z)$ at a zero or a pole of $f(z)$ is equal to the order of the zero or minus the order of the pole, respectively.*

The residue of a function $f(z)$ at infinity is defined as

$$\operatorname{res} f(\infty) = \frac{1}{2\pi i} \oint_{\Gamma} f(z)\, dz,$$

where Γ is a circle of sufficiently large radius $|z| = \rho$ and the integral is taken in the clockwise sense (so that the neighborhood of the point $z = \infty$ remains to the left of the contour, just as in the case of a finite point).

The residue of $f(z)$ at infinity is equal to minus the coefficient of z^{-1} in the Laurent expansion of $f(z)$ in a neighborhood of the point $z = \infty$,

$$\operatorname{res} f(\infty) = -c_{-1}.$$

THEOREM. *If a function $f(z)$ has finitely many singular points in the extended complex plane, then the sum of all its residues, including the residue at infinity, is zero:*

$$\operatorname{res} f(\infty) + \sum_{k=1}^{n} \operatorname{res} f(a_k) = 0, \qquad (\text{M9.2.5.5})$$

where $a_1, \ldots, a_n$ are finite singular points.

Example 1. Let us calculate the integral

$$\oint_{C} \frac{\ln(z+2)}{z^2}\, dz,$$

where C is the circle $|z| = \frac{1}{2}$.

In the disk $|z| \le \frac{1}{2}$, there is only one singular point of the integrand, $z = 0$, which is a second-order pole. The residue of $f(z)$ at $z = 0$ is calculated by the formula (M9.2.5.2)

$$\operatorname{res} f(0) = \lim_{z \to 0} \left[z^2 \frac{\ln(z+2)}{z^2} \right]'_z = \lim_{z \to 0} [\ln(z+2)]'_z = \lim_{z \to 0} \frac{1}{z+2} = \frac{1}{2}.$$

Using formula (M9.2.5.1), we obtain

$$\frac{1}{2} = \frac{1}{2\pi i} \oint_{C} \frac{\ln(z+2)}{z^2}\, dz, \qquad \oint_{C} \frac{\ln(z+2)}{z^2}\, dz = \pi i.$$

▶ **Jordan's lemma. Calculation of definite integrals using residues.** Suppose that we need to calculate the integral of a real function $f(x)$ over a (finite or infinite) interval (a, b). Let us supplement the interval (a, b) with a curve Γ that, together with (a, b), bounds a domain D, and then analytically continue the function $f(x)$ into $\overline{D}$. Then the residue theorem can be applied to this analytic continuation of $f(z)$, and by this theorem

$$\int_{a}^{b} f(x)\, dx + \int_{\Gamma} f(z)\, dz = 2\pi i \Lambda,$$

where Λ is the sum of residues of $f(z)$ in D. If $\int_{\Gamma} f(z)\, dz$ can be calculated or expressed in terms of the desired integral $\int_{a}^{b} f(x)\, dx$, then the problem will be solved.

When calculating integrals of the form $\int_{-\infty}^{\infty} f(x)\, dx$, one should apply (M9.2.5.4) to the contour C that consists of the interval $(-R, R)$ of the real axis and the arc C_R of the semicircle $|z| = R$ in the upper half-plane. Sometimes, it is only possible to find the limit as $R \to \infty$ of the integral over the contour C_R rather than to calculate it, and often it turns out that the limit of this integral is equal to zero.

LEMMA. *If a function $f(z)$ is analytic for $|z| > R_0$ and $z f(z) \to 0$ as $|z| \to \infty$ for $y \ge 0$, then*

$$\lim_{R \to \infty} \int_{C_R} f(z)\, dz = 0,$$

where C_R is the arc of the semicircle $|z| = R$ in the upper half-plane.

THEOREM. *Let a function $f(x)$ be defined on the whole real axis $-\infty < x < \infty$ and let it can be analytically continued to the upper half-plane $\operatorname{Im} z > 0$ so that the continuation $f(z)$ satisfies the conditions of the previous lemma. Then the improper integral $\int_{-\infty}^{\infty} f(x)\,dx$ exists and is equal to*

$$\int_{-\infty}^{\infty} f(x)\,dx = 2\pi i \sum_{k=1}^{n} \operatorname{res} f(a_k), \qquad (\text{M9.2.5.6})$$

where the a_k are singular points of $f(z)$ in the upper half-plane.

Example 2. Calculate the integral

$$I = \int_{-\infty}^{\infty} \frac{dx}{1 + x^4}.$$

The analytic continuation of the integrand into the upper half-plane is $f(z) = (1+z^4)^{-1}$; it satisfies the conditions of the above lemma. The function $f(z)$ has two singular points, $a_1 = e^{i\pi/4}$ and $a_2 = e^{i3\pi/4}$, in the upper half-plane (both points are first-order poles). Using formulas (M9.2.5.3) and (M9.2.5.6), one finds

$$I = 2\pi i \left(\frac{1}{4z^3}\Big|_{z=e^{i\pi/4}} + \frac{1}{4z^3}\Big|_{z=e^{i3\pi/4}} \right) = \frac{\pi\sqrt{2}}{2}.$$

JORDAN'S LEMMA. *If a function $g(z)$ tends to zero uniformly with respect to $\arg z$ along a sequence of circular arcs $C_{R_n} : |z| = R_n$, $\operatorname{Im} z > -a$ (where $R_n \to \infty$ and a is fixed), then*

$$\lim_{n\to\infty} \int_{C_{R_n}} g(z)e^{i\lambda z}\,dz = 0$$

for any positive number λ.

Example 3 (Laplace integral). To calculate the integral

$$\int_0^\infty \frac{\cos x}{x^2 + a^2}\,dx,$$

one uses the auxiliary function

$$f(z) = \frac{e^{iz}}{z^2 + a^2} = g(z)e^{iz}, \quad g(z) = \frac{1}{z^2 + a^2}$$

and the contour shown in Fig. M9.5. Since $g(z)$ satisfies the inequality $|g(z)| < (R^2 - a^2)^{-1}$ on C_R, it follows that this function uniformly tends to zero as $R \to \infty$, and by Jordan's lemma with $\lambda = 1$ we obtain

$$\int_{C_R} f(z)\,dz = \int_{C_R} g(z)e^{iz}\,dz \to 0$$

as $R \to \infty$.

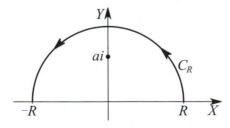

Figure M9.5. The contour for the calculation of the Laplace integral.

By the residue theorem

$$\int_{-R}^{R} \frac{e^{ix}}{x^2 + a^2}\, dx + \int_{C_R} f(z)\, dz = 2\pi i\, \frac{e^{-a}}{2ai}$$

for any $R > a$. (The residue at the singular point $z = ai$ of the function $f(z)$, which is a first-order pole and the only singular point of this function lying inside the contour, can be calculated by formula (M9.2.5.3).) In the limit as $R \to \infty$, we obtain

$$\int_{-\infty}^{\infty} \frac{e^{ix}}{x^2 + a^2}\, dx = \frac{\pi}{ae^a}.$$

Separating the real part and using the fact that the function is even, we obtain

$$\int_{0}^{\infty} \frac{\cos x}{x^2 + a^2}\, dx = \frac{\pi}{2ae^a}.$$

Bibliography for Chapter M9

Ablowitz, M. J. and Fokas, A. S., *Complex Variables: Introduction and Applications* (Cambridge Texts in Applied Mathematics), *2nd Edition*, Cambridge University Press, Cambridge, England, 2003.

Berenstein, C. A. and Roger Gay, R., *Complex Variables: An Introduction* (Graduate Texts in Mathematics), Springer, New York, 1997.

Bieberbach, L., *Conformal Mapping*, American Mathematical Society, Providence, Rhode Island, 2000.

Bronshtein, I. N., Semendyayev, K. A., Musiol, G., and Mühlig, H., *Handbook of Mathematics, 4th Edition*, Springer, New York, 2004.

Brown, J. W. and Churchill, R. V., *Complex Variables and Applications, 7th Edition*, McGraw-Hill, New York, 2003.

Caratheodory, C., *Conformal Representation*, Dover Publications, New York, 1998.

Carrier, G. F., Krock, M., and Pearson, C. E., *Functions of a Complex Variable: Theory and Technique* (Classics in Applied Mathematics), Society for Industrial & Applied Mathematics, University City Science Center, Philadelphia, 2005.

Cartan, H., *Elementary Theory of Analytic Functions of One or Several Complex Variables*, Dover Publications, New York, 1995.

Conway, J. B., *Functions of One Complex Variable I* (Graduate Texts in Mathematics), *2nd Edition*, Springer, New York, 1995.

Conway, J. B., *Functions of One Complex Variable II* (Graduate Texts in Mathematics), *2nd Edition*, Springer, New York, 1996.

Dettman, J. W., Applied Complex Variables (Mathematics Series), Dover Publications, New York, 1984.

England, A. H., *Complex Variable Methods in Elasticity, Dover Edition*, Dover Publications, New York, 2003.

Fisher, S. D., *Complex Variables* (Dover Books on Mathematics), *2nd Edition*, Dover Publications, New York, 1999.

Flanigan, F. J., *Complex Variables, Dover Edition*, Dover Publications, New York, 1983.

Greene, R. E. and Krantz, S. G., *Function Theory of One Complex Variable* (Graduate Studies in Mathematics), *Vol. 40, 2nd Edition*, American Mathematical Society, Providence, Rhode Island, 2002.

Ivanov, V. I. and Trubetskov, M. K., *Handbook of Conformal Mapping with Computer-Aided Visualization*, CRC Press, Boca Raton, Florida, 1995.

Korn, G. A and Korn, T. M., *Mathematical Handbook for Scientists and Engineers: Definitions, Theorems, and Formulas for Reference and Review*, Dover Edition, Dover Publications, New York, 2000.

Krantz, S. G., *Handbook of Complex Variables*, Birkhäuser, Boston, Massachusetts, 1999.

Lang, S., *Complex Analysis* (Graduate Texts in Mathematics), *4th Edition*, Springer, New York, 2003.

Lavrentiev, M. A. and Shabat, B. V., *Methods of the Theory of Functions of a Complex Variable, 5th Edition [in Russian]*, Nauka Publishers, Moscow, 1987.

LePage, W. R., *Complex Variables and the Laplace Transform for Engineers*, Dover Publications, New York, 1980.

Markushevich, A. I. and Silverman, R. A. (Editor), *Theory of Functions of a Complex Variable, 2nd Rev. Edition*, American Mathematical Society, Providence, Rhode Island, 2005.

Narasimhan, R. and Nievergelt, Y., *Complex Analysis in One Variable, 2nd Edition*, Birkhäuser, Boston, Basel, Stuttgart, 2000.

Needham, T., *Visual Complex Analysis, Rep. Edition*, Oxford University Press, Oxford, 1999.

Nehari, Z., *Conformal Mapping*, Dover Publications, New York, 1982.

Paliouras, J. D. and Meadows, D. S., *Complex Variables for Scientists and Engineers, Facsimile Edition*, Macmillan Coll. Div., New York, 1990.

Pierpont, J., *Functions of a Complex Variable (Phoenix Edition)*, Dover Publications, New York, 2005.

Schinzinger, R. and Laura, P. A. A., *Conformal Mapping: Methods and Applications*, Dover Publications, New York, 2003.

Silverman, R. A., *Introductory Complex Analysis*, Dover Publications, New York, 1984.

Spiegel, M. R., *Schaum's Outline of Complex Variables*, McGraw-Hill, New York, 1968.

Sveshnikov, A. G. and Tikhonov, A. N., The Theory of Functions of a Complex Variable, Mir Publishers, Moscow, 1982.

Wunsch, D. A., *Complex Variables with Applications, 2nd Edition*, Addison Wesley, Boston, Massachusetts, 1993.

Chapter M10
Integral Transforms

M10.1. General Form of Integral Transforms. Inversion Formulas

Normally an integral transform has the form

$$\widetilde{f}(\lambda) = \int_a^b \varphi(x, \lambda) f(x)\, dx.$$

The function $\widetilde{f}(\lambda)$ is called the *transform* of the function $f(x)$ and $\varphi(x, \lambda)$ is called the *kernel* of the integral transform. The function $f(x)$ is called the *inverse transform* of $\widetilde{f}(\lambda)$. The limits of integration a and b are real numbers (usually, $a = 0$, $b = \infty$ or $a = -\infty$, $b = \infty$). For brevity, we rewrite the integral transform as follows: $\widetilde{f}(\lambda) = \mathcal{L}\{f(x)\}$.

General properties of integral transforms (linearity):

$$\mathcal{L}\{kf(x)\} = k\mathcal{L}\{f(x)\},$$
$$\mathcal{L}\{f(x) \pm g(x)\} = \mathcal{L}\{f(x)\} \pm \mathcal{L}\{g(x)\}.$$

Here, k is an arbitrary constant; it is assumed that integral transforms of the functions $f(x)$ and $g(x)$ exist.

In Subsections M10.2–M10.4, the most popular (Laplace, Fourier, Mellin, etc.) integral transforms are described. These subsections also describe the corresponding inversion formulas, which normally have the form

$$f(x) = \int_C \psi(x, \lambda) \widetilde{f}(\lambda)\, d\lambda$$

and make it possible to recover $f(x)$ if $\widetilde{f}(\lambda)$ is given. The integration path C can lie either on the real axis or in the complex plane.

In many cases, to evaluate the integrals in the inversion formula—in particular, to find the inverse Laplace, Mellin, and Fourier transforms—methods of the theory of functions of a complex variable can be applied, including the theorems about residue and Jordan's lemma, which are outlined in Subsection M9.2.5.

M10.2. Laplace Transform
M10.2.1. Laplace Transform and the Inverse Laplace Transform

▶ **Laplace transform.** The *Laplace transform* of an arbitrary (complex-valued) function $f(x)$ of a real variable x ($x \geq 0$) is defined by

$$\widetilde{f}(p) = \int_0^\infty e^{-px} f(x)\, dx, \tag{M10.2.1.1}$$

where $p = s + i\sigma$ is a complex variable.

The Laplace transform exists for any continuous or piecewise-continuous function satisfying the condition $|f(x)| < Me^{\sigma_0 x}$ with some $M > 0$ and $\sigma_0 \geq 0$. In the following, σ_0 often means the greatest lower bound of the possible values of σ_0 in this estimate; this value is called the *growth exponent* of the function $f(x)$.

For any $f(x)$, the transform $\widetilde{f}(p)$ is defined in the half-plane $\mathrm{Re}\, p > \sigma_0$ and is analytic there.

For brevity, we shall write formula (M10.2.1.1) as follows:

$$\widetilde{f}(p) = \mathcal{L}\left\{f(x)\right\} \qquad \text{or} \qquad \widetilde{f}(p) = \mathcal{L}\left\{f(x),\, p\right\}.$$

▶ **Inverse Laplace transform.** Given the transform $\widetilde{f}(p)$, the function $f(x)$ can be found by means of the inverse Laplace transform

$$f(x) = \frac{1}{2\pi i} \int_{c-i\infty}^{c+i\infty} \widetilde{f}(p)e^{px}\, dp, \qquad i^2 = -1, \tag{M10.2.1.2}$$

where the integration path is parallel to the imaginary axis and lies to the right of all singularities of $\widetilde{f}(p)$, which corresponds to $c > \sigma_0$.

The integral in inversion formula (M10.2.1.2) is understood in the sense of the Cauchy principal value:

$$\int_{c-i\infty}^{c+i\infty} \widetilde{f}(p)e^{px}\, dp = \lim_{\omega \to \infty} \int_{c-i\omega}^{c+i\omega} \widetilde{f}(p)e^{px}\, dp.$$

In the domain $x < 0$, formula (M10.2.1.2) gives $f(x) \equiv 0$.

Formula (M10.2.1.2) holds for continuous functions. If $f(x)$ has a (finite) jump discontinuity at a point $x = x_0 > 0$, then the right-hand side of (M10.2.1.2) evaluates to $\frac{1}{2}[f(x_0 - 0) + f(x_0 + 0)]$ at this point (for $x_0 = 0$, the first term in the square brackets must be omitted).

For brevity, we write the Laplace inversion formula (M10.2.1.2) as follows:

$$f(x) = \mathcal{L}^{-1}\left\{\widetilde{f}(p)\right\} \qquad \text{or} \qquad f(x) = \mathcal{L}^{-1}\left\{\widetilde{f}(p),\, x\right\}.$$

There are tables of direct and inverse Laplace transforms (see Sections S2.1 and S2.2), which are handy in solving linear differential and integral equations.

M10.2.2. Main Properties of the Laplace Transform. Inversion Formulas for Some Functions

▶ **Convolution theorem. Main properties of the Laplace transform.**

$1°$. The *convolution* of two functions $f(x)$ and $g(x)$ is defined as an integral of the form $\int_0^x f(t)g(x-t)\, dt$, and is usually denoted by $f(x) * g(x)$, so that

$$f(x) * g(x) = \int_0^x f(t)\, g(x-t)\, dt.$$

By performing substitution $x - t = u$, we see that the convolution is symmetric with respect to the convolved functions: $f(x) * g(x) = g(x) * f(x)$.

The *convolution theorem* states that

$$\mathcal{L}\left\{f(x) * g(x)\right\} = \mathcal{L}\left\{f(x)\right\} \mathcal{L}\left\{g(x)\right\}$$

and is frequently applied to solve Volterra equations with kernels depending on the difference of the arguments.

$2°$. The main properties of the correspondence between functions and their Laplace transforms are gathered in Table M10.1.

$3°$. The Laplace transforms of some functions are listed in Table M10.2; for more detailed tables, see Section S2.1 and the list of references at the end of this chapter.

▶ **Inverse transforms of rational functions.** Consider the important case in which the transform is a rational function of the form

$$\widetilde{f}(p) = \frac{R(p)}{Q(p)}, \tag{M10.2.2.1}$$

where $Q(p)$ and $R(p)$ are polynomials in the variable p and the degree of $Q(p)$ exceeds that of $R(p)$.

Assume that the zeros of the denominator are simple, i.e.,

$$Q(p) \equiv \mathrm{const}\,(p - \lambda_1)(p - \lambda_2) \ldots (p - \lambda_n).$$

Then the inverse transform can be determined by the formula

$$f(x) = \sum_{k=1}^{n} \frac{R(\lambda_k)}{Q'(\lambda_k)} \exp(\lambda_k x), \tag{M10.2.2.2}$$

where the primes denote the derivatives.

If $Q(p)$ has multiple zeros, i.e.,

$$Q(p) \equiv \mathrm{const}\,(p - \lambda_1)^{s_1}(p - \lambda_2)^{s_2} \ldots (p - \lambda_m)^{s_m},$$

then

$$f(x) = \sum_{k=1}^{m} \frac{1}{(s_k - 1)!} \lim_{p \to s_k} \frac{d^{s_k - 1}}{dp^{s_k - 1}} \Big[(p - \lambda_k)^{s_k} \widetilde{f}(p) e^{px} \Big]. \tag{M10.2.2.3}$$

Example 1. The transform

$$\widetilde{f}(p) = \frac{b}{p^2 - a^2} \qquad \text{(a and b real numbers)}$$

can be represented as the fraction (M10.2.2.1) with $R(p) = b$ and $Q(p) = (p-a)(p+a)$. The denominator $Q(p)$ has two simple roots, $\lambda_1 = a$ and $\lambda_2 = -a$. Using formula (M10.2.2.2) with $n = 2$ and $Q'(p) = 2p$, we obtain the inverse transform in the form

$$f(x) = \frac{b}{2a} e^{ax} - \frac{b}{2a} e^{-ax} = \frac{b}{a} \sinh(ax).$$

Example 2. The transform

$$\widetilde{f}(p) = \frac{b}{p^2 + a^2} \qquad \text{(a and b real numbers)}$$

can be written as the fraction (M10.2.2.1) with $R(p) = b$ and $Q(p) = (p - ia)(p + ia)$, $i^2 = -1$. The denominator $Q(p)$ has two simple pure imaginary roots, $\lambda_1 = ia$ and $\lambda_2 = -ia$. Using formula (M10.2.2.2) with $n = 2$, we find the inverse transform:

$$f(x) = \frac{b}{2ia} e^{iax} - \frac{b}{2ia} e^{-iax} = \frac{b}{a} \sin(ax).$$

Example 3. The transform

$$\widetilde{f}(p) = a p^{-n},$$

where n is a positive integer, can be written as the fraction (M10.2.2.1) with $R(p) = a$ and $Q(p) = p^n$. The denominator $Q(p)$ has one root of multiplicity n, $\lambda_1 = 0$. By formula (M10.2.2.3) with $m = 1$ and $s_1 = n$, we find the inverse transform:

$$f(x) = \frac{a}{(n - 1)!} x^{n-1}.$$

Remark. Fairly detailed tables of inverse Laplace transforms can be found in Section S2.2.

TABLE M10.1
Main properties of the Laplace transform.

No.	Function	Laplace transform	Operation
1	$af_1(x) + bf_2(x)$	$a\widetilde{f}_1(p) + b\widetilde{f}_2(p)$	Linearity
2	$f(x/a),\ a > 0$	$a\widetilde{f}(ap)$	Scaling
3	$f(x - a),$ $f(x - a) \equiv 0$ for $x < a$	$e^{-ap}\widetilde{f}(p)$	Shift of the argument
4	$x^n f(x);\ n = 1, 2, \ldots$	$(-1)^n \widetilde{f}_p^{(n)}(p)$	Differentiation of the transform
5	$\dfrac{1}{x}f(x)$	$\displaystyle\int_p^\infty \widetilde{f}(q)\,dq$	Integration of the transform
6	$e^{ax}f(x)$	$\widetilde{f}(p - a)$	Shift in the complex plane
7	$f'_x(x)$	$p\widetilde{f}(p) - f(+0)$	Differentiation
8	$f_x^{(n)}(x)$	$p^n \widetilde{f}(p) - \displaystyle\sum_{k=1}^n p^{n-k} f_x^{(k-1)}(+0)$	Differentiation
9	$x^m f_x^{(n)}(x),\ m = 1, 2, \ldots$	$(-1)^m \dfrac{d^m}{dp^m}\left[p^n \widetilde{f}(p) - \displaystyle\sum_{k=1}^n p^{n-k} f_x^{(k-1)}(+0) \right]$	Differentiation
10	$\dfrac{d^n}{dx^n}\left[x^m f(x) \right],\ m \geq n$	$(-1)^m p^n \dfrac{d^m}{dp^m} \widetilde{f}(p)$	Differentiation
11	$\displaystyle\int_0^x f(t)\,dt$	$\dfrac{\widetilde{f}(p)}{p}$	Integration
12	$\displaystyle\int_0^x f_1(t)f_2(x - t)\,dt$	$\widetilde{f}_1(p)\widetilde{f}_2(p)$	Convolution

TABLE M10.2
The Laplace transforms of some functions.

No.	Function, $f(x)$	Laplace transform, $\widetilde{f}(p)$	Remarks
1	1	$1/p$	
2	x^n	$\dfrac{n!}{p^{n+1}}$	$n = 1, 2, \ldots$
3	x^a	$\Gamma(a + 1)p^{-a-1}$	$a > -1$
4	e^{-ax}	$(p + a)^{-1}$	
5	$x^a e^{-bx}$	$\Gamma(a + 1)(p + b)^{-a-1}$	$a > -1$
6	$\sinh(ax)$	$\dfrac{a}{p^2 - a^2}$	
7	$\cosh(ax)$	$\dfrac{p}{p^2 - a^2}$	
8	$\ln x$	$-\dfrac{1}{p}(\ln p + \mathcal{C})$	$\mathcal{C} = 0.5772\ldots$ is the Euler constant
9	$\sin(ax)$	$\dfrac{a}{p^2 + a^2}$	
10	$\cos(ax)$	$\dfrac{p}{p^2 + a^2}$	
11	$\operatorname{erfc}\left(\dfrac{a}{2\sqrt{x}}\right)$	$\dfrac{1}{p}\exp\left(-a\sqrt{p}\right)$	$a \geq 0$
12	$J_0(ax)$	$\dfrac{1}{\sqrt{p^2 + a^2}}$	$J_0(x)$ is the Bessel function

M10.2.3. Limit Theorems. Representation of Inverse Transforms as Convergent Series

▶ **Limit theorems.**

THEOREM 1. *Let $0 \le x < \infty$ and $\widetilde{f}(p) = \mathcal{L}\left\{f(x)\right\}$ be the Laplace transform of $f(x)$. If a limit of $f(x)$ as $x \to 0$ exists, then*

$$\lim_{x \to 0} f(x) = \lim_{p \to \infty} \left[p\widetilde{f}(p)\right].$$

THEOREM 2. *If a limit of $f(x)$ as $x \to \infty$ exists, then*

$$\lim_{x \to \infty} f(x) = \lim_{p \to 0} \left[p\widetilde{f}(p)\right].$$

▶ **Representation of inverse transforms as convergent series.**

THEOREM 1. *Suppose the transform $\widetilde{f}(p)$ can be expanded into series in negative powers of p,*

$$\widetilde{f}(p) = \sum_{n=1}^{\infty} \frac{a_n}{p^n},$$

convergent for $|p| > R$, where R is an arbitrary positive number; note that the transform tends to zero as $|p| \to \infty$. Then the inverse transform can be obtained by the formula

$$f(x) = \sum_{n=1}^{\infty} \frac{a_n}{(n-1)!} x^{n-1},$$

where the series on the right-hand side is convergent for all x.

THEOREM 2. *Suppose the transform $\widetilde{f}(p)$, $|p| > R$, is represented by an absolutely convergent series,*

$$\widetilde{f}(p) = \sum_{n=0}^{\infty} \frac{a_n}{p^{\lambda_n}}, \tag{M10.2.3.1}$$

where $\{\lambda_n\}$ is any positive increasing sequence, $0 < \lambda_0 < \lambda_1 < \cdots \to \infty$. Then it is possible to proceed termwise from series (M10.2.3.1) to the following inverse transform series:

$$f(x) = \sum_{n=0}^{\infty} \frac{a_n}{\Gamma(\lambda_n)} x^{\lambda_n - 1}, \tag{M10.2.3.2}$$

where $\Gamma(\lambda)$ is the Gamma function. Series (M10.2.3.2) is convergent for all real and complex values of x other than zero (if $\lambda_0 \ge 1$, the series is convergent for all x).

M10.3. Various Forms of the Fourier Transform

M10.3.1. Fourier Transform and the Inverse Fourier Transform

▶ **Standard form of the Fourier transform.** The *Fourier transform* is defined as follows:

$$\widetilde{f}(u) = \frac{1}{\sqrt{2\pi}} \int_{-\infty}^{\infty} f(x) e^{-iux} \, dx. \tag{M10.3.1.1}$$

For brevity, we rewrite formula (M10.3.1.1) as follows:

$$\widetilde{f}(u) = \mathfrak{F}\{f(x)\} \qquad \text{or} \qquad \widetilde{f}(u) = \mathfrak{F}\{f(x), u\}.$$

Given $\widetilde{f}(u)$, the function $f(x)$ can be found by means of the *inverse Fourier transform*

$$f(x) = \frac{1}{\sqrt{2\pi}} \int_{-\infty}^{\infty} \widetilde{f}(u)\, e^{iux}\, du. \tag{M10.3.1.2}$$

Formula (M10.3.1.2) holds for continuous functions. If $f(x)$ has a (finite) jump discontinuity at a point $x = x_0$, then the right-hand side of (M10.3.1.2) evaluates to $\frac{1}{2}\left[f(x_0 - 0) + f(x_0 + 0)\right]$ at this point.

For brevity, we rewrite formula (M10.3.1.2) as follows:

$$f(x) = \mathfrak{F}^{-1}\{\widetilde{f}(u)\} \qquad \text{or} \qquad f(x) = \mathfrak{F}^{-1}\{\widetilde{f}(u),\, x\}.$$

▶ **Convolution theorem. Main properties of the Fourier transform.**

$1°$. The *convolution* of two functions $f(x)$ and $g(x)$ is defined as

$$f(x) * g(x) \equiv \frac{1}{\sqrt{2\pi}} \int_{-\infty}^{\infty} f(x - t)g(t)\, dt.$$

By performing substitution $x - t = u$, we see that the convolution is symmetric with respect to the convolved functions: $f(x) * g(x) = g(x) * f(x)$.

The *convolution theorem* states that

$$\mathfrak{F}\{f(x) * g(x)\} = \mathfrak{F}\{f(x)\}\, \mathfrak{F}\{g(x)\}.$$

$2°$. The main properties of the correspondence between functions and their Fourier transforms are gathered in Table M10.3.

TABLE M10.3
Main properties of the Fourier transform.

No.	Function	Fourier transform	Operation
1	$a f_1(x) + b f_2(x)$	$a\widetilde{f}_1(u) + b\widetilde{f}_2(u)$	Linearity
2	$f(x/a),\ a > 0$	$a\widetilde{f}(au)$	Scaling
3	$x^n f(x);\ n = 1, 2, \ldots$	$i^n \widetilde{f}_u^{(n)}(u)$	Differentiation of the transform
4	$f''_{xx}(x)$	$-u^2 \widetilde{f}(u)$	Differentiation
5	$f_x^{(n)}(x)$	$(iu)^n \widetilde{f}(u)$	Differentiation
6	$\displaystyle\int_{-\infty}^{\infty} f_1(\xi) f_2(x - \xi)\, d\xi$	$\widetilde{f}_1(u)\widetilde{f}_2(u)$	Convolution

M10.3.2. Fourier Cosine and Sine Transforms

▶ **Fourier cosine transform.**

1°. Let a function $f(x)$ be integrable on the semiaxis $0 \le x < \infty$. The *Fourier cosine transform* is defined by

$$\widetilde{f}_c(u) = \sqrt{\frac{2}{\pi}} \int_0^\infty f(x) \cos(xu)\, dx, \qquad 0 < u < \infty. \tag{M10.3.2.1}$$

Given $\widetilde{f}_c(u)$, the function $f(x)$ can be found by means of the *Fourier cosine inversion formula*

$$f(x) = \sqrt{\frac{2}{\pi}} \int_0^\infty \widetilde{f}_c(u) \cos(xu)\, du, \qquad 0 < x < \infty. \tag{M10.3.2.2}$$

The Fourier cosine transform (M10.3.2.1) is denoted for brevity by $\widetilde{f}_c(u) = \mathfrak{F}_c\{f(x)\}$.

2°. Some other properties of the Fourier cosine transform:

$$\mathfrak{F}_c\{x^{2n} f(x)\} = (-1)^n \frac{d^{2n}}{du^{2n}} \mathfrak{F}_c\{f(x)\}, \qquad n = 1, 2, \dots ;$$
$$\mathfrak{F}_c\{f''(x)\} = -u^2 \mathfrak{F}_c\{f(x)\}.$$

The function $f(x)$ here is assumed to vanish sufficiently rapidly (exponentially) as $x \to \infty$. In the latter formula, the condition $f'(0) = 0$ is assumed to hold.

Parseval's relation for the Fourier cosine transform:

$$\int_0^\infty \mathfrak{F}_c\{f(x)\} \mathfrak{F}_c\{g(x)\}\, du = \int_0^\infty f(x)g(x)\, dx.$$

There are tables of the Fourier cosine transform (see Section S2.3 and the references listed at the end of the current chapter).

▶ **Fourier sine transform.**

1°. Let a function $f(x)$ be integrable on the semiaxis $0 \le x < \infty$. The *Fourier sine transform* is defined by

$$\widetilde{f}_s(u) = \sqrt{\frac{2}{\pi}} \int_0^\infty f(x) \sin(xu)\, dx, \qquad 0 < u < \infty. \tag{M10.3.2.3}$$

For given $\widetilde{f}_s(u)$, the function $f(x)$ can be found by means of the *inverse Fourier sine transform*

$$f(x) = \sqrt{\frac{2}{\pi}} \int_0^\infty \widetilde{f}_s(u) \sin(xu)\, du, \qquad 0 < x < \infty. \tag{M10.3.2.4}$$

The Fourier sine transform (M10.3.2.3) is briefly denoted by $\widetilde{f}_s(u) = \mathfrak{F}_s\{f(x)\}$.

$2°$. Some other properties of the Fourier sine transform:

$$\mathfrak{F}_s\big\{x^{2n}f(x)\big\} = (-1)^n \frac{d^{2n}}{du^{2n}}\mathfrak{F}_s\big\{f(x)\big\}, \qquad n = 1, 2, \dots ;$$

$$\mathfrak{F}_s\big\{f''(x)\big\} = -u^2 \mathfrak{F}_s\big\{f(x)\big\}.$$

The function $f(x)$ here is assumed to vanish sufficiently rapidly (exponentially) as $x \to \infty$. In the latter formula, the condition $f(0) = 0$ is assumed to hold.

Parseval's relation for the Fourier sine transform:

$$\int_0^\infty \mathfrak{F}_s\big\{f(x)\big\}\mathfrak{F}_s\big\{g(x)\big\}\, du = \int_0^\infty f(x)g(x)\, dx.$$

There are tables of the Fourier sine transform (see Section S2.4 and the references listed at the end of the current chapter).

M10.4. Mellin Transform and Other Transforms

M10.4.1. Mellin Transform and the Inversion Formula

▶ **Mellin transform.** Suppose that a function $f(x)$ is defined for positive x and satisfies the conditions

$$\int_0^1 |f(x)|\, x^{\sigma_1-1}\, dx < \infty, \qquad \int_1^\infty |f(x)|\, x^{\sigma_2-1}\, dx < \infty$$

for some real numbers σ_1 and σ_2, $\sigma_1 < \sigma_2$.

The Mellin transform of $f(x)$ is defined by

$$\hat{f}(s) = \int_0^\infty f(x)x^{s-1}\, dx, \qquad (\text{M10.4.1.1})$$

where $s = \sigma + i\tau$ is a complex variable ($\sigma_1 < \sigma < \sigma_2$).

For brevity, we rewrite formula (M10.4.1.1) as follows:

$$\hat{f}(s) = \mathfrak{M}\{f(x)\} \qquad \text{or} \qquad \hat{f}(s) = \mathfrak{M}\{f(x), s\}.$$

▶ **Inverse Mellin transform.** Given $\hat{f}(s)$, the function $f(x)$ can be found by means of the *inverse Mellin transform*

$$f(x) = \frac{1}{2\pi i}\int_{\sigma-i\infty}^{\sigma+i\infty} \hat{f}(s)x^{-s}\, ds \qquad (\sigma_1 < \sigma < \sigma_2), \qquad (\text{M10.4.1.2})$$

where the integration path is parallel to the imaginary axis of the complex plane s and the integral is understood in the sense of the Cauchy principal value.

Formula (M10.4.1.2) holds for continuous functions. If $f(x)$ has a (finite) jump discontinuity at a point $x = x_0 > 0$, then the right-hand side of (M10.4.1.2) evaluates to $\frac{1}{2}\big[f(x_0 - 0) + f(x_0 + 0)\big]$ at this point (for $x_0 = 0$, the first term in the square brackets must be omitted).

For brevity, we rewrite formula (M10.4.1.2) in the form

$$f(x) = \mathfrak{M}^{-1}\{\hat{f}(s)\} \qquad \text{or} \qquad f(x) = \mathfrak{M}^{-1}\{\hat{f}(s), x\}.$$

M10.4.2. Main Properties of the Mellin Transform. Relation Among the Mellin, Laplace, and Fourier Transforms

▶ **Main properties of the Mellin transform.** The main properties of the correspondence between the functions and their Mellin transforms are gathered in Table M10.4.

TABLE M10.4
Main properties of the Mellin transform.

No.	Function	Mellin transform	Operation
1	$af_1(x) + bf_2(x)$	$a\hat{f}_1(s) + b\hat{f}_2(s)$	Linearity
2	$f(ax), \ a > 0$	$a^{-s}\hat{f}(s)$	Scaling
3	$x^a f(x)$	$\hat{f}(s + a)$	Shift of the argument of the transform
4	$f(x^2)$	$\frac{1}{2}\hat{f}\left(\frac{1}{2}s\right)$	Squared argument
5	$f(1/x)$	$\hat{f}(-s)$	Inversion of the argument of the transform
6	$x^\lambda f(ax^\beta), \ a > 0, \beta \neq 0$	$\frac{1}{\beta} a^{-\frac{s+\lambda}{\beta}} \hat{f}\left(\frac{s+\lambda}{\beta}\right)$	Power law transform
7	$f'_x(x)$	$-(s-1)\hat{f}(s-1)$	Differentiation
8	$x f'_x(x)$	$-s\hat{f}(s)$	Differentiation
9	$f_x^{(n)}(x)$	$(-1)^n \dfrac{\Gamma(s)}{\Gamma(s-n)} \hat{f}(s-n)$	Multiple differentiation
10	$\left(x\dfrac{d}{dx}\right)^n f(x)$	$(-1)^n s^n \hat{f}(s)$	Multiple differentiation
11	$x^\alpha \displaystyle\int_0^\infty t^\beta f_1(xt) f_2(t)\,dt$	$\hat{f}_1(s+\alpha)\hat{f}_2(1-s-\alpha+\beta)$	Complicated integration
12	$x^\alpha \displaystyle\int_0^\infty t^\beta f_1\left(\dfrac{x}{t}\right) f_2(t)\,dt$	$\hat{f}_1(s+\alpha)\hat{f}_2(s+\alpha+\beta+1)$	Complicated integration

▶ **Relation among the Mellin, Laplace, and Fourier transforms.** There are tables of direct and inverse Mellin transforms (see the references listed at the end of the current chapter) that are useful in solving specific integral and differential equations. The Mellin transform is related to the Laplace and Fourier transforms by

$$\mathfrak{M}\{f(x), s\} = \mathfrak{L}\{f(e^x), -s\} + \mathfrak{L}\{f(e^{-x}), s\} = \mathfrak{F}\{f(e^x), is\},$$

which makes it possible to apply much more common tables of direct and inverse Laplace and Fourier transforms.

M10.4.3. Summary Table of Integral Transforms

Table M10.5 summarizes the integral transforms considered above and also lists some other integral transforms; for the constraints imposed on the functions and parameters occurring in the integrand, see the references given at the end of this section.

TABLE M10.5
Summary table of integral transforms.

Integral transform	Definition	Inversion formula
Laplace transform	$\tilde{f}(p) = \int_0^\infty e^{-px} f(x)\, dx$	$f(x) = \dfrac{1}{2\pi i} \int_{c-i\infty}^{c+i\infty} e^{px} \tilde{f}(p)\, dp$
Laplace–Carson transform	$\tilde{f}(p) = p \int_0^\infty e^{-px} f(x)\, dx$	$f(x) = \dfrac{1}{2\pi i} \int_{c-i\infty}^{c+i\infty} e^{px} \dfrac{\tilde{f}(p)}{p}\, dp$
Two-sided Laplace transform	$\tilde{f}_*(p) = \int_{-\infty}^\infty e^{-px} f(x)\, dx$	$f(x) = \dfrac{1}{2\pi i} \int_{c-i\infty}^{c+i\infty} e^{px} \tilde{f}_*(p)\, dp$
Fourier transform	$\tilde{f}(u) = \dfrac{1}{\sqrt{2\pi}} \int_{-\infty}^\infty e^{-iux} f(x)\, dx$	$f(x) = \dfrac{1}{\sqrt{2\pi}} \int_{-\infty}^\infty e^{iux} \tilde{f}(u)\, du$
Fourier sine transform	$\tilde{f}_s(u) = \sqrt{\dfrac{2}{\pi}} \int_0^\infty \sin(xu) f(x)\, dx$	$f(x) = \sqrt{\dfrac{2}{\pi}} \int_0^\infty \sin(xu) \tilde{f}_s(u)\, du$
Fourier cosine transform	$\tilde{f}_c(u) = \sqrt{\dfrac{2}{\pi}} \int_0^\infty \cos(xu) f(x)\, dx$	$f(x) = \sqrt{\dfrac{2}{\pi}} \int_0^\infty \cos(xu) \tilde{f}_c(u)\, du$
Hartley transform	$\tilde{f}_h(u) = \dfrac{1}{\sqrt{2\pi}} \int_{-\infty}^\infty (\cos xu + \sin xu) f(x)\, dx$	$f(x) = \dfrac{1}{\sqrt{2\pi}} \int_{-\infty}^\infty (\cos xu + \sin xu) \tilde{f}_h(u)\, du$
Mellin transform	$\hat{f}(s) = \int_0^\infty x^{s-1} f(x)\, dx$	$f(x) = \dfrac{1}{2\pi i} \int_{c-i\infty}^{c+i\infty} x^{-s} \hat{f}(s)\, ds$
Hankel transform	$\hat{f}_\nu(w) = \int_0^\infty x J_\nu(xw) f(x)\, dx$	$f(x) = \int_0^\infty w J_\nu(xw) \hat{f}_\nu(w)\, dw$
Y-transform	$F_\nu(u) = \int_0^\infty \sqrt{ux}\, Y_\nu(ux) f(x)\, dx$	$f(x) = \int_0^\infty \sqrt{ux}\, \mathbf{H}_\nu(ux) F_\nu(u)\, du$
Meijer transform (K-transform)	$\hat{f}(s) = \sqrt{\dfrac{2}{\pi}} \int_0^\infty \sqrt{sx}\, K_\nu(sx) f(x)\, dx$	$f(x) = \dfrac{1}{i\sqrt{2\pi}} \int_{c-i\infty}^{c+i\infty} \sqrt{sx}\, I_\nu(sx) \hat{f}(s)\, ds$
Kontorovich–Lebedev transform	$F(\tau) = \int_0^\infty K_{i\tau}(x) f(x)\, dx$	$f(x) = \dfrac{2}{\pi^2 x} \int_0^\infty \tau \sinh(\pi\tau) K_{i\tau}(x) F(\tau)\, d\tau$

NOTATION: $i = \sqrt{-1}$; $J_\mu(x)$ and $Y_\mu(x)$ are the Bessel functions of the first and the second kind, respectively; $I_\mu(x)$ and $K_\mu(x)$ are the modified Bessel functions of the first and the second kind, respectively; and $\mathbf{H}_\nu(x) = \displaystyle\sum_{j=0}^\infty \dfrac{(-1)^j (x/2)^{\nu+2j+1}}{\Gamma\left(j + \frac{3}{2}\right) \Gamma\left(\nu + j + \frac{3}{2}\right)}$ is the Struve function.

Bibliography for Chapter M10

Bateman, H. and Erdélyi, A., *Tables of Integral Transforms. Vols. 1 and 2*, McGraw-Hill, New York, 1954.

Beerends, R. J., ter Morschem, H. G., and van den Berg, J. C., *Fourier and Laplace Transforms*, Cambridge University Press, Cambridge, England, 2003.

Bellman, R. and Roth, R., *The Laplace Transform*, World Scientific Publishing Co., Singapore, 1984.

Ditkin, V. A. and Prudnikov, A. P., *Integral Transforms and Operational Calculus*, Pergamon Press, New York, 1965.

Oberhettinger, F., *Tables of Fourier Transforms and Fourier Transforms of Distributions*, Springer-Verlag, Berlin, 1980.

Oberhettinger, F. and Badii, L., *Tables of Laplace Transforms*, Springer-Verlag, New York, 1973.

Prudnikov, A. P., Brychkov, Yu. A., and Marichev, O. I., *Integrals and Series, Vol. 4, Direct Laplace Transform*, Gordon & Breach, New York, 1992.

Prudnikov, A. P., Brychkov, Yu. A., and Marichev, O. I., *Integrals and Series, Vol. 5, Inverse Laplace Transform*, Gordon & Breach, New York, 1992.

Sneddon, I., *Fourier Transforms*, Dover Publications, New York, 1995.

Chapter M11

Ordinary Differential Equations

M11.1. First-Order Differential Equations

M11.1.1. General Concepts. The Cauchy Problem. Uniqueness and Existence Theorems

▶ **Equations solved for the derivative. General solution.** A *first-order ordinary differential equation** solved for the derivative has the form

$$y'_x = f(x, y). \tag{M11.1.1.1}$$

Sometimes it is represented in terms of differentials as $dy = f(x, y)\, dx$.

A *solution of a differential equation* is a function $y(x)$ that, when substituted into the equation, turns it into an identity. The *general solution of a differential equation* is the set of all its solutions. In some cases, the general solution can be represented as a function $y = \varphi(x, C)$ that depends on one *arbitrary constant* C; specific values of C define specific solutions of the equation (*particular solutions*). In practice, the general solution more frequently appears in implicit form, $\Phi(x, y, C) = 0$, or parametric form, $x = x(t, C)$, $y = y(t, C)$.

Geometrically, the general solution (also called the general integral) of an equation is a family of curves in the xy-plane depending on a single parameter C; these curves are called *integral curves* of the equation. To each particular solution (particular integral) there corresponds a single curve that passes through a given point (x_0, y_0) in the plane.

For each point (x, y), the equation $y'_x = f(x, y)$ defines a value of y'_x, i.e., the slope of the integral curve that passes through this point. In other words, the equation generates a field of directions in the xy-plane. From the geometrical point of view, the problem of solving a first-order differential equation involves finding the curves, the slopes of which at each point coincide with the direction of the field at this point.

Figure M11.1 depicts the tangent to an integral curve at a point (x_0, y_0); the slope of the integral curve at this point is determined by the right-hand side of equation (M11.1.1.1): $\tan \alpha = f(x_0, y_0)$. The little segments show the field of tangents to the integral curves of the differential equation (M11.1.1.1) at other points.

▶ **Equations integrable by quadrature.** The process of finding a solution to a differential equation is called integration of this equation. The problem of integration of equation (M11.1.1.1) can often be reduced to the problem of finding indefinite integrals, or quadratures. A solution is expressed as a quadrature if it expressed in terms of elementary functions and the functions appearing in the equation using a finite set of the arithmetic operations, function compositions, and indefinite integrals. An equation is said to be integrable by quadrature if its general solution can be expressed in terms of quadratures.

---

* In what follows, we often call an ordinary differential equation a "differential equation" or, even shorter, an "equation."

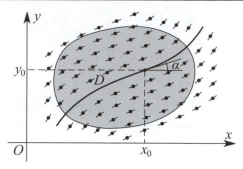

Figure M11.1. The direction field of a differential equation and the integral curve passing through a point (x_0, y_0).

▶ **Cauchy problem. The uniqueness and existence theorems.** The *Cauchy problem*: find a solution of equation (M11.1.1.1) that satisfies the *initial condition*

$$y = y_0 \quad \text{at} \quad x = x_0, \tag{M11.1.1.2}$$

where y_0 and x_0 are some numbers.

Geometrical meaning of the Cauchy problem: find an integral curve of equation (M11.1.1.1) that passes through the point (x_0, y_0); see Fig. M11.1.

Condition (M11.1.1.2) is alternatively written $y(x_0) = y_0$ or $y|_{x=x_0} = y_0$.

THEOREM (EXISTENCE, PEANO). *Let the function $f(x, y)$ be continuous in an open domain D of the xy-plane. Then there is at least one integral curve of equation (M11.1.1.1) that passes through a point $(x_0, y_0) \in D$; each of these curves can be extended at both ends up to the boundary of any closed domain $D_0 \subset D$ such that (x_0, y_0) belongs to the interior of D_0.*

THEOREM (UNIQUENESS). *Let the function $f(x, y)$ be continuous in an open domain D and have in D a bounded partial derivative with respect to y (or the Lipschitz condition holds: $|f(x, y) - f(x, z)| \le M|y - z|$, where M is some positive number and $(x, z) \in D$). Then there is a unique solution of equation (M11.1.1.1) satisfying condition (M11.1.1.2).*

▶ **Equations not solved for the derivative. The existence theorem.** A first-order differential equation not solved for the derivative can generally be written as

$$F(x, y, y'_x) = 0. \tag{M11.1.1.3}$$

THEOREM (EXISTENCE AND UNIQUENESS). *There exists a unique solution $y = y(x)$ of equation (M11.1.1.3) satisfying the conditions $y|_{x=x_0} = y_0$ and $y'_x|_{x=x_0} = t_0$, where t_0 is one of the real roots of the equation $F(x_0, y_0, t_0) = 0$ if the following conditions hold in a neighborhood of the point (x_0, y_0, t_0):*

1. *The function $F(x, y, t)$ is continuous in each of the three arguments.*
2. *The partial derivative F_t exists and is nonzero.*
3. *There is a bounded partial derivative with respect to y, $|F_y| \le M$.*

The solution surely exists if $|x - x_0| \le a$, where a is a (sufficiently small) positive number.

▶ **Singular solutions.** A point (x, y) at which the uniqueness of the solution to equation (M11.1.1.3) is violated is called a *singular* point. If conditions 1 and 3 of the existence and uniqueness theorem hold, then

$$F(x, y, t) = 0, \quad F_t(x, y, t) = 0 \tag{M11.1.1.4}$$

simultaneously at each singular point. Relations (M11.1.1.4) define a *t-discriminant curve* in parametric form. In some cases, the parameter t can be eliminated from (M11.1.1.4) to give an equation of this curve in implicit form, $\Psi(x, y) = 0$. If a branch $y = \psi(x)$ of the curve $\Psi(x, y) = 0$ consists of singular points and, at the same time, is an integral curve, then this branch is called a *singular integral curve* and the function $y = \psi(x)$ is a *singular solution* of equation (M11.1.1.3).

The singular solutions can be obtained by finding the *envelope of the family of integral curves*, $\Phi(x, y, C) = 0$, of equation (M11.1.1.3). The envelope is part of the *C-discriminant curve*, which is defined by the equations

$$\Phi(x, y, C) = 0, \qquad \Phi_C(x, y, C) = 0.$$

The branch of the *C*-discriminant curve at which

(a) there exist bounded partial derivatives, $|\Phi_x| < M_1$ and $|\Phi_y| < M_2$, and
(b) $|\Phi_x| + |\Phi_y| \neq 0$

is the envelope.

M11.1.2. Equations Solved for the Derivative. Simplest Techniques of Integration

▶ **Equations with separated or separable variables.**

$1°$. An *equation with separated variables* (a *separated equation*) has the form

$$f(y)y'_x = g(x).$$

Equivalently, the equation can be rewritten as $f(y)\, dy = g(x)\, dx$ (the right-hand side depends on x alone and the left-hand side on y alone). The general solution can be obtained by integration:

$$\int f(y)\, dy = \int g(x)\, dx + C,$$

where C is an arbitrary constant.

$2°$. An *equation with separable variables* (a *separable equation*) is generally represented by

$$f_1(y)g_1(x)y'_x = f_2(y)g_2(x).$$

Dividing the equation by $f_2(y)g_1(x)$, one obtains a separated equation. Integrating yields

$$\int \frac{f_1(y)}{f_2(y)}\, dy = \int \frac{g_2(x)}{g_1(x)}\, dx + C.$$

Remark. Solutions corresponding to $f_2(y) = 0$ may be lost when dividing the equation by $f_2(y)g_1(x)$. Therefore, the case of $f_2(y) = 0$, should be treated separately.

Many ordinary differential equations are reduced to separable equations.

Example. The equation

$$y'_x = yf(e^{\lambda x}y^k)$$

can be reduced, with the substitution $z = e^{\lambda x}y^k$, to a separable equation: $z'_x = \lambda z + kzf(z)$.

Some other equations reducible to separable equations are considered below.

▶ **Equation of the form** $y'_x = f(ax + by)$. For $b = 0$, it is an equation with separated variables. For $b \neq 0$, the substitution $z = ax + by$ brings the equation to a separable equation, $z'_x = bf(z) + a$.

▶ **Homogeneous equation.** A *homogeneous equation* remains the same under simultaneous scaling (dilation) of the independent and dependent variables in accordance with the rule $x \rightarrow \alpha x$, $y \rightarrow \alpha y$, where α is an arbitrary constant ($\alpha \neq 0$). Such equations can be represented in the form

$$y'_x = f\left(\frac{y}{x}\right).$$

The substitution $u = y/x$ brings a homogeneous equation to a separable one, $xu'_x = f(u) - u$.

Remark. The equations of the form

$$y'_x = f\left(\frac{a_1 x + b_1 y + c_1}{a_2 x + b_2 y + c_2}\right)$$

can be reduced to a homogeneous equation. To this end, for $a_1 x + b_1 y \neq k(a_2 x + b_2 y)$, one should use the change of variables $\xi = x - x_0$, $\eta = y - y_0$, where the constants x_0 and y_0 are determined by solving the linear algebraic system

$$a_1 x_0 + b_1 y_0 + c_1 = 0,$$
$$a_2 x_0 + b_2 y_0 + c_2 = 0.$$

As a result, one arrives at the following equation for $\eta = \eta(\xi)$:

$$\eta'_\xi = f\left(\frac{a_1 \xi + b_1 \eta}{a_2 \xi + b_2 \eta}\right).$$

On dividing the numerator and denominator of the argument of f by ξ, one obtains a homogeneous equation whose right-hand side is dependent on the ratio η/ξ only:

$$\eta'_\xi = f\left(\frac{a_1 + b_1 \eta/\xi}{a_2 + b_2 \eta/\xi}\right).$$

▶ **Generalized homogeneous equation.**

$1°$. A *generalized homogeneous equation* remains the same under simultaneous scaling of the independent and dependent variables in accordance with the rule $x \rightarrow \alpha x$, $y \rightarrow \alpha^k y$, where $\alpha \neq 0$ is an arbitrary constant and k is some number. Such equations can be represented in the form

$$y'_x = x^{k-1} f(yx^{-k}).$$

The substitution $u = yx^{-k}$ brings a generalized homogeneous equation to a separable equation, $xu'_x = f(u) - ku$.

Example. Consider the equation

$$y'_x = ax^2 y^4 + by^2. \tag{M11.1.2.1}$$

Let us perform the transformation $x = \alpha \bar{x}$, $y = \alpha^k \bar{y}$ and then multiply the resulting equation by α^{1-k} to obtain

$$\bar{y}'_{\bar{x}} = a\alpha^{3(k+1)} \bar{x}^2 \bar{y}^4 + b\alpha^{k+1} \bar{y}^2. \tag{M11.1.2.2}$$

It is apparent that if $k = -1$, the transformed equation (M11.1.2.2) is the same as the original one, up to notation. This means that equation (M11.1.2.1) is generalized homogeneous of degree $k = -1$. Therefore, the substitution $u = xy$ brings it to a separable equation: $xu'_x = au^4 + bu^2 + u$.

$2°$. Alternatively, a generalized homogeneous equation can be represented as

$$y'_x = \frac{y}{x} f(x^n y^m).$$

The substitution $z = x^n y^m$ leads to a separable equation: $xz'_x = nz + mzf(z)$.

▶ **Linear equation.** A first-order *linear equation* is written as

$$y'_x + f(x)y = g(x). \tag{M11.1.2.3}$$

The solution is sought in the product form $y = uv$, where $v = v(x)$ is any function that satisfies the "truncated" equation $v'_x + f(x)v = 0$ [as $v(x)$ one takes the particular solution $v = e^{-F}$, where $F = \int f(x)\,dx$]. As a result, one obtains the following separable equation for $u = u(x)$: $v(x)u'_x = g(x)$. Integrating it yields the general solution:

$$y(x) = e^{-F}\left(\int e^F g(x)\,dx + C\right), \qquad F = \int f(x)\,dx,$$

where C is an arbitrary constant.

▶ **Bernoulli equation.** A *Bernoulli equation* has the form

$$y'_x + f(x)y = g(x)y^a, \qquad a \neq 0, 1. \tag{M11.1.2.4}$$

(For $a = 0$ and $a = 1$, it is a linear equation; see above.) The substitution $z = y^{1-a}$ brings it to a linear equation, $z'_x + (1-a)f(x)z = (1-a)g(x)$. With this in view, one can obtain the general integral:

$$y^{1-a} = Ce^{-F} + (1-a)e^{-F}\int e^F g(x)\,dx, \qquad \text{where} \quad F = (1-a)\int f(x)\,dx.$$

▶ **Equation with exponential nonlinearity reducible to a linear equation.** Consider the equation

$$y'_x = f(x)e^{\lambda y} + g(x).$$

The substitution $u = e^{-\lambda y}$ leads to a linear equation: $u'_x = -\lambda g(x)u - \lambda f(x)$.

M11.1.3. Exact Differential Equations. Integrating Factor

▶ **Exact differential equations.** An *exact differential equation* has the form

$$f(x, y)\,dx + g(x, y)\,dy = 0, \quad \text{where} \quad \frac{\partial f}{\partial y} = \frac{\partial g}{\partial x}. \tag{M11.1.3.1}$$

The left-hand side of the equation is the total differential of a function of two variables $U(x, y)$. In this case, the general integral is given by

$$U(x, y) = C,$$

where C is an arbitrary constant and the function U is determined from the system

$$\frac{\partial U}{\partial x} = f, \qquad \frac{\partial U}{\partial y} = g.$$

Integrating the first equation yields $U = \int f(x, y)\,dx + \Psi(y)$ (while integrating, the variable y is treated as a parameter). On substituting this expression into the second equation, one identifies the function Ψ (and hence, U). As a result, the general integral of an exact differential equation can be represented in the form

$$\int_{x_0}^{x} f(\xi, y)\,d\xi + \int_{y_0}^{y} g(x_0, \eta)\,d\eta = C, \tag{M11.1.3.2}$$

where x_0 and y_0 are any numbers (from the domain of definition of the equation).

TABLE M11.1

An integrating factor $\mu = \mu(x, y)$ for some types of ordinary differential equations $f\,dx + g\,dy = 0$, where $f = f(x, y)$ and $g = g(x, y)$. The subscripts x and y indicate the corresponding partial derivatives.

No.	Conditions for f and g	Integrating factor	Remarks
1	$f = y\varphi(xy),\ g = x\psi(xy)$	$\mu = \frac{1}{xf-yg}$	$xf - yg \not\equiv 0$; $\varphi(z)$ and $\psi(z)$ are any functions
2	$f_x = g_y,\ f_y = -g_x$	$\mu = \frac{1}{f^2+g^2}$	$f + ig$ is an analytic function of the complex variable $x + iy$
3	$\frac{f_y-g_x}{g} = \varphi(x)$	$\mu = \exp\left[\int \varphi(x)\,dx\right]$	$\varphi(x)$ is any function
4	$\frac{f_y-g_x}{f} = \varphi(y)$	$\mu = \exp\left[-\int \varphi(y)\,dy\right]$	$\varphi(y)$ is any function
5	$\frac{f_y-g_x}{g-f} = \varphi(x+y)$	$\mu = \exp\left[\int \varphi(z)\,dz\right],\ z = x+y$	$\varphi(z)$ is any function
6	$\frac{f_y-g_x}{yg-xf} = \varphi(xy)$	$\mu = \exp\left[\int \varphi(z)\,dz\right],\ z = xy$	$\varphi(z)$ is any function
7	$\frac{x^2(f_y-g_x)}{yg+xf} = \varphi\left(\frac{y}{x}\right)$	$\mu = \exp\left[-\int \varphi(z)\,dz\right],\ z = \frac{y}{x}$	$\varphi(z)$ is any function
8	$\frac{f_y-g_x}{xg-yf} = \varphi(x^2+y^2)$	$\mu = \exp\left[\frac{1}{2}\int \varphi(z)\,dz\right],\ z = x^2+y^2$	$\varphi(z)$ is any function
9	$f_y - g_x = \varphi(x)g - \psi(y)f$	$\mu = \exp\left[\int \varphi(x)\,dx + \int \psi(y)\,dy\right]$	$\varphi(x)$ and $\psi(y)$ are any functions

Example. Consider the equation

$$(ay^n + bx)y_x' + by + cx^m = 0, \quad \text{or} \quad (by + cx^m)\,dx + (ay^n + bx)\,dy = 0,$$

defined by the functions $f(x, y) = by + cx^m$ and $g(x, y) = ay^n + bx$. Computing the derivatives, we have

$$\frac{\partial f}{\partial y} = b, \quad \frac{\partial g}{\partial x} = b \quad \Longrightarrow \quad \frac{\partial f}{\partial y} = \frac{\partial g}{\partial x}.$$

Hence the given equation is an exact differential equation. Its solution can be found using formula (M11.1.3.2) with $x_0 = y_0 = 0$:

$$\frac{a}{n+1}y^{n+1} + bxy + \frac{c}{m+1}x^{m+1} = C.$$

▶ **Integrating factor.** An *integrating factor* for the equation

$$f(x, y)\,dx + g(x, y)\,dy = 0$$

is a function $\mu(x, y) \neq 0$ such that the left-hand side of the equation, when multiplied by $\mu(x, y)$, becomes a total differential, and the equation itself becomes an exact differential equation.

An integrating factor satisfies the first-order partial differential equation,

$$g\frac{\partial \mu}{\partial x} - f\frac{\partial \mu}{\partial y} = \left(\frac{\partial f}{\partial y} - \frac{\partial g}{\partial x}\right)\mu,$$

which is not generally easier to solve than the original equation.

Table M11.1 lists some special cases where an integrating factor can be found in explicit form.

M11.1.4. Riccati Equation

▶ **General Riccati equation.** A *Riccati equation* has the general form

$$y'_x = f_2(x)y^2 + f_1(x)y + f_0(x). \tag{M11.1.4.1}$$

If $f_2 \equiv 0$, we have a linear equation, and if $f_0 \equiv 0$, we have a Bernoulli equation (see equation (M11.1.2.4) for $a = 2$), whose solutions were given previously. For arbitrary f_2, f_1, and f_0, the Riccati equation is not integrable by quadrature.

▶ **Use of particular solutions to construct the general solution.**

1°. Given a particular solution $y_0 = y_0(x)$ of the Riccati equation (M11.1.4.1), the general solution can be written as

$$y = y_0(x) + \Phi(x)\left[C - \int \Phi(x)f_2(x)\,dx\right]^{-1}, \tag{M11.1.4.2}$$

where C is an arbitrary constant and

$$\Phi(x) = \exp\left\{\int \left[2f_2(x)y_0(x) + f_1(x)\right]\,dx\right\}. \tag{M11.1.4.3}$$

To the particular solution $y_0(x)$ there corresponds $C = \infty$.

2°. Let $y_1 = y_1(x)$ and $y_2 = y_2(x)$ be two different particular solutions of equation (M11.1.4.1). Then the general solution can be expressed as

$$y = \frac{Cy_1 + U(x)y_2}{C + U(x)}, \quad \text{where} \quad U(x) = \exp\left[\int f_2(y_1 - y_2)\,dx\right].$$

To the particular solution $y_1(x)$, there corresponds $C = \infty$; and to $y_2(x)$, there corresponds $C = 0$.

3°. Let $y_1 = y_1(x)$, $y_2 = y_2(x)$, and $y_3 = y_3(x)$ be three distinct particular solutions of equation (M11.1.4.1). Then the general solution can be found without quadrature:

$$\frac{y - y_2}{y - y_1}\frac{y_3 - y_1}{y_3 - y_2} = C.$$

▶ **Some transformations.**

1°. The transformation (φ, ψ_1, ψ_2, ψ_3, and ψ_4 are arbitrary functions)

$$x = \varphi(\xi), \quad y = \frac{\psi_4(\xi)u + \psi_3(\xi)}{\psi_2(\xi)u + \psi_1(\xi)}$$

reduces the Riccati equation (M11.1.4.1) to a Riccati equation for $u = u(\xi)$.

2°. Let $y_0 = y_0(x)$ be a particular solution of equation (M11.1.4.1). Then the substitution $y = y_0 + 1/w$ leads to a first-order linear equation for $w = w(x)$:

$$w'_x + \left[2f_2(x)y_0(x) + f_1(x)\right]w + f_2(x) = 0.$$

For solution of linear equations, see Subsection M11.1.2.

▶ **Reduction of the Riccati equation to a second-order linear equation.** The substitution

$$u(x) = \exp\left(-\int f_2 y\,dx\right)$$

reduces the general Riccati equation (M11.1.4.1) to a second-order linear equation:

$$f_2 u''_{xx} - \left[(f_2)'_x + f_1 f_2\right]u'_x + f_0 f_2^2 u = 0,$$

which often may be easier to solve than the original Riccati equation.

M11.1.5. Equations Not Solved for the Derivative

▶ **Method of "integration by differentiation."** In the general case, a first-order equation not solved for the derivative,

$$F(x, y, y'_x) = 0, \tag{M11.1.5.1}$$

can be rewritten in the equivalent form

$$F(x, y, t) = 0, \qquad t = y'_x. \tag{M11.1.5.2}$$

We look for a solution in parametric form: $x = x(t)$, $y = y(t)$. In accordance with the first relation in (M11.1.5.2), the differential of F is given by

$$F_x \, dx + F_y \, dy + F_t \, dt = 0. \tag{M11.1.5.3}$$

Using the relation $dy = t \, dx$, we eliminate successively dy and dx from (M11.1.5.3). As a result, we obtain the system of two first-order ordinary differential equations:

$$\frac{dx}{dt} = -\frac{F_t}{F_x + tF_y}, \qquad \frac{dy}{dt} = -\frac{tF_t}{F_x + tF_y}. \tag{M11.1.5.4}$$

By finding a solution of this system, one thereby obtains a solution of the original equation (M11.1.5.1) in parametric form, $x = x(t)$, $y = y(t)$.

Remark 1. The application of the above method may lead to loss of individual solutions (satisfying the condition $F_x + tF_y = 0$); this issue requires further investigation.

Remark 2. One of the differential equations of system (M11.1.5.4) can be replaced by the algebraic equation $F(x, y, t) = 0$; see equation (M11.1.5.2). This technique is used further for solving some equations.

▶ **Equations of the form** $y = f(y'_x)$**.** This equation is a special case of equation (M11.1.5.1), with $F(x, y, t) = y - f(t)$. The method of "integration by differentiation" yields

$$\frac{dx}{dt} = \frac{f'(t)}{t}, \qquad y = f(t). \tag{M11.1.5.5}$$

Note the original equation is used here instead of the second equation in system (M11.1.5.4); this is convenient because the first equation in (M11.1.5.4) does not depend on y explicitly.

Integrating the first equation in (M11.1.5.5) yields the solution in parametric form,

$$x = \int \frac{f'(t)}{t} \, dt + C, \qquad y = f(t).$$

▶ **Equations of the form** $x = f(y'_x)$**.** This equation is a special case of equation (M11.1.5.1), with $F(x, y, t) = x - f(t)$. The method of "integration by differentiation" yields

$$x = f(t), \qquad \frac{dy}{dt} = tf'(t). \tag{M11.1.5.6}$$

Note the original equation is used here instead of the first equation in system (M11.1.5.4); this is convenient because the second equation in (M11.1.5.4) does not depend on x explicitly.

Integrating the second equation in (M11.1.5.6) yields the solution in parametric form,

$$x = f(t), \qquad y = \int tf'(t) \, dt + C.$$

▶ **Clairaut's equation** $y = xy'_x + f(y'_x)$. *Clairaut's equation* is a special case of equation (M11.1.5.1), with $F(x, y, t) = y - xt - f(t)$. It can be rewritten as

$$y = xt + f(t), \qquad t = y'_x. \tag{M11.1.5.7}$$

This equation corresponds to the degenerate case $F_x + tF_y \equiv 0$, where system (M11.1.5.4) cannot be obtained. One should proceed in the following way: the first relation in (M11.1.5.7) gives $dy = x \, dt + t \, dx + f'(t) \, dt$; performing the substitution $dy = t \, dx$, which follows from the second relation in (M11.1.5.7), one obtains

$$[x + f'(t)] \, dt = 0.$$

This equation splits into $dt = 0$ and $x + f'(t) = 0$. The solution of the first equation is obvious: $t = C$; it gives the general solution of Clairaut's equation,

$$y = Cx + f(C), \tag{M11.1.5.8}$$

which is a family of straight lines. The second equation generates a solution in parametric form,

$$x = -f'(t), \qquad y = -tf'(t) + f(t), \tag{M11.1.5.9}$$

which is a singular solution and is the envelope of the family of lines (M11.1.5.8).

Remark. There are also "compound" solutions of Clairaut's equation; they consist of part of the curve (M11.1.5.9) joined with the tangents at finite points; these tangents are defined by formula (M11.1.5.8).

▶ **Lagrange's equation** $y = xf(y'_x) + g(y'_x)$. *Lagrange's equation* is a special case of equation (M11.1.5.1), with $F(x, y, t) = y - xf(t) - g(t)$. In the special case $f(t) \equiv t$, it coincides with Clairaut's equation.

The method of "integration by differentiation" yields

$$\frac{dx}{dt} + \frac{f'(t)}{f(t) - t}x = \frac{g'(t)}{t - f(t)}, \qquad y = xf(t) + g(t). \tag{M11.1.5.10}$$

Here, the original equation is used instead of the second equation in system (M11.1.5.4); this is convenient because the first equation in (M11.1.5.4) does not depend on y explicitly.

The first equation of system (M11.1.5.10) is linear and can easily be integrated to obtain a solution to Lagrange's equation in parametric form.

Remark. With the above method, solutions of the form $y = t_k x + g(t_k)$, where the t_k are roots of the equation $f(t) - t = 0$, may be lost. These solutions can be particular or singular solutions of Lagrange's equation.

M11.1.6. Approximate Analytic Methods for Solution of Equations

▶ **Method of successive approximations (Picard's method).** The method of successive approximations consists of two stages. At the first stage, the Cauchy problem

$$y'_x = f(x, y) \qquad \text{(equation)}, \tag{M11.1.6.1}$$

$$y(x_0) = y_0 \qquad \text{(initial condition)} \tag{M11.1.6.2}$$

is reduced to the equivalent integral equation:

$$y(x) = y_0 + \int_{x_0}^{x} f(t, y(t)) \, dt. \tag{M11.1.6.3}$$

Then a solution of equation (M11.1.6.3) is sought using the formula of successive approximations:

$$y_{n+1}(x) = y_0 + \int_{x_0}^{x} f(t, y_n(t)) \, dt; \qquad n = 0, 1, 2, \ldots$$

The initial approximation $y_0(x)$ can be chosen arbitrarily; the simplest way is to take $y_0(x) = y_0$. The iterative process converges as $n \to \infty$, provided the conditions of the theorems in Subsection M11.1.1 are satisfied.

▶ **Method of Taylor series expansion in the independent variable.** A solution of the Cauchy problem (M11.1.6.1)–(M11.1.6.2) can be sought in the form of the Taylor series in powers of $(x - x_0)$:

$$y(x) = y(x_0) + y'_x(x_0)(x - x_0) + \frac{y''_{xx}(x_0)}{2!}(x - x_0)^2 + \cdots . \qquad \text{(M11.1.6.4)}$$

The first term $y(x_0)$ in solution (M11.1.6.4) is prescribed by the initial condition (M11.1.6.2). The values of the derivatives of $y(x)$ at $x = x_0$ are determined from equation (M11.1.6.1) and its derivative equations (obtained by successive differentiation), taking into account the initial condition (M11.1.6.2). In particular, setting $x = x_0$ in (M11.1.6.1) and substituting (M11.1.6.2), one obtains the value of the first derivative:

$$y'_x(x_0) = f(x_0, y_0). \qquad \text{(M11.1.6.5)}$$

Further, differentiating equation (M11.1.6.1) yields

$$y''_{xx} = f_x(x, y) + f_y(x, y)y'_x. \qquad \text{(M11.1.6.6)}$$

On substituting $x = x_0$, as well as the initial condition (M11.1.6.2) and the first derivative (M11.1.6.5), into the right-hand side of this equation, one calculates the value of the second derivative:

$$y''_{xx}(x_0) = f_x(x_0, y_0) + f(x_0, y_0)f_y(x_0, y_0).$$

Likewise, one can determine the subsequent derivatives of y at $x = x_0$.

Solution (M11.1.6.4) obtained by this method can normally be used in only some sufficiently small neighborhood of the point $x = x_0$.

Example. Consider the Cauchy problem for the equation

$$y'_x = e^y + \cos x$$

with the initial condition $y(0) = 0$.

Since $x_0 = 0$, we will be constructing a series in powers of x. It follows from the equation that $y'_x(0) = e^0 + \cos 0 = 2$. Differentiating the original equation yields $y''_{xx} = e^y y'_x - \sin x$. Using the initial condition and the condition $y'_x(0) = 2$ just obtained, we have $y''_{xx}(0) = e^0 \times 2 - \sin 0 = 2$. Similarly, we find that $y'''_{xxx} = e^y y''_{xx} + e^y(y'_x)^2 - \cos x$, whence $y'''_{xxx}(0) = e^0 \times 2 + e^0 \times 2^2 - \cos 0 = 5$.

Substituting the values of the derivatives at $x = 0$ into series (M11.1.6.4), we obtain the desired series representation of the solution: $y = 2x + x^2 + \frac{5}{6}x^3 + \cdots$.

M11.1.7. Numerical Integration of Differential Equations

▶ **Method of Euler polygonal lines.** Consider the Cauchy problem for the first-order differential equation

$$y'_x = f(x, y)$$

with the initial condition $y(x_0) = y_0$. Our aim is to construct an approximate solution $y = y(x)$ of this equation on an interval $[x_0, x_*]$.

Let us split the interval $[x_0, x_*]$ into n equal segments of length $\Delta x = \dfrac{x_* - x_0}{n}$. We seek approximate values $y_1, y_2, \ldots, y_n$ of the solution $y(x)$ at the partition points $x_1, x_2, \ldots, x_n = x_*$.

For a given initial value $y_0 = y(x_0)$ and a sufficiently small Δx, the values of the unknown function $y_k = y(x_k)$ at the other points $x_k = x_0 + k\Delta x$ are calculated successively by the formula

$$y_{k+1} = y_k + f(x_k, y_k)\Delta x \qquad \textit{(Euler polygonal line)},$$

where $k = 0, 1, \ldots, n - 1$. The Euler method is a single-step method of the first-order approximation (with respect to the step Δx).

▶ **Single-step methods of the second-order approximation.** Two single-step methods for solving the Cauchy problem in the second-order approximation are specified by the recurrence formulas

$$y_{k+1} = y_k + f\left(x_k + \tfrac{1}{2}\Delta x,\ y_k + \tfrac{1}{2}f_k\Delta x\right)\Delta x \quad \text{(first method)},$$

$$y_{k+1} = y_k + \tfrac{1}{2}\left[f_k + f(x_{k+1},\ y_k + f_k\Delta x)\right]\Delta x \quad \text{(second method)},$$

where $f_k = f(x_k, y_k)$; $k = 0,\ 1,\ \ldots,\ n-1$.

▶ **Runge–Kutta method of the fourth-order approximation.** This is one of the widely used methods. The unknown values y_k are successively found by the formulas

$$y_{k+1} = y_k + \tfrac{1}{6}(f_1 + 2f_2 + 2f_3 + f_4)\Delta x,$$

where

$$f_1 = f(x_k,\ y_k), \qquad f_2 = f(x_k + \tfrac{1}{2}\Delta x,\ y_k + \tfrac{1}{2}f_1\Delta x),$$

$$f_3 = f(x_k + \tfrac{1}{2}\Delta x,\ y_k + \tfrac{1}{2}f_2\Delta x), \qquad f_4 = f(x_k + \Delta x,\ y_k + f_3\Delta x).$$

Remark 1. All methods described in Subsection M11.1.7 are special cases of the Runge–Kutta method (a detailed description of this method can be found in the monographs listed at the end of the current chapter).

Remark 2. In practice, calculations are performed on the basis of any of the above recurrence formulas with two different steps Δx, $\tfrac{1}{2}\Delta x$ and an arbitrarily chosen small Δx. Then one compares the results obtained at common points. If these results coincide within the given order of accuracy, one assumes that the chosen step Δx ensures the desired accuracy of calculations. Otherwise, the step is halved and the calculations are performed with the steps $\tfrac{1}{2}\Delta x$ and $\tfrac{1}{4}\Delta x$, after which the results are compared again, etc. (Quite often, one compares the results of calculations with steps varying by a factor of ten or more.)

M11.2. Second-Order Linear Differential Equations

M11.2.1. Formulas for the General Solution. Some Transformations

▶ **Homogeneous linear equations. Formulas for the general solution.**

$1°$. Consider a second-order homogeneous linear equation in the general form

$$f_2(x)y''_{xx} + f_1(x)y'_x + f_0(x)y = 0. \tag{M11.2.1.1}$$

The *trivial solution*, $y = 0$, is a particular solution of the homogeneous linear equation.

Let $y_1(x)$, $y_2(x)$ be a fundamental system of solutions (nontrivial linearly independent particular solutions) of equation (M11.2.1.1). Then the general solution is given by

$$y = C_1 y_1(x) + C_2 y_2(x), \tag{M11.2.1.2}$$

where C_1 and C_2 are arbitrary constants.

$2°$. Let $y_1 = y_1(x)$ be any nontrivial particular solution of equation (M11.2.1.1). Then its general solution can be represented as

$$y = y_1\left(C_1 + C_2 \int \frac{e^{-F}}{y_1^2}\,dx\right), \quad \text{where} \quad F = \int \frac{f_1}{f_2}\,dx. \tag{M11.2.1.3}$$

▶ **Wronskian determinant and Liouville's formula.** The *Wronskian determinant* (or *Wronskian*) is defined by

$$W(x) = \begin{vmatrix} y_1(x) & y_2(x) \\ y'_1(x) & y'_2(x) \end{vmatrix} = y_1(y_2)'_x - y_2(y_1)'_x,$$

where $y_1(x)$, $y_2(x)$ is a fundamental system of solutions of equation (M11.2.1.1).

Liouville's formula:

$$W(x) = W(x_0) \exp\left[-\int_{x_0}^{x} \frac{f_1(t)}{f_2(t)}\,dt\right].$$

▶ **Simplest second-order linear equations and their solutions.**

1°. The *second-order linear equation with constant coefficients*

$$y''_{xx} + ay'_x + by = 0 \tag{M11.2.1.4}$$

has the following fundamental system of solutions:

$$
\begin{array}{lll}
y_1(x) = \exp\!\left(-\tfrac{1}{2}ax\right)\sinh\!\left(\tfrac{1}{2}x\sqrt{a^2-4b}\right), & y_2(x) = \exp\!\left(-\tfrac{1}{2}ax\right)\cosh\!\left(\tfrac{1}{2}x\sqrt{a^2-4b}\right) & \text{if } a^2 > 4b; \\[2mm]
y_1(x) = \exp\!\left(-\tfrac{1}{2}ax\right)\sin\!\left(\tfrac{1}{2}x\sqrt{4b-a^2}\right), & y_2(x) = \exp\!\left(-\tfrac{1}{2}ax\right)\cos\!\left(\tfrac{1}{2}x\sqrt{4b-a^2}\right) & \text{if } a^2 < 4b; \\[2mm]
y_1(x) = \exp\!\left(-\tfrac{1}{2}ax\right), & y_2(x) = x\exp\!\left(-\tfrac{1}{2}ax\right) & \text{if } a^2 = 4b.
\end{array}
$$

Remark. In physics equation (M11.2.1.4) is often called an *equation of damped oscillations.*

2°. The *Euler equation*

$$x^2 y''_{xx} + axy'_x + by = 0 \tag{M11.2.1.5}$$

is reduced by the change of variable $x = ke^t$ ($k \neq 0$) to the second-order linear equation with constant coefficients $y''_{tt} + (a-1)y'_t + by = 0$, which is treated in Item 1°.

Equation (M11.2.1.5) has the following fundamental system of solutions:

$$
\begin{array}{lll}
y_1(x) = |x|^{\frac{1-a}{2}+\mu}, & y_2(x) = |x|^{\frac{1-a}{2}-\mu} & \text{if } (1-a)^2 > 4b, \\[2mm]
y_1(x) = |x|^{\frac{1-a}{2}}, & y_2(x) = |x|^{\frac{1-a}{2}}\ln|x| & \text{if } (1-a)^2 = 4b, \\[2mm]
y_1(x) = |x|^{\frac{1-a}{2}}\sin(\mu\ln|x|), & y_2(x) = |x|^{\frac{1-a}{2}}\cos(\mu\ln|x|) & \text{if } (1-a)^2 < 4b,
\end{array}
$$

where $\mu = \tfrac{1}{2}|(1-a)^2 - 4b|^{1/2}$.

▶ **Bessel equation and related equations.**

1°. The *Bessel equation* has the form

$$x^2 y''_{xx} + xy'_x + (x^2 - \nu^2)y = 0.$$

It often arises in numerous applications.

Let ν be an arbitrary noninteger. Then the general solution of the Bessel equation is given by

$$y = C_1 J_\nu(x) + C_2 Y_\nu(x),$$

where $J_\nu(x)$ and $Y_\nu(x)$ are the *Bessel functions* of the first and second kind:

$$J_\nu(x) = \sum_{k=0}^{\infty} \frac{(-1)^k (x/2)^{2k+\nu}}{k!\,\Gamma(\nu+k+1)}, \qquad Y_\nu(x) = \frac{J_\nu(x)\cos\pi\nu - J_{-\nu}(x)}{\sin\pi\nu}.$$

In the case $\nu = n + \tfrac{1}{2}$, where $n = 0, 1, 2, \ldots$, the Bessel functions are expressed in terms of elementary functions:

$$J_{n+\frac{1}{2}}(x) = \sqrt{\frac{2}{\pi}}\, x^{n+\frac{1}{2}}\left(-\frac{1}{x}\frac{d}{dx}\right)^n \frac{\sin x}{x}, \qquad J_{-n-\frac{1}{2}}(x) = \sqrt{\frac{2}{\pi}}\, x^{n+\frac{1}{2}}\left(\frac{1}{x}\frac{d}{dx}\right)^n \frac{\cos x}{x},$$

$$Y_{n+\frac{1}{2}}(x) = (-1)^{n+1} J_{-n-\frac{1}{2}}(x).$$

The Bessel functions are described in Section M13.6 in detail.

TABLE M11.2
Some second-order linear equations whose solutions are expressed
in terms of Bessel functions and modified Bessel functions.

Equation	General solution	Remarks
$y''_{xx} - ax^k y = 0$	$y = C_1\sqrt{x}\,J_{\frac{1}{2q}}\left(\frac{\sqrt{\lvert a\rvert}}{q}x^q\right) + C_2\sqrt{x}\,Y_{\frac{1}{2q}}\left(\frac{\sqrt{\lvert a\rvert}}{q}x^q\right)$ if $a < 0$ $y = C_1\sqrt{x}\,I_{\frac{1}{2q}}\left(\frac{\sqrt{a}}{q}x^q\right) + C_2\sqrt{x}\,K_{\frac{1}{2q}}\left(\frac{\sqrt{a}}{q}x^q\right)$ if $a > 0$	$q = \frac{1}{2}(k+2)$
$y''_{xx} + ay'_x + (bx + c)y = 0$	$y = e^{-ax/2}\sqrt{\xi}\left[C_1 J_{1/3}\left(\frac{2}{3}\sqrt{b}\,\xi^{3/2}\right) + C_2 Y_{1/3}\left(\frac{2}{3}\sqrt{b}\,\xi^{3/2}\right)\right]$	$\xi = x + \dfrac{4c - a^2}{4b}$
$xy''_{xx} + ay'_x + by = 0$	$y = x^{\frac{1-a}{2}}\left[C_1 J_\nu\left(2\sqrt{bx}\right) + C_2 Y_\nu\left(2\sqrt{bx}\right)\right]$ if $bx > 0$ $y = x^{\frac{1-a}{2}}\left[C_1 I_\nu\left(2\sqrt{\lvert bx\rvert}\right) + C_2 K_\nu\left(2\sqrt{\lvert bx\rvert}\right)\right]$ if $bx < 0$	$\nu = \lvert 1 - a\rvert$
$xy''_{xx} + ay'_x + bxy = 0$	$y = x^{\frac{1-a}{2}}\left[C_1 J_\nu\left(\sqrt{b}\,x\right) + C_2 Y_\nu\left(\sqrt{b}\,x\right)\right]$ if $b > 0$ $y = x^{\frac{1-a}{2}}\left[C_1 I_\nu\left(\sqrt{\lvert b\rvert}\,x\right) + C_2 K_\nu\left(\sqrt{\lvert b\rvert}\,x\right)\right]$ if $b < 0$	$\nu = \frac{1}{2}\lvert 1 - a\rvert$
$xy''_{xx} + ay'_x + bx^k y = 0$	$y = x^{\frac{1-a}{2}}\left[C_1 J_\nu\left(\frac{2\sqrt{b}}{k+1}x^{\frac{k+1}{2}}\right) + C_2 Y_\nu\left(\frac{2\sqrt{b}}{k+1}x^{\frac{k+1}{2}}\right)\right]$	$\nu = \dfrac{\lvert 1 - a\rvert}{k+1}$
$x^2 y''_{xx} + xy'_x + (x^2 - \nu^2)y = 0$	$y = C_1 J_\nu(x) + C_2 Y_\nu(x)$	The Bessel equation
$x^2 y''_{xx} + xy'_x - (x^2 + \nu^2)y = 0$	$y = C_1 I_\nu(x) + C_2 K_\nu(x)$	The modified Bessel equation
$y''_{xx} + ae^{\lambda x}y = 0$	$y = C_1 J_0(z) + C_2 Y_0(z)$	$z = \dfrac{2\sqrt{a}}{\lambda}e^{\lambda x/2}$
$y''_{xx} + (ae^x - b)y = 0$	$y = C_1 J_{2\sqrt{b}}\left(2\sqrt{a}\,e^{x/2}\right) + C_2 Y_{2\sqrt{b}}\left(2\sqrt{a}\,e^{x/2}\right)$	
$y''_{xx} + ay'_x + (be^{\lambda x} + c)y = 0$	$y = e^{-ax/2}\left[C_1 J_\nu\left(2\lambda^{-1}\sqrt{b}\,e^{\lambda x/2}\right) + C_2 Y_\nu\left(2\lambda^{-1}\sqrt{b}\,e^{\lambda x/2}\right)\right]$	$\nu = \dfrac{1}{\lambda}\sqrt{a^2 - 4c}$

2°. The *modified Bessel equation* has the form

$$x^2 y''_{xx} + xy'_x - (x^2 + \nu^2)y = 0.$$

It can be reduced to the Bessel equation by means of the substitution $x = i\bar{x}$ ($i^2 = -1$).

The general solution of the modified Bessel equation is given by

$$y = C_1 I_\nu(x) + C_2 K_\nu(x),$$

where $I_\nu(x)$ and $K_\nu(x)$ are *modified Bessel functions* of the first and second kind:

$$I_\nu(x) = \sum_{k=0}^{\infty} \frac{(x/2)^{2k+\nu}}{k!\,\Gamma(\nu + k + 1)}, \qquad K_\nu(x) = \frac{\pi}{2}\frac{I_{-\nu}(x) - I_\nu(x)}{\sin \pi\nu}.$$

The modified Bessel functions are described in Subsection M13.7 in detail.

3°. Table M11.2. lists some second-order linear equations whose solutions are expressed in terms of Bessel functions and modified Bessel functions.

▶ **Nonhomogeneous linear equations. The existence theorem.** A second-order nonhomogeneous linear equation has the form

$$f_2(x)y''_{xx} + f_1(x)y'_x + f_0(x)y = g(x). \tag{M11.2.1.6}$$

THEOREM (EXISTENCE AND UNIQUENESS). *On an open interval $a < x < b$, let the functions f_2, f_1, f_0, and g be continuous and $f_2 \neq 0$. Also let*

$$y(x_0) = A, \quad y'_x(x_0) = B$$

be arbitrary initial conditions, where x_0 is any point such that $a < x_0 < b$, and A and B are arbitrary prescribed numbers. Then a solution of equation (M11.2.1.6) exists and is unique. This solution is defined for all $x \in (a, b)$.

▶ **Formulas for the general solution.**

$1°$. The general solution of the nonhomogeneous linear equation (M11.2.1.6) is the sum of the general solution of the corresponding homogeneous linear equation (M11.2.1.1) and any particular solution of the nonhomogeneous equation (M11.2.1.6).

$2°$. Let $y_1 = y_1(x)$, $y_2 = y_2(x)$ be a fundamental system of solutions of the corresponding homogeneous equation, with $g \equiv 0$. Then the general solution of nonhomogeneous equation (M11.2.1.6) can be represented as

$$y = C_1 y_1 + C_2 y_2 + y_2 \int y_1 \frac{g}{f_2} \frac{dx}{W} - y_1 \int y_2 \frac{g}{f_2} \frac{dx}{W}, \tag{M11.2.1.7}$$

where $W = y_1(y_2)'_x - y_2(y_1)'_x$ is the Wronskian determinant.

Remark. Given a nontrivial particular solution $y_1 = y_1(x)$ of the homogeneous equation (with $g \equiv 0$), a second particular solution $y_2 = y_2(x)$ of the homogeneous equation can be calculated from formula (M11.2.1.3). Then the general solution of equation (M11.2.1.6) can be constructed by (M11.2.1.3) and (M11.2.1.7).

$3°$. Let $\bar{y}_1$ and $\bar{y}_2$ be respective solutions of the nonhomogeneous differential equations $L[\bar{y}_1] = g_1(x)$ and $L[\bar{y}_2] = g_2(x)$, which have the same left-hand side but different right-hand sides, where $L[y]$ is the left-hand side of equation (M11.2.1.6). Then the function $\bar{y} = \bar{y}_1 + \bar{y}_2$ is also a solution of the equation $L[\bar{y}] = g_1(x) + g_2(x)$.

M11.2.2. Representation of Solutions as a Series in the Independent Variable

▶ **Equation coefficients are representable in power series form.** Let us consider a homogeneous linear differential equation of the general form

$$y''_{xx} + f(x)y'_x + g(x)y = 0. \tag{M11.2.2.1}$$

Assume that the functions $f(x)$ and $g(x)$ are representable, in a neighborhood of a point $x = x_0$, in power series form,

$$f(x) = \sum_{n=0}^{\infty} A_n(x - x_0)^n, \quad g(x) = \sum_{n=0}^{\infty} B_n(x - x_0)^n, \tag{M11.2.2.2}$$

on the interval $|x - x_0| < R$, where R stands for the minimum radius of convergence of the two series in (M11.2.2.2). In this case, the point $x = x_0$ is referred to as an *ordinary point*, and equation (M11.2.2.1) possesses two linearly independent solutions of the form

$$y_1(x) = \sum_{n=0}^{\infty} a_n(x - x_0)^n, \qquad y_2(x) = \sum_{n=0}^{\infty} b_n(x - x_0)^n. \qquad \text{(M11.2.2.3)}$$

The coefficients a_n and b_n are determined by substituting the series (M11.2.2.2) and (M11.2.2.3) into equation (M11.2.2.1) followed by matching the coefficients of like powers of $(x - x_0)$.*

▶ **Equation coefficients have poles at some point.** Assume that the functions $f(x)$ and $g(x)$ are representable, in a neighborhood of a point $x = x_0$, in the form

$$f(x) = \sum_{n=-1}^{\infty} A_n(x - x_0)^n, \qquad g(x) = \sum_{n=-2}^{\infty} B_n(x - x_0)^n, \qquad \text{(M11.2.2.4)}$$

on the interval $|x - x_0| < R$. In this case, the point $x = x_0$ is referred to as a *regular singular point*. Let λ_1 and λ_2 be roots of the quadratic equation

$$\lambda^2 + (A_{-1} - 1)\lambda + B_{-2} = 0.$$

Depending on the values of λ_1 and λ_2, three cases are possible. These cases are considered below.

1. If $\lambda_1 \neq \lambda_2$ and $\lambda_1 - \lambda_2$ is not an integer, equation (M11.2.2.1) has two linearly independent solutions of the form

$$y_1(x) = |x - x_0|^{\lambda_1}\left[1 + \sum_{n=1}^{\infty} a_n(x - x_0)^n\right],$$
$$\text{(M11.2.2.5)}$$
$$y_2(x) = |x - x_0|^{\lambda_2}\left[1 + \sum_{n=1}^{\infty} b_n(x - x_0)^n\right].$$

2. If $\lambda_1 = \lambda_2 = \lambda$, equation (M11.2.2.1) possesses two linearly independent solutions:

$$y_1(x) = |x - x_0|^{\lambda}\left[1 + \sum_{n=1}^{\infty} a_n(x - x_0)^n\right],$$
$$y_2(x) = y_1(x)\ln|x - x_0| + |x - x_0|^{\lambda}\sum_{n=0}^{\infty} b_n(x - x_0)^n.$$

3. If $\lambda_1 = \lambda_2 + N$, where N is a positive integer, equation (M11.2.2.1) has two linearly independent solutions of the form

$$y_1(x) = |x - x_0|^{\lambda_1}\left[1 + \sum_{n=1}^{\infty} a_n(x - x_0)^n\right],$$
$$y_2(x) = ky_1(x)\ln|x - x_0| + |x - x_0|^{\lambda_2}\sum_{n=0}^{\infty} b_n(x - x_0)^n,$$

where k is a constant to be determined (it may be equal to zero).

---

* Prior to that, the terms containing the same powers $(x - x_0)^k$, $k = 0, 1, \ldots$, should be collected.

To construct the solution in each of the three cases, the following procedure should be performed: substitute the above expressions of y_1 (resp., y_2) into the original equation (M11.2.2.1), taking into account (M11.2.2.4), and equate the coefficients of $(x - x_0)^n$ and (resp., $(x - x_0)^n \ln |x - x_0|$) for like values of n to obtain recurrence relations for the unknown coefficients. The desired solution can be found from these recurrence relations.

M11.2.3. Boundary Value Problems

▶ **First, second, third, and mixed boundary value problems** $(x_1 \leq x \leq x_2)$. We consider the second-order nonhomogeneous linear differential equation

$$y''_{xx} + f(x)y'_x + g(x)y = h(x). \tag{M11.2.3.1}$$

$1°$. *The first boundary value problem*: Find a solution of equation (M11.2.3.1) satisfying the boundary conditions

$$y = a_1 \quad \text{at} \quad x = x_1, \qquad y = a_2 \quad \text{at} \quad x = x_2. \tag{M11.2.3.2}$$

(The values of the unknown are prescribed at two distinct points x_1 and x_2.)

$2°$. *The second boundary value problem*: Find a solution of equation (M11.2.3.1) satisfying the boundary conditions

$$y'_x = a_1 \quad \text{at} \quad x = x_1, \qquad y'_x = a_2 \quad \text{at} \quad x = x_2. \tag{M11.2.3.3}$$

(The values of the derivative of the unknown are prescribed at two distinct points x_1 and x_2.)

$3°$. *The third boundary value problem*: Find a solution of equation (M11.2.3.1) satisfying the boundary conditions

$$
\begin{aligned}
y'_x + k_1 y = a_1 \quad \text{at} \quad x = x_1, \\
y'_x + k_2 y = a_2 \quad \text{at} \quad x = x_2.
\end{aligned}
\tag{M11.2.3.4}
$$

$4°$. *The mixed boundary value problem*: Find a solution of equation (M11.2.3.1) satisfying the boundary conditions

$$y = a_1 \quad \text{at} \quad x = x_1, \qquad y'_x = a_2 \quad \text{at} \quad x = x_2. \tag{M11.2.3.5}$$

(The unknown is prescribed at one point, and its derivative at another point.)

Conditions (M11.2.3.2), (M11.2.3.3), (M11.2.3.4), and (M11.2.3.5) are called *homogeneous* if $a_1 = a_2 = 0$.

▶ **Simplification of boundary conditions. The self-adjoint form of equations.**

$1°$. Nonhomogeneous boundary conditions can be reduced to homogeneous ones by the change of variable $z = A_2 x^2 + A_1 x + A_0 + y$ (the constants A_2, A_1, and A_0 are selected using the method of undetermined coefficients). In particular, the nonhomogeneous boundary conditions of the first kind (M11.2.3.2) can be reduced to homogeneous boundary conditions by the linear change of variable

$$z = y - \frac{a_2 - a_1}{x_2 - x_1}(x - x_1) - a_1.$$

$2°$. On multiplying by $p(x) = \exp\left[\int f(x)\, dx\right]$, one reduces equation (M11.2.3.1) to the self-adjoint form:

$$[p(x)y'_x]'_x + q(x)y = r(x). \tag{M11.2.3.6}$$

Without loss of generality, we can further consider equation (M11.2.3.6) instead of (M11.2.3.1). We assume that the functions p, p'_x, q, and r are continuous on the interval $x_1 \leq x \leq x_2$, and p is positive.

▶ **Green's function. Linear problems for nonhomogeneous equations.** The *Green's function* of the first boundary value problem for equation (M11.2.3.6) with homogeneous boundary conditions (M11.2.3.2) is a function of two variables $G(x, \xi)$ that satisfies the following conditions:

$1°$. $G(x, \xi)$ is continuous in x for fixed ξ, with $x_1 \leq x \leq x_2$ and $x_1 \leq \xi \leq x_2$.

$2°$. $G(x, \xi)$ is a solution of the homogeneous equation (M11.2.3.6), with $r = 0$, for all $x_1 < x < x_2$ exclusive of the point $x = \xi$.

$3°$. $G(x, \xi)$ satisfies the homogeneous boundary conditions $G(x_1, \xi) = G(x_2, \xi) = 0$.

$4°$. The derivative $G'_x(x, \xi)$ has a jump of $1/p(\xi)$ at the point $x = \xi$, that is,

$$G'_x(x, \xi)\big|_{x \to \xi, \, x > \xi} - G'_x(x, \xi)\big|_{x \to \xi, \, x < \xi} = \frac{1}{p(\xi)}.$$

For the second, third, and mixed boundary value problems, the Green's function is defined likewise except that in $3°$ one adopts, respectively, the homogeneous boundary conditions (M11.2.3.3), (M11.2.3.4), and (M11.2.3.5), with $a_1 = a_2 = 0$.

The solution of the nonhomogeneous equation (M11.2.3.6) subject to appropriate homogeneous boundary conditions is expressed in terms of the Green's function as follows:*

$$y(x) = \int_{x_1}^{x_2} G(x, \xi) r(\xi) \, d\xi.$$

▶ **Representation of the Green's function in terms of particular solutions.** We consider the first boundary value problem. Let $y_1(x)$ and $y_2(x)$ be linearly independent particular solutions of the homogeneous equation (M11.2.3.6), with $r = 0$, that satisfy the conditions

$$y_1(x_1) = 0, \qquad y_2(x_2) = 0.$$

(Each of the solutions satisfies one of the homogeneous boundary conditions.)

The Green's function is expressed in terms of solutions of the homogeneous equation as follows:

$$G(x, \xi) = \begin{cases} \dfrac{y_1(x) y_2(\xi)}{p(\xi) W(\xi)} & \text{for } x_1 \leq x \leq \xi, \\[3mm] \dfrac{y_1(\xi) y_2(x)}{p(\xi) W(\xi)} & \text{for } \xi \leq x \leq x_2, \end{cases} \tag{M11.2.3.7}$$

where $W(x) = y_1(x) y_2'(x) - y_1'(x) y_2(x)$ is the Wronskian determinant.

Remark. Formula (M11.2.3.7) can also be used to construct the Green's functions for the second, third, and mixed boundary value problems. To this end, one should find two linearly independent solutions, $y_1(x)$ and $y_2(x)$, of the homogeneous equation; the former satisfies the corresponding homogeneous boundary condition at $x = x_1$ and the latter satisfies the one at $x = x_2$.

M11.2.4. Eigenvalue Problems

▶ **Sturm–Liouville problem.** Consider the second-order homogeneous linear differential equation

$$[p(x) y_x']_x' + [\lambda \rho(x) - q(x)] y = 0 \tag{M11.2.4.1}$$

---

* The homogeneous boundary value problem, with $r(x) = 0$ and $a_1 = a_2 = 0$, is assumed to have only the trivial solution.

subject to linear boundary conditions of the general form

$$s_1 y'_x + k_1 y = 0 \quad \text{at} \quad x = x_1,$$
$$s_2 y'_x + k_2 y = 0 \quad \text{at} \quad x = x_2. \tag{M11.2.4.2}$$

It is assumed that the functions p, p'_x, ρ, and q are continuous, and p and ρ are positive on an interval $x_1 \le x \le x_2$. It is also assumed that $|s_1| + |k_1| > 0$ and $|s_2| + |k_2| > 0$.

The *Sturm–Liouville problem*: Find the values λ_n of the parameter λ at which problem (M11.2.4.1)–(M11.2.4.2) has a nontrivial solution. Such λ_n are called *eigenvalues* and the corresponding solutions $y_n = y_n(x)$ are called *eigenfunctions* of the Sturm–Liouville problem (M11.2.4.1)–(M11.2.4.2).

▶ **General properties of the Sturm–Liouville problem (M11.2.4.1)–(M11.2.4.2).**

$1°$. There are infinitely (countably) many eigenvalues. All eigenvalues can be ordered so that $\lambda_1 < \lambda_2 < \lambda_3 < \cdots$. Moreover, $\lambda_n \to \infty$ as $n \to \infty$; hence, there can only be a finite number of negative eigenvalues.

$2°$. The eigenfunctions are defined up to a constant factor. Each eigenfunction $y_n(x)$ has precisely $n - 1$ zeros on the open interval (x_1, x_2).

$3°$. Any two eigenfunctions $y_n(x)$ and $y_m(x)$, $n \ne m$, are orthogonal with weight $\rho(x)$ on the interval $x_1 \le x \le x_2$:

$$\int_{x_1}^{x_2} \rho(x) y_n(x) y_m(x) \, dx = 0 \quad \text{if} \quad n \ne m.$$

$4°$. An arbitrary function $F(x)$ that has a continuous derivative and satisfies the boundary conditions of the Sturm–Liouville problem can be decomposed into an absolutely and uniformly convergent series in the eigenfunctions

$$F(x) = \sum_{n=1}^{\infty} F_n y_n(x),$$

where the Fourier coefficients F_n of $F(x)$ are calculated by

$$F_n = \frac{1}{\|y_n\|^2} \int_{x_1}^{x_2} \rho(x) F(x) y_n(x) \, dx, \qquad \|y_n\|^2 = \int_{x_1}^{x_2} \rho(x) y_n^2(x) \, dx.$$

$5°$. If the conditions

$$q(x) \ge 0, \quad s_1 k_1 \le 0, \quad s_2 k_2 \ge 0 \tag{M11.2.4.3}$$

hold true, there are no negative eigenvalues. If $q \equiv 0$ and $k_1 = k_2 = 0$, the least eigenvalue is $\lambda_1 = 0$, to which there corresponds an eigenfunction $y_1 = \text{const}$. In the other cases where conditions (M11.2.4.3) are satisfied, all eigenvalues are positive.

$6°$. The following asymptotic formula is valid for eigenvalues as $n \to \infty$:

$$\lambda_n = \frac{\pi^2 n^2}{\Delta^2} + O(1), \qquad \Delta = \int_{x_1}^{x_2} \sqrt{\frac{\rho(x)}{p(x)}} \, dx. \tag{M11.2.4.4}$$

Remark 1. Equation (M11.2.4.1) can be reduced to the case where $p(x) \equiv 1$ and $\rho(x) \equiv 1$ by the change of variables

$$\zeta = \int \sqrt{\frac{\rho(x)}{p(x)}} \, dx, \qquad u(\zeta) = \left[p(x) \rho(x) \right]^{1/4} y(x).$$

In this case, the boundary conditions are transformed to boundary conditions of a similar form.

TABLE M11.3

Example estimates of the first eigenvalue λ_1 in Sturm–Liouville problems with boundary conditions of the first kind $y(0) = y(1) = 0$ obtained using the Rayleigh–Ritz principle [the right-hand side of relation (M11.2.4.6)].

Equation	Test function	λ_1, approximate	λ_1, exact
$y''_{xx} + \lambda(1 + x^2)^{-2}y = 0$	$z = \sin \pi x$	15.337	15.0
$y''_{xx} + \lambda(4 - x^2)^{-2}y = 0$	$z = \sin \pi x$	135.317	134.837
$y''_{xx} + \lambda(1 + \sin \pi x)y = 0$	$z = \sin \pi x$ $z = x(1 - x)$	$0.54105\,\pi^2$ $0.55204\,\pi^2$	$0.54032\,\pi^2$ $0.54032\,\pi^2$
$(\sqrt{1 + x}\, y'_x)'_x + \lambda y = 0$	$z = \sin \pi x$	11.9956	11.8985

Remark 2. The second-order linear equation

$$\varphi_2(x)y''_{xx} + \varphi_1(x)y'_x + [\lambda + \varphi_0(x)]y = 0$$

can be represented in the form of equation (M11.2.4.1) where $p(x)$, $\rho(x)$, and $q(x)$ are given by

$$p(x) = \exp\left[\int \frac{\varphi_1(x)}{\varphi_2(x)}\,dx\right], \quad \rho(x) = \frac{1}{\varphi_2(x)}\exp\left[\int \frac{\varphi_1(x)}{\varphi_2(x)}\,dx\right], \quad q(x) = -\frac{\varphi_0(x)}{\varphi_2(x)}\exp\left[\int \frac{\varphi_1(x)}{\varphi_2(x)}\,dx\right].$$

▶ **Problems with boundary conditions of the first kind.** Let us note some special properties of the Sturm–Liouville problem that is the first boundary value problem for equation (M11.2.4.1) with the boundary conditions

$$y = 0 \quad \text{at} \quad x = x_1, \qquad y = 0 \quad \text{at} \quad x = x_2. \tag{M11.2.4.5}$$

$1°$. For $n \to \infty$, the asymptotic relation (M11.2.4.4) can be used to estimate the eigenvalues λ_n. In this case, the asymptotic formula

$$\frac{y_n(x)}{\|y_n\|} = \left[\frac{4}{\Delta^2 p(x)\rho(x)}\right]^{1/4}\sin\left[\frac{\pi n}{\Delta}\int_{x_1}^{x}\sqrt{\frac{\rho(x)}{p(x)}}\,dx\right] + O\left(\frac{1}{n}\right), \qquad \Delta = \int_{x_1}^{x_2}\sqrt{\frac{\rho(x)}{p(x)}}\,dx$$

holds true for the eigenfunctions $y_n(x)$.

$2°$. If $q \geq 0$, the following upper estimate holds for the least eigenvalue (*Rayleigh–Ritz principle*):

$$\lambda_1 \leq \frac{\int_{x_1}^{x_2}\left[p(x)(z'_x)^2 + q(x)z^2\right]\,dx}{\int_{x_1}^{x_2}\rho(x)z^2\,dx}, \tag{M11.2.4.6}$$

where $z = z(x)$ is any twice differentiable function that satisfies the conditions $z(x_1) = z(x_2) = 0$. The equality in (M11.2.4.6) is attained if $z = y_1(x)$, where $y_1(x)$ is the eigenfunction corresponding to the eigenvalue λ_1. One can take $z = (x - x_1)(x_2 - x)$ or $z = \sin\left[\dfrac{\pi(x - x_1)}{x_2 - x_1}\right]$ in (M11.2.4.6) to obtain specific estimates.

It is significant to note that the left-hand side of (M11.2.4.6) usually gives a fairly precise estimate of the first eigenvalue (see Table M11.3).

$3°$. The extension of the interval $[x_1, x_2]$ results in decreasing the eigenvalues.

4°. Let the inequalities

$$0 < p_{\min} \le p(x) \le p_{\max}, \quad 0 < \rho_{\min} \le \rho(x) \le \rho_{\max}, \quad 0 < q_{\min} \le q(x) \le q_{\max}$$

be satisfied. Then the following bilateral estimates hold:

$$\frac{p_{\min}}{\rho_{\max}} \frac{\pi^2 n^2}{(x_2 - x_1)^2} + \frac{q_{\min}}{\rho_{\max}} \le \lambda_n \le \frac{p_{\max}}{\rho_{\min}} \frac{\pi^2 n^2}{(x_2 - x_1)^2} + \frac{q_{\max}}{\rho_{\min}}.$$

5°. In engineering calculations for eigenvalues, the approximate formula

$$\lambda_n = \frac{\pi^2 n^2}{\Delta^2} + \frac{1}{x_2 - x_1} \int_{x_1}^{x_2} \frac{q(x)}{\rho(x)}\, dx, \quad \Delta = \int_{x_1}^{x_2} \sqrt{\frac{\rho(x)}{p(x)}}\, dx \tag{M11.2.4.7}$$

may be quite useful. This formula provides an exact result if $p(x)\rho(x) = \text{const}$ and $q(x)/\rho(x) = \text{const}$ (in particular, for constant equation coefficients, $p = p_0$, $q = q_0$, and $\rho = \rho_0$) and gives a correct asymptotic behavior of (M11.2.4.4) for any $p(x)$, $q(x)$, and $\rho(x)$. In addition, relation (M11.2.4.7) gives two correct leading asymptotic terms as $n \to \infty$ if $p(x) = \text{const}$ and $\rho(x) = \text{const}$ [and also if $p(x)\rho(x) = \text{const}$].

6°. Suppose that $p(x) = \rho(x) = 1$ and the function $q(x)$ has a continuous derivative. The following asymptotic relations hold for eigenvalues λ_n and eigenfunctions $y_n(x)$ as $n \to \infty$:

$$\sqrt{\lambda_n} = \frac{\pi n}{x_2 - x_1} + \frac{1}{\pi n} Q(x_1, x_2) + O\left(\frac{1}{n^2}\right),$$

$$y_n(x) = \sin \frac{\pi n(x - x_1)}{x_2 - x_1} - \frac{1}{\pi n}\left[(x_1 - x)Q(x, x_2) + (x_2 - x)Q(x_1, x)\right] \cos \frac{\pi n(x - x_1)}{x_2 - x_1} + O\left(\frac{1}{n^2}\right),$$

where

$$Q(u, v) = \frac{1}{2} \int_u^v q(x)\, dx. \tag{M11.2.4.8}$$

7°. Let us consider the eigenvalue problem for the equation with a small parameter

$$y_{xx}'' + [\lambda + \varepsilon q(x)]y = 0 \qquad (\varepsilon \to 0)$$

subject to the boundary conditions (M11.2.4.5) with $x_1 = 0$ and $x_2 = 1$. We assume that $q(x) = q(-x)$.

This problem has the following eigenvalues and eigenfunctions:

$$\lambda_n = \pi^2 n^2 - \varepsilon A_{nn} + \frac{\varepsilon^2}{\pi^2} \sum_{k \ne n} \frac{A_{nk}^2}{n^2 - k^2} + O(\varepsilon^3), \quad A_{nk} = 2 \int_0^1 q(x) \sin(\pi n x) \sin(\pi k x)\, dx;$$

$$y_n(x) = \sqrt{2}\, \sin(\pi n x) - \varepsilon \frac{\sqrt{2}}{\pi^2} \sum_{k \ne n} \frac{A_{nk}}{n^2 - k^2} \sin(\pi k x) + O(\varepsilon^2).$$

Here, the summation is carried out over k from 1 to ∞. The next term in the expansion of y_n can be found in Nayfeh (1973).

M11.3. Second-Order Nonlinear Differential Equations
M11.3.1. Form of the General Solution. Cauchy Problem

▶ **Equations solved for the derivative. General solution.** A *second-order ordinary differential equation* solved for the highest derivative has the form

$$y''_{xx} = f(x, y, y'_x). \tag{M11.3.1.1}$$

The general solution of this equation depends on two arbitrary constants, C_1 and C_2. In some cases, the general solution can be written in explicit form, $y = \varphi(x, C_1, C_2)$, but more often implicit or parametric forms of the general solution are encountered.

▶ **Cauchy problem. The existence and uniqueness theorem.** *Cauchy problem*: Find a solution of equation (M11.3.1.1) satisfying the *initial conditions*

$$y(x_0) = y_0, \qquad y'_x(x_0) = y_1. \tag{M11.3.1.2}$$

(At a point $x = x_0$, the value of the unknown function, y_0, and its derivative, y_1, are prescribed.)

EXISTENCE AND UNIQUENESS THEOREM. *Let $f(x, y, z)$ be a continuous function in all its arguments in a neighborhood of a point (x_0, y_0, y_1) and let f have bounded partial derivatives f_y and f_z in this neighborhood, or the Lipschitz condition is satisfied: $|f(x, y, z) - f(x, \bar{y}, \bar{z})| \leq A(|y - \bar{y}| + |z - \bar{z}|)$, where A is some positive number. Then a solution of equation (M11.3.1.1) satisfying the initial conditions (M11.3.1.2) exists and is unique.*

M11.3.2. Equations Admitting Reduction of Order

▶ **Equations not containing y explicitly.** In the general case, a second-order equation that does not contain y explicitly has the form

$$F(x, y'_x, y''_{xx}) = 0. \tag{M11.3.2.1}$$

Such equations remain unchanged under an arbitrary translation of the dependent variable: $y \to y + \text{const}$. The substitution $y'_x = z(x)$, $y''_{xx} = z'_x(x)$ brings (M11.3.2.1) to a first-order equation: $F(x, z, z'_x) = 0$.

▶ **Equations not containing x explicitly (autonomous equations).** In the general case, a second-order equation that does not contain x explicitly has the form

$$F(y, y'_x, y''_{xx}) = 0. \tag{M11.3.2.2}$$

Such equations remain unchanged under an arbitrary translation of the independent variable: $x \to x + \text{const}$. Using the substitution $y'_x = w(y)$, where y plays the role of the independent variable, and taking into account the relations $y''_{xx} = w'_x = w'_y y'_x = w'_y w$, one can reduce (M11.3.2.2) to a first-order equation: $F(y, w, ww'_y) = 0$.

Example 1. Consider the autonomous equation

$$y''_{xx} = f(y),$$

which often arises in the theory of heat and mass transfer and combustion. The change of variable $y'_x = w(y)$ leads to a separable first-order equation: $ww'_y = f(y)$. Integrating yields $w^2 = 2F(y) + C_1$, where $F(y) = \int f(y)\,dy$. Solving for w and returning to the original variable, we obtain the separable equation $y'_x = \pm\sqrt{2F(y) + C_1}$. Its general solution is expressed as

$$\int \frac{dy}{\sqrt{2F(y) + C_1}} = \pm x + C_2, \quad \text{where} \quad F(y) = \int f(y)\,dy.$$

Remark. The equation $y''_{xx} = f(y + ax^2 + bx + c)$ is reduced by the change of variable $u = y + ax^2 + bx + c$ to an autonomous equation, $u''_{xx} = f(u) + 2a$.

▶ **Equations of the form** $F(ax + by, y'_x, y''_{xx}) = 0$. Such equations are invariant under simultaneous translations of the independent and dependent variables in accordance with the rule $x \to x + bc$, $y \to y - ac$, where c is an arbitrary constant.

For $b = 0$, see equation (M11.3.2.1). For $b \neq 0$, the substitution $bw = ax + by$ leads to the autonomous equation, $F(bw, w'_x - a/b, w''_{xx}) = 0$, that does not contain x explicitly.

▶ **Homogeneous equations.**

1°. The *equations homogeneous in the independent variable* remain unchanged under scaling of the independent variable, $x \to \alpha x$, where α is an arbitrary nonzero number. In the general case, such equations can be written in the form

$$F(y, xy'_x, x^2 y''_{xx}) = 0. \tag{M11.3.2.3}$$

The substitution $z(y) = xy'_x$ leads to a first-order equation, $F(y, z, zz'_y - z) = 0$.

2°. The *equations homogeneous in the dependent variable* remain unchanged under scaling of the variable sought, $y \to \alpha y$, where α is an arbitrary nonzero number. In the general case, such equations can be written in the form

$$F(x, y'_x/y, y''_{xx}/y) = 0. \tag{M11.3.2.4}$$

The substitution $z(x) = y'_x/y$ leads to a first-order equation, $F(x, z, z'_x + z^2) = 0$.

3°. The *equations homogeneous in both variables* are invariant under simultaneous scaling (dilatation) of the independent and dependent variables, $x \to \alpha x$ and $y \to \alpha y$, where α is an arbitrary nonzero number. In the general case, such equations can be written in the form

$$F(y/x, y'_x, xy''_{xx}) = 0. \tag{M11.3.2.5}$$

The transformation $t = \ln|x|$, $w = y/x$ leads to the autonomous equation, $F(w, w'_t + w, w''_{tt} + w'_t) = 0$, that does not contain t explicitly.

Example 2. The homogeneous equation

$$xy''_{xx} - y'_x = f(y/x)$$

is reduced by the transformation $t = \ln|x|$, $w = y/x$ to the autonomous form: $w''_{tt} = f(w) + w$. For solution of this equation, see Example 1 above (the notation of the right-hand side has to be changed there).

▶ **Generalized homogeneous equations.**

1°. The *generalized homogeneous equations* remain unchanged under simultaneous scaling of the independent and dependent variables in accordance with the rule $x \to \alpha x$ and $y \to \alpha^k y$, where α is an arbitrary nonzero number and k is some number. Such equations can be written in the form

$$F(x^{-k}y, x^{1-k}y'_x, x^{2-k}y''_{xx}) = 0. \tag{M11.3.2.6}$$

The transformation $t = \ln x$, $w = x^{-k}y$ leads to the autonomous equation,

$$F\left(w, w'_t + kw, w''_{tt} + (2k - 1)w'_t + k(k - 1)w\right) = 0,$$

that does not contain t explicitly.

$2°$. The most general form of representation of generalized homogeneous equations is as follows:

$$\mathcal{F}(x^n y^m, xy'_x/y, x^2 y''_{xx}/y) = 0. \tag{M11.3.2.7}$$

The transformation $z = x^n y^m$, $u = xy'_x/y$ reduces this equation to the first-order equation

$$\mathcal{F}\left(z, u, z(mu + n)u'_z - u + u^2\right) = 0.$$

Remark. For $m \neq 0$, equation (M11.3.2.7) is equivalent to equation (M11.3.2.6) in which $k = -n/m$. To the particular values $n = 0$ and $m = 0$ there correspond equations (M11.3.2.3) and (M11.3.2.4) homogeneous in the independent and dependent variables, respectively. For $n = -m \neq 0$, we have an equation homogeneous in both variables, which is equivalent to equation (M11.3.2.5).

▶ **Equations invariant under scaling–translation transformations.**

$1°$. The equations of the form

$$F(e^{\lambda x} y, e^{\lambda x} y'_x, e^{\lambda x} y''_{xx}) = 0 \tag{M11.3.2.8}$$

remain unchanged under simultaneous translation and scaling of variables, $x \to x + \alpha$ and $y \to \beta y$, where $\beta = e^{-\alpha \lambda}$ and α is an arbitrary number. The substitution $w = e^{\lambda x} y$ brings (M11.3.2.8) to the autonomous equation, $F(w, w'_x - \lambda w, w''_{xx} - 2\lambda w'_x + \lambda^2 w) = 0$, that does not contain x explicitly.

$2°$. The equation

$$F(e^{\lambda x} y^n, y'_x/y, y''_{xx}/y) = 0 \tag{M11.3.2.9}$$

is invariant under the simultaneous translation and scaling of variables, $x \to x + \alpha$ and $y \to \beta y$, where $\beta = e^{-\alpha \lambda/n}$ and α is an arbitrary number. The transformation $z = e^{\lambda x} y^n$, $w = y'_x/y$ brings (M11.3.2.9) to a first-order equation: $F\left(z, w, z(nw + \lambda)w'_z + w^2\right) = 0$.

$3°$. The equation

$$F(x^n e^{\lambda y}, xy'_x, x^2 y''_{xx}) = 0 \tag{M11.3.2.10}$$

is invariant under the simultaneous scaling and translation of variables, $x \to \alpha x$ and $y \to y + \beta$, where $\alpha = e^{-\beta \lambda/n}$ and β is an arbitrary number. The transformation $z = x^n e^{\lambda y}$, $w = xy'_x$ brings (M11.3.2.10) to a first-order equation: $F\left(z, w, z(\lambda w + n)w'_z - w\right) = 0$.

▶ **Equations of the form** $F(x, xy'_x - y, y''_{xx}) = 0$. The substitution $w(x) = xy'_x - y$ leads to a first-order equation: $F(x, w, w'_x/x) = 0$.

M11.3.3. Methods of Regular Series Expansions with Respect to the Independent Variable

A solution of the Cauchy problem

$$y''_{xx} = f(x, y, y'_x), \tag{M11.3.3.1}$$
$$y(x_0) = y_0, \quad y'_x(x_0) = y_1 \tag{M11.3.3.2}$$

can be sought in the form of a Taylor series in powers of the difference $(x - x_0)$, specifically:

$$y(x) = y(x_0) + y'_x(x_0)(x - x_0) + \frac{y''_{xx}(x_0)}{2!}(x - x_0)^2 + \frac{y'''_{xxx}(x_0)}{3!}(x - x_0)^3 + \cdots . \tag{M11.3.3.3}$$

The first two coefficients $y(x_0)$ and $y'_x(x_0)$ in solution (M11.3.3.3) are defined by the initial conditions (M11.3.3.2). The values of the subsequent derivatives of y at the point $x = x_0$ are determined from equation (M11.3.3.1) and its derivative equations (obtained by successive differentiation of the equation) taking into account the initial conditions (M11.3.3.2). In particular, setting $x = x_0$ in (M11.3.3.1) and substituting (M11.3.3.2), we obtain the value of the second derivative:

$$y''_{xx}(x_0) = f(x_0, y_0, y_1). \tag{M11.3.3.4}$$

Further, differentiating (M11.3.3.1) yields

$$y'''_{xxx} = f_x(x, y, y'_x) + f_y(x, y, y'_x)y'_x + f_{y'_x}(x, y, y'_x)y''_{xx}. \tag{M11.3.3.5}$$

On substituting $x = x_0$, the initial conditions (M11.3.3.2), and the expression of $y''_{xx}(x_0)$ of (M11.3.3.4) into the right-hand side of equation (M11.3.3.5), we calculate the value of the third derivative:

$$y'''_{xxx}(x_0) = f_x(x_0, y_0, y_1) + y_1 f_y(x_0, y_0, y_1) + f(x_0, y_0, y_1)f_{y'_x}(x_0, y_0, y_1).$$

The subsequent derivatives of the unknown are determined likewise.

The thus obtained solution (M11.3.3.3) can only be used in a small neighborhood of the point $x = x_0$.

Example. Consider the following Cauchy problem for a second-order nonlinear equation:

$$y''_{xx} = yy'_x + y^3; \tag{M11.3.3.6}$$
$$y(0) = y'_x(0) = 1. \tag{M11.3.3.7}$$

Substituting the initial values of the unknown and its derivative (M11.3.3.7) into equation (M11.3.3.6) yields the initial value of the second derivative:

$$y''_{xx}(0) = 2. \tag{M11.3.3.8}$$

Differentiating equation (M11.3.3.6) gives

$$y'''_{xxx} = yy''_{xx} + (y'_x)^2 + 3y^2 y'_x. \tag{M11.3.3.9}$$

Substituting here the initial values from (M11.3.3.7) and (M11.3.3.8), we obtain the initial condition for the third derivative:

$$y'''_{xxx}(0) = 6. \tag{M11.3.3.10}$$

Differentiating (M11.3.3.9) followed by substituting (M11.3.3.7), (M11.3.3.8), and (M11.3.3.10), we find that

$$y''''_{xxxx}(0) = 24. \tag{M11.3.3.11}$$

On substituting the initial data (M11.3.3.7), (M11.3.3.8), (M11.3.3.10), and (M11.3.3.11) into (M11.3.3.3), we arrive at the Taylor series expansion of the solution about $x = 0$:

$$y = 1 + x + x^2 + x^3 + x^4 + \cdots. \tag{M11.3.3.12}$$

This geometric series is convergent only for $|x| < 1$. In this case, summing up the series (M11.3.3.12) gives the exact solution of the Cauchy problem (M11.3.3.6)–(M11.3.3.7) of the form $y(x) = \dfrac{1}{1-x}$.

M11.3.4. Perturbation Methods in Problems with a Small Parameter

▶ **Preliminary remarks.** Perturbation methods are widely used in nonlinear mechanics and theoretical physics for solving problems that are described by differential equations with a small parameter ε. The primary purpose of these methods is to obtain an approximate solution that would be equally suitable at all (small, intermediate, and large) values of the independent variable as $\varepsilon \to 0$.

It is further assumed that the order of the equation remains unchanged at $\varepsilon = 0$.

In many problems of nonlinear mechanics and theoretical physics, the independent variable is dimensionless time t. Therefore, in this subsection we use the conventional t ($0 \le t < \infty$), instead of x.

▶ **Method of regular (direct) expansion in powers of the small parameter.** We consider an equation of general form with a parameter ε:

$$y_{tt}'' + f(t, y, y_t', \varepsilon) = 0. \tag{M11.3.4.1}$$

We assume that the function f can be represented as a series in powers of ε:

$$f(t, y, y_t', \varepsilon) = \sum_{n=0}^{\infty} \varepsilon^n f_n(t, y, y_t'). \tag{M11.3.4.2}$$

Solutions of the Cauchy problem and various boundary value problems for equation (M11.3.4.1) with $\varepsilon \to 0$ are sought in the form of a power series expansion:

$$y = \sum_{n=0}^{\infty} \varepsilon^n y_n(t). \tag{M11.3.4.3}$$

One should substitute expression (M11.3.4.3) into equation (M11.3.4.1) taking into account (M11.3.4.2). Then the functions f_n are expanded into a power series in the small parameter and the coefficients of like powers of ε are collected and equated to zero to obtain a system of equations for y_n:

$$y_0'' + f_0(t, y_0, y_0') = 0, \tag{M11.3.4.4}$$

$$y_1'' + F(t, y_0, y_0')y_1' + G(t, y_0, y_0')y_1 + f_1(t, y_0, y_0') = 0, \quad F = \frac{\partial f_0}{\partial y'}, \quad G = \frac{\partial f_0}{\partial y}. \tag{M11.3.4.5}$$

Only the first two equations are written out here. The prime denotes differentiation with respect to t. To obtain the initial (or boundary) conditions for y_n, the expansion (M11.3.4.3) should be taken into account.

The success in the application of this method is primarily determined by the possibility of constructing a solution of equation (M11.3.4.4) for the leading term y_0. It is significant to note that the other terms y_n with $n \geq 1$ are governed by linear equations with homogeneous initial conditions.

Example 1. The Duffing equation

$$y_{tt}'' + y + \varepsilon y^3 = 0 \tag{M11.3.4.6}$$

with initial conditions

$$y(0) = a, \quad y_t'(0) = 0$$

describes the motion of a cubic oscillator, i.e., oscillations of a point mass on a nonlinear spring. Here, y is the deviation of the point mass from the equilibrium and t is dimensionless time.

For $\varepsilon \to 0$, an approximate solution of the problem is sought in the form of the asymptotic expansion (M11.3.4.3). We substitute (M11.3.4.3) into equation (M11.3.4.6) and initial conditions and expand in powers of ε. On equating the coefficients of like powers of the small parameter to zero, we obtain the following problems for y_0 and y_1:

$$y_0'' + y_0 = 0, \qquad y_0(0) = a, \quad y_0'(0) = 0;$$
$$y_1'' + y_1 = -y_0^3, \qquad y_1(0) = 0, \quad y_1'(0) = 0.$$

The solution of the problem for y_0 is given by

$$y_0 = a \cos t.$$

Substituting this expression into the equation for y_1 and taking into account the identity $\cos^3 t = \frac{1}{4} \cos 3t + \frac{3}{4} \cos t$, we obtain

$$y_1'' + y_1 = -\tfrac{1}{4}a^3(\cos 3t + 3 \cos t), \qquad y_1(0) = 0, \quad y_1'(0) = 0.$$

Integrating yields

$$y_1 = -\tfrac{3}{8}a^3 t \sin t + \tfrac{1}{32}a^3(\cos 3t - 3\cos t).$$

Thus the two-term solution of the original problem is given by

$$y = a\cos t + \varepsilon a^3\left[-\tfrac{3}{8}t\sin t + \tfrac{1}{32}(\cos 3t - 3\cos t)\right] + O(\varepsilon^2).$$

Remark 1. The term $t\sin t$ causes $y_1/y_0 \to \infty$ as $t \to \infty$. For this reason, the solution obtained is unsuitable at large times. It can only be used for $\varepsilon t \ll 1$; this results from the condition of applicability of the expansion, $y_0 \gg \varepsilon y_1$.

This circumstance is typical of the method of regular series expansions with respect to the small parameter; in other words, the expansion becomes unsuitable at large values of the independent variable. Methods that allow avoiding this difficulty are discussed below.

Remark 2. Growing terms as $t \to \infty$, like $t\sin t$, that narrow down the domain of applicability of asymptotic expansions are called *secular*.

▶ **Method of scaled parameters (Lindstedt–Poincaré method).** This method is usually used for finding periodic solutions to equations of the form

$$y''_{tt} + \omega_0 y = \varepsilon f(y, y'_t), \tag{M11.3.4.7}$$

where $\varepsilon \ll 1$.

A solutions is sought in the form

$$t = z\left(1 + \sum_{k=0}^{n} \varepsilon^k \omega_k\right), \qquad y(t) = \sum_{k=0}^{n} \varepsilon^k y_k(z). \tag{M11.3.4.8}$$

The constants ω_k and functions $y_k(z)$ are determined; it is assumed that $y_{k+1}/y_k = O(1)$. By choosing appropriate ω_k, one removes the secular terms from the solution.

Example 2. Consider the Duffing equation (M11.3.4.6) once again. Following (M11.3.4.8), one performs the change of variable

$$t = z(1 + \varepsilon\omega_1 + \cdots)$$

to obtain

$$y''_{zz} + (1 + \varepsilon\omega_1 + \cdots)^2(y + \varepsilon y^3) = 0. \tag{M11.3.4.9}$$

The solution is sought in the series form (M11.3.4.8), $y = y_0(z) + \varepsilon y_1(z) + \cdots$. Substituting it into equation (M11.3.4.9) and matching the coefficients of like powers of ε, one arrives at the following system of equations for two leading terms of the series:

$$y''_0 + y_0 = 0, \tag{M11.3.4.10}$$

$$y''_1 + y_1 = -y_0^3 - 2\omega_1 y_0, \tag{M11.3.4.11}$$

where the prime denotes differentiation with respect to z.

The general solution of equation (M11.3.4.10) is given by

$$y_0 = a\cos(z + b), \tag{M11.3.4.12}$$

where a and b are constants of integration. Taking into account (M11.3.4.12) and rearranging terms, we reduce equation (M11.3.4.11) to

$$y''_1 + y_1 = -\tfrac{1}{4}a^3\cos\left[3(z+b)\right] - 2a\left(\tfrac{3}{8}a^2 + \omega_1\right)\cos(z + b). \tag{M11.3.4.13}$$

For $\omega_1 \neq -\tfrac{3}{8}a^2$, the particular solution of equation (M11.3.4.13) contains a secular term proportional to $z\cos(z + b)$. In this case, the condition of applicability of the expansion, $y_1/y_0 = O(1)$, cannot be satisfied at sufficiently large z. For this condition to be met, one should set

$$\omega_1 = -\tfrac{3}{8}a^2. \tag{M11.3.4.14}$$

In this case, the solution of equation (M11.3.4.13) is given by

$$y_1 = \tfrac{1}{32}a^3\cos\left[3(z+b)\right]. \tag{M11.3.4.15}$$

Subsequent terms of the expansion can be found likewise.

With (M11.3.4.12), (M11.3.4.14), and (M11.3.4.15), we obtain a solution of the Duffing equation in the form

$$y = a\cos(\omega t + b) + \tfrac{1}{32}\varepsilon a^3\cos\left[3(\omega t + b)\right] + O(\varepsilon^2),$$

$$\omega = \left[1 - \tfrac{3}{8}\varepsilon a^2 + O(\varepsilon^2)\right]^{-1} = 1 + \tfrac{3}{8}\varepsilon a^2 + O(\varepsilon^2).$$

▶ **Averaging method (Van der Pol–Krylov–Bogolyubov scheme).** The averaging method involves two stages. First, the second-order nonlinear equation (M11.3.4.7) is reduced with the transformation

$$y = a \cos \varphi, \quad y'_t = -\omega_0 a \sin \varphi, \quad \text{where} \quad a = a(t), \quad \varphi = \varphi(t),$$

to an equivalent system of two first-order differential equations:

$$a'_t = -\frac{\varepsilon}{\omega_0} f(a \cos \varphi, -\omega_0 a \sin \varphi) \sin \varphi,$$
$$\varphi'_t = \omega_0 - \frac{\varepsilon}{\omega_0 a} f(a \cos \varphi, -\omega_0 a \sin \varphi) \cos \varphi. \tag{M11.3.4.16}$$

The right-hand sides of equations (M11.3.4.16) are periodic in φ, with the amplitude a being a slow-varying function of time t. The amplitude and the oscillation character are changing little during the time the phase φ changes by 2π.

At the second stage, the right-hand sides of equations (M11.3.4.16) are being averaged with respect to φ. This procedure results in an approximate system of equations:

$$a'_t = -\frac{\varepsilon}{\omega_0} f_s(a),$$
$$\varphi'_t = \omega_0 - \frac{\varepsilon}{\omega_0 a} f_c(a), \tag{M11.3.4.17}$$

where

$$f_s(a) = \frac{1}{2\pi} \int_0^{2\pi} \sin \varphi \, f(a \cos \varphi, -\omega_0 a \sin \varphi) \, d\varphi,$$
$$f_c(a) = \frac{1}{2\pi} \int_0^{2\pi} \cos \varphi \, f(a \cos \varphi, -\omega_0 a \sin \varphi) \, d\varphi.$$

System (M11.3.4.17) is substantially simpler than the original system (M11.3.4.16)—the first equation in (M11.3.4.17), for the oscillation amplitude a, is a separable equation and, hence, can readily be integrated; then the second equation in (M11.3.4.17) can also be integrated.

M11.3.5. Galerkin Method and Its Modifications (Projection Methods)

▶ **General form of an approximate solution.** Consider a boundary value problem for the equation

$$\mathfrak{F}[y] - f(x) = 0 \tag{M11.3.5.1}$$

with linear homogeneous boundary conditions* at the points $x = x_1$ and $x = x_2$ ($x_1 \le x \le x_2$). Here, $\mathfrak{F}$ is a linear or nonlinear differential operator of the second order (or a higher order operator); $y = y(x)$ is the unknown function and $f = f(x)$ is a given function. It is assumed that $\mathfrak{F}[0] = 0$.

Let us choose a sequence of linearly independent functions (called *basis functions*)

$$\varphi = \varphi_n(x) \qquad (n = 1, 2, \dots, N) \tag{M11.3.5.2}$$

---

* Nonhomogeneous boundary conditions can be reduced to homogeneous ones by the change of variable $z = A_2 x^2 + A_1 x + A_0 + y$ (the constants A_2, A_1, and A_0 are selected using the method of undetermined coefficients).

satisfying the same boundary conditions as $y = y(x)$. According to all methods that will be considered below, an approximate solution of equation (M11.3.5.1) is sought as a linear combination

$$y_N = \sum_{n=1}^{N} A_n \varphi_n(x), \tag{M11.3.5.3}$$

with the unknown coefficients A_n to be found in the process of solving the problem.

The finite sum (M11.3.5.3) is called an *approximation function*. The remainder term R_N obtained after the finite sum has been substituted into the left-hand side of equation (M11.3.5.1),

$$R_N = \mathfrak{F}[y_N] - f(x). \tag{M11.3.5.4}$$

If the remainder R_N is identically equal to zero, then the function y_N is the exact solution of equation (M11.3.5.1). In general, $R_N \not\equiv 0$.

▶ **Galerkin method.** In order to find the coefficients A_n in (M11.3.5.3), consider another sequence of linearly independent functions

$$\psi = \psi_k(x) \qquad (k = 1, 2, \ldots, N). \tag{M11.3.5.5}$$

Let us multiply both sides of (M11.3.5.4) by ψ_k and integrate the resulting relation over the region $V = \{x_1 \le x \le x_2\}$, in which we seek the solution of equation (M11.3.5.1). Next, we equate the corresponding integrals to zero (for the exact solutions, these integrals are equal to zero). Thus, we obtain the following system of algebraic equations for the unknown coefficients A_n:

$$\int_{x_1}^{x_2} \psi_k R_N \, dx = 0 \qquad (k = 1, 2, \ldots, N). \tag{M11.3.5.6}$$

Relations (M11.3.5.6) mean that the approximation function (M11.3.5.3) satisfies equation (M11.3.5.1) "on the average" (i.e., in the integral sense) with weights ψ_k. Introducing the scalar product $\langle g, h \rangle = \int_{x_1}^{x_2} gh \, dx$ of arbitrary functions g and h, we can consider equations (M11.3.5.6) as the condition of orthogonality of the remainder R_N to all weight functions ψ_k.

The Galerkin method can be applied not only to boundary value problems, but also to eigenvalue problems (in the latter case, one takes $f = \lambda y$ and seeks eigenfunctions y_n, together with eigenvalues λ_n).

Mathematical justification of the Galerkin method for specific boundary value problems can be found in the literature listed at the end of Chapter M11. Below we describe some other methods that are in fact special cases of the Galerkin method.

Remark. Most often, one takes suitable sequences of polynomials or trigonometric functions as $\varphi_n(x)$ in the approximation function (M11.3.5.3).

▶ **Bubnov–Galerkin method, the moment method, the least squares method.**

$1°$. The sequences of functions (M11.3.5.2) and (M11.3.5.5) in the Galerkin method can be chosen arbitrarily. In the case of equal functions,

$$\varphi_k(x) = \psi_k(x) \qquad (k = 1, 2, \ldots, N), \tag{M11.3.5.7}$$

the method is often called the *Bubnov–Galerkin method*.

$2°$. The *moment method* is the Galerkin method with the weight functions (M11.3.5.5) being powers of x,

$$\psi_k = x^k. \tag{M11.3.5.8}$$

$3°$. Sometimes, the functions ψ_k are expressed in terms of φ_k by the relations

$$\psi_k = \mathfrak{F}[\varphi_k] \qquad (k = 1, 2, \ldots),$$

where $\mathfrak{F}$ is the differential operator of equation (M11.3.5.1). This version of the Galerkin method is called the *least squares method*.

▶ **Collocation method.** In the collocation method, one chooses a sequence of points x_k, $k = 1, \ldots, N$, and imposes the condition that the remainder (M11.3.5.4) be zero at these points,

$$R_N = 0 \quad \text{at} \quad x = x_k \qquad (k = 1, \ldots, N). \tag{M11.3.5.9}$$

When solving a specific problem, the points x_k, at which the remainder R_N is set equal to zero, are regarded as most significant. The number of collocation points N is taken equal to the number of the terms of the series (M11.3.5.3). This allows one to obtain a complete system of algebraic equations for the unknown coefficients A_n (for linear boundary value problems, this algebraic system is linear).

Note that the collocation method is a special case of the Galerkin method with the sequence (M11.3.5.5) consisting of the Dirac delta functions:

$$\psi_k = \delta(x - x_k).$$

In the collocation method, there is no need to calculate integrals, and this essentially simplifies the procedure of solving nonlinear problems (although usually this method yields less accurate results than other modifications of the Galerkin method).

Example. Consider the boundary value problem for the linear second-order ordinary differential equation with variable coefficients

$$y''_{xx} + g(x)y - f(x) = 0 \tag{M11.3.5.10}$$

subject to the boundary conditions of the first kind

$$y(-1) = y(1) = 0. \tag{M11.3.5.11}$$

Assume that the coefficients of equation (M11.3.5.10) are smooth even functions, so that $f(x) = f(-x)$ and $g(x) = g(-x)$. We use the collocation method for the approximate solution of problem (M11.3.5.10)–(M11.3.5.11).

$1°$. Take the polynomials

$$y_n(x) = x^{2n-2}(1 - x^2), \qquad n = 1, 2, \ldots N,$$

as the basis functions; they satisfy the boundary conditions (M11.3.5.11), $y_n(\pm 1) = 0$.

Let us consider three collocation points

$$x_1 = -\sigma, \quad x_2 = 0, \quad x_3 = \sigma \qquad (0 < \sigma < 1) \tag{M11.3.5.12}$$

and confine ourselves to two basis functions ($N = 2$), so that the approximation function is taken in the form

$$y(x) = A_1(1 - x^2) + A_2 x^2(1 - x^2). \tag{M11.3.5.13}$$

Substituting (M11.3.5.13) into the left-hand side of equation (M11.3.5.10) yields the remainder

$$R(x) = A_1\left[-2 + (1 - x^2)g(x)\right] + A_2\left[2 - 12x^2 + x^2(1 - x^2)g(x)\right] - f(x).$$

It must vanish at the collocation points (M11.3.5.12). Taking into account the properties $f(\sigma) = f(-\sigma)$ and $g(\sigma) = g(-\sigma)$, we obtain two linear algebraic equations for the coefficients A_1 and A_2:

$$A_1\left[-2 + g(0)\right] + 2A_2 - f(0) = 0 \qquad \text{(at } x = 0\text{)},$$
$$A_1\left[-2 + (1 - \sigma^2)g(\sigma)\right] + A_2\left[2 - 12\sigma^2 + \sigma^2(1 - \sigma^2)g(\sigma)\right] - f(\sigma) = 0 \qquad \text{(at } x = \pm\sigma\text{)}. \tag{M11.3.5.14}$$

$2°$. To be specific, let us take the following functions appearing in equation (M11.3.5.10):

$$f(x) = -1, \quad g(x) = 1 + x^2. \tag{M11.3.5.15}$$

On solving the corresponding system of algebraic equations (M11.3.5.14), we find the coefficients

$$A_1 = \frac{\sigma^4 + 11}{\sigma^4 + 2\sigma^2 + 11}, \quad A_2 = -\frac{\sigma^2}{\sigma^4 + 2\sigma^2 + 11}. \tag{M11.3.5.16}$$

In Fig. M11.2, the solid line depicts the numerical solution to problem (M11.3.5.10)–(M11.3.5.11), with the functions (M11.3.5.15), obtained by the shooting method (see Subsection M11.3.6). The dashed lines 1 and 2 show the approximate solutions obtained by the collocation method using the formulas (M11.3.5.13), (M11.3.5.16) with $\sigma = \frac{1}{2}$ (equidistant points) and $\sigma = \frac{\sqrt{2}}{2}$ (Chebyshev points*), respectively. It is evident that both cases provide good agreement between the approximate and numerical solutions; the use of Chebyshev points gives a more accurate result.

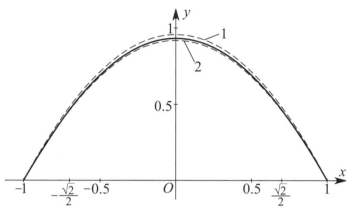

Figure M11.2. Comparison of the numerical solution of problem (M11.3.5.10), (M11.3.5.11), (M11.3.5.15) with the approximate analytical solution (M11.3.5.13), (M11.3.5.16) obtained with the collocation method.

▶ **Method of partitioning the domain.** The domain $V = \{x_1 \le x \le x_2\}$ is split into N subdomains: $V_k = \{x_{k1} \le x \le x_{k2}\}$, $k = 1, \dots, N$. In this method, the weight functions are chosen as follows:

$$\psi_k(x) = \begin{cases} 1 & \text{for } x \in V_k, \\ 0 & \text{for } x \notin V_k. \end{cases}$$

The subdomains V_k are chosen according to the specific properties of the problem under consideration and can generally be arbitrary (the union of all subdomains V_k may differ from the domain V, and some V_k and V_m may overlap).

▶ **Least squared error method.** Sometimes, in order to find the coefficients A_n of the approximation function (M11.3.5.3), one uses the least squared error method based on the minimization of the functional:

$$\Phi = \int_{x_1}^{x_2} R_N^2 \, dx \rightarrow \min. \tag{M11.3.5.17}$$

For given functions φ_n in (M11.3.5.3), the integral Φ is a function with respect to the coefficients A_n. The corresponding necessary conditions of minimum in (M11.3.5.17) have the form

$$\frac{\partial \Phi}{\partial A_n} = 0 \qquad (n = 1, \dots, N).$$

This is a system of algebraic equations for the coefficients A_n.

---

* Chebyshev nodes (points) are generally defined by $x_i = \cos\left(\frac{2i-1}{2m}\pi\right)$, $i = 1, \dots, m$. In this case, $m = 2$.

M11.3.6. Iteration and Numerical Methods

▶ **Method of successive approximations (Cauchy problem).** The method of successive approximations is implemented in two steps. First, the Cauchy problem

$$y''_{xx} = f(x, y, y'_x) \qquad \text{(equation)}, \qquad \text{(M11.3.6.1)}$$
$$y(x_0) = y_0, \quad y'_x(x_0) = y'_0 \qquad \text{(initial conditions)} \qquad \text{(M11.3.6.2)}$$

is reduced to an equivalent system of integral equations by the introduction of the new variable $u(x) = y'_x$. These integral equations have the form

$$u(x) = y'_0 + \int_{x_0}^x f\big(t, y(t), u(t)\big)\, dt, \qquad y(x) = y_0 + \int_{x_0}^x u(t)\, dt. \qquad \text{(M11.3.6.3)}$$

Then the solution of system (M11.3.6.3) is sought by means of successive approximations defined by the following recurrence formulas:

$$u_{n+1}(x) = y'_0 + \int_{x_0}^x f\big(t, y_n(t), u_n(t)\big)\, dt, \qquad y_{n+1}(x) = y_0 + \int_{x_0}^x u_n(t)\, dt; \qquad n = 0, 1, 2, \dots$$

As the initial approximation, one can take $y_0(x) = y_0$ and $u_0(x) = y'_0$.

▶ **Runge–Kutta method (Cauchy problem).** For the numerical integration of the Cauchy problem (M11.3.6.1)–(M11.3.6.2), one often uses the Runge–Kutta method.

Let Δx be sufficiently small. We introduce the following notation:

$$x_k = x_0 + k\Delta x, \quad y_k = y(x_k), \quad y'_k = y'_x(x_k), \quad f_k = f(x_k, y_k, y'_k); \qquad k = 0, 1, 2, \dots$$

The desired values y_k and y'_k are successively found by the formulas

$$y_{k+1} = y_k + y'_k \Delta x + \tfrac{1}{6}(f_1 + f_2 + f_3)(\Delta x)^2,$$
$$y'_{k+1} = y'_k + \tfrac{1}{6}(f_1 + 2f_2 + 2f_3 + f_4)\Delta x,$$

where

$$f_1 = f\big(x_k, y_k, y'_k\big),$$
$$f_2 = f\big(x_k + \tfrac{1}{2}\Delta x, \ y_k + \tfrac{1}{2}y'_k \Delta x, \ y'_k + \tfrac{1}{2}f_1 \Delta x\big),$$
$$f_3 = f\big(x_k + \tfrac{1}{2}\Delta x, \ y_k + \tfrac{1}{2}y'_k \Delta x + \tfrac{1}{4}f_1(\Delta x)^2, \ y'_k + \tfrac{1}{2}f_2 \Delta x\big),$$
$$f_4 = f\big(x_k + \Delta x, \ y_k + y'_k \Delta x + \tfrac{1}{2}f_2(\Delta x)^2, \ y'_k + f_3 \Delta x\big).$$

In practice, the step Δx is determined in the same way as for first-order equations (see Remark 2 in Subsection M11.1.7).

▶ **Shooting method (boundary value problems).** In order to solve the boundary value problem for equation (M11.3.6.1) with the boundary conditions

$$y(x_1) = y_1, \qquad y(x_2) = y_2, \qquad \text{(M11.3.6.4)}$$

one considers an auxiliary Cauchy problem for equation (M11.3.6.1) with the initial conditions

$$y(x_1) = y_1, \qquad y'_x(x_1) = a. \qquad \text{(M11.3.6.5)}$$

(The solution of this Cauchy problem can be obtained by the Runge–Kutta method or some other numerical method.) The parameter a is chosen so that the value of the solution

$y = y(x, a)$ at the point $x = x_2$ coincides with the value required by the second boundary condition in (M11.3.6.4):

$$y(x_2, a) = y_2.$$

First, one finds an a_1 and an a_2 $(a_1 < a_2)$ such that

$$[y(x_2, a_1) - y_2][y(x_2, a_2) - y_2] < 0.$$

This implies that the desired a in (M11.3.6.5) belongs to the interval (a_1, a_2). Then a sequence of numbers a_n such that

$$[y(x_2, a_{n-1}) - y_2][y(x_2, a_n) - y_2] < 0$$

is determined numerically, for example, by using the bisection method. The desired a is obtained as: $a = \lim\limits_{n \to \infty} a_n$.

In a similar way one constructs the solution of the boundary value problem with mixed boundary conditions

$$y(x_1) = y_1, \qquad y'_x(x_2) + ky(x_2) = y_2. \tag{M11.3.6.6}$$

In this case, one also considers the auxiliary Cauchy problem (M11.3.6.1), (M11.3.6.5). The parameter a is chosen so that the solution $y = y(x, a)$ satisfies the second boundary condition in (M11.3.6.6) at the point $x = x_2$.

M11.4. Linear Equations of Arbitrary Order and Linear Systems of Equations

M11.4.1. Linear Equations with Constant Coefficients

▶ **Homogeneous linear equations.** An nth-order homogeneous linear equation with constant coefficients has the general form

$$y_x^{(n)} + a_{n-1} y_x^{(n-1)} + \cdots + a_1 y'_x + a_0 y = 0. \tag{M11.4.1.1}$$

The general solution of this equation is determined by the roots of the characteristic equation

$$P(\lambda) = 0, \qquad \text{where} \quad P(\lambda) = \lambda^n + a_{n-1} \lambda^{n-1} + \cdots + a_1 \lambda + a_0. \tag{M11.4.1.2}$$

The following cases are possible:

1°. All roots $\lambda_1, \lambda_2, \ldots, \lambda_n$ of the characteristic equation (M11.4.1.2) are real and distinct. Then the general solution of the homogeneous linear differential equation (M11.4.1.1) has the form

$$y = C_1 \exp(\lambda_1 x) + C_2 \exp(\lambda_2 x) + \cdots + C_n \exp(\lambda_n x).$$

2°. There are m equal real roots $\lambda_1 = \lambda_2 = \cdots = \lambda_m$ $(m \le n)$, and the other roots are real and distinct. In this case, the general solution is given by

$$y = \exp(\lambda_1 x)(C_1 + C_2 x + \cdots + C_m x^{m-1})$$
$$+ C_{m+1} \exp(\lambda_{m+1} x) + C_{m+2} \exp(\lambda_{m+2} x) + \cdots + C_n \exp(\lambda_n x).$$

3°. There are m equal complex conjugate roots $\lambda_{1,2} = \alpha \pm i\beta$ $(2m \le n)$, and the other roots are real and distinct. In this case, the general solution is

$$y = \exp(\alpha x) \cos(\beta x)(A_1 + A_2 x + \cdots + A_m x^{m-1})$$
$$+ \exp(\alpha x) \sin(\beta x)(B_1 + B_2 x + \cdots + B_m x^{m-1})$$
$$+ C_{2m+1} \exp(\lambda_{2m+1} x) + C_{2m+2} \exp(\lambda_{2m+2} x) + \cdots + C_n \exp(\lambda_n x),$$

where $A_1, \ldots, A_m, B_1, \ldots, B_m, C_{2m+1}, \ldots, C_n$ are arbitrary constants.

4°. In the general case, where there are r different roots $\lambda_1, \lambda_2, \ldots, \lambda_r$ of multiplicities $m_1, m_2, \ldots, m_r$, respectively, the left-hand side of the characteristic equation (M11.4.1.2) can be represented as the product

$$P(\lambda) = (\lambda - \lambda_1)^{m_1}(\lambda - \lambda_2)^{m_2} \ldots (\lambda - \lambda_r)^{m_r},$$

where $m_1 + m_2 + \cdots + m_r = n$. The general solution of the original equation is given by the formula

$$y = \sum_{k=1}^{r} \exp(\lambda_k x)(C_{k,0} + C_{k,1}x + \cdots + C_{k,m_k-1}x^{m_k-1}),$$

where $C_{k,l}$ are arbitrary constants.

If the characteristic equation (M11.4.1.2) has complex conjugate roots $\lambda_s = \alpha_s + i\beta_s$ and $\lambda_{s+1} = \alpha_s - i\beta_s$, then in the above solution, the corresponding functions $\exp(\lambda_s x)$ and $\exp(\lambda_{s+1}x)$ should be replaced with $\exp(\alpha_s x)\cos(\beta_s x)$ and $\exp(\alpha_s x)\sin(\beta_s x)$, respectively, in a similar way to that in Item 3°.

Example 1. Find the general solution of the linear third-order equation

$$y''' + ay'' - y' - ay = 0.$$

Its characteristic equation is $\lambda^3 + a\lambda^2 - \lambda - a = 0$, or, in factorized form,

$$(\lambda + a)(\lambda - 1)(\lambda + 1) = 0.$$

Depending on the value of the parameter a, three cases are possible.

1. Case $a \neq \pm 1$. There are three different roots, $\lambda_1 = -a$, $\lambda_2 = -1$, and $\lambda_3 = 1$. The general solution of the differential equation is expressed as $y = C_1 e^{-ax} + C_2 e^{-x} + C_3 e^x$.

2. Case $a = 1$. There is a double root, $\lambda_1 = \lambda_2 = -1$, and a simple root, $\lambda_3 = 1$. The general solution of the differential equation has the form $y = (C_1 + C_2 x)e^{-x} + C_3 e^x$.

3. Case $a = -1$. There is a double root, $\lambda_1 = \lambda_2 = 1$, and a simple root, $\lambda_3 = -1$. The general solution of the differential equation is expressed as $y = (C_1 + C_2 x)e^x + C_3 e^{-x}$.

Example 2. Consider the linear fourth-order equation

$$y''''_{xxxx} - y = 0.$$

Its characteristic equation, $\lambda^4 - 1 = 0$, has four distinct roots, two real and two pure imaginary,

$$\lambda_1 = 1, \quad \lambda_2 = -1, \quad \lambda_3 = i, \quad \lambda_4 = -i.$$

Therefore, the general solution of the equation in question has the form (see Item 3°)

$$y = C_1 e^x + C_2 e^{-x} + C_3 \sin x + C_4 \cos x.$$

▶ **Nonhomogeneous linear equations. Forms of particular solutions.**

1°. An nth-order nonhomogeneous linear equation with constant coefficients has the general form

$$y_x^{(n)} + a_{n-1}y_x^{(n-1)} + \cdots + a_1 y_x' + a_0 y = f(x). \tag{M11.4.1.3}$$

The general solution of this equation is the sum of the general solution of the corresponding homogeneous equation with $f(x) \equiv 0$ (see equation M11.4.1.1) and any particular solution of the nonhomogeneous equation (M11.4.1.3).

If the roots $\lambda_1, \lambda_2, \ldots, \lambda_n$ of the characteristic equation (M11.4.1.2) are all real and distinct, equation (M11.4.1.3) has the general solution:

$$y = \sum_{\nu=1}^{n} C_\nu e^{\lambda_\nu x} + \sum_{\nu=1}^{n} \frac{e^{\lambda_\nu x}}{P'_\lambda(\lambda_\nu)} \int f(x)e^{-\lambda_\nu x}\, dx. \tag{M11.4.1.4}$$

In the general case, if the characteristic equation (M11.4.1.2) has complex and/or multiple roots, the solution to equation (M11.4.1.3) can be constructed using formula (M11.4.2.5).

TABLE M11.4

Forms of particular solutions to the nonhomogeneous linear equation with constant coefficients
$y_x^{(n)} + a_{n-1} y_x^{(n-1)} + \cdots + a_1 y_x' + a_0 y = f(x)$, that correspond to some special forms of the function $f(x)$.

Form of the function $f(x)$	Roots of the characteristic equation $\lambda^n + a_{n-1}\lambda^{n-1} + \cdots + a_1\lambda + a_0 = 0$	Form of particular solution
$P_m(x)$	Zero is not a root of the characteristic equation (i.e., $a_0 \neq 0$)	$\widetilde{P}_m(x)$
	Zero is a root of the characteristic equation (multiplicity r)	$x^r \widetilde{P}_m(x)$
$P_m(x)e^{\alpha x}$ (α is a real constant)	α is not a root of the characteristic equation	$\widetilde{P}_m(x)e^{\alpha x}$
	α is a root of the characteristic equation (multiplicity r)	$x^r \widetilde{P}_m(x)e^{\alpha x}$
$P_m(x)\cos\beta x + Q_k(x)\sin\beta x$	$i\beta$ is not a root of the characteristic equation	$\widetilde{P}_\nu(x)\cos\beta x + \widetilde{Q}_\nu(x)\sin\beta x$
	$i\beta$ is a root of the characteristic equation (multiplicity r)	$x^r[\widetilde{P}_\nu(x)\cos\beta x + \widetilde{Q}_\nu(x)\sin\beta x]$
$[P_m(x)\cos\beta x + Q_k(x)\sin\beta x]e^{\alpha x}$	$\alpha + i\beta$ is not a root of the characteristic equation	$[\widetilde{P}_\nu(x)\cos\beta x + \widetilde{Q}_\nu(x)\sin\beta x]e^{\alpha x}$
	$\alpha + i\beta$ is a root of the characteristic equation (multiplicity r)	$x^r[\widetilde{P}_\nu(x)\cos\beta x + \widetilde{Q}_\nu(x)\sin\beta x]e^{\alpha x}$

Notation: P_m and Q_k are polynomials of degrees m and k with given coefficients; $\widetilde{P}_m$, $\widetilde{P}_\nu$, and $\widetilde{Q}_\nu$ are polynomials of degrees m and ν whose coefficients are determined by substituting the particular solution into the basic equation; $\nu = \max\{m, k\}$; and α and β are real numbers, $i^2 = -1$.

2°. Table M11.4 lists the forms of particular solutions corresponding to some special forms of functions on the right-hand side of the linear nonhomogeneous equation.

3°. Consider the Cauchy problem for equation (M11.4.1.3) subject to the homogeneous initial conditions

$$y(0) = y_x'(0) = \cdots = y_x^{(n-1)}(0) = 0. \tag{M11.4.1.5}$$

Let $y(x)$ be the solution of problem (M11.4.1.3), (M11.4.1.5) for arbitrary $f(x)$ and let $u(x)$ be the solution of the auxiliary, simpler problem (M11.4.1.3), (M11.4.1.5) with $f(x) \equiv 1$, so that $u(x) = y(x)|_{f(x)\equiv 1}$. Then the formula

$$y(x) = \int_0^x f(t)u_x'(x - t)\, dt$$

holds. It is called the *Duhamel integral*.

▶ **Solution of the Cauchy problem using the Laplace transform.** Consider the Cauchy problem for equation (M11.4.1.3) with arbitrary initial conditions

$$y(0) = y_0, \quad y_x'(0) = y_1, \quad \ldots, \quad y_x^{(n-1)}(0) = y_{n-1}, \tag{M11.4.1.6}$$

where $y_0, y_1, \ldots, y_{n-1}$ are given constants.

Problem (M11.4.1.3), (M11.4.1.6) can be solved using the Laplace transform based on the formulas (for details, see Section M10.2)

$$\widetilde{y}(p) = \mathfrak{L}\{y(x)\}, \quad \widetilde{f}(p) = \mathfrak{L}\{f(x)\}, \quad \text{where} \quad \mathfrak{L}\{f(x)\} \equiv \int_0^\infty e^{-px} f(x)\, dx.$$

To this end, let us multiply equation (M11.4.1.3) by e^{-px} and then integrate with respect to x from zero to infinity. Taking into account the formula

$$\mathfrak{L}\{y_x^{(n)}(x)\} = p^n \widetilde{y}(p) - \sum_{k=1}^n p^{n-k} y_x^{(k-1)}(+0)$$

and the initial conditions (M11.4.1.6), we arrive at a linear algebraic equation for the transform $\widetilde{y}(p)$:

$$P(p)\widetilde{y}(p) - Q(p) = \widetilde{f}(p), \tag{M11.4.1.7}$$

where

$$P(p) = p^n + a_{n-1}p^{n-1} + \cdots + a_1 p + a_0, \quad Q(p) = b_{n-1}p^{n-1} + \cdots + b_1 p + b_0,$$
$$b_k = y_{n-k-1} + a_{n-1}y_{n-k-2} + \cdots + a_{k+2}y_1 + a_{k+1}y_0, \quad k = 0, 1, \ldots, n-1.$$

The polynomial $P(p)$ coincides with the characteristic polynomial (M11.4.1.2) at $\lambda = p$. The solution of equation (M11.4.1.7) is given by the formula

$$\widetilde{y}(p) = \frac{\widetilde{f}(p) + Q(p)}{P(p)}. \tag{M11.4.1.8}$$

On applying the Laplace inversion formula (see in Section M10.2) to (M11.4.1.8), we obtain a solution to problem (M11.4.1.3), (M11.4.1.6) in the form

$$y(x) = \frac{1}{2\pi i} \int_{c-i\infty}^{c+i\infty} \frac{\widetilde{f}(p) + Q(p)}{P(p)} e^{px}\, dp. \tag{M11.4.1.9}$$

Since the transform $\widetilde{y}(p)$ (M11.4.1.8) is a rational function, the inverse Laplace transform (M11.4.1.9) can be obtained using the formulas from Subsection M10.2.2 or the tables of Section S2.2.

Remark. In practice, the solution method for the Cauchy problem based on the Laplace transform leads to the solution faster than the direct application of general formulas like (M11.4.1.4), where one has to determine the coefficients $C_1, \ldots, C_n$.

Example 3. Consider the following Cauchy problem for a homogeneous fourth-order equation:

$$y_{xxxx}'''' + a^4 y = 0; \quad y(0) = y_x'(0) = y_{xxx}'''(0) = 0, \quad y_{xx}''(0) = b.$$

The Laplace transform reduces this problem to a linear algebraic equation for $\widetilde{y}(p)$: $(p^4 + a^4)\widetilde{y}(p) - bp = 0$. It follows that

$$\widetilde{y}(p) = \frac{bp}{p^4 + a^4}.$$

In order to invert this expression, let us use the table of inverse Laplace transforms S2.2.2 (see row 52) and take into account that a constant multiplier can be taken outside the transform operator to obtain the solution to the original Cauchy problem in the form

$$y(x) = \frac{b}{a^2} \sin\left(\frac{ax}{\sqrt{2}}\right) \sinh\left(\frac{ax}{\sqrt{2}}\right).$$

M11.4.2. Linear Equations with Variable Coefficients

▶ **Homogeneous linear equations. Structure of the general solution.** The general solution of the nth-order homogeneous linear differential equation

$$f_n(x)y_x^{(n)} + f_{n-1}(x)y_x^{(n-1)} + \cdots + f_1(x)y_x' + f_0(x)y = 0 \qquad \text{(M11.4.2.1)}$$

has the form

$$y = C_1 y_1(x) + C_2 y_2(x) + \cdots + C_n y_n(x), \qquad \text{(M11.4.2.2)}$$

where the functions $y_1(x)$, $y_2(x)$, $\ldots$, $y_n(x)$ are a fundamental system of solutions (the y_k are linearly independent particular solutions, $y_k \not\equiv 0$); C_1, C_2, $\ldots$, C_n are arbitrary constants.

▶ **Utilization of particular solutions for reducing the order of the equation.**

$1°$. Let $y_1 = y_1(x)$ be a nontrivial particular solution of equation (M11.4.2.1). The substitution

$$y = y_1(x) \int z(x)\,dx$$

results in a linear equation of order $n - 1$ for the function $z(x)$.

$2°$. Let $y_1 = y_1(x)$ and $y_2 = y_2(x)$ be two nontrivial linearly independent solutions of equation (M11.4.2.1). The substitution

$$y = y_1 \int y_2 w\,dx - y_2 \int y_1 w\,dx$$

results in a linear equation of order $n - 2$ for $w(x)$.

$3°$. Suppose that m linearly independent solutions $y_1(x)$, $y_2(x)$, $\ldots$, $y_m(x)$ of equation (M11.4.2.1) are known. Then one can reduce the order of the equation to $n - m$ by successive application of the following procedure. The substitution $y = y_m(x) \int z(x)\,dx$ leads to an equation of order $n - 1$ for the function $z(x)$ with known linearly independent solutions:

$$z_1 = \left(\frac{y_1}{y_m}\right)_x', \quad z_2 = \left(\frac{y_2}{y_m}\right)_x', \quad \ldots, \quad z_{m-1} = \left(\frac{y_{m-1}}{y_m}\right)_x'.$$

The substitution $z = z_{m-1}(x) \int w(x)\,dx$ yields an equation of order $n - 2$. Repeating this procedure m times, we arrive at a homogeneous linear equation of order $n - m$.

▶ **Wronskian determinant and Liouville formula.** The *Wronskian determinant* (or simply, *Wronskian*) is the function defined as

$$W(x) = \begin{vmatrix} y_1(x) & \cdots & y_n(x) \\ y_1'(x) & \cdots & y_n'(x) \\ \cdots & \cdots & \cdots \\ y_1^{(n-1)}(x) & \cdots & y_n^{(n-1)}(x) \end{vmatrix}, \qquad \text{(M11.4.2.3)}$$

where $y_1(x)$, $\ldots$, $y_n(x)$ is a fundamental system of solutions of the homogeneous equation (M11.4.2.1); $y_k^{(m)}(x) = \dfrac{d^m y_k}{dx^m}$, $m = 1, \ldots, n-1$; $k = 1, \ldots, n$.

The following *Liouville formula* holds:

$$W(x) = W(x_0) \exp\left[-\int_{x_0}^x \frac{f_{n-1}(t)}{f_n(t)}\,dt\right],$$

where x_0 is an arbitrary number.

▶ **Nonhomogeneous linear equations. Construction of the general solution.**

$1°$. The general nonhomogeneous nth-order linear differential equation has the form

$$f_n(x)y_x^{(n)} + f_{n-1}(x)y_x^{(n-1)} + \cdots + f_1(x)y_x' + f_0(x)y = g(x). \qquad \text{(M11.4.2.4)}$$

The general solution of the nonhomogeneous equation (M11.4.2.4) is obtained as the sum of the general solution of the corresponding homogeneous equation (M11.4.2.1) and any particular solution of equation (M11.4.2.4).

$2°$. Let $y_1(x), \ldots, y_n(x)$ be a fundamental system of solutions of the homogeneous equation (M11.4.2.1), and let $W(x)$ be the Wronskian determinant (M11.4.2.3). Then the general solution of the nonhomogeneous linear equation (M11.4.2.4) can be represented as

$$y = \sum_{\nu=1}^{n} C_\nu y_\nu(x) + \sum_{\nu=1}^{n} y_\nu(x) \int \frac{W_\nu(x)\,dx}{f_n(x)W(x)}, \qquad \text{(M11.4.2.5)}$$

where $W_\nu(x)$ is the determinant obtained by replacing the νth column of the matrix (M11.4.2.3) by the column vector with the elements $0, 0, \ldots, 0, g$.

$3°$. *Superposition principle.* A particular solution of a nonhomogeneous linear equation

$$\mathbf{L}[y] = \sum_{k=1}^{m} g_k(x), \qquad \mathbf{L}[y] \equiv f_n(x)y_x^{(n)} + f_{n-1}(x)y_x^{(n-1)} + \cdots + f_1(x)y_x' + f_0(x)y$$

is determined by adding together particular solutions,

$$y = \sum_{k=1}^{m} y_k,$$

of m (simpler) equations,

$$\mathbf{L}[y_k] = g_k(x), \qquad k = 1, 2, \ldots, m,$$

corresponding to respective nonhomogeneous terms in the original equation.

▶ **Euler equation.** The nonhomogeneous Euler equation has the form

$$x^n y_x^{(n)} + a_{n-1}x^{n-1}y_x^{(n-1)} + \cdots + a_1 x y_x' + a_0 y = f(x).$$

The substitution $x = be^t$ $(b \neq 0)$ leads to a constant coefficient linear equation of the form (M11.4.1.3).

Particular solutions of the homogeneous Euler equation [with $f(x) \equiv 0$] are sought in the form $y = x^k$. If all k are real and distinct, its general solution is expressed as

$$y(x) = C_1|x|^{k_1} + C_2|x|^{k_2} + \cdots + C_n|x|^{k_n}.$$

Remark. To a pair of complex conjugate values $k = \alpha \pm i\beta$ there corresponds a pair of particular solutions: $y = |x|^\alpha \sin(\beta|x|)$ and $y = |x|^\alpha \cos(\beta|x|)$.

M11.4.3. Systems of Linear Equations with Constant Coefficients

▶ **Systems of first-order linear homogeneous equations. The general solution.**

$1°$. In general, a homogeneous linear system of first-order ordinary differential equations with constant coefficients has the form

$$
\begin{aligned}
y_1' &= a_{11}y_1 + a_{12}y_2 + \cdots + a_{1n}y_n, \\
y_2' &= a_{21}y_1 + a_{22}y_2 + \cdots + a_{2n}y_n, \\
&\cdots\cdots\cdots\cdots\cdots\cdots\cdots\cdots\cdots \\
y_n' &= a_{n1}y_1 + a_{n2}y_2 + \cdots + a_{nn}y_n,
\end{aligned}
\tag{M11.4.3.1}
$$

where a prime stands for the derivative with respect to x. In the sequel, all the coefficients a_{ij} of the system are assumed to be real numbers.

The homogeneous system (M11.4.3.1) has the trivial particular solution $y_1 = y_2 = \cdots = y_n = 0$.

Superposition principle for a homogeneous system: any linear combination of particular solutions of system (M11.4.3.1) is also a solution of this system.

The general solution of the system of differential equations (M11.4.3.1) is the sum of its n linearly independent (nontrivial) particular solutions each multiplied by an arbitrary constant.

Remark. System (M11.4.3.1) can be reduced to a single homogeneous linear constant-coefficient nth-order equation.

$2°$. For brevity (and clarity), system (M11.4.3.1) is conventionally written in vector-matrix form:

$$
\mathbf{y}' = \mathbf{A}\mathbf{y},
\tag{M11.4.3.2}
$$

where $\mathbf{y} = (y_1, y_2, \ldots, y_n)^{\mathrm{T}}$ is the column vector of the unknowns and $\mathbf{A} = (a_{ij})$ is the matrix of the equation coefficients. The superscript T denotes the transpose of a matrix or a vector. So, for example, a row vector is converted into a column vector:

$$
(y_1, y_2)^{\mathrm{T}} \equiv \begin{pmatrix} y_1 \\ y_2 \end{pmatrix}.
$$

The right-hand side of equation (M11.4.3.2) is the product of the $n \times n$ square matrix $\mathbf{a}$ by the $n \times 1$ matrix (column vector) $\mathbf{y}$.

Let $\mathbf{y}_k = (y_{k1}, y_{k2}, \ldots, y_{kn})^{\mathrm{T}}$ be linearly independent particular solutions* of the homogeneous system (M11.4.3.1), where $k = 1, 2, \ldots, n$; the first subscript in $y_{km} = y_{km}(x)$ denotes the number of the solution and the second subscript ($m = 1, \ldots, n$) indicates the component of the vector solution. Then the general solution of the homogeneous system (M11.4.3.2) is expressed as

$$
\mathbf{y} = C_1\mathbf{y}_1 + C_2\mathbf{y}_2 + \cdots + C_n\mathbf{y}_n.
\tag{M11.4.3.3}
$$

A method for the construction of particular solutions that can be used to obtain the general solution by formula (M11.4.3.3) is presented below.

---

* This means that the condition $\det |y_{km}(x)| \neq 0$ holds.

▶ **Systems of first-order linear homogeneous equations. Particular solutions.** Particular solutions to system (M11.4.3.1) are determined by the roots of the characteristic equation

$$\Delta(\lambda) = 0, \quad \text{where} \quad \Delta(\lambda) \equiv \begin{vmatrix} a_{11} - \lambda & a_{12} & \dots & a_{1n} \\ a_{21} & a_{22} - \lambda & \dots & a_{2n} \\ \dots & \dots & \dots & \dots \\ a_{n1} & a_{n2} & \dots & a_{nn} - \lambda \end{vmatrix}. \qquad \text{(M11.4.3.4)}$$

The following cases are possible:

1°. Let λ be a simple real root of the characteristic equation (M11.4.3.4). The corresponding particular solution of the homogeneous linear system of equations (M11.4.3.1) has the exponential form

$$y_1 = A_1 e^{\lambda x}, \quad y_2 = A_2 e^{\lambda x}, \quad \dots, \quad y_n = A_n e^{\lambda x}, \qquad \text{(M11.4.3.5)}$$

where the coefficients $A_1, A_2, \dots, A_n$ are determined by solving the associated linear homogeneous system of equations:

$$\begin{aligned} (a_{11} - \lambda)A_1 + a_{12}A_2 + \cdots + a_{1n}A_n &= 0, \\ a_{21}A_1 + (a_{22} - \lambda)A_2 + \cdots + a_{2n}A_n &= 0, \\ \dots\dots\dots\dots\dots\dots\dots\dots\dots\dots\dots\dots\dots\dots& \\ a_{n1}A_1 + a_{n2}A_2 + \cdots + (a_{nn} - \lambda)A_n &= 0. \end{aligned} \qquad \text{(M11.4.3.6)}$$

The solution of this system is unique to within a constant factor.

If all roots of the characteristic equation $\lambda_1, \lambda_2, \dots, \lambda_n$ are real and distinct, then the general solution of system (M11.4.3.1) has the form

$$\begin{aligned} y_1 &= C_1 A_{11} e^{\lambda_1 x} + C_2 A_{12} e^{\lambda_2 x} + \cdots + C_n A_{1n} e^{\lambda_n x}, \\ y_2 &= C_1 A_{21} e^{\lambda_1 x} + C_2 A_{22} e^{\lambda_2 x} + \cdots + C_n A_{2n} e^{\lambda_n x}, \\ \dots\dots\dots\dots\dots\dots\dots\dots\dots\dots\dots\dots\dots\dots\dots\dots& \\ y_n &= C_1 A_{n1} e^{\lambda_1 x} + C_2 A_{n2} e^{\lambda_2 x} + \cdots + C_n A_{nn} e^{\lambda_n x}, \end{aligned} \qquad \text{(M11.4.3.7)}$$

where $C_1, C_2, \dots, C_n$ are arbitrary constants. The second subscript in A_{mk} indicates a coefficient corresponding to the root λ_k.

2°. For each simple complex root, $\lambda = \alpha + i\beta$, of the characteristic equation (M11.4.3.4), the corresponding particular solution is obtained in the same way as in the simple real root case; the associated coefficients $A_1, A_2, \dots, A_n$ in (M11.4.3.5) will be complex. Separating the real and imaginary parts in (M11.4.3.5) results in two real particular solutions to system (M11.4.3.1); the same two solutions are obtained if one takes the complex conjugate root, $\bar{\lambda} = \alpha - i\beta$.

3°. Let λ be a real root of the characteristic equation (M11.4.3.4) of multiplicity m. The corresponding particular solution of system (M11.4.3.1) is sought in the form

$$y_1 = P_m^1(x) e^{\lambda x}, \quad y_2 = P_m^2(x) e^{\lambda x}, \quad \dots, \quad y_n = P_m^n(x) e^{\lambda x}, \qquad \text{(M11.4.3.8)}$$

where the $P_m^k(x) = \sum_{i=0}^{m-1} B_{ki} x^i$ are polynomials of degree $m-1$. The coefficients of these polynomials result from the substitution of expressions (M11.4.3.8) into equations (M11.4.3.1); after dividing by $e^{\lambda x}$ and collecting like terms, one obtains n equations, each representing a polynomial equated to zero. By equating the coefficients of all resulting polynomials to zero, one arrives at a linear algebraic system of equations for the coefficients B_{ki}; the solution to this system will contain m free parameters.

4°. For a multiple complex root, $\lambda = \alpha + i\beta$, of multiplicity m, the corresponding particular solution is sought, just as in the case of a multiple real root, in the form (M11.4.3.8); here the coefficients B_{ki} of the polynomials $P_m^k(x)$ will be complex. Finally, in order to obtain real solutions of the original system (M11.4.3.1), one separates the real and imaginary parts in formulas (M11.4.3.8), thus obtaining two particular solutions with m free parameters each. The two solutions correspond to the complex conjugate roots $\lambda = \alpha \pm i\beta$.

5°. In the general case, where the characteristic equation (M11.4.3.4) has simple and multiple, real and complex roots (see Items 1°–4°), the general solution to system (M11.4.3.1) is obtained as the sum of all particular solutions multiplied by arbitrary constants.

Example 1. Consider the homogeneous system of two linear differential equations

$$y_1' = y_1 + 4y_2,$$
$$y_2' = y_1 + y_2.$$

The associated characteristic equation,

$$\begin{vmatrix} 1 - \lambda & 4 \\ 1 & 1 - \lambda \end{vmatrix} = \lambda^2 - 2\lambda - 3 = 0,$$

has distinct real roots:

$$\lambda_1 = 3, \qquad \lambda_2 = -1.$$

The system of algebraic equations (M11.4.3.6) for the solution coefficients becomes

$$(1 - \lambda)A_1 + 4A_2 = 0,$$
$$A_1 + (1 - \lambda)A_2 = 0. \tag{M11.4.3.9}$$

Substituting the first root, $\lambda = 3$, into system (M11.4.3.9) yields $A_1 = 2A_2$. We can set $A_1 = 2$ and $A_2 = 1$, since the solution is determined to within a constant factor. Thus the first particular solution of the homogeneous system of linear ordinary differential equations (M11.4.3.9) has the form

$$y_1 = 2e^{3x}, \qquad y_2 = e^{3x}. \tag{M11.4.3.10}$$

The second particular solution, corresponding to $\lambda = -1$, is found in the same way:

$$y_1 = -2e^{-x}, \qquad y_2 = e^{-x}. \tag{M11.4.3.11}$$

The sum of the two particular solutions (M11.4.3.10) and (M11.4.3.11) multiplied by arbitrary constants, C_1 and C_2, gives the general solution to the original homogeneous system of linear ordinary differential equations:

$$y_1 = 2C_1 e^{3x} - 2C_2 e^{-x}, \qquad y_2 = C_1 e^{3x} + C_2 e^{-x}.$$

Example 2. Consider the system of ordinary differential equations

$$y_1' = -y_2,$$
$$y_2' = 2y_1 + 2y_2. \tag{M11.4.3.12}$$

The characteristic equation

$$\begin{vmatrix} -\lambda & -1 \\ 2 & 2 - \lambda \end{vmatrix} = \lambda^2 - 2\lambda + 2 = 0$$

has complex conjugate roots:

$$\lambda_1 = 1 + i, \qquad \lambda_2 = 1 - i.$$

The algebraic system (M11.4.3.6) for the complex coefficients A_1 and A_2 becomes

$$-\lambda A_1 - A_2 = 0,$$
$$2A_1 + (2 - \lambda)A_2 = 0.$$

With $\lambda = 1 + i$, one nonzero solution is given by $A_1 = 1$ and $A_2 = -1 - i$. The corresponding complex solution to system (M11.4.3.12) has the form

$$y_1 = e^{(1+i)x}, \qquad y_2 = (-1 - i)e^{(1+i)x}.$$

Separating the real and imaginary parts, taking into account the formulas

$$e^{(1+i)x} = e^x(\cos x + i \sin x) = e^x \cos x + i e^x \sin x,$$

$$(-1-i)e^{(1+i)x} = -(1+i)e^x(\cos x + i \sin x) = e^x(\sin x - \cos x) - i e^x(\sin x + \cos x),$$

and making linear combinations of them, one arrives at the general solution to the original system (M11.4.3.12):

$$y_1 = C_1 e^x \cos x + C_2 e^x \sin x,$$
$$y_2 = C_1 e^x(\sin x - \cos x) - C_2 e^x(\sin x + \cos x).$$

▶ **Nonhomogeneous systems of linear first-order equations.**

$1°$. In general, a nonhomogeneous linear system of first-order differential equations with constant coefficients has the form

$$
\begin{aligned}
y_1' &= a_{11}y_1 + a_{12}y_2 + \cdots + a_{1n}y_n + f_1(x), \\
y_2' &= a_{21}y_1 + a_{22}y_2 + \cdots + a_{2n}y_n + f_2(x), \\
&\cdots\cdots\cdots\cdots\cdots\cdots\cdots\cdots\cdots\cdots\cdots\cdots\cdots\cdots\cdots \\
y_n' &= a_{n1}y_1 + a_{n2}y_2 + \cdots + a_{nn}y_n + f_n(x).
\end{aligned}
\qquad \text{(M11.4.3.13)}
$$

For brevity, the conventional vector notation will also be used:

$$\mathbf{y}' = \mathbf{A}\mathbf{y} + \mathbf{f}(x),$$

where $\mathbf{f}(x) = (f_1(x), f_2(x), \dots, f_n(x))^{\mathrm{T}}$.

The general solution of this system is the sum of the general solution to the corresponding homogeneous system with $f_k(x) \equiv 0$ [see system (M11.4.3.1)] and any particular solution of the nonhomogeneous system (M11.4.3.13).

$2°$. Let $\mathbf{y}_k = (D_{k1}(x), D_{k1}(x), \dots, D_{kn}(x))^{\mathrm{T}}$ be particular solutions to the homogeneous linear system of first-order constant-coefficient differential equations (M11.4.3.1) that satisfy the special initial conditions

$$y_k(0) = 1, \qquad y_m(0) = 0 \quad \text{for} \quad m \neq k; \qquad k, m = 1, \dots, n.$$

Then the general solution to the nonhomogeneous system (M11.4.3.13) is expressed as

$$y_m(x) = \sum_{k=1}^{n} \int_0^x f_k(t) D_{km}(x-t)\, dt + \sum_{k=1}^{n} C_k D_{km}(x), \qquad m = 1, \dots, n. \quad \text{(M11.4.3.14)}$$

The solution of the Cauchy problem for the nonhomogeneous system (M11.4.3.13) with arbitrary initial conditions,

$$y_1(0) = y_1^\circ, \qquad y_2(0) = y_2^\circ, \qquad \dots, \qquad y_n(0) = y_n^\circ, \qquad \text{(M11.4.3.15)}$$

is determined by formulas (M11.4.3.14) with $C_k = y_k^\circ$, $k = 1, \dots, n$.

M11.5. Nonlinear Equations of Arbitrary Order

M11.5.1. Structure of the General Solution. Cauchy Problem

▶ **Equations solved for the highest derivative. General solution.** An *nth-order differential equation solved for the highest derivative* has the form

$$y_x^{(n)} = f(x, y, y_x', \dots, y_x^{(n-1)}). \qquad \text{(M11.5.1.1)}$$

The general solution of this equation depends on n arbitrary constants $C_1, \ldots, C_n$. In some cases, the general solution can be written in explicit form as

$$y = \varphi(x, C_1, \ldots, C_n). \tag{M11.5.1.2}$$

▶ **Cauchy problem. The existence and uniqueness theorem.** The *Cauchy problem*: find a solution of equation (M11.5.1.1) with the *initial conditions*

$$y(x_0) = y_0, \quad y_x'(x_0) = y_0^{(1)}, \quad \ldots, \quad y_x^{(n-1)}(x_0) = y_0^{(n-1)}. \tag{M11.5.1.3}$$

(At a point x_0, the values of the unknown function $y(x)$ and all its derivatives of orders $\leq n-1$ are prescribed.)

EXISTENCE AND UNIQUENESS THEOREM. *Suppose the function $f(x, y, z_1, \ldots, z_{n-1})$ is continuous in all its arguments in a neighborhood of the point $(x_0, y_0, y_0^{(1)}, \ldots, y_0^{(n-1)})$ and has bounded derivatives with respect to y, z_1, $\ldots$, z_{n-1} in this neighborhood. Then a solution of equation (M11.5.1.1) satisfying the initial conditions (M11.5.1.3) exists and is unique.*

▶ **Reduction of an nth-order equation to a system of n first-order equations.** The differential equation (M11.5.1.1) is equivalent to the following system of n first-order equations:

$$y_0' = y_1, \quad y_1' = y_2, \quad \ldots, \quad y_{n-2}' = y_{n-1}, \quad y_{n-1}' = f(x, y_0, y_1, \ldots, y_{n-1}),$$

where the notation $y_0 \equiv y$ is adopted.

M11.5.2. Equations Admitting Reduction of Order

▶ **Equations not containing $y, y_x', \ldots, y_x^{(k)}$ explicitly.** An equation that does not explicitly contain the unknown function and its derivatives up to order k inclusive can generally be written as

$$F\left(x, y_x^{(k+1)}, \ldots, y_x^{(n)}\right) = 0 \qquad (1 \leq k+1 < n). \tag{M11.5.2.1}$$

Such equations are invariant under arbitrary translations of the unknown function, $y \to y + \text{const}$ (the form of such equations is also preserved under the transformation $u(x) = y + a_k x^k + \cdots + a_1 x + a_0$, where the a_m are arbitrary constants). The substitution $z(x) = y_x^{(k+1)}$ reduces (M11.5.2.1) to an equation whose order is by $k+1$ smaller than that of the original equation, $F\left(x, z, z_x', \ldots, z_x^{(n-k-1)}\right) = 0$.

▶ **Equations not containing x explicitly (autonomous equations).** In general, an equation that does not explicitly contain x has form

$$F\left(y, y_x', \ldots, y_x^{(n)}\right) = 0. \tag{M11.5.2.2}$$

Such equations are invariant under arbitrary translations of the independent variable, $x \to x + \text{const}$. The substitution $y_x' = w(y)$ (where y plays the role of the independent variable) reduces by one the order of an autonomous equation. Higher derivatives can be expressed in terms of w and its derivatives with respect to the new independent variable, $y_{xx}'' = ww_y'$, $y_{xxx}''' = w^2 w_{yy}'' + w(w_y')^2, \ldots$

▶ **Some other equations admitting reduction of order.** Table M11.5 lists the above nonlinear equations as well as some other equations admitting order reduction. The second column gives simple transformations that allow checking whether the equation is one of this type.

TABLE M11.5
Some ordinary differential equations that admit order reduction by one.

Equation	Transformation preserving form of equation (a is an arbitrary constant)	Transformation reducing order of equation, $u = u(z)$
$F\left(x, y'_x, y''_{xx}, \ldots, y_x^{(n)}\right) = 0$	$y = \bar{y} + a$	$z = x, \ u = y'_x$
$F\left(y, y'_x, y''_{xx}, \ldots, y_x^{(n)}\right) = 0$	$x = \bar{x} + a$	$z = y, \ u = y'_x$
$F\left(\alpha x + \beta y + \gamma, y'_x, y''_{xx}, \ldots, y_x^{(n)}\right) = 0$	$x = \bar{x} + a\beta, \ y = \bar{y} - a\alpha$	$z = \alpha x + \beta y + \gamma, \ u = y'_x$
$F\left(x, y'_x/y, y''_{xx}/y, \ldots, y_x^{(n)}/y\right) = 0$	$y = a\bar{y}$	$z = x, \ u = y'_x/y$
$F\left(y, xy'_x, x^2 y''_{xx}, \ldots, x^n y_x^{(n)}\right) = 0$	$x = a\bar{x}$	$z = y, \ u = xy'_x$
$F\left(y/x, y'_x, xy''_{xx}, \ldots, x^{n-1} y_x^{(n)}\right) = 0$	$x = a\bar{x}, \ y = a\bar{y}$	$z = y/x, \ u = y'_x$
$F\left(e^{\lambda x} y, y'_x/y, y''_{xx}/y, \ldots, y_x^{(n)}/y\right) = 0$	$x = \bar{x} - \frac{1}{\lambda} \ln a, \ y = a\bar{y}$	$z = e^{\lambda x} y, \ u = y'_x/y$
$F\left(xe^{\lambda y}, xy'_x, x^2 y''_{xx}, \ldots, x^n y_x^{(n)}\right) = 0$	$x = a\bar{x}, \ y = \bar{y} - \frac{1}{\lambda} \ln a$	$z = xe^{\lambda y}, \ u = xy'_x$
$F\left(x^k y, x^{k+1} y'_x, x^{k+2} y''_{xx}, \ldots, x^{k+n} y_x^{(n)}\right) = 0$	$x = a\bar{x}, \ y = a^{-k}\bar{y}$	$z = x^k y, \ u = x^{k+1} y'_x$

Bibliography for Chapter M11

Boyce, W. E. and DiPrima, R. C., *Elementary Differential Equations and Boundary Value Problems, 8th Edition*, John Wiley & Sons, New York, 2004.

Cole, G. D., *Perturbation Methods in Applied Mathematics*, Blaisdell Publishing Company, Waltham, Massachusetts, 1968.

Dormand, J. R., *Numerical Methods for Differential Equations: A Computational Approach*, CRC Press, Boca Raton, Florida, 1996.

El'sgol'ts, L. E., *Differential Equations*, Gordon & Breach, New York, 1961.

Ince, E. L., *Ordinary Differential Equations*, Dover Publications, New York, 1964.

Kamke, E., *Differentialgleichungen: Lösungsmethoden und Lösungen, I, Gewöhnliche Differentialgleichungen*, B. G. Teubner, Leipzig, 1977.

Marchenko, V. A., *Sturm–Liouville Operators and Applications*, Birkhäuser Verlag, Basel, 1986.

Murphy, G. M., *Ordinary Differential Equations and Their Solutions*, D. Van Nostrand, New York, 1960.

Nayfeh, A. H., *Perturbation Methods*, Wiley-Interscience, New York, 1973.

Nayfeh, A. H., *Introduction to Perturbation Techniques*, John Wiley & Sons, New York, 1981.

Polyanin, A. D. and Zaitsev, V. F., *Handbook of Exact Solutions for Ordinary Differential Equations, 2nd Edition*, Chapman & Hall/CRC Press, Boca Raton, 2003.

Schiesser, W. E., *Computational Mathematics in Engineering and Applied Science: ODEs, DAEs, and PDEs*, CRC Press, Boca Raton, Florida, 1993.

Tenenbaum, M. and Pollard, H., *Ordinary Differential Equations*, Dover Publications, New York, 1985.

Zwillinger, D., *Handbook of Differential Equations, 3rd Edition*, Academic Press, New York, 1997.

Chapter M12
Partial Differential Equations

M12.1. First-Order Quasilinear Partial Differential Equations

M12.1.1. Characteristic System. General Solution

▶ **Equations with two independent variables. General solution. Examples.** A *first-order quasilinear partial differential equation with two independent variables* has the general form

$$f(x, y, w)\frac{\partial w}{\partial x} + g(x, y, w)\frac{\partial w}{\partial y} = h(x, y, w). \qquad \text{(M12.1.1.1)}$$

Such equations are encountered in various applications (continuum mechanics, gas dynamics, hydrodynamics, heat and mass transfer, wave theory, acoustics, multiphase flows, chemical engineering, etc.).

If two independent integrals,

$$u_1(x, y, w) = C_1, \qquad u_2(x, y, w) = C_2, \qquad \text{(M12.1.1.2)}$$

of the *characteristic system*

$$\frac{dx}{f(x, y, w)} = \frac{dy}{g(x, y, w)} = \frac{dw}{h(x, y, w)} \qquad \text{(M12.1.1.3)}$$

are known, then the *general solution* of equation (M12.1.1.1) is given by

$$\Phi(u_1, u_2) = 0, \qquad \text{(M12.1.1.4)}$$

where $\Phi(u, v)$ is an arbitrary function of two variables. With equation (M12.1.1.4) solved for u_1 or u_2, we often specify the general solution in the form

$$u_k = \Psi(u_{3-k}),$$

where $k = 1$ or 2 and $\Psi(u)$ is an arbitrary function of one variable.

Remark. In the special case $h(x, y, w) \equiv 0$, one of the integrals of the characteristic system is $w = C_1$. Another integral may be determined from the first equation in (M12.1.1.3).

Example. Consider the linear constant coefficient equation

$$\frac{\partial w}{\partial x} + a\frac{\partial w}{\partial y} = b.$$

The characteristic system for this equation is

$$\frac{dx}{1} = \frac{dy}{a} = \frac{dw}{b}.$$

It has two independent integrals:

$$y - ax = C_1, \quad w - bx = C_2.$$

Hence, the general solution of the original equation is given by $\Phi(y - ax, w - bx) = 0$. On solving this equation for w, one obtains the general solution in explicit form

$$w = bx + \Psi(y - ax),$$

where $\Psi(u)$ is an arbitrary function.

▶ **Equations with n independent variables. General solution.** A *first-order quasilinear partial differential equation with n independent variables* has the general form

$$f_1(x_1, \ldots, x_n, w)\frac{\partial w}{\partial x_1} + \cdots + f_n(x_1, \ldots, x_n, w)\frac{\partial w}{\partial x_n} = g(x_1, \ldots, x_n, w). \quad \text{(M12.1.1.5)}$$

Let n independent integrals,

$$u_1(x_1, \ldots, x_n, w) = C_1, \quad \ldots, \quad u_n(x_1, \ldots, x_n, w) = C_n,$$

of the characteristic system

$$\frac{dx_1}{f_1(x_1, \ldots, x_n, w)} = \cdots = \frac{dx_n}{f_n(x_1, \ldots, x_n, w)} = \frac{dw}{g(x_1, \ldots, x_n, w)}$$

be known. Then the general solution of equation (M12.1.1.5) is given by

$$\Phi(u_1, \ldots, u_n) = 0,$$

where Φ is an arbitrary function of n variables.

M12.1.2. Cauchy Problem

▶ **Two formulations of the Cauchy problem.** Consider two formulations of the Cauchy problem.

1°. *Generalized Cauchy problem.* Find a solution $w = w(x, y)$ of equation (M12.1.1.1) satisfying the initial conditions

$$x = h_1(\xi), \quad y = h_2(\xi), \quad w = h_3(\xi), \quad \text{(M12.1.2.1)}$$

where ξ is a parameter ($\alpha \leq \xi \leq \beta$) and the $h_k(\xi)$ are given functions.

Geometric interpretation: find an integral surface of equation (M12.1.1.1) passing through the line defined parametrically by equations (M12.1.2.1).

2°. *Classical Cauchy problem.* Find a solution $w = w(x, y)$ of equation (M12.1.1.1) satisfying the initial condition

$$w = \varphi(y) \quad \text{at} \quad x = 0, \quad \text{(M12.1.2.2)}$$

where $\varphi(y)$ is a given function.

It is convenient to represent the classical Cauchy problem as a generalized Cauchy problem by rewriting condition (M12.1.2.2) in the parametric form

$$x = 0, \quad y = \xi, \quad w = \varphi(\xi). \quad \text{(M12.1.2.3)}$$

▶ **Procedure of solving the Cauchy problem.** The procedure of solving the Cauchy problem (M12.1.1.1), (M12.1.2.1) involves several steps. First, two independent integrals (M12.1.1.2) of the characteristic system (M12.1.1.3) are determined. Then, to find the constants of integration C_1 and C_2, the initial data (M12.1.2.1) must be substituted into the integrals (M12.1.1.2) to obtain

$$u_1\big(h_1(\xi), h_2(\xi), h_3(\xi)\big) = C_1, \quad u_2\big(h_1(\xi), h_2(\xi), h_3(\xi)\big) = C_2. \quad \text{(M12.1.2.4)}$$

Eliminating C_1 and C_2 from (M12.1.1.2) and (M12.1.2.4) yields

$$\begin{aligned}
u_1(x, y, w) &= u_1\big(h_1(\xi), h_2(\xi), h_3(\xi)\big), \\
u_2(x, y, w) &= u_2\big(h_1(\xi), h_2(\xi), h_3(\xi)\big).
\end{aligned} \quad \text{(M12.1.2.5)}$$

Formulas (M12.1.2.5) are a parametric form of the solution of the Cauchy problem (M12.1.1.1), (M12.1.2.1). In some cases, one may succeed in eliminating the parameter ξ from relations (M12.1.2.5), thus obtaining the solution in an explicit form.

Example 1. Consider the Cauchy problem for linear equation

$$\frac{\partial w}{\partial x} + a \frac{\partial w}{\partial y} = bw \tag{M12.1.2.6}$$

subjected to the initial condition (M12.1.2.2).

The corresponding characteristic system for equation (M12.1.2.6),

$$\frac{dx}{1} = \frac{dy}{a} = \frac{dw}{bw},$$

has two independent integrals

$$y - ax = C_1, \qquad we^{-bx} = C_2. \tag{M12.1.2.7}$$

Represent the initial condition (M12.1.2.2) in parametric form (M12.1.2.3) and then substitute the data (M12.1.2.3) into the integrals (M12.1.2.7). As a result, for the constants of integration we obtain $C_1 = \xi$ and $C_2 = \varphi(\xi)$. Substituting these expressions into (M12.1.2.7), we arrive at the solution of the Cauchy problem (M12.1.2.6), (M12.1.2.2) in parametric form:

$$y - ax = \xi, \qquad we^{-bx} = \varphi(\xi).$$

By eliminating the parameter ξ from these relations, we obtain the solution of the Cauchy problem (M12.1.2.6), (M12.1.2.2) in explicit form:

$$w = e^{bx}\varphi(y - ax).$$

Example 2. Consider the Cauchy problem for *Hopf's equation*

$$\frac{\partial w}{\partial x} + w \frac{\partial w}{\partial y} = 0 \tag{M12.1.2.8}$$

subject to the initial condition (M12.1.2.2).

First, we rewrite the initial condition (M12.1.2.2) in the parametric form (M12.1.2.3). Solving the characteristic system

$$\frac{dx}{1} = \frac{dy}{w} = \frac{dw}{0}, \tag{M12.1.2.9}$$

we find two independent integrals,

$$w = C_1, \quad y - wx = C_2. \tag{M12.1.2.10}$$

Using the initial conditions (M12.1.2.3), we find that $C_1 = \varphi(\xi)$ and $C_2 = \xi$. Substituting these expressions into (M12.1.2.10) yields the solution of the Cauchy problem (M12.1.2.8), (M12.1.2.2) in the parametric form

$$w = \varphi(\xi), \tag{M12.1.2.11}$$

$$y = \xi + \varphi(\xi)x. \tag{M12.1.2.12}$$

The straight lines defined by equation (M12.1.2.12) are called *characteristics*. They have the slope $\varphi(\xi)$ and intersect the y-axis at the points ξ. On each characteristic, the function w has the same value equal to $\varphi(\xi)$ (generally, w takes different values on different characteristics).

For $\varphi'(\xi) > 0$, different characteristics do not intersect and, hence, formulas (M12.1.2.11) and (M12.1.2.12) define a unique solution.

M12.2. Classification of Second-Order Linear Partial Differential Equations

M12.2.1. Equations with Two Independent Variables

▶ **Examples of equations encountered in applications.** There are three basic types of linear partial differential equations — *parabolic*, *hyperbolic*, and *elliptic*. The solutions of the equations pertaining to each of the types have their own characteristic qualitative differences.

The simplest example of a *parabolic equation* is the *heat equation*

$$\frac{\partial w}{\partial t} - \frac{\partial^2 w}{\partial x^2} = 0, \tag{M12.2.1.1}$$

where the variables t and x play the role of time and the spatial coordinate, respectively. Note that equation (M12.2.1.1) contains only one highest derivative term. Frequently encountered particular solutions of equation (M12.2.1.1) can be found at the end of Subsection M12.4.1.

The simplest example of a *hyperbolic equation* is the *wave equation*

$$\frac{\partial^2 w}{\partial t^2} - \frac{\partial^2 w}{\partial x^2} = 0, \tag{M12.2.1.2}$$

where the variables t and x play the role of time and the spatial coordinate, respectively. Note that the highest derivative terms in equation (M12.2.1.2) differ in sign. The general solution of equation (M12.2.1.2) can be found at the end of Subsection M12.4.1.

The simplest example of an *elliptic equation* is the *Laplace equation*

$$\frac{\partial^2 w}{\partial x^2} + \frac{\partial^2 w}{\partial y^2} = 0, \tag{M12.2.1.3}$$

where x and y play the role of the spatial coordinates. Note that the highest derivative terms in equation (M12.2.1.3) have like signs. Frequently encountered particular solutions of equation (M12.2.1.3) can be found at the end of Subsection M12.4.1.

Any linear partial differential equation of the second order with two independent variables can be reduced, by appropriate manipulations, to a simpler equation that has one of the three highest derivative combinations specified above in examples (M12.2.1.1), (M12.2.1.2), and (M12.2.1.3).

▶ **Types of equations. Characteristic equations.** Consider a second-order partial differential equation with two independent variables that has the general form

$$a(x,y)\frac{\partial^2 w}{\partial x^2} + 2b(x,y)\frac{\partial^2 w}{\partial x \partial y} + c(x,y)\frac{\partial^2 w}{\partial y^2} = F\left(x, y, w, \frac{\partial w}{\partial x}, \frac{\partial w}{\partial y}\right), \tag{M12.2.1.4}$$

where a, b, c are some functions of x and y that have continuous derivatives up to the second order inclusive.*

Given a point (x, y), equation (M12.2.1.4) is said to be

$$
\begin{array}{ll}
\textit{parabolic} & \text{if } b^2 - ac = 0, \\
\textit{hyperbolic} & \text{if } b^2 - ac > 0, \\
\textit{elliptic} & \text{if } b^2 - ac < 0
\end{array}
$$

at this point.

In order to reduce equation (M12.2.1.4) to a canonical form, one should first write out the characteristic equation

$$a\,(dy)^2 - 2b\,dx\,dy + c\,(dx)^2 = 0,$$

which splits into two equations

$$a\,dy - \left(b + \sqrt{b^2 - ac}\,\right) dx = 0 \tag{M12.2.1.5}$$

and

$$a\,dy - \left(b - \sqrt{b^2 - ac}\,\right) dx = 0, \tag{M12.2.1.6}$$

and then find their general integrals.

---

* The right-hand side of equation (M12.2.1.4) may be nonlinear. The classification and the procedure of reducing such equations to a canonical form are only determined by the left-hand side of the equation.

Remark. The characteristic equations (M12.2.1.5) and (M12.2.1.6) may be used if $a \not\equiv 0$. If $a \equiv 0$, the simpler equations

$$dx = 0,$$
$$2b \, dy - c \, dx = 0$$

should be used; the first equation has the obvious general solution $x = C$.

▶ **Canonical form of parabolic equations (case $b^2 - ac = 0$).** In this case, equations (M12.2.1.5) and (M12.2.1.6) coincide and have a common general integral,

$$\varphi(x, y) = C.$$

By passing from x, y to new independent variables ξ, η in accordance with the relations

$$\xi = \varphi(x, y), \qquad \eta = \eta(x, y),$$

where $\eta = \eta(x, y)$ is any twice differentiable function that satisfies the condition of nondegeneracy of the Jacobian $\frac{D(\xi, \eta)}{D(x, y)}$ in the given domain, we reduce equation (M12.2.1.4) to the canonical form

$$\frac{\partial^2 w}{\partial \eta^2} = F_1 \left(\xi, \eta, w, \frac{\partial w}{\partial \xi}, \frac{\partial w}{\partial \eta} \right). \tag{M12.2.1.7}$$

As η, one can take $\eta = x$ or $\eta = y$.

It is apparent that the transformed equation (M12.2.1.7) has only one highest-derivative term, just as the heat equation (M12.2.1.1).

Remark. In the degenerate case where the function F_1 does not depend on the partial derivative $\partial_\xi w$, equation (M12.2.1.7) is an ordinary differential equation for the variable η, in which ξ serves as a parameter.

▶ **Canonical forms of hyperbolic equations (case $b^2 - ac > 0$).** The general integrals

$$\varphi(x, y) = C_1, \qquad \psi(x, y) = C_2$$

of equations (M12.2.1.5) and (M12.2.1.6) are real and different. These integrals determine two different families of real characteristics.

By passing from x, y to new independent variables ξ, η in accordance with the relations

$$\xi = \varphi(x, y), \qquad \eta = \psi(x, y),$$

we reduce equation (M12.2.1.4) to

$$\frac{\partial^2 w}{\partial \xi \partial \eta} = F_2 \left(\xi, \eta, w, \frac{\partial w}{\partial \xi}, \frac{\partial w}{\partial \eta} \right).$$

This is the so-called *first canonical form of a hyperbolic equation.*

The transformation

$$\xi = t + z, \qquad \eta = t - z$$

brings the above equation to another canonical form,

$$\frac{\partial^2 w}{\partial t^2} - \frac{\partial^2 w}{\partial z^2} = F_3 \left(t, z, w, \frac{\partial w}{\partial t}, \frac{\partial w}{\partial z} \right),$$

where $F_3 = 4F_2$. This is the so-called *second canonical form of a hyperbolic equation.* Apart from notation, the left-hand side of the last equation coincides with that of the wave equation (M12.2.1.2).

In some cases, reduction of an equation to a canonical form allows finding its general solution.

▶ **Canonical form of elliptic equations (case $b^2 - ac < 0$).** In this case the general integrals of equations (M12.2.1.5) and (M12.2.1.6) are complex conjugate; these determine two families of complex characteristics.

Let the general integral of equation (M12.2.1.5) have the form

$$\varphi(x, y) + i\psi(x, y) = C, \qquad i^2 = -1,$$

where $\varphi(x, y)$ and $\psi(x, y)$ are real-valued functions.

By passing from x, y to new independent variables ξ, η in accordance with the relations

$$\xi = \varphi(x, y), \qquad \eta = \psi(x, y),$$

we reduce equation (M12.2.1.4) to the canonical form

$$\frac{\partial^2 w}{\partial \xi^2} + \frac{\partial^2 w}{\partial \eta^2} = F_4\left(\xi, \eta, w, \frac{\partial w}{\partial \xi}, \frac{\partial w}{\partial \eta}\right).$$

Apart from notation, the left-hand side of the last equation coincides with that of the Laplace equation (M12.2.1.3).

M12.2.2. Equations with Many Independent Variables

Let us consider a second-order partial differential equation with n independent variables $x_1, \ldots, x_n$ that has the form

$$\sum_{i,j=1}^{n} a_{ij}(\mathbf{x}) \frac{\partial^2 w}{\partial x_i \partial x_j} = F\left(\mathbf{x}, w, \frac{\partial w}{\partial x_1}, \ldots, \frac{\partial w}{\partial x_n}\right), \qquad \text{(M12.2.2.1)}$$

where the a_{ij} are some functions that have continuous derivatives with respect to all variables to the second order inclusive, and $\mathbf{x} = \{x_1, \ldots, x_n\}$. [The right-hand side of equation (M12.2.2.1) may be nonlinear. The left-hand side only is required for the classification of this equation.]

At a point $\mathbf{x} = \mathbf{x}_0$, the following quadratic form is assigned to equation (M12.2.2.1):

$$Q = \sum_{i,j=1}^{n} a_{ij}(\mathbf{x}_0)\xi_i\xi_j. \qquad \text{(M12.2.2.2)}$$

By an appropriate linear nondegenerate transformation

$$\xi_i = \sum_{k=1}^{n} \beta_{ik}\eta_k \qquad (i = 1, \ldots, n) \qquad \text{(M12.2.2.3)}$$

the quadratic form (M12.2.2.2) can be reduced to the canonical form

$$Q = \sum_{i=1}^{n} c_i\eta_i^2, \qquad \text{(M12.2.2.4)}$$

where the coefficients c_i assume the values 1, −1, and 0. The number of negative and zero coefficients in (M12.2.2.4) does not depend on the way in which the quadratic form is reduced to the canonical form.

TABLE M12.1
Classification of equations with many independent variables.

Type of equation (M12.2.2.1) at a point $\mathbf{x} = \mathbf{x}_0$	Coefficients of the canonical form (M12.2.2.4)
Parabolic (in the broad sense)	At least one coefficient of the c_i is zero
Hyperbolic (in the broad sense)	All c_i are nonzero and some c_i differ in sign
Elliptic	All c_i are nonzero and have like signs

Table M12.1 presents the basic criteria according to which the equations with many independent variables are classified.

Suppose all coefficients of the highest derivatives in (M12.2.2.1) are constant, $a_{ij} =$ const. By introducing the new independent variables $y_1, \dots, y_n$ in accordance with the formulas $y_i = \sum_{k=1}^{n} \beta_{ik} x_k$, where the β_{ik} are the coefficients of the linear transformation (M12.2.2.3), we reduce equation (M12.2.2.1) to the canonical form

$$\sum_{i=1}^{n} c_i \frac{\partial^2 w}{\partial y_i^2} = F_1\left(\mathbf{y}, w, \frac{\partial w}{\partial y_1}, \dots, \frac{\partial w}{\partial y_n}\right). \tag{M12.2.2.5}$$

The coefficients c_i here are the same as in the quadratic form (M12.2.2.4), and $\mathbf{y} = \{y_1, \dots, y_n\}$.

Specific equations of parabolic, hyperbolic, and elliptic types will be discussed further in Sections M12.3, M12.5, and M12.7–M12.9.

M12.3. Basic Problems for Linear Equations of Mathematical Physics

M12.3.1. Initial and Boundary Conditions. Cauchy Problem. Boundary Value Problems

Every equation of mathematical physics governs infinitely many qualitatively similar phenomena or processes. This follows from the fact that differential equations have infinitely many particular solutions. The specific solution that describes the physical phenomenon under study is isolated from the set of particular solutions of the given differential equation by means of initial and boundary conditions.

Throughout this section, we consider linear equations in the n-dimensional Euclidean space $\mathbb{R}^n$ or in an open domain $V \in \mathbb{R}^n$ (exclusive of the boundary) with a sufficiently smooth boundary $S = \partial V$.

▶ **Parabolic equations. Initial and boundary conditions.** In general, a linear second-order partial differential equation of the parabolic type with n independent variables can be written as

$$\frac{\partial w}{\partial t} - L_{\mathbf{x},t}[w] = \Phi(\mathbf{x}, t), \tag{M12.3.1.1}$$

where

$$L_{\mathbf{x},t}[w] \equiv \sum_{i,j=1}^{n} a_{ij}(\mathbf{x}, t) \frac{\partial^2 w}{\partial x_i \partial x_j} + \sum_{i=1}^{n} b_i(\mathbf{x}, t) \frac{\partial w}{\partial x_i} + c(\mathbf{x}, t) w, \tag{M12.3.1.2}$$

$$\mathbf{x} = \{x_1, \ldots, x_n\}, \quad \sum_{i,j=1}^{n} a_{ij}(\mathbf{x}, t)\xi_i\xi_j \geq \sigma \sum_{i=1}^{n} \xi_i^2, \quad \sigma > 0.$$

Parabolic equations govern unsteady thermal, diffusion, and other phenomena dependent on time t.

Equation (M12.3.1.1) is called homogeneous if $\Phi(\mathbf{x}, t) \equiv 0$.

Cauchy problem ($t \geq 0$, $\mathbf{x} \in \mathbb{R}^n$). Find a function w that satisfies equation (M12.3.1.1) for $t > 0$ and the initial condition

$$w = f(\mathbf{x}) \quad \text{at} \quad t = 0. \tag{M12.3.1.3}$$

Example 1. The solution to the Cauchy problem for the one-dimensional heat equation (M12.2.1.1) with the initial condition (M12.3.1.3) is given by

$$w(x, t) = \frac{1}{2\sqrt{\pi t}} \int_{-\infty}^{\infty} f(\xi) \exp\left[-\frac{(x - \xi)^2}{4t}\right] d\xi.$$

The derivation of this formula is shown in the example of Subsection M12.6.2.

*Boundary value problem** ($t \geq 0$, $\mathbf{x} \in V$). Find a function w that satisfies equation (M12.3.1.1) for $t > 0$, the initial condition (M12.3.1.3), and the boundary condition

$$\Gamma_{\mathbf{x},t}[w] = g(\mathbf{x}, t) \quad \text{at} \quad \mathbf{x} \in S \qquad (t > 0). \tag{M12.3.1.4}$$

In general, $\Gamma_{\mathbf{x},t}$ is a first-order linear differential operator in the space variables $\mathbf{x}$ with coefficients dependent on $\mathbf{x}$ and t. The basic types of boundary conditions are described in Subsection M12.3.2.

The initial condition (M12.3.1.3) is called homogeneous if $f(\mathbf{x}) \equiv 0$. The boundary condition (M12.3.1.4) is called homogeneous if $g(\mathbf{x}, t) \equiv 0$.

▶ **Hyperbolic equations. Initial and boundary conditions.** Consider a second-order linear partial differential equation of the hyperbolic type with n independent variables of the general form

$$\frac{\partial^2 w}{\partial t^2} + \varphi(\mathbf{x}, t)\frac{\partial w}{\partial t} - L_{\mathbf{x},t}[w] = \Phi(\mathbf{x}, t), \tag{M12.3.1.5}$$

where the linear differential operator $L_{\mathbf{x},t}$ is defined by (M12.3.1.2). Hyperbolic equations govern unsteady wave processes, which depend on time t.

Equation (M12.3.1.5) is said to be homogeneous if $\Phi(\mathbf{x}, t) \equiv 0$.

Cauchy problem ($t \geq 0$, $\mathbf{x} \in \mathbb{R}^n$). Find a function w that satisfies equation (M12.3.1.5) for $t > 0$ and the initial conditions

$$\begin{aligned} w &= f_0(\mathbf{x}) \quad \text{at} \quad t = 0, \\ \partial_t w &= f_1(\mathbf{x}) \quad \text{at} \quad t = 0. \end{aligned} \tag{M12.3.1.6}$$

Example 2. The solution to the Cauchy problem for the one-dimensional wave equation (M12.2.1.2) with the initial conditions (M12.3.1.6) is given by *d'Alembert's formula*:

$$w(x, t) = \frac{1}{2}[f_0(x + t) + f_0(x - t)] + \frac{1}{2}\int_{x-t}^{x+t} f_1(\xi)\,d\xi.$$

Boundary value problem ($t \geq 0$, $\mathbf{x} \in V$). Find a function w that satisfies equation (M12.3.1.5) for $t > 0$, the initial conditions (M12.3.1.6), and boundary condition (M12.3.1.4).

The initial conditions (M12.3.1.6) are called homogeneous if $f_0(\mathbf{x}) \equiv 0$ and $f_1(\mathbf{x}) \equiv 0$.

Goursat problem. On the characteristics of a hyperbolic equation with two independent variables, the values of the unknown function w are prescribed; for details, see Subsection M12.8.3.

---

* *Boundary value problems* for parabolic and hyperbolic equations are sometimes called *mixed* or *initial-boundary value problems*.

▶ **Elliptic equations. Boundary conditions.** In general, a second-order linear partial differential equation of elliptic type with n independent variables can be written as

$$-L_{\mathbf{x}}[w] = \Phi(\mathbf{x}), \tag{M12.3.1.7}$$

where

$$L_{\mathbf{x}}[w] \equiv \sum_{i,j=1}^{n} a_{ij}(\mathbf{x})\frac{\partial^2 w}{\partial x_i \partial x_j} + \sum_{i=1}^{n} b_i(\mathbf{x})\frac{\partial w}{\partial x_i} + c(\mathbf{x})w, \tag{M12.3.1.8}$$

$$\sum_{i,j=1}^{n} a_{ij}(\mathbf{x})\xi_i\xi_j \geq \sigma \sum_{i=1}^{n} \xi_i^2, \quad \sigma > 0.$$

Elliptic equations govern steady-state thermal, diffusion, and other phenomena independent of time t.

Equation (M12.3.1.7) is said to be homogeneous if $\Phi(\mathbf{x}) \equiv 0$.

Boundary value problem. Find a function w that satisfies equation (M12.3.1.7) and the boundary condition

$$\Gamma_{\mathbf{x}}[w] = g(\mathbf{x}) \quad \text{at} \quad \mathbf{x} \in S. \tag{M12.3.1.9}$$

In general, $\Gamma_{\mathbf{x}}$ is a first-order linear differential operator in the space variables $\mathbf{x}$. The basic types of boundary conditions are described below in Subsection M12.3.2.

The boundary condition (M12.3.1.9) is called homogeneous if $g(\mathbf{x}) \equiv 0$. The boundary value problem (M12.3.1.7)–(M12.3.1.9) is said to be homogeneous if $\Phi \equiv 0$ and $g \equiv 0$.

M12.3.2. First, Second, Third, and Mixed Boundary Value Problems

For any (parabolic, hyperbolic, and elliptic) second-order partial differential equations, it is conventional to distinguish four basic types of boundary value problems, depending on the form of the boundary conditions (M12.3.1.4) [see also the analogous condition (M12.3.1.9)]. For simplicity, here we confine ourselves to the case where the coefficients a_{ij} of equations (M12.3.1.1) and (M12.3.1.5), with the operator defined by (M12.3.1.2), have the special form

$$a_{ij}(\mathbf{x}, t) = a(\mathbf{x}, t)\delta_{ij}, \quad \delta_{ij} = \begin{cases} 1 & \text{if } i = j, \\ 0 & \text{if } i \neq j. \end{cases}$$

This situation is rather frequent in applications; such coefficients are used to describe various phenomena (processes) in isotropic media.

▶ **First boundary value problem.** The function $w(\mathbf{x}, t)$ takes prescribed values at the boundary S of the domain:

$$w(\mathbf{x}, t) = g_1(\mathbf{x}, t) \quad \text{for} \quad \mathbf{x} \in S. \tag{M12.3.2.1}$$

▶ **Second boundary value problem.** The derivative along the (outward) normal is prescribed at the boundary S of the domain:

$$\frac{\partial w}{\partial N} = g_2(\mathbf{x}, t) \quad \text{for} \quad \mathbf{x} \in S. \tag{M12.3.2.2}$$

In heat transfer problems, where w is temperature, the left-hand side of the boundary condition (M12.3.2.2) is proportional to the heat flux per unit area of the surface S.

TABLE M12.2

Boundary conditions for various boundary value problems specified by parabolic
and hyperbolic equations in two independent variables ($x_1 \leq x \leq x_2$).

Type of problem	Boundary condition at $x = x_1$	Boundary condition at $x = x_2$
First boundary value problem	$w = g_1(t)$	$w = g_2(t)$
Second boundary value problem	$\partial_x w = g_1(t)$	$\partial_x w = g_2(t)$
Third boundary value problem	$\partial_x w + \beta_1 w = g_1(t)$ $(\beta_1 < 0)$	$\partial_x w + \beta_2 w = g_2(t)$ $(\beta_2 > 0)$
Mixed boundary value problem	$w = g_1(t)$	$\partial_x w = g_2(t)$
Mixed boundary value problem	$\partial_x w = g_1(t)$	$w = g_2(t)$

▶ **Third boundary value problem.** A linear relationship between the unknown function
and its normal derivative is prescribed at the boundary S of the domain:

$$\frac{\partial w}{\partial N} + k(\mathbf{x}, t)w = g_3(\mathbf{x}, t) \quad \text{for} \quad \mathbf{x} \in S. \tag{M12.3.2.3}$$

It is usually assumed that $k(\mathbf{x}, t) = \text{const}$. In mass transfer problems, where w is concentration, the boundary condition (M12.3.2.3) with $g_3 \equiv 0$ describes a surface chemical reaction
of the first order.

▶ **Mixed boundary value problems.** Conditions of various types, listed above, are set at
different portions of the boundary S.

If $g_1 \equiv 0$, $g_2 \equiv 0$, or $g_3 \equiv 0$, the respective boundary conditions (M12.3.2.1), (M12.3.2.2),
and (M12.3.2.3) are said to be homogeneous.

Boundary conditions for various boundary value problems for parabolic and hyperbolic
equations in two independent variables x and t are displayed in Table M12.2. The equation
coefficients are assumed to be continuous, with the coefficients of the highest derivatives
being nonzero in the range $x_1 \leq x \leq x_2$ considered.

Remark 1. For elliptic equations, the coefficients a_{ij} as well as the right-hand sides of the boundary
conditions (M12.3.2.1)–(M12.3.2.3) are independent of time t.

Remark 2. For elliptic equations, the first boundary value problem is often called the *Dirichlet problem*,
and the second boundary value problem is called the *Neumann problem*.

M12.4. Properties and Exact Solutions of Linear Equations

M12.4.1. Homogeneous Linear Equations and Their Particular Solutions

▶ **Preliminary remarks.** For brevity, in this paragraph a homogeneous linear partial
differential equation will be written as

$$\mathcal{L}[w] = 0. \tag{M12.4.1.1}$$

For second-order linear parabolic and hyperbolic equations, the linear differential operator $\mathcal{L}[w]$ is defined by the left-hand side of equations (M12.3.1.1) and (M12.3.1.5), respectively. It is assumed that equation (M12.4.1.1) is an arbitrary homogeneous linear partial
differential equation of any order in the variables $t, x_1, \ldots, x_n$ with sufficiently smooth
coefficients.

A linear differential operator $\mathcal{L}$ possesses the properties

$$\mathcal{L}[w_1 + w_2] = \mathcal{L}[w_1] + \mathcal{L}[w_2],$$
$$\mathcal{L}[Aw] = A\mathcal{L}[w], \quad A = \text{const.}$$

An arbitrary homogeneous linear equation (M12.4.1.1) has a trivial solution, $w \equiv 0$.

A function w is called a *classical solution* of equation (M12.4.1.1) if w, when substituted into (M12.4.1.1), turns the equation into an identity and if all partial derivatives of w that occur in (M12.4.1.1) are continuous; the notion of a classical solution is directly linked to the range of the independent variables. In what follows, we usually write "solution" instead of "classical solution" for brevity.

▶ **Use of particular solutions for the construction of other solutions.** Below are some properties of particular solutions of homogeneous linear equations.

1°. Let $w_1 = w_1(\mathbf{x}, t)$, $w_2 = w_2(\mathbf{x}, t)$, ..., $w_k = w_k(\mathbf{x}, t)$ be any particular solutions of the homogeneous equation (M12.4.1.1). Then the linear combination

$$w = A_1 w_1 + A_2 w_2 + \cdots + A_k w_k \tag{M12.4.1.2}$$

with arbitrary constants A_1, A_2, ..., A_k is also a solution of equation (M12.4.1.1); in physics, this property is known as the *principle of linear superposition*.

Suppose $\{w_k\}$ is an infinite sequence of solutions of equation (M12.4.1.1). Then the series $\sum_{k=1}^{\infty} w_k$, irrespective of its convergence, is called a *formal solution* of (M12.4.1.1). If the solutions w_k are classical, the series is uniformly convergent, and the sum of the series has all the necessary partial derivatives, then the sum of the series is a classical solution of equation (M12.4.1.1).

2°. Let the coefficients of the linear differential operator $\mathcal{L}$ be independent of time t. If equation (M12.4.1.1) has a particular solution $\widetilde{w} = \widetilde{w}(\mathbf{x}, t)$, then the partial derivatives of $\widetilde{w}$ with respect to time,*

$$\frac{\partial \widetilde{w}}{\partial t}, \quad \frac{\partial^2 \widetilde{w}}{\partial t^2}, \quad \ldots, \quad \frac{\partial^k \widetilde{w}}{\partial t^k}, \quad \ldots,$$

are also solutions of equation (M12.4.1.1).

3°. Let the coefficients of the linear differential operator $\mathcal{L}$ be independent of the space variables $x_1, \ldots, x_n$. If equation (M12.4.1.1) has a particular solution $\widetilde{w} = \widetilde{w}(\mathbf{x}, t)$, then the partial derivatives of $\widetilde{w}$ with respect to the space coordinates

$$\frac{\partial \widetilde{w}}{\partial x_1}, \quad \frac{\partial \widetilde{w}}{\partial x_2}, \quad \frac{\partial \widetilde{w}}{\partial x_3}, \quad \ldots, \quad \frac{\partial^2 \widetilde{w}}{\partial x_1^2}, \quad \frac{\partial^2 \widetilde{w}}{\partial x_1 \partial x_2}, \quad \ldots, \quad \frac{\partial^{k+m} \widetilde{w}}{\partial x_2^k \partial x_3^m}, \quad \ldots$$

are also solutions of equation (M12.4.1.1).

If the coefficients of $\mathcal{L}$ are independent of only one space coordinate, say x_1, and equation (M12.4.1.1) has a particular solution $\widetilde{w} = \widetilde{w}(\mathbf{x}, t)$, then the partial derivatives

$$\frac{\partial \widetilde{w}}{\partial x_1}, \quad \frac{\partial^2 \widetilde{w}}{\partial x_1^2}, \quad \ldots, \quad \frac{\partial^k \widetilde{w}}{\partial x_1^k}, \quad \ldots$$

are also solutions of equation (M12.4.1.1).

---

* Here and in what follows, it is assumed that the particular solution $\widetilde{w}$ is differentiable sufficiently many times with respect to t and $x_1, \ldots, x_n$ (or the parameters).

4°. Let the coefficients of the linear differential operator $\mathcal{L}$ be constant and let equation (M12.4.1.1) have a particular solution $\widetilde{w} = \widetilde{w}(\mathbf{x}, t)$. Then any partial derivatives of $\widetilde{w}$ with respect to time and the space coordinates (including mixed derivatives)

$$\frac{\partial \widetilde{w}}{\partial t}, \quad \frac{\partial \widetilde{w}}{\partial x_1}, \quad \dots, \quad \frac{\partial^2 \widetilde{w}}{\partial x_2^2}, \quad \frac{\partial^2 \widetilde{w}}{\partial t \partial x_1}, \quad \dots, \quad \frac{\partial^k \widetilde{w}}{\partial x_3^k}, \quad \dots$$

are solutions of equation (M12.4.1.1).

5°. Suppose equation (M12.4.1.1) has a particular solution dependent on a parameter μ, $\widetilde{w} = \widetilde{w}(\mathbf{x}, t; \mu)$, and the coefficients of the linear differential operator $\mathcal{L}$ are independent of μ (but can depend on time and the space coordinates). Then, by differentiating $\widetilde{w}$ with respect to μ, one obtains other solutions of equation (M12.4.1.1),

$$\frac{\partial \widetilde{w}}{\partial \mu}, \quad \frac{\partial^2 \widetilde{w}}{\partial \mu^2}, \quad \dots, \quad \frac{\partial^k \widetilde{w}}{\partial \mu^k}, \quad \dots$$

Example 1. A linear *heat equation with a source* of the form

$$\frac{\partial w}{\partial t} = a \frac{\partial^2 w}{\partial x^2} + bw$$

has a particular solution

$$w(x, t) = \exp[\mu x + (a\mu^2 + b)t],$$

where μ is an arbitrary constant. Differentiating this solution with respect to μ yields another solution

$$w(x, t) = (x + 2a\mu t) \exp[\mu x + (a\mu^2 + b)t].$$

Let some constants $\mu_1, \dots, \mu_k$ belong to the range of the parameter μ. Then the sum

$$w = A_1 \widetilde{w}(\mathbf{x}, t; \mu_1) + \dots + A_k \widetilde{w}(\mathbf{x}, t; \mu_k), \tag{M12.4.1.3}$$

where $A_1, \dots, A_k$ are arbitrary constants, is also a solution of the homogeneous linear equation (M12.4.1.1). The number of terms in the sum (M12.4.1.3) can be both finite and infinite.

6°. Another effective way of constructing solutions involves the following considerations. The particular solution $\widetilde{w}(\mathbf{x}, t; \mu)$, which depends on the parameter μ (as before, it is assumed that the coefficients of the linear differential operator $\mathcal{L}$ are independent of μ), is first multiplied by an arbitrary function $\varphi(\mu)$. Then the resulting expression is integrated with respect to μ over some interval $[\alpha, \beta]$. Thus, one obtains a new function,

$$\int_\alpha^\beta \widetilde{w}(\mathbf{x}, t; \mu) \varphi(\mu) \, d\mu,$$

which is also a solution of the original homogeneous linear equation.

The properties listed in Items 1°–6° enable one to use known particular solutions to construct other particular solutions of homogeneous linear equations of mathematical physics.

▶ **Multiplicative separable solutions.** Many homogeneous linear partial differential equations have solutions that can be represented as the product of functions depending on different arguments. Such solutions are referred to as *multiplicative separable solutions*; very commonly these solutions are briefly, but less accurately, called just *separable solutions*.

TABLE M12.3
Homogeneous linear partial differential equations that admit multiplicative separable solutions.

No.	Form of equation (M12.4.1.1)	Form of particular solutions
1	Equation coefficients are constant	$w(\mathbf{x}, t) = A \exp(\lambda t + \beta_1 x_1 + \cdots + \beta_n x_n)$, $\lambda, \beta_1, \ldots, \beta_n$ are related by an algebraic equation
2	Equation coefficients are independent of time t	$w(\mathbf{x}, t) = e^{\lambda t} \psi(\mathbf{x})$, λ is an arbitrary constant, $\mathbf{x} = \{x_1, \ldots, x_n\}$
3	Equation coefficients are independent of the coordinates $x_1, \ldots, x_n$	$w(\mathbf{x}, t) = \exp(\beta_1 x_1 + \cdots + \beta_n x_n) \psi(t)$, $\beta_1, \ldots, \beta_n$ are arbitrary constants
4	Equation coefficients are independent of the coordinates $x_1, \ldots, x_k$	$w(\mathbf{x}, t) = \exp(\beta_1 x_1 + \cdots + \beta_k x_k) \psi(t, x_{k+1}, \ldots, x_n)$, $\beta_1, \ldots, \beta_k$ are arbitrary constants
5	$L_t[w] + L_{\mathbf{x}}[w] = 0$, operator L_t depends on only t, operator $L_{\mathbf{x}}$ depends on only $\mathbf{x}$	$w(\mathbf{x}, t) = \varphi(t) \psi(\mathbf{x})$, $\varphi(t)$ satisfies the equation $L_t[\varphi] + \lambda \varphi = 0$, $\psi(\mathbf{x})$ satisfies the equation $L_{\mathbf{x}}[\psi] - \lambda \psi = 0$
6	$L_t[w] + L_1[w] + \cdots + L_n[w] = 0$, operator L_t depends on only t, operator L_k depends on only x_k	$w(\mathbf{x}, t) = \varphi(t) \psi_1(x_1) \ldots \psi_n(x_n)$, $\varphi(t)$ satisfies the equation $L_t[\varphi] + \lambda \varphi = 0$, $\psi_k(x_k)$ satisfies the equation $L_k[\psi_k] + \beta_k \psi_k = 0$, $\lambda + \beta_1 + \cdots + \beta_n = 0$
7	$f_0(x_1) L_t[w] + \sum_{k=1}^{n} f_k(x_1) L_k[w] = 0$, operator L_t depends on only t, operator L_k depends on only x_k	$w(\mathbf{x}, t) = \varphi(t) \psi_1(x_1) \ldots \psi_n(x_n)$, $L_t[\varphi] + \lambda \varphi = 0$, $L_k[\psi_k] + \beta_k \psi_k = 0$, $k = 2, \ldots, n$, $f_1(x_1) L_1[\psi_1] - \left[\lambda f_0(x_1) + \sum_{k=2}^{n} \beta_k f_k(x_1)\right] \psi_1 = 0$

Table M12.3 presents the most commonly encountered types of homogeneous linear differential equations with many independent variables that admit exact separable solutions. Linear combinations of particular solutions that correspond to different values of the separation parameters, $\lambda, \beta_1, \ldots, \beta_n$, are also solutions of the equations in question. For brevity, the word "operator" is used below to denote "linear differential operator."

For a constant coefficient equation (see the first row in Table M12.3), the separation parameters must satisfy the algebraic equation

$$D(\lambda, \beta_1, \ldots, \beta_n) = 0, \qquad (\text{M12.4.1.4})$$

which results from substituting the solution into equation (M12.4.1.1). In physical applications, equation (M12.4.1.4) is usually referred to as a *dispersion equation*. Any n of the $n + 1$ separation parameters in (M12.4.1.4) can be treated as arbitrary.

Example 2. Consider the *telegraph equation*

$$\frac{\partial^2 w}{\partial t^2} + k \frac{\partial w}{\partial t} = a^2 \frac{\partial^2 w}{\partial x^2} + b \frac{\partial w}{\partial x} + cw.$$

A particular solution is sought in the form

$$w = A \exp(\beta x + \lambda t).$$

This results in the dispersion equation $\lambda^2 + k\lambda = a^2 \beta^2 + b\beta + c$, where one of the two parameters β or λ can be treated as arbitrary.

Remark 1. For stationary equations that do not depend on t, one should set $\lambda = 0$, $L_t[w] \equiv 0$, and $\varphi(t) \equiv 1$ in rows 1, 6, and 7 of Table M12.3.

Remark 2. Multiplicative separable solutions play an important role in the theory of linear partial differential equations; they are used for finding solutions to stationary and nonstationary boundary value problems; see Sections M12.5 and M12.7–M12.9.

▶ **Solutions of some linear equations encountered in applications.**

$1°$. Some particular solutions of the heat equation

$$\frac{\partial w}{\partial t} = a\frac{\partial^2 w}{\partial x^2}$$

are listed below:

$$w(x) = Ax + B,$$
$$w(x, t) = A(x^2 + 2at) + B,$$
$$w(x, t) = A(x^3 + 6atx) + B,$$
$$w(x, t) = A(x^4 + 12atx^2 + 12a^2t^2) + B,$$
$$w(x, t) = x^{2n} + \sum_{k=1}^{n} \frac{(2n)(2n-1)\dots(2n-2k+1)}{k!}(at)^k x^{2n-2k},$$
$$w(x, t) = x^{2n+1} + \sum_{k=1}^{n} \frac{(2n+1)(2n)\dots(2n-2k+2)}{k!}(at)^k x^{2n-2k+1},$$
$$w(x, t) = A\exp(a\mu^2 t \pm \mu x) + B,$$
$$w(x, t) = A\frac{1}{\sqrt{t}}\exp\left(-\frac{x^2}{4at}\right) + B,$$
$$w(x, t) = A\frac{x}{t^{3/2}}\exp\left(-\frac{x^2}{4at}\right) + B,$$
$$w(x, t) = A\exp(-a\mu^2 t)\cos(\mu x + B) + C,$$
$$w(x, t) = A\exp(-\mu x)\cos(\mu x - 2a\mu^2 t + B) + C,$$
$$w(x, t) = A\operatorname{erf}\left(\frac{x}{2\sqrt{at}}\right) + B,$$

where A, B, C, and μ are arbitrary constants, n is a positive integer, and $\operatorname{erf} z \equiv \frac{2}{\sqrt{\pi}}\int_0^z \exp(-\xi^2)\,d\xi$ is the error function (probability integral).

$2°$. The general solution of the wave equation

$$\frac{\partial^2 w}{\partial t^2} = a^2\frac{\partial^2 w}{\partial x^2}$$

is given by

$$w(x, t) = \varphi(x + at) + \psi(x - at),$$

where $\varphi(z_1)$ and $\psi(z_2)$ are arbitrary functions.

$3°$. Particular solutions of Laplace equation (M12.2.1.3) in the Cartesian coordinate system:

$$w(x, y) = Ax + By + C,$$
$$w(x, y) = A(x^2 - y^2) + Bxy,$$

$$w(x, y) = A(x^3 - 3xy^2) + B(3x^2 y - y^3),$$

$$w(x, y) = \frac{Ax + By}{x^2 + y^2} + C,$$

$$w(x, y) = \exp(\pm \mu x)(A \cos \mu y + B \sin \mu y),$$

$$w(x, y) = (A \cos \mu x + B \sin \mu x) \exp(\pm \mu y),$$

$$w(x, y) = (A \sinh \mu x + B \cosh \mu x)(C \cos \mu y + D \sin \mu y),$$

$$w(x, y) = (A \cos \mu x + B \sin \mu x)(C \sinh \mu y + D \cosh \mu y),$$

where A, B, C, D, and μ are arbitrary constants.

Particular solutions of Laplace equation (M12.2.1.3) in the polar coordinate system:

$$w(r) = A \ln r + B,$$

$$w(r, \varphi) = \left(Ar^m + \frac{B}{r^m} \right)(C \cos m\varphi + D \sin m\varphi),$$

where A, B, C, and D are arbitrary constants, and $m = 1, 2, \ldots$

M12.4.2. Nonhomogeneous Linear Equations and Their Particular Solutions

▶ **Simplest properties of nonhomogeneous linear equations.** For brevity, we write a nonhomogeneous linear partial differential equation in the form

$$\mathcal{L}[w] = \Phi(\mathbf{x}, t), \tag{M12.4.2.1}$$

where the linear differential operator $\mathcal{L}$ is defined above (see the beginning of Subsection M12.4.1).

Below are the simplest properties of particular solutions of the nonhomogeneous equation (M12.4.2.1).

1°. If $\widetilde{w}_\Phi(\mathbf{x}, t)$ is a particular solution of the nonhomogeneous equation (M12.4.2.1) and $\widetilde{w}_0(\mathbf{x}, t)$ is a particular solution of the corresponding homogeneous equation (M12.4.1.1), then the sum

$$A \widetilde{w}_0(\mathbf{x}, t) + \widetilde{w}_\Phi(\mathbf{x}, t),$$

with arbitrary constant A, is also a solution of the nonhomogeneous equation (M12.4.2.1). The following, more general statement holds: The general solution of the nonhomogeneous equation (M12.4.2.1) is the sum of the general solution of the corresponding homogeneous equation (M12.4.1.1) and any particular solution of the nonhomogeneous equation (M12.4.2.1).

2°. Suppose w_1 and w_2 are solutions of nonhomogeneous linear equations with the same left-hand side and different right-hand sides, i.e.,

$$\mathcal{L}[w_1] = \Phi_1(\mathbf{x}, t), \qquad \mathcal{L}[w_2] = \Phi_2(\mathbf{x}, t).$$

Then the function $w = w_1 + w_2$ is a solution of the equation

$$\mathcal{L}[w] = \Phi_1(\mathbf{x}, t) + \Phi_2(\mathbf{x}, t).$$

▶ **Fundamental and particular solutions of stationary equations.** Consider the second-order linear stationary (time-independent) nonhomogeneous equation

$$L_{\mathbf{x}}[w] = -\Phi(\mathbf{x}). \tag{M12.4.2.2}$$

Here, $L_{\mathbf{x}}$ is a linear differential operator of the second (or any) order of general form whose coefficients are dependent on $\mathbf{x}$, where $\mathbf{x} \in \mathbb{R}^n$.

A function $\mathscr{E} = \mathscr{E}(\mathbf{x}, \mathbf{y})$ that satisfies the equation with a special right-hand side

$$L_{\mathbf{x}}[\mathscr{E}] = -\delta(\mathbf{x} - \mathbf{y}) \tag{M12.4.2.3}$$

is called a *fundamental solution* corresponding to the operator $L_{\mathbf{x}}$. In (M12.4.2.3), $\delta(\mathbf{x})$ is an n-dimensional Dirac delta function and the vector quantity $\mathbf{y} = \{y_1, \ldots, y_n\}$ appears in equation (M12.4.2.3) as an n-dimensional free parameter. It is assumed that $\mathbf{y} \in \mathbb{R}^n$.

The n-dimensional Dirac delta function possesses the following basic properties:

$$1. \quad \delta(\mathbf{x}) = \delta(x_1)\delta(x_2)\ldots\delta(x_n),$$

$$2. \quad \int_{\mathbb{R}^n} \Phi(\mathbf{y})\delta(\mathbf{x} - \mathbf{y})\,d\mathbf{y} = \Phi(\mathbf{x}),$$

where $\delta(x_k)$ is the one-dimensional Dirac delta function, $\Phi(\mathbf{x})$ is an arbitrary continuous function, and $d\mathbf{y} = dy_1 \ldots dy_n$.

For constant coefficient equations, a fundamental solution always exists.

The fundamental solution $\mathscr{E} = \mathscr{E}(\mathbf{x}, \mathbf{y})$ can be used to construct a particular solution of the linear stationary nonhomogeneous equation (M12.4.2.2) for arbitrary continuous $\Phi(\mathbf{x})$; this particular solution is expressed as follows:

$$w(\mathbf{x}) = \int_{\mathbb{R}^n} \Phi(\mathbf{y})\mathscr{E}(\mathbf{x}, \mathbf{y})\,d\mathbf{y}. \tag{M12.4.2.4}$$

Remark 1. The fundamental solution $\mathscr{E}$ is not unique; it is defined up to an additive term $w_0 = w_0(\mathbf{x})$, which is an arbitrary solution of the homogeneous equation $L_{\mathbf{x}}[w_0] = 0$.

Remark 2. For constant coefficient differential equations, the fundamental solution possesses the property $\mathscr{E}(\mathbf{x}, \mathbf{y}) = \mathscr{E}(\mathbf{x} - \mathbf{y})$.

Remark 3. The right-hand sides of equations (M12.4.2.2) and (M12.4.2.3) are often prefixed with the plus sign. In this case, formula (M12.4.2.4) remains valid.

Example. For the two- and three-dimensional *Poisson equations*, the fundamental solutions have the forms

Equations		*Fundamental solutions*
$\dfrac{\partial^2 w}{\partial x_1^2} + \dfrac{\partial^2 w}{\partial x_2^2} = -\Phi(x_1, x_2)$	$\implies$	$\mathscr{E}(x_1, x_2, y_1, y_2) = \dfrac{1}{2\pi} \ln \dfrac{1}{\rho},$
$\dfrac{\partial^2 w}{\partial x_1^2} + \dfrac{\partial^2 w}{\partial x_2^2} + \dfrac{\partial^2 w}{\partial x_3^2} = -\Phi(x_1, x_2, x_3)$	$\implies$	$\mathscr{E}(x_1, x_2, x_3, y_1, y_2, y_3) = \dfrac{1}{4\pi r},$

where $\rho = \sqrt{(x_1 - y_1)^2 + (x_2 - y_2)^2}$ and $r = \sqrt{(x_1 - y_1)^2 + (x_2 - y_2)^2 + (x_3 - y_3)^2}$.

M12.5. Method of Separation of Variables (Fourier Method)

M12.5.1. Solution of Problems for Parabolic Equations

▶ **Mathematical statement of the problem.** Consider boundary value problems for a linear homogeneous parabolic equation of the form

$$\frac{\partial w}{\partial t} = a(x)\frac{\partial^2 w}{\partial x^2} + b(x)\frac{\partial w}{\partial x} + \big[c(x) + \gamma(t)\big]w \tag{M12.5.1.1}$$

with homogeneous linear boundary conditions,

$$s_1 \partial_x w + k_1 w = 0 \quad \text{at} \quad x = x_1,$$
$$s_2 \partial_x w + k_2 w = 0 \quad \text{at} \quad x = x_2, \tag{M12.5.1.2}$$

and an arbitrary initial condition,

$$w = f_0(x) \quad \text{at} \quad t = 0. \tag{M12.5.1.3}$$

We assume that the coefficients of equation (M12.5.1.1) and boundary conditions (M12.5.1.2) meet the following requirements:

$a(x)$, $b(x)$, $c(x)$, and $\gamma(t)$ are continuous functions; $a(x) > 0$, $|s_1| + |k_1| > 0$, $|s_2| + |k_2| > 0$.

Remark. In various applications, equations of the form (M12.5.1.1) may arise with the coefficient $b(x)$ going to infinity at the boundary, $b(x) \to \infty$ as $x \to x_1$, with the other coefficients being continuous. In this case, the first boundary condition in (M12.5.1.2) should be replaced with a condition of boundedness of the solution as $x \to x_1$. This may occur in spatial problems with central or axial symmetry where the solution depends only on the radial coordinate.

▶ **Derivation of equations and boundary conditions for particular solutions.** The method of separation of variables is based on searching for particular solutions of equation (M12.5.1.1) in the product form

$$w(x, t) = \varphi(x)\,\psi(t). \tag{M12.5.1.4}$$

Let us substitute (M12.5.1.4) into (M12.5.1.1) and divide all terms of the equation by $\varphi(x)\,\psi(t)$. Then we move the terms dependent on t to the left-hand side and those dependent on x to the right-hand side to obtain

$$\frac{\psi'_t}{\psi} - \gamma(t) = \frac{a(x)\varphi''_{xx} + b(x)\varphi'_x + c(x)\varphi}{\varphi}$$

This equality is achieved if both sides are equal to the same constant quantity, $-\lambda$. The free parameter λ is called the *separation constant*. Thus, we arrive at the following linear ordinary differential equations for $\varphi = \varphi(x)$ and $\psi = \psi(t)$:

$$a(x)\varphi''_{xx} + b(x)\varphi'_x + [\lambda + c(x)]\varphi = 0, \tag{M12.5.1.5}$$
$$\psi'_t + [\lambda - \gamma(t)]\psi = 0. \tag{M12.5.1.6}$$

Substituting (M12.5.1.4) into (M12.5.1.2) yields the boundary conditions for $\varphi = \varphi(x)$:

$$s_1 \varphi'_x + k_1 \varphi = 0 \quad \text{at} \quad x = x_1,$$
$$s_2 \varphi'_x + k_2 \varphi = 0 \quad \text{at} \quad x = x_2. \tag{M12.5.1.7}$$

The homogeneous linear ordinary differential equation (M12.5.1.5) in conjunction with the homogeneous linear boundary conditions (M12.5.1.7) makes up an eigenvalue problem.

▶ **Solution of eigenvalue problems. Orthogonality of eigenfunctions.** Let $\widetilde{\varphi}_1(x, \lambda)$ and $\widetilde{\varphi}_2(x, \lambda)$ be linearly independent particular solutions of equation (M12.5.1.5). Then the general solution of this equation can be represented as the linear combination

$$\varphi = C_1 \widetilde{\varphi}_1(x, \lambda) + C_2 \widetilde{\varphi}_2(x, \lambda), \tag{M12.5.1.8}$$

where C_1 and C_2 are arbitrary constants.

Substituting solution (M12.5.1.8) into the boundary conditions (M12.5.1.7) yields the following homogeneous linear algebraic system of equations for C_1 and C_2:

$$\sigma_{11}(\lambda)C_1 + \sigma_{12}(\lambda)C_2 = 0,$$
$$\sigma_{21}(\lambda)C_1 + \sigma_{22}(\lambda)C_2 = 0, \qquad \text{(M12.5.1.9)}$$

where $\sigma_{ij}(\lambda) = \left[s_i(\widetilde{\varphi}_j)'_x + k_i\widetilde{\varphi}_j\right]_{x=x_i}$. For system (M12.5.1.9) to have nontrivial solutions, its determinant must be zero; we have

$$\sigma_{11}(\lambda)\sigma_{22}(\lambda) - \sigma_{12}(\lambda)\sigma_{21}(\lambda) = 0. \qquad \text{(M12.5.1.10)}$$

Solving the transcendental equation (M12.5.1.10) for λ, one obtains the *eigenvalues* $\lambda = \lambda_n$, where $n = 1, 2, \ldots$ For these values of λ, equation (M12.5.1.5) has nontrivial solutions,

$$\varphi_n(x) = \sigma_{12}(\lambda_n)\widetilde{\varphi}_1(x, \lambda_n) - \sigma_{11}(\lambda_n)\widetilde{\varphi}_2(x, \lambda_n), \qquad \text{(M12.5.1.11)}$$

which are called *eigenfunctions* (these functions are defined up to a constant multiplier).

To facilitate the further analysis, we represent equation (M12.5.1.5) in the form

$$[p(x)\varphi'_x]'_x + [\lambda\rho(x) - q(x)]\varphi = 0, \qquad \text{(M12.5.1.12)}$$

where

$$p(x) = \exp\left[\int \frac{b(x)}{a(x)}\,dx\right], \quad q(x) = -\frac{c(x)}{a(x)}\exp\left[\int \frac{b(x)}{a(x)}\,dx\right], \quad \rho(x) = \frac{1}{a(x)}\exp\left[\int \frac{b(x)}{a(x)}\,dx\right].$$
$$\text{(M12.5.1.13)}$$

It follows from the adopted assumptions that $p(x)$, $p'_x(x)$, $q(x)$, and $\rho(x)$ are continuous functions, with $p(x) > 0$ and $\rho(x) > 0$.

The eigenvalue problem for equation (M12.5.1.12) subject to the boundary conditions (M12.5.1.7) is known to possess the following properties:

1. All eigenvalues $\lambda_1, \lambda_2, \ldots$ are real, and $\lambda_n \to \infty$ as $n \to \infty$; consequently, the number of negative eigenvalues is finite. Each eigenvalue is of multiplicity 1.

2. The system of eigenfunctions $\varphi_1(x)$, $\varphi_2(x)$, $\ldots$ is orthogonal on the interval $x_1 \leq x \leq x_2$ with weight $\rho(x)$, i.e.,

$$\int_{x_1}^{x_2} \rho(x)\varphi_n(x)\varphi_m(x)\,dx = 0 \quad \text{for} \quad n \neq m. \qquad \text{(M12.5.1.14)}$$

3. If

$$q(x) \geq 0, \quad s_1 k_1 \leq 0, \quad s_2 k_2 \geq 0, \qquad \text{(M12.5.1.15)}$$

there are no negative eigenvalues. If $q \equiv 0$ and $k_1 = k_2 = 0$, the least eigenvalue is $\lambda_1 = 0$ and the corresponding eigenfunction is $\varphi_1 = \text{const}$. Otherwise, all eigenvalues are positive, provided that conditions (M12.5.1.15) are satisfied; the first inequality in (M12.5.1.15) is satisfied if $c(x) \leq 0$.

Subsection M11.2.4 presents some estimates for the eigenvalues λ_n and eigenfunctions $\varphi_n(x)$.

▶ **Series solutions of boundary value problems for parabolic equations.** The solution of equation (M12.5.1.1) corresponding to the eigenvalues $\lambda = \lambda_n$ and satisfying the normalizing conditions $\psi_n(0) = 1$ has the form

$$\psi_n(t) = \exp\left[-\lambda_n t + \int_0^t \gamma(\xi)\, d\xi\right]. \tag{M12.5.1.16}$$

Then the solution of the nonstationary boundary value problem for equation (M12.5.1.1) with the boundary conditions (M12.5.1.2) and initial condition (M12.5.1.3) is sought in the series form

$$w(x,t) = \sum_{n=1}^{\infty} A_n \varphi_n(x)\, \psi_n(t); \tag{M12.5.1.17}$$

the A_n are some constant quantities, whose values are determined in the course of solving the problem. The functions $w_n(x,t) = \varphi_n(x)\psi_n(t)$ in (M12.5.1.17) are particular solutions of equation (M12.5.1.1) satisfying the boundary conditions (M12.5.1.2). By the principle of linear superposition, series (M12.5.1.17) is also a solution of the original partial differential equation that satisfies the boundary conditions.

To determine the coefficients A_n, we substitute series (M12.5.1.17) into the initial condition (M12.5.1.3), thus obtaining

$$\sum_{n=1}^{\infty} A_n \varphi_n(x) = f_0(x).$$

Multiplying this equation by $\rho(x)\varphi_n(x)$, where the weight function $\rho(x)$ is defined in (M12.5.1.13), then integrating the resulting relation with respect to x over the interval $x_1 \le x \le x_2$, and taking into account the properties (M12.5.1.14), we find

$$A_n = \frac{1}{\|\varphi_n\|^2} \int_{x_1}^{x_2} \rho(x)\varphi_n(x)f_0(x)\, dx, \quad \|\varphi_n\|^2 = \int_{x_1}^{x_2} \rho(x)\varphi_n^2(x)\, dx. \tag{M12.5.1.18}$$

Relations (M12.5.1.17), (M12.5.1.16), (M12.5.1.11), and (M12.5.1.18) give a formal solution of the nonstationary boundary value problem (M12.5.1.1)–(M12.5.1.3).

Example. Consider the first boundary value problem on the interval $0 \le x \le l$ for the heat equation

$$\frac{\partial w}{\partial t} = \frac{\partial^2 w}{\partial x^2} \tag{M12.5.1.19}$$

with the general initial condition (M12.5.1.3) and the homogeneous boundary conditions

$$w = 0 \quad \text{at} \quad x = 0, \qquad w = 0 \quad \text{at} \quad x = l. \tag{M12.5.1.20}$$

The functions $\psi_n(t)$ in solution (M12.5.1.17) are found from (M12.5.1.16) where $\gamma(t) = 0$:

$$\psi_n(t) = \exp(-\lambda_n t). \tag{M12.5.1.21}$$

The functions $\varphi_n(x)$ are determined by solving the eigenvalue problem (M12.5.1.5), (M12.5.1.7) with $a(x) = 1$, $b(x) = c(x) = 0$, $s_1 = s_2 = 0$, $k_1 = k_2 = 1$, $x_1 = 0$, and $x_2 = l$:

$$\varphi_{xx}'' + \lambda\varphi = 0; \qquad \varphi = 0 \quad \text{at} \quad x = 0, \qquad \varphi = 0 \quad \text{at} \quad x = l.$$

So we obtain the eigenfunctions and eigenvalues:

$$\varphi_n(x) = \sin\left(\frac{n\pi x}{l}\right), \qquad \lambda_n = \left(\frac{n\pi}{l}\right)^2, \qquad n = 1, 2, \dots \tag{M12.5.1.22}$$

The solution to problem (M12.5.1.19), (M12.5.1.20), (M12.5.1.3) is given by formulas (M12.5.1.17), (M12.5.1.18), (M12.5.1.21), and (M12.5.1.22). Taking into account that $\|\varphi_n\|^2 = l/2$, we obtain

$$w(x,t) = \sum_{n=1}^{\infty} A_n \sin\left(\frac{n\pi x}{l}\right) \exp\left(-\frac{n^2\pi^2 t}{l^2}\right), \qquad A_n = \frac{2}{l}\int_0^l f_0(\xi)\sin\left(\frac{n\pi\xi}{l}\right) d\xi. \qquad \text{(M12.5.1.23)}$$

If the function $f_0(x)$ is twice continuously differentiable and the compatibility conditions (see below) are satisfied, then series (M12.5.1.23) is convergent and admits termwise differentiation, once with respect to t and twice with respect to x. In this case, formula (M12.5.1.23) gives the classical smooth solution of problem (M12.5.1.19, (M12.5.1.20), (M12.5.1.3). [If $f_0(x)$ is not as smooth as indicated or if the compatibility conditions are not met, then series (M12.5.1.23) may converge to a discontinuous function, thus giving only a generalized solution.]

▶ **Conditions of compatibility of initial and boundary conditions.** Suppose the function w has a continuous derivative with respect to t and two continuous derivatives with respect to x and is a solution of problem (M12.5.1.1)–(M12.5.1.3). Then the boundary conditions (M12.5.1.2) and the initial condition (M12.5.1.3) must be consistent; namely, the following compatibility conditions must hold:

$$[s_1 f_0' + k_1 f_0]_{x=x_1} = 0, \qquad [s_2 f_0' + k_2 f_0]_{x=x_2} = 0. \qquad \text{(M12.5.1.24)}$$

If $s_1 = 0$ or $s_2 = 0$, then the additional compatibility conditions

$$\begin{aligned} [a(x)f_0'' + b(x)f_0']_{x=x_1} = 0 \quad &\text{if} \quad s_1 = 0, \\ [a(x)f_0'' + b(x)f_0']_{x=x_2} = 0 \quad &\text{if} \quad s_2 = 0 \end{aligned} \qquad \text{(M12.5.1.25)}$$

must also hold; the primes denote the derivatives with respect to x.

M12.5.2. Solution of Problems for Hyperbolic Equations

▶ **Solution of problems by the method of separation of variables.** Consider a linear homogeneous hyperbolic equation of the form

$$\alpha(t)\frac{\partial^2 w}{\partial t^2} + \beta(t)\frac{\partial w}{\partial t} = a(x)\frac{\partial^2 w}{\partial x^2} + b(x)\frac{\partial w}{\partial x} + \left[c(x) + \gamma(t)\right]w \qquad \text{(M12.5.2.1)}$$

with homogeneous linear boundary conditions (M12.5.1.2) and arbitrary initial conditions,

$$\begin{aligned} w &= f_0(x) \quad \text{at} \quad t = 0, & \text{(M12.5.2.2)} \\ \partial_t w &= f_1(x) \quad \text{at} \quad t = 0. & \text{(M12.5.2.3)} \end{aligned}$$

We assume that the coefficients of equation (M12.5.2.1) and boundary conditions (M12.5.1.2) meet the following requirements:

$$\begin{aligned} &\alpha(t), \ \beta(t), \ \gamma(t), \ a(x), \ b(x), \ \text{and} \ c(x) \ \text{are continuous functions}, \\ &\alpha(t) > 0, \quad a(x) > 0, \quad |s_1| + |k_1| > 0, \quad |s_2| + |k_2| > 0. \end{aligned}$$

Let us use the method of separation of variables to solve boundary value problems for homogeneous linear equations of the hyperbolic type (M12.5.2.1) with homogeneous boundary conditions (M12.5.1.2) and nonhomogeneous initial conditions (M12.5.2.2) and (M12.5.2.3).

The approach is based on searching for particular solutions of equation (M12.5.2.1) in the product form (M12.5.1.4). After separating the variables and rearranging (in much the same way as in Subsection M12.5.1 for parabolic equations), one arrives at the following linear ordinary differential equations for the functions $\varphi = \varphi(x)$ and $\psi = \psi(t)$:

$$a(x)\varphi''_{xx} + b(x)\varphi'_x + [\lambda + c(x)]\varphi = 0, \qquad (M12.5.2.4)$$
$$\alpha(t)\psi''_{tt} + \beta(t)\psi'_t + [\lambda - \gamma(t)]\psi = 0. \qquad (M12.5.2.5)$$

Note that equation (M12.5.2.4) coincides with (M12.5.1.5).

For hyperbolic equations, the solution of the boundary value problem (M12.5.2.1)–(M12.5.2.3), (M12.5.1.2) is sought in the series form

$$w(x, t) = \sum_{n=1}^{\infty} \varphi_n(x) \left[A_n\psi_{n1}(t) + B_n\psi_{n2}(t)\right]. \qquad (M12.5.2.6)$$

The A_n and B_n are some constant quantities, whose values are determined further in the course of solving the problem. The functions $\varphi_n(x)$ and the numbers λ_n are determined by solving the eigenvalue problem (M12.5.2.4), (M12.5.1.7). The functions $\psi_{n1}(t)$ and $\psi_{n2}(t)$ are particular solutions of the linear equation (M12.5.2.5) for ψ (with $\lambda = \lambda_n$) that satisfy the conditions

$$\psi_{n1}(0) = 1, \quad \psi'_{n1}(0) = 0; \qquad \psi_{n2}(0) = 0, \quad \psi'_{n2}(0) = 1. \qquad (M12.5.2.7)$$

Substituting solution (M12.5.2.6) into the initial conditions (M12.5.2.2) and (M12.5.2.3) yields

$$\sum_{n=1}^{\infty} A_n\varphi_n(x) = f_0(x), \qquad \sum_{n=1}^{\infty} B_n\varphi_n(x) = f_1(x).$$

Multiplying these equations by $\rho(x)\varphi_n(x)$, where the weight function $\rho(x)$ is defined in (M12.5.1.13), then integrating the resulting relations with respect to x over the interval $x_1 \leq x \leq x_2$, and taking into account the properties (M12.5.1.14), we obtain the coefficients of series (M12.5.2.6) in the form

$$A_n = \frac{1}{\|\varphi_n\|^2} \int_{x_1}^{x_2} \rho(x)\varphi_n(x)f_0(x)\,dx, \quad B_n = \frac{1}{\|\varphi_n\|^2} \int_{x_1}^{x_2} \rho(x)\varphi_n(x)f_1(x)\,dx.$$
$$(M12.5.2.8)$$

The quantity $\|\varphi_n\|$ is defined in (M12.5.1.18).

Relations (M12.5.2.6), (M12.5.1.11), and (M12.5.2.8) together with the solutions of the problems for equation (M12.5.2.5) subject to the initial conditions (M12.5.2.7) give a formal solution of the nonstationary boundary value problem for equation (M12.5.2.1) subject to the initial conditions (M12.5.2.2) and (M12.5.2.3) and boundary conditions (M12.5.1.2).

Example. Consider a mixed boundary value problem on the interval $0 \leq x \leq l$ for the wave equation

$$\frac{\partial^2 w}{\partial t^2} = \frac{\partial^2 w}{\partial x^2} \qquad (M12.5.2.9)$$

with the general initial conditions (M12.5.2.2) and (M12.5.2.3) and the homogeneous boundary conditions

$$w = 0 \quad \text{at} \quad x = 0, \qquad \partial_x w = 0 \quad \text{at} \quad x = l. \qquad (M12.5.2.10)$$

The solution of this mixed boundary value problem for equation (M12.5.2.9) is sought in the series form (M12.5.2.6). The functions $\varphi_n(x)$ are determined by solving the eigenvalue problem (M12.5.2.4), (M12.5.1.7) with $a(x) = 1$, $b(x) = c(x) = 0$, $s_1 = k_2 = 0$, $s_2 = k_1 = 1$, $x_1 = 0$, and $x_2 = l$:

$$\varphi''_{xx} + \lambda\varphi = 0; \qquad \varphi = 0 \quad \text{at} \quad x = 0, \qquad \varphi'_x = 0 \quad \text{at} \quad x = l.$$

So we obtain the eigenfunctions and eigenvalues:

$$\varphi_n(x) = \sin(\mu_n x), \quad \mu_n = \sqrt{\lambda_n} = \frac{\pi(2n-1)}{2l}, \quad n = 1, 2, \ldots \tag{M12.5.2.11}$$

The functions $\psi_{n1}(t)$ and $\psi_{n2}(t)$ are determined by the linear equation [see (M12.5.2.5) with $\alpha(t) = 1$, $\beta(t) = \gamma(t) = 0$, and $\lambda = \lambda_n$]

$$\psi_{tt}'' + \lambda\psi = 0$$

with the initial conditions (M12.5.2.7). We find

$$\psi_{n1}(t) = \cos\left(\sqrt{\lambda_n}\,t\right), \quad \psi_{n2}(t) = \frac{1}{\sqrt{\lambda_n}}\sin\left(\sqrt{\lambda_n}\,t\right). \tag{M12.5.2.12}$$

The solution to problem (M12.5.2.9), (M12.5.2.10), (M12.5.2.2), (M12.5.2.3) is given by formulas (M12.5.2.6), (M12.5.2.11), (M12.5.2.12), and (M12.5.2.8). Taking into account that $\|\varphi_n\|^2 = l/2$, we have

$$w(x,t) = \sum_{n=1}^{\infty}\left[A_n\cos(\mu_n t) + B_n\sin(\mu_n t)\right]\sin(\mu_n x), \quad \mu_n = \frac{\pi(2n-1)}{2l},$$

$$A_n = \frac{2}{l}\int_0^l f_0(x)\sin(\mu_n x)\,dx, \quad B_n = \frac{2}{l\mu_n}\int_0^l f_1(x)\sin(\mu_n x)\,dx. \tag{M12.5.2.13}$$

If $f_0(x)$ and $f_1(x)$ have three and two continuous derivatives, respectively, and the compatibility conditions are met (see below), then series (M12.5.2.13) is convergent and admits double termwise differentiation. In this case, formula (M12.5.2.13) gives the classical smooth solution of problem (M12.5.2.9), (M12.5.2.10), (M12.5.2.2), (M12.5.2.3).

▶ **Conditions of compatibility of initial and boundary conditions.** Suppose w is a twice continuously differentiable solution of the problem for equation (M12.5.2.1) subject to the initial conditions (M12.5.2.2) and (M12.5.2.3) and boundary conditions (M12.5.1.2). Then conditions (M12.5.1.24) and (M12.5.1.25) must hold. In addition, the following conditions of compatibility of the boundary conditions (M12.5.1.2) and initial condition (M12.5.2.3) must be satisfied:

$$[s_1 f_1' + k_1 f_1]_{x=x_1} = 0, \quad [s_2 f_1' + k_2 f_1]_{x=x_2} = 0.$$

M12.5.3. Solution of Problems for Elliptic Equations

▶ **Solution of a special problem for elliptic equations.** Now consider a boundary value problem for the elliptic equation

$$a(x)\frac{\partial^2 w}{\partial x^2} + \alpha(y)\frac{\partial^2 w}{\partial y^2} + b(x)\frac{\partial w}{\partial x} + \beta(y)\frac{\partial w}{\partial y} + \left[c(x) + \gamma(y)\right]w = 0 \tag{M12.5.3.1}$$

with homogeneous linear boundary conditions (M12.5.1.2) in x and the following mixed (homogeneous and nonhomogeneous) boundary conditions in y:

$$\begin{aligned} \sigma_1\partial_y w + \nu_1 w &= 0 && \text{at} \quad y = y_1, \\ \sigma_2\partial_y w + \nu_2 w &= f(x) && \text{at} \quad y = y_2. \end{aligned} \tag{M12.5.3.2}$$

We assume that the coefficients of equation (M12.5.3.1) and boundary conditions (M12.5.1.2) and (M12.5.3.2) meet the following requirements:

$a(x)$, $b(x)$, $c(x)$ $\alpha(y)$, $\beta(y)$, and $\gamma(y)$ are continuous functions,

$a(x) > 0$, $\alpha(y) > 0$, $|s_1| + |k_1| > 0$, $|s_2| + |k_2| > 0$, $|\sigma_1| + |\nu_1| > 0$, $|\sigma_2| + |\nu_2| > 0$.

The method of separation of variables is based on searching for particular solutions of equation (M12.5.3.1) in the product form

$$w(x, y) = \varphi(x)\,\psi(y). \tag{M12.5.3.3}$$

As before, we first arrive at the eigenvalue problem (M12.5.1.5), (M12.5.1.7) for the function $\varphi = \varphi(x)$; the solution procedure is detailed in Subsection M12.5.1. Further on, we assume that the λ_n and $\varphi_n(x)$ have been found. The functions $\psi_n = \psi_n(y)$ are determined by solving the linear ordinary differential equation

$$\alpha(y)\psi''_{yy} + \beta(y)\psi'_y + [\gamma(y) - \lambda_n]\psi = 0 \tag{M12.5.3.4}$$

subject to the homogeneous boundary condition

$$\sigma_1\partial_y\psi + \nu_1\psi = 0 \quad \text{at} \quad y = y_1, \tag{M12.5.3.5}$$

which follows from the first condition in (M12.5.3.2). The functions ψ_n are determined up to a constant factor.

Taking advantage of the principle of linear superposition, we seek the solution to the boundary value problem (M12.5.3.1), (M12.5.3.2), (M12.5.1.2) in the series form

$$w(x, y) = \sum_{n=1}^{\infty} A_n\varphi_n(x)\psi_n(y); \tag{M12.5.3.6}$$

The A_n are some constants, which are to be determined. By construction, series (M12.5.3.6) will satisfy equation (M12.5.3.1) with the boundary conditions (M12.5.1.2) and the first boundary condition (M12.5.3.2). In order to find the coefficients A_n, substitute (M12.5.3.6) into the second boundary condition (M12.5.3.2) to obtain

$$\sum_{n=1}^{\infty} A_n B_n\varphi_n(x) = f(x), \qquad B_n = \sigma_2\frac{d\psi_n}{dy}\bigg|_{y=y_2} + \nu_2\psi_n(y_2). \tag{M12.5.3.7}$$

Further, we follow the same procedure as in Subsection M12.5.1. Specifically, multiplying (M12.5.3.7) by $\rho(x)\varphi_n(x)$, then integrating the resulting relation with respect to x over the interval $x_1 \leq x \leq x_2$ and taking into account the properties (M12.5.1.14), we obtain

$$A_n = \frac{1}{B_n\|\varphi_n\|^2}\int_{x_1}^{x_2} \rho(x)\varphi_n(x)f(x)\,dx, \qquad \|\varphi_n\|^2 = \int_{x_1}^{x_2} \rho(x)\varphi_n^2(x)\,dx, \tag{M12.5.3.8}$$

where the weight function $\rho(x)$ is defined in (M12.5.1.13).

Example. Consider the first (Dirichlet) boundary value problem for the Laplace equation

$$\frac{\partial^2 w}{\partial x^2} + \frac{\partial^2 w}{\partial y^2} = 0 \tag{M12.5.3.9}$$

subject to the boundary conditions

$$\begin{array}{llll} w = 0 & \text{at} & x = 0, & w = 0 \quad \text{at} \quad x = l_1; \\ w = 0 & \text{at} & y = 0, & w = f(x) \quad \text{at} \quad y = l_2 \end{array} \tag{M12.5.3.10}$$

in a rectangular domain $0 \leq x \leq l_1, 0 \leq y \leq l_2$.

Particular solutions to equation (M12.5.3.9) are sought in the form (M12.5.3.3). We have the following eigenvalue problem for $\varphi(x)$:

$$\varphi_{xx}'' + \lambda\varphi = 0; \qquad \varphi = 0 \quad \text{at} \quad x = 0, \qquad \varphi = 0 \quad \text{at} \quad x = l_1.$$

On solving this problem, we find the eigenfunctions with respective eigenvalues

$$\varphi_n(x) = \sin(\mu_n x), \quad \mu_n = \sqrt{\lambda_n} = \frac{\pi n}{l_1}, \quad n = 1, 2, \ldots \tag{M12.5.3.11}$$

The functions $\psi_n = \psi_n(y)$ are determined by solving the following problem for a linear ordinary differential equation with homogeneous boundary conditions (at this stage, the functions ψ_n are determined up to a constant factor each):

$$\psi_{yy}'' - \lambda_n\psi = 0; \qquad \psi = 0 \quad \text{at} \quad y = 0. \tag{M12.5.3.12}$$

It is a special case of problem (M12.5.3.4), (M12.5.3.5) with $\alpha(y) = 1$, $\beta(y) = \gamma(y) = 0$, $\sigma_1 = 0$, and $\nu_1 = 1$. The nontrivial solutions of problem (M12.5.3.12) are expressed as

$$\psi_n(y) = \sinh(\mu_n y), \quad \mu_n = \sqrt{\lambda_n} = \frac{\pi n}{l_1}, \quad n = 1, 2, \ldots \tag{M12.5.3.13}$$

Using formulas (M12.5.3.6), (M12.5.3.8), (M12.5.3.11), (M12.5.3.13) and taking into account the relations $B_n = \psi_n(l_2) = \sinh(\mu_n l_2)$, $\rho(x) = 1$, and $\|\varphi_n\|^2 = l_1/2$, we find the solution of the original problem (M12.5.3.9), (M12.5.3.10) in the form

$$w(x, y) = \sum_{n=1}^{\infty} A_n \sin(\mu_n x) \sinh(\mu_n y), \quad A_n = \frac{2}{l_1 \sinh(\mu_n l_2)} \int_0^{l_1} f(x) \sin(\mu_n x)\, dx, \quad \mu_n = \frac{\pi n}{l_1}.$$

▶ **Generalization to the case of nonhomogeneous boundary conditions.** Now consider the linear boundary value problem for the elliptic equation (M12.5.3.1) with general nonhomogeneous boundary conditions

$$s_1\partial_x w + k_1 w = f_1(y) \quad \text{at} \quad x = x_1, \qquad s_2\partial_x w + k_2 w = f_2(y) \quad \text{at} \quad x = x_2,$$
$$\sigma_1\partial_y w + \nu_1 w = f_3(x) \quad \text{at} \quad y = y_1, \qquad \sigma_2\partial_y w + \nu_2 w = f_4(x) \quad \text{at} \quad y = y_2. \tag{M12.5.3.14}$$

The solution to this problem is the sum of solutions to four simpler auxiliary problems for equation (M12.5.3.1), each corresponding to one nonhomogeneous and three homogeneous boundary conditions in (M12.5.3.14); in these problems, all but one of the functions f_n are zero. Each auxiliary problem is solved using the procedure described above in this subsection, beginning with the search for solutions in the form of the product of functions with different arguments (M12.5.3.3), determined by equations (M12.5.1.5) and (M12.5.3.4). The solution to each of the auxiliary problems is sought in the series form (M12.5.3.6).

M12.6. Integral Transforms Method

Various integral transforms are widely used to solve linear problems of mathematical physics. The Laplace transform and the Fourier transform are in most common use (these and other integral transforms are considered in Chapter M10 in detail).

M12.6.1. Laplace Transform and Its Application to Linear Mathematical Physics Equations

▶ **Laplace and inverse Laplace transforms. Laplace transforms for derivatives.** The Laplace transform of an arbitrary (complex-valued) function $f(t)$ of a real variable t ($t \geq 0$) is defined by

$$\widetilde{f}(p) = \mathfrak{L}\{f(t)\}, \quad \text{where} \quad \mathfrak{L}\{f(t)\} \equiv \int_0^{\infty} e^{-pt} f(t)\, dt, \tag{M12.6.1.1}$$

where $p = s + i\sigma$ is a complex variable, $i^2 = -1$.

Given the transform $\widetilde{f}(p)$, the function $f(t)$ can be found by means of the inverse Laplace transform

$$f(t) = \mathcal{L}^{-1}\{\widetilde{f}(p)\}, \quad \text{where} \quad \mathcal{L}^{-1}\{\widetilde{f}(p)\} \equiv \frac{1}{2\pi i} \int_{c-i\infty}^{c+i\infty} \widetilde{f}(p)e^{pt}\, dp, \qquad \text{(M12.6.1.2)}$$

where the integration path is parallel to the imaginary axis and lies to the right of all singularities of $\widetilde{f}(p)$.

In order to solve nonstationary boundary value problems, the following Laplace transform formulas for derivatives will be required:

$$\mathcal{L}\{f'(t)\} = p\widetilde{f}(p) - f(0),$$
$$\mathcal{L}\{f''(t)\} = p^2\widetilde{f}(p) - pf(0) - f'(0), \qquad \text{(M12.6.1.3)}$$

where $f(0)$ and $f'(0)$ are the initial values.

More details on the properties of the Laplace transform and the inverse Laplace transform can be found in Section M10.2. The Laplace transforms of some functions are listed in Section S2.1. Tables of inverse Laplace transforms are listed in Section S2.2. Such tables are convenient to use when solving linear problems for partial differential equations.

▶ **Solution procedure for linear problems using the Laplace transform.** Figure M12.1 shows schematically how one can utilize the Laplace transforms to solve boundary value problems for linear parabolic or hyperbolic equations with two independent variables in the case where the equation coefficients are independent of t (the same procedure can be applied to solving linear problems characterized by higher-order equations). Here and henceforth, the short notation $\widetilde{w}(x, p) = \mathcal{L}\{w(x, t)\}$ will be used; the arguments of $\widetilde{w}$ may be omitted.

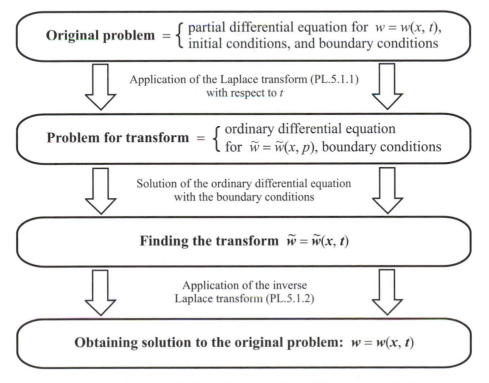

Figure M12.1. Solution procedure for linear boundary value problems using the Laplace transform.

It is significant that with the Laplace transform, the original problem for a partial differential equation is reduced to a simpler problem for an ordinary differential equation with parameter p; the derivatives with respect to t are replaced by appropriate algebraic expressions taking into account the initial conditions; see formulas (M12.6.1.3).

▶ **Solving linear problems for parabolic equations using the Laplace transform.** Consider a linear nonstationary boundary value problem for the parabolic equation

$$\frac{\partial w}{\partial t} = a(x)\frac{\partial^2 w}{\partial x^2} + b(x)\frac{\partial w}{\partial x} + c(x)w + \Phi(x,t) \tag{M12.6.1.4}$$

with the initial condition (M12.5.1.3) and the general nonhomogeneous boundary conditions

$$\begin{aligned} s_1\partial_x w + k_1 w &= g_1(t) \quad \text{at} \quad x = x_1, \\ s_2\partial_x w + k_2 w &= g_2(t) \quad \text{at} \quad x = x_2. \end{aligned} \tag{M12.6.1.5}$$

The application of the Laplace transform (M12.6.1.1) results in the problem determined by the ordinary differential equation for x (p is treated as a parameter)

$$a(x)\widetilde{w}''_{xx} + b(x)\widetilde{w}'_x + [c(x) - p]\widetilde{w} + f_0(x) + \widetilde{\Phi}(x,p) = 0 \tag{M12.6.1.6}$$

with the boundary conditions

$$\begin{aligned} s_1\widetilde{w}'_x + k_1\widetilde{w} &= \widetilde{g}_1(p) \quad \text{at} \quad x = x_1, \\ s_2\widetilde{w}'_x + k_2\widetilde{w} &= \widetilde{g}_2(p) \quad \text{at} \quad x = x_2. \end{aligned} \tag{M12.6.1.7}$$

Notation used: $\widetilde{\Phi}(x,p) = \mathcal{L}\left\{\Phi(x,t)\right\}$ and $\widetilde{g}_n(p) = \mathcal{L}\left\{g_n(t)\right\}$ ($n = 1, 2$). On solving problem (M12.6.1.6), (M12.6.1.7), one should apply to the resulting solution $\widetilde{w} = \widetilde{w}(x,p)$ the inverse Laplace transform (M12.6.1.2) to obtain the solution, $w = w(x,t)$, of the original problem.

Example. Consider the first boundary value problem for the heat equation:

$$\frac{\partial w}{\partial t} = \frac{\partial^2 w}{\partial x^2} \qquad (x > 0, \ t > 0),$$

$$w = 0 \quad \text{at} \quad t = 0 \qquad \text{(initial condition)},$$

$$w = w_0 \quad \text{at} \quad x = 0 \qquad \text{(boundary condition)},$$

$$w \to 0 \quad \text{at} \quad x \to \infty \quad \text{(boundary condition)}.$$

We apply the Laplace transform with respect to t. Let us multiply the equation, the initial condition, and the boundary conditions by e^{-pt} and then integrate with respect to t from zero to infinity. Taking into account the relations

$\mathcal{L}\{\partial_t w\} = p\widetilde{w} - w|_{t=0} = p\widetilde{w}$ (used are first property (M12.6.1.3) and the initial condition),

$\mathcal{L}\{w_0\} = w_0\mathcal{L}\{1\} = w_0/p$ (used are the first property of Subsection S2.1.1 and the relation $\mathcal{L}\{1\} = 1/p$),

we arrive at the following problem for a second-order linear ordinary differential equation with parameter p:

$$\widetilde{w}''_{xx} - p\widetilde{w} = 0,$$

$$\widetilde{w} = w_0/p \quad \text{at} \quad x = 0 \qquad \text{(boundary condition)},$$

$$\widetilde{w} \to 0 \qquad \text{at} \quad x \to \infty \quad \text{(boundary condition)}.$$

Integrating the equation yields the general solution $\widetilde{w} = A_1(p)e^{-x\sqrt{p}} + A_2(p)e^{x\sqrt{p}}$. Using the boundary conditions, we determine the constants, $A_1(p) = w_0/p$ and $A_2(p) = 0$. Thus, we have

$$\widetilde{w} = \frac{w_0}{p}e^{-x\sqrt{p}}.$$

Let us apply the inverse Laplace transform to both sides of this relation. We refer to the table in Subsection S2.2.5, row 20 (where x must be replaced by t and then a by x^2), to find the inverse transform of the right-hand side. Finally, we obtain the solution of the original problem:

$$w = w_0 \operatorname{erfc}\left(\frac{x}{2\sqrt{t}}\right), \qquad \text{where} \quad \operatorname{erfc} z = \frac{2}{\sqrt{\pi}}\int_z^\infty \exp\left(-\xi^2\right) d\xi.$$

M12.6.2. Fourier Transform and Its Application for Linear Equations of Mathematical Physics

▶ **Fourier transform and its properties.** The Fourier transform is defined as follows:

$$\tilde{f}(u) = \mathfrak{F}\{f(x)\}, \quad \text{where} \quad \mathfrak{F}\{f(x)\} \equiv \frac{1}{\sqrt{2\pi}} \int_{-\infty}^{\infty} f(x)e^{-iux}\, dx, \qquad i^2 = -1.$$

(M12.6.2.1)

This relation is meaningful for any function $f(x)$ absolutely integrable on the interval $(-\infty, \infty)$.

Given $\tilde{f}(u)$, the function $f(x)$ can be found by means of the inverse Fourier transform

$$f(x) = \mathfrak{F}^{-1}\{\tilde{f}(u)\}, \quad \text{where} \quad \mathfrak{F}^{-1}\{\tilde{f}(u)\} \equiv \frac{1}{\sqrt{2\pi}} \int_{-\infty}^{\infty} \tilde{f}(u)e^{iux}\, du, \qquad (\text{M12.6.2.2})$$

where the integral is understood in the sense of the Cauchy principal value.

The main properties of the correspondence between functions and their Fourier transforms are gathered in Table M10.4.

▶ **Solving linear problems of mathematical physics by the Fourier transform.** The Fourier transform is usually employed to solve problems for linear partial differential equations whose coefficients are independent of the space variable x, $-\infty < x < \infty$.

The scheme for solving linear boundary value problems with the help of the Fourier transform is similar to that used in solving problems with the help of the Laplace transform. With the Fourier transform, the derivatives with respect to x in the equation are replaced by appropriate algebraic expressions; see Property 4 or 5 in Table M10.4. In the case of two independent variables, the problem for a partial differential equation is reduced to a simpler problem for an ordinary differential equation with parameter u. On solving the latter problem, one determines the transform. After that, by applying the inverse Fourier transform, one obtains the solution of the original boundary value problem.

Example. Consider the following Cauchy problem for the heat equation:

$$\frac{\partial w}{\partial t} = \frac{\partial^2 w}{\partial x^2} \qquad (-\infty < x < \infty),$$

$$w = f(x) \quad \text{at} \quad t = 0 \qquad (\text{initial condition}).$$

We apply the Fourier transform with respect to the space variable x. Setting $\tilde{w} = \mathfrak{F}\{w(x,t)\}$ and taking into account the relation $\mathfrak{F}\{\partial_{xx}w\} = -u^2\tilde{w}$ (see Property 4 in Table M10.4), we arrive at the following problem for a linear first-order ordinary differential equation in t with parameter u:

$$\tilde{w}_t' + u^2\tilde{w} = 0,$$

$$\tilde{w} = \tilde{f}(u) \quad \text{at} \quad t = 0,$$

where $\tilde{f}(u)$ is defined by (M12.6.2.1). On solving this problem for the transform $\tilde{w}$, we find

$$\tilde{w} = \tilde{f}(u)e^{-u^2 t}.$$

Let us apply the inversion formula to both sides of this equation. After some calculations, we obtain the solution of the original problem in the form

$$w = \frac{1}{\sqrt{2\pi}} \int_{-\infty}^{\infty} \tilde{f}(u)e^{-u^2 t}e^{iux}\, du = \frac{1}{2\pi} \int_{-\infty}^{\infty} \left[\int_{-\infty}^{\infty} f(\xi)e^{-iu\xi}\, d\xi\right] e^{-u^2 t + iux}\, du$$

$$= \frac{1}{2\pi} \int_{-\infty}^{\infty} f(\xi)\, d\xi \int_{-\infty}^{\infty} e^{-u^2 t + iu(x-\xi)}\, du = \frac{1}{\sqrt{2\pi t}} \int_{-\infty}^{\infty} f(\xi)\exp\left[-\frac{(x-\xi)^2}{4t}\right] d\xi.$$

At the last stage we used the relation $\displaystyle\int_{-\infty}^{\infty} \exp\left(-a^2 u^2 + bu\right) du = \frac{\sqrt{\pi}}{|a|}\exp\left(\frac{b^2}{4a^2}\right).$

M12.7. Boundary Value Problems for Parabolic Equations with One Space Variable. Green's Function

M12.7.1. Representation of Solutions via the Green's Function

▶ **Statement of the problem** ($t \geq 0$, $x_1 \leq x \leq x_2$). In general, a nonhomogeneous linear differential equation of the parabolic type with variable coefficients with one spatial coordinate can be written as

$$\frac{\partial w}{\partial t} - L_{x,t}[w] = \Phi(x,t), \tag{M12.7.1.1}$$

where

$$L_{x,t}[w] \equiv a(x,t)\frac{\partial^2 w}{\partial x^2} + b(x,t)\frac{\partial w}{\partial x} + c(x,t)w, \qquad a(x,t) > 0. \tag{M12.7.1.2}$$

Consider the nonstationary boundary value problem for equation (M12.7.1.1) with an initial condition of general form

$$w = f(x) \quad \text{at} \quad t = 0, \tag{M12.7.1.3}$$

and arbitrary nonhomogeneous linear boundary conditions

$$s_1\frac{\partial w}{\partial x} + k_1 w = g_1(t) \quad \text{at} \quad x = x_1, \tag{M12.7.1.4}$$

$$s_2\frac{\partial w}{\partial x} + k_2 w = g_2(t) \quad \text{at} \quad x = x_2. \tag{M12.7.1.5}$$

By appropriately choosing the coefficients s_1, s_2, k_1, and k_2 in (M12.7.1.4) and (M12.7.1.5), one arrives at the first, second, third, and mixed boundary value problems for equation (M12.7.1.1).

▶ **Representation of the problem solution in terms of the Green's function.** The solution of the nonhomogeneous linear boundary value problem (M12.7.1.1)–(M12.7.1.5) can be represented as

$$w(x,t) = \int_0^t \int_{x_1}^{x_2} \Phi(y,\tau)G(x,y,t,\tau)\,dy\,d\tau + \int_{x_1}^{x_2} f(y)G(x,y,t,0)\,dy$$

$$+ \int_0^t g_1(\tau)a(x_1,\tau)\Lambda_1(x,t,\tau)\,d\tau + \int_0^t g_2(\tau)a(x_2,\tau)\Lambda_2(x,t,\tau)\,d\tau. \tag{M12.7.1.6}$$

Here, $G(x,y,t,\tau)$ is the Green's function that satisfies, for $t > \tau \geq 0$, the homogeneous equation

$$\frac{\partial G}{\partial t} - L_{x,t}[G] = 0 \tag{M12.7.1.7}$$

with the nonhomogeneous initial condition of a special form

$$G = \delta(x-y) \quad \text{at} \quad t = \tau \tag{M12.7.1.8}$$

TABLE M12.4
Expressions of the functions $\Lambda_1(x, t, \tau)$ and $\Lambda_2(x, t, \tau)$ involved
in the integrands of the last two terms in solution (M12.7.1.6).

Type of problem	Form of boundary conditions	Functions $\Lambda_m(x, t, \tau)$
First boundary value problem $(s_1 = s_2 = 0, \ k_1 = k_2 = 1)$	$w = g_1(t)$ at $x = x_1$ $w = g_2(t)$ at $x = x_2$	$\Lambda_1(x, t, \tau) = \partial_y G(x, y, t, \tau)\big|_{y=x_1}$ $\Lambda_2(x, t, \tau) = -\partial_y G(x, y, t, \tau)\big|_{y=x_2}$
Second boundary value problem $(s_1 = s_2 = 1, \ k_1 = k_2 = 0)$	$\partial_x w = g_1(t)$ at $x = x_1$ $\partial_x w = g_2(t)$ at $x = x_2$	$\Lambda_1(x, t, \tau) = -G(x, x_1, t, \tau)$ $\Lambda_2(x, t, \tau) = G(x, x_2, t, \tau)$
Third boundary value problem $(s_1 = s_2 = 1, \ k_1 < 0, \ k_2 > 0)$	$\partial_x w + k_1 w = g_1(t)$ at $x = x_1$ $\partial_x w + k_2 w = g_2(t)$ at $x = x_2$	$\Lambda_1(x, t, \tau) = -G(x, x_1, t, \tau)$ $\Lambda_2(x, t, \tau) = G(x, x_2, t, \tau)$
Mixed boundary value problem $(s_1 = k_2 = 0, \ s_2 = k_1 = 1)$	$w = g_1(t)$ at $x = x_1$ $\partial_x w = g_2(t)$ at $x = x_2$	$\Lambda_1(x, t, \tau) = \partial_y G(x, y, t, \tau)\big|_{y=x_1}$ $\Lambda_2(x, t, \tau) = G(x, x_2, t, \tau)$
Mixed boundary value problem $(s_1 = k_2 = 1, \ s_2 = k_1 = 0)$	$\partial_x w = g_1(t)$ at $x = x_1$ $w = g_2(t)$ at $x = x_2$	$\Lambda_1(x, t, \tau) = -G(x, x_1, t, \tau)$ $\Lambda_2(x, t, \tau) = -\partial_y G(x, y, t, \tau)\big|_{y=x_2}$

and the homogeneous boundary conditions

$$s_1 \frac{\partial G}{\partial x} + k_1 G = 0 \quad \text{at} \quad x = x_1, \tag{M12.7.1.9}$$

$$s_2 \frac{\partial G}{\partial x} + k_2 G = 0 \quad \text{at} \quad x = x_2. \tag{M12.7.1.10}$$

The quantities y and τ appear in problem (M12.7.1.7)–(M12.7.1.10) as free parameters, with $x_1 \leq y \leq x_2$, and $\delta(x)$ is the Dirac delta function.

The initial condition (M12.7.1.8) implies the limit relation

$$f(x) = \lim_{t \to \tau} \int_{x_1}^{x_2} f(y) G(x, y, t, \tau)\, dy$$

for any continuous function $f = f(x)$.

The functions $\Lambda_1(x, t, \tau)$ and $\Lambda_2(x, t, \tau)$ involved in the integrands of the last two terms in solution (M12.7.1.6) can be expressed in terms of the Green's function $G(x, y, t, \tau)$. The corresponding formulas for $\Lambda_m(x, t, \tau)$ are given in Table M12.4 for the basic types of boundary value problems.

It is significant that the Green's function G and the functions Λ_1, Λ_2 are independent of the functions Φ, f, g_1, and g_2 that characterize various nonhomogeneities of the boundary value problem.

If the coefficients of equation (M12.7.1.1)–(M12.7.1.2) are independent of time t, i.e., the conditions

$$a = a(x), \quad b = b(x), \quad c = c(x) \tag{M12.7.1.11}$$

hold, then the Green's function depends on only three arguments,

$$G(x, y, t, \tau) = G(x, y, t - \tau). \tag{M12.7.1.12}$$

In this case, the functions Λ_m depend on only two arguments, $\Lambda_m = \Lambda_m(x, t - \tau)$, $m = 1, 2$.

Formula (M12.7.1.6) also remains valid for the problem with boundary conditions of the third kind if $k_1 = k_1(t)$ and $k_2 = k_2(t)$. Here, the relations between Λ_m ($m = 1, 2$) and the Green's function G are the same as in the case of constant k_1 and k_2; the Green's function itself is now different.

Remark. In the first, second, and third boundary value problems that are considered on the interval $x_1 \leq x < \infty$, a condition of boundedness of the solution as $x \to \infty$ is set out. In this case, the solution is calculated by formula (M12.7.1.6) with $\Lambda_2 = 0$ and Λ_1 specified in Table M12.4.

▶ **Formulas for calculating Green's functions.** Consider the parabolic equation of a special form

$$\frac{\partial w}{\partial t} = a(x)\frac{\partial^2 w}{\partial x^2} + b(x)\frac{\partial w}{\partial x} + \left[c(x) + \gamma(t)\right]w + \Phi(x,t), \tag{M12.7.1.13}$$

which is a special case of equation (M12.7.1.1) with operator (M12.7.1.2) where $a(x,t) = a(x)$, $b(x,t) = b(x)$, and $c(x,t) = c(x) + \gamma(t)$. It is assumed that $a(x) > 0$.

The solution of the nonhomogeneous linear boundary value problem (M12.7.1.13), (M12.7.1.2)–(M12.7.1.5) is found by formula (M12.7.1.6), where the Green's function is given by

$$G(x,y,t,\tau) = \rho(y) \sum_{n=1}^{\infty} \frac{\varphi_n(x)\varphi_n(y)}{\|\varphi_n\|^2} \exp\left[-\lambda_n(t-\tau) + \int_{\tau}^{t} \gamma(\xi)\,d\xi\right], \tag{M12.7.1.14}$$

where the λ_n and $\varphi_n(x)$ are the eigenvalues and the corresponding eigenfunctions of the Sturm–Liouville problem for the linear ordinary differential equation (M12.5.1.5) with the homogeneous linear boundary conditions (M12.5.1.7), and

$$\rho(y) = \frac{1}{a(y)} \exp\left[\int \frac{b(y)}{a(y)}\,dy\right], \qquad \|\varphi_n\|^2 = \int_{x_1}^{x_2} \rho(x)\varphi_n^2(x)\,dx, \tag{M12.7.1.15}$$

Table M12.5 lists Green's functions for some problems for the nonhomogeneous heat equation that corresponds to $a(x) = a = \text{const}$ and $b(x) = c(x) = \gamma(t) \equiv 0$ in (M12.7.1.13). The function $G(x,y,t-\tau)$ from this table must be substituted for $G(x,y,t,\tau)$ in formula (M12.7.1.6).

M12.7.2. Problems for Parabolic Equation
$$\rho(x)\frac{\partial w}{\partial t} = \frac{\partial}{\partial x}\left[p(x)\frac{\partial w}{\partial x}\right] - q(x)w + \Phi(x,t)$$

▶ **General formulas for solving nonhomogeneous boundary value problems.** Consider a linear parabolic equation of the special form

$$\rho(x)\frac{\partial w}{\partial t} = \frac{\partial}{\partial x}\left[p(x)\frac{\partial w}{\partial x}\right] - q(x)w + \Phi(x,t). \tag{M12.7.2.1}$$

It is often encountered in heat and mass transfer theory and chemical engineering sciences. Throughout this subsection, we assume that the functions ρ, p, p'_x, and q are continuous and $\rho > 0$, $p > 0$, and $x_1 \leq x \leq x_2$.

The solution of equation (M12.7.2.1) under the initial condition (M12.7.1.3) and the arbitrary linear nonhomogeneous boundary conditions (M12.7.1.4) and (M12.7.1.5) can be represented as the sum

$$w(x,t) = \int_0^t \int_{x_1}^{x_2} \Phi(y,\tau)\mathcal{G}(x,y,t-\tau)\,dy\,d\tau + \int_{x_1}^{x_2} \rho(y)f(y)\mathcal{G}(x,y,t)\,dy$$
$$+ p(x_1)\int_0^t g_1(\tau)\Lambda_1(x,t-\tau)\,d\tau + p(x_2)\int_0^t g_2(\tau)\Lambda_2(x,t-\tau)\,d\tau. \tag{M12.7.2.2}$$

TABLE M12.5
The Green's functions for some boundary value problems for the heat equation.

Type of problem	Green's function, $G(x, y, t)$	Remarks
First boundary value problem $(0 \leq x < \infty)$	$\dfrac{1}{2\sqrt{\pi at}}\left\{\exp\left[-\dfrac{(x-y)^2}{4at}\right] - \exp\left[-\dfrac{(x+y)^2}{4at}\right]\right\}$	Looking for solution bounded as $x \to \infty$
Second boundary value problem $(0 \leq x < \infty)$	$\dfrac{1}{2\sqrt{\pi at}}\left\{\exp\left[-\dfrac{(x-y)^2}{4at}\right] + \exp\left[-\dfrac{(x+y)^2}{4at}\right]\right\}$	Looking for solution bounded as $x \to \infty$
First boundary value problem $(0 \leq x \leq l)$	$\dfrac{2}{l}\displaystyle\sum_{n=1}^{\infty}\sin\left(\dfrac{n\pi x}{l}\right)\sin\left(\dfrac{n\pi y}{l}\right)\exp\left(-\dfrac{an^2\pi^2 t}{l^2}\right)$ $=\dfrac{1}{2\sqrt{\pi at}}\displaystyle\sum_{n=-\infty}^{\infty}\left\{\exp\left[-\dfrac{(x-y+2nl)^2}{4at}\right] - \exp\left[-\dfrac{(x+y+2nl)^2}{4at}\right]\right\}$	The first series converges rapidly at large t and the second series at small t
Second boundary value problem $(0 \leq x \leq l)$	$\dfrac{1}{l} + \dfrac{2}{l}\displaystyle\sum_{n=1}^{\infty}\cos\left(\dfrac{n\pi x}{l}\right)\cos\left(\dfrac{n\pi y}{l}\right)\exp\left(-\dfrac{an^2\pi^2 t}{l^2}\right)$ $=\dfrac{1}{2\sqrt{\pi at}}\displaystyle\sum_{n=-\infty}^{\infty}\left\{\exp\left[-\dfrac{(x-y+2nl)^2}{4at}\right] + \exp\left[-\dfrac{(x+y+2nl)^2}{4at}\right]\right\}$	The first series converges rapidly at large t and the second series at small t
Mixed boundary value problem $(0 \leq x \leq l)$; w set at $x = 0$ and $\partial_x w$ set at $x = l$	$\dfrac{2}{l}\displaystyle\sum_{n=0}^{\infty}\sin\left[\dfrac{\pi(2n+1)x}{2l}\right]\sin\left[\dfrac{\pi(2n+1)y}{2l}\right]\exp\left[-\dfrac{a\pi^2(2n+1)^2 t}{4l^2}\right]$ $=\dfrac{1}{2\sqrt{\pi at}}\displaystyle\sum_{n=-\infty}^{\infty}(-1)^n\left\{\exp\left[-\dfrac{(x-y+2nl)^2}{4at}\right] - \exp\left[-\dfrac{(x+y+2nl)^2}{4at}\right]\right\}$	The first series converges rapidly at large t and the second series at small t

Here, the modified Green's function is given by

$$\mathcal{G}(x, y, t) = \sum_{n=1}^{\infty} \frac{\varphi_n(x)\varphi_n(y)}{\|\varphi_n\|^2}\exp(-\lambda_n t), \qquad \|\varphi_n\|^2 = \int_{x_1}^{x_2}\rho(x)\varphi_n^2(x)\,dx, \quad \text{(M12.7.2.3)}$$

where the λ_n and $\varphi_n(x)$ are the eigenvalues and corresponding eigenfunctions of the following Sturm–Liouville problem for a second-order linear ordinary differential equation:

$$[p(x)\varphi_x']_x' + [\lambda\rho(x) - q(x)]\varphi = 0,$$
$$s_1\varphi_x' + k_1\varphi = 0 \quad \text{at} \quad x = x_1, \qquad \text{(M12.7.2.4)}$$
$$s_2\varphi_x' + k_2\varphi = 0 \quad \text{at} \quad x = x_2.$$

The functions $\Lambda_1(x, t)$ and $\Lambda_2(x, t)$ that occur in the integrands of the last two terms in solution (M12.7.2.2) are expressed in terms of the Green's function (M12.7.2.3).

Remark. The corresponding formulas for $\Lambda_m(x, t)$ for the basic types of boundary value problems can be obtained from Table M12.4; to do so, one should set $G = \mathcal{G}$ in all formulas of the last column and omit the last argument τ.

▶ **Properties of Sturm–Liouville problem (M12.7.2.4). Heat equation with a source.** The basic properties of the Sturm–Liouville problem (M12.7.2.4), which coincides with problem (M12.5.1.12), (M12.5.1.7), are outlined in Subsections M12.5.1 and M11.2.4. In particular, there are infinitely many eigenvalues. All eigenvalues are real and different and can be ordered so that $\lambda_1 < \lambda_2 < \lambda_3 < \cdots$, with $\lambda_n \to \infty$ as $n \to \infty$. The different eigenfunctions $\varphi_n(x)$ and $\varphi_m(x)$ are orthogonal with weight $\rho(x)$ on the interval $x_1 \leq x \leq x_2$; see formulas (M12.5.1.14).

Example. Consider the first boundary value problem in the domain $0 \leq x \leq l$ for the heat equation with a source

$$\frac{\partial w}{\partial t} = a\frac{\partial^2 w}{\partial x^2} - bw$$

under the initial condition (M12.7.1.3) and boundary conditions

$$w = g_1(t) \quad \text{at} \quad x = 0,$$
$$w = g_2(t) \quad \text{at} \quad x = l.$$

The above equation is a special case of equation (M12.7.2.1) with $\rho(x) = 1$, $p(x) = a$, $q(x) = b$, and $\Phi(x, t) = 0$. The corresponding Sturm–Liouville problem (M12.7.2.4) has the form

$$a\varphi''_{xx} + (\lambda - b)\varphi = 0, \qquad \varphi = 0 \quad \text{at} \quad x = 0, \qquad \varphi = 0 \quad \text{at} \quad x = l.$$

The eigenfunctions and eigenvalues are found to be

$$\varphi_n(x) = \sin\left(\frac{\pi n x}{l}\right), \qquad \lambda_n = b + \frac{a\pi^2 n^2}{l^2}, \qquad n = 1, 2, \ldots$$

Using formula (M12.7.2.3) and taking into account that $\|\varphi_n\|^2 = l/2$, we obtain the Green's function

$$\mathcal{G}(x, y, t) = \frac{2}{l}e^{-bt}\sum_{n=1}^{\infty} \sin\left(\frac{\pi n x}{l}\right)\sin\left(\frac{\pi n y}{l}\right)\exp\left(-\frac{a\pi^2 n^2}{l^2}t\right).$$

Substituting this expression into (M12.7.2.2) with $p(x_1) = p(x_2) = \rho(y) = 1$, $x_1 = 0$, and $x_2 = l$ and taking into account the formulas

$$\Lambda_1(x, t) = \partial_y \mathcal{G}(x, y, t)\big|_{y=x_1}, \qquad \Lambda_2(x, t) = -\partial_y \mathcal{G}(x, y, t)\big|_{y=x_2}$$

(see the first row in Table M12.4 and Remark in Subsection M12.7.2), one obtains the solution to the problem in question.

M12.8. Boundary Value Problems for Hyperbolic Equations with One Space Variable. Green's Function. Goursat Problem

M12.8.1. Representation of Solutions in terms of the Green's Function

▶ **Statement of the problem** ($t \geq 0$, $x_1 \leq x \leq x_2$). In general, a one-dimensional nonhomogeneous linear differential equation of hyperbolic type with variable coefficients is written as

$$\frac{\partial^2 w}{\partial t^2} + \varphi(x, t)\frac{\partial w}{\partial t} - L_{x,t}[w] = \Phi(x, t), \qquad (\text{M12.8.1.1})$$

where the operator $L_{x,t}[w]$ is defined by (M12.7.1.2).

Consider the nonstationary boundary value problem for equation (M12.8.1.1) with the initial conditions

$$w = f_0(x) \quad \text{at} \quad t = 0,$$
$$\partial_t w = f_1(x) \quad \text{at} \quad t = 0 \qquad (\text{M12.8.1.2})$$

and arbitrary nonhomogeneous linear boundary conditions (M12.7.1.4) and (M12.7.1.5).

▶ **Representation of the problem solution in terms of the Green's function.** The solution of problem (M12.8.1.1), (M12.8.1.2), (M12.7.1.4), (M12.7.1.5) can be represented as the sum

$$
\begin{aligned}
w(x,t) = {} & \int_0^t \int_{x_1}^{x_2} \Phi(y,\tau) G(x,y,t,\tau)\, dy\, d\tau \\
& - \int_{x_1}^{x_2} f_0(y) \left[\frac{\partial}{\partial \tau} G(x,y,t,\tau) \right]_{\tau=0} dy + \int_{x_1}^{x_2} \big[f_1(y) + f_0(y)\varphi(y,0) \big] G(x,y,t,0)\, dy \\
& + \int_0^t g_1(\tau) a(x_1,\tau) \Lambda_1(x,t,\tau)\, d\tau + \int_0^t g_2(\tau) a(x_2,\tau) \Lambda_2(x,t,\tau)\, d\tau. \quad \text{(M12.8.1.3)}
\end{aligned}
$$

Here, the Green's function $G(x,y,t,\tau)$ is determined by solving the homogeneous equation

$$
\frac{\partial^2 G}{\partial t^2} + \varphi(x,t) \frac{\partial G}{\partial t} - L_{x,t}[G] = 0 \qquad \text{(M12.8.1.4)}
$$

with the semihomogeneous initial conditions

$$
\begin{aligned}
G &= 0 & \text{at} \quad t &= \tau, & \text{(M12.8.1.5)} \\
\partial_t G &= \delta(x-y) & \text{at} \quad t &= \tau, & \text{(M12.8.1.6)}
\end{aligned}
$$

and the homogeneous boundary conditions (M12.7.1.9), (M12.7.1.10). The quantities y and τ appear in problem (M12.8.1.4)–(M12.8.1.6), (M12.7.1.9), (M12.7.1.10) as free parameters ($x_1 \le y \le x_2$), and $\delta(x)$ is the Dirac delta function.

The functions $\Lambda_1(x,t,\tau)$ and $\Lambda_2(x,t,\tau)$ involved in the integrands of the last two terms in solution (M12.8.1.3) can be expressed in terms of the Green's function $G(x,y,t,\tau)$. The corresponding formulas for $\Lambda_m(x,t,\tau)$ are given in Table M12.4 for the basic types of boundary value problems.

It is significant that the Green's function G and Λ_1, Λ_2 are independent of the functions Φ, f_0, f_1, g_1, and g_2 that characterize various nonhomogeneities of the boundary value problem.

If the coefficients of equation (M12.8.1.1) are independent of time t, then the Green's function depends on only three arguments, $G(x,y,t,\tau) = G(x,y,t-\tau)$. In this case, one can set $\frac{\partial}{\partial \tau} G(x,y,t,\tau)\big|_{\tau=0} = -\frac{\partial}{\partial t} G(x,y,t)$ in solution (M12.8.1.3).

▶ **Formulas for the calculation of Green's functions.** Consider the hyperbolic equation of a special form

$$
\frac{\partial^2 w}{\partial t^2} + \beta(t) \frac{\partial w}{\partial t} = a(x) \frac{\partial^2 w}{\partial x^2} + b(x) \frac{\partial w}{\partial x} + \big[c(x) + \gamma(t) \big] w + \Phi(x,t), \qquad \text{(M12.8.1.7)}
$$

which is a special case of equation (M12.8.1.1) with operator (M12.7.1.2) where $\varphi(x,t) = \beta(t)$, $a(x,t) = a(x)$, $b(x,t) = b(x)$, and $c(x,t) = c(x) + \gamma(t)$. We assume that $a(x) > 0$.

The solution of the nonhomogeneous linear boundary value problem (M12.8.1.7), (M12.8.1.2), (M12.7.1.4), (M12.7.1.5) is found by formula (M12.8.1.3), where the Green's function is given by

$$
G(x,y,t,\tau) = \rho(y) \sum_{n=1}^{\infty} \frac{\varphi_n(x)\varphi_n(y)}{\|\varphi_n\|^2} \psi_n(t,\tau), \qquad \text{(M12.8.1.8)}
$$

TABLE M12.6
The Green's functions for some boundary value problems for the
wave equation $\frac{\partial^2 w}{\partial t^2} = a^2 \frac{\partial^2 w}{\partial x^2}$ in a bounded domain $0 \leq x \leq l$.

Type of problem	Green's function, $G(x, \xi, t)$	Remarks
First boundary value problem	$\dfrac{2}{a\pi} \displaystyle\sum_{n=1}^{\infty} \frac{1}{n} \sin\left(\frac{n\pi x}{l}\right) \sin\left(\frac{n\pi \xi}{l}\right) \sin\left(\frac{n\pi a t}{l}\right)$	
Second boundary value problem	$\dfrac{t}{l} + \dfrac{2}{a\pi} \displaystyle\sum_{n=1}^{\infty} \frac{1}{n} \cos\left(\frac{n\pi x}{l}\right) \cos\left(\frac{n\pi \xi}{l}\right) \sin\left(\frac{n\pi a t}{l}\right)$	
Mixed boundary value problem; w is set at $x = 0$ and $\partial_x w$ is set at $x = l$	$\dfrac{2}{al} \displaystyle\sum_{n=1}^{\infty} \frac{1}{\lambda_n} \sin(\lambda_n x) \sin(\lambda_n \xi) \sin(\lambda_n a t)$	$\lambda_n = \dfrac{\pi(2n+1)}{2l}$

where the λ_n and $\varphi_n(x)$ are the eigenvalues and the corresponding eigenfunctions of the Sturm–Liouville problem for the linear ordinary differential equation (M12.5.1.5) with the homogeneous linear boundary conditions (M12.5.1.7); $\rho(y)$ and $\|\varphi_n\|$ are determined by (M12.7.1.15), and $\psi = \psi_n(t, \tau)$ is the solution of equation (M12.5.2.5) with $\alpha(t) = 1$ and $\lambda = \lambda_n$ that satisfies the initial conditions

$$\psi = 0 \quad \text{at} \quad t = \tau, \qquad \psi'_t = 1 \quad \text{at} \quad t = \tau.$$

In the special case $\beta(t) = \gamma(t) = 0$, we have $\psi_n(t, \tau) = \lambda_n^{-1/2} \sin\left[\lambda_n^{1/2}(t - \tau)\right]$.

Table M12.6 lists Green's functions for some problems for a nonhomogeneous wave equation, which corresponds to $a(x, t) = a^2 = \text{const}$, $b(x) = c(x) = \beta(t) = \gamma(t) \equiv 0$ in (M12.8.1.7). One should substitute $G(x, y, t - \tau)$ from this table for $G(x, y, t, \tau)$ in formula (M12.8.1.3).

M12.8.2. Problems for Hyperbolic Equation

$$\rho(x)\frac{\partial^2 w}{\partial t^2} = \frac{\partial}{\partial x}\left[p(x)\frac{\partial w}{\partial x}\right] - q(x)w + \Phi(x, t)$$

▶ **General relations for solving nonhomogeneous boundary value problems.** Consider a linear hyperbolic equation of the special form

$$\rho(x)\frac{\partial^2 w}{\partial t^2} = \frac{\partial}{\partial x}\left[p(x)\frac{\partial w}{\partial x}\right] - q(x)w + \Phi(x, t). \tag{M12.8.2.1}$$

It is assumed that the functions ρ, p, p'_x, and q are continuous and the inequalities $\rho > 0$, $p > 0$ hold for $x_1 \leq x \leq x_2$.

The solution of equation (M12.8.2.1) under the general initial conditions (M12.8.1.2) and the arbitrary linear nonhomogeneous boundary conditions (M12.7.1.4)–(M12.7.1.5) can be represented as the sum

$$
\begin{aligned}
w(x, t) = &\int_0^t \int_{x_1}^{x_2} \Phi(\xi, \tau) \mathcal{G}(x, \xi, t - \tau) \, d\xi \, d\tau \\
&+ \frac{\partial}{\partial t} \int_{x_1}^{x_2} \rho(\xi) f_0(\xi) \mathcal{G}(x, \xi, t) \, d\xi + \int_{x_1}^{x_2} \rho(\xi) f_1(\xi) \mathcal{G}(x, \xi, t) \, d\xi \\
&+ p(x_1) \int_0^t g_1(\tau) \Lambda_1(x, t - \tau) \, d\tau + p(x_2) \int_0^t g_2(\tau) \Lambda_2(x, t - \tau) \, d\tau. \tag{M12.8.2.2}
\end{aligned}
$$

Here, the modified Green's function is determined by

$$\mathcal{G}(x,\xi,t) = \sum_{n=1}^{\infty} \frac{\varphi_n(x)\varphi_n(\xi)\sin\left(t\sqrt{\lambda_n}\right)}{\|\varphi_n\|^2\sqrt{\lambda_n}}, \qquad \|\varphi_n\|^2 = \int_{x_1}^{x_2} \rho(x)\varphi_n^2(x)\,dx, \quad \text{(M12.8.2.3)}$$

where the λ_n and $\varphi_n(x)$ are the eigenvalues and the corresponding eigenfunctions of the Sturm–Liouville problem for the second-order linear ordinary differential equation:

$$[p(x)\varphi'_x]'_x + [\lambda\rho(x) - q(x)]\varphi = 0,$$
$$s_1\varphi'_x + k_1\varphi = 0 \quad \text{at} \quad x = x_1, \qquad \text{(M12.8.2.4)}$$
$$s_2\varphi'_x + k_2\varphi = 0 \quad \text{at} \quad x = x_2.$$

The functions $\Lambda_1(x,t)$ and $\Lambda_2(x,t)$ that occur in the integrands of the last two terms in solution (M12.8.2.2) are expressed in terms of the Green's function of (M12.8.2.3).

Remark. The corresponding formulas for $\Lambda_m(x,t)$ for the basic types of boundary value problems can be obtained using Table M12.4; to do so, one should set $G = \mathcal{G}$ in all formulas of the last column and omit the last argument τ.

▶ **Properties of the Sturm–Liouville problem. The Klein–Gordon equation.** The general and special properties of the Sturm–Liouville problem (M12.8.2.4) are given in Subsections M12.5.1 and M11.2.5; various asymptotic and approximate formulas for eigenvalues and eigenfunctions can also be found there.

Example. Consider the second boundary value problem in the domain $0 \le x \le l$ for the Klein–Gordon equation

$$\frac{\partial^2 w}{\partial t^2} = a^2 \frac{\partial^2 w}{\partial x^2} - bw,$$

under the initial conditions (M12.8.1.2) and boundary conditions

$$\partial_x w = g_1(t) \quad \text{at} \quad x = 0,$$
$$\partial_x w = g_2(t) \quad \text{at} \quad x = l.$$

The Klein–Gordon equation is a special case of equation (M12.8.2.1) with $\rho(x) = 1$, $p(x) = a^2$, $q(x) = b$, and $\Phi(x,t) = 0$. The corresponding Sturm–Liouville problem (M12.8.2.4) has the form

$$a^2\varphi''_{xx} + (\lambda - b)\varphi = 0, \qquad \varphi'_x = 0 \quad \text{at} \quad x = 0, \qquad \varphi'_x = 0 \quad \text{at} \quad x = l.$$

The eigenfunctions and eigenvalues are found to be

$$\varphi_{n+1}(x) = \cos\left(\frac{\pi n x}{l}\right), \qquad \lambda_{n+1} = b + \frac{a\pi^2 n^2}{l^2}, \qquad n = 0, 1, \ldots$$

Using formula (M12.8.2.3) and taking into account that $\|\varphi_1\|^2 = l$ and $\|\varphi_n\|^2 = l/2$ $(n = 1, 2, \ldots)$, we find the Green's function:

$$\mathcal{G}(x,\zeta,t) = \frac{1}{l\sqrt{b}}\sin\left(t\sqrt{b}\right) + \frac{2}{l}\sum_{n=1}^{\infty}\cos\left(\frac{\pi n x}{l}\right)\cos\left(\frac{\pi n \xi}{l}\right)\frac{\sin\left(t\sqrt{(a\pi n/l)^2 + b}\right)}{\sqrt{(a\pi n/l)^2 + b}}.$$

Substituting this expression into (M12.8.2.2) with $p(x_1) = p(x_2) = \rho(\xi) = 1$, $x_1 = 0$, and $x_2 = l$ and taking into account the formulas

$$\Lambda_1(x,t) = -\mathcal{G}(x,x_1,t), \qquad \Lambda_2(x,t) = \mathcal{G}(x,x_2,t)$$

(see the second row in Table M12.4 and the above remark), one obtains the solution to the problem in question.

M12.8.3. Goursat Problem (a Problem with Initial Data of Characteristics)

▶ **Statement of the Goursat problem. Basic property of the solution.** Consider the general linear hyperbolic equation in two independent variables which is reduced to the first canonical form (see Subsection M12.2.1):

$$\frac{\partial^2 w}{\partial x \partial y} + a(x,y)\frac{\partial w}{\partial x} + b(x,y)\frac{\partial w}{\partial y} + c(x,y)w = f(x,y), \tag{M12.8.3.1}$$

where $a(x,y)$, $b(x,y)$, $c(x,y)$, and $f(x,y)$ are continuous functions.

The *Goursat problem* for equation (M12.8.3.1) is stated as follows: find a solution to equation (M12.8.3.1) that satisfies the conditions at characteristics

$$w(x,y)|_{x=x_1} = g(y), \qquad w(x,y)|_{y=y_1} = h(x), \tag{M12.8.3.2}$$

where $g(y)$ and $h(x)$ are given continuous functions that match each other at the point of intersection of the characteristics, so that

$$g(y_1) = h(x_1).$$

Basic properties of the Goursat problem: the value of the solution at any point $M(x_0, y_0)$ depends only on the values of $g(y)$ at the segment AN (which is part of the characteristic $x = x_1$), the values of $h(x)$ at the segment BN (which is part of the characteristic $y = y_1$), and the values of the functions $a(x,y)$, $b(x,y)$, $c(x,y)$, and $f(x,y)$ in the rectangle $NAMB$; see Fig. M12.2. The domain of influence on the solution at the point $M(x_0, y_0)$ is shaded for clarity.

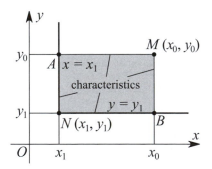

Figure M12.2. Domain of influence of the solution to the Goursat problem at a point M.

▶ **Riemann function.** A *Riemann function*, $\mathcal{R} = \mathcal{R}(x,y; x_0, y_0)$, corresponding to equation (M12.8.3.1) is defined as a solution to the equation

$$\frac{\partial^2 \mathcal{R}}{\partial x \partial y} - \frac{\partial}{\partial x}\big[a(x,y)\mathcal{R}\big] - \frac{\partial}{\partial y}\big[b(x,y)\mathcal{R}\big] + c(x,y)\mathcal{R} = 0 \tag{M12.8.3.3}$$

that satisfies the conditions

$$\mathcal{R} = \exp\left[\int_{y_0}^{y} a(x_0, \xi)\, d\xi\right] \quad \text{at} \quad x = x_0, \qquad \mathcal{R} = \exp\left[\int_{x_0}^{x} b(\xi, y_0)\, d\xi\right] \quad \text{at} \quad y = y_0 \tag{M12.8.3.4}$$

at the characteristics $x = x_0$ and $y = y_0$. Here, (x_0, y_0) is an arbitrary point from the domain of definition of equation (M12.8.3.1). The x_0 and y_0 appear in problem (M12.8.3.3), (M12.8.3.4) as parameters in the boundary conditions only.

THEOREM. *If the functions a, b, c and the partial derivatives a_x, b_y are all continuous, then the Riemann function $\mathcal{R}(x, y; x_0, y_0)$ exists. Moreover, the function $\mathcal{R}(x_0, y_0, x, y)$, obtained by swapping the parameters and the arguments, is a solution to the homogeneous equation (M12.8.3.1), with $f = 0$.*

Remark. It is significant that the Riemann function is independent of the initial data on characteristics (M12.8.3.2).

Example 1. The Riemann function for the equation $w_{xy} = 0$ is just $\mathcal{R} \equiv 1$.

Example 2. The Riemann function for the equation

$$w_{xy} + cw = 0 \qquad (c = \text{const})$$

is expressed in terms of the Bessel function $J_0(z)$ as

$$\mathcal{R} = J_0\left(\sqrt{4c(x_0 - x)(y_0 - y)}\right).$$

▶ **Solution representation for the Goursat problem in terms of the Riemann function.** Given a Riemann function, the solution to the Goursat problem (M12.8.3.1), (M12.8.3.2) at any point (x_0, y_0) can be written as

$$w(x_0, y_0) = (w\mathcal{R})_N + \int_N^A \mathcal{R}(g'_y + bg)\, dy + \int_N^B \mathcal{R}(h'_x + ah)\, dx + \iint_{NAMB} f\mathcal{R}\, dx\, dy.$$

The first term on the right-hand side is evaluated at the point of intersection of the characteristics (x_1, y_1). The second and third terms are integrals along the characteristics $y = y_1$ ($x_1 \leq x \leq x_0$) and $x = x_1$ ($y_1 \leq y \leq y_0$); these involve the initial data of (M12.8.3.2). The last integral is taken over the rectangular domain $NAMB$ defined by the inequalities $x_1 \leq x \leq x_0$, $y_1 \leq y \leq y_0$.

The Goursat problem for a hyperbolic equation reduced to the second canonical form (see Subsection M12.2.1) is treated similarly.

Example 3. Consider the Goursat problem for the wave equation

$$\frac{\partial^2 w}{\partial t^2} - a^2 \frac{\partial^2 w}{\partial x^2} = 0$$

with the boundary conditions prescribed on its characteristics

$$\begin{aligned}
w &= f(x) \quad \text{for} \quad x - at = 0 \quad (0 \leq x \leq b), \\
w &= g(x) \quad \text{for} \quad x + at = 0 \quad (0 \leq x \leq c),
\end{aligned} \tag{M12.8.3.5}$$

where $f(0) = g(0)$.

Substituting the values set on the characteristics (M12.8.3.5) into the general solution of the wave equation, $w = \varphi(x - at) + \psi(x + at)$, we arrive to a system of linear algebraic equations for $\varphi(x)$ and $\psi(x)$. As a result, the solution to the Goursat problem is obtained in the form

$$w(x, t) = f\left(\frac{x + at}{2}\right) + g\left(\frac{x - at}{2}\right) - f(0).$$

The propagation domain of the solution is the parallelogram bounded by the four straight lines

$$x - at = 0, \quad x + at = 0, \quad x - at = 2c, \quad x + at = 2b.$$

M12.9. Boundary Value Problems for Elliptic Equations with Two Space Variables

M12.9.1. Problems and the Green's Functions for Equation $a(x)\dfrac{\partial^2 w}{\partial x^2} + \dfrac{\partial^2 w}{\partial y^2} + b(x)\dfrac{\partial w}{\partial x} + c(x)w = -\Phi(x, y)$

▶ **Statements of boundary value problems.** Consider two-dimensional boundary value problems for the equation

$$a(x)\frac{\partial^2 w}{\partial x^2} + \frac{\partial^2 w}{\partial y^2} + b(x)\frac{\partial w}{\partial x} + c(x)w = -\Phi(x, y) \tag{M12.9.1.1}$$

with general boundary conditions in x,

$$\begin{aligned} s_1\frac{\partial w}{\partial x} - k_1 w &= f_1(y) \quad \text{at} \quad x = x_1, \\ s_2\frac{\partial w}{\partial x} + k_2 w &= f_2(y) \quad \text{at} \quad x = x_2, \end{aligned} \tag{M12.9.1.2}$$

and different boundary conditions in y. It is assumed that the coefficients of equation (M12.9.1.1) and the boundary conditions (M12.9.1.2) meet the requirements

$$a(x),\ b(x),\ c(x) \text{ are continuous } (x_1 \le x \le x_2); \quad a > 0, \quad |s_1| + |k_1| > 0, \quad |s_2| + |k_2| > 0.$$

▶ **Relations for the Green's function.** In the general case, the Green's function can be represented as

$$G(x, y, \xi, \eta) = \rho(\xi) \sum_{n=1}^{\infty} \frac{\varphi_n(x)\varphi_n(\xi)}{\|\varphi_n\|^2} \Psi_n(y, \eta; \lambda_n), \tag{M12.9.1.3}$$

where

$$\rho(x) = \frac{1}{a(x)} \exp\left[\int \frac{b(x)}{a(x)}\, dx\right], \quad \|\varphi_n\|^2 = \int_{x_1}^{x_2} \rho(x)\varphi_n^2(x)\, dx, \tag{M12.9.1.4}$$

and the λ_n and $\varphi_n(x)$ are the eigenvalues and eigenfunctions of the homogeneous boundary value problem for the ordinary differential equation

$$a(x)\varphi_{xx}'' + b(x)\varphi_x' + [\lambda + c(x)]\varphi = 0, \tag{M12.9.1.5}$$
$$s_1\varphi_x' - k_1\varphi = 0 \quad \text{at} \quad x = x_1, \tag{M12.9.1.6}$$
$$s_2\varphi_x' + k_2\varphi = 0 \quad \text{at} \quad x = x_2. \tag{M12.9.1.7}$$

The functions Ψ_n for various boundary conditions in y are specified in Table M12.7.

Example. Consider a boundary value problem for the Laplace equation

$$\frac{\partial^2 w}{\partial x^2} + \frac{\partial^2 w}{\partial y^2} = 0 \tag{M12.9.1.8}$$

in a strip $0 \le x \le l,\ -\infty < y < \infty$ with mixed boundary conditions

$$w = f_1(y) \quad \text{at} \quad x = 0, \qquad \frac{\partial w}{\partial x} = f_2(y) \quad \text{at} \quad x = l.$$

TABLE M12.7
The functions Ψ_n in (M12.9.1.3) for various boundary conditions. Notation: $\sigma_n = \sqrt{\lambda_n}$.

Domain	Boundary conditions	Function $\Psi_n(y, \eta; \lambda_n)$
$-\infty < y < \infty$	$|w| < \infty$ for $y \to \pm\infty$	$\dfrac{1}{2\sigma_n} e^{-\sigma_n |y - \eta|}$
$0 \le y < \infty$	$w = 0$ for $y = 0$	$\dfrac{1}{\sigma_n} \begin{cases} e^{-\sigma_n y} \sinh(\sigma_n \eta) & \text{for } y > \eta, \\ e^{-\sigma_n \eta} \sinh(\sigma_n y) & \text{for } \eta > y \end{cases}$
$0 \le y < \infty$	$\partial_y w = 0$ for $y = 0$	$\dfrac{1}{\sigma_n} \begin{cases} e^{-\sigma_n y} \cosh(\sigma_n \eta) & \text{for } y > \eta, \\ e^{-\sigma_n \eta} \cosh(\sigma_n y) & \text{for } \eta > y \end{cases}$
$0 \le y < \infty$	$\partial_y w - k_3 w = 0$ for $y = 0$	$\dfrac{1}{\sigma_n(\sigma_n + k_3)} \begin{cases} e^{-\sigma_n y}[\sigma_n \cosh(\sigma_n \eta) + k_3 \sinh(\sigma_n \eta)] & \text{for } y > \eta, \\ e^{-\sigma_n \eta}[\sigma_n \cosh(\sigma_n y) + k_3 \sinh(\sigma_n y)] & \text{for } \eta > y \end{cases}$
$0 \le y \le h$	$w = 0$ at $y = 0$, $w = 0$ at $y = h$	$\dfrac{1}{\sigma_n \sinh(\sigma_n h)} \begin{cases} \sinh(\sigma_n \eta) \sinh[\sigma_n(h-y)] & \text{for } y > \eta, \\ \sinh(\sigma_n y) \sinh[\sigma_n(h-\eta)] & \text{for } \eta > y \end{cases}$
$0 \le y \le h$	$\partial_y w = 0$ at $y = 0$, $\partial_y w = 0$ at $y = h$	$\dfrac{1}{\sigma_n \sinh(\sigma_n h)} \begin{cases} \cosh(\sigma_n \eta) \cosh[\sigma_n(h-y)] & \text{for } y > \eta, \\ \cosh(\sigma_n y) \cosh[\sigma_n(h-\eta)] & \text{for } \eta > y \end{cases}$
$0 \le y \le h$	$w = 0$ at $y = 0$, $\partial_y w = 0$ at $y = h$	$\dfrac{1}{\sigma_n \cosh(\sigma_n h)} \begin{cases} \sinh(\sigma_n \eta) \cosh[\sigma_n(h-y)] & \text{for } y > \eta, \\ \sinh(\sigma_n y) \cosh[\sigma_n(h-\eta)] & \text{for } \eta > y \end{cases}$

This equation is a special case of equation (M12.9.1.1) with $a(x) = 1$ and $b(x) = c(x) = \Phi(x, t) = 0$. The corresponding Sturm–Liouville problem (M12.9.1.5)–(M12.9.1.7) is written as

$$u''_{xx} + \lambda u = 0, \qquad u = 0 \text{ at } x = 0, \qquad u'_x = 0 \text{ at } x = l.$$

The eigenfunctions and eigenvalues are found as

$$\varphi_n(x) = \sin\left[\frac{\pi(2n-1)x}{l}\right], \qquad \lambda_n = \frac{\pi^2(2n-1)^2}{l^2}, \qquad n = 1, 2, \dots$$

Using formulas (M12.9.1.3) and (M12.9.1.4) and taking into account the identities $\rho(\xi) = 1$ and $\|y_n\|^2 = l/2$ ($n = 1, 2, \dots$) and the expression for Ψ_n from the first row in Table M12.7, we obtain the Green's function in the form

$$G(x, y, \xi, \eta) = \frac{1}{l} \sum_{n=1}^{\infty} \frac{1}{\sigma_n} \sin(\sigma_n x) \sin(\sigma_n \xi) e^{-\sigma_n |y - \eta|}, \qquad \sigma_n = \sqrt{\lambda_n} = \frac{\pi(2n-1)}{l}.$$

M12.9.2. Representation of Solutions to Boundary Value Problems via the Green's Functions

▶ **First boundary value problem.** The solution of the first boundary value problem for equation (M12.9.1.1) with the boundary conditions

$$w = f_1(y) \quad \text{at} \quad x = x_1, \qquad w = f_2(y) \quad \text{at} \quad x = x_2,$$
$$w = f_3(x) \quad \text{at} \quad y = 0, \qquad w = f_4(x) \quad \text{at} \quad y = h$$

is expressed in terms of the Green's function as

$$w(x, y) = a(x_1) \int_0^h f_1(\eta) \left[\frac{\partial}{\partial \xi} G(x, y, \xi, \eta)\right]_{\xi = x_1} d\eta - a(x_2) \int_0^h f_2(\eta) \left[\frac{\partial}{\partial \xi} G(x, y, \xi, \eta)\right]_{\xi = x_2} d\eta$$

$$+ \int_{x_1}^{x_2} f_3(\xi) \left[\frac{\partial}{\partial \eta} G(x, y, \xi, \eta)\right]_{\eta = 0} d\xi - \int_{x_1}^{x_2} f_4(\xi) \left[\frac{\partial}{\partial \eta} G(x, y, \xi, \eta)\right]_{\eta = h} d\xi$$

$$+ \int_{x_1}^{x_2} \int_0^h \Phi(\xi, \eta) G(x, y, \xi, \eta) \, d\eta \, d\xi.$$

▶ **Second boundary value problem.** The solution of the second boundary value problem for equation (M12.9.1.1) with the boundary conditions

$$\partial_x w = f_1(y) \quad \text{at} \quad x = x_1, \qquad \partial_x w = f_2(y) \quad \text{at} \quad x = x_2,$$
$$\partial_y w = f_3(x) \quad \text{at} \quad y = 0, \qquad \partial_y w = f_4(x) \quad \text{at} \quad y = h$$

is expressed in terms of the Green's function as

$$w(x, y) = -a(x_1) \int_0^h f_1(\eta) G(x, y, x_1, \eta) \, d\eta + a(x_2) \int_0^h f_2(\eta) G(x, y, x_2, \eta) \, d\eta$$
$$- \int_{x_1}^{x_2} f_3(\xi) G(x, y, \xi, 0) \, d\xi + \int_{x_1}^{x_2} f_4(\xi) G(x, y, \xi, h) \, d\xi$$
$$+ \int_{x_1}^{x_2} \int_0^h \Phi(\xi, \eta) G(x, y, \xi, \eta) \, d\eta \, d\xi.$$

M12.9.3. Solutions of Problems for the Laplace Equation in the Polar Coordinate System

In the polar coordinates, defined by $x = r \cos \varphi$ and $y = r \sin \varphi$, Laplace's equation (M12.9.1.8) becomes

$$\Delta w \equiv \frac{1}{r} \frac{\partial}{\partial r} \left(r \frac{\partial w}{\partial r} \right) + \frac{1}{r^2} \frac{\partial^2 w}{\partial \varphi^2} = 0. \tag{M12.9.3.1}$$

▶ **First boundary value problem ($0 \le r \le R$).** A circle is considered. A boundary condition is prescribed:

$$w = f(\varphi) \quad \text{at} \quad r = R. \tag{M12.9.3.2}$$

The solution to equation (M12.9.3.1) subject to the boundary condition (M12.9.3.2) is given by

$$w(r, \varphi) = \frac{1}{2\pi} \int_0^{2\pi} f(\psi) \frac{R^2 - r^2}{r^2 - 2Rr \cos(\varphi - \psi) + R^2} \, d\psi.$$

This formula is conventionally referred to as the *Poisson integral*.

▶ **Second boundary value problem ($0 \le r \le R$).** A circle is considered. A boundary condition is prescribed:

$$\partial_r w = f(\varphi) \quad \text{at} \quad r = R. \tag{M12.9.3.3}$$

The solution to equation (M12.9.3.1) subject to the boundary condition (M12.9.3.3) is expressed as

$$w(r, \varphi) = \frac{R}{2\pi} \int_0^{2\pi} f(\psi) \ln \frac{r^2 - 2Rr \cos(\varphi - \psi) + R^2}{R^2} \, d\psi + C,$$

where C is an arbitrary constant; this formula is known as the *Dini integral*.

Remark. The function $f(\varphi)$ must satisfy the solvability condition $\int_0^{2\pi} f(\varphi) \, d\varphi = 0$.

M12.10. Problems with Many Space Variables. Representation of Solutions via Green's Functions

M12.10.1. Boundary Value Problems for Parabolic Equations

▶ **Statement of the problem.** In general, a nonhomogeneous linear differential equation of the parabolic type in n space variables has the form

$$\frac{\partial w}{\partial t} - L_{\mathbf{x},t}[w] = \Phi(\mathbf{x}, t), \tag{M12.10.1.1}$$

where

$$L_{\mathbf{x},t}[w] \equiv \sum_{i,j=1}^{n} a_{ij}(\mathbf{x}, t)\frac{\partial^2 w}{\partial x_i \partial x_j} + \sum_{i=1}^{n} b_i(\mathbf{x}, t)\frac{\partial w}{\partial x_i} + c(\mathbf{x}, t)w,$$

$$\mathbf{x} = \{x_1, \dots, x_n\}, \quad \sum_{i,j=1}^{n} a_{ij}(\mathbf{x}, t)\xi_i\xi_j \ge \sigma \sum_{i=1}^{n} \xi_i^2, \quad \sigma > 0. \tag{M12.10.1.2}$$

Let V be some simply connected domain in $\mathbb{R}^n$ with a sufficiently smooth boundary S. We consider the nonstationary boundary value problem for equation (M12.10.1.1) in the domain V with an arbitrary initial condition,

$$w = f(\mathbf{x}) \quad \text{at} \quad t = 0, \tag{M12.10.1.3}$$

and nonhomogeneous linear boundary conditions,

$$\Gamma_{\mathbf{x},t}[w] = g(\mathbf{x}, t) \quad \text{for} \quad \mathbf{x} \in S. \tag{M12.10.1.4}$$

In the general case, $\Gamma_{\mathbf{x},t}$ is a first-order linear differential operator in the space coordinates with coefficients dependent on $\mathbf{x}$ and t.

▶ **Representation of the problem solution in terms of the Green's function.** The solution of the nonhomogeneous linear boundary value problem defined by (M12.10.1.1)–(M12.10.1.4) can be represented as the sum

$$w(\mathbf{x}, t) = \int_0^t \int_V \Phi(\mathbf{y}, \tau)G(\mathbf{x}, \mathbf{y}, t, \tau) \, dV_y \, d\tau + \int_V f(\mathbf{y})G(\mathbf{x}, \mathbf{y}, t, 0) \, dV_y$$

$$+ \int_0^t \int_S g(\mathbf{y}, \tau)H(\mathbf{x}, \mathbf{y}, t, \tau) \, dS_y \, d\tau, \tag{M12.10.1.5}$$

where $G(\mathbf{x}, \mathbf{y}, t, \tau)$ is the Green's function; for $t > \tau \ge 0$, it satisfies the homogeneous equation

$$\frac{\partial G}{\partial t} - L_{\mathbf{x},t}[G] = 0 \tag{M12.10.1.6}$$

with the nonhomogeneous initial condition of special form

$$G = \delta(\mathbf{x} - \mathbf{y}) \quad \text{at} \quad t = \tau \tag{M12.10.1.7}$$

and the homogeneous boundary condition

$$\Gamma_{\mathbf{x},t}[G] = 0 \quad \text{for} \quad \mathbf{x} \in S. \tag{M12.10.1.8}$$

The vector $\mathbf{y} = \{y_1, \ldots, y_n\}$ appears in problem (M12.10.1.6)–(M12.10.1.8) as an n-dimensional free parameter ($\mathbf{y} \in V$), and $\delta(\mathbf{x} - \mathbf{y}) = \delta(x_1 - y_1) \ldots \delta(x_n - y_n)$ is the n-dimensional Dirac delta function. The Green's function G is independent of the functions Φ, f, and g that characterize various nonhomogeneities of the boundary value problem. In (M12.10.1.5), the integration is performed everywhere with respect to $\mathbf{y}$, with $dV_y = dy_1 \ldots dy_n$.

The function $H(\mathbf{x}, \mathbf{y}, t, \tau)$ in the integrand of the last term in solution (M12.10.1.5) can be expressed via the Green's function $G(\mathbf{x}, \mathbf{y}, t, \tau)$. The corresponding formulas for $H(\mathbf{x}, \mathbf{y}, t, \tau)$ are given in Table M12.8 for the three basic types of boundary value problems; in the third boundary value problem, the coefficient k can depend on $\mathbf{x}$ and t. The boundary conditions of the second and third kinds, as well as the solution of the first boundary value problem, involve operators of differentiation along the conormal of operator (M12.10.1.2); these operators act as follows:

$$\frac{\partial G}{\partial M_x} \equiv \sum_{i,j=1}^{n} a_{ij}(\mathbf{x}, t) N_j \frac{\partial G}{\partial x_i}, \quad \frac{\partial G}{\partial M_y} \equiv \sum_{i,j=1}^{n} a_{ij}(\mathbf{y}, \tau) N_j \frac{\partial G}{\partial y_i}, \tag{M12.10.1.9}$$

where $\mathbf{N} = \{N_1, \ldots, N_n\}$ is the unit outward normal to the surface S. In the special case where $a_{ii}(\mathbf{x}, t) = 1$ and $a_{ij}(\mathbf{x}, t) = 0$ for $i \neq j$, operator (M12.10.1.9) coincides with the ordinary operator of differentiation along the outward normal to S.

TABLE M12.8
The form of the function $H(\mathbf{x}, \mathbf{y}, t, \tau)$ for the basic types of nonstationary boundary value problems.

Type of problem	Form of boundary condition (M12.10.1.4)	Function $H(\mathbf{x}, \mathbf{y}, t, \tau)$
First boundary value problem	$w = g(\mathbf{x}, t)$ for $\mathbf{x} \in S$	$H(\mathbf{x}, \mathbf{y}, t, \tau) = -\dfrac{\partial G}{\partial M_y}(\mathbf{x}, \mathbf{y}, t, \tau)$
Second boundary value problem	$\dfrac{\partial w}{\partial M_x} = g(\mathbf{x}, t)$ for $\mathbf{x} \in S$	$H(\mathbf{x}, \mathbf{y}, t, \tau) = G(\mathbf{x}, \mathbf{y}, t, \tau)$
Third boundary value problem	$\dfrac{\partial w}{\partial M_x} + kw = g(\mathbf{x}, t)$ for $\mathbf{x} \in S$	$H(\mathbf{x}, \mathbf{y}, t, \tau) = G(\mathbf{x}, \mathbf{y}, t, \tau)$

If the coefficient of equation (M12.10.1.6) and the boundary condition (M12.10.1.8) are independent of t, then the Green's function depends on only three arguments, $G(\mathbf{x}, \mathbf{y}, t, \tau) = G(\mathbf{x}, \mathbf{y}, t - \tau)$.

Remark. Let S_i ($i = 1, \ldots, p$) be different portions of the surface S such that $S = \sum\limits_{i=1}^{p} S_i$ and let boundary conditions of various types be set on the S_i,

$$\Gamma_{\mathbf{x},t}^{(i)}[w] = g_i(\mathbf{x}, t) \quad \text{for} \quad \mathbf{x} \in S_i, \qquad i = 1, \ldots, p. \tag{M12.10.1.10}$$

Then formula (M12.10.1.5) remains valid but the last term in (M12.10.1.5) must be replaced by the sum

$$\sum_{i=1}^{p} \int_0^t \int_{S_i} g_i(\mathbf{y}, \tau) H_i(\mathbf{x}, \mathbf{y}, t, \tau) \, dS_y \, d\tau. \tag{M12.10.1.11}$$

M12.10.2. Boundary Value Problems for Hyperbolic Equations

▶ **Statement of the problem.** The general nonhomogeneous linear differential hyperbolic equation in n space variables can be written as

$$\frac{\partial^2 w}{\partial t^2} + \varphi(\mathbf{x}, t)\frac{\partial w}{\partial t} - L_{\mathbf{x},t}[w] = \Phi(\mathbf{x}, t), \tag{M12.10.2.1}$$

where the operator $L_{\mathbf{x},t}[w]$ is defined explicitly in (M12.10.1.2).

We consider the nonstationary boundary value problem for equation (M12.10.2.1) in the domain V with arbitrary initial conditions,

$$w = f_0(\mathbf{x}) \quad \text{at} \quad t = 0, \tag{M12.10.2.2}$$

$$\partial_t w = f_1(\mathbf{x}) \quad \text{at} \quad t = 0, \tag{M12.10.2.3}$$

and the nonhomogeneous linear boundary condition (M12.10.1.4).

▶ **Representation of the problem solution in terms of the Green's function.** The solution of the nonhomogeneous linear boundary value problem defined by (M12.10.2.1)–(M12.10.2.3), (M12.10.1.4) can be represented as the sum

$$
w(\mathbf{x}, t) = \int_0^t \int_V \Phi(\mathbf{y}, \tau)G(\mathbf{x}, \mathbf{y}, t, \tau)\,dV_y\,d\tau - \int_V f_0(\mathbf{y})\left[\frac{\partial}{\partial \tau}G(\mathbf{x}, \mathbf{y}, t, \tau)\right]_{\tau=0} dV_y
$$
$$
+ \int_V \left[f_1(\mathbf{y}) + f_0(\mathbf{y})\varphi(\mathbf{y}, 0)\right] G(\mathbf{x}, \mathbf{y}, t, 0)\,dV_y
$$
$$
+ \int_0^t \int_S g(\mathbf{y}, \tau)H(\mathbf{x}, \mathbf{y}, t, \tau)\,dS_y\,d\tau. \tag{M12.10.2.4}
$$

Here, $G(\mathbf{x}, \mathbf{y}, t, \tau)$ is the Green's function; for $t > \tau \geq 0$, it satisfies the homogeneous equation

$$\frac{\partial^2 G}{\partial t^2} + \varphi(\mathbf{x}, t)\frac{\partial G}{\partial t} - L_{\mathbf{x},t}[G] = 0 \tag{M12.10.2.5}$$

with the semihomogeneous initial conditions

$$G = 0 \quad \text{at} \quad t = \tau,$$
$$\partial_t G = \delta(\mathbf{x} - \mathbf{y}) \quad \text{at} \quad t = \tau,$$

and the homogeneous boundary condition (M12.10.1.8).

If the coefficients of equation (M12.10.2.5) and the boundary condition (M12.10.1.8) are independent of time t, then the Green's function depends on only three arguments, $G(\mathbf{x}, \mathbf{y}, t, \tau) = G(\mathbf{x}, \mathbf{y}, t - \tau)$. In this case, one can set $\frac{\partial}{\partial \tau}G(\mathbf{x}, \mathbf{y}, t, \tau)\big|_{\tau=0} = -\frac{\partial}{\partial t}G(\mathbf{x}, \mathbf{y}, t)$ in solution (M12.10.2.4).

The function $H(\mathbf{x}, \mathbf{y}, t, \tau)$ in the integrand of the last term in solution (M12.10.2.4) can be expressed via the Green's function $G(\mathbf{x}, \mathbf{y}, t, \tau)$. The corresponding formulas for H are given in Table M12.8 for the three basic types of boundary value problems; in the third boundary value problem, the coefficient k can depend on $\mathbf{x}$ and t.

M12.10.3. Boundary Value Problems for Elliptic Equations

▶ **Statement of the problem.** In general, a nonhomogeneous linear elliptic equation can be written as

$$-L_{\mathbf{x}}[w] = \Phi(\mathbf{x}), \tag{M12.10.3.1}$$

where

$$L_{\mathbf{x}}[w] \equiv \sum_{i,j=1}^{n} a_{ij}(\mathbf{x}) \frac{\partial^2 w}{\partial x_i \partial x_j} + \sum_{i=1}^{n} b_i(\mathbf{x}) \frac{\partial w}{\partial x_i} + c(\mathbf{x})w. \tag{M12.10.3.2}$$

We consider equation (M12.10.3.1) with operator (M12.10.3.2) in a domain V and assume that the equation is subject to the general linear boundary condition

$$\Gamma_{\mathbf{x}}[w] = g(\mathbf{x}) \quad \text{for} \quad \mathbf{x} \in S. \tag{M12.10.3.3}$$

The solution of the stationary problem (M12.10.3.1)–(M12.10.3.3) can be obtained by passing in (M12.10.1.5) to the limit as $t \to \infty$. To this end, one should start with equation (M12.10.1.1), whose coefficients are independent of t, and take the homogeneous initial condition (M12.10.1.3), with $f(\mathbf{x}) = 0$, and the stationary boundary condition (M12.10.3.3).

▶ **Representation of the problem solution in terms of the Green's function.** The solution of the linear boundary value problem (M12.10.3.1)–(M12.10.3.3) can be represented as the sum

$$w(\mathbf{x}) = \int_V \Phi(\mathbf{y}) G(\mathbf{x}, \mathbf{y}) \, dV_y + \int_S g(\mathbf{y}) H(\mathbf{x}, \mathbf{y}) \, dS_y. \tag{M12.10.3.4}$$

Here, the Green's function $G(\mathbf{x}, \mathbf{y})$ satisfies the nonhomogeneous equation of special form

$$-L_{\mathbf{x}}[G] = \delta(\mathbf{x} - \mathbf{y}) \tag{M12.10.3.5}$$

with the homogeneous boundary condition

$$\Gamma_{\mathbf{x}}[G] = 0 \quad \text{for} \quad \mathbf{x} \in S. \tag{M12.10.3.6}$$

The vector $\mathbf{y} = \{y_1, \dots, y_n\}$ appears in problem (M12.10.3.5), (M12.10.3.6) as an n-dimensional free parameter ($\mathbf{y} \in V$). Note that G is independent of the functions Φ and g characterizing various nonhomogeneities of the original boundary value problem.

The function $H(\mathbf{x}, \mathbf{y})$ in the integrand of the second term in solution (M12.10.3.4) can be expressed via the Green's function $G(\mathbf{x}, \mathbf{y})$. The corresponding formulas for H for the three basic types of boundary value problems can be obtained using Table M12.8, where one should consider the functions to be independent of t and τ. The boundary conditions of the second and third kinds, as well as the solution of the first boundary value problem, involve operators of differentiation along the conormal of operator (M12.10.3.2); these operators are defined by (M12.10.1.9); in this case, the coefficients a_{ij} depend on $\mathbf{x}$ only.

M12.10.4. Construction of the Green's Functions. General Formulas and Relations

Table M12.9 lists the Green's functions of boundary value problems for second-order equations of various types in a bounded domain V. It is assumed that $L_{\mathbf{x}}$ is a second-order linear (self-adjoint) differential operator of the form

$$L_{\mathbf{x}}[w] \equiv \sum_{i,j=1}^{n} \frac{\partial}{\partial x_i} \left[a_{ij}(\mathbf{x}) \frac{\partial w}{\partial x_j} \right], \quad \text{where} \quad \sum_{i,j=1}^{n} a_{ij}(\mathbf{x}) \xi_i \xi_j \geq \sigma \sum_{i=1}^{n} \xi_i^2, \quad \sigma > 0,$$

TABLE M12.9
The Green's functions of boundary value problems for equations of various types in bounded domains. In all problems, the operators $L_\mathbf{x}$ and $\Gamma_\mathbf{x}$ are the same; $\mathbf{x} = \{x_1, \ldots, x_n\}$.

Equation	Initial and boundary conditions	Green's function
Elliptic equation $-L_\mathbf{x}[w] = \Phi(\mathbf{x})$	$\Gamma_\mathbf{x}[w] = g(\mathbf{x})$ for $\mathbf{x} \in S$ (no initial condition required)	$G(\mathbf{x}, \mathbf{y}) = \sum\limits_{k=1}^{\infty} \dfrac{u_k(\mathbf{x})u_k(\mathbf{y})}{\|u_k\|^2 \lambda_k}, \quad \lambda_k \neq 0$
Parabolic equation $\partial_t w - L_\mathbf{x}[w] = \Phi(\mathbf{x}, t)$	$w = f(\mathbf{x})$ at $t = 0$ $\Gamma_\mathbf{x}[w] = g(\mathbf{x}, t)$ for $\mathbf{x} \in S$	$G(\mathbf{x}, \mathbf{y}, t) = \sum\limits_{k=1}^{\infty} \dfrac{u_k(\mathbf{x})u_k(\mathbf{y})}{\|u_k\|^2} \exp(-\lambda_k t)$
Hyperbolic equation $\partial_{tt} w - L_\mathbf{x}[w] = \Phi(\mathbf{x}, t)$	$w = f_0(\mathbf{x})$ at $t = 0$ $w = f_1(\mathbf{x})$ at $t = 0$ $\Gamma_\mathbf{x}[w] = g(\mathbf{x}, t)$ for $\mathbf{x} \in S$	$G(\mathbf{x}, \mathbf{y}, t) = \sum\limits_{k=1}^{\infty} \dfrac{u_k(\mathbf{x})u_k(\mathbf{y})}{\|u_k\|^2 \sqrt{\lambda_k}} \sin(t\sqrt{\lambda_k})$

in the space variables $x_1, \ldots, x_n$, and $\Gamma_\mathbf{x}$ is a zeroth- or first-order linear boundary operator that can define a boundary condition of the first, second, or third kind; the coefficients of the operators $L_\mathbf{x}$ and $\Gamma_\mathbf{x}$ can depend on the space variables but are independent of time t. The coefficients λ_k and the functions $u_k(\mathbf{x})$ are determined by solving the homogeneous eigenvalue problem

$$L_\mathbf{x}[u] + \lambda u = 0, \tag{M12.11.1.1}$$

$$\Gamma_\mathbf{x}[u] = 0 \quad \text{for} \quad \mathbf{x} \in S. \tag{M12.11.1.2}$$

It is apparent from Table M12.9 that, given the Green's function in the problem for a parabolic (or hyperbolic) equation, one can easily construct the Green's functions of the corresponding problems for elliptic and hyperbolic (or parabolic) equations. In particular, the Green's function of the problem for an elliptic equation can be expressed via the Green's function of the problem for a parabolic equation as follows:

$$G_0(\mathbf{x}, \mathbf{y}) = \int_0^\infty G_1(\mathbf{x}, \mathbf{y}, t)\, dt. \tag{M12.11.1.3}$$

It is assumed here that $\lambda = 0$ is not an eigenvalue of problem (M12.11.1.1), (M12.11.1.2).

Remark 1. Formula (M12.11.1.3) can also be used if the domain V is infinite. In this case, one should make sure that the integral on the right-hand side is convergent.

Remark 2. Suppose the equations given in the first column of Table M12.9 contain $-L_\mathbf{x}[w] - \beta w$ instead of $-L_\mathbf{x}[w]$, with β being a free parameter. Then the λ_k in the expressions of the Green's function in the third column of Table M12.9 must be replaced by $\lambda_k - \beta$; just as previously, the λ_k and $u_k(\mathbf{x})$ were determined by solving the eigenvalue problem (M12.11.1.1), (M12.11.1.2).

M12.11. Duhamel's Principles in Nonstationary Problems

M12.11.1. Problems for Homogeneous Linear Equations

▶ **Parabolic equations with two independent variables.** Consider the problem for the homogeneous linear equation of parabolic type

$$\frac{\partial w}{\partial t} = a(x)\frac{\partial^2 w}{\partial x^2} + b(x)\frac{\partial w}{\partial x} + c(x)w \tag{M12.11.1.1}$$

with the homogeneous initial condition

$$w = 0 \quad \text{at} \quad t = 0 \tag{M12.11.1.2}$$

and the boundary conditions

$$s_1 \partial_x w + k_1 w = g(t) \quad \text{at} \quad x = x_1, \tag{M12.11.1.3}$$

$$s_2 \partial_x w + k_2 w = 0 \quad \text{at} \quad x = x_2. \tag{M12.11.1.4}$$

By appropriately choosing the values of the coefficients s_1, s_2, k_1, and k_2 in (M12.11.1.3) and (M12.11.1.4), one can obtain the first, second, third, and mixed boundary value problems for equation (M12.11.1.1).

The solution of problem (M12.11.1.1)–(M12.11.1.4) with the nonstationary boundary condition (M12.11.1.3) at $x = x_1$ can be expressed by the formula (*Duhamel's first principle*)

$$w(x, t) = \frac{\partial}{\partial t} \int_0^t u(x, t - \tau) \, g(\tau) \, d\tau = \int_0^t \frac{\partial u}{\partial t}(x, t - \tau) \, g(\tau) \, d\tau \tag{M12.11.1.5}$$

in terms of the solution $u(x, t)$ of the auxiliary problem for equation (M12.11.1.1) with the initial and boundary conditions (M12.11.1.2) and (M12.11.1.4), for u instead of w, and the following simpler stationary boundary condition at $x = x_1$:

$$s_1 \partial_x u + k_1 u = 1 \quad \text{at} \quad x = x_1. \tag{M12.11.1.6}$$

Remark. A similar formula also holds for the homogeneous boundary condition at $x = x_1$ and a nonhomogeneous nonstationary boundary condition at $x = x_2$.

Example. Consider the first boundary value problem for the heat equation

$$\frac{\partial w}{\partial t} = \frac{\partial^2 w}{\partial x^2} \tag{M12.11.1.7}$$

with the homogeneous initial condition (M12.11.1.2) and the boundary condition

$$w = g(t) \quad \text{at} \quad x = 0. \tag{M12.11.1.8}$$

(The second boundary condition is not required, since $0 \le x < \infty$ in this case.)

First consider the following auxiliary problem for the heat equation with the homogeneous initial condition and a simpler boundary condition:

$$\frac{\partial u}{\partial t} = \frac{\partial^2 u}{\partial x^2}, \quad u = 0 \quad \text{at} \quad t = 0, \quad u = 1 \quad \text{at} \quad x = 0.$$

This problem has a self-similar solution of the form

$$w = w(z), \quad z = xt^{-1/2},$$

where the function $w(z)$ is determined by the following ordinary differential equation and boundary conditions:

$$u''_{zz} + \tfrac{1}{2} z u'_z = 0, \quad u = 1 \quad \text{at} \quad z = 0, \quad u = 0 \quad \text{at} \quad z = \infty.$$

Its solution is expressed as

$$u(z) = \operatorname{erfc}\left(\frac{z}{2}\right) \quad \Longrightarrow \quad u(x, t) = \operatorname{erfc}\left(\frac{x}{2\sqrt{t}}\right),$$

where $\operatorname{erfc} z = \dfrac{2}{\sqrt{\pi}} \displaystyle\int_z^\infty \exp(-\xi^2) \, d\xi$ is the complementary error function. Substituting the obtained expression of $u(x, t)$ into (M12.11.1.5), we obtain the solution to the first boundary value problem for the heat equation (M12.11.1.7) with the initial condition (M12.11.1.2) and an arbitrary boundary condition (M12.11.1.8) in the form

$$w(x, t) = \frac{x}{2\sqrt{\pi}} \int_0^t \exp\left[-\frac{x^2}{4(t - \tau)}\right] \frac{g(\tau) \, d\tau}{(t - \tau)^{3/2}}.$$

▶ **Hyperbolic equations with two independent variables.** Consider the problem for the homogeneous linear hyperbolic equation

$$\frac{\partial^2 w}{\partial t^2} + \varphi(x)\frac{\partial w}{\partial t} = a(x)\frac{\partial^2 w}{\partial x^2} + b(x)\frac{\partial w}{\partial x} + c(x)w \qquad \text{(M12.11.1.9)}$$

with the homogeneous initial conditions

$$\begin{aligned} w &= 0 \quad \text{at} \quad t = 0, \\ \partial_t w &= 0 \quad \text{at} \quad t = 0, \end{aligned} \qquad \text{(M12.11.1.10)}$$

and the boundary conditions (M12.11.1.3) and (M12.11.1.4).

The solution of problem (M12.11.1.9), (M12.11.1.10), (M12.11.1.3), (M12.11.1.4) with the nonstationary boundary condition (M12.11.1.3) at $x = x_1$ can be expressed by formula (M12.11.1.5) in terms of the solution $u(x, t)$ of the auxiliary problem for equation (M12.11.1.9) with the initial conditions (M12.11.1.10) and boundary condition (M12.11.1.4), for u instead of w, and the simpler stationary boundary condition (M12.11.1.6) at $x = x_1$.

In this case, the remark made after the boundary condition (M12.11.1.6) remains valid.

M12.11.2. Problems for Nonhomogeneous Linear Equations

▶ **Parabolic equations.** The solution of the nonhomogeneous linear equation

$$\frac{\partial w}{\partial t} = \sum_{i,j=1}^{n} a_{ij}(\mathbf{x})\frac{\partial^2 w}{\partial x_i \partial x_j} + \sum_{i=1}^{n} b_i(\mathbf{x})\frac{\partial w}{\partial x_i} + c(\mathbf{x})w + \Phi(\mathbf{x}, t)$$

with the homogeneous initial condition (M12.11.1.2) and the homogeneous boundary condition

$$\Gamma_{\mathbf{x}}[w] = 0 \quad \text{for} \quad \mathbf{x} \in S \qquad \text{(M12.11.2.1)}$$

can be represented in the form (*Duhamel's second principle*)

$$w(\mathbf{x}, t) = \int_0^t U(\mathbf{x}, t - \tau, \tau)\, d\tau. \qquad \text{(M12.11.2.2)}$$

Here, $U(\mathbf{x}, t, \tau)$ is the solution of the auxiliary problem for the homogeneous equation

$$\frac{\partial U}{\partial t} = \sum_{i,j=1}^{n} a_{ij}(\mathbf{x})\frac{\partial^2 U}{\partial x_i \partial x_j} + \sum_{i=1}^{n} b_i(\mathbf{x})\frac{\partial U}{\partial x_i} + c(\mathbf{x})U$$

with the boundary condition (M12.11.2.1), in which w must be substituted by U, and the nonhomogeneous initial condition

$$U = \Phi(\mathbf{x}, \tau) \quad \text{at} \quad t = 0,$$

where τ is a parameter.

Note that (M12.11.2.1) can represent a boundary condition of the first, second, or third kind; the coefficients of the operator $\Gamma_{\mathbf{x}}$ are assumed to be independent of t.

▶ **Hyperbolic equations.** The solution of the nonhomogeneous linear equation

$$\frac{\partial^2 w}{\partial t^2} + \varphi(\mathbf{x})\frac{\partial w}{\partial t} = \sum_{i,j=1}^{n} a_{ij}(\mathbf{x})\frac{\partial^2 w}{\partial x_i \partial x_j} + \sum_{i=1}^{n} b_i(\mathbf{x})\frac{\partial w}{\partial x_i} + c(\mathbf{x})w + \Phi(\mathbf{x}, t)$$

with the homogeneous initial conditions (M12.11.1.10) and homogeneous boundary condition (M12.11.2.1) can be expressed by formula (M12.11.2.2) in terms of the solution $U = U(\mathbf{x}, t, \tau)$ of the auxiliary problem for the homogeneous equation

$$\frac{\partial^2 U}{\partial t^2} + \varphi(\mathbf{x})\frac{\partial U}{\partial t} = \sum_{i,j=1}^{n} a_{ij}(\mathbf{x})\frac{\partial^2 U}{\partial x_i \partial x_j} + \sum_{i=1}^{n} b_i(\mathbf{x})\frac{\partial U}{\partial x_i} + c(\mathbf{x})U$$

with the homogeneous initial and boundary conditions, (M12.11.1.2) and (M12.11.2.1), where w must be replaced by U, and the nonhomogeneous initial condition

$$\partial_t U = \Phi(\mathbf{x}, \tau) \quad \text{at} \quad t = 0,$$

where τ is a parameter.

Note that (M12.11.2.1) can represent a boundary condition of the first, second, or third kind.

Bibliography for Chapter M12

Akulenko, L. D. and Nesterov, S. V., *High Precision Methods in Eigenvalue Problems and Their Applications*, Chapman & Hall/CRC Press, Boca Raton, Florida, 2004.

Butkovskiy, A. G., *Green's Functions and Transfer Functions Handbook*, Halstead Press–John Wiley & Sons, New York, 1982.

Carslaw, H. S. and Jaeger, J. C., *Conduction of Heat in Solids*, Clarendon Press, Oxford, 1984.

Courant, R. and Hilbert, D., *Methods of Mathematical Physics, Vol. 2*, Wiley-Interscience, New York, 1989.

Ditkin, V. A. and Prudnikov, A. P., *Integral Transforms and Operational Calculus*, Pergamon Press, New York, 1965.

Duffy, D. G., *Transform Methods for Solving Partial Differential Equations, 2nd Edition*, Chapman & Hall/CRC Press, Boca Raton, Florida, 2004.

Farlow, S. J., *Partial Differential Equations for Scientists and Engineers*, John Wiley & Sons, New York, 1982.

Guenther, R. B. and Lee, J. W., *Partial Differential Equations of Mathematical Physics and Integral Equations*, Dover Publications, New York, 1996.

Haberman, R., *Elementary Applied Partial Differential Equations with Fourier Series and Boundary Value Problems*, Prentice-Hall, Englewood Cliffs, New Jersey, 1987.

Kamke, E., *Differentialgleichungen: Lösungsmethoden und Lösungen, II, Partielle Differentialgleichungen Erster Ordnung für eine gesuchte Funktion*, Akad. Verlagsgesellschaft Geest & Portig, Leipzig, 1965.

Leis, R., *Initial-Boundary Value Problems in Mathematical Physics*, John Wiley & Sons, Chichester, 1986.

Mikhlin, S. G. (Editor), *Linear Equations of Mathematical Physics*, Holt, Rinehart and Winston, New York, 1967.

Moon, P. and Spencer, D. E., *Field Theory Handbook, Including Coordinate Systems, Differential Equations and Their Solutions, 3rd Edition*, Springer-Verlag, Berlin, 1988.

Morse, P. M. and Feshbach, H., *Methods of Theoretical Physics, Vols. 1 and 2*, McGraw-Hill, New York, 1953.

Petrovsky, I. G., *Lectures on Partial Differential Equations*, Dover Publications, New York, 1991.

Polyanin, A. D., *Handbook of Linear Partial Differential Equations for Engineers and Scientists*, Chapman & Hall/CRC Press, Boca Raton, Florida, 2002.

Polyanin, A. D., Zaitsev, V. F., and Moussiaux, A., *Handbook of First Order Partial Differential Equations*, Taylor & Francis, London, 2002.

Sneddon, I. N., *Fourier Transformations*, Dover Publications, New York, 1995.

Stakgold, I., *Boundary Value Problems of Mathematical Physics. Vols. 1 and 2*, Society for Industrial & Applied Mathematics, Philadelphia, 2000.

Strauss, W. A., *Partial Differential Equations. An Introduction*, John Wiley & Sons, New York, 1992.
Tikhonov, A. N. and Samarskii, A. A., *Equations of Mathematical Physics*, Dover Publications, New York, 1990.
Vladimirov, V. S., *Equations of Mathematical Physics*, Dekker, New York, 1971.
Zauderer, E., *Partial Differential Equations of Applied Mathematics*, Wiley-Interscience, New York, 1989.
Zwillinger, D., *Handbook of Differential Equations, 3rd Edition*, Academic Press, New York, 1997.

Chapter M13
Special Functions and Their Properties

Throughout Chapter M13 it is assumed that n is a positive integer unless otherwise specified.

M13.1. Some Coefficients, Symbols, and Numbers
M13.1.1. Factorials. Binomial Coefficients. Pochhammer Symbol

▶ **Factorial.**
$$0! = 1! = 1, \quad n! = 1 \cdot 2 \cdot 3 \ldots (n-1)n, \quad n = 2, 3, 4, \ldots$$

▶ **Double factorial.**
$$0!! = 1!! = 1, \quad (2n)!! = 2 \cdot 4 \cdot 6 \ldots (2n), \quad (2n+1)!! = 1 \cdot 3 \cdot 5 \ldots (2n+1), \quad \text{where}$$
$n = 1, 2, 3, \ldots$

▶ **Binomial coefficients.**
$$C_n^k = \binom{n}{k} = \frac{n!}{k! \, (n-k)!}, \quad \text{where} \quad k = 1, \ldots, n;$$
$$C_a^0 = 1, \quad C_a^k = \binom{a}{k} = \frac{a(a-1)\ldots(a-k+1)}{k!}, \quad \text{where} \quad k = 1, 2, \ldots$$

Here, a is an arbitrary real number.

Remark. In various texts, the binomial coefficients are also denoted $C_r^n = {}_nC_r = C(n, r)$.

▶ **Generalization. Some properties.** General case:
$$C_a^b = \frac{\Gamma(a+1)}{\Gamma(b+1)\Gamma(a-b+1)}, \quad \text{where} \quad \Gamma(x) \text{ is the gamma function.}$$

Properties:
$$C_a^0 = 1, \quad C_n^k = 0 \quad \text{for} \quad k = -1, -2, \ldots \text{ or } k > n,$$
$$C_a^{b+1} = \frac{a}{b+1}C_{a-1}^b = \frac{a-b}{b+1}C_a^b, \quad C_a^b + C_a^{b+1} = C_{a+1}^{b+1},$$
$$C_{-1/2}^n = \frac{(-1)^n}{2^{2n}}C_{2n}^n = (-1)^n\frac{(2n-1)!!}{(2n)!!},$$
$$C_{1/2}^n = \frac{(-1)^{n-1}}{n2^{2n-1}}C_{2n-2}^{n-1} = \frac{(-1)^{n-1}}{n}\frac{(2n-3)!!}{(2n-2)!!},$$
$$C_{n+1/2}^{2n+1} = (-1)^n 2^{-4n-1}C_{2n}^n, \quad C_{2n+1/2}^n = 2^{-2n}C_{4n+1}^{2n},$$
$$C_n^{1/2} = \frac{2^{2n+1}}{\pi C_{2n}^n}, \quad C_n^{n/2} = \frac{2^{2n}}{\pi}C_n^{(n-1)/2}.$$

▶ **Pochhammer symbol.**
$$(a)_n = a(a+1)\ldots(a+n-1) = \frac{\Gamma(a+n)}{\Gamma(a)} = (-1)^n\frac{\Gamma(1-a)}{\Gamma(1-a-n)}.$$

M13.1.2. Bernoulli Numbers

▶ **Definition.** The *Bernoulli numbers* are defined by the recurrence relation

$$B_0 = 1, \quad \sum_{k=0}^{n-1} C_n^k B_k = 0, \quad n = 2, 3, \ldots$$

Numerical values:

$$B_0 = 1, \quad B_1 = -\tfrac{1}{2}, \quad B_2 = \tfrac{1}{6}, \quad B_4 = -\tfrac{1}{30}, \quad B_6 = \tfrac{1}{42}, \quad B_8 = -\tfrac{1}{30}, \quad B_{10} = \tfrac{5}{66}, \quad \ldots,$$
$$B_{2m+1} = 0 \quad \text{for} \quad m = 1, 2, \ldots$$

All odd-numbered Bernoulli numbers but B_1 are zero; all even-numbered Bernoulli numbers have alternating signs.

▶ **Generating function.**

$$\frac{x}{e^x - 1} = \sum_{n=0}^{\infty} B_n \frac{x^n}{n!}, \quad |x| < 2\pi.$$

This relation may be regarded as a definition of the Bernoulli numbers.

M13.2. Error Functions. Exponential and Logarithmic Integrals

M13.2.1. Error Function and Complementary Error Function

▶ **Integral representations.** Definitions:

$$\operatorname{erf} x = \frac{2}{\sqrt{\pi}} \int_0^x \exp(-t^2)\, dt \quad \text{(error function, also called probability integral)},$$

$$\operatorname{erfc} x = 1 - \operatorname{erf} x = \frac{2}{\sqrt{\pi}} \int_x^\infty \exp(-t^2)\, dt \quad \text{(complementary error function)}.$$

Properties:

$$\operatorname{erf}(-x) = -\operatorname{erf} x; \quad \operatorname{erf}(0) = 0, \quad \operatorname{erf}(\infty) = 1; \quad \operatorname{erfc}(0) = 1, \quad \operatorname{erfc}(\infty) = 0.$$

▶ **Expansions as $x \to 0$ and $x \to \infty$. Definite integral.** Expansion of $\operatorname{erf} x$ into series in powers of x as $x \to 0$:

$$\operatorname{erf} x = \frac{2}{\sqrt{\pi}} \sum_{k=0}^{\infty} (-1)^k \frac{x^{2k+1}}{k!\,(2k+1)} = \frac{2}{\sqrt{\pi}} \exp(-x^2) \sum_{k=0}^{\infty} \frac{2^k x^{2k+1}}{(2k+1)!!}.$$

Asymptotic expansion of $\operatorname{erfc} x$ as $x \to \infty$:

$$\operatorname{erfc} x = \frac{1}{\sqrt{\pi}} \exp(-x^2) \left[\sum_{m=0}^{M-1} (-1)^m \frac{\left(\frac{1}{2}\right)_m}{x^{2m+1}} + O\left(|x|^{-2M-1}\right) \right], \quad M = 1, 2, \ldots$$

Integral:

$$\int_0^x \operatorname{erf} t\, dt = x \operatorname{erf} x - \frac{1}{2} + \frac{1}{2} \exp(-x^2).$$

M13.2.2. Exponential Integral

▶ **Integral representations.** Definition:

$$\mathrm{Ei}(x) = \int_{-\infty}^{x} \frac{e^t}{t}\,dt = -\int_{-x}^{\infty} \frac{e^{-t}}{t}\,dt \qquad \text{for}\quad x < 0,$$

$$\mathrm{Ei}(x) = \lim_{\varepsilon \to +0}\left(\int_{-\infty}^{-\varepsilon} \frac{e^t}{t}\,dt + \int_{\varepsilon}^{x} \frac{e^t}{t}\,dt\right) \quad \text{for}\quad x > 0.$$

Other integral representations:

$$\mathrm{Ei}(-x) = -e^{-x}\int_0^\infty \frac{x\sin t + t\cos t}{x^2 + t^2}\,dt \quad \text{for}\quad x > 0,$$

$$\mathrm{Ei}(-x) = e^{-x}\int_0^\infty \frac{x\sin t - t\cos t}{x^2 + t^2}\,dt \quad \text{for}\quad x < 0.$$

▶ **Expansions as $x \to 0$ and $x \to \infty$.** Expansion into series in powers of x as $x \to 0$:

$$\mathrm{Ei}(x) = \begin{cases} \mathcal{C} + \ln(-x) + \displaystyle\sum_{k=1}^{\infty} \frac{x^k}{k!\,k} & \text{if } x < 0, \\[2mm] \mathcal{C} + \ln x + \displaystyle\sum_{k=1}^{\infty} \frac{x^k}{k!\,k} & \text{if } x > 0, \end{cases}$$

where $\mathcal{C} = 0.5772\ldots$ is the Euler constant.

Asymptotic expansion as $x \to \infty$:

$$\mathrm{Ei}(-x) = e^{-x}\sum_{k=1}^{n} (-1)^k \frac{(k-1)!}{x^k} + R_n, \qquad R_n < \frac{n!}{x^n}.$$

M13.2.3. Logarithmic Integral

▶ **Integral representations.** Definition:

$$\mathrm{li}(x) = \begin{cases} \displaystyle\int_0^x \frac{dt}{\ln t} & \text{if } 0 < x < 1, \\[3mm] \displaystyle\lim_{\varepsilon \to +0}\left(\int_0^{1-\varepsilon} \frac{dt}{\ln t} + \int_{1+\varepsilon}^{x} \frac{dt}{\ln t}\right) & \text{if } x > 1. \end{cases}$$

▶ **Limiting properties. Relation to the exponential integral.** For small x,

$$\mathrm{li}(x) \approx \frac{x}{\ln(1/x)}.$$

For large x,

$$\mathrm{li}(x) \approx \frac{x}{\ln x}.$$

Asymptotic expansion as $x \to 1$:

$$\mathrm{li}(x) = \mathcal{C} + \ln|\ln x| + \sum_{k=1}^{\infty} \frac{\ln^k x}{k!\,k}, \qquad \mathcal{C} = 0.5772\ldots$$

M13.3. Sine Integral and Cosine Integral. Fresnel Integrals

M13.3.1. Sine Integral

▶ **Integral representations. Properties.** Definition:

$$\text{Si}(x) = \int_0^x \frac{\sin t}{t}\, dt, \qquad \text{si}(x) = -\int_x^\infty \frac{\sin t}{t}\, dt = \text{Si}(x) - \frac{\pi}{2}.$$

Specific values:

$$\text{Si}(0) = 0, \quad \text{Si}(\infty) = \frac{\pi}{2}, \quad \text{si}(\infty) = 0.$$

Properties:

$$\text{Si}(-x) = -\text{Si}(x), \quad \text{si}(x) + \text{si}(-x) = -\pi, \quad \lim_{x \to -\infty} \text{si}(x) = -\pi.$$

▶ **Expansions as $x \to 0$ and $x \to \infty$.** Expansion into series in powers of x as $x \to 0$:

$$\text{Si}(x) = \sum_{k=1}^\infty \frac{(-1)^{k+1} x^{2k-1}}{(2k-1)(2k-1)!}.$$

Asymptotic expansion as $x \to \infty$:

$$\text{si}(x) = -\cos x \left[\sum_{m=0}^{M-1} \frac{(-1)^m (2m)!}{x^{2m+1}} + O\big(|x|^{-2M-1}\big) \right] + \sin x \left[\sum_{m=1}^{N-1} \frac{(-1)^m (2m-1)!}{x^{2m}} + O\big(|x|^{-2N}\big) \right],$$

where $M, N = 1, 2, \ldots$

M13.3.2. Cosine Integral

▶ **Integral representations.** Definition:

$$\text{Ci}(x) = -\int_x^\infty \frac{\cos t}{t}\, dt = \mathcal{C} + \ln x + \int_0^x \frac{\cos t - 1}{t}\, dt,$$

where $\mathcal{C} = 0.5772\ldots$ is the Euler constant.

▶ **Expansions as $x \to 0$ and $x \to \infty$.** Expansion into series in powers of x as $x \to 0$:

$$\text{Ci}(x) = \mathcal{C} + \ln x + \sum_{k=1}^\infty \frac{(-1)^k x^{2k}}{2k\,(2k)!}.$$

Asymptotic expansion as $x \to \infty$:

$$\text{Ci}(x) = \cos x \left[\sum_{m=1}^{M-1} \frac{(-1)^m (2m-1)!}{x^{2m}} + O\big(|x|^{-2M}\big) \right] + \sin x \left[\sum_{m=0}^{N-1} \frac{(-1)^m (2m)!}{x^{2m+1}} + O\big(|x|^{-2N-1}\big) \right],$$

where $M, N = 2, 3, \ldots$

M13.3.3. Fresnel Integrals

▶ **Integral representations.** Definitions:

$$
S(x) = \frac{1}{\sqrt{2\pi}} \int_0^x \frac{\sin t}{\sqrt{t}}\, dt = \sqrt{\frac{2}{\pi}} \int_0^{\sqrt{x}} \sin t^2\, dt,
$$

$$
C(x) = \frac{1}{\sqrt{2\pi}} \int_0^x \frac{\cos t}{\sqrt{t}}\, dt = \sqrt{\frac{2}{\pi}} \int_0^{\sqrt{x}} \cos t^2\, dt.
$$

▶ **Expansions as $x \to 0$ and $x \to \infty$.** Expansion into series in powers of x as $x \to 0$:

$$
S(x) = \sqrt{\frac{2}{\pi}}\, x \sum_{k=0}^{\infty} \frac{(-1)^k x^{2k+1}}{(4k+3)(2k+1)!},
$$

$$
C(x) = \sqrt{\frac{2}{\pi}}\, x \sum_{k=0}^{\infty} \frac{(-1)^k x^{2k}}{(4k+1)(2k)!}.
$$

Asymptotic expansion as $x \to \infty$:

$$
S(x) = \frac{1}{2} - \frac{\cos x}{\sqrt{2\pi x}} P(x) - \frac{\sin x}{\sqrt{2\pi x}} Q(x),
$$

$$
C(x) = \frac{1}{2} + \frac{\sin x}{\sqrt{2\pi x}} P(x) - \frac{\cos x}{\sqrt{2\pi x}} Q(x),
$$

$$
P(x) = 1 - \frac{1 \times 3}{(2x)^2} + \frac{1 \times 3 \times 5 \times 7}{(2x)^4} - \cdots, \quad Q(x) = \frac{1}{2x} - \frac{1 \times 3 \times 5}{(2x)^3} + \cdots.
$$

M13.4. Gamma Function, Psi Function, and Beta Function

M13.4.1. Gamma Function

▶ **Integral representations. Simplest properties.** The gamma function, $\Gamma(z)$, is an analytic function of the complex argument z everywhere except for the points $z = 0, -1, -2, \ldots$

For $\mathrm{Re}\, z > 0$,

$$
\Gamma(z) = \int_0^{\infty} t^{z-1} e^{-t}\, dt.
$$

For $-(n+1) < \mathrm{Re}\, z < -n$, where $n = 0, 1, 2, \ldots$,

$$
\Gamma(z) = \int_0^{\infty} \left[e^{-t} - \sum_{m=0}^{n} \frac{(-1)^m}{m!} \right] t^{z-1}\, dt.
$$

Simplest properties:

$$
\Gamma(z+1) = z\Gamma(z), \quad \Gamma(n+1) = n!, \quad \Gamma(1) = \Gamma(2) = 1.
$$

Fractional values of the argument:

$$\Gamma\left(\frac{1}{2}\right) = \sqrt{\pi}, \qquad \Gamma\left(n + \frac{1}{2}\right) = \frac{\sqrt{\pi}}{2^n}(2n-1)!!,$$

$$\Gamma\left(-\frac{1}{2}\right) = -2\sqrt{\pi}, \qquad \Gamma\left(\frac{1}{2} - n\right) = (-1)^n \frac{2^n \sqrt{\pi}}{(2n-1)!!}.$$

▶ **Euler, Stirling, and other formulas.** Euler formula

$$\Gamma(z) = \lim_{n\to\infty} \frac{n!\, n^z}{z(z+1)\ldots(z+n)} \qquad (z \neq 0, -1, -2, \ldots).$$

Symmetry formulas:

$$\Gamma(z)\Gamma(-z) = -\frac{\pi}{z \sin(\pi z)}, \qquad \Gamma(z)\Gamma(1-z) = \frac{\pi}{\sin(\pi z)},$$

$$\Gamma\left(\frac{1}{2} + z\right)\Gamma\left(\frac{1}{2} - z\right) = \frac{\pi}{\cos(\pi z)}.$$

Multiple argument formulas:

$$\Gamma(2z) = \frac{2^{2z-1}}{\sqrt{\pi}}\Gamma(z)\Gamma\left(z + \frac{1}{2}\right),$$

$$\Gamma(3z) = \frac{3^{3z-1/2}}{2\pi}\Gamma(z)\Gamma\left(z + \frac{1}{3}\right)\Gamma\left(z + \frac{2}{3}\right),$$

$$\Gamma(nz) = (2\pi)^{(1-n)/2} n^{nz-1/2} \prod_{k=0}^{n-1} \Gamma\left(z + \frac{k}{n}\right).$$

Asymptotic expansion (*Stirling formula*):

$$\Gamma(z) = \sqrt{2\pi}\, e^{-z} z^{z-1/2} \left[1 + \tfrac{1}{12}z^{-1} + \tfrac{1}{288}z^{-2} + O(z^{-3})\right] \qquad (|\arg z| < \pi).$$

M13.4.2. Psi Function (Digamma Function)

Definition:

$$\psi(z) = \frac{d \ln \Gamma(z)}{dz} = \frac{\Gamma_z'(z)}{\Gamma(z)}.$$

The psi function is the logarithmic derivative of the gamma function and is also called the *digamma function*.

Integral representations (Re $z > 0$):

$$\psi(z) = \int_0^\infty \left[e^{-t} - (1+t)^{-z}\right] t^{-1}\, dt,$$

$$\psi(z) = \ln z + \int_0^\infty \left[t^{-1} - (1 - e^{-t})^{-1}\right] e^{-tz}\, dt,$$

$$\psi(z) = -C + \int_0^1 \frac{1 - t^{z-1}}{1 - t}\, dt,$$

where $C = -\psi(1) = 0.5772\ldots$ is the Euler constant.

Values for integer argument:

$$\psi(1) = -\mathcal{C}, \qquad \psi(n) = -\mathcal{C} + \sum_{k=1}^{n-1} k^{-1} \quad (n = 2, 3, \dots).$$

Asymptotic expansion as $z \to \infty$ ($|\arg z| < \pi$):

$$\psi(z) = \ln z - \frac{1}{2z} - \frac{1}{12z^2} + \frac{1}{120z^4} - \frac{1}{252z^6} + \cdots = \ln z - \frac{1}{2z} - \sum_{n=1}^{\infty} \frac{B_{2n}}{2nz^{2n}},$$

where the B_{2n} are Bernoulli numbers.

M13.4.3. Beta Function

▶ **Integral representation. Relationship with the gamma function.** Definition:

$$B(x, y) = \int_0^1 t^{x-1}(1-t)^{y-1} \, dt,$$

where $\text{Re}\, x > 0$ and $\text{Re}\, y > 0$.

Relationship with the gamma function:

$$B(x, y) = \frac{\Gamma(x)\Gamma(y)}{\Gamma(x+y)}.$$

▶ **Some properties.**

$$B(x, y) = B(y, x);$$
$$B(x, y+1) = \frac{y}{x} B(x+1, y) = \frac{y}{x+y} B(x, y);$$
$$B(x, 1-x) = \frac{\pi}{\sin(\pi x)}, \quad 0 < x < 1;$$
$$\frac{1}{B(n, m)} = mC_{n+m-1}^{n-1} = nC_{n+m-1}^{m-1} = mC_{n+m-1}^{m} = nC_{n+m-1}^{n},$$

where n and m are positive integers.

M13.5. Incomplete Gamma Function

M13.5.1. Integral Representations. Recurrence Formulas

Definitions:

$$\gamma(\alpha, x) = \int_0^x e^{-t} t^{\alpha-1} \, dt, \qquad \text{Re}\, \alpha > 0,$$

$$\Gamma(\alpha, x) = \int_x^\infty e^{-t} t^{\alpha-1} \, dt = \Gamma(\alpha) - \gamma(\alpha, x).$$

Recurrence formulas:

$$\gamma(\alpha+1, x) = \alpha\gamma(\alpha, x) - x^\alpha e^{-x},$$
$$\gamma(\alpha+1, x) = (x+\alpha)\gamma(\alpha, x) + (1-\alpha)x\gamma(\alpha-1, x),$$
$$\Gamma(\alpha+1, x) = \alpha\Gamma(\alpha, x) + x^\alpha e^{-x}.$$

Special cases:

$$\gamma(n+1, x) = n! \left[1 - e^{-x} \left(\sum_{k=0}^{n} \frac{x^k}{k!}\right)\right], \qquad n = 0, 1, \ldots;$$

$$\Gamma(n+1, x) = n! \, e^{-x} \sum_{k=0}^{n} \frac{x^k}{k!}, \qquad n = 0, 1, \ldots;$$

$$\Gamma(-n, x) = \frac{(-1)^n}{n!} \left[\Gamma(0, x) - e^{-x} \sum_{k=0}^{n-1} (-1)^k \frac{k!}{x^{k+1}}\right], \quad n = 1, 2, \ldots$$

M13.5.2. Expansions as $x \to 0$ and $x \to \infty$. Relation to Other Functions

Asymptotic expansions as $x \to 0$:

$$\gamma(\alpha, x) = \sum_{n=0}^{\infty} \frac{(-1)^n x^{\alpha+n}}{n! \, (\alpha + n)},$$

$$\Gamma(\alpha, x) = \Gamma(\alpha) - \sum_{n=0}^{\infty} \frac{(-1)^n x^{\alpha+n}}{n! \, (\alpha + n)}.$$

Asymptotic expansions as $x \to \infty$:

$$\gamma(\alpha, x) = \Gamma(\alpha) - x^{\alpha-1} e^{-x} \left[\sum_{m=0}^{M-1} \frac{(1-\alpha)_m}{(-x)^m} + O\left(|x|^{-M}\right)\right], \qquad M = 1, 2, \ldots;$$

$$\Gamma(\alpha, x) = x^{\alpha-1} e^{-x} \left[\sum_{m=0}^{M-1} \frac{(1-\alpha)_m}{(-x)^m} + O\left(|x|^{-M}\right)\right] \qquad \left(-\tfrac{3}{2}\pi < \arg x < \tfrac{3}{2}\pi\right).$$

Representation of the error function, complementary error function, and exponential integral in terms of the gamma functions:

$$\operatorname{erf} x = \frac{1}{\sqrt{\pi}} \gamma\left(\frac{1}{2}, x^2\right), \quad \operatorname{erfc} x = \frac{1}{\sqrt{\pi}} \Gamma\left(\frac{1}{2}, x^2\right), \quad \operatorname{Ei}(-x) = -\Gamma(0, x).$$

M13.6. Bessel Functions (Cylinder Functions)

M13.6.1. Definitions and Basic Formulas

▶ **Bessel functions of the first and the second kind.** The *Bessel function of the first kind*, $J_\nu(x)$, and the *Bessel function of the second kind*, $Y_\nu(x)$ (also called the *Neumann function*), are solutions of the Bessel equation

$$x^2 y''_{xx} + x y'_x + (x^2 - \nu^2) y = 0$$

and are defined by the formulas

$$J_\nu(x) = \sum_{k=0}^{\infty} \frac{(-1)^k (x/2)^{\nu+2k}}{k! \, \Gamma(\nu + k + 1)}, \quad Y_\nu(x) = \frac{J_\nu(x) \cos \pi\nu - J_{-\nu}(x)}{\sin \pi\nu}. \tag{M13.6.1.1}$$

The formula for $Y_\nu(x)$ is valid for $\nu \neq 0, \pm 1, \pm 2, \ldots$ (the cases $\nu = 0, \pm 1, \pm 2, \ldots$ are discussed in what follows).

The general solution of the Bessel equation has the form $Z_\nu(x) = C_1 J_\nu(x) + C_2 Y_\nu(x)$ and is called a *cylinder function*.

▶ **Some formulas.**

$$2\nu Z_\nu(x) = x[Z_{\nu-1}(x) + Z_{\nu+1}(x)],$$

$$\frac{d}{dx}Z_\nu(x) = \frac{1}{2}[Z_{\nu-1}(x) - Z_{\nu+1}(x)] = \pm\left[\frac{\nu}{x}Z_\nu(x) - Z_{\nu\pm1}(x)\right],$$

$$\frac{d}{dx}[x^\nu Z_\nu(x)] = x^\nu Z_{\nu-1}(x), \qquad \frac{d}{dx}[x^{-\nu}Z_\nu(x)] = -x^{-\nu}Z_{\nu+1}(x),$$

$$\left(\frac{1}{x}\frac{d}{dx}\right)^n[x^\nu J_\nu(x)] = x^{\nu-n}J_{\nu-n}(x), \qquad \left(\frac{1}{x}\frac{d}{dx}\right)^n[x^{-\nu}J_\nu(x)] = (-1)^n x^{-\nu-n}J_{\nu+n}(x).$$

▶ **Bessel functions for $\nu = \pm n \pm \frac{1}{2}$, where $n = 0, 1, 2, \ldots$**

$$J_{1/2}(x) = \sqrt{\frac{2}{\pi x}}\sin x, \qquad\qquad J_{-1/2}(x) = \sqrt{\frac{2}{\pi x}}\cos x,$$

$$J_{3/2}(x) = \sqrt{\frac{2}{\pi x}}\left(\frac{1}{x}\sin x - \cos x\right), \qquad J_{-3/2}(x) = \sqrt{\frac{2}{\pi x}}\left(-\frac{1}{x}\cos x - \sin x\right),$$

$$J_{n+1/2}(x) = \sqrt{\frac{2}{\pi x}}\left[\sin\left(x - \frac{n\pi}{2}\right)\sum_{k=0}^{[n/2]}\frac{(-1)^k(n+2k)!}{(2k)!\,(n-2k)!\,(2x)^{2k}}\right.$$

$$\left. + \cos\left(x - \frac{n\pi}{2}\right)\sum_{k=0}^{[(n-1)/2]}\frac{(-1)^k(n+2k+1)!}{(2k+1)!\,(n-2k-1)!\,(2x)^{2k+1}}\right],$$

$$J_{-n-1/2}(x) = \sqrt{\frac{2}{\pi x}}\left[\cos\left(x + \frac{n\pi}{2}\right)\sum_{k=0}^{[n/2]}\frac{(-1)^k(n+2k)!}{(2k)!\,(n-2k)!\,(2x)^{2k}}\right.$$

$$\left. - \sin\left(x + \frac{n\pi}{2}\right)\sum_{k=0}^{[(n-1)/2]}\frac{(-1)^k(n+2k+1)!}{(2k+1)!\,(n-2k-1)!\,(2x)^{2k+1}}\right],$$

$$Y_{1/2}(x) = -\sqrt{\frac{2}{\pi x}}\cos x, \qquad\qquad Y_{-1/2}(x) = \sqrt{\frac{2}{\pi x}}\sin x,$$

$$Y_{n+1/2}(x) = (-1)^{n+1}J_{-n-1/2}(x), \qquad Y_{-n-1/2}(x) = (-1)^n J_{n+1/2}(x),$$

where $[A]$ stands for the integer part of the number A.

▶ **Bessel functions for $\nu = \pm n$, where $n = 0, 1, 2, \ldots$** Let $\nu = n$ be an arbitrary integer. The relations

$$J_{-n}(x) = (-1)^n J_n(x), \qquad Y_{-n}(x) = (-1)^n Y_n(x)$$

are valid. The function $J_n(x)$ is given by the first formula in (M13.6.1.1) with $\nu = n$, and $Y_n(x)$ can be obtained from the second formula in (M13.6.1.1) by passing to the limit $\nu \to n$. For nonnegative n, $Y_n(x)$ can be represented in the form

$$Y_n(x) = \frac{2}{\pi}J_n(x)\ln\frac{x}{2} - \frac{1}{\pi}\sum_{k=0}^{n-1}\frac{(n-k-1)!}{k!}\left(\frac{2}{x}\right)^{n-2k} - \frac{1}{\pi}\sum_{k=0}^{\infty}(-1)^k\left(\frac{x}{2}\right)^{n+2k}\frac{\psi(k+1) + \psi(n+k+1)}{k!\,(n+k)!},$$

where $\psi(1) = -\mathcal{C}$, $\psi(n) = -\mathcal{C} + \sum_{k=1}^{n-1}k^{-1}$, $\mathcal{C} = 0.5772\ldots$ is the Euler constant, and $\psi(x) = [\ln\Gamma(x)]'_x$ is the logarithmic derivative of the gamma function, also known as the digamma function (see Subsection M13.4.2).

▶ **Wronskians.**

$$W(J_\nu, J_{-\nu}) = -\frac{2}{\pi x}\sin(\pi\nu), \quad W(J_\nu, Y_\nu) = \frac{2}{\pi x},$$

Here, the notation $W(f, g) = fg'_x - f'_x g$ is used.

M13.6.2. Integral Representations and Asymptotic Expansions

▶ **Integral representations.** The functions $J_\nu(x)$ and $Y_\nu(x)$ can be represented in the form of definite integrals (for $x > 0$):

$$\pi J_\nu(x) = \int_0^\pi \cos(x\sin\theta - \nu\theta)\,d\theta - \sin\pi\nu \int_0^\infty \exp(-x\sinh t - \nu t)\,dt,$$

$$\pi Y_\nu(x) = \int_0^\pi \sin(x\sin\theta - \nu\theta)\,d\theta - \int_0^\infty (e^{\nu t} + e^{-\nu t}\cos\pi\nu)\,e^{-x\sinh t}\,dt.$$

For $|\nu| < \frac{1}{2}$, $x > 0$,

$$J_\nu(x) = \frac{2^{1+\nu}x^{-\nu}}{\pi^{1/2}\Gamma(\frac{1}{2} - \nu)}\int_1^\infty \frac{\sin(xt)\,dt}{(t^2 - 1)^{\nu+1/2}},$$

$$Y_\nu(x) = -\frac{2^{1+\nu}x^{-\nu}}{\pi^{1/2}\Gamma(\frac{1}{2} - \nu)}\int_1^\infty \frac{\cos(xt)\,dt}{(t^2 - 1)^{\nu+1/2}}.$$

For $\nu > -\frac{1}{2}$,

$$J_\nu(x) = \frac{2(x/2)^\nu}{\pi^{1/2}\Gamma(\frac{1}{2} + \nu)}\int_0^{\pi/2} \cos(x\cos t)\sin^{2\nu} t\,dt \quad (\textit{Poisson's formula}).$$

For $\nu = 0$, $x > 0$,

$$J_0(x) = \frac{2}{\pi}\int_0^\infty \sin(x\cosh t)\,dt, \qquad Y_0(x) = -\frac{2}{\pi}\int_0^\infty \cos(x\cosh t)\,dt.$$

For integer $\nu = n = 0, 1, 2, \ldots$,

$$J_n(x) = \frac{1}{\pi}\int_0^\pi \cos(nt - x\sin t)\,dt \quad (\textit{Bessel's formula}),$$

$$J_{2n}(x) = \frac{2}{\pi}\int_0^{\pi/2} \cos(x\sin t)\cos(2nt)\,dt,$$

$$J_{2n+1}(x) = \frac{2}{\pi}\int_0^{\pi/2} \sin(x\sin t)\sin[(2n + 1)t]\,dt.$$

▶ **Asymptotic expansions as $|x| \to \infty$.**

$$J_\nu(x) = \sqrt{\frac{2}{\pi x}}\left\{\cos\left(\frac{4x - 2\nu\pi - \pi}{4}\right)\left[\sum_{m=0}^{M-1}(-1)^m(\nu, 2m)(2x)^{-2m} + O(|x|^{-2M})\right]\right.$$

$$- \sin\left(\frac{4x - 2\nu\pi - \pi}{4}\right)\left[\sum_{m=0}^{M-1}(-1)^m(\nu, 2m+1)(2x)^{-2m-1} + O(|x|^{-2M-1})\right]\Bigg\},$$

$$Y_\nu(x) = \sqrt{\frac{2}{\pi x}}\Bigg\{\sin\left(\frac{4x - 2\nu\pi - \pi}{4}\right)\left[\sum_{m=0}^{M-1}(-1)^m(\nu, 2m)(2x)^{-2m} + O(|x|^{-2M})\right]$$

$$+ \cos\left(\frac{4x - 2\nu\pi - \pi}{4}\right)\left[\sum_{m=0}^{M-1}(-1)^m(\nu, 2m+1)(2x)^{-2m-1} + O(|x|^{-2M-1})\right]\Bigg\},$$

where $(\nu, m) = \dfrac{1}{2^{2m}m!}(4\nu^2 - 1)(4\nu^2 - 3^2)\dots[4\nu^2 - (2m-1)^2] = \dfrac{\Gamma(\frac{1}{2} + \nu + m)}{m!\,\Gamma(\frac{1}{2} + \nu - m)}.$

For nonnegative integer n and large x,

$$\sqrt{\pi x}\, J_{2n}(x) = (-1)^n(\cos x + \sin x) + O(x^{-2}),$$
$$\sqrt{\pi x}\, J_{2n+1}(x) = (-1)^{n+1}(\cos x - \sin x) + O(x^{-2}).$$

▶ **Zeros of Bessel functions.** Each of the functions $J_\nu(x)$ and $Y_\nu(x)$ has infinitely many real zeros (for real ν). All zeros are simple, except possibly for the point $x = 0$.

The zeros γ_m of $J_0(x)$, i.e., the roots of the equation $J_0(\gamma_m) = 0$, are approximately given by

$$\gamma_m = 2.4 + 3.13\,(m-1) \qquad (m = 1, 2, \dots),$$

with a maximum error of 0.2%.

M13.7. Modified Bessel Functions

M13.7.1. Definitions. Basic Formulas

▶ **Modified Bessel functions of the first and the second kind.** The *modified Bessel function of the first kind*, $I_\nu(x)$, and the *modified Bessel function of the second kind*, $K_\nu(x)$ (also called the *Macdonald function*), of order ν are solutions of the modified Bessel equation

$$x^2 y''_{xx} + x y'_x - (x^2 + \nu^2)y = 0$$

and are defined by the formulas

$$I_\nu(x) = \sum_{k=0}^{\infty} \frac{(x/2)^{2k+\nu}}{k!\,\Gamma(\nu + k + 1)}, \qquad K_\nu(x) = \frac{\pi}{2}\frac{I_{-\nu}(x) - I_\nu(x)}{\sin(\pi\nu)},$$

(see below for $K_\nu(x)$ with $\nu = 0, 1, 2, \dots$).

▶ **Some formulas.** The modified Bessel functions possess the following properties:

$$K_{-\nu}(x) = K_\nu(x); \qquad I_{-n}(x) = (-1)^n I_n(x), \quad n = 0, 1, 2, \dots$$
$$2\nu I_\nu(x) = x[I_{\nu-1}(x) - I_{\nu+1}(x)], \qquad 2\nu K_\nu(x) = -x[K_{\nu-1}(x) - K_{\nu+1}(x)],$$
$$\frac{d}{dx}I_\nu(x) = \frac{1}{2}[I_{\nu-1}(x) + I_{\nu+1}(x)], \qquad \frac{d}{dx}K_\nu(x) = -\frac{1}{2}[K_{\nu-1}(x) + K_{\nu+1}(x)].$$

▶ **Modified Bessel functions for $\nu = \pm n \pm \frac{1}{2}$, where $n = 0, 1, 2, \ldots$**

$$I_{1/2}(x) = \sqrt{\frac{2}{\pi x}} \sinh x, \qquad I_{-1/2}(x) = \sqrt{\frac{2}{\pi x}} \cosh x,$$

$$I_{3/2}(x) = \sqrt{\frac{2}{\pi x}} \left(-\frac{1}{x} \sinh x + \cosh x \right), \qquad I_{-3/2}(x) = \sqrt{\frac{2}{\pi x}} \left(-\frac{1}{x} \cosh x + \sinh x \right),$$

$$I_{n+1/2}(x) = \frac{1}{\sqrt{2\pi x}} \left[e^x \sum_{k=0}^{n} \frac{(-1)^k (n+k)!}{k!\,(n-k)!\,(2x)^k} - (-1)^n e^{-x} \sum_{k=0}^{n} \frac{(n+k)!}{k!\,(n-k)!\,(2x)^k} \right],$$

$$I_{-n-1/2}(x) = \frac{1}{\sqrt{2\pi x}} \left[e^x \sum_{k=0}^{n} \frac{(-1)^k (n+k)!}{k!\,(n-k)!\,(2x)^k} + (-1)^n e^{-x} \sum_{k=0}^{n} \frac{(n+k)!}{k!\,(n-k)!\,(2x)^k} \right],$$

$$K_{\pm 1/2}(x) = \sqrt{\frac{\pi}{2x}}\, e^{-x}, \qquad K_{\pm 3/2}(x) = \sqrt{\frac{\pi}{2x}} \left(1 + \frac{1}{x} \right) e^{-x},$$

$$K_{n+1/2}(x) = K_{-n-1/2}(x) = \sqrt{\frac{\pi}{2x}}\, e^{-x} \sum_{k=0}^{n} \frac{(n+k)!}{k!\,(n-k)!\,(2x)^k}.$$

▶ **Modified Bessel functions for $\nu = n$, where $n = 0, 1, 2, \ldots$** If $\nu = n$ is a nonnegative integer, then

$$K_n(x) = (-1)^{n+1} I_n(x) \ln \frac{x}{2} + \frac{1}{2} \sum_{m=0}^{n-1} (-1)^m \left(\frac{x}{2} \right)^{2m-n} \frac{(n-m-1)!}{m!}$$

$$+ \frac{1}{2}(-1)^n \sum_{m=0}^{\infty} \left(\frac{x}{2} \right)^{n+2m} \frac{\psi(n+m+1) + \psi(m+1)}{m!\,(n+m)!}; \qquad n = 0, 1, 2, \ldots,$$

where $\psi(z)$ is the logarithmic derivative of the gamma function; for $n = 0$, the first sum is omitted.

▶ **Wronskians.**

$$W(I_\nu, I_{-\nu}) = -\frac{2}{\pi x} \sin(\pi \nu), \qquad W(I_\nu, K_\nu) = -\frac{1}{x},$$

where $W(f, g) = f g'_x - f'_x g$.

M13.7.2. Integral Representations and Asymptotic Expansions

▶ **Integral representations.** The functions $I_\nu(x)$ and $K_\nu(x)$ can be represented in terms of definite integrals:

$$I_\nu(x) = \frac{x^\nu}{\pi^{1/2} 2^\nu \Gamma(\nu + \frac{1}{2})} \int_{-1}^{1} \exp(-xt)(1-t^2)^{\nu-1/2}\, dt \qquad (x > 0,\ \nu > -\tfrac{1}{2}),$$

$$K_\nu(x) = \int_{0}^{\infty} \exp(-x \cosh t) \cosh(\nu t)\, dt \qquad (x > 0),$$

$$K_\nu(x) = \frac{1}{\cos\left(\frac{1}{2}\pi\nu\right)} \int_{0}^{\infty} \cos(x \sinh t) \cosh(\nu t)\, dt \qquad (x > 0,\ -1 < \nu < 1),$$

$$K_\nu(x) = \frac{1}{\sin\left(\frac{1}{2}\pi\nu\right)} \int_{0}^{\infty} \sin(x \sinh t) \sinh(\nu t)\, dt \qquad (x > 0,\ -1 < \nu < 1).$$

For integer $\nu = n$,

$$I_n(x) = \frac{1}{\pi} \int_0^\pi \exp(x \cos t) \cos(nt)\, dt \qquad (n = 0, 1, 2, \dots),$$

$$K_0(x) = \int_0^\infty \cos(x \sinh t)\, dt = \int_0^\infty \frac{\cos(xt)}{\sqrt{t^2 + 1}}\, dt \qquad (x > 0).$$

▶ **Asymptotic expansions as $x \to \infty$:**

$$I_\nu(x) = \frac{e^x}{\sqrt{2\pi x}} \left\{ 1 + \sum_{m=1}^M (-1)^m \frac{(4\nu^2 - 1)(4\nu^2 - 3^2) \dots [4\nu^2 - (2m-1)^2]}{m!\, (8x)^m} \right\},$$

$$K_\nu(x) = \sqrt{\frac{\pi}{2x}}\, e^{-x} \left\{ 1 + \sum_{m=1}^M \frac{(4\nu^2 - 1)(4\nu^2 - 3^2) \dots [4\nu^2 - (2m-1)^2]}{m!\, (8x)^m} \right\}.$$

The terms of the order of $O(x^{-M-1})$ are omitted in the braces.

M13.8. Degenerate Hypergeometric Functions (Kummer Functions)

M13.8.1. Definitions and Basic Formulas

▶ **Degenerate hypergeometric functions $\Phi(a, b; x)$ and $\Psi(a, b; x)$.** The *degenerate hypergeometric functions (Kummer functions)* $\Phi(a, b; x)$ and $\Psi(a, b; x)$ are solutions of the degenerate hypergeometric equation

$$xy''_{xx} + (b - x)y'_x - ay = 0.$$

In the case $b \neq 0, -1, -2, -3, \dots$, the function $\Phi(a, b; x)$ can be represented as Kummer's series:

$$\Phi(a, b; x) = 1 + \sum_{k=1}^\infty \frac{(a)_k}{(b)_k} \frac{x^k}{k!},$$

where $(a)_k = a(a + 1) \dots (a + k - 1)$, $(a)_0 = 1$.

Table M13.1 presents some special cases where Φ can be expressed in terms of simpler functions.

The function $\Psi(a, b; x)$ is defined as follows:

$$\Psi(a, b; x) = \frac{\Gamma(1 - b)}{\Gamma(a - b + 1)} \Phi(a, b; x) + \frac{\Gamma(b - 1)}{\Gamma(a)} x^{1-b} \Phi(a - b + 1,\, 2 - b;\, x).$$

Table M13.2 presents some special cases where Ψ can be expressed in terms of simpler functions.

▶ **Kummer transformation and linear relations.** Kummer transformation:

$$\Phi(a, b; x) = e^x \Phi(b - a, b; -x), \qquad \Psi(a, b; x) = x^{1-b} \Psi(1 + a - b, 2 - b; x).$$

TABLE M13.1

Special cases of the Kummer function $\Phi(a, b; z)$.

a	b	z	Φ	Notation used
a	a	x	e^x	
1	2	$2x$	$\dfrac{1}{x} e^x \sinh x$	
a	$a+1$	$-x$	$a x^{-a} \gamma(a, x)$	Incomplete gamma function $\gamma(a, x) = \displaystyle\int_0^x e^{-t} t^{a-1}\, dt$
$\dfrac{1}{2}$	$\dfrac{3}{2}$	$-x^2$	$\dfrac{\sqrt{\pi}}{2} \operatorname{erf} x$	Error function $\operatorname{erf} x = \dfrac{2}{\sqrt{\pi}} \displaystyle\int_0^x \exp(-t^2)\, dt$
$-n$	$\dfrac{1}{2}$	$\dfrac{x^2}{2}$	$\dfrac{n!}{(2n)!} \left(-\dfrac{1}{2}\right)^{-n} H_{2n}(x)$	Hermite polynomials $H_n(x) = (-1)^n e^{x^2} \dfrac{d^n}{dx^n}\left(e^{-x^2}\right),$
$-n$	$\dfrac{3}{2}$	$\dfrac{x^2}{2}$	$\dfrac{n!}{(2n+1)!} \left(-\dfrac{1}{2}\right)^{-n} H_{2n+1}(x)$	$n = 0, 1, 2, \ldots$
$-n$	b	x	$\dfrac{n!}{(b)_n} L_n^{(b-1)}(x)$	Laguerre polynomials $L_n^{(\alpha)}(x) = \dfrac{e^x x^{-\alpha}}{n!} \dfrac{d^n}{dx^n}\left(e^{-x} x^{n+\alpha}\right),$ $\alpha = b-1,$ $(b)_n = b(b+1)\ldots(b+n-1)$
$\nu+\dfrac{1}{2}$	$2\nu+1$	$2x$	$\Gamma(1+\nu) e^x \left(\dfrac{x}{2}\right)^{-\nu} I_\nu(x)$	Modified Bessel functions $I_\nu(x)$
$n+1$	$2n+2$	$2x$	$\Gamma\left(n+\dfrac{3}{2}\right) e^x \left(\dfrac{x}{2}\right)^{-n-\frac{1}{2}} I_{n+\frac{1}{2}}(x)$	

Linear relations for Φ:

$$(b-a)\Phi(a-1, b; x) + (2a-b+x)\Phi(a, b; x) - a\Phi(a+1, b; x) = 0,$$

$$b(b-1)\Phi(a, b-1; x) - b(b-1+x)\Phi(a, b; x) + (b-a)x\Phi(a, b+1; x) = 0,$$

$$(a-b+1)\Phi(a, b; x) - a\Phi(a+1, b; x) + (b-1)\Phi(a, b-1; x) = 0,$$

$$b\Phi(a, b; x) - b\Phi(a-1, b; x) - x\Phi(a, b+1; x) = 0,$$

$$b(a+x)\Phi(a, b; x) - (b-a)x\Phi(a, b+1; x) - ab\Phi(a+1, b; x) = 0,$$

$$(a-1+x)\Phi(a, b; x) + (b-a)\Phi(a-1, b; x) - (b-1)\Phi(a, b-1; x) = 0.$$

Linear relations for Ψ:

$$\Psi(a-1, b; x) - (2a-b+x)\Psi(a, b; x) + a(a-b+1)\Psi(a+1, b; x) = 0,$$

$$(b-a-1)\Psi(a, b-1; x) - (b-1+x)\Psi(a, b; x) + x\Psi(a, b+1; x) = 0,$$

$$\Psi(a, b; x) - a\Psi(a+1, b; x) - \Psi(a, b-1; x) = 0,$$

$$(b-a)\Psi(a, b; x) - x\Psi(a, b+1; x) + \Psi(a-1, b; x) = 0,$$

$$(a+x)\Psi(a, b; x) + a(b-a-1)\Psi(a+1, b; x) - x\Psi(a, b+1; x) = 0,$$

$$(a-1+x)\Psi(a, b; x) - \Psi(a-1, b; x) + (a-c+1)\Psi(a, b-1; x) = 0.$$

TABLE M13.2
Special cases of the Kummer function $\Psi(a, b; z)$.

a	b	z	Ψ	Notation used
$1-a$	$1-a$	x	$e^x \Gamma(a, x)$	Incomplete gamma function $\Gamma(a, x) = \int_x^\infty e^{-t} t^{a-1}\, dt$
$\dfrac{1}{2}$	$\dfrac{1}{2}$	x^2	$\sqrt{\pi}\, \exp(x^2)\, \mathrm{erfc}\, x$	Complementary error function $\mathrm{erfc}\, x = \dfrac{2}{\sqrt{\pi}} \int_x^\infty \exp(-t^2)\, dt$
1	1	$-x$	$-e^{-x}\, \mathrm{Ei}(x)$	Exponential integral $\mathrm{Ei}(x) = \int_{-\infty}^x \dfrac{e^t}{t}\, dt$
1	1	$-\ln x$	$-x^{-1}\, \mathrm{li}\, x$	Logarithmic integral $\mathrm{li}\, x = \int_0^x \dfrac{dt}{t}$
$\dfrac{1}{2} - \dfrac{n}{2}$	$\dfrac{3}{2}$	x^2	$2^{-n} x^{-1} H_n(x)$	Hermite polynomials $H_n(x) = (-1)^n e^{x^2} \dfrac{d^n}{dx^n}\left(e^{-x^2}\right),$ $n = 0, 1, 2, \ldots$
$\nu + \dfrac{1}{2}$	$2\nu+1$	$2x$	$\pi^{-1/2}(2x)^{-\nu} e^x K_\nu(x)$	Modified Bessel functions $K_\nu(x)$

▶ **Differentiation formulas and Wronskian.** Differentiation formulas:

$$\frac{d}{dx}\Phi(a, b; x) = \frac{a}{b}\Phi(a+1, b+1; x), \qquad \frac{d^n}{dx^n}\Phi(a, b; x) = \frac{(a)_n}{(b)_n}\Phi(a+n, b+n; x),$$

$$\frac{d}{dx}\Psi(a, b; x) = -a\Psi(a+1, b+1; x), \qquad \frac{d^n}{dx^n}\Psi(a, b; x) = (-1)^n (a)_n \Psi(a+n, b+n; x).$$

Wronskian:

$$W(\Phi, \Psi) = \Phi\Psi'_x - \Phi'_x \Psi = -\frac{\Gamma(b)}{\Gamma(a)} x^{-b} e^x.$$

▶ **Degenerate hypergeometric functions for $n = 0, 1, 2, \ldots$**

$$\Psi(a, n+1; x) = \frac{(-1)^{n-1}}{n!\,\Gamma(a-n)} \left\{ \Phi(a, n+1; x) \ln x \right.$$

$$+ \sum_{r=0}^{\infty} \frac{(a)_r}{(n+1)_r}\left[\psi(a+r) - \psi(1+r) - \psi(1+n+r)\right]\frac{x^r}{r!} \left. \right\} + \frac{(n-1)!}{\Gamma(a)} \sum_{r=0}^{n-1} \frac{(a-n)_r}{(1-n)_r}\frac{x^{r-n}}{r!},$$

where $n = 0, 1, 2, \ldots$ (the last sum is omitted for $n = 0$), $\psi(z) = [\ln \Gamma(z)]'_z$ is the logarithmic derivative of the gamma function,

$$\psi(1) = -\mathcal{C}, \quad \psi(n) = -\mathcal{C} + \sum_{k=1}^{n-1} k^{-1},$$

where $\mathcal{C} = 0.5772\ldots$ is the Euler constant.

If $b < 0$, then the formula

$$\Psi(a, b; x) = x^{1-b}\Psi(a - b + 1,\ 2 - b;\ x)$$

is valid for any x.

For $b \neq 0, -1, -2, -3, \ldots$, the general solution of the degenerate hypergeometric equation can be represented in the form

$$y = C_1\Phi(a, b; x) + C_2\Psi(a, b; x),$$

and for $b = 0, -1, -2, -3, \ldots$, in the form

$$y = x^{1-b}\big[C_1\Phi(a - b + 1,\ 2 - b;\ x) + C_2\Psi(a - b + 1,\ 2 - b;\ x)\big].$$

M13.8.2. Integral Representations and Asymptotic Expansions

▶ **Integral representations.**

$$\Phi(a, b; x) = \frac{\Gamma(b)}{\Gamma(a)\,\Gamma(b - a)} \int_0^1 e^{xt} t^{a-1} (1 - t)^{b-a-1}\, dt \quad \text{(for } b > a > 0),$$

$$\Psi(a, b; x) = \frac{1}{\Gamma(a)} \int_0^\infty e^{-xt} t^{a-1} (1 + t)^{b-a-1}\, dt \qquad \text{(for } a > 0,\ x > 0),$$

where $\Gamma(a)$ is the gamma function.

▶ **Asymptotic expansion as $|x| \to \infty$.**

$$\Phi(a, b; x) = \frac{\Gamma(b)}{\Gamma(a)} e^x x^{a-b} \left[\sum_{n=0}^N \frac{(b - a)_n (1 - a)_n}{n!} x^{-n} + \varepsilon \right], \quad x > 0,$$

$$\Phi(a, b; x) = \frac{\Gamma(b)}{\Gamma(b - a)} (-x)^{-a} \left[\sum_{n=0}^N \frac{(a)_n (a - b + 1)_n}{n!} (-x)^{-n} + \varepsilon \right], \quad x < 0,$$

$$\Psi(a, b; x) = x^{-a} \left[\sum_{n=0}^N (-1)^n \frac{(a)_n (a - b + 1)_n}{n!} x^{-n} + \varepsilon \right], \quad -\infty < x < \infty,$$

where $\varepsilon = O(x^{-N-1})$.

M13.9. Elliptic Integrals

M13.9.1. Complete Elliptic Integrals

▶ **Definitions. Properties. Conversion formulas.** *Complete elliptic integral of the first kind*:

$$\mathsf{K}(k) = \int_0^{\pi/2} \frac{d\alpha}{\sqrt{1 - k^2 \sin^2 \alpha}} = \int_0^1 \frac{dx}{\sqrt{(1 - x^2)(1 - k^2 x^2)}}.$$

Complete elliptic integral of the second kind:

$$\mathsf{E}(k) = \int_0^{\pi/2} \sqrt{1 - k^2 \sin^2 \alpha}\, d\alpha = \int_0^1 \frac{\sqrt{1 - k^2 x^2}}{\sqrt{1 - x^2}}\, dx.$$

The argument k is called the *elliptic modulus* ($k^2 < 1$).
Notation:

$$k' = \sqrt{1 - k^2}, \quad K'(k) = K(k'), \quad E'(k) = E(k'),$$

where k' is the *complementary modulus*.

Properties:

$$K(-k) = K(k), \quad E(-k) = E(k);$$
$$K(k) = K'(k'), \quad E(k) = E'(k');$$
$$E(k) K'(k) + E'(k) K(k) - K(k) K'(k) = \frac{\pi}{2}.$$

Conversion formulas for complete elliptic integrals:

$$K\left(\frac{1 - k'}{1 + k'}\right) = \frac{1 + k'}{2} K(k),$$

$$E\left(\frac{1 - k'}{1 + k'}\right) = \frac{1}{1 + k'} \left[E(k) + k' K(k)\right],$$

$$K\left(\frac{2\sqrt{k}}{1 + k}\right) = (1 + k) K(k),$$

$$E\left(\frac{2\sqrt{k}}{1 + k}\right) = \frac{1}{1 + k} \left[2 E(k) - (k')^2 K(k)\right].$$

▶ **Representation of complete elliptic integrals in series form.** Representation of complete elliptic integrals in the form of series in powers of the modulus k:

$$K(k) = \frac{\pi}{2}\left\{1 + \left(\frac{1}{2}\right)^2 k^2 + \left(\frac{1 \times 3}{2 \times 4}\right)^2 k^4 + \cdots + \left[\frac{(2n - 1)!!}{(2n)!!}\right]^2 k^{2n} + \cdots\right\},$$

$$E(k) = \frac{\pi}{2}\left\{1 - \left(\frac{1}{2}\right)^2 \frac{k^2}{1} - \left(\frac{1 \times 3}{2 \times 4}\right)^2 \frac{k^4}{3} - \cdots - \left[\frac{(2n - 1)!!}{(2n)!!}\right]^2 \frac{k^{2n}}{2n - 1} - \cdots\right\}.$$

▶ **Differentiation formulas. Differential equations.** Differentiation formulas:

$$\frac{d K(k)}{dk} = \frac{E(k)}{k(k')^2} - \frac{K(k)}{k}, \quad \frac{d E(k)}{dk} = \frac{E(k) - K(k)}{k}.$$

The functions $K(k)$ and $K'(k)$ satisfy the second-order linear ordinary differential equation

$$\frac{d}{dk}\left[k(1 - k^2)\frac{d K}{dk}\right] - k K = 0.$$

The functions $E(k)$ and $E'(k) - K'(k)$ satisfy the second-order linear ordinary differential equation

$$(1 - k^2)\frac{d}{dk}\left(k\frac{d E}{dk}\right) + k E = 0.$$

M13.9.2. Incomplete Elliptic Integrals

▶ **Definitions. Properties.** *Elliptic integral of the first kind*:

$$F(\varphi, k) = \int_0^\varphi \frac{d\alpha}{\sqrt{1 - k^2 \sin^2 \alpha}} = \int_0^{\sin \varphi} \frac{dx}{\sqrt{(1 - x^2)(1 - k^2 x^2)}}.$$

Elliptic integral of the second kind:

$$E(\varphi, k) = \int_0^\varphi \sqrt{1 - k^2 \sin^2 \alpha} \, d\alpha = \int_0^{\sin \varphi} \frac{\sqrt{1 - k^2 x^2}}{\sqrt{1 - x^2}} \, dx.$$

Elliptic integral of the third kind:

$$\Pi(\varphi, n, k) = \int_0^\varphi \frac{d\alpha}{(1 - n \sin^2 \alpha)\sqrt{1 - k^2 \sin^2 \alpha}} = \int_0^{\sin \varphi} \frac{dx}{(1 - nx^2)\sqrt{(1 - x^2)(1 - k^2 x^2)}}.$$

The quantity k is called the *elliptic modulus* ($k^2 < 1$), $k' = \sqrt{1 - k^2}$ is the *complementary modulus*, and n is the *characteristic parameter*.

Complete elliptic integrals:

$$\mathsf{K}(k) = F\left(\frac{\pi}{2}, k\right), \qquad \mathsf{E}(k) = E\left(\frac{\pi}{2}, k\right),$$
$$\mathsf{K}'(k) = F\left(\frac{\pi}{2}, k'\right), \quad \mathsf{E}'(k) = E\left(\frac{\pi}{2}, k'\right).$$

Properties of elliptic integrals:

$$F(-\varphi, k) = -F(\varphi, k), \qquad F(n\pi \pm \varphi, k) = 2n\,\mathsf{K}(k) \pm F(\varphi, k);$$
$$E(-\varphi, k) = -E(\varphi, k), \qquad E(n\pi \pm \varphi, k) = 2n\,\mathsf{E}(k) \pm E(\varphi, k).$$

▶ **Conversion formulas.** Conversion formulas for elliptic integrals (first set):

$$F\left(\psi, \frac{1}{k}\right) = kF(\varphi, k),$$
$$E\left(\psi, \frac{1}{k}\right) = \frac{1}{k}\left[E(\varphi, k) - (k')^2 F(\varphi, k)\right],$$

where the arguments φ and ψ, which may be treated as angles, are related by $\sin \psi = k \sin \varphi$, $\cos \psi = \sqrt{1 - k^2 \sin^2 \varphi}$.

Conversion formulas for elliptic integrals (second set):

$$F\left(\psi, \frac{1 - k'}{1 + k'}\right) = (1 + k')F(\varphi, k),$$
$$E\left(\psi, \frac{1 - k'}{1 + k'}\right) = \frac{2}{1 + k'}\left[E(\varphi, k) + k' F(\varphi, k)\right] - \frac{1 - k'}{1 + k'} \sin \psi,$$

where φ and ψ are related by $\tan(\psi - \varphi) = k' \tan \varphi$.

Transformation formulas for elliptic integrals (third set):

$$F\left(\psi, \frac{2\sqrt{k}}{1 + k}\right) = (1 + k)F(\varphi, k),$$
$$E\left(\psi, \frac{2\sqrt{k}}{1 + k}\right) = \frac{1}{1 + k}\left[2E(\varphi, k) - (k')^2 F(\varphi, k) + 2k \frac{\sin \varphi \cos \varphi}{1 + k \sin^2 \varphi} \sqrt{1 - k^2 \sin^2 \varphi}\right],$$

where φ and ψ are related by $\sin \psi = \dfrac{(1 + k) \sin \varphi}{1 + k \sin^2 \varphi}.$

M13.10. Orthogonal Polynomials

All zeros of each of the orthogonal polynomials $\mathcal{P}_n(x)$ considered in this section are real and simple. The zeros of the polynomials $\mathcal{P}_n(x)$ and $\mathcal{P}_{n+1}(x)$ are alternating.

M13.10.1. Legendre Polynomials and Legendre Functions

▶ **Explicit and recurrence formulas for Legendre polynomials and functions.** The *Legendre polynomials* $P_n(x)$ and the *Legendre functions* $Q_n(x)$ are solutions of the second-order linear ordinary differential equation

$$(1 - x^2)y''_{xx} - 2xy'_x + n(n+1)y = 0.$$

The *Legendre polynomials* $P_n(x)$ and the Legendre functions $Q_n(x)$ are defined by the formulas

$$P_n(x) = \frac{1}{n!\,2^n} \frac{d^n}{dx^n} (x^2 - 1)^n,$$

$$Q_n(x) = \frac{1}{2} P_n(x) \ln \frac{1+x}{1-x} - \sum_{m=1}^{n} \frac{1}{m} P_{m-1}(x) P_{n-m}(x).$$

The polynomials $P_n(x)$ can be calculated using the formulas

$$P_0(x) = 1, \quad P_1(x) = x, \quad P_2(x) = \frac{1}{2}(3x^2 - 1),$$

$$P_3(x) = \frac{1}{2}(5x^3 - 3x), \quad P_4(x) = \frac{1}{8}(35x^4 - 30x^2 + 3),$$

$$P_{n+1}(x) = \frac{2n+1}{n+1} x P_n(x) - \frac{n}{n+1} P_{n-1}(x).$$

The first five functions $Q_n(x)$ have the form

$$Q_0(x) = \frac{1}{2} \ln \frac{1+x}{1-x}, \quad Q_1(x) = \frac{x}{2} \ln \frac{1+x}{1-x} - 1,$$

$$Q_2(x) = \frac{1}{4}(3x^2 - 1) \ln \frac{1+x}{1-x} - \frac{3}{2}x, \quad Q_3(x) = \frac{1}{4}(5x^3 - 3x) \ln \frac{1+x}{1-x} - \frac{5}{2}x^2 + \frac{2}{3},$$

$$Q_4(x) = \frac{1}{16}(35x^4 - 30x^2 + 3) \ln \frac{1+x}{1-x} - \frac{35}{8}x^3 + \frac{55}{24}x.$$

The polynomials $P_n(x)$ have the explicit representation

$$P_n(x) = 2^{-n} \sum_{m=0}^{[n/2]} (-1)^m C_n^m C_{2n-2m}^n x^{n-2m},$$

where $[A]$ stands for the integer part of a number A.

▶ **Zeros and orthogonality of the Legendre polynomials.** The polynomials $P_n(x)$ (with natural n) have exactly n real distinct zeros; all zeros lie on the interval $-1 < x < 1$. The zeros of $P_n(x)$ and $P_{n+1}(x)$ alternate with each other. The function $Q_n(x)$ has exactly $n+1$ zeros, which lie on the interval $-1 < x < 1$.

The functions $P_n(x)$ form an orthogonal system on the interval $-1 \le x \le 1$, with

$$\int_{-1}^{1} P_n(x)P_m(x)\,dx = \begin{cases} 0 & \text{if } n \ne m, \\ \dfrac{2}{2n+1} & \text{if } n = m. \end{cases}$$

▶ **Generating functions.** The generating function for Legendre polynomials is

$$\frac{1}{\sqrt{1-2sx+s^2}} = \sum_{n=0}^{\infty} P_n(x)s^n \qquad (|s| < 1).$$

The generating function for Legendre functions is

$$\frac{1}{\sqrt{1-2sx+s^2}} \ln\left[\frac{x-s+\sqrt{1-2sx+s^2}}{\sqrt{1-x^2}}\right] = \sum_{n=0}^{\infty} Q_n(x)s^n \qquad (|s| < 1,\ x > 1).$$

▶ **Associated Legendre functions with integer indices. Differential equation.** The *associated Legendre functions* $P_n^m(x)$ of order m are defined by the formulas

$$P_n^m(x) = (1-x^2)^{m/2}\frac{d^m}{dx^m}P_n(x), \qquad n = 1, 2, 3, \ldots, \qquad m = 0, 1, 2, \ldots$$

It is assumed by definition that $P_n^0(x) = P_n(x)$.
Properties:

$$P_n^m(x) = 0 \quad \text{if} \quad m > n, \qquad P_n^m(-x) = (-1)^{n-m}P_n^m(x).$$

The associated Legendre functions $P_n^m(x)$ have exactly $n - m$ distinct real zeros, all of which lie on the interval $-1 < x < 1$.

Some of the associated Legendre functions $P_n^m(x)$ with lower indices are

$$P_1^1(x) = (1-x^2)^{1/2}, \quad P_2^1(x) = 3x(1-x^2)^{1/2}, \quad P_2^2(x) = 3(1-x^2),$$
$$P_3^1(x) = \tfrac{3}{2}(5x^2-1)(1-x^2)^{1/2}, \quad P_3^2(x) = 15x(1-x^2), \quad P_3^3(x) = 15(1-x^2)^{3/2}.$$

The associated Legendre functions $P_n^m(x)$ with $n > m$ are solutions of the linear ordinary differential equation

$$(1-x^2)y''_{xx} - 2xy'_x + \left[n(n+1) - \frac{m^2}{1-x^2}\right]y = 0.$$

▶ **Orthogonality of the associated Legendre functions.** The functions $P_n^m(x)$ form an orthogonal system on the interval $-1 \le x \le 1$, with

$$\int_{-1}^{1} P_n^m(x)P_k^m(x)\,dx = \begin{cases} 0 & \text{if } n \ne k, \\ \dfrac{2}{2n+1}\dfrac{(n+m)!}{(n-m)!} & \text{if } n = k. \end{cases}$$

The functions $P_n^m(x)$ (with $m \ne 0$) are orthogonal on the interval $-1 \le x \le 1$ with weight $(1-x^2)^{-1}$, that is,

$$\int_{-1}^{1} \frac{P_n^m(x)P_n^k(x)}{1-x^2}\,dx = \begin{cases} 0 & \text{if } m \ne k, \\ \dfrac{(n+m)!}{m(n-m)!} & \text{if } m = k. \end{cases}$$

M13.10.2. Laguerre Polynomials and Generalized Laguerre Polynomials

▶ **Laguerre polynomials.** The Laguerre polynomials $L_n(x)$ satisfy the second-order linear ordinary differential equation

$$xy''_{xx} + (1 - x)y'_x + ny = 0$$

and are defined by the formulas

$$L_n(x) = \frac{1}{n!} e^x \frac{d^n}{dx^n} (x^n e^{-x}) = \frac{(-1)^n}{n!} \left[x^n - n^2 x^{n-1} + \frac{n^2(n-1)^2}{2!} x^{n-2} + \cdots \right].$$

The first four polynomials have the form

$$L_0(x) = 1, \quad L_1(x) = -x + 1, \quad L_2(x) = \tfrac{1}{2}(x^2 - 4x + 2), \quad L_3(x) = \tfrac{1}{6}(-x^3 + 9x^2 - 18x + 6).$$

To calculate $L_n(x)$ for $n \geq 2$, one can use the recurrence formulas

$$L_{n+1}(x) = \frac{1}{n + 1} \left[(2n + 1 - x)L_n(x) - nL_{n-1}(x) \right].$$

The functions $L_n(x)$ form an orthonormal system on the interval $0 < x < \infty$ with weight e^{-x}:

$$\int_0^\infty e^{-x} L_n(x) L_m(x) \, dx = \begin{cases} 0 & \text{if } n \neq m, \\ 1 & \text{if } n = m. \end{cases}$$

The generating function is

$$\frac{1}{1 - s} \exp\left(-\frac{sx}{1 - s} \right) = \sum_{n=0}^\infty L_n(x)s^n, \qquad |s| < 1.$$

▶ **Generalized Laguerre polynomials.** The generalized Laguerre polynomials $L_n^\alpha(x)$ $(\alpha > -1)$ satisfy the equation

$$xy''_{xx} + (\alpha + 1 - x)y'_x + ny = 0$$

and are defined by the formulas

$$L_n^\alpha(x) = \frac{1}{n!} x^{-\alpha} e^x \frac{d^n}{dx^n} (x^{n+\alpha} e^{-x}) = \sum_{m=0}^n C_{n+\alpha}^{n-m} \frac{(-x)^m}{m!} = \sum_{m=0}^n \frac{\Gamma(n + \alpha + 1)}{\Gamma(m + \alpha + 1)} \frac{(-x)^m}{m!\,(n - m)!}.$$

Notation: $L_n^0(x) = L_n(x)$.
Special cases:

$$L_0^\alpha(x) = 1, \quad L_1^\alpha(x) = \alpha + 1 - x, \quad L_n^{-n}(x) = (-1)^n \frac{x^n}{n!}.$$

To calculate $L_n^\alpha(x)$ for $n \geq 2$, one can use the recurrence formulas

$$L_{n+1}^\alpha(x) = \frac{1}{n + 1} \left[(2n + \alpha + 1 - x)L_n^\alpha(x) - (n + \alpha)L_{n-1}^\alpha(x) \right].$$

Other recurrence formulas:

$$L_n^\alpha(x) = L_{n-1}^\alpha(x) + L_n^{\alpha-1}(x), \qquad \frac{d}{dx} L_n^\alpha(x) = -L_{n-1}^{\alpha+1}(x), \qquad x\frac{d}{dx} L_n^\alpha(x) = nL_n^\alpha(x) - (n+\alpha)L_{n-1}^\alpha(x).$$

The functions $L_n^\alpha(x)$ form an orthogonal system on the interval $0 < x < \infty$ with weight $x^\alpha e^{-x}$:

$$\int_0^\infty x^\alpha e^{-x} L_n^\alpha(x) L_m^\alpha(x)\, dx = \begin{cases} 0 & \text{if } n \neq m, \\ \frac{\Gamma(\alpha+n+1)}{n!} & \text{if } n = m. \end{cases}$$

The generating function is

$$(1-s)^{-\alpha-1} \exp\left(-\frac{sx}{1-s}\right) = \sum_{n=0}^\infty L_n^\alpha(x)s^n, \qquad |s| < 1.$$

M13.10.3. Chebyshev Polynomials

▶ **Chebyshev polynomials of the first kind.** The *Chebyshev polynomials of the first kind* $T_n(x)$ satisfy the second-order linear ordinary differential equation

$$(1-x^2)y''_{xx} - xy'_x + n^2 y = 0$$

and are defined by the formulas

$$T_n(x) = \cos(n \arccos x) = \frac{(-2)^n n!}{(2n)!} \sqrt{1-x^2}\, \frac{d^n}{dx^n}\left[(1-x^2)^{n-\frac{1}{2}}\right]$$

$$= \frac{n}{2} \sum_{m=0}^{[n/2]} (-1)^m \frac{(n-m-1)!}{m!\,(n-2m)!} (2x)^{n-2m} \qquad (n = 0, 1, 2, \dots),$$

where $[A]$ stands for the integer part of a number A.

An alternative representation of the Chebyshev polynomials:

$$T_n(x) = \frac{(-1)^n}{(2n-1)!!} (1-x^2)^{1/2} \frac{d^n}{dx^n} (1-x^2)^{n-1/2}.$$

The first five Chebyshev polynomials of the first kind are

$$T_0(x) = 1, \quad T_1(x) = x, \quad T_2(x) = 2x^2 - 1, \quad T_3(x) = 4x^3 - 3x, \quad T_4(x) = 8x^4 - 8x^2 + 1.$$

The recurrence formulas:

$$T_{n+1}(x) = 2xT_n(x) - T_{n-1}(x), \qquad n \geq 2.$$

The functions $T_n(x)$ form an orthogonal system on the interval $-1 < x < 1$ with weight $(1-x^2)^{-1/2}$:

$$\int_{-1}^1 \frac{T_n(x)T_m(x)}{\sqrt{1-x^2}}\, dx = \begin{cases} 0 & \text{if } n \neq m, \\ \frac{1}{2}\pi & \text{if } n = m \neq 0, \\ \pi & \text{if } n = m = 0. \end{cases}$$

The generating function is

$$\frac{1-sx}{1-2sx+s^2} = \sum_{n=0}^\infty T_n(x)s^n \qquad (|s| < 1).$$

The functions $T_n(x)$ have only real simple zeros, all lying on the interval $-1 < x < 1$.

The normalized Chebyshev polynomials of the first kind, $2^{1-n}T_n(x)$, deviate from zero least of all. This means that among all polynomials of degree n with the leading coefficient 1, the maximum of the modulus $\max_{-1 \leq x \leq 1} |2^{1-n}T_n(x)|$ has the least value, with the maximum being equal to 2^{1-n}.

▶ **Chebyshev polynomials of the second kind.** The *Chebyshev polynomials of the second kind* $U_n(x)$ satisfy the second-order linear ordinary differential equation

$$(1 - x^2)y''_{xx} - 3xy'_x + n(n+2)y = 0$$

and are defined by the formulas

$$U_n(x) = \frac{\sin[(n+1)\arccos x]}{\sqrt{1-x^2}} = \frac{2^n(n+1)!}{(2n+1)!}\frac{1}{\sqrt{1-x^2}}\frac{d^n}{dx^n}(1-x^2)^{n+1/2}$$

$$= \sum_{m=0}^{[n/2]}(-1)^m\frac{(n-m)!}{m!\,(n-2m)!}(2x)^{n-2m} \quad (n = 0, 1, 2, \dots).$$

The first five Chebyshev polynomials of the second kind are

$$U_0(x) = 1, \quad U_1(x) = 2x, \quad U_2(x) = 4x^2 - 1, \quad U_3(x) = 8x^3 - 4x, \quad U_4(x) = 16x^4 - 12x^2 + 1.$$

The recurrence formulas:

$$U_{n+1}(x) = 2xU_n(x) - U_{n-1}(x), \quad n \geq 2.$$

The generating function is

$$\frac{1}{1 - 2sx + s^2} = \sum_{n=0}^{\infty}U_n(x)s^n \quad (|s| < 1).$$

The Chebyshev polynomials of the first and second kinds are related by

$$U_n(x) = \frac{1}{n+1}\frac{d}{dx}T_{n+1}(x).$$

M13.10.4. Hermite Polynomials

▶ **Various representations of the Hermite polynomials.** The *Hermite polynomials* $H_n(x)$ satisfy the second-order linear ordinary differential equation

$$y''_{xx} - 2xy'_x + 2ny = 0$$

and are defined by the formulas

$$H_n(x) = (-1)^n\exp(x^2)\frac{d^n}{dx^n}\exp(-x^2) = \sum_{m=0}^{[n/2]}(-1)^m\frac{n!}{m!\,(n-2m)!}(2x)^{n-2m}.$$

The first five polynomials are

$$H_0(x) = 1, \quad H_1(x) = 2x, \quad H_2(x) = 4x^2 - 2, \quad H_3(x) = 8x^3 - 12x, \quad H_4(x) = 16x^4 - 48x^2 + 12.$$

Recurrence formulas:

$$H_{n+1}(x) = 2xH_n(x) - 2nH_{n-1}(x), \quad n \geq 2;$$

$$\frac{d}{dx}H_n(x) = 2nH_{n-1}(x).$$

Integral representation:

$$H_{2n}(x) = \frac{(-1)^n 2^{2n+1}}{\sqrt{\pi}}\exp(x^2)\int_0^{\infty}\exp(-t^2)t^{2n}\cos(2xt)\,dt,$$

$$H_{2n+1}(x) = \frac{(-1)^n 2^{2n+2}}{\sqrt{\pi}}\exp(x^2)\int_0^{\infty}\exp(-t^2)t^{2n+1}\sin(2xt)\,dt,$$

where $n = 0, 1, 2, \dots$

▶ **Orthogonality. The generating function. An asymptotic formula.** The functions $H_n(x)$ form an orthogonal system on the interval $-\infty < x < \infty$ with weight e^{-x^2}:

$$\int_{-\infty}^{\infty} \exp(-x^2) H_n(x) H_m(x)\, dx = \begin{cases} 0 & \text{if } n \neq m, \\ \sqrt{\pi}\, 2^n n! & \text{if } n = m. \end{cases}$$

Generating function:

$$\exp(-s^2 + 2sx) = \sum_{n=0}^{\infty} H_n(x) \frac{s^n}{n!}.$$

Asymptotic behavior as $n \to \infty$:

$$H_n(x) \approx 2^{\frac{n+1}{2}} n^{\frac{n}{2}} e^{-\frac{n}{2}} \exp(x^2) \cos\left(\sqrt{2n+1}\, x - \tfrac{1}{2}\pi n\right).$$

▶ **Hermite functions.** The *Hermite functions* $h_n(x)$ are introduced by the formula

$$h_n(x) = \exp\left(-\frac{1}{2}x^2\right) H_n(x) = (-1)^n \exp\left(\frac{1}{2}x^2\right) \frac{d^n}{dx^n} \exp(-x^2), \qquad n = 0, 1, 2, \dots$$

The Hermite functions satisfy the second-order linear ordinary differential equation

$$h''_{xx} + (2n + 1 - x^2)h = 0.$$

The functions $h_n(x)$ form an orthogonal system on the interval $-\infty < x < \infty$:

$$\int_{-\infty}^{\infty} h_n(x) h_m(x)\, dx = \begin{cases} 0 & \text{if } n \neq m, \\ \sqrt{\pi}\, 2^n n! & \text{if } n = m. \end{cases}$$

Bibliography for Chapter M13

Abramowitz, M. and Stegun, I. A. (Editors), *Handbook of Mathematical Functions with Formulas, Graphs and Mathematical Tables*, National Bureau of Standards Applied Mathematics Series, Washington, D.C., 1964.

Bateman, H. and Erdélyi, A., *Higher Transcendental Functions, Vol. 1 and Vol. 2*, McGraw-Hill, New York, 1953.

Bateman, H. and Erdélyi, A., *Higher Transcendental Functions, Vol. 3*, McGraw-Hill, New York, 1955.

Gradshteyn, I. S. and Ryzhik, I. M., *Tables of Integrals, Series, and Products*, Academic Press, New York, 1980.

Magnus, W., Oberhettinger, F., and Soni, R. P., *Formulas and Theorems for the Special Functions of Mathematical Physics, 3rd Edition*, Springer-Verlag, Berlin, 1966.

McLachlan, N. W., *Bessel Functions for Engineers*, Clarendon Press, Oxford, 1955.

Polyanin, A. D. and Zaitsev, V. F., *Handbook of Exact Solutions for Ordinary Differential Equations, 2nd Edition*, Chapman & Hall/CRC Press, Boca Raton, Florida, 2003.

Slavyanov, S. Yu. and Lay, W., *Special Functions: A Unified Theory Based on Singularities*, Oxford University Press, Oxford, 2000.

Weisstein, E. W., *CRC Concise Encyclopedia of Mathematics, 2nd Edition*, CRC Press, Boca Raton, Florida, 2003.

Zwillinger, D., *CRC Standard Mathematical Tables and Formulae, 31st Edition*, CRC Press, Boca Raton, Florida, 2002.

Chapter M14

Probability Theory

M14.1. Basic Concepts and Simplest Probabilistic Models

M14.1.1. Rules and Formulas of Combinatorics

Below are rules and formulas of combinatorics that are useful in solving problems of probability theory.

▶ **Rule of sum and rule of product.**

Rule of sum. Suppose there is a collection of objects (of arbitrary nature). If an object α can be chosen from this collection in m different ways and another object β can be chosen in k ways, then one can choose *either object α or object β* in $m + k$ different ways.

Rule of product. If an object α can be chosen in m ways and after that an object β can be chosen in k ways, then one can choose the *ordered pair of objects (α, β)* in mk ways.

Example 1. How many two-digit numbers are there consisting of different digits?

The first digit can be one of 1, 2, ..., 9 and the other, one of the eight remaining plus 0 (i.e., one of nine digits). So, by the rule of product, there are $9 \times 9 = 81$ two-digit numbers consisting of different digits.

▶ **Permutations and combinations.**

Given a set of n (distinct) elements, an arrangement of r elements from this set ($r \leq n$) taken in a certain order is called a *permutation*. The total number of different permutations is equal to

$$P_n^r = n(n-1)(n-2)\ldots(n-r+1) = \frac{n!}{(n-r)!},$$

which is also denoted $P_r^n = {}_nP_r = P(n, r)$.

Example 2. How many three-digit numbers can be made up from the six digits 0, 1, 2, 3, 4, 5 without repetitions?

Three out of the six digits (in a certain order) can be selected in P_6^3 ways. The triplets that start with 0 (i.e., the number of two-digit numbers consisting of 1, ..., 5) must be subtracted from this. So the desired number of three-digit numbers is $P_6^3 - P_5^2 = 6 \cdot 5 \cdot 4 - 5 \cdot 4 = 100$.

Note that the number of different permutations on a set of n elements (i.e., arrangements that differ from each other by only the order of selection of the n elements) is equal to

$$P_n^n = n! = 1 \cdot 2 \cdot 3 \ldots n.$$

Note that $0! = 1$.

Example 3. How many ways can four people be seated at a table?

This can be done in $P_4^4 = 4! = 1 \cdot 2 \cdot 3 \cdot 4 = 24$ ways.

An arrangement of r elements selected from a set of n elements where the order is not important is called a *combination*. The number of combinations is equal to

$$C_n^r = \frac{P_n^r}{P_r^r} = \frac{n(n-1)(n-2)\ldots(n-r+1)}{r!} = \frac{n!}{r!\,(n-r)!}$$

and is also denoted $C_r^n = {}_nC_r = C(n, r) = \binom{n}{r}$. Note that the relation $C_n^r = C_n^{n-r}$ holds.

Example 4. How many ways can three balls be selected from a box containing six numbered balls?

The desired number equals the number of combinations of three elements from a set of six elements, or $C_6^3 = \dfrac{6 \cdot 5 \cdot 4}{3 \cdot 2 \cdot 1} = 20$.

M14.1.2. Probabilities of Random Events

▶ **Random events. Basic definitions.** The simplest mutually exclusive outcomes of an experiment are called *elementary events* ω_i. The set of all elementary outcomes, which we denote by the symbol Ω, is called the *space of elementary events* (or the *sample space*). Any subset of Ω is called a *random event* A (or simply an *event* A). Event A is said to occur when the outcome of the experiment is one of the elementary events that make up A.

Example 1. Suppose the experiment is rolling a six-sided die once. There are six possible mutually exclusive outcomes and the space of elementary events is $\Omega = \{\omega_1, \omega_2, \omega_3, \omega_4, \omega_5, \omega_6\}$, where elementary event ω_k stands for rolling a k on the die. If event A is rolling an odd number on the die, then A occurs when the outcome is one of the elementary events $\omega_1, \omega_3, \omega_5$.

An event A *implies* an event B ($A \subseteq B$) if B occurs when A occurs. Events A and B are said to be *equivalent* ($A = B$) if A implies B and B implies A, i.e., if both events A and B occur or do not occur simultaneously.

The *union* $C = A \cup B = A + B$ of events A and B is the event that at least one of the events A or B occurs. The elementary outcomes of the union $A + B$ are the elementary outcomes that belong to at least one of the events A and B.

The *intersection* $C = A \cap B = AB$ of events A and B is the event that both A and B occur. The elementary outcomes of the intersection AB are the elementary outcomes that simultaneously belong to A and B.

The *difference* $C = A \backslash B = A - B$ of events A and B is the event that A occurs and B does not occur. The elementary outcomes of the difference $A \backslash B$ are the elementary outcomes of A that do not belong to B.

The event that A does not occur is called the *complement* of A, or the *complementary event*, and is denoted by $\overline{A}$. The elementary outcomes of $\overline{A}$ are the elementary outcomes that do not belong to event A.

The set Ω is called a *sure event* or a *certain event* (it will always happen). The empty set $\varnothing$ is called an *impossible event* (it will never happen).

Two complementary events A and $\overline{A}$ are said to be *opposite*; they simultaneously satisfy the following two conditions:

$$A \cup \overline{A} = \Omega, \quad A \cap \overline{A} = \varnothing.$$

Events A and B are said to be *incompatible*, or *mutually exclusive*, if their simultaneous realization is impossible, i.e., if $A \cap B = \varnothing$.

Events $H_1, \ldots, H_n$ are said to form a *complete group of events*, or to be *collectively exhaustive*, if at least one of them necessarily occurs for each trial of the experiment, i.e., if

$$H_1 \cup \cdots \cup H_n = \Omega.$$

Events $H_1, \ldots, H_n$ form a *complete group of mutually exclusive* or *pairwise incompatible events* if exactly one of the events necessarily occurs for each trial of the experiment, i.e., if

$$H_1 \cup \cdots \cup H_n = \Omega \quad \text{and} \quad H_i \cap H_j = \varnothing \quad (i \neq j).$$

Main properties of random events:

1. $A \cup B = B \cup A$ and $A \cap B = B \cap A$ (commutativity).

2. $(A \cup B) \cap C = (A \cap C) \cup (B \cap C)$ and $(A \cap B) \cup C = (A \cup C) \cap (B \cup C)$ (distributivity).
3. $(A \cup B) \cup C = A \cup (B \cup C)$ and $(A \cap B) \cap C = A \cap (B \cap C)$ (associativity).
4. $A \cup A = A$ and $A \cap A = A$.
5. $A \cup \Omega = \Omega$ and $A \cap \Omega = A$.
6. $A \cup \overline{A} = \Omega$ and $A \cap \overline{A} = \varnothing$.
7. $\overline{\varnothing} = \Omega$, $\overline{\Omega} = \varnothing$, and $\overline{\overline{A}} = A$.
8. $A \setminus B = A \cap \overline{B}$.
9. $\overline{A \cup B} = \overline{A} \cap \overline{B}$ and $\overline{A \cap B} = \overline{A} \cup \overline{B}$ (*de Morgan's laws*).

▶ **Discrete probability space. Classical definition of probability.** Suppose that $\Omega = \{\omega_1, \ldots, \omega_n\}$ is a finite sample space. To each elementary event $\omega_i \in \Omega$ $(i = 1, 2, \ldots, n)$ there corresponds a number $p(\omega_i)$, called the probability of the elementary event ω_i. Thus a real function satisfying the following two conditions is defined on the set Ω:

1. *Nonnegativity condition:* $p(\omega_i) \geq 0$ for any $\omega_i \in \Omega$.
2. *Normalization condition:* $\sum_{i=1}^{n} p(\omega_i) = 1$.

The *probability* $P(A)$ of an event A for any subset $A \subset \Omega$ is defined to be the sum of the probabilities of the elementary events that form A; i.e.,

$$P(A) = \sum_{\omega_i \in A} p(\omega_i). \tag{M14.1.2.1}$$

Note that the relations $0 \leq P(A) \leq 1$, $P(\varnothing) = 0$, and $P(\Omega) = 1$ are always valid.

A special case of the definition of probability (M14.1.2.1) is the *classical definition of probability*, in which all elementary events are equally likely: $p(\omega_1) = \cdots = p(\omega_n) = 1/n$. Then the probability of the event A is

$$P(A) = \frac{m}{n}, \tag{M14.1.2.2}$$

where m is the number of elementary events making up A (the number *favorable* outcomes).

Example 2. In example 1, the probability of rolling an odd number is
$$P(A) = \tfrac{3}{6} = \tfrac{1}{2}.$$

Example 3. Let two dice be rolled. Under the assumption that the elementary events are equiprobable, find the probability of the event A that the sum of the numbers rolled is greater than 10. Obviously, the sample space can be represented as $\Omega = \{(i, j) : i, j = 1, 2, 3, 4, 5, 6\}$, where i is the number shown by the first die and j is that shown by the second die. The total number of elementary events is $|\Omega| = 36$. Event A is represented by the subset $A = \{(5, 6), (6, 5), (6, 6)\}$ of Ω. Since $|A| = 3$, formula (M14.1.2.2) gives $P(A) = |A|/|\Omega| = 3/36 = 1/12$.

▶ **Statistical definition of probability.** Suppose an experiment has been repeated N times and an event A has occurred k times. The ratio k/N is called the *relative frequency* of the event A in the given series of experiments. The number about which the relative frequency oscillates in long series of experiments is called the probability $P(A)$ of the event A.

▶ **Geometric definition of probability.** In the geometric approach to defining probability, one takes an arbitrary set on a straight line, in a plane, or in space to be the sample space Ω. A trial of an experiment is interpreted as a random selection of a point in Ω. The occurrence of an event A is treated as the occurrence of the point in a subdomain A of Ω. It is assumed that the probability of selecting a point in the domain A is proportional to the measure of A (i.e., to its length, area, or volume) and is independent of the position and shape of the domain. The probability of the event A is defined as

$$P(A) = \frac{\text{measure } A}{\text{measure } \Omega}.$$

Example 3. A point is randomly dropped inside a circle of radius $R = 1$. Find the probability of the event that the point lands in the circle of radius $r = \frac{1}{2}$ concentric with the larger one.

Let A be the event that the point lands in the smaller circle. We find the probability $P(A)$ as the ratio of the area of the smaller circle to that of the larger one:

$$P(A) = \frac{\pi r^2}{\pi R^2} = \frac{1}{4}.$$

▶ **Axiomatic definition of probability.** The *probability of an event* is defined to be a single-valued real function $P(A)$ satisfying the following three axioms:
1. *Nonnegativity*: $P(A) \geq 0$ for any $A \in \Omega$.
2. *Normalization*: $P(\Omega) = 1$.
3. *Additivity*: $P\left(\bigcup_n A_n\right) = \sum_n P(A_n)$, provided that $A_i \cap A_j = \varnothing$ whenever $i \neq j$.

Properties of probability:
1. The probability of an impossible event is zero; i.e., $P(\varnothing) = 0$.
2. The probability of the event $\overline{A}$ opposite to an event A is equal to $P(\overline{A}) = 1 - P(A)$.
3. Probability is a bounded function; i.e., $0 \leq P(A) \leq 1$.
4. If an event A implies an event B ($A \subseteq B$), then $P(A) \leq P(B)$.
5. If events $A_1, \ldots, A_n$ form a complete group of pairwise incompatible events, then
$$\sum_{i=1}^n P(A_i) = 1.$$

M14.1.3. Conditional Probability. Total Probability Formula

▶ **Probability of the union of events.** The probability of realization of at least one of two events H_1 and H_2 is given by the formula (addition theorem)

$$P(H_1 \cup H_2) = P(H_1) + P(H_2) - P(H_1 \cap H_2).$$

In particular, for $H_1 \cap H_2 = \varnothing$, we have

$$P(H_1 \cup H_2) = P(H_1) + P(H_2).$$

The probability of realization of at least one of n events is given by the formula

$$P(H_1 \cup \cdots \cup H_n) = \sum_{k=1}^n P(H_k) - \sum_{1 \leq k_1 < k_2 \leq n} P(H_{k_1} \cap H_{k_2})$$

$$+ \sum_{1 \leq k_1 < k_2 < k_3 \leq n} P(H_{k_1} \cap H_{k_2} \cap H_{k_3}) - \cdots + (-1)^{n-1} P(H_1 \cap \cdots \cap H_n). \quad \text{(M14.1.3.1)}$$

Note that $P(H_1 \cup H_2)$ is also written $P(H_1 + H_2)$.

For n pairwise incompatible events $H_1, H_2, \ldots, H_n$ ($H_i \cap H_j = \varnothing$ for $i \neq j$), formula (M14.1.3.1) simplifies to become

$$P(H_1 \cup \cdots \cup H_n) = P(H_1) + P(H_2) + \cdots + P(H_n).$$

It may often be convenient to calculate the probability of the sum of events via the probability of the product of opposite events:

$$P(H_1 \cup \cdots \cup H_n) = 1 - P(\bar{H}_1 \cap \bar{H}_2 \cap \cdots \cap \bar{H}_n).$$

▶ **Conditional probability.** The *conditional probability* $P(A|H)$ of an event A given the occurrence of some other event H is defined by the formula

$$P(A|H) = \frac{P(A \cap H)}{P(H)}, \quad P(H) > 0. \tag{M14.1.3.2}$$

Relation (M14.1.3.2) can be rewritten (multiplication theorem)

$$P(A \cap H) = P(H)\, P(A|H). \tag{M14.1.3.3}$$

The probability $P(A \cap H)$ is also written $P(AH)$.

The formula

$$P(A_1 \cap \cdots \cap A_n) = P(A_1)\, P(A_2|A_1)\, P(A_3|A_1 \cap A_2) \ldots P(A_n|A_1 \cap \cdots \cap A_{n-1})$$

is a generalization of (M14.1.3.3).

Example 1. Two guns shoot one and the same target independently, once each. The probability for the first gun to hit the target is 0.8 and that for the second gun is 0.9. It is required to find:

a) the probability that the target is hit only once and
b) the probability that the target is hit at least once.

a) Let A and B denote the events that the first and second guns hit the target, respectively. Then the probability that the target is hit only once is $P(A\bar{B} + \bar{A}B) = P(A\bar{B}) + P(\bar{A}B) = P(A)P(\bar{B}) + P(\bar{A})P(B) = 0.8\,(1 - 0.9) + (1 - 0.8)\,0.9 = 0.26$.

b) The probability of at least one hit is $P(A + B) = P(A) + P(B) - P(A)P(B) = 0.8 + 0.9 - 0.8 \cdot 0.9 = 0.98$. Note that this result is usually obtained as follows: $P(A+B) = 1 - P(\bar{A}\bar{B}) = 1 - P(\bar{A})P(\bar{B}) = 1 - 0.2 \cdot 0.1 = 0.98$.

▶ **Independence of events.** Two random events A and B are said to be *independent* if the conditional probability of A, given the occurrence of B, coincides with the unconditional probability of A,

$$P(A|B) = P(A).$$

In this case, $P(A \cap B) = P(A)P(B)$.

Random events $A_1, \ldots, A_n$ are jointly independent if the relation

$$P\left(\bigcap_{k=1}^{m} A_{i_k}\right) = \prod_{k=1}^{m} P(A_{i_k})$$

holds whenever $1 \le i_1 < \cdots < i_m \le n$ and $m \le n$.

The pairwise independence of the events A_i and A_j for all $i \ne j$ $(i, j = 1, 2, \ldots, n)$ does not imply that the events $A_1, \ldots, A_n$ are jointly independent.

Example 2. Suppose that the experiment is to draw one of four balls. Let three of them be labeled by the numbers 1, 2, and 3, and let the fourth ball bear all these numbers. By A_i $(i = 1, 2, 3)$ we denote the event that the chosen ball bears the number i. Are the events A_1, A_2, and A_3 dependent?

Since each number is encountered twice, $P(A_1) = P(A_2) = P(A_3) = 1/2$. Since any two distinct numbers are present only on one of the balls, we have $P(A_1 A_2) = P(A_2 A_3) = P(A_1 A_3) = 1/4$, and hence the events A_1, A_2, and A_3 are pairwise independent. All three distinct numbers are present only on one of the balls, and $P(A_1 A_2 A_3) = 1/4 \ne P(A_1)\, P(A_2)\, P(A_3) = 1/8$.

Thus we see that the events A_1, A_2, and A_3 are jointly dependent, even though they are pairwise independent.

▶ **Total probability formula. Bayes's formula.** Suppose that a group of pairwise incompatible events $H_1, \ldots, H_n$ is given and their probabilities $P(H_1), \ldots, P(H_n)$ as well as the conditional probabilities $P(A|H_1), \ldots, P(A|II_n)$ of an event A are known. Then the probability of A can be determined by the *total probability formula*

$$P(A) = \sum_{k=1}^{n} P(H_k)\, P(A|H_k). \tag{M14.1.3.4}$$

The events H_k are called *hypotheses* with respect to the event A.

Example 3. There are two urns: the first urn contains a white balls and b black ones, and the second urn contains c white balls and d black ones. We take one ball from the first urn and put it into the second urn. After this, we draw one ball from the second urn. Find the probability of the event that this ball is white.

Let A be the event of drawing a white ball. Consider the following group of events:

H_1, a white ball is taken from the first urn and put into the second urn.

H_2, a black ball is taken from the first urn and put into the second urn. Obviously,

$$P(H_1) = \frac{a}{a+b}, \quad P(H_2) = \frac{b}{a+b}; \quad P(A|H_1) = \frac{c+1}{c+d+1}, \quad P(A|H_2) = \frac{c}{c+d+1}.$$

Now by the total probability formula (M14.1.3.4) we obtain

$$P(A) = P(H_1)\,P(A|H_1) + P(H_2)\,P(A|H_2) = \frac{a}{a+b}\frac{c+1}{c+d+1} + \frac{b}{a+b}\frac{c}{c+d+1}.$$

If it is known that the event A has occurred as a result of one trial but it is unknown which of the events $H_1, \ldots, H_n$ has occurred, then the conditional probabilities of the hypotheses H_k (under the condition that the event A has occurred) are calculated using *Bayes's formula*:

$$P(H_k|A) = \frac{P(H_k)\,P(A|H_k)}{P(A)} \qquad (k = 1, 2, \ldots, n).$$

Example 4. A box contains same parts made by two automats, 40% produced by the first automat and the rest by the second. It is known that 3% of the first automat's products are faulty, while the percentage of faulty products made by the second automat is 2%. Find

a) the probability that a randomly chosen part will be faulty and

b) the probability that a randomly chosen part was made by the first automat if found faulty.

Let A denote the event that a randomly chosen part is faulty and let H_1 and H_2 denote the events that a part is made by the first and second automat, respectively. Then

a) the total probability formula gives

$$P(A) = P(H_1)\,P(A|H_1) + P(H_2)\,P(A|H_2) = 0.4 \cdot 0.03 + (1 - 0.4) \cdot 0.02 = 0.024;$$

b) Bayes's formula gives

$$P(H_1|A) = \frac{P(H_1)\,P(A|H_1)}{P(A)} = \frac{0.4 \cdot 0.03}{0.024} = 0.5.$$

M14.1.4. Sequence of Trials

▶ **Bernoulli process.** In this case, some event A occurs with probability $p = P(A)$ (the probability of "success") and does not occur with probability $q = P(\overline{A}) = 1 - P(A) = 1 - p$ (the probability of "failure") in each trial. The probability that in n independent trials, the event A ("success") occurs exactly k times is determined by the *Bernoulli formula (binomial distribution)*:

$$P_n(k) = C_n^k p^k (1-p)^{n-k} \qquad (k = 0, 1, \ldots, n). \tag{M14.1.4.1}$$

Note that the relation $\sum_{k=0}^{n} P_n(k) = 1$ holds.

Example 1. Find the probability that there will be 5 heads in tossing a coin 10 times.

We have $n = 10$, $k = 5$, and $p = \frac{1}{2}$. Then, by the Bernoulli formula, the desired probability is $P_{10}(5) = C_{10}^5 \left(\frac{1}{2}\right)^5 \left(\frac{1}{2}\right)^{10-5} = \frac{10 \cdot 9 \cdot 8 \cdot 7 \cdot 6}{5 \cdot 4 \cdot 3 \cdot 2 \cdot 1} \cdot \frac{1}{2^{10}} = \frac{63}{256}$.

Example 2 (Banach's problem).   A smoker mathematician has two matchboxes on him, each of which initially contains exactly n matches. Each time he needs to light a cigarette, he selects a matchbox at random. Find the probability of the event that as the mathematician takes out an empty box for the first time, precisely k matches will be left in the other box ($k \leq n$).

The mathematician has taken matches $2n - k$ times, n out of them from the box that is eventually empty. This scheme corresponds to the scheme of $2n - k$ independent Bernoulli trials with n "successes." The probability of a "success" in a single trial is equal to 0.5. The desired probability can be found by the formula

$$P_{2n-k}(n) = C_{2n-k}^n p^n (1-p)^{n-k} = C_{2n-k}^n \left(\tfrac{1}{2}\right)^{2n-k}.$$

The probability that the event occurs at least m times in n independent trials is calculated by the formula

$$P_n(k \geq m) = \sum_{k=m}^{n} P_n(k) = 1 - \sum_{k=0}^{m-1} P_n(k).$$

The probability that the event occurs at least once in n independent trials is calculated by the formula

$$P_n(k \geq 1) = 1 - (1-p)^n.$$

The number n of independent trials necessary for the event to occur at least once with probability at least P is given by the formula

$$n \geq \frac{\ln(1-P)}{\ln(1-p)}.$$

▶ **Limit formulas for Bernoulli process.** It is very difficult to use Bernoulli's formula (M14.1.4.1) for large n an k. In this case, one has to use approximate formulas for calculating $P_n(k)$ with desired accuracy.

Poisson formula. If the number of independent trials increases unboundedly ($n \to \infty$) and the probability p simultaneously goes to zero ($p \to 0$) so that their product np is a constant ($np = \lambda = \text{const}$), then the probability $P_n(k)$ satisfies the limit relation

$$\lim_{n \to \infty} P_n(k) = \frac{\lambda^k}{k!} e^{-\lambda}. \tag{M14.1.4.2}$$

Local de Moivre–Laplace theorem. Suppose that $n \to \infty$, $p = \text{const}$, $0 < p < 1$, and $0 < c_1 \leq x_{n,k} = (k - np)[np(1-p)]^{-1/2} \leq c_2 < \infty$; then

$$P_n(k) = \frac{1}{\sqrt{2\pi np(1-p)}} \exp\left[-\frac{(k-np)^2}{2np(1-p)}\right]\left[1 + O(1/\sqrt{n}\,)\right] \tag{M14.1.4.3}$$

uniformly with respect to $x_{n,k} \in [c_1, c_2]$.

Integral de Moivre–Laplace theorem. Suppose that $n \to \infty$ and $p = \text{const}$, $0 < p < 1$. The probability that the number of successes k in n independent trials is between k_1 and k_2 can be approximately calculated by the formula

$$P_n\{k_1 \leq k \leq k_2\} \approx \Phi\left(\frac{k_2 - np}{\sqrt{np(1-p)}}\right) - \Phi\left(\frac{k_1 - np}{\sqrt{np(1-p)}}\right). \tag{M14.1.4.4}$$

Here

$$\Phi(x) = \frac{1}{\sqrt{2\pi}} \int_{-\infty}^{x} \exp\left(-\frac{t^2}{2}\right) dt$$

is the cumulative distribution function of the standard normal distribution (also known as the Laplace function), which is tabulated.

The approximate formula (M14.1.4.2) is normally used for $n \geq 50$ and $np \leq 10$. The approximate formulas (M14.1.4.3) and (M14.1.4.4) normally are used for $np(1-p) > 9$.

▶ **Sequence of n independent trials.** In a series of trials, the trials are said to be *independent* if the occurring events are independent.

Let $p_k = P(A_k)$. Then the probability of the event that n_1 events A_1, n_2 events A_2, ..., and n_k events A_k occur in n independent trials is equal to

$$P_n(n_1, \ldots, n_k) = \frac{n!}{n_1! \ldots n_k!} \, p_1^{n_1} \ldots p_k^{n_k}. \tag{M14.1.4.5}$$

Remark. The probability (M14.1.4.5) is the coefficient of $x_1^{n_1} \ldots x_k^{n_k}$ in the expansion of the polynomial $(p_1 x_1 + \cdots + p_k x_k)^n$ in powers of $x_1, \ldots, x_k$.

M14.2. Random Variables and Their Characteristics

M14.2.1. One-Dimensional Random Variables

▶ **Notion of a random variable. The distribution function of a random variable.** Let $\Omega = \{\omega\}$ be the space of elementary events. A *random variable* X is a real number function $X = X(\omega)$ defined on the set Ω.

Any rule (table, function, graph, or otherwise) that permits one to find the probabilities of events is usually called the distribution law of a random variable. In general, random variables can be *discrete* or *continuous*.

The *cumulative distribution function* of a random variable X is the function $F(x)$ whose value at every point x is equal to the probability of the event $\{X < x\}$:

$$F(x) = P(X < x).$$

Properties of the cumulative distribution function:
1. $0 \le F(x) \le 1$.
2. $\lim\limits_{x \to -\infty} F(x) = F(-\infty) = 0$, $\lim\limits_{x \to +\infty} F(x) = F(+\infty) = 1$.
3. If $x_2 > x_1$, then $F(x_2) \ge F(x_1)$.
4. $P(x_1 \le X < x_2) = F(x_2) - F(x_1)$.
5. $F(x)$ is continuous from the left, i.e., $\lim\limits_{x \to x_0 - 0} F(x) = F(x_0)$.

▶ **Discrete random variables.** Let X be a *discrete* random variable assuming the values x_1, x_2, ..., x_n, ... with probabilities p_1, p_2, ..., p_n, ..., so its distribution law can be defined by the table

X	x_1	x_2	...	x_n	...
P	p_1	p_2	...	p_n	...

$$\left(\sum_i p_i = 1 \right).$$

(Here and in what follows, it is assumed that the values of a discrete random variable X are arranged in ascending order, so that $x_1 < x_2 < \cdots < x_{k-1} < x_k < \cdots$.)

In this case, the cumulative distribution function of a discrete random variable X is the step function defined as the sum

$$F(x) = P\{X < x\} = \sum_{x_n < x} p_n.$$

▶ **Continuous random variables. The probability density function.** A *continuous random variable* is defined by either a distribution function $F(x)$ or a *probability density function* $p(x)$, which are related by

$$p(x) = F'(x), \quad F(x) = \int_{-\infty}^{x} p(z) \, dz.$$

Sometimes, the distribution function and probability density function of a random variable X are denoted $F_X(x)$ and $p_X(x)$, respectively, rather than $F(x)$ and $p(x)$.

Properties of the probability density function:

1. $p(x) \geq 0$.
2. $p(-\infty) = p(+\infty) = 0$.
3. $\int_{-\infty}^{+\infty} p(x)\, dx = 1$.
4. $P\{a \leq X \leq b\} = \int_a^b p(x)\, dx$.

The differential $dF(x) = p(x)\, dx \approx P(x \leq X \leq x + dx)$ is called a *probability element*.

Remark. For continuous random variables, one always has $P(X = x_0) = 0$, but the event $\{X = x_0\}$ is not necessarily impossible.

▶ **Functions of random variables.** Suppose that a random variable Y is related to a random variable X by a functional dependence $Y = \varphi(X)$. If X is discrete, then, obviously, Y is also discrete. To find the distribution law of the random variable Y, it suffices to calculate the values $\varphi(x_i)$. If there are repeated values among $y_i = \varphi(x_i)$, then these repeated values are taken into account only once, the corresponding probabilities being added.

If X is a continuous random variable with probability density function $p(x)$, then, in general, the random variable Y is also continuous. The cumulative distribution function of Y is given by the formula

$$F_Y(y) = P(Y < y) = P(\varphi(X) < y) = \int_{\varphi(x) < y} p(x)\, dx. \tag{M14.2.1.1}$$

If the function $y = \varphi(x)$ is differentiable and monotone on the entire range of the argument x, then the probability density function $p_Y(y)$ of the random variable Y is given by the formula

$$p_Y(y) = p(g(y))\, |g_y'(y)|,$$

where g is the inverse of $\varphi(x)$.

Example 1. Suppose that a random variable X has the probability density

$$p(x) = \frac{1}{\sqrt{2\pi}} e^{-x^2/2}.$$

Find the distribution of the random variable $Y = X^2$.

In this case, $y = \varphi(x) = x^2$. According to (M14.2.1.1), we obtain

$$F_Y(y) = \int_{x^2 < y} \frac{1}{\sqrt{2\pi}} e^{-x^2/2}\, dx = \frac{1}{\sqrt{2\pi}} \int_{-\sqrt{y}}^{\sqrt{y}} e^{-x^2/2}\, dx = \frac{2}{\sqrt{2\pi}} \int_0^{\sqrt{y}} e^{-x^2/2}\, dx = \frac{1}{\sqrt{2\pi}} \int_0^y \frac{e^{-t/2}}{\sqrt{t}}\, dt.$$

Example 2. Suppose that a random variable X has the probability density

$$p(x) = \frac{1}{\sqrt{2\pi}\sigma} \exp\left[-\frac{(x-a)^2}{2\sigma^2}\right].$$

Find the probability density of the random variable $Y = e^X$.

For $y > 0$, the cumulative distribution function of the random variable $Y = e^X$ is determined by the relations

$$F_Y(y) = P(Y < y) = P(e^X < y) = P(X < \ln y) = F(\ln y).$$

We differentiate this relation and obtain

$$f_Y(y) = \frac{dF_Y(y)}{dy} = \frac{dF(\ln y)}{dy} = p(\ln y)\frac{1}{y} = \frac{1}{\sqrt{2\pi}\sigma y} \exp\left[-\frac{(\ln y - a)^2}{2\sigma^2}\right] \quad \text{for } y > 0.$$

The distribution of Y is called the *log-normal distribution*.

Example 3. Suppose that a random variable X has the probability density $p(x)$ for $x \in (-\infty, \infty)$. Then the probability density of the random variable $Y = |X|$ is given by the formula $p_Y(y) = p(x) + p(-x)\ (y \geq 0)$. In particular, if X is symmetric (i.e., the condition $P(X < -x) = P(X > x)$ holds for all x), then $p_Y(y) = 2p(x)$ $(y \geq 0)$.

M14.2.2. Expectation, Variance and Moments of a Random Variable

▶ **Expectation.** The *expectation* (*expected value*) $E\{X\}$ *of a discrete or continuous random variable* X is the expression given by the formula

$$E\{X\} = \begin{cases} \sum_i x_i p_i & \text{in the discrete case,} \\ \int_{-\infty}^{+\infty} x p(x)\, dx & \text{in the continuous case.} \end{cases} \tag{M14.2.2.1}$$

For the existence of the expectation (M14.2.2.1), it is necessary that the corresponding series or integral converge absolutely.

The expectation is the main characteristic defining the "position" of a random variable, i.e., the number near which its possible values are concentrated.

▶ **Expectation of function of random variable.** If a random variable Y is related to a random variable X by a functional dependence $Y = \varphi(X)$, then the expectation of the random variable $Y = \varphi(X)$ can be determined by two methods. The first method is to construct the distribution of the random variable Y and then use already known formulas to find $E\{Y\}$. The second method is to use the formulas

$$E\{Y\} = E\{\varphi(X)\} = \begin{cases} \sum_i \varphi(x_i) p_i & \text{in the discrete case,} \\ \int_{-\infty}^{+\infty} \varphi(x) p(x)\, dx & \text{in the continuous case} \end{cases}$$

if these expressions exist in the sense of absolute convergence.

Example. Suppose that a random variable X is uniformly distributed in the interval $(-\pi/2, \pi/2)$, i.e., $p(x) = 1/\pi$ for $x \in (-\pi/2, \pi/2)$ and $p(x) = 0$ for $|x| > \pi/2$. Then the expectation of the random variable $Y = \sin(X)$ is equal to

$$E\{Y\} = \int_{-\infty}^{+\infty} \varphi(x) p(x)\, dx = \int_{-\pi/2}^{\pi/2} \frac{1}{\pi} \sin x\, dx = 0.$$

Properties of the expectation:

1. $E\{a\} = a$ for any real number $a = \text{const.}$
2. $E\{\alpha X + \beta Y\} = \alpha E\{X\} + \beta E\{Y\}$ for any real α and β.
3. $E\{XY\} = E\{X\}E\{Y\}$ for independent random variables X and Y.
4. $|E\{X\}| \le E\{|X|\}$.
5. $E\{X\} \le E\{Y\}$ if $X(\omega) \le Y(\omega)$, $\omega \in \Omega$.
6. $g(E\{X\}) \le E\{g(X)\}$ for convex functions $g(X)$.
7. The *Cauchy–Schwarz inequality* $(E\{|XY|\})^2 \le (E\{X\})^2 (E\{Y\})^2$ holds.
8. $E\left\{\sum_{k=1}^{\infty} X_k\right\} = \sum_{k=1}^{\infty} E\{X_k\}$ if the series $\sum_{k=1}^{\infty} E\{|X_k|\}$ converges.

▶ **Moments.** The expectation $E\{(X - a)^k\}$ is called the *kth moment* of the random variable X about the number a. The moments about zero are usually referred to simply as the moments of a random variable. (Sometimes they are called *initial moments*.) The kth moment satisfies the relation

$$\alpha_k = E\{X^k\} = \begin{cases} \sum_i x_i^k p_i & \text{in the discrete case,} \\ \int_{-\infty}^{+\infty} x^k p(x)\, dx & \text{in the continuous case.} \end{cases}$$

If $a = E\{X\}$, then the kth moment of the random variable X about a is called the *kth central moment*. The kth central moment satisfies the relation

$$\mu_k = E\{(X - E\{X\})^k\} = \begin{cases} \displaystyle\sum_i (x_i - E\{X\})^k p_i & \text{in the discrete case,} \\ \displaystyle\int_{-\infty}^{+\infty} (x - E\{X\})^k p(x)\, dx & \text{in the continuous case.} \end{cases}$$

In particular, $\mu_0 = 1$ and $\mu_1 = 0$ for any random variable.

The number $m_k = E\{|X - a|^k\}$ is called the *kth absolute moment* of X about a.

The existence of a kth moment α_k or μ_k implies the existence of the moments α_m and μ_m of all orders $m < k$; if the integral (or series) for α_k or μ_k diverges, then all integrals (series) for α_m and μ_m of orders $m > k$ also diverge.

There is a simple relationship between the central and initial moments:

$$\mu_k = \sum_{m=0}^{k} C_k^m \alpha_m (\alpha_1)^{k-m}, \quad \alpha_0 = 1; \quad \alpha_k = \sum_{m=0}^{k} C_k^m \mu_m (\alpha_1)^{k-m}.$$

The probability distribution is uniquely determined by the moments $\alpha_0, \alpha_1, \dots$ provided that they all exist and the series $\sum_{m=0}^{\infty} |\alpha_m| t^m / m!$ converges for some $t > 0$.

▶ **Variance.** The *variance* of a random variable, $\text{Var}\{X\}$, is the measure of the deviation of a random variable X from its expectation $E\{X\}$, determined by the relation

$$\text{Var}\{X\} = E\{(X - E\{X\})^2\}.$$

The variance $\text{Var}\{X\}$ is the second central moment of the random variable X. The variance can be determined by the formulas

$$\text{Var}\{X\} = \begin{cases} \displaystyle\sum_i (x_i - E\{X\})^2 p_i & \text{in the discrete case,} \\ \displaystyle\int_{-\infty}^{+\infty} (x - E\{X\})^2 p(x)\, dx & \text{in the continuous case.} \end{cases}$$

The variance characterizes the spread in values of the random variable X about its expectation.

Properties of the variance:

1. $\text{Var}\{a\} = 0$ for any real number $a = \text{const}$.
2. The variance is nonnegative: $\text{Var}\{X\} \geq 0$.
3. $\text{Var}\{X\} = E\{X^2\} - (E\{X\})^2$.
4. $\text{Var}\{\alpha X + \beta\} = \alpha^2 \text{Var}\{X\}$ for any real numbers α and β.
5. $\text{Var}\{X \pm Y\} = \text{Var}\{X\} + \text{Var}\{Y\}$ for independent random variables X and Y.
6. If X and Y are independent random variables, then

$$\text{Var}\{XY\} = \text{Var}\{X\}\text{Var}\{Y\} + \text{Var}\{X\}(E\{Y\})^2 + \text{Var}\{Y\}(E\{X\})^2.$$

7. $\min_m E\{(X - m)^2\} = \text{Var}\{X\}$ and is attained for $m = E\{X\}$.

▶ **Other numerical characteristics of random variables.** The *standard deviation (root mean square deviation)* of a random variable X is the square root of its variance,

$$\sigma = \sqrt{\text{Var}\{X\}}.$$

The standard deviation has the same dimension as the random variable itself.

The *coefficient of variation* is the ratio of the standard deviation to the expected value,

$$v = \frac{\sigma}{E\{X\}}.$$

The *asymmetry coefficient*, or *skewness*, is defined by the formula

$$\gamma_1 = \frac{\mu_3}{(\mu_2)^{3/2}}.$$

If $\gamma_1 > 0$, then the distribution curve is more flattened to the right of the mode Mode$\{X\}$; if $\gamma_1 < 0$, then the distribution curve is more flattened to the left of the mode Mode$\{X\}$ (see Fig. M14.1).

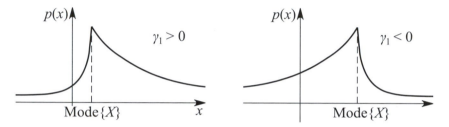

Figure M14.1. Relationship between the distribution curve and the asymmetry coefficient.

The *excess coefficient*, or *excess*, or *kurtosis*, is defined by the formula

$$\gamma_2 = \frac{\mu_4}{\mu_2^2} - 3.$$

One says that for $\gamma_2 = 0$ the distribution has a normal excess, for $\gamma_2 > 0$ the distribution has a positive excess, and for $\gamma_2 < 0$ the distribution has a negative excess.

Remark. The coefficients γ_1^2 and $\gamma_2 + 3$ or $(\gamma_2 + 3)/2$ are often used instead of γ_1 and γ_2.

A *mode* Mode$\{X\}$ *of a continuous probability distribution* is a point of local maximum of the probability density function $p(x)$. A *mode of a discrete probability distribution* is a value preceded and followed by values having probabilities smaller than $p(\text{Mode}\{X\})$.

Distributions with one, two, or more modes are said to be *unimodal*, *bimodal*, or *multimodal*, respectively.

M14.2.3. Main Discrete Distributions and Their Numerical Characteristics

▶ **Binomial distribution.** A random variable X has a *binomial distribution* with parameters n and p if

$$P_n(k) = P(X = k) = C_n^k p^k (1-p)^{n-k}, \quad k = 0, 1, \dots, n,$$

where $0 < p < 1$ and $n \geq 1$.

The cumulative distribution function has the form

$$
F(x) = \begin{cases} 1 & \text{for } x > n, \\ \sum_{k=1}^{m} C_n^k p^k (1-p)^{n-k} & \text{for } m \le x < m+1 \ (m = 1, 2, \dots, n-1), \\ 0 & \text{for } x < 0, \end{cases}
$$

and the numerical characteristics are given by the formulas

$$
E\{X\} = np, \quad \text{Var}\{X\} = np(1-p), \quad \gamma_1 = \frac{1-2p}{\sqrt{np(1-p)}}, \quad \gamma_2 = \frac{1-6p(1-p)}{np(1-p)}.
$$

The binomial distribution is a model of random experiments consisting of n independent identical Bernoulli trials. If $X_1, \dots, X_n$ are independent random variables, each of which can take only two values 1 or 0 with probabilities p and $q = 1-p$, respectively, then the random variable $X = \sum_{k=1}^{n} X_k$ has the binomial distribution with parameters (n, p).

See also the limit formulas (M14.1.4.2)–(M14.1.4.4) for the Bernoulli process.

▶ **Geometric distribution.** A random variable X has a *geometric distribution* with parameter p $(0 < p < 1)$ if

$$
P(X = k) = p(1-p)^k, \quad k = 0, 1, 2, \dots
$$

Numerical characteristics for the geometric distribution can be calculated by the formulas

$$
E\{X\} = \frac{1-p}{p}, \quad \text{Var}\{X\} = \frac{1-p}{p^2}, \quad \alpha_2 = \frac{(1-p)(2-p)}{p^2}, \quad \gamma_1 = \frac{2-p}{\sqrt{1-p}}, \quad \gamma_2 = 6 + \frac{p^2}{1-p}.
$$

The geometric distribution has a maximum at $k = 0$ and decreases monotonically with increasing k.

The geometric distribution describes a random variable X equal to the number of failures before the first success in a sequence of Bernoulli trials with probability p of success in each trial.

The geometric distribution is the only discrete distribution that is *memoryless*, i.e., satisfies the relation

$$
P(X > t + s \mid X > t) = P(X > s)
$$

for any $s > 0$ and $t > 0$. This property permits one to view the geometric distribution as the discrete analogue of the exponential distribution (see Subsection M14.2.4).

▶ **Hypergeometric distribution.** A random variable X has a *hypergeometric distribution* with parameters N, p, and n if

$$
P(X = k) = \frac{C_{Np}^k C_{N(1-p)}^{n-k}}{C_N^n}, \quad k = 0, 1, \dots, n,
$$

where $0 < p < 1$, $0 \le n \le N$, and $N > 0$.

The numerical characteristics are given by the formulas

$$
E\{X\} = np, \quad \text{Var}\{X\} = \frac{N-n}{N-1} np(1-p).
$$

A typical scheme in which the hypergeometric distribution arises is as follows: n elements are randomly drawn without replacement from a population of N elements containing exactly Np elements of type I and $N(1-p)$ elements of type II. The number of elements of type I in the sample is described by the hypergeometric distribution.

If $n \ll N$ (in practice, $n < 0.1N$), then

$$\frac{C_{Np}^k C_{N(1-p)}^{n-k}}{C_N^n} \approx C_n^k p^k (1-p)^{n-k};$$

i.e., the hypergeometric distribution tends to the binomial distribution.

▶ **Poisson distribution.** A random variable X has a *Poisson distribution* with parameter λ ($\lambda > 0$) if

$$P(X = k) = \frac{\lambda^k}{k!} e^{-\lambda}, \quad k = 0, 1, 2, \dots$$

The cumulative distribution function of the Poisson distribution at the points $k = 0, 1, 2, \dots$ is given by the formula

$$F(k) = \frac{1}{k!} \int_\lambda^\infty x^k e^{-x} \, dx,$$

and the numerical characteristics are given by

$$E\{X\} = \lambda, \quad \text{Var}\{X\} = \lambda, \quad \alpha_2 = \lambda^2 + \lambda, \quad \alpha_3 = \lambda(\lambda^2 + 3\lambda + 1),$$
$$\alpha_4 = \lambda(\lambda^3 + 6\lambda^2 + 7\lambda + 1), \quad \mu_3 = \lambda, \quad \mu_4 = 3\lambda^2 + \lambda, \quad \gamma_1 = \lambda^{-1/2}, \quad \gamma_2 = \lambda^{-1}.$$

The sum of independent random variables $X_1, \dots, X_n$ obeying the Poisson distributions with parameters $\lambda_1, \dots, \lambda_n$, respectively, has the Poisson distribution with parameter $\lambda_1 + \dots + \lambda_n$.

The Poisson distribution is the limit distribution for many discrete distributions such as the hypergeometric distribution, the binomial distribution, the negative binomial distribution, distributions arising in problems of arrangement of particles in cells, etc. The Poisson distribution is an acceptable model for describing the random number of occurrences of certain events on a given time interval in a given domain in space.

▶ **Negative binomial distribution.** A random variable X has a *negative binomial distribution* with parameters r and p if

$$P(X = k) = C_{r+k-1}^{r-1} p^r (1-p)^k, \quad k = 0, 1, \dots, r,$$

where $0 < p < 1$ and $r > 0$.

The numerical characteristics can be calculated by the formulas

$$E\{X\} = \frac{r(1-p)}{p}, \quad \text{Var}\{X\} = \frac{r(1-p)}{p^2}, \quad \gamma_1 = \frac{2-p}{\sqrt{r(1-p)}}, \quad \gamma_2 = \frac{6}{r} + \frac{p^2}{r(1-p)}.$$

The negative binomial distribution describes the number X of failures before the rth success in a Bernoulli process with probability p of success on each trial. For $r = 1$, the negative binomial distribution coincides with the geometric distribution.

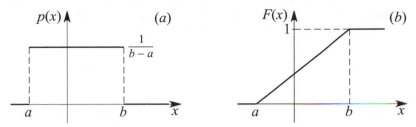

Figure M14.2. Probability density function (*a*) and cumulative distribution function (*b*) of a uniform distribution.

M14.2.4. Main Continuous Distributions and Their Numerical Characteristics

▶ **Uniform distribution.** A random variable X is *uniformly* distributed on an interval $[a, b]$ (Fig. M14.2 *a*) if

$$p(x) = \begin{cases} \dfrac{1}{b-a} & \text{for } x \in [a, b], \\ 0 & \text{for } x \notin [a, b]. \end{cases}$$

The cumulative distribution function (see Fig. M14.2 *b*) has the form

$$F(x) = \begin{cases} 0 & \text{for } x \le a, \\ \dfrac{x-a}{b-a} & \text{for } a < x \le b, \\ 1 & \text{for } x > b, \end{cases}$$

and the numerical characteristics are given by the expressions

$$E\{X\} = \frac{a+b}{2}, \quad \text{Var}\{X\} = \frac{(b-a)^2}{12}, \quad \gamma_1 = 0, \quad \gamma_2 = -1.2.$$

The uniform distribution does not have a mode.

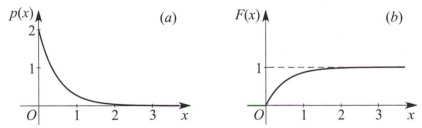

Figure M14.3. Probability density function (*a*) and cumulative distribution function (*b*) of an exponential distribution for $\lambda = 2$.

▶ **Exponential distribution.** A random variable X has the *exponential distribution* with parameter $\lambda > 0$ (Fig. M14.3 *a*) if

$$p(x) = \begin{cases} \lambda e^{-\lambda x} & \text{for } x > 0, \\ 0 & \text{for } x \le 0. \end{cases}$$

The cumulative distribution function (see Fig. M14.3 *b*) has the form

$$F(x) = \begin{cases} 1 - e^{-\lambda x} & \text{for } x > 0, \\ 0 & \text{for } x \le 0, \end{cases}$$

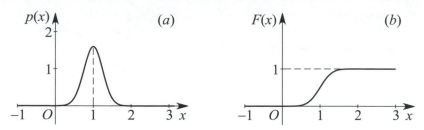

Figure M14.4. Probability density function (*a*) and cumulative distribution function (*b*) of a normal distribution for $a = 1$ and $\sigma = 1/4$.

and the numerical characteristics are given by the formulas

$$E\{X\} = \frac{1}{\lambda}, \quad \text{Var}\{X\} = \frac{1}{\lambda^2}, \quad \alpha_2 = \frac{2}{\lambda^2}, \quad \gamma_1 = 2, \quad \gamma_2 = 6.$$

The exponential distribution is the continuous analogue of the geometric distribution and is memoryless: $P(X > t + s \mid X > t) = P(X > s)$.

▶ **Normal distribution.** A random variable X has a *normal distribution* with parameters a and σ^2 (see Fig. M14.4 *a*) if its probability density function has the form

$$p(x) = \frac{1}{\sqrt{2\pi}\sigma} \exp\left[-\frac{(x-a)^2}{2\sigma^2}\right], \quad x \in (-\infty, \infty).$$

The cumulative distribution function (see Fig. M14.4 *b*) has the form

$$F(x) = \frac{1}{\sqrt{2\pi}\,\sigma} \int_{-\infty}^{x} \exp\left[-\frac{(t-a)^2}{2\sigma^2}\right] dt \equiv \frac{1}{2}\left[1 + \text{erf}\left(\frac{x-a}{\sqrt{2}\,\sigma}\right)\right],$$

where erf z is the error function (see Section M13.2).

The numerical characteristics are given by the formulas

$$E\{X\} = a, \quad \text{Var}\{X\} = \sigma^2, \quad \text{Mode}\{X\} = a, \quad \gamma_1 = 0, \quad \gamma_2 = 0,$$
$$\mu_k = \begin{cases} 0, & k = 2l - 1, \ l = 1, 2, \dots \\ (2k-1)!!\,\sigma^{2k}, & k = 2l, \ l = 1, 2, \dots \end{cases}$$

The linear transformation $Y = \frac{X-a}{\sigma}$ reduces the normal distribution with parameters (a, σ^2) and cumulative distribution function $F(x)$ to the standard normal distribution with parameters $(0, 1)$ and cumulative distribution function

$$\Phi(x) = \frac{1}{\sqrt{2\pi}} \int_{-\infty}^{x} e^{-t^2/2}\, dt. \tag{M14.2.4.1}$$

The probability that a random variable X normally distributed with parameters (m, σ^2) lies in the interval (a, b) is given by the formula

$$P(a < x < b) = \Phi\left(\frac{b-m}{\sigma}\right) - \Phi\left(\frac{a-m}{\sigma}\right).$$

A normally distributed random variable takes values close to its expectation with large probability; this is expressed by the *sigma rule*

$$P(|X - m| \geq k\sigma) = 2[1 - \Phi(k)] = \begin{cases} 0.3173 & \text{for } k = 1, \\ 0.0456 & \text{for } k = 2, \\ 0.0027 & \text{for } k = 3. \end{cases}$$

The three-sigma rule is most frequently used.

The fundamental role of the normal distribution is due to the fact that, under mild assumptions, the distribution of a sum of random variables is asymptotically normal as the number of terms increases. The corresponding conditions are specified in the central limit theorem.

▶ **Chi-square distribution.** A random variable $X = \chi^2(n)$ has a *chi-square distribution* with n degrees of freedom if its probability density function has the form

$$p(x) = \begin{cases} \dfrac{1}{2^{n/2}\,\Gamma(n/2)} x^{-1+n/2} e^{-x/2} & \text{for } x > 0, \\ 0 & \text{for } x \leq 0, \end{cases} \tag{M14.2.4.2}$$

where $\Gamma(z)$ is the Gamma function (see Subsection M13.4.1).

The cumulative distribution function can be written as

$$F(x) = \frac{1}{2^{n/2}\Gamma(n/2)} \int_0^x \xi^{-1+n/2} e^{-\xi/2}\, d\xi$$

and the numerical characteristics are given by the formulas

$$E\{\chi^2(n)\} = n, \quad \operatorname{Var}\{\chi^2(n)\} = 2n, \quad \alpha_k = n(n+2) \cdot \ldots \cdot [n + 2(k-1)],$$

$$\gamma_1 = 2\sqrt{\frac{2}{n}}, \quad \gamma_2 = \frac{12}{n}, \quad \operatorname{Mode}\{\chi^2(n)\} = n - 2 \quad (n \geq 2).$$

Main property of the chi-square distribution. For an arbitrary n, the sum

$$X = \sum_{k=1}^{n} X_k^2,$$

of squares of independent random variables obeying the standard normal distribution has the chi-square distribution with n degrees of freedom.

Relationship with other distributions:

1. For $n = 1$, formula (M14.2.4.2) gives the probability density function of the square X^2 of a random variable with the standard normal distribution.
2. For $n = 2$, formula (M14.2.4.2) gives the exponential distribution with parameter $\lambda = \frac{1}{2}$.
3. As $n \to \infty$, the random variable $X = \chi^2(n)$ has an asymptotically normal distribution with parameters $(n, 2n)$.
4. As $n \to \infty$, the random variable $\sqrt{2\chi^2(n)}$ has an asymptotically normal distribution with parameters $(\sqrt{2n-1}, 1)$.

M14.2.5. Two-dimensional and Multivariate Random Variables

▶ **Distribution of bivariate random variable.** Suppose that random variables X_1 and X_2 are defined in a probability space Ω_2; then one says that a *two-dimensional random vector* $\mathbf{X} = (X_1, X_2)$ or a *system of two random variables* is given.

The *distribution function* $F(x_1, x_2)$ (also denoted $F_{X_1, X_2}(x_1, x_2)$) of a two-dimensional *random vector* (X_1, X_2), or the *joint distribution function of the random variables* X_1 and X_2, is defined as the probability of the simultaneous occurrence (intersection) of the events $(X_1 < x_1)$ and $(X_2 < x_2)$; i.e.,

$$F(x_1, x_2) = P(X_1 < x_1, X_2 < x_2).$$

Properties of the joint distribution function of random variables X_1 and X_2:

1. The function $F(x_1, x_2)$ is a nondecreasing function of each of the arguments.
2. $F(x_1, -\infty) = F(-\infty, x_2) = F(-\infty, -\infty) = 0$.
3. $F(+\infty, +\infty) = 1$.
4. The probability that the random vector lies in a rectangle with sides parallel to the coordinate axes is

$$P(a_1 \leq X_1 < b_1, a_2 \leq X_2 < b_2) = F(b_1, b_2) - F(b_1, a_2) - F(a_1, b_2) + F(a_1, a_2).$$

5. The function $F(x_1, x_2)$ is left-continuous in either argument.

▶ **Discrete bivariate random variables.** A bivariate random variable (X_1, X_2) is said to be *discrete* if each of the random variables X_1 and X_2 is discrete.

If the random variable X_1 takes the values $x_{11}, \ldots, x_{1m}$ and the random variable X_2 takes the values $x_{21}, \ldots, x_{2n}$, then the random vector (X_1, X_2) can take only the pairs of values (x_{1i}, x_{2j}) $(i = 1, \ldots, m, j = 1, \ldots, n)$. The entries $p_{ij} = P(X_1 = x_{1i}, X_2 = x_{2j})$ are the probabilities of the simultaneous occurrence of the events $(X_1 = x_{1i})$ and $(X_2 = x_{2j})$; $P_{X_1, i} = p_{i1} + \cdots + p_{in}$ is the probability that the random variable X_1 takes the value x_{1i}; $P_{X_2, j} = p_{1j} + \cdots + p_{mj}$ is the probability that the random variable X_2 takes the value x_{2j}.

The distribution function of a discrete bivariate random variable can be determined by the formula

$$F(x_1, x_2) = \sum_{\substack{x_{1i} < x_1 \\ x_{2j} < x_2}} p_{ij}.$$

▶ **Continuous bivariate random variables.** A bivariate random variable (X_1, X_2) is said to be *continuous* if its joint distribution function $F(x_1, x_2)$ can be represented as

$$F(x_1, x_2) = \int_{-\infty}^{x_2} \int_{-\infty}^{x_1} p(y_1, y_2) \, dy_1 \, dy_2, \tag{M14.2.5.1}$$

where the *joint probability density function* $p(x_1, x_2)$ is piecewise continuous.

The joint probability density function can be expressed in terms of the joint distribution function as follows:

$$p(x_1, x_2) = \frac{\partial^2}{\partial x_1 \partial x_2} F(x_1, x_2). \tag{M14.2.5.2}$$

Formulas (M14.2.5.1) and (M14.2.5.2) establish a one-to-one correspondence (up to sets of probability zero) between the joint probability density functions and the joint distribution functions of continuous bivariate random variables. The differential $p(x_1, x_2) \, dx_1 \, dx_2$ is called a *probability element*. Up to higher-order infinitesimals, the probability element is equal to the probability for the random variable (X_1, X_2) to lie in the infinitesimal rectangle $(x_1, x_1 + \Delta x_1) \times (x_2, x_2 + \Delta x_2)$.

Properties of the joint probability density function of random variables X_1 and X_2:

1. The function $p(x_1, x_2)$ is nonnegative; i.e., $p(x_1, x_2) \geq 0$.

2. $\displaystyle\int_{-\infty}^{+\infty} \int_{-\infty}^{+\infty} p(x_1, x_2)\, dx_1\, dx_2 = 1.$

3. $P(a_1 < X_1 < b_1, a_2 < X_2 < b_2) = \displaystyle\int_{a_1}^{b_1} dx_1 \int_{a_2}^{b_2} p(x_1, x_2)\, dx_2 = \int_{a_2}^{b_2} dx_2 \int_{a_1}^{b_1} p(x_1, x_2)\, dx_1.$

4. The probability for a two-dimensional random variable (X_1, X_2) to lie in a domain $D \subset \mathbb{R}^2$ is numerically equal to the volume of the curvilinear cylinder with base D bounded above by the surface of the joint probability density function:

$$P[(X_1, X_2) \in D] = \iint_{(x_1, x_2) \in D} p_{X_1, X_2}(x_1, x_2)\, dx_1\, dx_2.$$

Random variables X_1 and X_2 are said to be *independent* if the relation

$$P(X_1 \in S_1, X_2 \in S_2) = P(X_1 \in S_1)\, P(X_2 \in S_2)$$

holds for any measurable sets S_1 and S_2.

THEOREM. *Random variables X_1 and X_2 are independent if and only if*

$$F(x_1, x_2) = F_1(x_1)\, F_2(x_2),$$

where F_1 and F_2 are the cumulative distribution functions of X_1 and X_2, respectively.

▶ **Numerical characteristics of bivariate random variables.** The *expectation* of a function $\varphi(X_1, X_2)$ of a bivariate random variable (X_1, X_2) is defined as the expression computed by the formula

$$E\{\varphi(X_1, X_2)\} = \begin{cases} \displaystyle\sum_i \sum_j \varphi(x_{1i}, x_{2j}) p_{ij} & \text{in the discrete case,} \\[2mm] \displaystyle\int_{-\infty}^{+\infty} \int_{-\infty}^{+\infty} \varphi(x_1, x_2) p(x_1, x_2)\, dx_1\, dx_2 & \text{in the continuous case,} \end{cases}$$

if these expressions exist in the sense of absolute convergence; otherwise, one says that $E\{\varphi(X_1, X_2)\}$ does not exist.

The *moment of order* $r_1 + r_2$ of a two-dimensional random variable (X_1, X_2) about a point (a_1, a_2) is defined as the expectation $E\{(X_1 - a_1)^{r_1}(X_2 - a_2)^{r_2}\}$.

If $a_1 = a_2 = 0$, then the moment of order $r_1 + r_2$ of a two-dimensional random variable (X_1, X_2) is called simply the moment, or the *initial moment*. The initial moment of order $r_1 + r_2$ is usually denoted by α_{r_1, r_2}; i.e., $\alpha_{r_1, r_2} = E\{X_1^{r_1} X_2^{r_2}\}$.

The first initial moments are the expectations of the random variables X_1 and X_2; i.e., $\alpha_{1,0} = E\{X_1^1 X_2^0\} = E\{X_1\}$ and $\alpha_{0,1} = E\{X_1^0 X_2^1\} = E\{X_2\}$. The point $(E\{X_1\}, E\{X_2\})$ on the OXY-plane characterizes the position of the random point (X_1, X_2); this position spreads about the point $(E\{X_1\}, E\{X_2\})$.

If $a_1 = E\{X_1\}$ and $a_2 = E\{X_2\}$, then the moment of order $r_1 + r_2$ of the bivariate random variable (X_1, X_2) is called the *central moment*. The central moment of order $r_1 + r_2$

is usually denoted by μ_{r_1,r_2}; i.e., $\mu_{r_1,r_2} = E\{(X_1 - E\{X_1\})^{r_1}(X_2 - E\{X_2\})^{r_2}\}$. Obviously, the first central moments are zero.

The second central moments are of special interest and have special names and notation:

$$\lambda_{11} = \mu_{2,0} = \text{Var}\{X_1\}, \quad \lambda_{22} = \mu_{0,2} = \text{Var}\{X_2\},$$

$$\lambda_{12} = \lambda_{21} = \mu_{1,1} = E\{(X_1 - E\{X_1\})(X_2 - E\{X_2\})\}.$$

The first two of these moments are the variances of the respective random variables, and the third one is called the *covariance* and will be considered below.

▶ **Covariance and correlation of two random variables.** The *covariance* (correlation moment, or mixed second moment) $\text{Cov}(X_1, X_2)$ of random variables X_1 and X_2 is defined as the central moment of order $(1 + 1)$:

$$\text{Cov}(X_1, X_2) = \alpha_{1,1} = E\{(X_1 - E\{X_1\})(X_2 - E\{X_2\})\}.$$

Properties of the covariance:

1. $\text{Cov}(X_1, X_2) = \text{Cov}(X_2, X_1)$.
2. $\text{Cov}(X, X) = \text{Var}\{X\}$.
3. If the random variables X_1 and X_2 are independent, then $\text{Cov}(X_1, X_2) = 0$ and if $\text{Cov}(X_1, X_2) \neq 0$, then the random variables X_1 and X_2 are dependent.
4. If $Y_1 = a_1 X_1 + b_1$ and $Y_2 = a_2 X_2 + b_2$, then $\text{Cov}(Y_1, Y_2) = a_1 a_2 \text{Cov}(X_1, X_2)$.
5. $\text{Cov}(X_1, X_2) = E\{X_1 X_2\} - E\{X_1\}E\{X_2\}$.
6. $|\text{Cov}(X_1, X_2)| \leq \sqrt{\text{Var}\{X_1\}\text{Var}\{X_2\}}$. Moreover, $\text{Cov}(X_1, X_2) = \pm\sqrt{\text{Var}\{X_1\}\text{Var}\{X_2\}}$ if and only if the random variables X_1 and X_2 are linearly dependent.
7. $\text{Var}\{X_1 + X_2\} = \text{Var}\{X_1\} + \text{Var}\{X_2\} + 2\text{Cov}(X_1, X_2)$.

If $\text{Cov}(X_1, X_2) = 0$, then the random variables X_1 and X_2 are said to be *uncorrelated*; if $\text{Cov}(X_1, X_2) \neq 0$, then they are *correlated*. Independent random variables are always uncorrelated, but uncorrelated random variables are not necessarily independent in general.

Example. Suppose that we roll two dice. Let X_1 be the number on the first die and let X_2 be the number on the second die. We consider the random variables $Y_1 = X_1 + X_2$ and $Y_2 = X_1 - X_2$ (the sum and difference of the points obtained). Then

$$\text{Cov}(Y_1, Y_2) = E\{(X_1 + X_2 - E\{X_1 + X_2\})(X_1 - X_2 - E\{X_1 - X_2\})\}$$
$$= E\{(X_1 - E\{X_1\})^2 - (X_2 - E\{X_2\})^2\}$$
$$= \text{Var}\{X_1\} - \text{Var}\{X_2\} = 0,$$

since X_1 and X_2 are identically distributed and hence $\text{Var}\{X_1\} = \text{Var}\{X_2\}$. But Y_1 and Y_2 are obviously dependent; for example, if $Y_1 = 2$ then one necessarily has $Y_2 = 0$.

The covariance of random variables X_1 and X_2 characterizes both their dependence on each other and their spread around the point $(E\{X_1\}, E\{X_2\})$. The covariance of X_1 and X_2 has the dimension equal to the product of dimensions of X_1 and X_2. Along with the covariance of X_1 and X_2, one often uses the coefficient of correlation $\rho(X_1, X_2)$, which is a dimensionless normalized quantity. The *correlation coefficient* of random variables X_1 and X_2 is the ratio of the covariance of X_1 and X_2 to the product of their standard deviations,

$$\rho(X_1, X_2) = \frac{\text{Cov}(X_1, X_2)}{\sigma_1 \sigma_2}.$$

The correlation coefficient of random variables X_1 and X_2 indicates the degree of linear dependence between the variables. If $\rho(X_1, X_2) = 0$, then there is no linear relation between the random variables, but there may well be some other relation between them.

Properties of the correlation coefficient:

1. $\rho(X_1, X_2) = \rho(X_2, X_1)$.
2. $\rho(X, X) = 1$.
3. If random variables X_1 and X_2 are independent, then $\rho(X_1, X_2) = 0$. If $\rho(X_1, X_2) \neq 0$, then the random variables X_1 and X_2 are dependent.
4. If $Y_1 = a_1 X_1 + b_1$ and $Y_2 = a_2 X_2 + b_2$, then $\rho(Y_1, Y_2) = \pm\rho(X_1, X_2)$.
5. $|\rho(X_1, X_2)| \leq 1$. Moreover, $\rho(X_1, X_2) = \pm 1$ if and only if the random variables X_1 and X_2 are linearly dependent.

The theory of distribution functions of multivariate random variables (random vectors) can be developed in a similar way.

M14.3. Limit Theorems

M14.3.1. Convergence of Random Variables

▶ **Convergence in probability.** A sequence of random variables $X_1, X_2, \ldots$ is said to *converge in probability* to a random variable $X (X_n \xrightarrow{P} X)$ if

$$\lim_{n \to \infty} P(|X_n - X| \geq \varepsilon) = 0$$

for any $\varepsilon > 0$, i.e., if for any $\varepsilon > 0$ and $\delta > 0$ there exists a number N, depending on ε and δ, such that the inequality

$$P(|X_n - X| \geq \varepsilon) < \delta$$

holds for $n > N$. A sequence of k-dimensional random variables $\mathbf{X}_n$ is said to converge in probability to a random variable $\mathbf{X}$ if each coordinate of the random variable $\mathbf{X}_n$ converges in probability to the respective coordinate of the random variable $\mathbf{X}$.

▶ **Convergence in the mean.** A sequence of random variables $X_1, X_2, \ldots$ with finite qth initial moments $(E\{X^q\} < \infty, q = 1, 2, \ldots)$ is said to *converge in the qth mean* to a random variable X if

$$\lim_{n \to \infty} E\{|X_n - X|^q\} = 0.$$

Convergence in the qth mean for $q = 2$ is called *convergence in mean square*. If $X_n \to X$ in the qth mean then $X_n \to X$ in the q_1th mean for all $q_1 \leq q$.

Convergence in the qth mean implies convergence in probability. The converse statement is generally not true.

▶ **Convergence in distribution.** Suppose that a sequence $F_1(x), F_2(x), \ldots$ of cumulative distribution functions converges to a distribution function $F(x)$,

$$\lim_{n \to \infty} F_n(x) = F(x),$$

for every point x. In this case, we say that the sequence $X_1, X_2, \ldots$ of the corresponding random variables converges to the random variable X *in distribution*. The random variables $X_1, X_2, \ldots$ can be defined on different probability spaces.

A sequence $F_1(x), F_2(x), \ldots$ of distribution functions *weakly converges* to a distribution function $F(x)$ $(F_n \to F)$ if

$$\lim_{n \to \infty} E\{h(X_n)\} = E\{h(X)\}$$

for any bounded continuous function h.

Convergence in distribution and weak convergence of distribution functions are equivalent.

The weak convergence $F_{X_n} \to F$ for random variables having a probability density function means the convergence

$$\int_{-\infty}^{+\infty} g(x) p_{X_n}(x) \, dx \to \int_{-\infty}^{+\infty} g(x) p(x) \, dx$$

for any bounded continuous function $g(x)$, where $p(x) = \lim_{n \to \infty} p_{X_n}(x)$.

M14.3.2. Limit Theorems

▶ **Law of large numbers.** The law of large numbers consists of several theorems establishing the stability of average results and revealing conditions for this stability to occur.

The notion of convergence in probability is most often used for the case in which the limit random variable X has the degenerate distribution concentrated at a point a ($P(X = a) = 1$) and

$$X_n = \frac{1}{n} \sum_{k=1}^{n} Y_k,$$

where $Y_1, Y_2, \dots$ are arbitrary random variables.

A sequence $Y_1, Y_2, \dots$ satisfies the *weak law of large numbers* if the limit relation

$$\lim_{n \to \infty} P\left(\left| \frac{1}{n} \sum_{k=1}^{n} Y_k - a \right| \ge \varepsilon \right) \equiv \lim_{n \to \infty} P(|X_n - a| \ge \varepsilon) = 0 \tag{M14.3.2.1}$$

holds for any $\varepsilon > 0$.

If the relation

$$P\left(\omega \in \Omega : \lim_{n \to \infty} \frac{1}{n} \sum_{k=1}^{n} Y_k = a \right) \equiv P\left(\omega \in \Omega : \lim_{n \to \infty} X_n = a \right) = 1$$

is satisfied instead of (M14.3.2.1), i.e., the sequence X_n converges to the number a with probability 1, then the sequence $Y_1, Y_2, \dots$ satisfies the *strong law of large numbers*.

Markov's inequality. For any nonnegative random variable X that has an expectation $E\{X\}$, the inequality

$$P(X \ge \varepsilon) \le \frac{E\{X\}}{\varepsilon} \tag{M14.3.2.2}$$

holds for any $\varepsilon > 0$. It follows from inequality (M14.3.2.2) that

$$P(X < \varepsilon) \ge 1 - \frac{E\{X\}}{\varepsilon}.$$

Chebyshev's inequality. For any random variable X with finite variance, the inequality

$$P(|X - E\{X\}| \ge \varepsilon) \le \frac{\mathrm{Var}\{X\}}{\varepsilon^2} \tag{M14.3.2.3}$$

holds for any $\varepsilon > 0$. It follows from inequality (M14.3.2.3) that

$$P(|X - E\{X\}| < \varepsilon) \ge 1 - \frac{\mathrm{Var}\{X\}}{\varepsilon^2}$$

CHEBYSHEV'S THEOREM. *If $X_1, X_2, \ldots$ is a sequence of pairwise independent random variables with uniformly bounded finite variances, $\text{Var}\{X_1\} \leq C$, $\text{Var}\{X_2\} \leq C, \ldots$, then the limit relation*

$$\lim_{n \to \infty} P\left(\left|\frac{1}{n}\sum_{k=1}^{n} X_k - \frac{1}{n}\sum_{k=1}^{n} E\{X_k\}\right| < \varepsilon\right) = 1$$

holds for any $\varepsilon > 0$.

BERNOULLI'S THEOREM. *Let μ_n be the number of occurrences of an event A (the number of successes) in n independent trials, and let $p = P(A)$ be the probability of the occurrence of the event A (the probability of success) in each of the trials. Then the sequence of relative frequencies μ_n/n of the occurrence of the event A in n independent trials converges in probability to $p = P(A)$ as $n \to \infty$; i.e., the limit relation*

$$\lim_{n \to \infty} P\left(\left|\frac{\mu_n}{n} - p\right| < \varepsilon\right) = 1$$

holds for any $\varepsilon > 0$.

POISSON'S THEOREM. *If in a sequence of independent trials the probability that an event A occurs in the kth trial is equal to p_k, then*

$$\lim_{n \to \infty} P\left(\left|\frac{\mu_n}{n} - \frac{p_1 + \cdots + p_n}{n}\right| < \varepsilon\right) = 1.$$

KOLMOGOROV'S THEOREM. *If a sequence of independent random variables $X_1, X_2, \ldots$ satisfies the condition*

$$\sum_{k=1}^{\infty} \frac{\text{Var}\{X_k\}}{k^2} < +\infty,$$

then it obeys the strong law of large numbers.

The existence of the expectation is a necessary and sufficient condition for the strong law of large numbers to apply to a sequence of independent identically distributed random variables.

▶ **Central limit theorems.** A random variable X_n with distribution function F_{X_n} is *asymptotically normally distributed* if there exists a sequence of pairs of real numbers m_n, σ_n^2 such that the random variables $(X_n - m_n)/\sigma_n$ converge in probability to a standard normal variable. This occurs if and only if the limit relation

$$\lim_{n \to \infty} P(m_n + a\sigma_n < X_n < m_n + b\sigma_n) = \Phi(b) - \Phi(a),$$

where $\Phi(x)$ is the distribution function of the standard normal law (M14.2.4.1), holds for any a and b $(b > a)$.

LYAPUNOV'S CENTRAL LIMIT THEOREM. *If $X_1, \ldots, X_n, \ldots$ is a sequence of independent random variables having finite expectation values $E\{X_k\}$ and finite variances $\text{Var}\{X_k\}$ and satisfying Lyapunov's condition*

$$\lim_{n \to \infty} \frac{\sum_{k=1}^{n} \alpha_3(X_k)}{\sqrt{\sum_{k=1}^{n} \text{Var}\{X_k\}}} = 0,$$

where $\alpha_3(X_k)$ is the third initial moment of the random variable X_k, then the sequence of random variables

$$Y_n = \frac{\sum_{k=1}^{n}(X_k - E\{X_k\})}{\sqrt{\sum_{k=1}^{n} \text{Var}\{X_k\}}}$$

converges in distribution to the normal law, i.e., the following limit exists:

$$\lim_{n \to \infty} P\left(\frac{\sum_{k=1}^{n}(X_k - E\{X_k\})}{\sqrt{\sum_{k=1}^{n} \mathrm{Var}\{X_k\}}} < t \right) = \frac{1}{\sqrt{2\pi}} \int_{-\infty}^{t} e^{-u^2/2} \, du = \Phi(t).$$

LINDEBERG'S CENTRAL LIMIT THEOREM. *Let $X_1, X_2, \ldots$ be a sequence of independent identically distributed random variables with finite expectation $E\{X_k\} = m$ and finite variance σ^2. Then, as $n \to \infty$, the random variable $\frac{1}{n} \sum_{k=1}^{n} X_k$ has an asymptotically normal probability distribution with parameters $(m, \sigma^2/n)$.*

Let μ_n be the number of occurrences of an event A (the number of successes) in n independent trials, and let $p = P(A)$ denote the probability of the occurrence of the event A (the probability of success) in each of the trials. Then the sequence of relative frequencies μ_n/n has an asymptotically normal probability distribution with parameters $(p, p(1 - p)/n)$.

Bibliography for Chapter M14

Bain, L. J. and Engelhardt, M., *Introduction to Probability and Mathematical Statistics, 2nd Edition* (Duxbury Classic), Duxbury Press, Boston, Massachusetts, 2000.

Bertsekas, D. P. and Tsitsiklis, J. N., *Introduction to Probability*, Athena Scientific, Belmont, Massachusetts, 2002.

Burlington, R. S. and May, D., *Handbook of Probability and Statistics With Tables, 2nd Edition*, McGraw-Hill, New York, 1970.

Chung, K. L., *A Course in Probability Theory Revised, 2nd Edition*, Academic Press, Boston, Massachusetts, 2000.

DeGroot, M. H. and Schervish, M. J., *Probability and Statistics, 3rd Edition*, Addison Wesley, Boston, Massachusetts, 2001.

Garcia, L., *Probability and Random Processes for Electrical Engineering: Student Solutions Manual, 2nd Edition*, Addison Wesley, Boston, Massachusetts, 1993.

Gnedenko, B. V. and Khinchin, A. Ya., *An Elementary Introduction to the Theory of Probability, 5th Edition*, Dover Publications, New York, 1962.

Goldberg, S., *Probability: An Introduction*, Dover Publications, New York, 1987.

Grimmett, G. R. and Stirzaker, D. R., *Probability and Random Processes, 3rd Edition*, Oxford University Press, Oxford, 2001.

Hsu, H., *Schaum's Outline of Probability, Random Variables, and Random Processes*, McGraw-Hill, New York, 1996.

Kokoska, S. and Zwillinger, D. (Editors), *CRC Standard Probability and Statistics Tables and Formulae, Student Edition*, Chapman & Hall/CRC, Boca Raton, Florida, 2000.

Lange, K., *Applied Probability*, Springer, New York, 2004.

Ledermann, W., *Probability* (Handbook of Applicable Mathematics), Wiley, New York, 1981.

Lipschutz, S., *Schaum's Outline of Probability, 2nd Edition*, McGraw-Hill, New York, 2000.

Lipschutz, S. and Schiller, J., *Schaum's Outline of Introduction to Probability and Statistics*, McGraw-Hill, New York, 1998.

Montgomery, D. C. and Runger, G. C., *Applied Statistics and Probability for Engineers, Student Solutions Manual, 4th Edition*, Wiley, New York, 2006.

Pitman, J., *Probability*, Springer, New York, 1993.

Ross, S. M., *Introduction to Probability Models, 8th Edition*, Academic Press, Boston, Massachusetts, 2002.

Ross, S. M., *A First Course in Probability, 7th Edition*, Prentice Hall, Englewood Cliffs, New Jersey, 2005.

Rozanov, Y. A., *Probability Theory: A Concise Course*, Dover Publications, New York, 1977.

Seely, J. A., *Probability and Statistics for Engineering and Science, 6th Edition*, Brooks Cole, Stanford, 2003.

Shiryaev, A. N., *Probability, 2nd Edition* (Graduate Texts in Mathematics), Springer, New York, 1996.

Walpole, R. E., Myers, R. H., Myers, S. L., and Ye, K., *Probability & Statistics for Engineers & Scientists, 8th Edition*, Prentice Hall, Englewood Cliffs, New Jersey, 2006.

Zwillinger, D. (Editor), *CRC Standard Mathematical Tables and Formulae, 31st Edition*, Chapman & Hall/CRC, Boca Raton, Florida, 2002.

Part II
Physics

Chapter P1
Physical Foundations of Mechanics

Preliminary remarks. *Mechanical motion* is change in the location of a body with respect to other bodies. This definition implies that mechanical motion is relative. In order to describe motion, one should specify a *frame of reference*, which includes a body of reference, a coordinate system fixed relative to the body, and a set of clocks synchronized with one another. Mechanics studies motions of model objects, a point particle (or a point mass) and a rigid body. The location of these objects is determined by a finite set of independent parameters; the objects are said to have finitely many *degrees of freedom*. *Kinematics* deals with the characterization of motion without finding out its reasons.

P1.1. Kinematics of a Point

P1.1.1. Basic Definitions. Velocity and Acceleration

▶ **Point particle. Law of motion. Path, distance and displacement.** A body whose dimensions can be neglected in studying its motion (compared to the distances of its movement) is called a *point particle* (or just a *particle*). The position of a point particle at an instant of time t is determined by the *position vector* $\mathbf{r}$ from the origin of some reference frame to the particle (see Fig. P1.1). As the particle moves, the end of the position vector traces a spatial curve, a *path* (also called a *trajectory*). In a rectangular Cartesian reference frame, the position vector is determined by its projections onto the coordinate axis, its x-, y-, and z-coordinates. The motion of a particle is completely determined by specifying its *law of motion*, a single vector function $\mathbf{r}(t)$ or three scalar functions $x(t)$, $y(t)$, $z(t)$. A position vector (or any other vector) can be conveniently written in terms of its projections using unit vectors, $\mathbf{i}$, $\mathbf{j}$, and $\mathbf{k}$, of the respective coordinate axes as follows: $\mathbf{r} = x\mathbf{i} + y\mathbf{j} + z\mathbf{k}$. The *distance* traveled by the particle in a given time interval is measured along the curvilinear path. Distance is a scalar quantity; it is nonnegative and nondecreasing with time. The *displacement* of a particle is the vector $\Delta\mathbf{r} = \mathbf{r} - \mathbf{r}_0$ that connects an original position of the particle with a final one and is equal to the difference of the position vectors at the initial and final time.

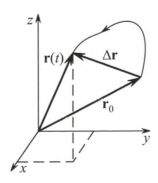

Figure P1.1. Path and displacement.

405

▶ **Velocity.** The (instantaneous) *velocity* of a particle is the derivative of its position vector with respect to time:

$$\mathbf{v} = \frac{d\mathbf{r}}{dt}.$$

The velocity is tangent to the path. The (time) average velocity over a finite time interval Δt is defined as the ratio of the displacement to the time interval, $\mathbf{v}_{\text{ave}} = \Delta \mathbf{r}/\Delta t$. (The mean speed is equal to the ratio of the distance traveled to the time interval.) Motion is called *uniform* if $\mathbf{v} = \text{const}$. A uniform motion is a motion along a straight line. A *uniform motion along a given curvilinear path* is a motion with a constant velocity magnitude. (An example of such a motion is a uniform motion in a circle.)

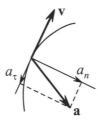

Figure P1.2. Normal and tangential acceleration.

▶ **Acceleration.** The *acceleration* of a particle is the derivative of the velocity with respect to time:

$$\mathbf{a} = \frac{d\mathbf{v}}{dt}.$$

The vector of acceleration lies in the same plane as the portion of the path where the motion takes places and is directed "inward" the path (if the path is a straight line, the acceleration is directed along it). The projection of the acceleration on the velocity direction is called the *tangential acceleration* and denoted a_τ; it defines the rate at which the velocity magnitude changes with time and equals its time derivative, $a_\tau = dv/dt$. The other component of the acceleration, which is perpendicular to the velocity, is called the *normal acceleration* (Fig. P1.2) and denoted a_n; it characterizes the rate at which the velocity direction changes and equals $a_n = v^2/R$, where R is the radius of curvature of the path (i.e., the radius of the circle that is the best approximation to the path at the given point). The average acceleration over a time interval Δt is defined as $\mathbf{a}_{\text{ave}} = \Delta \mathbf{v}/\Delta t$. A motion is said to be *uniformly accelerated* if $\mathbf{a} = \text{const}$. A *uniformly accelerated motion along a given trajectory* is a motion with $a_\tau = \text{const}$.

P1.1.2. The Direct and Inverse Problems of Kinematics

▶ **Direct problem.** The problem of determining the characteristics of a motion for a given law of motion is called the *direct problem of kinematics*.

Example 1. Let the law of motion of a particle be given by $x = R\cos\omega t$ and $y = R\sin\omega t$. Find the trajectory equation, velocity, and acceleration of the particle.

Solution. Eliminating time (by making use of the identity $\sin^2\omega t + \cos^2\omega t = 1$) gives the trajectory equation: $x^2 + y^2 = R^2$ (a circle of radius R). Further,

$$v_x = \frac{dx}{dt} = -\omega R\sin\omega t,$$
$$\text{(velocity components)},$$
$$v_y = \frac{dy}{dt} = \omega R\cos\omega t$$

$$a_x = \frac{d^2x}{dt^2} = -\omega^2 R \cos \omega t,$$
$$\qquad \text{(acceleration components).}$$
$$a_y = \frac{d^2y}{dt^2} = -\omega^2 R \sin \omega t$$

The magnitude of the velocity, $v = \sqrt{v_x^2 + v_y^2} = \omega R$, and magnitude of the acceleration, $a = \omega^2 R$, are independent of time and are related by $a = v^2/R$. The acceleration is perpendicular to the velocity and directed towards the center of the circle: $\mathbf{a} = -\omega^2 \mathbf{r}$.

▶ **Inverse problem.** The problem of determining the law of motion for a given acceleration $\mathbf{a}(t)$ is called the *inverse problem of kinematics*. For unique solvability of this problem, one needs to know the *initial conditions*—the position and the velocity of the particle at an initial instant of time.

Example 2. Uniformly accelerated motion. Given the acceleration $\mathbf{a} = \mathrm{const}$ of a particle and its initial velocity $\mathbf{v}_0$ and initial position $\mathbf{r}_0$, find the trajectory and the law of motion of the particle.

Solution. Integrating the relation $\frac{d}{dt}\mathbf{v} = \mathbf{a}$ gives the velocity: $\mathbf{v} = \mathbf{v}_0 + \int_0^t \mathbf{a}\, dt = \mathbf{v}_0 + \mathbf{a}t$. Integrating $\frac{d}{dt}\mathbf{r} = \mathbf{v}$ gives the position vector of the particle: $\mathbf{r} = \mathbf{r}_0 + \int_0^t \mathbf{v}\, dt = \mathbf{r}_0 + \mathbf{v}_0 t + \frac{1}{2}\mathbf{a}t^2$.

The particle moves in the plane of the vectors $\mathbf{v}_0$ and $\mathbf{a}$ in a parabola. This becomes apparent in the reference frame where the y-axis is directed along $\mathbf{a}$ and the x-axis is perpendicular to $\mathbf{a}$; we have $x = x_0 + v_{0x}t$ and $y = y_0 + v_{0y}t + \frac{1}{2}at^2$.

▶ **Other problems.** In the cases where the kinematical problem is neither direct nor inverse and not reduced to a direct or inverse problem, one has to consider a differential equation. Sometimes this equation can be solved by separation of variables.

Example 3. Quite often, when solving dynamical problems, one has to deal with an equation like $dv/dt = -\alpha v^2$ (the body slows down with an acceleration proportional to the velocity squared; α is a constant quantity). Find the speed of the body versus time if its initial value is v_0. Also find the speed versus the displacement s.

Solution. Separating the variables gives $dv/v^2 = -\alpha\, dt$. Integrating the left-hand side from v_0 to v and the right-hand side from 0 to t, one finds $v^{-1} - v_0^{-1} = \alpha t$, and subsequently $v = v_0/(1 + v_0\alpha t)$. Integrating further, one could find the law of motion $s(t)$ and then, having eliminated t, determine $v(s)$. However, the result can be obtained in an easier way: since $v\, dt = ds$, the original separable equation can be rewritten as $dv/v = -\alpha\, ds$. Integrating gives $\ln(v/v_0) = -\alpha s$, and hence $v = v_0 \exp(-\alpha s)$.

P1.1.3. Circular Motion

Motion in a circular path can be described using three angular variables: angle of rotation φ, *angular velocity* $\omega = d\varphi/dt$, and *angular acceleration* $\varepsilon = d\omega/dt$. If angles are measured in radians, then the length of the arc corresponding to the angle φ equals $s = \varphi R$. It follows that

$$v = \omega R, \quad a_\tau = \frac{dv}{dt} = \varepsilon R, \quad a_n = \frac{v^2}{R} = \omega^2 R.$$

In the case of a uniform circular motion, the normal acceleration a_n is sometimes called *centripetal*.

The above formulas for the normal and tangential accelerations can be derived by writing the velocity as $\mathbf{v} = v\boldsymbol{\tau}$, where $\boldsymbol{\tau}$ is the unit vector directed along the velocity. Then we have $\mathbf{a} = (dv/dt)\boldsymbol{\tau} + v(d\boldsymbol{\tau}/dt)$. Here we should take into account that $d\boldsymbol{\tau}/dt = \mathbf{n}(d\varphi/dt) = \mathbf{n}(v/R)$, where $\mathbf{n}$ is the unit vector perpendicular to the velocity and directed along the radius. This holds for an arbitrary path as well, provided that the radius of curvature R and the unit normal direction $\mathbf{n}$ are defined by the relation $d\boldsymbol{\tau} = \mathbf{n}(ds/R)$.

P1.1.4. Relativity of Motion. Addition of Velocities

If the motion of a particle is considered relative to two frames of reference, K and K', whose respective axes remain parallel to each other, then the velocities of the particle in these reference frames, $\mathbf{v}$ and $\mathbf{v}'$, satisfy the following velocity-addition formula at any instant of time:

$$\mathbf{v} = \mathbf{v}' + \mathbf{v}_K, \qquad (P1.1.4.1)$$

where $\mathbf{v}_K$ is the velocity of the reference frame K' relative to K. The accelerations satisfy a similar addition formula: $\mathbf{a} = \mathbf{a}' + \mathbf{a}_K$.

P1.2. Kinematics of a Rigid Body

P1.2.1. Translational and Rotational Motion. Plane Motion

▶ **Rigid body. Translational and rotational motion**. In mechanics, a *rigid body* is an idealized body the distance between any two points of which does not change (i.e., there is no deformation). Special attention is paid to the description of two simple kinds of motion of a rigid body: translational and rotational. In *translational motion*, the segment connecting any two points of the body moves parallel to itself at all times. Since all points of the body move in the same manner, it suffices to describe the motion of a single point only. In *rotational motion*, all points of the body move in circular paths whose centers lie on the same straight line, called the *axis of rotation*, and the velocities of all points are perpendicular to it. The angular velocities of all points are the same at any instant of time; therefore, it is convenient to introduce a common *vector of angular velocity* $\boldsymbol{\omega}$, directed along the axis of rotation following the right-hand rule. The distribution of linear velocities of points of the body is expressed using cross product*

$$\mathbf{v} = \boldsymbol{\omega} \times \mathbf{r}, \qquad (P1.2.1.1)$$

where the positions vectors are drawn from any point at the axis of rotation. If the axis of rotation is fixed, the vector of angular acceleration, $\boldsymbol{\varepsilon} = d\boldsymbol{\omega}/dt$, is also directed along the axis.

▶ **Plane motion of a rigid body.** Plane motion of a rigid body is defined as motion in which the velocities of all points of the body are parallel to some plane. If there is a translationally moving reference frame associated with an arbitrary point of the body (or its imaginary continuation), then the relative motion of any other point will be purely rotational about a fixed axis perpendicular to the plane of motion.

   Example. Wheel rolling without slipping on a flat surface with a speed v is convenient to represent as the combination of a translation motion with the speed v and a rotational motion with an angular velocity ω (Fig. P1.3). The velocity of any point of the wheel relative to the surface can be found by the velocity-addition law (P1.1.4.1). The speed of the lowest point O' must be zero; it follows that v and ω are related by $v = \omega R$. The accelerations of all points are directed toward the wheel center.

P1.2.2. Instantaneous Axis of Rotation. Addition of Angular Velocities. Motion Relative to a Rotational Frame

▶ **Instantaneous axis of rotation.** If some point of a body (or its imaginary continuation) is at rest at some instant, then there exists a straight line that passes through this point and is at rest at this instant. This straight line is called an *instantaneous axis of rotation*. The velocity distribution at the instant in question is given by (P1.2.1.1). The instantaneous axis and ω can change their position both in space and relative to the body. In particular, the velocities of all points of the wheel in the above example can be obtained as the result of pure rotation about the instantaneous axis through contact point O'. The angular acceleration ε can be nonparallel to the instantaneous axis.

---

* This equation is a special case of the important mathematical formula $d\mathbf{A}/dt = \boldsymbol{\omega} \times \mathbf{A}$ expressing the time derivative of any vector $\mathbf{A}$ rotating with an angular velocity $\boldsymbol{\omega}$.

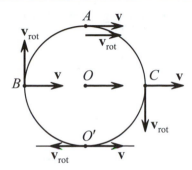

Figure P1.3. Rolling of a wheel can be represented as the sum of its translational and rotational motions or as a pure rotation about the instantaneous axis O'.

▶ **Motion of a rigid body with a fixed point. Addition of angular velocities.** The motion of a rigid body with a fixed point is, at any instant of time, pure rotation about an instantaneous axis. If this motion can be represented as rotation with an angular velocity $\boldsymbol{\omega}_2$ relative to a reference frame rotating with an angular velocity $\boldsymbol{\omega}_1$, then the resulting angular velocity equals $\boldsymbol{\omega} = \boldsymbol{\omega}_1 + \boldsymbol{\omega}_2$. To prove this, one should generalize the velocity-addition law (P1.1.4.1) to nontranslational motion of the reference frame:

$$\mathbf{v} = \mathbf{v}_{\text{rel}} + \mathbf{v}_{\text{transl}}, \tag{P1.2.2.1}$$

where $\mathbf{v}_{\text{transl}}$ is the *translatory velocity* of the moving reference frame at the point considered. If the reference is rotating, we have

$$\mathbf{v} = \mathbf{v}_{\text{rel}} + \boldsymbol{\omega}_1 \times \mathbf{r}. \tag{P1.2.2.2}$$

If the relative motion is pure rotation, $\mathbf{v}_{\text{rel}} = \boldsymbol{\omega}_2 \times \mathbf{r}$, then $\mathbf{v} = \boldsymbol{\omega}_1 \times \mathbf{r} + \boldsymbol{\omega}_2 \times \mathbf{r} = (\boldsymbol{\omega}_1 + \boldsymbol{\omega}_2) \times \mathbf{r}$ and so the resulting motion is rotation with the angular velocity $\boldsymbol{\omega}_1 + \boldsymbol{\omega}_2$.

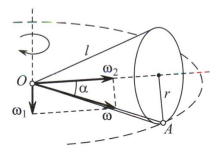

Figure P1.4. A cone rolls on a plane. The sum of the rotational motions is a pure rotation about the axis OA.

Example. A circular cone with a semi-apex angle α is put on its side and rolled without sliding so that its axis rotates with an angular velocity $\boldsymbol{\omega}_1$ (Fig. P1.4). Find the total angular velocity and angular acceleration of the cone.

Solution. In the rotating reference frame where the cone axis is at rest, the cone rotates about this axis with an angular velocity $\boldsymbol{\omega}_2$. The relation between ω_1 and ω_2 can be found from the condition that the point A is at rest; we have $\omega_1 l = \omega_2 r$, or $\omega_1 = \omega_2 \sin \alpha$. It follows that the vector $\boldsymbol{\omega} = \boldsymbol{\omega}_1 + \boldsymbol{\omega}_2$ is horizontal and directed along the line of contact between the cone and the surface; this line is, of course, the instantaneous axis of rotation. The angular acceleration is contributed by only the rotation of the vector $\boldsymbol{\omega}_2$ with the angular velocity $\boldsymbol{\omega}_1$, so that $\boldsymbol{\varepsilon} = d\boldsymbol{\omega}/dt = \boldsymbol{\omega}_1 \times \boldsymbol{\omega}_2$ and $\varepsilon = \omega_1 \omega_2 \cos \alpha$.

Formula (P1.2.2.2) is a special case of the formula for the time derivative of an arbitrary vector $\mathbf{A}$:

$$\frac{d\mathbf{A}}{dt} = \left(\frac{d\mathbf{A}}{dt}\right)_{\text{rel}} + \boldsymbol{\omega} \times \mathbf{A}, \tag{P1.2.2.3}$$

where the first term denotes the rate of change of **A** relative to the rotating reference frame (see also the footnote to formula (P1.2.1.1)).

In studying motion in noninertial reference frames (see Subsection P1.8.3), we will need the relation between the accelerations of a particle in a fixed reference frame and a rotating one. This relation can be obtained by applying (P1.2.2.3) to either term in (P1.2.2.2):

$$\frac{d\mathbf{v}}{dt} = \frac{d\mathbf{v}_{rel}}{dt} + \boldsymbol{\omega} \times \frac{d\mathbf{r}}{dt} = \left(\frac{d\mathbf{v}_{rel}}{dt}\right)_{rel} + \boldsymbol{\omega} \times \mathbf{v}_{rel} + \boldsymbol{\omega} \times (\mathbf{v}_{rel} + \boldsymbol{\omega} \times \mathbf{r}) = \mathbf{a}_{rel} + 2\boldsymbol{\omega} \times \mathbf{v}_{rel} + \boldsymbol{\omega} \times (\boldsymbol{\omega} \times \mathbf{r}).$$

The last term, $\boldsymbol{\omega} \times (\boldsymbol{\omega} \times \mathbf{r})$, is the translatory acceleration of the reference frame at the given point (radial or centripetal). Note that along with the relative and translatory accelerations, an additional term arises here. This term, $2\boldsymbol{\omega} \times \mathbf{v}_{rel}$, is called the *Coriolis acceleration*.

P1.3. Dynamics

P1.3.1. Newton's First Law. Mass. Momentum. Force

▶ **Newton's first law**. Newton's first law states that there are reference frames in which a body not interacting with other bodies is either at rest or in motion in a straight line with constant speed. The reference frames postulated by this law are called *inertial*.

It follows from the velocity-addition law that any reference frame that moves with a constant velocity relative to an inertial reference frame is also inertial. A heliocentric reference frame, related to the Sun and distant stars, can be considered inertial virtually exactly. The Earth can be treated as an inertial reference frame only approximately — the acceleration of the points at its surface amounts to $0.034\,\text{m/s}^2$ due to Earth's daily rotation. In addition, the Earth orbits the Sun but the associated acceleration is three orders of magnitude lower (check this as an exercise).

▶ **Mass. Momentum**. From Newton's first law it follows that in inertial reference frames, a body's acceleration results from interaction of the body with other bodies. This acceleration depends on the *inertia* of the body, or its ability to resist a change in velocity, and also on the intensity and direction of action of other bodies.

The *mass* of a body, m, is a positive scalar quantity that characterizes the body's inertia. It follows from experiments that when two bodies interact with each other, their accelerations in an inertial reference frame, $\mathbf{a}_1$ and $\mathbf{a}_2$, are opposite to each other in direction and the ratio of their magnitudes, a_1/a_2, is independent of the character and intensity of the interaction. This makes it possible to define the ratio of two masses as the reciprocal of the ratio of the accelerations resulting from the interaction of the bodies: $m_1/m_2 = a_2/a_1$. In order to express mass numerically, one has to choose a mass unit. In the International System of Units (SI), the unit of mass is the *kilogram* (kg). Mass possesses the property of additivity: if a body is divided into a number of parts, the sum of masses of all parts equals the mass of the whole body. However, the theory of relativity shows that this property is only valid approximately — it is violated if there is strong interaction between parts of the body. For example, the mass of a nucleus turns out to be less than the sum of masses of its constituting nucleons.

The *momentum* (also referred to as *linear momentum*) of a particle is a vector quantity that equals the product of the mass of the particle by its velocity:

$$\mathbf{p} = m\mathbf{v}. \tag{P1.3.1.1}$$

Definition (P1.3.1.1) applies only if $v \ll c$, where c is the speed of light. A more general definition that is valid for any $v < c$ is introduced in the theory of relativity.

▶ **Force.** A force acting on a body of mass m interacting with another body is the vector quantity $\mathbf{F} = m\mathbf{a}$, where $\mathbf{a}$ is the acceleration imparted by this force to the body m in an inertial reference frame. The more general definition is $\mathbf{F} = d\mathbf{p}/dt$. In Newtonian mechanics, both definitions are equivalent. However, in the theory of relativity, only the latter is valid. In SI, force is measured in *newtons* (N); $1\,\mathrm{N} = 1\,\mathrm{kg\,m/s^2}$.

The change of momentum in a time interval Δt, from time 0 to t, under the action of a force $\mathbf{F}$ equals the integral $\Delta\mathbf{p} = \mathbf{p}(t) - \mathbf{p}(0) = \int_0^t \mathbf{F}\,dt$. This quantity is called the *impulse* produced by the force F in the time interval Δt. The average force equals $\mathbf{F}_{\text{ave}} = \Delta\mathbf{p}/\Delta t$.

P1.3.2. Newton's Second and Third Laws

▶ **Newton's second law.** If a particle is acted upon by N bodies with forces $\mathbf{F}_1, \ldots, \mathbf{F}_N$, then the particle acquires an acceleration $\mathbf{a}$ determined by the *vector sum of the forces*:

$$\mathbf{a} = \frac{\mathbf{F}}{m} = \frac{1}{m}\sum_{j=1}^{N}\mathbf{F}_j \qquad \text{or} \qquad \frac{d\mathbf{p}}{dt} = \sum_{j=1}^{N}\mathbf{F}_j.$$

The vector $\mathbf{F} = \sum_{j=1}^{N}\mathbf{F}_j$ is called the *net force* (also known as the *resultant force*) of the forces $\mathbf{F}_1, \ldots, \mathbf{F}_N$. Newton's second law is an *equation of motion* of a particle, since it allows (in principle) the calculation of the motion of the particle if the net force applied is a known function of time and the particle position (and velocity). Likewise, given a system of particles, the set of Newton's second laws for each of them allows the calculation of the motion of the particles if the forces applied are known functions of time and mutual positions of the particles.

Outlined below are the properties of forces that are useful in solving mechanical problems.

▶ **Newton's third law** states that the forces $\mathbf{F}_{AB}$ and $\mathbf{F}_{BA}$ with which two bodies, A and B, act upon each other are directed along the same straight line, opposite in direction, and equal in magnitude:

$$\mathbf{F}_{AB} = -\mathbf{F}_{BA}.$$

P1.3.3. Action at a Distance. Causality. Natural Forces

▶ **Action at a distance. Causality. Field.** In Newtonian mechanics, Newton's third law is considered to hold regardless of the nature of interaction and of whether the bodies are in direct contact with each other or interact at a distance by means of gravitational or electromagnetic forces. However, the validity of Newton's third law for long-range forces would mean instantaneous transmission of information about changes in body positions. The notion of long-range action contradicts the postulates of the theory of relativity, which forbids information transmission at speeds faster than the speed of light. Modern physics has rejected the use of the concept of action at a distance. Instead, it has introduced a new participant—a *force field* (e.g., gravitational, electric, etc.) that fills the whole of space. A particle at a given point in space is acted upon not by distant bodies but by a field in the vicinity of the point; this field is produced by distant bodies (*field sources*). A change in the field is transmitted from point to point and propagates with a final speed (speed of light). Newtonian mechanics is valid in low-speed approximation, $v \ll c$, and in this, *nonrelativistic limit*, one can use both action at a distance and field.

It is convenient to study motion of a particle (or a system of particles) in an *external stationary field* generated by sources at rest (e.g., Earth's gravitational field). A force field is called *central* if the force exerted by the field on a particle placed in it is directed toward a single point called the *field origin* (*force center*) and the magnitude of the force depends only on the distance r to the origin, so that

$$\mathbf{F}(\mathbf{r}) = F_r(r)\frac{\mathbf{r}}{r},$$

where $F_r(r)$ is the projection of the force onto the radial direction; $F_r > 0$ corresponds to repulsive forces and $F_r < 0$ corresponds to attractive forces. A field is called *uniform* in a region of space if this force is the same (in direction and magnitude) at every point of space: $\mathbf{F} = \text{const}$.

▶ **Natural forces**. The whole diversity of naturally occurring forces can be reduced to a few fundamental types of interaction: gravitational, weak, electromagnetic, and strong. However, for practical purposes, this approach is unsuitable. Apart from the fundamental long-range forces (gravitational force in mechanics), one needs to know properties of various forces arising when macroscopic bodies are in contact with one another. Below is the list of forces arising in solving mechanical problems.

1. *Gravitational force* $m\mathbf{g}$ (the force with which a body at the Earth's surface is pulled to the Earth's center).

2. *Elastic force*. When a body is deformed in a certain direction, the restoring elastic force is proportional to the deformation (*Hooke's law*): $F_e = k|\Delta l|$, where k is the spring constant (body stiffness) and Δl is the deformation in this direction. The spring constant of an elastic rod is proportional to its cross-sectional area and inversely proportional to its length, so that

$$F_e = ES\frac{|\Delta l|}{l},$$

where E is *Young's modulus*, which depends on the rod material. In the limit of infinitely large stiffness, the (normal) reaction force N and the string tension force T arise; these forces are determined from the equations of motion. Such forces are called *constraint reaction forces* or *constraint reactions*.

3. *Dry friction force*. Sliding friction force: $F_{\text{fr}} = \mu_1 N$, where μ_1 is the sliding friction coefficient. Static friction force: $F_{\text{fr}} \leq \mu_2 N$, where μ_2 is the static friction coefficient. It is usually assumed that $\mu_1 = \mu_2$ in solving problems. The static friction force is also a constraint reaction.

4. *Drag force*. Drag refers to forces that oppose the relative motion of a body through a liquid or gas. It depends on the relative speed and the transverse size of the body. For low relative speeds and small sizes, the drag force is proportional to the speed (viscous drag). For large relative speeds, the drag force is proportional to the cross-sectional area and the speed squared (head-on drag or quadratic drag).

P1.3.4. Galilean Relativity Principle

Galilean relativity (also known as *Galilean invariance*) is a principle that states that all laws of mechanics are the same in all inertial reference frames. Suppose a frame K' moves relative to a frame K with a constant velocity $\mathbf{V}$. Let the respective axes of the two frames be parallel to each other and let the x- and x'-axes be directed along $\mathbf{V}$ (Fig. P1.5). Also let time be counted off from the instant when the frames coincide. Then the coordinates and time in the frame K' are expressed in terms of those in the frame K using the *Galileo transformations*

$$x' = x - Vt, \qquad y' = y, \qquad z' = z, \qquad t' = t. \tag{P1.3.4.1}$$

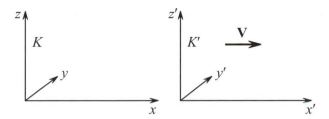

Figure P1.5. Two inertial reference frames. The mechanical laws have the same form in both frames.

Under the Galileo transformations, the difference of the velocities of two particles as well as the distance between them remain unchanged; this means that the forces acting between the two particles also remain the same. Furthermore, the Galileo transformations do not change the accelerations of the particles. Consequently, both sides of the equation of Newton's second law remain unchanged. So the equation of motion has the same form in various reference frames (*invariance under the Galileo transformations*).

The Galileo transformations are based on the assumption that the course of time and the lengths of segments are independent of the reference frame, which were considered to be inherent properties of space and time. The theory of relativity revises the traditional view of space and time and shows that the Galileo transformations are valid for $V \ll c$. The more general transformations are the Lorenz transformations, which are valid for any $V < c$.

P1.4. Law of Conservation of Momentum

P1.4.1. Center of Mass. Momentum of a System of Particles

▶ **System of particles. Center of mass.** Consider a system of N particles. The forces acting on a jth particle can be divided into *internal* ones, $\mathbf{F}^i_{jk}$, exerted by the other particles, and *external* ones, the resultant of which will be denoted $\mathbf{F}^e_j$. According to Newton's third law, the sum of the internal forces is zero. Hence, the sum of all forces acting on the system equals the sum of external forces. If the particles do not interact with external bodies, then the system is called *closed* (or *isolated*).

The *center of mass* (or *center of inertia*) of a system of N particles with masses $m_1, \ldots, m_N$ and position vectors $\mathbf{r}_1, \ldots, \mathbf{r}_N$ is the point whose position is defined by the vector

$$\mathbf{r}_{cm} = \frac{m_1 \mathbf{r}_1 + \cdots + m_N \mathbf{r}_N}{m_1 + \cdots + m_N}. \tag{P1.4.1.1}$$

A translationally moving reference frame with origin at the center of mass is called a *center-of-mass frame*.

If the mass of a system is distributed continuously (in space, on surface, or along a line), the system can be mentally divided into a large number of small regions to represent particles. Further, passing to a limit, one can replace summation by integration. The distribution of mass over volume is defined using *mass density* (or simply *density*), $\rho(\mathbf{r})$, such that $dm = \rho(\mathbf{r})\,dV$. Then, for continuously distributed mass, the definition of the center of mass (P1.4.1.1) becomes

$$\mathbf{r}_{cm} = \frac{\int \mathbf{r}\rho(\mathbf{r})\,dV}{\int \rho(\mathbf{r})\,dV}.$$

A body with constant density across the whole volume is called a *uniform body*.

▶ **Momentum.** The momentum of a system of particles is defined as the sum of momenta of the constituting particles:

$$\mathbf{p} = \sum \mathbf{p}_j.$$

Differentiating equation (P1.4.1.1) with respect to time, one can find that the momentum of the system is expressed in terms of the center-of-mass velocity as

$$\mathbf{p} = m\mathbf{v}_{\text{cm}}, \qquad\qquad (\text{P1.4.1.2})$$

where m is the total mass of the system. It follows that the momentum of a system of particles equals zero in a center-of-mass frame.

Summing up the Newton's Second Law equations for each particle, $d\mathbf{p}_j/dt = \mathbf{F}_j$, and taking into account that the internal forces are canceled out, one obtains

$$\frac{d\mathbf{p}}{dt} = \sum \mathbf{F}_j^{\text{e}} \qquad\qquad (\text{P1.4.1.3})$$

This means that the rate of change of the momentum of the system equals the net external force.

P1.4.2. Law of Conservation of Momentum. Motion of Center of Mass

▶ **Law of conservation of momentum.** From equation (P1.4.1.3) it follows that the *momentum of a closed system is conserved.* The law of conservation of (linear) momentum is a consequence of uniformity of space (there is no distinguished point in space) and is therefore a fundamental law of nature. Although derived within the scope of Newtonian mechanics, this law remains valid beyond it. Even with allowance for a finite speed of signal propagation, which leads to violation of Newton's third law, the law of conservation of momentum is obeyed exactly, provided that the momentum of the transmitter (force field) is taken into account.

The *momentum of an unclosed system* is conserved in the following cases:

1. The net external force is zero.

2. The net external force is perpendicular to some direction. Then the projection of momentum on this direction (not the whole momentum) is conserved.

3. If the interaction last a very short time Δt and the net external force $\mathbf{F}^{\text{e}}$ is bounded,* then the change in the momentum of the system, $\mathbf{F}^{\text{e}}\Delta t$, can be neglected (effectively assuming that $\Delta t \rightarrow 0$).

▶ **Motion of center of mass**. On differentiating equation (P1.4.1.2) with respect to time and taking into account (P1.4.1.3), we arrive at the equation of motion of the center of mass

$$m\mathbf{a}_{\text{cm}} = \sum \mathbf{F}_j^{\text{e}}. \qquad\qquad (\text{P1.4.2.1})$$

It follows that the center of mass moves in the same manner as though it was a single particle, whose mass is equal to the mass of the system, moving under the action of the net external force.

---

* Under these conditions, the momenta of particles can change considerably under the action of large internal forces (explosion, impact, etc.).

Example. A thin rod of length l and mass m rotates in a horizontal plane about one of its ends with an angular velocity ω. Find the tension force at the middle of the rod.

Solution. The center of mass of the outer half of the rod moves in a circle of radius $\frac{3}{4}l$ under the action of a single force, the unknown tension force. It follows from equation (P1.4.2.1) that $F = \left(\frac{1}{2}m\right)\omega^2\left(\frac{3}{4}l\right) = \frac{3}{8}m\omega^2 l$.

Note that the center of mass of a closed system moves with a constant velocity; hence, a center-of-mass frame is inertial.

▶ **Spacecraft propulsion.** For a spacecraft in outer space (far away from gravitating bodies), a change in its velocity is only possible by exhausting part of mass, rocket propellant. The equation of motion of a spacecraft under the action of an external force $\mathbf{F}$ and a propulsive jet can be easily obtained by writing down Newton's second law in the form (P1.4.1.3) in an inertial frame associated with the spacecraft at the current instant:

$$m\,\Delta\mathbf{v} + \Delta M\,\mathbf{u} = \mathbf{F}\,\Delta t,$$

where $\mathbf{u}$ is the jet velocity relative to the spacecraft and ΔM is the mass of the propellant exhausted in time Δt. Dividing by Δt, one arrives at the *Meshcherskii equation*:

$$m\mathbf{a} = \mathbf{F} - \mu\mathbf{u},$$

where $\mu = \Delta M/\Delta t = -\Delta m/\Delta t$ is the *propellant use* in the jet. The latter term on the right-hand side is called the *propulsive force*.

Let us write the Meshcherskii equation for the case of motion in a straight line with no external force applied to obtain $m(dv/dt) = -(dm/dt)u$. Assuming that u is constant and denoting the initial mass of the spacecraft by m_0, we find how the speed of the spacecraft depends on its mass: $v/u = -\ln(m/m_0)$ (*Tsiolkovsky equation*).

P1.5. Law of Conservation of Energy

P1.5.1. Work and Power. Kinetic Energy

▶ **Mechanical work** is a scalar quantity that is a measure of change in energy. The *work of external forces equals the change in energy*:

$$A = E_2 - E_1. \tag{P1.5.1.1}$$

Based on this equation, we will define different kinds of energy. (A refinement of this principle will be given in Chapter P2.)

The work of a force on a small portion of a path is defined as the dot product of the force by the displacement: $\delta A = \mathbf{F} \cdot d\mathbf{r}$. The work on the whole path from position 1 to position 2 equals

$$A = \int_1^2 \mathbf{F} \cdot d\mathbf{r} = \int_1^2 F_\tau\,|d\mathbf{r}| = \int_1^2 F\,|d\mathbf{r}|\cos\alpha,$$

where F_τ is the projection of the force on the direction of motion of the point where the force is applied and α is the angle between the force and this direction. In SI, work and energy are measured in *joules* (J) ($1\,\text{J} = 1\,\text{N m}$).

Example 1. The work of a constant force $\mathbf{F}$,

$$A = \int_1^2 \mathbf{F} \cdot d\mathbf{r} = \left(\mathbf{F} \cdot \int_1^2 d\mathbf{r}\right) = \mathbf{F} \cdot (\mathbf{r}_2 - \mathbf{r}_1),$$

equals the dot product of the force by the total displacement and is independent of the path of the point at which the force is applied.

Example 2. The work of a central force, the force exerted by a central field (see Subsection P1.3.3), equals

$$A = \int_1^2 \mathbf{F} \cdot d\mathbf{r} = \int_{r_1}^{r_2} F_r(r)\, dr.$$

It has been taken into account that the projection of $d\mathbf{r}$ onto the radial direction is equal to the change in the distance r to the origin. The work depends only on the initial, r_1, and final, r_2, distance to the origin and is independent of the path.

▶ **Mechanical power** is the rate at which work is performed. The average power is the ratio of the work done to the time interval. The instantaneous work equals the time derivative of the work done:

$$P = \frac{dA}{dt} = \frac{\mathbf{F} \cdot d\mathbf{r}}{dt} = \mathbf{F} \cdot \mathbf{v} = F_\tau v.$$

Power is measured in *watts* (W): $1\,\mathrm{W} = 1\,\mathrm{J/s}$.

▶ **Kinetic energy.** The kinetic energy of a particle is the energy it possesses due to its motion and depending on its speed. The velocity of a particle changes under the action of a net force $\mathbf{F}$, the work of which equals

$$A = \int_1^2 F_\tau\, |d\mathbf{r}| = m \int_1^2 a_\tau v\, dt = m \int_1^2 \frac{dv}{dt} v\, dt = m \int_{v_1}^{v_2} v\, dv = \tfrac{1}{2} m v_2^2 - \tfrac{1}{2} m v_1^2. \quad \text{(P1.5.1.2)}$$

It is apparent that, in accordance with the general principle (P1.5.1.1), kinetic energy can be defined as $E_k = mv^2/2$. The above equation, stating that the change in the kinetic energy equals the work of the net force, is known as the *work-energy theorem*.

The *kinetic energy of a system of particles* is defined as the sum of the kinetic energies of all constituting particles. The change in the kinetic energy of the system equals the work of all forces acting on its particles. The kinetic energy of a system of particles equals

$$E_k = \frac{m v_{cm}^2}{2} + E_{rel}, \quad \text{(P1.5.1.3)}$$

where m is the mass of the system, v_{cm} is the speed of its center of mass, and E_{rel} is the kinetic energy relative to a center-of-mass frame. Equation (P1.5.1.3) is known as *König's theorem*.

P1.5.2. Conservative Forces. Potential Energy

▶ **Conservative forces. Potential field.** A force of interaction between two particles is called *conservative* if the work of this force depends on only the initial and final positions of the particles and does not depend on their paths. Forces that do not satisfy this condition are called *nonconservative*.

An external stationary field is called *potential* if its work on moving a particle depends on only its initial and final positions and does not depend on its path. (It is equivalent to say that the work of the field on moving a particle through a closed path is zero.) A potential field is a field of conservative forces; it is generated by fixed external sources. As follows from example 1, a uniform field is potential. From example 2 it follows that any central field is potential. A superposition of several central fields (generated by several sources) is also a potential field. Hence, an electrostatic field and a stationary gravitational field are also potential. From example 2 it also follows that the elastic force produced by a light spring is conservative.

A system of particles with only conservative forces acting between them is referred to as a *conservative system*.

▶ **Potential energy** is energy stored within a system; it characterizes interaction between particles and depends on their mutual positions. The *potential energy of a particle in an external potential field* is defined as follows. Let us define the *difference between the potential energies* of the particle as the work performed by the field to move the particle from one position to another:

$$E_p(\mathbf{r}_1) - E_p(\mathbf{r}_2) = A_{12}. \tag{P1.5.2.1}$$

This work is independent of the path. Hence, the *change in the potential energies*, $E_2 - E_1$, equals the *negative* of the work done by the field. If the particle is moved very slowly by an *external* force, then the work performed by the external force is equal in magnitude and opposite in sign to the work of the field, so $A_{12}^e = E_2 - E_1$ in accordance with the general principle (P1.5.1.1).

Equation (P1.5.2.1) defines the difference of potential energies; so the potential energy is defined up to an additive constant. To make the definition unambiguous, one should specify the value (usually zero) of the potential energy at some point in space.

Example 1. The work done by the force of gravity mg on lifting a particle of mass m from height h_1 to height h_2 is equal to $mgh_1 - mgh_2$. Consequently, the potential energy of the particle in the gravitational field is $E_p = mgh$, where the height is counted off from a chosen zero level. The potential energy of a system of points in the field of gravity equals

$$E_p = \sum_j m_j g h_j = mg \frac{\sum_j m_j h_j}{m} = mgh_{cm},$$

where m is the mass of the system and h_{cm} is the height of its center of mass.

Example 2. The work done by the elastic force equals $\int_{x_1}^{x_2} (-kx)\, dx = \frac{1}{2} kx_1^2 - \frac{1}{2} kx_2^2$, where x_1 and x_2 are the initial and final points of the spring deformation. Consequently, the potential energy of the elastic spring is $E_p = \frac{1}{2} kx^2$, where zero energy corresponds to the undeformed spring.

Example 3. The work done by friction or resistive forces is negative on each segment of the path and also along a closed path. Consequently, these forces do not satisfy the condition of being conservative and are nonconservative.

The *potential energy of conservative interation* of two particles can be treated as the potential energy of one particle in the field generated by the other. The result is independent of which particle is considered to be the field source.

▶ **Relation between force and potential energy.** By writing equation (P1.5.2.1) for two close particles lying on the x-axis and separated by a distance dx, one obtains $E_p(x) - E_p(x + dx) = F_x\, dx$. It follows that the projection F_x of the force $\mathbf{F}$ onto the x-axis is determined by the derivative of the potential energy:

$$F_x = -\frac{\partial E_p}{\partial x}. \tag{P1.5.2.2}$$

The partial derivative means that E_p is treated as a function of only x here. The same is obviously valid for projections on any other direction. Consequently, the vector of force equals the opposite of the *gradient* of the potential energy:

$$\mathbf{F} = -\left(\frac{\partial E_p}{\partial x} \mathbf{i} + \frac{\partial E_p}{\partial y} \mathbf{j} + \frac{\partial E_p}{\partial z} \mathbf{k} \right) = -\operatorname{grad} E_p, \tag{P1.5.2.3}$$

where $\mathbf{i}$, $\mathbf{j}$, and $\mathbf{k}$ are unit vectors in the x-, y-, and z-axis, respectively.

For central field, formula (P1.5.2.2) becomes

$$F_r = -\frac{dE_p(r)}{dr}. \tag{P1.5.2.4}$$

P1.5.3. Mechanical Energy of a System of Particles. Law of Conservation of Mechanical Energy

▶ **Mechanical energy.** The mechanical energy of a system of particles is defined as the sum of its kinetic energy, the internal potential energy due to interaction between the particles, and the potential energy in an external field:

$$E_{\text{mech}} = \sum_j \frac{m_j v_j^2}{2} + \sum_{j<n} E_{jn} + \sum_j E_j^{\text{e}}. \tag{P1.5.3.1}$$

According to König's theorem (P1.5.1.3), the kinetic energy of a system equals the sum of two terms, the energy of its motion as a whole, $mv_{\text{cm}}^2/2$, and the kinetic energy in a center-of-mass frame, E_{rel}. The sum of E_{rel} and the potential energy of particle interactions is sometimes called the *internal mechanical energy* of the system.

▶ **Change in mechanical energy.** The change in the kinetic energy of a system equals the total work of all forces applied to the particles (see equation (P1.5.1.2)). The change in the potential energy equals the negative of the work of the conservative forces and external potential fields. Consequently, the change in the mechanical energy of the system equals the work done by the internal nonconservative forces and all external forces, except for the potential fields, whose contribution is included into E_{mech}. So we have

$$\Delta E_{\text{mech}} = A_{\text{nonconservative}} + A^{\text{e}}. \tag{P1.5.3.2}$$

▶ **Law of conservation of mechanical energy.** The mechanical energy of a closed conservative system does not change. This statement is a special case of the general fundamental principle of conservation of energy: the *total energy of a closed system is conserved*. Apart from the mechanical energy, the total energy also includes the different kinds of internal energy: thermal, chemical, and nuclear. The general principle of conservation of energy goes far beyond the scope of Newtonian mechanics, within the framework of which the law of conservation of the mechanical energy has been obtained. This principle is in close connection with the fundamental postulate of *uniformity of time* (there is no selected instant in time) and is one of the fundamental principles on which modern physics is based.

Note that if a conservative system is in an external potential field, the mechanical energy is also conserved, provided that the potential energy of the particles in the field is included in it.

The condition for a system to be conservative is equivalent to the requirement that both the mechanical and internal components of the total energy must be conserved individually. If, for example, there are friction forces acting in the system, whose work is negative, then the mechanical energy is reduced (see equation (P1.5.3.2)) being converted into the internal energy; in this case, heat is said to be produced in the system. The mechanical energy can change also in the case where there is a mechanism present in the system that can produce work at the expense of the internal energy (e.g., an internal combustion engine, a human being, etc.).

Example 1. Elastic collision. In a central elastic impact of two balls, both the momentum of the system and its mechanical energy are conserved. If m_1 and m_2 are the masses of the balls, v_1 and v_2 are speeds of the balls before the collision, and u_1 and u_2 are those after the collision, we have

$$m_1 v_1 + m_2 v_2 = m_1 u_1 + m_2 u_2,$$

$$\frac{m_1 v_1^2}{2} + \frac{m_2 v_2^2}{2} = \frac{m_1 u_1^2}{2} + \frac{m_2 u_2^2}{2}.$$

Instead of the second equation, it is more convenient to use the condition that the relative speed of the balls does not change in magnitude but changes in sign: $v_{1x} - v_{2x} = u_{2x} - u_{1x}$. This equation can be derived directly from the first two, but it becomes obvious if we consider the motion in a center-of-mass frame (the relative speed remains the same here). In this frame, the total momentum of the system is zero and the speed of either ball simply reverses after the impact; both conservation laws are obeyed. Then, solving the two linear equations, we find the final speeds of the balls:

$$u_1 = \frac{(m_1 - m_2)v_1 + 2m_2v_2}{m_1 + m_2}, \qquad u_2 = \frac{(m_2 - m_1)v_2 + 2m_1v_1}{m_1 + m_2}.$$

If we replace the second ball with a moving wall ($m_2 \gg m_1$), we get

$$u_1 \approx -v_1 + 2v_2, \quad u_2 \approx v_2.$$

Example 2. Inelastic collision. Two balls of masses m_1 and m_2 with velocities $\mathbf{v}_1$ and $\mathbf{v}_2$ undergo a perfectly inelastic collision such that the balls move as one with a velocity $\mathbf{u}$ after the collision. (No rotation occurs if the impact is central in a center-of-mass frame.) The velocities become equal as a result of nonconservative forces, which means that some heat is released in the impact. This can be verified by finding the final velocity using the law of conservation of momentum, $m_1\mathbf{v}_1 + m_2\mathbf{v}_2 = (m_1 + m_2)\mathbf{u}$, and calculating the decrease in the mechanical energy:

$$\frac{m_1 v_1^2}{2} + \frac{m_2 v_2^2}{2} - \frac{(m_1 + m_2)u^2}{2} = \frac{m_1 m_2}{m_1 + m_2} \frac{(\mathbf{v}_1 - \mathbf{v}_2)^2}{2}.$$

It is even easier to obtain this result considering the motion in a center-of-mass frame, where the balls are at rest after the collision.

P1.5.4. Potential Curves. Stability

If we know the potential energy $E_p(r)$ of a one-dimensional motion in a potential field (an example is shown in Fig. P1.6), then we can find out the following based on relation (P1.5.2.2):

a) The direction of the force ($F > 0$ for $0 < r < r_{\min}$ and $r > r_{\max}$; $F < 0$ for $r_{\min} < r < r_{\max}$).

b) The points of equilibrium ($F = 0$ at $r = r_{\min}$ and $r = r_{\max}$).

c) Stability of equilibrium. In the vicinity of the point $r = r_{\min}$, the force is directed toward the point, so the equilibrium is stable. At $r = r_{\max}$, the equilibrium is unstable. A stable equilibrium corresponds to a minimum of the potential energy.

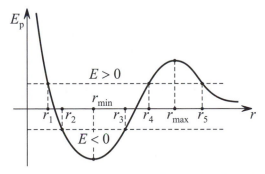

Figure P1.6. Given a potential energy curve, one can determine the direction of the force, the positions of the equilibrium points and their stability, and the character of motion.

d) The character of motion for a given value of the total mechanical energy E. Motion is possible only where $E > E_p$ (the kinetic energy is nonnegative). If $E < E_p(\infty)$ (in Fig. P1.6, $E_p(\infty) = 0$), motion is *finite*, so it can occur in a limited range of r. With $E < 0$ in Fig. P1.6, motion occurs only between the *turning points* r_2 and r_3. For $E > 0$, motion is either *infinite*, where the particle goes to infinity with kinetic energy E after reflecting from the turning point r_5, or locked by a potential barrier, where the particle moves between the turning points r_1 and r_4. In classical mechanics, a potential barrier is impenetrable. In quantum mechanics, there is a nonzero probability that a potential barrier can be penetrated (tunnel effect).

P1.6. Law of Conservation of Angular Momentum

P1.6.1. Moment of Force. Net Force

▶ **Moment of force.** The moment of a force $\mathbf{F}$ about a point O is defined as

$$\mathbf{M} = \mathbf{r} \times \mathbf{F}, \tag{P1.6.1.1}$$

where $\mathbf{r}$ is the vector from O to the point where the force is applied. A moment of force is also called a *torque*. The moment of force does not change if the force is moved along the straight line of its application. The magnitude of the moment of force equals $M = Fr \sin \alpha = Fd$, where α is the angle between the force vector and the radius vector $\mathbf{r}$, and d is the distance between the point O and the line of action of the force; the distance d is called the *moment arm* (Fig. P1.7).

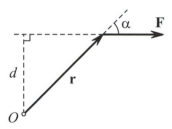

Figure P1.7. The moment of the force $\mathbf{F}$ is perpendicular to the plane of the figure and directed away from the reader. The magnitude of the moment equals the product of F by the arm d.

The most important property of moment of force is that the sum of the moments of internal forces in an arbitrary system of particles is zero. This fact follows from Newton's third law (see Subsection P1.3.2), since the forces acting between each pair of particles are equal in magnitude, opposite in direction, and directed along the straight line connecting the particles.

Note another property of moment of force: if the sum of forces is zero, then these forces produce the same net moment about any point in space.

▶ **Net force.** The *net force* of a system of forces is the force that equals the vector sum of them and is applied so that its moment about any point in space equals the net moment of this system of forces.

Example. Compute the net moment of the forces of gravity acting on the particles of a system:

$$\mathbf{M} = \sum \mathbf{r}_j \times m_j \mathbf{g} = \left(\sum m_j \mathbf{r}_j \right) \times \mathbf{g} = \mathbf{r}_{\mathrm{cm}} \times (m\mathbf{g}),$$

where m is the mass of the system, $\mathbf{r}_{\mathrm{cm}}$ is the radius vector of the center of mass. It is apparent that the net force passes through the center of mass; hence, *the center of mass is also the center of gravity.*

The *moment of force about an axis* z is the z-projection, M_z, of the moment of force about any point on the axis. If both $\mathbf{r}$ and $\mathbf{F}$ in (P1.6.1.1) are resolved into two components, parallel and perpendicular to the axis z, so that $\mathbf{r} = \mathbf{r}_\parallel + \mathbf{r}_\perp$ and $\mathbf{F} = \mathbf{F}_\parallel + \mathbf{F}_\perp$, then the components parallel to the axis will not contribute to the projection of the moment onto the axis. It follows that $M_z = (\mathbf{r}_\perp \times \mathbf{F}_\perp)_z$. In the plane perpendicular to the axis, we have the same picture as in Fig. P1.7, except that $\mathbf{r}$ and $\mathbf{F}$ must be replaced with $\mathbf{r}_\perp$ and $\mathbf{F}_\perp$. Accordingly, the magnitude of the moment about an axis is also equal to the product of the force by the moment arm, while the choice of the positive direction along the axis can be replaced with the easier choice of the positive direction of rotation about the axis.

P1.6.2. Angular Momentum of a Particle. Motion in Central Field

▶ **Angular momentum.** The angular momentum of a particle about a point O is the vector quantity

$$\mathbf{L} = \mathbf{r} \times \mathbf{p}, \tag{P1.6.2.1}$$

where $\mathbf{r}$ is the vector from O to the position of the particle and $\mathbf{p} = m\mathbf{v}$ is the (linear) momentum of the particle. Angular momentum is also known as *moment of momentum*. If the particle moves with a constant velocity, its angular momentum does not change. The magnitude of the angular momentum equals the product of the momentum by the arm. The angular momentum about an axis z is the z-projection, L_z, of the angular momentum $\mathbf{L}$ about any point at the axis. The angular momentum about an axis is determined by the projection of the momentum on a plane perpendicular to the axis: $L_z = (\mathbf{r}_\perp \times \mathbf{p}_\perp)_z$. These properties are similar to those of the moment of force.

The time derivative of the angular momentum of a particle about a point O,

$$\frac{d\mathbf{L}}{dt} = \frac{d\mathbf{r}}{dt} \times \mathbf{p} + \mathbf{r} \times \frac{d\mathbf{p}}{dt} = \mathbf{v} \times \mathbf{p} + \mathbf{r} \times \mathbf{F} = \mathbf{M}, \tag{P1.6.2.2}$$

equals the moment of the net force about O (since $\mathbf{v}$ and $\mathbf{p}$ are parallel to each other, we have $\mathbf{v} \times \mathbf{p} = \mathbf{0}$)

▶ **Motion in central field.** Since the force acting on a particle in a *central field* (also called a *central-force field*) is directed to the origin of the field, the moment of this force about the origin is identically zero. Hence, the vector angular momentum $\mathbf{L}$ is conserved. It follows from definition (P1.6.2.1) that the motion occurs in one plane perpendicular to $\mathbf{L}$ and the quantity $L = mvr \sin \alpha$ is conserved. The magnitude of the angular momentum is proportional to the rate at which the radius vector $\mathbf{r}$ sweeps out area (Fig. P1.8): $dS/dt = \frac{1}{2}r(v\,dt)\sin \alpha/dt = \frac{1}{2}L/m$. Consequently, the statement that angular momentum is conserved in a central field turns out to be equivalent to *Kepler's second law of planetary motion*.

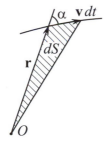

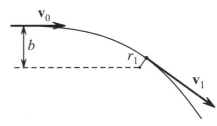

Figure P1.8. The rate at which the radius vector sweeps out area is expressed in terms of the angular momentum.

Figure P1.9. Fly-by of a particle near a force field center; b is the aiming parameter.

Example 1. Aiming parameter. Conservation of angular momentum in a central field allows one to relate the aiming parameter b of a particle approaching the field origin from far away (Fig. P1.9) to the distance of closest approach r_1 to the origin. (The *aiming parameter* is the distance from the field origin to the straight line of particle motion in the absence of interaction.) At the instant of closest approach to the origin, the particle velocity is perpendicular to the radius vector, so the law of conservation of angular momentum can be written as

$$mv_0 b = mv_1 r_1,$$

where v_0 is the initial speed of the particle far away from the origin and v_1 is the particle speed at the instant of closest approach. Expressing v_1 and substituting into the law of conservation of energy, $\frac{1}{2}mv_0^2 = \frac{1}{2}mv_1^2 + E_p(r_1)$, one can find r_1.

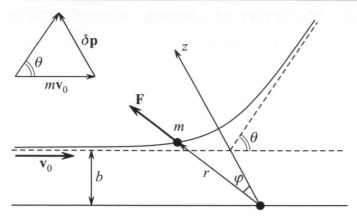

Figure P1.10. To the problem of determining the scattering angle θ in the field $F_r = A/r^2$.

Example 2. Scattering angle. The problem of particle scattering on a Coulomb (or gravitational) center is of significant importance in astrophysics and nuclear physics. Consider a particle of mass m moving toward a (repulsive) force center $F_r = A/r^2$ with the initial kinetic energy $E_k = \frac{1}{2}mv_0^2$ and aiming parameter b. Find the scattering angle θ.

Solution. From symmetry considerations it is clear that the change in momentum in the whole duration of the motion is directed along the z-axis (see Fig. P1.10) and is equal to

$$|\Delta \mathbf{p}| = \Delta p_z = 2mv_0 \sin(\theta/2).$$

The change in momentum (projected onto the z-axis) in the time dt is

$$dp_z = F_z \, dt = F \cos \varphi \, dt = \frac{A}{r^2} \frac{d\varphi}{v_\varphi/r} \cos \varphi = \frac{mA}{L} \cos \varphi \, d\varphi,$$

where $v_\varphi = r(d\varphi/dt)$ is the azimuthal (perpendicular to the radius vector) velocity component (see Fig. P1.11). The angular momentum of the particle is expressed in terms of v_φ as $L = mv_\varphi r$. Integrating with respect to φ from $-(\pi - \theta)/2$ to $(\pi - \theta)/2$ yields

$$\Delta p_z = \frac{2mA}{L} \cos(\theta/2).$$

Comparing the two expressions of Δp_z and taking into account that $L = mv_0 b$, we find that

$$\tan \frac{\theta}{2} = \frac{A}{2E_k b}.$$

This formula will be used subsequently in the discussion of Rutherford's experiments.

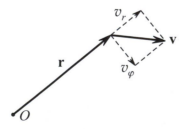

Figure P1.11. Radial and azimuthal components of velocity.

▶ **Effective potential energy.** With the laws of conservation of energy and angular momentum, one can establish the time dependence of the central distance r by reducing the problem to that of one-dimensional motion. To this end, the particle velocity should be resolved into two mutually perpendicular components (Fig. P1.11), a radial one, $v_r = dr/dt$, and an azimuthal one, v_φ. The angular momentum is expressed in terms of the azimuthal speed: $L = mrv_\varphi$. Hence, the mechanical energy of the particle can be written as

$$E = \frac{m(v_r^2 + v_\varphi^2)}{2} + E_{\mathrm{p}}(r) = \frac{m}{2}\left(\frac{dr}{dt}\right)^2 + \left[\frac{L^2}{2mr^2} + E_{\mathrm{p}}(r)\right].$$

It is apparent that the dependence $r(t)$ is the same as in a one-dimensional motion with the effective potential energy

$$E_{\mathrm{p}}^{\mathrm{eff}} = E_{\mathrm{p}} + \frac{L^2}{2mr^2}.$$

The second term is sometimes called the *centrifugal energy*. The graph of the effective potential curve is determined by the value of L, which can be evaluated from the initial conditions.

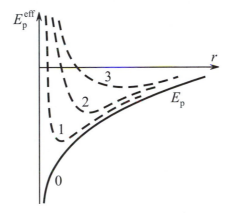

Figure P1.12. The form of the effective potential curve is determined by the angular momentum magnitude L.

Figure P1.12 depicts qualitatively the dependence of the effective potential energy $E_{\mathrm{p}}^{\mathrm{eff}}$ on the distance r for various L ($L_0 = 0$ and $L_1 < L_2 < L_3$) in the case that $E_{\mathrm{p}} = -b/r$. It is apparent that, for any L, the condition of finiteness of the motion remains the same: $E < 0$.

P1.6.3. Conservation of Angular Momentum for a System of Particles

▶ **Angular momentum of a system of particles.** The angular momentum of a system of particles about a point O is the vector sum of the angular momenta of the constituting particles. It is equal to

$$\mathbf{L} = \mathbf{r}_{\mathrm{cm}} \times \mathbf{p} + \mathbf{L}_{\mathrm{rel}}, \tag{P1.6.3.1}$$

where $\mathbf{L}_{\mathrm{rel}}$ is the angular momentum in a center-of-mass frame. One of the consequences of equation (P1.6.3.1) is that if the linear momentum of the system is zero, the angular momentum $\mathbf{L}$ does not depend on the choice of the point O. Summing up equation (P1.6.2.2) over all point of the system and taking into account that the net moment of internal forces is zero, we find that

$$\frac{d\mathbf{L}}{dt} = \mathbf{M}^{\mathrm{e}}, \tag{P1.6.3.2}$$

or *the time derivative of the angular momentum of the system equals the net moment of external forces.*

▶ **Law of conservation of angular momentum.** It follows from equation (P1.6.3.2) that *the angular momentum of a closed system is conserved.* The law of conservation of angular momentum is a fundamental law based on the postulate of isotropy of space (there is no selected direction in space). Just as with the laws of conservation of linear momentum and energy, the application of the law of conservation of angular momentum goes beyond the scope of Newtonian mechanics, where it was derived.

The angular momentum of a *unclosed* system is conserved in the following cases:

1. If the net moment of external forces $\mathbf{M}^e$ is zero. (Example: a system of interacting particles in an external central field.)

2. If the moment of external forces about an axis is zero, then the angular momentum about this axis is conserved.

Example. If a weight of mass m attached to one end of a weightless string moves in a horizontal circle, the other end of the string fixed, then the angular momentum about any point at the axis of rotation (except for the center of the circle) is not conserved. However, the angular momentum about the axis of rotation is conserved, since the moment of the force of gravity and that of the tension force about this axis are both zero.

3. If the external forces are bounded, then the change in angular momentum $\Delta \mathbf{L} = \mathbf{M}^e \Delta t$ in the short time of impact Δt can be neglected.

P1.6.4. Two-Body Problem. Reduced Mass

▶ **Reduced mass.** Consider a closed system consisting of two interacting particles. Solving the problem of motion of these particles (known as the *two-body problem*) means determining the particle positions at any time for given initial conditions. The positions of the particles are expressed in terms of the position of their center of mass $\mathbf{r}_{cm}(t)$ (see (P1.4.1.1)) and their relative displacement $\mathbf{r}_{12} = \mathbf{r}_1 - \mathbf{r}_2$ as follows:

$$\mathbf{r}_1 = \mathbf{r}_{cm} + \frac{m_2}{m_1 + m_2}\mathbf{r}_{12}, \qquad \mathbf{r}_2 = \mathbf{r}_{cm} - \frac{m_1}{m_1 + m_2}\mathbf{r}_{12}.$$

The center of mass moves with a constant velocity, with the initial position and velocity found from the initial conditions (see equations (P1.4.1.1) and (P1.4.1.2)). Thus, solving the two-body problem is reduced to determining $\mathbf{r}_{12}$.

From the conditions of uniformity and isotropy of space and uniformity of time, it follows that the particles interact with central forces such that $\mathbf{F}_{12} = -\mathbf{F}_{21} = \mathbf{F}$, where $\mathbf{F}$ is parallel to $\mathbf{r}_{12}$ and the magnitude F is dependent on $|\mathbf{r}_{12}|$ only. Let us write the equations of motion for each particle:

$$\mathbf{F} = m_1 \frac{d^2\mathbf{r}_1}{dt^2}, \qquad -\mathbf{F} = m_2 \frac{d^2\mathbf{r}_2}{dt^2}.$$

Subtracting the second equation divided by m_2 from the first equation divided by m_1, we get

$$\left(\frac{1}{m_1} + \frac{1}{m_2}\right)\mathbf{F} = \frac{d^2\mathbf{r}_{12}}{dt^2} \qquad \text{or} \qquad \mathbf{F} = \mu \frac{d^2\mathbf{r}_{12}}{dt^2},$$

where μ is the reduced mass determined by

$$\frac{1}{\mu} = \frac{1}{m_1} + \frac{1}{m_2}, \qquad \text{so} \qquad \mu = \frac{m_1 m_2}{m_1 + m_2}. \qquad (P1.6.4.1)$$

It is apparent that the vector $\mathbf{r}_{12}$ is determined by solving the problem of motion of a particle of mass μ in a central field.

▶ **Linear momenta and kinetic energies of two bodies in a center-of-mass frame.** The particle momenta in a center-of-mass frame, $\widetilde{\mathbf{p}}_1$ and $\widetilde{\mathbf{p}}_2$, are equal in magnitude (the total momentum is zero, with $\widetilde{\mathbf{p}}_2 = -\widetilde{\mathbf{p}}_1$ and $|\widetilde{\mathbf{p}}_1| = |\widetilde{\mathbf{p}}_2| = \widetilde{p}$) and are expressed in terms of the particle relative velocity (independent of the reference frame) as

$$|\mathbf{v}_{\text{rel}}| = \left| \left(\frac{\widetilde{\mathbf{p}}_1}{m_1} - \frac{\widetilde{\mathbf{p}}_2}{m_2} \right) \right| = \frac{\widetilde{p}}{\mu},$$

The kinetic energy of the system in the center-of-mass frame can also be expressed in terms of the relative velocity as

$$\widetilde{E}_k = \frac{\widetilde{p}^2}{2m_1} + \frac{\widetilde{p}^2}{2m_2} = \frac{\widetilde{p}^2}{2\mu} = \frac{\mu v_{\text{rel}}^2}{2}.$$

This formula was already encountered in considering inelastic collision (example 2, Subsection P1.5.3).

P1.7. Gravitational Field

P1.7.1. Law of Universal Gravitation. Superposition Principle

▶ **Newton's law of universal gravitation.** Two particles of masses m_1 and m_2 separated by a distance r are pulled to each other with a force of gravity (gravitational force) equal to

$$F = G\frac{m_1 m_2}{r^2}, \tag{P1.7.1.1}$$

where $G \approx 6.673 \times 10^{-11} \text{ N m}^2/\text{kg}^2$ is the gravitational constant. The gravitational force is a central force; it acts along the line connecting the particles.

The force exerted on a point mass m in a central field of gravity (gravitational field) produced by a fixed mass M can be represented as (see Subsection P1.3.3)

$$\mathbf{F} = -G\frac{mM}{r^2}\frac{\mathbf{r}}{r} \qquad \text{or} \qquad F_r(r) = -G\frac{mM}{r^2}. \tag{P1.7.1.2}$$

Thus, the role of the "charge" for a gravitational field is played by the inertial mass m. This property is sometimes formulated as the *equivalence of gravitational and inertial masses*. The potential energy of a particle in a central field of gravity can be found by using the relation between force and potential energy (formula (P1.5.2.4)); we have $-GmM/r^2 = -dE_p/dr$, so $E_p = -GmM/r + \text{const}$. The constant is usually set to zero, thus taking the potential energy to be zero at infinity:

$$E_p = -G\frac{mM}{r}. \tag{P1.7.1.3}$$

▶ **Principle of superposition.** If a field of gravity is generated by several point masses, $M_1, \ldots, M_N$, then the force acting on a particle m and its potential energy are expressed as

$$\mathbf{F} = \sum \mathbf{F}_j = -\sum G\frac{mM_j}{(\mathbf{r} - \mathbf{r}_j)^2}\frac{\mathbf{r} - \mathbf{r}_j}{|\mathbf{r} - \mathbf{r}_j|},$$
$$E_p = \sum E_{pj} = -\sum G\frac{mM_j}{|\mathbf{r} - \mathbf{r}_j|}, \tag{P1.7.1.4}$$

where $\mathbf{r}$ is the position vector of the particle m and $\mathbf{r}_j$ is the position vector of the mass point M_j. If a gravitational field is generated by a continuously distributed mass, then the summation in (P1.7.1.4) must be replaced by integration.

P1.7.2. Strength and Potential of Gravitational Field

▶ **Strength and potential of the gravitational field generated by a point mass.** It is apparent from equations (P1.7.1.4) that both the force acting on a particle of mass m in a gravitational field and the potential energy of the particle are proportional to m. Hence, the specific force, $\mathbf{F}/m$, and specific energy, E_p/m, are independent of m and so characterize the field itself. These are called the *strength* and *potential* of the gravitational field, respectively, and denoted

$$\mathbf{g} = \frac{\mathbf{F}}{m}, \qquad \varphi = \frac{E_p}{m}.$$

The field strength has a simple physical meaning: it is the *acceleration due to gravity*, or the acceleration acquired by any particle placed at a given point due to the action of the gravitational field only. The strength and potential created by a point mass M are expressed as

$$\mathbf{g} = -G\frac{M}{r^2}\frac{\mathbf{r}}{r}, \qquad \varphi = -G\frac{M}{r}. \tag{P1.7.2.1}$$

The strength and potential of the gravitation field generated by several point masses are calculated using the principle of superposition. For example, let us write equations (P1.7.2.1) for the case of a body with distributed mass:

$$\mathbf{g}(\mathbf{r}) = -\int G\frac{\rho(\mathbf{r}')}{(\mathbf{r}-\mathbf{r}')^2}\frac{\mathbf{r}-\mathbf{r}'}{|\mathbf{r}-\mathbf{r}'|}\,dV', \qquad \varphi(\mathbf{r}) = -\int G\frac{\rho(\mathbf{r}')}{|\mathbf{r}-\mathbf{r}'|}\,dV', \tag{P1.7.2.2}$$

where $\rho(\mathbf{r})$ is the mass density at the position $\mathbf{r}$ and the integration is performed over the volume occupied by the body.

Example 1. Show that the gravitational strength inside a hollow spherical layer is zero.

Solution. For a given point A, consider the contributions to the field strength of the small regions B and C cut out from the sphere by a thin cone with apex at A (Fig. P1.13). The ratio of the areas of these regions, and hence the ratio of their masses, equals the ratio of the squared distances of the regions to the point A. It follows that the strengths produced by the regions B and C at A are equal in magnitude.

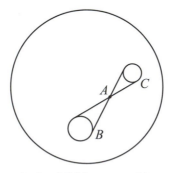

Figure P1.13. The strengths of the gravitational fields generated by opposite regions on a spherical surface cancel each other out.

The field strength produced by a thin hollow sphere of mass M outside the sphere turns out to be equal to the strength produced by a point mass M placed at the sphere center. Cumbersome integration is required to prove this result (Newton was first to do this). In Chapter P3, this statement will be proved using Gauss's theorem. The same result holds true for any spherically symmetric distributed mass and, in particular, for any spherical planet.

Example 2. Suppose a mass M is uniformly distributed along a segment of length l. Compute the gravitational strength and potential at the point lying outside on the continuation of the segment at a distance x from its center.

Solution. The mass of a segment of length dy equals $dm = M\,dy/l$. Integrating yields

$$g(x) = -G\int_{x-l/2}^{x+l/2}\frac{M\,dy}{ly^2} = -G\frac{M}{x^2-(l/2)^2},$$

$$\varphi(x) = -G\int_{x-l/2}^{x+l/2}\frac{M\,dy}{ly} = -G\frac{M}{l}\ln\frac{x+l/2}{x-l/2}.$$

It is apparent that a symmetric but nonspherical body cannot be replaced by a point mass placed at its center. This example illustrates also that the strength and potential are linked by the relation $g_x = -\partial\varphi/\partial x$, which is similar to (P1.5.2.2).

P1.7.3. Motion in a Central Field of Gravity. Kepler's Laws. Orbital and Escape Velocities

▶ **Motion in a central field of gravity. Kepler's laws.** Motion in a central field of gravity obeys the general laws of motion in a central field. However, it has some specific features reflected in Kepler's first and third laws initially formulated for planets of the solar system.

Kepler's first law states that a finite motion ($E < 0$) of a particle in a central gravitational field occurs in a closed path—an ellipse with the force center (the sun) located at one of its foci.

Kepler's second law affirms the constancy of the areal velocity, i.e., the rate at which the radius vector of the moving particle sweeps area. This law holds for any central field and follows directly from the law of conservation of momentum (see Subsection P1.6.2).

Kepler's third law states that, for any two particles moving in a central gravitational field, the ratio of the squares of their orbital periods equals the ratio of the cubes of the semimajor axes of their orbits: $T_1^2/T_2^2 = a_1^3/a_2^3$.

Kepler's first law should be supplemented with the statement that an infinite motion of a particle in a central gravitational field occurs in either a parabola ($E = 0$) or a hyperbola ($E > 0$). The third law can be supplemented by the relationship between the specific energy of orbital motion and the semimajor axis: $|E|/m = GM/(2a)$. It is apparent that the orbital period is uniquely determined by the specific energy of the moving particle.

▶ **Orbital and escape velocities.** The *orbital velocity* (also known as the *first cosmic velocity*) of a small body around a planet is the speed at which the body moves in a circular orbit around the planet near its surface, with the atmospheric drag neglected. The orbital speed is determined from the equation $mg = mv_0^2/R$ and equals $v_0 = \sqrt{gR} = \sqrt{GM/R}$, where M is the mass of the planet. (For Earth, $v_0 \approx 7.9\,\text{km/s}$.)

The *escape velocity* (also known as the *second cosmic velocity*) of a body is the minimum speed the body must have to "break free" from a planet's gravitational pull and go to infinity. As follows from Subsection P1.6.2, the condition for a motion to be infinite is the inequality $E \geq 0$; hence, the escape velocity is determined by the equation $mv_e^2/2 - GmM/R = 0$ and equals $v_e = \sqrt{2GM/R} = \sqrt{2gR}$. (For Earth, $v_e \approx 11.2\,\text{km/s}$.)

P1.8. Non-Inertial Reference Frames

P1.8.1. Definition of Fictitious Forces

In many cases, it is convenient to solve a dynamics problem in a non-inertial frame of reference (a reference frame that is not inertial) instead of recalculating results first obtained in an inertial frame. To this end, *fictitious forces* (also called *inertial forces*) are introduced as follows. In Newton's second law for a particle of mass m moving under the action of a net force $\mathbf{F}$, the acceleration is first represented as the sum of two components, $\mathbf{a} = \mathbf{a}_{rel} + \mathbf{a}^*$, where $\mathbf{a}_{rel}$ is the acceleration of the particle in the non-inertial frame. Then the other component, multiplied by m, is transferred to the other side of the equation and called the fictitious force. So we have

$$\mathbf{F} = m\mathbf{a},$$
$$\mathbf{F} = m(\mathbf{a}_{rel} + \mathbf{a}^*),$$
$$\mathbf{F} + (-m\mathbf{a}^*) = m\mathbf{a}_{rel},$$
$$\mathbf{F} + \mathbf{F}_{fict} = m\mathbf{a}_{rel}.$$

The fictitious forces are thus defined by

$$\mathbf{F}_{fict} = -m\mathbf{a}^*, \qquad (P1.8.1.1)$$

where the acceleration $\mathbf{a}^*$ is determined by the kinematic equation

$$\mathbf{a} = \mathbf{a}_{rel} + \mathbf{a}^* \qquad (P1.8.1.2)$$

and is dependent on the non-inertial frame itself and the position and velocity of the particle in it. Consider some specific cases.

P1.8.2. Translationally Moving Non-Inertial Frames

In this case, $\mathbf{a}^*$ equals the acceleration of the inertial frame K (see formula (P1.1.4.1)). Then the fictitious force is expressed as

$$\mathbf{F}_{fict} = -m\mathbf{a}_K.$$

It is apparent that the fictitious force here is equivalent to the force of gravity. So it is convenient to combine the two forces together in solving problems. With this done, the introduction of a fictitious force turns out to be equivalent to adjusting the acceleration due to gravity (see Subsection P1.7.2), so that $\mathbf{g}$ is effectively replaced by $\mathbf{g}^* = \mathbf{g} + \mathbf{a}^* = \mathbf{g} - \mathbf{a}_K$.

Example. A container filled with a liquid moves with a constant horizontal acceleration $\mathbf{a}$. Find the angle β between surface of the liquid and the horizontal plane.

Solution. In a reference frame associated with the container, the liquid surface is at rest and "horizontal," which means that it is perpendicular to the vector $\mathbf{g}^* = \mathbf{g} + \mathbf{a}^* = \mathbf{g} - \mathbf{a}$. So we need to find the angle between $\mathbf{g}$ and $\mathbf{g} - \mathbf{a}$. We have $\tan\beta = a/g$.

The principle of equivalence of inertial and gravitational forces was laid by Einstein at the basis of his *general theory of relativity*, which is a relativistic theory of gravitation and explains the appearance of gravitational forces by the fact that the space–time continuum is curved by external massive bodies.

P1.8.3. Uniformly Rotating Reference Frame

In this case the kinematic equation for the acceleration (P1.8.1.2) becomes: $\mathbf{a} = \mathbf{a}_{rel} + \boldsymbol{\omega} \times (\boldsymbol{\omega} \times \mathbf{r}) + 2\boldsymbol{\omega} \times \mathbf{v}_{rel}$ (see Subsection P1.2.2). The second term is associated with the rotation of the translatory velocity $\boldsymbol{\omega} \times \mathbf{r}$ together with the reference frame. This vector is directed toward the axis of rotation and its magnitude equals $\omega^2 R$ (R is the distance to the axis), and hence it is the normal (centripetal) acceleration of the reference frame at the given point.

The third term (Coriolis acceleration) is due to, first, the rotation of $\mathbf{v}_{\text{rel}}$ together with the reference frame and, second, the change in the translatory velocity $\boldsymbol{\omega} \times \mathbf{r}$ due to the movement of the particle from one point of the rotating non-inertial frame to another. Accordingly, the fictitious force (P1.8.1.1) is the sum of two components, the first called the *centrifugal force* and the second called the *Coriolis force*:

$$\mathbf{F}_{\text{fict}} = \mathbf{F}_{\text{cf}} + \mathbf{F}_{\text{Cor}} = m\omega^2 \mathbf{R} + 2m\mathbf{v}_{\text{rel}} \times \boldsymbol{\omega}. \tag{P1.8.3.1}$$

The centrifugal fictitious force is directed outward from the axis of rotation ($\mathbf{R}$ is directed from the axis and is perpendicular to it). Since it is independent of the particle velocity, its action is indistinguishable from a (nonuniform) gravitational field. For example, the force of gravity measured at the Earth's surface is the sum of the gravitational force and the centrifugal inertial force.

The Coriolis force is perpendicular to the particle velocity. For example, in the Northern hemisphere, for motions along the Earth's surface the horizontal component of the Coriolis force is directed to the right of the motion, which exhibits itself in the formation of cyclones (rotate counterclockwise), greater erosion of right river banks, etc.

Example. Find the deflecting action of the Coriolis force onto a body freely falling on the Earth's surface from a height h at the equator.

Solution. Since the deviation is small, the "undisturbed" velocity of free falling $\mathbf{v} = \mathbf{g}t$ can be substituted in the Coriolis force (P1.8.3.1). The vector $\boldsymbol{\omega}$ is perpendicular to $\mathbf{v}$, and hence the Coriolis force is directed eastward, its magnitude equal $2mv\omega$, and the horizontal acceleration imparted to the body is $a_{\text{east}} = 2\omega gt$. Integrating yields the horizontal speed $v_{\text{east}} = g\omega t^2$ and the horizontal displacement $s_{\text{east}} = \frac{1}{3}g\omega t^3$. Substituting the time of falling $t = \sqrt{2h/g}$ gives the final deviation. For example, if the initial height is $h = 300$ m, the final deviation is approximately 10 cm.

P1.9. Dynamics of Rigid Body

P1.9.1. Rotation About a Fixed Axis

For a rigid body consisting of many particles of mass m_j and rotating about a fixed axis z with an angular velocity ω, the magnitude of its angular momentum about the axis (see Subsection P1.6.2) equals

$$|L_z| = \sum_j R_j m_j v_j = \omega \sum_j m_j R_j^2 = I_z \omega, \tag{P1.9.1.1}$$

where R_j is the distance from the particle m_j to the axis; the relation $v_j = \omega R_j$ has been used here. The angular momentum direction coincides with that of $\boldsymbol{\omega}$ (determined by the right hand rule). The quantity

$$I_z = \sum_j m_j R_j^2 = \int R^2 \, dm \tag{P1.9.1.2}$$

is called the *moment of inertia* of the body about the axis z. Differentiating (P1.9.1.1) with respect to time and taking into account that $dL_z/dt = M_z$, where M_z is the moment of external forces about the axis of rotation (see equation (P1.6.3.2)), we get

$$M_z = I_z \varepsilon, \tag{P1.9.1.3}$$

where $\varepsilon = d\omega/dt$ is the angular acceleration. This equation is referred to as the *basic equation of rotational dynamics of a rigid body about a fixed axis*. Compute also the kinetic energy of the rotating body,

$$E_k = \sum_j \frac{m_j v_j^2}{2} = \tfrac{1}{2}\omega^2 \sum_j m_j R_j^2 = \tfrac{1}{2} I_z \omega^2, \tag{P1.9.1.4}$$

and the work done by an external force $\mathbf{F}$ on rotating the body,

$$A = \int d\mathbf{r} \cdot \mathbf{F} = \int (\omega \, dt \times \mathbf{r}) \cdot \mathbf{F} = \int (\omega \, dt) \cdot (\mathbf{r} \times \mathbf{F}) = \int \mathbf{M} \cdot d\varphi,$$

where $d\varphi = \omega \, dt$.

Example. A body of mass m is attached to one end of a weightless rope wound around a pulley of radius R and moment of inertia I. Find the acceleration of the body after it has been released.

Solution. Let us write Newton's second law for the body: $mg - T = ma$. The equation of motion of the pulley (P1.9.1.3) is $TR = I\varepsilon$. Using the kinematic relation $a = \varepsilon R$ (the rope does not slide against the pulley) and solving the above equations for a, we obtain $a = g/(1 + I/mR^2)$.

P1.9.2. Properties of Moment of Inertia

The moment of inertia (P1.9.1.2) is a scalar quantity that characterizes the distribution mass of the body relative to the axis of rotation. It can be seen from equations (P1.9.1.3) and (P1.9.1.4) that the moment of inertia is a measure of the body's inertia with respect to rotational motion; moment of inertia plays the same role as mass in translational motion.

Example. Compute the moment of inertia of a thin disk of mass m and radius R about its axis of symmetry.
Solution. Dividing the disk into thin circular rings and integrating, we obtain

$$I = \int r^2 \, dm = \int_0^R r^2 \left(\frac{m}{\pi R^2}\right) 2\pi r \, dr = \frac{mR^2}{2}.$$

The same result holds also for a homogeneous solid cylinder.

▶ **Parallel axis theorem.** The *parallel axis theorem* (also known as the *Huygens–Steiner theorem*) relates the moment of inertia I about an arbitrary axis to the moment of inertia I_0 about a parallel axis passing through the center of mass of the body:

$$I = I_0 + ma^2, \tag{P1.9.2.1}$$

where m is mass of the body and a is the distance between the axes. For example, the moment of inertia of a disk about an axis perpendicular to its plane and passing through its edge equals $\tfrac{1}{2}mR^2 + mR^2 = \tfrac{3}{2}mR^2$. The minimum moment of inertia among all parallel axes is for the axis passing the center of mass.

▶ **Perpendicular axis theorem.** The *perpendicular axis theorem* asserts that the moment of inertia of a plane body about an arbitrary axis z perpendicular to the body plane equals the sum of the moments of inertia about mutually perpendicular axes x and y lying in the body plane and intersecting z:

$$I_z = I_x + I_y.$$

For example, the moment of inertia of a thin disk about an axis of symmetry lying in its plane equals $I_x = I_y = \tfrac{1}{2}I_z = \tfrac{1}{4}mR^2$.

▶ **Moments of inertia of some bodies.** Listed below are moments of inertia of some bodies of various shape.

1) Thin ring of radius R (about its axis of symmetry): $I = mR^2$. The same result holds for a thin hollow cylinder (without end caps).

2) Thin rod of length l (about an axis through its center and perpendicular to its axis): $I = \frac{1}{12}ml^2$. A thin rectangular plate has the same moment of inertia about the axis through the centers of its opposite sides having the length l. The moment of inertia about a parallel axis through one of the ends of the rod equals $I = \frac{1}{3}ml^2$.

3) A thin rectangular plate (about the axis through its center and perpendicular to its plane): $I = \frac{1}{12}m(a^2 + b^2)$. A rectangular parallelepiped has the same moment of inertia about the axis through the centers of its opposite faces.

4) Thin spherical shell (about its axis of symmetry): $I = \frac{2}{3}mR^2$.

5) Homogeneous solid ball (about its axis of symmetry): $I = \frac{2}{5}mR^2$.

6) Hollow cylinder with inner radius R_1 and outer radius R_2: $I = \frac{1}{2}m(R_1^2 + R_2^2)$.

P1.9.3. Plane Motion of a Rigid Body

Plane motion of a rigid body is the superposition of translational motion of its center of mass and rotational motion in a center-of-mass frame (see Subsection P1.4.2). The motion of the center of mass is described by Newton's second law and determined by the net external force (equation (P1.4.2.1)), where only real external forces must be taken into account, since the moment of fictitious forces relative to the center of mass is zero (just like in the case of the forces of gravity in the example of Subsection P1.6.1). The kinetic energy of plane motion equals $E_k = \frac{1}{2}mv_{cm}^2 + \frac{1}{2}I_0\omega^2$ (see equation (P1.5.1.3)). The angular momentum about a fixed axis perpendicular to the plane of motion is calculated as (see equation (P1.6.1.3)): $L_z = \pm mv_{cm}d \pm I_0\omega$, where d is the arm of the center-of-mass velocity relative to the axis, and the plus or minus sign is determined by the positive direction of rotation selected.

Example 1. Find the acceleration of a round body that rolls down an inclined plane without slipping (Fig. P1.14). The radius of the body is R, its mass is m, and the moment of inertia about the central axis is I. The angle between the inclined plane and the horizontal surface is α.

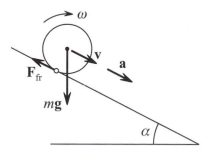

Figure P1.14. Forces acting on a body rolling down an inclined plane.

Solution. The equations of motion are: $mg\sin\alpha - F_{fr} = ma$ and $F_{fr}R = I\varepsilon$. The no-slipping condition ($v = \omega R$) leads to the equation $a = \varepsilon R$ (see the example in Subsection P1.2.1). Solving the above equations gives the acceleration $a = g\sin\alpha/(1 + I/mR^2)$. Since the static friction force does not produce any work, the mechanical energy is conserved.

Example 2. A thin rod of length l and mass M rests on a smooth horizontal plane. A ball of mass m moves in the plane and undergoes an elastic collision with an end of the rod perpendicularly to the line of the rod. The final speed of the ball, u, and the rod center, V, as well as angular velocity of the rod, ω, after the collision are determined from the following three equations:

1) law of conservation of linear momentum: $mv = mu + MV$,

2) law of conservation of energy: $mv^2/2 = mu^2/2 + MV^2/2 + (Ml^2/12)\omega^2/2$,

3) law of conservation of angular momentum (e.g., about the point of impact): $0 = MV(l/2) - (Ml^2/12)\omega$.

P1.9.4. Motion with a Fixed Point

For a rigid body rotating about a fixed point, its angular velocity is directed along the instantaneous axis of rotation and changes its direction both in space and relative to the body itself. The equation of motion

$$\frac{d\mathbf{L}}{dt} = \mathbf{M}^e,$$

which is referred to as the *basic equation of motion of a rigid body with a fixed point*, allows us to find out how the angular momentum $\mathbf{L}$ changes in time. Since $\mathbf{L}$ is generally non-parallel to the angular velocity ω, we need an equation that relates $\mathbf{L}$ and ω in order to close the system of equations of motion.

Example. As an inclined dumbbell rotates about a vertical axis (as shown in Fig. P1.15), its angular momentum $\mathbf{L} = \mathbf{r}_1 \times \mathbf{p}_1 + \mathbf{r}_2 \times \mathbf{p}_2$ is perpendicular to the straight line connecting the masses m_1 and m_2 of the dumbbell and directed at a non-right angle to the axis of rotation. Also the vector $\mathbf{L}$ itself rotates about the vertical axis with the angular velocity ω.

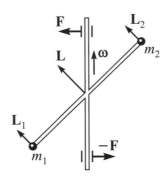

Figure P1.15. The vector of angular momentum $\mathbf{L}$ can be nonparallel to the axis of rotation (to ω).

▶ **Theorem of principal axes of inertia.** The relationship between $\mathbf{L}$ and ω can be found using the *theorem of principal axes of inertia* which states that for any rigid body and any point O there are three mutually perpendicular axes through O such that, when the body rotates about any of these axes, the vector $\mathbf{L}$ is parallel to the axis: $\mathbf{L}_i = I_i\omega_i$ ($i = 1, 2, 3$). The moments of inertia about these axes are called *principal moments of inertia*. If the rotation occurs about an arbitrary axis through O, the angular velocity can be resolved into its projections onto the principal axes, $\omega = \omega_1 + \omega_2 + \omega_3$, and the angular momentum can be expressed as

$$\mathbf{L} = I_1\omega_1 + I_2\omega_2 + I_3\omega_3. \tag{P1.9.4.1}$$

(Likewise, $\mathbf{L}$ can be resolved into its projections onto the principal axes and then ω can be expressed in terms of them.)

If the principal axes are through the center of mass (center of inertia) of the body, they are called *free axes*. When the body rotates about any of the free axes (free rotation), both the linear momentum and the angular momentum of the body are conserved, which means that there is no need to apply an external force or external moment to maintain rotation. (In the above example, the net force is zero; however, there are a couple of forces arising at points of support whose moments provide change of $\mathbf{L}$ with time.) It should be noted that free rotation is stable only about two of the three axes, the ones with the minimum and maximum principal moments of inertia.

Example 1. When is the choice of the free axes unique?

Solution. The free axes are uniquely determined only if the three principal moments of inertia are all different. If two moments are equal (e.g., $I_1 = I_2 = I$), then any axis that lies in the plane of the two free axes is also a free axis, with $\mathbf{L} = I\boldsymbol{\omega}_1 + I\boldsymbol{\omega}_2 = I\boldsymbol{\omega}$. If all three principal moments of inertia are equal (e.g., such is a homogeneous cube), then any axis through the center of mass is a free axis.

Example 2. Given three principal moments of inertia (I_1, I_2, and I_3) of a body, find the moment of inertia I about the axis that makes angles γ_1, γ_2, and γ_3 with the principal axes.

Solution. The kinetic energy of the body rotating with an angular velocity ω about the specified axis can be calculated in two ways. On the one hand, it is equal to

$$E = \frac{I\omega^2}{2}.$$

On the other hand,

$$E = \frac{L_\omega \omega}{2} = \frac{(\mathbf{L} \cdot \boldsymbol{\omega})}{2} = \frac{I_1\omega_1^2}{2} + \frac{I_2\omega_2^2}{2} + \frac{I_2\omega_2^2}{2},$$

where $L_\omega = I\omega$ is the projection of $\mathbf{L}$ onto $\boldsymbol{\omega}$. Since $\omega_i/\omega = \cos\gamma_i$, we find that

$$I = I_1 \cos^2\gamma_1 + I_2 \cos^2\gamma_2 + I_3 \cos^2\gamma_3.$$

Example 3. Free symmetric top. Nutation of the Earth's axis. Let is find out how the position of the axis of rotation changes relative to a rotating body (in a reference frame associated with the body). Let us confine ourselves to the case of a *free symmetric top*, or a freely rotating body with $I_1 = I_2 = I_\perp$ and $I_3 = I_\| \neq I_\perp$. (The Earth is an example of such a top.) For a symmetric top, the vector $\mathbf{L} = I_\|\boldsymbol{\omega}_\| + I_\perp\boldsymbol{\omega}_\perp$ lies in the same plane as the third axis (axis of symmetry) and the vector $\boldsymbol{\omega} = \boldsymbol{\omega}_\| + \boldsymbol{\omega}_\perp$. In an inertial reference frame, the angular momentum $\mathbf{L}$ is conserved. Consequently, in the rotating top-related reference frame, the vector $\mathbf{L}$ performs rotation with an angular velocity $\boldsymbol{\omega}$ at the current time, and hence it obeys the equation

$$\frac{d\mathbf{L}}{dt} = \boldsymbol{\omega} \times \mathbf{L}.$$

This equation is called *Euler's equation* (usually, a system of three scalar equations is written out which is obtained by projecting this vector equation onto axes 1, 2, and 3 taking into account relation (P1.9.4.1)). The complexity of Euler's equation is that the vector $\boldsymbol{\omega}$ is not constant but linked to $\mathbf{L}$ by the linear relation (P1.9.4.1). For a symmetric top, $\boldsymbol{\omega} \times \mathbf{L}$ is perpendicular to the plane $(\boldsymbol{\omega}, \mathbf{L})$; hence, $d\mathbf{L}_\|/dt = I_\| d\boldsymbol{\omega}_\|/dt = \mathbf{0}$ ($\boldsymbol{\omega}_\| = \mathrm{const}$). For $\boldsymbol{\omega}_\perp$, we have the equation

$$I_\perp \frac{d\boldsymbol{\omega}_\perp}{dt} = (\boldsymbol{\omega}_\| + \boldsymbol{\omega}_\perp) \times (I_\|\boldsymbol{\omega}_\| + I_\perp\boldsymbol{\omega}_\perp) = (I_\perp - I_\|)\,\boldsymbol{\omega}_\| \times \boldsymbol{\omega}_\perp$$

or

$$\frac{d\boldsymbol{\omega}}{dt} = \boldsymbol{\Omega} \times \boldsymbol{\omega}, \qquad \text{where} \quad \boldsymbol{\Omega} = \frac{(I_\perp - I_\|)}{I_\perp}\boldsymbol{\omega}_\|$$

This equation shows that in the top-related frame, the axis of rotation performs a circular motion about axis 3 with the angular velocity $\boldsymbol{\Omega}$. For the Earth, there is a small relative difference between the moments of inertia $I_\perp$ and $I_\|$, which leads to a slow periodic motion of the Earth's axis (*nutation of the Earth's axis*) with a period of tens of days.

▶ **Gyroscopes.** A gyroscope is a rigid body that rapidly spins about its axis of symmetry. The problem of motion of the gyroscope's axis can be solved in the so-called *gyroscope approximation*, $\mathbf{L} = I\boldsymbol{\omega}$, where both $\mathbf{L}$ and $\boldsymbol{\omega}$ are directed along the axis of symmetry. A perfectly balanced gyroscope is a gyroscope whose center of mass is fixed. It possesses the property of being *inertialess*: its axis stops moving as soon as the external action is removed ($\mathbf{M} = \mathbf{0}$). This property allows to use such gyroscopes to preserve orientation in space. A *heavy gyroscope* is a gyroscope whose center of mass is displaced by a distance d from the fixed point (Fig. P1.16). The moment of the force of gravity acting on a heavy gyroscope is perpendicular to $\mathbf{L}$. Since $d\mathbf{L}/dt \perp \mathbf{L}$, the angular momentum $\mathbf{L}$ and the gyroscope's axis perform regular rotation about the vertical axis (*precession of a gyroscope*). The end of the vector $\mathbf{L}$ moves in a horizontal circle of radius $L \sin\alpha$ with the angular velocity

$$\Omega = \frac{|d\mathbf{L}/dt|}{L \sin\alpha} = \frac{mgd \sin\alpha}{I\omega \sin\alpha} = \frac{mgd}{I\omega}.$$

The angular velocity of precession is independent of the angle α made by the axis of rotation with the vertical.

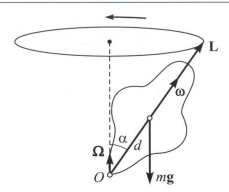

Figure P1.16. Gyroscope precession in the field of gravity.

P1.10. Special Theory of Relativity

P1.10.1. Basics of the Special Theory of Relativity

Einstein's *special theory of relativity* (STR), also known as *special relativity* (SR), extends the limits of classical Newtonian physics, valid for *nonrelativistic velocities* (small compared to the speed of light c), to any velocities, including *relativistic* (comparable with c). In the classical limit $v/c \to 0$, all results of special relativity reduce to results of classical, nonrelativistic physics.

▶ **Postulates of STR.** The special theory of relativity relies on two postulates:

The *principle of relativity*: all physical laws, both mechanical and electromagnetic, have the same form in all inertial reference frames. This implies that no inertial reference frame can be distinguished from any other by any experiments and taken to be the rest inertial frame. This postulate is an extension of the Galilean relativity principle (see Subsection P1.3.4) to electromagnetic phenomena.

The *principle of invariant light speed*: the speed of light in vacuum is the same for any inertial frame and equals $c \approx 3 \times 10^8$ m/s. This postulate contains two statements: (i) the speed of light is independent of the state of motion of the light source and (ii) the speed of light is independent of which inertial frame is the observer measuring this speed in (i.e., independent of the speed of the receiver).

The invariance of the light speed and its independence of the state of motion of the light source follow from Maxwell's equations of electromagnetic field. It seemed obvious that this statement could hold in only one reference frame. From the classical conceptions of space and time, any other observer moving with a speed v relative to this frame should measure the speed of light as $c + v$ if moving toward the source or $c - v$ if moving away from the source. This result would mean that Maxwell's equations hold in only one reference frame filled with quiescent "ether," through which electromagnetic waves travel. However, the attempt to detect a change in the light speed due to Earth's motion relative to the ether (Michelson–Morley experiment) failed to detect such a change. Einstein put forward the hypothesis that Maxwell's equations, as well as all physics laws, must have the same form in all inertial frames, and hence the light speed must be the same in all inertial frames. This hypothesis led to revision of the basic concepts of space and time.

▶ **Lorentz transformations.** The Lorentz transformations link the coordinates and time of an event measured in two inertial frames, one of which (K') moves relative to the other (K) with a constant velocity **V**. If the coordinate axes and the initial instant are chosen in the same way as in the Galilean transformations (see formulas (P1.3.4.1)), the Lorentz transformations have the form

$$x' = \frac{x - Vt}{\sqrt{1 - V^2/c^2}}, \quad y' = y, \quad z' = z, \quad t' = \frac{t - (V/c^2)x}{\sqrt{1 - V^2/c^2}}. \tag{P1.10.1.1}$$

It is often convenient to use the transformations for the differences of the coordinates and time instants of two events:

$$\Delta x' = \gamma(\Delta x - V\Delta t), \quad \Delta y' = \Delta y, \quad \Delta z' = \Delta z, \quad \Delta t' = \gamma\left[\Delta t - (V/c^2)\Delta x\right]. \quad \text{(P1.10.1.2)}$$

where the notation

$$\gamma = \frac{1}{\sqrt{1 - V^2/c^2}}, \qquad \gamma \geq 1, \qquad\qquad\qquad \text{(P1.10.1.3)}$$

is used for brevity. The Lorentz transformations reduce to the Galilean transformations for $V \ll c$. The Lorentz transformations are derived from the second pustulate of STR (invariance of light speed) and the requirement of linearity of the transformations (this requirement expresses the condition of uniformity of space). The *inverse transformations*, expressing the coordinates and time in K via those in K', can be obtained from (P1.10.1.1) and (P1.10.1.2) by substituting $-V$ for V:

$$\Delta x = \gamma(\Delta x' + V\Delta t'), \quad \Delta y = \Delta y', \quad \Delta z = \Delta z', \quad \Delta t = \gamma\left[\Delta t' + (V/c^2)\Delta x'\right]. \quad \text{(P1.10.1.4)}$$

P1.10.2. Consequences of the Lorentz Transformations. STR Kinematics

▶ **Length contraction.** The length of a moving segment is defined as the distance between the points at which the segment ends were at the same time (i.e., $\Delta t = 0$). Consider a rigid body that moves translationally with a speed v and let K' denote a reference frame associated with it. From equation (P1.10.1.2), where one should set $V = v$ and $\Delta t = 0$, one finds that the longitudinal dimension of a moving body is reduced:

$$l = l_0/\gamma = l_0\sqrt{1 - v^2/c^2}, \qquad\qquad\qquad \text{(P1.10.2.1)}$$

where l_0 is the proper longitudinal dimension measured in frame K', where the body is at rest. The transverse dimensions of a moving body do not change.

Example 1. A square plate moving at the speed $v = 0.8\,c$ along one of its sides becomes a rectangular with the angle between the diagonals equal to $\operatorname{arccot}\sqrt{1 - v^2/c^2} = \operatorname{arccot} 0.6 \approx 59°$.

▶ **Time dilation.** It can be seen from the Lorentz transformations that time passes differently in different inertial frames. In particular, two events occurring simultaneously ($\Delta t = 0$) but at different points in frame K may not be simultaneous in frame K', so that $\Delta t' = -\gamma V\Delta x/c^2$ can be positive or negative. This effect is known as *relativity of simultaneity*. A clock that moves together with frame K' (i.e., that rests relative to K', with $\Delta x' = 0$) shows the *proper time* of this frame. For an observer in frame K, this clock is slow (*time dilation*). Considering two readings of the moving clock as two events, we find from (P1.10.1.4) that

$$\Delta t = \gamma\Delta\tau = \frac{\Delta\tau}{\sqrt{1 - V^2/c^2}}, \qquad\qquad\qquad \text{(P1.10.2.2)}$$

where $\Delta\tau = \Delta t'$ is the proper time of the moving clock (to be precise, of the associated reference frame). The fact that all reference frames are indistinguishable manifests itself in that a clock resting relative to frame K appears to the observer in frame K' to lag behind the observer's clock in K'. (Note that, to control the moving clock, the stationary observer in K uses *different* clocks at different instants.)

The *twin paradox* consists in that special relativity predicts a change in the age of two twins, one of which stayed on the Earth and the other traveled in deep space at relativistic speeds and then returned to the Earth (the traveler will be younger). It may seem as though this thought experiment violates the principle of indistinguishability of reference frames. In fact, the Earth's twin stayed all the time in one and the same inertial frame but the astronaut changed his or her frame to return to the Earth (the astronaut-associated frame was not inertial).

Example 2. The mean proper lifetime of the muon (an unstable subatomic particle) is $\tau = 2.2 \times 10^{-6}$ s, with $c\tau \sim 660$ m (c is the speed of light). Due to time dilation, from the viewpoint of an Earth's observer, a space muon flying with a relativistic speed v with $\gamma \gg 1$ has a mean lifetime of $\gamma\tau$ and travels a distance $c\gamma\tau$ from its birth in upper atmosphere, which allows detecting the muon on the Earth's surface.

▶ **Velocity addition in special relativity.** If a particle moves with velocity $\mathbf{v}'$ relative to frame K', then its velocity relative to frame K can be found by expressing dx, dy, dz, and dt from (P1.10.1.4) and substituting into $v_x = dx/dt$, $v_y = dy/dt$, and $v_z = dz/dt$ to obtain

$$v_x = \frac{v_x' + V}{1 + v_x'V/c^2}, \quad v_y = \frac{v_y'\sqrt{1 - V^2/c^2}}{1 + v_x'V/c^2}, \quad v_z = \frac{v_z'\sqrt{1 - V^2/c^2}}{1 + v_x'V/c^2}. \tag{P1.10.2.3}$$

For nonrelativistic speeds, $v_x \ll c$ and $V \ll c$, the above formulas reduce to the nonrelativistic velocity-addition law (see equation (P1.1.4.1)). Formulas (P1.10.2.3) have the important property that if V and v' less than c, then v is also less than c. For example, if the particle is first accelerated to $V = 0.9\,c$ relative to the rest frame and then is further accelerated to $v' = 0.9\,c$ in the moving, particle-associated frame, the resulting speed will be $1.8\,c/1.81 < c$ rather than $1.8\,c$. It is apparent that the speed of light cannot be exceeded. The light speed is the maximum possible speed at which interactions can be transferred in nature.

▶ **Spacetime interval. Casuality.** The Lorentz transformations preserve neither the time interval nor the length of a spatial segment. However, it can be shown that the following quantity is preserved under the Lorentz transformations:

$$s_{12}^2 = (c\Delta t)^2 - (\Delta x)^2 - (\Delta y)^2 - (\Delta z)^2 = (c\Delta t)^2 - (\Delta \mathbf{r})^2. \tag{P1.10.2.4}$$

The quantity s_{12} is called the *spacetime interval* between events 1 and 2 ($\Delta t = t_2 - t_1$ and $\Delta \mathbf{r} = \mathbf{r}_2 - \mathbf{r}_1$).

If $s_{12}^2 > 0$, the spacetime interval is called *time-like*. In this case, there is an inertial frame where $\Delta \mathbf{r} = \mathbf{0}$, which means that the events occur at the same place but at different times. Such events can have a cause–effect relationship.

If $s_{12}^2 < 0$, the spacetime interval between events is called *space-like*. In this case, there is an inertial frame in which $\Delta t = 0$, so that the events occur simultaneously at different locations in space. Such events cannot have a causal relationship. The condition $c|\Delta t| < |\Delta \mathbf{r}|$, or $c|t_2 - T_1| < |\mathbf{r}_2 - \mathbf{v}_1|$, means that a light beam emitted at the instant of the earlier event (at t_1) from the point $\mathbf{r}_1$ cannot reach the point $\mathbf{r}_2$ by the instant t_2. Events separated from event 1 by time-like intervals represent its *absolute past* (if $t_2 - t_1 < 0$) or *absolute future* (if $t_2 - t_1 > 0$); the sequence of events is the same in all inertial frames. The sequence of events separated by a space-like interval can be different in different inertial frames.

▶ **Lorentz four-vectors.** A quadruple of scalar quantities $(A_x, A_y, A_z, A_\tau) = (\mathbf{A}, A_\tau)$ that are transformed, when passing from frame K to frame K', in the same way as (x, y, z, ct), or (see (P1.10.1.1))

$$A_x' = \frac{A_x - (V/c)A_\tau}{\sqrt{1 - V^2/c^2}}, \quad A_y' = A_y, \quad A_z' = A_z, \quad A_\tau' = \frac{A_\tau - (V/c)A_x}{\sqrt{1 - V^2/c^2}}, \tag{P1.10.2.5}$$

is called a *Lorentz four-dimensional vector* (or, for short, *four-vector* or *4-vector*). The numbers A_x, A_y, and A_z are called the spatial components of the 4-vector and A_τ is called its time component. The sum of two 4-vectors and the product of a 4-vector by a scalar number are also 4-vectors. The quantity $A^2 = A_\tau^2 - \mathbf{A}^2$, analogous to the spacetime interval, and also the scalar product $A_\tau B_\tau - \mathbf{A} \cdot \mathbf{B}$ are preserved in changing from one inertial frame to another. A physical equation written as an equality of two 4-vectors remains valid in all inertial frames.

P1.10.3. STR Dynamics

▶ **Momentum and energy in special relativity.** The velocity components are not transformed in the same way as the components of a 4-vector (cf. equations (P1.10.2.3) and (P1.10.2.5)), since both the numerator and the denominator are transformed in the expression $\mathbf{v} = d\mathbf{r}/dt$. Consequently, the quantity $\sum_j m_j \mathbf{v}_j$, corresponding to the classical definition of momentum, cannot be preserved in all inertial frames. The relativistic 4-vector of momentum is defined as

$$\mathbf{p} = m\frac{d\mathbf{r}}{d\tau} = \frac{m}{\sqrt{1-v^2/c^2}}\frac{d\mathbf{r}}{dt}, \qquad p_\tau = m\frac{d(ct)}{d\tau} = \frac{mc}{\sqrt{1-v^2/c^2}},$$

where $d\tau = dt\sqrt{1-v^2/c^2}$ is the infinitesimal change in the proper time of the particle (see (P1.10.2.2)), i.e., the time measured in an inertial frame whose velocity coincides with that of the particle at the given instant ($d\tau$ is independent of which inertial frame the particle is observed from). The spatial components of the 4-vector make up the *relativistic momentum*

$$\mathbf{p} = \frac{m\mathbf{v}}{\sqrt{1-v^2/c^2}}, \tag{P1.10.3.1}$$

while the time component p_τ turns out to be equal to E/c, where E is the *relativistic energy* of the particle:

$$E = \frac{mc^2}{\sqrt{1-v^2/c^2}}. \tag{P1.10.3.2}$$

The 4-vector $(\mathbf{p}, E/c)$ is called the *momentum-energy four-vector* or the *four-momentum*. It is noteworthy that the relativistic energy and momentum are linked by the simple relation

$$\mathbf{p} = \frac{E\mathbf{v}}{c^2}. \tag{P1.10.3.3}$$

In accordance with (P1.10.2.5), in changing from one inertial frame to another, the energy and momentum are transformed as

$$p_x{'} = \frac{p_x - VE/c^2}{\sqrt{1-V^2/c^2}}, \quad p_y{'} = p_y, \quad p_z{'} = p_z, \quad E' = \frac{E - Vp_x}{\sqrt{1-V^2/c^2}}. \tag{P1.10.3.4}$$

The relativistic energy of a particle is nonzero at $v = 0$; it consists of the rest energy, mc^2, and the kinetic energy,

$$E = \frac{mc^2}{\sqrt{1-v^2/c^2}} = mc^2 + E_k. \tag{P1.10.3.5}$$

Note that for $v/c \ll 1$, the relativistic kinetic energy reduces to the classical kinetic energy $E_k = mv^2/2$. Since the quantity $(E/c)^2 - p^2$ is conserved, it can be calculated in a frame where the particle is at rest at the given instant:

$$\frac{E^2}{c^2} - p^2 = m^2 c^2 \qquad \text{or} \qquad E^2 - p^2 c^2 = m^2 c^4. \tag{P1.10.3.6}$$

For particles with zero mass (e.g., photons), the energy–momentum relation becomes

$$E = pc \tag{P1.10.3.7}$$

(see also (P1.10.3.3)). Substituting $E = mc^2 + E_k$ into (P1.10.3.6) yields a convenient relation between momentum and kinetic energy:

$$p^2 c^2 = E_k(E_k + 2mc^2).$$

▶ **Inelastic collision. Relation between energy and mass.** Let us write down the law of conservation of relativistic energy for a perfectly inelastic collision of two particles of mass m each that moved toward each other with identical speeds v. We have

$$\frac{mc^2}{\sqrt{1 - v^2/c^2}} + \frac{mc^2}{\sqrt{1 - v^2/c^2}} = Mc^2.$$

It is apparent that the mass M of the combined particle is larger than the total mass of the initial particles. The increase in the internal energy at the inelastic collision by ΔE resulted in a mass increase of $\Delta E/c^2$. This example illustrates the general Einstein relationship between the relativistic energy of a resting body and its mass:

$$E = mc^2. \tag{P1.10.3.8}$$

Relativistic energy includes all kinds of energy.

Example 1. Suppose the energy of a body at rest equals E. Find the momentum of the body in a reference frame moving with a speed $-v$.

Solution. In accordance with the relativistic transformation formulas (P1.10.3.4), the momentum equals

$$\mathbf{p}' = (E'/c^2)\mathbf{v} = \frac{(E/c^2)\mathbf{v}}{\sqrt{1 - v^2/c^2}}.$$

It is seen that we have obtained the formula for relativistic momentum with mass $m = E/c^2$.

Example 2. Accelerators with colliding particle beams. Suppose that in order to achieve the threshold energy of a nuclear reaction, the particles in colliding beams must be accelerated to a kinetic energy αmc^2. To what kinetic energy must a single particle be accelerated if the other particle is in a stationary target?

Solution. In the nonrelativistic case, the energy in a center-of-mass frame equals half the energy of the oncoming particle. It is this energy that can be spent to the reaction, while the other half is inaccessible— it remains as kinetic energy of the system. In the relativistic (an especially ultrarelativistic) limit, the picture is cardinally different. Compute the energy of the moving particle (the other particle is at rest). Write the energy-momentum invariance condition for $E^2 - p^2 c^2$. In a center-of-mass frame, we have $E = 2\alpha mc^2 + 2mc^2$ and $p = 0$. In a moving inertial frame where one of the particles is at rest, we have $E = E_1 + mc^2$ and $p = p_1$, with E_1 and p_1 being the energy and momentum of the moving particle. On rearranging and taking into account the identity $E_1^2 - p_1^2 c^2 = m^2 c^4$, we find the kinetic energy of the oncoming particle as $E_1 - mc^2 = 2mc^2\alpha(\alpha + 2)$. For example, if $\alpha = 100$, we find that one particle must be accelerated to $20,400\,mc^2$. Approximately, $1/100$ of the particle energy is available for the reaction.

▶ **Basic law of relativistic dynamics.** The force exerted on a particle equals, just as in the case of classical mechanics, the time derivative of the momentum

$$\mathbf{F} = \frac{d\mathbf{p}}{dt}.$$ (P1.10.3.9)

However, the relativistic momentum (P1.10.3.1) is different from the classical one. Under the action of the load applied, the momentum can increase unboundedly. However, as seen from definition (P1.10.3.1), the speed will be less than c. The work done by the force (P1.10.3.9),

$$\int \mathbf{F} \cdot d\mathbf{r} = \int \frac{d\mathbf{p}}{dt} \cdot \mathbf{v}\, dt = \int \mathbf{v} \cdot d\mathbf{p}$$
$$= \int \sqrt{1 - v^2/c^2}\, \frac{\mathbf{p} \cdot d\mathbf{p}}{m} = \int \sqrt{1 - v^2/c^2}\, \frac{E\, dE}{mc^2} = \int dE,$$

equals the change in the kinetic energy. Here we have used formulas (P1.10.3.1) and (P1.10.3.2) and also $\mathbf{p} \cdot d\mathbf{p} = E\, dE/c^2$ (see (P1.10.3.6)).

Bibliography for Chapter P1

Alenitsyn, A. G., Butikov, E. I., and Kondratyev, A. S., *Concise Handbook of Mathematics and Physics*, CRC Press, Boca Raton, Florida, 1998.

Benenson, W., Harris, J. W., Stocker, H., and Lutz, H. (Eds.), *Handbook of Physics*, Springer-Verlag, New York, 2002.

French, A. P., *Special Relativity (M.I.T. Introductory Physics Series)*, W. W. Norton & Company, New York, 1968.

French, A. P., *Newtonian Mechanics (M.I.T. Introductory Physics Series)*, W. W. Norton & Company, New York, 1971.

Goldstein, H., Poole, C. P., and Safko, J. L., *Classical Mechanics (3rd Edition)*, Addison-Wesley, New York, 2002.

Irodov, I. E., *Mechanics. Basic Laws* [in Russian], Fizmatlit, Moscow, 2006

José J. V. and Saletan E. J., *Classical Dynamics: A Contemporary Approach*, Cambridge University Press, Cambridge, England, 1998.

Kibble, T. W. B. and Berkshire, F. H., *Classical Mechanics, 5th edition* , Imperial College Press, London, 2004.

Kittel, C., Knight, W. D., and Ruderman, M. A., *Mechanics (Berkeley Physics Course, Vol. 1), 2nd edition*, McGraw-Hill Book Company, New York, 1973.

Kleppner, D. and Kolenkow, R. J., *Introduction to Mechanics*, McGraw-Hill Inc., Boston, Massachusetts, 1973.

Landau, L. D. and Lifshitz, E. M., *Mechanics, Third Edition: Volume 1 (Course of Theoretical Physics)*, Butterworth-Heinemann, Oxford, England, 1976.

Morin, D., *Introduction to Classical Mechanics With Problems and Solutions*, Cambridge University Press, Cambridge, England, 2008.

Okun, L. B., *Energy and Mass in Relativity Theory*, World Scientific Publishing Company, London, 2009.

Prokhorov, A. M. (Ed.), *Encyclopedia of Physics, Volumes 1–5* [in Russian], Bolshaya Russkaya Encyclopedia, Moscow, 1998.

Prokhorov, A. M. (Ed.), *Encyclopedic Dictionary of Physics* [in Russian], Sovetskaya Encyclopedia, Moscow, 1984 (dic.academic.ru).

Resnick, R., *Introduction to Special Relativity*, John Wiley and Sons, Inc., New York, 1968.

Sivukhin, D. V., *General Physics, Volume 1, Mechanics* [in Russian], Fizmatlit, Moscow, 2005.

Sivukhin, D. V. and Yakovlev, I. A., *A Collection of Problems, Volume 1, Mechanics* [in Russian], Fizmatlit, Moscow, 2006.

Symon, K. R., *Mechanics, 3rd Edition*, Addison-Wesley, Reading, Massachusetts, 1971.

Taylor, J. R., *Classical Mechanics*, University Science Books, Sausalito, 2005.

Thornton, S. T. and Marion, J. B., *Classical Dynamics of Particles and Systems, 5th edition*, Brooks Cole, Belmont, California, 2003.

Woan, G., *The Cambridge Handbook of Physics Formulas*, Cambridge University Press, Cambridge, England, 2003.

Chapter P2

Molecular Physics and Thermodynamics

P2.1. Basic Principles and Definitions

P2.1.1. Two Approaches to Studying Matter

▶ **Thermodynamic approach.** The *thermodynamic approach* of studying substances consists of establishing links and relations between experimentally determined (phenomenological) parameters (called *thermodynamic parameters*) based on several postulates (*laws of thermodynmics*).

▶ **Statistical approach.** The *statistical approach* relies on kinetic-molecular postulates on the structure of substance (*fundamentals of kinetic-molecular theory*). These postulates include the following:

1. All bodies consist of a huge number of tiny particles—atoms and molecules.
2. These molecules are in constant random motion.
3. The molecules interact with one another: they experience attractive forces at large distances and repulsive forces at small enough distances.

The thermodynamic parameters are calculated within the framework of a specific model of internal structure of substance (i.e., the model of motion and interaction of atoms and molecules) by averaging over a huge number of states of the system. Statistical physics employs methods of probability theory and mathematical statistics. The classical theory relies on the classical laws of molecular motion, while quantum statistics relies on the laws of quantum mechanics.

▶ **Amount of substance. Mole.** The amount of substance in a system—or the number of constituting structural units, atoms and molecules—is measured in *moles*. One mole of any substance contains a certain number of molecules, called *Avogadro's number* (or the *Avogadro constant*) and equal to the number of atoms in 12 g of carbon-12. Avogadro's number equals $N_A \approx 6.02 \times 10^{23}$ mol^{-1}. The number of moles in the system is expressed as

$$\nu = \frac{N}{N_A} = \frac{m}{\mu}, \tag{P2.1.1.1}$$

where N is the number of molecules in the system, $m = m_0 N$ is the mass of the system, m_0 being the mass of a single molecule, and $\mu = m_0 N_A$ is the molar mass of the substance.

P2.1.2. Equation of State

▶ **Equilibrium states.** A thermodynamic system that is kept under unchanged external conditions comes to an *equilibrium state*, where there are no fluxes of any kind (e.g., mass or energy fluxes). The thermodynamic parameters of the equilibrium state (pressure p, temperature T, volume V, density ρ, molar mass μ, etc.) are related by an *equation of state*. For example, the *ideal gas law* (also known as the *Clapeyron–Mendeleev equation*), which is the equation of state of an ideal gas, has the form

$$pV = \nu RT \qquad \text{or} \qquad pV = NkT, \tag{P2.1.2.1}$$

441

where $R \approx 8.31\,\text{J/(mol K)}$ is the *gas constant* and $k = R/N_A \approx 1.38 \times 10^{-23}\,\text{J/K}$ is the *Boltzmann constant*. This equation was obtained empirically and it describes well the behavior of thin real gases (see Subsection P2.5.2). The *van der Waals equation*,

$$(V - \nu b)\left(p + \frac{a\nu^2}{V^2}\right) = \nu RT, \tag{P2.1.2.2}$$

is an equation of state that provides good approximation to the behavior of dense real gases (see Section P2.6).

A process of changing the state of a system is called *equilibrium* (also *quasi-equilibrium* or *quasistatic*) if it occurs so slowly that every intermediate state of the system can be treated as equilibrium. Equilibrium processes are typically *reversible* (i.e., can be made to pass through all their intermediate states in order); however, irreversible equilibrium processes are possible (e.g., for systems with memory). In what follows, when speaking of an equilibrium process we imply a reversible equilibrium process.

▶ **Simple thermodynamic system.** A homogeneous and isotropic thermodynamic system whose chemical composition does not change is called (*thermodynamically*) *simple*. Examples of simple systems are one-component (pure) gases and liquids. A simple system has two degrees of freedom, which means that only two of its thermodynamic parameters are independent. Equilibrium states of a simple system can be represented by points in a plane (e.g., in coordinates p and V) and equilibrium processes, by lines in this plane. The work done by a simple system against external forces equals

$$\delta A = p\,dV, \qquad A = \int_{V_1}^{V_2} p(V)\,dV, \tag{P2.1.2.3}$$

where the relationship $p(V)$ is called a *process equation*. Since the work is not a function of state, the infinitesimal work increment is denoted by δA, in order to distinguish it from the infinitesimal increment of a function (differential).

If any three parameters of a simple system are related to one another, due to an equation of state, their derivatives are also related by an equation. For example, expressing the infinitesimal increment of T in terms of those of p and V, we have $dT = (\partial T/\partial V)_p\,dV + (\partial T/\partial p)_V\,dp$. By applying this expression to isothermal process ($dT = 0$), we obtain

$$\left(\frac{\partial V}{\partial T}\right)_p \left(\frac{\partial T}{\partial p}\right)_V \left(\frac{\partial p}{\partial V}\right)_T = -1.$$

This equation relates three coefficients

$$\alpha = \frac{1}{V}\left(\frac{\partial V}{\partial T}\right)_p \qquad \textit{(thermal volume expansion coefficient)},$$

$$\beta = \frac{1}{p}\left(\frac{\partial p}{\partial T}\right)_V \qquad \textit{(thermal pressure coefficient)},$$

$$K = -V\left(\frac{\partial p}{\partial V}\right)_T \qquad \textit{(isothermal bulk modulus)}.$$

▶ **Nonequilibrium systems.** *Equilibrium thermodynamics* deals with equilibrium states and processes. It can judge the *direction of nonequilibrium processes* between the initial and final equilibrium states but cannot provide quantitative characterization of these processes. Nonequilibrium systems are studied by *physical kinetics* and *nonequilibrium thermodynamics*. Note that the time in which a system comes to equilibrium (*relaxation time*) decreases with decreasing system size, and therefore the term *local equilibrium* in a small portion of a nonequilibrium system makes sense. Consequently, it is meaningful to speak of local thermodynamic parameters (functions of state) that vary from point to point in a nonequilibrium system (e.g., temperature $T(\mathbf{r})$).

An important role in studying nonequilibrium systems is played by the notions of *flux* of a physical quantity (e.g., the number of molecules, mass, momentum, energy, charge, etc.) through a given surface and *flux density* at a given point in space.

For instance, energy flux W equals the amount of energy transferred through a surface per unit time (measured in J/s, or W). A *flux density* is a vector quantity that characterizes the direction and intensity of transfer of a physical quantity through a given point in space. For instance, energy flux density $\mathbf{w}$ equals in magnitude to the ratio of the energy flux ΔW through a small surface to the area ΔS of this surface (measured in W/m^2) in the limit $\Delta S \to 0$. Among all small surfaces through the given point, one must take one for which the above ratio is maximum. The vector $\mathbf{w}$ is taken to be perpendicular to this surface and pointing in the direction of the energy transfer. The energy flux through a given surface equals $W = \int \mathbf{w}\, d\mathbf{s}$, where $d\mathbf{s} = \mathbf{n} \cdot ds$, with $\mathbf{n}$ being the positive unit normal vector at the given point of the surface. Note that in electrodynamics (see Subsection P3.7.1), the flux of electric charge is called the electric current I and the flux density, the current density $\mathbf{j}$.

A state of a nonequilibrium system is called *stationary* if the flux densities of all physical quantities at all points are time independent. Recall that in an equilibrium state, there are no fluxes, that is, the flux densities at all points are zero.

P2.2. First Law of Thermodynamics

P2.2.1. Quantity of Heat. Internal Energy

▶ **Quantity of heat.** Unlike mechanical energy, which can only change through work, internal energy can change both due to work and due to contact with bodies having different temperatures (in the process of heat exchange). The amount of energy absorbed or lost by a system in a thermal contact is called the *quantity of heat* or just *heat* and denoted Q. Heat is considered positive if energy is received by the system and negative if it is given off.

In a thermal contact, energy is transferred from a body with a higher temperature to a body with a lower temperature. Any temperature scale must satisfy this property. Empirical temperature scales are based on indirect measurements, that is, on changes in parameters monotonically dependent on temperature. A gas scale of temperatures relies on the equation of state of the ideal gas (P2.1.2.1). A thermodynamic, or absolute scale, is based on the second law of thermodynamics.

▶ **Internal energy.** From the viewpoint of classical kinetic-molecular theory, the internal energy U of a thermodynamic system equals the sum of the kinetic energy of motion of all its molecules and the potential energy of their interaction. The internal energy, $U = U(V, T, \ldots)$, is a function of state of a thermodynamic system;* it is the most important characteristic of the system.

Thermodynamics must define any quantity phenomenologically, regardless of our knowledge about the internal structure of the substance. The concept of an *adiabatic envelope* can be used to see that this is possible in principle. The adiabatic envelope of thermodynamic system allows the shape of the system to be changed and work on the system to be done but excludes thermal exchange of the system with external bodies. In this case, the work done by the external forces, A^e, as the system performs a transition from state 1 to state 2 is independent of the transition process and is only dependent on the initial and final states. This allows us to define the internal energy difference as $U_2 - U_1 = A^e(1 \to 2)$.

P2.2.2. First Law of Thermodynamics

The general law of conservation of energy, which takes into account thermal exchange and internal energy, reads

$$\Delta E_{\mathrm{mech}} + \Delta U = A^e + Q. \tag{P2.2.2.1}$$

It states that the change in the total (mechanical plus internal) energy equals the work done by external forces and the quantity of heat received by the system in thermal exchange

---

* The relationship $U = U(V, T, \ldots)$ is called a *caloric equation of state*, in contract to $p = p(V, T)$, called a *thermal equation of state*.

with external bodies. Sometimes, the law of conservation of energy is formulated as the impossibility of creating a *perpetual motion machine of the first kind* (that would produce energy from nothing). The *first law of thermodynamics* is the special case of the law of conservation of energy as applied to thermodynamic systems, whose mechanical energy does not change. In addition, in thermodynamics it is customary to use the work done by the system against external forces: $A = -A^e$. We have

$$Q = \Delta U + A, \tag{P2.2.2.2}$$

which means that the heat supplied to a thermodynamic system is spent to change its internal energy and produce work against external forces.

▶ **First law of thermodynamics for a simple system. Heat capacity.** Using equation (P2.1.2.3), let us write out the first law of thermodynamics for a simple system that has received an infinitesimal quantity of heat:

$$\delta Q = dU + p\,dV. \tag{P2.2.2.3}$$

The ratio of δQ to the temperature increment dT is called the *heat capacity* of the system in the equilibrium process being considered. Heat capacity is measured in J/K and denoted by C. We have

$$C\,dT = dU + p\,dV. \tag{P2.2.2.4}$$

Considering U in (P2.2.2.4) to be a function of two variables (V, T), we can write

$$dU = \left(\frac{\partial U}{\partial T}\right)_V dT + \left(\frac{\partial U}{\partial V}\right)_T dV = C_V\,dT + \left(\frac{\partial U}{\partial V}\right)_T dV, \tag{P2.2.2.5}$$

where $C_V = \left(\dfrac{\partial U}{\partial T}\right)_V$ is the *heat capacity at constant volume* (see equation (P2.2.2.4)). So we have

$$C\,dT = C_V\,dT + \left[p + \left(\frac{\partial U}{\partial V}\right)_T\right]dV. \tag{P2.2.2.6}$$

It can be seen that the heat capacity for a process $V(T)$ depends on the value of the derivative dV/dT. Since the parameters (V, T) can be varied independently, this derivative and, hence, the heath capacity at a given point can take any value.

▶ **Enthalpy. The Joule–Thomson process.** If the state of a system is changed as an equilibrium process under a constant pressure, then the heat supplied can be expressed as

$$Q = \Delta U + p\Delta V = \Delta(U + pV) = \Delta H \qquad (p = \text{const}). \tag{P2.2.2.7}$$

The quantity $H = U + pV$ is called the *enthalpy* of the system. So the heat supplied equals the enthalpy difference at the final and initial states. Enthalpy is also used to describe the *Joule–Thomson process* (also called a *throttling process*), where a substance is forced through a porous wall without heat exchange with the environment. This process is used to achieve very low temperatures. Enthalpy is conserved in the Joule–Thomson process: $U_1 + p_1V_1 = U_2 + p_2V_2$.

▶ **Thermodynamic cycle.** A thermodynamic cycle is a series of thermodynamic processes in which a system eventually returns to its initial equilibrium state. At some stage, the system absorbs a quantity of heat Q_1 from higher-temperature bodies (*hot reservoir*) and at another stage, it gives up some heat Q_2 to lower-temperature bodies (*cold reservoir*). (This follows from the second law of thermodynamics; see Section P2.3.) For a heat engine, a device that runs on a thermodynamic cycle, its *thermal efficiency* is the number defined as

$$\eta = \frac{A}{Q_1} = \frac{Q_1 - Q_2}{Q_1} = 1 - \frac{Q_2}{Q_1}, \tag{P2.2.2.8}$$

where A is the net work done in one cycle. In a thermodynamic cycle, $\Delta U = 0$, and hence $Q_1 - Q_2 = A$.

P2.2.3. First Law of Thermodynamics for an Ideal Gas

▶ **Internal energy of an ideal gas**. The internal energy of an ideal gas depends only on its temperature, so that $U(p_1, V_1, T) = U(p_2, V_2, T)$. There is experimental evidence supporting this statement.

1. When a gas expands into a void space inside a rigid adiabatic container, both the work done by the gas, A, and the heat absorbed, Q, are zero, and hence $\Delta U = 0$. It was found (experiments by Gay-Lussac and Joule) that in this non-equilibrium process, the temperature of ideal gases does not change.

2. The temperature of an ideal gas does not change in the Joule–Thomson process.

The second law of thermodynamics allows us to prove that $(\partial U/\partial V)_T = 0$ based on the thermic equation of state (P2.1.2.1).

From equations (P2.2.2.4) and (P2.2.2.5) we find the increment in the internal energy and the first law of thermodynamics for an ideal gas:

$$dU = C_V\, dT, \qquad \delta Q = C_V\, dT + p\, dV. \tag{P2.2.3.1}$$

Experiments show that C_V is independent of T in a wide range of temperatures. Hence, we can write $U = C_V T + \text{const}$.

▶ **Processes in an ideal gas.** Consider a few examples to illustrate the application of the first law of thermodynamics to specific equilibrium processes in an ideal gas.

Example 1. Isothermal process ($dT = 0$). In this case, $dU = 0$. Hence,

$$\delta Q = p\, dV, \qquad Q = A = \int_{V_1}^{V_2} p(V)\, dV = \nu RT \ln\left(\frac{V_2}{V_1}\right). \tag{P2.2.3.2}$$

The heat capacity of this process can be considered infinitely large.

Example 2. Isobaric process ($dp = 0$). Introducing the *heat capacity at constant pressure* and taking into account the equation of state (P2.1.2.1), we get

$$C_p\, dT = C_V\, dT + p\, dV = C_V\, dT + \nu R\, dT.$$

Dividing by dT gives the relation between the heat capacities (*Mayer's relation*):

$$C_p = C_V + \nu R, \qquad c_p = c_V + \frac{R}{\mu}, \qquad C_p^{\mathrm{m}} = C_V^{\mathrm{m}} + R, \tag{P2.2.3.3}$$

where $c = C/m$ is the *specific heat capacity* and $C^{\mathrm{m}} = C/\nu$ is the *molar heat capacity* of the gas (measured in J/(kg K) and J/(mol K), respectively).

Example 3. Adiabatic process. If an equilibrium process occurs without heat supply ($\delta Q = 0$), then

$$C_V\, dT = -p\, dV \quad\Longrightarrow\quad \nu C_V^{\mathrm{m}}\, dT = -\frac{\nu RT}{V}\, dV \quad\Longrightarrow\quad \frac{dT}{T} = -\frac{R}{C_V^{\mathrm{m}}}\frac{dV}{V}.$$

Integrating yields the equation of the adiabatic process (*Poisson's equation*):

$$TV^{\gamma-1} = \text{const} \qquad \text{or} \qquad pV^{\gamma} = \text{const}, \tag{P2.2.3.4}$$

where the dimensionless parameter $\gamma = C_p/C_V = 1 + (R/C_V^{\mathrm{m}})$ is called the *adiabatic index*. Using Mayer's relation (P2.2.3.3), we can express the molar heat capacities in terms of γ as

$$C_V^{\mathrm{m}} = \frac{R}{\gamma - 1}, \qquad C_p^{\mathrm{m}} = \frac{\gamma R}{\gamma - 1}.$$

Example 4. Polytropic process. A thermodynamic process with constant heat capacity ($C = \text{const}$) is called polytropic. We have

$$C\, dT = C_V\, dT + p\, dV \quad\Longrightarrow\quad (C_V - C)\, dT = -p\, dV.$$

Proceeding in the same manner as for the adiabatic process, we arrive at the *polytropic process equation*

$$pV^n = \text{const},$$

where $n = 1 + \dfrac{R}{C_V^{\mathrm{m}} - C^{\mathrm{m}}} = \dfrac{C_p - C}{C_V - C}$ is the *polytropic index*.

It can be verified that the above thermodynamic processes are special cases of a polytropic process—isochoric with $n = \infty$, isobaric with $n = 0$, isothermal with $n = 1$, and adiabatic with $n = \gamma$.

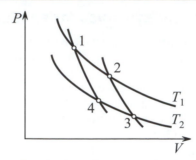

Figure P2.1 A Carnot cycle consists of two isothermal processes (1 to 2 and 3 to 4) and two adiabatic processes (2 to 3 and 4 to 1).

Example 5. Carnot cycle. A Carnot cycle is a thermodynamic cycle consisting of two isothermal processes (at temperatures T_1 and T_2 with $T_1 > T_2$) and two adiabatic processes as shown in Fig. P2.1.

From equation (P2.2.3.2) we have

$$Q_1 = \nu R T_1 \ln\left(\frac{V_2}{V_1}\right), \qquad Q_2 = \nu R T_2 \ln\left(\frac{V_3}{V_4}\right).$$

From Poisson's equation (P2.2.3.4) we have

$$T_1 V_1^{\gamma-1} = T_2 V_4^{\gamma-1},$$
$$T_1 V_2^{\gamma-1} = T_2 V_3^{\gamma-1}.$$

It follows that $V_2/V_1 = V_3/V_4$. Hence, the thermal efficiency is

$$\eta = 1 - \frac{T_2}{T_1}. \tag{P2.2.3.5}$$

This means that the thermal efficiency of a Carnot cycle depends only on the cold-to-hot reservoir temperature ratio.

P2.3. Second Law of Thermodynamics

P2.3.1. Formulations

Below are two most common formulations of the second law of thermodynamics:

1. It is impossible to construct an engine that would convert 100% of heat extracted from a heat reservoir into work in a cyclic process (formulation by William Thomson—Lord Kelvin). It is the same to say that it is impossible to create a *perpetual motion machine of the second kind* (that would produce work at the expense of solely the internal energy of a heat reservoir).

2. It is impossible to construct a refrigerator that would produce no other effect than the transfer energy from a cooler body to a hotter one (formulation by Rudolf Clausius).

The Thomson and Clausius formulations are equivalent.

P2.3.2. Carnot's Theorem

In general, a *Carnot cycle* is a thermodynamic cycle in which the working substance (of a heat engine) absorbs heat from a hot reservoir (at a constant temperature T_1) and gives up heat to a cold reservoir (at a constant temperature T_2).

Carnot's theorem states that the thermal efficiency of an arbitrary Carnot cycle cannot exceed that of any reversible Carnot cycle operating at the same temperatures T_1 and T_2. It immediately follows that the thermal efficiency of a reversible Carnot cycle depends on T_1 and T_2 only and is independent of the working substance nature.

Let us sketch the proof of Carnot's theorem. Assume that the thermal efficiency of the reversible heat engine is less than that of the irreversible one. Let us choose the amount of the working substance in the reversible engine such that the work done in one cycle is the same as the work done by the irreversible engine. Using (P2.2.2.8) and the above assumption, we can write the thermal efficiency inequality as $A/Q_1^{\text{rev}} < A/Q_1^{\text{irrev}}$, or $Q_1^{\text{rev}} > Q_1^{\text{irrev}}$. Now let us revert the operation of the reversible engine so that its work is consumed by the irreversible engine. In one cycle, the combined engine produces zero work and the hot reservoir receives the energy $Q_1^{\text{rev}} - Q_1^{\text{irrev}} > 0$ completely extracted from the cold reservoir. We have come to a conclusion that contradicts the Clausius formulation of the second law.

Since we know the thermal efficiency of a Carnot engine, the one that runs on an ideal gas (see (P2.2.3.5)), we can rewrite Carnot's theorem as

$$\frac{Q_1 - Q_2}{Q_1} \leq \frac{T_1 - T_2}{T_1} \qquad \text{or} \qquad \frac{Q_1}{T_1} - \frac{Q_2}{T_2} \leq 0. \qquad (\text{P2.3.2.1})$$

The equality corresponds to a reversible Carnot cycle.

▶ **Thermodynamic temperature scale.** Carnot's theorem allows us to introduce a temperature scale that does not depend on the properties of specific bodies involved. The temperature ratio of two bodies can be determined by connecting them with a reversible Carnot engine. Since the ratio Q_1/Q_2 depends only on the two temperatures, it can be taken to be equal to the ratio of the thermodynamic temperatures: $\theta_1/\theta_2 = Q_1/Q_2$. As seen from (P2.3.2.1), the ratio of the thermodynamic temperatures equals that of the ideal gas temperatures (in the range where the gas scale is defined).

P2.3.3. Calculation of the Internal Energy

▶ **Relation between U and V.** The second law of thermodynamics allows us to derive the important formula for the internal energy of a simple system

$$\left(\frac{\partial U}{\partial V}\right)_T = T\left(\frac{\partial p}{\partial T}\right)_V - p, \qquad (\text{P2.3.3.1})$$

which cannot be obtained based on the first law only.

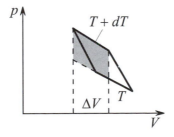

Figure P2.2. Infinitesimal Carnot cycle.

Let us derive formula (P2.3.3.1) from Carnot's theorem. Consider a very small (infinitesimal) reversible Carnot cycle for a simple system and plot it in the pV-coordinates. The work done by the system in one cycle, equal to the area of the small parallelogram (Fig. P2.2), will not change if we replace the adiabatic curves by vertical segments, whose lengths equal $\left(\frac{\partial p}{\partial T}\right)_V \Delta T$. Multiplying by the height ΔV gives $\delta A = \left(\frac{\partial p}{\partial T}\right)_V \Delta T \Delta V$. The heat absorbed at the top isotherm is equal to $\delta Q_1 = \Delta U + p\,dV = \left(\frac{\partial U}{\partial V}\right)_T \Delta V + p\,\Delta V,$

where ΔU has been substituted by its expression (P2.2.2.5) at constant temperature. From Carnot's theorem and equation (P2.3.2.1) we get

$$\frac{\delta A}{\delta Q_1} = \frac{\Delta T}{T} \implies \frac{\left(\frac{\partial p}{\partial T}\right)_V \Delta T \Delta V}{\left[\left(\frac{\partial U}{\partial V}\right)_T + p\right]\Delta V} = \frac{\Delta T}{T},$$

which immediately leads to (P2.3.3.1).

Below are a few examples illustrating the application of formula (P2.3.3.1).

Example 1. Internal energy of an ideal gas. Let us substitute the equation of state $p = \frac{1}{V}\nu RT$ into (P2.3.3.1) to obtain $\left(\frac{\partial U}{\partial V}\right)_T = 0$, which means that the internal energy is independent of volume.

Example 2. Internal energy of a van der Waals gas. Expressing pressure from the equation of state (P2.1.2.2) and substituting into (P2.3.3.1), we obtain $\left(\frac{\partial U}{\partial V}\right)_T = \frac{a\nu^2}{V^2}$. It follows that

$$dU = C_V\, dT + \frac{a\nu^2}{V^2}\, dV.$$

Furthermore, we have

$$\left(\frac{\partial C_V}{\partial V}\right)_T = \frac{\partial}{\partial V}\left[\left(\frac{\partial U}{\partial T}\right)_V\right] = \frac{\partial}{\partial T}\left[\left(\frac{\partial U}{\partial V}\right)_T\right] = \frac{\partial}{\partial T}\left(\frac{a\nu^2}{V^2}\right)_V = 0,$$

which means that C_V is independent of volume. For the temperature range where C_V is weakly dependent on T, we can write

$$U = C_V T - \frac{a\nu^2}{V}. \tag{P2.3.3.2}$$

Example 3. General formula for $C_p - C_V$. Substituting (P2.3.3.1) into (P2.2.2.6) and taking pressure to be constant, we find

$$C_p - C_V = T\left(\frac{\partial p}{\partial T}\right)_V\left(\frac{\partial V}{\partial T}\right)_p = -T\left(\frac{\partial p}{\partial V}\right)_T\left(\frac{\partial V}{\partial T}\right)_p^2 \geq 0.$$

(We have used the relation for the derivatives given at the end of Subsection P2.1.2.) This inequality follows from the condition of mechanical stability $\left(\frac{\partial p}{\partial V}\right)_T \leq 0$.

P2.4. Entropy

P2.4.1. Clausius Inequality

Inequality (P2.3.2.1) is a special case of the *Clausius inequality*, which holds for any thermodynamic cycle. If a system absorbs, in a thermodynamic process, some quantities of heat $Q_1, \ldots, Q_N$ from external reservoirs having temperatures $T_1^e, \ldots, T_N^e$, then the following inequality holds:

$$\sum_{i=1}^{N} \frac{Q_i}{T_i^e} \leq 0 \quad \text{or} \quad \oint \frac{\delta Q}{T^e} \leq 0. \tag{P2.4.1.1}$$

For a reversible process, the inequality becomes equality and the temperature of the reservoir with which the system is in thermal contact at the given point of the cycle is equal to the temperature of the system: $T^e = T$. In this case,

$$\oint_{\text{rev}} \frac{\delta Q}{T} = 0. \tag{P2.4.1.2}$$

Equation (P2.4.1.2) serves to define another function of state—entropy.

P2.4.2. Definition of Entropy

The Clausius equation for a reversible cycle (P2.4.1.2) can be reformulated as the condition for the integral

$$\int_A^B \frac{\delta Q}{T}$$

to be independent of the equilibrium path by which the system passes from state A to state B. This allows one to define a state variable, *entropy*, as

$$S_B - S_A = \int_A^B \frac{\delta Q}{T}, \qquad dS = \left(\frac{\delta Q}{T}\right)_{\text{rev}}. \qquad (P2.4.2.1)$$

To be precise, this relation defines the entropy difference between two states. If there is no heat supply in an equilibrium process, entropy does not change; so an adiabatic process can be called isentropic. For a simple system in an equilibrium process, the first law of thermodynamics (P2.2.2.3) acquires the form

$$T\, dS = dU + p\, dV. \qquad (P2.4.2.2)$$

Since $\delta Q = T\, dS$ for any equilibrium process, it is convenient to plot equilibrium processes and cycles in the TS-coordinates, where the area under the curve $T(S)$ equals the quantity of heat received. For example, a Carnot cycle is represented by a rectangle in TS-coordinates.

▶ **Examples of entropy calculations.**

Example 1. Process with constant heat capacity. The entropy change corresponding to the temperature change from T_1 to T_2 equals

$$S_2 - S_1 = \int_{T_1}^{T_2} \frac{C\, dT}{T} = C \ln\left(\frac{T_2}{T_1}\right). \qquad (P2.4.2.3)$$

Example 2. Entropy of an ideal gas. Substituting the internal energy equation for an ideal gas (P2.2.3.1) into (P2.4.2.2) and taking into account the equation of state $pV = \nu RT$, we obtain

$$dS = \nu C_V^m \frac{dT}{T} + \nu R \frac{dV}{V},$$
$$S_2 - S_1 = \nu \left[C_V^m \ln\left(\frac{T_2}{T_1}\right) + R \ln\left(\frac{V_2}{V_1}\right) \right]. \qquad (P2.4.2.4)$$

Sometimes, entropy is written as $S = \nu(C_V^m \ln T + R \ln V)$, bearing in mind that an entropy difference actually arises in applications.

Example 3. Entropy of a van der Waals gas. Substituting (P2.3.3.2) into (P2.4.2.2) and taking into account the equation of state (P2.1.2.2), we obtain

$$dS = \nu C_V^m \frac{dT}{T} + \frac{a\nu^2}{T} \frac{dV}{V^2} + \frac{p}{T} dV = \nu C_V^m \frac{dT}{T} + \frac{\nu R\, dV}{V - \nu b},$$
$$S = \nu \left[C_V^m \ln T + R \ln(V - \nu b) \right].$$

P2.4.3. Direction of Non-Equilibrium Processes in a Thermally Insulated System

Consider a non-equilibrium process that takes a system from an equilibrium state 1 to an equilibrium state 2. Let us set up a cyclic process by returning from 2 to 1 using any equilibrium process. Writing down the Clausius inequality (P2.4.1.1) for this cycle and taking into account the entropy definition (P2.4.2.1) for the equilibrium process, we get

$$\int_1^2 \frac{\delta Q}{T^e} \le S_2 - S_1. \qquad (P2.4.3.1)$$

If the non-equilibrium process occurs in an adiabatic envelope, then the left-hand side vanishes and then

$$S_2 \geq S_1, \tag{P2.4.3.2}$$

which means that entropy does not decrease in any process without heat supply. The equilibrium state of a thermally insulated system corresponds to the maximum of entropy.

▶ **Examples of non-equilibrium processes.**

Example 1. Heat exchange. Consider heat exchange between two bodies having initial temperatures T_1 and T_2 and identical heat capacities C. From the heat balance equation it follows that the bodies will reach the same equilibrium temperature $T_f = \frac{1}{2}(T_1 + T_2)$. Moreover, according to (P2.4.2.3), the entropy change of the system will be positive:

$$\Delta S = C \ln\left(\frac{T_f}{T_1}\right) + C \ln\left(\frac{T_f}{T_2}\right) = C \ln\left(\frac{T_f^2}{T_1 T_2}\right) > 0;$$

the argument of the logarithm is greater than one, since the arithmetic mean is greater than the geometric mean.

Example 2. Expansion of an ideal gas into a vacuum. The internal energy and, hence, temperature of an ideal gas remain constant when it expands into a vacuum. The entropy change can be found from (P2.4.2.4):

$$S_2 - S_1 = \nu R \ln\left(\frac{V_2}{V_1}\right). \tag{P2.4.3.3}$$

Entropy increase in adiabatic expansion into a vacuum (volume increase under constant internal energy) can be proved for any simple system. If S in (P2.4.2.2) is treated as a function of U and V, then $\left(\frac{\partial S}{\partial V}\right)_U = \frac{p}{T} > 0$.

Example 3. Gas mixing. Consider a thermally insulated container divided by a partition into two equal compartments each containing one mole of an ideal gas under the same temperatures and pressures. When the partition is removed, the gases mix with each other. As follows from (P2.4.3.3), the entropy of the system is increased by $2R \ln 2$ (*entropy of mixing*). If the gases are different, the initial state of the system is different from the final state; the gases can be separated again by doing work on special pistons letting through only one sort of molecules. However, if the gases are identical, the initial and final states of the system are the same, and hence the entropy must not change. The absence of a continuous transition when nearly identical molecules are replaced by completely identical ones is called the *Gibbs paradox*.

P2.4.4. Statistical Meaning of Entropy

Boltzmann's entropy formula relates the entropy S of a macroscopic state of a system to the *statistical weight* W of this macrostate (or the number of microstates corresponding to the given macrostate):

$$S = k \ln W, \tag{P2.4.4.1}$$

where k is the *Boltzmann constant*. The greater W, the higher the probability of the macrostate; in non-equilibrium processes, the systems goes from less probable states to more probable states. The logarithm provides for the additivity of entropy: the statistical weight of a system consisting of two independent subsystems equals the product of their statistical weights. (For more details about the statistical meaning of entropy, see Subsection P7.1.1.

Example. Identity of particles and the Gibbs paradox. To illustrate the application of formula (P2.4.4.1), let us calculate the spatial part of the entropy of an ideal gas. Imagine that the whole volume V of the system is divided into small cells of volume v. The corresponding number of independent states equals $W = (V/v)^N$, where N is the number of molecules in the gas. Then the entropy is $S = k \ln W = kN \ln(V/v) = \nu R \ln(V/v)$. Quantum statistics states that two identical particles are indistinguishable in principle, which means that W must be divided by the number of permutations $N!$. Finally, the spatial part of the entropy becomes $S = kN \ln(V/Nv)$. It can be seen that this approach allows us to resolve the Gibbs paradox.

P2.4.5. Third Law of Thermodynamics

The third law of thermodynamics (also known as *Nernst's theorem*) states that as a system approaches absolute zero, its entropy tends to a certain constant value independent of the values of other thermodynamic parameters. The entropy of a system at $T = 0$ is commonly taken to be zero. From the statistical viewpoint this is explained as follows: at $T = 0$, the system is in its lowest energy state whose statistical weight is very low (close to 1) and, by Boltzmann formula (P2.4.4.1), the entropy is negligibly small. It follows from this law that as $T \to 0$, the heat capacities C_V and C_p of the system as well as its thermal volume expansion and pressure coefficients, α and β, tend to zero.

P2.4.6. Thermodynamic Potentials

▶ **System in a thermostat.** Consider a change in the state of a system that is in a thermal contact with a thermostat having a constant temperature T. The system will have the same temperature at its initial and final equilibrium states. With $T^e = $ const and after a sequence of transformations

$$Q/T \leq S_2 - S_1 \quad \Longrightarrow \quad (U_2 - U_1) + A \leq TS_2 - TS_1,$$

inequality (P2.4.3.1) becomes

$$(U_1 - TS_1) - (U_2 - TS_2) \geq A \qquad \text{(P2.4.6.1)}$$

▶ **Helmholtz free energy.** With the new state variable

$$F = U - TS, \qquad \text{(P2.4.6.2)}$$

called the *Helmholtz free energy*, inequality (P2.4.6.1) becomes

$$F_1 - F_2 \geq A.$$

If the volume of the system does not change in a non-equilibrium process, then

$$F_2 \leq F_1,$$

and so the free energy does not increase. This fact can be rephrased as follows: an equilibrium state of a system of constant volume that is in thermal contact with a thermostat corresponds to the minimum of the free energy. For an equilibrium process, both inequalities above become equalities.

If a system placed in a thermostat and having constant volume is used to produce some "useful" work (e.g., by connecting internal moving parts of the system with external bodies), then the *maximum work* done by the system as it passes from state 1 to state 2 is equal to

$$A_{\max} = F_1 - F_2,$$

with the maximum work being done in a completely reversible equilibrium process. This is why this thermodynamic function is called free energy.

From formulas (P2.4.6.2) and (P2.4.2.2) it follows that

$$dF = -S\,dT - p\,dV. \qquad \text{(P2.4.6.3)}$$

Therefore, the natural variables to express free energy are volume and temperature: $F = F(V, T)$.

▶ **Gibbs free energy.** Suppose that a thermodynamic system is placed in a thermostat and is in mechanical contact (e.g., through a light piston or a flexible envelope) with the environment that has a fixed pressure p (the system will have the same pressure in the equilibrium state). Then the work done by the system to change its volume equals $A = p(V_2 - V_1)$. From (P2.4.6.1) we get

$$F_1 + pV_1 \geq F_2 + pV_2 \qquad \text{or} \qquad G_1 \geq G_2, \tag{P2.4.6.4}$$

with

$$G = F + pV = U - TS + pV. \tag{P2.4.6.5}$$

The new thermodynamic function G is called the *Gibbs free energy*. It does not increase in non-equilibrium processes and remains constant in completely reversible equilibrium processes. The minimum of $G(p, T)$ corresponds to an equilibrium state. The maximum "useful" work that the system can perform (i.e., the total work minus the work done to change the volume of the system) equals

$$A_{\max} = G_1 - G_2.$$

Since

$$dG = -S\,dT + V\,dp, \tag{P2.4.6.6}$$

the natural variables for the Gibbs free energy are pressure and temperature: $G = G(p, T)$.

▶ **Chemical potential.** If the pressure and temperature are kept constant, the Gibbs free energy is proportional to the quantity of substance: $G(p, T, m) = mg$ or $G(p, T, N) = N\mu$, where g is the *specific Gibbs free energy* and μ is the thermodynamic potential per molecule, or the *chemical potential*. (Sometimes, the chemical potential per mole is used.)

▶ **Canonic equations of state. Natural variables.** Although the functions U, H, F, and G, called *thermodynamic potentials*, can be expressed in terms of any independent variables, there are so-called *natural variables* associated with each of them. These are S and V for internal energy, S and p for enthalpy, V and T for Helmholtz free energy, and p and T for the Gibbs free energy. Below are the definitions and the differentials of the above functions listed together:

$$
\begin{aligned}
& & dU &= T\,dS - p\,dV + \mu\,dN; \\
H &= U + pV; & dH &= T\,dS + V\,dp + \mu\,dN; \\
F &= U - TS; & dF &= -S\,dT - p\,dV + \mu\,dN; \\
G &= U + pV - TS; & dG &= -S\,dT + V\,dp + \mu\,dN.
\end{aligned}
\tag{P2.4.6.7}
$$

Note that each differential has the additive term $\mu\,dN$, which shows the dependence of each function on the number of particles in the system. The expression of a thermodynamic potential in terms of its natural variables is called a *canonic equation of state*. Given a canonic equation, one can obtain both the thermic, $p = p(V, T)$, and caloric, $U(V, T)$, equations of state. For example, $-\left(\dfrac{\partial F}{\partial V}\right)_T$ gives the expression for $p(V, T)$ and also, having expressed $S(V, T) = -\left(\dfrac{\partial F}{\partial T}\right)_V$, one finds $U(V, T) = F + TS$.

▶ **Maxwell's identities.** By equating the cross derivatives of any thermodynamic function with respect to any two variables, one can obtain useful thermodynamic relations, called

Maxwell's identities. Below are Maxwell's identities for the thermodynamic potentials:

$$
\left(\frac{\partial T}{\partial V}\right)_S = -\left(\frac{\partial p}{\partial S}\right)_V;
$$
$$
\left(\frac{\partial T}{\partial p}\right)_S = \left(\frac{\partial V}{\partial S}\right)_p;
$$
$$
\left(\frac{\partial S}{\partial V}\right)_T = \left(\frac{\partial p}{\partial T}\right)_V;
$$
$$
\left(\frac{\partial S}{\partial p}\right)_T = -\left(\frac{\partial V}{\partial T}\right)_p.
$$

(P2.4.6.8)

Example 1. With Maxwell's identities, it is quite easy to obtain the formula for $\left(\frac{\partial U}{\partial V}\right)_T$ that was derived in Subsection P2.3.3 using an infinitesimally small Carnot cycle. The first equation in (P2.4.6.7) gives

$$
\left(\frac{\partial U}{\partial V}\right)_T = T\left(\frac{\partial S}{\partial V}\right)_T - p.
$$

Inserting the third identity in (P2.4.6.8), we arrive at the desired formula.

Example 2. Find the quantity of heat required for 1 kg of mercury to be isothermally compressed from 1 atm to 10 atm. The thermal volume expansion coefficient $\alpha = 1.8 \times 10^{-4}\,\mathrm{K}^{-1}$ and the mercury density $\rho = 13.55 \times 10^3\,\mathrm{kg\,m^{-3}}$ are assumed to be unchangeable.

Solution. It follows from the last Maxwell's identity in (P2.4.6.8) that

$$
Q = T\Delta S = T\left(\frac{\partial S}{\partial p}\right)_T \Delta p = -T\left(\frac{\partial V}{\partial T}\right)_p \Delta p = -TV\frac{1}{V}\left(\frac{\partial V}{\partial T}\right)_p \Delta p = -\frac{m}{\rho}T\alpha\Delta p = -3.6\,\mathrm{J}.
$$

P2.5. Kinetic Theory of Ideal Gases

P2.5.1. The Basic Equation of the Kinetic Theory

▶ **Pressure of an ideal gas.** From the viewpoint of kinetic-molecular theory, a gas is considered ideal if the potential energy of interaction among its molecules is negligible as compared to the kinetic energy and the molecule size is negligible as compared to the average distance between the molecules. The pressure on the walls of the container results from numerous elastic collisions of the molecules with them. (The collisions can be treated as elastic, on the average, since the gas is in thermal equilibrium with the walls.) The *basic equation of the kinetic theory of an ideal gas* expresses pressure in terms of the mean square velocity of the molecules:

$$
p = \tfrac{1}{3}n\langle\mathbf{p}\mathbf{v}\rangle = \tfrac{1}{3}nm_0\langle v^2\rangle = \tfrac{2}{3}n\langle\varepsilon_{\text{transl}}\rangle,
\tag{P2.5.1.1}
$$

where $n = N/V$ is the number of molecules per unit volume, m_0 is the mass of one molecule, $\mathbf{p} = m_0\mathbf{v}$ is its momentum, and $\langle\varepsilon_{\text{transl}}\rangle = \langle\tfrac{1}{2}m_0 v^2\rangle$ is the mean kinetic energy of translational motion of a molecule.

If the relation between energy and momentum is other than classical, then the expression in (P2.5.1.1) is different. For example, $|\mathbf{p}| = \varepsilon/c$ for a photon (see Subsection P1.10.3). As a result, pressure is expressed as

$$
p = \tfrac{1}{3}n\langle\varepsilon\rangle
\tag{P2.5.1.2}
$$

for a photon gas.

▶ **Number of collisions against the wall.** The number of molecules, ΔN, hitting a wall region of area ΔS in time Δt equals

$$\frac{\Delta N}{\Delta S \Delta t} = \frac{n\langle v \rangle}{4}. \qquad \text{(P2.5.1.3)}$$

Calculation formulas. Formulas for the transfer of the number of molecules, momentum, energy, etc. through a small area are obtained by dividing molecules into groups and calculating the contribution of each group. One of the most common approaches is to group molecules having nearly the same velocity, **v**. Consider the group of molecules having velocity magnitudes in the range from v to $v + dv$ in a solid angle $d\Omega$. The number of molecules in this group per unit volume equals $dn(v, v + dv; d\Omega) = dn(v, v + dv) \, d\Omega / 4\pi$. In order to find how many molecules in this group pass through a plane region of area ΔS in time Δt, we should construct an oblique cylinder based on this region (Fig. P2.3) whose generatrix has the length $v\Delta t$ and make an angle θ with the x-axis. The molecules from the group concerned that belong to this cylinder only will reach its base in time Δt. Integrating first with respect to angles ($d\Omega = \sin\theta \, d\theta \, d\varphi$) and then with respect to v, one can obtain formulas (P2.5.1.1), (P2.5.1.3), and others. For example, the energy transferred through area ΔS in time Δt is expressed as

$$\Delta E = \int \frac{m_0 v^2}{2} \, dn(v, v + dv) \frac{d\Omega}{4\pi} \left(\Delta S \, v\Delta t \cos\theta \right)$$

$$= \frac{m_0 n}{8} \Delta S \, \Delta t \int_0^\infty \frac{v^3 \, dn(v, v + dv)}{n} = \frac{m_0 n \langle v^3 \rangle}{8} \Delta S \, \Delta t. \qquad \text{(P2.5.1.4)}$$

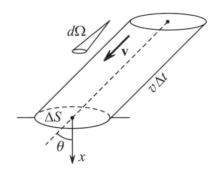

Figure P2.3. To the calculation of the number of molecules passing through a region of area ΔS in time Δt.

P2.5.2. Temperature and Energy of an Ideal Gas

▶ **Temperature definition in kinetic theory of gases.** It is proved in kinetic theory of gases that if two subsystems (consisting of identical or different molecules) can exchange energy, their mean kinetic energies of translational motion per molecule are equal to each other in equilibrium. Based on this fact, kinetic theory of gases defines temperature as a quantity proportional to the average kinetic energy of translational motion of molecules:

$$\langle \varepsilon_{\text{transl}} \rangle = \frac{3}{2} kT, \qquad \text{(P2.5.2.1)}$$

where k is the *Boltzmann constant*. The proportionality coefficient is chosen so that the equation of state of an ideal gas,

$$p = nkT = \frac{N}{V} kT, \qquad \text{(P2.5.2.2)}$$

obtained from the basic equation (P2.5.1.1) and the temperature definition (P2.5.2.1), coincides with equation (P2.1.2.1), where the gas temperature scale is used.

▶ **Root mean square velocity.** Using (P2.5.1.1) and (P2.5.2.1), one can calculate the root mean square velocity of translational motion of molecules:

$$v_{\text{rms}} = \sqrt{\langle v^2 \rangle} = \sqrt{\frac{3p}{\rho}} = \sqrt{\frac{3kT}{m_0}} = \sqrt{\frac{3RT}{\mu}}, \qquad \text{(P2.5.2.3)}$$

where $\rho = m_0 n$ is the gas density.

▶ **Internal energy of an ideal gas.** The *number of degrees of freedom* of a molecule, i, is an important characteristic of an ideal gas. A monatomic molecule has only three degrees of freedom; these correspond to translational motion in three spatial coordinates: $i = i_{\text{transl}} = 3$. A rigid diatomic molecule has two additional, rotational degrees of freedom, associated with the rotation about two axes perpendicular to the axis of the molecule: $i = i_{\text{transl}} + i_{\text{rot}} = 3 + 2 = 5$. A rigid multiatomic molecule (whose atoms are not collinear) has three rotational degrees of freedom, and therefore $i = 6$. Classical statistical physics proves the *equipartition theorem* (also known as *law of equipartition*). According to this theorem, an average energy of $\frac{1}{2}kT$ per molecule is associated with any degree of freedom to which there corresponds a term like αx^2 or $\beta \dot{x}^2$ in the expression of the molecule energy. Formula (P2.5.2.1) is in complete agreement with this theorem. Accordingly, the average energy per one molecule and the total internal energy of the gas are expressed as

$$\langle \varepsilon \rangle = \tfrac{1}{2} i_{\text{phys}} kT, \qquad U = N \langle \varepsilon \rangle = \tfrac{1}{2} i_{\text{phys}} \nu RT. \qquad \text{(P2.5.2.4)}$$

The physical number of degrees of freedom, i_{phys}, for rigid molecules coincides with the mathematical one. However, at sufficiently high temperatures ($T \sim 10^3$ K), when molecules can no longer be treated as rigid and vibrational energy must be taken into account, an average energy of kT will be associated with each vibrational degree of freedom. As a result, the total number of degrees of freedom becomes $i_{\text{phys}} = i_{\text{transl}} + i_{\text{rot}} + 2 i_{\text{vib}}$, where i_{vib} is the mathematical number of vibrational degrees of freedom in a molecule. For an s-atom molecule, $i_{\text{vib}} = 3s - (i_{\text{transl}} + i_{\text{rot}})$. In particular, for a nonrigid diatomic molecule, we have $i_{\text{vib}} = 6 - (3 + 2) = 1$ and $i_{\text{phys}} = 3 + 2 + 2 \times 1 = 7$.

▶ **Heat capacities of an ideal gas.** Using (P2.5.2.4) and the definitions of heat capacities at constant volume and constant pressure as well as adiabatic index (see Subsection P2.2.3), we find that

$$C_V^{\text{m}} = \frac{i_{\text{phys}}}{2} RT, \qquad C_p^{\text{m}} = \frac{i_{\text{phys}} + 2}{2} RT, \qquad \gamma = \frac{C_p}{C_V} = \frac{i_{\text{phys}} + 2}{i_{\text{phys}}}. \qquad \text{(P2.5.2.5)}$$

At room temperatures, the heat capacity follows the model of rigid molecules; the rotational degrees of freedom are not excited, or, as is often said, "frozen." However, at temperature increases to $\sim 10^3$ K, the heat capacity begins to increase, that is, vibrational degrees of freedom "unfreeze." Furthermore, as temperature decreases to several tens of kelvins, the rotational degrees of freedom "freeze out," which results in decreasing heat capacity. The phenomenon of *freezing-out of degrees of freedom* is explained by quantum mechanics: if the average energy of thermal motion, kT, is small compared to the distance to the nearest discrete energy level, the kind of motion concerned is not excited.

▶ **Mixture of ideal gases.** *Dalton's law* reads: the pressure of a mixture of ideal gases equals the sum of their partial pressures. In particular, for two gases,

$$p = p_1 + p_2 = (n_1 + n_2)kT = (\nu_1 + \nu_2)RT.$$

The internal energy of a mixture equals the sum of the internal energies of its components: $U = \frac{1}{2} i_1 \nu_1 RT + \frac{1}{2} i_2 \nu_2 RT$. This formula allows us to introduce the so-called effective number of degrees of freedom, $i(\nu_1 + \nu_2) = i_1 \nu_1 + i_2 \nu_2$, effective molar heat capacity, $(\nu_1 + \nu_2)C_\mu = \nu_1 C_{\mu 1} + \nu_2 C_{\mu 2}$, and effective molar mass, $(\nu_1 + \nu_2)\mu = \nu_1 \mu_1 + \nu_2 \mu_2$.

P2.5.3. Maxwell's Distribution

▶ **Definition and properties of distribution functions.** The distribution of molecules over their velocities is characterized by the following functions:

$$\varphi(v_x)\, dv_x = \frac{dn(v_x,\, v_x + dv_x)}{n},$$

$$f(v)\, dv = \frac{dn(v,\, v + dv)}{n}, \qquad (\text{P2.5.3.1})$$

$$\Phi(\mathbf{v})\, dv_x\, dv_y\, dv_z = \frac{dn(v_x,\, v_x + dv_x;\; v_y,\, v_y + dv_y;\; v_z,\, v_z + dv_z)}{n}.$$

The definition of each of the distribution functions is based on the assumption that the fraction of molecules that fall (on the average) within a very small interval of a given variable (e.g., velocity, its projection onto an axis, energy, etc.) is proportional to the width of this interval. Note that dv denotes a physically very narrow, rather than mathematically infinitesimal, interval but wide enough to contain a large number of molecules. The mean fraction of molecules having a certain property (e.g., having velocities within a given interval) can be treated as the probability that an arbitrary molecule possesses this property. Therefore, a distribution function is sometimes referred to as a *probability density function*.

Listed below are properties of a distribution function (exemplified by $f(v)$).

1. The fraction of particles (probability) having velocities within a finite interval (v_1, v_2):

$$\frac{\Delta n(v_1, v_2)}{n} = \int_{v_1}^{v_2} f(v)\, dv.$$

2. Normality:

$$\int_0^\infty f(v)\, dv = 1. \qquad (\text{P2.5.3.2})$$

3. Computation of the average of any function $\chi(v)$:

$$\langle \chi(v) \rangle = \int_0^\infty \chi(v)\, f(v)\, dv. \qquad (\text{P2.5.3.3})$$

The three distribution functions defined in (P2.5.3.1) are related by

$$f(v) = \Phi(v) 4\pi v^2, \qquad \Phi(v_x, v_y, v_z) = \varphi(v_x)\varphi(v_y)\varphi(v_z).$$

The function φ is even, and hence one can write $\varphi = \varphi(v_x^2)$. The function $\Phi(\mathbf{v})$ depends on $v^2 = v_x^2 + v_y^2 + v_z^2$ only. The relation between Φ and φ is satisfied only if $\varphi(v_x) = A \exp(-\zeta v_x^2)$. The coefficients A and ζ are determined from two conditions: (a) the normality of φ and (b) the requirement that $\langle v_x^2 \rangle = \frac{1}{3}\langle v^2 \rangle = kT/m_0$; see (P2.5.2.3). Thorough calculations give the answer

$$\varphi(v_x) = \left(\frac{m_0}{2\pi kT}\right)^{1/2} \exp\left(-\frac{m_0 v_x^2}{2kT}\right),$$

$$f(v) = \left(\frac{m_0}{2\pi kT}\right)^{3/2} \exp\left(-\frac{m_0 v^2}{2kT}\right) 4\pi v^2. \qquad (\text{P2.5.3.4})$$

The latter expression is referred to as *Maxwell's distribution* (Fig. P2.4). The function $f(v)$ attains its maximum at $v_{mp} = (2kT/m_0)^{1/2}$, which is called the *most probable velocity*. The value of $f(v)$ at this point is $f(v_{mp}) = 4e^{-1}(2\pi kT/m_0)^{-1/2}$. For example, it

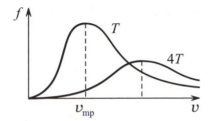

Figure P2.4. Maxwell's distribution function.

follows that if T increases by a factor of 4, the most probable velocity doubles and the value of the function, $f(v_{mp})$, is halved; note that the area under the curve $f(v)$ is equal to one.

The *average velocity* (arithmetic mean) of molecules is calculated by (P2.5.3.3):

$$\langle v \rangle = \int_0^\infty v f(v)\, dv = \sqrt{\frac{8kT}{\pi m_0}} = \sqrt{\frac{8RT}{\pi \mu}}.$$

The distribution of molecules over the energies of their translational motion is determined by

$$\psi(\varepsilon)\, d\varepsilon = \frac{dn(\varepsilon,\ \varepsilon + d\varepsilon)}{n} = \frac{2}{\sqrt{\pi}} \exp\left(-\frac{\varepsilon}{kT}\right) \sqrt{\frac{\varepsilon}{kT}}\, d\left(\frac{\varepsilon}{kT}\right).$$

Useful integrals. In calculating averages using Maxwell's distribution, the following integrals often arise: $I_n = \int_0^\infty x^n \exp(-\beta x^2)\, dx$ $(n = 0, 1, 2, \ldots)$. The first two integrals are $I_0 = \frac{1}{2} \sqrt{\pi}\, \beta^{-1/2}$ and $I_1 = \frac{1}{2}\beta^{-1}$. Other integrals can be calculated by differentiating I_0 or I_1 with respect to β. For example, $I_2 = -\dfrac{dI_0}{d\beta} = \frac{1}{4}\sqrt{\pi}\, \beta^{-3/2}$.

P2.5.4. Boltzmann's Distribution

▶ **Boltzmann's distribution.** If a gas is in an external force field, the concentration of molecules depends on the spatial position. It can be obtained from the condition of mechanical equilibrium of the gas that

$$n(\mathbf{r}) = n(\mathbf{r}_0) \exp\left[-\frac{\varepsilon_p(\mathbf{r}) - \varepsilon_p(\mathbf{r}_0)}{kT}\right], \tag{P2.5.4.1}$$

where $\varepsilon_p(\mathbf{r})$ is the potential energy of a molecule in the external field. Formula (P2.5.4.1) is known as *Boltzmann's distribution*. The barometric formula

$$p(h) = n(h)kT = p_0 \exp\left(-\frac{m_0 gh}{kT}\right), \quad \text{where} \quad n(h) = n_0 \exp\left(-\frac{m_0 gh}{kT}\right), \tag{P2.5.4.2}$$

is a special case of Boltzmann's distribution.

Let us illustrate the derivation of Boltzmann's distribution using the barometric formula as an example. The equilibrium condition for a vertical cylinder with the base area s and height dh in the field of gravity is: $s\, dp = -\rho(s\, dh)g$. Then, using the equations $p = nkT$ and $\rho = m_0 n$, we obtain $\dfrac{dn}{dh} = -\dfrac{m_0 g}{kT} n$. Integrating yields the barometric formula.

▶ **Maxwell–Boltzmann distribution.** Both Maxwell's and Boltzmann's distributions, (P2.5.3.4) and (P2.5.4.1), contain the expression $\exp(-\varepsilon/kT)$. The combined Maxwell–Boltzmann distribution expresses the probability that an arbitrary molecule from a container with N molecules is in the spatial region $(x, x+dx; y, y+dy; z, z+dz)$ and has the velocity within $(v_x, v_x + dv_x; v_y, v_y + dv_y; v_z, v_z + dv_z)$:

$$\frac{dN}{N} = \frac{dn(\mathbf{r})\, d^3\mathbf{r}}{N} = \frac{n(\mathbf{r})\Phi(\mathbf{v})}{N}\, d^3\mathbf{r}\, d^3\mathbf{v} = A\exp\left(-\frac{\varepsilon}{kT}\right) d^3\mathbf{r}\, d^3\mathbf{v}, \qquad \text{(P2.5.4.3)}$$

where $d^3\mathbf{r} = dx\, dy\, dz$, $d^3\mathbf{v} = dv_x\, dv_y\, dv_z$, $\varepsilon = \frac{1}{2}m_0 v^2 + \varepsilon_{\mathrm{p}}(\mathbf{r})$ is the mechanical energy of a molecule, A is a coefficient that can be found from the normality condition. The Maxwell–Boltzmann distribution applies to any kinds of molecule energy: rotational, vibrational, potential, angular orientation dependent, and others.

P2.6. Real Gases. Van Der Waals Equation

P2.6.1. Isotherms of a Real Gas

▶ **Low temperatures, $T < T_c$.** If the gas temperature is less than the critical temperature (see below), then the following processes occur in equilibrium isothermal volume contraction (Fig. P2.5):

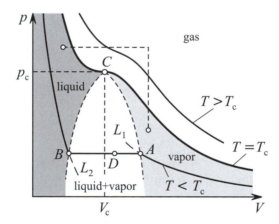

Figure P2.5. Isotherms of a real gas. ACB is the region of two-phase equilibrium.

1. Pressure increases to the point A. If temperature is well below the critical temperature, then the ideal gas law works well up to the point A.

2. Pressure does not change along AB. Some liquid emerges in the container; it is separated from the gas by a clear interphase surface. A gas which is in thermodynamic equilibrium with its liquid phase is called a *saturated vapor*. The liquid-to-gas mass ratio at a point D on the segment AB equals the ratio between the lengths of the segments AD and DB: $m_{\mathrm{liq}}/m_{\mathrm{vap}} = AD/DB$. The isotherm portions AL_1 and L_2B correspond to *metastable states*, or states at which the system can remain for some time but then quickly passes to a stable state on the segment AB. A gas and liquid in a metastable state are called a *supercooled vapor* and *superheated liquid* (at constant pressure). The existence of metastable states is accounted for by the fact that the formation of tiny nuclei of a new phase (liquid in vapor or vapor in liquid) requires a considerable amount of energy, known as surface energy (see Subsection P2.8.1).

3. After the point B, pressure sharply increases. There is only the liquid phase in the container.

▶ **Critical temperature,** $T = T_c$. The higher the temperature, the shorter the horizontal portion of the isotherm. At some *critical temperature*, T_c, it reduces to a mere point, C, the inflection point of the isotherm (Fig. P2.5). The pressure at the *critical point* is called the *critical pressure*, p_c, and the volume of a single mole of the gas is called the *critical molar volume*, V_c.

▶ **High temperatures,** $T > T_c$. For higher temperatures, $T > T_c$, the isotherm is a smooth monotonic curve. The region above the critical isotherm is called the true gas (it cannot be isothermally transferred to a liquid + vapor state). The region below the critical isotherm consists of three subregions: liquid, unsaturated vapor, and liquid + saturated vapor. It is possible to pass (non-isothermally) from the vapor region to the liquid region without entering the liquid + vapor region. So gas and liquid differ from each other by only quantitative characteristics.

The critical parameters of some substances are shown in the table below.

TABLE P2.1
Critical parameters of some substances.

Substance	T_c, K	p_c, MPa	V_c, cm³/mol
Helium, He	5.19	0.277	57.4
Hydrogen, H₂	33.24	1.30	65.0
Neon, Ne	44.4	2.654	41.7
Nitrogen, N₂	126.3	3.40	90.1
Oxygen, O₂	154.6	5.04	78
Methane, CH₄	190.66	4.626	99.38
Krypton, Kr	209.39	5.49	91.9
Carbon dioxide, CO₂	304.1	7.38	94.0
Hydrogen sulphide, H₂S	373.6	9.01	97.4
Water, H₂O	647.3	22.12	56
Mercury, Hg	1460	166.1	48
Lithium, Li	3200	68.9	66

P2.6.2. Van Der Waals Equation

▶ **Van der Waals equation.** The equation of state for one mole of a van der Waals gas is

$$(V_\mu - b)\left(p + \frac{a}{V_\mu^2}\right) = RT. \tag{P2.6.2.1}$$

Note that equation (P2.1.2.2) from Section P2.1 for ν moles is obtained by substituting $V_\mu = V/\nu$. The positive constant b reflects the fact that not all of the container volume is available for the motion of molecules due to their own volume. This constant is taken to be the total volume of all molecules multiplied by four: $b = 4N_A \times \frac{1}{6}\pi d^3$, where d is the diameter of a molecule. The term a/V_μ^2 reflects the fact that the total pressure decreases due to mutual interaction of molecules. This can be clearly seen from the expression of the internal energy of a van der Waals gas: $U = C_{\mu V}T - a/V_\mu$.

The average potential energy of interaction among the molecules is characterized by a model expression containing two terms, first corresponding to strong repulsion at small distances ($\sim 10^{-10}$ m) and second corresponding to van der Waals attractive forces at large distances (Fig. P2.6):

$$E_p = \frac{\alpha}{r^{12}} - \frac{\beta}{r^6}.$$

The minimum of E_p corresponds to the average inter-molecule distance in the absence of thermal motion. For $kT \ll E_{min}$, the motion of molecules reduces to vibrations about the bottom of the potential well (solid state). The case $kT \sim E_{min}$ corresponds to liquid state and the case $kT \gg E_{min}$ corresponds to gaseous state. Since the left branch of the potential curve, corresponding to repulsion, is nearly vertical, the minimum distance to which molecules approach each other before rebounding is weakly dependent on their average translational energy (see also Subsection P1.5.4).

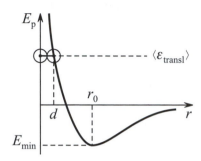

Figure P2.6. Potential energy of interaction of molecules.

▶ **Van der Waals isotherms.** Figure P2.7 depicts isotherms predicted by the van der Waals equation (P2.6.2.1). At temperatures below the critical value, the isotherms have two extrema, K_1 and K_2. However, for the path AK_1K_2B of a subcritical isotherm, the condition of mechanical stability, $(\partial p/\partial V)_T \leq 0$, is violated. In fact, the transition between the points A and B occurs along a horizontal line. According to *Maxwell's rule*, the position of the segment AB is determined by the condition that shaded areas must be equal. Otherwise, the isothermal cycle AK_1K_2BA would perform work, which contradicts the second law of thermodynamics. The portions AK_1 and K_2B correspond to metastable states.

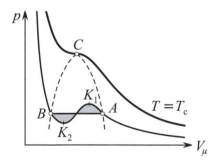

Figure P2.7. Van der Waals isotherms.

▶ **Calculation of critical parameters.** At the critical point, the following two conditions hold:

$$\left(\frac{\partial p}{\partial V}\right)_T = 0, \qquad \frac{\partial^2 p}{\partial V^2} = 0. \tag{P2.6.2.2}$$

Expressing p from (P2.6.2.1) and substituting into these equations, we find that

$$V_c = 3b, \quad p_c = \frac{1}{27}\frac{a}{b^2}, \quad T_c = \frac{8}{27}\frac{a}{bR}.$$

The critical parameters satisfy the relation $p_c V_c = \frac{3}{8} R T_c$. By introducing the reduced state variables $v = V/V_c$, $\pi = p/p_c$, and $\theta = T/T_c$, one can rewrite the van der Waals equation (P2.6.2.1) as

$$\left(v - \frac{1}{3}\right)\left(\pi + \frac{3}{v^2}\right) = \frac{8}{3}\theta.$$

This equation is the same for all gases. It follows that the van der Waals equation satisfies the experimentally established *principle of corresponding states*. This principle asserts that if two respective reduced state variables for two gases coincide, the third will also coincide. This law is obeyed very well by most real gases.

P2.7. Phase Equilibrium. Phase Transitions

P2.7.1. Conditions for Phase Equilibrium

▶ **Phase.** A *phase* is a part of a system which is homogeneous in its physical and chemical properties. The same substance can be in different phases, although its chemical composition does not change. Examples of a *two-phase system* are: liquid+saturated vapor, solid + liquid, and solid + vapor.

▶ **Three equilibrium conditions.** For the phases of a system to be in equilibrium, the following conditions must hold: (1) the pressures must be equal (mechanical equilibrium), (2) the temperatures must be equal (thermal equilibrium), and (3) the chemical potentials must be equal (equilibrium with respect to interphase transitions). The last condition corresponds to the minimum of the Gibbs free energy $G = \mu_1 N_1 + \mu_2(N - N_1)$ (see Subsection P2.4.6) with respect to the number of molecules in one of the phases, N_1. For a simple system, the condition $\mu_1(p, T) = \mu_2(p, T)$ determines the equation of the *two-phase coexistence curve* $p(T)$; a simple two-phase system has only one degree of freedom. In particular, the pressure of a saturated vapor, determined by the horizontal portion of the isotherm (see Subsection P2.6.1), is a single-valued function of temperature (the *evaporation–condensation curve*). The same applies to the dependence of the melting temperature on the external pressure (the *melting–freezing curve*) and also to the temperature dependence of the vapor pressure over a solid surface (the *sublimation–deposition curve*).

▶ **Triple point.** Figure P2.7 depicts a schematic phase diagram for water, which includes the three phase-equilibrium curves: evaporation–condensation (I), melting–freezing (II), and sublimation–deposition (III). The point of intersection of the three curves is called the *triple point* (for water, $T_3 = 0.01°C$ and $p_3 = 610.5$ Pa). The conditions $\mu_{vap}(p, T) = \mu_{liq}(p, T)$ and $\mu_{liq}(p, T) = \mu_{sol}(p, T)$ must hold simultaneously for a *three-phase system* to exist. The pressure and temperature, identifying the triple point, are determined uniquely by these conditions.

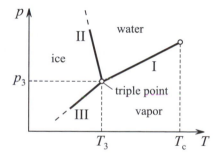

Figure P2.7. Phase diagram for water.

P2.7.2. First-Order Phase Transitions

▶ **Transition heat and entropy.** In order to transfer a portion of substance of mass m in a two-phase system from one phase to the other, the quantity of heat $Q = mq$ must be supplied (or removed) in an equilibrium isobar-isothermal process. This quantity of heat Q is called the *phase transition heat* or *latent heat*; the quantity q is the *specific phase transition heat* or *specific latent heat*. When a substance changes to a new phase, its entropy changes by $\Delta S = Q/T = mq/T$ and the internal energy, by $\Delta U = mq - mp\Delta v$, where $v = V/m$ is the *specific volume*. The quantity $\Delta s = q/T$ is called the specific entropy of the phase transition, and $\Delta u = q - p\Delta v$ is the change in the specific internal energy at the phase transition.

A transition between phases in a two-phase system in which some heat is absorbed or released and an abrupt change in internal energy, entropy, density, volume, etc. occurs is called a *first-order phase transition*. This name is due to the fact that the first derivatives of the Gibbs free energy, $S = -(\partial G/\partial T)_p$ and $V = (\partial G/\partial p)_T$, experience a jump discontinuity.

▶ **Clausius–Clapeyron relation.** If the temperature of a two-phase system is changed, then its pressure will adapt so that the chemical potentials (or the specific Gibbs free energies) of the phases remain the same: $dg_1 = dg_2$, or $-s_1\, dT + v_1\, dp = -s_2\, dT + v_2\, dp$ (see Subsection P2.4.6). It follows that the slope of the phase equilibrium curve equals

$$\frac{dp}{dT} = \frac{\Delta s}{\Delta v} = \frac{q}{T\Delta v}. \tag{P2.7.2.1}$$

This equation is referred to as the *Clausius–Clapeyron relation.*

Example 1. For the ice–water phase transition at atmospheric pressure, the melting temperature is 273 K, $q = 3.34 \times 10^5$ J/kg, and $v_w - v_i = -0.91 \times 10^{-4}$ m³/kg. Consequently $dp/dT = -1.1 \times 10^7$ Pa/K. The minus sign reflects the anomalous property of water: its density increases in melting.

Example 2. If heat is continuously supplied to a liquid, intensive evaporation begins across the whole volume when the temperature attains the value at which the saturation vapor pressure equals the external pressure. This process is called *boiling*. Equation (P2.7.2.1) determines the dependence between the *boiling temperature* and external pressure. Neglecting the specific volume of the liquid and expressing the specific volume of the vapor from the ideal gas law $v = \dfrac{RT}{\mu p}$, we arrive at the approximate equation $\dfrac{dp}{dT} = \dfrac{\mu q p}{RT^2}$. In particular, for water ($\mu = 18$ g/mol) at atmospheric pressure ($p = 10^5$ Pa), we have the boiling temperature $T = 373$ K and the specific latent heat of evaporation $q = 2.26 \times 10^6$ J/kg. If follows that $\dfrac{dp}{dT} = 3.5 \times 10^3$ Pa/K.

▶ **Second-order phase transitions.** In a second-order phase transition, density, internal energy, and entropy (the first derivatives of the Gibbs free energy) are changed continuously, while the various derivatives of these thermodynamic functions (such as heat capacity, compressibility, dielectric constant, magnetic permeability, etc.) experience a jump discontinuity or show critical behavior (tend to infinity when approaching the transition point). The name is due to the fact that the second derivatives of the Gibbs free energy experience a discontinuity. Examples of such transitions are the paramagnetic-to-ferromagnetic transition at the Curie point, transition of liquid helium to the superconducting state, superfluid transitions, etc. A second-order phase transition occurs at the whole volume at once and is associated with a qualitative change in the internal structure (such as appearance of additional symmetries or loss of some symmetries).

P2.8. Surface Tension

P2.8.1. Surface Energy

▶ **Surface tension coefficient.** Molecules at the surface of a liquid possess higher potential energy as compared to molecules in the bulk, since the former have fewer "neighbors," the energy of interaction with which is negative (see Fig. P2.6). At constant temperature, the additional energy is proportional to the surface area, Σ:

$$F_{\text{surf}} = \sigma(T)\Sigma, \tag{P2.8.1.1}$$

where σ is the surface free energy per unit area, called the *surface tension coefficient*.

In an equilibrium isothermal change of a rectangular area on the surface, the work done by external forces equals the change in the Helmholtz free energy (P2.4.6.2). So $A = \Delta F_{\text{surf}} = \sigma\Delta\Sigma = \sigma Lx$, where L is the rectangle side and x is its displacement. It follows that a *surface tension force* acts on the rectangle side which is perpendicular to the side and tangent to the liquid surface:

$$f_{\text{surf}} = \sigma L. \tag{P2.8.1.2}$$

In equilibrium, the Helmholtz free energy is minimum and, hence, a liquid of a liquid film tend to minimize their surface. For example, a drop of liquid tends to a spherical shape at zero gravity.

▶ **Thermodynamics of surface layer.** Since F_{surf} is a function of T and Σ, then $dF_{\text{surf}} = \sigma \, d\Sigma + \Sigma \dfrac{d\sigma}{dT} \, dT$. It follows, taking into account (P2.4.6.3), that the surface layer entropy and internal energy as well as the heat absorbed by the layer under isothermal area extension are expressed as

$$S = -\frac{d\sigma}{dT}\Sigma, \quad U = F + TS = \left(\sigma - T\frac{d\sigma}{dT}\right)\Sigma, \quad \delta Q = T\,dS = -T\frac{d\sigma}{dT}\,d\Sigma.$$

With increasing temperature T, the surface tension coefficient σ decreases and vanishes at a critical temperature.

P2.8.2. Additional Pressure Under a Curved Surface

▶ **Spherical surface.** It follows from the condition of mechanical equilibrium of the surface layer that the pressure inside a convex surface must be higher than outside. For a spherical surface of radius R, the additional pressure, Δp, can be found by increasing its radius by dR and equating the work done by external forces with the change in the Helmholtz energy: $\Delta p \, dV = \sigma \, d\Sigma$ and hence $\Delta p \, d(\frac{4}{3}\pi R^3) = \sigma \, d(4\pi R^2)$. Finally, we have

$$\Delta p = \frac{2\sigma}{R}. \tag{P2.8.2.1}$$

▶ **Young–Laplace formula.** The above equation can be generalized to include two different radii of curvature:

$$\Delta p = \sigma\left(\frac{1}{R_1} + \frac{1}{R_2}\right), \tag{P2.8.2.2}$$

where R_1 and R_2 are the principal radii of curvature of the surface (i.e., the minimum and maximum radii of curvature of the curves obtained by cutting the surface with two mutually perpendicular planes at a given point). Relation (P2.8.2.1) is called the *Young–Laplace formula* or *Young–Laplace equation.*

P2.8.3. Wetting. Capillary Effects

▶ **Contact angle.** When a liquid and solid are in contact with each other and both are in a gas environment, the liquid surface (liquid–gas interface) and the solid surface (solid–gas interface) meet at an angle, which is called the *contact angle*. The value of the contact angle, θ, depends on the surface tension coefficients at the liquid–gas ($\sigma_{\text{liq–gas}}$), liquid–solid ($\sigma_{\text{liq–sol}}$), and solid–gas ($\sigma_{\text{sol–gas}}$) interfaces. It follows from the condition of mechanical equilibrium at the line where the three phases meet (see Fig. P2.8) that $\sigma_{\text{sol–gas}} = \sigma_{\text{liq–sol}} + \sigma_{\text{liq–gas}} \cos \theta$ (per unit length). Hence,

$$\cos \theta = \frac{\sigma_{\text{sol–gas}} - \sigma_{\text{liq–sol}}}{\sigma_{\text{liq–gas}}}.$$

If $0 < \cos \theta < 1$, or $0 < \theta < 90°$, the liquid is said to *wet* the solid surface. If $-1 < \cos \theta \le 0$, or $90° \le \theta < 180°$, it is said that the liquid does not wet the surface. If the right-hand side of the equation is greater than or equal to 1 ($\theta = 0$), the liquid forms a thin molecular film of the solid surface (this phenomenon is known as *perfect wetting*); in a container, the liquid makes a zero angle with the solid wall. If the right-hand side of the equation is less than or equal to -1 ($\theta = 180°$), this situation is known as *perfect non-wetting*.

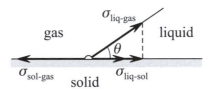

Figure P2.8. Contact angle.

▶ **Capillary effects.** If one end of a thin, capillary tube is immersed in a wetting liquid, the liquid will go up the capillary and reach a height h. This height is determined by the mechanical equilibrium condition.

If there is a capillary tube with inner radius r and the radius of curvature of the liquid meniscus is $R = r / \cos \theta$, the pressure under the meniscus (see (P2.8.2.1)) must be less than atmospheric pressure by the value $2\sigma/R = 2\sigma \cos \theta / r$. Equating the pressure difference with $\rho g h$, we find the height of the liquid column: $h = \dfrac{2\sigma \cos \theta}{\rho g r}$.

P2.9. Transfer Phenomena in Gases

P2.9.1. Collision of Molecules in a Gas

▶ **Collision frequency.** For a given molecule in a gas, its average number of collisions per unit time (collision frequency) with other molecules is equal to

$$z = n\sigma\langle v \rangle\sqrt{2}, \qquad\qquad (P2.9.1.1)$$

where n is the molecule concentration, $\langle v \rangle$ is the average speed of the molecules, and σ is the *effective cross-section* of elastic collisions. If the molecules are modeled as identical rigid balls of diameter d, we have $\sigma = \pi d^2$. The factor $\sqrt{2}$ accounts for the motion of oncoming molecules; this factor is usually omitted in the qualitative theory of transfer phenomena.

In order to estimate the collision frequency, the polygonal path of a molecule is usually surrounded by a tube of radius d. The moving molecule will collide with any resting molecule whose center is inside the tube. In a unit time, the molecule travels the distance $\langle v \rangle$, the corresponding tube volume equals $\pi d^2 \langle v \rangle$, and the number of molecules in the tube is $z = \pi d^2 \langle v \rangle n$.

▶ **Mean free path.** The average distance traveled by a molecule between successive collisions, the *mean free path*, is equal to

$$\lambda = \frac{\langle v \rangle}{z} = \frac{1}{n\sigma\sqrt{2}}. \tag{P2.9.1.2}$$

The total number of collisions among molecules in a unit volume per unit time is

$$\nu = \frac{zn}{2} = \frac{n\langle v \rangle}{2\lambda} = \frac{n^2\sigma\langle v \rangle\sqrt{2}}{2}. \tag{P2.9.1.3}$$

Effective cross-section. The effective cross-section of an interparticle collision process (with respect to a result of interest, such as scattering within a given solid angle, particle absorption, etc.) is determined as follows. Consider a fixed molecule (a scattering center) and an incident beam of molecules with flux $I = nv$ (the number of particles passing through a unit area per unit time, with the area perpendicular to the velocity vector). The effective cross-section of a process is the ratio between the number of collisions per unit time that have led to the desired result, ΔN, and the incident flux:

$$\sigma = \Delta N / I. \tag{P2.9.1.4}$$

The attenuation of the molecule beam traveling through a gas over a distance dx equals

$$dI = -I\sigma n\, dx = -(I/\lambda)\, dx,$$

where σ is the cross-section of elastic scattering, $n\, dx$ is the number of scattering centers along the path dx (per unit area), and $\lambda = 1/(n\sigma)$. Integrating yields the incident flux as a function of the distance traveled:

$$I = I_0 \exp(-n\sigma x) = I_0 \exp(-x/\lambda). \tag{P2.9.1.5}$$

It follows that the probability for a molecule to experience a collision on the interval $(x, x + dx)$ is $|dI|/I_0 = \exp(-x/\lambda)\, d(x/\lambda)$. The mean distance traveled by a molecule to the first collision equals

$$\langle x \rangle = \int_0^\infty x \exp\left(-\frac{x}{\lambda}\right) \frac{dx}{\lambda} = \lambda.$$

P2.9.2. Qualitative Model of Transfer Phenomena in Gases

The phenomena of viscosity, heat conduction, and diffusion are accounted for by the transfer of linear momentum, energy, and concentration due to chaotic thermal motion.

▶ **Viscosity. Fourier's law.** Viscosity is the appearance of frictional forces between parallel gas layers as they move relative to each other. If the directed flow velocity, $\mathbf{u}$, is parallel to the x-axis and only changes in the y-direction (Fig. P2.9), then a resistive force arises which is proportional to the area and the velocity gradient:

$$|F_x| = \eta S \frac{du}{dy}, \tag{P2.9.2.1}$$

where η is the *coefficient of viscosity* (or simply the *viscosity*). Equation (P2.9.2.1) is known as *Fourier's law*. The appearance of the resistive force is associated with the transfer of the momentum p_x along the y-axis by the molecules crossing the contact area due to their thermal motion. The force per unit area equals the momentum flux density.

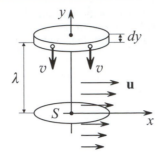

Figure P2.9. A frictional force arises between adjacent gas layers having different relative velocities.

▶ **Heat conduction. Newton's law of cooling.** If the gas temperature rather than the directed flow velocity changes along the y-axis, then the local heat flux density in the y-direction is expressed as

$$q = -\kappa \frac{dT}{dy}, \qquad (P2.9.2.2)$$

where κ is the *thermal conductivity*. This equation is known as Newton's law of cooling.

▶ **Diffusion. Fick's law.** If the concentration of selected gas molecules, n_1, changes along the y-axis (the total concentration n must be constant), then the flux density of the selected molecules, j, in the y-direction is given by

$$j = -D \frac{dn_1}{dy}, \qquad (P2.9.2.3)$$

where D is the *diffusion coefficient* (or *diffusivity*). This equation is known as *Fick's law of diffusion*. The quantity j is called the *diffusion flux density*.

▶ **Flux calculation formulas.** Let $g(y)$ denote a quantity whose transfer along the y-axis is of interest. Examples of such a quantity are the momentum of directed motion, $g(y) = m_0 u(y)$, for viscosity, the mean energy of thermal motion, $g(y) = \langle \varepsilon \rangle = c_V m_0 T$ (c_V is the specific heat capacity at constant volume), for heat conduction, and the relative concentration of selected molecules, $g(y) = n_1(y)/n$, for diffusion. Let us assume that each molecule transfers a certain (average) amount of the quantity g through a given surface; this amount of g was acquired by the molecule at its latest collision. Then the flux density of g in the y-direction can be calculated using the formula

$$G = -\frac{1}{3} n \langle v \rangle \lambda \frac{dg}{dy}. \qquad (P2.9.2.4)$$

For rough estimates, it can be assumed that the particles move with velocity $\langle v \rangle$ in one of the six directions determined by the coordinate axes. Then the particle flux through a unit area is equal to $n\langle v \rangle$. It can also be assumed that the particles crossing the area the y-level experienced their latest collision at an average distance λ (Fig. P2.9). Then we have $G \approx \frac{1}{6} n \langle v \rangle [g(y - \lambda) - g(y + \lambda)]$, and hence $G \approx \frac{1}{3} n \langle v \rangle \lambda (dg/dy)$.

▶ **Formulas for transfer coefficients in a gas.**

1. Substituting $g = m_0 v_x$ into (P2.9.2.4), gives the momentum flux, which is equal to the frictional force (P2.9.2.1) for a unit area. Then the coefficient of viscosity is expressed as

$$\eta = \tfrac{1}{3} n m_0 \langle v \rangle \lambda = \tfrac{1}{3} \rho \langle v \rangle \lambda. \qquad (P2.9.2.5)$$

2. Substituting $g = c_V m_0 T$ into (P2.9.2.4) and comparing with (P2.9.2.2), we find the thermal conductivity:

$$\kappa = \tfrac{1}{3} n m_0 \langle v \rangle c_V \lambda = \tfrac{1}{3} \rho \langle v \rangle c_V \lambda. \tag{P2.9.2.6}$$

3. Substituting $g = n_1/n$ into (P2.9.2.4) and comparing with (P2.9.2.3) yields the diffusivity:

$$D = \tfrac{1}{3} \langle v \rangle \lambda. \tag{P2.9.2.7}$$

There are two important conclusions that can be drawn from formulas (P2.9.2.5)–(P2.9.2.7):

1) The mean free path is inversely proportional to ρ, which can be seen from equation (P2.9.1.2); hence, the viscosity and thermal conductivity of gases at a given fixed temperature are pressure independent (Maxwell's law). This conclusion holds true as long as the mean free path is small compared to the container size.

2) The ratio between the viscosity and the thermal conductivity is temperature independent.

▶ **Extremely rarefied gas.** When the mean free path becomes commensurable with the container size (*rarefied gas*), the viscosity and thermal conductivity begin to depend on density. For example, consider the heat transfer between two walls having temperatures T_1 and T_2 and separated by a distance $d \ll \lambda$ (*extremely rarefied gas*). It can be assumed that, after its collision with the wall having the temperature T_1, each molecule acquires (on the average) the energy $c_V m_0 T_1$, then flies without collisions to the opposite wall and gives up the energy $c_V m_0 (T_2 - T_1)$. It follows from (P2.5.1.3) that the number of molecules reaching the wall in time Δt is equal to $\tfrac{1}{4} n \langle v \rangle S \Delta t$. Consequently, $q = \tfrac{1}{4} \rho \langle v \rangle c_V \Delta T$, and hence $\kappa = \tfrac{1}{4} \rho \langle v \rangle c_V d$.

Bibliography for Chapter P2

Alenitsyn, A. G., Butikov, E. I., and Kondratyev, A. S., *Concise Handbook of Mathematics and Physics*, CRC Press, Boca Raton, Florida, 1998.

Ansel'm, A. I., *Basics of Statistical Physics and Thermodynamics* [in Russian], Lan, St. Petersburg, 2007.

Benenson, W., Harris, J. W., Stocker, H., and Lutz, H. (Eds.), *Handbook of Physics*, Springer-Verlag, New York, 2002.

Borgnakke, C. and Sonntag, R. E., *Fundamentals of Thermodynamics, 7th edition*, Wiley, New York, 2008.

Callen, H. B., *Thermodynamics and an Introduction to Thermostatistics, 2nd edition*, John Wiley & Sons, New York, 1985.

Demtröder, W., *Molecular Physics: Theoretical Principles and Experimental Methods*, Wiley-VCH, New York, 2006.

Fermi, E., *Thermodynamics*, Dover Publications, New York, 1956.

Huang K., *Statistical Mechanics, 2nd edition*, Wiley, New York, 1987.

Irodov, I. E., *Physics of Macrosystems. Basic Laws* [in Russian], Fizmatlit, Moscow, 2006.

Kittel, C. and Kroemer, H., *Thermal Physics, 2nd Edition*, W. H. Freeman, San Francisco, California, 1980.

Landau, L. D. and Lifshitz, E. M., *Statistical Physics, Third Edition, Part 1: Volume 5 (Course of Theoretical Physics, Volume 5)*, Butterworth-Heinemann, Oxford, England, 1980.

Moran, J. M. and Shapiro, H. N., *Fundamentals of Engineering Thermodynamics, 6th Edition*, Wiley, New York, 2007.

Pathria, R. K., *Statistical Physics, 2nd edition*, Butterworth-Heinemann, Oxford, UK and Woburn, Massachusetts, 1996.

Pitaevskii, L. P. and Lifshitz, E. M., *Physical Kinetics: Volume 10 (Course of Theoretical Physics, Volume 10)*, Butterworth-Heinemann, Oxford, England, 1981.

Pitaevskii, L. P. and Lifshitz, E. M., *Statistical Physics, Part 2: Volume 9 (Course of Theoretical Physics, Volume 9)*, Butterworth-Heinemann, Oxford, England, 1980.

Plischke, M. and Bergersen, B., *Equilibrium Statistical Physics*, Prentice-Hall, New York, 1989.

Prokhorov, A. M. (Ed.), *Encyclopedia of Physics, Volumes 1–5* [in Russian], Bolshaya Russkaya Encyclopedia, Moscow, 1998.

Prokhorov, A. M. (Ed.), *Encyclopedic Dictionary of Physics* [in Russian], Sovetskaya Encyclopedia, Moscow, 1984 (dic.academic.ru).

Reichl, L. E., *A Modern Course on Statistical Physics*, Wiley-Interscience, New York, 1998.

Schroeder, D. V., *An Introduction to Thermal Physics*, Addison-Wesley, Reading, Massachusetts, 1999.

Sivukhin, D. V., *General Physics, Volume 2. Thermodynamics and Molecular Physics* [in Russian], Fizmatlit, Moscow, 2006.

Sivukhin, D. V. and Yakovlev, I. A., *A Collection of Problems, Volume 2. Thermodynamics and Molecular Physics* [in Russian], Fizmatlit, Moscow, 2006.

Van Ness, H. C., *Understanding Thermodynamics*, Dover Publications, New York, 1983.

Woan, G., *The Cambridge Handbook of Physics Formulas*, Cambridge University Press, Cambridge, England, 2003.

Chapter P3

Electrodynamics

P3.1. Electric Charge. Coulomb's Law

P3.1.1. Electric Charge and Its Properties

▶ **Electric charge** is a scalar physical quantity that determines the electromagnetic interaction of subatomic particles. In SI, the unit of charge is the *coulomb* (C); it is defined in terms of the unit of electric current, the *ampere* (A): $1\,\text{C} = 1\,\text{A s}$.

Properties of electric charge:

1. The carriers of electric charge are many subatomic particles such as protons and electrons (as well as antiprotons and positrons) and some metastable particles as π-mesons, K-mesons, muons, etc. Charged particles interact with each other with forces that decay with the distance as slowly as gravitational forces but are much stronger in magnitude.

2. Proton and electron and many other charged subatomic particles have the same electric charge in magnitude. This charge is called the *elementary charge* and denoted e. In SI, $e = 1.602 \times 10^{-19}$ C. As experiments have shown, the charge of subatomic particles is independent of their velocity.

3. The charge of subatomic particles can be either positive or negative. Particles with like charges repel one another and particles with unlike charges attract. By convention, the charge of a proton is positive $(+e)$ and the charge of an electron is negative $(-e)$.

If a macroscopic body includes different numbers of electrons, N_e, and protons, N_p, the body is charged and its charge is always multiple of the elementary charge: $q = e(N_p - N_e)$.

▶ **Law of conservation of electric charge.** A system is called *electrically closed* if it does not exchange charges with external bodies, which means that charges do not cross its boundary. The total charge of an electrically closed system, or the algebraic sum of all its charges, remains constant. This statement is obvious if there are no elementary particle reactions. However, the law of conservation of charge is fundamental—it holds even if charged particles are created or destroyed.

▶ **Charge distribution.** The spatial distribution of charge is characterized by its *volume density* $\rho(\mathbf{r})$, measured in C/m^3, *surface density* $\sigma(\mathbf{r})$, measured in C/m^2, and *linear density* $\lambda(\mathbf{r})$, measured in C/m. We have

$$dq = \rho\,dV, \qquad dq = \sigma\,dS, \qquad dq = \lambda\,dl. \qquad \text{(P3.1.1.1)}$$

▶ **Dipole. Dipole moment.** A system of charges is called *electrically neutral* if its net charge is zero. An *electric dipole* is a simple example of an electrically neutral system; it consists of two point charges, q and $-q$, separated by a distance l. The *electric dipole moment* is a vector quantity defined as

$$\mathbf{p} = q\mathbf{l}, \qquad \text{(P3.1.1.2)}$$

where $\mathbf{l} = \mathbf{r}_+ - \mathbf{r}_-$ is the displacement vector from the charge $-q$ to the charge q. The dipole moment is measured in C m.

The following definition is a generalization to the case of an electrically neutral system:

$$\mathbf{p} = \sum_i q_i \mathbf{r}_i = q(\mathbf{r}_+ - \mathbf{r}_-),$$

where $q = \sum_{q_i>0} q_i$ is the sum of all positive charges, $\mathbf{r}_+ = \sum_{q_i>0} q_i \mathbf{r}_i / q$ is the center of positive charges, and $\mathbf{r}_- = -\sum_{q_i<0} q_i \mathbf{r}_i / q$ is the center of negative charges.

P3.1.2. Coulomb's Law

Coulomb's law describes the electrostatic interaction of *point charges*, i.e., charged particles and bodies whose dimensions can be neglected as compared to the distances between them. Coulomb's law states that, in vacuum, the electrostatic force, F, acting on a point charge q_1 due to the presence of a point charge q_2 is directly proportional to each of the charges and inversely proportional to the square of the distance r between them:

$$F = k\frac{|q_1||q_2|}{r^2}, \tag{P3.1.2.1}$$

where k is the proportionality coefficient, dependent on the unit system chosen. The force is directed along the straight line connecting the point charges; like charges repel and unlike charges attract.

In SI, the coefficient in (P3.1.2.1) equals $k = 9.0 \times 10^9$ N m^2/C^2. Also $k = 1/(4\pi\varepsilon_0)$, where $\varepsilon_0 = 8.85 \times 10^{-12}$ F/m is the *electric constant*. In Gaussian units (CGS system), k is taken to be equal to one and the unit of charge is defined by Coulomb's law.

The electrostatic force acting on a charge q due to the presence of many point charges, $Q_1, Q_2, \ldots$, equals the linear vector superposition of forces

$$\mathbf{F} = \sum_i k\frac{qQ_i}{(\mathbf{r} - \mathbf{r}_i)^2} \frac{\mathbf{r} - \mathbf{r}_i}{|\mathbf{r} - \mathbf{r}_i|}, \tag{P3.1.2.2}$$

where $\mathbf{r}$ is the position vector of the charge q and $\mathbf{r}_i$ is the position vector of the charge Q_i ($i = 1, 2, \ldots$).

▶ **On unit conventions in electrostatics.** In what follows, all formulas will be given in SI. To convert to Gaussian units (CGS system), it will be sufficient to set $k = 1$ in most cases; if a formula contains ε_0, one should first replace ε_0 with $1/(4\pi k)$.

P3.2. Electric Field. Strength and Potential

P3.2.1. Strength of Electric Field

▶ **Electromagnetic field.** The interaction between charged particles is implemented through *electromagnetic field*. This means that: (a) a charged particle generates an electromagnetic field in the surrounding space and (b) the electromagnetic field exerts a force on other charged particles. The field generated by a point charge is proportional to its charge; the force exerted by the field on a particle is proportional to the charge of this particle.

▶ **Electric field.** An electromagnetic field is a combination of an *electric field* and a *magnetic field*, which are closely related to each other. The action of an electric field on a charged particle does not depend on the particle velocity and that of a magnetic field does. The sources of electric fields are any charged particles, while magnetic fields are generated by moving charges. The proportion between electric and magnetic fields changes when passing from one inertial frame to another. The properties of electric and magnetic fields are completely determined by Maxwell's equations (see Subsection P3.11.1). Maxwell's equations are invariant under relativistic, Lorentz transformations.

▶ **Test charge. Strength of an electric field.** The characteristics of an electromagnetic field are studied using the concept of a *test charged particle*, which is an idealized small object whose dimensions and charge are small enough so that, when placed into a field of interest, it does not distort the field (does not change the positions of the field sources). The force acting on a stationary test particle of charge q is proportional to the charge and is only determined by the electric field:

$$\mathbf{F}_q = q\mathbf{E}. \tag{P3.2.1.1}$$

This formula defines the *strength of an electric field*, $\mathbf{E}$. It also answers the question what force the field exerts on any charge q, moving or stationary. In SI, the strength of an electric field is measured in N/C or V/m. For brevity, the term electric field is used to mean the strength of an electric field whenever this does not lead to confusion.

▶ **Electrostatic field. Principle of superposition.** A stationary electric field generated by a system of stationary charges is called an *electrostatic field*. The strength of an electrostatic field can be determined using Coulomb's law (P3.1.2.2).

1. *Electric field of a point charge.* If a point charge q is placed in the field of a point charge Q, then it follows from (P3.1.2.2) and (P3.2.1.1) that

$$\mathbf{E} = k\frac{Q}{r^2}\frac{\mathbf{r}}{r} = E_r\frac{\mathbf{r}}{r}, \tag{P3.2.1.2}$$

where E_r is the projection of $\mathbf{E}$ onto the radial direction.

2. *Electric field of a system of point charges.* Using the principle of superposition, we get

$$\mathbf{E} = \mathbf{E}_1 + \mathbf{E}_2 + \cdots = \sum_i k\frac{Q_i}{(\mathbf{r} - \mathbf{r}_i)^2}\frac{\mathbf{r} - \mathbf{r}_i}{|\mathbf{r} - \mathbf{r}_i|}. \tag{P3.2.1.3}$$

For a continuously distributed charge, the summation must be replaced by integration (see (P3.1.1.1)). (Compare with the formulas for the gravitational field in Subsection P1.7.2.)

Example 1. There is a charge continuously distributed along an arc of a circle of radius R with a linear density λ (Fig. P3.1). The arc angle is $2\alpha_0$. Find the strength of the electric field generated by the charge, at the circle center.

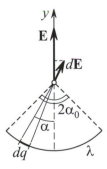

Figure P3.1. Strength of the electric field generated by a charged arc of a circle.

Solution. By symmetry, the field strength is directed along the y-axis, which passes through the midpoint of the arc and the circle center O. The strength of the field generated by the arc element $d\alpha$ at O equals $dE = k\,dq/R^2$, where $dq = \lambda R\,d\alpha$. Projecting onto y and integrating yields

$$E = E_y = \int_{-\alpha_0}^{\alpha_0} k\frac{\lambda R\,d\alpha}{R^2}\cos\alpha = k\frac{2\lambda\sin\alpha_0}{R}.$$

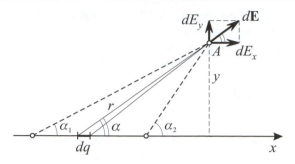

Figure P3.2. Strength of the electric field generated by a charged segment.

Example 2. Field of a segment. A charge is evenly distributed along a segment with a linear density λ. Find the field strength at the point A whose position is determined by the distance y to the straight line of the segment and two angles α_1 and α_2 as shown in Fig. P3.2.

Solution. The contribution of the element dx to the field strength at A equals $dE = k\lambda\,dx/r^2$, where r is the distance from dx to A. The respective projections on the x- and y-axes are $dE\cos\alpha$ and $dE\sin\alpha$. It is convenient to integrate with respect to the angle α. Substituting $r = y/\sin\alpha$ and $x = -y\cot\alpha$, with $dx = y\,d\alpha/\sin^2\alpha$, and integrating, we obtain the following expressions of E_x and E_y:

$$E_x = k\frac{\lambda}{y}(\sin\alpha_2 - \sin\alpha_1), \qquad E_y = k\frac{\lambda}{y}(\cos\alpha_1 - \cos\alpha_2).$$

If $\alpha_1 = \pi - \alpha_2 = \alpha$ (the point A lies on the straight line perpendicular to the segment through its midpoint), then $E_x = 0$ and $E_y = 2k\lambda\cos\alpha/y$. If $\alpha \to 0$ (the segment becomes a straight line), then $E_y = 2k\lambda/y$. If $\alpha_1 = 0$ and $\alpha_2 = \pi/2$ (a ray), then $E_x = E_y = k\lambda/y$.

P3.2.2. Potential of an Electrostatic Field

▶ **Definition of potential.** An electrostatic field is *potential* (see Subsection P1.5.3) just as any static central field. This means that the work done by the field to move a test charge from one point in space to another is independent of the path. This allows us to define the *potential of an electrostatic field*, $\varphi(\mathbf{r})$. We have

$$A_q(1 \to 2) = W_q(\mathbf{r}_1) - W_q(\mathbf{r}_2) = q[\varphi(\mathbf{r}_1) - \varphi(\mathbf{r}_2)], \qquad (P3.2.2.1)$$

where $A_q(1 \to 2)$ is the work done by the field to transfer the charge q from point $\mathbf{r}_1$ to point $\mathbf{r}_2$, $W_q(\mathbf{r}) = q\varphi(\mathbf{r})$ is the *potential energy of the charge in the electrostatic field* (throughout this chapter, energy will be denoted by W). The scalar quantity $\varphi(\mathbf{r}_1) - \varphi(\mathbf{r}_2)$ is called the *potential difference* between points $\mathbf{r}_1$ and $\mathbf{r}_2$. For $W_q(\mathbf{r})$ and $\varphi(\mathbf{r})$ to be uniquely defined, one has to choose a point where they are zero. In SI, potential is measure in volts (V).

▶ **Relation between field strength and potential.** Given the field strength $\mathbf{E}(\mathbf{r})$, one can determine the potential difference and potentials as

$$\varphi(\mathbf{r}_1) - \varphi(\mathbf{r}_2) = \int_{\mathbf{r}_1}^{\mathbf{r}_2} \mathbf{E}\,d\mathbf{r}, \qquad \varphi(\mathbf{r}) = \int_{\mathbf{r}}^{\mathbf{r}_0} \mathbf{E}\,d\mathbf{r}, \qquad (P3.2.2.2)$$

where $\mathbf{r}_0$ is the reference point where the potential is taken to be zero. Given the potential $\varphi(\mathbf{r})$, one can find the projection of the field strength onto any direction l (based on (P3.2.2.2) for two close points):

$$E_l = -\frac{\partial\varphi}{\partial l}. \qquad (P3.2.2.3)$$

Consequently, we can find the field strength vector:

$$\mathbf{E} = -\,\text{grad}\,\varphi = -\left(\mathbf{i}\frac{\partial\varphi}{\partial x} + \mathbf{j}\frac{\partial\varphi}{\partial y} + \mathbf{k}\frac{\partial\varphi}{\partial z}\right), \qquad (P3.2.2.4)$$

where $\mathbf{i}$, $\mathbf{j}$, and $\mathbf{k}$ are the unit vectors of the respective coordinate axes.

▶ **Principle of superposition for potential.**

1. The *potential of the field of a point charge Q* can be calculated using (P3.2.2.3):

$$E_r = -\frac{d\varphi}{dr} \quad \Longrightarrow \quad \frac{d\varphi}{dr} = -k\frac{Q}{r^2} \quad \Longrightarrow \quad \varphi(r) = k\frac{Q}{r} + \text{const.}$$

Usually, the potential is taken to be zero at infinity. In this case,

$$\varphi(r) = k\frac{Q}{r}. \tag{P3.2.2.5}$$

2. The *potential of the field of a system of point charges* equals the algebraic sum of the potentials of individual charges (principle of superposition for potential):

$$\varphi(\mathbf{r}) = \varphi_1(\mathbf{r}) + \varphi_2(\mathbf{r}) + \cdots, \qquad \varphi(\mathbf{r}) = \sum_i k\frac{Q_i}{|\mathbf{r} - \mathbf{r}_i|}. \tag{P3.2.2.6}$$

For a continuously distributed charge, the summation must be replaced by integration.

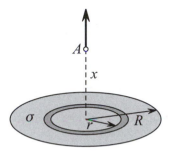

Figure P3.3. To the calculation of the electrostatic field of a charged disk.

Example 1. Field of a disk. A thin disk of radius R is evenly charged with a surface density σ. Find the potential and field strength at the point A located on the disk axis at a distance x from the center (Fig. P3.3).

Solution. The contribution of a thin ring bounded by the circles of radii r and $r + dr$ is

$$d\varphi = k\frac{dq}{\sqrt{x^2 + r^2}} = k\frac{\sigma\, 2\pi r\, dr}{\sqrt{x^2 + r^2}};$$

note that all points of the ring are essentially at the same distance to A. Integrating yields the desired potential

$$\varphi = \int_0^R k\frac{\sigma\, 2\pi r\, dr}{\sqrt{x^2 + r^2}} = 2k\pi\sigma\left(\sqrt{x^2 + R^2} - x\right).$$

Using (P3.2.2.3), we find the field strength (it is directed along x):

$$E_x = -\frac{d\varphi}{dx} = 2\pi k\sigma\left(1 - \frac{x}{\sqrt{x^2 + R^2}}\right).$$

In the limit $R \to \infty$, we arrive at the field strength of a plane: $E_x = 2\pi k\sigma = \sigma/(2\varepsilon_0)$.

Example 2. Field of a dipole. Find the potential and field strength far away from a dipole shown in Fig. P3.4.

Solution. The potential at a remote point determined by $\mathbf{r}$ ($r = |\mathbf{r}| \gg l$) equals

$$\varphi(\mathbf{r}) = \frac{kq}{r_1} - \frac{kq}{r_2} \approx \frac{kq}{r - s} - \frac{kq}{r + s} \approx k\frac{(\mathbf{p} \cdot \mathbf{r})}{r^3},$$

where $s = \left(\frac{1}{2}\mathbf{l} \cdot \frac{1}{r}\mathbf{r}\right)$ and $\mathbf{p} = q\mathbf{l}$ (see Subsection P3.1.1).

The field strength can be found using the identities $\text{grad}(1/r^3) = -3\mathbf{r}/r^5$ and $\text{grad}(\mathbf{p} \cdot \mathbf{r}) = \mathbf{p}$. We have

$$\mathbf{E} = -\text{grad}\,\varphi = k\left[\frac{3(\mathbf{p} \cdot \mathbf{r})\mathbf{r}}{r^5} - \frac{\mathbf{p}}{r^3}\right]. \tag{P3.2.2.7}$$

It follows that the magnitude of the field strength equals $E = kpr^{-3}\sqrt{3\cos^2\theta + 1}$.

The same expressions hold true for any electrically neutral system with a nonzero dipole moment (see Subsection P3.1.1) at large distances.

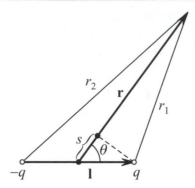

Figure P3.4. To the calculation of the field of a dipole.

▶ **Dipole in an external field.** A small dipole with dipole moment $\mathbf{p} = q\mathbf{l}$ in an external electrostatic field has the potential energy

$$W_{\mathrm{p}} = q\varphi(\mathbf{r}_+) - q\varphi(\mathbf{r}_-) = q\frac{\partial\varphi}{\partial l}l = -pE_l = -(\mathbf{p}\cdot\mathbf{E}). \qquad (\text{P3.2.2.8})$$

This energy changes if the dipole is turned by an angle—the field exerts a rotational moment on the dipole:

$$\mathbf{M} = \mathbf{l}\times(q\mathbf{E}) = \mathbf{p}\times\mathbf{E}, \qquad (\text{P3.2.2.9})$$

The energy also changes if the dipole moves in a nonuniform field—the field exerts the force

$$\mathbf{F} = q\mathbf{E}(\mathbf{r}_+) - q\mathbf{E}(\mathbf{r}_-) = q\frac{\partial\mathbf{E}}{\partial l}l = p\frac{\partial\mathbf{E}}{\partial l} = p_x\frac{\partial\mathbf{E}}{\partial x} + p_y\frac{\partial\mathbf{E}}{\partial y} + p_z\frac{\partial\mathbf{E}}{\partial z}. \qquad (\text{P3.2.2.10})$$

An equilibrium orientation with minimum potential energy corresponds to $\mathbf{p}\parallel\mathbf{E}$. A dipole oriented along the field strength vector is pulled into an area where the field is stronger.

P3.3. Gauss's Theorem

P3.3.1. Statement of the Theorem

▶ **Flux of electric field.** The flux of the electric field $\mathbf{E}$ through a plane area element ds is defined as

$$d\Phi = (\mathbf{E}\cdot d\mathbf{s}) = E_n ds = E\cos\theta\, ds, \qquad (\text{P3.3.1.1})$$

where $d\mathbf{s} = \mathbf{n}\, ds$, $\mathbf{n}$ is a normal vector to the plane, θ is the angle between $\mathbf{E}$ and $\mathbf{n}$, $E_n = E\cos\theta$ is the projection of $\mathbf{E}$ onto the normal $\mathbf{n}$. The sign of $d\Phi$ depends on how the normal direction is chosen. The flux of the field $\mathbf{E}$ through a finite surface S is determined as

$$\Phi = \int_S E_n\, ds. \qquad (\text{P3.3.1.2})$$

If the surface is closed, the normal is taken to point outward.

Flux additivity: if an electrostatic field is a superposition of several fields, $\mathbf{E}(\mathbf{r}) = \sum_i \mathbf{E}_i$, then $\Phi = \sum_i \Phi_i$ for any surface.

The flux of the electric field generated by point charge Q through an area ds is

$$d\Phi = E_r\cos\theta\, ds = \pm k\frac{Q}{r^2}\, ds_\perp = \pm kQ\, d\Omega,$$

where $ds_\perp = |ds\cos\theta|$ is the projection of ds onto the plane perpendicular to $\mathbf{r}$, and $d\Omega = ds_\perp/r^2$ is the solid angle at which the area ds is seen from the charge Q. The plus sign corresponds to an acute angle between $\mathbf{n}$ and $\mathbf{r}$, and the minus sign corresponds to an obtuse angle. The field flux through a closed surface is equal to $4\pi kQ$ if the charge is inside the surface, and zero if it is outside.

▶ **Gauss's theorem.** The flux of an electrostatic field through a closed surface is only determined by the net charge, q_{enc}, enclosed within the surface:

$$\Phi = 4\pi k q_{\text{enc}} = \frac{q}{\varepsilon_0}. \tag{P3.3.1.3}$$

▶ **Gauss's theorem and electric field lines.** Electric field lines are useful to visualize electric fields. These are drawn using the following rules: (a) the tangent to a field line is directed along the vector $\mathbf{E}$ at every point in space and (b) the density of field lines is proportional to the magnitude of the field at point of interest. The field flux through a surface is proportional to the number of field lines piercing the surface. (Lines of any vector field can be introduced likewise.) Gauss's theorem implies that the electric field lines must start and finish at charges and also they must be continuous in an empty space. It is noteworthy that Gauss's theorem and continuity of field lines are corollaries of the fact that the Coulomb force decays with distance as $1/r^2$.

Apart from field lines, *equipotential surfaces* are also used for visualization of electric fields. Equipotential surfaces are everywhere perpendicular to field lines (the work done by the field to transfer a test charge along such a surface must be zero).

P3.3.2. Application of Gauss's Theorem

▶ **Calculating the field using Gauss's theorem.** Suppose, from symmetry considerations, it is possible to construct an imaginary surface such that $\mathbf{E}$ is perpendicular to it and constant in magnitude on some part of the surface and is parallel to the surface ($\mathbf{E} \perp \mathbf{n}$) on the other part. Such a surface is called a *Gaussian surface*. Gauss's theorem can be applied to this surface to find the field magnitude E.

Example 1. Field of a charged string. Find the electric field generated by an infinitely long, evenly charged straight string (Fig. P3.5).

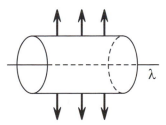

Figure P3.5. To the calculation of the electric field generated by an infinite string using Gauss's theorem.

Solution. The electric field is perpendicular to the string at any point in space. The field magnitude E_r is only dependent on the distance r to the string. Evidently, any cylinder of radius r with generatrix of length l and the axis lying on the string, can be taken as a Gaussian surface. Writing Gauss's theorem (P3.3.1.2), we find

$$E_r\, 2\pi rl = \frac{\lambda l}{\varepsilon_0} \quad \Longrightarrow \quad E_r = \frac{\lambda}{2\pi\varepsilon_0 r} = \frac{2k\lambda}{r}$$

(cf. example 2 from Subsection P3.2.1). The potential difference between two points equals

$$\varphi_1 - \varphi_2 = \frac{\lambda}{2\pi\varepsilon_0}\ln\frac{r_2}{r_1}.$$

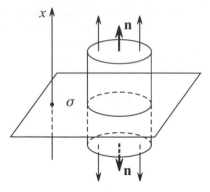

Figure P3.6. To the calculation of the electric field generated by a charged plane using Gauss's theorem.

Example 2. Field of a charged plane. Find the electric field generated by an evenly charged plane (Fig. P3.6).

Solution. The field is everywhere perpendicular to the plane and E_x is only dependent on the distance to the plane. A right cylinder whose bases of area S are symmetric with respect to the plane can be taken as a Gaussian surface. The field flux through this surface equals $E_x(x)S - E_x(-x)S = 2E_xS$. Writing Gauss's theorem, we find that

$$2E_xS = \frac{\sigma S}{\varepsilon_0} \quad \Longrightarrow \quad E_x(x) = -E_x(-x) = \frac{\sigma}{2\varepsilon_0}$$

(see example 1 from Subsection P3.2.2). Taking $\varphi = 0$ at $x = 0$, we have $\varphi(x) = \dfrac{\sigma}{2\varepsilon_0}|x|$.

Example 3. Field of a charged ball. Find the electric field of a uniformly charged ball of radius R with charge q.

Solution. The field of a uniformly charged ball is a central field. A concentric sphere of radius r can be taken as a Gaussian surface. If $r > R$, then the surface encloses the entire ball and so $q_{\text{enc}} = q$. Hence, $E_r \, 4\pi r^2 = q/\varepsilon_0$. If $r < R$, then $q_{\text{enc}} = qr^3/R^3$ and hence $E_r \, 4\pi r^2 = qr^3/(\varepsilon_0 R^3)$. Finally, we have

$$\mathbf{E} = \begin{cases} k\dfrac{q}{r^2}\dfrac{\mathbf{r}}{r} = \dfrac{\rho R^3}{3\varepsilon_0 r^2}\dfrac{\mathbf{r}}{r} & \text{for } r > R, \\[2mm] k\dfrac{q\mathbf{r}}{R^3} = \dfrac{\rho\mathbf{r}}{3\varepsilon_0} & \text{for } r < R, \end{cases}$$

where ρ is the charge density. The radial component of the field equals $E_r = -d\varphi/dr$ (see Subsection P3.2.2). Hence,

$$\varphi = \begin{cases} k\dfrac{q}{r} & \text{for } r > R, \\[2mm] -k\dfrac{qr^2}{2R^3} + \text{const} = -k\dfrac{qr^2}{2R^3} + \dfrac{3}{2}k\dfrac{q}{R} & \text{for } r < R. \end{cases}$$

The unknown constant is determined from the continuity condition for the potential at the ball surface.

▶ **Earnshaw's theorem.** Earnshaw's theorem is a corollary of Gauss's theorem. It states that a charged particle in vacuum cannot be in a stable equilibrium under the action of electric forces only. Indeed, if this were the case, the electric fields generated by the other charges would all have to point inward a small sphere enclosing the particle (if its charge is positive). But this would contradict Gauss's theorem (all other charges are outside the sphere).

▶ **Maxwell's equations in electrostatics.** The condition that an electrostatic field must be potential,

$$\oint \mathbf{E} \cdot d\mathbf{r} = 0 \quad \Longrightarrow \quad \text{curl } \mathbf{E} = \mathbf{0},$$

is one of Maxwell's equations (in integral and differential forms) written for the special case of electrostatics.

The second equation is Gauss's theorem (written for a spatially distributed charge):

$$\oint \mathbf{E} \cdot d\mathbf{s} = \frac{1}{\varepsilon_0} \int \rho \, dV \quad \Longrightarrow \quad \operatorname{div} \mathbf{E} = \frac{\rho}{\varepsilon_0}.$$

Substituting $\mathbf{E} = -\operatorname{grad} \varphi$ in the last equation yields *Poisson's equation for electric potential*:

$$\operatorname{div} \operatorname{grad} \varphi = -\frac{\rho}{\varepsilon_0}, \quad \text{or} \quad \Delta\varphi = -\frac{\rho}{\varepsilon_0}.$$

P3.4. Conductors and Capacitors
P3.4.1. Electrostatics of Conductors

▶ **Field in a conductor. The potential of a conductor.** A conductor is a material that is capable of carrying electric current. To this end, the material must contain movable electric charges that can travel across. Electrostatics deals with states where charges are at equilibrium, when there is no directed motion of charged particles—electric current. This means that the electric field must be zero everywhere within the conductor. It follows that:

1. The potential of all point within the conduction has the same value, which is called the *potential of the conductor*.

2. The conductor surface is an equipotential surface. The electric field lines are perpendicular to it.

3. The volume charge density in the conductor is zero (it follows from Gauss's theorem that it must be zero charge in any volume within the conductor). The uncompensated charges must be distributed on the surface.

4. The electric field near the conductor surface is related to the local surface charge density σ by

$$E = \frac{\sigma}{\varepsilon_0}. \tag{P3.4.1.1}$$

This formula can be obtained from Gauss's theorem (P3.3.1.3) by taking as a Gaussian surface a small cylinder whose one base is outside the conductor but near its surface and the other is inside. Then we get $ES = \sigma S/\varepsilon_0$.

▶ **Isolated conductor.** The potential of an isolated conductor is proportional to its charge:

$$\varphi = \frac{1}{C} q, \tag{P3.4.1.2}$$

where C is the *capacitance* of the conductor; it depends on the shape and size of the conductor and the dielectric permittivity of the surrounding medium. In SI, the unit of capacitance is the *farad* (F). In CGS, capacitance is measured in centimeters (cm). For example, the potential and capacitance of an isolated ball of radius R equal

$$\varphi = \frac{1}{4\pi\varepsilon_0} \frac{q}{R}, \quad C = 4\pi\varepsilon_0 R. \tag{P3.4.1.3}$$

▶ **Connection of conductors.** When two conductors having different potentials are connected to each other with a conducting wire, their potentials become equal. If the conductors are sufficiently far away from each other, their individual potentials can be calculated as the potential of an isolated conductor. Then the law of conservation of electric charge allows us to find the resulting common potential, φ':

$$C_1\varphi_1 + C_2\varphi_2 = C_1\varphi' + C_2\varphi' \quad \Longrightarrow \quad \varphi' = \frac{C_1\varphi_1 + C_2\varphi_2}{C_1 + C_2}$$

(the capacitance of the wire and its charge may be neglected). If the capacitance of one conductor is much larger than that of the other, its potential remains almost unchanged. For example, if a conductor is connected to the earth (it is said to be earthed), its potential becomes equal to the Earth's potential, which is constant and usually taken to be zero.

▶ **Conductor in an external electric field.** When any conductor is placed in an external electric field, its movable charges redistribute so that the resulting field (superposition of the external and internal fields) within the conductor is zero. For example, if a positive point charge is brought near an uncharged conductor, then negative charges will gather on the side that is closer to the charge and positive charges will move to the side that is further away. As a result, the charge and conductor will attract. (This phenomenon is called *electrostatic induction*.) Since the charge distribution on the conductor surface is usually unknown, the principle of superposition becomes unsuitable for the calculation of the resulting field. Sometimes, the problem may be solved by indirect techniques.

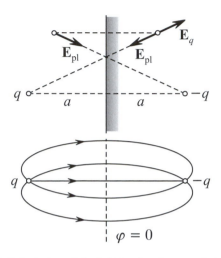

Figure P3.7. Calculation of the field by the electrostatic image method.

Example 1. Let us bring a positive point charge q near the surface of an earthed conductor to a small distance a compared to the conductor size. In this case, the conductor can be replaced by an infinite conducting half-space (Fig. P3.7). There will be induced negative charges at the plane surface. It is required to find the field $\mathbf{E}_{pl}$ produced by these charges at any point of the free space. Note that the charge distribution is unknown!

To begin with, let us find the field $\mathbf{E}_{pl}$ inside the conductor at the point symmetric to an outside point of interest. The field $\mathbf{E}_{pl}$ will compensate the field produced by the charge q at this point. Hence, the field $\mathbf{E}_{pl}$ inside will exactly coincide with that generated by an imaginary charge $-q$ placed at the same point as q. Now we see that the field $\mathbf{E}_{pl}$ produced by the surface charges of the conductor is symmetric relative to the plane. Consequently, $\mathbf{E}_{pl}$ *outside* will exactly coincide with the field produced by the imaginary charge $-q$ placed *symmetrically relative to the charge q*. This symmetric charge is called the *image* of the charge q and the method is known as the *electrostatic image method*. It is now fairly easy to find: (a) the attractive force between the charge and the conductor, $F = kq^2/(2a)^2$, (b) the electric field everywhere outside the conductor (it is equal to the field produced by the charge q and its image $-q$), and (c) the charge distribution of the surface—once the field near the surface is known, we can find the charge density σ using formula (P3.4.1.1).

▶ **Principle of uniqueness.** The principle of uniqueness is useful in solving problems with unknown charge distribution. This principle states that for a system of conductors, there exists a unique surface charge distribution and a unique field $\mathbf{E}(\mathbf{r})$ such that the net fields inside the conductors vanish and their potentials and charges acquire prescribed values. If one guesses a charge distribution or a field configuration that satisfies the above conditions, it will be the correct solution. For example, one can notice that the equipotential surface $\varphi = 0$ of the field produced by two charges, q and $-q$, coincides with their symmetry plane (see Fig. P3.7). Then it can be inferred that the fields produced by the charge q and it image $-q$ coincides with that produced by the charge q and the induced charge of the conducting plane (see the preceding example).

Example 2. A conducting ball in a uniform field. Consider an uncharged conducting ball of radius R in a unform field $\mathbf{E}_0$. The induced charges on the ball will redistribute so as to generate the field $-\mathbf{E}_0$ inside the ball. In order to obtain this distribution, consider two imaginary balls of the same radius but having a uniform volume charge, one of density ρ and the other of density $-\rho$. (The field of a uniformly charged ball was considered in example 3 from Subsection P3.3.2.) If the center of the positively charged ball is slightly displaced by $\mathbf{l}$ relative to the center of the negatively charged ball (Fig. P3.8), then the field in the overlapping region is

$$\mathbf{E} = -\frac{\rho\mathbf{r}}{3\varepsilon_0} + \frac{\rho(\mathbf{r}-\mathbf{l})}{3\varepsilon_0} = -\frac{\rho\mathbf{l}}{3\varepsilon_0},$$

and hence is uniform. The charge inside the overlapping region is zero. Now by letting $\mathbf{l} \to \mathbf{0}$ and $\rho \to \infty$ so that $\rho\mathbf{l}/(3\varepsilon_0) = \mathbf{E}_0$, we obtain the desired induced charge distribution: $\sigma = \sigma_0 \cos\theta$, where $\sigma_0 = \rho l = 3\varepsilon_0 E_0$. Also, we find the field the induced charges produce outside the ball. It coincides with that produced by a point dipole with the dipole moment $\mathbf{p} = (\rho V)\mathbf{l} = 4\pi\varepsilon_0 R^3 \mathbf{E}_0$ (see example 2, Subsection P3.2.2).

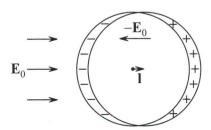

Figure P3.8. To the calculation of induced charges on a conducting ball in a uniform electric field.

P3.4.2. Capacitors

▶ **Capacitance of a capacitor.** A *capacitor* is an object consisting of a pair of insulated conductors, one of which having a charge $+q$ and the other the charge $-q$. The conductors are often called the *plates* of the capacitor. The charge q is called the *capacitor charge*. The potential difference between the positive and negative plates, U, is called the *capacitor voltage*. The voltage is proportional to the capacitor charge:

$$U = \frac{1}{C}q, \tag{P3.4.2.1}$$

where C is the *electric capacitance* (or just *capacitance*) of the capacitor.

▶ **Simple capacitors.** A capacitor is called simple if: (a) its electric field is concentrated within a limited region between the plates (so it can be assumed that all field lines start at the positive plate and finish at the negative one), (b) the whole region where the field is concentrated is filled with a dielectric material having a dielectric permittivity ε. (A dielectric weakens the electric field by a factor of ε; see Subsection P3.5.1.)

Example 1. Parallel-plate capacitor. A parallel-plate capacitor is one in which the conductors are parallel plates of area S each separated by a small distance b as compared to the plate size ($b \ll \sqrt{S}$). The electric field between the plates can be regarded as uniform everywhere except for the very edges. The field can be found using formula (P3.4.1.1): $E = q/(\varepsilon_0 \varepsilon S)$. The voltage of a parallel-plate capacitor equals $U = Eb$ and its capacitance is

$$C = \frac{\varepsilon_0 \varepsilon S}{b}. \tag{P3.4.2.2}$$

Example 2. Spherical capacitor. In a spherical capacitor, its plates are concentric spheres of radii R_1 and R_2. The field between the plates is $E = \dfrac{q}{4\pi\varepsilon_0 \varepsilon r^2}$, where r is the distance to the center. The voltage is expressed as $U = \int_{R_1}^{R_2} E(r)\,dr$ and the capacitance equals

$$C = \frac{4\pi\varepsilon_0 \varepsilon R_1 R_2}{R_2 - R_1}.$$

For $R_2 \to \infty$, this expression reduces to the capacitance of an isolated sphere (see formula (P3.4.1.3)).

Example 3. Cylindrical capacitor. In a cylindrical capacitor, its plates are concentric cylinders of radii R_1 and R_2 with length $l \gg R_2 - R_1$. The field between the plates is $E = \dfrac{q}{2\pi\varepsilon_0\varepsilon r l}$. The voltage is given by $U = \int_{R_1}^{R_2} E(r)\,dr$ and the capacitance equals is expressed as

$$C = \frac{2\pi\varepsilon_0\varepsilon l}{\ln(R_2/R_1)}.$$

▶ **Compound capacitors.** By connecting plates of several simple capacitors, one obtains a *compound capacitor*. The capacitance of a compound capacitor is determined by (P3.4.2.1) and is expressed in terms of the capacitances of its constituents.

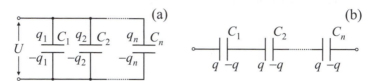

Figure P3.9. Parallel (a) and series (b) connection of capacitors.

1. *Parallel connection.* If n capacitors are connected in parallel with one another as shown in Fig. P3.9(a), they form a compound capacitor for which

$$U = U_i, \quad q = \sum_{i=1}^{n} q_i, \quad C = \sum_{i=1}^{n} C_i.$$

The voltage of all capacitors is the same.

2. *Series connection.* If n capacitors are connected in series as shown in Fig. P3.9(b), they form a compound capacitor for which

$$q = q_i, \quad U = \sum_{i=1}^{n} U_i, \quad \frac{1}{C} = \sum_{i=1}^{n} \frac{1}{C_i}.$$

The total charge at each pair of inner plates is zero.

3. The capacitance of a compound capacitor can often be found by replacing groups of capacitors connected in series or in parallel by an equivalent capacitor, eventually reducing the circuit to a series or parallel one. If the circuit cannot be reduced in this manner, the method of nodal potentials may be applied.

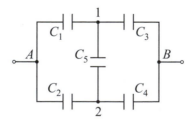

Figure P3.10. A capacitor connection irreducible to a serial or parallel circuit.

Example 4. Consider the circuit shown in Fig. P3.10. Let us take $\varphi_A = 0$. Then $\varphi_B = U$. The three connected plates of the capacitors C_1, C_3, and C_5 must have the same potential, φ_1, and their total charge is zero. The same holds for the capacitors C_2, C_4, and C_5. This allows us to write a system of equations for the potentials φ_1 and φ_2:

$$(\varphi_1 - U)C_3 + (\varphi_1 - \varphi_2)C_5 + \varphi_1 C_1 = 0,$$
$$(\varphi_2 - U)C_4 + (\varphi_2 - \varphi_1)C_5 + \varphi_2 C_2 = 0.$$

Solving the system, we find the charge of the compound capacitor, $q = C_1\varphi_1 + C_2\varphi_2$, and its capacitance $C = q/U$.

P3.5. Electrostatics of Dielectrics

P3.5.1. Polarization of Dielectrics

▶ **Bound charges. Polarization density.** A dielectric is a material that does not conduct electric current and, hence, has no free charges (or their number is negligibly small). Nevertheless, an electric field becomes weaker inside a dielectric. This indicates that some macroscopic charges arise in the volume and on the surface of a dielectric when placed in an electric field. These charges are due to *polarization* of the dielectric—the phenomenon where charges bound to individual molecules or the crystal lattice are displaced along the field; these charges are called *bound charges*. The average dipole moment of molecules becomes nonzero, which results in a nonzero dipole moment of any volume of the dielectric. The degree of polarization is characterized by the *polarization density vector*, $\mathbf{P}$, which is defined as the net dipole moment of the molecules per unit volume:

$$\mathbf{P} = \frac{\sum_i \mathbf{P}_i}{\Delta V} = n \langle \mathbf{p} \rangle,$$

where n is the molecule concentration.

Macroscopic equations of the electric field in a dielectric can be derived using the simplified assumption that all molecules in the small volume ΔV have the same dipole moment $\langle \mathbf{p} \rangle = \mathbf{P}/n = q_0 \langle \mathbf{l} \rangle$.

▶ **Formation of macroscopic bounded charges.** Macroscopic charges distributed over surface and volume can be used to describe the electric field produced by a huge number of oriented microscopic dipoles. The situation where polarization is uniform ($\mathbf{P}$ = const) is easiest to study. In this case, the volume of the dielectric remain electrically neutral and positive surface charges arise on one side of the dielectric and negative surface charges arise on the other side. The vector $\mathbf{P}$ is directed from the negative to positive charges and the electric field produced will be in *opposite* direction.

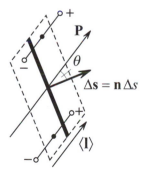

Figure P3.11. Some charges cross a selected area when the dielectric is polarized.

Polarized molecules in the volume of a dielectric are displaced from their original positions so that some charge, Δq, crosses an arbitrary area Δs in the perpendicular direction. This charge equals

$$\Delta q = n(\Delta s \langle l \rangle \cos \theta) q_0 = P_n \Delta s = (\mathbf{P} \cdot \Delta \mathbf{s}) \tag{P3.5.1.1}$$

where $\Delta s \langle l \rangle \cos \theta$ is the volume of a thin cylindrical layer whose molecules cross the area (Fig. P3.11). Integrating over any closed surface yields the net bound charge displaced from the enclosed volume (the net bound charge enclosed within the surface will have the opposite sign):

$$q_{enc}^b = -\oint \mathbf{P} \cdot d\mathbf{s} \quad \text{or} \quad \text{div}\,\mathbf{P} = -\rho^b, \tag{P3.5.1.2}$$

where $d\mathbf{s} = \mathbf{n}\,ds$ is the surface element vector directed along the normal toward the surrounding space and ρ^b is the volume density of the bound charge. The surface density of the bound charge of (P3.5.1.1) is given by

$$\sigma^b = P_n. \tag{P3.5.1.3}$$

▶ **Electric displacement.** The macroscopic electric field, $\mathbf{E}(\mathbf{r})$, in a polarized dielectric is determined by averaging the microscopic field, $\mathbf{E}_{micro}$ (highly varying at interatomic distances), over a small volume containing sufficiently many molecules: $\mathbf{E}(\mathbf{r}) = \langle \mathbf{E}_{micro} \rangle$. The total electric field is determined by both bound charges and free charges brought to the dielectric from outside:

$$\oint \mathbf{E} \cdot d\mathbf{s} = \frac{1}{\varepsilon_0}(q_{enc}^f + q_{enc}^b). \tag{P3.5.1.4}$$

Since the bound charge distribution is unknown in advance, it is convenient to introduce a vector quantity

$$\mathbf{D} = \varepsilon_0 \mathbf{E} + \mathbf{P}, \tag{P3.5.1.5}$$

which is called the *electric displacement field* or just the *electric displacement*. It follows from equations (P3.5.1.2) and (P3.5.1.4) that the electric displacement is determined by the free charges only:

$$\oint \mathbf{D} \cdot d\mathbf{s} = q_{enc}^f. \tag{P3.5.1.6}$$

This equation is known as *Gauss's law for the electric displacement*. In differential form, this law reads

$$\text{div}\,\mathbf{D} = \rho^f.$$

▶ **Isotropic dielectric.** The polarization density $\mathbf{P}$ at a given point in a dielectric is determined by the external electric field $\mathbf{E}$. For not too strong fields, $\mathbf{P}$ is linearly dependent on $\mathbf{E}$. Furthermore, $\mathbf{P} \parallel \mathbf{E}$ in an *isotropic* dielectric, and hence

$$\mathbf{P} = \varepsilon_0 \chi \mathbf{E}, \tag{P3.5.1.7}$$

where χ is the *electric susceptibility* of the dielectric material. Then the electric displacement can be expressed as

$$\mathbf{D} = \varepsilon_0 \mathbf{E} + \mathbf{P} = \varepsilon_0 \mathbf{E} + \varepsilon_0 \chi \mathbf{E} = \varepsilon_0 \varepsilon \mathbf{E}, \tag{P3.5.1.8}$$

where $\varepsilon = 1 + \chi$ is the *electric permittivity* of the material.

▶ **Boundary conditions at dielectric interface.** For two dielectric media meeting along a common interface, the following boundary conditions for tangential and normal components hold:

$$E_{1\tau} = E_{2\tau}, \qquad P_{2n} - P_{1n} = -\sigma^b, \qquad D_{2n} - D_{1n} = \sigma^f. \tag{P3.5.1.9}$$

The normal $\mathbf{n}$ points from medium 1 to medium 2. If there is no free charge at the interface, the normal components of the displacement are equal to each other. The first condition in (P3.5.1.9) results from the potentiality of the field $\mathbf{E}$ (the work along a closed path is zero). The second and third conditions follow from formulas (P3.5.1.2) and (P3.5.1.6), respectively. For an isotropic dielectric, we have

$$E_{1\tau} = E_{2\tau}, \quad \varepsilon_1 E_{1n} = \varepsilon_2 E_{2n}; \qquad D_{1n} = D_{2n}, \quad \frac{D_{1\tau}}{\varepsilon_1} = \frac{D_{2\tau}}{\varepsilon_2}. \tag{P3.5.1.10}$$

It can be seen that the field lines are refracted at the interface between dielectric media. Note that unlike the electric field, the electric displacement lines are continuous at the interface.

▶ **Calculation of fields in the presence of dielectrics.** Given the distribution of free charges, the field in the whole space is uniquely determined by the equations for $\mathbf{E}$ and $\mathbf{D}$ together with the relation between them (P3.5.1.8) and the boundary conditions (P3.5.1.10). The solution has the following important properties:

1. Using equations (P3.5.1.2) and (P3.5.1.6) and the formula $\mathbf{P} = -\dfrac{\varepsilon - 1}{\varepsilon}\mathbf{D}$, which follows from (P3.5.1.7) and (P3.5.1.8), we find for a homogeneous isotropic dielectric that

$$\rho^b = -\frac{\varepsilon - 1}{\varepsilon}\rho^f. \tag{P3.5.1.11}$$

Consequently, surface bound charge only will arise due to polarization in the absence of volume free charge. Note that for an inhomogeneous dielectric, for which ε depends on the coordinates, volume bound charge arises even if there is no volume free charge.

2. If the boundaries of a homogeneous dielectric coincide with equipotential surfaces of the field $\mathbf{E}_0$ produced by free charges in the absence of the dielectric, then the solution

$$\mathbf{E(r)} = \frac{\mathbf{E_0(r)}}{\varepsilon} \tag{P3.5.1.12}$$

satisfies the boundary conditions.

Example 1. Suppose a point charge q is surrounded by a hollow dielectric sphere with inner radius R_1 and outer radius R_2. The field in the cavity and outside the dielectric will remain the same. Inside the dielectric, the field will change so that

$$E = \frac{1}{4\pi\varepsilon_0\varepsilon}\frac{q}{r^2}, \quad P = \varepsilon_0(\varepsilon - 1)E = \frac{\varepsilon - 1}{\varepsilon}\frac{q}{4\pi r^2},$$

$$\sigma_2^b = P(R_2) = \frac{\varepsilon - 1}{\varepsilon}\frac{q}{4\pi R_2^2}, \quad \sigma_1^b = -\frac{\varepsilon - 1}{\varepsilon}\frac{q}{4\pi R_1^2}.$$

The bound charges at the inner and outer surfaces are equal to each other in magnitude: $q_2^b = -q_1^b = q\dfrac{\varepsilon - 1}{\varepsilon}$.

Example 2. Consider a ball made from a homogeneous dielectric and charged uniformly by a free charge with volume density ρ. The field in the absence of the dielectric was found in example 3, Subsection P3.3.2. The field in the dielectric, for $r < R$, will be weaker by a factor of ε:

$$E = \frac{\rho r}{3\varepsilon_0\varepsilon}, \quad P = \varepsilon_0(\varepsilon - 1)E = \frac{\varepsilon - 1}{\varepsilon}\frac{\rho r}{3},$$

$$\sigma^b = P(R) = \frac{\varepsilon - 1}{\varepsilon}\frac{\rho R}{3}, \quad \rho^b = -\frac{\varepsilon - 1}{\varepsilon}\rho.$$

It is not difficult to verify that $\sigma^b 4\pi R^2 = -\rho^b \frac{4}{3}\pi R^3$, so the total bound charge is zero, as one would expect.

Example 3. A dielectric slab is placed in a field $\mathbf{E}_0$ so that it enters the slab at the right angle. The electric permittivity of the slab varies from ε_1 at one side to ε_2 at the other side. Then we have

$$D = \varepsilon_0 E_0 = \text{const}, \quad E = \frac{E_0}{\varepsilon(x)}, \quad P = D\frac{\varepsilon - 1}{\varepsilon}, \quad \rho^b = -\frac{dP}{dx} = -\frac{D}{\varepsilon^2}\frac{d\varepsilon}{dx},$$

$$\sigma_2^b = P_2 = D\frac{\varepsilon_2 - 1}{\varepsilon_2}, \quad \sigma_1^b = -P_1 = -D\frac{\varepsilon_1 - 1}{\varepsilon_1}.$$

It is easy to verify that the total bound charge is zero.

Example 4. If a homogeneous dielectric ball is placed in a uniform field $\mathbf{E}_0$, the ball will polarize uniformly ($\mathbf{P} = \text{const}$). Indeed, uniform polarization corresponds to two uniformly charged balls whose centers are displaced by a small vector $\mathbf{l}$, as was shown in example 2, Subsection P3.4.1. The induced polarization field inside the balls is $\mathbf{E}_1 = -\rho\mathbf{l}/3\varepsilon_0 = -(q\mathbf{l}/V)/3\varepsilon_0 = -\mathbf{P}/3\varepsilon_0$. Then the total field in the dielectric is $\mathbf{E} = \mathbf{E}_0 + \mathbf{E}_1 = \mathbf{E}_0 - \mathbf{P}/3\varepsilon_0$. From this equation and the condition $\mathbf{P} = (\varepsilon - 1)\varepsilon_0\mathbf{E}$, we find that

$$\mathbf{E} = \frac{3\mathbf{E_0}}{\varepsilon + 2}, \quad \mathbf{P} = \frac{3\varepsilon_0(\varepsilon - 1)}{\varepsilon + 2}\mathbf{E_0}.$$

The boundary conditions are all satisfied.

P3.5.2. Types of Dielectric and Polarization Mechanisms

There are three main mechanisms of polarization. Accordingly, three basic types of dielectric are distinguished.

▶ **1. Dipolar polarization in polar dielectrics.** The molecules of a *polar dielectric* (*polar molecules*) have a permanent dipole moment, p_0, in the absence of electric field. Examples of such substances, are water (H_2O), hydrochloric acid (HCl), ammonia (NH_3), etc. Due to thermal motion, all dipoles are oriented chaotically in the absence of electric field, resulting in a zero average dipole moment. If there is an external field $\mathbf{E}$, dipoles tend to orient along the field (see Subsection P3.2.2). The competition between the chaotic thermal motion and the orienting action of the electric field results in an *average dipole moment* $\langle \mathbf{p} \rangle$. In a weak field ($p_0 E \ll kT$), the average dipole moment is proportional to $\mathbf{E}$ and decreases with increasing temperature.

In gaseous dielectrics, the probability for polar molecules to be oriented along $\mathbf{p}$ is proportional to $\exp(-W_p/kT)$, where $W_p = -(\mathbf{p} \cdot \mathbf{E})$, which follows from Boltzmann's distribution (see Subsection P2.5.4). It is apparent that, in an external electric field, the average dipole moment must be directed along $\mathbf{E}$ and proportional to $p_0 f(p_0 E/kT)$, where $f(x)$ is some function such that $f(0) = 0$ and $f(x) \to 1$ as $x \to \infty$. The case $f(x) \to 1$ corresponds to a strong field, where all dipoles are nearly parallel to $\mathbf{E}$. For small x (a weak field), we have $f(x) \approx \gamma x$ and $\langle \mathbf{p} \rangle \approx \gamma (p_0^2/kT)\,\mathbf{E}$, so the average dipole moment is proportional to $\mathbf{E}$ and inversely proportional to T. The analysis gives $\gamma = 1/3$. Then, $\mathbf{P} = \frac{1}{3}(np_0^2/kT)\,\mathbf{E}$, and hence the electric permittivity evaluates to

$$\varepsilon = 1 + \frac{1}{3}\frac{np_0^2}{\varepsilon_0 kT}.$$

This formula is known as the *Langevin–Debye law*.

▶ **2. Electronic polarization in nonpolar dielectrics.** In the absence of external electric fields, the dipole moment of a *nonpolar molecule* (e.g., H_2, O_2, N_2, etc.) is zero. So the minimum potential energy corresponds to the state where the center of the negative charge (electronic cloud) of the molecule coincides with that of the positive charge. In the presence of an electric field, the centers of the negative and positive charges are displaced and a force arises that tries to return them to an equilibrium. If the displacement, $\mathbf{l}$, is small, the restoring force is proportional to it: $\mathbf{F} = -\beta \mathbf{l}$ (*quasielastic force*). In equilibrium, $q\mathbf{E} = \beta \mathbf{l}$ and the *induced* dipole moment, $\mathbf{p} = q_0 \mathbf{l}$, turns out to be proportional to the field: $\mathbf{p} = \alpha \varepsilon_0 \mathbf{E}$. The coefficient α here is called the *molecular polarizability*. The polarization density is $\mathbf{P} = n\mathbf{p} = n\varepsilon_0 \alpha \mathbf{E}$, where $\mathbf{E}$ is the field acting on a single molecule. In gases, this field can be regarded as the average field in the dielectric. Consequently, the electric permittivity evaluates to

$$\varepsilon = 1 + \chi = 1 + \alpha n.$$

In dense gases as well as liquid and solid dielectrics, one should take into account that the field acting on a molecule is different from the average field. The molecular field includes the local field produced by the molecule itself, whereas the average field does not.

▶ **Difference between the local and average fields.** Consider a homogeneous dielectric crystal with nonpolar molecules. In order to find the field acting on a given molecule, consider a sphere with its center at the molecule. If the molecules are arranged in a regular manner (e.g., at the vertices of a cubic lattice), the field produced by the molecules included within the sphere is zero at the center. Hence, the local field at a given point is $\mathbf{E}_{loc} = \mathbf{E} + \mathbf{E}_1$, where $\mathbf{E}$ is the average macroscopic field (equal to the sum of the external field and the outer surface charge field) and $\mathbf{E}_1$ is the inner surface charge field of the spherical hollow. As follows from example 4 above, $\mathbf{E}_1 = \mathbf{P}/3\varepsilon_0$, and hence $\mathbf{P} = n\varepsilon_0 \alpha \left(\mathbf{E} + \frac{1}{3\varepsilon_0}\mathbf{P} \right)$, whence

$$\mathbf{P} = \frac{\varepsilon_0 n\alpha \mathbf{E}}{1 - \frac{n\alpha}{3}}.$$

Taking into account that $\mathbf{P} = \varepsilon_0(\varepsilon - 1)\mathbf{E}$, we get

$$\varepsilon - 1 = \frac{n\alpha}{1 - \frac{n\alpha}{3}} \qquad \text{or} \qquad \frac{\varepsilon - 1}{\varepsilon + 2} = \frac{n\alpha}{3}.$$

This result is known as the *Clausius–Mossotti relation*.

▶ **3. Ionic polarization in solid crystal dielectrics.** Examples of such dielectrics are NaCl, KCl, NaF, KI, etc. In the presence of an external electric field, the sublattice of positive ions (e.g., Na^+) is shifted, as a whole, along the field and the sublattice of negative ions (e.g., Cl^-) is displaced in the opposite direction.

▶ **Ferroelectrics.** Ferroelectric materials are substances that contain small regions (domains) where all dipoles are spontaneously polarized in the same direction. Ferroelectrics demonstrate this property only below a certain phase transition temperature, T_c, called the *Curie temperature*. These materials are electric analogues of ferromagnetic materials. They have anomalously large electric permittivity, dependent on the electric field and temperature, demonstrate hysteresis properties when polarized, and become usual dielectrics above the Curie temperature. (For details about ferromagnetics, see Subsection P3.9.2.)

P3.6. Energy of Electrostatic Field

P3.6.1. Energy of a System of Charges

▶ **Potential energy of interaction of charged particles.** The energy of interparticle interaction in a system of charged particles is defined as the work done by external forces to create this system or as the work done by the electric field to destroy this system. The energy of interaction between two charged particles equals the potential energy of one particle in the field produced by the other:

$$W_{12} = q_1\varphi_2(\mathbf{r}_1) = q_2\varphi_1(\mathbf{r}_2) = k\frac{q_1q_2}{r}.$$

The energy of a system of charged particles is

$$W = \sum_{i>j} W_{ij} = \frac{1}{2}\sum_{i \neq j} W_{ij} = \frac{1}{2}\sum_i q_i\varphi(\mathbf{r}_i), \tag{P3.6.1.1}$$

where $\mathbf{r}_i$ is the position vector of the ith particle and $\varphi(\mathbf{r}_i)$ is the potential created by all other particles at $\mathbf{r}_i$. For continuously distributed charges with volume density $\rho(\mathbf{r})$ and surface density $\sigma(\mathbf{r})$, the above formula becomes

$$W = \frac{1}{2}\int_V \varphi(\mathbf{r})\rho(\mathbf{r})\,dV + \frac{1}{2}\int_S \varphi(\mathbf{r})\rho(\mathbf{r})\,ds. \tag{P3.6.1.2}$$

Although formula (P3.6.1.2) is obtained as a formal generalization of formula (P3.6.1.1), there are essential differences between them. In the latter, $\varphi(\mathbf{r}_i)$ is the potential produced by all charges but the ith, whereas in the former, $\varphi(\mathbf{r})$ is the potential of *all* the charges at $\mathbf{r}$ (the contribution of a volume or surface element decreases to zero with the element size). However, the most important difference is this. Formula (P3.6.1.1) considers only the energy of interaction and if the point charges are moved far away from one another, the energy of interaction will be negligibly small but there will still remain the fields of individual charges. Unlike this, formula (P3.6.1.2) calculates the *total energy* of a system of continuously distributed charges and if the system is divided into increasingly smaller parts that are moved away to infinity, the total energy of all these parts will tend to zero.

Example 1. Calculate the energy of a uniformly charged ball of radius R with total charge q.

Solution. Substituting the ball field potential (see example 3, Subsection P3.3.2) into (P3.6.1.2) gives

$$W = \frac{1}{2} \int \rho\varphi \, dV = \frac{\rho}{2} \int_0^R \left(-k\frac{qr^2}{2R^3} + \frac{3}{2}k\frac{q}{R}\right) 4\pi r^2 dr = \frac{3}{5}k\frac{q^2}{R}.$$

▶ **Energy of an isolated conductor.** All charges are on the surface and have the same potential, which is the potential of the conductor. From (P3.6.1.2) we have

$$W = \frac{q\varphi}{2} = \frac{C\varphi^2}{2} = \frac{q^2}{2C}, \qquad (P3.6.1.3)$$

where C is the electric capacitance of the conductor (see Subsection P3.4.1). This result can also be obtained by directly calculating the work done by the field to take the whole charge to infinity: $W = \int \varphi(q) \, dq = C \int_0^\varphi \varphi \, d\varphi = \frac{1}{2}C\varphi^2$.

Example 2. The energy of an isolated conducting sphere equals

$$W = \frac{1}{2}q \cdot \frac{q}{4\pi\varepsilon_0 R} = \frac{q^2}{8\pi\varepsilon_0\varepsilon R},$$

where ε is the electric permittivity of the surrounding medium.

▶ **Capacitor energy.** From (P3.6.1.2) we get

$$W = \frac{q\varphi_1}{2} - \frac{q\varphi_2}{2} = \frac{qU}{2} = \frac{CU^2}{2} = \frac{q^2}{2C}. \qquad (P3.6.1.4)$$

The field in the capacitor can be destroyed by transferring the charge in small portions from one plate to the other. So $W = \int U(q) \, dq = C \int_0^U U \, dU = \frac{1}{2}CU^2$. For example, the energy of a parallel-plate capacitor is

$$W = \frac{\varepsilon_0\varepsilon S}{2b}(Eb)^2 = \frac{\varepsilon_0\varepsilon E^2}{2} Sb. \qquad (P3.6.1.5)$$

P3.6.2. Energy of an Electrostatic Field

▶ **Energy density of an electric field.** Within a field approach, one should deal with the energy of the field surrounding charged particles rather than the energy of interparticle interaction. The volume energy density of an electric field, $w = dW/dV$, in vacuum is only dependent on the field strength:

$$w = \frac{1}{2}\varepsilon_0 E^2. \qquad (P3.6.2.1)$$

This expression holds for any charge distribution (e.g., it can be obtained by considering the uniform field of a parallel-plate capacitor).

With the field equations (see Subsection P3.3.2), formula (P3.6.1.2) can be rewritten as

$$W = \frac{1}{2} \int \varepsilon_0 E^2 \, dV, \qquad (P3.6.2.2)$$

where the integration is performed over the whole space.

Formulas (P3.6.1.2) and (P3.6.2.2) allow one to calculate the total energy in the whole space only if: (a) there are no point or linear charges and (b) the charges are limited to a finite region of space. However, formula (P3.6.2.2) calculates not only field energy in the whole space but also energy limited to a finite region of space even if the total energy is infinite.

Example 1. When the charged ball energy was calculated (see example 1) using (P3.6.1.2), the integration was performed of the ball volume (since $\rho = 0$ outside the ball). However, formula (P3.6.2.2) reveals that not all of the field energy is stored within the ball:

$$W_1 = \int \frac{1}{2}\varepsilon_0 E^2 \, dV = \int_0^R \frac{1}{2}\varepsilon_0 \Big(\frac{qr}{4\pi\varepsilon_0 R^3}\Big)^2 (4\pi r^2 dr) = \frac{1}{10}k\frac{q^2}{R}.$$

It makes only 1/6 of the total energy while the 5/6 is stored in the surrounding void space:

$$W_2 = \int_R^\infty \frac{1}{2}\varepsilon_0 \Big(\frac{q}{4\pi\varepsilon_0 r^2}\Big)^2 (4\pi r^2 dr) = \frac{1}{2}k\frac{q^2}{R}.$$

In a dielectric, the energy density is larger by a factor of ε:

$$w = \tfrac{1}{2}\varepsilon_0\varepsilon E^2 = \tfrac{1}{2}ED = \tfrac{1}{2}\varepsilon_0 E^2 + \tfrac{1}{2}\varepsilon_0(\varepsilon - 1)E^2.$$

The first term in the last expression is the field energy in the dielectric itself while the second term, equal to $\frac{1}{2}\mathbf{E} \cdot \mathbf{P}$, is the work done by the field to polarize the dielectric. This expression can be generalized to the case where $\mathbf{P}$ (and $\mathbf{D}$) are not proportional to $\mathbf{E}$ (e.g., for ferroelectrics and strong fields). To this end, one should look at the process of charging a parallel-plate capacitor filled with such a dielectric:

$$w = \frac{W}{V} = \frac{1}{V}\int U \, dq = \frac{1}{V}\int (Eb)\, d(DS) = \int_0^D \mathbf{E} \cdot d\mathbf{D}.$$

▶ **Law of conservation of energy. Force calculation.** To calculate the forces, acting on a charged or polarized body in an electric field, one should look at the process of slow isothermal change of the body position. The work done by external forces together with work done by voltage sources are equal to the change in the field energy. Since the body is in mechanical equilibrium, the work done by external forces equals the work done by the force exerted by the field, taken with the opposite sign.

Example 2. A parallel-plate capacitor charged to a voltage U is filled with a liquid or gaseous dielectric with permittivity ε. Find the force acting on either plate. Also treat the case of a solid dielectric.

Solution. Assume that the capacitor is disconnected from the voltage source. When the distance between the plates is increased by dx, the energy increases by $\frac{1}{2}EDS\,dx$ and the work done is $F\,dx$. Hence, the force of attraction equals $F = \frac{1}{2}EDS = \frac{1}{2}Eq$, where $\frac{1}{2}E$ is the field strength in the dielectric produced by either plate ($E = U/b$). However, if the dielectric is a solid slab, then there is an air gap between the dielectric and either plate. Then the energy increment equals $\frac{1}{2}\varepsilon_0 E_0^2 S\,dx$ and the force is

$$F = \frac{1}{2}\varepsilon_0 E_0^2 S = \frac{1}{2}E_0 DS = \frac{1}{2}qE_0,$$

where $\frac{1}{2}E_0$ is the field strength produced by one plate within the air gap ($E_0 = \varepsilon E = \varepsilon U/b$).

Example 3. Find the force at which a dielectric slab is pulled into the gap between the plates of a parallel-plate capacitor charged to a voltage U (see Fig. P3.12).

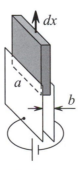

Figure P3.12. To example 3.

Solution. Assume that the respective dimensions of the slab coincide with those of the plates and the slab is slightly thinner than the interplate distance, b. We also assume that the capacitor is connected to a constant voltage source, U. Let us pull the slab out of the capacitor by a distance dx (Fig. P3.12). The field energy within the volume $dV = ab\,dx$ will change by $dW = \left(\frac{1}{2}\varepsilon_0 E^2 - \frac{1}{2}\varepsilon_0\varepsilon E^2\right) ab\,dx$ (a is the slab width and $E = U/b$ is the field). The charge at the plates will change by $dq = \Delta\sigma a\,dx = (\varepsilon_0 E - \varepsilon_0\varepsilon E)a\,dx$ and the work done by the source will be $\delta A_{\text{source}} = U\,dq$, which is twice the field energy increment. Writing the law of conservation of energy, $F\,dx + \delta A_{\text{source}} = dW$, we find that

$$F = \frac{1}{2}\varepsilon_0(\varepsilon - 1)abE^2.$$

Note that this force arises due to field distortion near the plate edges, although the energy calculation was performed without taking into account edge effects.

P3.7. Direct Current

P3.7.1. Electric Current in a Circuit

▶ **Electric current. Current density.** *Electric current* is an ordered motion of electric charges. *Convection current* is a motion of charges associated with the motion of a charged body. *Conduction current* is a drift motion of *free charges* (*charge carriers*) through a substance (conductor) due to an electric field applied. Examples of free charges are electrons in metals, ions in electrolytes, electrons and ions in plasma, and electrons and "holes" in semiconductors. The positive direction of electric current is conventionally taken to be the direction of positive moving charges.

The charge passing in time dt through an area $d\mathbf{s} = \mathbf{n}\,ds$ inside a conductor in the normal direction $\mathbf{n}$ is determined by the average velocity of free charges, $\langle \mathbf{v} \rangle$:

$$dq = q_0 n \langle \mathbf{v} \rangle \cdot d\mathbf{s}\,dt = \mathbf{j} \cdot \mathbf{s}\,dt, \tag{P3.7.1.1}$$

where q_0 is the charge of a single charge carrier, n is the carrier concentration, and $\mathbf{j} = q_0 n \langle \mathbf{v} \rangle$ is the *current density*. The charge passing through the cross-section, S, of a conductor per unit time is called the electric current (or the current) and denoted I:

$$dq = I\,dt = dt \int \mathbf{j} \cdot d\mathbf{s}, \tag{P3.7.1.2}$$

where the integration is performed over the cross-section. For a *linear* conductor ($S = \text{const}$), the current density is constant, $\mathbf{j} = \text{const}$, and formula (P3.7.1.2) gives $I = jS$. A current is said to be *constant* if $I = \text{const}$.

The decrease in the total charge within a closed surface equals $dq_{\text{enc}} = -dt \oint \mathbf{j} \cdot d\mathbf{s}$. In differential form, the law of conservation of charge is expressed as $\frac{d\rho}{dt} = -\operatorname{div}\mathbf{j}$. For a constant current, $\oint \mathbf{j} \cdot d\mathbf{s} = 0$, or $\operatorname{div}\mathbf{j} = 0$, so the current flows in continuous lines directed along the conductor.

▶ **Ohm's law in differential form.** To maintain current through a substance, free charges must be acted upon by an electromagnetic force $\mathbf{F}$, characterized by an electric field $\mathbf{E} = \mathbf{F}/q_0$. As soon as the force is removed, the average velocity of free charges vanishes in a very short time. For not too strong fields, the current density in an isotropic conductor must be proportional to $\mathbf{E}$:

$$\mathbf{j} = \sigma\mathbf{E} = \frac{1}{\rho}\mathbf{E}, \tag{P3.7.1.3}$$

where σ is the *conductivity* of the material and ρ is the *resistivity*. It follows from the law of conservation of charge that if $\rho = \text{const}$, then $\oint \mathbf{E} \cdot d\mathbf{s} = \rho \oint \mathbf{j} \cdot d\mathbf{s} = 0$, or $\operatorname{div}\mathbf{E} = 0$. Consequently, the field lines in a *homogeneous conductor* are continuous and in a wire of constant cross-section, they are parallel to the wire surface.

▶ **Joule–Lenz law in differential form.** As current flows through a conductor, the work done by the field **E** on free charges converts completely into thermal energy due to collisions with ions. The volume density of the thermal power of the current equals the work done by the field on charges per unit time in a unit volume:

$$w_{\text{therm}} = n\langle \mathbf{v} \rangle \cdot \mathbf{F} = q_0 n \langle \mathbf{v} \rangle \cdot \mathbf{E} = \mathbf{j} \cdot \mathbf{E} = \sigma E^2 = \rho j^2. \tag{P3.7.1.4}$$

▶ **Simple circuit segment. Resistance.** If current flows through a segment of an electrical circuit is due to an electrostatic (Coulomb) field only ($\rho \mathbf{j} = \mathbf{E}_{\text{coulomb}}$), this circuit segment is called *simple*. In a homogeneous conductor ($\rho = \text{const}$), $\oint \mathbf{E} \cdot d\mathbf{s} = \rho \oint \mathbf{j} \cdot d\mathbf{s} = 0$, and, according to Gauss's theorem (see Section P3.3), the field producing charge is only on the surface. In a wire of constant cross-section, the equipotential surfaces coincide with wire cross-sections. The current from a section 1 to section 2 is proportional to the potential difference $\varphi_1 - \varphi_2$:

$$\int_1^2 \mathbf{E}_{\text{coulomb}} \cdot d\mathbf{l} = \int_1^2 \rho \mathbf{j} \cdot d\mathbf{l} = I \int_1^2 \rho \frac{dl}{S} \quad \Longrightarrow \quad \varphi_1 - \varphi_2 = IR. \tag{P3.7.1.5}$$

This result is known as *Ohm's law for a simple circuit segment.* If $I < 0$, then current flows from section 2 to 1. The quantity $R = \int_1^2 (\rho/S)\, dl$ is the *electrical resistance of the conductor.* The SI unit of electrical resistance is the ohm (denoted Ω).

For a homogeneous linear conductor, $R = \rho l/S$, where ρ is the electrical resistivity, l is the conductor length, and S is the cross-sectional area. The electrical resistivity is temperature dependent; this dependence is practically linear in a wide range of temperatures: $\rho = \rho_0[1 + \alpha(T - T_0)]$, where ρ_0 is the resistivity at the reference temperature T_0 (e.g., $0°$ C), α is the *temperature coefficient of resistivity.* For many metals, $\alpha \approx 1/273$ (so $\rho \approx \rho_0 \alpha T$); for semiconductors, $\alpha < 0$.

▶ **Electromotive force.** Although electrostatic forces can maintain current through an open circuit segment, the total work done by these forces around a closed circuit is zero. Consequently, other, nonelectrostatic forces must be present in a closed circuit to compensate for thermal losses; the total work of these forces around a closed circuit is nonzero, and hence these are nonconservative. However, these forces are of electromagnetic nature, $\mathbf{F}_{\text{noncons}} = q_0 \mathbf{E}_{\text{noncons}}$, and their work done to transfer a test charge q from point 1 to point 2 is proportional to q:

$$A_q(1 \to 2) = q \int_1^2 \mathbf{E}_{\text{noncons}} \cdot d\mathbf{l} = q \mathscr{E}_{12} = \pm q \mathscr{E}, \tag{P3.7.1.6}$$

where $\mathscr{E}$ is the so-called *electromotive force* (or *emf* for short), which has the dimension of voltage and is measured in volts (V). A source of electromotive force is usually regarded as a source of current (such as a battery) and denoted in circuit diagrams by a circle with the minus and plus signs as shown in Fig. P3.13. The plus sign in formula (P3.7.1.6) is taken when the nonconservative forces act in the direction from point 1 to point 2, and the minus sign is taken otherwise. The power of the nonconservative forces (the power of the emf source) equals

$$P_{\text{noncons}} = \frac{A_q(1 \to 2)}{t} = I \mathscr{E}_{12}. \tag{P3.7.1.7}$$

The work done by the nonconservative forces in time dt on all charges in the conductor equals the work done to transfer the charge $I\, dt$ from one end of the conductor to the other.

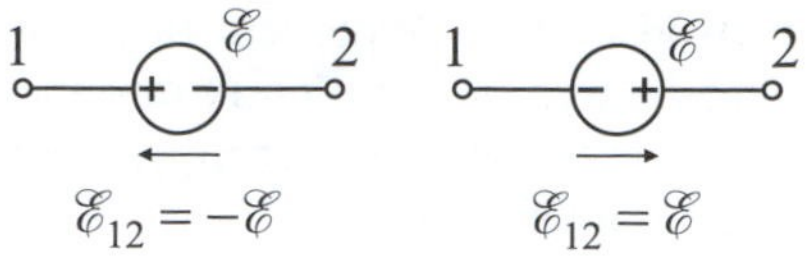

Figure P3.13. Sign rule for an emf (voltage) source.

▶ **Ohm's law for a circuit with a current source.** In a circuit containing a source of current, there are both electrostatic and nonelectrostatic forces acting on charges. In this case, Ohm's law in differential form (P3.7.1.3) reads

$$\mathbf{j} = \frac{1}{\rho}(\mathbf{E}_{\text{coulomb}} + \mathbf{E}_{\text{noncons}}), \tag{P3.7.1.8}$$

Ohm's law in integral form becomes

$$I\int_1^2 \rho\frac{dl}{S} = \int_1^2 (\mathbf{E}_{\text{coulomb}} + \mathbf{E}_{\text{noncons}}) \cdot d\mathbf{l} \implies IR = (\varphi_1 - \varphi_2) + \mathscr{E}_{12}. \tag{P3.7.1.9}$$

If $I < 0$, current flows from 2 to 1. The quantity $U_{12} = (\varphi_1 - \varphi_2) + \mathscr{E}_{12}$, equal to the work done by all forces to transfer a unit charge from cross-section 1 to 2, is called the *voltage* on the selected segment of the circuit. For a simple segment, the voltage equals the potential difference. Thus, Ohm's law states that $IR = U_{12}$ on any circuit segment.

▶ **Joule–Lenz law for a circuit segment.** The thermal power of electric current on a circuit segment can be found using formula (P3.7.1.4):

$$P_{\text{therm}} = \int_1^2 wS\,dl = \int_1^2 \rho j^2 S\,dl = I^2 \int_1^2 \rho\frac{dl}{S} = I^2 R. \tag{P3.7.1.10}$$

P3.7.2. Current in a Closed Circuit

▶ **Ohm's law for an unbranched circuit.** The current through an unbranched closed circuit is constant, $I = $ const. Adding up equations (P3.7.1.9) for all circuit segments, we get

$$I\sum R_i = \sum \mathscr{E}_{ij}. \tag{P3.7.2.1}$$

For example, if the current source has internal resistance r and the resistance of the remaining circuit is R, we have $I = \mathscr{E}/(R + r)$. The potential difference at the source terminals is $\varphi_+ - \varphi_- = IR = \mathscr{E} - Ir$. The short circuit current of the source ($R = 0$) equals $I_{\text{sc}} = \mathscr{E}/r$. Ohm's law (P3.7.2.1) expresses the law of conservation of energy for an unbranched closed circuit: the power of nonconservative forces equals the power of thermal losses at the resistors in the circuit.

▶ **Useful power of a current source.** If a current source is used to transmit energy into an external circuit, then the total (spent) power equals the power on nonconservative forces: $P_{\text{total}} = \mathscr{E}I$. The lost power equals the thermal losses at the internal resistance: $P_{\text{lost}} = I^2 r$. Then the useful power is $P_{\text{useful}} = \mathscr{E}I - I^2 r = I\Delta\varphi$, where $\Delta\varphi$ is the potential difference at the source terminals. It can be seen that the maximum of P_{useful} is attained at $I = \mathscr{E}/(2r)$ and is equal to $P_{\text{max}} = \mathscr{E}^2/(4r)$.

If the potential difference supplied at the ends of a circuit segment is $\varphi_1 - \varphi_2$ and the segment contains a device that performs work against the external forces (e.g., an electric motor), then the total power is $P_{\text{total}} = (\varphi_1 - \varphi_2)I$ and the lost power is $P_{\text{lost}} = I^2 R$, where R is the resistance of the motor winding. Hence, the useful power equals $P_{\text{useful}} = (\varphi_1 - \varphi_2)I - I^2 R = -\mathscr{E}_{12}I = \mathscr{E}I$. The quantities $\mathscr{E}$ and I depend on the revolution speed of the rotor and the maximum of P_{useful} is attained at $I = (\varphi_1 - \varphi_2)/(2R)$.

▶ **Analysis of branched circuits. Kirchhoff's circuit laws.** In order to determine the currents at various segments of a branched circuit, one should assign symbols to the unknown currents and specify the current directions (in a voluntary way). One of the following two methods should be used thereafter:

1. *Node potential method.* For a circuit involving N nodes, the nodal potentials are taken to be the unknowns (one of the potentials is set to zero). The currents are expressed in terms of the potentials using (P3.7.1.9). The law of conservation of charge is written for each of the $N-1$ unknown nodal potentials in the form $\sum_i I_i = 0$ (the algebraic sum of the currents flowing into and out of a node is zero; the summation is performed over all currents involved). Solving the resulting linear system, one finds the unknown potentials φ_i and then calculates the currents.

2. *Kirchhoff's laws.* In this method, the currents are taken to be the unknowns. First, the equation $\sum_i I_i = 0$ (known as *Kirchhoff's current law* or *Kirchhoff's first rule*) is written for $N-1$ nodes. Then, one chooses an arbitrary closed path and traces it (either clockwise or counterclockwise) summing up the equations of the form (P3.7.1.9) for each segment. This results in an equation of the form $\sum_i I_i R_i = \sum_{i,j} \mathscr{E}_{ij}$ (known as *Kirchhoff's voltage law* or *Kirchhoff's second rule*). This procedure is repeated for all independent closed paths to obtain the required number of equations. Finally, the resulting linear system of equations is solved for the unknown currents.

For more details and examples, see Chapter E6.

▶ **Equivalent resistances and sources.** Several resistances R_i connected with one another can be replaced by a single equivalent resistance R.

1) For resistances connected in series, $R = \sum_i R_i$.
2) For resistances connected in parallel, $R^{-1} = \sum_i R_i^{-1}$.
3) In the general case, the equivalent resistance is calculated by analyzing branched circuits.

Several connected current sources with emf $\mathscr{E}_i$ and internal resistance r_i can be replaced by a single equivalent sources with parameters $\mathscr{E}$ and r.

1) For sources connected in series, $\mathscr{E} = \sum_i \mathscr{E}_i$ and $r = \sum_i r_i$.
2) For sources connected in parallel, $\mathscr{E} = \sum_i \mathscr{E}_i r_i^{-1} / \sum_i r_i^{-1}$ and $r^{-1} = \sum_i r_i^{-1}$.
3) In the general case, the equivalent emf and resistance are calculated by analyzing branched circuits.

P3.7.3. Classical Free-Electron Theory of Metals

▶ **Basic postulates.** Conduction of metals is explained by the behavior of valence electrons. The free valence electrons are assumed to form an electron gas. In the classical free-electron theory of Drude–Lorentz, the electron gas is considered to be a classical gas whose concentration equals the atomic concentration of the metal and temperature equals the metal temperature. Conduction current is calculated by assuming that the electrons acquire the drift velocity as they freely move without collisions and that they lose this velocity completely at each collision with metal ions. In addition, it is assumed that $|\langle \mathbf{v} \rangle| \ll \langle v \rangle \sim \sqrt{kT}$. The average drift velocity is expressed as $\langle \mathbf{v} \rangle \sim a\tau/2 = e\mathbf{E}\tau/(2m_e)$, where the mean time between collisions τ is expressed in terms of the mean free path and the mean thermal speed: $\tau = \lambda/\langle v \rangle$. Using (P3.7.1.1), we get

$$\mathbf{j} = en\langle \mathbf{v} \rangle = \frac{e^2 n\lambda}{2m_e \langle v \rangle}\mathbf{E} \quad \text{or} \quad \rho = \frac{2m_e \langle v \rangle}{e^2 n\lambda}.$$

A thorough analysis leads to the same result except that the 2 is absent from the numerator; however, this is not essential for estimates.

▶ **Discussion of the result of the classical free-electron theory.**

1. The theory explains the increase of ρ with temperature but predicts the behavior $\rho \sim \sqrt{T}$, which is different from the observed relationship $\rho \sim T$.

2. The theory successfully explains the *Wiedemann–Franz law*: for all metals, the product of the thermal conductivity κ (see Subsection P2.9.2) by the electric resistivity ρ is proportional to temperature T (the product $\kappa\rho$ contains $\langle v \rangle^2$ only).

3. The mean free path of electrons predicted by the classical theory is several orders of magnitude larger than the interatomic distance, which contradicts the basic postulates.

4. The electron contribution to the molar heat capacity (see Subsection P2.5.2) is predicted to be $\frac{3}{2}R$. However, the experimentally found contribution is negligibly small.

P3.8. Constant Magnetic Field

P3.8.1. Lorentz Force and Ampère Force

▶ **Magnetic field. Lorentz force.** Magnetic field is one of the two components of the electromagnetic field (see Subsection P3.2.1). It acts on moving charged particles, currents, and magnetic moments. The sources of magnetic field are moving charged particles, currents, magnetic moments, and changing electric field. A magnetic field is characterized by a vector quantity **B** called the *magnetic induction* (also called the *magnetic field* or *magnetic flux density*). The vector **B** is determined by the equation expressing the force acting on a moving charged particle in an electromagnetic field:

$$\mathbf{F} = q\mathbf{E} + q\mathbf{v} \times \mathbf{B}. \tag{P3.8.1.1}$$

The force $\mathbf{F}_L = q\mathbf{v} \times \mathbf{B}$ exerted by the magnetic field on a moving charged particle is called the *Lorentz force*. (The electromagnetic force **F** is sometimes also called the Lorentz force or *generalized Lorentz force*.) The Lorentz force perpendicular to the particle velocity and, hence, does not produce work.

In order to determine the magnetic induction **B** by formula (P3.8.1.1), one should do the following:

1) measure the force acting on a static particle to isolate the action of the electric field;

2) find the direction of the velocity **v** for which the magnitude of the magnetic force is maximum at constant v;

3) calculate the magnitude of the magnetic induction as $B = F_{max}/(qv)$;

4) for known $\mathbf{F}_{max}$ and **v**, find the direction of **B** by the right-hand rule.

The magnetic induction can also be determined by the moment of force the magnetic field acts on a small loop of electric current. The electric field does not affect the electric current loop.

The SI unit of magnetic induction is the *tesla* (T): $1\,\text{T} = 1\,\text{N}/(\text{A}\,\text{m}) = 1\,\text{kg}/(\text{A}\,\text{s}^2)$.

Example 1. The Hall effect. A conductor with a current flowing through it is immersed in a transverse constant magnetic field (perpendicular to the current). An electrostatic field perpendicular to the magnetic induction arises within the conductor as shown in Fig. P3.14 (the Hall effect). This field results from displacement of the drifting free charges by the action of the Lorentz force. The displacement continues until the Lorentz force is equalized by the electric force: $q_0\langle\mathbf{v}\rangle \times \mathbf{B} = -q_0\mathbf{E}$. Multiplying by the concentration of free charges, n, we express **E** in terms of the current density to obtain $\mathbf{E} = -\frac{1}{q_0 n}\mathbf{j} \times \mathbf{B}$. It can be seen that the free charge concentration (for charges of the same type) and the sign of the charges can be obtained from the measurement of the magnitude and direction of the electric field **E**.

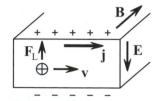

Figure P3.14. A conductor with a current immersed in a magnetic field (the Hall effect).

Example 2. Particle in a uniform magnetic field. Consider a nonrelativistic particle of mass m and charge q moving with a velocity $\mathbf{v}$ in a uniform magnetic field $\mathbf{B}$. Suppose that $\mathbf{v}$ makes an angle α with $\mathbf{B}$ at some instant. The Lorentz force is perpendicular to both $\mathbf{v}$ (hence, $v = \text{const}$) and $\mathbf{B}$ (hence, the velocity projections $v_\parallel$ and $v_\perp$ onto $\mathbf{B}$ and the plane perpendicular to $\mathbf{B}$ are conserved). In the projection onto the perpendicular plane, the particle moves in a circle, whose radius can be found from Newton's second law: $qv_\perp B = mv_\perp^2/R$. The period of revolutions, $T = 2\pi R/v_\perp = 2\pi m/qB$, is independent of the speed. So the resulting motion occurs in a helical path with radius R and pitch $h = v_\parallel T = 2\pi mv\cos\alpha/(qB)$.

▶ **Ampère's law.** The force acting on a current element in a magnetic field (*Ampère's force*) equals the sum of the forces acting on the moving free charges:

$$d\mathbf{F} = q_0[\langle \mathbf{v}\rangle \times \mathbf{B}]\,dN = nq_0[\langle \mathbf{v}\rangle \times \mathbf{B}]\,dV = \mathbf{j}\times\mathbf{B}\,dV = I\,d\mathbf{l}\times\mathbf{B}. \tag{P3.8.1.2}$$

This result is known as *Ampère's force law* (in differential form). Note the use of formula (P3.7.1.1), linking the current density to the mean velocity of free charges, to relate the force acting on a volume element dV with current density $\mathbf{j}$ and a line element $d\mathbf{l}$ with current I. The total force acting of a finite volume with distributed current $\mathbf{j(r)}$ of a linear current-carrying wire is obtained by integrating. For example, the force acting on a wire of length $\mathbf{l}$ with current I in a uniform magnetic field $\mathbf{B}$ equals $I\mathbf{l}\times\mathbf{B}$. The force acting on a current-carrying loop is zero: $\mathbf{F} = \oint I\,d\mathbf{l}\times\mathbf{B} = I\left(\oint d\mathbf{l}\right)\times\mathbf{B} = \mathbf{0}$.

▶ **Magnetic moment of a current-carrying loop.** The magnetic moment of a wire loop carrying current I is defined as the vector quantity

$$\mathbf{p}_{\mathrm{m}} = I\int \mathbf{n}\,ds, \tag{P3.8.1.3}$$

where the integration is carried out over any surface spanning the loop and the direction of the normal is determined by the right-hand rule. For a plane loop, we have

$$\mathbf{p}_{\mathrm{m}} = IS\mathbf{n} = I\mathbf{S}.$$

The magnetic moment of a loop is a magnetic analogue of the electric dipole moment (see Subsections P3.1.1 and P3.2.2); it determines the magnetic field of the loop at large distances from it and characterizes the behavior of the loop in a magnetic field.

Example 3. Consider a rectangular loop with sides a and b, suspended at the side a is a vertical uniform magnetic field $\mathbf{B}$ (Fig. P3.15). When current I is supplied, the loop will deflect by an angle β such that the moment of the force of gravity is equalized by the moment of the Ampère force: $mg\frac{1}{2}b\sin\beta = IBab\cos\beta$, or $\tan\beta = 2IBa/(mg)$. Note that the torque exerted by the magnetic field on the loop equals $\mathbf{M} = I\mathbf{S}\times\mathbf{B} = \mathbf{p}_{\mathrm{m}}\times\mathbf{B}$. A similar expression was obtained for the torque acting on an electric dipole in an electric field (see Subsection P3.2.2).

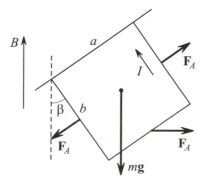

Figure P3.15. A rectangular current-carrying loop in a magnetic electric (to example 3).

P3.8.2. Calculation of Magnetic Induction

▶ **Biot–Savart–Laplace law.** The magnetic induction created by a current element $I\,d\mathbf{l}$ at a point A equals

$$d\mathbf{B} = \frac{\mu_0}{4\pi}\frac{I\,d\mathbf{l}\times\mathbf{r}}{r^3}, \tag{P3.8.2.1}$$

where $\mathbf{r}$ is the radius vector from the current element to the point A and $\mu_0 = 1/(\varepsilon_0 c^2) = 4\pi\times10^{-7}$ T m/A (or H/m) is the magnetic constant. This result is known as the *Biot–Savart law* or the *Biot–Savart–Laplace law*. Equation (P3.8.2.1) can be rewritten for the magnetic induction of a volume current element $\mathbf{j}\,dV$, a surface current element $\mathbf{i}\,ds$, or, on dividing by the number of free current carriers, a nonrelativistic particle $q\mathbf{v}$:

$$d\mathbf{B} = \frac{\mu_0}{4\pi}\frac{\mathbf{j}\times\mathbf{r}}{r^3}dV,\quad d\mathbf{B} = \frac{\mu_0}{4\pi}\frac{\mathbf{i}\times\mathbf{r}}{r^3}ds,\quad d\mathbf{B} = \frac{\mu_0}{4\pi}\frac{q\mathbf{v}\times\mathbf{r}}{r^3}. \tag{P3.8.2.2}$$

(The surface density of electric current is the current per unit length of the segment perpendicular to the current: $dI = i\,dl$.) In order to calculate the magnetic induction produced by a finite volume, surface, or wire, one has to perform integration (principle of superposition for magnetic induction).

▶ **On systems of units in electromagnetism.** Here and henceforth the formulas are given in SI. To convert units to the Gaussian system (CGS), it suffices in most cases to substitute $\mathbf{B}$ by $\mathbf{B}/c$ (and the flux Φ by Φ/c) and μ_0 by $4\pi/c^2$. If a formula involves the Coulomb constant k, it must be substituted by 1. The electric constant ε_0 is substituted by $1/(4\pi)$. It is noteworthy that $\mu_0\varepsilon_0 = 1/c^2$.

Example 1. Find the magnetic force acting between two particle of charges q_1 and q_2 having identical velocities, $\mathbf{v}_1 = \mathbf{v}_2 = \mathbf{v}$ ($v \ll c$), at some instant of time with $\mathbf{v}$ being perpendicular to the line connecting the charges.

Solution. The magnitude of the magnetic induction created by the first particle at the point where the second particle is located is $B = \frac{\mu_0}{4\pi}\frac{q_1 v}{r^2}$, where r is the interparticle distance. The field is perpendicular to $\mathbf{v}$ and $\mathbf{r}$. For charges of like sign, the Lorentz force acts in the direction of particle 1 and its magnitude equals $F = \frac{\mu_0}{4\pi}\frac{q_1 q_2 v^2}{r^2}$. The ratio of the magnetic force to the electrostatic force is $\mu_0\varepsilon_0 v^2 = \frac{v^2}{c^2}$.

Example 2. Magnetic field in the center of a circular current. Find the magnetic induction created by a circular loop of radius R at its center. The current flowing through the loop is I.

Solution. The magnetic induction created by an arc element $d\varphi$ equals

$$dB = \frac{\mu_0}{4\pi}\frac{IR\,d\varphi}{R^2} = \frac{\mu_0 I\,d\varphi}{4\pi R}$$

in magnitude and is directed along the axis (the right-hand rule). At its center, the whole loop creates the field

$$B = \frac{\mu_0 I}{2R}.$$

Example 3. Magnetic field at the axis of a circular current. Find the magnetic induction produced by a circular current (of radius R) on its axis at the point A located at a distance y from the loop center (Fig. P3.16).

Solution. The vector $\mathbf{B}$ is directed along the y-axis and the contribution of an arc element $d\varphi$ of the loop equals

$$dB_y = dB\sin\alpha = \frac{\mu_0}{4\pi}\frac{IR\,d\varphi}{R^2 + y^2}\frac{R}{\sqrt{R^2 + y^2}}.$$

Integrating with respect to φ yields

$$B = \frac{\mu_0 I R^2}{2(R^2 + y^2)^{3/2}}.$$

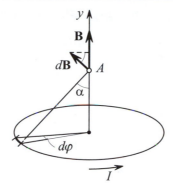

Figure P3.16. Magnetic field at the axis of a circular current.

At large distances from the loop ($y \gg R$), the magnetic induction

$$B \approx \frac{\mu_0 I R^2}{2y^3} = \frac{\mu_0}{4\pi} \frac{2I(\pi R^2)}{y^3}$$

is expressed in terms of the magnetic moment $\mathbf{p}_m = I\mathbf{S}$ in a similar manner as the electric field strength is expressed via the electric dipole moment (see example 4, Subsection P3.2.2, the case $\mathbf{r} \parallel \mathbf{p}$). I can be shown that at an arbitrary point $\mathbf{r}$, the magnetic induction of a small plane current-carrying loop equals

$$\mathbf{B} = \frac{\mu_0}{4\pi r^3} \left(\frac{3\mathbf{p}_m \cdot \mathbf{r}}{r^2} \mathbf{r} - \mathbf{p}_m \right). \tag{P3.8.2.3}$$

Example 4. Magnetic field of a linear current segment. Find the magnetic induction produced by a linear wire segment at the point A whose location is determined by its distance y to the straight line of the segment and two angles α_1 and α_2 as shown in Fig. P3.17.

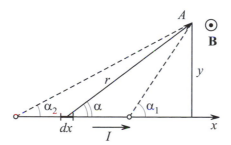

Figure P3.17. Magnetic field of a linear current segment.

Solution. The contribution of the element dx equals $dB = \frac{\mu_0}{4\pi} \frac{I\,dx}{r^2} \sin\alpha$; the field is perpendicular to the plane through the segment and the point A. It is convenient to integrate with respect to the angle α. Substituting $r = y/\sin\alpha$, changing the variable $x = -y \cot\alpha$, $dx = y\,d\alpha/\sin^2\alpha$, and integrating, we obtain

$$B = \frac{\mu_0 I}{4\pi y} (\cos\alpha_1 - \cos\alpha_2). \tag{P3.8.2.4}$$

Setting $\alpha_1 = 0$ and $\alpha_2 = \pi$, we find the magnetic induction of an infinitely long straight current-carrying wire:

$$B = \frac{\mu_0 I}{2\pi y}. \tag{P3.8.2.5}$$

Let us use (P3.8.2.4) to calculate the magnetic induction at the center of a rectangular loop to obtain

$$B = \frac{\mu_0}{4\pi} \frac{8I\sqrt{a^2 + b^2}}{ab},$$

where a and b are the sides of the rectangle.

P3.8.3. Ampère's Circuital Law and Flux of Magnetic Induction

▶ **Ampère's circuital law.** From the Biot–Savart–Laplace law it can be inferred that the *circulation of magnetic induction* around any closed path is determined by the algebraic sum of the currents enclosed by the path (i.e., by the current passing through a surface spanning the path):

$$\oint \mathbf{B} \cdot d\mathbf{l} = \mu_0 I_{\text{enc}} = \mu_0 \int \mathbf{j} \cdot d\mathbf{s}. \tag{P3.8.3.1}$$

The positive direction of the surface normal is determined by the right-hand rule.

▶ **Gauss's law for magnetism.** The *flux of magnetic induction* (*magnetic flux*) through any closed surface is zero:

$$\oint \mathbf{B} \cdot d\mathbf{s} = 0. \tag{P3.8.3.2}$$

This means the magnetic field lines are everywhere continuous, or there are no *magnetic charges*, at which magnetic field lines would start or terminate.

The magnetic flux through any surface S is defined as

$$\Phi = \int_S \mathbf{B} \cdot d\mathbf{s}.$$

The SI unit of magnetic flux is the *weber* (Wb), with $1 \, \text{Wb} = 1 \, \text{T m}^2$. It is noteworthy that magnetic flux through a surface spanning a closed path does not depend on the shape of the surface. Therefore, it is meaningful to speak about magnetic flux through a closed path.

Example 1. Magnetic field of a solenoid. A solenoid is a helical coil formed by a long thin wire densely wound on a cylinder. A solenoid is characterized by the number of turns per unit length, n. The ideal model of a solenoid is an infinitely long thin cylindrical surface carrying current $i = nI$, with the current perpendicular to the axis. It is not difficult to verify that the magnetic field inside an ideal solenoid is everywhere parallel to the axis (to this end, one should consider the contributions of any two symmetric current elements). There is no magnetic field outside the solenoid. It follows from Ampère's circuital law and Gauss's law that $\mathbf{B} = \text{const}$ inside the solenoid. In order to find the magnitude B, one should apply Ampère's circuital law to a rectangular loop one side of which is inside the solenoid and parallel to its axis and the opposite side is outside as shown in Fig. P3.18. One finds that $Bl = \mu_0 il$. So the magnetic field inside the solenoid is

$$B = \mu_0 i = \mu_0 n I. \tag{P3.8.3.3}$$

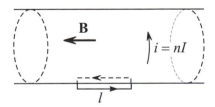

Figure P3.18. The magnetic field inside a solenoid can be found using Ampère's circuital law.

Example 2. Magnetic field of a rod. Find the magnetic field of an infinitely long straight rod of radius R carrying current I (Fig. P3.19). The magnetic lines are circles. Choose a circle of radius r as the closed path for Ampère's circuital law (P3.8.3.1). For $r > R$, we get $B \times 2\pi r = \mu_0 I$. For $r < R$, we find that $B \times 2\pi r = \mu_0 I(\pi r^2/\pi R^2)$. Hence,

$$B = \begin{cases} \dfrac{\mu_0 I}{2\pi r} & \text{for } r > R, \\[2mm] \dfrac{\mu_0 I r}{2\pi R^2} & \text{for } r < R. \end{cases}$$

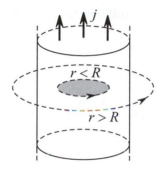
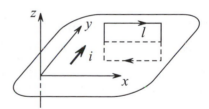

Figure P3.19. Magnetic field of a rod.   **Figure P3.20.** Magnetic field of a current plane.

Example 3. Magnetic field of a current plane. Consider an infinite plane (x, y) carrying current in the y-direction with a constant density i. The magnetic field above the plane, $z > 0$, is directed in the positive x-direction and that below the plane, $z < 0$, is in the negative x-direction (this can be verified by considering the contributions of two infinite straight currents symmetric relative to the point of interest). Choose a rectangular path with one side above the plane and in the x-direction and the other being symmetrically located below the plane (Fig. P3.20). Ampère's circuital law gives $2Bl = \mu_0 il$, and hence $B = \frac{1}{2}\mu_0 i$.

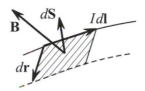

Figure P3.21. Displacement of a current loop element in a magnetic field.

▶ **Displacement of a current loop in a magnetic field.** When a loop carrying constant current I is moved or deformed in a constant magnetic field, the Ampère force performs work. The elementary work done by the Ampère force acting on a loop element equals

$$\delta A = d\mathbf{F} \cdot d\mathbf{r} = I[d\mathbf{l} \times \mathbf{B}] \cdot d\mathbf{r} = I\mathbf{B} \cdot [d\mathbf{r} \times d\mathbf{l}] = I\mathbf{B} \cdot d\mathbf{s},$$

where $d\mathbf{r}$ is the displacement of the loop element $d\mathbf{l}$, $d\mathbf{s}$ is the area element swept out by $d\mathbf{l}$ (Fig. P3.21). Integrating around the loop yields the elementary mechanical work done by the magnetic field in terms of the increment in the magnetic flux Φ through the loop:

$$\delta A_{\text{mech}} = I\, d\Phi. \tag{P3.8.3.4}$$

It follows that a *potential function*, $W_{\text{p}} = -I\Phi$, can be associated with a loop carrying current I in a constant magnetic field. A change in this function equals (with the opposite sign) the work done by the Ampère force:

$$A_{\text{mech}} = -\Delta W_{\text{p}}, \qquad W_{\text{p}} = -I\Phi. \tag{P3.8.3.5}$$

For a uniform field or a small plane loop, formula (P3.8.3.5) becomes

$$W_{\text{p}} = -\mathbf{p}_{\text{m}} \cdot \mathbf{B}, \tag{P3.8.3.6}$$

where $\mathbf{p}_{\text{m}}$ is the magnetic moment of the current loop (see Subsection P3.8.1). Since this expression coincides with the potential energy of an electric dipole (see Subsection P3.2.2), it is natural to assume that the torque acting on a magnetic dipole and the force acting on a magnetic dipole in a nonuniform field should have the same form:

$$\mathbf{M} = \mathbf{p}_{\text{m}} \times \mathbf{B}, \qquad \mathbf{F} = p_{\text{m}}\frac{\partial \mathbf{B}}{\partial n} = p_{\text{m}x}\frac{\partial \mathbf{B}}{\partial x} + p_{\text{m}y}\frac{\partial \mathbf{B}}{\partial y} + p_{\text{m}z}\frac{\partial \mathbf{B}}{\partial z}, \tag{P3.8.3.7}$$

where $\mathbf{n}$ is parallel $\mathbf{p}_{\text{m}}$. The equilibrium orientation of a magnetic dipole corresponds to the minimum potential energy, and hence it corresponds to $\mathbf{p}_{\text{m}} \parallel \mathbf{B}$. A current loop oriented in this manner is pulled into the region of a stronger magnetic field.

▶ **Maxwell's equations for a constant magnetic field.** In differential form, these equations are

$$\operatorname{curl} \mathbf{B} = \mu_0 \mathbf{j}, \qquad \operatorname{div} \mathbf{B} = 0.$$

In integral form, these are equations (P3.8.3.1) and (P3.8.3.2).

▶ **Vector potential.** Maxwell's equation div $\mathbf{B} = 0$ is satisfied if the magnetic field is sought in the form $\mathbf{B} = \operatorname{curl} \mathbf{A}$. The *vector potential* $\mathbf{A}$ is defined up to an arbitrary constant scalar field χ. The potential $\mathbf{A}' = \mathbf{A} + \operatorname{grad} \chi$ generates the same field $\mathbf{B}$ as $\mathbf{A}$; this expression is known as the *calibration transformation for a vector potential*. This transformation allows us to simplify the vector potential equation resulting from substituting $\mathbf{B} = \operatorname{curl} \mathbf{A}$ into the first Maxwell equation: $\operatorname{grad} \operatorname{div} \mathbf{A} - \Delta \mathbf{A} = \mu_0 \mathbf{j}$. By choosing a scalar field χ such that div $\mathbf{A} = 0$, we arrive at the equation

$$\Delta \mathbf{A} = -\mu_0 \mathbf{j}.$$

The equation for each projection of $\mathbf{A}$ onto the coordinate axes is mathematically identical with the equation for the electric scalar potential (see Subsection P3.3.2). Consequently, the solution of these equations is given by

$$\mathbf{A}(\mathbf{r}) = \frac{\mu_0}{4\pi} \int \frac{\mathbf{j}(\mathbf{r}')\, dV'}{|\mathbf{r} - \mathbf{r}'|}.$$

In particular, one can find the field of a small current loop (see (P3.8.2.3)) or prove the Biot–Savart law using this expression.

P3.9. Magnetic Field in Matter

P3.9.1. Magnetization. Molecular Currents

▶ **Magnetization of a material.** A material is regarded as magnetized if it creates a magnetic field inside itself and in the surrounding space in the absence of conduction currents. The source of magnetic field of a magnetized material are magnetic moments of atoms and molecules (atomic and molecular currents). In the absence of an external magnetic field, microscopic magnetic moments are either absent (in *diamagnetic* materials) or chaotically oriented (in *paramagnetic* materials). In both cases, the macroscopic magnetic field is zero. Constant magnets (made from *ferromagnetic* substances) are an exception; they can have a macroscopic magnetic field (see below for details). When a material is immersed in an external magnetic field, the material is magnetized and an induced volumetric magnetic moment arises. The degree to which a material is magnetized is characterized by the vector of *magnetization* $\mathbf{J}$, which is defined as the net magnetic moment per unit volume:

$$\mathbf{J} = \frac{\sum_i \mathbf{p}_{\mathrm{m}i}}{\Delta V} = n \langle \mathbf{p}_{\mathrm{m}} \rangle, \qquad\qquad (\text{P3.9.1.1})$$

where n is the molecule concentration. For the construction of macroscopic field equations for a magnetized material, it can be assumed for simplicity that all molecules in a small volume ΔV have the same magnetic moment $\langle \mathbf{p}_{\mathrm{m}} \rangle = \mathbf{J}/n = I_0 \mathbf{s}$, or they can be represented as identical microscopic current loops.

▶ **Macroscopic molecular currents.** In characterizing magnetic field in matter, the huge number of oriented microscopic currents (magnetic moments) may be replaced by macroscopic molecular currents distributed across surface and volume. For a simple example, consider a homogeneous magnetized cylinder whose molecular current loops are all oriented along the cylinder axis. The different currents will compensate for one another inside the cylinder and there will be uncompensated current flowing along the surface. The magnetic

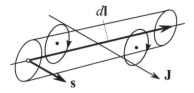

Figure P3.22. Magnetization creates a molecular current piercing the loop.

field of the surface current, $\mathbf{B}_{mol}$, will have, inside the cylinder, the *same direction* as the magnetization $\mathbf{J}$.

For an arbitrary closed path, its element $d\mathbf{l}$ inside the magnetized material pierces the current loops whose centers fall within the volume $dV = \mathbf{s} \cdot d\mathbf{l}$ (Fig. P3.22). The contribution into the molecular current enclosed by the path (i.e., the current crossing the surface that spans the path) equals $I_0 n\, dV = nI_0 \mathbf{s} \cdot d\mathbf{l} = \mathbf{J} \cdot d\mathbf{l}$. Integrating gives

$$\oint \mathbf{J} \cdot d\mathbf{l} = I_{enc}^{mol}. \tag{P3.9.1.2}$$

In differential form, this equation becomes curl $\mathbf{J} = \mathbf{j}_{mol}$. These two equations allow us to express the current density at the surface as

$$\mathbf{i}_{mol} = \mathbf{J} \times \mathbf{n}, \tag{P3.9.1.3}$$

where $\mathbf{n}$ is the outward normal to the surface of the magnetized material.

▶ **Magnetic field strength.** Macroscopic magnetic induction in a magnetized material results from averaging the microscopic field $\mathbf{B}$, highly varying at interatomic distances, over a small volume, though large enough to contain a large number of molecules: $\mathbf{B}(\mathbf{r}) = \langle \mathbf{B}_{micro} \rangle$. The magnetic induction is determined by both molecular and nonmolecular currents (such as conduction currents or convection currents):

$$\oint \mathbf{B} \cdot d\mathbf{l} = \mu_0 (I_{enc}^{nonmol} + I_{enc}^{mol}). \tag{P3.9.1.4}$$

Since the molecular current distribution is usually unknown, it is convenient to introduce a new vector quantity called the *magnetic field strength* and defined as

$$\mathbf{H} = \frac{\mathbf{B}}{\mu_0} - \mathbf{J}, \tag{P3.9.1.5}$$

which is only defined by nonmolecular currents:

$$\oint \mathbf{H} \cdot d\mathbf{l} = I_{enc}^{nonmol} \tag{P3.9.1.6}$$

(Ampère's circuital law for the magnetic field strength). In differential form, this expression becomes: curl $\mathbf{H} = \mathbf{j}^{nonmol}$. The SI unit for magnetic field strength is A/m.

▶ **Boundary conditions at the interface between different magnetic materials.** For the normal and tangential components of $\mathbf{B}$, $\mathbf{H}$, and $\mathbf{J}$, the following boundary conditions hold at the interface between two magnetic media:

$$B_{2n} = B_{1n}, \quad \mathbf{H}_1 \times \mathbf{n} - \mathbf{H}_2 \times \mathbf{n} = \mathbf{i}_{nonmol}, \quad \mathbf{J}_1 \times \mathbf{n} - \mathbf{J}_2 \times \mathbf{n} = \mathbf{i}_{mol}, \tag{P3.9.1.7}$$

where $\mathbf{n}$ is a normal vector pointing from medium 1 to medium 2 and $|\mathbf{H} \times \mathbf{n}| = H_\tau$. If there is no nonmolecular current at the interface, the second condition can be rewritten as $H_{1\tau} = H_{2\tau}$. The first condition follows from Gauss's law (P3.8.3.2) for $\mathbf{B}$, the second follows from (P3.9.1.6), and the third from (P3.9.1.3). There is a formal analogy in boundary conditions between $\mathbf{H}$ and $\mathbf{E}$ and also between $\mathbf{B}$ and $\mathbf{D}$ (see Subsection P3.5.1).

▶ **Isotropic magnetic material.** The magnetization $\mathbf{J}$ at a given point inside a magnetic material arises due to an external magnetic field and is determined by the vector $\mathbf{B}$. For not too strong fields, $\mathbf{J}$ depends on $\mathbf{B}$ linearly and in an *isotropic material*, in addition, $\mathbf{J} \parallel \mathbf{B}$. Then both vectors $\mathbf{J}$ and $\mathbf{H}$ are proportional to $\mathbf{B}$ and to each other. For historical reasons (due to the analogy between $\mathbf{E}$ and $\mathbf{H}$ in boundary conditions), the *magnetic susceptibility* is defined by

$$\mathbf{J} = \chi\mathbf{H}, \qquad \mathbf{J} = \frac{\chi}{\mu_0(1 + \chi)}\mathbf{B}. \tag{P3.9.1.8}$$

From (P3.9.1.5) we get

$$\mathbf{B} = \mu_0(1 + \chi)\mathbf{H} = \mu_0\mu\mathbf{H}, \tag{P3.9.1.9}$$

where μ is the *magnetic permeability* of the material. For paramagnetic materials, $\chi > 0$. For diamagnetic materials, $\chi < 0$. In both cases, $|\chi| \ll 1$ (weak magnetic properties). For ferromagnetic materials, $\chi \gg 1$.

The boundary conditions at the interface between two isotropic magnetic media are

$$B_{1n} = B_{2n}, \quad \frac{B_{1\tau}}{\mu_1} = \frac{B_{2\tau}}{\mu_2}; \qquad \mu_1 H_{1n} = \mu_2 H_{2n}, \quad H_{1\tau} = H_{2\tau}.$$

▶ **Calculation of fields in the presence of a magnetic material.** The equations for $\mathbf{B}$ and $\mathbf{H}$ together with the above boundary conditions determine magnetic field in the whole space if the nonmolecular currents are known. Below are some important properties of solutions:

1. The volume density of the molecular current is proportional to that of the nonmolecular current: $\mathbf{j}^{\text{mol}} = \chi\mathbf{j}^{\text{nonmol}}$. If there are no nonmolecular currents, then surface molecular currents only arise when the material is magnetized.

2. If the boundaries of a homogeneous magnetic material are everywhere tangent to the original ($\mathbf{B}_0$) magnetic lines, then $\mathbf{B} = \mu\mathbf{B}_0$ everywhere in the magnetic material.

Example 1. A long paramagnetic cylinder is immersed in a uniform magnetic field whose induction $\mathbf{B}_0$ is parallel to the cylinder axis. Find the magnetic induction, magnetic field strength, and magnetization inside the cylinder.

Solution. Everywhere inside the cylinder (except for its edges) we have: $\mathbf{B} = \mu\mathbf{B}_0$, $\mathbf{H} = \mathbf{H}_0 = \mathbf{B}_0/\mu_0$, and $\mathbf{J} = (\mu - 1)\mathbf{B}_0/\mu_0$.

Example 2. An infinitely long rod of circular cross-section, made from a material with magnetic permeability $\mu > 1$, carries a constant current with density $\mathbf{j}$. Find the magnetic field strength, magnetic induction, and magnetization. Also calculate the surface and volume density of the molecular current.

Solution. The molecular current within the rod volume has the density $\mathbf{j}^{\text{mol}} = (\mu-1)\mathbf{j}$. The field strength H can be found using Ampère's circuital law (P3.9.1.6). We get $H = I/(2\pi r)$ for $r > R$ and $H = Ir/(2\pi R^2)$ for $r < R$ (see example 2, Subsection P3.8.3). The magnetic induction equals $B = \mu_0 H$ for $r > R$ and $B = \mu_0\mu H$ for $r < R$; it has a jump discontinuity at the surface current. The magnetization equals $J = (\mu - 1)Ir/(2\pi R^2)$ ($J = 0$ outside the rod). The magnetic lines are concentric circles and the vectors $\mathbf{H}$ and $\mathbf{J}$ are parallel to $\mathbf{B}$. The surface molecular current is $\mathbf{i} = \mathbf{J} \times \mathbf{n}$; it equals $i_{\text{mol}} = (\mu - 1)I/(2\pi R)$ in magnitude and is opposite to $\mathbf{j}$ in direction. The total molecular current is zero, and hence $j_{\text{mol}}\pi R^2 = i_{\text{mol}}2\pi R$.

P3.9.2. Types of Magnetic Materials and Magnetization Mechanisms

Atomic and molecular magnetic moments are created by (a) the motion of electrons in closed orbits (orbital magnetic moments) and (b) the intrinsic magnetic moments of electrons related to the intrinsic mechanical angular momentum—spin (spin magnetic moments). Each type of magnetic moment is proportional to the corresponding mechanical angular momentum: $\mathbf{p}_{\text{m}} = -g(e/2m)\mathbf{L}$, where g is a dimensionless coefficient called the *gyromagnetic ratio*. For orbital moment, $g = 1$. For spin moment, $g = 2$.

▶ **Diamagnets. Larmor's mechanism of diamagnetic magnetization.** In the absence of external magnetic fields, the magnetic moments of atomic (molecular) electrons in a diamagnetic material (diamagnet) compensate for one another. When a magnetic field is applied, a torque starts to act on each magnetic moment. As a result, the magnetic moment rotates about the magnetic field direction (*Larmor precession*):

$$\frac{d\mathbf{L}}{dt} = \mathbf{p}_m \times \mathbf{B} = -g\frac{e}{2m}\mathbf{L}\times\mathbf{B} = \left(g\frac{e}{2m}\mathbf{B}\right)\times\mathbf{L}.$$

The vector $\mathbf{L}$ (and also $\mathbf{p}_m$) rotates with an angular velocity $\boldsymbol{\omega}_L = g\frac{e}{2m}\mathbf{B}$ (Fig. P3.23). This rotation gives rise to an additional magnetic moment of each electron, $\Delta\mathbf{p}_m$, opposing to $\mathbf{B}$ (independently of the original direction of $\mathbf{p}_m$).

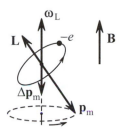

Figure P3.23. Larmor precession in diamagnetic materials.

▶ **Orientational mechanism of magnetization in paramagnetic materials.** In paramagnets, the magnetic moments of electrons do not compensate for one another even in the absence of external magnetic fields. The weak mechanism of diamagnetic magnetization is superimposed by a stronger orientational mechanism: in the presence of magnetic field, the orientation of the magnetic moment along the field is less costly energetically. In weak fields ($p_m B \ll kT$), the average magnetic moment is proportional to the magnetic induction. In strong fields, saturation is achieved, where all magnetic moments are oriented along the field and the magnetization virtually does not change. The orientational mechanism of paramagnetic magnetization is similar to that of polar dielectric polarization (see Subsection P3.5.2). In particular, the conclusion about inverse proportionality between the susceptibility and temperature remains valid.

▶ **Ferromagnets** are materials that show spontaneous magnetization in macroscopic (compared to interatomic distances) regions called *domains*. The domains are small ($\sim 10^{-5}$ to 10^{-4} m) as compared to the size of the sample under examination. Within each domain, spin magnetic moments are parallel to one another, which results from a special quantum interaction among them. In the absence of external magnetic fields, the domains are oriented in such a way that the average magnetization of the material is zero. When an external magnetic field is applied, the domains oriented along the field increase in size and the domains oriented against the field decrease. For stronger fields, the domains tend to turn as a whole to be oriented along the field. In strong enough fields, all domains are oriented along the field and saturation is achieved, when the magnetization reaches its maximum magnitude, J_{sat}, called the *saturation magnetization*.

The relationship between $\mathbf{J}$ and $\mathbf{H}$ is nonlinear (see Fig. P3.24). This means that the magnetic permeability, determined by the formula $\mathbf{J} = (\mu - 1)\mathbf{H}$, depends on H and attains very large values ($\mu \sim 10^5$–10^6). When the external field is removed, some *residual magnetization*, J_{res}, is observed. The field strength required to completely demagnetize the sample, H_C, is called the *coercive field* or *coercive force*. Ferromagnetic materials possessing a large coercive force are used to make permanent magnets. The dependence $J(H)$ for direct and reverse magnetization to saturation is called a *hysteresis loop*.

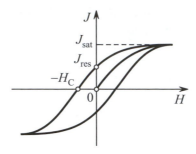

Figure P3.24. Hysteresis loop of a ferromagnetic material.

When heated above the *Curie temperature*, the material loses its ferromagnetic properties (spontaneous magnetization of domains disappears) and becomes a paramagnet. A second-order phase transition occurs at the Curie point.

P3.10. Electromagnetic Induction

P3.10.1. Faraday's Law of Electromagnetic Induction

▶ **Motion of a wire loop in a constant field.** When a current-carrying wire loop is moved or deformed in a constant magnetic field, the Ampère forces are everywhere perpendicular to the current and, according to (P3.8.3.4), perform mechanical work $\delta A_\perp = I\, d\Phi$. Since the total work done by the Lorentz force on any moving particle is zero, the same work (but opposite in sign) is performed by the magnetic field (i.e., the tangential component of the Lorentz force) on free charges moving through the wire. This gives rise to an additional electromotive force (*induced emf*) in the loop:

$$\delta A_\parallel = -I\, d\Phi = -(I\,dt)\frac{d\Phi}{dt} = \mathscr{E}\, dq, \qquad \mathscr{E} = -\frac{d\Phi}{dt}. \tag{P3.10.1.1}$$

▶ **Static loop. Solenoidal electric field.** It follows from Einstein's relativity principle that an induced emf must also arise when a static current-carrying loop is placed in a time varying magnetic field. Consider, for example, a permanent magnet approaching a closed loop at a constant velocity. In the magnet-associated inertial frame of reference, the field produced by the magnet is time independent and the induced emf in the loop is accounted for by the action of the Lorentz force. On the other hand, in the loop-associated inertial frame, the same emf is observed, but it cannot be caused by magnetic forces. Hence, the varying magnetic field must bring about an electric field. Since the work of this field around the loop must be nonzero, it follows that the field is not electrostatic—it is called a *solenoidal* (*eddy*) *electric field*. Thus, an induced emf arises in the loop regardless of the reason for which the magnetic flux through the loop changes. The induced emf equals

$$\mathscr{E} = -\frac{d\Phi}{dt}. \tag{P3.10.1.2}$$

This equation is known as *Faraday's law of electromagnetic induction*. The average induced emf in a finite time Δt equals $\mathscr{E}_{ave} = -\Delta\Phi/\Delta t$. The charge passed through the loop in this time is

$$q = \int I\, dt = \int \frac{\mathscr{E}}{R}\, dt = -\frac{\Delta\Phi}{R},$$

where R is the resistance of the loop.

▶ **Lenz's rule.** The direction of the emf is determined by its sign. It should be reminded (see Subsection P3.8.3) that the loop is *oriented*: the positive direction of the normal to the surface, in the formula for the calculation of Φ, and the positive direction of tracing the path, which determines the sign of the emf, are related by the right-hand rule. The minus sign in (P3.10.1.2) is in accordance with *Lenz's rule*. It states that the induced emf must be directed so that the magnetic field produced by the induced current flowing in the same direction, generates a magnetic flux (induced magnetic flux) of opposite sign as compared to the flux change that caused the emf. Lenz's rule is closely associated with the law of conservation of energy.

▶ **Motion of part of a loop.** When a loop element $d\mathbf{l}$ moves with a velocity $\mathbf{v}$, the work done by the Lorentz force to transfer a test charge q along $d\mathbf{l}$ equals $\delta A = q[\mathbf{v} \times \mathbf{B}] \cdot d\mathbf{l} = q\, d\mathscr{E}$. The electromotive forces induced by an arbitrarily moving portion of the loop in an arbitrary constant field and by a translationally moving portion of the loop in a uniform constant field are, respectively, equal to

$$\mathscr{E} = \int [\mathbf{v} \times \mathbf{B}] \cdot d\mathbf{l}, \qquad \mathscr{E} = [\mathbf{v} \times \mathbf{B}] \cdot \mathbf{l}, \tag{P3.10.1.3}$$

where $\mathbf{l}$ is the vector connecting the beginning and the end of the moving portion. If the wire loop is unclosed, a potential difference will arise between its ends, which is equal to the emf (since the electrostatic force is equalized by the Lorentz force at any point of the wire).

Example 1. A crosspiece of length l and resistance r moves with a speed v along two horizontal rails immersed in a vertical magnetic field (Fig. P3.25). Find the current through the crosspiece.

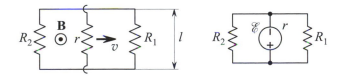

Figure P3.25. To example 1. An equivalent diagram is on the right.

Solution. Suppose that the rails are connected by a resistance R_1 on one side and a resistance R_2 on the other side. In order to find the current through the crosspiece as it moves with the speed v, we use (P3.10.1.3) to calculate the emf induced in the crosspiece: $\mathscr{E} = Bvl$. The fact that the emf is localized within the moving crosspiece enables us to draw an equivalent diagram, shown on the right, and eventually find the current: $I = \mathscr{E}/(r + R)$, where $R = R_1 R_2/(R_1 + R_2)$.

▶ **Equations of solenoidal electric field.** By writing down Faraday's law (P3.10.1.2) for a static loop in a varying magnetic field, we obtain a solenoidal electric field equation in integral form:

$$\oint \mathbf{E} \cdot d\mathbf{l} = -\frac{d}{dt} \int \mathbf{B} \cdot d\mathbf{s}. \tag{P3.10.1.4}$$

Since the circulation of the electrostatic field strength around a closed path is zero, the total electric field satisfies the same equation. The differential form of this equation is

$$\operatorname{curl} \mathbf{E} = -\frac{\partial \mathbf{B}}{\partial t}. \tag{P3.10.1.5}$$

Equations (P3.10.1.4) and (P3.10.1.5) represent one of Maxwell's equation (in integral and differential forms).

▶ **Quasistationary approximation.** If the currents and fields change sufficiently slowly, time delay of processes can be neglected. This means that the varying magnetic field is determined at any point (using the Biot–Savart–Laplace formulas) by the currents at the same instant of time. This also means that the electrostatic field is determined by the charge distribution at the same instant and the solenoidal electric field is calculated from the varying magnetic field using (P3.10.1.4) and (P3.10.1.5). It is this approximation that is used for the analysis of alternating current circuits. For this approximation to be valid it is required that the characteristic current variation time, τ, is large compared to (a) the local relaxation times (e.g., the time in which an equilibrium charge distribution is established in the conductor) and (b) the delay time $\tau_d = l/c$, where l is the characteristic size of the system and c is the speed of light. The latter condition suggests that we neglect the displacement current in Maxwell's equations (i.e., we assume that curl $\mathbf{B} = \mu_0 \mathbf{j}$).

Example 2. Let us estimate the time τ of charge redistribution in a conductor with resistivity ρ. We assume that there is a local volume charge somewhere in the conductor. The charge redistribution within an arbitrary volume is governed by the equation $\dfrac{dq}{dt} = -\oint j_n \, ds = -\dfrac{1}{\rho} \oint E_n \, ds = -\dfrac{q}{\varepsilon_0 \rho}$, whence $q = q_0 \exp\left(-\dfrac{t}{\varepsilon_0 \rho}\right)$, and so $\tau = \varepsilon_0 \rho$.

P3.10.2. Self-Induction and Mutual Induction

▶ **Self-induction.** Consider a stationary wire loop carrying current I changing in time. The magnetic induction produced by the current will change at a rate proportional to dI/dt. Hence, the solenoidal electric field arising at every point in space is also proportional to dI/dt. This field induces the emf

$$\mathscr{E}_{\text{self}} = -L\frac{dI}{dt} \tag{P3.10.2.1}$$

in the loop. This emf is called the *self-induction electromotive force*. The quantity L is called the *coefficient of self-inductance* or the *inductance* of the loop. The SI unit of inductance is the *henry* (H). The quantity $\Phi_{\text{self}} = LI$ is the self-induced magnetic flux through the area spanned by the loop. Using (P3.10.2.1), we get

$$\mathscr{E}_{\text{self}} = -\frac{d\Phi_{\text{self}}}{dt}. \tag{P3.10.2.2}$$

This formula remains valid not only if I changes but also if L does.

Note that the definition of L as the coefficient of proportionality between magnetic flux and current meets with difficulties: for a linear wire, the magnetic flux is infinitely large and for a wire of finite thickness, the usual definition of the flux is unsuitable, since it is unclear how to choose a path.

Example 1. Inductance of a solenoid. Consider a solenoid of length l and cross-sectional area S ($l \gg \sqrt{S}$) carrying current with surface density $i = (N/l)I$, where N is the number of turns and I is the current in the coil. The magnetic field in the solenoid is uniform and its magnetic induction (in the ideal solenoid approximation) is given by (P3.8.3.3): $B = \mu_0 \mu i$. For a changing current, the circulation of the solenoidal electric field is calculated as $\oint \mathbf{E} \cdot d\mathbf{l} = -S\dfrac{dB}{dt} = -\mu_0 \mu S \dfrac{di}{dt}$ and the emf induced along the whole length of the winding equals $\mathscr{E}_c = -N\mu_0 \mu S \dfrac{di}{dt} = -\dfrac{N^2 \mu_0 \mu S}{l} \dfrac{dI}{dt}$. The inductance and the magnetic flux are expressed as

$$L = \frac{N^2 \mu_0 \mu S}{l}, \qquad \Phi_{\text{self}} = LI = NBS. \tag{P3.10.2.3}$$

Example 2. Since the self-induction emf is proportional to the current change rate, the presence of inductance in a circuit impedes too rapid changes of the current. Suppose that a resistor R and an induction coil L are in a closed circuit with a source of constant emf $\mathscr{E}$. The current increases from zero to $I_0 = \mathscr{E}/R$ in a finite time and this process is governed by the equation $\mathscr{E} - L(dI/dt) = IR$, whose solution is $I = I_0[1 - \exp(-tR/L)]$. Hence, the characteristic time of current increasing equals $\tau = L/R$.

▶ **Mutual induction.** If there are two stationary loops, the current change in one of them will bring about in the other loop an electromotive force proportional to dI/dt, called the *mutual induction emf.* We have

$$\mathscr{E}_{21} = -L_{21}\frac{dI_1}{dt}, \qquad \mathscr{E}_{12} = -L_{12}\frac{dI_2}{dt}, \qquad \text{(P3.10.2.4)}$$

where L_{12} and L_{21} are the so-called *coefficients of mutual inductance*; the *mutuality theorem* (in a non-ferromagnetic medium) states that $L_{12} = L_{21}$. The magnetic flux through loop 1 produced by the current through loop 2 is given by $\Phi_{12} = L_{12}I_2$. This quantity is meaningful for two linear currents also.

In the quasistationary approximation, the vector potential can be calibrated in the same way as in magnetostatics (see Subsection P3.8.3). The magnetic flux can be written as

$$\Phi = \int \mathbf{B}\, ds = \int \text{curl}\,\mathbf{A}\, ds = \oint \mathbf{A}\, dl.$$

For Φ_{12}, we get

$$\Phi_{12} = \oint \mathbf{A}\, dl_1 = \oint\oint \frac{\mu_0}{4\pi}\frac{I_2\, dl_2}{r_{12}}\, dl_1.$$

It is apparent that the expression of L_{12} is symmetric with respect to the subscripts 1 and 2.

P3.10.3. Energy of Magnetic Field

▶ **Energy of a loop.** The energy of a current-carrying loop is the work done by external forces and sources to create the current (or, equivalently, the work of the solenoidal field to destroy the current). The work done by the current source against the self-induction emf of the loop is calculated as

$$W = A = \int_0^\infty L\frac{dI}{dt}(I\, dt) = \frac{LI^2}{2}. \qquad \text{(P3.10.3.1)}$$

Within a consistent field approach, the energy must be related with the magnetic field rather than the loop current. The easiest way to find the dependence between the energy density and the magnetic induction is to consider a long solenoid with a uniform field in it. We have

$$\frac{W}{V} = \frac{1}{Sl}\frac{\mu_0\mu N^2 S}{l}\frac{I^2}{2} = \left(\frac{\mu_0\mu IN}{l}\right)^2\frac{1}{2\mu_0\mu} = \frac{B^2}{2\mu_0\mu} = \frac{BH}{2}. \qquad \text{(P3.10.3.2)}$$

If the relationship between $\mathbf{B}$ and $\mathbf{H}$ is nonlinear, the energy density is expressed as

$$\frac{W}{V} = \int \mathbf{H}\cdot d\mathbf{B}.$$

Using the mutuality theorem ($L_{12} = L_{21}$), we can generalize (P3.10.3.1) to include two loops:

$$W = \int_0^\infty \left(L_1\frac{dI_1}{dt} + L_{12}\frac{dI_2}{dt}\right)(I_1\, dt) + \int_0^\infty \left(L_2\frac{dI_2}{dt} + L_{21}\frac{dI_1}{dt}\right)(I_2\, dt)$$

$$= \frac{L_1 I_1^2}{2} + \frac{L_2 I_2^2}{2} + \frac{L_{12} + L_{21}}{2}I_1 I_2 = \frac{I_1\Phi_1}{2} + \frac{I_2\Phi_2}{2}. \qquad \text{(P3.10.3.3)}$$

Likewise, for several loops, we have

$$W = \sum_i \frac{I_i \Phi_i}{2},$$

where Φ_i is the total magnetic flux through the ith loop. It can be seen that the current energy includes the energy of each individual current and the energy of their interaction. With the magnetic field equations, it can be shown that the above expression always converts to

$$W = \int \frac{B^2}{2\mu_0\mu}\, dV = \int \frac{BH}{2}\, dV. \tag{P3.10.3.4}$$

▶ **Law of conservation of energy and force calculation.** If the positions of the bodies and current-carrying loops change sufficiently slowly and isothermally, the total work done by external forces and current sources is spent on changing the magnetic field energy and releasing Joulean heat. Moreover, the work done by external forces on each body equals the work done by the magnetic forces on the body. If the currents are maintained constant, one has only to take into account the additional work done by the current sources against the induction emf; the work done by the original emf equals the heat released and, hence, they cancel each other out in the law of conservation of energy.

> **Example.** Consider two loops carrying currents I_1 and I_2 that are maintained constant. Let us calculate the work A^e done by external forces for a slow movement of the first loop. According to (P3.10.3.3), the changing part of the energy is only the energy of current interaction, which is equal to $W_{12} = L_{12}I_1I_2 = I_1\Phi_{12}$, where Φ_{12} is the magnetic flux created by the second current and passing through the first loop. At the same time, the potential function of the loop in an external field is $W_p = -I_1\Phi_{12}$ (see (P3.8.3.5)), and hence $\delta A^e = -I_1\, d\Phi_{12} = -I_1 I_2\, dL_{12}$. The apparent contradiction is resolved by taking into account the additional work done by the current sources against the induction emf in either loop. This work equals $\delta A^{\text{sources}} = I_1\, d\Phi_{12} + I_2\, d\Phi_{21} = I_1\, dL_{12}\, I_2 + I_2\, dL_{21}\, I_1$, which is twice the field energy change $dW_{12} = I_1 I_2\, dL_{12}$, and $\delta A^e + \delta A^{\text{sources}} = dW_{12}$.

P3.11. Maxwell's Equations

P3.11.1. Maxwell's Equations in Vacuum and Matter

▶ **Displacement current.** To generalize the electromagnetic field equations in vacuum to cover varying fields, only one of the four equations (see Subsections P3.3.2, P3.8.3, and P3.10.1) has to be changed. Three equations hold true in all cases but Ampère's circuital law (P3.8.3.1) turns out to be invalid for varying fields and currents. According to (P3.8.3.1), the current $\int \mathbf{j} \cdot d\mathbf{s}$ must be the same for any two surfaces spanning the closed path; if the charge within the volume between the two selected surfaces changes, this statement will contradict the principle of conservation of charge. For example, when a parallel-plate capacitor is being charged (Fig. P3.26), the current through one of the surfaces is $I = dq/dt$ and there in no current through the other surface (passing between the plates).

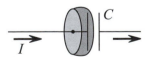

Figure P3.26. To the determination of the displacement current.

To remove the contradiction, Maxwell introduced a *displacement current* in the equation, with its density proportional to the rate of change of the electric field **D**:

$$\mathbf{j}_d = \frac{\partial \mathbf{D}}{\partial t}. \tag{P3.11.1.1}$$

In a dielectric medium, the displacement current density equals

$$\mathbf{j}_d = \varepsilon_0 \frac{\partial \mathbf{E}}{\partial t} + \frac{\partial \mathbf{P}}{\partial t}. \qquad (\text{P3.11.1.2})$$

The first term represents the displacement current density in vacuum and the second term is the real current density due to motion of bound charges in changing polarization. The displacement current through a surface equals $d\Phi_D/dt$, where Φ_D is the flux of $\mathbf{D}$ through the surface. The introduction of the displacement current removes the contradiction with the principle of conservation of charge. For example, when a parallel-plate capacitor is being charged, the displacement current through a surface between the plates, $I_d = S(dD/dt) = S(d\sigma/dt) = dq/dt$, equals the current flowing through the wires.

▶ **System of Maxwell's equations in vacuum.** Once the displacement current has been introduced, the system of Maxwell's equations in differential form becomes:

1) $\operatorname{curl} \mathbf{E} = -\dfrac{\partial \mathbf{B}}{\partial t}$ (Faraday's law of induction)

2) $\operatorname{curl} \mathbf{B} = \mu_0 \mathbf{j} + \mu_0 \varepsilon_0 \dfrac{\partial \mathbf{E}}{\partial t}$ (Ampère's circuital law with Maxwell's correction)

3) $\operatorname{div} \mathbf{E} = \dfrac{\rho}{\varepsilon_0}$ (Gauss's law)

4) $\operatorname{div} \mathbf{B} = 0$ (Gauss's law for magnetism)

The system of Maxwell's equations in integral form:

1) $\oint \mathbf{E} \cdot d\mathbf{l} = -\dfrac{d}{dt} \int \mathbf{B} \cdot d\mathbf{s}$,   2) $\oint \mathbf{B} \cdot d\mathbf{l} = \mu_0 \int \mathbf{j} \cdot d\mathbf{s} + \mu_0 \varepsilon_0 \dfrac{d}{dt} \int \mathbf{E} \cdot d\mathbf{s}$,

3) $\oint \mathbf{E} \cdot d\mathbf{s} = \dfrac{1}{\varepsilon_0} \int \rho \, dV$,   4) $\oint \mathbf{B} \cdot d\mathbf{s} = 0$.

The charge density and the current density satisfy the relation

$$\oint \mathbf{j} \cdot d\mathbf{s} = -\frac{d}{dt} \int \rho \, dV \qquad \left(\operatorname{div} \mathbf{j} = -\frac{\partial \rho}{\partial t} \right),$$

which expresses the law of conservation of charge and is a corollary of Maxwell's equations.

▶ **Solution of Maxwell's equations in the form of delayed potentials.** The first and fourth Maxwell equations can be satisfied by using the fields in the form

$$\mathbf{B} = \operatorname{curl} \mathbf{A}, \qquad \mathbf{E} = -\operatorname{grad} \varphi - \frac{\partial \mathbf{A}}{\partial t}.$$

The scalar potential φ and vector potential $\mathbf{A}$ are defined up to the *calibration transformation*:

$$\mathbf{A}' = \mathbf{A} + \operatorname{grad} \chi, \qquad \varphi' = \varphi - \frac{\partial \chi}{\partial t}.$$

The equations for the potentials resulting from the second and third Maxwell equations can be simplified using the calibration transformation. To this end, the potentials are required to satisfy *Lorentz's calibration condition*:

$$\operatorname{div} \mathbf{A} + \frac{1}{c^2} \frac{\partial \varphi}{\partial t} = 0.$$

In this case, the equations for the potentials satisfy *d'Alembert's equation*:

$$\Delta \mathbf{A} - \frac{1}{c^2} \frac{\partial^2 \mathbf{A}}{\partial t^2} = -\mu_0 \mathbf{j}, \qquad \Delta \varphi - \frac{1}{c^2} \frac{\partial^2 \varphi}{\partial t^2} = -\frac{\rho}{\varepsilon_0}.$$

If the right-hand side vanishes, d'Alembert's equation becomes the wave equation (see Chapter P4). In the presence of time-varying current and charge sources, d'Alembert's equations have solutions in the form of *delayed potentials*:

$$\mathbf{A}(\mathbf{r},t) = \frac{\mu_0}{4\pi} \int \frac{\mathbf{j}\left(\mathbf{r}',t-\frac{|\mathbf{r}-\mathbf{r}'|}{c}\right) dV'}{|\mathbf{r}-\mathbf{r}'|}, \qquad \varphi(\mathbf{r},t) = \frac{1}{4\pi\varepsilon_0} \int \frac{\rho\left(\mathbf{r}',t-\frac{|\mathbf{r}-\mathbf{r}'|}{c}\right) dV'}{|\mathbf{r}-\mathbf{r}'|}.$$

The potentials at a given point in space and a given time instant are determined by the charges and currents at all other points at preceding instants of time, taking into account the time required for a light signal to deliver information to the given point about the current and charge values at a remote point.

▶ **Maxwell's equations in a medium** are written as

differential form   integral form

1) $\quad \operatorname{curl} \mathbf{E} = -\dfrac{\partial \mathbf{B}}{\partial t}, \qquad \oint \mathbf{E}\cdot d\mathbf{l} = -\dfrac{d}{dt}\int \mathbf{B}\cdot d\mathbf{s},$

2) $\quad \operatorname{curl} \mathbf{H} = \mathbf{j} + \dfrac{\partial \mathbf{D}}{\partial t}, \qquad \oint \mathbf{H}\cdot d\mathbf{l} = \int \mathbf{j}\cdot d\mathbf{s} + \dfrac{d}{dt}\int \mathbf{D}\cdot d\mathbf{s},$

3) $\quad \operatorname{div} \mathbf{D} = \rho, \qquad \oint \mathbf{D}\cdot d\mathbf{s} = \int \rho\, dV,$

4) $\quad \operatorname{div} \mathbf{B} = 0, \qquad \oint \mathbf{B}\cdot d\mathbf{s} = 0.$

These serve to determine the four vector quantities **E**, **D**, **B**, and **H**. Maxwell's equations in a medium must be supplemented by the *material equations* relating **D** to **E** and **H** to **B** and characterizing the electric and magnetic properties of the medium. For isotropic linear media, the equations are

$$\mathbf{D} = \varepsilon_0\varepsilon\mathbf{E}, \qquad \mathbf{B} = \mu_0\mu\mathbf{H}.$$

Maxwell's equations can also be used to obtain boundary conditions for **E**, **D**, **B**, and **H** (see Subsections P3.5.1 and P3.9.1).

P3.11.2. Law of Conservation of Energy for Electromagnetic Fields

▶ **Poynting vector.** The following equation for any volume V bounded by a surface S can be derived from Maxwell's equations:

$$\frac{d}{dt}\int\left(\frac{\varepsilon_0 E^2}{2} + \frac{B^2}{2\mu_0}\right)dV + \int (\mathbf{j}\cdot\mathbf{E})\, dV + \frac{1}{\mu_0}\oint [\mathbf{E}\times\mathbf{B}]\cdot d\mathbf{s} = 0.$$

The first term characterizes the rate of change of electromagnetic field energy in the volume. In general, the formulas obtained before for electrostatic and magnetostatic fields remain valid for the electromagnetic field energy. The second term represents the power of the field work on the charged particles within the volume of interest; it is equal to the rate of change of their kinetic energy. It follows from the law of conservation of energy that the third term has the meaning of (outward) electromagnetic energy flux through the closed surface bounding the volume. The energy flux at a given point in space (the *Poynting vector*) is determined by the vectors **E** and **B** at this point:

$$\mathbf{S} = \frac{1}{\mu_0}\mathbf{E}\times\mathbf{B} = \mathbf{E}\times\mathbf{H}. \tag{P3.11.2.1}$$

The last expression represents the electromagnetic energy flux in a substance. The energy density in a substance is expressed as

$$w = \tfrac{1}{2}\mathbf{E} \cdot \mathbf{D} + \tfrac{1}{2}\mathbf{B} \cdot \mathbf{H}. \tag{P3.11.2.2}$$

Example. Consider the process of charging a parallel-plate capacitor with circular plates separated by a distance b. The rate of change of the energy in a cylinder of radius r (smaller than the plate radius) equals $\dfrac{d}{dt}\left(\tfrac{1}{2}c_0\varepsilon E^2\right)(\pi r^2 b) = \pi r^2 b E\dfrac{dD}{dt}$. The magnetic field strength can be found from the second Maxwell equations: $H \times 2\pi r = \dfrac{dD}{dt}\pi r^2$ (displacement current is on the right-hand side). We find that the rate of energy influx through the lateral surface of the cylinder, $S(2\pi rb) = HE(2\pi rb) = E\pi r^2 b\dfrac{dD}{dt}$, equals the energy change in the volume.

P3.11.3. Relativistic Properties of Fields

▶ **Transformation of charges and currents.** When changing from one inertial frame of reference to another, both the electromagnetic field sources (charge and current densities) and the fields themselves change. However, Maxwell's equations remain the same. The source transformation formulas for ρ (density of moving charge) and $\mathbf{j} = \rho\mathbf{v}$ have the simplest form. Denoting by ρ_0 the charge density in the inertial frame where $\mathbf{j} = \mathbf{v} = 0$ and taking into account length contraction (see Subsection P1.10.3), we get

$$\rho = \gamma\rho_0, \quad \mathbf{j} = \gamma\rho_0\mathbf{v}; \qquad \gamma = \frac{1}{\sqrt{1 - v^2/c^2}}.$$

Comparing with the momentum-energy 4-vector, we see that $(\mathbf{j}, \rho c)$ also forms a 4-vector, which means that $\mathbf{j}$ and ρc are transformed in the same manner as $(\mathbf{r}, ct)$, by the Lorentz formulas.

▶ **Transformation of fields.** Once we know the transformation formulas for field sources, it is not difficult to find out how $\mathbf{E}$ and $\mathbf{B}$ are transformed. The formulas are

$$\begin{aligned}
\mathbf{E}'_{\parallel} &= \mathbf{E}_{\parallel}, & \mathbf{E}'_{\perp} &= \gamma\left(\mathbf{E}_{\perp} + \mathbf{V} \times \mathbf{B}\right), \\
\mathbf{B}'_{\parallel} &= \mathbf{B}_{\parallel}, & \mathbf{B}'_{\perp} &= \gamma\left(\mathbf{B}_{\perp} - \frac{1}{c^2}\mathbf{V} \times \mathbf{E}\right),
\end{aligned} \tag{P3.11.3.1}$$

where $\mathbf{V}$ is the velocity of inertial frame K' relative to inertial frame K; the transformations are written for the field components parallel and perpendicular to $\mathbf{V}$. The scalar quantities $\mathbf{E} \cdot \mathbf{B}$ and $E^2 - c^2 B^2$ are invariant under these transformations:

$$\mathbf{E} \cdot \mathbf{B} = \mathbf{E}' \cdot \mathbf{B}', \qquad E^2 - c^2 B^2 = (E')^2 - c^2(B')^2. \tag{P3.11.3.2}$$

For $V \ll c$, formulas (P3.11.3.1) simplify to

$$\mathbf{E}' = \mathbf{E} + \mathbf{V} \times \mathbf{B}, \qquad \mathbf{B}' = \mathbf{B} - \frac{1}{c^2}\mathbf{V} \times \mathbf{E}. \tag{P3.11.3.3}$$

Example 1. Magnetic field of a nonrelativistic particle. Consider a particle with charge q moving relative to inertial frame K with a constant nonrelativistic velocity $\mathbf{v}$. In the particle associated inertial frame K', there is only the electric field $\mathbf{E}' = kq\mathbf{r}'/r'^3$. To find the fields in frame K, one has to write out the transformation formulas (P3.11.3.3) with $\mathbf{V} = -\mathbf{v}$. Since there is no length contraction in the nonrelativistic limit, we find that

$$\mathbf{E} = \mathbf{E}' = \frac{1}{4\pi\varepsilon_0}\frac{q\mathbf{r}}{r^3}, \qquad \mathbf{B} = \frac{\mathbf{v} \times \mathbf{E}'}{c^2} = \frac{\mu_0 q}{4\pi}\frac{\mathbf{v} \times \mathbf{r}}{r^3}$$

for the instant when the particle passes through the origin of coordinates in frame K. In deriving these formulas, we have used the relation $1/c^2 = \mu_0\varepsilon_0$.

Example 2. Polarization of a dielectric moving in a magnetic field. When a dielectric moves through a magnetic field with an nonrelativistic velocity perpendicular to the magnetic field lines, it becomes polarized. In the dielectric-associated inertial frame, there is transverse electric field: $\mathbf{E}' = \mathbf{v} \times \mathbf{B}$. The detailed polarization picture depends on the shape of the dielectric.

Example 3. Electric field of a relativistic particle. Suppose a particle with charge q moves relative to inertial frame K with a constant relativistic velocity $\mathbf{v}$. In the particle associated inertial frame K', there is only the electric field $\mathbf{E}' = kq\mathbf{r}'/r'^3$. To find the fields in frame K, one has to write out the transformation formulas (P3.11.3.1) with $\mathbf{V} = -\mathbf{v}$. Let us write out the formulas for the instant when the particle passes through the origin of coordinates in frame K. For simplicity, consider the plane (x, y). When changing from the coordinates (x', y') to (x, y), we must take into account that $x' = \gamma x$ and $y' = y$ (the coordinates of a point in frame K are measured at the same instant when the particle passes through the origin). We get

$$E_x = E_x' = k\frac{q\gamma x}{[(\gamma x)^2 + y^2]^{3/2}}, \qquad E_y = \gamma E_y' = k\frac{q\gamma y}{[(\gamma x)^2 + y^2]^{3/2}}.$$

It is seen that $\mathbf{E}$ is collinear with $\mathbf{r}$. However, the field at a point on the line of motion will be smaller than that at a point which is at the same distance from the charge but on a straight line perpendicular to $\mathbf{v}$. The magnetic field is obtained as

$$\mathbf{B} = \frac{\gamma}{c^2}\mathbf{v} \times \mathbf{E}' = \frac{1}{c^2}\mathbf{v} \times \mathbf{E} = \frac{\mu_0}{4\pi}\frac{\gamma q\,\mathbf{v} \times \mathbf{r}}{[(\gamma x)^2 + y^2]^{3/2}}.$$

Note that the above electric field is not potential.

Bibliography for Chapter P3

Alenitsyn, A. G., Butikov, E. I., and Kondratyev, A. S., *Concise Handbook of Mathematics and Physics*, CRC Press, Boca Raton, Florida, 1998.

Benenson, W., Harris, J. W., Stocker, H., and Lutz, H. (Eds.), *Handbook of Physics*, Springer-Verlag, New York, 2002.

Cottingham, W. N. and Greenwood, D. A., *Electricity and Magnetism*, Cambridge University Press, Cambridge, England, 1991.

Fitzpatrick, R., *Maxwell's Equations and the Principles of Electromagnetism*, Infinity Science Press, Hingham, Massachusetts, 2008.

Fleisch, D., *A Student's Guide to Maxwell's Equations*, Cambridge University Press, Cambridge, England, 2008.

Griffiths, D. J., *Introduction to Electrodynamics 3rd Edition*, Benjamin Cummings, New York, 1999.

Irodov, I. E., *Electromagnetism. Basic Laws* [in Russian], Fizmatlit, Moscow, 2006.

Jefimenko, O. D., *Electricity and Magnetism: An Introduction to the Theory of Electric and Magnetic Fields, 2nd Edition*, Electret Scientific Co., Star City, West Virginia, 1989.

Landau, L. D. and Lifshitz, E. M., *The Classical Theory of Fields, Fourth Edition: Volume 2 (Course of Theoretical Physics Series)*, Butterworth-Heinemann, Oxford, England, 1980.

Landau, L. D., Pitaevskii, L. P., and Lifshitz, E. M., *Electrodynamics of Continuous Media, Second Edition: Volume 8 (Course of Theoretical Physics)*, Butterworth-Heinemann, Oxford, England, 1984.

Melia F., *Electrodynamics (Chicago Lectures in Physics)*, University of Chicago Press, Chicago, Illinois, 2001.

Panofsky, W. K. H. and Phillips, M., *Classical Electricity and Magnetism: Second Edition*, Addison-Wesley, Reading, Massachusetts, 1962.

Pollack, G. and Stump, D., *Electromagnetism*, Addison-Wesley, Reading, Massachusetts, 2001.

Prokhorov, A. M. (Ed.), *Encyclopedia of Physics, Volumes 1–5* [in Russian], Bolshaya Russkaya Encyclopedia, Moscow, 1998.

Prokhorov, A. M. (Ed.), *Encyclopedic Dictionary of Physics* [in Russian], Sovetskaya Encyclopedia, Moscow, 1984 (dic.academic.ru).

Purcell, E. M., *Electricity and Magnetism, Berkeley Physics Course, Vol. II, 2nd Edition*, McGraw-Hill Science/Engineering/Math, New York, 1984.

Saslow, W. M., *Electricity, Magnetism, and Light*, Academic Press, New York, 2002.

Schwartz, M., *Principles of Electrodynamics*, Dover Publications, New York, 1987.

Sivukhin, D. V., *General Physics, Volume 3, Electricity* [in Russian], Fizmatlit, Moscow, 2004.

Sivukhin, D. V. and Yakovlev, I. A., *A Collection of Problems, Volume 3, Electricity and Magnetism* [in Russian], Fizmatlit, Moscow, 2006.

Tamm, I. E., *Basics of Theory of Electricity* [in Russian], Fizmatlit, Moscow, 2003.

Woan, G., *The Cambridge Handbook of Physics Formulas*, Cambridge University Press, Cambridge, England, 2003.

Chapter P4

Oscillations and Waves

P4.1. Oscillations

▶ **General definitions.** Oscillations are repetitive variations in the state of a system such that the state parameters vary in time according to a periodic or almost periodic law. If oscillations occur without external action due to system deviation from a stable equilibrium, the oscillations are said to be *free* or *natural*. If the oscillations occur under the action of an external periodic force, then they are said to be *forced*. The oscillations are characterized by their *period* T and *frequency* $\nu = 1/T$ (measured in *hertzs*: $1\,\text{Hz} = 1\,\text{s}^{-1}$). The term *vibrations* is often used in a narrower sense to mean mechanical oscillations, but sometimes is used synonymously with oscillations.

P4.1.1. Harmonic Oscillations. Composition of Oscillations

▶ **Simple harmonic oscillations.** An oscillation of a quantity x is said to be a *simple harmonic oscillation* (or *simple harmonic motion*) if x varies in time t by the law

$$x = A\cos(\omega t + \varphi_0), \tag{P4.1.1.1}$$

where A is the *amplitude*, $\varphi = \omega t + \varphi_0$ is the *phase*, φ_0 is the *initial phase*, and $\omega = 2\pi/T$ is the *angular* or *circular frequency* of the oscillation. The first and second time-derivatives of the quantity x,

$$\begin{aligned}
\dot{x} &= -A\omega\sin(\omega t + \varphi_0) = A\omega\cos\left(\omega t + \varphi_0 + \pi/2\right), \\
\ddot{x} &= -A\omega^2\cos(\omega t + \varphi_0) = A\omega^2\cos(\omega t + \varphi_0 + \pi),
\end{aligned} \tag{P4.1.1.2}$$

oscillate harmonically with the same frequency but with amplitudes ωA and $\omega^2 A$ and with the phase shifts $\pi/2$ and π, respectively.

Example. If the initial values (at $t = 0$) of the quantity x and its derivative, $x(0) = x_0$ and $\dot{x}(0) = v_0$, are known, then the amplitude and the initial phase of the oscillation can be determined. The equations $x_0 = A\cos\varphi_0$ and $v_0 = -\omega A\sin\varphi_0$ allow one to find $A = \sqrt{x_0^2 + (v_0/\omega)^2}$ and $\tan\varphi_0 = -v_0/(\omega x_0)$.

▶ **Equation of simple harmonic oscillations.** It follows from (P4.1.1.2) that if x varies by the harmonic law (P4.1.1.1), then x satisfies the *equation of simple harmonic oscillations*:

$$\ddot{x} + \omega^2 x = 0. \tag{P4.1.1.3}$$

The converse is also true: if the *equation of motion* of a physical system whose state is determined by a single quantity x can be reduced under certain conditions (usually, for small values of x) to the differential equation $\ddot{x} + \gamma x = 0$, where γ is a positive constant, then x varies according to the law (P4.1.1.1) with $\omega = \sqrt{\gamma}$ (the parameters A and φ_0 are determined by the initial conditions; see the example above).

▶ **Energy method for determining the frequency.** If the state of a physical system undergoes repetitive variations and the quadratic function

$$E = \tfrac{1}{2}\mu\dot{x}^2 + \tfrac{1}{2}\kappa x^2 \qquad (\mu \text{ and } \kappa \text{ are positive constants}) \tag{P4.1.1.4}$$

remains constant during these variations, then x changes according to the law (P4.1.1.1) with $\omega = \sqrt{\kappa/\mu}$. Indeed, differentiating (P4.1.1.4) with respect to time, we obtain the equation $\mu\ddot{x} + \kappa x = 0$, which is the equation of simple harmonic oscillations. Usually, the quantity E is proportional to the energy of the oscillatory system for small x; accordingly, such an approach is called the *energy method* for determining the oscillation frequency.

▶ **Complex exponential and vector diagram.** Using Euler's formula $e^{i\varphi} = \cos\varphi + i\sin\varphi$, one can treat the law of harmonic oscillations (P4.1.1.1) as the real part of a complex exponential function, $x(t) = \operatorname{Re}\widetilde{x}$, where

$$\widetilde{x} = \widetilde{A}\exp(i\omega t), \tag{P4.1.1.5}$$

and $\widetilde{A} = A\exp(i\varphi_0)$ is called the *complex amplitude* of oscillations. This technique is especially convenient for studying systems described by linear equations, since, in this case, the real and imaginary parts are transformed independently of each other.

The harmonic law (P4.1.1.1) can be obtained as the projection on the x-axis of a radius vector of length A that uniformly rotates counterclockwise with a constant angular velocity ω from an initial angular position φ_0; in this case, the angle made with the axis x varies by the law $\varphi = \varphi_0 + \omega t$. This technique is called the *method of vector diagrams*; it is especially convenient for the *composition of harmonic oscillations*. With this technique, the addition of functions is replaced by the graphical addition of vectors; note that the projection of the sum of vectors is equal to the sum of their individual projections.

▶ **Composition of harmonic oscillations of a single direction.** The sum of two harmonic oscillations of the same frequency with amplitudes A_1 and A_2 and the initial phases φ_1 and φ_2 is a harmonic oscillation of the same frequency, whose amplitude and initial phase can be found by the method of vector diagrams (Fig. P4.1):

$$A^2 = A_1^2 + A_2^2 + 2A_1A_2\cos(\varphi_2 - \varphi_1), \qquad \tan\varphi_0 = \frac{A_1\sin\varphi_1 + A_2\sin\varphi_2}{A_1\cos\varphi_1 + A_2\cos\varphi_2}.$$

The parallelogram spanned by the vectors rotates with angular velocity ω as a rigid body. The phase difference of oscillations with the same frequency does not vary in time; such oscillations are said to be *coherent*. If $\varphi_2 - \varphi_1 = \pm 2m\pi$, the amplitude is a maximum, $A = A_1 + A_2$, and if $\varphi_2 - \varphi_1 = \pm(2m + 1)\pi$, the amplitude is a minimum, $A = |A_1 - A_2|$.

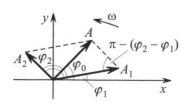

Figure P4.1. Composition of oscillations by the method of vector diagrams.

In the vector addition of incoherent oscillations with different frequencies, the parallelogram spanned by the vectors deforms with time and the magnitude of the resultant vector and its angular velocity vary, and hence the composite motion is not a harmonic oscillation. However, oscillations *with close frequencies* ($\Delta\omega = |\omega_1 - \omega_2| \ll \max\{\omega_1, \omega_2\}$) can be assumed to be approximately coherent on a sufficiently short time interval as compared to the *time of coherence* $\tau_{\mathrm{coh}} = 2\pi/\Delta\omega$. The composite oscillation will occur with the angular frequency $\omega = (\omega_1 + \omega_2)/2$ and an amplitude periodically varying between $A_1 + A_2$ and $|A_1 - A_2|$ with period $2\pi/\Delta\omega$. Such oscillations are called *beats*, $\omega_{\mathrm{b}} = \Delta\omega$ is the beat angular frequency, and the period of variation in the amplitude $T_{\mathrm{b}} = 2\pi/\omega_{\mathrm{b}}$ is the beat period. If $A_1 = A_2 = A$, then

$$x(t) = 2A \cos\left(\frac{\omega_2 - \omega_1}{2}t\right) \cos\left(\frac{\omega_2 + \omega_1}{2}t + \varphi_0\right).$$

The beats are an example of *modulated oscillations*, or oscillations obeying the harmonic law (P4.1.1.1) with one of the parameters (amplitude, frequency, or phase) varying periodically in time with a period significantly greater than that of the base oscillations. Amplitude, frequency, and phase modulation techniques are distinguished.

An arbitrary periodic oscillation with period T can be expanded into a *Fourier series* in simple harmonic oscillations with frequencies $\omega_n = n(2\pi/T)$. Such a representation is called the *harmonic analysis* of the periodic oscillation. The Fourier series terms are called the first (fundamental), second, third, etc. *harmonics* of the periodic oscillation. Aperiodic oscillations have, as a rule, a continuous spectrum of frequencies and can be represented as a *Fourier integral* in harmonic oscillations of all frequencies ranging from zero to infinity. All periodic oscillations have a line (discrete) spectrum of frequencies, but oscillations with a line spectrum are not necessarily periodic.

▶ **Composition of mutually perpendicular harmonic oscillations.** If a point moves in the plane so that its projections on the x- and y-axes oscillate harmonically, this motion is said to be the composition of mutually perpendicular simple harmonic oscillations. If the oscillations in either coordinate have *equal frequencies*, $x = A_1 \cos(\omega t + \varphi_1)$ and $y = A_2 \cos(\omega t + \varphi_2)$, the path of the point is an inclined ellipse:

$$\frac{x^2}{A_1^2} + \frac{y^2}{A_2^2} - \frac{2xy}{A_1 A_2} \cos(\varphi_2 - \varphi_1) = \sin^2(\varphi_2 - \varphi_1).$$

Such a motion is called an *elliptically polarized oscillation*. If $\varphi_2 - \varphi_1 = (2m+1)\pi/2$, the ellipse axes coincide with the coordinate axes. If $\varphi_2 - \varphi_1 = m\pi$, the ellipse degenerates into a straight line segment; such a motion is called a *linearly polarized oscillation*. The point can move clockwise (for $2m\pi < \varphi_2 - \varphi_1 < 2m\pi + \pi$) or counterclockwise; in these cases, one speaks about the right or left elliptic polarization.

Two-dimensional harmonic motions with frequencies ω_1 and ω_2 occur along curves called *Lissajous figures*. If the frequency ratio ω_1/ω_2 is a rational number, these figures are closed curves. Closed Lissajous figures can be used to determine the frequency ratio.

P4.1.2. Free Undamped Oscillations

▶ **Free mechanical oscillations with a single degree of freedom.** If the displacement of a conservative mechanical system from a stable equilibrium is described by a single parameter x, then for small x, the potential energy has the form

$$E_{\mathrm{p}}(x) = \frac{1}{2}x^2 \left(\frac{d^2 E_{\mathrm{p}}}{dx^2}\right)_{x=0} + \frac{1}{6}x^3 \left(\frac{d^3 E_{\mathrm{p}}}{dx^3}\right)_{x=0} + \cdots \approx \frac{\kappa x^2}{2}. \tag{P4.1.2.1}$$

It follows that small oscillations of the system about its stable equilibrium occur harmonically. The potential energy of an oscillatory system is usually reckoned from the stable equilibrium ($E_p = 0$ for $x = 0$). For small x, the *generalized force*, $F_x = -dE_p/dx$, is proportional to the displacement: $F_x = -\kappa x$. (If x is a linear coordinate, then the generalized force is the force projection on the displacement direction, and if x is an angle, then F_x is a torque.)

If a mechanical system is a point particle or a translationally moving body, then the kinetic energy of the system has the form $E_k = \frac{1}{2}m\dot{x}^2$. The equation of motion

$$m\ddot{x} = -\kappa x \tag{P4.1.2.2}$$

has the same form as that of a mass attached to a spring. For small x, the *restoring force* acting on the body (i.e., the force that tends to bring the body back to the equilibrium) is called the *quasielastic force*, $F_x = -\kappa x$, and the coefficient κ is the *effective spring constant*. The angular frequency and oscillation period are expressed as

$$\omega = \sqrt{\frac{\kappa}{m}}, \qquad T = 2\pi\sqrt{\frac{m}{\kappa}}. \tag{P4.1.2.3}$$

The particle oscillates harmonically, $x = A\cos(\omega t + \varphi_0)$, and its kinetic and potential energies oscillate harmonically with the frequency 2ω:

$$E_k = E\sin^2(\omega t + \varphi_0) = \frac{1}{2}E\left[1 - \cos(2\omega t + 2\varphi_0)\right],$$
$$E_p = E\cos^2(\omega t + \varphi_0) = \frac{1}{2}E\left[1 + \cos(2\omega t + 2\varphi_0)\right],$$

where $E = \frac{1}{2}m\omega^2 A^2 = \frac{1}{2}\kappa A^2$ is the total mechanical energy, which is proportional to the amplitude squared. The average kinetic energy is equal to the average potential energy.

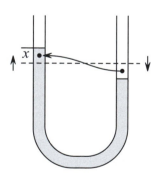

Figure P4.2. To example 1.

Example 1. If the level of a liquid in an U-shaped tube (Fig. P4.2) is displaced by a distance x from equilibrium, the potential energy of the liquid increases by $\Delta mgx = (\rho Sx)gx = (2\rho gS)x^2/2$ (where ρ is the liquid density and S is the tube cross-section) and the kinetic energy is equal to $m\dot{x}^2/2$. Hence the angular frequency of oscillations is equal to $\omega = \sqrt{2\rho gS/m}$.

Example 2. In one of the early models of the hydrogen atom (the Thomson model), it was assumed that the positive charge is uniformly distributed across a ball of radius R, equal to the atomic radius, and the electron was placed at the center of the ball. Let us find the frequency of electron oscillations when it is displaced from the equilibrium. If r denotes the displacement, the attraction force is equal to $F_r = -ke^2r/R^3$ (see Subsection P3.3.2). Hence, for small displacements, $r < R$, there occur harmonic oscillations with frequency $\omega = \sqrt{ke^2/(mR^3)}$.

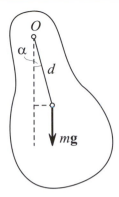

Figure P4.3. Physical pendulum.

Example 3. Physical pendulum. The *physical pendulum* is a rigid body whose horizontal immovable axis of rotation does not pass through its center of mass. For the pendulum deviation from equilibrium by a small angle α (Fig. P4.3), there arises a moment of the gravity force equal to $M = -mgd\sin\alpha$ (d is the distance from the center of mass to the axis of rotation), which tends to return the pendulum to equilibrium. The equation of rotational motion of the rigid body has the form: $I\ddot{\alpha} = -mgd\sin\alpha$, where I is the moment of inertia of the body about the axis of rotation. For small deviation angles $\alpha \ll 1$, this equation becomes the equation of harmonic oscillations with respect to the angle α: $\ddot{\alpha} + (mgd/I)\alpha = 0$. The pendulum angular frequency and oscillation period are equal to

$$\omega = \sqrt{mgd/I}, \qquad T = 2\pi\sqrt{I/(mgd)}.$$

For example, for a homogeneous rod of length l suspended by one of its ends, we obtain $I = ml^2/3$, $d = l/2$, $T = 2\pi\sqrt{2l/(3g)}$. The *reduced length* L of a physical pendulum is the length of a mathematical pendulum such that its period $T = 2\pi\sqrt{L/g}$ is equal to the period of the given physical pendulum: $L = I/(md)$. (For a homogeneous rod suspended by an end, we have $L = 2l/3$.) The point lying at a distance L from the suspension axis on the straight line passing through the center of mass perpendicularly to the suspension axis is called the *center of oscillation*. If a body is suspended on the axis passing through the center of oscillation parallel to the previous axis, then the oscillation period, and hence the pendulum reduced length do not change.

In this example, the generalized force is the moment of gravity force. The frequency can also be determined by the energy method if the dependence of the potential energy on the deviation angle is found: $E_p = mgd(1-\cos\alpha) \approx mgd\alpha^2/2$. For the total energy, we obtain $E = \frac{1}{2}I\dot{\alpha}^2 + \frac{1}{2}mgd\alpha^2$, which implies the oscillation frequency.

▶ **Isochronism of harmonic oscillations.** The harmonic oscillations are *isochronous*, which means that their period is independent of the amplitude. The point is that, in the equation of harmonic oscillations $\ddot{x} + \gamma x = 0$, the coefficient γ has dimension s^{-2} and it follows from the dimension considerations that the period cannot depend on anything else except for γ. The oscillations of large amplitude cease to be harmonic and hence isochronous; the period begins to depend on the amplitude. But in several special cases, even small oscillations may not be isochronous. This occurs if the second derivative in formula (P4.1.2.1) for the expansion of E_p in x is zero. Such a situation arises, for example, in the case where a load fixed at the middle of an *unstrained* spring with fixed ends deviates in the direction perpendicular to the spring. For small x, the potential energy is proportional to x^4, and the equation of motion has the form: $\ddot{x} + \gamma x^3 = 0$. In this case, the dimension γ is $m^{-2}s^{-2}$, and the oscillation period $T \sim \gamma^{-1/2}A^{-1}$ is inversely proportional to the amplitude.

▶ **Oscillatory systems with several degrees of freedom.** If the system displacement from the stable equilibrium is characterized by N independent parameters $x_1, \ldots, x_N$, then for an arbitrarily small initial deviation, the dependence $x_i(t)$ cannot be harmonic oscillations. But any motion can be represented as the sum of N special motions called *normal vibrational modes*, in each of which, all x_i harmonically oscillate with a angular

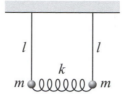

Figure P4.4. Coupled pendulums.

frequency characterizing this mode. In the general case, all the N frequencies can be distinct.

Example 4. Coupled pendulums. We consider two equal mathematical pendulums of length l whose loads are connected by an unstrained spring of rigidity k (Fig. P4.4). The normal modes of this system can be guessed because of the system symmetry. Harmonic oscillations arise for: (a) equal deviations of the pendulums to the same side ($x_1 = x_2$) and (b) for the equal deviations of the pendulums to opposite sides ($x_1 = -x_2$). In the first case: $\omega_1 = \sqrt{g/l}$, in the second case: $\omega_2 = \sqrt{(g/l) + (2k/m)}$. If only the first pendulum deviates to a side, which corresponds to the initial conditions $(x_{10}, x_{20}) = (A, 0)$, then such a deviation can be represented as the sum of two normal deviations: $(A, 0) = \left(\frac{1}{2}A, \frac{1}{2}A\right) + \left(\frac{1}{2}A, -\frac{1}{2}A\right)$. This means that the motion of the first pendulum obeys the law: $x_1 = \frac{1}{2}A\cos(\omega_1 t) + \frac{1}{2}A\cos(\omega_2 t)$, and the motion of the second pendulum obeys the law: $x_2 = \frac{1}{2}A\cos(\omega_1 t) - \frac{1}{2}A\cos(\omega_2 t)$. If $k/m \ll g/l$, then the frequencies ω_1 and ω_2 are close to each other: $\Delta\omega \approx \omega_1 \dfrac{k/m}{g/l}$, and, under superposition of two oscillations with close frequencies, one observes beats (see Subsection P4.1.1): in the time $T_b/2 = \pi/\Delta\omega$, the second pendulum is in full swing, and the first pendulum stops, and after the same time period, the energy is again transferred to the first pendulum, etc.

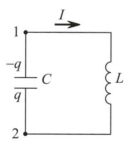

Figure P4.5. Electric oscillatory circuit.

▶ **Free oscillations in electric circuit.** The oscillatory circuit is a closed circuit consisting of a capacitor of capacitance C and a coil of inductance L (Fig. P4.5). The processes in an oscillatory circuit, just as in alternating current circuits, are studied in the domains of frequencies, where the *quasi-stationarity* condition is satisfied (see Subsection P3.10.1); that is, $\omega \ll c/l$, where l are typical dimensions of the system (in this approximation, in particular, one can neglect the circuit radiation). For the chosen rule of bypassing the circuit, the Ohm law for the subcircuit containing the inductance and for the capacitor charge has the form: $IR = (\varphi_1 - \varphi_2) + \mathscr{E}$, where R is the resistance of the wires and the coil. Taking the relations $I = \dot{q}$ (that holds the chosen signs), $q = (\varphi_2 - \varphi_1)/R$, and $\mathscr{E} = -L\dot{I}$ into account, we obtain $IR = -(q/C) - L\dot{I}$. This implies that

$$\ddot{q} + \frac{R}{L}\dot{q} + \frac{1}{LC}q = 0. \tag{P4.1.2.3}$$

If the circuit resistance is negligibly small (which corresponds to $R \ll \sqrt{L/C}$), then the capacitor charge varies according to the harmonic law $q = q_m\cos(\omega t + \varphi_0)$ with angular frequency and period equal to

$$\omega = \frac{1}{\sqrt{LC}} \quad \text{(Thomson's formula)}, \qquad T = 2\pi\sqrt{LC}. \tag{P4.1.2.4}$$

The current in the circuit varies according to the harmonic law with amplitude $I_m = \omega q_m$ with the phase lead by $\pi/2$.

Thomson's formula (P4.1.2.4) can be obtained by the energy method starting from the expression for the circuit energy:

$$W = \frac{LI^2}{2} + \frac{q^2}{2C},$$

taking into account that $I = \dot{q}$. The energy is transformed from electric into magnetic and conversely: at the moment of maxima capacitor charge, it is equal to $W = q_m^2/(2C)$, and when the capacitor charge is zero, it is equal to $W = LI_m^2/2$. The relations for the time-dependence of the magnetic and electric energy are the same as for the kinetic and potential energies in the case of mechanical oscillations.

P4.1.3. Damped and Forced Oscillations

▶ **Damped oscillations.** If the energy is lost in an oscillatory system, then the oscillation amplitude decreases with time. If the losses of energy of mechanical oscillations are determined by the force of viscous friction, which is proportional to the velocity for small velocities, then the equation of motion $m\ddot{x} = -\kappa x - b\dot{x}$ can be reduced to the linear differential equation

$$\ddot{x} + 2\beta\dot{x} + \omega_0^2 x = 0, \tag{P4.1.3.1}$$

which is called the *equation of damped oscillations*. Here $\beta = b/(2m)$ is the *damping coefficient*, ω_0 is the angular frequency of natural oscillations without damping (as $\beta \to 0$). It is convenient to seek the solution of this equation in exponential form: $\tilde{x} = \tilde{A}\exp(-\gamma t)$. Substituting this into (P4.1.3.1), we obtain the quadratic equation for γ: $\gamma^2 - 2\beta\gamma + \omega_0^2 = 0$. If $\beta < \omega_0$, then the equation has two complex conjugate solutions: $\gamma_{1,2} = \beta \pm i\sqrt{\omega_0^2 - \beta^2}$, which result in equal answers, and hence one can take any of them. Applying the Euler formula (see Subsection P4.1.1), we obtain

$$x(t) = A_0 e^{-\beta t}\cos(\omega t + \varphi_0), \qquad \omega = \sqrt{\omega_0^2 - \beta^2}, \tag{P4.1.3.2}$$

where the constants A_0 and φ_0 are determined by the initial conditions.

The function (P4.1.3.2) takes the zero value in equal time intervals $T = 2\pi/\omega$ (Fig. P4.6), and hence ω and T are conditionally called the frequency and period of damped oscillations. If $\beta \ll \omega_0$, then on each time interval $T \ll \tau \ll \beta^{-1}$, the oscillations can be assumed to be harmonic, and $A(t) = A_0\exp(-\beta t)$ has the meaning of the oscillation amplitude on this interval. Therefore (for any $\beta < \omega_0$), $A(t)$ is said to be time-dependent *amplitude of damped oscillations*. In time $\tau = \beta^{-1}$, the amplitude decreases by e times. The amplitude decrease per a period, equal to $A(t)/A(t + T) = \exp(\beta T)$, is called the *damping decrement*, and the

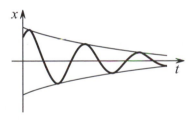

Figure P4.6. Damped oscillations.

logarithm of the decrement

$$\lambda = \ln \frac{A(t)}{A(t+T)} = \beta T \tag{P4.1.3.3}$$

is called the *logarithmic damping decrement*. The quantity $1/\lambda$ is the number of total oscillations in which the amplitude decreases by a factor of e.

If $\beta > \omega_0$ (strong damping), then the quadratic equation for γ has two real positive roots: $\gamma_{1,2} = \beta \pm \sqrt{\beta^2 - \omega_0^2}$, and the general solution

$$x(t) = A_1 \exp(-\gamma_1 t) + A_2 \exp(-\gamma_2 t) \tag{P4.1.3.4}$$

is of non-oscillatory (anharmonic) character (the parameters A_1 and A_2 are determined by the initial conditions).

▶ **Damped electric oscillations.** Comparing the equation of oscillations in an electric circuit (P4.1.2.4) with the equation of damped oscillations (P4.1.3.1), we find the expressions for the damping coefficient and the logarithmic damping decrement:

$$\beta = \frac{R}{2L}, \qquad \lambda = \beta T = \pi \frac{R}{\sqrt{L/C}}. \tag{P4.1.3.5}$$

The condition for weak damping $\lambda \ll 1$ has the form: $R \ll \sqrt{L/C}$ (the quantity $\sqrt{L/C}$ is called the *wave resistance of the circuit*).

▶ **Forced mechanical oscillations.** The motion of a system under the action of an external periodic force is called the *forced oscillations*, the external force itself is called the *driving force*. From the equation of motion $m\ddot{x} = -\kappa x - b\dot{x} + F_x(t)$, we obtain the *equation of forced oscillations*:

$$\ddot{x} + 2\beta\dot{x} + \omega_0^2 x = f(t), \tag{P4.1.3.6}$$

where the notation $f(t) = F_x(t)/m$ is used.

The general solution of such an *inhomogeneous* (with nonzero right-hand side) equation can be represented as the sum of a particular (that is, any) solution of the inhomogeneous equation and the general solution of the homogeneous equation. (The general solution must contain free parameters, which permit satisfying any initial conditions.) The homogeneous equation is the equation of damped oscillations, its general solution (formulas (P4.1.3.2) and (P4.1.3.4)) exponentially decays in time $\tau \sim 1/\beta$. The damping of natural oscillations means the termination of the regime of transient oscillations and the beginning of the regime of *steady-state* forced oscillations whose characteristics are determined by the functions $f(t)$ and the parameters β, ω_0, but are independent of the initial conditions.

We seek a particular solution of equation (P4.1.3.6) in the form of steady-state oscillations. Because any periodic force $F_x(t)$ can be expanded in the Fourier series, it is natural to study the steady-state forced oscillations under the action of a harmonic driving force $f(t) = F_x(t)/m = (F_0/m)\cos\Omega t$. We seek them in the form of harmonic oscillations of the same frequency but with a phase shift:

$$x = A\cos(\Omega t + \varphi). \tag{P4.1.3.7}$$

Substituting (P4.1.3.7) into equation (P4.1.3.6), we obtain

$$A = \frac{F_0}{m\sqrt{(\omega_0^2 - \Omega^2)^2 + 4\beta^2\Omega^2}},$$
$$\tan\varphi = -\frac{2\beta\Omega}{\omega_0^2 - \Omega^2}. \tag{P4.1.3.8}$$

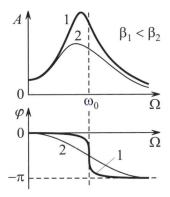

Figure P4.7. The amplitude and phase shift dependence on the frequency of the driving force.

For $\Omega = 0$, we obtain $\varphi = 0$ and $A = F_0/(m\omega_0^2) = F_0/\kappa$, which corresponds to the static displacement of the body following the slowly varying force. For $\Omega \to \infty$, we have $A \to 0$ and $\varphi \to -\pi$. The graphs of $A(\Omega)$ and $\varphi(\Omega)$ are given in Fig. P4.7.

▶ **Resonance.** The maximum value of the amplitude of steady-state oscillations is attained at the *resonance frequency* $\Omega_r = \sqrt{\omega_0^2 - 2\beta^2} = \sqrt{\omega^2 - \beta^2}$ and is equal to $A_{\max} = A(\Omega_r) = F_0/(2m\beta\omega)$, where $\omega = \sqrt{\omega_0^2 - \beta^2}$ is the angular frequency of damped oscillations. For $\beta \ll \omega_0$, the dependence $A(\Omega)$ contains a sharp and narrow maximum at the resonance frequency, which in this limit is close to the natural frequency of the system oscillations. This phenomenon is called *resonance*, and the curves of the dependence $A(\Omega)$ are the *resonance curves*. The maxima characteristics (for $\beta \ll \omega_0$) are: the ratio of $A_{\max}$ to the static deviation $A(0)$ is equal to $\omega/(2\beta) = \pi/\lambda$ (λ is the logarithmic damping decrement); the quantity $Q = \pi/\lambda$ is called the *quality factor* of the oscillatory system. The width of the resonance curve at the level $A_{\max}/\sqrt{2}$ is equal to the damping coefficient: $\Delta\Omega \approx \beta$.

The amplitude of steady-state oscillations of the velocity attains the maximum value $F_0/(2m\beta)$ at $\Omega = \omega_0$. At resonance, the velocity oscillations coincide in phase with the driving force oscillations.

We consider the process of establishing the oscillations in the case where the frequency driving force is equal to the resonance frequency (it is assumed that $\beta \ll \omega_0$). If, at the initial moment, the point displacement and velocity are zero, then in the limit under study the initial conditions are satisfied by the solution:

$$x \approx A_{\max}\sin\omega t - A_{\max}e^{-\beta t}\sin\omega t = A_{\max}\left(1 - e^{-\beta t}\right)\sin\omega t.$$

The obtained dependence is shown in Fig. P4.8. For $t \ll 1/\beta$, the amplitude increases proportionally to time: $A(t) \approx A_{\max}\beta t = = F_0 t/(2m\omega)$; the damping at this stage does not exhibit any influence. We note that the time of establishment of oscillations is large compared with the period: $1/\beta \gg 2\pi/\omega$. If the frequency Ω is close to ω, but differs from it, then the motion at the initial stage $t \ll 1/\beta$ is the sum of oscillations with close frequencies. If the condition $\omega \gg |\Omega - \omega| \gg 1/\beta$ is satisfied, then in the process of establishment of oscillations, profound beats can be observed (the oscillation amplitude increases almost up to $2A_{\max}$ and decreases almost to zero with period $T_b = 2\pi/|\Omega - \omega|$).

Resonance under arbitrary periodic action. If the period of external action $F_x(t)$ is multiple of the period of the system natural oscillations $T = 2\pi/\omega$, then the expansion of $F_x(t)$ in the Fourier series can contain the harmonic with frequency ω. If the Q-factor of an oscillatory system is large, then harmonic oscillations of a noticeable amplitude can arise under the action of this harmonic. For example, the swings can be slightly pushed at each swing (then the first harmonic has the resonance frequency), or at the next nearest swing (then the second harmonic will work), etc.

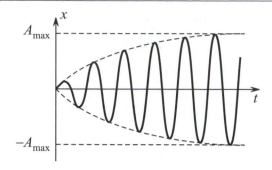

Figure P4.8. Establishment of oscillations at the resonance frequency.

▶ **Forced electric oscillations.** If the oscillatory circuit (see Subsection P4.2.1) contains alternating emf $\mathcal{E}(t)$, then the forced oscillations are described by the same differential equation as that for a mechanical system:

$$\ddot{q} + 2\beta\dot{q} + \omega_0^2 q = \mathcal{E}(t)/L, \tag{P4.1.3.9}$$

where q is capacitor charge, $\beta = R/(2L)$ is the damping coefficient, and $\omega_0^2 = 1/\sqrt{LC}$ is the angular frequency of the circuit natural oscillations. We consider the steady-state oscillations under the action of a harmonic force $\mathcal{E}(t) = \mathcal{E}_0 \cos \Omega t$. The charge oscillations obey the law $q(t) = q_0 \cos(\Omega t + \varphi)$, where the amplitude q_0 and the phase shift φ satisfy expressions (P4.1.3.8). However, in electric circuits, it is of interest to study not the charge oscillations but the oscillations of current $I = \dot{q}$, which occur according to the law $I = I_0 \cos(\Omega t - \varphi_1)$ with amplitude and phase shift equal to

$$I_0 = \mathcal{E}_0 \left[R^2 + \left(\Omega L - \frac{1}{\Omega C} \right)^2 \right]^{-1/2}, \qquad \tan \varphi_1 = \frac{1}{R} \left(\Omega L - \frac{1}{\Omega C} \right). \tag{P4.1.3.10}$$

The maximum of the current oscillation amplitude (resonance) is attained for $\Omega_r = \omega_0 = 1/\sqrt{LC}$, and it is equal to $I_0(\Omega_r) = \mathcal{E}_0/R$. At resonance, the current oscillations coincide in phase with emf oscillations.

The voltage oscillations on the resistor, capacitor, and the induction coil are described by the following formulas:

$$U_R = IR = I_0 R \cos\left(\Omega t - \varphi_1\right),$$
$$U_C = \frac{q}{C} = I_0 \frac{1}{\Omega C} \cos\left(\Omega t - \varphi_1 - \frac{\pi}{2}\right),$$
$$U_L = L\frac{dI}{dt} = I_0(\Omega L) \cos\left(\Omega t - \varphi_1 + \frac{\pi}{2}\right)$$

(the current amplitude is the same on all the elements, because they are connected in series).

The coefficients of proportionality between the current and voltage amplitudes are called as follows: $x_C = 1/(\Omega C)$ is the *capacitive reactance* and $x_L = \Omega L$ is *inductive reactance*. The quantity $x = x_L - x_C$ is called the *reactance*, R is the *resistance* (the energy is dissipated only at this resistance), and the quantity $z = \sqrt{R^2 + x^2}$ is the *impedance* of the circuit. Formula (P4.1.3.10) can be rewritten as $I_0 = \mathcal{E}_0/z$, $\tan \varphi_1 = x/R$. All these relations become rather obvious if the vector diagrams are used (Fig. P4.9). At resonance, the reactance vanishes, $x = 0$, that is, the voltage oscillations on the capacitance and inductance compensate for each other and the circuit impedance becomes purely *resistive*, $z = R$. If the resistance of a circuit is zero, $R = 0$, the impedance is called purely *reactive*, with $z = x$.

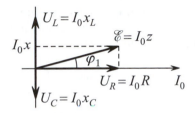

Figure P4.9. Vector diagram of currents and voltages for elements connected in series.

▶ **Power on a subcircuit.** The *active* or *effective value* of alternating current (voltage, emf) is the value of the constant current at which the quantity of heat released on the resistance in a period coincides with the quantity of heat released by the alternating current (that is, the mean square value over the period):

$$\int_0^T I^2 R \, dt = I_{\text{ef}}^2 R T.$$

For the sinusoidal current, $I_{\text{ef}} = I_0/\sqrt{2}$, and $U_{\text{ef}} = U_0/\sqrt{2}$. If the phase shift between the current and the voltage is φ, then the period-average power is equal to

$$P = \frac{1}{T} \int_0^T U_0 \cos(\Omega t) I_0 \cos(\Omega t - \varphi) \, dt = \frac{1}{2} U_0 I_0 \cos \varphi = I_{\text{ef}} U_{\text{ef}} \cos \varphi.$$

For purely reactive impedance, the average power is zero.

P4.2. Wave Processes

P4.2.1. Elastic and Acoustic Waves

▶ **Basic definitions.** If a continuum has elastic properties, then the motion of points at one place of the medium (at the *source*) results in propagation of this motion with a certain velocity in the form of an elastic wave. The medium wave motion is described by the function $\xi(\mathbf{r}, t)$ determining the *medium particle* displacement from equilibrium $\mathbf{r}$ at time t. (The medium particle is a sufficiently small *macroscopic* element of the medium, that is, an element containing rather many atoms and molecules.) An elastic wave can propagate in three-dimensional medium, two-dimensional medium (elastic membrane), and one-dimensional medium (elastic rod, stretched rigid fiber, air column). The simplest wave mode is the *plane wave*, where the function $\xi = \xi(x, t)$ varies only in one direction and depends on the coordinate x. If in this case the vector ξ is perpendicular to the wave propagation direction, then the wave is said to be *transverse*, and if it is parallel, then the wave is said to be *longitudinal*.

A medium is said to be *homogeneous* if all its points are equivalent, *isotropic* if all the directions in it are equivalent, *elastic* if all forces arising in it depend only on the displacement (strain), and *linear* if the forces are proportional to the strains. In an absolutely elastic medium, the wave mechanical energy is not dissipated (is not transformed into internal energy). If the medium has only volume elasticity, then only longitudinal waves can propagate in it (liquid, gas); if it also has elasticity of form, then transverse (shear) waves are also possible. In the linear medium, the wave propagation is described by linear differential equations, and the *superposition principle* holds: if two independent sources cause two different waves $\xi_1(\mathbf{r}, t)$ and $\xi_2(\mathbf{r}, t)$, then the common action of the sources results in the wave $\xi(\mathbf{r}, t) = \xi_1(\mathbf{r}, t) + \xi_2(\mathbf{r}, t)$. The same statement holds for a single source whose motion can be decomposed into two motions. Since any function of time can be represented as the Fourier integral in harmonic functions, it follows that the wave motion of a linear medium can be decomposed into *harmonic waves*. Through an arbitrary point of a harmonic wave, one can draw a single surface of constant phase, which is called the *wave surface*. The wave surface coinciding with the leading edge of the wave perturbation is called the *wave front*.

▶ **Equation of plane harmonic wave.** If the displacement in the source obeys the law $\xi(0, t) = \xi_m \cos \omega t$, then at the points with the coordinate x, the displacement occurs according to the same law, but with the delay by x/v, where v is the wave propagation velocity:

$$\xi(x, t) = \xi_m \cos\left[\omega\left(t - \frac{x}{v}\right)\right] = \xi_m \cos\left(\frac{2\pi}{T}t - \frac{2\pi}{\lambda}x\right) = \xi_m \cos(\omega t - kx), \quad \text{(P4.2.1.1)}$$

where $\lambda = 2\pi v/\omega = vT$ is *wavelength*, and $k = \omega/v = 2\pi/\lambda$ is the *wavenumber*. The wavelength is the shortest distance between points exhibiting synchronous oscillations (the phase difference is $\Delta\varphi = 2\pi$); the wavelength is the spatial period of the harmonic wave profile at any time moment. The phase difference between arbitrary points of the wave is equal to $\Delta\varphi = 2\pi(x_2 - x_1)/\lambda$. In equation (P4.2.1.1), it is assumed that the energy losses at the distance x are negligibly small; a small energy dissipation can be taken into account empirically introducing the factor $\exp(-\gamma x)$, where γ is *coefficient of wave damping*, in (P4.2.1.1). If we introduce the *wave vector* $\mathbf{k}$ directed perpendicularly to the wave front in the direction of wave propagation, then (P4.2.1.1) takes the form invariant under the choice of the coordinate system:

$$\xi(\mathbf{r}, t) = \xi_m \cos(\omega t - \mathbf{k} \cdot \mathbf{r}), \qquad \widetilde{\xi}(\mathbf{r}, t) = \widetilde{\xi}_m e^{i(\omega t - \mathbf{k} \cdot \mathbf{r})}. \quad \text{(P4.2.1.2)}$$

The last form of the traveling wave equation is based on the Euler formula; only the real part $\text{Re}\,\widetilde{\xi}$ has a physical meaning. If the wave velocity is independent of the frequency (there is no dispersion), then the plane wave perturbation of any form propagates without distortion.

▶ **Spherical and cylindrical diverging waves.** If the wave propagates from a point (spherical) or linear (cylindrical) source isotropically in all directions, then the wave surfaces have the respective shape of spheres or circular cylinders. In these cases, the equations of diverging waves without energy losses have the form:

$$\xi(r, t) = \frac{\xi_0}{r} \cos(\omega t - kr) \qquad \text{(spherical wave)},$$

$$\xi(\mathcal{R}, t) = \frac{\xi_0}{\sqrt{\mathcal{R}}} \cos(\omega t - k\mathcal{R}) \qquad \text{(cylindrical wave)},$$

where $r = \sqrt{x^2 + y^2 + z^2}$, $\mathcal{R} = \sqrt{x^2 + y^2}$. The amplitude decreases because the wave surface area increases; the energy conservation law demands that the energy flux remain constant (we recall that the energy of oscillations is proportional to the squared amplitude; see also below).

▶ **Wave equation.** A direct substitution shows that the displacement of the harmonic wave $\xi(\mathbf{r}, t)$ satisfies the equation

$$\frac{\partial^2 \xi}{\partial x^2} + \frac{\partial^2 \xi}{\partial y^2} + \frac{\partial^2 \xi}{\partial z^2} = \frac{1}{v^2}\frac{\partial^2 \xi}{\partial t^2}, \quad \text{or} \quad \Delta\xi = \frac{1}{v^2}\frac{\partial^2 \xi}{\partial t^2}, \quad \text{(P4.2.1.3)}$$

which is called the *wave equation*. If the harmonic wave velocity is independent of the frequency (the dependence of ω on k is linear, that is, the dispersion is absent), then any superposition of plane waves satisfies the same equation. The converse statement also holds: if the equation of motion of a particle of the medium can be reduced to (P4.2.1.3), then any wave perturbations of this medium propagate with the velocity v.

In the one-dimensional case, the general solution of equation (P4.2.1.3) has the form

$$\xi = f_1(x - vt) + f_2(x + vt), \tag{P4.2.1.4}$$

where f_1 and f_2 are arbitrary (differential) functions. The first term describes propagation of the plane signal $f_1(x)$ with velocity v without changes in the form in the direction of positive x, and the second term describes propagation of the signal $f_2(x)$ with velocity v in the direction of negative x.

Example 1. Elastic rod. The forces F arising in a thin rod of cross-sectional area S subjected to longitudinal elastic deformations obeys Hooke's law: $\sigma = E\varepsilon$, where $\sigma = F/S$ is the *stress* at a given point of the rod, ε is the strain, and E is Young's modulus. If x is the coordinate along the rod and $\xi(x,t)$ is the longitudinal displacement of the particle with coordinate x at the time t, the strain at the point x is expressed in terms of the partial derivative of the displacement as $\varepsilon = \dfrac{\xi(x + dx) - \xi(x)}{dx} \approx \dfrac{\partial \xi}{\partial x}$. The force acting on the particle of size Δx and mass $\Delta m = \rho\,\Delta x\,S$ is equal to $[\sigma(x + \Delta x) - \sigma(x)]S$. Then the equation of motion of the particle has the form

$$[\sigma(x + \Delta x) - \sigma(x)]S = \Delta m \frac{\partial^2 \xi}{\partial t^2} \quad \Longrightarrow \quad \frac{\partial^2 \xi}{\partial x^2} = \frac{\rho}{E} \frac{\partial^2 \xi}{\partial t^2}.$$

Hence, the wave propagation speed in an elastic rod is equal to $v = \sqrt{E/\rho}$. Longitudinal waves in elastic media propagate with the same speed.

Example 2. Stretched string. A transverse wave can propagate along a string whose tension is equal to T. Suppose that the string profile at a given time moment is described by the function $\xi(x,t)$ (Fig. P4.10). We obtain the equation of motion of the medium particle of dimension Δx under the assumption of low-slope wave: $\partial \xi/\partial x \ll 1$. The vertical projection of the tensile force is equal to $F_\perp(x) = T\sin\alpha \approx T(\partial \xi/\partial x)$ (α is the string inclination angle); the vertical force acting on the medium particle of dimension Δx and mass $\Delta m = \rho_{\text{lin}}\Delta x\,S$ is equal to $F_\perp(x + \Delta x) - F_\perp(x)$ (ρ_{lin} is the linear string density). The equation of motion of the medium particle has the form

$$\frac{\partial^2 \xi}{\partial x^2} = \frac{\rho_{\text{lin}}}{T} \frac{\partial^2 \xi}{\partial t^2},$$

hence the wave propagation velocity in the string is equal to $v = \sqrt{T/\rho_{\text{lin}}}$.

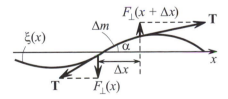

Figure P4.10. Element of a stretched string (to example 2).

Example 3. Air column in a cylinder. The volume elasticity of a gas (or any thermodynamically simple medium) is characterized by the bulk modulus of elasticity $K = -V(\partial p/\partial V) = \rho(\partial p/\partial \rho)$. The value of K depends on the condition under which bulk deformation of the gas occurs. For sound oscillations ($\nu > 16$ Hz), one can neglect the heat exchange between distinct medium particles and assume that K is the adiabatic modulus of elasticity: $K_{\text{ad}} = -V\left(\partial p/\partial V\right)_{\text{ad}}$. The variation in the pressure at a given point of the cylinder (for small strains) is $\Delta p(x) = -\lim_{V\to 0} K(\Delta V/V) = -\lim_{\Delta x\to 0} K[S\xi(x+\Delta x) - S\xi(x)]/(S\Delta x) = -K(\partial \xi/\partial x)$; the force acting on the air medium particle of dimension Δx and mass $\Delta m = \rho\,\Delta x\,S$ is equal to $\Delta F = [\Delta p(x) - \Delta p(x+\Delta x)]S$. The equation of motion of the medium particle has the form

$$\frac{\partial^2 \xi}{\partial x^2} = \frac{\rho}{K} \frac{\partial^2 \xi}{\partial t^2},$$

and hence the wave propagation velocity in the air cylinder is equal to $v = \sqrt{K/\rho} = \sqrt{\partial p/\partial \rho}$. (These expressions remain valid for any thermodynamically simple medium such as a real gas or a liquid.) For the sound waves in an ideal gas, we obtain: $K = \gamma p$, $v = \sqrt{\gamma p/\rho} = \sqrt{\gamma RT/\mu}$, where γ is the adiabatic exponent (see Subsection P2.2.3).

▶ **Volume density of energy. Energy flux density.** The densities of the kinetic and potential energies at a given point of the plane elastic longitudinal wave are equal to

$$w_k = \frac{\rho}{2}\left(\frac{\partial \xi}{\partial t}\right)^2, \qquad w_p = \frac{E}{2}\left(\frac{\partial \xi}{\partial x}\right)^2.$$

For a traveling wave of arbitrary form $\xi = f(x \pm vt)$, we have the relation $(\partial \xi/\partial t)^2 = v^2(\partial \xi/\partial x)^2$. Taking into account that $v^2 = E/\rho$ for the elastic longitudinal wave, we see that the densities of the kinetic and potential energies at each point of the traveling wave are equal to each other. This statement holds for traveling waves of any nature. In the case of a harmonic plane wave (P4.2.1.1), the wave energy per unit volume and its time average have the form:

$$w = w_k + w_p = \rho\omega^2\xi_m^2 \sin^2(\omega t - kx), \qquad \langle w \rangle = \tfrac{1}{2}\rho\omega^2\xi_m^2.$$

The energy flux density (the *Umov vector*) at a given point of the traveling wave is equal to

$$\mathbf{U} = w\mathbf{u}, \tag{P4.2.1.5}$$

where $\mathbf{u}$ is the rate of the energy transfer. In the dispersionless medium, $\mathbf{u} = \mathbf{v}$, where $\mathbf{v}$ is the wave propagation velocity contained in the wave equation (P4.2.1.4). For harmonic traveling wave (P4.2.1.1), $u = v$, where v is the phase velocity of the wave: $v = \omega/k$. The energy transferred by the traveling wave through the site $d\mathbf{s}$ per unit time (the energy flux) is equal to $d\Phi = \mathbf{U} \cdot d\mathbf{s}$. The magnitude of time-average value of the Umov vector (it is sometimes called the *wave intensity*) for a harmonic plane wave is equal to $I = |\langle \mathbf{U} \rangle| = v\langle w \rangle = \tfrac{1}{2}\rho v\omega^2\xi_m^2$. In the case of an arbitrary (including the standing) longitudinal elastic wave, one can use the expression for $\mathbf{U}$ in terms of the power of elastic forces (per unit area):

$$\mathbf{U} = -\sigma\frac{\partial \boldsymbol{\xi}}{\partial t},$$

where $\sigma = E\,\partial \xi/\partial x$ is the stress at a given point of the medium (see example 1). For the transverse wave in the string (example 2), the energy flux has a similar form:

$$P = -F_\perp\frac{\partial \xi}{\partial t} = -T\frac{\partial \xi}{\partial x}\frac{\partial \xi}{\partial t}.$$

▶ **Dispersion. Group velocity.** The dependence of the harmonic wave velocity on the frequency or the wavelength is called the *dispersion*, and the medium where this phenomenon is observed is called the *dispersion medium*. In the dispersion medium, the nonsinusoidal traveling pulse changes its shape in the process of motion, because the harmonic waves contained in its Fourier expansion move with different velocities. The energy transfer rate contained in expression (P4.2.1.5) for the Umov vector is determined in this case as the velocity of motion of the center of pulse (or as the velocity of the point at which the strain is maximum) and is called the *group velocity* of the wave packet (in contrast to the *phase velocity* of the sinusoidal wave $v = \omega/k$). A finite pulse of dimension Δx is called the *wave packet* because its expansion in the Fourier integral in harmonic waves contains a finite

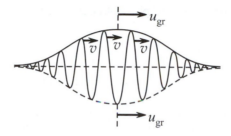

Figure P4.11. Phase and group velocities.

group of waves whose spectral width is determined by the relation: $\Delta x \, \Delta k \sim 1$. The group velocity of such a pulse is equal to

$$u_{\text{gr}} = \frac{d\omega}{dk} = v - \lambda \frac{dv}{d\lambda}, \qquad (\text{P4.2.1.6})$$

where $v = \omega/k$ is the phase velocity of the wave. The points of maxima (the "humps") inside the wave packet move with the phase velocity v, and the envelope of these maxima moves with the velocity u_{gr} (Fig. P4.11).

The notion of group velocity of a wave packet can be used only if its spectral width is small so that the dependence $\omega(k)$ can be assumed to be linear.

To illustrate this, we consider the wave that is the superposition of at most two traveling waves with close frequencies and equal amplitude:

$$\xi = \xi_m \cos[(\omega + \Delta\omega)t - (k + \Delta k)x] + \xi_m \cos[(\omega - \Delta\omega)t - (k - \Delta k)x]$$
$$= 2\xi_m \cos(\Delta\omega \, t - \Delta k \, x) \cos(\omega t - kx).$$

This shows that the wave envelope has the spatial period $\Delta x = 2\pi/\Delta k$ and propagates with velocity $u = \Delta\omega/\Delta k$.

Example 4. The group velocity can be both less and greater than the phase velocity. The phase velocity of waves on the surface of a fluid for long (gravitational) waves depends on the wavelength according to the law $v = \sqrt{g\lambda/2\pi}$, and for short (capillary) waves, according to the law $v = \sqrt{2\pi\sigma/\rho\lambda}$. Substituting this in formula (P4.2.1.6), we see that the group velocity in the first case is equal to $u_{\text{gr}} = v/2$, and in the second, to $u_{\text{gr}} = 3v/2$. We also see that the group velocity can be strongly different from the phase velocity.

▶ **Wave interference.** Two waves are called *coherent* if, at any point of the space, they generate coherent oscillations whose phase difference does not vary in time (see Subsection P4.1.1). The sources of coherent waves are called *coherent sources*. In the case of addition of incoherent waves, the time-average energy of the resultant oscillation is equal to the sum of their average energies. In the case of addition of coherent waves and in the case of particle oscillations that occur in the same or close directions, the phenomenon of *interference* can be observed. Interference of two or more coherent waves is characterized by increased amplitudes of the resulting wave at some points of space (where wave crests meet) and decreased amplitudes at other points (where a crest of a wave meets a trough of another wave). If a point lies at the distance r_1 from one of the coherent sources and at the distance r_2 from the other, then the phase difference between the oscillations at this point is equal to

$$\Delta\varphi = (\varphi_{10} - kr_1) - (\varphi_{20} - kr_2) = 2\pi\frac{r_2 - r_1}{\lambda} - (\varphi_{20} - \varphi_{10}) = 2\pi\frac{\Delta}{\lambda} - \Delta\varphi_0.$$

Here $\Delta\varphi_0$ is the phase difference between the source oscillations, $\Delta = (r_2 - r_1)$ is the *wave path difference*. The maximum conditions for the resultant oscillations is $\Delta\varphi = \pm 2\pi m$, the minimum condition is $\Delta\varphi = \pm 2\pi \left(m - \frac{1}{2}\right)$, where m is the order of the interference maximum (minimum). In the case of in-phase sources ($\Delta\varphi_0 = 0$) these conditions take the following especially simple form:

$$\Delta = \pm m\lambda \qquad \text{(maximum)},$$
$$\Delta = \pm\left(m - \tfrac{1}{2}\right)\lambda \qquad \text{(minimum)}.$$

Example 5. We consider two linear coherent sources of cylindrical waves that are at the distance $\lambda/2$ from each other. For the points at far distances from sources, the wave propagation difference is determined only by the direction of radiation and is equal to $\Delta = \frac{1}{2}\lambda\cos\theta$ (Fig. P4.12). In the case of in-phase sources, the radiation is weakened in the direction $\theta = 0$ and reinforced in the direction $\theta = \pi/2$, and in the case of sources oscillating in opposite phase ($\Delta\varphi_0 = \pi$), the situation is opposite.

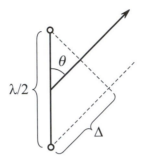

Figure P4.12. To example 5.

▶ **Standing waves. Natural oscillations.** An important example of interference is the *standing wave*, arising in addition of two equal plane traveling waves propagating towards each other:

$$\xi = \xi_m\cos(\omega t - kx) + \xi_m\cos(\omega t + kx) = 2\xi_m\cos kx\cos\omega t. \qquad \text{(P4.2.1.7)}$$

The oscillation amplitude $A(x) = 2\xi_m\cos kx$ of the wave points varies periodically from zero (at the wave *nodes*) to $2\xi_m$ (at the wave *antinodes*). The distance between neighboring nodes is equal to $\lambda/2$. The oscillations of points between two nodes occur in phase, but the oscillations of points on the opposite sides of the node occur in opposite phase. At the nodes, the velocity oscillation amplitude is zero, but the oscillation amplitude of the strain $\partial\xi/\partial x$ takes the maximum value. Conversely, at the antinodes, the strains are zero, but the velocity oscillations have the maximum amplitude. The time-average energy has the same value at all the points of the standing wave. The Umov vector is zero at both the nodes and antinodes and periodically changes its direction at the intermediate points.

If the end of the rod in oscillatory motion is rigidly fixed, then the displacement of the rod end points is zero; if the rod end is free, then the strain is zero. In the first case, the boundary condition has the form $\xi|_{x=0} = 0$, and in the second case, the form $\partial\xi/\partial x|_{x=0} = 0$, where the value $x = 0$ corresponds to the rod end. (The same conditions arise for the air oscillations in a cylindrical tube in the case of closed or open end of the tube. For transverse oscillations of the stretched string considered in example 2, only the first version in which the string end is fixed is possible.) If all the points of a finite rod oscillate harmonically with a single frequency, then such a motion is called a *natural* or *free oscillation* of the rod. If an infinitely long rod is mentally cut at the point where the standing wave node is located, then the boundary condition for the fixed end is satisfied; and if the cut is at the point of antinode, then the boundary condition for the free end is satisfied. Hence the free oscillations of a rod of length l in the case of two free or two fixed ends must satisfy the condition $l = \frac{1}{2}m\lambda$ or $\omega_m = \pi m v/l$ (the spectrum of natural oscillations, $m = 1, 2, \ldots$),

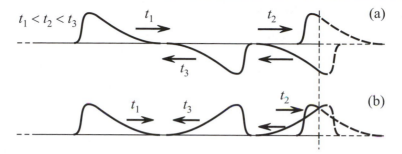

Figure P4.13. Reflection of a wave pulse from an end of a rod: (a) fixed end, (b) free end.

and in the case of one fixed end, the condition $l = \frac{1}{2}\left(m - \frac{1}{2}\right)\lambda$ or $\omega_m = \pi\left(m - \frac{1}{2}\right)v/l$. The natural oscillation with the least possible frequency is called the *fundamental oscillation*, all the other natural oscillations are *overtones* or *harmonics*.

▶ **Traveling wave reflection.** We consider reflection of traveling pulse from the rod end. If the rod end is fixed, then the boundary condition $\xi|_{x=0} = 0$ and the initial conditions (one incident pulse) are satisfied by the superposition of the incident pulse and the pulse of just the same form traveling towards it in which the point displacements have opposite direction (Fig. P4.13a). But if the rod end is free, then the boundary condition $\partial\xi/\partial x|_{x=0} = 0$ is satisfied by the pulse of displacements of the same sign traveling towards it (Fig. P4.13b). In the case of a harmonic traveling wave, the reflection from the free end occurs without changes in the phase, and a standing wave with antinode at the rod end is formed in this case. The reflection from the fixed end occurs with the phase shift by π, and a standing wave with node at the rod end is formed.

▶ **Doppler effect in acoustics.** In motion of the wave source and (or) the wave receiver with respect to the medium where the wave propagates, the registered frequency differs from the one under study (if the distance between the source and the receiver varies). We direct the axis x from the receiver to the source. If the source radiates sound with frequency ν_0 and moves along the axis x with velocity v_x^S (Fig. P4.14), then the distance between the

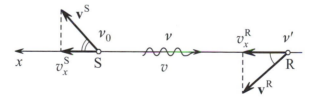

Figure P4.14. Doppler effect arises when the source (S) moves and also when the receiver (R) moves.

neighboring maxima of the wave propagating along x with velocity v in the direction of the receiver is equal to $\lambda = vT_0 + v_x^S T_0$, and its frequency is equal to v/λ:

$$\nu = \nu_0 \frac{v}{v + v_x^S}. \tag{P4.2.1.8}$$

If the receiver is fixed, then precisely this frequency is registered. If the receiver moves with velocity v_x^R, then the time interval between the arrival of the neighboring maxima is equal to $T' = \lambda/(v + v_x^R)$, and the frequency of the received signal is equal to $1/T'$:

$$\nu' = \nu \frac{v + v_x^R}{v} = \nu_0 \frac{v + v_x^R}{v + v_x^S}. \tag{P4.2.1.9}$$

The Doppler effect is observed if the distance between the source and the receiver varies.

P4.2.2. Electromagnetic Waves

▶ **Properties of electromagnetic waves.** The electromagnetic wave is the propagation of perturbations of an electromagnetic field in empty space or in a medium without sources. The existence of electromagnetic waves follows from the Maxwell theory (see Subsection P3.11.1), which states that an alternating electric field generates an alternating magnetic field and an alternating magnetic field generates an alternating electric field. In the case of a homogeneous isotropic medium, the Maxwell equation have the form

$$
\text{1) } \operatorname{curl} \mathbf{E} = -\frac{\partial \mathbf{B}}{\partial t}, \qquad \text{2) } \operatorname{curl} \mathbf{H} = \frac{\partial \mathbf{D}}{\partial t},
$$

$$
\text{3) } \operatorname{div} \mathbf{D} = 0, \qquad \text{4) } \operatorname{div} \mathbf{B} = 0, \tag{P4.2.2.1}
$$

where $\mathbf{D} = \varepsilon\varepsilon_0\mathbf{E}$ and $\mathbf{B} = \mu\mu_0\mathbf{H}$. For the electromagnetic field depending only on the coordinate x, we obtain the equations:

$$
-\mu\mu_0\frac{\partial H_z}{\partial t} = \frac{\partial E_y}{\partial x}, \qquad \mu\mu_0\frac{\partial H_y}{\partial t} = \frac{\partial E_z}{\partial x},
$$

$$
\varepsilon\varepsilon_0\frac{\partial E_y}{\partial t} = -\frac{\partial H_z}{\partial x}, \qquad \varepsilon\varepsilon_0\frac{\partial E_z}{\partial t} = \frac{\partial H_y}{\partial x}. \tag{P4.2.2.2}
$$

In this case, the equations for longitudinal components have the form $\partial E_x/\partial x = \partial E_x/\partial t = 0$, $\partial H_x/\partial x = \partial H_x/\partial t = 0$, that is, all longitudinal components of the field are constant quantities.

It is easy to see that the transverse components of the electromagnetic field satisfy the wave equation:

$$
\Delta E_y = \varepsilon\varepsilon_0\mu\mu_0\frac{\partial^2 E_y}{\partial t^2}, \qquad \Delta H_z = \varepsilon\varepsilon_0\mu\mu_0\frac{\partial^2 H_z}{\partial t^2}
$$

(the quantities E_z and H_y satisfy the same equations). It follows from these equations that the electromagnetic perturbations in a medium propagate with the velocity

$$
v = \frac{1}{\sqrt{\varepsilon\varepsilon_0\mu\mu_0}} = \frac{c}{\sqrt{\varepsilon\mu}} = \frac{c}{n}, \tag{P4.2.2.3}
$$

where $c = 1/\sqrt{\varepsilon_0\mu_0} \approx 3 \times 10^8$ m/c is the *velocity of electromagnetic waves in vacuum*, and $n = \sqrt{\varepsilon\mu}$ is the *index of refraction* of the medium.

We consider the plane electromagnetic wave of a certain frequency (the *monochromatic plane wave*) propagating along the axis x: $E_y = E_{0y}\cos(\omega t - kx)$, $E_z = E_{0z}\cos(\omega t - kx + \varphi)$. From (P4.2.2.2) we find that $H_y = -\sqrt{\varepsilon\varepsilon_0/\mu\mu_0}\,E_z$ and $H_z = \sqrt{\varepsilon\varepsilon_0/\mu\mu_0}\,E_y$. These formulas permit one to obtain the following properties of the plane traveling electromagnetic wave:

1) The electromagnetic wave is *transverse*.

2) Since $\mathbf{E} \cdot \mathbf{H} = 0$, it follows that $\mathbf{E} \perp \mathbf{H}$. The vectors $(\mathbf{E}, \mathbf{H}, \mathbf{k})$ form the right triple of vectors; that is, the direction of the vector $\mathbf{E} \times \mathbf{H}$ coincides with the direction of the wave propagation.

3) The values of the vectors $\mathbf{E}$ and $\mathbf{H}$ at each time moment are related as $\sqrt{\varepsilon\varepsilon_0}\,E = \sqrt{\mu\mu_0}\,H$ ($E = vB$).

The character of the wave polarization depends on the phase difference φ. For $\varphi = \pm m\pi$, the wave is *plane polarized*; that is, the vector $\mathbf{E}$ oscillates in one plane (this plane is called the *wave polarization plane*), and for another phase difference, the wave is *elliptically polarized*, or (for $\varphi = \pm(m + \frac{1}{2})\pi$ and for $E_{0x} = E_{0y}$), the wave is *circularly polarized* (see Subsection P4.1.1).

▶ **Energy of electromagnetic waves.** The volume density of the electromagnetic wave energy is calculated by the formula

$$w = \tfrac{1}{2}\varepsilon\varepsilon_0 E^2 + \tfrac{1}{2}\mu\mu_0 H^2 = \varepsilon\varepsilon_0 E^2 = \mu\mu_0 H^2 = \frac{1}{v}EH.$$

In the case of a monochromatic plane polarized wave, we have

$$w = \varepsilon\varepsilon_0 E_0^2 \cos^2(\omega t - kx), \qquad \langle w \rangle = \tfrac{1}{2}\varepsilon\varepsilon_0 E_0^2.$$

For the elliptically polarized wave:

$$w = \varepsilon\varepsilon_0 \left[E_{0y}^2 \cos^2(\omega t - kx) + E_{0z}^2 \cos^2(\omega t - kx + \varphi) \right],$$
$$\langle w \rangle = \tfrac{1}{2}\varepsilon\varepsilon_0 \left(E_{y0}^2 + E_{z0}^2 \right).$$

For the wave circularly polarized ($E_{0y} = E_{0z} = E_0$, $\varphi = \tfrac{1}{2}\pi \pm \pi m$), the energy density does not vary in time: $w = \varepsilon\varepsilon_0 E_0^2$.

The energy flux density in an electromagnetic wave is determined by the Poynting vector (see Subsection P3.11.2):

$$\mathbf{S} = w\mathbf{v} = \mathbf{E} \times \mathbf{H},$$

which points in the direction of the wave propagation. The *intensity of an electromagnetic wave* is the magnitude of the time-average of the Poynting vector:

$$I = |\langle \mathbf{S} \rangle| = v\langle w \rangle.$$

▶ **Wave pressure.** According to the theory of relativity, the energy flux density of electromagnetic field in a vacuum means that there exists a momentum of the electromagnetic field. The volume density of the electromagnetic field momentum is equal (in magnitude) to w/c, the momentum flux density is equal to $\mathbf{S}/c$, and its modulus is equal to w. In the case of wave reflection or absorption, there is a variation in the wave momentum, which must manifest itself in the wave pressure on an obstacle. In the case of normal reflection of a plane wave, the pressure is equal to $p = \langle w \rangle(1 + R)$, where R is the fraction of the reflected energy (the *coefficient of reflection*). Within the framework of electrodynamics, the wave pressure is explained by the action of the wave magnetic field on the current excited on reflecting surface by the wave electric field.

▶ **Radiation of plane current.** The simplest example of a system creating alternating fields and radiating a plane electromagnetic wave is the time-alternating plane current. We consider the current in the plane (x, z) directed along z and varying by the law $i_z = i_0 \cos \omega t$ (Fig. P4.15). It follows from the second of Maxwell's equations in integral form (see Subsection P3.11.1) that the displacement current vanishes, and hence, for a narrow loop surrounding the current, we have $B_x(0, t) = \pm\tfrac{1}{2}\mu_0 i_0 \cos \omega t$, where the minus sign corresponds to the field on the right of the plane (for small $y > 0$), and the plus sign corresponds to the field on the left of the plane. Using the relation between E and B in a plane wave, we obtain $E_z(0, t) = -A_E \cos \omega t$, where $A_E = \tfrac{1}{2}\mu_0 c i_0 = i_0/(2\varepsilon_0 c)$. The electric field is directed opposite to the current, its work over the current is negative; that is, the current gives energy to the wave. The field far from the plane has the form $E(y, t) = -A_E \cos(\omega t - ky)$ for $y > 0$ and $E(y, t) = -A_E \cos(\omega t + ky)$ for $y < 0$.

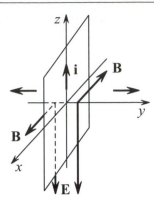

Figure P4.15. Radiation of plane current.

▶ **Wave radiation by moving charges.** The radiation of electromagnetic waves occurs in accelerated motion of charges. An electric dipole whose dipole moment rapidly varies in time (a *Hertz oscillator* or a *Hertz vibrator*) is a simple radiating system. Since the electromagnetic perturbations propagate with velocity c, the field of radiation at the distance r at time t is determined by the motion of charges in the dipole at time $t' = t - r/c$. The radiation field has a rather simple form only in the *wave zone*; that is, at distances that are large compared with both the dipole dimensions and $\lambda = c\tau$, where τ is the typical time of the dipole moment variation (in the case of harmonic oscillations, λ is the wavelength). The electric and magnetic fields in the wave zones are perpendicular to the radius vector $\mathbf{r}$ drawn from the dipole to the point of observation:

$$\mathbf{E} = \frac{1}{4\pi\varepsilon_0 c^2 r^3}\left[(\ddot{\mathbf{p}}\times\mathbf{r})\times\mathbf{r}\right]_{t-r/c}, \qquad \mathbf{B} = \frac{\mu_0}{4\pi c r^2}\left[\ddot{\mathbf{p}}\times\mathbf{r}\right]_{t-r/c}.$$

The mutual location of the vectors $\mathbf{E}$, $\mathbf{H}$, and $\mathbf{r}$ is shown in Fig. P4.16.

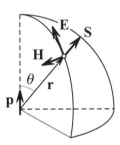

Figure P4.16. Radiation of a moving electric dipole.

The solution of the Maxwell equations in the form or retarded potentials (see Subsection P3.11.1) permits one to write the expressions for the potentials at the point $\mathbf{r}$ at time t. For example, the vector potential has the form:

$$\mathbf{A}(\mathbf{r}, t) = \frac{\mu_0}{4\pi}\frac{q\mathbf{v}(t-r/c)}{r} = \frac{\mu_0}{4\pi}\frac{\dot{\mathbf{p}}(t-r/c)}{r}.$$

In calculations of $\mathbf{B} = \text{curl}\,\mathbf{A}$, the principal role in the wave zone is played by the term arising in the argument differentiation $\dot{\mathbf{p}}$, which results in the above expression for $\mathbf{B}$. Similar but more cumbersome calculations can also be performed for the electric field.

The fields decrease with the distance according to the law $1/r$. The Poynting vector in the wave zone is directed along $\mathbf{r}$; in the direction making an angle θ with the vector $\mathbf{p}$, it is equal to

$$S = \frac{1}{4\pi\varepsilon_0}\frac{\sin^2\theta}{4\pi c^3 r^2}\left(\ddot{\mathbf{p}}\big|_{t-r/c}\right)^2.$$

The energy flux through a closed surface is precisely the power of the dipole radiation:

$$P = \int_0^\pi S(\theta) 2\pi r^2 \sin\theta \, d\theta = \frac{1}{4\pi\varepsilon_0} \frac{2\ddot{\mathbf{p}}^2}{3c^3}.$$

Assuming that only one of the dipole charges is in motion, we obtain the formula for the power of radiation of a particle in accelerated motion:

$$P = \frac{1}{4\pi\varepsilon_0} \frac{2q^2\ddot{\mathbf{r}}^2}{3c^3}, \qquad\qquad (\text{P4.2.2.4})$$

which is proportional to its squared acceleration.

An electric dipole whose moment varies by the harmonic law, $\mathbf{p} = \mathbf{p}_0 \cos\omega t$, is a *linear harmonic oscillator*. The dependence of the radiated wave intensity on the angle θ has the form

$$I = |\langle\mathbf{S}\rangle| = \frac{1}{4\pi\varepsilon_0} \frac{p_0^2 \omega^4 \sin^2\theta}{8\pi c^3 r^2}.$$

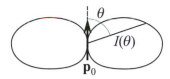

Figure P4.17. Directional pattern of a linear oscillator.

The dependence $I(\theta)$ for $r = \text{const}$ (the *directional pattern*) is shown in Fig. P4.17. The time-average power of radiation of such an oscillator

$$P = \frac{1}{4\pi\varepsilon_0} \frac{\omega^4 \mathbf{p}_0^2}{3c^3} \qquad\qquad (\text{P4.2.2.5})$$

is proportional to the fourth power of its frequency.

Example 1. A free electron (with mass m) in the field of a plane electromagnetic wave with amplitude E_0 exhibits forced oscillations with amplitude $A = eE_0/(m\omega^2)$ (the equation of motion of the electron has the form $m\ddot{x} = -eE_0 \cos\omega t$). One can see that the radiation power of a free electron in the field of an electromagnetic wave is independent of the frequency, and the scattering cross-section (the ratio of the radiation power to the incident wave intensity) is equal to

$$\sigma = \frac{P}{I} = \frac{8}{3} \frac{(ke^2)^2}{m^2 c^4} \sim 10^{-30} \, \text{m}^2.$$

Another situation arises if, in the electromagnetic wave field, there is an atomic electron the natural frequency ω_0 of whose oscillations is large compared to the frequency of the incident (light) wave. Then the oscillation amplitude (see Subsection P4.1.3) is equal to $A = eE_0/(m\omega_0^2)$; that is, it is almost independent of ω. In this case, the radiation power and the light scattering cross-section are proportional to ω^4, which qualitatively explains the blue color of the sky (the blue color scattering is stronger).

Example 2. In a hydrogen atom, let the electron move about the nucleus in a circular orbit with a decreasing radius r and speed v. The electron energy and radiation power are

$$W = \frac{mv^2}{2} - \frac{ke^2}{r} = -\frac{ke^2}{2r}, \qquad P = \frac{2ke^2\ddot{r}}{3c^3} = \frac{2}{3} \frac{(ke^2)^3}{m^2 c^3} \frac{1}{r^4},$$

since $m\ddot{r} = mv^2/r = ke^2/r^2$. It follows from the law of conservation of energy, $P = -dW/dt$, that

$$\frac{dr}{dt} = -\frac{4}{3} \frac{(ke^2)^2}{m^2 c^3} \frac{1}{r^2}.$$

Separating the variables and integrating, we find the time in which the radius decreases from its initial value r_0 to zero (i.e., the time in which the electron falls onto the nucleus, or the *deexcitation time*):

$$t = \frac{1}{4} \frac{m^2 c^3}{(ke^2)^2} r_0^3 \sim 10^{-10} \, \text{s},$$

where $r_0 \sim 10^{-10}$ m is the atomic radius.

▶ **Doppler effect for electromagnetic waves in vacuum.** The main distinction from the acoustic Doppler effect (see Subsection P4.2.1) is that there is no reference system associated with the medium, where the wave propagates. Therefore, the relation between the radiation frequency ν_0 (measured in the source reference system K_S) and the registered frequency ν (measured in the receiver reference system K_R) can depend only on their relative velocity. Assume that in K_R, the source moves with velocity v directed at the angle θ made with the radius vector drawn from the receiver to the source. Repeating the argument given in Subsection P4.2.1, we obtain

$$\nu = \frac{\widetilde{\nu}_0}{1 + (v/c)\cos\theta},$$

where $\widetilde{\nu}_0$ is the source frequency measured by the clock in the system K_R. Passing to the intrinsic time ($\widetilde{T}_0 = \gamma T_0$, or $\widetilde{\nu}_0 = \nu_0/\gamma$; see Subsection P1.10.2), we obtain

$$\nu = \frac{\nu_0}{\gamma[1 + (v/c)\cos\theta]} = \frac{\nu_0\sqrt{1 - v^2/c^2}}{1 + (v/c)\cos\theta}. \tag{P4.2.2.6}$$

One can see that, because of the time transformation, not only the longitudinal but also the transverse Doppler effect is observed (for the source motion perpendicular to the direction to the receiver).

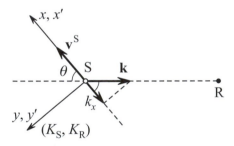

Figure P4.18. To the explanation of the Doppler effect for electromagnetic waves.

It is useful to consider the Doppler effect from a different viewpoint. Since the oscillation phase at a given point of the wave ($\omega t - \mathbf{k} \cdot \mathbf{r}$) must be invariant under the transition to another reference system, the ($\mathbf{k}, \omega/c$) is a four-vector of zero length ($k^2 - \omega^2/c^2 = 0$). If, in the frame K_R, the axis x is directed parallel to the source velocity $\mathbf{v}^S$ (Fig. P4.18), then, in this reference system, we have $k_x = -k\cos\theta = -(\omega/c)\cos\theta$. Using the Lorentz transformations to pass to the frame K_S, we obtain: $\omega_0/c = \gamma[(\omega/c) - k_x(v/c)]$, or $\omega_0 = \gamma[\omega + \omega(v/c)\cos\theta]$. Expressing ω, we obtain (P4.2.2.6). This method allows us, by writing the Lorentz transformations for the wave vector $\mathbf{k}$, to obtain formulas for the light aberration phenomenon, which is important in astronomy.

▶ **Scale of electromagnetic waves.** Depending on the way of radiation and registration of electromagnetic waves, several ranges with conditional (overlapping) boundaries are distinguished in the scale of the frequency ν or of the wavelengths in a vacuum $\lambda = c/\nu$:
1. The radio waves ($\lambda > 0.05$ mm).
2. The optical (light) radiation (10 nm $< \lambda <$ 1 mm):
    a) the infrared (IR) radiation (770 nm $< \lambda <$ 1 mm),
    b) the visible light (380 nm $< \lambda <$ 770 nm),
    c) the ultraviolet (UV) radiation (10 nm $< \lambda <$ 380 nm).
3. The X-rays radiation (0.01 pm $< \lambda <$ 100 nm).
4. The gamma radiation (gamma rays) ($\lambda < 0.1$ nm).

Bibliography for Chapter P4

Alenitsyn, A. G., Butikov, E. I., Kondratyev, A. S., *Concise Handbook of Mathematics and Physics*, CRC Press, Boca Raton, Florida, 1998.

Bekefi, G. and Barrett, A. H., *Electromagnetic Vibrations, Waves, and Radiation*, The MIT Press, Cambridge, Massachusetts, 1977.

Benenson, W., Harris, J. W., Stocker, and Lutz, H. (Eds.), *Handbook of Physics*, Springer-Verlag, New York, 2002.

Bloch, I., *The Physics of Oscillations and Waves: With Applications in Electricity and Mechanics*, Springer-Verlag, New York, 2004.

Crawford, F. S., Jr., *Waves and Oscillations. Berkeley Physics Course. Volume 3*, McGraw-Hill, New York, 1968.

Elmore, W. C. and Heald, M. A., *Physics of Waves*, Dover Publications, New York, 1985.

French, A. P., *Vibrations and Waves (M.I.T. Introductory Physics Series)*, W. W. Norton & Company, New York, 1971.

Gorelik, G. S., *Oscillations and Waves. Introduction to Acoustics, Radiophysics and Optics* [in Russian], Fizmatlit, Moscow, 2007.

Ingard, K. U., *Fundamentals of Waves and Oscillations*, Cambridge University Press, Cambridge, England, 1988.

Irodov, I. E., *Wave Processes. Basic Laws* [in Russian], Fizmatlit, Moscow, 2004.

Kneubühl, F. K., *Oscillations and Waves*, Springer, Berlin, 1997.

Pain, H. J., *The Physics of Vibrations and Waves, 6th Edition*, John Wiley & Sons, Chichester, England, 2005.

Prokhorov, A. M. (Ed.), *Encyclopedia of Physics, Volumes 1–5* [in Russian], Bolshaya Russkaya Encyclopedia, Moscow, 1998.

Prokhorov, A. M. (Ed.), *Encyclopedic Dictionary of Physics* [in Russian], Sovetskaya Encyclopedia, Moscow, 1984 (dic.academic.ru).

Sivukhin, D. V., *General Physics, Volume 4, Optics* [in Russian], Fizmatlit, Moscow, 2006.

Sivukhin, D. V., Yakovlev, I. A., *A Collection of Problems, Volume 4, Optics* [in Russian], Fizmatlit, Moscow, 2006.

Smith, W. F., *Waves and Oscillations: A Prelude to Quantum Mechanics* , Oxford University Press, Oxford, England, 2010.

Woan, G., *The Cambridge Handbook of Physics Formulas*, Cambridge University Press, Cambridge, England, 2003.

Chapter P5

Optics

P5.1. Geometric Optics. Radiometry

P5.1.1. Geometric Optics

▶ **Fundamental laws of geometric optics.** Optics studies the electromagnetic radiation of optic (light) range (see Subsection P4.2.2) as well as phenomena arising in its propagation in space and under interaction with matter. *Geometric optics* does not consider the wave character and polarization of light radiation and deals with notions of *light rays* pointing out the direction of light propagation and narrow *light beams* formed by light rays.

The fundamental laws of geometric optics are listed below.

1. *Law of rectilinear light propagation.*

2. *Law of independence of light beams.* The energy in each beam propagates independently of other beams; illumination of the surface on which several beams are incident, is equal to the sum of illuminations created by each beam separately.

3. *Law of light reflection.* The reflected ray lies in *incidence plane*, formed by the incident ray and the normal to the surface at the point of incidence; the angle of incidence is equal to the angle of reflection. All the angles are counted from the normal.

4. *Law of light refraction.* The refracted ray lies in the plane of incidence; the ratio of the sine of the angle of incidence α_1 to the sine of angle of refraction α_2 depends on the wavelength but is independent of the angle of incidence (Snell law):

$$\frac{\sin \alpha_1}{\sin \alpha_2} = n_{21} = \frac{n_2}{n_1}. \tag{P5.1.1.1}$$

The constant quantity n_{21} is called the *relative index of refraction* of the second medium with respect to the first one, which is equal to the ratio of (*absolute*) *indices of refraction* of each of the media (of indices of refraction of the medium with respect to a vacuum). From the viewpoint of wave optics, the absolute index of refraction shows by what factor the phase velocity of the light wave of a given frequency is less than the velocity of this wave in a vacuum:

$$v = c/n, \qquad n = \sqrt{\mu\varepsilon} \approx \sqrt{\varepsilon} \tag{P5.1.1.2}$$

(see Subsection P4.2.2). If $n_2 < n_1$, then for $\alpha_1 > \alpha_c$, where $\sin \alpha_c = n_{21} < 1$, one observes the *total reflection of light*; that is, the refracted ray is absent. If the light ray propagates in a medium with index of refraction gradually varying along an axis (in a layered medium), then it is convenient to use formula (P5.1.1.1) in the form $n \sin \alpha = \text{const}$, where α is the angle with this axis. One can see that α varies gradually, which means that the ray is bending.

The laws of geometric optics listed above must be supplemented with the *principle of reversibility of light beams*.

▶ **Limits of applicability of geometric optics.** The laws of geometric optics act where the phenomena of interference, diffraction, and polarization are insignificant. This happens when the wave amplitude and its first spatial derivatives vary just slightly over the wavelength. These conditions are violated at the shadow boundary, near the geometric point of convergence of light rays (focus), where light travels past narrow diaphragms, and also light propagates through media with rapidly varying index of refraction or with strong absorption. For example, when a light beam travels past a circular diaphragm of diameter d (for holes of irregular shape, d is understood as the minimum transverse dimension), it spreads out due to diffraction at a distance $l \sim d^2/\lambda$; one can use the laws of geometric optics at small distances compared to l.

▶ **Huygens principle and Fermat principle.** The laws of geometric optics are closely related to the *Huygens principle* (Huygens wave construction) and the *Fermat principle*. The Huygens principle is based on abstract-wave concepts and, at each point, permits constructing an auxiliary wave surface the normal to which shows the light ray direction. To construct the wave surface at time $t + \Delta t$, one considers all the points of the wave surface at time t as sources of secondary light waves. The envelope of spherical wave surfaces of secondary waves forms a new wave surface of the principal wave.

The Fermat principle is one of the examples of variational principles that play an important role in physics. It states that the time of light propagation along the true trajectory is extremal (usually, minimum or maximum) compared with all the virtual nearby trajectories. More precisely, this time does not vary (in the first order of magnitude) for a small distortion of the trajectory.

The Huygens and Fermat principles permit obtaining the fundamental laws of geometric optics but do not permit leaving its limits.

▶ **Optical path length.** It is convenient to express the time of light propagation through a medium with a variable index of refraction in terms of the *optical path length* L:

$$t = \int_1^2 \frac{dl}{c/n} = \frac{L}{c}, \qquad L = \int_1^2 n\, dl. \tag{P5.1.1.3}$$

This notion is also useful in studying the interference.

▶ **Images in optical systems.** One of the important problems of geometric optics is the construction of *images* formed by optical systems and the study of their properties. An image is the point of intersection of ray transmitted through an optical system. Images are *real* (the rays converge after leaving the system) and *virtual* (emergent rays diverge from the virtual point of intersection of their virtual continuations). The system of reflecting and refracting spherical (and plane) surfaces perpendicular to an axis forms the image of a point source by rays incident at small angles to the axis (*paraxial approximation*). The system axis is called the *principal optic axis*. The point of convergence of rays parallel to the axis is called the *focus of the system*; the plane perpendicular to the axis and drawn through the focus is called the *focal plane*. Like the images, the foci can be real and apparent. Additionally, one introduces the notion of "apparent source" denoting the point of intersection of imaginary continuations of the converging rays incident on the system.

In what follows, we use the important property of images: the optical path lengths of all rays from the source to the image are the same; that is, the optical system does not change the path difference of rays.

The basic elements of optical systems are the simple optical systems: plane and spherical mirrors, plane and spherical refracting surfaces, and thin lenses. The image formed by each of these elements is the source for the subsequent element.

When considering a simple system, one counts all the distances from the point of intersection of this system with the principal optical axis. The distance from this point to the source is denoted by s, the distance to the image, by s', and the distance to the focus, by f (*focal distance*). To real sources, images, and foci there correspond positive values, to apparent sources, negative values.

Below are formulas relating s, s', and f for simple optical systems.

1) **Spherical mirror** of radius R:

$$\frac{1}{s} + \frac{1}{s'} = \frac{1}{f}. \tag{P5.1.1.4}$$

The focal distance is equal to $f = R/2$, for a concave mirror, $R > 0$, for a convex mirror, $R < 0$.

2) **Plane mirror** forms the image of a point source located behind the mirror symmetrically to the source. Formally, (P5.1.1.4) holds for it as $R \to \infty$; that is, $s' = -s$ (to a real source there corresponds a virtual image, to an apparent source, there corresponds a real image).

3) **Spherical refracting surface** of radius R with the relative index of refraction $n_{21} = n_2/n_1$:

$$\frac{1}{s} + \frac{n_{21}}{s'} = \frac{n_{21} - 1}{R}.$$

For a convex surface, $R > 0$, for a concave surface, $R < 0$.

4) **Plane refracting surface** is obtained from the spherical surface in the limit as $|R| \to \infty$:

$$s' = -n_{21}s.$$

5) **Thin lens** bounded by spherical surfaces of radii R_1 and R_2. Relation (P5.1.1.4) holds, the focal distance of a thin lens is calculated by the formula:

$$\frac{1}{F} = (n_{21} - 1) \left(\frac{1}{R_1} + \frac{1}{R_2} \right).$$

The radii R_1, R_2 are assumed to be positive for convex surfaces and negative for concave surfaces. The quantity $D = 1/F$ is called the *focal power*; it is measured in *dioptres* (dpt), with 1 dpt = 1/m.

The theoretical limit of the image dimensions and the resolving power of optical systems is determined by the light diffraction.

P5.1.2. Radiometry and Photometry

Radiometry studies the measurement of electromagnetic radiation at all wavelengths, including visible light. Photometry studies the measurement of light within the wavelength range visible by the human eye.

▶ **Characteristics of radiation in space.** The radiation power emitted by a radiation source at a given point of space in a particular direction is characterized by the *radiant intensity*, I:

$$d\Phi = I \cos\theta \, ds \, d\Omega. \tag{P5.1.2.1}$$

The radiation power $d\Phi$ is called the *radiant flux*; it is the energy transferred per unit time through the area element ds by the light rays contained in the solid angle $d\Omega$. The angle θ is the angle between a selected direction and the normal to the area element. The product $\cos\theta \, ds$ is called the *apparent size* of the area element in the specified direction. The volume density of *radiant energy*, $u = dW/dV$, is calculated as

$$u = \frac{1}{c} \int I \, d\Omega. \tag{P5.1.2.2}$$

In the case of isotropic radiation, $u = 4\pi I/c$. For the intensity I and the density u, one can obtain the *spectral decomposition* in frequencies or wavelengths:

$$u = \int_0^\infty u_\nu \, d\nu = \int_0^\infty u_\omega \, d\omega = \int_0^\infty u_\lambda \, d\lambda.$$

Equating the energy contained in the mutually corresponding spectral intervals, one can find the relation between distinct spectral representations. For example, by setting $u_\nu \, |d\nu| = u_\lambda \, |d\lambda|$ with $\lambda = c/\nu$ taken into account, we obtain

$$u_\nu = \left(\frac{c}{\nu^2}\right) u_\lambda.$$

Similarly, we can obtain the decomposition in two mutually perpendicular polarizations.

▶ **Light and energy characteristics.** The above-introduced *energy characteristics* are used to obtain an objective description of the radiation. The *photometric* or *light* characteristics take account of the sensitivity of the eye to the optical radiation. The radiant power is measured in watts and the photometric unit corresponding to it is called the *luminous flux* and is measured in *lumens* (lm). To pass to the luminous flux, it is necessary to multiply the radiant power by the *luminous efficacy* η (measured in lm/W), which is equal to $\eta_0 = 625$ lm/W for the wavelength $\lambda_0 = 555$ nm and to $\eta_0 V_\lambda$ for other wavelengths, where V_λ is the curve of *visibility*, or *relative spectral sensitivity* of the eye ($V_\lambda = 1$ for $\lambda = 555$ nm and decreases to zero approaching the boundaries of the optical range). Recalculations from arbitrary radiant power to luminous flux can be performed by the formula $\Phi_{lu} = \eta_0 \int \Phi_\lambda V_\lambda \, d\lambda$. Recalculations from other energy characteristics to luminous characteristics and conversely can be performed similarly (for example, from the radiation intensity to the light intensity).

▶ **Characteristics of radiation source.** The point source is characterized by the *source energy power* determined as the radiant power in a given direction per unit solid angle: $d\Phi = J \, d\Omega$. The luminous characteristic—that is, the *source luminous intensity*—is determined in terms of the luminous flow. The radiant power unit, *candela*, is one of the basic SI units (1 lm = 1 cd × 1 sr, where "cd" stands for the *candela* and "sr" stands for the *steradian*). If there is no absorption or scattering of radiation, then the radiation intensity in the same direction at the distance r from the source is equal to

$$I = \frac{J}{r^2}.$$

For area sources—that is, for radiating surfaces—the notion of *energy radiance* in a given direction B_θ is introduced (the luminous characteristic is the *brightness*). The energy power emitted by a surface element ds is equal to

$$dJ = B_\theta \cos\theta \, ds \quad \text{or} \quad d\Phi = B_\theta \cos\theta \, ds \, d\Omega, \tag{P5.1.2.3}$$

where θ is the angle between the normal to the radiating surface and the radiation direction. The source whose B_θ is independent of the direction is called a *Lambertian source*. A Lambertian source of any shape appears uniformly bright (for example, the solar disk). The *radiant exitance* R (the luminous characteristic is the *luminous exitance*) is determined as the total radiant power in the solid angle 2π per unit area of a radiating surface:

$$R = \int B_\theta \cos\theta \, d\Omega. \tag{P5.1.2.4}$$

For a Lambertian source, $R = \pi B$.

▶ **Irradiance.** The radiant energy flux per unit area of illuminated surface is called the *irradiance* and denoted E. The corresponding light quantity is called the *illuminance* and measured in *luxes* (lx), with $1\,\mathrm{lx} = 1\,\mathrm{lm/m^2}$. If a point source radiation is incident at an angle θ with the normal, then

$$E = \frac{J}{r^2}\cos\theta.$$

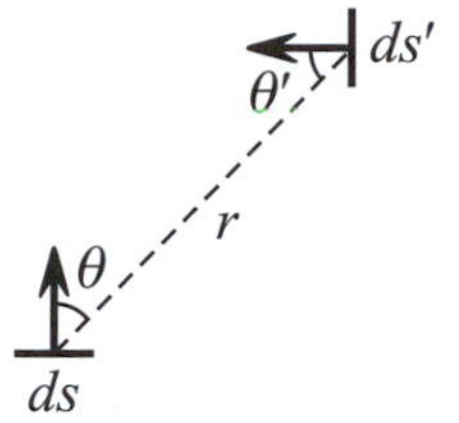

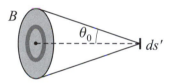

Figure P5.1. To calculations of the irradiance created by an area source.

The irradiance created by a luminous surface S on a site of the irradiated surface ds' is equal to the sum of contributions of all the sites ds. The solid angle from which ds' is viewed at ds is equal to $\cos\theta'\,ds'/r^2$ (Fig. P5.1). For the luminous flux radiated into this solid angle, we have $B_\theta \cos\theta\,ds(\cos\theta'\,ds'/r^2)$. Since $\cos\theta\,ds/r^2$ is equal to the solid angle $d\Omega$ from which the radiating site is viewed at the point of irradiance, it follows that $dE = B_\theta \cos\theta'\,d\Omega$.

Example. Calculate the irradiance created by a Lambertian source of disk shape irradiating a small area located on the disk axis oriented perpendicularly to the axis (Fig. P5.2).

Figure P5.2. Irradiance of a surface element ds' by a Lambertian disk.

Solution. The solid angle element bounded by θ' and $\theta' + d\theta'$ is equal to $d\Omega = 2\pi \sin\theta'd\theta'$. After integration, we obtain $E = \pi B \sin^2\theta_0$, where θ_0 is the cone half-opening angle from which the disk is viewed from the irradiated area element.

P5.2. Wave Properties of Light
P5.2.1. Interference of Light

▶ **Coherence of light waves.** In the superposition of light waves with a very high frequency of oscillations, one can observe only the time-average energy of oscillations, which is characterized by the *oscillation intensity* $I \sim \langle \mathbf{E}^2 \rangle$. When adding two oscillations $\mathbf{E} = \mathbf{E}_1 + \mathbf{E}_2$, we obtain $I = I_1 + I_2 + I_{12}$, where $I_{12} = 2\langle \mathbf{E}_1 \cdot \mathbf{E}_2 \rangle$ is called the interference term. If $I_{12} = 0$, then the light beams are said to be *incoherent*. The waves polarized in mutually perpendicular directions are always incoherent. Monochromatic waves are coherent only if their phase difference is constant (that is, their frequencies coincide) and if they are parallel polarized (or almost parallel polarized). In superposition of such waves,

$$I = I_1 + I_2 + 2\sqrt{I_1 I_2}\cos\Delta\varphi,$$

where $\Delta\varphi$ is the oscillation phase difference at a given point. In the special case for $I_1 = I_2 = I_0$, we have $I = 2I_0(1 + \cos\Delta\varphi)$. For in-phase oscillations, $I = (\sqrt{I_1} + \sqrt{I_2})^2$, for anti-phase oscillations, $I = (\sqrt{I_1} - \sqrt{I_2})^2$.

▶ **Interference of two waves.** In superposition of two plane waves $E_1 = E_{10} \cos(\omega t - \mathbf{k}_1 \cdot \mathbf{r})$ and $E_2 = E_{20} \cos(\omega t - \mathbf{k}_2 \cdot \mathbf{r})$, we obtain $\Delta \varphi = (\mathbf{k}_1 - \mathbf{k}_2) \cdot \mathbf{r}$. The surfaces of constant phase difference are planes perpendicular to the vector $\mathbf{k}_1 - \mathbf{k}_2$. The distance between neighboring planes with maximum intensity is equal to

$$\Delta x = \frac{2\pi}{|\mathbf{k}_1 - \mathbf{k}_2|} = \frac{\pi}{k \sin(\alpha/2)} = \frac{\lambda}{2 \sin(\alpha/2)}, \tag{P5.2.1.1}$$

where α is the angle between the vectors $\mathbf{k}_1$ and $\mathbf{k}_2$. For small α, we obtain $\Delta x \sim \lambda/\alpha$. If in the region of wave overlap there is a plane screen parallel to $\mathbf{k}_1 - \mathbf{k}_2$, then one can observe parallel alternating dark and light fringes on it.

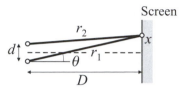

Figure P5.3. Interference of two closely located linear sources.

In superposition of spherical waves from two in-phase point sources, the maximum condition for the intensity of the mth order has the form $r_2 - r_1 = m\lambda$ (this is the equation of a hyperboloid of rotation whose axis passes through the sources). If a screen is placed parallel to this axis, we obtain light and dark fringes of hyperbolic shape. If the distance to the screen D is large compared to the distance d between the sources, then, at the screen center, we obtain equidistant almost parallel fringes. The distance between the fringes is the same as in the *Young experiment*, where the interference is created by two parallel linear sources of coherent light. If x is the distance from a screen point to its center (Fig. P5.3), then $r_1^2 = D^2 + (x + d/2)^2$, $r_2^2 = D^2 + (x - d/2)^2$, and for $d, x \ll D$, we obtain the optical path difference $\Delta = r_1 - r_2 \approx xd/D \approx \alpha x$, where $\alpha \approx d/D$ is the angle from which the source is viewed at the screen center. The intensity varies from $4I$ for $\Delta = \alpha x = m\lambda$ to zero for $\Delta = (m + \frac{1}{2})\lambda$. The distance between the fringes is equal to $\Delta x = \lambda/\alpha = \lambda D/d$.

▶ **Fraunhofer approximation.** At a large distance from two in-phase linear sources ($r_1, r_2 \gg d$), one can write $r_2 - r_1 \approx d \sin \theta$, where θ is the angle between the normal to the plane of sources and the direction of the interference observation (*the Fraunhofer approximation*). The oscillation amplitudes in interfering waves are assumed to be equal; that is, $I = 2I_0[1 + \cos(kd \sin \theta)]$. The interference maxima are observed at the angles satisfying the relation $\sin \theta = m\lambda/d$.

Let us specify at what distance from the sources the Fraunhofer approximation can be used. We express r_2 according to the cosine theorem:

$$r_2 = (r_1^2 + d^2 - 2dr_1 \sin \theta)^{1/2}$$

(the angle in the triangle is equal to $90° - \theta$), expand it up to the terms of the second order in d/r_1, and obtain

$$r_2 - r_1 \approx d \sin \theta + \frac{d^2}{2r_1} - \frac{d^2}{r_1} \sin^2 \theta.$$

The correction can be neglected if $d^2/r_1 \ll \lambda$. We see that the Fraunhofer approximation works in the case where the distance r from the observation point to the sources satisfies the relation

$$r \gg \frac{d^2}{\lambda}. \tag{P5.2.1.2}$$

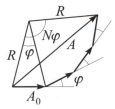

Figure P5.4. Composition of oscillations from several sources.

We note that this inequality is stronger than the simple condition $r \gg d$.

▶ **Interference of many waves.** We consider the interference of N equal in-phase sources located on the same straight line at the distance d from one another. At a large distance from the sources ($r \gg (Nd)^2/\lambda$) in the direction of θ, the composition of N oscillations of the same amplitude A_0 occurs, and the phase difference between the neighboring oscillations is equal to $\phi = (2\pi/\lambda)d\sin\theta$. The resultant amplitude A is obtained from the equations (see the vector diagram in Fig. P5.4): $A_0 = R\sin(\phi/2)$, $A = R\sin(N\phi/2)$. Eliminating the radius of the circumscribed circle R, we obtain

$$A = A_0 \frac{\sin(N\phi/2)}{\sin(\phi/2)}, \qquad I = I_0 \frac{\sin^2(N\phi/2)}{\sin^2(\phi/2)}. \tag{P5.1.2.3}$$

▶ **Interference from natural sources of light.** The radiation of natural (thermal) sources of light is the set of *wave trains* spontaneously radiated by excited atoms in their *luminescence*; that is, when they return to normal state. The train duration is $\sim 10^{-10}$–10^{-8} c, it contains 10^6–10^8 oscillations. Hence two different natural sources are incoherent even if a narrow spectral band is separated in their radiation, because the oscillation phase difference varies very fast and chaotically at each observation point. To observe the interference, it is necessary to split the radiation of one source into two or several beams and to make them to enter the observation point by different paths. In this case, each train interferes with itself, and the maximum or minimum condition is simultaneously satisfied for all the trains of the same frequency radiated from the same point of the source. It is convenient to introduce *optical path difference* of beams

$$\Delta = \int n_2 \, dl - \int n_1 \, dl,$$

where the integration is performed over the line passed by a given beam from the radiation point to the interference point. The maximum condition is $\Delta = m\lambda$; the minimum condition is $\Delta = \left(m + \frac{1}{2}\right)\lambda$, where m is called the *order of interference*.

We present the classical examples of obtaining two coherent sources.

1. *The Young experiment* (mentioned above). The sunlight is incident on a very narrow slit in the first screen, diverges because of diffraction, and is incident on the two narrow slits in the second screen. Again because of diffraction behind these slits, the light splits and forms overlapping coherent beams.

2. *Fresnel mirrors.* The light from the brightly illuminated slit is incident on two mirrors placed at an angle nearly equal to 180°. The closely located virtual images of the slit form two coherent sources.

3. *Fresnel biprism.* The light from a brightly illuminated slit refracts in two glass prisms with small refraction angles that contact by their bases. As the result of refraction, two closely located virtual images of the slit are formed.

4. *Billet split lens.* A collecting lens is cut into halves and the halves are slightly moved apart. The split lens is illuminated through a narrow split parallel to the cut line. Each of the lens halves forms its own real image of the slit.

5. *Lloyd mirror.* The light from a narrow slit is reflected from the mirror plane and forms the virtual image of the slit. The light from the slit itself and from its image interfere.

▶ **The influence of the source dimensions. The spatial coherence.** The optical path difference at a given point of the screen has a definite value only in the case of a point source. The path difference varies in transitions from point to point of the area source. If the path difference varies by $\lambda/2$, then the maximum condition becomes the minimum condition; that is, the interference patterns from different sites of the area source overlap and the total interference pattern is smeared.

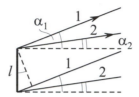

Figure P5.5. Influence of the source dimensions on the interference pattern.

For example, we consider a plane source. We assume that the first of the interfering rays is radiated by the source at an angle α_1 with its normal, and the second, at an angle α_2. We assume that these rays lie in the one plane with the normal and the angles of radiation are almost the same for all point of the source (Fig. P5.5). In transition from one end of the source to the other, the path length of the first ray varies by $l \sin \alpha_1$, and that of the second ray varies by $l \sin \alpha_2$ (l is the dimension of the source in the plane of rays), and the path difference varies by $l(\sin \alpha_2 - \sin \alpha_1)$. The interference pattern conservation condition (condition of *spatial coherence*) takes the form

$$l \left| \sin \alpha_2 - \sin \alpha_1 \right| < \lambda/2.$$

Example 1. In the Young experiment, l is the slit width in the first screen, and $\alpha_1 = -\alpha_2 = b/(2L)$, where b is the distance between the slits in the second screen and $L \gg b$ is the distance between the screens; we obtain: $2l(b/2L) < \lambda/2$; that is, $l < \lambda L/(2b)$.

We rewrite this formula as: $b < \lambda/(2\psi)$, where $\psi = l/L$ is the angular dimension of the source or the divergence angle of the rays incident on the slits. For example, in the case of the sun direct illumination of the slits, $\psi \approx 0.009$ rad, $\lambda \approx 500$ nm; that is, the distance between the slits must be less than 0.06 mm. Just therefore, the sun light must first be transmitted through a narrow slit.

▶ **Influence of the light nonmonochromaticity. Temporal coherence.** A violation of the monochromaticity of the combined waves may result in the interference pattern smearing. We assume that the frequencies of the radiated waves lie in a narrow spectral interval $\Delta \omega$. A strong distortion of the interference pattern occurs if the path difference Δ exceeds a certain critical value L_c, which is called the *coherence length*; L_c has the meaning of the *length of the wave train* radiated by the atom in a single act of radiation (we recall that stable interference occurs when the trains in one of the separated beams are combined with the trains corresponding to them in the other beam).

The time $\tau_c = L_c/c$ that is called the *coherence time* has the meaning of *duration of the wave train.* According to general properties of the Fourier transform, the spectral width of a wave packet is related to its duration as $\Delta \omega \, \Delta t \sim 2\pi$.

Example 2. The "truncated" sinusoid $A(t) \sin \omega_0 t$, where $A = \mathrm{const}$ in the interval $-\tau/2 < t < \tau/2$ and $A = 0$ outside this interval, has the Fourier transform $a(\omega) \sim \dfrac{\sin[\frac{1}{2}(\omega - \omega_0)\tau]}{(\omega - \omega_0)\tau}$ whose width is equal to $\Delta \omega \sim \dfrac{2\pi}{\tau}$.

We see that the spectral width of interference is related to the coherence duration as $\Delta\omega \sim 2\pi/\tau_c$. The maximum order of the spectrum, where one can observe the interference can be estimated as $N \sim \tau_c/T = \omega/\Delta\omega = \lambda/\Delta\lambda$. It is useful to note that the bands of neighboring orders merge just in this order:

$$(N + 1)\lambda = N(\lambda + \delta\lambda)$$

(the trailing edge of the band of order $N + 1$ coincides with the leading edge of the band of order N). The coherence length is often expressed in terms of $\delta\lambda$:

$$L_c = N\lambda = \frac{\lambda^2}{\delta\lambda}.$$

▶ **Interference in thin films.** We consider the interference of rays reflected from the leading and trailing surfaces of a thin film (Fig. P5.6). We assume that the wave front is plane; that is, the source is sufficiently remote. Since the front of refracted wave is

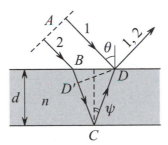

Figure P5.6. Interference in light reflection from a thin film.

perpendicular to the ray, it follows that ray 1 at point D and ray 2 at point D' have equal phases. Hence the optical path difference of rays at point D is equal to $n(D'C + CD)$. Moreover, $\lambda/2$ must be added to the path difference, which permits taking account of the phase variation on π in reflection from a medium with a greater index of refraction (on the air–film interface at point D). After transformations, we obtain:

$$\Delta = 2dn \cos\psi + \frac{\lambda}{2} = 2d\sqrt{n^2 - \sin^2\theta} + \frac{\lambda}{2},$$

where d is the film width, n is its index of refraction, and θ and ψ are the incidence and refraction angles. The maximum condition for observation in reflected light $\Delta = m\lambda$ is satisfied for certain wavelengths. For very thin films, the maximum condition is satisfied for one or two wavelengths from the visible region, and the film turns out to be colored. The maximum condition for observation in reflected light corresponds to the minimum condition for observation in transmitted light (no reflection occurs from the air–film interface). As always in interference, the energy does not increase but is redistributed.

Now we consider two important cases:

1) *Fringes of equal thickness.* If the rays are incident at an almost constant angle, for example, normally and the film width varies, then the lines of constant thickness are lines of constant path difference. In illumination by monochromatic light, these lines are visible as dark or light fringes. In observation in white light (for films of small thickness) the lines are colored. The interference occurs near the film surface (interference pattern is localized at the surface).

2) *Fringes of equal inclination.* The film thickness is constant, the illumination is by scattered light from a remote source. Changing the angle of observation, we alternatively

obtain the maximum condition and the minimum condition. The interference pattern is localized at infinity (or in the focal plane lens). For extremely thin films, the light can be nonmonochromatic; the observation at a given angle separates the wavelength for which the maximum condition is satisfied.

Example 3. Newton rings. If a plane-convex lens is put on the surface of a glass plate (Fig. P5.7) and illuminated by normally incident monochromatic light, then fringes of the shape of rings of equal thickness are observed in the air interval. The air interval thickness is equal to $d \approx \sqrt{R^2 + r^2} - R \approx r^2/(2R)$, where r is the circle radius. The minimum condition has the form $2d + (\lambda/2) = \left(m + \frac{1}{2}\right)\lambda$ or $r_m = \sqrt{m\lambda R}$. At the pattern center, there is a dark spot.

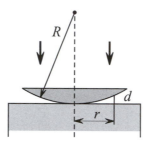

Figure P5.7. Newton rings.

▶ **Holography principles.** Holography is used to record the volume image of an object. The object is illuminated by laser light with a very high degree of coherence and after reflection goes to a photographic plate. The information about the object shape is carried by the dependence of the object wave phase on the coordinates along the photographic plate. If, simultaneously with light reflected from the object, the photographic plate is illuminated by a reference ray, the same as that used to illuminate the object but going directly to the plate after reflection from a mirror. Then, as the result of interference, a wave is generated whose amplitude and intensity will depend on the phase of the object wave. Since the photographic plate blackening is proportional to the intensity, it preserves the information about the wave phase. Illuminating the developed photographic plate by the light of the same laser, one can reconstruct the initial signal.

Recording the interference signal in the bulk of a thick transparent photographic plate, one obtains a *volume hologram*, which bears information about both the object shape and the laser signal wavelength. If the hologram is illuminated by white light, the waves of other frequencies cancel one another due to the interference, and the image of the object illuminated by the monochromatic laser light arises. If three holograms from lasers with different wavelengths are recorded in one photographic plate, then the volume colored image appears in illumination by white light.

▶ **Vavilov–Cherenkov radiation.** If a particle moves in a medium with velocity V exceeding the phase velocity $v = c/n$ of light in this medium, then the radiation of electromagnetic waves arises at a certain angle with the direction of the particle motion. The elementary explanation of this phenomenon is quite similar to the explanation of the appearance of shock wave in motion with supersonic speed. As a particle moves, the medium atoms excited by the particle begin to radiate. The path difference of waves radiated by excited atoms in the direction θ from positions A and B (Fig. P5.8) is equal to

$$\Delta = v\frac{L}{V} - L\cos\theta = L\left(\frac{v}{V} - \cos\theta\right).$$

If $V < v$, then for any angle θ, one can choose L such that the radiation of points A and B cancel each other ($\Delta = \lambda/2$). For $V > v$, the same concerns any angle other than

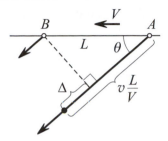

Figure P5.8. Vavilov–Cherenkov radiation.

$\theta_0 = \arccos(v/V)$ (one can attain $\Delta = \pm\lambda/2$). But in the direction θ_0, all the points radiate wave with the same phase; that is, the radiation is mutually reinforced.

P5.2.2. Diffraction

▶ **Huygens–Fresnel principle.** Deviations from the law of light rectilinear propagation is called diffraction. Approximate calculations of the light intensity behind obstacles or holes in the screen is based on the *Huygens–Fresnel principle* which supplements the geometric Huygens construction (see Subsection P5.1.1) with the condition of interference of secondary rays. According to this principle, all points of an arbitrary surface surrounding the wave sources are sources of coherent secondary waves, and the amplitude of the wave oscillations at an arbitrary point outside this surface can be obtained as the result of interference of secondary waves. Usually, for such a surface one takes the wave front, and then the sources of secondary waves are in phase. The secondary wave amplitude is proportional to the primary wave amplitude and to the area of the surface site:

$$dE = E_0 \frac{K(\alpha)}{r} \exp\left[i(\omega t - kr)\right] ds, \qquad (\text{P5.2.2.1})$$

where E_0 is the (complex) amplitude of the primary wave at the front points, α is the angle between the radiation direction and the normal to the surface, r is the distance from the front of secondary sources to the point of observation, and $K(\alpha)$ is a slowly decreasing function ($K(\pi) = 0$).

We consider the diffraction of a plane monochromatic wave incident normally on a flat screen with a hole of various shape.

Circular hole of radius R_h. The point of observation B lies on the hole axis at a distance l from its center (Fig. P5.9). For the surface of secondary waves we take a part of the front in the plane of the hole. We divide the front into thin circular segments, by cutting circles from point B on it at intervals whose lengths are $l+\delta r, l+2\delta r, \ldots$ ($\delta r \ll \lambda$). Since the optical path length acquires the constant increment δr in transition from segment to segment, it follows that the oscillation phase increments are also the same; moreover, the amplitudes of oscillations from separate circular segments, which are calculated by formula (P5.2.2.1), also coincide. Indeed, from $r^2 = R^2 + l^2$ we obtain $R\delta R = r\delta r$ and $\delta S/r = 2\pi\delta r = \text{const}$ ($\delta S = 2\pi R\delta R$). Hence the vector diagram of combined oscillations (Fig. P5.10) has the form of a slowly twisted spiral (with slow decrease in $K(\alpha)$ taken into account). From the symmetry consideration, one can conclude that the spiral converges to the center of the circle. The sum of the entire spiral corresponds to the free propagation of light in the absence of the screen.

The part of the surface from center to the points, where $r_1 = l + \lambda/2$, is called the *first Fresnel zone*, the part of the surface from the edge of the first zone to $r_2 = l + \lambda$ is the second Fresnel zone, etc. One can see that if only the first Fresnel zone is open, and all the others are closed by a screen, then the oscillation amplitude is twice greater than that without the screen, and the intensity is greater by the factor 4. If the first two zones are open, then the amplitude and intensity at point B are close to zero, etc. Conversely, if a small number of central Fresnel zones is closed by the screen, then this hardly affects the oscillation intensity at point B, it is almost the same as without the screen (this paradoxical fact is known as the "Poisson spot").

If an obstacle of complicated shape is prepared to close the next nearest Fresnel zones (all even zones or all odd zones), then the intensity increases many times at the point of observation. Such an instrument focusing the light like a lens is called a *zone plate* and finds numerous applications.

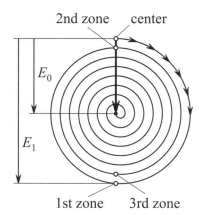

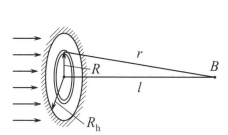

Figure P5.9. Wave front separation into thin circular segments.

Figure P5.10. Vector diagram for the wave front separation into circular segments.

We find the radius of the mth Fresnel zone from the Pythagorean theorem $r^2 = R^2 + l^2$, substituting $r = l + \Delta$ (where $\Delta \ll l$ is the path difference between the hole edge and the hole center):

$$R(\Delta) = \sqrt{2l\Delta}. \tag{P5.2.2.2}$$

If $\Delta = m\lambda/2$ is substituted, we obtain the radius of the mth Fresnel zone: $R_m = \sqrt{ml\lambda}$.

We also present the formula for the case of incidence at the hole of a spherical wave radiated by a point source lying on the hole axis at the distance a from its center:

$$R_m = \sqrt{\frac{2al\Delta}{a+l}} = \sqrt{\frac{mal\lambda}{a+l}}.$$

As $a \to \infty$, this expression becomes (P5.2.2.2).

The number of Fresnel zones opened by the hole is equal to $m = R_{\mathrm{h}}^2/l\lambda$. Hence for $l \ll R_{\mathrm{h}}^2/\lambda$, there are many open Fresnel zones and the light intensity at the point of observation slightly differs from the intensity of the incident wave; that is, the diffraction effects at the hole axis are weak (the geometric optics approximation works). In the converse limit $l \gg R_{\mathrm{h}}^2/\lambda$, only a small part of the first Fresnel zone is open; that is, all the oscillations at the point of observation are combined in one phase (the Fraunhofer approximation).

For a circular hole, it is more difficult to calculate the diffraction for off-axis points of observation. We will revisit this problem below considering it for the case of Fraunhofer diffraction. Unlike a circular hole, it turns out that the solution of the diffraction problem for all observation points is possible for a long slit-shaped hole.

Slit-shaped hole of width b. The point of observation B lies at a distance l from the screen. In this case, the front must be divided not into circular but into thin rectilinear segments (Fig. P5.11) of the same width δx. The main difference is that the segment areas are proportional not to $R\,\delta R$ but simply to δx. In this case, just as

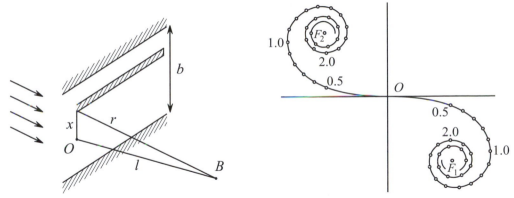

Figure P5.11. Wave front separation into thin recti-linear segments.

Figure P5.12. Vector diagram for the wave front sep-aration into thin rectilinear segments (Cornu spiral).

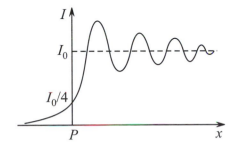

Figure P5.13. Intensity distribution on the screen in diffraction at the half-plane.

before, we have the identity $r^2 = x^2 + l^2$, with $x\,\delta x = r\,\delta r$, and hence $\delta r = x\,\delta s/r$, and the phase increment from segment to segment increases with the distance from point O. This results in that the radius of curvature of the vector diagram gradually decreases and it acquires the shape of spiral called the "Cornu spiral" (Fig. P5.12). The values of the parameter $v = x\sqrt{2/(l\lambda)}$ are marked on this spiral. For $x \ll l$, the path difference is equal to $\Delta = r - l \approx x^2/2l$, and hence $\Delta = v^2 \frac{\lambda}{4}$ ($v = 1$ corresponds to $\Delta = \lambda/4$ and $v = 2$ corresponds to $\Delta = \lambda$; see Fig. P5.12). The distance between finite points F_1 and F_2 of the spiral is equal to the amplitude A_0 of oscillations without the obstacle, or to the amplitude of the incident wave. If the slit edges are at the distances x_1 and x_2 from point O, then we calculate the corresponding values v_1 and v_2, obtain the corresponding points on the spiral, and, connecting them, determine the amplitude. Thus, one can obtain the diffraction pattern at the slit not only at the screen center but also aside the center. The case $b \ll \sqrt{l\lambda}$ corresponds to the Fraunhofer diffraction (a small part of the spiral is open); the case $b \gg \sqrt{l\lambda}$ corresponds to the geometric optics approximation. In this case, near the slit edges, one observes the diffraction pattern corresponding to diffraction at the half-plane (Fig. P5.13); precisely opposite the slit edge (point P), the amplitude OF_2 is equal to half the amplitude of the incident wave A_0 (intensity is four times less than I_0), and the width of the diffraction spreading is $\sim \sqrt{l\lambda}$.

Using the Fresnel zone method, one can study the diffraction pattern at a slit graphically using the Cornu spiral obtained numerically. A simple analytic description of diffraction at a slit can be obtained in the Fraunhofer approximation.

▶ **Fraunhofer diffraction at a slit.** Fraunhofer diffraction is diffraction in parallel rays, where the phase difference of secondary waves radiated from different points of the wave front segment under study can be found under the assumption that they are parallel. In this approximation, the diffraction pattern becomes simpler, and one can calculate the light intensity at different points of the screen. The Fraunhofer approximation acts in the following two cases: (i) if the diffracted light rays are collected by a lens in its focal plane or (ii) if the screen is separated from the diffraction hole by a sufficiently large distance $r \gg R^2/\lambda$, where R is the source dimension (see the discussion of the Fraunhofer condition in Subsection P5.2.1, formula (P5.2.1.2).

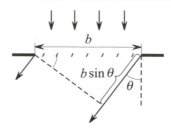

Figure P5.14. Fraunhofer diffraction at a thin slit.

In the Fraunhofer approximation, we calculate the diffraction of a plane monochromatic wave at a slit of width b. We consider only the normal incidence of a wave on a screen with a slit; in this case, all the slit points are sources of secondary in-phase waves. We divide the plane slit into a large number N of equal narrow strips (Fig. P5.14). The light amplitude from each of the strips is equal to E_0/N, where E_0 is the amplitude of light radiated by the slit in the direction $\theta = 0$ (in this case, all the strips radiate in-phase light). In the case of radiation in the direction at an angle θ with the normal, the radiation phase difference between the neighboring strips is equal to $\delta\varphi = (2\pi/\lambda)(b/N)\sin\theta$, the phase difference between the extreme strips is equal to $\varphi = (2\pi/\lambda)b\sin\theta = N\delta\varphi$. Using expression (P5.2.1.3) for the interference of N sources (Subsection P5.2.1), we obtain the amplitude for the slit radiation in the direction θ ($\sin\delta\varphi$ must be replaced by $\delta\varphi$):

$$E_1(\theta) = E_0 \frac{\sin(\varphi/2)}{\varphi/2}, \qquad \text{where} \quad \varphi = 2\pi\frac{b}{\lambda}\sin\theta. \tag{P5.2.2.3}$$

The radiation intensity is equal to $I_1 = I_0 \sin^2(\varphi/2)/(\varphi/2)^2$, the typical vector diagrams are shown in Fig. P5.15. The minima condition for the radiation has the form:

$$b\sin\theta = m\lambda \qquad (m = 0, \pm 1, \pm 2, \dots). \tag{P5.2.2.4}$$

the maxima condition is $b\sin\theta \approx (m + \frac{1}{2})\lambda$ ($m = 1, 2, \dots$). The maxima of intensity are to one another as $I_0 : I_1 : I_2 : \dots \approx 1 : (1.5\pi)^2 : (2.5\pi)^2 : \dots$. The maximum order of the spectrum is determined by the condition $m \le b/\lambda$.

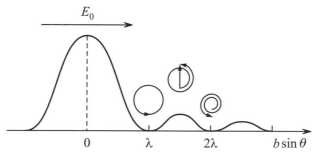

Figure P5.15. Angular distribution of radiation intensity and the vector diagrams for the Fraunhofer diffraction at a slit.

▶ **Fraunhofer diffraction at a circular hole.** In the case of diffraction at a circular hole, the intensity distribution has the form of concentric dark and light circles around the central light spot. The intensity distribution for small angles θ with hole axis is expressed in terms of the Bessel function of the first order: $I = I_0 J_1^2(\pi D\theta/\lambda)$, where D is the hole diameter. In outward appearance, the angular distribution of the intensity slightly differs from the case of diffraction at a slit; the first minimum corresponds to the angle $\theta_1 = 1.22\,\lambda/D$.

The angle θ_1 can be treated as the angle of diffraction expansion of the light beam. The beam spreading is weak at distances l satisfying the condition: $l\theta_1 \ll D$ or $l \ll D^2/\lambda$. This inequality is a condition of the geometric optics applicability (see Subsection P5.1.1).

▶ **Rayleigh criterion. Resolving power of optical instrument.** When two nearby points are observed in an optical instrument (lens, telescope), their images on the screen are small diffraction circles. It is conditionally assumed that the image points are indistinguishable if the distance between them is so small that the central maximum of one of the circles coincides with the first minimum of the other circle (*Rayleigh criterion*). Hence the angular resolution of such optical instruments (with an open diaphragm of diameter D) is given by the formula

$$\theta_{\min} = 1.22\frac{\lambda}{D}.$$

This limit of an instrument possible resolution is imposed by the wave nature of light and cannot be overcome by any technical improvements.

▶ **Diffraction grating.** The diffraction grating is a sequence of large number N of equal parallel slits (Fig. P5.16). The width of each slit is denoted by b, and the distance between the corresponding points (for example, centers) of neighboring slits, which is

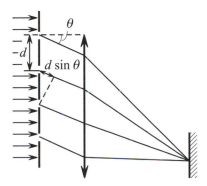

Figure P5.16. Diffraction grating.

called the *diffraction grating period*, is denoted by d ($d = l/N$, where l is the grating width). In the direction θ, each slit radiates light whose amplitude E_1 is determined by expression (P5.2.2.3). The phase difference between the neighboring slits is equal to $\delta = (2\pi/\lambda)d\sin\theta$, and to calculate the final amplitude, one can again use formula (P5.1.2.3) for the sum of oscillations of N independent sources:

$$E = E_1\frac{\sin(N\delta/2)}{\sin(\delta/2)}, \quad I = I_1\frac{\sin^2(N\delta/2)}{\sin^2(\delta/2)}, \quad \text{where} \quad \delta = 2\pi\frac{d}{\lambda}\sin\theta.$$

The *principal maxima of the diffraction grating* are determined by the condition

$$d\sin\theta = m\lambda. \tag{P5.2.2.5}$$

At the principal maxima, the phase difference between the neighboring slits equals $2\pi m$ and the oscillations from all slits are in phase. The vector diagram is a straight line of length NE_1. In the direction of the principal maxima, the radiated wave intensity $I = N^2 I_1$ is completely determined by the radiation intensity of a separate slit in this direction. If the principal maximum is near the minimum of $I_1(\theta)$ determined by the condition $b\sin\theta = m\lambda$, then this maximum is suppressed. The maximum order of the spectrum is determined by the condition $m \leq d/\lambda$.

The diffraction pattern has $N - 1$ minimum and $N - 2$ *additional maxima* between each two principal maxima. The minima are determined by the conditions $d \sin \theta = (m + n/N)\lambda$ ($n = 1, 2, \ldots, N - 1$). The first minimum is especially important because it determines the width of the principal maximum. In the vector diagram, the first minimum corresponds to the phase difference between the slits $2\pi/N$, and the vector diagram for N slits is a closed figure, namely, a regular N-gon. One can see that the width of the principal maxima decreases proportionally to $1/N$. The intensity of additional maxima is much less than that of the principal maximum, and it can be neglected in the case of a large number of slits.

If light is incident on the grating at an angle $\theta_0 \neq 0$, then the principal maxima condition takes the form:

$$d(\sin \theta - \sin \theta_0) = m\lambda. \tag{P5.2.2.6}$$

For coarse gratings ($d \gg \lambda$), it is efficient to use grazing rays ($\theta_0 \approx \pi/2$), for which the principal maxima condition takes the form: $d \cos \theta_0(\theta - \theta_0) = m\lambda$. One can see that the role of the effective period in this case is played by the small quantity $d \cos \theta_0$.

▶ **Diffraction grating as a spectral device.** The properties of any spectral device are determined by its *angular dispersion, free spectral range,* and *resolving power.*

The angular dispersion is the derivative $D_m = d\theta_m/d\lambda$, where θ_m is the position of the principal maximum of mth order. It follows from the principal maximum condition that

$$D_m = \frac{m}{d \cos \theta}.$$

The free spectral range is the maximum width $\Delta\lambda$ of the spectral interval for which there still is no overlapping of the spectra of neighboring orders. From the equations $d \sin \theta = m(\lambda + \Delta\lambda)$ and $d \sin \theta = (m + 1)\lambda$, we have $\Delta\lambda = \lambda/m$. One can see that to study wide parts of the spectrum, it is necessary to use lower orders.

The resolving power is the ratio $R = \lambda/\delta\lambda$, where $\delta\lambda$ is the least wavelength difference that can be resolved by the spectral device. Two lines are assumed to be resolved if the maximum of one line lies on the minimum of the other line (Rayleigh criterion). From the equations $d \sin \theta = (m + 1/N)\lambda$ and $d \sin \theta = m(\lambda + \delta\lambda)$ it follows that $\delta\lambda = \lambda/(mN)$; that is,

$$R = \frac{\lambda}{\delta\lambda} = mN.$$

▶ **Diffraction of X-rays at a crystal.** The wavelength of X-rays is comparable with the distance between the atoms of a crystal lattice, which is a *spatial diffraction grating* for incident rays. The condition of diffraction maxima consists in simultaneous satisfaction of three equations (P5.2.2.6) for the three mutually perpendicular crystal axes (*Laue conditions*):

$$d_1(\cos \alpha - \cos \alpha_0) = n_1\lambda, \quad d_2(\cos \beta - \cos \beta_0) = n_2\lambda, \quad d_3(\cos \gamma - \cos \gamma_0) = n_3\lambda,$$

where α, β, γ is angles with the axes such that $\cos^2\alpha + \cos^2\beta + \cos^2\gamma = 1$. These three equations cannot be satisfied simultaneously for an arbitrarily chosen direction of the incident ray and for a given wavelength. This means that in the case of the crystal irradiation by monochromatic but scattered X-rays (that is, in all possible directions), there arise diffraction maxima in quite certainly determined directions. An analysis of the *Laue patterns* thus obtained permits obtaining information about the crystal structure. The diffraction maxima are absent if $\lambda/2$ exceeds all the lattice periods; for visible light, the crystal can be treated as a homogeneous medium.

The reflection of X-rays from the crystal surface can also be considered as the interference of rays reflected from a system of successive atomic planes in the crystal. By analogy with the interference in thin films (see Subsection P5.2.1), the path difference between the rays reflected from the neighboring planes is equal to $2d \sin \theta$, where θ is the angle between the incident ray and the atomic plane (not the normal!), and d is the interplanar distance. (We note that the index of refraction of X-rays is almost equal to one; see Subsection P5.3.2.) The reflection is observed only in the directions of the diffraction maxima satisfying the *Wulf–Bragg condition*:

$$2d \sin \theta = m\lambda,$$

where $m = 1, 2, \ldots$ is the order of the diffraction maximum.

P5.3. Interaction Between Light and Matter

P5.3.1. Light Polarization. Fresnel Formulas

▶ **Polarized and natural light.** A plane wave is said to be *linearly polarized* or *plane polarized* if the vector **E** oscillates in one plane that is perpendicular to the wave front (it is called the *plane of polarization* of the wave). The monochromatic plane wave is either linearly polarized or elliptically polarized or circularly polarized (see Subsection P4.2.2). The elliptically polarized wave is the sum of two mutually perpendicular plane waves such that there is a phase difference between their oscillations. The *natural light* radiated by heated bodies is not polarized, because the direction of the vector **E** oscillations at each point varies rapidly and chaotically. A mixture of natural and polarized light is called *partially polarized* light.

The *polarizer* is a device at whose output the light is linearly polarized in a certain plane called the *transmission plane of polarizer*. The point is that the polarizer completely absorbs the light polarized perpendicularly to the transmission plane. If the natural light is transmitted through a polarizer, then it becomes linearly polarized and its intensity decreases twice (if there is no absorption in the transmission plane of the polarizer). If the linearly polarized light of intensity I_0 is transmitted through a polarizer whose transmission plane makes an angle α with the plane of oscillations of the light wave, then the intensity of the transmitted wave is equal to

$$I = I_0 \cos^2 \alpha$$

(*Malus's law*). This can be explained by the fact that the linearly polarized light with amplitude E_0 is the sum of two linearly polarized waves: the wave polarized in the transmission plane (its amplitude is equal to $E_0 \cos \alpha$) propagates through the polarizer without changes and the second wave is completely absorbed.

▶ **Reflection and refraction of waves. Fresnel formulas.** The intensity and polarization of the reflected and refracted waves depend on the incident wave polarization. We write the boundary conditions on the interface between two media:

$$E_\tau^i + E_\tau^r = E_\tau^d, \quad n_1^2(E_n^i + E_n^r) = n_2^2 E_n^d, \quad H_\tau^i + H_\tau^r = H_\tau^d, \quad H_n^i + H_n^r = H_n^d.$$

The subscripts τ and n denote the tangential and normal components, and the superscripts i, r, and d corresponds to the incident, reflected, and refracted (deflected) waves, respectively. Moreover, it is necessary to take account of the relations between the electric and magnetic fields in the plane of the electromagnetic wave (see Subsection P4.2.2). For a plane monochromatic wave, we have

$$\mathbf{E}^i = \mathbf{E}_0^i \exp\left[i(\omega t - \mathbf{k}^i \cdot \mathbf{r})\right],$$

$$\mathbf{E}^r = \mathbf{E}_0^r \exp\left[i(\omega t - \mathbf{k}^r \cdot \mathbf{r})\right],$$

$$\mathbf{E}^d = \mathbf{E}_0^d \exp\left[i(\omega t - \mathbf{k}^d \cdot \mathbf{r})\right]$$

The wave vectors (Fig. P5.17) are related by

$$k_\parallel^i = k_\parallel^r = k_\parallel^d,$$

or

$$k_1 \sin \alpha = k_1 \sin \beta = k_2 \sin \psi,$$

where $k_1 = \omega n_1/c$ and $k_2 = \omega n_2/c$. We obtain the reflection law $\beta = \alpha$ and refraction law $n_1 \sin \alpha = n_2 \sin \psi$.

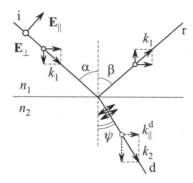

Figure P5.17. Decomposition in polarizations for incident, reflected, and refracted waves.

Special attention is drawn to the case where the light is incident from an optically denser medium ($n_1 > n_2$) at an angle α greater than the *limit angle* α_c *of total reflection* ($\sin \alpha_c = n_2/n_1$). In this case, $k_1 \sin \alpha > k_2$, and $k_\perp^d$ is imaginary: $k_\perp^d = \sqrt{k_2^2 - (k_\parallel^d)^2} = i|k_\perp^d|$, where $|k_\perp^d| = \frac{\omega}{c} \sqrt{n_1^2 \sin^2 \alpha - n_2^2}$. This means that the amplitude of the transmitted wave decays exponentially at the distance $\delta \sim 1/|k_\perp^d|$, and the amplitude of the reflected wave is equal to the amplitude of the incident wave.

The amplitudes of the transmitted and reflected waves depend on the polarization of the incident wave. We present the result of the reflected waves:

$$E_\parallel^r = -E_\parallel^i \frac{\tan(\alpha - \psi)}{\tan(\alpha + \psi)}, \qquad E_\perp^r = -E_\perp^i \frac{\sin(\alpha - \psi)}{\sin(\alpha + \psi)}$$

(*Fresnel formulas*). Here the first formula concerns the wave polarized in the incidence plane (it is convenient to derive it from the boundary conditions for **E**), and the second formula concerns the wave polarized in the perpendicular plane (it is convenient to derive it from the boundary conditions for **H**). One can see that, for an angle of incidence satisfying the condition $\alpha + \psi = \pi/2$, the wave polarized in the incidence plane does not reflect at all. Since $\sin \psi = \cos \alpha$ in this case, it follows that the angle of incidence at which the reflected wave is linearly polarized perpendicularly to the incidence plane (the *Brewster angle*) satisfies the relation:

$$\tan \alpha = n.$$

The qualitative explanation is that, in this case, the direction of the dipole oscillations (shown in the figure), generated in the second media by the wave polarized in the incidence plane turns out to be parallel to the direction of the reflected wave (reflected and refracted rays are mutually perpendicular). But the oscillator does not radiate any wave in the direction of its oscillations (see Subsection P4.2.2).

In the case of normal incidence, there is no difference between polarizations:

$$E^r = E^i \frac{n_1 - n_2}{n_1 + n_2}.$$

One can see that in the case of reflection from an optically denser medium ($n_2 > n_1$), the oscillation phase becomes opposite (more precisely, π is added to the phase). By the way, this property follows directly from the Fresnel formulas: for $\alpha > \psi$, the oscillations change the sign.

The ratio of the reflected energy to the flux of the incident energy is called the *reflection index*. In the case of normal incidence, it is equal to

$$R = \frac{(n_1 - n_2)^2}{(n_1 + n_2)^2}. \tag{P5.3.1.1}$$

The transmission coefficient is equal to $D = 1 - R$ (the energy conservation law).

We note that if the index of reflection is equal to the ratio of the energy volume densities of the reflected and incident waves: $R = (E^{\mathrm{r}}/E^{\mathrm{i}})^2$, then, calculating the transmission coefficient, one must take account of both the difference in the wave velocities in different media and (for incidence at an angle) the variation in the transverse area due to refraction: $S^{\mathrm{d}}/S^{\mathrm{i}} = \cos\psi / \cos\alpha$ (Fig. P5.18). Finally,

$$D = \frac{v_2 (E^{\mathrm{d}})^2 \cos\psi}{v_1 (E^{\mathrm{i}})^2 \cos\alpha}.$$

(The energy flux is equal to the product of the energy volume density by the wave velocity and the area of the transverse cross-section.)

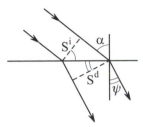

Figure P5.18. In refraction, the beam transverse cross-section is changed.

The coefficients R and D depend only on the *relative index of refraction* $n_{21} = n_2/n_1$ of the two media. The index of reflection is usually not large; for example, for $n_{21} = 1.5$, we obtain $R = 0.04$.

Example. Blooming of optical systems. The reflection index of glasses in optical instruments is not large (of several percents); nevertheless, it is an important problem to decrease the reflection for certain wavelengths. To this end, the surface is covered with a transparent film with the index of refraction $n' = \sqrt{n}$ (n is the index of refraction of glass) and of thickness $\lambda/(4n')$. The optical path difference of the rays reflected from the film surfaces is equal to $\lambda/2$ (the phase variation in reflection need not be taken into account, because it occurs for each of the rays), and the indices of reflection from these surfaces are close to each other (see formula (P5.3.1.1)). As a result, the light reflection is suppressed almost completely.

▶ **Optically anisotropic media.** When an electromagnetic wave travels through an anisotropic media, the vectors **E** and **D** are generally not parallel to each other. The linear relation between them is of tensor character, which means that each component of the vector **D** can be expressed as a linear combination of all three components of **E**. There exist three mutually perpendicular axes, called the *dielectric axes of the crystal*, for which **D** ∥ **E**: $D_i = \varepsilon_0 \varepsilon_i E_i$ ($i = 1, 2, 3$). The values ε_i are called the *principal dielectric permittivities* of the crystal. We consider only the case of *uniaxial crystals* for which two of the three ε_i coincide ($\varepsilon = \varepsilon_\perp$). The distinguished axis ($\varepsilon = \varepsilon_\parallel$) is called the *optical axis of the crystal*.

If a plane wave propagates in a uniaxial crystal, then the *principal plane of the crystal* is introduced, which is the plane passing through the optical axis and the vector **n** normal to the wave front. It turns out that the propagation of a linearly polarized light wave depends on the direction of its polarization. The wave polarized perpendicularly to the principal plane is said to be *ordinary*. The speed of such a wave propagation $v_\perp = c/\sqrt{\varepsilon_\perp}$ is independent of the direction; the oscillations of the vectors **E** and **D** have the same directions; the direction of energy propagation (i.e., the Poynting vector $\mathbf{S} = \mathbf{E} \times \mathbf{H}$) is perpendicular to the wave front. The wave polarized parallel to principal plane is said to be *extraordinary*. The velocity of its propagation depends on the angle between **n** and the optical axis (for the angle $\pi/2$ between them, it is equal to $v_\parallel = c/\sqrt{\varepsilon_\parallel}$). The vectors **E** and **D** oscillate in different directions, and the Poynting vector $\mathbf{S} = \mathbf{E} \times \mathbf{H}$ is not perpendicular to the wave front (the normal to the wave front is parallel to $\mathbf{D} \times \mathbf{H}$). The difference between the ordinary and extraordinary rays disappears only if the light propagation is parallel to the optical axis.

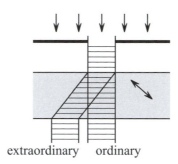

extraordinary   ordinary

Figure P5.19. Ordinary and extraordinary rays in the case of light normal incidence on the crystal cut at an angle with the optical axis.

If the light is incident on the crystal surface, then it splits into ordinary and extraordinary rays that are linearly polarized perpendicular to each other and have different indices of refraction (*phenomenon of birefringence*). For an extraordinary ray, only the direction of the front propagation obeys the refraction law (see Subsection P5.1.1), but the ray itself can leave the incidence plane. Even in the case of the ray normal incidence on a crystal cut at an angle with the optical axis, the spatial splitting of rays occurs (Fig. P5.19). The fronts are indicated by dashes, the optical axis is indicated by arrow. The extraordinary ray is polarized in the plane of the drawing, the ordinary ray is polarized perpendicularly to it.

To obtain and analyze polarized light, one uses the *polarization prisms (nicols)* cut at an angle with the ray propagation direction so that the ordinary ray is totally reflected from the plane of the cut and propagates sidewise, while the extraordinary propagates straightforwardly. Another method for obtaining polarized light is based on the difference in absorption of ordinary and extraordinary rays in some matters (*dichroism phenomenon*). When the light propagates through a *dichroic plate* (tourmaline plate, polaroid), the ordinary ray is absorbed and the linearly polarized extraordinary ray is transmitted.

To analyze the light polarization character, one studies the intensity dependence on the nicol orientation. If the intensity does not vary, then the light is either natural or circularly polarized. To distinguish these cases, one uses a *quarter wave plate* or a *compensator*. The plate thickness d is chosen so that the difference between the ordinary and extraordinary rays $\Delta = \Delta n\, d$ is equal to $\lambda/4$. The phase shift between mutually perpendicular oscillations becomes either zero or π, and the circular polarization becomes the linear polarization.

▶ **Rotation of the polarization plane.** In the case of propagation of linearly polarized light in several matters (they are said to be *optically active*) the *rotation of polarization plane* can occur. The rotation angle is proportional to the plate thickness: $\chi = \alpha l$, where α is *rotation per unit length*. Depending on the rotation direction, there are right- and left-handed matters. An example is given by a quartz plate cut perpendicularly to the optical axis (quartz can be either levorotary or dextrorotary). In solutions of optically active matter in an inactive solvent, α is proportional to the concentration. The molecules of active matters are antisymmetric with respect to the right- and left-handed rotation of spiral type. The phenomenon of rotation of the polarization plane can be characterized as the circular birefringence. The wave circularly polarized in opposite directions propagates with different velocities; that is, the phase difference between them varies. The sum of two such oscillations is a linear oscillation whose direction depends on the phase difference.

▶ **Artificial anisotropy.** If many isotropic bodies are placed in a homogeneous electric field, they acquire uniaxial anisotropy with the optical axis oriented parallel to the electric intensity field (the *electro-optical Kerr effect*). The phase difference between the ordinary and extraordinary rays in light propagation perpendicularly to $\mathbf{E}$ is proportional to the squared electric intensity:

$$\Delta\varphi = 2\pi BlE^2,$$

where l is the thickness of the matter layer and B is called the *Kerr constant*. Artificial anisotropy arises in the cases where the matter molecule polarizability depends on their orientation with respect to the field. A similar effect arise in the case of several matters placed in a magnetic field (the *Cotton–Mouton effect*). It is described by the relation $\Delta\varphi = 2\pi ClB^2$.

In the case of inactive matters placed in a strong magnetic field, the optical activity may arise for light propagating parallel to the vector $\mathbf{B}$ (*magnetic rotation of polarization plane*). The rotation angle per unit length in this case (for diamagnetic and paramagnetic materials) is proportional to the value of the magnetic induction: $\alpha = RB$, where R is called the *Verdet constant*.

P5.3.2. Light Dispersion and Absorption

▶ **Classical model of dispersion medium.** When an electromagnetic wave propagates through matter, the charged particles of the medium come into forced oscillatory motion. The amplitude of these oscillations and their phase shift with respect to the wave field strength oscillations depend on the relation between the wave frequency ω and the frequency ω_0 of natural oscillations of particles (see Subsection P4.1.3). The resultant wave perturbation can be considered as the result of interference of the original wave and the waves radiated by the medium particles (such an approach is called the *molecular optics*). But in the case of homogeneous media, it is possible to obtain the wave frequency characteristics semi-phenomenologically taking account of the polarization arising due to the particle displacement, introducing the dielectric susceptibility and permittivity depending on the frequency, and calculating the index of refraction. The wave damping—that is, the transformation of the energy oscillations into thermal energy—is taken into account by introducing the semi-empirical attenuation of oscillators; then the dielectric permittivity and the index of refraction become complex numbers.

First, we consider the medium of equal oscillators. The equation of motion of a charged particle has the form

$$\ddot{\boldsymbol{\xi}} + 2\beta\dot{\boldsymbol{\xi}} + \omega_0^2\boldsymbol{\xi} = \frac{e}{m}\mathbf{E},$$

where $\mathbf{E}$ is the field acting on the particle (in the optical range, only the electrons are significant). In rare gases, one can neglect the difference between the local and average fields; that is, one can assume that the electrons are under direct action of the wave field $\mathbf{E} = \mathbf{E}_0\exp\left[i(\omega t - \mathbf{k}\cdot\mathbf{r})\right]$. We seek the solution of the equation of motion in the form $\boldsymbol{\xi} = \boldsymbol{\xi}_0\exp(i\omega t)$ and, after the substitution, we obtain

$$\boldsymbol{\xi} = \frac{e}{m(\omega_0^2 - \omega^2 + 2i\beta\omega)}\mathbf{E}$$

(in complex representation, the phase shift is automatically taken into account). The particle displacement results in the molecules acquiring the dipole moments $\mathbf{p} = e\boldsymbol{\xi}$, thus leading to the polarization $\mathbf{P} = N\mathbf{p}$ (N is the particle concentration). From the relation $\mathbf{P} = \varepsilon_0(\varepsilon - 1)\mathbf{E}$ we derive the complex dielectric permittivity

$$\varepsilon = 1 + \frac{Ne^2}{\varepsilon_0 m(\omega_0^2 - \omega^2 + 2i\beta\omega)}. \tag{P5.3.2.1}$$

The index of refraction is also complex, $\sqrt{\varepsilon} = n - i\kappa$, as well as the wavenumber, $\widetilde{k} = \frac{\omega}{c}\widetilde{n} = \frac{\omega}{c}(n - i\kappa)$. Consequently, for the argument of the exponential function we have $i(\omega t - \widetilde{k}x) = i(\omega t - kx) - \kappa\frac{\omega}{c}x$, where $k = \frac{\omega n}{c} = \frac{\omega}{v}$, and the attenuation coefficient is equal to $\kappa\frac{\omega}{c}$. Hence,

$$E = E_0 \exp\left(-\frac{\kappa\omega}{c}x\right) \exp[i(\omega t - kx)]. \tag{P5.3.2.2}$$

To find $n(\omega)$ and $\kappa(\omega)$, it is necessary to equate the real and imaginary parts in $\varepsilon = (n - i\kappa)^2$. Far from the natural frequency (for $|\omega - \omega_0| \gg \beta$), we obtain

$$\varepsilon = n^2 = 1 + \frac{Ne^2}{\varepsilon_0 m(\omega_0^2 - \omega^2)}.$$

The dependencies $n(\omega)$ and $\kappa(\omega)$ are qualitatively shown in Fig. P5.20 (n_0 is the value of n as $\omega \to 0$, which is called the static value). In the region where the absorption is not large, the index of refraction increases with frequency (*normal dispersion*). In the narrow region of strong absorption, one observes the *anomalous dispersion*.

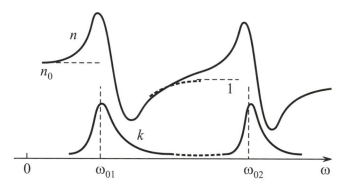

Figure P5.20. Dependence of the index of refraction (upper curve) and of the wave attenuation on the frequency.

A similar situation arises near each natural frequency (ω_{01} and ω_{02} in Fig. P5.20). For example, in the infrared region of the spectrum, the absorption and anomalous dispersion bands related to ion oscillations are observed. The absorption bands in the ultraviolet (sometimes, in the optical) regions of the spectrum can be explained by the electron oscillations on the atomic outer shells (the optical electrons). In the X-ray region of the spectrum, the wave frequency ω is large compared with all natural frequencies and the dependence $n(\omega)$ is determined by electron oscillations, which can be assumed to be free:

$$n^2 = 1 - \frac{Ne^2}{m\varepsilon_0\omega^2}. \tag{P5.3.2.3}$$

The index of refraction of X-rays is slightly different from one. The same formula holds for a wave propagating in rarefied plasma containing free electrons.

The phase velocity of a wave in plasma (and also to the right from the absorption band in dielectrics) turns out to be greater than the light speed in a vacuum ($n < 1$). But this does not contradict the theory of relativity, because the wave group velocity $u = d\omega/dk$ (see Subsection P4.2.1) is less than c in this case. Let us verify this fact for a wave in plasma. Using the relation $k^2 = \omega^2 n^2(\omega)/c^2$ and equation (P5.3.2.3), we obtain:

$$c^2 k \, dk = \omega \, d\omega \quad \Longrightarrow \quad \frac{\omega}{k} \frac{d\omega}{dk} = c^2.$$

Hence, in this case, $u = nc < c$.

In the case of polar molecules (for example, of water), a wide band of anomalous dispersion lies in the region of centimeter radio waves, where the amplitude of rotational oscillations of dipoles tending to rotate following the field strength strongly depends on the frequency. Just in this region, $n = \sqrt{\varepsilon}$ decreases from a large static value (for water, $n_0 \approx 9$) to the high-frequency value (for water, $n \approx 1.3$).

Formula (P5.3.2.1) is only valid for n close to unity, or in the regions where the difference of the field acting on the molecule from the average field in the matter can be neglected. A generalization to the case of dense fluids is the *Lorentz–Lorentz formula*:

$$\frac{n^2 - 1}{n^2 + 2} = \frac{Ne^2}{3\varepsilon_0 m(\omega_0^2 - \omega^2)},$$

which follows from the Clausius–Mossotti formula (Subsection P3.5.2). As the matter density varies, the quantity

$$r = \frac{1}{\rho} \frac{n^2 - 1}{n^2 + 2},$$

which is called the *specific refraction*, must remain constant.

▶ **Light scattering. Wave attenuation.** The wave intensity decreases in a medium not only because of the light absorption, but also because of its scattering. The scattering can be explained by the light emission by atomic oscillators, which occurs in all directions (see Subsection P4.2.2). But in an ideal homogeneous medium, the light scattered at molecules located at the distance $\lambda/2$ from each other would experience the total interference damping, and no attenuation due to scattering would occur in this case. The scattering is observed at small foreign particle (*Tyndall scattering in turbid media*) and at inhomogeneities arising due to the density fluctuations (the *Rayleigh scattering*).

The intensity of light scattered at inhomogeneities whose dimensions are small compared with the wavelength is proportional to λ^{-4} (*Rayleigh law*, also see Subsection P4.2.2). This explains the blue color of the sky (the scattered sunlight) and the yellow-red color of the sun (the transmitted light). The degree of polarization of the scattered natural light depends on the angle of scattering; the light scattered at the angle $\pi/2$ is completely polarized. The qualitative explanation is that only the oscillators whose oscillation direction is perpendicular to the scattering direction radiate in this direction. The scattering at inhomogeneities large compared with the wavelength weakly depend on the frequency; this explains the white color of clouds.

The Rayleigh scattering at the density fluctuations or concentration depend on the temperature. As the critical point is approached, the average values of fluctuations increase sharply and one observes white turbidity of the fluid, which is called the *critical opalescence*.

The light beam attenuation in the case of not large intensity occurs by the exponential law (the *Bouger law*):

$$I = I_0 e^{-\alpha x},$$

where the *wave attenuation coefficient* α is equal to the sum of the absorption coefficient, expressed in terms of the imaginary part of the index of refraction (see formula (P5.3.2.2)), and the scattering coefficient, which describes the wave attenuation due to scattering.

P5.4. Foundations of Quantum Optics

P5.4.1. Thermal Radiation

▶ **Equilibrium thermal radiation.** The radiation of electromagnetic (radiant) energy by a body due to the energy of chaotic (thermal) motion of its molecules is called the *thermal radiation*. The properties of thermal radiation are determined by the body material and its temperature. If one makes a closed cavity of any material and maintains the temperature of its walls to be constant, then the system (wall + radiation) comes to the state of thermodynamical equilibrium, and the *equilibrium thermal radiation* is attained in the bulk of the cavity. The most important characteristic of equilibrium thermal radiation is that its properties are completely determined by the wall temperature and are independent of their material. This statement follows from the second law of thermodynamics. Moreover, the equilibrium thermal radiation is homogeneous and isotropic.

The main characteristics of both the radiation from a body surface and the radiation in volume were introduced in Subsection P5.1.2. The radiation from a surface is characterized by the radiance B and the radiant exitance R equal to the radiant energy radiated from unit surface per unit time in all directions (that is, in the solid angle 2π). The spectral decompositions of the radiant exitance r_λ, r_ω, r_ν are also introduced, for example, $R = \int_0^\infty r_\lambda \, d\lambda$; the quantities r are called the *radiation capacities of a body*. The volume radiation is characterized by the luminous flux intensity I and the volume density of radiant energy u, as well as by their spectral decompositions. In the case of isotropic radiation, they are related as $u = 4\pi I/c$. The illumination E is determined as the total radiant power through unit site from all directions (from the solid angle 2π); in the case of isotropic radiation,

$$E = \pi I = \tfrac{1}{4} c u.$$

The spectral densities of illumination E are denoted by e_λ, e_ν and e_ω. The recalculation from one spectral characteristic to another is discussed in Subsection P5.1.2.

▶ **Absorption capacity. Kirchhoff law.** The *absorption capacity* of a body is the fraction of incident radiant energy absorbed by the body for a narrow interval of wavelengths or frequencies:

$$a_\lambda = \frac{d\Phi_{\text{abs}}}{d\Phi_{\text{i}}}.$$

A body for which $a_\lambda = 1$ in the entire spectral interval is called a *black body*. A black body can be modeled by a closed cavity with a small hole; almost all the rays incident into the cavity through the hole are absorbed as the result of multiple reflections from the internal walls. A body with $a_\lambda = \text{const} < 1$ (usually, $a_\lambda < 0.99$) is called a *grey body*.

Since the equilibrium thermal radiation is in equilibrium with the surface, it follows that for any spectral interval, the quantity of absorbed radiant energy, which is equal to $e_\lambda a_\lambda \, d\lambda$, must be equal to the quantity of the radiated energy, which is equal to $r_\lambda \, d\lambda$. Since the characteristics of the equilibrium bulk radiation are independent of the properties of a specific body, it follows that the ratio of the radiation capacity of any body to its absorption capacity is a universal function of the wavelength and temperature (*Kirchhoff law*):

$$\frac{r_\lambda}{a_\lambda} = e_\lambda(T) = \frac{1}{4} c u_\lambda(T). \tag{P5.4.1.1}$$

Since, for a black body, the absorption capacity is equal to one, it follows that the function on the right-hand side is precisely the *radiation capacity of a black body*, which we denote by r_λ^*:

$$\frac{r_\lambda}{a_\lambda} = r_\lambda^*(T).$$

One can see that the radiation capacity of a black body and its radiant exitance are independent of the method of its production; they are related to the energy volume density as

$$r_\lambda^*(T) = \tfrac{1}{4}cu_\lambda(T), \qquad R^*(T) = \tfrac{1}{4}cu(T). \tag{P5.4.1.2}$$

At a fixed temperature, a black body has the largest radiation capacity and radiant exitance. For example, for a grey body with $a_\lambda = \text{const} = \alpha$, we have $R = \alpha R^*$. We note that since the equilibrium thermal radiation is isotropic, it follows that a black body is a Lambertian source (see Subsection P5.1.2).

▶ **Stefan–Boltzmann and Wien laws.** The radiation capacity of a black body at a given temperature tends to zero for small and large λ and attains the maximum value for a certain wavelength λ_m, which depends on the temperature. The area under the curve r_λ^* is equal to the radiant exitance $R^*(T)$. The application of general thermodynamical relations to the equilibrium thermal radiation in a cavity allows one to obtain several general relations for it. (The temperature of the equilibrium thermal radiation is assumed to be equal to the temperature of the walls.) The *Stefan–Boltzmann law* states that the radiant exitance of a black body is proportional to the fourth power of temperature:

$$R^* = \sigma T^4, \tag{P5.4.1.3}$$

where $\sigma = 5.67 \times 10^{-8}$ W m^{-2} K^{-4} is the *Stefan–Boltzmann constant*.

To derive (P5.4.1.3), we use the expression for the pressure of isotropic radiation $p = u/3$ (see Subsection P2.5.1) and the formula $\left(\frac{\partial U}{\partial V}\right)_T = T\left(\frac{\partial p}{\partial T}\right)_V - p$, which is a consequence of the second law of thermodynamics (Subsection P2.3.1). Substituting $U = u(T)V$, we obtain the equation $4u = du/dT$, which implies $u \sim T^4$. Moreover, from the formula for pressure and from the first law of thermodynamics $\left(0 = d(uV) + \frac{1}{3}u\,dV\right)$, one can derive the adiabatic process equation for the equilibrium radiation: $uV^{4/3} = \text{const}$. Taking into account that $u \sim T^4$, we obtain $VT^3 = \text{const}$.

If we consider a slow adiabatic variation in a radiation volume contained in a vessel with mirror walls and apply the Doppler effect formula to the light reflection from a moving mirror (see Subsections P4.2.1 and P4.2.2), then we can prove the *Wien formula*:

$$r_\lambda^* = \lambda^{-5} f_1(\lambda T) \qquad \text{or} \qquad r_\lambda^* = T^5 f_2(\lambda T), \tag{P5.4.1.4}$$

where $f_i(x)$ are unknown functions whose form cannot be found in the framework of thermodynamics. Similar expressions for r_ω^* have the form

$$r_\omega^* = \omega^3 \varphi_1(\omega/T) \qquad \text{or} \qquad r_\omega^* = T^3 \varphi_2(\omega/T). \tag{P5.4.1.5}$$

From the Wien formula (P5.4.1.4) (or (P5.4.1.5)), one can derive the Stefan–Boltzmann law (P5.4.1.3). Moreover, these formulas imply the *Wien displacement law*, which presents the temperature dependence of the position of the maximum of the function r_λ^* (or r_ω^*):

$$\lambda_{\max} T = b \qquad (\omega_{\max}/T = b_1), \tag{P5.4.1.6}$$

where $b = 2.9 \times 10^{-3}$ m K is the Wien constant. For example, as the temperature decreases by the factor 2, then the position of the maximum of the function r_ω^* (or u_ω) becomes twice closely to the origin, and the maximum itself is eight times below (Fig. P5.21); in this case, the area under the graph decreases by the factor 16.

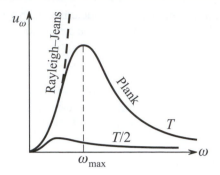

Figure P5.21. Frequency dependence of the energy volume density of the equilibrium thermal radiation.

▶ **Rayleigh–Jeans law.** Rayleigh and Jeans tried to obtain the form of the function u_ω in the framework of the classical statistical physics. They considered radiation in a cavity as an ensemble of standing electromagnetic waves randomly exchanging energy with the walls and between each other. From the viewpoint of statistics, each independent standing wave with a certain frequency of oscillations is equivalent to the oscillator with the same frequency. The energy calculation is reduced to the two independent problems:

1) What is the number dg of oscillators (standing waves) contained in the frequency interval $d\omega$? The answer must be expressed in the form of the function $G(\omega)$, which is called the *density of states*: $dg = VG(\omega)\,d\omega$, where V is the vessel volume.

To calculate $G(\omega)$, one can consider a vessel of the shape of a rectilinear parallelepiped with sides L_x, L_y, L_z. The boundary conditions (for example, the requirement that the nodes of standing waves lie on the boundaries) lead to the conditions $k_\xi L_\xi = m_\xi \pi$ ($\xi = x, y, z$). Therefore, in the space of wave vectors, the admissible states correspond to the nodes of the lattice with sides π/L_ξ and the cell volume $\delta = \pi^3/(L_x L_y L_z) = \pi^3/V$. The volume of the k-space, corresponding to the variation in the value of the wave vector from k to $k + dk$, is equal to $\frac{1}{8}(4\pi k^2\,dk)$ (the volume of the spherical layer cut by the first quadrant). Dividing by the cell volume, we obtain the number of spatially distinct oscillations in the interval dk: $dg = V\dfrac{k^2\,dk}{2\pi^2}$. It is also necessary to take account of the additional degrees of freedom (in the case of electromagnetic waves, there are two possible states of polarization), which we generally take into account by the additional factor γ.

Let us find the number of states per unit volume:

$$\frac{dg}{V} = \gamma \frac{k^2\,dk}{2\pi^2}. \qquad (P5.4.1.7)$$

This formula is obtained from the boundary conditions and has a very general character and numerous applications. To pass to ω, we must take the formula $k = \omega/c$ into account. Finally, we obtain ($\gamma = 2$)

$$G(\omega) = \frac{\omega^2}{\pi^2 c^3}. \qquad (P5.4.1.8)$$

2) What is the average energy of a single oscillator?

The classical physics gives the following answer (see Subsection P2.5.2): to each oscillator, independently of its frequency, one must assign two degrees of freedom and, according to the energy equipartition theorem, the average energy of each oscillator must be equal to kT, where k is the Boltzmann constant.

As a result of this reasoning, the classical physics gives the *Rayleigh–Jeans law*:

$$u_\omega = G(\omega)kT = \frac{\omega^2 kT}{\pi^2 c^3}. \qquad (P5.4.1.9)$$

Experiments show that the Rayleigh–Jeans law is well satisfied at small frequencies (for $\omega \ll \omega_m$) but is absolutely false at large frequencies (Fig. P5.21). Indeed, although (P5.4.1.9) satisfies the conditions imposed on any possible function u_ω by the Wien formula (P5.4.1.5), one can readily see that the function thus obtained does not have any maximum; it monotonically increases, and the integral $\int_0^\omega u_\omega d\omega$; that is, the total energy of radiation is equal to infinity! This situation is one of the symptoms of the deep crisis in the classical physics and the contemporaries called it the *ultraviolet catastrophe*.

▶ **Planck's law.** The energy *equipartition theorem* is a consequence of the fact that the oscillator classical energy is proportional to its squared amplitude and can take any even very small values. According to *Planck's quantization assumption*, the oscillator energy (counted from the minimum value) can take only discrete values multiple of a quantity depending on the oscillator frequency:

$$\varepsilon_n(\omega) = n\varepsilon_1(\omega), \qquad n = 0, 1, \ldots \qquad (\text{P5.4.1.10})$$

To calculate the average energy, one can use the Maxwell–Boltzmann formula (see Subsection P2.5.4), according to which the probability of a state with energy ε is proportional to $\exp\left[-\varepsilon/(kT)\right]$. We obtain

$$\langle \varepsilon(\omega) \rangle = \frac{\displaystyle\sum_{n=0}^{\infty} n\varepsilon_1 \exp\left[-n\varepsilon_1/(kT)\right]}{\displaystyle\sum_{n=0}^{\infty} \exp\left[-n\varepsilon_1/(kT)\right]}.$$

The series in the denominator is simply the sum of a geometric progression, and the numerator is obtained from the denominator by differentiation with respect to $1/(kT)$. After calculations, we obtain

$$\langle \varepsilon(\omega) \rangle = \varepsilon_1 \left[\exp\left(\frac{\varepsilon_1}{kT}\right) - 1\right]^{-1}, \qquad u_\omega = G(\omega)\langle \varepsilon(\omega) \rangle = \frac{\omega^2 \langle \varepsilon(\omega) \rangle}{\pi^2 c^3}. \qquad (\text{P5.4.1.11})$$

Comparing with Wien formula (P5.4.1.5), we see that $\varepsilon_1 = \varepsilon_1(\omega)$ must be proportional to ω:

$$\varepsilon_1(\omega) = \hbar\omega, \qquad (\text{P5.4.1.12})$$

where $\hbar = h/(2\pi) = 1.05 \times 10^{-34}$ J s is a universal constant. The constant h is called *Planck's constant* and $\hbar$ is the *reduced Planck constant*, often called "h-bar"; it is convenient to use the former constant in the case of frequency, and the latter, in the case of cyclic frequency: $\varepsilon_1 = h\nu = \hbar\omega$. Since the quantum energy is proportional to the oscillator frequency, it follows, that at a given temperature, the high-frequency oscillations are excited with a very small probability and their contribution to the energy radiation energy is negligibly small. This solves the problem of ultraviolet catastrophe.

After substitution of (P5.4.1.12) into the formula for the average energy, we obtain *Planck's law* for the spectral density of energy:

$$u_\omega = \frac{\hbar\omega^3}{\pi^2 c^3}\left[\exp\left(\frac{\hbar\omega}{kT}\right) - 1\right]^{-1}. \qquad (\text{P5.4.1.13})$$

The graph of this function is given in Fig. P5.21. We write Planck's law also in the variables ν and λ:

$$u_\nu = \frac{8\pi h\nu^3}{c^3}\left[\exp\left(\frac{h\nu}{kT}\right) - 1\right]^{-1}, \qquad u_\lambda = \frac{8\pi hc}{\lambda^5}\left[\exp\left(\frac{hc}{\lambda kT}\right) - 1\right]^{-1}. \qquad (\text{P5.4.1.14})$$

For $\hbar\omega \ll kT$, Planck's law becomes the Rayleigh–Jeans formula (P5.4.1.9). From Planck's law we can obtain the expressions for the Stefan–Boltzmann and Wien constants in terms of the universal constants:

$$\sigma = \frac{\pi^2 k^4}{60\, c^2 \hbar^3}, \qquad b \approx 0.20 \frac{hc}{k}.$$

Planck's law agrees well with experiments in the entire frequency range.

P5.4.2. Light Quanta

▶ **Photoeffect.** The *external photoelectric effect* (*photoeffect*) is the effect of light-induced emission of electrons from a matter. To study the photoeffect, one employs a vacuum tube with cold cathode (in this case, the thermoelectron emission can be neglected). Radiating the cathode by light of fixed frequency and intensity, one takes the volt-ampere characteristic of the tube (dependence of the current on the anode voltage). The volt-ampere characteristic (Fig. P5.22) permits finding: (a) the number of electrons emitted from the cathode per unit time (it is expressed in terms of the saturation current: $N = I_{sat}/e$) and (b) the maximum kinetic energy of emitted electron; it is expressed in terms of the *stopping voltage*, the anode voltage at which the current is zero:

$$\tfrac{1}{2}mv^2 = e|U_s|.$$

At this voltage, even the fastest electrons cannot reach the anode.

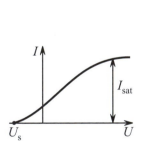

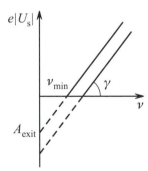

Figure P5.22. Dependence of the photoelectron current on the tube voltage.

Figure P5.23. The maximum energy of photoelectrons versus the light frequency for different metals.

The first law of photoeffect: the number of electrons emitted from metal by light per unit time is directly proportional to the light wave intensity.

The second law of photoeffect: the maximum kinetic energy of photoelectrons linearly increases with light frequency and is independent of light intensity. If the light frequency is less than a certain minimum frequency ν_{min} determined for a given material, then the photoeffect is not observed. It was experimentally found that the dependence of $e|U_s|$ on ν for a given metal has the form of an inclined straight line, and the inclination of the lines constructed for different metals turned out to be the same (Fig. P5.23). The classical wave theory of light could not explain the second law of photoeffect. Moreover, this theory cannot explain the fact that the *photoeffect is inertialess*; that is, absolutely no delay arises between the beginning of radiation and the appearance of current.

▶ **Light quanta.** The photoeffect laws were explained by Einstein. He used Planck's quantization assumption (Subsection P5.4.1), but proceeded much further assuming that the light energy quanta are completely absorbed by separate electrons. This means that in the process of absorption, light behaves like a localized particle (it was called *photon*) with energy

$$E = h\nu = \hbar\omega = hc/\lambda. \tag{P5.4.2.1}$$

Just as any massless particle moving with the speed of light, a *light quantum* (a *photon*) has the momentum

$$p = E/c = h\nu/c = h/\lambda. \tag{P5.4.2.2}$$

The relation between the energy and momentum of a massless particle is given by the theory of relativity (see Subsection P1.10.3).

The quantum properties of light manifest themselves in light radiation, absorption, and scattering. Its wave properties manifest themselves in the phenomena related to light propagation. The light is of dual nature (the wave-particle duality). The same properties are exhibited by all elementary particles.

The *photoeffect* (an act of photoeffect) is the absorption of photon by a particle, for example, by electron. As a result of photoeffect, the light quantum disappears, and the electron acquires additional energy. If a photoelectron is emitted from a matter, then the external photoeffect is observed; if it remains inside, then the *internal photoeffect* takes place. In the case of internal photoeffect, electrons can move from a bound state in a free state, which increases the number of current carriers and hence decreases the resistance.

Photoeffect is used to construct photoelements, photorelay, etc.

Example 1. Can photoeffect occur on a free electron?

Solution. The answer is negative, because, in this case, the energy and momentum conservation laws cannot be satisfied simultaneously. This becomes obvious if we pass to the inertial reference system in which the electron after the photoeffect is at rest. Before the photoeffect, the system contained a light quantum and a moving electron, and after the photoeffect, it contains only an immovable electron; that is, the energy is not conserved.

Absorbing a light quantum, the electron acquires the energy $h\nu$. After escape from a metal, the energy of each electron decreases by a certain value, which is called the *photoelectric work function* A_{exit} (the work required to remove an electron from metal; the photoelectric work function depends on the kind of the matter). The maximum energy of electrons after escape (if there are no other losses) has the form

$$\tfrac{1}{2}mv^2 = h\nu - A_{exit} \tag{P5.4.2.3}$$

(*Einstein equation*). If $h\nu < A_{exit}$, then the external photoeffect does not occur. Hence the threshold frequency is equal to

$$\nu_{min} = A_{exit}/h.$$

It follows from (P5.4.2.3) that the inclination of lines in the graph of the dependence of $e|U_s|$ on ν (Fig. P5.23) is equal to h and the segment cut by the straight line on the ordinate axis is equal to the photoelectric work.

The photon energy and the photoelectric work are usually expressed in *electronvolts* (eV). One electronvolt, which is a non-SI unit, equals the energy acquired by an electron as it accelerates through a potential difference of -1 V: $1\,\text{eV} = 1.6022 \times 10^{-19}$ J. If, for example, the stopping voltage is -3.5 V, then the maximum kinetic energy of electrons is equal to $3.5\,\text{eV}$.

Example 2. The existence of the photoelectric work function means that, on the metal boundary, there arise forces retaining the electron inside the metal. How can the electron attraction to an electrically neutral metal be explained?

Solution. A charge particle is attracted by charges of opposite sign induced on the conductor surface. The attraction force is calculated by using the method of electrostatic images (Subsection P3.4.1). The escaping and returning electrons form a charge cloud near the surface, and the charges on the metal surface form a positively charged layer. Between the charged layers, there exists a nonzero average field intensity directed outwards.

▶ **Boundary of the X-ray spectrum.** If electrons are accelerated in a vacuum tube whose electrodes are charges at several kilovolts, then on impact of electrons on anode, the effect of *slowing down the X-ray radiation* arises. The study of the spectrum of this radiation shows that it does not contain the wavelengths less than a certain value λ_c, which is inversely proportional to the voltage applied to the tube. This fact finds a natural explanation in quantum optics. The radiated photon energy cannot exceed the energy electron kinetic energy: $hc/\lambda \le eU$, which implies

$$\lambda \ge \lambda_c = \frac{hc}{eU}.$$

The theoretical value of the coefficient of proportionality between λ_c and $1/U$ is in good agreement with experiments.

▶ **Light pressure.** The light pressure was predicted by Maxwell on the basis of the electromagnetic theory and was measured by Lebedev. Lebedev's installation consisted of a light rod suspended in a vacuum on a thin filament. Two thin plates were fixed at the ends of the rod, one of them was a reflecting plate, the other was an absorbing plate. Illuminating the plate and measuring the filament twisting, he calculated the light pressure.

The electromagnetic theory gave the following explanation of the light pressure: the electric field of an electromagnetic wave induces current in the metal and this current is under the action of the Ampère force from the magnetic field of the wave; this force has the same direction as the wave propagation and is the cause of light pressure. The explanation of pressure in the language of light quanta looks much simpler: the photons with the momentum (P5.4.2.2) are absorbed or reflected transferring their momentum to the matter. In the case of photon reflection, the transferred momentum is twice larger than that in the case of absorption (also see Subsection P4.2.2).

▶ **Compton effect.** As a photon interacts with a free electron, the process of photon absorption is forbidden by conservation laws, but the photon scattering may occur. If the electron was originally at rest, then, as a result of interaction, it acquires a certain velocity. The energy conservation laws require that the photon energy decrease by the value of the electron kinetic energy, which means that its frequency must also decrease. At the same time, from the viewpoint of the wave theory, the frequency of scattered light must coincide with the frequency of incident light. This phenomenon is called the *Compton effect*, it was discovered in X-ray scattering and played an important role in the formation of quantum theory.

The photon scattering on an electron can be considered as an elastic collision of two particles obeying the energy and momentum conservation laws:

$$pc + mc^2 = p'c + E'_e, \qquad \mathbf{p} = \mathbf{p}' + \mathbf{p}'_e,$$

where $\mathbf{p}$ and $\mathbf{p}'$ is the initial and terminal photon momenta, $\mathbf{p}'_e$ and E'_e are the momentum and energy of the recoil electron (Fig. P5.24). We express the energy and momentum of the recoil electron and substitute them into the relation $E'^2_e - \mathbf{p}'^2_e c^2 = m^2 c^4$ (see Subsection P1.10.3). After transformations, we obtain

$$mc(p - p') = pp'(1 - \cos\theta),$$

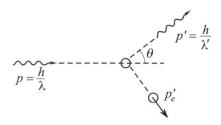

Figure P5.24. Photon scattering on an electron at rest (Compton effect).

where θ is angle of the photon scattering (the angle between the vectors **p** and **p**$'$). We express the momenta of incident and scattered photon from equation (P5.4.2.2): $p = h/\lambda$, $p' = h/\lambda'$, and obtain the formula for dependence of the wavelength increment on the angle of scattering:

$$\lambda' - \lambda = \lambda_C(1 - \cos \theta).$$

The quantity $\lambda_C = h/mc = 2.43 \times 10^{-12}$ m is called the electron *Compton wavelength*. The energy of photon with wavelength λ_C is equal to the energy of electron at rest mc^2. The maximum effect corresponds to the photon scattering at the angle $\theta = \pi$.

▶ **Number of photons in equilibrium thermal radiation.** Planck's formula (P5.4.1.13) for the density of energy of equilibrium thermal radiation can be written in the language of light quanta as follows:

$$u_\omega = \hbar\omega G(\omega)N_\omega,$$

where $\hbar\omega$ is the energy of a quantum, $G(\omega) = \omega^2/\pi^2 c^3$ is the density of states, and

$$N_\omega = \frac{1}{e^{\hbar\omega/kT} - 1} \tag{P5.4.2.4}$$

has the meaning of the number of photons in the state with a certain frequency ω.

P5.4.3. Principle of Laser Operation

▶ **Spontaneous and forced transitions. Einstein coefficients.** We consider an atom in the field of equilibrium radiation. If the atom is in the state with energy E_i, then, under the action of radiation, it can move into a state with larger energy E_j absorbing a light quantum

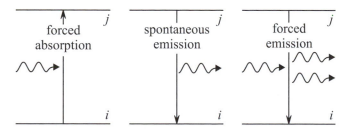

Figure P5.25. Schematic representation of processes of absorption and radiation of light quanta by atoms.

with frequency $\omega = (E_j - E_i)/\hbar$ (Fig. P5.25). Such a process is called the *forced (induced) absorption*. The probability of such a transition per unit time is proportional to the density of energy radiation at the frequency ω:

$$P_{i\to j} = B_{ij}u_\omega.$$

The reverse transition from a state with larger energy to a state with lesser energy is related to creation and emission of a quantum of frequency ω. It can occur in two ways. The first type of transitions is called the *spontaneous emission*, it occurs even in the absence of external radiation (for $u_\omega = 0$), and its probability is independent of u_ω:

$$P_{j \to i}^{\mathrm{sp}} = A_{ji}.$$

The second type of transitions is called the *forced emission*, it occurs under the action of a quantum of frequency ω. The transition probability per unit time, just as in the case of forced absorption, is proportional to u_ω:

$$P_{j \to i}^{\mathrm{ind}} = B_{ji} u_\omega.$$

The coefficients B_{ij}, B_{ji}, and A_{ji} thus derived are called *Einstein coefficients*, and they are independent of the radiation temperature.

▶ **Equilibrium between atoms and radiation.** We assume that a volume filled with equilibrium radiation at temperature T contains equal atoms. We also assume that the thermal equilibrium is established between atoms and radiation, and in this case the average number of atoms in state i is equal to N_i and that in state j is equal to N_j. These numbers at temperature T are related to each other by the *Boltzmann relation*:

$$N_j = N_i e^{-(E_j - E_i)/kT}.$$

In equilibrium, the number of transitions from i to j must be equal to the number of reverse transitions:

$$N_i B_{ij} u_\omega = N_j (A_{ji} + B_{ji} u_\omega). \tag{P5.4.3.1}$$

It follows from the Boltzmann relation that $N_i/N_j \to 1$ as $T \to \infty$. Taking into account that then $u_\omega \to \infty$, we obtain $B_{ij} = B_{ji}$.

Let us express u_ω from (P5.4.3.1):

$$u_\omega = \frac{A_{ji}/B_{ji}}{e^{\hbar\omega/kT} - 1} \tag{P5.4.3.2}$$

(we took into account that $N_i/N_j = \exp(\hbar\omega/kT)$). One can see that, in fact, we derived Planck's law for equilibrium radiation. The ratio A_{ji}/B_{ji} can be obtained from the condition that the classical Rayleigh–Jeans law must hold at high temperatures:

$$\frac{A_{ji}}{B_{ji}} = \frac{\hbar\omega^3}{\pi^2 c^3}.$$

We look at this relation from a different standpoint. Taking formula (P5.4.2.4) for the number of quanta with a certain frequency into account, we write (P5.4.3.2) as

$$\frac{B_{ji} u_\omega}{A_{ji}} = N_\omega.$$

Hence the ratio of the probability of forced radiation to the probability of spontaneous emission is equal to the number of quanta of frequency ω. In other words, the probability of forced emission calculated per one quantum is equal to the probability of spontaneous emission. One can see that the spontaneous and forced emissions must be of the same nature. The (qualitative) explanation is that even in the absence of light quanta in the system, the electromagnetic field has the so-called *zero-point oscillations* of a given frequency (the oscillator energy in quantum mechanics is equal to $\hbar\omega(N + \frac{1}{2})$). Precisely these oscillations cause spontaneous transitions.

▶ **Inverse population of levels.** In each act of forced emission, a new quantum is formed, and the created quantum has the same phase as the quantum generating this radiation. In other words, instead of one quantum, two *coherent quanta* appear in the system. Each of these quanta can in turn cause the appearance of a new coherent quantum, etc. Thus, coherent radiation may be amplified. But the newly created quanta are absorbed by atoms that are in the lower of the two energy states. In equilibrium, the number of atoms in the lower state (the population of the lower level) is higher than that in the upper level. Hence absorption prevails over the forced emission, and amplification is impossible.

To make a coherent ray amplification possible, it is necessary to create a metastable state with inverse population of levels. Then amplification of the quantum flux J occurs by the formula

$$dJ = \sigma(n_j - n_i)J\,dx, \qquad \text{or} \qquad J(x) = J(0)e^{\gamma x}, \qquad (P5.4.3.3)$$

where n_j and n_i are the atom concentrations, σ is the cross-section of forced absorption or emission, and $\gamma = \sigma(n_j - n_i)$ is the *quantum amplification coefficient* of the medium.

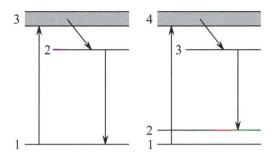

Figure P5.26. Three- and four-level schemes of laser operation.

The inverse population is created by using different methods of pumping. The most widely spread are the three- and four-level schemes (Fig. P5.26). In the three-level scheme (ruby laser), there is intense pumping from the lower level 1 to the wide short-living level 3 from which the long-living (metastable) level 2 is filled. It is necessary to ensure that the population of the upper operation level 2 exceeds the population of the lower operation level 1. The difficulty is that the initial population of level 2 is much less than the population of level 1, and intensive pumping is required. The four-level scheme does not have this drawback. In this scheme, the lower operation level 2 first contains as few atoms as level 3. The filling of level 3 occurs from level 4 which is pumped from the ground level 1. The neodymium laser operates according to this scheme.

▶ **Construction elements of lasers.** To use the phenomenon of quantum amplification of coherent light for creating the *quantum generator*, it is necessary to ensure positive feedback in order that a beam amplify itself. To this end, mirrors are arranged at the ends of the sample (or a mirror coating is applied). One of the mirrors is partially transparent, which ensures that part of light escapes from the system as a laser beam. The initial beam amplifies itself by multiple reflections from the two mirrors if the following condition is satisfied:

$$2Ln = m\lambda,$$

where n is index of refraction, L is the distance between the mirrors (the sample length), and m is an integer. To ensure coherent light generation, the total amplification coefficient on the entire closed path must be greater than 1:

$$R_1 R_2 e^{2\gamma L} > 1,$$

where γ is the quantum amplification coefficient (see (P5.4.3.3)) and R_1 and R_2 are the indices of reflection of the mirrors.

Bibliography for Chapter P5

Alenitsyn, A. G., Butikov, E. I., Kondratyev, A. S., *Concise Handbook of Mathematics and Physics*, CRC Press, Boca Raton, Florida, 1998.

Benenson, W., Harris, J. W., Stocker, and Lutz, H. (Eds.), *Handbook of Physics*, Springer-Verlag, New York, 2002.

Bennett, C. A., *Principles of Physical Optics*, John Wiley & Sons, New York, 2008.

Born, M. and Wolf, E., *Principles of Optics: Electromagnetic Theory of Propagation, Interference and Diffraction of Light (7th Edition)*, Cambridge University Press, Cambridge, England, 1999.

Fowles, G. R., *Introduction to Modern Optics, 2nd edition*, Dover Publications, New York, 1989.

Fox, M., *Quantum Optics: An Introduction (Oxford Master Series in Physics, 6)*, Oxford University Press, New York, 2006.

Hecht, E., *Optics, 4th Edition*, Addison Wesley, Reading, Massachusetts, 2001.

Irodov, I. E., *Wave Processes. Basic Laws* [in Russian], Fizmatlit, Moscow, 2004.

Landsberg, G. S., *Optics, 6th Edition* [in Russian], Fizmatlit, Moscow, 2006.

Pedrotti, F. L., Pedrotti, L. M. and Pedrotti, L. S., *Introduction to Optics, 3rd Edition*, Benjamin Cummings, New Jersey, 2006.

Prokhorov, A. M. (Ed.), *Encyclopedia of Physics, Volumes 1–5* [in Russian], Bolshaya Russkaya Encyclopedia, Moscow, 1998.

Prokhorov, A. M. (Ed.), *Encyclopedic Dictionary of Physics* [in Russian], Sovetskaya Encyclopedia, Moscow, 1984 (dic.academic.ru).

Saleh, B. E. A. and Teich, M. C., *Fundamentals of Photonics, 2nd edition (Wiley Series in Pure and Applied Optics)*, Wiley-Interscience, New York, 2007.

Saslow, W. M., *Electricity, Magnetism, and Light*, Academic Press, New York, 2002.

Sivukhin, D. V., *General Physics, Volume 4, Optics* [in Russian], Fizmatlit, Moscow, 2006.

Sivukhin, D. V., Yakovlev, I. A., *A Collection of Problems, Volume 4, Optics* [in Russian], Fizmatlit, Moscow, 2006.

Woan, G., *The Cambridge Handbook of Physics Formulas*, Cambridge University Press, Cambridge, England, 2003.

Chapter P6

Quantum Mechanics. Atomic Physics

By the beginning of the twentieth century, classical physics had accumulated many contradictions that could only be resolved by changing the main concepts of space and time (Einstein's theory of relativity, Chapter P1) and the laws of motion of particles on atomic scales (quantum mechanics). The first quantum concepts appeared at the very beginning of the century (Planck's formula for thermal radiation and Einstein's formula for photoeffect, Chapter P5), but quantum mechanics as a closed science was not constructed until the end of the 1920s.

P6.1. Atomic Structure. Bohr Model

P6.1.1. Nuclear Atomic Model

▶ **Experimental antecedents.** The atomic theory must, first of all, explain the accumulated experimental facts concerning the spectra of radiation of rarefied gases (that is, of separate atoms).

The atomic spectra consist of separate very narrow lines (*line spectra*) whose location is typical of each element. It was noted that any frequency is the difference of some two *terms*: $\nu_{mn} = T_m - T_n$, so that the element can be characterized not by a set of frequencies but by a set of terms. For hydrogen, we have $T_n = R_\nu/n^2$, where R_ν is the *Rydberg constant* (for frequency). The spectral lines for which one of the terms is fixed are called a *spectral series*. For example, in the hydrogen atom spectrum, the Lyman series ($n = 1$), the Balmer series ($n = 2$), etc. are distinguished.

But at the beginning of the twentieth century, there were no clear concepts about the atomic structure, the massive nucleus and electronic orbits. The *Thomson model*, in which the mass and the positive charge are uniformly distributed over the atomic volume and the electrons are contained inside this positive cloud as raisin in a pudding, seemed to be more attractive.

▶ **Rutherford experiments.** The nuclear atomic model was constructed in *Rutherford's experiments* on scattering of alpha particles at the atoms of thin goldfoil (performed by his collaborators Geiger and Marsden in 1911). It was found that some particles deviate by larger angles and even fly in the opposite direction. The number of such particles was small (approximately one of 8000 particles was scattered by an angle larger than 90°), but the mere fact of the existence of such particles contradicted the Thomson model.

Example 1. Estimate the maximum angle of deviation of alpha particles by the atom in the Thomson model.

Solution. To obtain this estimate, we assume that, in the Thomson model, the maximum deviation is exhibited by the particles flying near the edge of the atom (a ball of radius $R \sim 10^{-10}$ m), where the electric field strength is maximum (see example 3, Subsection P3.3.2). The formula for the angle of the particle deviation by the Coulomb field was derived in Subsection P1.6.2 (example 2):

$$\tan\frac{\theta}{2} = \frac{A}{2E_k b}, \qquad \text{(P6.1.1.1)}$$

where $A = 2Ze^2/4\pi\varepsilon_0$ ($Z = 79$), $E_k \sim 5$–$7\,$MeV is the kinetic energy of alpha particles, and b is the impact parameter. Taking $b \approx R \sim 10^{-10}\,$m for the Thomson model, and see that the maximum deviation of alpha particles is $\sim 0.01°$. If the multiple scattering is taken into account, then a greater angle of deviation is obtained, but for the film thickness $d \sim 10^{-7}$–$10^{-6}\,$m, the probability of scattering even by several degrees is negligibly small.

Rutherford showed that the experimental results are in good agreement with the formula for the angular distribution of α-particles scattered at a point Coulomb center with charge Ze.

Example 2. Rutherford's formula. We consider a thin target whose unit area contains $N_0 = nd$ nuclei (n is the bulk concentration of nuclei, d is the target width). Assuming that each particle interacts only with a single nucleus, we see that the particles scattered in the angle interval $(\theta, \theta + d\theta)$ are the particles whose impact parameter with respect to any nucleus lies in the interval $[b(\theta), b(\theta + d\theta)]$. The fraction of particles scattered in this angle is equal to the total area of such annuli around all the nuclei contained in the unit area of the target:

$$\frac{dN}{N} = N_0 2\pi b\, db = N_0 2\pi b \frac{db}{d\theta}\, d\theta.$$

Taking into account that the solid angle is equal to $d\Omega = 2\pi \sin\theta d\theta$, we obtain

$$\frac{dN}{N} = N_0 b \frac{db}{d\theta} \frac{d\Omega}{\sin\theta}.$$

Expressing b from formula (P6.1.1.1) for the scattering of alpha particles at the Coulomb center Ze, we obtain the *Rutherford formula*:

$$\frac{dN}{N} = \left(\frac{Ze^2}{8\pi\varepsilon_0 E_k} \right)^2 \frac{N_0\, d\Omega}{\sin^4(\theta/2)}. \tag{P6.1.1.2}$$

Numerous experiments confirmed that this formula is in good agreement with experimental data. (We note that the coefficient at $N_0 d\Omega$ is called the *differential scattering cross-section*; see Subsection P8.2.2 for details.)

▶ **Atomic structure.** An atom with atomic number Z consists of a heavy nucleus with charge $+Ze$ and electrons in the field of this nucleus. An atom whose number of electrons is not equal to its atomic number Z is called an *ion*. The minimum energy required to remove an electron and form an ion is called the *ionization energy*.

The nuclear model is in unresolvable contradiction with the fundamental concepts of classical physics. The electrons are attracted to the nucleus, and hence they cannot be at rest but must move around the nucleus (therefore, the nuclear model is sometimes called the *planetary model*). But in accelerated motion, the electron radiates electromagnetic waves, and within a time interval of the order of $10^{-10}\,$s, it must lose its entire energy and fall on the nucleus (see example 2, Subsection P4.2.2). Within the framework of classical physics, the atom cannot be stable.

P6.1.2. Bohr Model

▶ **Bohr postulates.** Bohr's semiclassical model describes the properties of the hydrogen atom and hydrogen-like ions (one electron in the field of the nucleus Ze). The model is based on the *Bohr postulates*:

1) The electrons can move only in certain stationary orbits according to the classical equations of motion but without energy radiation.

2) In this case, the angular momentum can take only discrete values $L_n = n\hbar$. For circular orbits, the quantization rule takes the form

$$mv_n r_n = n\hbar. \tag{P6.1.2.1}$$

The number n is called the *principal quantum number*.

3) The radiation or absorption of quanta of electromagnetic radiation occurs only in transition from one stationary orbit to another stationary orbit:

$$\hbar\omega = E_n - E_m. \tag{P6.1.2.2}$$

▶ **Hydrogen atom.** For the motion of an electron in a circular orbit, from the quantization rule (P6.1.2.1) and the equation of motion

$$\frac{ke^2}{r_n^2} = m\frac{v_n^2}{r_n},$$ (P6.1.2.3)

we find the electron velocity in the nth orbit and its radius:

$$v_n = \frac{ke^2}{n\hbar}, \quad r_n = \frac{n^2\hbar^2}{mke^2},$$

and the electron energy

$$E_n = \frac{mv_n^2}{2} - \frac{ke^2}{r_n} = -\frac{ke^2}{2r_n} = -\frac{R}{n^2},$$ (P6.1.2.4)

where

$$R = \frac{m(ke^2)^2}{2\hbar^2} \approx 13.6 \text{ eV}$$ (P6.1.2.5)

is the Rydberg constant (for the electron energy). The state with number $n = 1$ and energy $E_1 = -R = -13.6$ eV is called the *ground state of the atom*, the other values correspond to the excited states. The lifetime of excited states is of the order of 10^{-8} s. The radius of the orbit of electron in the ground state in the hydrogen atom is called the *Bohr radius* and is denoted by a_0:

$$a_0 = r_1 = \frac{\hbar^2}{mke^2} \approx 0.53 \text{ Å}.$$

The *energy of ionization* of the hydrogen atom is equal to $|E_1| = R$.

The radiation spectrum of the hydrogen atom is:

$$\frac{1}{\lambda} = R_\lambda \left(\frac{1}{n_2^2} - \frac{1}{n_1^2} \right),$$ (P6.1.2.6)

where $R_\lambda = \dfrac{m(ke^2)^2}{4\pi c\hbar^3} = 1.097 \times 10^7$ m^{-1} is the *Rydberg constant* (for the reciprocal wavelength), n_1 is the number of the initial state, n_2 is the number of the final state of electron (Fig. P6.1). The hydrogen lines with $n_2 = 1$ and $n_1 = 2, 3, \ldots$ form the Lyman series (ultraviolet region), the lines with $n_2 = 2$ and $n_1 = 3, 4, \ldots$ form the Balmer (visible region), the lines with $n_2 = 3$ form the Paschen series (infrared region), the further series are Brackett, Pfund, etc. series.

▶ **Hydrogen-like ions. Taking account of the nucleus motion**. The formulas obtained for the hydrogen atom can easily be generalized to the case of a hydrogen-like ion, or the case of one electron in the field of a charge Ze. It is simply required to replace e^2 by Ze^2 in all formulas. In this case, the Rydberg constant is multiplied by Z^2. For example, in the case of helium one-electron ion He$^+$, the electron energy takes the values $E_n = -4R/n^2$. Obviously, the energy with number $2n$ coincides with the electron energy in the hydrogen atom with number n. This means that the transition, for example, from the fourth level to the second level must correspond to the same frequency as the transition from the second level in the hydrogen atom to the first level. An analysis of the spectra shows that all the corresponding levels of the helium ion lie below the levels of the hydrogen atom

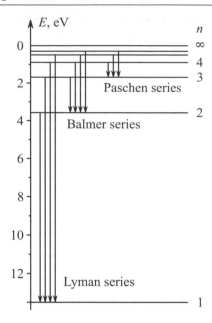

Figure P6.1. Hydrogen atom energy levels and spectral series.

approximately by 0.04 %. This difference can be explained by the fact that the nucleus mass is finite: both the electron and nucleus move about their common center of mass. The simplest way to take the nucleus motion into account is to replace the electron mass m by the reduced mass μ (see Subsection P1.6.4). This results in multiplication of all values of the energy and frequencies by the factor $M/(m + M)$, where M is the nucleus mass. The largest increase in the energy is for the hydrogen atom.

By the same reason, the energy levels of electrons in heavier isotopes lie below (*isotopic effect*). For example, the deuterium levels lie a little below the hydrogen levels, and the tritium levels lie a little below the deuterium levels.

▶ **The Bohr–Sommerfeld quantization condition.** A generalization of the orbit quantization rule is the *Bohr–Sommerfeld quantization condition*:

$$\oint p \, dq = 2\pi n\hbar, \tag{P6.1.2.7}$$

where q and p are the generalized coordinate and momentum. If for the generalized coordinates of motion in a central field we take the angle φ and the angular momentum L, then we obtain the Bohr quantization rule. In the case of one-dimensional motion, the quantization condition (P6.1.2.7) takes the form

$$2 \int_{x_1}^{x_2} \sqrt{2m(E_n - U(x))} \, dx = 2\pi n\hbar.$$

where x_1 and x_2 are the turning points, at which the momentum is zero. For example, in the case of oscillator, $U(x) = m\omega^2 x^2/2$, and calculating the integral, we obtain

$$E_n = n\hbar\omega,$$

which up to the term $\hbar\omega/2$ (zero-point oscillations) coincides with the exact answer given by quantum mechanics.

The Bohr theory is not a consistent quantum theory. It does not describe the fine structure of hydrogen-like atoms and the properties of many-electron atoms, starting from the helium atom.

P6.2. Elements of Quantum Mechanics
P6.2.1. Wave Function. Uncertainty Relation

▶ **The corpuscular-wave duality. De Broglie waves.** A moving particle of matter can exhibit wave-like properties; also a wave can exhibit particle-like properties. This behavior is known as *corpuscular-wave duality* or *wave-particle duality*. The corresponding wavelengths, wave vector, and cyclic frequency are given by the *de Broglie formulas*:

$$\lambda = \frac{h}{p}, \quad k = \frac{p}{\hbar}, \quad \omega = \frac{E}{\hbar}, \tag{P6.2.1.1}$$

where p is the particle linear momentum and E is its relativistic energy.

An important argument for de Broglie waves was the fact that, with definition (P6.2.1.1) taken into account, the Bohr quantization rule (formula (6.1.2.1)) takes the form of the existence condition for the closed standing wave: $2\pi r_n = n\lambda$ (an integer number of de Broglie waves can be put on the orbit length).

The phase velocity of the de Broglie wave of a free particle moving with velocity v is equal to c^2/v, its group velocity is equal to v. (The phase velocity is not an observable quantity.)

▶ **Wave function.** A particle is associated with a complex *wave function* $\psi(\mathbf{r})$, which has a statistical meaning. The squared modulus of the wave function is equal to the density of probability of finding a particle at a given point in space:

$$dP = |\psi(\mathbf{r})|^2 \, dV. \tag{P6.2.1.2}$$

The interference and diffraction phenomena in propagation of particles and in their detection arise due to *superposition principle*: if states ψ_1 and ψ_2 are realized, then the superposition $\psi = c_1\psi_1 + c_2\psi_2$ is also realized. The wave properties of particles were first discovered in experiments on diffraction of electrons at crystals in 1927. (*Davison–Germer experiments*).

▶ **Heisenberg uncertainty relation.** The measurement of a particle coordinates and projections of its momentum is of statistical character. The uncertainties in measurements of these quantities are related as

$$\Delta p_x \Delta x \geq \tfrac{1}{2}\hbar, \quad \Delta p_y \Delta y \geq \tfrac{1}{2}\hbar, \quad \Delta p_z \Delta z \geq \tfrac{1}{2}\hbar. \tag{P6.2.1.3}$$

The momentum and coordinates cannot simultaneously take exact values, and the notion of classical trajectory becomes meaningless.

The electron state with a definite momentum p_x is determined by the wave function in the plane wave form:

$$\psi(x) = \frac{1}{\sqrt{2\pi\hbar}} \exp\left(\frac{i}{\hbar} p_x x\right).$$

In this case, the electron is completely delocalized in space. To estimate the uncertainty in measuring the momentum for an electron with localized wave function $\psi(x)$, it is necessary to expand this function in the Fourier integral in wave functions with a definite momentum:

$$\psi(x) = \frac{1}{\sqrt{2\pi\hbar}} \int \exp\left(\frac{i}{\hbar} p_x x\right) \varphi(p_x) \, dp_x.$$

The probability of that the result of measuring of p_x lies in the interval $(p_x, p_x + dp_x)$ is determined by the squared Fourier transform $|\varphi(p_x)|^2 dp_x$. According to general properties of the Fourier transformation, the function width and its Fourier transform are related as $\Delta x \, \Delta(p_x/\hbar) \geq \tfrac{1}{2}$ (also see Subsection P5.2.1, discussion of the temporal coherence).

The uncertainty relation for time and energy has the form:

$$\Delta E \Delta t \geq \hbar, \tag{P6.2.1.4}$$

where ΔE is the state energy uncertainty and Δt is the particle lifetime in this state.

Example 1. Impossibility of electron localization in the nucleus.
If we assume that the electron is contained inside a nucleus of dimension $r \sim 10^{-15}$ m, then the uncertainty of its coordinate is of the order of the nucleus dimensions and the uncertainty of the momentum is $\Delta p \sim \hbar/r$. Assuming that the average momentum is of the order of its uncertainty, we obtain the following estimate for the energy: $E = p^2/2m \sim 10^{-8} J \approx 10^5$ MeV, which exceeds the electron binding energy in the atom by many orders of magnitude.

Example 2. Estimate of the ground state energy of the hydrogen atom. Assuming that the coordinate uncertainty is of the order of the atom average dimension and the momentum uncertainty is of the order of the average momentum, we obtain the following relation between momentum and radius:

$$p \sim \frac{\hbar}{r}.$$

Substituting $p(r)$ into expression (P6.1.2.4) for the electron energy, we obtain the energy as a function of one parameter, the orbit radius:

$$E \sim \frac{\hbar^2}{2mr^2} - k\frac{e^2}{r}.$$

This expression takes the minimum value equal (up to a dimensionless factor) to the ground state energy for r equal to the Bohr radius.

▶ **Applicability criterion for the classical description.** The notion of trajectory can be used only if the coordinate uncertainty Δx is small compared with the typical dimensions R of the motion region and the momentum uncertainty Δp is small compared to the momentum p. In other words, the inequality $pR \gg \hbar$ must be satisfied. In addition to this criterion, one can compare $\hbar$ with other combinations of the system parameters that are of dimension of angular momentum: (momentum × length) or (energy × time).

P6.2.2. The Schrödinger Equation

▶ **Time-dependent and time-independent Schrödinger equations.** The evolution of the wave function of a particle in the force field $U(\mathbf{r})$ is described by the *Schrödinger equation*:

$$i\hbar\frac{\partial\psi}{\partial t} = -\frac{\hbar^2}{2m}\Delta\psi + U\psi. \tag{P6.2.2.1}$$

In the state with a certain energy E (*stationary state*), the wave function has the form

$$\psi(\mathbf{r}, t) = \psi(\mathbf{r})\exp\left(-i\frac{E}{\hbar}t\right). \tag{P6.2.2.2}$$

The spatial part of the wave function of stationary state satisfies the *time-independent Schrödinger equation*:

$$\Delta\psi + \frac{2m}{\hbar^2}(E - U)\psi = 0. \tag{P6.2.2.3}$$

The wave function of a free particle ($U = 0$) in the state with a certain momentum $\mathbf{p}$ has the form

$$\psi(\mathbf{r}, t) = \frac{1}{(2\pi\hbar)^{3/2}}\exp\left[-\frac{i}{\hbar}(Et - \mathbf{p}\,\mathbf{r})\right].$$

▶ **The probability flux density** is determined by the expression

$$\mathbf{S} = -\frac{i\hbar}{2m}\left[\psi^*\nabla\psi - \psi\nabla\psi^*\right] \tag{P6.2.2.4}$$

and in the case of a free particle with momentum $\mathbf{p} = m\mathbf{v}$ is equal to $\mathbf{S} = \mathbf{v}|\psi(\mathbf{r})|^2$. The variation in the probability of finding a particle in a certain region is determined by the probability flux through the boundary of this region.

▶ **Properties of stationary states.** The solutions of the time-independent Schrödinger equation (P6.2.2.3) describing the *bound states* of a particle in a potential field must satisfy the following conditions:

1) The function and its first derivatives must be continuous.

2) The function must be normed: $\int |\psi|^2 \, dV = 1$. These two conditions can be satisfied only for definite values of energy $E_1, E_2, \ldots$ (*discreteness* of energy levels).

3) In the one-dimensional case, the solutions are real and *nondegenerate*; that is, to each energy value E_n there corresponds one function $\psi_n(x)$. The function ψ_n is zero at $n-1$ points (*nodes* of the wave function). If the potential energy is an even function ($U(-x) = U(x)$), then the wave functions with odd numbers (including the wave function of the ground state) are even ($\psi_{2n+1}(-x) = \psi_{2n+1}(x)$), and the wave functions with even numbers are odd ($\psi_{2n}(-x) = -\psi_{2n}(x)$).

The average value of the physical quantity $g(\mathbf{r})$, which is a function of coordinates, is calculated by the formula:

$$\langle g \rangle = \int g|\psi|^2 \, dV,$$

where $\psi(\mathbf{r})$ is the normed wave function.

Example 1. Well with infinite walls. If the potential energy has the form

$$U = \begin{cases} 0, & \text{for } 0 \le x \le L; \\ \infty, & \text{for } x < 0 \text{ or } x > L, \end{cases}$$

then the wave function vanishes at edges of the well, at $x = 0$ and $x = L$. The Schrödinger equation (P6.2.2.3) takes the form

$$\frac{d^2\psi}{dx^2} + \frac{2m}{\hbar^2} E\psi = 0$$

and has the solutions

$$\psi_n(x) = \sqrt{\frac{2}{L}} \sin k_n x, \quad k_n = \frac{n\pi}{L}.$$

The well width is equal to an integer number of de Broglie half-waves. The energy of the nth level is equal to $E_n = \dfrac{\hbar^2 k_n^2}{2m} = \dfrac{n^2\pi^2\hbar^2}{2mL^2}$. One can see that the wave functions with odd numbers are symmetric with respect to the well center, and the wave functions with even numbers are antisymmetric.

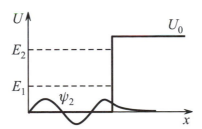

Figure P6.2. The wave function and energy levels in a one-dimensional potential well.

Example 2. Well with finite walls. If the potential energy has the form (Fig. P6.2)

$$U = \begin{cases} 0 & \text{for } 0 \le x \le L; \\ U_0 > 0 & \text{for } x > L; \\ \infty & \text{for } x < 0, \end{cases}$$

then the Schrödinger equation inside the well has the same form as in the preceding example and outside the well, for $x > L$, takes the form

$$\frac{d^2\psi}{dx^2} + \frac{2m}{\hbar^2}(E - U_0)\psi = 0.$$

The localized (bound) states must have energies in the interval $0 < E < U_0$. The wave function vanishes for $x = 0$ and is equal to $\psi_1(x) = A \sin(k_1 x)$ inside the well and to $\psi_2(x) = B \exp(-k_2 x) + C \exp(k_2 x)$ outside the well, where $k_1 = \sqrt{2mE}/\hbar$, $k_2 = \sqrt{2m(U_0 - E)}/\hbar$. It follows from the finiteness and integrability condition for the wave function that $C = 0$. Eliminating the coefficients A and B from the conditions of continuity, $\psi_1(L) = \psi_2(L)$, and smoothness, $\dfrac{d\psi_1}{dx}\Big|_{x=L} = \dfrac{d\psi_2}{dx}\Big|_{x=L}$, of the wave function, we obtain the equation for the allowed values of the energy: $k_2 = -k_1 \cot k_1 L$.

The roots must be sought graphically, as points of intersection of the graph $k_2 L = -k_1 L \cot k_1 L$ with the circle $(k_1 L)^2 + (k_2 L)^2 = 2mU_0 L^2/\hbar^2$ in the region $k_1 L \geq 0$, $k_2 L \geq 0$ (Fig. P6.3). The new bound states arise under the condition $E = U_0$ with $k_1 L = \pi(n + 1/2)$; for $U_0 L^2 < \pi^2 \hbar^2/(8m)$ (a very "narrow" or very "shallow" well), there are no bound states.

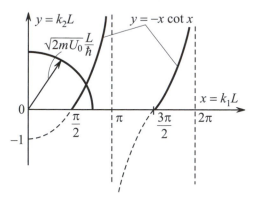

Figure P6.3. Graphical solution of the equations in example 2.

We note that the wave function of a bound state is nonzero for $x > L$ (Fig. P6.2). According to the classical concepts, the particle penetration under the potential barrier is impossible, because for $E < U_0$, the kinetic energy in this region is negative. In quantum mechanics, the probability of finding a particle under the barrier is nonzero, although it decreases exponentially with the distance from the barrier.

Example 3. Harmonic oscillator. The potential energy of harmonic oscillator is $U(x) = m\omega^2 x^2/2$. The energies of stationary states are equal to $E_n = (n + 1/2)\hbar\omega$, and the wave functions of stationary states are expressed in terms of special functions, Hermitian polynomials (see Subsection M13.10.4). The energy of the ground state is equal to $\hbar\omega/2$ (the energy of zero-point oscillations).

Passing to dimensionless variables by the change $E = \dfrac{\hbar\omega}{2}\varepsilon$, $x = \sqrt{\dfrac{\hbar}{m\omega}}\xi$, we obtain the Schrödinger equation for the oscillator in the form

$$\frac{d^2\psi}{d\xi^2} + (\varepsilon - \xi^2)\psi = 0.$$

As $\xi \to \infty$, the term ε in parentheses can be neglected, and we obtain the asymptotic for large ξ behavior of the wave function: $\psi \sim \exp(-\xi^2/2)$. We seek the wave function in the form $\psi = \exp(-\xi^2/2)\varphi(\xi)$, and for φ, we obtain the equation

$$\varphi'' - 2\xi\varphi' + (\varepsilon - 1)\varphi = 0.$$

Representing $\varphi(\xi)$ as a series, $\varphi = a_0 + a_1\xi + \cdots$, and substituting it into the equation, we obtain a series of recursive relations: $2a_2 + (\varepsilon - 1)a_0 = 0$, $6a_3 + (\varepsilon - 3)a_1 = 0$, $\ldots$, $(n + 1)(n + 2)a_{n+2} + (\varepsilon - 2n - 1)a_n = 0$, $\ldots$. The wave function $\psi(\xi)$ satisfies the finiteness condition only if the series for $\varphi(\xi)$ contains finitely many terms (otherwise, as $\xi \to \infty$, $\varphi(\xi)$ increases as $\exp(\xi^2)$). The series can be truncated under the condition that it contains either only odd powers (for this, we must set $a_0 = 0$) or only even powers ($a_1 = 0$). The series truncation condition permits obtaining allowed values of the energy: $\varepsilon_n = 2n + 1$ or $E_n = \hbar\omega(n + 1/2)$ ($n = 0, 1, \ldots$).

Just as in the preceding example, the wave functions of the oscillator stationary states are nonzero in the region, where the oscillator energy is less than its potential energy. The probability of finding a particle in the forbidden region decreases exponentially in the direction inside this region.

We note that in the ground state, the oscillator energy is equal not to zero but to $\hbar\omega/2$; this energy is called the *energy of zero-point oscillations*. The existence of such an energy is inevitable from the viewpoint of the uncertainty relation: a particle cannot be at the lowest point of the well and have zero momentum.

Just as any other quantum system, the oscillator can radiate or absorb photons in transitions from level to level. An analysis of probabilities of such transitions shows that they can occur only between neighboring levels; in other words, the transitions between the oscillator levels obey the *rule of selection*: $\Delta n = \pm 1$. This means that the oscillator radiates photons only of a single frequency ω.

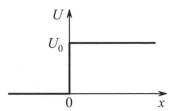

Figure P6.4. A step-like potential barrier (example 4).

Example 4. Reflection from a barrier. We show how to calculate the probability of reflection of the incident particles from an obstacle (barrier). We consider a barrier of the simplest shape, which is like a step of height U_0 (Fig. P6.4):

$$U = \begin{cases} 0 & \text{for } x \le 0; \\ U_0 & \text{for } x > 0. \end{cases}$$

We write the Schrödinger equation as (see example 22)

$$\psi'' + k_1^2 \psi = 0 \qquad \text{for } x < 0;$$
$$\psi'' + k_2^2 \psi = 0 \qquad \text{for } x > 0.$$

where $k_1 = \sqrt{2mE}/\hbar$, $k_2 = \sqrt{2m(E - U_0)}/\hbar$. The solution of this equation must describe the incident and reflected waves for $x < 0$ and the transmitted wave for $x > 0$:

$$\psi(x) = \begin{cases} A \exp(ik_1 x) + B \exp(-ik_1 x) & \text{for } x < 0; \\ C \exp(ik_2 x) & \text{for } x > 0. \end{cases}$$

(We omit the time-dependent part $\exp(-iEt/\hbar)$ of the wave function.) The densities of the probability flux (formula (P6.2.2.4)) of the incident, reflected, and transmitted waves are equal to $S_1 = |A|^2 (\hbar k_1/m)$, $S_2 = |B|^2 (\hbar k_1/m)$, and $S_3 = |C|^2 (\hbar k_2/m)$. Hence the reflection and transmission coefficients are equal to $R = S_2/S_1 = |B/A|^2$ and $D = S_3/S_1 = (k_2/k_1)|C/A|^2$. From the continuity conditions for the wave function and its derivatives for $x = 0$ we have

$$1 + \frac{B}{A} = \frac{C}{A}, \qquad k_1 \left(1 + \frac{B}{A}\right) = k_2 \frac{C}{A},$$

which implies

$$R = \left| \frac{k_1 - k_2}{k_1 + k_2} \right|^2, \qquad D = \left| \frac{4k_1 k_2}{(k_1 + k_2)^2} \right|. \tag{P6.2.2.5}$$

For $E > U_0$, the numbers k_1 and k_2 are real, the coefficients R and D are nonzero, and $R + D = 1$. We note that, in classical physics, there is no reflection of particles for $E > U_0$; that is, $R = 0$. In quantum physics, R is not only nonzero, but for $E - U_0 \ll U_0$, we obtain $k_2 \ll k_1$, and hence $D \to 0$ and $R \to 1$. Even more amazing purely quantum effect is obtained for $U_0 < 0$ (formally, (P6.2.2.5) remains valid, but $k_2 > k_1$): in the limit $E \ll |U_0|$, we obtain $k_1 \ll k_2$, and hence $R \to 1$ (!).

For $E < U_0$ ($U_0 > 0$), the solution remains formally true if we set $k_2 = i\kappa$, where $\kappa = \sqrt{2m(U_0 - E)}/\hbar$. In this case, $R = 1$, and the solution for $x > 0$ is the real exponential $|\psi|^2 = |C|^2 \exp(-2\kappa x)$; the depth of penetration of particles into the classically forbidden region is equal to $1/2\kappa$.

Example 5. Tunnel effect. In the case of incidence on a potential barrier of the de Broglie wave corresponding to a free particle with energy less than the barrier height, there is a nonzero probability of finding the particle on the other side of the barrier. The barrier transparency is the ratio the intensity (the density of the probability flux) of the transmitted wave to that of the incident wave. The potential barrier transparency (Fig. P6.5) is given by the formula

$$D = D_0 \exp\left(-\frac{2}{\hbar} \int_{x_1}^{x_2} \sqrt{2m[U(x) - E]} \, dx\right).$$

The particle can be found under the barrier in the region forbidden for motion in classical mechanics.

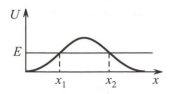

Figure P6.5. Tunnel effect (example 5).

P6.2.3. Operator Approach in Quantum Mechanics

▶ **Operators of physical quantities.** Any observable physical quantity in quantum mechanics is associated with a Hermitian (self-adjoint) operator acting on the wave function. A physical quantity G has a definite value λ_G only in the states corresponding to *eigenfunctions* of the operator $\widehat{G}$:

$$\widehat{G}\psi = \lambda_G \psi,$$

and the measurable values of the physical quantity in any of these states coincide with the *eigenvalues* λ_G of the operator $\widehat{G}$. The system of eigenfunctions of Hermitian operator is *complete*; that is, any function can be represented as a linear superposition of these functions. If a state is a superposition of eigenfunctions corresponding to different λ, $\psi = \sum_i C_i \psi_i$, then the probability of obtaining the value λ_i in measurements is equal to $|C_i|^2$, and the average value of G in this state is equal to

$$\langle G \rangle = \sum_i \lambda_i |C_i|^2 = \int \psi^* \widehat{G}\psi \, dV. \tag{P6.2.3.1}$$

The coordinate operator has the form $\widehat{x}\psi = x\psi$, the momentum projection operator has the form $\widehat{p}_x = -i\hbar\frac{\partial}{\partial x}$, and the energy operator (*Hamiltonian*) is determined by the expression:

$$\widehat{H} = \frac{\widehat{p}^2}{2m} + U = -\frac{\hbar^2}{2m}\Delta + U. \tag{P6.2.3.2}$$

The Schrödinger equations (P6.2.2.1) and (P6.2.2.3) become

$$i\hbar\frac{\partial \psi}{\partial t} = \widehat{H}\psi \quad \text{and} \quad \widehat{H}\psi = E\psi; \tag{P6.2.3.3}$$

that is, the stationary states are the eigenfunctions of the Hamiltonian, and the energy of these states are the corresponding eigenvalues.

▶ **Commensurability of physical quantities.** Two physical quantities, A and B, are said to be *commensurable* (can simultaneously have definite values) if they commute: $\widehat{A}\widehat{B} - \widehat{B}\widehat{A} = 0$. For the coordinate and momentum projections, we have

$$\widehat{x}\widehat{p}_x - \widehat{p}_x\widehat{x} = i\hbar. \tag{P6.2.3.4}$$

Example 1. Uncertainty principle. Operator approach. In the language of quantum operators, the uncertainty principle becomes an exact mathematical statement. The uncertainty of a physical quantity in the state ψ is $\Delta G = \sqrt{\langle (\widehat{G} - \langle \widehat{G}\rangle)^2 \rangle}$ (in the eigenstate of the operator $\widehat{G}$, the uncertainty is zero). The degree of commensurability of two physical quantities $\widehat{A}$ and $\widehat{B}$ is determined by their commutator $\widehat{C} = \widehat{A}\widehat{B} - \widehat{B}\widehat{A}$, which is expressed by the inequality

$$\Delta A \Delta B \geq \tfrac{1}{2}|\langle \widehat{C}\rangle|.$$

▶ **Conservation condition for a physical quantity.** An important statement of quantum mechanics is that the physical quantity is preserved in time if its operator commutes with the Hamiltonian. Indeed,

$$\frac{d}{dt}\langle \widehat{G}\rangle = \int \frac{\partial \psi^*}{\partial t}\widehat{G}\psi \, dV + \int \psi^*\widehat{G}\frac{\partial \psi}{\partial t}\, dV = \frac{i}{\hbar}\int \psi^*[\widehat{H}\widehat{G}]\psi \, dV = \frac{i}{\hbar}\langle[\widehat{H}\widehat{G}]\rangle,$$

where $\partial \psi/dt$ is expressed by using the Schrödinger equation (P6.2.3.3) (and $\partial\psi^*/dt$ is expressed by using the dual equation).

Example 2. Quantum oscillator. Operator approach. The commutation relations permit obtaining important results by pure algebraic methods. Let us find the eigenvalues and eigenfunctions of the Hamiltonian describing the quantum oscillator:

$$\widehat{H} = \frac{\widehat{p}_x^2}{2m} + \frac{m\omega^2\widehat{x}^2}{2},$$

starting only from the commutation relation (P6.2.3.4) between the coordinate and momentum. Further, we introduce dimensionless operators $\widehat{q}$, $\widehat{p}$, and $\widehat{h}$ using the formulas

$$\widehat{x} = \sqrt{\frac{\hbar}{m\omega}}\,\widehat{q}, \quad \widehat{p}_x = \sqrt{\hbar m\omega}\,\widehat{p}, \quad \widehat{H} = \frac{\hbar\omega}{2}\widehat{h},$$

and $\widehat{p} = -i\frac{d}{dq}$. Then, the dimensionless Hamiltonian acquires the form

$$\widehat{h} = \widehat{p}^2 + \widehat{q}^2,$$

and the dimensionless coordinate and momentum operators satisfy the relation

$$\widehat{q}\widehat{p} - \widehat{p}\widehat{q} = i.$$

We introduce the "raising" and "lowering" operators:

$$\widehat{a}^+ = \widehat{q} - i\widehat{p}, \qquad \widehat{a}^- = \widehat{q} + i\widehat{p},$$

which satisfy the identities

$$\widehat{a}^+\widehat{a}^- = \widehat{h} - 1, \quad \widehat{a}^-\widehat{a}^+ = \widehat{h} + 1, \quad [\widehat{a}^+\widehat{a}^-] = -2.$$

The eigenvalues λ and eigenfunctions ψ_λ of the operator $\widehat{A} = \widehat{a}^+\widehat{a}^-$ have the following properties:

1) The eigenvalues are nonnegative ($\lambda \geq 0$). An eigenvalue is zero if and only if $\widehat{a}^-\psi_0 = 0$.
2) If ψ_λ is a normed eigenfunction with $\lambda > 0$, then $\frac{1}{\sqrt{\lambda}}\widehat{a}^-\psi_\lambda$ is a normed eigenfunction with $\lambda' = \lambda - 2$, and $\frac{1}{\sqrt{\lambda+2}}\widehat{a}^+\psi_\lambda$ is a normed eigenfunction with $\lambda' = \lambda + 2$.

This implies that the eigenvalues of the operator $\widehat{A}$ are even integers, the eigenvalues of the operator $\widehat{h}$ are odd integers, and the eigenvalues of the energy operator are $E_n = \hbar\omega(n + \frac{1}{2})$. The ground state ($n = 0$) satisfies the equation $\widehat{a}^-\psi_0 = 0$ or $\frac{d\psi_0}{dq} = -q\psi_0$. We obtain $\psi_0(q) = \sqrt{\frac{2}{\pi}}\exp(-q^2/2)$. All the other functions can be obtained by successive application of the raising operator: $\psi_1 = \frac{1}{\sqrt{2}}(q - d/dq)\psi_0$, $\psi_2 = \frac{1}{\sqrt{3}}(q - d/dq)\psi_1$, etc.

▶ **Three-dimensional Schrödinger equation.** Now we consider the three-dimensional Schrödinger equation. This equation is much more complicated than the one-dimensional equation, but in several cases, it can be reduced to three one-dimensional equations by the method of separation of variables. We illustrate this method with an example of the three-dimensional oscillator.

Example 3. Three-dimensional quantum oscillator. Separation of variables. The Schrödinger equation for the three-dimensional oscillator has the form

$$\Delta\psi + \frac{2m}{\hbar^2}\left(E - \frac{m\omega^2 r^2}{2}\right)\psi = 0. \tag{P6.2.3.5}$$

The oscillator potential energy is unique: it is spherically symmetric and simultaneously the sum of three terms each of which depends only on one coordinate and coincides with the potential energy of the one-dimensional oscillator ($U_x = m\omega^2 x^2/2$ etc.). This permits seeking the solution in the form

$$\psi(\mathbf{r}) = \psi_1(x)\psi_2(y)\psi_3(z).$$

Substituting ψ in equation (P6.2.3.5) and dividing it by ψ, we obtain

$$\left(\frac{1}{\psi_1}\frac{\partial^2\psi_1}{\partial x^2} - \frac{2m}{\hbar^2}U_x\right) + \left(\frac{1}{\psi_2}\frac{\partial^2\psi_2}{\partial y^2} - \frac{2m}{\hbar^2}U_y\right) + \left(\frac{1}{\psi_3}\frac{\partial^2\psi_3}{\partial z^2} - \frac{2m}{\hbar^2}U_z\right) + \frac{2m}{\hbar^2}E = 0.$$

Since only the first term depends on x, it must be a constant that we denote by $2mE_1/\hbar^2$. The obtained equation coincides with the equation for the one-dimensional oscillator, and hence E_1 can take only the values $\hbar\omega(n_1+\frac{1}{2})$. The same concerns the other two terms. Since $E = E_1 + E_2 + E_3$, for the energy of the three-dimensional oscillator we obtain

$$E_{n_1,n_2,n_3} = \hbar\omega(n + \tfrac{3}{2}),$$

where $n = n_1 + n_2 + n_3$, and n_1, n_2, n_3 take the values $0, 1, 2, \ldots$. One can see that the ground state ($n = 0$) is nondegenerate, the level with $n = 1$ is triply degenerate, the level with $n = 2$ is sixfold degenerate, the level with $n = 3$ is tenfold degenerate, etc.

The operator representation permits writing the method of separation of variable more compactly. The Schrödinger equation (P6.2.3.5) takes the form

$$(\widehat{H}_1 + \widehat{H}_2 + \widehat{H}_3)\psi_1\psi_2\psi_3 = E\psi_1\psi_2\psi_3.$$

Dividing by $\psi_1\psi_2\psi_3$, we obtain

$$\frac{\widehat{H}_1\psi_1}{\psi_1} + \frac{\widehat{H}_2\psi_2}{\psi_2} + \frac{\widehat{H}_3\psi_3}{\psi_3} = E.$$

Only the first term depends on x; it is equal to E_1, etc.

P6.2.4. Motion in a Central Field. Hydrogen Atom

The general approach to solving the Schrödinger equation in a spherically symmetric field relies on using the conservation law for the moment of momentum (angular momentum). Such an approach permits separating the variables (in spherical coordinates) for any spherically symmetric potential.

▶ **Operator of angular momentum.** The *operator of orbital angular momentum* (moment of momentum) of an electron $\widehat{\mathbf{L}}$ with respect to point O is determined by the relations

$$\widehat{L}_x = \widehat{y}\widehat{p}_z - \widehat{z}\widehat{p}_y, \quad \widehat{L}_y = \widehat{z}\widehat{p}_x - \widehat{x}\widehat{p}_z, \quad \widehat{L}_z = \widehat{x}\widehat{p}_y - \widehat{y}\widehat{p}_x, \tag{P6.2.4.1}$$

and the operator of squared angular momentum is determined by the relation $\widehat{L}^2 = \widehat{L}_x^2 + \widehat{L}_y^2 + \widehat{L}_z^2$. The angular momentum projections do not commute with each other:

$$[\widehat{L}_x\widehat{L}_y] = i\hbar\widehat{L}_z, \quad [\widehat{L}_y\widehat{L}_z] = i\hbar\widehat{L}_x, \quad [\widehat{L}_z\widehat{L}_x] = i\hbar\widehat{L}_y, \tag{P6.2.4.2}$$

which means that they cannot simultaneously take definite values. But each of the projections commutes with the squared angular momentum, and hence we can speak about common eigenfunctions of the operators $\widehat{L}_z$ and $\widehat{L}^2$.

The triple of operators satisfying relations (P6.2.4.2) is called the vector of angular momentum even if it is impossible to write (P6.2.4.1) for them. The general notion of angular momentum in quantum mechanics is wider than the notion of classical orbital angular momentum of a particle, it also includes the notion of intrinsic angular momentum (spin) of subatomic particles (see Subsection P6.2.5).

The eigenvalues can be found purely algebraically from the commutation relations (P6.2.4.2). The common eigenfunctions of the operators $\widehat{L}^2$ and $\widehat{L}_z$ can correspond only to the following eigenvalues:

$$\widehat{L}^2\psi_{l,m} = l(l + 1)\hbar^2\psi_{l,m}, \qquad \widehat{L}_z\psi_{l,m} = m\hbar\psi_{l,m},$$

where l is an integer or half-integer nonnegative number, and for a given l, the number m can take $2l + 1$ values: $m = -l, -l + 1, \ldots, l - 1, l$.

We pass to dimensionless operators: $\widehat{L}_x = \hbar\widehat{l}_x$, $\widehat{L}_y = \hbar\widehat{l}_y$, $\widehat{L}_z = \hbar\widehat{l}_z$, satisfying the commutation relations $[\widehat{l}_x\widehat{l}_y] = i\widehat{l}_z$ etc. We introduce the "raising" and "lowering" operators:

$$\widehat{l}^+ = \widehat{l}_x + i\widehat{l}_y, \qquad \widehat{l}^- = \widehat{l}_x - i\widehat{l}_y,$$

satisfying the relations

$$[\widehat{l}_z\widehat{l}^+] = \widehat{l}^+, \quad [\widehat{l}_z\widehat{l}^-] = -\widehat{l}^-, \quad \widehat{l}^+\widehat{l}^- = \widehat{l}^2 - \widehat{l}_z^2 + \widehat{l}_z, \quad \widehat{l}^-\widehat{l}^+ = \widehat{l}^2 - \widehat{l}_z^2 - \widehat{l}_z.$$

If ψ is an eigenfunction of the operators $\widehat{l}^2$ and $\widehat{l}_z$ with the eigenvalues λ and μ, then $\widehat{l}^+\psi$ is an eigenfunction of these operators with the eigenvalues λ, $\mu + 1$, and $\widehat{l}^-\psi$ is an eigenfunction with the eigenvalues λ, $\mu - 1$. The minimum value μ_1 and maximum value μ_2 satisfy the equations:

$$\lambda - \mu_1^2 + \mu_1 = 0, \quad \lambda - \mu_2^2 - \mu_2 = 0,$$

which implies that $\mu_1 = -\mu_2$, and $\mu_2 - \mu_1 = 2\mu_2 = l$ is an integer. For the eigenvalue of the operator $\widehat{l}^2$, we obtain $\lambda = l(l+1)$.

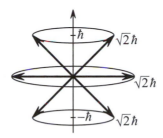

Figure P6.6. Representation of the angular momentum in the vector model.

One can see that the maximum value of the projection of the angular momentum $\hbar l$ is less than its magnitude $\hbar\sqrt{l(l+1)}$. In the *vector model*, the angular momentum is depicted as a vector making a fixed angle with the axis z and rotating about this axis. In this case, the projections L_x and L_y do not have fixed values, which shows that L_z is not commensurable with L_x and L_y. In Fig. P6.6, we present the case $l = 1$.

▶ **Eigenfunctions of the orbital angular momentum operator.** To find the eigenvalues of the *orbital* angular momentum determined by formulas (P6.2.4.1), it is convenient to pass to spherical coordinates. The orbital angular momentum is a function of only angular variables, and the commuting operators $\widehat{L}_z$ and $\widehat{L}^2$ take the form

$$\widehat{L}_z = -i\hbar\frac{\partial}{\partial\varphi},$$

$$\widehat{L}^2 = -\hbar^2\widehat{\Lambda}, \quad \text{where} \quad \widehat{\Lambda} = \frac{1}{\sin\theta}\frac{\partial}{\partial\theta}\left(\sin\theta\frac{\partial}{\partial\theta}\right) + \frac{1}{\sin^2\theta}\frac{\partial^2}{\partial\varphi^2}. \tag{P6.2.4.3}$$

In what follows, it is important that the operator $\widehat{\Lambda}$ coincides with the angular part of the Laplace operator written in spherical coordinates:

$$\Delta = \frac{1}{r^2}\frac{\partial}{\partial r}\left(r^2\frac{\partial}{\partial r}\right) + \frac{1}{r^2}\widehat{\Lambda}. \tag{P6.2.4.4}$$

The eigenfunctions can be represented as products of functions of different angular variables: $\psi_{l,m} = \Theta_{l,m}(\theta)\,\Phi_m(\varphi)$. The equation for Φ

$$-i\hbar\frac{d\Phi_m}{d\varphi} = m\hbar\Phi_m$$

has the solution $\Phi_m = A\exp(im\varphi)$, which is a unique function of the angle φ *only for integer values* of m. Hence, in the case of orbital angular momentum, only *integer* values of the numbers l and m are realized. The number l is called the *orbital quantum number* and the number m is called the *magnetic quantum number* (denoted by m_l). Half-integer values are realized in the case of intrinsic angular momentum—*spin*; the corresponding *spin quantum numbers* are denoted by s and m_s.

The equation for the eigenfunctions of the operator $\widehat{\Lambda}$ has solution in the form of *spherical functions* $Y_{lm}(\theta, \varphi) = \Phi_m(\varphi)\Theta_{ml}(\theta)$ with eigenvalues $-l(l+1)$, which are well known in mathematical physics. (The functions $\Theta(\theta)$ are expressed in terms of *associated Legendre functions*; see Subsection M13.10.1). For example,

$$\Phi_m(\phi) = \frac{1}{\sqrt{2\pi}}e^{im\phi}, \quad \Theta_{1,0}(\theta) = \sqrt{\frac{3}{2}}\cos\theta, \quad \Theta_{1,\pm1} = \mp\sqrt{\frac{3}{4}}\sin\theta, \quad \Theta_{2,0} = \sqrt{\frac{5}{8}}(3\cos^2\theta - 1).$$

Each of the functions is normalized as follows: $\int_0^{2\pi}\Phi^2\,d\varphi = 1$ and $\int_0^\pi\Theta^2\sin\theta\,d\theta = 1$. The spherical functions form a complete system of functions of angular variables; that is, any function of the variables θ and φ can be expanded in spherical functions.

▶ **Addition of angular momenta.** The operator of the sum of two angular momenta $\widehat{\mathbf{L}} = \widehat{\mathbf{L}}_1 + \widehat{\mathbf{L}}_2$, whose values are characterized by quantum numbers l_1 and l_2 (i.e., $\widehat{L}_1^2$ and $\widehat{L}_2^2$ have the eigenvalues $\hbar\sqrt{l_1(l_1+1)}$ and $\hbar\sqrt{l_2(l_2+1)}$), is characterized by the quantum number l, which can take the following values: $|l_1 - l_2| \leq l \leq l_1 + l_2$. The maximum value corresponds to the parallel orientation of the angular momenta, the minimum value corresponds to their antiparallel orientation. The operator $\widehat{L}^2$ does not commute with the operators L_{1z} and L_{2z}, and hence the eigenfunctions of the operators $\widehat{L}^2$ and $\widehat{L}_z$ are not eigenfunctions of L_{1z} and L_{2z} (but are eigenfunctions of $\widehat{L}_1^2$ and $\widehat{L}_2^2$).

▶ **Motion in a central field.** In this case, the potential energy $U(\mathbf{r})$ depends only on the modulus of $\mathbf{r}$. Since the orbital angular momentum operators $\widehat{L}_z$ and $\widehat{L}^2$ commute with $\widehat{r}^2$ and $\widehat{p}^2$, they also commute with the Hamiltonian. Hence, just as in classical mechanics, the value of the angular momentum and its projection onto the z-axis are preserved in the motion of a particle in a central field. The wave function of any state $\psi(r, \theta, \varphi)$ can be expanded in the common eigenfunctions ψ_{E,l,m_l} of the three commuting operators $\widehat{H}$, $\widehat{L}_z$ and $\widehat{L}^2$ corresponding to the three simultaneously measurable physical quantities: the energy, the orbital angular momentum projection, and its value. We use the separation of variables; that is, we seek the eigenfunctions in the form: $\psi_{E,l,m_l} = R_{E,l}(r)\Phi_{m_l}(\varphi)\Theta_{l,m_l}(\theta)$. Writing the Hamiltonian in spherical coordinates and taking (P6.2.4.3) and (P6.2.4.4) into account, we obtain

$$\widehat{H} = -\frac{\hbar^2}{2mr^2}\frac{\partial}{\partial r}\left(r^2\frac{\partial}{\partial r}\right) + \frac{\widehat{L}^2}{2mr^2} + U(r). \qquad (P6.2.4.5)$$

The Schrödinger equation in form (P6.2.3.5) after the action of the Hamiltonian (P6.2.4.5) on the function ψ_{E,l,m_l} becomes the equation for the radial part of the wave function $R_{E,l}(r)$:

$$-\frac{\hbar^2}{2m}\frac{1}{r^2}\frac{\partial}{\partial r}\left(r^2\frac{\partial}{\partial r}\right)R_{E,l} + \left(\frac{\hbar^2 l(l+1)}{2mr^2} + U(r)\right)R_{E,l} = ER_{E,l}. \qquad (P6.2.4.6)$$

Just as in classical mechanics (see Subsection P1.6.2), the potential energy is supplemented with the *centrifugal energy* $\hbar^2 l(l+1)/2mr^2$. One can see that the magnetic quantum number is not contained in the equation for allowed values of the energy. To each energy level there correspond two quantum numbers: the *orbital number* l and the *radial number* $n_r = 0, 1, 2, \ldots$ numbering the allowed discrete values of the energy. The level $E_{n_r,l}$ is $(2l+1)$-fold degenerate with respect to the values of the magnetic quantum number m_l.

▶ **Magnetic moment.** In classical physics, a charged particle moving in a closed trajectory has a magnetic moment. The value and direction of the magnetic moment completely determine both the magnetic field of the moving charge and its interaction with the external magnetic field (see Section P3.8). For orbital motion of an electron, the ratio of its magnetic moment to its mechanical angular momentum is equal to $-e/2m$. Hence, the projection of the orbital magnetic moment on the axis z takes the values $\mu_z = -(e/2m)\hbar m_l = -\mu_B m_l$, and the value of the orbital magnetic moment is equal to $\mu_B \sqrt{l(l+1)}$, where $\mu_B = e\hbar/2m \approx 9.274 \times 10^{-24}$ J/T is called the *Bohr magneton*. In the general case, the relation between the magnetic moment and mechanical angular momentum can be written as: $\mu_z = g(e/2m)L_z$ (or $(\mu_z/\mu_B) = g(L_z/\hbar)$). The dimensional multiplier g is called the *gyromagnetic ratio*; it is equal to 1 for orbital angular momentum (and denoted g_l) and 2 for spin (denoted g_s).

For an atom in a magnetic field with induction **B** directed along the axis z, the electron acquires additional energy of interaction between the magnetic moment and the magnetic field: $\Delta E = -\mu_z B = \mu_B m_l B$. One can see that in the magnetic field, the degeneration in the magnetic quantum number m_l must be removed; that is, instead of one level there arise $2l + 1$ equidistant levels the distance between which is proportional to the magnetic field. In this case, each spectral line must split into three equidistant lines (taking the *rules of selection* discussed below into account, we see that the variation in m_l in transitions between the levels can only be $\Delta m_l = 0, \pm 1$). This effect is called the *simple (normal) Zeeman effect*. The fact that the electron has an intrinsic (spin) magnetic moment strongly complicates the pattern of the line splitting in the magnetic field.

▶ **Hydrogen atom in quantum mechanics.** With the potential energy $U(r) = -ke^2/r$, equation (P6.2.4.6) has the following solutions:

1) The possible values of the energy are determined by the formula

$$E_{n_r,l} = -\frac{R}{(n_r + l + 1)^2},$$

where R is the Rydberg constant (see (P6.1.2.5)). The quantity $n = n_r + l + 1$ is called the *principal quantum number*. One can see that the electron energy is determined only by the principal quantum number. To one and the same principal quantum number n there correspond states with distinct values of l: $l = 0, 1, \ldots, n-1$. The total degeneracy of each energy level equals $\sum_{l=0}^{n-1}(2l + 1) = n^2$. The additional degeneracy is a special property of the Coulomb potential. We note that the energy levels coincide with those obtained in the semiclassical Bohr theory (Subsection P6.1.2), but there are profound important distinctions: in Bohr's theory, the orbital angular momentum in the ground state ($n = 1$) is $L = \hbar$, and in quantum theory, the angular momentum of the ground state is zero.

2) The radial functions have the form $R_{n,l} = r^l e^{-r/na_0} P_{n_r}(r)$, where $P_{n_r}(r)$ is a polynomial of degree $n_r = n - l - 1$, and $a_0 = \hbar^2/(mke^2)$ is the radius of the first electron orbit in Bohr's theory (the Bohr radius). For example,

$$R_{1,0} = \frac{2}{\sqrt{a_0^3}} e^{-r/a_0}, \quad R_{2,0} = \frac{1}{\sqrt{8a_0^3}} e^{-r/2a_0}\left(2 - \frac{r}{a_0}\right), \quad R_{2,1} = \frac{1}{2\sqrt{6a_0^3}} \frac{r}{a_0} e^{-r/2a_0}.$$

The radial functions are normed by unity: $\int_0^\infty R^2 r^2\, dr = 1$. The probability of finding the electron at a distance from the nucleus lying in the interval $(r, r + dr)$ is equal to $dP = \int_{(r,r+dr)} |\psi|^2\, dV = R^2(r)r^2\, dr$. In the ground state, the most probable distance from the electron to the nucleus is equal to the Bohr radius a_0.

The electron state is denoted by the Latin letter indicating the value of the orbital quantum number (s, p, d, and f correspond to $l = 0, 1, 2, 3$, respectively, and further in alphabetical order), before which the value of the principal quantum number is indicated. For example, $3d$ corresponds to the electron with $n = 3$ and $l = 2$.

▶ **The spectra of alkali metals**. In the case of alkali metals (Li, Na, …), the atomic state is determined by the electron state on the outer shell. This electron is in the centrally symmetric field of the nucleus and $Z - 1$ electrons on the inner shells. The outer shell radius is significantly larger than the inner shell radii ($r_n \sim n^2$), and hence, in the main region of electron existence, the electric field coincides with that of the charge $+e$. The field distortion by inner shell electrons exerts larger influence on a state with lesser orbital number l, because $\psi(r) \sim r^l$ for small r. For the energy levels $E_{n,l}$, where $n = n_r + l + 1$ is the principal quantum number, one usually employs the expression

$$E_{n,l} = -\frac{R}{(n + \alpha_l)^2},$$

where $\alpha_l < 0$ is the *Rydberg correction*. For example, for Li, $\alpha_0 = -0.41$ and $\alpha_1 = -0.04$, and for Na, $\alpha_1 = -0.88$.

The difference between the field where an electron moves from the Coulomb field can be taken into account in the first approximation by adding the term A/r^2 to the potential energy. This additional term takes account of two effects at once: first, the screening field of the electron "core," and second, the polarization of the core by the outer shell electron, which results in that the core acquires a small electric dipole moment directed at each time moment towards this electron. The term A/r^2 can be added to the centrifugal energy (see formula (P6.2.4.6)), and both terms can be written together as the centrifugal energy with redefined $l' = l + \alpha_l$.

To construct the spectrum, one must take account of the fact that the principal quantum number n of the lowest state of the outer shell electron is equal to the shell number in which this electron is located (for lithium, $n = 2$, for potassium, $n = 3$, etc.). Moreover, it is necessary to take account of the rule of selection of the orbital number l: $\Delta l = \pm 1$. In Fig. P6.7, we present the diagram of spectral lines of Li atom and give the names of the first principal series.

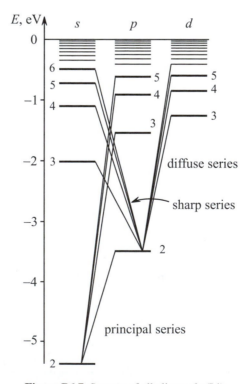

Figure P6.7. Spectra of alkali metals (Li).

P6.2.5. Electron Spin and Fine Structure of the Spectrum

▶ **Spin of electrons.** Several experimental facts show that to describe the state of electrons in an atom, it is insufficient to have three quantum numbers related to its spatial motion. These facts are:

1) The multiplet (fine) structure of spectral lines of alkali metals. For example, the most intensive line in the principal series; that is, the resonance line corresponding to the transition to the ground s-state from the first excited p-state, has the form of a pronounced doublet.

2) The complicated (anomalous) Zeeman effect. Splitting of atomic lines in a magnetic field cannot be described as the simple Zeeman effect related to the magnetic moment of the orbital motion.

3) The results of the Stern–Gerlach experiments on measuring magnetic moments of silver atoms (1921). A beam of silver atoms was directed into the region of a strongly inhomogeneous magnetic field, where the atoms were acted upon by the force $\mu_z(dB/dz)$. Since a silver atom has a single electron in the outermost shell, the atomic angular momentum and magnetic moment are equal to those of this electron. The orbital magnetic moment projection $\mu_z = m_l \mu_B$ can have $2l + 1$ distinct values. Atoms with electrons in the s-state do not deflect in magnetic field; if there is a fraction of atoms with electrons in the p-state, then they must split into three beams. It was discovered that a beam of atoms splits into two beams deflected by the magnetic field to opposite sides by the same distances.

The conclusion is that the electron has an intrinsic angular momentum (called *spin*) that is not related to the electron orbital motion. The spin projection on the axis z can take two values, which implies that the *spin quantum number s* must be equal to $1/2$ (because $2s + 1 = 2$). The value of the spin angular momentum is equal to $\hbar\sqrt{\frac{1}{2}(\frac{1}{2} + 1)} = \frac{\sqrt{3}}{2}\hbar$. The spin angular momentum projection on the axis z is equal to $\hbar m_s$, where $m_s = \pm\frac{1}{2}$ is the *magnetic spin quantum number* (also known as the *secondary spin quantum number*). We recall that half-integer values of the angular momentum quantum number do not contradict the commutation relations.

Since the spin projection can take two values, it follows that to the set of quantum numbers n and l, there correspond $2(2l + 1)$ distinct states and to the quantum number n, there correspond $2n^2$ states.

A variation in the value of deviation of atoms in a magnetic field permits finding the value of the intrinsic magnetic moment. It turned out that the intrinsic magnetic moment of an electron has anomalous gyromagnetic ratio $g_s = 2$: the ratio of the spin magnetic moment to the spin angular momentum is equal not to $e/2m$ but to e/m. The magnetic moment projection on the axis z takes the values $(e/m)\hbar m_s = \pm\mu_B$. The anomalous values of the gyromagnetic ratio are confirmed by the results of the *Einstein–de Haas experiments* (1914), in which magnetomechanical effects were studied (formation of a mechanical angular momentum in magnetization of an iron sample).

All subatomic particles have an intrinsic angular momentum associated—the spin. The spin quantum number of proton, neutron, and μ-mesons is equal to $1/2$, that of photon is equal to 1, and that of π-mesons is 0.

▶ **Fine structure of spectral lines.** The fine structure of lines of hydrogen-like atoms and alkali metals can be explained by interaction between the spin magnetic moment and the orbital angular momentum. The easiest way to understand the appearance of such an interaction is to use the reference frame associated with the electron. In this reference frame, the moving nucleus creates a magnetic field with which the spin magnetic moment interacts. With the *spin-orbit interaction* (also known as *spin-orbit coupling*) taken into account, the following operators are preserved (commute with the Hamiltonian): the total angular momentum $\widehat{J}^2$, its projection $\widehat{J}_z$, and the value of the orbital angular momentum $\widehat{L}^2$. The position of the levels $E_{n,l,j}$ is determined by three quantum numbers, n, l, and j. (In many-electron atoms, the quantum numbers of the angular momentum are denoted by the capital letters L, S, and J. In spectroscopy, the orbital angular momentum is denoted by the letters S, P, D, F, ... In alkali metals, the atomic state is determined by a single electron, and one can use both capital and small letters.) The degeneration of each level is given by $2j + 1$ states with distinct values of the projection of the total angular momentum. For a given l, the total angular momentum number can take two values: $j = (l \pm \frac{1}{2})$, for $l = 0$, there is only one value $j = \frac{1}{2}$. Hence the s-levels do not split (singlet levels), and the levels with $l > 0$ split into two near levels (doublet levels). When constructing the spectra, one must take the rules of selection into account: $\Delta l = \pm 1$, $\Delta j = 0, \pm 1$. For example, in transition from the doublet p-state to the lower singlet s-state, there arise two lines (doublet of the principal series), and the distance between them decreases with increasing n; and in transition from singlet s-levels to the lowest p-level, there arise doublets of the sharp series, which have the same splitting. In transition from the double d-state into the lower doublet p-state, there arise three lines (multiple doublets of diffusion series). The splitting values decreases with increase in both the orbital and principal quantum numbers, but it increases fast in transitions to atoms with larger serial numbers.

In the hydrogen atom and hydrogen-like ions, the spin-orbit interaction is partially compensated by relativistic corrections, and the final answer for the energy levels does not contain any dependence on the orbital quantum number:

$$\Delta E_{n,j} = \alpha^2 R \frac{Z}{n^3} \left(\frac{3}{4n} - \frac{1}{j + 1/2} \right),$$

where $\alpha = ke^2/\hbar c \approx 1/137$ is called the *fine structure constant*. Additional degeneration in the orbital quantum number is a specific characteristic of the Coulomb potential.

P6.3. Structure and Spectra of Complex Atoms

P6.3.1. Pauli Exclusion Principle. Structure of Complex Atoms

▶ **Self-consistent field.** For many-electron atoms, the problem of describing the atomic states becomes significantly more complicated. In the general case, it is necessary to consider the wave function depending on the coordinates and spin variables of all electrons and to take into account the interaction of electrons not only with the nucleus but also with one another. But a good description is given by the self-consistent field approximation, in which each electron moves in the average field of the nucleus and the other electrons. The self-consistent field potential $\varphi(r)$ is described by the Thomas–Fermi equation, which permits obtaining the energetically most profitable distribution of electrons in the space surrounding the atom.

Solving the Schrödinger equation for an electron in a self-consistent field one can obtain stationary states whose energy $E_{n,l}$ depends on the principal quantum number and the orbital quantum number. First, it is necessary to find the structure of the ground state (with the least energy) of the atom. What one-electron states are occupied by electrons in the ground state? At the first glance, it seems that it is most profitable to place all the electrons in the lowest state $1s$. But such a distribution of electrons is impossible because of the *Pauli exclusion principle* related to the identity property of particles.

▶ **Particle identity. Pauli principle.** The particle identity principle in quantum mechanics has a stronger meaning than that in classical mechanics. In classical physics, particles are identical but distinguishable from one another; for example, a permutation of two electrons results in a state that must be taken into account as a new state independent of the initial state. Unlike this, in quantum mechanics, particles are identical and indistinguishable, so that the state arising from permutation of particles absolutely identically coincides with the initial state. This means that $|\psi(1,2)|^2 = |\psi(2,1)|^2$. There are two possibilities: either the wave function is symmetric with respect to particle states, so that $\psi(1,2) = \psi(2,1)$, or it is antisymmetric, so that $\psi(1,2) = -\psi(2,1)$. The particles of the first type are called *bosons*, the particles of the second type are called *fermions*. In particle physics, it is proved that bosons are particles with integer spin and the fermions are particles with half-integer spin. If the wave function is constructed of one-particle wave functions, then it must be symmetrized in the case of bosons and antisymmetrized in the case of fermions. For example, in the case of two bosons with one-particle states ψ_1 and ψ_2, the wave function has the form $\psi(\mathbf{r}_1, \mathbf{r}_2) = \frac{1}{\sqrt{2}}[\psi_1(\mathbf{r}_1)\psi_2(\mathbf{r}_2) + \psi_2(\mathbf{r}_1)\psi_1(\mathbf{r}_2)]$, and in the case of two fermions, it has the form $\psi(\mathbf{r}_1, \mathbf{r}_2) = \frac{1}{\sqrt{2}}[\psi_1(\mathbf{r}_1)\psi_2(\mathbf{r}_2) - \psi_2(\mathbf{r}_1)\psi_1(\mathbf{r}_2)]$.

Electrons are fermions, and hence a system of several electrons is described by an antisymmetric (in particle states) wave function. In such a system (just as in a system of any identical fermions), the *Pauli exclusion principle* acts: *two electrons cannot be in absolutely equal quantum states*; indeed, the wave function of such a state must be identically zero.

▶ **Periodic table (Mendeleev's table).** To describe the electronic structure of the ground state of an atom with atomic number Z, one must place Z electrons in different quantum states with least possible energy. To calculate electronic states, one can choose any set of independent quantum numbers of a single electron; for example, just as in the case of alkali metals, these can be n, l, j, and m_j. But it is more convenient to neglect the spin-orbit interaction and use the numbers n, l, m_l, and m_s for an electron in a centrally symmetric self-consistent field. (To completely description the terms of the ground and excited states, one must choose a set of quantum numbers that describe the state of the entire atom and correspond to quantities that are conserved.)

All the electrons with the same principal quantum number n correspond to one *electron shell*. The maximum number of electrons in the nth shell is equal to $2n^2$. The shell with $n = 1$ is called the K-shell, the shell with $n = 2$ is called the L-shell, and further, by the alphabet. All electrons in a given shell with a fixed number l form an *electron subshell* with $2(2l + 1)$ electrons; $l = 0, \ldots, n - 1$. The electrons of a given shell with a larger value of l have a larger energy due to increasing contribution of the centrifugal energy. The electron state is denoted by the number of the principal quantum number followed by the letter corresponding to its orbital number. If an atom contains several electrons in the same subshell, then the superscript on the right is the number of electrons. We present the electron configurations for several atoms:

K-shell:

$$_1\text{H: } 1s, \quad _2\text{He: } 1s^2;$$

L-shell:

$$_3\text{Li: } 1s^2\, 2s, \qquad _4\text{Be: } 1s^2\, 2s^2,$$

$$_5\text{B: } 1s^2\, 2s^2\, 2p, \qquad _6\text{C: } 1s^2\, 2s^2\, 2p^2, \qquad _7\text{N: } 1s^2\, 2s^2\, 2p^3,$$

$$_8\text{O: } 1s^2\, 2s^2\, 2p^4, \qquad _9\text{F: } 1s^2\, 2s^2\, 2p^5, \qquad _{10}\text{Ne: } 1s^2\, 2s^2\, 2p^6;$$

two subshells (s and p) in the M-shell:

$$_{11}\text{Na: } [\text{Ne}]3s, \qquad _{12}\text{Mg: } [\text{Ne}]3s^2,$$

$$_{13}\text{Al: } [\text{Ne}]3s^2\, 3p, \qquad _{14}\text{Si: } [\text{Ne}]3s^2\, 3p^2, \qquad _{15}\text{P: } [\text{Ne}3s^2\, 3p^3,$$

$$_{16}\text{S: } [\text{Ne}]3s^2\, 3p^4, \qquad _{17}\text{Cl: } [\text{Ne}]3s^2\, 3p^5, \qquad _{18}\text{Ar: } [\text{Ne}]3s^2\, 3p^6;$$

s-subshell in the N-shell:

$$_{19}\text{K: [Ne]}3s^2\,3p^6\,4s, \quad _{20}\text{Ca: [Ne]}3s^2\,3p^6\,4s^2;$$

d-subshell in the M-shell:

$$_{21}\text{Sc: [Ne]}3s^2\,3p^6\,3d\,4s^2, \quad _{22}\text{Ti: [Ne]}3s^2\,3p^6\,3d^2\,4s^2,$$

$$_{23}\text{V: [Ne]}3s^2\,3p^6\,3d^3\,4s^2, \quad _{24}\text{Cr: [Ne]}3s^2\,3p^6\,3d^5\,4s;$$

and so on. The notation [Ne] stands for the electron configuration of $_{10}\text{Ne}$ $(1s^2\,2s^2\,2p^6)$.

One can see that, starting from number 19, the sequential filling of shells is violated: the energy of s-electrons in the N-shell is less than that of d-electrons in the M-shell and the $4s$-subshell is filled before the $3d$-subshell (see also Chapter S7). The chemical properties of elements are determined by the number of electrons in the outermost shell, and this results in the periodicity of chemical properties. For example, the noble gases— helium $_2\text{He}$, neon $_{10}\text{Ne}$, argon $_{18}\text{Ar}$, krypton $_{36}\text{Kr}$, xenon $_{54}\text{Xe}$, and radon $_{86}\text{Rn}$—differ in that their p-subshells are complete, and hence these elements have a large energy of ionization and tend to be chemically inert. The atoms with further numbers (lithium, sodium, potassium, rubidium, cesium, and francium) have complete p-subshells and one electron in the outermost s-subshell. These atoms form the group of alkali metals; they have a small energy of ionization and are all highly reactive.

We note that the total angular momentum and the total magnetic moment of any completely filled shell are zero.

P6.3.2. Spectra of Complex Atoms

▶ **Addition of orbital and spin angular momenta.** In light and medium atoms, the spin-orbit interaction is weak for separate electrons, and the addition of angular momenta occurs by the following scheme. The orbital angular momenta of electrons are added together to form the total orbital angular momentum $\widehat{\mathbf{L}}$, which is characterized by the quantum number L, and the electron spins are added up to make the total spin angular momentum $\widehat{\mathbf{S}}$ (quantum number S). In turn, the orbital and spin angular momenta add up to form the atom total angular momentum $\widehat{\mathbf{J}}$. Such a scheme is called the normal LS-coupling or the *Russell–Saunders coupling*. The number L is always integer, and the number S is integer if the shell contains an even number of electrons and is half-integer if the number of electrons is odd. We note that one must add up only the angular momenta of electrons in an unfilled shell.

In heavy atoms, the angular momenta are added together according to the scheme called the jj-coupling. First, one adds up the orbital and spin angular momenta of each individual electron ($\mathbf{j}_i = \mathbf{l}_i + \mathbf{s}_i$) and then the electron angular momenta are added up to form the total angular momentum of the atom ($\mathbf{J} = \sum_i \mathbf{j}_i$).

▶ **Classification of terms of complex atoms.** In the case of normal coupling, the total angular momentum is preserved (the value and the projection) and the values of the orbital and spin angular momenta. Each term is characterized by the quantum numbers L, S and J and, in the absence of a magnetic field, it is $(2J + 1)$-fold degenerate (the states that differ only in the value of the total angular momentum projection have the same energy). For given L and S, the quantum number J can take $2S + 1$ values from $L - S$ to $L + S$, and hence the number $2S + 1$ is called the term *mutliplicity*. The difference between the energies of terms in the same multiplet is determined by a comparatively weak spin-orbit interaction. The spectral notation of terms is: the orbital quantum number of an atom is denoted by the capital letter S, P, D, F, G, and further in alphabetical order (to distinguish from one-electron states, where the lowercase letters are used). The subscript on the right shows the number J, which distinguished this term in the multiplet, and the superscript on the left indicates the term multiplicity $2S + 1$. (It should be noted that, for $L < S$, the actual multiplicity of a term is equal to $2L + 1$, but it is conventional to show $2S + 1$.) For example, the term $^2P_{3/2}$ corresponds to the quantum numbers $L = 1$, $S = 1/2$, and $J = 3/2$. In addition, in the scheme of spectral transitions, where the excited states must be shown, the principal quantum number of the excited electron is given (the least number is equal to the number of the shell to which the unfilled subshell belongs).

▶ **The Hund rule.** In the addition of spin and orbital angular momenta of electrons, one can obtain distinct quantum numbers. To understand what quantum numbers correspond to the ground state of the atom, one can use the empirical *Hund rules*. The first rule says that the term with the least energy has the maximum S and, for this S, the value of L is a maximum. (First, one determines the maximum possible m_S and m_L and they are used to determine S and L.) The second rule: $J = |L - S|$ if less than half the subshell is filled, and $J = L + S$ otherwise. The meaning of the first Hund rule is that if electrons have the same spin states, then they have different spatial states, which corresponds to the lesser energy of interaction between the electrons. For example, for carbon with subshell $2p^2$, the total spin is equal to $S = 1$, and the orbital number cannot take the maximum value $L = 2$, because, in this case, the two electrons would have the same quantum numbers $m_l = 1$; hence $L = 1$, and since two electrons occupy less than half the p-subshell, where there are 6 places, it follows that $J = |L - S| = 0$. The principal term of the carbon atom is $^3P_0$. Verify that the principal term of the nitrogen atom is $^4S_{3/2}$.

▶ **Selection rules.** When constructing the scheme of spectral lines of an atom and determining the multiplicity of a radiation line, it is necessary to use both the term multiplicity and the *selection rules* for transitions between the terms. The selection rules for radiation transitions can be obtained by calculating the probability of transitions between the states. But the meaning of the selection rules can be understood using the momentum conservation law and taking into account the fact that the spin of an emitted photon is equal to 1.

The selection rule for the number J is: $\Delta J = 0$, ± 1, but the transition between two states with $J = 0$ is forbidden. The selection rule for the number m_J is: $\Delta m_J = 0, \pm 1$. The selection rule for the quantum number S is: $\Delta S = 0$. The selection rule for the quantum number L is: $\Delta L = 0$, ± 1, but the transition between two states with $L = 0$ is forbidden. In atoms with a single electron in an unfilled subshell, the transitions with $\Delta L = 0$ are forbidden.

The selection rule for the spin number is related to the fact that the magnetic field of an electromagnetic wave weakly interacts with the magnetic moment of the electron (the action on the orbital angular momentum is determined by electric forces). For example, in atoms with two electrons, one can observe independent series for states with $S = 0$ (singlet terms) and for states with $S = 1$ (triplet terms). In the case of helium, the data are described using the historical concept about two types of helium: parahelium ($S = 0$) and orthohelium ($S = 1$).

▶ **Characteristic X-ray spectra.** In the bombardment of the anode of an electron tube by energetic electrons (with energies $\sim 10^2$–10^4 eV), one can observe the X-ray radiation of two types. The slowing-down radiation is related to radiation of X-ray quanta directly by electrons. A typical characteristic of this radiation is a short-wavelength boundary of the spectrum (see Subsection P5.3.2) and a smooth distribution of the radiation intensity over the spectrum. Against this background, there are several intensive narrow lines of the characteristic radiation. Such a radiation appears in the case of liberation of an electron belonging to one of the inner electron subshells, after which the free place is occupied by an electron from a higher subshell and an X-ray quant is radiated. If one of the two electrons in the K-shell is liberated, then the following notation is used for the X-ray lines: K_α for the electron transition from the L-shell, K_β for the electron transition from the M-shell, etc. If an electron is liberated from the L-shell, then the spectral lines are denoted by L_α, L_β, etc.

For electrons in the inner subshells, the field of the nucleus is nearly unscreened. Therefore, in contrast to electrons in the outer subshells, the energy of inner subshell electrons must strongly depend on the serial number of the nucleus. One can expect that the energy of these electron levels must be described by the formula for hydrogen-like atoms with the screening correction: $E_n = R(Z - \delta)^2/n^2$. But such an approach does not give any good description of experimental facts and the empirical screening correction is introduced directly in the formula for spectral lines. For example, for the line K_α, we have

$$\frac{1}{\lambda} = \frac{3}{4}R_\lambda(Z - \sigma)^2,$$

where the screening correction σ for light atoms is approximately equal to 1. This law was discovered in 1913 by Moseley, who formulated it as the law of linear dependence of $\sqrt{\nu}$ for the characteristic line frequency on the serial number Z. The *Moseley law* allowed one to specify the serial numbers of many elements in the periodic table.

Since the X-ray spectrum is determined by the atomic inner subshell, it is independent of whether the atom is isolated or is contained in a molecule or a crystal. Therefore, the matter composition can be analyzed by using the characteristic X-ray spectra with a high universality.

Similarly to the optical spectra, the X-ray spectra have a fine structure because of the splitting of the levels of inner subshell electrons. But, in this case, one deals not with the normal LS-coupling but with the jj-coupling. For example, an electron in the L-shell can have $l = 0$ and then $j = 1/2$ or $l = 1$ and then the following two versions are possible: $j = 1/2$ or $j = 3/2$. Hence, the L-shell splits into three lines, and the difference in their energies is determined by the spin-orbit interaction. The K-shell contains only one level, the M-shell contains five levels, etc. In the analysis of spectra, one must follow the selection rules $\Delta l = \pm 1$ and $\Delta j = 0, \pm 1$. For example, the K_α line splits into two lines: $K_{\alpha 1}$ and $K_{\alpha 2}$, and there is no transition from the level with $l = 0$.

The study of absorption of the X-ray radiation directly allows one to see the position of energy levels of electrons in inner subshells. As the frequency increases, the absorption coefficients decreases smoothly, but for frequency equal to $\omega_i = E_i/\hbar$, where E_i is the energy of ionization of the electron from one of the inner levels, one observes a sharp increase in this coefficient, which is related to the fact that an addition absorption channel opens. According to the above, one can observe a single peak on the energy of the K-shell, three peaks on the energy of the L-shell, five peaks on the energy of the M-shell, etc.

▶ **Atom in an external magnetic field.** In sufficiently weak magnetic fields ($B < 1$ Tesla), the energy of the electron interaction with the magnetic field is small compared with the spin-orbit interaction, and the magnetic splitting of terms is small compared with the distance between the terms of the multiplet. In this case, one observes a *complicated (anomalous) Zeeman effect*; that is, the energy level splits into $2J + 1$ sublevels with distinct quantum numbers m_J, and the distance between the sublevels is proportional to the value of the magnetic field and depends on the quantum numbers L, S and J. The value of the sublevel displacement is given by the formula $\Delta E = g_J m_J \mu_B B$, where μ_B is the Bohr magneton, and

$$g_J = 1 + \frac{J(J + 1) + S(S + 1) - L(L + 1)}{2J(J + 1)}$$

is called the *Landé splitting factor*.

The complicated character of the Zeeman effect in weak fields can be explained by an anomalous value of the spin magnetic moment for which the gyromagnetic ratio is equal to 2. As a result, the total magnetic moment $\widehat{\mu} = -\frac{e}{2m}(\widehat{\mathbf{L}} + 2\widehat{\mathbf{S}})$ is not proportional to the total angular momentum $\widehat{\mathbf{J}}$. In the vector model, the magnetic moment can be represented as rapidly rotating about the preserved vector of total angular momentum, and, after averaging, only the magnetic moment projection onto the total angular momentum must remain:

$$\mu_J = \mathbf{J}\frac{(\mu \mathbf{J})}{J^2} = -\frac{e}{2m}\mathbf{J}\frac{(\mathbf{L} + 2\mathbf{S})(\mathbf{L} + \mathbf{S})}{J^2}.$$

We express $(\mathbf{LS}) = \frac{1}{2}(J^2 - L^2 - S^2)$, replace J^2 by $J(J + 1)$, L^2 by $L(L + 1)$, and S^2 by $S(S + 1)$ and obtain $\mu_J = -\frac{e}{2m}g_J\mathbf{J}$, where g_J is the Landé splitting factor, whose expression was given above.

We note that even for nonzero S, L, and J, the Landé splitting factor can be zero ($S = 3/2$, $L = 2$, $J = 1/2$) and even negative ($S = 5/2$, $L = 3$, $J = 1/2$). These effects are purely quantum and do not have classical analogs.

To obtain the splitting of spectral lines in a magnetic field, one must take the selection rule into account. For example, the level $^2P_{3/2}$ splits into 4 levels with the Landé splitting factor $\frac{4}{3}$, and the singlet level $^2S_{1/2}$ splits into two levels with the Landé splitting factor 2 (Fig. P6.8), but one observes not 8 lines but 6, because the transitions from $m_J = 3/2$ to $m_J = -1/2$ and from $m_J = -3/2$ to $m_J = 1/2$ are forbidden by the selection rule $\Delta m_J = 0, \pm 1$.

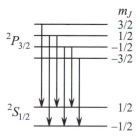

Figure P6.8. Anomalous Zeeman effect.

The simple Zeeman effect (splitting of one line into three) is observed in transition between the levels for which $g_1 = g_2 = g$. In this case, $\Delta\omega = g\Delta m_J B/\hbar$, where $\Delta m_J = 0, \pm 1$. Such a situation appears not only in transitions between two singlet levels ($S = 0$, $g = 1$) but also in some other cases.

If an atom is placed in a strong magnetic field (if the magnetic splitting of terms is much larger than the distance between the multiplet terms), then the fine structure of levels can be neglected—that is, the spin-orbit interaction can be neglected. The level with quantum numbers m_L, m_S is displaced in the magnetic field by $\Delta E = \mu_B B(m_L + 2m_S)$. The variation in the frequency of radiation is equal to $\hbar\Delta\omega = \mu_B B(\Delta m_L + \Delta m_S)$. With the selection rule $\Delta m_S = 0$, $\Delta m_L = 0, \pm 1$ taken into account, we see that the each line splits into three—two displaced lines and one undisplaced; hence, the Zeeman effect becomes simple in a strong magnetic field. This phenomenon is called the *Paschen–Buck effect*.

P6.4. Elements of Physics of Molecules

A molecule is a bound state of an electroneutral system consisting of nuclei of several atoms and of electrons surrounding them. The molecules of two atoms are said to be diatomic, the molecules of three and more atoms are said to be polyatomic. The molecule with insufficient or excessive number of electrons is called a molecular ion.

We consider only the simplest diatomic systems such as the hydrogen molecule H_2 and the hydrogen molecular ion H_2^+.

P6.4.1. Classification of Levels of a Molecule

▶ **Energy hierarchy.** Like an atom, a molecule is a microscopic system and has discrete energy levels corresponding to its ground and bound excited states. But, in contrast with an atom, in addition to the electron energy corresponding to possible states of electrons in the field of stationary nuclei, there also are two types of motion in a molecule: oscillatory motion related to variations in the distance between the nuclei, and rotation of the molecule as a whole about an axis. The molecule energy (counted from the ground state) can be assumed to be equal to the sum of energies corresponding to the above-listed motions:

$$E = E_{\text{el}} + E_{\text{vib}} + E_{\text{rot}}. \tag{P6.4.1.1}$$

There exists a pronounced hierarchy of these energies: $E_{\text{el}} \gg E_{\text{vib}} \gg E_{\text{rot}}$. For outermost shell electrons, E_{el} is of the order of several eV, E_{vib} is of the order of 10^{-2} eV, and E_{rot} is of the order of 10^{-4} eV. Just as $E_{\text{rot}}/E_{\text{vib}}$, the ratio $E_{\text{vib}}/E_{\text{el}}$ is equal, in the order of magnitude, to $\sqrt{m/M}$, where m is the mass of electron, and M is the mass of the nucleus; that is, for heavy molecules, the energy hierarchy increases.

In the first approximation, all three energies are quantized independently, and to each of them there corresponds its own system of discrete levels. The energy spectrum of a molecule is the set of distant electron levels, more closely located vibrational levels, and even more closely located rotational levels. The system of levels is schematically shown in Fig. P6.9 for two neighboring electron levels, where the letters v and J denote the vibrational and rotational quantum numbers.

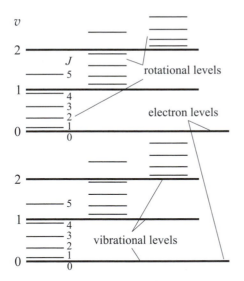

Figure P6.9. Structure of energy levels of a molecule.

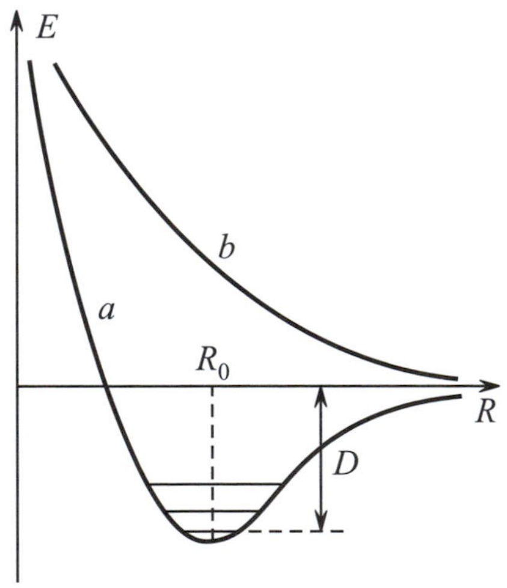

Figure P6.10. Dependence of electron energy of a molecule on the distance between the atoms.

▶ **The wave function of a molecule.** The wave function of a molecule can be represented as the product of wave functions of motions of three types. Each of these functions, just as the corresponding values of energies, can be obtained by solving the corresponding Schrödinger equation. First, one solves the Schrödinger equation for electrons for a given distance R between the nuclei; this energy depends on R as on a parameter. The energy of interaction between the nuclei is equal to the sum of the electron energy and the energy of the Coulomb repulsion of the nuclei. For some electron levels, this energy has the form of a potential well (curve a in Fig. P6.10), for the other levels (curve b), it has the form of a monotone dependence.

In the first case, there exists a bound state, in the second case, there is no bound state. For the electron ground state, the interaction energy has the form of a potential well, and the depth of this well has the meaning of the molecule dissipation energy. (More precisely, the dissipation energy D is less than the well depth by the energy of the zero-point vibrational level counted from the well depth equal to $\hbar\omega/2$.)

After the dependence of the energy of interaction between the nuclei on the distance R is calculated, one solves the problem about oscillations near the bottom of the obtained potential well and about rotations of a molecule with the distance between the nuclei equal to the average equilibrium distance R_0.

P6.4.2. Electron Energy. Theory of Chemical Bond

▶ **One electron in the field of two nuclei.** We begin with a molecular ion H_2^+. In this case, the potential of two protons for the motion of a single electron has a cylindrical symmetry with respect to the molecule axis but does not have the spherical symmetry intrinsic to the hydrogen atom. Moreover, the angular momentum projection on the molecule axis is preserved, but the angular momentum magnitude is not. The energy levels are characterized by the magnitude of the magnetic quantum number m_l, which is denoted by the letter λ: $\lambda = |m_l| = 0, 1, \ldots$. The electron states with $\lambda = 0, 1, 2, \ldots$ are denoted by the letters $\sigma, \pi, \delta, \ldots$ and are called *molecular orbitals*. With spin taken into account, the σ-state is twice degenerate, and the states with $\lambda > 0$ are fourfold degenerate. In the σ-state, the electron density is concentrated in the region between the nuclei, the nuclei are attracted to the electron cloud, and thus the attraction between the nuclei described by the potential well a in Fig. P6.10 is realized (for small R, repulsion between the nuclei predominates). The dissipation energy H_2^+ is equal to 2.65 eV. For π-electrons, the electron density has the form of two clouds located between the nuclei symmetrically with respect to a plane passing through the molecule axis, and the electron density in the plane itself is zero.

▶ **Hydrogen molecule.** In the hydrogen molecule, the chemical bond is realized not by one but by two electrons that are attracted to the nuclei, but repulse from one another. Just as in the case of any fermions, the wave function of two electrons must change the sign in their permutation. If the spin wave function is symmetric with respect to a particle permutation (the spins are parallel, and hence $S = 1$), then the coordinate wave function is antisymmetric and conversely. The state with antiparallel spins ($S = 0$) and a symmetric coordinate wave function has a lesser energy. In this state, the probability for an electron to stay in the space between the nuclei is sufficiently large and the nuclei are attracted to the electron cloud located between them.

Thus, the chemical bond is realized by two electrons with antiparallel spins. The dependence of the potential electrostatic energy on the distance between the nuclei has the form of curve a in Fig. P6.10. The hydrogen molecule dissociation energy is equal to 4.5 eV; this is less than the doubled energy of dissociation of the molecular ion H_2^+ because of the electrostatic repulsion between the electrons. For electrons with parallel spins, the dependence of the energy on the distance between the nuclei has the form of curve b in Fig. P6.10; that is, no bound state is formed.

This mechanism is called the *covalent (homeopolar) chemical bond*. It is realized not only in the hydrogen molecule but by any pair of outermost shell electrons in complex atoms combined in a molecule. For diatomic molecules consisting of different atoms, there exists another mechanism, called the *ionic (heteropolar) chemical bond*, in which one of the atoms gives part of the outermost shell electrons to another atom that has vacancies in the outermost subshell. As a result, two ions are formed that, at large distances, are attracted by the Coulomb law and, at small distances, begin to repulse. Intermediate cases between the covalent and ionic bonds are possible.

▶ **Exchange energy.** The difference in the energies of the symmetric and antisymmetric states has the general character and is a consequence of the so-called *exchange interaction*, which is a pure quantum phenomenon. The meaning of this phenomenon is reduced to the following. We construct a coordinate wave function of two electrons in the field of two nuclei from the one-electron wave functions $\psi_\alpha(\mathbf{r}_1)$ and $\psi_\beta(\mathbf{r}_2)$. This function can be either symmetric or antisymmetric (depending on the form of the spin function):

$$\varphi(\mathbf{r}_1, \mathbf{r}_2) = \frac{1}{\sqrt{2}} \left(\psi_\alpha(\mathbf{r}_1)\psi_\beta(\mathbf{r}_2) \pm \psi_\alpha(\mathbf{r}_2)\psi_\beta(\mathbf{r}_1) \right).$$

The average potential energy in these states is equal to

$$\langle U \rangle = \int \varphi^* \widehat{U} \varphi \, d^3\mathbf{r}_1 d^3\mathbf{r}_2$$

$$= \int |\psi_\alpha(\mathbf{r}_1)|^2 \widehat{U} |\psi_\beta(\mathbf{r}_2)|^2 d^3\mathbf{r}_1 d^3\mathbf{r}_2 \pm \int \psi_\alpha(\mathbf{r}_1)\psi_\beta^*(\mathbf{r}_1)\widehat{U}\psi_\beta(\mathbf{r}_2)\psi_\alpha^*(\mathbf{r}_2) d^3\mathbf{r}_1 d^3\mathbf{r}_2,$$

where $\widehat{U}$ is the potential energy of interaction of the electrons between each other and with the nuclei. The first term corresponds to the usual classical interaction of electron clouds and the nuclei. The second term is called the *exchange energy*, it is different from zero only in the case where the electron wave functions overlap. In the case of hydrogen molecule, the exchange energy is negative, and therefore, the state with a symmetric coordinate wave function has a lesser energy.

The exchange energy manifests itself in many quantum systems, in particular, it is responsible for the difference between the energies of the singlet and triplet states of electrons in the helium atom.

The chemical bond is usually approximately calculated by one of the two methods. In the *method of electron pairs*, for the initial one-particle wave functions $\psi(\mathbf{r}_1)$ and $\psi(\mathbf{r}_2)$ one takes the atomic wave functions of electrons localized near the first and second nuclei. As the nuclei approach each other, the wave functions begin to overlap, the exchange energy increase and a bound state is formed. In the *method of molecular orbitals*, the electrons successively fill the free σ, π, δ molecular states with different spins and thus form closed electron shells.

P6.4.3. Vibrational and Rotational Spectra

▶ **Vibrational levels.** The energy of interaction between the nuclei, shown in Fig. P6.10, has the minimum for $R = R_0$. Near this point, the energy can be expanded in a series and represented as

$$U(R) = U(R_0) + \tfrac{1}{2}k(R - R_0)^2 + \cdots,$$

where $k = d^2U/dR^2\big|_{R=R_0}$ is a quasi-elastic constant. The cyclic frequency of small oscillations of a molecule near the equilibrium is equal to $\omega_{\text{vib}} = \sqrt{k/\mu}$, where $\mu = m_1 m_2/(m_1 + m_2)$ is the reduced mass of a diatomic molecule (see Subsection P1.6.4). For permitted values of the oscillator energy, quantum mechanics gives values

$$E - U(R_0) = \hbar\omega_{\text{vib}}(v + \tfrac{1}{2}), \quad v = 0, 1, 2, \ldots.$$

For large v, the terms neglected in $U(R)$ become essential. If these are taken into account, the distance between the levels begins to decrease.

▶ **Rotational levels.** The rotational energy levels of a molecule are obtained as eigenvalues of the operator

$$\widehat{H} = \frac{\widehat{L}^2}{2I}, \tag{P6.4.3.1}$$

where $\widehat{L}$ is the operator of angular momentum and I is the moment of inertia of the molecule with respect to the center of mass. A rotating system with a constant moment of inertia, which is described by the Hamiltonian (P6.4.3.1), is called a *rotator*. The permitted values of the rotator energy are

$$E = \frac{\hbar^2}{2I} J(J+1), \quad J = 0, 1, 2, \dots.$$

The frequency of radiation or absorption is obtained from the relation

$$\hbar\omega = \Delta E_{el} + \Delta E_{rot} + \Delta E_{vib}, \quad (P6.4.3.2)$$

where the first term corresponds to the transition between distinct electron levels, the second term corresponds to the variation in the vibrational state, and the third term corresponds to the variation in the state of rotation.

If the electron state does not vary in the transition ($\Delta E_{el} = 0$), then one observes either the pure rotational spectrum (if $\Delta E_{vib} = 0$) or the vibrational-rotational spectrum. For the pure rotational and vibrational-rotational transitions, one has the selection rule $\delta v = \pm 1$ and $\Delta J = \pm 1$.

For the pure rotational transitions ($\Delta E_{el} = 0$ and $\Delta E_{vib} = 0$), the variation in the rotational energy is equal to

$$\Delta E_{rot} = E(J+1) - E(J) = \frac{\hbar^2}{I}(J+1).$$

One can see that the rotational transitions form a system of close equidistant spectral lines $\omega_{rot}, 2\omega_{rot}, \dots$, where $\omega_{rot} = \hbar^2/I$ (Fig. P6.11). The rotational lines can be resolved only by devices with a very high resolving power, but usually they are perceived as lines of finite width. The pure rotational transitions correspond to absorption in the far infrared region (hundreds and thousands of μm).

The vibrational transitions (for $\Delta E_{el} = 0$) are observed in absorption in the near infrared region (of several μm). Since, in this case, one usually has $\Delta E_{rot} \neq 0$, one observes a vibrational-rotational band of finite width rather than an infinitely thin line. With the selection rules taken into account, we obtain

$$\Delta E_{vib} + \Delta E_{rot} = \hbar(\omega_{vib} \pm n\omega_{rot}), \quad (n = 1, 2, \dots).$$

Hence, the distance between neighboring lines at the center of the band is equal to $2\omega_{rot}$ (Fig. P6.12).

The character of the vibrational-rotational spectra becomes significantly more complicated if the fact that the oscillations are anharmonic is taken into account. As $R - R_0$ increases, the potential energy curve (Fig. P6.10) becomes nonsymmetric. First of all, this results in displacement of vibrational levels; that is, as v increases, the vibrational levels approach one another. Moreover, in the state with larger v, the average distance between the atoms increases, and hence the moment of inertia increases with increasing v, while ω_{vib} decreases. This results in that the lines in the vibrational-rotational band cease to be equidistant.

But if $\Delta E_{el} \neq 0$, then one observes the electron spectra of emission and absorption in the visible and ultraviolet regions. To each electron transition, there corresponds a system of vibrational-rotational bands. In contrast to the vibrational-rotational transitions, in the case of transitions with variation in the electron state, the selection rules $\Delta v = \pm 1$ is not satisfied; that is, the transitions between levels with any values of vibrational quantum numbers are possible. Therefore, one can observe not a single vibrational-rotational band, but a series of bands separated from one another by ω_{vib}.

We note that diatomic molecules with zero dipole moment (for example, N_2 and O_2) do not have vibrational-rotational spectra. Just for this reason, such molecules do not absorb in the infrared (IR) region of the spectrum, and the absorption by air of the IR radiation occurs because the air contains the molecules H_2O and CO_2.

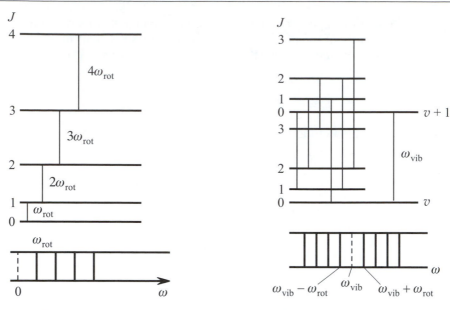

Figure P6.11. Rotational spectrum.   **Figure P6.12.** Vibrational-rotational spectrum.

Bibliography for Chapter P6

Alenitsyn, A. G., Butikov, E. I., and Kondratyev, A. S., *Concise Handbook of Mathematics and Physics*, CRC Press, Boca Raton, Florida, 1998.

Benenson, W., Harris, J. W., Stocker, H., and Lutz, H. (Eds.), *Handbook of Physics*, Springer-Verlag, New York, 2002.

Blokhintsev, D. I., *Basics of Quantum Mechanics* [in Russian], Lan, St. Petersburg, 2006.

Bohm, D., *Quantum Theory*, Dover Publications, New York, 1989.

Bransden, B. H. and Joachain, C. J., *Physics of Atoms and Molecules, 2nd Edition*, Benjamin Cummings, New York, 2003.

Budker, D., Kimball, D., and DeMille, D., *Atomic Physics: An Exploration through Problems and Solutions, 2nd Edition*, Oxford University Press, New York, 2008.

Demtröder, W., *An Introduction to Atomic and Molecular Physics*, Springer, Berlin, 2005.

Dirac, P. A. M., *Lectures on Quantum Mechanics*, Dover Publications, New York, 2001.

Foot, C. J., *Atomic Physics (Oxford Master Series in Atomic, Optical and Laser Physics)*, Oxford University Press, New York, 2005.

Griffiths, D. J., *Introduction to Quantum Mechanics (2nd Edition)*, Benjamin Cummings, San Francisco, California, 2004.

Heisenberg, W., *The Physical Principles of the Quantum Theory*, Dover Publications, New York, 1930.

Karlov, N. V. and Kirichenko, N. A., *Introductory Chapters of Quantum Mechanics* [in Russian], Fizmatlit, Moscow, 2006.

Landau, L. D. and Lifshitz, E. M., *Quantum Mechanics Non-Relativistic Theory, Third Edition: Volume 3, (Course of Theoretical Physics Series)*, Butterworth-Heinemann, Oxford, England, 1981.

Liboff, R., *Introductory Quantum Mechanics, 4th Edition*, Addison Wesley, San Francisco, California, 2002.

Mahan G. D., *Quantum Mechanics in a Nutshell*, Princeton University Press, Princeton, New York, 2008.

Prokhorov, A. M. (Ed.), *Encyclopedia of Physics, Volumes 1–5* [in Russian], Bolshaya Russkaya Encyclopedia, Moscow, 1998.

Prokhorov, A. M. (Ed.), *Encyclopedic Dictionary of Physics* [in Russian], Sovetskaya Encyclopedia, Moscow, 1984 (dic.academic.ru).

Sivukhin, D. V., *General Physics, Volume 5, Atomic and Nuclear Physics* [in Russian], Fizmatlit, Moscow, 2006.

Sivukhin, D. V. and Yakovlev, I. A., *A Collection of Problems, Volume 5, Atomic Physics. Nuclear and Particle Physics* [in Russian], Fizmatlit, Moscow, 2006.

Woan, G., *The Cambridge Handbook of Physics Formulas*, Cambridge University Press, Cambridge, England, 2003.

Chapter P7

Quantum Theory of Crystals

Although the laws of quantum mechanics act on an atomic scale, they also affect the macroscopic properties of gases, liquids, and rigid bodies. Quantum mechanics and quantum statistics—that is, statistics of indistinguishable particles—allowed one to obtain correct and consistent descriptions of thermal, magnetic, and many other properties of large ensembles of particles. Here a special place is occupied by quantum theory of crystals, which allowed one to remove several unsolvable contradictions of classical theory, to obtain quantitative description of electromagnetic properties of some classes of matter, to explain the superconductivity and superfluidity, etc.

P7.1. Quantum Statistical Distributions and Their Application

P7.1.1. Elements of Quantum Statistics

Quantum statistics studies the statistical properties of large ensembles of identical indistinguishable particles, fermions and bosons. Statistics of classical (i.e., distinguishable) particles is called the *Maxwell–Boltzmann statistics* (see Subsection P2.5.4). Statistics of *fermions*—that is, indistinguishable particles that cannot share identical states—is called the *Fermi–Dirac statistics*. Statistics of *bosons*—that is, indistinguishable particles that can share identical states—is called the *Bose–Einstein statistics*.

▶ **Statistical distributions.** Statistics of each type is characterized by its own *statistical distribution* of particles over states. A statistical distribution tells the average number of particles in a given state under the condition that the system is in thermal equilibrium. The most important property of all distributions is that they are functions of a single parameter, the particle energy:* $\langle N \rangle = f(\varepsilon)$.

In quantum mechanics, the notion of a state of a single particle is assumed to be quite definite, but only under certain conditions. For the energy levels of a particle system to be discrete, the system must be localized in a finite region of space. Usually, it is assumed that the system is closed in a box of a sufficiently large volume V. Moreover, the notion of a one-particle state with a certain energy assumes that there is no interparticle interaction. This last difficulty can be overcome by introducing a *self-consistent external field*, which includes the averaged field of the other particles.

For an ensemble of classical particles, one introduces the notion of an *elementary cell of phase space*. The phase space is the space of three coordinates and three momenta. In classical physics, to each state there corresponds one point of the phase space. In this case, the elementary cell corresponding to a single state is introduced for convenience in statistical description, and it can have an arbitrary (sufficiently small) volume. In quantum mechanics, a state of a particle cannot be represented by a point in the phase space, because

---

* In what follows, the energy of an individual particle is denoted by ε and the energy of the system, by E.

there are uncertainties in its position and momentum, which are related by the Heisenberg relations $\Delta x \Delta p_x \sim \hbar$, etc. One can see that the volume $\Delta x \Delta y \Delta z \Delta p_x \Delta p_y \Delta p_z \sim \hbar^3$ must correspond to each individual state. A more precise analysis shows that the volume of an elementary cell is equal to $(2\pi\hbar)^3$. To understand how many states there are contained in a finite region of the phase space, it is necessary to divide its volume by the volume of the elementary cell. Such an approach in which the elementary volume of a single state for a classical particle is chosen from quantum considerations is said to be *semiclassical*.

▶ **Microscopic and macroscopic states. Statistical weight.** To introduce the microscopic state (microstate) of a system means to define the state of all its particles at a given time instant. In classical physics, it is required to determine the coordinates and momenta of all particles—that is, the positions of all particles in the phase space. In the semiclassical approach, one finds the distribution of particles over elementary cells of volume $(2\pi\hbar)^3$ each. In quantum mechanics, one determines the distribution of particles over states with a certain energy (for the motion in a self-consistent field). For convenience of our presentation, we combine the last two cases and also call the states of particles with a certain energy *cells*. Due to interparticle interaction and external forces, the microstate of the system varies all the time and one can observe only the average characteristics in terms of which the macrostate can be determined.

To describe the macrostate quantitatively, let us divide the entire set of cells into *boxes*. Each box contains a large amount of cells with very close energy values lying within a narrow energy interval $(\varepsilon, \varepsilon + \Delta\varepsilon)$. Denote the number of particles in the ith box by N_i. We assume that the macrostate is determined by a set of numbers N_i, and hence all microstates with identical sets of these numbers belong to the same macrostate.

Without loss of generality, we consider a system of particles contained in a box with rigid thermally insulated walls, so that its total energy remains unchanged.

Why does this assumption not restrict the generality of consideration? First of all, we are interested in the properties of the equilibrium macrostate, and in this case, the energy of a large ensemble of particles can be determined with a very high accuracy under any external conditions. For example, for a system of N particles in a thermostat, the relative deviation from the average energy is $1/\sqrt{N}$. Let us explain this with an example of an ideal gas. If the average energy of a single particle is equal to $\langle \varepsilon \rangle$ and the root mean square deviation from the average energy is $\delta\varepsilon = \sqrt{\langle(\varepsilon - \langle\varepsilon\rangle)^2\rangle}$, then the average energy of the system is equal to $\langle E \rangle = N\langle\varepsilon\rangle$, and the root mean square deviation is

$$\sqrt{\langle(E - \langle E\rangle)^2\rangle} = \sqrt{\left\langle\left[\sum_i(\varepsilon_i - \langle\varepsilon\rangle)\right]^2\right\rangle} = \sqrt{\sum_i\langle(\varepsilon_i - \langle\varepsilon\rangle)^2\rangle} = \sqrt{N}\,\delta\varepsilon;$$

we have taken into account the fact that the deviations of the energies of distinct particles from the average value are mutually independent.

▶ **The principal postulate of statistical physics** for a system with a fixed energy states that all the microstates admissible for it (that is, those with a given value of total energy) are realized with equal probability. This postulate is closely related to the so-called *ergodic hypothesis* stating that, in the process of time evolution, the system passes through all admissible microstates.

The main conclusion from the principal postulate is the following: the greatest part of time the system is in the macrostates to which there corresponds the greatest number of microstates. The number of microstates corresponding to a given macrostate is called its *statistical weight* and is denoted by W. The macrostate with maximum W is called the *equilibrium macrostate*. According to the Boltzmann postulate (see Subsection P2.4.4), the system entropy in a given macrostate is related to its statistical weight as follows:

$$S = k\ln W.$$

The deviations from the equilibrium state are called *fluctuations*. We note that, because of an enormous number of particles, a large deviation in the system has an extremely small probability; that is, only small fluctuations are practically observable. If the system was initially in a highly non-equilibrium state, then, as the system evolves, it must move from macrostates with lesser statistical weights (lesser entropy) to a macrostate with a larger entropy approaching the equilibrium (the second law of thermodynamics).

▶ **Equilibrium macrostates.** In order to find out which particle distribution over boxes, $\{N_1, N_2, \dots\}$, corresponds to equilibrium, one must calculate $W(N_1, N_2, \dots)$ or $S/k = \ln W$ and take the maximum of this expression (taking into account the conditions that $\sum_i \varepsilon_i N_i = E$ and $\sum_i N_i = N$). The expressions for the statistical weight and entropy are significantly different for particles with different types of statistical behavior. Below are the expressions of the statistical weight for three different statistical models:

for the Maxwell–Boltzmann statistics,

$$W = \prod_i g_i^{N_i} \frac{N!}{N_1! \, N_2! \, \dots},$$

for the Bose–Einstein statistics,

$$W = \prod_i \frac{(N_i + g_i - 1)!}{N_i! \, (g_i - 1)!},$$

and for in the Fermi–Dirac statistics,

$$W = \prod_i \frac{g_i!}{N_i! \, (g_i - N_i)!},$$

where g_i stands for the number of cells in the ith box.

The equilibrium state $\{N_i\}$ corresponding to the maximum statistical weight under the condition that E and N are fixed can be found using the method of Lagrange multipliers. (It is necessary to find the maximum of the expression $(S/k) - \alpha \sum_i N_i - \beta \sum_i \varepsilon_i N_i$ under the assumption that all N_i are independent. Then the Lagrange multipliers α and β are determined from the equations $\sum_i N_i = N$ and $\sum_i \varepsilon_i N_i = E$.) We obtain the following expressions:

for the Maxwell–Boltzmann statistics,

$$N_i = g_i e^{-\alpha} e^{-\beta \varepsilon_i};$$

for the Bose–Einstein statistics,

$$N_i = \frac{g_i}{e^{\alpha} e^{\beta \varepsilon_i} - 1}$$

and for the Fermi–Dirac statistics,

$$N_i = \frac{g_i}{e^{\alpha} e^{\beta \varepsilon_i} + 1}.$$

The average number of particles per cell is obtained by dividing these expressions by g_i.

The physical meaning of the multipliers α and β can be clarified from thermodynamical considerations. We write the maximum condition for the expression $(S/k) - \alpha N - \beta E$

in the form $T\,dS/kT - \beta\,dE - \alpha\,dN = 0$ and compare with the thermodynamical identity $T\,dS - dU - \mu\,dN = 0$, which holds for variations in the state under constant volume. We obtain $\beta = 1/kT$ and $\alpha = \mu/kT$, where μ is the *chemical potential*.

Finally, the quantum distributions take the form:

Bose–Einstein distribution:

$$\langle N \rangle = f_B(\varepsilon) = \frac{1}{e^{(\varepsilon - \mu)/kT} - 1}. \tag{P7.1.1.1}$$

Fermi–Dirac distribution:

$$\langle N \rangle = f_F(\varepsilon) = \frac{1}{e^{(\varepsilon - \mu)/kT} + 1}. \tag{P7.1.1.2}$$

The chemical potential μ for a system with a fixed number of particles, N, is calculated as

$$N = \sum_i g_i f(\varepsilon_i),$$

where $f(\varepsilon_i)$ is the corresponding distribution (Bose–Einstein or Fermi–Dirac). But if the number of particles in the system is not fixed, then the chemical potential must be set to zero.

For $\varepsilon - \mu \gg kT$, both quantum distributions become the classical *Maxwell–Boltzmann distribution*:

$$\langle N \rangle = A e^{-\varepsilon/kT}. \tag{1}$$

This is related to the fact that the average number of particles in a cell is small, and the exclusion rules and the particle identity become insignificant.

Usually, the least possible value of energy (energy of the lowest level) is set to be zero.

▶ **Properties of the Bose–Einstein distribution.**

1) Bosons are particles with integer value of the intrinsic angular momentum (spin). Examples of bosons are π-mesons, photons, and phonons.

2) The chemical potential of bosons cannot be positive: $\mu \leq 0$.

3) As $T \to 0$, the chemical potential tends to zero and vanishes at $T = 0$.

4) At $T = 0$, all particles are in the lowest energy state with $\varepsilon = 0$. At finite but sufficiently low temperatures, a finite fraction of all particles is in the lowest energy state.

This phenomenon is called *Bose–Einstein condensation*, and the ensemble of particles in the lowest state is called a *Bose–Einstein condensate*. Condensation of bosons begins at a certain critical temperature (the condensation temperature). This phenomenon plays an important role in the explanation of phenomena such as superfluidity and superconductivity.

5) An example of the Bose–Einstein distribution is the Planck formula for the photon distribution over energies (Subsection P5.4.1). The number of photons is not fixed; therefore, $\mu = 0$.

▶ **Properties of the Fermi–Dirac distribution.**

1) Fermions are particles with half-integer spin. Examples of fermions are electrons, protons, neutrons, and μ-mesons.

2) The chemical potential of the Fermi gas can be either positive or negative. At low temperatures, the chemical potential is positive. The chemical potential of fermion gas for $T = 0$ is called the *Fermi energy* and is denoted by ε_F.

3) At $T = 0$, the energy-dependence of the average number of particles in a cell $\langle N \rangle$ has the form of a sharp step: for $\varepsilon < \varepsilon_F$, the cells are filled with particles ($\langle N \rangle = f_F(\varepsilon) = 1$), and for $\varepsilon > \varepsilon_F$, the cells are free ($f_F(\varepsilon) = 0$).

4) At low temperatures ($kT \ll \varepsilon_F$), the dependence $f_F(\varepsilon)$ has the form of a smoothed step (Fig. P7.1). The spreading width is equal to kT in the order of magnitude. For $\varepsilon = \mu$, the average number of particles is equal to 1/2. For $kT \ll \varepsilon_F$, the particle distribution over energies radically differs from the classical distribution; in this case, the gas is said to be *degenerate*. For $kT \gg \varepsilon_F$, the fermion gas does not differ noticeably from the classical gas.

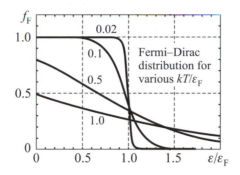

Figure P7.1. Fermi–Dirac distribution for various kT/ε_F (for ideal fermion gas).

▶ **Energy density of states.** The more specific properties of an ensemble of bosons or fermions depend on the physical properties of the system under study, mainly, of the form of the function $G(\varepsilon)$ characterizing the number of states Δg that correspond to a narrow energy interval $\Delta\varepsilon$ calculated per unit volume:

$$\frac{\Delta g}{V} = G(\varepsilon)\Delta\varepsilon. \tag{P7.1.1.4}$$

This function is called the *energy density of states* or simply the *density of states* (cf. Subsection P5.4.1). For example, the number of particles per unit volume is expressed in terms of the density of states:

$$n = \frac{N}{V} = \int_0^\infty f(\varepsilon)G(\varepsilon)\,d\varepsilon \tag{P7.1.1.5}$$

as well as the energy of particles per unit volume:

$$\frac{E}{V} = \int_0^\infty f(\varepsilon)\varepsilon G(\varepsilon)\,d\varepsilon. \tag{P7.1.1.6}$$

P7.1.2. Electron Gas in Metal

The classical electron theory of metal conductivity originates from the idea that the electrons in metal can be considered as the classical ideal gas (Subsection P3.7.3). One of the main difficulties encountered by this theory was that the electron contribution to the heat capacity predicted by this theory was equal to the heat capacity of a monatomic ideal gas $C_{el} = \frac{3}{2}kN$, where N is the number of free electrons. It is precisely this value that must be the difference between the heat capacities of a metal and a dielectric with similar crystal lattices. But experiment showed that the electron heat capacity at room temperatures is much less than kN (approximately, by a factor of 100) and varies proportionally to the temperature. These and some other facts can naturally be explained by the *Sommerfeld theory*, where the electrons in metal are considered as a quantum gas of fermions.

▶ **Density of states of electron gas.** We assume that, inside the metal, the electrons are free and the work required for an electron to leave the metal is infinitely large; that is, we treat them as particles in a potential well with infinitely high walls. The boundary conditions for the wave functions $\psi_x = A \sin k_x x$ have the form $k_x L_x = \pi n$, where L_x is the potential well dimension (that is, the metal sample dimension) along the axis x, k_x is the projection of the wave vector related to the electron momentum as $\mathbf{k} = \mathbf{p}/\hbar$, and n is a positive integer (see Subsection P6.2.2). The boundary conditions for the wave vector identically coincide with those arising in the case of consideration of electromagnetic oscillations in a closed cavity (Subsection P5.4.1). Therefore, the density of states in this case has the same form:

$$\frac{dg}{V} = \gamma \frac{k^2 \, dk}{2\pi^2},$$

where V is the vessel volume and γ is a multiplier taking the intrinsic degrees of freedom into account. In the case of electrons, $\gamma = 2$ takes account of the two possible projections of spin. To pass to the electron momentum, we must replace $k^2 \, dk$ by $p^2 \, dp/\hbar^3$:

$$\frac{dg}{V} = \frac{p^2 \, dp}{\pi^2 \hbar^3}.$$

We rewrite this formula as

$$dg = \gamma \frac{V \, 4\pi p^2 \, dp}{(2\pi \hbar)^3}.$$

One can see that to determine the number of states, it is necessary to divide the phase space volume by the elementary cell volume $(2\pi\hbar)^3$.

To pass to the energy $\varepsilon = p^2/2m$, it is necessary to make the change $p = \sqrt{2m\varepsilon}$, $p \, dp = m \, d\varepsilon$:

$$\frac{dg}{V} = \frac{m\sqrt{2m}\sqrt{\varepsilon} \, d\varepsilon}{\pi^2 \hbar^3}, \qquad G(\varepsilon) = \frac{m\sqrt{2m}\sqrt{\varepsilon}}{\pi^2 \hbar^3}.$$

▶ **Fermi energy of electrons.** At $T = 0$, all the energy levels from zero to $\varepsilon = \varepsilon_F$ are filled. To relate the Fermi energy to the electron concentration n, it is necessary to express the total number of electrons per unit volume and equate with the concentration

$$\int_0^{\varepsilon_F} G(\varepsilon) \, d\varepsilon = \frac{(2m)^{3/2} \varepsilon_F^{3/2}}{3\pi^2 \hbar^3} = n.$$

This implied the Fermi energy:

$$\varepsilon_F = \frac{\hbar^2 (3\pi^2 n)^{2/3}}{2m}. \tag{P7.1.2.1}$$

The momentum of an electron with Fermi energy is determined by the relation $\varepsilon_F = p_F^2/2m$ and is equal to

$$p_F = \hbar(3\pi^2 n)^{1/3}. \tag{P7.1.2.2}$$

The surface in momentum space formed by states with $\varepsilon = \varepsilon_F$ is called the *Fermi surface*. In the case of free electrons, the Fermi surface is a sphere of radius p_F.

The density of levels on the Fermi surface is equal to

$$G(\varepsilon_F) = \frac{dn}{d\varepsilon_F} = \frac{3}{2} \frac{n}{\varepsilon_F}.$$

The average energy of electrons at $T = 0$ is equal to

$$\langle \varepsilon \rangle = \frac{\int_0^{\varepsilon_F} \varepsilon G(\varepsilon)\, d\varepsilon}{\int_0^{\varepsilon_F} G(\varepsilon)\, d\varepsilon} = \frac{\int_0^{\varepsilon_F} \varepsilon \sqrt{\varepsilon}\, d\varepsilon}{\int_0^{\varepsilon_F} \sqrt{\varepsilon}\, d\varepsilon} = \tfrac{3}{5} \varepsilon_F.$$

For $T > 0$, the chemical potential μ is determined by the equation

$$n = \int_0^{\infty} f_F(\varepsilon) G(\varepsilon)\, d\varepsilon.$$

As the temperature increases, μ gradually decreases, but for $kT \ll \varepsilon_F$, its difference from ε_F is very small: $|\mu - \varepsilon_F|/\varepsilon_F \approx (\pi^2/12)(kT/\varepsilon_F)^2$. At a certain temperature T_0 ($kT_0 \approx \varepsilon_F$), the chemical potential is zero, and for higher temperatures, $\mu < 0$ (the chemical potential is determined from the graph in Fig. P7.1; for $\varepsilon = \mu$, we have $f_F(\varepsilon) = 0.5$).

To simplify the process of determining the temperature region where the electron gas is degenerate, one introduces the *Fermi temperature* by the relation $kT_F = \varepsilon_F$. For $T \ll T_F$, the electron gas is degenerate. For the majority of metals, T_F is equal to several tens of thousands of degrees, which means that the electron gas in metals is strongly degenerate at room temperatures.

▶ **Electron heat capacity of metals.** In the case of highly degenerate gas, only those electrons will participate in thermal motion whose energy differs from the Fermi energy by a value of the order of kT. Only such electrons can make transitions to a free state with energy greater than ε_F. The fraction of such electrons is $\sim kT/\varepsilon_F = T/T_F$ and the variation in their energy is $\sim kT$. Therefore, as the temperature increases from zero to T, the energy of degenerate electron gas increases by $\sim kNT^2/T_F$, and the contribution of electrons to the heat capacity $C_{el} \sim kN(T/T_F)$ varies proportionally to the temperature. (Precise calculations give the expression $C_{el} = (\pi^2/2)kN(T/T_F)$.) The number of free electrons per one mole of metal depends on the valency; for example, in the case of an univalent metal, $N \sim N_A$.

▶ **Electrical and thermal conduction of metals.** The transport theory gives the following expressions for the electrical and thermal conductivities of electrons (see also Subsections P3.7.3 and P2.9.2):

$$\sigma = \frac{ne^2\lambda}{m\langle v \rangle}, \qquad \kappa = \tfrac{1}{3}C_{el}\langle v \rangle \lambda.$$

Here λ is mean free path and C_{el} is the heat capacity of electrons per unit volume of metal. Participating in collisions with ions will be only those electrons that are located near the Fermi surface; therefore, $\langle v \rangle \approx v_F = p_F/m$, which means that the average speed is almost independent of the temperature. Nevertheless, the *Wiedemann–Franz law* remains true: the ratio $\kappa/\sigma \sim \langle v \rangle^2 C_{el}/n$ for metals is a linear function of the temperature. But in contrast with the classical electron gas, where C_{el}/n is independent of the temperature and $\langle v \rangle^2 \sim T$, the linear dependence on the temperature in the Sommerfeld theory is determined by the electron heat capacity.

The Sommerfeld theory cannot explain the dependence of the electrical conduction of pure metals on the temperature, because, in this theory, just as in the Drude theory of the classical electron gas, the mean free path is a pure empirical parameter. The dependence of the mean free path on the temperature is explained in the theory, where the periodicity of the crystal lattice field and the interaction of electrons with the lattice thermal vibrations (with phonons) are taken into account.

▶ **Richardson–Dashman formula.** The Sommerfeld theory permits explaining the temperature dependence for the current of thermoelectron emission from the metal surface. The metal surface can be left only by the electrons for which $p_z^2/2m > \varepsilon_F + A$, where A is *work of liberation* of electron from metal (the axis z is perpendicular to the surface). Omitting the calculations, we present the answer

$$j = CT^2 \exp(-A/kT),$$

where the constant $C = \dfrac{mek^2}{2\pi^2\hbar^3} = 120\,\mathrm{A}/(\mathrm{m^2\,K^2})$ is the same for all metals.

P7.1.3. Heat Capacity of Crystal Lattice. Phonons

▶ **Dulong–Petit law.** The classical statistical physics predicts that the contribution of the lattice vibrations to the molar heat capacity must be equal to $3R$. Indeed, each atom of the lattice exhibits three independent vibrations near the equilibrium and to each vibrational degree of freedom there corresponds the average energy kT. The energy of vibrations of all atoms is equal to $3NkT = 3\nu RT$, and the molar heat capacity of the lattice need not depend on the temperature and be equal to $3R$. Indeed, at room temperature, the molar heat capacity of many dielectrics is close to $3R$ (*Dulong–Petit law*). But as the temperature decreases, in all cases, one observes a significant decrease in the heat capacity, and at low temperatures, it tends to zero as T^3. Such a behavior of the lattice heat capacity is explained by quantum theory.

▶ **Einstein theory.** Einstein modeled vibrations of atoms at the nodes of the crystal lattice using $3N$ mutually independent equal quantum harmonic oscillators. The average energy of each such oscillator is equal to (see Subsection P5.4.1)

$$\langle\varepsilon\rangle = \frac{\hbar\omega}{2} + \frac{\hbar\omega}{e^{\hbar\omega/kT} - 1}. \tag{P7.1.3.1}$$

For $kT \gg \hbar\omega$, the molar heat capacity of a system of quantum oscillators is equal to $3R$, and as $T \to 0$ it exponentially tends to zero. The Einstein theory explained that the heat capacity decreases with decreasing temperature, but not by the law T^3.

▶ **Elastic waves.** The difficulties in the Einstein theory originate from the fact that the elastic vibrations of noninteracting oscillators of the same frequency were considered as elementary excitations (of normal oscillatory modes). In fact, due to interaction between neighboring atoms, elastic waves of different frequencies and wavelength propagate in crystal. The elementary excitations are waves with definite values of frequency ω and the wave vector $\mathbf{k}$. The average energy per unit wave with frequency ω is given by the formula (P7.1.3.1). To calculate the number of distinct waves in the interval $d\omega$, it is necessary to solve the following two problems:

1) To find the number of independent oscillations per unit volume element $d^3\mathbf{k}$ in the space of wave vectors. This problem can be solved starting from the boundary conditions. The total number of independent oscillations must be equal to the number of degrees of freedom $3N$. For traveling waves, we use *periodic boundary conditions*: $\xi(L_x) = \xi(0)$, with similar relations for y and z. This leads to

$$k_x L_x = 2\pi m, \quad \ldots, \quad \Delta k_x = \frac{2\pi}{L_x}, \quad \ldots, \quad \Delta k_x \Delta k_y \Delta k_z = \frac{(8\pi)^3}{V}.$$

Finally, for the density of states in the **k**-space, we have

$$dg = \gamma \frac{4\pi k^2\, dk}{\Delta k_x \Delta k_y \Delta k_z} = \gamma \frac{k^2\, dk}{2\pi^2} V. \tag{P7.1.3.2}$$

So we get the same formula as for standing waves (see Subsection P5.4.1). The multiplier γ is equal to three in the case of sound waves; it takes account of the two longitudinal waves and one transverse wave.

2) To find the law of dispersion, that is, the dependence $\omega(\mathbf{k})$. To solve this problem, it is necessary to write the equation of motion of separate atoms taking account of their interaction with neighboring atoms. The theoretical calculations are possible only for rather simple models, but in the general case, it is possible to find several important features.

For the wavelength larger than the interatomic distances ($ka \ll 1$), good results are given by the *continuum approximation* in which the medium is assumed to be continuous. In this case, the equations of motion of particles are reduced to the wave equation and the law of dispersion of sound waves has the form:

$$\omega = vk, \tag{P7.1.3.3}$$

where v is the velocity of the sound wave propagation (for simplicity, we neglect the difference between the longitudinal and transverse waves). The main contribution to the energy is given by the waves with $\hbar\omega \sim k_B T$, where k_B is the Boltzmann constant (the probability of formation a wave with $\hbar\omega \gg k_B T$ is exponentially small); hence the contribution of sound waves (for which $ka \ll 1$ or $\omega \ll v/a$) is the main contribution under the condition that $k_B T \ll \hbar v/a$, that is, at sufficiently low temperatures. Conversely, at high temperatures, the contribution to the energy from each of the elementary waves is equal to $k_B T$, and the energy is determined only by the common number of degrees of freedom.

▶ **Debye theory.** To obtain the correct value of the heat capacity at both high and low temperatures in the framework of a simple model, Debye assumed that the density of states (P7.1.3.2) and the dispersion law (P7.1.3.3) hold for all k up to some maximum value k_D and a corresponding frequency ω_D above which the density of states is zero. The *Debye frequency* ω_D is chosen so that the total number of state is equal to $3N$, which ensures that the Dulong–Petit law holds at high temperatures.

The density of the number of states in the Debye model is expressed as

$$G(\omega)\, d\omega = \begin{cases} \dfrac{3\,\omega^2\, d\omega}{2\pi^2 v^3} V & \text{for } \omega < \omega_D; \\[2mm] 0 & \text{for } \omega > \omega_D. \end{cases}$$

To take the difference between the transverse and longitudinal waves into account, we must define the average velocity by the relation

$$\frac{3}{v^3} = \left(\frac{1}{v_\parallel^3} + \frac{2}{v_\perp^3} \right).$$

We determine the Debye frequency from the equation

$$3N = \int_0^{\omega_D} \frac{3}{2}\, \frac{\omega^2\, d\omega}{\pi^2 v^3} V$$

to obtain

$$\omega_D = (6\pi^2 n)^{1/3} v, \qquad k_D = (6\pi^2 n)^{1/3}. \tag{P7.1.3.4}$$

The energy of the lattice vibrations in the Debye model is given by the expression

$$E_{cr} = \int_0^{\omega_D} \frac{\hbar \omega G(\omega)\, d\omega}{e^{\hbar \omega / k_B T} - 1}.$$

To separate the regions of high and low temperatures, it is convenient to introduce the *Debye temperature* by the relation

$$k_B T_D = \hbar \omega_D. \tag{P7.1.3.5}$$

Introducing the dimensionless variable $x = \hbar \omega / k_B T$, we obtain

$$E_{cr} = 9 N k_B T_D \left(\frac{T}{T_D} \right)^4 \int_0^{T_D/T} \frac{x^3\, dx}{e^x - 1}.$$

For $T \gg T_D$, expanding the exponential function, we obtain $E_{cr} = 3 k_B N T = 3\nu R T$, that is, the Dulong–Petit law. For $T \ll T_D$, the integration can be continued to infinity, and it becomes equal to $\pi^4/15$. We see that the energy of the lattice vibrations is proportional to T^4, and for the heat capacity, we obtain the law of T^3:

$$C_V = \frac{12}{5} \pi^4 k_B N \left(\frac{T}{T_D} \right)^3.$$

Example 1. Calculate the energy of zero-point vibrations of the crystal lattice in the Debye model.
Solution. This energy is equal to

$$E_0 = \int_0^{\omega_D} \frac{1}{2} \hbar \omega G(\omega) d\omega = \frac{3\hbar \omega_D^4}{16\pi^2 v^3} N = \frac{9}{8} N \hbar \omega_D = \frac{9}{8} N k_B T_D.$$

Example 2. Calculate the energy of the one-dimensional crystal lattice vibrations in the Debye model (only for longitudinal vibrations).
Solution. For the density of states, $G(k) = L/(2\pi)$, we obtain

$$2G(k)\, dk = G(\omega)\, d\omega = \begin{cases} \dfrac{d\omega}{\pi v} L & \text{for } \omega < \omega_D, \\ 0 & \text{for } \omega > \omega_D \end{cases}$$

(there are two traveling waves for one ω), where L is the length of the chain. Since the total number of states is equal to N, we have

$$\omega_D = \pi \frac{N}{L} v = \frac{\pi}{a} v. \tag{P7.1.3.6}$$

For the crystal energy, we obtain

$$E_{cr} = \int_0^{\omega_D} \frac{\hbar \omega G(\omega)\, d\omega}{e^{\hbar \omega / k_B T} - 1} = N k_B T_D \left(\frac{T}{T_D} \right)^2 \int_0^{T_D/T} \frac{x\, dx}{e^x - 1}.$$

For $T \ll T_D$, the energy of vibrations is proportional to T^2. For $T \gg T_D$, $E_{cr} = k_B N T$.

▶ **One-dimensional chain. Dispersion law.** To understand what consequences the discreteness in the mass distribution may lead to, we consider an infinitely long chain of atoms. We assume that the atomic mass is m, the distance between the neighboring atoms in equilibrium is a, and an elastic force with rigidity κ acts between the neighboring atoms. The equation of motion of the sth atom has the form

$$\xi_s'' = \omega_0^2 (\xi_{s+1} - 2\xi_s + \xi_{s-1}),$$

where ξ_s is the displacement of the sth atom from equilibrium, $\omega_0 = \sqrt{\kappa/m}$. We seek the solution in the form $\xi_s = A \exp[i(ksa - \omega t)]$. For the relation between ω and k, we obtain

$$\omega = 2\omega_0 \left| \sin \frac{ka}{2} \right|. \tag{P7.1.3.7}$$

The velocity of the sound wave ($ka \ll 1$) is equal to $v = \omega_0 a$, and the Debye frequency corresponding to this velocity is $\omega_D = \pi\omega_0 = (\pi/2)\omega_{max}$ ($\omega_{max} = 2\omega_0$). We note that we obtained a periodic function with period $2\pi/a$ (Fig. P7.2).

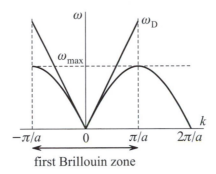

Figure P7.2. Dispersion law and the first Brillouin zone for a one-dimensional chain of atoms.

▶ **The first Brillouin zone.** We find what values of k correspond to physically distinct states. If we consider the values of the wave vector that differ by $2\pi n/a$: $k' = k + 2\pi n/a$, then we obtain not only equal values of ω but also the identical functions: $\xi_s(k') = \xi_s(k) \cdot \exp(2\pi i n s) = \xi_s(k)$. Hence all the physically distinct states are described by wave vectors lying in the interval $(-\pi/a < k \leq \pi/a)$. This interval is called the *first Brillouin zone*.

On the zone boundary, the traveling wave becomes a standing wave: $A \exp[i(\frac{\pi}{a}as - \omega t)] = A \exp(-i\omega t)(-1)^s$. The neighboring atoms oscillate in antiphase, the wave nodes lie between the atoms. We note that on the zone boundary, the wave group velocity is zero:

$$v_g = \frac{d\omega}{dk} = \omega_0 a \cos \frac{ka}{2} = \frac{a}{2}\sqrt{\omega_{max}^2 - \omega^2}, \tag{P7.1.3.8}$$

where $\omega_{max} = 2\omega_0$ is the frequency value on the Brillouin zone boundary.

▶ **The density of states in the Brillouin zone.** The first Brillouin zone contains exactly N distinct values of k (see example 2 above; N is the number of atoms in the chain), which means that all degrees of freedom are exhausted within the framework of the first Brillouin zone (we consider only the longitudinal traveling waves). In the one-dimensional Debye model, the density of states $G(\omega)$ is constant. However, from the dispersion law (P7.1.3.7) we find that

$$G(\omega)\,d\omega = 2G(k)\,dk = 2\frac{L}{2\pi}\frac{dk}{d\omega}\,d\omega = \frac{L}{\pi}\frac{d\omega}{v_g}.$$

Using (P7.1.3.7) and (P7.1.3.8), we obtain

$$G(\omega) = \frac{2L}{\pi a}(\omega_{max}^2 - \omega^2)^{-1/2}.$$

At the Brillouin zone boundary, the density of states $G(\omega)$ becomes infinitely large.

▶ **Diatomic chain. Optical branches.** If an elementary chain contains several distinct atoms, then in addition to *sound branches*, for which $\omega \sim k$ as $k \to 0$, there arise *optical branches*, on which ω remains finite for all k in the Brillouin zone. For example, for longitudinal oscillations in a chain of alternating atoms with masses m_1 and m_2, there are two branches shown in Fig. P7.3. For small k (near the zone center), the sound branch corresponds to deviations of distinct atoms in the same direction, and the optical branch corresponds to deviations in opposite directions. In the general case, where the cell contains p distinct atoms, there are three sound branches and $3p - 3$ optical branches.

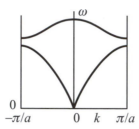

Figure P7.3. Sound and optical branches for a chain of distinct atoms.

▶ **Phonons.** Just as in the case where oscillations of an electromagnetic field and electromagnetic waves are associated with quantum particles, photons, the lattice elementary vibrations in the form of traveling waves are associated with particles whose momentum and energy are related to the wave vector and frequency by the relations $\mathbf{p} = \hbar\mathbf{k}$ and $\varepsilon = \hbar\omega$. These particles are called *phonons*. Since the phonon energy is a periodic function of their momentum, which is not typical of usual particles, they are called *quasiparticles*, and their momentum is called the *quasi-momentum*. The phonons obey the Bose–Einstein statistics with $\mu = 0$, because the number of phonons in a system is not fixed. The phonons can scatter at impurities and crystal defects and interact with other particles, for example, with electrons and photons. A phonon can also split into two phonons (because of nonlinear effects, which must be taken into account only for sufficiently energetic phonons).

In all these processes, the energy and momentum conservation laws are fulfilled. A nontrivial characteristic of phonons is that, in the final state, the momentum conservation law can contain a phonon that is not in the first Brillouin zone. If we pass to the corresponding phonon from the first zone, then an additional momentum $\hbar\mathbf{G}$, which describes the shift of this Brillouin zone with respect to the first, appears in the momentum conservation law. This momentum corresponds to the motion of the crystal as a whole (which cannot be discovered experimentally, because the mass of the crystal is very large). For example, in the one-dimensional case, $G = 2\pi n/a$, which corresponds to the case of equal displacement of all nodes of the lattice.

The processes of phonons scattering determine their mean free path in terms of which the phonon thermal conduction is expressed:

$$\kappa_{\mathrm{ph}} = \tfrac{1}{3}C_{\mathrm{cr}}\lambda_{\mathrm{ph}}v,$$

where $C_{\mathrm{cr}} \sim T^3$ is volume heat capacity of the lattice, v is the speed of sound. As the temperature increases, the mean free path decreases rapidly.

P7.2. Electrons in a Periodic Field. Energy Bands

Within the framework of the Sommerfeld theory, where electrons in metal are considered as a free fermion gas, it is possible to explain several facts, mainly, the behavior of the electron heat capacity. But many phenomena cannot be explained in this theory. First, this is the problem of origination of metals, semiconductors, and dielectrics and the description of their properties. To clarify these problems, it is necessary to take account of the interaction of electrons with the crystal lattice, which forms a periodic external field for the electrons. The interaction of electrons with the lattice periodic field results in the appearance in their spectrum of *allowed energy bands* separated by *forbidden energy bands*; that is, by the energy intervals that do not contain electron levels. The existence of crystals of different types corresponds to the different character of filling the energy bands by electrons.

P7.2.1. Qualitative Theory of Energy Bands

▶ **Origination of bands from atomic levels.** Let us track the behavior of electron energy levels in the case of gradual formation of a crystal from isolated atoms. First, we assume that N atoms are at a distance from one another such that the wave functions of electrons do not overlap. In this case, we can assume that each nondegenerate level of a separate atom in the system of N atoms is N-fold degenerate, but if an atomic level is p-fold degenerate, then its degree of degeneration in a system of N atoms is equal to pN. We begin to bring the atoms together to the distance corresponding to the crystal in equilibrium. As a result of interaction between the overlapping electron clouds related to separate atoms, one level yields N very closely located levels contained in a band of width ΔE (Fig. P7.4). The band width is determined by the energy of interaction between the corresponding electrons of neighboring atoms, and it is independent of the number of atoms. For the outer shells, the interaction is strong (the wave functions overlap significantly) and the band width is large, for inner shells, the band width is small. The number of levels in each band is large, so that the spectrum can be assumed to be continuous. Such energy bands are called *allowed bands*, and the energy intervals between the bands, where there are no electron levels, are called *forbidden bands*.

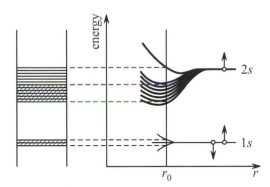

Figure P7.4. Formation of energy bands as the lithium atoms approach each other.

▶ **Classification of crystals.** The type of the crystal conductivity is defined by the character of filling by electrons (at $T = 0$) of the upper allowed band (above which all the allowed bands are empty). The bands below the upper allowed band are completely filled with electrons. The electrons in these bands at $T = 0$ cannot participate in conductivity, because there are no vacancies where they could move as the external field is switched on. The upper band originating from the atomic outer shell can be filled with electrons either partially or completely.

Metals. If this band is filled partially, then the crystal is a *metal* (or a *semimetal* if the number of electrons in this band is very small). This band is called the *conductivity band*. The Fermi energy of a metal lies in the interior of the conductivity band, the electrons located in this band form a set of free carriers participating in conductivity at any arbitrarily low temperature. The properties of these electrons have much in common with the electron gas in the Sommerfeld model (if the energy is counted from the lower boundary of the band), but there are important distinctions, which we discuss below.

The outer s-shell of atoms of alkali metals contain two electron states (corresponding to different projections of spin), and only one of them is occupied by the electron. Therefore, the conductivity band contains $2N$ free places, and only half of them is filled (see Fig. P7.4 for lithium). For rare-earth metals, both s-states of the outer shell are occupied by electrons, but there is another mechanism of filling the empty band: the energy band originating from the s-shell and containing $2N$ places is overlapped with the band originating from the p-shell containing $6N$ places. The allowed band thus formed contains $8N$ place only the $2N$ of which are filled with electrons (see Fig. P7.5 for beryllium).

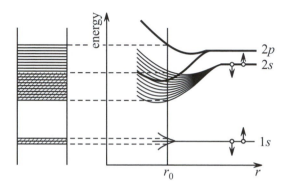

Figure P7.5. Formation of energy bands as beryllium atoms approach each other.

Dielectrics and semiconductors. If the upper band at $T = 0$ is completely filled with electrons, then at $T = 0$, the crystal is a dielectric. For $T > 0$, the conductivity of such a crystal is determined by the width ΔE of the forbidden band lying between the filled band (it is called the *valence band*) and the nearest free band (which in this case is called the *conductivity band*). Because of the thermal motion, the electrons can move from the valence band into the conductivity band. For $k_B T \ll \Delta E$, the electron concentration in the conductivity band is exponentially small ($\sim \exp(-\Delta E / k_B T)$). For $k_B T \sim \Delta E$, the electron concentration in the conductivity band becomes noticeable and rapidly increases with temperature (the conductivity becomes noticeable already for $k_B T \sim 0.1 \Delta E$). In this temperature region, the crystal behaves as a *pure semiconductor*. Usually, the semiconductors are crystals whose conductivity becomes noticeable at room temperature.

In a pure semiconductor, simultaneously with the appearance of negative charge carrier—electrons—in the conductivity band, the same number of positive charge carriers—vacant places—arise in the valence band; these positive charge carriers are called *holes*.

The conductivity of pure semiconductors related to the appearance of the equal quantities of electrons and holes is said to be *intrinsic*.

Extrinsic semiconductors. Addition of an impurity results in violation of the symmetry between positive and negative charges. If an impurity whose atom valency is greater than the crystal atom valency is added, there arise excessive electrons that are weakly related to the impurity atoms and become free at the temperatures at which the transitions between the bands are still absent. To these excessive electrons there correspond *donor energy levels* located in the forbidden band near the lower boundary of the conductivity band. The electrons from these levels move into the conductivity band, and since there is still no vacancy in the valence band, the conductivity in this temperature region is realized mainly by electrons. Such extrinsic semiconductors are called *semiconductors of n-type*.

In *semiconductors of p-type*, the valency of impurity is less that the valency of crystal atoms, which results in the appearance of vacant places to which the electrons begin to move at sufficiently low temperatures. To these vacancies there correspond empty energy levels (*acceptor levels*) located near the upper boundary of the valence band. The transition to these levels of electrons from the valence band leads to the formation of holes without appearance of electrons in the conductivity band.

P7.2.2. Quantum Mechanical Description of Energy Regions

Although it is appropriate to use quantum numbers of individual atoms to label energy bands, the state of an electron within an allowed band is described using a completely different quantum number associated with the so-called the *quasi-momentum* of the electron. The quasi-momentum varies quasi-continuously and reflects the translational properties of the Schrödinger equation and the wave function.

▶ **Bloch waves. Quasi-momentum of electron.** The wave function of a stationary state of an electron in an infinite periodic potential can be written as

$$\psi(\mathbf{r}) = e^{i\mathbf{k}\cdot\mathbf{r}}\varphi(\mathbf{r}), \tag{P7.2.2.1}$$

where $\varphi(\mathbf{r})$ is a periodic function of coordinates (with the same periodic properties as the potential energy of an electron). This statement is called the *Bloch theorem*, and the vector $\hbar\mathbf{k}$ is called the *quasi-momentum of an electron*. The vectors $\mathbf{k}$ and $\mathbf{p}$ for an electron in a crystal are in a sense similar to the wave vector and momentum of the free electron in the Sommerfeld model. Let us clarify the Bloch theorem and the meaning of these notions in the one-dimensional case.

Let the period of one-dimensional lattice be equal to a, and let the potential energy of electron have the same periodicity: $U(x + a) = U(x)$. It is easy to see that if $\psi(x)$ is a solution of the time-independent Schrödinger equation, then $\psi(x + a)$ is also a solution of this equation. Hence $\psi(x+na) = T(na)\psi(x)$, and the discrete function $T(x)$ (determined for $x = na$) has the following properties: $|T(na)| = 1$, $T(na) = [T(a)]^n$. We obtain $T(x) = e^{ikx}$, and the function $\varphi(x) = e^{-ikx}\psi(x)$ is a periodic function of x: $\varphi(x + a) = \varphi(x)$.

For each given value k, from the Schrödinger equation for $\psi(x)$ we obtain a second-order ordinary differential equation for $\varphi(x)$ with the boundary condition $\varphi(a) = \varphi(0)$. This equation has solution only for discrete values $E_n(k)$. If we pass from k to $k' = k + 2\pi/a$, then we obtain the same equation for $\varphi(x)$ and the same values of the energy $E_{k'}$ as for the number k. We conclude that the energy is a periodic function of k with period $G = 2\pi/a$ and the wave functions for the values of k that differ by G are identical. Hence we can reduce all the electron states to the first Brillouin zone. At the well boundaries, there are extrema (minima and maxima) of $E(k)$, which ensure the continuity and smoothness of this function in transitions from one Brillouin zone to another (Fig. P7.6). At the center of the zone (for $k = 0$), $E(k)$ usually also has an extremum.

The above-discussed separation of crystals into metals and dielectrics (at $T = 0$) can be well explained by the dependence $E(k)$ in the first Brillouin zone (Fig. P7.6). Figure P7.6a corresponds to a dielectric, Fig. P7.6b to a metal (or semimetal) in the case of overlapping bands, and Fig. P7.6c to a metal in the case of partially filled band.

▶ **The number of states in the Brillouin zone.** As k varies in the first Brillouin zone, the energy of the nth level $E_n(k)$ varies within the nth allowed energy band. If we take into account that the crystal is finite (by using, for example, periodic boundary conditions), then in the first Brillouin zone, k takes N values: $0, \pm 2\pi/L, \pm 4\pi/L, \ldots, \pi/a$, where $L = Na$. Hence the nth band contains N levels $E_n(k_i)$. With the spin taken into account, we obtain $2N$ electron states.

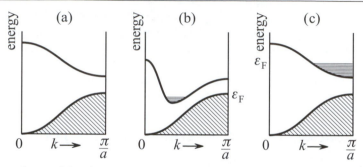

Figure P7.6. Dependence of the electron energy on the quasi-momentum in the first Brillouin zone for the upper energy bands: (a) dielectric, (b) metal or semimetal (overlapping bands), (c) metal (partially filled band).

P7.2.3. Dynamics of Electron in Crystal

▶ **Equation of motion of an electron in crystal.** The energy spectrum of electrons in the upper bands (the valence band and the conductivity band) bears information about interaction of electrons with the crystal lattice. If this spectrum is used to calculate the group velocity of an electron

$$v_g = \frac{d\omega}{dk} = \frac{1}{\hbar}\frac{d\varepsilon}{dk} = \frac{d\varepsilon}{dp}, \tag{P7.2.3.1}$$

then we can write the equations of motion in the form taking account of the action on electrons of *only external forces*. This equation has the form of usual Newton law

$$\frac{d\mathbf{p}}{dt} = \mathbf{F}, \tag{P7.2.3.2}$$

where **F** is an external force (that does not take account of the interaction with the lattice) and **p** is the quasi-momentum of electron.

For clarity, we present the expression for the rate of variation in the electron energy

$$\frac{d\varepsilon}{dt} = \frac{d\varepsilon}{dp}\frac{dp}{dt} = v_g\frac{dp}{dt},$$

which must be equal to the power of the external force $F v_g$.

As an example of the use of equations of motion (P7.2.3.1) and (P7.2.3.2), we consider the behavior of an electron in the empty energy band (where no other electrons are contained) under the action of a constant force. We assume that this electron does not collide with other electrons, phonons, and lattice defects and that, at the initial time, the electron is at rest. In the Sommerfeld model, the free electron would move with constant acceleration away from the point of the beginning of motion. In the band theory, the electron is first at the center of the band (for $p = 0$), then its momentum linearly increases, but the velocity varies differently in this case. It first increases, then decreases approaching the Brillouin zone boundary, and is zero at zone boundary. At this instant, the electron is "transferred" to the left boundary of the zone (if we want to stay in the framework of the first Brillouin zone), and the electron velocity changes the sign. At the time when the electron momentum is again at the center of the zone, its velocity again is zero, and the electron is transferred to the previous place. The electron motion has an oscillatory character. To avoid misunderstanding, we note that the electrons in metal cannot move in such a cycle, because they do not have time between the collisions to go far enough from the Fermi surface and approach the zone boundary.

▶ **Effective mass.** At the center and the band boundaries, we have $d\varepsilon/dp = 0$, and hence the energy near these points can approximately be written as

$$\varepsilon = \varepsilon_0 + \frac{1}{2}\frac{d^2\varepsilon}{dp^2}\bigg|_{p_0}(p - p_0)^2.$$

where ε_0, p_0 are the energy and quasi-momentum of the center (or boundary) of the band. If in all cases, the momentum and energy are counted from the parabola vertex, then this relation can be written as follows:

$$\varepsilon = \frac{p^2}{2m^*},$$

where

$$\frac{1}{m^*} = \frac{d^2\varepsilon}{dp^2}\bigg|_{p_0}, \tag{P7.2.3.3}$$

and the quantity m^* thus defined is called the *effective mass* of electron. For the electron velocity we obtain the usual relation

$$v_{\mathrm{g}} = \frac{p}{m^*}.$$

We note that the electron effective mass defined by equation (P7.2.3.3) is positive near the energy band bottom and is negative near the band upper boundary. As we saw, near the band upper boundary, under the action of an external force, the electron velocity does not increase but decreases.

Definition (P7.2.3.3) for m^* acts not only near the band boundaries. Indeed, at any point, we can write

$$\frac{dv_{\mathrm{g}}}{dt} = \frac{d}{dt}\frac{d\varepsilon}{dp} = \frac{d^2\varepsilon}{dp^2}\frac{dp}{dt},$$

or

$$F = \frac{dp}{dt} = m^*\frac{dv_{\mathrm{g}}}{dt}. \tag{P7.2.3.4}$$

The effective mass is an important dynamical characteristic of an electron in a crystal. The effective mass of an electron can either be greater than or less than its true mass (sometimes, by several dozens of times). The effective mass stands in the place of the usual mass in the expression for the electrical conduction: $\sigma = e^2 n\tau/m^*$, and in many other physical formulas. The effective mass is especially convenient to describe the properties of semiconductors, where the electron concentration in the conductivity band (and of holes in the valence band) is not large. In anisotropic crystals, the effective mass depends on the direction of motion of the electron.

Although an individual electron near the upper boundary of the band must be assigned a negative effective mass, the absence of such an electron in the completely filled valence band is equivalent to the presence of a single particle with positive mass and positive charge. As was already noted, such charge carriers are called *holes*.

▶ **Fermi surface for electrons in metals.** In the band theory, the Fermi surface of electrons is not a simple sphere of radius p_{F}, as it is the case for a free electron. Metals can have quite intricate Fermi surfaces. The determination of electronic Fermi surfaces is one of the most important problems in metal physics. Recall that at room temperatures, the electrons in metals are in a strongly degenerate state ($k_{\mathrm{B}}T \ll \varepsilon_{\mathrm{F}}$), and a special role is played by electrons near the Fermi surface. Fermi surfaces are usually depicted within the first

Brillouin zone; however, it is more graphic to display a periodic connected surface covering several Brillouin zones. Figure P7.7 depicts the Fermi surface and the boundaries of the first Brillouin zone for copper; the Fermi surfaces for silver, gold, and other metals of this group look quite similar. In Fig. P7.7, one can clearly see the transitions into neighboring Brillouin zones. (The three-dimensional presentations of the Fermi surfaces for most elements of the periodic table can be found at the Website `http://www.phys.ufl.edu/fermisurface`.)

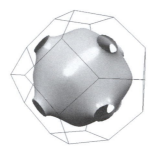

Figure P7.7. Fermi surface and boundaries of the first Brillouin zone for copper.

P7.2.4. Band Theory of Semiconductors

The difference between dielectrics and intrinsic semiconductors is determined by the width of the forbidden band Δ between the filled (at $T = 0$) valence band and the empty conductivity band. To semiconductors one usually refers the crystals for which $\Delta < 3\,\text{eV}$; for example, for germanium, $\Delta = 0.67\,\text{eV}$, and for silicon, $\Delta = 1.11\,\text{eV}$. The current carrier concentration is many orders of magnitude less for semiconductors than for metals and it varies quite strongly if impurities, even in negligible concentrations, are added.

▶ **Fermi energy and conductivity of an intrinsic semiconductor.** The position of the Fermi energy in intrinsic semiconductors is determined by the condition that the electron concentration in the conductivity band and the hole concentration in the valence band coincide. For $k_{\mathrm{B}}T \ll \Delta$, the Fermi energy is very close to the center of the forbidden band. (A deviation from the center of the band appears because the effective mass of electrons near the conductivity band bottom differs from the effective mass of holes near the top of the valence band). The concentration of charge carriers is calculated from the Fermi distribution, the dependence of the concentration on the temperature is mainly determined by the multiplier $\exp(-\Delta/2k_{\mathrm{B}}T)$, where $\Delta/2$ is the distance from the Fermi level to the band boundary.

The electron concentration in the conductivity band is calculated by the formula

$$\int_{\varepsilon_0}^{\infty} f_{\mathrm{F}}(\varepsilon) G(\varepsilon)\, d\varepsilon = \int_{\varepsilon_0}^{\infty} e^{-(\varepsilon - \varepsilon_{\mathrm{F}})/k_{\mathrm{B}}T} G(\varepsilon)\, d\varepsilon$$

$$= \frac{m^{*3/2}\sqrt{2}}{\pi^2 \hbar^3} e^{-(\varepsilon_0 - \varepsilon_{\mathrm{F}})/k_{\mathrm{B}}T} \int_0^{\infty} e^{-\varepsilon'/k_{\mathrm{B}}T} \sqrt{\varepsilon'}\, d\varepsilon',$$

where ε_0 is energy of the conductivity band bottom, $\varepsilon' = \varepsilon - \varepsilon_0$ is the energy counted from the band bottom (it is taken into account that $k_{\mathrm{B}}T \ll \varepsilon_0 - \varepsilon_{\mathrm{F}}$).

The intrinsic conductivity of semiconductors depends on the temperature similarly:

$$\sigma = \sigma_0 e^{-\Delta/2k_{\mathrm{B}}T}.$$

▶ **Extrinsic semiconductors.** In the case of extrinsic semiconductors, the Fermi energy is displaced to the boundary of the band, where the main charge carriers are located. In the case of semiconductors of n-type with donor conductivity, the Fermi level is located near the conductivity band bottom, in the case of semiconductors of p-type with acceptor conductivity, near the upper boundary of the valence band (Fig. P7.8). At lower temperatures, the number of impurity carriers in the band strongly depends on the temperature, and the conductivity rapidly increases with increasing temperature. Then the impurity levels become "impoverished"; that is, no electrons remain at donor levels (or no free places at acceptor levels), and the conductivity slowly decreases. (Just as in the case of metals, the mean free path decreases because the number of phonons in the system increases.) At higher temperatures, the number of charge carriers begins to increase again, because the mechanism of intrinsic conductivity begins to act. The Fermi energy in this case approaches the center of the forbidden band.

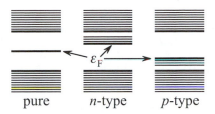

Figure P7.8. Position of the Fermi level in the intrinsic and extrinsic semiconductors.

▶ **Contact phenomena. p-n transition.** If a semiconductor of n-type comes in contact with a semiconductor of p-type, then a *p-n transition* arises. As a result of diffusion of the majority carriers in the neighboring semiconductor in the transition region, there arises a contact difference of potentials, and the potential of the semiconductor of n-type turns out to be higher. The contact difference of potentials prevents the charge carriers from motion. The potential difference between the semiconductors may attain such a value (Fig. P7.9) that the Fermi energy of the semiconductors becomes equal (the condition that the chemical potentials be equal). In equilibrium, the diffusion current of the majority carrier through the potential barrier is equal to the drift current of minority carriers, to which the contact difference of potentials is no obstacle.

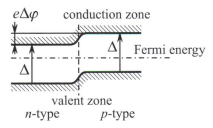

Figure P7.9. Position of energy bands in the region of p-n transition.

If the external field is applied in the forward direction (the positive pole) to the semiconductor of p-type, then the majority carrier current increases in $\exp(eU/k_{\mathrm{B}}T)$ times, because the barrier decreases by eU (U is the applied voltage). If the field is applied in the backward direction, then the forward carrier current decreases (the barrier increases). The applied voltage does not affect the minority carrier current. The volt-ampere characteristic of the p-n transition is determined by the expression

$$I = I_0\left(e^{eU/k_{\mathrm{B}}T} - 1\right),$$

where I_0 is the majority carrier current in the absence of the field (equal to the opposed minority carrier current).

Bibliography for Chapter P7

Alenitsyn, A. G., Butikov, E. I., and Kondratyev, A. S., *Concise Handbook of Mathematics and Physics,* CRC Press, Boca Raton, Florida, 1998.

Anselm, A. I., *Introduction to Physics of Semiconductors* [in Russian], Lan, St. Petersburg, 2008.

Ashcroft, N. W. and Mermin, N. D., *Solid State Physics,* Brooks Cole, Belmont, California, 1976.

Benenson, W., Harris, J. W., Stocker, H., and Lutz, H. (Eds.), *Handbook of Physics,* Springer-Verlag, New York, 2002.

Blakemore, J. S., *Solid State Physics 2nd Edition,* Cambridge University Press, Cambridge, England, 1985.

Chaikin, P. M. and Lubensky, T. C., *Principles of Condensed Matter Physics,* Cambridge University Press, Cambridge, England, 2000.

Ginsburg, I. F., *Introduction to Solid State Physics. Fundamentals of Quantum Mechanics and Statistical Physics with Selected Problems from Solid State Physics* [in Russian], Lan, St. Petersburg, 2007.

Grosso, G. and Parravicini, G. P., *Solid State Physics,* Academic Press, London, 2000.

Hofmann, P., *Solid State Physics: An Introduction,* Wiley-VCH, Berlin, 2008.

Hook, J. R. and Hall, H. E., *Solid State Physics, 2nd Edition,* John Wiley & Sons, Chichester, England, 1991.

Kittel, C., *Introduction to Solid State Physics, 8th Edition,* Wiley, New York, 2004.

Mihaly, L. and Martin, M. C., *Solid State Physics: Problems and Solutions, 2nd Edition,* Wiley-VCH, Berlin, 2009.

Omar, M. A., *Elementary Solid State Physics: Principles and Applications, 4th Edition,* Addison-Wesley, New York, 1994.

Prokhorov, A. M. (Ed.), *Encyclopedia of Physics, Volumes 1–5* [in Russian], Bolshaya Russkaya Encyclopedia, Moscow, 1998.

Prokhorov, A. M. (Ed.), *Encyclopedic Dictionary of Physics* [in Russian], Sovetskaya Encyclopedia, Moscow, 1984 (dic.academic.ru).

Rosenberg, H. M., *The Solid State: An Introduction to the Physics of Crystals for Students of Physics, Materials Science, and Engineering (Oxford Physics Series, 9), 3rd Edition,* Oxford University Press, New York, 1988.

Snoke, D. W., *Solid State Physics: Essential Concepts,* Addison Wesley, New York, 2008.

Woan, G., *The Cambridge Handbook of Physics Formulas,* Cambridge University Press, Cambridge, England, 2003.

Ziman, J. M., *Principles of the Theory of Solids, 2nd Edition,* Cambridge University Press, Cambridge, England, 1979.

Chapter P8

Elements of Nuclear Physics

The idea that practically the entire mass of an atom is concentrated in its positively charged nucleus of infinitesimally small dimensions is due to Rutherford's experiments (see Chapter P6). Since the dimensions of nuclei turned out to be by five orders less than those of atoms, it can be assumed, in the framework of atomic physics, that the nucleus is a point Coulomb center. Actually, the nucleus is a complex structure formed by strongly interacting particles (several ones or hundreds) obeying the laws of quantum mechanics and quantum statistics. The nuclei can undergo radioactive transformations, participate in nuclear reactions, disintegrate, and merge with other nuclei. The characteristic energies in that nuclear world are measured in millions of electron-volts, which explains why nuclei appear as stable objects in atomic processes with energies up to several hundred electron-volts.

P8.1. Basic Properties of Nuclei

P8.1.1. Characteristics of Nuclei

▶ **Nuclear Composition.** The atomic nucleus consists of protons and neutrons—particles collectively named *nucleons*.

The proton is a subatomic particle with positive charge e and mass $m_p = 1836.15\, m_e$, where m_e is the electron mass. The *neutron* is an electrically neutral particle whose mass is slightly greater than that of the proton: $m_n = 1838.68\, m_e$. In its free state, the neutron is unstable in the sense that within 15.5 minutes, on the average, it turns into a proton after emitting an electron and an antineutrino: $n \rightarrow p + e + \tilde{\nu}$. The spin* of both the proton and neutron is equal to $1/2$, which means that they belong to the class of fermions. Both the proton and the neutron possess nonzero magnetic moments: $\mu_p = 2.793\, \mu_N$ and $\mu_n = -1.91\, \mu_N$, where $\mu_N = e\hbar/2m_p$ is the so-called *nuclear magneton* ($\mu_N = \mu_B/1836.15$). The nonzero magnetic moment of the neutron suggests that nucleons must have an internal structure involving more elementary charged particles.

The nuclear composition is described by two integers: the *atomic number* Z, equal to the number of protons in the nucleus (and so identical to the *charge number* of the nucleus), and the *mass number* A, equal to the total number of nucleons in the nucleus. The number of neutrons (the *neutron number*) is $N = A - Z$. The charge number of an element coincides with its ordinal number in Mendeleev's table (see Chapter S7) and determines its physical-chemical properties, since these depend on the charge of the nucleus and are almost independent of its mass. The charge and the mass numbers of an element X are specified as indices to the left of its symbol: $^A_Z X$; for example, $^3_2 He$, $^{13}_6 C$. Since the element symbol uniquely determines its number in the periodic table, which coincides with the charge number Z, the left lower index is sometimes dropped ($^3 He$, $^{13} C$).

Particular atoms with A nucleons in their nucleus and Z protons, together with their electronic shell containing Z electrons, are called *nuclides*. All nuclides with the same Z

---

* Here and henceforth, we often use the word "spin" for short to mean the "spin quantum number."

are called *isotopes* of a given element. The isotopes of hydrogen ^{1}H, ^{2}H, and ^{3}H, whose nuclei are essentially distinct, are respectively called the common hydrogen (sometimes the term *protium* is used), *deuterium* (D), and *tritium* (T), and the nuclei of deuterium and tritium are called *deuteron* (d) and *triton* (t).

There are stable and radioactive isotopes. Radioactive isotopes are characterized by their type of radioactivity (α-decay or β-decay) and their lifetime. (In more detail, radioactive transformations are considered below.) There are many radioactive isotopes occurring in nature, provided that their lifetime is very long (^{235}U, ^{232}Th, and others) or they are constantly produced as a result of decay of such long-lived nuclei (for example, ^{223}Fr). There are no stable isotopes for $Z = 43$ (technetium), $Z = 61$ (promethium) and for all elements with $Z > 83$ (starting from polonium). The other elements possess from 1 to 9 (tin Sn) stable isotopes, and on the average, the number of these is equal to 3. Different stable (or long-lived) isotopes have different occurrence, the observed *atomic weight* is an average over several isotopes. For instance, the atomic weight of natural chlorine is equal to 35.5, and it contains 76% of ^{35}Cl and 24% of ^{37}Cl.

Nuclides with the same mass number A are called *isobars* (for instance, $^{40}_{18}$Ar is $^{40}_{20}$Ca). In the presence of β-decay, isobars can transform into one another. Note that for a stable nuclide, its closest isobars are β-active.

Among stable and long-lived nuclei, the most frequent are even-even ones (even number of protons and even number of neutrons) and the rarest are odd-odd ones (only four stable ones: $^2_1$H, $^6_3$Li, $^{10}_5$B and $^{14}_7$N).

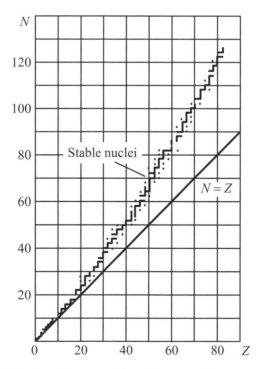

Figure P8.1. Number of neutrons in a nucleus versus its consecutive number.

Figure P8.1 shows the dependence of N on Z for stable nuclides. It can be seen that for the first 15 to 20 elements, the number of protons is approximately the same as that of neutrons ($Z \sim A/2$). With the increase of the consecutive number, there is a clear tendency to the number of neutrons being greater than that of protons (thus, the only stable isotope of bismuth is $^{209}_{83}$Bi).

▶ **Nuclear spin and magnetic moments.** Nuclear spin I is defined as the total angular momentum of the nucleus, which is the composition of spins and orbital angular momenta of separate nucleons. Nuclei with evenly many nucleons have integer spin, while the spin of nuclei with an odd number of nucleons is half-integer. For nuclei with an even number of both protons and neutrons, the spin of the ground state is equal to zero. This means that angular momenta of separate nucleons compensate one another. It should be mentioned, however, that $I = 1$ for the ground state of deuteron.

As a rule, nuclei with nonzero spin have also a nonzero magnetic moment. Nuclear magnetic moments are usually very small (of the order of nuclear magneton), but their interaction with orbital motion of electrons gives rise to a *hyperfine structure* of electronic levels, which can be observed.

▶ **Nuclear dimensions and shape.** Nucleon distribution over the nuclear volume has been established in many experiments with scattering of fast electrons and nucleons. (The de Broglie wavelength of scattered particles should be small compared with nuclear dimensions.) In electronic scattering experiments aimed at testing only the electromagnetic interaction one finds spatial distribution of the nuclear charge (i.e., proton distribution), and in neutron scattering experiments one finds spatial distribution of all nucleons. It turns out that these distributions differ but slightly, and in what follows we speak only of the nucleon distribution.

As it happens, the shape of most nuclei is nearly spherical. For nuclei with a fairly large number of nucleons ($A > 12$), one can identify an interior region in which nucleon concentration is constant, and there is a small transition layer over which the concentration drops to zero. If the nuclear radius is defined as the distance from the center where the concentration drops by a factor of two, one arrives at the following dependence on the mass number:

$$R \sim R_0 A^{1/3} \, \text{fm}, \tag{P8.1.1.1}$$

where $R_0 = 1.2$–1.4 fm and fm (*fermi*) is a non-SI unit of distance (1 fm = 10^{-15} m = 1 femtometre), common in nuclear physics. It can be seen that nucleon concentration within the nucleus is nearly the same for all nuclei.

Example. Estimate the average nucleon concentration in the nucleus and the average nuclear substance density.

Solution. The average nucleon concentration is equal to $n = A/(\frac{4}{3}\pi R^3) \sim 1.0$–$1.5 \times 10^{44}$ m^{-3}. This concentration corresponds to the density $\rho = Mn \sim 2 \times 10^{17}$ kg/m^3, where M is the nucleon mass.

Note that for many nuclei, one observes a considerable deviation of particle distribution from spherical symmetry (this deviation is especially pronounced in excited states, but also happens in the ground state). Thus, for the simplest compound nucleus, the deuteron, charge distribution in the ground state is nonspherical. (This fact is indicative of a noncentral character of nuclear forces.)

▶ **Nuclear binding energy.** Nuclear masses measured with a fair degree of accuracy happen to be smaller than the sum of the nucleon masses involved. The nonadditive character of mass in this situation is a direct consequence of the Einstein formula relating mass and energy at rest (see Subsection P1.10.3). Between nucleons, there exist attraction forces by which they are kept within the nucleus in spite of the strong (due to the small distances) Coulomb repulsion, and positive work is required in order to split the nucleus into nucleons. The amount of that work is called *nuclear binding energy*:

$$E_\text{b} = (Zm_p + Nm_n - m_\text{nuc})c^2, \tag{P8.1.1.2}$$

where $N = A - Z$ is the number of neutrons in the nucleus. The huge magnitude of nuclear interactions (compared with, say, atomic or chemical interactions that have a purely electromagnetic nature) made it possible to measure directly the said mass decrease.

Let us make a few remarks with regard to formula (P8.1.1.2).

1. In nuclear physics, nuclear mass and particle mass are expressed either in energy units or in atomic mass units. In the first case, one gives the value of the rest energy mc^2 expressed in electronvolts. Thus, the masses of electron, proton, and neutron in these units are respectively equal to $m_e = 0.511$ MeV, $m_p = 938.28$ MeV, $m_n = 939.55$ MeV. When energy units are used, many formulas become shorter. Thus, formula (P8.1.1.2) becomes

$$E_b = Zm_p + Nm_n - m_{\text{nuc}}. \tag{P8.1.1.3}$$

Note that $1\,\text{kg} = 5.61 \times 10^{29}$ MeV, $1\,\text{MeV} = 1.78 \times 10^{-30}$ kg.

By definition, the atomic mass unit (AMU) is $1/12$ of the mass of carbon nuclide (not nucleus) ^{12}C. Reference tables usually give nuclide passes in AMUs. For example, $m_e = 5.486 \times 10^{-4}$ AMU, $m_p = 1.007276$ AMU, $m_n = 1.00898$ AMU. Note that 1 AMU $= 931.50$ MeV $= 1.660 \times 10^{-27}$ kg.

2. Since reference tables usually list masses of nuclides rather than nuclear masses, it is convenient to rewrite formula (P8.1.1.3) as follows:

$$E_b = ZM_H + Nm_n - M_a, \tag{P8.1.1.4}$$

where M_H is the mass of the hydrogen atom and M_a is the mass of the atom (nuclide) corresponding to the said nucleus. The first term of this formula differs from that of (P8.1.1.3) by Zm_e; and the nuclide mass M_a differs from the nuclear mass m_n by the same amount, if one neglects the energy of electric interaction between electrons and the nucleus.

3. Instead of nuclide mass, tables often list the so-called *mass defect* $\Delta_a = M_a - A$ (where M_a is expressed in AMUs). Thus, the mass defect of the nuclide ^{12}C is equal to zero, the mass defect of the hydrogen atom is equal to 0.007825 AMU, the mass defect of the nuclide ^{3}He is equal to 0.016030 AMU. Binding energy can be expressed in terms of mass defects as follows:

$$E_b = Z\Delta_H + N\Delta_n - \Delta_a. \tag{P8.1.1.5}$$

▶ **Dependence of binding energy on nuclear composition.** It turns out that in the first rough approximation, binding energy grows in direct proportion to the number of nucleons in the nucleus, i.e., its mass number A. For qualitative and numerical analysis, it is convenient to use *specific binding energy* defined as $\varepsilon = E_b/A$. Figure P8.2 shows an approximate dependence of ε on the mass number A. Being approximately equal to 8 MeV/nucleon, specific energy first (for $A < 16$) undergoes a sharp increase, then continues to grow smoothly and for $A \sim 60$ (in the region of iron) attains its maximum, 8.8 MeV/nucleon, after which it smoothly decreases to 7.6 MeV/nucleon for the last natural nuclide ^{238}U. Note that due to the growth of ε for small A, fusion of light nuclei is energetically efficient, and due to the decrease of ε for large A, fission of heavy nuclei is energetically efficient. The energy released by fission of heavy nuclei is called *atomic*, and that released by fusion of light nuclei is called *thermonuclear*. Thermonuclear energy per nucleon is several times greater than atomic energy.

Nuclear forces and nuclear models are considered in the next section, but here we make a few brief remarks.

1. The fact that specific binding energy, for most nuclei, is approximately constant and equal to its average value indicates that nuclear forces have a *short-range* character and have the property of *saturation* (each nucleon interacts only with its several closest neighbors).

2. The decrease of ε for small A is explained by an increase of surface effects that depend on the relative fraction of surface particles.

3. The decrease of specific binding energy for large A is explained by an increased relative contribution of Coulomb energy.

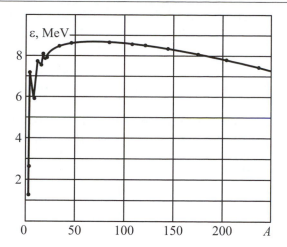

Figure P8.2. Specific binding energy versus mass number.

4. For light elements, the maximum binding energy for nuclides with fixed A (isobars) corresponds to $Z \sim A/2$. (For this reason, such nuclides are usually called β-stable.) This phenomenon is referred to as the *symmetry effect*. The shift to a larger percentage of neutrons observed with the growth of A is explained by an increased role of Coulomb repulsion of protons.

5. Specific binding energy of even-even nuclei is, on the average, greater than that of even-odd nuclei, and specific binding energy of even-odd nuclei is greater than that of odd-odd nuclei. This is explained by *pairing* of identical nucleons in nuclei. The union of two identical nucleons that form a zero-spin pair results in a gain approximately equal to 1 MeV.

6. Most stable (i.e., having the largest ε) among all even-even nuclei are the so-called *magic nuclei* whose number of protons or that of neutrons are equal to one of the *magic numbers*: 2, 8, 20, 28, 50, 82, 126. Especially stable are *double-magic nuclei*; i.e., those with the magic number of both protons and neutrons (e.g., ^{4}He, ^{16}O, ^{40}Ca, ^{208}Pb, etc.). The existence of magic nuclei indicates that nuclei have a shell structure similar to that of atoms.

P8.1.2. Nuclear Forces

▶ **Properties of nuclear forces.** Between the nucleons in a nucleus there exist nuclear attraction forces which are hundreds of times greater than electromagnetic repulsion forces between protons. These attraction forces ensure nuclear stability. Let us mention some (most important) features of nuclear forces.

1. Nuclear forces have a short-range character with the effective range $r_0 \sim 2.0$–2.5 fm. As the distance between nucleons becomes greater than the effective range, nuclear interaction rapidly decays. If that distance becomes several times smaller than the effective range, attraction is replaced by strong repulsion.

Example. Knowing the deuteron mass, one can calculate binding energy of the simplest nucleus, deuteron. This energy is equal to 2.22 MeV. On the basis of this binding energy, estimate the potential well depth corresponding to nuclear interaction between two nucleons.

Solution. First of all, let us replace the system of two interacting particles of mass M by a single particle with reduced mass $m = M/2$. For simplicity, we assume that potential energy is represented by a well of radius $R = 2.3$ fm and depth U_0. The quantity U_0 can be found from the condition that the ground-state energy in this potential well should be equal to $E_1 = -2.2$ MeV. For the spherically symmetric wave function of the ground state, the Schrödinger equation has the form

$$\frac{1}{r^2} \frac{d}{dr}\left(r^2 \frac{d\psi}{dr}\right) + \frac{2m}{\hbar^2}(E - U)\psi = 0,$$

where $U = -U_0$ for $r < R$ and $U = 0$ for $r > R$. Setting $\Psi(r) = r\psi(r)$, we come to the equation

$$\frac{d^2\Psi}{dr^2} + \frac{M}{\hbar^2}(E - U)\Psi = 0,$$

whose solution was obtained in Subsection P6.2.2. Omitting further calculations, we just state the answer: to get the correct ground-state energy, one should have the depth of the potential well about 29 MeV. (For a well of radius 2 fm, the value obtained for the depth is about 35 MeV.)

Note that this attraction value corresponds only to the state of a two-nucleon system in which their total spin is equal to 1. In the state with zero spin, attraction is nearly twice weaker and there is no bound state. This is why there is no stable formation consisting either of two neutrons or two protons (according to Pauli's principle, two identical particles cannot have parallel spins).

There is no final answer with regard to the dependence of nuclear forces on the distance between nucleons. With equal success, different models of that dependence are used for processing experimental data. The best theoretical justification has been given to the Yukawa formula

$$U(r) = A\frac{e^{-r/r_0}}{r}. \tag{P8.1.2.1}$$

2. The short-range nature of nuclear forces is closely connected with their *saturation property*. Every nucleon inside the nuclear substance is interacting with a certain number of its closest neighbors. (The nucleons near the surface of a nucleus have a smaller number of neighbors.) For this reason, nuclear binding energy grows in proportion to the number of nucleons, while the average nuclear density remains constant. If nuclear forces were long-range (as, for instance, the Coulomb force), then every nucleon would interact with all the others and the binding energy would grow faster. For example, electric energy of a ball with charge Ze and radius R is proportional to Z^2/R (see Subsection P3.6.1).

3. Nuclear forces have the property of *charge independence*: they do not depend on the type of interacting nucleons (p-p, p-n, or n-n).

4. Nuclear forces are non-central. This is confirmed, in particular, by the non-spherical charge distribution in the deuteron ground state. If the interaction was exactly central, the orbital angular momentum would be conserved and would be strictly equal to zero in the ground state, which corresponds to a spherically symmetric wave function.

5. Nuclear forces depend on spin orientations. (See above with regard to nucleon interaction in the deuteron.)

▶ **Nature of nuclear forces.** In modern physics, any fundamental interaction is regarded as a result of particle exchange (particles are agents of interaction). Nuclear interaction between nucleons is implemented through the exchange of *π-mesons (pions)*, which are particles with zero spin and mass m_π about 200 times greater than that of the electron.

Around a single nucleon, the so-called *virtual pions* constantly are born and die. The term "virtual" is used here, because these particles exist for such a short time that due to the time-energy uncertainty relation ($\Delta t \, \Delta E \sim \hbar$), it is possible for the energy of the system to change by $\sim m_\pi c^2$. For virtual particles, the common relation between energy and momentum does not hold. These particles have different impulses (of different directions) with the average value $m_\pi c$ and in their lifetime travel the average distance $c\Delta t \sim \hbar/(m_\pi c)$ equal to the Compton wavelength of a π-meson. Sometimes in this situation one speaks of a single nucleon surrounded by a coat of π-mesons whose radius, r_0, is equal to their Compton length. As two nucleons approach one another by the distance r_0, their meson coats begin to overlap and some virtual pions born of one nucleon are absorbed by the other. In 1935, Yukawa showed that upon averaging over the momenta of virtual pions, there arises attraction with potential energy described by (P8.1.2.1) with $r_0 = \hbar/(m_\pi c)$. At that time, π-mesons were still unknown. Yukawa predicted their existence and estimated their mass in terms of the nuclear interaction radius. (It was not until 1947 that π-mesons were discovered in cosmic rays.)

Let it be mentioned here that the carriers of electromagnetic interaction are virtual photons. Photons have zero mass, and therefore, electromagnetic interaction has long-range character ($r_0 = \infty$, and the Yukawa formula (P8.1.2.1) results in the Coulomb interaction). Weak interaction is implemented through the exchange of very massive W-bosons, and therefore, its radius is much smaller than that of nuclear interaction. Note also that the contemporary theory assumes that the basic carriers of strong interaction are *gluons* implementing the interaction between *quarks*, of which all strongly interacting particles are built (see Subsection P8.4.1).

P8.1.3. Nuclear Models

▶ **Nuclear models.** It should be observed from the start that there is still no universal model of the nucleus that could provide a unified theoretical approach to the calculation of various nuclear characteristics. Each of the existing models is capable of successfully explaining some properties of nuclei, while the other properties remain unaccounted for or their explanation is too complex or questionable. Among these models, a special place is occupied by the *microscopic nuclear model* proposed by A. B. Migdal, who called it the *theory of finite Fermi systems*. This model is based on the theory of quantum Fermi fluid which applies the methods of the quantum field theory to a statistical ensemble of interacting fermions. The major achievements of the microscopic model are that it provides a comparison and a unified justification of outwardly distinct theoretical models (sometimes, it is called a "model of nuclear models").

The other models can be provisionally divided into two groups: collective models and single-particle models. The first group covers, in particular, the theory of infinite nuclear matter and the liquid-drop nuclear model. The second group contains the theories dealing with the motion of separate nucleons either in a plane potential well of finite depth (nuclear Fermi gas) or in some effective potential (shell models). Some sophisticated shell models take into account pairing of identical nucleons.

There are some new models that combine certain features of the said two groups. The interior part of the nucleus corresponding to completely filled up shells is regarded as a drop of nuclear fluid in whose field external nucleons are moving.

▶ **Liquid-drop (hydrodynamic) nuclear model** treats the nucleus as a charged drop of nuclear liquid with surface tension. This model describes basic properties of the dependence of binding energy on A and Z and characterizes surface vibrations of nuclei and some properties of fission of heavy nuclei.

▶ **The nuclear Fermi gas model** is used for the explanation of the symmetry effect (Z is approximately equal to N in light nuclei) and its violation in heavy nuclei, and this model allows one to calculate the depth of the effective potential well.

In this model, both protons and neutrons are regarded as two subsystems of identical fermions moving freely in a potential well. On each level, two identical particles with different spins can exist. The more nucleons of a given type are present in the nucleus, the more their Fermi energy. If there is a considerable divergence between the Fermi energies of neutrons and protons, then one nucleon will transform into another with total energy decrease (either $n \rightarrow p + e^- + \widetilde{\nu}$ or $p \rightarrow n + e^+ + \nu$). Therefore, in a stable nucleus, the Fermi energies should be approximately the same. With the Coulomb interaction neglected, the total (common) energy minimum corresponds to approximately the same number of neutrons and protons. This situation corresponds to light nuclei. In heavy nuclei, the Coulomb interaction between protons is increasingly significant and leveling of the Fermi energies occurs when the number of neutrons exceeds that of protons.

Example. Using the mean concentration of nucleons in the nucleus calculated in the example of Subsection P8.1.1, estimate the depth of the potential well occupied by these nucleons.

Solution. To obtain such an estimate, we neglect the Coulomb interaction. The mean kinetic energy of nucleons of one type is equal to (see Subsection P7.1.2)

$$K_{\text{mean}} = \frac{3}{5} E_F = \frac{3}{5} \frac{\hbar^2 (3\pi^2 n_1)^{2/3}}{2M} \sim 20\,\text{MeV/nucleon},$$

where $n_1 \sim 0.6 \times 10^{44}\,\text{m}^{-3}$ is the concentration of nucleons of one type equal to one half of the nucleon concentration. The mean specific binding energy is about 8 MeV/nucleon. This energy can be calculated as the difference between the well depth and the mean kinetic energy. We find that the well depth should be about 30 MeV. As will be shown below, the estimate just obtained is well below the actual value.

▶ **Empirical formula for binding energy.** On the basis of the liquid-drop and the gas models, one can write a fairly simple empirical formula (Weizsäcker's formula) for nuclear binding energy (1935):

$$E_{\text{b}} = \varepsilon A = \alpha_1 A - \alpha_2 A^{2/3} - \alpha_3 \frac{Z^2}{A^{1/3}} - \alpha_4 \frac{(A - 2Z)^2}{A} - \alpha_5 \frac{\delta(Z, N)}{A^{3/4}}, \qquad \text{(P8.1.3.1)}$$

where $\delta(Z, N)$ is equal to -1 for even-even nuclei, 0 for even-odd nuclei, and 1 for odd-odd nuclei. The coefficients α_i are chosen from experimental data and at present the following values are adopted: $\alpha_1 = 15.75\,\text{MeV}$, $\alpha_2 = 17.8\,\text{MeV}$, $\alpha_3 = 0.71\,\text{MeV}$, $\alpha_4 = 23.7\,\text{MeV}$, and $\alpha_5 = 34\,\text{MeV}$.

The first three terms correspond to the concepts of the liquid-drop model. The first term reflects approximate proportionality of binding energy to the number of nucleons; the coefficient α_1 can be understood as binding energy associated with a single nucleon in an imaginary nuclear substance (matter) with electric interaction "turned off." The second term describes "surface tension" of a nuclear drop (the nucleons on the surface of the nucleus are in special conditions), and the third term describes its electrostatic energy.

The fourth term is due to the symmetry effect explained in the framework of the nuclear Fermi gas. This term attains its minimum for $N = Z$, since the divergence from total symmetry observed with an increase of A is connected with the Coulomb energy, and this has been accounted for in the third term. The fifth term reflects stepwise change of binding energy at the transition from a given nucleus to nuclei with adjacent values of A or Z.

It is of interest to note a considerable divergence of the coefficient α_1 from the mean value of specific binding energy $\varepsilon \sim 8\,\text{MeV}$. The point is that ε already contains the Coulomb and the surface energies. If ε is replaced by α_1 in example (when estimating the well depth), then the binding energy acquires the value $\sim 40\,\text{MeV}$.

On the basis of Weizsäcker's formula one can perform fairly precise ($\sim 10^{-4}$) calculations of binding energy for many nuclides, especially heavy ones, α-decay energy, fission and fusion energies, and many other characteristics. However, to explain the existence of magic nuclei, to calculate spins and magnetic moments, one cannot do without the shell model.

▶ **Shell model of the nucleus** (Goeppert-Mayer and Jensen, 1949). This model assumes that each nucleon is moving in a self-consistent field produced by other nucleons. The potential energy of this field is chosen so as to give the best agreement with experimental data for a wide range of nuclei.

In contrast with electrons in an atom, spin-orbit interaction plays a much greater role for nucleons and and can reach 10–15%. Furthermore, the Russel–Saunders coupling common for electrons (i.e., LS-coupling, in which the spins and orbital angular momenta of all particles are first summed up separately and then the total spin is added to the total orbital angular momentum) is suitable for the lightest nuclei only, whereas for most other nuclei the jj-coupling is more appropriate, in which the spin of each individual nucleon is added to its orbital angular momentum. The nucleon energy levels are characterized by the quantum numbers l (denoted by the letters $s, p, d, f, g, \dots$), j (equal to $l - \frac{1}{2}$ or $l + \frac{1}{2}$

and specified as the right subscript), and m_j, and also the number n that enumerates the levels with a given l and can take the values 1, 2, 3, ... To each level, there correspond $2j + 1$ states. The nucleon energy levels are grouped into shells; the distances between these groups are significantly greater than the interlevel distances within a shell. The magic numbers correspond to shells being completely filled up:

I: $1s_{1/2}$ (2 nucleons);
II: $2p_{3/2}, 2p_{1/2}$ (6 nucleons, altogether 8);
III: $1d_{5/2}, 2s_{1/2}, 1d_{3/2}$ (12 nucleons, altogether 20);
IV: $1f_{7/2}$ (8 nucleons, altogether 28);
V: $2p_{3/2}, 1f_{5/2}, 2p_{1/2}, 1g_{9/2}$ (22 nucleons, altogether 50);
VI: $1g_{7/2}, 2d_{5/2}, 2d_{3/2}, 3s_{1/2}, 1h_{11/2}$ (32 nucleons, altogether 82);
VII: $1h_{9/2}, 2f_{7/2}, 2f_{5/2}, 3p_{3/2}, 3p_{1/2}, 1i_{13/2}$ (44 nucleons, altogether 126).

Nucleon states are filled up in the order indicated in the above scheme. Using this, one can obtain correct values for the spins of almost all nuclei.

P8.2. Nuclear Transformations

P8.2.1. Radioactivity

Radioactivity is a spontaneous emission of one or several particles from a nucleus accompanied by its transformation into another nucleus (or its transition to another state). The original radioactive nucleus (nuclide) is called *parent nucleus*, and product nuclei are called *daughter nuclei*.

Basic types of radioactivity are the following: α-decay, β-decay, and γ-decay. In spite of different physical processes associated with these types of decay, their time evolution is described by one and the same law.

▶ **Radioactive decay law.** Since radioactive decay has a random nature, it can be assumed that on a small time-interval dt the fraction of split nuclei is proportional to dt. Denoting by $N(t)$ the number of intact nuclei by the time instant t, we come to the equation

$$\frac{dN}{N} = -\lambda \, dt. \tag{P8.2.1.1}$$

The coefficient λ is called the *decay constant*. The solution of equation (P8.2.1.1) has the form

$$N = N_0 e^{-\lambda t}, \tag{P8.2.1.2}$$

where N_0 is the number of nuclei for $t = 0$. This relation is called the *radioactive decay law*.

The number of nuclei decaying per unit time is called *radioactive sample activity*:

$$A = \left| \frac{dN}{dt} \right| = \lambda N = A_0 e^{-\lambda t}, \tag{P8.2.1.3}$$

where A_0 is the initial activity of the a sample. Activity per unit mass is called *specific activity*. Activity is measured in *becquerels* (1 Bq corresponds to one decay per second) or *curies* (1 Ci = 3.7×10^{10} Bq).

Apart from the constant λ, decay is characterized by the *half-life period* $T_{1/2}$ (often called *half-life*) equal to the time in which half of the nuclei undergo decay ($N = N_0/2$). From (P8.2.1.2), we find that

$$T_{1/2} = \frac{\ln 2}{\lambda} = \frac{0.693}{\lambda}. \tag{P8.2.1.4}$$

It is natural to introduce the *mean lifetime* of a nucleus, τ. The probability of the event that the nucleus decays on the time interval $(t, t+dt)$ is equal to $|dN|/N_0 = \lambda N dt/N_0 = \lambda e^{-\lambda t} dt$, and therefore,

$$\tau = \int_0^\infty t \lambda e^{-\lambda t} dt = \frac{1}{\lambda} = \frac{T_{1/2}}{\ln 2} = 1.44 T_{1/2}. \tag{P8.2.1.5}$$

Example. Calculating sample activity. The half-life of the isotope $_{38}^{90}\text{Sr}$ is about 28 years. What is the activity of its $m = 1$ g sample?

Solution. The activity is equal to the product of the decay constant $\lambda = \ln 2/T_{1/2}$ and the number of nuclei $N = N_A(m/\mu)$, where $\mu = 90$ g/mol. We obtain $A = 5.2 \times 10^{12}$ c^{-1} = 140 curie.

▶ **Alpha decay.** The alpha decay (or α-decay) of a nucleus X with atomic number Z and mass number A results in the nucleus ^4He (known as the *alpha particle* or α-*particle*) and a daughter nucleus Y with atomic number $Z - 2$ and mass number $A - 4$:

$$_Z^A\text{X} \rightarrow {}_{Z-2}^{A-4}\text{Y} + {}_2^4\text{He}. \tag{P8.2.1.6}$$

Alpha decay is observed only for heavy nuclei (mostly with $A > 209$).

The energy released during the α-decay is equal to

$$Q = m_\text{X} - m_\alpha - m_\text{Y}, \tag{P8.2.1.7}$$

where the nuclear masses are expressed in energy units. Note that instead of nuclear masses one can substitute masses of nuclides (electron masses are mutually canceled) or their mass defects.

The fact that one observes decay with the emission of the nucleus ^4He, and not other light nuclei or single nucleons, is explained by the abnormally large binding energy of the nucleus ^4He. For example, the α-decay of the nucleus ^{232}U is accompanied by the emission of 5.4 MeV energy, and for the emission of a proton by this nucleus it should be supplied with 6.1 MeV energy.

Due to the momentum conservation law, the nucleus experiences recoil and carries away part of the energy. For this reason, the registered energy of α-particles is less than the reaction energy Q. Assuming that the parent nucleus were at rest, let us write the momentum conservation law (for) $p_\alpha = p_\text{Y}$ in terms of kinetic energies of the daughter nucleus and the α-particle:

$$m_\alpha K_\alpha = m_\text{Y} K_\text{Y}$$

Expressing K_Y from the above formula and inserting it into the energy conservation law

$$K_\alpha + K_\text{Y} = Q,$$

we obtain

$$K_\alpha = \frac{m_\text{Y}}{m_\text{Y} + m_\alpha} Q = \frac{A - 4}{A} Q,$$

where A is the mass number of the parent nucleus. As a rule, we have $A > 210$, and therefore, the kinetic energy of the α-particle is greater than 98% of the decay energy.

Since the average lifetime of most α-active nuclei is great, they are subject to decay from the ground state. Moreover, due to a very "sharp" (strong) dependence of the decay constant on the energy of the outgoing particles, one observes the most intensive transition to the ground state of the parent nucleus. Therefore, in most cases, all α-particles of a given decay have the same energy. In some cases, one observes α-particles that correspond to low (rotational) levels, and this leads to a slight blurring of the α-spectrum. Moreover, for some extremely short-lived nuclei, decay is possible from excited states of the parent nucleus and one observes α-particles with energies considerably larger than the principal value (for example, polonium isotopes $_{84}^{212}\text{Po}$ and $_{84}^{214}\text{Po}$ emit not only "common" α-particles but those with energies up to 10.5 MeV).

▶ **Theory of alpha decay.** For a long time, some properties of α-decay looked quite mysterious. In the first place, it is an unusually strong dependence of the half-life on the energy of α-particles. In the principal region of α-activity ($A \geq 210$), α-particles with energies significantly smaller than 4 MeV are never observed (their half-lives are too great), for $^{204}_{82}\text{Pb}$, $K_\alpha \approx 4$ MeV and $T_{1/2} = 1.4 \times 10^{17}$ years, and for $^{212}_{84}\text{Po}$, $K_\alpha \approx 9$ MeV and $T_{1/2} = 3 \times 10^{-7}$ s. Moreover, one could not understand how it was possible for an α-particle to leave the nucleus at all. Outside the nucleus, its Coulomb potential energy

$$U_{\text{col}} = \frac{2(Z-2)e^2}{4\pi\varepsilon_0 r}$$

attains its maximum on the boundary of the nucleus for $r = R = r_0 A^{1/3}$ (see (P8.1.1.1)), and inside the nucleus it enters a negative potential well of nuclear forces. Hence, the α-particle is locked by a potential barrier (Fig. P8.3) whose height $U_0 \sim 25$–30 MeV is considerably greater than the energy of α-particles 4–9 MeV. The alpha particle can penetrate this barrier only through quantum tunneling (see Subsection P6.2.2). The theory of α-decay as a tunneling transition was developed in 1928 by Gamow, Condon, and Gurney and explained all its major properties.

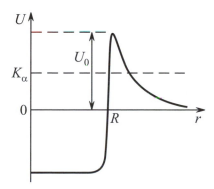

Figure P8.3. Potential energy of an α-particle in the nucleus has the shape of a potential barrier.

In order to calculate the decay constant λ, the probability p of the production of an α-particle inside the nucleus (it is assumed that $p \sim 1$) should be multiplied by the number of its collisions with the barrier $\nu \sim v/2R$ (v is speed of the α-particle) and the transmission coefficient (of passing through the barrier)

$$D \sim \exp\left\{ -\frac{2}{\hbar} \int_R^{r_1} \sqrt{2m_\alpha \left(U(r) - K_\alpha \right)}\, dr \right\}.$$

This transmission coefficient is responsible for the sharp (exponential) dependence of the decay constant (and the half-life) on the energy of the outgoing particle K_α.

▶ **Beta decay.** Beta decay (or β-decay) is the term applied to radioactive nuclear transformations in which a proton turns into a neutron (or vice versa) and involving electrons (or positrons) and neutrinos (or antineutrinos). Basic types of β-decay are the following:

a) Electron β^--decay (beta-minus):

$$^A_Z\text{X} \rightarrow \,^A_{Z+1}\text{Y} + e^- + \widetilde{\nu} \tag{P8.2.1.8}$$

(for example, decay of tritium $^3_1\text{H} \rightarrow \,^3_2\text{He} + e^- + \widetilde{\nu}$).

b) Positron β^+-decay (beta-plus):

$$\,^A_Z X \rightarrow \,^A_{Z-1}Y + e^+ + \nu \tag{P8.2.1.9}$$

(for example, decay of the artificial radioactive isotope of nitrogen $\,^{13}_7 N \rightarrow \,^{13}_6 C + e^+ + \nu$).

c) Electron capture:

$$e^- + \,^A_Z X \rightarrow \,^A_{Z-1}Y + \nu \tag{P8.2.1.10}$$

(for example, the reaction $e^- + \,^7_4 B \rightarrow \,^7_3 Li + \nu$). In electron capture, the nucleus absorbs an electron from an atomic shell, usually from the K-shell, and therefore, this process is called K-capture (one also observes L-capture). Electron capture is accompanied by the characteristic X-ray radiation.

An important distinction of β-decays from other types of radioactivity is that the former are intranucleon (pertaining a nucleon) processes rather than intranuclear (pertaining to the entire nucleus) ones. This means that an elementary act of such a process consists in the transformation of one nucleon into another. The schemes of the transformations listed above are the following:

a) $n \rightarrow p + e^- + \tilde{\nu}$;
b) $p \rightarrow n + e^+ + \nu$;
c) $e^- + p \rightarrow n + \nu$.

Such transformations (and therefore, nuclear β-decay) are possible only because of weak interaction. In free space, only the first process (neutron decay) can happen spontaneously. However, within the nucleus the new nucleon turns out surrounded by other nucleons, in the process of its birth the entire nucleus changes its structure, and the energy condition required for the decay is determined by masses of the nuclei (the original and the produced) rather than the nucleon masses.

Spontaneous decay is energetically allowed, if the sum of the masses of the original particles is greater than the sum of the masses of the produced particles. For β-decay, this condition has the form:

$m_X > m_Y + m_e$ (β^- and β^+-decay);
$m_X + m_e > m_Y$ (electron capture).

It is convenient to rewrite these formulas in terms of nuclide masses. Thus, for β^--decay, one should take $m_X = M_X - Z m_e$, $m_Y = M_Y - (Z+1)m_e$ and the decay condition becomes

$M_X > M_Y$ (β^--decay).

For the other types of decay, we have:

$M_X > M_Y + 2m_e$ (β^+-decay);
$M_X > M_Y$ (electron capture).

Note that the last two types of decay result in identical daughter nuclei, i.e., are competitive; however, electronic capture is more energetically advantageous.

Weizsäcker's formula (P8.1.3.1) can be utilized to find the parameters (Z, A) for the nuclei most stable with respect to the above types of β-decay. Since the mass number A is unchanged in these types of decay, one should compare the masses (binding energies) of all isobars. Consider, for instance, an odd A (even-odd nuclei, $\Delta(A, Z) = 0$). Minimizing the nuclear energy $E = Z m_p + (A - Z)m_n - E_b$ with respect to Z, we find Z_β corresponding to the maximum β-stability (Fig. P8.4a). The isobars with $Z < Z_\beta$ should be examined with regard to β^--decay, and those with $Z > Z_\beta$ should be examined with regard to β^+-decay and electron capture. For even A, we obtain two branches of $E(Z)$ (Fig. P8.4b): the upper one

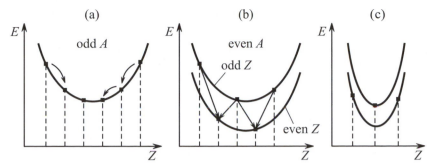

Figure P8.4. Isobaric nuclei energy versus Z and energetically allowed types of β-decay for nuclei with even A (a) and nuclei with odd A (b); the right figure (c) corresponds to a β-stable nucleus with an even A.

for odd-odd nuclei and the lower one for even-even nuclei. The nuclei on the upper branch are usually unstable with respect to transitions to the lower branch, but there are 4 β-stable nuclei corresponding to the situation represented in Fig. P8.4c.

The number of β-active nuclides is much larger than that of α-active nuclides (each element has a β-active isotope). The half-life depends on energy (the greater the energy gain, the smaller $T_{1/2}$), but this dependence is not as strong as in the case of α-decay. Moreover, the greater the divergence between the structures of the final and the original nuclei (for instance, the difference of their spins), the slower the decay. Sometimes, the transition to the ground state of the daughter nucleus has small probability (is a rare event), and as a result of β-decay a nucleus in an excited state is produced.

The neutrino (more precisely, electron-type neutrino, which is one of its three modifications) is a very light (possibly, massless) neutral particle with spin 1/2, which participates only in weak interactions. For this reason, its interaction with matter is extremely weak and its detection is very difficult. Thus, for a neutrino with energy 100 MeV the mean free path in water is of the order 10^{19} m (about 100 light years). A direct experimental proof of the existence of neutrino was not obtained until 1956, when it became possible to utilize a powerful flux of neutrino resulting from uranium fission in nuclear reactor. (Reactions of antineutrino capture $\tilde{\nu} + p \rightarrow n + e^{+}$ were registered.) The existence of neutrino was hypothesized by W. Pauli in 1930, in a rather dramatic situation due to its "invisibility."

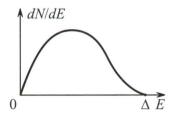

Figure P8.5. Distribution of β-electrons with respect to energies.

It was considered a proven fact that β^{-}-decay results only in two particles: the daughter nucleus and the electron. In that case, as in the case of any two-particle decay (say, α-decay) of a fixed nucleus, the decay products should have certain energies. With the kinetic energy of the daughter nucleus neglected, the electron energy should be equal to $\Delta = m_{X} - m_{Y} - m_{e} = M_{X} - M_{Y}$. However, in 1914, J. Chadwick established the continuity of β-spectra. It turned out that the distribution of electrons with respect to energies is like that in Fig. P8.5, and the electron energy averaged with respect to that distribution is as small as $\Delta/3$. The situation looked so paradoxical that in 1924 N. Bohr guessed that the energy conservation law could be violated in micro-processes! As noticed in 1929, the angular momentum conservation law is violated for some types of β-decay. Thus, for $_{83}^{210}\text{Bi} \rightarrow _{84}^{210}\text{Po} + e^{-}$, the spins of the nuclei are equal to 1 and 0 and the electron carries away a half-integer spin. This deadlock was broken by Pauli's hypothesis about the existence of neutrino. On the basis of this hypothesis, in 1934, E. Fermi developed the theory of weak interactions and calculated the activities of β-decays and the spectra of outgoing electrons. In 1936, N. Bohr accepted the Fermi theory and rejected the idea of non-conservation of energy.

▶ **Radioactive families.** Most natural radioactive elements belong to four radioactive families. All elements of these families are obtained by consecutive decay from a single parent nuclide and end with some stable element. The decay family of thorium has mass numbers $4n$, starts with $^{232}_{90}\text{Th}$ and ends with $^{208}_{82}\text{Pb}$. The decay family of neptunium ($A = 4n+1$) starts with $^{237}_{93}\text{Ne}$ and ends with $^{209}_{83}\text{Bi}$. The decay family of uranium ($A = 4n+2$) starts with $^{238}_{92}\text{U}$ and ends with $^{206}_{82}\text{Pb}$. Finally, the actinium family ($A = 4n + 3$) starts with $^{235}_{92}\text{U}$ and ends with $^{207}_{82}\text{Pb}$. (The first three terms of the last family were discovered later.) As an example, let us give the sequence of elements of the thorium family:

$$^{232}_{90}\text{Th} \xrightarrow{\alpha} {}^{228}_{88}\text{Ra} \xrightarrow{\beta} {}^{228}_{89}\text{Ac} \xrightarrow{\beta} {}^{228}_{90}\text{Th} \xrightarrow{\alpha} {}^{224}_{88}\text{Ra} \xrightarrow{\alpha} {}^{220}_{86}\text{Rn} \xrightarrow{\alpha} {}^{216}_{84}\text{Po}$$

$$\xrightarrow{\alpha} {}^{212}_{82}\text{Pb} \xrightarrow{\beta} {}^{212}_{83}\text{Bi} \xrightarrow{\alpha} {}^{208}_{81}\text{Tl} \xrightarrow{\beta} {}^{208}_{82}\text{Pb} \quad \left({}^{212}_{83}\text{Bi} \xrightarrow{\beta} {}^{212}_{84}\text{Po} \xrightarrow{\alpha} {}^{208}_{82}\text{Pb} \right).$$

The element $^{212}_{83}\text{Bi}$ can decay by way of α- and β-radiation, and therefore, on this element the decay sequence branches.

▶ **Gamma decay.** Gamma radiation is observed in transitions of an excited nucleus to lower energy levels, with unchanged A and Z and emitted γ-quanta. After emitting one or several photons, the nucleus comes to the ground state. Since nuclear energy values are discrete, the spectrum of γ-radiation is also discrete. Excited nuclei are produced in β-decay, if the decay to the ground state is forbidden. For example, β^--decay of $^{24}_{11}\text{Na}$ with spin $I = 4$ leads neither to the ground state $^{24}_{12}\text{Mg}$ with spin $I = 0$, nor to the first excited level ($I = 2$), but leads to the second level with $I = 4$, upon which the nucleus consecutively emits two γ-quanta with energies 2.7 MeV and 1.4 MeV and comes to the ground state.

A free nucleon cannot emit γ-quanta, since this would contradict the energy and the momentum conservation laws. However, inside the nucleus, the momentum is redistributed between all nucleons and radiation becomes possible. Thus, in contrast to β-decay, the emission of γ-quanta is an intranuclear rather than intranucleon process.

The Coulomb barrier does not obstruct γ-radiation, and therefore, the lifetime of γ-active nuclei is usually small, (10^{-7}–10^{-15} s). However, in the cases when the transition to the ground state is strongly hindered, one observes long-lived excited states called *isomers*. The lifetime of isomers may reach several hours and even more. Some properties of isomers (in particular, probabilities of β-decay) may be observably different from those of the ground states.

A competitor of γ-decay is a process called *internal conversion*, in which an excited nucleus transmits the energy surplus to one of the electrons of the atomic shell. The spectrum of the outgoing electrons, in contrast to that of β-electrons, is discrete. Moreover, internal conversion is accompanied by the characteristic X-ray radiation.

▶ **Mössbauer effect.** If the energy of an incident photon coincides with the energy of the transition of a system from the ground state to an excited state, then the probability of such a photon being absorbed drastically increases. This phenomenon is called *resonant absorption*. The dependence of the absorption probability on energy has the shape of a sharp peak whose width is equal to that of the excited level ΔE ($\Delta E \sim \hbar/\Delta t$, where Δt is the lifetime of the excited state). The natural scheme of resonant absorption utilizes identical atoms as the emitter and the absorber. Such a scheme is used in optics, where the emission and the absorption are determined by electron transitions. However, observation of resonant absorption of nuclear γ-quanta encounters serious obstacles. The point is that when emitting a γ-quantum the atom experiences a much greater recoil and carries away more energy than in the case of emitting an optical photon. This leads to a more significant *decrease* of the frequency of the emitted photon and its absorption becomes a rare event (has small probability) (moreover, during absorption part of the photon energy is lost on the kinetic energy of the atomic recoil and the energy of the absorbed photon must be not *smaller* but *greater* than the energy of the excited level).

The momentum of nuclear recoil is equal to that of the photon: $p = \hbar\omega/c$, and the nuclear energy is equal to

$$E_{\text{nuc}} = \frac{p^2}{2m_{\text{nuc}}} = \frac{(\hbar\omega)^2}{2m_{\text{nuc}}c^2}. \tag{P8.2.1.11}$$

For resonance absorption, the photon energy shift (coinciding with nuclear energy E_{nuc}) should not exceed the level width ΔE. However, we have $\hbar\omega > 10^4$ eV, and $m_{\text{nuc}}c^2 < 2 \times 10^{11}$ eV; therefore, $E_{\text{nuc}} > 10^{-3}$ eV. At the same time, the common value of the excited level width is $\Delta E \sim 10^{-8}$–10^{-7} eV. We see that nuclear recoil causes a decrease of the γ-quantum energy by a quantity that exceeds the width of excited nuclear levels by many orders.

In 1958, Mössbauer discovered that at low temperatures resonant absorption is observed on the isotope $^{191}_{77}\text{Ir}$. The essence of the phenomenon known as the Mössbauer effect is the following. During both emission and absorption of γ-quanta by atoms forming a crystal lattice, the recoil momentum can be imparted to the entire crystal as a whole (more precisely, to a group of $N \sim 10^8$ atoms). This means that the recoil energy becomes negligibly small (m_{nuc} in (P8.2.1.11) should be replaced by the crystal mass) and the conditions of resonance absorption hold. There appears the possibility to fix changes of energy of γ-quanta of the level width order, i.e., $\Delta E \sim 10^{-8}$ eV. The important quantity here is not ΔE itself (for electron levels it is of the same order) but its *ratio* to the energy of photons (this ratio characterizes relative measurement errors). Since the energy of nuclear γ-quanta is 10^4–10^8 eV, the precision of measurements based on the Mössbauer effect reaches 10^{-12}–10^{-16}.

One of the first applications of the Mössbauer effect was the measurement of the gravitational red shift in the Earth's field of gravity (Pound and Rebka, 1960). The essence of the phenomenon is that as a photon moves in the gravitational field, its energy should change. The relative variation of the frequency and the energy should be equal to

$$\frac{\Delta E}{E} = \frac{(E/c^2)\Delta\varphi}{E} = \frac{\Delta\varphi}{c^2},$$

where $\Delta\varphi$ is the gravitational potential. If the Mössbauer emitter and receiver are placed on the same vertical line with the distance h between them, then $\Delta\varphi = gh$. For $h = 30$ m, the relative frequency variation $\Delta\omega/\omega = \Delta E/E = 3 \times 10^{-15}$ is within the precision range of the method.

The change of the frequency shift is measured on the basis of the Doppler effect. The emitter and the receiver being fixed, resonance absorption is absent because of the gravitational shift. Moving the receiver in the vertical direction, one finds the speed v at which there again appears resonant absorption. Then $\Delta\omega/\omega = v/c$.

▶ **Radiation dose measurement.** The *absorbed dose* of ionizing radiation is defined as the energy absorbed by a medium per unit mass and expressed in *greys* (Gy) (1 Gy = 1 J/kg).

The ionizing effect of X-rays and gamma rays on dry air is characterized by the *exposure dose*, defined as the net charge of all ions of like sign produced in one kilogram of irradiated air and measured in *roentgens* (R); 1 R = 2.58×10^{-4} C/kg. Alternatively, 1 R is defined as the amount of radiation required to generate 2.08×10^9 ion pairs in 1 cm^3 of dry air at standard temperature and pressure.

To evaluate the *biological effect* of radiation on living organisms, one uses the so-called *radiation quality factor* Q (also known as the *radiation weighting factor*), which indicates how many times a particular type of radiation is more dangerous than X-rays or gamma rays for the same dose absorbed. For beta radiation, $Q = 1$; for protons with energies less than 10 MeV, $Q = 10$; for thermal neutrons with energies less than 20 keV, $Q = 3$; for neutrons with energies 0.5–10 MeV, $Q = 10$; for alpha radiation with energy less than 10 MeV, $Q = 20$; for heavy nuclei, $Q = 20$. The *equivalent radiation dose* is the product of the absorbed dose and the quality factor. The equivalent dose is measured in *sieverts* (1 Sv is equal to 1 Gy for $Q = 1$). The unit *rem* is also used (roentgen equivalent in man or mammal, a biological equivalent of roentgen): 1 rem = 0.01 Sv.

P8.2.2. Nuclear Reactions

Nuclear reaction is a process of the type

$$a + {}^A_Z X \rightarrow {}^{A'}_{Z'} Y + b, \tag{P8.2.2.1}$$

where a is the original (incident) particle, ${}^A_Z X$ is the original nucleus (target nucleus), ${}^{A'}_{Z'} Y$ is the final nucleus, b is the final (outgoing) particle. One often uses the following symbolic notation:

$$^A X(a, b)^{A'} Y, \tag{P8.2.2.2}$$

and the type of the process is briefly denoted by (a, b). For $b = a$ (and $X = Y$), the corresponding process is called *scattering*. If, in addition, the final nucleus Y is in the same state as the original X, then one speaks of *elastic scattering*.

For the reaction (P8.2.2.1) to be possible, it is necessary that conservation laws hold for the electric charge and some other charges (lepton, baryon, etc.), which is discussed in Subsection P8.4.2. Moreover, the states of the resulting particles and the original ones are related by the conservation laws for energy, linear momentum, and angular momentum.

Nuclear reactions are accompanied by changes in the nuclear structure, more precisely, these reactions are possible only if the distances between the reacting particles become smaller than the nuclear interaction radius $r_0 \sim 10^{-15}$ m.

▶ **Yield and cross-section of nuclear reaction.** Consider a flux of incident particles (a) of given energy hitting a fixed nuclear target X. We start with a thin target of width Δx and the number of nuclei per unit area being $N_0 = n_0 \Delta x$ (n_0 is volume concentration of target nuclei). By definition, the *yield of nuclear reaction* is the fraction of incident particles (a) taking part in a given reaction (that is, the probability of the reaction):

$$w = \frac{\Delta N}{N}. \tag{P8.2.2.3}$$

Clearly, the reaction yield w is proportional to N_0. The proportionality coefficient has dimensionality of area and is called the *cross-section the nuclear reaction*:

$$w = \sigma N_0 \qquad \sigma = \frac{\Delta N}{N N_0}. \tag{P8.2.2.4}$$

Let us make a few remarks about the definition of the cross-section.

1. In Subsection P2.9.1, when discussing transfer phenomena, we introduced the effective cross-section of molecular scattering, whose meaning is similar to that of the nuclear reaction cross-section. Besides, the notion of the differential cross-section of elastic scattering was required in our analysis of Rutherford's experiments (see Subsection P6.1.1).

2. In (P8.2.2.4), the quantity N can be replaced by the particle flux (number of particles per unit time) or by the flux density J (flux per unit area), and the structure of the formula remains the same, since it involves the *relative number of particles*.

3. In the case of a thin target, it can be assumed that each incident particle interacts with only one target nucleus, and therefore, one can consider interaction of incident particles with a single nucleus. Since the number of particles hitting the target is equal to $N = JSt$ (J is flux density, S is target area) and $N_0 S$ is the total number of target nuclei, we find that the cross-section can be defined as the ratio of the number of particles reacting with a single nucleus per unit time to the particle flux density.

4. The yield and the cross-section are introduced for any selected interaction result. For a given input channel $(a + X)$ there are usually several output channels $(Y + b)$. Thus, the result of proton bombardment of nuclei ${}_3^7\text{Li}$ can be ${}_3^7\text{Li} + p$ (elastic and inelastic scattering), $\alpha + \alpha$, ${}_4^8\text{Be} + \gamma$, etc. Some channels are closed for low energies. It should be kept in mind that one of the general statements of quantum mechanics is that any process which is not forbidden by any of the conservation laws is always realized (even if with small probability). In particular, among possible channels, there is always elastic scattering.

5. In order to describe the dependence of the final result of scattering on continuous parameters (such as the exit angle or energy of a particle), one introduces various types of *differential cross-section*. These are defined as the ratio of the number of reactions resulting in the said parameter (say, the angle θ) belonging to a certain narrow interval (say, between θ and $\theta + d\theta$) to the number of incident particles N multiplied by the number of target particles per unit area N_0 and the interval width:

$$\frac{d\sigma}{d\theta} = \frac{dN}{N N_0 \, d\theta}$$

or

$$\frac{d\sigma}{d\Omega} = \frac{dN}{N N_0 \, d\Omega},$$

where Ω is solid angle, with $d\Omega = \sin\theta \, d\theta \, d\varphi$. Integrating the differential cross-section over all possible values of the parameter (parameters), one obtains the total cross-section of the reaction.

6. To have a more clear idea of the cross-section, one can imagine that the particles are classical and their interaction has such a short range that they can be replaced by balls of radii R_1 and R_2, and the reaction occurs as soon as the particles come into contact. Then, $\sigma = \pi(R_1 + R_2)^2$ (such a model is used in the theory of transfer phenomena). This model has some bearing to reality in the cases of high energy of incident particles, if the de Broglie wavelength λ is small compared with the radius R of the nucleus (and there is no Coulomb interaction). In this situation,

$$\sigma \sim \pi R^2 \sim 10^{-30}\text{--}10^{-28} \text{ m}^2. \tag{P8.2.2.5}$$

For slow particles $(\lambda \gg R)$, the cross-section drastically increases, since it is determined not by the nuclear radius but by the de Broglie wavelength:

$$\sigma \sim \pi\lambda^2.$$

However, this estimate pertains only to neutrons, since a slow charged particle cannot overcome Coulomb repulsion and approach the nucleus.

The estimate (P8.2.2.5) shows why nuclear cross-sections are expressed in *barns* (b) ($1 \text{ b} = 10^{-28} \text{ m}^2$).

7. If the target is not thin, then the beam intensity decreases inside the target and the cross-section of the reaction is expressed through its yield by a more complex formula than (P8.2.2.4). This can be done, if one knows the total cross-section of the processes leading to beam attenuation. The simplest situation is that of the reaction under consideration being the only possible one. Then, for each layer one can write

$$\frac{dN}{N} = -\sigma n\, dx, \tag{P8.2.2.6}$$

where N is the number of beam particles that have reached the layer with the coordinate x, $n\,dx$ is the number of target atoms in a layer of thickness dx (per unit area), n is volume concentration of atoms, the symbol $(-)$ indicates that reacted particles abandon the beam. The solution of equation (P8.2.2.6) has the form

$$N = N_0 e^{-\sigma n x}, \tag{P8.2.2.7}$$

where N_0 is the original number of beam particles. The quantity $\alpha = \sigma n$ is called the *absorption coefficient*. It is easy to check that the mean free path is equal to

$$l = \frac{1}{\sigma n} \tag{P8.2.2.8}$$

(see Subsection P2.9.1).

Example. It is required to calculate the free path of neutrino in iron if the neutrino interaction cross-section is $\sigma_\nu \approx 10^{-19}\,\mathrm{b} = 10^{-47}\,\mathrm{m}^2$.

Solution. The iron atomic concentration is equal to $n = \rho N_A/\mu$, where N_A is Avogadro's number, $\rho = 7.8 \times 10^3\,\mathrm{kg/m^3}$ is iron density, and $\mu = 58 \times 10^{-3}\,\mathrm{kg/mol}$ is molar (atomic) mass of iron. We obtain $l = 1/(n\sigma) = 1.2 \times 10^{18}\,\mathrm{m}$, which is about 130 light years.

▶ **Reaction energy.** The energy conservation law for an arbitrary nuclear reaction can be written in the form

$$\sum_i m_i + \sum_i K_i = \sum_i m'_i + \sum_i K'_i \tag{P8.2.2.9}$$

(the masses are expressed in energy units). The *reaction energy* Q is defined as the kinetic energy increment (or mass decrease):

$$Q = \sum_i K'_i - \sum_i K_i = \sum_i m_i - \sum_i m'_i = \sum_i \Delta_i - \sum_i \Delta'_i, \tag{P8.2.2.10}$$

where Δ is the mass defect (see Subsection P8.1.1). Note also that reaction energy can be expressed through nuclear binding energy ($Q = \sum_i E'_{bi} - \sum_i E_{bi}$), but only if the number of protons is preserved (if charges are carried only by nuclear protons). Reactions with energy release ($Q > 0$) are called *exothermic*, and those with $Q < 0$ are called *endothermic*. Sometimes, reaction energy is indicated in its notation, for example, $n + {}^{10}\mathrm{B} \to {}^{7}\mathrm{Li} + {}^{4}\mathrm{He} + 2.79\,\mathrm{MeV}$ or ${}^{4}_{2}\mathrm{He} + {}^{14}_{7}\mathrm{N} \to {}^{17}_{8}\mathrm{O} + {}^{1}_{1}\mathrm{H} - 1.2\,\mathrm{MeV}$.

Since the last expression in (P8.2.2.10) does not depend on the reference frame, the reaction energy is equal to the kinetic energy increment in any reference frame, in particular, that of the center of mass

$$Q = \sum_i \widetilde{K}'_i - \sum_i \widetilde{K}_i. \tag{P8.2.2.11}$$

If the reaction involves an incident (non-relativistic) particle of mass m_a and a *stationary* target particle of mass M_X, then the kinetic energy of the system in its center-of-mass frame can be linked to the kinetic energy of the incident particle K:

$$\widetilde{K} = \frac{M_X}{m_a + M_X} K. \tag{P8.2.2.12}$$

The kinetic energy in the center-of-mass frame is expressed via the reduced mass $\mu = m_a M_X / (m_a + M_X)$ and the relative particle velocity $\mathbf{v}_{rel} = \mathbf{v}_1 - \mathbf{v}_2$:

$$\widetilde{K} = \frac{\mu \mathbf{v}_{rel}^2}{2} = \frac{M_X}{m_a + M_X} \frac{m_a \mathbf{v}_a^2}{2}$$

(see Subsection P1.6.4). It has been taken into account that the relative velocity does not depend on the reference frame and the target particle is at rest in the laboratory reference frame.

▶ **Threshold of nuclear reaction.** Endothermic reaction can proceed only if the kinetic energy of the incident particle is greater than some minimum value called *threshold of reaction*. The threshold of reaction K_{thr} does not coincide with $|Q|$, since due to the momentum conservation law the final kinetic energy in the laboratory reference frame differs from zero and $K_{thr} = |Q| + K'$ (see (P8.2.2.10)). However, in the center-of-mass frame, on the reaction threshold the particles in the final state are at rest ($\widetilde{K}' = 0$) and the initial kinetic energy is equal to $|Q|$ (see (P8.2.2.11)). From (P8.2.2.12), we get

$$K_{thr} = \frac{m_a + M_X}{M_X} \widetilde{K} = \frac{m_a + M_X}{M_X} |Q|. \tag{P8.2.2.13}$$

In the situation when the incident particle and decay products cannot be assumed non-relativistic, the threshold of endothermic reaction $m_a + M_X \to m_1 + m_2 + \cdots$ can be expressed through its energy $|Q| = (m_1 + m_2 + \cdots) c^2 - (m_a + M_X) c^2$ as follows. The quantity $E^2 - p^2 c^2$ (here, E is the energy of the system and p is its momentum) is preserved during the reaction and is also unchanged, if one passes to another reference frame. Let us equate the initial value of this quantity in the laboratory reference system to its final value calculated in the center-of-mass frame (where $p = 0$):

$$[K_{thr} + (m_a + M_X) c^2]^2 - p^2 c^2 = (m_1 + m_2 + \cdots)^2 c^4.$$

Taking into account that the momentum of the target M_X is zero, for particle m_a we can write

$$p^2 c^2 = (K_{thr} + m_a c^2)^2 - m_a^2 c^4 = K_{thr}(K_{thr} + 2 m_a c^2).$$

After suitable rearrangements, we get

$$(m_a + M_X)^2 c^2 + 2 M_X K_b = (m_1 + m_2 + \cdots)^2 c^2,$$

and therefore,

$$K_{thr} = \frac{(m_1 + m_2 + \cdots)^2 - (m_a + M_X)^2}{2 M_X} c^2 = |Q| \left(\frac{m_a + M_X}{M_X} + \frac{|Q|}{2 M c^2} \right). \tag{P8.2.2.14}$$

This formula can be applied also in the case of the incident particle being the γ-quantum, in which case one should take $m_a = 0$. The photon threshold energy is equal to

$$\varepsilon_{thr} = \frac{(m_1 + m_2 + \cdots)^2 - M_X^2}{2 M_X} c^2 = |Q| \left(1 + \frac{|Q|}{2 M_X c^2} \right). \tag{P8.2.2.15}$$

Example. Calculate the threshold energy of γ-quantum for the production of an electron-positron pair on a proton and an electron at rest.

Solution. Note that the production of an electron-positron pair by a photon is impossible in empty space (see above), but it is possible in the field of a third particle (e.g., a proton). In this example, we are dealing with the reaction $\gamma + p \to e^- + e^+ + p$. Setting $m_a = 0$, $M_X = m_p$, and $m_1 + m_2 + \cdots = 2 m_e + m_p$ in (P8.2.2.14), we obtain

$$\varepsilon_{thr} = \frac{(2 m_e + m_p)^2 - m_p^2}{2 m_p} c^2 = 2 m_e c^2 \left(1 + \frac{m_e}{m_p} \right) \approx 1.02 \text{ MeV};$$

cf. formula (P8.2.2.15) with $|Q| = 2 m_e c^2$.

In the case where electron-positron pair production is due photon-electron interaction, we substitute m_p by m_e to obtain $\varepsilon_{thr} = 4 m_e c^2 = 2.04 \text{ MeV}$.

▶ **Mechanisms of nuclear reactions.** Just as there is still no universal theory accounting for all properties of nuclei, there is no unified theory of nuclear reactions. The features of different types of reactions in different energy intervals are described in terms of different models (or *mechanisms*, as they are also called) of nuclear reactions. Let us briefly dwell on some of those.

The mechanism of direct reactions (S. Butler, 1953) is aimed at explaining some features of reactions occurring at high energies, for instance, high anisotropy (maximum probability of forward emission), almost complete energy transfer to the outgoing fragment, etc. Such processes include, in particular, the *stripping reaction* (d, p), (d, n) and the *pickup reaction* (p, d), (n, d). In direct reaction, the incident particle reacts not with the entire nucleus but with its fragment — a nucleon or a small group of nucleons — transmitting energy to that fragment in a short (nuclear) time 10^{-22}–10^{-21} s. (This time is typical of reactions determined exclusively by nuclear forces. *Nuclear time* is estimated as $\tau_{\text{nucl}} = 2R/v \sim 10^{-23}$–$10^{-21}$ s, where $v \sim 10^{7}$–10^{8} m/s is the nucleon speed.)

The mechanism of compound nucleus (N. Bohr, 1936) accounts for many features of reactions occurring at low energies. In the first place, it is the long duration (on a nuclear scale) of many reactions ($\tau \sim 10^{-16}$–10^{-12} s, which is by many orders greater than nuclear time $\tau_{\text{nucl}} \sim 10^{-22}$ s). Secondly, it is the symmetry (in the center-of-mass frame) of forward-backward particle emission (sometimes, even complete isotropy). Thirdly, it is the resonance character of reactions: their cross-sections have sharp maxima for some energy values. The essence of the mechanism of compound nucleus consists in that the reaction occurs in two stages:

$$a + X \to C^* \to Y + b. \tag{P8.2.2.16}$$

On the first stage, the parent nucleus X captures the incident particle and a compound nucleus C (in an excited state, which is indicated by the superscript $*$) is formed. The excitation energy is calculated as the sum of the binding energy of the particle a in the nucleus C (equal to $E_{aC} = m_X + m_a - m_C$) and the initial kinetic energy $\widetilde{K}$ (in center-of-mass frame):

$$E_{\text{exc}} = E_{aC} + \widetilde{K}. \tag{P8.2.2.17}$$

The process of particle capture and distribution of surplus energy between nucleons of the compound nucleus goes on for a short time $\sim \tau_{\text{nucl}}$. Then, the compound nucleus disintegrates in one of the possible ways (with different probabilities): either the surplus energy is randomly concentrated on a single particle (nucleon or a group of nucleons) and it abandons the nucleus, or a γ-quantum is emitted, or there is nuclear fission. All these processes are much slower than nuclear ones, which accounts for the slowness of the reactions occurring with compound nucleus formation.

The symmetry of emission of decay products is explained by the fact that the compound nucleus is capable of "forgetting" the direction of the incident particle. The resonance character of the reaction cross-section is connected with the fact that the probability of particle capture (and therefore, the cross-section of the entire process) increases sharply, if E_{exc} in (P8.2.2.17) coincides with one of the discrete energy levels of the compound nucleus. These values correspond to discrete values of kinetic energy $\widetilde{K}$, through which the energy of the incident particle is expressed (as shown above).

The mechanism of compound nucleus gives an acceptable explanation of the fact that reactions with different initial particles may have similar final results. Thus, the reactions

$$n + {}^{13}_{7}\text{N}, \quad {}^{1}_{1}\text{H} + {}^{13}_{6}\text{C}, \quad {}^{2}_{1}\text{H} + {}^{12}_{6}\text{C}, \quad {}^{3}_{1}\text{H} + {}^{11}_{6}\text{C}, \quad {}^{3}_{2}\text{He} + {}^{11}_{5}\text{B}, \quad {}^{4}_{2}\text{He} + {}^{10}_{5}\text{B}$$

occur with the formation of one and the same compound nucleus ${}^{14}_{7}\text{N}^*$, and if their excitation energies turn out to be equal, then they will have identical output channels.

▶ **Statistical model.** For large excitation energies, the spectrum of the compound nucleus becomes practically continuous, and the *statistical model of decay*, or the *nuclear evaporation model*, becomes applicable. In this model, the compound nucleus is regarded as a drop of hot liquid with temperature $T = E_{exc}/k$ (k is the Boltzmann constant) and particle emission as evaporation from its surface. In the framework of this model, it is clear why within this energy range the angular distribution of emitted particles becomes isotropic (in the center-of-mass frame).

P8.3. Nuclear Power Engineering

Since maximum specific binding energy is characteristic of medium nuclei (with $A \sim$ 50–60), it is natural to expect that fission of heavy nuclei, as well as fusion (synthesis) of light nuclei is accompanied by energy release. The energy released in fission is called *atomic* and that released in fusion *thermonuclear*. The construction in the middle of the twentieth century of atomic and thermonuclear (hydrogen) bombs, in which the said reactions occur uncontrollably, put the very existence of humankind at risk. At the same time, controllable generation of energy at nuclear power plants is of great importance for power industry of many countries (in spite of the many ecological problems involved). The solution of the problem of controlled thermonuclear fusion would give a practically inexhaustible, ecologically safe source of energy. Moreover, reactions of fusion are the main source of energy of the sun and other stars.

P8.3.1. Nuclear Fission

▶ **Spontaneous fission.** In order to have a qualitative picture of fission phenomena and obtain some quantitative estimates, one can use the liquid-drop nuclear model and the corresponding Weizsäcker's formula (P8.1.3.1). For qualitative understanding, one can consider fission resulting in two identical fragments. Fissure leads to Coulomb energy drop and an increase of surface energy. The ratio of the corresponding terms in (P8.1.3.1) depends on the parameter $\kappa = Z^2/A$, and fission probability also depends on this parameter. It is not difficult to check that fission becomes energetically efficient for

$$\frac{Z^2}{A} > \kappa_1 \approx 18. \tag{P8.3.1.1}$$

The substitution of actual values of Z and A shows that fission starts to be efficient for nuclei in the niobium–palladium range, i.e., starting from $A \sim 100$. However, spontaneous fission is observed only for very heavy nuclei (with $A \sim 250$).

This is due to the fact that although the two-fragment state has a smaller energy, the transition to that state requires overcoming a substantial potential barrier. Subject to gradual deformation, the nucleus passes through intermediate states whose energy is greater than that of the initial state. Figure P8.6 schematically represents the dependence of the energy of the deformed nucleus on a conditional parameter γ describing the degree of deformation. The barrier height equal to the difference between the maximum intermediate energy and the initial energy of the nucleus is called *activation energy* E_{act}. If excitation energy received by the nucleus exceeds E_{act}, it quickly splits into two fragments. Activation energy decreases with the growth of Z^2/A and vanishes for

$$\frac{Z^2}{A} > \kappa_2 \approx 50. \tag{P8.3.1.2}$$

In the absence of a barrier, the nucleus instantly splits into two fragments without receiving any additional energy. Due to quantum tunneling, spontaneous fission from the ground state is also possible for $Z^2/A < \kappa_2$, but the barrier should be not too high; otherwise, fission probability would be very small. Thus, for $^{238}_{92}U$, we have $Z^2/A = 35.5$, which is considerably less than κ_2, and the and half-life of $^{238}_{92}U$ with respect to spontaneous fission is $T_{1/2} \sim 10^{16}$ years, which is by 6 orders greater than $T_{1/2}$ of this element with respect to α-fission. On the other hand, for the artificial nucleus with $Z = 107$ and $A = 261$, we have $Z^2/A = 43.9$, and its half-life with respect to spontaneous fission is $T_{1/2} \sim 10^{-3}$, which is comparable with $T_{1/2}$ with respect to α-fission.

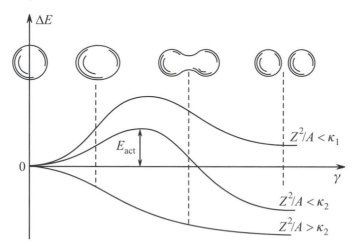

Figure P8.6. Energy of intermediate states of a nucleus in the process if its fission.

Specific binding energy of nuclei with $A \sim 240$ is by 0.9 MeV less than that of fission fragments. This means that in every instance of fission, about 200 MeV energy is released. More than 80% of that energy is carried away as kinetic energy of the fragments. The remaining energy goes for the emission of neutrons and γ- and β-particles by the fragments.

As a rule, fission results in fragments with different masses. Light fragments are grouped around Kr, and heavy ones around Xe. This is explained by the fact that the numbers of neutrons in these nuclei are close to magic numbers 50 and 82.

Since the fraction of neutrons in heavy nuclei is considerably greater than in light nuclei, daughter nuclei have an excess number of neutrons. Immediately after fission, daughter nuclei emit 2 or 3 neutrons with kinetic energy ~ 2 MeV, after which there occur several β-decays that bring the fragments to a stable state. Neutron emission by fission products plays the key role in ensuring the chain reaction of fission.

Of great importance for the *controlled* chain reaction are the so-called *delayed neutrons*. These are produced in the case when a sufficiently slow β-decay of a fission product results in excited nuclear state with small energy of separation of a neutron, upon which the neutron is instantly emitted. The delay is determined by the β-decay time and may reach tens of seconds.

▶ **Fission of excited nuclei.** When a nucleus receives a sufficient amount of excitation energy (several MeV), it starts to oscillate with its shape periodically changing from oblate to prolate. Even if the excitation energy is less than the activation energy, the probability of fission of an excited nucleus is substantially higher than in the ground state. A competitive process is that of emission of excitation energy in the form of γ-quanta.

An excited nucleus with a considerable fission probability can be obtained by exposing it to different types of particle radiation. The most effective is exposure to neutrons; there is no Coulomb barrier for neutrons and they can approach the nucleus even if their kinetic energy is very small (of the order of 1 keV).

▶ **Neutron capture.** The principal mechanism of fission of natural uranium isotopes, ^{235}U and ^{238}U, is neutron capture with the formation of compound nuclei ^{236}U and ^{239}U in excited state. It is important that ^{235}U is broken down by any, in particular, thermal neutrons (with energies less than 0.5 keV), while ^{238}U only by fast neutrons (with energies of the order of 2 MeV). The point is that the compound nucleus ^{236}U is of the even-even type and for the same amount of neutron energy, its excitation energy turns out greater than that of the even-odd nucleus ^{239}U.

Excitation energy of a compound nucleus can be found from the equation

$$m_n + K_n + M_A = M_{A+1} + E_{\text{exc}},$$

where K_n is the kinetic energy of the neutron (the kinetic energy of the compound nucleus is neglected). We get

$$E_{\text{exc}} = (m_n + M_A - M_{A+1}) + K_n = \varepsilon_{A+1} + K_n,$$

where ε_{A+1} is the neutron binding energy in the compound nucleus. A compact nucleus undergoes fission, if E_{exc} becomes greater than activation energy. For ^{239}U, we have $E_{\text{act}} = 7.1$ MeV, and $\varepsilon_{A+1} = 5.5$ MeV; therefore, when exposed to neutron radiation, the nucleus ^{238}U requires for fission K_n not less than 1.6 MeV. For ^{236}U, we have $\varepsilon_{A+1} = 6.8$ MeV, which is greater than $E_{\text{act}} = 6.5$ MeV, and thus the nucleus ^{235}U will disintegrate when exposed to any neutrons.

The cross-section of capture of slow neutrons is determined by their de Broglie wavelength, which can be much larger than the nuclear radius. The cross-section of capture of thermal neutrons with energy 0.025 keV is hundreds of times greater than the cross-section of capture of fast neutrons with energy 1 MeV, whence the great importance of fission caused by thermal neutrons.

▶ **Chain reaction.** The possibility of self-sustained reaction of fission (chain reaction) is due to the fact that upon nuclear fission by a *single neutron*, fission fragments emit *two or three* neutrons, which, in turn, can cause fission of other nuclei. The problem is that some of the produced neutrons drop out of the reaction: are either absorbed without causing fission or just leave the system through its boundary.

One defines the *neutron multiplication coefficient* as the ratio of the number of neutrons in the next generation to that in the preceding one:

$$k = N_{i+1}/N_i.$$

For the chain reaction to start, it is necessary that k be greater than unity. In order to keep the reactor steady, k should be maintained equal to unity.

With the increase of the system's dimensions, the fraction of neutrons abandoning the system through its boundary tends to zero. Therefore, whereas $k_\infty > 1$ for an infinite nuclear medium, the condition $k > 1$ for a finite system of a certain shape will hold, if the mass of that system exceeds some critical value called *critical mass*. The critical mass can be decreased if the active object is surrounded by a shell that reflects neutrons. Thus, for the pure element ^{235}U the critical mass is 47 kg (ball of radius 17 cm), but if ^{235}U is layered by polyethylene films and coated by a beryllium shell, the critical mass reduces to 250 g (ball of radius 3 cm). Next we consider the chain reaction in infinite nuclear medium.

Any nuclear fuel with reasonable production cost contains a high percentage of ^{238}U. Thus, natural uranium contains 99.3% ^{238}U and only 0.7% ^{235}U. Although ^{238}U is also capable of neutron fission, its probability is too small (and for neutrons with energies less than 1.6 MeV this probability is equal to zero), and the rest of the neutrons are absorbed by ^{238}U without fission and with emission of γ-quanta (*radiation capture* (n, γ)). At the same time, the probability of nuclear fission of ^{235}U (and also ^{233}U and $^{239}_{94}Pu$) is fairly high, 85–90%, and these elements can be split by both fast and slow neutrons.

High enrichment of uranium (i.e., a substantial increase of ^{235}U concentration) is a very costly process and is used only for the production of nuclear bomb charges (where economical factors are inessential) and the construction of breeders—fast neutron reactors (discussed below).

Chain reactions are mainly obtained with the use of slow (thermal) neutrons. Fission neutrons have mean energy $\sim 2\,$MeV. For this energy, the cross-sections of neutron capture by nuclei of ^{235}U and ^{238}U are of the same order. However, for thermal neutrons the situation is quite different. For ^{238}U, the capture cross-section remains small ($\approx 2.8\,$b), and for ^{235}U, it sharply increases and is $\approx 650\,$b (which is much greater than the geometrical cross-section); moreover, the greater part (84%) of absorbed neutrons causes fission (fission cross-section is 550 b).

It is easy to calculate that for thermal neutrons the cross-section of fission of a "medium" nucleus of natural uranium is $550 \times 0.007 = 3.9\,$b, and the cross-section of radiation capture is $2.8 + 100 \times 0.007 = 3.5\,$b. We see that the store of neutrons is not very large (2.5 neutrons are born, on the average, in each fission event, but half of these are lost); therefore, for the realization of chain reaction it is important to decrease the loss of neutrons in the process of their moderation. The best moderator, from the mechanical standpoint, should be hydrogen (contained in common water molecules), since in elastic collision the greatest energy transfer occurs in the case of equal masses of the colliding particles (see Subsection P1.5.3). However, hydrogen atoms easily absorb neutrons ($^1H + n \rightarrow {}^2H + \gamma$), and therefore, deuterium (contained in heavy water), carbon, or beryllium are used. In order to reduce the absorption of moderated neutrons by nuclei of ^{238}U, the nuclear fuel, instead of being mixed with the moderator to form a homogeneous structure, is set in layers that alternate with layers of the moderator to form a heterogeneous structure. Getting into a moderator layer, the neutron slows down without the risk of being absorbed, and then goes back to the region of reaction.

Had the chain reaction been due to merely fission neutrons, such a reaction could hardly be controlled. Indeed, the lifetime of a single generation of fission neutrons is $\sim 10^{-3}\,$s. If the multiplication coefficient randomly increases, say, to $k = 1.005$, then in 1 s the number of neutrons increases 150 times, which means that the reaction turns into explosion. The stability of controlled reactions is achieved by way of *delayed neutrons* mentioned above. Even though their fraction is not great ($\sim 0.75\%$), the average lifetime of their generation is $\sim 0.1\,$s, and in the above example the number of neutrons would increase only 1.5 times instead of 150.

▶ **Transuranic elements.** The only natural fissile element is ^{235}U, but its resources are not too great. Though reactors on fast neutrons require a substantial enrichment of uranium (up to 15%) and are not very powerful, they can be used for a special purpose—to produce a new type of nuclear fuel (this is why they are called *breeders*). Thus, radiation capture of neutrons by ^{238}U, which is so harmful for chain reactions, leads to the production of the artificial transuranic element $^{239}_{94}Pu$:

$$^{238}_{92}U + n \rightarrow {}^{239}_{92}U + \gamma;$$

$$^{239}_{92}U \rightarrow {}^{239}_{93}Ne + e^- + \widetilde{\nu}; \qquad T_{1/2} = 2.3 \text{ min};$$

$$^{239}_{93}Ne \rightarrow {}^{239}_{94}Pu + e^- + \widetilde{\nu}; \qquad T_{1/2} = 2.3 \text{ days}.$$

Plutonium is practically stable (its half-life is 24,000 years) and is easily fissionable by neutrons with any energy. Plutonium can be more easily separated from ^{238}U, since their chemical properties are different.

P8.3.2. Fusion Reactions

As mentioned above, for light nuclei, the binding energy ε increases with the growth of A (see Fig. P8.2). Thus, reactions of fusion of two light nuclei with the formation of a heavier nucleus (reactions of synthesis) should occur with energy release. Since the growth of $\varepsilon(A)$ for small A is steeper than its drop for large A, much more energy (per one nucleon) is released in fusion than in fission.

▶ **Conditions of fusion.** Fusion of two nuclei is possible only if they approach one another by a distance at which nuclear forces start to be active, which is hindered by Coulomb repulsion. To overcome the Coulomb barrier the nuclei must have large kinetic energies. If the substance is heated up to a high temperature, then conditions of self-sustained reaction may arise, and energy release will compensate its loss (for instance, by radiation). For this reason, reactions of fusion are called thermonuclear reactions. Such reactions are the main source of energy of the sun and other stars.

Note that the temperatures at which fusion is possible are so high that the substance acquires the state of fully ionized plasma.

Example. Estimate the temperature at which fusion of two protons to deuteron can occur.

Solution. For this reaction to take place, the protons must approach one another by the distance equal to two effective radii of nuclear forces $r_0 \sim 2 \times 10^{-15}$ m (see Subsection P8.1.1). From the energy conservation law

$$2E_{\text{kin}} = \frac{1}{4\pi\varepsilon_0}\frac{e^2}{2r_0}$$

we find the kinetic energy $E_{\text{kin}} \sim 0.1\,\text{MeV}$, and then, using the classical formula $E_{\text{kin}} = \frac{3}{2}k_\text{B}T$, estimate the temperature, $T \sim 10^9$ K.

▶ **Thermonuclear energy of the sun and the stars.** The temperature in the center of the sun is by two orders less than the threshold value obtained above. This means that the *mean* nuclear energy is insufficient for fusion to take place. Still, the reaction occurs, and the energy released by the sun is huge: $\sim 4 \times 10^{26}$ W. The point is that, according to the Maxwellian distribution, a ceratin (very small) part of particles possesses energy much larger than the mean value. Moreover, due to quantum tunneling, nuclei with energies less than the Coulomb barrier can participate in the reaction. All this means that there is only a trace amount of colliding nuclei that react. For this reason the specific energy released by the sun per unit time is so small, $\sim 10^{-4}$ W/kg, but owing to its huge mass, the total energy release is so great.

Thermonuclear energy of the sun is released in two cycles of thermonuclear reactions: the *carbon cycle* and the *hydrogen cycle*. Both these cycles result in the transformation of four protons into a helium nucleus (with the emission of positrons, photons, and neutrino), with the total energy release $\sim 25\,\text{MeV}$. The carbon cycle requires the presence of carbon nuclei, which play the role of a catalyst, since they are restored as a result of the cycle. The reactions of this cycle are the following:

$$^1_1\text{H} + {}^{12}_6\text{C} \rightarrow {}^{13}_7\text{N} + \gamma;$$

$$^{13}_7\text{N} \rightarrow {}^{13}_6\text{C} + e^+ + \nu;$$

$$^1_1\text{H} + {}^{13}_6\text{C} \rightarrow {}^{14}_7\text{N} + \gamma;$$

$$^1_1\text{H} + {}^{14}_7\text{N} \rightarrow {}^{15}_8\text{O} + \gamma;$$

$$^{15}_8\text{O} \rightarrow {}^{15}_7\text{N} + e^+ + \nu;$$

$$^1_1\text{H} + {}^{15}_7\text{N} \rightarrow {}^{12}_6\text{C} + {}^4_2\text{He}.$$

The second cycle contains three different reactions: the first and the second take place twice (altogether five reactions):

$$_1^1\text{H} + {}_1^1\text{H} \rightarrow {}_1^2\text{H} + e^+ + \gamma;$$

$$_1^1\text{H} + {}_1^2\text{H} \rightarrow {}_2^3\text{He} + \gamma;$$

$$_2^3\text{He} + {}_2^3\text{He} \rightarrow {}_2^4\text{He} + 2{}_1^1\text{H}.$$

At temperatures up to 1.5×10^7 K, the hydrogen cycle makes a larger contribution to energy release; at higher temperatures, a larger contribution is due to the carbon cycle. For the sun, both cycles are equally important. In less bright stars, the hydrogen cycle dominates and in brighter stars, the carbon cycle does. For giant stars, other cycles are crucial (e.g., the helium and neon cycles).

▶ **Controlled fusion.** Fusion control would give mankind a practically inexhaustible energy source. In contrast to reactions occurring inside stars, energy should be obtained here from a fairly small amount of nuclear fuel with maximum efficiency. To that end, the substance should be heated up to a temperature at which the probability of nuclear fusion at collisions is sufficiently large. Moreover, one has to find a way to confine incandescent plasma for a sufficiently long time and extract the released energy, transforming it, say, to electricity.

The major technological problem is plasma confinement. In this connection, the most promising method is to place plasma in magnetic field of a certain structure (magnetic trap). In a *tokamak*, plasma is confined by magnetic field produced by the current in the plasma itself (the same current is first used for heating plasma). A fairly long-time plasma confinement is hindered by various types of instabilities that arise and quickly develop in hot plasma placed in magnetic field.

For each reaction, there is a characteristic temperature T_0 to which plasma should be heated in order to ensure the most efficient extraction of energy. The second characteristic parameter is the minimum time τ of plasma confinement necessary for energy gain (more precisely, the product of time and plasma concentration, $n\tau$).

Proton–proton fusion reaction fuels the sun and could be most advantageous as regards energy release; furthermore, protons are abundant on Earth (hydrogen is contained in water). Unfortunately, the cross-section of this reaction is so small that it cannot be considered as a basis for controlled fusion. Equally attractive seem *deuterium–deuterium* reactions

$$_1^2\text{H} + {}_1^2\text{H} \rightarrow {}_2^3\text{He} + n + 3.25\,\text{MeV},$$

$$_1^2\text{H} + {}_1^2\text{H} \rightarrow {}_1^3\text{H} + {}_1^1\text{H} + 4.03\,\text{MeV},$$

since deuterium constitutes 0.015% of all hydrogen resources. This reaction is characterized by the parameters $T_0 \approx 10^9$ K and $n\tau > 10^{22}$ s/m^3. However, the next most promising candidate is the *deuterium–tritium* reaction

$$_1^2\text{H} + {}_1^3\text{H} \rightarrow {}_2^4\text{He} + n + 17.6\,\text{MeV},$$

which has a considerably greater power efficiency and better critical parameters: $T_0 \approx 2 \times 10^8$ K and $n\tau > 10^{20}$ s/m^3, which compensates for the difficulties connected with the production of tritium.

P8.4. Subatomic Particles

In this section, we give some brief reference information with regard to classification of subatomic particles, quantum numbers, conservation laws, and quark structure of hadrons.

P8.4.1. Classes of Subatomic Particles

In modern particle physics, two main types of subatomic particles are distinguished: elementary particles, which are not known consist of other particles, and composite particles. Elementary particles include quarks, leptons, gauge bosons, photons, and gluons. Composite subatomic particles consist of two or more elementary particles and include baryons (e.g., protons and neutrons) and mesons.

Remark. For historical reasons, all subatomic particles are very often called elementary particles and the structureless particles are called truly elementary particles or truly fundamental particles.

▶ **Fundamental interactions.** Transformations and decay of subatomic particles are governed by the following fundamental interactions: strong, electromagnetic, and weak (gravitational interaction is inessential here). The weaker the interaction, the slower the associated processes. *Strong interaction* has the range 10^{-15} m and the characteristic time of the associated processes is 10^{-23} s. Strong interaction binds nucleons in atomic nuclei. *Electromagnetic interaction* is long-range (it has an infinite range), the characteristic time of the associated processes is not less than 10^{-20} s. A testing ground of this most well-known type interaction is the control of atomic electrons, i.e., of all diversity of substances and chemical reactions. *Weak interaction* has the range 10^{-18} m, the characteristic time of the associated processes (involving neutrino, as a rule) is not less than 10^{-10} s. It is weak interaction that controls fusion of two protons, which is crucial for energy processes in the sun, and it also controls various types of β-decay. Weak interaction is the most universal of the three; it is always present, but its determining character is manifested only in the absence of the other two.

Fundamental interactions are realized by the exchange of special particles that belong to the class of *fundamental vector bosons* (all these particles have unit spin). The interaction carriers are produced by one of the interacting particles and are absorbed by the other. The produced particles are *virtual*, they do not obey the relation $E^2 - p^2c^2 = m^2c^4$, which is allowed by the uncertainty principle $\Delta E \Delta t \sim \hbar$, provided that their lifetime is small. If the mass of the carriers differs from zero, then the interaction radius is equal to their Compton wave-length $\hbar/mc$ (for details, see Subsection P8.1.2). The electromagnetic interaction carrier is a massless particle called a *photon*, the well-known quantum of electromagnetic field. Weak interaction carriers are heavy *intermediate bosons* W^+, W^- and Z, whose masses are almost equal to 100 GeV. Strong interaction carriers are called *gluons*—the quanta of the eight gluon fields that realize interaction between quarks. (In the early π-meson theory of nuclear forces described in Subsection P8.1.2, π-mesons were regarded as strong interaction carriers). The gluon mass is equal to zero, but in contrast to photons, gluons interact with one another and with quarks and cannot go away farther than 10^{-15} m.

▶ **General properties of particles.** We are going to describe different classes of subatomic particles. First, we list some general properties.

1. Particles can be stable (electron, proton, neutrino) or unstable. In the last case, one indicates the particle lifetime (or width of the energy interval $\Delta E = \hbar/\Delta t$), as well as decay channels. Particles with lifetime $\sim 10^{-23}$ s are called *resonance particles*. Particles whose lifetime is much greater than the said value are called quasi-stable.

2. Particles can be either composite (consisting of other particles; e.g., nucleons consist of quarks) or, for presently accessible energy levels, structureless (photon, electron, neutrino, quarks).

3. A particle can be either a boson (integer spin) or fermion (half-integer spin).

4. Almost any particle has its counterpart—an antiparticle whose charges (all its charges, not only the electric one) have opposite signs. Thus, electron corresponds to positron, proton to antiproton, neutron to antineutron. A particle that coincides with its antiparticle is called *truly neutral* (for example, photon or π^0-meson).

5. All particles are divided into strongly interacting hadrons (baryons and baryon resonances, mesons and meson resonances), leptons (fermions not involved in strong interaction) and fundamental bosons (interaction carriers).

▶ **Hadrons and quarks.** The term *hadron* refers to all particles involved in strong interaction. Hadrons are divided into two large classes: *baryons* (hadrons with half-integer spin, i.e., fermions) and *mesons* (hadrons with integer spin, i.e., bosons). The class of baryons contains, in particular, nucleons, and that of mesons contains π-mesons.

Hadrons are quite numerous (more than 400 hadrons have been discovered). A systematic classification of hadrons and understanding of many of their properties were not possible until the appearance (in the 1960s) of the hypothesis of *quark structure of hadrons*, according to which all hadrons consist of elementary particles called *quarks*. All quarks have spin 1/2, and therefore, are fermions. There are six varieties of quarks (called *quark flavors*): u, d, s, c, b, and t (from *up, down, strange, charm, bottom,* and *top*). The masses of the quarks are: $m_u \approx 5$ MeV, $m_d \approx 7$ MeV, $m_s \approx 150$ MeV, $m_c \approx 1.3$ GeV, $m_b \approx 4.3$ GeV, and $m_t \approx 175$ GeV. The most extraordinary property of quarks is that they possess fractional electric charges. The quarks u, c, and t are called upper quarks and their charge is equal to $+2/3$; the quarks d, s, and b are called lower quarks and their charge is equal to $-1/3$ (in elementary charge units).

All baryons consist of three quarks, and all mesons consist of a quark and an antiquark. Nucleons consist of light quarks u and d: proton contains two u-quarks and one d-quark ($p = uud$), neutron contains two d-quarks and one u-quark ($n = ddu$). These quarks also form π-mesons: $\pi^+ = u\widetilde{d}$ and $\pi^- = d\widetilde{u}$ (a tilde over a quark symbol denotes the corresponding antiquark); π^0 is a quantum superposition of two states: $\pi^0 = (u\widetilde{u} + d\widetilde{d})/\sqrt{2}$. We see that π^+ and π^- are mutual antiparticles, and π^0 coincides with its antiparticle. Particles containing an s-quark are called strange and those with a c-quark are called charmed particles.

It should be immediately observed that, at first sight, the masses of hadrons are much greater than the sum of the masses of their quarks. But the mass of any bound system should be smaller than the sum of the masses of its constituents by the quantity equal to the binding energy. The point is that the quarks in the hadrons are surrounded by a cloud of virtual gluons, sometimes called the "gluon coat." The masses of "coated" quarks are about 350 MeV greater than the masses of "naked" quarks.

Baryons that contain more heavy quarks than u and d are called "hyperons." From quarks u, d, and s, six hyperons with spin 1/2 can be composed: Λ^0 (uds, $m \approx 1120$ MeV), Σ^+, Σ^0, Σ^- (uus, uds, dds, $m \approx 1200$ MeV), Ξ^0, Ξ^- (uss, dss, $m \approx 1320$ MeV). Decay of all these hyperons is determined by weak interaction, and therefore, their lifetime is $\tau \sim 10^{-10}$ s (except for the Σ^0-hyperon whose decay is determined by electromagnetic interaction, its lifetime is $\tau \sim 10^{-20}$ s). Together with proton and neutron, these hyperons form an octet of baryons with spin 1/2. (The difference between Λ^0 and Σ^0 is explained below.)

Next, we have the decuplet of baryons with spin 3/2, among which there are baryons consisting of three identical quarks, for example, Δ^-, Δ^0, Δ^+, Δ^{++} (ddd, udd, uud, uuu, $m \approx 1230$ MeV). The heaviest in the decuplet is the Ω^--hyperon (sss, $m \approx 1670$ MeV). Note that subatomic particles consisting of the same quarks and differing only in spin (or other quantum numbers) often bear different names (e.g., neutron and Δ^0-hyperon).

Further, we briefly describe mesons. In addition to the three π-mesons, there are six mesons having zero spin and orbital momentum and containing an s-quark: the four K^+, K^0, $\widetilde{K}^0$, K^- ($u\widetilde{s}$, $d\widetilde{s}$, $s\widetilde{d}$, $s\widetilde{u}$, $m \approx 500$ MeV) and two truly neutral mesons η and η' (different quantum superpositions of the states $u\widetilde{u}$, $d\widetilde{d}$ and $s\widetilde{s}$). Next, we have nine mesons with spin $S = 1$, among which there are three ρ, four K^*, ω^0, and φ^0. All these mesons (as well as η') are of resonance type $\tau \sim 10^{-23}$ s).

An important event of 1974 was the discovery of the first charmed particle, J/ψ-meson ($m = 3.1\,\text{GeV}$); this particle represents one of the states of *charmonium* $c\widetilde{c}$ (charmonium has "hidden charm"). Thereafter, other charmed mesons were found (for example, $D^+ = c\widetilde{d}$, $D^0 = c\widetilde{u}$, and $D_s^+ = c\widetilde{s}$) and charmed baryons (for example, $\Lambda_c^+ = udc$), and also more heavy "beauty" hadrons containing a b-quark.

▶ **Quark color and gluons.** The existence of baryons consisting of three identical quarks is in contradiction with the Pauli principle, since two of them must have the same spin direction. Moreover, the Ω^--hyperon has spin 3/2; i.e., all its three s-quarks can be in the same state (when the hyperon spin projection is maximum). How can this be explained? It turned out that there are *three* quarks with the same flavor and not just a single such quark. The distinction between these is described by an additional quantum number that can take three values. It was proposed to use the term *color* for that number (or color charge) and ascribe to it three values: R (red), G (green), and B (blue). Accordingly, the colors of antiquarks are denoted by $\widetilde{R}$, $\widetilde{G}$, $\widetilde{B}$ (anti-red, anti-green, and anti-blue). Thus, there are altogether 18 quarks and 18 antiquarks. The mixture of the three colors is regarded as colorless (white). A quark–antiquark pair possesses a "hidden" color.

The color charges of quarks are sources of gluon fields (just as the electric charge is the source of electromagnetic field). The quanta of these fields, called *gluons* (from "glue"), realize strong interaction. (Electromagnetic interaction is realized by the exchange of photons, i.e., quanta of electromagnetic field.) Both gluons and photons are massless particles with spin 1 and negative parity (see below). Photons, however, are electrically neutral, while gluons are "colored"; i.e., take part in strong interactions. Each gluon carries two color charges: one color and one anti-color. There are eight different gluons: six with explicit color, ($g_{R\widetilde{G}}$, $g_{R\widetilde{B}}$, $g_{G\widetilde{R}}$, $g_{G\widetilde{B}}$, $g_{B\widetilde{R}}$, $g_{B\widetilde{G}}$) and two with hidden color.

According to the color charge conservation law, a quark must change its color, when emitting or absorbing a gluon. Thus, when a red R-quark emits a gluon $R\widetilde{G}$, it becomes a G-quark, but its flavor does not change:

$$u_R \rightarrow u_G + g_{R\widetilde{G}}.$$

(Quark flavor can change only in weak interactions; i.e., when W- or Z-bosons are emitted.)

Since gluons possess color charges, they can emit and absorb gluons and be involved in mutual strong interaction. Strong interaction, regarded as a result of emission and absorption of gluons by quarks and gluons, is studied by *quantum chromodynamics* (by analogy with *quantum electrodynamics*, which studies electromagnetic interaction as a result exchange of photons). One of the basic statements of quantum chromodynamics is the nonexistence of free colored objects. Free existence is allowed only for such color charge combinations that carry no color charge as a whole. For example, free hadrons can only exist in the "white" state, its three quarks being of different colors at each time instant. (Still, each quark continuously changes its color, emitting or absorbing virtual gluons that have been either emitted by another quark or produced in its "gluon coat"—cloud of virtual gluons. But the total color charge of the hadron remains white.) The same principle applies to mesons. For example, π^+-meson is a quantum superposition of three color states: $u_R\widetilde{d}_{\widetilde{R}}$, $u_G\widetilde{d}_{\widetilde{G}}$, $u_B\widetilde{d}_{\widetilde{B}}$. Each of these states has a hidden color, but their superposition is "white."

▶ **Confinement.** Why is it impossible to divide a white object into two colored ones (for instance, to remove one quark from a meson or hadron)? First of all, the exchange of colored gluons leads to an interaction whose potential grows with the distance not slower than the linear function (the interaction force does not decrease). A possible explanation is that the gluon field is not scattered in all directions (this would correspond to the Coulomb potential), but is concentrated in a narrow "pipe" (string) joining the quarks. Therefore, infinite energy would be required to split the quarks. In fact, as the quarks are separated from one another by a distance greater than 10^{-15} m, the gluon string is "broken," a pair quark–antiquark is produced, and the removed quark turns into a white meson. Attempting to obtain two colored objects, we again obtain two white objects. (Note that for large energies of scattered quarks, each of these transforms not into a single meson but into several hadrons, and the so-called "hadron jet" is produced. Observation of hadron jets was an important evidence supporting the existence of quarks.) In this way, quantum chromodynamics explains the failure of all attempts to detect free quarks or knock them out of hadrons, in spite of the fact that the existence of quarks within hadrons was confirmed by experiments with scattering of high energy electrons on hadrons. This behavior of quarks is referred to as "confinement."

▶ **Leptons.** Leptons are fermions that do not participate in strong interactions. There are three pairs of leptons: electron e^- and electron-type neutrino ν_e, muon μ^- and muon-type neutrino ν_μ, tau-lepton τ^- and tau-neutrino ν_τ. The spin of all leptons is equal to 1/2. Each charged lepton has its antiparticle: e^+, μ^+ and τ^+. There is no final answer to the question whether neutrinos are truly neutral particles or have antiparticles—antineutrinos, $\widetilde{\nu}_e$, $\widetilde{\nu}_\mu$ and $\widetilde{\nu}_\tau$. Leptons are fundamental particles—no structure has been detected in them. As shown by experiments, each charged lepton takes part in weak interactions coupled to its neutrino, for example:

$$n \to p e^- \widetilde{\nu}_e, \qquad \pi^+ \to \mu^+ \nu_\mu, \qquad \tau^+ \to \widetilde{\nu}_\tau e^+ \nu_e.$$

The masses of muon and tau-lepton are respectively equal to 106 MeV and 1784 MeV. These particles are unstable: muon decays along the channel $\mu^- \to e^- \widetilde{\nu}_e \nu_\mu$, its lifetime is 2×10^{-6} s; the heavier tau-lepton decays along many channels, its lifetime is 5×10^{-13} s. For the neutrino masses, upper bounds can be given: $m_{\nu_e} < 30$ eV, $m_{\nu_\mu} < 0.5$ MeV, and $m_{\nu_\tau} < 150$ MeV.

▶ **Fundamental fermions.** Six leptons and six quarks form the class of *fundamental fermions* of which all other subatomic particles are "made." A special role is played by the "first generation" of fundamental fermions u, d, ν_e, e^-, which compose nucleons and atoms and determine almost all processes observed in nature. The "second generation" of fundamental fermions (c, s, ν_μ, μ^-) and their "third generation" (t, d, ν_τ, τ^-) are artificially obtained in powerful accelerators. However, only the investigation of all fundamental particles in their entirety may lead to the understanding of the structure and the evolution of our universe.

All processes in the world of subatomic particles amount to interactions between fundamental fermions realized by their exchanging several *fundamental vector bosons* (photons, gluons, three intermediate bosons). All these processes are governed by conservation laws.

Recall that the electromagnetic interaction carrier—photon—can be emitted or absorbed only by charged particles, and in this process neither the charge, nor the color, nor the flavor of the particle can change. The carriers of strong interaction—gluons—can be emitted or absorbed by quarks or gluons, and in this process the quark may change its color, but its charge and its flavor remain the same.

Weak interactions are realized in terms of emission and absorption of intermediate vector bosons W^+, W^-, and Z^0. As a charged boson is emitted or absorbed by a charged lepton, the latter transforms into the corresponding neutrino (and vice versa, a neutrino transforms into the corresponding charged lepton). The universal character of weak interactions is manifested in that vector bosons are emitted and absorbed not only by leptons, but also by quarks. (Note that in contrast to gluons, intermediate bosons have no color, i.e., take no part in strong interactions.) After emitting or absorbing a charged boson, the upper quark turns into the lower one and vice versa. It is essential that any upper quark (u, c, t) can transform into *any* lower quark (d, s, b).

For example, neutron decay $n \to pe^-\widetilde{\nu}_e$ amounts to d-quark decay $d \to ue^-\widetilde{\nu}_e$, which is described as the transformation of the d-quark into an u-quark with emission of a W^+-boson, which then turns into the pair $e^-\widetilde{\nu}_e$.

An important event in particle physics was the discovery in 1974 of intermediate bosons, whose existence and properties had been predicted by the so-called "standard theory of electroweak interaction" of Glashow, Weinberg, and Salam. The discovery was made in specially constructed proton–antiproton colliders (colliding beam accelerators). Upon collision, one quark and one antiquark turned into a powerful hadron jets. In time $\sim 10^{-25}$ s, the produced boson either decayed to quark + antiquark or to two leptons (e.g., electron + antineutrino). It was possible to detect the electrons produced in the W^--boson decay (only those processes were chosen in which the momentum conservation law was violated by the value of the momentum carried away by neutrino, which is equal to the electron momentum). To give an idea of the complexity of the events, it can be mentioned that in the first series of experiments, six reliable events were chosen out of one billion.

P8.4.2. Characteristics of Particles

▶ **Particle characteristics and conservation laws.** All particle characteristics can be divided into two groups: geometrical characteristics connected with space-time properties (mass, spin, spatial parity) and internal quantum numbers reflecting the symmetry of fundamental interactions (charge, baryon charge, lepton charge, isospin, strangeness, charm, and some others).

Each of these characteristics is associated with some conservation law and it makes sense to discuss them jointly. Some conservation laws are regarded as absolutely precise, others are approximate, i.e., valid for a certain class of reactions. Some approximate conservation laws are violated only by weak interactions (in processes determined by weak interactions), others are violated by both weak and electromagnetic interactions. Conservation laws allow us to predict the possibility or the impossibility of decay processes and reactions, understand the corresponding time lengths, predict the existence of new particles and describe the conditions of their birth. Recall the principle adopted in quantum theory, "what is not forbidden is possible," which makes the application of conservation laws especially effective.

▶ **Particle mass and law of conservation of energy-momentum.** Particle mass m is its most important characteristic. All subatomic particles have different masses. The Einstein formula relates the mass of a particle to its energy at rest.

The laws of conservation of mass and linear momentum follow from the homogeneity of time and space. These laws are crucial in the examination of reactions. The law of conservation of energy implies, in particular, that decay of an unstable particle can proceed only along such channels in which the sum of the masses of the produced particles is less than the mass of the original particle. Another example: the momentum conservation law forbids annihilation of an electron–positron pair with the emission of one γ-quantum (in center-of-mass frame the pair momentum is zero). The laws of conservation of energy and momentum can be used to calculate the threshold of any reaction (not forbidden by other conservation laws). Such calculations were carried out in Subsection P8.2.2; see formulas (P8.2.2.13)–(P8.2.2.15).

▶ **Particle spin I and angular momentum conservation law.** The spin of a particle determines whether the particle is subject to the Fermi–Dirac statistics (half-integer spin) or the Bose–Einstein statistics (integer spin). The spin of a hadron adds up from spins and orbital angular momenta of its constituent quarks. The spin of a particle determines the unique selected direction in a reference frame in which it is at rest.

The law of conservation of angular momentum is a consequence of the isotropy of space. The application of this law to the analysis of reactions is hindered by the fact that spins are added up by the rule of addition of angular momenta. For example, the total angular momentum of a system of two particles with spin 1 may be zero. However, it is easy to find out whether the total spin is integer or half-integer. This result often suffices for establishing the impossibility of some reaction. Thus, neutrinoless decay of a neutron into two particles—proton and electron—is forbidden by the angular momentum conservation law.

▶ **Parity.** Parity P of a particle (more precisely, its *spatial parity*) is a purely quantum notion. It indicates the behavior of the particle wave function under *spatial inversion*, i.e., the transformation of coordinates $x \rightarrow -x$, $y \rightarrow -y$, and $z \rightarrow -z$. There are two alternatives: the wave function does not change (positive parity, $P = +1$); the wave function changes its sign (negative parity, $P = -1$). Parity is an internal characteristic of a particle (just as spin, charge, etc.) Electron, proton, and neutron have parity $P = 1$, while photon and π-mesons have parity $P = -1$. Parity of any quark is equal to +1, and that of antiquark is −1. Parity of a compound particle (in the state with zero orbital angular momentum) is equal to the product of parities of its constituent particles.

For a long time, it was assumed that spatial parity is preserved in all reactions. The point is that the parity conservation law is a consequence of the statement that claims the invariance of processes with respect to spatial inversion or, equivalently, with respect to mirror reflection, since inversion reduces to a reflection and a rotation (for instance, reflection with respect to the plane XY and rotation by the angle 180° with respect to the axis Z). It was considered evident that all processes in the common and the "mirror" world should obey the same physical laws. However, later it was found that this statement is incorrect for processes determined by weak interactions. Violation of parity was predicted by Tsung-Dao Lee and Chen-Ning Yang in 1956, and shortly confirmed experimentally by Chien-Shiung Wu. As shown by these experiments, the electrons emitted in β-decay mostly move in the direction opposite to that of nuclear spin. Since nuclear spin changes its direction under reflection, electrons are emitted mostly in the direction of spin in the mirror world.

In the same experiment, one observed violation of another symmetry, the so-called charge symmetry or C-symmetry, when all particles are replaced by the corresponding antiparticles. It turned out that in the "antiworld," positrons would be emitted mostly in the direction of spin. However, under both transformations (i.e., spatial reflection together with the replacement of particles by their antiparticles), the process again transforms into itself. For some time it was assumed that this "combined" CP-symmetry is absolutely precise. But later weak processes were discovered for which the CP-symmetry is violated. At present, only one symmetry is accepted as absolutely precise—the CPT-symmetry, where the above two transformations are supplemented with time reflection. This symmetry is sufficient for proving that a particle and its antiparticle have equal masses and lifetimes.

▶ **Lepton charges L_e, L_μ, and L_τ.** Experiments show that in any processes the difference between the number of leptons and that of antileptons of a given type is preserved. Accordingly, three types of lepton charges have been introduced: L_e, L_μ, and L_τ. For e and ν_e, it is assumed that $L_e = 1$, for their antiparticles, $L_e = -1$, and for all other particles, $L_e = 0$. In the same way, the other charges are treated.

Lepton charges are preserved in all processes (for energies accessible at present).

The remaining conservation laws pertain to the world of hadrons. These laws are introduced empirically and can be naturally accounted for in the framework of the quark model.

▶ **Baryon charge B.** As shown by experiments, the difference between the number of baryons and that of antibaryons is preserved in all processes (for energies accessible at present). Accordingly, each baryon is ascribed baryon charge $B = 1$; for any antibaryon, $B = -1$, and for any meson $B = 0$.

If all quarks are ascribed charge $B = 1/3$ (and one takes $B = -1/3$ for antiquarks), then the baryon charge conservation law reduces to the preservation of the number of quarks (more precisely, the difference between the number of quarks and that of antiquarks).

▶ **Strangeness.** When first hyperons were discovered, some strange properties of their behavior were noticed. First, it was found that although hyperon emission occurred in high energy nucleon collisions (i.e., with strong interactions involved), their decay time was $\sim 10^{-10}$ s, which corresponds to weak interactions. The second strange property was that a hyperon could never be born alone. Thus, in proton collision, a Λ^0-hyperon is born only together with a K^+-meson or Σ^+-hyperon, but never with K^--meson or Σ^--hyperon.

In order to explain these features, particles were ascribed a new quantum number, *strangeness* S, which is preserved in strong and electromagnetic interactions and may change by ± 1 in weak interactions. To account for experimental data, the particles Λ^0, Σ^0, Σ^- and K^- were ascribed strangeness $S = -1$, Ξ-hyperons strangeness $S = -2$, and Ω^--hyperon strangeness $S = -3$.

In the quark model of hadrons, the adopted carrier of strangeness is the flavor quark s (for this one takes $S = -1$, and for all other quarks, $S = 0$). The slow rate of decay processes connected with variation of strangeness is explained by that quark flavor change $s \to u$ may occur only in weak interactions with emission of a W^--boson or absorption of a W^+-boson.

▶ **Charm C, beauty b, and truth t.** In exactly the same way, one introduces the quantum numbers of *charm* C (the c-quark is ascribed $C = 1$) and *beauty* b (the b-quark is ascribed $b = 1$). The last and the heaviest quark, the t-quark, corresponds to the quantum number t, *truth*. Just as in the case of strangeness, the quantum numbers C, b, and t are preserved in strong reactions, but may change in weak reactions.

▶ **Isotopic spin T.** Isotopic spin was introduced in nuclear physics and played an important role in the classification of the ground and the excited nuclear states. The introduction of this unusual quantum number, which has no analogue in the classical physics, reflects the assumption of *charge independence* of nuclear forces, i.e., the identical coincidence of these forces in the pairs n-n, n-p, p-p in the same spatial and spin states. To describe charge independence, one introduces the *isotopic spin vector* **T**: its magnitude for both nucleons is equal to $T = 1/2$, and its projections are $T_z = 1/2$ for proton and $T_z = -1/2$ for neutron. Thus, the two nucleons form an *isotopic doublet*. Nuclear interaction is assumed independent of the "direction" of the vector **T**, i.e., in the absence of electromagnetic interaction, the isotopic spin space is isotropic (properties of the system do not change under rotations of the vector **T**).

Summation of isotopic spins in a system of several nucleons is carried out by the same rules as that of angular momenta. Nuclei with the same number of nucleons (isobars) and the same values of T form a *nuclear multiplet* of $2T + 1$ nuclei (with different T_z, i.e., different charges), whose members have close properties. For example, the nuclei $^{10}_{6}C$, $^{10}_{5}B$ (in excited state), and $^{10}_{4}Be$ form an isotopic triplet with $T = 1$, and $^{4}_{2}He$ is an isotopic singlet ($T = 0$). A nucleus in its ground state has the minimum possible value of T for a given projection T_z (calculated from its composition); for example, the ground state of $^{10}_{5}B$ has $T = 0$.

Isotopic spin plays an important role in the world of subatomic particles, in particular, for the classification of hadrons. Baryons and mesons are split into multiplets with close properties, and the number of particles in a multiplet is $2T + 1$. Thus, three π-mesons form a triplet with $T = 1$; K^0- and K^+-mesons form a doublet with $T = 1/2$; three Σ-hyperons form a triplet with $T = 1$; four Δ-hyperons have $T = 3/2$; the Λ^0-hyperon constitutes a singlet with $T = 0$. The members of a multiplet differ by their isotopic spin projections T_z.

In the quark model, the presence of multiplets is explained by the fact that the u- and d-quarks have close masses and differ only by their charges. These quarks are ascribed the value $T = 1/2$; for the u-quark, one takes $T_z = 1/2$, and for the d-quark $T_z = -1/2$. For the other quarks, one takes $T = 0$. Isotopic spin of hadrons is calculated by the summation rule for angular momenta. Thus, the doublet of K-mesons has quark composition $u\tilde{s}$ and $d\tilde{s}$. Summation of two isotopic spins $1/2$ can result in either $T = 0$ ($T_z = 0$) or $T = 1$ ($T_z = -1$, $T_z = 0$, or $T_z = 1$). The first state corresponds to a hyperon, $\Lambda^0 = uds$ (isotopic singlet), and an isotopic triplet is formed by the hyperons $\Sigma^- = dds$, $\Sigma^0 = uds$, and $\Sigma^+ = uus$.

In reactions determined by strong interactions, one observes the *isotopic spin conservation law*: the quantities T and T_z are preserved. Electromagnetic and weak interactions violate the conservation law for T, but T_z is preserved in electromagnetic interactions and may change only in weak interactions. The point is that T_z can change only if a u- or a d-quark transform into another quark, which is possible only in weak interactions (when a charged W-boson is absorbed or emitted).

P8.4.3. Interaction Unification

▶ **Electroweak interaction.** In the *standard theory of electroweak interaction*, both interactions, the weak and the electromagnetic, together with their vector bosons, appear in the framework of a unified approach. As the original particles (also called *seed particles*) supposed to realize the unified electroweak interaction one adopts four vector massless particles, two neutral particles, and two charged ones. In the real world, the symmetry between these particles becomes spontaneously broken and three of them acquire mass (intermediate bosons), and one particle remains massless (photon). Many conclusions of the theory (in particular, mass values obtained for intermediate bosons) have had experimental support. The value of the constant of weak interaction happens to be greater than that of electromagnetic interaction, and the slow rate of weak processes (compared with electromagnetic processes) is due to the large mass of W- and Z-bosons ($\sim 100\,\text{GeV}$). For energies much larger than these masses, the symmetry is restored (masses can be neglected), and the unified symmetric electroweak interaction becomes active.

▶ **Grand unification.** The brilliant success of the theory of electroweak interactions opened the way to the creation of new theories that allow for the unification of the three fundamental interactions—strong, weak, and electromagnetic. These theories were called the "Grand Unification." The boundary energy for this unification is unimaginable, $E \sim 10^{15}\,\text{GeV}$. At this energy, the constants of the three interactions become equal. For larger energies, the three interactions come forth together as a universal symmetric interaction. The carriers of this interaction are 24 vector bosons: eight gluons, three intermediate bosons, photon, and superheavy X- and Y-bosons with masses $\sim 10^{15}\,\text{GeV}$. These bosons carry both an electric charge (4/3 for X and 1/3 for Y) and a color charge (taking into account three colors and antiparticles, we have 12 particles). Upon emission or absorption of an X- or a Y-boson, a quark may turn into a lepton and vice versa. Therefore, participation of these bosons in interactions leads to a violation of the laws of conservation of baryon and lepton charges (the electric charge conservation law remains unquestionable). As a result, there is a very small probability (because of the great masses of X and Y) of proton decay into a positron and several π-mesons. The proton lifetime predicted by the theory lies in the interval $\tau \sim 10^{30}$–10^{35} years. Current attempts to detect proton decay (in great masses of water placed in deep mines for the exclusion of background noise) give the estimate $\tau > 10^{32}$ years.

Bibliography for Chapter P8

Alenitsyn, A. G., Butikov, E. I., and Kondratyev, A. S., *Concise Handbook of Mathematics and Physics*, CRC Press, Boca Raton, Florida, 1998.

Benenson, W., Harris, J. W., Stocker, and Lutz, H. (Eds.), *Handbook of Physics*, Springer-Verlag, New York, 2002.

Bertulani, C. A., *Nuclear Physics in a Nutshell*, Princeton University Press, Princeton, New York, 2007.

Bohr, N., Neutron capture and nuclear constitution, *Nature*, Vol. 137, 1936, pp. 344–348.

Butler, S. T. and Hittmair, O. H., *Nuclear Stripping Reactions*, John Wiley and Sons Inc., New York, 1959.

Cottingham, W. N. and Greenwood, D. A., *An Introduction to Nuclear Physics*, Cambridge University Press, Cambridge, England, 2001.

Goeppert-Mayer, M., On closed shells in nuclei, *Physical Review*, Vol. 74, 1948, pp. 235–239.

Goeppert-Mayer, M., On closed shells in nuclei. II, *Physical Review*, Vol. 75, 1949, pp. 1969–1970.

Goeppert-Mayer, M. and Jensen, J. H. D., *Elementary Theory of Nuclear Shell Structure*, John Wiley and Sons, New York, 1955.

Griffiths, D., *Introduction to Elementary Particles, 2nd Edition*, Wiley-VCH Verlag GmbH & Co. KgaA, Weinheim, 2008.

Haxel, O., Jensen, J. H. D., and Suess, H. E., On the "magic numbers" in nuclear structure, *Physical Review*, Vol. 75, 1949, pp. 1766–1766.

Krane, K. S., *Introductory Nuclear Physics*, John Wiley and Sons, New York, 1987.

Lilley, J. S., *Nuclear Physics: Principles and Applications*, Wiley, New York, 2001.

Martin, B. R., *Nuclear and Particle Physics: An Introduction*, John Wiley and Sons, Chichester, England, 2006.

Meier, C. A. (Ed.), *Atom and Archetype: The Pauli/Jung Letters, 1932–1958*, Princeton University Press, Princeton, New Jersey, 2001.

Mukhin, K. N., *Experimental Nuclear Physics, Volumes 1-3, 7th Edition* [in Russian], Lan, Moscow, 2009.

Okun, L. B., *Elementary Introduction to Physics of Elementary Particles* [in Russian], Fizmatlit, Moscow, 2009.

Okun, L. B., *Particle Physics: The Quest for the Substance of Substance (Contemporary Concepts in Physics Series)*, OPA Ltd., Amsterdam, for Harwood Academic Publishers GmbH, 1985.

Okun, L. B., *Physics of Elementary Particles* [in Russian], LKI, Moscow, 2008.

Perkins, D. H., *Introduction to High Energy Physics, 4th edition*, Cambridge University Press, Cambridge, England, 2000.

Pound, R. V. and Rebka, Jr. G. A., Apparent weight of photons, *Physical Review Letters*, Vol. 4, No. 7, 1960, pp. 337–341.

Pound, R. V. and Rebka, Jr. G. A., Gravitational red-shift in nuclear resonance, *Physical Review Letters*, Vol. 3, No. 9, 1959, pp. 439–441.

Prokhorov, A. M. (Ed.), *Encyclopedia of Physics, Volumes 1-5* [in Russian], Bolshaya Russkaya Encyclopedia, Moscow, 1998.

Prokhorov, A. M. (Ed.), *Encyclopedic Dictionary of Physics* [in Russian], Sovetskaya Encyclopedia, Moscow, 1984 (dic.academic.ru).

Sivukhin, D. V., *General Physics, Volume 5, Atomic and Nuclear Physics* [in Russian], Fizmatlit, Moscow, 2006.

Sivukhin, D. V. and Yakovlev, I. A., *A Collection of Problems, Volume 5, Atomic Physics. Nuclear and Particle Physics* [in Russian], Fizmatlit, Moscow, 2006.

Williams, W. S. C., *Nuclear and Particle Physics*, Oxford University Press, New York, 1991.

Woan, G., *The Cambridge Handbook of Physics Formulas*, Cambridge University Press, Cambridge, England, 2003.

Part III

Elements of Applied and Engineering Sciences

Chapter E1

Dimensions and Similarity

E1.1. Dimensions

E1.1.1. Basic and Derived Units of Measurement. Systems of Units of Measurement

▶ **Basic and derived units of measurement.** Mechanical, physical, and physicochemical quantities are characterized by numbers that are determined by direct or indirect comparison with the accepted *units of measurement*. The units of measurement are divided into the *basic* and *derived* units of measurement. The basic units of measurement are defined as artificial and/or natural standards that are chosen from convenience considerations (i.e., rather arbitrarily). The derived units of measurement can be obtained from the basic units in the following two ways:

(i) starting from the definition of the corresponding physical quantity, which prompts the method for its (direct or mental) measurement, or

(ii) by a direct use of formulas relating different quantities by some quantitative relations.

Example 1. The velocity (in uniform motion), by definition, is the ratio of the distance passed in a certain time interval to the value of this time interval. Therefore, in this case, for the unit of velocity it is convenient to take the ratio of the length unit to the time unit.

▶ **Systems of units of measurement and classes of systems of units of measurement.** A *system of units of measurement* is a set of basic units of measurement that are sufficient to measure numerical characteristics of the class of phenomena under study.

To describe mechanical quantities, it suffices to introduce three independent basic units of measurement, namely, units of mass, length, and time.

Example 2. Nowadays, the most widely used system is the system of units of measurement named the International System of Units (SI), where the unit of mass is 1 kilogram (kg), i.e., the mass of some specially manufactured and thoroughly preserved standard, the unit of length is 1 meter (m), i.e., the length of another standard, and the unit of time is 1 second (s), i.e., the 1/86400 fraction of the mean solar day. (There is also a more precise definition of the second.)

A *class of systems of units of measurement* is a set of systems of units of measurement that differ from one another only in the value of the basic units of measurement.

Example 3. The SI system mentioned in Example 2 belongs to the class of systems of units of measurement denoted by MLT (mass — length — time). The MLT class contains the sometimes used CGS system, where the unit of mass is 1 gram (g) = 0.001 kg (one thousandth of the kilogram), the unit of length is 1 centimeter (cm) = 0.01 m (one hundredth of the metre), and the unit of time is 1 s.

E1.1.2. Dimensional and Dimensionless Quantities. Dimensional Formula

▶ **Dimensional and dimensionless quantities.** In any practical calculations and in the processing of experimental data whose results are represented as sets of numbers characterizing the properties of phenomena and processes, two types of quantities are encountered.

The quantities whose numerical value depends on the system of units of measurement used within a given class of systems of units of measurement are called *dimensional*. The quantities whose numerical values are independent of the system of units of measurement used within a given class are called *dimensionless*.

▶ **Dimensional formula. Dimensions of mechanical quantities.** The *dimension of a physical quantity* (in other words, the *dimensional formula*) is a function determining by how many times the numerical value of this quantity varies when passing from the original system of units of measurement to another system contained in the given class. In a dimensional formula, the symbol of unit length is used to be denoted by L, of unit mass, by M, and of unit time, by T; the dimension of any derived quantity A is denoted by the symbol $[A]$. The dimension of a dimensionless quantity is equal to 1.

Example. If the length unit is decreased by a factor of L and the time unit is decreased by a factor of T, then the numerical value of any measured acceleration changes by a factor of LT^{-2}. Thus, for the dimension of acceleration in the class of MLT systems, we have

$$[a] = LT^{-2}.$$

Similarly, the dimension of force in the MLT class is determined by the formula: $[F] = MLT^{-2}$. (This formula can also be obtained from Newton's second law $ma = F$.)

Table E1.1 presents the dimensions of basic mechanical quantities in the class of MLT systems.

One can see that for all mechanical quantities given in Table E1.1, the dimension is a power monomial of the form

$$[A] = L^{\alpha}M^{\beta}T^{\gamma}, \tag{E1.1.2.1}$$

where α, β, and γ are some constants.

One can prove the following more general statement: *the dimension is always a power monomial*. Such a simple structure of a dimensional formula is necessarily determined by the following physical condition: the ratio of two numerical values of any derived quantity must be independent of the choice of scales for the basic units of measurement.

▶ **Properties of dimensions.**
1. The dimension of a physical quantity does not vary if it is multiplied by a dimensionless quantity.
2. The dimension of the product of physical quantities is equal to the product of dimensions of these quantities:
$$[A_1 A_2 \ldots A_m] = [A_1][A_2] \ldots [A_m].$$
3. The exponent can be brought outside the symbol of dimension:
$$[A^p] = [A]^p.$$
4. The physical quantities can be added (subtracted) only if they have equal dimensions.
5. The exponents contained in the right-hand sides of dimension formulas like (E1.1.2.1) are rational numbers (i.e., they can be represented as the fractions p/q, where p and q are integers).

E1.2. The π-Theorem and Its Practical Use

E1.2.1. The π-Theorem

▶ **Passing from dimensional quantities to dimensionless quantities. The π-theorem.** Assume that there is a functional dependence between dimensional physical quantities,

$$f(A_1, \ldots, A_k, A_{k+1}, \ldots, A_n) = 0. \tag{E1.2.1.1}$$

TABLE E1.1
Dimensions of mechanical quantities in the class of MLT systems.

Quantity	Dimensional formula	Dimension in the SI system	Dimension in the CGS system
Length	L	m	cm
Area	L^2	m^2	cm^2
Volume	L^3	m^3	cm^3
Curvature	L^{-1}	m^{-1}	cm^{-1}
Time	T	s	s
Velocity	LT^{-1}	m/s	cm/s
Acceleration	LT^{-2}	m/s^2	cm/s^2
Angular velocity, frequency	T^{-1}	s^{-1}	s^{-1}
Angular acceleration	T^{-2}	s^{-2}	s^{-2}
Mass	M	kg	g
Density	ML^{-3}	kg/m^3	g/cm^3
Force	MLT^{-2}	$kg{\cdot}m/s^2$	$g{\cdot}cm/s^2$
Moment of force, torque	ML^2T^{-2}	$kg \cdot m^2/s^2$	$g \cdot cm^2/s^2$
Momentum	MLT^{-1}	$kg{\cdot}m/s$	$g{\cdot}cm/s$
Angular momentum	ML^2T^{-1}	$kg{\cdot}m^2/s$	$g{\cdot}cm^2/s$
Pressure, tension	$ML^{-1}T^{-2}$	$kg/(m{\cdot}s^2)$	$g/(cm{\cdot}s^2)$
Work, energy	ML^2T^{-2}	$kg{\cdot}m^2/s^2$	$g{\cdot}cm^2/s^2$
Power	ML^2T^{-3}	$kg{\cdot}m^2/s^3$	$g{\cdot}cm^2/s^3$
Modulus of elasticity	$ML^{-1}T^{-2}$	$kg/(m{\cdot}s^2)$	$g/(cm{\cdot}s^2)$
Kinematic viscosity	L^2T^{-1}	m^2/s	cm^2/s
Dynamic viscosity	$ML^{-1}T^{-1}$	$kg/(m{\cdot}s)$	$g/(cm{\cdot}s)$
Angle	dimensionless quantity	dimensionless quantity	dimensionless quantity

(In the process under study, some of these quantities can be variable, and the other, constant; moreover, here we do not distinguish between dependent and independent variables.) Let $A_1, \ldots, A_k$ $(k \le n-1)$ be dimensional quantities of independent dimensions (these quantities are called *constitutive parameters*), and suppose that the dimensions of the other quantities can be expressed via the dimensions of the constitutive parameters:

$$[A_{k+1}] = [A_1]^{m_1\,k+1}[A_2]^{m_2\,k+1}\ldots[A_k]^{m_k\,k+1},$$
$$\ldots\ldots\ldots\ldots\ldots\ldots\ldots\ldots\ldots\ldots\ldots\ldots\ldots\ldots\ldots,\qquad\text{(E1.2.1.2)}$$
$$[A_n] = [A_1]^{m_1\,n}[A_2]^{m_2\,n}\ldots[A_k]^{m_k\,n},$$

where the m_{ij} are constants. Set

$$\Pi_1 = \frac{A_{k+1}}{A_1^{m_1\,k+1}A_2^{m_2\,k+1}\ldots A_k^{m_k\,k+1}},\quad\ldots,\quad\Pi_{n-k} = \frac{A_n}{A_1^{m_1\,n}A_2^{m_2\,n}\ldots A_k^{m_k\,n}}.\qquad\text{(E1.2.1.3)}$$

It follows from relations (E1.2.1.2) that the quantities $\Pi_1, \Pi_2, \ldots, \Pi_{n-k}$ are dimensionless. Each of the dimensionless quantities Π_i has the form of a power monomial consisting of $\leq k+1$ dimensional quantities. (The inequality takes into account the fact that some of the exponents m_{ij} can be zero.)

THE π-THEOREM. *The original relation* (E1.2.1.1) *among* n *dimensional physical quantities can be represented as a relation among* $n-k$ *dimensionless quantities* (E1.2.1.3),

$$F(\Pi_1, \Pi_2, \ldots, \Pi_{n-k}) = 0.\qquad\text{(E1.2.1.4)}$$

It follows from the above that every physical relation between dimensional quantities can be reformulated as a relation between fewer dimensionless quantities. This important fact is a source of useful applications of the π-theorem for the analysis of various phenomena and processes. In particular, for $k = n - 1$, the same n dimensional parameters can form only a single dimensionless combination, and hence, by the π-theorem, the initial functional dependence can be represented in the following simple form:

$$A_1^{m_1} A_2^{m_2} \ldots A_n^{m_n} = \text{const}.\qquad\text{(E1.2.1.5)}$$

Here the exponents $m_1, m_2, \cdots, m_n$ can readily be determined by dimensional formulas, where it is taken into account that the left-hand side of (E1.2.1.5) is a dimensionless complex and the unknown dimensionless constant on the right-hand side in (E1.2.1.5) can be found either experimentally or theoretically by solving the corresponding mathematical problem (see examples in Section E1.2.2 below).

Remark 1. Without loss of generality, one can assume that one of the exponents m_i in (E1.2.1.5) is equal to 1 (in particular, if $m_1 \neq 0$, then one can set $m_1 = 1$).

Remark 2. If $\Pi_1, \Pi_2, \ldots, \Pi_s$ are dimensionless quantities, then the complex $\Pi = \Pi_1^{\beta_1}\Pi_2^{\beta_2}\ldots\Pi_s^{\beta_s}$, where $\beta_1, \beta_2, \ldots, \beta_s$ are arbitrary constants, is also a dimensionless quantity. This allows one to compose new $n - k$ functionally independent dimensionless quantities from the dimensionless quantities (E1.2.1.3) and write out a relation of the form (E1.2.1.4) for them. A successful choice of the form of the dimensionless parameters permits reducing the number of experiments and simplifying their analysis.

▶ **Experimental data must be processed in dimensionless variables.** The π-theorem leads to a very important conclusion that experimental data must be processed in dimensionless variables, thus reducing the number of measured and varied quantities.

In practical experiments, it is usually assumed that establishing the dependence of a quantity on some constitutive parameter (argument) means measuring this quantity for at least 10 values of the given argument (of course, the number 10 is here taken at random). Thus, to experimentally determine the quantity A as a function of n constitutive parameters $A_1, \ldots, A_n$, one must perform 10^n experiments. But, using the π-theorem, one can reduce the problem to determining a function of $n - k$ dimensionless arguments $\Pi_1, \ldots, \Pi_{n-k}$; to this end, one must perform 10^{n-k} experiments, i.e., 10^k times fewer. In other words, the labouriousness of determining the desired function is reduced by as many orders of magnitude as the number of quantities of independent dimensions contained in the set of constitutive parameters.

E1.2.2. Simplest Examples of the Use of the π-Theorem

In what follows, all dimensions of constitutive parameters are written in the class of MLT systems of units of measurement.

▶ **The Pythagorean theorem.** Let us describe an amusing version of proving the Pythagorean theorem by analyzing the dimensions.

The area S of a rectangular triangle is determined by the length c of the hypotenuse and, for example, the smaller of the acute angles, β. Using the three quantities S, c, and β whose dimensions have the form $[S] = L^2$, $[c] = L$, and $[\beta] = 1$ (the angle β is a dimensionless quantity), one can compose two dimensionless combinations:

$$\Pi_1 = S/c^2, \quad \Pi_2 = \beta. \tag{E1.2.2.1}$$

We set $n = 3$ and $k = 1$ in formula (E1.2.1.4) and take the explicit form of the dimensionless complexes (E1.2.2.1) into account. We have $F(S/c^2, \beta) = 0$. Resolving the dependence thus obtained for the first argument, we obtain

$$S = c^2\Phi(\beta),$$

where $\Phi(\beta)$ is a function. (The explicit form of this function is not required in the following.)

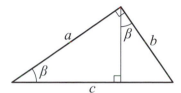

Figure E1.1. A right triangle.

The height perpendicular to the hypotenuse (see Fig. E1.1) divides the main triangle into two right triangles similar to it whose hypotenuses are the corresponding legs a and b of the original triangle. The areas of the new triangles are $S_a = a^2\Phi(\beta)$ and $S_b = b^2\Phi(\beta)$, respectively, where $\Phi(\beta)$ is the same function as above. Since the sum of S_a and S_b is the area of the original triangle, $S = S_a + S_b$, we have $c^2\Phi(\beta) = a^2\Phi(\beta) + b^2\Phi(\beta)$. By canceling $\Phi(\beta)$, we obtain

$$c^2 = a^2 + b^2,$$

as desired.

▶ **Shock wave front propagation in nuclear explosion.** In nuclear explosion, considerable energy E_0 is released very rapidly (we assume that it is released instantaneously) in a rather small area (which is approximately assumed to be a point). A powerful shock wave propagates from the explosion center, and the pressure behind it initially equals hundreds of thousands of atmospheres. This pressure is much larger than the initial air pressure, whose influence can be neglected at the first stage of the explosion. Thus, after the time interval t, the shock wave front radius r_f depends on E_0, t, and the initial air density ρ_0.

This problem contains four dimensional quantities r_f, E_0, t, and ρ_0, which have the following dimensions:

$$[r_f] = L, \quad [E_0] = ML^2T^{-2}, \quad [t] = T, \quad [\rho_0] = ML^{-3}. \tag{E1.2.2.2}$$

In the set of these four quantities, one can choose three quantities with independent dimensions, which corresponds to $k = 3$ and $n = 4$ in the statement of the π-theorem. Therefore, we have $k + 1 = n$ and can form a single dimensionless complex of the quantities (E1.2.2.2), and the desired functional dependence can be represented by analogy with (E1.2.1.5) as

$$r_f E_0^{m_2} t^{m_3} \rho_0^{m_4} = C = \text{const},$$ (E1.2.2.3)

where, without loss of generality, the first exponent m_1 was set equal to 1. The exponents m_2, m_3, and m_4 are determined by the condition that the dimension of the left-hand side of (E1.2.2.3) should be equal to 1 (because the right-hand side is a dimensionless constant). Therefore, we obtain the chain of relations

$$[r_f E_0^{m_2} t^{m_3} \rho_0^{m_4}] = [r_f][E_0]^{m_2}[t]^{m_3}[\rho_0]^{m_4} = L[ML^2T^{-2}]^{m_2} T^{m_3}[ML^{-3}]^{m_4}$$
$$= M^{m_2+m_4} L^{2m_2-3m_4+1} T^{-2m_2+m_3} = 1.$$

For the last relation to be satisfied, it is necessary that all exponents be zero:

$$m_2 + m_4 = 0,$$
$$2m_2 - 3m_4 + 1 = 0,$$
$$-2m_2 + m_3 = 0.$$

The solution of this linear system of algebraic equations has the form

$$m_2 = -\tfrac{1}{5}, \quad m_3 = -\tfrac{2}{5}, \quad m_4 = \tfrac{1}{5}.$$ (E1.2.2.4)

Substituting the exponents (E1.2.2.4) into (E1.2.2.3) and solving the resulting relation for r_f, we obtain the shock wave propagation law

$$r_f = C\left(\frac{E_0 t^2}{\rho_0}\right)^{1/5}.$$ (E1.2.2.5)

If the quantities E_0 and ρ_0 are known, then to find the constant C in formula (E1.2.2.5) it suffices to know the shock wave front radius r_f for a single value $t = t_1$ of the time interval. A more detailed study (beyond the framework of the theory of dimensions) shows that $C \approx 1$.

From formula (E1.2.2.5), one can obtain the time dependence of the shock wave front radius:

$$r_f = r_{f1}\left(t/t_1\right)^{2/5},$$

where $r_{f1} = r_f|_{t=t_1}$.

▶ **Pendulum oscillations in the field of gravity forces.** We consider a mathematical pendulum, which is a heavy material point of mass m suspended on a weightless unstretchable string of length l; it is assumed that the other end of the string is rigidly fixed. We are interested in the oscillation period T_p of the pendulum that makes an angle φ_0 with the vertical axis at the initial time instant. (The weight velocity is zero at the initial time instant.)

This problem is characterized by the five quantities m, l, T_p, φ_0, and g (g is the acceleration due to gravity), which have the dimensions

$$[m] = \text{M}, \quad [l] = \text{L}, \quad [T_p] = T, \quad [\varphi_0] = 1, \quad [g] = \text{LT}^{-2};$$ (E1.2.2.6)

three of these quantities have independent dimensions. Thus, in this case we have $n = 5$ and $k = 3$, and one can compose two dimensionless complexes of these quantities (E1.2.2.6):

$$\Pi_1 = \varphi_0, \quad \Pi_2 = T_p\sqrt{g/l}.$$ (E1.2.2.7)

Therefore, by the π-theorem, we have the dependence $F(\varphi_0, T_p \sqrt{g/l}) = 0$. By solving it for the oscillation period T_p, we obtain

$$T_p = f_1(\varphi_0)\sqrt{l/g}. \tag{E1.2.2.8}$$

where $f_1(\varphi_0)$ is a function.

It is important to note that the dimensionless parameters (E1.2.2.7) are independent of the load mass m. Therefore, the oscillation period (E1.2.2.8) is independent of the pendulum mass.

The function $f_1(\varphi)$ can be determined both theoretically and experimentally.

Assuming that the function $f_1(\varphi)$ can be expanded in a Taylor series for small φ, we have

$$f_1(\varphi_0) = a + b\varphi_0 + c\varphi_0^2 + \cdots.$$

Assume that $a \neq 0$. Then, for small pendulum oscillations, by retaining only the leading term of the expansion in (E1.2.2.8), we obtain the simple formula

$$T_p = a\sqrt{l/g}, \tag{E1.2.2.9}$$

where a is a constant.

Thus, for small pendulum oscillations, we can use the theory of dimensions to obtain formula (E1.2.2.9) for the oscillation period up to a constant factor. A theoretical analysis (independent of the theory of dimensions) permits finding the unknown constant $a = 2\pi$.

▶ **Interphase transport.** When describing the interphase transport in gas–liquid systems, the Higbie–Danckwerts surface renewal model has been widely used. According to this model, turbulent pulsations with frequency $\omega = 1/t_\omega$ in the liquid phase constantly renew the interphase surface. The contact time t_ω is determined experimentally. The diffusion flow J through an element of the renewed surface is assumed to be purely molecular. It depends on the molecular diffusion coefficient D of the gas dissolved in the liquid. The mass transfer coefficient is defined as follows:

$$k = \frac{J}{S(C_s - C_\infty)},$$

where S is the interphase surface area, C_∞ is the concentration of the transferred component in the depth of the liquid, and C_s is the concentration on the interphase surface.

Experiments show that the mass transfer coefficient k mainly depends only on the contact time t_ω and the diffusion coefficient D. The dimensions of the three above-listed quantities have the form

$$[k] = \frac{L}{T}, \quad [D] = \frac{L^2}{T}, \quad [t_\omega] = T.$$

There relations contain only two independent dimensions L and T. Therefore, by the π-theorem, one can construct only one dimensionless complex of the form (E1.2.1.5) from the quantities k, D, and t_ω, which results in the dependence

$$k = C\sqrt{D/t_\omega},$$

where C is a constant.

According to the theoretical results obtained by Higbie and Danckwerts, $C = 2/\sqrt{\pi}$. In the Danckwerts model, it is proposed to determine the value of t_ω by the spectrum of the time of residence of liquid elements on the interphase surface. The exponential form of this spectrum is accepted starting from the concept of statistical independence of turbulent vortices arising in the depth of the liquid and reaching the interphase surface.

▶ **Equation of state of an ideal gas.** An ideal gas is understood as a gas whose static state is completely determined by the values of pressure P, density ρ, absolute temperature T_a, and the heat capacity c_v.

The dimensions of these quantities have the form

$$[P] = \frac{M}{LT^2}, \quad [\rho] = \frac{M}{L^3}, \quad [T_a] = K, \quad [c_v] = \frac{L^2}{T^2 K}. \tag{E1.2.2.10}$$

We take into account the fact that the dimension K of temperature (in degrees Kelvin) is a derived unit of measurement, which can be written in terms of dimensions of the basic units of measurement as follows: $K = ML^2/T^2$. Therefore, relations (E1.2.2.10) for four quantities contain three independent dimensions. Hence, by the π-theorem, the desired functional dependence for P, ρ, T_a, and c_v can be represented in a simple form like (E1.2.1.5). In the case under study, this leads to the Clapeyron—Mendeleev equation

$$\frac{P}{c_v \rho T_a} = \text{const},$$

which has the more usual form $P = \rho R T_a$, where R is the gas constant.

▶ **Flow of a viscous fluid about a solid particle.** Consider a homogenous flow of a viscous incompressible fluid about a solid spherical particle of diameter d. The particle drag force F caused by viscosity depends on the four parameters d, U_i, ν, and ρ, where U_i is the unperturbed flow velocity far from the particle and ν and ρ are the kinematic viscosity and the fluid density, respectively.

The dimensions of all five above-listed quantities (including the force) have the form

$$[F] = \frac{ML}{T^2}, \quad [d] = L, \quad [U_i] = \frac{L}{T}, \quad [\nu] = \frac{L^2}{T}, \quad [\rho] = \frac{M}{L^3}.$$

Out of these quantities, one can choose three with independent dimensions and compose two dimensionless complexes

$$\Pi_1 = \frac{F}{d^2 \rho U_i^2}, \quad \Pi_2 = \frac{dU_i}{\nu}, \tag{E1.2.2.11}$$

which, by the π-theorem, satisfy the functional relation $F(\Pi_1, \Pi_2) = 0$. By solving this relation for Π_1, we find the general structure of the dependence of the drag force on the other parameters:

$$\frac{F}{d^2 \rho U_i^2} = \Phi\left(\frac{dU_i}{\nu}\right). \tag{E1.2.2.12}$$

Thus, dimension analysis permits significantly simplifying experimental studies by decreasing the number of varied variables in the problem from five to two. The complex $\mathsf{Re} = dU_i/\nu$ on the right-hand side of the expression (E1.2.2.12) is called the *Reynolds number*. (One often takes the particle radius, $a = d/2$, to be the characteristic length when determining the Reynolds number.)

Experiments show that $\Phi \to A$ as $\mathsf{Re} \to \infty$, where A is a constant. This situation in which the drag force is independent of viscosity and quadratically depends on velocity is called the Newton flow mode.

As $\mathsf{Re} \to 0$, we have $\Phi \to B/\mathsf{Re}$. (This result can be obtained theoretically.) In this case, the drag force linearly depends on both the viscosity and the velocity. Such a flow is called a Stokes (or creeping) flow.

Remark. In the case of a homogeneous translational flow of a viscous incompressible fluid about a body of nonspherical shape with typical linear dimension d (for example, for d one can take the diameter of a spherical body with equivalent volume), the dependence (E1.2.2.12) also holds, where the function $\Phi(z)$ depends on the shape of the body.

▶ **Stirring a liquid.** The power N required to stir a liquid with density ρ and kinematic viscosity ν depends on the mixer diameter d, the angular velocity of stirring ω, and the acceleration due to gravity g. The problem complexity does not allow one to calculate the power by solving the hydrodynamic equations analytically. Therefore, we use dimensional analysis.

We write out the dimensions of the main quantities:

$$[N] = \frac{\mathrm{ML}^2}{\mathrm{T}^3}, \quad [\rho] = \frac{\mathrm{M}}{\mathrm{L}^3}, \quad [\nu] = \frac{\mathrm{L}^2}{\mathrm{T}}, \quad [d] = \mathrm{L}, \quad [\omega] = \frac{1}{\mathrm{T}}, \quad [g] = \frac{\mathrm{L}}{\mathrm{T}^2}.$$

We use the π-theorem to determine the number of dimensionless complexes of the problem:

$$3 = 6(\textit{the number of parameters of the problem}) - 3(\textit{the number of primary dimensions}).$$

Under the assumption that the quantities d, ρ, and ω can enter any dimensionless combination, we have

$$\frac{N}{\rho d^5 \omega^3} = \Phi\left(\frac{d^2 \omega}{\nu}, \frac{d\omega^2}{g}\right).$$

The structure of the obtained functional dependence allows us to simplify the experimental study significantly by decreasing the number of varied variables in the problem from six to three. Experiments show that for $d^2\omega/\nu \le 10^{-2}$, when the influence of the viscosity is large, the mixer power consumption ceases to depend on the dimensionless complex $d^2\omega/\nu$.

▶ **Turbulent flow in a tube.** Consider a steady turbulent flow in a circular smooth tube of radius a. Assume that r is the radial coordinate counted off from the tube axis; τ_w is the friction stress* on the tube wall; $\overline{V}$ is the average component of the fluid velocity; $\overline{V}_{\max}$ is the maximum velocity on the tube axis; and ν and ρ are the liquid kinematic viscosity and density, respectively.

It is easily seen that the average velocities $\overline{V}$ and $\overline{V}_{\max}$ of the fluid motion are determined by the following set of parameters:

$$a, \quad \rho, \quad \nu, \quad \tau_w, \quad r = a - y.$$

All the dimensionless quantities depend on the two dimensionless complexes

$$\mathrm{Re}_* = \frac{a U_*}{\nu}, \quad \frac{r}{a} = 1 - \frac{y}{a},$$

where $U_* = \sqrt{\tau_w/\rho}$ is the velocity of the tangential friction stress on the tube wall.

It follows from the similarity theory that

$$\frac{\overline{V}_{\max} - \overline{V}}{U_*} = F\left(\mathrm{Re}_*, \frac{a - y}{a}\right). \tag{E1.2.2.13}$$

The difference $\overline{V}_{\max} - \overline{V}$ is called the velocity defect. It characterizes the fluid velocity distribution over the tube cross-section with respect to the motion on the axis.

---

* The dependence $\tau_w = \frac{a}{2}\frac{\Delta P}{L}$ holds, where $\Delta P > 0$ is the pressure difference on the tube length L.

For very large Reynolds numbers Re_*, formula (E1.2.2.13) implies that

$$\frac{\overline{V}_{\max} - \overline{V}}{U_*} = F_1\left(\frac{a-y}{a}\right), \tag{E1.2.2.14}$$

where $F_1(z) = F(\infty, z)$.

The expression (E1.2.2.14) is called the universal law of velocity distribution for the velocity defect. Experiments show that the law (E1.2.2.14) holds in the entire flow region inside smooth and rough tubes except for a narrow near-wall region. In this case, the function $F_1(z)$ on the right-hand side in formula (E1.2.2.14) is well approximated by the Prandtl expression $F_1(z) = 2{,}5 \ln z$.

E1.3. Similarity and Simulation. General Remarks
E1.3.1. Similarity and Simulation

▶ **Physical similarity and simulation.** Prior to manufacturing a large and expensive structure (such as ship, aircraft, cooler, chemical reactor, etc.), one often turns to simulation, i.e., model testing, so as to obtain the best characteristics of the structure under its expected operation conditions. In simulation, it is necessary to know how to recalculate the results obtained in experiments with models to the case of actual conditions: if this is unknown, then simulation is useless. This also applies to complicated phenomena and processes. The theory of simulation is based on the notion of similar phenomena.

Consider a class of qualitatively similar physical phenomena or processes that differ only in the numerical values of the constitutive parameters that are used to compose s dimensionless complexes $\Pi_1, \Pi_2, \ldots, \Pi_s$. (Here $s = n - k$; see the π-theorem.) Two phenomena or processes from a given class are said to be physically *similar* if they are described by the same dimensionless complexes:

$$\Pi_1' = \Pi_1'', \quad \Pi_2' = \Pi_2'', \quad \ldots, \quad \Pi_s' = \Pi_s'', \tag{E1.3.1.1}$$

where the single prime refers to the first phenomenon (called the model phenomenon), and the double prime refers to the second phenomenon (called the full-scale phenomenon). In this definition, the dimensionless complexes $\Pi_1, \Pi_2, \ldots, \Pi_s$ are called the *similarity parameters*.

Physical similarity is a generalization of geometric similarity concerned with bodies (objects) whose characteristic dimensions are proportional.

▶ **Rules for recalculating the measurement results from the model to the case of actual conditions.** If the similarity parameters are determined by formulas (E1.2.1.3), then relations (E1.3.1.1) become

$$\frac{A_q'}{(A_1')^{m_{1q}}(A_2')^{m_{2q}}\ldots(A_k')^{m_{kq}}} = \frac{A_q''}{(A_1'')^{m_{1q}}(A_2'')^{m_{2q}}\ldots(A_k'')^{m_{kq}}}, \quad q = k+1, \ldots, n \tag{E1.3.1.2}$$

and determine simple rules for recalculating the measurement results from the model to the case of actual conditions (under which it is difficult to perform direct measurements).

Conditions (E1.3.1.2) show how to choose the model constitutive parameters $A_{k+1}', \ldots, A_n'$ to ensure the similarity of the model to the case of actual conditions:

$$A_q' = A_q''(A_1'/A_1'')^{m_{1q}}\ldots(A_k'/A_k'')^{m_{kq}}, \quad q = k+1, \ldots, n; \tag{E1.3.1.3}$$

the model parameters $A_1', \ldots, A_k'$ can be chosen arbitrarily. The quantities $A_1'', \ldots, A_n''$ in (E1.3.1.3) are introduced based on required technical (or actually existing) parameters of the process under study.

Example. Suppose a streamlined ship moves on the surface of a liquid. The main contribution to the drag of the rapidly moving ship is due to the surface waves created by it. We consider an ideal situation in which the contribution of the streamlined ship viscous drag can be neglected as the first approximation. The ship drag force F is determined by its characteristic length l, its velocity U, the liquid density ρ, and the acceleration due to gravity g. (The last parameter is essential, because the force of gravity has a determining effect on waves.) The quantities under study have the following dimensions:

$$[F] = \text{MLT}^{-2}, \quad [l] = \text{L}, \quad [U] = \text{LT}^{-1}, \quad [\rho] = \text{ML}^{-3}, \quad [g] = \text{LT}^{-2}.$$

Here $n = 5$ and $k = 3$, and therefore, one can compose two dimensionless complexes

$$\Pi_1 = \frac{U}{\sqrt{lg}}, \quad \Pi_2 = \frac{F}{\rho(lU)^2}. \tag{E1.3.1.4}$$

The dynamic similarity parameter Π_1 is called the *Froude number*, which is denoted by the special symbol Fr. Thus, for the model and in the case of actual conditions, the quantity U^2/l must be the same (with g remaining unchanged), which results in the following rule for recalculating the model velocities to the case of actual conditions:

$$U' = U'' \sqrt{\frac{l'}{l''}}. \tag{E1.3.1.5}$$

Using the second similarity parameter in (E1.3.1.4) and assuming that the liquid is the same for the model and the actual conditions (more precisely, $\rho' = \rho''$), we obtain the following rule for recalculating the drag force:

$$F' = F'' \left(\frac{l'U'}{l''U''} \right)^2 = F'' \left(\frac{l'}{l''} \right)^3, \tag{E1.3.1.6}$$

i.e., the drag force is proportional to the cubed linear scale of simulation. Formula (E1.3.1.6) was derived using relation (E1.3.1.5).

E1.3.2. General Concluding Remarks

The use of dimensionless parameters simplifies analysis and interpretation of experimental data to a large extent and, in numerous cases, allows successful simulation of phenomena and processes that cannot be studied under laboratory conditions. Combining the theory of dimensions and the similarity theory with the conclusions obtained experimentally or by mathematical analysis of the corresponding equations, one can often obtain rather significant and nontrivial results.

Although undoubtedly useful, the theory of dimensions and similarity is in fact a restricted theory, because it does not permit obtaining the explicit form of the functional relations between dimensionless quantities (it only reveals the structure of these dependencies and the number of dimensionless parameters). The desired functional dependencies describing the phenomenon or process under study can be determined both experimentally and theoretically by using the existing investigation methods. For the problems that are stated mathematically as appropriate equations and boundary conditions, the role of methods of the theory of dimensions and similarity is mostly of auxiliary character. (After the problem has been stated mathematically in dimensional variables, it is often reasonable to reformulate it in terms of dimensionless variables.)

Bibliography for Chapter E1

Barenblatt, G. I., *Dimensional Analysis*, Gordon & Breach, New York, 1989.
Bridgman, P. W., *Dimensional Analysis*, Yale Univ. Press, New Haven, Connecticut, 1931.
Danckwerts, P. V., Gas–Liquid Reactions, McGraw-Hill, New York, 1970.
Sedov, L. I., *Similarity and Dimensional Methods in Mechanics*, CRC Press, Boca Raton, Florida, 1993.
Taylor, G. I., The formation of a blast wave by a very intense explosion, *Proc. Roy. Soc., Ser. A,* 1950, Vol. 201, No. 1065, pp. 159–186.

Chapter E2

Mechanics of Point Particles and Rigid Bodies

E2.1. Kinematics

Kinematics is the field of mechanics where the geometric properties of motion of a point or a rigid body are studied independently of their mass and causes of their motion.

E2.1.1. Kinematics of a Point

To specify the motion of a point M is to introduce a rule determining the point position with respect to a reference frame $Oxyz$ at each time instant. The most routine and convenient methods for specifying the motion of a point are the vector, coordinate, and natural methods.

▶ **The vector method for specifying the motion of a point.** In this method, the position of point M is specified by the *position vector* $\mathbf{r}(t)$ drawn to this point from a stationary center (pole, point). For the pole one can take any point, for example, the origin O of a coordinate system (Fig. E2.1). The position vector is a vector function of a scalar argument—time t—and in the course of time the endpoint of the vector $\mathbf{r}$ traces a curve in space, which is called the *trajectory* (or *path*) of the point.

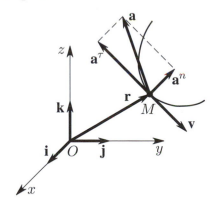

Figure E2.1. Specifying the motion of point M in a Cartesian coordinate system. The trajectory of motion of the point, its velocity, and acceleration.

Along with the position vector, the kinematic state of point M is characterized by the velocity vector $\mathbf{v}(t)$ and the acceleration vector $\mathbf{a}(t)$.

The *velocity* $\mathbf{v} = \dfrac{d\mathbf{r}}{dt} = \dot{\mathbf{r}}$, which is the derivative of the position vector with respect to time,* characterizes the rate and direction of variation in the position of the point in space. The velocity vector is directed along the tangent to the trajectory at point M.

---

   * In this chapter, the time derivatives are usually denoted by a dot (which is traditional notation in classical mechanics).

The *acceleration* $\mathbf{a} = \dot{\mathbf{v}} = \ddot{\mathbf{r}}$ is the first time derivative of the velocity vector, or the second derivative of the position vector. It is convenient to represent the acceleration as the sum of two vectors $\mathbf{a}^\tau$ and $\mathbf{a}^n$ lying in the plane spanned by the vectors $\mathbf{v}$ and $\mathbf{a}$ (see Fig. E2.1):

$$\mathbf{a} = \mathbf{a}^\tau + \mathbf{a}^n.$$

The vector $\mathbf{a}^\tau$ is parallel or antiparallel to $\mathbf{v}$ and is called the *tangential* acceleration, and the vector $\mathbf{a}^n$ is perpendicular to $\mathbf{a}^\tau$, points to the center of curvature of the trajectory, and is called the *normal* acceleration.

The projection $\text{comp}_\mathbf{v}\,\mathbf{a}$ of the vector $\mathbf{a}$ on the direction of the vector $\mathbf{v}$, the tangential acceleration magnitude a^τ, and the normal acceleration magnitude a^n can be found by the formulas

$$\text{comp}_\mathbf{v}\,\mathbf{a} = \frac{\mathbf{a} \cdot \mathbf{v}}{v} = \dot{v}, \quad a^\tau = |\dot{v}|, \quad a^n = \sqrt{a^2 - (a^\tau)^2},$$

where $\mathbf{a} \cdot \mathbf{v}$ is the scalar product of the vectors $\mathbf{a}$ and $\mathbf{v}$.

The sign of $\text{comp}_\mathbf{v}\,\mathbf{a}$ determines the character of motion of the point at the time instant under study: if this quantity is positive, then the velocity magnitude v increases, and the motion is said to be *accelerated*; otherwise, the velocity magnitude decreases, and the motion is said to be *decelerated*.

▶ **The coordinate method for specifying the motion of a point.** The position of point M in space is specified by three functions of time,

$$x = x(t), \qquad y = y(t), \qquad z = z(t), \tag{E2.1.1.1}$$

where x, y, z are the Cartesian coordinates of M. This method is another representation of the vector method, because the position vector $\mathbf{r}$ issuing from the origin can be represented as the sum

$$\mathbf{r} = x(t)\mathbf{i} + y(t)\mathbf{j} + z(t)\mathbf{k},$$

where $\mathbf{i}$, $\mathbf{j}$, $\mathbf{k}$ are the unit vectors (basis vectors) directed along the coordinate x-, y-, and z-axes (Fig. E2.1).

The projections of the velocity on the coordinate v_x-, v_y-, and v_z-axes can be found by differentiation with respect to time of the corresponding functions (E2.1.1.1):

$$v_x = \dot{x}, \qquad v_y = \dot{y}, \qquad v_z = \dot{z}. \tag{E2.1.1.2}$$

The velocity vector can be found as the vector sum of three vectors,

$$\mathbf{v} = v_x\mathbf{i} + v_y\mathbf{j} + v_z\mathbf{k},$$

and the velocity magnitude is given by the formula

$$v = \sqrt{v_x^2 + v_y^2 + v_z^2}. \tag{E2.1.1.3}$$

The projections a_x, a_y, and a_z of the acceleration onto the coordinate axes, the acceleration $\mathbf{a}$, and its magnitude a can be found by the formulas

$$a_x = \dot{v}_x = \ddot{x}, \quad a_y = \dot{v}_y = \ddot{y}, \quad a_z = \dot{v}_z = \ddot{z},$$
$$\mathbf{a} = a_x\mathbf{i} + a_y\mathbf{j} + a_z\mathbf{k}, \quad a = \sqrt{a_x^2 + a_y^2 + a_z^2}. \tag{E2.1.1.4}$$

The tangential acceleration is calculated by the formulas

$$\text{comp}_{\mathbf{V}}\, \mathbf{a} = \dot{v} = \frac{\dot{x}\ddot{x} + \dot{y}\ddot{y} + \dot{z}\ddot{z}}{\sqrt{\dot{x}^2 + \dot{y}^2 + \dot{z}^2}}; \qquad a^\tau = |\dot{v}|. \tag{E2.1.1.5}$$

▶ **The natural method for specifying the motion of a point.** In this case, the position of point M in space is specified by specifying

(a) The trajectory of the point.

(b) The origin (point O on the trajectory from which the angular position s is counted).

(c) The positive sense of the arc coordinate.

(d) The law of motion $s = s(t)$ of the point, i.e., the time dependence of the arc coordinate.

The arc coordinate s is the length of the trajectory arc counted from point O to point M. If the length is counted in the positive sense, then the arc coordinate is positive; otherwise, it is negative.

The natural coordinate axes are constructed at point M. For a given trajectory of the point, the position vector $\mathbf{r}$ connecting the fixed pole with point M becomes a function of the arc coordinate, $\mathbf{r} = \mathbf{r}(s)$. Its derivative $\boldsymbol{\tau} = \mathbf{r}'_s$ with respect to the arc coordinate is the unit vector *tangent* to the trajectory at point M in the sense of increasing arc coordinate.

The second derivative $\mathbf{r}''_{ss} = \boldsymbol{\tau}'_s = \mathbf{n}/\rho$ gives the second unit vector $\mathbf{n}$ of the *principal normal*, which is perpendicular to $\boldsymbol{\tau}$ and points to the center of curvature of the trajectory. The geometric meaning of the factor $1/\rho$ will be explained later.

The last, third unit vector $\mathbf{b}$ of the *binormal* is the cross product of the first two unit vectors, $\mathbf{b} = \boldsymbol{\tau} \times \mathbf{n}$.

The right trihedron of unit vectors thus constructed bears the name of the *natural coordinate axes*. The planes passing through them are called as follows: $(\mathbf{nb})$ is the *normal plane*, $(\mathbf{b}\boldsymbol{\tau})$ is the *rectifying plane*, and $(\boldsymbol{\tau}\mathbf{n})$ is the *osculating plane*.

If on the trajectory near the point M we choose a point M' whose arc coordinate $s + \Delta s$ differs by a small value Δs, them, up to terms of higher order of magnitude, the distances from M' to the above-listed planes are as follows: Δs to the normal plane, $(\Delta s)^2/(2\rho)$ to the rectifying plane, and $\beta(\Delta s)^3$ to the osculating plane, where β is a quantity of the order of unity. This means that the part of the trajectory in a small neighborhood of point M "practically" lies in the osculating plane.

Let us construct a family of circles tangent to the trajectory at point M so that their centers are located on the principal normal $\mathbf{n}$. These circles lie in the osculating plane. The circle that is tangent to the trajectory most closely is called the *osculating circle*. The center of this circle, its radius ρ, and the quantity $k = 1/\rho$ are called the *center of curvature*, the *radius of curvature*, and the *curvature* of the trajectory at point M (Fig. E2.2). On flat parts of the trajectory, the values of ρ are larger and the values of k are less than on the curved parts.

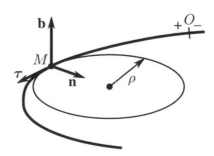

Figure E2.2. Osculating circle, the center of curvature, and the radius of curvature.

Using the natural coordinate axes, one can write the velocity and acceleration vectors of point M as

$$\mathbf{v} = \dot{s}\boldsymbol{\tau}, \qquad \mathbf{a} = \mathbf{a}^\tau + \mathbf{a}^n = \ddot{s}\boldsymbol{\tau} + (v^2/\rho)\mathbf{n}. \qquad \text{(E2.1.1.6)}$$

In the last expressions, one must take into account the following:

First, the vector $\mathbf{v}$ is directed along the tangent to the trajectory and $v = |\dot{s}|$; the point velocity $\mathbf{v}$ is directed in the sense of increasing s if $\dot{s} > 0$ and decreasing s if $\dot{s} < 0$.

Second, the vector $\mathbf{a}$ lies in the osculating plane and points in the direction of the trajectory concavity; $\ddot{s} = \text{comp}_\mathbf{v}\,\mathbf{a}$; $|\ddot{s}| = a^\tau$; the tangential acceleration vector $\mathbf{a}^\tau$ is directed in the sense of increasing s if $\ddot{s} > 0$ and decreasing s if $\ddot{s} < 0$.

Third, if $\dot{s}\ddot{s} > 0$, then, at the time instant under study, the velocity magnitude v increases and the motion is accelerated; for $\dot{s}\ddot{s} < 0$, the velocity magnitude v decreases and the motion is decelerated.

The normal acceleration is zero if the point moves along a rectilinear part of the trajectory, where $k = 1/\rho = 0$. The tangential acceleration of the point is zero if the velocity magnitude remains constant.

A *uniformly varying* motion of a point is a motion where $\ddot{s} = a_0 = \text{const}$. In this case, the values of the velocity and the arc coordinate can be found by the formulas

$$\dot{s} = \dot{s}_0 + a_0(t - t_0),$$
$$s = s_0 + \dot{s}_0(t - t_0) + \tfrac{1}{2}a_0(t - t_0)^2. \qquad \text{(E2.1.1.7)}$$

Here $\dot{s}_0$ and s_0 are the values of the velocity and the arc coordinate corresponding to the time instant t_0.

A *uniform motion* is a motion of a point in which $\ddot{s} = a_0 = 0$.

In the case of an arbitrary motion, the arc coordinate s and the path L passed over by the point can be found by the formulas

$$s = \int_{t_0}^{t} \dot{s}\,dt, \qquad L = \int_{t_0}^{t} v\,dt.$$

This clearly illustrates the difference between the arc coordinate and the path passed over by the point.

Example. An ellipsograph rod AB of length $2l = 2$ m (Fig. E2.3) moves so that the angle φ (in radians) varies according to the law $\varphi = -\tfrac{1}{2}t^2 + 2t + \tfrac{\pi}{6} - \tfrac{3}{2}$. Furthermore, point B of the rod moves along the horizontal line, and point A moves along the vertical line. Find the trajectory of point K lying at the middle of the rod, its velocity, acceleration, and the radius of curvature of the trajectory for $t = 1$ s.

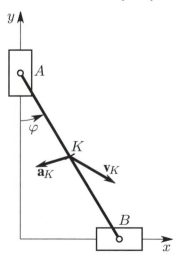

Figure E2.3. Ellipsograph (thin rod) whose endpoints can move along perpendicular straight lines.

Solution. The method used here to specify the motion of the point differs from the three above-mentioned ones. Therefore, we pass to the coordinate method and use the corresponding formulas. We introduce the coordinate axes as is shown in the figure and then find the coordinates of point K: $x = l \sin \varphi$, $y = l \cos \varphi$ (for convenience, the subscript is omitted). Squaring the right and left sides of the equations of motion and adding them termwise, we obtain the trajectory equation

$$x^2 + y^2 = l^2,$$

which implies that point K moves in a circle of radius l. Then, by using formulas (E2.1.1.2)—(E2.1.1.6) and by taking into account the fact that the quantities $\varphi = \pi/6$, $\dot{\varphi} = 1 \, \text{s}^{-1}$, and $\ddot{\varphi} = -1 \, \text{s}^{-2}$ correspond to time $t = 1$ s, we obtain

$$
\begin{aligned}
&v_x = \dot{x} = l \cos \varphi \, \dot{\varphi}; \quad v_x(1) = 0.87 \, \text{m/s}, \\
&v_y = \dot{y} = -l \sin \varphi \, \dot{\varphi}; \quad v_y(1) = -0.5 \, \text{m/s}, \\
&v = \sqrt{v_x^2 + v_y^2}; \quad v(1) = 1 \, \text{m/s}, \\
&a_x = \dot{v}_x = l(-\sin \varphi \, \dot{\varphi}^2 + \cos \varphi \, \ddot{\varphi}); \quad a_x(1) = -1.37 \, \text{m/s}^2, \\
&a_y = \dot{v}_y = l(-\cos \varphi \, \dot{\varphi}^2 - \sin \varphi \, \ddot{\varphi}); \quad a_y(1) = -0.37 \, \text{m/s}^2, \\
&a = \sqrt{a_x^2 + a_y^2}; \quad a(1) = 1.41 \, \text{m/s}^2, \\
&\dot{v} = \frac{v_x a_x + v_y a_y}{v}; \quad \dot{v}(1) = -1.0 \, \text{m/s}^2, \quad a^\tau(1) = 1.0 \, \text{m/s}^2, \\
&a^n = \sqrt{a^2 - (a^\tau)^2}; \quad a^n(1) = 1.0 \, \text{m/s}^2, \\
&\rho = \frac{v^2}{a^n}; \quad \rho(1) = 1.0 \, \text{m},
\end{aligned}
\tag{E2.1.1.8}
$$

and the sign of $\dot{v}$ means that, for $t = 1$ s, the motion of the point is decelerated; this result is obvious in advance, as follows from the equation of the trajectory.

E2.1.2. Kinematics of a Rigid Body

▶ **Some notions, definitions, and theorems.** An *absolutely rigid body* (or a *rigid body*, or a *body*) is a set of points moving in space so that the distance between any two of them remains constant. (In other words, the mutual position of points of a rigid body does not vary in time.)

The problem of kinematics of a rigid body is to obtain formulas for determining the velocity vector $\mathbf{v}$ and the acceleration vector $\mathbf{a}$ for any point of the rigid body.

THEOREM 1. *If a vector $\mathbf{b}$ belongs to a moving rigid body (connects two of its points), then the vector $\dot{\mathbf{b}}$ (the time derivative) is either zero or orthogonal to $\mathbf{b}$.*

Indeed, by differentiating the relation $\mathbf{b} \cdot \mathbf{b} = \text{const}$ with respect to time, we obtain $\mathbf{b} \cdot \dot{\mathbf{b}} = 0$, which proves the above statement.

THEOREM 2. *The projections of the velocities of any two points of a rigid body onto the straight line connecting them are equal,* $\text{comp}_{\overrightarrow{AB}} \mathbf{V}_A = \text{comp}_{\overrightarrow{AB}} \mathbf{V}_B$ *(Fig. E2.4).*

This theorem remains valid for an arbitrary motion of the body and states that the projections must be equal in both magnitude and direction.

THEOREM 3. *If the time derivatives $\dot{\mathbf{b}}_1$ and $\dot{\mathbf{b}}_2$ of two noncollinear vectors $\mathbf{b}_1$ and $\mathbf{b}_2$ belonging to a rigid body are known, then, for any vector $\mathbf{b}$ belonging to this rigid body, its time derivative can be found as the cross product $\dot{\mathbf{b}} = \boldsymbol{\omega} \times \mathbf{b}$, where*

$$
\begin{aligned}
&\boldsymbol{\omega} = 0 \text{ if } \dot{\mathbf{b}}_1 = \dot{\mathbf{b}}_2 = 0; \\
&\boldsymbol{\omega} = \eta \cdot \mathbf{b}_2, \text{ where } \eta = (\dot{\mathbf{b}}_1 \cdot \dot{\mathbf{b}}_1)/(\dot{\mathbf{b}}_1 \cdot (\mathbf{b}_2 \times \mathbf{b}_1)) \text{ if } \dot{\mathbf{b}}_1 \neq 0; \ \dot{\mathbf{b}}_2 = 0; \\
&\boldsymbol{\omega} = (\dot{\mathbf{b}}_1 \times \dot{\mathbf{b}}_2)/(\dot{\mathbf{b}}_1 \cdot \mathbf{b}_2) \text{ if } \dot{\mathbf{b}}_1 \neq 0; \ \dot{\mathbf{b}}_2 \neq 0 \quad (\textit{Markuzon's formula}).
\end{aligned}
\tag{E2.1.2.1}
$$

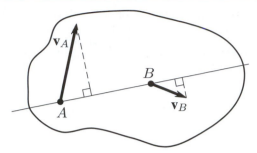

Figure E2.4. Projections of the velocities of two points of a rigid body onto the straight line passing through them.

The vector $\boldsymbol{\omega}$ is called the *vector of angular velocity of the rigid body*; it characterizes the kinematic state of the body at the time instant under study.

Although formulas (E2.1.2.1) make sense for any values of the vectors $\mathbf{b}_1$, $\mathbf{b}_2$, $\dot{\mathbf{b}}_1$, and $\dot{\mathbf{b}}_2$, one has to bear in mind that these vectors in problems of kinematics cannot be introduced arbitrarily; they must satisfy the conditions of the above-mentioned theorems of kinematics of a rigid body.

The vector $\boldsymbol{\varepsilon} = \dot{\boldsymbol{\omega}}$ is called the *vector of angular acceleration* of the rigid body.

We assume that the body moves relative to a fixed reference frame $O\xi\eta\zeta$, and at the time instant under study we know the position of point P of the body, its velocity $\mathbf{v}_P$ and acceleration $\mathbf{a}_P$, and the vectors of the body angular velocity $\boldsymbol{\omega}$ and angular acceleration $\boldsymbol{\varepsilon}$.

In this case, the velocity and acceleration of any point M of the body can be found by differentiating the vector equation $\overrightarrow{OM} = \overrightarrow{OP} + \mathbf{r}$ with respect to time (Fig. E2.5 *a*).

After the first differentiation, we obtain $\overrightarrow{OM} = \overrightarrow{OP} + \dot{\mathbf{r}}$, or

$$\mathbf{v}_M = \mathbf{v}_P + \boldsymbol{\omega} \times \mathbf{r}; \tag{E2.1.2.2}$$

after the repeated differentiation, we obtain $\dot{\mathbf{v}}_M = \dot{\mathbf{v}}_P + \dot{\boldsymbol{\omega}} \times \mathbf{r} + \boldsymbol{\omega} \times \dot{\mathbf{r}}$, or

$$\mathbf{a}_M = \mathbf{a}_P + \boldsymbol{\varepsilon} \times \mathbf{r} + \boldsymbol{\omega} \times (\boldsymbol{\omega} \times \mathbf{r}). \tag{E2.1.2.3}$$

Formulas (E2.1.2.2) and (E2.1.2.3) in general form solve the problem of kinematics of a rigid body, but they can hardly be used to solve problems frequently encountered in practice. In what follows, we consider special cases of motion of a rigid body.

▶ **Translational motion.** A motion of a rigid body is said to be *translational* if, in the course of motion, any straight line drawn in the body remains parallel to the initial direction of itself. In this case, $\boldsymbol{\omega} = 0$ and $\boldsymbol{\varepsilon} = 0$, which implies the properties of translational motion: being superimposed on each other, the trajectories of all points of a rigid body coincide (are congruent), and the velocities and accelerations of all points of the body are the same at each time instant:

$$\mathbf{v}_A = \mathbf{v}_B = \mathbf{v}, \qquad \mathbf{a}_A = \mathbf{a}_B = \mathbf{a},$$

where A and B are any points of the body.

Thus, the kinematics of translational motion of a body is reduced to the kinematics of one of its points.

▶ **Rotation about a fixed axis.** In *rotation of a rigid body about a fixed axis*, which is called the axis of rotation, the points lying on the axis remain fixed. The other points move in circles lying in the planes orthogonal to the rotation axis.

We introduce two reference frames, the fixed reference frame $O\xi\eta\zeta$ whose axis $O\zeta$ coincides with the rotation axis of the body and the moving frame $Oxyz$ rigidly fixed to

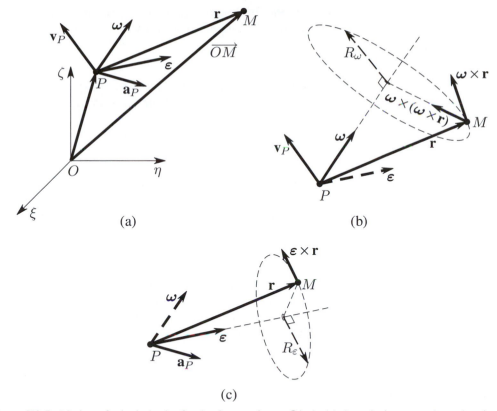

(a)   (b)

(c)

Figure E2.5. Motion of a body in the fixed reference frame $O\xi\eta\zeta$: (a) the velocity $\mathbf{v}_P$ and acceleration $\mathbf{a}_P$ of point P of the body, and the body angular velocity $\boldsymbol{\omega}$ and angular acceleration $\boldsymbol{\varepsilon}$; (b) the directions of the vectors $\boldsymbol{\omega} \times \mathbf{r}$ and $\boldsymbol{\omega} \times (\boldsymbol{\omega} \times \mathbf{r})$; (c) the direction of the vector $\boldsymbol{\varepsilon} \times \mathbf{r}$.

the body with axis Oz also directed along the rotation axis. The origin O of both frames is placed at the point of intersection of the rotation axis with the plane containing the point M whose velocity and acceleration are the desired variables.

The unit vectors $\boldsymbol{\xi}^\circ, \boldsymbol{\eta}^\circ, \boldsymbol{\zeta}^\circ$ of the fixed reference frame and $\mathbf{i}^\circ, \mathbf{j}^\circ, \mathbf{k}^\circ$ of the moving frame are shown in Fig. E2.6.

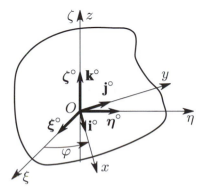

Figure E2.6. Rotation of a body about a fixed axis.

The dihedral angle φ between the moving plane $\mathbf{i}^\circ\mathbf{k}^\circ$ and the fixed plane $\boldsymbol{\xi}^\circ\boldsymbol{\zeta}^\circ$ is called the angle of rotation of the rigid body; it is measured in radians. The rotation angle is assumed to be positive if it is counted off from the fixed plane towards the moving plane anticlockwise if we look in the positive sense of the axis of rotation. To know the position of the body (and of any of its points) at any time instant, it is necessary to know the dependence of the angle φ on time t,

$$\varphi = \varphi(t).$$

This equation expresses the *law of rotation of a rigid body about a fixed axis.*

Along with the function $\varphi(t)$, its first and second time derivatives characterize the kinematic state of the rigid body at the time instant under study:

$$\omega = \dot{\varphi}, \qquad \varepsilon = \dot{\omega} = \ddot{\varphi},$$

where ω is the angular velocity of the body and ε is its angular acceleration.

The unit vectors of the moving frame are written as functions of the rotation angle φ as follows: $\mathbf{i}^\circ = \boldsymbol{\xi}^\circ \cos\varphi + \boldsymbol{\eta}^\circ \sin\varphi, \mathbf{j}^\circ = -\boldsymbol{\xi}^\circ \sin\varphi + \boldsymbol{\eta}^\circ \cos\varphi, \mathbf{k}^\circ = \boldsymbol{\zeta}^\circ = \overrightarrow{\text{const}}$. By differentiating these formulas with respect to time, we obtain $\mathbf{i}^\circ = \mathbf{j}^\circ \dot{\varphi}, \mathbf{j}^\circ = -\mathbf{i}^\circ \dot{\varphi}$, after which the angular velocity vector can be found: $\boldsymbol{\omega} = (\mathbf{i}^\circ \times \mathbf{j}^\circ)/(\mathbf{i}^\circ \cdot \mathbf{j}^\circ) = -(\mathbf{j}^\circ \times \mathbf{i}^\circ)\dot{\varphi}/(\mathbf{j}^\circ \cdot \mathbf{j}^\circ) = \boldsymbol{\zeta}^\circ \dot{\varphi}$.

After the differentiation of the last relation, we obtain $\boldsymbol{\varepsilon} = \dot{\boldsymbol{\omega}} = \boldsymbol{\zeta}^\circ \ddot{\varphi}$.

Thus, the vectors of angular velocity and angular acceleration of a body rotating about a fixed axis are directed along this axis. In this case, if we look in the direction opposite to the vector $\boldsymbol{\omega}$, then it seems that the rotation is anticlockwise (the right-hand screw rule). In these cases, the vector $\boldsymbol{\varepsilon}$ is parallel to the vector $\boldsymbol{\omega}$ if the signs of $\dot{\varphi}$ and $\ddot{\varphi}$ coincide. Otherwise, $\boldsymbol{\omega}$ and $\boldsymbol{\varepsilon}$ are antiparallel.

To determine the velocity $\mathbf{v}_M$ and the acceleration $\mathbf{a}_M$ of an arbitrary point M of the body, one can use the vector formulas (E2.1.2.2) and (E2.1.2.3), but here we perform this in a different, more illustrative way.

In the case of rotation of a body about a fixed axis, the trajectory of its arbitrary point M is a circle whose radius R is called the radius of rotation of a point. In Fig. E2.7, this circle is shown from the side of positive sense of the rotation axis. The sign of ω indicates the direction of rotation: if it is positive, then the rotation is anticlockwise. It is convenient to represent the angular velocity and the angular acceleration by using curved arrows with their sign taken into account (in Fig. E2.7, $\omega < 0$ and $\varepsilon > 0$). If ω and ε have the same signs, then one says that the rotation is accelerated, and if the signs are opposite, then the rotation is decelerated.

In the case of rotation of a body about a fixed axis, the magnitude of the velocity of point M can be found by the formula

$$v = |\omega|R; \tag{E2.1.2.4}$$

in this case the velocity is directed along the tangent to the circle, and so is perpendicular to the radius, and its sense is consistent with the sense of the curved arrow ω (Fig. E2.7). The tangential, normal, and total acceleration can be found by the following formulas, which are a special case of formulas (E2.1.2.3):

$$a^\tau = |\varepsilon|R, \quad a^n = \omega^2 R, \quad a = R\sqrt{\varepsilon^2 + \omega^4}. \tag{E2.1.2.5}$$

The tangential acceleration vector $\mathbf{a}^\tau$ is directed along the tangent to the circle, and its sense is consistent with that of the curved arrow ε; the vector $\mathbf{a}^n$ is directed along the radius R towards the center of the circle. The acute angle μ between $\mathbf{a}$ and R can be found by the formula

$$\tan\mu = \frac{a^\tau}{a^n} = \frac{|\varepsilon|}{\omega^2};$$

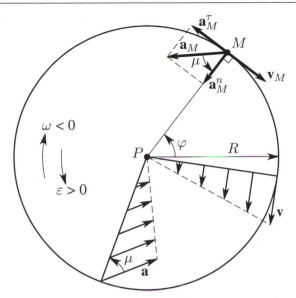

Figure E2.7. The trajectory of an arbitrary point M of a body rotating about a fixed axis; R is the radius of rotation of the point. The curved arrows denote the angular velocity and the angular acceleration.

in this case, the sense of the nearest rotation from $\mathbf{a}$ to R is consistent with the sense of the curved arrow ε.

It follows from the last two formulas that the vectors $\mathbf{v}$ and $\mathbf{a}$ of different points of the body that lie on the same radius are parallel to each other and linearly depend on the distance from the axis of rotation (Fig. E2.7):

$$\omega = \frac{v_B}{BP} = \frac{v_C}{CP} = \frac{v_D}{DP}, \qquad \varepsilon = \frac{a_B}{BP} = \frac{a_C}{CP} = \frac{a_D}{DP}. \tag{E2.1.2.6}$$

A rotation of a rigid body is said to be *uniformly varying* if $\varepsilon = \varepsilon_0 = \text{const}$. In this case, the time dependencies of the rotation angle and the angular velocity are given by the formulas

$$\omega = \omega_0 + \varepsilon_0(t - t_0),$$
$$\varphi = \varphi_0 + \omega_0(t - t_0) + \tfrac{1}{2}\varepsilon_0(t - t_0)^2,$$

which are similar to the formulas for the uniformly varying motion of a point (E2.1.1.7).

The *uniform* rotation of a rigid body is a special case of uniformly varying rotation, namely, the case in which $\varepsilon_0 = \text{const} = 0$.

E2.1.3. Plane-Parallel (Plane) Motion of a Rigid Body

▶ **Basic notions.** A *plane-parallel* (*plane*) motion is a motion of a body in which the distance from any of its points to some fixed plane remains unchanged. (Each point of the body moves in a plane, but the planes for different points of the body can be different.) The rotation of a body about a fixed axis is a special case of plane motion.

Studying a plane-parallel motion can be reduced to studying the motion of a plane figure in its plane. The rotation angle $\varphi(t)$ of a plane figure is counted off from a fixed straight line to a straight line rigidly fixed to the figure.

The position of a plane figure is completely determined if the coordinates of some of its points (pole, center) P and the rotation angle are known. The relations

$$x_P = x(t), \quad y_P = y(t), \quad \varphi = \varphi(t)$$

determining three functional dependencies on time are called the *equations of motion of a plane figure*.

The angular velocity vector $\boldsymbol{\omega}$ and the angular acceleration vector $\boldsymbol{\varepsilon}$ are introduced by analogy with the case of rotation of a body about a fixed axis; they are orthogonal to the base plane. One has $\omega = \dot{\varphi}$ and $\varepsilon = \dot{\omega} = \ddot{\varphi}$.

Just as in the case of rotation about a fixed axis, when solving practical problems it is convenient to represent the angular velocity and acceleration by curved arrows.

▶ **Theorem on velocities of points of a plane figure. Instantaneous center of velocities.**

THEOREM (ON VELOCITIES OF POINTS OF A PLANE FIGURE). *The velocity of any point B of a plane figure is equal to the vector sum of the velocity $\mathbf{v}_P$ of the pole and the velocity $\mathbf{v}_{BP}$ that point B would have in the imaginary rotation of the figure about the fixed pole P at the angular velocity ω (Fig. E2.8):*

$$\mathbf{v}_B = \mathbf{v}_P + \mathbf{v}_{BP}. \tag{E2.1.3.1}$$

The velocity $\mathbf{v}_{BP}$ is perpendicular to segment BP, its magnitude is calculated by the formula $v_{BP} = |\omega|BP$, and the sense is consistent with the sense of the curved arrow ω.

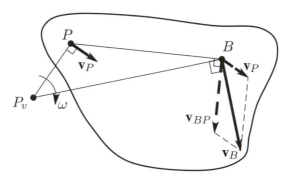

Figure E2.8. Determining the velocity of an arbitrary point of a plane figure (to the theorem on velocities of points of a plane figure).

The point P_v of a plane figure whose velocity at the time instant under study is zero is called the *instantaneous center of velocities*. If this point is taken for the pole, then the velocity of an arbitrary point B of the plane figure is determined by the formula $\mathbf{v}_B = \mathbf{v}_{BP_v}$. The direction of $\mathbf{v}_B$ is consistent with the direction of the curved arrow of the angular velocity, and its magnitude can be found by the formula $v_B = |\omega|BP_v$. The velocities of points of the plane figure at the time instant under study are distributed by analogy with the case of rotation of the figure about a fixed axis passing through the instantaneous center of velocities. Therefore, one can use formula (E2.1.2.6) for rotational motion:

$$\omega = \frac{v_B}{BP_v} = \frac{v_C}{CP_v} = \frac{v_D}{DP_v}. \tag{E2.1.3.2}$$

In Fig. E2.9a,b,c, we show several versions of location of the instantaneous center of velocities:

(a) In rolling motion without slip on a fixed surface, the instantaneous center of velocities lies at the point of tangency (Fig. E2.9a).

(b) If $\mathbf{v}_B \parallel \mathbf{v}_C$ and $\mathbf{v}_B \perp BC$, then the position of P_v is determined by the construction shown in Fig. E2.9b.

(c) If $\mathbf{v}_B \parallel \mathbf{v}_C$ and the segment BC is not perpendicular to $\mathbf{v}_B$, then there is no instantaneous center of velocities, $\omega = 0$, and the velocities of all points of the body are the same at a given time instant (Fig. E2.9c).

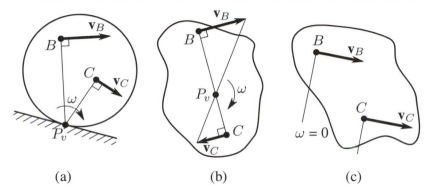

(a)   (b)   (c)

Figure E2.9. Possible positions of the instantaneous center of velocities: (a) in rolling motion without slip on a fixed surface; (b) the vectors at two points of the body are parallel and perpendicular to the line connecting them; (c) the vectors at two points of the body are parallel but not perpendicular to the line connecting them.

Example. At time $t = 1$ s, find the velocities of points A, B, and K of the ellipsograph rod whose motion is described in the example in Subsection E2.1.1.

Solution. The rod AB is in plane-parallel motion. Drawing perpendiculars to the directions of velocities of points A and B, we find the position of the instantaneous center P_v of velocities (Fig. E2.10). We represent the angular velocity $\omega = \dot{\varphi} = 1\,\text{s}^{-1}$ by a curved arrow with sign taken into account. We direct the velocities of points A, B, and K perpendicular to the straight lines connecting them with P_v so that their senses are consistent with the curved arrow of the angular velocity. The values of the velocities can be found from the relation

$$\omega = \frac{v_A}{AP_v} = \frac{v_B}{BP_v} = \frac{v_K}{KP_v},$$

which implies $v_A = v_K = 1.0\,\text{m/s}$ and $v_B = \sqrt{3} = 1.73\,\text{m/s}$.

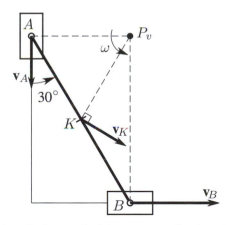

Figure E2.10. Ellipsograph (thin rod) whose endpoints move according to a given law along two perpendicular straight lines.

▶ **Theorem on accelerations of points of a plane figure. Instantaneous center of accelerations.**

THEOREM (ON ACCELERATIONS OF POINTS OF A PLANE FIGURE). *The acceleration of any point B of a plane figure is equal to the vector sum of the acceleration of the pole $\mathbf{a}_P$ and the acceleration $\mathbf{a}_{BP}$ that point B would have in the imaginary rotation of the figure about the fixed pole P:*

$$\mathbf{a}_B = \mathbf{a}_P + \mathbf{a}_{BP} = \mathbf{a}_P + \mathbf{a}_{BP}^\tau + \mathbf{a}_{BP}^n. \tag{E2.1.3.3}$$

The magnitudes of the vectors $\mathbf{a}_{BP}^\tau$ and $\mathbf{a}_{BP}^n$ and their orientation on the plane are as

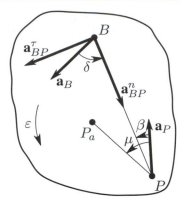

Figure E2.11. Determining the acceleration of an arbitrary point of a plane figure (to the theorem on accelerations of points of a plane figure).

follows: $\mathbf{a}^n_{BP} \parallel \overrightarrow{BP}$ and is directed to the pole P; $a^n_{BP} = \omega^2 BP$; $\mathbf{a}^\tau_{BP} \perp \overrightarrow{BP}$ and is consistent with the sense of the curved arrow ε, $a^\tau_{BP} = |\varepsilon| BP$ (Fig. E2.11).

If the vector relation (E2.1.3.3) is projected onto two directions, the parallel direction BP and the perpendicular direction BP, then we obtain the system of two scalar equations

$$a_B \cos \delta = -a_P \cos \beta + \omega^2 BP,$$
$$a_B \sin \delta = -a_P \sin \beta + |\varepsilon| BP$$

containing seven variables a_B, a_P, δ, β, BP, ω, and ε. If five of them are known, then the remaining two can be found by solving the system.

The *instantaneous center of accelerations* of a plane figure is a point P_a of this figure whose acceleration at the time instant under study is zero. The distribution of accelerations of points of the plane figure is the same as in the case of its rotation about a fixed axis passing through the point P_a.

Let us indicate the positions of the instantaneous center of velocities and the instantaneous center of accelerations of a disk that uniformly rolls without slip on a horizontal fixed straight line.

Since the rolling motion is without slip, it follows that the velocities of points of tangency of the disk and of the straight line are the same. And since the straight line is fixed, we conclude that the instantaneous center of velocities of the disk lies at the point of tangency.

The center of the disk moves uniformly, and hence its tangential acceleration is zero. Further, the trajectory of the center of the disk is rectilinear, and hence its normal acceleration is zero. Therefore, the total acceleration of the center, which is equal to the vector sum of the tangential and normal accelerations, is also zero; i.e., the instantaneous center of accelerations of the disk lies at the center of the disk.

The above example shows that the instantaneous center of velocities and the instantaneous center of accelerations are in general different points of a plane figure. But in special cases they can coincide; for example, in the case of rotation of a body about a fixed axis, both P_v and P_a always lie on the axis of rotation.

E2.1.4. Arbitrary Motion of a Rigid Body

The velocity $\mathbf{v}_M$ and the acceleration $\mathbf{a}_M$ of any point of a rigid body in an arbitrary motion can be found if, at the time instant under study, we know the velocity $\mathbf{v}_P$ and the acceleration $\mathbf{a}_P$ of a point P of the body and the vectors $\boldsymbol{\omega}$ of angular velocity and $\boldsymbol{\varepsilon}$ of angular acceleration of the body. The corresponding formulas have the form (see Eqs. (E2.1.2.2) and (E2.1.2.3))

$$\mathbf{v}_M = \mathbf{v}_P + \boldsymbol{\omega} \times \mathbf{r},$$
$$\mathbf{a}_M = \mathbf{a}_P + \boldsymbol{\varepsilon} \times \mathbf{r} + \boldsymbol{\omega} \times (\boldsymbol{\omega} \times \mathbf{r}). \tag{E2.1.4.1}$$

In the special case $\mathbf{v}_P = 0$, $\mathbf{a}_P = 0$, which takes place in *spherical motion*, formulas (E2.1.4.1) become

$$\mathbf{v}_M = \boldsymbol{\omega} \times \mathbf{r} \qquad \text{(Euler formula)},$$
$$\mathbf{a}_M = \boldsymbol{\varepsilon} \times \mathbf{r} + \boldsymbol{\omega} \times (\boldsymbol{\omega} \times \mathbf{r}) = \boldsymbol{\varepsilon} \times \mathbf{r} + \boldsymbol{\omega} \times \mathbf{v}_M.$$

In this case, the distribution of velocities of points of the body is the same as in the case of rotation about the fixed axis passing through P parallel to the vector $\boldsymbol{\omega}$; therefore, it is called the *instantaneous axis of rotation*. The velocities of all points of the body that lie on this axis at the time instant under study are zero, and a small displacement of the body can be treated as a rotation by a small angle about the instantaneous axis of rotation.

To find the specific features of the spatial orientation of the vectors contained in relations (E2.1.4.1), we rewrite them as

$$\mathbf{v}_M = \mathbf{v}_P + \mathbf{v}_\omega,$$
$$\mathbf{a}_M = \mathbf{a}_P + \mathbf{a}_\varepsilon^\tau + \mathbf{a}_\omega^n, \tag{E2.1.4.2}$$

where we introduce the notation $\mathbf{v}_\omega = \boldsymbol{\omega} \times \mathbf{r}$, $\mathbf{a}_\varepsilon^\tau = \boldsymbol{\varepsilon} \times \mathbf{r}$, and $\mathbf{a}_\omega^n = \boldsymbol{\omega} \times (\boldsymbol{\omega} \times \mathbf{r})$.

These vectors can be interpreted physically as follows (see Fig. E2.5b,c):

The vector $\mathbf{v}_\omega$ is the velocity of point M of the body in the case of imaginary rotation about the fixed axis directed along the vector $\boldsymbol{\omega}$ and passing through point P. In this imaginary motion, the point moves in a circle of radius R_ω lying in the plane orthogonal to the vector $\boldsymbol{\omega}$. The velocity is directed along the tangent to this circle, and its value can be calculated by formula $v_\omega = \omega R_\omega$.

$\mathbf{a}_\omega^n$ is the vector of normal acceleration of point M in the same imaginary motion of the body. The vector is directed along the radius of the circle towards its center, and its magnitude can be calculated by the formula $a_\omega^n = \omega^2 R_\omega$.

$\mathbf{a}_\varepsilon^\tau$ is the vector of tangential acceleration of point M in the case of *another* imaginary rotation about the fixed axis directed along the vector $\boldsymbol{\varepsilon}$ and passing through point P. In this imaginary motion, the point moves in a circle of radius R_ε lying in the plane orthogonal to the vector $\boldsymbol{\varepsilon}$. The vector of tangential acceleration is directed along the tangent to this circle, and its magnitude can be calculated by the formula $a_\varepsilon^\tau = \varepsilon R_\varepsilon$.

It follows from the above that any small displacement of a rigid body at the time instant under study can be represented as composed of three simultaneous imaginary small motions:

(1) The translational motion with velocity $\mathbf{v}_P$ and acceleration $\mathbf{a}_P$ and with zero magnitudes of ω and ε.

(2) The rotation about a fixed axis directed along the vector $\boldsymbol{\omega}$ and passing through point P, with zero magnitudes of $\mathbf{v}_P, \mathbf{a}_P$, and ε.

(3) The rotation about a fixed axis directed along the vector $\boldsymbol{\varepsilon}$ and passing through point P, with zero magnitudes of $\mathbf{v}_P, \mathbf{a}_P, \omega$, and t.

We take the scalar products of both sides of Eq. (E2.1.4.1) by the unit vector $\boldsymbol{\omega}/\omega$ and obtain the equation

$$\text{comp}_{\boldsymbol{\omega}}\, \mathbf{v}_P = \text{comp}_{\boldsymbol{\omega}}\, \mathbf{v}_M.$$

This means that the projection of the velocity of any point of the body onto the direction of the angular velocity vector is independent of the choice of the point.

Now we find the body points that have the minimum velocity. Without loss of generality, we direct the vector $\boldsymbol{\omega}$ along the axis Pz of the reference frame $Pxyz$, place the velocity vector $\mathbf{v}$ of point P in the plane xPz, and expand it into two orthogonal components, $\mathbf{v}_{\|\omega}$ parallel to the axis Pz and $\mathbf{v}_{\perp\omega}$ parallel to the axis Px. On the axis Py, we choose a point S so that $PS = v_{\perp\omega}/\omega$. A straightforward verification shows that $\mathbf{v}_S \parallel \boldsymbol{\omega}$ and $v_S = v_{\min} = v_{\|\omega}$. All the points lying on the axis SZ have the same property.

A system of vectors $\boldsymbol{\omega}, \mathbf{v}$ located on the same straight line is called a *screw* or a *wrench*. The scalar quantity $p = v/\omega$ is called the *parameter* of the screw.

The constant values accompanying some phenomenon or process are usually called *invariants*. Therefore, one can say that, in kinematics of a rigid body, there are two invariants with respect to the choice of the pole P, namely, the vector of angular velocity of a rigid body and the projection of the velocity of a point of the body on the direction of the angular velocity vector.

Thus, in an arbitrary motion of a rigid body, the distribution of velocities of its points at the time instant under study can be one of the following: (1) instantaneous rest, (2) instantaneous translational motion, (3) rotation about the instantaneous axis, and (4) instantaneous screw motion.

Example. A right circular cone with angle 2α at the vertex performs a spherical motion and rolls without slip on the surface of a fixed right circular cone with angle 2β at the vertex (Fig. E2.12). At the time instant under study, the axis of the moving cone, rotating about the vertical axis, has the angular velocity $\boldsymbol{\omega}_1$ and the angular acceleration $\boldsymbol{\varepsilon}_1$.

Find the angular velocity $\boldsymbol{\omega}$ and the angular acceleration $\boldsymbol{\varepsilon}$ of the moving cone at the time instant under study.

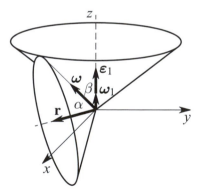

Figure E2.12. Rolling of a circular cone on the surface of another circular cone.

Solution. We introduce fixed Cartesian coordinate axes with origin at the vertex of the cone: the z-axis is directed along the axis of the fixed cone, the x-axis is perpendicular to the z-axis in the plane formed by the cone axes at the initial time instant $t = 0$, and the y-axis is orthogonal to the plane xz, so that the axes thus constructed form a right reference frame.

On the axis of the moving cone, we choose an arbitrary point M and denote the position vector connecting this point with the origin O by $\mathbf{r}$.

The moving cone is in spherical motion. The angular velocity vector $\boldsymbol{\omega}$ is directed along the common generator of the cones, because it is an instantaneous axis of rotation. The plane $\boldsymbol{\omega}\mathbf{r}$ rotates about the fixed z-axis, and the vectors $\boldsymbol{\omega}_1$ and $\boldsymbol{\varepsilon}_1$ are directed along it.

The velocity v_M can be found if we first refer it to the cone and then to the plane $\boldsymbol{\omega}\mathbf{r}$:

$$\boldsymbol{\omega} \times \mathbf{r} = \boldsymbol{\omega}_1 \times \mathbf{r} \quad \text{or} \quad (\boldsymbol{\omega} - \boldsymbol{\omega}_1) \times \mathbf{r} = \mathbf{0},$$

which implies that $(\boldsymbol{\omega} - \boldsymbol{\omega}_1) \parallel \mathbf{r} = 0$. This means that the straight line connecting the endpoints of the vectors $\boldsymbol{\omega}$ and $\boldsymbol{\omega}_1$ is parallel to the vector $\mathbf{r}$.

Considering the obtained triangle, we obtain $\omega = \omega_1 \sin(\alpha + \beta)/\sin\beta$ from the law of sines.

To find the vector ε, we represent the vector $\boldsymbol{\omega} = \boldsymbol{\omega}(t)$ as the sum of three components parallel to the coordinate axes:

$$\boldsymbol{\omega} = \omega_1 \frac{\sin(\alpha+\beta)\sin\alpha}{\sin\beta}\sin\left(\omega_1 t + \frac{\varepsilon_1 t^2}{2}\right)\mathbf{i} - \omega_1 \frac{\sin(\alpha+\beta)\sin\alpha}{\sin\beta}\cos\left(\omega_1 t + \frac{\varepsilon_1 t^2}{2}\right)\mathbf{j} + \omega_1 \frac{\sin(\alpha+\beta)\cos\alpha}{\sin\beta}\mathbf{k},$$

where $\varphi_1 = (\omega_1 t + \varepsilon_1 t^2/2)$ is the rotation angle of the plane $\boldsymbol{\omega}\mathbf{r}$ in time t. The angular acceleration can be obtained by differentiating the resulting angular velocity vector $\varepsilon = \dot{\boldsymbol{\omega}}$ with respect to time.

Now if the vectors $\boldsymbol{\omega}$ and ε are known, then one can find the velocity and the acceleration of any point of the moving cone.

The expression for the angular acceleration can be obtained in a simpler way by using the theorems presented below.

E2.1.5. Compound Motion of a Point

▶ **Basic notions.** Very often, two observers trace one and the same point M, one of them is in the laboratory reference frame $O\xi\eta\zeta$, which is assumed to be fixed, and the other is in a moving reference frame $Pxyz$, which moves relative to the fixed frame according to a given law (Fig. E2.13). In this case, if the law of motion of a point in the moving reference frame is given, then the point position in the fixed reference frame can also be determined at any time instant. The motion of a point defined in this way is called the *compound motion of a point*.

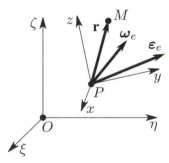

Figure E2.13. Compound motion of a point; the fixed reference frame $O\xi\eta\zeta$ and the moving reference frame $Pxyz$.

This method for specifying the motion of a point differs from those studied above. The velocity of a point and its acceleration can be determined as follows: use appropriate transformations to pass to one of the above-studied methods for determining the motion of a point and then proceed accordingly. But the same problem can be solved differently by using the notions and theorems on the velocities and accelerations in the compound motion of a point, which will be given below.

The *absolute motion of point* M is its motion relative to the fixed (absolute) reference frame. The trajectory of the point, its velocity, and acceleration are called the *absolute trajectory*, *absolute velocity* $\mathbf{v}_a$ and *absolute acceleration* $\mathbf{a}_a$, respectively.

The *relative motion of point* M is its motion relative to the moving reference frame. The trajectory of the point, its velocity, and acceleration relative to the moving reference frame are called the *relative trajectory*, *relative velocity* $\mathbf{v}_r$, and *relative acceleration* $\mathbf{a}_r$, respectively.

The *frame motion* is the motion of the moving reference frame relative to the fixed reference frame. In this motion, the moving frame "transfers" the set of points rigidly fixed to it.

The *frame velocity* $\mathbf{v}_e$ and the *frame acceleration* $\mathbf{a}_e$ of point M are the velocity and acceleration of the point B that is permanently attached to the moving frame and coincides with the moving point M at the given time instant.

If the velocity $\mathbf{v}_P$ and the acceleration $\mathbf{a}_P$ of a point P belonging to the moving frame, as well as the angular velocity $\boldsymbol{\omega}_e$ and the angular acceleration $\boldsymbol{\varepsilon}_e$ of the moving frame, are known, then $\mathbf{v}_e$ and $\mathbf{a}_e$ can be found by the formulas

$$\mathbf{v}_e = \mathbf{v}_B = \mathbf{v}_P + \boldsymbol{\omega}_e \times \mathbf{r}, \qquad \mathbf{a}_e = \mathbf{a}_B = \mathbf{a}_P + \boldsymbol{\varepsilon}_e \times \mathbf{r} + \boldsymbol{\omega}_e \times (\boldsymbol{\omega}_e \times \mathbf{r}),$$

where $\mathbf{r}$ is the position vector $\overrightarrow{PB}$ of point B of the moving frame at which the point M occurs at the time instant under study.

▶ **Main theorems.**

THEOREM ON VELOCITIES IN COMPOUND MOTION OF A POINT. *The absolute velocity vector is equal to the vector sum of the relative and frame velocity vectors,*

$$\mathbf{v}_a = \mathbf{v}_r + \mathbf{v}_e. \tag{E2.1.5.1}$$

THEOREM ON ACCELERATIONS IN COMPOUND MOTION OF A POINT. *The absolute acceleration vector is equal to the vector sum of the relative, frame, and Coriolis acceleration vectors,*

$$\mathbf{a}_a = \mathbf{a}_r + \mathbf{a}_e + \mathbf{a}_C. \tag{E2.1.5.2}$$

The Coriolis acceleration vector is determined by the formula

$$\mathbf{a}_C = 2(\boldsymbol{\omega}_e \times \mathbf{v}_r) \tag{E2.1.5.3}$$

and is consistent with the rules for calculating the cross product.

To find the vectors $\mathbf{a}_r$ and $\mathbf{a}_e$, one has to bear in mind that each of them can be the vector sum of several components, for example, of the tangential and normal accelerations.

Example. Point M moves according to the law $AM = 0.5\,t^2 - 2t + 2.5$ along the ellipsograph rod whose motion was described in the Example in Subsection E2.1.1. Find its velocity and acceleration at time $t = 1$ s.

Solution. We find the velocity and acceleration by applying the theorems on compound motion of a point. Let us introduce a moving reference frame attached to the rod AB. Under such a choice, the relative motion of point M can be introduced in a natural way. For $t = 1$ s, this point is at the middle of the segment.

We use formulas (E2.1.1.6) to obtain the relative velocity and acceleration

$$v_r = \dot{L} = t - 2; \quad v_r(1) = -1 \text{ m/s},$$
$$a_r^\tau = \ddot{L}; \quad a_r^\tau(1) = 1 \text{ m/s}^2, \quad a_r^n = v_r^2/\rho_r = 0 \quad (\text{because } 1/\rho_r = 0),$$

where the notation $L = AM$ is used.

In Fig. E2.14, we present the relative velocity and acceleration with signs taken into account.

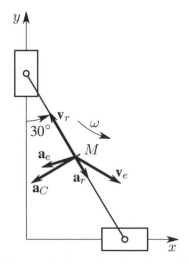

Figure E2.14. Relative and frame velocities and accelerations of point M of the ellipsograph.

The frame velocity and acceleration of point M are equal to the velocity and acceleration of the point on the ellipsograph rod where the point M occurs at time $t = 1\ s$. This point is the rod point K whose velocity and acceleration were obtained by solving the example in Subsection E2.1.1 (see formulas (E2.1.1.8)); we show them in the figure.

Now let us find the Coriolis acceleration. The angular velocity vector of the moving reference frame is directed towards the observer perpendicular to the figure plane, and the relative velocity vector lies in the figure plane. Therefore, the vector $\mathbf{a}_C$ lies in the figure plane and is perpendicular to $\mathbf{v}_r$, so that if we look towards $\mathbf{a}_C$, then the nearest rotation from $\boldsymbol{\omega}$ to $\mathbf{v}_r$ seems to be anticlockwise (Fig. E2.14). The magnitude of $\mathbf{a}_C$ is equal to

$$a_C = 2\omega v_r \sin 90° = 2\ \text{m/s}^2.$$

Let us find the absolute velocity of point M:

$$\mathbf{v}_a = \mathbf{v}_r + \mathbf{v}_e.$$

We project both sides of the vector equation onto the x- and y-axes:

$$v_{ax} = v_{rx} + v_{ex} = -1 \sin 30° + 1 \cos 30° = 0.37\ \text{m/s},$$
$$v_{ay} = v_{ry} + v_{ey} = 1 \cos 30° - 1 \sin 30° = 0.37\ \text{m/s},$$

which implies that $v_a = \sqrt{v_{ax}^2 + v_{ay}^2} = 0.52\ \text{m/s}$.

Let us find the absolute acceleration of point M:

$$\mathbf{a}_a = \mathbf{a}_r + \mathbf{a}_e + \mathbf{a}_C.$$

We project both sides of the last vector relation onto the x- and y-axes:

$$a_{ax} = 1 \sin 30° - 1.37 - 2 \cos 30° = -2.60\ \text{m/s}^2,$$
$$a_{ay} = -1 \cos 30° - 0.37 - 2 \sin 30° = -2.24\ \text{m/s}^2,$$

which implies that $a_a = \sqrt{a_{ax}^2 + a_{ay}^2} = 3.43\ \text{m/s}^2$.

When solving problems of kinematics by using theorems on the compound motion of a point, the kinematic characteristics of the relative and frame motions can be found only after the moving reference frame is chosen. This choice is not unique, because it depends on the researcher alone. Therefore, the problem on the direction of the Coriolis acceleration makes no sense until the fixed frame, the moving frame, and the law of motion of the latter are chosen.

The theorem on the velocities of a point in compound motion of a point can be stated differently. Indeed, for the point M, the theorem on velocities can be written as (Fig. E2.13)

$$\mathbf{v}_a = \mathbf{v}_r + \mathbf{v}_e \quad \text{or} \quad \frac{d\overrightarrow{OM}}{dt} = \mathbf{v}_r + \frac{d\overrightarrow{OP}}{dt} + \boldsymbol{\omega} \times \mathbf{r} \quad \text{or} \quad \frac{d\mathbf{r}}{dt} = \mathbf{v}_r + \boldsymbol{\omega} \times \mathbf{r}.$$

Taking into account the fact that $\mathbf{v}_r$ is the velocity of point M relative to the moving reference frame, we call it the *local (relative) derivative* of the vector $\mathbf{r}$ and denote by $\frac{\widetilde{d\mathbf{r}}}{dt}$; the quantity $\frac{d\mathbf{r}}{dt}$ will be called the *absolute derivative*. As a result, the last relation implies that

$$\frac{d\mathbf{r}}{dt} = \frac{\widetilde{d\mathbf{r}}}{dt} + \boldsymbol{\omega} \times \mathbf{r}.$$

This formula is sometimes called the *theorem on the relation between the absolute and local (relative) derivatives of a vector*.

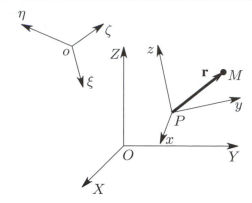

Figure E2.15. Compound motion of a body determined by three reference frames.

E2.1.6. Compound Motion of a Body

▶ **Absolute, relative, and frame motion.** One often encounters situations in which the position of points of a rigid body is determined by three reference frames: the frame $Pxyz$ rigidly fixed to the body, the moving frame $o\xi\eta\zeta$ in which the motion of the body is given, and the absolute (conventionally, fixed) frame $OXYZ$ relative to which the motion of the moving reference frame is given (Fig. E2.15).

The motion of a body in a moving frame is said to be relative, in the fixed frame, to be absolute, and the motion of the moving frame relative to the fixed frame is called the frame motion. The rigid body motion itself, when given as described above, is said to be *compound*.

▶ **Addition of translational motions.** Assume that, in a moving frame, a body is in instantaneous translational motion with velocity $\mathbf{v_r}$, while the moving frame itself is in instantaneous translational motion with velocity $\mathbf{v_e}$. Using the theorem on the addition of velocities in the compound motion of a point, we obtain the velocity of an arbitrary point M of the body,

$$\mathbf{v}_M = \mathbf{v_a} = \mathbf{v_r} + \mathbf{v_e}.$$

The point M in no way participates in the final result, and this means that the velocity vectors of all points of the body are the same at the time instant under study; that is, in the case of two instantaneous translational motions, the body is in *instantaneous translational motion* with velocity equal to the vector sum of the velocities of these two instantaneous translational motions.

This result can be generalized to the case of a larger number of translational motions: the resulting motion is an instantaneous translational motion with velocity equal to the vector sum of the velocities of the original instantaneous translational motions.

▶ **Addition of rotations about intersecting axes.** Assume that in the moving frame the body is in instantaneous rotation about an axis with angular velocity ω_1, while the moving frame itself is in instantaneous rotation about another axis with angular velocity ω_2; moreover, the axes intersect at point P (Fig. E2.16).

Using the theorem on the addition of velocities in compound motion of a point, we obtain the velocity of an arbitrary point M of the body:

$$\mathbf{v}_M = \mathbf{v_a} = \mathbf{v_r} + \mathbf{v_e} = \omega_1 \times \mathbf{r} + \omega_2 \times \mathbf{r} = (\omega_1 + \omega_2) \times \mathbf{r} = \mathbf{\Omega} \times \mathbf{r}, \quad \text{where} \quad \mathbf{\Omega} = \omega_1 + \omega_2.$$

In other words, if the instantaneous angular velocities intersect, then the resulting motion of the body is an instantaneous rotation, with angular velocity equal to their vector sum,

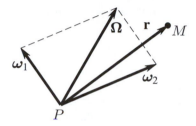

Figure E2.16. Addition of rotations about intersecting axes.

about an axis passing through the point of intersection. This statement holds for any number of instantaneous rotations about axes intersecting at one point.

▶ **Couple of rotations.** Assume that a body in the moving reference frame is in instantaneous rotation about an axis with angular velocity ω', while the moving frame itself is in instantaneous rotation about an axis with angular velocity ω''. Moreover, the axes are parallel, and the angular velocities are equal and opposite, $\omega' = \omega'' = \omega$ and $\omega' = -\omega''$ (Fig. E2.17).

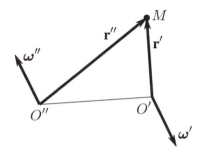

Figure E2.17. Addition of rotations about parallel axes (couple of rotations).

Using the theorem on the addition of velocities in compound motion of a point, we obtain the velocity of an arbitrary point M of the body:

$$\mathbf{v}_M = \mathbf{v}_a = \mathbf{v}_r + \mathbf{v}_e = \omega' \times \mathbf{r}' + \omega'' \times \mathbf{r}'' = \omega' \times (\mathbf{r}' - \mathbf{r}'') = \omega' \times \overrightarrow{O''O'} = \omega'' \times \overrightarrow{O'O''} = \overline{\mathrm{mom}}_{\omega',\omega''},$$

where the symbol $\overline{\mathrm{mom}}_{\omega',\omega''}$ denotes the vector of *moment of the couple of rotations*.

The arbitrary chosen point M of the body does not occur anywhere in the final result, and this means that the velocity vectors of all points of the body are the same at the time instant under study; i.e., in the case of a couple of rotations, the body is in *instantaneous translational motion* with velocity equal to the moment of the couple of rotations.

The result thus obtained can be generalized to the case of several couples of rotations: the resulting motion of the body is an instantaneous translational motion with velocity equal to the vector sum of the moments of couples of rotations.

One can show that

1. A couple of instantaneous rotations can be replaced by another equivalent couple. In this change, it is necessary that the vector of the moment of the new couple of rotations be equal to the vector of the moment of the initial couple.

2. Two couples of rotations are equivalent to one couple whose moment is equal to the vector sum of moments of the initial couples.

It follows from the above that any instantaneous motion of a rigid body can be described by using a *system of sliding angular velocity vectors*.

▶ **Theorems on sliding vectors.** In the theory of sliding vectors, the following theorems hold:

THEOREM 1. *An arbitrary system of sliding vectors can be reduced to an equivalent system consisting of a sliding vector applied at an arbitrary chosen point P and a couple of vectors; here the vector is equal to the vector sum of vectors of the system (the resultant vector of the system), and the vector of the couple moment is equal to the vector sum of moments of all vectors of the system about the point P (the resultant moment of the system about the point P).*

THEOREM 2. *Two systems of sliding vectors are equivalent if, after being referred to the same point P, their resultant vectors and their resultant moments coincide.*

THEOREM 3. *If a system of sliding vectors is first referred to point P' and then to point P'', then the resultant vectors Ω' and Ω'' and the moments $\mathbf{M}'$ and $\mathbf{M}''$ of the resultant couples are related as*

$$\Omega'' = \Omega' = \Omega,$$

$$\mathbf{M}'' = \mathbf{M}' + \overrightarrow{P''P'} \times \Omega'.$$

THEOREM 4. *It follows from Theorem 3 that the following quantities are independent of the choice of the reference point P:*

$$\Omega'' = \Omega' = \Omega,$$

$$\Omega' \cdot \mathbf{M}' = \Omega'' \cdot \mathbf{M}''.$$

Therefore, the resultant vector of the system and the scalar product of the resultant vector by the resultant moment are called *invariants of the system of sliding vectors* about the reference center.

The second invariant with the first taken into account can be written in another form:

$$\mathrm{comp}_\Omega\, \mathbf{M}' = \mathrm{comp}_\Omega\, \mathbf{M}'';$$

i.e., the projection of the resultant moment onto the direction of the resultant vector is independent of the choice of the center of reference.

THEOREM 5. *The locus of points of reference with coordinates x, y, z at which the vectors Ω and $\mathbf{M}$ are parallel is the straight line*

$$\frac{M_{Ox} - (y\Omega_z - z\Omega_y)}{\Omega_x} = \frac{M_{Oy} - (z\Omega_x - x\Omega_z)}{\Omega_y} = \frac{M_{Oz} - (x\Omega_y - y\Omega_x)}{\Omega_z}, \qquad \text{(E2.1.6.1)}$$

where $\Omega_x, \Omega_y, \Omega_z$ and M_{Ox}, M_{Oy}, M_{Oz} are the projections onto the Cartesian coordinate axes $Oxyz$ of the resultant vector Ω and the resultant moment $\mathbf{M}_O$ about the origin O.

The straight line (E2.1.6.1) is called the *screw axis* or the *central axis*, and the pair consisting of the resultant vector and the parallel vector of the resultant couple is called the *screw*.

Thus, just as in Subsection E2.1.4, we conclude that for $\Omega \cdot \mathbf{M} \neq 0$ the motion of the body can be reduced to instantaneous screw motion.

In special cases ($\Omega \cdot \mathbf{M} = 0$), there may be states of instantaneous rest ($\Omega = 0$ and $M = 0$), of instantaneous translational motion ($\Omega = 0$ and $M \neq 0$), and of rotation about instantaneous axis ($\Omega \neq 0$ and $M = 0$ or $\mathbf{M} \perp \Omega$).

E2.2. Basic Notions and Laws of Mechanics

E2.2.1. Basic Notions of Mechanics

▶ **Mechanical systems. Forces.** A *mechanical system* is defined to be a set of point particles in which the position, velocity, and acceleration of any point depend on the positions, velocities, and accelerations of the other points.

A *constrained mechanical system* is a system of points with constraints imposed on their positions and velocities. The material bodies realizing these constraints are called the *mechanical constraints* imposed on the mechanical system. The relations between the coordinates and velocities of points of the system and time are called *constraint equations*. A material system is said to be *free* if there are no mechanical constraints in the system.

The kinematic state of interacting points (bodies) of a mechanical system is changed. A measure of such interaction is the *force*, depicted in the figures as a directed rectilinear segment. The *line of action of the force* is the straight line on which this segment lies.

A set of forces distinguished according to some property is called a *system of forces* and is denoted by $\{\mathbf{F}_i\}$.

A system of forces is said to be *balanced* if, being applied to a free rigid body at rest, it does not take the body out of this state.

Systems of forces $\{\mathbf{F}_i\}$ and $\{\mathbf{G}_j\}$ are said to be *equivalent* if, being separately applied to a free rigid body, they cause the same changes in its kinematic state.

If there exists a force equivalent to a system of forces applied to a rigid body, then it is called the *resultant* of the system of forces.

The forces of mechanical constraints acting on points of a system are called *constraint reaction forces*. The forces that act on points of a system and are not reaction forces are called *active forces*.

▶ **Moment of force.** *The moment of a force* $\mathbf{F}$ *about a center* (*point*) P is the vector $\mathbf{M}_P(\mathbf{F})$ equal to the cross product of the position vector $\mathbf{r}$ drawn from the center to the force application point by the vector of force, $\mathbf{M}_P(\mathbf{F}) = \mathbf{r} \times \mathbf{F}$ (Fig. E2.18).

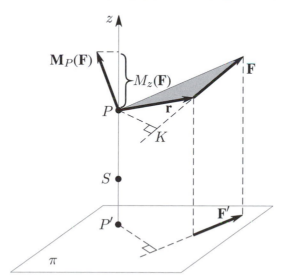

Figure E2.18. Moment of the force $\mathbf{F}$ about the center (point) P.

The magnitude of the moment of force is equal to the product of the magnitude of force by its moment arm about the center. The *moment arm of a force* $\mathbf{F}$ about a center P is the segment of the perpendicular PK drawn from the center to the force action line. The vector $\mathbf{M}_P(\mathbf{F})$ is perpendicular to the plane where the force and the center are located and is directed so that if we look in the direction opposite to this vector, then it seems that rotation of the force about the center is anticlockwise (the right-hand screw rule). If the force is translated along the line of its action, then its moment about the same center does not change.

If two arbitrary centers P and S are chosen on an axis (for example, on the z-axis), then

$$\mathrm{comp}_z \, \mathbf{M}_S(\mathbf{F}) = \mathrm{comp}_z(\overrightarrow{SP} + \mathbf{r}) \times \mathbf{F} = \mathrm{comp}_z \, \mathbf{M}_P(\mathbf{F});$$

i.e., the projection of the moment of a force about a center lying on an axis onto this axis is independent of the choice of the center. This scalar quantity is called the *moment of the force* $\mathbf{F}$ *about the z-axis* and is denoted by $M_z(\mathbf{F})$.

To calculate the moment of force about a given axis, one should

(1) Draw any plane π perpendicular to this axis and mark the point P' of their intersection (Fig. E2.18).

(2) Find the projection $\mathbf{F}'$ of the force onto this plane.

(3) Find the magnitude of the moment of the vector $\mathbf{F}'$ about the center P', which is equal to the product $F'h'$.

(4) Supply the obtained quantity with a sign observing the rotation of $\mathbf{F}'$ about P' from the side of positive sense of the axis, so that the positive sign corresponds to anticlockwise rotation.

The moment of a force about an axis does not vary if the force is transferred along its line of action. It is zero if the force and the axis are in the same plane; in this case, the line of action of the force is either parallel to the axis or intersects it.

The moment of a force $\mathbf{F}$ about the origin O of a reference frame and the moments of the force about the coordinate axes are calculated by the formulas

$$\mathbf{M}_O(\mathbf{F}) = \mathbf{r} \times \mathbf{F} = \begin{vmatrix} \mathbf{i} & \mathbf{j} & \mathbf{k} \\ x & y & z \\ F_x & F_y & F_z \end{vmatrix} = M_x(\mathbf{F})\,\mathbf{i} + M_y(\mathbf{F})\,\mathbf{j} + M_z(\mathbf{F})\,\mathbf{k};$$

$$M_x(\mathbf{F}) = yF_z - zF_y, \quad M_y(\mathbf{F}) = zF_x - xF_z, \quad M_z(\mathbf{F}) = xF_y - yF_x,$$

where F_x, F_y, and F_z are projections of the force onto the coordinate axes and x, y, and z are the coordinates of the force application point.

The *resultant vector of a system of forces* $\{\mathbf{F}_k\}$ is the vector $\mathbf{R}$ equal to the vector sum of all forces in the system:

$$\mathbf{R} = \sum_{k=1}^{n} \mathbf{F}_k.$$

The *resultant moment of a system of forces about a center* P is the vector $\mathbf{M}_P$ equal to the vector sum of the vectors of moments of all forces in the system about the center P:

$$\mathbf{M}_P = \sum_{k=1}^{n} \mathbf{M}_P(\mathbf{F}_k) = \sum_{k=1}^{n} \mathbf{r}_k \times \mathbf{F}_k.$$

Here $\mathbf{r}_k$ is the position vector drawn from the center P to the point of application of the force $\mathbf{F}_k$.

▶ **Couple of forces and its properties.** A couple of forces is a system of two forces $\mathbf{F}'$ and $\mathbf{F}''$ applied to the same rigid body so that they are equal in magnitude, parallel, and have opposite senses: $\mathbf{F}' = -\mathbf{F}''$ (Fig. E2.19).

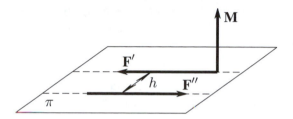

Figure E2.19. Couple of forces.

The shortest distance h between the lines of action of the forces in a couple is called the *arm of the couple*, and the plane where the forces lie is called the *plane of action* of the couple.

Properties of a couple of forces:

1. The resultant vector of a couple of forces is zero.

2. The resultant moment $\mathbf{M}(\mathbf{F}', \mathbf{F}'') = \mathbf{M}$ of a couple of forces is independent of the choice of the center and is called the *torque of the couple of forces*. It is equal to the moment of one of the forces in the couple about the point of application of the other force. The torque of a couple of forces is the measure of its mechanical action on a rigid body. Under the action of a couple of forces, a free rigid body being originally at rest begins to rotate about the axis parallel to the torque of the couple.

E2.2.2. Basic Laws of Mechanics

▶ **1. Law of inertia** (*Newton's first law*): if no forces act on a free point particle, then it preserves the rest state or a uniform rectilinear motion.

▶ **2. Law of proportionality between force and acceleration** (*Newton's second law*) can be written as the vector equation

$$m\mathbf{a} = \mathbf{F},$$

according to which a force acting on a free point particle causes point acceleration whose vector is parallel to the force and proportional to its magnitude.

The *point mass* m enters the second law as a scalar coefficient of proportionality. In mechanics, the mass is assumed to be constant; it is a measure of inertia and gravitational properties of a point particle.

The reference frames in which Newton's first and second laws are satisfied are called *inertial reference frames*. If these laws are not satisfied in a reference frame, then it is said to be *noninterial*.

▶ **3. Law of equality of action and reaction** (*Newton's third law*): to each action, there corresponds an equal and oppositely directed reaction.

The force actions of some material bodies on other bodies cannot be one-sided; they are always reciprocal. The forces do not arise by themselves but are the results of contact or spatial interaction of bodies. The appearance of a force acting on a body assumes that there is another body, which, in turn, is under the action of a force acting from the first body. The law states that these forces are equal in magnitude and opposite in direction, and the lines of their action coincide.

▶ **4.** A system of forces applied to a point has the resultant $\mathbf{R}^*$ equal to their vector sum. A consequence of this law is the following theorem.

THEOREM (VARIGNON) ON THE MOMENT OF THE RESULTANT OF A SYSTEM OF CON-
VERGING FORCES. *The moment of the resultant of a system of converging forces about an
arbitrary center is equal to the vector sum of the moments of forces of the system about the
same center.*

▶ **5.** The kinematic state of a constrained mechanical system does not vary if the mechanical constraints imposed on the system are replaced by the constraint reaction forces.

E2.3. Statics

E2.3.1. Basic Laws and Theorems of Statics

▶ **The problems of statics** include
(1) Transforming systems of forces into equivalent systems of forces.
(2) Determining equilibrium conditions for systems of forces acting on a rigid body.

▶ **Axioms of statics.** The general laws of mechanics are supplemented in statics with the
following axioms.
1. *Axiom about the balance of two forces applied to a rigid body*: two forces applied
to a rigid body are mutually balanced only if the forces have opposite directions along the
common line of action and their magnitudes are equal to each other.
2. *Axiom on inclusion and exclusion of balanced systems of forces*: the action of a
system of forces on a rigid body does not vary if a balanced system of forces is added to or
removed from the system.
It follows from the last two axioms that a force applied to a rigid body can be transferred
along its line of action, whereby the action of the force on the body remains the same.
3. *Axiom on preservation of equilibrium of a mechanical system under the action of
additionally imposed mechanical constraints*: if a mechanical system is in equilibrium,
then the equilibrium remains preserved if an additional mechanical constraint is imposed
on the system. In particular, this implies that the equilibrium of a deformed body under the
action of a system of forces is preserved in hardening, because the body hardening process
is equivalent to imposing additional mechanical constraints on the system of point particles
forming the deformed body.

▶ **Basic theorems of statics. Several formulas.** The axioms of statics define systems of
forces applied to a rigid body as systems of sliding vectors, and therefore, all results of the
theory of sliding vectors remain true for systems of forces.

THEOREM ON EQUIVALENCE OF COUPLES OF FORCES. *Two couples of forces are equivalent if their torques coincide.*

This means that, without changing the action of a couple on a rigid body, it can be
transferred to any place in the plane of its action; the couple can be turned and the force
magnitudes and the arm can be varied so that the rotation direction and the magnitude of
the couple torque remain constant. A couple can be transferred into another plane parallel
to the initial plane of the couple action. Thus, the main characteristic of a couple is the
vector of its torque, and hence it is more convenient to represent the couple by this vector
rather than by two antiparallel forces.

THEOREM ON THE RESULTANT COUPLE OF FORCES. *A system of couples of forces applied to a rigid body is equivalent to a single couple of forces whose torque is equal to the
vector sum of torques of the original couples.*

THEOREM ON THE PARALLEL TRANSFER OF A FORCE. *One can transfer a force to any point of a rigid body without changing its action on the body provided that one simultaneously adds a couple of forces (an associated couple) whose torque is equal to the moment of the force to be transferred about the point where the force is to be transferred.*

THEOREM ON REDUCING A SYSTEM OF FORCES TO A CENTER. *An arbitrary system of forces applied to a rigid body is equivalent to a single force applied at an a priori chosen center P and a single couple of forces. The force is equal to the resultant vector* $\mathbf{R}$ *of the system of forces, and the torque of the couple is equal to the resultant moment* $\mathbf{M}_P$ *of the system of forces about the chosen center.*

THEOREM ON EQUIVALENCE OF SYSTEMS OF FORCES APPLIED TO A RIGID BODY. *For two systems of forces to be equivalent, it is necessary and sufficient that they have equal resultant vectors and resultant moments about the same center.*

If a system of forces is reduced first to a point P' and then to a point P'', then the resultant vectors $\mathbf{R}'$ and $\mathbf{R}''$ and the resultant moments $\mathbf{M}'$ and $\mathbf{M}''$ satisfy the relations

$$\mathbf{R}'' = \mathbf{R}' = \mathbf{R},$$
$$\mathbf{M}'' = \mathbf{M}' + \overrightarrow{P''P'} \times \mathbf{R}'.$$

The invariants of a system of forces are the resultant vector and the scalar product of the resultant vector by the resultant moment of the system; they are independent of the choice of the reduction center:

$$\mathbf{R}'' = \mathbf{R}' = \mathbf{R},$$
$$\mathbf{R}' \cdot \mathbf{M}' = \mathbf{R}'' \cdot \mathbf{M}''.$$

The second invariant, with the first invariant taken into account, can be represented as

$$\text{comp}_\mathbf{R}\, \mathbf{M}' = \text{comp}_\mathbf{R}\, \mathbf{M}''.$$

For $\mathbf{R} \cdot \mathbf{M} \neq 0$, the system of forces can be reduced to a screw (wrench), and the equation of the screw axis has the form

$$\frac{M_{Ox} - (yR_z - zR_y)}{R_x} = \frac{M_{Oy} - (zR_x - xR_z)}{R_y} = \frac{M_{Oz} - (xR_y - yR_x)}{R_z},$$

where R_x, R_y, R_z and M_{Ox}, M_{Oy}, M_{Oz} are the projections on the Cartesian coordinate axes $Oxyz$ of the resultant vector $\mathbf{R}$ and of the vector $\mathbf{M}_O$ of the resultant moment about the origin O.

Thus, just as in Subsection E2.1.4, we conclude that for $\mathbf{R} \cdot \mathbf{M} \neq 0$ the system of forces can be reduced to a wrench. The special cases where $\mathbf{R} \cdot \mathbf{M} = 0$ include the case of a balanced system of forces ($R = 0$ and $M = 0$), the case of reduction to the resultant couple of forces ($R = 0$ and $\mathbf{M} \neq 0$), and the case of nonzero resultant force $\mathbf{R} \neq 0$.

THEOREM (VARIGNON) ON THE MOMENT OF THE RESULTANT. *If a system of forces has a resultant* $\mathbf{R}^*$*, then its moment about any center P is equal to the vector sum of the moments of all forces of the system about the same center:*

$$\mathbf{M}_P(\mathbf{R}^*) = \sum_{k=1}^{n} \mathbf{M}_P(\mathbf{F}_k).$$

Corollary: the moment of the resultant about an arbitrary axis is equal to the algebraic sum of moments of all forces of the system about the same axis.

E2.3.2. Balance Conditions for a System of Forces Applied to a Rigid Body

▶ **Equations of equilibrium of a rigid body.** For a system of forces applied to a rigid body to be balanced, it is necessary and sufficient that its resultant vector and resultant moment about a center, for example, about the origin O, be zero:

$$\mathbf{R} = \sum_k \mathbf{F}_k = 0; \qquad \mathbf{M}_O = \sum_k \mathbf{M}_O(\mathbf{F}_k) = 0. \tag{E2.3.2.1}$$

In coordinate form, the *balance conditions* for a system of forces applied to a rigid body are obtained from the vector equations and have the form of the system of six equations

$$
\begin{aligned}
&\sum_k F_{kx} = 0, &&\sum_k F_{ky} = 0, &&\sum_k F_{kz} = 0, \\
&\sum_k M_x(\mathbf{F}_k) = 0, &&\sum_k M_y(\mathbf{F}_k) = 0, &&\sum_k M_z(\mathbf{F}_k) = 0.
\end{aligned}
\tag{E2.3.2.2}
$$

This system is called the *primary system of equations of equilibrium of a rigid body* under the action of an arbitrary system of forces.

The upper three equations are called the *equations of projections* of forces onto the coordinate axes; they reflect the fact that if a rigid body is in equilibrium, then the algebraic sum of projections of all forces applied to the body onto each of the coordinate axes must be zero. The lower three equations are called the *equations of moments of forces* about the coordinate axes. These equations show that if a body is in equilibrium, then the algebraic sum of moments of all forces acting on the body about each of the coordinate axes must be zero.

There are other forms of the system of equilibrium equations, different from the primary system, each of which consists of six equations. When using them, one should verify conditions constraining the choice of axes about which the sums of moments of the forces are calculated.

In each problem on the equilibrium of a rigid body or a structure consisting of several bodies, in addition to some given quantities, there are quantities that must be determined when solving the problem. An equilibrium problem is said to be *statically determinate* if it can be solved completely by methods of statics. In this case, it is necessary that the number of unknowns does not exceed the number of equilibrium equations. A problem is said to be *statically indeterminate* it if cannot be solved by using the equations of statics alone. Thus, any equilibrium problem with seven or more unknowns for a single body is a priori statically indeterminable.

▶ **Special cases of equilibrium equations for a rigid body.** In special cases of arrangement of forces, one or several equations in the primary system can become identities. For example, (i) if all the forces are perpendicular to the coordinate z-axis, then the equation for the projections onto the z-axis becomes an identity; (ii) if the lines of action of all forces intersect the y-axis, then the equation for the moments about the y-axis becomes an identity. (In these cases, the number of equilibrium equations decreases to five.)

In what follows, we consider systems of forces whose equilibrium is described by three equilibrium equations.

A system of forces is said to be *converging* if the lines of action of the forces intersect at a single point. We choose the origin at this point. The equations of moments of the main system become identities, and the remaining three equations

$$\sum_k F_{kx} = 0, \quad \sum_k F_{ky} = 0, \quad \sum_k F_{kz} = 0 \tag{E2.3.2.3}$$

form the system of equilibrium equations.

A system of forces is said to be *plane* if all the forces lie in the same plane. We direct the coordinate axes so that all the forces lie in the plane xOy; in this case, the third, fourth, and fifth equations of the main system become identities, and the remaining three equations

$$\sum_k F_{kx} = 0, \quad \sum_k F_{ky} = 0, \quad \sum_k M_O(\mathbf{F}_k) = 0 \tag{E2.3.2.4}$$

form a system of equilibrium equations of a plane system of forces. Here the equation of moments about the z-axis is written in a different equivalent form: the moment vectors are perpendicular to the plane of action of the forces if the center lies in the same plane. Therefore, the moment of a force about the center can be viewed as a scalar quantity. In a right reference frame, the moment is assumed to be positive if the force tends to rotate anticlockwise about the center.

In the case of a plane system of forces, one can use other forms of equilibrium equations. One of them is

$$\sum_k F_{kx} = 0, \quad \sum_k M_P(\mathbf{F}_k) = 0, \quad \sum_k M_S(\mathbf{F}_k) = 0,$$

where P and S are any points of the plane for which the segment PS is not perpendicular to the x-axis.

Another form of equilibrium equations is

$$\sum_k M_P(\mathbf{F}_k) = 0, \quad \sum_k M_S(\mathbf{F}_k) = 0, \quad \sum_k M_E(\mathbf{F}_k) = 0,$$

where P, S, and E are any points on the plane that do not lie on the same straight line.

A system of forces is called a *system of parallel forces* if the lines of action of the forces are parallel. Let the z-axis be parallel to the lines of action. The first, second, and sixth equations of the main system become identities, and the remaining equations

$$\sum_k F_{kz} = 0, \quad \sum_k M_x(\mathbf{F}_k) = 0, \quad \sum_k M_y(\mathbf{F}_k) = 0$$

form the system of equilibrium equations for a body under the action of parallel forces.

E2.3.3. Solution of the Problems of Statics

▶ **The general scheme of solution of the equilibrium problem for a body** (or for structures consisting of several bodies) includes several stages. One should

1. Choose a body (or a structure) the study of whose equilibrium allows one to determine the desired quantities. Draw an *assumption diagram*, i.e., a simplified figure that contains only linear dimensions and angles necessary to solve the problem but does not contain insignificant details.

2. In the diagram, draw the active forces given in the assumptions of the problem.

3. In the case of a constrained body, omit the mechanical constraints imposed on it and replace their action by constraint reactions. After such a change, the body becomes free.

4. Verify whether the necessary condition for the problem to be statically determinate is satisfied: the number of unknowns in the assumption diagram should not exceed the number of equilibrium equations for the system of forces under study.

5. If this condition is satisfied, write out the system of equilibrium equations, solve it, and study the results.

When solving the problem, it is desirable to follow the above scheme strictly.

Experience shows that errors most frequently creep in when replacing the constraints by their reaction forces. Consider this issue in more detail.

▶ **Rules for the alignment of mechanical constraint reaction forces**. Every mechanical constraint is either a body or a mechanical device that imposes some constraints on the body displacements in space. Depending on the type of the constraint, some displacements are prohibited and some displacements are allowed. This permits one to predict some qualitative characteristics even before calculating the numerical values of the constraint forces. The rule that one should stick to when replacing the constraints by reaction forces is the following: the constraint reaction can generally consist of two force factors, namely, a force applied at the point where the constraint is imposed and a couple of forces.

If a constraint prohibits a translational displacement of the body, then there appears a reaction force whose direction is opposite to that of the prohibited displacement. If a constraint prohibits a rotation of the body, then there appears a couple of constraint reaction forces whose torque is directed along the axis of the prohibited rotation.

▶ **Some types of mechanical constraints**.

1. *Perfectly flexible unstretchable massless string*. In Fig. E2.20, a string attached to a rigid body at point B is shown. The constraint in question is imposed at point B, and since this is a perfectly flexible string, the body is allowed to rotate about any axis passing through this point. This means that no couple of reaction forces appears in this case.

For this type of constraints, small translational displacements of a body are allowed in which point B moves on the surface of a sphere of radius BK centered at point K. Therefore, no reaction force can arise in the direction of allowed displacements. The translational displacement of the body in direction KB is prohibited, because the string is unstretchable. This means that there arises a reaction force $\mathbf{N}_B$ applied to the body at point B in the direction of the straight line BK.

2. One aligns reaction forces in a similar way if the constraint is realized either by *free support of two bodies* such that the surface of one of them is absolutely smooth (Fig. E2.21) or by a *moving hinge B* (Fig. E2.22).

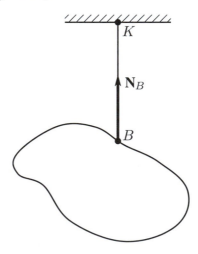

Figure E2.20. Perfectly flexible unstretchable massless string.

3. A *fixed hinge A* (Fig. E2.22) does not prevent beam AB from rotation about point A; therefore, no couple of reaction forces appears at point A. The device under study prohibits any translational displacement; therefore, neither the magnitude nor the direction of the reaction force $\mathbf{R}_A$ is known in advance. In this case, $\mathbf{R}_A$ is usually represented by two components $\mathbf{x}_A$ and $\mathbf{y}_A$ parallel to the coordinate axes.

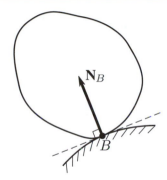

Figure E2.21. Free support of two bodies.

4. *Rigid clamping.* The constraint prohibits any translational displacement of the body; therefore, there arises a constraint reaction force $\mathbf{R}_B$ whose direction is unknown. It is usually represented by three components $\mathbf{x}_B, \mathbf{y}_B, \mathbf{z}_B$ parallel to the coordinate axes. Rigid clamping prohibits rotation about any axis passing through point B; therefore, there arises a couple of reaction forces whose torque $\mathbf{M}_B$ is known in advance neither in magnitude nor in direction. The unknown couple is usually represented by an equivalent system of three couples whose torques $\mathbf{M}_{Bx}, \mathbf{M}_{By}, \mathbf{M}_{Bz}$ are directed along the coordinate axes (Fig. E2.23).

Figure E2.22. Fixed hinge.

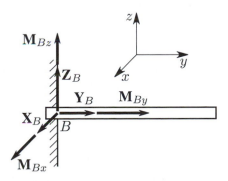

Figure E2.23. Rigid clamping.

If several constraints are imposed on a body, then each of them can be investigated independently of the other constraints and the forces applied to the body.

The unknowns in problems of statics can be not only the constraint reactions but also angles, linear dimensions of structures, and other parameters.

Example 1. Find the reaction of the rigid clamping B of a bent massless beam BDE (Fig. E2.24a) subjected to a force $F = 20$ kN and a couple of forces with torque $M = 2$ kN m assuming that $\alpha = 30°$, $BD = 2$ m, and $DE = 1$ m.

Solution. We follow the general scheme for solving body equilibrium problems.

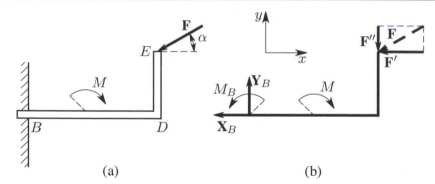

(a)   (b)

Figure E2.24. Bent beam with rigidly clamped left end is subjected to a force F and a couple of forces with torque M: (a) active forces; (b) constraint reaction forces.

1. The reaction of the rigid clamping B can be found by studying the beam equilibrium; therefore, for the rigid body we take the beam BDE. We separately draw the assumption diagram (Fig. E2.24b).

2. On the diagram, we align the active force factors acting on the beam, i.e., the force $\mathbf{F}$ and the couple of forces $\mathbf{M}$, which we present as a curved arrow in the direction of rotation of the couple. (In our case, the rotation is clockwise.)

3. The beam is not free, because a constraint is imposed on it; i.e., it is rigidly clamped at point B. We make the beam free by replacing the constraint with constraint reaction forces. The clamping reaction consists of a force $\mathbf{R}_B$ and a couple of forces with torque M_B, for both of which neither the magnitude nor the direction is known.

We introduce the reference frame xy shown in the figure. Since the direction of the force $\mathbf{R}_B$ is unknown, we represent it by two components $\mathbf{X}_B$ and $\mathbf{Y}_B$. The unknown couple M_B of reaction forces is represented by the curved arrow. Note that we need not guess the true directions of the components and the curved arrow of the torque of the couple, because all this will be clear after the problem is solved. (In the assumption diagram, $\mathbf{X}_B$ is shown as if it were directed to the left only because this way it is more noticeable in the figure.)

We replace $\mathbf{F}$ by an equivalent system of forces consisting of two components $\mathbf{F}'$ and $\mathbf{F}''$ parallel to the coordinate axes,

$$\mathbf{F} = \mathbf{F}' + \mathbf{F}'', \quad F' = F\cos\alpha, \quad F'' = F\sin\alpha.$$

4. Now the assumption diagram represents a free beam subjected to forces lying in the same plane. The system of equilibrium equations consists of three equations. The number of unknowns X_B, Y_B, M_B is also equal to three. This means that the number of unknowns and the number of equations coincide and the necessary conditions for the problem to be statically determinate are satisfied.

5. We use the basic form of equilibrium equations for a plane system of forces, taking the center B as the moment point.

The mishap most frequently encountered when writing out the equilibrium equations is to forget some force or couple of forces; therefore, it is recommended to arrange all forces and all couples of forces applied to the body in a row and then write out the equilibrium equations beneath it:

Forces, couples of forces	X_B	Y_B	M_B	F'	F''	M

$$\sum\nolimits_k F_{kx} = -X_B + 0 + 0 - F' + 0 + 0 = 0,$$
$$\sum\nolimits_k F_{ky} = 0 + Y_B + 0 + 0 - F'' + 0 = 0,$$
$$\sum\nolimits_k M_B(\mathbf{F}_k) = 0 + 0 + M_B + F'\,DE - F''\,BD - M = 0.$$

By substituting the original data, we obtain the solution of the problem:

$$X_B = -17.1 \text{ kN}, \quad Y_B = 10 \text{ kN}, \quad M_B = 4.9 \text{ kN m}.$$

The signs in the answers show that the direction of the component $\mathbf{X}_B$ is opposite to that shown in the assumption diagram, while the directions of the component $\mathbf{Y}_B$ and of the curved arrow M_B correspond to those in the diagram.

It is not recommended to redraw $\mathbf{X}_B$ in the diagram (Fig. E2.24b) with the sign of the answer taken into account, because should this be done, it would be impossible to verify whether the equations of equilibrium are composed correctly and interpret the results.

The clamping reaction force $\mathbf{R}_B$ is the vector sum of the orthogonal components $\mathbf{X}_B$ and $\mathbf{Y}_B$; therefore, its magnitude can be obtained by the formula

$$R_B = \sqrt{X_B^2 + Y_B^2} = 19.8 \text{ kN}.$$

Example 2. A structure consisting of two massless rods BD and DE is shown in Fig. E2.25. The rods are of the same length 4 m and are hinged to each other and to the base so that $BE = BD = 2a$. The left rod BD is loaded by the horizontal force $F = 20$ kN at the midpoint, and the right rod DE is loaded by the couple of forces $M = 10$ kN m. Find the support reactions and the pressure in the intermediate hinge D.

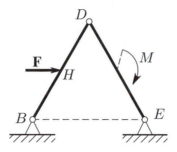

Figure E2.25. Structure consisting of two rods hinged to each other and to the base.

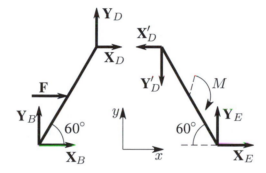

Figure E2.26. Assumption diagram for the two-rod problem.

Solution. We consider the equilibrium of each rod separately, for which we divide the structure into two parts, namely, the rods BD and DE, and align the active forces and the constraint reactions.

After this, the assumption diagram of the problem acquires the form shown in Fig. E2.26, where the reference frame is the same for both rods. The force $\mathbf{R}_D$ acting on the rod BD from the rod ED is represented by two components X_D and Y_D, and the force $\mathbf{R}'_D$ acting from the rod DB on the rod ED is represented by two components X'_D and Y'_D. Since the action force $\mathbf{R}_D$ and the reaction force $\mathbf{R}'_D$ must satisfy the action–reaction axiom, it follows that their components also satisfy the relations $X_D = -X'_D$ and $Y_D = -Y'_D$; i.e., the components are equal and oppositely directed, and this is already taken into account in the diagram. The magnitudes of the components satisfy the relations

$$X_D = X'_D, \quad Y_D = Y'_D. \tag{E2.3.3.1}$$

All in all, there are eight unknowns X_B, Y_B, X_D, Y_D, X'_D, Y'_D, X_E, and Y_E. The equilibrium of each rod is described by three equilibrium equations; together with the last two equations, they form a system of eight equations with eight unknowns. (The necessary conditions for the problem to be statically determinate are satisfied.)

The equilibrium equations read

for rod BD

$$\sum_k F_{kx} = X_B + X_D + F = 0,$$
$$\sum_k F_{ky} = Y_B + Y_D = 0,$$
$$\sum_k M_B(\mathbf{F}_k) = -Fa \sin 60°$$
$$- X_D 2a \sin 60° + Y_D 2a \cos 60° = 0,$$

for rod ED

$$\sum_k F_{kx} = X'_D + X_E = 0,$$
$$\sum_k F_{ky} = -Y'_D + Y_E = 0,$$
$$\sum_k M_E(\mathbf{F}_k) = X'_D 2a \sin 60°$$
$$+ Y'_D 2a \cos 60° - M = 0.$$

By solving this system with (E2.3.3.1) taken into account, we obtain $X_B = -16.44$ kN, $Y_B = -11.16$ kN, $X_D = X'_D = -3.56$ kN, $Y_D = Y'_D = 11.16$ kN, $X_E = -3.56$ kN, and $Y_E = 11.16$ kN.

E2.3.4. Center of Parallel Forces. Center of Gravity of a Rigid Body

▶ **Center of parallel forces.** Assume that a rigid body is under the action of a system $\{\mathbf{F}_k\}$ of parallel forces. The forces can rotate about the points of their application remaining parallel. Since $\mathbf{R} \cdot \mathbf{M}_P = 0$, it follows that the system has a resultant $\mathbf{R}^*$ provided that $\mathbf{R} \neq 0$. One can prove that the line of action of $\mathbf{R}^*$ always passes through the same point C of the body regardless of the force direction. This point is called the *center of parallel forces*, and its position in space is determined by the position vector $\mathbf{r}_C$,

$$\mathbf{r}_C = \frac{\sum_k F_k \mathbf{r}_k}{\sum_k F_k}, \tag{E2.3.4.1}$$

where the $\mathbf{r}_k$ are the position vectors of the points where the forces $\mathbf{F}_k$ are applied and the F_k are the force magnitudes.

The coordinates x_C, y_C, z_C of the center of parallel forces can be determined by the formulas

$$x_C = \frac{\sum_k F_k x_k}{\sum_k F_k}, \quad y_C = \frac{\sum_k F_k y_k}{\sum_k F_k}, \quad z_C = \frac{\sum_k F_k z_k}{\sum_k F_k}. \tag{E2.3.4.2}$$

▶ **Center of gravity of a body.** The center of gravity of a rigid body is the center of parallel forces that represent elementary forces of gravity of the material particles comprising the body. If the body is on Earth's surface and its dimensions are small compared to Earth's radius, then one can assume that the lines of action of the forces of gravity are parallel and their magnitudes depend only on the body volume V, the material density ρ, and the free fall acceleration g.

The formulas for finding the center of gravity read

$$\mathbf{r}_{\text{cg}} = \frac{1}{G} \int_V \gamma \mathbf{r} \, dV, \quad x_{\text{cg}} = \frac{1}{G} \int_V \gamma x \, dV, \quad y_{\text{cg}} = \frac{1}{G} \int_V \gamma y \, dV, \quad z_{\text{cg}} = \frac{1}{G} \int_V \gamma z \, dV,$$

where $\gamma = \gamma(x, y, z) = g\rho$ is the specific weight and $G = \int_V \gamma \, dV$ is the weight of the body.

Similar formulas hold for the center of gravity of a body that has the shape of a surface or a line.

▶ **Methods for finding the center of gravity.**

Symmetric bodies. If a body has a plane (axis, center) of material symmetry, then its center of gravity lies in this plane (on the axis, at the center).

Method of partition. Assume that a body consists, for example, of three parts (Fig. E2.27) for each of which we know the weight G_1, G_2, G_3 and the position $\mathbf{r}_1$, $\mathbf{r}_2$, $\mathbf{r}_3$ of the center of gravity. The position vector $\mathbf{r}_{123}$ of the center of gravity of the body consisting of three parts and its coordinates are determined by the formulas

$$\begin{aligned} \mathbf{r}_{123} &= \frac{G_1 \mathbf{r}_1 + G_2 \mathbf{r}_2 + G_3 \mathbf{r}_3}{G_1 + G_2 + G_3}, \quad x_{123} = \frac{G_1 x_1 + G_2 x_2 + G_3 x_3}{G_1 + G_2 + G_3}, \\ y_{123} &= \frac{G_1 y_1 + G_2 y_2 + G_3 y_3}{G_1 + G_2 + G_3}, \quad z_{123} = \frac{G_1 z_1 + G_2 z_2 + G_3 z_3}{G_1 + G_2 + G_3}. \end{aligned} \tag{E2.3.4.3}$$

Method of negative masses. Now assume that we need to find the position of the center of gravity of a new body consisting of parts *1* and *2* (see Fig. E2.27).

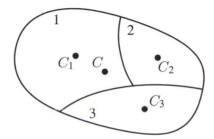

Figure E2.27. Determining the center of gravity of a body consisting of three parts.

The solution can be obtained either by the above formulas if one sets $G_3 = 0$ or by the formulas that, under certain conditions, can be more convenient:

$$\mathbf{r}_{12} = \frac{G\mathbf{r}_{123} - G_3\mathbf{r}_3}{G - G_3}, \qquad x_{12} = \frac{Gx_{123} - G_3x_3}{G - G_3},$$

$$y_{12} = \frac{Gy_{123} - G_3y_3}{G - G_3}, \qquad z_{12} = \frac{Gz_{123} - G_3z_3}{G - G_3},$$

where $G = G_{123} = G_1 + G_2 + G_3$ is the weight of the body consisting of three parts. The weight G_3 of the cutaway part enters the formulas with negative sign, and just this fact underlies the name of the method.

In Table E4.1 (Section E4.1), we present formulas for the coordinates of the center of gravity for homogeneous bodies of simplest shapes.

E2.3.5. Distributed Forces

In practice, one often encounters the cases in which the body is subjected not to lumped forces but to a load distributed over a volume, surface, or line. The special case of bulk distribution of the force of gravity was considered in the preceding section. The vector quantity $\mathbf{q}$ characterizing the load is called the *load intensity* and is measured in N/m^3, N/m^2, or N/m.

When solving problems of statics, a distributed load is usually replaced by a simpler statically equivalent force (or a system of forces).

1. A *uniformly distributed load* of intensity $\mathbf{q}$ (Fig. E2.28) in a plane problem of statics has the resultant $\mathbf{Q} = \mathbf{q}\,BD$ whose line of action passes through the midpoint of the interval where the load is applied, $BE = ED$.

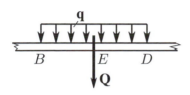

Figure E2.28. Uniformly distributed load.

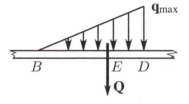

Figure E2.29. Linearly distributed load.

2. A *linearly distributed load* in plane problems of statics (Fig. E2.29) has the resultant $\mathbf{Q} = \frac{1}{2}\mathbf{q}_{max}\,BD$ passing through point E, and $BE = \frac{2}{3}BD$.

E2.3.6. Friction Laws (Coulomb Laws)

▶ **Law of sliding friction.** Consider a body of weight G at rest on a rough horizontal plane (Fig. E2.30). An attempt to move the body by applying a horizontal force $\mathbf{F}$ to it remains unsuccessful until the force attains a certain magnitude $\mathbf{F}^*$. The resultant of the support reaction forces can be represented as the sum of two components, the *normal pressure force* $\mathbf{N}$ and the *static friction force* $\mathbf{F}_{fr}$. The equation for the projections of the forces onto the horizontal axis gives $F_{fr} = F$.

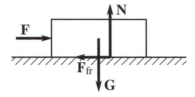

Figure E2.30. Friction force arising in an attempt to move a fixed body on a horizontal plane.

Experiments show that F^* and N satisfy the relation (*Coulomb's dry friction law*)

$$F^* = f_0 N, \tag{E2.3.6.1}$$

where f_0 is the *coefficient of static friction*, which depends on the materials of the contacting bodies and the state of their surfaces.

Until $F \leq F^*$, the body is at rest. But if a force larger than F^* is applied to it, the body begins to move. In motion, the resistance force can be found by the formula

$$F_{fr} = f N, \tag{E2.3.6.2}$$

where f is the *coefficient of kinetic friction* and F_{fr} is the force of kinetic friction.

Note that the coefficient of kinetic friction is always less than the coefficient of static friction, $f < f_0$.

▶ **Laws of rolling friction.** Consider a disk of radius R at rest on a nonsmooth horizontal plane (Fig. E2.31). An attempt to roll the disk by applying a horizontal force $\mathbf{F}$ to its center remains unsuccessful if the force magnitude is less than a certain limit value F^{**}.

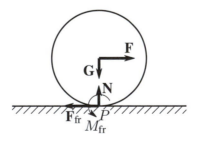

Figure E2.31. Friction force arising in an attempt to roll a circular disk.

According to the theorem on the reduction of a system of forces to a center, the support reaction forces distributed over a small surface near the contact point P can be replaced by an equivalent system, namely, by the normal pressure force $\mathbf{N}$, the static friction force $\mathbf{F}_{fr}$ applied at the contact point P, and a couple of friction forces with torque M_{fr}.

If the disk is in equilibrium, then the equations of moments about the center P imply $FR = M_{fr}$. Experiments show that F^{**} and N are related by $F^{**} = kN/R$ and

$$M^{**} = kN. \tag{E2.3.6.3}$$

The dimensional coefficient k [m] is called the *coefficient of rolling friction*. A rolling surface is said to be *absolutely rough* if $f \neq 0$ and $k = 0$.

Experiments show that, other conditions being equal, F^* is much larger than F^{**} (usually by two or three orders of magnitude), and hence, in technology, if it is necessary to decrease the friction losses, then one tries to replace sliding by rolling.

E2.4. Dynamics of a Point Particle

E2.4.1. Equations of Motion of a Point Particle

The motion of a point particle with respect to an inertial reference frame is described by Newton's second law:

$$m\mathbf{a} = \sum_{j=1}^{n} \mathbf{F}_j + \sum_{k=1}^{l} \mathbf{R}_k, \tag{E2.4.1.1}$$

where m is the mass of the point, $\mathbf{a}$ is its acceleration, and the right-hand side of this equation is the vector sum of all forces applied to the point. The causes of all these forces may be different. Here we distinguish between *active forces* $\mathbf{F}_j$ and *constraint reaction forces* $\mathbf{R}_k$. The active forces can depend on time t and on the position and velocity of the point. The active forces include the forces of gravity, elasticity, viscous friction, aerohydrodynamic drag, etc.

The constraint reaction forces act on a nonfree point particle whose motion is subjected to some mechanical constraints. These forces can only be determined in the course of solution of the problem of dynamics.

We take the Cartesian axes of an inertial reference frame x, y, z, project the vector equality (E2.4.1.1) onto these axes, and obtain

$$m\frac{d^2x}{dt^2} = \sum_j F_{jx} + \sum_k R_{kx},$$

$$m\frac{d^2y}{dt^2} = \sum_j F_{jy} + \sum_k R_{ky},$$

$$m\frac{d^2z}{dt^2} = \sum_j F_{jz} + \sum_k R_{kz}.$$

These three equations are called the *differential equations of motion of a point particle in Cartesian coordinates*. (From now on, summation indices are sometimes omitted.)

The *differential equations of motion of a point particle in projections onto the natural coordinate axes* have the form

$$m\frac{dv}{dt} = \sum F_{j\tau} + \sum R_{k\tau},$$

$$\frac{mv^2}{\rho} = \sum F_{jn} + \sum R_{kn},$$

$$0 = \sum F_{jb} + \sum R_{kb}.$$

Here we have taken into account the fact that $a_\tau = \dfrac{dv}{dt}$, $a_n = \dfrac{v^2}{\rho}$, and $a_b = 0$.

E2.4.2. First and Second Problems of Dynamics

In the equations of motion (see Subsection E2.4.1), unknowns can occur on both left- and right-hand sides. Depending on this, the problems of dynamics are divided into two types, which we consider below.

▶ **The first problem of dynamics.** The law of motion and the active forces are given, and one needs to find the constraint reaction forces.

Example 1. A load of weight G to which a certain initial velocity was imparted at time $t_0 = 0$ ascends an inclined rough plane (Fig. E2.32). Find the friction force F_{fr} and the normal pressure force N acting on the body under the assumption that the coefficient f of friction of the plane and the angle α of inclination are known.

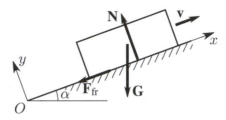

Figure E2.32. Motion of a load on an inclined rough plane.

Solution. We introduce Cartesian coordinate axes and place the origin O at the load position at time $t = 0$. Let us draw the load in an arbitrary position and the forces acting on it. Under the assumption that the load is a point particle, we write out Newton's second law

$$m\mathbf{a} = \mathbf{G} + \mathbf{N} + \mathbf{F}_{\text{fr}}.$$

By projecting both sides of this vector equation onto the y-axis, we obtain $0 = -G\cos\alpha + N$ (because the load acceleration is parallel to the x-axis), whence we find $N = G\cos\alpha$. Further, using the Coulomb law, we obtain the friction force $F_{\text{fr}} = fN = fG\cos\alpha$.

▶ **The second problem of dynamics.** The active forces, the mechanical constraint equations, the initial position of the point, and its initial velocity are given, and one needs to find the law of motion of the point and the constraint reaction.

It is recommended to solve the second problem of dynamics successively in several stages listed below.

1. The supposed trajectory of motion is drawn with the point particle shown on it.
2. The forces applied to the point are drawn.
3. Newton's second law is written in vector form.
4. A convenient reference frame is chosen.
5. The equations of motion of the point are written in projections on either the axes of the Cartesian coordinate system or the axes of the natural trihedron. In the first case, it is necessary to express all active forces in terms of t, x, y, z, $\dot{x}$, $\dot{y}$, and $\dot{z}$; in the second case, in terms of t, s, $\dot{s}$.
6. The obtained differential equations are supplemented with the initial conditions, i.e., the values of the coordinates and projections of the velocity of the point at the initial time instant. (They are taken from the conditions of the problem with the introduced reference frame taken into account.)
7. The problem thus posed is solved numerically or analytically.

It is recommended to perform these stages of solution without changing their order.

Example 2. In addition to the conditions of the problem in Example 1, it is assumed that the velocity of the load at time t^* is equal to half the initial velocity. Find the initial velocity v_0 of the load and the path L traveled in time t^*.

Solution. Continuing the process of solving the problem in Example 1, we project both sides of the vector equation onto the x-axis. Using the formulas $G = mg$ and $F_{\text{fr}} = fG\cos\alpha$, we obtain

$$\ddot{x} = -g\,(\sin\alpha + f\cos\alpha).$$

The general solution of the obtained differential equation and the expression for the velocity of the load are given by the formulas

$$x = -\tfrac{1}{2}g\,(\sin\alpha + f\cos\alpha)t^2 + C_1 t + C_2,$$
$$\dot{x} = -g\,(\sin\alpha + f\cos\alpha)t + C_1.$$

(The details of their derivation are omitted, and their correctness can be verified by differentiation.)

The last two relations must hold at any time instant t and hence at the initial time $t_0 = 0$ and at time t^*; i.e., the relations are satisfied if, instead of t, x, and $\dot{x}$, we first substitute the values 0, 0, v_0 and then the values t^*, L, $v_0/2$. After the substitutions, we obtain the system of four equations

$$0 = C_2,$$
$$v_0 = C_1,$$
$$L = -\tfrac{1}{2}g\,(\sin\alpha + f\cos\alpha)(t^*)^2 + C_1 t^* + C_2,$$
$$\tfrac{1}{2}v_0 = -g\,(\sin\alpha + f\cos\alpha)t^* + C_1$$

with four unknowns C_1, C_2, v_0, L. By solving this system, we obtain the desired quantities

$$v_0 = 2g\,(\sin\alpha + f\cos\alpha)t^*,$$
$$L = \tfrac{3}{2}g\,(\sin\alpha + f\cos\alpha)(t^*)^2.$$

E2.5. General Theorems of Dynamics of a Mechanical System

It is often possible to find important characteristics of motion of a mechanical system without integrating the system of differential equations of motion. This can be done by using general theorems of dynamics.

E2.5.1. Basic Notions and Definitions

▶ **Internal and external forces.** Any force acting on a point of a mechanical system is necessarily either an active force or a constraint reaction. The entire set of forces acting on the points of the system can also be divided into two classes in a different way: one distinguishes between the *external forces* $\mathbf{F}^e$ and the *internal forces* $\mathbf{F}^i$. The external forces are the forces acting on the points of the system from points and bodies that are not contained in the system under study. The internal forces are the forces of interaction between the points and bodies contained in the system under study.

This distinction depends on what point particles and bodies are included by the researcher into the mechanical system under study. If the system is extended by including additional points and bodies, then some forces that were external for the original system can become internal for the extended system.

▶ **Properties of internal forces.** Since these forces are forces of interaction between parts of the system, they are contained in the total system of internal forces in "pairs" formed according to the action–reaction axiom. For each of such "pairs" of forces, the resultant vector and the resultant moment about an arbitrary point are zero. Since the complete system of internal forces consists only of "pairs," it follows that

1. The resultant of the system of internal forces is zero.
2. The resultant moment of the system of internal forces about an arbitrary point is zero.

The *mass m of the system* is the arithmetic sum of masses m_k of all points and bodies that form the system,

$$m = \sum_k m_k.$$

▶ The **center of mass** (center of inertia) of a mechanical system is the geometric point C whose position vector $\mathbf{r}_C$ and whose coordinates x_C, y_C, z_C are determined by the following formulas similar to the formulas for the center of parallel forces (see Subsection E2.3.4):

$$\mathbf{r}_C = \frac{\sum_k m_k \mathbf{r}_k}{m}, \quad x_C = \frac{\sum_k m_k x_k}{m}, \quad y_C = \frac{\sum_k m_k y_k}{m}, \quad z_C = \frac{\sum_k m_k z_k}{m},$$

where $\mathbf{r}_k, x_k, y_k, z_k$ are the position vectors and the coordinates of the points comprising the system.

For a rigid body in a homogeneous field of gravity, the positions of the center of mass and the center of gravity coincide; in other cases, these are different geometric points.

Along with an inertial reference frame, one often considers a noninertial reference frame moving translationally. Its coordinate axes $Cx^*y^*z^*$ (the *König axes*) are chosen so that the origin C constantly coincides with the center of mass of the mechanical system. By definition, the center of mass is stationary in the König axes and is located at the origin.

▶ **Moments of inertia.** In the space, we choose a plane Π, an axis l, and a point O. The distances from a point of mass m_k to Π, l, and O are denoted by the symbols δ_k, Δ_k, and r_k, respectively. Consider the positive expressions

$$I_\Pi = \sum_k m_k \delta_k^2, \quad I_{ll} = \sum_k m_k \Delta_k^2, \quad I_O = \sum_k m_k r_k^2,$$

where the sum is taken over all the points of the system. These expressions are called the *moments of inertia about the plane Π, the axis l, and the point O*. In the Cartesian coordinate axes x, y, z with origin at point O, the moments of inertia satisfy the relations

$$I_{xx} = I_{yOx} + I_{xOz}, \quad I_{yy} = I_{zOy} + I_{yOx}, \quad I_{zz} = I_{xOz} + I_{zOy},$$
$$I_O = I_{xOy} + I_{yOz} + I_{zOx} = \tfrac{1}{2}(I_{xx} + I_{yy} + I_{zz}).$$

Consider a ray l passing through the origin O with direction cosines α, β, γ (Fig. E2.33). The squared distance Δ_k from the point of mass m_k with coordinates x_k, y_k, z_k to the ray is equal to

$$\Delta^2 = (x^2 + y^2 + z^2)(\alpha^2 + \beta^2 + \gamma^2) - (x\alpha + y\beta + z\gamma)^2.$$

(To be concise, we omit the subscript k.)

The moment of inertia I_{ll} of a material system about the ray is equal to

$$I_{ll} = \sum m\Delta^2 = I_{xx}\alpha^2 + I_{yy}\beta^2 + I_{zz}\gamma^2 - 2I_{yz}\beta\gamma - 2I_{zx}\gamma\alpha - 2I_{xy}\alpha\beta. \qquad \text{(E2.5.1.1)}$$

The quantities

$$I_{yz} = \sum myz, \quad I_{zx} = \sum mzx, \quad I_{xy} = \sum mxy,$$

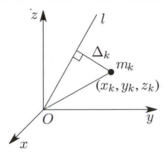

Figure E2.33. Moments of inertia of a point particle with mass m_k about the ray l.

are called the *centrifugal moments of inertia*; they can take both positive and negative values.

On the ray l at a distance $1/\sqrt{I_{ll}}$ from the origin, we mark a point with coordinates $X = \alpha/\sqrt{I_{ll}}$, $Y = \beta/\sqrt{I_{ll}}$, and $Z = \gamma/\sqrt{I_{ll}}$. The equation of the geometric locus of such points with (E2.5.1.1) taken into account has the form

$$I_{xx}X^2 + I_{yy}Y^2 + I_{zz}Z^2 - 2I_{yz}YZ - 2I_{zx}ZX - 2I_{xy}XY = 1.$$

Since the points of the system occupy a bounded domain in space, it follows that the last relation is the equation of an ellipsoid. It is called the *ellipsoid of inertia* of the system with respect to the point O. The principal axes of this ellipsoid are called the *principal axes of inertia* with respect to point O. In these axes, the centrifugal moments of inertia are zero.

If the principal axes of inertia are taken for the coordinate axes, then the expression for the moment of inertia acquires the form

$$I_{ll} = A\alpha^2 + B\beta^2 + C\gamma^2,$$

where $A = I_{xx}$, $B = I_{yy}$, and $C = I_{zz}$.

The principal axes of inertia passing through the center of mass of the system are called the *principal central* axes of inertia.

If a mechanical system is a rigid body, then the summation in the above formulas should be replaced by the integration over the volume V occupied by the body. For example,

$$I_{xOy} = \iiint_V \rho z^2 \, dx \, dy \, dz,$$

where $\rho = \rho(x, y, z)$ is the body density.

Table E2.1 presents formulas for calculating the moments of inertia of several figures.

Theorem (Huygens). *The moments of inertia about parallel axes one of which passes through the center of mass C are related by*

$$I_{zz} = I_{Cz} + md^2,$$

where I_{zz} and I_{Cz} *are moments of inertia about parallel axes z and Cz, the axis Cz passes through the center of mass, d is the distance between the axes, and m is the mass of the system.*

It follows from the Huygens theorem that $I_{Cz} \le I_{zz}$.

TABLE E2.1

The moments of inertia of the simplest homogeneous figures of mass m.

Type of figure	Description of figure	The moment of inertia is calculated about	Moment of inertia, I
Line	Straight line segment of length $2a$	the perpendicular passing through the midpoint of the segment	$\frac{1}{3}ma^2$
Line	Circle arc of radius R supported by angle α	the diameter dividing the arc in half	$\frac{1}{2}mR^2\left(1 - \frac{\sin\alpha}{\alpha}\right)$
Surface	Rectangle with sides $2a$ and $2b$	(i) the axis lying in its plane and passing through the center perpendicular to the side $2a$ (ii) the axis perpendicular to the plane of the rectangle and passing through its center	(i) $\frac{1}{3}ma^2$ (ii) $\frac{1}{3}m(a^2 + b^2)$
Surface	Ellipse with semiaxes a and b	(i) the axis a (ii) the axis b (iii) the axis perpendicular to the plane of the ellipse and passing through the center	(i) $\frac{1}{4}mb^2$ (ii) $\frac{1}{4}ma^2$ (iii) $\frac{1}{4}m(a^2 + b^2)$
Surface	Triangle	the axis lying in its plane and passing through one of the vertices	$\frac{1}{6}m(h_1^2 + h_1 h_2 + h_2^2)$, h_1, h_2 are distances from the other two vertices to the same axis
Surface	Hollow thin-walled cylinder of radius R	the cylinder axis	mR^2
Surface	Thin spherical shell of radius R	the axis passing through the center of the sphere	$\frac{2}{3}mR^2$
Three-dimensional body	Rectangular parallelepiped with sides $2a, 2b, 2c$	the axis passing through the center perpendicular to the face with sides $2a$ and $2b$	$\frac{1}{3}m(a^2 + b^2)$
Three-dimensional body	Rectangular pyramid of height h whose base is a rectangle with sides $2a$ and $2b$	(i) the height (ii) the axis passing through the center of gravity and parallel to the side $2a$	(i) $\frac{1}{5}m(a^2 + b^2)$ (ii) $\frac{1}{20}m(\frac{3}{4}h^2 + 4b^2)$
Three-dimensional body	Right circular cylinder of radius R and height h	(i) the cylinder axis (ii) the straight line passing through the center of gravity perpendicular to the cylinder axis	(i) $\frac{1}{2}mR^2$ (ii) $\frac{1}{4}m(\frac{1}{3}h^2 + R^2)$
Three-dimensional body	Ball of radius R	the axis passing through the center of the ball	$\frac{2}{5}mR^2$
Three-dimensional body	Triaxial ellipsoid with semiaxes a, b, c	the axis $2a$	$\frac{1}{5}m(b^2 + c^2)$
Three-dimensional body	Spherical sector of radius R and height h	the symmetry axis	$\frac{1}{5}m(3Rh - h^2)$

E2.5.2. Theorem on the Motion of the Center of Mass

▶ **Statement of the theorem and some comments.**

THEOREM. *The center of mass of a mechanical system moves as a point particle with mass m equal to the mass of the system would move under the action of external forces applied to the system:*

$$m\mathbf{a}_c = \sum_k \mathbf{F}_k^e, \tag{E2.5.2.1}$$

where m is the mass of the system and $\mathbf{a}_c$ is the acceleration of the center of mass.

The mathematical statement of the theorem is similar to that of Newton's second law. Let us explain the statement of the theorem in more detail. In motion of the system, its center of mass moves along a certain trajectory. We assume, for example, that at time t_0 the system is in position B and has a velocity $\mathbf{v}_{c0}$. Now if a point of mass m is placed in position B at time t_0, the velocity $\mathbf{v}_{c0}$ is imparted to it, and forces equal to the external forces acting on the system are applied to this point, then the point will move together with the center of mass of the system along the same trajectory with the same velocity and the same acceleration.

The vector equation implies the differential equations of motion of the center of mass in projections onto the axes of the Cartesian coordinate system:

$$m\ddot{x}_c = \sum_k F_{kx}^e, \quad m\ddot{y}_c = \sum_k F_{ky}^e, \quad m\ddot{z}_c = \sum_k F_{kz}^e.$$

▶ **Law of conservation of the velocity of the center of mass of a mechanical system:** if the resultant vector of external forces acting on a system is zero, then the center of mass of the system moves at a constant velocity; i.e., if $\sum_k \mathbf{F}_k^e = 0$, then $\mathbf{v}_c = \text{const}$.

Note that in this case the velocity vector itself (rather than its magnitude) is constant, and hence the center of mass will move uniformly and rectilinearly.

If the projection of the resultant vector of external forces of the system onto an axis is zero, then the projection of the velocity of the center of mass of the system onto this axis remains constant. For example, if $\sum_k F_{kx} = 0$, then $v_{cx} = \text{const}$.

Example. Under the action of the force of gravity, a homogeneous disk of mass m_2 rolls down the lateral face of a prism of mass m_1 located on a smooth horizontal plane (Fig. E2.34). The angle between the lateral face and the prism base is α. At the initial time, the velocities of the prism and the disk are zero. Find the distance s_1 traveled by the prism as the disk center travels the distance s_2 along the face.

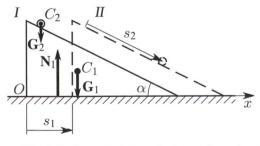

Figure E2.34. Rolling of a disk on the lateral face of a prism.

Solution. Consider the mechanical system consisting of the prism and the disk and draw the external forces, i.e., the active forces of gravity $\mathbf{G}_1$ and $\mathbf{G}_2$ and the reaction force $\mathbf{N}_1$ of the smooth plane. A typical characteristic of the system of external forces is that all of them are perpendicular to the horizontal axis, and hence the sum of their projections onto this axis is zero.

We direct the x-axis horizontally from left to right and place its origin at the point O. In Fig. E2.34, we draw the system in two positions, the initial position I and position II (drawn by dashed lines) of the system at the time by which the disk has traveled the distance s_2 along the prism face. Since the direction and magnitude of the prism displacement are unknown in advance, we put position II on the right of the initial position I without hesitation.

Since $\sum_k F_{kx}^e = 0$, it follows from the conservation law for the projection of the velocity of the system center of mass onto the x-axis that $v_{cx} = $ const. Since all velocities are zero in position I, we have $v_{cx} = 0$. It follows that $x_c = $ const, or $x_c^I = x_c^{II}$. We use the formulas for the coordinates of the center of mass to rewrite the last relation as

$$\frac{m_1 x_1^I + m_2 x_2^I}{m_1 + m_2} = \frac{m_1 x_1^{II} + m_2 x_2^{II}}{m_1 + m_2}.$$

Here x_1^I and x_2^I are the coordinates of the centers of mass of the prism and the disk in position I, and x_1^{II} and x_2^{II} are the respective quantities in position II.

It follows from Fig. E2.34 that $x_1^{II} = x_1^I + s_1$ and $x_2^{II} = x_2^I + s_1 + s_2 \cos \alpha$. By substituting these relations into the formula and by performing algebraic manipulations, we obtain $s_1 = -\dfrac{m_2 s_2 \cos \alpha}{m_1 + m_2}$.

The sign of the answer shows that the prism displacement is opposite to that shown in the figure.

E2.5.3. Theorem on the Momentum

▶ **Momentum.** The *momentum of a point* of mass m moving at a velocity $\mathbf{v}$ is the vector

$$\mathbf{Q} = m\mathbf{v}.$$

The *momentum of a mechanical system* is the resultant of momenta of all points of the system,

$$\mathbf{Q} = \sum_k \mathbf{Q}_k.$$

One can prove that the momentum of a system is equal to the momentum of a fictitious point particle with mass equal to the mass of the system and velocity equal to the velocity of the center of mass, $\mathbf{Q} = m\mathbf{v}_c$.

▶ **Impulse of force.** Suppose that a force $\mathbf{F}$ is applied to a moving point particle. (Along with this force, any other forces can act on the particle, but for now we consider only one of them.)

The *elementary impulse of force* $\mathbf{F}$ in an elementary time interval dt is the vector

$$d\mathbf{S} = \mathbf{F}\, dt.$$

The *impulse of force* $\mathbf{F}$ in a finite time interval from t_0 to t is the vector

$$\mathbf{S} = \int_{t_0}^{t} d\mathbf{S} = \int_{t_0}^{t} \mathbf{F}\, dt.$$

The projections of the impulse of force onto the coordinate axes can be calculated by the formulas

$$S_x = \int_{t_0}^{t} F_x\, dt, \quad S_y = \int_{t_0}^{t} F_y\, dt, \quad S_z = \int_{t_0}^{t} F_z\, dt.$$

► **Various statements of the theorem on the momentum.**

THEOREM ON THE MOMENTUM IN DIFFERENTIAL FORM. *The time derivative of the momentum of a mechanical system is equal to the resultant vector of external forces acting on the system,*

$$\frac{d\mathbf{Q}}{dt} = \sum_k \mathbf{F}_k^e; \qquad (E2.5.3.1)$$

in other words, the differential of the momentum of a system is equal to the vector sum of elementary impulses of all external forces acting on the points of the system,

$$d\mathbf{Q} = \sum_k d\mathbf{S}_k^e.$$

In projections onto the coordinate axes (e.g., onto the x-axis), we obtain

$$\frac{dQ_x}{dt} = \sum_k F_{kx}^e \quad \text{or} \quad dQ_x = \sum_k dS_{kx}^e.$$

THEOREM ON THE MOMENTUM IN INTEGRAL FORM. *The variation in the momentum of a mechanical system in a time interval is equal to the sum of impulses of external forces acting on the system in the same time interval,*

$$\mathbf{Q} - \mathbf{Q}_0 = \sum_k \mathbf{S}_k^e, \qquad (E2.5.3.2)$$

where the vectors $\mathbf{Q}$ *and* $\mathbf{Q}_0$ *correspond to time instants* t *and* t_0 *and the* $\mathbf{S}_k^e$ *are the impulses of external forces acting on the system in the time interval from* t_0 *to* t.

In projections onto the coordinate axes (e.g., onto the x-axis), we have

$$Q_x - Q_{x0} = \sum_k S_{kx}^e.$$

If the theorem is used in the case of a single point particle, one should remember that any force applied to the point is external.

► **The momentum conservation law.**
The law of conservation of the momentum of a mechanical system has the form

$$\text{if} \quad \sum_k \mathbf{F}_k^e = 0, \quad \text{then} \quad \mathbf{Q} = \text{const.}$$

The law of conservation of the momentum projection onto any axis (for example, onto the x-axis) has the form

$$\text{if} \quad \sum_k F_{kx}^e = 0, \quad \text{then} \quad Q_x = \text{const.}$$

Example. On a smooth horizontal plane, a rectangular parallelepiped of mass m_1 moves at a velocity v_0, and on the upper face of this parallelepiped, there are two propulsion devices with masses m_2 and m_3 (Fig. E2.35). At some time instant, the devices begin to move towards each other, and the laws of their motion with respect to the parallelepiped are determined by the functions $s_2 = h_2 t^3$ and $s_3 = h_3 t^3$. Find how the velocity of the parallelepiped depends on time.

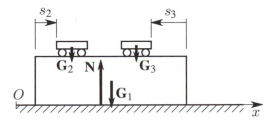

Figure E2.35. Motion of two propulsion devices on the surface of a moving rectangular parallelepiped.

Solution. Consider the system consisting of the three bodies (the parallelepiped and the propulsion devices). Let us draw the external forces, namely, the active forces of gravity $\mathbf{G}_1$, $\mathbf{G}_2$, and $\mathbf{G}_3$ and the plane reaction $\mathbf{N}$. All of them are vertical, and hence we can apply the law of conservation of the system momentum projection onto the x-axis; i.e., we write out Q_{x0} at time t_0 and Q_x at an arbitrary time t and equate the obtained quantities with each other:

$$Q_{x0} = m_1 v_0 + m_2 v_0 + m_3 v_0,$$
$$Q_x = m_1 v_1 + m_2(v_1 + \dot{s}_2) + m_3(v_1 - \dot{s}_3).$$

In the last expression, the velocities of the propulsion devices were calculated according to the theorem on the velocities in the compound motion of a point. By equating the right-hand sides of these equations and by performing algebraic manipulations, we obtain

$$v_1 = v_0 - \frac{3t^2(m_2 h_2 - m_3 h_3)}{m_1 + m_2 + m_3}.$$

E2.5.4. Theorem on the Angular Momentum

▶ **Moment of momentum and the angular momentum.** The *moment of momentum of a point particle about a fixed center* P is the vector $\mathbf{K}_P$ equal to the cross product of the position vector connecting the center with the point by the momentum of the point:

$$\mathbf{K}_P = \mathbf{M}_P(\mathbf{Q}) = \mathbf{r} \times m\mathbf{v}.$$

The *angular momentum (the net moment of momentum) of a mechanical system about the center* P is the vector sum of moments of momenta of all points of the system about the center:

$$\mathbf{K}_P = \sum_j \mathbf{K}_{Pj}.$$

The *moment of momentum of a point about the x-axis* is the quantity K_x equal to the projection onto this axis of the moment of momentum of the point about any center P belonging to the axis,

$$K_x = M_x(\mathbf{Q}).$$

The *angular momentum of a system about the x-axis* is the projection onto this axis of the angular momentum of the system about any center P belonging to the axis:

$$K_x = \sum_j K_{jx}.$$

The analytic expression for the angular momentum of a system about the coordinate x-axis has the form

$$K_x = \sum_j m_j(y_j \dot{z}_j - \dot{y}_j z_j).$$

The formulas for K_y and K_z are similar.

One can show that the angular momentum of a system about a fixed center P is equal to the sum of the moment of momentum of the center of mass about the center P and the angular momentum of the system about the center of mass C in its relative motion in the König system,

$$\mathbf{K}_P = \mathbf{K}_P + \mathbf{K}_C^*.$$

Here $\mathbf{K}_P = \mathbf{M}_P(m\mathbf{v}_C)$ and $\mathbf{K}_C^*$ is the angular momentum of the system in its motion with respect to the König reference frame.

The angular momentum K_z of a rigid body rotating about the fixed z-axis with angular velocity ω is calculated by the formula

$$K_z = I_{zz}\omega,$$

where I_{zz} is the moment of inertia of the rigid body about the z-axis.

The angular momentum of a rigid body is discussed in more detail in Section E2.8.

▶ **Theorem on angular momentum of a mechanical system.**

THEOREM. *The time derivative of the angular momentum about any fixed center P is equal to the resultant moment of external forces of the system about the same center,*

$$\frac{d\mathbf{K}_P}{dt} = \sum_j \mathbf{M}_P(\mathbf{F}_j^e). \qquad (E2.5.4.1)$$

Projecting the vector expression onto the coordinate axes (e.g., on the x-axis), we obtain the theorem on the angular momentum of the system about a fixed axis:

$$\frac{dK_x}{dt} = \sum_j M_x(\mathbf{F}_j^e).$$

If this theorem is used to study the motion of a rigid body rotating about a fixed axis z, then we obtain the *differential equation of rotational motion of a rigid body about a fixed axis*:

$$I_{zz}\ddot{\varphi} = \sum_j M_z(\mathbf{F}_j^e),$$

where φ is the rotation angle.

▶ **Law of conservation of angular momentum.** The law is stated as follows: if the resultant moment of external forces acting on a system about a center P is zero, then the net moment of momentum about this center is constant. For example,

$$\text{if} \quad \sum_j \mathbf{M}_P(\mathbf{F}_j^e) = \mathbf{0}, \quad \text{then} \quad \mathbf{K}_P = \text{const.}$$

The right-hand side of this relation contains a vector constant; i.e., neither the magnitude nor the direction of the vector depends on time.

If the sum of moments of external forces acting on the system about some fixed axis is zero, then the angular momentum of the system about this axis remains constant. For example,

$$\text{if} \quad \sum_j M_x(\mathbf{F}_j^e) = 0, \quad \text{then} \quad K_x = \overrightarrow{\text{const.}}$$

Example 1. A point particle moves under the action of a system of forces $\{\mathbf{F}_k\}$ so that the line of action of the resultant $\mathbf{R}^*$ passes through a fixed center O. (We take it for the origin of the reference frame.) Let us find several characteristics of motion of the point.

Since $\sum_k \mathbf{M}_O(\mathbf{F}_k) = \mathbf{M}_O(\mathbf{R}^*) = \mathbf{0}$, it follows that $\mathbf{K}_O = \overrightarrow{\text{const}}$, or $\mathbf{r} \times m\mathbf{v} = \overrightarrow{\text{const}}$.

Taking the scalar products of both sides of the last relation by $\mathbf{r}$, we obtain $\mathbf{r} \cdot \overrightarrow{\text{const}} = 0$, or $C_1 x + C_2 y + C_3 z = 0$, where x, y, z are the coordinates of the moving point and C_1, C_2, C_3 are the coordinates of the vector constant.

The last expression is the equation of a plane passing through the center O, and this means that the trajectory of the point lies in that plane.

Example 2. A massless string with a load of weight G_2 at the end is wound on a homogeneous drum of weight G_1 and radius R (Fig. E2.36). Find the acceleration of the load neglecting the friction forces in the rotation of the drum and assuming that the string unwinds from the drum without slip.

Solution. Let the system consist of the drum, the load, and the string. We draw the external forces, namely, the active forces of gravity $\mathbf{G}_1$ and $\mathbf{G}_2$ and the reaction force $\mathbf{N}_O$ passing through the rotation axis O. The direction of the force $\mathbf{N}_O$ is unknown in advance, therefore we draw it arbitrarily. There is no couple of friction forces on the axis, which follows from the assumptions of the problem.

We apply the theorem on the variations in the angular momentum of a system about the drum rotation axis:

$$\frac{dK_O}{dt} = \sum_j M_O(\mathbf{F}_j^e).$$

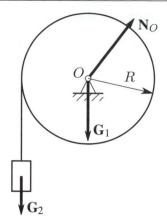

Figure E2.36. Unwinding of a massless string with a load from a rotating drum.

We calculate K_O as the sum of the angular momenta of the drum and of the load, take into account the equation $v_2 = \omega_1 R$, and obtain

$$K_O = v_2 R\left(\tfrac{1}{2}m_1 + M_2\right).$$

We calculate the sum of moments of external forces about the axis:

$$\sum_j M_O(\mathbf{F}_j^e) = m_2 g R.$$

Substituting the right-hand sides of the last formulas into the statement of the theorem, we obtain

$$a_2 = \frac{2m_2}{m_1 + 2m_2}\, g.$$

E2.5.5. Theorem on the Kinetic Energy

▶ **Elementary work.** Consider a point B moving under the action of a system of forces. A small displacement of the point along the trajectory is characterized by the vector $d\mathbf{r}$ (Fig. E2.37). We distinguish one force $\mathbf{F}$ in the system.

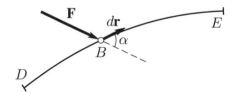

Figure E2.37. Elementary work of the force $\mathbf{F}$ on the displacement $d\mathbf{r}$.

The *elementary work of the force* $\mathbf{F}$ on the displacement $d\mathbf{r}$ is the scalar quantity $d'A$ equal to the scalar product of the vectors $\mathbf{F}$ and $d\mathbf{r}$:

$$d'A = \mathbf{F} \cdot d\mathbf{r} = F\, dr \cos \alpha.$$

In coordinate form, the elementary work is calculated by the formula

$$d'A = F_x\, dx + F_y\, dy + F_z\, dz,$$

where F_x, F_y, F_z and dx, dy, dz are the coordinates of the vectors $\mathbf{F}$ and $d\mathbf{r}$, respectively, in the Cartesian coordinate system.

Note that the elementary work $d'A$ need not be the total differential of a function depending on the coordinates. (This fact is also reflected in the above representation.)

The sign of the elementary work is determined by the cosine of the angle α: it is positive for $0 \le \alpha < \pi/2$, negative for $\pi/2 < \alpha \le \pi$, and zero for $\alpha = \pi/2$.

▶ **Calculation of elementary work in special cases.**

1. The elementary work of a force applied to a rigid body rotating about a fixed axis z is calculated by the formula

$$d'A = \pm M_z(\mathbf{F})\, d\varphi.$$

2. The sum of elementary works of a couple of forces applied to a body in arbitrary motion is calculated by the formula $d'A = (\mathbf{M} \cdot \boldsymbol{\omega})\, dt$, and in the cases of rotation about a fixed axis and of plane motion, it can be calculated as follows:

$$d'A = \pm M\, d\varphi.$$

Here M is the torque of a couple of forces, and $d\varphi$ is the angle of rotation of the body. The plus sign is taken if the sense of the curved arrows of the couple torque coincides with the sense of rotation, and the minus sign is taken if they are opposite. (It is assumed that the plane in which the couple acts is parallel to the base plane.)

3. When calculating the elementary work of friction forces applied to a body rolling without slip, it is necessary to take into account the fact that the following forces act at the point of tangency P (Fig. E2.31): the normal pressure force $\mathbf{N}$, the friction force $\mathbf{F}_{\mathrm{fr}}$, and the couple of rolling friction forces with torque $M_{\mathrm{fr}} = Nk$. Since there is no slip, so that the point of tangency is the instantaneous center of velocities and its velocity $\mathbf{v}_P$ is zero, it follows that $d\mathbf{r} = \mathbf{v}_P\, dt = 0$, which implies that

$$d'A_N = dA_{\mathrm{fr}} = 0, \quad d'A_M = -M_{\mathrm{fr}}\, d\varphi = -Nk\, d\varphi.$$

In rolling on an ideally rough surface, $d'A_M = 0$, because $M_{\mathrm{fr}} = 0$.

4. One can prove that the sum of elementary works of forces applied to a rigid body is equal to the sum of elementary works of a statically equivalent system of forces. By the theorem on reducing a system of forces to a given center, an arbitrary system of forces can be replaced by an equivalent system consisting of a force $\mathbf{R}$ applied at an a priori chosen point P and a couple of forces with torque $\mathbf{M}_P$. Therefore, instead of cumbersome calculations of the sum of elementary works of many forces applied to a body, one usually calculates the sum of elementary works of a single force and a single couple.

Example 1. A system of elementary forces of gravity acting on a rigid body always has the resultant equal to the body weight G applied at the center of gravity C. Therefore, the sum of elementary works of the forces of gravity is equal to the elementary work of the force of weight on the displacement of the center of gravity of the body.

Example 2. The sum of elementary works of internal forces applied to points of a rigid body is zero, because the resultant vector and the resultant moment of the system of internal forces are zero.

▶ **Work of a force. Potential force.** The *work of a force* $\mathbf{F}$ on a finite displacement of a point along a trajectory DE (Fig. E2.37) is given by the curvilinear integral

$$A = \int_{DE} d'A = \int_{DE} \mathbf{F} \cdot d\mathbf{r} = \int_{DE} F_x\, dx + F_y\, dy + F_z\, dz.$$

A *force field* is a part of space where a point particle is under the action of a force $\mathbf{F}$ depending on the position of the point and the time, $\mathbf{F} = \mathbf{F}(\mathbf{r}, t)$, where $\mathbf{r}$ is the position vector of the point. A force field is said to be *nonstationary* (transient) if the force explicitly depends on time and *stationary* if the force is independent of time.

In what follows, we deal only with stationary force fields, where $\mathbf{F} = \mathbf{F}(\mathbf{r})$. In this case, the projections of the force on the coordinate axes are functions of the coordinates of the point, $F_x = F_x(x, y, z)$, $F_y = F_y(x, y, z)$, and $F_z = F_z(x, y, z)$.

In the general case, the work of a force depends on the shape of the trajectory along which the point moves. But there are force fields in which the work depends only on the endpoints of the trajectory. One can show that in these cases the elementary work is the total differential of a function, $d'A = dU(x, y, z)$, and

$$F_x = \frac{\partial U}{\partial x}, \quad F_y = \frac{\partial U}{\partial y}, \quad F_z = \frac{\partial U}{\partial z}.$$

Such a force field is said to be *potential*, and the function U is called the *force function* or the *potential* of the field.

Examples of potential forces are as follows:

1. The force of gravity G; the expression for its work is given by the formula $A = G \Delta H$, where G is the weight of a point particle and ΔH is the height difference between the initial and final points of the trajectory.

2. The expression for the sum of works of forces applied to the ends of a weightless linearly elastic spring of rigidity γ is given by the formula $A = -\gamma(\lambda_e^2 - \lambda_s^2)/2$, where L is the spring current length, l is its length in the unstrained state, $\lambda = L - l$ is the length variation, λ_e corresponds to the terminal state of the spring, and λ_s corresponds to the initial state.

The function $\Pi(x, y, z) = -U + \text{const}$ is called the *potential energy* of the field, and the function $E = T + \Pi$ is called the *total mechanical energy*. The potential energy of a point particle is equal to the work of forces of the potential field in the transition from the current state of the point into the zero state. For the zero state we can take any state because of the constant contained in the definition.

▶ **Kinetic energy.** The *kinetic energy of a point* of mass m moving at a velocity **v** is the scalar quantity T determined by the formula

$$T = \frac{mv^2}{2}.$$

The *kinetic energy of a mechanical system* is the sum of the kinetic energies of all its points:

$$T = \sum_k T_k = \sum_k \frac{m_k v_k^2}{2}.$$

One can prove that the kinetic energy of a system is equal to the sum of the kinetic energy of the center of mass and the kinetic energy of the system in its relative motion in the König reference frame:

$$T = \frac{mv_C^2}{2} + \sum_k \frac{m_k(v_k^*)^2}{2} = T_C + T^*,$$

where $m = \sum_k m_k$ and the v_k^* are the relative velocities of the points.

The formulas for the kinetic energy of a rigid body in the simplest motions are as follows:

(a) $T = \frac{1}{2}mv_C^2$ in translational motion.

(b) $T = \frac{1}{2}I_{zz}\omega^2$ in rotation about the fixed z-axis.

(c)

$$T = \frac{1}{2}mv_C^2 + \frac{1}{2}I_{CC}\omega^2 \tag{E2.5.5.1}$$

in plane-parallel motion, where I_{CC} is the moment of inertia of the body about the axis perpendicular to the base plane and passing through the center of mass C.

The kinetic energy of a rigid body is discussed in more detail in Section E2.8.

▶ **Various statements of the theorem on the kinetic energy.**

THEOREM ON THE KINETIC ENERGY IN DIFFERENTIAL FORM. *The differential of the kinetic energy of a mechanical system is equal to the sum of elementary works of the forces applied to the points of the system on the elementary displacements of these points:*

$$dT = \sum d'A_j. \qquad (E2.5.5.2)$$

THEOREM ON THE KINETIC ENERGY IN INTEGRAL FORM. *The variation in the kinetic energy of a mechanical system on a certain displacement is equal to the sum of works of all forces applied to the points of the system on the displacements of these points:*

$$T - T_0 = \sum A_j.$$

Remark. In contrast to the three above-considered theorems on the dynamics of a system, the last theorem is characterized by the following specific features:

(1) The theorem on the variation in the kinetic energy relates scalar quantities rather than vector quantities.

(2) Both statements of the theorem contain works of all forces, not only external but also internal. (It is also possible to divide the sum of works into the sum of works of active forces and constraint reaction forces.)

(3) The sum of works of internal forces applied to the points of a rigid body is zero.

Example 3. A homogeneous disk under the action of the force of weight G rolls down without slip on an ideally rough plane inclined at an angle α to the horizon (Fig. E2.38). Find the acceleration of the disk center, the friction force magnitude, and the minimum value f^* of the coefficient of friction at which a rolling motion without slip is possible.

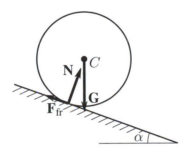

Figure E2.38. Rolling of a disk down an inclined plane under the action of the force of weight.

Solution. For the mechanical system we take the disk itself and study the motion by using the theorem on the kinetic energy in differential form. Calculating the kinetic energy of the disk, we obtain

$$T = \tfrac{1}{2}mv_C^2 + \tfrac{1}{2}I_{CC}\omega^2 = \tfrac{3}{4}mv_C^2.$$

Here we have taken into account the fact that $I_{CC} = \tfrac{1}{2}mr^2$ and $v_C = \omega r$. Since the system under study is a rigid body, it follows that $\sum d'A^i = 0$. We calculate the sum of elementary works of external forces, substitute the obtained expressions for T and $\sum d'A^e$ into the statement of the theorem, and obtain

$$d\left(\tfrac{3}{4}3mv_C^2\right) = mgr\sin\alpha\,d\varphi.$$

Dividing both sides of this relation by dt and taking into account the kinematic relations $\dot{v}_C = a_C$ and $\dot{\varphi} = \omega$, we obtain

$$a_C = \tfrac{2}{3}g\sin\alpha.$$

Let us write out the statement of the theorem on the motion of the center of mass of a system as applied to the problem under study:

$$ma_C = G + N + F_{fr}.$$

We project the obtained vector equation onto two perpendicular axes the first of which is parallel to the normal reaction force **N**:

$$0 = -G\cos\alpha + N,$$
$$ma_C = G\sin\alpha - F_{fr}.$$

Taking into account the value a_C obtained earlier, we obtain the solution of the system of equations: $F_{fr} = \frac{1}{3}G \sin \alpha$, $N = G \cos \alpha$. By substituting these quantities into the Coulomb friction law, we determine the minimum value f^* of the coefficient of friction for which a rolling motion without slip is possible: $f^* = \frac{1}{3} \mathrm{tg}\,\alpha$.

A mechanical system whose points are only subjected to potential forces is said to be *conservative*.

The *law of conservation of the total mechanical energy* states that the total mechanical energy of a conservative system remains constant, $E = T + \Pi = T_0 + \Pi_0 = \mathrm{const}$.

E2.6. Elements of Analytical Mechanics

E2.6.1. D'Alembert's Principle

▶ **Forces of inertia.** Assume that a point of a material system moves under the action of some system of forces (these forces can be divided either into external and internal forces or into active forces and constraint reaction forces). The resultant of this system of converging forces will be denoted by **F**.

The *force of inertia* of a point is the vector **Φ** equal and opposite to the product of the mass of the point by its acceleration,

$$\mathbf{\Phi} = -m\mathbf{a}.$$

The force **Φ** is fictitious; it is not an actual force acting on the point.

D'Alembert's principle: in the motion of a mechanical system (point), any of its states can be considered as an equilibrium if the real forces acting on each point of the system are supplemented with the fictitious forces of inertia.

According to this principle, if each point of the system is supplemented with the force $\mathbf{\Phi}_j = -m_j \mathbf{a}_j$, then the system of forces consisting of the real forces $\mathbf{F}_j$ and the fictitious forces $\mathbf{\Phi}_j$ satisfies all the equations of statics; i.e., the resultant vector of the system of forces and its resultant moment about an arbitrary center P are zero:

$$\sum \mathbf{F}_j + \sum \mathbf{\Phi}_j = 0,$$
$$\sum \mathbf{M}_P(\mathbf{F}_j) + \sum \mathbf{M}_P(\mathbf{\Phi}_j) = 0.$$

In coordinate form, these equations can be written as

$$\sum F_{jx} + \sum \Phi_{jx} = 0, \qquad \sum M_x(\mathbf{F}_j) + \sum M_x(\mathbf{\Phi}_j) = 0,$$
$$\sum F_{jy} + \sum \Phi_{jy} = 0, \qquad \sum M_y(\mathbf{F}_j) + \sum M_y(\mathbf{\Phi}_j) = 0,$$
$$\sum F_{jz} + \sum \Phi_{jz} = 0, \qquad \sum M_z(\mathbf{F}_j) + \sum M_z(\mathbf{\Phi}_j) = 0.$$

D'Alembert's principle allows one to transfer the methods for solving problems of statics to problems of dynamics.

Example 1. A load of weight G is in an elevator cabin ascending in decelerated motion with acceleration a (Fig. E2.39). Find the pressure exerted by the elevator cabin floor on the load.

Solution. We assume that the load is a point particle and draw the real forces acting on it: the active force of weight **G** and the force **N** of floor pressure on the load. We supplement these forces with the fictitious force of inertia $\mathbf{\Phi} = -m\mathbf{a}$. (Note that the force **Φ** in the figure is opposite not to the elevator displacement but to the acceleration vector.)

The obtained system of three forces **G**, **N**, and **Φ** is in equilibrium according to d'Alembert's principle. The lines of action of all the forces are directed along a single straight line, and hence the equilibrium of the system of forces is described by the single equation

$$-G + N + \Phi = 0.$$

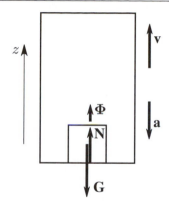

Figure E2.39. Ascent of an elevator with a load.

We substitute the magnitude $ma = \frac{G}{g}a$ of the force of inertia (where g is the free fall acceleration) for Φ and obtain $N = G(g - a)$. One can see that the floor pressure force is less than the weight of the load.

The system of forces of inertia can be very cumbersome in the case of numerous point particles or distributed masses. Using the theorem of statics on reducing a system of forces to a center, one can replace the system of forces Φ_j of inertia by an equivalent system consisting of a single force Φ applied at a center P given in advance (it is equal to the *resultant vector of forces of inertia*, $\Phi = \sum \Phi_j$, and is independent of the choice of the center) and by a single couple of forces whose torque $\mathbf{M}_P^\Phi$ is equal to the *net moment of forces of inertia* about the center, $\mathbf{M}_P^\Phi = \sum \mathbf{M}_P(\Phi_j)$.

One can show that Φ is calculated by the formula

$$\Phi = -m\mathbf{a}_C,$$

where m is the mass of the system and $\mathbf{a}_C$ is the acceleration of the center of mass.

▶ **Formulas for the moment of forces of inertia of a body.** Some useful formulas for the net moment of forces of inertia of a rigid body and its projections onto the coordinate axes are given below:

1. $\mathbf{M}_C^\Phi = 0$ in translational motion.
2. $M_z^\Phi = -I_{zz}\varepsilon$ in rotation about the fixed z-axis.
3. $M_C^\Phi = -I_{CC}\varepsilon$ in plane-parallel motion.

Here ε is the angular acceleration of the body, and I_{zz} and I_{CC} are the moments of inertia of the body about the z-axis and the axis passing through the center of mass and perpendicular to the base plane. (The minus signs in the formula mean that the senses of the angular acceleration and the torque of the couple of forces of inertia are opposite.)

Example 2. A homogeneous disk of radius r rolls without slip up a circular arc of radius R (Fig. E2.40). The coefficient of rolling friction is k. Find the acceleration of the disk center and the force of disk pressure on the support at the time instant when the velocity of the disk center is v_0 and the angle between the vertical line and the straight line connecting the centers of the disk and of the arc is α.

Solution. The acceleration of the disk center consists of two components a_C^τ and a_C^n, where the direction of a_C^τ is unknown in advance and $a_C^n = v_0^2/(R - r)$. In the no-slip case, we have $a_C^\tau = \varepsilon r$, where ε is the angular acceleration of the disk.

We reduce the forces of inertia of the disk to the center of mass and decompose the force of inertia into two components, $\Phi = \Phi^\tau + \Phi^n$, where $\Phi^\tau = -m\mathbf{a}_C^\tau$ and $\Phi^n = -m\mathbf{a}_C^n$. The torque of the couple of forces of inertia has the magnitude $M_C^\Phi = I_{cc}\varepsilon = mra_C^\tau/2$, and the corresponding curved arrow is opposite to the curved arrow of the assumed angular acceleration.

The system of forces in Fig. E2.40 is in equilibrium because of d'Alembert's principle. Let us write out the equations of equilibrium for a plane system of forces:

$$F_{\text{fr}} + \Phi^\tau - G \sin \alpha = 0,$$

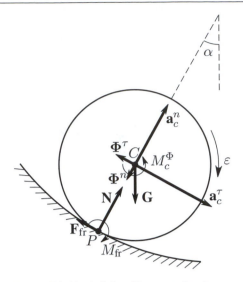

Figure E2.40. A disk rolling on a circular arc.

$$N - \Phi^n - G \cos \alpha = 0,$$
$$-M_{\text{fr}} + M_C^\Phi + \Phi^\tau r - Gr \sin \alpha = 0.$$

Here the first of the equations is the equation for the projections onto the direction of $\mathbf{F}_{\text{fr}}$, the second is the equation for the projections onto the direction of $\mathbf{N}$, and the last is the equation of moments about the center P.

We supplement these equations with Coulomb's law $M_{\text{fr}} = kN$ (see Subsection E2.3.6) and obtain the definitive system of four equations with four unknowns a_C^τ, N, F_{fr}, and M_{fr}. By solving this system, we obtain

$$a_C^\tau = \frac{2}{3r} \left[gr \sin \alpha + k \left(g \cos \alpha + \frac{v_0^2}{R-r} \right) \right], \quad N = m \left(g \cos \alpha + \frac{v_0^2}{R-r} \right).$$

E2.6.2. Classification of Mechanical Constraints. Generalized Coordinates

▶ **Classification of mechanical constraints.** *Mechanical constraints* are devices (bodies) imposing restrictions on the positions and velocities of points of a mechanical system. These restrictions are always satisfied regardless of the given forces and are written as relations called the *constraint equations*.

Stationary constraints are constraints independent of time; constraints depending on time are said to be *nonstationary*.

Constraints whose equations contain the coordinates of points and time are said to be *geometric*; constraints are said to be *kinematic* (*differential*) if the constraint equations contain not only the coordinates and time, but also the velocities of points.

If a kinematic constraint can be represented as an equivalent geometric constraint, then it is said to be *kinematic integrable*. Otherwise, if a constraint equation cannot be represented as a geometric constraint in principle, then it is called a *nonintegrable constraint*.

Geometric and kinematic integrable constraints are said to be *holonomic*, and kinematic nonintegrable constraints are said to be *nonholonomic*. A mechanical system is said to be *holonomic* if only holonomic constraints are imposed on it and *nonholonomic* if there is at least one nonholonomic constraint.

Constraints are said to be *bilateral* if the restrictions imposed by them on the positions of points, their velocity, and time, can be written as equalities. *Unilateral constraints* are written as inequalities.

▶ **Virtual displacement of a point.** A *virtual displacement of a point* of a mechanical system is any displacement $\delta\mathbf{r}$ admitted by the imposed constraints from the position occupied by the point at a given time instant. (When constructing such displacements, one should "freeze" the time, so that the nonstationary constraints become fixed, i.e., stationary.) The point does not actually perform virtual displacements but could do so without violating the constraints at a given time instant.

A *virtual displacement of a system* is an arbitrary set of virtual displacements $\delta\mathbf{r}_j$ of the points of the system admitted by all constraints imposed on it. By way of example, consider a point subjected to a nonstationary constraint given by a plane translationally moving at a velocity $\mathbf{v}$ (Fig. E2.41). According to the theorem on the compound motion of a point, its actual displacement $d\mathbf{r}$ is equal to the vector sum of the relative displacement $\delta\mathbf{r}$ and the translational displacement $\mathbf{v}\,dt$. The figure illustrates the difference between the actual and virtual displacements of a point.

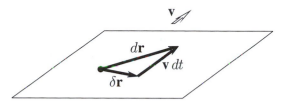

Figure E2.41. Virtual displacement of a point on a moving plane.

For stationary constraints, the actual displacements of points are contained in the set of virtual displacements.

A mechanical system can have quite a few various virtual displacements. But for systems consisting of material rigid bodies and finitely many point particles there exist several mutually independent virtual displacements in terms of which one can express any other virtual displacement. The number of independent displacements is called the *number of degrees of freedom of the mechanical system.*

▶ **Generalized coordinates.** *Generalized coordinates* are independent parameters that uniquely determine the position of each point of a mechanical system. There are as many degrees of freedom in a holonomic system as there are generalized coordinates, but this is not true for nonholonomic systems, which have fewer degrees of freedom than generalized coordinates.

Consider several specific examples.

1. A free point particle in space is a system with three degrees of freedom.

2. A free rigid body has six degrees of freedom. Indeed, the position of any point of the body in space can be determined if the positions of three points B_1, B_2, B_3 of the body that do not lie on a single straight line are known. The position of each of these points can be determined by three parameters, for example, the coordinates x_j, y_j, z_j ($j = 1, 2, 3$). All in all, there are nine coordinates, but they cannot be specified arbitrarily, because there are three equations saying that the distances s_{12}, s_{23}, s_{31} between the points must remain constant. (The body is rigid.) If, say, the six coordinates x_1, y_1, z_1, x_2, y_2, x_3 are known, then the remaining three coordinates z_2, y_3, z_3 can be found from the constraint equations.

3. A body rotating about a fixed axis has a single degree of freedom, and for the generalized coordinate one can take the rotation angle φ.

4. A rigid body in plane-parallel motion has three degrees of freedom; for generalized coordinates one can take, for example, the rotation angle and two Cartesian coordinates of any point of the body.

5. A rigid body in translational motion has three degrees of freedom; for generalized coordinates one can take three Cartesian coordinates of any point of the body.

6. A system consisting of a prism lying on a plane and a disk rolling without slip on the lateral face of the prism has two degrees of freedom (Fig. E2.42).

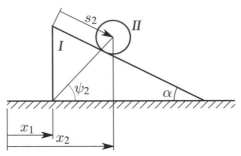

Figure E2.42. System consisting of a prism on a plane and a disk rolling without slip on the lateral face of the prism.

7. A system consisting of two free points has six degrees of freedom.

8. A sewing machine mechanism consisting of numerous rigid bodies has a single degree of freedom.

9. A thin rectilinear rod moving on a plane in such a way that the velocity of the rod center must be parallel to the rod axis has two degrees of freedom.

Of the above examples of mechanical systems, only one system (the last) is nonholonomic, while the others are holonomic.

We again return to the notion of generalized coordinates, which we illustrate by the system in Example 6 (Fig. E2.42). The position of each point of the disk and the prism is known as soon as we specify the values for one of the following pairs of variables: either (x_1, x_2), or (x_1, s_2), or (x_2, ψ_2), or (x_1, ψ_2), or $q_1 = \frac{1}{3}(x_1 + x_2)$, or $q_2 = \frac{1}{2}x_1 - x_2$, etc.

Let us summarize. For the system under study, there are infinitely many choices of generalized coordinates, but each set of these always contains two independent variables. Since the system is holonomic, it follows that the number of generalized coordinates is 2 (i.e., coincides with the number of degrees of freedom). The coordinates are referred to as generalized because they may fail to have a clear geometric meaning. (For example, this is the case for the coordinates (q_1, q_2).)

▶ **Ideal constraints.** Constraints are said to be *ideal* if the sum of works of their reactions $\mathbf{R}_j$ is zero on any virtual displacement of the system:

$$\sum \delta A_j^{\mathbf{R}} = 0.$$

An example of a system with ideal constraints is a free rigid body. Any complicated mechanism consisting of several rigid bodies can be treated as a mechanical system with ideal constraints if some bodies are connected absolutely rigidly and the other bodies are connected by ideal hinges (without friction) and also by weightless unstretchable perfectly flexible strings. Moreover, the contact surfaces must be either absolutely smooth or perfectly rough when one of the bodies rolls on another without slip so that the moment of rolling friction forces at the point of their tangency is zero and the static friction force is nonzero.

E2.6.3. Principle of Virtual Displacements

Principle of virtual displacements is stated as follows: for a mechanical system with ideal constraints to be in equilibrium, it is necessary and sufficient that the sum of elementary works of active forces $\mathbf{F}_i$ on any virtual displacement of the system be zero.

The mathematical representation of the principle of virtual displacements is

$$\sum \delta A_i^{\mathbf{F}} = 0. \tag{E2.6.3.1}$$

If a system has n degrees of freedom and its position is determined by generalized coordinates $q_j(j = 1, 2, \ldots, n)$, then the last expression can be rewritten as

$$\sum \delta A_i^{\mathbf{F}} = \sum \mathbf{F}_i \cdot \delta \mathbf{r}_i = \sum_{j=1}^{n} Q_j \, \delta q_j = 0,$$

where the δq_j are variations of the generalized coordinates.

The coefficients Q_j are called *generalized forces*. Since the variations of generalized coordinates are independent, we see that for the equation to be satisfied it is necessary that each of the factors multiplying δq_j be zero. Thus, if a system with ideal constraints is in equilibrium, then all generalized forces are zero.

Example. Find the angle φ of deviation from the vertical axis of a heavy homogeneous rod of weight G with a horizontal force F applied to the lower end B (Fig. E2.43).

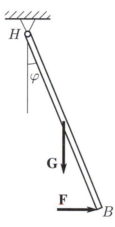

Figure E2.43. Deviation from the vertical axis of a heavy rod under the action of a horizontal force.

Solution. In the mechanical system we include a rod, which is a rigid body. Neglecting the friction in the hinge, we conclude that the constraints imposed on the system are ideal, and its equilibrium can be studied by the principle of virtual displacements.

In the equilibrium characterized by the angle φ, consider a virtual displacement of the system; i.e., we deflect the rod by a small angle $\delta\varphi$ about the hinge H, calculate the sum of elementary works of active forces **G** and **F**, and equate it with zero:

$$-GL \sin \varphi \, \delta\varphi + F2L \cos \varphi \, \delta\varphi = 0$$

(the rod length is $2L$), or, after transformations,

$$\delta\varphi \, (-GL \sin \varphi + F2L \cos \varphi) = 0.$$

Since the virtual displacement $\delta\varphi$ is arbitrary, it follows that, for the product to be zero, it is necessary to equate the expression in parentheses with zero:

$$-GL \sin \varphi + F2L \cos \varphi = 0.$$

This implies the desired angle $\varphi = \arctan(2F/G)$.

E2.6.4. General Equation of Dynamics (d'Alembert–Lagrange Principle)

Statement of the general equation of dynamics: A mechanical system with imposed ideal constraints moves in such a way that, at each time instant, the sum of elementary works of all active forces $\mathbf{F}_j$ and forces of inertia $\boldsymbol{\Phi}_j$ on any virtual displacement of the system is zero. The mathematical statement of the d'Alembert–Lagrange principle is

$$\sum_j \delta A_j^{\mathbf{F}} + \sum_j \delta A_j^{\Phi} = 0. \tag{E2.6.4.1}$$

Remark. Problems of system dynamics can generally be solved by various methods. The general equation of dynamics has the unquestionable advantage that once it is used, the solution process does not involve the unknown constraint reaction forces, which significantly decreases the order of the system of equations of motion.

Example. A rectangular parallelepiped of mass m_1 moves on a smooth horizontal surface, a homogeneous disk of mass m_2 and radius r_2 rolls on the upper ideally rough surface of the parallelepiped, and a constant horizontal force F_2 is applied to the center of the disk (Fig. E2.44a). Find the accelerations a_1 of the parallelepiped and a_2 of the disk center.

Solution. Consider the system consisting of the parallelepiped and the disk. The constraints imposed on the system are ideal, and we can use the general equation of dynamics to study the motion of the system. For the coordinates determining the system position we take the absolute coordinate x_1 of the parallelepiped and the coordinate x_2 characterizing the disk position with respect to the parallelepiped.

The acceleration vectors a_1 of the parallelepiped and a_2 of the disk center are horizontal, their magnitudes are $a_1 = \ddot{x}_1$ and $a_2 = \ddot{x}_1 + \ddot{x}_2$, and the disk angular acceleration is $\varepsilon_2 = \ddot{x}_2/r_2$. The expected directions of the vectors $\mathbf{a}_1$ and $\mathbf{a}_2$ and the corresponding direction of the curved arrow ε_2 are shown in Fig. E2.44b. We supplement the active forces $\mathbf{F}_2, \mathbf{G}_1, \mathbf{G}_2$ (where $\mathbf{G}_1$ and $\mathbf{G}_2$ are the forces of weight of the parallelepiped and of the disk) with the forces of inertia $\boldsymbol{\Phi}_1 = -m_1\mathbf{a}_1$, $\boldsymbol{\Phi}_2 = -m_2\mathbf{a}_2$ and the couple of forces of inertia with torque $M_2^{\Phi} = -I_{2c}\varepsilon_2$ (see Fig. E2.44b).

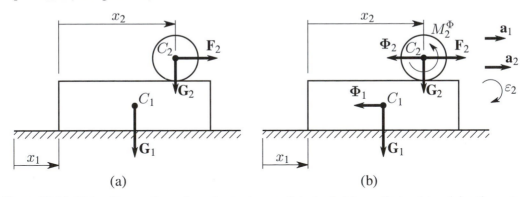

(a)   (b)

Figure E2.44. Disk rolling on the surface of a moving parallelepiped: (a) coordinates determining the system position and the active forces; (b) vectors of accelerations and forces of inertia.

Consider a virtual displacement of the system increasing the coordinates x_1, x_2 by some values $\delta x_1, \delta x_2$. Then the disk rotates by the angle $\delta\varphi_2 = \delta x_2/r_2$.

Now we calculate the sum of elementary works of active forces and forces of inertia on the virtual displacement and equate it with zero:

$$F_2(\delta x_1 + \delta x_2) - m_1 a_1\,\delta x_1 - m_2 a_2(\delta x_1 + \delta x_2) - I_{2c}\varepsilon_2\,\delta\varphi_2 = 0.$$

On rearranging, we obtain

$$\delta x_1\left[F_2 - m_1\ddot{x}_1 - m_2(\ddot{x}_1 + \ddot{x}_2)\right] + \delta x_2\left[F_2 - m_2(\ddot{x}_1 + \ddot{x}_2) - \tfrac{1}{2}m_2\ddot{x}_2\right] = 0.$$

Here the virtual displacements δx_1 and δx_2 can take arbitrary mutually independent values. For the equation to be always satisfied, it is necessary that both factors in square brackets multiplying δx_1 and δx_2 be zero. Thus, the last relation becomes a system of two equations with two unknowns,

$$F_2 - m_1\ddot{x}_1 - m_2(\ddot{x}_1 + \ddot{x}_2) = 0,$$
$$F_2 - m_2(\ddot{x}_1 + \ddot{x}_2) - \tfrac{1}{2}m_2\ddot{x}_2 = 0.$$

By solving the system, we obtain

$$a_1 = \ddot{x}_1 = \frac{F_2}{3m_1 + m_2}, \quad a_2 = \ddot{x}_1 + \ddot{x}_2 = \frac{F_2(2m_1 + m_2)}{m_2(3m_1 + m_2)}.$$

E2.6.5. Lagrange Equations of the Second Kind

To describe the motion of a holonomic mechanical system with n degrees of freedom with imposed ideal constraints, one often uses the *Lagrange equations of the second kind*

$$\frac{d}{dt}\left(\frac{\partial T}{\partial \dot{q}_j}\right) - \frac{\partial T}{\partial q_j} = Q_j, \quad j = 1, 2, \ldots, n. \tag{E2.6.5.1}$$

Here the q_j are *generalized coordinates* the number of which is equal to the number n of degrees of freedom, the $\dot{q}_j$ are the *generalized velocities* equal to the time derivatives of the generalized coordinates, T is the kinetic energy of the system, and the Q_j are *generalized forces*.

The quantities T and Q_j must be represented as functions of generalized velocities, generalized coordinates, and time:

$$T = T(\dot{q}_1, \ldots, \dot{q}_n, q_1, \ldots, q_n, t),$$
$$Q_j = Q_j(\dot{q}_1, \ldots, \dot{q}_n, q_1, \ldots, q_n, t).$$

The generalized forces are found from the expression for the sum of elementary works of active forces $\mathbf{F}_i$ on a virtual displacement of the system transformed to the form

$$\sum \delta A_i^{\mathbf{F}} = \sum \mathbf{F}_i \cdot \delta \mathbf{r}_i = \sum_{j=1}^{n} Q_j \, \delta q_j.$$

The number of generalized forces is equal to the number of degrees of freedom.

The physical dimension of the generalized force Q_j depends on the dimension of the corresponding generalized coordinate q_j, because the dimension of their product $Q_j \, \delta q_j$ must coincide with the dimension of work of force. Therefore, Q_j may have no clear physical meaning, and just this fact underlies the name "generalized force."

After the functions T and Q_j are substituted, the Lagrange equations acquire the form of a system of second-order ordinary differential equations, which should be integrated with the initial conditions taken into account.

Example. To visualize the advantages of the above method, we construct the differential equations of motion of the mechanical system considered in the example in the preceding section in the form of Lagrange equations of the second kind.

Solution. Consider a system consisting of a parallelepiped and a disk (Fig. E2.44). The imposed constraints are holonomic and ideal, and hence the Lagrange equations of the second kind can be used. The system has two degrees of freedom; for the generalized coordinates determining the system position we choose the absolute coordinates $q_1 = x_1$ of the parallelepiped and $q_2 = x_1 + x_2$ of the disk center (see Fig. E2.44a).

We write out the expression for the kinetic energy of the system and reduce it to the form of a function of $\dot{q}_1, \dot{q}_2, q_1, q_2,$ and t:

$$T = T_1 + T_2 = \frac{m_1 v_1^2}{2} + \frac{m_2 v_{2c}^2}{2} + \frac{I_{2c}\omega_2^2}{2} = \frac{m_1 \dot{q}_1^2}{2} + \frac{m_2 \dot{q}_2^2}{2} + \frac{m_2(\dot{q}_2 - \dot{q}_1)^2}{2}.$$

In these transformations, we have used the formula for the kinetic energy of the body in plane motion and the kinematic relation $\omega_2 = (\dot{q}_2 - \dot{q}_1)/r_2$.

Because of the simple kinematics and the fortunate choice of generalized coordinates, the expression for the kinetic energy (explicitly) contains neither the generalized coordinates nor time.

Let us calculate the generalized forces. To this end, we calculate the sum of elementary works of active forces $\mathbf{G}_1$, $\mathbf{G}_2$, $\mathbf{F}_2$ on a virtual displacement of the system determined by variations δq_1, δq_2 directed so that the coordinates q_1 and q_2 increase:

$$\sum \delta A_i^{\mathbf{F}} = F_2(\delta x_1 + \delta x_2) = F_2\,\delta q_2 = 0\,\delta q_1 + F_2\,\delta q_2.$$

The coefficients multiplying the variations of generalized coordinates are the desired expressions for generalized forces, whence

$$Q_1 = 0, \quad Q_2 = F_2.$$

It remains to write out the system of equations of motion:

$$\frac{d}{dt}\left(\frac{\partial T}{\partial \dot{q}_1}\right) - \frac{\partial T}{\partial q_1} = Q_1,$$

$$\frac{d}{dt}\left(\frac{\partial T}{\partial \dot{q}_2}\right) - \frac{\partial T}{\partial q_2} = Q_2.$$

In view of the obtained expressions for T, Q_1, and Q_2, after appropriate transformations the equations of motion acquire the form

$$m_1\ddot{q}_1 - \tfrac{1}{2}m_2(\ddot{q}_2 - \ddot{q}_1) = 0,$$

$$m_2\ddot{q}_2 + \tfrac{1}{2}m_2(\ddot{q}_2 - \ddot{q}_1) = F_2.$$

The solution thus obtained coincides with the solution of this problem obtained earlier by a different method in the preceding section.

In contrast to the Lagrange equations of the second kind, the general equation of dynamics can also be used to study nonholonomic systems and thus has a wider scope. At the same time, the Lagrange equations of the second kind are more convenient when studying holonomic systems, because they require fewer manipulations, and these manipulations are simpler.

E2.7. Small Oscillations of Mechanical Systems

E2.7.1. Preliminaries

In what follows, we consider a holonomic conservative mechanical system with n degrees of freedom whose position is determined by generalized coordinates $q_1, q_2, \ldots, q_n$. Without loss of generality, we assume that the coordinate values $q_1 = q_2 = \ldots = q_n = 0$ correspond to an equilibrium of the system. In this case, all generalized forces are zero, $Q_1 = Q_2 = \ldots = Q_n = 0$.

The equilibrium $q_j = 0, \dot{q}_j = 0; j = 1, 2, \ldots, n$ of the system is said to be *stable* if, for each $\varepsilon > 0$, there exists a $\delta = \delta(\varepsilon) > 0$ such that if the initial deviations for $t = 0$ are in the δ-neighborhood $|q_j^0| < \delta, |\dot{q}_j^0| < \delta$, then, for the entire time of motion, the system does not leave the ε-neighborhood of the equilibrium; i.e., $|q_j| < \varepsilon$ and $|\dot{q}_j| < \varepsilon$ for $t > 0$.

Example. A physical pendulum moving in the vertical plane (Fig. E2.45) has two equilibria corresponding to the angles $\varphi = 0$ and $\varphi = \pi$. It is intuitively clear that the first equilibrium is stable and the second equilibrium is unstable.

The character of equilibrium can be determined by studying the exact solution of the equations of motion of the system. But there exist stability criteria for the system equilibrium, which permit solving this problem more rationally.

One such criterion is given by the following theorem.

THEOREM (LAGRANGE). *If, in a certain position of a conservative system, the potential energy has a strict minimum, then this position is a stable equilibrium of the system.*

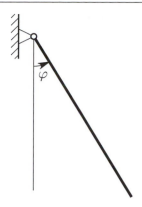

Figure E2.45. Oscillations of a physical pendulum in the vertical plane.

E2.7.2. Small Oscillations of a Conservative System

The kinetic and potential energy of a conservative system with n degrees of freedom can be expressed via the generalized coordinates q_j and generalized velocities $\dot{q}_j$ ($i = 1, \dots, n$):

$$T = \frac{1}{2} \sum_{i,j=1}^{n} a_{ij}(q_1, \dots, q_n)\dot{q}_i\dot{q}_j, \qquad \Pi = \Pi(q_1, \dots, q_n).$$

We assume that:

(1) The state of stable equilibrium of the system corresponds to the coordinates $q_1 = \cdots = q_n = 0$. By the Lagrange theorem, the potential energy of the system has a strict minimum at this point; i.e., $\frac{\partial \Pi}{\partial q_j}(0, 0, \dots, 0) = 0$, $j = 1, 2, \dots, n$.

(2) $\Pi(0, 0, \dots, 0) = 0$, which can always be assumed because of the arbitrary constant contained in the definition of the potential energy (see Subsection E2.5.5).

(3) In the motion of the system, the values of the coordinates q_j and velocities $\dot{q}_j$ are small.

We expand the functions T and Π in a neighborhood of $q_1 = \cdots = q_n = 0$ and $\dot{q}_1 = \cdots = \dot{q}_n = 0$ into series in powers of the coordinates and velocities preserving the terms of the second order of magnitude in q_j and $\dot{q}_j$:

$$T = \frac{1}{2} \sum_{i,j=1}^{n} a_{ij}\dot{q}_i\dot{q}_j, \qquad \Pi = \frac{1}{2} \sum_{i,j=1}^{n} c_{ij}q_iq_j,$$

where the constants a_{ij} and c_{ij} satisfy the conditions $a_{ij} = a_{ji}$ and $c_{ij} = c_{ji} = \left. \frac{\partial^2 \Pi}{\partial q_i \partial q_j} \right|_{q_k=0}$.

As a result, both functions become quadratic forms whose positive definiteness follows from the physical meaning of the kinetic energy $T \geq 0$ and the fact that the potential energy Π at a point of stable equilibrium has a strict minimum, $\Pi \geq 0$.

The Lagrange equations with the above expressions for T and Π taken into account acquire the form of a system of n second-order ordinary linear equations

$$\sum_{j=1}^{n}(a_{ij}\ddot{q}_j + c_{ij}q_j) = 0 \quad (i = 1, \dots, n).$$

The above-described process is called the linearization of equations of motion. The problem of closeness of the solutions of the linearized system to those of the original nonlinear problem is important, but we do not discuss it here.

We seek a particular solution of the above system in the form $q_j = u_j \sin(\omega t + \alpha)$ with $j = 1, \ldots, n$. By substituting this expression into the equations of motion and by cancelling $\sin(\omega t + \alpha)$, we obtain a system of linear algebraic equations for the amplitudes u_j. Since some of the amplitudes must be nonzero, it follows that the determinant of the system of homogeneous equations must be zero:

$$\det |c_{ij} - \lambda a_{ij}| = 0, \quad \lambda = \omega^2.$$

The obtained equation is called the *secular equation* or the *frequency equation*. By expanding the determinant, we obtain a polynomial of degree n in λ.

One can prove that in the case under study the secular equation has only real positive roots λ_j associated with real positive frequencies $\omega_j = \sqrt{\lambda_j}$. In this case, the n obtained particular solutions of the system are linearly independent (including the case of multiple roots of the secular equation); they are called the *normal oscillations* of the system.

Since the system of differential equations is linear, we see that any linear combination of normal oscillations with constant coefficients is also a solution of the system.

Example. A double mathematical pendulum consists of two weightless rods of equal length l with heavy point particles of equal mass m fixed at the ends of the rods. The construction moves in the plane of the figure, the upper rod can rotate about the fixed hinge O_1 adjusted to the base, and the second pendulum can rotate about the moving hinge O_2 connecting the rods (Fig. E2.46a). Study the small motions of the system near the stable equilibrium.

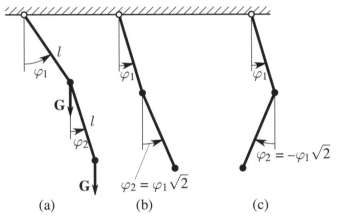

Figure E2.46. Oscillations of the double mathematical pendulum: (a) sketch of the device; (b) the angles made by the rods with the vertical axis for $C_1 = 0$; (c) the angles made by the rods with the vertical axis for $C_2 = 0$.

Solution. For the generalized coordinates we take φ_1 and φ_2, i.e., the rod rotation angles about the vertical. The expressions for the kinetic and potential energy of the system have the form

$$T = ml^2 \dot{\varphi}_1^2 + 2ml^2 \cos(\varphi_2 - \varphi_1) \dot{\varphi}_1 \dot{\varphi}_2 + ml^2 \dot{\varphi}_2^2,$$
$$\Pi = -2mgl \cos \varphi_1 - mgl \cos \varphi_2.$$

It follows from the expression for the potential energy that the stable equilibrium of the system corresponds to the coordinate values $\varphi_1 = \varphi_2 = 0$. Assuming that the oscillations are small, we rewrite the expressions for T and Π omitting the terms of the third and higher orders of magnitude:

$$T = ml^2 \dot{\varphi}_1^2 + 2ml^2 \dot{\varphi}_1 \dot{\varphi}_2 + ml^2 \dot{\varphi}_2^2,$$
$$\Pi = mgl\varphi_1^2 + \tfrac{1}{2}mgl\varphi_2^2.$$

We write out the equations of motion of the system as

$$2l\ddot{\varphi}_1 + l\ddot{\varphi}_2 + 2g\varphi_1 = 0,$$
$$l\ddot{\varphi}_1 + l\ddot{\varphi}_2 + g\varphi_2 = 0.$$

We seek a particular solution of this system in the form

$$\varphi_1 = u_1 \sin(\omega t + \alpha), \quad \varphi_2 = u_2 \sin(\omega t + \alpha).$$

The secular equation is reduced to the form

$$l^2 \omega^4 - 4lg\omega^2 + 2g^2 = 0,$$

whose roots are the two angular natural oscillation frequencies

$$\omega_1 = \sqrt{\left(2 - \sqrt{2}\right)\frac{g}{l}}, \quad \omega_2 = \sqrt{\left(2 + \sqrt{2}\right)\frac{g}{l}}.$$

It follows from the second equation of motion that $u_2 = u_1 \frac{l\omega^2}{g - l\omega^2}$, which implies that $u_2 = u_1\sqrt{2}$ for the frequency ω_1 and $u_2 = -u_1\sqrt{2}$ for the frequency ω_2. Therefore, the general solution of the linearized system of differential equations describing the motion of the double mathematical pendulum has the form

$$\varphi_1 = C_1 \sin(\omega_1 t + \alpha_1) + C_2 \sin(\omega_2 t + \alpha_2),$$

$$\varphi_2 = \sqrt{2}\, C_1 \sin(\omega_1 t + \alpha_1) - \sqrt{2}\, C_2 \sin(\omega_2 t + \alpha_2).$$

Here C_1, C_2, α_1, α_2 are arbitrary constants depending on the initial conditions, i.e., on the values φ_1^0, φ_2^0, $\dot{\varphi}_1^0$, $\dot{\varphi}_2^0$ at the initial time instant $t = 0$.

The angles φ_1 and φ_2 made by the rods with the vertical axis for $C_1 = 0$, $C_2 \neq 0$ and $C_2 = 0$, $C_1 \neq 0$ are shown in Fig. E2.46, b and c.

E2.7.3. Normal Coordinates

Obviously, the difficulties encountered when seeking the natural oscillation frequencies of a mechanical system increase with the number of degrees of freedom. Moreover, finding the general characteristics of the system behavior is also difficult, because the characteristic equation is cumbersome. The existence of *normal* or *principal coordinates* in a system permits one to make the analysis of the system less laborious.

THEOREM. *Two quadratic forms at least one of which is positive definite can be simultaneously reduced to "sums of squares" by a nonsingular linear transformation of the variables.*

In the case of oscillations of a conservative system, there are two positive definite quadratic forms: the kinetic energy, which is a quadratic form of the generalized velocities $\dot{q}_i$, and the potential energy, which is a quadratic form of the generalized coordinates q_i. Since the generalized coordinates and generalized velocities behave in the same way under a linear transformation,

$$q_k = \sum_{s=1}^{n} \gamma_{ks}\theta_s, \quad \dot{q}_k = \sum_{s=1}^{n} \gamma_{ks}\dot{\theta}_s, \quad k = 1, \ldots, n; \quad \det(\gamma_{ks}) \neq 0,$$

we see that the kinetic and potential energy of the system in the new variables become

$$T = \frac{1}{2}\sum_{j=1}^{n}\dot{\theta}_j^2, \quad \Pi = \frac{1}{2}\sum_{j=1}^{n}\lambda_j\theta_j^2.$$

The Lagrange equations in normal coordinates θ_j have the form

$$\ddot{\theta}_j + \lambda_j\theta_j = 0, \quad j = 1, \ldots, n.$$

Each of these equations contains only one unknown function. The general solutions of these equations determine the harmonic oscillations

$$\theta_j = C_j \sin(\omega_j t + \alpha_j), \quad \omega_j = \sqrt{\lambda_j}, \quad j = 1, \ldots, n,$$

where the C_j and α_j are arbitrary constants.

Thus, we have shown that all the roots of the secular equation for a conservative mechanical system are real and positive. In addition, to n frequencies $\omega_j = \sqrt{\lambda_j}$ there correspond n linearly independent amplitude vectors, and the general solution of the system does not contain secular terms of the form $\theta_j = (C_0 + C_1 t + C_2 t^2 + \cdots) \sin(\omega_j t + \alpha_j)$ in the case of multiple frequencies.

E2.7.4. Influence of Small Dissipative Forces on System Oscillations

Assume that each point of the system is subjected not only to potential forces but also to resistance forces depending on the velocity. Such forces are said to be *dissipative*. In this case, the total mechanical energy of the system does not remain constant but decays with time.

Let us estimate the influence of dissipative forces on system oscillations under the assumption that these forces are small and linearly depend on the velocities of points of the system. In addition, we assume that the conservative system has distinct oscillation frequencies. The expressions for the generalized forces acquire the form

$$Q_i = -\sum_{k=1}^{n} b_{ik} \dot{q}_k - \sum_{k=1}^{n} c_{ik} q_k, \qquad i = 1, \ldots, n.$$

By analogy with the potential energy of the system, we introduce the *dissipative Rayleigh function*, specifying it by a positive definite quadratic form R in the generalized velocities $\dot{q}_i$:

$$R = \frac{1}{2} \sum_{i,k=1}^{n} b_{ik} \dot{q}_i \dot{q}_k > 0 \quad \text{for} \quad \sum_{i=1}^{n} \dot{q}_i^2 > 0.$$

In the case under study, the system is *definitely dissipative*, and the equilibrium of the system is asymptotically stable; i.e., $q_i(t) \to 0$ as $t \to \infty$.

In normal coordinates, the equations of motion of the system acquire the form

$$\ddot{\theta}_i + \beta_i \dot{\theta}_i + \omega_i^2 \theta_i + \sum_{\substack{j=1 \\ j \neq i}}^{n} \beta_{ij} \dot{\theta}_j = 0 \quad (i, j = 1, \ldots, n).$$

Using the fact that the dissipative forces are small—i.e., the coefficients β_i and β_{ij} are small—in the first approximation, one can obtain a system of linearly independent particular solutions ("normal oscillations") in the form (Gantmakher, 1975)

$$\theta_i = A_i e^{-\frac{\beta_i}{2} t} \sin(\omega_i t + \alpha_i), \quad \theta_j = \varepsilon_j A_i e^{-\frac{\beta_i}{2} t} \sin\left(\omega_i t + \alpha_i + \frac{\pi}{2}\right),$$

$$\varepsilon_j = -\frac{\beta_{ji} \omega_i}{\omega_j^2 - \omega_i^2}, \quad j = 1, \ldots, i-1, i+1, \ldots, n,$$

where $i = 1, \ldots, n$. This implies that:

(1) The system frequencies are close to those of the corresponding conservative system in the absence of dissipative forces.

(2) The system oscillations are damped as $t \to \infty$.

(3) In the ith normal oscillation, all coordinates θ_j ($j \neq i$) are small compared with the ith coordinate θ_i and differ from it in phase by a quarter period.

E2.7.5. Forced Oscillations

Assume that the system considered in the preceding section is also subjected to *perturbing forces* Q_i. The equations of motion of such a system acquire the form

$$\sum_{j=1}^{n}(a_{ij}\ddot{q}_j + b_{ij}\dot{q}_j + c_{ij}q_j) = Q_i(t), \qquad i = 1,\ldots,n.$$

The general solution $q_i(t)$ of this inhomogeneous system of linear equations is the sum of the general solution $q_i^{\circ}(t)$ of the corresponding homogeneous system and a particular solution $\widetilde{q}_i(t)$ of the inhomogeneous system,

$$q_i(t) = q_i^{\circ}(t) + \widetilde{q}_i(t).$$

Since $q_i^{\circ}(t) \to 0$ as $t \to \infty$, it follows that the system behavior for large values t is determined by the particular solution, $q_i(t) \to \widetilde{q}_i(t)$ as $t \to \infty$.

The system under study is linear, and, because of the linear superposition of particular solutions, the general case of finding the forced oscillations is reduced to studying the solution of the system for the case in which only one of the perturbing forces is nonzero, say, $Q_1 \neq 0$.

We restrict ourselves to the case of a harmonic driving force $Q_1(t) = Ae^{i\Omega t}$, $Q_j = 0$ ($j = 2,\ldots,n$), where i is the imaginary unit ($i^2 = -1$). We seek the forced oscillations in the form $q_k = B_k e^{i\Omega t}$ ($k = 1,\ldots,n$). By substituting these expressions into the equations of motion, we obtain a system of algebraic equations for B_k:

$$\sum_{j=1}^{n}\left[a_{1j}(i\Omega)^2 + b_{1j}(i\Omega) + c_{1j}q_j\right]B_j = A,$$

$$\sum_{j=1}^{n}\left[a_{kj}(i\Omega)^2 + b_{kj}(i\Omega) + c_{kj}q_j\right]B_j = 0, \qquad k = 2,\ldots,n.$$

By solving this system, we obtain $B_k = W_{1k}(i\Omega)A$, where $W_{1k}(i\Omega) = \Delta_{1k}(i\Omega)/\Delta(i\Omega)$ is a proper rational function of $i\Omega$ with real coefficients. The hodograph of this function in the complex plane is called the *frequency* or *amplitude-phase characteristic*.

By setting $W_{1k}(i\Omega) = R_{1k}(\Omega)e^{i\Psi_{1k}(\Omega)}$, we rewrite the solution as

$$q_k = R_{1k}(\Omega)Ae^{i[\Omega t+\Psi_{1k}(\Omega)]}, \qquad k = 1,\ldots,n,$$

where $R_{1k}(\Omega)$ is the *amplitude* characteristic and $\Psi_{1k}(\Omega)$ is the *phase* characteristic.

In the case of the sinusoidal perturbing force

$$Q_1 = \sin(\Omega t) = \frac{A}{2i}(e^{i\Omega t} - e^{-i\Omega t}),$$

the solution becomes

$$q_k = R_{1k}(\Omega)A \sin[\Omega t + \Psi_{1k}(\Omega)] \quad (k = 1,\ldots,n);$$

i.e., a sinusoidal driving force causes a sinusoidal response, where the forces are multiplied by the amplitude characteristic $R_{1k}(\Omega)$ and the phase shift is determined by the phase characteristic $\Psi_{1k}(\Omega)$.

By adjusting the system characteristics, one can suppress oscillations at some "useless" frequencies and increase the amplitudes of these oscillations at the other "required" frequencies. This idea underlies the design of frequency filters.

Example. The equation of forced oscillations of a system with a single degree of freedom and its amplitude characteristic R and phase characteristic Ψ has the form

$$\ddot{q} + b\dot{q} + \omega^2 q = A\sin(\Omega t), \quad R = \frac{1}{\sqrt{(\omega^2 - \Omega^2)^2 + b^2\Omega^2}}, \quad \Psi = \arctan\left(\frac{b\Omega}{\omega^2 - \Omega^2}\right).$$

Here we note the following specific features of the obtained solution: the forced oscillation amplitude $B = AR$ becomes unboundedly large if the dissipative forces are very small and the driving force frequency tends to the natural oscillation frequencies ($B \to \infty$ as $b \to 0$ and $\Omega \to \omega$).

E2.8. Dynamics of an Absolutely Rigid Body

E2.8.1. Rotation of a Rigid Body about a Fixed Axis

Assume that a body of mass m rotates about the vertical z-axis under the action of a system of active forces $\{\mathbf{F}_k\}$. At points O_1 and O_2 of the axis, cylindrical hinges are located and a thrust bearing* is placed at the point O_1 (Fig. E2.47).

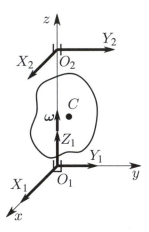

Figure E2.47. Rotation of a rigid body about a vertical axis under the action of active forces.

Theorems on the momentum and angular momentum with respect to the center O_1 imply the following system of six equations with six unknowns:

$$-mx_C\omega^2 - my_C\dot{\omega} = X_1 + X_2 + \sum F_{kx},$$

$$-my_C\omega^2 + mx_C\dot{\omega} = Y_1 + Y_2 + \sum F_{ky},$$

$$0 = Z_1 + \sum F_{kz},$$

$$I_{yz}\omega^2 - I_{zx}\dot{\omega} = -LY_2 + \sum M_x(\mathbf{F}_k),$$

$$-I_{zx}\omega^2 - I_{zx}\dot{\omega} = LX_2 + \sum M_y(\mathbf{F}_k),$$

$$I_{zz}\dot{\omega} = \sum M_z(\mathbf{F}_k).$$

Here x_C and y_C are the coordinates of the center of mass C of the body, X_1, X_2, Y_1, Y_2, Z_1 are the projections of the constraint reaction forces onto the coordinate axes, and $L = O_1O_2$.

---

* A thrust bearing is a device preventing the body from vertical displacements.

The last equation does not contain reactions; it is called the *equation of motion of a rigid body about a fixed axis*. By integrating this equation, one can find the angular velocity and the angle of rotation as a function of time: $\omega = \omega(t, \omega_0)$, $\varphi = \varphi(t, \omega_0, \varphi_0)$.

If the body is at rest $\omega = \dot{\omega} = 0$, then the constraint reactions are said to be *static*. In the case of body rotation, the reactions are said to be *dynamic*; they depend on the angular velocity and the angular acceleration of the body.

The conditions under which the dynamic reactions are equal to static reactions have the form

$$x_C = y_C = 0, \quad I_{yz} = I_{zx} = 0.$$

This implies that, as a rigid body rotates about a fixed axis, the dynamic constraint reactions are equal to static ones if the rotation axis is one of the principal central axes of inertia.

E2.8.2. Plane-Parallel (Plane) Motion of a Rigid Body

Each motion of a body can be viewed as a motion of the center of mass and the motions of the body with respect to the center of mass.

In the case of plane motion, the body points, including the center of mass, move in planes parallel to a fixed (base) plane. The motion about the center of mass is the rotation about an axis perpendicular to the base plane and passing through the center of mass.

We draw a plane parallel to the base plane through the center of mass C and introduce fixed axes $O\xi\eta$ in this plane (Fig. E2.48). In addition, we introduce two moving reference frames whose common origin is the center of mass C. One of them, Cx^*y^* (the König axes), moves translationally: its axes are parallel to the axes of the fixed system $O\xi\eta$; the second system, Cxy, is rigidly fixed to the body and rotates about C.

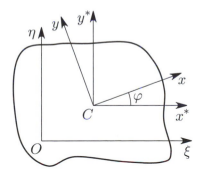

Figure E2.48. Plane-parallel motion of a rigid body.

The body position is determined by the following three parameters: the coordinates x_C, y_C of the center of mass in the fixed system $O\xi\eta$ and the rotation angle φ between the axes Cx^* and Cx.

The theorem on the motion of the center of mass of the body in the fixed system $O\xi\eta$ and the theorem on the angular momentum about the axis perpendicular to the base plane and passing through the center of mass give the following three equations of plane motion of a rigid body:

$$m\frac{d^2 x_C}{dt^2} = \sum_k F_{kx}^{\mathrm{e}}, \quad m\frac{d^2 y_C}{dt^2} = \sum_k F_{ky}^{\mathrm{e}}, \quad I_{CC}\frac{d^2\varphi}{dt^2} = \sum_k M_C \mathbf{F}_k^{\mathrm{e}}.$$

Here m is the mass of the body and I_{CC} is the moment of inertia of the body about the axis passing through the center of mass and perpendicular to the base plane; the expressions on the right-hand side of the relations contain the projections and moments of the external forces acting on the body.

By integrating the obtained system of equations with the initial conditions taken into account, one can determine x_C, y_C, and φ as functions of time t.

E2.8.3. Motion of a Rigid Body about a Fixed Point

▶ **Preliminary remarks and definitions.**

An arbitrary motion of a rigid body is the combination of two simultaneous motions: the translational motion together with an arbitrary point of the body and the spherical motion about this point treated as a fixed point.

In several problems of motion of a rigid body, the differential equations of motion determining the translational motion are independent of the equations of motion about the chosen point. Therefore, both systems of equations can be integrated independently of each other. As a rule, the difficulties of analytical integration arise in the second system. Therefore, the problem of motion of a body about a fixed point is most important in mechanics.

Any body with a single fixed point has three degrees of freedom, and its position can be determined by introducing three independent generalized coordinates.

We introduce two reference frames whose origins coincide with the fixed point O of the rigid body. The system $O\xi\eta\zeta$ is fixed, and the system $Oxyz$ is a moving system rigidly fixed to the body. The position of the moving system (i.e., the position of the rigid body) with respect to the fixed system is determined by the *Euler angles* φ, ψ, and θ.

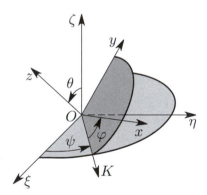

Figure E2.49. Motion of a rigid body about a fixed point: the fixed and moving reference frames $O\xi\eta\zeta$ and $Oxyz$, and the Euler angles φ, ψ, and θ.

The straight line OK of intersection of the planes $O\xi\eta$ and Oxy is called the *nodal line*. Its positive direction is chosen so that the nearest rotation from $O\zeta$ to Oz seems to be anticlockwise if it is observed from the side of the point K.

The *angle φ of proper rotation* is the angle between OK and Ox; it is assumed to be positive if, being observed from the positive sense of the axis, the rotation from OK to Ox seems to be anticlockwise. The axis Oz is called the *axis of proper rotation*. If only the angle φ varies, then the body rotates about the fixed axis Oz with the *angular velocity* $\dot{\varphi} = \dot{\varphi}\mathbf{z}^\circ$ *of proper rotation* directed along the axis Oz, where $\mathbf{z}^\circ$ is the unit vector of the axis Oz. (We do not show all unit vectors of coordinate axes in Fig. E2.49 so as not to overload it.)

The *precession angle ψ* is the angle between $O\xi$ and OK; it is assumed to be positive if, being observed from the positive sense of the axis $O\zeta$, the rotation from $O\xi$ to OK seems to be anticlockwise. The axis $O\zeta$ is called the *axis of precession*. If only the angle ψ varies, then the body rotates about the fixed axis $O\zeta$ with the *angular precession velocity* $\dot{\boldsymbol{\psi}} = \dot{\psi}\boldsymbol{\zeta}^\circ$ directed along the axis $O\zeta$, where $\boldsymbol{\zeta}^\circ$ is the unit vector of the axis $O\zeta$.

The *nutation angle* θ is the angle between $O\zeta$ and Oz. The nodal line OK is called the *axis of nutation*. If only the angle θ varies, then the body rotates about the fixed axis OK with the *angular nutation velocity* $\dot{\boldsymbol{\theta}} = \dot{\theta}\,\overrightarrow{OK}^\circ$ directed along the axis OK, where $\overrightarrow{OK}^\circ$ is the unit vector of the axis OK.

The superposition of the system $O\xi\eta\zeta$ with the system $Oxyz$ can be realized by three rotations: first, by the angle ψ about the axis $O\zeta$, then by the angle θ about OK, and finally, by the angle φ about the axis Oz.

▶ **Kinematic Euler equations.** If all three angles vary as the body moves, then the angular velocity of the body $\boldsymbol{\omega}$ according to the theorems on the compound motion of a rigid body is equal to the vector sum of the vectors of angular velocities of the proper rotation, precession, and nutation:

$$\boldsymbol{\omega} = \dot{\boldsymbol{\varphi}} + \dot{\boldsymbol{\psi}} + \dot{\boldsymbol{\theta}}.$$

We denote the projections of the angular velocity $\boldsymbol{\omega}$ on the axes of the moving reference frame $Oxyz$ by p, q, r. The expressions of these quantities via the Euler angles φ, ψ, θ and their derivatives $\dot{\varphi}, \dot{\psi}, \dot{\theta}$ have the form

$$
\begin{aligned}
p &= \dot{\psi}\sin\theta\sin\varphi + \dot{\theta}\cos\varphi, \\
q &= \dot{\psi}\sin\theta\cos\varphi - \dot{\theta}\sin\varphi, \\
r &= \dot{\psi}\cos\theta + \dot{\varphi}.
\end{aligned}
\tag{E2.8.3.1}
$$

For p', q', r', i.e., the projections of the angular velocity $\boldsymbol{\omega}$ on the axes of the fixed system $O\xi\eta\zeta$, the expressions via the Euler angles and their derivatives have the form

$$
\begin{aligned}
p' &= \dot{\varphi}\sin\theta\sin\psi + \dot{\theta}\cos\psi, \\
q' &= -\dot{\varphi}\sin\theta\cos\psi + \dot{\theta}\sin\psi, \\
r' &= \dot{\varphi}\cos\theta + \dot{\psi}.
\end{aligned}
\tag{E2.8.3.2}
$$

The last two groups of equations are called the *kinematic Euler equations*.

▶ **Formulas for the angular momentum. Dynamic Euler equations.** The expressions for the angular momentum $\mathbf{K}_O$ of a rigid body about a fixed point O and its projections K_{Ox}, K_{Oy}, K_{Oz} onto the axes of the moving reference frame have the form (in what follows, the subscript O is omitted for convenience):

$$
\mathbf{K} = \sum_j (\mathbf{r}_j \times m_j \mathbf{v}_j) = \sum_j [\mathbf{r}_j \times m_j(\boldsymbol{\omega} \times \mathbf{r}_j)] = \boldsymbol{\omega}\sum_j m_j r_j^2 - \sum_j m_j \mathbf{r}_j(\mathbf{r}_j \cdot \boldsymbol{\omega}),
$$

$$
\begin{aligned}
K_x &= pI_{xx} - qI_{xy} - rI_{xz}, \\
K_y &= -pI_{yx} + qI_{yy} - rI_{yz}, \\
K_z &= -pI_{zx} - qI_{zy} + rI_{zz}.
\end{aligned}
$$

$$\tag{E2.8.3.3}$$

The expression for the kinetic energy T of a rigid body moving about a fixed point has the form

$$
2T = \sum_j m_j v_j^2 = \sum_j m_j \mathbf{v}_j \cdot (\boldsymbol{\omega} \times \mathbf{r}_j) = \boldsymbol{\omega} \cdot \sum_j (\mathbf{r}_j \times m_j \mathbf{v}_j) = \boldsymbol{\omega} \cdot \mathbf{K},
$$

or, in coordinate form,

$$2T = \boldsymbol{\omega} \cdot \mathbf{K} = I_{xx}p^2 + I_{yy}q^2 + I_{zz}r^2 - 2I_{yz}qr - 2I_{zx}rp - 2I_{xy}pq.$$

The *dynamic Euler equations* of motion of a rigid body about a fixed point are obtained from the theorem on the angular momentum about a fixed point O:

$$\frac{d\mathbf{K}_O}{dt} = \sum \mathbf{M}_O \mathbf{F}_j^e \qquad \text{(in the fixed reference frame)}$$

or

$$\frac{\widetilde{d\mathbf{K}_O}}{dt} + \boldsymbol{\omega} \times \mathbf{K}_O = \mathbf{M}_O \mathbf{F}_j^e \quad \text{(in the moving reference frame)}.$$

It is inconvenient to project the vector relations on the axes of a fixed reference frame, because the coefficients $I_{\xi\xi}, I_{\eta\eta}, I_{\zeta\zeta}, \dots, I_{\xi\eta}$ contained in the expression for K_O depend on time, and the equations thus obtained are rather cumbersome. The equations of motion become concise if

1. The projections onto the axes of the moving reference frame are taken.

2. For the moving axes one takes the principal axes of inertia of the body for the point O, so that

$$K_{Ox} = I_{xx}p = Ap, \quad K_{Oy} = I_{yy}q = Bq, \quad K_{Oz} = I_{zz}r = Cr.$$

(Here the traditional notation $A = I_{xx}$, $B = I_{yy}$, $C = I_{zz}$ is used.)

As a result, the dynamic Euler equations acquire the form

$$A\frac{dp}{dt} + (C - B)qr = M_x,$$

$$B\frac{dq}{dt} + (A - C)rp = M_y, \qquad \text{(E2.8.3.4)}$$

$$A\frac{dr}{dt} + (B - A)pq = M_z.$$

Along with the kinematic Euler equations (E2.8.3.1), they form a system of six first-order ordinary nonlinear equations for the six unknown functions $p, q, r, \varphi, \psi, \theta$ of time. The general integrals of the system must contain six arbitrary constants, which are determined by the initial conditions $p_0, q_0, r_0, \varphi_0, \psi_0, \theta_0$ for $t = t_0$.

By eliminating p, q, r, one can pass from the system of six equations to a system of three second-order ordinary differential equations for the Euler angles.

The difficulties of analytic integration of the system of nonlinear equations of motion are obvious, because, in the general case, M_x, M_y, and M_z are functions of $t, p, q, r, \varphi, \psi$, and θ.

▶ **Statement of the problem on the motion of a rigid body about a fixed point.** Consider the motion of a heavy rigid body about a fixed point under the action of a single active force of weight $\mathbf{G}$. We neglect the Earth's rotation, assuming that the Earth is fixed. The axis $O\zeta$ of the stationary system $O\xi\eta\zeta$ fixed to the Earth is directed vertically upwards and opposite to the force of gravity. The position vector $\overrightarrow{OC}$ of the center of gravity of a body is denoted by $\mathbf{r}_c(x_c, y_c, z_c)$. The expressions for the direction cosines of the angles made by the axis $O\zeta$ with the axes of the moving reference frame become

$$\cos(\widehat{\zeta, x}) = \gamma_1 = \sin\theta\sin\varphi, \quad \cos(\widehat{\zeta, y}) = \gamma_2 = \sin\theta\cos\varphi, \quad \cos(\widehat{\zeta, z}) = \gamma_3 = \cos\theta.$$

Here $\gamma_1, \gamma_2, \gamma_3$ are the projections of the unit vector ζ° on the moving axes; i.e., $\zeta^\circ = \gamma_1\mathbf{i} + \gamma_2\mathbf{j} + \gamma_3\mathbf{k}$, where $\mathbf{i}, \mathbf{j}, \mathbf{k}$ are the unit vectors of the moving axes. Since the direction of the force of gravity does not vary, it follows that

$$\frac{d\zeta^\circ}{dt} \equiv \frac{\widetilde{d\zeta^\circ}}{dt} + \omega \times \zeta^\circ = 0,$$

which implies the *Poisson equations*:

$$\frac{d\gamma_1}{dt} = r\gamma_2 - q\gamma_3,$$

$$\frac{d\gamma_2}{dt} = p\gamma_3 - r\gamma_1,$$

$$\frac{d\gamma_3}{dt} = q\gamma_1 - p\gamma_2.$$

In the case under study, with the relations $\mathbf{G} = -G\zeta^\circ$ and $\mathbf{M}_0 = \mathbf{r} \times \mathbf{G}$ taken into account, the dynamic Euler equations acquire the form

$$A\frac{dp}{dt} + (C - B)qr = G(\gamma_2 z_\mathrm{c} - \gamma_3 y_\mathrm{c}),$$

$$B\frac{dq}{dt} + (A - C)rp = G(\gamma_3 x_\mathrm{c} - \gamma_1 z_\mathrm{c}),$$

$$A\frac{dr}{dt} + (B - A)pq = G(\gamma_1 y_\mathrm{c} - \gamma_2 x_\mathrm{c}).$$

Here the quantities A, B, C, x_c, y_c, z_c are constants.

The last equations, together with the Poisson equations, form a closed system of six equations with six unknown functions of time $\gamma_1, \gamma_2, \gamma_3, p, q, r$. If they are obtained, then to determine the Euler angles φ, ψ, θ as functions of time it is necessary to find $\psi(t)$ by a single quadrature of any of the kinematic Euler equations, because $\theta(t)$ and $\varphi(t)$ can be found from the expressions for the direction cosines of the angles made by the axis $O\zeta$ with the axes of the moving system.

One can show that the solution of the problem on the motion of a body is reduced to finding only a single integral.

If the system of equations of motion is represented in the canonical form

$$\frac{dp}{P} = \frac{dq}{Q} = \frac{dr}{R} = \frac{\gamma_1}{\Gamma_1} = \frac{\gamma_2}{\Gamma_2} = \frac{\gamma_3}{\Gamma_3} = dt,$$

where

$$P = G(\gamma_2 z_\mathrm{c} - \gamma_3 y_\mathrm{c}) - (C - B)qr, \quad \Gamma_1 = r\gamma_2 - q\gamma_3,$$
$$Q = G(\gamma_3 x_\mathrm{c} - \gamma_1 z_\mathrm{c}) - (A - C)rp, \quad \Gamma_2 = p\gamma_3 - r\gamma_1,$$
$$R = G(\gamma_1 y_\mathrm{c} - \gamma_2 x_\mathrm{c}) - (B - A)pq, \quad \Gamma_3 = q\gamma_1 - p\gamma_2,$$

then one can see that

(1) It is possible to integrate the system of the first five equations

$$\frac{dp}{P} = \frac{dq}{Q} = \frac{dr}{R} = \frac{\gamma_1}{\Gamma_1} = \frac{\gamma_2}{\Gamma_2} = \frac{\gamma_3}{\Gamma_3}$$

alone, because they do not contain the time t explicitly.

(2) The structure of this system of five equations is such that, according to the theory of the last Jacobi multiplier for the canonical system of differential equations (the details of this theory are beyond the scope of this book), to reduce the problem to quadratures it suffices to know four rather than five integrals of the system.

(3) One of the integrals, $\gamma_1^2 + \gamma_2^2 + \gamma_3^2 = 1$, is obvious from geometric considerations.

(4) The theorem on the angular momentum of the system about the axis $O\zeta$, $\frac{dK_\zeta}{dt} = M_\zeta = 0$, leads to the integral

$$Ap\gamma_1 + Bq\gamma_2 + Cr\gamma_3 = \text{const.}$$

(5) The theorem on the kinetic energy $dT = -Gd\zeta_c$ gives one more integral,

$$\frac{1}{2}(Ap^2 + Bq^2 + Cr^2) = -G(x_c\gamma_1 + y_c\gamma_2 + z_c\gamma_3) + \text{const.}$$

Thus, it remains to find one last integral. This problem can be solved only in the following three special cases:

(i) The *Euler–Poinsot case*: motion by inertia, when the moment of external forces about the fixed point is zero: $\mathbf{M}_O = 0$.

(ii) The *Lagrange–Poisson case*: the ellipsoid of inertia of a body about the fixed point is an ellipsoid of rotation, $A = B$, and the center of gravity of the body lies on the axis of rotation of the ellipsoid of inertia.

(iii) The *Kowalewski case* determined by the following two conditions: (a) $A = B = 2C$ and (b) the center of gravity lies in the equatorial plane of the ellipsoid of inertia.

The studies showed that it is only in these three cases that there exists an algebraic integral that holds for any initial conditions, and this reduces the problem to quadratures. The cases of integration described in the literature and differing from the above cases hold only under certain specifically chosen initial conditions.

Example. Under the conditions $A = B$ (the ellipsoid of inertia is an ellipsoid of rotation) and $\mathbf{M}_O = 0$, the solution of the problem on the motion of a body can be solved completely in elementary functions.

The dynamic Euler equations become

$$A\frac{dp}{dt} + (C - A)qr = 0,$$

$$B\frac{dq}{dt} + (A - C)rp = 0,$$

$$A\frac{dr}{dt} = 0.$$

The solution of this system gives the three integrals

$$A(p^2 + q^2) + Cr^2 = h, \quad A^2(p^2 + q^2) + C^2r^2 = K_O^2, \quad r = r_0 = \text{const.}$$

The first of them is the energy integral, the second follows from the law of conservation of the angular momentum $\mathbf{K}_O$, and the third follows directly from the last equation.

To solve the problem completely, it is necessary to express the Euler angles as functions of time. To simplify the solution, we direct the axis $O\zeta$ along the vector $\mathbf{K}_O$, which remains constant in magnitude and direction. The projections of $\mathbf{K}_O$ onto the axes Ox, Oy, and Oz are

$$K_x = Ap = G\sin\theta\sin\varphi, \quad K_y = Aq = G\sin\theta\cos\varphi, \quad K_z = Cr = G\cos\varphi.$$

It follows from the last equation that $\theta = \text{const} = \theta_0$, because $\cos\theta = Cr/G = Cr_0/G = \text{const.}$

After the corresponding substitutions into the kinematic Euler equations, we integrate them and finally obtain the expressions for the Euler angles:

$$\varphi = \left(r_0 - \frac{K_O}{A}\cos\theta_0\right)t + \varphi_0, \quad \psi = \frac{K_O}{A}t + \psi_0, \quad \theta = \theta_0.$$

The motion determined by these equations is called the *regular precession*.

E2.8.4. Elementary Theory of Gyroscope

A gyroscope is a homogeneous body of rotation rapidly rotating about its symmetry axis; the gyro axis can change its direction in space. Usually, the gyro is made as a massive symmetry body fixed so that one of the points of its axis remains immovable. This, for example, can be done by using the gimbal suspension (Fig. E2.50).

Figure E2.50. Gyroscope.

The exact analytical study of the gyro motion is a very complicated problem. The solution is significantly simplified if the angular velocity of the gyro rotation is sufficiently large.

If $M_O = 0$ for a gyro rotating about the symmetry axis, then the theorem on the angular momentum $d\mathbf{K}_O/dt = \mathbf{M}_O$ implies that $\mathbf{K}_O = I\boldsymbol{\omega}_1 = \textbf{const}$. This means that the gyro axis preserves its original direction with respect to inertial reference frame, and the angular velocity $\boldsymbol{\omega}_1$ is constant.

If $M_O \neq 0$, then, in addition to proper rotation, the gyro performs precession and nutation motions. Exact studies show that, for a rapidly rotating gyro, the variations in the axis direction due to the nutation oscillations, $\Delta\theta = (\theta_{\max} - \theta_{\min})$, remain in a very small range, and the angular nutation velocity $\dot{\theta}$ is also very small. Therefore, in the approximate theory of gyros, it is assumed that $\theta = \theta_0 = \text{const}$.

The angular precession velocity $\boldsymbol{\omega}_2$ is also small, but it is necessary to take this velocity into account, because it has a constant sign and the gyro axis changes its direction rather noticeably with time.

With the above assumptions taken into account, we have $\boldsymbol{\omega} = \boldsymbol{\omega}_1 + \boldsymbol{\omega}_2 \approx \boldsymbol{\omega}_1$, because $\omega_2 \ll \omega_1$. In what follows, we assume that, at any time instant, the angular velocity and angular momentum vectors are directed along the gyro axis, $\mathbf{K}_O = I\boldsymbol{\omega}_1$ (Fig. E2.51).

Now the theorem on the angular momentum $d\mathbf{K}_O/dt = \mathbf{M}_O$ can be represented as $d(I\boldsymbol{\omega}_1)/dt = \mathbf{M}_O$. The left-hand side of the equation contains the velocity of the endpoint of the vector $I\boldsymbol{\omega}_1$ rotating about the vertical axis at the angular velocity ω_2. Using the Euler formula, we write out the final result:

$$I(\boldsymbol{\omega}_2 \times \boldsymbol{\omega}_1) = \mathbf{M}_O.$$

Figure E2.51. Directions of the angular velocity of the gyro proper rotation ω_1 (along the gyro axis) and the angular precession velocity ω_2.

Since the intrinsic angular momentum $I\omega_1$ is known, the last expression allows one to solve the following two problems:

(1) If $\mathbf{M}_O$ is known, find ω_2, i.e., the angular precession velocity.

(2) Conversely, if ω_2 is known, find $\mathbf{M}_O$, i.e., the moment of forces responsible for the precession.

E2.9. Elementary Theory of Impact

As a rule, in the motion of a mechanical system, in each successive small time interval τ, the variations in the velocities of points of the system are also small and have the same order of magnitude. But in several cases, one can observe the phenomenon of *impact* in which the velocities of several points can vary by a constant value in the time interval from t_0 to $t_0 + \tau$. This occurs either in the case of instantaneous imposition of a mechanical constraint (for example, when a body falls on a massive plate) or in the case of instantaneous removal of a constraint (for example, when a flying missile bursts).

For such a point, we write the theorem on the momentum in the form $d\mathbf{Q} = \sum_k d\mathbf{S}_k$ or $d(m\mathbf{V}) = \sum_k \mathbf{F}_k \, dt$, perform integration, and obtain

$$\int_{\mathbf{V}_0}^{\mathbf{V}_1} d(m\mathbf{V}) = \int_{t_0}^{t_0+\tau} \sum_k \mathbf{F}_k \, dt \qquad \text{or} \qquad m(\mathbf{V}_1 - \mathbf{V}_0) = \sum_k \widetilde{\mathbf{F}}_k \tau.$$

Here $\mathbf{V}_0$ and $\mathbf{V}_1$ are the velocities of the point before and after the impact, and $\widetilde{\mathbf{F}}_k$ are the average forces acting on the point on impact.

Since the left-hand side of the last equation is a finite quantity, it follows that at least one of the quantities $\widetilde{\mathbf{F}}_k$ must be of the order of $1/\tau$ on impact; i.e., since τ is small, it must be much larger than the "ordinary" forces acting on the point. In what follows, such forces are called *impact forces*.

Since the impact processes are very cumbersome, the law of variation of the impact force is, as a rule, unknown. It follows from general considerations that the graph of the impact force magnitude has the shape shown in Fig. E2.52, where the average impact force magnitude is given.

Example 1. Approximate estimates of the impact force in the case of fall of a stone of weight 1 kg on a plate from the height of 10 m give a quantity of the of order of 1.5 tons under the assumptions that the stone

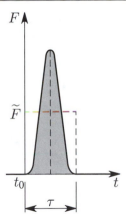

Figure E2.52. Qualitative graph of the impact force magnitude and computation of the average impact force.

after fall does not bounce and $\tau = 0.001$ s. Judging from the figure, the maximum magnitude of the impact force in the time of impact is at least twice larger, i.e., attains the value of the order of 3 tons and exceeds the magnitude of the "ordinary" force of weight approximately by a factor of 3000.

With the above explanations taken into account, the theorem on the momentum of a point can be stated as

$$\mathbf{Q}_1 - \mathbf{Q}_0 = \sum \mathbf{S}_k;$$

i.e., the variation of the momentum of a point in the time of impact is equal to the sum of impact momenta.

The *impact momentum* is the vector $\mathbf{S}_k = \int_{t_0}^{t_0+\tau} \mathbf{F}_k \, dt$. The statement of the theorem contains only the momenta of impact forces, because the momenta of "ordinary" forces can be neglected as being of small order of magnitude.

At the time of impact, the velocity of the point varies according to the relation

$$m\dot{\mathbf{r}} - m\mathbf{V}_0 = \sum \mathbf{S}_k,$$

where $\mathbf{r}$ is the position vector of the point with respect to a fixed center O.

Separating the variables and integrating over the time of impact, we obtain

$$\int_{\mathbf{r}^\circ}^{\mathbf{r}} m \, d\mathbf{r} = \int_{t_0}^{t_0+\tau} \left(m\mathbf{V}_0 + \sum \mathbf{S}_k \right) dt,$$

or

$$m(\mathbf{r} - \mathbf{r}^\circ) = \left(m\mathbf{V}_0 + \sum \widetilde{\mathbf{S}}_k \right) \tau,$$

where $\mathbf{r}^\circ$, $\mathbf{r}$, $\widetilde{\mathbf{S}}_k$ are the position vectors of the point at the beginning and at the end of the impact and the average value of the impact momentum. The right-hand side of the equation is a small value of the order of τ, which implies that $\mathbf{r} = \mathbf{r}^\circ$; i.e., the displacement of the point in the time of impact can be neglected.

The last fact allows us to obtain the following theorem.

THEOREM ON THE VARIATION IN THE MOMENT OF MOMENTUM OF A POINT ON IMPACT. *In the time of impact, the variation in the moment of momentum of a point about any fixed center is equal to the sum of moments of impact momenta about the same center:*

$$\mathbf{r} \times \mathbf{Q}_1 - \mathbf{r}^\circ \times \mathbf{Q}_0 = \mathbf{r}^\circ \times \sum \mathbf{S}_k \qquad or \qquad \mathbf{K}_o - \mathbf{K}_o^\circ = \sum \mathbf{M}_o \mathbf{S}_k.$$

THEOREM ON THE MOMENTUM OF A MECHANICAL SYSTEM ON IMPACT. *In the time of impact, the variation in the momentum of a system is equal to the sum of external impact momenta:*

$$\mathbf{Q} - \mathbf{Q}^\circ = \sum \mathbf{S}_k^{\mathrm{e}}.$$

THEOREM ON THE ANGULAR MOMENTUM OF A MECHANICAL SYSTEM ON IMPACT. *In the time of impact, the variation in the angular momentum of the system about any fixed center O is equal to the sum of moments of external impact momenta about the same center:*

$$\mathbf{K}_o - \mathbf{K}_o^\circ = \sum \mathbf{M}_o \mathbf{S}_k^{\mathrm{e}}.$$

In collision of two bodies, at the point of contact, there arise impact forces applied to each body (Fig. E2.53). These forces (and hence the impact momenta) are equal and opposite, $\mathbf{S}_1 = -\mathbf{S}_2$.

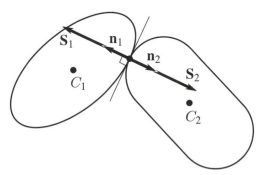

Figure E2.53. Collision of two bodies.

We assume that the forces of friction are negligibly small; in this case, the impact momenta are directed along the common normal to the surfaces of colliding bodies.

As observations show, the hypothesis on the absolute rigidity of bodies under the conditions in question is not satisfied, and therefore, it is necessary to take into account some additional properties.

The impact process can be divided into two phases, the initial and the final. In the initial phase, some constraints are imposed instantaneously, and the deformable bodies approach each other along the line of the common normal. In the final phase of the impact, the constraints are removed instantaneously, the bodies remove from each other and completely or partially restore their original shape.

The complete investigation of the phenomenon under study is far beyond the framework of theoretical mechanics. But here it is also possible to obtain results that are in good agreement with practice if the *Newton's hypothesis* is accepted.

We assume that, in the time of impact, the site of contact of bodies is small and can be assumed to be a point. We measure the relative velocity of approach of the points of contact of the bodies directly before impact and calculate the absolute value of its projection V' on the common normal to the surfaces of the colliding bodies. We perform similar actions when the points of contact separate immediately after impact, and obtain the values of V''.

According to Newton's hypothesis, the quantity $\varepsilon = V''/V'$ characterizes the properties of colliding bodies and is independent of their masses and relative velocity.

This hypothesis can be stated in a different way: the ratio of the magnitudes of momenta in the first and second phases of impact is a constant value for colliding bodies and is a quantity independent of masses and velocities of the bodies.

The physical constant ε is called the *coefficient of elasticity*. It follows from physical considerations that $0 \le \varepsilon \le 1$. For $\varepsilon = 1$, the impact is said to be *absolutely elastic*, for $\varepsilon = 0$, *absolutely inelastic*.

Example 2. The values of the coefficient of elasticity can be obtained experimentally. A heavy ball raised to a height h_0 over a massive horizontal plate is lowered without initial velocity, and the maximal height h_1 of its raising after rebound is measured (Fig. E2.54).

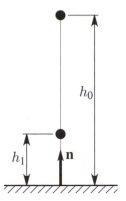

Figure E2.54. Experimental determination of the coefficient of elasticity by measuring the height of the ball rebound from the surface.

Neglecting the air resistance, we determine the velocity of fall on the plate V_0 and the rebound velocity V_1 from the relations $V_0 = \sqrt{2gh_0}$ and $V_1 = \sqrt{2gh_1}$. As a result, the formula for the coefficient of elasticity acquires the form

$$\varepsilon = \frac{V_1}{V_0} = \sqrt{\frac{h_1}{h_2}}.$$

Example 3 (Direct central impact of two bodies). The impact of two bodies in which the vectors of internal impact momenta and the velocities of the centers of mass are directed along the straight line connecting the centers of mass is called a *direct central impact* (Fig. E2.55). We assume that the velocities of the centers of mass of the bodies before impact and the coefficient of elasticity are known. It is required to find the velocities of centers of mass of the bodies after impact and the impact momenta.

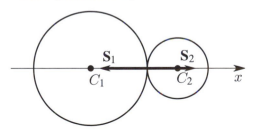

Figure E2.55. Direct central impact of two moving bodies (of different masses).

Solution. We denote the velocities of the centers of mass of the bodies before impact by V_1 and V_2 and after impact by V_1' and V_2'. Since the first body overtakes the second body, we have $V_1 > V_2$, $V_1' \leq V_2'$ before impact and $\frac{V_2' - V_1'}{V_1 - V_2} = \varepsilon$ after impact.

Since the acting impact momenta are internal, the law of conservation of momentum on impact holds, which, in projections onto the direction of the x-axis, gives the first equation

$$m_1 V_1 + m_2 V_2 = m_1 V_1' + m_2 V_2'.$$

In addition, for a known coefficient of elasticity, there is a second equation relating the unknowns V_1' and V_2':

$$\varepsilon(V_1 - V_2) = V_2' - V_1'.$$

By solving the resulting system of two equations, we obtain

$$V_1' = V_1 + \frac{m_2(1+\varepsilon)(V_2 - V_1)}{m_1 + m_2}, \quad V_2' = V_2 + \frac{m_1(1+\varepsilon)(V_1 - V_2)}{m_2 + m_1}.$$

Now we find the impact momentum S_1 applied to the first body:

$$S_1 = m_1(V_1' - V_1) = -\frac{m_1 m_2(1+\varepsilon)(V_1 - V_2)}{m_1 + m_2}.$$

The impact momentum S_2 applied to the second body is related to S_1 as $S_2 = -S_1$.

Example 4 (Action of the impact on a body rotating about a fixed axis). To a body rotating with angular velocity ω about the vertical fixed axis AB, the impact momentum $\mathbf{S}$ is applied at point M (Fig. E2.56). In cylindrical hinges A, B and the thrust bearing, there arise impact momenta $\mathbf{S}_{Ax}, \mathbf{S}_{Ay}, \mathbf{S}_{Az}, \mathbf{S}_{Bx}, \mathbf{S}_{By}$, which must be found under given dynamic characteristics of the body.

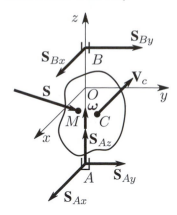

Figure E2.56. Impact on a body rotating about a fixed axis.

Solution. To make the further calculations more convenient, we place the coordinate axes so that the axis Oz coincides with the axis of rotation ($x_A = y_A = x_B = y_B = 0$; $\omega_x = \omega_y = 0$) and the center of mass of the body C lie in the plane yAz ($y_C = 0$); in addition, we choose the origin, i.e., point O, so that the point of application of the impact momentum M lies in the plane xOy ($z_M = 0$).

The theorem on the variation in the momentum of the body and the theorem on the variation in its angular momentum about the center O acquire the form

$$m(\mathbf{V}'_C - \mathbf{V}_C) = \mathbf{S} + \mathbf{S}_A + \mathbf{S}_B,$$

$$\mathbf{K}'_O - \mathbf{K}_O = \mathbf{r}_M \times \mathbf{S} + \mathbf{r}_A \times \mathbf{S}_A + \mathbf{r}_B \times \mathbf{S}_B,$$

where $\mathbf{S}_A$ and $\mathbf{S}_B$ are the momenta of the constraint reaction forces, $\mathbf{r}_M$ is the position vector of point M, and $\mathbf{V}_C$, $\mathbf{K}$ and $\mathbf{V}'_C$, $\mathbf{K}'_A$ are the velocity of the center of mass and the angular momentum of the body before and after the impact.

To two vector equations, there corresponds a system of six scalar equations

$$S_x + S_{Ax} + S_{Bx} = m y_C(\omega - \omega'),$$

$$S_y + S_{Ay} + S_{By} = 0,$$

$$S_z + S_{Az} = 0,$$

$$-I_{xz}(\omega' - \omega) = y_M S_z - z_A S_{Ay} - z_B S_{By},$$

$$-I_{yz}(\omega' - \omega) = -x_M S_z + z_A S_{Ax} + z_B S_{Bx},$$

$$I_{zz}(\omega' - \omega) = x_M S_y - y_M S_x,$$

with six unknowns ω', S_{Ax}, S_{Ay}, S_{Az}, S_{Bx}, S_{By}.

From the sixth equation, we obtain the body angular velocity after the impact:

$$\omega' = \omega + \frac{1}{I_{zz}}(x_M S_y - y_M S_x),$$

and then find the impact momenta of the constraint reactions.

Now we obtain the conditions under which there are no impact momenta of the constraint reactions. In this case, the point of application of the impact momentum is called the *center of impact*.

If there are no impact momenta of the supports $S_{Ax} = S_{Ay} = S_{Az} = S_{Bx} = S_{By} = 0$, then the system of the above-obtained equations becomes

$$S_x = m y_{\tilde{N}}(\omega' - \omega),$$

$$S_y = 0,$$

$$S_z = 0,$$

$$-I_{xz}(\omega' - \omega) = y_M S_z,$$

$$-I_{yz}(\omega' - \omega) = -x_M S_z,$$

$$I_{zz}(\omega' - \omega) = x_M S_y - y_M S_x.$$

This implies that the supports do not experience impact loads if

1. The external impact momentum is perpendicular to the plane passing through the center of mass and the axis of rotation of the body, because it follows from the second and third equations that $S_y = 0$ and $S_z = 0$.

2. The axis of the body rotation is a principal axis of inertia for the point O at which it intersects the orthogonal plane containing the point of application of the impact momentum, because it follows from the fourth and fifth equations that $I_{xz} = I_{yz} = 0$.

3. The position of the impact center is determined by the formula $y_M = -I_{zz}.my_C$, which follows from the first and sixth equations.

The theorem on the variation in the kinetic energy of a material system on impact (the Carnot theorem) is formulated for two cases: instantaneous imposition of constraints and instantaneous removal of constraints.

THEOREM (CARNOT).

(i) *The kinetic energy lost by the system in instantaneous imposition of absolutely inelastic constraints on it is equal to the kinetic energy that the system would have if its points moved with the lost velocities:*

$$T - T' = \sum_\nu \frac{1}{2} m_\nu (\mathbf{v}_\nu - \mathbf{v}'_\nu)^2;$$

this is the case in which the impact has only the first phase.

(ii) *The kinetic energy acquired by the system in instantaneous removal of constraints is equal to the kinetic energy that the system would have if its points moved with acquired velocities:*

$$T' - T = \sum_\nu \frac{1}{2} m_\nu (\mathbf{v}'_\nu - \mathbf{v}_\nu)^2;$$

this is the case where the impact has only the second phase.

Bibliography for Chapter E2

Arnold, V. I., *Mathematical Methods in Classical Mechanics, 2nd Edition*, Springer-Verlag, New York, 1989.

Bat, M. J., Ganelidze, G. U., and Kelson, A.S., *Theoretical Mechanics in Examples and Problems, Vols. 1 and 2* [in Russian], Nauka, Moscow, 1990 and 1991.

Bottema, O. and Roth, B., *Theoretical Kinematics*, Dover Publications, New York, 1990.

Bukhgolts, N. N., *Basic Course of Theoretical Mechanics, Parts 1 and 2* [in Russian], Nauka, Moscow, 1972.

Butenin, N. V. and Fufaev, N. A., *Introduction to Analytical Mechanics* [in Russian], Nauka, Moscow, 1991.

Gantmakher, F. R., *Lectures in Analytical Mechanics*, Mir Publ., Moscow, 1975.

Greenwood, D. T., *Classical Mechanics* [reprint of 1977 ed.], Courier Dover Publ., New York, 1997.

Gregory, D., *Classical Mechanics: An Undergraduate Text*, Cambridge University Press, Cambridge, England, 2006.

Hibbeler, R. C., *Engineering Mechanics, Static, 11th Edition*, Prentice Hall, Upper Saddle River, New Jersey, 2006.

Kibble, T. W. B. and Berkshire, F. H., *Classical Mechanics*, Imperial College Press, England, 2004.

Landau, L. D. and Lifshitz, E. M., *Mechanics, 3rd Edition*, Butterworth-Heinemann, Oxford, England, 1976.

Lurie, A. I., *Analytical Mechanics*, Springer, Berlin, 2002.

Marion, J. B. and Thornton, S. T., *Classical Dynamics of Particles and Systems, 5th Edition*, Brooks Cole, Pacific Grove, California, 2003.

Mescherskii, I. V., *A Collection of Problems of Mechanics* [in Russian], Nauka, Moscow, 1986.

Morin, D., *Introduction to Classical Mechanics: with Problems and Solutions*, Cambridge University Press, Cambridge, England, 2008.

Targ, S., *Theoretical Mechanics: a Short Course*, Gordon and Breach, New York, 1967.

Taylor, J. R., *Classical Mechanics*, University Science Books, Sausalito, California, 2005.

Chapter E3
Elements of Strength of Materials

E3.1. Basic Notions

E3.1.1. Introduction. External and Internal Forces

▶ **Deformation, strength, and rigidity.** The strength of materials is the field of mechanics where design and analysis of structural members for strength, rigidity, and stability are considered.

The strength of materials is based on the knowledge accumulated in theoretical mechanics. The difference is that the object in theoretical mechanics is an absolutely rigid body, but the strength of materials deals with deformable rigid bodies.

In practice, the actual members of machines and structures experience the action of different forces. Under the action of these forces, the bodies are *deformed*; i.e., the mutual position of the material particles is changed. If the forces are sufficiently large, then the body fracture may occur.

The ability of a body to take loads without fracture and large deformations is called the *strength* and *rigidity*, respectively.

Several equilibria of bodies and structures are *unstable*; i.e., in these states, any negligible actions of random character, as a rule, may lead to significant deviations from these states. But if the deviations are also small, then such equilibria are said to be *stable*.

▶ **External forces.** The external forces acting on a structure comprise the active forces (loads) and the reactions of external constraints. Several types of loads are distinguished.

Lumped force applied at a point. They are introduced instead of actual forces acting on a small site of the surface of a structural member whose dimensions can be neglected.

Distributed forces. For example, the forces of fluid pressure on the vessel bottom are loads distributed over the surface and are measured in N/m^2, and the weight forces are loads distributed over the volume and measured in N/m^3. In several cases, a load distributed over a line is introduced; the intensity of such a load is measured in N/m.

One version of the load is the *lumped torque* (couple of forces).

▶ **Internal forces in a rod.** The rod is the most widespread structural member; therefore, the main attention in the strength of materials is paid to the rod.

The longitudinal axis and the transverse cross-section are the basic geometric elements of the rod. It is assumed that the rod transverse cross-sections are perpendicular to the longitudinal axis, and the longitudinal axis passes through the centers of gravity of the transverse cross-sections.

The *internal forces* of a rod are the forces of interaction between its separate parts arising under the action of external forces (it is assumed that, in the absence of external forces, the internal forces are zero).

We consider a rod in equilibrium under the action of a system of external forces (Fig. E3.1a). We mentally draw an arbitrary transverse cross-section dividing the rod into two parts L and R. The right part R is subjected to a system of forces exerted the left part L and distributed over the surface of the transverse cross-section, i.e., a system of internal forces with respect to the entire rod. This system can be reduced to the resultant vector $\mathbf{R}$ and the resultant moment $\mathbf{M}$ if the center of gravity of the cross-section, i.e., point O, is taken to be the center of reduction.

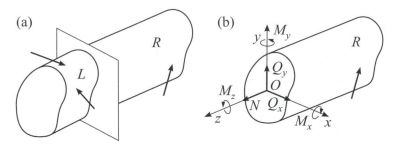

Figure E3.1. Internal forces in a rod: (a) (imaginary) cross-section of the rod, (b) internal forces and moments at the cross-section.

▶ **Internal force factors.** We choose the coordinate system so that the axes x, y lie in the transverse cross-section and the axis z is perpendicular to it and decompose $\mathbf{R}$ and $\mathbf{M}$ into components in these axes: Q_x, Q_y, N and M_x, M_y, M_z (Fig. E3.1b).

These six quantities are called *internal force factors of the rod (or internal forces)* in the cross-section under study. Each of these forces has its own name corresponding to its direction or the specific form of the rod deformation caused by these forces. The forces Q_x and Q_y are called the *transverse (cross-axis) forces*, and N is called the *normal (longitudinal) force*. The moments M_x and M_y are called the *bending moments*, and M_z is called the *twisting moment*.

▶ **Method of sections.** Since the cut-off part of the rod R is in equilibrium, it is possible to compose six static equations for the forces acting on this part, which determine all six internal force factors:

$$\sum F_x = 0, \qquad \sum F_y = 0, \qquad \sum F_z = 0,$$
$$\sum m_x(\mathbf{F}) = 0, \qquad \sum m_y(\mathbf{F}) = 0, \qquad \sum m_z(\mathbf{F}) = 0.$$

This method for determining the internal forces is called the *method of sections*.

On the basis of the law of action and reaction, the right-hand part of the rod acts on the left-hand part with equal but opposite forces; therefore, they can also be determined starting from the fact that the rod part L is in equilibrium.

E3.1.2. Stresses and Strains at a Point

▶ **Stresses.** The *stress vector p* is the intensity of internal forces distributed over the cross-section, at a point of the cross-section (Fig. E3.2). Its components lying in the plane of the cross-section are called *tangential (shear) stresses* τ_x, τ_y, and the component perpendicular to the cross-section is called the *normal stress* σ. The stresses are measured in N/m^2 (Pa) and depend not only on the choice of the point but also on the orientation of the cross-section (or of the site dA, Fig. E3.2) passing through this point. The complete set of stresses at a given point for different sites is called the *stress state* at this point.

If the stresses in the rod cross-section are known, then its internal force factors can be found by the formulas

$$N = \int_A \sigma \, dA, \qquad Q_x = \int_A \tau_x \, dA, \qquad Q_y = \int_A \tau_y \, dA,$$

$$M_x = -\int_A \sigma y \, dA, \qquad M_y = -\int_A \sigma x \, dA, \qquad M_z = \int_A (\tau_y x - \tau_x y) \, dA,$$

$$\text{(E3.1.2.1)}$$

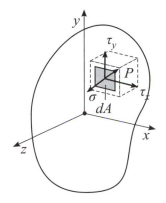

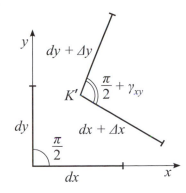

Figure E3.2. Tangential and normal stresses.   **Figure E3.3.** Deformation at a point K.

where the integrals are taken over the entire cross-section area A.

▶ **Strains.** Consider an arbitrary point K of the body in the initial unstrained state and draw two infinitely small segments of lengths dx and dy through this point in the direction of the axes x and y (Fig. E3.3).

Under the action of the load, the body is strained, the point K moves to the point K', the segments dx and dy change their lengths by Δx and Δy, and the angle $\pi/2$ between them varies by γ_{xy}.

The ratios

$$\varepsilon_x = \frac{\Delta x}{dx}, \quad \varepsilon_y = \frac{\Delta y}{dy}$$

are called the *linear strains* at point K in the direction of the axes x and y, respectively, and the quantity γ_{xy} is called the *angular strain* (shear angle) at point K between the axes x and y. The complete set of linear and angular strains in different directions passing through the point under study is called the *strain state* at this point.

E3.1.3. Basic Notions and Hypotheses

▶ **Elasticity** is the property of a body that is expressed as a unique dependence between the forces acting on the body and the body strains. In particular, an elastic body returns to the initial state after the loads are removed.

In several cases, after the loads are removed, only part of the total strain disappears, and this part is said to be *elastic*. The remaining part, the so-called *residual* (*plastic*) strain, is related to the body property called *plasticity*.

▶ **Basic hypotheses accepted in the strength of materials:**

1. The bodies are assumed to be solid (without hollows) and homogeneous; i.e., the properties of the body material are the same at different points.

2. The body material is *isotropic*; i.e., its properties are the same in all directions. (In several cases, it is necessary to abandon this assumption for *anisotropic* bodies, whose material properties are different in different directions. For example, the properties of wood in the directions along the fibers and across them are different.)

3. The body strains at each point are directly proportional to the stresses at this point. This property is called the *linear elasticity law* or *Hooke's law*.

4. It is assumed that the strains in the body, as well as the displacements of its particles, are small compared with the geometric dimensions of the body itself.

5. The *principle of independence of forces* (principle of superposition) holds, according to which any quantity a depending on the action of several forces is equal to the sum of

quantities a_i determined for each separate force. This principle follows from hypotheses 3 and 4.

E3.2. Stress–Strain State at a Point

E3.2.1. Types of Stress State

▶ **Principal stresses and principal directions.** An analysis of the stress state at a point K of the body becomes more convenient if we mentally separate an elementary parallelepiped around this point and consider the stresses acting on its faces. Since the parallelepiped is small, we can assume that the stress state is the same at all of its points (is homogeneous) and coincides with the stress state at point K under study. Varying the orientation of this parallelepiped, we can make all its faces free of shear stresses. The corresponding normal stresses are called the *principal stresses*, and their directions are called the *principal directions* at point K.

The *linear, plane, volumetric (uniaxial, biaxial, and triaxial)* stress states at a point are distinguished depending on whether the parallelepiped oriented in the principal directions experiences extension (or compression) respectively in one, two, or three mutually perpendicular directions.

▶ **Plane stress state. Reciprocity law.** In what follows, because of its special importance, we consider only the case of plane stress state in which there are no stresses on two opposite faces of an elementary parallelepiped. The other faces, in the general case of their orientation, are under the action of the shear and normal stresses (Fig. E3.4). Since the parallelepiped stress state is homogenous, it follows that the similar stresses on opposite faces are numerically equal. Obviously, the normal forces are mutually balanced on the parallelepiped faces. The tangential forces on the same faces form two couples of forces with moments $(\tau_y\, dy\, dx)\, dz$ and $(\tau_z\, dz\, dx)\, dy$ of opposite direction, where dx, dy, dz—are the parallelepiped dimensions. These moments must be balanced, which implies $\tau_z = \tau_y$.

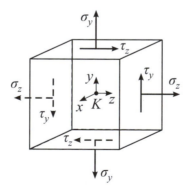

Figure E3.4. Plane stress state at a point.

The last equality expresses the *reciprocity law for shear stresses*: on any mutually perpendicular sites, the shear stresses are equal in value and directed so that they tend to rotate the element in opposite directions (Fig. E3.4). The aforesaid allows one to introduce the unique notation: $\tau = \tau_z = \tau_y$.

E3.2.2. Uniaxial Tension and Compression

▶ **Modulus of elasticity of the first kind.** We consider an elementary parallelepiped in the state of uniaxial tension or compression with only normal stresses acting on its two opposite faces (Fig. E3.5).

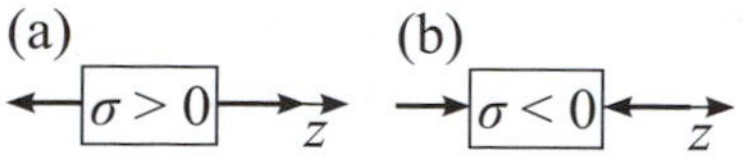

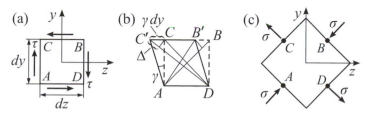

Figure E3.5. (a) Uniaxial tension and (b) uniaxial compression.

Rule of signs: the tensile stresses are assumed to be positive, and the compressing stresses are assumed to be negative. According to Hooke's law, the stress and the linear strain in the direction of the axis z are directly proportional:

$$\sigma = E\varepsilon_z. \tag{E3.2.2.1}$$

The proportionality coefficient E is called the *modulus of elasticity of the first kind* and has the dimension of stress, because ε_z is a dimensionless variable. For $\sigma > 0$ (tension), ε_z is also greater than zero, which corresponds to the parallelepiped elongation in the direction of the axis z. For $\sigma < 0$ (compression), we have $\varepsilon_z < 0$, which corresponds to the parallelepiped shortening.

▶ **Poisson coefficient.** Because of symmetry, the linear strains in the directions of the axes x and y are equal to each other and also proportional to the stress, or, with (E3.2.2.1) taken into account, to the strain ε_z:

$$\varepsilon_x = \varepsilon_y = -\nu\varepsilon_z = -\frac{\nu\sigma}{E}. \tag{E3.2.2.2}$$

The constant dimensionless proportionality coefficient ν is called the *Poisson ratio*, and the minus sign takes into account the opposite signs of the longitudinal strain ε_z and transverse strains ε_x and ε_y. In particular, for $\sigma > 0$ and $\varepsilon_z > 0$, the quantities ε_x and ε_y are less than zero, and hence the parallelepiped transverse dimensions decrease.

The constant quantities E and ν characterize the elastic properties of a specific material; for example, for steel we have $E = 210$ GPa, $0.25 \leq \nu \leq 0.30$.

E3.2.3. Simple Shear

▶ **Simple shear.** Consider a specific form of plane stress state in which the lateral faces of an elementary parallelepiped are only under the action of shear stresses τ (Fig. E3.6a). By the reciprocity law, the stresses on neighboring faces are the same. Such a form of stress state is called the *simple shear*.

Figure E3.6. Pure shear: (a) initial state (tangential stresses are only acting on the lateral face), (b) deformation of the elementary parallelepiped, (c) normal stresses acting on the elementary parallelepiped rotated by 45°.

The character of the parallelepiped strain is shown in Fig. E3.6b. Initially, the right angles between the lateral faces vary by the quantity γ, which is called the shear angle, and the linear strains ε_x, ε_y, ε_z are zero.

▶ **Shear modulus.** In this case, Hooke's law has the form

$$\tau = G\gamma. \tag{E3.2.3.1}$$

The proportionality coefficient G is called the *shear modulus* or the *modulus of elasticity of the second kind*. The modulus dimension coincides with the dimension of stresses.

It is easy to see that, for $dy = dz$, the diagonal segment AB in the parallelepiped becomes shorter and the segment CD elongates by the same value $\Delta = \gamma\, dy \sin 45°$ (Fig. E3.6b); therefore, the other elementary parallelepiped, distinguished in a neighborhood of the same point and rotated by the angle $45°$ with respect to the first parallelepiped, must be under the action of the corresponding normal stresses σ (Fig. E3.6c). In this case, there are already no shear stresses, and hence the normal stresses σ are the principal stresses, and their directions are the principal directions for a given stress state.

▶ **Relation between G, E, and ν.** Consider the element formed by the diagonal cut of the parallelepiped (see Fig. E3.7). Its faces are under the action of forces whose values are equal to $\tau\, dx\, dy$, $\tau\, dx\, dz$, and $\sigma\, dx\, dz/\cos 45°$.

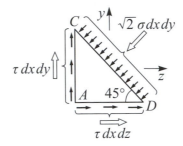

Figure E3.7. Equilibrium of the element formed by the diagonal cut of the parallelepiped.

Since this element is in equilibrium, we obtain

$$\sigma = \tau. \tag{E3.2.3.2}$$

With (E3.2.3.1) and (E3.2.3.2) taken into account, the strain of segment CD (Fig. E3.6b) is equal to $\varepsilon_{CD} = \Delta/CD = \frac{1}{2}\gamma = \frac{1}{2}\tau/G = \frac{1}{2}\sigma/G$.

On the other hand, treating the state in Fig. E3.6c as a combination of two uniaxial stress states (extension and compression in perpendicular directions), we obtain $\varepsilon_{CD} = \dfrac{\sigma}{E} + \dfrac{\nu\sigma}{E}$ form Hooke's law (E3.2.2.1), (E3.2.2.2). This implies $\dfrac{\sigma}{2G} = \dfrac{\sigma}{E} + \dfrac{\nu\sigma}{E}$ or $G = \dfrac{E}{2(1+\nu)}$.

Thus, the shear modulus G is determined via the already introduced constants E and ν.

E3.3. Central Tension and Compression

E3.3.1. Longitudinal Force

The *central tension (compression)* is a type of a rod deformation such that only a longitudinal force N arises in the rod transverse cross-sections and all the other internal forces are zero.

Rule of signs: the tensile longitudinal forces are assumed to be positive, and the compressing longitudinal forces are assumed to be negative.

Example. In Fig. E3.8a we show a rod loaded by two forces F_1 and F_2 lying on its axis. The longitudinal force will be determined by the method of sections (see Subsection E3.1.1). We draw an arbitrary cross-section I on segment "a" of the rod and consider the equilibrium of the free cut-off part (Fig. E3.8b). We replace the action of the cut-off part by an unknown force N_1 assuming that it is positive. The equilibrium equation for this cut-off part, stating that the sum of projections of all forces on the axis z is zero, i.e., $N_1 - F_1 = 0$, implies

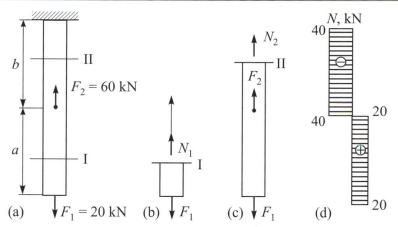

Figure E3.8. Central tension (compression) of a rod: (a) the rod with two forces applied, (b) equilibrium of the cut-off part to cross-section I, (c) equilibrium of the cut-off part to cross-section II, (d) longitudinal force distribution diagram.

that $N_1 = F_1 = 20$ kN. Similarly, for the cross-section II on segment "b" of the rod, we obtain (Fig. E3.8c) $N_2 + F_2 - F_1 = 0$, which implies $N_2 = F_1 - F_2 = -40$ kN.

Thus, the longitudinal force on segments "a" and "b" of the rod is different. Segment "a" of the rod experiences tension ($N > 0$), and segment "b" is under compression ($N < 0$). In transition from segment "a" to segment "b," the force N experiences a jump by the value of the force F_2.

To illustrate the character of variation in the longitudinal force along the rod, one usually constructs the graph of the function $N(z)$, which is called the axial force *diagram*. This graph presents numerical values of the longitudinal (axial) force in typical cross-sections and their signs. For the example under study, the diagram of $N(z)$ is shown in Fig. E3.8d.

E3.3.2. Stresses and Strains under Tension or Compression

▶ **Flat cross-section hypothesis.** The stresses under tension or compression will be determined by using the *flat cross-section hypothesis*: after strain, the rod transverse cross-sections remain plane and perpendicular to the rod axis. Experiments show that this hypothesis is violated only in the regions of the so-called local stresses, immediately near the points of application of external forces or sites of sharp variation in the area of the transverse cross-section, where the so-called *stress concentration* occurs. Taking them into account is a special problem, and we do not consider it here.

▶ **Formulas for stresses and strains.** Consider a rod under the action of two tensile forces (Fig. E3.9a). According to the flat cross-section hypothesis and the fact that the longitudinal force is constant along the rod ($N = F$), the stress–strain state of the rod points outside a neighborhood of the endpoints is homogeneous, and the normal stresses are uniformly distributed over the transverse cross-section (Fig. E3.9b).

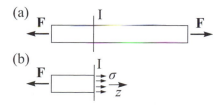

Figure E3.9. (a) A rod under tension by two forces and (b) uniform stress distribution across the cross-section.

It follows from formula (E3.1.2.1) that $N = \int_A \sigma\, dA = \sigma A$, which implies

$$\sigma = \frac{N}{A}. \tag{E3.3.2.1}$$

Formula (E3.3.2.1) permits calculating the normal stresses in the rod transverse cross-sections in the case of central tension or compression from the known longitudinal force N and the known area A of the transverse cross-section.

The elementary parallelepiped separated in the rod is with the conditions of uniaxial tension (Fig. E3.5a). Its strains are determined by Hooke's law (see Subsection E3.2.2). In particular, the longitudinal strain is

$$\varepsilon_z = \frac{\sigma}{E} = \frac{N}{EA}.$$

Summing the elongations of small elements over the entire length of the rod, we obtain its *absolute elongation*

$$\Delta l = \int_0^l \varepsilon_z\, dz = \varepsilon_z l = \frac{Nl}{EA},$$

where l is the rod length.

The quantity EA is called the *rigidity of the rod cross-section under tension and compression*.

E3.3.3. Strength Analysis under Tension and Compression

▶ **Admissible stress.** To ensure the strength of a rod under tension or compression, it is necessary to bound the maximum stresses of this rod by a value called the *admissible stress* $[\sigma]$. Its value is chosen on the basis of the unsafe stress value, which is introduced in advance for a given material, with the safety margin taken into account. For materials that differently resist tension and compression, the two different admissible stresses $[\sigma^t]$ and $[\sigma^c]$ are introduced.

▶ **Problems for strength analysis.** Thus, the strength condition in this case has the form

$$\sigma_{\max} = \frac{|N|}{A} \le [\sigma].$$

Using this condition, one can solve the following problems:

1. From given loads, cross-section dimensions, and the value of admissible stress, verify the rod strength.

2. From given loads and a known value of admissible stress, find the dimensions of the transverse cross-section:

$$A \ge \frac{|N|}{[\sigma]}.$$

3. From given cross-section dimensions and a known value of admissible stress, find the value of admissible longitudinal force:

$$N \le [\sigma]A.$$

Example. For a cast-iron rod (Fig. E3.8a) with cross-section area $A = 5\ \text{cm}^2$, it is necessary to verify the strength conditions for $[\sigma^t] = 30\ \text{MPa}$, $[\sigma^c] = 100\ \text{MPa}$.

Solution. On segment "*a*," the rod is under the action of the tensile force $N_1 = 20\ \text{kN}$ (see Fig. E3.8d); therefore, the normal stress in the cross-sections on this segment is $\sigma_1 = N_1/A = 40\ \text{MPa}$.

On segment "*b*," the rod is under compression, $N_2 = -40\ \text{kN}$. The corresponding stress is $\sigma_2 = |N_2|/A = 80\ \text{MPa}$.

Since the admissible stresses under tension and compression for cast iron are different, it is necessary to verify two conditions, $\sigma_1 \le [\sigma^t]$ and $\sigma_2 \le [\sigma^c]$. The first condition is not satisfied, and hence the strength condition is not satisfied in the whole.

E3.4. Torsion

E3.4.1. Twisting Moment

The *torsion* is a type of rod deformation such that only the twisting moment M_z arises in the rod transverse cross-sections and all the other internal force factors are zero.

Rule of signs: if observed on the side of the external normal to the cross-section, the twisting moment seems to be anticlockwise, then it is assumed to be positive.

Example. Consider a rod loaded by two moments m_1 and m_2 such that the planes of their action are perpendicular to the rod axis (Fig. E3.10a). According to the method of sections, we draw an arbitrary transverse cross-section I on segment "a" and consider the free cut-off part of the rod in equilibrium (Fig. E3.10b). We replace the action of the cut-off part by an unknown twisting moment M_{z1} assuming that its direction is positive. The equilibrium equation for the cut-off part, stating that the sum of moments of all forces about the axis z is zero, i.e., $M_{z1} - m_1 = 0$, implies that $M_{z1} = m_1 = 30\,\text{kN m}$. Similarly, for cross-section II we have (Fig. E3.10c) $M_{z2} + m_2 - m_1 = 0$, which implies $M_{z2} = m_1 - m_2 = -10\,\text{kN m}$.

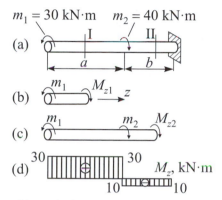

Figure E3.10. Twisting moment: (a) a rod with two moments applied, (b) equilibrium of the cut-off part to cross-section I, (c) equilibrium of the cut-off part to cross-section II, (d) twisting moment diagram.

In Fig. E3.10d, we present the twisting moment diagram, which illustrates the character of variation in M_z along the rod axis. On the boundary between segments "a" and "b" of the rod, the twisting moment experiences a jump by the value of the moment m_2.

E3.4.2. Stresses and Strains in Torsion

▶ **Stress state in torsion.** We consider only rods whose transverse cross-sections are of a disk or annulus shape.

To find the law of stress distribution over the cross-section, we assume, which is confirmed experimentally, that the result of deformation is that the rod transverse cross-sections in torsion rotate about the longitudinal axis as rigid disks (annuli). The applicability conditions for this assumption are similar to the applicability conditions of the flat cross-section hypothesis under tension or compression (see Subsection E3.3.2).

Consider a rod experiencing torsion under the action of two moments applied at its ends (Fig. E3.11a). We divide it into coaxial tubes (Fig. E3.11b) and distinguish one of them with the internal radius ρ and infinitely small thickness $d\rho$ (Fig. E3.11c).

By the above assumption about the rod strain character, we assume that an infinitely small element $ACBD$ of the tube (Fig. E3.11c) experiences a strain corresponding to simple shear (see Subsection E3.2.3). This implies that only the shear stresses τ act on the lateral faces of the element (see Fig. E3.6a).

▶ **Polar moment of inertia.** The shear angle of an element is $\gamma = BB'/dz = \rho\,d\varphi/dz$ (Fig. E3.11c), where $d\varphi/dz$ is the *running angle of torsion* of the tube or the angle of torsion

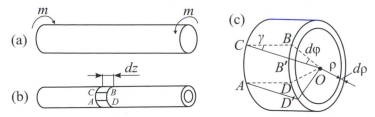

Figure E3.11. Stresses and strains in torsion: (a) a rod with two moments applied, (b) a tube element, (c) deformation of the tube element.

per unit length. It follows from formula (E3.2.3.1) (Hooke's law) that

$$\tau = G\gamma = G\rho\frac{d\varphi}{dz}. \tag{E3.4.2.1}$$

The value of the running angle of torsion $d\varphi/dz$ is the same for all coaxial tubes comprising the rod. Therefore, it follows from formula (E3.4.2.1) that the shear stresses in the rod cross-sections are directly proportional to the distance ρ to its axis (Fig. E3.12).

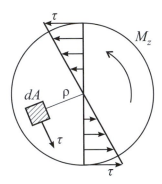

Figure E3.12. Tangential stresses in rod cross-sections.

The twisting moment M_z can be obtained by summing the moments of all stresses distributed over the cross-section about the axis z:

$$M_z = \int_A \tau\rho\, dA = GI_{\mathrm{p}}\frac{d\varphi}{dz}, \quad \text{where} \quad I_{\mathrm{p}} = \int_A \rho^2\, dA. \tag{E3.4.2.2}$$

The integral I_{p} over the cross-section surface is called the *polar moment of inertia of the cross-section* and is a geometric characteristic of this cross-section.

For a disk of diameter D, we have $I_{\mathrm{p}} = \frac{\pi}{32}D^4$.

For an annulus, we have $I_{\mathrm{p}} = \frac{\pi}{32}(D^4 - d^4)$, where d and D are the outer and inner diameters, respectively.

▶ **Shear stresses. Angle of torsion.** Using (E3.4.2.2), we obtain

$$\frac{d\varphi}{dz} = \frac{M_z}{GI_{\mathrm{p}}}. \tag{E3.4.2.3}$$

Substituting this expression into formula (E3.4.2.1) for τ, we obtain the basic formula for shear stresses in the rod cross-sections in torsion

$$\tau = \frac{M_z\rho}{I_{\mathrm{p}}}. \tag{E3.4.2.4}$$

The *angle of torsion* of the entire rod in Fig. E3.11a—i.e., the angle of mutual rotation of its end cross-sections—can be found if we know the running angle of torsion (E3.4.2.3) and take into account the fact that the value of M_z is constant along the rod axis ($M_z = m$):

$$\Delta\varphi = \int_0^l \frac{d\varphi}{dz} = \frac{M_z l}{G I_p},$$

where l is the rod length.

The quantity $G I_p$ is called the *cross-section rigidity in torsion* or *torsional rigidity*.

E3.4.3. Strength Analysis in Torsion

The value of the maximum shear stress in torsion is found from formula (E3.4.2.4) for $\rho = \rho_{max}$:

$$\tau_{max} = \frac{|M_z|\rho_{max}}{I_p} = \frac{|M_z|}{W_p}. \tag{E3.4.3.1}$$

The quantity $W_p = I_p/\rho_{max}$ is called the *polar moment of resistance of the cross-section*. For a disk, we have $W_p = \frac{\pi}{16} D^3$.

For an annulus, we have $W_p = \frac{\pi}{16} D^3 (1 - d^4/D^4)$.

The strength condition is reduced to the inequality $\tau_{max} = |M_z|/W_p \leq [\tau]$, where $[\tau]$ is the admissible stress in torsion.

Just as in case of strength calculated under tension and compression, the following three types of problems are possible, which differ in the form of the strength condition.

1. Checking calculation: $|M_z|/W_p \leq [\tau]$.
2. Choosing the cross-section: $W_p \geq |M_z|/[\tau]$.
3. Determining the admissible load: $|M_z| \leq [\tau]W_p$.

Example. For a steel rod of circular cross-section (Fig. E3.10a), choose the diameter from the strength condition for $[\tau] = 100$ MPa.

Solution. Since the cross-section is constant along the rod, the cross-sections are unsafe on segment "a," where the maximum twisting moment $M_z = 30$ kN m arises (see Fig. E3.10d).

From the strength condition

$$\tau_{max} = \frac{M_z}{W_p} = \frac{16 M_z}{\pi D^3} \leq [\tau]$$

we obtain $D \geq \sqrt[3]{16 M_z/(\pi[\tau])} = 0.119$ m.

Rounding up, we finally obtain $D = 12$ cm.

E3.5. Symmetrical Bending

E3.5.1. Bending Moment and Transverse Force

▶ **Definitions.** The *bending* is a type of rod deformation such that the bending moments M_x and M_y and, perhaps, the transverse forces Q_x and Q_y arise in the rod cross-sections and the other internal force factors (N and M_z) are zero.

If the transverse forces are zero, the bending is said to be *pure*, and if there are transverse forces, then we have the so-called *lateral bending*.

Consider the rods whose cross-sections have the axis of symmetry y. If such a rod is subjected to a system of external forces lying in the plane yz and perpendicular to the rod longitudinal axis (Fig. E3.13), then only two internal force factors—i.e., the bending

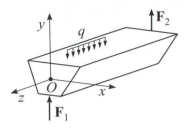

Figure E3.13. Symmetric bending.

moment M_x and the transverse force Q_y—arise in its cross-section. Such a bending is said to be *symmetric* (for the oblique bending, see Subsection E3.6.1).

Rule of signs: the bending moment is assumed to be positive if it results in extension of the lower fibers and compression of the upper fibers; the transverse force is positive if it is directed downwards acting on the cross-section from the right and is directed upwards acting on the cross-section from the left.

As was already noted, the rod internal force factors depend on the longitudinal coordinate of the cross-section under study; hence, they are functions of the variable z.

▶ **Differential equations of equilibrium of a rod.** Consider a rod under the action of a load q distributed along its axis (Fig. E3.14a). We distinguish a small element by two cross-sections (Fig. E3.14b) and replace the action of the cut-off parts of the rod by the corresponding internal forces with positive directions.

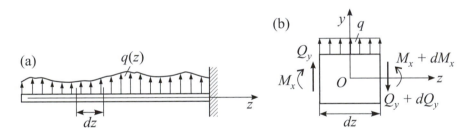

Figure E3.14. Rod equilibrium: (a) a rod under a distributed load (b) equilibrium of a small element.

Calculating the sum of projections of all forces on the axis y and the sum of moments of forces about the point O, we obtain the corresponding equations of equilibrium of the element:

$$q\,dz + Q_y - (Q_y + dQ_y) = 0,$$
$$-M_x + (M_x + dM_x) - Q_y\,dz = 0.$$

These equations imply the relations

$$\frac{dQ_y}{dz} = q, \qquad \frac{dM_x}{dz} = Q_y, \tag{E3.5.1.1}$$

which are called the *differential equations of equilibrium of a rod under bending*. Equation (E3.5.1.1), which must be satisfied by the functions $Q_y(z)$ and $M_x(z)$, is often used to construct diagrams of bending moments and transverse forces, and to verify this construction.

The rods that mainly work under bending are usually called *beams*.

▶ **Examples of diagram construction.** Consider examples of constructing diagrams of $Q_y(z)$ and $M_x(z)$ for several beams by using the method of sections.

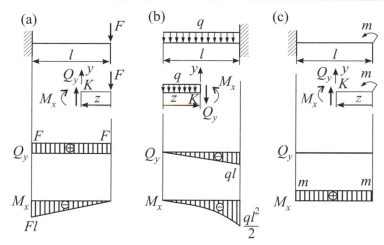

Figure E3.15. Cantilever beam (with clamped end): (a) loaded with a point force, (b) loaded with an evenly distributed force, (c) loaded with a moment.

In Fig. E3.15a, we show a cantilever beam (i.e., a beam with a fixed end) loaded by a lumped force. According to the method of sections, we draw an arbitrary cross-section K at a distance z from the free end of the beam and consider the free cut-off part in equilibrium. We replace the action of the cut-off part by the unknown forces Q_y and M_x assuming that their directions are positive. Calculating the sum of projections of all forces on the axis y and the sum of moments of forces about the point K, we compose the corresponding equations of equilibrium,

$$Q_y - F = 0, \quad -M_x - Fz = 0,$$

and whence obtain $Q_y = F$, $M_x = -Fz$. Thus, the transverse force is constant and the bending moment varies directly proportional to the distance z taking the values 0 and $-Fl$ at the free ($z = 0$) and fixed ($z = l$) ends of the beam, respectively. In Fig. E3.15a, we present the corresponding diagrams of Q_y and M_x.

The same procedure for a beam uniformly loaded by a distributed load (Fig. E3.15b) gives

$$-qz - Q_y = 0, \quad \tfrac{1}{2}qz^2 + M_x = 0,$$

which implies $Q_y = -qz$ and $M_x = -\tfrac{1}{2}qz^2$. In this case, the transverse force is a linear function of z, and the bending moment depends on z according to the quadratic parabola law, which agrees with the diagrams of Q_y and M_x in Fig. E3.15b.

Similarly, for a cantilever beam loaded by a moment (Fig. E3.15c), we obtain $Q_y = 0$, $M_x = m$.

For two-support beams (Fig. E3.16) that do not have free ends, it is necessary first to find the support reactions from the equilibrium conditions for the beam as a whole. After this, one can determine the internal forces by analogy with the above procedure.

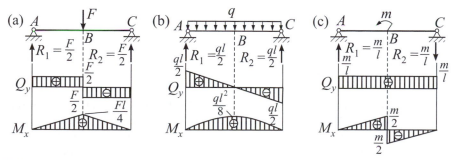

Figure E3.16. A beam hinged at both ends: (a) loaded with a point force, (b) loaded with an evenly distributed force, (c) loaded with a moment.

For an arbitrary cross-section on segment AB of a beam loaded at the center ($AB = BC = l/2$) by a lumped force F (Fig. E3.16a), we obtain $Q_y = R_1 = F/2$, $M_x = R_1 z = Fz/2$, where z is the distance from the left end A. For a cross-section on segment BC, we find $Q_y = R_1 - F = -F/2$, $M_x = R_1 z - F(z - l/2) = F(l - z)/2$. The corresponding diagrams are given at the bottom of Fig. E3.16a.

For the beam in Fig. E3.16b: $Q_y = R_1 - qz = q(l/2 - z)$, $M_x = R_1z - qz^2/2 = qz(l - z)/2$. At the middle of the beam, we have $M_x = ql^2/8$.

For segment AB of the beam in Fig. E3.16c, we have $Q_y = R_1 = m/l$, $M_x = R_1z = mz/l$, and for segment BC, we have $Q_y = R_1 = m/l$, $M_x = R_1z - m = m(z/l - 1)$.

▶ **Specific characteristics of diagrams of M_x and Q_y.** We briefly formulate the diagram qualitative characteristics, which follows from relations (E3.5.1.1).

1. On the beam part free from any load ($q = 0$), Q_y is constant, and M_x is a linear function of z.

2. On the segment of uniformly distributed load ($q = $ const), Q_y is a linear function, and M_x is a quadratic function (see Figs. E3.15b and E3.16b).

3. At the points of application of the lumped force, Q_y experiences a jump by the value of this force, and the diagram of M_x has a break point (Fig. E3.16a).

4. At the points of application of the lumped moment, M_x experiences a jump by the value of this moment (Fig. E3.16c).

At the extrema points of the function M_x, $Q_y = 0$ (Fig. E3.16b).

The knowledge of qualitative characteristics of diagrams allows one to construct diagrams without finding the functional dependencies $Q_y(z)$ and $M_x(z)$ but calculating the values of Q_y and M_x only on the boundaries of typical segments. The extremum values of M_x are determined after finding the extrema points from the conditions $Q_y = 0$.

E3.5.2. Stresses and Strains under Symmetric Pure Bending

▶ **Pure bending hypotheses.** Consider a beam in the state of pure bending (Fig. E3.17). In its cross-sections, there arises a single nonzero force factor, i.e., the constant bending moment $M_x = m$.

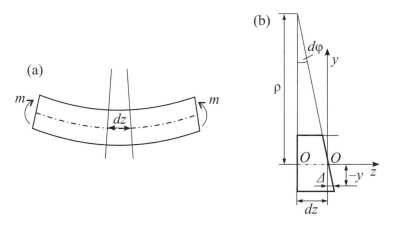

Figure E3.17. Simple pure bending of a beam: (a) deformation in bending, (b) deformation of a small element.

We use the *flat cross-section hypothesis*[*]: after the strain, the beam cross-sections plane before the strain remain plane and perpendicular to the strained longitudinal axis.

According to this hypothesis, two close cross-sections rotate about each other by an angle $d\varphi$ (Fig. E3.17b). As a result of such strain, there arises a *neutral layer* whose longitudinal fibers $O - O$ do not change their length. In the lower part of the cross-section, the longitudinal fibers elongate (in the upper part, become shorter) by the value $\Delta = -y\,d\varphi$. The longitudinal strain of a fiber is equal to $\varepsilon = \Delta/dz = -y\,d\varphi/(\rho\,d\varphi) = -y/\rho$, where ρ is the radius of curvature of the neutral layer.

---

[*] A similar hypothesis was used in Subsection E3.3.2.

In addition to the flat cross-section hypothesis, we assume that the beam longitudinal fibers do not exert their weight upon one another. This means that they are in the state of uniaxial tension or compression (see Fig. E3.5).

▶ **Axis curvature. Formula for normal stresses.** According to Hooke's law, we have

$$\sigma = E\varepsilon = -\frac{Ey}{\rho}. \tag{E3.5.2.1}$$

This implies that the normal stresses σ in the beam cross-section vary directly proportionally to the distance y from the neutral layer (Fig. E3.18).

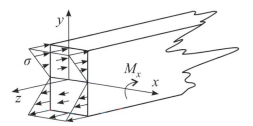

Figure E3.18. Normal stresses in the beam cross-section.

Using formulas (E3.1.2.1) and the fact that the longitudinal force is zero, we obtain

$$N = \int_A \sigma \, dA = -\frac{E}{\rho} \int_A y \, dA = 0,$$

$$M_x = -\int_A \sigma y \, dA = \frac{E}{\rho} \int_A y^2 \, dA.$$

It follows from the first relation that the *neutral axis* x of the cross-section passes through the center of gravity, and hence the quantity ρ is the radius of curvature of the deflected axis of the beam. From the second relation, we derive the formula for the axis curvature

$$k = \frac{1}{\rho} = \frac{M_x}{EI_x}, \quad \text{where} \quad I_x = \int_A y^2 \, dA. \tag{E3.5.2.2}$$

The quantity I_x is called the *moment of inertia of the cross-section about the axis x* (see Subsection E2.5.1 in the chapter "Mechanics of point particles and rigid bodies"), and the product EI_x is called the *cross-section rigidity in bending* or *flexural rigidity*.

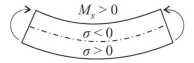

Figure E3.19. Signs of normal stresses in simple bending.

From (E3.5.2.1) and (E3.5.2.2) we obtain the basic formula for the rod normal stresses in the case of symmetric bending

$$\sigma = -\frac{M_x y}{I_x}, \tag{E3.5.2.3}$$

where the "minus" sign is necessary to match the rules of sign for the quantities σ and M_x. For $M_x > 0$ (Fig. E3.19), the upper fibers are compressed ($y > 0$, $\sigma < 0$), and the lower fibers are extended ($y < 0$, $\sigma > 0$); for $M_x < 0$, the signs of stresses in the upper and lower parts of the cross-section are changed to the opposite.

E3.5.3. Stresses and Strains in Symmetric Lateral Bending

▶ **Shear stresses. Zhuravskii formula.** In the case of symmetric lateral bending, in the rod cross-sections there arises not only a bending moment M_x but also a transverse force Q_y related to the shear stresses in the cross-section by the formula

$$Q_y = \int_A \tau_y \, dA.$$

Although there are shear stresses, formulas (E3.5.2.2) and (E3.5.2.3) obtained for the pure bending can also be used in this case.

To determine the shear stresses, consider the beam in Fig. E3.20a. We cut a small element from it by two neighboring cross-sections and one longitudinal cross-section (Fig. E3.20b). The normal stresses calculated by formula (E3.5.2.3) act on its lateral faces. Since the bending moment on the right is greater than the bending moment on the left by the value dM_x, the normal stress on the right is also greater by $d\sigma = dM_x y/I_x$.

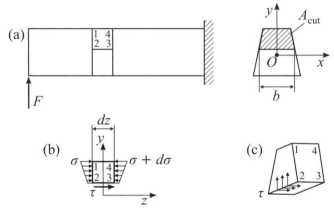

Figure E3.20. Symmetric lateral bending of a cantilever beam: (a) small element and its deformation, (b) normal and tangential stresses at the faces of the small element, (c) normal and tangential stresses in the cross-section.

For this element to be in equilibrium, the shear stresses τ must be applied to its lower face (Fig. E3.20b).

We assume that they are uniformly distributed over the width of the cross-section. From the equilibrium condition for an element stating that the sum of projections of all forces on the axis z is zero,

$$\tau b \, dz - \int_{A_{\text{cut}}} d\sigma \, dA = 0,$$

where the integral is extended to the cut-off part of the rod cross-section, we obtain

$$\tau = \frac{\int_{A_{\text{cut}}} d\sigma \, dA}{b \, dz} = \frac{dM_x}{dz} \frac{\int_{A_{\text{cut}}} y \, dA}{b I_x}.$$

Taking into account the second equation in (E3.5.1.1), we obtain the *Zhuravskii formula* for shear stresses:

$$\tau = \frac{Q_y S_x^{\text{cut}}}{I_x b}, \quad \text{where} \quad S_x^{\text{cut}} = \int_{A_{\text{cut}}} y \, dA. \qquad \text{(E3.5.3.1)}$$

The quantity S_x^{cut} is called the *static moment of the cross-section cut-off part about the axis x*.

According to the reciprocity law for shear stresses (see Subsection E3.2.1), formula (E3.5.3.1) also determines the shear stresses in the rod cross-section (Fig. E3.20c).

▶ **Cross-section of rectangular shape.** In the special case of the cross-section of rectangular shape (Fig. E3.21), we have

$$I_x = \tfrac{1}{12} b h^3, \quad S_x^{\text{cut}} = y_C \, A_{\text{cut}} = \tfrac{1}{2} b \left(\tfrac{1}{4} h^2 - y^2 \right).$$

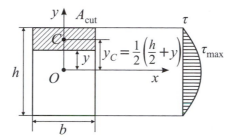

Figure E3.21. Rectangular cross-section and tangential stress diagram.

Substituting these expressions into formula (E3.5.3.1), we obtain

$$\tau = \frac{Q_y S_x^{\text{cut}}}{I_x b} = \frac{6 Q_y}{b h^3} \left(\frac{h^2}{4} - y^2 \right).$$

One can see that the quantity τ varies over the cross-section height according to the quadratic parabola law and attains the maximum value for $y = 0$ (i.e., on the axis x):

$$\tau_{\max} = \frac{3 Q_y}{2 b h}. \tag{E3.5.3.2}$$

In Fig. E3.21, we present the diagram of shear stresses over the height of the rectangular cross-section.

E3.5.4. Strength Analysis in the Case of Symmetric Bending

▶ **Maximal stresses.** It follows from formula (E3.5.2.3) that the maximum (in magnitude) normal stresses are attained at the cross-section points most remote from the neutral axis x, i.e., for $|y| = y_{\max}$:

$$\sigma_{\max} = \frac{|M_x| y_{\max}}{I_x} = \frac{|M_x|}{W_x}, \quad \text{where} \quad W_x = \frac{I_x}{y_{\max}}. \tag{E3.5.4.1}$$

The quantity W_x is called the *moment of resistance of the cross-section in bending* (axial moment of resistance).

For the cross-section of rectangular shape (Fig. E3.21), we have

$$W_x = \frac{b h^2}{6} \quad \text{and} \quad \sigma_{\max} = \frac{6 |M_x|}{b h^2}. \tag{E3.5.4.2}$$

We compare the quantities σ_{max} and τ_{max} for a beam of rectangular cross-section. In the order of magnitude, $Q_y \sim F$ and $M_x \sim Fl$, where F is the characteristic force acting on the beam and l is the beam length. Then it follows from (E3.5.3.2) and (E3.5.4.2) that

$$\frac{\tau_{max}}{\sigma_{max}} \sim \frac{h}{l} \ll 1.$$

Thus, the shear stresses are small compared with the normal stresses. This conclusion also holds for rods with cross-sections of different shapes (an exception is thin-walled rods, in which the appearance of large shear stresses is possible).

▶ **Beam strength conditions.** As a rule, the strength analysis for beams is performed for the normal stresses using the strength condition in the form

$$\sigma_{max} = \frac{|M_x|}{W_x} \le [\sigma].$$

But if the rod material (for example, wood) badly resists the shear strains (cut off), then the calculations for the shear stresses are also necessary:

$$\tau_{max} = \frac{|Q_y|}{I_x}\left(\frac{S_x^{cut}}{b}\right)_{max} \le [\tau].$$

In the cases where the rod material has different admissible stresses under tension $[\sigma^t]$ and compression $[\sigma^c]$, the strength analysis is performed separately for the tensile and compressing stresses:

$$\sigma_{max}^t = \frac{|M_x|y_{max}^t}{I_x} \le [\sigma^t],$$

$$\sigma_{max}^c = \frac{|M_x|y_{max}^c}{I_x} \le [\sigma^c],$$

where y_{max}^t and y_{max}^c are the distances from the axis x to the respective points of the extended and compressed parts of the cross-section that are most remote from this axis.

Just as in the case of tension and compression or torsion, in the strength analysis for rods under bending, one solves the following three problems:

1) checking calculation,
2) choosing the cross-section,
3) determining the admissible load.

All geometric characteristics required in strength analysis of rolled cross-sections (double tee, channel bar, etc.) are given in special tables called the range of rolled steel products.

Example. For a steel double tee beam (No. 55) of length $l = 3.2$ m (see Fig. E3.16b), determine the admissible load $[q]$ for $[\sigma] = 160$ MPa from the strength condition.

Solution. For the beam under study, the unsafe cross-section is that in the middle of the beam span, where the maximum bending moment $M_x = ql^2/8$ arises.

In the range of rolled steel products, we find the resistance moment $W_x = 2000$ cm^3 for the double tee. It follows from the strength condition

$$\sigma_{max} = \frac{M_x}{W_x} = \frac{ql^2}{8W_x} \le [\sigma]$$

that $q \le 8W_x[\sigma]/l^2 = 250$ N/m. Thus, $[q] = 250$ kN/m.

E3.6. Combined Stress

The *combined stress* includes the types of a rod strain such that at least two nonzero internal force factors arise simultaneously in the rod cross-sections. An exception is the symmetric lateral bending, which is considered as a simple type of strain, although there are two force factors in this case, the bending moment and the transverse force, because in the overwhelming majority of cases, the strength and rigidity analysis are performed without taking into account the transverse forces, i.e., for a single force factor, which is the bending moment.

E3.6.1. Oblique Bending

▶ **Definition.** When studying the symmetric lateral bending, we assume that the cross-section is symmetric with respect to the axis y. It turns out that all the results of Section E3.5 also hold for cross-sections of arbitrary shape if the axes x and y are the *principal central axes of inertia*; i.e., they are mutually perpendicular axes passing through the cross-section center of gravity and the axial moments of inertia about these axes take extremum values. In what follows, for the axes x and y we take the principal central axes.

The *oblique bending* is the general case of the rod bending (see Subsection E3.5.1) in which two bending moments M_x and M_y and perhaps the transverse forces Q_x and Q_y appear in the rod cross-section.

Using the force superposition principle, we can determine the normal stresses at any point of the cross-section with coordinates x, y by the formula

$$\sigma = -\frac{M_x y}{I_x} - \frac{M_y x}{I_y}, \qquad (E3.6.1.1)$$

which follows from formula (E3.5.2.3).

▶ **Neutral axis in the case of oblique bending.** Just as in the case of symmetric bending, in the cross-section there exists a neutral axis at whose points the stresses are zero. It is determined by the equation

$$\frac{M_x y}{I_x} + \frac{M_y x}{I_y} = 0. \qquad (E3.6.1.2)$$

Listed below are several properties of the neutral axis in oblique bending.
1. The neutral axis passes through the center of gravity of the cross-section.

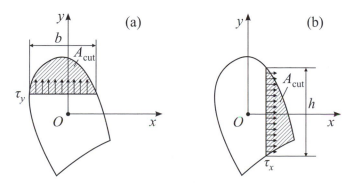

Figure E3.22. Tangential stresses τ_y (a) and τ_x (b) in the cross-section in oblique bending.

2. The neutral axis divides the cross-section into two parts: in one of them, $\sigma > 0$, in the other, $\sigma < 0$.

3. The normal stresses in the cross-section vary directly proportionally to the distance from the neutral axis attaining the maximum values (in magnitude) at the cross-section points most remote from the neutral axis.

The shear stresses in the cross-section (Fig. E3.22) can be determined by formula (E3.5.3.1):

$$\tau_y = \frac{Q_y S_x^{\text{cut}}}{I_x b}, \qquad \tau_x = \frac{Q_x S_y^{\text{cut}}}{I_y h}.$$

Just as in the case of symmetric lateral bending, the shear stresses, as a rule, play a secondary role, and hence the strength conditions impose constraints on the maximum normal stresses.

Example. Calculate the maximum normal stresses in the unsafe cross-section of the beam (Fig. E3.23) under oblique bending. The cross-section is of rectangular shape, $b = 6\,\text{cm}$, $h = 12\,\text{cm}$.

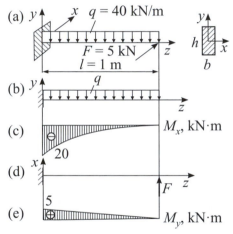

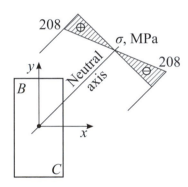

Figure E3.23. Oblique bending: (a) a beam in oblique bending, (b) vertical loading, (c) bending moment diagram for vertical loading, (d) horizontal loading, (e) bending moment diagram for horizontal loading.

Figure E3.24. Neutral axis and the normal stress diagram in the transversal direction.

Solution. In Fig. E3.23b and d, we show the beam loading diagram in the vertical and horizontal planes, and in Fig. E3.23c and e, the corresponding diagrams of the bending moments M_x, M_y. Judging from these diagrams, the unsafe cross-section is the fixation cross-section, where $M_x = -20\,\text{kN m}$, $M_y = 5\,\text{kN m}$.

For the rectangular cross-section, $I_x = bh^3/12$, $I_y = hb^3/12$, and the neutral axis equation (E3.6.1.2) for the unsafe cross-section takes the form $-y + x = 0$.

It follows from the position of the neutral line (Fig. E3.24) that the most unsafe points of the cross-section are points B and C; i.e., the points most remote from the neutral axis. Substituting their coordinates into (E3.6.1.1), we calculate the stresses at these points

$$\sigma_B = -\frac{M_x y_B}{I_x} - \frac{M_y x_B}{I_y} = 208\,\text{MPa},$$

$$\sigma_C = -\frac{M_x y_C}{I_x} - \frac{M_y x_C}{I_y} = -208\,\text{MPa}.$$

In Fig. E3.24, we show the diagram for σ, which characterized the variations in the normal stresses in the direction perpendicular to the neutral axis.

E3.6.2. Eccentric Extension or Compression

▶ The **eccentric extension or compression** is a type of a rod deformation such that the longitudinal force N and bending moments M_x, M_y (and perhaps the transverse forces Q_x, Q_y) arise in the rod cross-section.

The longitudinal force and the bending moments can be treated as the result of action of an eccentrically applied force $F = N$ on the rod (Fig. E3.25). Therefore, such a type of combined stresses is called eccentric extension or compression.

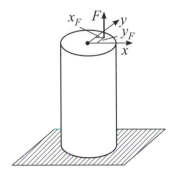

Figure E3.25. Eccentric extension of a rod.

The bending moments are related to the coordinates of the point of application of the force F by the relations $M_x = -Fy_F$, $M_y = -Fx_F$. Therefore, from (E3.6.1.1), formulas (E3.3.2.1), and the force superposition principle for normal stresses at an arbitrary point of any cross-section with coordinates x, y, we obtain

$$\sigma = \frac{N}{A} - \frac{M_x y}{I_x} - \frac{M_y x}{I_y} = F\left(\frac{1}{A} + \frac{y_F y}{I_x} + \frac{x_F x}{I_y}\right). \tag{E3.6.2.1}$$

▶ **Neutral axis in the case of eccentric extension or compression.** In this case, the equation of the cross-section neutral axis at whose points the stresses are zero becomes

$$\frac{1}{A} + \frac{y_F y}{I_x} + \frac{x_F x}{I_y} = 0. \tag{E3.6.2.2}$$

It is easy to see that the neutral axis does not pass through the center of gravity of the cross-section. The other properties are the same as in the case of oblique bending. In addition, we mention one more property of the neutral axis in the case of eccentric extension or compression: the neutral axis does not intersect the quarter of the cross-section where the force F is applied.

▶ **Core of a cross-section.** The neutral axis position, as follows from Eq. (E3.6.2.2), depends on the coordinates of the point of application of the force F. If the point of application of the force F is sufficiently close to the center of gravity of the cross-section, namely, is in the region which is called the *core of the cross-section*, then the neutral axis passes beyond the cross-section; i.e., all the points of the cross-section experience normal stresses of the same sign. In Fig. E3.26, we show the cores for a rectangular and circular cross-sections. The strength conditions in the case of eccentric extension or compression have the form of constraints on the maximum normal stresses.

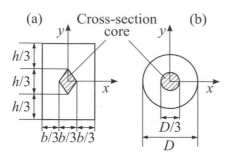

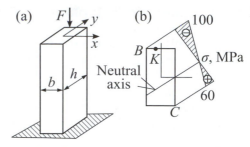

Figure E3.26. (a) Rectangular cross-section core, (b) circular cross-section core.

Figure E3.27. (a) Eccentrically compressed beam of rectangular cross-section, (b) force application point and the normal stress diagram.

Example. Calculate the maximum normal stresses in the cross-section of an eccentrically compressed rod of rectangular cross-section for $b = 6$ cm, $h = 12$ cm (Fig. E3.27). The point K of application of the force $F = 144$ kN has the coordinates $x_F = -1$ cm, $y_F = 6$ cm (Fig. E3.27b).

Solution. We calculate the geometric characteristics of the cross-section:

$$A = bh = 72 \, \text{cm}^2,$$

$$I_x = bh^3/12 = 864 \, \text{cm}^4,$$

$$I_y = hb^3/12 = 216 \, \text{cm}^4.$$

The neutral axis equation (E3.6.2.2) becomes $1 + y/2 - x/3 = 0$. Its position (Fig. E3.27b) shows that B and C are the most stressed points of the cross-section. Substituting their coordinates into (E3.6.2.1), we find the stresses at these points (the force F must be taken with "minus" sign, because its direction is opposite to that shown in Fig. E3.25):

$$\sigma_B = -F \left(\frac{1}{A} + \frac{y_F y_B}{I_x} + \frac{x_F x_B}{I_y} \right) = -100 \, \text{MPa},$$

$$\sigma_C = -F \left(\frac{1}{A} + \frac{y_F y_C}{I_x} + \frac{x_F x_C}{I_y} \right) = 60 \, \text{MPa}.$$

In Fig. E3.27b, we construct the diagram of σ, which shows the law of normal stress distribution in the direction perpendicular to the neutral axis.

E3.6.3. Bending with Torsion

▶ The **bending with torsion** is a type of rod deformation such that the twisting moment M_z, the bending moments M_x and M_y, and, perhaps, the transverse forces Q_x and Q_y arise in its cross-sections.

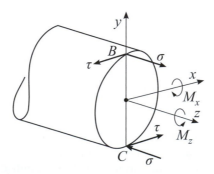

Figure E3.28. Combined bending and torsion of circular beam.

We consider rods of a circular cross-section. Because it is symmetric, any two mutually perpendicular central axes of the cross-section are principal axes. Therefore, they can always be chosen so that, in the cross-section under study, the total bending moment has only a single component, for example, M_x. The maximum (in magnitude) normal stresses in the cross-section are determined by formula (E3.5.4.1) for the symmetric bending

$$\sigma_{\max} = \frac{|M_x|}{W_x} \qquad (E3.6.3.1)$$

and are attained at points B and C that are most remote from the neutral axis x (Fig. E3.28).

By formula (E3.4.3.1), we obtain the maximum shear stresses in torsion:

$$\tau_{\max} = \frac{|M_z|}{W_p}.$$

They appear at the points of the cross-section boundary, including points B and C.

Thus, points B and C are the most unsafe points. In their neighborhood, the plane stress state is realized, which is a combination of uniaxial extension (or compression) and simple shear.

▶ **The Saint-Venant and von Mises strength conditions.** In contrast to all the strain types considered above, in this case, both σ and τ may be significant, and therefore, the problem of the strength condition arises for such a stress state. There exist different theories of strength answering this question. Most often, one applies the Saint-Venant strength condition

$$\sqrt{\sigma^2 + 4\tau^2} \leq [\sigma]$$

or the von Mises strength condition

$$\sqrt{\sigma^2 + 3\tau^2} \leq [\sigma].$$

Using the relation between the resistance moments $W_p = 2W_x$ and taking into account the fact that, in the case of arbitrary orientation of the axes x and y, the quantity M_x in formula (E3.6.3.1) must be replaced by the total bending moment $M_b = \sqrt{M_x^2 + M_y^2}$, we reduce the strength conditions to the final form:

$$\frac{\sqrt{M_x^2 + M_y^2 + M_z^2}}{W_x} \leq [\sigma] \qquad \text{(Saint-Venant)},$$

$$\frac{\sqrt{M_x^2 + M_y^2 + 0.75M_z^2}}{W_x} \leq [\sigma] \qquad \text{(von Mises)},$$

where $W_x = \pi D^3/32$ is the axial moment of resistance for a disk of diameter D.

Example. Using the von Mises strength condition, find the diameter of a steel shaft under bending with torsion (Fig. E3.29). The admissible stress is $[\sigma] = 150\,\text{MPa}$.

Solution. The diagram of the shaft loading in the vertical plane and the corresponding diagram of the bending moment M_x are given in Fig. E3.29b and c.

The diagram of the shaft loading by the external moment and the diagram of the arising twisting moments M_z are given in Fig. E3.29d and e.

The most unsafe cross-section is the fixation cross-section, where the maximum bending $M_x = 13\,\text{kN m}$ and twisting $M_z = 6\,\text{kN m}$ moments arise.

From the strength condition

$$\frac{32\sqrt{M_x^2 + 0.75M_z^2}}{\pi D^3} \leq [\sigma]$$

we obtain: $D \geq 0.098\,\text{m}$. Thus, we can round up: $D = 10\,\text{cm}$.

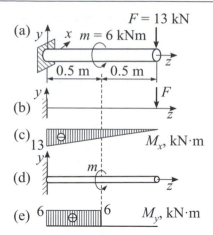

Figure E3.29. A shaft in combined bending and torsion: (a) dimensions and loads, (b) vertical loading, (c) bending moment diagram, (d) torsion loading, (e) twisting moment diagram.

E3.7. Stability of Compressed Rods

E3.7.1. Critical Force

In Subsection E3.1.1, it was already noted that some equilibria of bodies or systems turn out to be unstable, i.e., such that any negligible mechanical actions may lead to significant deviations from these states. Each of these equilibria is characterized by a certain value of the load which separates the regions of stable and unstable equilibria.

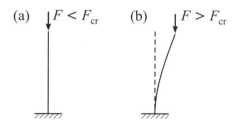

Figure E3.30. A compressed rod: (a) stable state, (b) unstable state.

In the case of a compressed rod (Fig. E3.30), it turns out that if the compressing force F does not exceed a certain value called the *critical force* F_{cr}, then the rectilinear form of the rod equilibrium is stable. But if $F > F_{cr}$, then the rectilinear form of the rod equilibrium is unstable, and the rod tends to take another stable form of equilibrium with a curvilinear axis.

The compressing stresses $\sigma_{cr} = F_{cr}/A$ corresponding to the critical force are called *critical stresses*.

E3.7.2. Euler Formula

Thus, for $F > F_{cr}$, the solution of the problem of the rod equilibrium is not unique. An analysis of the conditions of this ambiguity permits obtaining the formula for the value of F_{cr}.

Consider a hinged centrally compressed rod in a slightly deflected state (Fig. E3.31). The bending moment in the rod cross-section is equal to $M = -Fv$, where the deflection $v(z)$

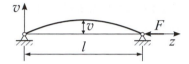

Figure E3.31. A hinged, centrally compressed rod.

determines the rod curvilinear axis. For small deflections of the rod, the curvature k of the deflected axis ($v/l \ll 1, v' \ll 1$) is equal to

$$k = \frac{1}{\rho} = \frac{v''}{(1 + v'^2)^{3/2}} \approx \frac{d^2 v}{dz^2}.$$

An analysis of the bending strain leads to formula (E3.7.2.3) in Section E3.5 relating the curvature and the bending moment,

$$k = \frac{1}{\rho} = \frac{M}{EI},$$

where the rules of signs agree well for the curvature $k = 1/\rho$ and the moment M.

Thus, we obtain the differential equation for the rod deflected axis:

$$v'' = -a^2 v, \tag{E3.7.2.1}$$

where $a^2 = F/(EI)$.

The boundary conditions take into account the fact that the rod ends are fixed:

$$v(0) = v(l) = 0. \tag{E3.7.2.2}$$

Equation (E3.7.2.1) with boundary conditions (E3.7.2.2) has the obvious trivial solution $v = 0$ corresponding to the rectilinear form of equilibrium. We are interested in nonzero solutions.

The general integral of Eq. (E3.7.2.1) has the form

$$v = C_1 \sin(az) + C_2 \cos(az).$$

It follows from the first boundary condition that $C_2 = 0$, and from the second, that $C_1 \sin(al) = 0$. Eliminating $C_1 \neq 0$, we obtain the relation

$$\sin(al) = 0,$$

which is satisfied if $al = \pi n$ or

$$F = \frac{\pi^2 n^2 EI}{l^2} \quad (n = 1, 2, \dots). \tag{E3.7.2.3}$$

Thus, it was found that if the compression force F takes discrete values determined by formula (E3.7.2.3), then, along with the rectilinear equilibrium, the rod curved equilibria of the sine curve form are possible;

$$v = C_1 \sin\left(\frac{n\pi z}{l}\right).$$

From (E3.7.2.3) with $n = 1$, we obtain the *Euler formula* for the critical force

$$F_{\mathrm{cr}} = \frac{\pi^2 EI}{l^2}. \tag{E3.7.2.4}$$

A more exact theory shows that the ambiguity of the equilibrium forms occurs for all forces F exceeding the Euler force (E3.7.2.4).

E3.7.3. Influence of Methods for Fixation of the Rod Ends on the Value of the Critical Force

Similar problems for other cases of rod ends fixation (Fig. E3.32) can also be considered. Their solution shows that the critical force can be determined by the formula similar to (E3.7.2.4):

$$F_{cr} = \frac{\pi^2 EI}{(\mu l)^2}, \qquad (E3.7.3.1)$$

introducing the notions of *reduced length* μl and of *reduced length coefficient* μ.

The quantity μl can be treated as the rod length part on which the half-wave of the sine curve corresponding to the rod curved axis can be placed. The values of the coefficient μ in several cases of the rod fixation are given in Fig. E3.32.

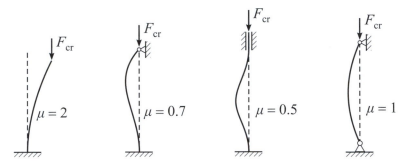

Figure E3.32. Reduced length coefficients μ for various fixing conditions of a rod.

Example. Find the critical force for a steel rod of length $l = 4$ m (Fig. E3.32) one of whose ends is rigidly fixed. The rod cross-section is the double tee No. 22.

Solution. In the range of rolled steel products, we find the moments of inertia $I_x = 2550$ cm^4, $I_y = 157$ cm^4 for this double tee. Obviously, the rod becomes unstable in plane of minimum rigidity, and therefore, we must take the minimum moment of inertia $I_{min} = I_y = 157$ cm^4.

For the given fixation coefficient $\mu = 2$, the modulus of elasticity for steel is $E = 210$ GPa. By the Euler formula (E3.7.3.1), we obtain $F_{cr} = \dfrac{\pi^2 E I_{min}}{(\mu l)^2} = 50.8$ kN.

E3.7.4. Scope of the Euler Formula

The critical stresses corresponding to the critical force (E3.7.3.1) are equal to

$$\sigma_{cr} = \frac{F_{cr}}{A} = \frac{\pi^2 E}{(\mu l / i)^2},$$

where $i = \sqrt{I/A}$ is the radius of inertia of the cross-section.

We introduce the notation $\lambda = \mu l / i$. Then the preceding formula takes the form

$$\sigma_{cr} = \frac{\pi^2 E}{\lambda^2}. \qquad (E3.7.4.1)$$

The quantity λ is called the *rod flexibility*.

It follows from (E3.7.4.1) that as the flexibility decreases, the value of σ_{cr} increases unboundedly. It is clear that formula (E3.7.4.1) cannot be used for stresses that are too large, because the rod material under these conditions does not obey Hooke's law.

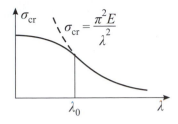

Figure E3.33. Dependence of the critical stress on the rod flexibility.

The value of the maximum stress at which Hooke's law is still satisfied is called the *proportionality limit* σ_{prop}.

Thus, the range of applicability for the Euler formula is determined by the inequality $\sigma_{\text{cr}} < \sigma_{\text{prop}}$ or $\pi^2 E/\lambda^2 < \sigma_{\text{prop}}$. As a result, we obtain $\lambda > \lambda_0 = \pi\sqrt{E/\sigma_{\text{prop}}}$. The dependence of σ_{cr} on λ for $\lambda < \lambda_0$ is usually determined experimentally. In Fig. E3.33, the solid line schematically shows the complete graph of the dependence $\sigma_{\text{cr}}(\lambda)$. For very small flexibilities, the rod loses strength under compression, and therefore, the critical stresses are practically equal to the corresponding strength loss stresses.

E3.7.5. Stability Analysis of Compressed Rods

In practice, it is necessary to ensure the stable operation of compressed rods, and therefore, the stresses in the rod cannot exceed the critical stresses with a certain margin:

$$\sigma = \frac{F}{A} \leq \frac{\sigma_{\text{cr}}}{n}, \tag{E3.7.5.1}$$

where n is the safety factor.

For convenience, the *coefficient of longitudinal bending* $\varphi = \sigma_{\text{cr}}/(n[\sigma])$ is introduced. Then condition (E3.7.5.1) takes the form

$$\frac{F}{A} \leq [\sigma]\varphi. \tag{E3.7.5.2}$$

The dependencies $\varphi(\lambda)$ for different materials are given in reference books in the form of tables.

Using condition (E3.7.5.2), one can solve the same problems as those in the strength analysis (see Subsection E3.3.3).

Bibliography for Chapter E3

Beer, F.P., Johnston, E.R. et al., *Mechanics of Materials, 3rd Edition*, McGraw-Hill, New York, 2001.

Birger, I. A. and Mavljutov, R. R., *Strength of Materials* [in Russian], Nauka, Moscow, 1986.

Case, J., Chilver, H., and Ross, C. T. F., *Strength of Materials and Structures, 4th Edition*, Elsevier, Amsterdam, 1999.

Darkov, A. V. and Shapiro, G. S., *Strength of Materials* [in Russian], Vysshaya Shkola, Moscow, 1989.

Den Hartog, J. P., *Strength of Materials*, Dover Publications, New York, 1961.

Drucker, D. C., *Introduction to Mechanics of Deformable Solids*, McGraw-Hill, New York, 1967.

Feodos'ev, V. I., *Strength of Materials* [in Russian], Nauka, Moscow, 1986.

Gastev, V. A., *A Short Course on Strength of Materials* [in Russian], Nauka, Moscow, 1977.

Mott, R. L., *Applied Strength of Materials, 4th Edition*, Prentice-Hall, New York, 2002.

Nash, W., *Schaum's Outline of Strength of Materials, 4th Edition*, McGraw-Hill, New York, 1998.

Popov, E. P., *Engineering Mechanics of Solids*, Prentice Hall, Englewood Cliffs, New Jersey, 1990.

Rabotnov, Yu. N., *Mechanics of Solids* [in Russian], Nauka, Moscow, 1988.

Smith, W. F. and Hashemi, J., *Foundations of Materials Science and Engineering, 4th Edition*, McGraw-Hill, New York, 2006.

Timoshenko, S. P., *Strength of Materials, 3rd Edition*, Krieger Publishing Co., Malabar, Florida, 1983.

Timoshenko, S. P. and Young, D. H., *Elements of Strength of Materials, 5th Edition*, D. Van Nostrand Co., New York, 1968.

Chapter E4

Hydrodynamics*

E4.1. Hydrostatics

E4.1.1. Properties of Incompressible Fluid at Rest. Basic Law of Hydrostatics

▶ **Preliminaries.** *Fluids* differ from solids in the absence of shape and in light mobility of small volumes. The ability of fluids to take any shape without separation into fractions under the action of small forces is called *fluidity*.

In their mechanical properties, the fluids are separated into the two classes, *liquids* and *gases*.

Liquids are fluids that can form a free surface, namely, a surface of contact (interface) between the liquid and the surrounding gas (another liquid). Fluids have low compressibility. This class of liquids comprises water, gasoline, oil, petroleum, mercury, etc.

Gases are fluids occupying the entire possible volume. Gases are easily compressed and expanded under the action of applied forces.

▶ **Properties of an incompressible liquids at rest.**

1. *Pascal's law*: pressure applied to a liquid is transferred to any point of the liquid equally in all directions (the hydrostatic pressure on a small plane site is independent of its orientation in space).

   Corollary: for a vessel of an arbitrary shape and any dimensions, the liquid pressure is the same at the same depth.

   Remark. Pascal's law also holds for gases.

2. The interface between two immiscible liquids (or a liquid and a gas) is a horizontal plane.

3. The *law of communicating vessels*: the level of a homogeneous liquid in communicating vessels is the same.

Example 1. A hydraulic press consists of two cylindrical communicating vessels of different diameters equipped with pistons whose areas are S_1 and S_2. The cylinders are filled with liquid oil. The unloaded pistons are at the same level. A force F_1 acting on the piston with area S_1 creates the additional pressure $p = F_1/S_1$ in the liquid. According to Pascal's law, this pressure is transferred by the liquid in all directions without change. Therefore, the pressure force $F_2 = pS_2 = F_1S_2/S_1$ acts on the second piston. From this relation, we obtain

$$\frac{F_2}{F_1} = \frac{S_2}{S_1}. \tag{E4.1.1.1}$$

Hence the forces acting on the hydraulic press pistons are proportional to the areas of these pistons. Therefore, using a hydraulic press, one can win much in the force by choosing $S_2 \gg S_1$.

---

* In this chapter, we consider several problems of hydrodynamics which are often encountered in chemical technology.

Remark. Deriving formula (E4.1.1.1), we neglected the term $\rho g h$ determining the pressure in the liquid due to the gravity force (see below).

▶ **Basic law of hydrostatics.** Consider the equilibrium of an incompressible homogeneous liquid in the field of gravity force. We introduce a rectangular Cartesian coordinate system whose Z-axis is directed vertically upwards (the gravity force is directed vertically downwards). *Basic law of hydrostatics*: the pressure distribution in a given volume of liquid is determined by the formula

$$p = p_0 - \rho g(z - z_0),   \text{(E4.1.1.2)}$$

where p_0 is the pressure for $z = z_0$, ρ is the liquid density, and g is the free fall acceleration.

Example 2. Assume that the coordinate z_0 corresponds to the free surface of the liquid. Then for the liquid pressure at the depth $h = z_0 - z$, we have $p = p_0 + \rho g h$, where p_0 is the atmospheric pressure (the gas pressure) on the free surface.

▶ **Rotation of a liquid in a cylindrical vessel.** Consider a liquid in a cylindrical vessel rotating about the vertical axis at a constant angular velocity ω. We introduce a rectangular Cartesian coordinate system whose Z-axis is directed along the axis of rotation.

The pressure distribution in the liquid is determined by the formula

$$p = p_0 + \frac{1}{2}\rho\omega^2 r^2 - \rho g(z - z_0),   \text{(E4.1.1.3)}$$

where $r = \sqrt{x^2 + y^2}$ is the radial distance to the cylinder axis. For $\omega = 0$, formula (E4.1.1.3) becomes (E4.1.1.2).

Assume that the coordinate z_0 corresponds to the free surface of the liquid on the axis of rotation. The surface of the liquid becomes a paraboloid of rotation and is described by the equation

$$\frac{1}{2}\omega^2 r^2 = g(z - z_0),   \text{(E4.1.1.4)}$$

which is obtained by substituting the pressure $p = p_0$ into (E4.1.1.3). In (E4.1.1.4), we set $r = R$, where R is the radius of the rotating vessel, and obtain the paraboloid height

$$H = \frac{\omega^2 R^2}{2g},$$

where $H = z_R - z_0$.

E4.1.2. Force of Pressure on a Plane and on a Curved Wall

▶ **Force of pressure on a plane wall.** Consider the wetted part of a plane wall of area S immersed in a quiescent liquid at an angle α with the horizon. The resultant pressure force P exerted by the liquid on this wall is directed along the normal to its surface and is numerically equal to

$$P = p_c S.   \text{(E4.1.2.1)}$$

Here p_c is the excess pressure at the center of gravity of the wetted part of the wall, which is calculated by the formula

$$p_c = p_0 - p_a + \rho g h_c^\circ,   \text{(E4.1.2.2)}$$

where p_0 is the pressure on the free surface of the liquid, p_a is the external pressure (for example, the atmospheric pressure) on the outer dry side of the wall, and h_c° is depth of immersion of the wall center of gravity with respect to the free surface.

TABLE E4.1

The area, the coordinate of the center of gravity, and the moment of inertia of plane symmetric figures about the axis passing through the center of gravity (the figures are symmetric with respect to the vertical axis and the coordinate z is counted downwards from the upper point of the figure along the axis of symmetry).

Name and description of the figure	Coordinate of the center of gravity, z_0	Area of the figure, S	Moment of inertia, I_c
Rectangle with sides a and b; the side a is parallel to the Z-axis	$\dfrac{1}{2}a$	ab	$\dfrac{1}{12}a^3 b$
Isosceles triangle with base a and height h	$\dfrac{2}{3}h$	$\dfrac{1}{2}ah$	$\dfrac{1}{36}ah^3$
Isosceles trapezium with bases a and b and height h (a is the upper base)	$\dfrac{1}{3}h\dfrac{a+2b}{a+b}$	$\dfrac{1}{2}h(a+b)$	$\dfrac{1}{36}h^3\dfrac{a^2+4ab+b^2}{a+b}$
Disk of radius R	$\dfrac{1}{2}R$	πR^2	$\dfrac{1}{4}\pi R^4$
Half-disk of radius R; rectilinear base is above	$\dfrac{4}{3\pi}R$	$\tfrac{1}{2}\pi R^2$	$\dfrac{9\pi^2-64}{72\pi}R^4$
Annular region with radii R and r ($R > r$)	R	$\pi(R^2 - r^2)$	$\dfrac{1}{4}\pi(R^4 - r^4)$
Ellipse with semiaxes a and b; semiaxis a is directed vertically	a	πab	$\dfrac{1}{4}\pi a^3 b$

Formula (E4.1.2.2) is often written as $p_c = \rho g(H + h_c^\circ)$, where $H = \dfrac{p_0 - p_a}{\rho g}$ is the *piezometric head*. The surface $z = H$ is called the *piezometric plane*. The sign of the difference $(p_0 - p_a)$ determines the sign of H.

The vertical coordinate of the point of pressure force application to the wetted part of the plane wall of area S is determined by the formula

$$h_p = h_c + \frac{I_c}{h_c S}\sin^2\alpha, \qquad (\text{E4.1.2.3})$$

where h_p and h_c are the vertical distances from the center of pressure and the center of gravity to the piezometric plane, α is the angle of inclination of the wall to the horizon, and I_c is the moment of inertia of the wall area about the horizontal axis passing through the wall center of gravity. In the case of a horizontal wall (for $\alpha = 0$), the center of gravity and center of pressure of the wall coincide.

Table E4.1 presents formulas for calculation of the moment of inertia of several plane figures about the axis passing through the center of gravity.

For $h_c > 0$, the piezometric plane lies above the center of gravity of the wetted surface, and the center of pressure lies below the center of gravity. For $h_c < 0$, the piezometric plane lies below the center of gravity of the wetted surface, and the center of pressure may be above the center of gravity (if $h_p < 0$). For $h_c = 0$, the piezometric plane passes through the center of gravity of the wetted surface, and the pressure forces are reduced to a couple of forces.

▶ **Force of pressure on a curved wall.** In the general case of an arbitrary curvilinear surface, the distributed load of the liquid pressure normal at each point of the surface can be reduced to the resultant vector and the principal moment.

For curvilinear walls symmetric with respect to the vertical plane (the most common situation in applications), the sum of elementary pressure forces can be reduced to a single resultant force lying in the plane of symmetry or to a couple of forces lying in the same plane. The value and the direction of the resultant force P are determined by two components, usually, horizontal and vertical.

The horizontal component of the pressure force on the curvilinear wall is equal to the pressure force on the projection of this wall onto the vertical plane of symmetry and is determined by the formula

$$P_1 = \rho g h_c S_1, \tag{E4.1.2.4}$$

where h_c is the depth of immersion of the center of gravity of the above-mentioned wall projection counted from the piezometric plane and S_1 is the area of the wall projection onto the vertical plane. The line of action of the force horizontal component P_1 passes through the center of pressure of the wall projection onto the vertical plane, lies in the plane of symmetry, and is displaced (downwards if $h_c > 0$ or upwards if $h_c < 0$) with respect to the center of gravity of this projection of the wall by the distance

$$\Delta h = \frac{I_c}{h_c S_1},$$

where I_c is the moment of inertia of the above-mentioned projection area of the wall about the horizontal axis passing through the center of gravity of this projection.

The vertical component of the force of pressure on the curvilinear wall is equal to the weight of liquid in the volume V bounded by the wall, the piezometric plane, and the vertical projected surface constructed on the wall contour; it is determined by the formula

$$P_2 = \rho g V. \tag{E4.1.2.5}$$

The vertical component of the pressure force passes through the center of gravity of the volume V (this volume is often called the volume of the body of pressure).

In formulas (E4.1.2.4) and (E4.1.2.5), it is assumed that the liquid is on one side of the wall and the pressure is constant (for example, equal to the atmospheric pressure) on the other (dry) side of the wall. The total pressure force on the wall is the geometric sum of the forces P_1 and P_2, and its absolute value is equal to

$$P = \sqrt{P_1^2 + P_2^2}.$$

The line of action of the total pressure force P passes through the point of intersection of the lines of action of the force components P_1 and P_2. The angle φ of inclination of the resultant force to the horizon is determined by the formula

$$\tan \varphi = \frac{P_2}{P_1}.$$

In the case of walls of constant curvature (cylindrical and spherical walls), the total pressure force passes trough the wall center of curvature.

In the case of liquid excess pressures on the side of the wetted side of the wall, all components and the total force are directed from the liquid to the wall (outside from the inside).

In the case of two-sided action of liquids on the wall, one first determines the horizontal and vertical components on each side of the wall under the assumption of one-sided action of the liquid, and then the total horizontal and vertical components of action of both liquids are determined.

If the curvilinear surface under study is intersected by the vertical axis at more than one point, it is necessary to divide this surface into simple parts (a simple part of a surface is its part intersected by the vertical axis only at a single point) and calculate the vertical component for each of the parts. Then the obtained vertical components are summed algebraically. The same process is used if the surface under study is intersected by the horizontal axis at more than one point.

E4.1.3. Archimedes Principle. Stability of Floating Bodies

▶ **Archimedes principle.** The body (completely or partly) immersed in a liquid is under the action of the buoyancy force F_A numerically equal to the weight of the liquid in the volume displaced by the body (*Archimedes principle*). Thus, we have

$$F_A = \rho g V_b,$$

where V_b is the volume of the liquid displaced by the body. The force F_A passes through the center of gravity of the volume displaced by the body, which is called the *center of buoyancy*.

▶ **Stability of floating bodies.** If a floating body is in equilibrium, then its center of gravity and the center of buoyancy lie on the common vertical line called the *axis of buoyancy* (for a body symmetric with respect to a plane, the axis of buoyancy lies in this plane). For a floating body completely immersed in a liquid (underwater floating) to be in stable equilibrium, it is necessary that the body center of gravity be below the center of buoyancy.

If a body is floating on the surface of a liquid (surface floating), then stable equilibrium is possible in several cases where the body center of gravity lies above the center of buoyancy. For a body to be in stable equilibrium in the case of surface floating, it is necessary that, when heeled (the body axis of buoyancy is inclined by an angle θ, see Fig. E4.1), the body *metacenter M* (i.e., the point of intersection of the line of the Archimedes force action with the axis of buoyancy) lie above the center of gravity C. In Fig. E4.1, point B' is the center of buoyancy when the body is heeled, and point B corresponds to the original position of the center of buoyancy on the axis of buoyancy in equilibrium.

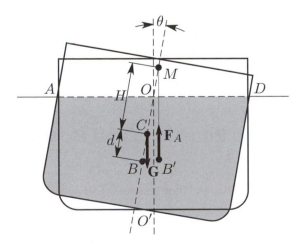

Figure E4.1. Stability of floating bodies. The center of gravity C lies above the center of buoyancy B.

The distance H between points M and C is called the *metacentric height*. For small angles of heeling, the metacentric height is determined by the formula

$$H = \frac{I}{V} - d \qquad (\theta \ll 1),$$

where I is the moment of inertia of the floating section AD about the axis of suspension OO' (see Fig. E4.1), V is the body volume immersed in the liquid, and d is the excess of the center of gravity over the center of buoyancy in equilibrium. For a floating body to be in stable equilibrium, it is necessary that the following condition be satisfied:

$$H = \frac{I}{V} - d > 0. \qquad (E4.1.3.1)$$

The stability of a floating body must be verified for the axis about which the moment of inertia of the floatation section is minimal.

Example. A wooden bar of square cross-section $a \times a$ and height h and of density ρ_b is floating in a liquid of density ρ_l ($\rho_l > \rho_b$). Find the maximum value of the bar height h_{max} for which the floating is still stable.

To solve this problem, we use formula (E4.1.3.1) and substitute the minimum value of the metacentric height into this formula:

$$\frac{I}{V} - d = 0. \qquad (E4.1.3.2)$$

The bar immersion depth x is determined by the Archimedes principle: the body weight must be equal to the weight of the liquid displaced by the body, $\rho_b g a^2 h = \rho_f g a^2 x$. We have

$$x = \frac{\rho_b}{\rho_f} h.$$

Now we calculate the quantities occurring on the left-hand side of (E4.1.3.2):

$$I = \frac{1}{12} a^4, \quad V = a^2 x = \frac{\rho_b}{\rho_f} a^2 h, \quad d = \frac{h}{2} - \frac{x}{2} = \frac{1}{2} h \left(1 - \frac{\rho_b}{\rho_f}\right).$$

(To find the moment of inertia I, we used the data for the rectangle given in Table E4.1 for $a = b$.) Substituting the values thus obtained into formula (E4.1.3.2), we have

$$\frac{1}{12} \frac{\rho_f}{\rho_b} \frac{a^2}{h} - \frac{1}{2} h \left(1 - \frac{\rho_b}{\rho_f}\right) = 0.$$

Resolving this equation for h, we obtain the maximal height of the bar:

$$h_{max} = \frac{a}{\sqrt{6\delta(1-\delta)}}, \qquad \delta = \frac{\rho_b}{\rho_f}.$$

E4.2. Hydrodynamic Equations and Boundary Conditions

In this section, we present equations and boundary conditions used in solving hydrodynamic problems. Their detailed derivation, as well as an analysis of scope, various physical statements and solutions of related problems, and applied issues can be found, e.g., in the books presented in the literature at the end of this chapter. We consider incompressible fluids with constant density ρ and dynamic viscosity μ.

E4.2.1. Laminar Flows. Navier–Stokes Equations. Euler Equations

▶ **Preliminaries.** Experiments show that two qualitatively different forms of fluid motion are possible. In particular, at small velocities of the flow in a circular tube, the fluid moves in straight lines parallel to one another and to the tube axis. Such a "smooth" flow without mixing is said to be *laminar*. In the general case of laminar steady-state flow, the fluid elements move along stream lines (a stream line is a line such that the tangent at each of its points coincides with the direction of the fluid particle velocity at this point). For large velocities of motion, an intensive mixing of the fluid is observed in the tube, and this mixing is characterized by unsteady chaotic motion of the fluid elements in the transverse direction (there appear velocity components perpendicular to the tube axis). Such a flow regime is said to be *turbulent*.

In Subsection E4.2.1, equations for laminar flows of fluids are considered, and in Subsection E4.2.3, equations for turbulent flows.

For brevity, in what follows, we often refer to "laminar flows" simply as "flows."

▶ **Navier–Stokes equations for a viscous incompressible fluid.** The closed system of equations of motion for a viscous incompressible Newtonian fluid consists of the continuity equation

$$\frac{\partial V_X}{\partial X} + \frac{\partial V_Y}{\partial Y} + \frac{\partial V_Z}{\partial Z} = 0 \tag{E4.2.1.1}$$

and three Navier–Stokes equations

$$\frac{\partial V_X}{\partial t} + V_X \frac{\partial V_X}{\partial X} + V_Y \frac{\partial V_X}{\partial Y} + V_Z \frac{\partial V_X}{\partial Z}$$
$$= -\frac{1}{\rho} \frac{\partial P}{\partial X} + \nu \left(\frac{\partial^2 V_X}{\partial X^2} + \frac{\partial^2 V_X}{\partial Y^2} + \frac{\partial^2 V_X}{\partial Z^2} \right) + g_X,$$

$$\frac{\partial V_Y}{\partial t} + V_X \frac{\partial V_Y}{\partial X} + V_Y \frac{\partial V_Y}{\partial Y} + V_Z \frac{\partial V_Y}{\partial Z} \tag{E4.2.1.2}$$
$$= -\frac{1}{\rho} \frac{\partial P}{\partial Y} + \nu \left(\frac{\partial^2 V_Y}{\partial X^2} + \frac{\partial^2 V_Y}{\partial Y^2} + \frac{\partial^2 V_Y}{\partial Z^2} \right) + g_Y,$$

$$\frac{\partial V_Z}{\partial t} + V_X \frac{\partial V_Z}{\partial X} + V_Y \frac{\partial V_Z}{\partial Y} + V_Z \frac{\partial V_Z}{\partial Z}$$
$$= -\frac{1}{\rho} \frac{\partial P}{\partial Z} + \nu \left(\frac{\partial^2 V_Z}{\partial X^2} + \frac{\partial^2 V_Z}{\partial Y^2} + \frac{\partial^2 V_Z}{\partial Z^2} \right) + g_Z,$$

Equations (E4.2.1.1) and (E4.2.1.2) are written in an orthogonal Cartesian system X, Y, Z in physical space; t is time; g_X, g_Y, and g_Z are the mass force (e.g., the gravity force) density components; $\nu = \mu/\rho$ is the kinematic viscosity of the fluid. The three components V_X, V_Y, V_Z of the fluid velocity and the pressure P are the unknowns.

By introducing the fluid velocity vector $\mathbf{V} = \mathbf{i}_X V_X + \mathbf{i}_Y V_Y + \mathbf{i}_Z V_Z$, where $\mathbf{i}_X$, $\mathbf{i}_Y$, and $\mathbf{i}_Z$ are the unit vectors of the Cartesian coordinate system, and by using the symbolic differential operators

$$\nabla \equiv \mathbf{i}_X \frac{\partial}{\partial X} + \mathbf{i}_Y \frac{\partial}{\partial Y} + \mathbf{i}_Z \frac{\partial}{\partial Z}, \qquad \Delta \equiv \frac{\partial^2}{\partial X^2} + \frac{\partial^2}{\partial Y^2} + \frac{\partial^2}{\partial Z^2},$$

one can rewrite system (E4.2.1.1), (E4.2.1.2) in the concise vector form

$$\nabla \cdot \mathbf{V} = 0, \tag{E4.2.1.3}$$

$$\frac{\partial \mathbf{V}}{\partial t} + (\mathbf{V} \cdot \nabla) \mathbf{V} = -\frac{1}{\rho} \nabla P + \nu \, \Delta \mathbf{V} + \mathbf{g}. \tag{E4.2.1.4}$$

The continuity and Navier–Stokes equations in the cylindrical and spherical coordinate systems can be obtained using the expressions of the operators ∇ and Δ given in Section S3.2.

▶ **Dimensionless form of equations.** To analyze the hydrodynamic equations (E4.2.1.3) and (E4.2.1.4), it is convenient to introduce dimensionless variables and unknown functions as follows:

$$\tau = \frac{Ut}{a}, \quad x = \frac{X}{a}, \quad y = \frac{Y}{a}, \quad z = \frac{Z}{a}, \quad \mathbf{v} = \frac{\mathbf{V}}{U}, \quad p = \frac{P}{\rho U^2},$$

where a and U are the characteristic length and the characteristic velocity, respectively. As a result, we obtain

$$\nabla \cdot \mathbf{v} = 0, \tag{E4.2.1.5}$$

$$\frac{\partial \mathbf{v}}{\partial t} + (\mathbf{v} \cdot \nabla)\mathbf{v} = -\nabla p + \frac{1}{\mathsf{Re}}\Delta \mathbf{v} + \frac{1}{\mathsf{Fr}}\frac{\mathbf{g}}{g}. \tag{E4.2.1.6}$$

In Eq. (E4.2.1.6), the following basic dimensionless state-geometric parameters of the flow are used:

$$\mathsf{Re} = \frac{aU}{\nu} \text{ is the } Reynolds\ number, \qquad \mathsf{Fr} = \frac{U^2}{ga} \text{ is the } Froude\ number.$$

Small values of Reynolds numbers correspond to slow ("creeping") flows and high Reynolds numbers, to rapid flows. Since these limit cases contain a small or large dimensionless parameter, one can efficiently use various modifications of the perturbation method.

▶ **Euler equations for an ideal incompressible fluid.** In the special case $V_X = V_Y = V_Z = 0$, the Navier–Stokes equations (E4.2.1.2) become the *equilibrium equations of a quiescent incompressible fluid*, which are independent of the fluid viscosity. From the equilibrium equations of a quiescent incompressible fluid, one can derive the laws of hydrostatics described in Section E4.1.

In the special case $\nu = 0$, the Navier–Stokes equations (E4.2.1.2) become the *Euler equations*,

$$\frac{\partial V_X}{\partial t} + V_X\frac{\partial V_X}{\partial X} + V_Y\frac{\partial V_X}{\partial Y} + V_Z\frac{\partial V_X}{\partial Z} = -\frac{1}{\rho}\frac{\partial P}{\partial X} + g_X,$$

$$\frac{\partial V_Y}{\partial t} + V_X\frac{\partial V_Y}{\partial X} + V_Y\frac{\partial V_Y}{\partial Y} + V_Z\frac{\partial V_Y}{\partial Z} = -\frac{1}{\rho}\frac{\partial P}{\partial Y} + g_Y, \tag{E4.2.1.7}$$

$$\frac{\partial V_Z}{\partial t} + V_X\frac{\partial V_Z}{\partial X} + V_Y\frac{\partial V_Z}{\partial Y} + V_Z\frac{\partial V_Z}{\partial Z} = -\frac{1}{\rho}\frac{\partial P}{\partial Z} + g_Z,$$

which, together with the continuity equation (E4.2.1.1), describe the flows of ideal (inviscid) incompressible fluid.

We assume that the flow is stationary and of the mass forces only the gravity force is acting, which is characterized by the values $g_X = g_Y = 0$, $g_Z = -g$ (the coordinate Z is counted off from the Earth's surface, and g is the free fall acceleration). In this case, the Euler equations admit the *Bernoulli integral*; namely, the following conservation law is satisfied on any stream line:

$$\frac{V^2}{2} + \frac{P}{\rho} + gZ = \text{const}, \tag{E4.2.1.8}$$

where $V = \sqrt{V_X^2 + V_Y^2 + V_Z^2}$ is the modulus of the fluid velocity. In the general case, the Bernoulli integral constant on the right-hand side in (E4.2.1.8) is different for different stream lines.

Example. Assume that there is a small hole in the wall of a wide vessel at a depth h. Find the velocity of the liquid flow through the hole.

Consider the stream line issuing from the liquid free surface in the vessel and passing through the hole. For this stream line, we write the Bernoulli integral (E4.2.1.8) in the form

$$\frac{V_1^2}{2} + \frac{P_1}{\rho} + g_1 Z = \frac{V_2^2}{2} + \frac{P_2}{\rho} + g Z_2, \tag{E4.2.1.9}$$

where subscript 1 denotes the quantities on the liquid surface and the subscript 2 denotes the quantities on the hole output. The coordinate Z is counted off from the hole. Substituting the values $V_1 \approx 0$, $P_1 = P_2 = P_a$, $Z_1 = h$, $Z_2 = 0$ into (E4.2.1.9), we obtain the velocity of the liquid flow from the hole:

$$V_0 = \sqrt{2gh}.$$

This formula is called the *Torricelli formula*.

The motion of an ideal fluid is said to be *potential* if the fluid velocity vector admits the representation

$$\mathbf{V} = \operatorname{grad} \varphi, \tag{E4.2.1.10}$$

where the function φ is called the *potential*. Because of the continuity equation (E4.2.1.1), the potential satisfies the Laplace equation

$$\Delta\varphi = 0,$$

and the pressure is determined by the Cauchy–Lagrange integral

$$\frac{\partial\varphi}{\partial t} + \frac{V^2}{2} + \frac{P}{\rho} + gZ = f(t). \tag{E4.2.1.11}$$

where $f(t)$ is an arbitrary function. When writing this integral, it was assumed that of the mass forces only the gravity force is acting. Relation (E4.2.1.11) is satisfied at all points of the region of potential flow of an ideal fluid; to find the function $f(t)$, it suffices to know the left-hand side of the integral as a function of time at any point of the flow.

In the special case of a steady-state potential motion of a fluid, the Cauchy–Lagrange integral (E4.2.1.11) coincides with the Bernoulli integral (E4.2.1.8), where the constant on the right-hand side is the same for all stream lines.

Note that the Euler equations (E4.2.1.7) have a very restricted domain of application (especially in problems of chemical technology).

E4.2.2. Initial Conditions and the Simplest Boundary Conditions

For the solution of system (E4.2.1.1), (E4.2.1.2) to determine the velocity and pressure fields uniquely, we must impose initial and boundary conditions.

▶ **Initial conditions.** In nonstationary problems, where the terms with partial derivatives with respect to time are retained in the equations of motion, the initial velocity field must be given in the entire flow region and satisfy the continuity equation (E4.2.1.1) there. The initial pressure field need not be given, since the equations do not contain the derivative of pressure with respect to time.

▶ **Boundary conditions on the surface of a solid body.** As a rule, the region occupied by a moving fluid is not the entire space but only part of it bounded by some surfaces. According to whether the point at infinity belongs to the flow region or not, the problem of finding the unknown functions is called the exterior or interior problem of hydrodynamics, respectively.

On the surface S of a solid body moving in a flow of a viscous fluid, the no-slip condition is imposed. This condition says that the vector $\mathbf{V}|_S$ of the fluid velocity on the surface of the solid is equal to the vector $\mathbf{V}_0$ of the solid velocity. If the solid is at rest, then $\mathbf{V}|_S = \mathbf{0}$. In the projections on the normal $\mathbf{n}$ and the tangent τ to the surface S, this condition reads

$$V_n|_S = 0, \qquad V_\tau|_S = 0.$$

More complicated boundary conditions are posed on an interface between two fluids (for example, see boundary conditions (E4.7.1.1)–(E4.7.1.4)).

▶ **Boundary condition far from the body for the translational flow.** To solve the exterior hydrodynamic problem, one must impose a condition at infinity (that is, far from the body, the drop, or the bubble).

For uniform translational flow with velocity $\mathbf{U}_i$ around a finite body, the boundary condition far from the body has the form

$$\mathbf{V} \to \mathbf{U}_i \qquad \text{as} \quad R \to \infty, \tag{E4.2.2.1}$$

where $R = \sqrt{X^2 + Y^2 + Z^2}$.

▶ **Shear flows.** An arbitrary stationary velocity field $\mathbf{V}(\mathbf{R})$ in an incompressible medium can be approximated near the point $\mathbf{R} = \mathbf{0}$ by two terms of the Taylor series:

$$V_k(\mathbf{R}) = V_k(\mathbf{0}) + G_{km}X_m,$$
$$G_{km} \equiv (\partial V_k/\partial X_m)_{\mathbf{R}=0}, \quad G_{11} + G_{22} + G_{33} = 0. \tag{E4.2.2.2}$$

Here V_k and G_{km} are the fluid velocity and the shear tensor components in the Cartesian coordinates X_1, X_2, X_3. The sum is taken over the repeated index m; since the fluid is incompressible, it follows that the sum of the diagonal entries G_{mm} is zero.

For viscous flows around particles whose size is much less than the characteristic size of flow inhomogeneities, the velocity distribution (1.1.15) can be viewed as the velocity field far from the particle. The special case $G_{km} = 0$ corresponds to a uniform translational flow. For $V_k(\mathbf{0}) = 0$, Eq. (E4.2.2.2) describes the velocity field in an arbitrary linear shear flow.

Example (Simple shear flow). For the simple shear (Couette) flow, we have

$$V_X = GY, \qquad V_Y = 0, \qquad V_Z = 0. \tag{E4.2.2.3}$$

In this case, G is called the gradient of the flow rate or the shear rate. The Couette flow occurs between two planes one of which is immovable and the other moves at a constant velocity parallel to the first plane.

The expressions (E4.2.2.3) give the exact solution of the Navier–Stokes equations (E4.2.1.1)–(E4.2.1.2) in the case of potential mass forces.

E4.2.3. Turbulent Flows. Reynolds Equations

▶ **Reynolds equations.** Formally, stationary solutions of the Navier–Stokes equations are possible for any Reynolds numbers. But practically, only stable flows with respect to small perturbations, always present in the flow, can exist. For sufficiently high Reynolds numbers, the stationary solutions become unstable; i.e., the amplitude of small perturbations increases with time. For this reason, stationary solutions can only describe real flows at not too high Reynolds numbers.

The flow in the boundary layer on a flat plate is laminar up to $\mathsf{Re}_X = U_iX/\nu \approx 3.5\times10^5$ (U_i is the unperturbed fluid velocity far from the plate, and the coordinate X is counted along the plate from the front edge), and that in a circular smooth tube is laminar for $\mathsf{Re} = a\langle V\rangle/\nu < 1500$ (a is the radius of a tube, and $\langle V\rangle$ is the mean flow rate velocity).

For higher Reynolds numbers, the laminar flow loses its stability and a transient regime of development of unstable modes takes place. For $\mathrm{Re}_X > 10^7$ and $\mathrm{Re} > 2500$, a fully developed regime of turbulent flow is established, which is characterized by chaotic variations in the basic macroscopic flow parameters in time and space.

When mathematically describing a fully developed turbulent motion of fluid, it is common to represent the velocity components and pressure in the form

$$V_i = \overline{V_i} + V_i', \quad P = \overline{P} + P', \tag{E4.2.3.1}$$

where the bar and prime denote the time-average and fluctuating components, respectively. The averages of the fluctuations are zero, $\overline{V_i'} = \overline{P'} = 0$.

The representation (E4.2.3.1) of the hydrodynamic parameters of turbulent flow as the sum of the average and fluctuating components followed by the averaging process made it possible, on the basis of the continuity equation (E4.2.1.3) and the Navier–Stokes equations (E4.2.1.4), to obtain (under some assumptions) the Reynolds equations

$$\nabla \cdot \overline{\mathbf{V}} = 0,$$
$$\frac{\partial \overline{\mathbf{V}}}{\partial t} + (\overline{\mathbf{V}} \cdot \nabla)\overline{\mathbf{V}} = -\frac{1}{\rho}\nabla\overline{P} + \nu\Delta\overline{\mathbf{V}} + \mathbf{g} + \frac{1}{\rho}\nabla \cdot \boldsymbol{\sigma}^{\mathrm{t}} \tag{E4.2.3.2}$$

for the averaged pressure and velocity fields. These equation contain the Reynolds turbulent shear stress tensor $\boldsymbol{\sigma}^{\mathrm{t}}$, whose components are defined as

$$\sigma_{ij}^{\mathrm{t}} = -\rho\overline{V_i'V_j'}. \tag{E4.2.3.3}$$

The variable $\rho\overline{V_i'V_j'}$ is the average rate at which the turbulent fluctuations transfer the jth momentum component along the ith axis.

▶ **The closure problem. Turbulent viscosity.** Unlike the Navier–Stokes equations completed by the continuity equation, the Reynolds equations form an unclosed system of equations, since these contain the *a priori* unknown turbulent stress tensor $\boldsymbol{\sigma}^{\mathrm{t}}$ with components (E4.2.3.3). Additional hypotheses must be invoked in order to close system (E4.2.3.2).

So far the closure problem for the system of Reynolds equations has not been solved theoretically in a conclusive way. In engineering calculations, various assumptions that the Reynolds stresses depend on the average turbulent flow parameters are often adopted as closure conditions. These conditions are usually formulated on the basis of experimental data, dimensional considerations, analogies with molecular rheological models, etc.

One traditional approach to the closure of the Reynolds equations is outlined below. This approach is based on Boussinesq's model of turbulent viscosity completed by Prandtl's hypothesis. For simplicity, we confine our consideration to the case of simple shear flow, where the transverse coordinate $Y = X_2$ is measured from the wall (the results are also applicable to turbulent boundary layers). According to Boussinesq's model, the only nonzero component of the Reynolds turbulent shear stress tensor and the divergence of this tensor are defined as

$$\sigma_{1,2}^{\mathrm{t}} = \rho\nu_{\mathrm{t}}\frac{\partial\overline{V}}{\partial Y}, \quad \nabla \cdot \boldsymbol{\sigma}^{\mathrm{t}} = \rho\frac{\partial}{\partial Y}\left(\nu_{\mathrm{t}}\frac{\partial\overline{V}}{\partial Y}\right), \tag{E4.2.3.4}$$

where $\overline{V}$ stands for the longitudinal average velocity component. Formulas (E4.2.3.4) contain the turbulent (eddy) viscosity ν_{t}, which is not a physical constant but is a function of geometric and kinematic flow parameters. It is necessary to specify this function to close the Reynolds equations.

Following Prandtl, we have

$$\nu_t = \kappa^2 Y^2 \left| \frac{\partial \overline{V}}{\partial Y} \right|, \tag{E4.2.3.5}$$

where $\kappa = 0.4$ is the von Kármán empirical constant (in the literature, this constant is most frequently taken to be $\kappa = 0.40$ or 0.41, although other values can sometimes be encountered).

There are a number of other methods for closing the Reynolds equations, also based on the notion of turbulent viscosity. These methods, as well as other models and methods of turbulence theory, can be found in the literature listed at the end of this chapter.

E4.3. Hydrodynamics of Thin Films

E4.3.1. Preliminary Remarks. Different Regimes of Film Type Flow

Film type flows are widely used in chemical technology (in contact devices of absorption, chemosorption, and rectification columns as well as evaporators, dryers, heat exchangers, film chemical reactors, extractors, and condensers).

As a rule, the liquid and the gas phase are simultaneously fed into an apparatus where the fluids undergo physical and chemical treatment. Therefore, generally speaking, there is a dynamic interaction between the phases until the flooding mode sets in the countercurrent flows of gas and liquid. However, for small values of gas flow rate one can neglect the dynamic interaction and assume that the liquid flow in a film is due to the gravity force alone.

The value of the Reynolds number $\mathrm{Re} = Q/\nu$, where Q is the volume flow rate per unit film width, determines whether the flow in the gravitational film is laminar, wave, or turbulent. It is well known that laminar flow becomes unstable at the critical value $\mathrm{Re}_* = 2$ to 6. However, the point starting from which the waves actually occur is noticeably shifted downstream. Even in the range $6 \leq \mathrm{Re} \leq 400$, corresponding to wave flows, a considerable part of the film remains wave-free. Since this part is much larger than the initial part where the velocity profile and the film width reach their steady-state values, we see that for films in which viscous and gravity forces are in balance, the hydrodynamic laws of steady-state laminar flow virtually determine the rate of mass exchange in various apparatuses, like packed absorbing and fractionating columns, widely used in chemical and petroleum industry. In these columns, the films flow over the packing surface whose linear dimensions do not exceed a few centimeters (Raschig rings, Palle rings, Birle seats, etc.).

Paradoxically, the range of flow rates (or Reynolds numbers) for which the assumption of laminar flow can be used in practice is bounded below (rather than above). Indeed, there is a threshold value $Q_{\min}$ of the volume flow rate per unit film width such that for $Q < Q_{\min}$ the flow in separate jets is energetically favorable. It was theoretically established that

$$Q_{\min} = 2.15 \left(\frac{\nu \sigma^3}{\rho^3 g} \right)^{1/5} (1 - \cos \theta)^{3/5},$$

where σ is the surface tension of the liquid and θ is the wetting angle of the wall material and the liquid (see Figure E4.2), determined by Young's fundamental relation

$$\sigma_{gw} = \sigma \cos \theta + \sigma_{fw},$$

Figure E4.2. Definition of the wetting angle.

where σ_{gw} and σ_{fw} are the specific excess surface energies for the gas–wall and liquid–wall interfaces.

Recently, the criterion of nonbreaking film flow was thermodynamically substantiated with the aid of Prigogine's principle of minimum entropy production, including the case of a double film flow.

In practice, Q_{min} can be reduced by wall hydrophilization—that is, by treating the surface with alcohol, which decreases the wetting angle.

E4.3.2. Films on an Inclined Plane and on a Cylindrical Surface

▶ **Film flowing by gravity on an inclined plane.** Consider a thin liquid film flowing by gravity on a solid plane surface (Figure E4.3). Let α be the angle of inclination. We assume that the motion is sufficiently slow, so that we can neglect inertial forces (that is, convective terms) compared with the viscous friction and the gravity force. Let the film thickness h (which is assumed to be constant) be much less than the film length. In this case, in the first approximation, the normal component of the liquid velocity is small compared with the longitudinal component, and we can neglect the derivatives along the film surface compared with the normal derivatives.

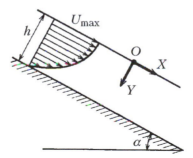

Figure E4.3. Steady waveless laminar flow in thin film on an inclined plane.

These assumptions result in the one-dimensional velocity and pressure profiles $V = V(Y)$ and $P = P(Y)$, where Y is the coordinate measured along the normal to the film surface. The corresponding hydrodynamic equations of thin films expressing the balance of viscous and gravity forces have the form

$$\mu \frac{d^2 V}{dY^2} + \rho g \sin \alpha = 0,$$
$$\frac{dP}{dY} - \rho g \cos \alpha = 0. \tag{E4.3.2.1}$$

To these equations one should add the boundary conditions

$$\frac{dV}{dY} = 0, \quad P = P_0 \qquad \text{at} \quad Y = 0,$$
$$V = 0 \qquad\qquad\qquad \text{at} \quad Y = h, \tag{E4.3.2.2}$$

which show that the tangential stress is zero, the pressure is equal to the atmosphere pressure at the free surface, and the no-slip condition is satisfied at the surface of the plane.

The solution of problem (E4.3.2.1), (E4.3.2.2) has the form

$$V = U_{\max}(1 - y^2),$$
$$P = P_0 + \rho g h \cos \alpha \, y, \tag{E4.3.2.3}$$

where $U_{\max} = \frac{1}{2}(g/\nu)h^2 \sin \alpha$ is the maximum flow velocity (the velocity at the free boundary) and $y = Y/h$ is the dimensionless transverse coordinate.

The volume flow rate per unit width is given by the formula

$$Q = \int_0^h V(Y)\,dY = \frac{g h^3 \sin \alpha}{3\nu} = \tfrac{2}{3} U_{\max} h. \tag{E4.3.2.4}$$

The mean flow rate velocity $\langle V \rangle$ is equal to $2/3$ of the maximum velocity,

$$\langle V \rangle = \tfrac{2}{3} U_{\max}.$$

Let us find the Reynolds number for the film flow:

$$\mathrm{Re} = \frac{Q}{\nu} = \frac{g h^3 \sin \alpha}{3\nu^2}.$$

This allows us to express the film thickness via the Reynolds number and the volume flow rate per unit width:

$$h = \left(\frac{3\nu^2}{g \sin \alpha}\mathrm{Re}\right)^{1/3} = \left(\frac{3\nu}{g \sin \alpha}Q\right)^{1/3}.$$

▶ **Film on a cylindrical surface.** Consider a thin liquid film of thickness h flowing by gravity on the surface of a vertical circular cylinder of radius a. In the cylindrical coordinates $\mathcal{R}, \varphi, Z$, the only nonzero component of the liquid velocity satisfies the equation

$$\mu\left(\frac{d^2 V_Z}{d\mathcal{R}^2} + \frac{1}{\mathcal{R}}\frac{\partial V_Z}{\partial \mathcal{R}}\right) + \rho g = 0. \tag{E4.3.2.5}$$

The boundary conditions on the wall and on the free surface can be written as

$$V_Z = 0 \quad \text{at} \quad \mathcal{R} = a, \qquad \frac{dV_Z}{d\mathcal{R}} = 0 \quad \text{at} \quad \mathcal{R} = a + h. \tag{E4.3.2.6}$$

The solution of problem (E4.3.2.5), (E4.3.2.6) is given by the formula

$$V_Z(\mathcal{R}) = \frac{\rho g}{4\mu}\left\{a^2 - \mathcal{R}^2 + \left[(a+h)^2 - a^2\right]\frac{\ln(\mathcal{R}/a)}{\ln(1 + h/a)}\right\}.$$

E4.4. Fluid Flows in Tubes

E4.4.1. Laminar Flows in Tubes of Various Cross-Sections

Laminar steady-state fluid flows in tubes are often encountered in practice (water-, gas- and oil pipelines, heat exchangers, etc.). It is worth noting that in these cases the corresponding hydrodynamic equations admit an exact closed-form solution. In what follows, we describe the most important results in that direction.

▶ **Statement of the problem.** Consider a laminar steady-state fluid flow in a rectilinear tube of constant cross-section. The fluid streamlines in such systems are strictly parallel (we neglect the influence of the tube endpoints on the flow). We shall use the Cartesian coordinates X, Y, Z with the Z-axis directed along the flow. Let us take into account the fact that the transverse velocity components of the fluid are zero and the longitudinal component depends only on the transverse coordinates. In this case, the continuity equation (E4.2.1.1) and the first two Navier–Stokes equations in (E4.2.1.2) are satisfied automatically, and it follows from the third equation in (E4.2.1.2) that

$$\frac{\partial^2 V}{\partial X^2} + \frac{\partial^2 V}{\partial Y^2} = \frac{1}{\mu} \frac{dP}{dZ}, \tag{E4.4.1.1}$$

where $V \equiv V_Z$ is the longitudinal velocity component.

Equation (E4.4.1.1) must be supplemented by the no-slip condition

$$V = 0 \qquad \text{on the tube surface.} \tag{E4.4.1.2}$$

The pressure gradient dP/dZ in the steady state is constant along the tube and can be represented in the form

$$\frac{dP}{dZ} = -\frac{\Delta P}{L}, \tag{E4.4.1.3}$$

where $\Delta P > 0$ is the total pressure drop along a tube part of length L.

The main flow characteristics are the volume flow rate

$$Q = \int_S V \, dS \tag{E4.4.1.4}$$

and the mean flow rate velocity

$$\langle V \rangle = \frac{Q}{S_*}, \tag{E4.4.1.5}$$

where S_* is the area of the tube cross-section.

▶ **Plane channel.** First, consider the flow between two infinite parallel planes at a distance $2h$ from each other. The coordinate X is measured from one of the planes along the normal. Since the fluid velocity is independent of the coordinate Y, we can rewrite (E4.4.1.1) in the form

$$\frac{d^2 V}{dX^2} = -\frac{\Delta P}{\mu L}.$$

The solution of this equation under the no-slip boundary conditions on the planes ($V = 0$ for $X = 0$ and $X = 2h$) has the form

$$V = \frac{\Delta P}{2\mu L} X(2h - X). \tag{E4.4.1.6}$$

Formula (E4.4.1.6) describes the parabolic velocity field in a *plane Poiseuille flow* symmetric with respect to the midplane $X = h$ of the channel.

The volume flow rate per unit width of the channel can be found by integrating (E4.4.1.6) over the cross-section:

$$Q = \frac{2h^3 \Delta P}{3\mu L}. \tag{E4.4.1.7}$$

The mean flow rate velocity is

$$\langle V \rangle = \frac{h^2 \Delta P}{3\mu L}.$$
(E4.4.1.8)

The maximum velocity is attained on the midplane of the channel:

$$U_{\max} = \frac{h^2 \Delta P}{2\mu L} \qquad (\text{at} \quad X = h).$$

▶ **Circular tube.** In the case of a circular tube, Eq. (E4.4.1.1), with regard to (E4.4.1.3), acquires the form

$$\frac{1}{\mathcal{R}} \frac{d}{d\mathcal{R}} \left(\mathcal{R} \frac{dV}{d\mathcal{R}} \right) = -\frac{\Delta P}{\mu L}, \qquad \mathcal{R} = \sqrt{X^2 + Y^2}.$$
(E4.4.1.9)

The solution of this equation under the no-slip condition on the surface of a tube of radius a ($V = 0$ for $\mathcal{R} = a$) describes an *axisymmetric Poiseuille flow* with parabolic velocity profile:

$$V = \frac{\Delta P}{4\mu L} \left(a^2 - \mathcal{R}^2 \right).$$
(E4.4.1.10)

The volume flow rate can be obtained by integrating over the cross-section:

$$Q = 2\pi \int_0^a \mathcal{R} V \, d\mathcal{R} = \frac{\pi a^4 \Delta P}{8\mu L}.$$
(E4.4.1.11)

By using (E4.4.1.5), we obtain the mean flow rate velocity

$$\langle V \rangle = \frac{a^2 \Delta P}{8\mu L}.$$
(E4.4.1.12)

The maximum fluid velocity is attained at the tube axis:

$$U_{\max} = \frac{a^2 \Delta P}{4\mu L} \qquad (\text{at} \ \mathcal{R} = 0).$$
(E4.4.1.13)

▶ **Annular channel between two coaxial circular cylinders.** Now consider the flow in an annular channel between two coaxial circular cylinders of radii a_1 and a_2 ($a_1 < a_2$). In this case, Eq. (E4.4.1.9) remains valid. The solution of this equation satisfying the no-slip conditions on the cylinder surfaces,

$$V = 0 \quad \text{at} \quad \mathcal{R} = a_1, \qquad V = 0 \quad \text{at} \quad \mathcal{R} = a_2,$$

has the form

$$V = \frac{\Delta P}{4\mu L} \left[a_2^2 - \mathcal{R}^2 + \frac{a_2^2 - a_1^2}{\ln(a_2/a_1)} \ln \frac{\mathcal{R}}{a_2} \right].$$
(E4.4.1.14)

The volume flow rate is given by the formula

$$Q = \frac{\pi \Delta P}{8\mu L} \left[a_2^4 - a_1^4 - \frac{(a_2^2 - a_1^2)^2}{\ln(a_2/a_1)} \right].$$
(E4.4.1.15)

▶ **Tube of elliptic cross-section.** Now consider a tube whose cross-section is an ellipse with semiaxes a and b. The surface of this tube is given by the equation

$$\left(\frac{X}{a}\right)^2 + \left(\frac{Y}{b}\right)^2 = 1. \tag{E4.4.1.16}$$

The solution of Eq. (E4.4.1.1) under the no-slip condition on the surface (E4.4.1.16) has the form

$$V = \frac{a^2 b^2 \Delta P}{2\mu L(a^2 + b^2)}\left(1 - \frac{X^2}{a^2} - \frac{Y^2}{b^2}\right). \tag{E4.4.1.17}$$

The volume flow rate is

$$Q = \frac{\pi \Delta P}{4\mu L}\frac{a^3 b^3}{a^2 + b^2}. \tag{E4.4.1.18}$$

By using formula (E4.4.1.5), we find the mean flow rate velocity

$$\langle V \rangle = \frac{\Delta P}{4\mu L}\frac{a^2 b^2}{a^2 + b^2}. \tag{E4.4.1.19}$$

The maximum velocity is attained at the tube axis

$$U_{\max} = \frac{a^2 b^2 \Delta P}{2\mu L(a^2 + b^2)} \qquad \text{(at } X = Y = 0). \tag{E4.4.1.20}$$

In the special case $a = b$, formulas (E4.4.1.17)–(E4.4.1.20) are reduced to the corresponding formulas (E4.4.1.10)–(E4.4.1.13) for a circular tube.

▶ **Tube of rectangular cross-section.** Now consider a tube of rectangular cross-section with sides a and b. We assume that the flow region is described by the inequalities $0 \leq X \leq a$ and $0 \leq Y \leq b$. The solution of Eq. (E4.4.1.1) under the no-slip conditions on the tube surface has the form

$$V = \frac{\Delta P}{2\mu L}X(a-X) + \sum_{m=1}^{\infty} \sin\left(\frac{\pi m X}{a}\right)\left(A_m \cosh\frac{\pi m Y}{a} + B_m \sinh\frac{\pi m Y}{a}\right),$$

$$A_m = \frac{a^2 \Delta P}{\pi^3 m^3 \mu L}[\cos(\pi m) - 1], \quad B_m = -A_m \frac{\cosh(\pi m k) - 1}{\sinh(\pi m k)}, \quad k = \frac{b}{a}. \tag{E4.4.1.21}$$

By integrating the expression for V, we obtain the volume flow rate

$$Q = \frac{\Delta P}{24\mu L}ab(a^2 + b^2)$$

$$- \frac{8\Delta P}{\pi^5 \mu L}\sum_{m=1}^{\infty}\frac{1}{(2m-1)^5}\left[a^4 \tanh\left(\pi b\frac{2m-1}{2a}\right) + b^4 \tanh\left(\pi a\frac{2m-1}{2b}\right)\right]. \tag{E4.4.1.22}$$

For a tube of square section with side a, this formula acquires the form

$$Q = \frac{a^4 \Delta P}{12\mu L}\left[1 - \frac{192}{\pi^5}\sum_{m=1}^{\infty}\frac{1}{(2m-1)^5}\tanh\left(\pi\frac{2m-1}{2}\right)\right], \tag{E4.4.1.23}$$

or, after summing the series,

$$Q = 0.0351 \frac{a^4 \Delta P}{\mu L}. \tag{E4.4.1.23a}$$

It is useful to rewrite the last expression as

$$\frac{Q}{Q_0} = 0.883,$$

where Q_0 is the volume flow rate for a circular tube with the same cross-section area. The volume flow rate for a square tube is smaller, because the cross-section has corners near which the velocity of a viscous fluid decreases noticeably.

▶ **Tube of triangular cross-section.** Now suppose that the cross-section of the tube is an equilateral triangle with side b. We place the origin at the center of the cross-section and measure the coordinate X along one of the sides of the triangle. In this case, the solution of Eq. (E4.4.1.1) under the boundary condition (E4.4.1.2) has the form

$$V = \frac{\sqrt{3} \Delta P}{6 \mu b L} \left(Y - \frac{b}{2\sqrt{3}} \right) \left(Y + \sqrt{3} X - \frac{b}{\sqrt{3}} \right) \left(Y - \sqrt{3} X - \frac{b}{\sqrt{3}} \right).$$

The volume flow rate of this flow is given by the formula

$$Q = \frac{\sqrt{3}}{320} \frac{b^4 \Delta P}{\mu L}.$$

It is useful to compare Q with the volume flow rate for a circular tube of the same cross-section area:

$$\frac{Q}{Q_0} = 0.726.$$

This expression shows that the volume flow rate for a tube whose cross-section is an equilateral triangle is substantially lower than for tubes of square or circular cross-section of the same area.

▶ **Rectilinear tube of arbitrary cross-section.** By using dimensional considerations, one can obtain the following formulas for the volume flow rate and the maximum velocity in a rectilinear tube of constant cross-section of arbitrary shape:

$$Q = \mathcal{K} \frac{S_*^2 \Delta P}{\mu L}, \qquad U_{\max} = \mathcal{K}_v \frac{S_* \Delta P}{\mu L}, \tag{E4.4.1.24}$$

where S_* is the tube cross-section area and $\mathcal{K}$ and $\mathcal{K}_v$ are dimensionless coefficients depending on the shape of the cross-section. The coefficients $\mathcal{K}$ and $\mathcal{K}_v$ can be obtained either experimentally or theoretically.

The most important dimensionless geometric parameter characterizing the cross-section shape is the ratio $\sqrt{S_*}/\mathcal{P}$, where $\mathcal{P}$ is the cross-section perimeter. In calculations, it is convenient to use the shape parameter

$$\xi = 2\sqrt{\pi} \frac{\sqrt{S_*}}{\mathcal{P}}, \tag{E4.4.1.25}$$

which always lies in the range $0 \le \xi \le 1$. The value $\xi = 1$ corresponds to a circular tube. This condition is ensured by choosing the proportionality factor $2\sqrt{\pi}$ in formula (E4.4.1.25).

For tubes of convex cross-section that is nearly circular or at least does not deviate very much from the circular shape, it is natural to assume that the coefficients $\mathcal{K}$ and $\mathcal{K}_v$ in (E4.4.1.24) depend only on ξ:

$$\mathcal{K} = \mathcal{K}(\xi), \qquad \mathcal{K}_v = \mathcal{K}_v(\xi). \tag{E4.4.1.26}$$

Then in many cases the function $\mathcal{K} = \mathcal{K}(\xi)$ (which determines the volume flow rate Q in (E4.4.1.24)) is well approximated by the linear function

$$\mathcal{K} = \mathcal{K}_0 \xi, \qquad \mathcal{K}_0 = \frac{1}{8\pi} \approx 0.0398. \tag{E4.4.1.27}$$

For $\xi = 1$, the approximate formula (E4.4.1.27) gives the exact value of $\mathcal{K}_0$ corresponding to a circular tube. For example, formula (E4.4.1.27) can be used for tubes whose cross-section is a regular N-gon ($N = 4, 5, \dots$). In particular, for a tube of square cross-section, Eq. (E4.4.1.25) gives $\xi = \frac{1}{2}\sqrt{\pi} \approx 0.886$. Substituting this into (E4.4.1.27), we obtain $\mathcal{K} = 0.0353$. This differs from the exact value $\mathcal{K} = 0.0351$ only by 0.6% (see (E4.4.1.23a)). For a tube of elliptic cross-section with axial ratio $a/b = 1.5$, the error in (E4.4.1.27) is about 5%.

▶ **Hydrodynamic drag coefficient.** Consider the hydrodynamic drag for the laminar flow of a fluid in tubes of various shape. The drag coefficient λ relating the pressure drop and the characteristic pressure head is introduced by the relation

$$\lambda = \frac{d_e}{\frac{1}{2}\rho\langle V\rangle^2} \frac{\Delta P}{L}, \tag{E4.4.1.28}$$

where d_e is equivalent (or "hydraulic") diameter and $\langle V \rangle$ is the mean flow rate velocity.

Let us introduce the equivalent (or "hydraulic") diameter d_e by the formula

$$d_e = \frac{4S_*}{\mathcal{P}},$$

where S_* is the area of the tube cross-section and $\mathcal{P}$ is the cross-section perimeter. For tubes of circular cross-section, d_e coincides with the diameter, and for a plane channel, d_e is twice the height of the channel.

Table E4.2 presents the values of the drag coefficients for tubes with various shapes of the cross-section. The Reynolds number $\mathsf{Re}_d = d_e\langle V\rangle/\nu$ can be calculated from the mean flow rate velocity and the equivalent diameter.

For a tube whose cross-section is a regular N-gon, the value $\lambda\,\mathsf{Re}_d$ is given by the approximate formula

$$\lambda\,\mathsf{Re}_d = \frac{64\,N - 82}{N - 0.95}. \tag{E4.4.1.29}$$

The comparison with Table E4.2 shows that the maximum error in (E4.4.1.29) is less than 0.6% at $N = 3, 4, 6, \infty$.

E4.4.2. Turbulent Flows in Tubes

▶ **Tangential stress. Turbulent viscosity.** A flow of a fluid through a smooth tube of circular cross-section remains laminar while $\mathsf{Re} = a\langle V\rangle/\nu < 1500$, where a is the tube radius and $\langle V \rangle$ the mean flow rate velocity of the fluid. For higher Reynolds numbers, the loss of stability of the laminar flow is observed and an intermediate regime occurs. For

TABLE E4.2
Values of the drag coefficients for laminar flow in tubes of various shape.

Tube profile		$\lambda \, \mathrm{Re}_d$	Equivalent diameter d_e
Circular tube of diameter d		64.000	d
Flat tube of width $2h$		96.000	$4h$
Elliptic tube with semiaxes a and b	$b/a =$ 1.00 0.80 0.50 0.25 0.125 0.0625 0	64.000 64.392 67.292 72.960 76.584 78.144 78.956	$\dfrac{\pi b}{E\left(\sqrt{1 - b^2/a^2}\,\right)}$, where $E(\vartheta)$ is the complete elliptic integral of the second kind
Tube of rectangular cross-section with sides a and b	$b/a =$ 1.00 0.714 0.50 0.25 0.125 0.05 0	58.008 58.260 62.192 72.932 82.336 89.908 96.000	$\dfrac{2ab}{a+b}$
Equilateral triangle with side a		53.348	$\dfrac{a\sqrt{3}}{3}$
Regular hexagon with side a		60.216	$a\sqrt{3}$
Semi-circle of diameter d		63.068	$\dfrac{\pi d}{\pi + 2}$

Re > 2500, a fully developed regime of turbulent flow is established, which is characterized by a chaotic variation of the velocity and pressure in time and space.

In a turbulent flow in a tube, there are two significantly different flow regions. In the first, entry region, the average velocity profile $\overline{V}$ changes dramatically with the distance from the entry cross-section. In the second, stabilized flow region, the average velocity profile is the same at each cross-section. The length of the entry (stabilization) region depends on the Reynolds number and the roughness of the walls and occupies a few dozen diameters (from 25 to 100). For rough estimates, this length is frequently taken to be 50 tube diameters.

In the stabilized flow region, the average fluid velocity $\overline{V}$ is directed along the tube axis and depends only on the distance Y from the tube wall. The integration of the Reynolds equations (E4.2.3.2) yields the following expression for the shear stress:

$$\tau = \tau_s(1 - Y/a), \tag{E4.4.2.1}$$

where τ_s is the friction stress at the wall. Near the tube axis, as $Y/a \to 1$, it follows from (E4.4.2.1) that $\tau/\tau_s \to 0$. Near the wall, as $Y/a \to 0$, we have $\tau/\tau_s \to 1$.

The friction stress at the wall for a circular tube is calculated by the formula $\tau_s = \frac{1}{2}a(\Delta P/L)$, where $\Delta P > 0$ is the total pressure drop along a tube part of length L.

In accordance with Boussinesq's model (E4.2.3.4), the shear stress can be represented as

$$\tau = \rho(\nu + \nu_t)\frac{\partial \overline{V}}{\partial Y}, \tag{E4.4.2.2}$$

where ν is the kinematic viscosity and ν_t is the turbulent viscosity.

Formulas (E4.4.2.1) and (E4.4.2.2) have formed the basis of most theoretical studies of the average fluid velocity and the drag coefficient in the stabilized region of turbulent flow in a circular tube (and a plane channel of width $2a$). The corresponding results obtained on the basis of Prandtl's relation (E4.2.3.4) and von Kármán's relation for turbulent viscosity can be found in the literature given at the end of this chapter. In what follows, major attention will be paid to empirical and semi-empirical formulas that approximate numerous experimental data quite well.

▶ **Structure of the flow. Velocity profile in a circular tube.** Experiments show that in the stabilized flow region, two characteristic subregions can be singled out, namely,

$$0 \le Y/a \le \sigma \quad \text{(wall region)},$$
$$\sigma \le Y/a \le 1 \quad \text{(core of turbulent flow)},$$

where Y is the transverse coordinate measured from the wall and $\sigma = 0.1$ to 0.2.

To describe the turbulent flow in the wall region, one introduces the so-called friction velocity U_* and the dimensionless internal coordinate y^+ according to the formulas

$$U_* = \sqrt{\tau_s/\rho}, \quad y^+ = Y U_*/\nu, \tag{E4.4.2.3}$$

where τ_s is the shear stress at the wall and ρ is the fluid density. According to von Kármán, it is convenient to single out three subdomains in the wall region:

$$\text{wall region} = \begin{cases} \text{viscous sublayer } (\nu_t \ll \nu), \\ \text{buffer layer } (\nu_t \sim \nu), \\ \text{logarithmic layer } (\nu_t \gg \nu). \end{cases}$$

In the viscous sublayer, the turbulent viscosity tends to zero near the wall (ν_t is proportional to Y^3). In the logarithmic layer, the turbulent viscosity depends on the transverse coordinate linearly, $\nu_t = 0.4 U_* Y$.

The approximate ranges of the above subdomains in terms of y^+ are as follows:

$$\begin{array}{ll} \text{viscous sublayer} & 0 \le y^+ \le 5, \\ \text{buffer layer} & 5 \le y^+ \le 30, \\ \text{logarithmic layer} & 30 \le y^+. \end{array}$$

In the viscous sublayer, the average velocity distribution is linear,

$$\frac{\overline{V}(Y)}{U_*} = y^+ \qquad \text{for} \qquad 0 \le y^+ \le 5. \tag{E4.4.2.4}$$

Note that turbulent fluctuations can penetrate into the viscous sublayer, although the turbulent friction is small there.

The average velocity profile in the thickest, logarithmic layer can be described by the formula

$$\frac{\overline{V}(Y)}{U_*} = 2.5 \ln y^+ + 5 \qquad \text{for} \qquad 30 \le y^+ \tag{E4.4.2.5}$$

quite well. This relation is referred to as Prandtl's law of the wall. It is worth noting that the factor 2.5 occurring in Eq. (E4.4.2.5) comes from $1/\kappa$, where $\kappa = 0.4$ is the von Kármán constant. The value 5 of the constant term was obtained on the basis of numerous experimental data.

It should be pointed out that formulas (E4.4.2.4) and (E4.4.2.5) are not only supported by experimental data quite well but also have certain theoretical justification.

For the velocity distribution in the buffer layer, one can use the simple interpolation formula

$$\frac{\overline{V}(Y)}{U_*} = y^+ \cos^{3.24}\xi + (2.5 \ln y^+ + 5)\sin^{2.24}\xi, \quad \xi = \frac{\pi}{2}\frac{y^+}{18 + y^+}. \tag{E4.4.2.6}$$

This formula well agrees with the existing experimental data and defines a continuous and smooth profile of the average velocity over the entire wall domain (including the viscous sublayer, the buffer layer, and the logarithmic layer).

Numerous experimental data provide evidence that the average velocity $\overline{V}$ in the core of turbulent flow can be approximated as

$$\frac{U_{\max} - \overline{V}(Y)}{U_*} = 2.5 \ln \frac{a}{Y}, \tag{E4.4.2.7}$$

where $U_{\max}$ is the velocity at the flow axis. Equation (E4.4.2.7) is universal and is referred to as von Kármán's velocity defect law;* it is applicable for smooth and rough tubes and any Reynolds numbers corresponding to turbulent flows. Formula (E4.4.2.7) can be somewhat refined by adding the term $0.6(1 - Y/a)^2$ to the right-hand side, thus extending the formula to the range $0.01 \leq Y/a < 1$.

Darcy's formula

$$\frac{U_{\max} - \overline{V}(Y)}{U_*} = 5.08\left(1 - \frac{Y}{a}\right)^{3/2} \tag{E4.4.2.8}$$

is worth mentioning. It provides a more accurate prediction of the turbulent flow near the flow axis $(Y/a \approx 1)$ compared with Eq. (E4.4.2.7) but has a narrower scope, $0.25 \leq Y/a \leq 1$.

In engineering calculations, it is not uncommon to approximate the average fluid velocity in a turbulent flow by the power law

$$\frac{\overline{V}(Y)}{U_{\max}} = \left(\frac{Y}{a}\right)^{1/n}, \tag{E4.4.2.9}$$

where the parameter n slowly increases with the Reynolds number. The value $n = 7$ suggested by Blasius is most frequently used. In this case, formula (E4.4.2.9) agrees well with experimental data within the range $3 \times 10^3 \leq \mathsf{Re}_d \leq 10^5$, where $\mathsf{Re}_d = d\langle V\rangle/\nu$ is the Reynolds number for a tube of diameter $d = 2a$.

The average velocity profile for the entire cross-section of a circular tube can be calculated using the unified interpolation formula

$$\frac{\overline{V}(Y)}{U_*} = 2.5 \ln(1 + 4\zeta) + 7.5\left(1 - e^{-\zeta} - \zeta e^{-3\zeta}\right) + 2.5 \ln\frac{1.5(2 - \eta)}{1 + 2(1 - \eta)^2}, \tag{E4.4.2.10}$$

where $\zeta = 0.1\,YU_*/\nu$ and $\eta = Y/a$. This formula is a bit simpler than Reichardt's formula. The former is obtained from the latter by a slight change in numerical coefficients, which provides a better agreement with the experimental data in the logarithmic layer, with the same accuracy in the other domains.

---

* It is remarkable that Eq. (E4.4.2.7) was first suggested for the logarithmic layer, but it turned out that it can well be extrapolated to almost the entire domain of turbulent core.

▶ **Drag coefficient of a circular tube.** The drag coefficient λ is expressed via other hydrodynamic parameters as follows:

$$\lambda = \frac{\Delta P}{L}\frac{4a}{\rho\langle V\rangle^2} = \frac{8\tau_{\mathrm{s}}}{\rho\langle V\rangle^2} = 8\left(\frac{U_*}{\langle V\rangle}\right)^2. \tag{E4.4.2.11}$$

In the region of stabilized turbulent flow in a smooth circular tube, the drag coefficient can be estimated by the Prandtl–Nikuradze implicit formula

$$\frac{1}{\sqrt{\lambda}} = 0.88\ln\left(\mathsf{Re}_d\sqrt{\lambda}\right) - 0.82, \tag{E4.4.2.12}$$

where Re_d is the Reynolds number determined by the diameter. Within the range $3\times10^3 \leq \mathsf{Re}_d \leq 3\times10^6$, the maximum deviation of the result predicted by Eq. (E4.4.2.12) from experimental data is about 2%. In practice, it is more convenient to use simpler explicit formulas of Blasius and Nikuradze:

$$\lambda = \begin{cases} 0.3164\mathsf{Re}_d^{-0.25} & \text{for } 3\times10^3 \leq \mathsf{Re}_d \leq 10^5, \\ 0.0032 + 0.221\,\mathsf{Re}_d^{-0.237} & \text{for } 10^5 \leq \mathsf{Re}_d. \end{cases} \tag{E4.4.2.13}$$

These formulas are also accurate within 2%. The first line in Eq. (E4.4.2.13) follows from the assumption that the average velocity profile is given by the power law (E4.4.2.9) with $n = 7$.

In the transient zone, one can calculate the drag coefficient by the formula

$$\lambda = 6.3\times10^{-4}\sqrt{\mathsf{Re}_d} \qquad (2200 \leq \mathsf{Re}_d \leq 4000).$$

▶ **Turbulent flow in a plane channel.** Qualitatively, the picture of stabilized turbulent flow in a plane channel is similar to that in a circular tube. Indeed, in the viscous sublayer adjacent to the channel walls, the velocity distribution increases linearly with the distance from the wall: $\overline{V}(Y)/U_* = y^+$. In the logarithmic layer, the average velocity profile can be described by the expression

$$\frac{\overline{V}(Y)}{U_*} = 2.5\ln y^+ + 5.2.$$

In the flow core, the average velocity distribution in a plane channel of width $2h$ can be approximately described by formulas of the form

$$\frac{U_{\max} - \overline{V}(Y)}{U_*} = A\left(1 - \frac{Y}{h}\right)^m,$$

where $A = 6.5$ and $m = 1.9$.

In the region of stabilized turbulent flow, the drag coefficient can be determined from the implicit relation

$$\frac{1}{\sqrt{\lambda}} = 0.86\ln\left(\mathsf{Re}\sqrt{\lambda}\right) - 0.35,$$

where $\lambda = 8\left(U_*/\langle\overline{V}\rangle\right)^2$ and $\mathsf{Re} = 2h\langle\overline{V}\rangle/\nu$. This relation is in good agreement with experimental data for all $\mathsf{Re} \leq 10^4$.

▶ **Drag coefficient for tubes of other shape.** The drag coefficient λ for turbulent flows in rectilinear tubes of noncircular cross-section can also be computed using relations (E4.4.2.12) and (E4.4.2.13), where the equivalent diameter

$$d_e = \frac{4S_*}{\mathcal{P}} \tag{E4.4.2.14}$$

should be regarded as the characteristic length used to calculate the Reynolds number. In Eq. (E4.4.2.14), S_* is the cross-section area of the tube and $\mathcal{P}$ the cross-section perimeter. The values of the drag coefficient predicted by this approach fairly well agree with experimental data for tubes of rectangular and triangular cross-section.

More detailed information about the structure of turbulent flows in a circular (or noncircular) tube and a plane channel, as well as various relations for determining the average velocity profile and the drag coefficient, can be found in the literature given at the end of this chapter.

E4.5. Hydrodynamic Boundary Layers on a Flat Plate

E4.5.1. Laminar Boundary Layer

▶ **Preliminary remarks. Flow for large Reynolds numbers.** In practice, one often deals with outside flows around stationary extended equipment elements such as plates, guiding elements, or tubes. In this case, the action of external mass forces can often be neglected, and the hydrodynamic laws are determined by the relation between pressure, viscous, and inertial forces. Then the system of dimensionless steady-state hydrodynamic equations becomes

$$\nabla \cdot \mathbf{v} = 0,$$
$$(\mathbf{v} \cdot \nabla)\mathbf{v} = -\nabla p + \frac{1}{\mathsf{Re}} \, \Delta \mathbf{v}. \tag{E4.5.1.1}$$

The system contains a single parameter, the Reynolds number. The solution in the general case is very complicated (the system is nonlinear), but it can be simplified if we consider the passage to the limit as $\mathsf{Re} \to 0$ or $\mathsf{Re} \to \infty$. In this section, we solve the problem on a longitudinal flat-plate flow assuming that $\mathsf{Re} \to \infty$, that is, for a liquid with "vanishing viscosity." The solution is not straightforward: one cannot just disregard the term $\mathsf{Re}^{-1}\Delta\mathbf{v}$, thus obtaining the equation of an ideal fluid. The mathematical difficulty is that the small parameter Re^{-1} occurs in the term with higher derivatives. By neglecting this term, we change the order and the type of the equation, and the solution of the original system need not converge as $\mathsf{Re}^{-1} \to 0$ to the solution of the limit system with $\mathsf{Re}^{-1} = 0$. Here we deal with a *singular perturbation*. Moreover, it is clear from physical considerations that an ideal fluid flow past a body cannot satisfy the no-slip condition on the surface. Actually, the tangential velocity varies from zero on the surface of the body to the velocity of the undisturbed flow remote from the body.

For fluids with low viscosity, this change of velocity occurs in a thin fluid layer adjacent to the surface of the body. Prandtl termed this layer a *boundary layer*. The magnitude of $\Delta\mathbf{v}$ is very large in this layer. Thus, although Re^{-1} is a small parameter, one cannot disregard the term $\mathsf{Re}^{-1}\Delta\mathbf{v}$ in the boundary layer. Nevertheless, since the longitudinal and the transverse coordinates play different roles in the boundary layer, we can simplify the equations. The corresponding formal estimate of terms in the second equation in (E4.5.1.1) can be found in the books listed at the end of this chapter.

▶ **Statement of the Blasius problem.** Consider the steady-state problem on the longitudinal zero-pressure-gradient flow ($\nabla P \equiv 0$) past a half-infinite flat plate ($0 \leq X < \infty$). We assume that the coordinates X and Y are directed along the plate and transverse to the plate, respectively, and the origin is placed at the front edge of the plate. The velocity of the incoming flow is U_i.

Let us write out the final system of boundary layer equations for an incompressible fluid

$$V_X \frac{\partial V_X}{\partial X} + V_Y \frac{\partial V_X}{\partial Y} = \nu \frac{\partial^2 V_X}{\partial Y^2},$$
$$\frac{\partial V_X}{\partial X} + \frac{\partial V_Y}{\partial Y} = 0. \tag{E4.5.1.2}$$

The natural boundary conditions have the form

$$\begin{array}{lll} Y = 0, & V_X = V_Y = 0 & \text{(no-slip condition)}, \\ Y \to \infty, & V_X \to U_i & \text{(condition far from the plate)}. \end{array} \tag{E4.5.1.3}$$

▶ **Blasius solution. Friction coefficient.** We express the fluid velocity components via the stream function Ψ by the formulas

$$V_X = \frac{\partial \Psi}{\partial Y}, \qquad V_Y = -\frac{\partial \Psi}{\partial X} \tag{E4.5.1.4}$$

and substitute them into (E4.5.1.2).

Then we seek the stream function in the form

$$\Psi(X,Y) = \sqrt{\nu X U_i}\, f(\eta), \qquad \eta = Y \sqrt{\frac{U_i}{\nu X}}, \tag{E4.5.1.5}$$

where η is a self-similar variable.

We obtain the following boundary value problem for the function $f(\eta)$:

$$\begin{array}{ll} 2f''' + f f'' = 0; \\ \eta = 0, & f = f' = 0; \\ \eta \to \infty, & f' \to 1. \end{array} \tag{E4.5.1.6}$$

Detailed tables containing the numerical solution of this problem can be found, for example, in Schlichting's book (1981).

By using (E4.5.1.4), we can calculate the fluid velocity components as follows:

$$V_X = U_i f'(\eta), \qquad V_Y = \frac{1}{2}\sqrt{\frac{\nu U_i}{X}} \left[\eta f'(\eta) - f(\eta) \right]. \tag{E4.5.1.7}$$

The obtained solution also allows us to calculate some variables that are of practical interest. For example, the local frictional stress on the wall is

$$\tau_w(X) = \mu \left(\frac{\partial V_X}{\partial Y} \right)_{Y=0} = \mu U_i \sqrt{\frac{U_i}{\nu X}}\, f''(0) = 0.332\, \mu U_i \sqrt{\frac{U_i}{\nu X}}, \tag{E4.5.1.8}$$

and the local friction coefficient is given by

$$c_f(X) = \frac{\tau_w(X)}{\frac{1}{2}\rho U_i^2} = 0.664 \sqrt{\frac{\nu}{U_i X}}. \tag{E4.5.1.9}$$

The total friction coefficient for a plate of length L is given by the formula

$$\langle c_{\mathrm{f}} \rangle = \frac{1}{L} \int_0^L c_{\mathrm{f}}(X) \, dX = \frac{1.328}{\sqrt{\mathsf{Re}_L}}, \tag{E4.5.1.10}$$

where $\mathsf{Re}_L = U_{\mathrm{i}} L / \nu$ is the Reynolds number for the plate in the flow.

Formula (E4.5.1.9) is known as the Blasius law for the drag in longitudinal flat-plate flow. This formula can be used in laminar flow, that is, for $\mathsf{Re}_L < 3.5 \times 10^5$.

Although the boundary layer in this statement of the problem is asymptotic—that is, extends infinitely along the Y-coordinate—one can approximately estimate its thickness if we adopt the convention that the velocity on the boundary of the layer differs from the undisturbed flow velocity at most by 1%. Then the boundary layer thickness is

$$\delta(X) \approx 5 \sqrt{\nu X / U_{\mathrm{i}}}.$$

The Blasius solution shows that the longitudinal velocity profiles are affinely similar to each other for all cross-sections of the boundary layer.

▶ **Reversed statement of the Blasius problem.** In applications, one often deals with the "reversed" statement of the Blasius problem, in which a half-infinite plate moves in its plane at a velocity U_{i}. In this case, the stream function is also sought in the form (E4.5.1.5), but the boundary value problem (E4.5.1.6) for the function $f = f(\eta)$ is replaced by

$$2f''' + f f'' = 0;$$
$$\eta = 0, \qquad f = 0, \quad f' = 1;$$
$$\eta \to \infty, \qquad f' \to 0,$$

In this case, in contrast with (E4.5.1.8), the local friction stress on the wall is given by

$$\tau_w(X) = 0.444 \, \mu U_{\mathrm{i}} \sqrt{\frac{U_{\mathrm{i}}}{\nu X}}.$$

E4.5.2. Turbulent Boundary Layer

▶ **Structure of the flow. Velocity profile.** The flow in the boundary layer on a flat plate is laminar until $\mathsf{Re}_X = U_{\mathrm{i}} X / \nu \approx 3 \times 10^5$. On a longer plate, the boundary layer becomes turbulent; that is, its thickness increases sharply and the longitudinal velocity profile alters.

In accordance with Boussinesq's model the turbulence boundary layer on a flat plate is described by the equations

$$\overline{V}_X \frac{\partial \overline{V}_X}{\partial X} + \overline{V}_Y \frac{\partial \overline{V}_X}{\partial Y} = \frac{\partial}{\partial Y} \left[(\nu + \nu_t) \frac{\partial \overline{V}_X}{\partial Y} \right],$$
$$\frac{\partial \overline{V}_X}{\partial X} + \frac{\partial \overline{V}_Y}{\partial Y} = 0,$$

where ν_t is the turbulent viscosity.

These equations have formed the basis of most theoretical investigations on the determination of the average fluid velocity and the drag coefficient in the stabilized region of turbulent flow on a flat plate. In what follows, major attention will be paid to empirical and semiempirical formulas that approximate numerous experimental data quite well.

Experiments show that the turbulent boundary layer on a flat plate includes two qualitatively different regions, namely, the wall region (adjacent to the plate surface) and the outer region (bordering the unperturbed stream). By analogy with the flow through a circular tube, it is common to subdivide the thin wall region into three subdomains (von Kármán's scheme):

$$\boxed{\text{wall region}} \;=\; \boxed{\text{viscous sublayer}} \;+\; \boxed{\text{buffer layer}} \;+\; \boxed{\text{logarithmic layer}}$$

The friction velocity U_* and the dimensionless internal coordinate y^+ defined by relations (E4.4.2.3) are introduced to describe the turbulent flow of the fluid in the wall region. The transverse coordinate Y in Eq. (E4.4.2.3) is measured from the plate surface.

The velocity profile in the viscous sublayer is linear, $\overline{V}_X(Y)/U_* = y^+$.

In the logarithmic layer and the outer region, the average velocity profile is quite well described by the relations (Monin & Yaglom, 1992)

$$\frac{U_i - \overline{V}_X(Y)}{U_*} = \begin{cases} -2.5\ln(Y/\delta) + 2.2 & \text{for } Y/\delta \le 0.15, \\ 9.6\,[1 - (Y/\delta)]^2 & \text{for } 0.15 \le Y/\delta \le 1, \end{cases}$$

where $\delta = \delta(X)$ is the boundary layer thickness.

Note the Spalding implicit interpolation formula

$$y^+ = 2.5\,V^+ + 0.135\left[\exp(-V^+) - 1 - V^+ - \tfrac{1}{2}(V^+)^2 - \tfrac{1}{6}(V^+)^3\right],$$

where $y^+ = YU_*/\nu$ and $V^+ = 0.4\,\overline{V}(Y)/U_*$. This formula quite well describes the average velocity profile within the entire wall region $0 \le Y/\delta \le 0.15$.

The turbulent boundary layer thickness can be calculated from the formula

$$\delta = 0.33\,X\sqrt{\tfrac{1}{2}c_f}. \tag{E4.5.2.1}$$

▶ **The local coefficient of friction.** In the case of a two-sided flow past a flat plate, the local coefficient of friction $c_f = c_f(X)$ is expressed via other hydrodynamic parameters as

$$c_f = \frac{\tau_s}{\tfrac{1}{2}\rho U_i^2} = 2\left(\frac{U_*}{U_i}\right)^2.$$

For the turbulent boundary layer on a flat plate, von Kármán's friction law with modified numerical coefficients

$$\frac{1}{\sqrt{c_f}} = 1.77\ln\left(c_f \mathsf{Re}_X\right) + 2.4, \tag{E4.5.2.2}$$

is typically used, which quite well agrees with experimental data. In Eq. (E4.5.2.2), $\mathsf{Re}_X = U_i X/\nu$ is the local Reynolds number. Here it is assumed that the flow turbulization in the boundary layer starts from the front edge of the plate.

Relation (E4.5.2.2) defines the friction coefficient versus the Reynolds number implicitly. For $\mathsf{Re}_X > 10^9$, it is more convenient to use the simpler explicit Falkner's formula

$$c_f = 0.0262\,\mathsf{Re}_X^{-1/7}, \tag{E4.5.2.3}$$

The corresponding mean coefficient of friction for a plate of length L is given by

$$\langle c_f \rangle = \frac{1}{L}\int_0^L c_f\,dX = 0.0306\,\mathsf{Re}_L^{-1/7}, \tag{E4.5.2.4}$$

where $\mathsf{Re}_L = U_i L/\nu$.

The comparison of Eq. (E4.5.2.4) with Eq. (E4.5.1.10) reveals that the resistance of a flat plate to turbulent flow is much greater than to laminar flow and decreases with increasing Reynolds number considerably slower.

Note also the Schlichting formulas

$$c_{\mathrm{f}} = \left(2\log_{10}\mathrm{Re}_X - 0.65\right)^{-2.3}, \qquad \langle c_{\mathrm{f}}\rangle = 0.455\left(\log_{10}\mathrm{Re}_L\right)^{-2.58},$$

which are accurate to within few percent for $10^5 \le \mathrm{Re}_L \le 10^9$.

More detailed information about the structure of turbulent flows on a flat plate, as well as various relations for determining the average velocity profile and the local coefficient of friction, can be found in the bibliography given at the end of this chapter.

E4.6. Spherical Particles and Circular Cylinders in Translational Flow

▶ **Preliminary remarks.** Solving the problem on the interaction of a solid particle, drop, or bubble with the surrounding continuous phase underlies the design and analysis of many technological processes. The industrial applications of such interaction include classification of suspensions in hydrocyclones, sedimentation of colloids, pneumatic conveyers, fluidization, heterogeneous catalysis in suspension, dissolving solid particles, extraction from drops, absorption, and evaporation into bubbles.

The description of a large variety of meteorological phenomena is also based on the analysis of motion of a collection of drops in air. The recent increase in atmosphere pollution is a serious problem, which requires understanding the transfer of mechanical, chemical, and radioactive particles in the atmosphere.

In rarefied systems of particles, drops, or bubbles, the particle–particle interaction can be neglected in the first approximation; then one deals with the behavior of a single particle moving in fluid. In this case, the streamline pattern depends on the particle shape, the flow type (translational or shear), and a number of other geometric factors.

E4.6.1. Stokes Equations and Their Solution for the Axisymmetric Case

▶ **Stokes equations.** One of the main approaches to the analysis and simplification of the Navier–Stokes equations is as follows. One assumes that the nonlinear inertia term $(\mathbf{V} \cdot \nabla)\mathbf{V}$ is small compared with the linear viscous term $\nu\Delta\mathbf{V}$ and hence can be neglected altogether or taken into account in some special way. This method is well-founded for $\mathrm{Re} = LU/\nu \ll 1$ and is widely used for studying the motion of particles, drops, and bubbles in fluids. Low Reynolds numbers are typical of the following three cases: slow (creeping) flows, highly viscous fluids, and small dimensions of particles.

For steady-state flows of viscous incompressible fluid, by neglecting the inertia terms in (E4.2.1.4) and by including all conservative mass forces in the pressure P, we arrive at the Stokes equations

$$\begin{aligned} \nabla \cdot \mathbf{V} &= 0, \\ \mu\Delta\mathbf{V} &= \nabla P. \end{aligned} \qquad (\text{E4.6.1.1})$$

The Stokes equations (E4.6.1.1) are linear and hence much simpler than the nonlinear Navier–Stokes equations. For any two solutions $\{\mathbf{V}_1, P_1\}$ and $\{\mathbf{V}_2, P_2\}$ of (E4.6.1.1), the sum $\{\alpha\mathbf{V}_1 + \beta\mathbf{V}_2, \alpha P_1 + \beta P_2\}$ satisfies the same equations for any constants α and β.

In axisymmetric problems, all the variables in the spherical coordinates R, θ, φ are independent of φ, and the third component of the fluid velocity is zero, $V_\varphi = 0$. The fluid velocity components V_R and V_θ can be expressed via the stream function Ψ as follows:

$$V_R = \frac{1}{R^2 \sin\theta} \frac{\partial\Psi}{\partial\theta}, \quad V_\theta = -\frac{1}{R\sin\theta} \frac{\partial\Psi}{\partial R}. \tag{E4.6.1.2}$$

Then the first equation in (E4.6.1.1) (the continuity equation) is satisfied automatically. On eliminating the terms containing the pressure from the second vector equation (E4.6.1.1) and replacing the velocity components by the right-hand side of (E4.6.1.2), we obtain the following equation for the stream function:

$$E^2\left(E^2\Psi\right) = 0, \quad E^2 \equiv \frac{\partial^2}{\partial R^2} + \frac{\sin\theta}{R^2}\frac{\partial}{\partial\theta}\left(\frac{1}{\sin\theta}\frac{\partial}{\partial\theta}\right). \tag{E4.6.1.3}$$

▶ **General solution of Stokes equations for the axisymmetric case.** The general solution of Eq. (E4.6.1.3), which gives the finite fluid velocities on the flow axis for $\theta = 0$ and $\theta = \pi$, has the form

$$\Psi(R,\theta) = \sum_{n=2}^{\infty}\left(A_n R^n + B_n R^{1-n} + C_n R^{n+2} + D_n R^{3-n}\right)\mathcal{J}_n(\cos\theta). \tag{E4.6.1.4}$$

Here A_n, B_n, C_n, and D_n are arbitrary constants, and $\mathcal{J}_n(\zeta)$ are the Gegenbauer functions of the first kind, which can be represented in the form

$$\mathcal{J}_n(\zeta) = \frac{P_{n-2}(\zeta) - P_n(\zeta)}{2n-1} = -\frac{1}{(n-1)!}\left(\frac{d}{d\zeta}\right)^{n-2}\left(\frac{\zeta^2-1}{2}\right)^{n-1},$$

$$P_n(\zeta) = \frac{1}{n!\,2^n}\frac{d^n}{d\zeta^n}(\zeta^2-1)^n,$$

where $P_n(\zeta)$ are the Legendre functions. The Gegenbauer functions of the first kind are polynomials; in particular,

$$\mathcal{J}_0(\zeta)=1, \quad \mathcal{J}_1(\zeta)=-\zeta, \quad \mathcal{J}_2(\zeta)=\tfrac{1}{2}(1-\zeta^2), \quad \mathcal{J}_3(\zeta)=\tfrac{1}{2}\zeta(1-\zeta^2), \quad \mathcal{J}_4(\zeta)=\tfrac{1}{8}(1-\zeta^2)(5\zeta^2-1).$$

The corresponding fluid velocity components and pressure are given by the formulas

$$V_R = -\sum_{n=2}^{\infty}\left(A_n R^{n-2} + B_n R^{-n-1} + C_n R^n + D_n R^{1-n}\right)P_{n-1}(\cos\theta),$$

$$V_\theta = \sum_{n=2}^{\infty}\left[nA_n R^{n-2} - (n-1)B_n R^{-n-1} + (n+2)C_n R^n - (n-3)D_n R^{1-n}\right]\frac{\mathcal{J}_n(\cos\theta)}{\sin\theta},$$

$$P = -2\mu\sum_{n=2}^{\infty}\left(\frac{2n+1}{n-1}C_n R^{n-1} + \frac{2n-3}{n}D_n R^{-n}\right)P_{n-1}(\cos\theta) + \text{const}.$$

The force exerted by the fluid on any spherical boundary described by the equation $R = \text{const}$ is given by

$$F = 4\pi\mu D_2.$$

It is of interest that this force is completely determined by only one coefficient of the series (E4.6.1.4).

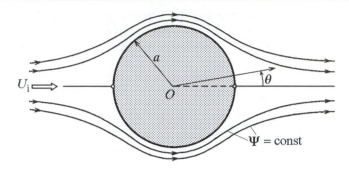

Figure E4.4. Translational Stokes flow past a spherical particle.

E4.6.2. Spherical Particles in Translational Stokes Flow (at Re → 0)

▶ **Flow past a spherical particle.** Consider a solid spherical particle of radius a in a translational Stokes flow with velocity U_i and dynamic viscosity μ (Figure E4.4). We assume that the fluid has dynamic viscosity μ. We use the spherical coordinate system R, θ, φ with origin at the center of the particle and with angle θ measured from the direction of the incoming flow (that is, from the rear stagnation point on the particle surface). In view of the axial symmetry, only two components of the fluid velocity, namely, V_R and V_θ, are nonzero, and all the unknowns are independent of the third coordinate φ.

The fluid velocity distribution is given by the Stokes equations (E4.6.1.1) with the no-slip boundary conditions on the surface of the solid sphere,

$$V_R = V_\theta = 0 \qquad \text{at} \quad R = a, \tag{E4.6.2.1}$$

and the boundary conditions at infinity of the form

$$V_R \to U_i \cos \theta, \quad V_\theta \to -U_i \sin \theta \qquad \text{as} \quad R \to \infty, \tag{E4.6.2.2}$$

which correspond to condition (E4.2.2.1) of translational unperturbed flow far from the particle.

By passing from the fluid velocity components V_R, V_θ to the stream function Ψ according to formulas (E4.6.1.2), we arrive at Eq. (E4.6.1.3). It follows from the remote boundary conditions (E4.6.2.2) that in the general solution (E4.6.1.4) it suffices to retain only the first term (corresponding to the case $n = 2$). The no-slip conditions (E4.6.2.1) allow us to find the unknown constants A_2, B_2, C_2, and D_2. The resulting expression for the stream function,

$$\Psi = \frac{1}{2} U_i R^2 \left(1 - \frac{3}{2} \frac{a}{R} + \frac{1}{2} \frac{a^3}{R^3} \right) \sin^2 \theta, \tag{E4.6.2.3}$$

allows us to find the fluid velocity and pressure in the form

$$V_R = U_i \left(1 - \frac{3}{2} \frac{a}{R} + \frac{1}{2} \frac{a^3}{R^3} \right) \cos \theta,$$

$$V_\theta = -U_i \left(1 - \frac{3}{4} \frac{a}{R} - \frac{1}{4} \frac{a^3}{R^3} \right) \sin \theta, \tag{E4.6.2.4}$$

$$P = P_i - \frac{3\mu U_i a \cos \theta}{2R^2},$$

where P_i is the unperturbed pressure far from the particle.

The viscous drag acting on the spherical particle is given by the Stokes formula

$$F = 6\pi\mu a U_i. \tag{E4.6.2.5}$$

▶ **Steady-state motion of spherical particles in a fluid.** In chemical technology, one often meets the problem of a steady-state motion of a spherical particle with velocity U_i in a quiescent fluid. Since the Stokes equations are linear, the solution of this problem can be obtained from formulas (E4.6.2.4) by adding the terms $\bar{V}_R = -U_i\cos\theta$ and $\bar{V}_\theta = U_i\sin\theta$, which describe a translational flow with velocity U_i in the direction opposite to the incoming flow.

The drag coefficient for a solid spherical particle is calculated by the formula (the Stokes law)

$$c_f = \frac{F}{\frac{1}{2}\rho U_i^2 \pi a^2} = \frac{12}{\mathsf{Re}}, \tag{E4.6.2.6}$$

where $\mathsf{Re} = \rho U_i a/\mu$ and ρ is the density of the fluid. The Stokes law is confirmed by experiments for $\mathsf{Re} < 0.1$.

By equating the drag force F of the sphere with the difference $\frac{4}{3}\pi a^3 g\Delta\rho$ between the gravity and buoyancy forces, one can estimate the steady-state velocity of relative motion of phases (the velocity at which a spherical particle falls or rises) as

$$U = \frac{2}{9}\frac{ga^2\Delta\rho}{\mu}, \tag{E4.6.2.7}$$

where $\Delta\rho$ is the difference between the densities of the surrounding fluid and the particle and g is the free fall acceleration.

E4.6.3. Spherical Particles in Translational Flow in a Wide Range of Re

▶ **Flow past spherical particles in a wide range of Re.** Neglecting the inertia term for the flow past a sphere is adequate to experiments only in the limit case $\mathsf{Re} \to 0$. For $\mathsf{Re} = 0.05$, the error in estimating the drag by formula (E4.6.2.6) is 1.5 to 2%, and for $\mathsf{Re} = 0.5$ the error becomes 10.5 to 11%. Therefore, one can use the estimate $c_f = 12/\mathsf{Re}$ for the drag coefficient only for $\mathsf{Re} < 0.2$ (in this case, the maximum error does not exceed 5%).

More exact than the Stokes law (E4.6.2.6) is the two-term Oseen's expansion of the drag coefficient as $\mathsf{Re} \to 0$:

$$c_f = \frac{12}{\mathsf{Re}}\left(1 + \frac{3}{8}\mathsf{Re}\right). \tag{E4.6.3.1}$$

The difference between Oseen's approximation (E4.6.3.1) and experimental data is 0 to 1.0% for $\mathsf{Re} \le 0.05$ and 4 to 6% for $\mathsf{Re} = 0.5$.

For $\mathsf{Re} > 0.5$, asymptotic solutions no longer give an adequate description of translational flow of a viscous fluid past a spherical particle.

Numerous available numerical solutions for the Navier–Stokes equations, as well as experimental data, provide a detailed analysis of the flow pattern for increasing Reynolds numbers. For $0.5 < \mathsf{Re} < 10$, there is no flow separation, although the fore-and-aft symmetry typical of inertia-free Stokes flow past a sphere is more and more distorted. Finally, at $\mathsf{Re} \approx 10$, flow separation occurs at the rear of the particle.

The range $10 < \mathsf{Re} < 65$ is characterized by the existence of a closed stable area at the rear, in which there is an axisymmetric recirculating wake. As Re increases, the wake lengthens, and the separation ring θ_s moves forward from the rear point ($\theta_s = 0°$ at $\text{Re} = 10$) to $\theta_s = 72°$ at $\text{Re} = 200$ according to the law

$$\theta_s = 42.5 \left(\ln \frac{\text{Re}}{10} \right)^{0.483} \qquad \text{for} \quad 10 < \text{Re} < 200. \qquad \text{(E4.6.3.2)}$$

Here, as well as in (E4.6.3.3), θ_s is measured in degrees.

At $\text{Re} > 65$, the vorticity region in the rear area ceases to be stable and becomes unsteady. At $65 < \text{Re} < 200$, a long oscillating wake is formed behind the particle, which gradually becomes turbulent for $200 < \text{Re} < 1.5 \times 10^5$. Simultaneously, the separation point moves upstream according to the law

$$\theta_s = 102 - 213 \, \text{Re}^{-0.37} \qquad \text{for} \quad 200 < \text{Re} < 1.5 \times 10^5. \qquad \text{(E4.6.3.3)}$$

▶ **Formulas for the drag coefficient in a wide range of Re.** We give two simple approximate formulas for the drag coefficient of a spherical particle

$$
\begin{aligned}
c_f &= \frac{12}{\text{Re}} \left(1 + 0.241 \, \text{Re}^{0.687} \right), & 0 \le \text{Re} \le 400, \\
c_f &= \frac{12}{\text{Re}} \left(1 + 0.0811 \, \text{Re}^{0.879} \right), & 200 \le \text{Re} \le 2500,
\end{aligned}
\qquad \text{(E4.6.3.4)}
$$

where the Reynolds number is determined with respect to the radius. In formulas (E4.6.3.4), the maximum error does not exceed 5% in the given ranges.

In a wide range of Reynolds numbers, one can use the following more complicated approximation to the drag coefficient:

$$c_f = \frac{12}{\text{Re}} \left(1 + 0.241 \, \text{Re}^{0.687} \right) + 0.42 \left(1 + 1.902 \times 10^4 \, \text{Re}^{-1.16} \right)^{-1},$$

whose maximum error does not exceed 6% for $\text{Re} < 1.5 \times 10^5$.

At $\text{Re} \approx 1.5 \times 10^5$, one can observe the "drag crisis" characterized by a sharp decrease in the drag coefficient; the boundary layer becomes more and more turbulent; the separation point shifts abruptly to the aft area.

For $\text{Re} \ge 1.7 \times 10^5$, the drag coefficient can be calculated by the formulas

$$
c_f = \begin{cases}
28.18 - 5.3 \log_{10} \text{Re} & \text{for } 1.7 \times 10^5 \le \text{Re} \le 2 \times 10^5, \\
0.1 \log_{10} \text{Re} - 0.46 & \text{for } 2 \times 10^5 < \text{Re} \le 5 \times 10^5, \\
0.19 - 4 \times 10^4 \text{Re}^{-1} & \text{for } 5 \times 10^5 < \text{Re}.
\end{cases}
$$

E4.6.4. Translational Flow Past a Cylinder (the Plane Problem)

In chemical technology and power engineering, equipment containing heat exchanging pipes and various cylindrical links immersed in a moving fluid is often used. The estimation of the hydrodynamic action on these elements is based on the solution of the plane problem on the flow past a cylinder.

▶ **Flow past a cylinder at low Reynolds numbers.** Consider the translational flow of a viscous incompressible fluid with velocity U_i about a circular cylinder of radius a. The drag coefficient for a circular cylinder at low Reynolds numbers is calculated by the asymptotic formula

$$c_f = \frac{F}{aU_i^2 \rho} = \frac{4\pi}{\mathsf{Re}}(\Delta - 0.87\Delta^3),$$

$$\Delta = \left(\ln \frac{3.703}{\mathsf{Re}} \right)^{-1}, \quad \mathsf{Re} = \frac{aU_i\rho}{\mu},$$

(E4.6.4.1)

where F is the force per unit length of the cylinder and ρ is the density of the fluid.

Comparison with experimental data shows that formula (E4.6.4.1) can be used for $0 < \mathsf{Re} < 0.4$.

▶ **Flow past a cylinder at moderate Reynolds numbers.** According to experimental data, a nonseparating flow past a circular cylinder is realized at $\mathsf{Re} \le 2.5$. At such Reynolds numbers, one can use the following approximate formula for calculating the drag coefficient:

$$c_f = 5.65\,\mathsf{Re}^{-0.78}\left(1 + 0.26\,\mathsf{Re}^{0.82}\right) \qquad \text{for} \quad 0.05 \le \mathsf{Re} \le 2.5;$$

this formula was obtained from experimental and numerical results.

If the Reynolds number becomes larger than the critical value $\mathsf{Re} \approx 2.5$, then the vortex counterflow with closed streamlines arises near the rear point, that is, separation occurs. As the Reynolds number increases, the separation point gradually moves from the axis upward along the cylinder surface. The drag coefficient for a separated flow past a cylinder at moderate Reynolds numbers can be calculated by the empirical formulas

$$c_f = 5.65 \times 10^{-0.78}\left(1 + 0.333\,\mathsf{Re}^{0.55}\right) \qquad \text{for} \quad 2.5 < \mathsf{Re} \le 20,$$

$$c_f = 5.65 \times 10^{-0.78}\left(1 + 0.148\,\mathsf{Re}^{0.82}\right) \qquad \text{for} \quad 20 < \mathsf{Re} \le 200.$$

▶ **Flow past a cylinder at high Reynolds numbers.** With further increase of Re, the rear vortices become longer and then alternative vortex separation occurs (the Kármán vortex street is formed). Simultaneously, the separation point moves closer to the equatorial section. The frequency ν_f of vortex shedding from the rear area is an important characteristic of the flow past a cylinder. It can be determined from the empirical formula

$$\mathsf{St} = \frac{0.13}{c_f}\left[1 - \exp(-2.38\,c_f)\right],$$

where $\mathsf{St} = a\nu_f/U_i$ is the Strouhal number.

We also present another useful formula for the vortex separation frequency: $\nu_f = 0.08\,U_i/b$, where b is the half-width of the wake at the point of destruction.

Starting from $\mathsf{Re} \approx 0.5 \times 10^3$, one can speak of a developed hydrodynamic boundary layer. The flow remains laminar in a considerable part of this layer. If the Reynolds number varies in the range $0.5 \times 10^3 < \mathsf{Re} < 0.5 \times 10^5$, the separation point θ_s of the laminar boundary layer gradually moves from $71.2°$ to $95°$.

For $\mathsf{Re} > 2000$, the wake becomes totally turbulent at large distances from the body.

The curve $c_f(\mathsf{Re})$ contains two straight-line segments (self-similarity areas), where the drag coefficient is practically constant,

$$c_f = 1.0 \qquad \text{for} \quad 3 \times 10^2 < \mathsf{Re} < 3 \times 10^3,$$

$$c_f = 1.1 \qquad \text{for} \quad 4 \times 10^3 < \mathsf{Re} < 10^5.$$

In the intermediate region between these segments, the drag coefficient monotonically increases with Reynolds number.

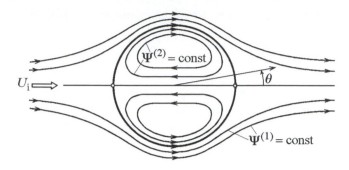

Figure E4.5. Translational Stokes flow past a spherical drop.

▶ **Developed turbulence in the boundary layer on a cylinder.** Developed turbulence within the boundary layer takes place at higher Reynolds numbers $\mathrm{Re} \approx 10^5$ and is accompanied by the "drag crisis." The cylinder drag first decreases sharply to $c_f \approx 0.3$ at $\mathrm{Re} = 3.5 \times 10^5$ and then begins to increase and again enters the self-similar regime characterized by the constant value

$$c_f = 0.9 \qquad \text{for} \quad \mathrm{Re} > 5 \times 10^5.$$

E4.7. Spherical Drops and Bubbles in Translational Flow

E4.7.1. Spherical Drops and Bubbles in Translational Stokes Flow (at Re → 0)

▶ **Flow past a spherical drop or bubble.** Now consider a spherical drop of radius a in a translational Stokes flow of another fluid with velocity U_i (Figure E4.5). We assume that the dynamic viscosities of the outer and inner fluids are equal to μ_1 and μ_2, respectively. The unknown variables outside and inside the drop are indicated by the superscripts (1) and (2), respectively.

To obtain the velocity and pressure profiles for the fluid in each phase, we shall use the Stokes equations (E4.6.1.1). As previously, the condition that the flow is uniform far from the drop has the form (E4.6.2.2).

Let us write out four conditions that must be satisfied on the boundary of a spherical drop.

There is no flow across the interface:

$$V_R^{(1)} = V_R^{(2)} = 0 \qquad \text{at} \quad R = a. \tag{E4.7.1.1}$$

The tangential velocity is continuous across the interface:

$$V_\theta^{(1)} = V_\theta^{(2)} \qquad \text{at} \quad R = a. \tag{E4.7.1.2}$$

The normal stress jump across the interface is equal to the pressure increment due to interfacial tension:

$$P^{(1)} - 2\mu_1 \frac{\partial V_R^{(1)}}{\partial R} + \frac{2\sigma}{a} = P^{(2)} - 2\mu_2 \frac{\partial V_R^{(2)}}{\partial R} \qquad \text{at} \quad R = a, \tag{E4.7.1.3}$$

where σ is the interfacial tension.

The tangential stress is continuous across the interface:

$$\mu_1 \left(\frac{\partial V_\theta^{(1)}}{\partial R} - \frac{V_\theta^{(1)}}{R} \right) = \mu_2 \left(\frac{\partial V_\theta^{(2)}}{\partial R} - \frac{V_\theta^{(2)}}{R} \right) \qquad \text{at} \quad R = a. \tag{E4.7.1.4}$$

We also use the boundedness of the solution at the drop center:

$$V_R^{(2)} < \infty, \quad V_\theta^{(2)} < \infty \quad \text{at} \quad R = 0. \tag{E4.7.1.5}$$

We introduce the stream function $\Psi^{(m)}$ in each of the phases ($m = 1, 2$) according to formulas (E4.6.1.2). Conditions (E4.7.1.1)–(E4.7.1.5) allow us to determine the constants in the general solutions (E4.7.1.5) inside and outside the drop. As a result, we obtain the Hadamard–Rybczynski solution

$$\Psi^{(1)} = \frac{1}{4} U_i R^2 \left(2 - \frac{2 + 3\beta}{1 + \beta} \frac{a}{R} + \frac{\beta}{1 + \beta} \frac{a^3}{R^3} \right) \sin^2 \theta,$$

$$\Psi^{(2)} = -\frac{U_i}{4(1 + \beta)} R^2 \left(1 - \frac{R^2}{a^2} \right) \sin^2 \theta, \quad \beta = \frac{\mu_2}{\mu_1}. \tag{E4.7.1.6}$$

By using formulas (E4.6.1.2), we calculate the velocity and the pressure outside the drop:

$$V_R^{(1)} = U_i \left[1 - \frac{2 + 3\beta}{2(1 + \beta)} \frac{a}{R} + \frac{\beta}{2(1 + \beta)} \frac{a^3}{R^3} \right] \cos \theta,$$

$$V_\theta^{(1)} = -U_i \left[1 - \frac{2 + 3\beta}{4(1 + \beta)} \frac{a}{R} - \frac{\beta}{4(1 + \beta)} \frac{a^3}{R^3} \right] \sin \theta, \tag{E4.7.1.7}$$

$$P^{(1)} = P_0^{(1)} - \frac{\mu_1 U_i a (2 + 3\beta)}{2(1 + \beta)} \frac{\cos \theta}{R^2}, \quad \beta = \frac{\mu_2}{\mu_1}.$$

The velocity and the pressure inside the drop are given by

$$V_R^{(2)} = -\frac{U_i}{2(1 + \beta)} \left(1 - \frac{R^2}{a^2} \right) \cos \theta,$$

$$V_\theta^{(2)} = \frac{U_i}{2(1 + \beta)} \left(1 - 2\frac{R^2}{a^2} \right) \sin \theta, \tag{E4.7.1.8}$$

$$P^{(2)} = P_0^{(2)} + \frac{5\mu_2 U_i R \cos \theta}{a^2 (1 + \beta)}.$$

The constants $P_0^{(1)}$ and $P_0^{(2)}$ in the expressions (E4.7.1.7) and (E4.7.1.8) for the pressure fields are related by

$$P_0^{(2)} - P_0^{(1)} = \frac{2\sigma}{a}. \tag{E4.7.1.9}$$

The drag force acting on the spherical drop is

$$F = 2\pi a U_i \frac{2\mu_1 + 3\mu_2}{\mu_1 + \mu_2}. \tag{E4.7.1.10}$$

As $\beta = \mu_2/\mu_1 \to \infty$, Eq. (E4.7.1.10) becomes the Stokes formula (E4.6.2.5) for a solid particle. The passage to the limit as $\beta \to 0$ corresponds to a gas bubble.

▶ **Steady-state motion of drops and bubbles in a fluid.** Consider the problem of a steady-state motion of a spherical drop or bubble with velocity U_i in a quiescent fluid. Since the Stokes equations are linear, the solution of this problem can be obtained from formulas (E4.7.1.7) and (E4.7.1.8) by adding the terms $\bar{V}_R = -U_i \cos \theta$ and $\bar{V}_\theta = U_i \sin \theta$, which describe a translational flow with velocity U_i in the direction opposite to the incoming flow.

The drag coefficient for a drop is calculated by the formula

$$c_f = \frac{F}{\frac{1}{2}\rho_1 U_i^2 \pi a^2} = \frac{4}{Re}\left(\frac{2+3\beta}{1+\beta}\right), \quad \text{where} \quad Re = \frac{\rho_1 U_i a}{\mu_1}. \tag{E4.7.1.11}$$

By equating the drag force F of the sphere with the difference $\frac{4}{3}\pi a^3 g \Delta\rho$ between the gravity and buoyancy forces, one can estimate the steady-state velocity of relative motion of phases (the velocity at which a spherical drop falls or rises) as

$$U = \frac{2}{3}\frac{ga^2\Delta\rho}{\mu_1}\left(\frac{1+\beta}{2+3\beta}\right), \tag{E4.7.1.12}$$

where $\Delta\rho$ is the difference between the densities of the outer and the inner fluid and g is the free fall acceleration.

Relations (E4.7.1.11) and (E4.7.1.12) cover the entire range $0 \le \beta \le \infty$ of the phase viscosity ratio. In the limit case $\beta \to \infty$, formulas (E4.7.1.11) and (E4.7.1.12) become formulas (E4.6.2.6) and (E4.6.2.7) for a solid spherical particle. In the other limit case $\beta = 0$ corresponding to a spherical gas bubble in a highly viscous liquid, formulas (E4.7.1.11) and (E4.7.1.12) become

$$c_f = \frac{8}{Re}, \quad U = \frac{1}{3}\frac{ga^2\Delta\rho}{\mu_1} \quad \text{(gas bubble)}. \tag{E4.7.1.13}$$

These formulas hold only for extremely pure liquids without any surfactants.

It is known that even very small quantities of surfactants are adsorbed on the bubble surface and lead to its "solidification." This results in eliminating internal circulation, and hence the bubble rises according to the Stokes law (E4.7.1.11) for solid particles.

E4.7.2. Spherical Drops and Bubbles in Translational Flow at Various Reynolds Numbers

▶ **Bubble in a translational flow.** For determining the drag coefficient of a spherical bubble in a translational flow in the entire range of Reynolds numbers, one can use the interpolation formula

$$c_f = \frac{8}{Re} + \frac{16}{Re+16}, \quad Re = \frac{aU_i}{\nu}, \tag{E4.7.2.1}$$

where a is the bubble radius and ν is the kinematic viscosity of the surrounding fluid. In both limit cases $Re \to 0$ and $Re \to \infty$, formula (E4.7.2.1) gives correct asymptotic results; its maximum error for intermediate Reynolds numbers is less than 4.5%. The drag coefficient of a spherical bubble monotonically decreases as the Reynolds number increases.

▶ **Drop in a translational liquid flow.** For the drag coefficient of a spherical drop in a translational flow at small Reynolds numbers, we have the two-term expansion

$$c_f = \frac{3\beta+2}{\beta+1}\left(\frac{4}{Re} + \frac{Re}{2}\right), \quad Re = \frac{aU_i}{\nu}, \tag{E4.7.2.2}$$

where ν is the kinematic viscosity of the fluid surrounding the drop.

The spherical form of a drop or a bubble in a Stokes flow follows from the fact that the flow is inertia-free. However, even for the case in which the inertia forces dominate viscous forces and the Reynolds number cannot be considered small, the drop remains undeformed if the inertia forces are small compared with the capillary forces. The ratio of inertial to capillary forces is measured by the Weber number $We = \rho_1 U_i^2 a / \sigma$, where σ is the surface tension at the drop boundary. For small We, a deformable drop will preserve the spherical form.

At the end of Subsection E4.7.1, it was already noted that even a small amount of surfactants in any of the adjacent phases may lead to the "solidification" of the interface, so that the laws of flow around a drop become close to those for a solid particle. This effect often occurs in practice.

The drag coefficient of a spherical drop can be determined by the formula

$$c_f(\beta, Re) = \frac{1}{\beta + 1} c_f(0, Re) + \frac{\beta}{\beta + 1} c_f(\infty, Re). \tag{E4.7.2.3}$$

Here $c_f(0, Re)$ is the drag coefficient of the spherical bubble, which can be calculated by the formula (E4.7.2.1), and $c_f(\infty, Re)$ is the drag coefficient of a solid spherical particle, which can be calculated by (E4.6.3.4). The approximate expression (E4.7.2.3) gives three correct terms of the expansion for small Reynolds numbers; for $0 \le Re \le 50$, the maximum error is less than 5%.

E4.8. Flow Past Nonspherical Particles

E4.8.1. Translational Stokes Flow Past Ellipsoidal Particles

The axisymmetric problem about a translational Stokes flow past an ellipsoidal particle admits an exact closed-form solution. In what follows, we present a brief summary of the corresponding results (see Happel & Brenner, 1965).

▶ **Oblate ellipsoid of revolution.** Consider an oblate ellipsoid of revolution with semiaxes a and b ($a > b$) in a translational Stokes flow with velocity U_i. We assume that the fluid viscosity is equal to μ.

We pass from the Cartesian coordinates X, Y, Z to the reference frame σ, τ, φ fixed to the oblate ellipsoid of revolution by using the transformations

$$X^2 = m^2(1 + \sigma^2)(1 - \tau^2)\cos^2\varphi, \quad Y^2 = m^2(1 + \sigma^2)(1 - \tau^2)\sin^2\varphi, \quad Z = m\sigma\tau,$$

$$\text{where} \quad m = \sqrt{a^2 - b^2} \quad (\sigma \ge 0, \, -1 \le \tau \le 1).$$

As a result, the ellipsoid surface is given by a constant value of the coordinate σ:

$$\sigma = \sigma_0, \quad \text{where} \quad \sigma_0 = \left[(a/b)^2 - 1\right]^{-1/2}.$$

Since the problem is axisymmetric, we introduce the stream function as

$$V_\sigma = \frac{1}{m^2\sqrt{(1 + \sigma^2)(\sigma^2 + \tau^2)}} \frac{\partial\Psi}{\partial\tau}, \quad V_\tau = -\frac{1}{m^2\sqrt{(1 - \tau^2)(\sigma^2 + \tau^2)}} \frac{\partial\Psi}{\partial\sigma}.$$

Then the Stokes equations (E4.6.1.1) are reduced to one equation for Ψ, which can be solved by the separation of variables. By satisfying the boundary condition for a uniform flow remote from the particle and the no-slip conditions at the particle boundary, we obtain

$$\Psi = \frac{1}{2}m^2 U_i (1 - \tau^2) \left[\sigma^2 + 1 - \frac{(\sigma_0^2 + 1)\sigma - (\sigma_0^2 - 1)(\sigma^2 + 1)\operatorname{arccot}\sigma}{\sigma_0 - (\sigma_0^2 - 1)\operatorname{arccot}\sigma_0}\right].$$

In the similar problem on the motion of an oblate ellipsoid with velocity U_i in a quiescent fluid, we have

$$\Psi = -\frac{1}{2}m^2 U_i(1-\tau^2)\frac{(\sigma_0^2+1)\sigma - (\sigma_0^2-1)(\sigma^2+1)\operatorname{arccot}\sigma}{\sigma_0 - (\sigma_0^2-1)\operatorname{arccot}\sigma_0}. \qquad \text{(E4.8.1.1)}$$

The force exerted on the ellipsoid by the fluid is

$$F = \frac{8\pi\mu U_i\sqrt{a^2-b^2}}{\sigma_0 - (\sigma_0^2-1)\operatorname{arccot}\sigma_0}.$$

As $\sigma_0 \to 0$, an oblate ellipsoid degenerates into an infinite thin disk of radius a. By passing to the limit in (E4.8.1.1), we can obtain the following expression for the stream function:

$$\Psi = -\frac{1}{\pi}a^2 U_i(1-\tau^2)\big[\sigma + (\sigma^2+1)\operatorname{arccot}\sigma\big].$$

The disk moving in the direction perpendicular to its plane with velocity U_i in a quiescent fluid experiences the drag force

$$F = 16\mu a U_i, \qquad \text{(E4.8.1.2)}$$

which is less than the force acting on a sphere of the same radius (for the sphere, we have $F = 6\pi\mu a U_i$). Formula (E4.8.1.2) is confirmed by experimental data.

▶ **Prolate ellipsoid of revolution.** To solve the corresponding problem about an ellipsoidal particle in a translational Stokes flow, we use the reference frame σ, τ, φ fixed to the prolate ellipsoid of revolution. The transformation to the coordinates (σ, τ, φ) is determined by the formulas

$$X^2 = m^2(\sigma^2-1)(1-\tau^2)\cos^2\varphi, \quad Y^2 = m^2(\sigma^2-1)(1-\tau^2)\sin^2\varphi, \quad Z = m\sigma\tau,$$

$$\text{where} \quad m = \sqrt{a^2-b^2} \quad (\sigma \geq 1 \geq \tau \geq -1).$$

Here, as previously, the larger semiaxis is denoted by a. In this case, the ellipsoid surface is given by the equation

$$\sigma = \sigma_0, \qquad \text{where} \quad \sigma_0 = \big[1-(b/a)^2\big]^{-1/2}.$$

The fluid velocity is given by

$$V_\sigma = \frac{1}{m^2\sqrt{(\sigma^2-1)(\sigma^2-\tau^2)}}\frac{\partial\Psi}{\partial\tau}, \quad V_\tau = -\frac{1}{m^2\sqrt{(1-\tau^2)(\sigma^2-\tau^2)}}\frac{\partial\Psi}{\partial\sigma}$$

in terms of the stream function

$$\Psi = \frac{1}{2}m^2 U_i(1-\tau^2)\left[\sigma^2 - 1 - \frac{(\sigma_0^2+1)(\sigma^2-1)\operatorname{arctanh}\sigma - (\sigma_0^2-1)\sigma}{(\sigma_0^2+1)\operatorname{arctanh}\sigma_0 - \sigma_0}\right],$$

where $\operatorname{arctanh}\sigma = \dfrac{1}{2}\ln\dfrac{\sigma+1}{\sigma-1}$.

In the problem about a prolate ellipsoid of revolution moving at a velocity U_i in a quiescent fluid, the corresponding stream function has the form

$$\Psi = -\frac{1}{2}m^2 U_i(1-\tau^2)\frac{(\sigma_0^2+1)(\sigma^2-1)\operatorname{arctanh}\sigma - (\sigma_0^2-1)\sigma}{(\sigma_0^2+1)\operatorname{arctanh}\sigma_0 - \sigma_0}.$$

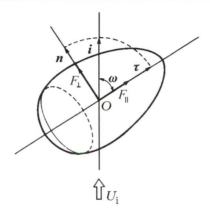

Figure E4.6. Body of revolution in translational flow (arbitrary orientation)

The drag force is calculated as

$$F = \frac{8\pi\mu U_i \sqrt{a^2 - b^2}}{(\sigma_0^2 + 1)\operatorname{arctanh}\sigma_0 - \sigma_0}.$$

If $a \gg b$, then the prolate ellipsoid degenerates into a needle-like rod. In this case, the force acting on the needle of length a and radius b which moves in the direction of its axis at a velocity U_i has the form

$$F = \frac{4\pi\mu a U_i}{\ln(a/b) + 0.193}.$$

E4.8.2. Translational Stokes Flow Past Bodies of Revolution

Consider bodies of revolution of any shape with arbitrary orientation in a translational flow at low Reynolds numbers. We assume that the axis of the body of revolution forms an angle ω with the direction of the fluid velocity at infinity (Figure E4.6). The unit vector $\mathbf{i}$ directed along the flow can be represented as the sum $\mathbf{i} = \boldsymbol{\tau}\cos\omega + \mathbf{n}\sin\omega$, where $\boldsymbol{\tau}$ is the unit vector directed along the body axis and $\mathbf{n}$ the unit vector in the plane of rotation of the body. In the Stokes approximation, the drag force is given by the following expression in the general case:

$$\mathbf{F} = \boldsymbol{\tau}F_{\|}\cos\omega + \mathbf{n}F_{\perp}\sin\omega, \tag{E4.8.2.1}$$

where $F_{\|}$ and $F_{\perp}$ are the drag forces of the body of revolution for its parallel ($\omega = 0$) and perpendicular ($\omega = \pi/2$) positions in the translational flow.

The projection of the drag force onto the incoming flow direction is equal to the inner product

$$(\mathbf{F} \cdot \mathbf{i}) = F_{\|}\cos^2\omega + F_{\perp}\sin^2\omega. \tag{E4.8.2.2}$$

It follows from (E4.8.2.1) and (E4.8.2.2) that to calculate the drag force of a body of revolution of any shape with arbitrary orientation in a Stokes flow, it suffices to know the value of this force only for two special positions of the body in space. The "axial" ($F_{\|}$) and "transversal" ($F_{\perp}$) drags can be obtained both theoretically and experimentally. In what follows, we present the expressions for $F_{\|}$ and $F_{\perp}$ for some bodies of revolution of nonspherical shape.

For a thin circular disk of radius a, one has

$$F_{\|} = 16\mu a U_i, \qquad F_{\perp} = \tfrac{32}{3}\mu a U_i.$$

For a dumbbell-like particle consisting of two adjacent spheres of equal radius a, one has

$$F_\parallel = 12\pi\mu a U_i \lambda_\parallel, \qquad \lambda_\parallel \approx 0.645,$$
$$F_\perp = 12\pi\mu a U_i \lambda_\perp, \qquad \lambda_\perp \approx 0.716.$$

In these formulas, the product $12\pi\mu a U_i$ is equal to the sum of drag forces for two isolated spheres of radius a.

For oblate ellipsoids of revolution with semiaxes a and b, one has

$$F_\parallel = 3.77\,(4a + b), \qquad F_\perp = 3.77\,(3a + 2b), \tag{E4.8.2.3}$$

where a is the equatorial radius ($a > b$).

For prolate ellipsoids of revolution with semiaxes a and b, one has

$$F_\parallel = 3.77\,(a + 4b), \qquad F_\perp = 3.77\,(2a + 3b), \tag{E4.8.2.4}$$

where b is the equatorial radius ($b > a$).

Formulas (E4.8.2.3) and (E4.8.2.4) are approximate. They work well for slightly deformed ellipsoids of revolution. In (E4.8.2.3), the maximum error is less than 6% for any ratio of the semiaxes.

E4.8.3. Translational Stokes Flow Past Particles of Arbitrary Shape

▶ **Hydrodynamic force and angular momentum acting on a particle.** A particle of arbitrary shape moving in an infinite fluid that is at rest at infinity is subject to the action of the hydrodynamic force and angular momentum due to its translational motion and rotation, respectively:

$$\mathbf{F} = \mu(\mathbf{K}\,\mathbf{U} + \mathbf{S}\,\boldsymbol{\omega}), \tag{E4.8.3.1}$$
$$\mathbf{M} = \mu(\mathbf{S}\,\mathbf{U} + \boldsymbol{\Omega}\,\boldsymbol{\omega}), \tag{E4.8.3.2}$$

where $\mathbf{K}$, $\mathbf{S}$, and $\boldsymbol{\Omega}$ are tensors of rank two depending on the particle geometry.

The symmetric tensor $\mathbf{K} = [K_{ij}]$ is called translational. It characterizes the drag of a body under translational motion and depends only on the size and shape of the body. In the principal axes, the translation tensor is reduced to the diagonal form

$$\mathbf{K} = \begin{bmatrix} K_1 & 0 & 0 \\ 0 & K_2 & 0 \\ 0 & 0 & K_3 \end{bmatrix}, \tag{E4.8.3.3}$$

where K_1, K_2, and K_3 are the principal drags acting on the body as it moves along the major axes. For orthotropic bodies (with three symmetry planes orthogonal to each other), the principal axes of the translational tensor are perpendicular to the corresponding symmetry planes. For axisymmetric bodies,, one of the axes (say, the first) is a major axis, and $K_2 = K_3$. For a sphere of radius a, any three pairwise perpendicular axes are major, and $K_1 = K_2 = K_3 = 6\pi a$.

A symmetric tensor $\boldsymbol{\Omega}$ is called a rotational tensor. It depends both on the shape and size of the particle and on the choice of the origin. The rotational tensor characterizes the drag under rotation of the body and has diagonal form with entries Ω_1, Ω_2, Ω_3 in the principal axes (the positions of the principal axes of the rotational and translational tensors in space are different). For axisymmetric bodies, one of the major axes (for instance, the first) is parallel to the symmetry axis, and in this case $\Omega_2 = \Omega_3$. For a spherical particle, we have $\Omega_1 = \Omega_2 = \Omega_3$.

The tensor **S** is symmetric only at a point O unique for each body; this point is called the center of hydrodynamic reaction. This tensor is called the conjugate tensor and characterizes the combined reaction of the body under translational and rotational motion (the drag moment in the translational motion and the drag force in the rotational motion). For bodies with orthotropic, axial, or spherical symmetry, the conjugate tensor is zero. However, it is necessary to take this tensor into account for bodies with helicoidal symmetry (propeller-like bodies).

In problems of gravity settling of particles, the translational tensor is most important.

▶ **Principal drags of some nonspherical bodies.**
For a thin circular disk of radius a, we have

$$K_1 = 16a, \qquad K_2 = \tfrac{32}{3}a, \qquad K_3 = \tfrac{32}{3}a. \qquad \text{(E4.8.3.4)}$$

For needle-like ellipsoids of length l and radius a, one has

$$K_1 = \frac{4\pi l}{2\ln(l/a) - 1}, \quad K_2 = \frac{4\pi l}{\ln(l/a) + 0.5}, \quad K_3 = \frac{4\pi l}{\ln(l/a) + 0.5}. \qquad \text{(E4.8.3.5)}$$

For thin circular cylinders of length l and radius a, one has

$$K_1 = \frac{4\pi l}{\ln(l/a) - 0.72}, \quad K_2 = \frac{4\pi l}{\ln(l/a) + 0.5}, \quad K_3 = \frac{4\pi l}{\ln(l/a) + 0.5}. \qquad \text{(E4.8.3.6)}$$

For a dumbbell-like particle consisting of two adjacent spheres of equal radius a, one has

$$K_1 = 24.3\,a, \qquad K_2 = 27.0\,a, \qquad K_3 = 27.0\,a. \qquad \text{(E4.8.3.7)}$$

▶ **Axisymmetric bodies of arbitrary shape.** For axisymmetric bodies of arbitrary shape, we introduce the notion of perimeter-equivalent sphere. To this end, we project all points of the surface of the body on a plane perpendicular to its axis. The projection is a circle of radius $a_\perp$. The perimeter-equivalent sphere has the same radius.

We let $\bar{c}_\parallel$ denote the relative coefficient of the axial drag equal to the ratio of the axial drag of an axisymmetric body to the drag of a sphere with equivalent perimeter. The relative coefficients of transversal drag $\bar{c}_\perp$ equal to the ratio of the transversal drag of an axisymmetric body to the drag of a sphere with equivalent perimeter are introduced similarly.

Experimental data and numerical results for principal values of the translational tensor for some axisymmetric and orthotropic bodies (cylinders, doubled cones, parallelepipeds) are well approximated by the following dependence for the relative coefficient of the axial drag:

$$\bar{c}_\parallel = 0.244 + 1.035\,\Sigma - 0.712\,\Sigma^2 + 0.441\,\Sigma^3 \qquad (0.4 \le \Sigma \le 1.2).$$

The values of the perimeter-equivalent factor Σ are equal to the ratio of the surface area of the particle to the surface area of the perimeter-equivalent sphere.

To determine the relative coefficients of transversal drag, one can use the formula

$$\bar{c}_\perp = 0.392 + 0.621\,\Sigma - 0.04\,\Sigma^2,$$

which is well consistent with experimental data for $0.5 \le \Sigma \le 4.5$.

E4.8.4. Sedimentation of Isotropic and Nonisotropic Particles

▶ **Sedimentation of isotropic particles.** The steady-state rate U_i of particle settling in mass force fields, primarily, in the gravitational field, is an important hydrodynamic characteristic of processes in chemical technology such as settling and sedimentation. Any spherically isotropic body homogeneous with respect to density experiences the same drag under translational motion regardless of its orientation. Such a body is also isotropic with respect to any pair of forces that arise when it rotates around an arbitrary axis passing though its center. If at the initial time such a body has some orientation in fluid and can fall without initial rotation, then this body falls vertically without rotation, preserving its initial orientation.

It is convenient to describe the free fall of nonspherical isotropic particles by using the sphericity parameter

$$\psi = \frac{S_e}{S_*},$$

where S_* is the surface area of the particle and S_e is the surface area of the volume-equivalent sphere. If the motion is slow, one can calculate the settling rate by the empirical formula

$$\mathbf{U}_i = \frac{2}{9} \frac{Q \Delta \rho a_e^2}{\mu} \mathbf{g},$$

where a_e is the radius of the volume-equivalent sphere and

$$Q = 0.843 \ln \frac{\psi}{0.065}.$$

We present some values of the sphericity factor ψ for some particles: sphere, 1.000; octahedron, 0.846; cube, 0.806; tetrahedron, 0.670.

▶ **Sedimentation of nonisotropic particles.** If the velocity of a spherical particle in Stokes settling is always codirected with the gravity force, even for homogeneous axisymmetric particles the velocity is directed vertically if and only if the vertical coincides with one of the principal axes of the translational tensor **K**. If the angle between the symmetry axis and the vertical is φ, then the velocity direction is given by the angle

$$\theta = \pi + \arctan\left(\frac{K_2}{K_1} \tan \varphi\right),$$

where K_1 and K_2 are the axial and transversal principal drags to the translational motion. The velocity is given by

$$U_i = \frac{V \Delta \rho g}{\mu} (K_1^2 \cos^2 \theta + K_2^2 \sin^2 \theta)^{-1/2},$$

where V is the particle volume.

If the settling direction is not vertical, this means that a falling particle is subject to the action of a transverse force, which leads to its horizontal displacement. An additional complication is that the center of hydrodynamic reaction (including the buoyancy force) does not coincide with the particle center of mass. In this case, in addition to the translational motion, the particle is subject to rotation under the action of the arising moment of forces (e.g., the "somersault" of a bullet with displaced center of mass). For axisymmetric particles, this rotation stops when the system "the mass center + the reaction center" becomes stable, that is, the mass center is ahead of the reaction center. In this case, the settling trajectory becomes stable and rectilinear.

However, in the more general case of an asymmetric particle, the combined action of the lateral and rotational forces may lead to motion along a 3D (for instance, spiral, trajectory). At the same time, a steady-state settling trajectory with helicoidal (propeller-like) symmetry remains rectilinear, notwithstanding the body rotation.

Two theorems are useful for estimating the steady-state rate of settling of Stokes non-spherical particles. One theorem, proved by Hill & Power (1956), states that the Stokes drag of an arbitrary body moving in a viscous fluid is larger than the Stokes drag of any inscribed body. Thus, to determine the upper and lower bounds for the Stokes drag of a body of an exotic shape, one can suggest a reasonable set of inscribed and circumscribed bodies with known drags. The other theorem, proved by Weinberger (1972), states that of all particles of different shape but equal volume, the spherical particle has the maximum Stokes settling rate.

▶ **Mean velocity of nonisotropic particles falling in a fluid.** The mean flow rate velocity $\langle \mathbf{U} \rangle$ of a particle, which is obtained in a large number of experiments when an arbitrarily oriented particle falls in a fluid, is determined for the Stokes flow by the formula:

$$\langle \mathbf{U} \rangle = \frac{V \Delta \rho}{\mu \overline{K}} \mathbf{g}, \qquad (E4.8.4.1)$$

where V is the particle volume, $\Delta \rho$ is the difference between the densities of the particle and the fluid, $\mathbf{g}$ is the free fall acceleration, and $\overline{K}$ is the average drag expressed via the principal drags as

$$\frac{1}{\overline{K}} = \frac{1}{3} \left(\frac{1}{K_1} + \frac{1}{K_2} + \frac{1}{K_3} \right). \qquad (E4.8.4.2)$$

The average drag force acting on an arbitrarily oriented particle falling in fluid is given by

$$\langle \mathbf{F} \rangle = -\mu \overline{K} \langle \mathbf{U} \rangle. \qquad (E4.8.4.3)$$

Formulas (E4.8.4.1)–(E4.8.4.3) are important in view of some problems of Brownian motion.

In the special case of a spherical particle, one must set $V = \frac{4}{3}\pi a^3$ and $\overline{K} = 6\pi a$ in (E4.8.4.1).

Let us calculate the average drag for a thin disk of radius a. To this end, we substitute the principal drags (E4.8.3.4) into (E4.8.4.2). As a result, we obtain

$$\overline{K} = 12a.$$

Substituting the coefficients K_1, K_2, and K_3 from (E4.8.3.5)–(E4.8.3.7) into (E4.8.4.2) and using (E4.8.4.1), one obtains the average settling rate for the above-mentioned non-spherical bodies.

E4.8.5. Flow Past Nonspherical Particles at Various Reynolds Numbers

The Stokes flow around particles of any shape is separation-free—that is, the stream lines come from infinity—bend round the body everywhere closely attaining the body surface, and return to infinity. However, for higher Reynolds numbers, separation occurs, which leads to wake formation behind the body. As the Reynolds number increases, the size of the wake region (the wake length) grows differently for different bodies. With further increase in the Reynolds number, the wake becomes unsteady and completely turbulent and goes to infinity. The force action of the flow on the body is closely related to the wake size and state. The limit asymptotic cases of this action are the Stokes (as $\mathrm{Re} \to 0$) and the Newtonian (as $\mathrm{Re} \to \infty$) regimes of flow. We have already considered the characteristics of the Stokes flow. The Newtonian regime of flow is characterized by the fact that the drag coefficient c_f of the body is constant.

In axial flow past disks, which are the limit cases of axisymmetric bodies of small length, the drag coefficients for the entire range of Reynolds numbers calculated with respect to the radius can be determined by approximate formulas, which agree well with numerical results and experimental data:

$$
\begin{aligned}
c_f &= 10.2 \, \mathrm{Re}^{-1}(1 + 0.318 \, \mathrm{Re}) && \text{for} \quad \mathrm{Re} \le 0.005, \\
c_f &= 10.2 \, \mathrm{Re}^{-1}(1 + 10^s) && \text{for} \quad 0.005 < \mathrm{Re} \le 0.75, \\
c_f &= 10.2 \, \mathrm{Re}^{-1}(1 + 0.239 \, \mathrm{Re}^{0.792}) && \text{for} \quad 0.75 < \mathrm{Re} \le 66.5, \\
c_f &= 1.17 && \text{for} \quad \mathrm{Re} > 66.5,
\end{aligned}
$$

where $s = -0.61 + 0.906 \log_{10} \mathrm{Re} - 0.025 \, (\log_{10} \mathrm{Re})^2$.

The steady-state settling rate of a particle of an arbitrary shape (for Newtonian regime of a flow at high Reynolds numbers) can be obtained by the formula

$$
U = 0.69 \, \gamma^{1/36} [g a_e (\gamma - 1)(1.08 - \psi)]^{1/2} \quad \text{for} \quad 1.1 < \gamma < 8.6,
$$

where γ is the particle-fluid density ratio, a_e is the radius of the volume-equivalent sphere, and ψ is the ratio of the area of the volume-equivalent sphere to the particle surface area.

E4.9. Motion of Several Particles. Effective Viscosity of Suspensions

The motion of a particle in infinite fluid creates some velocity and pressure fields. Neighboring particles move in already perturbed hydrodynamic fields. Simultaneously, the first particle itself experiences hydrodynamic interaction with the neighboring particles and neighboring moving or fixed surfaces. Since in a majority of actual disperse systems, the existence of an ensemble of particles and the apparatus walls is inevitable, the consideration of the hydrodynamic interaction between these objects is very important. One method for obtaining the required information about the interaction is based on the construction of exact closed-form solutions. However, even within the framework of Stokes hydrodynamics, to describe the motion of an ensemble of particles is a very complicated problem, which admits an exact closed-form solution only in exceptional cases.

E4.9.1. Motion of Two Spheres. Gravitational Sedimentation of Several Spheres

▶ **Motion of two spheres along a line passing through their centers.** Consider the axisymmetric problem about the motion of two spheres at the same velocity U. In the Stokes approximation, the force acting on each of the spheres is described by the formula

$$
F = 6\pi \mu a U \lambda, \tag{E4.9.1.1}
$$

where a is the sphere radius and λ is the correction coefficient depending on the radii and the distance l between the centers of the spheres. In the case of spheres of equal radii, the expression for λ has the form

$$\lambda = \frac{4}{3} \sinh k \sum_{n=1}^{\infty} \frac{n(n+1)}{(2n-1)(2n+3)} \left\{ 1 - \frac{4 \sinh^2[(n+\frac{1}{2})k] - (2n+1)^2 \sinh^2 k}{2 \sinh[(2n+1)k] + (2n+1) \sinh 2k} \right\}, \quad \text{(E4.9.1.2)}$$

where $k = \ln\left[\frac{1}{2}(l/a) + \sqrt{\frac{1}{4}(l/a)^2 - 1}\right]$.

For numerical calculations, it is convenient to use the approximate formula

$$\lambda = \frac{0.88\,a + l}{2.5\,a + l}, \quad \text{(E4.9.1.3)}$$

whose maximal error is less than 1.3% for any a and l.

Since $\lambda \leq 1$, it follows from (E4.9.1.1) that the velocity of the steady-state motion of each of the spheres in the ensemble is greater than the velocity of a single sphere.

▶ **Gravitational sedimentation of several spheres.** Some relations between the drag force F and the sedimentation velocity U averaged with respect to various orientations of particles of equal radius in space were obtained by Happel & Brenner (1965). It was assumed that the maximum distance l between the sphere centers is much larger than their radius a. In all considered cases, the drag force is described by formula (E4.9.1.1), where λ is the correction coefficient depending on the configuration of the system of particles. In what follows, we write out the correction coefficient for some typical configurations of particles.

For a system of two spheres, one has

$$\lambda = \frac{2}{1 + (a/l)}.$$

For a system of three spheres arranged in a line,

$$\lambda = \frac{3}{1 + \frac{10}{3}(a/l) - \frac{1}{4}(a/l)^2}.$$

For a system of four spheres arranged in a line,

$$\lambda = \frac{4}{1 + \frac{13}{2}(a/l) - \frac{9}{8}(a/l)^2}.$$

For a system of four spheres arranged in vertices of a square,

$$\lambda = \frac{4}{1 + 2.7\,(a/l) - 0.04\,(a/l)^2}.$$

For a system of eight spheres arranged in vertices of a cube,

$$\lambda = \frac{8}{1 + 5.7\,(a/l) - 0.34\,(a/l)^2}.$$

E4.9.2. Rate of Suspension Precipitation. Effective Viscosity of Suspensions

▶ **Cellular model. Rate of suspension precipitation.** The two-phase media with large number of particles are described by a widely used approximate model, namely, the cell model. To each particle of the disperse phase, this model assigns the corresponding volume of the free liquid. Thus, the entire suspension (or emulsion) is divided into a collection of spherical cells of radius b whose centers coincide with the centers of particles of radius a. The geometric parameters of these cells are related to the volume concentration ϕ of the disperse phase as follows:

$$b = a\phi^{-1/3}.$$

The definition of a particle velocity $\mathbf{U}$ determines the axial symmetry of the problem, which is convenient to consider in spherical coordinates. In the Stokes approximation, the general solution of such a problem is determined by formula (E4.6.1.4), where arbitrary constants must be determined from the conditions that the solution is bounded at infinity and the velocity is known on the particle surface and from some other conditions on the cell boundary (for $R = b$). The condition that the normal velocity on this boundary is zero—that is, there is no flow across the cell—is undoubtedly obvious. The second condition, which is necessary for the complete identification of the solution, leads to different opinions. Cunningham postulated that the tangential velocity is zero, thus considering a cell as a container with rigid boundary. Happel proposed to use the condition that the tangential stress is zero, thus postulating that the cell is an isolated force unit (is in equilibrium under the action of forces). Kuwabara proposed to use the condition that the vortex intensity is zero on the cell boundary. Finally, Slobodov and Chepura assumed that the vortex intensity flux vanishes on the cell boundary.

The choice of the boundary condition essentially determines the model of force interaction between the particle at the center of the cell and the other particles. The steady-state velocities of the gravitational sedimentation in suspensions obtained with the help of cell models were compared with numerous experimental data and it was shown that the most precise results can be obtained by using the Slobodov–Chepura model. This model, under the assumption that the particle at the cell center is a drop of a liquid with different viscosity, leads to formula (E4.9.1.1) for the drag forces, where the correction coefficient can be calculated as

$$\lambda = \frac{\beta + \frac{2}{3}}{1 - \frac{3}{5}\phi^{1/3} - \frac{2}{5}\phi^2 + \beta(1 - \frac{9}{10}\phi^{1/3} + \frac{1}{2}\phi + \frac{2}{5}\phi^2)}, \qquad \beta = \frac{\mu_2}{\mu_1}.$$

As $\phi \to 0$ and $\beta \to \infty$, we have $\lambda \to 1$, which corresponds to the Stokes drag law.

▶ **Effective viscosity of suspensions.** Suspensions of particles in fluid are widely used in various processes of chemical technology. If the dimensions of particles in the suspension are essentially smaller than those of the apparatus, then one can consider the suspension as some continuous medium with properties other than properties of the disperse phase.

Very often, this medium remains Newtonian from the viewpoint of its rheological properties, but the viscosity of this medium is somewhat larger than that of the continuous phase. This viscosity μ_{ef} is called the effective viscosity. In practice, it is convenient to relate the effective viscosity to the viscosity μ of the continuous phase and consider the dimensionless effective viscosity $\bar{\mu} = \mu_{\text{ef}}/\mu$.

The value $\bar{\mu}$ depends primarily on the volume concentration of the disperse phase ϕ. The well-known Einstein formula

$$\bar{\mu} = 1 + 2.5\,\phi, \tag{E4.9.2.1}$$

holds for strongly rarefied suspensions of solid spherical particles.

For $0 \le \phi \le 0.4$, the results of experimental and numerical calculations are well approximated by the formula

$$\bar{\mu} = 1 + 2.5\,\phi + 12.5\,\phi^2,$$

which becomes the Einstein formula (E4.9.2.1) as $\phi \to 0$.

The dimensionless effective viscosity of a rarefied emulsion of spherical drops and bubbles moving in a fluid can be determined by the formula

$$\bar{\mu} = 1 + \frac{5\beta + 2}{2\beta + 2}\,\phi, \qquad \text{(E4.9.2.2)}$$

where β is the ratio of the drop viscosity to the fluid viscosity. By passing to the limit as $\beta \to \infty$ in (E4.9.2.2), one obtains the Einstein formula (E4.9.2.1). The value $\beta = 0$ corresponds to gas bubbles.

Bibliography for Chapter E4

Batchelor, G. K., *An Introduction to Fluid Dynamics*, Cambridge Univ. Press, Cambridge, England, 1967.

Blasius, H., Crenzschichten in Flussigkeiten mit Kleiner Reibung, *Zeitschr. für Math. und Phys.,* Bd. 56, Ht. 1, S. 1–37, 1908.

Bradshaw, P., *Turbulence*, Springer-Verlag, Berlin, 1978.

Chhabra, R. P., *Bubbles, Drops, and Particles in Non-Newtonian Fluids*, CRC Press, Boca Raton, Florida, 1993.

Clift, R., Grace, J. R., and Weber, M. E., *Bubbles, Drops and Particles*, Acad. Press, New York, 1978.

Emanuel, G., *Analytical Fluid Dynamics*, CRC Press, Boca Raton, Florida, 1993.

Faber, T. E., *Fluid Dynamics for Physicists*, Cambridge Univ. Press, Cambridge, England, 1995.

Frost, W. and Moulden, T. H., *Handbook of Turbulence*, Plenum Press, New York, 1977.

Hadamard, J. S., Mouvement permanent lent d'une sphere liquide et visqueuse dans un liquide visqueux, *Comp. Rend. Acad. Sci. Paris*, Vol. 152, No. 25, pp. 1735–1739, 1911 and Vol. 154, No. 3, p. 109, 1912.

Happel, J. and Brenner, H., *Low Reynolds Number Hydrodynamics*, Prentice-Hall, Englewood Cliffs, New Jersey, 1965.

Hetsroni, G., *Handbook of Multiphase Systems*, Hemisphere Publ. Corp., Washington, 1982.

Hill, R. and Power, G., Extremum principles for slow viscous flow and approximate calculation of drag, *Quarterly J. Mech. Appl. Math.*, 1956, Vol. 9, No. 3, pp. 313–319.

Hinze, J. O., *Turbulence*, McGraw-Hill, New York, 1975.

Idelchik, I. E., *Handbook of Hydraulic Resistance*, Begell House, New York, 1994.

Kaplun, S. and Lagerstrom, P. A., Asymptotic expansions of Navier–Stokes solutions for small Reynolds numbers, *J. Math. Mech.,* Vol. 6, pp. 585–593, 1957.

Kármán, Th., von, *Mechanische Ahnlichkeit und Turbulenz*, Verhandlg. d. III Intern. Kongress fur Techn. Mechanik, Stockholm, Bd. 1, S. 85, 1930.

Kutateladze, S. S., *Heat Transfer and the Hydrodynamical Resistance. A Reference Book* [in Russian], Energoatomizdat, Moscow, 1990.

Lamb, H., *Hydrodynamics*, Dover Publ., New York, 1945.

Landau, L. D. and Lifschitz, E. M., *Fluid Mechanics. Course of Theoretical Physics*, Pergamon Press, Oxford, 1987.

Lesieur, M., *Turbulence in Fluids*, Kluwer Acad. Publ., Dordrecht, 1997.

Levich, V. G., *Physicochemical Hydrodynamics*, Prentice-Hall, Englewood Cliffs, New Jersey, 1962.

Loitsyanskiy, L. G., *Mechanics of Liquids and Gases*, Begell House, New York, 1996.

Mobius, D. and Miller, R., *Drops and Bubbles in Interfacial Research*, Elsevier Science Ltd., Oxford, 1997.

Monin, A. S. and Yaglom, A. M., *Statistical Fluid Mechanics: Theory of Turbulence, Vols. 1 and 2* [in Russian], Gidrometeoizdat, St. Petersburg, 1992 and 1996.

Nikitin N. V., Statistical characteristics of wall turbulence, *Fluid Dynamics*, Vol. 31, No. 3, pp. 361–370, 1996.

Oseen, C. W., Über die Stokes'sche Formel, und uber eine verwandte Aufgabe in der Hydrodynamik, *Ark. Math. Astronom. Fys.*, Bd. 6, Ht. 29, 1910.

Paneli, D. and Gutfinger, C., *Fluid Mechanics*, Cambridge Univ. Press, Cambridge, England, 1997.

Petukhov, B. S., *Heat Transfer and Drag in Laminar Flow of Liquids in Tubes* [in Russian], Energiya, Moscow, 1967.

Polyanin, A. D., Kutepov, A. M., Vyazmin, A. V., and Kazenin, D. A., *Hydrodynamics, Mass and Heat Transfer in Chemical Engineering*, Taylor & Francis, London, 2002.

Prandtl, L., The mechanics of viscous fluids, In: *Aerodynamics Theory*, Vol. 3, pp. 34–208; Springer-Verlag, Berlin, 1935.

Proudman, I. and Pearson, J. R. A., Expansions at small Reynolds number for the flow past a sphere and circular cylinder, *J. Fluid Mech.,* Vol. 2, No. 3, pp. 237–262, 1957.

Reid, R. C., Prausnitz, J. M., and Sherwood, T. K., *The Properties of Gases and Liquids*, McGraw-Hill Book Comp., New York, 1977.

Reynolds, A. J., *Turbulent Flows in Engineering*, Wiley, London, 1974.

Rybczynski M. W., Über die fortschreitende Bewegung einer flüssigen Kugel in einem zähen Medium, *Bull. Acad. Sci. Cracovie, Ser. A, Sci. Math.,* Bd. 1, S. 40–46, 1911.

Schlichting, H., *Boundary Layer Theory*, McGraw-Hill, New York, 1981.

Shivamoggi, B., *Theoretical Fluid Dynamics*, Wiley VCH, Chichester, New York, 1998.

Slobodov, E. B. and Chepura I. V., A cellular model of biphasal media, *Theor. Found. Chem. Engng.,* 1982, Vol. 16, No. 3, pp. 235–239.

So, R. M. C., Speciale, C. G., and Launder, B. E., *Near-Wall Turbulent Flows*, Elsevier, Amsterdam, 1993.

Soo, S. L., *Fluid Dynamics of Multiphase Systems*, Blaisdell Publ. Comp., Waltham, 1968.

Stimson, M. and Jeffrey, G. B., The motion of two spheres in a viscous flow, *Proc. Roy. Soc. London,* Vol. A111, No. 757, p. 110, 1926.

Stokes, G. G., On the effect of the internal friction of fluids on the motion of pendulums, *Trans. Camb. Phil. Soc.,* Vol. 9, No. 2, pp. 8–106, 1851.

Taylor, G. I., Viscosity of a fluid, containing small drops of another fluid, *Proc. Roy. Soc. London,* Vol. A138, No. 834, pp. 41–48, 1932.

Townsend, A. A., *The Structure of Turbulent Shear Flow*, Cambridge Univ. Press, London, 1976.

Van Dyke, M. D., *Perturbation Methods in Fluid Mechanics*, Parabolic Press, Stanford California, 1975.

Vargaftic, N. V., Vinogradov, Y. K., and Yargin, V. S., *Handbook of Physical Properties of Liquids and Gases*, Begell House Inc., New York, 1996.

Warsi, Z. U. A., *Fluid Dynamics*, CRC Press, Boca Raton, Florida, 1993.

Weinberger, H. F., Variational properties of steady fall in Stokes flow, *J. Fluid Mech.,* 1972, Vol. 52, Pt. 2, pp. 321–344.

Zapryanov, Z. and Tabakova, S., *Dynamics of Bubbles, Drops and Rigid Particles*, Kluwer Acad. Publ., Dordrecht, 1999.

Zdravkovich, M. M., *Flow Around Circular Cylinders*, Oxford Sci. Publ., Oxford, 1997.

Chapter E5

Mass and Heat Transfer

E5.1. Convective Mass and Heat Transfer. Equations and Boundary Conditions

E5.1.1. Mass and Heat Transfer in Laminar Flows

▶ **Mass concentration.** So far, we have considered the motion of fluids of homogeneous physical and chemical composition. In practice, one often meets more complicated situations in which the fluid contains dissolved substances (contaminants or reactants) and is a solution or a mixture.

Water solutions of common salt or sugar and water-alcohol mixtures are the simplest examples of such systems.

The composition of solutions and mixtures is usually characterized by the mass density of the substance (the mass of dissolved substance per unit volume) or the dimensionless mass concentration C (the ratio of the mass density of a substance to the total density of the mixture). The latter is normally used in this chapter; for brevity, we refer to it simply as the concentration.

▶ **Convective mass transfer equation.** Let us write out the main equations and boundary conditions used in the mathematical statement of mass transfer and chemical hydrodynamic problems. We assume that the medium density and viscosity are independent of the concentration and that the concentration distribution does not affect the flow field. This allows one to analyze the hydrodynamic problem about the fluid motion and the diffusion problem of finding the concentration field independently. It is assumed that the information about the fluid velocity field necessary for the solution of the diffusion problem is known. We also assume that the diffusion coefficient is independent of the concentration. For simplicity, we restrict our considerations to the case of two-component solutions.

In Cartesian coordinates X, Y, Z, the solute transfer in absence of homogeneous transformations is described by the equation

$$\frac{\partial C}{\partial t} + V_X \frac{\partial C}{\partial X} + V_Y \frac{\partial C}{\partial Y} + V_Z \frac{\partial C}{\partial Z} = D \left(\frac{\partial^2 C}{\partial X^2} + \frac{\partial^2 C}{\partial Y^2} + \frac{\partial^2 C}{\partial Z^2} \right), \qquad \text{(E5.1.1.1)}$$

where C is the concentration, D is the diffusion coefficient, and V_X, V_Y, and V_Z are the fluid velocity components, which are assumed to be given.

Equation (E5.1.1.1) reflects the fact that the transfer of a substance in a moving medium is due to two distinct physical mechanisms. First, there is molecular diffusion due to concentration difference in a liquid or gas, which tends to equalize the concentrations. Second, the solute is carried along by the moving medium. The combination of these two processes is usually called *convective diffusion*.

▶ **Initial condition and the simplest boundary conditions.** To complete the statement of the problem, it is necessary to supplement equation (E5.1.1.1) with initial and boundary conditions. As the initial condition, one usually takes the concentration profile in the flow at time $t = 0$. The boundary conditions are, as a rule, given on some surface and far from it, in the bulk of the solution. The latter condition corresponds to prescribing the unperturbed concentration C_i at infinity:

$$\xi_* \to \infty, \quad C \to C_i, \tag{E5.1.1.2}$$

where ξ_* is the distance measured along the normal to the surface.

In solid dissolution problems, it is usually assumed in the boundary conditions that the concentration is zero in the bulk of the fluid, $C_i = 0$, and constant on the crystal surface,

$$\xi_* = 0, \quad C = C_s, \tag{E5.1.1.3}$$

where C_s is given. The boundary conditions (E5.1.1.2) (with $C_i = 0$) and (E5.1.1.3) are also used in liquid drop evaporation problems.

The boundary conditions on a surface where a chemical reaction occurs depend on the specific physical statement of the problem. In the special case of an "infinitely rapid" heterogeneous chemical reaction, the corresponding boundary condition has the form

$$\xi_* = 0, \quad C = 0 \tag{E5.1.1.4}$$

and means that the reagent is completely taken up in the reaction on the interface. Such a situation is often called the diffusion regime of reaction. Condition (E5.1.1.4) has the following meaning: the chemical reaction on the interface proceeds vigorously, so that all the substance that approaches the interface takes part in the reaction. Note that condition (E5.1.1.4) is a special case of (E5.1.1.3) for $C_s = 0$.

▶ **Dimensionless mass transfer equation and boundary conditions.** Let us introduce a characteristic length a (for example, the particle or tube radius) and a characteristic velocity U (for example, the unperturbed flow velocity far from a particle or the fluid velocity on the tube axis). First, consider the boundary conditions (E5.1.1.2) and (E5.1.1.3). Then it is convenient to rewrite equation (E5.1.1.1) for the convective mass transfer in the following dimensionless form. We introduce dimensionless variables by setting

$$\tau = \frac{Dt}{a^2}, \quad x = \frac{X}{a}, \quad y = \frac{Y}{a}, \quad z = \frac{Z}{a}, \quad \xi = \frac{\xi_*}{a},$$
$$v_x = \frac{V_X}{U}, \quad v_y = \frac{V_Y}{U}, \quad v_z = \frac{V_Z}{U}, \quad c = \frac{C_i - C}{C_i - C_s} \tag{E5.1.1.5}$$

and substitute them into (E5.1.1.1). As a result, we obtain

$$\frac{\partial c}{\partial \tau} + \mathsf{Pe}\left(v_x \frac{\partial c}{\partial x} + v_y \frac{\partial c}{\partial y} + v_z \frac{\partial c}{\partial z} \right) = \frac{\partial^2 c}{\partial x^2} + \frac{\partial^2 c}{\partial y^2} + \frac{\partial^2 c}{\partial z^2}. \tag{E5.1.1.6}$$

Here the Peclet number $\mathsf{Pe} = aU/D$ is a dimensionless parameter characterizing the ratio of convective transfer to diffusion transfer and a is a characteristic length (tube radius, film thickness, etc.).

In the new variables (E5.1.1.5), the boundary condition far from the surface (E5.1.1.2) has the form

$$\xi \to \infty, \quad c \to 0. \tag{E5.1.1.7}$$

In a similar way, taking into account (E5.1.1.3) and (E5.1.1.5), we obtain the boundary condition

$$\xi = 0, \quad c = 1. \tag{E5.1.1.8}$$

For brevity, we rewrite the convective diffusion equation (E5.1.1.6) in the following frequently used form:

$$\frac{\partial c}{\partial \tau} + \text{Pe}\,(\mathbf{v}\cdot\nabla)\,c = \Delta c, \tag{E5.1.1.9}$$

where ∇ is the Hamilton operator and Δ is the Laplace operator; the explicit form of these operators in the Cartesian coordinates x, y, z follows by comparing (E5.1.1.6) and (E5.1.1.9).

▶ **Heat transfer equation. Boundary conditions.** We assume that the medium density and viscosity and the thermal conductivity coefficient are independent of temperature and that the temperature distribution does not affect the flow field. This allows one to analyze the hydrodynamic problem about the fluid motion and the heat problem of finding the temperature field independently. It is assumed that the information about the fluid velocity field necessary for the solution of the heat problem is known.

The equation of heat transfer in a moving medium is similar to the convective diffusion equation (E5.1.1.1) and has the form

$$\frac{\partial T_*}{\partial t} + V_X\frac{\partial T_*}{\partial X} + V_Y\frac{\partial T_*}{\partial Y} + V_Z\frac{\partial T_*}{\partial Z} = \chi\left(\frac{\partial^2 T_*}{\partial X^2} + \frac{\partial^2 T_*}{\partial Y^2} + \frac{\partial^2 T_*}{\partial Z^2}\right), \tag{E5.1.1.10}$$

where T_* is temperature and χ is the thermal diffusivity. The dissipative heating of fluid is neglected in Eq. (E5.1.1.10).

For nonstationary problems, the temperature distribution in the flow at the initial instant of time must be given.

Far from the surface, one usually has the condition that temperature is constant in the flow region:

$$\xi_* \to \infty, \quad T_* \to T_\text{i}. \tag{E5.1.1.11}$$

For body–medium heat exchange problems, if constant temperature is maintained on the body surface, the second boundary condition is

$$\xi_* = 0, \quad T_* = T_\text{s}. \tag{E5.1.1.12}$$

By using the new dimensionless variables

$$\bar{\tau} = \frac{\chi t}{a^2}, \quad x = \frac{X}{a}, \quad y = \frac{Y}{a}, \quad z = \frac{Z}{a}, \quad \text{Pe}_T = \frac{aU}{\chi},$$
$$v_x = \frac{V_X}{U}, \quad v_y = \frac{V_Y}{U}, \quad v_z = \frac{V_Z}{U}, \quad T = \frac{T_\text{i}-T_*}{T_\text{i}-T_\text{s}}, \tag{E5.1.1.13}$$

one can rewrite Eq. (E5.1.1.10) and the boundary conditions (E5.1.1.11) and (E5.1.1.12) in the form

$$\frac{\partial T}{\partial \bar{\tau}} + \text{Pe}_T\,(\mathbf{v}\cdot\nabla)\,T = \Delta T; \tag{E5.1.1.14}$$

$$\xi \to \infty, \quad T \to 0; \qquad \xi = 0, \quad T = 1. \tag{E5.1.1.15}$$

Obviously, from the mathematical viewpoint, the problem for equation (E5.1.1.14) with conditions (E5.1.1.15) describing the body–medium heat exchange is identical to problem (E5.1.1.6)–(E5.1.1.8) describing the flow–particle mass exchange in the case of a diffusion regime of reaction on the particle surface. By virtue of this analogy, in what follows we usually consider only mass transfer problems.

E5.1.2. Basic Dimensionless Parameters. Diffusion Fluxes and the Sherwood Number

▶ **Basic dimensionless parameters.** The diffusion and thermal Peclet numbers in the convective mass and heat transfer equations (E5.1.1.6) and (E5.1.1.14) are related to the Reynolds number $\mathsf{Re} = aU/\nu$ (where ν is the kinematic viscosity of the fluid) on the right-hand side in the Navier–Stokes equations (1.1.12) by the formulas

$$\mathsf{Pe} = \mathsf{Re}\,\mathsf{Sc}, \qquad \mathsf{Pe}_T = \mathsf{Re}\,\mathsf{Pr}. \tag{E5.1.2.1}$$

Here the Schmidt number $\mathsf{Sc} = \nu/D$ and the Prandtl number $\mathsf{Pr} = \nu/\chi$ are dimensionless values depending only on the physical properties of the continuous phase.

For ordinary gases, the diffusion coefficient and the kinematic viscosity are of the same order of magnitude, which corresponds to Schmidt numbers of the order of one ($\mathsf{Sc} \sim 1$).

For ordinary liquids like water, the kinematic viscosity is several orders of magnitude larger than the diffusion coefficient ($\mathsf{Sc} \sim 10^3$). In extremely viscous liquids like glycerin, the Schmidt number is of the order of 10^6.

The range of the Prandtl number is narrower than that of the Schmidt number. In gases such as air, $\mathsf{Pr} \sim 1$, and in liquids like water, $\mathsf{Pr} \sim 10$. In extremely viscous liquids like glycerin, the Prandtl number is of the order of 10^3. Liquid metals (sodium, lithium, mercury, etc.) are characterized by low Prandtl numbers: $5 \times 10^{-3} \le \mathsf{Pr} \le 5 \times 10^{-2}$.

The Reynolds number $\mathsf{Re} = aU/\nu$ is not a physical constant of a medium and depends on geometric and kinematic factors. Therefore, the range of variation of this number may be arbitrary.

It follows from the considered examples and relation (E5.1.2.1) that the Peclet numbers in problems of physicochemical hydrodynamics can vary in a wide range.

Since diffusion processes in fluids are characterized by very large Schmidt numbers, we point out that the Peclet number in convective mass transfer problems in fluid media is also large even for low Reynolds numbers at which the Stokes flow law applies (for a creeping flow).

▶ **Local and total diffusion fluxes and the Sherwood number.** A local (or differential) diffusion flux of a solute to the surface is determined by

$$j_* = D\rho\left(\frac{\partial C}{\partial \xi_*}\right)_{\xi_*=0} \tag{E5.1.2.2}$$

and in general varies along the surface.

The total (or integral) diffusion flux can be obtained by integrating (E5.1.2.2) over the entire surface S,

$$I_* = \iint_S j_*\, dS. \tag{E5.1.2.3}$$

The total diffusion flux I_* is the amount of substance that reacts on the entire surface per unit time.

In mass transfer problems with the boundary conditions (E5.1.1.2) and (E5.1.1.3), one often replaces (E5.1.2.2) and (E5.1.2.3) by the dimensionless diffusion fluxes

$$j = \frac{aj_*}{D\rho\,(C_{\mathrm{i}} - C_{\mathrm{s}})}, \qquad I = \frac{I_*}{aD\rho\,(C_{\mathrm{i}} - C_{\mathrm{s}})}. \tag{E5.1.2.4}$$

For the diffusion regime of reaction, which corresponds to the boundary conditions (E5.1.1.2) and (E5.1.1.4), as well as for the finite rate of surface chemical reaction in the case of the boundary conditions (E5.1.1.2) and (E5.1.1.5), one should set $C_s = 0$ in (E5.1.2.4).

The main quantity of practical interest is the mean Sherwood number

$$\text{Sh} = \frac{I}{S}, \tag{E5.1.2.5}$$

where $S = S_*/a^2$ is the dimensionless surface area and S_* is the corresponding dimensional area.

The calculation of the diffusion fluxes and the mean Sherwood number is usually carried out in three steps. First, the convective mass transfer problem is solved and the concentration field is determined. Second, the normal derivative $\left(\partial C/\partial \xi_*\right)_{\xi_*=0}$ on the surface is evaluated. Finally, one applies formulas (E5.1.2.2)–(E5.1.2.5).

For brevity, throughout the book (where it cannot lead to a misunderstanding) we use the terms "concentration" and "diffusion flux" instead of "dimensionless concentration" and "dimensionless diffusion flux."

E5.1.3. Mass and Heat Transfer in Turbulent Flows

▶ **Equation of mass and heat transfer.** When mathematically describing the transfer of a passive admixture in a turbulent flow, the admixture concentration and the fluid velocity components are represented as

$$C = \overline{C} + C', \quad V_i = \overline{V}_i + V'_i, \tag{E5.1.3.1}$$

where the bar and prime denote the time-average and fluctuating components, respectively. The averages of the fluctuations are zero, $\overline{C'} = \overline{V'_i} = 0$.

The introduction of the average and fluctuating components by relations (E5.1.3.1) followed by the application of the averaging operation made it possible, under some assumptions, to obtain the following equation for the average concentration of the admixture from Eq. (E5.1.1.1):

$$\frac{\partial \overline{C}}{\partial t} + (\overline{\mathbf{V}} \cdot \nabla)\overline{C} = D\Delta\overline{C} + \nabla \cdot (-\overline{C'\mathbf{V}'}). \tag{E5.1.3.2}$$

Here $-\overline{C'\mathbf{V}'}$ is the turbulent (eddy) flux of the admixture. This vector can be determined using some closure hypothesis, based, as a rule, on empirical information.

By analogy with Boussinesq's ideas about transfer of momentum, it is often assumed for transfer of a substance that

$$-\overline{C'\mathbf{V}'} = \mathbf{D}_t \cdot \nabla\overline{C}, \tag{E5.1.3.3}$$

where $\mathbf{D}_t$ is the turbulent diffusion tensor.

▶ **Turbulent Prandtl number.** In a number of problems, it is important to know, as a rule, only one component of turbulent transfer, namely, the component normal to the wall. In this case, the turbulent diffusion strength is characterized by a single scalar quantity D_t, just as the intensity of turbulent transfer of momentum is characterized by a single scalar quantity ν_t. The ratio of the turbulent viscosity coefficient to the turbulent diffusion coefficient,

$$\text{Pr}_t = \frac{\nu_t}{D_t}, \tag{E5.1.3.4}$$

is referred to as the turbulent Prandtl (Schmidt) number.

The simplest way to close equation (E5.1.3.2) is to use the hypothesis that the turbulent Prandtl number for the process examined is constant. Then it readily follows from formula (E5.1.3.4) that the turbulent diffusion coefficient is proportional to the turbulent viscosity, $D_t = \nu_t/Pr_t$. By using the expression for ν_t borrowed from the corresponding hydrodynamic model, one can obtain the desired value of D_t. In particular, following Prandtl's or von Kármán's model, one can use formula (1.1.21) or (1.1.22) for ν_t.

For a long time, it was assumed without sufficient justification that $Pr_t = 1$. With this value of Pr_t, the mechanism of turbulent transfer of momentum turns out to be identical to that of any passive scalar substance (the Reynolds analogy). According to up-to-date knowledge, the Reynolds analogy can be used in some cases for rough estimation of parameters of real flows, but the scope of the predicted results is highly restricted. The turbulent Prandtl number, as well as ν_t and D_t, depends on physical, geometric, and kinematic properties of the turbulent flow. For wall flows, numerous experimental measurements available give the value

$$Pr_t \approx 0.85. \tag{E5.1.3.5}$$

For jet flows and mixing layers, there are various estimates for Pr_t ranging from 0.5 to 0.75. These estimates are helpful for approximate calculations of turbulent heat and mass transfer and can be used for closing equation (E5.1.3.2).

E5.2. Mass Transfer in Liquid Films

E5.2.1. Mass Exchange Between Gases and Liquid Films

▶ **Statement of the problem.** Dissolution of gases in flowing liquid films is one of the most important methods for dissolving gases, which is widely used in technology. Film absorbers with irrigated walls are used for obtaining water solutions of gases (e.g., absorption of HCl vapor by water), for separating gas mixtures (e.g., absorption of benzene in the cake and by-product process), for purifying gases from detrimental effluents (e.g., purification of coke-oven from H_2S), etc.

Let us consider the absorption of weakly soluble gases on the free surface of a liquid film in a laminar flow on an inclined plane. It follows from the results of Section E4.3 that for moderate velocities of motion, the steady-state distribution of the velocity inside the film has the form of a semiparabola with maximum velocity U_{max} on the free surface, which is one and a half the mean flow velocity $\langle V \rangle$:

$$U_{max} = \frac{3}{2}\langle V \rangle = \frac{gh^2}{2\nu}\sin\alpha.$$

Here g is the gravitational acceleration, α the angle between the plane and the horizon, and h the film thickness, given by the expression

$$h = \left(\frac{3\nu^2}{g}\,Re\right)^{1/3},$$

where $Re = Q/\nu$ is the Reynolds number and Q is the irrigation density (that is, the volume rate flow of the liquid per unit width of the film).

The liquid velocity inside the film has a parabolic profile and is given by the formula

$$V = U_{max}(1 - y^2), \qquad y = Y/h,$$

where the Y-axis is normal to the film surface (Figure E4.3).

Assume that in the cross-section $X = 0$ the fluid flow contacts with a gas, so that a constant concentration $C = C_s$ of the absorbed component is attained on the free boundary ($Y = 0$), while no solute is contained in the irrigating liquid. Furthermore, we assume that there is no flow across the wall. We restrict ourselves to the case of high Peclet numbers, so that one can ignore the molecular diffusion along the film.

It follows from these assumptions that the concentration distribution inside the film is described by the equation and the boundary conditions

$$(1 - y^2)\frac{\partial c}{\partial x} = \frac{1}{\mathsf{Pe}}\frac{\partial^2 c}{\partial y^2}; \tag{E5.2.1.1}$$

$$x = 0, \quad c = 0 \qquad (0 \leq y \leq 1); \tag{E5.2.1.2}$$

$$y = 0, \quad c = 1 \qquad (x > 0); \tag{E5.2.1.3}$$

$$y = 1, \quad \partial c/\partial y = 0 \quad (x > 0), \tag{E5.2.1.4}$$

where the following dimensionless variables have been used:

$$x = \frac{X}{h}, \quad y = \frac{Y}{h}, \quad c = \frac{C}{C_s}, \quad \mathsf{Pe} = \frac{hU_{\max}}{D}. \tag{E5.2.1.5}$$

Note that for small $x \leq O(\mathsf{Pe}^{-1/2})$, near the input section one should consider the complete mass transfer equation with the term $\partial^2 c/\partial y^2$ replaced by Δc.

▶ **Diffusion boundary layer approximation.** For $x = O(1)$, the concentration mostly varies on the initial interval in a thin diffusion boundary layer near the free boundary of the film. In this region, we expand the transverse coordinate according to the rule

$$y = \zeta/\sqrt{\mathsf{Pe}}. \tag{E5.2.1.6}$$

By substituting (E5.2.1.6) into (E5.2.1.1) and by passing to the limit as $\mathsf{Pe} \to \infty$ (it is assumed that the variables x and ζ and the corresponding derivatives are of the order of 1), we obtain the equation

$$\frac{\partial c}{\partial x} = \frac{\partial^2 c}{\partial \zeta^2}. \tag{E5.2.1.7}$$

In view of (E5.2.1.6), the distance to the wall determined by the coordinate $y = 1$ corresponds to $\zeta = \sqrt{\mathsf{Pe}}$. Therefore, as $\mathsf{Pe} \to \infty$, the value $y = 1$ in the boundary condition (E5.2.1.4) corresponds to $\zeta \to \infty$. This implies that we can rewrite the boundary conditions (E5.2.1.2)–(E5.2.1.4) as follows:

$$x = 0, \quad c = 0; \qquad \zeta = 0, \quad c = 1; \qquad \zeta \to \infty, \quad \partial c/\partial \zeta \to 0. \tag{E5.2.1.8}$$

The solution of problem (E5.2.1.7), (E5.2.1.8) is given by the formula

$$c = \mathrm{erfc}\left(\frac{\zeta}{2\sqrt{x}}\right), \tag{E5.2.1.9}$$

where $\mathrm{erfc}\, z = \dfrac{2}{\sqrt{\pi}} \displaystyle\int_z^\infty \exp(-t^2)\, dt$ is the complementary error function.

By differentiating (E5.2.1.9), we obtain the dimensionless local diffusion flux to the film surface:

$$j = -\left(\frac{\partial c}{\partial y}\right)_{y=0} = \left(\frac{\mathsf{Pe}}{\pi x}\right)^{1/2}. \tag{E5.2.1.10}$$

TABLE E5.1
The eigenvalues λ_m and the coefficients A_m in the expansion of the solution (E5.2.1.12) for
the concentration distribution inside the film with gas absorption on the film surface.

m	λ_m	A_m	m	λ_m	A_m
1	2.2631	1.3382	6	22.3181	−0.1873
2	6.2977	−0.5455	7	26.3197	0.1631
3	10.3077	0.3589	8	30.3209	−0.1449
4	14.3128	−0.2721	9	34.3219	0.1306
5	18.3159	0.2211	10	38.3227	−0.1191

The dimensionless total diffusion flux on the part of the film lying in the interval from 0 to x is equal to

$$I = \int_0^x j \, dx = 2 \left(\frac{\mathsf{Pe}}{\pi} x \right)^{1/2}. \tag{E5.2.1.11}$$

Formulas (E5.2.1.10) and (E5.2.1.11) cannot be used for sufficiently large x, where the diffusion boundary layer "grows" through the entire film. To estimate the scope of these formulas, consider the original problem (E5.2.1.1)–(E5.2.1.4).

▶ **Exact solution.** For $0 \le x < \infty$, the exact solution of problem (E5.2.1.1)–(E5.2.1.4) can be represented in the form of the series

$$c = 1 - \sum_{m=0}^{\infty} A_m \exp\left(-\frac{\lambda_m^2}{\mathsf{Pe}} x \right) H_m(y), \tag{E5.2.1.12}$$

where the functions $H_m(y)$ and the coefficients A_m and λ_m are independent of the Peclet number. The functions $H_m(y)$ are defined by the formulas

$$H_m(y) = y \exp\left(-\tfrac{1}{2}\lambda_m y^2 \right) \Phi\left(a_m + \tfrac{1}{2}, \tfrac{3}{2}; \lambda_m y^2 \right), \quad a_m = \tfrac{1}{4}(1 - \lambda_m), \tag{E5.2.1.13}$$

where $\Phi(a, b, \xi) = 1 + \sum_{m=1}^{\infty} \dfrac{a(a+1)\ldots(a+m-1)}{b(b+1)\ldots(b+m-1)} \dfrac{\xi^m}{m!}$ is the degenerate hypergeometric function. The eigenvalues λ_m can be found by solving the transcendental equation

$$\lambda_m \Phi\left(a_m + \tfrac{1}{2}, \tfrac{3}{2}; \lambda_m \right) - \Phi\left(a_m + \tfrac{1}{2}, \tfrac{1}{2}; \lambda_m \right) = 0,$$

and the coefficients A_m are computed by the formulas

$$A_m = \frac{\displaystyle\int_0^1 (1 - y^2) H_m(y) \, dy}{\displaystyle\int_0^1 (1 - y^2) H_m^2(y) \, dy}, \quad \text{where} \quad m = 1, 2, \ldots \tag{E5.2.1.14}$$

Table E5.1 contains the first ten eigenvalues λ_m and the coefficients A_m calculated by Rotem & Neilson (1966).

Taking into account (E5.2.1.12), we obtain the dimensional total diffusion flux to the part of the film boundary from 0 to x in the form

$$I = -\int_0^x \left(\frac{\partial c}{\partial y} \right)_{y=0} dx = \mathsf{Pe} \sum_{m=1}^{\infty} \frac{A_m}{\lambda_m^2} \left(\frac{dH_m}{dy} \right)_{y=0} \left[1 - \exp\left(-\frac{\lambda_m^2}{\mathsf{Pe}} x \right) \right]. \tag{E5.2.1.15}$$

The substitution of the function (E5.2.1.13) into (E5.2.1.15) yields

$$I = \text{Pe} \sum_{m=1}^{\infty} \frac{A_m}{\lambda_m^2} \left[1 - \exp\left(-\frac{\lambda_m^2}{\text{Pe}} x \right) \right].$$

The comparison of this formula with (E5.2.1.11) shows that the diffusion boundary layer approximation can be used in the region $x \le 0.1 \, \text{Pe}$.

E5.2.2. Dissolution of a Plate by a Laminar Liquid Film

▶ **Statement of the problem.** Now consider mass transfer from a solid wall to a liquid film at high Peclet numbers. Such a problem is of serious interest in dissolution, crystallization, corrosion, anodic dissolution of metals in some electrochemical processes, etc. In many practical cases, dissolution processes are rather rapid compared with diffusion. Therefore, we assume that the concentration on the plate surface is equal to the constant C_s and the incoming liquid is pure. As before, we introduce dimensionless variables according to formulas (E5.2.1.5). In this case, the convective mass transfer in the liquid film is described by Eq. (E5.2.1.1), the boundary condition (E5.2.1.2) imposed on the longitudinal variable x, and the following boundary conditions with respect to the transverse coordinate:

$$y = 0, \quad \frac{\partial c}{\partial y} = 0 \quad (x > 0); \tag{E5.2.2.1}$$

$$y = 1, \quad c = 1 \quad (x > 0). \tag{E5.2.2.2}$$

Although this problem differs from the previously studied problem (E5.2.1.1)–(E5.2.1.4) only by a rearrangement of the boundary conditions (E5.2.1.3) and (E5.2.1.4), there is a substantial difference in the solutions of these problems.

▶ **Diffusion boundary layer approximation.** The concentration mostly varies on the initial interval $x = O(1)$, that is, in the diffusion boundary layer near the film surface. In this region, the asymptotic solution can be obtained by substituting the expanded coordinate

$$\xi = (1 - y) \, \text{Pe}^{1/3} \tag{E5.2.2.3}$$

into Eq. (E5.2.1.1) and by taking the leading term of the concentration expansion as $\text{Pe} \to \infty$. As a result, we arrive at the diffusion boundary layer equation

$$2\xi \frac{\partial c}{\partial x} = \frac{\partial^2 c}{\partial \xi^2}. \tag{E5.2.2.4}$$

By analogy with the already considered problem about gas absorption on the film surface, using (E5.2.1.2), (E5.2.2.1), and (E5.2.2.2), we obtain the boundary conditions for Eq. (E5.2.2.4), which coincide with (E5.2.1.8) up to the replacement $\xi \to \zeta$. The corresponding problem has the solution

$$c = \frac{1}{\Gamma(1/3)} \Gamma\left(\frac{1}{3}, \frac{2\xi^3}{9x} \right), \tag{E5.2.2.5}$$

where $\Gamma(1/3, z)$ is the incomplete gamma function (see Section M13.4).

By differentiating (E5.2.2.5), we calculate the dimensionless local diffusion flux

$$j = \frac{6^{1/3}}{\Gamma(1/3)} \frac{\text{Pe}^{1/3}}{x^{1/3}} \approx 0.678 \frac{\text{Pe}^{1/3}}{x^{1/3}}. \tag{E5.2.2.6}$$

The corresponding dimensionless total diffusion flux to the film boundary is

$$I = \int_0^x j \, dx = 1.02 \, \text{Pe}^{1/3} x^{2/3}. \tag{E5.2.2.7}$$

By comparing formulas (E5.2.1.9)–(E5.2.1.11) with (E5.2.2.5)–(E5.2.2.7), we see that for $x \sim 1$ the thickness of the diffusion boundary layer near the free surface of the film, $\delta_0 \sim \text{Pe}^{-1/2}$, is considerably less than that of the boundary layer near the solid surface, $\delta_{\text{sol}} \sim \text{Pe}^{-1/3}$. Accordingly, the diffusion flux to the free surface is larger than that to the solid surface. Moreover, the diffusion flux decreases more rapidly on the free surface than on the solid boundary with the increase of the distance from the input cross-section. These effects are due to the fact that the fluid moves much more rapidly near the free surface than near the solid boundary, where the no-slip condition is satisfied.

All facts established for a fluid film remain valid for a majority of problems on the diffusion boundary layer. Namely, near a gas–fluid or fluid–fluid interface, the dimensionless thickness of the layer is proportional to $\text{Pe}^{-1/2}$ (for the diffusion flux we have $j \sim \text{Pe}^{1/2}$), and near the fluid–solid interface the thickness of the boundary layer is proportional to $\text{Pe}^{-1/3}$ (the diffusion flux is $j \sim \text{Pe}^{1/3}$).

▶ **Exact solution.** The exact solution of the problem for equation (E5.2.1.1) with conditions (E5.2.1.2), (E5.2.2.1), and (E5.2.2.2) in the entire region $0 \leq x < \infty$ can be represented in the form of the series (E5.2.1.12), where the eigenfunctions $H_m(y)$ are expressed via the hypergeometric functions by the formulas

$$H_m(y) = \exp\left(-\tfrac{1}{2}\lambda_m y^2\right) \Phi\left(\tfrac{1}{4} - \tfrac{1}{4}\lambda_m, \tfrac{1}{2}; \lambda_m y^2\right) \tag{E5.2.2.8}$$

and the transcendental equation for the eigenvalues λ_m is

$$\Phi\left(a_m, \tfrac{1}{2}; \lambda_m\right) = 0, \qquad \text{where} \quad a_m = \tfrac{1}{4} - \tfrac{1}{4}\lambda_m. \tag{E5.2.2.9}$$

The roots of Eq. (E5.2.2.9) are positive and increase monotonically, so that $\lambda_m \to \infty$ as $m \to \infty$. To determine the eigenvalues λ_m, one can use the approximate formula

$$\lambda_m = 4m + 1.68 \qquad (m = 0, 1, 2, \ldots), \tag{E5.2.2.10}$$

whose error is less than 0.2%.

As before, the coefficients A_m in the series (E5.2.1.12) are determined by (E5.2.1.14), where the eigenfunctions $H_m(y)$ are given in (E5.2.2.8). To compute the coefficients A_m, one can use the approximate formulas

$$A_0 = 1.2; \quad A_m = 2.27 \, (-1)^m \lambda_m^{-7/6} \qquad \text{for} \quad m = 1, 2, 3, \ldots, \tag{E5.2.2.11}$$

where the λ_m are given in (E5.2.2.10). The maximum error of (E5.2.2.11) is less than 0.1%.

The total diffusion flux on the film surface is given by (E5.2.1.15) with $(dH_m/dy)_{y=0}$ replaced by $(dH_m/dy)_{y=1}$. One can use the expressions (E5.2.2.8), (E5.2.2.10), and (E5.2.2.11) for calculating the eigenfunctions $H_m(y)$ and the numbers λ_m and A_m.

E5.3. Heat Transfer to a Flat Plate

E5.3.1. Heat Transfer in Laminar Flow

▶ **Statement of the problem. Thermal boundary layer.** Consider heat transfer to a flat plate in a longitudinal translational flow of a viscous incompressible fluid with velocity U_i at high Reynolds numbers. We assume that the temperature on the plate surface and far from it is equal to the constants T_s and T_i, respectively. The origin of the rectangular coordinates X, Y is at the front edge of the plate, the X-axis is tangent to the plate, and the Y-axis is normal to it.

Numerous experiments and numerical calculations show that the laminar hydrodynamic boundary layer occurs for $5 \times 10^2 \leq \mathsf{Re}_X \leq 5 \times 10^5$ to 10^6. In this region, the thermal Peclet number $\mathsf{Pe}_X = \mathsf{Pr}\,\mathsf{Re}_X$ is large for gases and common liquids. For liquid metals, there is a range of Reynolds numbers, $10^4 \leq \mathsf{Re}_X \leq 10^6$, where the Peclet numbers are also large.

Taking into account the previous discussion, we restrict ourselves to the case of high Peclet numbers, for which the longitudinal molecular heat transfer may be neglected. The corresponding thermal boundary layer equations and the boundary conditions have the form

$$v_x \frac{\partial T}{\partial x} + v_y \frac{\partial T}{\partial y} = \frac{1}{\mathsf{Pr}} \frac{\partial^2 T}{\partial y^2}; \tag{E5.3.1.1}$$

$$x = 0, \quad T = 0; \qquad y = 0, \quad T = 1; \qquad y \to \infty, \quad T \to 0.$$

Here the dimensionless variables are introduced by formulas (E5.1.1.13) with characteristic length $a = \nu / U_i$ and ν is the kinematic viscosity of the fluid.

The fluid velocity in (E5.3.1.1) is given by the Blasius solution

$$v_x = f'(\eta), \quad v_y = \frac{\eta f' - f}{2\sqrt{x}}, \qquad \text{where} \quad \eta = \frac{y}{\sqrt{x}}. \tag{E5.3.1.2}$$

The function $f = f(\eta)$ was previously described in Section E4.5.1, and the prime stands for the derivative with respect to η.

▶ **Temperature field.** The solution of problem (E5.3.1.1) in conjunction with relations (E5.3.1.2) will be sought in the self-similar form $T = T(\eta)$. As a result, we obtain the ordinary differential equation

$$\frac{d^2 T}{d\eta^2} + \frac{1}{2} \mathsf{Pr} f(\eta) \frac{dT}{d\eta} = 0; \tag{E5.3.1.3}$$

$$\eta = 0, \quad T = 1; \qquad \eta \to \infty, \quad T \to 0.$$

By taking into account the relation $f = -f'''/f''$, which follows from the equation for f, we can write out the solution of problem (E5.3.1.3) as follows:

$$T = \frac{\displaystyle\int_\eta^\infty \left[f''(\eta) \right]^{\mathsf{Pr}} d\eta}{\displaystyle\int_0^\infty \left[f''(\eta) \right]^{\mathsf{Pr}} d\eta}. \tag{E5.3.1.4}$$

For $\mathsf{Pr} = 1$, this formula implies the simple relation $T(\eta) = 1 - f'(\eta) = 1 - v_x$ between temperature and the longitudinal velocity.

▶ **Heat flux and the Nusselt number.** By differentiating (E5.3.1.4) and by using the numerical value $f''(0) = 0.332$, we obtain the dimensionless local heat flux to the plate surface:

$$j_T = -\left(\frac{\partial T}{\partial y}\right)_{y=0} = \frac{\varphi(\text{Pr})}{\sqrt{x}}, \quad \text{where} \quad \varphi(\text{Pr}) = \frac{(0.332)^{\text{Pr}}}{\int_0^\infty \left[f''(\eta)\right]^{\text{Pr}} d\eta}. \quad (\text{E5.3.1.5})$$

It is convenient to seek the asymptotics of the function $\varphi(\text{Pr})$ at low and high Prandtl numbers starting from Eq. (E5.3.1.3) with the extended variable $\eta = \zeta/\text{Pr}$. As a result, we obtain the equation $T''_{\zeta\zeta} + f(\zeta/\text{Pr})T'_{\zeta} = 0$. As $\text{Pr} \to 0$, the argument of the function $f(\zeta/\text{Pr})$ tends to infinity, which corresponds to a constant velocity inside the thermal boundary layer and $f(\eta) \approx \eta$. In the other limit case as $\text{Pr} \to \infty$, the argument of the function $f(\zeta/\text{Pr})$ tends to zero, which corresponds to the linear approximation of the velocity inside the boundary layer and $f(\eta) \approx 0.166\,\eta^2$. By substituting the above-mentioned leading terms of the asymptotic expansion of f into Eq. (E5.3.1.3) and by solving the corresponding problems, we obtain the following expressions for the heat flux (E5.3.1.5):

$$\begin{aligned} \varphi(\text{Pr}) &\to \left(\text{Pr}/\pi\right)^{1/2} & (\text{Pr} \to 0), \\ \varphi(\text{Pr}) &\to 0.339\,\text{Pr}^{1/3} & (\text{Pr} \to \infty). \end{aligned} \quad (\text{E5.3.1.6})$$

Both considered limit situations occur in numerous problems of convective heat transfer. In the case $\text{Pr} \to 0$, which approximately takes place for liquid metals (e.g., mercury), one can neglect the dynamic boundary layer in the calculation of the temperature boundary layer and replace the velocity profile $v(x, y)$ by the velocity $v_\infty(x)$ of the inviscid outer flow. As $\text{Pr} \to \infty$, which corresponds to the case of strongly viscous fluids (e.g., glycerin), the temperature boundary layer is very thin and lies inside the dynamic boundary layer, where the velocity increases linearly with distance from the plate surface.

In the entire range of the Prandtl number, the function $\varphi(\text{Pr})$ in formula (E5.3.1.5) is well approximated by the expression

$$\varphi(\text{Pr}) = 0.0817 \left[(1 + 72\,\text{Pr})^{2/3} - 1\right]^{1/2}, \quad (\text{E5.3.1.7})$$

whose maximum error is about 0.5%.

Let us write out the local Nusselt number:

$$\text{Nu}_X = -\frac{X}{T_s - T_i}\left(\frac{\partial T_*}{\partial Y}\right)_{Y=0} = \sqrt{\text{Re}_X}\,\varphi(\text{Pr}), \quad (\text{E5.3.1.8})$$

where $\text{Re}_X = XU_i/\nu$ is the local Reynolds number.

E5.3.2. Heat Transfer in Turbulent Flow

▶ **Temperature profile.** Let us discuss qualitative specific features of convective heat and mass transfer in turbulent flow past a flat plate. Experimental evidence indicates that several characteristic regions with different temperature profiles can be distinguished in the thermal boundary layer on a flat plate. At moderate Prandtl numbers ($0.5 \le \text{Pr} \le 2.0$), it can be assumed for rough estimates that the characteristic sizes of these regions are of the same order of magnitude as those of the wall layer and the core of the turbulent stream, see Section E4.5.2.

For the description of heat and mass transfer in the wall layer, it is common to introduce the friction velocity U_*, friction temperature Θ_*, and dimensionless internal coordinate y^+ by the relations

$$U_* = \sqrt{\frac{\tau_s}{\rho}}, \qquad \Theta_* = \frac{q_s}{\rho c_p U_*}, \qquad y^+ = \frac{Y U_*}{\nu}, \tag{E5.3.1.9}$$

where τ_s is the shear stress at the wall, ρ the fluid density, q_s the heat flux at the wall, c_p the specific heat of the fluid, ν the kinematic viscosity, and Y the distance to the plate surface.

In the molecular thermal conduction layer, adjacent to the plate surface, the deviation of the average temperature $\overline{T}$ from the wall temperature T_s depends linearly on the transverse coordinate:

$$\frac{T_s - \overline{T}(Y)}{\Theta_*} = \mathsf{Pr}\, y^+. \tag{E5.3.1.10}$$

For $\mathsf{Pr} > 1$, the linear law (E5.3.1.10) is satisfied for $0 \leq y^+ \leq 9\,\mathsf{Pr}^{-1/3}$.

In the logarithmic layer, the average temperature can be determined with the use of the relations*

$$\frac{T_s - \overline{T}(Y)}{\Theta_*} = 2.12 \ln y^+ + \beta(\mathsf{Pr}), \tag{E5.3.1.11}$$

$$\beta(\mathsf{Pr}) = (3.85\,\mathsf{Pr}^{1/3} - 1.3)^2 + 2.12 \ln \mathsf{Pr}.$$

These formulas are valid within a wide range of Prandtl numbers, $6 \times 10^{-3} \leq \mathsf{Pr} \leq 10^4$, for $y^+ \geq y^+_{\log}$. The lower bound of the logarithmic layer, $y^+_{\log}$, depends on the Prandtl number as follows:

$$y^+_{\log} \simeq \begin{cases} 12\,\mathsf{Pr}^{-1/3} & \text{if } \mathsf{Pr} \gg 1 \quad \text{(for liquids)}, \\ 30 & \text{if } \mathsf{Pr} \sim 1 \quad \text{(for gases)}, \\ 2/\mathsf{Pr} & \text{if } \mathsf{Pr} \ll 1 \quad \text{(for liquid metals)}. \end{cases}$$

In the stream core and a major part of the logarithmic layer, the average temperature profile can be described by the single formula

$$\frac{\overline{T}(Y) - \overline{T}_m}{\Theta_*} = -2.12 \ln \eta + 1.5\,(1 - \eta), \tag{E5.3.1.12}$$

where $\overline{T}_m$ is the average temperature at the stream axis, $\eta = Y/\delta_T$ the dimensionless distance to the plate surface, and δ_T the thermal boundary layer thickness. The results predicted by Eq. (E5.3.1.12) fairly well agree with experimental data provided by various authors for turbulent flows of air, water, and transformer oil for $10^{-2} < Y/\delta_T < 1$.

The thermal boundary layer thickness can be estimated from the relation

$$\delta_T(X) = 0.45\, X \sqrt{\tfrac{1}{2} c_\mathrm{f}}, \tag{E5.3.1.13}$$

where c_f is the local friction coefficient determined by formula (E4.5.2.2) or (E4.5.2.3). The empirical coefficient 0.45 was obtained by processing experimental data for $0.7 \leq \mathsf{Pr} \leq 64$. It is greater than the similar coefficient in relation (E4.5.2.1) for the hydrodynamic turbulent boundary layer thickness. (Qualitatively, this is accounted for by the fact that the turbulent Prandtl number is less than unity; that is, the hydrodynamic boundary layer lies within the thermal boundary layer.)

---

* The universal constant 2.12 multiplying the logarithm is defined as Pr_t/κ, where $\mathsf{Pr}_t = 0.85$ is the turbulent Prandtl number for wall flows and $\kappa = 0.4$ the von Kármán constant.

▶ **Nusselt number.** For determining the local Nusselt number, it is recommended to use the Kader–Yaglom formula

$$\mathsf{Nu}_X = \frac{\mathsf{Pr}\,\mathsf{Re}_X(\tfrac{1}{2}c_f)^{1/2}}{2.12\ln(\tfrac{1}{2}c_f\,\mathsf{Pr}\,\mathsf{Re}_X) + (3.85\,\mathsf{Pr}^{1/3} - 1.3)^2 + 1.5}, \tag{E5.3.1.14}$$

which quite well agrees with numerous experimental data within a wide range of Prandtl numbers $(0.5 \le \mathsf{Pr} \le 100)$.

For rough estimates, the simpler formula

$$\mathsf{Nu}_X = 0.0288\,\mathsf{Pr}^{0.4}\mathsf{Re}_X^{0.8}$$

can be used, which is valid for $0.5 \le \mathsf{Pr} \le 50$.

More detailed information about heat and mass transfer in turbulent flows past a flat plate, as well as various relations for determining the temperature profile and Nusselt (Sherwood) numbers, and a lot of other useful information can be found in the references given at the end of this chapter.

E5.4. Convective Heat and Mass Transfer in Tubes and Channels

Many convective mass and heat transfer processes in chemical industry, petroleum chemistry, gas, nuclear, and other branches of industry occur in pipes (water, gas, and oil pipelines, heat exchangers, etc.).

Starting from the classical works by Graetz (1883) and Nusselt (1910), many authors considered the problem about the temperature distribution in a fluid moving in a tube under various assumptions on the type of flow, the tube shape, the form of boundary conditions, the value of the Peclet number, and some other simplifications. In this section, we outline the most important results obtained in this field.

E5.4.1. Heat and Mass Transfer in a Circular Tube with Constant Temperature of the Wall for Laminar Flows

▶ **Statement of the problem.** Consider a laminar steady-state fluid flow in a circular tube of radius a with Poiseuille velocity profile (see Section E4.4.1). We introduce cylindrical coordinates $\mathcal{R}$, Z with the Z-axis in the flow direction. We assume that for $Z > 0$ the temperature on the wall is equal to the constant T_2. In the entry area $Z < 0$, the temperature on the wall is also constant but takes another value T_1.

The convective heat transfer in a tube is described by the equation

$$\mathsf{Pe}_T(1 - \varrho^2)\frac{\partial T}{\partial z} = \frac{\partial^2 T}{\partial \varrho^2} + \frac{1}{\varrho}\frac{\partial T}{\partial \varrho} + \frac{\partial^2 T}{\partial z^2} \tag{E5.4.1.1}$$

and the boundary conditions

$$\varrho = 0, \quad \frac{\partial T}{\partial \varrho} = 0; \qquad \varrho = 1, \quad T = \begin{cases} 0 & \text{for } z < 0; \\ 1 & \text{for } z > 0; \end{cases} \tag{E5.4.1.2}$$

$$z \to -\infty, \quad T \to 0; \qquad z \to \infty, \quad T \to 1. \tag{E5.4.1.3}$$

The following dimensionless variables have been used:

$$\varrho = \frac{\mathcal{R}}{a}, \quad z = \frac{Z}{a}, \quad T = \frac{T_* - T_1}{T_2 - T_1}, \quad \mathsf{Pe}_T = \frac{aU_{\max}}{\chi},$$

where T_* is the fluid temperature, χ the thermal diffusivity, $U_{\max} = a^2 \Delta P / (4\mu L)$ the maximum velocity at the center of the tube, ΔP the pressure increment along the distance L, and μ the dynamic viscosity of the fluid.

▶ **High Peclet numbers (initial region).** As $\mathsf{Pe}_T \to \infty$, the fluid temperature in the region $z < 0$ is constant and equal to the temperature $T \approx 0$ on the wall. In the region $z > 0$ at $z = O(1)$, a thin boundary layer is being formed near the wall of the tube. In this region, on the left-hand side of Eq. (E5.4.1.1) one may retain only the leading term of the velocity expansion as $\varrho \to 1$ and write $v = 1 - \varrho^2 \approx 2\xi$, where $\xi = 1 - \varrho$. Moreover, in contrast with the first term on the right-hand side in (E5.4.1.1), one can neglect the last two terms; that is, $\Delta T \approx \partial^2 T / \partial \xi^2$. Thus, we arrive at an equation that coincides with (E5.2.2.4) up to notation. Taking into account the boundary conditions (E5.4.1.2), we obtain the temperature distribution in the boundary layer,

$$T = \frac{1}{\Gamma(1/3)} \Gamma\left(\frac{1}{3}, \frac{2\,\mathsf{Pe}_T (1-\varrho)^3}{9z}\right), \tag{E5.4.1.4}$$

where $\Gamma(1/3, z)$ is the incomplete gamma function (see Section M13.4).

The corresponding dimensionless local and total heat fluxes have the form

$$j_T = -\left(\frac{\partial T}{\partial \varrho}\right)_{\varrho=1} = \frac{1}{\Gamma(1/3)} \left(\frac{6\,\mathsf{Pe}_T}{z}\right)^{1/3}, \tag{E5.4.1.5}$$

$$I_T = \int_0^z j_T\, dz = \frac{3(6\,\mathsf{Pe}_T)^{1/3}}{2\Gamma(1/3)} z^{2/3}. \tag{E5.4.1.6}$$

The scope of formulas (E5.4.1.4)–(E5.4.1.6) is bounded by the condition $z \ll \mathsf{Pe}_T$. This restriction practically always holds in the similar problem about the diffusion boundary layer.

▶ **Arbitrary Peclet numbers.** In the general case $0 \le \mathsf{Pe}_T < \infty$, we seek the fluid temperature distribution separately on both sides of the inlet cross-section of the tube in the form of series:

$$T = 1 - \sum_{m=0}^{\infty} A_m \exp\left(-\frac{\lambda_m^2}{\mathsf{Pe}_T} z\right) f_m(\varrho) \qquad \text{for} \quad z > 0, \tag{E5.4.1.7}$$

$$T = \sum_{k=0}^{\infty} B_k \exp\left(\frac{\eta_k^2}{\mathsf{Pe}_T} z\right) g_k(\varrho) \qquad \text{for} \quad z < 0. \tag{E5.4.1.8}$$

The eigenfunctions f_m and g_k and the eigenvalues λ_m and η_k are determined from two spectral problems that are obtained by substituting the expansions (E5.4.1.7) and (E5.4.1.8) into Eq. (E5.4.1.1) and the boundary conditions (E5.4.1.2) and (E5.4.1.3).

The temperature $T = T(\varrho, z)$ and its derivative must be continuous across the section $z = 0$:

$$T(\varrho, -0) = T(\varrho, +0); \qquad \frac{\partial T}{\partial z}(\varrho, -0) = \frac{\partial T}{\partial z}(\varrho, +0). \tag{E5.4.1.9}$$

The consistency conditions (E5.4.1.9) allow one to find the coefficients A_m and B_k in the series (E5.4.1.7) and (E5.4.1.8).

In what follows, we give the results of the analysis of this problem only for the region $z > 0$.

The eigenfunctions $f_m(\rho)$ are expressed via the degenerate hypergeometric function $\Phi(a, b; \xi)$ as

$$f_m(\rho) = \exp\left(-\tfrac{1}{2}\lambda_m \varrho^2\right) \Phi(a_m, 1; \lambda_m \varrho^2), \qquad (E5.4.1.10)$$

$$a_m = \frac{1}{2} - \frac{1}{4}\lambda_m - \frac{\lambda_m^3}{4\,\mathsf{Pe}_T^2},$$

and the eigenvalues λ_m satisfy the transcendental equation

$$\Phi(a_m, 1; \lambda_m) = 0. \qquad (E5.4.1.11)$$

The coefficients A_m can be computed by the formula

$$A_m = -\frac{2}{\lambda_m \left(\dfrac{\partial f}{\partial \lambda}\right)_{\varrho=1,\,\lambda=\lambda_m}} \qquad (E5.4.1.12)$$

with the function $f(\rho) = f(\rho; \lambda)$ determined by relations (E5.4.1.10) where the indices m are omitted.

Consider the asymptotic behavior of the eigenvalues λ_m and coefficients A_m in some limit cases.

As $\mathsf{Pe}_T \to 0$, we have

$$\lambda_m = (\gamma_m\,\mathsf{Pe}_T)^{1/2}, \quad A_m = -\left[\gamma_m J_1(\gamma_m)\right]^{-1}, \quad f_m = J_0(\gamma_m \varrho), \qquad (E5.4.1.13)$$

where $J_0 = J_0(\xi)$ and $J_1 = J_1(\xi)$ are the Bessel functions and the γ_m are the roots of J_0, $J_0(\gamma_m) = 0$.

It is convenient to calculate approximate values of γ_m by using the approximate expression

$$\gamma_m = 2.4 + 3.13\,m \qquad (m = 0,\ 1,\ 2,\ \ldots), \qquad (E5.4.1.14)$$

whose maximum error is less than 0.2%.

As $\mathsf{Pe} \to \infty$, the asymptotics of the eigenvalues λ_m for large m has the form

$$\lambda_m = 4m + \tfrac{8}{3}. \qquad (E5.4.1.15)$$

This formula constructed under the assumption that $m \gg 1$ can be used for all m. The maximum error in (E5.4.1.15) is attained at $m = 0$ and is equal to 1.4%.

Instead of (E5.4.1.15), it is convenient to use the more precise formula

$$\lambda_m = 4m + 2.7 \qquad (m = 0,\ 1,\ 2,\ \ldots), \qquad (E5.4.1.16)$$

whose maximum error is 0.3%.

The coefficients A_m of the series (E5.4.1.7) can be calculated by the formula

$$A_m = 2.85\,(-1)^m \lambda_m^{-2/3}, \qquad (E5.4.1.17)$$

whose maximum error is 0.5%.

For large but finite Peclet numbers, the expressions (E5.4.1.16) and (E5.4.1.17) can be used only for a bounded set of eigenvalues such that $\lambda_m \ll \mathsf{Pe}_T$.

The dependence of the ground eigenvalue λ_0 on the Peclet number is nicely approximated by the formula

$$\lambda_0 = 2.7 \sqrt{\frac{\exp(0.27\,\mathsf{Pe}_T) - 1}{\exp(0.27\,\mathsf{Pe}_T) - 0.18}},$$

whose maximum error is about 1%.

▶ **Bulk temperature. Heat flux.** Taking into account the fact that the dimensionless fluid velocity distribution in a tube is $u(\varrho) = 1 - \varrho^2$, we obtain the bulk temperature in an arbitrary cross-section:

$$\langle T \rangle = \frac{\displaystyle\int_0^1 T u(\varrho) 2\pi \varrho \, d\varrho}{\displaystyle\int_0^1 u(\varrho) 2\pi \varrho \, d\varrho} = 4 \int_0^1 T(1 - \varrho^2) \varrho \, d\varrho.$$

By substituting the series (E5.4.1.7) into this formula, we obtain

$$\langle T \rangle = 1 - \sum_{m=0}^{\infty} E_m \exp\left(-\frac{\lambda_m^2}{\mathsf{Pe}_T} z\right), \qquad (E5.4.1.18)$$

where

$$E_m = 4 A_m \int_0^1 f_m (1 - \varrho^2) \varrho \, d\varrho.$$

By differentiating (E5.4.1.7), we arrive at the dimensionless local heat flux to the tube wall,

$$j_T = \left(\frac{\partial T}{\partial \varrho}\right)_{\varrho=1} = -\sum_{m=0}^{\infty} A_m f_m'(1) \exp\left(-\frac{\lambda_m^2}{\mathsf{Pe}_T} z\right), \qquad (E5.4.1.19)$$

where

$$f_m'(1) = 2 a_m \lambda_m \exp(-\tfrac{1}{2}\lambda_m) \Phi(a_m + 1, 2; \lambda_m).$$

▶ **Nusselt number.** The most practically important variable is the Nusselt number

$$\mathsf{Nu} = \frac{2 j_T}{1 - \langle T \rangle}. \qquad (E5.4.1.20)$$

Here $1 - \langle T \rangle$ is the temperature head equal to the difference between the wall temperature and the mean flow rate temperature.

It follows from (E5.4.1.18)–(E5.4.1.20) that at large distances from the inlet cross-section (as $z \to +\infty$) the Nusselt number tends to the constant value

$$\mathsf{Nu}_\infty = \frac{-f_0'(1)}{2 \displaystyle\int_0^1 f_0(\varrho)(1 - \varrho^2) \varrho \, d\varrho}. \qquad (E5.4.1.21)$$

For low Peclet numbers, the limit Nusselt number is

$$\mathsf{Nu}_\infty = \frac{\gamma_0^3}{4} \frac{J_1(\gamma_0)}{J_2(\gamma_0)} \approx 4.16 \qquad (\text{as } \mathsf{Pe}_T \to 0). \qquad (E5.4.1.22)$$

For high Peclet numbers, the limit Nusselt number is

$$\mathsf{Nu}_\infty = \tfrac{1}{2}\lambda_0^2 \approx 3.66 \qquad (\text{as } \mathsf{Pe}_T \to \infty). \qquad (E5.4.1.23)$$

In the entire range of the Peclet number, the limit Nusselt number is nicely approximated by the formula

$$\text{Nu}_\infty = \frac{4.16 + 1.15 \, \text{Pe}_T}{1 + 0.315 \, \text{Pe}_T}, \tag{E5.4.1.28}$$

whose maximum error is about 0.6%.

Calculations according to (E5.4.1.20) show that for high Peclet numbers one can conventionally divide the entire length of a heated (cooled) tube into two parts. On the first part, a temperature profile is formed with radial temperature distribution varying from the initial value (at $z = 0$) to some limit value $f_0(\varrho)$. In this region, the number Nu decreases near the inlet cross-section as a power-law function $\text{Nu} \approx 2j_T$, where j_T is described by (E5.4.1.5). On the second part of the tube, the radial distribution of the excess temperature $\delta T = 1 - T$ does not vary along the tube (though the absolute values of temperature do vary), and the number Nu preserves the constant value 3.66. The first part of the tube is called the thermal initial region, and the second one, the region of steady-state heat exchange.

It is conventional to define the length of the thermal initial region as the distance from the inlet cross-section to the point at which the Nusselt number differs from its limit value (E5.4.1.23) by 1%. Calculations show that the dimensional length of the thermal initial region is $l = 0.11a \, \text{Pe}_T$.

E5.4.2. Heat and Mass Transfer in a Circular Tube with Constant Heat Flux at the Wall for Laminar Flows

▶ **Statement of the problem.** Now let us study the case in which a constant heat flux $q = \kappa (\partial T / \partial \mathcal{R})_{\mathcal{R}=a} = \text{const}$, where κ is the thermal conductivity coefficient of the fluid, is given on the surface of a circular tube for $Z > 0$. The entry part is modeled by the region $Z < 0$, where there is no heat flux across the tube surface and the temperature tends to the constant value T_1 as $Z \to -\infty$.

In this case, it is convenient to introduce the dimensionless temperature as

$$\widetilde{T} = \frac{\kappa(T - T_1)}{aq}, \tag{E5.4.2.1}$$

and the other dimensionless variables are determined in the same way as in problem (E5.4.1.1)–(E5.4.1.3).

The considered heat transfer process is described by Eq. (E5.4.1.1) with T replaced by $\widetilde{T}$ and with the boundary conditions

$$\varrho = 0, \quad \frac{\partial \widetilde{T}}{\partial \varrho} = 0; \qquad \varrho = 1, \quad \frac{\partial \widetilde{T}}{\partial \varrho} = \begin{cases} 0 & \text{for } z < 0; \\ 1 & \text{for } z > 0; \end{cases} \tag{E5.4.2.2}$$

$$z \to -\infty, \quad \widetilde{T} \to 0. \tag{E5.4.2.3}$$

Note that the temperature distribution as $z \to +\infty$ is not known in advance.

▶ **Temperature field far from the inlet cross-section.** The temperature distribution far from the inlet cross-section has the form

$$\widetilde{T} = 4\frac{z}{\text{Pe}_T} + \varrho^2 - \frac{\varrho^4}{4} + \frac{8}{\text{Pe}_T^2} - \frac{7}{24} \qquad (z \gg 1). \tag{E5.4.2.4}$$

The mean flow rate temperature of the fluid and the temperature head in the region of heat stabilization are, respectively, equal to

$$\langle \widetilde{T} \rangle_m = \frac{4}{\mathsf{Pe}_T} z + \frac{8}{\mathsf{Pe}_T^2}, \qquad \widetilde{T}_s - \langle \widetilde{T} \rangle_m = \frac{11}{24},$$

where $\widetilde{T}_s$ is the dimensionless temperature on the tube wall.

The limit Nusselt number is

$$\mathsf{Nu}_\infty = \frac{2(d\widetilde{T}/d\varrho)_{\varrho=1}}{\widetilde{T}_s - \langle \widetilde{T} \rangle} = \frac{48}{11} \approx 4.36.$$

Obviously, Nu_∞ is independent of the Peclet number.

▶ **Temperature field in the initial region of the tube.** For $z < 0$, we seek the solution of the complete problem (E5.4.1.1), (E5.4.2.2), (E5.4.2.3) in the form of the series (E5.4.1.8) (with T replaced by $\widetilde{T}$). For $z > 0$, the temperature field is constructed on the basis of the asymptotic distribution (E5.4.2.4) as follows:

$$\widetilde{T} = 4\frac{z}{\mathsf{Pe}_T} + \varrho^2 - \frac{\varrho^4}{4} + \frac{8}{\mathsf{Pe}_T^2} - \frac{7}{24} - \sum_{m=0}^{\infty} A_m \exp\left(-\lambda_m^2 \frac{z}{\mathsf{Pe}_T}\right) f_m(\varrho).$$

By substituting these series into (E5.4.1.1), (E5.4.2.2), and (E5.4.2.3) and by separating the variables, we obtain spectral problems for the eigenfunctions f_m and g_k and the eigenvalues λ_m and η_k.

The solution of the problem for f_m is given by (E5.4.1.10), where the numbers λ_m satisfy the transcendental equation

$$\Phi(a_m, 1; \lambda_m) = 2a_m \Phi(a_m + 1, 2; \lambda_m).$$

The coefficients of the expansions A_m and B_m are determined by the condition that the temperature and its derivative are continuous across the section $z = 0$ (E5.4.1.9).

If we determine the length l of the thermal initial region on the basis of the formula $\mathsf{Nu} = 1.01\,\mathsf{Nu}_\infty$, then we obtain $l = 0.14\,\mathsf{Pe}\,a$.

E5.4.3. Heat and Mass Transfer in Plane Channels for Laminar Flows

▶ **Plane channel with constant temperature of the wall.** We shall study the heat exchange in a laminar fluid flow with parabolic velocity profile in a plane channel of width $2h$. Let us introduce rectangular coordinates X, Y with the X-axis codirected with the flow and lying at equal distances from the channel walls. We assume that the temperature on the walls (at $Y = \pm h$) is constant and is equal to T_1 for $X < 0$ and to T_2 for $X > 0$. Since the problem is symmetric with respect to the X-axis, it suffices to consider half of the flow region, $0 \leq Y \leq h$.

The temperature distribution T_* is described by the following equation and boundary conditions:

$$\mathsf{Pe}_T(1 - y^2)\frac{\partial T}{\partial x} = \frac{\partial^2 T}{\partial x^2} + \frac{\partial^2 T}{\partial y^2}; \tag{E5.4.3.1}$$

$$y = 0, \quad \frac{\partial T}{\partial y} = 0; \qquad y = 1, \quad T = \begin{cases} 0 & \text{for } x < 0; \\ 1 & \text{for } x > 0; \end{cases} \tag{E5.4.3.2}$$

$$x \to -\infty, \quad T \to 0; \qquad x \to +\infty, \quad T \to 1, \tag{E5.4.3.3}$$

with the dimensionless variables

$$x = \frac{X}{h}, \quad y = \frac{Y}{h}, \quad T = \frac{T_* - T_1}{T_2 - T_1}, \quad \mathsf{Pe}_T = \frac{hU_{\max}}{\chi}, \quad U_{\max} = \frac{3}{2}\langle V \rangle,$$

where $U_{\max}$ is the maximum fluid velocity on the flow axis and $\langle V \rangle$ is the mean flow velocity over the cross-section.

By analogy with the case of a circular tube, we seek the solution of problem (E5.4.3.1)–(E5.4.3.3) in the form of a series for separated variables:

$$T = 1 - \sum_{m=0}^{\infty} A_m \exp\left(-\frac{\lambda_m^2}{\mathsf{Pe}_T} x\right) f_m(y) \qquad \text{for} \quad x > 0, \qquad (E5.4.3.4)$$

$$T = \sum_{k=0}^{\infty} B_k \exp\left(\frac{\eta_k^2}{\mathsf{Pe}_T} x\right) g_k(y) \qquad \text{for} \quad x < 0. \qquad (E5.4.3.5)$$

The eigenvalues λ_m, η_k, the eigenfunctions f_m, g_k, and the coefficients A_m and B_k can be obtained by analogy with the case of a circular tube in Section E5.4.1.

In what follows, we present only the basic solutions of the problem for $x > 0$.

The eigenfunctions $f_m(y)$ can be written as

$$f_m(y) = \exp\left(-\frac{1}{2}\lambda_m y^2\right) \Phi\left(\frac{1}{4} - \frac{1}{4}\lambda_m - \frac{\lambda_m^3}{4\,\mathsf{Pe}_T^2}, \frac{1}{2}; \lambda_m y^2\right). \qquad (E5.4.3.6)$$

Here $\Phi(a, b; \xi)$ is the degenerate hypergeometric function and the λ_m satisfy the transcendental equation

$$\Phi\left(a_m, \frac{1}{2}; \lambda_m\right) = 0, \quad \text{where} \quad a_m = \frac{1}{4} - \frac{\lambda_m}{4} - \frac{\lambda_m^3}{4\,\mathsf{Pe}_T^2}. \qquad (E5.4.3.7)$$

The coefficients A_m are calculated by the formula (E5.4.1.12) with ϱ replaced by y, where the auxiliary function f can be obtained from (E5.4.3.6) after the indices m are omitted.

In the limit case as $\mathsf{Pe}_T \to 0$, we have

$$\lambda_m = \sqrt{\mathsf{Pe}_T\left(\frac{\pi}{2} + \pi m\right)}, \quad A_m = \frac{4(-1)^m}{(\pi + 2\pi m)^2}, \quad f_m = \cos\left[\left(\frac{\pi}{2} + \pi m\right)y\right], \quad (E5.4.3.8)$$

where $m = 0, 1, 2, \ldots$

As $\mathsf{Pe}_T \to \infty$, the eigenvalues λ_m and the coefficients A_m can be obtained from formulas (E5.2.2.10) and (E5.2.2.11) (since formulas (E5.4.3.6) and (E5.4.3.7) pass as $\mathsf{Pe}_T \to \infty$ into formulas (E5.2.2.8) and (E5.2.2.9), respectively).

▶ **Mean flow rate temperature. Nusselt number.** The flow rate temperature for a plane channel is given by the formula

$$\langle T \rangle = \frac{3}{2} \int_0^1 T(1 - y^2)\, dy. \qquad (E5.4.3.9)$$

The local heat flux can be found by using (E5.4.1.19) with z replaced by x and the derivative $f_m'(1)$ calculated from (E5.4.3.6)

Let us substitute formulas (E5.4.3.4), (E5.4.3.9), and (E5.4.1.19) into (E5.4.1.20), and let x tend to infinity. As a result, we obtain the limit Nusselt number

$$\mathrm{Nu}_\infty = \frac{-4 f_0'(1)}{3 \int_0^1 f_0(y)(1 - y^2)\, dy}. \tag{E5.4.3.10}$$

The eigenfunction $f_0(y) = \cos(\pi y/2)$ follows from (E5.4.3.8) for low Peclet numbers. By applying (E5.4.3.10), we obtain

$$\mathrm{Nu}_\infty = \frac{\pi^4}{24} \approx 4.06 \qquad (\text{as } \mathrm{Pe}_T \to 0). \tag{E5.4.3.11}$$

For high Peclet numbers, the limit Nusselt number is

$$\mathrm{Nu}_\infty = \frac{4}{3}\lambda_0^2 \approx 3.77 \qquad (\text{as } \mathrm{Pe}_T \to \infty). \tag{E5.4.3.12}$$

Over the entire range of Peclet numbers, Nu_∞ is nicely approximated by the formula

$$\mathrm{Nu}_\infty = \frac{4.06 + 3.66\,\mathrm{Pe}_T}{1 + 0.97\,\mathrm{Pe}_T}, \tag{E5.4.3.13}$$

whose maximum error is about 0.5%.

▶ **Plane channel with constant heat flux at the wall.** Now consider the case in which a constant heat flux q is given on the walls of a plane channel for $X > 0$. We assume that for $X < 0$ the walls are thermally insulated and the temperature tends to a constant T_1 as $X \to -\infty$.

For $a \equiv h$, we introduce the dimensionless temperature $\widetilde{T}$ according to (E5.4.2.1). The heat exchange in a plane channel is described by Eq. (E5.4.3.1) (as $T \to \widetilde{T}$) and the boundary conditions (E5.4.2.2), (E5.4.2.3), where z and ϱ are replaced by x and y, respectively.

The temperature distribution far from the inlet cross-section (in the heat stabilization region) has the form

$$\widetilde{T} = \frac{3}{2}\frac{x}{\mathrm{Pe}_T} + \frac{3}{4}y^2 - \frac{1}{8}y^4 + \frac{9}{4\,\mathrm{Pe}_T^2} - \frac{39}{280} \qquad (\text{for } x \gg 1). \tag{E5.4.3.14}$$

From the temperature distribution (E5.4.3.14) far from the input cross-section, one can find the limit Nusselt number

$$\mathrm{Nu}_\infty = 4.12.$$

E5.4.4. Turbulent Heat Transfer in Circular Tube and Plane Channel

▶ **Temperature profile.** Let us discuss qualitative specific features of convective heat and mass transfer in a turbulent flow through a circular tube and plane channel in the stabilized flow region. Experimental evidence indicates that several characteristic regions with different temperature profiles can be distinguished. At moderate Prandtl numbers ($0.5 \le \mathrm{Pr} \le 2.0$), the structure and sizes of these regions are similar to those of the wall layer and the core of the turbulent stream considered in Section E4.4.2.

In the molecular thermal conduction layer, adjacent to the tube wall, the deviation of the average temperature $\overline{T}$ from the wall temperature T_s satisfies the linear dependence (E5.3.1.10). In the logarithmic layer, the average temperature can be estimated using relations (E5.3.1.11), which are valid for liquids, gases, and liquid metals within a wide range of Prandtl numbers, $6 \times 10^{-3} \leq \mathrm{Pr} \leq 10^4$.

In the stream core and a major part of the logarithmic layer, the average temperature profile can be described by the single formula

$$\frac{\overline{T}(Y) - \overline{T}_\mathrm{m}}{\Theta_*} = -2.12 \ln \eta + 0.3 \, (1 - \eta^2), \tag{E5.4.4.1}$$

where $\overline{T}_\mathrm{m}$ is the average temperature at the stream axis, $\eta = Y/a$ the dimensionless distance to the tube wall, and a the tube radius. The results predicted by Eq. (E5.4.4.1) fairly well agree with experimental data provided by various authors for turbulent flows of water, air, and liquid metals through circular tubes and plane channels for $3 \times 10^{-2} < Y/a < 1$.

▶ **Nusselt number for the thermal stabilized region.** In the thermal stabilization region of a turbulent flow through a smooth tube, for computing the limiting Nusselt number one can use the formula

$$\mathrm{Nu}_\infty = \frac{\mathrm{Pr}\,\mathrm{Re}_d \xi}{2.12 \ln(\mathrm{Pr}\,\mathrm{Re}_d \xi) + (3.85\,\mathrm{Pr}^{1/3} - 1.3)^2 - 4.3 + 6.7\,\xi}. \tag{E5.4.4.2}$$

Here the following notation is used:

$$\mathrm{Nu}_\infty = \frac{\alpha d}{c_p \rho \chi}, \quad \alpha = \frac{q_s}{T_s - \overline{T}_\mathrm{m}}, \quad \mathrm{Re}_d = \frac{d\langle V\rangle}{\nu}, \quad \mathrm{Pr} = \frac{\nu}{\chi}, \quad \xi = \sqrt{\tfrac{1}{8}\lambda},$$

where α is the heat transfer coefficient, $d = 2a$ the tube diameter, c_p the specific heat at constant pressure, ρ the fluid density, χ the thermal diffusivity, q_s the heat flux at the wall, ν the kinematic viscosity, and $\langle V\rangle$ the mean flow rate velocity. The drag coefficient λ can be found from Eq. (E4.4.2.12) or (E4.4.2.13). Formula (E5.4.4.2) unifies the results of more than fifty experimental studies and applies to liquids and gases within wide ranges of Reynolds and Prandtl numbers, $5 \times 10^3 \leq \mathrm{Re}_d \leq 2 \times 10^6$ and $0.6 \leq \mathrm{Pr} \leq 4 \times 10^4$. In the case of mass exchange, the Nusselt number must be replaced by the Sherwood number and the thermal Prandtl number, by the diffusion one.

In engineering calculations, the Nusselt number is often determined from simple two-term (or one-term) formulas like

$$\mathrm{Nu}_\infty = A + B\,\mathrm{Pr}^n \mathrm{Re}_d^m \tag{E5.4.4.3}$$

valid in a limited range of Prandtl and Reynolds numbers. For liquid metals, in the case of $10^4 < \mathrm{Re}_d < 10^6$ and constant temperature or heat flux at the wall, in Eq. (E5.4.4.3) one should set

$$\begin{aligned} A &= 5, & B &= 0.025, & n &= m = 0.8 & (\text{if } T_s = \text{const}), \\ A &= 6.8, & B &= 0.044, & n &= m = 0.75 & (\text{if } q_s = \text{const}). \end{aligned} \tag{E5.4.4.4}$$

▶ **Intermediate domain and the entry region of the tube.** In the intermediate domain ($2200 \leq \mathrm{Re}_d \leq 4000$) at moderate Prandtl numbers, the Nusselt number can be estimated using the formula

$$\mathrm{Nu} = 3 \times 10^{-4}\mathrm{Pr}^{0.35}\mathrm{Re}_d^{1.5} \qquad (\text{for } 0.5 \leq \mathrm{Pr} \leq 5).$$

For the entry region of the tube in the case of simultaneous development of hydrodynamic and thermal boundary layers, the local Nusselt number $\mathsf{Nu} = \mathsf{Nu}(X)$ at constant temperature or heat flux at the wall can be calculated from the formulas

$$
\mathsf{Nu} = \begin{cases} 0.02\,\mathsf{Re}_d^{0.75}\mathsf{Pr}^{0.4}\zeta(\zeta-1)^{-0.2} & \text{if } T_s = \text{const}, \\ 0.021\,\mathsf{Re}_d^{0.75}\mathsf{Pr}^{0.4}\zeta^{0.8}(\ln\zeta)^{-0.2} & \text{if } q_s = \text{const}. \end{cases} \tag{E5.4.4.5}
$$

Here the following notation is used:

$$
\mathsf{Nu} = \frac{\alpha d}{c_p \rho \chi}, \quad \mathsf{Re}_d = \frac{dU_i}{\nu}, \quad \zeta = (1.18)^b, \quad b = \mathsf{Re}_d^{-0.25}\frac{X}{a},
$$

where U_i is the inlet velocity and $\alpha = \alpha(X)$ the local heat transfer coefficient (based on the inlet temperature). Formulas (E5.4.4.5) are valid for $0.5 \le \mathsf{Pr} \le 200$ in the entire entry region, whose length x_L is estimated as $x_L = 1.3\,\mathsf{Re}^{0.25}$.

More detailed information about heat and mass transfer in turbulent flows through a circular tube or plane channel, as well as various relations for determining the temperature profile and Nusselt (Sherwood) numbers, and other useful information can be found in the references given at the end of this chapter.

E5.4.5. Limit Nusselt Numbers for Tubes of Various Cross-Sections

▶ **Laminar flows.** Below we give some results for the limit Nusselt numbers corresponding to the heat stabilization region for fully developed laminar flow of fluids in tubes of various cross-sections in the case of high Peclet numbers.

Let us introduce the equivalent (or "hydraulic") diameter d_e by the formula

$$
d_e = \frac{4S_*}{\mathcal{P}}, \tag{E5.4.5.1}
$$

where S_* is the tube cross-section area and $\mathcal{P}$ is the cross-section perimeter. For tubes of circular cross-section, d_e coincides with the diameter, and for a plane channel, d_e is twice the height of the channel.

Consider a tube of arbitrary shape and denote the contour of the cross-section by Γ. Generally speaking, the Nusselt number varies along the contour Γ. We define the perimeter-average Nusselt number $\overline{\mathsf{Nu}}$ as follows:

$$
\overline{\mathsf{Nu}} = \frac{q_s}{T_s - \langle T_* \rangle_m}\frac{d_e}{\kappa}. \tag{E5.4.5.2}
$$

Here T_s is the temperature on the wall of the tube, $\langle T_* \rangle_m$ is the mean flow rate temperature of the fluid, κ is the thermal conductivity coefficient, and q_s is the perimeter-average heat flux given by the formula

$$
q_s = -\frac{1}{\mathcal{P}}\kappa\int_\Gamma \left(\frac{\partial T_*}{\partial\xi}\right)_\Gamma d\Gamma, \tag{E5.4.5.3}
$$

where $\partial T_*/\partial\xi$ is the derivative of the temperature T_* along the normal to the contour of the tube cross-section.

TABLE E5.2
The values of the limit numbers $\overline{\mathrm{Nu}}_\infty$ for fully developed flow in tubes of
various shape for high Peclet numbers (the subscript T corresponds to the
constant temperature of the wall and the subscript q, to the constant heat flux).

Tube profile		$\overline{\mathrm{Nu}}_{\infty T}$	$\overline{\mathrm{Nu}}_{\infty q}$	Equivalent diameter d_e
Circular tube of diameter d		3.658	4.364	d
Flat tube of width $2h$		7.541	8.235	$4h$
Elliptic tube with semiaxes a and b	$b/a =$ 1.00	3.658	4.364	$\dfrac{\pi b}{E\left(\sqrt{1-b^2/a^2}\right)}$, where $E(\vartheta)$ is the complete elliptic integral of the second kind
	0.80	3.669	4.387	
	0.50	3.742	4.558	
	0.25	3.792	4.880	
	0.125	3.725	5.085	
	0.0625	3.647	5.176	
	0	3.488	5.225	
Tube of rectangular cross-section with sides a and b	$b/a =$ 1.00	2.976	3.608	$\dfrac{2ab}{a+b}$
	0.714	3.077	3.734	
	0.50	3.391	4.123	
	0.25	4.439	5.331	
	0.125	5.597	6.490	
	0.05		7.451	
	0	7.541	8.235	
Equilateral triangle with side a		2.47	3.111	$\dfrac{a\sqrt{3}}{3}$
Regular hexagon with side a		3.34	4.002	$a\sqrt{3}$
Semi-circle of diameter d			4.089	$\dfrac{\pi d}{\pi + 2}$

Table E5.2 presents the perimeter-average Nusselt numbers for tubes with various cross-section shapes.

At constant temperature on the wall of a tube of rectangular cross-section with sides a and b, the value $\overline{\mathrm{Nu}}_\infty$ for $a \geq b$ is nicely approximated by the formula

$$\overline{\mathrm{Nu}}_\infty = 7.5 - 17.5\,\epsilon + 23\,\epsilon^2 - 10\,\epsilon^3, \qquad \epsilon = b/a, \tag{E5.4.5.5}$$

whose maximum error is 3%.

At constant temperature on the wall of a tube whose cross-section is a regular N-gon, the limit Nusselt number is given by the approximate formula

$$\overline{\mathrm{Nu}}_\infty = 3.65 - 0.18\,N^{-1} - 10\,N^{-2}. \tag{E5.4.5.6}$$

The comparison with Table E5.2 shows that the maximum error in (E5.4.5.6) is less than 0.5% at $N = 3,\ 4,\ 6,\ \infty$.

▶ **Turbulent flows.** The strength of heat transfer in turbulent flows of fluids through tubes of noncircular cross-section can be estimated using relations (E5.4.4.2)–(E5.4.4.4) suggested for circular tubes. In this case, the characteristic length on which the Nusselt numbers Nu_∞ are based is taken to be the equivalent diameter, d_e, defined by Eq. (E5.4.5.1).

The quantity $\mathcal{P}$ in this formula is the wettable perimeter, regardless of what part of the perimeter exchanges heat with the fluid.

Detailed information about heat transfer in turbulent flows in tubes and channels, as well as various relations for determining the mean flow rate temperature and the Nusselt number, can be found in the references given at the end of this chapter.

E5.5. Interior Heat Problems for Solid Bodies of Various Shapes

E5.5.1. Statement of the Problem

Consider a class of problems concerning transient heat exchange between convex solid bodies (or a cavity filled with a quiescent fluid) of various shapes and the environment. At the initial time $t = 0$, the temperature is the same throughout the body and is equal to T_i, and for $t > 0$, the temperature on the surface Γ of the body is maintained constant and is equal to T_s. The temperature distribution inside the body is described by the heat equation

$$\frac{\partial T}{\partial \bar{\tau}} = \Delta T, \tag{E5.5.1.1}$$

the initial condition

$$T = 0 \quad \text{at} \quad \bar{\tau} = 0, \tag{E5.5.1.2}$$

and the boundary condition

$$T = 1 \quad \text{on} \quad \Gamma. \tag{E5.5.1.3}$$

The dimensionless variables are introduced according to formulas (E5.1.1.13), where the characteristic length can be taken arbitrarily.

The problem of transient diffusion into a cavity filled with a quiescent medium can be stated in a similar manner.

In this section, attention is chiefly paid to the study of the bulk body temperature

$$\langle T \rangle = \frac{1}{V} \int_v T \, dv, \tag{E5.5.1.4}$$

where $V = \int_v dv$ is the dimensionless volume of the body.

E5.5.2. Bulk Temperature for Bodies of Various Shapes

▶ **Bulk temperature for a sphere, a parallelepiped, and a cylinder of finite length.** First, consider the simplest one-dimensional (with respect to spatial coordinates) heat exchange problem for a solid sphere of radius a. The exact solution of this problem is well known and results in the following expression for the bulk temperature:

$$\langle T \rangle = 1 - \frac{6}{\pi^2} \sum_{k=1}^{\infty} \frac{1}{k^2} \exp(-\pi^2 k^2 \bar{\tau}), \qquad \bar{\tau} = \frac{\chi t}{a^2}. \tag{E5.5.2.1}$$

The exact solution of the corresponding three-dimensional problem (E5.5.1.1)–(E5.5.1.3) for a parallelepiped with sides L_1, L_2, and L_3 can be constructed by separation of variables and results in the following formula for bulk temperature:

$$\langle T \rangle = 1 - \left(\frac{8}{\pi^2}\right)^3 \sum_{k=1}^{\infty} \sum_{m=1}^{\infty} \sum_{l=1}^{\infty} \frac{1}{(2k-1)^2(2m-1)^2(2l-1)^2}$$

$$\times \exp\left\{-\pi^2 \left[\frac{(2k-1)^2}{L_1^2} + \frac{(2m-1)^2}{L_2^2} + \frac{(2l-1)^2}{L_3^2}\right] \chi t\right\}. \quad \text{(E5.5.2.2)}$$

Now consider heat exchange for a cylinder of finite length. Let a be the radius and L the length of the cylinder. By solving problem (E5.5.1.1)–(E5.5.1.3), we obtain the following expression for bulk temperature:

$$\langle T \rangle = 1 - \frac{32}{\pi^2} \sum_{k=1}^{\infty} \sum_{m=1}^{\infty} \frac{1}{\vartheta_k^2 (2m-1)^2} \exp\left\{-\left[\frac{\vartheta_k^2}{a^2} + \frac{\pi^2(2m-1)^2}{L^2}\right] \chi t\right\}, \quad \text{(E5.5.2.3)}$$

where the ϑ_k are the roots of the Bessel function of the first kind of index zero: $J_0(\vartheta_k) = 0$. Note that to compute the roots of this transcendental equation one can use the approximate formula

$$\vartheta_k = 2.4 + 3.13\,(k-1) \qquad (k = 1, 2, \dots)$$

with a maximum error of 0.2%.

▶ **Bulk temperature for a bounded body of arbitrary shape.** For a bounded body of arbitrary shape, the solution of problem (E5.5.1.1)–(E5.5.1.3) tends as $\bar{\tau} \to \infty$ to the limit value (equal to 1) determined by the boundary condition on the surface of the body. By setting $T = 1$ in (E5.5.1.4), we find the asymptotics of bulk temperature for large $\bar{\tau}$:

$$\langle T \rangle \to 1 \quad \text{as} \quad \bar{\tau} \to \infty. \quad \text{(E5.5.2.4)}$$

One can show that the bulk temperature of a bounded body of arbitrary shape has the following asymptotics for small $\bar{\tau}$:

$$\langle T \rangle \to 2\frac{S}{V}\sqrt{\frac{\bar{\tau}}{\pi}} \quad \text{at} \quad \bar{\tau} \to 0, \quad \text{(E5.5.2.5)}$$

where S and V are the dimensionless surface area and volume of the body.

For approximately determining the bulk temperature of a bounded convex body of arbitrary shape in the entire time range $0 \le \bar{\tau} < \infty$, one can use the formula

$$\langle T \rangle = 1 - \frac{6}{\pi^2} \sum_{k=1}^{\infty} \frac{1}{k^2} \exp\left(-\frac{\pi^2 k^2 S^2}{9V^2}\bar{\tau}\right).$$

This expression can be rewritten as follows:

$$\langle T \rangle = 1 - \frac{6}{\pi^2} \sum_{k=1}^{\infty} \frac{1}{k^2} \exp\left(-\frac{\pi^2}{9}k^2 \frac{S_*^2 \chi t}{V_*^2}\right), \quad \text{(E5.5.2.6)}$$

where S_* and V_* are, respectively, the dimensional surface area and volume of the body.

Formula (E5.5.2.6) gives the correct asymptotics (E5.5.2.4) and (E5.5.2.5) for small and large times and is exact for a spherical particle. (Formula (E5.5.2.6) for $S_* = 4\pi a$ and $V_* = \frac{4}{3}\pi a^3$ coincides with (E5.5.2.1).)

The approximation (E5.5.2.6) is compared with the exact bulk temperature (E5.5.2.2) for the parallelepiped in view of the relations $S_* = 2(L_1 L_2 + L_1 L_3 + L_2 L_3)$ and $V_* = L_1 L_2 L_3$. The maximum error of formula (E5.5.2.6) is about 5% for $0.25 \le L_3/L_1 \le 4.0$ and $L_2/L_1 = 1$.

The exact value (E5.5.2.3) is compared with the approximation (E5.5.2.6), where $S_* = 2\pi a(a + L)$ and $V_* = \pi a^2 L$, for various ratios of the cylinder dimensions. The maximum error in formula (E5.5.2.6) is about 3.5% for $0.25 \le 2a/L \le 4.0$.

For practical calculations, it is expedient to replace the infinite series (E5.5.2.6) by the simpler formula

$$\langle T \rangle = \sqrt{1 - e^{-1.27\,\omega}} + 0.6 \left(e^{-1.5\,\omega} - e^{-1.1\,\omega} \right), \qquad \omega = \frac{S_*^2 \chi t}{V_*^2},$$

whose maximum deviation from (E5.5.2.6) is about 1.7%.

E5.6. Mass and Heat Exchange Between Particles of Various Shapes and a Quiescent Medium

E5.6.1. Stationary Mass and Heat Exchange

▶ **Statement of the problem.** Consider stationary diffusion to a particle of finite size in a quiescent medium, which corresponds to the case $\mathsf{Pe} = 0$. We assume that the concentration on the surface of the particle and far from it is constant and equal to C_s and C_i, respectively. The concentration field outside the particle is described by the Laplace equation

$$\Delta c = 0 \qquad\qquad (E5.6.1.1)$$

and the boundary conditions

$$c = 1 \quad \text{on the surface } \Gamma \text{ of the particle,} \qquad (E5.6.1.2)$$
$$c = 0 \quad \text{remote from the particle,} \qquad (E5.6.1.3)$$

where $c = \dfrac{C_i - C}{C_i - C_s}$ is the dimensionless concentration.

The unknown quantity of most practical interest in these problems is the mean Sherwood number, which is determined by (E5.1.2.5) and is related to the mass transfer coefficient α_c by

$$\mathsf{Sh} = \frac{a\alpha_c}{D}, \qquad\qquad (E5.6.1.4)$$

where a is the characteristic length.

From the mathematical viewpoint, the diffusion problem (E5.6.1.1)–(E5.6.1.3) is equivalent to the problem on the electric field of a charged conductive body in a homogeneous charge-free dielectric medium. Therefore, the mean Sherwood number in a quiescent fluid coincides with the dimensionless electrostatic capacitance of the body and can be calculated or measured by methods of electrostatics.

▶ **Shape factor.** In what follows, it is convenient to introduce a shape factor Π, which has the dimension of length, as follows:

$$\Pi = \frac{\alpha_c S_*}{D} = \mathsf{Sh}\frac{S_*}{a}, \qquad\qquad (E5.6.1.5)$$

where S_* is the dimensional surface area of the particle. Note that sometimes Π is referred to as "conductance."

TABLE E5.3
Shape factor for particles in quiescent medium.

No	Shape of particle	Shape factor $\Pi = \mathrm{Sh}\dfrac{S_*}{a}$
1	Sphere of radius a	$4\pi a$
2	Oblate ellipsoid of revolution with semiaxes a and b, $\chi = b/a < 1$	$\dfrac{4\pi a\sqrt{1-\chi^2}}{\arccos\chi}$
3	Prolate ellipsoid of revolution with semiaxes a and b, $\chi = b/a > 1$	$\dfrac{4\pi a\sqrt{\chi^2-1}}{\ln\left(\chi+\sqrt{\chi^2-1}\,\right)}$
4	Circular cylinder of radius a and length L $(0 \le L/a \le 16)$	$\left[8 + 4.1\,(L/a)^{0.76}\right]a$
5	Tangent spheres of equal radius a	$2\ln 2\,(4\pi a)$
6	Tangent spheres of radii a_1 and a_2	$-\dfrac{4\pi a_1 a_2}{a_1 + a_2}\left[\psi\left(\dfrac{a_1}{a_1 + a_2}\right) + \psi\left(\dfrac{a_2}{a_1 + a_2}\right) + 2\ln\gamma\right],$ where $\psi(x) = \dfrac{d}{dx}\Gamma(x)$ is the logarithmic derivative of the gamma function and $\ln\gamma = -\psi(1) = 0.5772\ldots$ is the Euler constant
7	Orthogonally intersecting spheres with radii a_1 and a_2	$4\pi\left(a_1 + a_2 - \dfrac{a_1 a_2}{\sqrt{a_1^2 + a_2^2}}\right)$
8	Cube with edge a	$0.654\,(4\pi a)$
9	Thin rectangular plate with sides L_1 and L_2 $(L_1 \ge L_2)$	$\dfrac{2\pi L_1}{\ln(4L_1/L_2)}$

Table E5.3 shows the values of Π for particles of various shape. It follows from (E5.6.1.5) that the mean Sherwood number can be obtained from the data in this table by dividing the shape factor by the surface area of the particle and then by multiplying by the characteristic length.

One can interpret the table data as follows. Let us project a body of revolution onto a plane perpendicular to the symmetry axis. The projection is a disk of radius a_{p}. The sphere of radius a_{p} is called a perimeter-equivalent sphere. We introduce the perimeter-equivalent factor

$$\Sigma = \frac{S_*}{4\pi a_{\mathrm{p}}^2} = \frac{\text{surface area of particle}}{\text{surface area of perimeter-equivalent sphere}} \qquad (E5.6.1.6)$$

and consider the corresponding shape factor ratio

$$\widetilde{\Pi} = \frac{\Pi}{4\pi a_{\mathrm{p}}} = \frac{\text{shape factor of the particle}}{\text{shape factor of the perimeter-equivalent sphere}}. \qquad (E5.6.1.7)$$

The dimensionless variables (E5.6.1.6) and (E5.6.1.7) are invariant with respect to the choice of the characteristic length. The dependence of $\widetilde{\Pi}$ on Σ for particles of various geometric shapes is fairly well approximated by the formula

$$\widetilde{\Pi} = 0.637 + 0.327\,(2\Sigma - 1)^{0.76} \qquad (0.5 \le \Sigma \le 8.5). \qquad (E5.6.1.8)$$

It is expedient to use this approximate formula for the analysis of mass exchange between particles of complicated shape and a quiescent medium if the solution of problem (E5.6.1.1)–(E5.6.1.3) is unknown.

To calculate the particle shape factor, it is also convenient to use the simple approximate formula

$$\Pi = 5.25\, S_*^{1/4} V_*^{1/6}, \tag{E5.6.1.9}$$

where S_* is the surface area and V_* the volume of the particle.

For spheres and cubes, the error in formula (E5.6.1.9) is at most 0.13%. For circular cylinders of radius a and length $L = a$, $L = 2a$, or $L = 3a$, the error in formula (E5.6.1.9) is respectively 1.1%, 0.6%, and 2.1%.

It is useful to rewrite formula (E5.6.1.9) as follows:

$$\Pi = \sqrt{S_*}\, f(\xi), \qquad f(\xi) = A_s \sqrt{\xi}, \qquad A_s = 2\sqrt{\pi} \approx 3.545,$$

where ξ is a dimensionless geometric parameter,

$$\xi = \lambda \frac{V_*^{1/3}}{S_*^{1/3}}, \qquad \lambda = (36\pi)^{1/6} \approx 2.199.$$

The parameter ξ characterizes the particle shape and ranges in the interval $0 \le \xi \le 1$. The value $\xi = 1$ corresponds to a spherical particle. Formula (E5.6.1.9) gives good results for weakly and moderately deformed convex particles with $0.88 \le \xi \le 1.0$. (For example, this condition is satisfied for cubic particles, regular polyhedra, and circular cylinders of radius a and length L with $a \le L \le 3a$.)

Let us write out some lower and upper bounds for the shape factor. The lower bound for an arbitrary particle is determined by the shape factor of the sphere of the same volume V_*:

$$\Pi \ge (48\pi^2 V_*)^{1/3}. \tag{E5.6.1.10}$$

Another lower bound has the form

$$\Pi \ge 8(S_{max}/\pi)^{1/2},$$

where S_{max} is the maximum projected area of the particle. The last formula becomes an equality for a disk. An upper bound is given by the shape factor of any surface (say, a sphere or an ellipsoid of revolution) surrounding the particle.

E5.6.2. Transient Mass and Heat Exchange

Suppose that at the initial time $t = 0$ the concentration in the continuous medium is constant and is equal to C_i and that a constant surface concentration C_s is maintained for $t > 0$. The transient mass exchange between a particle and a quiescent medium is described by the equation

$$\frac{\partial c}{\partial \tau} = \Delta c \tag{E5.6.2.1}$$

with the boundary conditions (E5.6.1.2) and (E5.6.1.3) and the initial condition

$$\tau = 0, \quad c = 0, \tag{E5.6.2.2}$$

where $\tau = tD/a^2$ is dimensionless time.

For a spherical particle, the solution of problem (E5.6.2.1), (E5.6.2.2), (E5.6.1.2), (E5.6.1.3) can be expressed via the complementary error function and has the form

$$c = \frac{1}{r} \, \mathrm{erfc}\left(\frac{r-1}{2\sqrt{\tau}}\right), \tag{E5.6.2.3}$$

where r is the dimensionless radial coordinate normalized to the radius of the particle.

The mean Sherwood number for a sphere is

$$\mathsf{Sh} = 1 + \frac{1}{\sqrt{\pi \tau}}. \tag{E5.6.2.4}$$

For nonspherical particles, the mean Sherwood number can be approximated by the expression

$$\mathsf{Sh} = \mathsf{Sh}_{\mathrm{st}} + \frac{1}{\sqrt{\pi \tau}}, \tag{E5.6.2.5}$$

where $\mathsf{Sh}_{\mathrm{st}}$ is the Sherwood number corresponding to the solution of the stationary problem (E5.6.1.1)–(E5.6.1.3). In (E5.6.2.5), Sh, $\mathsf{Sh}_{\mathrm{st}}$, and τ are assumed to be reduced to dimensionless form by using the same characteristic length.

For nonspherical particles, one can obtain $\mathsf{Sh}_{\mathrm{st}}$ by using Table E5.3 and the expression (E5.6.1.8).

E5.7. Mass and Heat Transfer to a Spherical Particle in a Translational Flow

E5.7.1. Statement of the Problem

▶ **Equation and the boundary conditions.** Consider steady-state diffusion to a solid spherical particle of radius a in a translational flow with velocity U_i. We assume that the concentration on the particle surface and far from it is constant and equal to C_s and C_i, respectively. The mass transfer process in the continuous medium is described by the equation (E5.1.1.9), which in the stationary case in the spherical coordinate system R, θ, φ (in view of the fact that all variables in this axisymmetric problem are independent of φ) has the form

$$\mathsf{Pe}\left(v_r \frac{\partial c}{\partial r} + \frac{v_\theta}{r} \frac{\partial c}{\partial \theta}\right) = \frac{1}{r^2} \frac{\partial}{\partial r}\left(r^2 \frac{\partial c}{\partial r}\right) + \frac{1}{r^2 \sin\theta} \frac{\partial}{\partial \theta}\left(\sin\theta \frac{\partial c}{\partial \theta}\right), \tag{E5.7.1.1}$$

and the boundary conditions

$$\begin{aligned}
r = 1, &\quad c = 1 &\quad \text{(on the surface of the particle),} &\tag{E5.7.1.2}\\
r \to \infty, &\quad c \to 0 &\quad \text{(far from the particle),} &\tag{E5.7.1.3}
\end{aligned}$$

where the dimensionless variables are introduced according to the formulas

$$r = \frac{R}{a}, \quad v_r = \frac{V_R}{U_i}, \quad v_\theta = \frac{V_\theta}{U_i}, \quad c = \frac{C_i - C}{C_i - C_s}, \quad \mathsf{Pe} = \frac{aU_i}{D}. \tag{E5.7.1.4}$$

▶ **Mean Sherwood number.** In this section, attention is mainly paid to formulas for the calculation of the mean Sherwood number for a particle in a translational flow at various Reynolds and Peclet numbers. The Reynolds number and the mean Sherwood number, as well as the Peclet number, are determined by the particle radius: $\mathsf{Re} = aU_i/\nu$ and

$$\mathsf{Sh} = -\frac{1}{2} \int_0^\pi \sin\theta \left(\frac{\partial c}{\partial r}\right)_{r=1} d\theta. \tag{E5.7.1.5}$$

E5.7.2. Mass Transfer at Low Reynolds Numbers and Various Peclet Numbers

For a solid spherical particle in a translational Stokes flow (low Reynolds numbers, $Re \to 0$) the fluid velocity components V_R and V_θ in Eq. (E5.7.1.4) are determined by formulas (E4.6.2.4).

▶ **Low Peclet numbers.** For $Pe = 0$, which corresponds to a particle in a quiescent fluid, the exact solution of the problem (E5.7.1.1)–(E5.7.1.3) has the form

$$c = \frac{1}{r}. \tag{E5.7.2.1}$$

The corresponding mean Sherwood number is equal to unity,

$$Sh = 1 \qquad (Pe = 0). \tag{E5.7.2.2}$$

The three-term expansion of the mean Sherwood number as $Pe \to 0$ has the form

$$Sh = 1 + \tfrac{1}{2} Pe + \tfrac{1}{2} Pe^2 \ln Pe, \tag{E5.7.2.3}$$

whose error is of the order of Pe^2. The asymptotic formula (E5.7.2.3) with the last term omitted gives a good approximation to the Sherwood number for $Pe < 0.4$.

▶ **High Peclet numbers (the diffusion boundary layer).** For high Peclet numbers, a thin diffusion boundary layer is formed near the surface of the sphere. The ratio of the thickness of this layer to the particle radius is of the order of $Pe^{-1/3}$. In this region, the radial component of molecular diffusion to the particle surface is essential, and tangential diffusion may be neglected. Convective mass transfer due to the motion of the fluid should also be taken into account.

The concentration distribution in the diffusion boundary layer can be expressed as follows via the incomplete gamma function:

$$c = \frac{1}{\Gamma(1/3)} \Gamma\left(\frac{1}{3}, \frac{Pe}{3} \frac{(r-1)^3 \sin^3 \theta}{\pi - \theta + \frac{1}{2} \sin 2\theta} \right), \tag{E5.7.2.4}$$

where the angle θ is counted off as shown in Figure E4.4.

By differentiating equation (E5.7.2.4), we obtain the dimensionless local diffusion flux to the surface of the sphere in the form

$$j = -\left(\frac{\partial c}{\partial r} \right)_{r=1} = 0.766 \sin \theta \left(\pi - \theta + \frac{1}{2} \sin 2\theta \right)^{-1/3} Pe^{1/3}. \tag{E5.7.2.5}$$

One can see that the local diffusion flux attains its maximum at the front stagnation point on the surface of the sphere (at $\theta = \pi$) and monotonically decreases with the angular coordinate to the minimum value, which is zero and is attained at $\theta = 0$.

The corresponding mean Sherwood number is determined by Levich's formula

$$Sh = 0.625 \, Pe^{1/3} \qquad (Pe \to \infty). \tag{E5.7.2.6}$$

Let us also present the two-term expansion for the mean Sherwood number,

$$Sh = 0.625 \, Pe^{1/3} + 0.461 \qquad (Pe \to \infty). \tag{E5.7.2.7}$$

which refines formula (E5.7.2.6). Formula (E5.7.2.7) can be used for practical calculations if $Pe \geq 10$.

▶ **Arbitrary Peclet numbers.** The mean Sherwood number for a solid spherical particle in a translational Stokes flow in the entire range of Peclet numbers can be computed with the use of the Clift–Grace–Weber approximate formula

$$\mathrm{Sh} = 0.5 + (0.125 + 0.243\,\mathrm{Pe})^{1/3}. \tag{E5.7.2.8}$$

The interpolation formula (E5.7.2.8) gives exact asymptotic results both as $\mathrm{Pe} \to 0$ (E5.7.2.2) and as $\mathrm{Pe} \to \infty$ (E5.7.2.6). The maximum error of formula (E5.7.2.8) in the entire range of Peclet numbers is about 2%. (Comparison was made with available results of finite-difference numerical solutions of problem (E5.7.1.1)–(E5.7.1.3).)

E5.7.3. Mass Transfer at Moderate and High Reynolds Numbers

▶ **Spherical particle at various Reynolds numbers.** In the case of a spherical particle in a translational flow at $0.5 \le \mathrm{Re} \le 200$ and $0.125 \le \mathrm{Sc} \le 50$, a number of numerical results concerning the mean Sherwood number can be described by the Clift–Grace–Weber approximate formula

$$\mathrm{Sh} = 0.5 + 0.527\,\mathrm{Re}^{0.077}(1 + 2\,\mathrm{Re}\,\mathrm{Sc})^{1/3}, \tag{E5.7.3.1}$$

whose error is about 3%.

The analysis of available experimental data on heat and mass transfer to a solid sphere in a translational flow results in the following Clift–Grace–Weber correlations.

Heat exchange with air at $\mathrm{Pr} = 0.7$:

$$\mathrm{Nu} = 0.5 + 0.47\,\mathrm{Re}^{0.47} \qquad \text{for} \quad 50 \le \mathrm{Re} \le 2 \times 10^3,$$
$$\mathrm{Nu} = 0.5 + 0.2\,\mathrm{Re}^{0.58} \qquad \text{for} \quad 2 \times 10^3 \le \mathrm{Re} \le 5 \times 10^4.$$

Mass transfer with a liquid for large Schmidt numbers ($\mathrm{Sc} > 100$):

$$\mathrm{Sh} = 0.5 + 0.5\,\mathrm{Re}^{0.48}\,\mathrm{Sc}^{1/3} \qquad \text{for} \quad 50 \le \mathrm{Re} \le 10^3,$$
$$\mathrm{Sh} = 0.5 + 0.31\,\mathrm{Re}^{0.55}\,\mathrm{Sc}^{1/3} \qquad \text{for} \quad 10^3 \le \mathrm{Re} \le 5 \times 10^4.$$

The experimental data for $0.5 < \mathrm{Re} < 50$ are well described by the expression (E5.7.3.1).

▶ **General correlation for the Sherwood number.** For the calculation of the mean Sherwood number in laminar flows of various types without closed streamlines past a solid spherical particle, one can use the approximate formula

$$\mathrm{Sh} = 0.5 + (0.125 + \mathrm{Sh}_\infty^3)^{1/3}, \tag{E5.7.3.2}$$

where the auxiliary value Sh_∞ must be chosen equal to the leading term of the asymptotic expansion of the Sherwood number as $\mathrm{Pe} \to \infty$ (for given $\mathrm{Re} = \mathrm{const}$). Formula (E5.7.3.2) after the substitution of the value of Sh_∞ can be used for the calculation of the Sherwood number for various flows (one can consider, say, shear flows) in the entire range of Peclet numbers. For a solid spherical particle in a translational Stokes flow (as $\mathrm{Re} \to 0$), formula (E5.7.3.2) becomes (E5.7.2.8).

E5.8. Mass and Heat Transfer to Spherical Drops and Bubbles in Translational Flow

E5.8.1. Statement of the Problem. Some Remarks

In this section, attention is mainly paid to formulas for the calculation of the mean Sherwood number for spherical drops and bubbles of radius a in a translational flow with velocity U_i at various Reynolds and Peclet numbers. Consider the exterior problem with limiting diffusion resistance of the continuous phase. We assume that the concentration on the drop or bubble surface and far from it is constant and equal to C_s and C_i, respectively.

The mathematical statement of the problem on a steady-state mass transfer process in a continuous medium outside a drop is described by equation (E5.7.1.1) and the boundary conditions (E5.7.1.2)–(E5.7.1.3), where the fluid velocity components V_R and V_θ in Eq. (E5.7.1.4) are determined from the corresponding hydrodynamic problem. The mean Sherwood number, as in the case of a solid spherical particle, is computed by formula (E5.7.1.5).

The process of convective diffusion to the liquid–liquid (liquid–gas) interface substantially differs from the process of convective diffusion to the fluid–solid interface, which was considered in Section E5.7. This is due to the difference between the hydrodynamic conditions on the interfaces. The fluid velocity on the surface of a solid is always zero by virtue of the no-slip condition. On the contrary, the interface between two fluids can move, and the tangential velocity on the interface differs from zero.

Convective transfer of a substance by a moving fluid to the fluid–solid interface takes place under the condition that the flow is somewhat retarded, so that the transfer rate close to the surface is considerably lower than that in the bulk of the solution. On the contrary, diffusion to a liquid–liquid (liquid–gas) interface takes place under the more favorable conditions of nonretarded flow. That is why convective diffusion of a substance to a liquid–liquid interface is much more intensive than that to a fluid–solid interface.

We denote the mean Sherwood number for a gas bubble by Sh_b and for a solid sphere by Sh_p.

E5.8.2. Mass Transfer at Low Reynolds Numbers and Various Peclet Numbers

For spherical drops and bubbles in a translational Stokes flow (low Reynolds numbers, $\mathsf{Re} \to 0$) the fluid velocity components V_R and V_θ in Eq. (E5.7.1.4) are determined by formulas (E4.7.1.7), where $\beta = \mu_2/\mu_1$ is the drop–ambient medium dynamic viscosity ratio. (The value $\beta = 0$ corresponds to a gas bubble.)

▶ **Drops and bubbles at low Peclet numbers.** For $\mathsf{Pe} = 0$, which corresponds to a drop (bubble) in a quiescent fluid, the exact solution of problem (E5.7.1.1)–(E5.7.1.3), as in the case of a solid particle, has the form (E5.7.2.1), and the corresponding mean Sherwood number is unity; see Eq. (E5.7.2.2).

The three-term expansion of the mean Sherwood number as $\mathsf{Pe} \to 0$ has the form

$$\mathsf{Sh} = 1 + \frac{1}{2}\,\mathsf{Pe} + \frac{2+3\beta}{6(1+\beta)}\,\mathsf{Pe}^2\ln\mathsf{Pe} + O(\mathsf{Pe}^2). \tag{E5.8.2.1}$$

For $\beta \to \infty$, formula (E5.8.2.1) passes into formula (E5.7.2.3) for a solid particle.

▶ **Drops and bubbles at high Peclet numbers (the diffusion boundary layer).** For high Peclet numbers, a thin diffusion boundary layer is formed near the drop (bubble) surface.

The ratio of the thickness of this layer to the drop radius is of the order of $\mathrm{Pe}^{-1/2}$. In this region, the radial component of molecular diffusion to the drop surface is essential, and tangential diffusion may be neglected. Convective mass transfer due to the motion of the fluid should also be taken into account.

For $0 \le \beta \ll \mathrm{Pe}^{1/3}$, the concentration distribution in the diffusion boundary layer can be expressed via the complementary error function as follows:

$$c = \mathrm{erfc}\left(\frac{1}{4}\sqrt{\frac{6\,\mathrm{Pe}}{\beta+1}}\,(r-1)\,\frac{1-\cos\theta}{\sqrt{2-\cos\theta}}\right),\tag{E5.8.2.2}$$

where the angle θ is counted off as shown in Figure E4.5.

By differentiating equation (E5.8.2.2), we obtain the dimensionless local diffusion flux to the surface of the spherical drop,

$$j = -\left(\frac{\partial c}{\partial r}\right)_{r=1} = \sqrt{\frac{3\,\mathrm{Pe}}{\pi(\beta+1)}}\,\frac{1-\cos\theta}{\sqrt{2-\cos\theta}}.\tag{E5.8.2.3}$$

One can see that the local diffusion flux attains its maximum at the front stagnation point on the drop surface (at $\theta = \pi$) and monotonically decreases with the angular coordinate to the minimum value, which is zero and is attained at $\theta = 0$.

The mean Sherwood number for a spherical drop is given by the Levich formula

$$\mathrm{Sh} = \sqrt{\frac{2\,\mathrm{Pe}}{3\pi(\beta+1)}} = 0.461\left(\frac{\mathrm{Pe}}{\beta+1}\right)^{1/2}\qquad(\mathrm{Pe}\to\infty).\tag{E5.8.2.4}$$

Let us also give the two-term expansion for the mean Sherwood number,

$$\mathrm{Sh} = 0.461\left(\frac{\mathrm{Pe}}{\beta+1}\right)^{1/2} + 0.41\left(\frac{3}{4}\beta+1\right)\qquad(\mathrm{Pe}\to\infty),\tag{E5.8.2.5}$$

which refines formula (E5.8.2.4).

Formula (E5.8.2.5) can be used for practical calculations at $\mathrm{Pe} \ge 100$ for $0 \le \beta \le 0.82\,\mathrm{Pe}^{1/3} - 1$. The value $\beta = 0$ in (E5.8.2.4) and (E5.8.2.5) corresponds to a gas bubble.

▶ **Spherical bubbles at any Peclet numbers.** The results of the numerical solution of the problem of mass transfer to a spherical bubble in a translational flow is well approximated by the following expression for the mean Sherwood number:

$$\mathrm{Sh} = 0.6 + (0.16 + 0.213\,\mathrm{Pe})^{1/2}\qquad(\beta = 0),\tag{E5.8.2.6}$$

whose maximum error is about 3% in the entire range of Peclet numbers.

▶ **Spherical drops at any Peclet numbers.** In the range $0 \le \mathrm{Pe} \le 200$, the results of numerical calculations of mean Sherwood numbers for a spherical drop in a translational flow under a limiting resistance of the continuous phase is well described by the approximate formula

$$\mathrm{Sh} = \frac{1}{\beta+1}\mathrm{Sh_b} + \frac{\beta}{\beta+1}\mathrm{Sh_p},\tag{E5.8.2.7}$$

where β is the ratio of dynamic viscosities of the drop and the ambient fluid ($\beta = 0$ corresponds to a gas bubble, and $\beta \to \infty$, to a solid sphere), and $\mathrm{Sh_b}$ and $\mathrm{Sh_p}$ are the

Sherwood numbers for a bubble and for a solid particle, which can be calculated by formulas (E5.8.2.6) and (E5.7.2.8), respectively.

It is important to note that the expression (E5.8.2.7) gives three valid terms of the asymptotic expansion of Sh as $\mathsf{Pe} \to 0$ for any β, see formula (E5.8.2.1).

In the interval $200 \le \mathsf{Pe} < \infty$, for any values of the phase viscosities, the mean Sherwood number for a drop can be calculated by solving the cubic equation

$$\mathsf{Sh}^3 - 0.212 \frac{\mathsf{Pe}}{\beta + 1} \mathsf{Sh} - (0.624)^3 \, \mathsf{Pe} = 0. \tag{E5.8.2.8}$$

E5.8.3. Mass Transfer at Moderate and High Reynolds Numbers

▶ **Spherical bubble at any Peclet numbers for $\mathsf{Re} \ge 35$.** For a spherical bubble in a translational flow at moderate and high Reynolds numbers and high Peclet numbers, the mean Sherwood number can be calculated by the approximate formula

$$\mathsf{Sh} = \left(\frac{2}{\pi} \mathsf{Pe} \right)^{1/2} \left(1 - \frac{2}{\sqrt{\mathsf{Re}}} \right)^{1/2}, \tag{E5.8.3.1}$$

whose error is less than 7% for $\mathsf{Re} \ge 35$.

For $0 \le \mathsf{Pe} < \infty$ and $\mathsf{Re} \ge 35$, to calculate the mean Sherwood number for a spherical bubble, one can use the formula

$$\mathsf{Sh} = 0.6 + \left[0.16 + 0.637 \left(1 - \frac{2}{\sqrt{\mathsf{Re}}} \right) \mathsf{Pe} \right]^{1/2}, \tag{E5.8.3.2}$$

which is exact for $\mathsf{Pe} = 0$ and passes into (E5.8.3.1) as $\mathsf{Pe} \to \infty$. For $\mathsf{Re} = \infty$, the error of formula (E5.8.3.2) is about 3%.

▶ **Spherical drop at high Peclet numbers for $\mathsf{Re} \ge 35$.** For a spherical drop in a translational flow at moderate and high Reynolds numbers and high Peclet numbers, the mean Sherwood number is well approximated by the formula

$$\mathsf{Sh} = \left(\frac{2}{\pi} \mathsf{Pe} \right)^{1/2} \left(1 - \frac{2 + 1.49 \, \beta^{0.64}}{\sqrt{\mathsf{Re}}} \right)^{1/2}, \tag{E5.8.3.3}$$

which passes into (E5.8.3.1) for $\beta = 0$. Formula (E5.8.3.3) can be applied for $0 \le \beta \le 2$ and $\mathsf{Re} \ge 35$.

E5.8.4. General Correlations for the Sherwood Number

▶ **Bubbles.** For the calculation of the mean Sherwood number in laminar flows of various types past a spherical bubble, one can use the approximate formula

$$\mathsf{Sh} = 0.6 + (0.16 + \mathsf{Sh}_\infty^2)^{1/2}, \tag{E5.8.4.1}$$

where Sh_∞ is the asymptotic value of the mean Sherwood number as $\mathsf{Pe} \to \infty$, which can be calculated by solving the diffusion boundary layer equation in a given flow field. (It is assumed that there are no closed streamlines in the continuous phase.)

Formula (E5.8.4.1) after the substitution of the value of Sh_∞ can be used for the calculation of the Sherwood number in the entire range of Peclet numbers. For a spherical bubble in a translational Stokes flow (as $Re \to 0$), formula (E5.8.4.1) passes into (E5.8.2.6). The general formula (E5.8.4.1) can also be used in the case of linear shear flows.

▶ **Drops in the entire range of phase viscosities.** For low and moderate Peclet numbers in an arbitrary laminar flow past a spherical drop under limiting resistance of the continuous phase, it is expedient to calculate the mean Sherwood number by using formula (E5.8.2.7), where Sh_b and Sh_p are the Sherwood numbers for the limit cases of a bubble and a solid particle. These quantities can be calculated by formulas (E5.8.4.1) and (E5.7.3.2). For high Peclet numbers, in the entire range of phase viscosities, the mean Sherwood number can be found by solving the cubic equation

$$Sh^3 - Sh_\beta^2 Sh - Sh_p^3 = 0, \tag{E5.8.4.2}$$

where Sh_β is the asymptotic value of the mean Sherwood number obtained in the diffusion boundary layer approximation for a drop of moderate viscosity $\beta = O(1)$ as $Pe \to \infty$, and Sh_p is the corresponding asymptotic value for a solid particle ($\beta = \infty$) as $Pe \to \infty$. For the special case of a translational Stokes flow past a spherical drop, Eq. (E5.8.4.2) passes into (E5.8.2.8).

For Stokes flows ($Re \to 0$) at high Peclet numbers, one can use the approximate formula

$$Sh_\beta = \frac{Sh_0}{\sqrt{\beta + 1}}, \tag{E5.8.4.3}$$

where Sh_0 is the asymptotic value of the Sherwood number for a gas bubble ($\beta = 0$) as $Pe \to \infty$. For a translational or arbitrary linear straining Stokes flow past a spherical drop, formula (E5.8.4.3) is exact.

E5.9. Mass Transfer to Nonspherical Particles in Translational Flow

E5.9.1. Mass and Heat Transfer to a Particle of Arbitrary Shape at Low Peclet Numbers

▶ **Asymptotic formulas for the Sherwood number and diffusion flux.** For low Peclet numbers, the problem of mass exchange between a particle of arbitrary shape and a uniform translational flow were studied by Brenner (1963). The following expression was obtained for the mean Sherwood number up to first-order infinitesimals with respect to Pe:

$$\frac{Sh}{Sh_0} = 1 + \frac{1}{8\pi} Pe_M, \qquad Pe_M = \frac{\Pi U_i}{D}, \tag{E5.9.1.1}$$

where Sh_0 is the Sherwood number corresponding to a quiescent medium. The influence of the fluid motion is characterized by the modified Peclet number, in which the shape factor Π of the particle plays the role of the characteristic length.

Formula (E5.9.1.1) is quite general and holds for solid particles, drops, and bubbles of arbitrary shape in a uniform translational flow at any Re as $Pe \to 0$.

It gives a good approximation to the Sherwood number ratio for $Pe_M < 5$. In the special case of a spherical particle of radius a, in (E5.9.1.1) one should set $\Pi = 4\pi a$. For nonspherical particles, in (E5.9.1.1) one must use the values of Π from Table E5.3.

For a particle of arbitrary shape in a translational Stokes flow, the first three terms of the asymptotic expansion of the dimensionless total diffusion flux as $\mathsf{Pe} \to 0$ have the form

$$I = I_0 + \frac{1}{8\pi}\mathsf{Pe}\,I_0^2 + \frac{1}{8\pi}\mathsf{Pe}^2 \ln \mathsf{Pe}\,I_0^2(\mathbf{f} \cdot \mathbf{e}) + O(\mathsf{Pe}^2). \tag{E5.9.1.2}$$

Here $I_0 = \Pi/a$ is the total flux to the particle in a quiescent fluid, $\mathbf{f}$ is the dimensionless vector equal to the ratio of the drag of the particle to the Stokes drag of a solid sphere of radius a (a is the characteristic length that appears in the definitions of the dimensionless variables Pe, I, and I_0), and $\mathbf{e}$ is the unit vector codirected with the fluid velocity at infinity. The Sherwood number is given by $\mathsf{Sh} = I/S$, where S is the dimensionless surface area of the particle.

To calculate $I_0 = \Pi/a$, one can use the results of Section E5.6.

▶ **Some special cases.** For an ellipsoid of revolution with semiaxes a and b (a is the equatorial radius) whose symmetry axis is directed along the flow, the Stokes drag is given by the formula

$$(\mathbf{f} \cdot \mathbf{e}) = \begin{cases} \dfrac{4}{3}(\chi^2 + 1)^{-1/2}\left[\chi - (\chi^2 - 1)\operatorname{arccot}\chi\right]^{-1} & \text{for } a \geq b, \\[3mm] \dfrac{8}{3}(\chi^2 - 1)^{-1/2}\left[(\chi^2 + 1)\ln\dfrac{\chi + 1}{\chi - 1} - 2\chi\right]^{-1} & \text{for } a \leq b, \end{cases}$$

where $\chi = \left|(a/b)^2 - 1\right|^{-1/2}$.

For a body of revolution whose axis is inclined at an angle ω to the incoming flow direction (see Figure E4.6), the following formula holds in the Stokes approximation (as $\mathsf{Re} \to 0$):

$$(\mathbf{f} \cdot \mathbf{e}) = f_\| \cos^2\omega + f_\perp \sin^2\omega, \tag{E5.9.1.3}$$

where $f_\|$ and $f_\perp$ are the values of the dimensionless drag of the body for the cases in which its axis is parallel ($\omega = 0$) and perpendicular ($\omega = \pi/2$), respectively, to the flow direction.

In particular, for a thin circular disk, one should set $f_\| = 8/(3\pi)$ and $f_\perp = 16/(9\pi)$ in (E5.9.1.3); for a dumbbell-like particle consisting of two adjacent spheres of equal radius, $f_\| \approx 0.645$ and $f_\perp \approx 0.716$.

The logarithmic term sharply restricts the practical value of the expansion (E5.9.1.2). The two-term expression (E5.9.1.1) holds in a wider range of the Peclet number (although this expression is less accurate for very small Pe).

E5.9.2. Mass Transfer to a Ellipsoidal Particle in Translational Flow

▶ **High Peclet numbers (the diffusion boundary layer approximations).** Consider diffusion to the surface of a solid ellipsoidal particle in a homogeneous translational Stokes flow ($\mathsf{Re} \to 0$). The particle is an ellipsoid of revolution with semiaxes a and b oriented along and across the flow, respectively. (Here b is the equatorial radius.) We introduce the following notation:

$$\chi = b/a, \quad a_e = a\chi^{2/3}, \quad \mathsf{Pe}_e = a_e U_i/D, \tag{E5.9.2.1}$$

where a_e is the radius of the volume-equivalent sphere serving as the characteristic length.

The dimensionless total diffusion flux to the surface of an ellipsoidal particle at high Peclet numbers is determined by the formula

$$I = 7.85\,K(\chi)\mathsf{Pe}_c^{1/3}, \tag{E5.9.2.2}$$

where the shape coefficient K is given by

$$K(\chi) = \left(\frac{4}{3}\right)^{1/3} \chi^{-2/9} (\chi^2 - 1)^{1/3} \left(1 + \frac{\chi^2 - 2}{\sqrt{\chi^2 - 1}} \arctan \sqrt{\chi^2 - 1}\right)^{-1/3} \quad \text{for} \quad \chi \geq 1,$$

$$K(\chi) = \left(\frac{4}{3}\right)^{1/3} \chi^{-2/9} (1 - \chi^2)^{1/3} \left(\frac{2 - \chi^2}{2\sqrt{1 - \chi^2}} \ln \frac{1 + \sqrt{1 - \chi^2}}{1 - \sqrt{1 - \chi^2}} - 1\right)^{-1/3} \quad \text{for} \quad \chi \leq 1.$$

$$(E5.9.2.3)$$

For $\chi = 1$, we have $K = 1$, and formula (E5.9.2.2) after the division by 4π gives the Sherwood number (4.6.8) for the solid sphere.

For $0.5 \leq \chi \leq 3.0$, the shape coefficient is well approximated by the expression

$$K(\chi) = 1 + \tfrac{2}{45}(\chi - 1), \tag{E5.9.2.4}$$

whose maximum error is 0.8%.

The mean Sherwood number is given by the formula $\mathsf{Sh} = I/S$, where S is the dimensionless surface area of the ellipsoid of revolution,

$$S = \frac{2\pi}{\chi^{1/3}} \left(\chi + \frac{1}{2\sqrt{\chi^2 - 1}} \ln \frac{\chi + \sqrt{\chi^2 - 1}}{\chi - \sqrt{\chi^2 - 1}}\right) \quad \text{for} \quad \chi \geq 1,$$

$$S = \frac{2\pi}{\chi^{1/3}} \left(\chi + \frac{1}{\sqrt{1 - \chi^2}} \arcsin \sqrt{1 - \chi^2}\right) \quad \text{for} \quad \chi \leq 1.$$

$$(E5.9.2.5)$$

The dimensionless quantity S is related to the dimensional surface area S_* of the ellipsoid by $S = S_*/a_{\mathrm{e}}^2$.

▶ **Arbitrary Peclet numbers.** For arbitrary Peclet numbers, the mean Sherwood number (corresponding to the characteristic length a_{e}) for a translational Stokes flow past an ellipsoidal particle can be approximated by the formula

$$\mathsf{Sh} = 0.5\frac{1}{S}\left(\frac{\Pi}{a_{\mathrm{e}}}\right) + \frac{1}{S}\left\{0.125\left(\frac{\Pi}{a_{\mathrm{e}}}\right)^3 + \left[7.85\,K(\chi)\right]^3 \mathsf{Pe}_{\mathrm{e}}\right\}^{1/3}, \tag{E5.9.2.6}$$

where the shape factor Π is given in the second and third rows of Table E5.3 and the quantities Pe_{e}, K, and S are defined in Eqs. (E5.9.2.1), (E5.9.2.3), and (E5.9.2.5), respectively.

For $0.2 \leq \chi \leq 5.0$, the maximum error of formula (E5.9.2.6) for an ellipsoidal particle does not exceed 10%. (The comparison used the results of the numerical solution of the corresponding axisymmetric mass and heat transfer problem by the finite-difference method.)

Formulas (E5.9.2.2) and (E5.9.2.4) cannot be used in the case of a strongly oblate ($\chi \gg 1$) or a strongly prolate ($\chi \ll 1$) ellipsoid of revolution.

E5.9.3. Mass Transfer to Particles of Arbitrary Shape

▶ **Bodies of revolution.** Consider a bounded body of revolution whose axis is inclined at an angle ω to the translation flow velocity at infinity (see Figure E4.6). In this case, the mean Sherwood number can be computed by the approximate formula

$$\mathsf{Sh} = \mathsf{Sh}_\| \cos^2 \omega + \mathsf{Sh}_\perp \sin^2 \omega, \tag{E5.9.3.1}$$

where $Sh_\parallel$ and $Sh_\perp$ are the mean Sherwood numbers corresponding to the cases of parallel ($\omega = 0$) and perpendicular ($\omega = \pi/2$) positions of the body in the flow, respectively.

At low Peclet numbers, for the translational Stokes flow past an arbitrarily shaped body of revolution, formula (E5.9.3.1) coincides with the exact asymptotic expression in the first three terms of the expansion. Since (E5.9.3.1) holds identically for a spherical particle at all Peclet numbers, one can expect that for particles whose shape is nearly spherical, the approximate formula (E5.9.3.1) will give good results for low as well as moderate or high Peclet numbers.

▶ **Particles of arbitrary shape.** For a steady-state viscous flow (without closed streamlines) past arbitrarily shaped smooth particles, one can calculate the mean Sherwood number by the approximate formula

$$Sh = 0.5\, Sh_0 + (0.125\, Sh_0^3 + Sh_\infty^3)^{1/3}. \tag{E5.9.3.2}$$

Here the auxiliary variables Sh_0 and Sh_∞ in (E5.9.3.2) are the leading terms of the asymptotic expansions of the mean Sherwood number at small and large Peclet numbers, respectively. (In (E5.9.3.2), Sh, Sh_0, and Sh_∞ are defined on the basis of the same characteristic length.)

For spherical particles, we have $Sh_0 = 1$ (the radius is taken as the characteristic length), and (E5.9.3.2) turns into (E5.7.3.2). The substitution of the values of Sh_0 and Sh_∞ corresponding to ellipsoidal particles in a translational Stokes flow into (E5.9.3.2) results in (E5.9.2.6).

For a translational Stokes flow past a convex body of revolution of sufficiently smooth shape with symmetry axis parallel to the flow, the error $\mathcal{E}$ (in percent) in formula (E5.9.3.2) for the mean Sherwood number can be approximately estimated as follows:

$$\mathcal{E} < 2\left(\frac{a}{b} + \frac{b}{a}\right),$$

where a and b are the maximum longitudinal and transverse dimensions of the particle. This estimate agrees well with the results previously described for an ellipsoidal particle.

For a particle of a given shape, the auxiliary quantities Sh_0 and Sh_∞ occurring in (E5.9.3.2) can be determined either theoretically or experimentally. In the last case, the parameter Sh_0 must be found from experiments on diffusion to the particle in a quiescent fluid. (Recall that the value Sh_0 corresponds to the dimensionless capacity of the body; the electrostatic method for measuring this capacity is widely used in electrical engineering.) For a solid particle, the asymptotics of the mean Sherwood number as $Pe \to \infty$ has the form $Sh_\infty = B\, Pe^{1/3}$, where B is a constant. Therefore, to find the parameter B and hence Sh_∞, it suffices to carry out a single experiment at high Peclet numbers. (High Peclet numbers at low Reynolds numbers, $Re < 0.5$, can easily be achieved in water solutions of glycerin.) Thus, to find Sh_0 and Sh_∞, it suffices to carry out two fairly simple experiments.

E5.9.4. Mass and Heat Transfer to a Circular Cylinder in Translational Flow

Consider diffusion to the surface of a circular cylinder of radius a in a flow with velocity U_i directed along the normal to the cylinder axis. This is a model problem used in chemical engineering for calculating mass transfer to prolate particles; it is used even more widely in mechanics of aerosols for analyzing diffusion sedimentation of aerosols on fibrous filters.

At low Reynolds numbers $\mathsf{Re} = aU_i/\nu$, the analytical solution of this problem results in the following two-term expansion of the mean Sherwood number (based on the radius of the cylinder) with respect to the high Peclet number $\mathsf{Pe} = aU_i/D$:

$$\mathsf{Sh} = \frac{0.580}{(2.00 - \ln 2\,\mathsf{Re})^{1/3}}\mathsf{Pe}^{1/3} + 0.0993.$$

If $\mathsf{Sc} > 0.5$, then the mean Sherwood number for cylinders of various cross-sections perpendicular to the flow direction in a wide range of Reynolds numbers can be determined by using the following formula, derived from experimental data (Kutateladze, 1990):

$$\mathsf{Sh} = A\,\mathsf{Sc}^{0.37}\mathsf{Re}^m,$$

where the coefficients A and m are given below:

The range of Re	A	m
0.05 to 2	0.640	0.305
2 to 4	0.556	0.41
4 to 500	0.381	0.47
500 to 2.5×10^3	0.430	0.47
2.5×10^3 to 2.5×10^4	0.142	0.60
2.5×10^4 to 10^5	0.0168	0.80

E5.10. Transient Mass Transfer to Particles, Drops, and Bubbles in Steady-State Flows

E5.10.1. Statement of the Problem

We consider a laminar steady-state translational flow past a solid spherical particle (drop or bubble) of radius a and study transient mass transfer to the particle surface. At the initial time $t = 0$, the concentration in the continuous phase is constant and equal to C_i, whereas for $t > 0$ a constant concentration C_s is maintained on the particle surface.

In the particle-centered spherical coordinate system (R, θ, φ), the nonstationary problem for the concentration C in dimensionless variables comprises the convective diffusion equation

$$\frac{\partial c}{\partial \tau} + \mathsf{Pe}(\mathbf{v} \cdot \nabla)c = \Delta c \tag{E5.10.1.1}$$

and the initial and boundary conditions

$$\tau = 0, \quad c = 0; \qquad r = 1, \quad c = 1; \qquad r \to \infty, \quad c \to 0, \tag{E5.10.1.2}$$

where $c = (C_i - C)/(C_i - C_s)$, $\tau = Dt/a^2$, $r = R/a$, $\mathsf{Pe} = aU/D$, and U is the characteristic flow velocity (for the translational flow $U = U_i$, where U_i is the fluid velocity at infinity). The steady-state flow field $\mathbf{v}$ is given.

Table E5.4
The expressions for Sh_{st} in Eq. (E5.10.2.1) for spherical particles, drops, and bubbles; $\mathsf{Pe} = aU_i/D$.

No	Disperse phase	Flow mode	Sh_{st}
1	Bubble	Translational Stokes flow ($\mathsf{Re} \ll 1$)	$\left(\dfrac{2\,\mathsf{Pe}}{3\pi}\right)^{1/2}$
2	Bubble	Laminar translational flow at high Reynolds numbers ($\mathsf{Re} \gg 1$)	$\left(\dfrac{2\,\mathsf{Pe}}{\pi}\right)^{1/2}$
3	Drop	Translational Stokes flow ($\mathsf{Re} \ll 1$)	$\left[\dfrac{2\,\mathsf{Pe}}{3\pi(\beta + 1)}\right]^{1/2}$
4	Solid particle	Translational Stokes flow ($\mathsf{Re} \ll 1$)	$0.624\,\mathsf{Pe}^{1/3}$

E5.10.2. General Correlations for the Sherwood Number

▶ **Spherical particles and drops at high Peclet numbers.** In what follows we restrict ourselves to the case of high Peclet numbers and assume that there are no closed streamlines in the flow.

The diffusion boundary layer in problem (E5.10.1.1), (E5.10.1.2) is first adjacent to the particle surface and then rapidly spreads over the flow region with the subsequent exponential relaxation to a steady state. The characteristic relaxation time τ_r is of the order of $\mathsf{Pe}^{-2/3}$ for a solid particle and of the order of Pe^{-1} for bubbles and drops of moderate viscosity.

The time dependence of the mean Sherwood number for an arbitrary steady-state flow past spherical particles, drops, and bubbles can be determined by the approximate formula

$$\frac{\mathsf{Sh}}{\mathsf{Sh}_{st}} = \sqrt{\coth(\pi\,\mathsf{Sh}_{st}^2\tau)}, \qquad (E5.10.2.1)$$

where $\mathsf{Sh}_{st} = \lim\limits_{\tau \to \infty} \mathsf{Sh}$ is the Sherwood number for the steady-state diffusion mode; Sh_{st} depends on the Peclet number and can be determined by solving the corresponding stationary problem. (For the translational flow, the stationary problem is determined by the equation and the boundary conditions (E5.7.1.1)–(E5.7.1.3).)

Equation (E5.10.2.1) gives a valid asymptotic result for any flow field in both limit cases $\tau \to 0$ and $\tau \to \infty$.

Table E5.4 presents Sh_{st} for the translational flow past spherical particles, drops, and bubbles of radius a. The parameter β is the ratio of the dynamic viscosity of the drop to that of the ambient fluid and varies in the range $0 \le \beta \le 2$ (the value $\beta = 0$ corresponds to a gas bubble).

For example, in the case of nonstationary mass transfer in a steady-state translational Stokes flow past a spherical drop with limiting resistance of the continuous phase, the steady-state value Sh_{st} is presented in the third row of Table E5.4. By substituting this value into (E5.10.2.1), we obtain

$$\mathsf{Sh} = \left[\frac{2\,\mathsf{Pe}}{3\pi(\beta + 1)}\coth\left(\frac{2}{3}\frac{\mathsf{Pe}\tau}{\beta + 1}\right)\right]^{1/2}. \qquad (E5.10.2.2)$$

The maximum error of Eq. (E5.10.2.2) for $0 \le \tau < \infty$ does not exceed 0.7%.

Formula (E5.10.2.1) gives good results for the mean Sherwood number for linear shear flows past spherical particles, drops, and bubbles at high Peclet numbers. (In the diffusion boundary layer approximation, the maximum error of Eq. (E5.10.2.1) for all τ does not exceed 1.8% in any case.)

▶ **Spherical particles and drops at arbitrary Peclet numbers.** To calculate the mean Sherwood number for an arbitrary laminar flow past spherical particles, drops, and bubbles in the entire range of Peclet numbers, one can use the interpolation formula

$$\text{Sh} = (\text{Sh}_{\text{st}} - 1)\sqrt{\coth\left[\pi(\text{Sh}_{\text{st}} - 1)^2\tau\right]} + 1. \qquad (E5.10.2.3)$$

Consider the behavior of this function in various limit cases. Since $\text{Sh}_{\text{st}} \to 1$ as $\text{Pe} \to 0$, we see that Eq. (E5.10.2.3) yields the exact result (E5.6.2.4) for a quiescent medium. As $\text{Pe} \to \infty$, we have $\text{Sh}_{\text{st}} \to \infty$, and the expression (E5.10.2.3) passes into (E5.10.2.1). For small τ, Eq. (E5.10.2.3) yields the exact result $\text{Sh} \approx (\pi\tau)^{-1/2}$. As $\tau \to \infty$, we have $\text{Sh} \to \text{Sh}_{\text{st}}$, which follows from (E5.10.2.3).

Bibliography for Chapter E5

Acrivos, A. and Goddard, J. D., Asymptotic expansions for laminar forced-convection heat and mass transfer. Part 1. Low speed flows, *J. Fluid Mech.,* Vol. 23, No. 2, pp. 273–291, 1965.

Bird, R. B., Stewart, W. E., and Lightfoot, E. N., *Transport Phenomena*, Wiley, New York, 1965.

Brenner, H., Forced convection-heat and mass transfer at small Peclet numbers from particle of arbitrary shape, *Chem. Eng. Sci.,* Vol. 18, No. 2, pp. 109–122, 1963.

Carslow, H. S. and Jaeger, J. C., *Conduction of Heat in Solids*, Pergamon Press, New York, 1959.

Clift, R., Grace, J. R., and Weber, M. E., *Bubbles, Drops and Particles*, Acad. Press, New York, 1978.

Deavours, C. A., An exact solution for the temperature distribution in parallel plate Poiseuille flow, *Trans. ASME, J. Heat Transfer,* Vol. 96, No. 4, 1974.

De Kee, D. and Chhabra, R. P., *Transport Processes in Bubbles, Drops and Particles*, Hemisphere, New York, 1992.

Eckert, E. R. G. and Drake, R. M., *Analysis of Heat and Mass Transfer*, Hemisphere Publ., New York, 1987.

Friedlander, S. K., Mass and heat transfer to single spheres and cylinders at low Reynolds numbers, *AIChE J.,* Vol. 3, No. 1, pp. 43–48, 1957.

Graetz, L., Über die Warmeleitungsfähigkeit von Flüssigkeiten, *Annln. Phys.,* Bd. 18, S. 79–84, 1883.

Grigull, U. and Sandner, H., *Heat Conduction*, Hemisphere Publ., New York, 1984.

Gupalo, Yu. P., Polyanin, A. D., and Ryazantsev, Yu. S., *Mass and Heat Exchange Between Reacting Particles and Flow* [in Russian], Nauka, Moscow, 1985.

Hewitt, G. F., Shires, G. L., and Bott, T. R., *Process Heat Transfer*, Begell House, New York, 1994.

Hewitt, G. F. and Spalding, D. B., *Encyclopedia of Heat and Mass Transfer*, Hemisphere, New York, 1986.

Incropera, F. P. and Dewitt, D. P., *Fundamentals of Heat and Mass Transfer*, Wiley, New York, 1996.

Kader, B. A. and Yaglom, A. M., Heat and mass transfer laws for fully turbulent wall flows, *Int. J. Heat Mass Transfer*, Vol. 15, No. 12, pp. 2329–2351, 1972.

Kutateladze, S. S., *Heat Transfer and the Hydrodynamical Resistance. A Reference Book* [in Russian], Energoatomizdat, Moscow, 1990.

Levich, V. G., *Physicochemical Hydrodynamics*, Prentice-Hall, Englewood Cliffs, New Jersey, 1962.

Nusselt, W., Abhängigkeit der Wärmeübergangzahl con der Rohrlänge, *VDI Zeitschrift*, Bd. 54, Ht. 28, S. 1154–1158, 1910.

Oellrich, L., Schmidt-Traub, H., and Brauer, H., Theoretische Berechnung des Stofftransport in der Umgebung einer Einzelblase, *Chem. Eng. Sci.,* Vol. 28, No. 3, pp. 711–721, 1973.

Patankar, S. V. and Spalding, D. B., *Heat and Mass Transfer in Boundary Layers*, Morgan-Grampian, London, 1967.

Petukhov, B. S., *Heat Transfer and Drag in Laminar Flow of Liquids in Tubes* [in Russian], Energiya, Moscow, 1967.

Polyanin, A. D. and Dilman, V. V., *Methods of Modeling Equations and Analogies in Chemical Engineering*, CRC Press–Begell House, Boca Raton, Florida, 1994.

Polyanin, A. D., Kutepov, A. M., Vyazmin, A. V., and Kazenin, D. A., *Hydrodynamics, Mass and Heat Transfer in Chemical Engineering*, Taylor & Francis, London, 2002.

Reid, R. C., Prausnitz, J. M., and Sherwood, T. K., *The Properties of Gases and Liquids*, McGraw-Hill Book Comp., New York, 1977.

Rohsenow, W. M., Hartnett, J. P., and Ganic, E. N., *Handbook of Heat Transfer Fundamentals*, McGraw-Hill, New York, 1985.

Rotem, Z. and Neilson, J. E., Exact solution for diffusion to flow down an incline, *Can. J. Chem. Engng.*, Vol. 47, pp. 341–346, 1966.

Sherwood, T. K., Pigford, R. L., and Wilke, C. R., *Mass Transfer*, McGraw-Hill, New York, 1975.

Sparrow, E. M. and Ohadi, M. M., Numerical and experimental studies of turbulent heat transfer in a tube, *Numerical Heat Transfer*, Vol. 11, No. 4, pp. 461–476, 1987.

Vargaftic, N. V., Vinogradov, Y. K., and Yargin, V. S., *Handbook of Physical Properties of Liquids and Gases*, Begell House Inc., New York, 1996.

Zapryanov, Z. and Tabakova, S., *Dynamics of Bubbles, Drops and Rigid Particles*, Kluwer Acad. Publ., Dordrecht, 1999.

Chapter E6

Electrical Engineering

E6.1. Generalities

E6.1.1. Introduction. Basic Notions, Definitions, and Notation

▶ **Preliminary remarks.** *Electrical engineering* is a technical science studying physical phenomena related to directed motion of charged particles in conducting media at the level of macroprocesses and the problems of practical use of such phenomena. Electrical engineering is traditionally divided into the following fields: foundations of circuit theory, foundations of magnetic circuits, electric devices and transformers, electric machines, and foundations of electronics.

The basic notions of electrical engineering coincide with similar notions in the course of physics. Electrical engineering employs the following physical quantities in the sense explained below.

▶ **Notions, definitions, and notation.** The *potential* is the work done by electric forces to transfer the unit charge from a given point in space, in particular, a given point of the circuit to the point at infinity (the zero potential point). The potential is denoted by the symbol φ and is measured in volts (V). The potential can be determined for a given point in space (circuit) with respect to the zero potential.

The *voltage* (*potential difference*) is the work done by electric forces to transfer the unit charge from a point in space (circuit) to another point. The voltage is also measured in volts (V) and denoted by the symbol U. The voltage can be determined only between two points in space (circuit).

The *current strength* (*current*) is the electric charge passing through the conductor transverse cross-section per unit time. The current strength is measured in amperes (A) and is denoted by I. The current strength can be determined for a given element of the electric circuit.

The *electromotive force* (EMF) is the work done by forces of nonelectrostatical nature (extraneous forces) to transfer the unit charge from a point in space (circuit) to another point. The EMF is also measured in volts (V) and denoted by E. The EMF can be determined only for two points in space (circuit).

The *resistance* (*electric resistance*) is a physical quantity characterizing the property of bodies or media with current flowing in them to transform the electric energy into the heat energy of particles of these bodies or media. The resistance is measured in ohms (Ω) and is denoted by R. The reciprocal of resistance is called the conductance and is denoted by g. The SI unit of conductance is the siemens (S), $1\,\text{S} = 1/\Omega$.

The *capacitance* is the physical quantity characterizing the property of bodies to accumulate electric field energy. The SI unit of capacitance is the farad (F).

The *inductance* is the physical quantity characterizing the property of bodies with current flowing in them to accumulate the magnetic field energy. The SI unit of inductance is the henry (H).

The *power* is the work done in electric circuits by force of electrical and nonelectrical nature per unit time. The SI unit of power is the watt (W).

E6.1.2. Perfect Elements. Active and Passive Elements

▶ **Perfect elements. Notation conventions for perfect elements.** Electrical engineering operates with ideal or perfect elements. The idealization consists in that each of the perfect elements has only one of the properties inherent in actual elements. Any perfect element is an abstraction, i.e., a convenient model that with certain accuracy reflects the processes and phenomena occurring in actual elements.

In electrical engineering, the following perfect elements are studied.

A *conductor* is a perfect element in which the electric energy is neither transformed nor dissipated nor accumulated. The conductor has neither resistance nor inductance. The voltage at the ends of a perfect conductor is always zero. The connection of two points of a perfect conductor is called the *short circuit*.

A *disconnection* is a perfect element in which the electric energy is neither transformed nor dissipated nor accumulated. The disconnection has no capacitance, and the disconnection resistance and inductance can be assumed to be infinitely large. The current through a disconnection is always zero. The disconnection between two points of an electric circuit is called the *open circuit*.

A *switch* is a perfect element having two stable states in one of which it is a perfect conductor and in the other it is a disconnection. The transition from one state into another occurs in an infinitely small time interval.

A *resistor* is a perfect element that has only a finite value of resistance. The inductance of a perfect resistor is zero. In the resistor, the electric energy is dissipated; i.e., it is irreversibly transformed into the heat energy.

A *capacitor* is a perfect element that has only a finite value of capacitance. The resistance of a perfect capacitor is infinitely large, and its inductance is zero. In the capacitor, the electric field energy is accumulated.

A *coil* (*induction coil*) is a perfect element that has only a finite value of inductance. The resistance of ideal induction coil is zero. In the induction coil, the magnetic field energy is accumulated.

An *ideal EMF source* is an active element such that the voltage at its terminals is equal to the value of its electromotive force. This quantity is, in particular, independent of the current flowing through the EMF source. The short circuit problem for an ideal EMF source is ill posed.

The *ideal source of current* is an active element such that the strength of the current flowing through it is always equal to the current of the source. The current of the source is, in particular, independent of its voltage. The open circuit problem for the current source is ill posed.

By default, all the elements are assumed to be perfect. Any perfect element in electrical circuits (networks) is indicated by the symbol of its typical parameter. If it is necessary to distinguish several elements, then the parameters are equipped with corresponding indices.

The notation conventions for perfect elements are shown in Fig. E6.1.

Unless otherwise specified, it is assumed that the geometric dimensions of perfect elements are zero.

The direction from minus to plus at the ideal EMF source shows the direction of extraneous forces or the direction of increase in the potential. The voltage on such a source is equal to E.

A perfect element is said to be linear if its characteristic elements are independent of the current flowing in the element and of its voltage. By default, it is assumed that the elements are linear.

The mathematical model of an actual element can be constructed from perfect elements with any desired accuracy of details.

 —\/\/\/— R	Resistor of resistance R
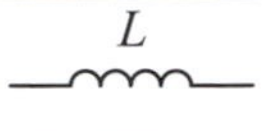 —റ്റ്റ— L	Ideal induction coil of inductance L
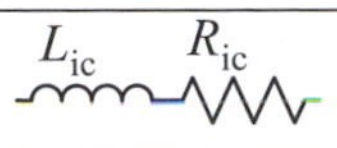 —റ്റ്റ—\/\/\/— L_{ic} R_{ic}	Nonideal, real induction coil of inductance L_{ic} and resistance R_{ic}
 —||— C	Capacitor of capacitance C
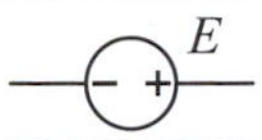 —⊕— E	Ideal source of voltage, electromotive force E
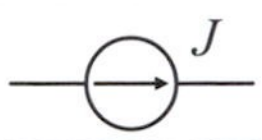 —⊖— J	Ideal source of direct current
 —/°— S	Switch (SPST = Single Pole, Single Throw)
 —\/\/\/— R	Variable resistor (rheostat)
 —\/\/\/— R	Nonlinear element (resistor)
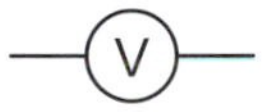 —(V)—	Voltmeter
 —(A)—	Ammeter

Figure E6.1. Notation conventions for perfect elements.

▶ **Active and passive elements.** The perfect elements are divided into active and passive elements according to the character of the energy transformation in them.

In passive elements, only the energy accumulation (inductance and capacitance) or its dissipation (irreversible transformation of electric energy into heat in resistors) may occur. In several passive elements, energy is neither accumulated nor transformed at all (conductor, disconnection, switch).

In active elements, energy is transformed from one type into another. An EMF source and a current source are active elements. The energy transformation in active elements is always reversible.

E6.1.3. Electric Circuits and Networks. Classification of Electric Circuits and Their Operation Modes

▶ **Electric circuits and networks.** An *electric circuit* is an electrical engineering device that exists or may exist in reality. An electric circuit consists of actual elements.

A *network (electrical network)* is a certain graphical representation of an actual device, which is constructed only of perfect elements.

An *equivalent circuit* of an electric circuit consists of a set of different idealized elements chosen so that it were possible to describe the processes in the circuit with a prescribed or desired accuracy.

▶ **Classification of electric circuits and their operation modes.** The electric circuits are classified according to the characteristics of the elements composing them as follows:

linear and nonlinear circuits, circuits with lumped parameters, and circuits with distributed parameters. For a circuit to be referred to as nonlinear, it suffices to contain at least one nonlinear element. For a circuit to be referred to as a circuit with distributed parameters, it suffices that the linear dimensions of at least one element are commensurable with the wavelength of the electromagnetic field arising in this circuit.

The operation modes of electric circuits are classified according to the character of variations in the input and output signals (currents and voltages) in time. The operation modes are divided into *stationary*, *quasistationary* (*periodic*), and *nonstationary* (*transient*).

E6.1.4. Structure Elements of Network. Notion of the Graph of a Network

▶ **Structure elements of network.** In any electrical network, the following structural elements can be distinguished: branch, node, loop.

An *branch* is a series connection of several perfect elements. A branch can consist of a single element if it is not a conductor or a disconnection. A series connection is a connection of perfect elements in which the "output" of one element is the "input" of one and only one subsequent element. The branch origin is the "input" of the first of the elements, and the branch end is the "output" of the last of the perfect elements. The origin and the end of a branch are, as a rule, conditional notions. It is convenient to denote all perfect elements belonging to the same branch by the same indices. It follows from the charge conservation law that the current strength is the same for all elements of the branch.

A *node* is a point common for three or more branches. For each of the branches belonging to (incident on) a given node, the node can be only the origin or the end of the branch. It follows from the definition of electric potential that the potentials of all points forming a given node are the same.

An *loop* is any closed path in an electric circuit network such that none of the branches and none of the nodes are encountered twice when going around this path. A loop may be dependent and independent.

▶ **Graph of a network.** The *graph of a network* is a graph such that the set of its vertices coincides with the set of the network nodes and the set of its edges coincides with the network branches. The set of independent loops in a network is the set of fundamental cycles in the graph of the network. The elements of a graph are branches, vertices, nodes, and loops. A branch is the current path between two vertices, and the current strength in all transverse cross-sections of the branch at a given time moment is constant. A node is a vertex at which three or more branches meet. A loop is a closed path beginning and terminating at the same vertex. The number of loops in a network may be rather large, but still, there is a number N of linearly independent loops, which can be determined by the formula

$$N = B - K + 1,$$

where B is the number of branches and K is the number of nodes.

E6.1.5. Physical Laws for Structure Elements

For the basic structure elements of any electrical network, there exists its own physical law:

Ohm's law   (for branches),
Kirchhoff's first law   (for nodes),
Kirchhoff's second law   (for closed loops).

▶ **Ohm's law.** *Ohm's law* is formulated as follows: the current strength on a part of the circuit is directly proportional to the voltage applied to this part and is inversely proportional to the resistance of this part.

Ohm's law can be interpreted for a part of the circuit containing an EMF source E (the generalized Ohm law) as follows: the current of a part of the circuit is equal to the algebraic sum of its voltage and EMF divided by the resistance of the part. The EMF and the voltage are taken with positive sign if their directions coincide with the current direction and are taken with negative sign if their directions are opposite to the current direction. The four situations shown in Fig. E6.2 are possible.

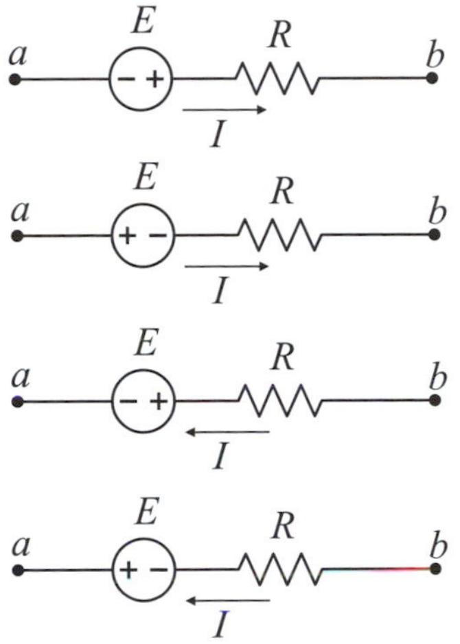

$$I = \frac{U_{ab} + E}{R}$$

$$I = \frac{U_{ab} - E}{R}$$

$$I = \frac{-U_{ab} - E}{R}$$

$$I = \frac{-U_{ab} + E}{R}$$

Figure E6.2. Generalized Ohm law for different parts of the circuit.

▶ **Kirchhoff's laws.** *Kirchhoff's first law* is formulated as follows:

$$\sum_{k=1}^{n} I_k = 0 \qquad \text{(the algebraic sum of currents at a node is zero).}$$

Here it is assumed that the currents flowing into a node are taken with the sign "+", and the currents flowing out of a node, with the sign "−".

Kirchhoff's second law can be formulated as follows:

$$\sum_{k=1}^{n} E_k = \sum_{l=1}^{m} R_l I_l,$$

where $\sum_{k=1}^{n} E_k$ is the algebraic sum of EMF E_k in a closed loop, and a term in this sum is taken with sign "+" if the directions of the source EMF and of an arbitrary chosen path tracing coincide with each other, otherwise, the sign "−" is taken; $\sum_{l=1}^{m} R_l I_l$ is the algebraic sum of voltages in the same loop, and a term of this sum is taken with sign "+" if the directions of current voltage and of an arbitrary chosen path tracing coincide with each other; otherwise, the sign "−" is taken.

E6.1.6. Problems of Analysis and Synthesis of a Network. An Analysis of a Network by Physical Laws. Potential Diagram

▶ **Problems of analysis and synthesis of a network.** In electrical engineering, the problems of network analysis and synthesis are distinguished. The analysis problem is a problem of calculating currents in branches of a network, of voltages between nodes, of powers, etc. from the well-known structure of the circuit and the parameters of its elements. The analysis problem is generally solved by constructing an equivalent circuit that, with a desired accuracy, reflects the circuit element properties important in the problem under study and the circuit structure. Further, the problem is solved by one of the methods for analyzing the circuits, which are considered below. The choice of any method is based on the following factors: the dimension of the obtained system of equations, the simplicity of algorithmization, and the restricted possibilities of using the method.

The circuit synthesis problem involves the construction of a network ensuring the prescribed values of one or several parameters for one or several elements under certain restrictions.

▶ **An analysis of a network by Ohm's and Kirchhoff's laws.** An analysis of any correctly described electric circuit can be performed by using only the three physical laws listed in Subsection E6.1.5. In general, to solve the problem, it is necessary to write out the equations of Kirchhoff's second law for all independent loops and of Kirchhoff's first law for all nodes except one arbitrary node. If necessary, the generalized Ohm law is written for separate branches. The system of linear equations thus obtained can be solved by any method.

The order of calculations may be as follows:

1. The positive directions of currents in all branches are chosen and indicated by arrows.
2. The positive directions of path tracing are chosen.
3. The system of equations according to Kirchhoff's laws is constructed: (i) by the first law, we obtain $N_1 = K - 1$ equations, (ii) by the second law, we have $N_2 = B - K + 1$ equations, where B is the number of branches and K is the number of nodes (in the whole, we obtain $N_1 + N_2 = B$ equations).

Example. As an example, consider the electrical network shown in Fig. E6.3. We choose the directions of the currents and the directions of the path tracing as is shown in Fig. E6.3. Then we obtain the system of equations:

$$I_1 - I_2 + I_3 = 0,$$
$$E_1 + E_2 = I_1 R_1 + I_2 R_2,$$
$$E_2 + E_3 = I_2 R_2 + I_3 R_3.$$

The unknowns in this system of equations may be any three quantities, and the others must be prescribed. Usually, the unknowns are the currents in the circuit branches. If it is necessary to introduce the voltage U_{AB} into the system of equations, then it is also required to write out the generalized Ohm law for any of the three branches.

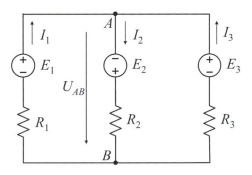

Figure E6.3. Electrical network.

Using Ohm's and Kirchhoff's laws, one can calculate the operation mode of any electric circuit. But the order of the system of equations can be large. The system can be simplified by different methods, and all of them are based on Kirchhoff's and Ohm's laws. A method for solving the problems is chosen starting from the complexity of the network and of given parameters.

▶ **Distribution of the potential along an unbranched (unramified) electric circuit.** The *potential diagram* is the distribution of electric potentials along parts of the circuit of a closed loop, namely, $\varphi(R)$.

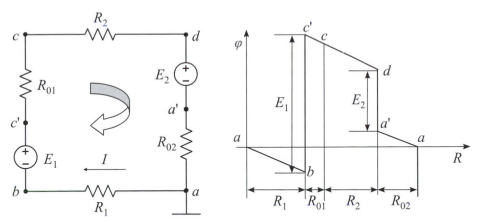

Figure E6.4. Electrical network and the potential diagram; R_{01} and R_{02} stand for the internal resistances of the EMF E_1 and E_2, respectively.

We illustrate the diagram construction with an example of a simple unbranched circuit shown in Fig. E6.4. Assume that the direction of the path tracing coincides with the direction of the current. To construct the potential diagram, we assume that one (any) of the potentials of the closed loop is zero. If the EMF E_1 exceeds the EMF E_2, then, in this case, we have

$$I = \frac{E_1 - E_2}{R_1 + R_{01} + R_2 + R_{02}}.$$

Let $\varphi_a = 0$; then

$$\varphi_b = \varphi_a - R_1 I, \quad \varphi_c = \varphi_b + E_1 - R_{01} I, \quad \varphi_d = \varphi_c - R_2 I, \quad \varphi_a = \varphi_d - E_2 - R_{01} I = 0.$$

In these expressions, the value of the EMF source voltage E is taken with sign "+" if the sense of the path coincides with the direction of E; otherwise, the sign "−" is taken. The value of the resistor voltage is taken with the sign "+" if the sense of the path is opposite to that of the current; otherwise, the sign "−" is taken.

E6.2. Linear Circuits of Direct Current

E6.2.1. Power of Direct Current

To estimate the energy conditions, it is important to know how fast the work is done; i.e, it is important to determine the power: $P = UI$.

For resistive elements, this expression can be transformed using Ohm's law as follows: $P_R = UI = I^2 R = U^2/R$.

For the EMF source whose direction coincides with the direction of the current, the power of extraneous forces is determined as $P_E = EI$. But if the current and EMF directions are opposite (for example, in the case of battery charging), then the power of extraneous forces is: $P_E = -EI$. In this case, the EMF source obtains energy from the external circuit.

The power of extraneous forces of the current source is equal to $P_J = U_{ab}I = U_{ab}J$. If $U_{ab} < 0$, then the source gains energy from the external circuit.

In any electric circuit, the energy balance—i.e., the power balance—must be satisfied, which means that the algebraic sum of powers of all energy sources must be equal to the algebraic sum of powers of all energy receivers:

$$\sum P_E = \sum P_R.$$

Example. Let us compose the energy balance for the network shown in Fig. E6.5. We obtain

$$U_{ab}J + E_1 I_1 = I_1^2 R_1 + I_2^2 R_2.$$

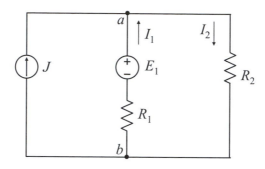

Figure E6.5. Electrical network illustrating the energy balance.

E6.2.2. Notion of Nonideal Source. Equivalent Circuits and External Characteristics

▶ **Nonideal source. Equivalent circuits.** A nonideal source of energy with known EMF E and internal resistance R_0 can be represented by two basic equivalent networks (equivalent circuits): the voltage source (Fig. E6.6a) or the current source (Fig. E6.6b).

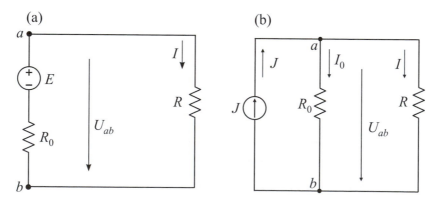

Figure E6.6. (a) Equivalent circuit of voltage source and (b) current source.

The representation of actual electric energy sources in the form of two equivalent circuits is an equivalent representation with respect to the exterior part of the circuit (receiver); i.e., in both cases, the voltages between the source terminals are the same (U_{ab}). But the energy relations in the two equivalent circuits are not the same. There is no equality between the power of the EMF sources and the current source $EI \neq U_{ab}I$ and the loss power $I^2 R_0 \neq I_0^2 R_0$.

▶ **External characteristic and operation modes of an EMF source.** According to Ohm's law, the EMF source terminal voltage takes the form

$$U_{ab} = E - IR_0.$$

The dependence between the voltage U and the current I flowing through the EMF source is called the *external characteristic* of the EMF source or the volt-ampere characteristic (VACh). The external characteristic of the source $U(I)$ completely determines the properties of this source and, for a majority of actual sources, this characteristic, can be represented as a straight line (see Fig. E6.7).

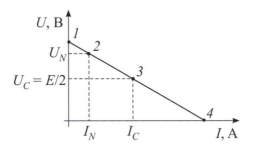

Figure E6.7. External characteristic of the EMF source.

Of all possible operation modes of an EMF source, we note the following four most important modes:

point 1: open circuit mode,
point 2: nominal (typical) operation mode,
point 3: consistent operation mode,
point 4: short circuit mode.

The *open circuit mode* (point 1) is an operation mode in which the receiver resistance is infinitely large ($R = \infty$). In practice, this corresponds to disconnection of an electric circuit, hence the open-circuit current is zero ($I_* = 0$), and hence $U_* = E$. In the open circuit tests, the EMF E can be determined from the readings of a voltmeter connected directly to the open-circuit terminals of the EMF source E.

The *nominal mode* (point 2) of a source is characterized by the fact that its voltage, current, and power correspond to the values calculated by the manufacturing departments. In this case, the best operation conditions are guaranteed for the EMF source (efficiency, operating life, etc.). The quantities determining the nominal mode are usually indicated in the certificate, in the catalog, or on the device panel. In the nominal mode, $U_N = E - R_0 I_N$.

In practice, the nominal mode is attained if U_N lies in the limits $0.90E \leq U_N \leq 0.95E$.

The *consistent mode* (point 3) is an operation mode in which the source transfers the maximal power P_{max} to the receiver (to the external circuit). To achieve this operation mode, it is necessary to choose a value of the receiver resistance R equal to the internal resistance of the source R_0, i.e., $R = R_0$.

The *short circuit mode* (point 4) is characterized by the fact that the receiver resistance becomes zero $R = 0$. As a rule, in practice, this is related to closing the electric energy receiver by a conductor of a very small resistance, the so-called industrial short circuit, often caused by a lot of abuse of electrical engineering devices (electric motors, transformers, home equipment, etc.).

For $R = 0$, we have $U' = RI' = 0$, and then the short circuit current is equal to $I' = E/R_0$. Because of a small internal resistance of the EMF source E, the short circuit current (I') can attain very large values significantly exceeding the value of nominal currents $(I' \gg I_N)$. Therefore, the short circuit mode, as a rule, is very dangerous and undesirable for both the source and the receiver. But in several cases, the short circuit mode is nominal for electrical engineering devices such as the welding transformers and generators, current transformers, etc.

The parameters of the EMF source E can be determined graphically by using its external characteristic.

E6.2.3. Problem of Maximal Power Transfer. The Source Efficiency Factor

For the receiver to deliver the maximal power P_{max}, it is necessary that the receiver operate with the EMF source E in the consistent operation mode. To achieve the consistent mode, it is necessary to choose the receiver resistance R so that it is equal to the internal resistance R_0 of the EMF source E.

Indeed, the power delivered by the receiver is equal to $P = RI^2$, where $I = \dfrac{E}{R + R_0}$. Therefore,

$$P = \frac{RE^2}{(R + R_0)^2}.$$

Let us analyze this formula. If $R = 0$, then $P = 0$; if $R = \infty$, then $P = 0$. Therefore, the function $P = P(R)$ has an extremum. One can show that this function attains its maximum for $R = R_0$. The corresponding power delivered by the receiver is calculated as

$$P_{max} = RI_C^2 = \frac{R_0 E^2}{(R_0 + R_0)^2} = \frac{E^2}{4R_0},$$

and the power delivered by the source is equal to

$$P_E = EI_C = E\frac{E}{R_0 + R_0} = \frac{E^2}{2R_0}.$$

The source efficiency factor in this operation mode is equal to

$$\eta = \frac{P_{max}}{P_E} = 0.5;$$

i.e., 50% of energy the source is transferred to the receiver, and the other 50% are lost in the source itself (due to its internal resistance).

The consistent mode is admissible and desirable in systems of telemechanics, automatics, telephony, etc., where it is necessary to distinguish the maximal possible power P_{max} in the receiver; the 50% losses in the source can be neglected, because the power consumption in such systems is small of the order of several watts.

In power-supply systems, such an operation mode is impossible, because the powers of the energy sources is of the order of megawatts (MW) and the 50% losses in the sources are extremely large.

E6.2.4. Theorem on Active Impedor (Two-Terminal Network) and Its Application

▶ **Theorem on active impedor.** The electric circuit (Fig. E6.8) can be divided into two parts: the first contains a branch (e.g., resistance R) in which it is required to find the current I, the second—i.e., the other part of the circuit—contains an active impedor.

The theorem on an active impedor can be formulated as follows: if the active circuit with a branch connected to it is replaced by a source with EMF E equal to the voltage at the terminals of the opened branch and with the resistance equal to the input resistance of active circuit, then the current in this branch docs not change.

An active impedor with respect to the resistance R can be treated as an equivalent generator with EMF E and the internal resistance R_0, which permits determining the current flowing through the resistance R as follows: $I = E/(R + R_0)$.

Sometimes, the theorem on active impedor is also called the equivalent generator method or the open-circuit and short-circuit method, the latter name explains the essence of this method.

The unknown values of EMF E and the internal resistance R_0 of an equivalent generator can easily be determined by open circuit and short circuit tests. From the open circuit test we determined U_* and the equivalent generator EMF $E = U_*$.

The impedor internal resistance can be determined in the following two ways. First, one can calculate the resistance of the entire electrical circuit without the resistor R with respect to the terminals of the active impedor for the bridged EMF sources E with their internal resistances taken into account and for the open current sources. Second, one can first find the short circuit current by setting that the resistance R is zero and then determine the internal resistance by the formula $R_0 = E/I'$.

▶ **Example of applying the theorem on active impedor.** We illustrate the use of the theorem on active impedor by an example for the network shown in Fig. E6.8a. In this network, we choose the positive direction of the desired current I_1. Then we consider the network part connected to the first branch under study (enclosed in the dashed line) as an active impedor with EMF E_* and internal resistance R_0 (see Fig. E6.8b). In the network, the direction of EMF E_* to point A is chosen arbitrarily; then $U_{AB} = E_*$. The full-scale network in open circuit mode is shown in Fig. E6.8c.

In the impedor internal branches, the current is calculated by the formula

$$I_* = \frac{E_2}{R_2 + R_3 + R_{02}}.$$

The open-circuit voltage determines the EMF of the active impedor as follows:

$$U_{AB} = E_* = R_3 I_*.$$

To calculate the impedor internal resistance, we transform the network replacing the source E_2 by a short-circuited part (see Fig. E6.8d). The network input resistance is the internal resistance of the active impedor:

$$R_0 = \frac{(R_2 + R_{02})R_3}{R_2 + R_3 + R_{02}}.$$

Returning to the original network, we determine the desired current by Ohm's law:

$$I_1 = \frac{E_* + E_1}{R_1 + R_{01} + R_0}.$$

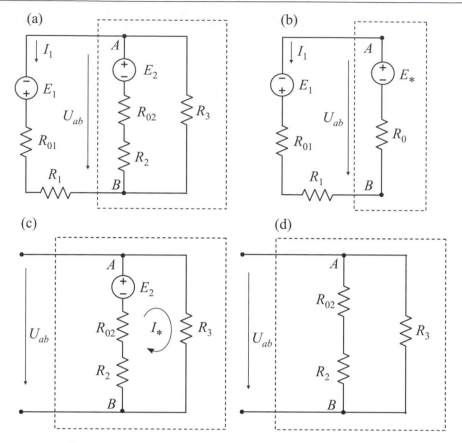

Figure E6.8. Electrical networks illustrating the active impedor method.

E6.2.5. Two Compensation Theorems. Properties of Linear Circuits: Linearity Theorem and Reciprocity Theorem

▶ **Compensation theorems.** The *voltage compensation theorem* (*voltage compensation principle*) says that the network part ab with voltage U_{ab} can be replaced by an equivalent EMF source $E = U_{ab}$ whose direction is opposite to the positive direction of the voltage U_{ab} (see Fig. E6.9). In this case, the current distribution in the network does not change.

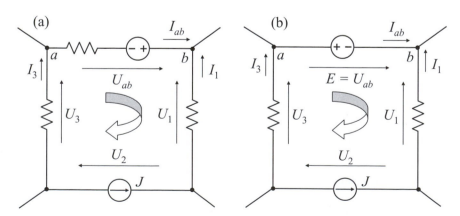

Figure E6.9. (a) Initial network and (b) transformed network.

The *current compensation theorem* (*current compensation principle*) says that the network part ab with current I_{ab} can be replaced by an equivalent current source $J = I_{ab}$ whose direction coincides with the positive direction of the current I_{ab}.

▶ **Linearity theorem and reciprocity theorem.** *Linearity theorem*: if, in any linear electric circuit, there are variations in the EMF E or in the resistance, then any two quantities (currents or voltages) for any two branches of this circuit are related by the linear expression $y = a + bx$, where x is the current or voltage in one branch, and y, in the other branch. If, in the network, the EMF or the resistance vary simultaneously in two branches, then any three quantities in the network (currents or voltages) are related by the linear expression: $y = a + bx + cz$.

The coefficients a, b, and c can be determined experimentally or can be calculated.

So, for example, as the EMF E_m varies in any mth branch, the currents I_k and I_p in any kth and any pth branches are related by the linear expression $I_k = a_k + b_k I_p$. According to the compensation theorem, any branch can be replaced by an appropriately chosen EMF; hence the variation in the resistance in the mth branch also results in variations in the currents and voltages in the branches, and there exists linear relations between these currents. The coefficients a_k and b_k can be determined if the values of the voltages and currents are known in the branches under study for two operation modes in the circuit.

Reciprocity theorem: in any electric circuit, the current I_k in the kth branch caused by the EMF E_m located in the mth branch is equal to the current I_m in the mth branch caused by the EMF E_k located in the kth branch if $E_m = E_k$.

E6.2.6. Equivalent Transformations in Electrical Networks

▶ **Preliminary remarks.** The analysis of a complicated electric circuit can often be simplified by replacing a group of resistive elements in its equivalent circuit with an equivalent group where the resistive elements are connected in a different way. An expedient transformation of electrical networks leads to a decrease in the number of its branches or nodes and hence in the number of equations determining its electric state. In all cases of replacement of electrical networks by equivalent networks of different form, it is necessary to ensure that the currents and voltages remain unchanged in the network parts that remain untransformed.

▶ **Series and parallel connection of resistors.** The *series connection of resistors* is a connection such that the current flowing through the resistors is one and the same. For one and the same current to flow through the receivers connected in series, it is necessary that there be no electrical nodes at the points of their connection. Then the equivalent resistance R for N resistors connected in series is determined as

$$R = R_1 + R_2 + \cdots + R_N. \tag{E6.2.6.1}$$

The *parallel connection of resistors* is a connection such that the voltage at the resistor terminals (points of connection) is one and the same. For one and the same voltage to be at the terminals of the resistors, it is necessary that they have a single common input (node) and a single common output (node).

For N resistors connected in parallel, we have

$$\frac{1}{R} = \frac{1}{R_1} + \frac{1}{R_2} + \cdots + \frac{1}{R_N}. \tag{E6.2.6.2}$$

▶ **Useful formulas for several resistors.** From formula (E6.2.6.2), one can obtain the equivalent resistances of two or three resistors connected in parallel:

$$R = \frac{R_1 R_2}{R_1 + R_2} \qquad \text{(two parallel resistors)},$$

$$R = \frac{R_1 R_2 R_3}{R_1 R_2 + R_1 R_3 + R_2 R_3} \qquad \text{(three parallel resistors)}.$$

A Y-network (also know as a star) of resistive elements can be transformed to an equivalent △-network; see Fig. E6.10. The inverse transformation is also possible.

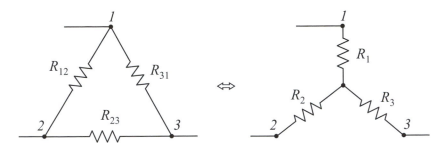

Figure E6.10. Invertible transformation between a △-network and a Y-network.

△-to-Y transformation formulas:

$$R_1 = \frac{R_{12} R_{31}}{R_{12} + R_{23} + R_{31}}, \qquad R_2 = \frac{R_{23} R_{12}}{R_{12} + R_{23} + R_{31}}, \qquad R_3 = \frac{R_{31} R_{23}}{R_{12} + R_{23} + R_{31}}.$$

The resistance of each ray in the equivalent Y-network is equal to the product of resistances of the two △-network branches adjacent to the corresponding ray divided by the sum of resistances of all branches in the △-network.

Y-to-△ transformation formulas:

$$R_{12} = R_1 + R_2 + \frac{R_1 R_2}{R_3}, \qquad R_{23} = R_2 + R_3 + \frac{R_2 R_3}{R_1}, \qquad R_{31} = R_3 + R_1 + \frac{R_3 R_1}{R_2}.$$

We also present similar formulas for conductances:

$$g_{12} = \frac{g_1 g_2}{g_1 + g_2 + g_3}, \qquad g_{23} = \frac{g_2 g_3}{g_1 + g_2 + g_3}, \qquad g_{23} = \frac{g_3 g_1}{g_1 + g_2 + g_3},$$

Thus, the conductance of each branch in the equivalent △-network is equal to the product of conductances of the Y-network rays adjacent to the same terminals divided by the sum of conductances of all rays.

E6.2.7. Application of Equivalent Transformations in Network Analysis. Convolution Method

It is convenient to use the method of equivalent transformations mainly for networks with a single energy source, provided that the energy source voltage does not change after the transformation. The convolution method is reduced to the transformation of the electric circuit into the simplest network with a single source and a single equivalent receiver; then the backward calculations are performed.

We illustrate the convolution method by an example of the network given in Fig. E6.11. First, we arbitrarily choose the direction of unknown currents I_1, I_2, I_{34}. The resistors with resistances R_3 and R_4 connected in series are replaced by a resistor with equivalent resistance R_{34} by the formula $R_{34} = R_3 + R_4$. We transform the original network (Fig. E6.11a) into an equivalent network (Fig. E6.11b). The resistors with resistances R_2 and R_{34} connected in parallel are replaced by a resistor with equivalent resistance R_{234} by the formula

$$R_{234} = \frac{R_2 R_{34}}{R_2 + R_{34}}.$$

We replace the transformed network (Fig. E6.11b) by an equivalent network (Fig. E6.11c). It follows from Fig. E6.11c that the resistors with resistances R_1 and R_{234} are connected in series. Therefore, the resistance equivalent to them is calculated by the formula

$$R' = R_1 + R_{234}.$$

Thus, after the above transformations, the original network (Fig. E6.11a) is transformed into an equivalent network (Fig. E6.11d).

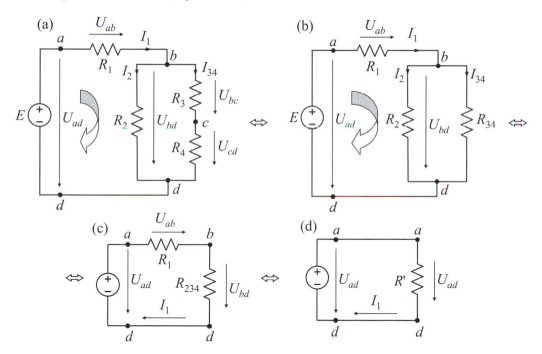

Figure E6.11. (a) Original electrical network and (b, c, d) the transformed networks connected in series.

The current I_1 at the circuit input (Fig. E6.11d) is calculated by Ohm's law:

$$I_1 = \frac{E}{R'} = \frac{U_{ab}}{R'}.$$

To determine the currents in parallel branches, it is necessary to calculate the voltage U_{bd} at them (Fig. E6.11c): $U_{bd} = R_{234} I_1$.

We determine the currents in parallel branches (Fig. E6.11b) by the formula

$$I_2 = \frac{U_{bd}}{R_2}, \qquad I_{34} = \frac{U_{bd}}{R_{34}}.$$

E6.2.8. Superposition Principle. The Notion of Input and Transfer Conductances (Admittances)

▶ **Superposition principle. Input and transfer conductances.** The following superposition principle holds for linear electric circuits: the current in any branch is equal to the algebraic sum of currents in this branch (partial currents) under the separate action of each source if the other sources are replaced by resistors with resistances equal to the internal resistances of the corresponding sources. An analog of such a principle can be found in mechanics, where the motion of a body under the action of several forces can be considered as a result of addition of actions caused by each separate force.

In the equivalent circuit with m branches, the current in each nth branch is equal to the algebraic sum of partial currents caused by each of the EMF and each of the current sources.

For a network without any current source, one can write an equation for the current in any nth branch of the linear electric circuit in the form

$$I_n = E_1 g_{1n} + E_2 G_{2n} + \cdots + E_m g_{mn} \qquad (n \le m).$$

In this equation, the factors multipying the EMF have the dimension of conductances. Therefore, each of the factors with two equal indices g_{nn} can be called the input conductance of branch n, and any of the factors with two distinct indices g_{mn} can be called the transfer conductance of the branches n and m.

The numerical values of the input and transfer conductances of branches can be calculated as follows. We equate all the EMF E_m with zero except for E_n.

Then $I_n = E_n g_{nn}$, whence $g_{nn} = I_n / E_n$. Therefore, the input conductance of any branch is determined by the current to EMF ratio in this branch provided that the EMF are zero in the other branches. The value of the current in branch m can be determined if we equate all the EMF with zero except for E_n. Then $I_m = E_m n g_{mn}$, whence $g_{mn} = I_m / E_n$. Therefore, the transfer conductance of any two branches is determined by the ratio of the current in one branch to the EMF in the other branch provided that the EMF in the other branches are zero.

The input conductance of a branch has a positive value, because it was assumed that we choose the same direction for the current and for the source EMF. The transfer conductance of two branches can take positive and negative values, and moreover, $g_{mn} = g_{nm}$, which means that the reciprocity principle is satisfied.

▶ **Electrical networks illustrating the superposition principle.** As an example, consider the network in Fig. E6.12a.

The currents in branches are equal to the sum of partial currents in the networks; see Fig. E6.12b and c.

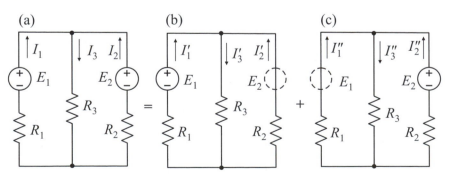

Figure E6.12. Electrical networks illustrating the superposition principle.

We determine the currents in the original network as the sum of partial currents:

$$I_1 = I_1' + I_1'' = g_{11}E_1 + g_{12}E_2 = \frac{R_1 + R_3}{R^2}E_1 - \frac{R_3}{R^2}E_2,$$

$$I_2 = I_2' + I_2'' = g_{21}E_1 + g_{22}E_2 = -\frac{R_3}{R^2}E_1 + \frac{R_1 + R_3}{R^2}E_2,$$

$$I_3 = I_3' + I_3'' = g_{31}E_1 + g_{32}E_2 = \frac{R_2}{R^2}E_1 + \frac{R_1}{R^2}E_2,$$

where $R = \sqrt{R_1 R_2 + R_1 R_3 + R_2 R_3}$.

In the equivalent circuits with current sources, the partial currents of branches are determined from each of them by eliminating the other current sources by disconnecting the branches containing them.

E6.2.9. Concise Methods for Analyzing Electric Circuits

▶ **Mesh-current method.** The mesh-current method is based on Kirchhoff's laws and is most convenient for analyzing electric circuits where the number of independent loops is less than the number of electrical nodes.

When solving the problems, it is assumed that, in the independent loops of the electric circuit, there are their own currents, which are denoted for the first, second, etc. loops, respectively.

Then the linear equations for the chosen unknown loop (fictitious) currents can be composed according to Kirchhoff's second law. The number of the composed independent linear equations must correspond to the number of unknown loop currents. When constructing the equations, first, one must compose the simplest independent equations, and second, each subsequent equation must contain at least one branch that does not enter the loops already used. Solving the obtained system of equations for the unknown loop currents, one can determine the unknown (actual) currents.

In the general case, for a network with n independent loops, the system of equations of the mesh-current method can be written as

$$
\begin{aligned}
R_{11}I_{11} + R_{12}I_{12} + \cdots + R_{1n}I_{nn} &= E_{11}, \\
R_{21}I_{11} + R_{22}I_{12} + \cdots + R_{2n}I_{nn} &= E_{22}, \\
&\cdots\cdots\cdots\cdots\cdots\cdots\cdots\cdots\cdots\cdots\cdots, \\
R_{n1}I_{11} + R_{n2}I_{12} + \cdots + R_{nn}I_{nn} &= E_{nn}.
\end{aligned}
\tag{E6.2.9.1}
$$

The solution of this system for loop currents can, for example, be found by the Cramer method (see Section M5.3).

Example. We illustrate the use of the mesh-current method by the example of the electrical network shown in Fig. E6.13.

We have the system of equations

$$
\begin{aligned}
(R_1 + R_4)I_{11} - R_4 I_{22} &= E_1, \\
-R_4 I_{11} + (R_3 + R_4 + R_5)I_{22} - R_3 I_{33} &= E_3 + E_5, \\
-R_3 I_{22} + (R_2 + R_3)I_{33} &= -E_2 - E_3.
\end{aligned}
\tag{E6.2.9.2}
$$

We denote

$$R_{11} = R_1 + R_4, \quad R_{22} = R_3 + R_4 + R_5, \quad R_{33} = R_2 + R_3, \quad R_{12} = R_{21} = -R_4, \quad R_{32} = R_{23} = -R_3,$$

$$E_{11} = E_1, \quad E_{22} = E_3 + E_5, \quad E_{33} = -E_2 - E_3.$$

As a result, system (E6.2.9.2) takes the standard form (E6.2.9.1):

$$R_{11}I_{11} + R_{12}I_{22} = E_{11},$$
$$R_{21}I_{11} + R_{22}I_{22} + R_{23}I_{33} = E_{22},$$
$$R_{32}I_{22} + R_{33}I_{33} = E_{33}.$$

Determining the loop currents from this system, we can readily find the actual currents in the circuit by the formulas

$$I_1 = I_{11}, \quad I_2 = -I_{33}, \quad I_3 = I_{22} - I_{33}, \quad I_4 = I_{22} - I_{11}, \quad I_5 = I_{22}.$$

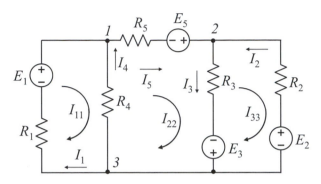

Figure E6.13. Electrical network illustrating the mesh-current method.

▶ **Node-potential method.** The operation mode of any circuit is completely characterized by the equations constructed on the basis of Kirchhoff's first and second laws, and to determine the currents in all branches, it is necessary to compose and solve a system with the same number of unknowns (unknown node potentials). But the number of equations to be solved can be decreased by using the node-potential method based on the application of Kirchhoff's first law and the generalized Ohm law.

To explain the essence of the method, consider the electrical network shown in Fig. E6.14. Assume that the potential of one of the nodes, for example, of the third, is zero, $\varphi_3 = 0$. Then for nodes 1 and 2 we compose the following equations according to Kirchhoff's first law:

$$I_5 - I_4 - I_1 + I_6 = 0, \quad -I_5 - I_6 - I_2 + I_3 = 0. \tag{E6.2.9.3}$$

The currents in the branches can be determined by using Ohm's law:

$$
\begin{aligned}
I_6 &= (\varphi_1 - \varphi_2)g_6, & I_1 &= (-\varphi_1 + E_1)g_1, \\
I_4 &= (-\varphi_1)g_4, & I_5 &= (\varphi_1 - \varphi_2 + E_5)g_5, \\
I_2 &= (-\varphi_2 + E_2)g_2, & I_3 &= (\varphi_2 + E_3)g_3,
\end{aligned}
\tag{E6.2.9.4}
$$

where φ_1 and φ_2 are the potentials of the first and second nodes. Substituting the expressions (E6.2.9.4) for the currents into (E6.2.9.3), we obtain

$$\varphi_1(g_1 + g_4 + g_5 + g_6) - \varphi_2(g_5 + g_6) = E_1g_1 - E_5g_5,$$
$$-\varphi_1(g_5 + g_6) + \varphi_2(g_2 + g_3 + g_5 + g_6) = E_1g_1 - E_3g_3 + E_5g_5,$$

or

$$\varphi_1 g_{11} - \varphi_2 g_{12} = E_1 g_1 - E_5 g_5,$$
$$-\varphi_1 g_{21} + \varphi_2 g_{22} = E_1 g_1 - E_3 g_3 + E_5 g_5.$$

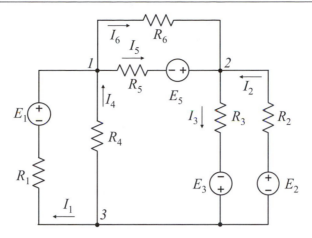

Figure E6.14. Electrical network illustrating the node-potential method.

Here $g_{11} = g_1 + g_4 + g_5 + g_6$ and $g_{22} = g_2 + g_3 + g_5 + g_6$ are the sums of conductances of the branches adjacent to the first and second nodes, respectively, and $g_{12} = g_{21} = g_5 + g_6$ is the sum of conductances of the branches connecting these nodes.

The right-hand side of the last equation is an algebraic sum of products of the EMF E of sources by the conductances of the branches connected to the node of interest. This product is taken with the plus sign if the EMF E is directed towards the node and with the minus sign if the EMF E is directed from the node. (The positive direction of an EMF source is from minus to plus.)

It is remarkable that the current directions in the branches need not be taken into account when composing the equations.

▶ **Nodal-pair method.** The nodal-pair method is based on the generalized Ohm law and Kirchhoff's first law and is used to calculate the electric circuits with two nodes (i.e., electric circuits with branches connected in parallel). When solving the problems by this method, one first determines the voltage between the two nodes by the formula:

$$U_{12} = \frac{\sum g_i E_i + \sum J_k}{\sum g_j}$$

where g_i are the electric conductances of the corresponding branches, J_k are the values of currents from the current sources, and $g_i E_i$ is the product of electric conductance of a branch by the EMF E of the same branch, which is taken with sign "+" if the EMF E is directed towards node 1; otherwise, the sign "−" is taken.

After the voltage drop between two nodes is determined, one calculates the unknown currents by the generalized Ohm law.

Example. We illustrate the use of the nodal-pair method by anthe example of the electrical network shown in Fig. E6.15.

The voltage between two nodes is equal to

$$U_{12} = \frac{E_1 g_1 - E_2 g_4 + J}{g_1 + g_2 + g_4},$$

where $g_1 = 1/R_1$, $g_2 = 1/(R_2 + R_3)$, and $g_3 = 1/R_4$.

▶ **Proportional parts method.** Sometimes, the currents in a circuit containing only one source can be found in a relatively simple way by using the proportional parts method. The essence of this method is that in one of the branches, usually in the most remote

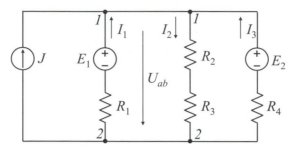

Figure E6.15. Electrical network illustrating the nodal-pair method.

from the energy source, an arbitrary value of the current strength is assumed, usually equal to 1 A. Then, starting from this current, one calculates the currents and voltages in the other branches, as a rule, by using Ohm's law; these calculations are performed until the energy source voltage corresponding to the accepted values of currents in branches is found. After this, the coefficient of proportionality between the actual voltage U_F and the calculated source voltage U' is derived: $k = U_F/U'$. To determine the true values of currents in all the branches, one must multiply the current values obtained earlier by the proportionality coefficient k: $I_m = kI'_m$.

E6.3. Nonlinear Electric Circuits of Direct Current

E6.3.1. Notion of a Nonlinear Element. Several Typical Characteristics of Nonlinear Elements

▶ **Nonlinear elements and their characteristics.** The electric circuits used in automatics, telemechanics, measurement and computation technology often contain separate passive elements whose electric resistance depends on the value and direction of the flowing current or the applied voltage. Such elements and the electric circuits containing them are said to be *nonlinear* and are used to determine certain functional dependencies between different electric quantities.

The nonlinear elements can be *uncontrolled* or *controlled*. The first ones, impedors, are intended to operate without any actions on them from a controlled factor; the second ones, multiterminal devices, are used under the action of controlled factor on them (transistors, thyristors, etc.).

The electric properties of nonlinear elements are represented by their volt-ampere characteristics (VACh) $I = f(U)$. In the uncontrolled nonlinear elements, the VACh is shown by a single curve, and in the controlled, by a family of curves whose parameter is the control factor. The VACh of nonlinear elements have different shapes and, according to this, can be divided into *symmetric* and *nonsymmetric*. The electric resistance of linear elements is constant and the VACh is a straight line passing through the origin (Fig. E6.16.).

In symmetric nonlinear elements, a change in the direction of the applied voltage U does not change the value of the current I, but changes its direction, and hence the relation $f(U) = -f(-U)$ holds for their VACh (Fig. E6.17a).

In nonsymmetric nonlinear elements, for the same magnitude of the voltage U applied in the opposite directions, the currents I_1 and I_2 are different in values and the inequality $f(U) \neq -f(-U)$ holds for VACh (Fig. E6.17b).

▶ **Static and differential resistances of a nonlinear element.** For each nonlinear element, the static resistance R_C corresponding to a given VACh point can be distinguished.

It follows from Fig. E6.18 that the value of this resistance for point A is proportional to the tangent of the angle β between the straight line connecting point A with the origin and

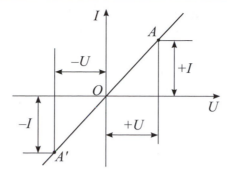

Figure E6.16. The volt-ampere characteristic of a linear element.

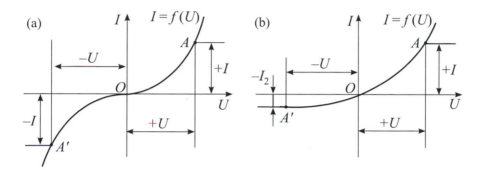

Figure E6.17. The volt-ampere characteristic of a nonlinear symmetric (a) and a nonsymmetric (b) element.

the axis of currents, i.e.,

$$R_S = \frac{U}{I} = \frac{m_U \, OB}{m_I \, BA} = m_R \tan \beta,$$

where m_U, m_I, and m_R are the voltage, current, and resistance scales, respectively.

The limit of the ratio of the voltage increment on the circuit part to the current increment on it, or the derivative of the voltage with respect to the current, determines the differential resistance. The value of this resistance is proportional to the tangent of the angle α between the tangent to VACh at the same point A and the axis of currents, i.e.,

$$R_D = \frac{dU}{dI} = \frac{m_U \, OB}{m_I \, CA} = m_R \tan \alpha.$$

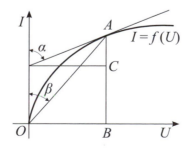

Figure E6.18. Determining the static and differential resistances of a nonlinear element.

The static resistance characterizes the behavior of a nonlinear element in the constant current operation mode, and the differential resistance, in the case of the current small deviations from the steady state.

Consider a closed circuit consisting of a nonlinear element and an active impedor with the external characteristic $U = E - R_0 I$. Point A of intersection of the external characteristic of an active impedor (line 1) with the VACh nonlinear resistive element (curve 2) determined the operation mode circuit (Fig. E6.19).

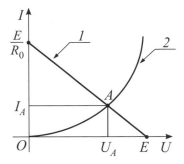

Figure E6.19. The volt-ampere characteristics of an active impedor and a nonlinear resistive element.

E6.3.2. Equivalent Transformations of Characteristics of Nonlinear Elements

If the VACh of a nonlinear element on the operation segment is practically linear, then, in calculation, the nonlinear element can replaced by an equivalent network consisting of a voltage source and a linear resistance R_D. So, for example, the VACh of a nonlinear element shown in Fig. E6.20 on small parts near the operation point A can be replaced by the straight line whose equation has the form

$$U_1 = E_1 + m_R \tan \alpha_1 I \qquad \text{or} \qquad U_1 = E_1 + R_{D1} I. \qquad \text{(E6.3.2.1)}$$

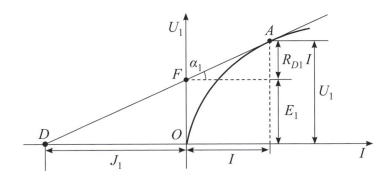

Figure E6.20. The volt-ampere characteristic of a nonlinear element.

At the operating point A, the voltage U_1 and the current I on the nonlinear element (Fig. E6.21a) are related by the expression (E6.3.2.1). The equivalent networks of such an element are shown in Figs. E6.21b and c. In this case, the EMF E_1 is opposite to the current I, because only in the case of this EMF direction the potential of point 1 exceeds

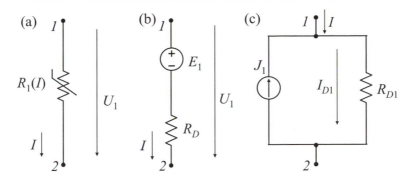

Figure E6.21. Equivalent circuits for replacing a nonlinear element.

the potential of point 2 by U_1. The value of J_1 is equal to the segment OD cut off on the axis of currents by the continuation of the tangent.

If a nonlinear element has the VACh shown in Fig. E6.22, then Eq. (E6.3.2.1) has the form

$$U_2 = -E_2 + m_R \tan \alpha_2 \, I \quad \text{or} \quad U_2 = -E_2 + R_{D2}I.$$

If the current and voltage of a nonlinear element have the same positive directions (Fig. E6.21a), then the directions of the EMF and of the current in the current source are changed to the opposite in the equivalent networks.

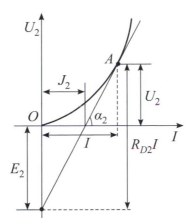

Figure E6.22. The volt-ampere characteristic of a nonlinear element.

E6.3.3. Graphical Method for Analyzing Circuits with Nonlinear Elements

▶ **Preliminary remarks.** Nonlinear electric circuits are often analyzed by the graphical method based on Kirchhoff's laws and the VACh of some elements contained in the electric circuit. The initial data for calculations (VACh) are given in the form of graphs and tables.

Remark. In the case of nonlinear circuits, one cannot use the theorems and methods based on the assumption that all the elements contained in these circuits are linear. So the superposition principle cannot be used in nonlinear circuits. The linearity theorem, the reciprocity theorem, and the proportional parts method also cannot be used.

We consider several problems with different versions of connection of nonlinear elements in nonlinear circuits whose networks are shown in Fig. E6.23.

▶ **Series connection of nonlinear elements.** Figure E6.23a shows the series connection of two nonlinear elements $R_1(I_1)$ and $R_2(I_2)$ whose VACh are given in Fig. E6.24. It is required to determine the current I and the voltages U_1 and U_2 of elements for a given voltage U at the circuit terminals.

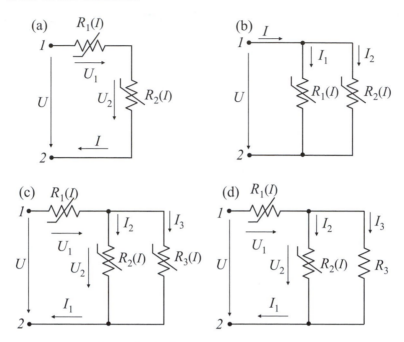

Figure E6.23. Examples of nonlinear circuits.

To calculate the current I and the voltages U_1 and U_2, one must construct an auxiliary characteristic, namely, the dependence of the current on the total voltage $U_1 + U_2$. Since, just as in an unbranched circuit, the current is the same in both nonlinear elements, it follows that, to construct the characteristic $I(U_1 + U_2)$, it is necessary to sum the voltages U_1 and U_2 for equal values of the current I.

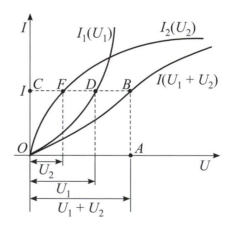

Figure E6.24. The volt-ampere characteristics of nonlinear resistive elements connected in series.

We put the voltage U on the abscissa axis and, from point A, draw the straight line AB parallel to the ordinate axis till the intersection with the curve $I(U_1 + U_2)$; the segment AB thus obtained is equal in scale to the current I. Then, from point B, we draw the straight line BC parallel to the ordinate axis. As a result, we obtain segments CD and CF equal in scale to U_1 and U_2, respectively. Similarly, we can solve the problem of analyzing a circuit consisting of three or more elements with nonlinear characteristics connected in series.

The graphical constructions for analyzing a series circuit can be performed differently. We assume that, in Fig. E6.23a, the ends of a circuit part are connected by the EMF source $E = U$. The voltage U_2 at the terminals of a nonlinear element $R_2(I_2)$ is determined, on one hand, by the VACh of this element, and, on the other hand, by the difference between the EMF $E = U$ and the voltage U_1 at the terminals of the nonlinear element $R_1(I_1)$.

Figure E6.25 presents the characteristic and the curves $I_2(U_2)$ and $I_1(U - U_2)$ at whose point of intersection the currents in the elements coincide.

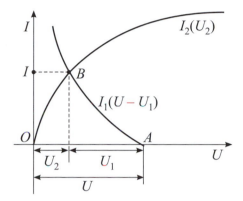

Figure E6.25. Curve $I_2(U_2)$ and curve $I_1(U - U_2)$.

▶ **Parallel connection of nonlinear elements.** In Fig. E6.23b, we show the parallel connection of two nonlinear elements $R_1(I_1)$ and $R_2(I_2)$ whose VACh are shown in Fig. E6.26. Assume that the voltage at the terminals of the circuit is equal to U. Then, by the VACh $I_1(U_1)$ and $I_2(U_2)$, it is easy to determine the currents I_1 and I_2 in the nonlinear elements and, by the formula $I_1 + I_2 = I$, the current in the unbranched part of the circuit.

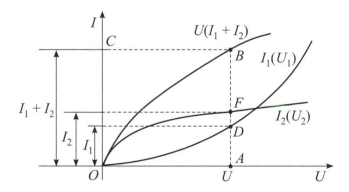

Figure E6.26. The volt-ampere characteristics of nonlinear resistive elements connected in parallel.

If a current I is given, then, to determine the voltage U and the currents I_1 and I_2 in nonlinear elements, one must construct the auxiliary quantity $(I_1 + I_2)U$, i.e., the dependence of the total current on the voltage U. Since, in parallel connection, $U_1 = U_2 = U$, it follows that, to construct this quantity, it is necessary to sum the ordinates of the curves $I_1(U_1)$ and $I_2(U_2)$ for the same values of the voltage. Then, on the ordinate axis, we put the interval OC equal in scale to the current I and, from point C, draw the straight line CB parallel to the abscissa axis till the intersection with the characteristic $(I_1 + I_2)U$. The segment $CB = OA$ thus obtained is equal in scale to the voltage U. Finally, the segments AD and AF are equal in scale to the currents I_1 and I_2.

▶ **Mixed connection of nonlinear elements.** In Figs. E6.23c and d, we show the mixed connection of nonlinear elements (c) and of nonlinear and linear elements (d). To determine the desired currents and voltages, one can also use the graphical method by analogy with calculations of series and parallel circuits.

E6.4. Magnetic Circuits

A set of devices containing ferromagnetic bodies and forming a closed circuit in which a magnetic flux is formed in the presence of a magnetizing force and along which the magnetic flux lines close up, is called a magnetic circuit. The magnetic circuits, just as the electric circuits, can be branched and unbranched.

E6.4.1. Magnetic Field and Its Parameters

▶ **Magnetic induction vector.** The magnetic field arising when a direct current flows through a conducting medium is graphically shown by magnetic lines of force and is characterized at each point by the magnetic induction vector **B** directed along the tangent to the field lines (Fig. E6.27).

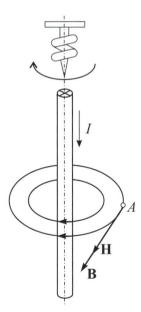

Figure E6.27. Magnetic field of a rectilinear conductor and an induction coil.

The direction of this vector is determined by the right-hand screw rule (corkscrew rule), which means that if the translational motion of such a screw is aligned in direction with the electric current I, then the direction of its rotational motion coincides with that of the magnetic induction vector **B**.

If the magnetic induction **B** has the same value and direction at all points of the field, then such a field is said to be *uniform* or *homogeneous*.

In the international system of units, or the SI system, the magnetic induction (flux density) B is measured in teslas (T) or webers per square meter (Wb/m^2).

▶ **Magnetic flux.** The second important quantity characterizing the magnetic field is the magnetic flux Φ, i.e., the flux of the magnetic induction vector **B** through a surface S (Fig. E6.28a):

$$\Phi = \int \mathbf{B} \cdot d\mathbf{S} = \int B \cos(\mathbf{B}, d\mathbf{S}) \, dS.$$

The quantity Φ can take either positive or negative value depending on the value of the angle between the directions of the vectors **B** and $d\mathbf{S}$.

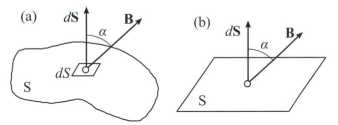

Figure E6.28. Determining the magnetic flux through (a) an arbitrary surface or (b) a plane surface in a uniform magnetic field.

If a magnetic field is uniform (homogeneous), then, for all points of space, the magnetic induction vector is constant, **B** = **const**, and the magnetic flux is determined by the formula (Fig. E6.28b):

$$\Phi = BS \cos \alpha. \tag{E6.4.1.1}$$

In SI, the magnetic flux is measured in webers (Wb).

▶ **Magnetic field strength vector. Absolute and relative magnetic permeability.** When studying magnetic fields and analyzing magnetic devices, one uses the magnetic field strength (intensity) vector **H**, which coincides in direction with the magnetic induction vector **B**:

$$\mathbf{B} = \mu_a \mathbf{H} = \mu \mu_0 \mathbf{H},$$

where μ_a is the absolute magnetic permeability of the medium, $\mu_0 = 4\pi \cdot 10^{-7}$ H/m (henry/meter) is the magnetic constant (permeability of vacuum), and μ is the (relative) magnetic permeability.

For nonferromagnetic materials and media (wood, paper, copper, aluminum, air), μ_a does not practically differ from μ_0 ($\mu = 1$).

For ferromagnetic materials (steel, cast iron, etc.), μ_a is many times larger than μ_0 and depends on the value of the magnetic field strength H.

In SI system, the magnetic field strength H is measured in amperes per meter (A/m).

E6.4.2. Ampère's Circuital Law

The calculations of magnetic circuits are based on Ampère's circuital law, which was obtained on the basis of numerous experiments. This law states that the linear integral of the magnetic field strength vector along any loop is equal to the total current surrounded by this loop (algebraic sum of currents related to the loop):

$$\oint \mathbf{H} \cdot d\mathbf{l} = \oint H \cos \alpha \, dl = \sum I_k,$$

where **H** is the magnetic field strength at a given point in space, $H = |\mathbf{H}|$, dl is a length element of the closed loop, and $\cos \alpha$ is the angle between the directions of the vectors **H** and $d\mathbf{l}$.

The currents are assumed to be positive if their direction corresponds to the corkscrew rule (is clockwise).

In particular, according to Ampère's circuital law, for the loop in Fig. E6.29, we have

$$\oint \mathbf{H} \cdot d\mathbf{l} = \oint H \cos \alpha \, dl = -I_1 + I_2 + I_3.$$

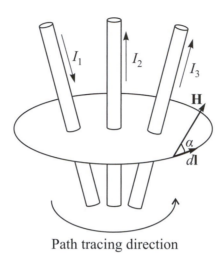

Path tracing direction

Figure E6.29. Ampère's circuital law.

The quantity $\sum I_k$ is called the magnetomotive force (MMF); in the SI system, the main unit of MMF is ampere (A). The main unit of the magnetic field strength in SI system is ampere per meter (A/m) and does not have any special name. One often uses the "ampere per centimeter" unit, $1\,\text{A/cm} = 100\,\text{A/m}$.

E6.4.3. Ohm's Law for Magnetic Circuit

To obtain the dependence between the magnetic flux and the MMF, we consider an example of unbranched circuit shown in Fig. E6.30.

We assume that the parts l_1 and l_2 of the magnetic circuit are made of the same ferromagnetic material but have different transverse cross-sections S_1 and S_2.

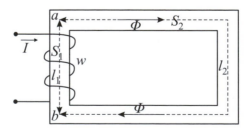

Figure E6.30. Unbranched circuit.

If the lengths of the magnetic circuit parts (l_1 and l_2) significantly exceed the transverse dimensions, then we can assume that the magnetic induction at all points of each part is the same.

The magnetic induction for two parts of the circuit is determined by the formulas

$$B_1 = \frac{\Phi}{S_1}, \qquad B_2 = \frac{\Phi}{S_2}.$$

The magnetic field strength on these parts is calculated as

$$H_1 = \frac{B_1}{\mu_0 \mu_1} = \frac{\Phi}{\mu_0 \mu_1 S_1}, \qquad H_2 = \frac{B_2}{\mu_0 \mu_2} = \frac{\Phi}{\mu_0 \mu_2 S_2}. \tag{E6.4.3.1}$$

Let us apply Ampère's circuital law to the loop coinciding with the center magnetic line of length $l = l_1 + l_2$. We obtain

$$F = \oint H \, dl = H_1 l_1 + H_2 l_2 = I w, \tag{E6.4.3.2}$$

where w is the number of turns of the current loop. In this equation, by analogy with the EMF of an electric circuit, $F = I w$ is called the magnetomotive force or MMF.

Substituting H_1 and H_2 from (E6.4.3.1) into (E6.4.3.2), we find

$$F = I w = \frac{\Phi l_1}{\mu_0 \mu_1 S_1} + \frac{\Phi l_2}{\mu_0 \mu_2 S_2} = \Phi R_{M1} + \Phi R_{M2}.$$

The quantity $l/(\mu_a S) = R_m$ is usually called the *magnetic resistance* of a magnetic circuit (by analogy with the electric resistance $R = l/\gamma S$).

The magnetic resistance of air (gaps) is linear, because $\mu_a = \mu_0 = \text{const}$. The magnetic resistance of the core is nonlinear, because μ_a depends on B.

The product of the magnetic flux Φ by the magnetic resistance R_M is called, by analogy with electric circuit, the *magnetic voltage*:

$$U_{Mab} = \Phi R_{M1},$$

whence we obtain the magnetic flux

$$\Phi = \frac{I w}{R_{M1} + R_{M2}} = \frac{I w - U_{Mab}}{R_{M1}} = \frac{U_{Mab}}{R_{M2}}.$$

This equation is usually called *Ohm's law for a magnetic circuit*. It should be noted that this analogy is purely formal, and the physical essence of processes in electric and magnetic circuits is different.

E6.4.4. Ferromagnetic Materials and Their Properties

▶ **Ferromagnetic materials. Hysteresis loop.** It is known that the magnetic permeability of ferromagnetic materials is a variable quantity and depends on B, and hence the magnetic resistance R_M is also nonconstant, which significantly complicates the calculations of magnetic circuits. Therefore, to calculate the magnetic circuits with ferromagnetic parts, it is necessary to know the magnetization curves, which present the dependence $B = f(H)$. These dependencies are obtained experimentally by testing closed magnetic circuits with distributed winding.

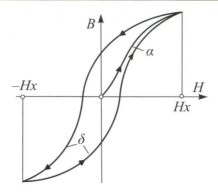

Figure E6.31. Dependence $B = B(H)$ and the hysteresis loop.

The primary magnetization of a specimen is described by the curve μ, which is called the curve of primary magnetization (Fig. E6.31).

If a specimen is cyclically magnetized with the magnetic field strength varying in the limits from Hx to $-Hx$, then the graph is a closed curve, which is known as the hysteresis loop.

If the process of cyclic magnetization is repeated for gradually increasing values of the magnetic field strength, then one can obtain a family of hysteresis loops and the so-called limit hysteresis loop.

To the limit loop there corresponds a variation in the magnetic field strength in the limits from H_{max} to $-H_{max}$. Any increase in H beyond H_{max} does not increase the hysteresis loop area. Such a hysteresis loop determines the value of the residual magnetic induction. The curve connecting the vertices of a hysteresis loops is called the normal magnetization curve. These curves are given in reference books and are used to analyze magnetic circuits.

▶ **Soft magnetic and hard magnetic materials.** The process of cyclic magnetization reversal requires some energy, which is proportional to the hysteresis loop area.

In this connection, for the magnetic circuits in electrical engineering devices operating under conditions of uninterrupted magnetization reversal (for example, transformers) it is expedient to use ferromagnetic materials with a narrow hysteresis loop (see curves α in Fig. E6.32). Such ferromagnetic materials are said to be *soft magnetic* (electrical grade sheet and several special alloys, for example, permalloy consisting of nickel, iron, and other components).

In production of permanent magnets, it is recommended to use ferromagnetic materials with a wide hysteresis loop (curves δ), which have a large residual induction. Such ferromagnetic materials are said to be *hard magnetic* (several alloys of iron with wolfram, chrome, and aluminum).

E6.4.5. Kirchhoff's Laws for Magnetic Circuits

▶ **Kirchhoff's first law for magnetic circuits.** The magnetic flux Φ completely closed by a magnetic circuit is called the *main flux*, and the magnetic flux closed partially by parts of a magnetic circuit and partially by the environment is called the *leakage flux*.

It is well known that the lines of the magnetic induction vector are closed. Therefore, the magnetic flux through any closed surface is always zero. Hence, the following equation holds for nodes of magnetic circuits:

$$\sum \Phi_k = 0. \qquad\qquad (E6.4.5.1)$$

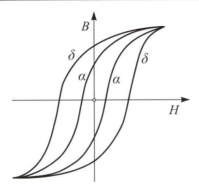

Figure E6.32. Hysteresis loops of soft magnetic (curves α) and hard magnetic (curves δ) materials.

In this equation, the positive signs are taken for the fluxes with the positive direction away from the node, and negative signs are taken for the fluxes with direction towards the node of the magnetic circuit (or vice versa), just by analogy with Kirchhoff's first law for direct current circuits.

Equation (E6.4.5.1) expresses Kirchhoff's first law for magnetic circuits: the algebraic sum of magnetic fluxes at a node of a magnetic circuit is zero.

▶ **Kirchhoff's second law for magnetic circuits.** Using Ampère's circuital law, one can obtain an equation by analogy with Kirchhoff's second law for electric circuits. In Fig. E6.33, we show a branched magnetic circuit. This circuit consists of six unbranched parts and has the same number of distinct magnetic fluxes. Each part of the magnetic circuit is made of the same material with constant transverse cross-section on its entire length. But different parts of this circuit can be produced of different ferromagnetic materials and can have different transverse cross-sections. Neglecting the leakage fluxes and assuming that the lengths of the parts significantly exceed their transverse dimensions, we can determine the magnetic inductions.

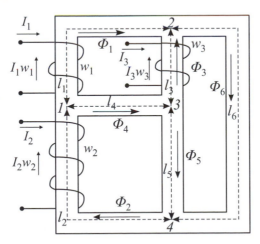

Figure E6.33. Branched magnetic circuit.

The magnetic field strengths on these parts of the magnetic circuit can be found from the magnetization curves under the obtained values of the magnetic induction. We apply Ampère's circuital law, for example, to close the loop 1-2-3-1 of the magnetic circuit (bypassed clockwise) and obtain

$$\sum F_k = \sum I_k w_k = I_1 w_1 - I_3 w_3 = H_1 l_1 - H_3 l_3 - H_4 l_4.$$

Taking into account the fact that

$$H_1 = \frac{\Phi_1}{\mu_0\mu_1 S_1}, \quad H_3 = \frac{\Phi_3}{\mu_0\mu_3 S_3}, \quad H_4 = \frac{\Phi_4}{\mu_0\mu_4 S_4},$$

we have

$$\sum F_k = \frac{\Phi_1 l_1}{\mu_0\mu_1 S_1} - \frac{\Phi_3 l_3}{\mu_0\mu_3 S_3} - \frac{\Phi_4 l_4}{\mu_0\mu_4 S_4} = \Phi_1 R_{M1} - \Phi_3 R_{M3} - \Phi_4 R_{M4}.$$

Generalizing the last formula to any closed loop, we obtain *Kirchhoff's second law for a magnetic circuit*:

$$\sum F_k = \sum I_n w_n = \sum \Phi_p R_{Mp} = \sum H_q l_q;$$

i.e., the algebraic sum of MMF in a loop of a magnetic circuit is equal to the algebraic sum of magnetic voltages in the same loop.

Here it should be specially noted that the resistance of any part of the magnetic circuit cannot usually be considered as a constant variable because of the dependence of the magnetic permeability of ferromagnetic materials on the induction. Therefore, the analysis of magnetic circuits is similar to the analysis of electric circuits with nonlinear VACh.

E6.4.6. Analysis of an Unbranched Magnetic Circuit

When analyzing an unbranched magnetic circuits, we encounter problems of two types.

In some problems, it is necessary to determine the magnetizing current from a given magnetic flux, in other problems, the flux from a given current or a given MMF. In both cases, as a rule, the geometric dimensions of all parts of magnetic circuits, the material of which they are produced, the normal magnetization curves or hysteresis loops, and the number of turns of the induction coil are given. In magnetic circuits composed of soft magnetic materials, hysteresis can usually be neglected; i.e., one can assume that the dependence of the induction on the magnetic field strength is unique and determined by the normal magnetization curve.

▶ **First problem for unbranched magnetic circuits.** Consider the unbranched magnetic circuit shown in Fig. E6.34, which consists of a π-shaped electromagnetic core and a steel plate closing its terminals. There is an air gap between the electromagnetic core terminals and the plate. It is required to determine the magnetizing current at which the magnetic flux in the air gap has a prescribed value. The electromagnet cross-section is S_1, and the plate cross-section is S_2.

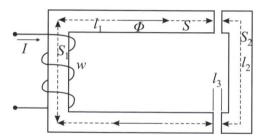

Figure E6.34. Unbranched magnetic circuit.

Under these conditions, we represent an entire magnetic circuit as three parts connected in series: the electromagnetic core, two air gaps, and the steel plate with the same flux Φ. To calculate the magnetic circuit, we determine the lengths of separate parts of the magnetic circuit: l_1, l_2, $2l_3$. For small dimensions of the core transverse cross-sections compared with their length, small inaccuracies in determining the center lengths of the circuit parts do not lead to significant calculational errors. The exact calculations of the magnetic flux distribution in the air gap is a difficult problem, but if the gap length is small and the plane of the ferromagnetic bodies surrounding the gap are parallel, then one can assume that the magnetic field in the gap is homogeneous and its cross-section S_3 is equal to the core cross-section S_1.

We represent the unbranched magnetic circuit as an equivalent network consisting of three series connected resistances R_{M1}, R_{M2}, R_{M3}, and MMF $F = Iw$ (Fig. E6.35a). In this scheme, the resistances R_{M1} and R_{M2} depend on the magnetic flux, and the resistance R_{M3} is a constant quantity. According to Kirchhoff's second law for magnetic circuit, we have

$$F = Iw = H_1 l_1 + H_2 l_2 + 2H_3 l_3.$$

The magnetic inductions B_1, B_2, B_3 are determined by a given value of the magnetic flux by formula (E6.4.1.1)

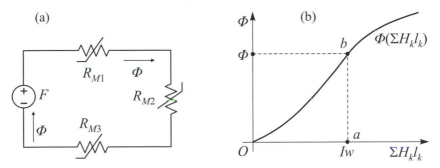

Figure E6.35. (a) Equivalent network of an unbranched magnetic circuit and (b) the auxiliary characteristic $\Phi = f(\sum H_k l_k)$.

From the obtained values B_1 and B_2 of the magnetic induction and the normal magnetization curves for the corresponding materials, we determine the magnetic field strengths H_1 and H_2. For the air gap, we have $H_3 = B_3/\mu_0$. The magnetizing current is calculated as

$$I = \frac{Iw}{w} = \frac{H_1 l_1 + H_2 l_2 + 2H_2 l_3}{w}.$$

▶ **Second problem for unbranched magnetic circuits.** Now we consider the same magnetic circuit for which it is required to determine the magnetic flux from a given value of MMF. In contrast to the preceding problem, this problem does not have a "direct" solution because of the nonlinear relation between the flux and the magnetizing current.

Such a solution can be obtained, for example, by the following method. First, we prescribe a value of the magnetic flux, for example, Φ'. Then, just as in the preceding problem, we find the MMF $F' = I'w = \sum (H_k l_k)'$. If the obtained value of MMF coincides with the given $F = Iw$, then the problem is solved. But such a coincidence is usually not obtained after the first attempt. Therefore, we must prescribe other values of the magnetic flux Φ'', Φ''', etc. and then find the corresponding values F'', F''', etc. and construct the auxiliary characteristic $\Phi = \Phi(\sum H_k l_k)$ shown in Fig. E6.35b.

Then we put the value of the given MMF $F = Iw$ (point a) on the abscissa axis and draw a straight line from this point parallel to the ordinate axis till the intersection with the curve $\Phi(\sum H_k l_k)$ at point b. As a result, we obtain segment ab determining the desired value of the magnetic flux.

It is necessary to point out that, in practical calculations, it does not make any sense to construct the entire curve $\Phi(\sum H_k l_k)$ starting from the zero value of the magnetic flux. To obtain the first point of the curve, we must equate the given MMF with the magnetic voltage on the site with maximal magnetic resistance $R_M \Phi' = I w$ and, using the magnetization curve, determine the maximal flux Φ' from this equation.

If there is an air gap in the magnetic circuit, then most often the gap is the site with maximal magnetic resistance. For the air gap, in the case of the circuit in Fig. E6.34, the last equation can be written as

$$\frac{2l_3}{\mu_0 S_3} \Phi' = I w \qquad \text{or} \qquad \Phi' = \frac{\mu_0 S_3 I w}{2l_3}.$$

Since the other parts of the same unbranched magnetic circuit, just as the part with maximal magnetic resistance, bound the magnetic flux, its subsequent values necessary to construct the curve $\Phi(\sum H_k l_k)$ must be taken less than Φ'.

Since the magnetic flux is the same in all parts of the unbranched magnetic circuit, the quantity $\Phi(\sum H_k l_k)$ can be constructed by analogy with the unbranched electric circuit by graphical summation of the abscissas of the straight line $\Phi(2H_3 l_3)$ and the curves $\Phi(H_1 l_1)$ and $\Phi(H_2 l_2)$ for the same values of the magnetic flux. The quantity $\Phi(2H_3 l_3)$ is described by the straight line passing through the origin. Its construction becomes easier if the magnetic field strength H_3 is found for each value of Φ, $H_3 = \Phi/(\mu_0 S_3)$.

The quantities $\Phi(H_1 l_1)$ and $\Phi(H_2 l_2)$ are constructed by using the normal magnetization curves for the materials of the first and second parts of the magnetizing circuit. For this, it is necessary to multiply the ordinates of the magnetization curves by the respective cross-sections of the first and second parts ($\Phi = BS$) and to multiply the abscissas by their lengths (Hl).

E6.4.7. Analysis of a Branched Magnetic Circuit

Analysis of branched magnetic circuits is based on the use of Kirchhoff's laws for magnetic circuits. Because of a nonlinear relation between the induction and the magnetic field strength for ferromagnetic materials, the analysis of such circuits is often performed by graphic-analytical methods by analogy with methods used to analyze nonlinear electric circuits.

When analyzing a magnetic circuit, just as in the case of an electric circuit, it is necessary to prescribe the positive direction of MMF and of magnetic fluxes (if they are not already given).

In Fig. E6.36, we present an example of a branched magnetic circuit with a single MMF and its equivalent network.

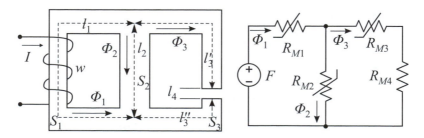

Figure E6.36. Branched magnetic circuit.

For such a magnetic circuit, "direct" computations are possible if the center rod has the same cross-section along the entire length and is produced of the same material as the other rods. The magnetic flux Φ_3 in the air gap must also be given.

From the already known magnetic flux Φ_3, we calculate the induction $B_1 = \Phi/S_1'$; from the magnetization curve, we find the magnetic field strength H_3 and the magnetic field strength in the air gap. The magnetic voltage in the third branch, i.e., between the nodes a and b, is determined as

$$U_{Mab} = \sum H_k l_k = H_3(l_3' + l_3'') + H_4 l_4.$$

Since the second and third branches are connected in parallel, it follows that $U_{Mab} = H_2 l_2$. After H_2 is computed, we find B_2 from the magnetization curve. The fluxes are calculated as follows: $\Phi_2 = B_2 S_2$, $\Phi_1 = \Phi_2 + \Phi_3$. We determine the flux Φ_1, calculate the magnetic induction B_1, and then find the magnetic field strength H_1 from the magnetization curve. According to Kirchhoff's second law, we have

$$F = Iw = H_1 l_1 + U_{Mab}.$$

A more general problem of calculating all fluxes for a given MMF can be solved by a graphical method by analogy with the calculations of direct current circuits with nonlinear elements. With the magnetic leakage fluxes taken into account, such a problem can be solved by the successive approximation method.

E6.4.8. Analysis of a Branched Homogeneous Magnetic Circuit with Permanent Magnet

The analysis of such a circuit cam be reduced to determining the value of magnetic induction in air gap (Fig. E6.37), which depends on the shape of the magnetization curve $B = f(H_M)$ and the relation between the air gap length l_a and the magnetic circuit ferromagnetic part length l_M, i.e., on the value $N = l_a/l_M$, which is the demagnetization factor if the air gap cross-section coincides with the ferromagnetic part cross-section.

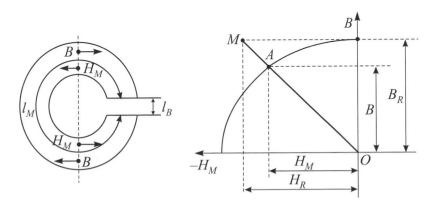

Figure E6.37. Magnetic induction in a magnetic gap and the magnetization curve.

Applying Ampère's circuital law along the center magnetic line of a permanent magnet and taking into account the fact that there is no winding with magnetizing current, we obtain

$$H_M l_M + H_a \delta = 0,$$

where H_M and H_a are the magnetic field strengths in a ferromagnetic material and in the air gap, respectively; and l_M and δ are the respective lengths of the center magnetic line on the same parts.

It follows from the above that

$$H_M = -\frac{\delta}{l_M} H_a = -\frac{\delta}{\mu_0 l_m} B = -\frac{NB}{\mu_0}, \qquad \text{(E6.4.8.1)}$$

where B is the value of the magnetic induction in the air gap and in the magnet body, provided that the air gap cross-section is equal in area to magnet cross-section and there is no magnetic leakage flux.

Since the quantities H_M and B are related to each other by the magnetization curve $B = f(H_M)$ and Eq. (E6.4.8.1) of the straight line OM, it follows that the ordinate of point A, which is the point of intersection of these lines, determines the value of the desired magnetic induction B. To construct the straight line OM, it suffices to connect the origin with point M whose ordinate is B_R, which is the residual magnetic induction, and whose abscissa is H_R.

The same calculations can be performed for a small length of the air gap. Otherwise, the pattern of the magnetic field distribution becomes more complicated, and it is necessary to pass from analytical calculations to experimental studies.

E6.5. Single-Phase Harmonic Current Circuits

E6.5.1. Effective and Average Values of Sinusoidal Quantities

▶ **Effective values of sinusoidal quantities.** The electromotive force e varying in time by a sinusoidal law (see Fig. E6.38) can be obtained by rotating an open rectangular wire loop in a homogeneous magnetic field with constant angular velocity about an axis perpendicular to the field lines.

The equation of sinusoidal EMF has the form

$$e = E_{\mathrm{m}} \sin(\omega t),$$

where E_{m} is its maximal value or the *amplitude*.

Under the action of a sinusoidal EMF, in an electric circuit, there arises a sinusoidal or harmonic current whose instantaneous value is determined by the formula

$$i = I_{\mathrm{m}} \sin(\omega t + \psi).$$

The *sinusoidal current* is a special case of alternating current whose value and direction vary periodically by the sinusoidal law.

The sinusoidal current has significant advantages over the alternating current of any other shape. The derivative and the integral of a sinusoidal function are again sinusoidal functions. Therefore, in the process of transformation of electric energy transferred over a distance, the shape of the curve of sinusoidal current remains the same. The power losses in sinusoidal current circuits are minimal, and the theoretical analysis of sinusoidal current electric circuits are simpler.

A sinusoid is characterized by the following three parameters:

1. The maximal value or amplitude E_{m}.

2. The *period* T, which is the time of one full wave. The number of waves per second is called the *frequency*. The frequency is the inverse of the period, it is denoted by letter f and is measured in hertzs (Hz). Since the rotation of the wire loop by the angle 2π corresponds to

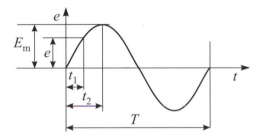

Figure E6.38. Graph of a sinusoidal electromotive force.

the time equal to one period T, it follows that the angular velocity or the *angular frequency* is $\omega = 2\pi f$. The angular frequency is measured in rad/s or s^{-1}.

3. The *phase*, $\omega t + \psi$, is the sine argument. The phase characterizes the state of oscillations at a given time moment t. The angle ψ is equal to the phase at the initial time moment $(t = 0)$ and hence is called the *initial phase*. The *phase displacement* between two sinusoidal quantities with equal frequencies is the difference between their initial phases (in other words, the phase shifting, phase shift, phase difference).

▶ **Average values of sinusoidal quantities.** The average value of a sinusoidally varying quantity is its value averaged over a half-period. For example, the average value of current is determined as

$$\langle I \rangle = \frac{2}{T} \int_0^{T/2} I_{\mathrm{m}} \sin(\omega t)\, dt = \frac{2}{\pi} T_{\mathrm{m}} \approx 0.638\, I_{\mathrm{m}}.$$

The notion of *effective value* of a sinusoidally varying quantity is used very widely. Electric energy transformed into heat energy is proportional to the squared current (the Joule(–Lenz) law). To evaluate the effects of alternating currents, EMF, and voltages, the notion of the quadratic mean over a period, called the *active* or *effective value*, is introduced. For example, the effective value of current is

$$I = \sqrt{\frac{1}{T} \int_0^T i^2\, dt} = \sqrt{\frac{1}{T} \int_0^T I_{\mathrm{m}}^2 \sin^2(\omega t)\, dt} = \frac{I_{\mathrm{m}}}{\sqrt{2}} \approx 0.707\, I_{\mathrm{m}}. \qquad \text{(E6.5.1.1)}$$

As a rule, the electrical measuring instruments of direct estimation show the effective values of the measured quantities.

The effective value of alternating current is numerically equal to the direct current for which, in the time interval equal to one period, the same quantity of heat is released in the resistance R as for the alternating current.

E6.5.2. Representation of Sinusoidal Quantities by Vectors on Complex Plane

▶ **Sum of two sinusoidal quantities. The vector diagram.** In alternating current circuits, Kirchhoff's laws hold for instantaneous values. Their direct application to sinusoidal current circuits leads to addition and transformation of trigonometric functions, and it is convenient to perform these operations by using complex numbers.

The instantaneous values of any sinusoidal quantities, for example, of the currents $i_1 = I_{\mathrm{m}1} \sin(\omega t + \psi_1)$ and $i_2 = I_{\mathrm{m}2} \sin(\omega t + \psi_2)$, can be represented at each time instant t by the projections of the amplitude vectors $I_{\mathrm{m}1}$ and $I_{\mathrm{m}2}$ onto the vertical axis as is shown

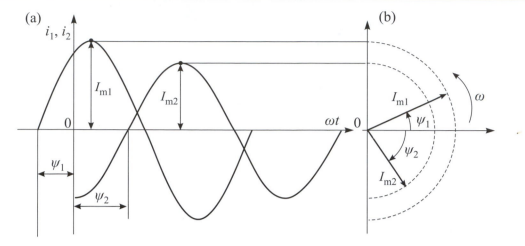

Figure E6.39. Determining the instantaneous value of the sum of two sinusoidal currents (a) by summing the projections of the amplitude vectors (b).

in Fig. E6.39. In this case, the vectors rotate in the plane with constant angular velocity ω about the point 0 anticlockwise. It should be noted that the vectors I_{m1} and I_{m2} rotate about the origin with the same angular velocity ω, and their mutual disposition remains unchanged. Therefore, they are usually represented at time $t = 0$.

The sum of instantaneous values of two sinusoidal currents i_1 and i_2 of the same frequency f can be determined as the projection of the amplitude vector I_m equal to the geometric sum of the summand vectors I_{m1} and I_{m2} and is presented in Fig. E6.40. The entire set of rotating vectors, constructed with preservation of the mutual orientation, is called the vector diagram, permits finding the sum or difference of several harmonic quantities, and reflects the processes in a sinusoidal current electric circuit.

▶ **Vector of current on complex plane. Complex effective value.** A complex number can be represented as a vector on plane with two mutually perpendicular reference axes for quantities: the real axis (denoted by $+1$) and the imaginary axis (denoted by $+j$, where $j^2 = -1$). The length of a vector in a chosen scale characterizes the absolute value (or modulus) of the complex number, and the angle of rotation of the vector about the real axis is the argument of the complex number. The positive angle is counted counterclockwise, the negative angle is counted clockwise.

Remark. In electrical engineering, the imaginary unit is denoted by the symbol j, which differs from its common notation i in mathematics (see Chapter M9). This is due to the fact that the symbol i in electrical engineering usually denotes the current.

A complex number can be written in algebraic, exponential, or trigonometric form. For example, to the current $i_m = I_m \sin(\omega t + \psi)$ represented in Fig. E6.41 by a rotating vector there corresponds the complex number

$$a + jb = I_m e^{j(\omega t + \psi)} = I_m \cos(\omega t + \psi) + j I_m \sin(\omega t + \psi),$$

written by using the Euler formula $e^{j\alpha} = \cos\alpha + j\sin\alpha$. The imaginary part b of this complex number is equal to the current i_m.

The amplitude is determined by the formula $I_m = \sqrt{a^2 + b^2}$, and the phase angle can be found from the projections of the vector onto the axes:

$$\sin(\omega t + \psi) = \frac{b}{\sqrt{a^2 + b^2}}, \qquad \cos(\omega t + \psi) = \frac{a}{\sqrt{a^2 + b^2}}.$$

In what follows, the vectors of various sinusoidal quantities are denoted by bold letters.

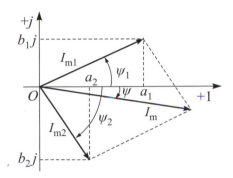

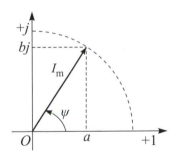

Figure E6.40. Vector summation of current vectors.　**Figure E6.41.** Vector of current on complex plane.

Example 1. Let us determine the resultant current i in Fig. E6.40. To this end, we represent each of the given currents in terms of complex amplitudes

$$\mathbf{I}_1 = I_{m1}e^{j\psi_1}, \quad \mathbf{I}_2 = I_{m2}e^{j\psi_2}.$$

The geometric sum of vectors $\mathbf{I}_1$ and $\mathbf{I}_2$ gives the complex amplitude of the total current, $\mathbf{I} = I_m e^{j\psi}$. The amplitude of the current I_m is determined by the length of the sum vector, and the initial phase ψ is determined by the angle made by this vector with the axis +1.

This example can also be solved analytically by summing the corresponding complex numbers represented in algebraic form:

$$\mathbf{I} = \mathbf{I}_1 + \mathbf{I}_2 = a_1 + jb_1 + a_2 + jb_2 = (a_1 + a_2) + j(b_1 + b_2).$$

The amplitude and the phase angle are determined by the formulas

$$I_m = \sqrt{(a_1 + a_2)^2 + (b_1 + b_2)^2}, \quad \tan\psi = \frac{b_1 + b_2}{a_1 + a_2}.$$

One can see that, in summation and subtraction of complex numbers, it is convenient to use their algebraic representation, and in multiplication (division), it is more convenient to use the exponential representation.

In the analysis of electric circuits, it is often necessary to pass from the algebraic representation of a complex number to the exponential representation or vice versa. In addition to the complex amplitude, the *complex effective value* is also used, which is introduced as follows:

$$\mathbf{I} = \frac{I_m}{\sqrt{2}}e^{j\psi}.$$

Here $\sqrt{2}$ in the denominator arises by analogy with the expression for the effective current (E6.5.1.1).

We consider two numerical examples of passing from the instantaneous value of current to the complex current and conversely.

Example 2. Express the complex effective value of the following current:

$$i = 3\sin\left(\omega t + \frac{\pi}{4}\right).$$

In this example, $I_m = 3$, and the initial phase is $\psi = \pi/4$. Therefore,

$$\mathbf{I} = \frac{I_m}{\sqrt{2}}e^{j\psi} = \frac{3}{\sqrt{2}}e^{j\pi/4} = 2.1213\left(\cos\frac{\pi}{4} + j\sin\frac{\pi}{4}\right) = 1.5 + 1.5\,j.$$

Example 3. Write the expression of the instantaneous value of current if its complex effective value has the form

$$\mathbf{I} = 2e^{-j\pi/3}.$$

We find the amplitude value $I_m = 2 \cdot \sqrt{2} = 2.8284$ and the initial phase $\psi = -\pi/3$. The instantaneous value of the current is $i = 2.8284\sin\left(\omega t - \frac{\pi}{3}\right)$.

Remark. In alternating current circuits, one can sum sinusoidal quantities (currents and voltages) that have the same frequency and different initial phases by using the trigonometric formulas

$$\sum_{k=1}^{n} A_k \sin(\omega t + \psi_k) = \sum_{k=1}^{n} A_k[\cos(\psi_k)\sin(\omega t) + \sin(\psi_k)\cos(\omega t)] = A_c \sin(\omega t) + A_s \cos(\omega t) = A\sin(\omega t + \psi),$$

where

$$A_c = \sum_{k=1}^{n} A_k \cos\psi_k, \quad A_s = \sum_{k=1}^{n} A_k \sin\psi_k, \quad A = \sqrt{A_c^2 + A_s^2}, \quad \sin\psi = \frac{A_s}{A}, \quad \cos\psi = \frac{A_c}{A}.$$

E6.5.3. Active Resistance in a Sinusoidal Current Circuit

▶ **Circuit with active resistance. Vector diagram.** Figure E6.42 presents a circuit with active resistance to which the sinusoidal voltage is applied:

$$u = U_m \sin(\omega t + \psi_U).$$

The current can be determined using Ohm's law for instantaneous values

$$i = \frac{u}{R} = \frac{U_m}{R}\sin(\omega t + \psi_U) = I_m \sin(\omega t + \psi_U).$$

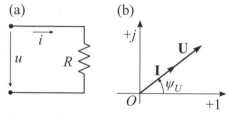

Figure E6.42. (a) A sinusoidal current circuit with active resistance and (b) the vector diagram for such a circuit.

Comparing the expressions for the instantaneous values of voltage and current, one can see that there is no phase shift between voltage and current (Fig. E6.42a), which is determined by the angle equal to the difference of the voltage and current initial phases, i.e.,

$$\varphi = \psi_U - \psi_I = 0.$$

Thus, the current and the voltage have the same initial phase (coincide in phase).

▶ **Instantaneous and average value of power. Ohm's law in complex form.** The instantaneous value of the power of this circuit is equal to the product of instantaneous values of its voltage and current:

$$p = ui = U_m I_m \sin^2(\omega t + \psi_U) = UI[1 - \cos(2\omega t + 2\psi_U)],$$

where $I = I_m/\sqrt{2}$ and $U = U_m/\sqrt{2}$ are the effective current and effective voltage. The instantaneous power is measured by doubled angular frequency 2ω and is all the time positive in the limits from zero to the amplitude value.

We calculate the average value of power over a period:

$$P = \frac{1}{T}\int_0^T p\,dt = UI.$$

Thus, the average value of power in this circuit is determined by the product of the effective values of voltage and current, is called the *active power*, and is measured in watts (W).

Ohm's law in complex form for a circuit with active resistance has the form

$$\mathbf{U} = R\mathbf{I}.$$

E6.5.4. Inductance in a Sinusoidal Current Circuit

▶ **Circuit with inductance. Vector diagram. Inductive reactance.** Consider a circuit (Fig. E6.43) with induction coil of inductance L, whose active resistance is negligibly small ($R_{\text{ric}} \approx 0$). Assume that sinusoidal current flows in this circuit:

$$i = I_m \sin(\omega t - \pi/2) = -I_m \cos(\omega t).$$

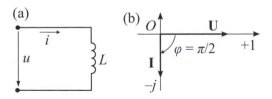

Figure E6.43. (a) A sinusoidal current circuit with inductance and (b) the vector diagram for such a circuit.

The EMF of self-induction induced in the inductor is calculated as

$$e_L = -L\frac{di}{dt} = -\omega L I_m \sin(\omega t) = E_{Lm} \sin(\omega t).$$

Here $E_{Lm} = -\omega L I_m$ is the amplitude.

According to Kirchhoff's second law, we have

$$e_L + u = 0,$$

which implies the instantaneous value of voltage applied to the inductor terminals

$$u_L = -e_L = \omega L I_m \sin(\omega t) = U_{Lm} \sin(\omega t).$$

The effective voltage U and effective current I are related according to Ohm's law as follows:

$$I = \frac{U}{\omega L} = \frac{U}{X_L},$$

where the quantity $X_L = \omega L = 2\pi f L$ is called the *inductive reactance*. The inductive reactance is directly proportional to the coil inductance and the frequency of the current flowing through it.

The angle of phase shift between voltage and current is determined by the difference of their initial phases:

$$\varphi = \psi_U - \psi_I = 0 - (-\pi/2) = \pi/2.$$

It is clear that the inductor voltage phase vector is $\pi/2$ ahead of the current vector.

▶ **Instantaneous value of power. Ohm's law for a circuit with inductance.** The instantaneous power in a circuit with inductance can be found as follows:

$$p_L = ui = -\frac{1}{2}U_m I_m \sin(2\omega t) = -UI \sin(2\omega t).$$

In the first quarter of the period, where the current i and the voltage u are positive (of the same sign), the power is also positive. The energy is transferred from source to circuit and is consumed to create a magnetic field. At the end of the first quarter of the period, the field energy is maximal. In the second quarter of the period, the current i remains positive but decreases, and the voltage u is negative; then the power p_L is negative. The magnetic field energy is transferred back into the circuit. The average value of power in such a circuit is zero:

$$P_L = \frac{1}{T} \int_0^T p_L \, dt = 0.$$

Thus, in a circuit with inductance, the permanent energy oscillation (exchange) between the circuit and the magnetic field occurs without the source energy consumption.

The amplitude power in an inductive circuit is determined by the formula

$$Q_L = UI = I^2 \omega L = 2\pi f L I^2,$$

is called the *reactive inductive power*, is determined by the product of the effective voltage and effective current, and is measured in reactive volt-amperes (RVA).

Ohm's law in complex form for a circuit with inductance has the form

$$\mathbf{U} = j\omega L \mathbf{I} = jX_L\mathbf{I}.$$

E6.5.5. Capacitance in a Sinusoidal Current Circuit

▶ **Circuit with capacitance. Vector diagram. Capacitive resistance.** Assume that the sinusoidal voltage is applied to a capacitor of capacitance C (Fig. E6.44):

$$u = U_{\mathrm{m}} \sin(\omega t).$$

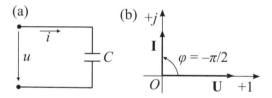

Figure E6.44. (a) A sinusoidal current circuit with capacitance and (b) the vector diagram for such a circuit.

The current flowing through the capacitor is equal to

$$i = C\frac{dU_C}{dt} = \omega C U_{Cm} \cos(\omega t) = I_{\mathrm{m}} \sin(\omega t + \pi/2),$$

where I_{m} is the current amplitude,

$$I_{\mathrm{m}} = \omega C U_{Cm} = \frac{U_{Cm}}{X_C}.$$

The quantity $X_C = \dfrac{1}{\omega C} = \dfrac{1}{2\pi f C}$ is called the *capacitive reactance*. The capacitive resistance is inversely proportional to the frequency of the feeding source and the capacitor capacitance.

The phase shift angle between voltage and current is determined by the difference of their initial phases:

$$\varphi = \psi_U - \psi_I = 0 - \pi/2 = -\pi/2.$$

It is clear that the capacitor voltage phase vector is $\pi/2$ behind the current vector.

▶ **Instantaneous value of power. Ohm's law for a circuit with capacitance.** The instantaneous power in a circuit with capacitance:

$$p_C = ui = UI \sin(2\omega t).$$

In the first quarter of the period, where the voltage u and the current i have the same sign, the capacitor consumes energy from the feeding source, which created the electric field of the capacitor. In the second quarter of the period, the current i is negative, and the capacitor voltage u, still remaining positive, decreases from its maximum to zero, and the energy stored in the electric field is transferred back to the source, i.e., the instantaneous power is negative. Then the processes repeat. Thus, the period-average in a sinusoidal current circuit with capacitance is zero:

$$P_C = \frac{1}{T} \int_0^T p_C \, dt = 0.$$

The power amplitude in a capacitive circuit is determined by the product of the effective voltage and effective current

$$Q_C = UI = U^2 \omega C = 2\pi f C U^2$$

and is called the *reactive capacitive power*; Q_C is measured in reactive volt-amperes (RVA).

Ohm's law in complex form for a circuit with capacitance has the form:

$$\mathbf{U}_C = -jX_C\mathbf{I}.$$

E6.5.6. Unbranched Electric Circuit

▶ **Kirchhoff's second law. Reactive resistance. Voltage and impedance triangles.** If the elements in an unbranched electric circuit do not affect one another, then the circuit can always be reduced to an equivalent circuit with three elements each of which is characterized by one of the parameters R, L, and C, as is shown in Fig. E6.45.

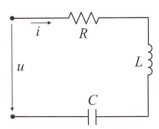

Figure E6.45. Unbranched sinusoidal current circuit.

In the analysis of such circuits, just as in the case of direct current circuits, it is necessary to introduce arbitrary positive directions of voltages and currents.

The instantaneous values of voltages in unbranched circuits always satisfy Kirchhoff's second law equation:

$$u = u_R + u_C + u_L.$$

Kirchhoff's second law equation can be written in complex form:

$$\mathbf{U} = \mathbf{U}_R + \mathbf{U}_C + \mathbf{U}_L. \tag{E6.5.6.1}$$

The voltages on separate elements are calculated by the formulas

$$\mathbf{U}_R = R\mathbf{I}, \quad \mathbf{U}_C = -jX_C\mathbf{I}, \quad \mathbf{U}_L = jX_L\mathbf{I}. \tag{E6.5.6.2}$$

Substituting expressions (E6.5.6.2) into (E6.5.6.1), we obtain the complex voltage at the terminals of a sinusoidal current circuit:

$$\mathbf{U} = R\mathbf{I} - jX_C\mathbf{I} + jX_L\mathbf{I}.$$

This formula can be represented as

$$\mathbf{U} = (R + jX)\mathbf{I},$$

where $X = X_L - X_C$ is the *reactance of the circuit* (reactive resistance), which can be both positive and negative.

The quantity $Z = R + jX$ is called the *circuit impedance* (total complex resistance); it can be represented in exponential form:

$$Z = ze^{j\varphi},$$

where $z = \sqrt{R^2 + X^2}$ is the modulus of the complex number Z and φ is the argument of the complex number Z, which is equal to the phase shift angle in the circuit between voltage and current ($\tan\varphi = X/R$).

In Fig. E6.46, we present the vector diagrams of an unbranched sinusoidal current circuit, which are constructed according to Kirchhoff's second law equation, and in Fig. E6.46a, we consider the case of reactance $X > 0$; i.e., the inductive load predominates in the circuit and the current is by a positive angle φ behind the voltage; in Fig. E6.46b, we consider the case of reactance $X < 0$; i.e., the capacitive load predominates in the circuit and the current is by a negative angle φ ahead of the applied voltage.

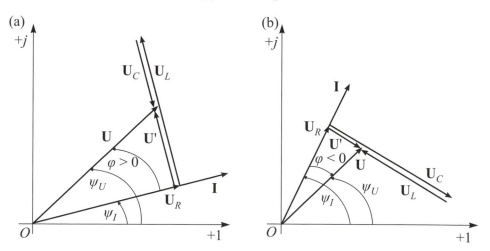

Figure E6.46. Vector diagrams of unbranched sinusoidal current circuit with predominating loads: (a) inductive load; (b) capacitive load.

The rectangular triangle with legs U_R and $U' = U_L + U_C$ and hypotenuse U is called the *voltage triangle*. One can readily pass from the voltage triangle to the similar *impedance triangle* shown in Fig. E6.47 with legs R and jX and hypotenuse Z.

The position of the impedance triangle is independent of the initial phases ψ_U and ψ_I, because the active resistance R is always put on the complex plane in the positive direction

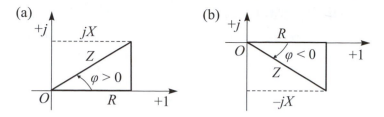

Figure E6.47. Impedance triangles with predominating loads: (a) inductive load; (b) capacitive load.

of the real axis, and the reactance X is put, depending on the sign, in the positive or negative direction of the imaginary axis.

▶ **Instantaneous value of power, active power, total power, and reactive power.** Assume that the voltage initial phase ψ_U is zero and the instantaneous values of voltage and current are respectively described by the formulas

$$u = U_m \sin(\omega t), \qquad i = I_m \sin(\omega t - \varphi).$$

We obtain the following expression for the instantaneous value of power

$$p = iu = U_m I_m \sin(\omega t) \sin(\omega t - \varphi) = \frac{1}{2} U_m I_m [\cos \varphi - \cos(2\omega t - \varphi)]$$
$$= U I [\cos \varphi - \cos(2\omega t - \varphi)].$$

We find the period-average power (or the active power):

$$P = \frac{1}{T} \int_0^T iu \, dt = U I \cos \varphi,$$

which is determined by half the product of the effective values of voltage, current, and power factor.

The power factor is determined by the formula

$$\cos \varphi = \frac{P}{UI} = \frac{P}{S},$$

where the quantity $S = UI$ is called the *total power*; it is measured in volt-amperes (VA) and is the power for given effective values of voltage and current, when the circuit power factor is equal to one, $\cos \varphi = 1$.

In the analysis of electric circuits, the notion of *reactive power* is also used:

$$Q = U I \sin \varphi,$$

which can be both positive (for $\varphi > 0$) and negative (for $\varphi < 0$), and its absolute value is determined by the formula

$$|Q| = \sqrt{S^2 - P^2}.$$

The *complex total power* is determined as

$$\mathbf{S} = \mathbf{U}\bar{\mathbf{I}} = UI e^{j(\psi_U - \psi_I)} = UI e^{j\varphi} = UI \cos \varphi + jUI \sin \varphi = P + jQ,$$

where $\bar{\mathbf{I}} = I e^{-j\psi_I}$ is the complex conjugate current.

E6.5.7. Branched Electric Circuit

▶ **Complex values of currents. The circuit susceptance and the complex admittance.**
In the case of parallel connection of elements with parameters L, C, R attached to a source
of sinusoidal voltage (Fig. E6.48), the following equation following from Kirchhoff's first
law holds:

$$\mathbf{I} = \mathbf{I}_R + \mathbf{I}_C + \mathbf{I}_L;$$

i.e., the complex value of the current in the unbranched part of the circuit is equal to the
algebraic sum of complex values of currents in separate branches.

Since the current in the branch with active resistance R coincides in phase with the
applied voltage, the current in the branch with inductance L is $\pi/2$ behind the voltage, and
the current in the branch with capacitance C is $\pi/2$ ahead of the voltage, we obtain:

$$\mathbf{I}_R = \frac{\mathbf{U}}{R} = g\mathbf{U}, \quad \mathbf{I}_L = -j\frac{\mathbf{U}}{X_L} = -jb_L\mathbf{U}, \quad \mathbf{I}_C = j\frac{\mathbf{U}}{X_C} = jb_C\mathbf{U},$$

where g, b_L, and b_C are, respectively, the (active) conductance, inductive susceptance,
and capacitive susceptance of separate branches of the branched sinusoidal current electric
circuit under study.

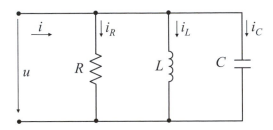

Figure E6.48. Branched alternating current electric circuit.

Let us determine the current in the unbranched part of the circuit:

$$\mathbf{I} = g\mathbf{U} - j(b_L - b_C)\mathbf{U} = (g - jb)\mathbf{U},$$

where $b = b_L - b_C$ is the *circuit susceptance*, which can be both positive and negative.

The quantity

$$Y = g - j(b_L - b_C) = g - jb$$

is called the *complex admittance of a circuit* and can be written in exponential form

$$Y = ye^{-j\varphi},$$

where $y = \sqrt{g^2 + b^2}$ is the modulus of the complex number Y and $\varphi = \arctan(b/g)$ is the
angle of phase shift between the applied voltage and the current in the unbranched part of
the circuit.

The complex admittance Y of any part of the circuit with parameters R, L, C is related
to the complex impedance Z by the inverse relationship:

$$Y = \frac{1}{Z} = \frac{1}{R + jX} = \frac{R}{R^2 + X^2} - j\frac{X}{R^2 + X^2} = \frac{R}{z^2} - j\frac{X}{z^2} = g - jb,$$

where $z = \sqrt{R^2 + X^2}$ is the impedance, $g = R/z^2$ is the conductance, $b = X/z^2$ is the
susceptance.

▶ **Vector diagrams and admittance triangles.** In Fig. E6.49, we present the vector diagrams constructed on the complex plane for an electric circuit with parallel connection of elements with the parameters R, L, C; the first of them corresponds to the case of susceptance $b > 0$, i.e., the inductive load predominates in the circuit and the current in the unbranched part of the circuit is by a positive angle φ behind the voltage, and the second corresponds to the case of susceptance $b < 0$; i.e., the capacitance load predominates in the circuit and the current in the unbranched part of the circuit is by a negative angle φ ahead of the voltage.

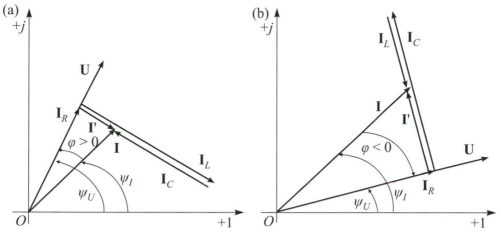

Figure E6.49. Vector diagrams of branched sinusoidal current circuit with predominating loads: (a) inductive load, (b) capacitive load.

Dividing the sides of a rectangular current triangle by the complex voltage, it is easy to obtain the admittance triangle (Fig. E6.50), which is similar to the first but rotated clockwise with respect to it by the angle ψ_I. The admittance triangle position is independent of the initial phases ψ_U and ψ_I, because the conductance g is always put on the complex plane in the positive direction of the real axis, and the susceptance b is put, depending on the sign, in the positive (for $b < 0$) or negative (for $b > 0$) direction of the imaginary axis.

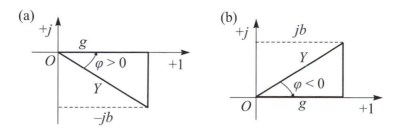

Figure E6.50. Admittance triangles with predominating loads: (a) inductive load, (b) capacitive load.

E6.5.8. Topographic Diagram

In direct current circuits, one constructs graphs of distribution of the potential, which graphically interprets Kirchhoff's second law. In the case of a sinusoidal current circuit, one constructs the topographic vector diagram, which is a set of points on the complex plane representing the complex potentials of the electric network points of the same name with respect to a single point whose potential is conditionally zero. Thus, the order of location of the voltage drop vectors on the diagram strictly corresponds to the order of location of elements in the network.

In Fig. E6.51a, we present the network of an unbranched electric circuit and construct its topographic diagram shown in Fig. E6.51b.

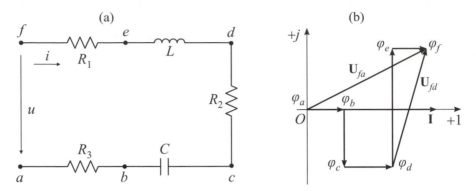

Figure E6.51. Unbranched electric circuit (a) and the topographic diagram (b).

The construction of a topographic diagram starts from calculating the current vector. It is convenient to direct the current along the real axis $\psi_i = 0$. Then one arbitrarily chooses a point whose potential is zero, for example, $\varphi_a = 0$. Thus, point a is at the origin.b

The directions of the voltage vectors on the topographic diagram are shown with arbitrarily chosen direction of the current vector $\mathbf{I}$. The network is traced towards the positive direction of the current $\mathbf{I}$. As we pass to point b, the potential increases by the value of the voltage drop on the resistor R_1 and is equal to

$$\varphi_b = \varphi_a + R_3\mathbf{I} = \mathbf{U}_{ba}.$$

The vector of this voltage drop coincides in phase with the current vector $\mathbf{I}$.

The potential of point c is equal to

$$\varphi_c = \varphi_b + (-jX_C)\mathbf{I}.$$

The voltage $\mathbf{U}_{cb} = (-jX_C)\mathbf{I}$ is by an angle $\pi/2$ behind the current. Similarly, the voltage vectors are constructed for other parts of the circuit. The vector of resultant voltage lies between points a and f.

From the topographic vector diagram, one can readily determine the vector of voltage difference between two arbitrary points of the circuit. For example, to determine the vector of voltage difference between points d and f of the network, it suffices to draw the voltage vector between these points, because

$$\mathbf{U}_{fd} = \varphi_f - \varphi_d.$$

This corresponds to the well-known rule of vector subtraction.

E6.5.9. Resonance Phenomena

▶ **Preliminary remarks.** The operation mode of an electric circuit containing induction coils and capacitors in which the equivalent (input) reactance or the equivalent (input) susceptance is zero is called the *resonance operation mode*. This operation mode is characterized by the relations

$$X_e = 0, \quad B_e = 0.$$

The electric circuit in resonance behaves as the *pure active resistance* with respect to the external circuit; i.e., the voltage and the current at the circuit input are in phase (coincide in phase).

The following two types of resonance are distinguished: the voltage resonance and the current resonance.

▶ **Voltage resonance.** The resonance arising in a circuit (Fig. E6.52), where the real induction coil and capacitor connected in series, is called the voltage resonance.

The complex equivalent impedance of such a circuit is

$$Z_e = R_e + jX_e = R_{ic} + jX_{ic} - jX_C = R_{ic} + j(X_{ic} - X_C) = R_{ic} + j\left(\omega L - \frac{1}{\omega C}\right).$$

Equating the imaginary part with zero ($\operatorname{Im} Z_e = 0$), we find the voltage resonance condition

$$X_e = X_{ic} - X_C = \omega_r L - 1/(\omega_r C) = 0 \implies X_{ic} = X_C \quad \text{or} \quad \omega_r L = 1/(\omega_r C),$$

where $\omega_r = 1/\sqrt{LC}$ is the resonance angular frequency.

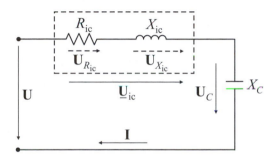

Figure E6.52. Electric network used to study the voltage resonance.

The voltage resonance can be obtained by changing either the frequency of the power supply voltage or the value of the inductance or capacitance. The first case is called a frequency resonance, and the other two are called parametric resonances.

The inductive (or capacitive) susceptance at the instant of the frequency resonance is called the *characteristic impedance*, is denoted by letter ρ, and is determined as

$$\rho = \omega_r L = \frac{1}{\omega_r C} = \sqrt{\frac{L}{C}}.$$

The strength of the current flowing in the circuit at the resonance instant is maximal:

$$I_r = \max I = \frac{U}{Z_e}\bigg|_{X_e=0} = \frac{U}{\sqrt{R_{ic}^2 + (X_{ic} - X_C)^2}}\bigg|_{X_{ic}=X_L} = \frac{U}{R_{ic}}.$$

For certain values of capacitance, the capacitor and coil voltages attain their maximal values $\max U_C$ and $\max U_{ic}$, which are determined from the condition that the derivatives are zero:

$$\frac{dU_C}{dC} = 0, \quad \frac{dU_{ic}}{dC} = 0.$$

If the induction coil is assumed to be ideal (i.e., $R_{ic} = 0$), then the coil and capacitor voltages are equal to each other:

$$X_{ic}I_r = X_C I_r, \quad U_{ic} = U_C.$$

The ideal voltage resonance is equivalent to the case of shorted input terminals of the circuit (since $I_r = U/R_r = U/0 = \infty$).

The ratios of the induction coil (or capacitor) voltage to the input voltage in resonance are called the *circuit quality factor* (Q factor):

$$q = \frac{U_{X_{ic}}}{U} = \frac{U_C}{U} = \frac{\rho I_r}{R_{ic}I_r} = \frac{\rho}{R_{ic}} \quad \text{or}$$

$$q = \frac{U_{X_{ic}}}{U} = \frac{U_C}{U} = \frac{X_{ic}I_r}{R_{ic}I_r} = \frac{X_C I_r}{R_{ic}I_r} = \frac{X_{ic}}{R_{ic}} = \frac{X_C}{R_{ic}}.$$

The quality factor q shows to what extent the induction coil (or capacitor) voltage exceeds the input voltage; it depends on the coil and capacitor parameters.

We write Kirchhoff's second law for this circuit (Fig. E6.52) in complex form:

$$\mathbf{U} = \mathbf{U}_{ic} + \mathbf{U}_C = \mathbf{U}_{R_{ic}} + \mathbf{U}_{X_{ic}} + \mathbf{U}_C = R_{ic}\mathbf{I} + jX_{ic}\mathbf{I} - jX_C\mathbf{I}.$$

We construct the vector diagrams (Fig. E6.53) on the basis of Kirchhoff's second law on the complex plane in the three cases: (a) before resonance ($U_{ic} < U_C$), (b) at the resonance instant ($U_{ic} \cong U_C$), and (c) after resonance ($U_{ic} > U_C$).

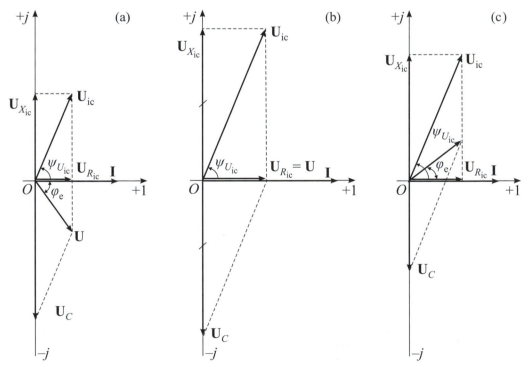

Figure E6.53. Vector diagrams of voltage: (a) before resonance ($U_{ic} < U_C$, $\varphi < 0$), (b) at the resonance instant ($U_{ic} \cong U_C$, $\varphi = 0$), and (c) after resonance ($U_{ic} > U_C$, $\varphi > 0$).

As follows from the vector voltage diagram in the voltage resonance mode, the input current vector I_r coincides in phase with the input voltage vector U; i.e., the phase difference between these vectors is zero:

$$\varphi_e = \varphi_U - \varphi_I = 0.$$

Here ψ_U is the initial phase of the input voltage, and ψ_I is the initial phase of the input current. In resonance, the equivalent reactive power of the entire circuit is zero:

$$Q_e = Q_{ic} - Q_C = X_{ic}I_r^2 - X_C I_r^2 = 0,$$

and the equivalent total power of the circuit becomes the pure active power

$$S_e = \sqrt{P_e^2 + Q_e^2} = \sqrt{P_e^2} = P_e = R_{ic}I_r^2.$$

In analyzing the parametric voltage resonance, as the capacitance value varies, the resonance curves $I(C)$, $U_{ic}(C)$, $U_C(C)$, $\varphi_e(C)$, $\cos\varphi_e(C)$ are constructed by the formulas

$$I(C) = \frac{U}{Z_e} = \frac{U}{\sqrt{R_{ic}^2 + (\omega L - 1/\omega C)^2}},$$

$$U_{ic}(C) = Z_{ic}I = \frac{U\sqrt{R_{ic}^2 + (\omega L)^2}}{\sqrt{R_{ic}^2 + (\omega L - 1/\omega C)^2}},$$

$$U_C(C) = X_C I = \frac{U}{\omega C\sqrt{R_{ic}^2 + (\omega L - 1/\omega C)^2}},$$

$$\varphi_e(C) = \arctan\frac{X_e}{R_e} = \arctan\frac{\omega L - 1/\omega C}{R_{ic}},$$

$$\cos\varphi_e(C) = \frac{R_e}{Z_e} = \frac{R_{ic}}{\sqrt{R_{ic}^2 + (\omega L - 1/\omega C)^2}}.$$

▶ **Current resonance.** The resonance arising in a circuit (Fig. E6.54) where a real induction coil and a capacitor are connected in parallel is called the current resonance.

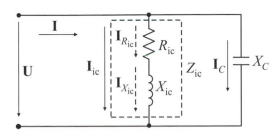

Figure E6.54. Electric network used to study the current resonance.

We find the complex equivalent admittance of such a circuit:

$$\mathbf{Y}_e = \mathbf{Y}_{ic} + \mathbf{Y}_C = \frac{1}{R_{ic} + jX_{ic}} + \frac{1}{-jX_C} = \frac{R_{ic} - jX_{ic}}{(R_{ic} + jX_{ic})(R_{ic} - jX_{ic})} + \frac{j}{j(-jX_C)}$$

$$= \frac{R_{ic}}{R_{ic}^2 + X_{ic}^2} - j\frac{X_{ic}}{R_{ic}^2 + X_{ic}^2} + j\frac{1}{X_C} = G_{ic} - jB_{ic} + jB_C = G_{ic} - j(B_{ic} - B_C),$$

where $G_{ic} = \dfrac{R_{ic}}{R_{ic}^2 + X_{ic}^2}$, $B_{ic} = \dfrac{X_{ic}}{R_{ic}^2 + X_{ic}^2}$, and $B_C = \dfrac{1}{X_C}$ are, respectively, the conductance and the susceptance of the coil and capacitor.

Equating the imaginary part of the obtained expression with zero, $B_{ic} - B_C = 0$, we obtain the current resonance condition

$$\frac{X_{ic}}{R_{ic}^2 + X_{ic}^2} - \frac{1}{X_C} = \frac{\omega_r L_{ic}}{R_{ic}^2 + (\omega_r L_{ic})^2} - \omega_r C = 0,$$

where ω_r is the resonance angular frequency.

The current strength at the resonance circuit input at the resonance instant is imaginary

$$I_r = \min I = Y_e U = \sqrt{G_{ic}^2 + (B_{ic} - B_C)^2}\, U = G_{ic} U.$$

If the induction coil is assumed to be ideal ($R_{ic} = 0$), then the currents in the coil and capacitor are equal to each other, $I'_{ic} = I'_C$, because $Y'_{ic} = B_C U$ and $B'_{ic} U = B_C U$, where

$$Y'_{ic} = \sqrt{(G'_{ic})^2 + (B'_{ic})^2} = \sqrt{0 + (B'_{ic})^2} = B'_{ic} = \frac{X_{ic}}{0 + X_{ic}^2} = \frac{1}{X_{ic}}.$$

The ideal current resonance is equivalent to the circuit break (open-circuit operation mode), because $I_r = G'_{ic} U = 0 \cdot U = 0$.

We write Kirchhoff's first law for the circuit under study in complex form:

$$\begin{aligned}
\mathbf{I} = \mathbf{I}_{ic} + \mathbf{I}_C &= \mathbf{Y}_{ic}\mathbf{U} + \mathbf{Y}_C\mathbf{U} = (G_{ic} - jB_{ic})\mathbf{U} + jB_C\mathbf{U} \\
&= G_{ic}\mathbf{U} - jB_{ic}\mathbf{U} + jB_C\mathbf{U} = \mathbf{I}_{R_{ic}} + \mathbf{I}_{X_{ic}} + \mathbf{I}_C,
\end{aligned}$$

where $\mathbf{I}_{R_{ic}} = G_{ic}\mathbf{U}$ is the active component of the coil current and $\mathbf{I}_{X_{ic}} = -jB_{ic}\mathbf{U}$ is the reactive component of the coil current.

We write Kirchhoff's first law on the complex plane in three cases: (a) before resonance ($I_{ic} > I_C$), (b) at the resonance instant ($I_{ic} \cong I_C$), and (c) after resonance ($I_{ic} < I_C$); i.e., construct the vector diagrams of currents (Fig. E6.55).

As follows from the current diagram (Fig. E6.55), in the current resonance operation mode, the vector of the input voltage coincides in phase with the vector of the input current; i.e., there is no phase shift between these vectors

$$\varphi_e = \psi_U - \psi_I = 0.$$

In resonance, the equivalent reactive power of the entire circuit is zero

$$Q_e = Q_{ic} - Q_C = B_{ic}U^2 - B_C U^2 = 0,$$

and the equivalent total power of the circuit S_e is purely active,

$$S_e = \sqrt{P_e^2 + Q_e^2} = \sqrt{P_e^2} = P_e = P_{ic} = G_{ic}U^2.$$

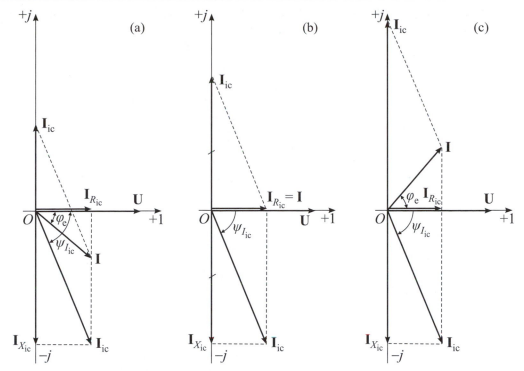

Figure E6.55. Vector diagrams of currents: (a) before resonance ($I_{ic} > I_C$, $\varphi_e > 0$); (b) at the resonance instant ($I_{ic} \cong I_C$, $\varphi_e = 0$); (c) after resonance ($I_{ic} < I_C$, $\varphi_e < 0$).

In the analysis of the parametric current resonance, the resonance curves $I_C(C)$, $I_{ic}(C)$, $I(C)$, $\varphi_e(C)$, $\cos \varphi_e(C)$ are constructed by the formulas

$$I_C(C) = B_C U = \omega C U,$$

$$I_{ic}(C) = Y_{ic} U = \sqrt{G_{ic}^2 + B_{ic}^2}\, U = U \sqrt{\left(\frac{R_{ic}}{R_{ic}^2 + (\omega L)^2}\right)^2 + \left(\frac{\omega L}{R_{ic}^2 + (\omega L)^2}\right)^2} = \text{const},$$

$$I(C) = Y_e U = \sqrt{G_{ic}^2 + (B_{ic} - B_C)^2}\, U = U \sqrt{\left(\frac{R_{ic}}{R_{ic}^2 + (\omega L)^2}\right)^2 + \left(\frac{\omega L}{R_{ic}^2 + (\omega L)^2} - \omega C\right)^2},$$

$$\varphi_e(C) = \arctan \frac{B_e}{G_e} = \arctan\left(\frac{\omega L - \omega C[R_{ic}^2 + (\omega L)^2]}{R_{ic}}\right),$$

$$\cos \varphi_e(C) = \frac{G_e}{Y_e} = \frac{R_{ic}}{\sqrt{R_{ic}[R_{ic}^2 + (\omega L)^2] + \{\omega L - \omega C[R_{ic}^2 + (\omega L)^2]\}^2}}.$$

The current resonance, in contrast to the voltage resonance (which causes the overvoltage in electrical devices) is safe for electrical plants and, in particular, can be used to compensate for the reactive power in them. In current resonance, large currents arise in circuits only if the circuit branches have large susceptance; i.e., the circuits contain large capacitor batteries or powerful induction coil.

E6.5.10. Improvement of the Power Factor

The power factor ($\cos \varphi$) is one of the fundamental energy characteristics of electrical engineering devices.

An increase in $\cos\varphi$ decreases the losses in the electric energy transfer from source to receivers and increases the efficiency factor of devices.

As is known, the power factor can be determined from the power triangle by the formula

$$\cos\varphi = \frac{P}{S} = \frac{P}{\sqrt{P^2 + Q^2}} = \frac{P}{\sqrt{P^2 + (Q_L - Q_C)^2}},$$

where φ is the phase shift between the voltage U and current I vectors, S is the total power of the circuit, P is the active power of the circuit, Q_L is the inductive power, Q_C is the capacitive power, $Q = Q_L - Q_C = UI\sin\varphi = UI_X$ is the reactive power of the circuit, and $I_X = I\sin\varphi$ is the reactive component of the actual current I (this current is called the *reactive current*).

As follows from the last formula, to increase $\cos\varphi$, it is necessary to decrease the reactive power Q, but in alternating current motors (asynchronous motors), the torque is obtained by using the stator and conductors rotating field interaction with the rotor current. Hence such machines need an alternating magnetizing reactive current I_X (i.e., the reactive power $Q = UI_X$) to create the torque moment of the field, and this worsens the power factor $\cos\varphi$ in industrial plant devices. A low value of the power factor results in an incomplete use of power in generators, power transmission lines, and transformers. They are uselessly loaded by the reactive current I_X. The reactive current I_X increases the losses ΔP_{lin} in wires as the electric energy is transferred. These losses consist of the losses arising due to the transfer of the active current I_R and the useless losses due to the transfer of the reactive current I_X. The latter are caused by the energy transmission from magnetic fields of motors to electrical stations and generators and by the reverse transmissions.

The use of the current resonance permits unloading the energy source and the transmitting devices from these useless oscillations of electric energy and hence from the reactive current I_X by closing the electric energy oscillations in the resonance circuit formed by capacitors of capacitance C and a induction coil of inductance L. Practically, this unloading is realized by parallel connection of motors with parameters (R_p, X_p) equivalent to those of a battery of capacitors of capacitance C (Fig. E6.56a).

To compensate for the phase shift completely, the reactive (capacitive) power Q_C of the batteries must be equal to the reactive (inductive) power of motors $Q_L = UI_p\sin\varphi$.

In a majority of cases, the phase shift is compensated for incompletely, because the presence of a small reactive current I_X is of no importance for $\cos\varphi \geq 0.95$, because $I = \sqrt{I_R^2 + I_X^2}$, and the complete compensation requires additional installation of devices with significant conductance (additional batteries of capacitors), which is often economically unprofitable.

Usually, the value of $\cos\varphi$ that the electrical engineering device must have after compensation is prescribed; if the initial value of the current receiver I_p and its $\cos\varphi_p$ are known (as a rule, these values are given in the certificates of electrical engineering devices), then the required value of the capacitance C of the battery of capacitors is determined as follows.

To decrease the phase shift φ_p to the value φ, it is necessary, as the vector diagram shows (Fig. E6.56b), to decrease the resultant reactive current of the device by the quantity $I_{X_p} - I_X$; here I_{X_p} is the device reactive current before compensation, and I_X is the reactive current after compensation.

The active current I_R is related to the reactive current I_X simply as $I_X = I_R\tan\varphi$; moreover, the active current can be expressed in terms of the active power P and the voltage U of the device (receiver) as

$$I_R = \frac{P}{U} = \frac{UI\cos\varphi}{U} = I\cos\varphi.$$

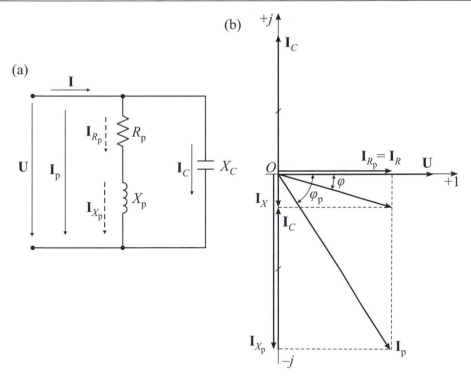

Figure E6.56. Phase shift compensation: (a) network and the vector diagram (b).

The active current does not vary in the course of compensation, $I_{R_p} = I_R = $ const. After the required substitutions, one can express the desired difference of reactive currents as follows:

$$I_{X_p} - I_X = I_R(\tan \varphi_p - \tan \varphi) = \frac{P}{U}(\tan \varphi_p - \tan \varphi).$$

The capacitive current I_C, necessary to perform the compensation must be numerically equal to this difference:

$$I_C = I_{X_p} - I_X = \frac{P}{U}(\tan \varphi_p - \tan \varphi). \tag{E6.5.10.1}$$

By Ohm's law, the capacitive current is related to the capacitance as

$$I_C = \frac{U}{X_C} = \frac{U}{1/\omega C} = \omega C U. \tag{E6.5.10.2}$$

From (E6.5.10.1) and (E6.5.10.2) we obtain the capacitance necessary to perform the compensation:

$$C = \frac{P}{\omega U^2}(\tan \varphi_p - \tan \varphi).$$

The improvement of the power factor ($\cos \varphi$) by including capacitors (batteries of capacitors) is called the *artificial improvement of the power factor* in contrast to the *natural improvement of the power factor*, which is attained by complete use of the power of motors and by installing motors that do not consume reactive current (synchronous motors).

Bibliography for Chapter E6

Atabekov, G. I., *Theoretical Foundations of Electrical Engineering. Linear Electrical Networks* [in Russian], Lan, Moscow, 2009.

Bird, J., *Electrical Circuit Theory and Technology, 2nd Edition*, Newnes, 2003.

Bessonov, L. A., *Theoretical Foundations of Electrical Engineering. Electrical Networks* [in Russian], Vysshaya Shkola, Moscow, 2006.

Bobrow, L. S., *Fundamentals of Electrical Engineering*, Oxford University Press, Oxford, England, 1996.

Chen, W. (Editor), *The Electrical Engineering Handbook*, Academic Press, New York, 2004.

Demirchan, K. S., Neiman, L. R., and Korovkin, N. V., *Theoretical Foundations of Electrical Engineering* [in Russian], Piter, Saint-Petersburg, 2009.

Dorf, R. C. (Editor), *The Electrical Engineering Handbook, Third Edition – 6 Volume Set, 3rd Edition*, CRC Press, Boca Raton, Florida, 2006.

Dorf, R. C. and Tallarida, R. J., *Pocket Book of Electrical Engineering Formulas*, CRC Press, Boca Raton, Florida, 1993.

Hambley, A. R., *Electrical Engineering: Principles and Applications, 5rd Edition*, Prentice Hall, New Jersey, 2010.

Kasatkin, A. S., *Electrical Engineering* [in Russian], Vysshaya Shkola, Moscow, 2000.

Kothari, D. P. and Nagrath, I. J., *Basic Electrical Engineering, 3rd Edition*, McGraw-Hill, New York, 2009.

Kubala, T., *Electricity 1, Devices, Circuits & Materials, 9th Edition*, Delmar Cengage Learning, New York, 2008.

Kubala, T., *Electricity 2, Devices, Circuits & Materials, 9th Edition*, Delmar Cengage Learning, New York, 2008.

Mittle, V. N. and Mittal, A., *Basic Electrical Engineering, 2nd Edition*, McGraw-Hill, New York, 2006.

Nagsarkar, T. K. and Sukhija, M. S., *Basic Electrical Engineering*, Oxford University Press, Oxford, England, 2005.

Nilsson, J. W. and Riedel, S. A., *Electric Circuits, 8th Edition*, Prentice Hall, New Jersey, 2008.

Phadke, A. G., *Handbook of Electrical Engineering Calculations*, CRC Press, Boca Raton, Florida, 1999.

Rizzoni, G., *Fundamentals of Electrical Engineering*, McGraw-Hill, New York, 2008.

Stanley W. D., Hackworth, J. R., and Jones, R. L., *Fundamentals of Electrical Engineering and Technology*, Delmar Cengage Learning, New York, 2006.

Chapter E7

Empirical and Engineering Formulas and Criteria for Their Applicability

E7.1. Empirical, Engineering and Interpolation Formulas. Least-Squares Method

E7.1.1. Empirical and Engineering Formulas

▶ **Preliminaries. Requirements to empirical formulas.** In science and engineering applications, functional dependences are often studied experimentally. The measurement results form a finite set of points, which is usually written as in the form of a table

$$
\begin{array}{|c|c|c|c|c|}
\hline
x_1 & x_2 & x_3 & \cdots & x_n \\
\hline
y_1 & y_2 & y_3 & \cdots & y_n \\
\hline
\end{array}
\tag{E7.1.1.1}
$$

Here x_1, x_2, x_3, ..., x_n are the values of the argument, and y_1, y_2, y_3, ..., y_n are the corresponding values of the function.

If the analytic dependence of y on x is unknown (or is known but very cumbersome), then the following practically important problem arises: find an *empirical formula*

$$
y = f(x)
\tag{E7.1.1.2}
$$

such that for all x_i ($i = 1, 2, \ldots, n$) the values provided by this formula only slightly differ from the experimental data y_i,

$$
|y_i - f(x_i)| < \varepsilon \qquad (i = 1, 2, \ldots, n).
$$

The empirical formula must satisfy the following two fundamental requirements:

(i) It must be sufficiently accurate.

(ii) It must be sufficiently simple. (Of two formulas with the same accuracy, the simpler one is preferable.)

If there is no information about intermediate data, then it is usually assumed that the empirical function is continuous, smooth, and monotone on any interval $[x_i, x_{i+1}]$ (unless $y_i = y_{i+1}$).

Remark. Any experimental data are approximate, because measurements always have certain errors. Therefore, the accuracy of any empirical formula can never be better than that of the experimental data used to construct it.

▶ **Choice of the form of an empirical formula.** The general form of the desired functional dependence (E7.1.1.2) is usually taken a priori from various considerations. At the initial stage, the basic functional dependence must contain a set of free parameters (indeterminate coefficients) whose values are further determined from the condition that the empirical formula is consistent with the experimental data. Linear, rational-fractional rational, power-law, exponential, and (in the case of periodic processes) trigonometric functions are used most frequently. In what follows, we consider the most widely used types of empirical formulas in Sections E7.1.2 and E7.2.

In many cases, if the argument ranges in a bounded interval, one can use a polynomial

$$y = \sum_{i=0}^{m} a_i x^i$$

to construct an empirical formula. The degree m of the polynomial and the indeterminate coefficients a_i can be found from various considerations, which can take into account a priori known boundary or initial conditions.

THEOREM (WEIERSTRASS). *Let $f(x)$ be a continuous function on a finite interval $[a, b]$. Then, for any $\varepsilon > 0$, there exists a polynomial $P(x)$ such that the inequality $|f(x) - P(x)| < \varepsilon$ holds at every point of $[a, b]$.*

Assume that there is a class of qualitatively similar problems or phenomena that differ from each other by geometric characteristics (for example, by the shape of the region under study, the flow structure, etc.) and by the numerical values of physical-chemical parameters. If the dependence (E7.1.1.2) of the desired quantity y on the argument x is known for some specific (for example, the simplest) problem (this dependence may be approximate and may be obtained theoretically or experimentally), then the empirical formula describing a similar quantity for another problem of the same class can be sought in the form

$$y = a_1 f(a_2 x + a_3) + a_4, \tag{E7.1.1.3}$$

where the numerical values of the parameters a_i are determined, for example, by the least-squares method (see Section E7.1.2). Formula (E7.1.1.3) is obtained from (E7.1.1.2) by combining the translation and scaling operations in both variables x and y. In several cases, it suffices to use two-parameter dependences (E7.1.1.3) in which either $a_2 = 1$ and $a_3 = 0$ or $a_3 = a_4 = 0$.

In certain cases, the form of the empirical formula can be chosen on the basis of additional theoretical considerations concerning the character of the dependence under study. (Such formulas are often said to be *semiempirical*.)

▶ **Engineering formulas.** The notion of *engineering formula* is wider than that of *empirical formula*. Engineering formulas are approximate formulas used in applications and obtained by either theoretical (approximate, analytic, or numerical) or experimental methods. Various combinations of theoretical and experimental methods can be used as well.

Ideally, the engineering formula must be sufficiently accurate and sufficiently simple and have a wide scope of applicability.

E7.1.2. Least-Squares Method

▶ **General case.** The *least-squares method* allows approximately determining a function presented in table form (E7.1.1.1) by measurement results.

Consider a functional dependence of the form

$$y = F(x, a_1, a_2, \ldots, a_m), \tag{E7.1.2.1}$$

where $a_1, a_2, \ldots, a_m$ are the desired parameters.

Let us write the sum

$$S = \sum_{i=1}^{n} \left[y_i - F(x_i, a_1, a_2, \ldots, a_m) \right]^2 \tag{E7.1.2.2}$$

of squared deviations, where x_i and y_i are the experimental data presented in the table of Eq. (E7.1.1.1).

The numerical values of the parameters a_i are determined by minimizing the sum (E7.1.2.2). Necessary conditions for the minimum of the expression (E7.1.2.2) lead to the following system of m algebraic (or transcendental) equations for the parameters a_i:

$$\frac{\partial S}{\partial a_1} = 0, \quad \frac{\partial S}{\partial a_2} = 0, \quad \ldots, \quad \frac{\partial S}{\partial a_m} = 0. \tag{E7.1.2.3}$$

These equations may be nonlinear in general.

If system (E7.1.2.3) has a unique solution, then this is the desired solution. If there are several solutions, then one should compare the corresponding values of the sum (E7.1.2.2) and choose the parameter values that provide the minimum of S.

Example 1. For the desired empirical dependence we take the linear function

$$y = ax + b.$$

To calculate the indeterminate coefficients a and b, one minimizes the sum

$$S = \sum_{i=1}^{n} (y_i - ax_i - b)^2.$$

By using the necessary conditions for the minimum of the function $S = S(a, b)$ (i.e., by equating the partial derivatives of S with respect to the parameters a and b with zero), we obtain the system of two linear equations

$$a \sum_{i=1}^{n} x_i^2 + b \sum_{i=1}^{n} x_i = \sum_{i=1}^{n} x_i y_i,$$

$$a \sum_{i=1}^{n} x_i + bn = \sum_{i=1}^{n} y_i.$$

for a and b. Here the values $x_1, x_2, \ldots, x_n$ and $y_1, y_2, \ldots, y_n$ are assumed to be taken from table (E7.1.1.1).

Example 2. If for the desired dependence we take the quadratic function

$$y = ax^2 + bx + c,$$

then we obtain the system of three linear equations

$$a \sum_{i=1}^{n} x_i^4 + b \sum_{i=1}^{n} x_i^3 + c \sum_{i=1}^{n} x_i^2 = \sum_{i=1}^{n} x_i^2 y_i,$$

$$a \sum_{i=1}^{n} x_i^3 + b \sum_{i=1}^{n} x_i^2 + c \sum_{i=1}^{n} x_i = \sum_{i=1}^{n} x_i y_i,$$

$$a \sum_{i=1}^{n} x_i^2 + b \sum_{i=1}^{n} x_i + cn = \sum_{i=1}^{n} y_i.$$

for the coefficients a, b, and c.

▶ **Special case** (a generalization of Examples 1 and 2). System (E7.1.2.3) is simplified dramatically if the empirical function (E7.1.2.1) is taken to be linear in the parameters a_i,

$$y = \varphi_0(x) + a_1 \varphi_1(x) + \cdots + a_m \varphi_m(x). \tag{E7.1.2.4}$$

Then system (E7.1.2.3) is the linear algebraic system that can be written as

$$a_1(\varphi_1, \varphi_1) + a_2(\varphi_1, \varphi_2) + \cdots + a_m(\varphi_1, \varphi_m) = (\varphi_1, Y),$$
$$a_1(\varphi_2, \varphi_1) + a_2(\varphi_2, \varphi_2) + \cdots + a_m(\varphi_2, \varphi_m) = (\varphi_2, Y),$$
$$\dots\dots\dots\dots\dots\dots\dots\dots\dots\dots\dots\dots\dots\dots\dots\dots\dots\dots\dots$$
$$a_1(\varphi_m, \varphi_1) + a_2(\varphi_m, \varphi_2) + \cdots + a_m(\varphi_m, \varphi_m) = (\varphi_m, Y),$$

where

$$(\varphi_j, \varphi_k) = \sum_{i=1}^{n} \varphi_j(x_i)\varphi_k(x_i), \quad (\varphi_j, Y) = \sum_{i=1}^{n} \varphi_j(x_i)Y_i, \quad Y_i = y_i - \varphi_0(x_i).$$

E7.1.3. Refinement of Empirical Formulas

▶ **Statement of the problem.** Assume that the empirical formula (E7.1.1.2) does not agree well with the experimental data (E7.1.1.1). It is required to modify this formula so as to satisfy the experimental data.

▶ **The simplest method for refining empirical formulas.** Here we describe the simplest method for refining an empirical formula if it the sum of squared deviations provided by this formula is larger than desired.

Consider the auxiliary function

$$y = f(x) + a, \tag{E7.1.3.1}$$

where a is an unknown constant. We determine a from the condition that the sum

$$S = \sum_{i=1}^{n} [y_i - f(x_i) - a]^2 \tag{E7.1.3.2}$$

of squared deviations be minimal. A necessary condition for the minimum of the sum (E7.1.3.2) can be found from the condition $S'_a = 0$ and gives

$$a = \frac{1}{n} \sum_{i=1}^{n} \varepsilon_i, \quad \varepsilon_i = y_i - f(x_i), \tag{E7.1.3.3}$$

which ensures the minimum of the sum under study. Here ε_i is the deviation of the experimental data from the results provided by the empirical formula (E7.1.1.2) at the point x_i.

Thus, if $a \neq 0$, then the addition of the constant a determined by formula (E7.1.3.3) to the right-hand side of the empirical formula (E7.1.1.2) results in a refinement of this formula in the sense that the sum of squared deviations decreases. If $a = 0$ or $a \approx 0$, then this method does not give the desired effect. In this case, one can use the more general method described below.

▶ **A more general method for refining empirical formulas.** Instead of (E7.1.3.1), we take a more general auxiliary function

$$y = f(x) + a\varphi(x), \tag{E7.1.3.4}$$

where $\varphi(x)$ is a given function and a is an unknown constant.

Consider the sum

$$S = \sum_{i=1}^{n} [y_i - f(x_i) - a\varphi(x_i)]^2. \tag{E7.1.3.5}$$

A necessary condition for the minimum of this function implies the equation

$$\sum_{i=1}^{n} \varphi(x_i)[y_i - f(x_i) - a\varphi(x_i)] = 0,$$

which permits finding the desired constant a,

$$a = \frac{\displaystyle\sum_{i=1}^{n} \varepsilon_i \varphi(x_i)}{\displaystyle\sum_{i=1}^{n} \varphi^2(x_i)}, \qquad \varepsilon_i = y_i - f(x_i). \tag{E7.1.3.6}$$

Remark 1. Instead of formula (E7.1.3.4), one can refine the empirical formula (E7.1.1.2) by using the more complicated two-parameter dependence

$$y = f(x) + a_1\varphi_1(x) + a_2\varphi_2(x).$$

Remark 2. To refine the empirical formula (E7.1.1.2), one can use the four-parameter dependence (E7.1.1.3).

E7.1.4. Interpolation Formulas

▶ **Linear interpolation. Lagrange interpolation polynomial.** Interpolation formulas are formulas that give the exact values y_m of a function on a given discrete set of values x_m the argument.

Linear interpolation formulas

$$y = \sum_{i=1}^{n} a_i\varphi_i(x), \tag{E7.1.4.1}$$

where $\varphi_1(x), \ldots, \varphi_n(x)$ is a given system of functions, are used most frequently. The coefficients a_i are determined from the condition that the function y coincides with the given values y_m at the respective points x_m,

$$y_m = \sum_{i=1}^{n} a_i\varphi_i(x_m), \qquad m = 1, 2, \ldots, n.$$

For the sequence $\varphi_m(x) = x^{m-1}$ of polynomials, we obtain the function

$$y = L_{n-1}(x) = \sum_{i=1}^{n} y_i \prod_{j \neq i} \frac{x - x_j}{x_i - x_j}, \tag{E7.1.4.2}$$

which is called the *Lagrange interpolation polynomial*. The $(n-1)$st-order polynomial (E7.1.4.2) exactly satisfied the data of table (E7.1.1.1).

Let a function $f(x)$ have an nth-order derivative on a finite interval $[a, b]$. Then the error in replacing $f(x)$ with the Lagrange polynomial $L_{n-1}(x)$ such that $f(x_m) = L_{n-1}(x_m)$ ($m = 1, \ldots, n$) is estimated as

$$|f(x) - L_{n-1}(x)| \leq \frac{1}{n!} \max_{a \leq x \leq b} |f^{(n)}(x)| \, |(x - x_1)(x - x_2) \ldots (x - x_n)|$$

$$\leq \frac{1}{n!} \max_{a \leq x \leq b} |f^{(n)}(x)| \left(\frac{b - a}{n} \right)^n.$$

▶ **Two-point linear interpolation.** In applications, the linear interpolation formula with two points x_k and x_{k+1} is used most frequently. This formula reads

$$y = \frac{y_{k+1} - y_k}{x_{k+1} - x_k}(x - x_k) + y_k \qquad (x_k \leq x \leq x_{k+1}) \tag{E7.1.4.3}$$

and corresponds to the case of $n = 2$ in formula (E7.1.4.2) with x_1 and x_2 redenoted by x_k and x_{k+1}, respectively. Formula (E7.1.4.3) describes the straight line segment passing with endpoints (x_k, y_k) and (x_{k+1}, y_{k+1}) on the (x, y)-plane.

Remark. The interpolation formulas (E7.1.4.2) with $n > 2$ are rather cumbersome and are not used practically to construct empirical formulas on the basis of experimental data (because it does not make any sense to take a function exactly coinciding with the measurement results, which always contain certain errors).

E7.2. Criteria for Applicability of Empirical Formulas

E7.2.1. Direct Method for Verifying Two-Parameter Formulas

▶ **Method of complete elimination of free parameters**. For simplicity and clarity, we illustrate the method of complete elimination of free parameters by an example of two-parameter functions explicitly defined as

$$y = F(x, a, b). \tag{E7.2.1.1}$$

It is required to examine whether formula (E7.2.1.1) can be used to approximate given experimental data by an appropriate choice of the constants a and b.

By specifying the values of the function at three arbitrary (but distinct) points x_i, x_j, and x_k, we obtain the three relations

$$y_i = F(x_i, a, b), \quad y_j = F(x_j, a, b), \quad y_k = F(x_k, a, b). \tag{E7.2.1.2}$$

From now on, we use the short notation $y_i = y(x_i)$, $y_j = y(x_j)$, and $y_k = y(x_k)$. By eliminating the free parameters a and b from (E7.2.1.2), we obtain the criterion relation. It is convenient to write it out as

$$\Theta = 1, \tag{E7.2.1.3}$$

where the left-hand side contains a function

$$\Theta = \Theta(x_i, x_j, x_k, y_i, y_j, y_k). \tag{E7.2.1.4}$$

It is necessary to satisfy relation (E7.2.1.3) for any values of i, j, k ($1 \leq i < j < k \leq n$, where n is the total number of experimental points) for the dependence (E7.2.1.1) to hold. For the applicability of formula (E7.2.1.1) as an empirical formula, it is necessary to satisfy relation (E7.2.1.3) with prescribed accuracy.

In practice, it is convenient to use the special case of formula (E7.2.1.4) with $i = j - 1$ and $k = j + 1$. In this case, it is useful to represent the results of experimental data processing obtained by calculating the function (E7.2.1.4) by points (x_j, Θ_j) on the (x, Θ)-plane and estimate their deviations from the straight line $\Theta = 1$ (see the criterion (E7.2.1.3)), which is parallel to the abscissa axis.

We use the above-described method in the case of two-parameter empirical dependences of special form

$$y = af(x) + b, \qquad (E7.2.1.5)$$

where $f(x)$ is a given function and a and b are free parameters. We take three distinct points x_i, x_j, and x_k to which there correspond three values y_i, y_j, and y_k of the desired quantity. Under the assumption that the functional dependence (E7.2.1.5) takes place, we have

$$y_i = af(x_i) + b, \quad y_j = af(x_j) + b, \quad y_k = af(x_k) + b. \qquad (E7.2.1.6)$$

By eliminating the parameters a and b from these relations, we obtain

$$\frac{y_k - y_i}{y_j - y_i} = \frac{f(x_k) - f(x_i)}{f(x_j) - f(x_i)}. \qquad (E7.2.1.7)$$

By introducing the new auxiliary variable

$$\Theta = \frac{y_k - y_i}{y_j - y_i} \frac{f(x_j) - f(x_i)}{f(x_k) - f(x_i)}, \qquad (E7.2.1.8)$$

we can verify the applicability of the two-parameter approximations of the form (E7.2.1.5), because condition (E7.2.1.3) must be satisfied for them, which follows from relation (E7.2.1.7).

Formula (E7.2.1.8) readily admits computer implementation, and by choosing the function $f(x)$ in (E7.2.1.5) in various ways, one can compare the corresponding approximation errors and then take the dependence with the least error.

Example 1. For the logarithmic dependence

$$y = a \ln x + b \equiv a \ln(cx), \qquad c = \exp(b/a), \qquad (E7.2.1.9)$$

formula (E7.2.1.8) becomes

$$\Theta = \frac{y_k - y_i}{y_j - y_i} \frac{\ln x_j - \ln x_i}{\ln x_k - \ln x_i}. \qquad (E7.2.1.10)$$

For the applicability of two-parameter approximations of the form (E7.2.1.9), relation (E7.2.1.3) must be satisfied under condition (E7.2.1.10).

Example 2. Consider the two-parameter power-law dependence

$$y = ax^m \qquad (x \geq 0, \ y \geq 0). \qquad (E7.2.1.11)$$

Taking the logarithms of both sides of this formula, we obtain a dependence of the form (E7.2.1.5) with $f(x) = \ln x$, $b = \ln a$, and $a = m$ and with y replaced by $\ln y$ on the left-hand side. By substituting $f(x) = \ln x$ into (E7.2.1.8) and by replacing y by $\ln y$, we obtain

$$\Theta = \frac{\ln y_k - \ln y_i}{\ln y_j - \ln y_i} \frac{\ln x_j - \ln x_i}{\ln x_k - \ln x_i}. \qquad (E7.2.1.12)$$

Formula (E7.2.1.12) permits verifying the applicability of power-law functions of the form (E7.2.1.11).

In Table E7.1, we present direct criteria (necessary conditions) for the applicability of two-parameter monotone dependences of the simplest form obtained by the method of complete elimination of free parameters.

TABLE E7.1
Some two-parameter dependences and direct criteria of their applicability of the form
(E7.2.1.3) obtained by the method of complete elimination of free parameters.

No.	Form of empirical formula	Form of criterion function Θ	Remarks
1	$y = ax + b$	$\dfrac{(x_j - x_i)(y_k - y_i)}{(x_k - x_i)(y_j - y_i)}$	
2	$y = \dfrac{a}{x} + b$	$\dfrac{x_j(x_k - x_i)(y_j - y_i)}{x_k(x_j - x_i)(y_k - y_i)}$	
3	$y = \dfrac{1}{ax + b}$	$\dfrac{(x_j - x_i)y_j(y_k - y_i)}{(x_k - x_i)y_k(y_j - y_i)}$	
4	$y = \dfrac{x}{ax + b}$	$\dfrac{x_j(x_k - x_i)y_k(y_j - y_i)}{x_k(x_j - x_i)y_j(y_k - y_i)}$	
5	$y = ax^m$	$\dfrac{\ln(x_k/x_i)\ln(y_j/y_i)}{\ln(x_j/x_i)\ln(y_k/y_i)}$	$x > 0$
6	$y = ae^{\lambda x}$	$\dfrac{(x_k - x_i)\ln(y_j/y_i)}{(x_j - x_i)\ln(y_k/y_i)}$	
7	$y = a \ln x + b$	$\dfrac{(y_k - y_i)\ln(x_j/x_i)}{(y_j - y_i)\ln(x_k/x_i)}$	$x > 0$
8	$y = af(x) + b$	$\dfrac{y_k - y_i}{y_j - y_i}\dfrac{f(x_j) - f(x_i)}{f(x_k) - f(x_i)}$	$f(x)$ is given
9	$y = a[f(x)]^m$	$\dfrac{\ln(y_k/y_i)}{\ln(y_j/y_i)}\dfrac{\ln[f(x_j)/f(x_i)]}{\ln[f(x_k)/f(x_i)]}$	$f(x) \geq 0$ is given
10	$y = a\exp[\lambda f(x)]$	$\dfrac{\ln(y_k/y_i)}{\ln(y_j/y_i)}\dfrac{f(x_j) - f(x_i)}{f(x_k) - f(x_i)}$	$f(x)$ is given

Remark 1. For a majority of two-parameter functions (E7.2.1.1), it is impossible to eliminate free parameters from relations (E7.2.1.2); i.e., it is impossible to obtain the criterion relation (E7.2.1.3) explicitly. This fact does not permit using this method efficiently for more general multi-parameter dependences.

Remark 2. The direct method of elimination of free parameters can be used in the case of multi-parameter dependences of the special form

$$y = h(x)F\left(\frac{\sum_{i=1}^{N} a_i f_i(x)}{\sum_{j=1}^{M} b_j g_j(x)}\right), \tag{E7.2.1.13}$$

where $f_i(x)$, $g_j(x)$, $h(x)$, and $F(z)$ are given functions and a_i and b_j are free parameters. To this end, we first must rewrite (E7.2.1.13) in the equivalent form

$$F^{-1}\left(\frac{y}{h(x)}\right)\sum_{j=1}^{M} b_j g_j(x) = \sum_{i=1}^{N} a_i f_i(x),$$

where F^{-1} is the inverse function of F. Further, taking the required number of points x_k ($k = 1, 2, \ldots, N+M+1$) and the values y_k corresponding to them, one can eliminate the parameters a_i and b_j from the resulting system of linear algebraic equations.

▶ **Method of partial elimination of free parameters.** We describe the method of partial elimination of free parameters for two-parameter functions $y = F(x, a, b)$. Prescribing the values of the function at two arbitrary points, we obtain two relations, $y_i = F(x_i, a, b)$ and

$y_j = F(x_j, a, b)$. By eliminating one of the parameters from them and then by solving the obtained relation for the other parameter, we obtain a criterion relation of the form $\varphi(x_i, x_j, y_i, y_j) = a$ or $\psi(x_i, x_j, y_i, y_j) = b$.

Let us apply this method to the two-parameter family of functions (E7.2.1.5). By eliminating the parameter b from the first two relations in (E7.2.1.6), we obtain the relation

$$\frac{y_j - y_i}{f(x_j) - f(x_i)} = a, \qquad (E7.2.1.14)$$

which can also be used to verify the applicability of functions of the form (E7.2.1.5). If any two points x_i and x_j and the values y_i and y_j corresponding to them are substituted into the left-hand side of (E7.2.1.14), then one and the same constant a must be obtained. (If Eq. (E7.2.1.14) is satisfied approximately with satisfactory accuracy, then the constant a can readily be verified by the least-squares method.) Another criterion relation for the family of functions (E7.2.1.6) can be derived in a similar way by eliminating the parameter a:

$$\frac{y_i f(x_j) - y_j f(x_i)}{f(x_j) - f(x_i)} = b. \qquad (E7.2.1.15)$$

In the special case of the logarithmic law (E7.2.1.9), the first criterion relation (E7.2.1.14) acquires the form

$$\frac{y_j - y_i}{\ln x_j - \ln x_i} = a, \qquad (E7.2.1.16)$$

and the second criterion relation (E7.2.1.15) becomes

$$\frac{y_i \ln x_j - y_j \ln x_i}{\ln x_j - \ln x_i} = b. \qquad (E7.2.1.17)$$

Now we consider the two-parameter power-law dependence (E7.2.1.11). We take logarithms of both sides of this formula and then apply the resulting equation to two points x_i and x_j. After the elimination of the parameter a, we obtain the first criterion relation

$$\frac{\ln y_j - \ln y_i}{\ln x_j - \ln x_i} = m. \qquad (E7.2.1.18)$$

For the power-law dependence (E7.2.1.11), we can also obtain the second criterion relation

$$\frac{\ln x_i \ln y_j - \ln x_j \ln y_i}{\ln x_i - \ln x_j} = \ln a \qquad (E7.2.1.19)$$

by eliminating the parameter m.

In Table E7.2, we present the direct criteria (necessary conditions) for the applicability of two-parameter monotone dependences of the simplest form obtained by the method of partial elimination of free parameters (usually, one sets $j = i + 1$).

The method of partial elimination of free parameters leads to simpler criteria than the method of complete elimination of free parameters; in addition, it permits obtaining one of the parameters readily (cf. Table E7.1 and Table E7.2).

TABLE E7.2
Several two-parameter dependences and direct criteria of their applicability obtained by
the method of partial elimination of free parameters; f^{-1} is the inverse function of f.

No.	Form of empirical formula	Applicability criterion	Remarks
1	$y = ax + b$	$\dfrac{y_i - y_j}{x_i - x_j} = a$	
2	$y = \dfrac{a}{x} + b$	$\dfrac{x_i x_j (y_j - y_i)}{x_i - x_j} = a$	
3	$y = \dfrac{1}{ax + b}$	$\dfrac{y_j - y_i}{(x_i - x_j) y_i y_j} = a$	
4	$y = \dfrac{x}{ax + b}$	$\dfrac{x_i x_j (y_i - y_j)}{(x_i - x_j) y_i y_j} = b$	
5	$y = ax^m$	$\dfrac{\ln(y_i / y_j)}{\ln x_i - \ln x_j} = m$	$x > 0$
6	$y = ae^{\lambda x}$	$\dfrac{\ln(y_i / y_j)}{x_i - x_j} = \lambda$	
7	$y = a \ln x + b$	$\dfrac{y_i - y_j}{\ln x_i - \ln x_j} = a$	$x > 0$
8	$y = a f(x) + b$	$\dfrac{y_i - y_j}{f(x_i) - f(x_j)} = a$	$f(x)$ is given
9	$y = a[f(x)]^m$	$\dfrac{\ln(y_i / y_j)}{\ln[f(x_i)/f(x_j)]} = m$	$f(x) \geq 0$ is given
10	$y = a \exp[\lambda f(x)]$	$\dfrac{\ln(y_i / y_j)}{f(x_i) - f(x_j)} = \lambda$	$f(x)$ is given
11	$y = f(ax + b)$	$\dfrac{f^{-1}(y_i) - f^{-1}(y_j)}{x_i - x_j} = a$	$f(z)$ is given
12	$y = a f(x) + b g(x)$	$\dfrac{y_i g(x_j) - y_j g(x_i)}{f(x_i)g(x_j) - f(x_j)g(x_i)} = a$	$f(x), g(x)$ is given

▶ **Several generalizations.** The method of partial elimination of free parameters can be
used in the case of multi-parameter dependences of special form (E7.2.1.13). For example,
for three-parameter dependence in the form of a quadratic polynomial

$$y = ax^2 + bx + c,$$

we obtain the criterion

$$\frac{(x_k - x_i)(y_j - y_i) - (x_j - x_i)(y_k - y_i)}{(x_k - x_i)(x_j^2 - x_i^2) - (x_j - x_i)(x_k^2 - x_i^2)} = a.$$

For a more general three-parameter dependence

$$y = a f(x) + b g(x) + c h(x),$$

where $f(x)$, $g(x)$, and $h(x)$ are given functions, we have the criterion

$$\frac{(y_j h_i - y_i h_j)(g_k h_i - g_i h_k) - (y_k h_i - y_i h_k)(g_j h_i - g_i h_j)}{(f_j h_i - f_i h_j)(g_k h_i - g_i h_k) - (f_k h_i - f_i h_k)(g_j h_i - g_i h_j)} = a.$$

Here we use the shorthand notation $f_p = f(x_p)$, $g_p = g(x_p)$, and $h_p = h(x_p)$, where $p = i, j, k$. The other two criteria can be obtained by cyclic permutations of the constants a, b, and c and the functions f, g, and h.

Remark. Instead of the explicit dependence (E7.2.1.1), one can consider a more general dependence $F(x, y, a, b) = 0$ and, in particular, an implicit dependence of the special form

$$g(x, y) = a f(x, y) + b,$$

which generalizes formula (E7.2.1.5). In this case, instead of (E7.2.1.14) we have the criterion

$$\frac{g(x_j, y_j) - g(x_i, y_i)}{f(x_j, y_j) - f(x_i, y_i)} = a.$$

E7.2.2. Three-Parameter Empirical Formulas

▶ **Three-parameter formula of power-law form.** Consider the power-law dependence

$$y = ax^m + b \tag{E7.2.2.1}$$

containing three parameters a, b, and m. We compose the geometric mean for the two extreme points: $x_{\mathrm{gm}} = \sqrt{x_1 x_n}$. Let $x_k \leq x_{\mathrm{gm}} \leq x_{k+1}$. We use the linear interpolation formula (E7.1.4.3) to obtain the approximate value y_{gm}. Assuming that the points (x_1, y_1), $(x_{\mathrm{gm}}, y_{\mathrm{gm}})$, (x_n, y_n) lie on the curve (E7.2.2.1), we obtain three equations

$$y_1 = ax_1^m + b, \quad y_{\mathrm{gm}} = ax_{\mathrm{gm}}^m + b, \quad y_n = ax_n^m + b.$$

By raising $x_{\mathrm{gm}} = \sqrt{x_1 x_n}$ to the power m and by multiplying by a, we obtain

$$ax_{\mathrm{gm}}^m = \sqrt{ax_1^m \, ax_n^m} \implies y_{\mathrm{gm}} - b = \sqrt{(y_1 - b)(y_n - b)}.$$

By squaring the last relation, we obtain the constant b,

$$b = \frac{y_1 y_n - y_{\mathrm{gm}}^2}{y_1 + y_n - 2y_{\mathrm{gm}}}. \tag{E7.2.2.2}$$

Since the coefficient b is now known, the change $Y = y - b$ takes formula (E7.2.2.1) to a formula of the form (E7.2.1.11) with two parameters. To verify its applicability, we can use the criterion relation (E7.2.1.12) or (E7.2.1.18), where y_s must be replaced by $y_s - b$, $s = i, j, k$.

▶ **Three-parameter formula of exponential form.** Consider the exponential dependence

$$y = ae^{\lambda x} + b \tag{E7.2.2.3}$$

containing three parameters a, b, and λ. We compose the arithmetic mean for the two extreme points, $x_{\mathrm{am}} = \frac{1}{2}(x_1 + x_n)$. Let $x_k \leq x_{\mathrm{am}} \leq x_{k+1}$. We use the linear interpolation formula (E7.1.4.3) to find the approximate value y_{am}. Assuming that the three points (x_1, y_1), $(x_{\mathrm{am}}, y_{\mathrm{am}})$, (x_n, y_n) lie on the curve (E7.2.2.3), we obtain

$$y_1 = ae^{\lambda x_1} + b, \quad y_{\mathrm{am}} = ae^{\frac{1}{2}\lambda(x_1 + x_n)} + b, \quad y_n = ae^{\lambda x_n} + b.$$

This implies that $y_1 - b = ae^{\lambda x_1}$ and $y_n - b = ae^{\lambda x_n}$, and hence $(y_1 - b)(y_n - b) = (y_{\mathrm{am}} - b)^2$. Now we determine the constant b:

$$b = \frac{y_1 y_n - y_{\mathrm{am}}^2}{y_1 + y_n - 2y_{\mathrm{am}}}. \tag{E7.2.2.4}$$

Since the coefficient b is now known, the change $Y = y - b$ takes formula (E7.2.2.3) to the formula $Y = ae^{\lambda x}$ with two parameters. To verify its applicability, we can use the criterion dependence given in Tables E7.1 and E7.2 (the 6th row in both cases), where y_s must be replaced by $y_s - b$, $s = i, j, k$.

Remark 1. An exact analytic criterion for the applicability of formula (E7.2.2.3) in the case of equally spaced argument, i.e., for

$$\Delta x_i = x_{i+1} - x_i = h \qquad (i = 1, 2, \ldots, n-1),$$

has the form

$$\lambda = \frac{1}{h} \ln \left(\frac{y_{i+2} - y_{i+1}}{y_{i+1} - y_i} \right) = \text{const}.$$

Remark 2. In Subsection E7.2.1, we presented criteria for verifying the applicability of a quadratic polynomial with three parameters. In Subsection E7.2.3, we describe more complicated criteria for verifying three-parameter formulas.

E7.2.3. Combined Method for Verifying Empirical Formulas

▶ **Description of the method. Three-parameter dependences of power-law form.** In the cases where there are sufficiently many experimental points, empirical formulas can be verified by a combined method. This method is based on the differentiation of the dependence under study, after which, using the original dependence and its derivative calculated at several points, the required number of free parameters is eliminated.

We illustrate the above by an example of the three-parameter power-law dependence

$$y = a(x + b)^m. \tag{E7.2.3.1}$$

By successively setting $x = x_i$, $x = x_j$, $x = x_k$ in (E7.2.3.1), we cannot eliminate the free parameters from the obtained relations; i.e., the direct method cannot be used.

Therefore, we proceed as follows. First, we differentiate (E7.2.3.1) with respect to x,

$$y' = am(x + b)^{m-1}, \tag{E7.2.3.2}$$

and then eliminate the parameter a from (E7.2.3.1) and (E7.2.3.2). We have

$$\frac{y}{y'} = \frac{x + b}{m}. \tag{E7.2.3.3}$$

Relation (E7.2.3.3) contains two parameters m and b and already admits the application of the direct method. By setting $x = x_i$ and $x = x_j$ in (E7.2.3.3) and then by eliminating one of the free parameters, we obtain two criterion relations containing the first derivatives:

$$\frac{(x_i - x_j)y_i'y_j'}{y_i y_j' - y_j y_i'} = m, \tag{E7.2.3.4}$$

$$\frac{x_i y_j y_i' - x_j y_i y_j'}{y_i y_j' - y_j y_i'} = b. \tag{E7.2.3.5}$$

In a similar way, consider the three-parameter dependence

$$y = ax^m + b, \tag{E7.2.3.6}$$

for which it is also impossible to derive criterion relations by the direct method. By differentiating (E7.2.3.6), we obtain

$$y' = amx^{m-1}. \tag{E7.2.3.7}$$

By eliminating the parameter a from (E7.2.3.6) and (E7.2.3.7), we obtain

$$\frac{y - b}{y'} = \frac{x}{m}. \tag{E7.2.3.8}$$

By setting $x = x_i$ and $x = x_j$ in (E7.2.3.8) and then by eliminating the parameter b, we obtain the criterion relation

$$\frac{x_i y_i' - x_j y_j'}{y_i - y_j} = m \tag{E7.2.3.9}$$

containing the first derivatives.

▶ **Dependences of exponential and logarithmic type.** We present several other simple useful criteria for widely used three-parameter functions.

For the exponential dependence

$$y = ae^{\lambda x} + b,$$

we have the criterion

$$\frac{y_i' - y_j'}{y_i - y_j} = \lambda. \qquad (E7.2.3.10)$$

For the logarithmic dependence

$$y = a\ln(bx + c) \equiv a\ln(x + D) + E \qquad (D = c/b, \ E = a\ln b),$$

we have the criterion

$$\frac{y_j - y_i}{\ln(y_i'/y_j')} = a. \qquad (E7.2.3.11)$$

▶ **Formula for calculation of the derivatives.** Criteria (E7.2.3.4), (E7.2.3.5), (E7.2.3.9), (E7.2.3.10), and (E7.2.3.11) contain derivatives, which can be calculated approximately by using appropriate approximation formulas.

In what follows, we present three difference schemes most widely used in numerical mathematics to calculate the derivatives at the point x_k:

$$y_k' = \frac{y_{k+1} - y_{k-1}}{x_{k+1} - x_{k-1}} \qquad \text{(central difference)},$$

$$y_k' = \frac{y_{k+1} - y_k}{x_{k+1} - x_k} \qquad \text{(right difference)},$$

$$y_k' = \frac{y_k - y_{k-1}}{x_k - x_{k-1}} \qquad \text{(left difference)}.$$

Numerical experiments show that the first scheme based on the central difference provides the most accurate results. It is expedient to use just this scheme for computations.

E7.3. Construction of Engineering Formulas by the Method of Asymptotic Analogies

E7.3.1. Description of the Method of Asymptotic Analogies

▶ **Preliminary remarks.** The most important stage in the study of specific physical and engineering problems is to find general quantitative laws valid for a class of qualitatively similar problems. In many cases, general results of this type can be obtained by the method of asymptotic analogies. The method is based on the passage from the usual dimensionless variables to special asymptotic coordinates and can be used for constructing wide-scope approximate formulas. (For example, in problems of mass and heat transfer one and the same formula can be used for describing a variety of qualitatively similar problems that differ in surface shape and flow structure.)

Suppose that there is a class of problems that differ in geometric characteristics and depend on a dimensionless parameter x $(0 \leq x < \infty)$. We assume that the dependence of the basic desired variable w on the parameter x is known for some specific (say, the simplest) geometry:

$$y = f(x), \qquad (E7.3.1.1)$$

where $f(x)$ is a monotone function.

In problems of mass and heat transfer, y is usually the Sherwood (Nusselt) number or the volume-average concentration; the parameter x is dimensionless time, the Peclet number, or the dimensionless rate constant of a chemical reaction.

▶ **Transition to asymptotic coordinates.** Let us transform (E7.3.1.1) as follows. Let the leading terms of the asymptotic expansions of y for small and large x have the form

$$y \to y_0 \quad (\text{as } x \to 0), \qquad y \to y_\infty \quad (\text{as } x \to \infty), \tag{E7.3.1.2}$$

where y_0 and y_∞ depend in certain way on x,

$$y_0 = \varphi(x), \quad y_\infty = \psi(x), \tag{E7.3.1.3}$$

and are determined from the analysis of formula (E7.3.1.1). We also assume that the condition $\varphi/\psi \neq \text{const}$ is satisfied.

Note that the original dependence (E7.3.1.1), as well as the asymptotics (E7.3.1.2) and (E7.3.1.3), can be determined either theoretically or experimentally.

Using (E7.3.1.1)–(E7.3.1.3), we write the following two relations:

$$\frac{y}{y_0} = \frac{f(x)}{\varphi(x)}, \qquad \frac{y_\infty}{y_0} = \frac{\psi(x)}{\varphi(x)}. \tag{E7.3.1.4}$$

By expressing x from second relation and by substituting it in the first equation (E7.3.1.4), we find the explicit form of y in terms of the asymptotics y_0 and y_∞. As follows from (E7.3.1.4), the structure of this relation is generally written as

$$\frac{y}{y_0} = F\left(\frac{y_\infty}{y_0}\right). \tag{E7.3.1.5}$$

The function f contained in the original formula (E7.3.1.4) and the function F contained in formula (E7.3.1.5) are related as

$$f(x) \equiv \varphi(x) F\left(\frac{\psi(x)}{\varphi(x)}\right).$$

It is seen that, in contrast to formula (E7.3.1.1), formula (E7.3.1.5) remains invariant with respect to the choice of the method for determining the dimensionless value y. We suggest that the range of each of the ratios y/y_0 and y_∞/y_0 be identical for all problems of the class under consideration. The variables y/y_0 and y_∞/y_0 are called *asymptotic coordinates*.

Example. In applications, one usually encounters power-law asymptotics,

$$y_0 = A x^k \quad \left(\lim_{x\to 0} y/y_0 = 1\right), \qquad y_\infty = B x^m \quad \left(\lim_{x\to\infty} y/y_\infty = 1\right), \tag{E7.3.1.6}$$

where A, B, k, and m are some constants and $k \neq m$ (for the entire class of considered problems, we assume that the constants k and m remain the same but the parameters A and B can vary).

In the case of power-law asymptotics (E7.3.1.6), formula (E7.3.1.5) can be expressed explicitly via the function f:

$$\frac{y}{y_0} = \frac{1}{A}\left(\frac{A}{B}\frac{y_\infty}{y_0}\right)^{\frac{k}{k-m}} f\left(\left(\frac{A}{B}\frac{y_\infty}{y_0}\right)^{\frac{1}{m-k}}\right), \tag{E7.3.1.7}$$

or, equivalently,

$$\frac{y}{y_\infty} = \frac{1}{B}\left(\frac{A}{B}\frac{y_\infty}{y_0}\right)^{\frac{m}{k-m}} f\left(\left(\frac{A}{B}\frac{y_\infty}{y_0}\right)^{\frac{1}{m-k}}\right). \tag{E7.3.1.8}$$

▶ **Description of the method.** The basic idea of the method of asymptotic analogies is to use the expression (E7.3.1.5) (or formulas (E7.3.1.7) and (E7.3.1.8)) to approximate similar

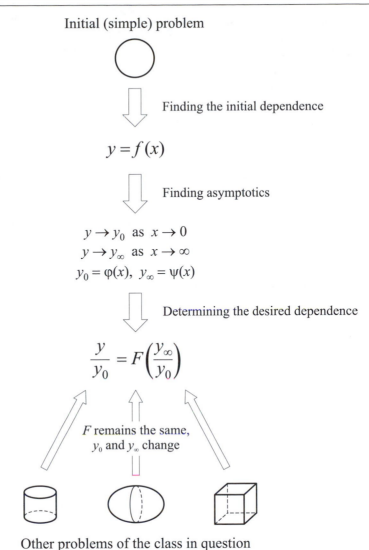

Initial (simple) problem

Finding the initial dependence

$$y = f(x)$$

Finding asymptotics

$$y \to y_0 \text{ as } x \to 0$$
$$y \to y_\infty \text{ as } x \to \infty$$
$$y_0 = \varphi(x), \quad y_\infty = \psi(x)$$

Determining the desired dependence

$$\frac{y}{y_0} = F\left(\frac{y_\infty}{y_0}\right)$$

F remains the same,
y_0 and y_∞ change

Other problems of the class in question

Figure E7.1. Scheme of application of the method of asymptotic analogies for a class of qualitatively similar problems.

characteristics for a wider class of problems describing qualitatively similar phenomena or processes. Specifically, after the relation (E7.3.1.5) has been constructed with the help of (E7.3.1.1) for some specific (say, the simplest) case, we can evaluate y for other problems of this class by finding the asymptotics y_0 (as $x \to 0$) and y_∞ (as $x \to \infty$) and then by substituting these asymptotics into (E7.3.1.5). The approximate formulas thus obtained are asymptotically sharp in both limit cases $x \to 0$ and $x \to \infty$.

Figure E7.1 presents a principle scheme of applying the method of asymptotic analogies.

In the books by Polyanin and Dilman (1994) and Polyanin et al. (2002), the formulas obtained by the method of asymptotic analogies were compared with the already known formulas obtained by exact, numerical, and approximate methods for a large number of specific cases. These investigations confirmed the high accuracy and wide capabilities of the method of asymptotic analogies. In other words, the final functional relation (E7.3.1.5) between y and its asymptotics remains the same (or varies slightly) in a wide class of problems of the same type, and the specific features of geometric distinctions between these problems (like the surface shape and the flow structure) are sufficiently well taken into account by the corresponding asymptotic parameters y_0 and y_∞.

As a result, the scope of the final formula (E7.3.1.5) (or formulas (E7.3.1.7) and (E7.3.1.8)) is substantially wider than that of the original formula (E7.3.1.1). In this sense, one can say that formulas of the type of (E7.3.1.6) are more informative than the original formula (E7.3.1.1).

Remark. The original dependence (E7.3.1.1), as well as the asymptotics (E7.3.1.3) can be determined either theoretically or experimentally.

E7.3.2. Illustrating Examples: Interior Heat Exchange Problems for Bodies of Various Shape

▶ **Statement of the problem.** Consider a class of problems concerning transient heat exchange between convex bodies of various shape and the ambient medium. At the initial time $t = 0$, the temperature is the same throughout the body and is equal to T_i, and for $t > 0$, the temperature on the surface Γ of the body is maintained constant and is equal to T_s. The general mathematical statement of the problem about the temperature distribution inside the body is given in Section E5.5.1 (see formulas (E5.5.1.1)–(E5.5.1.3)).

In this section, attention is paid to the study of the bulk body temperature

$$\langle T \rangle = \frac{1}{V} \int_v T \, dv, \tag{E7.3.2.1}$$

where T is the dimensionless temperature and $V = \int_v dv$ is the dimensionless volume of the body.

▶ **General formulas for the bulk temperature of the body.** To approximate the dependence of the bulk temperature on time, we use the method of asymptotic analogies. The simplest original problem is taken to be the one-dimensional (with respect to spatial coordinates) heat exchange problem for a sphere of radius a. The exact solution of this problem is well known and results in the following expression for the bulk temperature:

$$\langle T \rangle = 1 - \frac{6}{\pi^2} \sum_{k=1}^{\infty} \frac{1}{k^2} \exp(-\pi^2 k^2 \bar{\tau}), \qquad \tau = \frac{\chi t}{a^2}, \tag{E7.3.2.2}$$

where χ is the thermal diffusivity.

The asymptotic expressions (for small and large $\bar{\tau}$) of (E7.3.2.2) have the form

$$\langle T \rangle_0 = 6\pi^{-1/2} \sqrt{\bar{\tau}} \quad (\bar{\tau} \to 0); \qquad \langle T \rangle_\infty = 1 \quad (\bar{\tau} \to \infty) \tag{E7.3.2.3}$$

and are a special case of (E7.3.1.6) for $y_0 = \langle T \rangle_0$ and $y_\infty = \langle T \rangle_\infty$, where $A = 6\pi^{-1/2}$, $B = 1$, $k = \frac{1}{2}$, and $m = 0$. By substituting these values into (E7.3.1.7) with $f = \langle T \rangle$, we can rewrite (E7.3.2.2) as follows:

$$\frac{\langle T \rangle}{\langle T \rangle_\infty} = 1 - \frac{6}{\pi^2} \sum_{k=1}^{\infty} \frac{1}{k^2} \exp\left[-\frac{\pi^3}{36} k^2 \left(\frac{\langle T \rangle_0}{\langle T \rangle_\infty} \right)^2 \right]. \tag{E7.3.2.4}$$

Following the method of asymptotic analogies, we shall use formula (E7.3.2.4) for the calculation of bulk temperature for nonspherical bodies. To this end, for a body of a given shape, we must first calculate the asymptotics of bulk temperature for small and large $\bar{\tau}$ and then substitute these asymptotics into (E7.3.2.4).

For a bounded body of arbitrary shape, the solution of problem (E5.5.1.1)–(E5.5.1.3) tends as $\bar{\tau} \to \infty$ to the limit value (equal to 1) determined by the boundary condition on the surface of the body. By setting $T = 1$ in (E7.3.2.1), we find the asymptotics of the bulk temperature for large $\bar{\tau}$:

$$\langle T \rangle_\infty = 1. \tag{E7.3.2.5}$$

Now let us consider the initial stage of the process, corresponding to small values of dimensionless time. Let us integrate Eq. (E5.5.1.1) over the volume v occupied by the body. Taking into account the identity $\Delta T = \text{div}\,(\text{grad}\,T)$ and applying the Gauss divergence theorem, we replace the volume integral on the right-hand side by a surface integral. As a result, we obtain

$$\frac{\partial}{\partial \bar{\tau}} \int_v T\, dv = -\int_\Gamma \frac{\partial T}{\partial \xi}\, d\Gamma, \tag{E7.3.2.6}$$

where ξ is the coordinate along the inward normal on Γ.

For small $\bar{\tau}$, the temperature mainly varies in a thin region adjacent to the surface. In this region, the derivatives along the surface can be neglected compared with the normal derivatives. Therefore, the temperature distribution as $\bar{\tau} \to 0$ is described by the equation

$$\frac{\partial T}{\partial \bar{\tau}} = \frac{\partial^2 T}{\partial \xi^2} \tag{E7.3.2.7}$$

with the initial and boundary conditions,

$$\bar{\tau} = 0, \;\; T = 0; \qquad \xi = 0, \;\; T = 1,$$

where the value $\xi = 0$ corresponds to the surface of the body.

The solution of problem (E7.3.2.7) is given by the complementary error function

$$T = \text{erfc}\left(\frac{\xi}{2\sqrt{\bar{\tau}}}\right). \tag{E7.3.2.8}$$

By differentiating this formula with respect to ξ and by setting $\xi = 0$, we obtain the asymptotics as $\bar{\tau} \to 0$ of local heat flux to the surface of the body,

$$\left(\frac{\partial T}{\partial \xi}\right)_\Gamma = -\frac{1}{\sqrt{\pi \bar{\tau}}}. \tag{E7.3.2.9}$$

Let us substitute (E7.3.2.9) into (E7.3.2.6). After the integration, we obtain

$$\frac{\partial}{\partial \bar{\tau}} \int_v T\, dv = \frac{1}{\sqrt{\pi \bar{\tau}}} S, \tag{E7.3.2.10}$$

where S is the dimensionless surface area of the body.

Let us integrate both sides of (E7.3.2.10) with respect to $\bar{\tau}$ from 0 to $\bar{\tau}$. With regard to the initial condition (E5.5.1.2) and relation (E7.3.2.1), we obtain the desired asymptotic expression for bulk temperature as $\bar{\tau} \to 0$:

$$\langle T \rangle_0 = 2\frac{S}{V}\sqrt{\frac{\bar{\tau}}{\pi}}. \tag{E7.3.2.11}$$

The substitution of (E7.3.2.5) and (E7.3.2.11) into (E7.3.2.4) gives an approximate dependence of bulk temperature of an arbitrarily shaped body on time:

$$\langle T \rangle = 1 - \frac{6}{\pi^2} \sum_{k=1}^{\infty} \frac{1}{k^2} \exp\left(-\frac{\pi^2 k^2 S^2}{9 V^2} \bar{\tau}\right).$$

This expression can be rewritten as follows

$$\langle T \rangle = 1 - \frac{6}{\pi^2} \sum_{k=1}^{\infty} \frac{1}{k^2} \exp\left(-\frac{\pi^2}{9} k^2 \frac{S_*^2 \chi t}{V_*^2}\right), \tag{E7.3.2.12}$$

where S_* and V_* are, respectively, the dimensional surface area and volume of the body.

For practical calculations, it is expedient to replace the infinite series by the simpler formula

$$\langle T \rangle = \sqrt{1 - e^{-1.27\,\omega}} + 0.6\left(e^{-1.5\,\omega} - e^{-1.1\,\omega}\right), \qquad \omega = \frac{S_*^2 \chi t}{V_*^2}, \tag{E7.3.2.13}$$

whose maximum deviation from (E7.3.2.12) is about 1.7% (see Table E7.3).

▶ **Bulk temperature for bodies of various shape.** Let us compare the approximate dependence (E7.3.2.12) with some well-known exact results on heat exchange for nonspherical bodies.

First, consider a parallelepiped with sides L_1, L_2, and L_3. The solution of the corresponding three-dimensional problem (E5.5.1.1)–(E5.5.1.3) can be constructed by separation of variables and results in the following formula for bulk temperature:

$$\langle T \rangle = 1 - \left(\frac{8}{\pi^2}\right)^3 \sum_{k=1}^{\infty} \sum_{m=1}^{\infty} \sum_{l=1}^{\infty} \frac{1}{(2k-1)^2 (2m-1)^2 (2l-1)^2}$$
$$\times \exp\left\{-\pi^2 \left[\frac{(2k-1)^2}{L_1^2} + \frac{(2m-1)^2}{L_2^2} + \frac{(2l-1)^2}{L_3^2}\right] \chi t\right\}. \tag{E7.3.2.14}$$

Since the surface area and the volume of the parallelepiped are given by the formulas $S_* = 2(L_1 L_2 + L_1 L_3 + L_2 L_3)$ and $V_* = L_1 L_2 L_3$, we can rewrite (E7.3.2.14) as

$$\langle T \rangle = 1 - \left(\frac{8}{\pi^2}\right)^3 \sum_{k=1}^{\infty} \sum_{m=1}^{\infty} \sum_{l=1}^{\infty} \frac{1}{(2k-1)^2 (2m-1)^2 (2l-1)^2}$$
$$\times \exp\left[-\frac{\pi^2}{4} \frac{\left(\frac{2k-1}{L_1}\right)^2 + \left(\frac{2m-1}{L_2}\right)^2 + \left(\frac{2l-1}{L_3}\right)^2}{\left(\frac{1}{L_1} + \frac{1}{L_2} + \frac{1}{L_3}\right)^2} \frac{S_*^2 \chi t}{V_*^2}\right]. \tag{E7.3.2.15}$$

In Table E7.3, the approximation (E7.3.2.12) is compared with the exact bulk temperature (E7.3.2.15) for six distinct values of L_1, L_2, and L_3. The maximum error of formulas (E7.3.2.12) and (E7.3.2.13) is about 5% for $0.25 \le L_3/L_1 \le 4.0$ and $L_2/L_1 = 1$.

TABLE E7.3
Comparison of exact and approximate values of bulk temperature $\langle T \rangle$ for bodies of various shape.

Bodies of various shape		Dimensionless time $S_*^2 \chi t / V_*^2$							
		0.05	0.1	0.2	0.3	0.5	1.0	1.5	2.0
Sphere, formula (E7.3.2.12)		0.236	0.323	0.438	0.518	0.631	0.795	0.882	0.932
Approximate formula (E7.3.2.13)		0.237	0.324	0.437	0.514	0.623	0.782	0.870	0.923
Paralle-lepiped, formula (E7.3.2.15); $E_i = L_i/L_1$	$E_2 = 1, E_3 = 0.25$	0.237	0.326	0.443	0.527	0.647	0.821	0.907	0.951
	$E_2 = 1, E_3 = 0.5$	0.233	0.318	0.429	0.506	0.615	0.774	0.862	0.915
	$E_2 = 1, E_3 = 1$	0.232	0.316	0.425	0.499	0.604	0.757	0.843	0.897
	$E_2 = 1, E_3 = 2$	0.232	0.318	0.427	0.503	0.610	0.767	0.854	0.920
	$E_2 = 1, E_3 = 4$	0.234	0.320	0.432	0.510	0.620	0.782	0.871	0.952
	$E_2 = 2, E_3 = 4$	0.234	0.321	0.435	0.514	0.628	0.794	0.882	0.932
Cylinder, formula (E7.3.2.17); $E = 2a/L$	$E = 0.25$	0.236	0.325	0.440	0.522	0.638	0.807	0.894	0.942
	$E = 0.5$	0.234	0.321	0.434	0.513	0.624	0.787	0.875	0.926
	$E = 1$	0.233	0.319	0.429	0.506	0.613	0.770	0.857	0.910
	$E = 2$	0.234	0.320	0.431	0.509	0.619	0.780	0.868	0.920
	$E = 4$	0.237	0.326	0.444	0.528	0.649	0.823	0.909	0.952

Now consider heat exchange for a cylinder of finite length. Let a be the radius and L the length of the cylinder. By solving problem (E5.5.1.1)–(E5.5.1.3), we obtain the following expression for bulk temperature:

$$\langle T \rangle = 1 - \frac{32}{\pi^2} \sum_{k=1}^{\infty} \sum_{m=1}^{\infty} \frac{1}{\vartheta_k^2 (2m-1)^2} \exp \left\{ -\left[\frac{\vartheta_k^2}{a^2} + \frac{\pi^2 (2m-1)^2}{L^2} \right] \chi t \right\}, \qquad \text{(E7.3.2.16)}$$

where the ϑ_k are the roots of the Bessel function of the first kind of index zero: $J_0(\vartheta_k) = 0$. (The first sixty roots ϑ_k are tabulated in the book by Janke et al. (1960).)

Formula (E7.3.2.16) can be rewritten in the form

$$\langle T \rangle = 1 - \frac{32}{\pi^2} \sum_{k=1}^{\infty} \sum_{m=1}^{\infty} \frac{1}{\vartheta_k^2 (2m-1)^2} \exp \left[-\frac{L^2 \vartheta_k^2 + \pi^2 a^2 (2m-1)^2}{4(a+L)^2} \frac{S_*^2 \chi t}{V_*^2} \right], \qquad \text{(E7.3.2.17)}$$

where $S_* = 2\pi a(a+L)$ is the surface area and $V_* = \pi a^2 L$ the volume of the cylinder.

In Table E7.3, the exact value of the bulk temperature given by (E7.3.2.17) is compared with the approximation (E7.3.2.12) for various ratios of the cylinder dimensions. The maximum error in formula (E7.3.2.12) is about 3.5% for $0.25 \leq 2a/L \leq 4.0$.

E7.3.3. Illustrating Examples: Mass Exchange Between Bubbles/Particles and Flow

▶ **Mass transfer to a spherical bubble at low Reynolds numbers.** The problem of mass transfer to a spherical bubble in a translational flow as $\text{Re} \to 0$ was studied in the entire range of Peclet numbers by finite-difference methods. The results for the mean Sherwood number are well approximated by the expression (E5.8.2.6)

$$\text{Sh} = 0.6 + (0.16 + 0.213\,\text{Pe})^{1/2}, \qquad \text{(E7.3.3.1)}$$

where the bubble radius is taken as the characteristic scale.

The maximum deviation of (E7.3.3.1) in the entire range of Peclet numbers is about 3%.

We use formula (E7.3.3.1) as a basis for obtaining a more general dependence, which can be used for nonspherical bubbles and more complicated flows. To this end, we use the method of asymptotic analogies, by setting $x = \text{Pe}$ and $y = \text{Sh}$.

From formula (E7.3.3.1) we find its asymptotics:

$$\text{Sh}_0 = 1 \quad (\text{Pe} \to 0), \qquad \text{Sh}_\infty = (0.213\,\text{Pe})^{1/2} \quad (\text{Pe} \to \infty), \qquad (\text{E7.3.3.2})$$

which corresponds to the values of the exponents $k = 0$ and $m = 1/2$ in formulas (E7.3.1.6)–(E7.3.1.8).

Passing from the original variables Pe and Sh to the asymptotic coordinates Sh/Sh_0 and $\text{Sh}/\text{Sh}_\infty$ in (E7.3.3.1), we obtain the dependence

$$\frac{\text{Sh}}{\text{Sh}_0} = 0.6 + \left[0.16 + \left(\frac{\text{Sh}_\infty}{\text{Sh}_0}\right)^2\right]^{1/2}, \qquad (\text{E7.3.3.3})$$

where Sh_∞ is the asymptotic value of the mean Sherwood number as $\text{Pe} \to \infty$, which can be calculated by solving the diffusion boundary layer equation in a given flow field or determined experimentally.

Now we consider an *axisymmetric shear flow*, where the dimensional fluid velocity components remote from the bubble have the following form in the Cartesian coordinates X_1, X_2, X_3:

$$\mathbf{V} = (V_1, V_2, V_3) = (-GX_1, -GX_2, 2GX_3), \qquad (\text{E7.3.3.4})$$

where G is the shear coefficient.

For a spherical bubble in a shear Stokes flow (E7.3.3.4), the asymptotics of the mean Sherwood number has the form (Polyanin et al., 2002)

$$\text{Sh}_0 = 1 \quad (\text{Pe} \to 0), \qquad \text{Sh}_\infty = \left(\frac{3}{\pi}\,\text{Pe}\right)^{1/2} \quad (\text{Pe} \to \infty), \qquad (\text{E7.3.3.5})$$

where $\text{Pe} = a^2|G|/D$. By substituting the asymptotics (E7.3.3.5) into formula (E7.3.3.3), we obtain the formula

$$\text{Sh} = 0.6 + (0.16 + 0.955\,\text{Pe})^{1/2},$$

whose error in the entire range of the Peclet numbers does not exceed 3%.

▶ **Mass transfer to a solid particle at low Reynolds numbers.** The problem of mass transfer to a solid spherical particle in a translational Stokes flow ($\text{Re} \to 0$) was studied in the entire range of Peclet numbers by finite-difference methods. To find the mean Sherwood number for a spherical particle, it is convenient to use the following approximate formula (E5.7.2.8):

$$\text{Sh} = 0.5 + (0.125 + 0.243\,\text{Pe})^{1/3}, \qquad (\text{E7.3.3.6})$$

where the particle radius is taken as the characteristic scale.

The maximum deviation of (E7.3.3.6) in the entire range of Peclet numbers is about 2%.

From formula (E7.3.3.6), we find its asymptotics:

$$\text{Sh}_0 = 1 \qquad \left(\lim_{\text{Pe}\to 0} \text{Sh}/\text{Sh}_0 = 1\right),$$
$$\text{Sh}_\infty = (0.243\,\text{Pe})^{1/3} \qquad \left(\lim_{\text{Pe}\to\infty} \text{Sh}/\text{Sh}_\infty = 1\right), \qquad (\text{E7.3.3.7})$$

which corresponds to the exponents $k = 0$ and $m = 1/3$ in formulas (E7.3.1.6)–(E7.3.1.8).

By passing from the original variables to the asymptotic coordinates in (E7.3.3.6) with the use of (E7.3.3.7), one can derive the approximate formula

$$\frac{\text{Sh}}{\text{Sh}_0} = 0.5 + \left[0.125 + \left(\frac{\text{Sh}_\infty}{\text{Sh}_0}\right)^3\right]^{1/3}, \qquad (\text{E7.3.3.8})$$

which is equivalent to formula (E5.9.3.2). The approximate formula (E7.3.3.8) can be used to calculate the mean Sherwood number for a steady-state viscous flow (without closed streamlines) past smooth particles of nonspherical shape (for example, ellipsoidal particles). The auxiliary value Sh_∞ must be chosen equal to the leading term of the asymptotic expansion of the Sherwood number as $\mathsf{Pe} \to \infty$.

For a particle of a given shape, the auxiliary quantities Sh_0 and Sh_∞ occurring in (E7.3.3.8) can be determined either theoretically or experimentally (see Section E5.9.3).

Remark. For a solid spherical particle in an axisymmetric shear flow (E7.3.3.4), the maximum error of formula (E7.3.3.8) in the entire region of Peclet numbers is about 3%.

▶ **Transient mass transfer in steady-state flows.** We consider a laminar steady-state flow past a solid spherical particle (drop or bubble) of radius a and study transient mass transfer to the particle surface. At the initial time $t = 0$, the concentration in the continuous phase is constant and equal to C_i, whereas for $t > 0$ a constant concentration C_s is maintained on the particle surface. The mathematical statement of the problem about the concentration distribution in the flow is given in Section E5.10.1 (see formulas (E5.10.1.1) and (E5.10.1.2)).

In what follows, we restrict ourselves to the case of high Peclet numbers, in which there are no closed streamlines in the flow.

For an axisymmetric shear flow past a bubble (the fluid velocity field far from the bubble is determined by formula (E7.3.3.4)), the mean Sherwood number can be calculated by the formula

$$\mathsf{Sh} = \sqrt{\frac{3\,\mathsf{Pe}}{\pi} \coth(3\,\mathsf{Pe}\,\tau)}, \tag{E7.3.3.9}$$

which was obtained in the diffusion boundary layer approximation. In formula (E7.3.3.9), we use the notation $\tau = Dt/a^2$ and $\mathsf{Pe} = a^2 G/D$.

From formula (E7.3.3.9), we can find the asymptotics

$$\mathsf{Sh}_0 = (\pi\tau)^{-1/2} \quad (\tau \to 0), \qquad \mathsf{Sh}_\infty = (3\,\mathsf{Pe}/\pi)^{1/2} \quad (\tau \to \infty). \tag{E7.3.3.10}$$

By passing to asymptotic coordinates in (E7.3.3.9) with the use of (E7.3.3.10), we obtain

$$\frac{\mathsf{Sh}}{\mathsf{Sh}_\infty} = \sqrt{\coth\left(\frac{\mathsf{Sh}_\infty}{\mathsf{Sh}_0}\right)^2}. \tag{E7.3.3.11}$$

For an arbitrary steady-state flow past spherical particles, drops, and bubbles, we have $\mathsf{Sh}_0 = (\pi\tau)^{-1/2}$, and formula (E7.3.3.11) becomes

$$\frac{\mathsf{Sh}}{\mathsf{Sh}_\infty} = \sqrt{\coth(\pi\mathsf{Sh}_\infty^2\tau)}, \tag{E7.3.3.12}$$

where $\mathsf{Sh}_\infty = \lim_{\tau \to \infty} \mathsf{Sh}$ is the Sherwood number for the steady-state diffusion mode and $\tau = Dt/a^2$.

Equation (E7.3.3.12) gives a valid asymptotic result for any steady-state flow field in both limit cases $\tau \to 0$ and $\tau \to \infty$.

Table E7.4 presents a comparison of the mean Sherwood numbers calculated according to Eq. (E7.3.3.12) with available data for various flows past spherical drops, bubbles, and solid particles at high Peclet numbers (in this table, we use the abbreviation DBLA for "diffusion boundary layer approximation"). The parameter β is the ratio of the dynamic viscosity of the drop to that of the ambient fluid and varies in the range $0 \leq \beta \leq 2$. (The value $\beta = 0$ corresponds to a gas bubble.)

Remark. One can find other examples of using the method of asymptotic analogies in the books by Polyanin and Dilman (1994) and Polyanin et al. (2002).

TABLE E7.4
Maximum error of Eq. (E7.3.3.12) for various types of flow past spherical drops,
bubbles, and solid particles (according to the data by Polyanin and Dilman, 1994).

No.	Dispersed phase	Flow type	Solution method	Error, %
1	Drop, bubble	Axisymmetric shear Stokes flow	Analytical, DBLA	0
2	Drop, bubble	Translational Stokes flow	Analytical, DBLA	0.7
3	Drop, bubble	Two-dimensional shear Stokes flow	Analytical, DBLA	1.8
4	Bubble	Laminar translational flow at high Reynolds numbers	Analytical, DBLA	0.7
5	Bubble	Axisymmetric shear flow at high Reynolds numbers	Analytical, DBLA	0
6	Drop, bubble	Flow caused by an electric field	Analytical, DBLA	0
7	Solid particle	Translational flow of an ideal (inviscid) liquid	Analytical, DBLA	0.7
8	Solid particle	Translational Stokes flow	Interpolation of numerical and analytical results	1.4
9	Solid particle	Translational Stokes flow	Finite-difference numerical method (at $Pe = 500$)	4

Bibliography for Chapter E7

Björck, A., *Numerical Methods for Least Squares Problems*, Society for Ind. & Appl. Math., Philadelphia, 1996.

Demidovich, B. P., Maron, I. A., and Shuvalova, E., Z., *Numerical Methods of Analysis* [in Russian], Fizmatlit, Moscow, 1963.

Demidovich, B. P. and Maron, I. A., *Foundations of Comutational Mathematics* [in Russian], Nauka, Moscow, 1970.

Gieck, K. and Gieck, R., *Engineering Formulas, 8th Edition*, McGraw-Hill, New York, 2006.

Janke E., Emde F., Lösch F., *Tafeln Höherer Funktionen*, Teubner Verlogsgesellschaft, Stuttgart, 1960.

Kariya, T. and Kurata, H., *Generalized Least Squares*, Wiley, Hoboken, New Jersey, 2004.

Lawson, C. and Hanson, R., *Solving Least Squares Problems*, Prentice-Hall, Englewood Cliffs, New Jersey, 1974.

Merriman, M., *A Textbook on the Method of Least Squares, 8th Edition*, Cornell University Library, 2009.

Polyanin, A. D. and Dilman, V. V., *Methods of Modeling Equations and Analogies in Chemical Engineering*, CRC Press–Begell House, Boca Raton, Florida, 1994.

Polyanin, A. D., Kutepov, A. M., Vyazmin, A. V., and Kazenin, D. A., *Hydrodynamics, Mass and Heat Transfer in Chemical Engineering*, Taylor & Francis, London, 2002.

Polyanin, A. D., Vyazmina, E. A., and Dilman, V. V., New methods for processing experimental data: Applicability tests for empirical formulas, *Theor. Found. Chem. Eng.*, Vol. 42, No. 4, pp. 354–365, 2008.

Running, T. R., *Empirical Formulas*, Cornell University Library, 2009.

Semendyaev, K. A., *Empirical Formulas* [in Russian], Gos. Tech.-Teor. Izdat., Moscow, 1933.

Weisstein, E. W., *CRC Concise Encyclopedia of Mathematics, 2nd Edition*, CRC Press, Boca Raton, Florida, 2003.

Wolberg, J., *Data Analysis Using the Method of Least Squares: Extracting the Most Information from Experiments*, Springer-Verlag, Berlin, 2006.

Part IV

Supplements

Chapter S1

Integrals

S1.1. Indefinite Integrals

Throughout Section S1.1, the integration constant C is omitted for brevity.

S1.1.1. Integrals Involving Rational Functions

▶ **Integrals involving** $a + bx$.

1. $\displaystyle \int \frac{dx}{a + bx} = \frac{1}{b} \ln |a + bx|.$

2. $\displaystyle \int (a + bx)^n dx = \frac{(a + bx)^{n+1}}{b(n + 1)}, \qquad n \neq -1.$

3. $\displaystyle \int \frac{x\,dx}{a + bx} = \frac{1}{b^2} \left(a + bx - a \ln |a + bx| \right).$

4. $\displaystyle \int \frac{x^2\,dx}{a + bx} = \frac{1}{b^3} \left[\frac{1}{2}(a + bx)^2 - 2a(a + bx) + a^2 \ln |a + bx| \right].$

5. $\displaystyle \int \frac{dx}{x(a + bx)} = -\frac{1}{a} \ln \left| \frac{a + bx}{x} \right|.$

6. $\displaystyle \int \frac{dx}{x^2(a + bx)} = -\frac{1}{ax} + \frac{b}{a^2} \ln \left| \frac{a + bx}{x} \right|.$

7. $\displaystyle \int \frac{x\,dx}{(a + bx)^2} = \frac{1}{b^2} \left(\ln |a + bx| + \frac{a}{a + bx} \right).$

8. $\displaystyle \int \frac{x^2\,dx}{(a + bx)^2} = \frac{1}{b^3} \left(a + bx - 2a \ln |a + bx| - \frac{a^2}{a + bx} \right).$

9. $\displaystyle \int \frac{dx}{x(a + bx)^2} = \frac{1}{a(a + bx)} - \frac{1}{a^2} \ln \left| \frac{a + bx}{x} \right|.$

10. $\displaystyle \int \frac{x\,dx}{(a + bx)^3} = \frac{1}{b^2} \left[-\frac{1}{a + bx} + \frac{a}{2(a + bx)^2} \right].$

▶ **Integrals involving** $a + x$ **and** $b + x$.

1. $\displaystyle \int \frac{a + x}{b + x}\,dx = x + (a - b) \ln |b + x|.$

2. $\displaystyle \int \frac{dx}{(a + x)(b + x)} = \frac{1}{a - b} \ln \left| \frac{b + x}{a + x} \right|, \qquad a \neq b.$ For $a = b$, see Integral 2 with $n = -2$ in the previous item.

3. $\displaystyle \int \frac{x\,dx}{(a+x)(b+x)} = \frac{1}{a-b}\left(a\ln|a+x| - b\ln|b+x|\right).$

4. $\displaystyle \int \frac{dx}{(a+x)(b+x)^2} = \frac{1}{(b-a)(b+x)} + \frac{1}{(a-b)^2}\ln\left|\frac{a+x}{b+x}\right|.$

5. $\displaystyle \int \frac{x\,dx}{(a+x)(b+x)^2} = \frac{b}{(a-b)(b+x)} - \frac{a}{(a-b)^2}\ln\left|\frac{a+x}{b+x}\right|.$

6. $\displaystyle \int \frac{x^2\,dx}{(a+x)(b+x)^2} = \frac{b^2}{(b-a)(b+x)} + \frac{a^2}{(a-b)^2}\ln|a+x| + \frac{b^2-2ab}{(b-a)^2}\ln|b+x|.$

7. $\displaystyle \int \frac{dx}{(a+x)^2(b+x)^2} = -\frac{1}{(a-b)^2}\left(\frac{1}{a+x} + \frac{1}{b+x}\right) + \frac{2}{(a-b)^3}\ln\left|\frac{a+x}{b+x}\right|.$

8. $\displaystyle \int \frac{x\,dx}{(a+x)^2(b+x)^2} = \frac{1}{(a-b)^2}\left(\frac{a}{a+x} + \frac{b}{b+x}\right) + \frac{a+b}{(a-b)^3}\ln\left|\frac{a+x}{b+x}\right|.$

9. $\displaystyle \int \frac{x^2\,dx}{(a+x)^2(b+x)^2} = -\frac{1}{(a-b)^2}\left(\frac{a^2}{a+x} + \frac{b^2}{b+x}\right) + \frac{2ab}{(a-b)^3}\ln\left|\frac{a+x}{b+x}\right|.$

▶ **Integrals involving $a^2 + x^2$.**

1. $\displaystyle \int \frac{dx}{a^2+x^2} = \frac{1}{a}\arctan\frac{x}{a}.$

2. $\displaystyle \int \frac{dx}{(a^2+x^2)^2} = \frac{x}{2a^2(a^2+x^2)} + \frac{1}{2a^3}\arctan\frac{x}{a}.$

3. $\displaystyle \int \frac{dx}{(a^2+x^2)^3} = \frac{x}{4a^2(a^2+x^2)^2} + \frac{3x}{8a^4(a^2+x^2)} + \frac{3}{8a^5}\arctan\frac{x}{a}.$

4. $\displaystyle \int \frac{dx}{(a^2+x^2)^{n+1}} = \frac{x}{2na^2(a^2+x^2)^n} + \frac{2n-1}{2na^2}\int \frac{dx}{(a^2+x^2)^n}; \quad n = 1,\,2,\,\ldots$

5. $\displaystyle \int \frac{x\,dx}{a^2+x^2} = \frac{1}{2}\ln(a^2+x^2).$

6. $\displaystyle \int \frac{x\,dx}{(a^2+x^2)^2} = -\frac{1}{2(a^2+x^2)}.$

7. $\displaystyle \int \frac{x\,dx}{(a^2+x^2)^3} = -\frac{1}{4(a^2+x^2)^2}.$

8. $\displaystyle \int \frac{x\,dx}{(a^2+x^2)^{n+1}} = -\frac{1}{2n(a^2+x^2)^n}; \quad n = 1,\,2,\,\ldots$

9. $\displaystyle \int \frac{x^2\,dx}{a^2+x^2} = x - a\arctan\frac{x}{a}.$

10. $\displaystyle \int \frac{x^2\,dx}{(a^2+x^2)^2} = -\frac{x}{2(a^2+x^2)} + \frac{1}{2a}\arctan\frac{x}{a}.$

11. $\displaystyle \int \frac{x^2\,dx}{(a^2+x^2)^3} = -\frac{x}{4(a^2+x^2)^2} + \frac{x}{8a^2(a^2+x^2)} + \frac{1}{8a^3}\arctan\frac{x}{a}.$

12. $\displaystyle \int \frac{x^2\,dx}{(a^2+x^2)^{n+1}} = -\frac{x}{2n(a^2+x^2)^n} + \frac{1}{2n}\int \frac{dx}{(a^2+x^2)^n}; \quad n = 1,\,2,\,\ldots$

13. $\displaystyle\int \frac{x^3\,dx}{a^2 + x^2} = \frac{x^2}{2} - \frac{a^2}{2}\ln(a^2 + x^2).$

14. $\displaystyle\int \frac{x^3\,dx}{(a^2 + x^2)^2} = \frac{a^2}{2(a^2 + x^2)} + \frac{1}{2}\ln(a^2 + x^2).$

15. $\displaystyle\int \frac{x^3\,dx}{(a^2 + x^2)^{n+1}} = -\frac{1}{2(n-1)(a^2 + x^2)^{n-1}} + \frac{a^2}{2n(a^2 + x^2)^n}; \quad n = 2,\,3,\,\ldots$

16. $\displaystyle\int \frac{dx}{x(a^2 + x^2)} = \frac{1}{2a^2}\ln\frac{x^2}{a^2 + x^2}.$

17. $\displaystyle\int \frac{dx}{x(a^2 + x^2)^2} = \frac{1}{2a^2(a^2 + x^2)} + \frac{1}{2a^4}\ln\frac{x^2}{a^2 + x^2}.$

18. $\displaystyle\int \frac{dx}{x(a^2 + x^2)^3} = \frac{1}{4a^2(a^2 + x^2)^2} + \frac{1}{2a^4(a^2 + x^2)} + \frac{1}{2a^6}\ln\frac{x^2}{a^2 + x^2}.$

19. $\displaystyle\int \frac{dx}{x^2(a^2 + x^2)} = -\frac{1}{a^2 x} - \frac{1}{a^3}\arctan\frac{x}{a}.$

20. $\displaystyle\int \frac{dx}{x^2(a^2 + x^2)^2} = -\frac{1}{a^4 x} - \frac{x}{2a^4(a^2 + x^2)} - \frac{3}{2a^5}\arctan\frac{x}{a}.$

21. $\displaystyle\int \frac{dx}{x^3(a^2 + x^2)^2} = -\frac{1}{2a^4 x^2} - \frac{1}{2a^4(a^2 + x^2)} - \frac{1}{a^6}\ln\frac{x^2}{a^2 + x^2}.$

22. $\displaystyle\int \frac{dx}{x^2(a^2 + x^2)^3} = -\frac{1}{a^6 x} - \frac{x}{4a^4(a^2 + x^2)^2} - \frac{7x}{8a^6(a^2 + x^2)} - \frac{15}{8a^7}\arctan\frac{x}{a}.$

23. $\displaystyle\int \frac{dx}{x^3(a^2 + x^2)^3} = -\frac{1}{2a^6 x^2} - \frac{1}{a^6(a^2 + x^2)} - \frac{1}{4a^4(a^2 + x^2)^2} - \frac{3}{2a^8}\ln\frac{x^2}{a^2 + x^2}.$

▶ **Integrals involving $a^2 - x^2$.**

1. $\displaystyle\int \frac{dx}{a^2 - x^2} = \frac{1}{2a}\ln\left|\frac{a + x}{a - x}\right|.$

2. $\displaystyle\int \frac{dx}{(a^2 - x^2)^2} = \frac{x}{2a^2(a^2 - x^2)} + \frac{1}{4a^3}\ln\left|\frac{a + x}{a - x}\right|.$

3. $\displaystyle\int \frac{dx}{(a^2 - x^2)^3} = \frac{x}{4a^2(a^2 - x^2)^2} + \frac{3x}{8a^4(a^2 - x^2)} + \frac{3}{16a^5}\ln\left|\frac{a + x}{a - x}\right|.$

4. $\displaystyle\int \frac{dx}{(a^2 - x^2)^{n+1}} = \frac{x}{2na^2(a^2 - x^2)^n} + \frac{2n - 1}{2na^2}\int \frac{dx}{(a^2 - x^2)^n}; \quad n = 1,\,2,\,\ldots$

5. $\displaystyle\int \frac{x\,dx}{a^2 - x^2} = -\frac{1}{2}\ln|a^2 - x^2|.$

6. $\displaystyle\int \frac{x\,dx}{(a^2 - x^2)^2} = \frac{1}{2(a^2 - x^2)}.$

7. $\displaystyle\int \frac{x\,dx}{(a^2 - x^2)^3} = \frac{1}{4(a^2 - x^2)^2}.$

8. $\displaystyle\int \frac{x\,dx}{(a^2 - x^2)^{n+1}} = \frac{1}{2n(a^2 - x^2)^n}; \quad n = 1,\,2,\,\ldots$

9. $\displaystyle\int \frac{x^2\,dx}{a^2-x^2} = -x + \frac{a}{2}\ln\left|\frac{a+x}{a-x}\right|.$

10. $\displaystyle\int \frac{x^2\,dx}{(a^2-x^2)^2} = \frac{x}{2(a^2-x^2)} - \frac{1}{4a}\ln\left|\frac{a+x}{a-x}\right|.$

11. $\displaystyle\int \frac{x^2\,dx}{(a^2-x^2)^3} = \frac{x}{4(a^2-x^2)^2} - \frac{x}{8a^2(a^2-x^2)} - \frac{1}{16a^3}\ln\left|\frac{a+x}{a-x}\right|.$

12. $\displaystyle\int \frac{x^2\,dx}{(a^2-x^2)^{n+1}} = \frac{x}{2n(a^2-x^2)^n} - \frac{1}{2n}\int \frac{dx}{(a^2-x^2)^n}; \quad n = 1, 2, \ldots$

13. $\displaystyle\int \frac{x^3\,dx}{a^2-x^2} = -\frac{x^2}{2} - \frac{a^2}{2}\ln|a^2-x^2|.$

14. $\displaystyle\int \frac{x^3\,dx}{(a^2-x^2)^2} = \frac{a^2}{2(a^2-x^2)} + \frac{1}{2}\ln|a^2-x^2|.$

15. $\displaystyle\int \frac{x^3\,dx}{(a^2-x^2)^{n+1}} = -\frac{1}{2(n-1)(a^2-x^2)^{n-1}} + \frac{a^2}{2n(a^2-x^2)^n}; \quad n = 2, 3, \ldots$

16. $\displaystyle\int \frac{dx}{x(a^2-x^2)} = \frac{1}{2a^2}\ln\left|\frac{x^2}{a^2-x^2}\right|.$

17. $\displaystyle\int \frac{dx}{x(a^2-x^2)^2} = \frac{1}{2a^2(a^2-x^2)} + \frac{1}{2a^4}\ln\left|\frac{x^2}{a^2-x^2}\right|.$

18. $\displaystyle\int \frac{dx}{x(a^2-x^2)^3} = \frac{1}{4a^2(a^2-x^2)^2} + \frac{1}{2a^4(a^2-x^2)} + \frac{1}{2a^6}\ln\left|\frac{x^2}{a^2-x^2}\right|.$

▶ **Integrals involving $a^3 + x^3$.**

1. $\displaystyle\int \frac{dx}{a^3+x^3} = \frac{1}{6a^2}\ln\frac{(a+x)^2}{a^2-ax+x^2} + \frac{1}{a^2\sqrt{3}}\arctan\frac{2x-a}{a\sqrt{3}}.$

2. $\displaystyle\int \frac{dx}{(a^3+x^3)^2} = \frac{x}{3a^3(a^3+x^3)} + \frac{2}{3a^3}\int \frac{dx}{a^3+x^3}.$

3. $\displaystyle\int \frac{x\,dx}{a^3+x^3} = \frac{1}{6a}\ln\frac{a^2-ax+x^2}{(a+x)^2} + \frac{1}{a\sqrt{3}}\arctan\frac{2x-a}{a\sqrt{3}}.$

4. $\displaystyle\int \frac{x\,dx}{(a^3+x^3)^2} = \frac{x^2}{3a^3(a^3+x^3)} + \frac{1}{3a^3}\int \frac{x\,dx}{a^3+x^3}.$

5. $\displaystyle\int \frac{x^2\,dx}{a^3+x^3} = \frac{1}{3}\ln|a^3+x^3|.$

6. $\displaystyle\int \frac{dx}{x(a^3+x^3)} = \frac{1}{3a^3}\ln\left|\frac{x^3}{a^3+x^3}\right|.$

7. $\displaystyle\int \frac{dx}{x(a^3+x^3)^2} = \frac{1}{3a^3(a^3+x^3)} + \frac{1}{3a^6}\ln\left|\frac{x^3}{a^3+x^3}\right|.$

8. $\displaystyle\int \frac{dx}{x^2(a^3+x^3)} = -\frac{1}{a^3x} - \frac{1}{a^3}\int \frac{x\,dx}{a^3+x^3}.$

9. $\displaystyle\int \frac{dx}{x^2(a^3+x^3)^2} = -\frac{1}{a^6x} - \frac{x^2}{3a^6(a^3+x^3)} - \frac{4}{3a^6}\int \frac{x\,dx}{a^3+x^3}.$

▶ **Integrals involving $a^3 - x^3$.**

1. $\displaystyle\int \frac{dx}{a^3 - x^3} = \frac{1}{6a^2} \ln \frac{a^2 + ax + x^2}{(a-x)^2} + \frac{1}{a^2\sqrt{3}} \arctan \frac{2x+a}{a\sqrt{3}}.$

2. $\displaystyle\int \frac{dx}{(a^3 - x^3)^2} = \frac{x}{3a^3(a^3 - x^3)} + \frac{2}{3a^3} \int \frac{dx}{a^3 - x^3}.$

3. $\displaystyle\int \frac{x\,dx}{a^3 - x^3} = \frac{1}{6a} \ln \frac{a^2 + ax + x^2}{(a-x)^2} - \frac{1}{a\sqrt{3}} \arctan \frac{2x+a}{a\sqrt{3}}.$

4. $\displaystyle\int \frac{x\,dx}{(a^3 - x^3)^2} = \frac{x^2}{3a^3(a^3 - x^3)} + \frac{1}{3a^3} \int \frac{x\,dx}{a^3 - x^3}.$

5. $\displaystyle\int \frac{x^2\,dx}{a^3 - x^3} = -\frac{1}{3} \ln |a^3 - x^3|.$

6. $\displaystyle\int \frac{dx}{x(a^3 - x^3)} = \frac{1}{3a^3} \ln \left| \frac{x^3}{a^3 - x^3} \right|.$

7. $\displaystyle\int \frac{dx}{x(a^3 - x^3)^2} = \frac{1}{3a^3(a^3 - x^3)} + \frac{1}{3a^6} \ln \left| \frac{x^3}{a^3 - x^3} \right|.$

8. $\displaystyle\int \frac{dx}{x^2(a^3 - x^3)} = -\frac{1}{a^3 x} + \frac{1}{a^3} \int \frac{x\,dx}{a^3 - x^3}.$

9. $\displaystyle\int \frac{dx}{x^2(a^3 - x^3)^2} = -\frac{1}{a^6 x} - \frac{x^2}{3a^6(a^3 - x^3)} + \frac{4}{3a^6} \int \frac{x\,dx}{a^3 - x^3}.$

▶ **Integrals involving $a^4 \pm x^4$.**

1. $\displaystyle\int \frac{dx}{a^4 + x^4} = \frac{1}{4a^3\sqrt{2}} \ln \frac{a^2 + ax\sqrt{2} + x^2}{a^2 - ax\sqrt{2} + x^2} + \frac{1}{2a^3\sqrt{2}} \arctan \frac{ax\sqrt{2}}{a^2 - x^2}.$

2. $\displaystyle\int \frac{x\,dx}{a^4 + x^4} = \frac{1}{2a^2} \arctan \frac{x^2}{a^2}.$

3. $\displaystyle\int \frac{x^2\,dx}{a^4 + x^4} = -\frac{1}{4a\sqrt{2}} \ln \frac{a^2 + ax\sqrt{2} + x^2}{a^2 - ax\sqrt{2} + x^2} + \frac{1}{2a\sqrt{2}} \arctan \frac{ax\sqrt{2}}{a^2 - x^2}.$

4. $\displaystyle\int \frac{dx}{a^4 - x^4} = \frac{1}{4a^3} \ln \left| \frac{a+x}{a-x} \right| + \frac{1}{2a^3} \arctan \frac{x}{a}.$

5. $\displaystyle\int \frac{x\,dx}{a^4 - x^4} = \frac{1}{4a^2} \ln \left| \frac{a^2 + x^2}{a^2 - x^2} \right|.$

6. $\displaystyle\int \frac{x^2\,dx}{a^4 - x^4} = \frac{1}{4a} \ln \left| \frac{a+x}{a-x} \right| - \frac{1}{2a} \arctan \frac{x}{a}.$

S1.1.2. Integrals Involving Irrational Functions

▶ **Integrals involving $x^{1/2}$.**

1. $\displaystyle\int \frac{x^{1/2}\,dx}{a^2 + b^2 x} = \frac{2}{b^2} x^{1/2} - \frac{2a}{b^3} \arctan \frac{bx^{1/2}}{a}.$

2. $\displaystyle \int \frac{x^{3/2}\,dx}{a^2 + b^2 x} = \frac{2x^{3/2}}{3b^2} - \frac{2a^2 x^{1/2}}{b^4} + \frac{2a^3}{b^5}\arctan\frac{bx^{1/2}}{a}.$

3. $\displaystyle \int \frac{x^{1/2}\,dx}{(a^2 + b^2 x)^2} = -\frac{x^{1/2}}{b^2(a^2 + b^2 x)} + \frac{1}{ab^3}\arctan\frac{bx^{1/2}}{a}.$

4. $\displaystyle \int \frac{x^{3/2}\,dx}{(a^2 + b^2 x)^2} = \frac{2x^{3/2}}{b^2(a^2 + b^2 x)} + \frac{3a^2 x^{1/2}}{b^4(a^2 + b^2 x)} - \frac{3a}{b^5}\arctan\frac{bx^{1/2}}{a}.$

5. $\displaystyle \int \frac{dx}{(a^2 + b^2 x)x^{1/2}} = \frac{2}{ab}\arctan\frac{bx^{1/2}}{a}.$

6. $\displaystyle \int \frac{dx}{(a^2 + b^2 x)x^{3/2}} = -\frac{2}{a^2 x^{1/2}} - \frac{2b}{a^3}\arctan\frac{bx^{1/2}}{a}.$

7. $\displaystyle \int \frac{dx}{(a^2 + b^2 x)^2 x^{1/2}} = \frac{x^{1/2}}{a^2(a^2 + b^2 x)} + \frac{1}{a^3 b}\arctan\frac{bx^{1/2}}{a}.$

8. $\displaystyle \int \frac{x^{1/2}\,dx}{a^2 - b^2 x} = -\frac{2}{b^2}x^{1/2} + \frac{2a}{b^3}\ln\left|\frac{a + bx^{1/2}}{a - bx^{1/2}}\right|.$

9. $\displaystyle \int \frac{x^{3/2}\,dx}{a^2 - b^2 x} = -\frac{2x^{3/2}}{3b^2} - \frac{2a^2 x^{1/2}}{b^4} + \frac{a^3}{b^5}\ln\left|\frac{a + bx^{1/2}}{a - bx^{1/2}}\right|.$

10. $\displaystyle \int \frac{x^{1/2}\,dx}{(a^2 - b^2 x)^2} = \frac{x^{1/2}}{b^2(a^2 - b^2 x)} - \frac{1}{2ab^3}\ln\left|\frac{a + bx^{1/2}}{a - bx^{1/2}}\right|.$

11. $\displaystyle \int \frac{x^{3/2}\,dx}{(a^2 - b^2 x)^2} = \frac{3a^2 x^{1/2} - 2b^2 x^{3/2}}{b^4(a^2 - b^2 x)} - \frac{3a}{2b^5}\ln\left|\frac{a + bx^{1/2}}{a - bx^{1/2}}\right|.$

12. $\displaystyle \int \frac{dx}{(a^2 - b^2 x)x^{1/2}} = \frac{1}{ab}\ln\left|\frac{a + bx^{1/2}}{a - bx^{1/2}}\right|.$

13. $\displaystyle \int \frac{dx}{(a^2 - b^2 x)x^{3/2}} = -\frac{2}{a^2 x^{1/2}} + \frac{b}{a^3}\ln\left|\frac{a + bx^{1/2}}{a - bx^{1/2}}\right|.$

14. $\displaystyle \int \frac{dx}{(a^2 - b^2 x)^2 x^{1/2}} = \frac{x^{1/2}}{a^2(a^2 - b^2 x)} + \frac{1}{2a^3 b}\ln\left|\frac{a + bx^{1/2}}{a - bx^{1/2}}\right|.$

▶ **Integrals involving $(a + bx)^{p/2}$.**

1. $\displaystyle \int (a + bx)^{p/2}\,dx = \frac{2}{b(p + 2)}(a + bx)^{(p+2)/2}.$

2. $\displaystyle \int x(a + bx)^{p/2}\,dx = \frac{2}{b^2}\left[\frac{(a + bx)^{(p+4)/2}}{p + 4} - \frac{a(a + bx)^{(p+2)/2}}{p + 2}\right].$

3. $\displaystyle \int x^2(a + bx)^{p/2}\,dx = \frac{2}{b^3}\left[\frac{(a + bx)^{(p+6)/2}}{p + 6} - \frac{2a(a + bx)^{(p+4)/2}}{p + 4} + \frac{a^2(a + bx)^{(p+2)/2}}{p + 2}\right].$

▶ **Integrals involving $(x^2 + a^2)^{1/2}$.**

1. $\displaystyle \int (x^2 + a^2)^{1/2}\,dx = \frac{1}{2}x(a^2 + x^2)^{1/2} + \frac{a^2}{2}\ln\left[x + (x^2 + a^2)^{1/2}\right].$

2. $\displaystyle\int x(x^2 + a^2)^{1/2}\,dx = \frac{1}{3}(a^2 + x^2)^{3/2}.$

3. $\displaystyle\int (x^2 + a^2)^{3/2}\,dx = \frac{1}{4}x(a^2 + x^2)^{3/2} + \frac{3}{8}a^2 x(a^2 + x^2)^{1/2} + \frac{3}{8}a^4 \ln\big|x + (x^2 + a^2)^{1/2}\big|.$

4. $\displaystyle\int \frac{1}{x}(x^2 + a^2)^{1/2}\,dx = (a^2 + x^2)^{1/2} - a\ln\left|\frac{a + (x^2 + a^2)^{1/2}}{x}\right|.$

5. $\displaystyle\int \frac{dx}{\sqrt{x^2 + a^2}} = \ln\big[x + (x^2 + a^2)^{1/2}\big].$

6. $\displaystyle\int \frac{x\,dx}{\sqrt{x^2 + a^2}} = (x^2 + a^2)^{1/2}.$

7. $\displaystyle\int (x^2 + a^2)^{-3/2}\,dx = a^{-2}x(x^2 + a^2)^{-1/2}.$

▶ **Integrals involving $(x^2 - a^2)^{1/2}$.**

1. $\displaystyle\int (x^2 - a^2)^{1/2}\,dx = \frac{1}{2}x(x^2 - a^2)^{1/2} - \frac{a^2}{2}\ln\big|x + (x^2 - a^2)^{1/2}\big|.$

2. $\displaystyle\int x(x^2 - a^2)^{1/2}\,dx = \frac{1}{3}(x^2 - a^2)^{3/2}.$

3. $\displaystyle\int (x^2 - a^2)^{3/2}\,dx = \frac{1}{4}x(x^2 - a^2)^{3/2} - \frac{3}{8}a^2 x(x^2 - a^2)^{1/2} + \frac{3}{8}a^4 \ln\big|x + (x^2 - a^2)^{1/2}\big|.$

4. $\displaystyle\int \frac{1}{x}(x^2 - a^2)^{1/2}\,dx = (x^2 - a^2)^{1/2} - a\arccos\left|\frac{a}{x}\right|.$

5. $\displaystyle\int \frac{dx}{\sqrt{x^2 - a^2}} = \ln\big|x + (x^2 - a^2)^{1/2}\big|.$

6. $\displaystyle\int \frac{x\,dx}{\sqrt{x^2 - a^2}} = (x^2 - a^2)^{1/2}.$

7. $\displaystyle\int (x^2 - a^2)^{-3/2}\,dx = -a^{-2}x(x^2 - a^2)^{-1/2}.$

▶ **Integrals involving $(a^2 - x^2)^{1/2}$.**

1. $\displaystyle\int (a^2 - x^2)^{1/2}\,dx = \frac{1}{2}x(a^2 - x^2)^{1/2} + \frac{a^2}{2}\arcsin\frac{x}{a}.$

2. $\displaystyle\int x(a^2 - x^2)^{1/2}\,dx = -\frac{1}{3}(a^2 - x^2)^{3/2}.$

3. $\displaystyle\int (a^2 - x^2)^{3/2}\,dx = \frac{1}{4}x(a^2 - x^2)^{3/2} + \frac{3}{8}a^2 x(a^2 - x^2)^{1/2} + \frac{3}{8}a^4 \arcsin\frac{x}{a}.$

4. $\displaystyle\int \frac{1}{x}(a^2 - x^2)^{1/2}\,dx = (a^2 - x^2)^{1/2} - a\ln\left|\frac{a + (a^2 - x^2)^{1/2}}{x}\right|.$

5. $\displaystyle\int \frac{dx}{\sqrt{a^2 - x^2}} = \arcsin\frac{x}{a}.$

6. $\displaystyle\int \frac{x\,dx}{\sqrt{a^2 - x^2}} = -(a^2 - x^2)^{1/2}.$

7. $\displaystyle\int (a^2 - x^2)^{-3/2}\,dx = a^{-2}x(a^2 - x^2)^{-1/2}.$

▶ **Integrals involving arbitrary powers. Reduction formulas.**

1. $\displaystyle\int \frac{dx}{x(ax^n + b)} = \frac{1}{bn} \ln\left|\frac{x^n}{ax^n + b}\right|.$

2. $\displaystyle\int \frac{dx}{x\sqrt{x^n + a^2}} = \frac{2}{an} \ln\left|\frac{x^{n/2}}{\sqrt{x^n + a^2} + a}\right|.$

3. $\displaystyle\int \frac{dx}{x\sqrt{x^n - a^2}} = \frac{2}{an} \arccos\left|\frac{a}{x^{n/2}}\right|.$

4. $\displaystyle\int \frac{dx}{x\sqrt{ax^{2n} + bx^n}} = -\frac{2\sqrt{ax^{2n} + bx^n}}{bnx^n}.$

The parameters a, b, p, m, and n below in Integrals 5–8 can assume arbitrary values, except for those at which denominators vanish in successive applications of a formula. Notation: $w = ax^n + b$.

5. $\displaystyle\int x^m(ax^n + b)^p \, dx = \frac{1}{m + np + 1}\left(x^{m+1}w^p + npb \int x^m w^{p-1} \, dx\right).$

6. $\displaystyle\int x^m(ax^n + b)^p \, dx = \frac{1}{bn(p + 1)}\left[-x^{m+1}w^{p+1} + (m + n + np + 1)\int x^m w^{p+1} \, dx\right].$

7. $\displaystyle\int x^m(ax^n + b)^p \, dx = \frac{1}{b(m + 1)}\left[x^{m+1}w^{p+1} - a(m + n + np + 1)\int x^{m+n} w^p \, dx\right].$

8. $\displaystyle\int x^m(ax^n + b)^p \, dx = \frac{1}{a(m + np + 1)}\left[x^{m-n+1}w^{p+1} - b(m - n + 1)\int x^{m-n} w^p \, dx\right].$

S1.1.3. Integrals Involving Exponential Functions

1. $\displaystyle\int e^{ax} \, dx = \frac{1}{a}e^{ax}.$

2. $\displaystyle\int a^x \, dx = \frac{a^x}{\ln a}.$

3. $\displaystyle\int xe^{ax} \, dx = e^{ax}\left(\frac{x}{a} - \frac{1}{a^2}\right).$

4. $\displaystyle\int x^2 e^{ax} \, dx = e^{ax}\left(\frac{x^2}{a} - \frac{2x}{a^2} + \frac{2}{a^3}\right).$

5. $\displaystyle\int x^n e^{ax} \, dx = e^{ax}\left[\frac{1}{a}x^n - \frac{n}{a^2}x^{n-1} + \frac{n(n - 1)}{a^3}x^{n-2} - \cdots + (-1)^{n-1}\frac{n!}{a^n}x + (-1)^n \frac{n!}{a^{n+1}}\right],$
$n = 1, 2, \ldots$

6. $\displaystyle\int P_n(x)e^{ax} \, dx = e^{ax}\sum_{k=0}^{n} \frac{(-1)^k}{a^{k+1}}\frac{d^k}{dx^k}P_n(x),$ where $P_n(x)$ is an arbitrary polynomial of degree n.

7. $\displaystyle\int \frac{dx}{a + be^{px}} = \frac{x}{a} - \frac{1}{ap}\ln|a + be^{px}|.$

8. $\displaystyle \int \frac{dx}{ae^{px} + be^{-px}} = \begin{cases} \dfrac{1}{p\sqrt{ab}} \arctan\left(e^{px}\sqrt{\dfrac{a}{b}}\right) & \text{if } ab > 0, \\[4mm] \dfrac{1}{2p\sqrt{-ab}} \ln\left(\dfrac{b + e^{px}\sqrt{-ab}}{b - e^{px}\sqrt{-ab}}\right) & \text{if } ab < 0. \end{cases}$

9. $\displaystyle \int \frac{dx}{\sqrt{a + be^{px}}} = \begin{cases} \dfrac{1}{p\sqrt{a}} \ln \dfrac{\sqrt{a + be^{px}} - \sqrt{a}}{\sqrt{a + be^{px}} + \sqrt{a}} & \text{if } a > 0, \\[4mm] \dfrac{2}{p\sqrt{-a}} \arctan \dfrac{\sqrt{a + be^{px}}}{\sqrt{-a}} & \text{if } a < 0. \end{cases}$

S1.1.4. Integrals Involving Hyperbolic Functions

▶ **Integrals involving $\cosh x$.**

1. $\displaystyle \int \cosh(a + bx)\, dx = \frac{1}{b} \sinh(a + bx).$

2. $\displaystyle \int x \cosh x\, dx = x \sinh x - \cosh x.$

3. $\displaystyle \int x^2 \cosh x\, dx = (x^2 + 2) \sinh x - 2x \cosh x.$

4. $\displaystyle \int x^{2n} \cosh x\, dx = (2n)! \sum_{k=1}^{n} \left[\frac{x^{2k}}{(2k)!} \sinh x - \frac{x^{2k-1}}{(2k-1)!} \cosh x \right].$

5. $\displaystyle \int x^{2n+1} \cosh x\, dx = (2n+1)! \sum_{k=0}^{n} \left[\frac{x^{2k+1}}{(2k+1)!} \sinh x - \frac{x^{2k}}{(2k)!} \cosh x \right].$

6. $\displaystyle \int x^p \cosh x\, dx = x^p \sinh x - px^{p-1} \cosh x + p(p-1) \int x^{p-2} \cosh x\, dx.$

7. $\displaystyle \int \cosh^2 x\, dx = \frac{1}{2}x + \frac{1}{4} \sinh 2x.$

8. $\displaystyle \int \cosh^3 x\, dx = \sinh x + \frac{1}{3} \sinh^3 x.$

9. $\displaystyle \int \cosh^{2n} x\, dx = C_{2n}^n \frac{x}{2^{2n}} + \frac{1}{2^{2n-1}} \sum_{k=0}^{n-1} C_{2n}^k \frac{\sinh[2(n-k)x]}{2(n-k)}, \qquad n = 1, 2, \dots$

10. $\displaystyle \int \cosh^{2n+1} x\, dx = \frac{1}{2^{2n}} \sum_{k=0}^{n} C_{2n+1}^k \frac{\sinh[(2n-2k+1)x]}{2n-2k+1} = \sum_{k=0}^{n} C_n^k \frac{\sinh^{2k+1} x}{2k+1},$

    $n = 1, 2, \dots$

11. $\displaystyle \int \cosh^p x\, dx = \frac{1}{p} \sinh x \cosh^{p-1} x + \frac{p-1}{p} \int \cosh^{p-2} x\, dx.$

12. $\displaystyle \int \cosh ax \cosh bx\, dx = \frac{1}{a^2 - b^2} (a \cosh bx \sinh ax - b \cosh ax \sinh bx).$

13. $\displaystyle \int \frac{dx}{\cosh ax} = \frac{2}{a} \arctan\left(e^{ax}\right).$

14. $\displaystyle \int \frac{dx}{\cosh^{2n} x} = \frac{\sinh x}{2n-1} \left[\frac{1}{\cosh^{2n-1} x} \right.$

$$\left. + \sum_{k=1}^{n-1} \frac{2^k(n-1)(n-2)\ldots(n-k)}{(2n-3)(2n-5)\ldots(2n-2k-1)} \frac{1}{\cosh^{2n-2k-1} x} \right], \qquad n = 1, 2, \ldots$$

15. $\displaystyle \int \frac{dx}{\cosh^{2n+1} x} = \frac{\sinh x}{2n} \left[\frac{1}{\cosh^{2n} x} \right.$

$$\left. + \sum_{k=1}^{n-1} \frac{(2n-1)(2n-3)\ldots(2n-2k+1)}{2^k(n-1)(n-2)\ldots(n-k)} \frac{1}{\cosh^{2n-2k} x} \right] + \frac{(2n-1)!!}{(2n)!!} \arctan \sinh x,$$

$$n = 1, 2, \ldots$$

16. $\displaystyle \int \frac{dx}{a + b \cosh x} = \begin{cases} -\dfrac{\operatorname{sign} x}{\sqrt{b^2 - a^2}} \arcsin \dfrac{b + a \cosh x}{a + b \cosh x} & \text{if } a^2 < b^2, \\[3mm] \dfrac{1}{\sqrt{a^2 - b^2}} \ln \dfrac{a + b + \sqrt{a^2 - b^2}\,\tanh(x/2)}{a + b - \sqrt{a^2 - b^2}\,\tanh(x/2)} & \text{if } a^2 > b^2. \end{cases}$

▶ **Integrals involving $\sinh x$.**

1. $\displaystyle \int \sinh(a + bx)\, dx = \frac{1}{b} \cosh(a + bx).$

2. $\displaystyle \int x \sinh x\, dx = x \cosh x - \sinh x.$

3. $\displaystyle \int x^2 \sinh x\, dx = (x^2 + 2) \cosh x - 2x \sinh x.$

4. $\displaystyle \int x^{2n} \sinh x\, dx = (2n)! \left[\sum_{k=0}^{n} \frac{x^{2k}}{(2k)!} \cosh x - \sum_{k=1}^{n} \frac{x^{2k-1}}{(2k-1)!} \sinh x \right].$

5. $\displaystyle \int x^{2n+1} \sinh x\, dx = (2n+1)! \sum_{k=0}^{n} \left[\frac{x^{2k+1}}{(2k+1)!} \cosh x - \frac{x^{2k}}{(2k)!} \sinh x \right].$

6. $\displaystyle \int x^p \sinh x\, dx = x^p \cosh x - p x^{p-1} \sinh x + p(p-1) \int x^{p-2} \sinh x\, dx.$

7. $\displaystyle \int \sinh^2 x\, dx = -\frac{1}{2}x + \frac{1}{4} \sinh 2x.$

8. $\displaystyle \int \sinh^3 x\, dx = -\cosh x + \frac{1}{3} \cosh^3 x.$

9. $\displaystyle \int \sinh^{2n} x\, dx = (-1)^n C_{2n}^n \frac{x}{2^{2n}} + \frac{1}{2^{2n-1}} \sum_{k=0}^{n-1} (-1)^k C_{2n}^k \frac{\sinh[2(n-k)x]}{2(n-k)}, \qquad n = 1, 2, \ldots$

10. $\displaystyle \int \sinh^{2n+1} x\, dx = \frac{1}{2^{2n}} \sum_{k=0}^{n} (-1)^k C_{2n+1}^k \frac{\cosh[(2n-2k+1)x]}{2n-2k+1}$

$$= \sum_{k=0}^{n} (-1)^{n+k} C_n^k \frac{\cosh^{2k+1} x}{2k+1}, \qquad n = 1, 2, \ldots$$

11. $\displaystyle\int \sinh^p x\, dx = \frac{1}{p} \sinh^{p-1} x \cosh x - \frac{p-1}{p} \int \sinh^{p-2} x\, dx.$

12. $\displaystyle\int \sinh ax \sinh bx\, dx = \frac{1}{a^2 - b^2}\left(a \cosh ax \sinh bx - b \cosh bx \sinh ax\right).$

13. $\displaystyle\int \frac{dx}{\sinh ax} = \frac{1}{a} \ln\left|\tanh \frac{ax}{2}\right|.$

14. $\displaystyle\int \frac{dx}{\sinh^{2n} x} = \frac{\cosh x}{2n-1}\left[-\frac{1}{\sinh^{2n-1} x}\right.$

$$\left. + \sum_{k=1}^{n-1} (-1)^{k-1} \frac{2^k (n-1)(n-2)\ldots(n-k)}{(2n-3)(2n-5)\ldots(2n-2k-1)} \frac{1}{\sinh^{2n-2k-1} x}\right], \qquad n = 1, 2, \ldots$$

15. $\displaystyle\int \frac{dx}{\sinh^{2n+1} x} = \frac{\cosh x}{2n}\left[-\frac{1}{\sinh^{2n} x}\right.$

$$\left. + \sum_{k=1}^{n-1} (-1)^{k-1} \frac{(2n-1)(2n-3)\ldots(2n-2k+1)}{2^k (n-1)(n-2)\ldots(n-k)} \frac{1}{\sinh^{2n-2k} x}\right] + (-1)^n \frac{(2n-1)!!}{(2n)!!} \ln\tanh\frac{x}{2},$$

$n = 1, 2, \ldots$

16. $\displaystyle\int \frac{dx}{a + b\sinh x} = \frac{1}{\sqrt{a^2 + b^2}} \ln \frac{a\tanh(x/2) - b + \sqrt{a^2 + b^2}}{a\tanh(x/2) - b - \sqrt{a^2 + b^2}}.$

17. $\displaystyle\int \frac{Ax + B\sinh x}{a + b\sinh x}\, dx = \frac{B}{b} x + \frac{Ab - Ba}{b\sqrt{a^2 + b^2}} \ln \frac{a\tanh(x/2) - b + \sqrt{a^2 + b^2}}{a\tanh(x/2) - b - \sqrt{a^2 + b^2}}.$

▶ **Integrals involving** $\tanh x$ **or** $\coth x$.

1. $\displaystyle\int \tanh x\, dx = \ln \cosh x.$

2. $\displaystyle\int \tanh^2 x\, dx = x - \tanh x.$

3. $\displaystyle\int \tanh^3 x\, dx = -\tfrac{1}{2} \tanh^2 x + \ln \cosh x.$

4. $\displaystyle\int \tanh^{2n} x\, dx = x - \sum_{k=1}^{n} \frac{\tanh^{2n-2k+1} x}{2n - 2k + 1}, \qquad n = 1, 2, \ldots$

5. $\displaystyle\int \tanh^{2n+1} x\, dx = \ln \cosh x - \sum_{k=1}^{n} \frac{(-1)^k C_n^k}{2k \cosh^{2k} x} = \ln \cosh x - \sum_{k=1}^{n} \frac{\tanh^{2n-2k+2} x}{2n - 2k + 2},$

$n = 1, 2, \ldots$

6. $\displaystyle\int \tanh^p x\, dx = -\frac{1}{p-1} \tanh^{p-1} x + \int \tanh^{p-2} x\, dx.$

7. $\displaystyle\int \coth x\, dx = \ln|\sinh x|.$

8. $\displaystyle\int \coth^2 x\, dx = x - \coth x.$

9. $\displaystyle\int \coth^3 x\, dx = -\tfrac{1}{2}\coth^2 x + \ln|\sinh x|.$

10. $\displaystyle\int \coth^{2n} x\, dx = x - \sum_{k=1}^{n}\frac{\coth^{2n-2k+1} x}{2n-2k+1}, \quad n = 1, 2, \ldots$

11. $\displaystyle\int \coth^{2n+1} x\, dx = \ln|\sinh x| - \sum_{k=1}^{n}\frac{C_n^k}{2k\,\sinh^{2k} x} = \ln|\sinh x| - \sum_{k=1}^{n}\frac{\coth^{2n-2k+2} x}{2n-2k+2},$

 $n = 1, 2, \ldots$

12. $\displaystyle\int \coth^p x\, dx = -\frac{1}{p-1}\coth^{p-1} x + \int \coth^{p-2} x\, dx.$

S1.1.5. Integrals Involving Logarithmic Functions

1. $\displaystyle\int \ln ax\, dx = x\ln ax - x.$

2. $\displaystyle\int x\ln x\, dx = \tfrac{1}{2}x^2\ln x - \tfrac{1}{4}x^2.$

3. $\displaystyle\int x^p \ln ax\, dx = \begin{cases} \dfrac{1}{p+1}x^{p+1}\ln ax - \dfrac{1}{(p+1)^2}x^{p+1} & \text{if } p \neq -1, \\ \tfrac{1}{2}\ln^2 ax & \text{if } p = -1. \end{cases}$

4. $\displaystyle\int (\ln x)^2\, dx = x(\ln x)^2 - 2x\ln x + 2x.$

5. $\displaystyle\int x(\ln x)^2\, dx = \tfrac{1}{2}x^2(\ln x)^2 - \tfrac{1}{2}x^2\ln x + \tfrac{1}{4}x^2.$

6. $\displaystyle\int x^p(\ln x)^2\, dx = \begin{cases} \dfrac{x^{p+1}}{p+1}(\ln x)^2 - \dfrac{2x^{p+1}}{(p+1)^2}\ln x + \dfrac{2x^{p+1}}{(p+1)^3} & \text{if } p \neq -1, \\ \tfrac{1}{3}\ln^3 x & \text{if } p = -1. \end{cases}$

7. $\displaystyle\int (\ln x)^n\, dx = \frac{x}{n+1}\sum_{k=0}^{n}(-1)^k(n+1)n\ldots(n-k+1)(\ln x)^{n-k}, \quad n = 1, 2, \ldots$

8. $\displaystyle\int (\ln x)^q\, dx = x(\ln x)^q - q\int (\ln x)^{q-1}\, dx,\ q \neq -1.$

9. $\displaystyle\int x^n(\ln x)^m\, dx = \frac{x^{n+1}}{m+1}\sum_{k=0}^{m}\frac{(-1)^k}{(n+1)^{k+1}}(m+1)m\ldots(m-k+1)(\ln x)^{m-k},$

 $n, m = 1, 2, \ldots$

10. $\displaystyle\int x^p(\ln x)^q\, dx = \frac{1}{p+1}x^{p+1}(\ln x)^q - \frac{q}{p+1}\int x^p(\ln x)^{q-1}\, dx, \quad p, q \neq -1.$

11. $\displaystyle\int \ln(a+bx)\, dx = \frac{1}{b}(ax+b)\ln(ax+b) - x.$

12. $\displaystyle\int x\ln(a+bx)\,dx = \frac{1}{2}\left(x^2 - \frac{a^2}{b^2}\right)\ln(a+bx) - \frac{1}{2}\left(\frac{x^2}{2} - \frac{a}{b}x\right).$

13. $\displaystyle\int x^2\ln(a+bx)\,dx = \frac{1}{3}\left(x^3 - \frac{a^3}{b^3}\right)\ln(a+bx) - \frac{1}{3}\left(\frac{x^3}{3} - \frac{ax^2}{2b} + \frac{a^2x}{b^2}\right).$

14. $\displaystyle\int \frac{\ln x\,dx}{(a+bx)^2} = -\frac{\ln x}{b(a+bx)} + \frac{1}{ab}\ln\frac{x}{a+bx}.$

15. $\displaystyle\int \frac{\ln x\,dx}{(a+bx)^3} = -\frac{\ln x}{2b(a+bx)^2} + \frac{1}{2ab(a+bx)} + \frac{1}{2a^2b}\ln\frac{x}{a+bx}.$

16. $\displaystyle\int \frac{\ln x\,dx}{\sqrt{a+bx}} = \begin{cases} \dfrac{2}{b}\left[(\ln x - 2)\sqrt{a+bx} + \sqrt{a}\,\ln\dfrac{\sqrt{a+bx}+\sqrt{a}}{\sqrt{a+bx}-\sqrt{a}}\right] & \text{if } a>0, \\[3mm] \dfrac{2}{b}\left[(\ln x - 2)\sqrt{a+bx} + 2\sqrt{-a}\,\arctan\dfrac{\sqrt{a+bx}}{\sqrt{-a}}\right] & \text{if } a<0. \end{cases}$

17. $\displaystyle\int \ln(x^2+a^2)\,dx = x\ln(x^2+a^2) - 2x + 2a\arctan(x/a).$

18. $\displaystyle\int x\ln(x^2+a^2)\,dx = \frac{1}{2}\left[(x^2+a^2)\ln(x^2+a^2) - x^2\right].$

19. $\displaystyle\int x^2\ln(x^2+a^2)\,dx = \frac{1}{3}\left[x^3\ln(x^2+a^2) - \frac{2}{3}x^3 + 2a^2x - 2a^3\arctan(x/a)\right].$

S1.1.6. Integrals Involving Trigonometric Functions

▶ **Integrals involving** $\cos x$ ($n = 1, 2, \ldots$).

1. $\displaystyle\int \cos(a+bx)\,dx = \frac{1}{b}\sin(a+bx).$

2. $\displaystyle\int x\cos x\,dx = \cos x + x\sin x.$

3. $\displaystyle\int x^2\cos x\,dx = 2x\cos x + (x^2-2)\sin x.$

4. $\displaystyle\int x^{2n}\cos x\,dx = (2n)!\left[\sum_{k=0}^{n}(-1)^k\frac{x^{2n-2k}}{(2n-2k)!}\sin x + \sum_{k=0}^{n-1}(-1)^k\frac{x^{2n-2k-1}}{(2n-2k-1)!}\cos x\right].$

5. $\displaystyle\int x^{2n+1}\cos x\,dx = (2n+1)!\sum_{k=0}^{n}\left[(-1)^k\frac{x^{2n-2k+1}}{(2n-2k+1)!}\sin x + \frac{x^{2n-2k}}{(2n-2k)!}\cos x\right].$

6. $\displaystyle\int x^p\cos x\,dx = x^p\sin x + px^{p-1}\cos x - p(p-1)\int x^{p-2}\cos x\,dx.$

7. $\displaystyle\int \cos^2 x\,dx = \frac{1}{2}x + \frac{1}{4}\sin 2x.$

8. $\displaystyle\int \cos^3 x\,dx = \sin x - \frac{1}{3}\sin^3 x.$

9. $\displaystyle\int \cos^{2n} x \, dx = \frac{1}{2^{2n}} C_{2n}^n x + \frac{1}{2^{2n-1}} \sum_{k=0}^{n-1} C_{2n}^k \frac{\sin[(2n-2k)x]}{2n-2k}.$

10. $\displaystyle\int \cos^{2n+1} x \, dx = \frac{1}{2^{2n}} \sum_{k=0}^n C_{2n+1}^k \frac{\sin[(2n-2k+1)x]}{2n-2k+1}.$

11. $\displaystyle\int \frac{dx}{\cos x} = \ln\left|\tan\left(\frac{x}{2} + \frac{\pi}{4}\right)\right|.$

12. $\displaystyle\int \frac{dx}{\cos^2 x} = \tan x.$

13. $\displaystyle\int \frac{dx}{\cos^3 x} = \frac{\sin x}{2\cos^2 x} + \frac{1}{2} \ln\left|\tan\left(\frac{x}{2} + \frac{\pi}{4}\right)\right|.$

14. $\displaystyle\int \frac{dx}{\cos^n x} = \frac{\sin x}{(n-1)\cos^{n-1} x} + \frac{n-2}{n-1} \int \frac{dx}{\cos^{n-2} x}, \quad n > 1.$

15. $\displaystyle\int \frac{x \, dx}{\cos^{2n} x} = \sum_{k=0}^{n-1} \frac{(2n-2)(2n-4)\dots(2n-2k+2)}{(2n-1)(2n-3)\dots(2n-2k+3)} \frac{(2n-2k)x \sin x - \cos x}{(2n-2k+1)(2n-2k)\cos^{2n-2k+1} x}$

$$+ \frac{2^{n-1}(n-1)!}{(2n-1)!!}\left(x\tan x + \ln|\cos x|\right).$$

16. $\displaystyle\int \cos ax \cos bx \, dx = \frac{\sin\left[(b-a)x\right]}{2(b-a)} + \frac{\sin\left[(b+a)x\right]}{2(b+a)}, \quad a \neq \pm b.$

17. $\displaystyle\int \frac{dx}{a + b\cos x} = \begin{cases} \dfrac{2}{\sqrt{a^2-b^2}} \arctan \dfrac{(a-b)\tan(x/2)}{\sqrt{a^2-b^2}} & \text{if } a^2 > b^2, \\[4mm] \dfrac{1}{\sqrt{b^2-a^2}} \ln\left|\dfrac{\sqrt{b^2-a^2} + (b-a)\tan(x/2)}{\sqrt{b^2-a^2} - (b-a)\tan(x/2)}\right| & \text{if } b^2 > a^2. \end{cases}$

18. $\displaystyle\int \frac{dx}{(a + b\cos x)^2} = \frac{b\sin x}{(b^2-a^2)(a+b\cos x)} - \frac{a}{b^2-a^2}\int \frac{dx}{a+b\cos x}.$

19. $\displaystyle\int \frac{dx}{a^2 + b^2\cos^2 x} = \frac{1}{a\sqrt{a^2+b^2}} \arctan \frac{a\tan x}{\sqrt{a^2+b^2}}.$

20. $\displaystyle\int \frac{dx}{a^2 - b^2\cos^2 x} = \begin{cases} \dfrac{1}{a\sqrt{a^2-b^2}} \arctan \dfrac{a\tan x}{\sqrt{a^2-b^2}} & \text{if } a^2 > b^2, \\[4mm] \dfrac{1}{2a\sqrt{b^2-a^2}} \ln\left|\dfrac{\sqrt{b^2-a^2} - a\tan x}{\sqrt{b^2-a^2} + a\tan x}\right| & \text{if } b^2 > a^2. \end{cases}$

21. $\displaystyle\int e^{ax} \cos bx \, dx = e^{ax}\left(\frac{b}{a^2+b^2}\sin bx + \frac{a}{a^2+b^2}\cos bx\right).$

22. $\displaystyle\int e^{ax} \cos^2 x \, dx = \frac{e^{ax}}{a^2+4}\left(a\cos^2 x + 2\sin x \cos x + \frac{2}{a}\right).$

23. $\displaystyle\int e^{ax} \cos^n x \, dx = \frac{e^{ax}\cos^{n-1} x}{a^2+n^2}(a\cos x + n\sin x) + \frac{n(n-1)}{a^2+n^2}\int e^{ax} \cos^{n-2} x \, dx.$

▶ **Integrals involving $\sin x$ $(n = 1, 2, \ldots)$.**

1. $\displaystyle\int \sin(a + bx)\, dx = -\frac{1}{b}\cos(a + bx).$

2. $\displaystyle\int x \sin x \, dx = \sin x - x \cos x.$

3. $\displaystyle\int x^2 \sin x \, dx = 2x \sin x - (x^2 - 2) \cos x.$

4. $\displaystyle\int x^3 \sin x \, dx = (3x^2 - 6) \sin x - (x^3 - 6x) \cos x.$

5. $\displaystyle\int x^{2n} \sin x \, dx = (2n)! \left[\sum_{k=0}^{n} (-1)^{k+1} \frac{x^{2n-2k}}{(2n - 2k)!} \cos x + \sum_{k=0}^{n-1} (-1)^k \frac{x^{2n-2k-1}}{(2n - 2k - 1)!} \sin x \right].$

6. $\displaystyle\int x^{2n+1} \sin x \, dx = (2n+1)! \sum_{k=0}^{n} \left[(-1)^{k+1} \frac{x^{2n-2k+1}}{(2n - 2k + 1)!} \cos x + (-1)^k \frac{x^{2n-2k}}{(2n - 2k)!} \sin x \right].$

7. $\displaystyle\int x^p \sin x \, dx = -x^p \cos x + px^{p-1} \sin x - p(p - 1) \int x^{p-2} \sin x \, dx.$

8. $\displaystyle\int \sin^2 x \, dx = \frac{1}{2}x - \frac{1}{4}\sin 2x.$

9. $\displaystyle\int x \sin^2 x \, dx = \frac{1}{4}x^2 - \frac{1}{4}x \sin 2x - \frac{1}{8}\cos 2x.$

10. $\displaystyle\int \sin^3 x \, dx = -\cos x + \frac{1}{3}\cos^3 x.$

11. $\displaystyle\int \sin^{2n} x \, dx = \frac{1}{2^{2n}} C_{2n}^n x + \frac{(-1)^n}{2^{2n-1}} \sum_{k=0}^{n-1} (-1)^k C_{2n}^k \frac{\sin[(2n - 2k)x]}{2n - 2k},$

   where $C_m^k = \dfrac{m!}{k!\,(m - k)!}$ are binomial coefficients $(0! = 1).$

12. $\displaystyle\int \sin^{2n+1} x \, dx = \frac{1}{2^{2n}} \sum_{k=0}^{n} (-1)^{n+k+1} C_{2n+1}^k \frac{\cos[(2n - 2k + 1)x]}{2n - 2k + 1}.$

13. $\displaystyle\int \frac{dx}{\sin x} = \ln \left| \tan \frac{x}{2} \right|.$

14. $\displaystyle\int \frac{dx}{\sin^2 x} = -\cot x.$

15. $\displaystyle\int \frac{dx}{\sin^3 x} = -\frac{\cos x}{2 \sin^2 x} + \frac{1}{2} \ln \left| \tan \frac{x}{2} \right|.$

16. $\displaystyle\int \frac{dx}{\sin^n x} = -\frac{\cos x}{(n - 1) \sin^{n-1} x} + \frac{n - 2}{n - 1} \int \frac{dx}{\sin^{n-2} x}, \quad n > 1.$

17. $\displaystyle\int \frac{x \, dx}{\sin^{2n} x} = -\sum_{k=0}^{n-1} \frac{(2n-2)(2n-4)\ldots(2n-2k+2)}{(2n-1)(2n-3)\ldots(2n-2k+3)} \frac{\sin x + (2n-2k)x \cos x}{(2n-2k+1)(2n-2k)\sin^{2n-2k+1} x}$

$$+ \frac{2^{n-1}(n-1)!}{(2n-1)!!} \left(\ln |\sin x| - x \cot x \right).$$

18. $\displaystyle\int \sin ax \sin bx \, dx = \frac{\sin[(b-a)x]}{2(b-a)} - \frac{\sin[(b+a)x]}{2(b+a)}, \quad a \neq \pm b.$

19. $\displaystyle\int \frac{dx}{a + b \sin x} = \begin{cases} \dfrac{2}{\sqrt{a^2 - b^2}} \arctan \dfrac{b + a \tan x/2}{\sqrt{a^2 - b^2}} & \text{if } a^2 > b^2, \\[2mm] \dfrac{1}{\sqrt{b^2 - a^2}} \ln \left| \dfrac{b - \sqrt{b^2 - a^2} + a \tan x/2}{b + \sqrt{b^2 - a^2} + a \tan x/2} \right| & \text{if } b^2 > a^2. \end{cases}$

20. $\displaystyle\int \frac{dx}{(a + b \sin x)^2} = \frac{b \cos x}{(a^2 - b^2)(a + b \sin x)} + \frac{a}{a^2 - b^2} \int \frac{dx}{a + b \sin x}.$

21. $\displaystyle\int \frac{dx}{a^2 + b^2 \sin^2 x} = \frac{1}{a\sqrt{a^2 + b^2}} \arctan \frac{\sqrt{a^2 + b^2} \, \tan x}{a}.$

22. $\displaystyle\int \frac{dx}{a^2 - b^2 \sin^2 x} = \begin{cases} \dfrac{1}{a\sqrt{a^2 - b^2}} \arctan \dfrac{\sqrt{a^2 - b^2} \, \tan x}{a} & \text{if } a^2 > b^2, \\[2mm] \dfrac{1}{2a\sqrt{b^2 - a^2}} \ln \left| \dfrac{\sqrt{b^2 - a^2} \, \tan x + a}{\sqrt{b^2 - a^2} \, \tan x - a} \right| & \text{if } b^2 > a^2. \end{cases}$

23. $\displaystyle\int \frac{\sin x \, dx}{\sqrt{1 + k^2 \sin^2 x}} = -\frac{1}{k} \arcsin \frac{k \cos x}{\sqrt{1 + k^2}}.$

24. $\displaystyle\int \frac{\sin x \, dx}{\sqrt{1 - k^2 \sin^2 x}} = -\frac{1}{k} \ln \left| k \cos x + \sqrt{1 - k^2 \sin^2 x} \right|.$

25. $\displaystyle\int \sin x \sqrt{1 + k^2 \sin^2 x} \, dx = -\frac{\cos x}{2} \sqrt{1 + k^2 \sin^2 x} - \frac{1 + k^2}{2k} \arcsin \frac{k \cos x}{\sqrt{1 + k^2}}.$

26. $\displaystyle\int \sin x \sqrt{1 - k^2 \sin^2 x} \, dx$
$$= -\frac{\cos x}{2} \sqrt{1 - k^2 \sin^2 x} - \frac{1 - k^2}{2k} \ln \left| k \cos x + \sqrt{1 - k^2 \sin^2 x} \right|.$$

27. $\displaystyle\int e^{ax} \sin bx \, dx = e^{ax} \left(\frac{a}{a^2 + b^2} \sin bx - \frac{b}{a^2 + b^2} \cos bx \right).$

28. $\displaystyle\int e^{ax} \sin^2 x \, dx = \frac{e^{ax}}{a^2 + 4} \left(a \sin^2 x - 2 \sin x \cos x + \frac{2}{a} \right).$

29. $\displaystyle\int e^{ax} \sin^n x \, dx = \frac{e^{ax} \sin^{n-1} x}{a^2 + n^2} (a \sin x - n \cos x) + \frac{n(n-1)}{a^2 + n^2} \int e^{ax} \sin^{n-2} x \, dx.$

▶ **Integrals involving $\sin x$ and $\cos x$.**

1. $\displaystyle\int \sin ax \cos bx \, dx = -\frac{\cos[(a+b)x]}{2(a+b)} - \frac{\cos\left[(a-b)x\right]}{2(a-b)}, \quad a \neq \pm b.$

2. $\displaystyle\int \frac{dx}{b^2 \cos^2 ax + c^2 \sin^2 ax} = \frac{1}{abc} \arctan \left(\frac{c}{b} \tan ax \right).$

3. $\displaystyle\int \frac{dx}{b^2 \cos^2 ax - c^2 \sin^2 ax} = \frac{1}{2abc} \ln \left| \frac{c \tan ax + b}{c \tan ax - b} \right|.$

4. $\displaystyle\int \frac{dx}{\cos^{2n} x \sin^{2m} x} = \sum_{k=0}^{n+m-1} C_{n+m-1}^{k} \frac{\tan^{2k-2m+1} x}{2k - 2m + 1}, \quad n, m = 1, 2, \dots$

5. $\displaystyle\int \frac{dx}{\cos^{2n+1} x \sin^{2m+1} x} = C_{n+m}^m \ln|\tan x| + \sum_{k=0}^{n+m} C_{n+m}^k \frac{\tan^{2k-2m} x}{2k - 2m}$, $\quad n, m = 1, 2, \ldots$

▶ **Reduction formulas.**
The parameters p and q below can assume any values, except for those at which the denominators on the right-hand side vanish.

1. $\displaystyle\int \sin^p x \cos^q x \, dx = -\frac{\sin^{p-1} x \cos^{q+1} x}{p + q} + \frac{p - 1}{p + q} \int \sin^{p-2} x \cos^q x \, dx.$

2. $\displaystyle\int \sin^p x \cos^q x \, dx = \frac{\sin^{p+1} x \cos^{q-1} x}{p + q} + \frac{q - 1}{p + q} \int \sin^p x \cos^{q-2} x \, dx.$

3. $\displaystyle\int \sin^p x \cos^q x \, dx = \frac{\sin^{p-1} x \cos^{q-1} x}{p + q} \left(\sin^2 x - \frac{q - 1}{p + q - 2} \right)$
$$+ \frac{(p - 1)(q - 1)}{(p + q)(p + q - 2)} \int \sin^{p-2} x \cos^{q-2} x \, dx.$$

4. $\displaystyle\int \sin^p x \cos^q x \, dx = \frac{\sin^{p+1} x \cos^{q+1} x}{p + 1} + \frac{p + q + 2}{p + 1} \int \sin^{p+2} x \cos^q x \, dx.$

5. $\displaystyle\int \sin^p x \cos^q x \, dx = -\frac{\sin^{p+1} x \cos^{q+1} x}{q + 1} + \frac{p + q + 2}{q + 1} \int \sin^p x \cos^{q+2} x \, dx.$

6. $\displaystyle\int \sin^p x \cos^q x \, dx = -\frac{\sin^{p-1} x \cos^{q+1} x}{q + 1} + \frac{p - 1}{q + 1} \int \sin^{p-2} x \cos^{q+2} x \, dx.$

7. $\displaystyle\int \sin^p x \cos^q x \, dx = \frac{\sin^{p+1} x \cos^{q-1} x}{p + 1} + \frac{q - 1}{p + 1} \int \sin^{p+2} x \cos^{q-2} x \, dx.$

▶ **Integrals involving $\tan x$ and $\cot x$.**

1. $\displaystyle\int \tan x \, dx = -\ln|\cos x|.$

2. $\displaystyle\int \tan^2 x \, dx = \tan x - x.$

3. $\displaystyle\int \tan^3 x \, dx = \tfrac{1}{2} \tan^2 x + \ln|\cos x|.$

4. $\displaystyle\int \tan^{2n} x \, dx = (-1)^n x - \sum_{k=1}^{n} \frac{(-1)^k (\tan x)^{2n-2k+1}}{2n - 2k + 1}$, $\quad n = 1, 2, \ldots$

5. $\displaystyle\int \tan^{2n+1} x \, dx = (-1)^{n+1} \ln|\cos x| - \sum_{k=1}^{n} \frac{(-1)^k (\tan x)^{2n-2k+2}}{2n - 2k + 2}$, $\quad n = 1, 2, \ldots$

6. $\displaystyle\int \frac{dx}{a + b \tan x} = \frac{1}{a^2 + b^2} \left(ax + b \ln|a \cos x + b \sin x| \right).$

7. $\displaystyle\int \frac{\tan x \, dx}{\sqrt{a + b \tan^2 x}} = \frac{1}{\sqrt{b - a}} \arccos \left(\sqrt{1 - \frac{a}{b}} \cos x \right)$, $\quad b > a, \, b > 0.$

8. $\displaystyle\int \cot x \, dx = \ln|\sin x|.$

9. $\int \cot^2 x\,dx = -\cot x - x.$

10. $\int \cot^3 x\,dx = -\frac{1}{2}\cot^2 x - \ln|\sin x|.$

11. $\int \cot^{2n} x\,dx = (-1)^n x + \sum_{k=1}^{n} \frac{(-1)^k(\cot x)^{2n-2k+1}}{2n-2k+1}, \qquad n = 1, 2, \ldots$

12. $\int \cot^{2n+1} x\,dx = (-1)^n \ln|\sin x| + \sum_{k=1}^{n} \frac{(-1)^k(\cot x)^{2n-2k+2}}{2n-2k+2}, \qquad n = 1, 2, \ldots$

13. $\int \dfrac{dx}{a+b\cot x} = \dfrac{1}{a^2+b^2}\left(ax - b\ln|a\sin x + b\cos x|\right).$

S1.1.7. Integrals Involving Inverse Trigonometric Functions

1. $\int \arcsin \dfrac{x}{a}\,dx = x\arcsin \dfrac{x}{a} + \sqrt{a^2 - x^2}.$

2. $\int \left(\arcsin \dfrac{x}{a}\right)^2 dx = x\left(\arcsin \dfrac{x}{a}\right)^2 - 2x + 2\sqrt{a^2 - x^2}\,\arcsin \dfrac{x}{a}.$

3. $\int x\arcsin \dfrac{x}{a}\,dx = \dfrac{1}{4}(2x^2 - a^2)\arcsin \dfrac{x}{a} + \dfrac{x}{4}\sqrt{a^2 - x^2}.$

4. $\int x^2 \arcsin \dfrac{x}{a}\,dx = \dfrac{x^3}{3}\arcsin \dfrac{x}{a} + \dfrac{1}{9}(x^2 + 2a^2)\sqrt{a^2 - x^2}.$

5. $\int \arccos \dfrac{x}{a}\,dx = x\arccos \dfrac{x}{a} - \sqrt{a^2 - x^2}.$

6. $\int \left(\arccos \dfrac{x}{a}\right)^2 dx = x\left(\arccos \dfrac{x}{a}\right)^2 - 2x - 2\sqrt{a^2 - x^2}\,\arccos \dfrac{x}{a}.$

7. $\int x\arccos \dfrac{x}{a}\,dx = \dfrac{1}{4}(2x^2 - a^2)\arccos \dfrac{x}{a} - \dfrac{x}{4}\sqrt{a^2 - x^2}.$

8. $\int x^2 \arccos \dfrac{x}{a}\,dx = \dfrac{x^3}{3}\arccos \dfrac{x}{a} - \dfrac{1}{9}(x^2 + 2a^2)\sqrt{a^2 - x^2}.$

9. $\int \arctan \dfrac{x}{a}\,dx = x\arctan \dfrac{x}{a} - \dfrac{a}{2}\ln(a^2 + x^2).$

10. $\int x\arctan \dfrac{x}{a}\,dx = \dfrac{1}{2}(x^2 + a^2)\arctan \dfrac{x}{a} - \dfrac{ax}{2}.$

11. $\int x^2 \arctan \dfrac{x}{a}\,dx = \dfrac{x^3}{3}\arctan \dfrac{x}{a} - \dfrac{ax^2}{6} + \dfrac{a^3}{6}\ln(a^2 + x^2).$

12. $\int \operatorname{arccot} \dfrac{x}{a}\,dx = x\operatorname{arccot} \dfrac{x}{a} + \dfrac{a}{2}\ln(a^2 + x^2).$

13. $\int x\operatorname{arccot} \dfrac{x}{a}\,dx = \dfrac{1}{2}(x^2 + a^2)\operatorname{arccot} \dfrac{x}{a} + \dfrac{ax}{2}.$

14. $\int x^2 \operatorname{arccot} \dfrac{x}{a}\,dx = \dfrac{x^3}{3}\operatorname{arccot} \dfrac{x}{a} + \dfrac{ax^2}{6} - \dfrac{a^3}{6}\ln(a^2 + x^2).$

S1.2. Tables of Definite Integrals

Throughout Section S1.2 it is assumed that n is a positive integer, unless otherwise specified.

S1.2.1. Integrals Involving Power-Law Functions

▶ **Integrals over a finite interval.**

1. $$\int_0^1 \frac{x^n\,dx}{x+1} = (-1)^n\left[\ln 2 + \sum_{k=1}^{n}\frac{(-1)^k}{k}\right].$$

2. $$\int_0^1 \frac{dx}{x^2 + 2x\cos\beta + 1} = \frac{\beta}{2\sin\beta}.$$

3. $$\int_0^1 \frac{\left(x^a + x^{-a}\right)dx}{x^2 + 2x\cos\beta + 1} = \frac{\pi\sin(a\beta)}{\sin(\pi a)\sin\beta}, \qquad |a| < 1,\ \beta \neq (2n+1)\pi.$$

4. $$\int_0^1 x^a(1-x)^{1-a}\,dx = \frac{\pi a(1-a)}{2\sin(\pi a)}, \qquad -1 < a < 1.$$

5. $$\int_0^1 \frac{dx}{x^a(1-x)^{1-a}} = \frac{\pi}{\sin(\pi a)}, \qquad 0 < a < 1.$$

6. $$\int_0^1 \frac{x^a\,dx}{(1-x)^a} = \frac{\pi a}{\sin(\pi a)}, \qquad -1 < a < 1.$$

7. $$\int_0^1 x^{p-1}(1-x)^{q-1}\,dx \equiv B(p,q) = \frac{\Gamma(p)\Gamma(q)}{\Gamma(p+q)}, \qquad p,q > 0.$$

8. $$\int_0^1 x^{p-1}(1-x^q)^{-p/q}\,dx = \frac{\pi}{q\sin(\pi p/q)}, \qquad q > p > 0.$$

9. $$\int_0^1 x^{p+q-1}(1-x^q)^{-p/q}\,dx = \frac{\pi p}{q^2\sin(\pi p/q)}, \qquad q > p.$$

10. $$\int_0^1 x^{q/p-1}(1-x^q)^{-1/p}\,dx = \frac{\pi}{q\sin(\pi/p)}, \qquad p > 1,\ q > 0.$$

11. $$\int_0^1 \frac{x^{p-1} - x^{-p}}{1-x}\,dx = \pi\cot(\pi p), \qquad |p| < 1.$$

12. $$\int_0^1 \frac{x^{p-1} - x^{-p}}{1+x}\,dx = \frac{\pi}{\sin(\pi p)}, \qquad |p| < 1.$$

13. $$\int_0^1 \frac{x^p - x^{-p}}{x-1}\,dx = \frac{1}{p} - \pi\cot(\pi p), \qquad |p| < 1.$$

14. $$\int_0^1 \frac{x^p - x^{-p}}{1+x}\,dx = \frac{1}{p} - \frac{\pi}{\sin(\pi p)}, \qquad |p| < 1.$$

15. $$\int_0^1 \frac{x^{1+p} - x^{1-p}}{1-x^2}\,dx = \frac{\pi}{2}\cot\left(\frac{\pi p}{2}\right) - \frac{1}{p}, \qquad |p| < 1.$$

16. $\displaystyle\int_0^1 \frac{x^{1+p} - x^{1-p}}{1 + x^2}\, dx = \frac{1}{p} - \frac{\pi}{2\sin(\pi p/2)}, \quad |p| < 1.$

17. $\displaystyle\int_0^1 \frac{dx}{\sqrt{(1 + a^2 x)(1 - x)}} = \frac{2}{a}\arctan a.$

18. $\displaystyle\int_0^1 \frac{dx}{\sqrt{(1 - a^2 x)(1 - x)}} = \frac{1}{a}\ln\frac{1 + a}{1 - a}.$

19. $\displaystyle\int_{-1}^1 \frac{dx}{(a - x)\sqrt{1 - x^2}} = \frac{\pi}{\sqrt{a^2 - 1}}, \quad 1 < a.$

20. $\displaystyle\int_0^1 \frac{x^n\, dx}{\sqrt{1 - x}} = \frac{2\,(2n)!!}{(2n + 1)!!}, \quad n = 1,\,2,\,\dots$

21. $\displaystyle\int_0^1 \frac{x^{n-1/2}\, dx}{\sqrt{1 - x}} = \frac{\pi\,(2n - 1)!!}{(2n)!!}, \quad n = 1,\,2,\,\dots$

22. $\displaystyle\int_0^1 \frac{x^{2n}\, dx}{\sqrt{1 - x^2}} = \frac{\pi}{2}\,\frac{1 \times 3 \times \dots \times (2n - 1)}{2 \times 4 \times \dots \times (2n)}, \quad n = 1,\,2,\,\dots$

23. $\displaystyle\int_0^1 \frac{x^{2n+1}\, dx}{\sqrt{1 - x^2}} = \frac{2 \times 4 \times \dots \times (2n)}{1 \times 3 \times \dots \times (2n + 1)}, \quad n = 1,\,2,\,\dots$

24. $\displaystyle\int_0^1 \frac{x^{\lambda-1}\, dx}{(1 + ax)(1 - x)^\lambda} = \frac{\pi}{(1 + a)^\lambda \sin(\pi\lambda)}, \quad 0 < \lambda < 1, \quad a > -1.$

25. $\displaystyle\int_0^1 \frac{x^{\lambda-1/2}\, dx}{(1 + ax)^\lambda (1 - x)^\lambda} = 2\pi^{-1/2}\Gamma\!\left(\lambda + \tfrac{1}{2}\right)\Gamma\!\left(1 - \lambda\right)\cos^{2\lambda} k\,\frac{\sin[(2\lambda - 1)k]}{(2\lambda - 1)\sin k},$
$k = \arctan\sqrt{a}, \quad -\tfrac{1}{2} < \lambda < 1, \quad a > 0.$

▶ **Integrals over an infinite interval.**

1. $\displaystyle\int_0^\infty \frac{dx}{ax^2 + b} = \frac{\pi}{2\sqrt{ab}}.$

2. $\displaystyle\int_0^\infty \frac{dx}{x^4 + 1} = \frac{\pi\sqrt{2}}{4}.$

3. $\displaystyle\int_0^\infty \frac{x^{a-1}\, dx}{x + 1} = \frac{\pi}{\sin(\pi a)}, \quad 0 < a < 1.$

4. $\displaystyle\int_0^\infty \frac{x^{\lambda-1}\, dx}{(1 + ax)^2} = \frac{\pi(1 - \lambda)}{a^\lambda \sin(\pi\lambda)}, \quad 0 < \lambda < 2.$

5. $\displaystyle\int_0^\infty \frac{x^{\lambda-1}\, dx}{(x + a)(x + b)} = \frac{\pi(a^{\lambda-1} - b^{\lambda-1})}{(b - a)\sin(\pi\lambda)}, \quad 0 < \lambda < 2.$

6. $\displaystyle\int_0^\infty \frac{x^{\lambda-1}(x + c)\, dx}{(x + a)(x + b)} = \frac{\pi}{\sin(\pi\lambda)}\left(\frac{a - c}{a - b}a^{\lambda-1} + \frac{b - c}{b - a}b^{\lambda-1}\right), \quad 0 < \lambda < 1.$

7. $\displaystyle\int_0^\infty \frac{x^\lambda\, dx}{(x + 1)^3} = \frac{\pi\lambda(1 - \lambda)}{2\sin(\pi\lambda)}, \quad -1 < \lambda < 2.$

8. $\displaystyle\int_0^\infty \frac{x^{\lambda-1}\,dx}{(x^2+a^2)(x^2+b^2)} = \frac{\pi\left(b^{\lambda-2}-a^{\lambda-2}\right)}{2\left(a^2-b^2\right)\sin(\pi\lambda/2)}, \quad 0 < \lambda < 4.$

9. $\displaystyle\int_0^\infty \frac{x^{p-1}-x^{q-1}}{1-x}\,dx = \pi[\cot(\pi p)-\cot(\pi q)], \quad p, q > 0.$

10. $\displaystyle\int_0^\infty \frac{x^{\lambda-1}\,dx}{(1+ax)^{n+1}} = (-1)^n\,\frac{\pi C_{\lambda-1}^n}{a^\lambda \sin(\pi\lambda)}, \quad 0 < \lambda < n+1, \quad C_{\lambda-1}^n = \frac{(\lambda-1)(\lambda-2)\ldots(\lambda-n)}{n!}.$

11. $\displaystyle\int_0^\infty \frac{x^m\,dx}{(a+bx)^{n+1/2}} = 2^{m+1}m!\,\frac{(2n-2m-3)!!}{(2n-1)!!}\,\frac{a^{m-n+1/2}}{b^{m+1}},$

$a, b > 0, \quad n, m = 1, 2, \ldots, \quad m < b - \tfrac{1}{2}.$

12. $\displaystyle\int_0^\infty \frac{dx}{(x^2+a^2)^n} = \frac{\pi}{2}\,\frac{(2n-3)!!}{(2n-2)!!}\,\frac{1}{a^{2n-1}}, \quad n = 1, 2, \ldots$

13. $\displaystyle\int_0^\infty \frac{(x+1)^{\lambda-1}}{(x+a)^{\lambda+1}}\,dx = \frac{1-a^{-\lambda}}{\lambda(a-1)}, \quad a > 0.$

14. $\displaystyle\int_0^\infty \frac{x^{a-1}\,dx}{x^b+1} = \frac{\pi}{b\sin(\pi a/b)}, \quad 0 < a \le b.$

15. $\displaystyle\int_0^\infty \frac{x^{a-1}\,dx}{(x^b+1)^2} = \frac{\pi(a-b)}{b^2\sin[\pi(a-b)/b]}, \quad a < 2b.$

16. $\displaystyle\int_0^\infty \frac{x^{\lambda-1/2}\,dx}{(x+a)^\lambda(x+b)^\lambda} = \sqrt{\pi}\left(\sqrt{a}+\sqrt{b}\right)^{1-2\lambda}\frac{\Gamma(\lambda-1/2)}{\Gamma(\lambda)}, \quad \lambda > 0.$

17. $\displaystyle\int_0^\infty \frac{1-x^a}{1-x^b}\,x^{c-1}\,dx = \frac{\pi\sin A}{b\sin C\sin(A+C)}, \quad A = \frac{\pi a}{b}, \quad C = \frac{\pi c}{b}; \quad a+c < b, \quad c > 0.$

18. $\displaystyle\int_0^\infty \frac{x^{a-1}\,dx}{(1+x^2)^{1-b}} = \tfrac{1}{2}B\left(\tfrac{1}{2}a, 1-b-\tfrac{1}{2}a\right), \quad \tfrac{1}{2}a+b < 1, \quad a > 0.$

19. $\displaystyle\int_0^\infty \frac{x^{2m}\,dx}{(ax^2+b)^n} = \frac{\pi(2m-1)!!\,(2n-2m-3)!!}{2\,(2n-2)!!\,a^m b^{n-m-1}\sqrt{ab}}, \quad a, b > 0, \quad n > m+1.$

20. $\displaystyle\int_0^\infty \frac{x^{2m+1}\,dx}{(ax^2+b)^n} = \frac{m!\,(n-m-2)!}{2(n-1)!a^{m+1}b^{n-m-1}}, \quad ab > 0, \quad n > m+1 \ge 1.$

21. $\displaystyle\int_0^\infty \frac{x^{\mu-1}\,dx}{(1+ax^p)^\nu} = \frac{1}{pa^{\mu/p}}B\left(\frac{\mu}{p}, \nu-\frac{\mu}{p}\right), \quad p > 0, \quad 0 < \mu < p\nu.$

22. $\displaystyle\int_0^\infty \left(\sqrt{x^2+a^2}-x\right)^n\,dx = \frac{na^{n+1}}{n^2-1}, \quad n = 2, 3, \ldots$

23. $\displaystyle\int_0^\infty \frac{dx}{\left(x+\sqrt{x^2+a^2}\right)^n} = \frac{n}{a^{n-1}(n^2-1)}, \quad n = 2, 3, \ldots$

24. $\displaystyle\int_0^\infty x^m\left(\sqrt{x^2+a^2}-x\right)^n\,dx = \frac{m!\,na^{n+m+1}}{(n-m-1)(n-m+1)\ldots(n+m+1)},$

$n, m = 1, 2, \ldots, \quad 0 \le m \le n-2.$

25. $\displaystyle\int_0^\infty \frac{x^m\,dx}{\left(x+\sqrt{x^2+a^2}\right)^n} = \frac{m!\,n}{(n-m-1)(n-m+1)\ldots(n+m+1)a^{n-m-1}}, \quad n = 2, 3, \ldots$

S1.2.2. Integrals Involving Exponential Functions

1. $\displaystyle\int_0^\infty e^{-ax}\,dx = \frac{1}{a}, \quad a > 0.$

2. $\displaystyle\int_0^1 x^n e^{-ax}\,dx = \frac{n!}{a^{n+1}} - e^{-a}\sum_{k=0}^n \frac{n!}{k!}\frac{1}{a^{n-k+1}}, \quad a > 0, \ \ n = 1, 2, \ldots$

3. $\displaystyle\int_0^\infty x^n e^{-ax}\,dx = \frac{n!}{a^{n+1}}, \quad a > 0, \ \ n = 1, 2, \ldots$

4. $\displaystyle\int_0^\infty \frac{e^{-ax}}{\sqrt{x}}\,dx = \sqrt{\frac{\pi}{a}}, \quad a > 0.$

5. $\displaystyle\int_0^\infty x^{\nu-1} e^{-\mu x}\,dx = \frac{\Gamma(\nu)}{\mu^\nu}, \quad \mu, \nu > 0.$

6. $\displaystyle\int_0^\infty \frac{dx}{1 + e^{ax}} = \frac{\ln 2}{a}.$

7. $\displaystyle\int_0^\infty \frac{x^{2n-1}\,dx}{e^{px} - 1} = (-1)^{n-1}\left(\frac{2\pi}{p}\right)^{2n}\frac{B_{2n}}{4n}, \quad n = 1, 2, \ldots;$

   the B_m are Bernoulli numbers.

8. $\displaystyle\int_0^\infty \frac{x^{2n-1}\,dx}{e^{px} + 1} = (1 - 2^{1-2n})\left(\frac{2\pi}{p}\right)^{2n}\frac{|B_{2n}|}{4n}, \quad n = 1, 2, \ldots$

9. $\displaystyle\int_{-\infty}^\infty \frac{e^{-px}\,dx}{1 + e^{-qx}} = \frac{\pi}{q\sin(\pi p/q)}, \quad q > p > 0 \ \text{ or } \ 0 > p > q.$

10. $\displaystyle\int_0^\infty \frac{e^{ax} + e^{-ax}}{e^{bx} + e^{-bx}}\,dx = \frac{\pi}{2b\cos\left(\dfrac{\pi a}{2b}\right)}, \quad b > a.$

11. $\displaystyle\int_0^\infty \frac{e^{-px} - e^{-qx}}{1 - e^{-(p+q)x}}\,dx = \frac{\pi}{p+q}\cot\frac{\pi p}{p+q}, \quad p, q > 0.$

12. $\displaystyle\int_0^\infty \left(1 - e^{-\beta x}\right)^\nu e^{-\mu x}\,dx = \frac{1}{\beta}B\left(\frac{\mu}{\beta}, \nu + 1\right).$

13. $\displaystyle\int_0^\infty \exp\left(-ax^2\right)\,dx = \frac{1}{2}\sqrt{\frac{\pi}{a}}, \quad a > 0.$

14. $\displaystyle\int_0^\infty x^{2n+1}\exp\left(-ax^2\right)\,dx = \frac{n!}{2a^{n+1}}, \quad a > 0, \ \ n = 1, 2, \ldots$

15. $\displaystyle\int_0^\infty x^{2n}\exp\left(-ax^2\right)\,dx = \frac{1\times 3\times\ldots\times(2n-1)\sqrt{\pi}}{2^{n+1}a^{n+1/2}}, \quad a > 0, \ \ n = 1, 2, \ldots$

16. $\displaystyle\int_{-\infty}^\infty \exp\left(-a^2x^2 \pm bx\right)\,dx = \frac{\sqrt{\pi}}{|a|}\exp\left(\frac{b^2}{4a^2}\right).$

17. $\displaystyle\int_0^\infty \exp\left(-ax^2 - \frac{b}{x^2}\right)\,dx = \frac{1}{2}\sqrt{\frac{\pi}{a}}\exp\left(-2\sqrt{ab}\right), \quad a, b > 0.$

18. $\displaystyle\int_0^\infty \exp\left(-x^a\right)\,dx = \frac{1}{a}\Gamma\left(\frac{1}{a}\right), \quad a > 0.$

S1.2.3. Integrals Involving Hyperbolic Functions

1. $\displaystyle\int_0^\infty \frac{dx}{\cosh ax} = \frac{\pi}{2|a|}.$

2. $\displaystyle\int_0^\infty \frac{dx}{a + b\cosh x} = \begin{cases} \dfrac{2}{\sqrt{b^2 - a^2}}\arctan\dfrac{\sqrt{b^2 - a^2}}{a + b} & \text{if } |b| > |a|, \\[4mm] \dfrac{1}{\sqrt{a^2 - b^2}}\ln\dfrac{a + b + \sqrt{a^2 - b^2}}{a + b - \sqrt{a^2 + b^2}} & \text{if } |b| < |a|. \end{cases}$

3. $\displaystyle\int_0^\infty \frac{x^{2n}\,dx}{\cosh ax} = \left(\frac{\pi}{2a}\right)^{2n+1}|B_{2n}|, \quad a > 0; \quad \text{the } B_m \text{ are Bernoulli numbers.}$

4. $\displaystyle\int_0^\infty \frac{x^{2n}}{\cosh^2 ax}\,dx = \frac{\pi^{2n}(2^{2n} - 2)}{a(2a)^{2n}}|B_{2n}|, \quad a > 0.$

5. $\displaystyle\int_0^\infty \frac{\cosh ax}{\cosh bx}\,dx = \frac{\pi}{2b\cos\left(\dfrac{\pi a}{2b}\right)}, \quad b > |a|.$

6. $\displaystyle\int_0^\infty x^{2n}\frac{\cosh ax}{\cosh bx}\,dx = \frac{\pi}{2b}\frac{d^{2n}}{da^{2n}}\frac{1}{\cos\left(\frac{1}{2}\pi a/b\right)}, \quad b > |a|, \quad n = 1, 2, \ldots$

7. $\displaystyle\int_0^\infty \frac{\cosh ax\cosh bx}{\cosh(cx)}\,dx = \frac{\pi}{c}\frac{\cos\left(\dfrac{\pi a}{2c}\right)\cos\left(\dfrac{\pi b}{2c}\right)}{\cos\left(\dfrac{\pi a}{c}\right) + \cos\left(\dfrac{\pi b}{c}\right)}, \quad c > |a| + |b|.$

8. $\displaystyle\int_0^\infty \frac{x\,dx}{\sinh ax} = \frac{\pi^2}{2a^2}, \quad a > 0.$

9. $\displaystyle\int_0^\infty \frac{dx}{a + b\sinh x} = \frac{1}{\sqrt{a^2 + b^2}}\ln\frac{a + b + \sqrt{a^2 + b^2}}{a + b - \sqrt{a^2 + b^2}}, \quad ab \neq 0.$

10. $\displaystyle\int_0^\infty \frac{\sinh ax}{\sinh bx}\,dx = \frac{\pi}{2b}\tan\left(\frac{\pi a}{2b}\right), \quad b > |a|.$

11. $\displaystyle\int_0^\infty x^{2n}\frac{\sinh ax}{\sinh bx}\,dx = \frac{\pi}{2b}\frac{d^{2n}}{dx^{2n}}\tan\left(\frac{\pi a}{2b}\right), \quad b > |a|, \quad n = 1, 2, \ldots$

12. $\displaystyle\int_0^\infty \frac{x^{2n}}{\sinh^2 ax}\,dx = \frac{\pi^{2n}}{a^{2n+1}}|B_{2n}|, \quad a > 0.$

S1.2.4. Integrals Involving Logarithmic Functions

1. $\displaystyle\int_0^1 x^{a-1}\ln^n x\,dx = (-1)^n n!\,a^{-n-1}, \quad a > 0, \quad n = 1, 2, \ldots$

2. $\displaystyle\int_0^1 \frac{\ln x}{x + 1}\,dx = -\frac{\pi^2}{12}.$

3. $\displaystyle\int_0^1 \frac{x^n\ln x}{x + 1}\,dx = (-1)^{n+1}\left[\frac{\pi^2}{12} + \sum_{k=1}^n \frac{(-1)^k}{k^2}\right], \quad n = 1, 2, \ldots$

4. $\displaystyle\int_0^1 \frac{x^{\mu-1}\ln x}{x+a}\,dx = \frac{\pi a^{\mu-1}}{\sin(\pi\mu)}\big[\ln a - \pi\cot(\pi\mu)\big], \quad 0 < \mu < 1.$

5. $\displaystyle\int_0^1 |\ln x|^\mu\,dx = \Gamma(\mu+1), \quad \mu > -1.$

6. $\displaystyle\int_0^\infty x^{\mu-1}\ln(1+ax)\,dx = \frac{\pi}{\mu a^\mu \sin(\pi\mu)}, \quad -1 < \mu < 0.$

7. $\displaystyle\int_0^1 x^{2n-1}\ln(1+x)\,dx = \frac{1}{2n}\sum_{k=1}^{2n}\frac{(-1)^{k-1}}{k}, \quad n = 1, 2, \ldots$

8. $\displaystyle\int_0^1 x^{2n}\ln(1+x)\,dx = \frac{1}{2n+1}\left[\ln 4 + \sum_{k=1}^{2n+1}\frac{(-1)^k}{k}\right], \quad n = 0, 1, \ldots$

9. $\displaystyle\int_0^1 x^{n-1/2}\ln(1+x)\,dx = \frac{2\ln 2}{2n+1} + \frac{4(-1)^n}{2n+1}\left[\pi - \sum_{k=0}^{n}\frac{(-1)^k}{2k+1}\right], \quad n = 1, 2, \ldots$

10. $\displaystyle\int_0^\infty \ln\frac{a^2+x^2}{b^2+x^2}\,dx = \pi(a-b), \quad a, b > 0.$

11. $\displaystyle\int_0^\infty \frac{x^{p-1}\ln x}{1+x^q}\,dx = -\frac{\pi^2\cos(\pi p/q)}{q^2\sin^2(\pi p/q)}, \quad 0 < p < q.$

12. $\displaystyle\int_0^\infty e^{-\mu x}\ln x\,dx = -\frac{1}{\mu}(\mathcal{C}+\ln\mu), \quad \mu > 0, \quad \mathcal{C} = 0.5772\ldots$

S1.2.5. Integrals Involving Trigonometric Functions

▶ **Integrals over a finite interval.**

1. $\displaystyle\int_0^{\pi/2}\cos^{2n} x\,dx = \frac{\pi}{2}\,\frac{1\times 3\times\ldots\times(2n-1)}{2\times 4\times\ldots\times(2n)}, \quad n = 1, 2, \ldots$

2. $\displaystyle\int_0^{\pi/2}\cos^{2n+1} x\,dx = \frac{2\times 4\times\ldots\times(2n)}{1\times 3\times\ldots\times(2n+1)}, \quad n = 1, 2, \ldots$

3. $\displaystyle\int_0^{\pi/2} x\cos^n x\,dx = -\sum_{k=0}^{m-1}\frac{(n-2k+1)(n-2k+3)\ldots(n-1)}{(n-2k)(n-2k+2)\ldots n}\,\frac{1}{n-2k}$

$$+\begin{cases}\dfrac{\pi}{2}\dfrac{(2m-2)!!}{(2m-1)!!} & \text{if } n = 2m-1, \\[2mm] \dfrac{\pi^2}{8}\dfrac{(2m-1)!!}{(2m)!!} & \text{if } n = 2m, \end{cases} \qquad m = 1, 2, \ldots$$

4. $\displaystyle\int_0^\pi \frac{dx}{(a+b\cos x)^{n+1}} = \frac{\pi}{2^n(a+b)^n\sqrt{a^2-b^2}}\sum_{k=0}^{n}\frac{(2n-2k-1)!!\,(2k-1)!!}{(n-k)!\,k!}\left(\frac{a+b}{a-b}\right)^k, \quad a > |b|.$

5. $\displaystyle\int_0^{\pi/2}\sin^{2n} x\,dx = \frac{\pi}{2}\,\frac{1\times 3\times\ldots\times(2n-1)}{2\times 4\times\ldots\times(2n)}, \quad n = 1, 2, \ldots$

6. $\int_0^{\pi/2} \sin^{2n+1} x \, dx = \dfrac{2 \times 4 \times \ldots \times (2n)}{1 \times 3 \times \ldots \times (2n+1)}, \qquad n = 1, 2, \ldots$

7. $\int_0^{\pi} x \sin^{\mu} x \, dx = \dfrac{\pi^2}{2^{\mu+1}} \dfrac{\Gamma(\mu+1)}{\left[\Gamma\left(\mu+\frac{1}{2}\right)\right]^2}, \qquad \mu > -1.$

8. $\int_0^{\pi/2} \dfrac{\sin x \, dx}{\sqrt{1 - k^2 \sin^2 x}} = \dfrac{1}{2k} \ln \dfrac{1+k}{1-k}.$

9. $\int_0^{\pi/2} \sin^{2n+1} x \cos^{2m+1} x \, dx = \dfrac{n! \, m!}{2(n+m+1)!}, \qquad n, m = 1, 2, \ldots$

10. $\int_0^{\pi/2} \sin^{p-1} x \cos^{q-1} x \, dx = \frac{1}{2} B\left(\frac{1}{2}p, \frac{1}{2}q\right).$

11. $\int_0^{2\pi} (a \sin x + b \cos x)^{2n} \, dx = 2\pi \dfrac{(2n-1)!!}{(2n)!!} \left(a^2 + b^2\right)^n, \qquad n = 1, 2, \ldots$

12. $\int_0^{\pi} \dfrac{\sin x \, dx}{\sqrt{a^2 + 1 - 2a \cos x}} = \begin{cases} 2 & \text{if } 0 \le a \le 1, \\ 2/a & \text{if } 1 < a. \end{cases}$

13. $\int_0^{\pi/2} (\tan x)^{\pm \lambda} \, dx = \dfrac{\pi}{2 \cos\left(\frac{1}{2}\pi\lambda\right)}, \qquad |\lambda| < 1.$

▶ **Integrals over an infinite interval.**

1. $\int_0^{\infty} \dfrac{\cos ax}{\sqrt{x}} \, dx = \sqrt{\dfrac{\pi}{2a}}, \qquad a > 0.$

2. $\int_0^{\infty} \dfrac{\cos ax - \cos bx}{x} \, dx = \ln \left| \dfrac{b}{a} \right|, \qquad ab \ne 0.$

3. $\int_0^{\infty} \dfrac{\cos ax - \cos bx}{x^2} \, dx = \frac{1}{2}\pi(b - a), \qquad a, b \ge 0.$

4. $\int_0^{\infty} x^{\mu-1} \cos ax \, dx = a^{-\mu} \Gamma(\mu) \cos\left(\frac{1}{2}\pi\mu\right), \qquad a > 0, \quad 0 < \mu < 1.$

5. $\int_0^{\infty} \dfrac{\cos ax}{b^2 + x^2} \, dx = \dfrac{\pi}{2b} e^{-ab}, \qquad a, b > 0.$

6. $\int_0^{\infty} \dfrac{\cos ax}{b^4 + x^4} \, dx = \dfrac{\pi\sqrt{2}}{4b^3} \exp\left(-\dfrac{ab}{\sqrt{2}}\right) \left[\cos\left(\dfrac{ab}{\sqrt{2}}\right) + \sin\left(\dfrac{ab}{\sqrt{2}}\right)\right], \qquad a, b > 0.$

7. $\int_0^{\infty} \dfrac{\cos ax}{(b^2 + x^2)^2} \, dx = \dfrac{\pi}{4b^3}(1 + ab)e^{-ab}, \qquad a, b > 0.$

8. $\int_0^{\infty} \dfrac{\cos ax \, dx}{(b^2 + x^2)(c^2 + x^2)} = \dfrac{\pi \left(be^{-ac} - ce^{-ab}\right)}{2bc\left(b^2 - c^2\right)}, \qquad a, b, c > 0.$

9. $\int_0^{\infty} \cos\left(ax^2\right) dx = \dfrac{1}{2} \sqrt{\dfrac{\pi}{2a}}, \qquad a > 0.$

10. $\int_0^{\infty} \cos\left(ax^p\right) dx = \dfrac{\Gamma(1/p)}{pa^{1/p}} \cos \dfrac{\pi}{2p}, \qquad a > 0, \quad p > 1.$

11. $\displaystyle\int_0^\infty \frac{\sin ax}{x}\,dx = \frac{\pi}{2}\,\mathrm{sign}\,a.$

12. $\displaystyle\int_0^\infty \frac{\sin^2 ax}{x^2}\,dx = \frac{\pi}{2}|a|.$

13. $\displaystyle\int_0^\infty \frac{\sin ax}{\sqrt{x}}\,dx = \sqrt{\frac{\pi}{2a}}, \quad a > 0.$

14. $\displaystyle\int_0^\infty x^{\mu-1}\sin ax\,dx = a^{-\mu}\Gamma(\mu)\sin\left(\tfrac{1}{2}\pi\mu\right), \quad a > 0, \quad 0 < \mu < 1.$

15. $\displaystyle\int_0^\infty \sin\left(ax^2\right)dx = \frac{1}{2}\sqrt{\frac{\pi}{2a}}, \quad a > 0.$

16. $\displaystyle\int_0^\infty \sin\left(ax^p\right)dx = \frac{\Gamma(1/p)}{pa^{1/p}}\sin\frac{\pi}{2p}, \quad a > 0, \quad p > 1.$

17. $\displaystyle\int_0^\infty \frac{\sin x \cos ax}{x}\,dx = \begin{cases} \frac{\pi}{2} & \text{if } |a| < 1, \\ \frac{\pi}{4} & \text{if } |a| = 1, \\ 0 & \text{if } 1 < |a|. \end{cases}$

18. $\displaystyle\int_0^\infty \frac{\tan ax}{x}\,dx = \frac{\pi}{2}\,\mathrm{sign}\,a.$

19. $\displaystyle\int_0^\infty e^{-ax}\sin bx\,dx = \frac{b}{a^2 + b^2}, \quad a > 0.$

20. $\displaystyle\int_0^\infty e^{-ax}\cos bx\,dx = \frac{a}{a^2 + b^2}, \quad a > 0.$

21. $\displaystyle\int_0^\infty \exp\left(-ax^2\right)\cos bx\,dx = \frac{1}{2}\sqrt{\frac{\pi}{a}}\,\exp\left(-\frac{b^2}{4a}\right).$

22. $\displaystyle\int_0^\infty \cos(ax^2)\cos bx\,dx = \sqrt{\frac{\pi}{8a}}\left[\cos\left(\frac{b^2}{4a}\right) + \sin\left(\frac{b^2}{4a}\right)\right], \quad a, b > 0.$

23. $\displaystyle\int_0^\infty (\cos ax + \sin ax)\cos(b^2x^2)\,dx = \frac{1}{b}\sqrt{\frac{\pi}{8}}\,\exp\left(-\frac{a^2}{2b}\right), \quad a, b > 0.$

24. $\displaystyle\int_0^\infty \left[\cos ax + \sin ax\right]\sin(b^2x^2)\,dx = \frac{1}{b}\sqrt{\frac{\pi}{8}}\,\exp\left(-\frac{a^2}{2b}\right), \quad a, b > 0.$

S1.3. General Reduction Formulas for the Evaluation of Definite Integrals

▶ *Below are some general formulas, involving arbitrary functions and parameters, that could facilitate the evaluation of integrals.*

S1.3.1. Integrals Involving Functions of a Linear or Rational Argument

1. $\displaystyle\int_{-a}^a f(x)\,dx = 0 \quad \text{if } f(x) \text{ is odd.}$

2. $\displaystyle\int_{-a}^{a} f(x)\,dx = 2\int_{0}^{a} f(x)\,dx \quad$ if $f(x)$ is even.

3. $\displaystyle\int_{0}^{a} [f(x) + f(a-x)]\,dx = 2\int_{0}^{a} f(x)\,dx.$

4. $\displaystyle\int_{0}^{a} [f(x) - f(a-x)]\,dx = 0.$

5. $\displaystyle\int_{0}^{1} f\left(2x\sqrt{1-x^2}\right) dx = \int_{0}^{1} f\left(1 - x^2\right) dx.$

S1.3.2. Integrals Involving Functions with Trigonometric Argument

1. $\displaystyle\int_{0}^{\pi} f(\sin x)\,dx = 2\int_{0}^{\pi/2} f(\sin x)\,dx.$

2. $\displaystyle\int_{0}^{\pi/2} f(\sin x)\,dx = \int_{0}^{\pi/2} f(\cos x)\,dx.$

3. $\displaystyle\int_{0}^{\pi/2} f(\sin x, \cos x)\,dx = 0 \quad$ if $f(x, y) = -f(y, x).$

4. $\displaystyle\int_{0}^{\pi/2} f(\sin 2x)\cos x\,dx = \int_{0}^{\pi/2} f(\cos^2 x)\cos x\,dx.$

5. $\displaystyle\int_{0}^{n\pi} x f(\sin x)\,dx = \pi n^2 \int_{0}^{\pi/2} f(\sin x)\,dx \quad$ if $f(x) = f(-x).$

6. $\displaystyle\int_{0}^{n\pi} x f(\sin x)\,dx = (-1)^{n-1}\pi n \int_{0}^{\pi/2} f(\sin x)\,dx \quad$ if $f(-x) = -f(x).$

7. $\displaystyle\int_{0}^{2\pi} f(a\sin x + b\cos x)\,dx = \int_{0}^{2\pi} f\left(\sqrt{a^2+b^2}\,\sin x\right) dx = 2\int_{0}^{\pi} f\left(\sqrt{a^2+b^2}\,\cos x\right) dx.$

8. $\displaystyle\int_{0}^{\pi} f\left(\frac{\sin^2 x}{1 + 2a\cos x + a^2}\right) dx = \int_{0}^{\pi} f(\sin^2 x)\,dx \quad$ if $|a| \geq 1.$

9. $\displaystyle\int_{0}^{\pi} f\left(\frac{\sin^2 x}{1 + 2a\cos x + a^2}\right) dx = \int_{0}^{\pi} f\left(\frac{\sin^2 x}{a^2}\right) dx \quad$ if $0 < |a| < 1.$

S1.3.3. Integrals Involving Logarithmic Functions

1. $\displaystyle\int_{a}^{b} f(x)\ln^n x\,dx = \left[\left(\frac{d}{d\lambda}\right)^n \int_{a}^{b} x^\lambda f(x)\,dx\right]_{\lambda=0}.$

2. $\displaystyle\int_{a}^{b} f(x)\ln^n g(x)\,dx = \left[\left(\frac{d}{d\lambda}\right)^n \int_{a}^{b} f(x)[g(x)]^\lambda\,dx\right]_{\lambda=0}.$

3. $\displaystyle\int_{a}^{b} f(x)[g(x)]^\lambda \ln^n g(x)\,dx = \left(\frac{d}{d\lambda}\right)^n \int_{a}^{b} f(x)[g(x)]^\lambda\,dx.$

S1.3.4. General Reduction Formulas for the Calculation of Improper Integrals

▶ **Improper integrals involving power functions.**

1. $\displaystyle\int_0^\infty f\left(\frac{a+bx}{1+x}\right)\frac{dx}{(1+x)^2} = \frac{1}{b-a}\int_a^b f(x)\,dx.$

2. $\displaystyle\int_0^\infty \frac{f(ax)-f(bx)}{x}\,dx = \left[f(0)-f(\infty)\right]\ln\frac{b}{a}$ if $a>0$, $b>0$, $f(x)$ is continuous on $[0,\infty)$, and $f(\infty) = \lim\limits_{x\to\infty} f(x)$ is a finite quantity.

3. $\displaystyle\int_0^\infty \frac{f(ax)-f(bx)}{x}\,dx = f(0)\ln\frac{b}{a}$ if $a>0$, $b>0$, $f(x)$ is continuous on $[0,\infty)$, and the integral $\displaystyle\int_c^\infty \frac{f(x)}{x}\,dx$ exists; $c>0$.

4. $\displaystyle\int_0^\infty f\left(\left|ax-\frac{b}{x}\right|\right)dx = \frac{1}{a}\int_0^\infty f(|x|)\,dx$ if $a>0$, $b>0$.

5. $\displaystyle\int_0^\infty x^2 f\left(\left|ax-\frac{b}{x}\right|\right)dx = \frac{1}{a^3}\int_0^\infty (x^2+ab)f(|x|)\,dx$ if $a>0$, $b>0$.

6. $\displaystyle\int_0^\infty f\left(\left|ax-\frac{b}{x}\right|\right)\frac{dx}{x^2} = \frac{1}{b}\int_0^\infty f(|x|)\,dx$ if $a>0$, $b>0$.

7. $\displaystyle\int_0^\infty f\left(x,\frac{1}{x}\right)\frac{dx}{x} = 2\int_0^1 f\left(x,\frac{1}{x}\right)\frac{dx}{x}$ if $f(x,y)=f(y,x)$.

8. $\displaystyle\int_0^\infty f\left(x,\frac{a}{x}\right)\frac{dx}{x} = 0$ if $f(x,y)=-f(y,x)$, $a>0$ (the integral is assumed to exist).

▶ **Improper integrals involving logarithmic functions.**

1. $\displaystyle\int_0^\infty f\left(\frac{x}{a}+\frac{a}{x}\right)\frac{\ln x}{x}\,dx = \ln a\int_0^\infty f\left(\frac{x}{a}+\frac{a}{x}\right)\frac{dx}{x}$ if $a>0$.

2. $\displaystyle\int_0^\infty f\left(\frac{x^p}{a}+\frac{a}{x^p}\right)\frac{\ln x}{x}\,dx = \frac{\ln a}{p}\int_0^\infty f\left(\frac{x^p}{a}+\frac{a}{x^p}\right)\frac{dx}{x}$ if $a>0$, $p>0$.

3. $\displaystyle\int_0^\infty f(x^a+x^{-a})\frac{\ln x}{1+x^2}\,dx = 0$ (a special case of the integral below).

4. $\displaystyle\int_0^\infty f\left(x,\frac{1}{x}\right)\frac{\ln x}{1+x^2}\,dx = 0$ if $f(x,y)=f(y,x)$ (the integral is assumed to exist).

5. $\displaystyle\int_0^\infty f\left(x,\frac{1}{x}\right)\frac{\ln x}{x}\,dx = 0$ if $f(x,y)=f(y,x)$ (the integral is assumed to exist).

▶ **Improper integrals involving trigonometric functions.**

1. $\displaystyle\int_0^\infty f(x)\frac{\sin x}{x}\,dx = \int_0^{\pi/2} f(x)\,dx$ if $f(x)=f(-x)$ and $f(x+\pi)=f(x)$.

2. $\displaystyle\int_0^\infty f(x)\frac{\sin x}{x}\,dx = \int_0^{\pi/2} f(x)\cos x\,dx$ if $f(x)=f(-x)$ and $f(x+\pi)=-f(x)$.

3. $\displaystyle\int_0^\infty \frac{f(\sin x)}{x}\,dx = \int_0^{\pi/2} \frac{f(\sin x)}{\sin x}\,dx$ if $f(-x) = -f(x)$.

4. $\displaystyle\int_0^\infty \frac{f(\sin x)}{x^2}\,dx = \int_0^{\pi/2} \frac{f(\sin x)}{\sin^2 x}\,dx$ if $f(x) = f(-x)$.

5. $\displaystyle\int_0^\infty \frac{f(\sin x)}{x}\cos x\,dx = \int_0^{\pi/2} \frac{f(\sin x)}{\sin x}\cos^2 x\,dx$ if $f(-x) = -f(x)$.

6. $\displaystyle\int_0^\infty \frac{f(\sin x)}{x}\tan x\,dx = \int_0^{\pi/2} f(\sin x)\,dx$ if $f(-x) = f(x)$.

7. $\displaystyle\int_0^\infty \frac{f(\sin x)}{x^2 + a^2}\,dx = \frac{\sinh(2a)}{2a}\int_0^{\pi/2} \frac{f(\sin x)\,dx}{\cosh^2 a - \cos^2 x}$ if $f(-x) = f(x)$.

8. $\displaystyle\int_0^\infty f\left(x + \frac{1}{x}\right)\frac{\arctan x}{x}\,dx = \frac{\pi}{4}\int_0^\infty f\left(x + \frac{1}{x}\right)\frac{dx}{x}$.

▶ **Calculation of improper integrals using analytic functions.** Suppose

$$F(z) = f(r, x) + ig(r, x), \quad z = r(\cos x + i\sin x), \quad i^2 = -1,$$

where $F(z)$ is a function analytic in a circle of radius r, $f(r, x) = \operatorname{Re} F(z)$, and $g(r, x) = \operatorname{Im} F(z)$. Then the following formulas hold:

1. $\displaystyle\int_0^\infty \frac{f(r, x)}{x^2 + a^2}\,dx = \frac{\pi}{2a}F(re^{-a})$.

2. $\displaystyle\int_0^\infty \frac{xg(r, x)}{x^2 + a^2}\,dx = \frac{\pi}{2}[F(re^{-a}) - F(0)]$.

3. $\displaystyle\int_0^\infty \frac{g(r, x)}{x}\,dx = \frac{\pi}{2}[F(r) - F(0)]$.

4. $\displaystyle\int_0^\infty \frac{g(r, x)}{x(x^2 + a^2)}\,dx = \frac{\pi}{2a^2}[F(r) - F(re^{-a})]$.

Bibliography for Chapter S1

Dwight, H. B., *Tables of Integrals and Other Mathematical Data*, Macmillan, New York, 1961.

Gradshteyn, I. S. and Ryzhik, I. M., *Tables of Integrals, Series, and Products, 6th Edition*, Academic Press, New York, 2000.

Prudnikov, A. P., Brychkov, Yu. A., and Marichev, O. I., *Integrals and Series, Vol. 1, Elementary Functions*, Gordon & Breach, New York, 1986.

Prudnikov, A. P., Brychkov, Yu. A., and Marichev, O. I., *Integrals and Series, Vol. 2, Special Functions*, Gordon & Breach, New York, 1986.

Prudnikov, A. P., Brychkov, Yu. A., and Marichev, O. I., *Integrals and Series, Vol. 3, More Special Functions*, Gordon & Breach, New York, 1988.

Zwillinger, D., *CRC Standard Mathematical Tables and Formulae, 31st Edition*, CRC Press, Boca Raton, 2002.

Chapter S2
Integral Transforms*

S2.1. Tables of Laplace Transforms

S2.1.1. General Formulas

No.	Original function, $f(x)$	Laplace transform, $\widetilde{f}(p) = \int_0^\infty e^{-px} f(x)\, dx$
1	$af_1(x) + bf_2(x)$	$a\widetilde{f}_1(p) + b\widetilde{f}_2(p)$
2	$f(x/a),\ \ a > 0$	$a\widetilde{f}(ap)$
3	$\begin{cases} 0 & \text{if } 0 < x < a, \\ f(x-a) & \text{if } a < x \end{cases}$	$e^{-ap}\widetilde{f}(p)$
4	$x^n f(x);\ \ n = 1, 2, \ldots$	$(-1)^n \dfrac{d^n}{dp^n}\widetilde{f}(p)$
5	$\dfrac{1}{x} f(x)$	$\displaystyle\int_p^\infty \widetilde{f}(q)\, dq$
6	$e^{ax} f(x)$	$\widetilde{f}(p-a)$
7	$\sinh(ax)f(x)$	$\frac{1}{2}\left[\widetilde{f}(p-a) - \widetilde{f}(p+a)\right]$
8	$\cosh(ax)f(x)$	$\frac{1}{2}\left[\widetilde{f}(p-a) + \widetilde{f}(p+a)\right]$
9	$\sin(\omega x)f(x)$	$-\frac{i}{2}\left[\widetilde{f}(p-i\omega) - \widetilde{f}(p+i\omega)\right],\ \ i^2 = -1$
10	$\cos(\omega x)f(x)$	$\frac{1}{2}\left[\widetilde{f}(p-i\omega) + \widetilde{f}(p+i\omega)\right],\ \ i^2 = -1$
11	$f(x^2)$	$\dfrac{1}{\sqrt{\pi}} \displaystyle\int_0^\infty \exp\left(-\dfrac{p^2}{4t^2}\right)\widetilde{f}(t^2)\, dt$
12	$x^{a-1}f\left(\dfrac{1}{x}\right),\ \ a > -1$	$\displaystyle\int_0^\infty (t/p)^{a/2} J_a\left(2\sqrt{pt}\right)\widetilde{f}(t)\, dt$
13	$f(a \sinh x),\ \ a > 0$	$\displaystyle\int_0^\infty J_p(at)\widetilde{f}(t)\, dt$
14	$f(x+a) = f(x)$ (periodic function)	$\dfrac{1}{1 - e^{ap}} \displaystyle\int_0^a f(x)e^{-px}\, dx$
15	$f(x+a) = -f(x)$ (antiperiodic function)	$\dfrac{1}{1 + e^{-ap}} \displaystyle\int_0^a f(x)e^{-px}\, dx$
16	$f'_x(x)$	$p\widetilde{f}(p) - f(+0)$
17	$f_x^{(n)}(x)$	$p^n\widetilde{f}(p) - \displaystyle\sum_{k=1}^n p^{n-k} f_x^{(k-1)}(+0)$

---

* For definitions and properties of the special functions occurring in this chapter, such as $J_a(z)$, erfc z, $\Gamma(z)$, etc., see Chapter M13.

No.	Original function, $f(x)$	Laplace transform, $\widetilde{f}(p) = \displaystyle\int_0^\infty e^{-px} f(x)\,dx$
18	$x^m f_x^{(n)}(x), \quad m \geq n$	$\left(-\dfrac{d}{dp}\right)^m \left[p^n \widetilde{f}(p) \right]$
19	$\dfrac{d^n}{dx^n}\left[x^m f(x) \right], \quad m \geq n$	$(-1)^m p^n \dfrac{d^m}{dp^m} \widetilde{f}(p)$
20	$\displaystyle\int_0^x f(t)\,dt$	$\dfrac{\widetilde{f}(p)}{p}$
21	$\displaystyle\int_0^x (x-t) f(t)\,dt$	$\dfrac{1}{p^2} \widetilde{f}(p)$
22	$\displaystyle\int_0^x (x-t)^\nu f(t)\,dt, \qquad \nu > -1$	$\Gamma(\nu+1) p^{-\nu-1} \widetilde{f}(p)$
23	$\displaystyle\int_0^x e^{-a(x-t)} f(t)\,dt$	$\dfrac{1}{p+a} \widetilde{f}(p)$
24	$\displaystyle\int_0^x \sinh\left[a(x-t)\right] f(t)\,dt$	$\dfrac{a\widetilde{f}(p)}{p^2 - a^2}$
25	$\displaystyle\int_0^x \sin\left[a(x-t)\right] f(t)\,dt$	$\dfrac{a\widetilde{f}(p)}{p^2 + a^2}$
26	$\displaystyle\int_0^x f_1(t) f_2(x-t)\,dt$	$\widetilde{f}_1(p) \widetilde{f}_2(p)$
27	$\displaystyle\int_0^x \dfrac{1}{t} f(t)\,dt$	$\dfrac{1}{p} \displaystyle\int_p^\infty \widetilde{f}(q)\,dq$
28	$\displaystyle\int_x^\infty \dfrac{1}{t} f(t)\,dt$	$\dfrac{1}{p} \displaystyle\int_0^p \widetilde{f}(q)\,dq$
29	$\displaystyle\int_0^\infty \dfrac{1}{\sqrt{t}} \sin\left(2\sqrt{xt}\right) f(t)\,dt$	$\dfrac{\sqrt{\pi}}{p\sqrt{p}} \widetilde{f}\left(\dfrac{1}{p}\right)$
30	$\dfrac{1}{\sqrt{x}} \displaystyle\int_0^\infty \cos\left(2\sqrt{xt}\right) f(t)\,dt$	$\dfrac{\sqrt{\pi}}{\sqrt{p}} \widetilde{f}\left(\dfrac{1}{p}\right)$
31	$\displaystyle\int_0^\infty \dfrac{1}{\sqrt{\pi x}} \exp\left(-\dfrac{t^2}{4x}\right) f(t)\,dt$	$\dfrac{1}{\sqrt{p}} \widetilde{f}\left(\sqrt{p}\right)$
32	$\displaystyle\int_0^\infty \dfrac{t}{2\sqrt{\pi x^3}} \exp\left(-\dfrac{t^2}{4x}\right) f(t)\,dt$	$\widetilde{f}\left(\sqrt{p}\right)$
33	$f(x) - a \displaystyle\int_0^x f\left(\sqrt{x^2 - t^2}\right) J_1(at)\,dt$	$\widetilde{f}\left(\sqrt{p^2 + a^2}\right)$
34	$f(x) + a \displaystyle\int_0^x f\left(\sqrt{x^2 - t^2}\right) I_1(at)\,dt$	$\widetilde{f}\left(\sqrt{p^2 - a^2}\right)$

S2.1.2. Expressions Involving Power Functions

No.	Original function, $f(x)$	Laplace transform, $\widetilde{f}(p) = \int_0^\infty e^{-px} f(x)\, dx$
1	1	$\dfrac{1}{p}$
2	$\begin{cases} 0 & \text{if } 0 < x < a, \\ 1 & \text{if } a < x < b, \\ 0 & \text{if } b < x \end{cases}$	$\dfrac{1}{p}\left(e^{-ap} - e^{-bp}\right)$
3	x	$\dfrac{1}{p^2}$
4	$\dfrac{1}{x+a}$	$-e^{ap}\,\mathrm{Ei}(-ap)$
5	$x^n, \qquad n = 1, 2, \dots$	$\dfrac{n!}{p^{n+1}}$
6	$x^{n-1/2}, \qquad n = 1, 2, \dots$	$\dfrac{1 \times 3 \times \dots \times (2n-1)\sqrt{\pi}}{2^n p^{n+1/2}}$
7	$\dfrac{1}{\sqrt{x+a}}$	$\sqrt{\dfrac{\pi}{p}}\,e^{ap}\,\mathrm{erfc}\left(\sqrt{ap}\right)$
8	$\dfrac{\sqrt{x}}{x+a}$	$\sqrt{\dfrac{\pi}{p}} - \pi\sqrt{a}\,e^{ap}\,\mathrm{erfc}\left(\sqrt{ap}\right)$
9	$(x+a)^{-3/2}$	$2a^{-1/2} - 2(\pi p)^{1/2} e^{ap}\,\mathrm{erfc}\left(\sqrt{ap}\right)$
10	$x^{1/2}(x+a)^{-1}$	$(\pi/p)^{1/2} - \pi a^{1/2} e^{ap}\,\mathrm{erfc}\left(\sqrt{ap}\right)$
11	$x^{-1/2}(x+a)^{-1}$	$\pi a^{-1/2} e^{ap}\,\mathrm{erfc}\left(\sqrt{ap}\right)$
12	$x^\nu, \qquad \nu > -1$	$\Gamma(\nu+1)p^{-\nu-1}$
13	$(x+a)^\nu, \qquad \nu > -1$	$p^{-\nu-1} e^{-ap}\Gamma(\nu+1, ap)$
14	$x^\nu(x+a)^{-1}, \qquad \nu > -1$	$k e^{ap}\Gamma(-\nu, ap), \qquad k = a^\nu\Gamma(\nu+1)$
15	$(x^2 + 2ax)^{-1/2}(x+a)$	$a e^{ap} K_1(ap)$

S2.1.3. Expressions Involving Exponential Functions

No.	Original function, $f(x)$	Laplace transform, $\widetilde{f}(p) = \int_0^\infty e^{-px} f(x)\, dx$
1	e^{-ax}	$(p+a)^{-1}$
2	xe^{-ax}	$(p+a)^{-2}$
3	$x^{\nu-1} e^{-ax}, \qquad \nu > 0$	$\Gamma(\nu)(p+a)^{-\nu}$
4	$\dfrac{1}{x}\left(e^{-ax} - e^{-bx}\right)$	$\ln(p+b) - \ln(p+a)$
5	$\dfrac{1}{x^2}\left(1 - e^{-ax}\right)^2$	$(p+2a)\ln(p+2a) + p\ln p - 2(p+a)\ln(p+a)$
6	$\exp\left(-ax^2\right), \qquad a > 0$	$(\pi b)^{1/2}\exp\left(bp^2\right)\mathrm{erfc}(p\sqrt{b}), \qquad a = \dfrac{1}{4b}$
7	$x\exp\left(-ax^2\right)$	$2b - 2\pi^{1/2} b^{3/2} p\,\mathrm{erfc}(p\sqrt{b}), \qquad a = \dfrac{1}{4b}$

No.	Original function, $f(x)$	Laplace transform, $\widetilde{f}(p) = \int_0^\infty e^{-px} f(x)\,dx$
8	$\exp(-a/x), \qquad a \geq 0$	$2\sqrt{a/p}\,K_1\big(2\sqrt{ap}\,\big)$
9	$\sqrt{x}\,\exp(-a/x), \qquad a \geq 0$	$\frac{1}{2}\sqrt{\pi/p^3}\big(1 + 2\sqrt{ap}\,\big)\exp\big(-2\sqrt{ap}\,\big)$
10	$\dfrac{1}{\sqrt{x}}\exp(-a/x), \qquad a \geq 0$	$\sqrt{\pi/p}\,\exp\big(-2\sqrt{ap}\,\big)$
11	$\dfrac{1}{x\sqrt{x}}\exp(-a/x), \qquad a > 0$	$\sqrt{\pi/a}\,\exp\big(-2\sqrt{ap}\,\big)$
12	$x^{\nu-1}\exp(-a/x), \qquad a > 0$	$2(a/p)^{\nu/2}K_\nu\big(2\sqrt{ap}\,\big)$
13	$\exp\big(-2\sqrt{ax}\,\big)$	$p^{-1} - (\pi a)^{1/2}p^{-3/2}e^{a/p}\,\mathrm{erfc}\big(\sqrt{a/p}\,\big)$
14	$\dfrac{1}{\sqrt{x}}\exp\big(-2\sqrt{ax}\,\big)$	$(\pi/p)^{1/2}e^{a/p}\,\mathrm{erfc}\big(\sqrt{a/p}\,\big)$

S2.1.4. Expressions Involving Hyperbolic Functions

No.	Original function, $f(x)$	Laplace transform, $\widetilde{f}(p) = \int_0^\infty e^{-px} f(x)\,dx$
1	$\sinh(ax)$	$\dfrac{a}{p^2 - a^2}$
2	$\sinh^2(ax)$	$\dfrac{2a^2}{p^3 - 4a^2 p}$
3	$\dfrac{1}{x}\sinh(ax)$	$\dfrac{1}{2}\ln\dfrac{p+a}{p-a}$
4	$x^{\nu-1}\sinh(ax), \qquad \nu > -1$	$\frac{1}{2}\Gamma(\nu)\big[(p-a)^{-\nu} - (p+a)^{-\nu}\big]$
5	$\sinh\big(2\sqrt{ax}\,\big)$	$\dfrac{\sqrt{\pi a}}{p\sqrt{p}}e^{a/p}$
6	$\sqrt{x}\,\sinh\big(2\sqrt{ax}\,\big)$	$\pi^{1/2}p^{-5/2}\big(\frac{1}{2}p + a\big)e^{a/p}\,\mathrm{erf}\big(\sqrt{a/p}\,\big) - a^{1/2}p^{-2}$
7	$\dfrac{1}{\sqrt{x}}\sinh\big(2\sqrt{ax}\,\big)$	$\pi^{1/2}p^{-1/2}e^{a/p}\,\mathrm{erf}\big(\sqrt{a/p}\,\big)$
8	$\dfrac{1}{\sqrt{x}}\sinh^2\big(\sqrt{ax}\,\big)$	$\frac{1}{2}\pi^{1/2}p^{-1/2}\big(e^{a/p} - 1\big)$
9	$\cosh(ax)$	$\dfrac{p}{p^2 - a^2}$
10	$\cosh^2(ax)$	$\dfrac{p^2 - 2a^2}{p^3 - 4a^2 p}$
11	$x^{\nu-1}\cosh(ax), \qquad \nu > 0$	$\frac{1}{2}\Gamma(\nu)\big[(p-a)^{-\nu} + (p+a)^{-\nu}\big]$
12	$\cosh\big(2\sqrt{ax}\,\big)$	$\dfrac{1}{p} + \dfrac{\sqrt{\pi a}}{p\sqrt{p}}e^{a/p}\,\mathrm{erf}\big(\sqrt{a/p}\,\big)$
13	$\sqrt{x}\,\cosh\big(2\sqrt{ax}\,\big)$	$\pi^{1/2}p^{-5/2}\big(\frac{1}{2}p + a\big)e^{a/p}$
14	$\dfrac{1}{\sqrt{x}}\cosh\big(2\sqrt{ax}\,\big)$	$\pi^{1/2}p^{-1/2}e^{a/p}$
15	$\dfrac{1}{\sqrt{x}}\cosh^2\big(\sqrt{ax}\,\big)$	$\frac{1}{2}\pi^{1/2}p^{-1/2}\big(e^{a/p} + 1\big)$

S2.1.5. Expressions Involving Logarithmic Functions

No.	Original function, $f(x)$	Laplace transform, $\widetilde{f}(p) = \int_0^\infty e^{-px} f(x)\, dx$
1	$\ln x$	$-\dfrac{1}{p}(\ln p + \mathcal{C})$, $\mathcal{C} = 0.5772\ldots$ is the Euler constant
2	$\ln(1 + ax)$	$-\dfrac{1}{p} e^{p/a} \operatorname{Ei}(-p/a)$
3	$\ln(x + a)$	$\dfrac{1}{p}\left[\ln a - e^{ap} \operatorname{Ei}(-ap)\right]$
4	$x^n \ln x, \qquad n = 1, 2, \ldots$	$\dfrac{n!}{p^{n+1}}\left(1 + \frac{1}{2} + \frac{1}{3} + \cdots + \frac{1}{n} - \ln p - \mathcal{C}\right)$, $\mathcal{C} = 0.5772\ldots$ is the Euler constant
5	$\dfrac{1}{\sqrt{x}} \ln x$	$-\sqrt{\pi/p}\left[\ln(4p) + \mathcal{C}\right]$
6	$x^{n-1/2} \ln x, \qquad n = 1, 2, \ldots$	$\dfrac{k_n}{p^{n+1/2}}\left[2 + \frac{2}{3} + \frac{2}{5} + \cdots + \frac{2}{2n-1} - \ln(4p) - \mathcal{C}\right]$, $k_n = 1 \times 3 \times 5 \times \ldots \times (2n-1)\dfrac{\sqrt{\pi}}{2^n}, \quad \mathcal{C} = 0.5772\ldots$
7	$x^{\nu-1} \ln x, \quad \nu > 0$	$\Gamma(\nu)p^{-\nu}\left[\psi(\nu) - \ln p\right], \quad \psi(\nu)$ is the logarithmic derivative of the gamma function
8	$(\ln x)^2$	$\dfrac{1}{p}\left[(\ln x + \mathcal{C})^2 + \frac{1}{6}\pi^2\right], \quad \mathcal{C} = 0.5772\ldots$
9	$e^{-ax} \ln x$	$-\dfrac{\ln(p + a) + \mathcal{C}}{p + a}, \qquad \mathcal{C} = 0.5772\ldots$

S2.1.6. Expressions Involving Trigonometric Functions

No.	Original function, $f(x)$	Laplace transform, $\widetilde{f}(p) = \int_0^\infty e^{-px} f(x)\, dx$
1	$\sin(ax)$	$\dfrac{a}{p^2 + a^2}$
2	$|\sin(ax)|, \qquad a > 0$	$\dfrac{a}{p^2 + a^2} \coth\left(\dfrac{\pi p}{2a}\right)$
3	$\sin^{2n}(ax), \qquad n = 1, 2, \ldots$	$\dfrac{a^{2n}(2n)!}{p\left[p^2 + (2a)^2\right]\left[p^2 + (4a)^2\right]\ldots\left[p^2 + (2na)^2\right]}$
4	$\sin^{2n+1}(ax), \qquad n = 1, 2, \ldots$	$\dfrac{a^{2n+1}(2n+1)!}{\left[p^2 + a^2\right]\left[p^2 + 3^2 a^2\right]\ldots\left[p^2 + (2n+1)^2 a^2\right]}$
5	$x^n \sin(ax), \qquad n = 1, 2, \ldots$	$\dfrac{n!\, p^{n+1}}{(p^2 + a^2)^{n+1}} \displaystyle\sum_{0 \le 2k \le n} (-1)^k C_{n+1}^{2k+1}\left(\dfrac{a}{p}\right)^{2k+1}$
6	$\dfrac{1}{x} \sin(ax)$	$\arctan\left(\dfrac{a}{p}\right)$
7	$\dfrac{1}{x} \sin^2(ax)$	$\frac{1}{4}\ln\left(1 + 4a^2 p^{-2}\right)$

No.	Original function, $f(x)$	Laplace transform, $\widetilde{f}(p) = \int_0^\infty e^{-px} f(x)\, dx$
8	$\dfrac{1}{x^2}\sin^2(ax)$	$a\arctan(2a/p) - \tfrac{1}{4}p\ln\left(1 + 4a^2 p^{-2}\right)$
9	$\sin\left(2\sqrt{ax}\,\right)$	$\dfrac{\sqrt{\pi a}}{p\sqrt{p}}e^{-a/p}$
10	$\dfrac{1}{x}\sin\left(2\sqrt{ax}\,\right)$	$\pi\,\mathrm{erf}\left(\sqrt{a/p}\,\right)$
11	$\cos(ax)$	$\dfrac{p}{p^2 + a^2}$
12	$\cos^2(ax)$	$\dfrac{p^2 + 2a^2}{p\left(p^2 + 4a^2\right)}$
13	$x^n\cos(ax), \qquad n = 1, 2, \ldots$	$\dfrac{n!\,p^{n+1}}{\left(p^2 + a^2\right)^{n+1}}\displaystyle\sum_{0 \le 2k \le n+1}(-1)^k C_{n+1}^{2k}\left(\dfrac{a}{p}\right)^{2k}$
14	$\dfrac{1}{x}\left[1 - \cos(ax)\right]$	$\tfrac{1}{2}\ln\left(1 + a^2 p^{-2}\right)$
15	$\dfrac{1}{x}\left[\cos(ax) - \cos(bx)\right]$	$\dfrac{1}{2}\ln\dfrac{p^2 + b^2}{p^2 + a^2}$
16	$\sqrt{x}\cos\left(2\sqrt{ax}\,\right)$	$\tfrac{1}{2}\pi^{1/2}p^{-5/2}(p - 2a)e^{-a/p}$
17	$\dfrac{1}{\sqrt{x}}\cos\left(2\sqrt{ax}\,\right)$	$\sqrt{\pi/p}\,e^{-a/p}$
18	$\sin(ax)\sin(bx)$	$\dfrac{2abp}{\left[p^2 + (a+b)^2\right]\left[p^2 + (a-b)^2\right]}$
19	$\cos(ax)\sin(bx)$	$\dfrac{b\left(p^2 - a^2 + b^2\right)}{\left[p^2 + (a+b)^2\right]\left[p^2 + (a-b)^2\right]}$
20	$\cos(ax)\cos(bx)$	$\dfrac{p\left(p^2 + a^2 + b^2\right)}{\left[p^2 + (a+b)^2\right]\left[p^2 + (a-b)^2\right]}$
21	$\dfrac{ax\cos(ax) - \sin(ax)}{x^2}$	$p\arctan\dfrac{a}{x} - a$
22	$e^{bx}\sin(ax)$	$\dfrac{a}{(p-b)^2 + a^2}$
23	$e^{bx}\cos(ax)$	$\dfrac{p-b}{(p-b)^2 + a^2}$
24	$\sin(ax)\sinh(ax)$	$\dfrac{2a^2 p}{p^4 + 4a^4}$
25	$\sin(ax)\cosh(ax)$	$\dfrac{a\left(p^2 + 2a^2\right)}{p^4 + 4a^4}$
26	$\cos(ax)\sinh(ax)$	$\dfrac{a\left(p^2 - 2a^2\right)}{p^4 + 4a^4}$
27	$\cos(ax)\cosh(ax)$	$\dfrac{p^3}{p^4 + 4a^4}$

S2.1.7. Expressions Involving Special Functions

No.	Original function, $f(x)$	Laplace transform, $\widetilde{f}(p) = \int_0^\infty e^{-px} f(x)\, dx$
1	$\mathrm{erf}(ax)$	$\dfrac{1}{p} \exp\!\left(b^2 p^2\right) \mathrm{erfc}(bp), \qquad b = \dfrac{1}{2a}$
2	$\mathrm{erf}\!\left(\sqrt{ax}\,\right)$	$\dfrac{\sqrt{a}}{p\sqrt{p+a}}$
3	$e^{ax}\,\mathrm{erf}\!\left(\sqrt{ax}\,\right)$	$\dfrac{\sqrt{a}}{\sqrt{p}\,(p-a)}$
4	$\mathrm{erf}\!\left(\tfrac{1}{2}\sqrt{a/x}\,\right)$	$\dfrac{1}{p}\left[1 - \exp\!\left(-\sqrt{ap}\,\right)\right]$
5	$\mathrm{erfc}\!\left(\sqrt{ax}\,\right)$	$\dfrac{\sqrt{p+a}-\sqrt{a}}{p\sqrt{p+a}}$
6	$e^{ax}\,\mathrm{erfc}\!\left(\sqrt{ax}\,\right)$	$\dfrac{1}{p+\sqrt{ap}}$
7	$\mathrm{erfc}\!\left(\tfrac{1}{2}\sqrt{a/x}\,\right)$	$\dfrac{1}{p}\exp\!\left(-\sqrt{ap}\,\right)$
8	$\mathrm{Ci}(x)$	$\dfrac{1}{2p}\ln(p^2+1)$
9	$\mathrm{Si}(x)$	$\dfrac{1}{p}\,\mathrm{arccot}\,p$
10	$\mathrm{Ei}(-x)$	$-\dfrac{1}{p}\ln(p+1)$
11	$J_0(ax)$	$\dfrac{1}{\sqrt{p^2+a^2}}$
12	$J_\nu(ax), \qquad \nu > -1$	$\dfrac{a^\nu}{\sqrt{p^2+a^2}\,\left(p+\sqrt{p^2+a^2}\,\right)^\nu}$
13	$x^n J_n(ax), \qquad n = 1, 2, \ldots$	$1\times 3\times 5\times\ldots\times(2n-1)a^n\left(p^2+a^2\right)^{-n-1/2}$
14	$x^\nu J_\nu(ax), \qquad \nu > -\tfrac{1}{2}$	$2^\nu \pi^{-1/2}\Gamma\!\left(\nu+\tfrac{1}{2}\right)a^\nu\left(p^2+a^2\right)^{-\nu-1/2}$
15	$x^{\nu+1} J_\nu(ax), \qquad \nu > -1$	$2^{\nu+1}\pi^{-1/2}\Gamma\!\left(\nu+\tfrac{3}{2}\right)a^\nu p\left(p^2+a^2\right)^{-\nu-3/2}$
16	$J_0\!\left(2\sqrt{ax}\,\right)$	$\dfrac{1}{p}e^{-a/p}$
17	$\sqrt{x}\,J_1\!\left(2\sqrt{ax}\,\right)$	$\dfrac{\sqrt{a}}{p^2}e^{-a/p}$
18	$x^{\nu/2} J_\nu\!\left(2\sqrt{ax}\,\right), \qquad \nu > -1$	$a^{\nu/2}p^{-\nu-1}e^{-a/p}$
19	$J_0\!\left(a\sqrt{x^2+bx}\,\right)$	$\dfrac{1}{\sqrt{p^2+a^2}}\exp\!\left(bp - b\sqrt{p^2+a^2}\,\right)$
20	$I_0(ax)$	$\dfrac{1}{\sqrt{p^2-a^2}}$
21	$I_\nu(ax), \qquad \nu > -1$	$\dfrac{a^\nu}{\sqrt{p^2-a^2}\,\left(p+\sqrt{p^2-a^2}\,\right)^\nu}$

No.	Original function, $f(x)$	Laplace transform, $\widetilde{f}(p) = \int_0^\infty e^{-px} f(x)\, dx$
22	$x^\nu I_\nu(ax), \qquad \nu > -\tfrac{1}{2}$	$2^\nu \pi^{-1/2} \Gamma\left(\nu + \tfrac{1}{2}\right) a^\nu \left(p^2 - a^2\right)^{-\nu-1/2}$
23	$x^{\nu+1} I_\nu(ax), \qquad \nu > -1$	$2^{\nu+1} \pi^{-1/2} \Gamma\left(\nu + \tfrac{3}{2}\right) a^\nu p \left(p^2 - a^2\right)^{-\nu-3/2}$
24	$I_0\left(2\sqrt{ax}\,\right)$	$\dfrac{1}{p} e^{a/p}$
25	$\dfrac{1}{\sqrt{x}} I_1\left(2\sqrt{ax}\,\right)$	$\dfrac{1}{\sqrt{a}} \left(e^{a/p} - 1\right)$
26	$x^{\nu/2} I_\nu\left(2\sqrt{ax}\,\right), \qquad \nu > -1$	$a^{\nu/2} p^{-\nu-1} e^{a/p}$
27	$Y_0(ax)$	$-\dfrac{2}{\pi} \dfrac{\operatorname{arcsinh}(p/a)}{\sqrt{p^2 + a^2}}$
28	$K_0(ax)$	$\dfrac{\ln\left(p + \sqrt{p^2 - a^2}\,\right) - \ln a}{\sqrt{p^2 - a^2}}$

S2.2. Tables of Inverse Laplace Transforms

S2.2.1. General Formulas

No.	Laplace transform, $\widetilde{f}(p)$	Inverse transform, $f(x) = \dfrac{1}{2\pi i} \int_{c-i\infty}^{c+i\infty} e^{px} \widetilde{f}(p)\, dp$
1	$\widetilde{f}(p + a)$	$e^{-ax} f(x)$
2	$\widetilde{f}(ap), \quad a > 0$	$\dfrac{1}{a} f\left(\dfrac{x}{a}\right)$
3	$\widetilde{f}(ap + b), \quad a > 0$	$\dfrac{1}{a} \exp\left(-\dfrac{b}{a} x\right) f\left(\dfrac{x}{a}\right)$
4	$\widetilde{f}(p - a) + \widetilde{f}(p + a)$	$2 f(x) \cosh(ax)$
5	$\widetilde{f}(p - a) - \widetilde{f}(p + a)$	$2 f(x) \sinh(ax)$
6	$e^{-ap} \widetilde{f}(p), \quad a \geq 0$	$\begin{cases} 0 & \text{if } 0 \leq x < a, \\ f(x - a) & \text{if } a < x \end{cases}$
7	$p \widetilde{f}(p)$	$\dfrac{df(x)}{dx}, \quad \text{if } f(+0) = 0$
8	$\dfrac{1}{p} \widetilde{f}(p)$	$\displaystyle\int_0^x f(t)\, dt$
9	$\dfrac{1}{p + a} \widetilde{f}(p)$	$e^{-ax} \displaystyle\int_0^x e^{at} f(t)\, dt$
10	$\dfrac{1}{p^2} \widetilde{f}(p)$	$\displaystyle\int_0^x (x - t) f(t)\, dt$
11	$\dfrac{\widetilde{f}(p)}{p(p + a)}$	$\dfrac{1}{a} \displaystyle\int_0^x \left[1 - e^{a(x-t)}\right] f(t)\, dt$
12	$\dfrac{\widetilde{f}(p)}{(p + a)^2}$	$\displaystyle\int_0^x (x - t) e^{-a(x-t)} f(t)\, dt$
13	$\dfrac{\widetilde{f}(p)}{(p + a)(p + b)}$	$\dfrac{1}{b - a} \displaystyle\int_0^x \left[e^{-a(x-t)} - e^{-b(x-t)}\right] f(t)\, dt$

No.	Laplace transform, $\widetilde{f}(p)$	Inverse transform, $f(x) = \dfrac{1}{2\pi i}\displaystyle\int_{c-i\infty}^{c+i\infty} e^{px}\widetilde{f}(p)\,dp$
14	$\dfrac{\widetilde{f}(p)}{(p+a)^2 + b^2}$	$\dfrac{1}{b}\displaystyle\int_0^x e^{-a(x-t)}\sin\big[b(x-t)\big]f(t)\,dt$
15	$\dfrac{1}{p^n}\widetilde{f}(p),\quad n = 1, 2, \ldots$	$\dfrac{1}{(n-1)!}\displaystyle\int_0^x (x-t)^{n-1} f(t)\,dt$
16	$\widetilde{f}_1(p)\widetilde{f}_2(p)$	$\displaystyle\int_0^x f_1(t) f_2(x-t)\,dt$
17	$\dfrac{1}{\sqrt{p}}\widetilde{f}\Big(\dfrac{1}{p}\Big)$	$\displaystyle\int_0^\infty \dfrac{\cos\big(2\sqrt{xt}\,\big)}{\sqrt{\pi x}}f(t)\,dt$
18	$\dfrac{1}{p\sqrt{p}}\widetilde{f}\Big(\dfrac{1}{p}\Big)$	$\displaystyle\int_0^\infty \dfrac{\sin\big(2\sqrt{xt}\,\big)}{\sqrt{\pi t}}f(t)\,dt$
19	$\dfrac{1}{p^{2\nu+1}}\widetilde{f}\Big(\dfrac{1}{p}\Big)$	$\displaystyle\int_0^\infty (x/t)^\nu J_{2\nu}\big(2\sqrt{xt}\,\big)f(t)\,dt$
20	$\dfrac{1}{p}\widetilde{f}\Big(\dfrac{1}{p}\Big)$	$\displaystyle\int_0^\infty J_0\big(2\sqrt{xt}\,\big)f(t)\,dt$
21	$\dfrac{1}{p}\widetilde{f}\Big(p + \dfrac{1}{p}\Big)$	$\displaystyle\int_0^x J_0\big(2\sqrt{xt-t^2}\,\big)f(t)\,dt$
22	$\dfrac{1}{p^{2\nu+1}}\widetilde{f}\Big(p + \dfrac{a}{p}\Big),\quad -\dfrac{1}{2} < \nu \le 0$	$\displaystyle\int_0^x \Big(\dfrac{x-t}{at}\Big)^\nu J_{2\nu}\big(2\sqrt{axt-at^2}\,\big)f(t)\,dt$
23	$\widetilde{f}\big(\sqrt{p}\,\big)$	$\displaystyle\int_0^\infty \dfrac{t}{2\sqrt{\pi x^3}}\exp\Big(-\dfrac{t^2}{4x}\Big)f(t)\,dt$
24	$\dfrac{1}{\sqrt{p}}\widetilde{f}\big(\sqrt{p}\,\big)$	$\dfrac{1}{\sqrt{\pi x}}\displaystyle\int_0^\infty \exp\Big(-\dfrac{t^2}{4x}\Big)f(t)\,dt$
25	$\widetilde{f}\big(p + \sqrt{p}\,\big)$	$\dfrac{1}{2\sqrt{\pi}}\displaystyle\int_0^x \dfrac{t}{(x-t)^{3/2}}\exp\Big[-\dfrac{t^2}{4(x-t)}\Big]f(t)\,dt$
26	$\widetilde{f}\big(\sqrt{p^2 + a^2}\,\big)$	$f(x) - a\displaystyle\int_0^x f\big(\sqrt{x^2-t^2}\,\big)J_1(at)\,dt$
27	$\widetilde{f}\big(\sqrt{p^2 - a^2}\,\big)$	$f(x) + a\displaystyle\int_0^x f\big(\sqrt{x^2-t^2}\,\big)I_1(at)\,dt$
28	$\dfrac{\widetilde{f}\big(\sqrt{p^2 + a^2}\,\big)}{\sqrt{p^2 + a^2}}$	$\displaystyle\int_0^x J_0\big(a\sqrt{x^2-t^2}\,\big)f(t)\,dt$
29	$\dfrac{\widetilde{f}\big(\sqrt{p^2 - a^2}\,\big)}{\sqrt{p^2 - a^2}}$	$\displaystyle\int_0^x I_0\big(a\sqrt{x^2-t^2}\,\big)f(t)\,dt$
30	$\widetilde{f}\big(\sqrt{(p+a)^2 - b^2}\,\big)$	$e^{-ax}f(x) + be^{-ax}\displaystyle\int_0^x f\big(\sqrt{x^2-t^2}\,\big)I_1(bt)\,dt$
31	$\widetilde{f}(\ln p)$	$\displaystyle\int_0^\infty \dfrac{x^{t-1}}{\Gamma(t)}f(t)\,dt$
32	$\dfrac{1}{p}\widetilde{f}(\ln p)$	$\displaystyle\int_0^\infty \dfrac{x^t}{\Gamma(t+1)}f(t)\,dt$
33	$\widetilde{f}(p - ia) + \widetilde{f}(p + ia),\ \ i^2 = -1$	$2f(x)\cos(ax)$
34	$i\big[\widetilde{f}(p - ia) - \widetilde{f}(p + ia)\big],\ i^2 = -1$	$2f(x)\sin(ax)$

No.	Laplace transform, $\widetilde{f}(p)$	Inverse transform, $f(x) = \dfrac{1}{2\pi i}\displaystyle\int_{c-i\infty}^{c+i\infty} e^{px}\widetilde{f}(p)\,dp$
35	$\dfrac{d\widetilde{f}(p)}{dp}$	$-xf(x)$
36	$\dfrac{d^n\widetilde{f}(p)}{dp^n}$	$(-x)^n f(x)$
37	$p^n\dfrac{d^m\widetilde{f}(p)}{dp^m}, \quad m \geq n$	$(-1)^m\dfrac{d^n}{dx^n}\left[x^m f(x)\right]$
38	$\displaystyle\int_p^\infty \widetilde{f}(q)\,dq$	$\dfrac{1}{x}f(x)$
39	$\dfrac{1}{p}\displaystyle\int_0^p \widetilde{f}(q)\,dq$	$\displaystyle\int_x^\infty \dfrac{f(t)}{t}\,dt$
40	$\dfrac{1}{p}\displaystyle\int_p^\infty \widetilde{f}(q)\,dq$	$\displaystyle\int_0^x \dfrac{f(t)}{t}\,dt$

S2.2.2. Expressions Involving Rational Functions

No.	Laplace transform, $\widetilde{f}(p)$	Inverse transform, $f(x) = \dfrac{1}{2\pi i}\displaystyle\int_{c-i\infty}^{c+i\infty} e^{px}\widetilde{f}(p)\,dp$
1	$\dfrac{1}{p}$	1
2	$\dfrac{1}{p+a}$	e^{-ax}
3	$\dfrac{1}{p^2}$	x
4	$\dfrac{1}{p(p+a)}$	$\dfrac{1}{a}\left(1 - e^{-ax}\right)$
5	$\dfrac{1}{(p+a)^2}$	xe^{-ax}
6	$\dfrac{p}{(p+a)^2}$	$(1 - ax)e^{-ax}$
7	$\dfrac{1}{p^2 - a^2}$	$\dfrac{1}{a}\sinh(ax)$
8	$\dfrac{p}{p^2 - a^2}$	$\cosh(ax)$
9	$\dfrac{1}{(p+a)(p+b)}$	$\dfrac{1}{a-b}\left(e^{-bx} - e^{-ax}\right)$
10	$\dfrac{p}{(p+a)(p+b)}$	$\dfrac{1}{a-b}\left(ae^{-ax} - be^{-bx}\right)$
11	$\dfrac{1}{p^2 + a^2}$	$\dfrac{1}{a}\sin(ax)$
12	$\dfrac{p}{p^2 + a^2}$	$\cos(ax)$
13	$\dfrac{1}{(p+b)^2 + a^2}$	$\dfrac{1}{a}e^{-bx}\sin(ax)$
14	$\dfrac{p}{(p+b)^2 + a^2}$	$e^{-bx}\left[\cos(ax) - \dfrac{b}{a}\sin(ax)\right]$

No.	Laplace transform, $\widetilde{f}(p)$	Inverse transform, $f(x) = \dfrac{1}{2\pi i}\displaystyle\int_{c-i\infty}^{c+i\infty} e^{px}\,\widetilde{f}(p)\,dp$
15	$\dfrac{1}{p^3}$	$\frac{1}{2}x^2$
16	$\dfrac{1}{p^2(p+a)}$	$\dfrac{1}{a^2}\left(e^{-ax} + ax - 1\right)$
17	$\dfrac{1}{p(p+a)(p+b)}$	$\dfrac{1}{ab(a-b)}\left(a - b + be^{-ax} - ae^{-bx}\right)$
18	$\dfrac{1}{p(p+a)^2}$	$\dfrac{1}{a^2}\left(1 - e^{-ax} - axe^{-ax}\right)$
19	$\dfrac{1}{(p+a)(p+b)(p+c)}$	$\dfrac{(c-b)e^{-ax} + (a-c)e^{-bx} + (b-a)e^{-cx}}{(a-b)(b-c)(c-a)}$
20	$\dfrac{p}{(p+a)(p+b)(p+c)}$	$\dfrac{a(b-c)e^{-ax} + b(c-a)e^{-bx} + c(a-b)e^{-cx}}{(a-b)(b-c)(c-a)}$
21	$\dfrac{p^2}{(p+a)(p+b)(p+c)}$	$\dfrac{a^2(c-b)e^{-ax} + b^2(a-c)e^{-bx} + c^2(b-a)e^{-cx}}{(a-b)(b-c)(c-a)}$
22	$\dfrac{1}{(p+a)(p+b)^2}$	$\dfrac{1}{(a-b)^2}\left[e^{-ax} - e^{-bx} + (a-b)xe^{-bx}\right]$
23	$\dfrac{p}{(p+a)(p+b)^2}$	$\dfrac{1}{(a-b)^2}\left\{-ae^{-ax} + \left[a + b(b-a)x\right]e^{-bx}\right\}$
24	$\dfrac{p^2}{(p+a)(p+b)^2}$	$\dfrac{1}{(a-b)^2}\left[a^2 e^{-ax} + b(b - 2a - b^2 x + abx)e^{-bx}\right]$
25	$\dfrac{1}{(p+a)^3}$	$\frac{1}{2}x^2 e^{-ax}$
26	$\dfrac{p}{(p+a)^3}$	$x\left(1 - \frac{1}{2}ax\right)e^{-ax}$
27	$\dfrac{p^2}{(p+a)^3}$	$\left(1 - 2ax + \frac{1}{2}a^2 x^2\right)e^{-ax}$
28	$\dfrac{1}{p(p^2+a^2)}$	$\dfrac{1}{a^2}\left[1 - \cos(ax)\right]$
29	$\dfrac{1}{p\left[(p+b)^2 + a^2\right]}$	$\dfrac{1}{a^2+b^2}\left\{1 - e^{-bx}\left[\cos(ax) + \dfrac{b}{a}\sin(ax)\right]\right\}$
30	$\dfrac{1}{(p+a)(p^2+b^2)}$	$\dfrac{1}{a^2+b^2}\left[e^{-ax} + \dfrac{a}{b}\sin(bx) - \cos(bx)\right]$
31	$\dfrac{p}{(p+a)(p^2+b^2)}$	$\dfrac{1}{a^2+b^2}\left[-ae^{-ax} + a\cos(bx) + b\sin(bx)\right]$
32	$\dfrac{p^2}{(p+a)(p^2+b^2)}$	$\dfrac{1}{a^2+b^2}\left[a^2 e^{-ax} - ab\sin(bx) + b^2\cos(bx)\right]$
33	$\dfrac{1}{p^3+a^3}$	$\dfrac{1}{3a^2}e^{-ax} - \dfrac{1}{3a^2}e^{ax/2}\left[\cos(kx) - \sqrt{3}\sin(kx)\right],$ $k = \frac{1}{2}a\sqrt{3}$
34	$\dfrac{p}{p^3+a^3}$	$-\dfrac{1}{3a}e^{-ax} + \dfrac{1}{3a}e^{ax/2}\left[\cos(kx) + \sqrt{3}\sin(kx)\right],$ $k = \frac{1}{2}a\sqrt{3}$

No.	Laplace transform, $\widetilde{f}(p)$	Inverse transform, $f(x) = \dfrac{1}{2\pi i}\displaystyle\int_{c-i\infty}^{c+i\infty} e^{px}\widetilde{f}(p)\,dp$
35	$\dfrac{p^2}{p^3 + a^3}$	$\frac{1}{3}e^{-ax} + \frac{2}{3}e^{ax/2}\cos(kx), \quad k = \frac{1}{2}a\sqrt{3}$
36	$\dfrac{1}{(p+a)\left[(p+b)^2 + c^2\right]}$	$\dfrac{e^{-ax} - e^{-bx}\cos(cx) + ke^{-bx}\sin(cx)}{(a-b)^2 + c^2}, \quad k = \dfrac{a-b}{c}$
37	$\dfrac{p}{(p+a)\left[(p+b)^2 + c^2\right]}$	$\dfrac{-ae^{-ax} + ae^{-bx}\cos(cx) + ke^{-bx}\sin(cx)}{(a-b)^2 + c^2},$ $k = \dfrac{b^2 + c^2 - ab}{c}$
38	$\dfrac{p^2}{(p+a)\left[(p+b)^2 + c^2\right]}$	$\dfrac{a^2 e^{-ax} + (b^2 + c^2 - 2ab)e^{-bx}\cos(cx) + ke^{-bx}\sin(cx)}{(a-b)^2 + c^2},$ $k = -ac - bc + \dfrac{ab^2 - b^3}{c}$
39	$\dfrac{1}{p^4}$	$\frac{1}{6}x^3$
40	$\dfrac{1}{p^3(p+a)}$	$\dfrac{1}{a^3} - \dfrac{1}{a^2}x + \dfrac{1}{2a}x^2 - \dfrac{1}{a^3}e^{-ax}$
41	$\dfrac{1}{p^2(p+a)^2}$	$\dfrac{1}{a^2}x\left(1 + e^{-ax}\right) + \dfrac{2}{a^3}\left(e^{-ax} - 1\right)$
42	$\dfrac{1}{p^2(p+a)(p+b)}$	$-\dfrac{a+b}{a^2 b^2} + \dfrac{1}{ab}x + \dfrac{1}{a^2(b-a)}e^{-ax} + \dfrac{1}{b^2(a-b)}e^{-bx}$
43	$\dfrac{1}{(p+a)^2(p+b)^2}$	$\dfrac{1}{(a-b)^2}\left[e^{-ax}\left(x + \dfrac{2}{a-b}\right) + e^{-bx}\left(x - \dfrac{2}{a-b}\right)\right]$
44	$\dfrac{1}{(p+a)^4}$	$\frac{1}{6}x^3 e^{-ax}$
45	$\dfrac{p}{(p+a)^4}$	$\frac{1}{2}x^2 e^{-ax} - \frac{1}{6}ax^3 e^{-ax}$
46	$\dfrac{1}{p^2(p^2 + a^2)}$	$\dfrac{1}{a^3}\left[ax - \sin(ax)\right]$
47	$\dfrac{1}{p^4 - a^4}$	$\dfrac{1}{2a^3}\left[\sinh(ax) - \sin(ax)\right]$
48	$\dfrac{p}{p^4 - a^4}$	$\dfrac{1}{2a^2}\left[\cosh(ax) - \cos(ax)\right]$
49	$\dfrac{p^2}{p^4 - a^4}$	$\dfrac{1}{2a}\left[\sinh(ax) + \sin(ax)\right]$
50	$\dfrac{p^3}{p^4 - a^4}$	$\dfrac{1}{2}\left[\cosh(ax) + \cos(ax)\right]$
51	$\dfrac{1}{p^4 + a^4}$	$\dfrac{1}{a^3\sqrt{2}}\left(\cosh\xi\sin\xi - \sinh\xi\cos\xi\right), \quad \xi = \dfrac{ax}{\sqrt{2}}$
52	$\dfrac{p}{p^4 + a^4}$	$\dfrac{1}{a^2}\sin\left(\dfrac{ax}{\sqrt{2}}\right)\sinh\left(\dfrac{ax}{\sqrt{2}}\right)$
53	$\dfrac{p^2}{p^4 + a^4}$	$\dfrac{1}{a\sqrt{2}}\left(\cos\xi\sinh\xi + \sin\xi\cosh\xi\right), \quad \xi = \dfrac{ax}{\sqrt{2}}$

No.	Laplace transform, $\widetilde{f}(p)$	Inverse transform, $f(x) = \dfrac{1}{2\pi i}\displaystyle\int_{c-i\infty}^{c+i\infty} e^{px}\widetilde{f}(p)\,dp$
54	$\dfrac{1}{(p^2 + a^2)^2}$	$\dfrac{1}{2a^3}\big[\sin(ax) - ax\cos(ax)\big]$
55	$\dfrac{p}{(p^2 + a^2)^2}$	$\dfrac{1}{2a}x\sin(ax)$
56	$\dfrac{p^2}{(p^2 + a^2)^2}$	$\dfrac{1}{2a}\big[\sin(ax) + ax\cos(ax)\big]$
57	$\dfrac{p^3}{(p^2 + a^2)^2}$	$\cos(ax) - \tfrac{1}{2}ax\sin(ax)$
58	$\dfrac{1}{\big[(p+b)^2 + a^2\big]^2}$	$\dfrac{1}{2a^3}e^{-bx}\big[\sin(ax) - ax\cos(ax)\big]$
59	$\dfrac{1}{(p^2 - a^2)(p^2 - b^2)}$	$\dfrac{1}{a^2 - b^2}\Big[\dfrac{1}{a}\sinh(ax) - \dfrac{1}{b}\sinh(bx)\Big]$
60	$\dfrac{p}{(p^2 - a^2)(p^2 - b^2)}$	$\dfrac{\cosh(ax) - \cosh(bx)}{a^2 - b^2}$
61	$\dfrac{p^2}{(p^2 - a^2)(p^2 - b^2)}$	$\dfrac{a\sinh(ax) - b\sinh(bx)}{a^2 - b^2}$
62	$\dfrac{p^3}{(p^2 - a^2)(p^2 - b^2)}$	$\dfrac{a^2\cosh(ax) - b^2\cosh(bx)}{a^2 - b^2}$
63	$\dfrac{1}{(p^2 + a^2)(p^2 + b^2)}$	$\dfrac{1}{b^2 - a^2}\Big[\dfrac{1}{a}\sin(ax) - \dfrac{1}{b}\sin(bx)\Big]$
64	$\dfrac{p}{(p^2 + a^2)(p^2 + b^2)}$	$\dfrac{\cos(ax) - \cos(bx)}{b^2 - a^2}$
65	$\dfrac{p^2}{(p^2 + a^2)(p^2 + b^2)}$	$\dfrac{-a\sin(ax) + b\sin(bx)}{b^2 - a^2}$
66	$\dfrac{p^3}{(p^2 + a^2)(p^2 + b^2)}$	$\dfrac{-a^2\cos(ax) + b^2\cos(bx)}{b^2 - a^2}$
67	$\dfrac{1}{p^n},\quad n = 1, 2, \ldots$	$\dfrac{1}{(n-1)!}x^{n-1}$
68	$\dfrac{1}{(p+a)^n},\quad n = 1, 2, \ldots$	$\dfrac{1}{(n-1)!}x^{n-1}e^{-ax}$
69	$\dfrac{1}{p(p+a)^n},\quad n = 1, 2, \ldots$	$a^{-n}\big[1 - e^{-ax}e_n(ax)\big],\quad e_n(z) = 1 + \dfrac{z}{1!} + \cdots + \dfrac{z^n}{n!}$
70	$\dfrac{1}{p^{2n} + a^{2n}},\quad n = 1, 2, \ldots$	$-\dfrac{1}{na^{2n}}\displaystyle\sum_{k=1}^{n}\exp(a_k x)\big[a_k\cos(b_k x) - b_k\sin(b_k x)\big],$ $a_k = a\cos\varphi_k,\ b_k = a\sin\varphi_k,\ \varphi_k = \dfrac{\pi(2k-1)}{2n}$
71	$\dfrac{1}{p^{2n} - a^{2n}},\quad n = 1, 2, \ldots$	$\dfrac{1}{na^{2n-1}}\sinh(ax) + \dfrac{1}{na^{2n}}\displaystyle\sum_{k=2}^{n}\exp(a_k x)$ $\times\big[a_k\cos(b_k x) - b_k\sin(b_k x)\big],$ $a_k = a\cos\varphi_k,\ b_k = a\sin\varphi_k,\ \varphi_k = \dfrac{\pi(k-1)}{n}$

No.	Laplace transform, $\widetilde{f}(p)$	Inverse transform, $f(x) = \dfrac{1}{2\pi i}\displaystyle\int_{c-i\infty}^{c+i\infty} e^{px}\,\widetilde{f}(p)\,dp$
72	$\dfrac{1}{p^{2n+1} + a^{2n+1}}$, $\quad n = 0, 1, \ldots$	$\dfrac{e^{-ax}}{(2n+1)a^{2n}} - \dfrac{2}{(2n+1)a^{2n+1}}\displaystyle\sum_{k=1}^{n}\exp(a_k x)$ $\times \big[a_k\cos(b_k x) - b_k\sin(b_k x)\big]$, $a_k = a\cos\varphi_k,\ b_k = a\sin\varphi_k,\ \varphi_k = \dfrac{\pi(2k-1)}{2n+1}$
73	$\dfrac{1}{p^{2n+1} - a^{2n+1}}$, $\quad n = 0, 1, \ldots$	$\dfrac{e^{ax}}{(2n+1)a^{2n}} + \dfrac{2}{(2n+1)a^{2n+1}}\displaystyle\sum_{k=1}^{n}\exp(a_k x)$ $\times \big[a_k\cos(b_k x) - b_k\sin(b_k x)\big]$, $a_k = a\cos\varphi_k,\ b_k = a\sin\varphi_k,\ \varphi_k = \dfrac{2\pi k}{2n+1}$
74	$\dfrac{Q(p)}{P(p)}$, $P(p) = (p-a_1)\ldots(p-a_n)$; $Q(p)$ is a polynomial of degree $\leq n-1$; $a_i \neq a_j$ if $i \neq j$	$\displaystyle\sum_{k=1}^{n}\dfrac{Q(a_k)}{P'(a_k)}\exp\big(a_k x\big)$, (the prime stands for the differentiation)
75	$\dfrac{Q(p)}{P(p)}$, $P(p) = (p-a_1)^{m_1}\ldots(p-a_n)^{m_n}$; $Q(p)$ is a polynomial of degree $< m_1 + m_2 + \cdots + m_n - 1$; $a_i \neq a_j$ if $i \neq j$	$\displaystyle\sum_{k=1}^{n}\sum_{l=1}^{m_k}\dfrac{\Phi_{kl}(a_k)}{(m_k-l)!\,(l-1)!}\,x^{m_k-l}\exp\big(a_k x\big)$, $\Phi_{kl}(p) = \dfrac{d^{l-1}}{dp^{l-1}}\left[\dfrac{Q(p)}{P_k(p)}\right]$, $\quad P_k(p) = \dfrac{P(p)}{(p-a_k)^{m_k}}$
76	$\dfrac{Q(p) + pR(p)}{P(p)}$, $P(p) = (p^2 + a_1^2)\ldots(p^2 + a_n^2)$; $Q(p)$ and $R(p)$ are polynomials of degree $\leq 2n-2$; $a_l \neq a_j$, $l \neq j$	$\displaystyle\sum_{k=1}^{n}\dfrac{Q(ia_k)\sin(a_k x) + a_k R(ia_k)\cos(a_k x)}{a_k P_k(ia_k)}$, $P_m(p) = \dfrac{P(p)}{p^2 + a_m^2}$, $\quad i^2 = -1$

S2.2.3. Expressions Involving Square Roots

No.	Laplace transform, $\widetilde{f}(p)$	Inverse transform, $f(x) = \dfrac{1}{2\pi i}\displaystyle\int_{c-i\infty}^{c+i\infty} e^{px}\,\widetilde{f}(p)\,dp$
1	$\dfrac{1}{\sqrt{p}}$	$\dfrac{1}{\sqrt{\pi x}}$
2	$\sqrt{p-a} - \sqrt{p-b}$	$\dfrac{e^{bx} - e^{ax}}{2\sqrt{\pi x^3}}$
3	$\dfrac{1}{\sqrt{p+a}}$	$\dfrac{1}{\sqrt{\pi x}}e^{-ax}$
4	$\sqrt{\dfrac{p+a}{p}} - 1$	$\tfrac{1}{2}ae^{-ax/2}\big[I_1\big(\tfrac{1}{2}ax\big) + I_0\big(\tfrac{1}{2}ax\big)\big]$
5	$\dfrac{\sqrt{p+a}}{p+b}$	$\dfrac{e^{-ax}}{\sqrt{\pi x}} + (a-b)^{1/2}e^{-bx}\,\mathrm{erf}\big[(a-b)^{1/2}x^{1/2}\big]$
6	$\dfrac{1}{p\sqrt{p}}$	$2\sqrt{\dfrac{x}{\pi}}$
7	$\dfrac{1}{(p+a)\sqrt{p+b}}$	$(b-a)^{-1/2}e^{-ax}\,\mathrm{erf}\big[(b-a)^{1/2}x^{1/2}\big]$

No.	Laplace transform, $\widetilde{f}(p)$	Inverse transform, $f(x) = \dfrac{1}{2\pi i}\displaystyle\int_{c-i\infty}^{c+i\infty} e^{px}\widetilde{f}(p)\,dp$
8	$\dfrac{1}{\sqrt{p}\,(p-a)}$	$\dfrac{1}{\sqrt{a}}\,e^{ax}\,\mathrm{erf}\!\left(\sqrt{ax}\,\right)$
9	$\dfrac{1}{p^{3/2}(p-a)}$	$a^{-3/2}e^{ax}\,\mathrm{erf}\!\left(\sqrt{ax}\,\right) - 2a^{-1}\pi^{-1/2}x^{1/2}$
10	$\dfrac{1}{\sqrt{p}+a}$	$\pi^{-1/2}x^{-1/2} - ae^{a^2x}\,\mathrm{erfc}\!\left(a\sqrt{x}\,\right)$
11	$\dfrac{a}{p\left(\sqrt{p}+a\right)}$	$1 - e^{a^2x}\,\mathrm{erfc}\!\left(a\sqrt{x}\,\right)$
12	$\dfrac{1}{p+a\sqrt{p}}$	$e^{a^2x}\,\mathrm{erfc}\!\left(a\sqrt{x}\,\right)$
13	$\dfrac{1}{\left(\sqrt{p}+\sqrt{a}\,\right)^2}$	$1 - \dfrac{2}{\sqrt{\pi}}(ax)^{1/2} + (1-2ax)e^{ax}\left[\mathrm{erf}\!\left(\sqrt{ax}\,\right)-1\right]$
14	$\dfrac{1}{p\left(\sqrt{p}+\sqrt{a}\,\right)^2}$	$\dfrac{1}{a} + \left(2x-\dfrac{1}{a}\right)e^{ax}\,\mathrm{erfc}\!\left(\sqrt{ax}\,\right) - \dfrac{2}{\sqrt{\pi a}}\sqrt{x}$
15	$\dfrac{1}{\sqrt{p}\left(\sqrt{p}+a\right)^2}$	$2\pi^{-1/2}x^{1/2} - 2axe^{a^2x}\,\mathrm{erfc}\!\left(a\sqrt{x}\,\right)$
16	$\dfrac{1}{\left(\sqrt{p}+a\right)^3}$	$\dfrac{2}{\sqrt{\pi}}(a^2x+1)\sqrt{x} - ax(2a^2x+3)e^{a^2x}\,\mathrm{erfc}\!\left(a\sqrt{x}\,\right)$
17	$p^{-n-1/2}, \quad n=1,2,\ldots$	$\dfrac{2^n}{1\times 3\times\ldots\times(2n-1)\sqrt{\pi}}\,x^{n-1/2}$
18	$(p+a)^{-n-1/2}$	$\dfrac{2^n}{1\times 3\times\ldots\times(2n-1)\sqrt{\pi}}\,x^{n-1/2}e^{-ax}$
19	$\dfrac{1}{\sqrt{p^2+a^2}}$	$J_0(ax)$
20	$\dfrac{1}{\sqrt{p^2-a^2}}$	$I_0(ax)$
21	$\dfrac{1}{\sqrt{p^2+ap+b}}$	$\exp\!\left(-\tfrac{1}{2}ax\right)J_0\!\left[\left(b-\tfrac{1}{4}a^2\right)^{1/2}x\right]$
22	$\left(\sqrt{p^2+a^2}-p\right)^{1/2}$	$\dfrac{1}{\sqrt{2\pi x^3}}\sin(ax)$
23	$\dfrac{1}{\sqrt{p^2+a^2}}\left(\sqrt{p^2+a^2}+p\right)^{1/2}$	$\dfrac{\sqrt{2}}{\sqrt{\pi x}}\cos(ax)$
24	$\dfrac{1}{\sqrt{p^2-a^2}}\left(\sqrt{p^2-a^2}+p\right)^{1/2}$	$\dfrac{\sqrt{2}}{\sqrt{\pi x}}\cosh(ax)$
25	$\left(\sqrt{p^2+a^2}+p\right)^{-n}$	$na^{-n}x^{-1}J_n(ax)$
26	$\left(\sqrt{p^2-a^2}+p\right)^{-n}$	$na^{-n}x^{-1}I_n(ax)$
27	$\left(p^2+a^2\right)^{-n-1/2}$	$\dfrac{(x/a)^n J_n(ax)}{1\times 3\times 5\times\ldots\times(2n-1)}$
28	$\left(p^2-a^2\right)^{-n-1/2}$	$\dfrac{(x/a)^n I_n(ax)}{1\times 3\times 5\times\ldots\times(2n-1)}$

S2.2.4. Expressions Involving Arbitrary Powers

No.	Laplace transform, $\widetilde{f}(p)$	Inverse transform, $f(x) = \dfrac{1}{2\pi i}\displaystyle\int_{c-i\infty}^{c+i\infty} e^{px}\widetilde{f}(p)\,dp$
1	$(p+a)^{-\nu},\quad \nu>0$	$\dfrac{1}{\Gamma(\nu)}x^{\nu-1}e^{-ax}$
2	$\left[(p+a)^{1/2}+(p+b)^{1/2}\right]^{-2\nu},\quad \nu>0$	$\dfrac{\nu}{(a-b)^{\nu}}x^{-1}\exp\!\left[-\tfrac{1}{2}(a+b)x\right]I_{\nu}\!\left[\tfrac{1}{2}(a-b)x\right]$
3	$\left[(p+a)(p+b)\right]^{-\nu},\quad \nu>0$	$\dfrac{\sqrt{\pi}}{\Gamma(\nu)}\left(\dfrac{x}{a-b}\right)^{\nu-1/2}\exp\!\left(-\dfrac{a+b}{2}x\right)I_{\nu-1/2}\!\left(\dfrac{a-b}{2}x\right)$
4	$\left(p^2+a^2\right)^{-\nu-1/2},\quad \nu>-\tfrac{1}{2}$	$\dfrac{\sqrt{\pi}}{(2a)^{\nu}\Gamma\!\left(\nu+\tfrac{1}{2}\right)}x^{\nu}J_{\nu}(ax)$
5	$\left(p^2-a^2\right)^{-\nu-1/2},\quad \nu>-\tfrac{1}{2}$	$\dfrac{\sqrt{\pi}}{(2a)^{\nu}\Gamma\!\left(\nu+\tfrac{1}{2}\right)}x^{\nu}I_{\nu}(ax)$
6	$p\left(p^2+a^2\right)^{-\nu-1/2},\quad \nu>0$	$\dfrac{a\sqrt{\pi}}{(2a)^{\nu}\Gamma\!\left(\nu+\tfrac{1}{2}\right)}x^{\nu}J_{\nu-1}(ax)$
7	$p\left(p^2-a^2\right)^{-\nu-1/2},\quad \nu>0$	$\dfrac{a\sqrt{\pi}}{(2a)^{\nu}\Gamma\!\left(\nu+\tfrac{1}{2}\right)}x^{\nu}I_{\nu-1}(ax)$
8	$\begin{aligned}&\left[(p^2+a^2)^{1/2}+p\right]^{-\nu}=\\&a^{-2\nu}\left[(p^2+a^2)^{1/2}-p\right]^{\nu},\quad \nu>0\end{aligned}$	$\nu a^{-\nu}x^{-1}J_{\nu}(ax)$
9	$\begin{aligned}&\left[(p^2-a^2)^{1/2}+p\right]^{-\nu}=\\&a^{-2\nu}\left[p-(p^2-a^2)^{1/2}\right]^{\nu},\quad \nu>0\end{aligned}$	$\nu a^{-\nu}x^{-1}I_{\nu}(ax)$
10	$p\left[(p^2+a^2)^{1/2}+p\right]^{-\nu},\quad \nu>1$	$\nu a^{1-\nu}x^{-1}J_{\nu-1}(ax)-\nu(\nu+1)a^{-\nu}x^{-2}J_{\nu}(ax)$
11	$p\left[(p^2-a^2)^{1/2}+p\right]^{-\nu},\quad \nu>1$	$\nu a^{1-\nu}x^{-1}I_{\nu-1}(ax)-\nu(\nu+1)a^{-\nu}x^{-2}I_{\nu}(ax)$
12	$\dfrac{\left(\sqrt{p^2+a^2}+p\right)^{-\nu}}{\sqrt{p^2+a^2}},\quad \nu>-1$	$a^{-\nu}J_{\nu}(ax)$
13	$\dfrac{\left(\sqrt{p^2-a^2}+p\right)^{-\nu}}{\sqrt{p^2-a^2}},\quad \nu>-1$	$a^{-\nu}I_{\nu}(ax)$

S2.2.5. Expressions Involving Exponential Functions

No.	Laplace transform, $\widetilde{f}(p)$	Inverse transform, $f(x) = \dfrac{1}{2\pi i}\displaystyle\int_{c-i\infty}^{c+i\infty} e^{px}\widetilde{f}(p)\,dp$
1	$p^{-1}e^{-ap},\quad a>0$	$\begin{cases}0 & \text{if } 0<x<a,\\ 1 & \text{if } a<x\end{cases}$
2	$p^{-1}\!\left(1-e^{-ap}\right),\quad a>0$	$\begin{cases}1 & \text{if } 0<x<a,\\ 0 & \text{if } a<x\end{cases}$
3	$p^{-1}\!\left(e^{-ap}-e^{-bp}\right),\quad 0\le a<b$	$\begin{cases}0 & \text{if } 0<x<a,\\ 1 & \text{if } a<x<b,\\ 0 & \text{if } b<x\end{cases}$
4	$p^{-2}\!\left(e^{-ap}-e^{-bp}\right),\quad 0\le a<b$	$\begin{cases}0 & \text{if } 0<x<a,\\ x-a & \text{if } a<x<b,\\ b-a & \text{if } b<x\end{cases}$
5	$(p+b)^{-1}e^{-ap},\quad a>0$	$\begin{cases}0 & \text{if } 0<x<a,\\ e^{-b(x-a)} & \text{if } a<x\end{cases}$

No.	Laplace transform, $\widetilde{f}(p)$	Inverse transform, $f(x) = \dfrac{1}{2\pi i}\displaystyle\int_{c-i\infty}^{c+i\infty} e^{px}\widetilde{f}(p)\,dp$
6	$p^{-\nu}e^{-ap}, \quad \nu > 0$	$\begin{cases} 0 & \text{if } 0 < x < a, \\ \dfrac{(x-a)^{\nu-1}}{\Gamma(\nu)} & \text{if } a < x \end{cases}$
7	$p^{-1}\left(e^{ap}-1\right)^{-1}, \quad a > 0$	$f(x) = n \ \text{ if } \ na < x < (n+1)a; \quad n = 0,\,1,\,2,\,\ldots$
8	$e^{a/p}-1$	$\sqrt{\dfrac{a}{x}}\,I_1\!\left(2\sqrt{ax}\right)$
9	$p^{-1/2}e^{a/p}$	$\dfrac{1}{\sqrt{\pi x}}\cosh\!\left(2\sqrt{ax}\right)$
10	$p^{-3/2}e^{a/p}$	$\dfrac{1}{\sqrt{\pi a}}\sinh\!\left(2\sqrt{ax}\right)$
11	$p^{-5/2}e^{a/p}$	$\sqrt{\dfrac{x}{\pi a}}\cosh\!\left(2\sqrt{ax}\right) - \dfrac{1}{2\sqrt{\pi a^3}}\sinh\!\left(2\sqrt{ax}\right)$
12	$p^{-\nu-1}e^{a/p}, \quad \nu > -1$	$(x/a)^{\nu/2}I_\nu\!\left(2\sqrt{ax}\right)$
13	$1 - e^{-a/p}$	$\sqrt{\dfrac{a}{x}}\,J_1\!\left(2\sqrt{ax}\right)$
14	$p^{-1/2}e^{-a/p}$	$\dfrac{1}{\sqrt{\pi x}}\cos\!\left(2\sqrt{ax}\right)$
15	$p^{-3/2}e^{-a/p}$	$\dfrac{1}{\sqrt{\pi a}}\sin\!\left(2\sqrt{ax}\right)$
16	$p^{-5/2}e^{-a/p}$	$\dfrac{1}{2\sqrt{\pi a^3}}\sin\!\left(2\sqrt{ax}\right) - \sqrt{\dfrac{x}{\pi a}}\cos\!\left(2\sqrt{ax}\right)$
17	$p^{-\nu-1}e^{-a/p}, \quad \nu > -1$	$(x/a)^{\nu/2}J_\nu\!\left(2\sqrt{ax}\right)$
18	$\exp\!\left(-\sqrt{ap}\right), \quad a > 0$	$\dfrac{\sqrt{a}}{2\sqrt{\pi}}\,x^{-3/2}\exp\!\left(-\dfrac{a}{4x}\right)$
19	$p\exp\!\left(-\sqrt{ap}\right), \quad a > 0$	$\dfrac{\sqrt{a}}{8\sqrt{\pi}}\,(a-6x)x^{-7/2}\exp\!\left(-\dfrac{a}{4x}\right)$
20	$\dfrac{1}{p}\exp\!\left(-\sqrt{ap}\right), \quad a \geq 0$	$\mathrm{erfc}\!\left(\dfrac{\sqrt{a}}{2\sqrt{x}}\right)$
21	$\sqrt{p}\exp\!\left(-\sqrt{ap}\right), \quad a > 0$	$\dfrac{1}{4\sqrt{\pi}}\,(a-2x)x^{-5/2}\exp\!\left(-\dfrac{a}{4x}\right)$
22	$\dfrac{1}{\sqrt{p}}\exp\!\left(-\sqrt{ap}\right), \quad a \geq 0$	$\dfrac{1}{\sqrt{\pi x}}\exp\!\left(-\dfrac{a}{4x}\right)$
23	$\dfrac{1}{p\sqrt{p}}\exp\!\left(-\sqrt{ap}\right), \quad a \geq 0$	$\dfrac{2\sqrt{x}}{\sqrt{\pi}}\exp\!\left(-\dfrac{a}{4x}\right) - \sqrt{a}\,\mathrm{erfc}\!\left(\dfrac{\sqrt{a}}{2\sqrt{x}}\right)$
24	$\dfrac{\exp\!\left(-k\sqrt{p^2+a^2}\right)}{\sqrt{p^2+a^2}}, \quad k > 0$	$\begin{cases} 0 & \text{if } 0 < x < k, \\ J_0\!\left(a\sqrt{x^2-k^2}\right) & \text{if } k < x \end{cases}$
25	$\dfrac{\exp\!\left(-k\sqrt{p^2-a^2}\right)}{\sqrt{p^2-a^2}}, \quad k > 0$	$\begin{cases} 0 & \text{if } 0 < x < k, \\ I_0\!\left(a\sqrt{x^2-k^2}\right) & \text{if } k < x \end{cases}$

S2.2.6. Expressions Involving Hyperbolic Functions

No.	Laplace transform, $\widetilde{f}(p)$	Inverse transform, $f(x) = \dfrac{1}{2\pi i}\displaystyle\int_{c-i\infty}^{c+i\infty} e^{px}\widetilde{f}(p)\,dp$
1	$\dfrac{1}{p\sinh(ap)}, \quad a>0$	$f(x) = 2n$ if $a(2n-1)<x<a(2n+1)$; $n=0,1,2,\dots$ $(x>0)$
2	$\dfrac{1}{p^2\sinh(ap)}, \quad a>0$	$f(x) = 2n(x-an)$ if $a(2n-1)<x<a(2n+1)$; $n=0,1,2,\dots$ $(x>0)$
3	$\dfrac{\sinh(a/p)}{\sqrt{p}}$	$\dfrac{1}{2\sqrt{\pi x}}\left[\cosh\left(2\sqrt{ax}\right) - \cos\left(2\sqrt{ax}\right)\right]$
4	$\dfrac{\sinh(a/p)}{p\sqrt{p}}$	$\dfrac{1}{2\sqrt{\pi a}}\left[\sinh\left(2\sqrt{ax}\right) - \sin\left(2\sqrt{ax}\right)\right]$
5	$p^{-\nu-1}\sinh(a/p), \quad \nu>-2$	$\tfrac{1}{2}(x/a)^{\nu/2}\left[I_\nu\left(2\sqrt{ax}\right) - J_\nu\left(2\sqrt{ax}\right)\right]$
6	$\dfrac{1}{p\cosh(ap)}, \quad a>0$	$f(x) = \begin{cases} 0 & \text{if } a(4n-1)<x<a(4n+1), \\ 2 & \text{if } a(4n+1)<x<a(4n+3), \end{cases}$ $n=0,1,2,\dots$ $(x>0)$
7	$\dfrac{1}{p^2\cosh(ap)}, \quad a>0$	$x - (-1)^n(x-2an)$ if $2n-1<x/a<2n+1$; $n=0,1,2,\dots$ $(x>0)$
8	$\dfrac{\cosh(a/p)}{\sqrt{p}}$	$\dfrac{1}{2\sqrt{\pi x}}\left[\cosh\left(2\sqrt{ax}\right) + \cos\left(2\sqrt{ax}\right)\right]$
9	$\dfrac{\cosh(a/p)}{p\sqrt{p}}$	$\dfrac{1}{2\sqrt{\pi a}}\left[\sinh\left(2\sqrt{ax}\right) + \sin\left(2\sqrt{ax}\right)\right]$
10	$p^{-\nu-1}\cosh(a/p), \quad \nu>-1$	$\tfrac{1}{2}(x/a)^{\nu/2}\left[I_\nu\left(2\sqrt{ax}\right) + J_\nu\left(2\sqrt{ax}\right)\right]$
11	$\dfrac{1}{p}\tanh(ap), \quad a>0$	$f(x) = (-1)^{n-1}$ if $2a(n-1)<x<2an$; $n=1,2,\dots$
12	$\dfrac{1}{p}\coth(ap), \quad a>0$	$f(x) = (2n-1)$ if $2a(n-1)<x<2an$; $n=1,2,\dots$
13	$\operatorname{arccoth}(p/a)$	$\dfrac{1}{x}\sinh(ax)$

S2.2.7. Expressions Involving Logarithmic Functions

No.	Laplace transform, $\widetilde{f}(p)$	Inverse transform, $f(x) = \dfrac{1}{2\pi i}\displaystyle\int_{c-i\infty}^{c+i\infty} e^{px}\widetilde{f}(p)\,dp$
1	$\dfrac{1}{p}\ln p$	$-\ln x - \mathcal{C}$, $\mathcal{C} = 0.5772\dots$ is the Euler constant
2	$p^{-n-1}\ln p$	$\left(1 + \tfrac{1}{2} + \tfrac{1}{3} + \cdots + \tfrac{1}{n} - \ln x - \mathcal{C}\right)\dfrac{x^n}{n!}$, $\mathcal{C} = 0.5772\dots$ is the Euler constant
3	$p^{-n-1/2}\ln p$	$k_n\left[2 + \tfrac{2}{3} + \tfrac{2}{5} + \cdots + \tfrac{2}{2n-1} - \ln(4x) - \mathcal{C}\right]x^{n-1/2}$, $k_n = \dfrac{2^n}{1\times 3\times 5\times\dots\times(2n-1)\sqrt{\pi}}, \quad \mathcal{C} = 0.5772\dots$
4	$p^{-\nu}\ln p, \quad \nu>0$	$\dfrac{1}{\Gamma(\nu)}x^{\nu-1}\left[\psi(\nu) - \ln x\right], \quad \psi(\nu)$ is the logarithmic derivative of the gamma function
5	$\dfrac{1}{p}(\ln p)^2$	$(\ln x + \mathcal{C})^2 - \tfrac{1}{6}\pi^2, \quad \mathcal{C} = 0.5772\dots$

No.	Laplace transform, $\widetilde{f}(p)$	Inverse transform, $f(x) = \dfrac{1}{2\pi i} \displaystyle\int_{c-i\infty}^{c+i\infty} e^{px}\widetilde{f}(p)\,dp$
6	$\dfrac{1}{p^2}(\ln p)^2$	$x\left[(\ln x + \mathcal{C} - 1)^2 + 1 - \tfrac{1}{6}\pi^2\right]$
7	$\dfrac{\ln(p+b)}{p+a}$	$e^{-ax}\left\{\ln(b-a) - \mathrm{Ei}\big[(a-b)x\big]\right\}$
8	$\dfrac{\ln p}{p^2 + a^2}$	$\dfrac{1}{a}\cos(ax)\,\mathrm{Si}(ax) + \dfrac{1}{a}\sin(ax)\big[\ln a - \mathrm{Ci}(ax)\big]$
9	$\dfrac{p\ln p}{p^2 + a^2}$	$\cos(ax)\big[\ln a - \mathrm{Ci}(ax)\big] - \sin(ax)\,\mathrm{Si}(ax)$
10	$\ln\dfrac{p+b}{p+a}$	$\dfrac{1}{x}\left(e^{-ax} - e^{-bx}\right)$
11	$\ln\dfrac{p^2 + b^2}{p^2 + a^2}$	$\dfrac{2}{x}\big[\cos(ax) - \cos(bx)\big]$
12	$p\ln\dfrac{p^2 + b^2}{p^2 + a^2}$	$\dfrac{2}{x}\big[\cos(bx) + bx\sin(bx) - \cos(ax) - ax\sin(ax)\big]$
13	$\ln\dfrac{(p+a)^2 + k^2}{(p+b)^2 + k^2}$	$\dfrac{2}{x}\cos(kx)(e^{-bx} - e^{-ax})$
14	$p\ln\left(\dfrac{1}{p}\sqrt{p^2 + a^2}\right)$	$\dfrac{1}{x^2}\big[\cos(ax) - 1\big] + \dfrac{a}{x}\sin(ax)$
15	$p\ln\left(\dfrac{1}{p}\sqrt{p^2 - a^2}\right)$	$\dfrac{1}{x^2}\big[\cosh(ax) - 1\big] - \dfrac{a}{x}\sinh(ax)$

S2.2.8. Expressions Involving Trigonometric Functions

No.	Laplace transform, $\widetilde{f}(p)$	Inverse transform, $f(x) = \dfrac{1}{2\pi i} \displaystyle\int_{c-i\infty}^{c+i\infty} e^{px}\widetilde{f}(p)\,dp$
1	$\dfrac{\sin(a/p)}{\sqrt{p}}$	$\dfrac{1}{\sqrt{\pi x}}\sinh\big(\sqrt{2ax}\big)\sin\big(\sqrt{2ax}\big)$
2	$\dfrac{\sin(a/p)}{p\sqrt{p}}$	$\dfrac{1}{\sqrt{\pi a}}\cosh\big(\sqrt{2ax}\big)\sin\big(\sqrt{2ax}\big)$
3	$\dfrac{\cos(a/p)}{\sqrt{p}}$	$\dfrac{1}{\sqrt{\pi x}}\cosh\big(\sqrt{2ax}\big)\cos\big(\sqrt{2ax}\big)$
4	$\dfrac{\cos(a/p)}{p\sqrt{p}}$	$\dfrac{1}{\sqrt{\pi a}}\sinh\big(\sqrt{2ax}\big)\cos\big(\sqrt{2ax}\big)$
5	$\dfrac{1}{\sqrt{p}}\exp\big(-\sqrt{ap}\big)\sin\big(\sqrt{ap}\big)$	$\dfrac{1}{\sqrt{\pi x}}\sin\left(\dfrac{a}{2x}\right)$
6	$\dfrac{1}{\sqrt{p}}\exp\big(-\sqrt{ap}\big)\cos\big(\sqrt{ap}\big)$	$\dfrac{1}{\sqrt{\pi x}}\cos\left(\dfrac{a}{2x}\right)$
7	$\arctan\dfrac{a}{p}$	$\dfrac{1}{x}\sin(ax)$
8	$\dfrac{1}{p}\arctan\dfrac{a}{p}$	$\mathrm{Si}(ax)$
9	$p\arctan\dfrac{a}{p} - a$	$\dfrac{1}{x^2}\big[ax\cos(ax) - \sin(ax)\big]$
10	$\arctan\dfrac{2ap}{p^2 + b^2}$	$\dfrac{2}{x}\sin(ax)\cos\big(x\sqrt{a^2 + b^2}\big)$

S2.2.9. Expressions Involving Special Functions

No.	Laplace transform, $\widetilde{f}(p)$	Inverse transform, $f(x) = \dfrac{1}{2\pi i}\displaystyle\int_{c-i\infty}^{c+i\infty} e^{px}\widetilde{f}(p)\,dp$
1	$\exp\left(ap^2\right)\operatorname{erfc}\left(p\sqrt{a}\right)$	$\dfrac{1}{\sqrt{\pi a}}\exp\left(-\dfrac{x^2}{4a}\right)$
2	$\dfrac{1}{p}\exp\left(ap^2\right)\operatorname{erfc}\left(p\sqrt{a}\right)$	$\operatorname{erf}\left(\dfrac{x}{2\sqrt{a}}\right)$
3	$\operatorname{erfc}\left(\sqrt{ap}\right), \quad a>0$	$\begin{cases} 0 & \text{if } 0<x<a, \\ \dfrac{\sqrt{a}}{\pi x\sqrt{x-a}} & \text{if } a<x \end{cases}$
4	$\dfrac{1}{\sqrt{p}}\operatorname{erfc}\left(\sqrt{ap}\right), \quad a>0$	$\begin{cases} 0 & \text{if } 0<x<a, \\ \dfrac{1}{\sqrt{\pi x}} & \text{if } x>a \end{cases}$
5	$e^{ap}\operatorname{erfc}\left(\sqrt{ap}\right)$	$\dfrac{\sqrt{a}}{\pi\sqrt{x}\,(x+a)}$
6	$\dfrac{1}{\sqrt{p}}e^{ap}\operatorname{erfc}\left(\sqrt{ap}\right)$	$\dfrac{1}{\sqrt{\pi(x+a)}}$
7	$\dfrac{1}{\sqrt{p}}\operatorname{erf}\left(\sqrt{ap}\right), \quad a>0$	$\begin{cases} \dfrac{1}{\sqrt{\pi x}} & \text{if } 0<x<a, \\ 0 & \text{if } x>a \end{cases}$
8	$\operatorname{erf}\left(\sqrt{a/p}\right)$	$\dfrac{1}{\pi x}\sin\left(2\sqrt{ax}\right)$
9	$\dfrac{1}{\sqrt{p}}\exp(a/p)\operatorname{erf}\left(\sqrt{a/p}\right)$	$\dfrac{1}{\sqrt{\pi x}}\sinh\left(2\sqrt{ax}\right)$
10	$\dfrac{1}{\sqrt{p}}\exp(a/p)\operatorname{erfc}\left(\sqrt{a/p}\right)$	$\dfrac{1}{\sqrt{\pi x}}\exp\left(-2\sqrt{ax}\right)$
11	$p^{-a}\gamma(a,bp), \quad a,b>0$	$\begin{cases} x^{a-1} & \text{if } 0<x<b, \\ 0 & \text{if } b<x \end{cases}$
12	$\gamma(a,b/p), \quad a>0$	$b^{a/2}x^{a/2-1}J_a\left(2\sqrt{bx}\right)$
13	$a^{-p}\gamma(p,a)$	$\exp\left(-ae^{-x}\right)$
14	$K_0(ap), \quad a>0$	$\begin{cases} 0 & \text{if } 0<x<a, \\ (x^2-a^2)^{-1/2} & \text{if } a<x \end{cases}$
15	$K_\nu(ap), \quad a>0$	$\begin{cases} 0 & \text{if } 0<x<a, \\ \dfrac{\cosh\left[\nu\arccos h(x/a)\right]}{\sqrt{x^2-a^2}} & \text{if } a<x \end{cases}$
16	$K_0\left(a\sqrt{p}\right)$	$\dfrac{1}{2x}\exp\left(-\dfrac{a^2}{4x}\right)$
17	$\dfrac{1}{\sqrt{p}}K_1\left(a\sqrt{p}\right)$	$\dfrac{1}{a}\exp\left(-\dfrac{a^2}{4x}\right)$

S2.3. Tables of Fourier Cosine Transforms

S2.3.1. General Formulas

No.	Original function, $f(x)$	Cosine transform, $\check{f}_c(u) = \int_0^\infty f(x)\cos(ux)\,dx$
1	$af_1(x) + bf_2(x)$	$a\check{f}_{1c}(u) + b\check{f}_{2c}(u)$
2	$f(ax), \quad a > 0$	$\dfrac{1}{a}\check{f}_c\left(\dfrac{u}{a}\right)$
3	$x^{2n}f(x), \quad n = 1, 2, \dots$	$(-1)^n\dfrac{d^{2n}}{du^{2n}}\check{f}_c(u)$
4	$x^{2n+1}f(ax), \quad n = 0, 1, \dots$	$(-1)^n\dfrac{d^{2n+1}}{du^{2n+1}}\check{f}_s(u), \quad \check{f}_s(u) = \int_0^\infty f(x)\sin(xu)\,dx$
5	$f(ax)\cos(bx), \quad a, b > 0$	$\dfrac{1}{2a}\left[\check{f}_c\left(\dfrac{u+b}{a}\right) + \check{f}_c\left(\dfrac{u-b}{a}\right)\right]$

S2.3.2. Expressions Involving Power Functions

No.	Original function, $f(x)$	Cosine transform, $\check{f}_c(u) = \int_0^\infty f(x)\cos(ux)\,dx$
1	$\begin{cases} 1 & \text{if } 0 < x < a, \\ 0 & \text{if } a < x \end{cases}$	$\dfrac{1}{u}\sin(au)$
2	$\begin{cases} x & \text{if } 0 < x < 1, \\ 2-x & \text{if } 1 < x < 2, \\ 0 & \text{if } 2 < x \end{cases}$	$\dfrac{4}{u^2}\cos u \, \sin^2\dfrac{u}{2}$
3	$\dfrac{1}{a+x}, \quad a > 0$	$-\sin(au)\,\mathrm{si}(au) - \cos(au)\,\mathrm{Ci}(au)$
4	$\dfrac{1}{a^2 + x^2}, \quad a > 0$	$\dfrac{\pi}{2a}e^{-au}$
5	$\dfrac{1}{a^2 - x^2}, \quad a > 0$	$\dfrac{\pi\sin(au)}{2u}$ (the integral is understood in the sense of Cauchy principal value)
6	$\dfrac{a}{a^2 + (b+x)^2} + \dfrac{a}{a^2 + (b-x)^2}$	$\pi e^{-au}\cos(bu)$
7	$\dfrac{b+x}{a^2 + (b+x)^2} + \dfrac{b-x}{a^2 + (b-x)^2}$	$\pi e^{-au}\sin(bu)$
8	$\dfrac{1}{a^4 + x^4}, \quad a > 0$	$\tfrac{1}{2}\pi a^{-3}\exp\left(-\dfrac{au}{\sqrt{2}}\right)\sin\left(\dfrac{\pi}{4} + \dfrac{au}{\sqrt{2}}\right)$
9	$\dfrac{1}{(a^2 + x^2)(b^2 + x^2)}, \quad a, b > 0$	$\dfrac{\pi}{2}\dfrac{ae^{-bu} - be^{-au}}{ab(a^2 - b^2)}$
10	$\dfrac{x^{2m}}{(x^2 + a)^{n+1}},$ $n, m = 1, 2, \dots; \ n + 1 > m \geq 0$	$(-1)^{n+m}\dfrac{\pi}{2n!}\dfrac{\partial^n}{\partial a^n}\left(a^{1/\sqrt{m}}e^{-u\sqrt{a}}\right)$
11	$\dfrac{1}{\sqrt{x}}$	$\sqrt{\dfrac{\pi}{2u}}$
12	$\begin{cases} \dfrac{1}{\sqrt{x}} & \text{if } 0 < x < a, \\ 0 & \text{if } a < x \end{cases}$	$2\sqrt{\dfrac{\pi}{2u}}\,C(au), \quad C(u)$ is the Fresnel integral

No.	Original function, $f(x)$	Cosine transform, $\check{f}_c(u) = \int_0^\infty f(x)\cos(ux)\,dx$
13	$\begin{cases} 0 & \text{if } 0 < x < a, \\ \dfrac{1}{\sqrt{x}} & \text{if } a < x \end{cases}$	$\sqrt{\dfrac{\pi}{2u}}\,[1 - 2C(au)]$, $C(u)$ is the Fresnel integral
14	$\begin{cases} 0 & \text{if } 0 < x < a, \\ \dfrac{1}{\sqrt{x-a}} & \text{if } a < x \end{cases}$	$\sqrt{\dfrac{\pi}{2u}}\,[\cos(au) - \sin(au)]$
15	$\dfrac{1}{\sqrt{a^2 + x^2}}$	$K_0(au)$
16	$\begin{cases} \dfrac{1}{\sqrt{a^2 - x^2}} & \text{if } 0 < x < a, \\ 0 & \text{if } a < x \end{cases}$	$\dfrac{\pi}{2}\,J_0(au)$
17	$(a^2 + x^2)^{-1/2}\big[(a^2 + x^2)^{1/2} + a\big]^{1/2}$	$(2u/\pi)^{-1/2}e^{-au}, \quad a > 0$
18	$x^{-\nu}, \quad 0 < \nu < 1$	$\sin\big(\tfrac{1}{2}\pi\nu\big)\Gamma(1 - \nu)u^{\nu-1}$

S2.3.3. Expressions Involving Exponential Functions

No.	Original function, $f(x)$	Cosine transform, $\check{f}_c(u) = \int_0^\infty f(x)\cos(ux)\,dx$
1	e^{-ax}	$\dfrac{a}{a^2 + u^2}$
2	$\dfrac{1}{x}\big(e^{-ax} - e^{-bx}\big)$	$\dfrac{1}{2}\ln\dfrac{b^2 + u^2}{a^2 + u^2}$
3	$\sqrt{x}\,e^{-ax}$	$\tfrac{1}{2}\sqrt{\pi}\,(a^2 + u^2)^{-3/4}\cos\big(\tfrac{3}{2}\arctan\dfrac{u}{a}\big)$
4	$\dfrac{1}{\sqrt{x}}e^{-ax}$	$\sqrt{\dfrac{\pi}{2}}\left[\dfrac{a + (a^2 + u^2)^{1/2}}{a^2 + u^2}\right]^{1/2}$
5	$x^n e^{-ax}, \quad n = 1, 2, \ldots$	$\dfrac{a^{n+1}n!}{(a^2 + u^2)^{n+1}}\sum_{0 \le 2k \le n+1}(-1)^k C_{n+1}^{2k}\left(\dfrac{u}{a}\right)^{2k}$
6	$x^{n-1/2}e^{-ax}, \quad n = 1, 2, \ldots$	$k_n u\dfrac{\partial^n}{\partial a^n}\dfrac{1}{r\sqrt{r-a}}$, where $\quad r = \sqrt{a^2 + u^2}, \; k_n = (-1)^n\sqrt{\pi/2}$
7	$x^{\nu-1}e^{-ax}$	$\Gamma(\nu)(a^2 + u^2)^{-\nu/2}\cos\big(\nu\arctan\dfrac{u}{a}\big)$
8	$\dfrac{x}{e^{ax} - 1}$	$\dfrac{1}{2u^2} - \dfrac{\pi^2}{2a^2\sinh^2\big(\pi a^{-1}u\big)}$
9	$\dfrac{1}{x}\left(\dfrac{1}{2} - \dfrac{1}{x} + \dfrac{1}{e^x - 1}\right)$	$-\dfrac{1}{2}\ln\big(1 - e^{-2\pi u}\big)$
10	$\exp(-ax^2)$	$\dfrac{1}{2}\sqrt{\dfrac{\pi}{a}}\exp\left(-\dfrac{u^2}{4a}\right)$
11	$\dfrac{1}{\sqrt{x}}\exp\left(-\dfrac{a}{x}\right)$	$\sqrt{\dfrac{\pi}{2u}}\,e^{-\sqrt{2au}}\big[\cos\big(\sqrt{2au}\,\big) - \sin\big(\sqrt{2au}\,\big)\big]$
12	$\dfrac{1}{x\sqrt{x}}\exp\left(-\dfrac{a}{x}\right)$	$\sqrt{\dfrac{\pi}{a}}\,e^{-\sqrt{2au}}\cos\big(\sqrt{2au}\,\big)$

S2.3.4. Expressions Involving Hyperbolic Functions

No.	Original function, $f(x)$	Cosine transform, $\check{f}_c(u) = \int_0^\infty f(x)\cos(ux)\,dx$
1	$\dfrac{1}{\cosh(ax)}, \quad a > 0$	$\dfrac{\pi}{2a\cosh\left(\frac{1}{2}\pi a^{-1}u\right)}$
2	$\dfrac{1}{\cosh^2(ax)}, \quad a > 0$	$\dfrac{\pi u}{2a^2\sinh\left(\frac{1}{2}\pi a^{-1}u\right)}$
3	$\dfrac{\cosh(ax)}{\cosh(bx)}, \quad |a| < b$	$\dfrac{\pi}{b}\left[\dfrac{\cos\left(\frac{1}{2}\pi ab^{-1}\right)\cosh\left(\frac{1}{2}\pi b^{-1}u\right)}{\cos\left(\pi ab^{-1}\right)+\cosh\left(\pi b^{-1}u\right)}\right]$
4	$\dfrac{1}{\cosh(ax)+\cos b}$	$\dfrac{\pi\sinh\left(a^{-1}bu\right)}{a\sin b\,\sinh\left(\pi a^{-1}u\right)}$
5	$\exp(-ax^2)\cosh(bx), \quad a > 0$	$\dfrac{1}{2}\sqrt{\dfrac{\pi}{a}}\,\exp\left(\dfrac{b^2-u^2}{4a}\right)\cos\left(\dfrac{abu}{2}\right)$
6	$\dfrac{x}{\sinh(ax)}$	$\dfrac{\pi^2}{4a^2\cosh^2\left(\frac{1}{2}\pi a^{-1}u\right)}$
7	$\dfrac{\sinh(ax)}{\sinh(bx)}, \quad |a| < b$	$\dfrac{\pi}{2b}\dfrac{\sin\left(\pi ab^{-1}\right)}{\cos\left(\pi ab^{-1}\right)+\cosh\left(\pi b^{-1}u\right)}$
8	$\dfrac{1}{x}\tanh(ax), \quad a > 0$	$\ln\left[\coth\left(\frac{1}{4}\pi a^{-1}u\right)\right]$

S2.3.5. Expressions Involving Logarithmic Functions

No.	Original function, $f(x)$	Cosine transform, $\check{f}_c(u) = \int_0^\infty f(x)\cos(ux)\,dx$
1	$\begin{cases} \ln x & \text{if } 0 < x < 1, \\ 0 & \text{if } 1 < x \end{cases}$	$-\dfrac{1}{u}\operatorname{Si}(u)$
2	$\dfrac{\ln x}{\sqrt{x}}$	$-\sqrt{\dfrac{\pi}{2u}}\left[\ln(4u)+C+\dfrac{\pi}{2}\right],$ $C = 0.5772\ldots$ is the Euler constant
3	$x^{\nu-1}\ln x, \quad 0 < \nu < 1$	$\Gamma(\nu)\cos\left(\dfrac{\pi\nu}{2}\right)u^{-\nu}\left[\psi(\nu)-\dfrac{\pi}{2}\tan\left(\dfrac{\pi\nu}{2}\right)-\ln u\right]$
4	$\ln\left|\dfrac{a+x}{a-x}\right|, \quad a > 0$	$\dfrac{2}{u}\left[\cos(au)\operatorname{Si}(au)-\sin(au)\operatorname{Ci}(au)\right]$
5	$\ln\left(1+a^2/x^2\right), \quad a > 0$	$\dfrac{\pi}{u}\left(1-e^{-au}\right)$
6	$\ln\dfrac{a^2+x^2}{b^2+x^2}, \quad a,b > 0$	$\dfrac{\pi}{u}\left(e^{-bu}-e^{-au}\right)$
7	$e^{-ax}\ln x, \quad a > 0$	$-\dfrac{aC+\frac{1}{2}a\ln(u^2+a^2)+u\arctan(u/a)}{u^2+a^2}$, see row 2 for C
8	$\ln\left(1+e^{-ax}\right), \quad a > 0$	$\dfrac{a}{2u^2}-\dfrac{\pi}{2u\sinh\left(\pi a^{-1}u\right)}$
9	$\ln\left(1-e^{-ax}\right), \quad a > 0$	$\dfrac{a}{2u^2}-\dfrac{\pi}{2u}\coth\left(\pi a^{-1}u\right)$

S2.3.6. Expressions Involving Trigonometric Functions

No.	Original function, $f(x)$	Cosine transform, $\check{f}_c(u) = \int_0^\infty f(x)\cos(ux)\,dx$
1	$\dfrac{\sin(ax)}{x}, \quad a>0$	$\begin{cases} \frac{1}{2}\pi & \text{if } u<a, \\ \frac{1}{4}\pi & \text{if } u=a, \\ 0 & \text{if } u>a \end{cases}$
2	$x^{\nu-1}\sin(ax), \quad a>0,\ |\nu|<1$	$\pi\,\dfrac{(u+a)^{-\nu}-|u+a|^{-\nu}\,\mathrm{sign}(u-a)}{4\Gamma(1-\nu)\cos\left(\frac{1}{2}\pi\nu\right)}$
3	$\dfrac{x\sin(ax)}{x^2+b^2}, \quad a,b>0$	$\begin{cases} \frac{1}{2}\pi e^{-ab}\cosh(bu) & \text{if } u<a, \\ -\frac{1}{2}\pi e^{-bu}\sinh(ab) & \text{if } u>a \end{cases}$
4	$\dfrac{\sin(ax)}{x(x^2+b^2)}, \quad a,b>0$	$\begin{cases} \frac{1}{2}\pi b^{-2}\left[1-e^{-ab}\cosh(bu)\right] & \text{if } u<a, \\ \frac{1}{2}\pi b^{-2}e^{-bu}\sinh(ab) & \text{if } u>a \end{cases}$
5	$e^{-bx}\sin(ax), \quad a,b>0$	$\dfrac{1}{2}\left[\dfrac{a+u}{(a+u)^2+b^2}+\dfrac{a-u}{(a-u)^2+b^2}\right]$
6	$\dfrac{1}{x}\sin^2(ax), \quad a>0$	$\dfrac{1}{4}\ln\left|1-4\dfrac{a^2}{u^2}\right|$
7	$\dfrac{1}{x^2}\sin^2(ax), \quad a>0$	$\begin{cases} \frac{1}{4}\pi(2a-u) & \text{if } u<2a, \\ 0 & \text{if } u>2a \end{cases}$
8	$\dfrac{1}{x}\sin\left(\dfrac{a}{x}\right), \quad a>0$	$\dfrac{\pi}{2}J_0\left(2\sqrt{au}\right)$
9	$\dfrac{1}{\sqrt{x}}\sin\left(a\sqrt{x}\right)\sin\left(b\sqrt{x}\right), \ a,b>0$	$\sqrt{\dfrac{\pi}{u}}\,\sin\left(\dfrac{ab}{2u}\right)\sin\left(\dfrac{a^2+b^2}{4u}-\dfrac{\pi}{4}\right)$
10	$\sin\left(ax^2\right), \quad a>0$	$\sqrt{\dfrac{\pi}{8a}}\left[\cos\left(\dfrac{u^2}{4a}\right)-\sin\left(\dfrac{u^2}{4a}\right)\right]$
11	$\exp\left(-ax^2\right)\sin\left(bx^2\right), \quad a>0$	$\dfrac{\sqrt{\pi}}{(A^2+B^2)^{1/4}}\exp\left(-\dfrac{Au^2}{A^2+B^2}\right)\sin\left(\varphi-\dfrac{Bu^2}{A^2+B^2}\right),$ $A=4a,\ B=4b,\ \varphi=\frac{1}{2}\arctan(b/a)$
12	$\dfrac{1-\cos(ax)}{x}, \quad a>0$	$\dfrac{1}{2}\ln\left|1-\dfrac{a^2}{u^2}\right|$
13	$\dfrac{1-\cos(ax)}{x^2}, \quad a>0$	$\begin{cases} \frac{1}{2}\pi(a-u) & \text{if } u<a, \\ 0 & \text{if } u>a \end{cases}$
14	$x^{\nu-1}\cos(ax), \quad a>0,\ 0<\nu<1$	$\dfrac{1}{2}\Gamma(\nu)\cos\left(\frac{1}{2}\pi\nu\right)\left[|u-a|^{-\nu}+(u+a)^{-\nu}\right]$
15	$\dfrac{\cos(ax)}{x^2+b^2}, \quad a,b>0$	$\begin{cases} \frac{1}{2}\pi b^{-1}e^{-ab}\cosh(bu) & \text{if } u<a, \\ \frac{1}{2}\pi b^{-1}e^{-bu}\cosh(ab) & \text{if } u>a \end{cases}$
16	$e^{-bx}\cos(ax), \quad a,b>0$	$\dfrac{b}{2}\left[\dfrac{1}{(a+u)^2+b^2}+\dfrac{1}{(a-u)^2+b^2}\right]$
17	$\dfrac{1}{\sqrt{x}}\cos\left(a\sqrt{x}\right)$	$\sqrt{\dfrac{\pi}{u}}\,\sin\left(\dfrac{a^2}{4u}+\dfrac{\pi}{4}\right)$
18	$\dfrac{1}{\sqrt{x}}\cos\left(a\sqrt{x}\right)\cos\left(b\sqrt{x}\right)$	$\sqrt{\dfrac{\pi}{u}}\,\cos\left(\dfrac{ab}{2u}\right)\sin\left(\dfrac{a^2+b^2}{4u}+\dfrac{\pi}{4}\right)$
19	$\exp\left(-bx^2\right)\cos(ax), \quad b>0$	$\dfrac{1}{2}\sqrt{\dfrac{\pi}{b}}\exp\left(-\dfrac{a^2+u^2}{4b}\right)\cosh\left(\dfrac{au}{2b}\right)$
20	$\cos\left(ax^2\right), \quad a>0$	$\sqrt{\dfrac{\pi}{8a}}\left[\cos\left(\frac{1}{4}a^{-1}u^2\right)+\sin\left(\frac{1}{4}a^{-1}u^2\right)\right]$
21	$\exp\left(-ax^2\right)\cos\left(bx^2\right), \quad a>0$	$\dfrac{\sqrt{\pi}}{(A^2+B^2)^{1/4}}\exp\left(-\dfrac{Au^2}{A^2+B^2}\right)\cos\left(\varphi-\dfrac{Bu^2}{A^2+B^2}\right),$ $A=4a,\ B=4b,\ \varphi=\frac{1}{2}\arctan(b/a)$

S2.3.7. Expressions Involving Special Functions

No.	Original function, $f(x)$	Cosine transform, $\check{f}_c(u) = \int_0^\infty f(x)\cos(ux)\,dx$
1	$\mathrm{Ei}(-ax)$	$-\dfrac{1}{u}\arctan\left(\dfrac{u}{a}\right)$
2	$\mathrm{Ci}(ax)$	$\begin{cases} 0 & \text{if } 0 < u < a, \\ -\dfrac{\pi}{2u} & \text{if } a < u \end{cases}$
3	$\mathrm{si}(ax)$	$-\dfrac{1}{2u}\ln\left\lvert\dfrac{u+a}{u-a}\right\rvert, \quad u \neq a$
4	$J_0(ax), \quad a > 0$	$\begin{cases} (a^2 - u^2)^{-1/2} & \text{if } 0 < u < a, \\ 0 & \text{if } a < u \end{cases}$
5	$J_\nu(ax), \quad a > 0,\ \nu > -1$	$\begin{cases} \dfrac{\cos\left[\nu\arcsin(u/a)\right]}{\sqrt{a^2 - u^2}} & \text{if } 0 < u < a, \\ -\dfrac{a^\nu \sin(\pi\nu/2)}{\xi(u+\xi)^\nu} & \text{if } a < u, \end{cases}$ where $\xi = \sqrt{u^2 - a^2}$
6	$\dfrac{1}{x}J_\nu(ax), \quad a > 0,\ \nu > 0$	$\begin{cases} \nu^{-1}\cos\left[\nu\arcsin(u/a)\right] & \text{if } 0 < u < a, \\ \dfrac{a^\nu \cos(\pi\nu/2)}{\nu\left(u + \sqrt{u^2 - a^2}\right)^\nu} & \text{if } a < u \end{cases}$
7	$x^{-\nu}J_\nu(ax), \quad a > 0,\ \nu > -\tfrac{1}{2}$	$\begin{cases} \dfrac{\sqrt{\pi}\,(a^2 - u^2)^{\nu-1/2}}{(2a)^\nu \Gamma\left(\nu + \tfrac{1}{2}\right)} & \text{if } 0 < u < a, \\ 0 & \text{if } a < u \end{cases}$
8	$x^{\nu+1}J_\nu(ax),$ $a > 0,\ -1 < \nu < -\tfrac{1}{2}$	$\begin{cases} 0 & \text{if } 0 < u < a, \\ \dfrac{2^{\nu+1}\sqrt{\pi}\,a^\nu u}{\Gamma\left(-\nu - \tfrac{1}{2}\right)(u^2 - a^2)^{\nu+3/2}} & \text{if } a < u \end{cases}$
9	$J_0\left(a\sqrt{x}\right), \quad a > 0$	$\dfrac{1}{u}\sin\left(\dfrac{a^2}{4u}\right)$
10	$\dfrac{1}{\sqrt{x}}J_1\left(a\sqrt{x}\right), \quad a > 0$	$\dfrac{4}{a}\sin^2\left(\dfrac{a^2}{8u}\right)$
11	$x^{\nu/2}J_\nu\left(a\sqrt{x}\right),\ a > 0,\ -1 < \nu < \tfrac{1}{2}$	$\left(\dfrac{a}{2}\right)^\nu u^{-\nu-1}\sin\left(\dfrac{a^2}{4u} - \dfrac{\pi\nu}{2}\right)$
12	$J_0\left(a\sqrt{x^2 + b^2}\right)$	$\begin{cases} \dfrac{\cos\left(b\sqrt{a^2 - u^2}\right)}{\sqrt{a^2 - u^2}} & \text{if } 0 < u < a, \\ 0 & \text{if } a < u \end{cases}$
13	$Y_0(ax), \quad a > 0$	$\begin{cases} 0 & \text{if } 0 < u < a, \\ -(u^2 - a^2)^{-1/2} & \text{if } a < u \end{cases}$
14	$x^\nu Y_\nu(ax), \quad a > 0,\ \lvert\nu\rvert < \tfrac{1}{2}$	$\begin{cases} 0 & \text{if } 0 < u < a, \\ -\dfrac{(2a)^\nu\sqrt{\pi}}{\Gamma\left(\tfrac{1}{2} - \nu\right)(u^2 - a^2)^{\nu+1/2}} & \text{if } a < u \end{cases}$
15	$K_0\left(a\sqrt{x^2 + b^2}\right), \quad a, b > 0$	$\dfrac{\pi}{2\sqrt{u^2 + a^2}}\exp\left(-b\sqrt{u^2 + a^2}\right)$

S2.4. Tables of Fourier Sine Transforms

S2.4.1. General Formulas

No.	Original function, $f(x)$	Sine transform, $\check{f}_s(u) = \int_0^\infty f(x)\sin(ux)\,dx$
1	$af_1(x) + bf_2(x)$	$a\check{f}_{1s}(u) + b\check{f}_{2s}(u)$
2	$f(ax), \quad a > 0$	$\dfrac{1}{a}\check{f}_s\left(\dfrac{u}{a}\right)$
3	$x^{2n}f(x), \quad n = 1, 2, \ldots$	$(-1)^n \dfrac{d^{2n}}{du^{2n}}\check{f}_s(u)$
4	$x^{2n+1}f(ax), \quad n = 0, 1, \ldots$	$(-1)^{n+1}\dfrac{d^{2n+1}}{du^{2n+1}}\check{f}_c(u), \quad \check{f}_c(u) = \int_0^\infty f(x)\cos(xu)\,dx$
5	$f(ax)\cos(bx), \quad a, b > 0$	$\dfrac{1}{2a}\left[\check{f}_s\left(\dfrac{u+b}{a}\right) + \check{f}_s\left(\dfrac{u-b}{a}\right)\right]$

S2.4.2. Expressions Involving Power Functions

No.	Original function, $f(x)$	Sine transform, $\check{f}_s(u) = \int_0^\infty f(x)\sin(ux)\,dx$
1	$\begin{cases} 1 & \text{if } 0 < x < a, \\ 0 & \text{if } a < x \end{cases}$	$\dfrac{1}{u}\left[1 - \cos(au)\right]$
2	$\begin{cases} x & \text{if } 0 < x < 1, \\ 2 - x & \text{if } 1 < x < 2, \\ 0 & \text{if } 2 < x \end{cases}$	$\dfrac{4}{u^2}\sin u \sin^2\dfrac{u}{2}$
3	$\dfrac{1}{x}$	$\dfrac{\pi}{2}$
4	$\dfrac{1}{a+x}, \quad a > 0$	$\sin(au)\,\mathrm{Ci}(au) - \cos(au)\,\mathrm{si}(au)$
5	$\dfrac{x}{a^2 + x^2}, \quad a > 0$	$\dfrac{\pi}{2}e^{-au}$
6	$\dfrac{1}{x(a^2 + x^2)}, \quad a > 0$	$\dfrac{\pi}{2a^2}\left(1 - e^{-au}\right)$
7	$\dfrac{a}{a^2 + (x-b)^2} - \dfrac{a}{a^2 + (x+b)^2}$	$\pi e^{-au}\sin(bu)$
8	$\dfrac{x+b}{a^2 + (x+b)^2} - \dfrac{x-b}{a^2 + (x-b)^2}$	$\pi e^{-au}\cos(bu)$
9	$\dfrac{x}{(x^2 + a^2)^n}, \quad a > 0, \ n = 1, 2, \ldots$	$\dfrac{\pi u e^{-au}}{2^{2n-2}(n-1)!\,a^{2n-3}}\sum_{k=0}^{n-2}\dfrac{(2n-k-4)!}{k!\,(n-k-2)!}(2au)^k$
10	$\dfrac{x^{2m+1}}{(x^2 + a)^{n+1}},$ $n, m = 0, 1, \ldots;\ 0 \le m \le n$	$(-1)^{n+m}\dfrac{\pi}{2n!}\dfrac{\partial^n}{\partial a^n}\left(a^m e^{-u\sqrt{a}}\right)$
11	$\dfrac{1}{\sqrt{x}}$	$\sqrt{\dfrac{\pi}{2u}}$
12	$\dfrac{1}{x\sqrt{x}}$	$\sqrt{2\pi u}$

No.	Original function, $f(x)$	Sine transform, $\check{f}_s(u) = \int_0^\infty f(x) \sin(ux)\,dx$
13	$x(a^2 + x^2)^{-3/2}$	$uK_0(au)$
14	$\dfrac{\left(\sqrt{a^2 + x^2} - a\right)^{1/2}}{\sqrt{a^2 + x^2}}$	$\sqrt{\dfrac{\pi}{2u}}\,e^{-au}$
15	$x^{-\nu}, \quad 0 < \nu < 2$	$\cos\left(\tfrac{1}{2}\pi\nu\right)\Gamma(1-\nu)u^{\nu-1}$

S2.4.3. Expressions Involving Exponential Functions

No.	Original function, $f(x)$	Sine transform, $\check{f}_s(u) = \int_0^\infty f(x) \sin(ux)\,dx$
1	$e^{-ax}, \quad a > 0$	$\dfrac{u}{a^2 + u^2}$
2	$x^n e^{-ax}, \quad a > 0,\, n = 1, 2, \ldots$	$n!\left(\dfrac{a}{a^2 + u^2}\right)^{n+1} \displaystyle\sum_{k=0}^{[n/2]} (-1)^k C_{n+1}^{2k+1}\left(\dfrac{u}{a}\right)^{2k+1}$
3	$\dfrac{1}{x}e^{-ax}, \quad a > 0$	$\arctan\dfrac{u}{a}$
4	$\sqrt{x}\,e^{-ax}, \quad a > 0$	$\dfrac{\sqrt{\pi}}{2}(a^2 + u^2)^{-3/4}\sin\left(\dfrac{3}{2}\arctan\dfrac{u}{a}\right)$
5	$\dfrac{1}{\sqrt{x}}e^{-ax}, \quad a > 0$	$\sqrt{\dfrac{\pi}{2}}\,\dfrac{\left(\sqrt{a^2 + u^2} - a\right)^{1/2}}{\sqrt{a^2 + u^2}}$
6	$\dfrac{1}{x\sqrt{x}}e^{-ax}, \quad a > 0$	$\sqrt{2\pi}\left(\sqrt{a^2 + u^2} - a\right)^{1/2}$
7	$x^{n-1/2}e^{-ax}, \quad a > 0,\, n = 1, 2, \ldots$	$(-1)^n\sqrt{\dfrac{\pi}{2}}\,\dfrac{\partial^n}{\partial a^n}\left[\dfrac{\left(\sqrt{a^2 + u^2} - a\right)^{1/2}}{\sqrt{a^2 + u^2}}\right]$
8	$x^{\nu-1}e^{-ax}, \quad a > 0,\, \nu > -1$	$\Gamma(\nu)(a^2 + u^2)^{-\nu/2}\sin\left(\nu\arctan\dfrac{u}{a}\right)$
9	$x^{-2}\left(e^{-ax} - e^{-bx}\right), \quad a, b > 0$	$\dfrac{u}{2}\ln\left(\dfrac{u^2 + b^2}{u^2 + a^2}\right) + b\arctan\left(\dfrac{u}{b}\right) - a\arctan\left(\dfrac{u}{a}\right)$
10	$\dfrac{1}{e^{ax} + 1}, \quad a > 0$	$\dfrac{1}{2u} - \dfrac{\pi}{2a\sinh(\pi u/a)}$
11	$\dfrac{1}{e^{ax} - 1}, \quad a > 0$	$\dfrac{\pi}{2a}\coth\left(\dfrac{\pi u}{a}\right) - \dfrac{1}{2u}$
12	$\dfrac{e^{x/2}}{e^x - 1}$	$-\tfrac{1}{2}\tanh(\pi u)$
13	$x\exp\left(-ax^2\right)$	$\dfrac{\sqrt{\pi}}{4a^{3/2}}\,u\exp\left(-\dfrac{u^2}{4a}\right)$
14	$\dfrac{1}{x}\exp\left(-ax^2\right)$	$\dfrac{\pi}{2}\operatorname{erf}\left(\dfrac{u}{2\sqrt{a}}\right)$
15	$\dfrac{1}{\sqrt{x}}\exp\left(-\dfrac{a}{x}\right)$	$\sqrt{\dfrac{\pi}{2u}}\,e^{-\sqrt{2au}}\left[\cos\left(\sqrt{2au}\right) + \sin\left(\sqrt{2au}\right)\right]$
16	$\dfrac{1}{x\sqrt{x}}\exp\left(-\dfrac{a}{x}\right)$	$\sqrt{\dfrac{\pi}{a}}\,e^{-\sqrt{2au}}\sin\left(\sqrt{2au}\right)$

S2.4.4. Expressions Involving Hyperbolic Functions

No.	Original function, $f(x)$	Sine transform, $\check{f}_s(u) = \int_0^\infty f(x)\sin(ux)\,dx$
1	$\dfrac{1}{\sinh(ax)}, \quad a > 0$	$\dfrac{\pi}{2a}\tanh\left(\tfrac{1}{2}\pi a^{-1}u\right)$
2	$\dfrac{x}{\sinh(ax)}, \quad a > 0$	$\dfrac{\pi^2 \sinh\left(\tfrac{1}{2}\pi a^{-1}u\right)}{4a^2 \cosh^2\left(\tfrac{1}{2}\pi a^{-1}u\right)}$
3	$\dfrac{1}{x}e^{-bx}\sinh(ax), \quad b > |a|$	$\tfrac{1}{2}\arctan\left(\dfrac{2au}{u^2 + b^2 - a^2}\right)$
4	$\dfrac{1}{x\cosh(ax)}, \quad a > 0$	$\arctan\left[\sinh\left(\tfrac{1}{2}\pi a^{-1}u\right)\right]$
5	$1 - \tanh\left(\tfrac{1}{2}ax\right), \quad a > 0$	$\dfrac{1}{u} - \dfrac{\pi}{a\sinh\left(\pi a^{-1}u\right)}$
6	$\coth\left(\tfrac{1}{2}ax\right) - 1, \quad a > 0$	$\dfrac{\pi}{a}\coth\left(\pi a^{-1}u\right) - \dfrac{1}{u}$
7	$\dfrac{\cosh(ax)}{\sinh(bx)}, \quad |a| < b$	$\dfrac{\pi}{2b}\dfrac{\sinh\left(\pi b^{-1}u\right)}{\cos\left(\pi ab^{-1}\right) + \cosh\left(\pi b^{-1}u\right)}$
8	$\dfrac{\sinh(ax)}{\cosh(bx)}, \quad |a| < b$	$\dfrac{\pi}{b}\dfrac{\sin\left(\tfrac{1}{2}\pi ab^{-1}\right)\sinh\left(\tfrac{1}{2}\pi b^{-1}u\right)}{\cos\left(\pi ab^{-1}\right) + \cosh\left(\pi b^{-1}u\right)}$

S2.4.5. Expressions Involving Logarithmic Functions

No.	Original function, $f(x)$	Sine transform, $\check{f}_s(u) = \int_0^\infty f(x)\sin(ux)\,dx$
1	$\begin{cases} \ln x & \text{if } 0 < x < 1, \\ 0 & \text{if } 1 < x \end{cases}$	$\dfrac{1}{u}\left[\text{Ci}(u) - \ln u - \mathcal{C}\right],$ $\mathcal{C} = 0.5772\ldots$ is the Euler constant
2	$\dfrac{\ln x}{x}$	$-\tfrac{1}{2}\pi(\ln u + \mathcal{C})$
3	$\dfrac{\ln x}{\sqrt{x}}$	$-\sqrt{\dfrac{\pi}{2u}}\left[\ln(4u) + \mathcal{C} - \dfrac{\pi}{2}\right]$
4	$x^{\nu-1}\ln x, \quad |\nu| < 1$	$\dfrac{\pi u^{-\nu}\left[\psi(\nu) + \tfrac{\pi}{2}\cot\left(\tfrac{\pi\nu}{2}\right) - \ln u\right]}{2\Gamma(1-\nu)\cos\left(\tfrac{\pi\nu}{2}\right)}$
5	$\ln\left|\dfrac{a+x}{a-x}\right|, \quad a > 0$	$\dfrac{\pi}{u}\sin(au)$
6	$\ln\dfrac{(x+b)^2 + a^2}{(x-b)^2 + a^2}, \quad a,b > 0$	$\dfrac{2\pi}{u}e^{-au}\sin(bu)$
7	$e^{-ax}\ln x, \quad a > 0$	$\dfrac{a\arctan(u/a) - \tfrac{1}{2}u\ln(u^2 + a^2) - e^{\mathcal{C}}u}{u^2 + a^2}, \quad \mathcal{C} = 0.5772\ldots$
8	$\dfrac{1}{x}\ln\left(1 + a^2x^2\right), \quad a > 0$	$-\pi\,\text{Ei}\left(-\dfrac{u}{a}\right)$

S2.4.6. Expressions Involving Trigonometric Functions

No.	Original function, $f(x)$	Sine transform, $\check{f}_s(u) = \int_0^\infty f(x)\sin(ux)\,dx$
1	$\dfrac{\sin(ax)}{x}, \quad a > 0$	$\dfrac{1}{2}\ln\left\|\dfrac{u+a}{u-a}\right\|$
2	$\dfrac{\sin(ax)}{x^2}, \quad a > 0$	$\begin{cases} \frac{1}{2}\pi u & \text{if } 0 < u < a, \\ \frac{1}{2}\pi a & \text{if } u > a \end{cases}$
3	$x^{\nu-1}\sin(ax), \quad a > 0,\ -2 < \nu < 1$	$\pi\,\dfrac{\|u-a\|^{-\nu} - \|u+a\|^{-\nu}}{4\Gamma(1-\nu)\sin\left(\frac{1}{2}\pi\nu\right)}, \quad \nu \neq 0$
4	$\dfrac{\sin(ax)}{x^2+b^2}, \quad a,b > 0$	$\begin{cases} \frac{1}{2}\pi b^{-1}e^{-ab}\sinh(bu) & \text{if } 0 < u < a, \\ \frac{1}{2}\pi b^{-1}e^{-bu}\sinh(ab) & \text{if } u > a \end{cases}$
5	$\dfrac{\sin(\pi x)}{1-x^2}$	$\begin{cases} \sin u & \text{if } 0 < u < \pi, \\ 0 & \text{if } u > \pi \end{cases}$
6	$e^{-ax}\sin(bx), \quad a > 0$	$\dfrac{a}{2}\left[\dfrac{1}{a^2+(b-u)^2} - \dfrac{1}{a^2+(b+u)^2}\right]$
7	$x^{-1}e^{-ax}\sin(bx), \quad a > 0$	$\dfrac{1}{4}\ln\dfrac{(u+b)^2+a^2}{(u-b)^2+a^2}$
8	$\dfrac{1}{x}\sin^2(ax), \quad a > 0$	$\begin{cases} \frac{1}{4}\pi & \text{if } 0 < u < 2a, \\ \frac{1}{8}\pi & \text{if } u = 2a, \\ 0 & \text{if } u > 2a \end{cases}$
9	$\dfrac{1}{x^2}\sin^2(ax), \quad a > 0$	$\frac{1}{4}(u+2a)\ln\|u+2a\| + \frac{1}{4}(u-2a)\ln\|u-2a\| - \frac{1}{2}u\ln u$
10	$\exp(-ax^2)\sin(bx), \quad a > 0$	$\dfrac{1}{2}\sqrt{\dfrac{\pi}{a}}\,\exp\left(-\dfrac{u^2+b^2}{4a}\right)\sinh\left(\dfrac{bu}{2a}\right)$
11	$\dfrac{1}{x}\sin(ax)\sin(bx),\ a \geq b > 0$	$\begin{cases} 0 & \text{if } 0 < u < a-b, \\ \frac{\pi}{4} & \text{if } a-b < u < a+b, \\ 0 & \text{if } a+b < u \end{cases}$
12	$\sin\left(\dfrac{a}{x}\right), \quad a > 0$	$\dfrac{\pi\sqrt{a}}{2\sqrt{u}}J_1\left(2\sqrt{au}\right)$
13	$\dfrac{1}{\sqrt{x}}\sin\left(\dfrac{a}{x}\right), \quad a > 0$	$\sqrt{\dfrac{\pi}{8u}}\left[\sin\left(2\sqrt{au}\right) - \cos\left(2\sqrt{au}\right) + \exp\left(-2\sqrt{au}\right)\right]$
14	$\exp\left(-a\sqrt{x}\right)\sin\left(a\sqrt{x}\right),\ a > 0$	$a\sqrt{\dfrac{\pi}{8}}\,u^{-3/2}\exp\left(-\dfrac{a^2}{2u}\right)$
15	$\dfrac{\cos(ax)}{x}, \quad a > 0$	$\begin{cases} 0 & \text{if } 0 < u < a, \\ \frac{1}{4}\pi & \text{if } u = a, \\ \frac{1}{2}\pi & \text{if } a < u \end{cases}$
16	$x^{\nu-1}\cos(ax), \quad a > 0,\ \|\nu\| < 1$	$\dfrac{\pi(u+a)^{-\nu} - \text{sign}(u-a)\|u-a\|^{-\nu}}{4\Gamma(1-\nu)\cos\left(\frac{1}{2}\pi\nu\right)}$
17	$\dfrac{x\cos(ax)}{x^2+b^2}, \quad a,b > 0$	$\begin{cases} -\frac{1}{2}\pi e^{-ab}\sinh(bu) & \text{if } u < a, \\ \frac{1}{2}\pi e^{-bu}\cosh(ab) & \text{if } u > a \end{cases}$
18	$\dfrac{1-\cos(ax)}{x^2}, \quad a > 0$	$\dfrac{u}{2}\ln\left\|\dfrac{u^2-a^2}{u^2}\right\| + \dfrac{a}{2}\ln\left\|\dfrac{u+a}{u-a}\right\|$
19	$\dfrac{1}{\sqrt{x}}\cos\left(a\sqrt{x}\right)$	$\sqrt{\dfrac{\pi}{u}}\cos\left(\dfrac{a^2}{4u} + \dfrac{\pi}{4}\right)$
20	$\dfrac{1}{\sqrt{x}}\cos\left(a\sqrt{x}\right)\cos\left(b\sqrt{x}\right),\ a,b > 0$	$\sqrt{\dfrac{\pi}{u}}\cos\left(\dfrac{ab}{2u}\right)\cos\left(\dfrac{a^2+b^2}{4u} + \dfrac{\pi}{4}\right)$

S2.4.7. Expressions Involving Special Functions

No.	Original function, $f(x)$	Sine transform, $\check{f}_s(u) = \int_0^\infty f(x)\sin(ux)\,dx$
1	$\operatorname{erfc}(ax)$, $\quad a > 0$	$\dfrac{1}{u}\left[1 - \exp\left(-\dfrac{u^2}{4a^2}\right)\right]$
2	$\operatorname{ci}(ax)$, $\quad a > 0$	$-\dfrac{1}{2u}\ln\left\lvert 1 - \dfrac{u^2}{a^2}\right\rvert$
3	$\operatorname{si}(ax)$, $\quad a > 0$	$\begin{cases} 0 & \text{if } 0 < u < a, \\ -\frac{1}{2}\pi u^{-1} & \text{if } a < u \end{cases}$
4	$J_0(ax)$, $\quad a > 0$	$\begin{cases} 0 & \text{if } 0 < u < a, \\ \dfrac{1}{\sqrt{u^2 - a^2}} & \text{if } a < u \end{cases}$
5	$J_\nu(ax)$, $\quad a > 0,\ \nu > -2$	$\begin{cases} \dfrac{\sin\left[\nu\arcsin(u/a)\right]}{\sqrt{a^2 - u^2}} & \text{if } 0 < u < a, \\ \dfrac{a^\nu\cos(\pi\nu/2)}{\xi(u + \xi)^\nu} & \text{if } a < u, \end{cases}$ where $\xi = \sqrt{u^2 - a^2}$
6	$\dfrac{1}{x}J_0(ax)$, $\quad a > 0,\ \nu > 0$	$\begin{cases} \arcsin(u/a) & \text{if } 0 < u < a, \\ \pi/2 & \text{if } a < u \end{cases}$
7	$\dfrac{1}{x}J_\nu(ax)$, $\quad a > 0,\ \nu > -1$	$\begin{cases} \nu^{-1}\sin\left[\nu\arcsin(u/a)\right] & \text{if } 0 < u < a, \\ \dfrac{a^\nu\sin(\pi\nu/2)}{\nu\left(u + \sqrt{u^2 - a^2}\right)^\nu} & \text{if } a < u \end{cases}$
8	$x^{-1}e^{-ax}J_0(bx)$, $\quad a > 0$	$\arcsin\left(\dfrac{2u}{\sqrt{(u+b)^2 + a^2} + \sqrt{(u-b)^2 + a^2}}\right)$
9	$J_0\left(a\sqrt{x}\right)$, $\quad a > 0$	$\dfrac{1}{u}\cos\left(\dfrac{a^2}{4u}\right)$
10	$\dfrac{1}{\sqrt{x}}J_1\left(a\sqrt{x}\right)$, $\quad a > 0$	$\dfrac{2}{a}\sin\left(\dfrac{a^2}{4u}\right)$
11	$x^{\nu/2}J_\nu\left(a\sqrt{x}\right)$, $a > 0,\quad -2 < \nu < \frac{1}{2}$	$\dfrac{a^\nu}{2^\nu u^{\nu+1}}\cos\left(\dfrac{a^2}{4u} - \dfrac{\pi\nu}{2}\right)$
12	$Y_0(ax)$, $\quad a > 0$	$\begin{cases} \dfrac{2\arcsin(u/a)}{\pi\sqrt{a^2 - u^2}} & \text{if } 0 < u < a, \\ \dfrac{2\left[\ln\left(u - \sqrt{u^2 - a^2}\right) - \ln a\right]}{\pi\sqrt{u^2 - a^2}} & \text{if } a < u \end{cases}$
13	$K_0(ax)$, $\quad a > 0$	$\dfrac{\ln\left(u + \sqrt{u^2 + a^2}\right) - \ln a}{\sqrt{u^2 + a^2}}$
14	$x^{\nu+1}K_\nu(ax)$, $\quad a > 0,\ \nu > -\frac{3}{2}$	$\sqrt{\pi}\,(2a)^\nu\Gamma\left(\nu + \tfrac{3}{2}\right)u(u^2 + a^2)^{-\nu - 3/2}$

Bibliography for Chapter S2

Bateman, H. and Erdélyi, A., *Tables of Integral Transforms. Vols. 1 and 2*, McGraw-Hill, New York, 1954.

Ditkin, V. A. and Prudnikov, A. P., *Integral Transforms and Operational Calculus*, Pergamon Press, New York, 1965.

Oberhettinger, F., *Tables of Fourier Transforms and Fourier Transforms of Distributions*, Springer-Verlag, Berlin, 1980.

Oberhettinger, F. and Badii, L., *Tables of Laplace Transforms*, Springer-Verlag, New York, 1973.

Prudnikov, A. P., Brychkov, Yu. A., and Marichev, O. I., *Integrals and Series, Vol. 4, Direct Laplace Transforms*, Gordon & Breach, New York, 1992.

Prudnikov, A. P., Brychkov, Yu. A., and Marichev, O. I., *Integrals and Series, Vol. 5, Inverse Laplace Transforms*, Gordon & Breach, New York, 1992.

Chapter S3

Orthogonal Curvilinear Systems of Coordinates

S3.1. Arbitrary Curvilinear Coordinate Systems

S3.1.1. General Nonorthogonal Curvilinear Coordinates

▶ **Metric tensor. Arc length and volume elements in curvilinear coordinates.** The curvilinear coordinates x^1, x^2, x^3 are defined as functions of the rectangular Cartesian coordinates x, y, z:

$$x^1 = x^1(x, y, z), \quad x^2 = x^2(x, y, z), \quad x^3 = x^3(x, y, z).$$

Using these formulas, one can express x, y, z in terms of the curvilinear coordinates x^1, x^2, x^3 as follows:

$$x = x(x^1, x^2, x^3), \quad y = y(x^1, x^2, x^3), \quad z = z(x^1, x^2, x^3).$$

The *metric tensor components* g_{ij} are determined by the formulas

$$g_{ij}(x^1, x^2, x^3) = \frac{\partial x}{\partial x^i} \frac{\partial x}{\partial x^j} + \frac{\partial y}{\partial x^i} \frac{\partial y}{\partial x^j} + \frac{\partial z}{\partial x^i} \frac{\partial z}{\partial x^j};$$

$$g_{ij}(x^1, x^2, x^3) = g_{ji}(x^1, x^2, x^3); \qquad i, j = 1, 2, 3.$$

The arc length dl between close points $(x, y, z) \equiv (x^1, x^2, x^3)$ and $(x + dx,\ y + dy,\ z + dz) \equiv (x^1 + dx^1,\ x^2 + dx^2,\ x^3 + dx^3)$ is expressed as

$$(dl)^2 = (dx)^2 + (dy)^2 + (dz)^2 = \sum_{i=1}^{3} \sum_{j=1}^{3} g_{ij}(x^1, x^2, x^3)\, dx^i\, dx^j.$$

The volume of the elementary parallelepiped with vertices at the eight points (x^1, x^2, x^3), $(x^1 + dx^1, x^2, x^3)$, $(x^1, x^2 + dx^2, x^3)$, $(x^1, x^2, x^3 + dx^3)$, $(x^1 + dx^1, x^2 + dx^2, x^3)$, $(x^1 + dx^1, x^2, x^3 + dx^3)$, $(x^1, x^2 + dx^2, x^3 + dx^3)$, $(x^1 + dx^1, x^2 + dx^2, x^3 + dx^3)$ is given by

$$dV = \frac{\partial(x, y, z)}{\partial(x^1, x^2, x^3)}\, dx^1\, dx^2\, dx^3 = \pm\sqrt{\det |g_{ij}|}\, dx^1\, dx^2\, dx^3.$$

Here the plus sign corresponds to the standard situation where the tangent vectors to the coordinate lines x^1, x^2, x^3, pointing in the direction of growth of the respective coordinate, form a right-handed triple, just as unit vectors $\mathbf{i}, \mathbf{j}, \mathbf{k}$ of a right-handed rectangular Cartesian coordinate system.

▶ **Vector components in Cartesian and curvilinear coordinate systems.** The unit vectors **i**, **j**, **k** of a rectangular Cartesian coordinate system* x, y, z and the unit vectors $\mathbf{i}_1, \mathbf{i}_2, \mathbf{i}_3$ of a curvilinear coordinate system x^1, x^2, x^3 are connected by the linear relations

$$\mathbf{i}_n = \frac{1}{\sqrt{g_{nn}}} \left(\frac{\partial x}{\partial x^n} \mathbf{i} + \frac{\partial y}{\partial x^n} \mathbf{j} + \frac{\partial z}{\partial x^n} \mathbf{k} \right), \quad n = 1, 2, 3;$$

$$\mathbf{i} = \sqrt{g_{11}} \frac{\partial x^1}{\partial x} \mathbf{i}_1 + \sqrt{g_{22}} \frac{\partial x^2}{\partial x} \mathbf{i}_2 + \sqrt{g_{33}} \frac{\partial x^3}{\partial x} \mathbf{i}_3;$$

$$\mathbf{j} = \sqrt{g_{11}} \frac{\partial x^1}{\partial y} \mathbf{i}_1 + \sqrt{g_{22}} \frac{\partial x^2}{\partial y} \mathbf{i}_2 + \sqrt{g_{33}} \frac{\partial x^3}{\partial y} \mathbf{i}_3;$$

$$\mathbf{k} = \sqrt{g_{11}} \frac{\partial x^1}{\partial z} \mathbf{i}_1 + \sqrt{g_{22}} \frac{\partial x^2}{\partial z} \mathbf{i}_2 + \sqrt{g_{33}} \frac{\partial x^3}{\partial z} \mathbf{i}_3.$$

In the general case, the vectors $\mathbf{i}_1, \mathbf{i}_2, \mathbf{i}_3$ are not orthogonal and change their direction from point to point.

The components v_x, v_y, v_z of a vector **v** in a rectangular Cartesian coordinate system x, y, z and the components v_1, v_2, v_3 of the same vector in a curvilinear coordinate system x^1, x^2, x^3 are related by

$$\mathbf{v} = v_x \mathbf{i} + v_y \mathbf{j} + v_z \mathbf{k} = v_1 \mathbf{i}_1 + v_2 \mathbf{i}_2 + v_3 \mathbf{i}_3,$$

$$v_n = \sqrt{g_{nn}} \left(\frac{\partial x^n}{\partial x} v_x + \frac{\partial x^n}{\partial y} v_y + \frac{\partial x^n}{\partial z} v_z \right), \quad n = 1, 2, 3;$$

$$v_x = \frac{\partial x}{\partial x^1} \frac{v_1}{\sqrt{g_{11}}} + \frac{\partial x}{\partial x^2} \frac{v_2}{\sqrt{g_{22}}} + \frac{\partial x}{\partial x^3} \frac{v_3}{\sqrt{g_{33}}};$$

$$v_y = \frac{\partial y}{\partial x^1} \frac{v_1}{\sqrt{g_{11}}} + \frac{\partial y}{\partial x^2} \frac{v_2}{\sqrt{g_{22}}} + \frac{\partial y}{\partial x^3} \frac{v_3}{\sqrt{g_{33}}};$$

$$v_z = \frac{\partial z}{\partial x^1} \frac{v_1}{\sqrt{g_{11}}} + \frac{\partial z}{\partial x^2} \frac{v_2}{\sqrt{g_{22}}} + \frac{\partial z}{\partial x^3} \frac{v_3}{\sqrt{g_{33}}}.$$

S3.1.2. General Orthogonal Curvilinear Coordinates

▶ **Orthogonal coordinates. Length, area, and volume elements.** A system of coordinates is orthogonal if

$$g_{ij}(x^1, x^2, x^3) = 0 \quad \text{for} \quad i \neq j.$$

In this case the third invariant of the metric tensor is given by

$$g = \det |g_{ij}| = g_{11} g_{22} g_{33}.$$

The *Lamé coefficients* L_k of orthogonal curvilinear coordinates are expressed in terms of the components of the metric tensor as

$$L_i = \sqrt{g_{ii}} = \sqrt{\left(\frac{\partial x}{\partial x^i} \right)^2 + \left(\frac{\partial y}{\partial x^i} \right)^2 + \left(\frac{\partial z}{\partial x^i} \right)^2}, \quad i = 1, 2, 3.$$

---

* Here and henceforth the coordinate axes and the respective coordinates of points in space are denoted by the same letters.

Arc length element:

$$dl = \sqrt{(L_1\,dx^1)^2 + (L_2\,dx^2)^2 + (L_3\,dx^3)^2} = \sqrt{g_{11}(dx^1)^2 + g_{22}(dx^2)^2 + g_{33}(dx^3)^2}.$$

The area elements ds_i of the respective coordinate surfaces $x^i = \text{const}$ are given by

$$ds_1 = dl_2\,dl_3 = L_2 L_3\,dx^2\,dx^3 = \sqrt{g_{22}g_{33}}\,dx^2\,dx^3,$$
$$ds_2 = dl_1\,dl_3 = L_1 L_3\,dx^1\,dx^3 = \sqrt{g_{11}g_{33}}\,dx^1\,dx^3,$$
$$ds_3 = dl_1\,dl_2 = L_1 L_2\,dx^1\,dx^2 = \sqrt{g_{11}g_{22}}\,dx^1\,dx^2.$$

Volume element:

$$dV = L_1 L_2 L_3\,dx^1\,dx^2\,dx^3 = \sqrt{g_{11}g_{22}g_{33}}\,dx^1\,dx^2\,dx^3.$$

▶ **Basic differential relations in orthogonal curvilinear coordinates.** In what follows, we present the basic differential operators in the orthogonal curvilinear coordinates x^1, x^2, x^3. The corresponding unit vectors are denoted by $\mathbf{i}_1$, $\mathbf{i}_2$, $\mathbf{i}_3$.

The gradient of a scalar f is expressed as

$$\operatorname{grad} f \equiv \nabla f = \frac{1}{\sqrt{g_{11}}}\frac{\partial f}{\partial x^1}\mathbf{i}_1 + \frac{1}{\sqrt{g_{22}}}\frac{\partial f}{\partial x^2}\mathbf{i}_2 + \frac{1}{\sqrt{g_{33}}}\frac{\partial f}{\partial x^3}\mathbf{i}_3.$$

Divergence of a vector $\mathbf{v} = v_1\mathbf{i}_1 + v_2\mathbf{i}_2 + v_3\mathbf{i}_3$:

$$\operatorname{div} \mathbf{v} \equiv \nabla \cdot \mathbf{v} = \frac{1}{\sqrt{g}}\left[\frac{\partial}{\partial x^1}\left(v_1\sqrt{\frac{g}{g_{11}}}\right) + \frac{\partial}{\partial x^2}\left(v_2\sqrt{\frac{g}{g_{22}}}\right) + \frac{\partial}{\partial x^3}\left(v_3\sqrt{\frac{g}{g_{33}}}\right)\right].$$

Gradient of a scalar f along a vector $\mathbf{v}$:

$$(\mathbf{v} \cdot \nabla)f = \frac{v_1}{\sqrt{g_{11}}}\frac{\partial f}{\partial x^1} + \frac{v_2}{\sqrt{g_{22}}}\frac{\partial f}{\partial x^2} + \frac{v_3}{\sqrt{g_{33}}}\frac{\partial f}{\partial x^3}.$$

Gradient of a vector $\mathbf{w}$ along a vector $\mathbf{v}$:

$$(\mathbf{v} \cdot \nabla)\mathbf{w} = \mathbf{i}_1(\mathbf{v} \cdot \nabla)w_1 + \mathbf{i}_2(\mathbf{v} \cdot \nabla)w_2 + \mathbf{i}_3(\mathbf{v} \cdot \nabla)w_3.$$

Curl of a vector $\mathbf{v}$:

$$\begin{aligned}
\operatorname{curl}\mathbf{v} \equiv \nabla \times \mathbf{v} = {} & \mathbf{i}_1\frac{\sqrt{g_{11}}}{\sqrt{g}}\left[\frac{\partial}{\partial x^2}\left(v_3\sqrt{g_{33}}\right) - \frac{\partial}{\partial x^3}\left(v_2\sqrt{g_{22}}\right)\right] \\
& + \mathbf{i}_2\frac{\sqrt{g_{22}}}{\sqrt{g}}\left[\frac{\partial}{\partial x^3}\left(v_1\sqrt{g_{11}}\right) - \frac{\partial}{\partial x^1}\left(v_3\sqrt{g_{33}}\right)\right] \\
& + \mathbf{i}_3\frac{\sqrt{g_{33}}}{\sqrt{g}}\left[\frac{\partial}{\partial x^1}\left(v_2\sqrt{g_{22}}\right) - \frac{\partial}{\partial x^2}\left(v_1\sqrt{g_{11}}\right)\right].
\end{aligned}$$

Remark. Sometimes curl $\mathbf{v}$ is denoted by curl $\mathbf{v}$.

Laplace operator of a scalar f:

$$\Delta f \equiv \nabla^2 f = \frac{1}{\sqrt{g}}\left[\frac{\partial}{\partial x^1}\left(\frac{\sqrt{g}}{g_{11}}\frac{\partial f}{\partial x^1}\right) + \frac{\partial}{\partial x^2}\left(\frac{\sqrt{g}}{g_{22}}\frac{\partial f}{\partial x^2}\right) + \frac{\partial}{\partial x^3}\left(\frac{\sqrt{g}}{g_{33}}\frac{\partial f}{\partial x^3}\right)\right].$$

S3.2. Cylindrical and Spherical Coordinate Systems

S3.2.1. Cylindrical Coordinates

▶ **Transformations of coordinates and vectors. The metric tensor components.** The Cartesian coordinates are expressed in terms of the cylindrical ones as

$$x = \rho \cos \varphi, \quad y = \rho \sin \varphi, \quad z = z$$
$$(0 \le \rho < \infty, \quad 0 \le \varphi < 2\pi, \quad -\infty < z < \infty).$$

The cylindrical coordinates are expressed in terms of the Cartesian ones as

$$\rho = \sqrt{x^2 + y^2}, \quad \tan \varphi = y/x, \quad z = z \quad (\sin \varphi = y/\rho).$$

Coordinate surfaces:

$x^2 + y^2 = \rho^2$ (right circular cylinders with their axis coincident with the z-axis),

$y = x \tan \varphi$ (half-planes through the z-axis),

$z = z$ (planes perpendicular to the z-axis).

Direct and inverse transformations of the components of a vector $\mathbf{v} = v_x \mathbf{i} + v_y \mathbf{j} + v_z \mathbf{k} = v_\rho \mathbf{i}_\rho + v_\varphi \mathbf{i}_\varphi + v_z \mathbf{i}_z$:

$$v_\rho = v_x \cos \varphi + v_y \sin \varphi, \qquad\qquad v_x = v_\rho \cos \varphi - v_\varphi \sin \varphi,$$
$$v_\varphi = -v_x \sin \varphi + v_y \cos \varphi, \qquad\qquad v_y = v_\rho \sin \varphi + v_\varphi \cos \varphi,$$
$$v_z = v_z; \qquad\qquad\qquad\qquad\qquad\quad v_z = v_z.$$

Metric tensor components:

$$g_{\rho\rho} = 1, \quad g_{\varphi\varphi} = \rho^2, \quad g_{zz} = 1, \quad \sqrt{g} = \rho.$$

▶ **Basic differential relations.** Gradient of a scalar f:

$$\nabla f = \frac{\partial f}{\partial \rho} \mathbf{i}_\rho + \frac{1}{\rho} \frac{\partial f}{\partial \varphi} \mathbf{i}_\varphi + \frac{\partial f}{\partial z} \mathbf{i}_z.$$

Divergence of a vector $\mathbf{v}$:

$$\nabla \cdot \mathbf{v} = \frac{1}{\rho} \frac{\partial(\rho v_\rho)}{\partial \rho} + \frac{1}{\rho} \frac{\partial v_\varphi}{\partial \varphi} + \frac{\partial v_z}{\partial z}.$$

Gradient of a scalar f along a vector $\mathbf{v}$:

$$(\mathbf{v} \cdot \nabla)f = v_\rho \frac{\partial f}{\partial \rho} + \frac{v_\varphi}{\rho} \frac{\partial f}{\partial \varphi} + v_z \frac{\partial f}{\partial z}.$$

Gradient of a vector $\mathbf{w}$ along a vector $\mathbf{v}$:

$$(\mathbf{v} \cdot \nabla)\mathbf{w} = (\mathbf{v} \cdot \nabla)w_\rho \mathbf{i}_\rho + (\mathbf{v} \cdot \nabla)w_\varphi \mathbf{i}_\varphi + (\mathbf{v} \cdot \nabla)w_z \mathbf{i}_z.$$

Curl of a vector $\mathbf{v}$:

$$\nabla \times \mathbf{v} = \left(\frac{1}{\rho} \frac{\partial v_z}{\partial \varphi} - \frac{\partial v_\varphi}{\partial z} \right) \mathbf{i}_\rho + \left(\frac{\partial v_\rho}{\partial z} - \frac{\partial v_z}{\partial \rho} \right) \mathbf{i}_\varphi + \frac{1}{\rho} \left[\frac{\partial(\rho v_\varphi)}{\partial \rho} - \frac{\partial v_\rho}{\partial \varphi} \right] \mathbf{i}_z.$$

Laplacian of a scalar f:

$$\Delta f = \frac{1}{\rho} \frac{\partial}{\partial \rho} \left(\rho \frac{\partial f}{\partial \rho} \right) + \frac{1}{\rho^2} \frac{\partial^2 f}{\partial \varphi^2} + \frac{\partial^2 f}{\partial z^2}.$$

Remark. The cylindrical coordinates ρ, φ are also used as *polar coordinates* on the plane xy.

S3.2.2. Spherical Coordinates

▶ **Transformations of coordinates and vectors. The metric tensor components.** The Cartesian coordinates are expressed in terms of the spherical ones as

$$x = r\sin\theta\cos\varphi, \quad y = r\sin\theta\sin\varphi, \quad z = r\cos\theta$$
$$(0 \le r < \infty, \ 0 \le \theta \le \pi, \ 0 \le \varphi < 2\pi).$$

The spherical coordinates are expressed in terms of the Cartesian ones as

$$r = \sqrt{x^2 + y^2 + z^2}, \quad \theta = \arccos\frac{z}{r}, \quad \tan\varphi = \frac{y}{x} \quad \left(\sin\varphi = \frac{y}{\sqrt{x^2 + y^2}}\right).$$

Coordinate surfaces:

$$x^2 + y^2 + z^2 = r^2 \qquad \text{(spheres)},$$
$$x^2 + y^2 - z^2\tan^2\theta = 0 \quad \text{(circular cones)},$$
$$y = x\tan\varphi \qquad\qquad \text{(half-planes through the } z\text{-axis)}.$$

Direct and inverse transformations of the components of a vector $\mathbf{v} = v_x\mathbf{i} + v_y\mathbf{j} + v_z\mathbf{k} = v_r\mathbf{i}_r + v_\theta\mathbf{i}_\theta + v_\varphi\mathbf{i}_\varphi$:

$$v_r = v_x\sin\theta\cos\varphi + v_y\sin\theta\sin\varphi + v_z\cos\theta, \qquad v_x = v_r\sin\theta\cos\varphi + v_\theta\cos\theta\cos\varphi - v_\varphi\sin\varphi,$$
$$v_\theta = v_x\cos\theta\cos\varphi + v_y\cos\theta\sin\varphi - v_z\sin\theta, \qquad v_y = v_r\sin\theta\sin\varphi + v_\theta\cos\theta\sin\varphi + v_\varphi\cos\varphi,$$
$$v_\varphi = -v_x\sin\varphi + v_y\cos\varphi; \qquad v_z = v_r\cos\theta - v_\theta\sin\theta.$$

The metric tensor components are

$$g_{rr} = 1, \quad g_{\theta\theta} = r^2, \quad g_{\varphi\varphi} = r^2\sin^2\theta, \quad \sqrt{g} = r^2\sin\theta.$$

▶ **Basic differential relations.** Gradient of a scalar f:

$$\nabla f = \frac{\partial f}{\partial r}\mathbf{i}_r + \frac{1}{r}\frac{\partial f}{\partial\theta}\mathbf{i}_\theta + \frac{1}{r\sin\theta}\frac{\partial f}{\partial\varphi}\mathbf{i}_\varphi.$$

Divergence of a vector $\mathbf{v}$:

$$\nabla\cdot\mathbf{v} = \frac{1}{r^2}\frac{\partial}{\partial r}\left(r^2 v_r\right) + \frac{1}{r\sin\theta}\frac{\partial}{\partial\theta}\left(\sin\theta\,v_\theta\right) + \frac{1}{r\sin\varphi}\frac{\partial v_\varphi}{\partial\varphi}.$$

Gradient of a scalar f along a vector $\mathbf{v}$:

$$(\mathbf{v}\cdot\nabla)f = v_r\frac{\partial f}{\partial r} + \frac{v_\theta}{r}\frac{\partial f}{\partial\theta} + \frac{v_\varphi}{r\sin\theta}\frac{\partial f}{\partial\varphi}.$$

Gradient of a vector $\mathbf{w}$ along a vector $\mathbf{v}$:

$$(\mathbf{v}\cdot\nabla)\mathbf{w} = (\mathbf{v}\cdot\nabla)w_r\mathbf{i}_r + (\mathbf{v}\cdot\nabla)w_\theta\mathbf{i}_\theta + (\mathbf{v}\cdot\nabla)w_\varphi\mathbf{i}_\varphi.$$

Curl of a vector $\mathbf{v}$:

$$\nabla\times\mathbf{v} = \frac{1}{r\sin\theta}\left[\frac{\partial(\sin\theta\,v_\varphi)}{\partial\theta} - \frac{\partial v_\theta}{\partial\varphi}\right]\mathbf{i}_r + \frac{1}{r}\left[\frac{1}{\sin\theta}\frac{\partial v_r}{\partial\varphi} - \frac{\partial(rv_\varphi)}{\partial r}\right]\mathbf{i}_\theta + \frac{1}{r}\left[\frac{\partial(rv_\theta)}{\partial r} - \frac{\partial v_r}{\partial\theta}\right]\mathbf{i}_\varphi.$$

Laplacian of a scalar f:

$$\Delta f = \frac{1}{r^2}\frac{\partial}{\partial r}\left(r^2\frac{\partial f}{\partial r}\right) + \frac{1}{r^2\sin\theta}\frac{\partial}{\partial\theta}\left(\sin\theta\frac{\partial f}{\partial\theta}\right) + \frac{1}{r^2\sin^2\theta}\frac{\partial^2 f}{\partial\varphi^2}.$$

Bibliography for Chapter S3

Arfken, G. B. and Weber, H. J., *Mathematical Methods for Physicists, 4th Edition,* Academic Press, New York, 1995.

Korn, G. A. and Korn, T. M., *Mathematical Handbook for Scientists and Engineers, 2nd Edition,* Dover Publications, New York, 2000.

Menzel, D. H., *Mathematical Physics,* Dover Publications, New York, 1961.

Moon, P. and Spencer, D. E., *Field Theory Handbook, Including Coordinate Systems, Differential Equations and Their Solutions, 3rd Edition,* Springer-Verlag, Berlin, 1988.

Morse, P. M. and Feshbach, H., *Methods of Theoretical Physics, Part I,* McGraw-Hill, New York, 1953.

Weisstein, E. W., *CRC Concise Encyclopedia of Mathematics, 2nd Edition,* CRC Press, Boca Raton, Florida, 2003.

Zwillinger, D., *CRC Standard Mathematical Tables and Formulae, 31st Edition,* CRC Press, Boca Raton, Florida, 2002.

Chapter S4
Ordinary Differential Equations

▶ *In this chapter we shall often use the term "solution" to mean "general solution"; C, C_1, and C_2 are arbitrary constants. Notation like S4.2.1.12 stands for equation 12 from Subsection S4.2.1.*

S4.1. First-Order Equations

S4.1.1. Simplest Equations Integrable in Closed Form

1. $y'_x = f(y)$.

Autonomous equation.

Solution: $\quad x = \displaystyle\int \frac{dy}{f(y)} + C$.

Particular solutions: $y = A_k$, where the A_k are roots of the algebraic (transcendental) equation $f(A_k) = 0$.

2. $y'_x = f(x)g(y)$.

Separable equation.

Solution: $\quad \displaystyle\int \frac{dy}{g(y)} = \int f(x)\,dx + C$.

Particular solutions: $y = A_k$, where the A_k are roots of the algebraic (transcendental) equation $g(A_k) = 0$.

3. $g(x)y'_x = f_1(x)y + f_0(x)$.

Linear equation.
Solution:

$$y = Ce^F + e^F \int e^{-F} \frac{f_0(x)}{g(x)}\,dx, \quad \text{where} \quad F(x) = \int \frac{f_1(x)}{g(x)}\,dx.$$

4. $g(x)y'_x = f_1(x)y + f_0(x)y^k$.

Bernoulli equation. Here, k is an arbitrary number. For $k \ne 1$, the substitution $w(x) = y^{1-k}$ leads to a linear equation: $g(x)w'_x = (1-k)f_1(x)w + (1-k)f_0(x)$.
Solution:

$$y^{1-k} = Ce^F + (1-k)e^F \int e^{-F} \frac{f_0(x)}{g(x)}\,dx, \quad \text{where} \quad F(x) = (1-k)\int \frac{f_1(x)}{g(x)}\,dx.$$

5. $y'_x = f(y/x)$.

Homogeneous equation. The substitution $u(x) = y/x$ leads to a separable equation: $xu'_x = f(u) - u$.

6. $y'_x = f(ax + by + c)$.

If $b \neq 0$, the substitution $u(x) = ax + by + c$ leads to a separable equation: $u'_x = bf(u) + a$.

7. $y'_x = x^{k-1} f(yx^{-k})$.

Generalized homogeneous equation. The substitution $u = yx^{-k}$ leads to a separable equation, $xu'_x = f(u) - ku$.

8. $y'_x = \dfrac{y}{x} f(x^n y^m)$.

Generalized homogeneous equation. The substitution $z = x^n y^m$ leads to a separable equation: $xz'_x = nz + mzf(z)$.

9. $y'_x = f\left(\dfrac{ax + by + c}{\alpha x + \beta y + \gamma}\right)$.

$1°$. For $\Delta = a\beta - b\alpha \neq 0$, the transformation $x = u + \dfrac{b\gamma - c\beta}{\Delta}$, $y = v(u) + \dfrac{c\alpha - a\gamma}{\Delta}$ leads to an equation:

$$v'_u = f\left(\frac{au + bv}{\alpha u + \beta v}\right).$$

Dividing both the numerator and denominator of the fraction on the right-hand side by u, we obtain a homogeneous equation of the form S4.1.1.5.

$2°$. For $\Delta = 0$ and $b \neq 0$, the substitution $v(x) = ax + by + c$ leads to a separable equation of the form S4.1.1.2:

$$v'_x = a + bf\left(\frac{bv}{\beta v + b\gamma - c\beta}\right).$$

$3°$. For $\Delta = 0$ and $\beta \neq 0$, the substitution $v(x) = \alpha x + \beta y + \gamma$ also leads to a separable equation:

$$v'_x = \alpha + \beta f\left(\frac{bv + c\beta - b\gamma}{\beta v}\right).$$

S4.1.2. Riccati Equation

Preliminary remarks. A Riccati equation has the general form

$$y'_x = f_2(x)y^2 + f_1(x)y + f_0(x). \tag{S4.1.2.1}$$

$1°$. Given a particular solution $y_0 = y_0(x)$ of the Riccati equation (S4.1.2.1), the general solution can be written as:

$$y = y_0(x) + \Phi(x)\left[C - \int \Phi(x)f_2(x)\, dx\right]^{-1},$$

$$\Phi(x) = \exp\left\{\int \left[2f_2(x)y_0(x) + f_1(x)\right] dx\right\}. \tag{S4.1.2.2}$$

To the particular solution $y_0(x)$ there corresponds $C = \infty$.

     Often only particular solutions will be given for the specific equations presented below in Subsection S4.1.2. The general solutions of these equations can be obtained by formulas (S4.1.2.2).

2°. The substitution

$$u(x) = \exp\left(-\int f_2 y\, dx\right)$$

reduces the general Riccati equation to a second-order linear equation:

$$f_2 u''_{xx} - \left[(f_2)'_x + f_1 f_2\right] u'_x + f_0 f_2^2 u = 0,$$

which often may be easier to solve than the original Riccati equation. Specific second-order linear equations are outlined in Section S4.2.

3°. Many solvable equations of this form can be found in the book by Polyanin and Zaitsev (2003).

1. $y'_x = ay^2 + bx^n.$

Special Riccati equation, n is an arbitrary number.

1°. Solution for $n \neq -2$:

$$y = -\frac{1}{a}\frac{w'_x}{w}, \qquad w(x) = \sqrt{x}\left[C_1 J_{\frac{1}{2k}}\left(\frac{1}{k}\sqrt{ab}\,x^k\right) + C_2 Y_{\frac{1}{2k}}\left(\frac{1}{k}\sqrt{ab}\,x^k\right)\right],$$

where $k = \frac{1}{2}(n+2)$; $J_m(z)$ and $Y_m(z)$ are Bessel functions (see Subsection M13.6).

2°. Solution for $n = -2$:

$$y = \frac{\lambda}{x} - x^{2a\lambda}\left(\frac{ax}{2a\lambda+1}x^{2a\lambda} + C\right)^{-1},$$

where λ is a root of the quadratic equation $a\lambda^2 + \lambda + b = 0$.

2. $y'_x = y^2 + f(x)y - a^2 - af(x).$

Particular solution: $y_0 = a$. The general solution can be obtained by formulas (S4.1.2.2).

3. $y'_x = f(x)y^2 + ay - ab - b^2 f(x).$

Particular solution: $y_0 = b$. The general solution can be obtained by formulas (S4.1.2.2).

4. $y'_x = y^2 + xf(x)y + f(x).$

Particular solution: $y_0 = -1/x$. The general solution can be obtained by formulas (S4.1.2.2).

5. $y'_x = f(x)y^2 - ax^n f(x)y + anx^{n-1}.$

Particular solution: $y_0 = ax^n$. The general solution can be obtained by formulas (S4.1.2.2).

6. $y'_x = f(x)y^2 + anx^{n-1} - a^2 x^{2n} f(x).$

Particular solution: $y_0 = ax^n$. The general solution can be obtained by formulas (S4.1.2.2).

7. $y'_x = -(n+1)x^n y^2 + x^{n+1}f(x)y - f(x).$

Particular solution: $y_0 = x^{-n-1}$. The general solution can be obtained by formulas (S4.1.2.2).

8. $xy'_x = f(x)y^2 + ny + ax^{2n}f(x).$

Solution: $y = \begin{cases} \sqrt{a}\, x^n \tan\left[\sqrt{a} \int x^{n-1} f(x)\, dx + C\right] & \text{if } a > 0, \\ \sqrt{|a|}\, x^n \tanh\left[-\sqrt{|a|} \int x^{n-1} f(x)\, dx + C\right] & \text{if } a < 0. \end{cases}$

9. $xy'_x = x^{2n} f(x)y^2 + [ax^n f(x) - n]y + bf(x).$

The substitution $z = x^n y$ leads to a separable equation: $z'_x = x^{n-1} f(x)(z^2 + az + b).$

10. $y'_x = f(x)y^2 + g(x)y - a^2 f(x) - ag(x).$

Particular solution: $y_0 = a$. The general solution can be obtained by formulas (S4.1.2.2).

11. $y'_x = f(x)y^2 + g(x)y + anx^{n-1} - a^2 x^{2n} f(x) - ax^n g(x).$

Particular solution: $y_0 = ax^n$. The general solution can be obtained by formulas (S4.1.2.2).

12. $y'_x = ae^{\lambda x}y^2 + ae^{\lambda x} f(x)y + \lambda f(x).$

Particular solution: $y_0 = -\dfrac{\lambda}{a}e^{-\lambda x}$. The general solution can be obtained by formulas (S4.1.2.2).

13. $y'_x = f(x)y^2 - ae^{\lambda x} f(x)y + a\lambda e^{\lambda x}.$

Particular solution: $y_0 = ae^{\lambda x}$. The general solution can be obtained by formulas (S4.1.2.2).

14. $y'_x = f(x)y^2 + a\lambda e^{\lambda x} - a^2 e^{2\lambda x} f(x).$

Particular solution: $y_0 = ae^{\lambda x}$. The general solution can be obtained by formulas (S4.1.2.2).

15. $y'_x = f(x)y^2 + \lambda y + ae^{2\lambda x} f(x).$

Solution: $y = \begin{cases} \sqrt{a}\, e^{\lambda x} \tan\left[\sqrt{a} \int e^{\lambda x} f(x)\, dx + C\right] & \text{if } a > 0, \\ \sqrt{|a|}\, e^{\lambda x} \tanh\left[-\sqrt{|a|} \int e^{\lambda x} f(x)\, dx + C\right] & \text{if } a < 0. \end{cases}$

16. $y'_x = y^2 - f^2(x) + f'_x(x).$

Particular solution: $y_0 = f(x)$. The general solution can be obtained by formulas (S4.1.2.2).

17. $y'_x = f(x)y^2 - f(x)g(x)y + g'_x(x).$

Particular solution: $y_0 = g(x)$.

S4.1.3. Other Equations

1. $yy'_x = f(x)y + g(x).$

Abel equation of the second kind. Many solvable equations of this form can be found in the books by Zaitsev and Polyanin (1994) and Polyanin and Zaitsev (2003).

2. $yy'_x = f(x)y^2 + g(x)y + h(x)$.

Abel equation of the second kind. Many solvable equations of this form can be found in the books by Zaitsev and Polyanin (1994) and Polyanin and Zaitsev (2003).

▶ *In equations S4.1.3.3–S4.1.3.21 below, the functions f, g, and h are arbitrary composite functions whose arguments can depend on both x and y.*

3. $y'_x = f(y + ax^n + b) - anx^{n-1}$.

The substitution $u = y + ax^n + b$ leads to a separable equation: $u'_x = f(u)$.

4. $y'_x = -\dfrac{n}{m}\dfrac{y}{x} + y^k f(x)g(x^n y^m)$.

The substitution $z = x^n y^m$ leads to a separable equation: $z'_x = mx^{\frac{n-nk}{m}} f(x) z^{\frac{k+m-1}{m}} g(z)$.

5. $y'_x = x^{n-1} y^{1-m} f(ax^n + by^m)$.

The substitution $w = ax^n + by^m$ leads to a separable equation: $w'_x = x^{n-1}[an + bmf(w)]$.

6. $[x^n f(y) + xg(y)]y'_x = h(y)$.

This is a Bernoulli equation with respect to $x = x(y)$ (see equation S4.1.1.4).

7. $x[f(x^n y^m) + mx^k g(x^n y^m)]y'_x = y[h(x^n y^m) - nx^k g(x^n y^m)]$.

The transformation $t = x^n y^m$, $z = x^{-k}$ leads to a linear equation with respect to $z = z(t)$:
$t[nf(t) + mh(t)]z'_t = -kf(t)z - kmg(t)$.

8. $x[f(x^n y^m) + my^k g(x^n y^m)]y'_x = y[h(x^n y^m) - ny^k g(x^n y^m)]$.

The transformation $t = x^n y^m$, $z = y^{-k}$ leads to a linear equation with respect to $z = z(t)$:
$t[nf(t) + mh(t)]z'_t = -kh(t)z + kng(t)$.

9. $x[sf(x^n y^m) - mg(x^k y^s)]y'_x = y[ng(x^k y^s) - kf(x^n y^m)]$.

The transformation $t = x^n y^m$, $w = x^k y^s$ leads to a separable equation: $tf(t)w'_t = wg(w)$.

10. $[f(y) + amx^n y^{m-1}]y'_x + g(x) + anx^{n-1}y^m = 0$.

Solution: $\displaystyle\int f(y)\,dy + \int g(x)\,dx + ax^n y^m = C$.

11. $y'_x = e^{-\lambda x} f(e^{\lambda x} y)$.

The substitution $u = e^{\lambda x} y$ leads to a separable equation: $u'_x = f(u) + \lambda u$.

12. $y'_x = e^{\lambda y} f(e^{\lambda y} x)$.

The substitution $u = e^{\lambda y} x$ leads to a separable equation: $xu'_x = \lambda u^2 f(u) + u$.

13. $y'_x = yf(e^{\alpha x} y^m)$.

The substitution $z = e^{\alpha x} y^m$ leads to a separable equation: $z'_x = \alpha z + mzf(z)$.

14. $y'_x = \dfrac{1}{x} f(x^n e^{\alpha y}).$

The substitution $z = x^n e^{\alpha y}$ leads to a separable equation: $xz'_x = nz + \alpha z f(z).$

15. $y'_x = f(x)e^{\lambda y} + g(x).$

The substitution $u = e^{-\lambda y}$ leads to a linear equation: $u'_x = -\lambda g(x)u - \lambda f(x).$

16. $y'_x = -\dfrac{n}{x} + f(x)g(x^n e^y).$

The substitution $z = x^n e^y$ leads to a separable equation: $z'_x = f(x)zg(z).$

17. $y'_x = -\dfrac{\alpha}{m} y + y^k f(x)g(e^{\alpha x}y^m).$

The substitution $z = e^{\alpha x}y^m$ leads to a separable equation:

$$z'_x = m \exp\left[\frac{\alpha}{m}(1-k)x\right] f(x)z^{\frac{k+m-1}{m}} g(z).$$

18. $y'_x = e^{\alpha x - \beta y} f(ae^{\alpha x} + be^{\beta y}).$

The substitution $w = ae^{\alpha x} + be^{\beta y}$ leads to a separable equation: $w'_x = e^{\alpha x}[a\alpha + b\beta f(w)].$

19. $[e^{\alpha x}f(y) + a\beta]y'_x + e^{\beta y}g(x) + a\alpha = 0.$

Solution: $\displaystyle\int e^{-\beta y}f(y)\,dy + \int e^{-\alpha x}g(x)\,dx - ae^{-\alpha x - \beta y} = C.$

20. $x[f(x^n e^{\alpha y}) + \alpha yg(x^n e^{\alpha y})]y'_x = h(x^n e^{\alpha y}) - nyg(x^n e^{\alpha y}).$

The substitution $t = x^n e^{\alpha y}$ leads to a linear equation with respect to $y = y(t)$:
$t[nf(t) + \alpha h(t)]y'_t = -ng(t)y + h(t).$

21. $[f(e^{\alpha x}y^m) + mxg(e^{\alpha x}y^m)]y'_x = y[h(e^{\alpha x}y^m) - \alpha xg(e^{\alpha x}y^m)].$

The substitution $t = e^{\alpha x}y^m$ leads to a linear equation with respect to $x = x(t)$:
$t[\alpha f(t) + mh(t)]x'_t = mg(t)x + f(t).$

S4.2. Second-Order Linear Equations

Preliminary remarks. A homogeneous linear equation of the second order has the general form

$$f_2(x)y''_{xx} + f_1(x)y'_x + f_0(x)y = 0.$$

Let $y_0 = y_0(x)$ be a nontrivial particular solution ($y_0 \not\equiv 0$) of this equation. Then the general solution of this equation can be found from the formula

$$y = y_0\left(C_1 + C_2\int \frac{e^{-F}}{y_0^2}\,dx\right), \quad \text{where} \quad F = \int \frac{f_1}{f_2}\,dx.$$

For specific equations described below, sometimes only particular solutions are given, while the general solutions can be obtained with the above formula.

S4.2.1. Equations Involving Power Functions

1. $y''_{xx} + ay = 0.$

Equation of free oscillations.

$$\text{Solution:} \quad y = \begin{cases} C_1 \sinh(x\sqrt{|a|}) + C_2 \cosh(x\sqrt{|a|}) & \text{if } a < 0, \\ C_1 + C_2 x & \text{if } a = 0, \\ C_1 \sin(x\sqrt{a}) + C_2 \cos(x\sqrt{a}) & \text{if } a > 0. \end{cases}$$

2. $y''_{xx} - ax^n y = 0.$

$1°$. For $n = -2$, this is the Euler equation S4.2.1.12 (the solution is expressed in terms of elementary function).

$2°$. Assume $2/(n+2) = 2m + 1$, where m is an integer. Then the solution is

$$y = \begin{cases} x(x^{1-2q}D)^{m+1}\left[C_1 \exp\left(\dfrac{\sqrt{a}}{q}x^q\right) + C_2 \exp\left(-\dfrac{\sqrt{a}}{q}x^q\right)\right] & \text{if } m \geq 0, \\[4mm] (x^{1-2q}D)^{-m}\left[C_1 \exp\left(\dfrac{\sqrt{a}}{q}x^q\right) + C_2 \exp\left(-\dfrac{\sqrt{a}}{q}x^q\right)\right] & \text{if } m < 0, \end{cases}$$

where $D = \dfrac{d}{dx}$, $q = \dfrac{n+2}{2} = \dfrac{1}{2m+1}$.

$3°$. For any n, the solution is expressed in terms of Bessel functions and modified Bessel functions:

$$y = \begin{cases} C_1\sqrt{x}\,J_{\frac{1}{2q}}\left(\dfrac{\sqrt{-a}}{q}x^q\right) + C_2\sqrt{x}\,Y_{\frac{1}{2q}}\left(\dfrac{\sqrt{-a}}{q}x^q\right) & \text{if } a < 0, \\[4mm] C_1\sqrt{x}\,I_{\frac{1}{2q}}\left(\dfrac{\sqrt{a}}{q}x^q\right) + C_2\sqrt{x}\,K_{\frac{1}{2q}}\left(\dfrac{\sqrt{a}}{q}x^q\right) & \text{if } a > 0, \end{cases}$$

where $q = \frac{1}{2}(n+2)$. The functions $J_\nu(z)$, $Y_\nu(z)$ and $I_\nu(z)$, $K_\nu(z)$ are described in Sections M13.6 and M13.7 in detail; see also equations S4.2.1.13 and S4.2.1.14.

3. $y''_{xx} + ay'_x + by = 0.$

Second-order constant coefficient linear equation. In physics this equation is called an *equation of damped vibrations.*

$$\text{Solution:} \quad y = \begin{cases} \exp\left(-\frac{1}{2}ax\right)\left[C_1 \exp\left(\frac{1}{2}\lambda x\right) + C_2 \exp\left(-\frac{1}{2}\lambda x\right)\right] & \text{if } \lambda^2 = a^2 - 4b > 0, \\ \exp\left(-\frac{1}{2}ax\right)\left[C_1 \sin\left(\frac{1}{2}\lambda x\right) + C_2 \cos\left(\frac{1}{2}\lambda x\right)\right] & \text{if } \lambda^2 = 4b - a^2 > 0, \\ \exp\left(-\frac{1}{2}ax\right)\left(C_1 x + C_2\right) & \text{if } a^2 = 4b. \end{cases}$$

4. $y''_{xx} + ay'_x + (bx + c)y = 0.$

$1°$. Solution with $b \neq 0$:

$$y = \exp\left(-\tfrac{1}{2}ax\right)\sqrt{\xi}\left[C_1 J_{1/3}\left(\tfrac{2}{3}\sqrt{b}\,\xi^{3/2}\right) + C_2 Y_{1/3}\left(\tfrac{2}{3}\sqrt{b}\,\xi^{3/2}\right)\right], \quad \xi = x + \frac{4c - a^2}{4b},$$

where $J_{1/3}(z)$ and $Y_{1/3}(z)$ are Bessel functions.

$2°$. For $b = 0$, see equation S4.2.1.3.

5. $y''_{xx} + (ax + b)y'_x + (\alpha x^2 + \beta x + \gamma)y = 0.$

The substitution $y = u \exp(sx^2)$, where s is a root of the quadratic equation $4s^2 + 2as + \alpha = 0$, leads to an equation of the form S4.2.1.11: $u''_{xx} + [(a+4s)x+b]u'_x + [(\beta+2bs)x+\gamma+2s]u = 0.$

6. $xy''_{xx} + ay'_x + by = 0.$

$1°$. The solution is expressed in terms of Bessel functions and modified Bessel functions:

$$y = \begin{cases} x^{\frac{1-a}{2}} \left[C_1 J_\nu\left(2\sqrt{bx}\right) + C_2 Y_\nu\left(2\sqrt{bx}\right) \right] & \text{if } bx > 0, \\ x^{\frac{1-a}{2}} \left[C_1 I_\nu\left(2\sqrt{|bx|}\right) + C_2 K_\nu\left(2\sqrt{|bx|}\right) \right] & \text{if } bx < 0, \end{cases}$$

where $\nu = |1 - a|$.

$2°$. For $a = \frac{1}{2}(2n + 1)$, where $n = 0, 1, \ldots$, the solution is

$$y = \begin{cases} C_1 \dfrac{d^n}{dx^n} \cos\sqrt{4bx} + C_2 \dfrac{d^n}{dx^n} \sin\sqrt{4bx} & \text{if } bx > 0, \\ C_1 \dfrac{d^n}{dx^n} \cosh\sqrt{4|bx|} + C_2 \dfrac{d^n}{dx^n} \sinh\sqrt{4|bx|} & \text{if } bx < 0. \end{cases}$$

7. $xy''_{xx} + ay'_x + bxy = 0.$

$1°$. The solution is expressed in terms of Bessel functions and modified Bessel functions:

$$y = \begin{cases} x^{\frac{1-a}{2}} \left[C_1 J_\nu\left(\sqrt{b}\,x\right) + C_2 Y_\nu\left(\sqrt{b}\,x\right) \right] & \text{if } b > 0, \\ x^{\frac{1-a}{2}} \left[C_1 I_\nu\left(\sqrt{|b|}\,x\right) + C_2 K_\nu\left(\sqrt{|b|}\,x\right) \right] & \text{if } b < 0, \end{cases}$$

where $\nu = \frac{1}{2}|1 - a|$.

$2°$. For $a = 2n$, where $n = 1, 2, \ldots$, the solution is

$$y = \begin{cases} C_1 \left(\dfrac{1}{x}\dfrac{d}{dx}\right)^n \cos\left(x\sqrt{b}\right) + C_2 \left(\dfrac{1}{x}\dfrac{d}{dx}\right)^n \sin\left(x\sqrt{b}\right) & \text{if } b > 0, \\ C_1 \left(\dfrac{1}{x}\dfrac{d}{dx}\right)^n \cosh\left(x\sqrt{-b}\right) + C_2 \left(\dfrac{1}{x}\dfrac{d}{dx}\right)^n \sinh\left(x\sqrt{-b}\right) & \text{if } b < 0. \end{cases}$$

8. $xy''_{xx} + ny'_x + bx^{1-2n}y = 0.$

For $n = 1$, this is the Euler equation S4.2.1.12. For $n \neq 1$, the solution is

$$y = \begin{cases} C_1 \sin\left(\dfrac{\sqrt{b}}{n-1}x^{1-n}\right) + C_2 \cos\left(\dfrac{\sqrt{b}}{n-1}x^{1-n}\right) & \text{if } b > 0, \\ C_1 \exp\left(\dfrac{\sqrt{-b}}{n-1}x^{1-n}\right) + C_2 \exp\left(\dfrac{-\sqrt{-b}}{n-1}x^{1-n}\right) & \text{if } b < 0. \end{cases}$$

9. $xy''_{xx} + ay'_x + bx^n y = 0.$

If $n = -1$ and $b = 0$, we have the Euler equation S4.2.1.12. If $n \neq -1$ and $b \neq 0$, the solution is expressed in terms of Bessel functions:

$$y = x^{\frac{1-a}{2}} \left[C_1 J_\nu\left(\frac{2\sqrt{b}}{n+1}x^{\frac{n+1}{2}}\right) + C_2 Y_\nu\left(\frac{2\sqrt{b}}{n+1}x^{\frac{n+1}{2}}\right) \right], \qquad \text{where} \quad \nu = \frac{|1-a|}{n+1}.$$

10. $xy''_{xx} + (b - x)y'_x - ay = 0.$

Degenerate hypergeometric equation.

1°. If $b \neq 0, -1, -2, -3, \ldots$, Kummer's series is a particular solution:

$$\Phi(a, b; x) = 1 + \sum_{k=1}^{\infty} \frac{(a)_k}{(b)_k} \frac{x^k}{k!},$$

where $(a)_k = a(a + 1) \ldots (a + k - 1)$, $(a)_0 = 1$. If $b > a > 0$, this solution can be written in terms of a definite integral:

$$\Phi(a, b; x) = \frac{\Gamma(b)}{\Gamma(a)\,\Gamma(b - a)} \int_0^1 e^{xt} t^{a-1} (1 - t)^{b-a-1} \, dt,$$

where $\Gamma(z) = \int_0^{\infty} e^{-t} t^{z-1} \, dt$ is the gamma function.

If b is not an integer, then the general solution has the form:

$$y = C_1 \Phi(a, b; x) + C_2 x^{1-b} \Phi(a - b + 1, \; 2 - b; \; x).$$

2°. For $b \neq 0, -1, -2, -3, \ldots$, the general solution of the degenerate hypergeometric equation can be written in the form

$$y = C_1 \Phi(a, b; x) + C_2 \Psi(a, b; x),$$

while for $b = 0, -1, -2, -3, \ldots$, it can be represented as

$$y = x^{1-b} \big[C_1 \Phi(a - b + 1, \; 2 - b; \; x) + C_2 \Psi(a - b + 1, \; 2 - b; \; x) \big].$$

The functions $\Phi(a, b; x)$ and $\Psi(a, b; x)$ are described in Subsection M13.8 in detail.

11. $(a_2 x + b_2)y''_{xx} + (a_1 x + b_1)y'_x + (a_0 x + b_0)y = 0.$

Let the function $\mathcal{J}(a, b; x)$ be an arbitrary solution of the degenerate hypergeometric equation $xy''_{xx} + (b - x)y'_x - ay = 0$ (see equation S4.2.1.10), and let the function $Z_\nu(x)$ be an arbitrary solution of the Bessel equation $x^2 y''_{xx} + xy'_x + (x^2 - \nu^2)y = 0$ (see S4.2.1.13). The results of solving the original equation are presented in Table S4.1.

TABLE S4.1
Solutions of equation S4.2.1.11 for different values of the determining parameters.

Solution: $y = e^{kx} w(z)$, where $z = \dfrac{x - \mu}{\lambda}$					
Constraints	k	λ	μ	w	Parameters
$a_2 \neq 0$, $a_1^2 \neq 4a_0 a_2$	$\dfrac{\sqrt{D} - a_1}{2a_2}$	$-\dfrac{a_2}{2a_2 k + a_1}$	$-\dfrac{b_2}{a_2}$	$\mathcal{J}(a, b; z)$	$a = B(k)/(2a_2 k + a_1)$, $b = (a_2 b_1 - a_1 b_2)a_2^{-2}$
$a_2 = 0$, $a_1 \neq 0$	$-\dfrac{a_0}{a_1}$	1	$-\dfrac{2b_2 k + b_1}{a_1}$	$\mathcal{J}\big(a, \tfrac{1}{2}; \beta z^2\big)$	$a = B(k)/(2a_1)$, $\beta = -a_1/(2b_2)$
$a_2 \neq 0$, $a_1^2 = 4a_0 a_2$	$-\dfrac{a_1}{2a_2}$	a_2	$-\dfrac{b_2}{a_2}$	$z^{\nu/2} Z_\nu\big(\beta \sqrt{z}\big)$	$\nu = 1 - (2b_2 k + b_1)a_2^{-1}$, $\beta = 2\sqrt{B(k)}$
$a_2 = a_1 = 0$, $a_0 \neq 0$	$-\dfrac{b_1}{2b_2}$	1	$\dfrac{b_1^2 - 4b_0 b_2}{4a_0 b_2}$	$z^{1/2} Z_{1/3}\big(\beta z^{3/2}\big)$; see also 2.1.2.12	$\beta = \dfrac{2}{3}\left(\dfrac{a_0}{b_2}\right)^{1/2}$
Notation: $D = a_1^2 - 4a_0 a_2$, $B(k) = b_2 k^2 + b_1 k + b_0$					

12. $x^2 y''_{xx} + axy'_x + by = 0.$

Euler equation. Solution:

$$y = \begin{cases} |x|^{\frac{1-a}{2}} \left(C_1 |x|^{\mu} + C_2 |x|^{-\mu} \right) & \text{if } (1-a)^2 > 4b, \\ |x|^{\frac{1-a}{2}} (C_1 + C_2 \ln |x|) & \text{if } (1-a)^2 = 4b, \\ |x|^{\frac{1-a}{2}} \left[C_1 \sin(\mu \ln |x|) + C_2 \cos(\mu \ln |x|) \right] & \text{if } (1-a)^2 < 4b, \end{cases}$$

where $\mu = \frac{1}{2} |(1-a)^2 - 4b|^{1/2}$.

13. $x^2 y''_{xx} + xy'_x + (x^2 - \nu^2)y = 0.$

Bessel equation.

$1°$. Let ν be an arbitrary noninteger. Then the general solution is given by

$$y = C_1 J_\nu(x) + C_2 Y_\nu(x), \tag{1}$$

where $J_\nu(x)$ and $Y_\nu(x)$ are the Bessel functions of the first and second kind:

$$J_\nu(x) = \sum_{k=0}^{\infty} \frac{(-1)^k (x/2)^{\nu+2k}}{k! \, \Gamma(\nu+k+1)}, \qquad Y_\nu(x) = \frac{J_\nu(x) \cos \pi\nu - J_{-\nu}(x)}{\sin \pi\nu}. \tag{2}$$

Solution (1) is denoted by $y = Z_\nu(x)$, which is referred to as the cylindrical function.

The functions $J_\nu(x)$ and $Y_\nu(x)$ can be expressed in terms of definite integrals (with $x > 0$):

$$\pi J_\nu(x) = \int_0^\pi \cos(x \sin\theta - \nu\theta) \, d\theta - \sin\pi\nu \int_0^\infty \exp(-x \sinh t - \nu t) \, dt,$$

$$\pi Y_\nu(x) = \int_0^\pi \sin(x \sin\theta - \nu\theta) \, d\theta - \int_0^\infty (e^{\nu t} + e^{-\nu t} \cos\pi\nu) e^{-x \sinh t} \, dt.$$

$2°$. In the case $\nu = n + \frac{1}{2}$, where $n = 0, 1, 2, \ldots$, the Bessel functions are expressed in terms of elementary functions:

$$J_{n+\frac{1}{2}}(x) = \sqrt{\frac{2}{\pi}} \, x^{n+\frac{1}{2}} \left(-\frac{1}{x} \frac{d}{dx} \right)^n \frac{\sin x}{x}, \qquad J_{-n-\frac{1}{2}}(x) = \sqrt{\frac{2}{\pi}} \, x^{n+\frac{1}{2}} \left(\frac{1}{x} \frac{d}{dx} \right)^n \frac{\cos x}{x},$$

$$Y_{n+\frac{1}{2}}(x) = (-1)^{n+1} J_{-n-\frac{1}{2}}(x).$$

The Bessel functions are described in Section M13.6 in detail.

14. $x^2 y''_{xx} + xy'_x - (x^2 + \nu^2)y = 0.$

Modified Bessel equation. It can be reduced to equation S4.2.1.13 by means of the substitution $x = i\bar{x}$ ($i^2 = -1$).

Solution:

$$y = C_1 I_\nu(x) + C_2 K_\nu(x),$$

where $I_\nu(x)$ and $K_\nu(x)$ are modified Bessel functions of the first and second kind:

$$I_\nu(x) = \sum_{k=0}^{\infty} \frac{(x/2)^{2k+\nu}}{k! \, \Gamma(\nu+k+1)}, \qquad K_\nu(x) = \frac{\pi}{2} \frac{I_{-\nu}(x) - I_\nu(x)}{\sin \pi\nu}.$$

The modified Bessel functions are described in Section M13.7 in detail.

15. $x^2 y''_{xx} + ax y'_x + (bx^n + c)y = 0, \qquad n \neq 0.$

The case $b = 0$ corresponds to the Euler equation S4.2.1.12.
    For $b \neq 0$, the solution is

$$y = x^{\frac{1-a}{2}} \left[C_1 J_\nu \left(\frac{2}{n} \sqrt{b}\, x^{\frac{n}{2}} \right) + C_2 Y_\nu \left(\frac{2}{n} \sqrt{b}\, x^{\frac{n}{2}} \right) \right],$$

where $\nu = \frac{1}{n}\sqrt{(1-a)^2 - 4c}$; $J_\nu(z)$ and $Y_\nu(z)$ are the Bessel functions of the first and second kind.

16. $x^2 y''_{xx} + ax y'_x + x^n(bx^n + c)y = 0.$

The substitution $\xi = x^n$ leads to an equation of the form S4.2.1.11:

$$n^2 \xi y''_{\xi\xi} + n(n - 1 + a)y'_\xi + (b\xi + c)y = 0.$$

17. $x^2 y''_{xx} + (ax + b)y'_x + cy = 0.$

The transformation $x = z^{-1}$, $y = z^k e^z w$, where k is a root of the quadratic equation $k^2 + (1-a)k + c = 0$, leads to an equation of the form S4.2.1.11:

$$zw''_{zz} + [(2-b)z + 2k + 2 - a]w'_z + [(1-b)z + 2k + 2 - a - bk]w = 0.$$

18. $(1 - x^2)y''_{xx} - 2xy'_x + n(n+1)y = 0, \qquad n = 0, 1, 2, \ldots$

Legendre equation.
    The solution is given by

$$y = C_1 P_n(x) + C_2 Q_n(x),$$

where the Legendre polynomials $P_n(x)$ and the Legendre functions of the second kind $Q_n(x)$ are given by the formulas

$$P_n(x) = \frac{1}{n!\, 2^n} \frac{d^n}{dx^n}(x^2 - 1)^n, \quad Q_n(x) = \frac{1}{2} P_n(x) \ln \frac{1+x}{1-x} - \sum_{m=1}^{n} \frac{1}{m} P_{m-1}(x) P_{n-m}(x).$$

The functions $P_n = P_n(x)$ can be conveniently calculated using the recurrence relations

$$P_0(x) = 1, \quad P_1(x) = x, \quad P_2(x) = \frac{1}{2}(3x^2 - 1), \quad \ldots, \quad P_{n+1}(x) = \frac{2n+1}{n+1} x P_n(x) - \frac{n}{n+1} P_{n-1}(x).$$

Three leading functions $Q_n = Q_n(x)$ are

$$Q_0(x) = \frac{1}{2} \ln \frac{1+x}{1-x}, \quad Q_1(x) = \frac{x}{2} \ln \frac{1+x}{1-x} - 1, \quad Q_2(x) = \frac{3x^2 - 1}{4} \ln \frac{1+x}{1-x} - \frac{3}{2}x.$$

The Legendre polynomials and the Legendre functions are described in Subsection M13.10.1 in more detail.

19. $(1 - x^2)y''_{xx} - 2xy'_x + \nu(\nu + 1)y = 0.$

Legendre equation; ν is an arbitrary number. The case $\nu = n$ where n is a nonnegative integer is considered in S4.2.1.18.

The substitution $z = x^2$ leads to the hypergeometric equation. Therefore, with $|x| < 1$ the solution can be written as

$$y = C_1 F\left(-\frac{\nu}{2}, \frac{1 + \nu}{2}, \frac{1}{2}; x\right) + C_2 x F\left(\frac{1 - \nu}{2}, 1 + \frac{\nu}{2}, \frac{3}{2}; x\right),$$

where $F(\alpha, \beta, \gamma; x)$ is the hypergeometric series (see equation S4.2.1.22).

20. $(ax^2 + b)y''_{xx} + axy'_x + cy = 0.$

The substitution $z = \int \dfrac{dx}{\sqrt{ax^2 + b}}$ leads to a constant coefficient linear equation: $y''_{zz} + cy = 0.$

21. $(1 - x^2)y''_{xx} + (ax + b)y'_x + cy = 0.$

$1°.$ The substitution $2z = 1 + x$ leads to the hypergeometric equation S4.2.1.22:

$$z(1 - z)y''_{zz} + [az + \tfrac{1}{2}(b - a)]y'_z + cy = 0.$$

$2°.$ For $a = -2m - 3$, $b = 0$, and $c = \lambda$, the Gegenbauer functions are solutions of the equation.

22. $x(x - 1)y''_{xx} + [(\alpha + \beta + 1)x - \gamma]y'_x + \alpha\beta y = 0.$

Gaussian hypergeometric equation. For $\gamma \neq 0, -1, -2, -3, \ldots$, a solution can be expressed in terms of the hypergeometric series:

$$F(\alpha, \beta, \gamma; x) = 1 + \sum_{k=1}^{\infty} \frac{(\alpha)_k (\beta)_k}{(\gamma)_k} \frac{x^k}{k!}, \qquad (\alpha)_k = \alpha(\alpha + 1) \ldots (\alpha + k - 1),$$

which, *a fortiori*, is convergent for $|x| < 1$.

For $\gamma > \beta > 0$, this solution can be expressed in terms of a definite integral:

$$F(\alpha, \beta, \gamma; x) = \frac{\Gamma(\gamma)}{\Gamma(\beta)\Gamma(\gamma - \beta)} \int_0^1 t^{\beta-1}(1 - t)^{\gamma-\beta-1}(1 - tx)^{-\alpha} \, dt,$$

where $\Gamma(\beta)$ is the gamma function.

If γ is not an integer, the general solution of the hypergeometric equation has the form:

$$y = C_1 F(\alpha, \beta, \gamma; x) + C_2 x^{1-\gamma} F(\alpha - \gamma + 1, \beta - \gamma + 1, 2 - \gamma; x).$$

In the degenerate cases $\gamma = 0, -1, -2, -3, \ldots$, a particular solution of the hypergeometric equation corresponds to $C_1 = 0$ and $C_2 = 1$. If γ is a positive integer, another particular solution corresponds to $C_1 = 1$ and $C_2 = 0$. In both these cases, the general solution can be constructed by means of the last formula given in the preliminary remarks at the beginning of Section S4.2.

Table S4.2 gives the general solutions of the hypergeometric equation for some values of the determining parameters.

TABLE S4.2
General solutions of the hypergeometric equation for some values of the determining parameters.

α	β	γ	Solution: $y = y(x)$
0	β	γ	$C_1 + C_2 \int \|x\|^{-\gamma}\|x-1\|^{\gamma-\beta-1}\,dx$
α	$\alpha+\frac{1}{2}$	$2\alpha+1$	$C_1\left(1+\sqrt{1-x}\right)^{-2\alpha} + C_2 x^{-2\alpha}\left(1+\sqrt{1-x}\right)^{2\alpha}$
α	$\alpha-\frac{1}{2}$	$\frac{1}{2}$	$C_1\left(1+\sqrt{x}\right)^{1-2\alpha} + C_2\left(1-\sqrt{x}\right)^{1-2\alpha}$
α	$\alpha+\frac{1}{2}$	$\frac{3}{2}$	$\frac{1}{\sqrt{x}}\left[C_1\left(1+\sqrt{x}\right)^{1-2\alpha} + C_2\left(1-\sqrt{x}\right)^{1-2\alpha}\right]$
1	β	γ	$\|x\|^{1-\gamma}\|x-1\|^{\gamma-\beta-1}\left(C_1 + C_2 \int \|x\|^{\gamma-2}\|x-1\|^{\beta-\gamma}\,dx\right)$
α	β	α	$\|x-1\|^{-\beta}\left(C_1 + C_2 \int \|x\|^{-\alpha}\|x-1\|^{\beta-1}\,dx\right)$
α	β	$\alpha+1$	$\|x\|^{-\alpha}\left(C_1 + C_2 \int \|x\|^{\alpha-1}\|x-1\|^{-\beta}\,dx\right)$

23. $(1-x^2)^2 y''_{xx} - 2x(1-x^2)y'_x + [\nu(\nu+1)(1-x^2) - \mu^2]y = 0.$

Legendre equation, ν and μ are arbitrary parameters.

The transformation $x = 1 - 2\xi$, $y = |x^2 - 1|^{\mu/2}w$ leads to the hypergeometric equation S4.2.1.22:

$$\xi(\xi-1)w''_{\xi\xi} + (\mu+1)(1-2\xi)w'_\xi + (\nu-\mu)(\nu+\mu+1)w = 0$$

with parameters $\alpha = \mu - \nu$, $\beta = \mu + \nu + 1$, $\gamma = \mu + 1$.

In particular, the original equation is integrable by quadrature if $\nu = \mu$ or $\nu = -\mu - 1$.

24. $(x-a)^2(x-b)^2 y''_{xx} - cy = 0,$ $a \neq b.$

The transformation $\xi = \ln\left|\dfrac{x-a}{x-b}\right|$, $y = (x-b)\eta$ leads to a constant coefficient linear equation: $(a-b)^2(\eta''_{\xi\xi} - \eta'_\xi) - c\eta = 0$. Therefore, the solution is as follows:

$$y = C_1|x-a|^{(1+\lambda)/2}|x-b|^{(1-\lambda)/2} + C_2|x-a|^{(1-\lambda)/2}|x-b|^{(1+\lambda)/2},$$

where $\lambda^2 = 4c(a-b)^{-2} + 1 \neq 0$.

25. $(ax^2 + bx + c)^2 y''_{xx} + Ay = 0.$

The transformation $\xi = \int \dfrac{dx}{ax^2 + bx + c}$, $w = \dfrac{y}{\sqrt{|ax^2 + bx + c|}}$ leads to a constant coefficient linear equation of the form S4.2.1.1: $w''_{\xi\xi} + (A + ac - \frac{1}{4}b^2)w = 0$.

26. $x^2(ax^n - 1)y''_{xx} + x(apx^n + q)y'_x + (arx^n + s)y = 0.$

Find the roots A_1, A_2 and B_1, B_2 of the quadratic equations

$$A^2 - (q+1)A - s = 0, \quad B^2 - (p-1)B + r = 0$$

and define parameters c, α, β, and γ by the relations

$$c = A_1, \quad \alpha = (A_1 + B_1)n^{-1}, \quad \beta = (A_1 + B_2)n^{-1}, \quad \gamma = 1 + (A_1 - A_2)n^{-1}.$$

Then the solution of the original equation has the form $y = x^c u(ax^n)$, where $u = u(z)$ is the general solution of the hypergeometric equation S4.2.1.22: $z(z-1)u''_{zz} + [(\alpha+\beta+1)z-\gamma]u'_z + \alpha\beta u = 0$.

S4.2.2. Equations Involving Exponential and Other Functions

1. $y''_{xx} + ae^{\lambda x}y = 0, \qquad \lambda \neq 0.$

Solution: $y = C_1 J_0(z) + C_2 Y_0(z)$, where $z = 2\lambda^{-1}\sqrt{a}\,e^{\lambda x/2}$; $J_0(z)$ and $Y_0(z)$ are Bessel functions.

2. $y''_{xx} + (ae^x - b)y = 0.$

Solution: $y = C_1 J_{2\sqrt{b}}(2\sqrt{a}\,e^{x/2}) + C_2 Y_{2\sqrt{b}}(2\sqrt{a}\,e^{x/2})$, where $J_\nu(z)$ and $Y_\nu(z)$ are Bessel functions.

3. $y''_{xx} - (ae^{2\lambda x} + be^{\lambda x} + c)y = 0.$

The transformation $z = e^{\lambda x}$, $w = z^{-k}y$, where $k = \sqrt{c}/\lambda$, leads to an equation of the form S4.2.1.11: $\lambda^2 z w''_{zz} + \lambda^2(2k + 1)w'_z - (az + b)w = 0$.

4. $y''_{xx} + ay'_x + be^{2ax}y = 0.$

The transformation $\xi = e^{ax}$, $u = ye^{ax}$ leads to a constant coefficient linear equation of the form S4.2.1.1: $u''_{\xi\xi} + ba^{-2}u = 0$.

5. $y''_{xx} - ay'_x + be^{2ax}y = 0.$

The substitution $\xi = e^{ax}$ leads to a constant coefficient linear equation of the form S4.2.1.1: $y''_{\xi\xi} + ba^{-2}y = 0$.

6. $y''_{xx} + ay'_x + (be^{\lambda x} + c)y = 0.$

Solution: $y = e^{-ax/2}\left[C_1 J_\nu\left(2\lambda^{-1}\sqrt{b}\,e^{\lambda x/2}\right) + C_2 Y_\nu\left(2\lambda^{-1}\sqrt{b}\,e^{\lambda x/2}\right)\right]$, where $\nu = \frac{1}{\lambda}\sqrt{a^2 - 4c}$; $J_\nu(z)$ and $Y_\nu(z)$ are Bessel functions.

7. $y''_{xx} - (a - 2q\cosh 2x)y = 0.$

Modified Mathieu equation. The substitution $x = i\xi$ leads to the Mathieu equation S4.2.2.8:

$$y''_{\xi\xi} + (a - 2q\cos 2\xi)y = 0.$$

For eigenvalues $a = a_n(q)$ and $a = b_n(q)$, the corresponding solutions of the modified Mathieu equation are

$$\mathrm{Ce}_{2n+p}(x, q) = \mathrm{ce}_{2n+p}(ix, q) = \sum_{k=0}^{\infty} A_{2k+p}^{2n+p}\cosh[(2k + p)x],$$

$$\mathrm{Se}_{2n+p}(x, q) = -i\,\mathrm{se}_{2n+p}(ix, q) = \sum_{k=0}^{\infty} B_{2k+p}^{2n+p}\sinh[(2k + p)x],$$

where p can be either 0 or 1, and the coefficients A_{2k+p}^{2n+p} and B_{2k+p}^{2n+p} are specified in S4.2.2.8.

The modified Mathieu functions are discussed in the books by Bateman & Erdélyi (1955, vol. 3) and Abramowitz & Stegun (1964) in more detail.

8. $y''_{xx} + (a - 2q \cos 2x)y = 0$.

Mathieu equation.

1°. Given numbers a and q, there exists a general solution $y(x)$ and a characteristic index μ such that

$$y(x + \pi) = e^{2\pi\mu}y(x).$$

For small values of q, an approximate value of μ can be found from the equation

$$\cosh(\pi\mu) = 1 + 2\sin^2\left(\tfrac{1}{2}\pi\sqrt{a}\right) + \frac{\pi q^2}{(1-a)\sqrt{a}}\sin\left(\pi\sqrt{a}\right) + O(q^4).$$

If $y_1(x)$ is the solution of the Mathieu equation satisfying the initial conditions $y_1(0) = 1$ and $y'_1(0) = 0$, the characteristic index can be determined from the relation

$$\cosh(2\pi\mu) = y_1(\pi).$$

The solution $y_1(x)$, and hence μ, can be determined with any degree of accuracy by means of numerical or approximate methods.

The general solution differs depending on the value of $y_1(\pi)$ and can be expressed in terms of two auxiliary periodical functions $\varphi_1(x)$ and $\varphi_2(x)$ (see Table S4.3).

TABLE S4.3
The general solution of the Mathieu equation S4.2.2.8 expressed
in terms of auxiliary periodical functions $\varphi_1(x)$ and $\varphi_2(x)$.

Constraint	General solution $y = y(x)$	Period of φ_1 and φ_2	Index
$y_1(\pi) > 1$	$C_1 e^{2\mu x}\varphi_1(x) + C_2 e^{-2\mu x}\varphi_2(x)$	π	μ is a real number
$y_1(\pi) < -1$	$C_1 e^{2\rho x}\varphi_1(x) + C_2 e^{-2\rho x}\varphi_2(x)$	2π	$\mu = \rho + \tfrac{1}{2}i, \; i^2 = -1,$ ρ is the real part of μ
$\|y_1(\pi)\| < 1$	$(C_1 \cos\nu x + C_2 \sin\nu x)\varphi_1(x)$ $+(C_1 \cos\nu x - C_2 \sin\nu x)\varphi_2(x)$	π	$\mu = i\nu$ is a pure imaginary number, $\cos(2\pi\nu) = y_1(\pi)$
$y_1(\pi) = \pm1$	$C_1\varphi_1(x) + C_2 x\varphi_2(x)$	π	$\mu = 0$

2°. In applications, of major interest are periodical solutions of the Mathieu equation that exist for certain values of the parameters a and q (those values of a are referred to as eigenvalues). The most important solutions are listed in Table S4.4.

The Mathieu functions possess the following properties:

$$\mathrm{ce}_{2n}(x, -q) = (-1)^n \, \mathrm{ce}_{2n}\left(\frac{\pi}{2} - x, \, q\right), \qquad \mathrm{ce}_{2n+1}(x, -q) = (-1)^n \, \mathrm{se}_{2n+1}\left(\frac{\pi}{2} - x, \, q\right),$$

$$\mathrm{se}_{2n}(x, -q) = (-1)^{n-1} \, \mathrm{se}_{2n}\left(\frac{\pi}{2} - x, \, q\right), \qquad \mathrm{se}_{2n+1}(x, -q) = (-1)^n \, \mathrm{ce}_{2n+1}\left(\frac{\pi}{2} - x, \, q\right).$$

Selecting a sufficiently large m and omitting the term with the maximum number in the recurrence relations (indicated in Table S4.4), we can obtain approximate relations for the eigenvalues a_n (or b_n) with respect to parameter q. Then, equating the determinant of the corresponding homogeneous linear system of equations for coefficients A_m^n (or B_m^n) to zero, we obtain an algebraic equation for finding $a_n(q)$ (or $b_n(q)$).

TABLE S4.4

Periodical solutions of the Mathieu equation $ce_n = ce_n(x, q)$ and $se_n = se_n(x, q)$ (for odd n, the functions ce_n and se_n are 2π-periodical, and for even n, they are π-periodical); certain eigenvalues $a = a_n(q)$ and $b = b_n(q)$ correspond to each value of the parameter q; $n = 0, 1, 2, \ldots$.

Mathieu functions	Recurrence relations for coefficients	Normalization conditions
$ce_{2n}(x, q) = \sum\limits_{m=0}^{\infty} A_{2m}^{2n} \cos(2mx)$	$qA_2^{2n} = a_{2n} A_0^{2n}$; $qA_4^{2n} = (a_{2n}-4)A_2^{2n} - 2qA_0^{2n}$; $qA_{2m+2}^{2n} = (a_{2n}-4m^2)A_{2m}^{2n}$ $\quad - qA_{2m-2}^{2n}, \quad m \geq 2$	$(A_0^{2n})^2 + \sum\limits_{m=0}^{\infty}(A_{2m}^{2n})^2$ $= \begin{cases} 2 & \text{if } n=0 \\ 1 & \text{if } n \geq 1 \end{cases}$
$ce_{2n+1}(x, q) = \sum\limits_{m=0}^{\infty} A_{2m+1}^{2n+1} \cos[(2m+1)x]$	$qA_3^{2n+1} = (a_{2n+1}-1-q)A_1^{2n+1}$; $qA_{2m+3}^{2n+1} = [a_{2n+1}-(2m+1)^2]$ $\quad \times A_{2m+1}^{2n+1} - qA_{2m-1}^{2n+1}, \quad m \geq 1$	$\sum\limits_{m=0}^{\infty}(A_{2m+1}^{2n+1})^2 = 1$
$se_{2n}(x, q) = \sum\limits_{m=0}^{\infty} B_{2m}^{2n} \sin(2mx),$ $se_0 = 0$	$qB_4^{2n} = (b_{2n}-4)B_2^{2n}$; $qB_{2m+2}^{2n} = (b_{2n}-4m^2)B_{2m}^{2n}$ $\quad - qB_{2m-2}^{2n}, \quad m \geq 2$	$\sum\limits_{m=0}^{\infty}(B_{2m}^{2n})^2 = 1$
$se_{2n+1}(x, q) = \sum\limits_{m=0}^{\infty} B_{2m+1}^{2n+1} \sin[(2m+1)x]$	$qB_3^{2n+1} = (b_{2n+1}-1-q)B_1^{2n+1}$; $qB_{2m+3}^{2n+1} = [b_{2n+1}-(2m+1)^2]$ $\quad \times B_{2m+1}^{2n+1} - qB_{2m-1}^{2n+1}, \quad m \geq 1$	$\sum\limits_{m=0}^{\infty}(B_{2m+1}^{2n+1})^2 = 1$

For fixed real $q \neq 0$, the eigenvalues a_n and b_n are all real and different, while:

$$\text{if} \quad q > 0 \quad \text{then} \quad a_0 < b_1 < a_1 < b_2 < a_2 < \cdots;$$
$$\text{if} \quad q < 0 \quad \text{then} \quad a_0 < a_1 < b_1 < b_2 < a_2 < a_3 < b_3 < b_4 < \cdots$$

The eigenvalues possess the following properties:

$$a_{2n}(-q) = a_{2n}(q), \quad b_{2n}(-q) = b_{2n}(q), \quad a_{2n+1}(-q) = b_{2n+1}(q).$$

The solution of the Mathieu equation corresponding to eigenvalue a_n (or b_n) has n zeros on the interval $0 \leq x < \pi$ (q is a real number).

Listed below are two leading terms of asymptotic expansions of the Mathieu functions $ce_n(x, q)$ and $se_n(x, q)$, as well as of the corresponding eigenvalues $a_n(q)$ and $b_n(q)$, as $q \to 0$:

$$ce_0(x, q) = \frac{1}{\sqrt{2}}\left(1 - \frac{q}{2}\cos 2x\right), \quad a_0(q) = -\frac{q^2}{2} + \frac{7q^4}{128};$$

$$ce_1(x, q) = \cos x - \frac{q}{8}\cos 3x, \quad a_1(q) = 1 + q;$$

$$ce_2(x, q) = \cos 2x + \frac{q}{4}\left(1 - \frac{\cos 4x}{3}\right), \quad a_2(q) = 4 + \frac{5q^2}{12};$$

$$ce_n(x, q) = \cos nx + \frac{q}{4}\left[\frac{\cos(n+2)x}{n+1} - \frac{\cos(n-2)x}{n-1}\right], \quad a_n(q) = n^2 + \frac{q^2}{2(n^2-1)} \quad (n \geq 3);$$

$$se_1(x, q) = \sin x - \frac{q}{8}\sin 3x, \quad b_1(q) = 1 - q;$$

$$se_2(x, q) = \sin 2x - \frac{q}{12}\sin 4x, \quad b_2(q) = 4 - \frac{q^2}{12};$$

$$\mathrm{se}_n(x, q) = \sin nx - \frac{q}{4}\left[\frac{\sin(n+2)x}{n+1} - \frac{\sin(n-2)x}{n-1}\right], \quad b_n(q) = n^2 + \frac{q^2}{2(n^2-1)} \qquad (n \geq 3).$$

The Mathieu functions are discussed in the books by Bateman & Erdélyi (1955) and Abramowitz & Stegun (1964) in more detail.

9. $y''_{xx} + a \tan x\, y'_x + by = 0.$

1°. The substitution $\xi = \sin x$ leads to a linear equation of the form S4.2.1.21: $(\xi^2-1)y''_{\xi\xi} + (1-a)\xi y'_\xi - by = 0.$

2°. Solution for $a = -2$:

$$y \cos x = \begin{cases} C_1 \sin(kx) + C_2 \cos(kx) & \text{if } b+1 = k^2 > 0, \\ C_1 \sinh(kx) + C_2 \cosh(kx) & \text{if } b+1 = -k^2 < 0. \end{cases}$$

3°. Solution for $a = 2$ and $b = 3$: $\quad y = C_1 \cos^3 x + C_2 \sin x\, (1 + 2\cos^2 x).$

S4.2.3. Equations Involving Arbitrary Functions

▶ *Notation:* $f = f(x)$ and $g = g(x)$ are arbitrary functions; a, b, and λ are arbitrary parameters.

1. $y''_{xx} + fy'_x + a(f-a)y = 0.$
Particular solution: $y_0 = e^{-ax}.$

2. $y''_{xx} + xfy'_x - fy = 0.$
Particular solution: $y_0 = x.$

3. $xy''_{xx} + (xf+a)y'_x + (a-1)fy = 0.$
Particular solution: $y_0 = x^{1-a}.$

4. $xy''_{xx} + [(ax+1)f + ax - 1]y'_x + a^2xfy = 0.$
Particular solution: $y_0 = (ax+1)e^{-ax}.$

5. $xy''_{xx} + [(ax^2+bx)f + 2]y'_x + bfy = 0.$
Particular solution: $y_0 = a + b/x.$

6. $x^2y''_{xx} + xfy'_x + a(f-a-1)y = 0.$
Particular solution: $y_0 = x^{-a}.$

7. $y''_{xx} + (f + ae^{\lambda x})y'_x + ae^{\lambda x}(f+\lambda)y = 0.$
Particular solution: $y_0 = \exp\left(-\frac{a}{\lambda}e^{\lambda x}\right).$

8. $y''_{xx} - (f^2 + f'_x)y = 0.$
Particular solution: $y_0 = \exp\left(\int f\, dx\right).$

9. $y''_{xx} + 2fy'_x + (f^2 + f'_x)y = 0.$

Solution: $y = (C_2 x + C_1)\exp\left(-\int f\,dx\right).$

10. $y''_{xx} + (1-a)fy'_x - a(f^2 + f'_x)y = 0.$

Particular solution: $y_0 = \exp\left(a\int f\,dx\right).$

11. $y''_{xx} + fy'_x + (fg - g^2 + g'_x)y = 0.$

Particular solution: $y_0 = \exp\left(-\int g\,dx\right).$

12. $fy''_{xx} - af'_x y'_x - bf^{2a+1}y = 0.$

Solution: $y = C_1 e^u + C_2 e^{-u}$, where $u = \sqrt{b}\int f^a\,dx.$

13. $f^2 y''_{xx} + f(f'_x + a)y'_x + by = 0.$

The substitution $\xi = \int f^{-1}dx$ leads to a constant coefficient linear equation: $y''_{\xi\xi} + ay'_\xi + by = 0.$

14. $y''_{xx} - f'_x y'_x + a^2 e^{2f}y = 0.$

Solution: $y = C_1 \sin\left(a\int e^f\,dx\right) + C_2\cos\left(a\int e^f\,dx\right).$

15. $y''_{xx} - f'_x y'_x - a^2 e^{2f}y = 0.$

Solution: $y = C_1\exp\left(a\int e^f\,dx\right) + C_2\exp\left(-a\int e^f\,dx\right).$

S4.3. Second-Order Nonlinear Equations

S4.3.1. Equations of the Form $y''_{xx} = f(x, y)$

1. $y''_{xx} = f(y).$

Autonomous equation.

     Solution: $\int\left[C_1 + 2\int f(y)\,dy\right]^{-1/2} dy = C_2 \pm x.$

     Particular solutions: $y = A_k$, where A_k are roots of the algebraic (transcendental) equation $f(A_k) = 0.$

2. $y''_{xx} = Ax^n y^m.$

Emden–Fowler equation.

$1°$. With $m \neq 1$, the Emden–Fowler equation has a particular solution:

$$y = \lambda x^{\frac{n+2}{1-m}}, \quad \text{where } \lambda = \left[\frac{(n+2)(n+m+1)}{A(m-1)^2}\right]^{\frac{1}{m-1}}.$$

$2°$. The transformation $z = x^{n+2}y^{m-1}$, $w = xy'_x/y$ leads to a first-order (Abel) equation: $z[(m-1)w + n + 2]w'_z = -w^2 + w + Az.$

$3°$. The transformation $y = w/t$, $x = 1/t$ leads to the Emden–Fowler equation with the independent variable raised to a different power: $w''_{tt} = At^{-n-m-3}w^m$.

$4°$. The books by Zaitsev and Polyanin (1994) and Polyanin and Zaitsev (2003) present 28 solvable cases of the Emden–Fowler equation (corresponding to some pairs of n and m).

3. $y''_{xx} + f(x)y = ay^{-3}.$

Ermakov's equation. Let $w = w(x)$ be a nontrivial solution of the second-order linear equation $w''_{xx} + f(x)w = 0$. The transformation $\xi = \int \dfrac{dx}{w^2}$, $z = \dfrac{y}{w}$ leads to an autonomous equation of the form S4.3.1: $z''_{\xi\xi} = az^{-3}$.

Solution: $C_1 y^2 = aw^2 + w^2 \left(C_2 + C_1 \int \dfrac{dx}{w^2} \right)^2.$

▶ *Further on, f, g, h, and ψ are arbitrary composite functions of their arguments indicated in parentheses after the function name (the arguments can depend on x, y, y'_x).*

4. $y''_{xx} = f(ay + bx + c).$

The substitution $w = ay + bx + c$ leads to an equation of the form S4.3.1.1: $w''_{xx} = af(w)$.

5. $y''_{xx} = f(y + ax^2 + bx + c).$

The substitution $w = y + ax^2 + bx + c$ leads to an equation of the form S4.3.1.1: $w''_{xx} = f(w) + 2a$.

6. $y''_{xx} = x^{-1}f(yx^{-1}).$

Homogeneous equation. The transformation $t = -\ln|x|$, $z = y/x$ leads to an autonomous equation: $z''_{tt} - z'_t = f(z)$.

7. $y''_{xx} = x^{-3}f(yx^{-1}).$

The transformation $\xi = 1/x$, $w = y/x$ leads to the equation of the form S4.3.1.1: $w''_{\xi\xi} = f(w)$.

8. $y''_{xx} = x^{-3/2}f(yx^{-1/2}).$

Having set $w = yx^{-1/2}$, we obtain $\dfrac{d}{dx}(xw'_x)^2 = \tfrac{1}{2}ww'_x + 2f(w)w'_x$. Integrating the latter equation, we arrive at a separable equation.

Solution: $\int \left[C_1 + \tfrac{1}{4}w^2 + 2 \int f(w)\,dw \right]^{-1/2} dw = C_2 \pm \ln x.$

9. $y''_{xx} = x^{k-2}f(x^{-k}y).$

Generalized homogeneous equation. The transformation $z = x^{-k}y$, $w = xy'_x/y$ leads to a first-order equation: $z(w - k)w'_z = z^{-1}f(z) + w - w^2$.

10. $y''_{xx} = yx^{-2}f(x^n y^m).$

Generalized homogeneous equation. The transformation $z = x^n y^m$, $w = xy'_x/y$ leads to a first-order equation: $z(mw + n)w'_z = f(z) + w - w^2$.

11. $y''_{xx} = y^{-3} f \left(\dfrac{y}{\sqrt{ax^2 + bx + c}} \right).$

Setting $u(x) = y(ax^2 + bx + c)^{-1/2}$ and integrating the equation, we obtain a first-order separable equation:

$$(ax^2 + bx + c)^2 (u'_x)^2 = (\tfrac{1}{4}b^2 - ac)u^2 + 2 \int u^{-3} f(u) \, du + C_1.$$

12. $y''_{xx} = e^{-ax} f(e^{ax} y).$

The transformation $z = e^{ax} y$, $w = y'_x / y$ leads to a first-order equation: $z(w + a)w'_z = z^{-1} f(z) - w^2.$

13. $y''_{xx} = y f(e^{ax} y^m).$

The transformation $z = e^{ax} y^m$, $w = y'_x / y$ leads to a first-order equation: $z(mw + a)w'_z = f(z) - w^2.$

14. $y''_{xx} = x^{-2} f(x^n e^{ay}).$

The transformation $z = x^n e^{ay}$, $w = xy'_x$ leads to a first-order equation: $z(aw + n)w'_z = f(z) + w.$

15. $y''_{xx} = \dfrac{\psi''_{xx}}{\psi} y + \psi^{-3} f \left(\dfrac{y}{\psi} \right), \qquad \psi = \psi(x).$

The transformation $\xi = \int \dfrac{dx}{\psi^2}$, $w = \dfrac{y}{\psi}$ leads to an equation of the form S4.3.1.1: $w''_{\xi\xi} = f(w).$

Solution: $\displaystyle \int \left[C_1 + 2 \int f(w) \, dw \right]^{-1/2} dw = C_2 \pm \int \dfrac{dx}{\psi^2(x)}.$

S4.3.2. Equations of the Form $f(x, y)y''_{xx} = g(x, y, y'_x)$

1. $y''_{xx} - y'_x = f(y).$

Autonomous equation. The substitution $w(y) = y'_x$ leads to a first-order equation. For solvable equations of the form in question, see the book by Polyanin and Zaitsev (2003).

2. $y''_{xx} + f(y)y'_x + g(y) = 0.$

Lienard equation. The substitution $w(y) = y'_x$ leads to a first-order equation. For solvable equations of the form in question, see the book by Polyanin and Zaitsev (2003).

3. $y''_{xx} + [ay + f(x)]y'_x + f'_x(x)y = 0.$

Integrating yields a Riccati equation: $y'_x + f(x)y + \tfrac{1}{2}ay^2 = C.$

4. $y''_{xx} + [2ay + f(x)]y'_x + af(x)y^2 = g(x).$

On setting $u = y'_x + ay^2$, we obtain a first-order linear equation: $u'_x + f(x)u = g(x).$

5. $y''_{xx} = ay'_x + e^{2ax}f(y)$.

Solution: $\int \left[C_1 + 2 \int f(y)\,dy \right]^{-1/2} dy = C_2 \pm \dfrac{1}{a} e^{ax}$.

6. $y''_{xx} = f(y)y'_x$.

Solution: $\int \dfrac{dy}{F(y) + C_1} = C_2 + x$, where $F(y) = \int f(y)\,dy$.

7. $y''_{xx} = \left[e^{\alpha x} f(y) + \alpha \right] y'_x$.

The substitution $w(y) = e^{-\alpha x} y'_x$ leads to a first-order separable equation: $w'_y = f(y)$.

Solution: $\int \dfrac{dy}{F(y) + C_1} = C_2 + \dfrac{1}{\alpha} e^{\alpha x}$, where $F(y) = \int f(y)\,dy$.

8. $xy''_{xx} = ny'_x + x^{2n+1}f(y)$.

1°. Solution for $n \neq -1$:

$$\int \left[C_1 + 2 \int f(y)\,dy \right]^{-1/2} dy = \pm \dfrac{x^{n+1}}{n+1} + C_2.$$

2°. Solution for $n = -1$:

$$\int \left[C_1 + 2 \int f(y)\,dy \right]^{-1/2} dy = \pm \ln |x| + C_2.$$

9. $xy''_{xx} = f(y)y'_x$.

The substitution $w(y) = xy'_x/y$ leads to a first-order linear equation: $yw'_y = -w + 1 + f(y)$.

10. $xy''_{xx} = \left[x^k f(y) + k - 1 \right] y'_x$.

Solution: $\int \dfrac{dy}{F(y) + C_1} = C_2 + \dfrac{1}{k} x^k$, where $F(y) = \int f(y)\,dy$.

11. $x^2 y''_{xx} + xy'_x = f(y)$.

The substitution $x = \pm e^t$ leads to an autonomous equation of the form S4.3.1.1: $y''_{tt} = f(y)$.

12. $(ax^2 + b)y''_{xx} + axy'_x + f(y) = 0$.

The substitution $\xi = \int \dfrac{dx}{\sqrt{ax^2 + b}}$ leads to an autonomous equation of the form S4.3.1.1: $y''_{\xi\xi} + f(y) = 0$.

13. $y''_{xx} = f(y)y'_x + g(x)$.

Integrating yields a first-order equation: $y'_x = \int f(y)\,dy + \int g(x)\,dx + C$.

14. $xy''_{xx} + (n+1)y'_x = x^{n-1}f(yx^n)$.

The transformation $\xi = x^n$, $w = yx^n$ leads to an autonomous equation of the form S4.3.1.1: $n^2 w''_{\xi\xi} = f(w)$.

15. $gy''_{xx} + \frac{1}{2}g'_x y'_x = f(y), \qquad g = g(x).$

Integrating yields a first-order separable equation: $g(x)(y'_x)^2 = 2 \int f(y)\,dy + C_1$.
Solution for $g(x) \geq 0$:

$$\int \left[C_1 + 2 \int f(y)\,dy \right]^{-1/2} dy = C_2 \pm \int \frac{dx}{\sqrt{g(x)}}.$$

16. $y''_{xx} = -ay'_x + e^{ax} f(y e^{ax}).$

The transformation $\xi = e^{ax}$, $w = y e^{ax}$ leads to the equation $w''_{\xi\xi} = a^{-2} f(w)$, which is of the form of S4.3.1.1.

17. $xy''_{xx} = f(x^n e^{ay})y'_x.$

The transformation $z = x^n e^{ay}$, $w = xy'_x$ leads to the following first-order separable equation:
$z(aw + n)w'_z = [f(z) + 1]w.$

18. $x^2 y''_{xx} + xy'_x = f(x^n e^{ay}).$

The transformation $z = x^n e^{ay}$, $w = xy'_x$ leads to the following first-order separable equation:
$z(aw + n)w'_z = f(z).$

19. $yy''_{xx} + (y'_x)^2 + f(x)yy'_x + g(x) = 0.$

The substitution $u = y^2$ leads to a linear equation, $u''_{xx} + f(x)u'_x + 2g(x) = 0$, which can be reduced by the change of variable $w(x) = u'_x$ to a first-order linear equation.

20. $yy''_{xx} - (y'_x)^2 + f(x)yy'_x + g(x)y^2 = 0.$

The substitution $u = y'_x/y$ leads to a first-order linear equation: $u'_x + f(x)u + g(x) = 0.$

21. $yy''_{xx} - n(y'_x)^2 + f(x)y^2 + ay^{4n-2} = 0.$

$1°$. For $n = 1$, this is an equation of the form S4.3.2.22.

$2°$. For $n \neq 1$, the substitution $w = y^{1-n}$ leads to Ermakov's equation S4.3.1.5: $w''_{xx} + (1-n)f(x)w + a(1-n)w^{-3} = 0.$

22. $yy''_{xx} - n(y'_x)^2 + f(x)y^2 + g(x)y^{n+1} = 0.$

The substitution $w = y^{1-n}$ leads to a nonhomogeneous linear equation: $w''_{xx} + (1-n)f(x)w + (1-n)g(x) = 0.$

23. $yy''_{xx} + a(y'_x)^2 + f(x)yy'_x + g(x)y^2 = 0.$

The substitution $w = y^{a+1}$ leads to a linear equation: $w''_{xx} + f(x)w'_x + (a+1)g(x)w = 0.$

24. $yy''_{xx} = f(x)(y'_x)^2.$

The substitution $w(x) = xy'_x/y$ leads to a Bernoulli equation S4.1.4: $xw'_x = w + [f(x)-1]w^2.$

25. $y''_{xx} - a(y'_x)^2 + f(x)e^{ay} + g(x) = 0.$

The substitution $w = e^{-ay}$ leads to a nonhomogeneous linear equation: $w''_{xx} - ag(x)w = af(x)$.

26. $y''_{xx} - a(y'_x)^2 + be^{4ay} + f(x) = 0.$

The substitution $w = e^{-ay}$ leads to Ermakov's equation S4.3.1.5: $w''_{xx} - af(x)w = abw^{-3}$.

27. $y''_{xx} + a(y'_x)^2 - \frac{1}{2}y'_x = e^x f(y).$

The substitution $w(y) = e^{-x}(y'_x)^2$ leads to a first-order linear equation: $w'_y + 2aw = 2f(y)$.

28. $y''_{xx} + \alpha(y'_x)^2 = \left[e^{\beta x}f(y) + \beta\right]y'_x.$

Solution:
$$\int \frac{e^{\alpha y}\,dy}{F(y) + C_1} = C_2 + \frac{1}{\beta}e^{\beta x}, \qquad \text{where} \quad F(y) = \int e^{\alpha y}f(y)\,dy.$$

29. $y''_{xx} + f(y)(y'_x)^2 + g(y) = 0.$

The substitution $w(y) = (y'_x)^2$ leads to a first-order linear equation: $w'_y + 2f(y)w + 2g(y) = 0$.

30. $y''_{xx} + f(y)(y'_x)^2 - \frac{1}{2}y'_x = e^x g(y).$

The substitution $w(y) = e^{-x}(y'_x)^2$ leads to a first-order linear equation: $w'_y + 2f(y)w = 2g(y)$.

31. $y''_{xx} = xf(y)(y'_x)^3.$

Taking y to be the independent variable, we obtain a linear equation with respect to $x = x(y)$: $x''_{yy} = -f(y)x$.

32. $y''_{xx} = f(y)(y'_x)^2 + g(x)y'_x.$

Dividing by y'_x, we obtain an exact differential equation. Its solution follows from the equation:
$$\ln|y'_x| = \int f(y)\,dy + \int g(x)\,dx + C.$$

Solving the latter for y'_x, we arrive at a separable equation. In addition, $y = C_1$ is a singular solution, with C_1 being an arbitrary constant.

33. $y''_{xx} = f(x)g(xy'_x - y).$

The substitution $w = xy'_x - y$ leads to a first-order separable equation: $w'_x = xf(x)g(w)$.

34. $y''_{xx} = \frac{y}{x^2}f\left(\frac{xy'_x}{y}\right).$

The substitution $w(x) = xy'_x/y$ leads to a first-order separable equation: $xw'_x = f(w) + w - w^2$.

35. $gy''_{xx} + \frac{1}{2}g'_x y'_x = f(y)h(y'_x\sqrt{g}), \qquad g = g(x).$

The substitution $w(y) = y'_x\sqrt{g}$ leads to a first-order separable equation: $ww'_y = f(y)h(w)$.

36. $y''_{xx} = f(y'^2_x + ay).$

The substitution $w(y) = (y'_x)^2 + ay$ leads to a first-order separable equation: $w'_y = 2f(w) + a$.

Bibliography for Chapter S4

Abramowitz, M. and Stegun, I. A. (Editors), *Handbook of Mathematical Functions with Formulas, Graphs and Mathematical Tables*, National Bureau of Standards Applied Mathematics, Washington, 1964.

Bateman, H. and Erdélyi, A., *Higher Transcendental Functions*, Vol. 3, McGraw-Hill, New York, 1955.

Kamke, E., *Differentialgleichungen: Lösungsmethoden und Lösungen, I, Gewöhnliche Differentialgleichungen*, B. G. Teubner, Leipzig, 1977.

Murphy, G. M., *Ordinary Differential Equations and Their Solutions*, D. Van Nostrand, New York, 1960.

Polyanin, A. D. and Zaitsev, V. F., *Handbook of Exact Solutions for Ordinary Differential Equations, 2nd Edition*, Chapman & Hall/CRC Press, Boca Raton, Florida, 2003.

Zaitsev, V. F. and Polyanin, A. D., *Discrete-Group Methods for Integrating Equations of Nonlinear Mechanics*, CRC Press, Boca Raton, Florida, 1994.

Chapter S5
Some Useful Electronic Mathematical Resources

arXiv.org (http://arxiv.org). A service of automated e-print archives of articles in the fields of mathematics, nonlinear science, computer science, and physics.

Catalog of Mathematics Resources on the WWW and the Internet (http://mthwww.uwc.edu/wwwmahes/files/math01.htm).

CFD Codes List (http://www.fges.demon.co.uk/cfd/CFD_codes_p.html). Free software.

CFD Resources Online (http://www.cfd-online.com/Links). Software, modeling and numerics, etc.

Computer Handbook of ODEs (http://www.scg.uwaterloo.ca/ ecterrab/handbook_odes.html). An online computer handbook of methods for solving ordinary differential equations.

Deal.II (http://www.dealii.org). Finite element differential equations analysis library.

Dictionary of Algorithms and Data Structures—NIST (http://www.nist.gov/dads/). The dictionary of algorithms, algorithmic techniques, data structures, archetypical problems, and related definitions.

DOE ACTS Collection (http://acts.nersc.gov). The Advanced CompuTational Software (ACTS) Collection is a set of software tools for computation sciences.

EEVL: Internet Guide to Engineering, Mathematics and Computing (http://www.eevl.ac.uk). Cross-search 20 databases in engineering, mathematics, and computing.

EqWorld: World of Mathematical Equations (http://eqworld.ipmnet.ru). Extensive information on algebraic, ordinary differential, partial differential, integral, functional, and other mathematical equations.

FOLDOC—Computing Dictionary (http://foldoc.doc.ic.ac.uk/foldoc/index.html). The free on-line dictionary of computing is a searchable dictionary of terms from computing and related fields.

Free Software (http://www.wseas.com/software). Download free software packages for scientific-engineering purposes.

FSF/UNESCO Free Software Directory (http://directory.fsf.org).

GAMS: Guide to Available Mathematical Software (http://gams.nist.gov). A cross-index and virtual repository of mathematical and statistical software components of use in computational science and engineering.

Google—Mathematics Websites (http://directory.google.com/Top/Science/Math/). A directory of more than 11,000 mathematics Websites ordered by type and mathematical subject.

Google — Software (http://directory.google.com/Top/Science/Math/Software). A directory of software.

Mathcom—PDEs (http://www.mathcom.com/corpdir/techinfo.mdir/scifaq/q260.html). Partial differential equations and finite element modeling.

Mathematical Atlas (http://www.math-atlas.org). A collection of short articles designed to provide an introduction to the areas of modern mathematics.

Mathematical Constants and Numbers (http://numbers.computation.free.fr/Constants/constants.html). Mathematical, historical, and algorithmic aspects of some classical mathematical constants; easy and fast programs are also included.

Mathematical WWW Virtual Library (http://www.math.fsu.edu/Virtual/index.php). A directory of mathematics-related Websites ordered by type and mathematical subject.

Mathematics Archives (http://archives.math.utk.edu). Combined archive and directory of mathematics Websites, mailing lists, and teaching materials.

Mathematics Genealogy Project (http://www.genealogy.ams.org). A large biographical database of mathematicians.

Mathematics Websites (http://www.math.psu.edu/MathLists/Contents.html). A directory of mathematics Websites ordered by type and mathematical subject.

Math Forum: Internet Mathematics Library (http://mathforum.org). A directory of mathematics Websites ordered by the mathematical subject.

MathGuide: SUB Gottingen (http://www.mathguide.de). An Internet-based subject gateway to mathematics-related Websites.

MathWorld: World of Mathematics (http://www.mathworld.com). An online encyclopedia of mathematics, focusing on classical mathematics. The Web's most extensive mathematical resource.

MGNet (http://www.mgnet.org/mgnet-codes.html). Free software.

Netlib (http://www.netlib.org). A collection of mathematical software, papers, and databases.

Numerical Solutions (http://www.numericalmathematics.com/numerical_solutions.htm). A library of mathematical programs.

PlanetMath.Org (http://planetmath.org). An online mathematics encyclopedia.

Probability Web (http://www.mathcs.carleton.edu/probweb/probweb.html). A collection of probability resources on the World Wide Web; the pages are designed to be especially helpful to researchers and teachers.

Science Oxygen—Mathematics (http://www.scienceoxygen.com/math.html). Topics from various sections of mathematics.

Scilab (http://scilabsoft.inria.fr). A free scientific software package.

Software — Differential Equations (http://www.scicomp.uni-erlangen.de/SW/diffequ.html). General resources and methods for ODEs and PDEs.

S.O.S. Mathematics (http://www.sosmath.com). A free resource for math review material from algebra to differential equations.

Statistics Online Computational Resources (http://socr.stat.ucla.edu). Interactive distributions, statistical analysis, virtual probability-related experiments and demonstrations, computer games, and others.

Stat/Math Center (http://www.indiana.edu/statmath/bysubject/numerics.html). Numerical computing resources on the Internet.

UW-L Math Calculator (http://www.compute.uwlax.edu/index.php). Calculus, differential equations, numerical methods, statistics, and others.

Wikipedia: Free Encyclopedia—List of Open Source Software Packages (http://en.wikipedia.org/wiki/List_of_open-source_software_packages).

Wikipedia: Free Encyclopedia—Mathematics (http://en.wikipedia.org/wiki/Mathematics). A collection of short articles from various sections of mathematics.

Wolfram Functions Site (http://functions.wolfram.com). More than 87,000 formulas used by mathematicians, computer scientists, physicists, and engineers; more than 10,000 graphs and animations of the functions.

Yahoo—Mathematics Websites (http://dir.yahoo.com/science/mathematics/). A directory of mathematics Websites ordered by type and mathematical subject.

Yahoo—Software (http://dir.yahoo.com/Science/Mathematics/Software). A directory of software.

Chapter S6
Physical Tables

S6.1. Symbols and Units

S6.1.1. Greek Letters

Uppercase	Lowercase	Name	Uppercase	Lowercase	Name
A	α	Alpha	N	ν	Nu
B	β	Beta	Ξ	ξ	Xi
Γ	γ	Gamma	O	o	Omicron
Δ	δ	Delta	Π	π	Pi
E	ε, ϵ	Epsilon	P	ρ, ϱ	Rho
Z	ζ	Zeta	Σ	σ, ς	Sigma
H	η	Eta	T	τ	Tau
Θ	θ, ϑ	Theta	Y, Υ	υ	Upsilon
I	ι	Iota	Φ	ϕ, φ	Phi
K	$\kappa, \varkappa$	Kappa	X	χ	Chi
Λ	λ	Lambda	Ψ	ψ	Psi
M	μ	Mu	Ω	ω, ϖ	Omega

S6.1.2. SI Prefixes

Factor	Prefix	Symbol	Factor	Prefix	Symbol
10^1	deca	da	10^{-1}	deci	d
10^2	hecto	h	10^{-2}	centi	c
10^3	kilo	k	10^{-3}	milli	m
10^6	mega	M	10^{-6}	micro	μ
10^9	giga	G	10^{-9}	nano	n
10^{12}	tera	T	10^{-12}	pico	p
10^{15}	peta	P	10^{-15}	femto	f
10^{18}	exa	E	10^{-18}	atto	a
10^{21}	zetta	Z	10^{-21}	zepto	z
10^{24}	yotta	Y	10^{-24}	yocto	y

S6.1.3. SI Base Units

Name	Symbol	Quantity
metre	m	length
kilogram	kg	mass
second	s	time
ampere	A	electric current
kelvin	K	thermodynamic temperature
mole	mol	amount of substance
candela	cd	luminous intensity

S6.1.4. SI Derived Units with Special Names

Name	Symbol	Quantity	Expression in other units
hertz	Hz	frequency	$1/s$
radian	rad	angle	m/m (dimensionless)
steradian	sr	solid angle	m^2/m^2 (dimensionless)
newton	N	force, weight	$kg\,m/s^2$
pascal	Pa	pressure, stress	$N/m^2 = m^{-1}\,kg\,s^{-2}$
joule	J	energy, work, heat	$N\,m = W\,s = C\,V = m^2\,kg\,s^{-2}$
watt	W	power, radiant flux	$J/s = V\,A = m^2\,kg\,s^{-3}$
coulomb	C	electric charge, electric flux	$A\,s$
volt	V	voltage, electromotive force	$W/A = J/C = m^2\,kg\,s^{-3}\,A^{-1}$
farad	F	electric capacitance	$C/V = m^{-2}\,kg^{-1}\,s^4\,A^2$
ohm	Ω	electric resistance, impedance, reactance	$V/A = m^2\,kg\,s^{-3}\,A^{-2}$
siemens	S	electrical conductance	$1/\Omega = m^{-2}\,kg^{-1}\,s^3\,A^2$
weber	Wb	magnetic flux	$J/A = m^2\,kg\,s^{-2}\,A^{-1}$
tesla	T	magnetic field strength, magnetic flux density	$V\,s/m^2 = N/(A\,m) = Wb/m^2 = kg\,s^{-2}\,A^{-1}$
henry	H	inductance	$V\,s/A = Wb/A = m^2\,kg\,s^{-2}\,A^{-2}$
Celsius	°C, C	temperature	$K - 273.15$
lumen	lm	luminous flux	$lx\,m^2 = cd\,sr$
lux	lx	illuminance	$lm/m^2 = m^{-2}\,cd\,sr$
becquerel	Bq	radioactivity (decays per unit time)	$1/s$
gray	Gy	absorbed dose of ionizing radiation	$J/kg = m^2/s^2$
sievert	Sv	equivalent dose (of ionizing radiation)	$J/kg = m^2/s^2$
katal	kat	catalytic activity	mol/s

S6.1.5. Non-SI Units Accepted for Use with SI

Name	Symbol	Quantity	SI equivalent
minute	min	time	$1\,\text{min} = 60\,\text{s}$
hour	h	time	$1\,\text{h} = 60\,\text{min} = 3600\,\text{s}$
day	d	time	$1\,\text{d} = 24\,\text{h} = 1440\,\text{min} = 86400\,\text{s}$
degree of arc	°	angle	$1° = \frac{\pi}{180}\,\text{rad}$
minute of arc	′	angle	$1' = \frac{1}{60}° = \frac{\pi}{10800}\,\text{rad}$
second of arc	″	angle	$1'' = \frac{1}{60}' = \frac{1}{3600}° = \frac{\pi}{648000}\,\text{rad}$
hectare	ha	area	$1\,\text{ha} = 10000\,\text{m}^2$
litre	l, L	volume	$1\,\text{l} = 1\,\text{dm}^3 = 0.001\,\text{m}^3$
tonne	t	mass	$1\,\text{t} = 1000\,\text{kg}$

S6.1.6. Other Common Non-SI Units

Name	Symbol	Quantity	SI equivalent
fermi	fm	length	$1\,\text{fm} = 10^{-15}\,\text{m}$
ångström	Å	length	$1\,\text{Å} = 0.1\,\text{nm} = 10^{-10}\,\text{m}$
nautical mile	M	length	$1\,\text{M} = 1852\,\text{m}$
astronomical unit	AU	length	$1\,\text{AU} = 1.49598 \times 10^{11}\,\text{m}$
light-year	ly	length	$1\,\text{ly} = 9.46073 \times 10^{15}\,\text{m}$
parsec	pc	length	$1\,\text{pc} = 3.26156\,\text{ly} = 3.08568 \times 10^{16}\,\text{m}$
are	a	area	$1\,\text{a} = 1\,\text{dam}^2 = 100\,\text{m}^2$
barn	b	area	$1\,\text{b} = 10^{-28}\,\text{m}^2$
electronvolt	eV	energy	$1\,\text{eV} = 1.60218 \times 10^{-19}\,\text{J}$
electronvolt	eV	mass	$1\,\text{eV} = 1.78266 \times 10^{-36}\,\text{kg}$
atomic mass unit	AMU, u	mass	$1\,\text{AMU} = 1.66054 \times 10^{-27}\,\text{kg}$
dalton	Da	mass	$1\,\text{Da} = 1\,\text{AMU} = 1.66054 \times 10^{-27}\,\text{kg}$
bar	bar	pressure	$1\,\text{bar} = 10^5\,\text{Pa}$
millibar	mbar	pressure	$1\,\text{mbar} = 1\,\text{hPa} = 100\,\text{Pa}$
atmosphere	atm	pressure	$1\,\text{atm} = 1.01325 \times 10^5\,\text{Pa}$
millimetre of mercury	mmHg	pressure	$1\,\text{mmHg} = 133.322\,\text{Pa}$
decibel	dB	logarithmic ratio	

S6.1.7. Base Planck Units

The base Planck units are those of length, mass, time, electric charge, and temperature; these are derived from five fundamental physical constants: speed of light in vacuum (c), gravitational constant (G), reduced Planck constant ($\hbar$), Coulomb constant ($k_e = 1/(4\pi\varepsilon_0)$), and Boltzmann constant (k_B).

Name	Expression	Value in SI units
Planck length	$l_P = \sqrt{\hbar G/c^3}$	1.616×10^{-35} m
Planck mass	$m_P = \sqrt{\hbar c/G}$	2.176×10^{-8} kg
Planck time	$t_P = l_P/c = \sqrt{\hbar G/c^5}$	5.391×10^{-44} s
Planck charge	$q_P = \sqrt{4\pi\varepsilon_0 \hbar c}$	$1.875\,546 \times 10^{-18}$ C
Planck temperature	$T_P = m_P c^2/k_B = \sqrt{\hbar c^5/(Gk_B^2)}$	1.417×10^{32} K

S6.2. Physical Constants

S6.2.1. Universal Constants

Quantity	Symbol	Value
Gravitational constant	G, γ	$6.674(28 \pm 67) \times 10^{-11}$ N m^2/kg^2
Speed of light in vacuum	c	$299{,}792{,}458$ m/s
Planck constant	h	$6.626\,069 \times 10^{-34}$ J s
Reduced Planck constant	$\hbar = h/(2\pi)$	$1.054\,572 \times 10^{-34}$ J s

S6.2.2. Astronomical Constants

Name	Symbol	Value
Earth's standard surface gravity	g, g_n, g_0	9.80665 m/s^2
Earth mass	$M_\oplus, M_E$	5.9737×10^{24} kg
Mean Earth density	$\rho_\oplus, \rho_E$	5.515×10^3 kg/m^3
Mean Earth radius	$R_\oplus, R_E$	6.371×10^6 m
Mean Earth–Moon distance		3.844×10^8 m
Mean Earth–Sun distance		1.496×10^{11} m = 1 AU
Solar mass	$M_\odot, M_S$	1.989×10^{30} kg
Mean Solar radius	$R_\odot, R_S$	6.96×10^8 m
Lunar mass	$M_{\mathbb{C}}, M_L$	7.348×10^{22} kg
Mean Lunar radius	$R_{\mathbb{C}}, R_L$	1.737×10^6 m

S6.2.3. Planetary, Lunar, and Solar Parameters

Body	Equatorial radius, km	Mass, $\times 10^{24}$ kg	Mean density, g/cm^3	Mean distance from Sun, AU	Equatorial surface gravity, m/s^2
Mercury	2,440 (0.383 $R_\oplus$)	0.330 (0.055 $M_\oplus$)	5.427 (0.984 $\rho_\oplus$)	0.3871	3.7 (0.38 g)
Venus	6,052 (0.950 $R_\oplus$)	4.869 (0.815 $M_\oplus$)	5.243 (0.951 $\rho_\oplus$)	0.7233	8.87 (0.91 g)
Earth	6,378 (1.001 $R_\oplus$)	5.9737 ($M_\oplus$)	5.515 ($\rho_\oplus$)	1.0000	9.766 (0.996 g)
Earth's Moon	1,737 (0.273 $R_\oplus$)	0.073 (0.012 $M_\oplus$)	3.341 (0.606 $\rho_\oplus$)	0.002570*	1.622 (0.165 g)
Mars	3,397 (0.533 $R_\oplus$)	0.642 (0.107 $M_\oplus$)	3.933 (0.714 $\rho_\oplus$)	1.5237	3.693 (0.378 g)
Jupiter	71,492 (11.221 $R_\oplus$)	1,898 (317.8 $M_\oplus$)	1.326 (0.241 $\rho_\oplus$)	5.2034	20.87 (2.128 g)
Saturn	60,268 (9.460 $R_\oplus$)	568.5 (95.16 $M_\oplus$)	0.70 (0.127 $\rho_\oplus$)	9.5371	10.40 (1.060 g)
Uranus	25,559 (4.012 $R_\oplus$)	86.85 (14.37 $M_\oplus$)	1.30 (0.236 $\rho_\oplus$)	19.191	8.43 (0.860 g)
Neptune	24,764 (3.887 $R_\oplus$)	102.4 (17.15 $M_\oplus$)	1.76 (0.317 $\rho_\oplus$)	30.069	10.71 (1.092 g)
Pluto**	1,151 (0.181 $R_\oplus$)	0.013 (0.002 $M_\oplus$)	2 (0.4 $\rho_\oplus$)	39.482	0.81 (0.083 g)
Sun	695,500 (109 $R_\oplus$)	332,900 $M_\oplus$	1.409 (0.255 $\rho_\oplus$)		274.0 (27.9 g)

* This is the average Moon–Earth distance. The Moon's average distance to the Sun is 1 AU, the same as the Earth's.
** In 2006, Pluto was reclassified as a dwarf planet after the discovery of similar objects.

S6.2.4. Electromagnetic Constants

Quantity	Symbol	Value
Magnetic constant	μ_0	$4\pi \times 10^{-7}$ N/A^2 = $1.256\,637 \times 10^{-6}$ H/m
Electric constant	$\varepsilon_0 = 1/(\mu_0 c^2)$	$8.854\,188 \times 10^{-12}$ F/m
Coulomb's law coefficient	$k_e = 1/(4\pi\varepsilon_0)$, k	$8.987\,552 \times 10^9$ m/F
Biot–Savart law coefficient	$\mu_0/(4\pi)$	10^{-7} H/m
Elementary charge	e	$1.602\,176 \times 10^{-19}$ C
Bohr magneton	$\mu_B = e\hbar/(2m_e)$	$9.274\,009 \times 10^{-24}$ J/T
Nuclear magneton	$\mu_N = e\hbar/(2m_p)$	$5.050\,783 \times 10^{-27}$ J/T

S6.2.5. Atomic and Nuclear Constants

Quantity	Symbol	Value
Bohr radius	$a_0 = \hbar^2/(m_e k_e e^2)$	$5.291\,772 \times 10^{-11}$ m
Classical electron radius	$r_e = e^2/(4\pi\varepsilon_0 m_e c^2)$	$2.817\,940 \times 10^{-15}$ m
Electron rest mass	m_e	$9.109\,382 \times 10^{-31}$ kg
Proton rest mass	m_p	$1.672\,622 \times 10^{-27}$ kg
Neutron rest mass	m_n	$1.674\,927 \times 10^{-27}$ kg

Quantity	Symbol	Value
Electron rest energy	$m_e c^2$	510.999 keV
Proton rest energy	$m_p c^2$	938.272 MeV
Neutron rest energy	$m_n c^2$	939.566 MeV
Proton–electron mass ratio	m_p/m_e	1836.153
Specific charge of electron	$-e/m_e$	$-1.758\,820 \times 10^{11}$ C/kg
Specific charge of proton	e/m_p	$9.578\,834 \times 10^7$ C/kg
Magnetic moment of electron	μ_e	$-1.001\,160\,\mu_B$
Magnetic moment of proton	μ_p	$2.792\,847\,\mu_N$
Fine-structure constant	$\alpha = e^2/(4\pi\varepsilon_0\hbar c)$	$7.297\,353 \times 10^{-3}$
Rydberg constant (reciprocal wavelength)	$R_\infty = m_e e^4/(8\varepsilon_0^2 h^3 c)$	$1.097\,373 \times 10^7$ m^{-1}
Rydberg constant (electron energy)	$R_E = hcR_\infty$	13.605 69 eV = 1 Ry
Rydberg constant (frequency)	$R_\nu = cR_\infty$	$3.289\,842 \times 10^{15}$ s^{-1}
Rydberg constant (angular frequency)	$R_\omega = 2\pi cR_\infty$	$2.067\,069 \times 10^{16}$ s^{-1}
Compton wavelength of electron	$\lambda_e = h/(m_e c)$	$2.426\,310 \times 10^{-12}$ m
Compton wavelength of proton	$\lambda_p = h/(m_p c)$	$1.321\,410 \times 10^{-15}$ m

S6.2.6. Other Constants

Quantity	Symbol	Value
Avogadro's number	N_A	$6.022\,14 \times 10^{23}$ mol^{-1}
Gas constant	R	8.3144 J/(mol K)
Boltzmann constant	$k = R/N_A,\ k_B$	$1.380\,65 \times 10^{-23}$ J/K
Stefan–Boltzmann constant	$\sigma = \pi^2 k^4/(60\,\hbar^3 c^2)$	5.6704×10^{-8} W/(m^2 K^4)
Wien constant	b	$2.897\,76 \times 10^{-3}$ m K
Hubble constant	H_0	70.8 ± 4.0 (km/s)/Mpc

S6.3. Nuclear Physics

S6.3.1. Masses of Light Nuclides

Z	Nuclide	$M - A$, AMU	Z	Nuclide	$M - A$, AMU
0	n	0.00867	6	^{11}C	0.01143
1	^1H	0.00783	6	^{12}C	0
1	^2H	0.01410	6	^{13}C	0.00335
1	^3H	0.01605	7	^{13}N	0.00574
2	^3He	0.01603	7	^{14}N	0.0037
2	^4He	0.00260	7	^{15}N	0.00011
3	^6Li	0.01513	8	^{15}O	0.00307
3	^7Li	0.01601	8	^{16}O	−0.00509
4	^7Be	0.01693	8	^{17}O	−0.00087
4	^8Be	0.00531	9	^{19}F	−0.00160
4	^9Be	0.01219	10	^{20}Ne	−0.00756
4	^{10}Be	0.01354	11	^{23}Na	−0.01023
5	^{10}B	0.01294	11	^{24}Na	−0.00903
5	^{11}B	0.00930	12	^{24}Mg	−0.01496

Notation: Z charge number, M particle mass, A mass number.

S6.3.2. Leptons

Leptons are fermions, i.e., spin-$1/2$ elementary particles. They all have baryon number 0 and form three generations: electronic, comprising electrons (e^-) and electron neutrinos (ν_e), muonic, comprising muons (μ^-) and muon neutrinos (ν_μ), and tauonic, comprising tauons (τ^-) and tauon neutrinos (ν_τ). Every lepton has a corresponding antilepton. The main properties of leptons are listed below.

Lepton	Mass, MeV/c^2	Q	L_ν	L_μ	L_τ	Mean lifetime, s	Common decay modes
$\nu_e, \widetilde{\nu}_e$	$< 8 \times 10^{-6}$	0	±1	0	0	unknown	
$\nu_\mu, \widetilde{\nu}_\mu$	< 0.27	0	0	±1	0	unknown	
$\nu_\tau, \widetilde{\nu}_\tau$	< 31	0	0	0	±1	unknown	
e^-, e^+	0.51100	∓1	±1	0	0	stable	stable
μ^-, μ^+	105.658	∓1	0	±1	0	2.1970×10^{-6}	$\mu^- \to e^- + \widetilde{\nu}_e + \nu_\mu$ $\mu^+ \to e^+ + \nu_e + \widetilde{\nu}_\mu$
τ^-, τ^+	1,777	∓1	0	0	±1	2.91×10^{-13}	$\tau^- \to e^- + \widetilde{\nu}_e + \nu_\tau,$ $\mu^- + \widetilde{\nu}_\mu + \nu_\tau,$ $\pi^- + \nu_\tau, \; \ldots$ $\tau^+ \to e^+ + \nu_e + \widetilde{\nu}_\tau, \; \ldots$

Notation: Q electric charge (in e), $L_{\nu,\mu,\tau}$ lepton numbers. The tilde denotes an antiparticle.

S6.3.3. Quarks

Quarks combine to form composite particles—hadrons (baryons and mesons). All quarks are fermions and have spin $1/2$ and baryon number $1/3$; they have six flavors and form three generations: up and down (first), charm and strange (second), and top and bottom (third). Every quark flavor has a corresponding antiquark.

Quark	Q	I	I_z	C	S	T	B	Mass, GeV/c^2
u (up)	$+2/3$	$1/2$	$+1/2$	0	0	0	0	0.0015–0.0033
d (down)	$-1/3$	$1/2$	$-1/2$	0	0	0	0	0.0035–0.0060
c (charm)	$+2/3$	0	0	1	0	0	0	1.160–1.340
s (strange)	$-1/3$	0	0	0	-1	0	0	0.0899–0.0949
t (top)	$+2/3$	0	0	0	0	1	0	169.1–173.3
b (bottom)	$-1/3$	0	0	0	0	0	-1	4.130–4.370

Notation: Q electric charge (in e), I isospin, I_z isospin projection, C charm, S strangeness, T topness, B bottomness.

S6.3.4. Elementary Bosons

Elementary bosons are bosonic (integer-spin) elementary particles that act as carriers of the fundamental forces: electromagnetic (photons), weak (Z and W bosons), strong (gluons), and gravitational (gravitons, currently hypothetical). Also the existence of the Higgs boson is theoretically predicted, which is thought to be the mediator of mass.

Particle	Q, e	Mass, GeV/c^2	S	Width, GeV
Photon, γ	0	0	1	stable
Z^0	0	91.19	1	2.5
W^+, W^-	± 1	80.4	1	2.1
Gluon, g	0	0	1	
Graviton, g, G	0	0	2	stable
Higgs boson, H^0	0	115–150	0	

Notation: Q electric charge, S spin quantum number.

S6.3.5. Some Baryons

Baryons are the family of composite particles made of three quarks. Baryons have half-integer spins and, hence, are fermions. They have lepton number 0 and baryon number 1. Each baryon has a corresponding antibaryon, where each quark is replaced by the corresponding antiquark.

Particle	Quark content	Mass, MeV/c^2	Q	J^P	S	C	B	I	I_z	Mean lifetime
p	uud	938.272	+1	$1/2^+$	0	0	0	$1/2$	$1/2$	$> 2.1 \times 10^{29}$ yr
n	udd	939.566	0	$1/2^+$	0	0	0	$1/2$	$-1/2$	886 s
Λ^0	uds	1115.7	0	$1/2^+$	-1	0	0	0	0	2.6×10^{-10} s
Σ^+	uus	1189.4	+1	$1/2^+$	-1	0	0	1	1	8.0×10^{-11} s
Σ^0	uds	1192.6	0	$1/2^+$	-1	0	0	1	0	7×10^{-20} s
Σ^-	dds	1197.4	-1	$1/2^+$	-1	0	0	1	-1	1.5×10^{-10} s
Ξ^0	uss	1315	0	$1/2^+$	-2	0	0	$1/2$	$1/2$	2.9×10^{-10} s
Ξ^-	dss	1322	-1	$1/2^+$	-2	0	0	$1/2$	$-1/2$	1.6×10^{-10} s
Δ^-	ddd	1232	-1	$3/2^+$	0	0	0	$3/2$	$-3/2$	5.6×10^{-24} s
Δ^0	udd	1232	0	$3/2^+$	0	0	0	$3/2$	$-1/2$	5.6×10^{-24} s
Δ^+	uud	1232	+1	$3/2^+$	0	0	0	$3/2$	$1/2$	5.6×10^{-24} s
Δ^{++}	uuu	1232	+2	$3/2^+$	0	0	0	$3/2$	$3/2$	5.6×10^{-24} s
Ω^-	sss	1672	-1	$3/2^+$	-3	0	0	0	0	8×10^{-11} s
Λ_c^+	udc	2286	+1	$1/2^+$	0	1	0	0	0	2.0×10^{-13} s
Σ_c^0	ddc	2454	0	$1/2^+$	0	1	0	1	-1	3×10^{-13} s
Ξ_c^0	dsc	2471	0	$1/2^+$	-1	1	0	$1/2$	$-1/2$	10^{-13} s
Ω_c^0	ssc	2697	0	$1/2^+$	-2	1	0	0	0	7×10^{-14} s
Λ_b^0	udb	5620	0	$1/2^+$	0	0	-1	0	0	1.4×10^{-12} s

Notation: Q electric charge (e), J total angular momentum number, S strangeness, C charm, B bottomness, I isospin, I_z isospin projection.

S6.3.6. Some Mesons

Mesons are the family of composite particles made of a quark and an antiquark. Mesons have integer spins and, hence, are bosons. They have lepton number 0 and baryon number 0. Each meson has a corresponding antimeson, where each quark is replaced by the corresponding antiquark.

Particle	Quark content	Mass, MeV/c^2	Q	J^P	S	C	B	I	I_z	Mean lifetime, s
π^+, π^-	$u\widetilde{d}, d\widetilde{u}$	139.57	± 1	0^-	0	0	0	1	± 1	2.60×10^{-8}
π^0	$\frac{1}{\sqrt{2}}(u\widetilde{u} - d\widetilde{d})$	134.98	0	0^-	0	0	0	1	0	8.4×10^{-17}
η	$\frac{1}{\sqrt{6}}(u\widetilde{u} + d\widetilde{d} - 2s\widetilde{s})$	547.8	0	0^-	0	0	0	0	0	5.0×10^{-19}
η'	$\frac{1}{\sqrt{3}}(u\widetilde{u} + d\widetilde{d} + s\widetilde{s})$	957.7	0	0^-	0	0	0	0	0	3.2×10^{-21}

Particle	Quark content	Mass, MeV/c^2	Q	J^P	S	C	B	I	I_z	Mean lifetime, s
K^+, K^-	$u\widetilde{s}$, $\widetilde{u}s$	493.7	±1	0^-	±1	0	0	$1/2$	±$1/2$	1.24×10^{-8}
K^0, $\widetilde{K}^0$	$d\widetilde{s}$, $\widetilde{d}s$	497.7	0	0^-	±1	0	0	$1/2$	±$1/2$	8.9×10^{-11}, 5.2×10^{-8}
ρ^+, ρ^-	$u\widetilde{d}$, $d\widetilde{u}$	775.4	±1	1^-	0	0	0	1	±1	$\sim 4.5 \times 10^{-24}$
ρ^0	$\frac{1}{\sqrt{2}}(u\widetilde{u}-d\widetilde{d})$	775.5	0	1^-	0	0	0	1	0	$\sim 4.5 \times 10^{-24}$
ω	$\frac{1}{\sqrt{2}}(u\widetilde{u}+d\widetilde{d})$	782.6	0	1^-	0	0	0	0	0	7.7×10^{-23}
φ	$s\widetilde{s}$	1019.4	0	1^-	0	0	0	0	0	1.5×10^{-22}
K^{*+}, K^{*-}	$u\widetilde{s}$, $\widetilde{u}s$	891.7	±1	1^-	±1	0	0	$1/2$	±$1/2$	$\sim 7.3 \times 10^{-20}$
K^{*0}, $\widetilde{K}^{*0}$	$d\widetilde{s}$, $\widetilde{d}s$	896.0	0	1^-	±1	0	0	$1/2$	±$1/2$	7.3×10^{-20}
D^+, D^-	$\widetilde{d}c$, $d\widetilde{c}$	1869.6	±1	0^-	0	±1	0	$1/2$	$\mp 1/2$	1.04×10^{-12}
D^0, $\widetilde{D}^0$	$\widetilde{u}c$, $u\widetilde{c}$	1864.8	0	0^-	0	±1	0	$1/2$	$\mp 1/2$	4.10×10^{-13}
D_s^+, D_s^-	$c\widetilde{s}$, $\widetilde{c}s$	1968.5	±1	0^-	±1	±1	0	0	0	5.0×10^{-13}
B^+, B^-	$u\widetilde{b}$, $\widetilde{u}b$	5279	±1	0^-	0	0	±1	$1/2$	±$1/2$	1.64×10^{-12}
B^0, $\widetilde{B}^0$	$d\widetilde{b}$, $\widetilde{d}b$	5280	0	0^-	0	0	±1	$1/2$	±$1/2$	1.53×10^{-12}
η_c	$c\widetilde{c}$	2980	0	0^-	0	0	0	0	0	2.5×10^{-23}
J/ψ	$c\widetilde{c}$	3096.9	0	1^-	0	0	0	0	0	7.1×10^{-21}
Υ	$b\widetilde{b}$	9460	0	1^-	0	0	0	0	0	1.2×10^{-20}

Notation: Q electric charge (e), J total angular momentum number, P parity, S strangeness, C charm, B bottomness, I isospin, I_z isospin projection.

Bibliography for Chapter S6

Benenson, W., Harris, John W., Stocker, H., and Lutz, H. (Eds.), *Handbook of Physics*, Springer, New York, 2002.

Forsythe, W. E., *Smithsonian Physical Tables, 9th Revised Edition*, Smithsonian Institution Press, Washington, 1954; Knovel, New York, 2003.

Griffiths, D., *Introduction to Elementary Particles, 2nd Edition*, Wiley-VCH Verlag GmbH & Co. KgaA, Weinheim, 2008.

Haynes, W. M. (Ed.), *CRC Handbook of Chemistry and Physics, 91st Edition*, CRC Press, Boca Raton, Florida, 2010.

Nordling, C. and Osterman, J.,, *Physics Handbook for Science and Engineering*, Studentlitteratur, Lund, Sweden, 2003.

Prokhorov, A. M. (Ed.), *Encyclopedic Dictionary of Physics* [in Russian], Sovetskaya Encyclopedia, Moscow, 1984 (dic.academic.ru).

Prokhorov, A. M. (Ed.), *Encyclopedia of Physics, Volumes 1–5* [in Russian], Bolshaya Russkaya Encyclopedia, Moscow, 1998.

Chapter S7

Periodic Table

The *periodic table of the chemical elements* (also known as *Mendeleev's table*) illustrates recurring ("periodic") trends in the properties of the elements. Currently (as of May 2010), the standard table contains 118 elements; see next page.

The elements are listed in order of increasing *atomic number* (the number of protons in the atomic nucleus). Each element has a *name* and a *symbol* and is listed together with its *standard atomic weight*—the relative atomic mass, or the ratio of the average mass per atom of the element to 1/12 of the mass of an atom of ^{12}C, calculated for all naturally occurring isotopes of the element. If an element has no stable isotopes, its standard atomic weight is taken to be the atomic mass number (the total number of protons and neutrons in the atomic nucleus) of the longest-lived isotope.

The elements are arranged in 18 groups (vertical columns) and 7 periods (horizontal rows). In most groups, the elements have very similar properties that exhibit a clear trend. Many groups are given unsystematic names such as the *alkali metals* (group 1, except hydrogen), *alkaline earth metals* (group 2), *transition metals* (groups 3–11), *pnictogens* (group 15), *chalcogens* (group 16), *halogens* (group 17), and *noble gases* (group 18). Some regions of the periodic table show horizontal trends in properties, which is true for the transition metals and especially for the *lanthanides* (group 3, period 6) and *actinides* (group 3, period 7).

The chemical properties of an element are primarily determined by its *electron config-uration*—the distribution of electrons in atomic orbitals (allowed quantum states). The set of orbitals corresponding to the same principal quantum number, n, is called an *electron shell*. The shells corresponding to $n = 1, 2, 3, \ldots$ are denoted by K, L, M, $\ldots$ (K is the innermost shell). Each shell is divided into *subshells* corresponding to the azimuthal quantum number, l ($l = 0, \ldots, n-1$), and the subshells are labeled s, p, d, f, g, h, etc. Each subshell can accommodate up to a maximum of $2(2l+1)$ electrons and the maximum number of electrons in the nth shell is $2n^2$. The electrons in an atom occupy orbitals starting from the lowest-energy orbital.

The number of electrons in an atom determines its electron configuration. For example, sodium Na has 11 electrons and its electron configuration is written as $1s^2 2s^2 2p^6 3s$, which means that there are 2 electrons in the K-shell, 8 electrons in the L-shell with 2 in the s-subshell and 6 in the p-subshell, and 1 electron in the s-subshell of the M-shell. The electron configuration of Na can be written as [Ne] $3s$ for short, where [Ne] stands for that of Ne ($1s^2 2s^2 2p^6$).

The order in which orbitals are filled by the electrons is given by *Madelung's rule* (also known as the $n + l$ rule): (i) orbitals are filled in order of increasing $n + l$, (ii) if two orbitals have the same value of $n + l$, they are filled in order of increasing n. This rule is graphically illustrated in Fig. S7.1. According to the rule, the orbitals are filled in the order $1s\, 2s\, 2p\, 3s\, 3p\, 4s\, 3d\, 4p\, 5s\, 4d\, 5p\, 6s\, 4f\, 5d\, 6p\, 7s\, 5f\, 6d\, 7p$ for the 118 elements of the periodic table. Madelung's rule was established experimentally, it applies only to neutral atoms in their ground state, and it works very well for most of the elements. However, there are a few exceptions; for example, Cr should have four electrons in the $3d$ orbital but has five,

Periodic Table of the Chemical Elements

Legend:

Atomic number — Symbol — Atomic weight — Element name

Example:
4 — 9.0122 — **Be** — BERYLLIUM

Atomic mass number of the most stable isotope:
43 — (98) — **Tc** — TECHNETIUM

Physical state (300 K, 101 kPa):
- **Ti** solid
- **Br** liquid
- **Ne** gas

Natural occurrence:
- **Mn** primordial
- **Tc** from decay
- **Bh** synthetic

Group 1	2	3	4	5	6	7	8	9	10	11	12	13	14	15	16	17	18
1 1.0079 **H** HYDROGEN																	2 4.0026 **He** HELIUM
3 6.941 **Li** LITHIUM	4 9.0122 **Be** BERYLLIUM											5 10.811 **B** BORON	6 12.011 **C** CARBON	7 14.007 **N** NITROGEN	8 15.999 **O** OXYGEN	9 18.998 **F** FLUORINE	10 20.180 **Ne** NEON
11 22.990 **Na** SODIUM	12 24.305 **Mg** MAGNESIUM											13 26.982 **Al** ALUMINIUM	14 28.086 **Si** SILICON	15 30.974 **P** PHOSPHORUS	16 32.065 **S** SULPHUR	17 35.453 **Cl** CHLORINE	18 39.948 **Ar** ARGON
19 39.098 **K** POTASSIUM	20 40.078 **Ca** CALCIUM	21 44.956 **Sc** SCANDIUM	22 47.867 **Ti** TITANIUM	23 50.942 **V** VANADIUM	24 51.996 **Cr** CHROMIUM	25 54.938 **Mn** MANGANESE	26 55.845 **Fe** IRON	27 58.933 **Co** COBALT	28 58.693 **Ni** NICKEL	29 63.546 **Cu** COPPER	30 65.39 **Zn** ZINC	31 69.723 **Ga** GALLIUM	32 72.64 **Ge** GERMANIUM	33 74.922 **As** ARSENIC	34 78.96 **Se** SELENIUM	35 79.904 **Br** BROMINE	36 83.80 **Kr** KRYPTON
37 85.468 **Rb** RUBIDIUM	38 87.62 **Sr** STRONTIUM	39 88.906 **Y** YTTRIUM	40 91.224 **Zr** ZIRCONIUM	41 92.906 **Nb** NIOBIUM	42 95.94 **Mo** MOLYBDENUM	43 (98) **Tc** TECHNETIUM	44 101.07 **Ru** RUTHENIUM	45 102.91 **Rh** RHODIUM	46 106.42 **Pd** PALLADIUM	47 107.87 **Ag** SILVER	48 112.41 **Cd** CADMIUM	49 114.82 **In** INDIUM	50 118.71 **Sn** TIN	51 121.76 **Sb** ANTIMONY	52 127.60 **Te** TELLURIUM	53 126.90 **I** IODINE	54 131.29 **Xe** XENON
55 132.91 **Cs** CAESIUM	56 137.33 **Ba** BARIUM	57-71 **La-Lu** *	72 178.49 **Hf** HAFNIUM	73 180.95 **Ta** TANTALUM	74 183.84 **W** TUNGSTEN	75 186.21 **Re** RHENIUM	76 190.23 **Os** OSMIUM	77 192.22 **Ir** IRIDIUM	78 195.08 **Pt** PLATINUM	79 196.97 **Au** GOLD	80 200.59 **Hg** MERCURY	81 204.38 **Tl** THALLIUM	82 207.2 **Pb** LEAD	83 208.98 **Bi** BISMUTH	84 (209) **Po** POLONIUM	85 (210) **At** ASTATINE	86 (222) **Rn** RADON
87 (223) **Fr** FRANCIUM	88 (226) **Ra** RADIUM	89-103 **Ac-Lr** **	104 (267) **Rf** RUTHERFORDIUM	105 (268) **Db** DUBNIUM	106 (271) **Sg** SEABORGIUM	107 (270) **Bh** BOHRIUM	108 (269) **Hs** HASSIUM	109 (278) **Mt** MEITNERIUM	110 (281) **Ds** DARMSTADTIUM	111 (281) **Rg** ROENTGENIUM	112 (285) **Cn** COPERNICIUM	113 (286) **Uut** UNUNTRIUM	114 (289) **Uuq** UNUNQUADIUM	115 (289) **Uup** UNUNPENTIUM	116 (293) **Uuh** UNUNHEXIUM	117 (294) **Uus** UNUNSEPTIUM	118 (294) **Uuo** UNUNOCTIUM

* Lantanoids

57 138.91 **La** LANTHANIUM	58 140.12 **Ce** CERIUM	59 140.91 **Pr** PRASEODYMIUM	60 144.24 **Nd** NEODYMIUM	61 (145) **Pm** PROMETHIUM	62 150.36 **Sm** SAMARIUM	63 151.96 **Eu** EUROPIUM	64 157.25 **Gd** GADOLINIUM	65 158.93 **Tb** TERBIUM	66 162.50 **Dy** DYSPROSIUM	67 164.93 **Ho** HOLMIUM	68 167.26 **Er** ERBIUM	69 168.93 **Tm** THULIUM	70 173.04 **Yb** YTTERBIUM	71 174.97 **Lu** LUTETIUM

** Actinoids

89 (227) **Ac** ACTINIUM	90 232.04 **Th** THORIUM	91 231.04 **Pa** PROTACTINIUM	92 238.03 **U** URANIUM	93 (237) **Np** NEPTUNIUM	94 (244) **Pu** PLUTONIUM	95 (243) **Am** AMERICIUM	96 (247) **Cm** CURIUM	97 (247) **Bk** BERKELIUM	98 (251) **Cf** CALIFORNIUM	99 (252) **Es** EINSTEINIUM	100 (257) **Fm** FERMIUM	101 (258) **Md** MENDELEVIUM	102 (259) **No** NOBELIUM	103 (262) **Lr** LAWRENCIUM

Period: 1, 2, 3, 4, 5, 6, 7

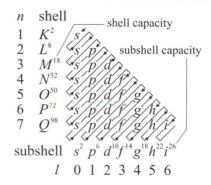

Figure S7.1. Approximate order in which atomic orbitals are filled by electrons (Madelung's rule).

with the electronic configuration $[Ar]\, 3d^5 4s^1$ rather than $[Ar]\, 3d^4 4s^2$ (the former turns out to have a lower energy than the latter). Similarly, Cu takes the configuration $[Ar]\, 3d^{10} 4s^1$ instead of $[Ar]\, 3d^9 4s^2$. There are 18 more known exceptions in periods 5, 6, and 7.

The chemical properties of an element are chiefly determined by the outmost (valence) electrons in the electronic configuration. These electrons have the highest energies and travel farthest from the nucleus and usually belong to the outmost (valence) shell, which is especially true for light atoms. Atoms with complete valence shells (all noble gases) are the most chemically inert, while those with only one electron in their valence shells (alkali metals) or just missing one electron from having a complete shell (halogens) are the most reactive. However, for heavier atoms, electrons in inner subshells can have higher energy than those in the outer subshell; for example, the $3d$ electrons have more energy than the $4s$ electrons and are, therefore, most important in chemical reactions.

The number of valence shell electrons determines the group to which the element belongs. The type of orbital in which the atom's outermost electrons reside determines the *block* to which it belongs. The s-block comprises groups 1 and 2 (plus helium), the p-block comprises the last six groups (13 through 18), the d-block contains all transition metals (groups 3 to 12). The lanthanides (elements 57–71) and actinides (elements 89–103) form the f-block. Elements within each block exhibit some similarity in their properties.

For more details about the periodic table, its applications, and the element properties, see the bibliography below.

Bibliography for Chapter S7

Atkins, P. W., *The Periodic Kingdom: A Journey in the Land of the Chemical Elements*, Basic Books, New York, 1995.

Ball, P., *The Ingredients: A Guided Tour of the Elements*, Oxford University Press, New York, 2002.

Griffiths, D. J., *Introduction to Quantum Mechanics, 2nd Edition*, Prentice Hall, Upper Saddle River, New Jersey, 2004.

Haynes, W. M. (Ed.), *CRC Handbook of Chemistry and Physics, 91st Edition*, CRC Press, Boca Raton, Florida, 2010.

Makarenya, A. A. and Trifonov, D. N., *The Periodic Law of D. I. Mendeleev* [in Russian], Prosveshchenie, Moscow, 1969.

Scerri, E., *The Periodic System, Its Story and Its Significance*, Oxford University Press, New York, 2007.

INDEX

A

F